FINANCIAL TIMES

WORLD DESK REFERENCE

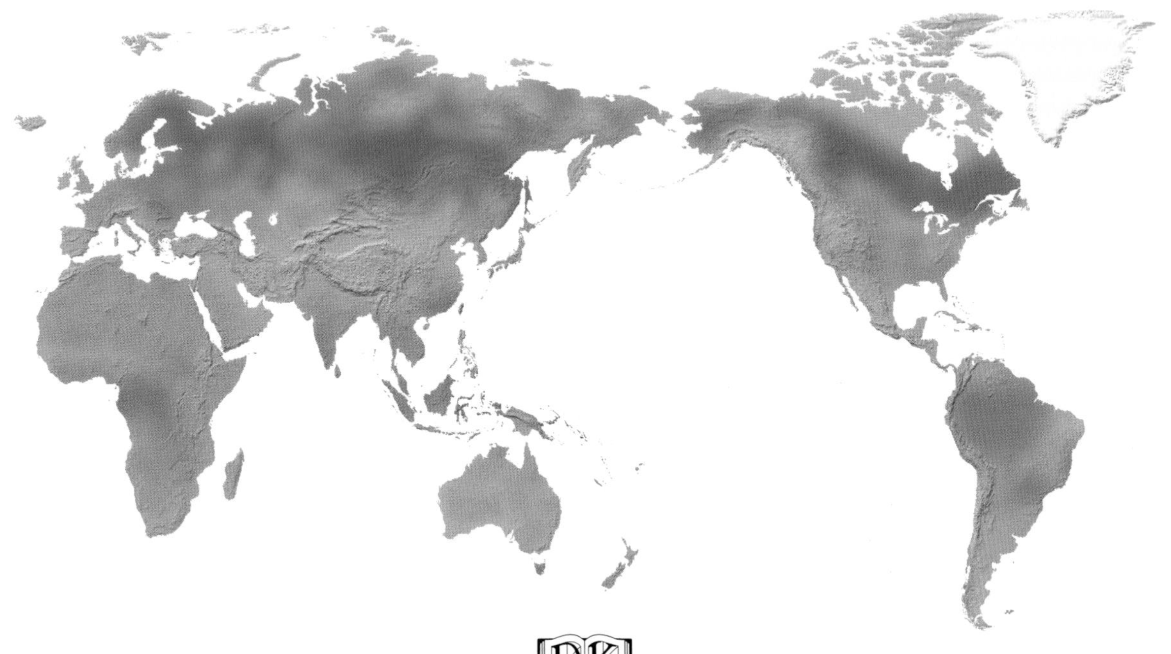

DORLING KINDERSLEY
LONDON • NEW YORK • MUNICH • MELBOURNE • DELHI

For the very latest information, visit:
www.dk.com and click on the Maps & Atlases icon

A DORLING KINDERSLEY BOOK
www.dk.com

FOR THE FIFTH EDITION

EDITOR-IN-CHIEF
Andrew Heritage

SENIOR CARTOGRAPHIC MANAGER
David Roberts

SENIOR CARTOGRAPHIC EDITOR
Simon Mumford

SYSTEMS COORDINATOR
Philip Rowles

FIFTH EDITION UPDATED AND EDITED BY
Cambridge International Reference on Current Affairs (CIRCA)

PROJECT MANAGER
Catherine Jagger

EDITORIAL SUPERVISION
Roger East

DATABASE
Carolyn Postgate, Kate Yedigaroff

GRAPHICS
Alex Yedigaroff, Elizabeth Postgate

EDITORS
Richard J. Thomas, Philippa Youngman

EDITORIAL AND RESEARCH
Joceline Bury, John Coggins, Ian Gorvin, Helen Hawkins
Lawrence Joffe, Frances Nicholson, Farzana Shaikh

PICTURE RESEARCH
Louise Thomas, ilumi

DORLING KINDERSLEY CARTOGRAPHY

EDITORIAL DIRECTION
Andrew Heritage

MANAGING EDITORS
Ian Castello-Cortes, Wim Jenkins

PROJECT EDITORS
Debra Clapson, Catherine Day,
Jo Edwards, Jane Oliver

EDITORS
Alastair Dougall, Ailsa Heritage,
Nicholas Kynaston, Lisa Thomas,
Susan Turner, Chris Whitwell, Elizabeth Wyse

ADDITIONAL EDITORIAL ASSISTANCE
Sam Atkinson, Louise Keane, Zoë Ellinson,
Caroline Lucas, Sophie Park, Laura Porter,
Jo Russ, Crispian Martin St. Valery,
Sally Wood, Ulrike Fritz-Weltz

READERS
Jane Bruton, Reg Grant, Ann Kramer, Lesley Riley

ART DIRECTION
Chez Picthall, Philip Lord

PROJECT DESIGNERS
Martin Biddulph, Scott David,
Carol Ann Davis, David Douglas,
Yahya El-Droubie, Karen Gregory

DESIGNERS
Tony Cutting, Rhonda Fisher,
Nicola Liddiard, Katy Wall

ADDITIONAL DESIGN ASSISTANCE
Paul Bayliss, Carol Ann Davis,
Adam Dobney, Kenny Laurenson,
Paul Williams

SYSTEMS COORDINATOR
Phil Rowles

PRODUCTION
Michelle Thomas, Wendy Penn

PROJECT CARTOGRAPHERS
Caroline Bowie, Ruth Duxbury,
James Mills-Hicks, John Plumer, Julie Turner

CARTOGRAPHERS
James Anderson, Dale Buckton,
Roger Bullen, Tony Chambers,
Jan Clark, Tom Coulson, Martin Darlison,
Claire Ellam, Julia Lunn, Michael Martin,
Alka Ranger, Peter Winfield, Claudine Zante

PICTURE RESEARCH
Alison McKittrick, Sarah Moule,
Christine Rista, Louise Thomas

DATABASE MANAGER
Simon Lewis

INDEX GAZETTEER
Margaret Hynes, Julia Lynch,
Barbara Nash, Jayne Parsons, Janet Smy

Printed and bound in Italy by Graphicom

First published in Great Britain in 1994
by Dorling Kindersley Limited,
80 Strand, London WC2R 0RL

A Penguin Company

Previously published as the DK World Reference Atlas

Reprinted with revisions 1995.Second Edition 1996. Revised 1998.
Third Edition (revised) 2000. Fourth Edition (revised) 2002, Fifth Edition (revised 2003).

A catalogue record for this book is available from the British Library.

ISBN: 0-7513-4889-9

Foreword

THIS DESK REFERENCE is presented to the public in the full knowledge that the world is in a state of continual flux. Political fashions and personalities come and go, while the ebb and flow of peoples and ideas across the face of the planet creates constant shifts in the cultural landscape. All the material assembled for this book has been researched from the most up-to-date and authoritative sources; our team of consultants and contributors, designers, editors, and cartographers have endeavored not only to explain the meaning of this material, to place it in a useful and clear context, but also to present it in a way that has a lasting value and relevance, regardless of the turmoil of daily events. This new edition, bearing the imprimatur of the *Financial Times*, has been completely revised and updated, to reflect the global changes of the past few years. It includes the latest statistical data, and over 60 new photographs.

The publishers would like to thank the many consultants and contributors whose diligence, perseverance, and attention to detail made this book possible.

General Consultants

Anthony Goldstone, Senior Editor Asia-Pacific, *The Economist* Intelligence Unit, London
Professor Jack Spence, Director of Studies, The Royal Institute of International Affairs, London

Regional Consultants

Asia
Anthony Goldstone, London

Africa
James Hammill, Lecturer in African Politics, University of Leicester
Kaye Whiteman, Editor-in-Chief, *West Africa Magazine*, London

Russia and CIS
Martin McCauley, Senior Lecturer, School of Slavonic and East European Studies, University of London

Central and South America
Nick Caistor, Producer, Latin American Section, BBC World Service

USA
Michael Elliot, Diplomatic Editor, *Newsweek*, Washington DC

Europe
John Ardagh, London
Rory Clarke, Senior Editor Europe, *The Economist* Intelligence Unit, London
Charles Powell, Centre for European Studies, St Antony's College, Oxford

Middle East
John Whelan, Ex Editor-in-Chief, *Middle East Economic Digest*

Pacific
Jim Boutilier, Professor in History, Royal Roads Military College, Victoria, Canada

Caribbean
Canute James, *Financial Times*, Kingston, Jamaica

Contributors

Janice Bell, School of Slavonic and East European Studies, University of London
Gerry Bourke, Asia Correspondent, *The Guardian*, Islamabad
Vincent Cable, Director, International Economics Programme
P K Clark, MA, Former Chief Map Research Officer, Ministry of Defence
Ken Davies, Senior Editor, *The Economist* Intelligence Unit, London
Roger Dunn, Analyst, Control Risks Group, London
Aidan Foster-Carter, Senior Lecturer in Sociology, University of Leeds
Professor Murray Forsyth, Centre for Federal Studies, University of Leicester
Natasha Franklin, School of Slavonic and East European Studies, London
Adam Hannestad, *Blomberg Business News*, Copenhagen
Peter Holden, *The Economist* Research Department, London
Tim Jones, Knight Ritter, Brussels
Angella Johnstone, Home Affairs Correspondent, *The Guardian*, London
Oliver Keserü, International Chamber of Commerce, Paris
Robert Macdonald, *The Economist* Intelligence Unit
William Mader, Former Europe Bureau Chief, *Time Magazine*, Washington DC
Professor Brian Matthews, Institute of Commonwealth Studies, London
Nick Middleton, Oriel College, Oxford
Professor Mya Maung, Department of Finance, Boston College, Massachusetts
Judith Nordby, Leeds University
Simon Orme, London
Professor Richard Overy, Department of History, King's College, London
Steve Percy, East Asia Service, BBC World Service
Douglas Rimmer, Honorary Senior Research Fellow, Centre for West African Studies, University of Birmingham
Donna Rispoli, Linacre College, Oxford
Ian Rodger, *The Financial Times*, Zürich
The Royal Institute of International Affairs, London
Struan Simpson, St. James Research, London
Julie Smith, Brasenose College, Oxford
Elizabeth Spencer, London
Michiel Van Kuyen, Erasmus University, Rotterdam
Steven Whitefield, Pembroke College, Oxford
Georgina Wilde, Regional Director, Asia-Pacific, *The Economist* Intelligence Unit, London
H P Willmott, Visiting Professor, Dept. of Military Strategy & Operations, The National War College, Washington DC
Andrew Wilson, Sydney Sussex College, Cambridge
Tom Wingfield, *Reuters*, Bangkok
The World Conservation Monitoring Centre, Cambridge
Cambridge International Reference on Current Affairs (CIRCA)

CONTENTS

1
WORLD FACTFILE

2
THE NATIONS OF THE WORLD

OVERSEAS TERRITORIES & DEPENDENCIES

3
INDEX ~ GAZETTEER

COVER FLAPS

Key to Charts and Icons

ICONS AND TREND INDICATORS vary. Not all variations are shown in the key below, but where they do occur the symbols have been "stacked."

COUNTRY PROFILES

Date of country's independence, or formation.

Date when the country's current borders were established.

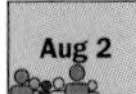
National Day

Vehicle country identifying code

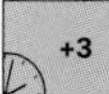
Time zone(s) of country (hours plus or minus from GMT)

International telephone dialling code

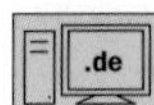
Internet country identifying code

CLIMATE

Indication of the climatic types and zones found in each country.

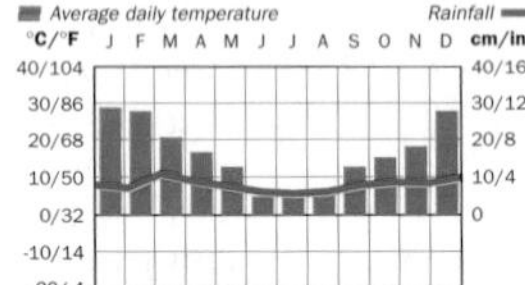
Statistics are given for the national capital. They represent maximum summer and minimum winter averages.

TRANSPORTATION

Indicates on which side of the road vehicles are driven in each country.

The country's principal international airport with annual passenger numbers.

Total size of national merchant or cargo fleet.

THE TRANSPORTATION NETWORK

National communications infrastructure given in kilometers and miles.

Extent of national paved road network

Extent of expressways, freeways, or major highways

Extent of commercial rail network

Extent of inland waterways navigable by commercial craft

TOURISM

The ratio of foreign visitors to population.

Number of visitors per year, including business travelers.

Indicators showing trend in recent visitor numbers (up/level/down).

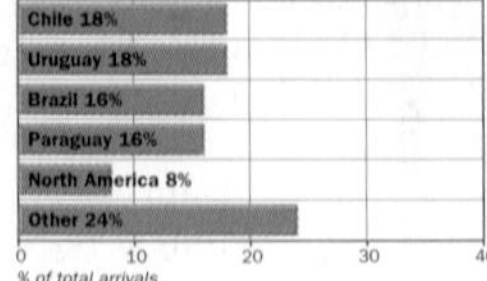
Chile 18%
Uruguay 18%
Brazil 16%
Paraguay 16%
North America 8%
Other 24%
0
10
20
30
40
% of total arrivals
The state of each nation's tourism is explained, with reasons given when there is no significant tourist industry. The chart shows the percentage of total visitors by country of origin.

PEOPLE

An easy indication of the population density in each country (high/medium/low).

Main languages spoken, including official languages.

Population density. This is an average over the whole country.

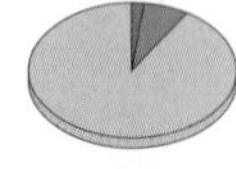
The pie chart proportions show the religious affiliations of those who profess a belief.

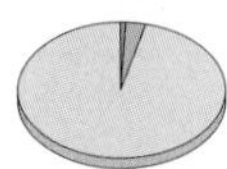
This pie chart illustrates the ethnic origin of the country's population.

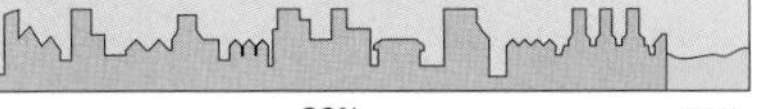
89%
11%
This graph represents the proportion of the population living in urban areas (gray) and rural areas (green).

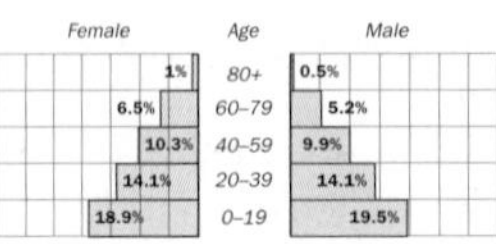
Female
Age
Male
80+
60–79
40–59
20–39
0–19
1%
6.5%
10.3%
14.1%
18.9%
0.5%
5.2%
9.9%
14.1%
19.5%
% of population by age group
This chart shows the breakdown of the population by age groupings, providing an interesting insight into the country's demography.

POLITICS

Indicates the type of elections held within each country.

Dates of last and next legislative elections for Lower (L.) and Upper (U.) Houses.

Name of head of state. In many cases this is a nominal position and does not indicate that this is the country's most powerful person.

A graphic representation of the political makeup of the country's government, based on each party's showing at the last election. Where there are two houses, the more important elected body is shown first.

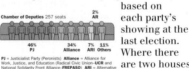

Chamber of Deputies 257 seats

2% AR

46% PJ | 34% Alliance | 7% ARI | 11% Others

PJ = Justicialist Party (Peronists) Alliance = Alliance for Work, Justice, and Education (Radical Civic Union–UCR and National Solidarity Front Alliance–FREPASO) ARI = Alternative for a Republic of Equals AR = Action for the Republic

Senate 72 seats

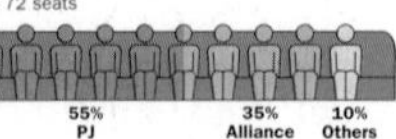
55% PJ
35% Alliance
10% Others

WORLD AFFAIRS

Indication of membership of the UN (United Nations), and date of entry.

Abbreviations indicate membership of international organizations.

Nonmembership of additional international organizations.

AID

Indication as to which countries are aid givers (donors) or aid recipients.

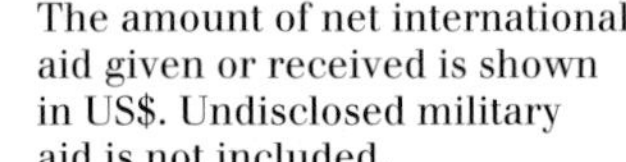
The amount of net international aid given or received is shown in US$. Undisclosed military aid is not included.

Symbols indicate whether aid payments or receipts are rising, level, or declining.

DEFENSE

An indication of the status of conscription and mandatory military service.

The defense budget, the country's annual expenditure (in US$) on arms and military personnel.

Symbols indicate if the trend in defense spending is rising, level, or declining.

THE ARMED FORCES

Icons represent the main branches of the national armed forces.

Army: equipment and personnel

Navy: equipment and personnel

Air force: equipment and personnel

Nuclear capability: armaments

ECONOMICS

An indication of the average rate of inflation per annum, over the past decade.

Gross National Product (GNP) – the total value of goods and services produced by a country.

Exchange rates against the US$ at the start and end of the last year.

❑ World GNP Ranking	24th
❑ GNP per Capita	$24,388
❑ Balance of Payments	$2.2bn
❑ Inflation	5.6%
❑ Unemployment	10.6%

The score cards are intended to give a broad picture of the country's economy. Gross National Product (GNP), unlike GDP, includes income from investments and businesses held abroad. Balance of payments is the difference between a country's payments to and receipts from abroad.

Consumer Price Index
GDP
160
140
120
100
80
1996
1997
1998
1999
2000
Consumer price index 1995=100
GDP 1996=100

This graph shows year-on-year variations in GDP and consumer prices.

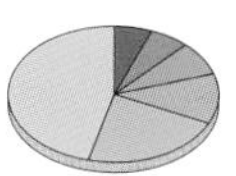

This pie chart gives a broad picture of the country's principal import trading partners.

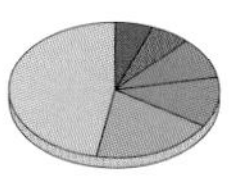

This pie chart gives a broad picture of the country's principal export trading partners.

RESOURCES

▷ Indicates the capacity of the combined national electricity generating sources (in kilowatts).

Oil produced in barrels per day (b/d). Refining capacity and oil reserves are given where applicable.

Estimated livestock resources.

Main mineral reserves are listed in descending order of economic importance.

Fish catch per year (where fishing is a major industry).

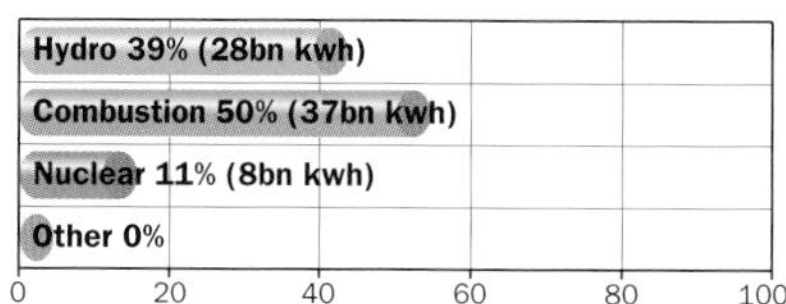

Hydro 39% (28bn kwh)
Combustion 50% (37bn kwh)
Nuclear 11% (8bn kwh)
Other 0%
0
20
40
60
80
100
% of total generation by type

Percentages of the different energy sources used for the generation of electricity are represented graphically ("Combustion" indicates the burning of fossil fuels, wood etc.). An account of the country's resource base is given in the text.

ENVIRONMENT

▷ The 2002 Index of progress toward environmental sustainability, based on 22 core indicators. Compiled by the World Economic Forum taskforce.

Percentage of land which is protected or conserved by law. Protection is often only theoretical.

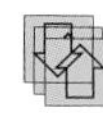

Trend in total CO_2 emissions since 1990 (up/level/down) and current emissions per capita.

ENVIRONMENTAL TREATIES

National parties to international environmental treaties.

Ramsar: (wetlands)

Montreal Protocol: (CFC emissions)

CITES: (endangered species)

CBD: (biological diversity)

Basel: (hazardous wastes)

Kyoto: (greenhouse gases)

MEDIA

▷ Indicates the average rates of television ownership across the country.

Media free to express critical views.

Partial controls or constraints on media freedom.

Severe restrictions on media freedom.

PUBLISHING AND BROADCAST MEDIA

National broadcast and print media, by size and ownership.

Main national newspapers

Television stations: state-owned/independent

Radio: state-owned/ independent

CRIME

▷ An indication of the status of capital punishment and the death penalty.

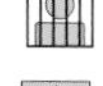

Prison population statistics

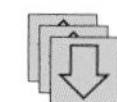

Symbols show general trend in crime figures.

Murders
4 per 100,000 population

Rapes
5 per 100,000 population

Thefts
431 per 100,000 population

This section records official crime figures only. Reported statistics are normally lower than the actual figures.

CHRONOLOGY

Beginning at a significant date in the recent history of the country, the outline chronology continues through to the present day, and highlights key dates and turning points.

EDUCATION

▷ Displays the age until which children are legally required to attend school.

Literacy rate. UNESCO defines as literate anyone who can read and write a short statement.

The number of students in all forms of tertiary education within that country.

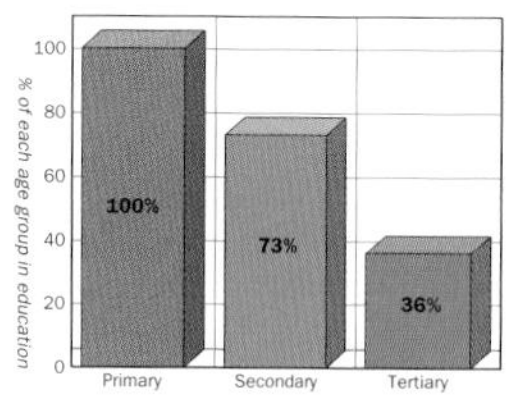

100
80
60
40
20
0
% of each age group in education
100%
73%
36%
Primary
Secondary
Tertiary

This graph shows, for each level of education, the total enrollment, regardless of age, as a percentage of the population of the age group that officially corresponds to that level.

HEALTH

▷ An indication of the existence of health benefits provided by the state.

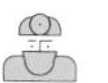

Ratio of the number of people per doctor is given as a national average.

Major causes of death are listed.

SPENDING

▷ Indicates the trend in GDP per capita over the previous decade.

Levels of car ownership (per 1000 head of population)

Rates of telephone connectivity (per 1000 head of population)

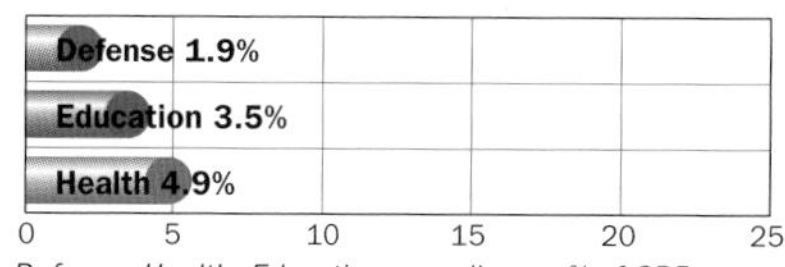

Defense 1.9%
Education 3.5%
Health 4.9%
0
5
10
15
20
25
Defense, Health, Education spending as % of GDP

Percentage of the country's GDP that is spent on defense, education, and health.

WORLD RANKING

Schooling, educational attainment, and human development rankings are based on the UN Human Development Index (which covers 172 countries and Hong Kong).

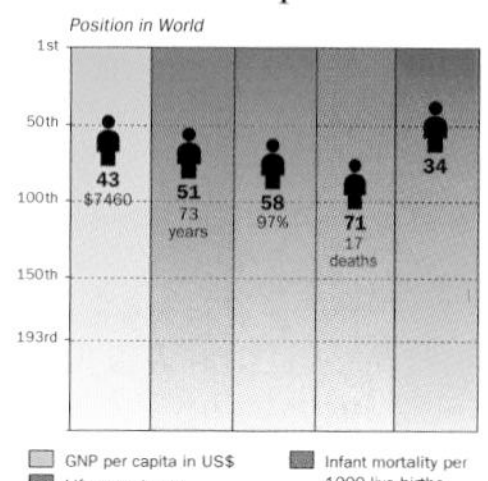

Position in World
1st
50th
100th
150th
193rd
43
$7460
51
73 years
58
97%
71
17 deaths
34
GNP per capita in US$
Life expectancy
Literacy
Infant mortality per 1000 live births
Human dev. index

SOURCES OF STATISTICAL DATA USED IN THIS BOOK

Airports Council International

Amnesty International

Automobile Association (AA)

British Petroleum:
World Energy Data

Cambridge International Reference on Current Affairs (CIRCA)

Commonwealth Secretariat:
Small States Economic Review and Basic Statistics

Dorling Kindersley

Europa World Yearbook

European Bank for Reconstruction and Development (EBRD)

Financial Times

Fischer Weltalmanach

Food and Agriculture Organization (FAO)

International Atomic Energy Agency (IAEA)

International Institute for Strategic Studies (IISS):
The Military Balance

International Labor Organization (ILO):
World Labor Report

International Monetary Fund (IMF):
Balance of Payments Statistics Yearbook,
Direction of Trade Statistics Yearbook,
Government Financial Statistics Yearbook,
International Financial Statistics,
World Economic Outlook

International Road Federation

International Union for Conservation of Nature (IUCN)

International Union of Railways

INTERPOL International Crime Statistics

Lloyd's Register of Shipping

Organization for Economic Cooperation Development (OECD):
Economic surveys

OECD Development Assistance Committee (DAC):
Development Cooperation Report

Organization of Petroleum Exporting Countries (OPEC)

Ramsar Convention Bureau

Royal Automobile Club (RAC)

United Nations Demographic Yearbook

United Nations Energy Statistics Yearbook

United Nations Industrial Commodity Statistics Yearbook

United Nations International Trade Statistics Yearbook

United Nations Statistical Yearbook

United Nations Statistical Yearbook of Asia and the Pacific

United Nations Children's Fund (UNICEF)

United Nations Development Program (UNDP):
Human Development Report

United Nations Educational, Scientific, and Cultural Organization (UNESCO): Statistical Yearbook

United Nations Environment Program (UNEP):
Ozone Secretariat
Secretariat of the Basel Convention
Secretariat of the Convention on Biological Diversity (CBD)
Secretariat of the Convention on International Trade in Endangered Species (CITES)

United Nations Framework Convention on Climate Change

United Nations Population Fund (UNFPA):
The State of World Population

United States Central Intelligence Agency (CIA)

World Bank: World Development Indicators,
World Development Report, World Bank Atlas

World Conservation Monitoring Center (WCMC):
Biodiversity Data Sourcebook

World Economic Forum

World Health Organization (WHO)

World Prison Population List (Research, Development, and Statistics Directorate, UK Home Office)

World Tourist Organization (WTO)

Worldwide Fund for Nature (WWF)

1

WORLD FACTFILE

The Physical World

The Earth's surface is constantly being transformed: it is uplifted, folded, and faulted by tectonic forces; weathered and eroded by wind, water, and ice. Sometimes change is dramatic, the spectacular results of earthquakes or floods. More often it is a slow process lasting millions of years. A physical map of the world represents a snapshot of the ever-evolving architecture of the Earth. This terrain map shows the whole surface of the Earth, both above and below the sea. The size of the Earth can be measured in different ways. When taken from the Equator, the diameter of the Earth measures 12,756 km (7927 miles); when taken from pole to pole, the diameter measures 12,714 km (7900 miles). Two-thirds of the Earth's surface is covered by oceans. The landscape of the ocean floor, like the surface of the land, has been shaped by movements of the Earth's crust over millions of years to form volcanic mountain ranges, deep trenches, basins, and plateaus. Ocean currents constantly redistribute warm and cold water around the world. The largest ocean in the world is the Pacific, which covers an area of over 181 million sq. km (70 million sq. miles).

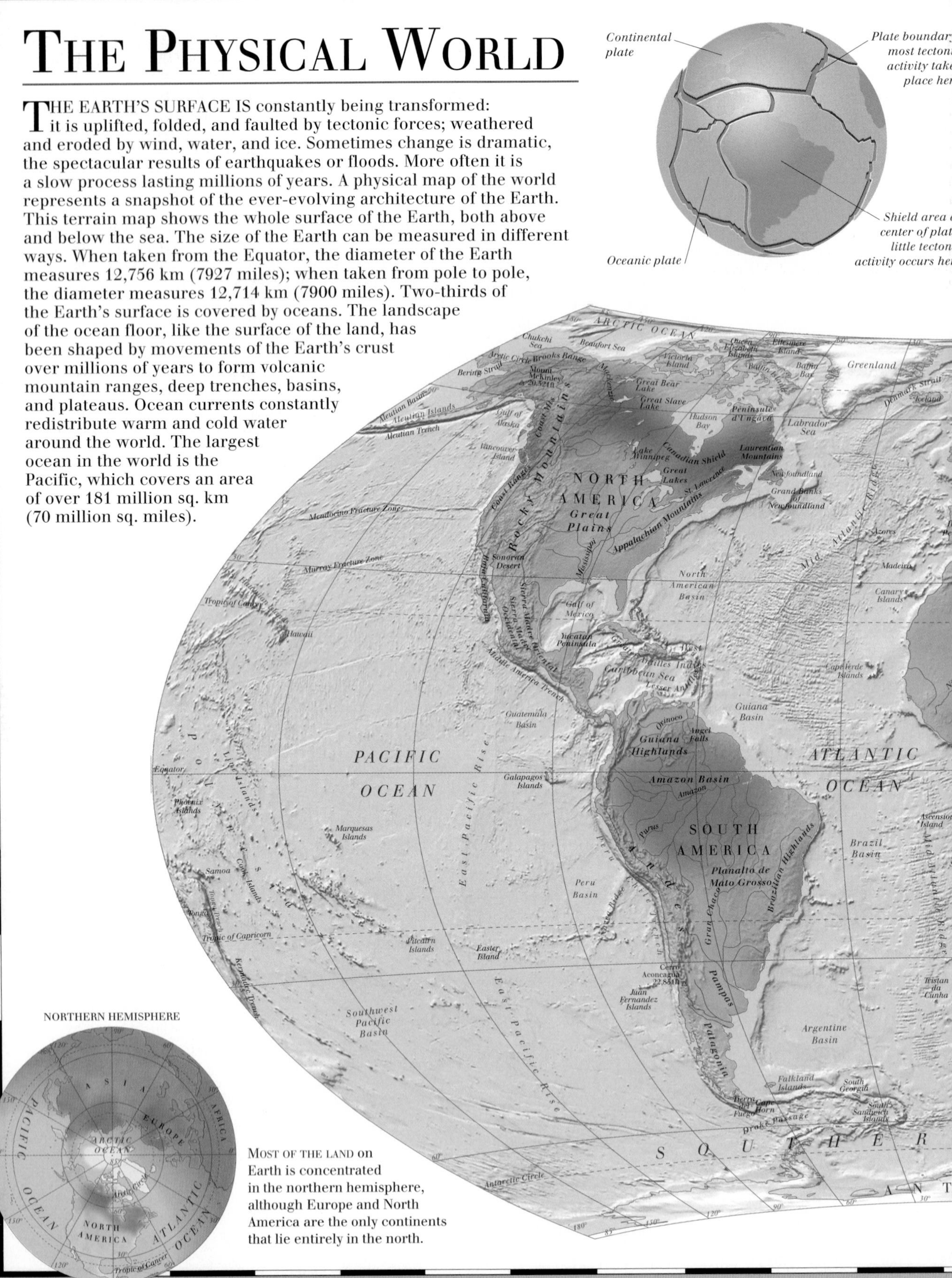
Continental plate
Oceanic plate
ARCTIC OCEAN
Chukchi Sea
Beaufort Sea
Bering Strait
Brooks Range
Mount McKinley 20,321ft
Aleutian Basin
Aleutian Islands
Aleutian Trench
Gulf of Alaska
Vancouver Island
Coast Mts
Victoria Island
Great Bear Lake
Great Slave Lake
Queen Elizabeth Islands
Ellesmere Island
Baffin Island
Baffin Bay
Greenland
Denmark Strait
Iceland
Hudson Bay
Peninsule d'Ungava
Labrador Sea
Lake Winnipeg
Canadian Shield
Laurentian Mountains
Great Lakes
St. Lawrence
Newfoundland
Grand Banks of Newfoundland
NORTH AMERICA
Rocky Mountains
Coast Ranges
Great Plains
Appalachian Mountains
Mississippi
Mendocino Fracture Zone
Murray Fracture Zone
Sonoran Desert
Baja California
Sierra Madre Occidental
Sierra Madre Oriental
Azores
Madeira
Canary Islands
North American Basin
Mid-Atlantic Ridge
Tropic of Cancer
Hawaii
Gulf of Mexico
Yucatan Peninsula
Caribbean Sea
Lesser Antilles
Middle America Trench
Cape Verde Islands
Guatemala Basin
Guiana Basin
Orinoco
Angel Falls
Guiana Highlands
PACIFIC OCEAN
ATLANTIC OCEAN
Equator
Line Islands
Phoenix Islands
Galapagos Islands
Amazon Basin
Amazon
Marquesas Islands
East Pacific Rise
SOUTH AMERICA
Purus
Ascension Island
Brazil Basin
Samoa
Cook Islands
Planalto de Mato Grosso
Brazilian Highlands
Peru Basin
Andes
Gran Chaco
Tonga
Tonga Trench
Tropic of Capricorn
Pitcairn Islands
Easter Island
Cerro Aconcagua 22,831ft
Pampas
Juan Fernandez Islands
Tristan da Cunha
Kermadec Trench
Southwest Pacific Basin
Patagonia
Argentine Basin
Falkland Islands
South Georgia
Cape Horn
Tierra del Fuego
Drake Passage
South Sandwich Islands
SOUTHER
Antarctic Circle
NORTHERN HEMISPHERE
ASIA
EUROPE
AFRICA
ARCTIC OCEAN
PACIFIC OCEAN
ATLANTIC OCEAN
Arctic Circle
NORTH AMERICA
Tropic of Cancer

Most of the land on Earth is concentrated in the northern hemisphere, although Europe and North America are the only continents that lie entirely in the north.

THE DYNAMIC EARTH

THE EARTH'S CRUST is made up of eight major (and several minor) rigid continental and oceanic ectonic plates, which constantly move relative to one another. It is this movement which causes volcanic eruptions, earthquakes, and sometimes tsunamis along the plate boundaries. The largest volcanoes formed by this process are Aconcagua in Argentina at 6959 m (22,831 ft) and Kilimanjaro in Tanzania at 5895 m (19,340 ft), both of which are now extinct. Plate tectonics are responsible for the formation of the Himalayas – which were created by two colliding plates – and the Hawaiian Islands, created by the Pacific plate's movement over a "hot spot" of magma.

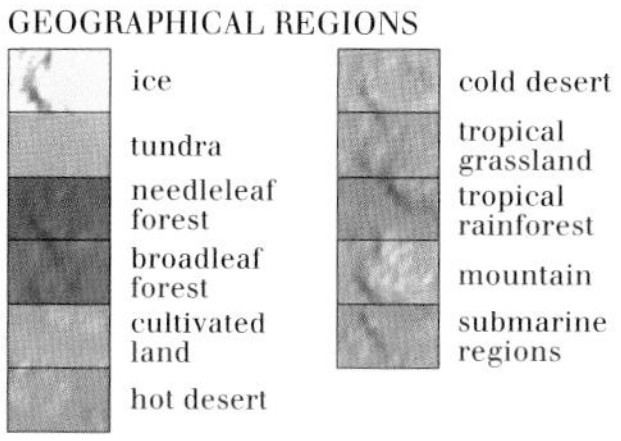

GEOGRAPHICAL REGIONS

- ice
- tundra
- needleleaf forest
- broadleaf forest
- cultivated land
- hot desert
- cold desert
- tropical grassland
- tropical rainforest
- mountain
- submarine regions

PHYSICAL WORLD FACTFILE

HIGHEST MOUNTAINS		
1 **Everest**	8848 m	(29,028 ft)
2 **K2**	8611 m	(28,251 ft)
3 **Kangchenjunga I**	8590 m	(28,169 ft)
4 **Makalu I**	8463 m	(27,766 ft)
5 **Cho Oyu**	8201 m	(26,906 ft)
LONGEST RIVERS		
1 **Nile**	6695 km	(4160 mi.)
2 **Amazon**	6516 km	(4048 mi.)
3 **Yangtze**	6380 km	(3964 mi.)
4 **Mississippi /Missouri**	6019 km	(3740 mi.)
5 **Ob'-Irtysh**	5570 km	(3461 mi.)
LARGEST DESERTS		
1 **Sahara**	9,065,000 km^2	(3,263,400 mi^2)
2 **Australian**	3,750,000 km^2	(1,350,000 mi^2)
3 **Gobi**	1,295,000 km^2	(466,200 mi^2)
4 **Arabian**	750,000 km^2	(270,000 mi^2)
5 **Sonoran**	311,000 km^2	(111,960 mi^2)

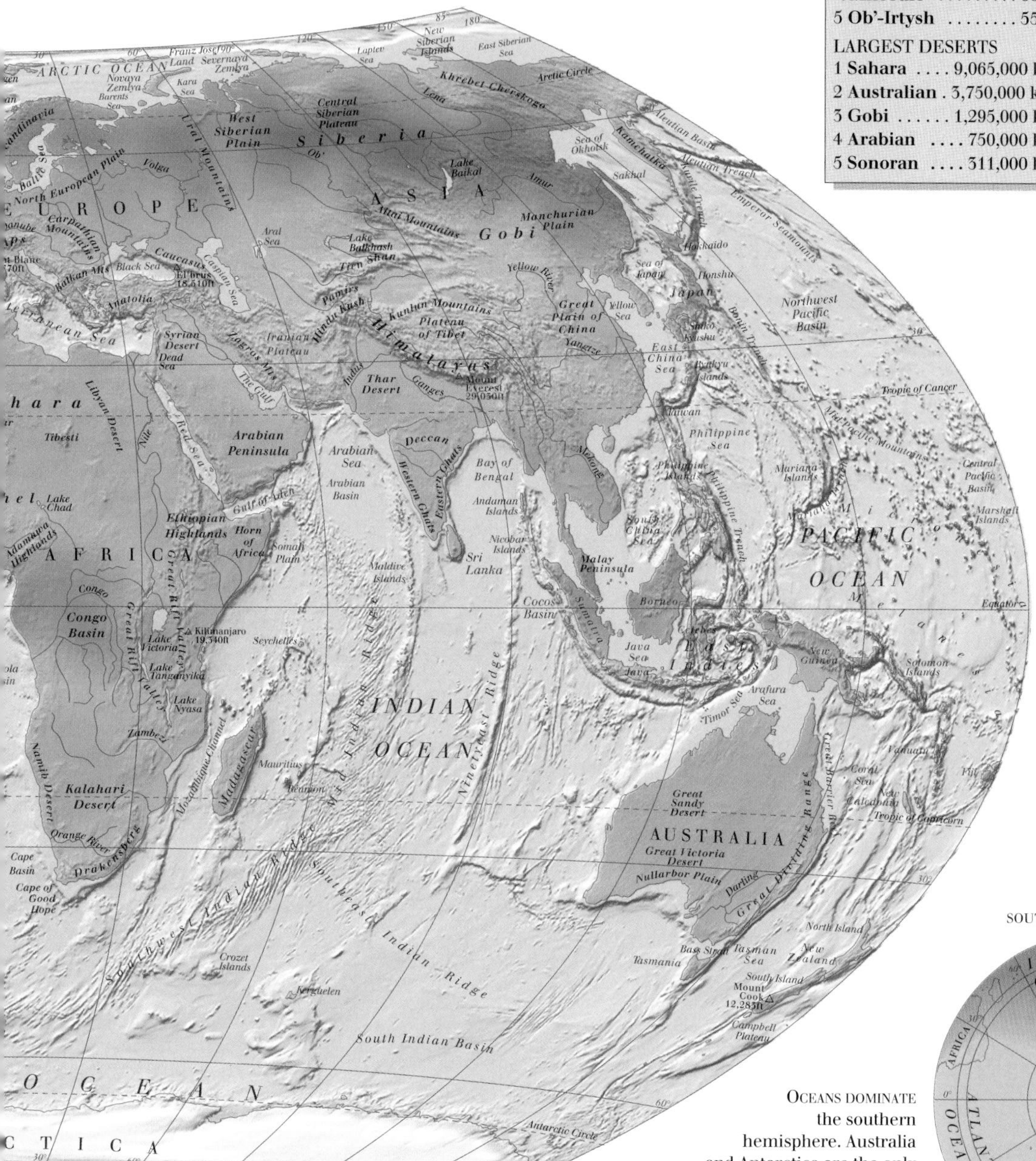

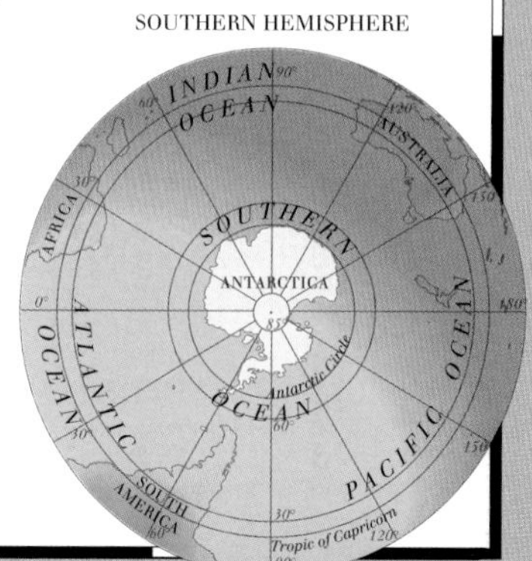

SOUTHERN HEMISPHERE

OCEANS DOMINATE the southern hemisphere. Australia and Antarctica are the only continental landmasses that lie entirely in the south.

THE POLITICAL WORLD

IN 2002, EAST TIMOR joined the international community, becoming the world's 193rd recognized independent state. In 1950 there were only 82. With the exception of Antarctica, where territorial claims have been deferred by international treaty, every land area of the Earth's surface either belongs to, or is claimed by, one country or another. Some 60 overseas dependent territories remain, administered variously by Australia, Denmark, France, the Netherlands, New Zealand, Norway, the UK, and the US. Over the last half-century, national self-determination has been a driving force for many states with a history of colonialism or oppression. While some new states on gaining independence moved peacefully to establish a democracy, many others have been torn by religious or ethnic conflicts or became submerged in power struggles resulting in dictatorship by a military regime or an individual despot.

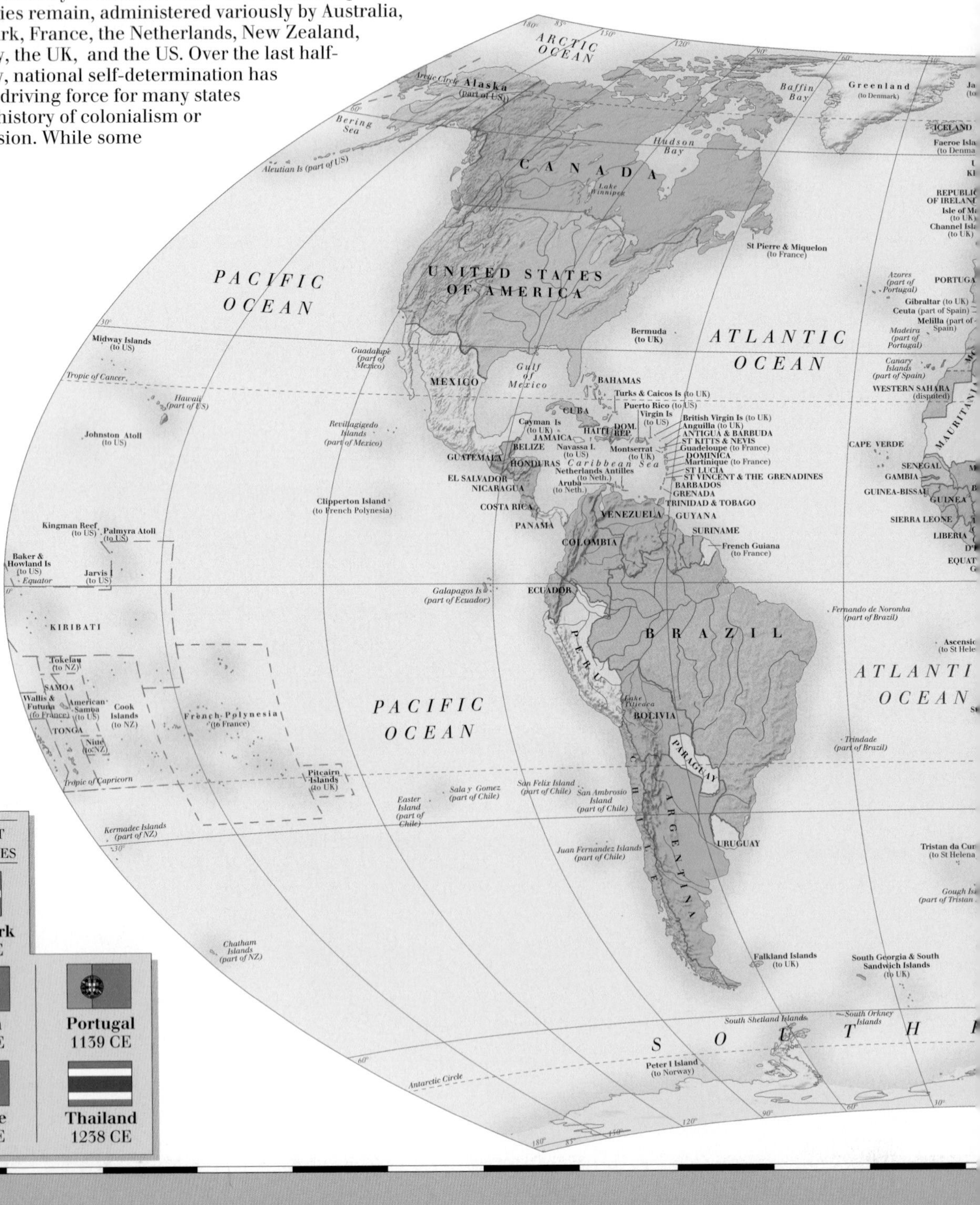

OLDEST COUNTRIES

Denmark
950 CE

China
960 CE

Portugal
1139 CE

France
987 CE

Thailand
1238 CE

KEY

———	Full borders
- - - - - - - -	Disputed borders
- - - - - -	Undefined borders
— — — —	Extent of dependent island territories
—— —— ——	Extent of country boundaries for island territories
Tristan da Cunha (to St Helena)	Dependent territory with self-government
Gough Island (part of Tristan da Cunha)	Territory without self-government (the state it belongs to is given in brackets)

International Borders

BOUNDARIES BETWEEN states fall into three categories. Full borders are internationally recognized territorial boundaries. Undefined borders exist where no fixed boundary has been demarcated. A disputed border is where a de facto boundary exists which is not agreed upon or is subject to arbitration. Disputed borders exist throughout the world, such as the land borders between India and China and between Ethiopia and Eritrea, and the maritime border between Samoa and American Samoa.

Countries with the Most Land Borders

1 **China:** *14* (Afghanistan, Bhutan, Burma, India, Kazakhstan, North Korea, Kyrgyzstan, Laos, Mongolia, Nepal, Pakistan, Russian Federation, Tajikistan, Vietnam)

Russian Federation: *14* (Azerbaijan, Belarus, China, Estonia, Finland, Georgia, Kazakhstan, North Korea, Latvia, Lithuania, Mongolia, Norway, Poland, Ukraine)

2 **Brazil:** *10* (Argentina, Bolivia, Colombia, French Guiana, Guyana, Paraguay, Peru, Suriname, Uruguay, Venezuela)

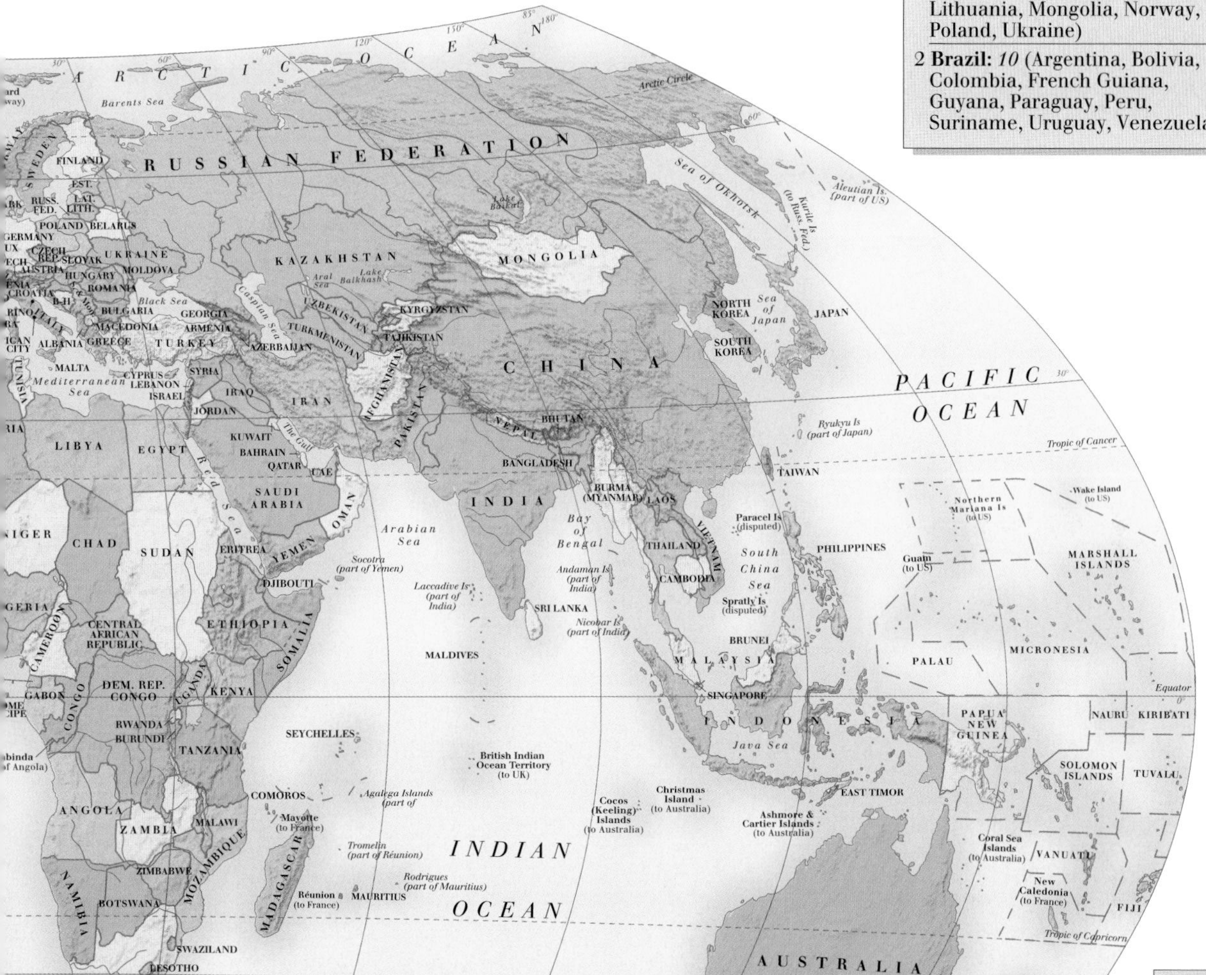

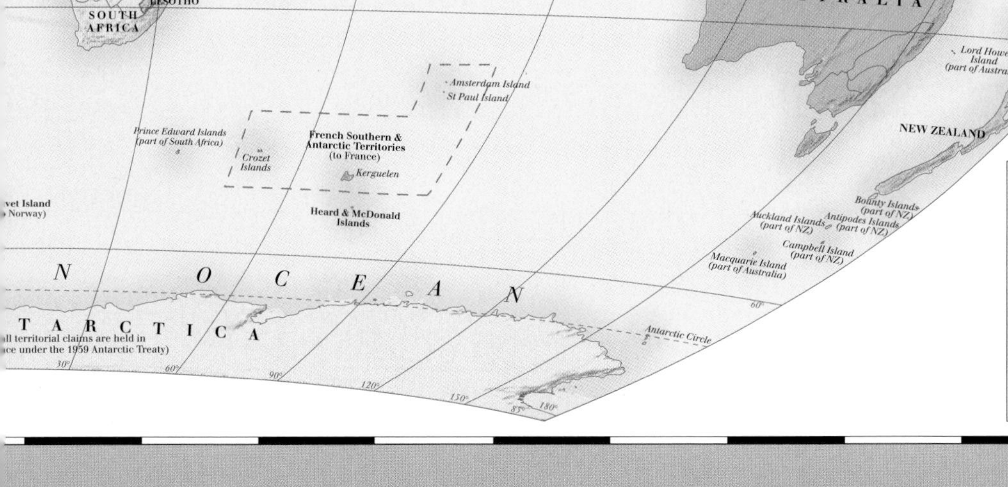

Youngest Countries

East Timor
2002

Slovakia
1993

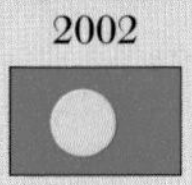

Palau
1994

Czech Rep.
1993

Eritrea
1993

MARS
- *Diameter: 6786 km*
- *Mass: 642 billion billion tons*
- *Temperature: –137 to 37°C*
- *Distance from Sun: 228 million km*
- *Length of year: 1.88 years*
- *Surface gravity: 1 kg = 0.38 kg*

EARTH
- *Diameter: 12,756 km*
- *Mass: 5976 billion billion tons*
- *Temperature: –70 to 55°C*
- *Distance from Sun: 150 million km*
- *Length of year: 365.25 days*
- *Surface gravity: 1kg = 1 kg*

THE EARTH

GASES SUCH AS CARBON dioxide are known as "greenhouse gases" because they prevent shortwave solar radiation from entering the Earth's atmosphere, but help to stop longwave radiation from escaping. This traps heat, raising the Earth's temperature. An excess of these gases traps more heat and can lead to global warming.

Incoming shortwave solar radiation

Greenhouse gases prevent the escape of longwave radiation

Longwave radiation deflected by the Earth heats the atmosphere

VENUS
- *Diameter: 12,102 km*
- *Mass: 4870 billion billion tons*
- *Temperature: 457°C*
- *Distance from Sun: 108 million km*
- *Length of year: 224.7 days*
- *Surface gravity: 1 kg = 0.88 kg*

MERCURY
- *Diameter: 4878 km*
- *Mass: 330 billion billion tons*
- *Temperature: –173 to 427°C*
- *Distance from Sun: 58 million km*
- *Length of year: 87.97 days*
- *Surface gravity: 1 kg = 0.38 kg*

THE SOLAR SYSTEM

THE SOLAR SYSTEM CONSISTS of the nine major planets, their moons, the asteroids, and the comets that orbit around the Sun. The Sun itself is composed of 70% hydrogen and 30% helium, and at its core nuclear fusion reactions turning hydrogen into helium produce the heat and light which make life possible on Earth. Of the planets, the inner four (Mercury, Venus, Earth, and Mars) are termed terrestrial, while the next four (Jupiter, Saturn, Uranus, and Neptune) are termed gas giants. Pluto, at the edge of the solar system, is much smaller, and made of rock. The largest natural satellite in the Solar System is Ganymede (5262 km – 3270 miles – in diameter), which orbits around Jupiter, the largest planet. Halley's comet is the brightest comet when seen from Earth, and orbits the Sun once every 76 years. The largest asteroid is named Ceres (940 km – 584 miles – in diameter), which is found in the asteroid belt between Mars and Jupiter. The planet Earth is unique within the solar system (and possibly the universe), being the only planet capable of sustaining life.

JUPITER
- *Diameter: 142,984 km*
- *Mass: 1,900,000,000 billion billion tons*
- *Temperature: –153°C*
- *Distance from Sun: 778 million km*
- *Length of year: 11.86 years*
- *Surface gravity: 1 kg = 2.53 kg*

Highest high tides
Lowest high tides
Earth
Moon
Sun
Tidal bulge created by gravitational pull

Saturn

- *Diameter: 120,660 km*
- *Mass: 570,000 billion billion tons*
- *Temperature: –185°C*
- *Distance from Sun: 1427 million km*
- *Length of year: 29.46 years*
- *Surface gravity: 1 kg = 1.07 kg*

Uranus

- *Diameter: 51,118 km*
- *Mass: 102,000 billion billion tons*
- *Temperature: –214°C*
- *Distance from Sun: 2870 million km*
- *Length of year: 84.01 years*
- *Surface gravity: 1 kg = 0.92 kg*

Moon and Tides

TIDES ARE CREATED by the pull of the Sun and the Moon's gravity on the surface of the oceans. Waves are formed by wind blowing over the surface of the oceans. The highest tides occur when the Earth, the Moon, and the Sun are aligned (*below left*). The lowest tides are experienced when the Sun and Moon align at right angles to one another (*below right*).

NEAR SIDE OF THE MOON

FAR SIDE OF THE MOON

Neptune

- *Diameter: 49,528 km*
- *Mass: 13 billion billion tons*
- *Temperature: –225°C*
- *Distance from Sun: 4497 million km*
- *Length of year: 164.79 years*
- *Surface gravity: 1 kg = 1.18 kg*

Pluto

- *Diameter: 2300 km*
- *Mass: 13 billion billion tons*
- *Temperature: –236°C*
- *Distance from Sun: 5900 million km*
- *Length of year: 248.54 years*
- *Surface gravity: 1 kg = 0.30 kg*

Timeline of Space Exploration

1955 · 1960 · 1970 · 1980 · 1990 · 2000 · 2010

- **1957:** USSR launches *Sputnik I* - first artificial satellite
- **Oct 10, 1959:** *Luna 3* sends back first pictures of dark side of the Moon
- **Apr 12, 1961:** Yuri Gagarin (USSR) first person in space
- **Jul 10, 1962:** Launch of *Telstar I*, first commercial communications satellite
- **Feb 13, 1966:** *Luna 9* first probe to land on Moon
- **Jul 21, 1969:** Neil Armstrong and Buzz Aldrin first people to land on Moon
- **1976:** Missions of *Viking 1* and *2* analyze surface of Mars
- **Jan 28, 1986:** *Challenger* shuttle explodes; all seven crew members killed
- **Feb 20, 1986:** Launch of space station *Mir*
- **Aug 25, 1989:** *Voyager 2* probe passes Neptune on way out of Solar System
- **Apr 24, 1990:** Launch of Hubble Space Telescope
- **Nov 20, 1998:** Launch of first part of International Space Station
- **2000:** *Mir* brought to earth. Dennis Tito is first space tourist. 100th shuttle mission completed
- **2001:** *Near* probe lands on Eros asteroid
- **2004:** Provisional date for a probe landing on Titan (Saturn's largest moon)
- **2005:** Provisional date for the launch of the first Chinese manned space flight

The Climate

THE EARTH'S CLIMATIC REGIONS consist of stable patterns of weather conditions averaged out over a long period of time. Different climates are categorized according to particular combinations of temperature and humidity. By contrast, weather consists of short-term fluctuations in wind, temperature, and humidity conditions. Different climates are determined by latitude, altitude, the prevailing wind, and circulation of ocean currents. Longer-term changes in climate, such as global warming or the onset of ice ages, are punctuated by shorter-term events which comprise the day-to-day weather of a region, such as frontal depressions, hurricanes, and blizzards.

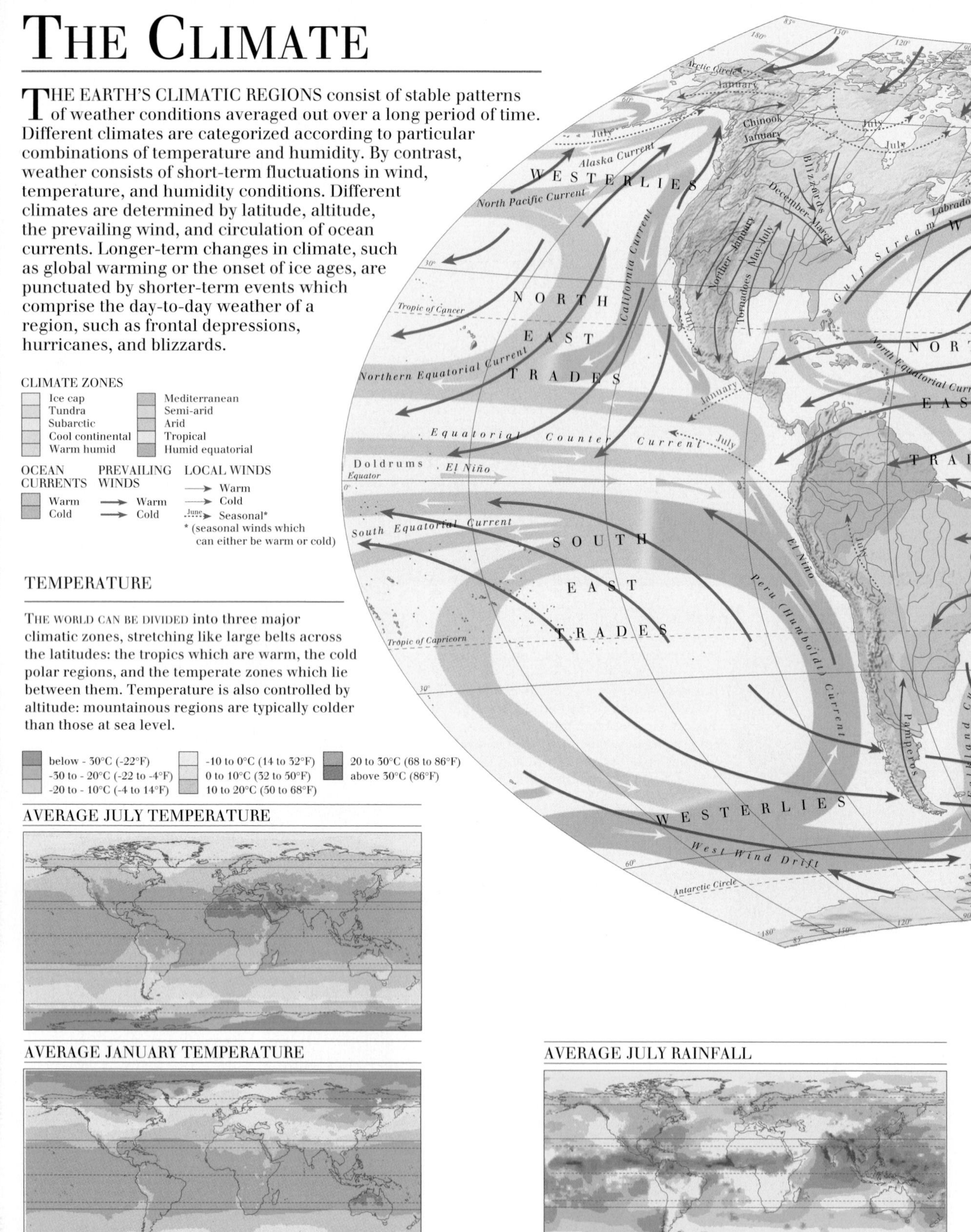
Arctic Circle
January
Chinook
January
July
July
July
Alaska Current
WESTERLIES
North Pacific Current
Blizzards
December–March
Northers January
Tornadoes May–July
California Current
Labrador Current
Gulf Stream
NORTH EAST TRADES
Tropic of Cancer
Northern Equatorial Current
North Equatorial Current
January
Equatorial Counter Current
July
Doldrums
El Niño
Equator
South Equatorial Current
SOUTH EAST TRADES
El Niño
July
Peru (Humboldt) Current
Tropic of Capricorn
Pamperos
Falkland Current
WESTERLIES
West Wind Drift
Antarctic Circle

CLIMATE ZONES

- Ice cap
- Tundra
- Subarctic
- Cool continental
- Warm humid
- Mediterranean
- Semi-arid
- Arid
- Tropical
- Humid equatorial

OCEAN CURRENTS

- Warm
- Cold

PREVAILING WINDS

- Warm
- Cold

LOCAL WINDS

- Warm
- Cold
- Seasonal*

* (seasonal winds which can either be warm or cold)

TEMPERATURE

THE WORLD CAN BE DIVIDED into three major climatic zones, stretching like large belts across the latitudes: the tropics which are warm, the cold polar regions, and the temperate zones which lie between them. Temperature is also controlled by altitude: mountainous regions are typically colder than those at sea level.

- below -30°C (-22°F)
- -30 to -20°C (-22 to -4°F)
- -20 to -10°C (-4 to 14°F)
- -10 to 0°C (14 to 32°F)
- 0 to 10°C (32 to 50°F)
- 10 to 20°C (50 to 68°F)
- 20 to 30°C (68 to 86°F)
- above 30°C (86°F)

AVERAGE JULY TEMPERATURE

AVERAGE JANUARY TEMPERATURE

AVERAGE JULY RAINFALL

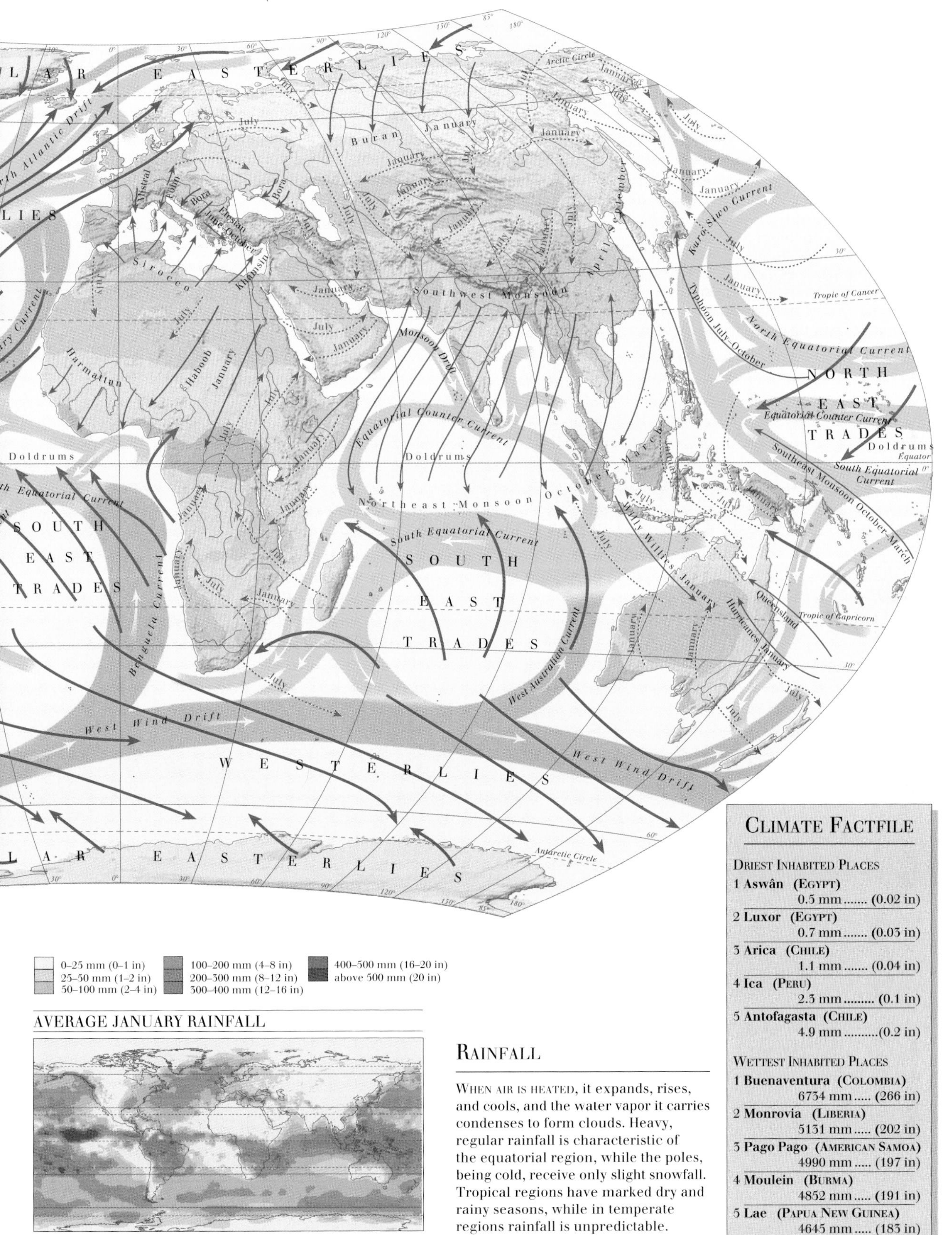
POLAR EASTERLIES
North Atlantic Drift
Buran
January
July
Mistral
Föhn
Bora
Etesian June-October
Sirocco
Khamsin
Southwest Monsoon
Monsoon Drift
April-September
Kuro Siwo Current
Typhoon July-October
Tropic of Cancer
North Equatorial Current
NORTH EAST TRADES
Equatorial Counter Current
Doldrums
Equator
South Equatorial Current
Southeast Monsoon October-March
Harmattan
Haboob
Northeast Monsoon
October-March
Willy Willies January
Queensland
Hurricanes January
Tropic of Capricorn
SOUTH EAST TRADES
Benguela Current
West Australian Current
West Wind Drift
WESTERLIES
POLAR EASTERLIES
Antarctic Circle
Arctic Circle
0–25 mm (0–1 in)
25–50 mm (1–2 in)
50–100 mm (2–4 in)
100–200 mm (4–8 in)
200–300 mm (8–12 in)
300–400 mm (12–16 in)
400–500 mm (16–20 in)
above 500 mm (20 in)
AVERAGE JANUARY RAINFALL

RAINFALL

WHEN AIR IS HEATED, it expands, rises, and cools, and the water vapor it carries condenses to form clouds. Heavy, regular rainfall is characteristic of the equatorial region, while the poles, being cold, receive only slight snowfall. Tropical regions have marked dry and rainy seasons, while in temperate regions rainfall is unpredictable.

CLIMATE FACTFILE

DRIEST INHABITED PLACES

1 **Aswân** (EGYPT)	0.5 mm	(0.02 in)
2 **Luxor** (EGYPT)	0.7 mm	(0.03 in)
3 **Arica** (CHILE)	1.1 mm	(0.04 in)
4 **Ica** (PERU)	2.3 mm	(0.1 in)
5 **Antofagasta** (CHILE)	4.9 mm	(0.2 in)

WETTEST INHABITED PLACES

1 **Buenaventura** (COLOMBIA)	6734 mm	(266 in)
2 **Monrovia** (LIBERIA)	5131 mm	(202 in)
3 **Pago Pago** (AMERICAN SAMOA)	4990 mm	(197 in)
4 **Moulein** (BURMA)	4852 mm	(191 in)
5 **Lae** (PAPUA NEW GUINEA)	4645 mm	(183 in)

THE ENVIRONMENT

THE EARTH CAN BE DIVIDED into a series of biogeographic regions, or biomes – ecological communities where certain species of plant and animal coexist within particular climatic conditions. Within these broad classifications, other factors affect the local distribution of species in each biome, such as soil richness, altitude, and human activities such as urbanization, intensive agriculture, and deforestation. Apart from the polar ice caps, there are few areas which have not been colonized by animals or plants over the course of the Earth's history. Because of all animals' reliance on plants for survival, plants are known as primary producers. The availability of nutrients and the temperature of an area define its primary productivity, which affects the number and type of animals which are able to live there; the level of humidity or aridity is also a determining factor.

BIODIVERSITY

THE NUMBER OF PLANT AND ANIMAL SPECIES, and the range of genetic diversity within the populations of each species, make up the Earth's biodiversity. The plants and animals which are endemic to a region – that is, those which are found nowhere else in the world – are also important in determining levels of biodiversity. Human settlement and intervention have encroached on many areas of the world once rich in endemic plant and animal species. Increasing international efforts are being made to monitor and conserve the biodiversity of the Earth's remaining wild places.

Arctic Circle
Rocky Mountains
Great Plains
Canadian Shield
North European Plain
Mediterranean Sea
ATLANTIC OCEAN
Tropic of Cancer
Sahara
Caribbean Sea
Sahel
PACIFIC OCEAN
Equator
Amazon Basin
Congo Basin
ATLANTIC OCEAN
Gran Chaco
Kalahari Desert
Tropic of Capricorn
Pampas
SOUTHERN OCE
Antarctic Circle
ANTARCT

ANIMALS

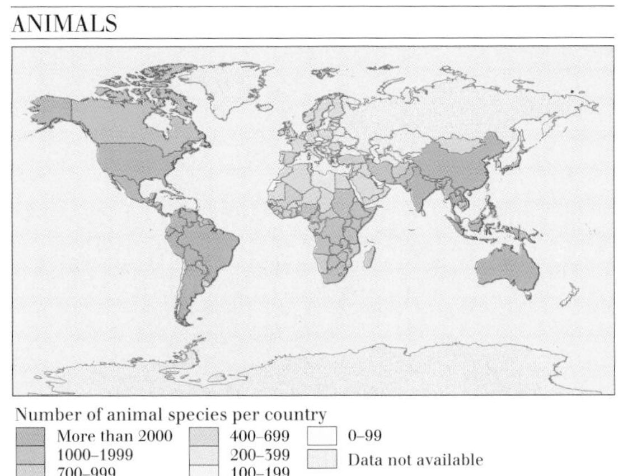

Number of animal species per country

- More than 2000
- 1000–1999
- 700–999
- 400–699
- 200–399
- 100–199
- 0–99
- Data not available

Animal Adaptation

The degree of an animal's adaptability to different climates and conditions is extremely important in ensuring its success as a species. Many animals, particularly the largest mammals, are becoming restricted to ever-smaller regions as human development and modern agricultural practices reduce their natural habitats. In contrast, humans have been responsible – both deliberately and accidentally – for the spread of some of the world's most successful species. Many of these introduced species are now more numerous than the indigenous animal populations.

PLANTS

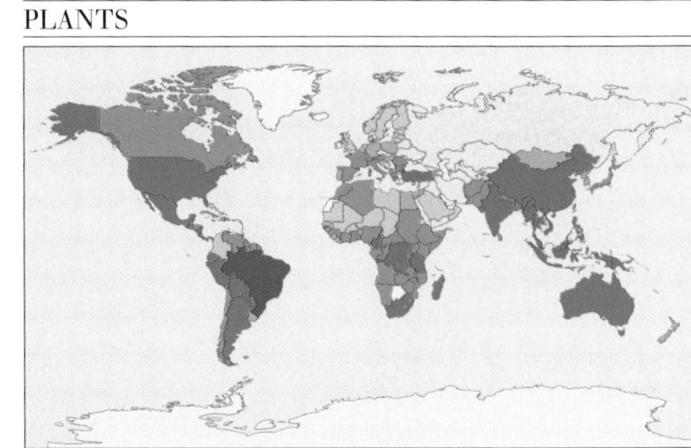

Number of plant species per country

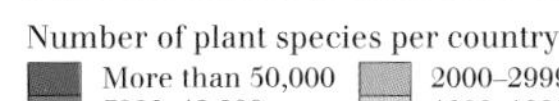

- More than 50,000
- 7000–49,999
- 3000–6999
- 2000–2999
- 1000–1999
- 600–999
- 0–599
- Data not available

Plant Adaptation

Environmental conditions, such as climate, soil type, and competition with other organisms, influence the development of plants into distinctive forms. Similar conditions in different parts of the world create similar adaptations in the plants, which may then be modified by other, local, factors specific to the region.

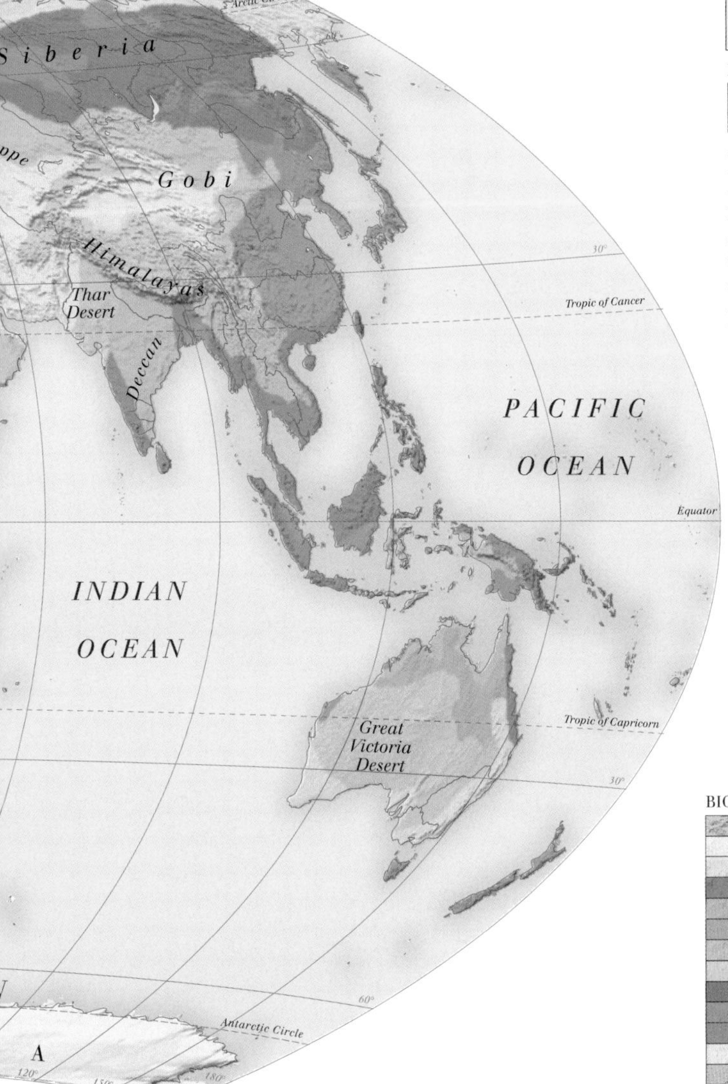
O C E A N
Arctic Circle
S i b e r i a
G o b i
Himalayas
Thar Desert
Deccan
Tropic of Cancer
PACIFIC
OCEAN
Equator
INDIAN
OCEAN
Tropic of Capricorn
Great Victoria Desert
Antarctic Circle
A

BIOME TYPES

- Mountains
- Polar regions
- Tundra
- Tropical rainforests
- Dry woodlands
- Savanna
- Temperate grasslands
- Mediterranean
- Coniferous forests
- Temperate rainforests
- Broadleaf forests
- Cold deserts
- Hot deserts
- Wetlands

Environment Factfile

LARGEST PROTECTED AREAS

Seychelles 95%
Ecuador 44%
Kiribati 39%
Venezuela 37%
Denmark 33%

HIGHEST ANNUAL DEFORESTATION

Brazil	23,090 km²	(8936 mi²)
China	18,060 km²	(6989 mi²)
Indonesia	13,120 km²	(5077 mi²)
Sudan	9590 km²	(3711 mi²)
Zambia	8510 km²	(3293 mi²)

IDENTIFIED SPECIES

Microorganisms 5800
Invertebrates 1,021,000
Plants 322,500
Fish 19,100
Reptiles & amphibians .. 12,000
Mammals 4000

ENDANGERED SPECIES

Mammals 484
Birds 403
Reptiles 100
Amphibians 49
Fish 291
Invertebrates 763

POPULATION
over 5 000 000
over 1 000 000
over 500 000
over 100 000
over 50 000
under 50 000
LAND HEIGHT
4000m/13 124ft
3000m/9843ft
2000m/6562ft
1000m/3281ft
200m/656ft
Sea Level
Line of cross-section
RUSSIAN FEDERATION
ARCTIC
OCEAN
North Pole
Arctic Circle
Bering Sea
Bering Strait
St Lawrence I.
Aleutian Islands
Attu
Atka
Umnak
Unalaska
Unimak
Alaska Peninsula
Kodiak I.
Gulf of Alaska
ALASKA (to US)
Yukon
Brooks Range
Alaska Range
Mt McKinley (Denali) 6194m
Fairbanks
Anchorage
Dawson
YUKON TERRITORY
Mackenzie Mts
Juneau
Alexander Archipelago
Coast Mountains
Queen Charlotte Is
Vancouver I.
Vancouver
BRITISH COLUMBIA
Rocky Mountains
Beaufort Sea
Inuvik
Great Bear Lake
NORTHWEST TERRITORIES
Yellowknife
Great Slave Lake
Banks I.
Victoria I.
Queen Elizabeth Islands
Ellesmere I.
Devon I.
Somerset I.
Prince of Wales I.
Baffin Island
NUNAVUT
Southampton I.
Iqaluit
Hudson Strait
Péninsula D'ungava
GREENLAND
(Danish province)
NUUK
Baffin Bay
Davis Strait
Denmark Strait
ICELAND
ALBERTA
Athabasca
Edmonton
Calgary
Medicine Hat
CANADA
Great Plains
SASKATCHEWAN
Saskatoon
Regina
MANITOBA
Churchill
Lake Winnipeg
Winnipeg
Hudson Bay
James Bay
ONTARIO
Thunder Bay
L. Superior
Great Lakes
Sault Saint Marie
L. Huron
L. Michigan
Toronto
L. Ontario
L. Erie
OTTAWA
QUEBEC
Québec
Montréal
St Lawrence
Labrador City
Sept Iles
NEWFOUNDLAND & LABRADOR
Labrador
Newfoundland
St John's
ST PIERRE & MIQUELON (to France)
PRINCE EDWARD I.
NEW BRUNSWICK
NOVA SCOTIA
Halifax
PACIFIC
WASHINGTON
Seattle
Spokane
Portland
Cascade Range
Coast Ranges
OREGON
Great Basin
IDAHO
Boise
Snake
MONTANA
Missouri
WYOMING
NORTH DAKOTA
Prairies
SOUTH DAKOTA
MINNESOTA
St Paul
Minneapolis
WISCONSIN
Milwaukee
Chicago
Mississippi
IOWA
NEBRASKA
Omaha
Denver
COLORADO
KANSAS
Kansas City
MISSOURI
St Louis
ILLINOIS
INDIANA
OHIO
Ohio
Cincinnati
Detroit
Cleveland
Buffalo
NEW YORK
New York
Hudson
PENNSYLVANIA
Pittsburgh
Philadelphia
NEW JERSEY
DELAWARE
MARYLAND
Baltimore
WASHINGTON DC
WEST VIRGINIA
VERMONT
NEW HAMPSHIRE
MAINE
MASSACHUSETTS
Boston
RHODE ISLAND
CONNECTICUT
UNITED STATES OF AMERICA
Sacramento
San Francisco
Reno
NEVADA
Sierra Nevada
CALIFORNIA
Salt Lake City
UTAH
Las Vegas
Death Valley
Los Angeles
San Diego
Colorado
ARIZONA
Phoenix

ATLANTIC OCEAN
Gulf of Mexico
Caribbean Sea
Golfo de California
Bahía de Campeche
Península de Yucatán
Sierra Madre Occidental
Sierra Madre Oriental
Altiplano de Mexicana
Rio Grande
Mississippi
Greater Antilles
Lesser Antilles
Tropic of Cancer
Equator
MEXICO
TEXAS
LOUISIANA
MISSISSIPPI
ALABAMA
GEORGIA
SOUTH CAROLINA
FLORIDA
ARKANSAS
CUBA
BAHAMAS
JAMAICA
HAITI
DOMINICAN REPUBLIC
BELIZE
GUATEMALA
HONDURAS
EL SALVADOR
NICARAGUA
COSTA RICA
PANAMA
COLOMBIA
VENEZUELA
Hermosillo
Ciudad Juárez
El Paso
Chihuahua
Culiacán
Torreón
Monterrey
San Antonio
Dallas
Houston
Baton Rouge
New Orleans
Mobile
Birmingham
Atlanta
Charleston
Jacksonville
Tampa
Miami
NASSAU
HAVANA (La Habana)
Tampico
Aguascalientes
León
Guadalajara
Morelia
MEXICO CITY
Puebla
Acapulco
Mérida
Campeche
BELMOPAN
GUATEMALA CITY
SAN SALVADOR
TEGUCIGALPA
MANAGUA
SAN JOSÉ
PANAMA CITY
KINGSTON
PORT-AU-PRINCE
SANTO DOMINGO
SAN JUAN
BERMUDA (to UK)
TURKS & CAICOS ISLANDS (to UK)
CAYMAN ISLANDS (to UK)
NAVASSA I. (to US)
PUERTO RICO (to US)
BRITISH VIRGIN IS (to UK)
VIRGIN IS (to US)
ANGUILLA (to UK)
ST KITTS & NEVIS
ANTIGUA & BARBUDA
MONTSERRAT (to UK)
GUADELOUPE (to France)
DOMINICA
MARTINIQUE (to France)
ST LUCIA
ST VINCENT & THE GRENADINES
BARBADOS
GRENADA
TRINIDAD & TOBAGO
ARUBA (to Neth.)
NETHERLANDS ANTILLES (to Neth.)
Islas Revillagigedo (to Mexico)
Guadalupe (to Mexico)
Galapagos (to Ecuador)
Scale : 1:32 500 000
0 250 500 km
0 250 500 miles

NORTH AMERICA

NORTH AMERICA'S climate is as varied as its topography: much of Canada is snowbound or clothed in forest, its sparse population congregating along the US border. Along the continent's western flank are the spectacular Rocky Mountains. To the east lie the older, wooded Appalachians. Between these are the Great Plains – grazed by herds of livestock or sown with cereals. These plains were once home to tribes of native Americans, supplanted by incoming white settlers. The population and industry of the US are concentrated in the temperate northeast, while the drier south and west are rural and thinly populated. North America is rich in minerals and oil. Mexico has the world's largest Spanish-speaking population. Central America and the Caribbean contain some 20 countries and numerous small islands. The climate is tropical and prone to storms, the landscape mountainous and volcanic.

CROSS-SECTION THROUGH NORTH AMERICA: 43° 30'N, 126°W–65°W

8000m
4000m
Sea Level
-4000m
-8000m
Temperate Rainforest
Mountain Vegetation
Cold Desert
Mountain Vegetation
Temperate Grassland
Broadleaf Forest
PACIFIC OCEAN
Cascade Range
Great Basin
Snake
Rocky Mountains
Great Plains
Missouri
Prairies
Mississippi
Lake Michigan
Great Lakes
Lake Huron
Lake Ontario
Appalachian Mountains
Hudson
ATLANTIC OCEAN
126°W
114°W
102°W
90°W
78°W
66°W

CONTINENTAL FACTS

PHYSICAL FEATURES

- LARGEST LAKE: Lake Superior, Canada/ US 83,270 km² (32,140 mi²)
- LONGEST RIVER: Mississippi-Missouri, US 6019 km (3740 miles)
- HIGHEST POINT: Mt. McKinley (Denali), Alaska, US 6194 m (20,322 ft)
- LOWEST POINT: Death Valley, California, US 86 m (282 ft) below sea level

POLITICAL FEATURES

- TOTAL POPULATION: 482.9 million
- LARGEST CITY AND ITS POPULATION: Mexico City, Mexico 18.1 million
- COUNTRY WITH HIGHEST POPULATION DENSITY: Barbados 623 people/km² (1614 people/mi²)
- LARGEST COUNTRY: Canada 9,976,140 km² (3,851,788 mi²)
- SMALLEST COUNTRY: Grenada 340 km² (131 mi²)

SOUTH AMERICA

THE WORLD'S fourth-largest continent includes one of its most important resources – the Amazonian rainforest. It is a major source of oxygen and includes half of all known living species, while the Amazon – the world's second-longest river – contains one-fifth of the world's fresh water. The Andes mountain chain reaches down South America's western flank, sheltering the prairies of the Gran Chaco, the Pampas, and the wastes of the far south. Most South Americans are *mestizo* – of mixed European and Amerindian descent – and live in the coastal regions. Spanish is the most widely spoken language, and over 90% of South Americans are Roman Catholic. South America has massive mineral resources, many exploited by US and European multinationals.

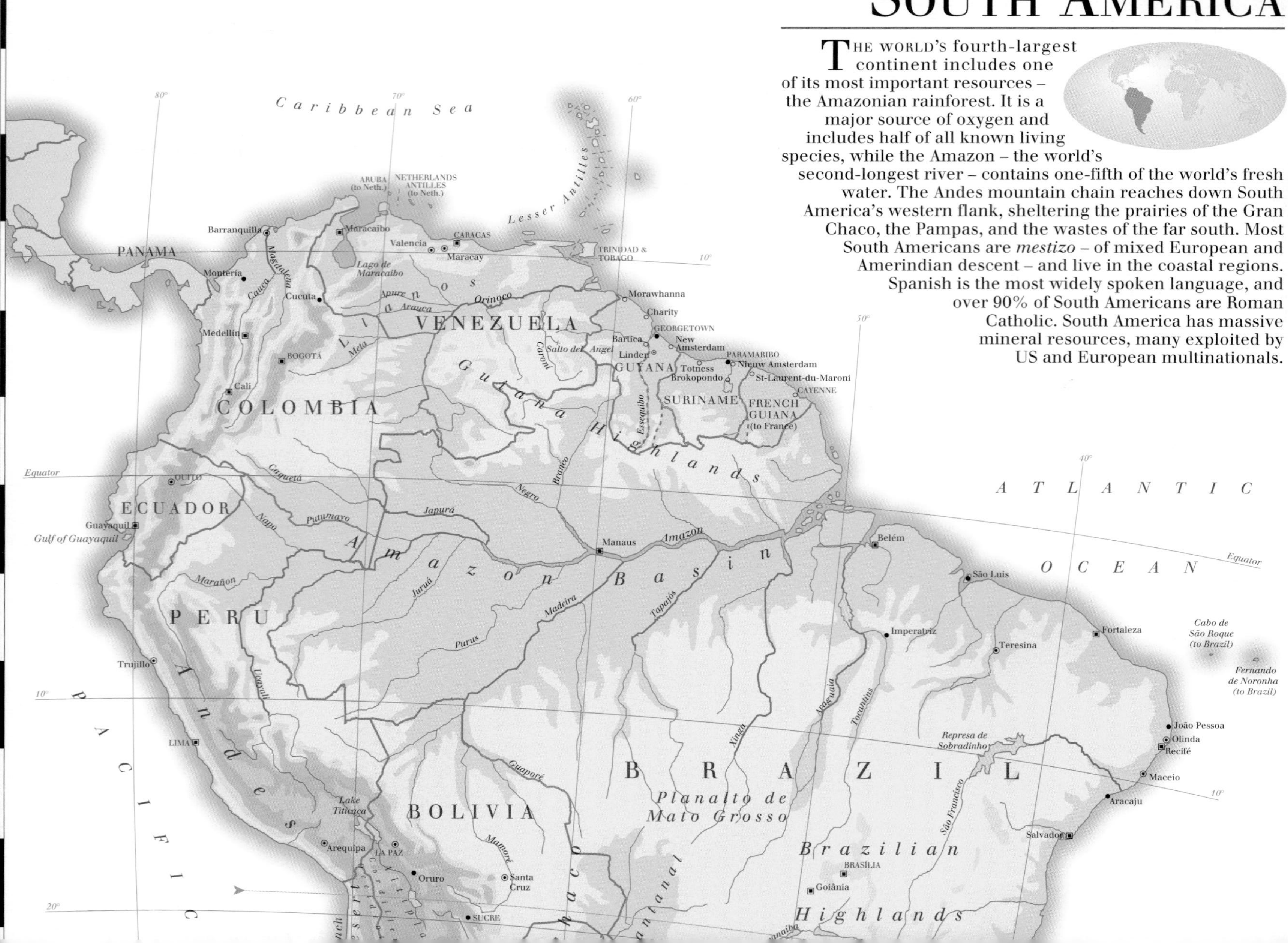

Caribbean Sea
Lesser Antilles
ARUBA (to Neth.)
NETHERLANDS ANTILLES (to Neth.)
TRINIDAD & TOBAGO
PANAMA
Barranquilla
Maracaibo
Lago de Maracaibo
Valencia
CARACAS
Maracay
Montería
Magdalena
Cauca
Cucuta
Apure
Arauca
Orinoco
Llanos
VENEZUELA
Medellín
BOGOTÁ
Meta
Cali
COLOMBIA
Caroní
Salto del Angel
Guiana Highlands
Morawhanna
Charity
GEORGETOWN
New Amsterdam
Bartica
Linden
GUYANA
Totness
Brokopondo
PARAMARIBO
Nieuw Amsterdam
St-Laurent-du-Maroni
CAYENNE
Essequibo
SURINAME
FRENCH GUIANA (to France)
Equator
QUITO
ECUADOR
Guayaquil
Gulf of Guayaquil
Caquetá
Napo
Putumayo
Japurá
Negro
Branco
Manaus
Amazon
Amazon Basin
Marañon
Juruá
Purus
Madeira
Tapajós
PERU
Trujillo
LIMA
Andes
Ucayali
PACIFIC
ATLANTIC OCEAN
Belém
São Luis
Imperatriz
Teresina
Fortaleza
Cabo de São Roque (to Brazil)
Fernando de Noronha (to Brazil)
João Pessoa
Olinda
Recifé
Maceio
Aracaju
Salvador
Represa de Sobradinho
São Francisco
Tocantins
Araguaia
Xingu
BRAZIL
Planalto de Mato Grosso
Brazilian Highlands
BRASÍLIA
Goiânia
Guaporé
BOLIVIA
Lake Titicaca
Arequipa
LA PAZ
Oruro
Mamoré
Santa Cruz
SUCRE
80°
70°
60°
50°
40°
10°
20°

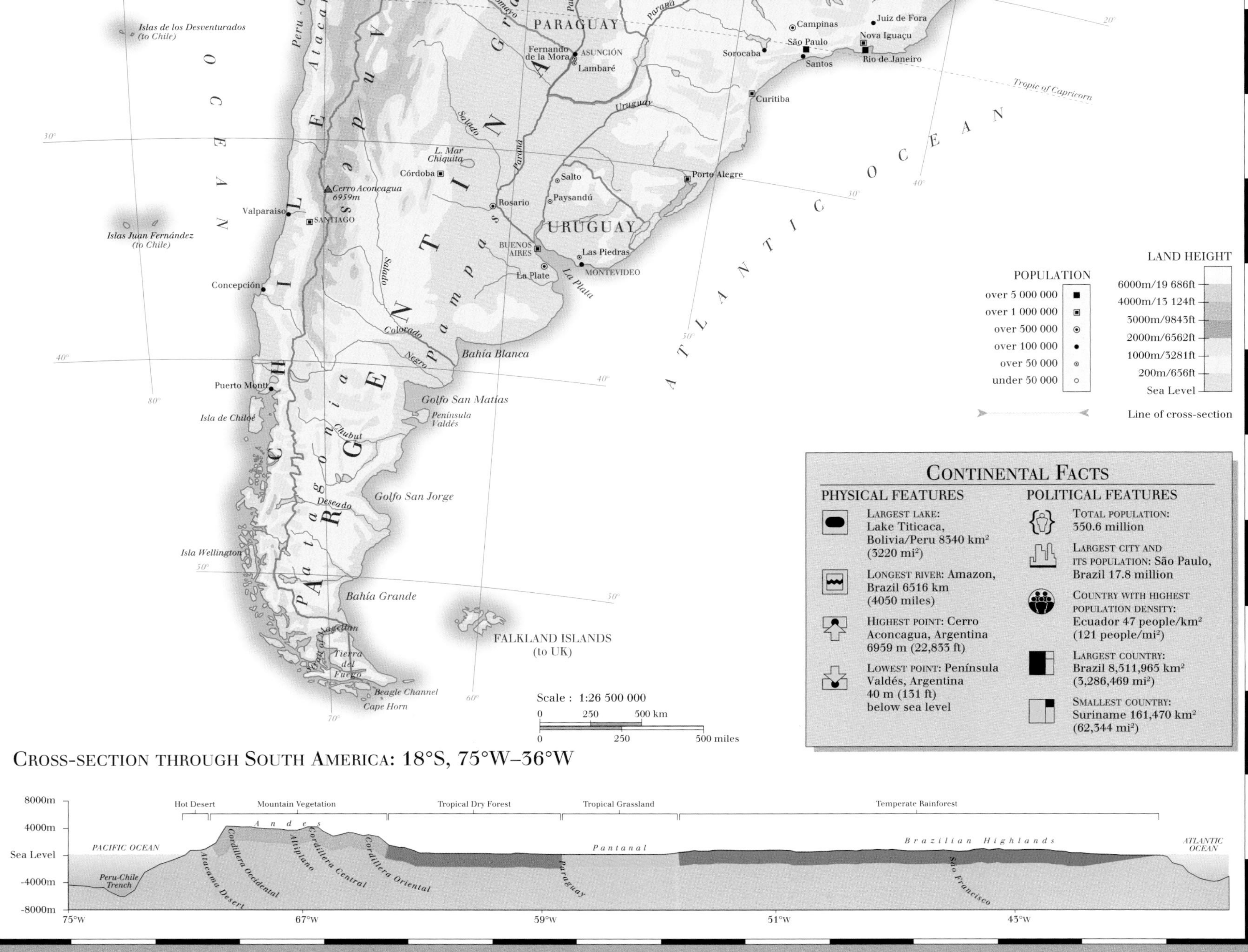
PARAGUAY
URUGUAY
ARGENTINA
CHILE
Patagonia
Pampas
Andes
Peru - Chile
Atacama
ATLANTIC OCEAN
OCEAN
Tropic of Capricorn
Islas de los Desventurados (to Chile)
Islas Juan Fernández (to Chile)
Campinas
Juiz de Fora
Nova Iguaçu
São Paulo
Sorocaba
Rio de Janeiro
Santos
Curitiba
Fernando de la Mora
ASUNCIÓN
Lambaré
Uruguay
Paraná
Salado
L. Mar Chiquita
Córdoba
Cerro Aconcagua 6959m
Valparaíso
SANTIAGO
Salto
Paysandú
Rosario
Porto Alegre
BUENOS AIRES
Las Piedras
MONTEVIDEO
La Plate
La Plata
Concepción
Colorado
Negro
Bahía Blanca
Puerto Montt
Isla de Chiloé
Golfo San Matías
Península Valdés
Chubut
Golfo San Jorge
Deseado
Isla Wellington
Bahía Grande
FALKLAND ISLANDS (to UK)
Strait of Magellan
Tierra del Fuego
Beagle Channel
Cape Horn
Scale : 1:26 500 000
0 250 500 km
0 250 500 miles
LAND HEIGHT
6000m/19 686ft
4000m/13 124ft
3000m/9843ft
2000m/6562ft
1000m/3281ft
200m/656ft
Sea Level
POPULATION
over 5 000 000
over 1 000 000
over 500 000
over 100 000
over 50 000
under 50 000
Line of cross-section
CONTINENTAL FACTS
PHYSICAL FEATURES
LARGEST LAKE: Lake Titicaca, Bolivia/Peru 8340 km² (3220 mi²)
LONGEST RIVER: Amazon, Brazil 6516 km (4050 miles)
HIGHEST POINT: Cerro Aconcagua, Argentina 6959 m (22,833 ft)
LOWEST POINT: Península Valdés, Argentina 40 m (131 ft) below sea level
POLITICAL FEATURES
TOTAL POPULATION: 350.6 million
LARGEST CITY AND ITS POPULATION: São Paulo, Brazil 17.8 million
COUNTRY WITH HIGHEST POPULATION DENSITY: Ecuador 47 people/km² (121 people/mi²)
LARGEST COUNTRY: Brazil 8,511,965 km² (3,286,469 mi²)
SMALLEST COUNTRY: Suriname 161,470 km² (62,344 mi²)
CROSS-SECTION THROUGH SOUTH AMERICA: 18°S, 75°W–36°W
8000m
4000m
Sea Level
-4000m
-8000m
Hot Desert
Mountain Vegetation
Tropical Dry Forest
Tropical Grassland
Temperate Rainforest
PACIFIC OCEAN
Peru-Chile Trench
Atacama Desert
Cordillera Occidental
Altiplano
Cordillera Central
Cordillera Oriental
Andes
Pantanal
Paraguay
Brazilian Highlands
São Francisco
ATLANTIC OCEAN
75°W
67°W
59°W
51°W
43°W

EUROPE

THE SMALLEST CONTINENT AFTER AUSTRALIA, Europe has a wide variety of climates and landscapes. The tundra of the far north gives way to a cool, wet, heavily forested region. The North European Plain is well-drained, fertile, and rich in oil, coal, and natural gas. The shores of the Mediterranean are generally warm, dry, and hilly, ideal for cultivating olives, citrus fruit, and grapes. A great curve of mountain ranges, including the Pyrenees, Alps, and Carpathians, divides north from south. To the east, the rolling plains of European Russia and the Ukraine, clad in coniferous forests or cultivated for wheat, run up to the Ural Mountains. Europeans are mainly Christian – Catholic, Orthodox, or Protestant – and speak a variety of languages, most of which spring from Latin (Romance), Germanic, or Slavic roots.

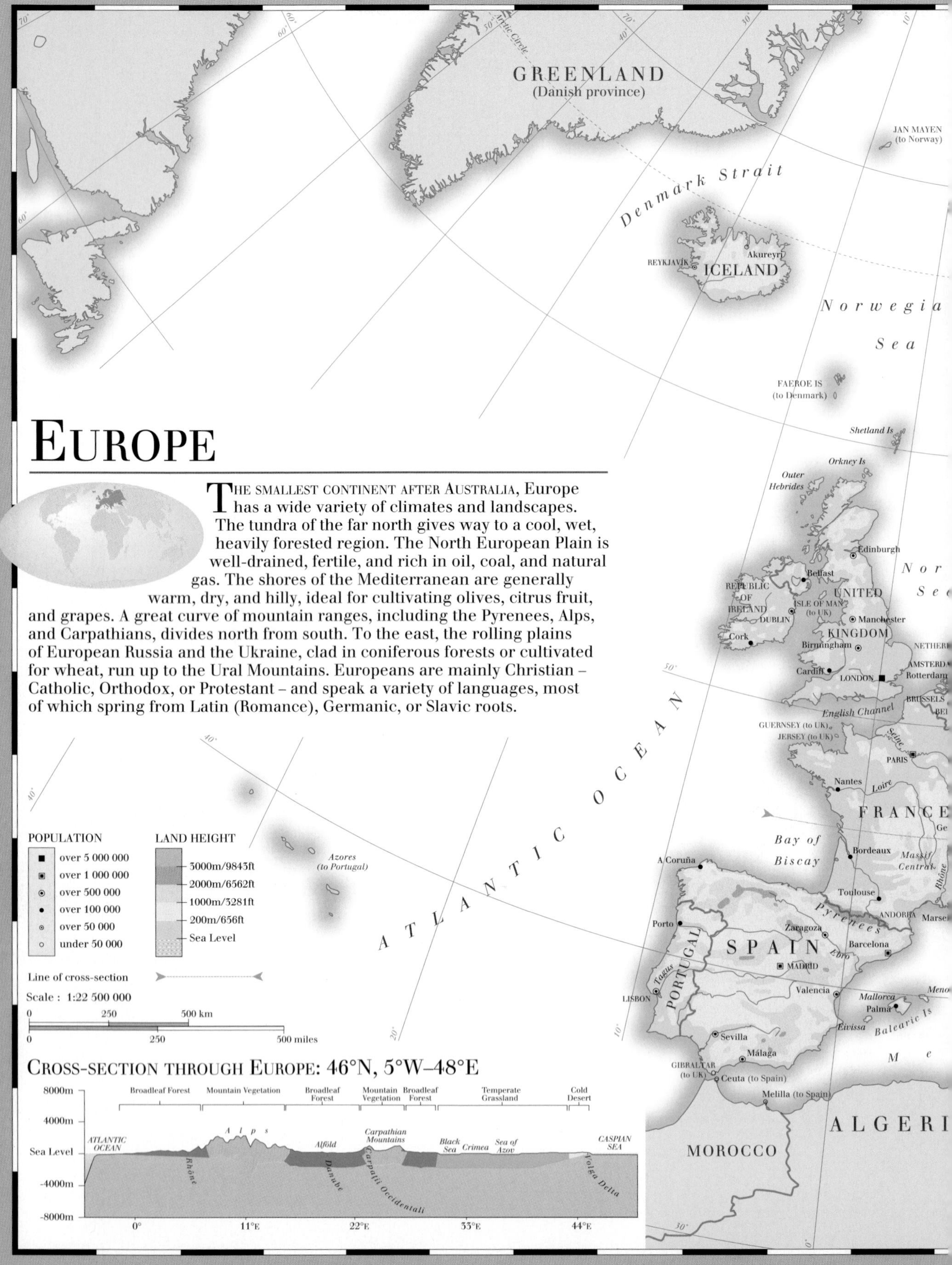
GREENLAND
(Danish province)
Arctic Circle
JAN MAYEN
(to Norway)
Denmark Strait
REYKJAVIK
Akureyri
ICELAND
Norwegian Sea
FAEROE IS
(to Denmark)
Shetland Is
Orkney Is
Outer Hebrides
Edinburgh
Belfast
REPUBLIC OF IRELAND
ISLE OF MAN (to UK)
DUBLIN
UNITED KINGDOM
Manchester
Cork
Birmingham
Cardiff
LONDON
AMSTERDAM
Rotterdam
BRUSSELS
English Channel
GUERNSEY (to UK)
JERSEY (to UK)
Seine
PARIS
Nantes
Loire
FRANCE
ATLANTIC OCEAN
Bay of Biscay
Bordeaux
Massif Central
Rhône
Toulouse
ANDORRA
Pyrenees
A Coruña
Azores
(to Portugal)
Porto
Zaragoza
Barcelona
SPAIN
Ebro
MADRID
PORTUGAL
Tagus
LISBON
Valencia
Mallorca
Palma
Eivissa
Balearic Is
Sevilla
Málaga
GIBRALTAR
(to UK)
Ceuta (to Spain)
Melilla (to Spain)
MOROCCO
ALGERI
POPULATION
over 5 000 000
over 1 000 000
over 500 000
over 100 000
over 50 000
under 50 000
LAND HEIGHT
3000m/9843ft
2000m/6562ft
1000m/3281ft
200m/656ft
Sea Level
Line of cross-section
Scale : 1:22 500 000
0
250
500 km
0
250
500 miles
CROSS-SECTION THROUGH EUROPE: 46°N, 5°W–48°E
8000m
4000m
Sea Level
-4000m
-8000m
Broadleaf Forest
Mountain Vegetation
Broadleaf Forest
Mountain Vegetation
Broadleaf Forest
Temperate Grassland
Cold Desert
ATLANTIC OCEAN
Alps
Rhône
Alföld
Danube
Carpathian Mountains
Carpatii Occidentali
Black Sea
Crimea
Sea of Azov
CASPIAN SEA
Volga Delta
0°
11°E
22°E
33°E
44°E

CONTINENTAL FACTS
PHYSICAL FEATURES
LARGEST LAKE: Ladoga, European Russia 18,390 km² (7100 mi²)
LONGEST RIVER: Volga, European Russia 3688 km (2290 miles)
HIGHEST POINT: El' brus, Caucasus Mountains, European Russia 5642 m (18,510 ft)
LOWEST POINT: Volga Delta, Caspian Sea, European Russia 28 m (92 ft) below sea level
POLITICAL FEATURES
TOTAL POPULATION: 701.9 million
COUNTRY WITH HIGHEST POPULATION DENSITY: Monaco 16,253 people/km² (42,095 people/mi²)
LARGEST CITY AND ITS POPULATION: Moscow, European Russia 9.3 million
LARGEST COUNTRY: European Russia 3,955,818 km² (1,527,341 mi²)
SMALLEST COUNTRY: Vatican City, Italy 0.44 km² (0.17 mi²)
SVALBARD (to Norway)
Novaya Zemlya
Karskoye More
Barents Sea
Arctic Circle
Pechora
Murmansk
Tromsø
Arkhangel'sk
Severnaya Dvina
Kemi
Gulf of Bothnia
Vaasa
Onezhskoye Ozero
Ladozhskoye Ozero
NORWAY
SWEDEN
FINLAND
Trondheim
Tampere
Turku
HELSINKI
Sankt Peterburg
RUSSIAN FEDERATION
Yaroslavl'
Kazan'
Nizhniy Novgorod
Ufa
Ural
Orenburg
Samara
Uppsala
Åland
OSLO
Örebro
STOCKHOLM
TALLINN
Tartu
ESTONIA
Vänern
Vättern
Kristiansand
Göteborg
Gotland
Baltic Sea
RIGA
LATVIA
MOSCOW
EUROPEAN RUSSIA
Tula
Saratov
Volga
KAZAKHSTAN
Šiauliai
Daugavpils
LITHUANIA
Orsha
DENMARK
Malmö
COPENHAGEN
KALININGRAD (to Russian Fed.)
VILNIUS
MINSK
BELARUS
Voronezh
North European Plain
Gdańsk
Homyel'
Don
Volgograd
Astrakhan'
Volga Delta
Hamburg
Elbe
BERLIN
Wisła
WARSAW
Łódź
Kharkiv
KIEV
Dnieper
GERMANY
POLAND
UKRAINE
Donets'k
Dnipropetrovs'k
Rostov-na-Donu
Caspian Sea
Frankfurt am Main
PRAGUE
Kraków
CZECH REPUBLIC
SLOVAKIA
Dniester
Carpathian Mts
Makhachkala
Sea of Azov
Krasnodar
Caucasus
Danube
Strasbourg
München
VIENNA
BRATISLAVA
Miskolc
MOLDOVA
CHIŞINĂU
Iaşi
Odesa
Crimea
El'brus 5642m
GEORGIA
TBILISI
AZERBAIJAN
Salzburg
AUSTRIA
BUDAPEST
HUNGARY
Alföld
Carpaţii Occidentali
ROMANIA
Braşov
Black Sea
ARMENIA
LJUBLJANA
SLOVENIA
ZAGREB
CROATIA
Timişoara
Novi Sad
BUCHAREST
Trabzon
Milano
Po
Torino
ITALY
SAN MARINO
BOSNIA & HERZEGOVINA
SARAJEVO
BELGRADE
SERBIA & MONTENEGRO (YUGOSLAVIA)
Danube
BULGARIA
Varna
MONACO
Corsica
Split
Podgorica
SOFIA
Plovdiv
Istanbul
ANKARA
Irmak
IRAN
Adriatic Sea
Appennino
SKOPJE
MACEDONIA
Bursa
TURKEY
ROME
VATICAN CITY
TIRANA
ALBANIA
Thessaloníki
Napoli
Tyrrhenian Sea
Gaziantep
Konya
Adana
Lárisa
GREECE
İzmir
Sardinia
Cagliari
ATHENS
Pátra
Aegean Sea
SYRIA
IRAQ
Palermo
Sicily
Ionian Sea
NICOSIA
CYPRUS
LEBANON
Iráklejo
Crete
MALTA
ISRAEL
TUNISIA
Mediterranean Sea
LIBYA
EGYPT

ATLANTIC OCEAN
INDIAN OCEAN
Mediterranean Sea
Black Sea
Red Sea
Gulf of Aden
Gulf of Guinea
SPAIN
PORTUGAL
FRANCE
ITALY
GREECE
TURKEY
SYRIA
LEBANON
ISRAEL
JORDAN
CYPRUS
MALTA
SAUDI ARABIA
YEMEN
MOROCCO
ALGERIA
TUNISIA
LIBYA
EGYPT
SUDAN
CHAD
NIGER
MALI
MAURITANIA
WESTERN SAHARA (Occupied by Morocco)
SENEGAL
GAMBIA
GUINEA-BISSAU
GUINEA
SIERRA LEONE
LIBERIA
IVORY COAST (COTE D'IVOIRE)
BURKINA
GHANA
TOGO
BENIN
NIGERIA
CAMEROON
CENTRAL AFRICAN REPUBLIC
EQUATORIAL GUINEA
GABON
CONGO
DEM. REP. CONGO
ANGOLA
UGANDA
KENYA
RWANDA
BURUNDI
TANZANIA
ETHIOPIA
ERITREA
DJIBOUTI
SOMALIA
SAO TOME & PRINCIPE
SEYCHELLES
VICTORIA
Madeira (to Portugal)
Islas Canarias (to Spain)
ASCENSION (to St Helena)
Sahara
Sahel
Atlas Mts
Ahaggar
Nubian Desert
Sudd
Ethiopian Highlands
Horn of Africa
Adamawa Highlands
Great Rift Valley
Kilimanjaro 5895m
Nile
Blue Nile
White Nile
Niger
Congo
L. Chad
Lake Volta
L. Victoria
L. Tanganyika
Lac'Assal
Tropic of Cancer
Equator
GIBRALTAR (to UK)
Tanger
Ceuta (to Spain)
Melilla (to Spain)
RABAT
Casablanca
Fez
Marrakech
Oran
ALGIERS
Constantine
TUNIS
Sfax
TRIPOLI
Banghāzī
Sabhā
Alexandria
CAIRO
Aswān
Port Sudan
KHARTOUM
El Obeid
Juba
Abéché
N'DJAMENA
Agadez
NIAMEY
Kano
ABUJA
Lagos
Ibadan
PORTO-NOVO
Port Harcourt
LOMÉ
ACCRA
OUAGADOUGOU
BAMAKO
YAMOUSSOUKRO
Abidjan
MONROVIA
FREETOWN
CONAKRY
BISSAU
BANJUL
DAKAR
NOUAKCHOTT
Nouâdhibou
Laâyoune
YAOUNDÉ
Douala
MALABO
Bata
LIBREVILLE
SÃO TOMÉ
BANGUI
BRAZZAVILLE
KINSHASA
Pointe-Noire
CABINDA (to Angola)
LUANDA
Kisangani
Kananga
Mbuji-Mayi
BUJUMBURA
KIGALI
KAMPALA
Mbale
Kisumu
Mwanza
NAIROBI
Mombasa
DODOMA
Zanzibar
Dar es Salaam
Kismaayo
MOGADISHU
Hargeysa
ADDIS ABABA
Dire Dawa
ASMARA

ST HELENA (to UK)
OCEAN
Namibe
Walvis Bay
WINDHOEK
NAMIBIA
BOTSWANA
Kalahari Desert
GABORONE
Francistown
ZIMBABWE
HARARE
Bulawayo
LUSAKA
Zambezi
Blantyre
Quelimane
Beira
MOZAMBIQUE
Mozambique Channel
Limpopo
PRETORIA
Johannesburg
MAPUTO
MBABANE
SWAZILAND
LESOTHO
MASERU
Bloemfontein
Orange
SOUTH AFRICA
Drakensberg
Durban
East London
Port Elizabeth
Cape Town
Cape of Good Hope
MADAGASCAR
Mahajanga
ANTANANARIVO
Fianarantsoa
Farafangana
MAURITIUS
PORT LOUIS
RÉUNION (to France)
Tropic of Capricorn

Scale : 1:36 000 000
0 500 1000 km
0 500 1000 miles

POPULATION
over 5 000 000
over 1 000 000
over 500 000
over 100 000
over 50 000
under 50 000

LAND HEIGHT
4000m/13 124ft
3000m/9843ft
2000m/6562ft
1000m/3281ft
200m/656ft
Sea Level

Line of cross-section

AFRICA

AFRICA IS THE SECOND-LARGEST CONTINENT after Asia. It is dominated by the Sahara in the north and the Great Rift Valley in the east. The Mediterranean climate of the extreme north and south enables cultivation of grapes and other fruit. A belt of tropical rainforest lies along the Equator, while Africa's great tropical grasslands provide grazing for herds of wild animals and domestic livestock. A narrow strip of Egypt is watered by the world's longest river, the Nile, which has sustained communities from prehistoric times. The center and south of the continent are rich in minerals. Just over one-tenth of the world's population lives in Africa – a wide variety of peoples with their own distinctive languages and cultures. Although Islam and Christianity are widespread, many Africans adhere to their own local customs and religious beliefs.

CONTINENTAL FACTS

PHYSICAL FEATURES

LARGEST LAKE: Lake Victoria, Kenya/Tanzania/Uganda 68,880 km² (26,560 mi²)

LONGEST RIVER: Nile, Uganda/Sudan/Egypt 6695 km (4160 miles)

HIGHEST POINT: Kilimanjaro, Tanzania 5895 m (19,341 ft)

LOWEST POINT: Lac' Assal, Djibouti 156 m (512 ft) below sea level

POLITICAL FEATURES

TOTAL POPULATION: 811.6 million

LARGEST CITY AND POPULATION: Lagos, Nigeria, 13.4 million

COUNTRY WITH HIGHEST POPULATION DENSITY: Mauritius 645 people/km² (1671 people/mi²)

LARGEST COUNTRY: Algeria 2,505,810 km² (967,493 mi²)

SMALLEST COUNTRY: Seychelles 455 km² (175 mi²)

CROSS-SECTION THROUGH AFRICA 7°N, 15°W–55°E

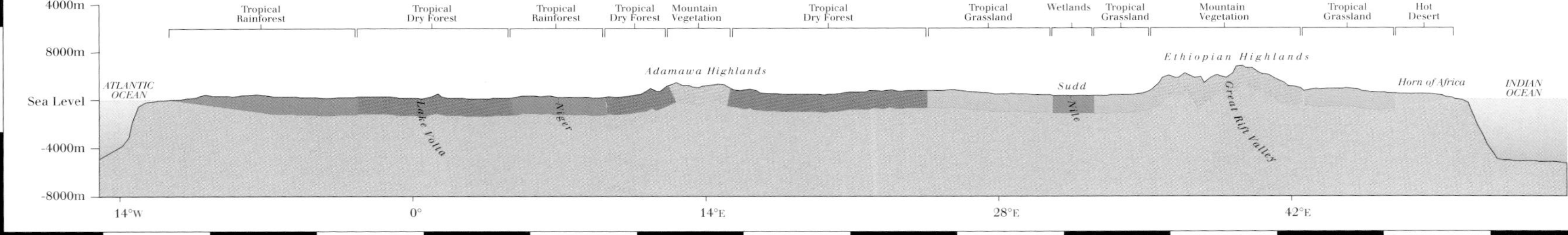
4000m
8000m
Sea Level
-4000m
-8000m
Tropical Rainforest
Tropical Dry Forest
Tropical Rainforest
Tropical Dry Forest
Mountain Vegetation
Tropical Dry Forest
Tropical Grassland
Wetlands
Tropical Grassland
Mountain Vegetation
Tropical Grassland
Hot Desert
ATLANTIC OCEAN
Lake Volta
Niger
Adamawa Highlands
Sudd
Nile
Ethiopian Highlands
Great Rift Valley
Horn of Africa
INDIAN OCEAN
14°W
0°
14°E
28°E
42°E

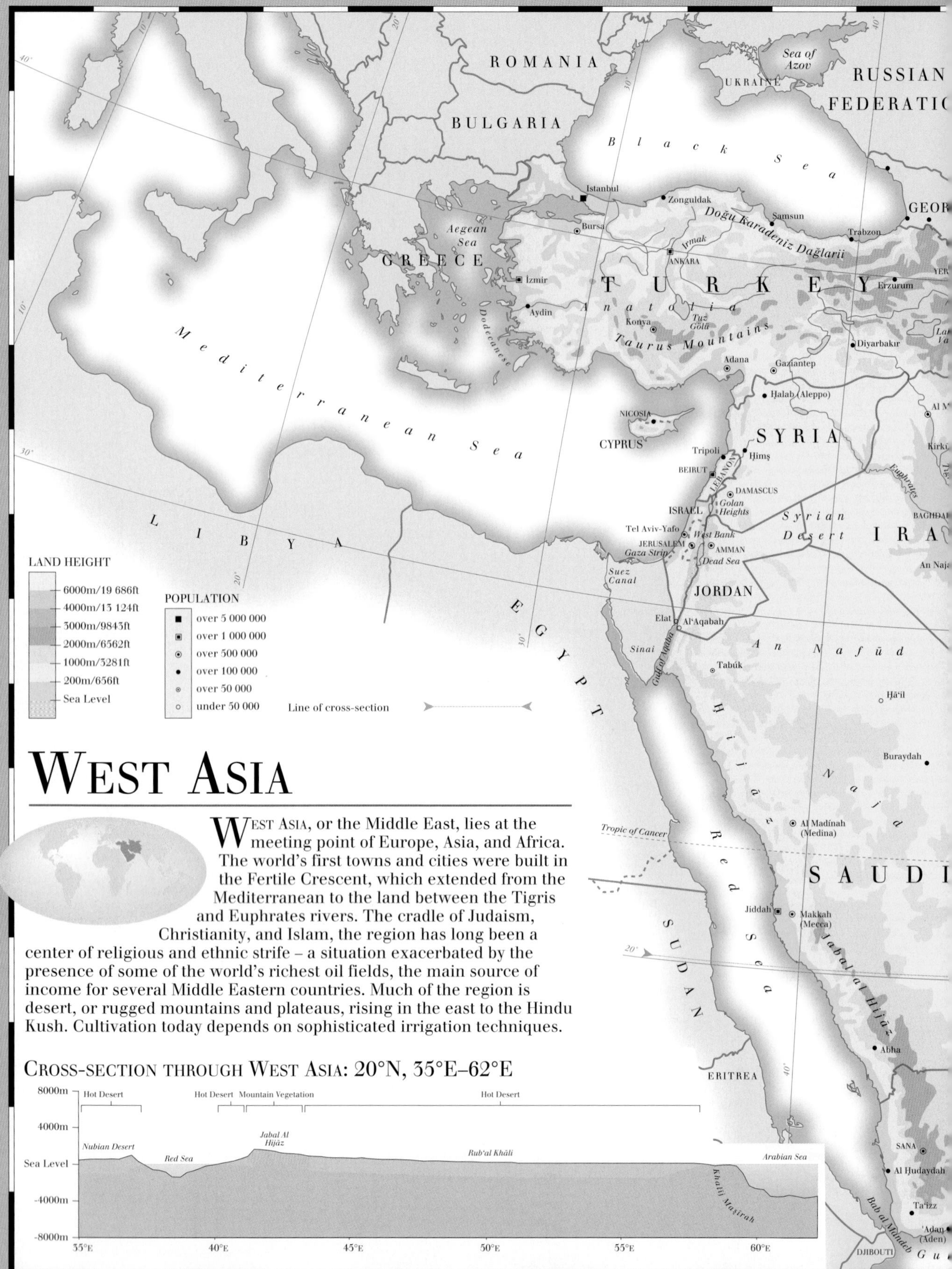
ROMANIA
BULGARIA
UKRAINE
Sea of Azov
RUSSIAN FEDERATIO
Black Sea
Istanbul
Zonguldak
Samsun
Trabzon
GEOR
Doğu Karadeniz Dağlarii
Bursa
Aegean Sea
Irmak
ANKARA
GREECE
İzmir
TURKEY
Erzurum
Aydin
Anatolia
Tuz Gölü
Konya
Dodecanese
Taurus Mountains
Diyarbakır
Mediterranean Sea
Adana
Gaziantep
Ḥalab (Aleppo)
NICOSIA
CYPRUS
SYRIA
Tripoli
Ḥimṣ
BEIRUT
LEBANON
DAMASCUS
ISRAEL
Golan Heights
Syrian Desert
Tel Aviv-Yafo
West Bank
JERUSALEM
AMMAN
Gaza Strip
Dead Sea
IRA
BAGHDA
Euphrates
Kirkū
Al M
An Naja
LIBYA
Suez Canal
JORDAN
Elat
Al'Aqabah
Sinai
Gulf of Aqaba
An Nafūd
EGYPT
Tabūk
Ḥā'il
Ḥijāz
Buraydah
Najd
Al Madīnah (Medina)
Tropic of Cancer
SAUDI
Red Sea
Jiddah
Makkah (Mecca)
SUDAN
Jabal al Ḥijāz
Abhā
ERITREA
SANA
Al Ḥudaydah
Ta'izz
'Adan (Aden)
Bab al Mandeb
DJIBOUTI
LAND HEIGHT
6000m/19 686ft
4000m/13 124ft
3000m/9843ft
2000m/6562ft
1000m/3281ft
200m/656ft
Sea Level
POPULATION
over 5 000 000
over 1 000 000
over 500 000
over 100 000
over 50 000
under 50 000
Line of cross-section

West Asia

West Asia, or the Middle East, lies at the meeting point of Europe, Asia, and Africa. The world's first towns and cities were built in the Fertile Crescent, which extended from the Mediterranean to the land between the Tigris and Euphrates rivers. The cradle of Judaism, Christianity, and Islam, the region has long been a center of religious and ethnic strife – a situation exacerbated by the presence of some of the world's richest oil fields, the main source of income for several Middle Eastern countries. Much of the region is desert, or rugged mountains and plateaus, rising in the east to the Hindu Kush. Cultivation today depends on sophisticated irrigation techniques.

Cross-section through West Asia: 20°N, 35°E–62°E

8000m
4000m
Sea Level
-4000m
-8000m
Hot Desert
Hot Desert
Mountain Vegetation
Hot Desert
Nubian Desert
Red Sea
Jabal Al Hijāz
Rub'al Khāli
Arabian Sea
Khalīj Maṣīrah
35°E
40°E
45°E
50°E
55°E
60°E

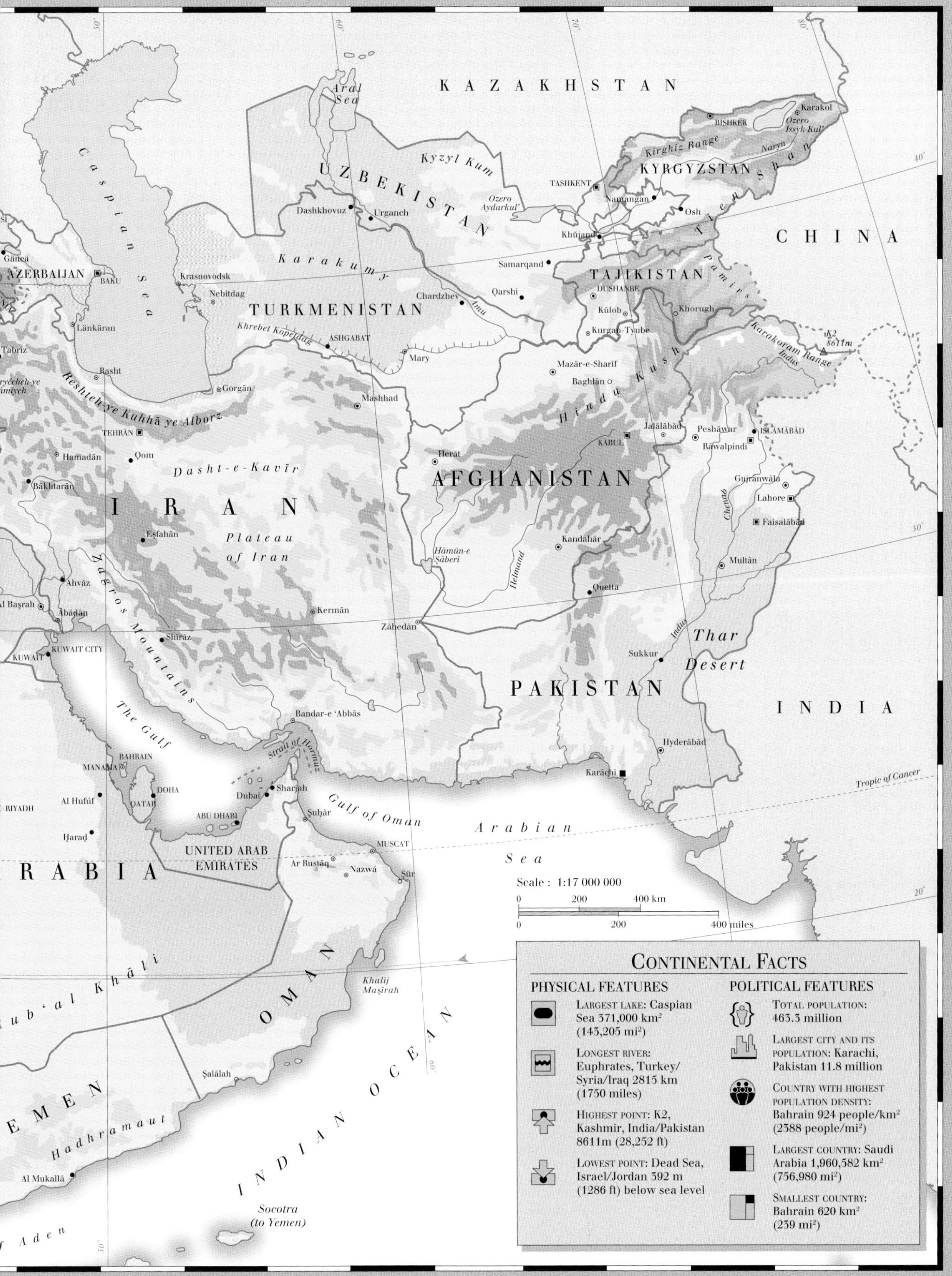
KAZAKHSTAN
Aral Sea
Caspian Sea
UZBEKISTAN
Kyzyl Kum
Ozero Aydarkul'
TASHKENT
BISHKEK
Karakol
Ozero Issyk-Kul'
Kirghiz Range
Naryn
KYRGYZSTAN
Tien Shan
Namangan
Osh
Dashkhovuz
Urganch
Khŭjand
CHINA
Samarqand
TAJIKISTAN
DUSHANBE
Pamirs
Gäncä
AZERBAIJAN
BAKU
Krasnovodsk
Nebitdag
Karakumy
TURKMENISTAN
Chardzhev
Amu
Qarshi
Kŭlob
Khorugh
Kurgan-Tyube
Länkäran
Tabriz
Khrebet Kopetdag
ASHGABAT
Mary
K2 8611m
Karakoram Range
Indus
Rasht
Mazār-e-Sharīf
Baghlān
Hindu Kush
Reshteh-ye Kuhhā ye Alborz
Gorgān
Mashhad
TEHRĀN
Jalālābād
KĀBUL
Peshāwar
ISLAMĀBĀD
Rawalpindi
Hamadān
Qom
Dasht-e-Kavīr
Herāt
AFGHANISTAN
Gujrānwāla
Lahore
Bākhtarān
IRAN
Chenab
Faisalābād
Eşfahān
Plateau of Iran
Kandahār
Hāmūn-e Şāberī
Helmand
Multān
Ahvāz
Quetta
Al Başrah
Ābādān
Kermān
Zāhedān
Zagros Mountains
Shīrāz
Thar Desert
KUWAIT
KUWAIT CITY
Sukkur
PAKISTAN
INDIA
The Gulf
Bandar-e 'Abbās
Strait of Hormuz
BAHRAIN
MANAMA
Hyderābād
Karāchi
Tropic of Cancer
DOHA
QATAR
Dubai
Sharjah
Al Hufūf
RIYADH
ABU DHABI
Şuḩār
Gulf of Oman
Arabian Sea
Ḩaraḑ
MUSCAT
UNITED ARAB EMIRATES
Ar Rustāq
Nazwá
Şūr
SAUDI ARABIA
Scale : 1:17 000 000
0 200 400 km
0 200 400 miles
Rub' al Khālī
OMAN
Khalīj Maşīrah
YEMEN
Şalālah
Hadhramaut
INDIAN OCEAN
Al Mukallā
Socotra (to Yemen)
Gulf of Aden
CONTINENTAL FACTS
PHYSICAL FEATURES
LARGEST LAKE: Caspian Sea 371,000 km² (143,205 mi²)
LONGEST RIVER: Euphrates, Turkey/Syria/Iraq 2815 km (1750 miles)
HIGHEST POINT: K2, Kashmir, India/Pakistan 8611m (28,252 ft)
LOWEST POINT: Dead Sea, Israel/Jordan 392 m (1286 ft) below sea level
POLITICAL FEATURES
TOTAL POPULATION: 463.3 million
LARGEST CITY AND ITS POPULATION: Karachi, Pakistan 11.8 million
COUNTRY WITH HIGHEST POPULATION DENSITY: Bahrain 924 people/km² (2388 people/mi²)
LARGEST COUNTRY: Saudi Arabia 1,960,582 km² (756,980 mi²)
SMALLEST COUNTRY: Bahrain 620 km² (239 mi²)

UKRAINE
EUROPEAN RUSSIA
Arctic Circle
Karskoye More
Ostrov Komsomolets
Ostrov Oktyabr'skoy Revolyutsii
Ostrov Bol'shevik
Severnaya Zemlya
Poluostrov Taymyr
Ozero Taymyr
Poluostrov Yamal
Gydanskiy Poluostrov
Severo-Sibirskaya Nizmennost'
Salekhard
Nadym
Noril'sk
Ural Mountains
Ob'
Serov
Ural'sk
Ural
Yekaterinburg
Khanty-Mansiyk
RUSSIA
ASIATIC RUSSIA
Chelyabinsk
Tyumen'
Zapadno-Sibirskaya Ravnina
Nizhnevartovsk
Yenisey
Nizhnyaya Tunguska
Sredne Sibirskoye Ploskogor'ye
Atyrau
Aktobe
Caspian Sea
Aktau
Kostanay
Irtysh
KAZAKHSTAN
Ustyurt Plateau
Ishim
Omsk
Podkamennaya Tunguska
Aral'sk
Aral Sea
Syr Darya
ASTANA
Tomsk
Novosibirsk
TURKMENISTAN
UZBEKISTAN
Zhezkazgan
Karaganda
Barnaul
Krasnoyarsk
Bratsk
Bodaybo
Lena
Kyzylorda
Rubtsovsk
Semipalatinsk
Abakan
Ozero Balkhash
Ozero Baykal
Shymkent
Taraz
Ili
Ozero Zaysan
Uvs Nuur
Kyzyl
Irkutsk
Ulan-Ude
Vitim
Taldykorgan
Ölgiy
Almaty
Hovd
Hyargas Nuur
Hövsgöl Nuur
Chita
Yablonovyy Khrebet
AFGHANISTAN
TAJIKISTAN
KYRGYZSTAN
Yining
Altai Mountains
Erdenet
Darhan
Kashi
Tien Shan
Ürümqi
Altay
ULAN BATOR
Choybalsan
Kerulen
Tarim He
Turpan Hami
MONGOLIA
PAKISTAN
Hami
Bayanhongor
Taklimakan Shamo
Hotan
Saynshand
Aksai Chin
Gobi
Kunlun Shan
Altun Shan
Yumen
Qilian Shan
Qaidam Pendi
Huang He
Baotou
Datong
BEIJING (PEKING)
Qing Zang Gaoyuan
Golmud
Xining
Tianjin
INDIA
Himalaya
Qinghai Hu
Lanzhou
Taiyuan
Shijiazhuang
TIBET
CHINA
Mekong
Salween
Xi'an
Huang He (Yellow R.)
Zhengzhou
NEPAL
Brahmaputra
Xixabangma Feng 8012m
Lhasa
Tropic of Cancer
BHUTAN
INDIA
Chengdu
Nanjing
Chang Jiang (Yangtze)
Wuhan
Chongqing
Dongting Hu
Poyang Hu
Changsha
Nanchang
Guiyang
Kunming
Xi Jiang
Fuzhou
BURMA (MYANMAR)
Shantou
Nanning
Guangzhou
Hong Kong (Xianggang)
Macao (Aomen)
VIETNAM
LAOS
Gulf of Tongking
Haikou
Hainan Dao
South China Sea

Continental Facts

PHYSICAL FEATURES

Largest lake: Aral Sea, Kazakhstan/Uzbekistan 66,500 km² (25,700 mi²)

Longest river: Chang Jiang (Yangtze), China 6380 km (3965 miles)

Highest point: Xixabangma Feng, China 8012 m (26,286 ft)

Lowest point: Turpan Hami (Turfan Basin), China 154 m (505 ft) below sea level

POLITICAL FEATURES

Total population: 1552 million

Largest city and its population: Tokyo, Japan 26.4 million

Country with highest population density: Taiwan 688 people/km² (1782 people/mi²)

Largest country: Asiatic Russia 13,119,582 km² (5,065,471 mi²)

Smallest country: Taiwan 32,260 km² (12,456 mi²)

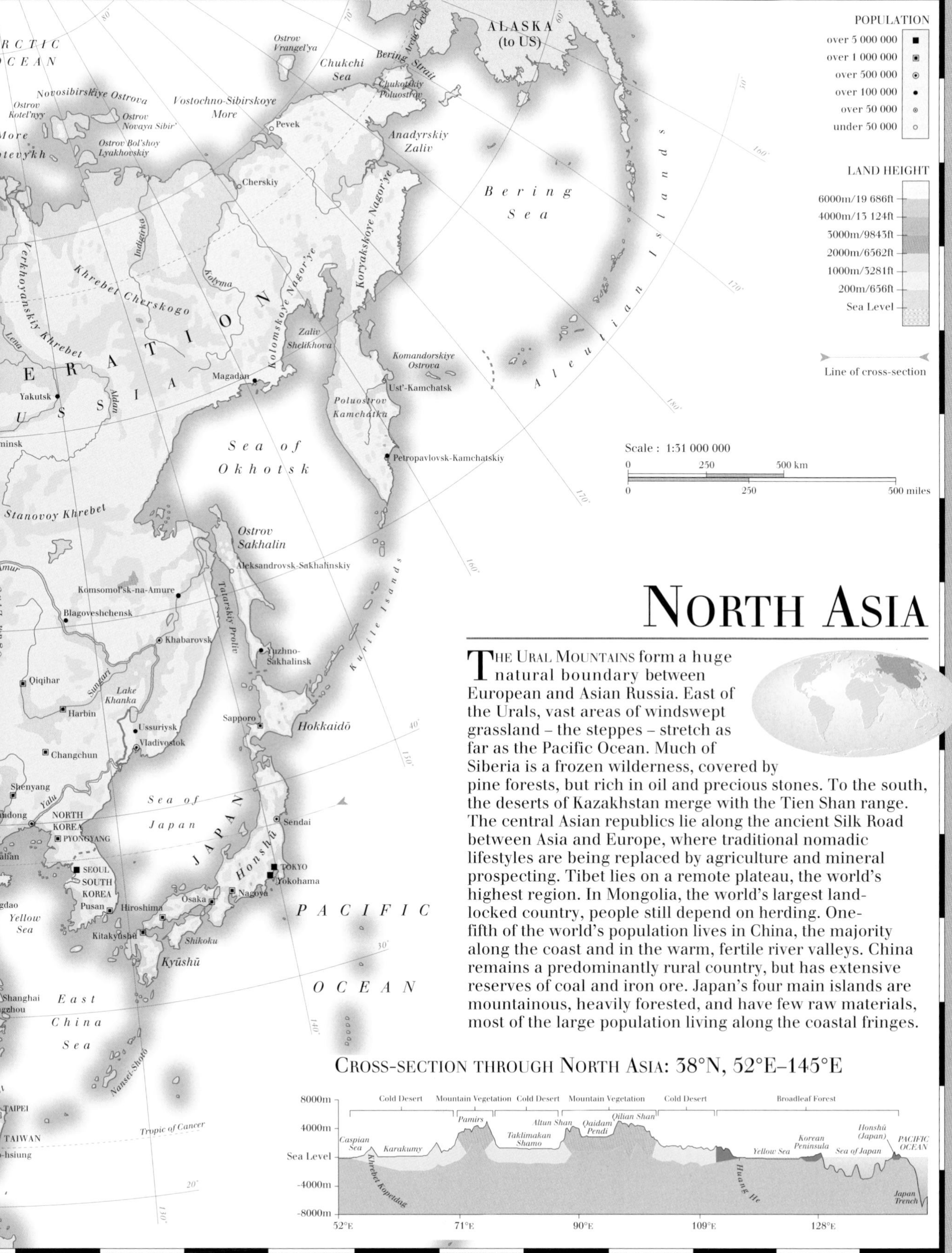
POPULATION
over 5 000 000
over 1 000 000
over 500 000
over 100 000
over 50 000
under 50 000
LAND HEIGHT
6000m/19 686ft
4000m/13 124ft
3000m/9843ft
2000m/6562ft
1000m/3281ft
200m/656ft
Sea Level
Line of cross-section
Scale : 1:31 000 000
0 250 500 km
0 250 500 miles
ALASKA (to US)
Ostrov Vrangel'ya
Chukchi Sea
Bering Strait
Arctic Circle
Chukotskiy Poluostrov
Novosibirskiye Ostrova
Ostrov Kotel'nyy
Ostrov Novaya Sibir'
Vostochno-Sibirskoye More
Ostrov Bol'shoy Lyakhovskiy
Pevek
Anadyrskiy Zaliv
Bering Sea
Cherskiy
Koryakskoye Nagor'ye
Aleutian Islands
Indigirka
Kolyma
Verkhoyanskiy Khrebet
Khrebet Cherskogo
Kolomskoye Nagor'ye
Zaliv Shelikhova
Komandorskiye Ostrova
Ust'-Kamchatsk
Lena
Aldan
Magadan
Yakutsk
Poluostrov Kamchatka
Sea of Okhotsk
Petropavlovsk-Kamchatskiy
Stanovoy Khrebet
Ostrov Sakhalin
Amur
Aleksandrovsk-Sakhalinskiy
Komsomol'sk-na-Amure
Blagoveshchensk
Tatarskiy Proliv
Khabarovsk
Yuzhno-Sakhalinsk
Kurile Islands
Qiqihar
Sungari
Lake Khanka
Harbin
Ussuriysk
Vladivostok
Sapporo
Hokkaidō
Changchun
Shenyang
Yalu
Sea of Japan
JAPAN
Sendai
NORTH KOREA
PYONGYANG
SEOUL
SOUTH KOREA
Pusan
Honshū
TOKYO
Yokohama
Nagoya
Osaka
Hiroshima
Kitakyūshū
Shikoku
Kyūshū
Yellow Sea
PACIFIC OCEAN
Shanghai
East China Sea
Nansei-Shotō
Tropic of Cancer
TAIPEI
TAIWAN

NORTH ASIA

The Ural Mountains form a huge natural boundary between European and Asian Russia. East of the Urals, vast areas of windswept grassland – the steppes – stretch as far as the Pacific Ocean. Much of Siberia is a frozen wilderness, covered by pine forests, but rich in oil and precious stones. To the south, the deserts of Kazakhstan merge with the Tien Shan range. The central Asian republics lie along the ancient Silk Road between Asia and Europe, where traditional nomadic lifestyles are being replaced by agriculture and mineral prospecting. Tibet lies on a remote plateau, the world's highest region. In Mongolia, the world's largest land-locked country, people still depend on herding. One-fifth of the world's population lives in China, the majority along the coast and in the warm, fertile river valleys. China remains a predominantly rural country, but has extensive reserves of coal and iron ore. Japan's four main islands are mountainous, heavily forested, and have few raw materials, most of the large population living along the coastal fringes.

CROSS-SECTION THROUGH NORTH ASIA: 38°N, 52°E–145°E

Cold Desert
Mountain Vegetation
Cold Desert
Mountain Vegetation
Cold Desert
Broadleaf Forest
8000m
4000m
Sea Level
-4000m
-8000m
Caspian Sea
Karakumy
Khrebet Kopetdag
Pamirs
Taklimakan Shamo
Altun Shan
Qaidam Pendi
Qilian Shan
Huang He
Yellow Sea
Korean Peninsula
Sea of Japan
Honshū (Japan)
PACIFIC OCEAN
Japan Trench
52°E
71°E
90°E
109°E
128°E

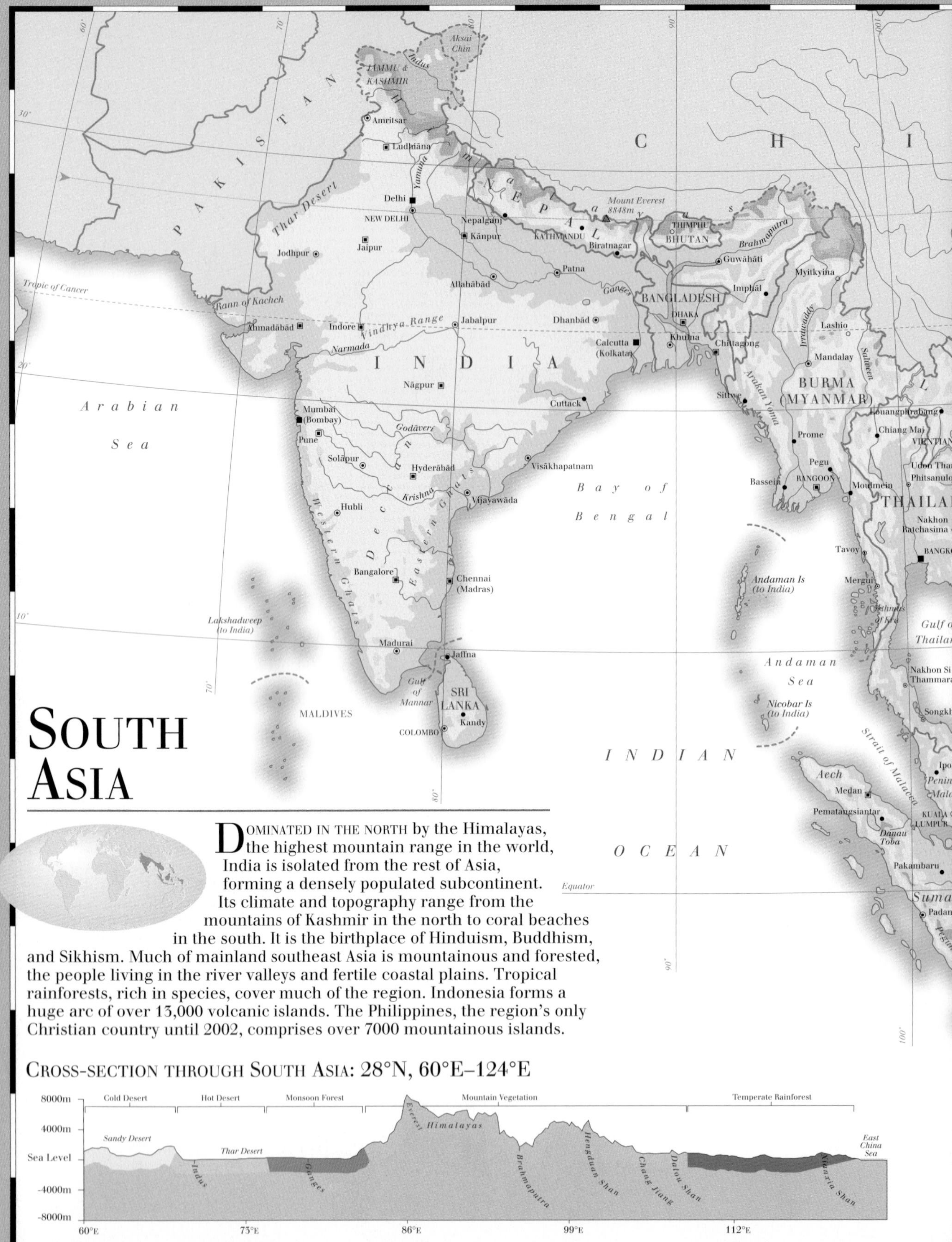

PAKISTAN
JAMMU & KASHMIR
Aksai Chin
Indus
Amritsar
Ludhiāna
Thar Desert
Delhi
NEW DELHI
Yamuna
NEPAL
Nepalganj
Kānpur
KATHMANDU
Mount Everest 8848m
Biratnagar
THIMPHU
BHUTAN
CHINA
Brahmaputra
Guwāhāti
Jodhpur
Jaipur
Allahābād
Patna
Ganges
Tropic of Cancer
Rann of Kachch
Ahmadābād
Indore
Vindhya Range
Jabalpur
Dhanbād
BANGLADESH
DHAKA
Khulna
Chittagong
Imphāl
Myitkyina
Irrawaddy
Lashio
Narmada
INDIA
Calcutta (Kolkata)
Mandalay
Salween
BURMA (MYANMAR)
Nāgpur
Cuttack
Sittwe
Arakan Yoma
Arabian Sea
Mumbai (Bombay)
Pune
Godāvari
Solāpur
Hyderābād
Deccan
Visākhapatnam
Bay of Bengal
Prome
Chiang Mai
VIENTIANE
Pegu
RANGOON
Bassein
Moulmein
Krishna
Vijayawāda
Hubli
Western Ghats
Eastern Ghats
Bangalore
Chennai (Madras)
Tavoy
BANGKOK
Andaman Is (to India)
Mergui
Isthmus of Kra
Gulf of Thailand
Lakshadweep (to India)
Madurai
Jaffna
Andaman Sea
Nakhon Si Thammarat
Gulf of Mannar
SRI LANKA
Kandy
COLOMBO
MALDIVES
Nicobar Is (to India)
Songkhla
Strait of Malacca
INDIAN OCEAN
Aceh
Medan
Ipoh
Pematangsiantar
Danau Toba
KUALA LUMPUR
Pakanbaru
Equator
Padang
Sumatra

SOUTH ASIA

DOMINATED IN THE NORTH by the Himalayas, the highest mountain range in the world, India is isolated from the rest of Asia, forming a densely populated subcontinent. Its climate and topography range from the mountains of Kashmir in the north to coral beaches in the south. It is the birthplace of Hinduism, Buddhism, and Sikhism. Much of mainland southeast Asia is mountainous and forested, the people living in the river valleys and fertile coastal plains. Tropical rainforests, rich in species, cover much of the region. Indonesia forms a huge arc of over 13,000 volcanic islands. The Philippines, the region's only Christian country until 2002, comprises over 7000 mountainous islands.

CROSS-SECTION THROUGH SOUTH ASIA: 28°N, 60°E–124°E

8000m
4000m
Sea Level
-4000m
-8000m
Cold Desert
Hot Desert
Monsoon Forest
Mountain Vegetation
Temperate Rainforest
Sandy Desert
Thar Desert
Indus
Ganges
Everest
Himalayas
Brahmaputra
Hengduan Shan
Chang Jiang
Dalou Shan
Xianxia Shan
East China Sea
60°E
75°E
86°E
99°E
112°E

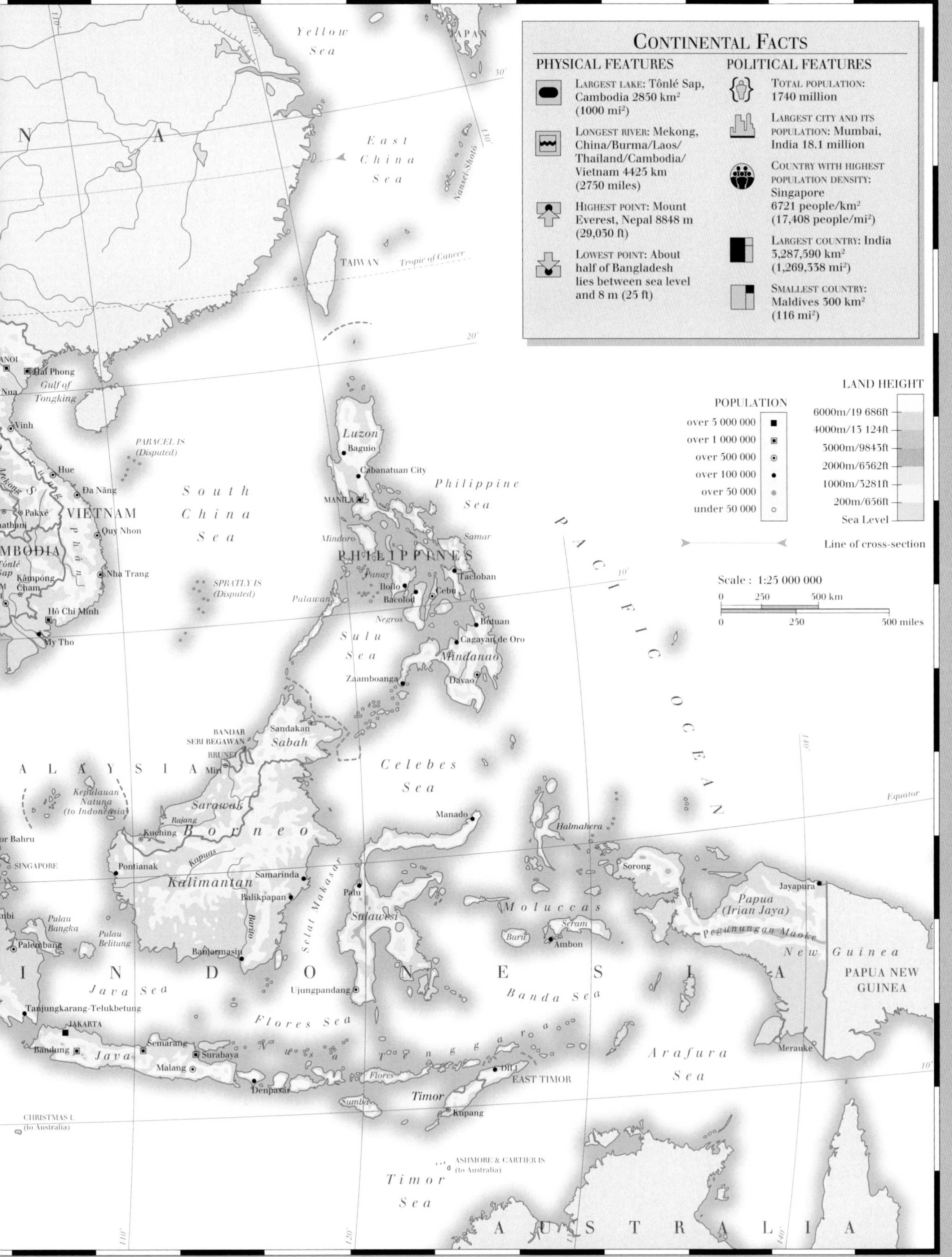
CONTINENTAL FACTS
PHYSICAL FEATURES
LARGEST LAKE: Tônlé Sap, Cambodia 2850 km² (1000 mi²)
LONGEST RIVER: Mekong, China/Burma/Laos/Thailand/Cambodia/Vietnam 4425 km (2750 miles)
HIGHEST POINT: Mount Everest, Nepal 8848 m (29,030 ft)
LOWEST POINT: About half of Bangladesh lies between sea level and 8 m (25 ft)
POLITICAL FEATURES
TOTAL POPULATION: 1740 million
LARGEST CITY AND ITS POPULATION: Mumbai, India 18.1 million
COUNTRY WITH HIGHEST POPULATION DENSITY: Singapore 6721 people/km² (17,408 people/mi²)
LARGEST COUNTRY: India 3,287,590 km² (1,269,338 mi²)
SMALLEST COUNTRY: Maldives 300 km² (116 mi²)
POPULATION
over 5 000 000
over 1 000 000
over 500 000
over 100 000
over 50 000
under 50 000
LAND HEIGHT
6000m/19 686ft
4000m/13 124ft
3000m/9843ft
2000m/6562ft
1000m/3281ft
200m/656ft
Sea Level
Line of cross-section
Scale : 1:25 000 000
0 250 500 km
0 250 500 miles
Yellow Sea
JAPAN
East China Sea
Nansei-Shoto
TAIWAN
Tropic of Cancer
Hai Phong
Gulf of Tongking
Vinh
Hue
Đa Nãng
PARACEL IS (Disputed)
VIETNAM
Pakxé
Quy Nhon
MBODIA
Nha Trang
Kâmpóng Cham
Hô Chi Minh
My Tho
South China Sea
SPRATLY IS (Disputed)
Luzon
Baguio
Cabanatuan City
MANILA
Philippine Sea
Mindoro
Samar
PHILIPPINES
Panay
Iloilo
Bacolod
Cebu
Tacloban
Palawan
Negros
Butuan
Cagayan de Oro
Sulu Sea
Mindanao
Zamboanga
Davao
PACIFIC OCEAN
Sandakan
Sabah
BANDAR SERI BEGAWAN
BRUNEI
Miri
Celebes Sea
Kepulauan Natuna (to Indonesia)
Sarawak
Rajang
Kuching
Borneo
Manado
Halmahera
SINGAPORE
Pontianak
Kapuas
Kalimantan
Samarinda
Balikpapan
Selat Makasar
Palu
Sulawesi
Sorong
Jayapura
Papua (Irian Jaya)
Moluccas
Seram
Buru
Ambon
Pegunungan Maoke
Pulau Bangka
Pulau Belitung
Palembang
Barito
Banjarmasin
New Guinea
INDONESIA
PAPUA NEW GUINEA
Java Sea
Ujungpandang
Banda Sea
Tanjungkarang-Telukbetung
Flores Sea
JAKARTA
Tenggara
Bandung
Java
Semarang
Surabaya
Malang
Flores
DILI
EAST TIMOR
Arafura Sea
Merauke
Denpasar
Timor
Sumba
Kupang
CHRISTMAS I. (to Australia)
ASHMORE & CARTIER IS (to Australia)
Timor Sea
AUSTRALIA
Equator

Tropic of Cancer
TAIWAN
NORTHERN MARIANA ISLANDS (to US)
WAKE I. (to US)
P A C I F I C
South China Sea
Philippine Sea
Saipan
GUAM (to US)
MARSHALL ISLANDS
Ratak Chain
Ralik Chain
P H I L I P P I N E S
Yap
Pohnpei
KOLONIA
KOROR
Babelthuap
Chuuk Is
Kosrae
Majuro
Sulu Sea
MALAYSIA
BRUNEI
PALAU
M I C R O N E S I A
BAIRIKI
Tarawa
Celebes Sea
Bismarck Archipelago
NAURU
Equator
Bismarck Sea
Rabaul
New Britain
TUVALU
Nanumea
I N D O N E S I A
Mt Wilhelm 4509m
PAPUA NEW GUINEA
New Guinea
Bougainville
SOLOMON ISLANDS
Santa Isabel
Malaita
Nukufetau
FONGAFALE
Solomon Sea
Guadalcanal
HONIARA
Santa Cruz Is
Nukulae
PORT MORESBY
Banda Sea
San Cristobal
Arafura Sea
Torres Strait
Rennell
EAST TIMOR
Timor Sea
Darwin
Gulf of Carpentaria
Coral Sea
VANUATU
Espiritu Santo
Vanua L
Malekula
Viti Levu
ASHMORE & CARTIER ISLANDS (to Australia)
CORAL SEA ISLANDS (to Australia)
PORT-VILA
Éfaté
SUVA
Cairns
NEW CALEDONIA (to France)
Great Barrier Reef
INDIAN OCEAN
Townsville
Îles Loyauté
FIJI
New Caledonia
NOUMÉA
MacKay
NORTHERN TERRITORY
Broome
Great Sandy Desert
Rockhampton
QUEENSLAND
A U S T R A L I A
Great Dividing Range
Alice Springs
Brisbane
Gold Coast
Toowoomba
NORFOLK ISLAND (to Australia)
Simpson Desert
Gibson Desert
Musgrave Ranges
Lake Eyre
Grey Range
Lord Howe I. (to Australia)
Ball's Pyramid (to Australia)
WESTERN AUSTRALIA
Great Victoria Desert
Darling
Tropic of Capricorn
SOUTH AUSTRALIA
Lake Torrens
NEW SOUTH WALES
Newcastle
Sydney
Wollongong
Murray
CANBERRA
Auckland
Hamilton
Geraldton
Kalgoorlie
Great Australian Bight
Adelaide
Port Lincoln
VICTORIA
Tasman Sea
Bendigo
Melbourne
Geelong
AUSTRALIAN CAPITAL TERRITORY
NEW ZEALAND
WELLINGTON
Perth
Esperance
Bunbury
Bass Strait
Albany
Launceston
Christchurch
TASMANIA
Hobart
Dunedin
Bounty Islands (to NZ)
Antipodes Islands (to NZ)
Auckland Islands (to NZ)
LAND HEIGHT
3000m/9843ft
2000m/6562ft
1000m/3281ft
200m/656ft
Sea Level
POPULATION
over 5 000 000
over 1 000 000
over 500 000
over 100 000
over 50 000
under 50 000
Scale : 1:40 000 000
0
500
1000 km
0
500
1000 miles

AUSTRALASIA & OCEANIA

OCEANIA, A CONTINENT OF ISLANDS stretching across a vast area of the Pacific Ocean, is home to only 0.5 per cent of the world's population. Dominated by Australia, it includes few other countries with significant land mass apart from New Zealand, Papua New Guinea, and Fiji, but a myriad volcanic and coral islands in three main groups, Micronesia, Melanesia, and Polynesia. Australia, flat and dry, is sparsely populated, most people living along the coastal lowlands, especially in the southeast. Its first inhabitants, the Aboriginal peoples, retain some of their original lands in the interior, but the European and Asian settlers of recent centuries form most of the population. Australia is rich in minerals, such as gold, uranium, and iron ore, which are the basis of its prosperity. Mountainous Papua New Guinea is covered in tropical rainforest, while New Zealand is temperate, rugged, and volcanic in the north. Owing to their isolation, these countries' flora and fauna have evolved many unique species. The peoples of Oceania colonized the Pacific by 1100 CE, and the many insular farming and fishing communities have developed distinctive cultures, the Maoris of New Zealand being among the most notable.

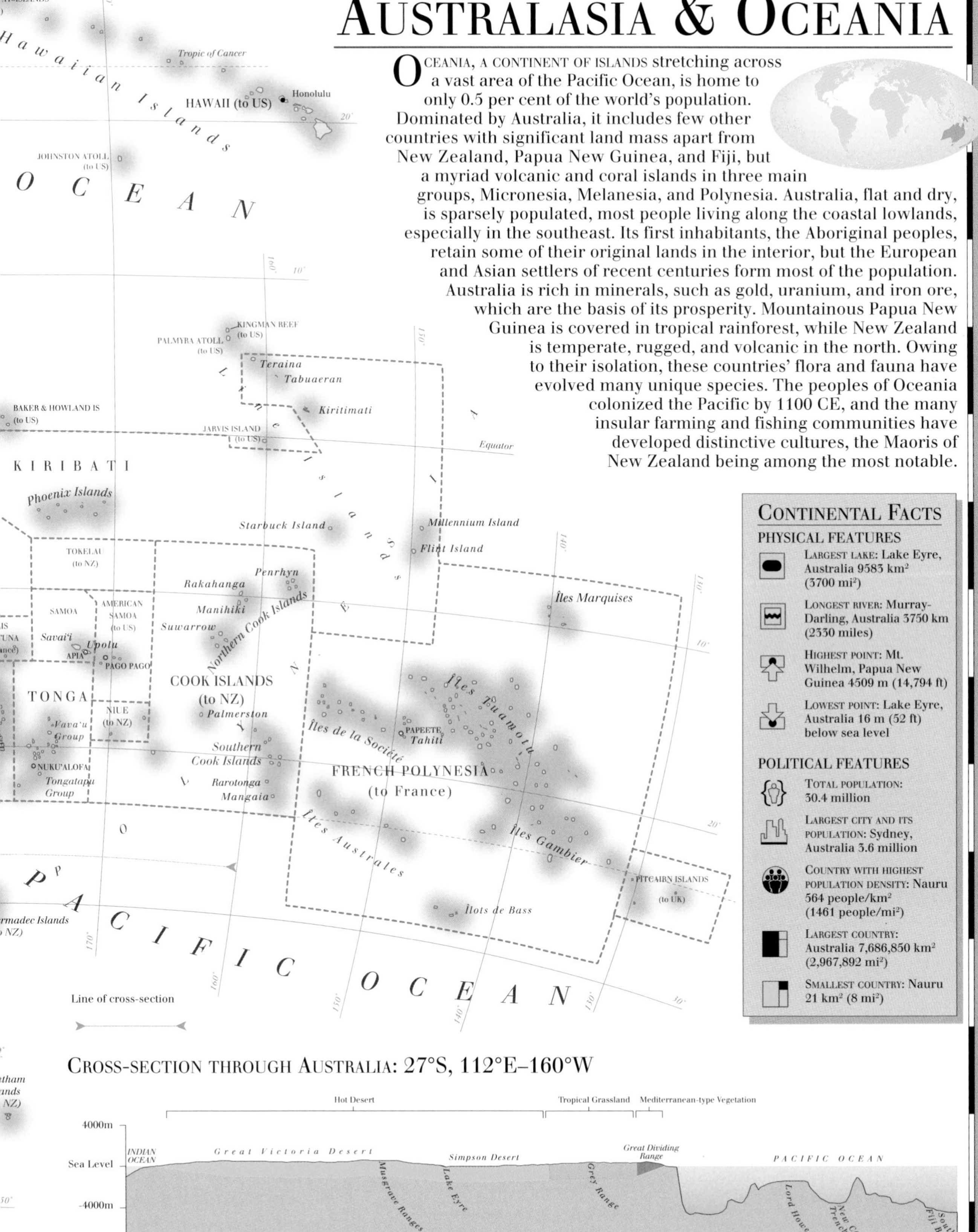
Hawaiian Islands
Tropic of Cancer
HAWAII (to US)
Honolulu
JOHNSTON ATOLL (to US)
OCEAN
KINGMAN REEF (to US)
PALMYRA ATOLL (to US)
Teraina
Tabuaeran
Kiritimati
Line Islands
BAKER & HOWLAND IS (to US)
JARVIS ISLAND (to US)
Equator
KIRIBATI
Phoenix Islands
Starbuck Island
Millennium Island
Flint Island
TOKELAU (to NZ)
Penrhyn
Rakahanga
Manihiki
Northern Cook Islands
Suwarrow
Îles Marquises
SAMOA
AMERICAN SAMOA (to US)
Savai'i
Upolu
APIA
PAGO PAGO
TONGA
NIUE (to NZ)
Vava'u Group
NUKU'ALOFA
Tongatapu Group
COOK ISLANDS (to NZ)
Palmerston
Southern Cook Islands
Rarotonga
Mangaia
Îles Tuamotu
Îles de la Société
PAPEETE
Tahiti
FRENCH POLYNESIA (to France)
Îles Australes
Îles Gambier
PITCAIRN ISLANDS (to UK)
Îlots de Bass
PACIFIC OCEAN
Line of cross-section

CONTINENTAL FACTS

PHYSICAL FEATURES

LARGEST LAKE: Lake Eyre, Australia 9583 km² (3700 mi²)

LONGEST RIVER: Murray-Darling, Australia 3750 km (2330 miles)

HIGHEST POINT: Mt. Wilhelm, Papua New Guinea 4509 m (14,794 ft)

LOWEST POINT: Lake Eyre, Australia 16 m (52 ft) below sea level

POLITICAL FEATURES

TOTAL POPULATION: 30.4 million

LARGEST CITY AND ITS POPULATION: Sydney, Australia 3.6 million

COUNTRY WITH HIGHEST POPULATION DENSITY: Nauru 564 people/km² (1461 people/mi²)

LARGEST COUNTRY: Australia 7,686,850 km² (2,967,892 mi²)

SMALLEST COUNTRY: Nauru 21 km² (8 mi²)

CROSS-SECTION THROUGH AUSTRALIA: 27°S, 112°E–160°W

Hot Desert
Tropical Grassland
Mediterranean-type Vegetation
4000m
Sea Level
-4000m
-8000m
INDIAN OCEAN
Great Victoria Desert
Musgrave Ranges
Lake Eyre
Simpson Desert
Grey Range
Great Dividing Range
PACIFIC OCEAN
Lord Howe Rise
New Caledonia Trench
South Fiji Basin
112°E
125°E
138°E
151°E
164°E

Chronology of World History

THIS TABLE PRESENTS A SUMMARY of the world's crucial historical events, from the first evidence of settlement and agriculture until 2000 CE. Each of the six columns is shaded a different color, with each color representing a particular continent. Reading across the columns, one can follow the development of cultures across the major landmasses of the world. By reading downwards, each continent's particular cultural history can be seen, from its first steps toward civilization, through periods of migration, empire, and revolution, to its involvement in the global wars and political diplomacy of the late 20th century.

North America

- **15,000 BCE** Evidence of human settlement in North America
- **4000 BCE** Earliest cultivation of corn in Central America
- **100 CE** Teotihuacan becomes capital of largest state in Mesoamerica
- **c.300 CE** Start of classic Mayan civilization in Yucatan
- **900** Toltecs rise to power after Teotihuacan and Mayan states collapse
- **c.1000** Vikings colonize Greenland and discover America (Vinland)
- **1200** Aztecs enter Valley of Mexico
- **1325** Tenochtitlan founded by Aztecs
- **1492** Columbus reaches Caribbean
- **c.1500** Inuit peoples found throughout Arctic region
- **1502** Introduction of African slaves to Caribbean
- **1519** Cortes begins conquest of Aztec empire
- **1565** First African slaves arrive on mainland North America
- **1607** First permanent English settlement in North America (Jamestown, Virginia)
- **1608** French colonists found Québec
- **1620** Puritans on *Mayflower* land in New England
- **1759** British capture Québec
- **1776** American Declaration of Independence
- **1789** George Washington becomes first president of US
- **1791** Revolution in Haiti
- **1803** Louisiana Purchase nearly doubles size of US
- **1810** Revolution in Mexico
- **1819** US buys Florida from Spain
- **1821** Mexico gains independence
- **1828** Federalist–Centralist wars in Mexico (to 1859)
- **1845** Texas annexed by US
- **1846** US–Mexican War (to 1848)
- **1848** Californian Gold Rush

South America

- **c.20,000 BCE** First settlers arrive
- **c.11,000 BCE** Evidence of settlement at Monte Verde in present-day Chile
- **c.4500 BCE** Evidence of agriculture and herding in central Andes
- **c.2500 BCE** Masonry building and temple architecture on Pacific coast
- **c.1800 BCE** Ceremonial center of La Florida built in Peru
- **c.450 CE** Nazca culture flourishing; lines and giant figures drawn in desert
- **c.900 CE** Wari and Tiwanaku flourish as capitals of first competing empires
- **1100** Emergence of Chimu state on north coast of Peru
- **1380** Beginning of Inca empire in central highlands
- **1475** Chimu conquered by Incas
- **1494** Treaty of Tordesillas divides western hemisphere between Spain and Portugal
- **1500** Cabral sights Brazilian coast
- **1502** First expedition sent from Portugal to exploit coast of Brazil
- **c.1510** First African slaves to South America
- **1525** Civil war in Inca empire
- **1532** Pizarro begins defeat of Incas (to 1540)
- **1562** War and disease kill much of Amerindian population of Brazil (to 1563)
- **1630** Dutch establish New Holland, covering much of northern Brazil
- **1654** Portuguese regain control of Brazil
- **1663** Brazil becomes viceroyalty
- **1695** Gold discovered in Brazil
- **1739** Viceroyalty of New Granada established to defend Spanish interests on Caribbean coast
- **1750** Treaty of Madrid defines boundary between Spanish colonies and Brazil

Europe

- **c.6500 BCE** Farming spreads rapidly into central Europe
- **c.2800 BCE** Building of Stonehenge begins
- **c.1600 BCE** Minoan Palace civilization on Crete
- **c.750 BCE** Beginnings of Greek city-states
- **510 BCE** Roman Republic founded
- **431 BCE** Outbreak of Peloponnesian War between Sparta and Athens
- **218 BCE** Carthaginians invade Italy under command of Hannibal
- **49 BCE** Julius Caesar conquers Gaul
- **43 CE** Roman invasion of Britain
- **238 CE** Goths begin to invade borders of Roman Empire
- **330** Constantinople becomes new capital of Roman Empire
- **410** Invasion and pillage of Rome by Visigoths
- **711** Spain invaded by Muslims
- **793** Viking raids across Europe
- **800** Charlemagne becomes first Holy Roman Emperor
- **950** Harold Bluetooth consolidates the unification of Denmark
- **987** Feudal lords elect first Capetian king of France
- **1066** Norman conquest of England
- **1236** Russia invaded by Mongols
- **1337** Onset of Hundred Years War
- **1453** Byzantine Empire collapses as Ottoman Turks capture Constantinople
- **1478** Ivan III first czar of Russia
- **c.1500** Italian Renaissance
- **1521** Beginning of Protestant Reformation
- **1534** Henry VIII of England breaks with Rome
- **1588** Spanish Armada defeated by English
- **1618** Onset of Thirty Years War
- **1642** English Civil War (to 1649)
- **1756** Onset of Seven Years War
- **1789** French Revolution

AFRICA

- ❑ **c.400,000** BCE First evidence of *Homo sapiens* (modern humans) in Rift Valley
- ❑ **3100** BCE King Narmer unifies Upper and Lower Egypt and becomes first pharaoh
- ❑ **c.2650** BCE Start of great pyramid building in Egypt
- ❑ **2040** BCE Beginning of Middle Kingdom in Egypt
- ❑ **1352** BCE Pharaoh Akhenaten promotes Aten (sun) worship in Egypt
- ❑ **814** BCE Foundation of Phoenician colony of Carthage
- ❑ **146** BCE Rome conquers Carthage
- ❑ **31** BCE Cleopatra's death marks end of Ptolemaic dynasty in Egypt
- ❑ **c.600** CE Kingdom of Ghana founded
- ❑ **641** CE Muslims conquer Egypt
- ❑ **c.900** CE Emergence of Great Zimbabwean state
- ❑ **1067** Almoravids destroy kingdom of Ghana
- ❑ **c.1300** Emergence of empire of Benin (Nigeria)
- ❑ **1390** Formation of kingdom of Kongo
- ❑ **1443** Portuguese begin mass export of slaves from Atlantic coast to Europe
- ❑ **1498** Vasco da Gama rounds Cape of Good Hope
- ❑ **1502** First slaves taken to New World
- ❑ **1570** Establishment of Portuguese colony of Angola
- ❑ **1652** Dutch establish colony at Cape of Good Hope
- ❑ **1787** British establish Sierra Leone for freed slaves
- ❑ **1795** British capture Cape of Good Hope from the Dutch
- ❑ **1798** Occupation of Egypt by Napoléon Bonaparte
- ❑ **1816** Shaka leads expansion of Zulu
- ❑ **1822** Freed black slaves found colony of Liberia
- ❑ **1830** French invasion of Algeria
- ❑ **1836** Start of Boer Great Trek
- ❑ **1848** Boers found Orange Free State
- ❑ **1853** Livingstone finds Victoria Falls (Musi-o-Tunya)
- ❑ **1869** Opening of Suez Canal
- ❑ **1875** Stanley establishes source of Nile

ASIA AND THE MIDDLE EAST

- ❑ **c.12,000** BCE Beginnings of farming in Palestine
- ❑ **c.4800** BCE 'Ubaid culture builds towns in Mesopotamia
- ❑ **c.3500** BCE Fortified towns built throughout northern China
- ❑ **c.2800** BCE Emergence of city states in Indus Valley
- ❑ Wheel used in Mesopotamia
- ❑ **c.1750** BCE Foundation of Old Babylonian Empire under Hammurabi
- ❑ **c.1200** BCE Traditional date for exodus of the Jews from Egypt
- ❑ **c.1100** BCE Phoenician civilization spreads throughout Mediterranean
- ❑ **c.660** BCE Japanese empire founded
- ❑ **550** BCE Persian Empire founded
- ❑ **c.480** BCE Death – or *parinibbana* – of Gautama Buddha
- ❑ **334** BCE Alexander the Great invades Asia Minor
- ❑ **332** BCE Foundation of Mauryan empire in India
- ❑ **202** BCE Han dynasty begins in China
- ❑ **c.112** BCE "Silk Road" links China to West
- ❑ **c.30** CE Crucifixion of Jesus of Nazareth, founder of Christianity
- ❑ **c.350** CE Huns invade Persia and India
- ❑ **622** Mohammed, founder of Islam, flees Mecca; start of Muslim calendar
- ❑ **960** China united under Sung dynasty
- ❑ **1044** Foundation of Burma
- ❑ **1099** Jerusalem sacked in First Crusade
- ❑ **1185** Minamoto shoguns rule Japan
- ❑ **1206** Mongols begin to conquer Asia under Genghis Khan
- ❑ **1238** Foundation of first Thai kingdom
- ❑ **1258** Baghdad sacked by Mongols
- ❑ **1264** Yuan dynasty founded in China by Kublai Khan
- ❑ **1275** Marco Polo arrives in China
- ❑ **1333** Civil war in Japan
- ❑ **1368** Ming dynasty begins in China
- ❑ **1392** Korea proclaims independence
- ❑ **1498** Vasco da Gama completes first European voyage to India
- ❑ **1526** Foundation of Mughal empire in India
- ❑ **1600** Charter granted to East India Company
- ❑ **1609** Beginning of Tokugawa shogunate in Japan
- ❑ **1619** Dutch found Batavia (Jakarta)

AUSTRALASIA AND OCEANIA

- ❑ **c.60,000** BCE First people arrive in Australia
- ❑ **c.30,000** BCE Aboriginal rock art begins to appear
- ❑ **c.8000-6000** BCE Rising sea level covers New Guinea land bridge
- ❑ **c.6000** BCE Migrations from southeast Asia give rise to Austronesian culture
- ❑ **c.4000** BCE Austronesians reach southwestern Pacific islands
- ❑ **c.1000** BCE Emergence of archaic Polynesian society in Fiji, Tonga, and Samoa
- ❑ **c.300** CE Easter Island settled
- ❑ **c.600** CE Polynesians arrive in Hawaii
- ❑ **1520** Magellan enters Pacific
- ❑ **1526** Jorge de Meneses first European to sight New Guinea
- ❑ **1606** Torres sails through strait that now bears his name; proves New Guinea is an island
- ❑ **1642** Tasman, searching for a southern continent, finds Tasmania and New Zealand
- ❑ **1688** Dampier first Englishman to visit Australia
- ❑ **1768** Cook's first voyage
- ❑ **1773** Cook crosses Antarctic Circle and explores Southern Ocean (to 1775)
- ❑ **1779** Cook killed in Hawaii on third voyage
- ❑ **1788** First penal settlement established at Port Jackson (Sydney)
- ❑ **1802** Flinders circumnavigates Australia (to 1803)
- ❑ **1818** Start of Maori "Musket Wars" in New Zealand
- ❑ **1819** Bellingshausen's expedition sights Antarctica
- ❑ **1829** Britain annexes western and final third of Australian continent
- ❑ **1830** A mere 200 foreigners, mostly British, permanently resident in New Zealand
- ❑ **1840** Treaty of Waitangi grants sovereignty over New Zealand to British
- ❑ **1841** New Zealand becomes a separate Crown colony
- ❑ **1845** Northern War in New Zealand (to 1846)

see pages 68–69 for CHRONOLOGY 2001–2002

North America (continued)

❑ **1861** US Civil War (to 1865)

❑ **1863** Emancipation Proclamation

❑ **1865** Lee surrenders to Grant at Appomattox
❑ Assassination of President Lincoln
❑ Slavery abolished in US

❑ **1867** US buys Alaska from Russia
❑ Dominion of Canada established

❑ **1869** 15th Amendment gives vote to freed slaves in US

❑ **1871** Start of Apache Wars

❑ **1876** Battle of Little Big Horn: Sioux warriors kill 250 US soldiers

❑ **1890** Massacre of Sioux warriors at Wounded Knee ends Amerindian wars

❑ **1896** Klondike Gold Rush, Alaska

❑ **1898** Spanish–American War

❑ **1899** Spain cedes Cuba and Puerto Rico to US

❑ **1910** Mexican Revolution begins

❑ **1921** US restricts immigration

❑ **1929** Wall Street Crash

❑ **1933** President Roosevelt introduces New Deal

❑ **1940s** Race riots in Harlem, Los Angeles, Detroit, and Chicago

❑ **1941** US enters war against Germany and Japan

❑ **1945** End of World War II

❑ **1949** Formation of NATO
❑ Cold War begins

❑ **1950** US supports south in Korean War (to 1953)

❑ **1959** Cuban Revolution

❑ **1962** Cuban missile crisis

❑ **1963** Martin Luther King leads march on Washington D.C.
❑ Assassination of President Kennedy

❑ **1964** US Congress approves sending first troops to Vietnam

❑ **1968** Assassination of Martin Luther King sparks riots in 124 US cities

❑ **1969** Neil Armstrong becomes first person on moon

❑ **1973** US withdraws from Vietnam

❑ **1974** President Nixon resigns over Watergate scandal

❑ **1979** Civil war in Nicaragua (to 1990)
❑ Civil war in El Salvador (to 1992)

❑ **1991** US leads alliance in Gulf War to end Iraq's invasion of Kuwait

❑ **1994** North American Free Trade Agreement (NAFTA) established

❑ **1999** President Clinton survives impeachment

❑ **2000** George W. Bush wins presidential elections but does not receive a majority of votes

South America (continued)

❑ **1811** Bolívar starts fight to liberate Venezuela
❑ Paraguay independent

❑ **1817** San Martin wins decisive victory over Spanish and liberates Chile

❑ **1821** Peru independent

❑ **1822** Brazil independent

❑ **1823** Slavery abolished in Chile

❑ **1825** Bolivia independent

❑ **1828** Uruguay independent

❑ **1830** Ecuador, Colombia, and Venezuela (formerly Gran Colombia) become separate states

❑ **1851** Slavery abolished in Colombia

❑ **1853** Slavery abolished in Ecuador, Argentina, and Uruguay

❑ **1854** Slavery abolished in Bolivia and Venezuela

❑ **1864** Paraguayan War: Brazil, Argentina, and Uruguay defeat Paraguay

❑ **1870** Slavery abolished in Paraguay

❑ **1888** Slavery abolished in Brazil

❑ **1900** Major Italian migration to Argentina

❑ **1914** Panama Canal opens

❑ **1930** Military revolution in Brazil

❑ **1932** Chaco War between Bolivia and Paraguay (to 1935); Paraguay defeats Bolivia

❑ **1937** "New State" in Brazil launched by Vargas

❑ **1946** Peron comes to power in Argentina

❑ **1955** Peron ousted by military coup; returns to power in 1973

❑ **1968** Tupamaros urban guerrilla group founded in Uruguay
❑ Military junta takes over Peru

❑ **1970** Allende elected president of Chile

❑ **1973** US backs Pinochet coup against elected government in Chile; Allende assassinated

❑ **1976** "Dirty War" of right-wing death squads in Argentina

❑ **1982** Falklands War between Argentina and UK

❑ **1983** Democracy restored in Argentina

❑ **1985** Democracy restored in Brazil and Uruguay

❑ **1989** Democracy restored in Chile

❑ **1999** Panama takes control of Panama Canal

❑ **2000** President Fujimori ousted amid massive corruption scandal in Peru

Europe (continued)

❑ **1804** Napoléon becomes emperor of France

❑ **1815** Napoléon defeated
❑ Treaty of Vienna

❑ **1845** Beginning of Irish potato famine

❑ **1854** Crimean War (to 1856)

❑ **1861** Italy unified
❑ Emancipation of serfs in Russia

❑ **1870** Franco-Prussian War

❑ **1871** Germany unified

❑ **1914** World War I (to 1918)

❑ **1917** Russian Revolution

❑ **1922** Mussolini comes to power in Italy after Fascist "March on Rome"

❑ **1933** Nazis take power in Germany; Hitler is elected chancellor

❑ **1936** Spanish Civil War (to 1939)

❑ **1939** Germany invades Poland precipitating World War II

❑ **1941** German forces invade Russia

❑ **1944** British and US troops land in Normandy; Russians advance into eastern Europe

❑ **1945** Defeat of Germany

❑ **1949** Formation of NATO
❑ Cold War begins

❑ **1957** Treaties of Rome establish European Economic Community

❑ **1961** Building of the Berlin Wall
❑ Yuri Gagarin first person in space

❑ **1968** Troubles in Northern Ireland

❑ **1973** UK and Ireland join European Communities

❑ **1975** End of dictatorship in Spain with death of Gen. Franco

❑ **1986** Explosion at Chernobyl nuclear power reactor
❑ Soviet launch of *Mir* space station

❑ **1989** Democratic revolutions in eastern Europe
❑ Berlin Wall demolished

❑ **1990** Reunification of Germany

❑ **1991** The Soviet Union splits into its component countries
❑ Slovenia and Croatia claim their independence

❑ **1992** Civil war in Bosnia & Herzegovina (to 1995)

❑ **1993** Velvet Divorce: separation of Czech Republic and Slovakia

❑ **1994** Outbreak of war in Chechnya

❑ **1995** European Union expands to 15 members as Austria, Finland, and Sweden join

❑ **1999** "Ethnic cleansing" of Albanians in Kosovo leads to NATO airstrikes against Yugoslavia

❑ **2000** President Milošević is ousted in Yugoslavia (Serbia & Montenegro) in a popular revolution

AFRICA (CONTINUED)

- ❑ **1879** British defeat Zulus
- ❑ **1881** French occupy Tunisia
- ❑ **1882** Britain occupies Egypt
- ❑ **1883** France begins conquest of Madagascar
- ❑ **1889** Colonization of "Rhodesia"
- ❑ **1890** Land connection between Angola and Mozambique ended
- ❑ **1894** Britain occupies Uganda
- ❑ **1896** Ethiopian emperor Menelik II defeats Italians at Adawa
- ❑ **1899** Boer War (to 1902)
- ❑ **1910** Formation of Union of South Africa
- ❑ **1911** Italian conquest of Libya
- ❑ **1935** Second Italian invasion of Ethiopia
- ❑ **1942** British halt German advance at El Alamein
- ❑ **1948** Pro-apartheid National Party wins power in South Africa
- ❑ **1956** UK fails to block Egypt's nationalization of Suez Canal
- ❑ **1960** Outbreak of civil war in Belgian Congo
  ❑ Fifteen countries gain independence
- ❑ **1962** Algeria gains independence
- ❑ **1963** Zambia and Malawi granted independence
  ❑ Organization of African Unity (OAU) founded
- ❑ **1964** Nelson Mandela sentenced to life imprisonment
- ❑ **1974** Emperor Haile Selassie of Ethiopia deposed
- ❑ **1975** Angola and Mozambique gain independence; civil wars ensue
- ❑ **1980** Black majority rule established in Zimbabwe
- ❑ **1981** President Sadat of Egypt assassinated
- ❑ **1984** Worst recent famine in Ethiopia
- ❑ **1990** Mandela released: apartheid begins to be dismantled
  ❑ Namibia becomes independent
- ❑ **1991** Civil war in Sierra Leone (to 2001)
- ❑ **1994** South Africa holds first multiracial election; Mandela wins presidency
  ❑ Attempted genocide of Tutsis by Hutu in Rwanda
- ❑ **1997** Overthrow of Mobutu in Zaire (Democratic Republic of the Congo – DRC)
- ❑ **1998** Start of civil war in DRC
- ❑ **1999** Nigeria returns to democracy
- ❑ **2000** Ethiopia–Eritrea conflict

ASIA AND THE MIDDLE EAST (CONTINUED)

- ❑ **1757** East India Company defeats Nawab of Bengal's forces at Plassey
- ❑ **1842** Opium Wars (to 1854), Britain compels China to open Treaty Ports and annexes Hong Kong
- ❑ **1851** Taiping rebellion in China, 20 million killed (to 1864)
- ❑ **1868** Meiji Restoration in Japan
- ❑ **1877** Queen Victoria proclaimed empress of India
- ❑ **1911** Manchu dynasty overthrown in China, republic declared
- ❑ **1922** The last Ottoman sultan is deposed; Turkey proclaimed a republic
- ❑ **1932** Kingdom of Saudi Arabia founded
- ❑ **1937** Japanese forces invade China
- ❑ **1941** Pearl Harbor attacked by Japan
- ❑ **1945** Atom bombs dropped on Hiroshima and Nagasaki; c.210,000 killed, Japan surrenders
- ❑ **1947** Partition of India: Pakistan and India independent
- ❑ **1948** Burma and Ceylon (Sri Lanka) proclaim their independence
  ❑ Establishment of Israel
- ❑ **1949** People's Republic of China proclaimed
  ❑ Indonesia independent
- ❑ **1950** Korean War (until 1953)
- ❑ **1954** Laos, Cambodia, and Vietnam proclaim their independence
- ❑ **1959** China occupies Tibet
- ❑ **1965** US combat troops in Vietnam
- ❑ **1966** Cultural Revolution in China
- ❑ **1971** East Pakistan (Bangladesh) claims independence
- ❑ **1975** Fall of Saigon ends Vietnam War
  ❑ Civil war in Lebanon (to 1989)
- ❑ **1979** Overthrow of shah in Iran, Islamic Republic founded
  ❑ Vietnam pushes Khmer Rouge from Cambodia
- ❑ **1980** Iran–Iraq War (to 1988)
- ❑ **1982** Israeli invasion of Lebanon
- ❑ **1986** Marcos deposed in Philippines
- ❑ **1989** Massacre in Tiananmen Square
- ❑ **1990** Invasion of Kuwait by Iraq
- ❑ **1991** Gulf War
- ❑ **1996** *Taliban* take over in Afghanistan
- ❑ **1997** Hong Kong is returned to China
- ❑ **1997** Asian financial crisis
- ❑ **1998** Suharto regime collapses in Indonesia
  ❑ India and Pakistan test nuclear weapons
- ❑ **2000** Palestinians begin new *intifada*

AUSTRALASIA AND OCEANIA (CONTINUED)

- ❑ **c.1850** Migrant workers from China, Japan, and Philippines start arriving in Hawaii
- ❑ **1851** Gold discovered in New South Wales
- ❑ **1858** King Movement demands Maori state and opposes further land sales
- ❑ **1860** European settlers outnumber Maoris in New Zealand
- ❑ **1862** Second Maori War
- ❑ **1864** First French convict settlers in New Caledonia
- ❑ **1865** 1000 Chinese brought to Tahiti to work cotton plantation (to 1866)
- ❑ **1869** Last convict ship arrives in Australia
- ❑ **1870** Maori resistance crushed
  ❑ Germans start to buy up large tracts of Western Samoa
- ❑ **1874** Indian sugarcane workers arrive in Fiji
- ❑ **1888** Chile starts colonization of Easter Island
- ❑ **1890** Gold discovered in Western Australia
- ❑ **1898** US annexes Hawaii and seizes Guam from Spain
- ❑ **1901** Australia is self-governing federation within British Empire
- ❑ **1912** Amundsen's expedition reaches South Pole
- ❑ **1914** Over 60,000 Australian troops and more than 15,000 New Zealanders lose their lives in World War I (to 1918)
- ❑ **1930s** Australia hit hard by global Depression
- ❑ **1942** Australia under threat of invasion as Japanese bomb Darwin
- ❑ **1946** US begins nuclear tests at Eniwetok and Bikini atolls in Micronesia
- ❑ **1952** UK begins nuclear tests (to 1991) on mainland Australia
- ❑ **1966** France begins nuclear tests in Tuamotu Islands
- ❑ **1975** Restrictions imposed on immigrants to Australia
- ❑ **1985** South Pacific Forum declares nuclear-free Pacific
- ❑ **1988** Bicentennial celebrations in Australia occasion Aboriginal protests
- ❑ **1996** France halts nuclear testing in the Pacific
- ❑ **1998** Ethnic conflict in Solomon Islands (to 2000)
- ❑ **1999** Australian referendum rejects proposal on becoming republic
- ❑ **2000** Coup in Fiji ousts first ethnic Indian government

THE FORMATION OF THE MODERN WORLD

THE WORLD AS WE KNOW IT today, like all of the species that inhabit it, is the product of many thousands of years of evolution. The political and cultural map of the globe bears the hallmark of many varied courses of human development the world over. Nevertheless, much of the modern human geography of the planet can be traced to developments in the relatively recent past. The following pages chart the rise and fall of the various states and empires of the early modern and modern ages. Beginning with the first great achievement of European exploration, the "discovery" of the Americas in 1492, the maps show the way in which various European and Asian powers expanded their cultural and political influence and control down to the present day. This process left indelible cultural imprints in the form of language, religion, education, and systems of government on every part of the planet.

MAJOR MIGRATIONS SINCE 1500

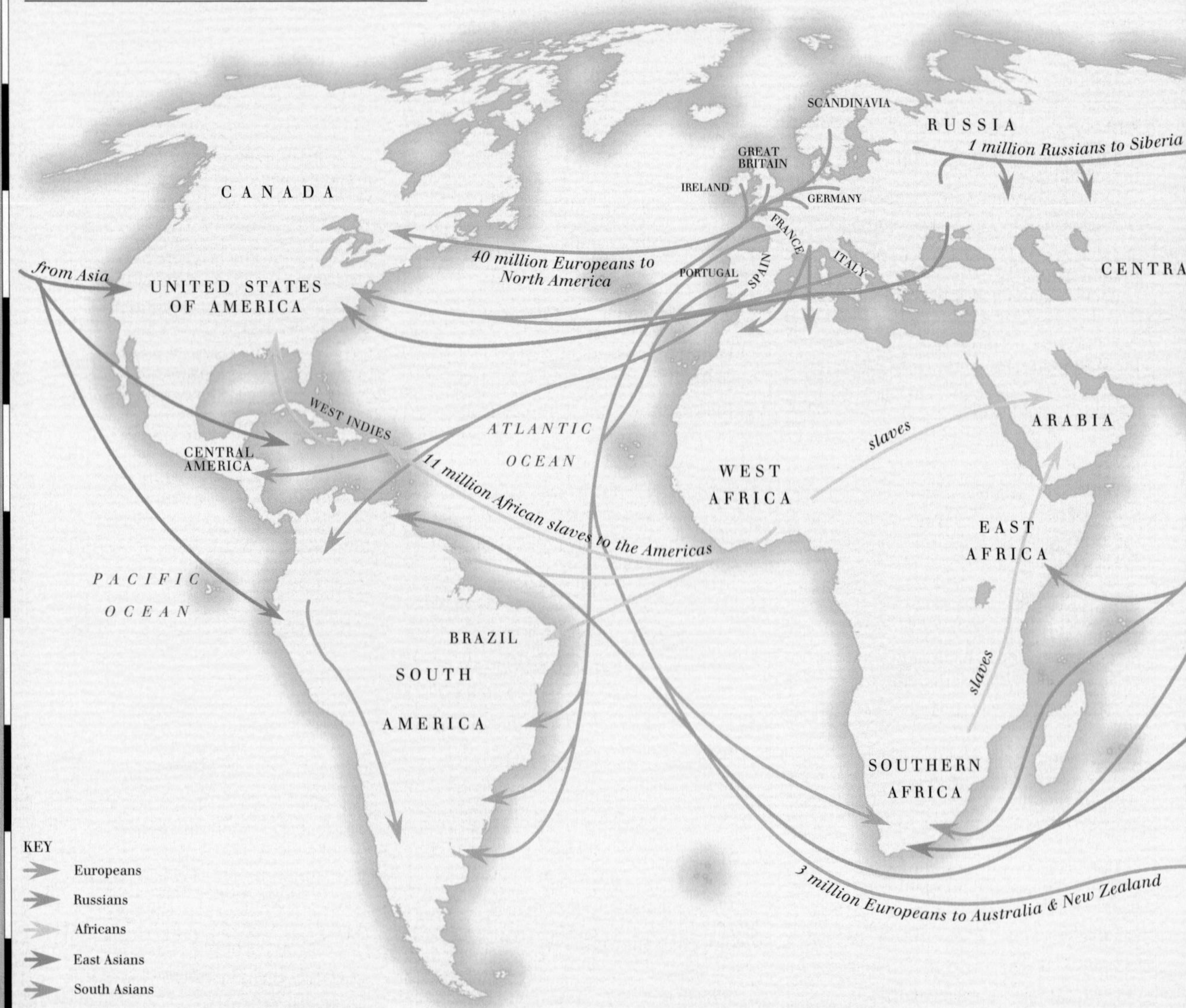
SCANDINAVIA
RUSSIA
1 million Russians to Siberia
GREAT BRITAIN
IRELAND
GERMANY
CANADA
FRANCE
ITALY
SPAIN
PORTUGAL
40 million Europeans to North America
from Asia
UNITED STATES OF AMERICA
CENTRA
WEST INDIES
ATLANTIC OCEAN
slaves
ARABIA
CENTRAL AMERICA
11 million African slaves to the Americas
WEST AFRICA
EAST AFRICA
PACIFIC OCEAN
BRAZIL
SOUTH AMERICA
slaves
SOUTHERN AFRICA
3 million Europeans to Australia & New Zealand
KEY
Europeans
Russians
Africans
East Asians
South Asians

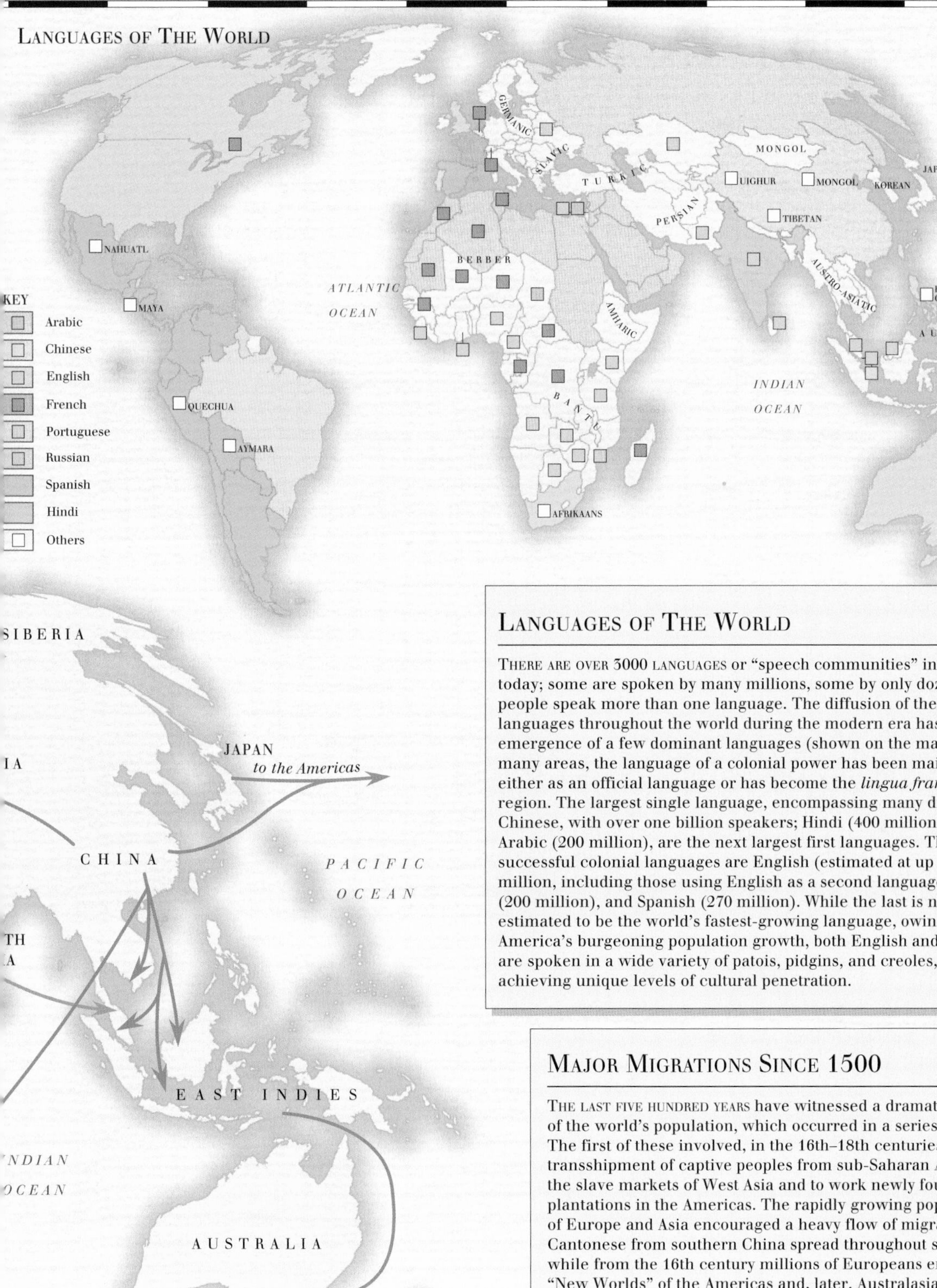
LANGUAGES OF THE WORLD
KEY
Arabic
Chinese
English
French
Portuguese
Russian
Spanish
Hindi
Others
NAHUATL
MAYA
QUECHUA
AYMARA
ATLANTIC OCEAN
GERMANIC
SLAVIC
TURKIC
BERBER
AMHARIC
BANTU
AFRIKAANS
MONGOL
UIGHUR
MONGOL
KOREAN
JAPANESE
PERSIAN
TIBETAN
AUSTRO-ASIATIC
FILIPINO, CEBUANO
AUSTRONESIAN
PACIFIC OCEAN
INDIAN OCEAN
SIBERIA
JAPAN
to the Americas
CHINA
PACIFIC OCEAN
EAST INDIES
INDIAN OCEAN
AUSTRALIA
NEW ZEALAND

LANGUAGES OF THE WORLD

THERE ARE OVER 3000 LANGUAGES or "speech communities" in the world today; some are spoken by many millions, some by only dozens. Many people speak more than one language. The diffusion of the major languages throughout the world during the modern era has seen the emergence of a few dominant languages (shown on the map). In many areas, the language of a colonial power has been maintained either as an official language or has become the *lingua franca* of the region. The largest single language, encompassing many dialects, is Chinese, with over one billion speakers; Hindi (400 million) and Arabic (200 million), are the next largest first languages. The most successful colonial languages are English (estimated at up to 1500 million, including those using English as a second language), French (200 million), and Spanish (270 million). While the last is now estimated to be the world's fastest-growing language, owing to Latin America's burgeoning population growth, both English and French are spoken in a wide variety of patois, pidgins, and creoles, thus achieving unique levels of cultural penetration.

MAJOR MIGRATIONS SINCE 1500

THE LAST FIVE HUNDRED YEARS have witnessed a dramatic redistribution of the world's population, which occurred in a series of waves. The first of these involved, in the 16th–18th centuries, the mass transshipment of captive peoples from sub-Saharan Africa to supply the slave markets of West Asia and to work newly founded European plantations in the Americas. The rapidly growing populations of Europe and Asia encouraged a heavy flow of migration. The Cantonese from southern China spread throughout southeast Asia, while from the 16th century millions of Europeans emigrated to the "New Worlds" of the Americas and, later, Australasia. This European diaspora reached a peak at the end of the 19th century. Then, as the colonial empires coalesced in the early years of the 20th century, there was a final wave of global movement within them, when south and east Asians migrated to fill labor markets and exploit opportunities in Africa and the Americas. While homogeneous societies have developed in North America and Australia, many diverse ethnic communities remain scattered across the world.

The World in 1492

When Christopher Columbus sailed west from Europe, seeking a quicker route to Asia, he launched a process of discovery that was eventually to bring the disparate regions of the world into closer contact, to form the global map we know today. The largest political entity in the world at that time was the Chinese Ming empire. Culturally, the Islamic faith had forged a bond of religious unity which extended in a broad swathe from southeast Asia to the Atlantic coast of north Africa. Europe was a mêlée of rival monarchies; sub-Saharan Africa a patchwork of trading kingdoms; the Americas, a separate world of rich tribal cultures, with empires established only in Central America and the central Andes.

Global States and Territories

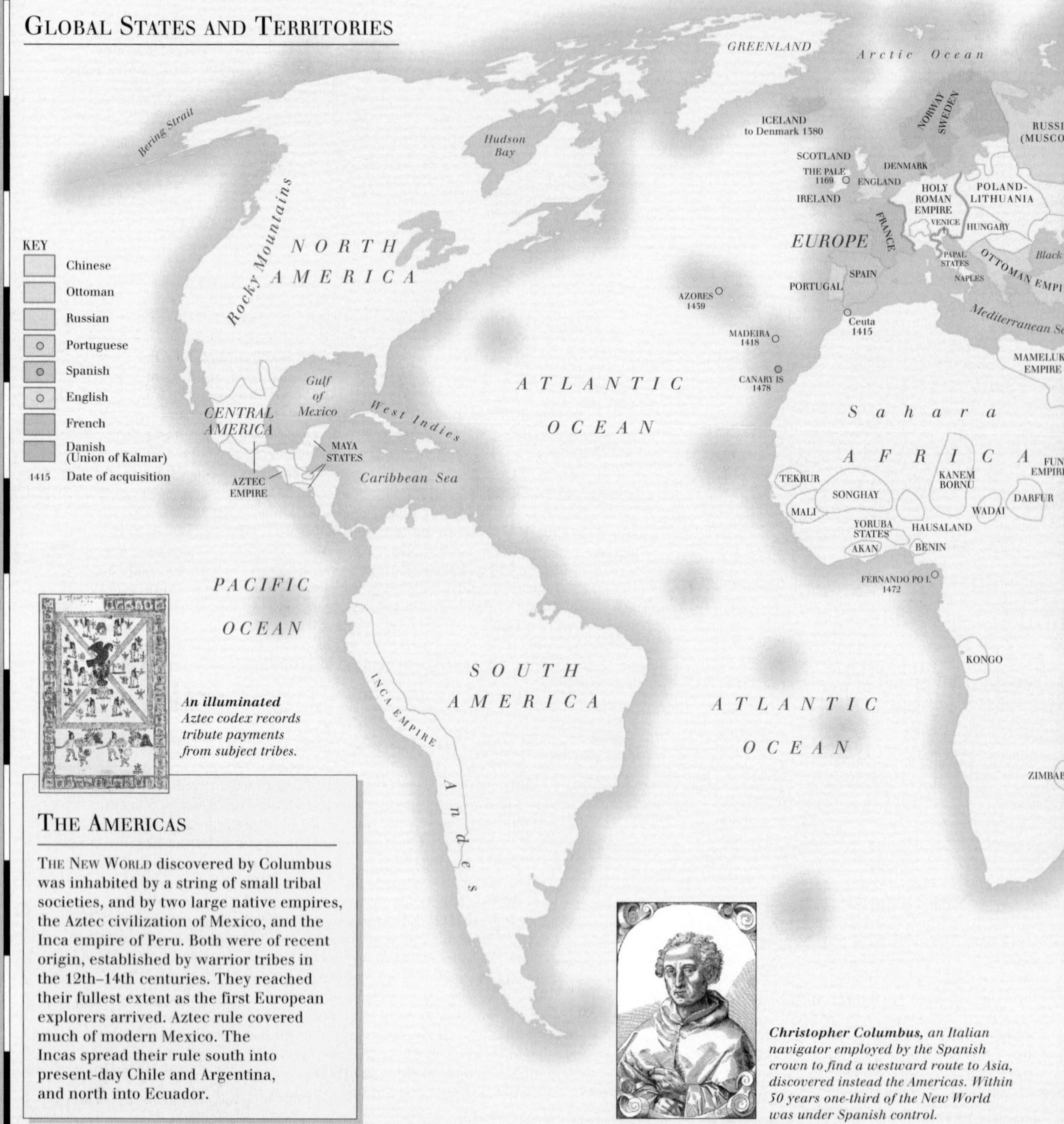
GREENLAND
Arctic Ocean
Bering Strait
Hudson Bay
ICELAND
to Denmark 1380
NORWAY
SWEDEN
RUSSIA (MUSCO
SCOTLAND
DENMARK
THE PALE 1169
ENGLAND
IRELAND
HOLY ROMAN EMPIRE
POLAND-LITHUANIA
VENICE
HUNGARY
FRANCE
EUROPE
PAPAL STATES
NAPLES
OTTOMAN EMPI
Black
SPAIN
PORTUGAL
Rocky Mountains
NORTH AMERICA
AZORES 1439
Ceuta 1415
Mediterranean Se
MADEIRA 1418
MAMELUK EMPIRE
KEY
Chinese
Ottoman
Russian
Portuguese
Spanish
English
French
Danish (Union of Kalmar)
1415 Date of acquisition
Gulf of Mexico
West Indies
ATLANTIC OCEAN
CANARY IS 1478
Sahara
CENTRAL AMERICA
MAYA STATES
AFRICA
FUN. EMPIRE
AZTEC EMPIRE
Caribbean Sea
TEKRUR
KANEM BORNU
SONGHAY
DARFUR
MALI
WADAI
YORUBA STATES
HAUSALAND
AKAN
BENIN
FERNANDO PO I. 1472
PACIFIC OCEAN
KONGO
SOUTH AMERICA
INCA EMPIRE
ATLANTIC OCEAN
Andes
ZIMBAB

An illuminated *Aztec codex records tribute payments from subject tribes.*

The Americas

The New World discovered by Columbus was inhabited by a string of small tribal societies, and by two large native empires, the Aztec civilization of Mexico, and the Inca empire of Peru. Both were of recent origin, established by warrior tribes in the 12th–14th centuries. They reached their fullest extent as the first European explorers arrived. Aztec rule covered much of modern Mexico. The Incas spread their rule south into present-day Chile and Argentina, and north into Ecuador.

Christopher Columbus, *an Italian navigator employed by the Spanish crown to find a westward route to Asia, discovered instead the Americas. Within 50 years one-third of the New World was under Spanish control.*

EUROPE

THOUGH CHRISTIAN EUROPE later transformed the exploration and settlement of the world, the Europe from which Columbus sailed was an unstable, violent continent, threatened by invaders from Asia to the east, and from the Ottoman Empire to the south. Civil wars and dynastic conflict resulted in shifting frontiers and small, militarily weak states. Only France, united by the late 15th century, Spain, a single monarchy from the 1490s, Portugal, and England were close to their modern forms.

The Portuguese caravel, *buoyant, sturdy, and lateen-rigged, was an ideal ocean-going vessel.*

EAST ASIA

THE MOST POWERFUL STATE in the world in 1492 was Ming China. Set up in 1386 after the collapse of Mongol power, the Ming dynasty ruled an area from Manchuria in the north to the borders of Vietnam in the south. Based on a traditional structure of bureaucratic control, the Ming emperors controlled their vast empire from Peking (Beijing), from where they launched punitive wars against the Mongols and Japanese pirates along the coast. Chinese culture and trade spread throughout east and southeast Asia, and Chinese navigators reached the Red Sea and the east African coast.

Chinese junks *plied the China seas, and traded as far as the East Indies, Ceylon (Sri Lanka), and east Africa.*

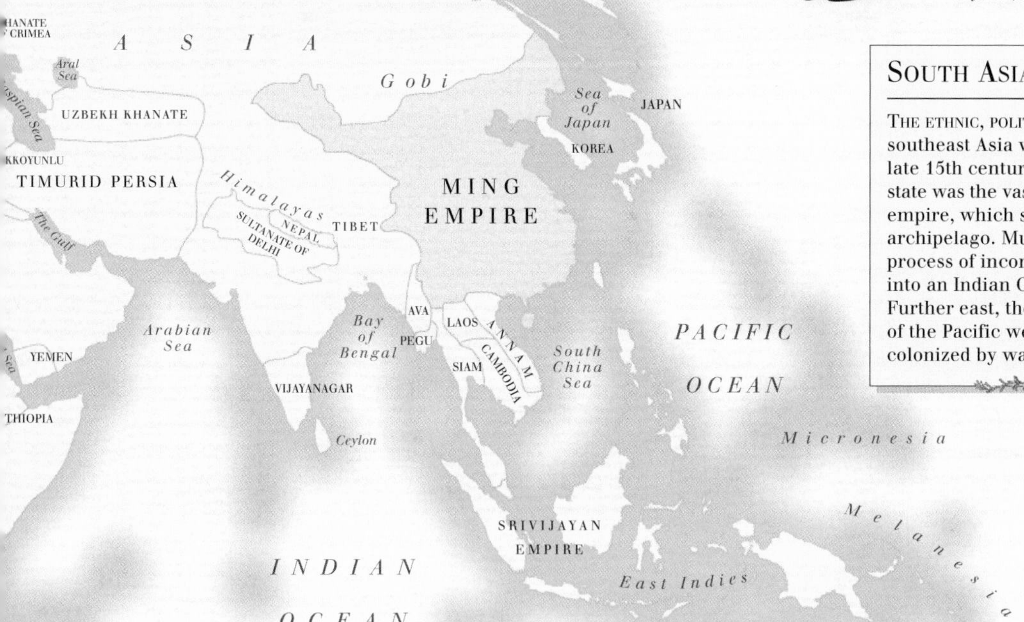

SOUTH ASIA AND OCEANIA

THE ETHNIC, POLITICAL, and religious map of southeast Asia was largely in place by the late 15th century. However, the largest state was the vast Srivijayan Hindu–Buddhist empire, which spanned the East Indies archipelago. Muslim traders were in the process of incorporating this rich region into an Indian Ocean trading empire. Further east, the scattered island groups of the Pacific were being successively colonized by waves of Melanesians.

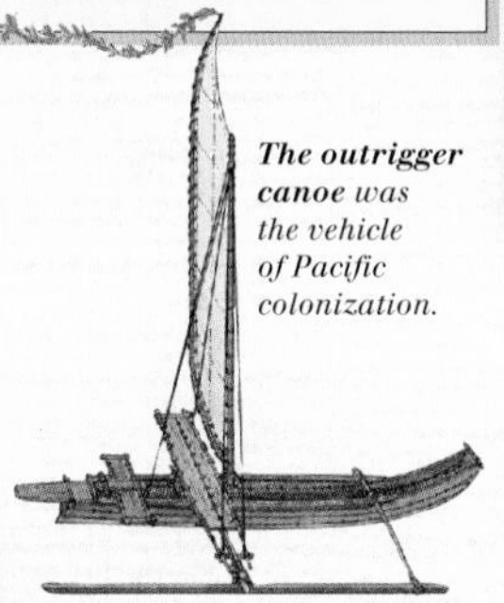

The outrigger canoe *was the vehicle of Pacific colonization.*

Arab dhows *built a trading network around the Indian Ocean.*

MIDDLE EAST AND AFRICA

AFTER CENTURIES OF INVASION from the Christian West and Asian nomadic empires, the Middle Eastern world stabilized around a revival of the Ottoman Empire. Vassal states extended across north Africa to Morocco, which linked the trading kingdoms of sub-Saharan Africa with the markets of Asia. The great cities of the Middle East surpassed those of Europe in wealth and learning.

The magnetic compass, *in use since the 13th century, was a primary navigational tool for the first ocean-going explorers, although early compasses were not always reliable, and ships often went astray. Accurate navigation only came later with the invention of the chronometer.*

THE AGE OF DISCOVERY: 1492–1648

THE FIRST STATE to take advantage of the new age of exploration was Spain. By the middle of the 16th century, under the Emperor Charles V, Spain was established as the foremost European colonial power, and one of the richest and most powerful kingdoms in Europe. Spanish rule was extended over the whole of Central America, much of South America, Florida, and the Caribbean; in Asia, Spanish rule was established in the Philippines. Spain led the way in establishing European settler colonies overseas. By the middle of the 17th century, British, Dutch, and French colonists began to challenge Spanish dominance in the Americas and east Asia, while pirates around the world plundered Spain's wealthy merchant convoys.

GLOBAL STATES AND TERRITORIES

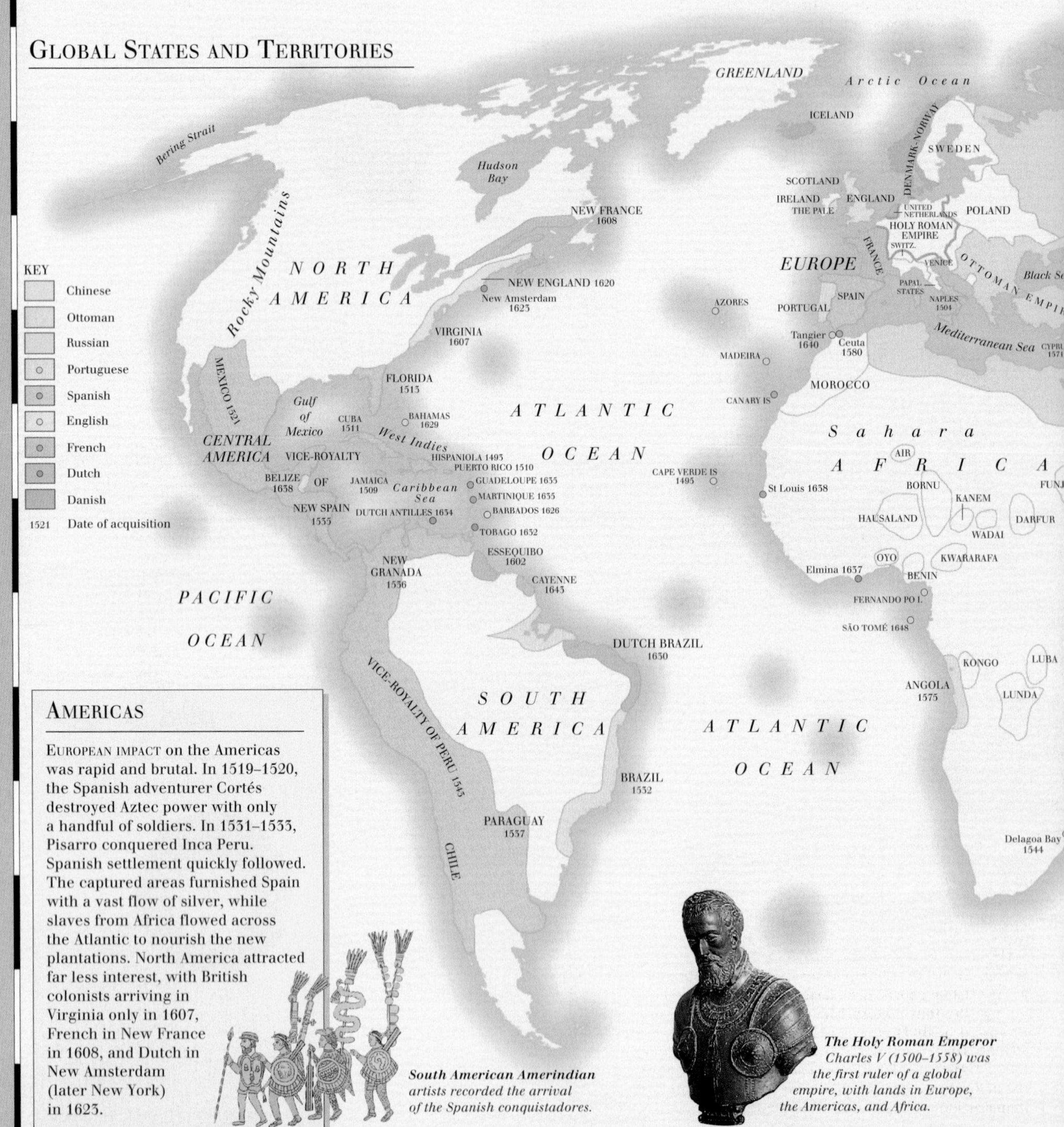
KEY
Chinese
Ottoman
Russian
Portuguese
Spanish
English
French
Dutch
Danish
1521 Date of acquisition
GREENLAND
Arctic Ocean
ICELAND
SWEDEN
DENMARK-NORWAY
SCOTLAND
IRELAND
THE PALE
ENGLAND
UNITED NETHERLANDS
POLAND
HOLY ROMAN EMPIRE
SWITZ.
VENICE
FRANCE
EUROPE
OTTOMAN EMPIRE
Black Sea
PAPAL STATES
NAPLES 1504
SPAIN
PORTUGAL
AZORES
Mediterranean Sea
CYPRUS 1571
Tangier 1640
Ceuta 1580
MADEIRA
MOROCCO
CANARY IS
Sahara
AFRICA
AIR
BORNU
KANEM
FUNJ
HAUSALAND
DARFUR
WADAI
OYO
KWARARAFA
BENIN
Elmina 1637
FERNANDO PO I.
SÃO TOMÉ 1648
KONGO
LUBA
ANGOLA 1575
LUNDA
Delagoa Bay 1544
St Louis 1638
CAPE VERDE IS 1495
Bering Strait
Hudson Bay
Rocky Mountains
NORTH AMERICA
NEW FRANCE 1608
NEW ENGLAND 1620
New Amsterdam 1625
VIRGINIA 1607
FLORIDA 1513
MEXICO 1521
Gulf of Mexico
CUBA 1511
BAHAMAS 1629
West Indies
CENTRAL AMERICA
VICE-ROYALTY OF NEW SPAIN 1535
HISPANIOLA 1493
PUERTO RICO 1510
BELIZE 1638
JAMAICA 1509
Caribbean Sea
GUADELOUPE 1635
MARTINIQUE 1635
BARBADOS 1626
DUTCH ANTILLES 1634
TOBAGO 1632
ESSEQUIBO 1602
CAYENNE 1643
NEW GRANADA 1536
ATLANTIC OCEAN
PACIFIC OCEAN
DUTCH BRAZIL 1630
SOUTH AMERICA
VICE-ROYALTY OF PERU 1543
BRAZIL 1532
PARAGUAY 1537
CHILE
ATLANTIC OCEAN

AMERICAS

EUROPEAN IMPACT on the Americas was rapid and brutal. In 1519–1520, the Spanish adventurer Cortés destroyed Aztec power with only a handful of soldiers. In 1531–1533, Pisarro conquered Inca Peru. Spanish settlement quickly followed. The captured areas furnished Spain with a vast flow of silver, while slaves from Africa flowed across the Atlantic to nourish the new plantations. North America attracted far less interest, with British colonists arriving in Virginia only in 1607, French in New France in 1608, and Dutch in New Amsterdam (later New York) in 1623.

***South American Amerindian** artists recorded the arrival of the Spanish conquistadores.*

***The Holy Roman Emperor** Charles V (1500–1558) was the first ruler of a global empire, with lands in Europe, the Americas, and Africa.*

EUROPE

FOR MORE THAN A CENTURY after Martin Luther inspired the Protestant Reformation in the 1520s, Europe was torn by religious wars. Scandinavia, England, and Scotland adopted the new beliefs, but elsewhere bitter civil conflicts led to the prolonged warfare and persecution known as the Thirty Years War. This ended in 1648; it destroyed wide areas of central Europe and decimated the German population, but resulted in a religious settlement which continues to the present. The Dutch Republic and northern Germany became Protestant while southern Germany, Poland, and southwest Europe remained Roman Catholic.

Printing*, using movable type, was a key development in the dissemination of ideas, knowledge, and commerce in early modern Europe.*

ASIA

IN 1480, THE SMALL PRINCIPALITY OF MUSCOVY (Moscow) threw off Mongol control, and proceeded to expand Muscovite power over the whole of the area from the Arctic Ocean to the Caspian Sea. In the 1550s, the conquest of Kazan brought Russian power to the Urals, and over the next century it spread across Siberia, reaching the Pacific coast by 1649. Much of the area remained uninhabited, but to the south this new empire jostled uneasily with a string of central Asian Muslim khanates, and with the newly established Manchurian Ch'ing dynasty, which wrested control of China from the Ming in 1644.

European navigators *and surveyors produced accurate maps and charts of their voyages.*

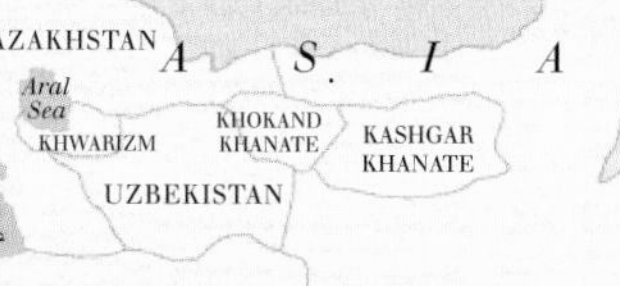

USSIAN EMPIRE
Siberia
Bering Strait
KAZAKHSTAN
ASIA
Aral Sea
KHIWARIZM
KHOKAND KHANATE
KASHGAR KHANATE
UZBEKISTAN
SAFAVID PERSIA
The Gulf
Himalayas
TIBET
NEPAL
MUGHAL EMPIRE
MANCHU (CH'ING) EMPIRE
Sea of Japan
JAPAN
KOREA
Deshima 1641
OMAN 1508
Diu 1535
Surat 1608
Daman 1559
Bombay 1534
Hooghly 1640
ARAKAN
BURMA
Bay of Bengal
Arabian Sea
Goa 1510
Masulipatam 1611
Madras 1639
CEYLON 1505
Galle 1640
LAOS
ANNAM
SIAM
Macao 1557
FORMOSA 1624
South China Sea
PHILIPPINES from 1565
PACIFIC OCEAN
Micronesia
Melanesia
Malacca 1641
MOLUCCAS from 1605
HIOPIA
RTUGUESE ST AFRICA rom 1505
INDIAN OCEAN
Makassar 1607
Batavia 1619
East Indies
1610 TIMOR 1618
Madagascar
AUSTRALIA
NEW ZEALAND

The Indian Mughal *ruler Shahjahan (1592–1648), builder of the Taj Mahal.*

SOUTH ASIA AND OCEANIA

THE PORTUGUESE and the Spanish were the first European powers to open trade with the powerful Asian states of Mughal India and Ch'ing China. The Spanish opened trans-Pacific routes between Central America, the Philippines, and China. But the establishment of the Dutch and British East India companies in the early 17th century announced the advent of two new maritime powers.

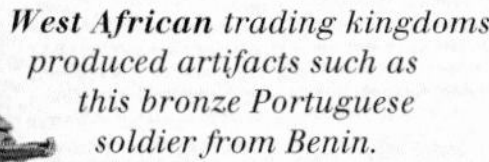

West African *trading kingdoms produced artifacts such as this bronze Portuguese soldier from Benin.*

AFRICA AND THE MIDDLE EAST

WHILE EUROPE WAS DIVIDED by the Reformation, Islam experienced a remarkable resurgence in the 16th century. The revival of the Ottoman Empire brought Islamic rule over much of southeast Europe. Islam spread along trade routes to sub-Saharan Africa. In east Africa, it spread south along the coast. Further east, Muslim rulers established new imperial states in Persia (Iran) and India.

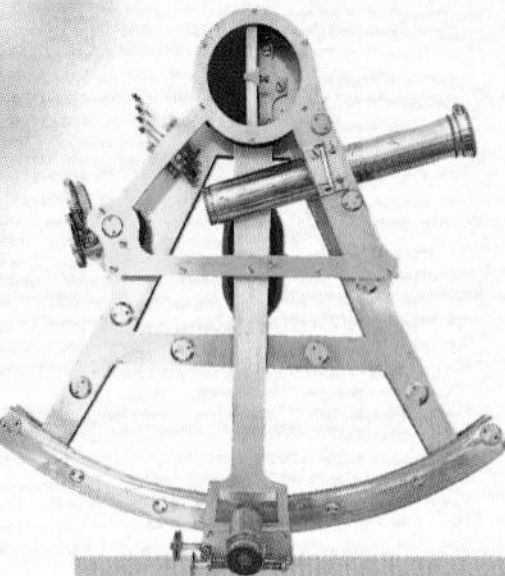

The sextant *allowed navigators to take accurate measurements of heavenly bodies in relation to the horizon, thus allowing latitude to be calculated correctly. Early sextants had to be hand-held and were often used on shore rather than on board ship.*

The Age of Expansion: 1648–1789

The years from the middle of the 17th century to the end of the 18th century saw a massive consolidation of European discovery and exploration, which took the form of colonial settlement and political expansion. This period also witnessed the beginning of a sharp rise in European population and in its economic strength, accompanied by rapid developments in the arts and sciences. All these factors powered European expansion – a process that would bring European culture to every part of the globe, gradually filling in the world map, and bringing it into often fatal contact with less robust indigenous cultures. By the last quarter of the 18th century, with Europe poised on the brink of political turmoil, only Africa and Australasia remained largely unmolested by European attentions.

Global States and Territories

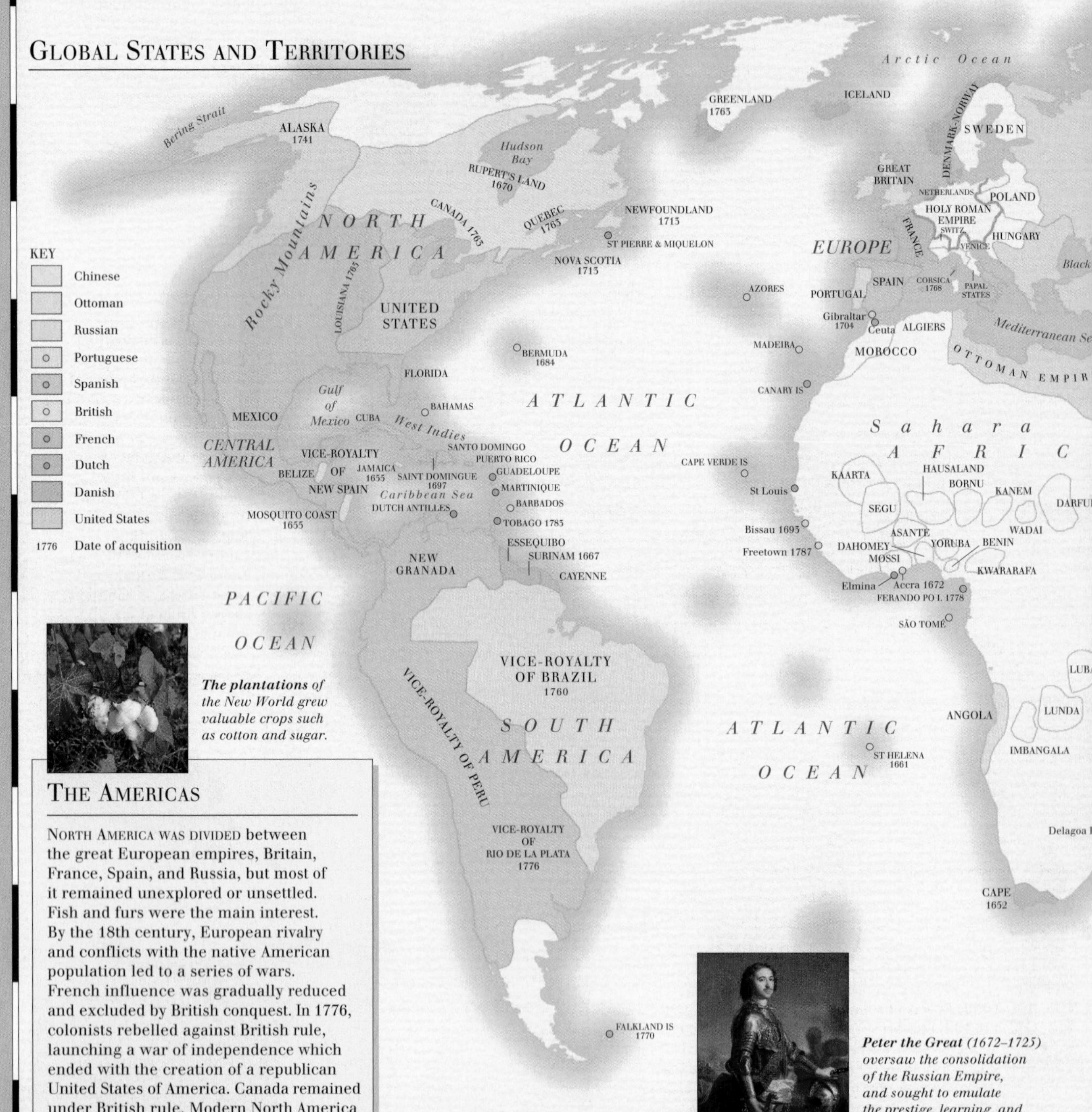
KEY
Chinese
Ottoman
Russian
Portuguese
Spanish
British
French
Dutch
Danish
United States
1776 Date of acquisition
Arctic Ocean
Bering Strait
ALASKA 1741
Hudson Bay
RUPERT'S LAND 1670
GREENLAND 1763
ICELAND
DENMARK-NORWAY
SWEDEN
GREAT BRITAIN
NETHERLANDS
POLAND
HOLY ROMAN EMPIRE
SWITZ
HUNGARY
VENICE
Black S
FRANCE
EUROPE
SPAIN
CORSICA 1768
PAPAL STATES
PORTUGAL
Gibraltar 1704
Ceuta
ALGIERS
Mediterranean Sea
MOROCCO
OTTOMAN EMPIRE
Rocky Mountains
NORTH AMERICA
CANADA 1763
QUEBEC 1763
NEWFOUNDLAND 1713
ST PIERRE & MIQUELON
NOVA SCOTIA 1713
LOUISIANA 1763
UNITED STATES
AZORES
MADEIRA
BERMUDA 1684
FLORIDA
CANARY IS
Gulf of Mexico
BAHAMAS
MEXICO
CUBA
West Indies
ATLANTIC OCEAN
Sahara
AFRICA
CENTRAL AMERICA
VICE-ROYALTY OF NEW SPAIN
BELIZE
JAMAICA 1655
SANTO DOMINGO
PUERTO RICO
SAINT DOMINGUE 1697
GUADELOUPE
MARTINIQUE
BARBADOS
TOBAGO 1783
Caribbean Sea
DUTCH ANTILLES
MOSQUITO COAST 1655
CAPE VERDE IS
St Louis
KAARTA
HAUSALAND
BORNU
KANEM
DARFUR
SEGU
WADAI
Bissau 1693
ASANTE
YORUBA
BENIN
DAHOMEY
MOSSI
Freetown 1787
KWARARAFA
Elmina
Accra 1672
FERANDO PO I. 1778
SÃO TOMÉ
ESSEQUIBO
SURINAM 1667
CAYENNE
NEW GRANADA
PACIFIC OCEAN
VICE-ROYALTY OF BRAZIL 1760
VICE-ROYALTY OF PERU
SOUTH AMERICA
LUBA
LUNDA
ANGOLA
IMBANGALA
ATLANTIC OCEAN
ST HELENA 1661
VICE-ROYALTY OF RIO DE LA PLATA 1776
Delagoa Ba
CAPE 1652
FALKLAND IS 1770

***The plantations** of the New World grew valuable crops such as cotton and sugar.*

The Americas

North America was divided between the great European empires, Britain, France, Spain, and Russia, but most of it remained unexplored or unsettled. Fish and furs were the main interest. By the 18th century, European rivalry and conflicts with the native American population led to a series of wars. French influence was gradually reduced and excluded by British conquest. In 1776, colonists rebelled against British rule, launching a war of independence which ended with the creation of a republican United States of America. Canada remained under British rule. Modern North America was gradually taking shape.

***Peter the Great** (1672–1725) oversaw the consolidation of the Russian Empire, and sought to emulate the prestige, learning, and sophistication of the western European monarchies.*

EUROPE

AFTER THE CRISIS of the Thirty Years War, Europe began to develop a more settled state system as successful dynastic houses imposed more centralized rule. The Habsburgs acquired control over Hungary and much of central Europe. Russia's frontiers pushed into Poland and the Ukraine. The French Bourbon monarchy became the most powerful in Europe. Its material wealth and culture made it a rival to the older empires of Asia. French became the common language of educated Europeans, and French philosophy led to the intellectual "Enlightenment."

***Isaac Newton** (1642–1727), the leading scientist of Europe's Age of Reason.*

ASIA

THE CH'ING DYNASTY forged the shape of modern China. By 1658 the whole of southern China was under Manchu control. Formosa (Taiwan) was occupied in 1683, outer Mongolia in 1697. A protectorate was established over Tibet in 1751. Over the course of this expansion, the population of China tripled and the economy boomed through trade in tea, porcelain, and silk with Russia and the West. Manchu China was powerful enough to resist incursions by the European empires, avoiding the fate of the crumbling Mughal empire in India, where Britain and France competed for trade and territory.

***Dutch and British** East Indiamen carried the vast European trade with Asia.*

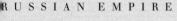

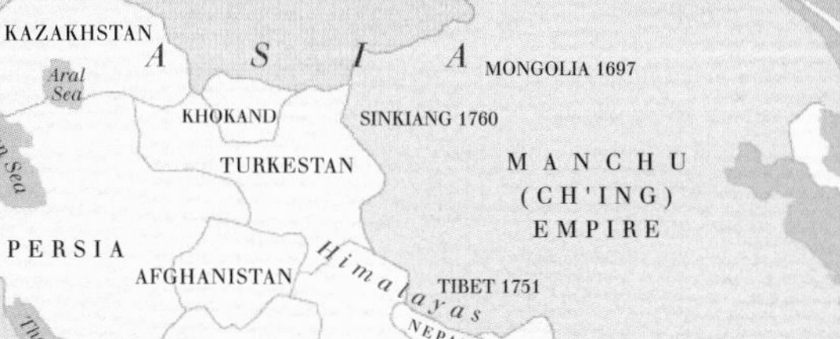
Bering Strait
RUSSIAN EMPIRE
KAZAKHSTAN
ASIA
Aral Sea
Caspian Sea
MONGOLIA 1697
KHOKAND
SINKIANG 1760
TURKESTAN
MANCHU (CH'ING) EMPIRE
Sea of Japan
JAPAN
KOREA
PERSIA
AFGHANISTAN
Himalayas
TIBET 1751
NEPAL
The Gulf
BALUCHISTAN
Deshima
PACIFIC OCEAN
BENGAL 1757
MARATHA CONFEDERACY
BURMA
Chandernagore 1688
Surat
Diu
Daman
Bombay 1661
Bay of Bengal
FORMOSA 1683
Macao
Arabian Sea
NORTHERN CIRCARS 1756
SIAM
ANNAM
South China Sea
PHILIPPINES
Goa
ANDAMAN IS 1789
MARIANAS 1668
Mahé 1725
MADRAS
Pondicherry 1674
Karikal 1738
CAROLINE IS 1686
ETHIOPIA
Galle
CEYLON 1658
Penang 1786
MALAYA
Micronesia
INDIAN OCEAN
MOLUCCAS
Melanesia
CHAGOS IS 1784
DUTCH EAST INDIES
PORTUGUESE EAST AFRICA
TIMOR
Madagascar
RÉUNION 1662
Fort Dauphin 1766
AUSTRALIA
LORD HOWE I. 1788
NEW SOUTH WALES 1788
NEW ZEALAND

***Maori New Zealand** was one of the few indigenous cultures to remain untouched by European contact until the 19th century.*

OCEANIA

SOUTHEAST ASIA AND OCEANIA were areas of small, warring kingdoms, increasingly prey to the ambitions of European traders, first Spanish and Portuguese, then Dutch and British. Yet, by the late 18th century, there was still little formal colonization. Officially discovered by Europeans in the early 17th century, most of Australasia was still unexplored and unsettled, except for a number of small penal colonies set up by the British in New South Wales (1788) and Tasmania (1804).

***African slavers** marched their human cargo from the interior to the coast for transshipment.*

AFRICA

DURING THE 17TH AND 18TH CENTURIES Africa was regarded by the rest of the world as a source of two things: gold and slaves. Some 13.5 million slaves were shipped in the 1700s, from the west coast and from Portuguese Angola. African dealers sold to European middlemen, who in turn sold on the surviving slaves. In northern and northeastern Africa, Arab slavers traded with the Ottoman Empire. But the rest of Africa remained isolated from the outside world.

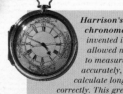

***Harrison's chronometer,** invented in 1762, allowed navigators to measure time accurately, and thus calculate longitude correctly. This greatly reduced the risk of shipwreck and heralded the beginning of accurate mapping of the world.*

The Age of Revolution: 1789–1830

In 1789 royal power was shattered by the French Revolution. The collapse of the most powerful monarchy in Europe reverberated worldwide. The revolutions in France and America ushered in the idea of the modern nation state, and of popular representative government. Revolutionary outbreaks occurred elsewhere in Europe, and overseas colonies in Latin America won their independence. At the same time, an industrial revolution was taking place in Europe, transforming the old trading economy into a manufacturing base which would require a global supply of raw materials and a global market to fuel it. The revolutionary years thus marked the beginning of the modern political and economic world order.

Global States and Territories

KEY
Chinese
Ottoman
Russian
Portuguese
Spanish
British
French
Dutch
Danish
United States
1790 Date of acquisition
[1820] Date of independence
GREENLAND
Arctic Ocean
ICELAND
NORWAY
SWEDEN
(Union 1815)
FINLAND to Russia 1809
UNITED KINGDOM
DENMARK
NETHERLANDS
GERMAN CONFED.
BELGIUM [1830]
FRANCE
AUSTRIA-HUNGARY
EUROPE
PORTUGAL
SPAIN
NAPLES
OTTOMAN
Black S
Gibraltar
Algiers 1830
MALTA 1800
Mediterranean Se
AZORES
MADEIRA
CANARY IS
MOROCCO
EGYP
(autonomo
Bering Strait
ALASKA
NORTH-WESTERN TERRITORIES
Hudson Bay
CANADA
OREGON COUNTRY
(US/Britain 1818-46)
Rocky Mountains
NORTH AMERICA
Louisiana Purchase 1803
UNITED STATES
ST PIERRE & MIQUELON
BERMUDA
MEXICO
[1821]
Gulf of Mexico
FLORIDA to US 1819
BAHAMAS
CUBA
West Indies
ATLANTIC OCEAN
CENTRAL AMERICA
BRITISH HONDURAS
JAMAICA
HAITI [1804]
Puerto Rico
ANGUILLA
GUADELOUPE
MARTINIQUE
ST LUCIA 1814
BARBADOS
TOBAGO
TRINIDAD
Caribbean Sea
DUTCH ANTILLES (to Br. 1807-15)
UNITED PROVINCES OF CENTRAL AMERICA
[1825-38]
MOSQUITO COAST
VENEZUELA [1830]
BRITISH GUIANA
DUTCH GUIANA
FRENCH GUIANA
REPUBLIC OF GREATER COLOMBIA
[1819-30]
NEW GRANADA [1831]
ECUADOR [1830]
PACIFIC OCEAN
SOUTH AMERICA
PERU [1821]
BOLIVIA [1825]
EMPIRE OF BRAZIL [1822]
PARAGUAY [1811]
ARGENTINA [1816]
CHILE [1818]
URUGUAY [1828]
Patagonia
FALKLAND IS (to Argentina 1820-33)
Sahara
AFRICA
SUDA
(to Egypt 18
CAPE VERDE IS
SENEGAL
Bathurst 1816
PORTUGUESE GUINEA
FUTA JALLON
MASSINA
DAHOMEY
BORNU
DARFUR
MOSSI
FULANI
WADAI
SIERRA LEONE
ASANTE
Assinie
LIBERIA (founded 1822)
Elmina
YORUBA
IBO
BENIN
Accra
FERNANDO PO I.
PRINCIPE
SÃO TOME
BUGAN
MANYEMA
NYAM
LUBA
ASCENSION I. 1815
ATLANTIC OCEAN
ANGOLA
LUNDA
ST HELENA
IMBANGALA
OVIMBUNDU
MATABELELAND
Delagoa B
SWAZI
SOTHO
ZULU
CAPE COLONY
TRISTAN DA CUNHA 1815

The Americas

The fledgling United States of America began to expand rapidly, purchasing the Midwest territories from France in 1803, and Florida from Spain in 1819. Revolutionary fervor both here and in Europe weakened the control of France, Spain, and Portugal throughout Latin America. From 1810 there followed 20 years of violent revolt, with native armies fighting their European masters and each other. The new states were prey to political violence and instability, but they never again came under European rule.

Simón Bolívar (1783–1830) led armies of liberation in Peru, Bolivia, and Venezuela.

Napoléon Bonaparte (1769–1821) began his career as a commander in the French Revolutionary wars. By 1804 he had become emperor of France, which dominated much of western Europe.

EUROPE

UNDER NAPOLÉON BONAPARTE, France subordinated a large part of Europe and destroyed the old feudal order. Napoléon helped to shape the new nation states that emerged in 19th-century Europe – Belgium, Italy, and Germany. He gave much of Europe its modern legal code and systems of measurement, education, and local government.

Steam-powered engines *transformed the European industrial economy.*

ASIA

THE PRINCIPAL COLONIAL POWER in Asia was Russia, whose consolidation of its empire in northern and central Asia continued throughout the 19th century. But now the Dutch began to extend their control of the East Indies, while a bitter struggle between the British and the French was conducted in and around the Indian Ocean. France was gradually forced to concede many of its footholds in India, where the British East India Company rapidly extended its interests by a mixture of diplomacy and military force. But the elusive key to Asia's largest markets remained the slumbering giant of Ch'ing China, whose Manchu rulers, like the shoguns of Japan, remained unimpressed by European overtures.

The spices of the East Indies, *such as pepper, were among the most highly valued trade commodities from Asia.*

RUSSIAN EMPIRE
Bering Strait
ASIA
Aral Sea
Caspian Sea
MONGOLIA
MANCHU (CH'ING) CHINA
Sea of Japan
JAPAN
KOREA
TIBET (Chinese protectorate from 1750)
Himalayas
NEPAL
BHUTAN
PERSIA
AFGHANISTAN
The Gulf
ARABIA
OMAN
INDIA
Diu
Daman
Arabian Sea
Goa
Mahé
Pondicherry
Karikal
Bay of Bengal
BURMA
TENASSERIM 1826
ANDAMAN IS
ANNAM
SIAM
Macao
FORMOSA
South China Sea
PACIFIC OCEAN
PHILIPPINES
MARIANAS
CAROLINE IS
Micronesia
Melanesia
ETHIOPIA
LACCADIVE IS 1791
Ceylon
MALAYA
Malacca 1824
SINGAPORE 1819
MALDIVE IS 1887
ZANZIBAR (to Oman)
SEYCHELLES 1794
CHAGOS IS
DUTCH EAST INDIES
New Guinea
Timor
INDIAN OCEAN
PORTUGUESE EAST AFRICA
Madagascar
HOVA KINGDOM
MAURITIUS 1810
RÉUNION
WESTERN AUSTRALIA 1829
NEW SOUTH WALES
AUSTRALIA
LORD HOWE I.
NEW ZEALAND
CHATHAM IS 1791
TASMANIA (Van Diemen's Land)
AUCKLAND IS 1806
MACQUARIE IS 1811

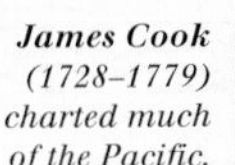

James Cook *(1728–1779) charted much of the Pacific.*

OCEANIA

THOUGH PORTUGUESE and Dutch explorers had confirmed the existence of Australasia in the 17th century, it was not until the voyages of Captain Cook in the 1770s that the geography of the Pacific was established, and the fertile eastern coast of Australia was explored and charted. Over the next 30 years, small settlements were established around the coast; by 1829, Britain had brought the whole of Australia under the British flag.

The first European *migrants to Africa settled in Cape Colony.*

AFRICA

THE NORTHERN REGIONS OF AFRICA were part of the vast Islamic Ottoman Empire; from here Islam spread south to west Africa and the Horn of Africa. Holy wars (or *jihads*) in the late 18th and early 19th centuries completed the conversion to Islam of much of Saharan and sub-Saharan Africa. Large tribal kingdoms flourished in the Congo basin and southern Africa.

The development *during the European industrial revolution of mechanized manufacturing plant and machinery, such as power looms, gave Europe effective control of a booming global trade in raw materials and mass-manufactured commodities.*

THE AGE OF EMPIRE: 1830–1914

THE 19TH CENTURY was dominated by the spread of modern industry and transportation, and the expansion of European trade and influence worldwide. Industry made Europe rich and powerful; its capital cities were monuments to the self-confidence of the new European age. Railroads and steamships revolutionized communications, bringing a stream of industrial goods, technical know-how, and European settlers across America, Africa, and Asia. Modern industry and weapons brought Europe to the summit of global influence. In these developments lay the origins of the division of the world into rich and poor regions; a developed, prosperous north and an underdeveloped, dependent south.

GLOBAL STATES AND TERRITORIES

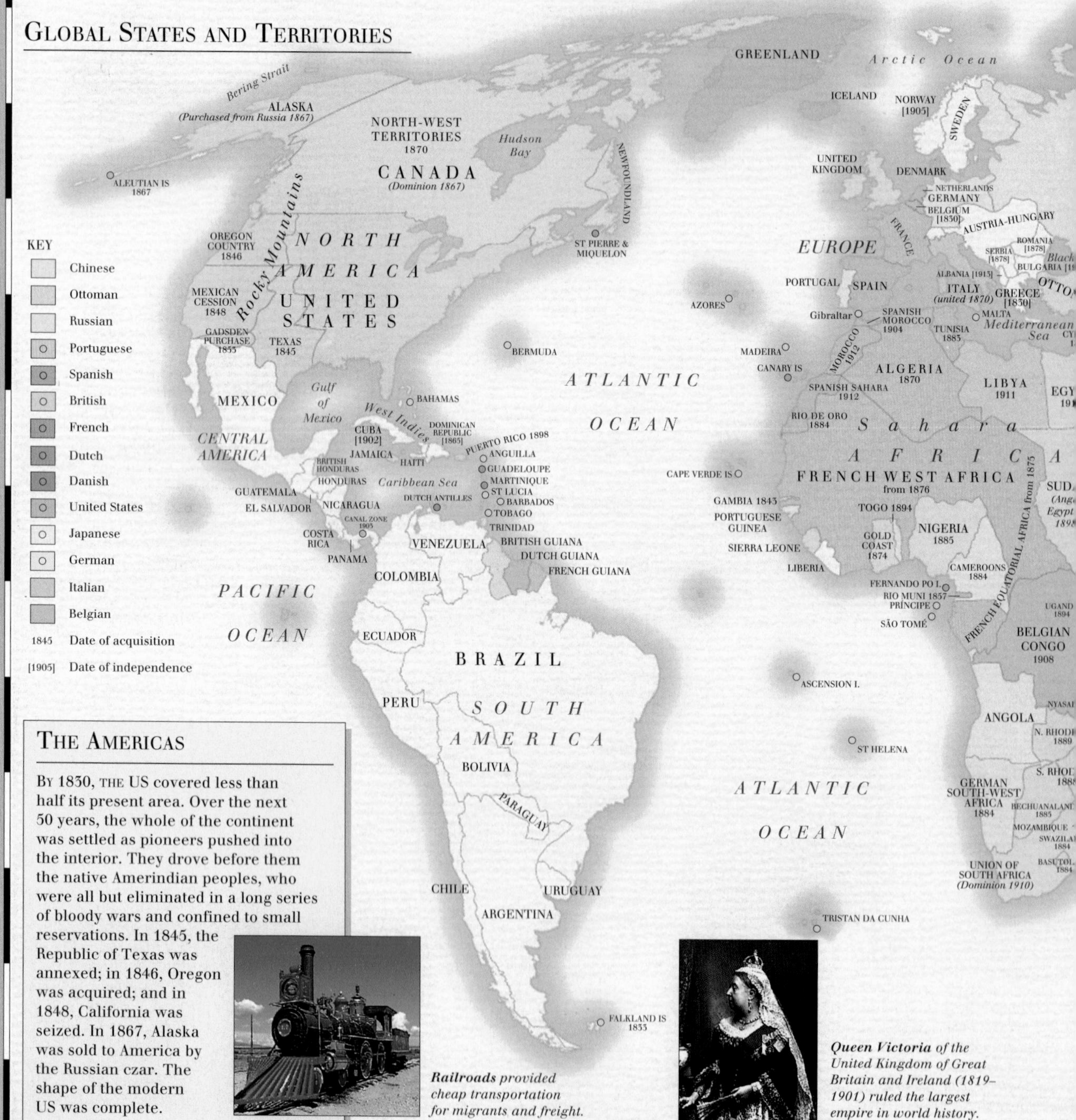
KEY
Chinese
Ottoman
Russian
Portuguese
Spanish
British
French
Dutch
Danish
United States
Japanese
German
Italian
Belgian
1845 Date of acquisition
[1905] Date of independence
Bering Strait
ALASKA
(Purchased from Russia 1867)
ALEUTIAN IS 1867
NORTH-WEST TERRITORIES 1870
CANADA
(Dominion 1867)
Hudson Bay
NEWFOUNDLAND
ST PIERRE & MIQUELON
GREENLAND
Arctic Ocean
ICELAND
NORWAY [1905]
SWEDEN
UNITED KINGDOM
DENMARK
NETHERLANDS
GERMANY
BELGIUM [1830]
AUSTRIA-HUNGARY
FRANCE
EUROPE
ROMANIA [1878]
SERBIA [1878]
BULGARIA
PORTUGAL
SPAIN
ALBANIA [1913]
ITALY (united 1870)
GREECE [1830]
MALTA
Gibraltar
SPANISH MOROCCO 1904
TUNISIA 1883
Mediterranean Sea
MOROCCO 1912
ALGERIA 1870
LIBYA 1911
MADEIRA
CANARY IS
SPANISH SAHARA 1912
RIO DE ORO 1884
Sahara
AFRICA
FRENCH WEST AFRICA from 1876
CAPE VERDE IS
GAMBIA 1843
PORTUGUESE GUINEA
SIERRA LEONE
LIBERIA
TOGO 1894
GOLD COAST 1874
NIGERIA 1885
CAMEROONS 1884
FERNANDO PO I.
RIO MUNI 1857
PRÍNCIPE
SÃO TOMÉ
FRENCH EQUATORIAL AFRICA from 1875
BELGIAN CONGO 1908
ANGOLA
GERMAN SOUTH-WEST AFRICA 1884
UNION OF SOUTH AFRICA (Dominion 1910)
ASCENSION I.
ST HELENA
TRISTAN DA CUNHA
AZORES
BERMUDA
ATLANTIC OCEAN
OREGON COUNTRY 1846
Rocky Mountains
NORTH AMERICA
UNITED STATES
MEXICAN CESSION 1848
GADSDEN PURCHASE 1853
TEXAS 1845
MEXICO
Gulf of Mexico
CENTRAL AMERICA
West Indies
BAHAMAS
CUBA [1902]
DOMINICAN REPUBLIC [1865]
JAMAICA
HAITI
PUERTO RICO 1898
ANGUILLA
GUADELOUPE
MARTINIQUE
ST LUCIA
BARBADOS
TOBAGO
TRINIDAD
BRITISH HONDURAS
HONDURAS
Caribbean Sea
DUTCH ANTILLES
GUATEMALA
EL SALVADOR
NICARAGUA
CANAL ZONE 1903
COSTA RICA
PANAMA
VENEZUELA
BRITISH GUIANA
DUTCH GUIANA
FRENCH GUIANA
COLOMBIA
PACIFIC OCEAN
ECUADOR
BRAZIL
PERU
SOUTH AMERICA
BOLIVIA
PARAGUAY
CHILE
URUGUAY
ARGENTINA
FALKLAND IS 1833

THE AMERICAS

BY 1830, THE US covered less than half its present area. Over the next 50 years, the whole of the continent was settled as pioneers pushed into the interior. They drove before them the native Amerindian peoples, who were all but eliminated in a long series of bloody wars and confined to small reservations. In 1845, the Republic of Texas was annexed; in 1846, Oregon was acquired; and in 1848, California was seized. In 1867, Alaska was sold to America by the Russian czar. The shape of the modern US was complete.

***Railroads** provided cheap transportation for migrants and freight.*

***Queen Victoria** of the United Kingdom of Great Britain and Ireland (1819–1901) ruled the largest empire in world history.*

EUROPE

IN THE 19TH CENTURY, Europe was transformed into an industrial economy. In the new industrial cities, pressure developed for liberal reforms and parliamentary politics. Nationalists created new states in Germany, Italy, Greece, Serbia, and Belgium. While the modern map of Europe gradually began to take shape, European imperialists brought still more areas of the world under their control.

***Sailing ships** carried most oceanic trade until 1900.*

ASIA

BUILDING ON COLONIAL INTERESTS that stretched back into the 18th century, Britain and France transformed the political world of south Asia. Britain extended its rule in India and, in 1885, Burma was brought under British control. The Vietnamese and Chinese Empires were pressured by Europeans anxious to trade and to spread Christianity: the Ch'ing empire conceded areas of influence; the Vietnamese empire resisted and was brought under French domination by force. By the 1890s the whole of southern Asia except for Siam was dominated by Europe, which created the modern state structure of the region.

***The Japanese emperor Meiji** (1852–1912) opened Japan to Western trade and influence.*

RUSSIAN EMPIRE
Bering Strait
KAZAKHSTAN 1854
Aral Sea
Caspian Sea
MONGOLIA (autonomous 1912)
MANCHURIA
AMUR 1858
1853 SAKHALIN 1905
USSURI 1860
KURILE IS 1875
TURKESTAN 1895
BUKHARA 1868
TURKMENISTAN 1885
ASIA
Sea of Japan
JAPAN
Port Arthur 1905
Weihaiwei 1898
Tsingtao 1898
KOREA 1905
AFGHANISTAN
TIBET [1912]
Himalayas
CHINA
PERSIA
EMPIRE
The Gulf
NEPAL
BHUTAN
RYUKYU IS 1874
BAHRAIN 1861
ARABIA
OMAN
Chandernagore
INDIA
BURMA
Macao
Hong Kong 1841
FORMOSA 1895
PACIFIC OCEAN

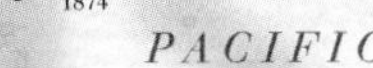

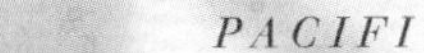

Diu
Daman
Arabian Sea
Bay of Bengal
FRENCH INDO-CHINA 1887
South China Sea
MARIANAS 1899
ERITREA 1889
HADHRAMAUT 1888
Aden 1839
Goa
SIAM
PHILIPPINES 1898
SOCOTRA 1886
Mahé
Pondicherry
GUAM 1898
BRITISH SOMALILAND 1884
FRENCH SOMALILAND 1884
LACCADIVE IS
Karikal
ANDAMAN IS
CAROLINE IS 1899
CEYLON
ETHIOPIA
ITALIAN SOMALILAND 1889
NICOBAR IS 1869
BRITISH NORTH BORNEO 1881
Micronesia
MALAYA
SARAWAK 1888
MALDIVE IS
BISMARCK ARCHIPELAGO 1884
NAURU 1888
BRITISH EAST AFRICA 1888
SEYCHELLES
Melanesia
ZANZIBAR 1890
CHAGOS IS
DUTCH EAST INDIES
NEW GUINEA
GERMAN EAST AFRICA 1885
PAPUA 1906
SOLOMON IS 1893
CHRISTMAS I. 1888
TIMOR
COMORO IS 1886
COCOS IS 1857
INDIAN OCEAN
MADAGASCAR 1882
MAURITIUS
RÉUNION
NEW CALEDONIA 1853
AUSTRALIA (Commonwealth 1901)
NORFOLK ISLAND
NEW ZEALAND 1840 (Dominion 1907)
CHATHAM IS.
TASMANIA
AUCKLAND IS
MACQUARIE IS

***The colonization** of Australia and New Zealand was based on sheep farming.*

OCEANIA

DURING THE 19TH century, Australia and New Zealand remained closely tied to the British homeland. British settlers came to farm and later to prospect for gold and other valuable minerals. In 1840, New Zealand came under British rule and the native Maoris were forced off the land. Not until 1872 was the continent of Australia traversed, and not until 1901 was a single state, the Commonwealth of Australia, proclaimed.

***Quinine** – the cure for malaria.*

***New medicines** made the colonization of Africa possible.*

AFRICA

THE POLITICAL STRUCTURE of independent Africa was torn up by encroaching European empires. As native societies reacted violently to European intrusion, so European military and political power was increased to secure European interests. In 1884, in Berlin, the European powers divided Africa between them. The "Partition of Africa" established the modern frontiers of many states.

***The European** imperial powers maintained control of their often far-flung colonies by military superiority. Native forces were rarely a match for the large, highly trained armies, powerful navies, and technically advanced weaponry which the Europeans had at their disposal.*

***The Gatling gun,** the most successful of the hand-crank-operated machine guns of the 19th century.*

THE AGE OF GLOBAL WAR: 1914–1945

IN 1914, IMPERIAL AND MILITARY rivalry in Europe provoked the first of two world wars, the largest and most destructive wars in human history. At the end of the first war, in 1918, the old international order was dead. The Russian Empire collapsed in revolution and was transformed by a communist minority into the Soviet Union. The German, Habsburg, and Ottoman empires were dismembered. A fragile peace ensued but the old equilibrium was gone. The rise of strident nationalism in Germany, Japan, and Italy destroyed the peace once again in 1939. The second war cost the lives of 50 million people and ravaged Europe and Asia. At its end, in 1945, the US and the Soviet Union had emerged as the new superpowers.

GLOBAL STATES AND TERRITORIES

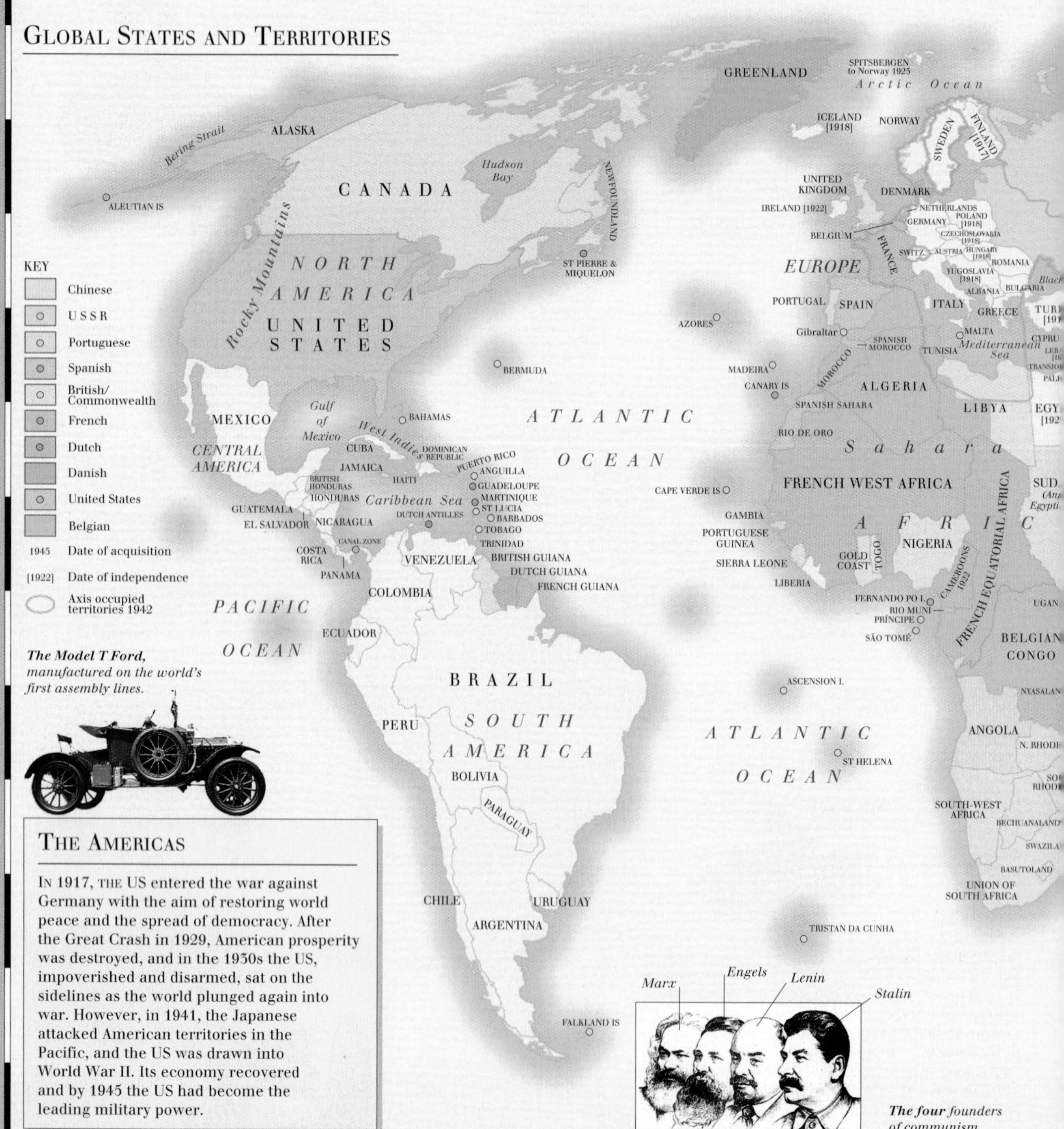
KEY
Chinese
USSR
Portuguese
Spanish
British/ Commonwealth
French
Dutch
Danish
United States
Belgian
1945 Date of acquisition
[1922] Date of independence
Axis occupied territories 1942
GREENLAND
SPITSBERGEN to Norway 1925
Arctic Ocean
ICELAND [1918]
NORWAY
SWEDEN
FINLAND [1917]
Bering Strait
ALASKA
ALEUTIAN IS
Hudson Bay
CANADA
NEWFOUNDLAND
UNITED KINGDOM
DENMARK
IRELAND [1922]
NETHERLANDS
POLAND [1918]
GERMANY
BELGIUM
CZECHOSLOVAKIA [1918]
FRANCE
SWITZ.
AUSTRIA
HUNGARY [1918]
ROMANIA
YUGOSLAVIA [1918]
ALBANIA
BULGARIA
Rocky Mountains
NORTH AMERICA
UNITED STATES
ST PIERRE & MIQUELON
EUROPE
PORTUGAL
SPAIN
ITALY
GREECE
AZORES
Gibraltar
SPANISH MOROCCO
MALTA
TUNISIA
Mediterranean Sea
BERMUDA
MADEIRA
CANARY IS
MOROCCO
ALGERIA
SPANISH SAHARA
LIBYA
MEXICO
Gulf of Mexico
BAHAMAS
ATLANTIC OCEAN
RIO DE ORO
CENTRAL AMERICA
West Indies
CUBA
DOMINICAN REPUBLIC
PUERTO RICO
Sahara
JAMAICA
HAITI
ANGUILLA
BRITISH HONDURAS
GUADELOUPE
FRENCH WEST AFRICA
HONDURAS
Caribbean Sea
MARTINIQUE
ST LUCIA
CAPE VERDE IS
GUATEMALA
DUTCH ANTILLES
BARBADOS
EL SALVADOR
NICARAGUA
TOBAGO
GAMBIA
AFRICA
CANAL ZONE
TRINIDAD
PORTUGUESE GUINEA
NIGERIA
COSTA RICA
VENEZUELA
BRITISH GUIANA
SIERRA LEONE
GOLD COAST
TOGO
FRENCH EQUATORIAL AFRICA
PANAMA
DUTCH GUIANA
FRENCH GUIANA
LIBERIA
CAMEROONS 1922
COLOMBIA
PACIFIC OCEAN
FERNANDO PO I.
RIO MUNI
PRÍNCIPE
SÃO TOMÉ
ECUADOR
BELGIAN CONGO
BRAZIL
ASCENSION I.
NYASALAND
PERU
SOUTH AMERICA
ATLANTIC OCEAN
ANGOLA
ST HELENA
BOLIVIA
PARAGUAY
SOUTH-WEST AFRICA
BECHUANALAND
SWAZILAND
BASUTOLAND
UNION OF SOUTH AFRICA
CHILE
URUGUAY
ARGENTINA
TRISTAN DA CUNHA
FALKLAND IS
Marx
Engels
Lenin
Stalin

The Model T Ford, manufactured on the world's first assembly lines.

The four founders of communism.

THE AMERICAS

IN 1917, THE US entered the war against Germany with the aim of restoring world peace and the spread of democracy. After the Great Crash in 1929, American prosperity was destroyed, and in the 1930s the US, impoverished and disarmed, sat on the sidelines as the world plunged again into war. However, in 1941, the Japanese attacked American territories in the Pacific, and the US was drawn into World War II. Its economy recovered and by 1945 the US had become the leading military power.

EUROPE

BOTH WORLD WARS had their origins in Europe. In 1914, Germany invaded Belgium; Britain, France, and Russia combined to defeat it, with US help. In 1918 new nation states were established in eastern Europe. But, by 1939, revived German nationalism started a second world war; much of western Europe came under a German "New Order" until the Soviet Union, Britain, and the US developed sufficient military strength to reconquer Europe and defeat Germany.

World War II *was decided by mechanical and industrial superiority.*

ASIA

THE COLLAPSE OF THE CHINESE EMPIRE in 1911, followed in 1917 by the disappearance of the Russian Empire, produced instability across Asia. Full-scale war broke out between Japan and China in 1937, with Japan trying to conquer China. The Soviet Union was the victim of German aggression from 1941. Both Japan and Germany were held at bay by communist forces which eventually succeeded in imposing stable politics on Asia. By 1945, the Soviet Union had reconquered its lost territories and dominated eastern Europe. In China, communist armies filled the vacuum left by the Japanese defeat.

Mahatma Gandhi *(1868–1948) led India to independence through peaceful noncooperation and protest.*

U S S R
Bering Strait
SAKHALIN 1945
KURILE IS 1945
MONGOLIA [1924]
Aral Sea
Caspian Sea
A S I A
Sea of Japan
JAPAN
KOREA [1945]
CHINA
IRAN (Persia)
AFGHANISTAN
Himalayas
TIBET
IRAQ [1932]
NEPAL
BHUTAN
The Gulf
BAHRAIN
PACIFIC OCEAN
RYUKYU IS 1945
Chandernagore
INDIA
Macao
Hong Kong
TAIWAN (Formosa) 1945
SAUDI ARABIA [1932]
OMAN
Diu
Daman
BURMA
FRENCH INDO-CHINA
YEMEN [1918]
Red Sea
Arabian Sea
Bay of Bengal
THAILAND (Siam)
South China Sea
PHILIPPINES
MARIANAS 1945
ERITREA 1941
HADHRAMAUT
Aden
Goa
SOCOTRA
Mahé
Pondicherry
GUAM
BRITISH SOMALILAND
FRENCH SOMALILAND
LACCADIVE IS
Karikal
ANDAMAN IS
CAROLINE IS 1945
CEYLON
ETHIOPIA
ITALIAN SOMALILAND 1941
NICOBAR IS
Micronesia
BRITISH NORTH BORNEO
MALAYA
SARAWAK
MALDIVE IS
KENYA
BISMARCK ARCHIPELAGO 1945
NAURU 1945
Melanesia
ZANZIBAR
SEYCHELLES
NEW GUINEA
CHAGOS IS
DUTCH EAST INDIES
PAPUA
SOLOMON IS
TANGANYIKA
MOZAMBIQUE
COMORO IS
INDIAN OCEAN
CHRISTMAS I. 1888
COCOS IS
TIMOR
MADAGASCAR
MAURITIUS
RÉUNION
NEW CALEDONIA
AUSTRALIA (Dominion 1926)
LORD HOWE I.
NEW ZEALAND
CHATHAM IS
TASMANIA
AUCKLAND IS
MACQUARIE IS

OCEANIA

FOR THE ONLY TIME in its history, Australia was faced with the very real prospect of invasion. In World War II, Japanese armies reached the island of New Guinea, and bombed towns in northern Australia. Japanese submarines attacked Sydney harbor. The Battle of the Coral Sea, in May 1942, saved Australia, but it took almost three years to clear Japanese forces from the South Pacific, where they hung on grimly to the rich oil and mineral resources they had captured.

Japan promoted *itself as the liberator of Asia from the chains of European colonialism.*

Haile Selassie *(1892–1975), ruler of Ethiopia, the only independent empire in Africa.*

MIDDLE EAST

IN 1918, THE OTTOMAN EMPIRE disappeared after being in existence for 400 years. The modern map of north Africa and the Middle East was carved out of its ruins by the victors of World War I. The genocide of Europe's Jews by Nazi Germany during World War II accelerated the foundation of a new state of Israel in 1948, leading inexorably to conflict between displaced native Arabs and Jewish migrants.

The conquest *of the air was the most important technological achievement of the period. It added a devastating dimension to warfare, in the form of aerial bombing, while transforming civil transportation.*

A German Zeppelin *airship of the 1930s.*

The Modern Age: 1945–Present Day

Two nuclear explosions in Japan finally ended World War II and ushered in a new era of global rivalry. Total destruction by nuclear weapons was a real possibility as the two superpowers, the Soviet Union and the US, became locked with their allies in a fearful Cold War. Various side conflicts were staged in the ruins of the old empires where new nation states, with old colonial borders, were emerging. By 1991, communist power in the Soviet Union and Eastern Europe had crumbled, while the capitalism of the West had gained in strength.

The US is now the sole global superpower and has used its status to push Western goods and ideals to every corner of the globe, provoking often violent reactions among proud, conservative cultures. The age of globalization has begun.

Global States and Territories

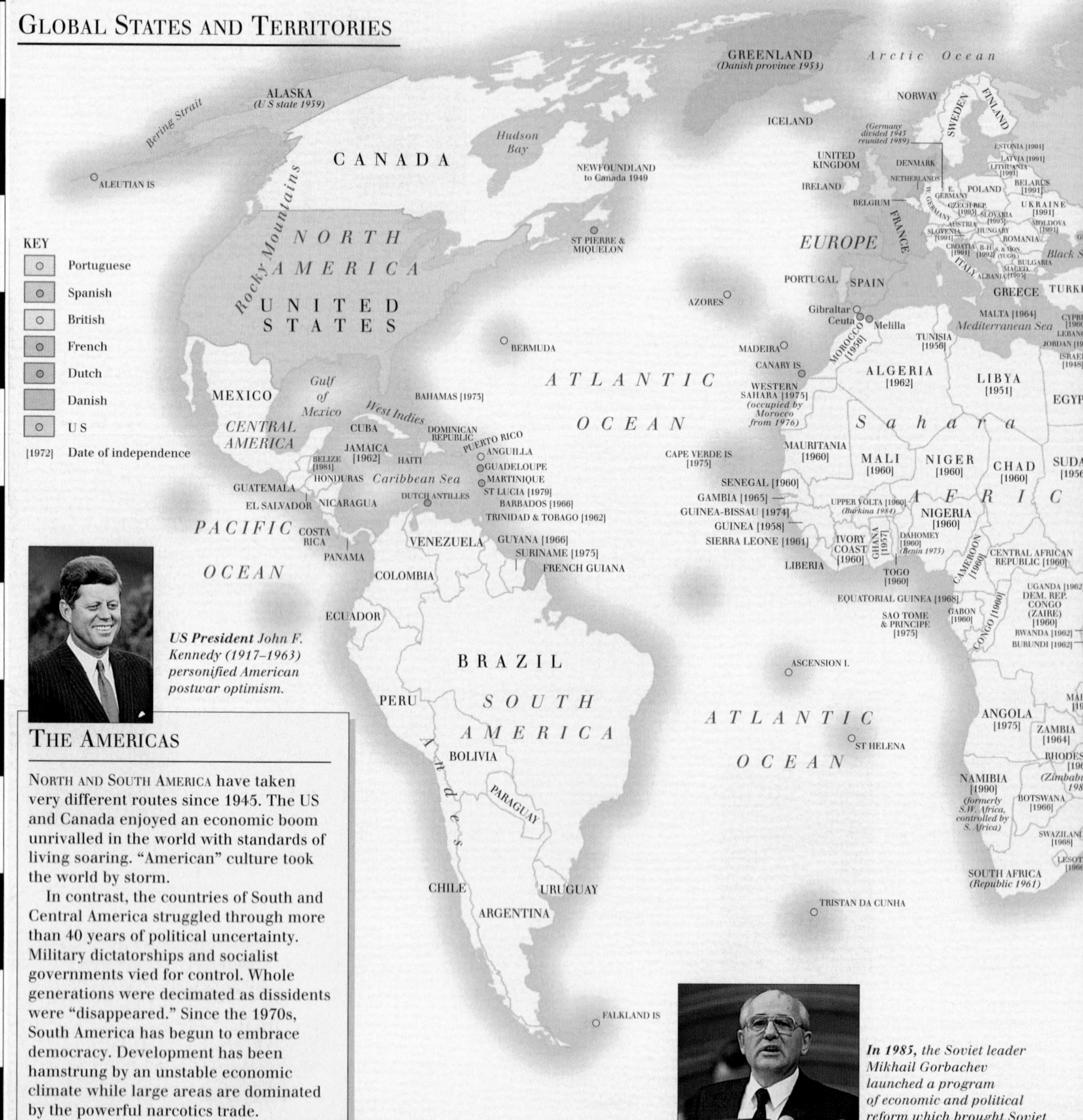

KEY
Portuguese
Spanish
British
French
Dutch
Danish
US
[1972] Date of independence
GREENLAND
(Danish province 1953)
Arctic Ocean
ALASKA
(U S state 1959)
Bering Strait
ALEUTIAN IS
Hudson Bay
CANADA
NEWFOUNDLAND to Canada 1949
ST PIERRE & MIQUELON
Rocky Mountains
NORTH AMERICA
UNITED STATES
BERMUDA
AZORES
MEXICO
Gulf of Mexico
BAHAMAS [1975]
West Indies
CENTRAL AMERICA
CUBA
DOMINICAN REPUBLIC
PUERTO RICO
JAMAICA [1962]
HAITI
BELIZE [1981]
HONDURAS
Caribbean Sea
ANGUILLA
GUADELOUPE
MARTINIQUE
ST LUCIA [1979]
BARBADOS [1966]
TRINIDAD & TOBAGO [1962]
DUTCH ANTILLES
GUATEMALA
EL SALVADOR
NICARAGUA
COSTA RICA
PANAMA
PACIFIC OCEAN
VENEZUELA
GUYANA [1966]
SURINAME [1975]
FRENCH GUIANA
COLOMBIA
ECUADOR
BRAZIL
SOUTH AMERICA
PERU
BOLIVIA
PARAGUAY
Andes
CHILE
URUGUAY
ARGENTINA
FALKLAND IS
ATLANTIC OCEAN
ATLANTIC OCEAN
ASCENSION I.
ST HELENA
TRISTAN DA CUNHA
ICELAND
NORWAY
SWEDEN
FINLAND
(Germany divided 1945 reunited 1989)
UNITED KINGDOM
IRELAND
DENMARK
NETHERLANDS
BELGIUM
W. GERMANY
E. GERMANY
POLAND
ESTONIA [1991]
LATVIA [1991]
LITHUANIA [1991]
BELARUS [1991]
UKRAINE [1991]
MOLDOVA [1991]
CZECH REP. [1993]
SLOVAKIA [1993]
AUSTRIA
HUNGARY
SLOVENIA [1991]
CROATIA [1991]
ROMANIA
BULGARIA
ALBANIA
ITALY
FRANCE
EUROPE
Black Sea
PORTUGAL
SPAIN
GREECE
TURKEY
Gibraltar
Ceuta
Melilla
MOROCCO [1956]
MALTA [1964]
Mediterranean Sea
TUNISIA [1956]
MADEIRA
CANARY IS
WESTERN SAHARA [1975] (occupied by Morocco from 1976)
ALGERIA [1962]
LIBYA [1951]
EGYPT
Sahara
CAPE VERDE IS [1975]
MAURITANIA [1960]
MALI [1960]
NIGER [1960]
CHAD [1960]
SUDAN [1956]
SENEGAL [1960]
GAMBIA [1965]
GUINEA-BISSAU [1974]
GUINEA [1958]
SIERRA LEONE [1961]
UPPER VOLTA [1960] (Burkina 1984)
AFRICA
NIGERIA [1960]
IVORY COAST [1960]
GHANA [1957]
DAHOMEY [1960] (Benin 1975)
LIBERIA
TOGO [1960]
CAMEROON [1960]
CENTRAL AFRICAN REPUBLIC [1960]
EQUATORIAL GUINEA [1968]
SAO TOME & PRINCIPE [1975]
GABON [1960]
CONGO [1960]
UGANDA [1962]
DEM. REP. CONGO (ZAIRE) [1960]
RWANDA [1962]
BURUNDI [1962]
ANGOLA [1975]
ZAMBIA [1964]
NAMIBIA [1990] (formerly S.W. Africa, controlled by S. Africa)
BOTSWANA [1966]
SWAZILAND [1968]
LESOTHO [1966]
SOUTH AFRICA (Republic 1961)

US President John F. Kennedy (1917–1963) *personified American postwar optimism.*

The Americas

North and South America have taken very different routes since 1945. The US and Canada enjoyed an economic boom unrivalled in the world with standards of living soaring. "American" culture took the world by storm.

In contrast, the countries of South and Central America struggled through more than 40 years of political uncertainty. Military dictatorships and socialist governments vied for control. Whole generations were decimated as dissidents were "disappeared." Since the 1970s, South America has begun to embrace democracy. Development has been hamstrung by an unstable economic climate while large areas are dominated by the powerful narcotics trade.

In 1985, the Soviet leader Mikhail Gorbachev launched a program of economic and political reform which brought Soviet communism to an end.

The Berlin ***Wall****, symbol of the Cold War division of Europe, was demolished in 1989.*

EUROPE

IN 1945, EUROPE LAY IN RUINS. The iron curtain descended to divide capitalist West from communist East. The West underwent an economic boom which restored widespread prosperity and political stability. It progressed toward economic and political unity under the EU. The East, meanwhile, labored under planned economies and quickly sought to catch up with the West after 1989, with most countries now aspiring to join the EU.

ASIA

IN SOUTHERN ASIA, popular nationalist movements came to power in India, Burma, Malaya, and Indonesia; in China and Indo-China, power passed to native communist movements whose roots went back to the 1920s. After 1949, China under Mao Zedong became, with its vast population and large military forces, a second communist superpower. Japan, meanwhile, was Asia's capitalist "miracle." Its economy and cities laid waste by bombing in 1945, it rebuilt with US aid so successfully that by the 1980s, it was the world's second-largest economy. China's potential economic growth, however, could put Japan's past achievements in the shade.

兵民是胜利之本

Chinese communism, *based on the mobilization of peasants and workers, has nevertheless recognized the need for economic reforms.*

RUSSIAN FEDERATION
Bering Strait
KAZAKHSTAN [1991]
ASIA
MONGOLIA
Aral Sea
UZBEKISTAN [1991]
KYRGYZSTAN [1991]
TURKMENISTAN [1991]
TAJIKISTAN [1991]
CHINA
Sea of Japan
JAPAN
S. KOREA [1948]
IRAN
AFGHANISTAN
Himalayas
PAKISTAN [1947]
NEPAL
BHUTAN
The Gulf
QATAR [1971]
BANGLADESH [1971] (formerly E. Pakistan)
INDIA [1947]
OMAN
Arabian Sea
Bay of Bengal
(Vietnam united 1976)
PACIFIC OCEAN
MARIANAS
(Yemen united 1990)
THAILAND
South China Sea
GUAM
DJIBOUTI [1977]
MICRONESIA [1991]
BRUNEI [1984]
PALAU [1995]
INDONESIA [1949]
Melanesia
SEYCHELLES [1976]
CHAGOS IS
E. TIMOR
SOLOMON IS [1978]
INDIAN OCEAN
MADAGASCAR [1960]
RÉUNION
NEW CALEDONIA
AUSTRALIA
NEW ZEALAND
AUCKLAND IS (to N Z)
MACQUARIE IS (to Australia)

A ***treaty*** *banning the testing of nuclear bombs in the Pacific was signed in 1986.*

OCEANIA

THE POSTWAR economies of Japan, the US, and Australia had by the 1990s created a new industrial and trading network around the Pacific Rim. Cheap labor and low overheads drew younger states – South Korea, Taiwan, Singapore, Malaysia, Indonesia – into the system and much of the world's manufacturing is now concentrated there, creating a consequent shift in the balance of the global economy.

Gamal Abd al-Nasser *(1918–1970) of Egypt, galvanized the Arab states to resist the West.*

AFRICA AND THE MIDDLE EAST

THE COLONIAL POWERS, weakened by war, faced an irresistible wave of demands for self-determination. Between 1958 and 1975, 41 African countries gained independence. In north Africa and throughout the Middle East a new form of anti-imperialism emerged in the 1970s in the form of Islamic fundamentalism. In South Africa, white rule and the apartheid system ended in 1994.

From the 1950s *to the 1970s, superpower rivalry focused on space exploration. The Soviets put the first man in space in 1961, and the Americans landed on the moon in 1969. Since then, both manned and unmanned missions have become almost everyday events.*

POPULATION

THE WORLD'S POPULATION – 6.2 billion in 2002 – is likely to reach nearly 10 billion by 2050. Better nutrition, health care, and sanitation mean fewer infant deaths and longer life expectancy, although around 800 million people in the developing world are malnourished and over one billion live in extreme poverty. In much of Africa in particular, the AIDS epidemic is so severe that the population is set to fall significantly. Elsewhere it is lower birthrates, already familiar in most industrialized countries, that have slowed the rate of growth. The result is a rapidly aging population: it is thought that by 2050 there will be around two billion people over the age of 60. The distribution of population is very uneven, dependent on climate, terrain, natural resources, and economic factors. The great majority of people live in coastal zones and along river valleys. Urbanization is on the increase, and by 2002 just under half of the world's population lived in cities – most of them in Asia. The mass migration of people from rural areas in search of work has resulted in the growth of huge sprawling squatter camps on the edge of many Third World cities.

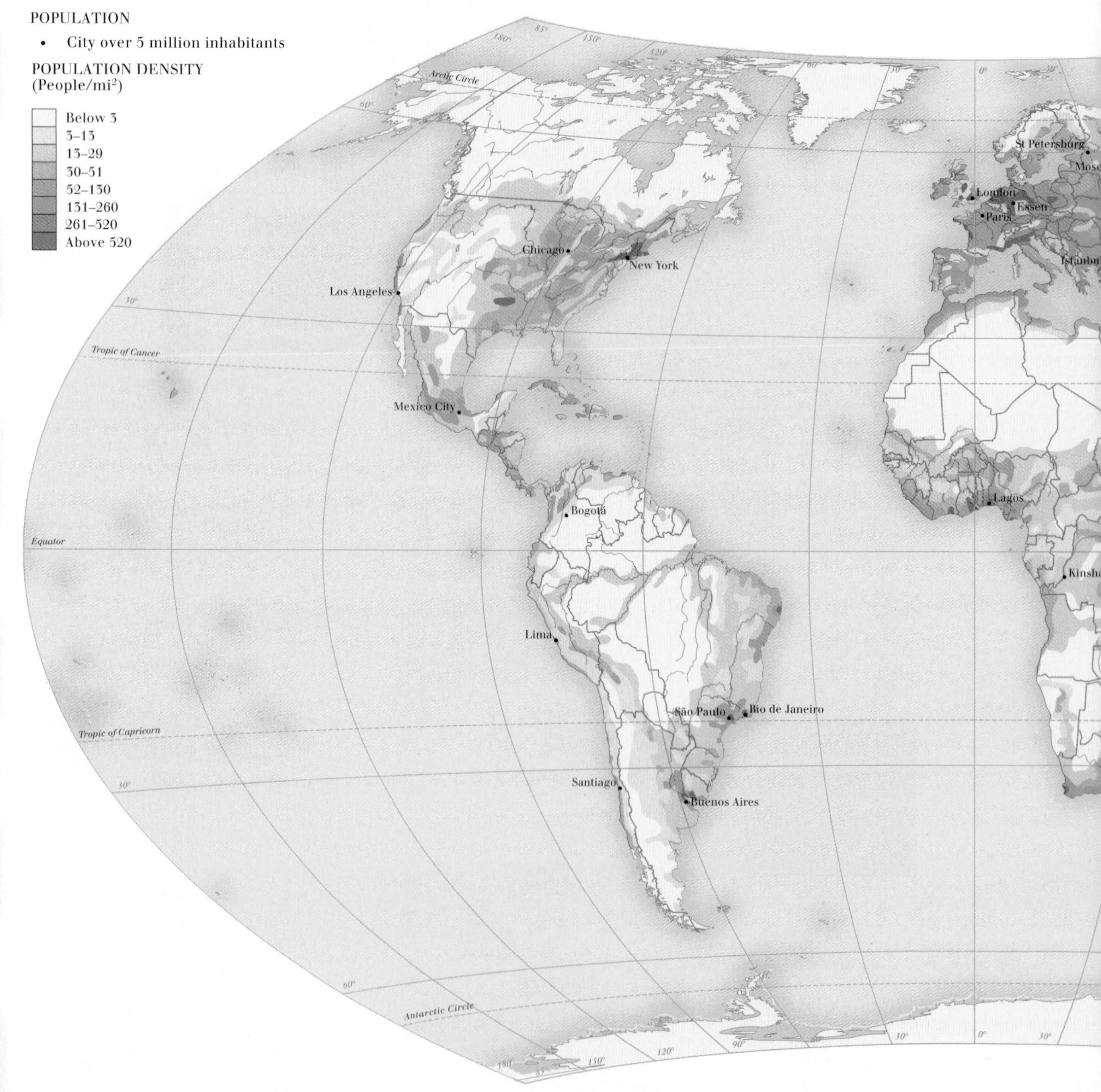
POPULATION
• City over 5 million inhabitants
POPULATION DENSITY
(People/mi²)
Below 3
3–13
13–29
30–51
52–130
131–260
261–520
Above 520
Arctic Circle
Tropic of Cancer
Equator
Tropic of Capricorn
Antarctic Circle
St Petersburg
London
Essen
Paris
Chicago
New York
Los Angeles
Mexico City
Bogotá
Lagos
Lima
São Paulo
Rio de Janeiro
Santiago
Buenos Aires

INFANT MORTALITY

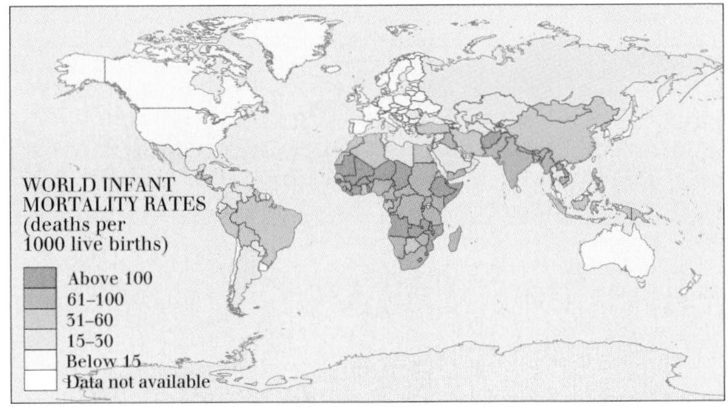
WORLD INFANT MORTALITY RATES (deaths per 1000 live births)
Above 100
61–100
31–60
15–30
Below 15
Data not available

Infant Mortality

INFANT MORTALITY RATES are highest in Africa, South America, and south Asia, where poverty and disease are rife, and where average standards of health care are not as high as in North America or Europe. The country with the highest infant mortality rate is Afghanistan, where years of conflict have devastated communities.

LIFE EXPECTANCY

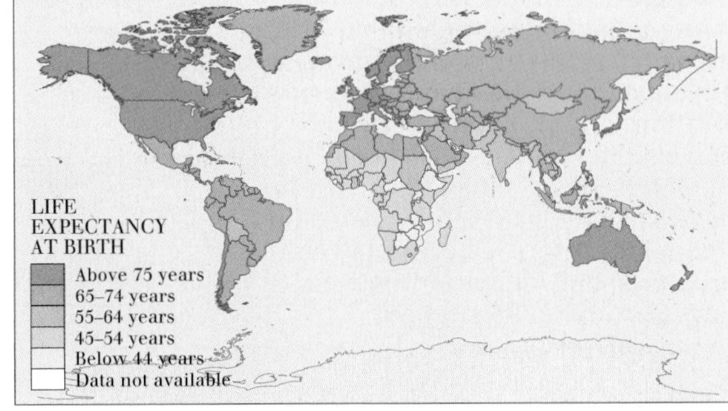
LIFE EXPECTANCY AT BIRTH
Above 75 years
65–74 years
55–64 years
45–54 years
Below 44 years
Data not available

Life Expectancy

LIFE EXPECTANCY IS poorest in Africa, for reasons similar to those noted above. In western Europe and North America, life expectancy is increasing at such a rate that each successive generation may expect to live longer than the last. In the developed world, people can now expect to live twice as long as they did a century ago.

Population Factfile

LARGEST POPULATIONS

1 **China**	1285 million people
2 **India**	1025 million people
3 **USA**	281 million people

LARGEST COUNTRIES BY AREA

1 **Russian Federation**	16,995,800 km² (6,562,100 mi²)
2 **China**	9,326,410 km² (3,600,927 mi²)
3 **Canada**	9,220,970 km² (3,560,217 mi²)

MOST DENSELY POPULATED COUNTRIES

1 **Monaco**	16,253 people/km² (42,095 people/mi²)
2 **Singapore**	6721 people/km² (17,408 people/mi²)
3 **Malta**	1225 people/km² (3173 people/mi²)

MOST SPARSELY POPULATED COUNTRIES

1 **Mongolia**	2 people/km² (4 people/mi²)
2 **Namibia**	2 people/km² (6 people/mi²)
3 **Australia**	3 people/km² (7 people/mi²)

World Economy

THE WEALTHY COUNTRIES of the developed world, with their aggressive, market-led economies and their access to productive new technologies and international markets, dominate the world economic system. At the other extreme, many of the countries of the developing world are locked in a cycle of unrepayable debt, rising populations, and unemployment. The state-managed economies of the former communist bloc began to be dismantled during the 1990s, and China is emerging as a major economic power following decades of isolation. Since the late 1980s, technological advances have enabled transactions between financial centers to occur at even greater speed, and new markets have sprung up throughout the world.

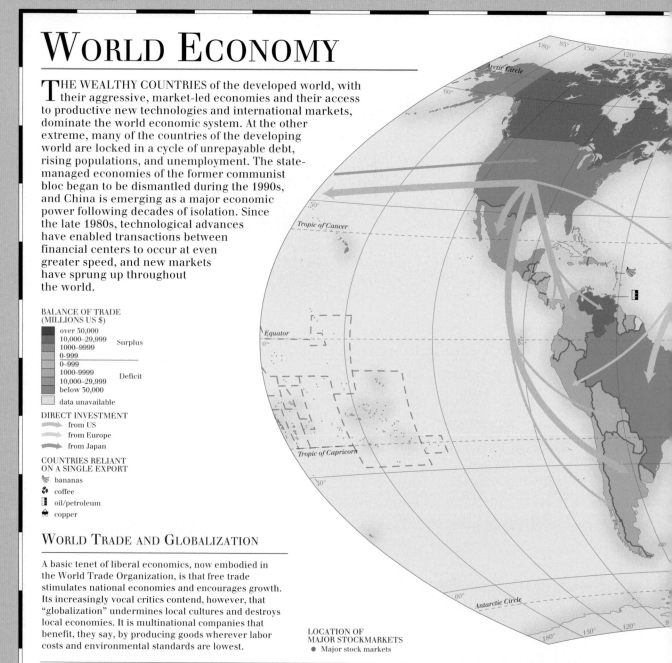

BALANCE OF TRADE (MILLIONS US $)

Surplus:
- over 30,000
- 10,000–29,999
- 1000–9999
- 0–999

Deficit:
- 0–999
- 1000-9999
- 10,000–29,999
- below 30,000

- data unavailable

DIRECT INVESTMENT
- from US
- from Europe
- from Japan

COUNTRIES RELIANT ON A SINGLE EXPORT
- bananas
- coffee
- oil/petroleum
- copper

World Trade and Globalization

A basic tenet of liberal economics, now embodied in the World Trade Organization, is that free trade stimulates national economies and encourages growth. Its increasingly vocal critics contend, however, that "globalization" undermines local cultures and destroys local economies. It is multinational companies that benefit, they say, by producing goods wherever labor costs and environmental standards are lowest.

LOCATION OF MAJOR STOCKMARKETS
- Major stock markets

INTERNATIONAL TRADE

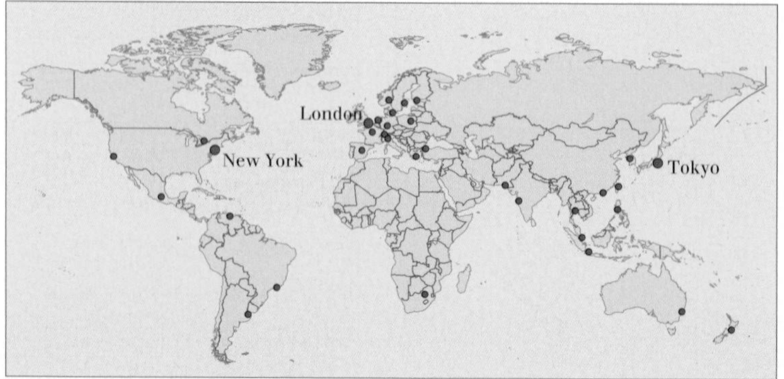
London
New York
Tokyo

World Economies

HIGHEST GNP PER CAPITA

1 Liechtenstein	$50,000
2 Luxembourg	$42,060
3 Switzerland	$38,140
4 Japan	$33,470
5 Norway	$32,050

LOWEST GNP PER CAPITA

1= Somalia	$100
1= Ethiopia	$100
3= Burundi	$110
3= Congo, Dem. Rep.	$110
5 Sierra Leone	$130

Trade Blocs

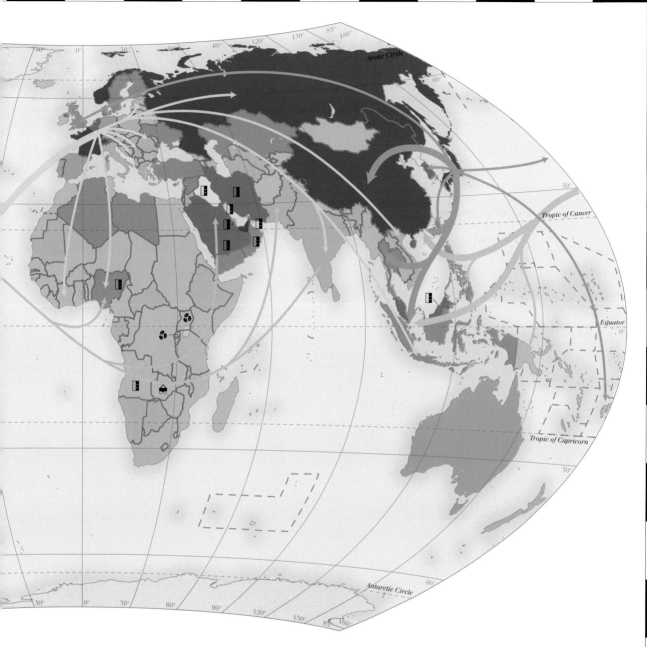

International trade blocs are formed when groups of countries, often already enjoying close military and political ties, join together to offer mutually preferential terms of trade for both imports and exports. Increasingly, global trade is dominated by three main blocs: the EU, NAFTA, and ASEAN. They are supplanting older trade blocs such as the Commonwealth, a legacy of colonialism.

TRADE BLOCS

- EU
- ASEAN
- NAFTA
- SADC
- MERCOSUR
- ECOWAS

TRADE BLOCS

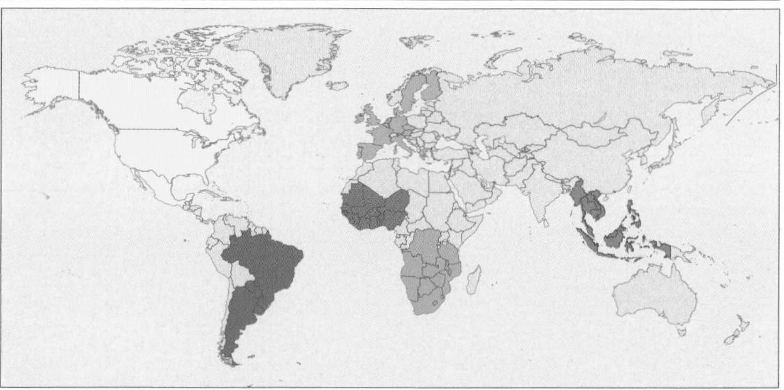

WORLD ECONOMY

THE SIZE OF A COUNTRY'S economy does not relate directly to its population or even its resources. Japan, for example, has a much "bigger" economy than China, India, Russia, or Latin America as a whole. Such imbalances usually occur because countries differ enormously in their living standards, the education and skills of their workforces, the productivity of their agriculture, and the value of their markets. A country's economic performance can be evaluated by calculating its gross national product (GNP). This is the total value of both the goods and the services (including so-called "invisible exports" – financial services, tourism, and so on) that it produces. Most trade (62% of the global total by value) is in manufactured goods, but during the last three decades the most rapidly growing sector has been services – banking, insurance, tourism, consultancy, accountancy, films, music and other cultural services, airlines, and shipping. Accounting for 20% of the total, services now exceed the value of trade in food and raw materials.

COMPARATIVE WORLD WEALTH

A global assessment of GNP by country reveals great disparities. The developed world, with only a quarter of the world's population, has 80% of the world's manufacturing income. This imbalance is maintained as war and political instability undermine poor countries' prospects.

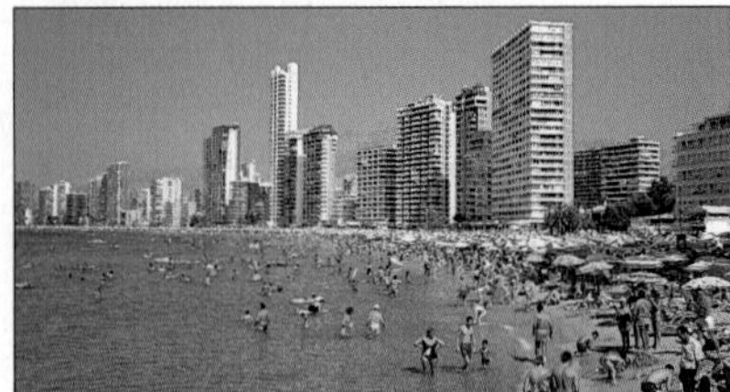

***Mass-market tourism** is now an all-important source of revenue in many countries.*

AVERAGE GDP PER CAPITA (IN $US)

- Above 10,000
- 2000–10,000
- 500–1999
- Below 500
- Data unavailable

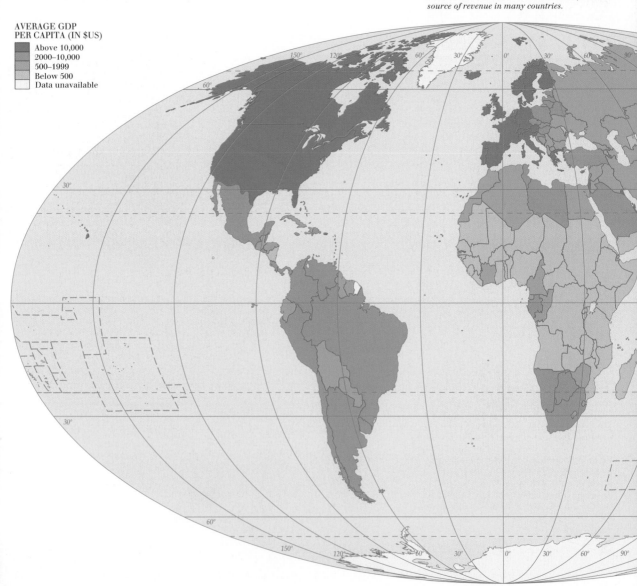

DEBT OF POOR AND MIDDLE-INCOME COUNTRIES

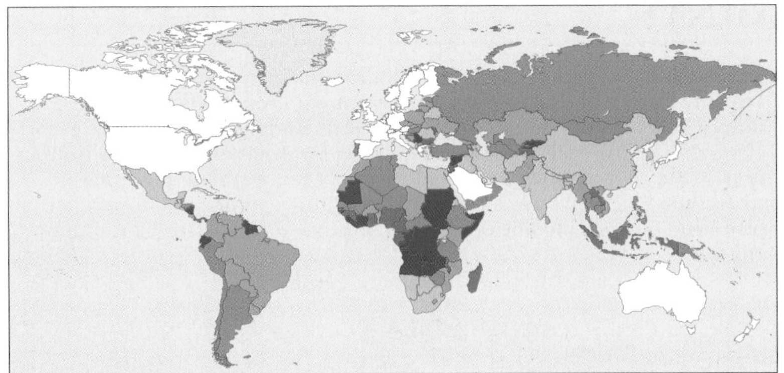

INTERNATIONAL DEBT (AS PERCENTAGE OF GNP)

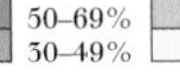
over 100%
70–100%

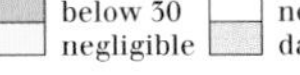
50–69%
30–49%

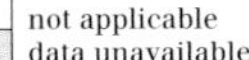
below 30
negligible

not applicable
data unavailable

INTERNATIONAL DEBT

In response to unsustainable levels of debt in the developing world, the IMF and World Bank have introduced a program to help heavily indebted poor countries (HIPCs) manage their repayments. The Jubilee 2000 campaign, advocating debt cancellation, has won some qualified support among creditor countries.

WORLD'S 20 LARGEST CORPORATIONS

		2001, $ millions
1	**Wal-Mart Stores (US)**	**$219,812**
2	**Exxon Mobil (US)**	**$191,581**
3	**General Motors (US)**	**$177,260**
4	**BP (UK)**	**$174,218**
5	**Ford Motor Company (US)**	**$162,412**
6	**Enron (US)**	**$138,718**
7	**DaimlerChrysler (US/Germany)**	**$136,897**
8	**Royal Dutch/Shell Group (UK/Neth.)**	**$135,211**
9	**General Electric (US)**	**$125,913**
10	**Toyota Motor Company (Japan)**	**$120,814**
11	**Citigroup (US)**	**$112,022**
12	**Mitsubishi (Japan)**	**$105,814**
13	**Mitsui (Japan)**	**$101,206**
14	**ChevronTexaco (US)**	**$99,699**
15	**Total Fina Elf (France)**	**$94,312**
16	**Nippon Telegraph and Telephone (Japan)**	**$93,425**
17	**Itochu (Japan)**	**$91,177**
18	**Allianz (Germany)**	**$85,929**
19	**Intl. Business Machines (US)**	**$85,866**
20	**ING Group (Neth.)**	**$82,999**

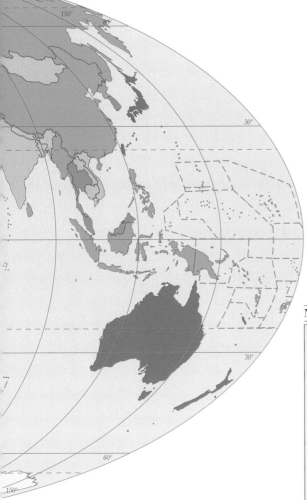

NEWLY INDUSTRIALIZED COUNTRIES

In the 1990s, the fast-growing export-oriented "Asian tiger" economies, such as Singapore, South Korea, and Taiwan, offered exciting prospects for foreign investors. The 1997–1998 Asian financial crisis came as a severe shock. Some countries quickly returned to growth, but investors became more cautious about their exposure in NICs around the world.

NEWLY INDUSTRIALIZED AND INDUSTRIALIZING COUNTRIES

MEXICO
VENEZUELA
COLUMBIA
BRAZIL
CHILE
ARGENTINA
SOUTH AFRICA
THAILAND
MALAYSIA
SINGAPORE
INDONESIA
SOUTH KOREA
TAIWAN
PHILIPPINES

Global Tourism

TOURISM IS THE WORLD'S biggest industry. In 2000 there were a record 697 million tourists worldwide, a number expected to rise to over one billion by 2010, despite the setbacks to the industry in 2001. France is the world's most popular destination country, with 77 million visitors annually, but with cheaper flights, improved transportation, and increased leisure time, many of the countries of the developing world are rapidly becoming tourist meccas. Since the 1960s, mass tourism has become increasingly specialized, encompassing sporting and adventure holidays as well as ecological tours. Tourism employs 200 million people worldwide – 8% of the working population. However, the benefits of tourism are not always felt at a local level, where jobs are often low-paid and menial.

TOURIST ARRIVALS
Data not available
Less than 50,000
50,000 - 99,999
100,000 - 399,999
400,000 - 1 million
1 million - 2.5 million
2.5 million - 10 million
More than 10 million
CARIBBEAN
BARBADOS
GRENADA
ST LUCIA
TRINIDAD & TOBAGO
DOMINICA
ANTIGUA & BARBUDA
GUADELOUPE (Fr.)
ST KITTS & NEVIS
BERMUDA (UK)
MARTINIQUE (Fr.)
ST VINCENT & THE GRENADINES
BAHAMAS
EUROPE
MONACO
SAN MARINO
LIECHTENSTEIN
GIBRALTAR (to UK)
LUXEMBOURG
MALTA
ANDORRA
AFRICA
CAPE VERDE
SAO TOME & PRINCIPE
Alaska (to USA)
CANADA
UNITED STATES OF AMERICA
GREENLAND (to Denmark)
ICELAND
Faeroe Islands (to Denmark)
NORWAY
SWEDEN
UNITED KINGDOM
REP. OF IRELAND
DENMARK
NETH.
BELG.
LUX.
GERMANY
FRANCE
SWITZ.
AUSTRIA
SLOVENIA
CROATIA
SAN MARINO
PORTUGAL
SPAIN
Gibraltar (to UK)
MOROCCO
ALGERIA
TUNISIA
WESTERN SAHARA
MAURITANIA
MALI
NIGER
CAPE VERDE
SENEGAL
GAMBIA
GUINEA-BISSAU
GUINEA
BURKINA
SIERRA LEONE
LIBERIA
IVORY COAST (CÔTE D'IVOIRE)
GHANA
TOGO
BENIN
NIGERIA
CAMEROON
EQUATORIAL GUINEA
SAO TOME & PRINCIPE
GABON
Bermuda (to UK)
MEXICO
BAHAMAS
CUBA
DOMINICAN REPUBLIC
Puerto Rico (to USA)
JAMAICA
HAITI
BELIZE
HONDURAS
GUATEMALA
EL SALVADOR
NICARAGUA
COSTA RICA
PANAMA
ST KITTS & NEVIS
ANTIGUA & BARBUDA
Guadeloupe (to Fr.)
DOMINICA
Martinique (to Fr.)
ST LUCIA
BARBADOS
GRENADA
ST VINCENT & THE GRENADINES
TRINIDAD & TOBAGO
VENEZUELA
GUYANA
SURINAME
French Guiana (to France)
COLOMBIA
ECUADOR
PERU
BRAZIL
BOLIVIA
PARAGUAY
CHILE
ARGENTINA
URUGUAY
PACIFIC OCEAN
ATLANTIC OCEAN

Even in the remote ***Himalayas,*** *rubbish discarded by trekkers and mountaineers pollutes the landscape.*

"Eco-tourism"

Spectacular and ecologically fascinating natural features such as tropical rainforests and coral reefs are becoming increasingly popular as tourist attractions. Holidays in such destinations can put serious pressure on the very ecological systems that bring in the visitors in the first place. Strenuous efforts are being made to limit the environmental impact of this so-called "eco-tourism."

***The beautiful island** of Phuket, Thailand, has been taken over by tourist developments.*

A TOURIST PARADISE?

THE MOST REMOTE CORNERS of the world are now being penetrated by tourists in their quest for the exotic. In many parts of the developing world, tourism can be described as a form of "neocolonialism;" hotels and beaches are owned by multinational companies, and most of the profits are taken outside the country. Tourism frequently alienates local people from their own land, and has a negative impact on the local culture and environment.

PACIFIC OCEAN
- FIJI
- MICRONESIA
- NAURU
- SOLOMON ISLANDS
- VANUATU
- SAMOA
- TONGA
- KIRIBATI

MIDDLE EAST
- BAHRAIN

INDIAN OCEAN
- COMOROS
- MALDIVES
- MAURITIUS
- SEYCHELLES

ASIA
- SINGAPORE

***Eco-tourists** travel to the distant Antarctic, where they observe its rich wildlife.*

Global Security

THE ENDING OF THE COLD WAR in 1989 greatly reduced the risk of another global (and possibly nuclear) war, but did little to end localized tensions and conflicts. Since then territorial disputes, and particularly ethnic tensions, have undermined peace and security around the world. Conflict is now more frequent within states than between them. The first post-Cold War international war, following the invasion of Kuwait by Iraq in 1990, set the tone for joint action by more powerful states to punish apparent injustices elsewhere. Economic sanctions are usually employed as the first stage of punitive measures. The UN, with its laborious decision-making process, has tried to fill the role of global peacemaker wherever possible, without becoming involved in military action. On the other hand, regional organizations such as NATO (with the US and the UK to the fore), the Russian-dominated Commonwealth of Independent States, or ECOWAS in west Africa have taken direct action against the forces of the supposed aggressor. Attacks on the US in 2001 underlined that ideological fanaticism is the greatest threat to global security.

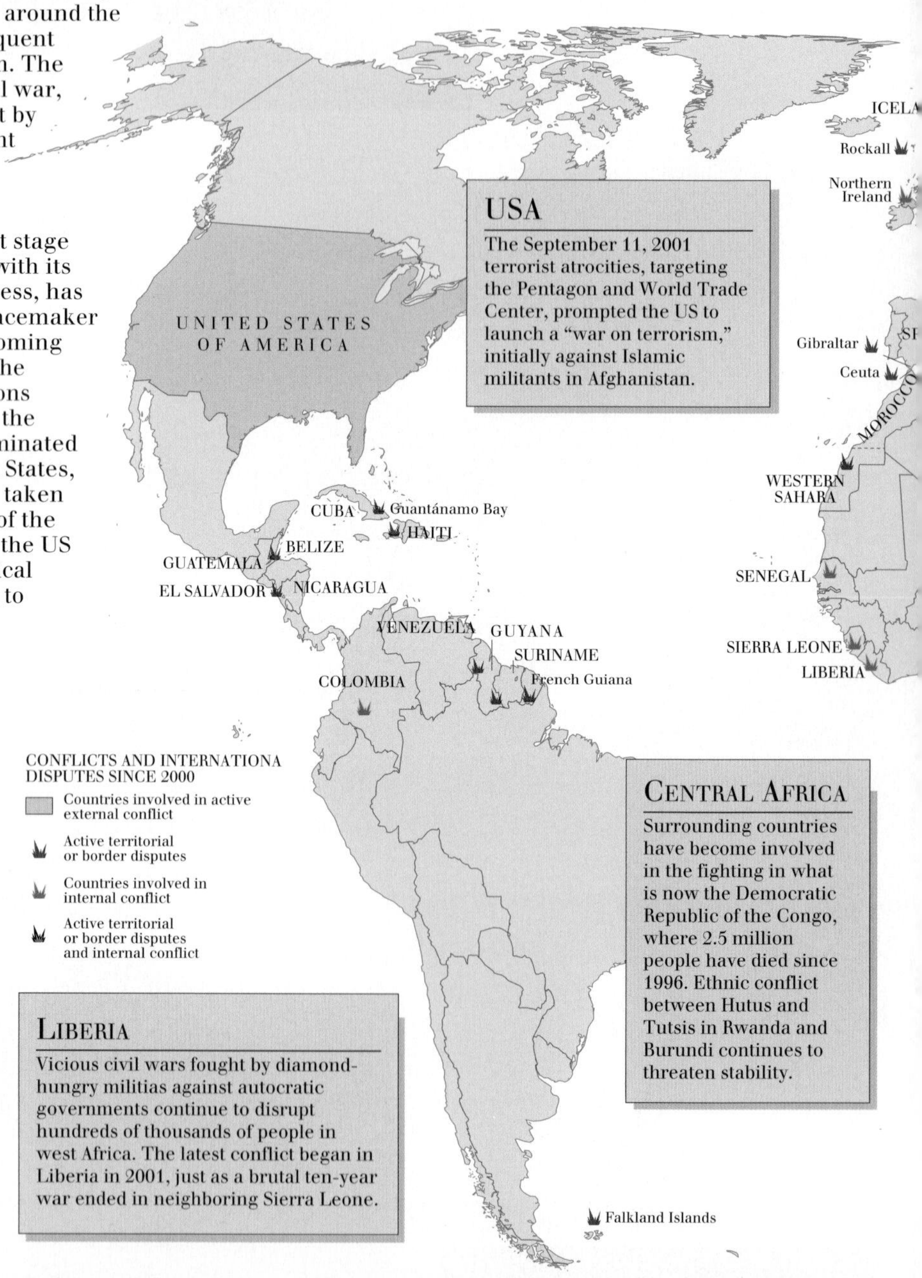

USA

The September 11, 2001 terrorist atrocities, targeting the Pentagon and World Trade Center, prompted the US to launch a "war on terrorism," initially against Islamic militants in Afghanistan.

UNITED STATES OF AMERICA
ICELA
Rockall
Northern Ireland
Gibraltar
Ceuta
MOROCCO
WESTERN SAHARA
CUBA
Guantánamo Bay
HAITI
BELIZE
GUATEMALA
EL SALVADOR
NICARAGUA
SENEGAL
VENEZUELA
GUYANA
SURINAME
French Guiana
COLOMBIA
SIERRA LEONE
LIBERIA
Falkland Islands

Colombia

For more than 40 years left- and right-wing guerrillas have fought a vicious civil war against government troops and one another. Civilians have all too often been caught in the crossfire. Idealistic notions of revolution have been superseded by a simple desire to defend the massive profits accumulated through the narcotics trade. The US has taken an active role in aiding the Colombian government's efforts to regain control and restore peace.

CONFLICTS AND INTERNATIONA DISPUTES SINCE 2000

- Countries involved in active external conflict
- Active territorial or border disputes
- Countries involved in internal conflict
- Active territorial or border disputes and internal conflict

Liberia

Vicious civil wars fought by diamond-hungry militias against autocratic governments continue to disrupt hundreds of thousands of people in west Africa. The latest conflict began in Liberia in 2001, just as a brutal ten-year war ended in neighboring Sierra Leone.

Central Africa

Surrounding countries have become involved in the fighting in what is now the Democratic Republic of the Congo, where 2.5 million people have died since 1996. Ethnic conflict between Hutus and Tutsis in Rwanda and Burundi continues to threaten stability.

Conflict in the Modern Era

1899–1902: Boer War in South Africa

1904–1905: Russo-Japanese War; series of Russian defeats

1910: Start of Mexican Revolution

1911: Start of Chinese revolution

Jun 1914: Assassination of Archduke Franz Ferdinand, heir to the Austro-Hungarian throne, at Sarajevo, precipitates start of World War I

1917: Russian revolution

1918: End of World War I

1932–1935: Paraguay defeats Bolivia in Gran Chaco War

1936–1939: Spanish Civil War

1939: Germany invades Poland; start of World War II

1940: Germany invades Norway, Denmark, Belgium, the Netherlands, and France

1941: Japan attacks US at Pearl Harbor; Germany attempts to invade Russia

1945: German surrender e war in Europe; atomic bor Japan end war in Pacific

1949: Mao Zec Chinese comn win civil w

1900 1910 1920 1930 1940 195

NEPAL

Maoist rebels launched a republican insurgency in 1999, gathering support in rural areas. Attempts to engage the rebels in the political process collapsed in 2001 and fighting intensified. In response the government has launched a determined effort to defeat the rebels, leaving heavy casualties on both sides and ruining the country's all-important tourist industry.

CHECHNYA

Russian troops first entered the breakaway republic of Chechnya in 1994, launching a bloody war to crush Islamic separatists there. Russia claimed to have definitively taken control of the republic in 2002, but Muslim guerrillas continue their struggle.

IRAQ

The regime of President Saddam Hussein has been besieged by the US and its allies since they failed to topple the dictator in their liberation of Kuwait in 1991. A strict economic embargo has been in place since the so-called Gulf War, and US and UK warplanes have carried out punitive airstrikes within strictly policed "no-fly" zones across the north and south of the country since 1998.

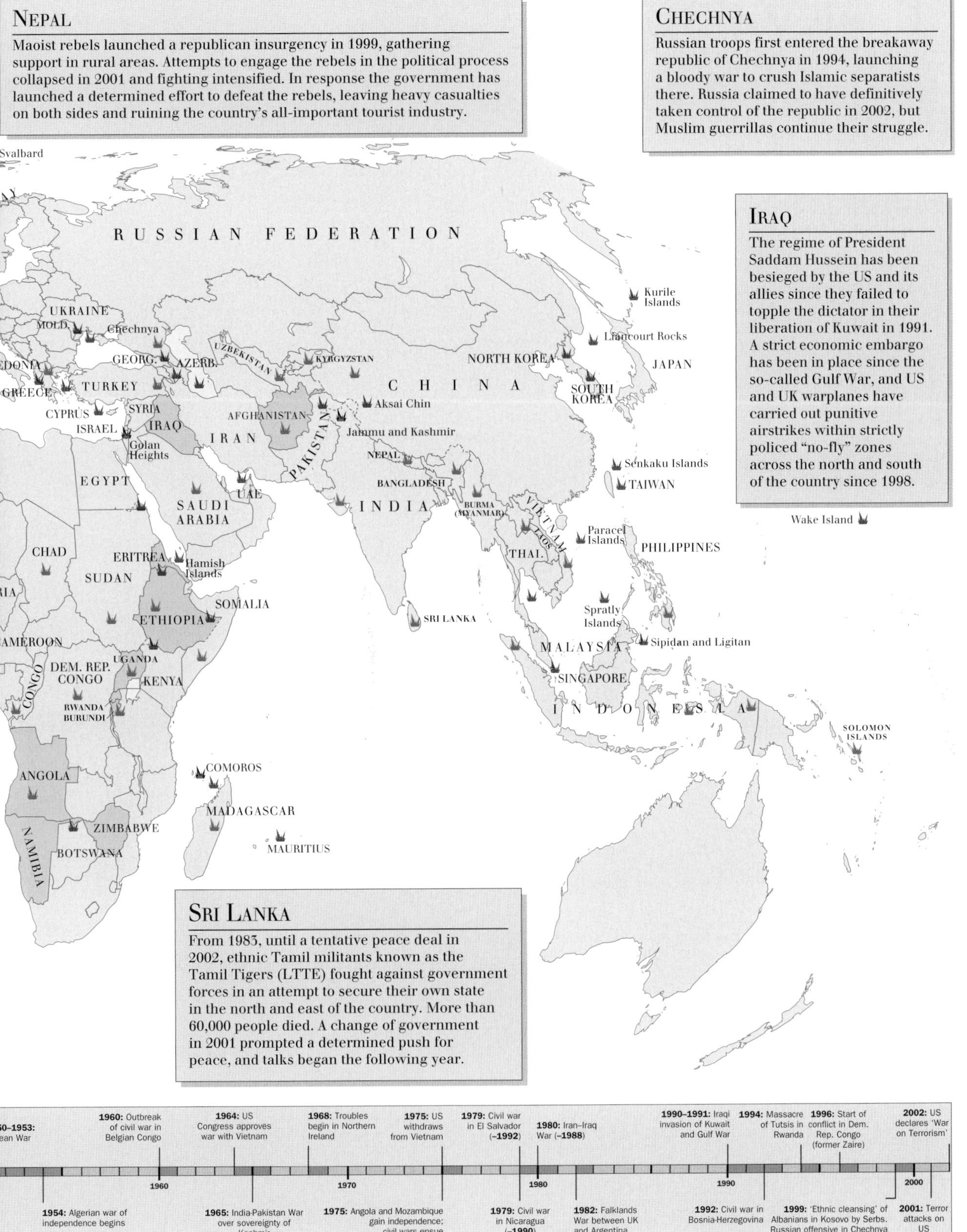
Svalbard
RUSSIAN FEDERATION
UKRAINE
MOLD.
Chechnya
GEORG.
AZERB.
UZBEKISTAN
KYRGYZSTAN
TURKEY
GREECE
CYPRUS
SYRIA
ISRAEL
IRAQ
Golan Heights
IRAN
AFGHANISTAN
PAKISTAN
Aksai Chin
Jammu and Kashmir
NEPAL
CHINA
NORTH KOREA
SOUTH KOREA
JAPAN
Kurile Islands
Liancourt Rocks
Senkaku Islands
TAIWAN
Wake Island
EGYPT
SAUDI ARABIA
UAE
BANGLADESH
INDIA
BURMA (MYANMAR)
VIETNAM
LAOS
THAI.
Paracel Islands
PHILIPPINES
CHAD
SUDAN
ERITREA
Hamish Islands
ETHIOPIA
SOMALIA
SRI LANKA
Spratly Islands
MALAYSIA
Sipidan and Ligitan
SINGAPORE
INDONESIA
SOLOMON ISLANDS
CAMEROON
CONGO
DEM. REP. CONGO
UGANDA
KENYA
RWANDA
BURUNDI
ANGOLA
COMOROS
MADAGASCAR
MAURITIUS
ZIMBABWE
NAMIBIA
BOTSWANA

SRI LANKA

From 1983, until a tentative peace deal in 2002, ethnic Tamil militants known as the Tamil Tigers (LTTE) fought against government forces in an attempt to secure their own state in the north and east of the country. More than 60,000 people died. A change of government in 2001 prompted a determined push for peace, and talks began the following year.

…50–1953: …rean War

1954: Algerian war of independence begins

1960: Outbreak of civil war in Belgian Congo

1964: US Congress approves war with Vietnam

1965: India-Pakistan War over sovereignty of Kashmir

1968: Troubles begin in Northern Ireland

1975: US withdraws from Vietnam

1975: Angola and Mozambique gain independence; civil wars ensue

1979: Civil war in El Salvador (**–1992**)

1979: Civil war in Nicaragua (**–1990**)

1980: Iran–Iraq War (**–1988**)

1982: Falklands War between UK and Argentina

1990–1991: Iraqi invasion of Kuwait and Gulf War

1992: Civil war in Bosnia-Herzegovina

1994: Massacre of Tutsis in Rwanda

1996: Start of conflict in Dem. Rep. Congo (former Zaire)

1999: 'Ethnic cleansing' of Albanians in Kosovo by Serbs. Russian offensive in Chechnya

2001: Terror attacks on US

2002: US declares 'War on Terrorism'

1960 1970 1980 1990 2000

TIME ZONES

–2
–1
0
+1
+2
+3
+4
+5
+6
+7
+8
+9
SVALBARD (to Norway)
JAN MAYEN (to Norway)
Franz Josef Land
Severnaya Zemlya
New Siberian Isla
Novaya Zemlya
ICELAND
FAEROE ISLANDS (to Denmark)
NORWAY
SWEDEN
FINLAND
RUSSIAN FEDERATION
European Russia
Asian Russia
UNITED KINGDOM
REPUBLIC OF IRELAND
CHANNEL ISLANDS (to UK)
DENMARK
EST.
LAT.
LITH.
RUS. FED.
POLAND
BELA.
NETH.
GERMANY
BELG.
LUX.
CZ.REP.
SLVK.
UKRAINE
MOLD.
FRANCE
LIECH.
SWITZ.
AUT.
HUNG.
ROM.
MONACO
SLVN.
CRO.
S.& MON. (YUGO.)
ANDORRA
B.&H.
BULG.
PORTUGAL
SPAIN
VAT. CITY
S.M.
ALB.
MACED.
ITALY
GREECE
TURKEY
GEORGIA
ARMENIA
AZERB.
KAZAKHSTAN
MONGOLIA
UZBEK.
KYRG.
TURKMEN.
TAJ.
CHINA
N. KOREA
S. KOREA
JAPAN
Azores (to Portugal)
GIBRALTAR (to UK)
Madeira (to Portugal)
Canary Islands (to Spain)
MALTA
TUNISIA
CYPRUS
SYRIA
LEB.
ISRAEL
IRAQ
IRAN
AFGH.
PAKISTAN
NEPAL
BHUTAN
JORDAN
KUWAIT
BAHRAIN
QATAR
U.A.E.
MOROCCO
ALGERIA
LIBYA
EGYPT
WESTERN SAHARA
SAUDI ARABIA
OMAN
BANGLADESH
INDIA
BURMA (MYANMAR)
LAOS
TAIWAN
MAURITANIA
MALI
NIGER
CHAD
SUDAN
ERITREA
YEMEN
CAPE VERDE
SENEGAL
GAMBIA
GUINEA-BISSAU
GUINEA
BURKINA
NIGERIA
DJIBOUTI
ETHIOPIA
SOMALIA
THAI.
VIETNAM
CAMB.
Paracel Islands (disputed)
PHILIPPINES
GUAM (to US)
SIERRA LEONE
LIBERIA
IVORY COAST (CÔTE D'IVOIRE)
GHANA
TOGO
BENIN
CAMEROON
EQ. GUINEA
C.A.R.
UGANDA
KENYA
Laccadive Islands (to India)
Andaman Islands (to India)
SRI LANKA
MALDIVES
Spratly Islands (disputed)
BRUNEI
MALAYSIA
SINGAPORE
MICRO
PALAU
ASCENSION ISLAND (to St Helena)
SAO TOME & PRINCIPE
GABON
CONGO
RWANDA
BURUNDI
TANZANIA
DEM. REP. CONGO
ST HELENA (to UK)
SEYCHELLES
BRITISH INDIAN OCEAN TERRITORY (to UK)
CHRISTMAS ISLAND (to Australia)
INDONESIA
EAST TIMOR
COMOROS
MAYOTTE (to France)
ANGOLA
MALAWI
ZAMBIA
MOZAMBIQUE
MADAGASCAR
COCOS (KEELING) ISLANDS (to Australia)
ASHMORE & CARTIER ISLANDS (to Australia)
ATLANTIC OCEAN
NAMIBIA
ZIMB.
BOTS.
MAURITIUS
REUNION (to France)
SWAZILAND
LESOTHO
SOUTH AFRICA
INDIAN OCEAN
AUSTRAL
TRISTAN DA CUNHA (to St Helena)
Gough Island (to Tristan da Cunha)
Greenwich Meridian
FRENCH SOUTHERN & ANTARCTIC TERRITORIES (to France)
Prince Edward Islands (to South Africa)
HEARD & MCDONALD ISLANDS (to Australia)
11:00
12:00
13:00
14:00
15:00
16:00
17:00
18:00
19:00
20:00
21:00
ANTARCTICA
(except for Graham Land, Antarctica is not subject to any time zone)

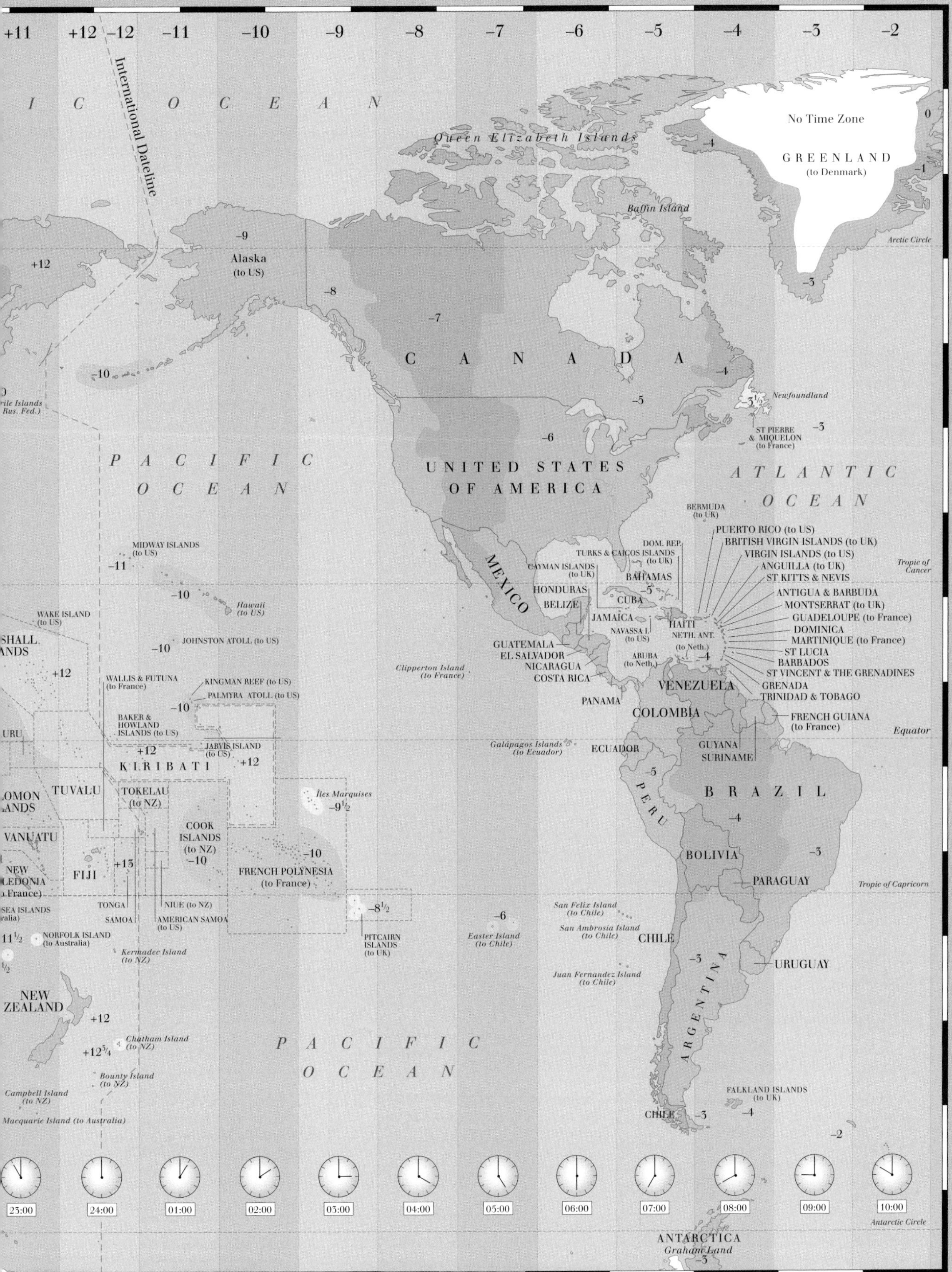
+11
+12
–12
–11
–10
–9
–8
–7
–6
–5
–4
–3
–2
International Dateline
PACIFIC OCEAN
Queen Elizabeth Islands
No Time Zone
GREENLAND
(to Denmark)
Baffin Island
Arctic Circle
Alaska
(to US)
CANADA
Newfoundland
ST PIERRE & MIQUELON
(to France)
UNITED STATES OF AMERICA
ATLANTIC OCEAN
BERMUDA
(to UK)
MIDWAY ISLANDS
(to US)
Hawaii
(to US)
WAKE ISLAND
(to US)
JOHNSTON ATOLL (to US)
MEXICO
PUERTO RICO (to US)
BRITISH VIRGIN ISLANDS (to UK)
VIRGIN ISLANDS (to US)
ANGUILLA (to UK)
ST KITTS & NEVIS
ANTIGUA & BARBUDA
MONTSERRAT (to UK)
GUADELOUPE (to France)
DOMINICA
MARTINIQUE (to France)
ST LUCIA
BARBADOS
ST VINCENT & THE GRENADINES
GRENADA
TRINIDAD & TOBAGO
DOM. REP.
TURKS & CAICOS ISLANDS
(to UK)
CAYMAN ISLANDS
(to UK)
BAHAMAS
HONDURAS
BELIZE
CUBA
JAMAICA
NAVASSA I.
(to US)
HAITI
NETH. ANT.
(to Neth.)
ARUBA
(to Neth.)
GUATEMALA
EL SALVADOR
NICARAGUA
COSTA RICA
PANAMA
Tropic of Cancer
Clipperton Island
(to France)
WALLIS & FUTUNA
(to France)
KINGMAN REEF (to US)
PALMYRA ATOLL (to US)
BAKER & HOWLAND ISLANDS (to US)
JARVIS ISLAND
(to US)
KIRIBATI
VENEZUELA
COLOMBIA
FRENCH GUIANA
(to France)
Equator
Galápagos Islands
(to Ecuador)
ECUADOR
GUYANA
SURINAME
TUVALU
TOKELAU
(to NZ)
Îles Marquises
–9½
PERU
BRAZIL
VANUATU
FIJI
COOK ISLANDS
(to NZ)
FRENCH POLYNESIA
(to France)
BOLIVIA
PARAGUAY
Tropic of Capricorn
NEW CALEDONIA
(to France)
+13
TONGA
SAMOA
NIUE (to NZ)
AMERICAN SAMOA
(to US)
–8½
PITCAIRN ISLANDS
(to UK)
Easter Island
(to Chile)
San Felix Island
(to Chile)
San Ambrosia Island
(to Chile)
CHILE
NORFOLK ISLAND
(to Australia)
Kermadec Island
(to NZ)
Juan Fernandez Island
(to Chile)
ARGENTINA
URUGUAY
NEW ZEALAND
Chatham Island
(to NZ)
+12¾
Bounty Island
(to NZ)
Campbell Island
(to NZ)
Macquarie Island (to Australia)
FALKLAND ISLANDS
(to UK)
CHILE
23:00
24:00
01:00
02:00
03:00
04:00
05:00
06:00
07:00
08:00
09:00
10:00
Antarctic Circle
ANTARCTICA
Graham Land

CHRONOLOGY 2001–2002

THE 20th CENTURY came to a definitive end on September 11, 2001. The momentum of the terrorist attacks in the US revolutionized relations between old enemies, helped to boost right-wing politics around the Western world – including the new US administration – and opened the 21st century with a global conflict. However, in some places peace prospects did improve, with agreements ending a 19-year war in Sri Lanka and the 27-year civil conflict in Angola. Into this altered landscape came the world's newest country, East Timor, and out of it crept "Yugoslavia," as the name was abandoned for good.

NORTH AMERICA

❑ **February 12, 2001** The human genome is mapped by a team of international scientists.

❑ **April 28, 2001** Dennis Tito from the US becomes the first space tourist, aboard the International Space Station.

❑ **May 1, 2001** US president George W. Bush declares the 1972 antiballistic missile (ABM) treaty a thing of the past, opening the way for development of the controversial missile defense shield.

❑ **July 23, 2001** The US is the only industrialized country not to sign up to a new global environmental treaty on greenhouse gas emissions.

❑ **September 11, 2001** Around 3000 people die in a concerted attack against US targets by Islamic militants: hijacked aircraft are flown into New York's celebrated World Trade Center (WTC) and the Pentagon in Washington D.C.

❑ **October 31. 2001** Figures show that the longest economic boom in the US has come to an end.

❑ **December 19, 2001** Fires in the ruins of the WTC are extinguished.

❑ **February 1, 2002** President Bush labels Iran, Iraq, and North Korea an "axis of evil."

❑ **June 17, 2002** Asteroid 2002MN whistles past the earth but is only detected three days later.

❑ **July 21, 2002** Just seven months after the collapse of the Enron corporation, telecommunications firm WorldCom files the largest corporate bankruptcy in US history amid revelations about false accounting.

CENTRAL AND SOUTH AMERICA

❑ **January 13, 2001** The first in a month of massive earthquakes hits El Salvador, killing over 1000.

❑ **July 9, 2001** Charges against former Chilean dictator Gen. Augusto Pinochet are dropped due to his deteriorating mental condition.

❑ **November 1, 2001** The first direct flights for 40 years between Cuba and the US are inaugurated.

❑ **November 11, 2001** Hurricane Michelle devastates Cuba, prompting trade with the US for essential supplies.

❑ **January 1, 2002** Eduard Duhalde is sworn in as Argentina's fifth president in 12 days. The country continues to suffer economic chaos.

❑ **February 20, 2002** Peace talks in Colombia are abandoned: it returns to full-scale civil war.

❑ **April 14, 2002** President Hugo Chávez is returned to power in Venezuela after a 48-hour coup.

❑ **June 30, 2002** Brazil wins the soccer World Cup for a record fifth time.

EUROPE

❑ **February 21, 2001** The first registered case of foot-and-mouth precipitates a one-year epidemic which devastates livestock across the UK.

❑ **April 10, 2001** The Netherlands becomes the first country in the world to legalize euthanasia.

❑ **May 13, 2001** Right-wing businessman Silvio Berlusconi returns as prime minister of Italy, bringing "postfascists" and northern separatists back into government.

❑ **July 12, 2001** A movement led by ex-king Simeon II wins elections in Bulgaria. He becomes prime minister, taking the surname Saxecoburggotski.

❑ **September 6, 2001** A peace deal in Macedonia ends a seven-month insurrection by ethnic Albanian rebels.

❑ **September 23, 2001** The historic Solidarity party is wiped out of the Polish Sejm (parliament) in elections.

❑ **November 10, 2001** A right-wing coalition wins elections in Norway. Ten days later a conservative coalition triumphs in Denmark.

❑ **January 1, 2002** The euro is fully introduced at midnight in 12 EU countries. Six months later the franc, deutsche mark, drachma, and other old currencies are consigned to the melting pot of history.

❑ **February 12, 2002** The trial begins in The Hague of former Serbian dictator Slobodan Milošević.

❑ **March 3, 2002** Switzerland votes to apply to become the 190th member of the UN, but pledges to maintain its neutrality.

❑ **March 14, 2002** The constituent governments of Yugoslavia agree to dissolve the union and replace it with "Serbia and Montenegro."

❑ **April 18, 2002** Russian president Vladimir Putin declares the Chechen war over. Guerrilla fighting continues.

❑ **April 21, 2002** Far-right leader Jean-Marie Le Pen shocks the world by coming second in the first round of the French presidential elections.

❑ **May 6, 2002** Controversial Dutch anti-immigrant politician Pim Fortuyn is shot dead. His party goes on to form part of the new government.

❑ **May 28, 2002** A Russia–NATO council is established, giving Russia more input into NATO policy.

❑ **August 14, 2002** Floodwaters recede from Prague. Rising river levels cause havoc across central Europe.

AFRICA

❑ **January 16, 2001** President Laurent Kabila of the Democratic Republic of Congo (DRC) is assassinated by a bodyguard. He is succeeded by his son Joseph, raising hopes of peace.

❑ **January 31, 2001** Abdelbaset al-Megrahi of Libya is convicted of the 1988 bombing of a PanAm airplane over Lockerbie, Scotland.

❑ **April 2, 2001** Ten years of brutal civil war are brought to an end in Sierra Leone.

❑ **April** 18, **2001** Violence breaks out in the ethnic Berber-dominated region of Kabylia in northeastern Algeria.

❑ **May** 28, **2001** A coup attempt in the Central African Republic turns into a pitched battle; thousands flee.

❑ **November** 20, **2001** A previously unknown rebel group launches an insurrection in Liberia.

❑ **January** 28, **2002** Over 1000 people die trying to escape a fire at a munitions dump in Lagos, Nigeria.

❑ **February** 22, **2002** Marc Ravalomanana declares himself the elected president of Madagascar, prompting a tense and increasingly violent four-month standoff with the incumbent, Didier Ratsiraka.

❑ **April** 4, **2002** The death of UNITA leader Joseph Savimbi prompts a cease-fire, ending 27 years of civil war in Angola.

❑ **July 9, 2002** The African Union is created as the OAU's successor in Durban, South Africa.

❑ **July** 22, **2002** Moroccan soldiers leave the disputed Spanish islet of Perejil after a ten-day standoff.

WEST ASIA/MIDDLE EAST

❑ **February** 6, **2001** Right-wing Likud leader Ariel Sharon is elected prime minister of Israel in the last direct elections for the post. Under his leadership, confrontation with the Palestinians intensifies.

❑ **May** 6, **2001** Pope John Paul II becomes the first pope to pray in a mosque, during his visit to Syria.

❑ **December** 6, **2001** Kandahar, the spiritual capital of the *taliban* in Afghanistan, is "liberated" by the Northern Alliance, signaling the collapse of the Islamic regime.

❑ **February** 14, **2002** Bahrain is declared a constitutional monarchy.

❑ **March** 19, **2002** US forces complete Operation Anaconda in Afghanistan, leaving more than 500 al-Qaida and *taliban* fighters dead.

❑ **May** 10, **2002** The five-week siege in Bethlehem of the Church of the Nativity is finally lifted; 39 Palestinian militants involved go into exile.

❑ **June 19, 2002** Israeli forces begin the construction of a wall between Israel proper and the Palestinian West Bank.

NORTH AND EAST ASIA

❑ **March** 21, **2001** Amnesty International declares that China executes more people than the rest of the world put together. Up to 40 people a week are killed as part of the anticorruption "Strike Hard" campaign.

❑ **April** 1, **2001** A US reconnaissance plane collides with a Chinese fighter near Hainan, causing a major diplomatic crisis.

❑ **July** 16, **2001** China and Russia sign their first friendship treaty since the Sino-Soviet split of the early 1960s. They both oppose the proposed US missile defense system.

❑ **August** 31, **2001** New populist Japanese prime minister Junichiro Koizumi inflames regional tensions by visiting the controversial Yasakuni war shrine.

❑ **September 26, 2001** The Communist Party of China (CCP) adopts the "Three Represents" doctrine encouraging the inclusion of capitalists in the Maoist party.

❑ **November 10, 2001** The World Trade Organization (WTO) ratifies China's membership. The next day Taiwan's accession is also ratified .

❑ **April** 11, **2002** The Chinese government admits that there are around 850,000 people in China who are HIV-positive. The UN estimates that the figure is already likely to be more than one million.

❑ **June** 7, **2002** Figures show that a six-month recession has come to an end in Japan.

❑ **August** 23, **2002** Reclusive North Korean leader Kim Jong Il meets Russian president Vladimir Putin in Vladivostok in only his third foreign visit in eight years.

SOUTH ASIA

❑ **January** 26, **2001** A powerful earthquake in Gujarat, India, kills an estimated 30,000 people and obliterates infrastructure.

❑ **June** 1,**2001** King Birendra of Nepal, and nine other members of the royal family, are murdered by Crown Prince Dipendra in a drunken rage. Dipendra also kills himself. His uncle Gyanendra becomes king.

❑ **November** 26, **2001** A state of emergency is declared in Nepal as the government launches a counteroffensive against Maoist rebels.

❑ **February** 28, **2002** The murder of 58 Hindus in Gujarat sparks sectarian violence across the state, leaving over 500 people – mostly Muslims – dead.

❑ **May** 6, **2002** Aung San Suu Kyi, the leader of the democratic opposition in Burma, is released from house arrest.

❑ **May** 14, **2002** Kashmiri separatists kill 34 people in a terrorist attack, raising tensions between India and Pakistan and bringing them to the brink of possible nuclear war.

SOUTHEAST ASIA

❑ **July** 23, **2001** President Wahid of Indonesia is ousted by parliament and replaced by his deputy, Megawati Sukarnoputri.

❑ **January** 1, **2002** The Indonesian provinces of Aceh and Papua (formerly Irian Jaya) are granted limited autonomy.

❑ **February** 21, **2002** A cease-fire in Sri Lanka promise peace after 19 years of civil war.

❑ **May** 20, **2002** East Timor becomes an independent country.

❑ **June** 22, **2002** Prime Minister Mahathir Mohamed of Malaysia announces he will resign in 2003 after 22 years in power.

AUSTRALIA AND OCEANIA

❑ **March** 23, **2001** The 15-year-old *Mir* space station, the last remnant of the Cold-War era of space exploration, is brought down into the Pacific.

❑ **August** 29, **2001** Australian troops board the *Tampa*, a Norwegian ship carrying over 400 mainly Afghan refugees, whose forcible relocation to camps in other Pacific countries heralds the start of the controversial "Pacific Solution."

❑ **August** 31, **2001** A final peace agreement is reached in Bougainville, Papua New Guinea, ending a 13-year independence struggle.

❑ **November** 10, **2001** The right-wing government of Prime Minister John Howard is unexpectedly reelected in Australia.

❑ **May** 14, **2002** AusAid predicts that 40% of Papua New Guinea's adults could die of AIDS by 2020.

❑ **July** 2, **2002** US adventurer Steve Fosset lands in Australia after completing the first solo circumnavigation of the Earth in a hot air balloon.

see pages 36–39 for CHRONOLOGY OF WORLD HISTORY

International Organizations

THIS LISTING GIVES the full names of all international organizations referred to, often by acronym, in the World Desk Reference (political parties are to be found under the Politics heading within each country entry). The full names are followed by the date of the establishment or foundation, an indication of membership, where appropriate, and a summary of the organization's aims and functions.

ACC
Arab Cooperation Council
established 1989
members – Egypt, Iraq, Jordan, Yemen
Promotes Arab economic cooperation

ACP
African, Caribbean, and Pacific Countries
established 1976
members – 78 developing countries and territories
Preferential economic and aid relationship with the EU under the Lomé Convention

ACS
Association of Caribbean States
established 1994
members – 25 Caribbean countries
Promotes economic, scientific, and cultural cooperation in the region

ADB
Asian Development Bank
established 1966
members – 43 Asia–Pacific countries and territories, 17 nonregional countries
Encourages regional development

AfDB
African Development Bank
established 1964
members – 53 African countries, 24 non-African countries
Encourages African economic and social development

AFESD
Arab Fund for Economic and Social Development
established 1968
members – 21 Arab countries (including Palestine)
Promotes social and economic development in Arab states

AL
League of Arab States (Arab League)
established 1945
members – 22 Arab countries (including Palestine)
Forum to promote Arab cooperation on social, political, and military issues

ALADI
Latin American Integration Association
established 1960
members – 12 Central and South American countries
Promotes trade and regional integration

AMCC
Amazonian Cooperation Council
established 1978
members – Bolivia, Brazil, Colombia, Ecuador, Guyana, Peru, Suriname, Venezuela
Promotes the harmonious development of the Amazon region

AMF
Arab Monetary Fund
established 1977
members – 21 Arab countries (including Palestine)
Promotes monetary and economic cooperation

AMU
Arab Maghreb Union
established 1989
members – Algeria, Libya, Mauritania, Morocco, Tunisia
Promotes integration and economic cooperation among north African Arab states

ANZUS
Australia–New Zealand–United States Security Treaty
established 1951
members – Australia, New Zealand, US
Trilateral security agreement. Security relations between the US and New Zealand were suspended in 1984 over the issue of US nuclear-powered or potentially nuclear-armed naval vessels visiting New Zealand ports. High-level contacts between the US and New Zealand were resumed in 1994

AP
Andean Pact (Acuerdo de Cartegena), also known as Andean Community
established 1969
members – Bolivia, Colombia, Ecuador, Peru, Venezuela
Promotes development through integration

APEC
Asia–Pacific Economic Cooperation
established 1989
members – 20 Pacific Rim countries
Promotes regional economic cooperation

ASEAN
Association of Southeast Asian Nations
established 1967
members – Brunei, Burma, Cambodia, Indonesia, Laos, Malaysia, Philippines, Singapore, Thailand, Vietnam
Promotes economic, social, and cultural cooperation

AU
African Union
established 2002
members – 52 African countries and Western Sahara
Promotes unity and cooperation in Africa. (The African Union was, until 2002, the Organization of African Unity, which had been founded in 1963)

BADEA
Arab Bank for Economic Development in Africa
established 1973
members – 18 Arab countries (including Palestine)
Established as an agency of the Arab League to promote economic development in Africa

BDEAC
Central African States Development Bank
established 1975
members – Cameroon, Central African Republic, Chad, Congo, Equatorial Guinea, France, Gabon, Germany, Kuwait
Furthers economic development

Benelux
Benelux Economic Union
established 1960
members – Belgium, Luxembourg, Netherlands
Develops economic ties between member countries

BOAD
West African Development Bank
established 1973
members – Benin, Burkina, Guinea-Bissau, Ivory Coast, Mali, Niger, Senegal, Togo
Promotes economic development and integration in West Africa

BSEC
Organization of the Black Sea Economic Cooperation
established 1992
members – Albania, Armenia, Azerbaijan, Bulgaria, Georgia, Greece, Moldova, Romania, Russia, Turkey, Ukraine
Furthers regional stability through economic cooperation

CAEU
Council of Arab Economic Unity
established 1957
members – 12 Arab countries (including Palestine)
Encourages economic integration

Caricom
Caribbean Community and Common Market
established 1973
members – 13 Caribbean countries and Montserrat
Fosters economic ties in the Caribbean

CBSS
Council of the Baltic Sea States
established 1992
members – Denmark, Estonia, Finland, Germany, Iceland, Latvia, Lithuania, Norway, Poland, Russia, Sweden
Promotes cooperation among Baltic Sea states

CDB
Caribbean Development Bank
established 1969
members – 17 Caribbean countries/dependencies, 8 non-Caribbean countries
Promotes regional development

CE
Council of Europe
established 1949
members – 44 European countries
Promotes unity and quality of life in Europe

CEFTA
Central European Free Trade Agreement
established 1992
members – Bulgaria, Czech Republic, Hungary, Poland, Romania, Slovakia, Slovenia
Promotes trade and cooperation

CEI
Central European Initiative
established 1989
members – 17 eastern and central European countries: Albania, Austria, Belarus, Bosnia and Herzegovina, Bulgaria, Croatia, Czech Republic, Hungary, Italy, Macedonia, Moldova, Poland, Romania, Serbia and Montenegro, Slovakia, Slovenia, Ukraine
Evolved from the Hexagonal Group; promotes economic and political cooperation, within the OSCE

CEMAC
Central African Economic and Monetary Community
established 1994
members – Cameroon, Central African Republic, Chad, Congo, Equatorial Guinea, Gabon
Aims to promote subregional integration, by economic and monetary union (replaced UDEAC)

CEPGL
Economic Community of the Great Lakes Countries
established 1976
members – Burundi, Democratic Republic of the Congo, Rwanda
Promotes regional economic cooperation

CERN
European Organization for Nuclear Research
established 1954
members – 20 European countries
Provides for collaboration in nuclear research for peaceful purposes

CILSS
Permanent Interstate Committee for Drought Control in the Sahel
established 1973
members – 9 African countries in the Sahel region
Promotes prevention of drought and crop failure in the region

CIS
Commonwealth of Independent States
established 1991
members – Armenia, Azerbaijan, Belarus, Georgia, Kazakhstan, Kyrgyzstan, Moldova, Russia, Tajikistan, Turkmenistan, Ukraine, Uzbekistan
Promotes interstate relationships among former republics of the Soviet Union

CMCA
Central American Monetary Council
established 1960
members – Costa Rica, El Salvador, Guatemala, Honduras, Nicaragua
Now a subsystem of SICA. Furthers economic ties between members; one of its institutions is the BCIE – Central American Bank for Economic Integration

COI
Indian Ocean Commission
established 1982
members – Comoros, France (representing Réunion), Madagascar, Mauritius, Seychelles
Promotes regional cooperation

COMESA
Common Market for Eastern and Southern Africa
established 1993
members – 20 African countries
Promotes economic development and cooperation (replaced PTA)

Comm
Commonwealth
established 1931
members – 54 countries (although Pakistan and Zimbabwe are currently suspended). Members are chiefly former members of the British Empire
Develops relationships and contacts between members

CP
Colombo Plan
established 1950
members – Four donor countries: Australia, Japan, New Zealand, US; and 20 Asia–Pacific countries
Encourages economic and social development in Asia–Pacific region

CPLP
Community of Portuguese-speaking Countries
established 1996
members – Portugal, Brazil, and five Portuguese-speaking African countries – Angola, Cape Verde, Guinea-Bissau, Mozambique, São Tomé and Príncipe
To promote political and diplomatic links between member states, and cooperation on economic, social, cultural, judicial, and scientific development among Portuguese-speaking countries

Damasc
Damascus Declaration
established 1991
members – Bahrain, Egypt, Kuwait, Oman, Qatar, Saudi Arabia, Syria, United Arab Emirates
A loose association, formed after the Gulf War, which aims to secure the stability of the region

EAC
East African Community
established 2001
members – Kenya, Tanzania, Uganda
Promotes economic cooperation

EAPC
Euro-Atlantic Partnership Council
established 1991
members – The 19 members of NATO plus 27 eastern European countries
Forum for cooperation on political and security issues (successor to the NACC, North Atlantic Cooperation Council)

EBRD
European Bank for Reconstruction and Development
established 1991
members – 60 countries
Helps transition of former communist European states to market economies

ECO
Economic Cooperation Organization
established 1985
members – Afghanistan, Azerbaijan, Iran, Kazakhstan, Kyrgyzstan, Pakistan, Tajikistan, Turkey, Turkmenistan, Uzbekistan
Aims at cooperation in economic, social, and cultural affairs

ECOWAS
Economic Community of West African States
established 1975
members – 15 west African countries
Promotes regional economic cooperation

EEA
European Economic Area
established 1994
members – The 15 members of the EU, and Iceland, Liechtenstein, and Norway
Aims to include EFTA members in the EU single market

EEC
Eurasian Economic Community
established 2001
members – Belarus, Kazakhstan, Kyrgyzstan, Russia, Tajikistan
Coordinates regional trade

EFTA
European Free Trade Association
established 1960
members – Iceland, Liechtenstein, Norway, Switzerland
Promotes economic cooperation

ESA
European Space Agency
established 1973
members – 15 European countries
Promotes cooperation in space research for peaceful purposes

EU
European Union
established 1992
members – 15 European countries
Aims to integrate the economies of member states and promote cooperation and coordination of policies. Previously the European Communities (EC), established in 1957 by the Treaties of Rome

FZ
Franc zone
established Not applicable
members – France (including overseas departments and territories), Monaco, and 15 African countries
Aims to form monetary union among countries whose currencies are linked to that of France

G3
Group of 3
established 1987
members – Colombia, Mexico, Venezuela
Aims to ease trade restrictions

G7
Group of 7
established 1975
members – The seven major industrialized countries: Canada, France, Germany, Italy, Japan, UK, US
Summit meetings of the seven major industrialized countries, originally for economic purposes. For political purposes summit meetings are now held as the G8, including Russia

G8
Group of 8
established 1994
members – Canada, France, Germany, Italy, Japan, Russia, UK, US
Global forum of world's major powers, which holds regular summit meetings

G10
Group of 10
established 1962
members – 11 members: G7, plus Belgium, the Netherlands, Sweden, and Switzerland
Ministers meet to discuss monetary issues

G15
Group of 15
established 1989
members – 19 developing countries
Meets annually to further cooperation among developing countries

G24
Group of 24
established Not applicable
members – 24 developing countries within the IMF
Promotes the interests of developing countries on monetary and development issues.

GCC
Gulf Cooperation Council
established 1981
members – Bahrain, Kuwait, Oman, Qatar, Saudi Arabia, UAE
Promotes cooperation in social, economic, and political affairs

GEPLACEA
Latin American and Caribbean Sugar Exporting Countries
established 1974
members – 23 countries
A forum for consultation on the production and sale of sugar

GGC
Gulf of Guinea Commission
established 2001
members – Angola, Cameroon, Congo, Democratic Republic of the Congo, Equatorial Guinea, Gabon, Nigeria, São Tomé and Príncipe
Promotes regional cooperation

IAEA
International Atomic Energy Agency
established 1957
members – 134 countries
Promotes and monitors peaceful use of atomic energy

IBRD
International Bank for Reconstruction and Development
(also known as the World Bank)
established 1945
members – 184 countries
UN agency providing economic development loans

ICRC
International Committee of the Red Cross
established 1863
members – Up to 25 Swiss nationals form the international committee. Red Cross or Red Crescent societies exist in 175 countries
Coordinates all international humanitarian activities of the International Red Cross and Red Crescent Movement, giving legal and practical assistance to the victims of wars and disasters. It works through national committees of Red Cross or Red Crescent societies

IDB
Inter-American Development Bank
established 1959
members – 28 American countries and 18 nonregional countries
Promotes development in Latin America and the Caribbean through the financing of economic and social development projects and the provision of technical assistance

IGAD
Intergovernmental Authority on Development
established 1996
members – Djibouti, Eritrea, Ethiopia, Kenya, Somalia, Sudan, Uganda
Promotes cooperation on food security, infrastructure, and other development issues (supersedes IGADD, founded 1986, to promote cooperation on drought-related matters)

IMF
International Monetary Fund
established 1945
members – 184 countries. The voting rights of the Democratic Republic of the Congo are currently suspended.
Promotes international monetary cooperation, the balanced growth of trade, and exchange rate stability; provides credit resources to members experiencing balance-of-trade difficulties.

ISDB
Islamic Development Bank
established 1975
members – 53 countries (including Palestine)
Promotes economic development on Islamic principles among Muslim communities
(agency of the OIC)

IWC
International Whaling Commission
established 1946
members – 48 countries
Reviews conduct of whaling throughout world; coordinates and funds whale research

LCBC
Lake Chad Basin Commission
established 1964
members – Cameroon, Central African Republic, Chad, Niger, Nigeria
Encourages economic and environmental development in Lake Chad region

Mekong River
Mekong River Commission
established 1995
members – Cambodia, Laos, Thailand, Vietnam
Accord on the sustainable development of Mekong River basin (replacing the 1958 interim Mekong Secretariat)

Mercosur
Southern Common Market
established 1991
members – Argentina, Brazil, Paraguay, Uruguay
Promotes economic integration, free trade, and common external tariffs

MRU
Mano River Union
established 1973
members – Guinea, Liberia, Sierra Leone
Aims to create customs and economic union in order to promote development

NAFTA
North American Free Trade Agreement
established 1994
members – Canada, Mexico, US
Free trade zone

NAM
Non-Aligned Movement
established 1961
members – 115 countries (including Palestine). Serbia and Montenegro has been suspended since 1992
Fosters political and military cooperation away from traditional Eastern or Western blocs

NATO
North Atlantic Treaty Organization
established 1949
members – 19 countries
Promotes mutual defense cooperation. Since January 1994, NATO's Partnerships for Peace program has provided a loose framework for cooperation with former members of the Warsaw Pact and the ex-Soviet republics. A historic Founding Act signed between Russia and NATO in May 1997 allowed for the organization's eastward expansion, under which the Czech Republic, Hungary, and Poland were the first three countries to join

NC
Nordic Council
established 1952
members – Denmark, Finland, Iceland, Norway, Sweden
Promotes cultural and environmental cooperation in Scandinavia

OAPEC
Organization of Arab Petroleum Exporting Countries
established 1968
members – 10 Arab countries: Algeria, Bahrain, Egypt, Iraq, Kuwait, Libya, Qatar, Saudi Arabia, Syria, UAE
Aims to promote the interests of member countries and increase cooperation in the petroleum industry

OAS
Organization of American States
established 1948
members – 35 American countries (although Cuba has been suspended since 1962)
Promotes security, economic, and social development in the Americas

OAU
Organization of African Unity
Predecessor of the AU.

OECD
Organization for Economic Cooperation and Development
established 1961
members – 30 industrialized democracies
Forum for coordinating economic policies among industrialized countries

OECS
Organization of Eastern Caribbean States
established 1981
members – 7 Caribbean countries/dependencies: Antigua and Barbuda, Dominica, Grenada, Montserrat, St. Kitts and Nevis, St. Lucia, St. Vincent and the Grenadines
Promotes political, economic, and defense cooperation

OIC
Organization of the Islamic Conference
established 1971
members – 57 countries (including Palestine)
Furthers Islamic solidarity and cooperation

OIF
International Organization of Francophony
established 1970
members – 48 countries and the governments of Québec, New Brunswick, and the French Community of Belgium were represented at the 1999 summit. The Intergovernmental Agency of Francophony has 49 members.
To promote cooperation and cultural and technical links among French-speaking countries and communities

OMVG
Gambia River Development Organization
established 1978
members – Gambia, Guinea, Guinea-Bissau, Senegal
Promotes integrated development of the Gambia River basin

Opanal
Agency for the Prohibition of Nuclear Weapons in Latin America and the Caribbean
established 1969
members – 32 countries (plus Cuba, who has signed but not ratified the Treaty of Tlatelolco)
Aims to ensure compliance with the Treaty of Tlatelolco (banning nuclear weapons from the region)

OPEC
Organization of the Petroleum Exporting Countries
established 1960
members – 11 oil producers: Algeria, Indonesia, Iran, Iraq, Kuwait, Libya, Nigeria, Qatar, Saudi Arabia, United Arab Emirates, Venezuela
Aims to coordinate oil policies to ensure fair and stable prices

OSCE
Organization for Security and Cooperation in Europe
established 1972
members – 55 countries
Aims to strengthen democracy and human rights, and settle disputes peacefully (formerly CSCE; renamed 1994)

Partnerships for Peace (PfP)
see NATO
established Not applicable
members – 27 members: eastern European and former Soviet countries, Sweden, Finland, Malta, Austria, and Switzerland

PC
Pacific Community (formerly South Pacific Commission)
established 1948
members – 27 countries and territories
A forum for dialogue between Pacific countries and powers administering Pacific territories

PIF
Pacific Islands Forum (formerly the South Pacific Forum)
established 1971
members – 16 countries and self-governing territories
Develops regional political cooperation

RG
Rio Group
established 1987
members – 19 Latin American and Caribbean countries
Forum for Latin American and Caribbean issues (evolved from Contadora Group, established 1948)

SAARC
South Asian Association for Regional Cooperation
established 1985
members – Bangladesh, Bhutan, India, Maldives, Nepal, Pakistan, Sri Lanka
Encourages economic, social, and cultural cooperation

SACU
Southern African Customs Union
established 1969
members – 5 southern African countries: Botswana, Lesotho, Namibia, South Africa, Swaziland
Promotes cooperation in trade and customs matters among southern African states

SADC
Southern African Development Community
established 1992
members – 14 southern African countries
Promotes economic integration

San José
San José Group
established 1988
members – Costa Rica, El Salvador, Guatemala, Honduras, Nicaragua, Panama
A 'complementary, voluntary, and gradual' economic union

SCO
Shanghai Cooperation Organization (formerly Shanghai Five)
established 1996
members – China, Kazakhstan, Kyrgyzstan, Russia, Tajikistan, Uzbekistan
Promotes regional security and cooperation

SELA
Latin American Economic System
established 1975
members – 28 countries
Promotes economic and social development through regional cooperation

SICA
Central American Integration System
established 1991
members – 6 countries: Costa Rica, El Salvador, Guatemala, Honduras, Nicaragua, Panama
Coordinates the political, economic, social, and environmental integration of the region

UEMOA
West African Economic and Monetary Union
established 1994
members – 8 West African countries
Aims for convergence of monetary policies and economic union

UN
United Nations
established 1945
members – 191 countries; permanent members of the Security Council – China, France, Russia, UK, US
Aims to maintain international peace and security and to promote cooperation over economic, social, cultural, and humanitarian problems

Agencies include the regional commissions of the UN's Economic and Social Council:
ECA (Economic Commission for Africa – established 1958);
ECE (Economic Commission for Europe – established 1947);
ECLAC (Economic Commission for Latin America and the Caribbean – established 1948);
ESCAP (Economic and Social Commission for Asia and the Pacific – established 1947);
ESCWA (Economic and Social Commission for Western Asia – established 1973).

Other bodies of the UN, in which most members participate, include
UNICEF (UN Children's Fund);
UNCTAD (UN Conference on Trade and Development);
UNDP (UN Development Program);
UNHCR (UN High Commissioner for Refugees);
UNFPA (UN Population Fund);
IDA (International Development Association)

Taiwan and the Vatican City do not belong to the UN.

WEU
Western European Union
established 1955
members – 10 countries
A forum for European military cooperation

WTO
World Trade Organization
established 1995
members – 143 countries (including Hong Kong and Macao) and the EU
Aims to liberalize trade through multilateral trade agreements (as the successor to GATT (the General Agreement on Tariffs and Trade))

2

THE NATIONS OF THE WORLD

THE NATIONS OF THE WORLD

• AFGHANISTAN ~ ZIMBABWE

OVERSEAS TERRITORIES & DEPENDENCIES

A

ASIA

Turkmenistan Uzb. Tajikistan China AFGHANISTAN Iran Pakistan

Asia

AFGHANISTAN

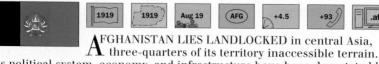

OFFICIAL NAME: Islamic State of Afghanistan **CAPITAL:** Kabul
POPULATION: 22.5 million **CURRENCY:** Afghani **OFFICIAL LANGUAGES:** Pashtu and Dari

1919

1919

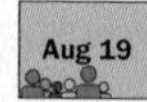
Aug 19

AFG

+4.5

+93

.af

AFGHANISTAN LIES LANDLOCKED in central Asia, three-quarters of its territory inaccessible terrain. Its political system, economy, and infrastructure have been devastated by decades of armed conflict. In the 1980s Islamic *mujahideen* factions defeated the Soviet-backed communist regime, but rivalries undermined their fragile power-sharing agreement and the hard-line *taliban* militia swept to power in 1996. Islamic dress codes and behavior were vigorously enforced and women left with few rights or opportunities. The *taliban* regime crumbled in the face of the US-led "war on terrorism" launched in late 2001.

***The Band-i-Amir River,** in the Hindu Kush. Afghanistan is mountainous and arid. Many Afghans are nomadic sheep farmers.*

CLIMATE

Mountain/cold desert

WEATHER CHART FOR KABUL

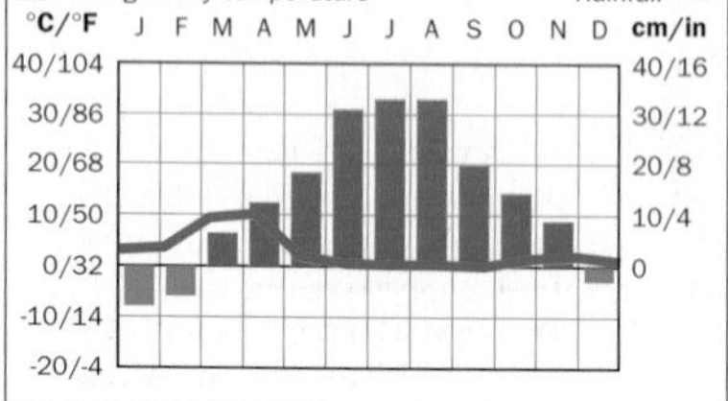
Average daily temperature
Rainfall
°C/°F
J F M A M J J A S O N D
cm/in
40/104
30/86
20/68
10/50
0/32
-10/14
-20/-4
40/16
30/12
20/8
10/4
0

Afghanistan has the world's widest temperature range, with lows of –50°C (–58°F) and highs of 53°C (127°F). Severe drought, a frequent problem, affected half the population in 2000.

TRANSPORTATION

Drive on right

Kābul International

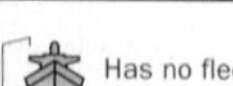

Has no fleet

THE TRANSPORTATION NETWORK

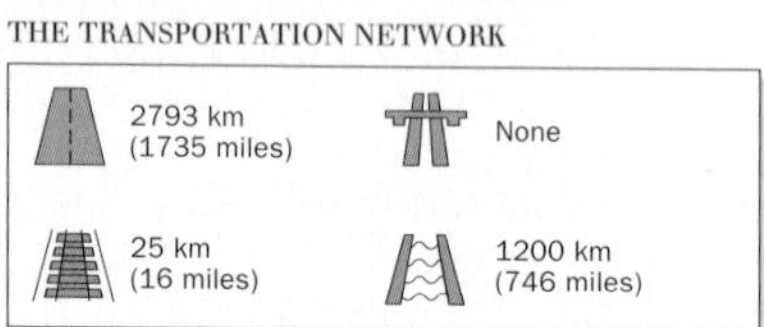
2793 km
(1735 miles)
None
25 km
(16 miles)
1200 km
(746 miles)

The repair and reconstruction of the roads, severely damaged by war, and the modernization of the air traffic control system are the most urgent priorities. The rebuilding of roads is usually carried out by local communities. However, neighboring Pakistan has undertaken to rebuild a number of key routes, including the Kabul–Peshawar link, which will benefit its own trade with central Asia.

Securing key supply routes was a crucial factor in intra-*mujahideen* feuding and in the *taliban*'s efforts to gain control over the whole country. Anti-*taliban* forces relied heavily on supplies from the north. Much of Afghanistan's outlying territory is sown with land mines.

TOURISM

Visitors : Population 1:5625

4000 visitors

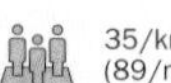

No change in 1995–1998

MAIN TOURIST ARRIVALS

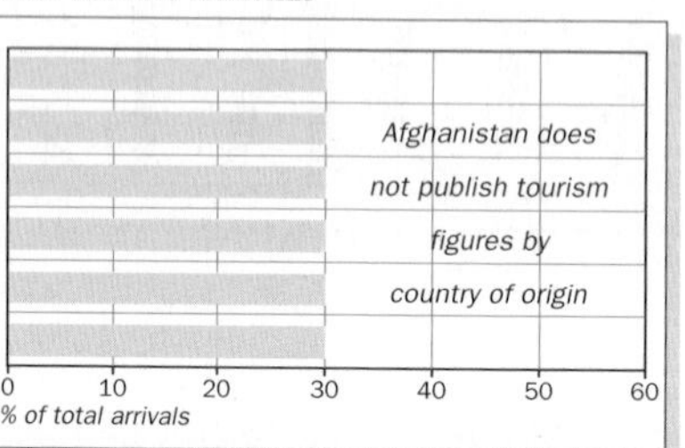
Afghanistan does
not publish tourism
figures by
country of origin
0
10
20
30
40
50
60
% of total arrivals

Afghanistan is a war zone. Restrictions on the activities of UN and aid agency personnel were lifted in 2001 following the overthrow of the *taliban* regime. Few hotels are open in Kabul and travel is extremely dangerous. Air Ariana, the Afghan national airline, lost six of its eight aircraft in the 2001 US bombing.

The lack of a formal economy means that Afghanistan gets few visits from businessmen, and any expatriates who were previously in Kabul left during the *taliban* period.

PEOPLE

Pop. density low

Pashtu, Tajik, Dari, Farsi, Uzbek, Turkmen

35/km² (89/mi²)

THE URBAN/RURAL POPULATION SPLIT

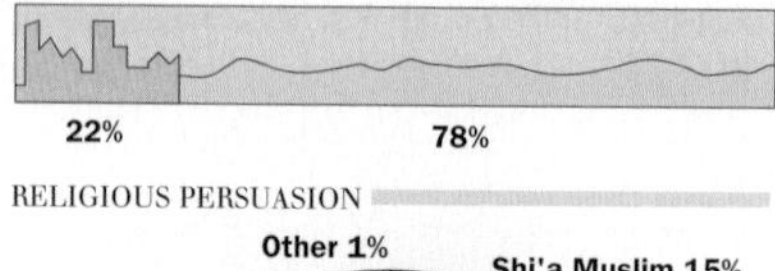
22%
78%

RELIGIOUS PERSUASION

Other 1%
Shi'a Muslim 15%
Sunni Muslim 84%

ETHNIC MAKEUP

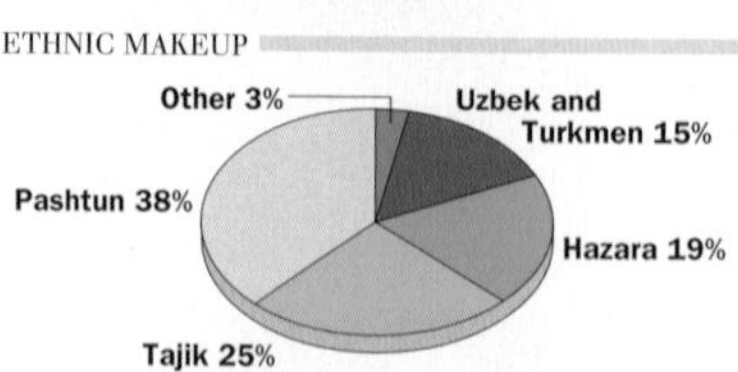
Other 3%
Uzbek and Turkmen 15%
Pashtun 38%
Hazara 19%
Tajik 25%

Ethnic divisions largely determined intra-*mujahideen* feuding after 1992. Pashtuns have traditionally been the rulers of Afghanistan, and dominated the *taliban*. The fall of that regime in 2001 provided the opportunity for the Tajik–Uzbek alliance to enforce a power-sharing agreement. Pashtuns have since faced reprisal attacks, particularly in the north, where they form a minority. Religious differences between Sunnis and Shi'as became particularly acute under the Sunni *taliban* regime.

Some two million of the country's population were killed in the ten-year conflict which followed the invasion by Soviet Union forces in 1979 and in the post-1992 civil war. As many people again were maimed. A further six million people were forced to flee to neighboring Pakistan and Iran; many returned, but the fighting in 2001 created a fresh wave of refugees and left hundreds of thousands more people internally displaced.

Women had few rights under the rigid Islamic regime of the *taliban*. The interim government has lifted the ban on women in employment and allowed girls to resume schooling.

POPULATION AGE BREAKDOWN

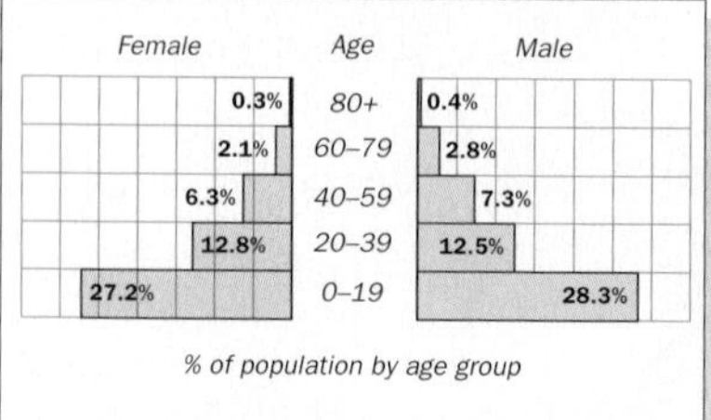
Female
Age
Male
0.3%
80+
0.4%
2.1%
60–79
2.8%
6.3%
40–59
7.3%
12.8%
20–39
12.5%
27.2%
0–19
28.3%
% of population by age group

POLITICS

In transition

1988/2004

President Hamid Karzai

AT THE LAST ELECTION

House of Representatives (dissolved)

Following the downfall of Najibullah's regime in April 1992, both houses were dissolved and an interim *mujahideen* legislature formed.

Senate (dissolved)

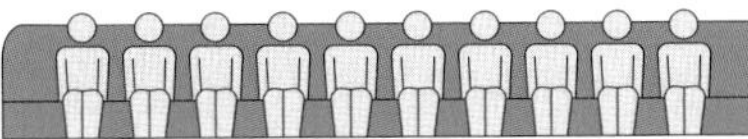

The political system had virtually collapsed in Afghanistan prior to the *taliban* takeover in 1996. Rival *mujahideen* factions had been in control since April 1992, when the communist President Najibullah, whose regime had been backed until 1990 by the Soviet Union, was forced to step down.

According to the 1993 Islamabad peace accord, elections were to be held by the end of the year. These were indefinitely postponed owing to factional fighting, which led in early 1995 to the emergence of a potent new force, the extremist Islamic *taliban*. Initially the product of fundamentalist Islamic schools established in the refugee camps on the Pakistani border, the *taliban* were mostly ethnic Pashtuns, and their support was strongest around Kandahar in the south. Kabul fell to them in 1996, and an opposition Northern Alliance headed by the former president, Burhanuddin Rabbani of the Jamiat-i-Islami, failed to reverse the situation. On assuming power the *taliban* declared Afghanistan a "complete" Islamic state, and imposed a strict Islamic code.

The US-led intervention in late 2001 helped replace the *taliban* regime with groups allied to the Northern Alliance, which succeeded in gaining control of most major cities, including Kabul. Later that year UN-sponsored talks led to the formation of an interim government headed by Pashtun leader Hamid Karzai, who is committed to implementing the decisions of a Loya Jirga (grand council), pending elections in 2004. Ex-king Zahir Shah returned in April 2002 to Kabul, where he endorsed Karzai's election as head of state by the Loya Jirga in June. The fragility of the political situation was underlined by the assassination in July of Vice President Haji Abdul Qadir.

Hamid Karzai *was confirmed as head of state in 2002 by the traditional Loya Jirga council.*

WORLD AFFAIRS

Joined UN in 1946

CP

ECO

IBRD

NAM

OIC

Under the Islamic extremist *taliban* government Afghanistan was a pariah state, with only three Islamic countries maintaining diplomatic ties. The regime became the first target in the US-led "war on terrorism" in October 2001, leading to its destruction. The new interim government is consequently heavily dependent on the countries which effectively installed it, making good relations with them essential for future aid. Pakistan was the last country to break off links with the *taliban* and remains Afghanistan's most delicately placed partner, with a history of religious as well as ethnic sympathies with the ousted, Pashtun-dominated, Islamic regime.

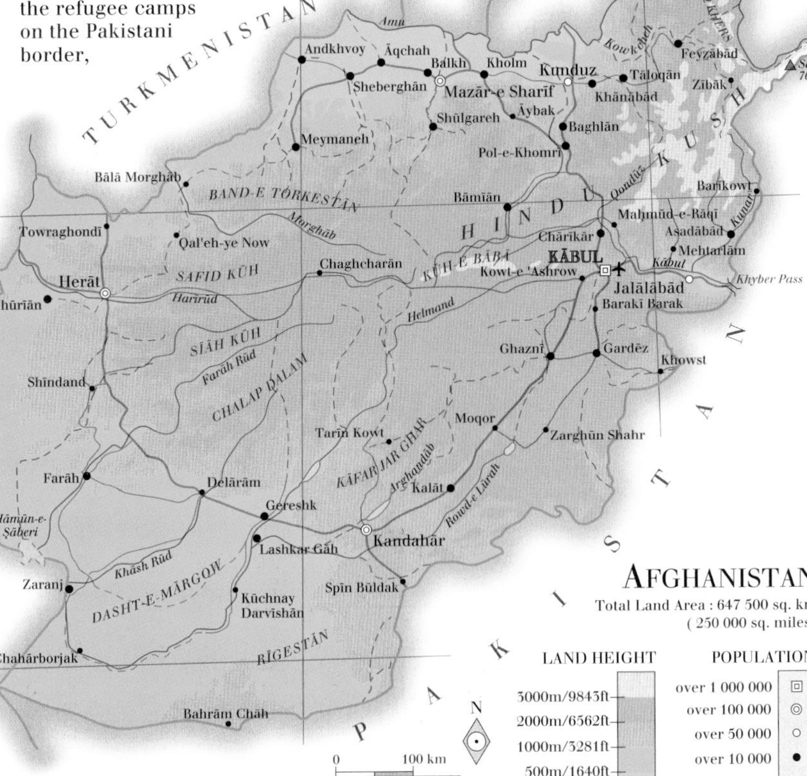
TAJIKISTAN
UZBEKISTAN
TURKMENISTAN
CHINA
PAKISTAN
KŪH-E VĀKHĀN
Sad Istragh 7037m
Feyzābād
Kunduz
Tāloqān
Khānābād
Zībāk
Andkhvoy
Āqchah
Balkh
Kholm
Mazār-e Sharīf
Sheberghān
Shūlgareh
Āybak
Baghlān
Pol-e-Khomrī
Meymaneh
Bālā Morghāb
BAND-E TORKESTĀN
Bāmīān
HINDU KUSH
Barikowt
Mahmūd-e-Rāqī
Asadābād
Mehtarlām
Chārīkār
KĀBUL
Kowt-e 'Ashrow
Jalālābād
Khyber Pass
Barakī Barak
Towraghondī
Qal'eh-ye Now
SAFID KŪH
Herāt
Harīrūd
Ghūrīān
Chaghcharān
Helmand
SIĀH KŪH
Farāh Rūd
CHALAP DALAN
Ghaznī
Gardēz
Khowst
Shīndand
Moqor
Tarīn Kowt
Zarghūn Shahr
Farāh
Delārām
Kalāt
Gereshk
Hāmūn-e-Şāberī
Lashkar Gāh
Kandahār
Khāsh Rūd
DASHT-E-MĀRGOW
Zaranj
Kūchnay Darvīshān
Spīn Būldak
RĪGESTĀN
Chahārborjak
Bahrām Chāh
AFGHANISTAN
Total Land Area : 647 500 sq. km
(250 000 sq. miles)
LAND HEIGHT
5000m/9843ft
2000m/6562ft
1000m/3281ft
500m/1640ft
200m/656ft
POPULATION
over 1 000 000
over 100 000
over 50 000
over 10 000
under 10 000
N
0 100 km
0 100 miles

CHRONOLOGY

The foundations of an Afghan state of Pashtun peoples were laid in the mid-18th century, when Durrani Ahmad Shah became paramount chief of the Abdali Pashtun peoples.

- ❑ **1838–1842** First Anglo-Afghan war.
- ❑ **1878** Second British invasion of Afghan territory.
- ❑ **1879** Under Treaty of Gandmak signed with Amir Yaqub Ali Khan, various Afghan areas annexed by Britain. Yaqub Ali Khan later exiled. New treaty signed with Amir Abdul Rahman, establishing the Durand line, a contentious boundary between Afghanistan and Pakistan.
- ❑ **1919** Independence declared.
- ❑ **1933** Mohammed Zahir Shah ascends throne.
- ❑ **1953–1963** Mohammed Daud Khan, prime minister, resigns after king rejects proposals for democratic reforms.

A

CHRONOLOGY *continued*

- ❑ **1965** Elections held, but monarchy retains power. Marxist Party of Afghanistan (PDPA) formed and banned. PDPA splits into the Parcham and Khalq factions.
- ❑ **1973** Daud mounts a coup, abolishes monarchy, and declares republic. *Mujahideen* rebellion begins. Refugees flee to Pakistan.
- ❑ **1978** Opposition to Daud from PDPA culminates in Saur revolution. Revolutionary Council under Mohammad Taraki takes power. Daud assassinated.
- ❑ **1979** Taraki ousted. Hafizullah Amin takes power. Amin killed in December coup backed by USSR. 80,000 Soviet Army troops invade Afghanistan. *Mujahideen* rebellion stepped up into full-scale guerrilla war, with US backing.
- ❑ **1980** Babrak Karmal, leader of Parcham PDPA, installed as head of Marxist regime.
- ❑ **1986** Najibullah replaces Karmal as head of government.
- ❑ **1989** Soviet Army withdraws. Najibullah remains in office.
- ❑ **1992** Najibullah hands over power to *mujahideen* factions.
- ❑ **1993** *Mujahideen* agree on formation of government.
- ❑ **1994** Power struggle between Rabbani and Hekmatyar.
- ❑ **1996** *Taliban* take power and impose strict Islamic regime.
- ❑ **1998** Earthquake in northern regions kills thousands.
- ❑ **1999** Power-sharing agreement between *taliban* and Northern Alliance breaks down.
- ❑ **2000** Worst drought in 30 years. UN imposes sanctions in response to *taliban* support for Osama bin Laden.
- ❑ **2001** *Taliban* government falls after intense US-led air strikes from October – first campaign in "war on terrorism." Interim government formed under Hamid Karzai; peacekeepers deployed in Kabul.
- ❑ **2002** Earthquakes kill thousands. Ex-king Zahir Shah returns from exile. Loya Jirga convenes, elects Karzai head of state.

AID

Recipient

$141m (receipts)

Down 1% in 2000

The fall of the *taliban* regime opened the floodgates for massive inflows of aid to rebuild the country, notably from the EU, Japan, Saudi Arabia, and the US. However, relations between foreign aid agencies and the authorities can be fraught, and working conditions for aid personnel are extremely hazardous.

DEFENSE

Compulsory military service

$245m

Down 2% in 2000

AFGHAN ARMED FORCES

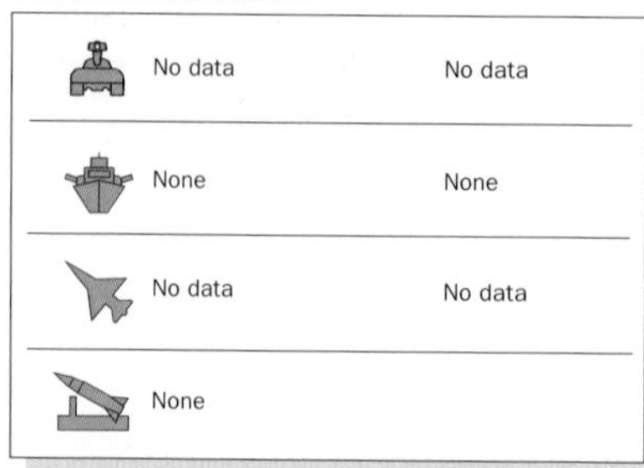
No data
No data
None
None
No data
No data
None

The new regime is establishing a national army; private armies have been banned. An international peacekeeping force maintains security in Kabul. Heavy ground offensives against the remaining al-Qaida and *taliban* forces were launched by the US in March 2002; hundreds were killed, including non-combatants.

The US–Russian agreement in 1991 to suspend military supplies to rival Afghan groups marked an end to active superpower involvement for a decade. The Kabul communists had been almost totally dependent on Moscow for arms, even after the Soviet military withdrawal in 1989. The covert arms trade expanded with the activity of Islamic militants from abroad. The bulk of these arms originate in eastern Europe and the former Soviet Union. The movement of Islamist militants and weapons between Tajikistan and Afghanistan is tackled by CIS troops.

Afghanistan holds hundreds of *Stinger* missiles given to the *mujahideen* by the US in the 1980s. The US, worried that they might be used against civilian airliners, has offered to buy them back, but none have so far been returned.

ECONOMICS

Not available

$5.87bn

4750 afghanis

SCORE CARD

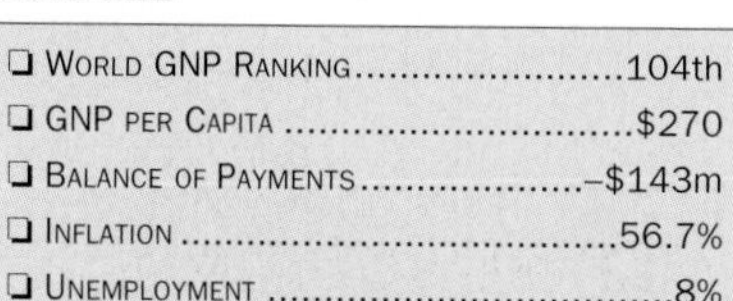

- ❑ WORLD GNP RANKING........................104th
- ❑ GNP PER CAPITA $270
- ❑ BALANCE OF PAYMENTS....................–$143m
- ❑ INFLATION 56.7%
- ❑ UNEMPLOYMENT 8%

EXPORTS

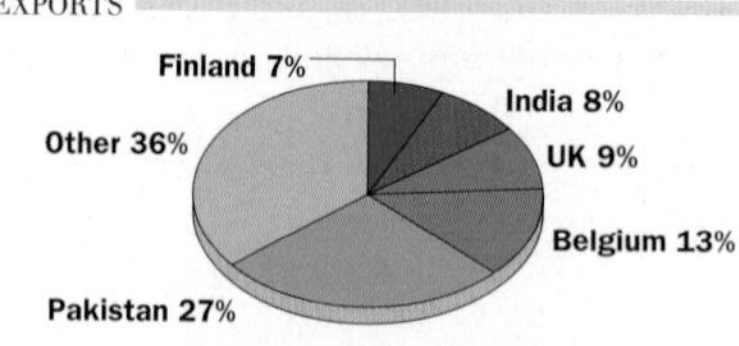
Finland 7%
India 8%
Other 36%
UK 9%
Belgium 13%
Pakistan 27%

IMPORTS

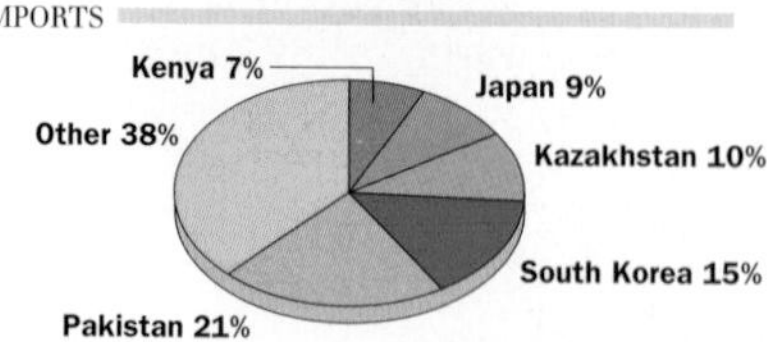
Kenya 7%
Japan 9%
Other 38%
Kazakhstan 10%
South Korea 15%
Pakistan 21%

STRENGTHS

Very few, apart from illicit opium trade. Agriculture still the largest sector. Overseas assets unfrozen from 2002.

WEAKNESSES

Decades of fighting; agriculture, industry, and infrastructure in ruins. Communication links damaged by earthquakes and devastated by bombing in 2001.

PROFILE

The protracted fighting has left Afghanistan one of the poorest and least developed countries in the world. Estimates suggest that $4 billion is needed to rebuild the country and that over 80% of infrastructure has been destroyed. Agricultural activity has fallen back from pre-1979 levels; the Soviets' "scorched earth" policy laid waste large areas, and much of the rural population fled to the cities. Many farmers turned back to growing poppies for opium production, but saw little profit from the trade, despite Afghanistan being one of the world's largest opium sources. Poppy cultivation was eventually banned by the *taliban*, and remains prohibited under the new regime.

ECONOMIC PERFORMANCE INDICATOR

Consumer Price Index
GDP

200
150
100
50
0
Consumer price index 1990=100
1988
1989
1990
1991
1992
GDP unavailable

Much of Afghanistan's infrastructure, already scarred by decades of civil war, was reduced to ruins by heavy US bombardment in 2001.

RESOURCES

Electric power 494,000 kw

1200 tonnes

Not an oil producer and has no refineries

14.3m sheep, 2.2m goats, 1.5m cattle, 7.2m chickens

Natural gas, salt, coal, copper, lapis lazuli, barytes, talc

ELECTRICITY GENERATION

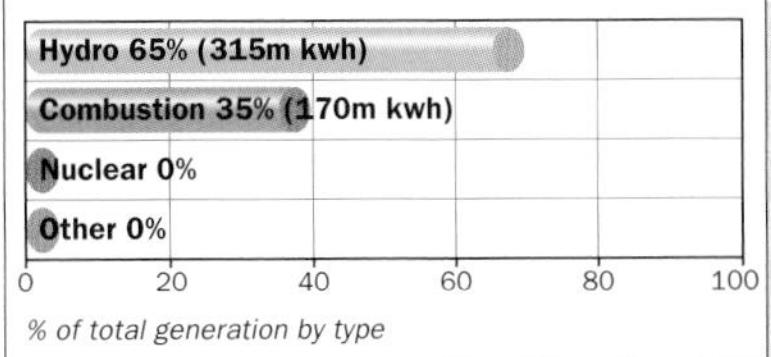
Hydro 65% (315m kwh)
Combustion 35% (170m kwh)
Nuclear 0%
Other 0%
0
20
40
60
80
100
% of total generation by type

Natural gas and coal are the most important strategic resources. In 2002 Afghanistan signed a gas agreement with Pakistan and Turkmenistan, which could generate revenue estimated at $300 million. Restoring the power generation system is a government priority. The construction of dams on the Kunar and Laghman rivers is being considered. Coal production has fallen from prewar levels and mines are in urgent need of rehabilitation.

AFGHANISTAN : LAND USE

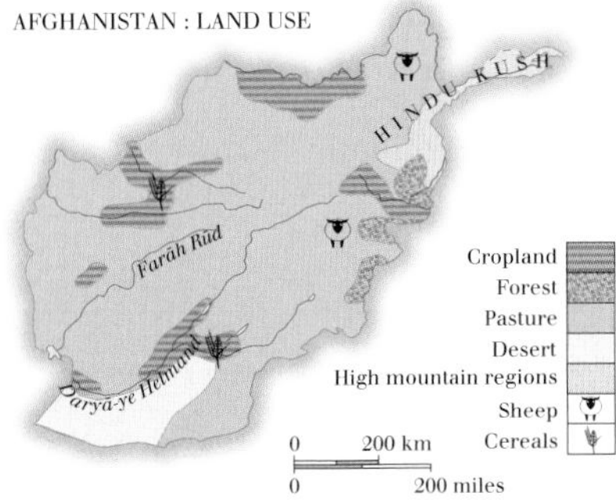
HINDU KUSH
Farah Rud
Darya-ye Helmand
Cropland
Forest
Pasture
Desert
High mountain regions
Sheep
Cereals
0
200 km
0
200 miles

ENVIRONMENT

Not available

0.3% (0.2% partially protected)

0.05 tonnes per capita

ENVIRONMENTAL TREATIES

No
Yes
No
No
No
No

Environmental priorities are low, given Afghanistan's relatively anarchic conditions. However, the country's lack of industry, even in Kabul, means that industrial pollution is minimal. The biggest problem facing Afghanistan is land mines: over ten million have been laid, and the UN estimates that it will take 100 years to make the country safe for civilians.

MEDIA

TV ownership low

Daily newspaper circulation 6 per 1000 people

PUBLISHING AND BROADCAST MEDIA

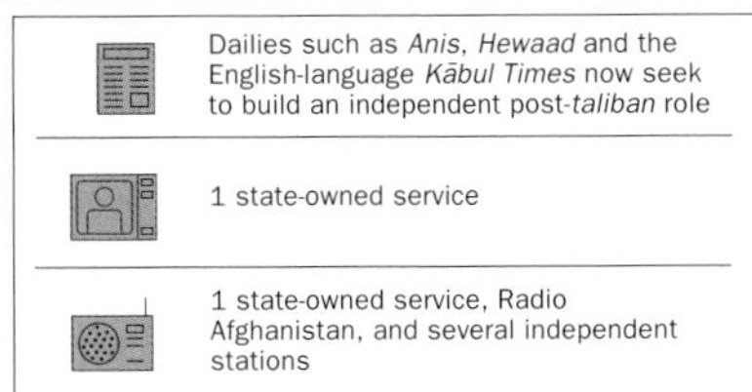
Dailies such as *Anis, Hewaad* and the English-language *Kābul Times* now seek to build an independent post-*taliban* role
1 state-owned service
1 state-owned service, Radio Afghanistan, and several independent stations

Rural Afghan factions run newspapers and radio stations. Television and the Internet were banned under the *taliban*, as were video cassette recorders and satellite dishes, but regular television broadcasts restarted in November 2001 in Kabul. The BBC, which broadcasts in Pashtu and Dari, is popular, especially for its soap operas, which convey information on welfare issues.

CRIME

Death penalty in use

Afghanistan does not publish prison figures

Levels of all crimes remain very high

CRIME RATES

No statistics for murders, rapes, or thefts are published due to the war situation

Gun law operates widely. Anyone traveling in rural areas is vulnerable to armed robbery. Cities vary according to which faction is dominant and the level of disruption which has been caused by war.

EDUCATION

School leaving age: 13

37%

12,800 students

THE EDUCATION SYSTEM

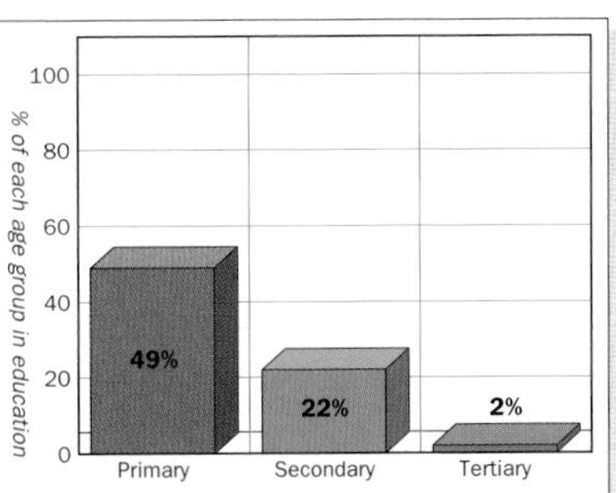
100
80
60
40
20
0
% of each age group in education
49%
22%
2%
Primary
Secondary
Tertiary

Under the *taliban* regime, education for women was strictly limited and segregation rigidly enforced; the literacy rate for women is now the lowest in the world. The Northern Alliance announced the lifting of restrictions on female education in November 2001.

Kabul University, which was closed in 1992, has been partially reopened.

HEALTH

No welfare state health benefits

1 per 10,000 people

Infectious, parasitic, respiratory, and digestive diseases

The health service has collapsed completely and almost all medical professionals have left the country. Infant and maternal mortality rates are among the highest in the world, and life expectancy is very low.

Parasitic diseases and infections are a particular problem. The UN organized a program for the chlorination of well water, following an outbreak of cholera in Kabul, and launched a mass measles vaccination program in January 2002.

Under the *taliban* regime most women in Afghanistan had very little access to health care; their admission to hospital was strongly discouraged, as was the employment of female medical staff.

SPENDING

GDP/cap. decrease

CONSUMPTION AND SPENDING

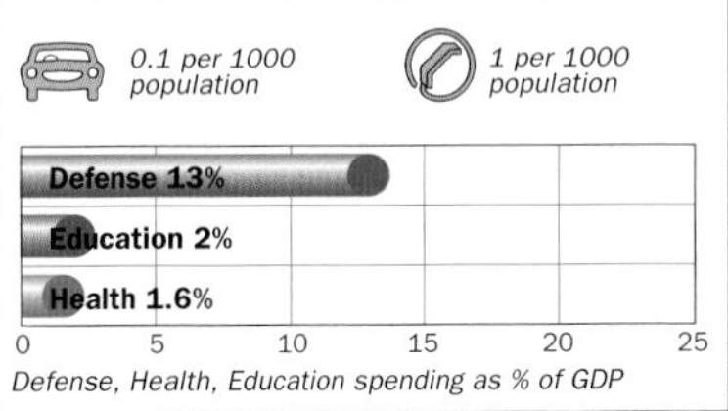
0.1 per 1000 population
1 per 1000 population
Defense 13%
Education 2%
Health 1.6%
0
5
10
15
20
25
Defense, Health, Education spending as % of GDP

The vast majority of Afghans live in conditions of extreme poverty. The country does not have the resources to feed its people at present – a situation exacerbated by the severe drought of 2000, the 2001 fighting, and the 2002 earthquakes. The return of refugees from neighboring Pakistan and Iran would make Afghanistan even more dependent on outside assistance for its rehabilitation.

A number of *mujahideen* leaders accumulated personal fortunes during the civil war. These derive in part from the substantial foreign aid that was once available and, in some cases, from the trafficking of opium.

WORLD RANKING

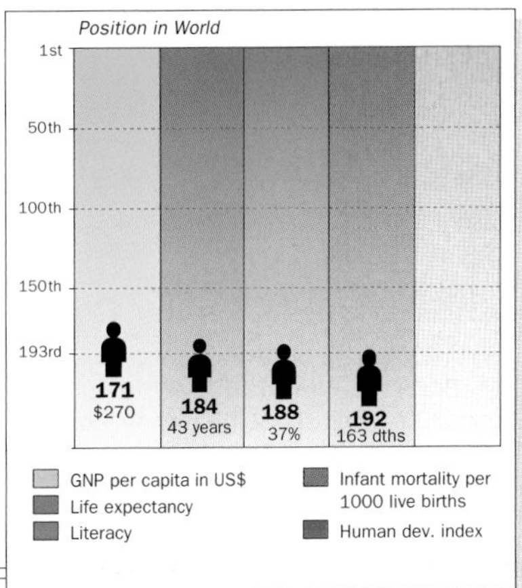
Position in World
1st
50th
100th
150th
193rd
171
$270
184
43 years
188
37%
192
163 dths
GNP per capita in US$
Life expectancy
Literacy
Infant mortality per 1000 live births
Human dev. index

A

ALBANIA

OFFICIAL NAME: Republic of Albania **CAPITAL:** Tirana
POPULATION: 3.1 million **CURRENCY:** Lek **OFFICIAL LANGUAGE:** Albanian

EUROPE

Serb. & Mon.
ALBANIA
Maced.
Italy
Greece
Europe

1912

1921

Nov 28
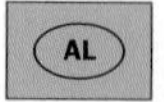
AL
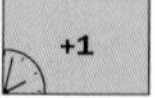
+1

+355

.al

LYING AT THE SOUTHEASTERN end of the Adriatic Sea, opposite the heel of Italy, Albania is a poverty-stricken, mountainous country which became a one-party communist state in 1944. The "land of the eagles," as it is known locally, first held multiparty elections in 1991. Economic collapse provoked uprisings and virtual anarchy in 1997, and the country was only stabilized by the intervention of OSCE troops.

CLIMATE

Mediterranean/ continental

WEATHER CHART FOR TIRANA

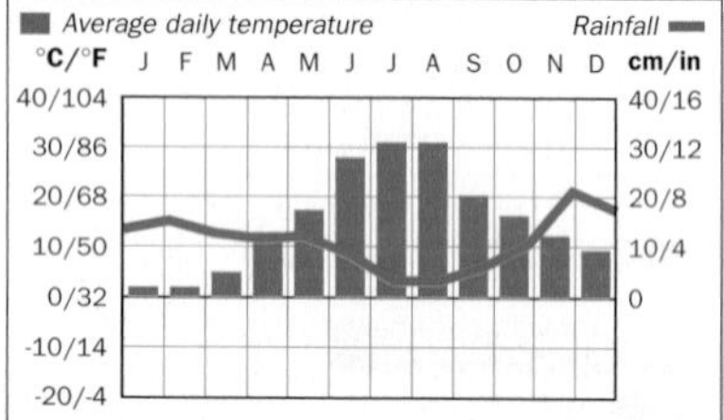
Average daily temperature
Rainfall
°C/°F
J F M A M J J A S O N D
cm/in
40/104
30/86
20/68
10/50
0/32
-10/14
-20/4
40/16
30/12
20/8
10/4
0

The coastal climate is Mediterranean, but rather wet in winter. Heavy rain or snow falls in winter in the mountains.

TRANSPORTATION

Drive on right

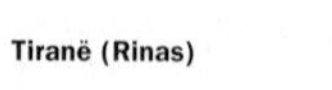
Tiranë (Rinas)

34 ships
23,928 grt

THE TRANSPORTATION NETWORK

5400 km (3355 miles)

None

440 km (273 miles)

43 km (27 miles)

The transportation infrastructure is poor. The rail network is limited and roads are in disrepair. Private cars were first allowed in 1991. Buses and private vans are the main means of transportation.

TOURISM

Visitors : Population 1:79

39,000 visitors
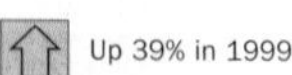
Up 39% in 1999

MAIN TOURIST ARRIVALS

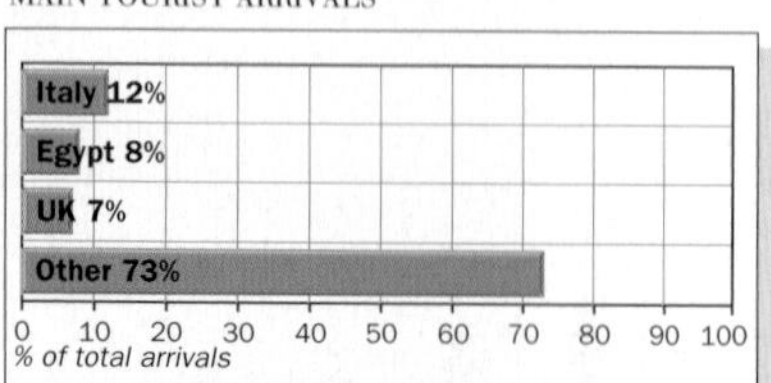
Italy 12%
Egypt 8%
UK 7%
Other 73%
0 10 20 30 40 50 60 70 80 90 100
% of total arrivals

Instability and then the war in Kosovo stalled plans to exploit Albania's scenic beauty. Facilities remain very limited, especially outside Tirana.

PEOPLE

Pop. density medium

Albanian, Greek

113/km²
(293/mi²)

THE URBAN/RURAL POPULATION SPLIT

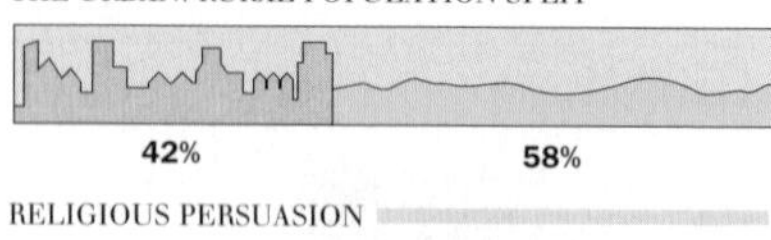
42%
58%

RELIGIOUS PERSUASION

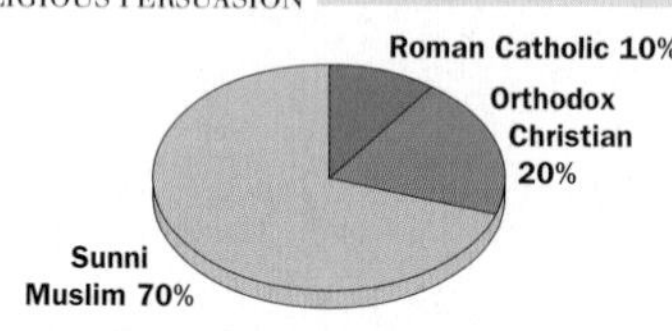
Roman Catholic 10%
Orthodox Christian 20%
Sunni Muslim 70%

The existence of ethnic minorities in Albania was only officially acknowledged in 1989. The Greek minority strongly contests statistics which state that 98% of the population are Albanian. Located mainly in the south and identifying with Athens rather than Tirana, the Greeks claim to make up 10% of the population. They suffer considerable discrimination.

Under communism, Albania was the only officially atheist state in the world. Many Albanians maintained their beliefs in private. Religious worship is now permitted and mosques have reopened. Society is traditional and male-dominated. The extended family remains a strong influence.

In 1999 the country temporarily sheltered nearly half a million ethnic Albanian refugees fleeing "ethnic cleansing" in neighboring Kosovo.

City of a thousand windows. *Berat was preserved as a museum while a new town was built further down the valley.*

POLITICS

Multiparty elections

2001/2005

President Alfred Moisiu

AT THE LAST ELECTION

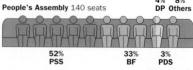

People's Assembly 140 seats

52% PSS | 33% BF | 3% PDS | 4% DP | 8% Others

PSS = Socialist Party of Albania **BF** = Union for Victory (led by the Democratic Party – **PD**) **DP** = Democrat Party (splinter from PD) **PDS** = Social Democratic Party

Albania was dominated for more than 40 years by communist ruler Enver Hoxha, who died in 1985. An exodus of Albanians in 1991 finally persuaded the regime to call multiparty elections. The resulting center-right coalition failed, however, to create a Western-style liberal state.

Many people were ruined in 1997 by the collapse of the "pyramid" savings schemes, prompting a rebellion in the south and forcing the government to resign. The new coalition led by the socialist PSS won elections later that year, and a further term in 2001, as Prime Minister Ilir Meta claimed credit for restoring a measure of hope and security. The poll, though disputed by the PD, was praised internationally as the fairest yet. In 2002 the PSS was wracked by internal disputes. By July, party leader Fatos Nano had won through to become prime minister.

WORLD AFFAIRS

Joined UN in 1955

WTO

CE

OSCE

OIC

PfP

Foreign policy in the late 1990s was dominated by the fate of Kosovo, the predominately ethnic Albanian province in neighboring Serbia. Ethnic Albanian separatism also erupted in Macedonia in 2001, although Albania was not directly involved. Membership of NATO and the EU is a long-term goal. In 2001 an interim EU "stabilization and association agreement" was offered.

AID

Recipient

$319m (receipts)

Down 35% in 2000

Since 1991 the West has provided aid. Food aid was stepped up in 1997, when anarchy swept the country, and again in 1999 to help cope with the hundreds of thousands of refugees arriving from Kosovo, in neighboring Serbia. EU aid now focuses on helping structural reforms.

DEFENSE

Compulsory military service

 $111m

 Down 21% in 2000

Officer ranks were reestablished in 1991, and a ten-year reconstruction program was launched in 2000. Military service of 18 months is mandatory. The Kosovo crisis prompted Albania in 1999 to make its airspace available to NATO.

ECONOMICS

Inflation 39% p.a. (1990–2000)

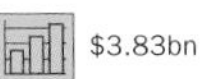 $3.83bn

 142.75–136.80 lekë

SCORE CARD

- ❏ World GNP Ranking 122nd
- ❏ GNP per Capita $1120
- ❏ Balance of Payments ....–$156m
- ❏ Inflation 0.1%
- ❏ Unemployment.................................16%

Strengths

Oil and gas reserves. Significant economic growth achieved since 1997 collapse. Progress with privatization.

Weaknesses

Rudimentary public services and energy, transportation, and water networks. Pyramid schemes wiped out savings.

EXPORTS

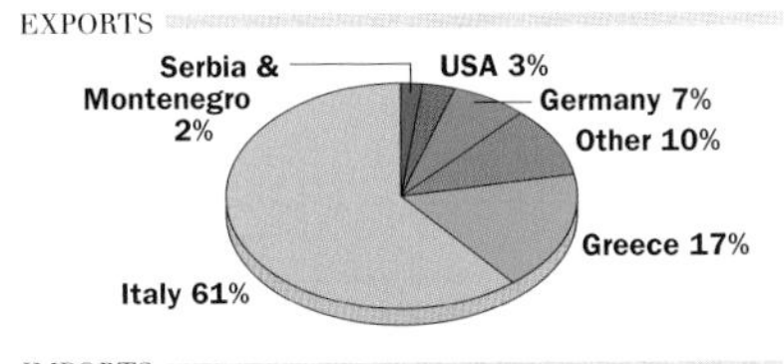
Serbia & Montenegro 2%
USA 3%
Germany 7%
Other 10%
Greece 17%
Italy 61%

IMPORTS

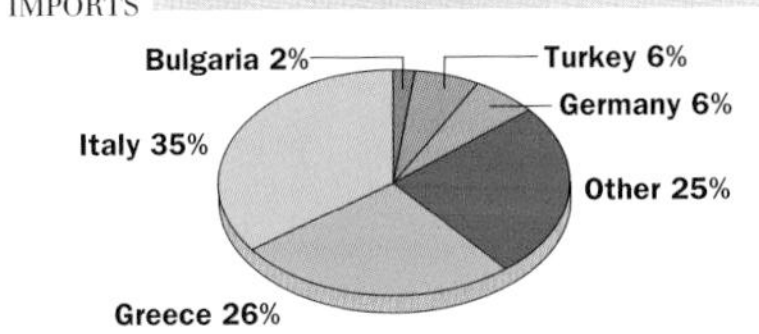
Bulgaria 2%
Turkey 6%
Germany 6%
Italy 35%
Other 25%
Greece 26%

ALBANIA

Total Land Area : 28 748 sq. km (11 100 sq. miles)

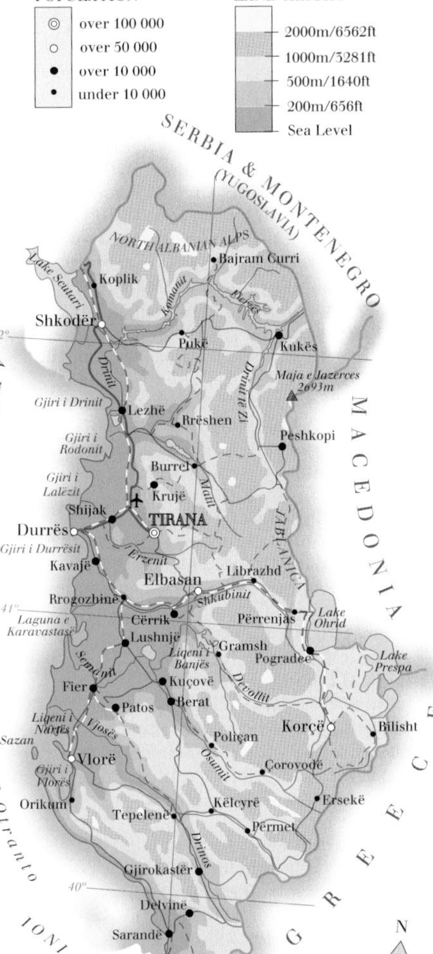
POPULATION
over 100 000
over 50 000
over 10 000
under 10 000
LAND HEIGHT
2000m/6562ft
1000m/3281ft
500m/1640ft
200m/656ft
Sea Level
SERBIA & MONTENEGRO (YUGOSLAVIA)
NORTH ALBANIAN ALPS
Bajram Curri
Koplik
Lake Scutari
Shkodër
Pukë
Kukës
Maja e Jazerces 2693m
Gjiri i Drinit
Lezhë
Rrëshen
Peshkopi
Gjiri i Rodonit
Burrel
Gjiri i Lalëzit
Krujë
Shijak
TIRANA
Durrës
Gjiri i Durrësit
Kavajë
Elbasan
Librazhd
Rrogozhinë
Laguna e Karavastasë
Cërrik
Lushnjë
Përrenjas
Lake Ohrid
Gramsh
Pogradec
Lake Prespa
Fier
Kuçovë
Patos
Berat
Ligeni i Nartës
Sazan
Vlorë
Poliçan
Korçë
Bilisht
Gjiri i Vlorës
Corovodë
Orikum
Tepelenë
Këlcyrë
Ersekë
Përmet
Gjirokastër
Delvinë
Sarandë
Konispol
ADRIATIC SEA
MACEDONIA
GREECE
Strait of Otranto
IONIAN SEA
42°
41°
40°
20°
N
0 40 km
0 40 miles

RESOURCES

Electric power 1.9m kw

3055 tonnes

7378 b/d (reserves 183m barrels)

1.94m sheep, 1.12m goats, 4m chickens

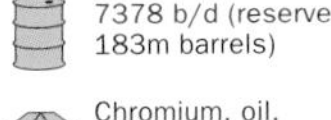 Chromium, oil, coal, natural gas, copper, nickel

Albania needs huge capital investment to develop its minerals and to create a modern electricity supply system.

ENVIRONMENT

Sustainability rank: 24th

 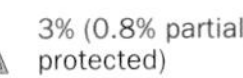 3% (0.8% partially protected)

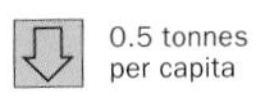 0.5 tonnes per capita

Toxic waste pollution from communist-era heavy industry is among the worst in Europe. Years of shortages mean that most materials are recycled.

MEDIA

TV ownership medium

 Daily newspaper circulation 37 per 1000 people

PUBLISHING AND BROADCAST MEDIA

There are 4 daily newspapers, including *Rilindja Demokratike*, *Zëri i Popullit*, and *Koha Jonë*, the best-selling newspaper

1 state-run service, 75 private stations

1 state-run service, 30 private stations

Media freedom is improving. A 1998 law banned political or religious control of TV stations. Newspaper sales are falling; some dailies are still party-run.

CRIME

Death penalty not used in practice

 2922 prisoners

 Down 6% in 1999

Lawlessness is widespread; guns are easily available after the anarchy of 1997. Cannabis is widely grown.

CHRONOLOGY

Albania gained independence in 1912 for the first time in its history.

- ❏ **1924–1939** Ahmet Zogu in power; crowned King Zog in 1928.
- ❏ **1939–1943** Occupied by Italy.
- ❏ **1944** Communist state; led by Enver Hoxha until 1985.
- ❏ **1991** First multiparty elections.
- ❏ **1997** Economic chaos as failure of pyramid schemes causes revolt.
- ❏ **1999** Refugee influx from Kosovo.
- ❏ **2001** PSS wins second term.

EDUCATION

School leaving age: 14

 85%

 34,257 students

The system is derived from the Soviet, Chinese, and Italian models. Albania has eight universities.

HEALTH

Welfare state health benefits

 1 per 769 people

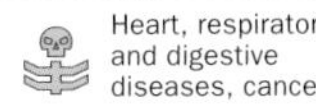 Heart, respiratory, and digestive diseases, cancers

The health service is rudimentary, and dependent on Western aid for most drugs and medical supplies.

SPENDING

GDP/cap. increase

CONSUMPTION AND SPENDING

29 per 1000 population

39 per 1000 population

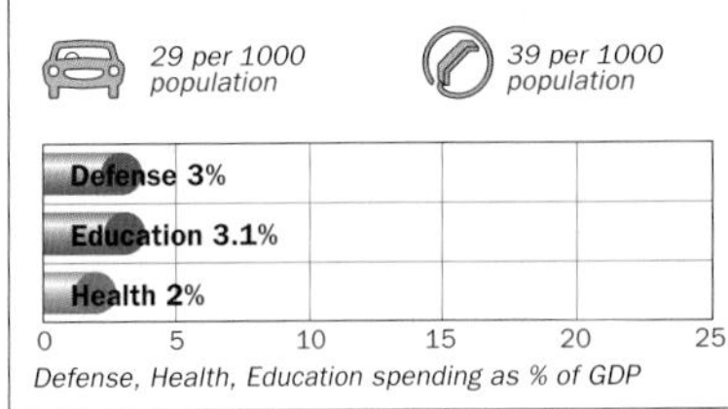
Defense 3%
Education 3.1%
Health 2%
0
5
10
15
20
25

Defense, Health, Education spending as % of GDP

Wealth is limited to a few private-sector entrepreneurs. Poverty is worst in northern rural areas but also acute in slum settlements around Tirana and other cities.

WORLD RANKING

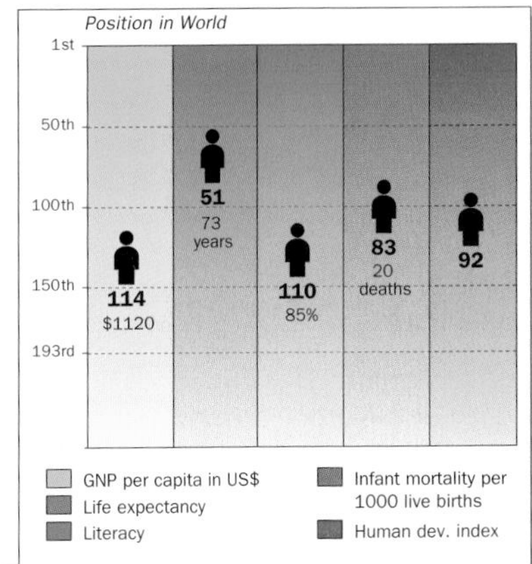
Position in World
1st
50th
100th
150th
193rd
114
$1120
51
73 years
110
85%
83
20 deaths
92
GNP per capita in US$
Life expectancy
Literacy
Infant mortality per 1000 live births
Human dev. index

A

ALGERIA

NORTH AFRICA

Morocco
Tunisia
ALGERIA
Western Sahara
Libya
Mauritania
Mali
Niger
Africa

OFFICIAL NAME: People's Democratic Republic of Algeria **CAPITAL:** Algiers
POPULATION: 30.8 million **CURRENCY:** Algerian dinar **OFFICIAL LANGUAGES:** Arabic and Tamazight

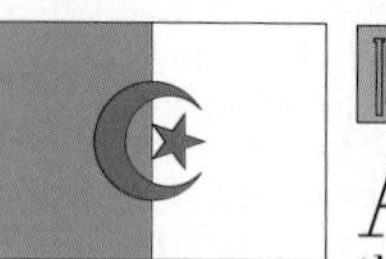

1962

1962

Nov 1

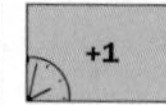
DZ

+1

+213

.dz

AFRICA'S SECOND-LARGEST country, which extends from a densely populated Mediterranean coast to the empty northern Sahara, Algeria won independence from France in 1962. The military blocked radical Islamists from taking power after winning elections in 1991, setting up a new civilian regime and fighting a bloody terrorist conflict ever since. Algeria has one of the youngest populations, and highest birthrates, in the region.

CLIMATE

Hot desert/ Mediterranean

WEATHER CHART FOR ALGIERS

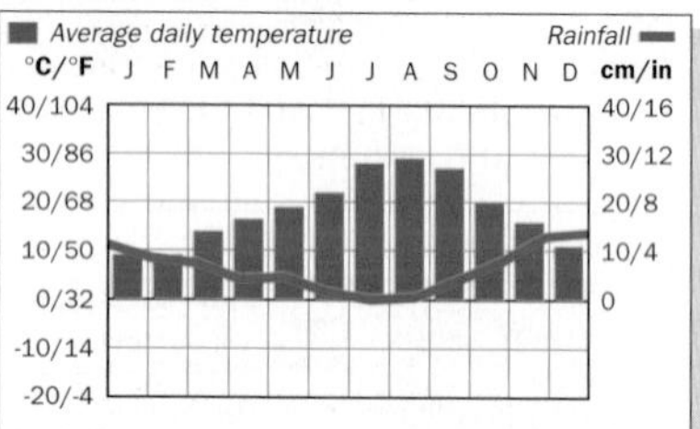
Average daily temperature
Rainfall
°C/°F
J F M A M J J A S O N D
cm/in
40/104
30/86
20/68
10/50
0/32
-10/14
-20/-4
40/16
30/12
20/8
10/4
0

Coastal areas have a warm, temperate climate. The whole area to the south of the Atlas Mountains is hot desert.

TRANSPORTATION

Drive on right

Houari Boumedienne, Algiers
2.78m passengers

142 ships
960,824 grt

THE TRANSPORTATION NETWORK

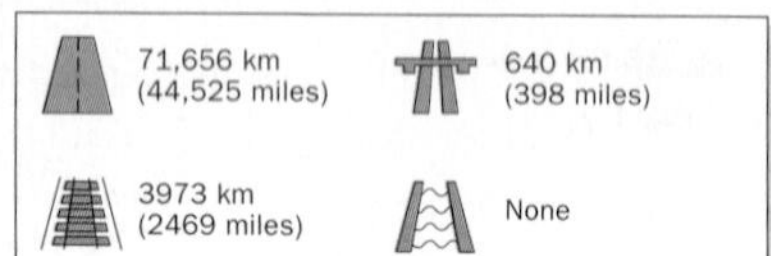
71,656 km
(44,525 miles)
640 km
(398 miles)
3973 km
(2469 miles)
None

There are five international airports. Rail is the quickest way to travel between the main urban centers.

TOURISM

Visitors : Population 1:36

866,000 visitors

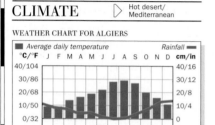

Up 16% in 2000

MAIN TOURIST ARRIVALS

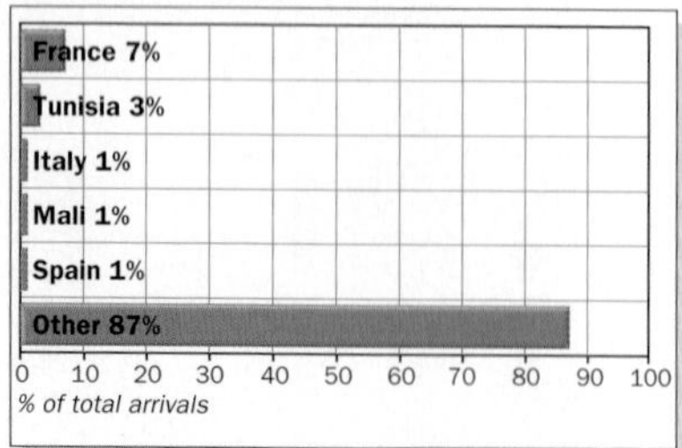
France 7%
Tunisia 3%
Italy 1%
Mali 1%
Spain 1%
Other 87%
0 10 20 30 40 50 60 70 80 90 100
% of total arrivals

The once-popular desert safaris are now rare. Tourists are a target for militant Islamist groups.

PEOPLE

Pop. density low

Arabic, Tamazight (Kabyle, Shawia, Tamashek), French

13/km² (33/mi²)

THE URBAN/RURAL POPULATION SPLIT

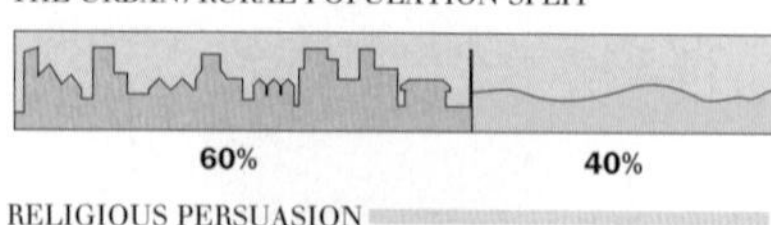
60%
40%

RELIGIOUS PERSUASION

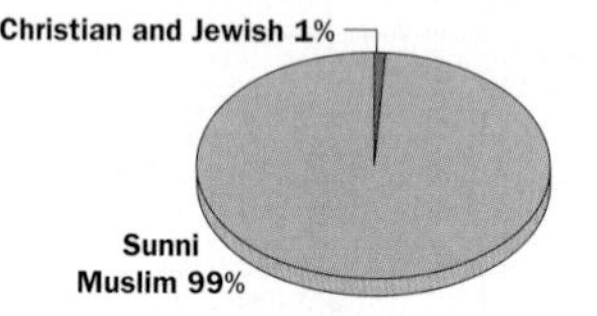
Christian and Jewish 1%
Sunni Muslim 99%

ETHNIC MAKEUP

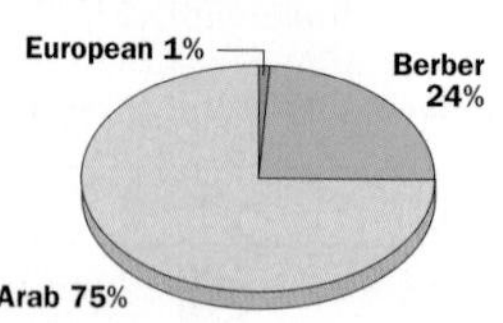
European 1%
Berber 24%
Arab 75%

Algeria's population is predominantly Arab, under 30 years of age, and urban; around a quarter are Berber. More than 85% speak Arabic and 99% are Sunni Muslim. Mosques also provide social and medical services. Of the million or so French who settled in Algeria before independence, only about 6000 remain. Most Berbers consider the mountainous Kabylia region their homeland. Demonstrations there have met with violent police crackdowns, particularly in the Berber Spring of 1980, and since its anniversary in 2001. The Berber language, Tamazight, was recognized as an official language in 2002.

POPULATION AGE BREAKDOWN

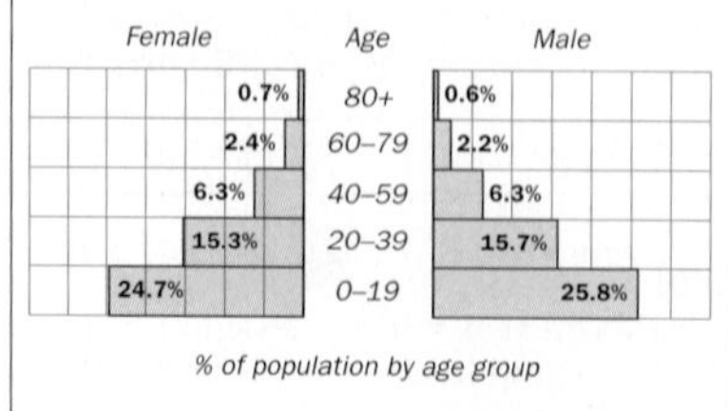
Female
Age
Male
0.7%
80+
0.6%
2.4%
60–79
2.2%
6.3%
40–59
6.3%
15.3%
20–39
15.7%
24.7%
0–19
25.8%
% of population by age group

POLITICS

Multiparty elections

L. House 2002/2007
U. House 2000/2003

President Abdelaziz Bouteflika

AT THE LAST ELECTION

National People's Assembly 389 seats

8% Ind **3% Others**

51% FLN **12% RND** **11% MRN** **10% MSP** **5% PT**

FLN = National Liberation Front **RND** = National Democratic Rally **MRN** = Movement for National Reform **MSP** = Movement for a Peaceful Society **Ind** = Independents **PT** = Workers' Party **App** = Appointed **FFS** = Front of Socialist Forces

Council of the Nation 144 seats

3% FFS

53% RND **33% App** **9% FLN** **2% MSP**

96 seats are elected, 48 are appointed by the president

Algeria has a multiparty parliamentary system.

PROFILE

Until 1988, Algeria was a single-party socialist-style regime. The aging ruling elite then adopted privatization policies strongly opposed by radical Islamists. The Islamic Salvation Front (FIS) won elections in 1991, but the results were canceled. Since 1992 the country has been scarred by appalling violence and tens of thousands of people have been killed in a terrorist campaign and by ruthless state counterterrorism. Hopes for peace rose when Abdelaziz Bouteflika became president in 1999, although the election was marred by the withdrawal of all other candidates. Parliamentary elections in 2002, won by the FLN, also suffered from a low turnout and a partial opposition boycott. Violence continues, with an annual death toll which is still in the thousands.

MAIN POLITICAL ISSUES

Islamic fundamentalism

Radical Islamists want Algeria to be a theocracy. The imposition of military rule to prevent them taking office in 1992 unleashed violence spearheaded by the extremist Armed Islamic Group (GIA).

The market economy

The FIS's 1991 election victory was in part due to popular reaction against economic reforms. After a brief suspension following the army takeover in 1992, the liberalization program was revived under pressure from the IMF and the World Bank.

WORLD AFFAIRS

Joined UN in 1962

AL

AMU

OIC

NAM

OPEC

Algeria's struggle for independence from France lasted from 1954 until 1962. Throughout the 1960s and 1970s, Algeria's success in rejecting a colonial power made it a champion for the developing world. It had a leading voice within the UN, the Arab League, and the Organization of African Unity. However, relations with the West remained essentially stable. Algeria was increasingly seen by the diplomatic community as a useful bridge between the West and Iran. In 1981, Algerian diplomats helped to secure the release of US hostages being held in Tehran. Algeria also attempted to act in a mediating role during the 1980–1988 Iran–Iraq War.

Algeria's influence overseas has diminished as the country has become increasingly unstable politically. A victory for the fundamentalist FIS in Algeria would greatly encourage Islamist militants in neighboring Morocco and Tunisia, and further undermine Egypt's embattled government.

France fears the spill-over of terrorism and has been shocked by the killings, especially those of seven French priests and of the French Roman Catholic bishop of Oran.

European governments are anxious to help stabilize the regime to avoid refugees seeking entry into France, Spain, and Italy.

Saharan town, *showing the wide range of Algeria's scenery, from lush, irrigated gardens near water sources to barren dunes beyond. 80% of Algeria is desert.*

AID

Recipient

$162m (receipts)

Up 82% in 2000

As a major oil producer, Algeria receives relatively small quantities of aid. During the 1980s, its economy became dependent on eastern European manufactures, which were swapped for oil. The collapse of this trade in the 1990s led Algeria to turn to the West for loans. The growing weight of Western economic involvement in turn fortified the regime against criticism of its hard-line methods against Islamic opponents. The IMF has provided loans to help Algeria meet payments on its debt, on condition that it move toward a market-oriented economy.

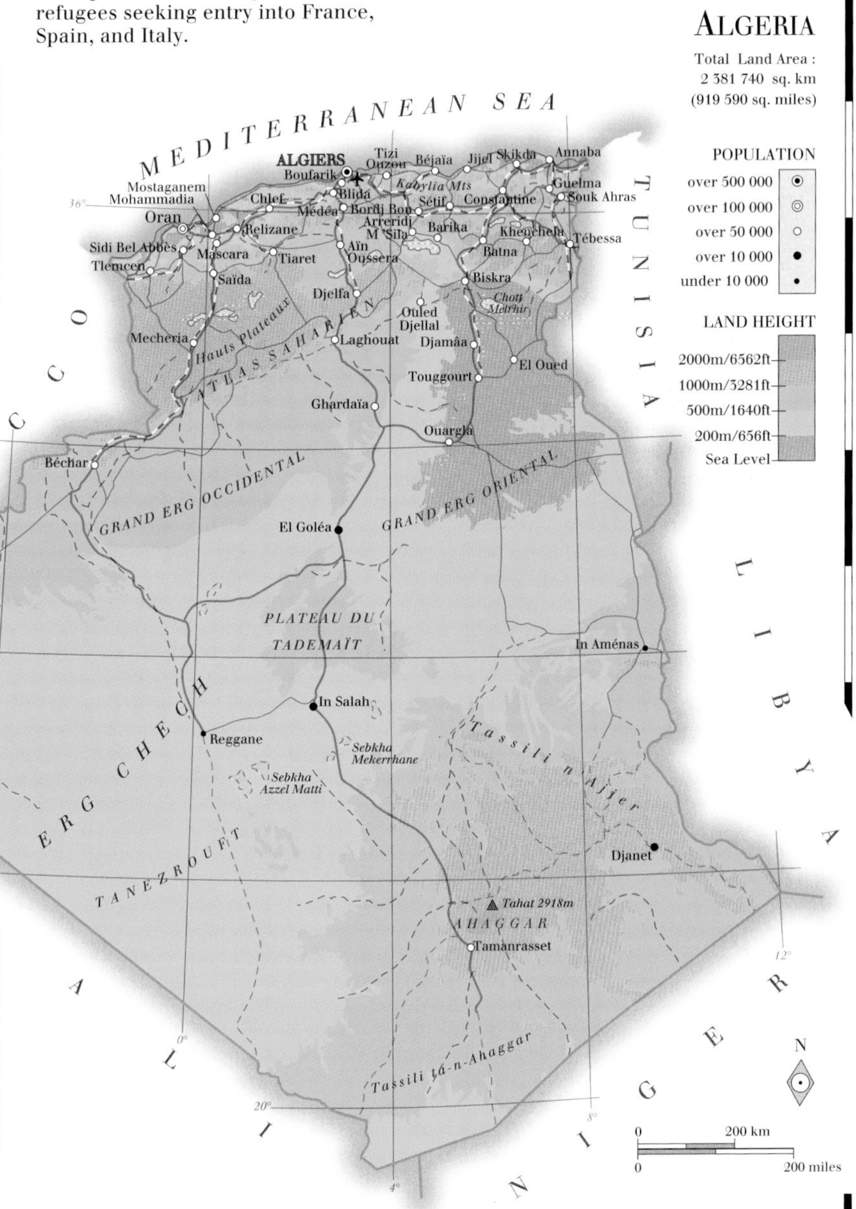

ALGERIA

Total Land Area : 2 381 740 sq. km (919 590 sq. miles)

POPULATION
over 500 000
over 100 000
over 50 000
over 10 000
under 10 000

LAND HEIGHT
2000m/6562ft
1000m/3281ft
500m/1640ft
200m/656ft
Sea Level

0 200 km
0 200 miles

Abdelaziz Bouteflika, *who was elected president in 1999.*

Abassi Madani, *leader of the Islamic Salvation Front (FIS).*

CHRONOLOGY

The conquest of Algeria by France began in 1830. By 1900, French settlers occupied most of the best land. In 1954, war was declared on the colonial administration by the National Liberation Front (FLN).

- ❑ **1962** Cease-fire agreed, followed by independence of Algerian republic.
- ❑ **1965** Military junta topples government of Ahmed Ben Bella. Revolutionary council set up.
- ❑ **1966** Judiciary "Algerianized." Tribunals try "economic crimes."
- ❑ **1971** Oil industry nationalized. President Boumedienne continues with land reform, a national health service, and "socialist" management.
- ❑ **1976** National Charter establishes a socialist state.
- ❑ **1980** Ben Bella released after 15 years' detention. Agreement with France whereby latter gives incentives for return home of 800,000 Algerian immigrants.
- ❑ **1981** Algeria helps to negotiate release of hostages from US embassy in Tehran, Iran.
- ❑ **1985** Two most popular Kabyle (Berber) singers given three-year jail sentences for opposing regime.
- ❑ **1987** Limited economic liberalization. Cooperation agreement with Soviet Union.
- ❑ **1988** Anti-FLN violence; state of emergency. Algeria negotiates release of Kuwaiti hostages from aircraft; Shi'a hijackers escape.
- ❑ **1989** Constitutional reforms diminish power of FLN. New political parties founded, including Islamic Salvation Front (FIS). AMU established.
- ❑ **1990** Political exiles permitted to return. FIS is victorious in municipal elections.
- ❑ **1991** FIS leaders Abassi Madani and Ali Belhadj arrested. FIS wins most seats in National People's Assembly.
- ❑ **1992** President Chadli overthrown by military. President Boudiaf assassinated. Madani and Belhadj given 12 years in jail.
- ❑ **1994** Political violence led by GIA.
- ❑ **1995** Democratic presidential elections won by Liamine Zéroual.
- ❑ **1996** Murders continue, notably of Catholic clergy and GIA leader.
- ❑ **1997** Madani released from jail but debarred from active politics.
- ❑ **1999** Abdelaziz Bouteflika elected president in poll boycotted by opposition candidates.
- ❑ **2001** Fresh investment in oil and gas benefits economy. Resurgence of Berber protests.
- ❑ **2002** Berber language Tamazight recognized as national language. FLN election victory.

DEFENSE

▷ Compulsory military service

 $2.93bn

 Down 5% in 2000

The National Liberation Army (NLA), equipped with Russian weapons, is the dominant power in politics. There have been fears that parts of the army would forge an alliance with Muslim militants; the extreme rebel Armed Islamic Group, which has split from the FIS, is led by former army officers. However, the military are also suspected of taking part in reprisal killings of large numbers of Islamists.

ALGERIAN ARMED FORCES

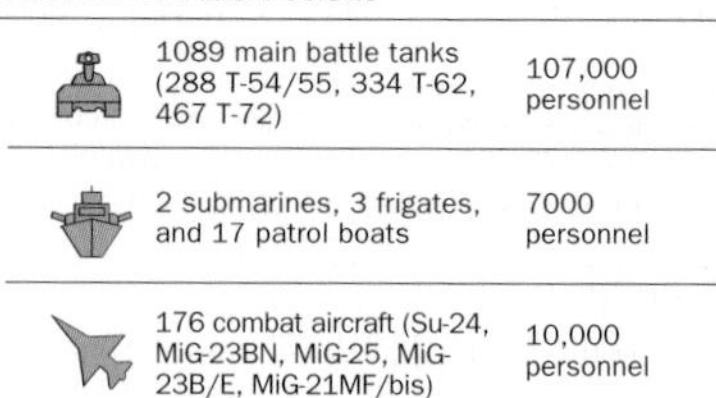

Army	1089 main battle tanks (288 T-54/55, 334 T-62, 467 T-72)	107,000 personnel
Navy	2 submarines, 3 frigates, and 17 patrol boats	7000 personnel
Air force	176 combat aircraft (Su-24, MiG-23BN, MiG-25, MiG-23B/E, MiG-21MF/bis)	10,000 personnel
Missiles	None	

ECONOMICS

▷ Inflation 18% p.a. (1990–2000)

 $47.9bn

 73.42–76.86 Algerian dinars

SCORE CARD

- ❑ World GNP Ranking..........................50th
- ❑ GNP per Capita..............................$1580
- ❑ Balance of Payments.......................$8.9bn
- ❑ Inflation..2.6%
- ❑ Unemployment..................................30%

EXPORTS

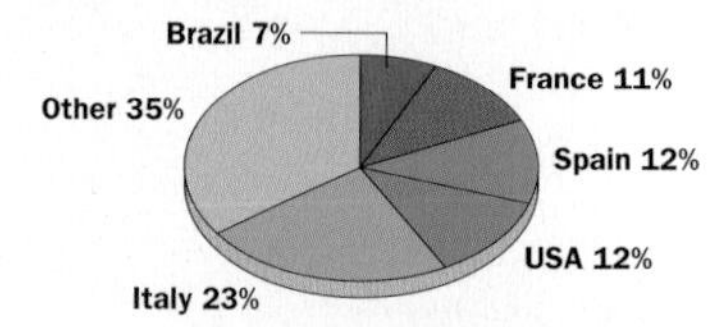

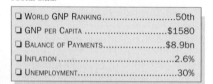

Brazil 7%
France 11%
Other 35%
Spain 12%
USA 12%
Italy 23%

IMPORTS

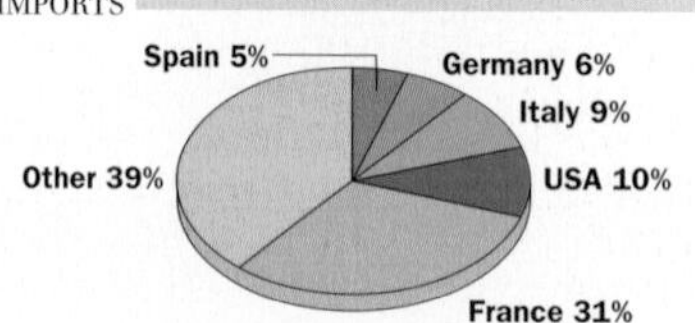

Spain 5%
Germany 6%
Italy 9%
Other 39%
USA 10%
France 31%

Strengths

Oil and gas. Recent collaboration with Western oil companies should see improvements in productivity. Natural gas is supplied to Europe, with plans in hand for the construction of a third undersea pipeline.

Weaknesses

Political turmoil threatens many new projects and has led to an exodus of European and other expatriate workers important to the economy. Lack of skilled labor coupled with high unemployment. Limited agriculture. Shortages of basic foodstuffs. Thriving black market.

Profile

Centralized socialist planning dominated the Algerian economy until the late 1980s, when the economic collapse of the Soviet Union led to a change in policy, and Algeria began moving toward a market economy. Reforms were frozen, however, following the military takeover in 1992, although many have since been resumed under pressure from the IMF and the World Bank. The majority of the economy's most productive sectors remain under state control, although private investment is encouraged in the oil industry and, since early 2001, in telecommunications. A number of Western oil companies have signed exploration contracts with Algeria since it has accepted more competitive production-sharing agreements. However, Western investment levels are likely to remain small as long as the political situation is unstable.

ECONOMIC PERFORMANCE INDICATOR

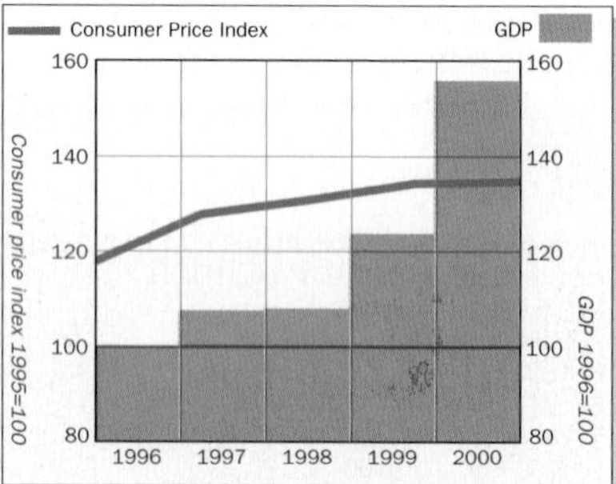

Consumer Price Index
GDP
160
140
120
100
80
Consumer price index 1995=100
GDP 1996=100
1996
1997
1998
1999
2000

ALGERIA : MAJOR BUSINESSES

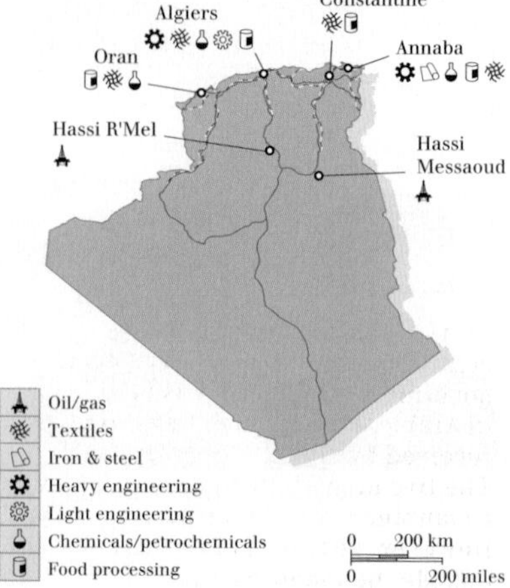

Algiers
Constantine
Oran
Annaba
Hassi R'Mel
Hassi Messaoud
Oil/gas
Textiles
Iron & steel
Heavy engineering
Light engineering
Chemicals/petrochemicals
Food processing
0 200 km
0 200 miles

RESOURCES

Electric power 6m kw

105,943 tonnes

1.56m b/d (reserves 9.2bn barrels)

19.3m sheep, 3.5m goats, 1.7m cattle, 110m chickens

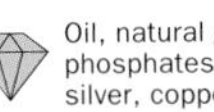
Oil, natural gas, iron, phosphates, lead, zinc, silver, copper, gold

ELECTRICITY GENERATION

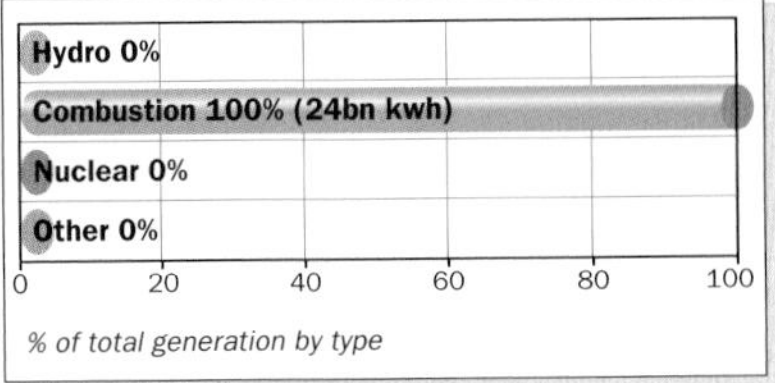
Hydro 0%
Combustion 100% (24bn kwh)
Nuclear 0%
Other 0%
0
20
40
60
80
100
% of total generation by type

Crude oil and natural gas, Algeria's main resources, have been produced since the 1950s. Algeria also has diverse minerals, including iron ore, zinc, silver, copper ore, lead, gold, and phosphates. In the 1960s and 1970s, Algeria sought to become a major manufacturer, with investments in building materials, refined products, and steel; none of these sectors is competitive on world markets. Agriculture employs one-quarter of Algeria's workforce, but its importance to the economy is diminishing. State forests cover some 2% of Algeria's land. Most are brushwood, but some areas include cork oak trees, Aleppo pine, evergreen oak, and cedar. Algeria has a large fishing fleet. Sardines, anchovies, tuna, and shellfish are the major species caught commercially.

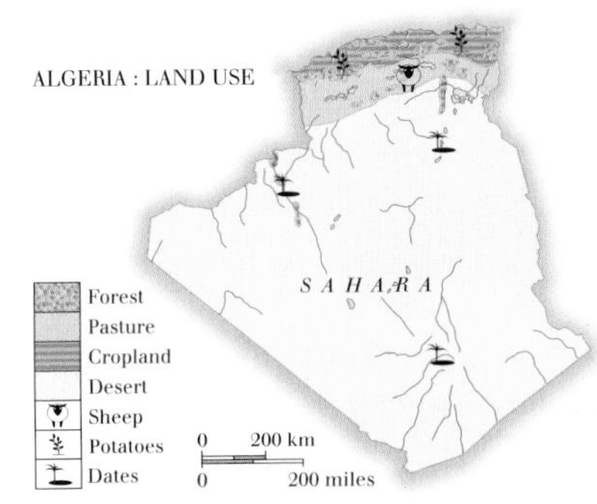
ALGERIA : LAND USE
SAHARA
Forest
Pasture
Cropland
Desert
Sheep
Potatoes
Dates
0 200 km
0 200 miles

ENVIRONMENT

Sustainability rank: 70th

3% (0.1% partially protected)

3.6 tonnes per capita

ENVIRONMENTAL TREATIES

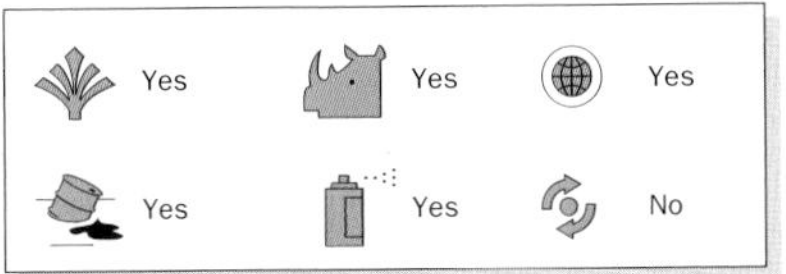
Yes
Yes
Yes
Yes
Yes
No

Since most of Algeria is desert or semidesert, over 90% of the population is forced to live on what remains – some 20% of the land. The desert is moving northward. Vegetation has been stripped for use as firewood and animal fodder, leaving fragile soils exposed which then require expensive specialist care to conserve them. Techniques for water purification are substandard, and rivers are being increasingly contaminated by untreated sewage, industrial effluent, and wastes from petroleum refining.

MEDIA

TV ownership medium

Daily newspaper circulation 38 per 1000 people

PUBLISHING AND BROADCAST MEDIA

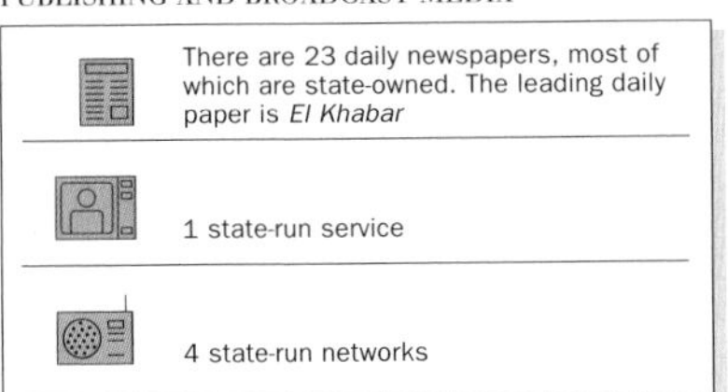
There are 23 daily newspapers, most of which are state-owned. The leading daily paper is El Khabar
1 state-run service
4 state-run networks

Newspapers, TV, and radio are state-controlled and permit no criticism of government actions. TV is broadcast in Arabic, French, and Tamazight, received by about two million sets. The five main daily newspapers have a combined circulation of 1.3 million. However, distribution is limited outside the major cities.

CRIME

Death penalty in use

35,737 prisoners

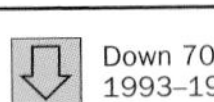
Down 70% 1993–1999

CRIME RATES

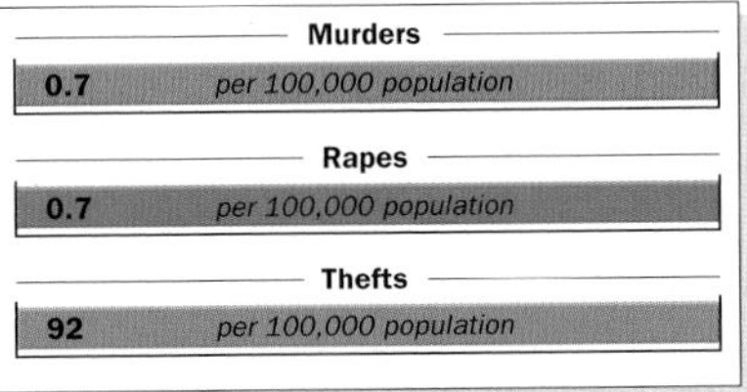
Murders
0.7 per 100,000 population
Rapes
0.7 per 100,000 population
Thefts
92 per 100,000 population

Thousands of people have been killed by radical Islamists since 1992, while human rights groups have accused pro-government death squads of brutal reprisal killings and of persecuting suspected Islamist militants.

EDUCATION

School leaving age: 15

68%

347,410 students

THE EDUCATION SYSTEM

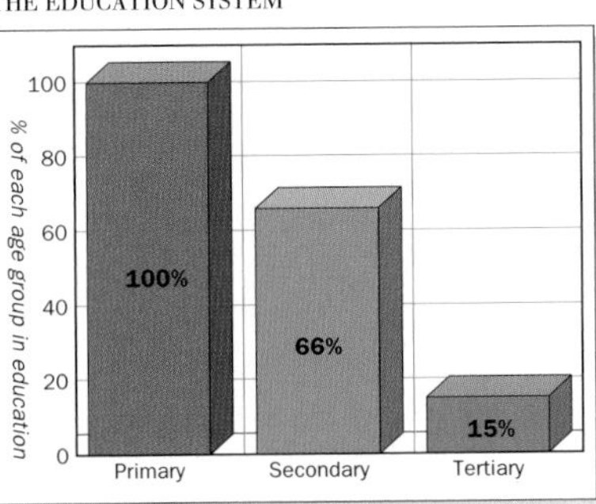
100
80
60
40
20
0
% of each age group in education
100%
66%
15%
Primary
Secondary
Tertiary

Over three-quarters of the school-age population receive a formal education, and the literacy rate is rising.

Since 1973, the curriculum has been Arabicized and the teaching of French has been restricted. New legislation in 1996 enforced the use of Arabic in public life, including for teaching.

Ten universities, seven polytechnics, and several technical colleges provide higher education.

HEALTH

Welfare state health benefits

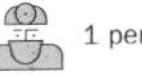
1 per 1000 people

Respiratory, heart, and cerebrovascular diseases, malaria

Since 1974 all Algerians have had the right to free health care. Primary health care is rudimentary outside main cities. Because the formal health care system is overburdened, many people turn to alternative forms of medicine. The infant mortality rate is well below the average for north Africa, and life expectancy is just above the regional average.

SPENDING

GDP/cap. increase

CONSUMPTION AND SPENDING

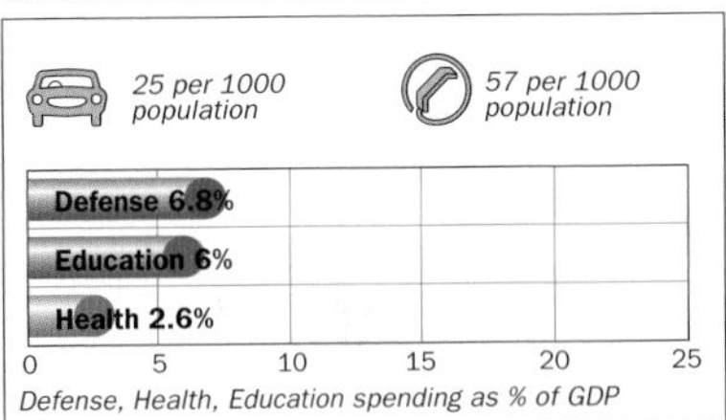
25 per 1000 population
57 per 1000 population
Defense 6.8%
Education 6%
Health 2.6%
0
5
10
15
20
25
Defense, Health, Education spending as % of GDP

There is great disparity in wealth between the political elite and the rest of the population. Those with connections in the military are the wealthiest group. Most Algerians have had to contend with soaring prices for basic necessities.

WORLD RANKING

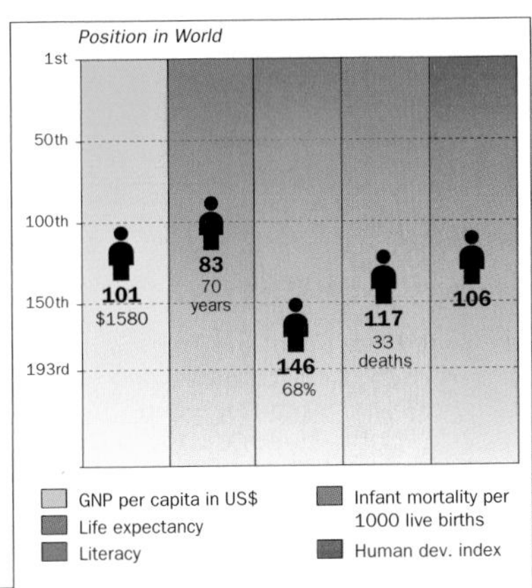
Position in World
1st
50th
100th
150th
193rd
101
$1580
83
70
years
146
68%
117
33
deaths
106
GNP per capita in US$
Life expectancy
Literacy
Infant mortality per 1000 live births
Human dev. index

A

ANDORRA

OFFICIAL NAME: Principality of Andorra **CAPITAL:** Andorra la Vella
POPULATION: 66,800 **CURRENCY:** Euro (French franc & Spanish peseta until 2002) **OFFICIAL LANGUAGE:** Catalan

1278

1278

Sept 8
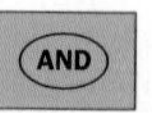
AND
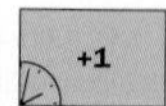
+1

+376

.ad

A TINY, LANDLOCKED principality between France and Spain, Andorra lies high in the eastern Pyrenees. From the 13th century, French and Spanish co-princes (today the President of France and the Bishop of Urgell) have ruled Andorra. In December 1993, the principality held its first full elections. Andorra's spectacular scenery, alpine climate, and duty-free shopping have made tourism, especially skiing, its main source of income.

Andorra's outstanding mountain scenery attracts skiers in winter, walkers in summer.

CLIMATE

Mountain

WEATHER CHART FOR ANDORRA LA VELLA

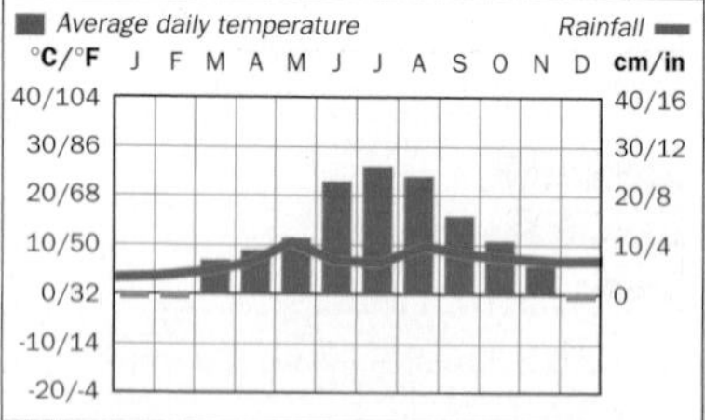

Average daily temperature
Rainfall
°C/°F
J F M A M J J A S O N D
cm/in
40/104
30/86
20/68
10/50
0/32
-10/14
-20/-4
40/16
30/12
20/8
10/4
0

Springs are cool and wettest in May; summers are relatively dry and warm. Snowfalls in December and January provide snow for good skiing up to March. Andorra's climate supports an abundance of wild flowers.

TRANSPORTATION

Drive on right

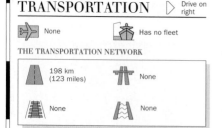

None

Has no fleet

THE TRANSPORTATION NETWORK

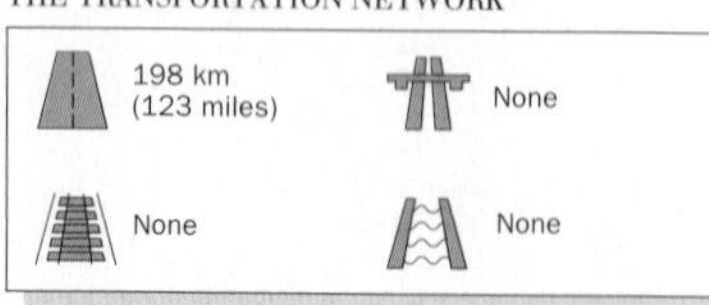

198 km (123 miles)
None
None
None

The road from France to Spain climbs to 2704 m (8875 ft) through one of the most dramatic mountain passes in Europe. Traffic congestion is a major problem in Andorra la Vella, especially in the summer. In 2001 plans for an overhead rail system were announced.

TOURISM

Visitors : Population 44:1

2.95m visitors

Up 26% in 2000

MAIN TOURIST ARRIVALS

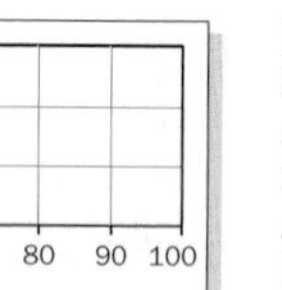

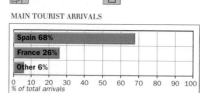

Spain 68%
France 26%
Other 6%
0 10 20 30 40 50 60 70 80 90 100
% of total arrivals

Most tourists visit Andorra to ski or shop. There is a traditional trade in day trippers from France and Spain, drawn by the many tax-free designer-label boutiques.

Five resorts offer Alpine skiing facilities, and specialize in Nordic skiing. In summer they cater instead for mountain hikers; Andorra's wide variety of wild flowers attract many visitors, but there is also much for the birdwatcher to see.

Although not strongly promoted, hunting of wild boar is popular, and the goat-like chamois can be hunted under special license.

EUROPE
France
ANDORRA
Spain

Europe

PEOPLE

Pop. density medium

Spanish, Catalan, French, Portuguese

144/km² (372/mi²)

THE URBAN/RURAL POPULATION SPLIT

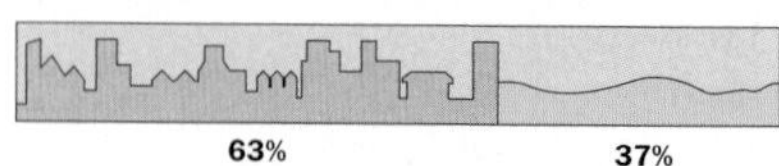

63%
37%

RELIGIOUS PERSUASION

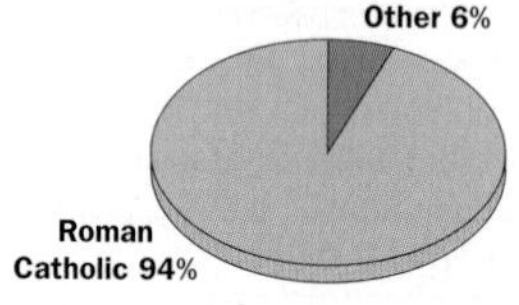
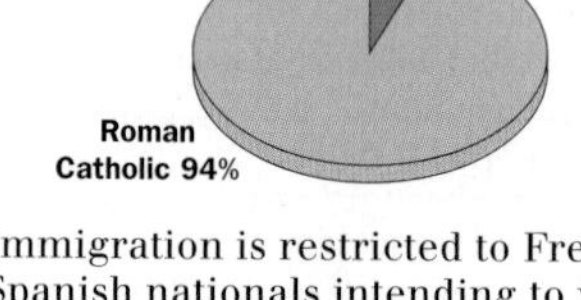

Other 6%
Roman Catholic 94%

Immigration is restricted to French and Spanish nationals intending to work in Andorra. Divorce is illegal and the marriage rate is low.

POLITICS

Multiparty elections

2001/2005

Co-Princes Jacques Chirac and Joan Martí Alanis

AT THE LAST ELECTION

General Council of the Valleys 28 seats

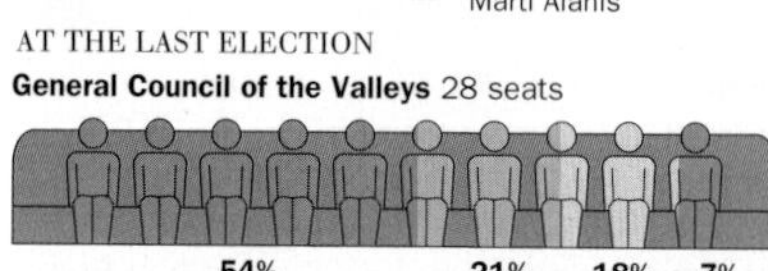

54% PLA
21% PSD
18% PD
7% UL

PLA = Liberal Party of Andorra **PSD** = Social Democratic Party **PD** = Democratic Party **UL** = Unió Laurediana

14 members are elected on a national list and 14 are elected in 7 dual-member parishes.

Andorra was a semifeudal state until 1993, when a referendum approved measures which legalized political parties and the right to strike, and altered relations with the co-princes. The ruling PLA, led by Marc Forné, has twice been returned to power.

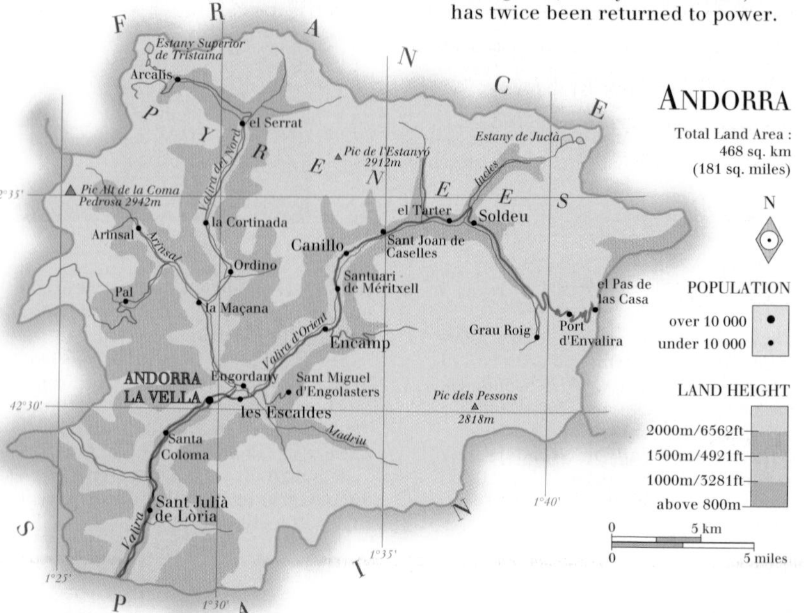

FRANCE
SPAIN
PYRENEES
Estany Superior de Tristaina
Arcalís
el Serrat
Pic de l'Estanyó 2912m
Estany de Juclà
Pic Alt de la Coma Pedrosa 2942m
Valira del Nord
Jucles
el Tarter
Soldeu
la Cortinada
Arinsal
Canillo
Sant Joan de Caselles
Ordino
Santuari de Meritxell
el Pas de las Casa
Pal
la Maçana
Grau Roig
Port d'Envalira
Valira d'Orient
Encamp
ANDORRA LA VELLA
Engordany
Sant Miguel d'Engolasters
les Escaldes
Pic dels Pessons 2818m
Madriu
Santa Coloma
Sant Julià de Lòria
Valira
42°35'
42°30'
1°25'
1°30'
1°35'
1°40'
ANDORRA
Total Land Area : 468 sq. km (181 sq. miles)
N
POPULATION
over 10 000
under 10 000
LAND HEIGHT
2000m/6562ft
1500m/4921ft
1000m/3281ft
above 800m
0 5 km
0 5 miles

A

WORLD AFFAIRS

Joined UN in 1993

CE

OSCE
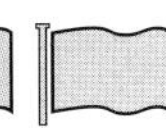
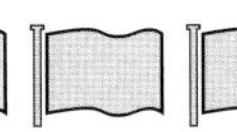

In 1991 Andorra became a member of the EU customs union and adopted the euro as its official currency in 2002. It joined the UN in 1993. Andorra's status as a tax haven has prompted criticism from the OECD.

AID

Not applicable

 Andorra has no aid receipts or donations

 Not applicable

The principality of Andorra neither receives nor provides aid, and has no plans to do so.

DEFENSE

No compulsory military service

 Andorra has no defense budget

 Not applicable

Andorra has no defense budget; France and Spain provide protection. The last military action was intervention by French *gendarmes* to restore order after a royalist coup in 1933.

ECONOMICS

Not available

$1.02bn

 1.0651–1.1231 euros

SCORE CARD

- ❑ WORLD GNP RANKING........................154th
- ❑ GNP PER CAPITA $15,600
- ❑ BALANCE OF PAYMENTS ......Included in Spanish total
- ❑ INFLATION 1.62%
- ❑ UNEMPLOYMENT ....Low unemployment

STRENGTHS

Tourism is the basis of the economy. Strict banking secrecy laws make Andorra an important tax haven. Healthy luxury retail sector. Farming: cereals, potatoes, and tobacco are the major products.

WEAKNESSES

France and Spain effectively decide economic policy. Dependence on imported food and raw materials.

EXPORTS

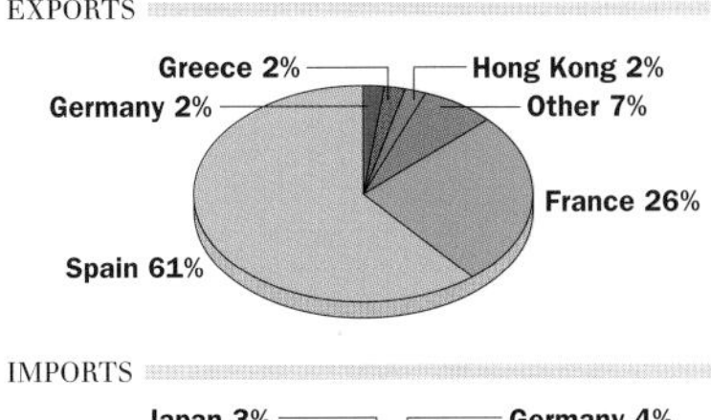
Greece 2%
Hong Kong 2%
Germany 2%
Other 7%
France 26%
Spain 61%

IMPORTS

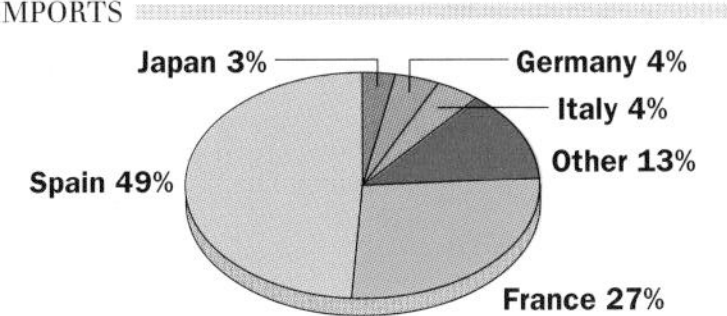
Japan 3%
Germany 4%
Italy 4%
Other 13%
Spain 49%
France 27%

RESOURCES

Not available

 None

 Not an oil producer and has no refineries

1586 sheep, 1181 cattle, 1037 horses, 854 goats

None

Water is a major resource, hydropower providing most energy needs. However, Andorra has to import twice as much electricity as it produces, and plans to develop wind power. A third of the country is designated forest.

ENVIRONMENT

Not available

 None

 Not available

The impact of millions of visitors each year on Andorra's alpine environment is of great concern. The development of hotels, ski resorts, and transportation links threaten to despoil the country's picturesque mountain landscape. It is also endangering the remarkable flora, and creates pressure to clear forested areas.

Hunting is no longer promoted but remains a popular attraction; the wild boar and the Pyrenean chamois are particularly targeted. Some restrictions have been introduced to preserve rarer animal species.

MEDIA

TV ownership high

 Daily newspaper circulation 60 per 1000 people

PUBLISHING AND BROADCAST MEDIA

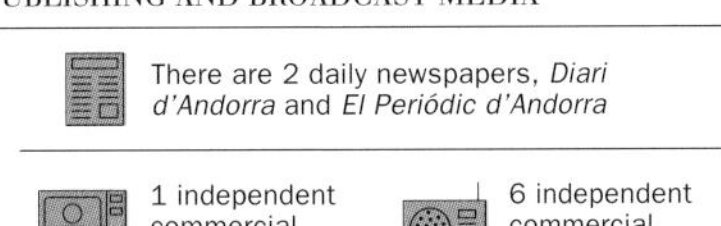
There are 2 daily newspapers, *Diari d'Andorra* and *El Periódic d'Andorra*
1 independent commercial channel
6 independent commercial stations

Andorra has one full domestic station, Radio i Televisio d'Andorra, and receives TV broadcasts from France and Spain. A Spanish TV company broadcasts one hour a day of programs for Andorra.

CRIME

No death penalty

 20 prisoners

 Little change in 1999

Tourists are natural targets for thieves, most of whom are not Andorran. Thefts of expensive cars for resale in France and Spain are on the increase.

Andorra's two criminal courts are known as the *Tribunals de Corts*.

EDUCATION

School leaving age: 14

 99%

 3186 students

There are around 30 schools in Andorra, with instruction in Catalan, French, and Spanish. The University of Andorra specializes in distance learning using the Internet.

CHRONOLOGY

Since 1278, Andorra has been autonomous, ruled by French and Spanish co-princes.

- ❑ **1970** Women get the vote.
- ❑ **1982** First constitution enshrines popular sovereignty.
- ❑ **1983** General Council votes in favor of income tax.
- ❑ **1984** Government resigns over attempt to introduce indirect taxes.
- ❑ **1991** EU customs union comes into effect.
- ❑ **1992** Political demonstrations demanding constitutional reform. Government resigns.
- ❑ **1993** Referendum approves new constitution.
- ❑ **1994** Government falls; replaced by center-right Liberal cabinet, which is reelected in 1997 and 2001.

HEALTH

Welfare state health benefits

 1 per 455 people

 Heart and cerebrovascular diseases

Andorra has one public hospital. Health spas are popular, and the hot springs at les Escaldes attract rheumatism sufferers.

SPENDING

GDP/cap. increase

CONSUMPTION AND SPENDING

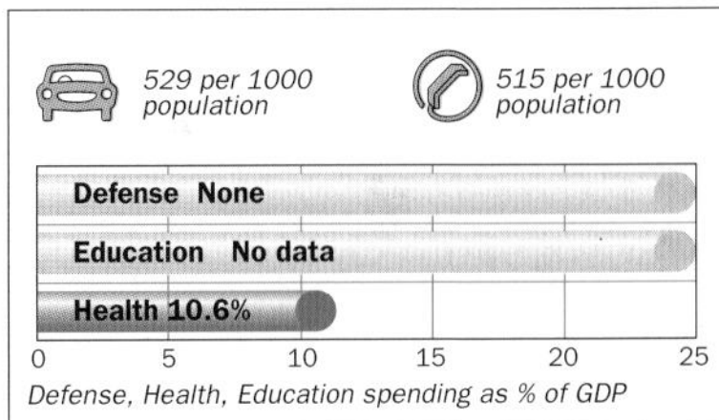
529 per 1000 population
515 per 1000 population
Defense None
Education No data
Health 10.6%
0
5
10
15
20
25
Defense, Health, Education spending as % of GDP

Hotel owners are the wealthiest group of citizens in Andorran society; many choose to live across the border in neighboring Spain.

WORLD RANKING

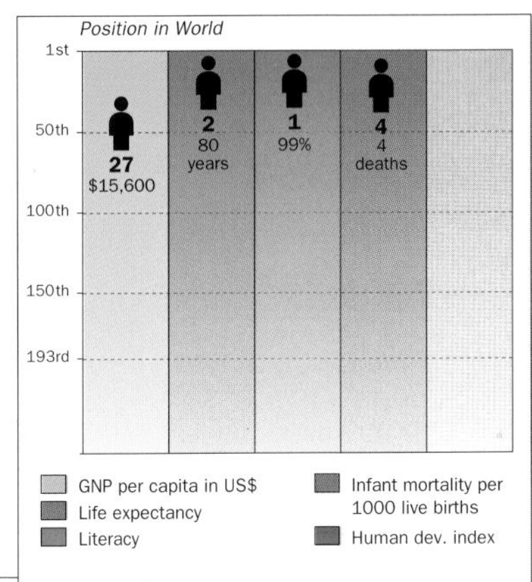
Position in World
1st
50th
100th
150th
193rd
27
$15,600
2
80
years
1
99%
4
4
deaths
GNP per capita in US$
Life expectancy
Literacy
Infant mortality per 1000 live births
Human dev. index

A

ANGOLA

OFFICIAL NAME: Republic of Angola **CAPITAL:** Luanda
POPULATION: 13.5 million **CURRENCY:** Readjusted kwanza **OFFICIAL LANGUAGE:** Portuguese

SOUTHERN AFRICA
Congo · Dem. Rep. Congo · ANGOLA · Zambia · Namibia · Africa

1975

1975

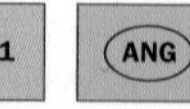
Nov 11

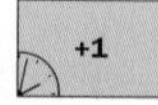
ANG

+1

+244

.ao

AN OIL- AND DIAMOND-RICH country in southwest Africa, Angola has suffered almost continuous civil war since independence from Portugal in 1975. During the Cold War the West supported UNITA rebels against the Soviet-backed MPLA government. After many failed peace initiatives, the latest, in 2002, has raised hopes yet again of a more permanent end to the violence.

Angola's capital, Luanda. Founded in 1575 by the Portuguese, it became a transshipment point for slaves en route to Brazil.

CLIMATE

Tropical/steppe

WEATHER CHART FOR LUANDA

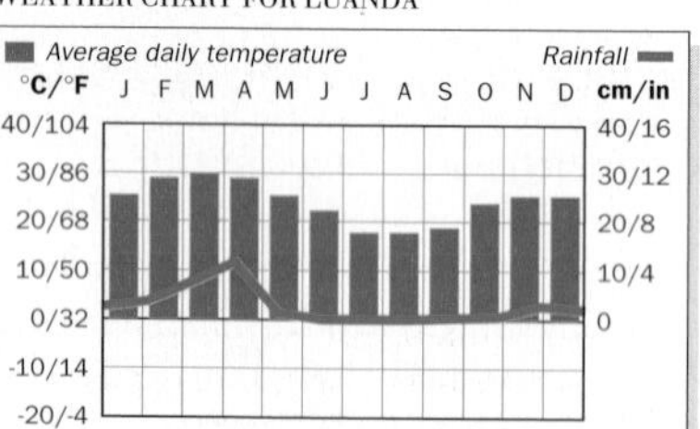

Average daily temperature
Rainfall
°C/°F
J F M A M J J A S O N D
cm/in
40/104
30/86
20/68
10/50
0/32
-10/14
-20/-4
40/16
30/12
20/8
10/4
0

The climate varies from temperate to tropical. Rainfall decreases from north to south. The Benguela Current makes the coast unusually cool and dry.

TRANSPORTATION

Drive on right

Luanda International 766,077 passengers

125 ships 66,335 grt

THE TRANSPORTATION NETWORK

19,156 km (11,903 miles)

Much of this infrastructure has been destroyed by civil war.

2952 km (1834 miles)

1295 km (805 miles)

War has destroyed infrastructure, restricted movement of people and goods, and devastated port traffic. UN peacekeepers have tried to clear mines and repair roads, bridges, and railroads.

TOURISM

Visitors : Population 1:265

51,000 visitors

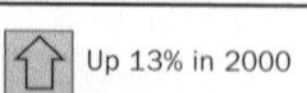

Up 13% in 2000

MAIN TOURIST ARRIVALS

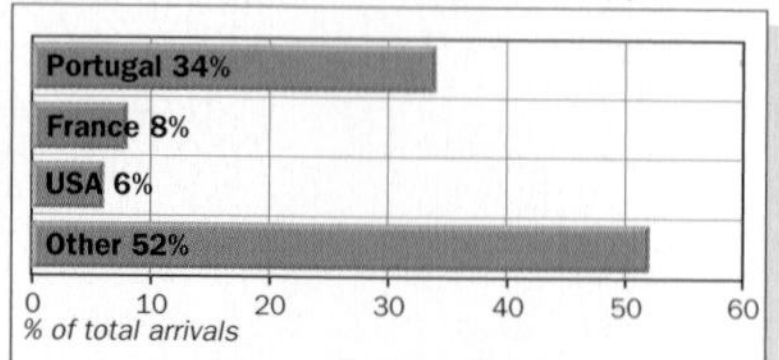

Portugal 34%
France 8%
USA 6%
Other 52%
0 10 20 30 40 50 60
% of total arrivals

Since independence Angola has been a war zone. Until the late 1990s the few visitors were journalists or employees of the oil multinationals in Cabinda.

PEOPLE

Pop. density low

Portuguese, Umbundu, Kimbundu, Kikongo

11/km² (28/mi²)

THE URBAN/RURAL POPULATION SPLIT

34% 66%

ETHNIC MAKEUP

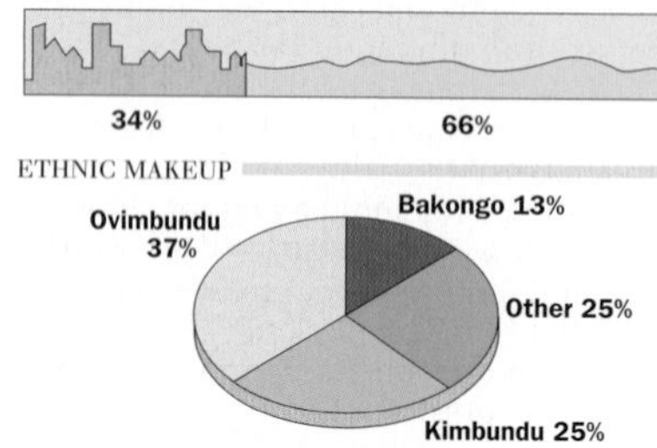

Ovimbundu 37%
Bakongo 13%
Other 25%
Kimbundu 25%

UNITA gave itself the role of sole representative of the predominantly rural-dwelling Ovimbundu, in order to attack the mainly urban-based, largely Kimbundu or mixed race (Portuguese–African) MPLA. Religion has revived since the MPLA abandoned its Marxist tenets in the 1980s. Some 20% of the population are internal refugees.

POLITICS

In transition

1992/1998 (postponed)

President José Eduardo dos Santos

AT THE LAST ELECTION

National Assembly 223 seats

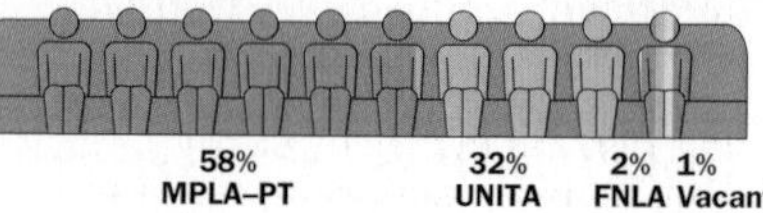

3% PRS
1% PLD
3% Others
58% MPLA–PT
32% UNITA
2% FNLA
1% Vacant

MPLA–PT = Popular Movement for the Liberation of Angola –Workers' Party **UNITA** = National Union for the Total Independence of Angola **PRS** = Social Renovated Party **FNLA** = Angolan National Liberation Front **PLD** = Liberal Democratic Party
The seats allotted to members from abroad remained vacant

In power since 1975, the MPLA abandoned one-party rule in 1991 and won Angola's first multiparty elections in 1992. Jonas Savimbi's defeated UNITA responded by reopening the civil war. The rivals signed a peace protocol in Lusaka, Zambia, in 1994, which stipulated the formation of a government of national unity (GURN). UNITA ministers eventually joined the GURN in 1997, but were expelled in 1998 as fighting escalated. Savimbi's death in February 2002 led to a renewed peace initiative, and a "definitive cease-fire" in April.

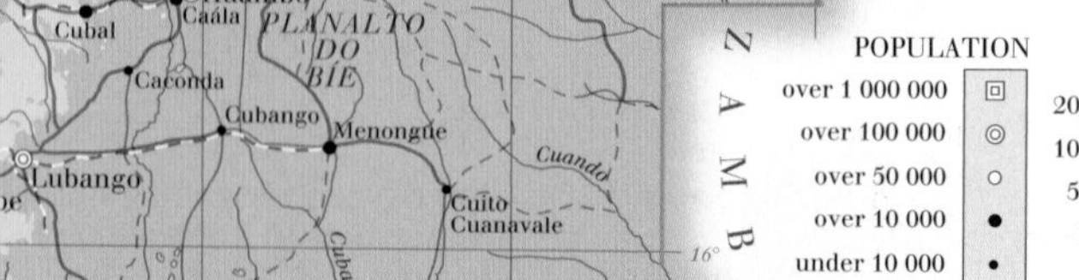

CONGO
CABINDA
Cabinda
DEM. REP. CONGO
ATLANTIC OCEAN
M'Banza Congo
Uíge
Ambriz
Caxito
LUANDA
Viana
Golungo Alto
Camabatela
Calandula
N'Dalatando
Lucala
Dondo
Malanje
Lóvua
Cambulo
Lucapa
Saurimo
Gabela
Sumbe
Uaco Cungo
Lobito
Catumbela
Benguela
Bailundo
Camacupa
Luena
Moco 2620m
Kuito
Huambo
Caála
PLANALTO DO BIE
Cubal
Caconda
Cubango
Menongue
Namibe
Lubango
Tômbua
Cuito Cuanavale
N'Giva
ZAMBIA
NAMIBIA
Caprivi Strip
BOTSWANA
Cuanza
Cuango
Cassai
Lumbezi
Cuando
Cubango
Cuito
Luiana
Cunene

ANGOLA
Total Land Area : 1 246 700 sq. km (481 351 sq. miles)

POPULATION
over 1 000 000
over 100 000
over 50 000
over 10 000
under 10 000

LAND HEIGHT
2000m/6562ft
1000m/3281ft
500m/1640ft
200m/656ft
Sea Level

0 200 km
0 200 miles

WORLD AFFAIRS

Joined UN in 1976

COMESA

IBRD

NAM

AU

SADC

Angola was a key Cold War frontier in the 1980s, but after 1992 UNITA lost international support. Post-apartheid South Africa joined the US, Russia, and Portugal as guarantors of Angola's peace process after 1994. Relations with Zambia are tense, as fighting has on occasion spilled across the border. Controls on the diamond trade were tightened in early 2000 in support of UN trade sanctions on UNITA-held regions. UNITA moved to secure territory and influence in the DRC in 2001.

AID

Recipient

 $307m (receipts)

 Down 21% in 2000

Donors from the World Bank, the EU, the USA, Japan, and Norway pledged aid in the mid-1990s to rebuild the shattered economy. However, resumed fighting strained their resources. Attacks on aid convoys prompted many NGOs to leave the country altogether. UN humanitarian flights ceased in 1999.

DEFENSE

No compulsory military service

 $1.25bn

 Up 24% in 2000

By 2002 government forces had asserted control over more than 90% of the country. Under the cease-fire, 5000 UNITA troops were integrated into the regular army. The remaining 80,000 were demobilized. The government has agreed to ratify the 1997 Ottawa treaty on the elimination of antipersonnel mines.

ECONOMICS

Inflation 741% p.a. (1990–2000)

$3.85bn

16.70–31.86 readjusted kwanza

SCORE CARD

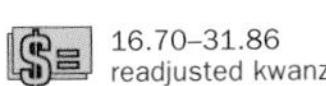

❑ WORLD GNP RANKING	121st
❑ GNP PER CAPITA	$290
❑ BALANCE OF PAYMENTS	–$4m
❑ INFLATION	325%
❑ UNEMPLOYMENT	50%

STRENGTHS

Oil an important foreign exchange earner; hopes for advanced offshore drilling. Some of the richest mineral deposits in Africa.

WEAKNESSES

Severe drought. Civil war: destruction of infrastructure. Ten million land mines laid nationwide.

EXPORTS

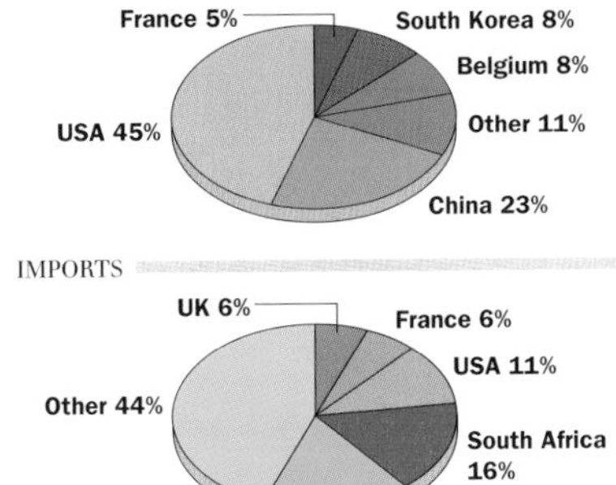
France 5%
South Korea 8%
Belgium 8%
USA 45%
Other 11%
China 23%

IMPORTS

UK 6%
France 6%
USA 11%
Other 44%
South Africa 16%
Portugal 17%

RESOURCES

Electric power 462,000 kw

177,497 tonnes

731,000 b/d (reserves 5.4bn barrels)

 4.04m cattle, 2.15m goats, 800,000 pigs, 6.8m chickens

 Oil, diamonds, iron, copper, lead, zinc, gold, manganese

Deepwater oil fields have been found. The rich alluvial diamond deposits were controlled by UNITA during the civil war.

ENVIRONMENT

Sustainability rank: 111th

 7% (1% partially protected)

 0.5 tonnes per capita

War has damaged the water supply. UNITA has been accused of widescale ivory poaching.

MEDIA

TV ownership low

 Daily newspaper circulation 11 per 1000 people.

PUBLISHING AND BROADCAST MEDIA

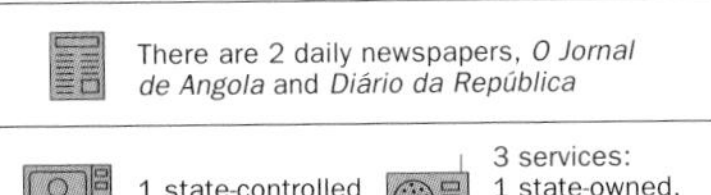

There are 2 daily newspapers, *O Jornal de Angola* and *Diário da República*

1 state-controlled service

3 services: 1 state-owned, 2 independent

Many foreign journalists have been arrested. UN officials accuse the MPLA of spreading "hostile war propaganda."

CRIME

No death penalty

 4884 prisoners

 Up 92% 1997–1999

Murder, theft, corruption, and diamond smuggling are commonplace. Rural areas are effectively controlled by gangs. Both the MPLA and UNITA have poor human rights records. Global companies assisted UNITA in sanctions-busting.

EDUCATION

School leaving age: 10

 40%

 8337 students

A government-backed initiative, Adra, is making progress in reviving schooling in cities.

CHRONOLOGY

The Portuguese first established coastal forts in 1482.

- ❑ **1975** Independence. Civil war between MPLA and UNITA.
- ❑ **1979** José Eduardo dos Santos (MPLA) becomes president.
- ❑ **1991** UN-brokered peace.
- ❑ **1992** MPLA election victory provokes UNITA to resume fighting.
- ❑ **1994** Lusaka peace agreement.
- ❑ **1998** Civil war reerupts.
- ❑ **2000** Fighting spreads as UNITA increases guerrilla activity.
- ❑ **2002** UNITA leader Jonas Savimbi killed. April, cease-fire signed.

HEALTH

No welfare state health benefits

 1 per 10,000 people

 Malaria, diarrheal and respiratory diseases, severe malnutrition

Angola's health system is barely able to cope with casualties of war and the threat of epidemics. Angola has the greatest number of amputees (caused by land mines) in the world.

Immunization of three million children against polio began in 2002.

SPENDING

GDP/cap. increase

CONSUMPTION AND SPENDING

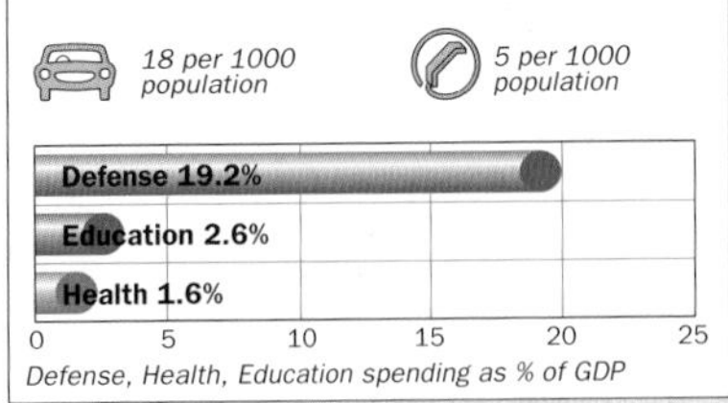
18 per 1000 population
5 per 1000 population
Defense 19.2%
Education 2.6%
Health 1.6%
0
5
10
15
20
25

Defense, Health, Education spending as % of GDP

State officials enjoy various luxuries, such as access to cars and other consumer goods, while the majority struggles to survive. The MPLA accuses its own generals of illicit diamond mining.

WORLD RANKING

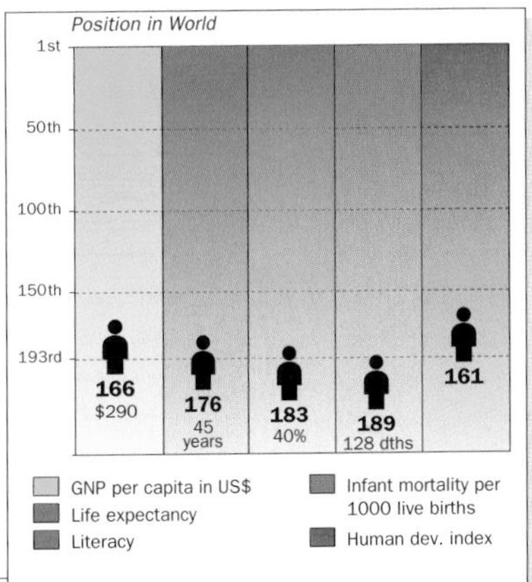
Position in World
1st
50th
100th
150th
193rd
166
$290
176
45 years
183
40%
189
128 dths
161
GNP per capita in US$
Life expectancy
Literacy
Infant mortality per 1000 live births
Human dev. index

A

ANTARCTICA

ANTARCTICA
Argentina
Chile
South America
Africa
Australia

OFFICIAL NAME: Antarctica **CAPITAL:** None
POPULATION: None **CURRENCY:** None **OFFICIAL LANGUAGE:** None

1961

1961

n/a
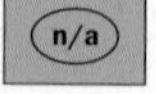
n/a

n/a

n/a
.aq

THE WORLD'S fifth-largest continent, Antarctica is almost entirely covered by ice over 2000 m (6560 ft) thick. The area sustains a varied wildlife, including seals, whales, and penguins. The Antarctic Treaty, signed in 1959 and in force in 1961, provides for international governance of Antarctica. To gain Consultative Status, countries have to set up a program of scientific research into the continent. Following a 1994 international agreement, a whale sanctuary was established around Antarctica.

PEOPLE

Not applicable

English, Spanish, French, Norwegian, Chinese, Polish, Russian, German, Japanese

Not applicable

ETHNIC MAKEUP

Antarctica has a transient population of Americans, English, French, Norwegians, Argentinians, Chileans, Chinese, Russians, Poles, and Japanese. Most are involved in research. Few stay more than two years.

Antarctica has no indigenous population. Around 80 Chilean settlers live at any one time in the continent's only permanent community on King George Island. The rest of the population are scientists and logistical staff working at the 40 permanent, and as many as 100 temporary, research stations. Most stations are too far apart for direct contact between different nationalities.

CLIMATE

Freezing

WEATHER CHART

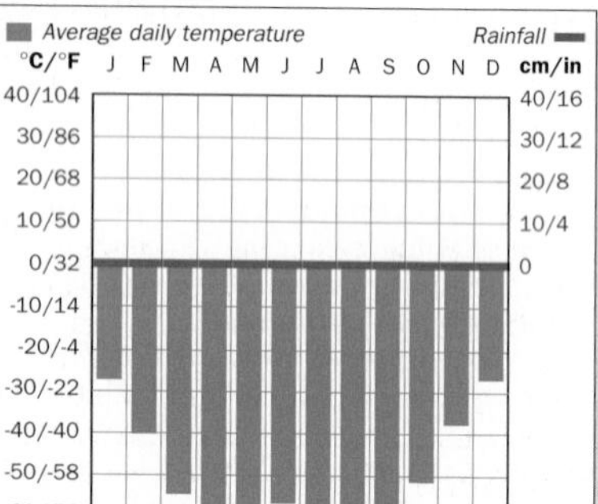
Average daily temperature
Rainfall
°C/°F
J F M A M J J A S O N D
cm/in
40/104
30/86
20/68
10/50
0/32
-10/14
-20/-4
-30/-22
-40/-40
-50/-58
-60/-76
40/16
30/12
20/8
10/4
0

Antarctica is the windiest as well as the coldest continent. Powerful winds create a narrow storm belt around the continent, which brings cloud, fog, and severe blizzards. Icebergs barricade more than 90% of the coastline, and climate change has seen an increase in their number and size in recent years. Antarctica contains over 80% of the world's fresh water.

TOURISM

Not applicable

9400 visitors

No change in 1999

Tourism is mainly by cruise ship to the Antarctic Peninsula, the Ross Sea, and the sub-Antarctic islands. In 1983, the Chileans began flights to King George Island, where an 80-bed hotel has been built. Main attractions are the wildlife, skiing, and visits to scientific stations and historic huts. The growth of tourism has disrupted scientific programs and official regulation of tourism is now essential.

TRANSPORTATION

Not applicable

Airstrips to some stations

Has no fleet

Ships are the main mode of transportation to Antarctica. They are also used for marine research projects. Air traffic from Chile is growing, and France and the UK are building new airstrips. Most planes have to be equipped with skis.

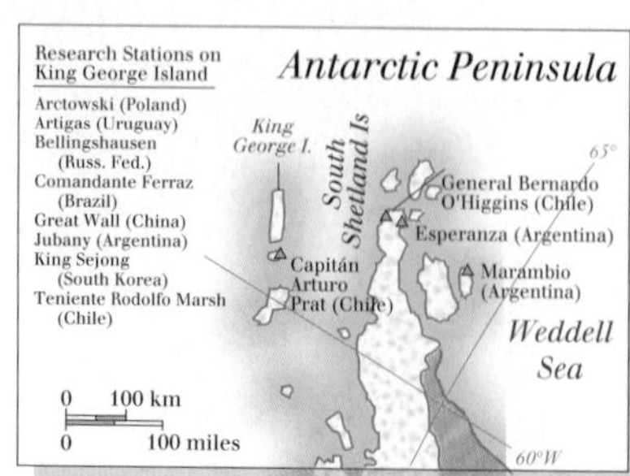
Research Stations on King George Island
Arctowski (Poland)
Artigas (Uruguay)
Bellingshausen (Russ. Fed.)
Comandante Ferraz (Brazil)
Great Wall (China)
Jubany (Argentina)
King Sejong (South Korea)
Teniente Rodolfo Marsh (Chile)
Antarctic Peninsula
King George I.
South Shetland Is
General Bernardo O'Higgins (Chile)
Esperanza (Argentina)
Capitán Arturo Prat (Chile)
Marambio (Argentina)
Weddell Sea
65°
60°W
0 100 km
0 100 miles

ANTARCTICA

Total Land Area : 14 000 000 sq. km (5 405 000 sq. miles)

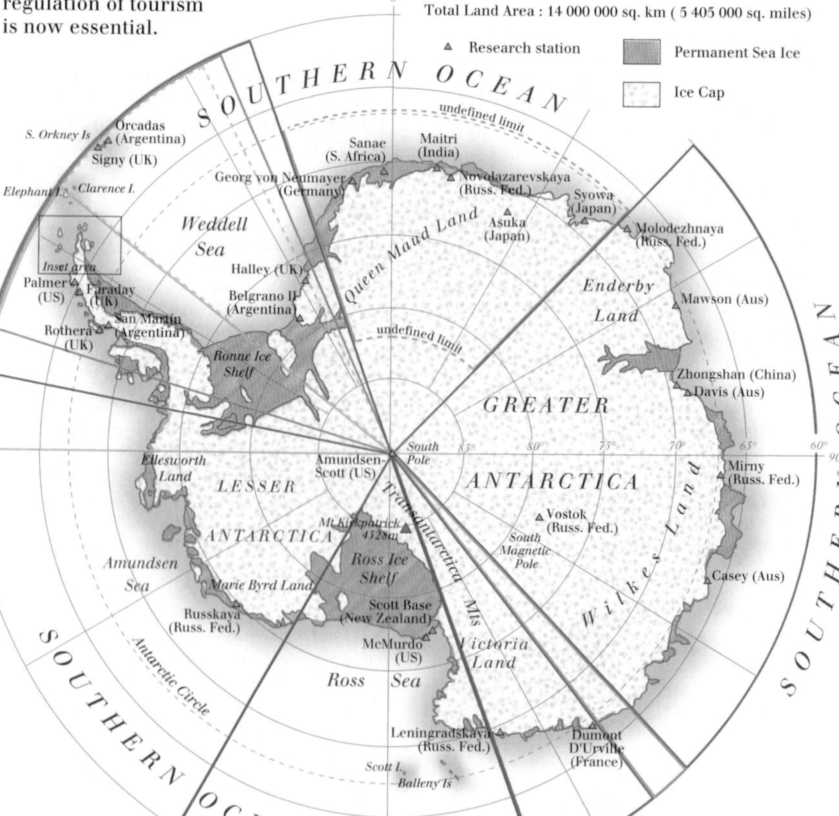
Research station
Permanent Sea Ice
Ice Cap
SOUTHERN OCEAN
undefined limit
S. Orkney Is
Orcadas (Argentina)
Signy (UK)
Elephant I.
Clarence I.
Inset area
Palmer (US)
Faraday (UK)
Rothera (UK)
San Martín (Argentina)
Weddell Sea
Georg von Neumayer (Germany)
Sanae (S. Africa)
Maitri (India)
Novolazarevskaya (Russ. Fed.)
Syowa (Japan)
Asuka (Japan)
Queen Maud Land
Halley (UK)
Belgrano II (Argentina)
Molodezhnaya (Russ. Fed.)
Enderby Land
Mawson (Aus)
Ronne Ice Shelf
Zhongshan (China)
Davis (Aus)
GREATER ANTARCTICA
South Pole
Amundsen-Scott (US)
Ellesworth Land
LESSER ANTARCTICA
Mirny (Russ. Fed.)
Vostok (Russ. Fed.)
South Magnetic Pole
Mt Kirkpatrick 4528m
Transantarctic Mts
Ross Ice Shelf
Wilkes Land
Casey (Aus)
Amundsen Sea
Marie Byrd Land
Russkaya (Russ. Fed.)
Scott Base (New Zealand)
McMurdo (US)
Victoria Land
Ross Sea
Antarctic Circle
Leningradskaya (Russ. Fed.)
Dumont D'Urville (France)
Scott I.
Balleny Is
0°
90°W
90°E
180°
85° 80° 75° 70° 65° 60°
0 1000 km
0 1000 miles
TERRITORIAL CLAIMS
Australian claim
Argentinian claim
British claim
Chilean claim
French claim
New Zealand claim
Norwegian claim
Brazilian zone of interest

Neumayer Channel, Antarctica. Many states are pressing for the whole of Antarctica to be protected as an international park.

POLITICS

Not applicable

 Not applicable

 Consultative Parties to Antarctic Treaty

NO LEGISLATIVE OR ADVISORY BODIES

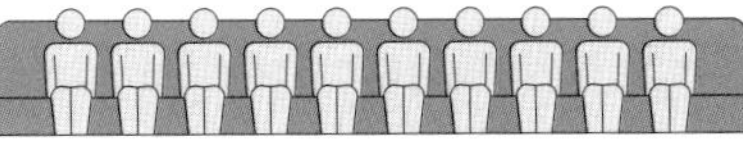

The Antarctic Treaty of 1959 was signed by 12 countries. Consultative meetings are held most years to discuss scientific, environmental, and political matters.

There are 27 parties to the Antarctic Treaty and 17 nations with observer status. There are territorial claims by Australia, France, New Zealand, and Norway, and overlapping claims in the Antarctic Peninsula by Argentina, Chile, and the UK. Other states do not recognize these claims.

Of main concern is the adoption of a wide range of environmental protection measures. Proposals include the monitoring of all scientific activities and also the prosecution of any country if it were demonstrated that its research would lead to detrimental global change.

WORLD AFFAIRS

Not a UN member

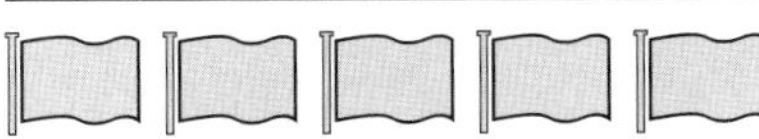

Rivalries exist between nations wishing to preserve Antarctica as a world park and those pursuing territorial claims.

AID

Recipient

 Research is funded by governments

 Subject to individual government budgets

Scientific programs in the Antarctic are almost entirely funded by government agencies in the home countries. Funding is occasionally provided by scientific institutions and universities.

DEFENSE

Not applicable

 No defense force

 Not applicable

Under the Antarctic Treaty, Antarctica can be used only for peaceful purposes. Any military personnel present perform purely scientific or logistic roles.

ECONOMICS

Not applicable

 Not applicable

 Antarctica has no currency

Research is government-funded and is therefore subject to reductions. The exploitation of marine stocks provides no income for Antarctica.

RESOURCES

Not applicable

 Included in national fish catch totals

 Not an oil producer and has no refineries

 None

Mineral extraction is banned

Antarctica's main resources are its marine stocks, including fin fish, seals, and whales. A campaign by environmental groups, supported by Australia and France, to ban mining and declare Antarctica a world park was rewarded with an agreement in 1991 to impose a 50-year ban on mining, and in 1994 by the approval of a whale sanctuary. Prospects for energy sources alternative to fossil fuels, such as solar power and wind generators, are being explored.

ENVIRONMENT

Not applicable

 Most of Antarctica is protected

 Not available

Antarctica is one of the Earth's last great wildernesses. Its layer of ice, 4000 m (13,120 ft) thick in places, has formed over thousands of years. Its ecosystem is such that a "footprint" will leave its mark for many years. Several species are unique to the continent, including king penguins. The blood of polar fish contains antifreeze agents. Ecological concerns include overfishing, particularly of krill, cod, and squid; the disintegration of ice shelves; the depletion of the ozone layer over Antarctica; and the various knock-on effects of global warming. In 1994 the IWC agreed to a French proposal to create an Antarctic whale sanctuary which, together with the Indian Ocean sanctuary, protects the feeding grounds of 90% of the world's whales.

MEDIA

Not applicable

 There are no daily newspapers produced in Antarctica

A few bases publish newssheets for local consumption. Local radio stations are found at some of the larger bases.

CRIME

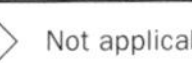

Not applicable

There are no prisons in Antarctica

 Crime is negligible

Each person in Antarctica is subject to their national laws. Occasional petty theft from stations is linked to visits from tourists.

CHRONOLOGY

The Russian explorer, Thaddeus von Bellingshausen, was the first to sight Antarctica, in 1820. The South Pole was first reached by the Norwegian, Roald Amundsen, in December 1911.

- **1957–1958** International Geophysical Year launches scientific exploration of Antarctica.
- **1959** Antarctic Treaty signed by 12 countries. Territorial claims frozen.
- **1978** Convention limiting seal hunting comes into force.
- **1985** Ozone depletion disclosed.
- **1994** Establishment of Antarctic whale sanctuary.
- **1998** Agreement on 50-year ban on mineral extraction comes into force.

EDUCATION

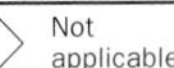

Not applicable

Not applicable

None

Schoolhouses exist on the Chilean base, Villa Las Estrellas, and the Argentinian base, Esperanza. Teaching is geared to the relevant national system. Some researchers' studies contribute to higher degrees.

Antarctic-based research has resulted in a number of scientific breakthroughs, including the discovery of the depletion of the ozone layer.

HEALTH

Not applicable

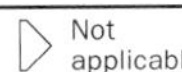

 1 medical officer per station

 Deaths are extremely rare in Antarctica

Each station has its own medical officer. The problems usually associated with polar conditions, such as frostbite and snow blindness, are very rare. All personnel are medically screened before arrival. If serious illnesses develop, patients have to be evacuated by air, including in recent years an acute case of gallstones and a doctor who self-diagnosed breast cancer.

SPENDING

Not applicable

Any evidence of wealth disparities reflects the different levels of base funding. US bases are the best-funded, while the budgets of other bases are subject to domestic politics. Most stations have TVs and video recorders. Telephone systems operate only within stations. Computers are supplied for scientific research.

WORLD RANKING

The UN Human Development Index conditions are not applicable to Antarctica.

A

ANTIGUA & BARBUDA

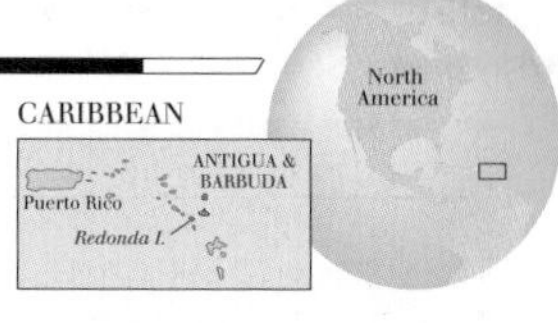
CARIBBEAN
North America
ANTIGUA & BARBUDA
Puerto Rico
Redonda I.

OFFICIAL NAME: Antigua and Barbuda **CAPITAL:** St. John's
POPULATION: 66,400 **CURRENCY:** Eastern Caribbean dollar **OFFICIAL LANGUAGE:** English

1981

1981

Nov 1
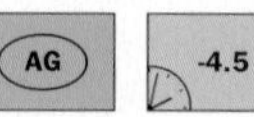
AG
-4.5

+1268

.ag

PART OF THE Leeward Islands chain, Antigua was in turn a Spanish, French, and British colony. British influence is still strong and most clearly revealed in the Antiguans' passion for cricket. Antigua has two dependencies: Barbuda, 50 km (30 miles) to the north, sporting a magnificent beach; and Redonda, 40 km (25 miles) west, an uninhabited rock with its own king.

CLIMATE

▷ Tropical oceanic

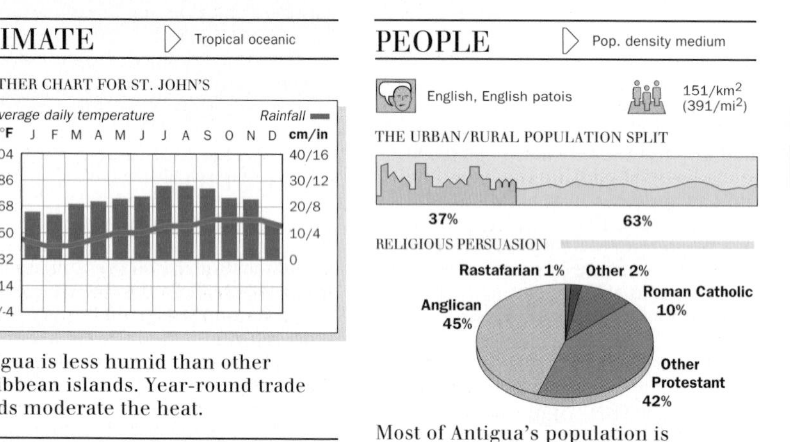
WEATHER CHART FOR ST. JOHN'S
Average daily temperature
Rainfall
°C/°F J F M A M J J A S O N D cm/in
40/104 40/16
30/86 30/12
20/68 20/8
10/50 10/4
0/32 0
-10/14
-20/-4

Antigua is less humid than other Caribbean islands. Year-round trade winds moderate the heat.

TRANSPORTATION

▷ Drive on left

V. C. Bird International, St. John's
727,292 passengers

759 ships
4.22m grt

THE TRANSPORTATION NETWORK

250 km (155 miles)

None

77 km (48 miles)

None

Recent multimillion EC$ projects have expanded the international airport and provided 140 km (90 miles) of roads with all-weather surfaces.

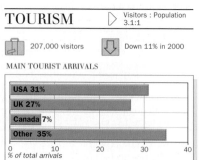

TOURISM

▷ Visitors : Population 3.1:1

207,000 visitors

Down 11% in 2000

MAIN TOURIST ARRIVALS

USA 31%
UK 27%
Canada 7%
Other 35%
0 10 20 30 40
% of total arrivals

Antigua is especially popular with US cruise ship tourists. Among other draws are the annual international tennis championship, the islands' three golf courses, and the attraction of large duty-free shopping malls.

PEOPLE

▷ Pop. density medium

English, English patois

151/km^2 (391/mi^2)

THE URBAN/RURAL POPULATION SPLIT

37% 63%

RELIGIOUS PERSUASION

Rastafarian 1%
Other 2%
Roman Catholic 10%
Anglican 45%
Other Protestant 42%

Most of Antigua's population is descended from Africans, brought over between the 16th and 19th centuries. There are, in addition, a few Europeans and South Asians. Racial tensions are few. Life is based around the extended family. Since the 1960s the status of women has risen as a result of their greater access to education, and many are now entering the legal, financial, and medical professions. By Caribbean standards, wealth disparities are small.

ANTIGUA & BARBUDA

Total Land Area : 442 sq. km (170 sq. miles)

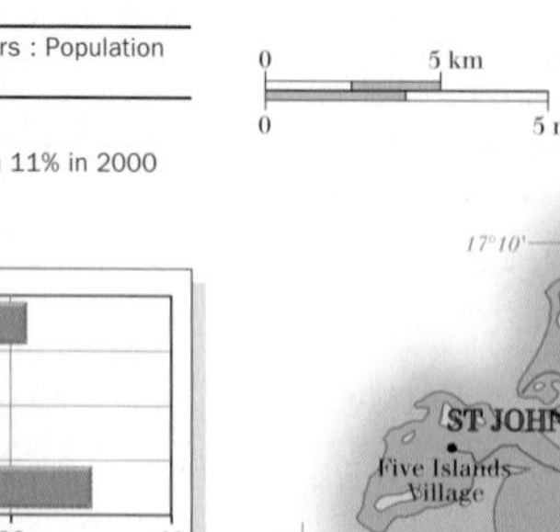
0 5 km
0 5 miles
N

POLITICS

▷ Multiparty elections

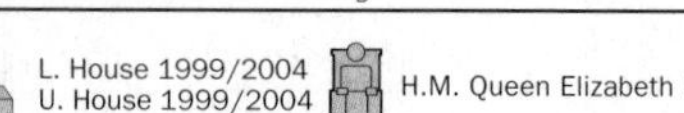
L. House 1999/2004
U. House 1999/2004
H.M. Queen Elizabeth II

AT THE LAST ELECTION

House of Representatives 17 seats

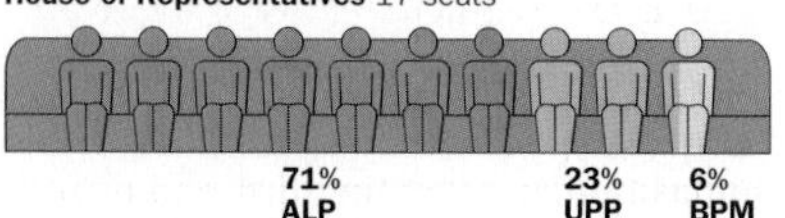
71% ALP
23% UPP
6% BPM

ALP = Antigua Labour Party **UPP** = United Progressive Party **BPM** = Barbuda People's Movement

Senate 17 seats

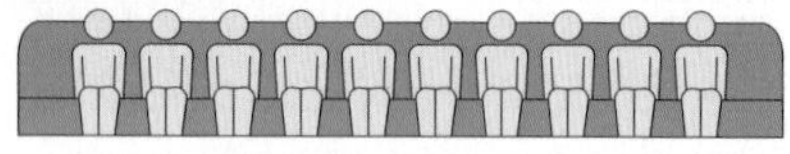

The members of the Senate are appointed. 11 are chosen by the prime minister, 4 by the leader of the opposition, 1 by the governor-general, and 1 by the Barbuda Council

Antigua's multiparty democracy has been dominated for four decades by the Bird family. Vere Bird Sr., the veteran prime minister and ALP leader, retired in 1994, and a battle between his two sons to succeed him was won by Lester Bird; his elder brother Vere Jr. had in 1990 been removed from public office, accused of involvement in gun-running. Lester Bird led the ALP into its sixth consecutive term in 1999.

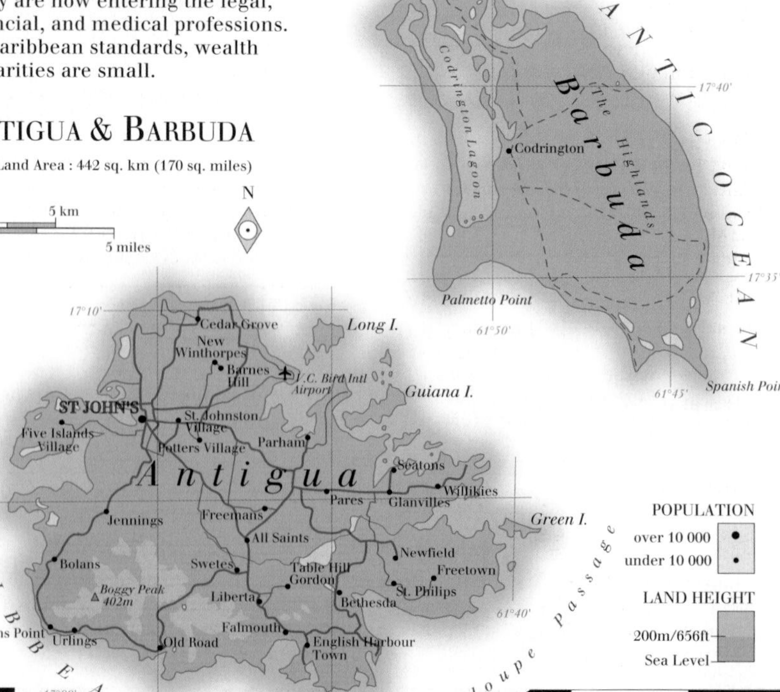
ATLANTIC OCEAN
Goat I.
Codrington Lagoon
Barbuda
The Highlands
Codrington
Palmetto Point
Spanish Point
Long I.
Cedar Grove
New Winthorpes
Barnes Hill
V.C. Bird Intl Airport
Guiana I.
ST JOHN'S
St. Johnston Village
Five Islands Village
Potters Village
Parham
Antigua
Seatons
Willikies
Pares
Glanvilles
Jennings
Freemans
All Saints
Green I.
Bolans
Swetes
Table Hill Gordon
Newfield
Freetown
St. Philips
Boggy Peak 402m
Liberta
Bethesda
Johnsons Point
Urlings
Falmouth
Old Road
English Harbour Town
CARIBBEAN SEA
Guadeloupe Passage
POPULATION
over 10 000
under 10 000
LAND HEIGHT
200m/656ft
Sea Level

A

WORLD AFFAIRS

Joined UN in 1981

ACS

Caricom

Comm

OECS
OAS

Antigua has tended to back US policy in the region, supporting both the US invasion of Grenada in 1983 and economic sanctions against Cuba.

AID

Recipient

 US$10m (receipts)

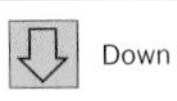 Down 9% in 2000

Donors include the US, the UK, France, Kuwait, Japan, and the EU. In 1998 the opposition accused the Bird regime of manipulating the distribution of hurricane relief for electoral gain.

DEFENSE

No compulsory military service

 US$4m

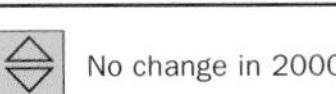 No change in 2000

There is a 170-strong defense force. Reports that the government had privately imported weaponry from the US in 1998 were denied. Two military bases are leased to the US.

ECONOMICS

Inflation 2.5% p.a. (1990–2000)

 US$642m

 2.7 Eastern Caribbean dollars

SCORE CARD

- ❑ WORLD GNP RANKING........................165th
- ❑ GNP PER CAPITA.........................US$9440
- ❑ BALANCE OF PAYMENTS..................–US$79m
- ❑ INFLATION.......................................1.6%
- ❑ UNEMPLOYMENT.................................7%

STRENGTHS

Tourism and construction of tourist hotels and infrastructure. Financial and communications services linked to offshore financial sector.

WEAKNESSES

Very little diversification makes Antigua vulnerable to downturns in the world tourism market. Evidence of offshore money laundering harms the country's reputation.

EXPORTS

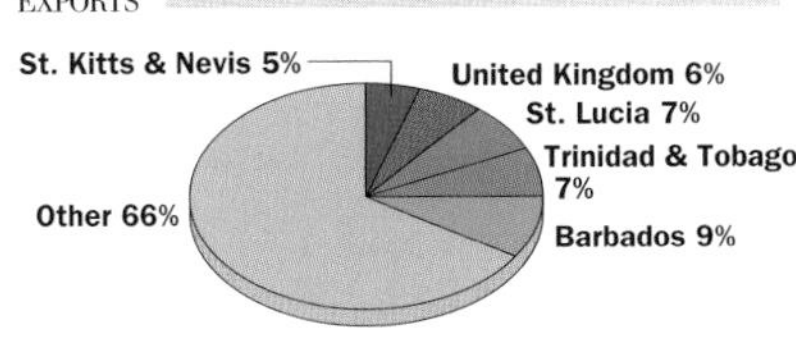
St. Kitts & Nevis 5%
United Kingdom 6%
St. Lucia 7%
Trinidad & Tobago 7%
Barbados 9%
Other 66%

IMPORTS

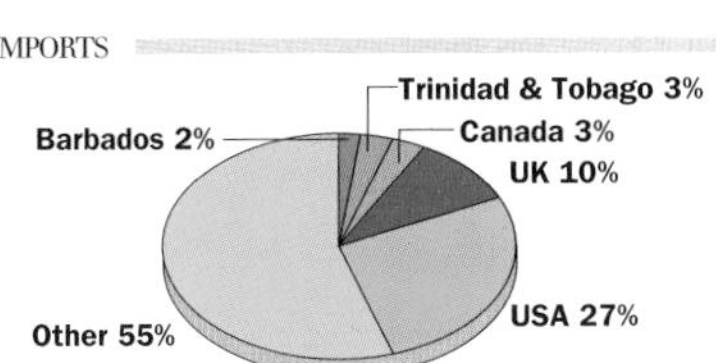
Trinidad & Tobago 3%
Barbados 2%
Canada 3%
UK 10%
USA 27%
Other 55%

Nelson's Dockyard. *Luxury yachts fitted with state-of-the-art gadgetry contrast with the 18th-century St. John's harbor.*

RESOURCES

Electric power 26,000 kw

 3185 tonnes

 Not an oil producer

15,700 cattle, 12,200 sheep, 90,000 chickens

 None

Antigua has no strategic or commodity resources and has to import almost all its energy requirements.

ENVIRONMENT

Not available

 9%

5 tonnes per capita

Sewage from hotels causes major problems. Untreated effluent pollutes the sea, while uncontrolled disposal has killed valuable inshore fish stocks in the mangrove swamps, whose whole ecosystem is under threat from poorly planned hotel development.

MEDIA

TV ownership hgh

Daily newspaper circulation 91 per 1000 people

PUBLISHING AND BROADCAST MEDIA

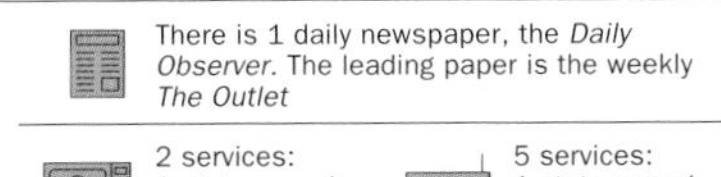
There is 1 daily newspaper, the *Daily Observer*. The leading paper is the weekly *The Outlet*
2 services: 1 state-owned, 1 independent
5 services: 1 state-owned, 4 independent

Arson was alleged when the printing equipment of opposition weekly *The Outlet* was destroyed in 1998. Of the privately owned broadcasting services, one is affiliated to the opposition.

CRIME

Death penalty in use

 186 prisoners

 Down 2% 1992–1996

Murder is rare. Rape, armed robbery, and burglary are main concerns, as is offshore money laundering.

EDUCATION

School leaving age: 16

 95%

 631 students

Education is based on the former British selective system. Students go on to the University of the West Indies, or to study in the UK or the US.

CHRONOLOGY

In 1667, Antigua became a British colony. Barbuda, formerly owned privately by the Codrington family, was annexed in 1860.

- ❑ **1951** Universal adult suffrage introduced.
- ❑ **1981** Independence from Britain; opposed by Barbudan secessionist movement.
- ❑ **1983** Supports US invasion of Grenada.
- ❑ **1994** Lester Bird elected prime minister.
- ❑ **1995** New taxes provoke protests.
- ❑ **1999** ALP wins sixth consecutive elections; Lester Bird remains prime minister.

HEALTH

Welfare state health benefits

1 per 877 people

 Heart and respiratory diseases, cancers

By Caribbean standards, the health system is efficient, with easy access to the state-run clinics and hospitals. A new hospital has been built in St. John's.

SPENDING

GDP/cap. increase

CONSUMPTION AND SPENDING

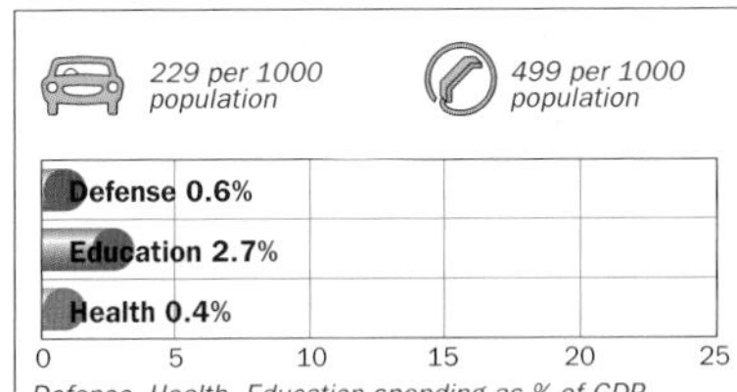
229 per 1000 population
499 per 1000 population
Defense 0.6%
Education 2.7%
Health 0.4%
0
5
10
15
20
25
Defense, Health, Education spending as % of GDP

Wealthy Antiguans are active in running the thriving tourist industry; some are allegedly also involved in money laundering. Unemployment is relatively low and the average per capita income is among the highest in the Caribbean.

WORLD RANKING

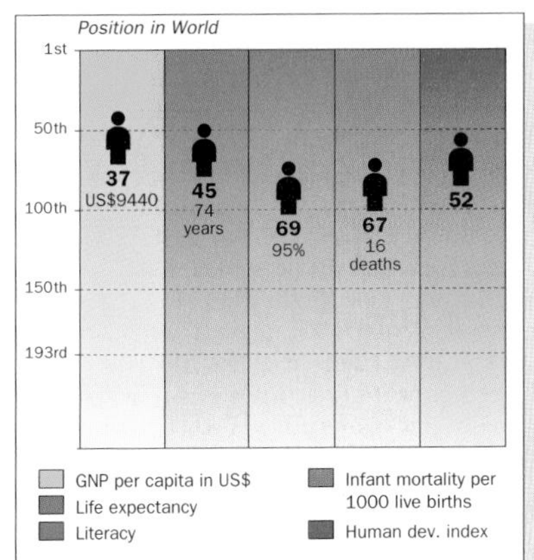
Position in World
1st
50th
100th
150th
193rd
37
US$9440
45
74 years
69
95%
67
16 deaths
52
GNP per capita in US$
Life expectancy
Literacy
Infant mortality per 1000 live births
Human dev. index

A

ARGENTINA

SOUTH AMERICA
Bolivia
Brazil
Paraguay
Uruguay
Chile
ARGENTINA
South America

OFFICIAL NAME: Republic of Argentina **CAPITAL:** Buenos Aires
POPULATION: 37.5 million **CURRENCY:** Argentine peso **OFFICIAL LANGUAGE:** Spanish

1816 | 1816 | May 25 | RA | -3 | +54 | .ar

OCCUPYING MOST OF THE southern half of South America, Argentina extends 3460 km (2150 miles) from Bolivia to Cape Horn. The Andes mountains in the west run north–south, forming a natural border with Chile. To the east they slope down to the fertile central pampas, the region known as Entre Ríos. Agriculture, especially beef, wheat, and fruit, and energy resources are Argentina's main sources of wealth. Politics in Argentina was characterized in the past by periods of military rule, but in 1983, Argentina returned to a system of multiparty democracy.

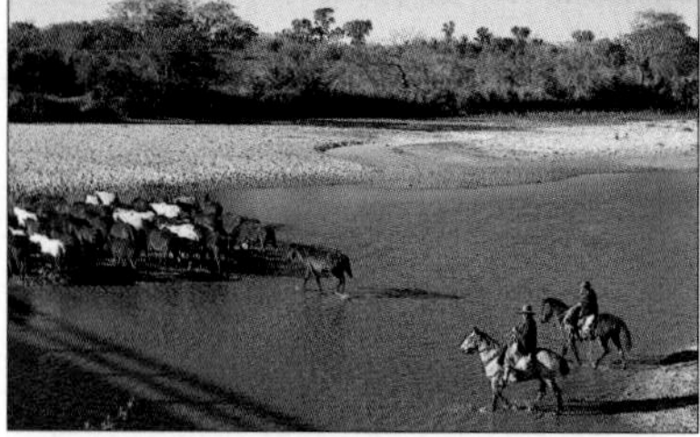

***Herding cattle** in the northeast, near Corrientes. Beef, Argentina's initial source of wealth, remains a major export.*

CLIMATE

Mountain/steppe/subtropical

WEATHER CHART FOR BUENOS AIRES

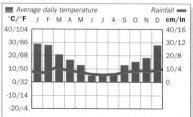

Average daily temperature — Rainfall
°C/°F J F M A M J J A S O N D cm/in
40/104, 30/86, 20/68, 10/50, 0/32, -10/14, -20/-4
40/16, 30/12, 20/8, 10/4, 0

The northeast is near-tropical. The Andes are semiarid in the north and snowy in the south. The western lowlands are desert, while the pampas have a mild climate with heavy summer rains.

TRANSPORTATION

Drive on right

Jorge Newberry, Buenos Aires 7.04m passengers

493 ships 464,300 grt

THE TRANSPORTATION NETWORK

63,553 km (39,490 miles)

734 km (456 miles)

33,000 km (20,506 miles)

11,000 km (6835 miles)

Air travel is expensive, and inadequate connections between provinces frustrate business and tourism. The national airline, Aerolineas Argentinas, was privatized in 1990, but in 2000 was the object of a rescue plan; the 37 airports are privately operated. The privatized railroad, one of the largest in the world, is primarily used for freight, but Buenos Aires' subway and commuter lines have attracted strong investment and heavy use. Some 9500 km (6000 miles) of roads are privatized, and tolls are among the highest worldwide. The six main terminals in the port of Buenos Aires are privately run. A $20 billion national infrastructure program announced in 2001 has been stalled by the economic crisis.

TOURISM

Visitors : Population 1:13

2.95m visitors

Up 2% in 2000

MAIN TOURIST ARRIVALS

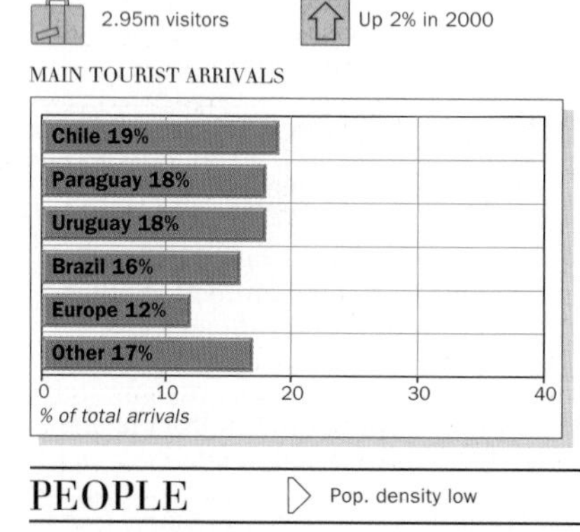
Chile 19%
Paraguay 18%
Uruguay 18%
Brazil 16%
Europe 12%
Other 17%
0
10
20
30
40
% of total arrivals

Tourism has been undersold, and the government, working with business, has launched a massive international marketing campaign. Visitors, still mostly from neighboring countries, are attracted by Buenos Aires' rich city life, the fashionable Atlantic coastal resort of Mar del Plata, ski stations such as Bariloche and Las Leñas in the Andes, and wineries around Mendoza. Other major attractions are the Iguazú National Park, Antarctic cruises, and the renowned whale-watching hot spot off Peninsula Valdés.

PEOPLE

Pop. density low

Spanish, Italian, Amerindian languages

14/km² (35/mi²)

THE URBAN/RURAL POPULATION SPLIT

90% 10%

RELIGIOUS PERSUASION

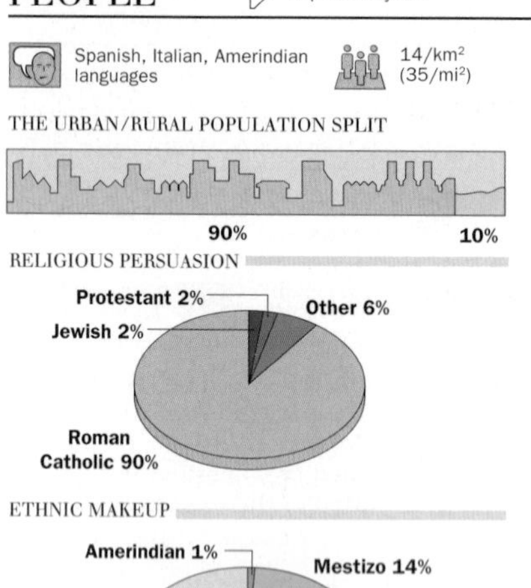
Protestant 2%
Other 6%
Jewish 2%
Roman Catholic 90%
ETHNIC MAKEUP
Amerindian 1%
Mestizo 14%
Indo-European 85%

Most Argentinians of European descent are from recent 20th-century migrations; over one-third are of Italian origin. Indigenous peoples now form a tiny minority, living mainly in Andean regions or in the Gran Chaco. Argentina also has communities of Welsh, Lebanese, Syrians, Armenians, Japanese, and Koreans.

POPULATION AGE BREAKDOWN

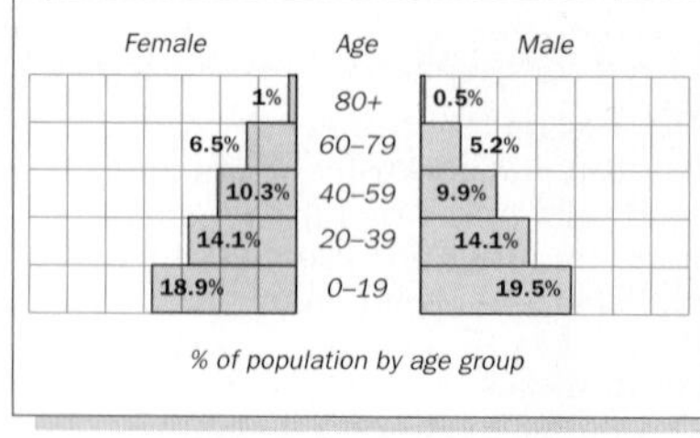
Female
Age
Male
1%
80+
0.5%
6.5%
60–79
5.2%
10.3%
40–59
9.9%
14.1%
20–39
14.1%
18.9%
0–19
19.5%
% of population by age group

The vast majority of Argentinians are urban dwellers, with some 40% of the population living in Buenos Aires, one of the largest cities in Central or South America.

Catholicism and the extended family remain strong in Argentina. In addition, the family forms the basis of many successful businesses.

Women have a higher profile than in most Latin American states, and were enfranchised in 1947. Today, many enter the professions and rise to positions of influence in service businesses such as the media. The exception is party politics. Eva Perón, who inspired the musical *Evita*, did help to push women into a more active political role in the 1940s and 1950s, but this trend was reversed under military rule.

POLITICS

Multiparty elections

L. House 1999/2003
U. House 2001/2003

President Eduardo Duhalde

Argentina is a multiparty democracy; the president is head of government.

PROFILE

The Peronists dominated politics from the 1940s. The party, founded on mass working-class and left-wing intellectual support, was inimical to the military; coups were staged in 1955, 1966, and 1976. President Carlos Menem won elections in 1989 on a populist platform but steered the Peronists toward a right-wing free-market agenda. The UCR tended to stay in opposition, only winning when the electorate registered a protest vote, as in 1983. In 1999 it captured the presidency as part of a center-left alliance with FREPASO. However, President Fernando de la Rúa was brought down by the 2001 economic crisis, and the Peronists, the largest party in the legislature since the mid-term elections, got their candidate Eduardo Duhalde installed as interim president.

MAIN POLITICAL ISSUES

The legitimacy of the political system

A shaky coalition of the main parties charged the Duhalde caretaker administration in January 2002 with rescuing Argentina from the worst economic and social crisis in a generation before handing power to an elected successor in 2003. From the outset his stabilization project was threatened by the unprecedented rejection of the political system, manifested in the waves of protest that had deposed the de la Rúa government and resulted in three stopgap presidents prior to Duhalde. Public opinion pointed to a major shake-up to restore public confidence in politicians, the judiciary, and business.

The economic crisis

Keeping the peso pegged to the dollar was blamed for prolonging economic recession and was ditched by Duhalde. The severe effects of industrial collapse and fiscal austerity, coupled with hated restrictions on bank withdrawals to defend the peso and head off inflation, hit domestic consumption and reduced a near cashless population to bartering goods. The IMF, however, made the release of funds dependent on fiscal, debt, and exchange rate performance.

AT THE LAST ELECTION

Chamber of Deputies 257 seats

46% **PJ** | **34%** **Alliance** | **7%** **ARI** | **11%** **Others** | **2%** **AR**

PJ = Justicialist Party (Peronists) **Alliance** = Alliance for Work, Justice, and Education (Radical Civic Union–**UCR** and National Solidarity Front Alliance–**FREPASO**) **ARI** = Alternative for a Republic of Equals **AR** = Action for the Republic

Senate 72 seats

55% **PJ** | **35%** **Alliance** | **10%** **Others**

ARGENTINA

Total Land Area : 2 766 890 sq. km (1 068 296 sq. miles)

POPULATION: over 1 000 000; over 500 000; over 100 000; over 50 000; over 10 000

LAND HEIGHT: 4000m/13124ft; 2000m/6562ft; 1000m/3281ft; 200m/656ft; Sea Level

0 200 km; 0 200 miles

Carlos Menem: authoritarian rule overshadowed free-market reforms.

Eduardo Duhalde, elected in 2002, became the fifth president in 12 days.

WORLD AFFAIRS

Joined UN in 1945

SELA

Mercsr

OAS

RG

G15

Argentina takes a pro-Western stance and has deployed its armed forces in a series of UN actions. Solid relations with potential aid donors and trade partners were made all the more essential following the collapse of the Argentine economy in 2001. Ties to the US are strong, although US pressure for an open-skies agreement and demands for royalty payments on patented drugs create some tensions.

Friction with Brazil over trade rules complicates Argentina's membership of Mercosur. It wants Mercosur to be strengthened, and widened to include Chile, currently an associate member.

The normalizing of relations with the UK in 1998 sidelined Argentina's claim to sovereignty over the nearby Falklands Islands (known locally as Las Islas Malvinas), the focus of the 1982 war between the two countries.

In 2002, Argentina requested the extradition of former Bolivian president Hugo Banzer (who later died) in relation to Plan Cóndor, which saw thousands kidnapped and murdered under military regimes in Argentina, Chile, and Bolivia in the 1970s.

A

AID

Recipient

 $76m (receipts)  Down 24% in 2000

Although one massive "financial shield" was agreed in 2001, the IMF's refusal of a further rescue package precipitated the crash in December.

CHRONOLOGY

The Spanish first established settlements in the Andean foothills in 1543. The indigenous Amerindians, who had stopped any Inca advance into their territory, also prevented the Spaniards from settling in the east until the 1590s.

- ❑ **1816** United Provinces of Río de la Plata declare independence; 70 years of civil war follow.
- ❑ **1835–1852** Dictatorship of Juan Manuel Rosas.
- ❑ **1853** Federal system set up.
- ❑ **1857** Europeans start settling the pampas; six million by 1930.
- ❑ **1877** First refrigerated ship starts frozen beef trade to Europe.
- ❑ **1878–1883** War against pampas Amerindians (almost exterminated).
- ❑ **1916** Hipólito Yrigoyen wins first democratic presidential elections.
- ❑ **1930** Military coup.
- ❑ **1943** New military coup. Gen. Juan Perón organizes trade unions.
- ❑ **1946** Perón elected president, with military and labor backing.
- ❑ **1952** Eva Perón, charismatic wife of Juan Perón, dies of leukemia.
- ❑ **1955** Military coup ousts Perón. Inflation, strikes, unemployment.
- ❑ **1973** Perón reelected president.
- ❑ **1974** Perón dies; succeeded by his third wife "Isabelita," who is unable to exercise control.
- ❑ **1976** Military junta seizes power. Political parties are banned. Brutal repression during "dirty war" sees "disappearance" of over 15,000 "left-wing suspects."
- ❑ **1981** Gen. Galtieri president.
- ❑ **1982** Galtieri orders invasion of Falkland Islands. UK retakes them.
- ❑ **1983** Pro-human rights candidate Raúl Alfonsín (UCR) elected president. Hyperinflation.
- ❑ **1989** Carlos Menem (Peronist) president.
- ❑ **1992** Inflation down to 25%.
- ❑ **1995** Economy enters recession.
- ❑ **1998–1999** Argentina weathers financial crisis in Brazil.
- ❑ **1999** Fernando de la Rúa of UCR–FREPASO alliance elected president.
- ❑ **2001** Weakened by corruption scandals, government is brought down by economic crisis.
- ❑ **2002** January, Eduardo Duhalde appointed president by Peronist-dominated Congress.

DEFENSE

No compulsory military service

 $4.69bn  Down 14% in 2000

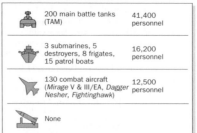

ARGENTINIAN ARMED FORCES

200 main battle tanks (TAM)	41,400 personnel
3 submarines, 5 destroyers, 8 frigates, 15 patrol boats	16,200 personnel
130 combat aircraft (*Mirage V & III/EA, Dagger Nesher, Fightinghawk*)	12,500 personnel
None	

The end of dictatorship led to trials and prison for the top brass, but subsequent immunity laws were meant to placate the military and close the chapter on the "dirty war" (1976–1983), during which some 15,000 to 30,000 were killed or "disappeared." The military made public admissions of guilt in 1995. A 2001 ruling in a kidnap and murder case, however, said that such immunity was unconstitutional and cleared the way for the trial of many more military personnel for human rights crimes. Former junta leader Gen. Leopoldo Galtieri was arrested in 2002. The armed forces now see themselves as modernized, participating in UN peacekeeping and cooperating on defense within the framework of the Mercosur trade bloc (Brazil, Argentina, Uruguay, and Paraguay).

ECONOMICS

Inflation 5.2% p.a. (1990–2000)

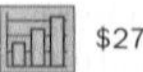 $276bn  0.9984–1 Argentine pesos

SCORE CARD

- ❑ World GNP Ranking.......................17th
- ❑ GNP per Capita..........................$7460
- ❑ Balance of Payments...................–$8.9bn
- ❑ Inflation.....................................–0.9%
- ❑ Unemployment..............................16%

EXPORTS

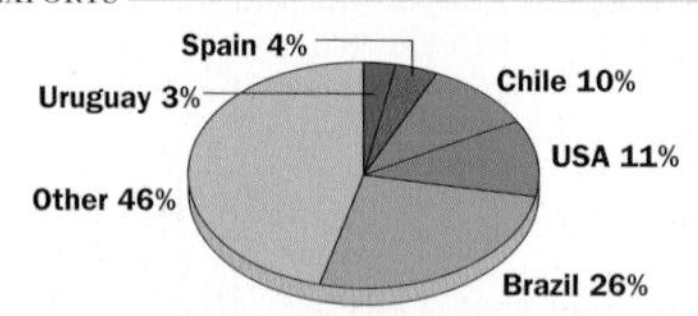
Spain 4%
Uruguay 3%
Chile 10%
USA 11%
Other 46%
Brazil 26%

IMPORTS

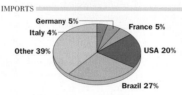

Germany 5%
Italy 4%
France 5%
Other 39%
USA 20%
Brazil 27%

ECONOMIC PERFORMANCE INDICATOR

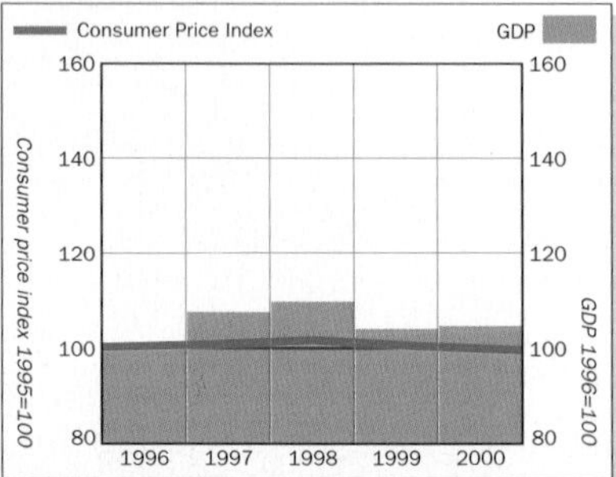
Consumer Price Index
GDP
160
140
120
100
80
Consumer price index 1995=100
GDP 1996=100
1996
1997
1998
1999
2000

Profile

The "miracle" recovery of the 1990s was based on stabilizing the peso (by pegging it to the US dollar) and on a combination of neoliberal reforms accompanied by privatization. Argentina rode out the Mexican crisis of 1995, but was hit by damage to foreign investor confidence and a shrinking Brazilian market in 1998–1999. Regional recession in 2001 brought economic crisis, and the world's largest default on international debt.

Strengths

Rich and varied agricultural base. Powerful agribusiness (mainly beef, soya, wheat, fruit, and wine) and wealth of energy resources. Net exporter of oil.

Weaknesses

End of currency stability based on pegging of peso to dollar. Vulnerability to external shocks and downturns in Brazil (largest single export market). Investors scared by changes in assessment of country risk in emerging markets. Heavy debts, public and private, to refinance. Weak banking sector. Global fluctuations in prices of vital non-oil commodities. High unemployment and risk of unrest. Endemic tax evasion. Subsidies and trade barriers bar agricultural produce from US and EU.

ARGENTINA : MAJOR BUSINESSES

Salta
San Salvador de Jujuy
Corrientes
Córdoba
Santa Fé
Mendoza
Buenos Aires
Viedma
Wine
Textiles
Agribusiness
Metals
Oranges
Tobacco
Vehicle assembly
Light engineering
Cattle/Meat packing
Heavy engineering
0 400 km
0 400 miles
* significant multinational ownership

RESOURCES

Electric power 23m kw

1.03m tonnes

822,000 b/d (reserves 3bn barrels)

50.2m cattle, 13.5m sheep, 4.2m pigs, 110m chickens

Oil, natural gas, coal, iron, zinc, lead, uranium, tin, silver, copper, gold

ELECTRICITY GENERATION

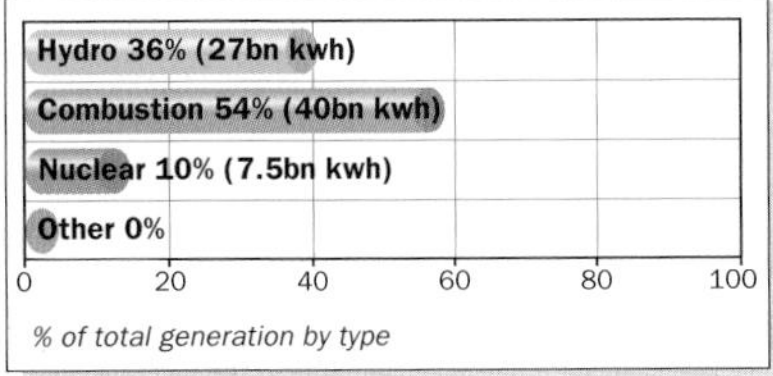
Hydro 36% (27bn kwh)
Combustion 54% (40bn kwh)
Nuclear 10% (7.5bn kwh)
Other 0%
0
20
40
60
80
100
% of total generation by type

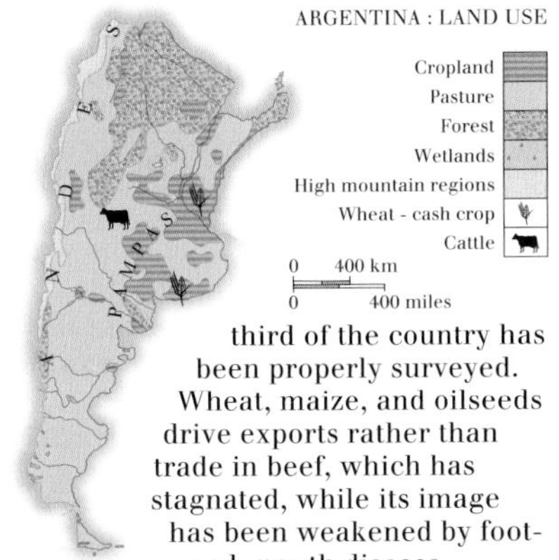
ARGENTINA : LAND USE
Cropland
Pasture
Forest
Wetlands
High mountain regions
Wheat - cash crop
Cattle
0 400 km
0 400 miles
ANDES
PAMPAS

Known oil and gas reserves are still underexploited, and copper and gold mining are just taking off; only one-third of the country has been properly surveyed. Wheat, maize, and oilseeds drive exports rather than trade in beef, which has stagnated, while its image has been weakened by foot-and-mouth disease.

ENVIRONMENT

Sustainability rank: 15th

2% (2% partially protected)

3.8 tonnes per capita

ENVIRONMENTAL TREATIES

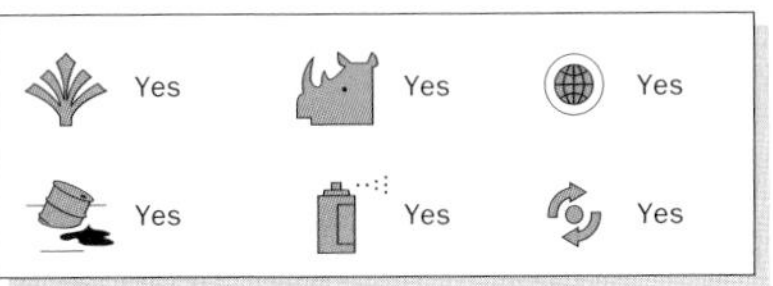
Yes
Yes
Yes
Yes
Yes
Yes

Environmental protection has low governmental priority. Legislation is weak and largely ignored by states which retain a good deal of autonomy. Political parties typically shy away from the level of public spending needed to tackle major environmental problems, and a corrupt judiciary had meant poor enforcement of existing laws. Key problems are hazardous waste, poor urban water and air quality and inadequate sewers, pesticide contamination due to agribusiness, deforestation, and illegal hunting.

MEDIA

TV ownership high

Daily newspaper circulation 56 per 1000 people

PUBLISHING AND BROADCAST MEDIA

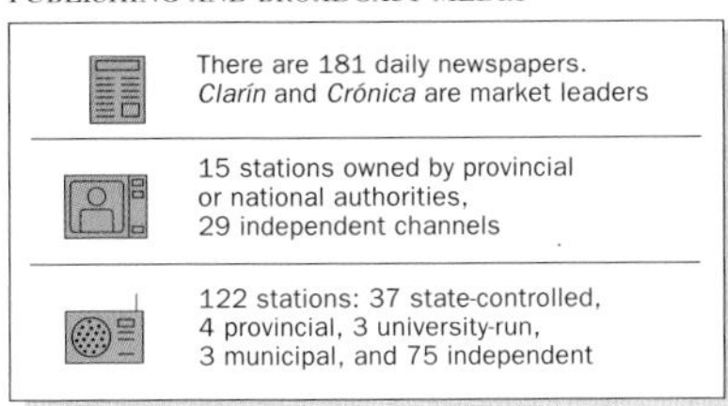
There are 181 daily newspapers.
Clarín and *Crónica* are market leaders
15 stations owned by provincial or national authorities, 29 independent channels
122 stations: 37 state-controlled, 4 provincial, 3 university-run, 3 municipal, and 75 independent

Many journalists were murdered by the military in the 1970s, but the press was liberated under the UCR (1983–1989) and harries governments, especially on corruption. Investigative journalists, however, still face intimidation. The Internet had reached over 10% of the population by 2001.

CRIME

Death penalty not used in practice

38,604 prisoners

Up 4% in 1999

CRIME RATES

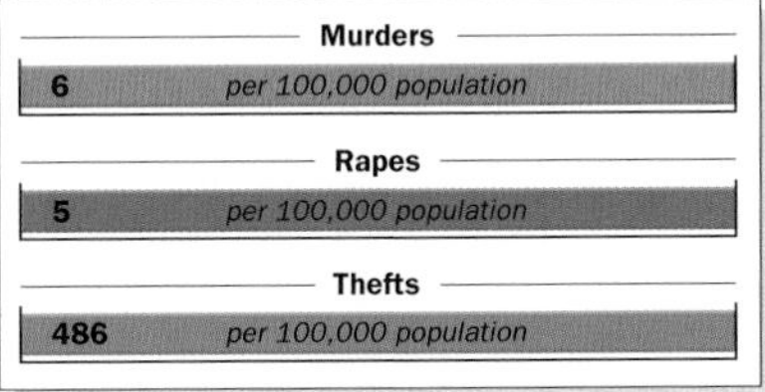
Murders
6 per 100,000 population
Rapes
5 per 100,000 population
Thefts
486 per 100,000 population

Half of Buenos Aires households own a gun for fear of violent crime. Law enforcement in the interior is weak. The judiciary and the police command little respect. Overcrowded prisons lead to frequent riots and criminal cases can take over a year to reach court.

EDUCATION

School leaving age: 14

97%

1.34m students

THE EDUCATION SYSTEM

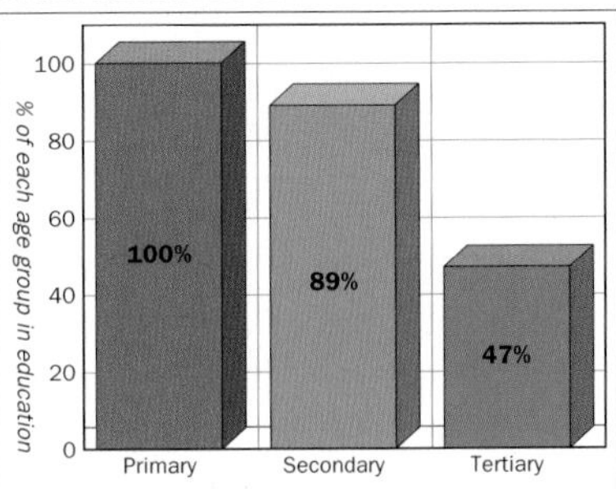
100
80
60
40
20
0
% of each age group in education
100%
89%
47%
Primary
Secondary
Tertiary

Public schooling is free and compulsory to the age of 14. Huge numbers of poor students drop out of the education system near or after this age, and 60–70% of middle-class students drop out by the close of five-year courses at cash-strapped free state universities. A fifth of government spending on education finds its way to private institutions.

HEALTH

Welfare state health benefits

1 per 370 people

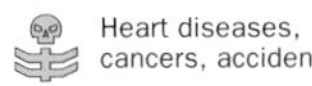
Heart diseases, cancers, accidents

Buenos Aires has 33 hospitals, but free state provision suffers from underfunding, poorly paid staff, and long queues. Government-sponsored vaccination, mother-and-child schemes, feeding programs, and rural health projects barely tackle such problems in the poorest provinces as malnutrition, lack of decent water and sewers, and threadbare medical cover. A health care deregulation bill was decreed in 2001 to dismantle trade unions' monopoly of health insurance schemes.

SPENDING

GDP/cap. increase

CONSUMPTION AND SPENDING

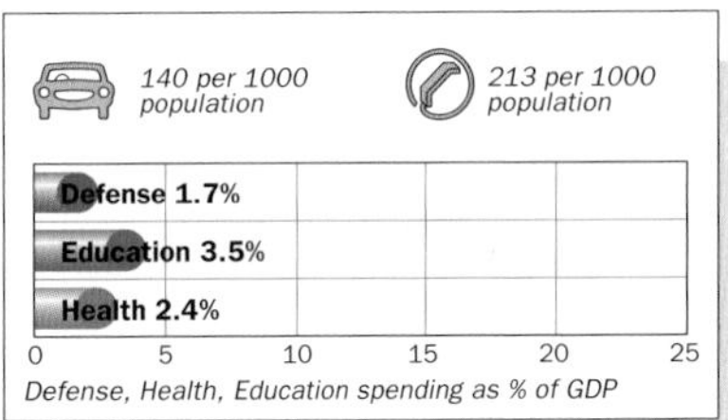
140 per 1000 population
213 per 1000 population
Defense 1.7%
Education 3.5%
Health 2.4%
0
5
10
15
20
25
Defense, Health, Education spending as % of GDP

Members of the wealthy elite, who travel in private jets to *estancias* (country estates), vacation in Europe and the US, and hold dollar accounts offshore to avoid tax, escaped the worst of the economic collapse in 2001–2002. Middle-income groups squeezed after years of free-market reforms lost out in the crisis which forced some 15 million below the poverty line, with four million in extreme poverty. Emergency government aid offered in 2002 to one million unemployed people with children under 18 was at a level below half the legal minimum salary. The precrisis figure of some 40% of workers in the low-wage black economy was set to balloon. Millions of cashless and poor people resorted to exchanging goods at barter clubs.

WORLD RANKING

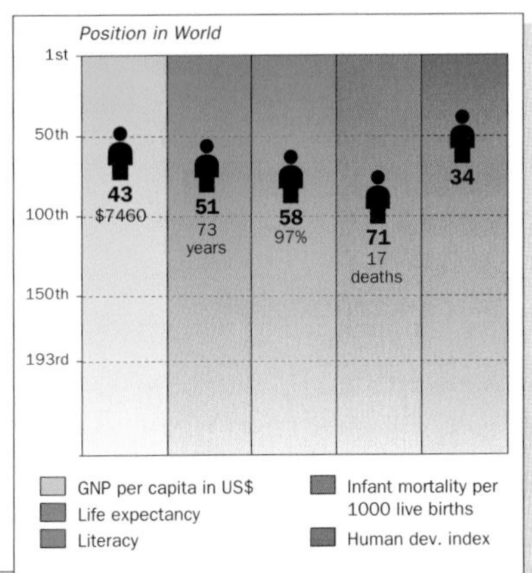
Position in World
1st
50th
100th
150th
193rd
43
$7460
51
73 years
58
97%
71
17 deaths
34
GNP per capita in US$
Life expectancy
Literacy
Infant mortality per 1000 live births
Human dev. index

A

ARMENIA

EUROPE

Europe

Georgia

ARMENIA

Azerbaijan

Turkey

Iran

OFFICIAL NAME: Republic of Armenia **CAPITAL:** Yerevan
POPULATION: 3.8 million **CURRENCY:** Dram **OFFICIAL LANGUAGE:** Armenian

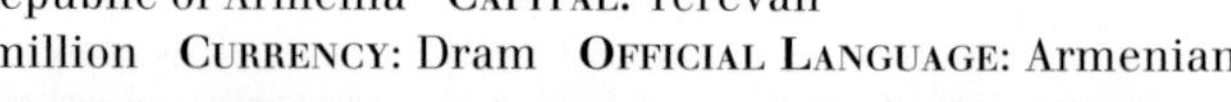
1991
1991
Sept 21
ARM
+4
+374
.am

LANDLOCKED IN THE Lesser Caucasus Mountains, Armenia is the smallest of the former Soviet Union's republics. It was the first to adopt Christianity as its state religion. It is bordered by Muslim states to the south, east, and west. Keen to strengthen links with the rest of the CIS, Armenia has kept to a path of radical economic reform, including privatization. The confrontation with Azerbaijan over the enclave of Nagorno Karabakh has dominated national life since 1988.

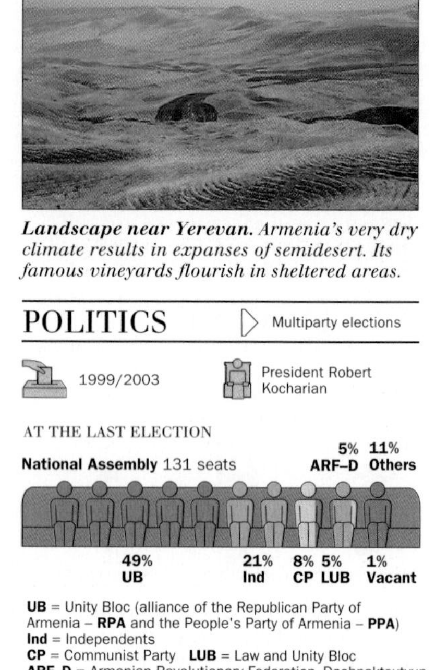

Landscape near Yerevan. *Armenia's very dry climate results in expanses of semidesert. Its famous vineyards flourish in sheltered areas.*

CLIMATE

Mountain

WEATHER CHART FOR YEREVAN

Average daily temperature — Rainfall

°C/°F J F M A M J J A S O N D cm/in

40/104, 30/86, 20/68, 10/50, 0/32, -10/14, -20/-4

40/16, 30/12, 20/8, 10/4, 0

Armenia has a continental climate, with little rainfall in the lowlands. Winters can be very cold.

TRANSPORTATION

Drive on right
Yerevan International
Has no fleet

THE TRANSPORTATION NETWORK

15,998 km (9941 miles)

7567 km (4702 miles)

796 km (495 miles)

None

Public transportation was badly hit by a war-induced fuel crisis. Road and rail links with Georgia, connecting with the main east–west corridor, need upgrading.

TOURISM

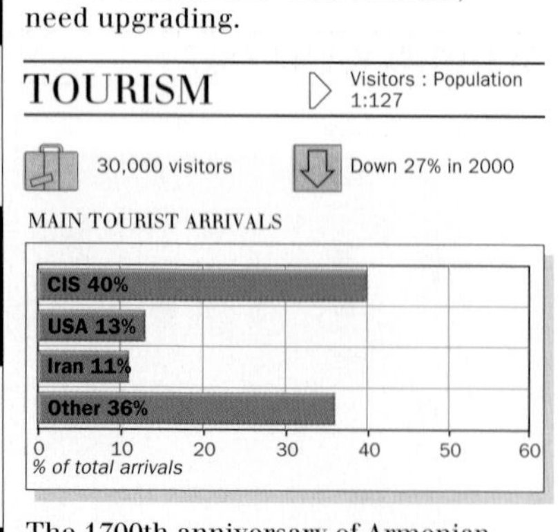
Visitors : Population 1:127
30,000 visitors
Down 27% in 2000
MAIN TOURIST ARRIVALS
CIS 40%
USA 13%
Iran 11%
Other 36%
0
10
20
30
40
50
60
% of total arrivals

The 1700th anniversary of Armenian Christianity in 2001 boosted the war-damaged tourist industry. Most visitors are diaspora Armenians.

PEOPLE

Pop. density medium

Armenian, Russian

128/km² (330/mi²)

THE URBAN/RURAL POPULATION SPLIT

70% — 30%

ETHNIC MAKEUP

Other 2% — Azeri 3% — Russian 2% — Armenian 93%

Minority nationalities are well integrated in Armenia. There are strong contacts with the many Armenian emigrants, numbering nine million in the US, France, and Syria.

Conflict with Azerbaijan forced 350,000 Armenians in Azerbaijan to return to Armenia and 190,000 Azeris in Armenia to return to Azerbaijan.

POLITICS

Multiparty elections

1999/2003

President Robert Kocharian

AT THE LAST ELECTION

National Assembly 131 seats

49% UB — 21% Ind — 8% CP — 5% LUB — 1% Vacant — 5% ARF–D — 11% Others

UB = Unity Bloc (alliance of the Republican Party of Armenia – **RPA** and the People's Party of Armenia – **PPA**)
Ind = Independents
CP = Communist Party **LUB** = Law and Unity Bloc
ARF–D = Armenian Revolutionary Federation–Dashnaktsutyun

Armenia became an independent multiparty democracy in 1991 and held its first parliamentary elections in July 1995. A new constitution approved by referendum set up a presidential republic, and in 1996 President Levon Ter-Petrossian was reelected for a five-year term. The war with Azerbaijan, over the issue of whether the Armenian enclave of Nagorno Karabakh inside Azerbaijan should become part of Armenia, has simmered since a 1994 cease-fire. Ter-Petrossian resigned in 1998, after parliament opposed his softer line in search of peace. He was succeeded by Robert Kocharian, a former premier and ex-governor of Nagorno Karabakh. In 1999 the prime minister, RPA leader Vazgen Sarkissian, was shot dead in a dramatic attack on parliament. He was succeeded first by his brother Aram, then in 2000 by Andranik Markarian.

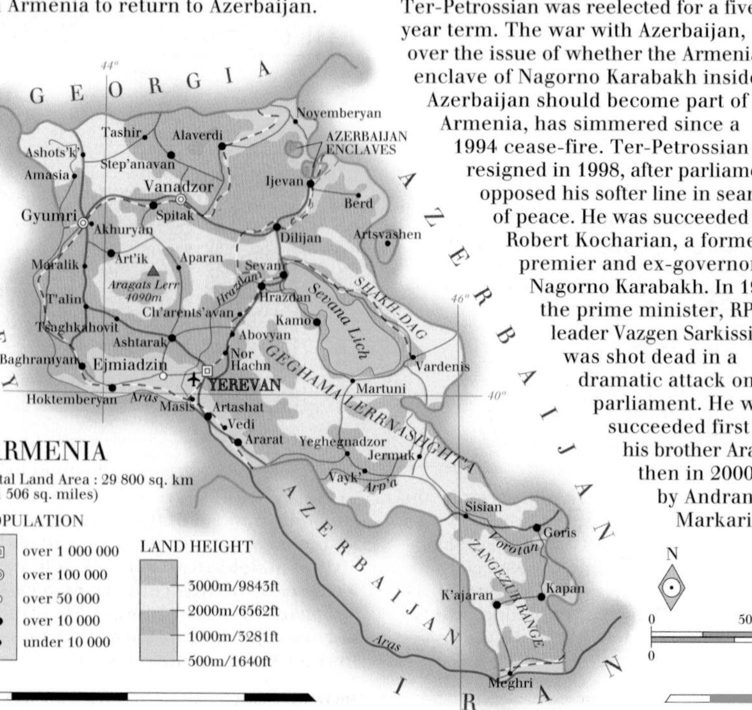
GEORGIA
TURKEY
AZERBAIJAN
IRAN
Noyemberyan
AZERBAIJAN ENCLAVES
Tashir
Alaverdi
Ashots'k'
Step'anavan
Amasia
Vanadzor
Ijevan
Berd
Gyumri
Spitak
Akhuryan
Dilijan
Artsvashen
Maralik
Art'ik
Aparan
Sevan
Aragats Lerr 4090m
T'alin
Hrazdan
Kamo
Ch'arents'avan
Sevana Lich
SHAKH-DAG
T'saghkahovit
Abovyan
Ashtarak
Nor Hachn
GEGHAMA LERRNASHGHT'A
Vardenis
Baghramyan
Ejmiadzin
YEREVAN
Martuni
Hoktemberyan
Aras
Masis
Artashat
Vedi
Ararat
Yeghegnadzor
Jermuk
Vayk'
Arp'a
Sisian
Vorotan
Goris
ZANGEZUR RANGE
K'ajaran
Kapan
Meghri
ARMENIA
Total Land Area : 29 800 sq. km (11 506 sq. miles)
POPULATION
over 1 000 000
over 100 000
over 50 000
over 10 000
under 10 000
LAND HEIGHT
3000m/9843ft
2000m/6562ft
1000m/3281ft
500m/1640ft
N
0
50 km

WORLD AFFAIRS

Joined UN in 1992

BSEC

CIS

CE

EAPC

OSCE

Armenia joined the Council of Europe in 2001. The continuing conflict with Azerbaijan is diplomatically damaging.

A Turkish trade embargo has been in place since 1988. Russia signed a treaty of friendship in 1997 but has been less pro-Armenian since the Azeris rejoined the CIS. Work began on a gas pipeline between Iran and Armenia in 1998.

AID

Recipient

 $216m (receipts)

 Up 3% in 2000

The EBRD and the World Bank back infrastructure projects. Expatriate Armenians such as US billionaire Kirk Kerkorian are a major source of funds.

DEFENSE

Compulsory military service

 $149m

 Down 6% in 2000

Successes in the fighting over Nagorno Karabakh have increased the profile and autonomy of the army, which includes conscripts on 24-month national service. A cease-fire has broadly held since 1994, but peace talks, revived in 1999–2000, remain inconclusive.

ECONOMICS

Inflation 213% p.a. (1990–2000)

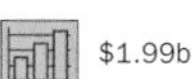 $1.99bn

 551.4–562.2 drams

SCORE CARD

- ❏ WORLD GNP RANKING....................137th
- ❏ GNP PER CAPITA..........................$520
- ❏ BALANCE OF PAYMENTS..................–$278m
- ❏ INFLATION....................................–0.8%
- ❏ UNEMPLOYMENT..............................12%

STRENGTHS

Strong ties with Armenian emigrants. Deposits of rare metals, currently unexploited. Machine building and manufacturing – includes textiles and bottling of mineral water.

WEAKNESSES

Dependent on imported energy, raw materials, and semi-finished goods. High unemployment and economic stagnation. Widespread corruption.

EXPORTS

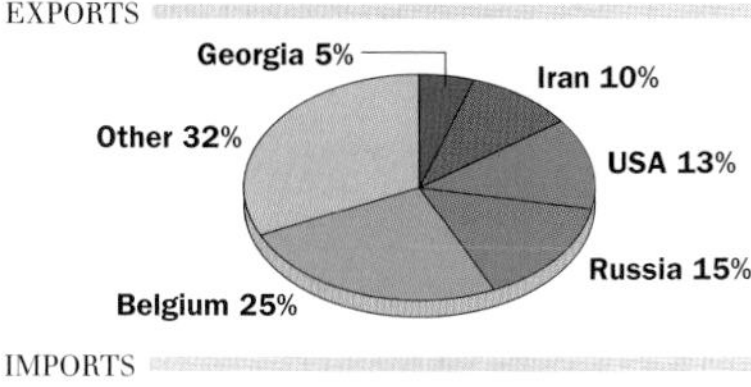
Georgia 5%
Iran 10%
Other 32%
USA 13%
Russia 15%
Belgium 25%

IMPORTS

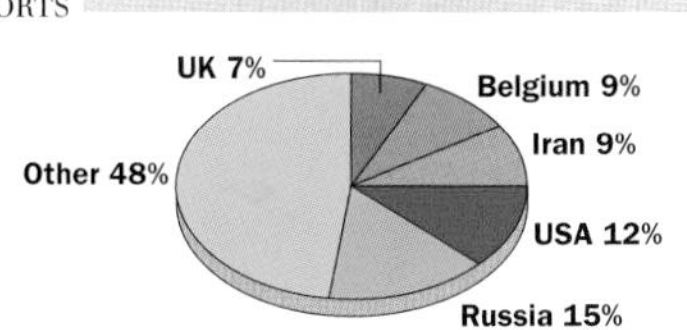
UK 7%
Belgium 9%
Iran 9%
Other 48%
USA 12%
Russia 15%

RESOURCES

Electric power 3m kw

877 tonnes

 Minimal oil production

497,155 sheep, 485,000 cattle, 4.3m chickens

 Coal, oil, natural gas, rare metals

Energy resources are negligible, and mismanagement of the energy industry was deemed to have cost the country $200 million in the 1990s. Vegetables and fruit are grown in fertile lowlands, and grains in the hills; agriculture accounts for as much as a third of GDP.

ENVIRONMENT

Sustainability rank: 38th

8%

 0.9 tonnes per capita

Environmental groups, backed by the EU, demand the closure of the Medzamor nuclear power station, declared unsafe after the 1988 earthquake, but reopened in 1995 owing to the energy crisis. HEP generation near Lake Sevan has seriously lowered its water level.

MEDIA

TV ownership medium

 Daily newspaper circulation 24 per 1000 people

PUBLISHING AND BROADCAST MEDIA

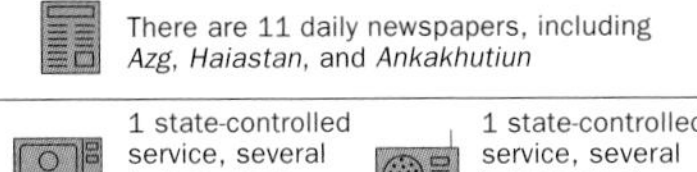

There are 11 daily newspapers, including *Azg, Haiastan,* and *Ankakhutiun*

1 state-controlled service, several independent stations

1 state-controlled service, several independent stations

Numerous TV and broadcasting stations assist media freedom. Independent journals and newspapers depend on the government-controlled paper industry.

CRIME

Death penalty in use

 7428 prisoners

 Down 7% in 1999

Reforms to the legal system introduced in 1999 included the replacement of the Supreme Court by an appeals court. Assassinations of political figures are common.

EDUCATION

School leaving age: 14

 98%

 39,770 students

The education system, previously conforming to that of the USSR, now emphasizes Armenian history and culture; 12% of the population have received higher education.

CHRONOLOGY

Armenia lost its autonomy in the 14th century. In 1639, Turkey took the west and Persia the east; Persia ceded its part to Russia in 1828.

- ❏ **1877–1878** Massacre of Armenians during Russo-Turkish war.
- ❏ **1915** Ottomans exile 1.75 million Turkish Armenians; most die.
- ❏ **1920** Independence.
- ❏ **1922** Becomes a Soviet republic.
- ❏ **1988** Earthquake kills 25,000. Conflict with Azerbaijan over Nagorno Karabakh begins.
- ❏ **1991** Independence from USSR.
- ❏ **1994** Cease-fire with Azerbaijan.
- ❏ **1995** First parliamentary elections.
- ❏ **1998** Kocharian elected president.
- ❏ **1999** Shooting of prime minister in attack on parliament.

HEALTH

No welfare state health benefits

 1 per 313 people

 Circulatory diseases, cancers, accidents, violence

Hospitals suffer from the erratic electricity supply. Poor sewerage and other services have led to a rise in hepatitis, tuberculosis, and cholera.

SPENDING

GDP/cap. decrease

CONSUMPTION AND SPENDING

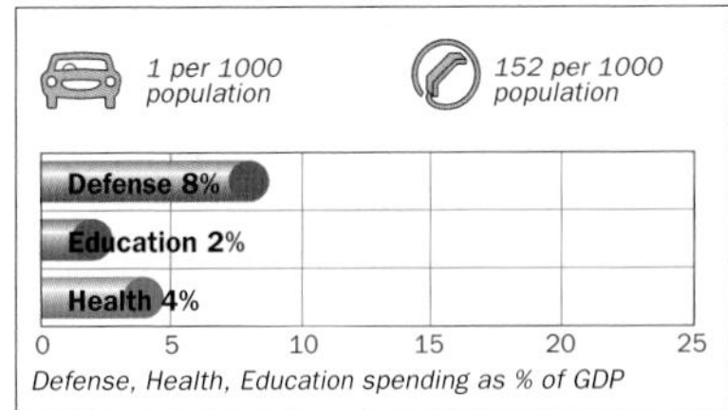
1 per 1000 population
152 per 1000 population
Defense 8%
Education 2%
Health 4%
0
5
10
15
20
25
Defense, Health, Education spending as % of GDP

The richest Armenian people are those living away from Armenia itself, particularly in the US and France. The many refugees from Baku, Azerbaijan, are the poorest.

WORLD RANKING

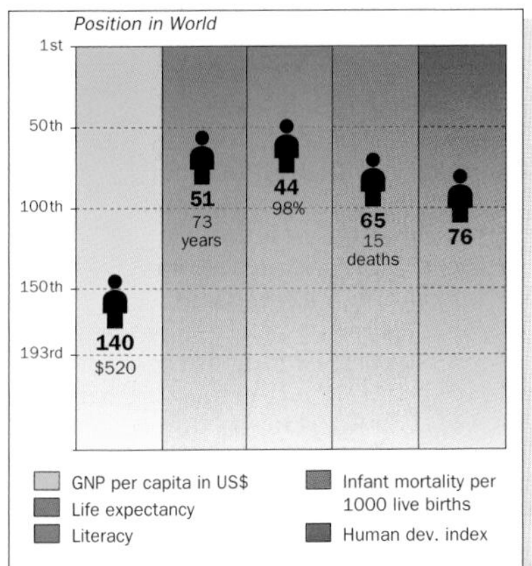
Position in World
1st
50th
100th
150th
193rd
140
$520
51
73
years
44
98%
65
15
deaths
76
GNP per capita in US$
Life expectancy
Literacy
Infant mortality per 1000 live births
Human dev. index

A

AUSTRALIA

AUSTRALASIA
Asia
Pacific Ocean
Indonesia
Papua New Guinea
AUSTRALIA
New Zealand
Australia

OFFICIAL NAME: Commonwealth of Australia **CAPITAL:** Canberra
POPULATION: 19.3 million **CURRENCY:** Australian dollar **OFFICIAL LANGUAGE:** English

1901

1901

Jan 26

AUS
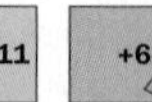
+8 to +11

+61

.au

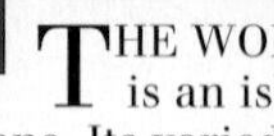

THE WORLD'S SIXTH-LARGEST country, Australia is an island continent located between the Indian and Pacific Oceans. Its varied landscapes include tropical rainforests, the deserts of the arid "red center," snowcapped mountains, rolling tracts of pastoral land, and magnificent beaches. Famous natural features include Uluru (Ayers Rock) and the Great Barrier Reef. Most Australians live on the coast, and all the state capitals, including Sydney, host of the 2000 Olympics, are coastal cities. Only Canberra, the national capital, lies inland. The vast interior is dotted with large reserves, sparsely inhabited by communities from the small Aboriginal population.

Uluru (Ayers Rock), Northern Territory. *The renaming of Ayers Rock reflects growing Aboriginal influence in Australia.*

CLIMATE

Hot desert/steppe/ tropical/Mediterranean

WEATHER CHART FOR CANBERRA

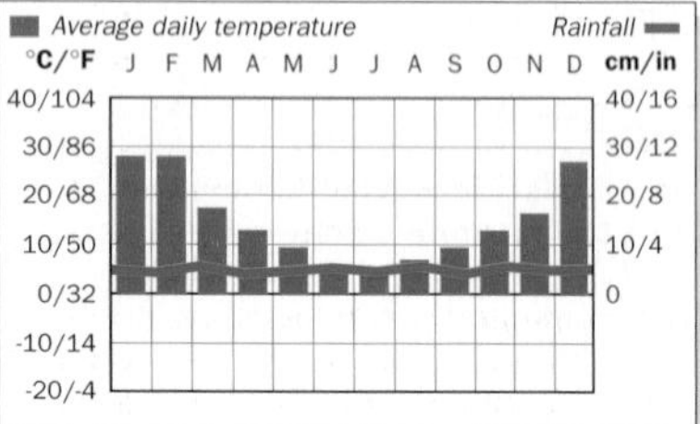
Average daily temperature
Rainfall
°C/°F
J F M A M J J A S O N D
cm/in
40/104
30/86
20/68
10/50
0/32
-10/14
-20/-4
40/16
30/12
20/8
10/4
0

The interior, west, and south are arid or semiarid and very hot in summer; central desert temperatures can reach 50°C (122°F). The north, around Darwin and Cape York Peninsula, is hot all year and humid during the summer monsoon. Only the east and southeast, within 400 km (250 miles) of the coast, and the southwest, around Perth, are temperate: most Australians live in these areas.

TRANSPORTATION

Drive on left

Kingsford Smith, Sydney 25.8m passengers

631 ships 1.91m grt

THE TRANSPORTATION NETWORK

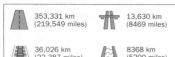

353,331 km (219,549 miles)

13,630 km (8469 miles)

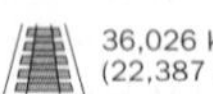
36,026 km (22,387 miles)

8368 km (5200 miles)

Air transportation is well developed and vital to Australia's sparsely populated center and west. Sydney suffers from air congestion, but proposals for a new West Sydney airport remain controversial. A proposed high-speed train network on the east coast has been shelved due to cost. Most long-distance freight in Australia travels in massive trucks known as "road trains." Improvements in urban transportation are a priority and gained impetus in Sydney from the 2000 Olympic Games.

TOURISM

Visitors : Population 1:3.9

4.95m visitors

Up 11% in 2000

MAIN TOURIST ARRIVALS

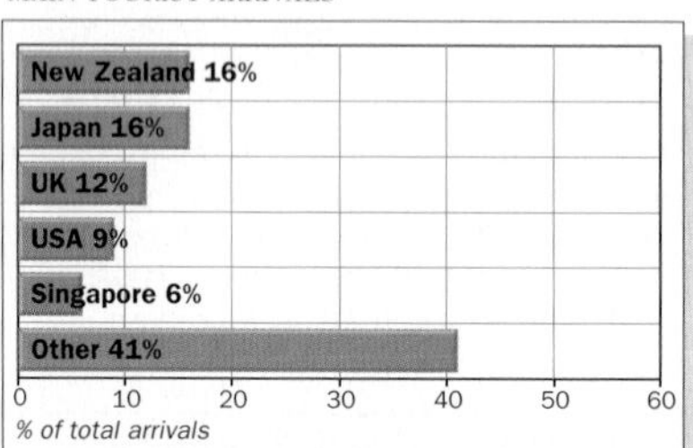
New Zealand 16%
Japan 16%
UK 12%
USA 9%
Singapore 6%
Other 41%
0
10
20
30
40
50
60
% of total arrivals

Tourism is now Australia's largest single foreign exchange earner. Faster, cheaper air travel and highly successful government marketing campaigns, both on a national and a state level, attract tourists in increasing numbers. The focus during the 1990s on drawing tourists from nearby Asian countries left the Australian tourist industry vulnerable after the Asian financial crisis of 1997. In recent years New Zealanders have equaled the Japanese as the largest single category of visitors to Australia. However, tourists from southeast Asia continue to arrive in significant numbers.

The country's many attractions include wildlife, swimming and surfing off Pacific and Indian Ocean beaches, skin diving along the Great Barrier Reef, and skiing in the Australian Alps. Uluru, Aboriginal culture, and the town of Alice Springs are among the outback's attractions. The far north has tropical resorts, the northwest, pearl fishing. The vineyards of the south and southeast attract many visitors, as do the cultural life of Melbourne and Sydney and the arts festivals held in state capitals. Sydney's famous landmarks and cosmopolitan feel, as well as the world-renowned Bondi Beach, make it a favorite.

The mid-1980s saw a phenomenal boom; tourist arrivals almost tripled in five years to reach two million in 1990. Even though growth slowed during the early 1990s, by 2000 the number of visitors had reached almost five million, boosted greatly by the celebrated Sydney Olympic Games.

Bonaparte Archipelago
Derby
Broome
INDIAN OCEAN
Port Hedland
GREAT SA
DESERT
Onslow
Hamersley Range
Newman
Paraburdoo
Lake Disappointment
WESTERN AUSTRALIA
Lake Macleod
Carnarvon
Shark Bay
Dirk Hartog Island
Lake Carnegie
Meekatharra
Mount Magnet
Geraldton
Kalgoorlie
Moora
Perth
Fremantle
Northam
Merredin
Norseman
Rockingham
Bunbury
Collie
Wagin
Bridgetown
Cape Leeuwin
Manjimup
Ravensthorpe
Esperance
Albany

AUSTRALIA

Total Land Area : 7 686 850 sq. km (2 967 893 sq. miles)

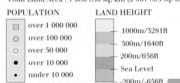

POPULATION

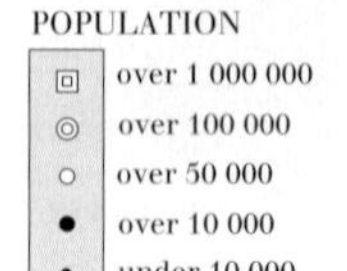
over 1 000 000
over 100 000
over 50 000
over 10 000
under 10 000

LAND HEIGHT

1000m/3281ft
500m/1640ft
200m/656ft
Sea Level
-200m/-656ft

PEOPLE

Pop. density low

English, Italian, Cantonese, Greek, Arabic, Vietnamese, Aboriginal languages

3/km² (7/mi²)

THE URBAN/RURAL POPULATION SPLIT

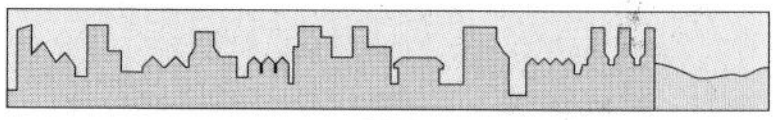

85% 15%

The first settlers arrived in Australia at least 100,000 years ago. Their modern descendants, the Aborigines, today make up less than 3% of the population. European settlement began in 1788 and was dominated by British and Irish immigrants – some of whom were convicts – until the gold rushes of the 1850s. Immigrants of other nationalities – including many Chinese – arrived to prospect for gold, then settled in the cities, especially Melbourne and Sydney. When the new federal government was installed in 1901, one of its first acts was to prevent further Chinese immigration. The act set out the "White Australia" policy, which conditioned attitudes to immigration for almost 70 years.

A massive immigration drive after World War II brought many more British settlers to Australia in the 1950s. Further government initiatives to "populate or perish" saw the arrival of large numbers of Italians and Greeks.

From the late 1960s, the "White Australia" policy was progressively wound down. It was officially ended during the 1972–1975 Whitlam administration. Ever since, up to 50% of immigrants each year have come from Asia, transforming Australia from an almost exclusively European enclave into a multicultural society in which immigrant groups are encouraged to maintain connections with their own cultures and languages.

Aborigines, the exception in an otherwise integrated society, number around 410,000. Economically and socially marginalized, they face considerable discrimination. Until the mid-1960s, they were denied the vote and full social benefits. Their land had been occupied as *terra nullius* – belonging to no one. Since the 1970s, Aborigines have made a more organized stand on land and civil rights. Native title to land was recognized in 1993, although controversies continue over the extent of its application. Civil rights campaigns have moved on from the initial phase of antiracist protests to demand greater equality in areas such as health, housing, and education.

Average life expectancy is still 20 years lower than the rest of the population. Alcoholism is a pervasive problem both in towns and rural areas. Aborigines in urban areas may be relatively better housed but face particular problems in asserting their cultural identity.

During the 1950s and 1960s, Catholic–Protestant differences were sufficient to cause a rift in the ALP. However, a subsequent policy encouraging mixed denomination schooling, coupled with a decline in religious observance, has largely neutralized the issue.

RELIGIOUS PERSUASION

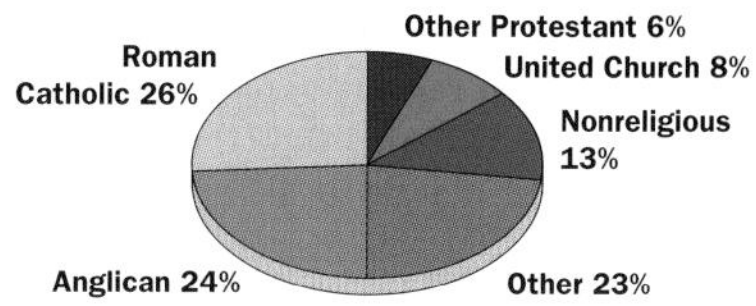

Roman Catholic 26%
Other Protestant 6%
United Church 8%
Nonreligious 13%
Other 23%
Anglican 24%

ETHNIC MAKEUP

Aboriginal and Other 3%
Asian 5%
European 92%

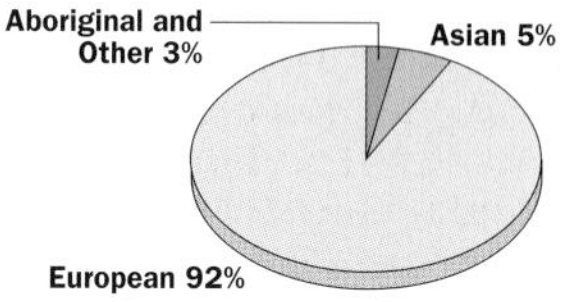

POPULATION AGE BREAKDOWN

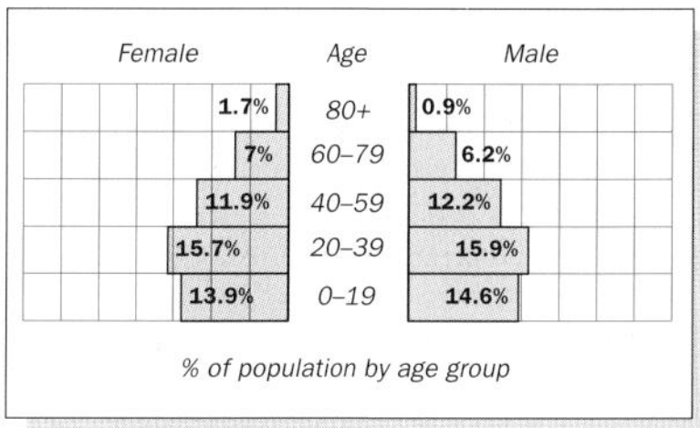

Female	Age	Male
1.7%	80+	0.9%
7%	60–79	6.2%
11.9%	40–59	12.2%
15.7%	20–39	15.9%
13.9%	0–19	14.6%

% of population by age group

A

CHRONOLOGY

Dutch, Portuguese, French, and – decisively – British incursions throughout the 17th and 18th centuries signaled the end of a millennia of Aboriginal isolation. Governor Arthur Philip raised the British Union Flag at Sydney Cove on 26 January 1788.

- ❑ **1901** Inauguration of Commonwealth of Australia.
- ❑ **1915** Australian troops suffer heavy casualties at Gallipoli.
- ❑ **1939** Prime Minister Menzies announces Australia will follow Britain into war with Germany.
- ❑ **1942** Fall of Singapore to Japanese army. Japanese invasion of Australia seems imminent. Government turns to US for help.
- ❑ **1950** Australian troops committed to UN/US Korean War against North Korean communists.
- ❑ **1962** Menzies government commits Australian aid to war in Vietnam.
- ❑ **1966** Adopts decimal currency.
- ❑ **1972** Whitlam government elected. Aid to South Vietnam ceases.
- ❑ **1975** Whitlam government dismissed by Governor-General Sir John Kerr. Malcolm Fraser forms coalition government.
- ❑ **1983** Bob Hawke becomes prime minister at the head of an ALP administration.
- ❑ **1985** Corporate boom followed by deepening recession.
- ❑ **1992** Paul Keating defeats Hawke in leadership vote, becomes prime minister; announces "Turning toward Asia" policy. High Court's "Mabo Judgment" recognizes Aboriginal land rights.
- ❑ **1993** Against most predictions, Keating's ALP government reelected. Native Title Act provides compensation for Aboriginal rights extinguished by existing land title.
- ❑ **1996** Defeat of Keating government. Liberal John Howard becomes prime minister. Shooting of 35 people by gunman in Port Arthur, Tasmania, prompts tightening of gun control laws. First death under Northern Territory's controversial euthanasia legislation; legislation later overruled at federal level.
- ❑ **1998** Elections: Howard's Liberal and National coalition retains power with reduced majority; fears of right-wing One Nation party breakthrough prove unfounded.
- ❑ **1999** Referendum rejects proposals to replace Queen as head of state by indirectly elected president.
- ❑ **2000** Olympic Games in Sydney.
- ❑ **2001** Surprise reelection of Liberal–National coalition.

POLITICS

Multiparty elections

L. House 2001/2004
U. House 2001/2004

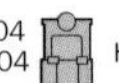
H.M. Queen Elizabeth II

AT THE LAST ELECTION

House of Representatives 150 seats

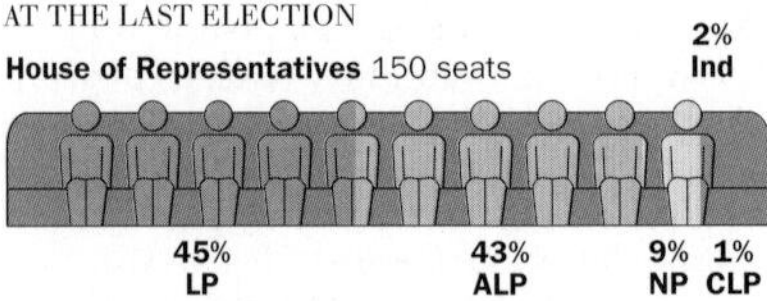

2% Ind
45% LP
43% ALP
9% NP
1% CLP

LP = Liberal Party **ALP** = Australian Labor Party **NP** = National Party **Ind** = Independents **CLP** = County–Liberal Party **AD** = Australian Democrats **G** = Greens

Senate 76 seats

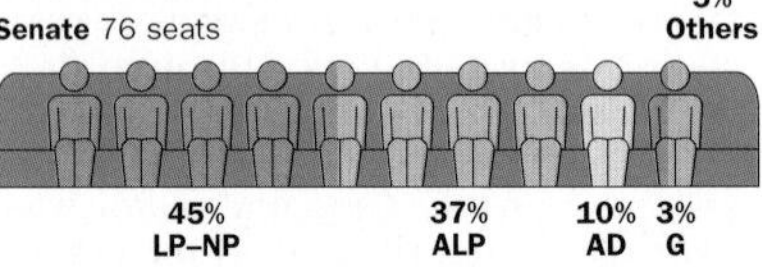

5% Others
45% LP–NP
37% ALP
10% AD
3% G

12 seats in the Senate are apportioned to each of the country's constituent states and two each to the Northern Territory and the Australian Capital Territory

Australia is a parliamentary democracy on the British model. There are six state governments, all but one (Queensland) bicameral. The Northern Territory became self-governing in 1978.

Profile

The ALP and the Liberal and National parties have dominated Australian politics since 1945. The latter two, politically to the right, work together in coalition and broadly represent big business and agricultural interests. The ALP gained some of this support in the 1980s, adopting free-market policies and blurring the differences between parties, but 13 years of ALP rule ended in 1996, when a Liberal–National coalition took office under John Howard. It retained power with a much reduced majority in the 1998 poll, and again in 2001, due largely to popularity for its hard-line stance on immigration and terrorism. However, by March 2002 it faced ALP regional governments in all states and territories, highlighting the fragility of its hold on power.

Main Political Issues

Immigration

Public concern at images of boatloads of would-be immigrants entering Australian waters has boosted right-wing policies. Dramatic attempts by asylum seekers to protest over the conditions in which they are kept have only served further to alienate public opinion.

The Howard government has courted public mistrust of refugees, focusing on "unacceptable" behavior. Its allegations that immigrants had thrown children overboard in a desperate attempt to secure asylum were proved groundless in 2002, and its notorious "Pacific solution" – encouraging tiny Pacific states to house asylum applicants temporarily in return for financial aid – has been roundly criticized by the international community.

Vineyards in South Australia. *Wine-making has been one of Australia's greatest agricultural success stories in recent years.*

The far right

Although consistently unsuccessful at the federal level (it failed to win a single seat in the 2001 general election), the local success of Pauline Hanson's One Nation party has done much to legitimize many right-wing policies. In response conservative parties have adopted a notably anti-immigrant position while the ALP has conspicuously toned down its support for immigration and embraced market economics.

Republicanism

Despite international press coverage, the republican issue is not of major importance to most Australians. The former ALP government played down the debate in order to avoid inflaming monarchist groups. In his 1996 oath of office, Prime Minister Howard swore allegiance to Queen Elizabeth II, but not her successors. His government's Constitutional Convention of February 1998 discussed various republican models, including the popular election of a head of state. The alternative to monarchy which was eventually chosen to put to referendum in 1999 – where the president would be chosen by parliament, requiring a two-thirds majority – was rejected by the voters.

Paul Keating, *resigned as leader of the ALP after his 1996 election defeat.*

John Howard, *leader of the LP, was elected prime minister in 1996.*

Aden Ridgeway, *Aboriginal senator and deputy leader of the AD.*

WORLD AFFAIRS

Joined UN in 1945

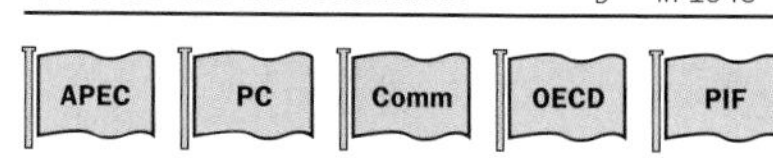
APEC
PC
Comm
OECD
PIF

Australia's international focus has shifted from Europe and the US toward Asia. Geopolitically it is in an ambiguous position. It lost its place as a major UK trading partner after the UK joined the EU, but was still regarded as a Western outsider by the Asian nations with which it wanted closer links. Australia took practical steps to redefine its role, backing the 1989 Asia–Pacific Economic Cooperation forum (APEC) to create a multilateral regional trading bloc similar to the EU and NAFTA. The US was a strong supporter, seeing APEC as a means of promoting free-market economics in Asia. Australia's ambition is for APEC to become the leading association in the region. However, the move toward market liberalization slowed, following the 1997 Asian financial crisis.

Relations with the US are tense on questions of trade. Australia objects to subsidized US wheat undercutting its own exports, particularly in the key Chinese market. It sees the EU and the US as its main competitors in selling to southeast Asia.

On security issues Australia still supports the West. It sent troops to the 1991 Gulf War and to assist in the "war on terrorism" in Afghanistan in 2001. Its commitment to the Pacific region also remains strong. In the post-Cold War era this is expressed mainly through development aid rather than defense arrangements, but Australian troops led the UN force in East Timor in 1999.

Australia objects to longline fishing, which kills large numbers of dolphins, and employs submarine patrols in an effort to regulate the industry. There have been a number of minor skirmishes with Indonesian and Japanese boats.

Australia has been strongly criticized by the international community and the UN for its treatment of the Aboriginal population and asylum seekers. The "Pacific Solution" to the immigration problem prompted condemnation on its adoption in late 2001.

AID

Donor

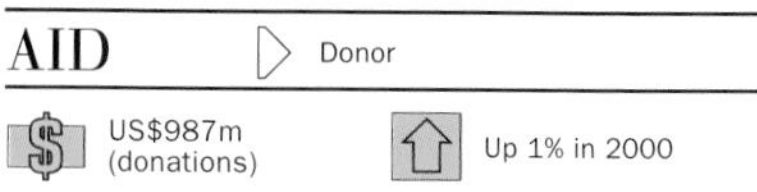
US$987m (donations)
Up 1% in 2000

Australia spends only 0.25% of its GNP on aid programs. Most is spent in the Asia–Pacific region. Particular areas of focus are those of HIV/AIDS programs and nongovernmental organizations. The recipient of by far the greatest amount is Papua New Guinea, where Australian companies such as Broken Hill Proprietary have major mining operations.

DEFENSE

No compulsory military service

US$6.95bn

Down 11% in 2000

Strategic ties with the US remain important. Australia has defense arrangements with the Philippines, Brunei, and Thailand among others. It dramatically increased expenditure in 2001–2002 to promote its position as a regional power. Updating military equipment and exploiting information technologies are priorities. 1550 soldiers were despatched to assist in the "war on terrorism" in late 2001.

AUSTRALIAN ARMED FORCES

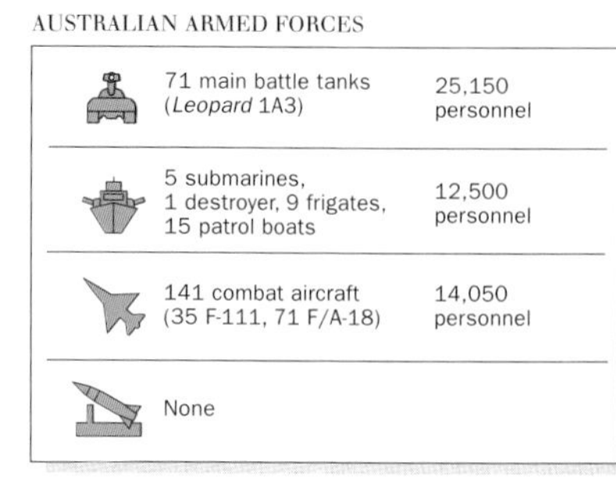

71 main battle tanks (*Leopard* 1A3)	25,150 personnel
5 submarines, 1 destroyer, 9 frigates, 15 patrol boats	12,500 personnel
141 combat aircraft (35 F-111, 71 F/A-18)	14,050 personnel
None	

ECONOMICS

Inflation 1.5% p.a. (1990–2000)

US$388bn

1.7997–1.9535 Australian dollars

SCORE CARD

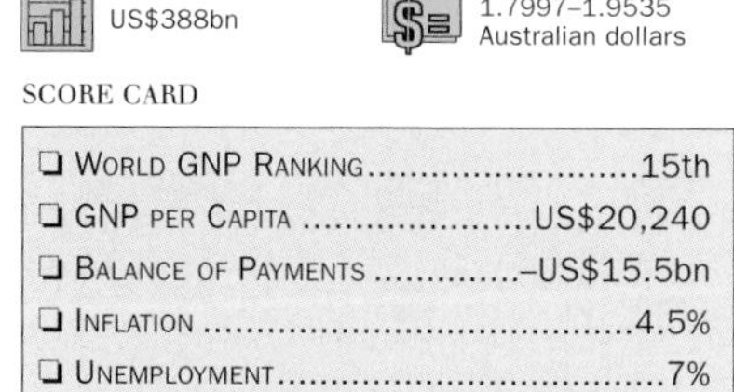

❑ WORLD GNP RANKING	15th
❑ GNP PER CAPITA	US$20,240
❑ BALANCE OF PAYMENTS	–US$15.5bn
❑ INFLATION	4.5%
❑ UNEMPLOYMENT	7%

EXPORTS

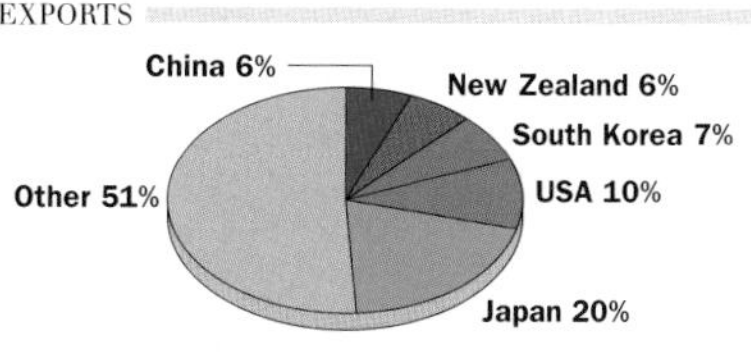
China 6%
New Zealand 6%
South Korea 7%
USA 10%
Japan 20%
Other 51%

IMPORTS

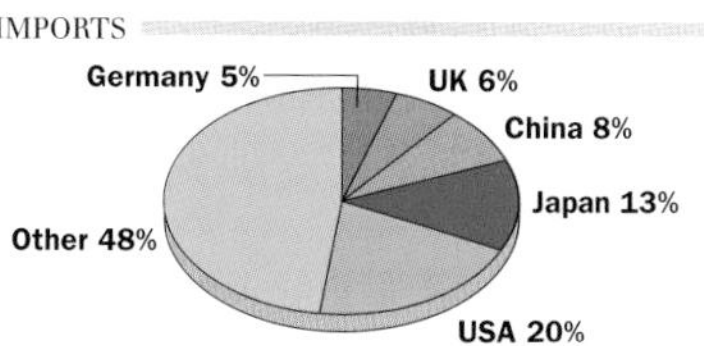
Germany 5%
UK 6%
China 8%
Japan 13%
USA 20%
Other 48%

STRENGTHS

Efficient agricultural and mining industries. Vast mineral deposits. Highly profitable tourist industry with record of dramatic growth. Good history regarding both economic growth and inflation.

ECONOMIC PERFORMANCE INDICATOR

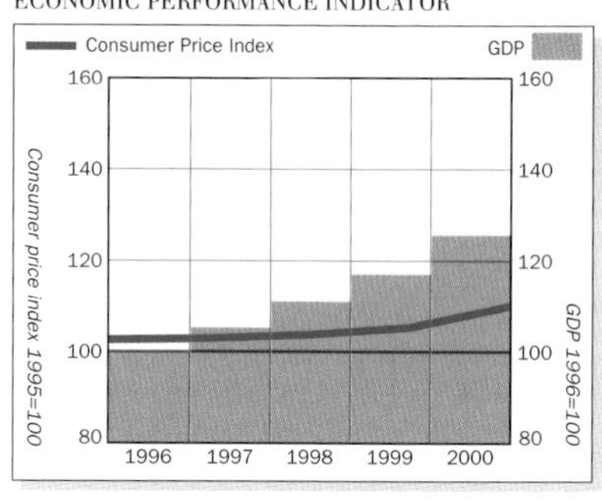
Consumer Price Index
GDP
Consumer price index 1995=100
GDP 1996=100
160
140
120
100
80
1996
1997
1998
1999
2000

WEAKNESSES

May suffer from EU and NAFTA protectionist policies. Political and financial instability in export markets in southeast Asia. Competition from Asian economies with lower wage rates and poorer working conditions. Unemployment likely to remain high. Balance-of-payments deficit.

PROFILE

Companies concentrated during the 1990s on the Asian market, which grew to represent 60% of Australia's trade. They were hit hard when the 1997 Asian financial crisis tipped the region into recession. In order to compete in Asia, Australia's economy has been undergoing massive structural adjustment. The Howard government, like its ALP predecessor, has been dismantling the tariffs that had made Australia one of the most protected OECD economies. Higher unemployment and the collapse of many businesses accompanied this change. However, the economy appeared to have weathered the global economic slowdown in 2001, posting positive growth and lower unemployment by the end of that year.

AUSTRALIA : MAJOR BUSINESSES

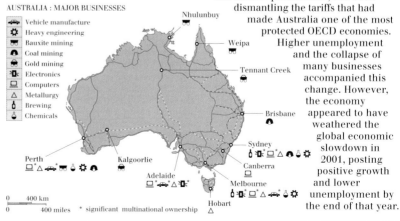

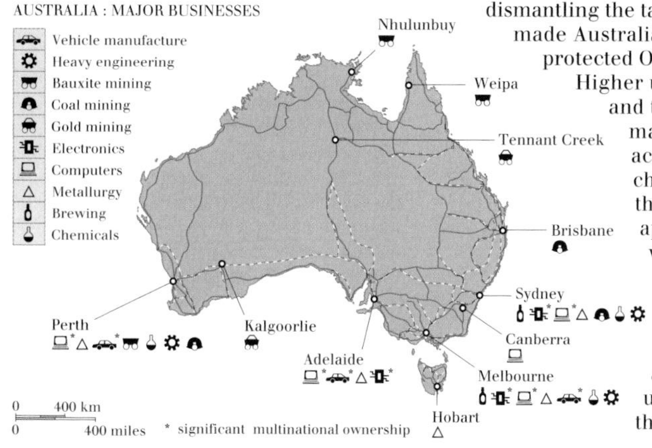
Vehicle manufacture
Heavy engineering
Bauxite mining
Coal mining
Gold mining
Electronics
Computers
Metallurgy
Brewing
Chemicals
Nhulunbuy
Weipa
Tennant Creek
Brisbane
Sydney
Canberra
Melbourne
Hobart
Adelaide
Kalgoorlie
Perth
0 400 km
0 400 miles
* significant multinational ownership

RESOURCES

Electric power 39.7m kw

250,075 tonnes

733,000 b/d (reserves 3.5bn barrels)

120m sheep, 27.6m cattle, 96m chickens

Coal, iron, bauxite, zinc, lead, copper, nickel, opals, gold, uranium

ELECTRICITY GENERATION

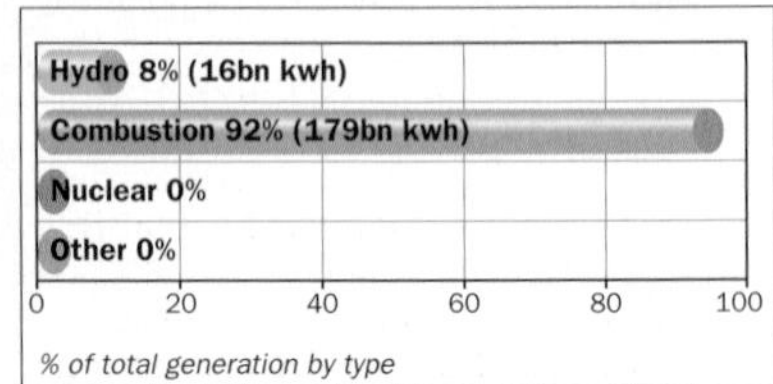
Hydro 8% (16bn kwh)
Combustion 92% (179bn kwh)
Nuclear 0%
Other 0%
0 20 40 60 80 100
% of total generation by type

Australia has one of the world's most important mining industries. It is a world leader in exports of coal, iron ore, gold, bauxite, and copper. Minerals account for a tenth of Australia's GDP and over half of its merchandise export earnings. Since the first discoveries of coal in 1798, mineral production in Australia has risen every year; in the decade to 1992 it doubled. Growth slowed but continued during the 1990s: many new projects are planned. The share of minerals in the total economy is expected to continue growing, but, having benefited from Australia's location close to the markets of southeast Asia, it was left vulnerable following the regional crisis of 1997.

While minerals underpin much of Australia's wealth, there is growing concern at the environmental cost of extraction. There is also ongoing uncertainty over the possibility of Aboriginal claims to land holding valuable minerals. The 1992 "Mabo Judgment" recognized Aboriginal land rights predating European settlement. The 1993 Native Title Act confirmed these rights, and in 1996 the High Court's historic "Wik decision" enabled claims to be made over land which was subject to a "pastoral" lease. But legislation passed in 1998 cut back Aborigines' rights to make such claims.

AUSTRALIA : LAND USE

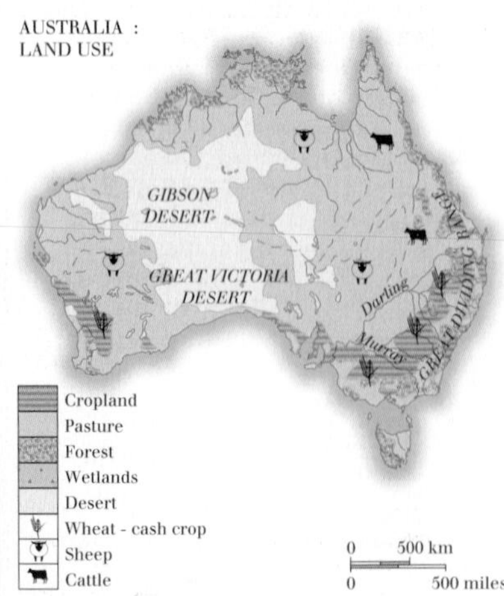
GIBSON DESERT
GREAT VICTORIA DESERT
Darling
Murray
GREAT DIVIDING RANGE
Cropland
Pasture
Forest
Wetlands
Desert
Wheat - cash crop
Sheep
Cattle
0 500 km
0 500 miles

Green Island in the far north of Queensland. It is part of the Great Barrier Reef which stretches 1995 km (1240 miles) down the coast.

ENVIRONMENT

Sustainability rank: 16th

7% (2% partially protected)

17.7 tonnes per capita

ENVIRONMENTAL TREATIES

Yes	Yes	Yes
Yes	Yes	No

Australians are among the most environmentally conscious in the world, but the government has strongly resisted any commitment to cutting greenhouse gas emissions to help limit global warming. It has not even endorsed the 1997 Kyoto protocol, which would in fact allow Australia to increase emissions by up to 8% until 2010, whereas most industrialized countries had to commit to cuts.

Green issues are dominated by Greenpeace and the less radical Australian Conservation Foundation (ACF), which has been highly critical of Australia's record of destroying natural habitats.

MEDIA

TV ownership high

Daily newspaper circulation 296 per 1000 people

PUBLISHING AND BROADCAST MEDIA

There are 65 daily newspapers, mostly state circulation. The leading papers are the *Sydney Morning Herald* and *The Australian*

1 state-owned service, 44 independent commercial stations

6 state-owned networks, 166 independent commercial stations

The Australian press is firmly in the grip of press barons such as Rupert Murdoch and Kerry Packer, although cross-media rules prevent the ownership of newspapers and TV channels in the same city. Public-sector broadcasting remains dominated by the politically neutral Australian Broadcasting Corporation (ABC), which receives complaints about its coverage from all main parties.

CRIME

No death penalty

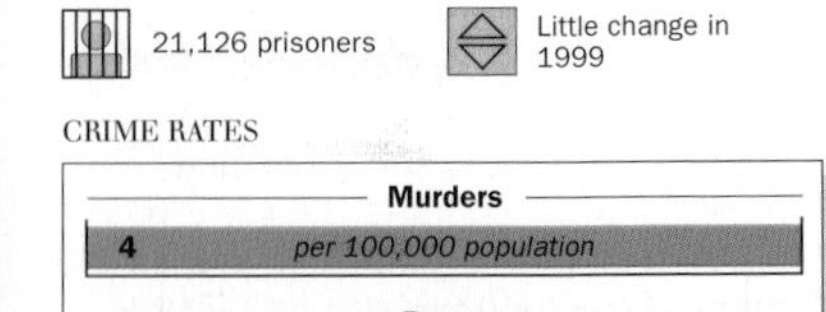

21,126 prisoners

Little change in 1999

CRIME RATES

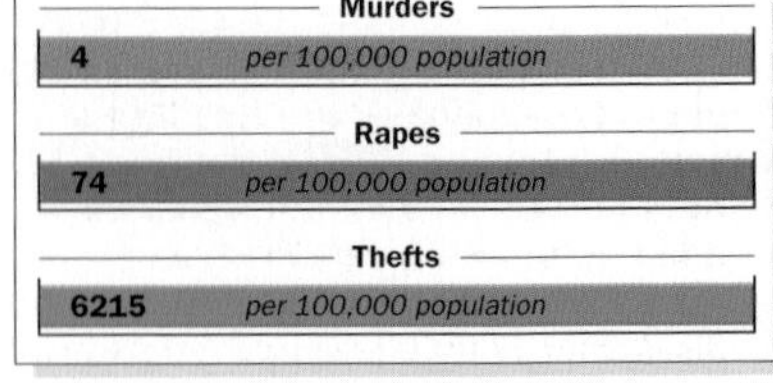
Murders
4 per 100,000 population
Rapes
74 per 100,000 population
Thefts
6215 per 100,000 population

Each state has its own police force and court system. The High Court and Family Court both have national jurisdiction. Since the 1970s, the legal system has been placing greater emphasis on individual rights. The disproportionate number of Aboriginal deaths in custody is of concern, as are rising narcotics-related offenses. Australia is active in narcotics control throughout southeast Asia. Gun control laws were strengthened following the 1996 Port Arthur shooting, when a lone gunman killed 35 people. In 1997 the Wood inquiry uncovered widespread police corruption in New South Wales and led to major reforms.

EDUCATION

School leaving age: 15

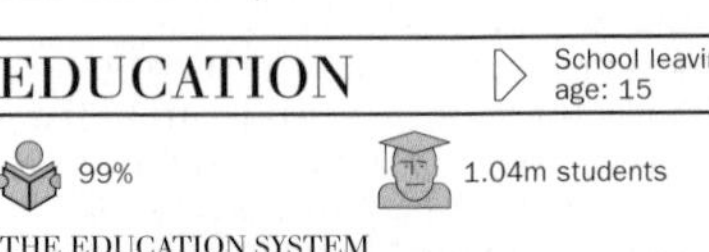

99%

1.04m students

THE EDUCATION SYSTEM

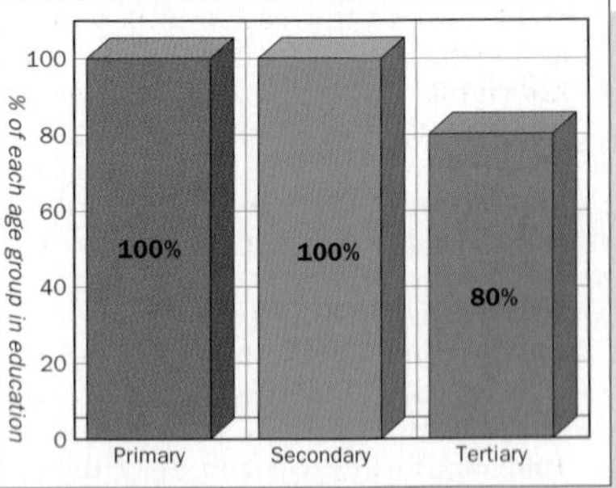
100
80
60
40
20
0
% of each age group in education
100%
100%
80%
Primary
Secondary
Tertiary

Education in Australia is a state responsibility, except in Canberra (where it is funded by the federal government). State education departments run the government schools and set the policies for educational practice and standards for all schools. Nongovernment schools, run by religious and other groups, exist in all states. Special provision is made for inaccessible outback areas, with recent moves to bring new technologies to the bush.

Schooling is compulsory from age 5–6 to age 15–16 in all states. After their final year at school, students sit for the Higher School Certificate. Universities are independent of state control and are funded by the federal government.

ABORIGINAL RIGHTS

Central to Aboriginal culture and beliefs is the relationship with tribal lands. The Aborigines' land rights campaign, like their civil rights campaign, began gathering momentum in the 1960s, when they were first included in the census and won the right to vote. Now a key political issue, it has far-reaching implications for national identity, for the country's most disadvantaged people, and for powerful mining and farming interests.

Background

In the millennia before the arrival of Europeans, Aboriginal peoples ranged widely over their land. Their number was then greatly reduced by exposure to disease and by conflict. Settlers acquired land without reference to any preexisting rights of tribal peoples, based on the legal doctrine that before 1788 the land was *terra nullius* – belonging to no one. Those Aborigines still living a traditional lifestyle were largely confined to reserves.

Land rights campaigns won an initial breakthrough when land councils were set up, first in the Northern Territory in 1976, then in other states. These provided a structure for holding freehold title to land in trust for its tribal inhabitants. However, in the absence of a uniform national land rights policy conservative states such as Queensland – with the largest Aboriginal population – resisted moves to consolidate Aborigine ownership. Particular flashpoints arose when mining companies were granted concessions to exploit sacred sites.

Native Title Acts in the 1990s

The first nationwide Native Title Act was introduced in 1993. It recognized the new situation created by a crucial 1992 court ruling. The so-called Mabo Judgment established that rights to

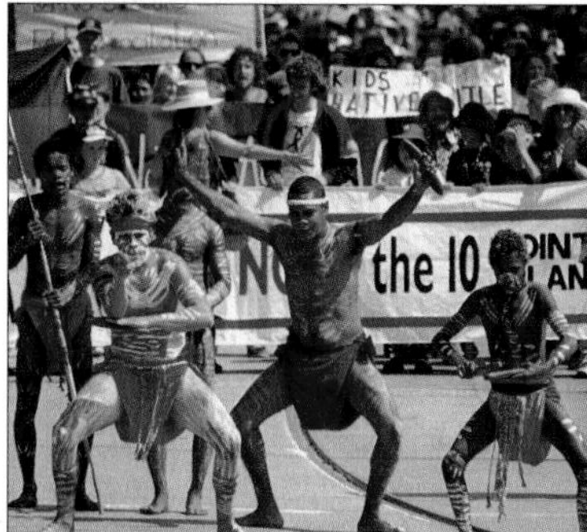
KIDS
the 10

Native dancers from Kuranda, *northern Queensland, perform a traditional dance in front of protesters marching for increased Aboriginal land rights.*

An Aboriginal elder *surveys his native land in Western Australia.*

land ownership did indeed exist in common law based on native title, effectively reversing the concept of *terra nullius* and recognizing the pre-1788 Aboriginal occupancy.

The 1993 law specified that native title existed for all Crown land held by federal or by state government, unless it had specifically been extinguished. Native title met powerful resistance from mining companies, especially in Western Australia. In 1996, a court case brought by the Wik people of the Cape York Peninsula in Queensland took the matter further. The Wik ruling said that native title still coexisted with the rights of farmers who had long leases from the Crown on huge tracts of grazing land, granted earlier in the century. Changes of use on such land (to allow mining, cash crops, and tourist developments) would therefore require consultation with Aborigines.

The Liberal–National government responded with a plan to protect leaseholders. Its own Native Title Amendment Bill, tightly restricting Aborigine claims, split the country but was eventually passed in 1998.

As Australia's indigenous population (410,000 and rising) develops a growing sense of pride and assertiveness, court battles continue over mining developments on sensitive tribal sites. Meanwhile, Liberal prime minister John Howard has resisted pressure to go beyond expressions of "regret" and make a formal government apology for the way in which white settlers mistreated Aboriginal people. Campaigners for such an apology see it as central to a national reconciliation agreement, especially after the recent controversy over the "stolen generation" – Aboriginal children forcibly removed from their families and brought up by whites under an assimilation policy which continued until the late 1960s.

HEALTH

Welfare state health benefits

1 per 400 people

Heart, cerebrovascular, and respiratory diseases, cancers

Australia's extensive public health service has among the highest standards in the world. Hospital waiting lists are short. Outback areas are serviced by the efficient Royal Flying Doctor Service. While vigilance continues in the areas of hygiene, nutrition, and general living standards, health authorities have targeted heart disease, injury prevention, personal fitness, Aboriginal health, and the prevention of cancers – particularly lung, cervical, breast, and skin cancers – as current priorities. Incentives to encourage private health insurance, introduced during the 1990s, sparked fears over public health funding and quality.

SPENDING

GDP/cap. increase

CONSUMPTION AND SPENDING

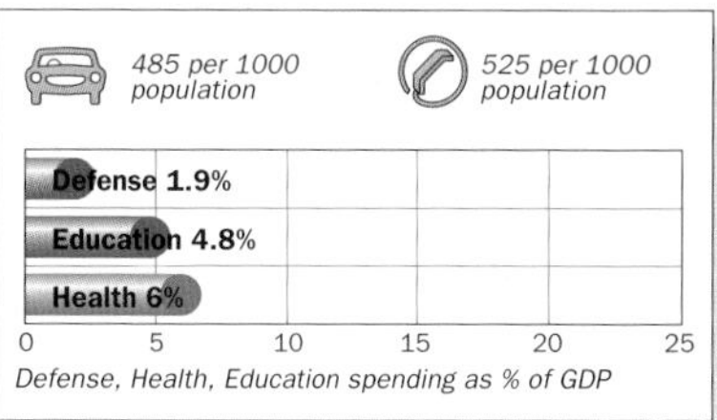
485 per 1000 population
525 per 1000 population
Defense 1.9%
Education 4.8%
Health 6%
0
5
10
15
20
25
Defense, Health, Education spending as % of GDP

Australians traditionally enjoyed reasonable equality of wealth distribution. A large proportion of families own two cars and have relatively high disposable incomes, and a benign climate helps most people to live comfortably. However, high unemployment during the 1990s recession widened the gap between rich and poor, and Australia slipped down the world standard of living list for a few years. The incidence of homelessness, critical poverty, and child neglect due to poverty have increased in recent years.

WORLD RANKING

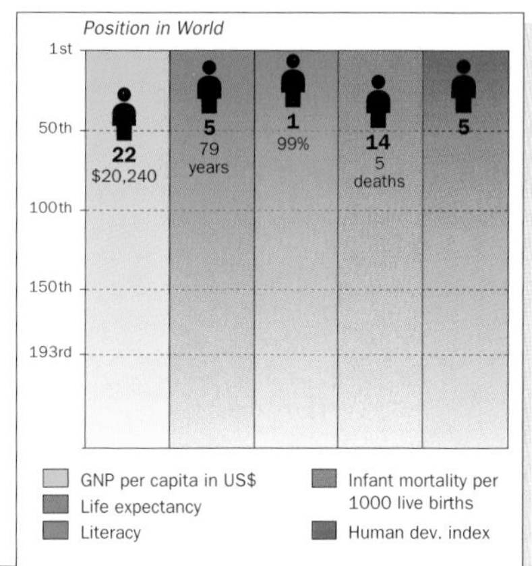
Position in World
1st
50th
100th
150th
193rd
22
$20,240
5
79
years
1
99%
14
5
deaths
5
GNP per capita in US$
Life expectancy
Literacy
Infant mortality per 1000 live births
Human dev. index

A

AUSTRIA

Europe
EUROPE
Germany
Czech Rep.
AUSTRIA
Hungary
Switz.
Italy
Slovenia

OFFICIAL NAME: Republic of Austria **CAPITAL:** Vienna **POPULATION:** 8.1 million
CURRENCY: Euro (Schilling until 2002) **OFFICIAL LANGUAGE:** German

1918 1919

Oct 26

A

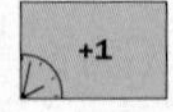
+1

+43

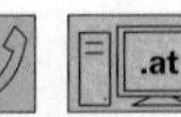
.at

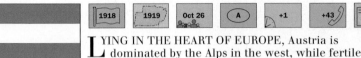

LYING IN THE HEART OF EUROPE, Austria is dominated by the Alps in the west, while fertile plains make up the east and north. A separate republic after the collapse of the Austro-Hungarian Empire, Austria was absorbed into Hitler's Germany in 1938. It regained independence in 1955 after the departure of the last Soviet troops from the Allied Occupation Force. Its economy encompasses successful high-tech sectors, a tourist industry which attracts wealthier visitors, and a strong agricultural base. Joining the EU in 1995, in 2002 it was one of 12 EU states to adopt the euro.

CLIMATE

Mountain/continental

WEATHER CHART FOR VIENNA

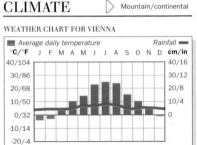

Average daily temperature · Rainfall
°C/°F J F M A M J J A S O N D cm/in
40/104 30/86 20/68 10/50 0/32 -10/14 -20/-4
40/16 30/12 20/8 10/4 0

Austria has a temperate continental climate. Alpine areas experience colder temperatures and higher precipitation.

TRANSPORTATION

Drive on right

Schwechat, Vienna 11.9m passengers

24 ships 89,600 grt

THE TRANSPORTATION NETWORK

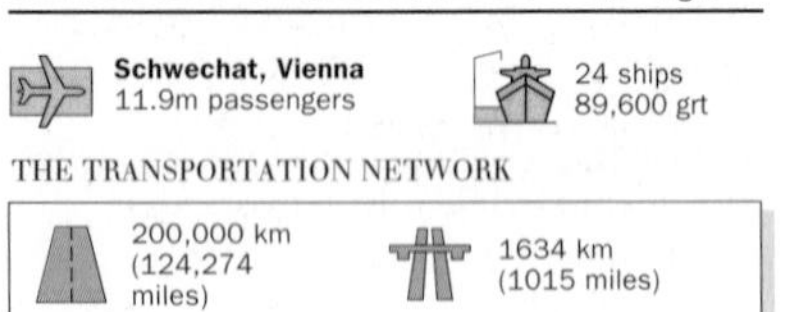
200,000 km (124,274 miles)
1634 km (1015 miles)
5740 km (3567 miles)
358 km (222 miles)

Austria's central location in Europe has encouraged the development of a sophisticated communications and transportation network.

The Tirol is situated in the heart of Austria's Alps. It is the most mountainous region of all and attracts both winter and summer visitors.

TOURISM

Visitors : Population 2.2:1

18.2m visitors

Up 1% in 2001

MAIN TOURIST ARRIVALS

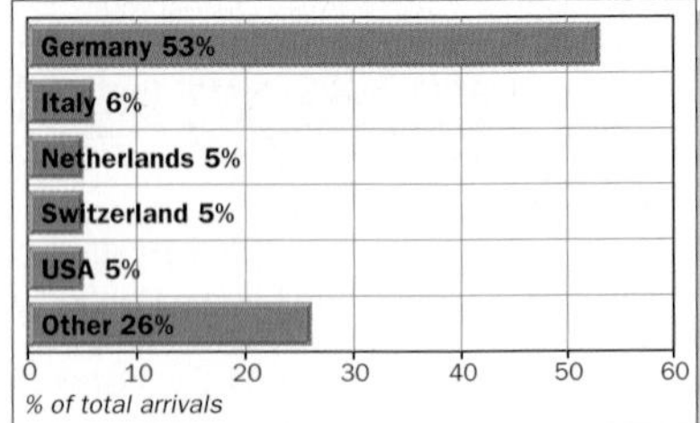
Germany 53%
Italy 6%
Netherlands 5%
Switzerland 5%
USA 5%
Other 26%
0 10 20 30 40 50 60
% of total arrivals

Well-developed Alpine skiing and winter sports resorts account for almost one-third of the country's total tourist earnings. Many resorts, such as St. Anton and Kitzbühel, cater for the top end of the market. In the summer season, which peaks in July and August, tourists visit the scenic Tirol and the lakes around Bad Ischl. Year-round major attractions are Vienna, with its coffee houses and the Prater park (whose Ferris wheel was immortalized in *The Third Man*), and Salzburg, Austria's second city. The latter is internationally famous for its summer music festival and as the birthplace of Mozart.

AUSTRIA

Total Land Area : 83 858 sq. km (32 378 sq. miles)

LAND HEIGHT: 3000m/9843ft · 2000m/6562ft · 1000m/3281ft · 500m/1640ft · 200m/656ft · Sea Level

POPULATION: over 1 000 000 · over 500 000 · over 100 000 · over 50 000 · over 10 000

0 50 km · 0 50 miles

PEOPLE

Pop. density medium

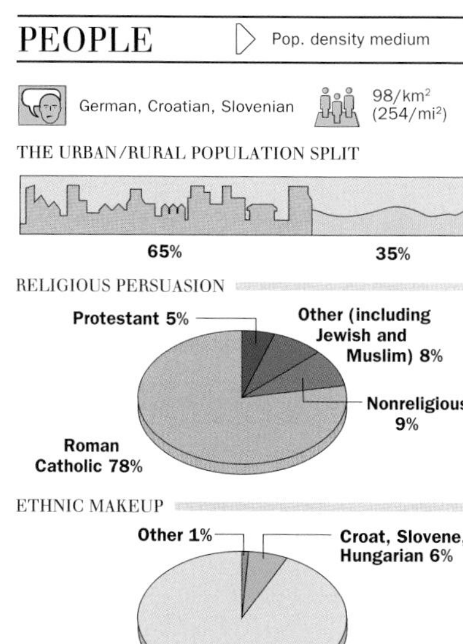
German, Croatian, Slovenian
98/km2 (254/mi2)
THE URBAN/RURAL POPULATION SPLIT
65%
35%
RELIGIOUS PERSUASION
Protestant 5%
Other (including Jewish and Muslim) 8%
Nonreligious 9%
Roman Catholic 78%
ETHNIC MAKEUP
Other 1%
Croat, Slovene, Hungarian 6%
German 93%

Austrian society is homogeneous. Almost all Austrians are German speakers, although Austrians like to stress their distinctive identity in relation to Germany. Minorities are few; there are some ethnic Slovenes, Croats, and Hungarians in the south and east, as well as some Roma communities. These minorities have been augmented by large numbers of immigrants from eastern Europe and one-time refugees from the former Yugoslavia. The result has been a perceptible increase in ethnic tension, particularly as the far right claims that migrants are taking jobs from the local population.

The nuclear family is the norm in Austria. It is common for both parents to work. While gender equality is enshrined in the constitution, in practice society is still strongly patriarchal. Compared with the rest of Europe, few women enter politics.

Young Austrians tend to live in their parental home until they marry. This practice reflects the long time taken to complete university degrees, for which students do not receive maintenance grants. Austrians marry at a younger age than the European average. Nominally a Catholic country, Austria is socially less conservative than some Catholic German *Länder*.

POPULATION AGE BREAKDOWN

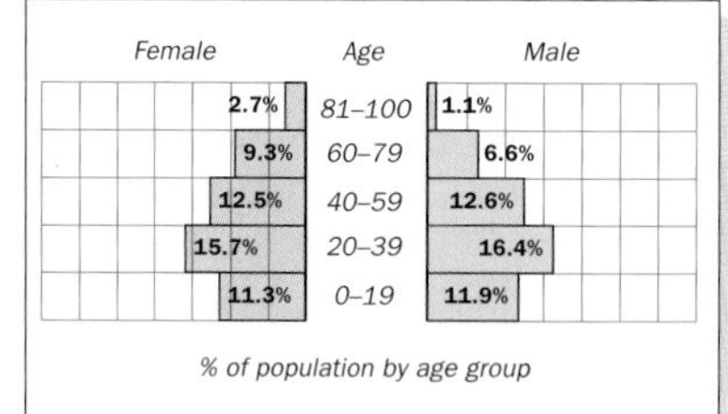
Female
Age
Male
2.7% 81–100 1.1%
9.3% 60–79 6.6%
12.5% 40–59 12.6%
15.7% 20–39 16.4%
11.3% 0–19 11.9%
% of population by age group

POLITICS

Multiparty elections

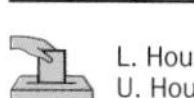

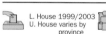

L. House 1999/2003
U. House varies by province

President Thomas Klestil

AT THE LAST ELECTION

National Council 183 seats

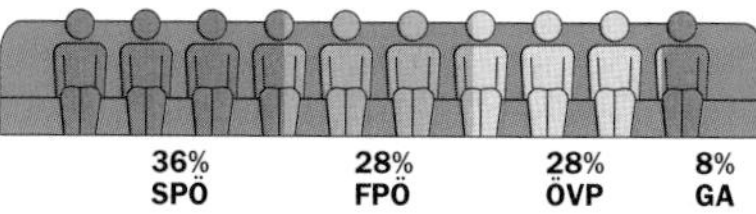

36% **SPÖ** **28%** **FPÖ** **28%** **ÖVP** **8%** **GA**

SPÖ = Social Democratic Party of Austria **FPÖ** = Freedom Party of Austria **ÖVP** = Austrian People's Party **GA** = Green Alternative

Federal Council 64 seats

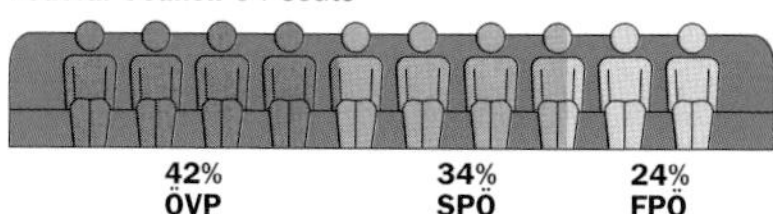

42% **ÖVP** **34%** **SPÖ** **24%** **FPÖ**

Austria is a federal, multiparty democracy. The chancellor (premier) holds real executive power.

PROFILE

The long-established "grand coalition" between the socialist SPÖ and the conservative ÖVP governed Austria almost without interruption from the 1950s until 1999. Their hold on power, with no real alternative, reached into all areas of public life. The far-right FPÖ achieved a breakthrough in elections in October 1999, when it came equal second with the ÖVP. A new right-wing coalition of the ÖVP and the FPÖ was met with regional and international criticism. FPÖ leader Jörg Haider drew strong condemnation, having openly expressed admiration for some of Hitler's policies. He remained a driving force behind the party despite resigning formal leadership in May 2000. Later in that year, local elections showed dwindling support for the FPÖ, prompting talk of Haider's return to the forefront of Austrian politics. By 2002 the ruling coalition was under strain. The FPÖ adopted a far more skeptical approach to EU enlargement, calling for a veto of the Czech Republic's membership bid unless the Czech Temelín nuclear energy plant was closed.

The nine provincial assemblies and governments have considerable powers. Vienna, with provincial status, has long been dominated by the SPÖ.

MAIN POLITICAL ISSUE
Relations with EU partners

Austria's population is divided over the merits of EU membership. While the farming lobby remains apprehensive of EU agricultural policies, the country as a whole has benefited from lower food prices and greater consumer choice. Fears that membership is eroding national identity, as well as encouraging an influx of cheap east European labor, provided an electoral basis for the nationalist FPÖ. In turn, EU states responded to the FPÖ's membership of the ruling coalition in 2000 by imposing diplomatic sanctions for seven months.

Dr. Thomas Klestil, *the ÖVP candidate, was elected president in 1992.*

Wolfgang Schüssel, *ÖVP chancellor since 2000, in coalition with the far right.*

WORLD AFFAIRS

Joined UN in 1955

EU CE

PfP

OECD

OSCE

Despite the importance of relations with Germany, Austria's powerful northern neighbor and main trading partner, there has been a conscious policy of stressing Austria's independence and creating some diplomatic distance. Austria is keen to maintain its relationship with the US, which is reinforced by Austria's role as supplier of small arms to the US Army. The inclusion of the far right in the Austrian government in 2000 provoked the imposition of diplomatic sanctions by EU states for seven months.

Austria's status as a neutral state has begun to be questioned since it joined the EU and NATO's Partnerships for Peace program. Austria is part of the Schengen Convention, ending border controls with other participating EU members. Its geopolitical position gives it considerable influence in eastern Europe, and exports to the region trebled in the 1990s. The ÖVP has strongly supported the eastward enlargement of the EU, but the FPÖ is more hostile.

AID

Donor

$423m (donations)

Down 20% in 2000

New projects are now assessed for their impact on the environment and on gender issues. Poland is the largest recipient of official aid and Bosnia and Herzegovina the second. A major exporter to the former Yugoslavia before the wars there in the 1990s, Austria has a key role in regional reconstruction.

A

CHRONOLOGY

Austria came under the control of the Habsburgs in 1273. In 1867, the Dual Monarchy of Austria-Hungary was formed under Habsburg rule. Defeat in World War I in 1918 led to the breakup of the Habsburg empire and the formation of the Republic of Austria.

- ❑ **1934** Chancellor Dollfuss dismisses parliament and starts imprisoning social democrats, communists, and National Socialist (Nazi) Party members. Nazis attempt coup.
- ❑ **1938** The Anschluss – Austria incorporated into Germany by Hitler.
- ❑ **1945** Austria occupied by Soviet, British, US, and French forces. Elections result in ÖVP–SPÖ coalition.
- ❑ **1950** Attempted coup by Communist Party fails. Marshall Aid helps economic recovery.
- ❑ **1955** Occupying troops withdrawn. Austria recognized as a neutral sovereign state.
- ❑ **1971** SPÖ government formed under Chancellor Bruno Kreisky who dominates Austrian politics for 12 years.
- ❑ **1983** Socialists and the FPÖ form a coalition government under Fred Sinowatz.
- ❑ **1986** Kurt Waldheim, former UN secretary-general, elected president, despite war crimes allegations. Franz Vranitzky replaces Sinowatz as federal chancellor. Nationalist Jörg Haider becomes FPÖ leader, prompting the SPÖ to pull out of government. Elections produce stalemate. Return to "grand coalition" of SPÖ–ÖVP.
- ❑ **1990** ÖVP loses support in elections.
- ❑ **1992** Thomas Klestil (ÖVP) elected president. Elections confirm some traditional ÖVP supporters defecting to FPÖ.
- ❑ **1995** Austria joins EU. Elections after coalition disagreement over budget; SPÖ and ÖVP increase representation; "grand coalition" re-forms in early 1996.
- ❑ **1997** Vranitzky resigns; replaced by Viktor Klima.
- ❑ **1998** Klestil reelected president.
- ❑ **1999** Haider's FPÖ wins 40% of votes in Carinthia regional poll, is equal second with ÖVP in general election in October; SPÖ remains as largest party.
- ❑ **2000** ÖVP accepts FPÖ into coalition, with Wolfgang Schüssel as chancellor; political crisis. EU imposes diplomatic sanctions, lifted after seven months.
- ❑ **2002** Euro fully adopted.

DEFENSE

▷ Compulsory military service

 $1.61bn

 Down 5% in 2000

Under the 1955 State Treaty, which granted Austria its full independence, Austria is neutral, although it has participated in NATO's Partnerships for Peace program since 1995.

Despite the small size of its own forces, the Austrian arms industry is strong. It not only meets most of the hardware needs of its own army, but also exports arms to the US and other countries.

AUSTRIAN ARMED FORCES

Army	277 main battle tanks (163 M-60A3, 114 *Leopard* 2A4)	34,600 personnel
Navy	None	
Air force	52 combat aircraft (23 SAAB J-350e)	6500 personnel
Missiles	None	

ECONOMICS

▷ Inflation 2% p.a. (1990–2000)

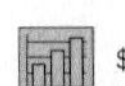 $205bn

 1.0651–1.1231 euros

SCORE CARD

❑ World GNP Ranking	22nd
❑ GNP per Capita	$25,220
❑ Balance of Payments	–$5.21bn
❑ Inflation	2.4%
❑ Unemployment	6%

EXPORTS

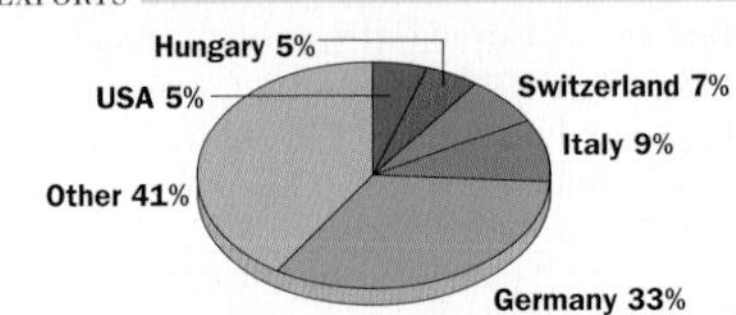

Hungary 5%
USA 5%
Switzerland 7%
Italy 9%
Other 41%
Germany 33%

IMPORTS

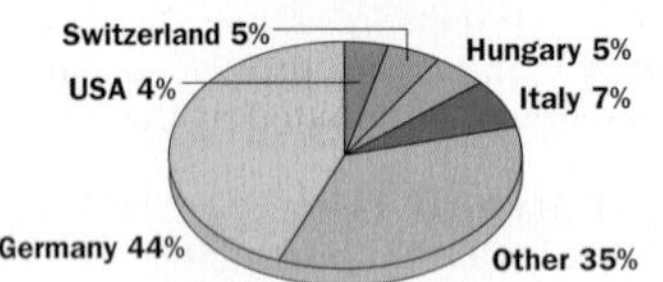

Switzerland 5%
USA 4%
Hungary 5%
Italy 7%
Germany 44%
Other 35%

Strengths

Large manufacturing base. Strong chemical and petrochemical industries. Electrical engineering sector, textiles, and wood processing industries. Highly skilled labor force. Tourism is an important foreign currency earner.

Weaknesses

Lacks natural resources. Reliant on imported raw materials, particularly oil and gas. Process of introducing greater competition and deregulation has been slow.

Profile

Austria's industrial and high-tech sector is highly developed and contributes over a quarter of GDP. Some services, notably tourism, are highly sophisticated and profitable, although tourism receipts have been down in recent years.

A recession in the early 1990s was reversed by a rapid increase in exports to eastern Europe and Germany and by increased domestic demand.

ECONOMIC PERFORMANCE INDICATOR

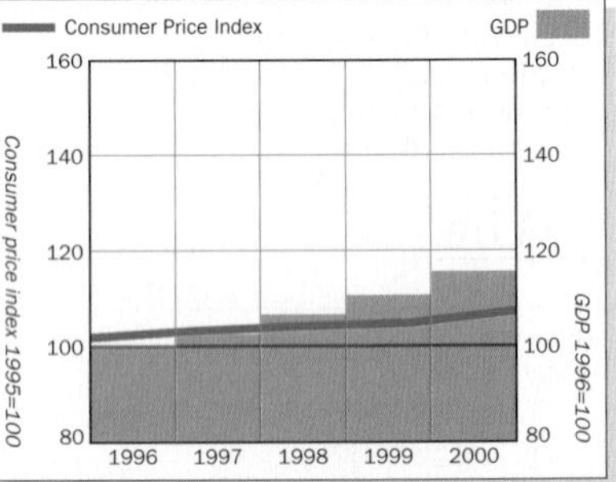

Consumer Price Index
GDP
160
140
120
100
80
1996
1997
1998
1999
2000
Consumer price index 1995=100
GDP 1996=100

There have been benefits from EU membership since 1995. Prices for many products, particularly food and books, have fallen. The Austrian labor market has also seen an influx of migrant labor more willing to accept flexible working arrangements and lower wages. Foreign investment has increased, as more multinationals locate their headquarters for east European operations in Austria. A far-reaching fiscal stabilization program instituted after accession enabled Austria to meet the economic convergence criteria necessary for it to join the final stage of economic and monetary union, fully adopting the euro in January 2002.

AUSTRIA : MAJOR BUSINESSES

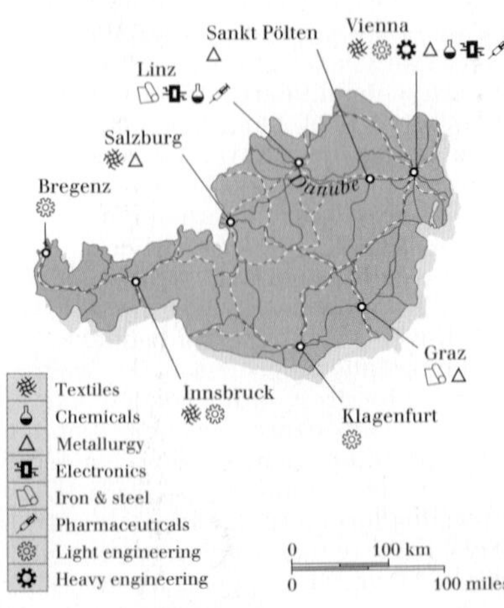

Sankt Pölten
Vienna
Linz
Salzburg
Bregenz
Danube
Graz
Innsbruck
Klagenfurt
Textiles
Chemicals
Metallurgy
Electronics
Iron & steel
Pharmaceuticals
Light engineering
Heavy engineering
0 100 km
0 100 miles

RESOURCES

Electric power 18.3m kw

3502 tonnes

19,228 b/d (reserves 77m barrels)

3.43m pigs, 2.16m cattle, 11.1m chickens

Iron, coal, magnesite, zinc, lead

ELECTRICITY GENERATION

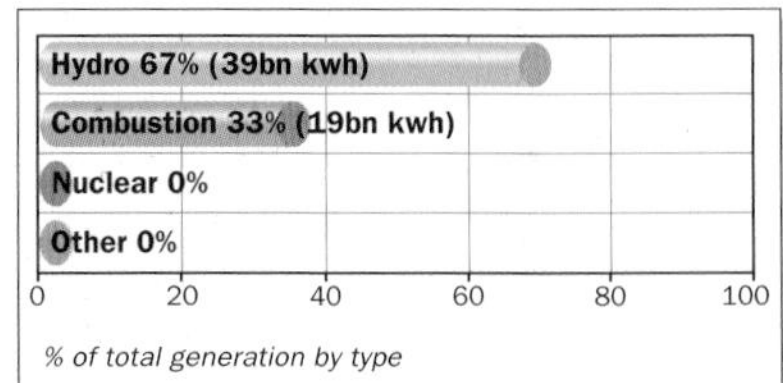
Hydro 67% (39bn kwh)
Combustion 33% (19bn kwh)
Nuclear 0%
Other 0%
0
20
40
60
80
100
% of total generation by type

AUSTRIA : LAND USE

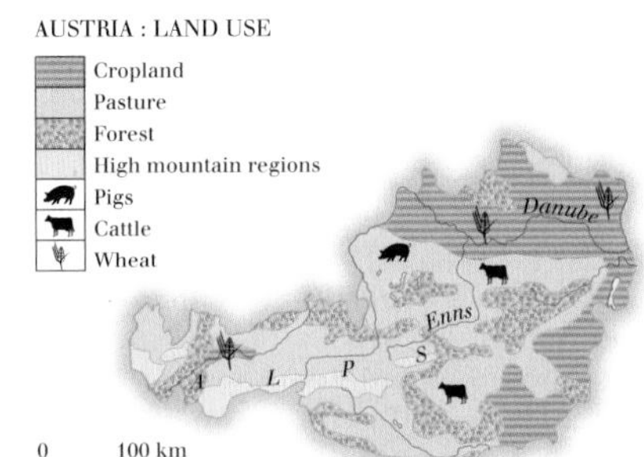
Cropland
Pasture
Forest
High mountain regions
Pigs
Cattle
Wheat
Danube
Enns
I
L
P
S
0
100 km
0
100 miles

Austria has few resources. It lacks significant oil, coal, and gas deposits and has to import a large amount of its energy. Russia is a key energy supplier, and gas is provided via pipelines running through the Czech Republic and Slovakia. Oil is imported up the Danube. Russia and Germany are the major suppliers of iron ore and raw steel for Austria's industry.

ENVIRONMENT

Sustainability rank: 7th

30% (24% partially protected)

7.9 tonnes per capita

ENVIRONMENTAL TREATIES

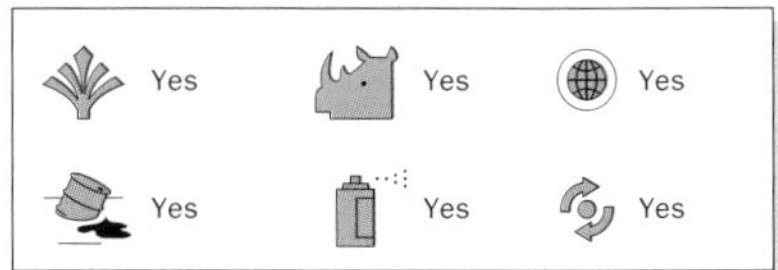
Yes
Yes
Yes
Yes
Yes
Yes

Environmental awareness is high and the government invests nearly 3% of GDP in environmental protection. Roughly half of all domestic waste is separated for recycling, with heavy fines for failing to observe regulations. The safety of nuclear reactors in the neighboring Czech Republic, Slovakia, and Slovenia is a major concern.

MEDIA

TV ownership high

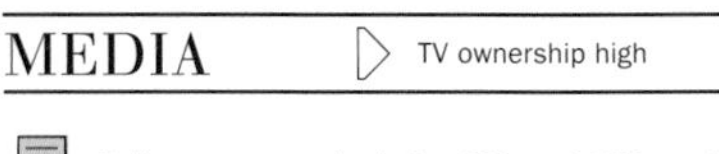

Daily newspaper circulation 296 per 1000 people

PUBLISHING AND BROADCAST MEDIA

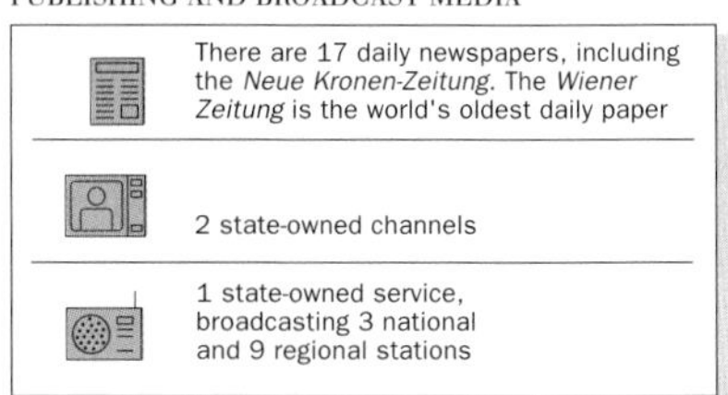
There are 17 daily newspapers, including the *Neue Kronen-Zeitung*. The *Wiener Zeitung* is the world's oldest daily paper
2 state-owned channels
1 state-owned service, broadcasting 3 national and 9 regional stations

TV and radio are controlled by the Austrian Broadcasting Company (ÖRF), which has a politically appointed general director. Private radio stations are restricted to regional broadcasting. Cable TV is licensed by ÖRF to prevent it taking viewers from existing stations. A German-language satellite channel is run jointly with German and Swiss TV.

CRIME

No death penalty

6896 prisoners

Up 3% in 1999

CRIME RATES

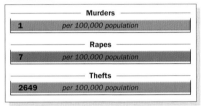

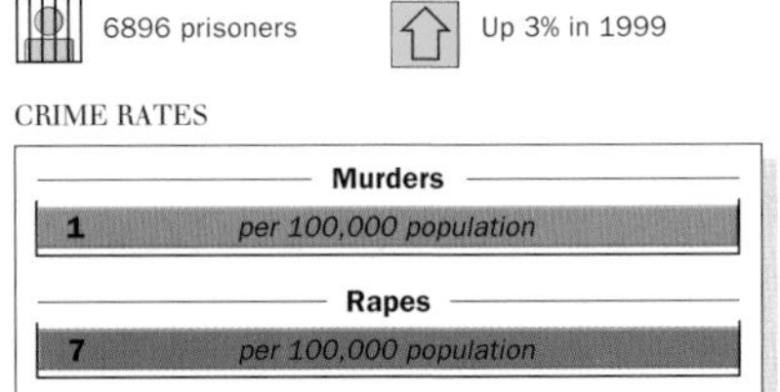
Murders
1 per 100,000 population
Rapes
7 per 100,000 population
Thefts
2649 per 100,000 population

Austria's crime rate is below Europe's average. However, the number of burglaries is rising. The arrival of the Russian mafia in Vienna has led to an increase in money laundering.

EDUCATION

School leaving age: 15

99%

230,030 students

THE EDUCATION SYSTEM

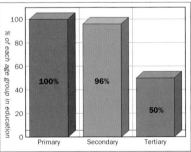

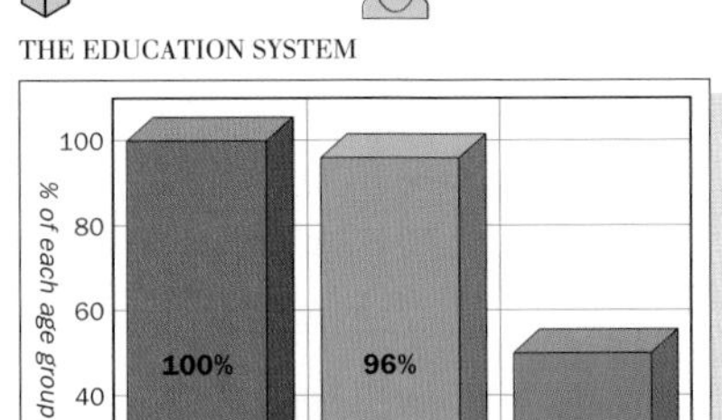
100
80
60
40
20
0
% of each age group in education
100%
96%
50%
Primary
Secondary
Tertiary

Of total government expenditure some 10% is spent on education. Children are streamed into two types of secondary school according to their academic ability. Those in a *Gymnasium* (11–18) take the *Reifeprüfung* or *Matura* which determines entry to university, but children in a *Hauptschule* (11–15) do not. The universities are oversubscribed; students take six years or more to finish their first degrees.

HEALTH

Welfare state health benefits

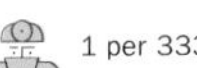

1 per 333 people

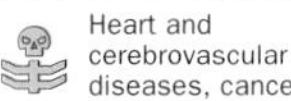

Heart and cerebrovascular diseases, cancers

Austria has relatively high levels of spending on health, accounting for 13% of total government expenditure. However, private spending on health is increasing and accounts for nearly one-third of the total, the highest proportion of any EU state, as patients increasingly choose to use the private health sector to avoid waiting lists for operations.

SPENDING

GDP/cap. increase

CONSUMPTION AND SPENDING

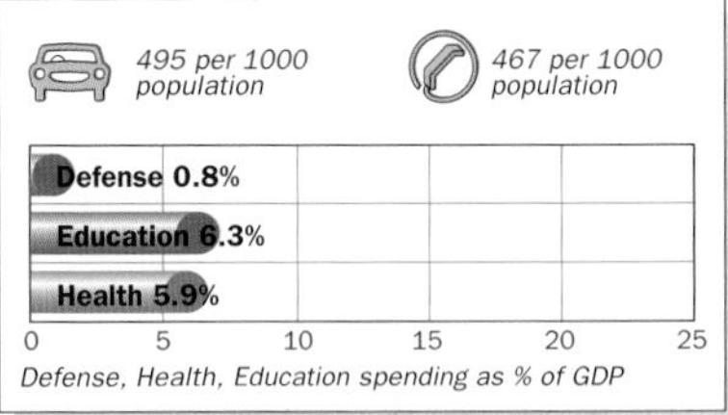
495 per 1000 population
467 per 1000 population
Defense 0.8%
Education 6.3%
Health 5.9%
0
5
10
15
20
25
Defense, Health, Education spending as % of GDP

Despite four decades of mainly left-of-center government, Austria has retained many of its traditional social divisions. Inherited wealth is still respected above earned wealth, and social mobility is lower than in neighboring Germany. Austrians have the highest savings rate of any OECD country. Relatively few Austrians own stocks and shares, and most companies are dominated by a single shareholder. Limited amounts are invested in property. Legislation in 2000 banned anonymous savings accounts, a system unique in the EU to Austria which, it had been argued, encouraged money laundering and insider dealing. Government bonds offer low rates of interest and the property market is weak; many people, particularly in Vienna, tend to rent rather than buy their apartments. Refugees from conflicts in the former Yugoslavia form the poorest group in Austrian society.

WORLD RANKING

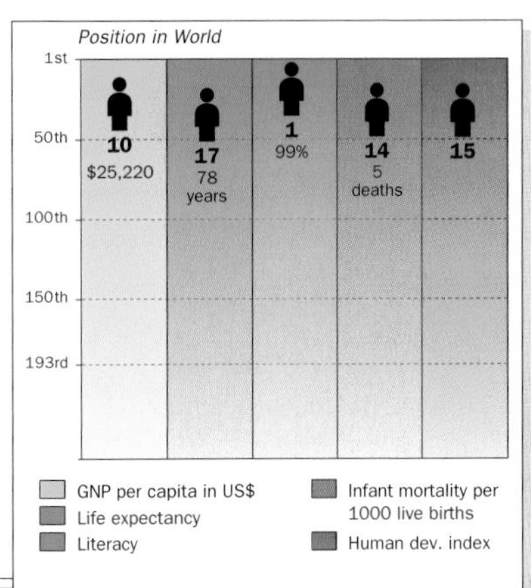
Position in World
1st
50th
100th
150th
193rd
10
$25,220
17
78 years
1
99%
14
5 deaths
15
GNP per capita in US$
Life expectancy
Literacy
Infant mortality per 1000 live births
Human dev. index

A

AZERBAIJAN

ASIA

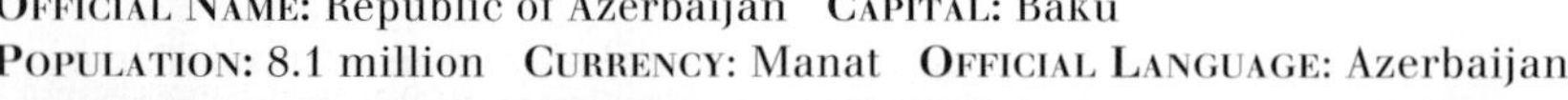

OFFICIAL NAME: Republic of Azerbaijan **CAPITAL:** Baku
POPULATION: 8.1 million **CURRENCY:** Manat **OFFICIAL LANGUAGE:** Azerbaijani

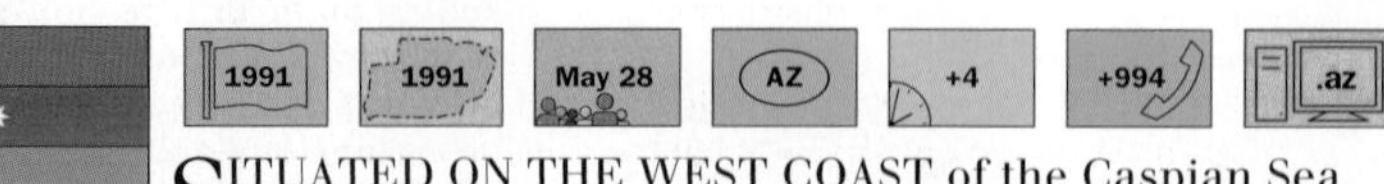
1991
1991
May 28
AZ
+4
+994
.az

SITUATED ON THE WEST COAST of the Caspian Sea, Azerbaijan was the first Soviet republic to declare independence. The issue of the disputed enclave of Nagorno Karabakh, whose Armenian population seeks secession, led to full-scale war (1988–1994) and is still a dominant concern. Over 200,000 refugees, and more than twice as many internally displaced, added to the problems of the troubled economy. Azerbaijan's oil wealth, however, gives it long-term potential.

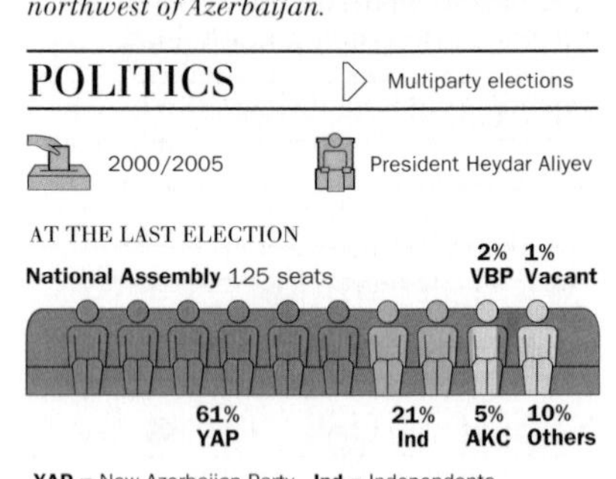

Landscape typical of the Lesser Caucasus *mountains near Qazax in the extreme northwest of Azerbaijan.*

CLIMATE

Mountain/steppe

WEATHER CHART FOR BAKU

Average daily temperature — Rainfall
°C/°F J F M A M J J A S O N D cm/in
40/104, 30/86, 20/68, 10/50, 0/32, -10/14, -20/-4 — 40/16, 30/12, 20/8, 10/4, 0

Coastal areas are subtropical, but bitter winters inland have become a life-or-death issue for thousands of refugees.

TRANSPORTATION

Drive on right

Baku

284 ships 647,022 grt

THE TRANSPORTATION NETWORK

23,057 km (14,327 miles)

None

2116 km (1315 miles)

None

Buses provide the most efficient public transportation. Access to the Naxçivan enclave is by air or via Iran.

TOURISM

Visitors : Population 1:12

681,000 visitors

Up 13% in 2000

MAIN TOURIST ARRIVALS

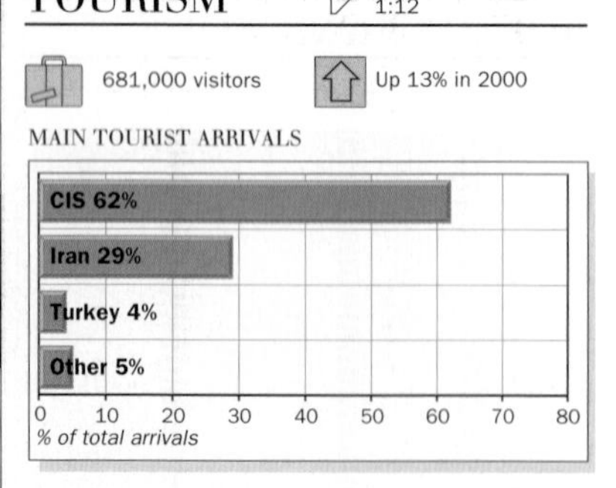
CIS 62%
Iran 29%
Turkey 4%
Other 5%
0 10 20 30 40 50 60 70 80
% of total arrivals

Because of the Nagorno Karabakh conflict, and strong anti-Western feelings (Azerbaijan perceives the West as taking the Armenian side), there are few visitors, most of them on business.

PEOPLE

Pop. density medium

Azerbaijani, Russian

94/km² (242/mi²)

THE URBAN/RURAL POPULATION SPLIT

57% 43%

ETHNIC MAKEUP

Armenian 2%
Russian 3%
Other 2%
Dagestani 3%
Azeri 90%

Before independence Armenians and Russians accounted for 11% of the population. The violence of the early 1990s and accompanying Azeri nationalism caused thousands to flee. Racial hostility against those who remain is increasing.

Women, once prominent in the ruling party, have lost their political status, and their general status is declining.

The once effective social security system is now under great strain.

POLITICS

Multiparty elections

2000/2005

President Heydar Aliyev

AT THE LAST ELECTION

National Assembly 125 seats

61% YAP, **21% Ind**, **5% AKC**, **10% Others**, **2% VBP**, **1% Vacant**

YAP = New Azerbaijan Party **Ind** = Independents
AKC = Azerbaijan Popular Front **VBP** = Civic Solidarity Party
Vacant = Seat reserved for member from Nagorno Karabakh

The determination of the Nagorno Karabakh enclave to secede from Azerbaijan led to war from 1988 to 1994, with Armenian forces gaining control over 20% of Azeri territory. Peace talks have yet to reach an accord.

The YAP replaced the communists in 1995 and again dominated elections in 2000, which observers criticized as seriously flawed. It supports Heydar Aliyev, first elected president in 1993 (at the age of 70) and reelected in 1998.

AZERBAIJAN

Total Land Area : 86 600 sq. km (33 436 sq. miles)

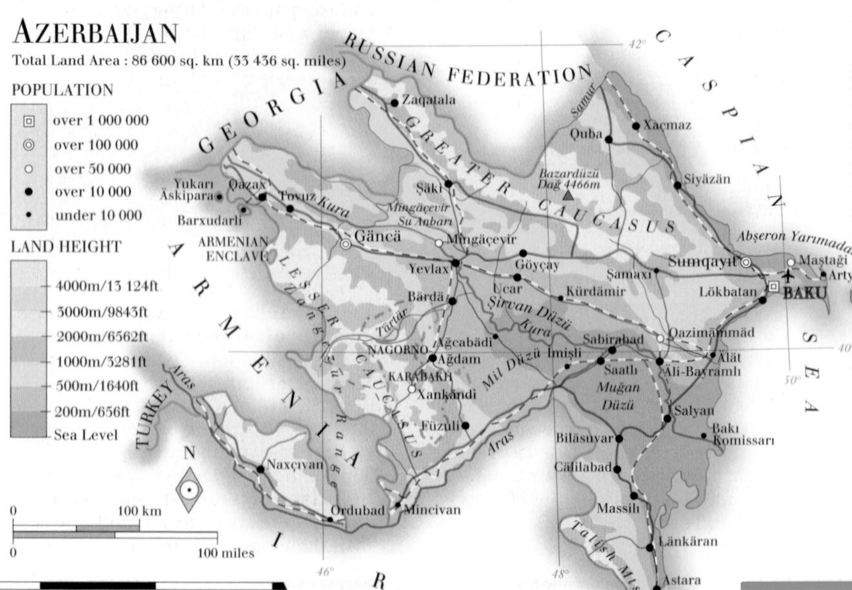
RUSSIAN FEDERATION
CASPIAN SEA
GEORGIA
ARMENIA
TURKEY
IRAN
GREATER CAUCASUS
Zaqatala
Quba
Xaçmaz
Siyäzän
Bazardüzü Dag 4466m
Şäki
Qazax
Yukarı Äskipara
Tovuz
Barxudarli
ARMENIAN ENCLAVE
Gäncä
Mingäçevir
Mingäçevir Su Anbari
Yevlax
Göyçay
Şamaxı
Sumqayıt
Abşeron Yarımadası
Maştağı
BAKU
Lökbatan
Kürdämir
Bärdä
Ucar
Şirvan Düzü
Ağcabädi
Ağdam
NAGORNO KARABAKH
Xankändi
Füzuli
Imişli
Sabirabad
Qazimämmäd
Älät
Ali-Bayramlı
Saatlı
Muğan Düzü
Salyan
Bakı Komissarı
Biläsuvar
Cälilabad
Massılı
Länkäran
Astara
Talish Mts
Naxçıvan
Ordubad
Mincivan
LESSER CAUCASUS
Zangezur Range
Mil Düzü
Kura
Aras
Tärtär
POPULATION
over 1 000 000
over 100 000
over 50 000
over 10 000
under 10 000
LAND HEIGHT
4000m/13 124ft
3000m/9843ft
2000m/6562ft
1000m/3281ft
500m/1640ft
200m/656ft
Sea Level
0 100 km
0 100 miles

WORLD AFFAIRS

Joined UN in 1992

CIS

CE

EAPC

OIC

OSCE

Relations with Armenia remain the central foreign policy issue. Turkey, with its common history and culture, is a natural ally. The West, neighboring Iran (with a large Azeri population), and Russia are interested in Azeri oil fields in the Caspian Sea. Azerbaijan joined the Council of Europe in 2001.

AID

Recipient

 $139m (receipts)

 Down 18% in 2000

World Bank aid has grown steadily. From 1992 to 2002 a pro-Armenian US Congress allowed limited humanitarian aid only.

DEFENSE

Compulsory military service

 $213m

 Up 5% in 2000

Azerbaijan has been a member of NATO's Partnership for Peace program since 1994. Its naval forces operate under CIS control.

ECONOMICS

Inflation 199% p.a. (1990–2000)

 $4.85bn

 4456–4475 manats

SCORE CARD

- ❑ WORLD GNP RANKING........................113th
- ❑ GNP PER CAPITA..............................$600
- ❑ BALANCE OF PAYMENTS....................–$168m
- ❑ INFLATION..1.8%
- ❑ UNEMPLOYMENT....................................1%

STRENGTHS

Extensive oil and natural gas reserves starting to come on stream. Iron, copper, lead, and salt deposits. Cotton and silk.

WEAKNESSES

Antiquated Soviet-era industry. Poor infrastructure and corruption threaten development. Oil pipeline to Turkey not yet built. Fallout from war in Nagorno Karabakh still drains state resources.

EXPORTS

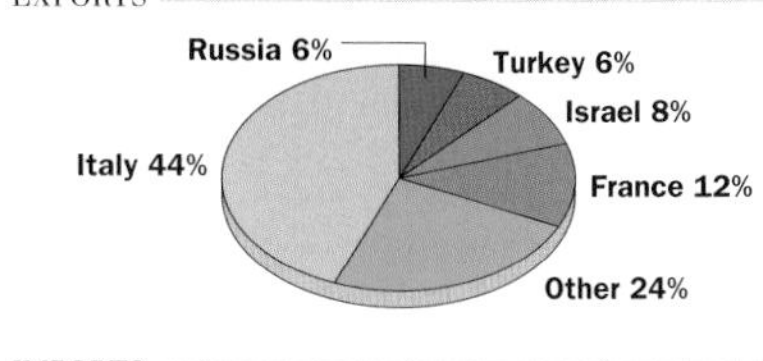
Russia 6%
Turkey 6%
Israel 8%
Italy 44%
France 12%
Other 24%

IMPORTS

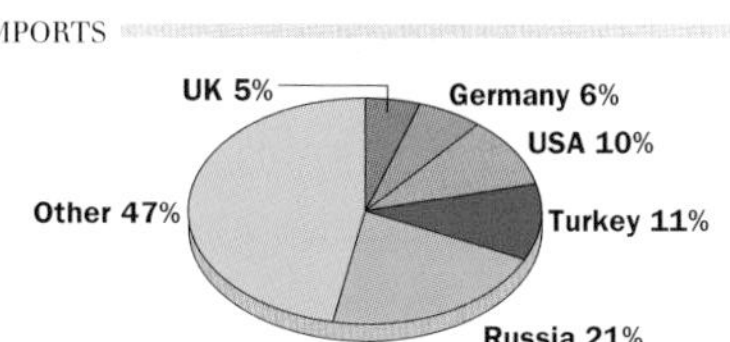
UK 5%
Germany 6%
USA 10%
Other 47%
Turkey 11%
Russia 21%

RESOURCES

Electric power 5.2m kw

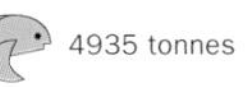 4935 tonnes

 300,000 b/d (reserves 7bn barrels)

5.56m sheep, 2m cattle, 14.5m chickens

 Iron, bauxite, copper, lead, zinc, limestone, salt, oil, gas

Relatively neglected in the Soviet period, Azerbaijan's Caspian Sea oil fields have attracted international interest. The shallowwater Guneshli field alone has over four million barrels of reserves. Offshore natural gas is also plentiful.

ENVIRONMENT

Sustainability rank: 114th

 6%

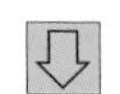 4.9 tonnes per capita

Under the Soviet regime oil pollution devastated the Caspian Sea, and pesticides were massively overused in agriculture. Major rivers suffer heavy pollution from Georgia and Armenia. Lack of funds restricts action.

MEDIA

TV ownership high

 Daily newspaper circulation 10 per 1000 people

PUBLISHING AND BROADCAST MEDIA

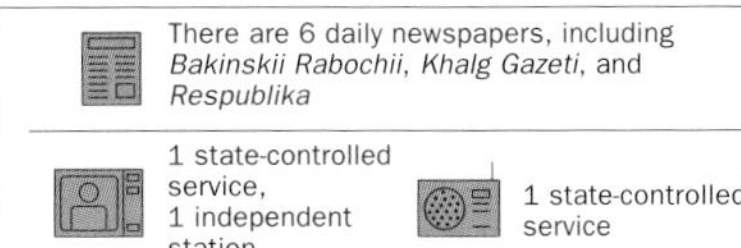

There are 6 daily newspapers, including *Bakinskii Rabochii*, *Khalg Gazeti*, and *Respublika*

1 state-controlled service, 1 independent station

1 state-controlled service

A 1998 decree abolished censorship, but freedom is limited by newsprint controls, license restrictions, and intimidation.

CRIME

No death penalty

 24,881 prisoners

 Down 6% in 1999

The judicial system returned to political control in 1993. Criminality is a particular problem in camps for those displaced in the Nagorno Karabakh conflict. Elsewhere, there is a low rate of violent crime, but assaults in the street have become less rare.

EDUCATION

School leaving age: 17

 96%

 156,800 students

When it came to power in the mid-1990s, the YAP began reversing communist control over education policy, which had been particularly noticeable in the teaching of history. Baku State, the largest of eight universities, is investing in modernizing its facilities.

HEALTH

Welfare state health benefits

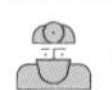 1 per 278 people

 Heart, cerebrovascular and respiratory diseases, cancers

The already poor health care system effectively collapsed as a result of war and the transition to a market economy.

CHRONOLOGY

Under consecutive Persian, Ottoman, and Russian influences, Azerbaijan, one of the world's major oil producers in 1900, attained independence in 1918.

- ❑ **1920** Red Army invades. Soviet republic established.
- ❑ **1922** Incorporated in Transcaucasian Soviet Federative Socialist Republic (TSFSR).
- ❑ **1930** Forced collectivization of agriculture.
- ❑ **1936** TSFSR disbanded; Azerbaijan a full union republic (ASSR).
- ❑ **1945** Attempted annexation of Azeri region of Iran.
- ❑ **1985** Gorbachev tackles corruption in Communist Party of Azerbaijan.
- ❑ **1988** Nagorno Karabakh seeks unification with Armenia.
- ❑ **1990** Nagorno Karabakh attempts secession. Soviet troops move in.
- ❑ **1991** Independence.
- ❑ **1993** Heydar Aliyev president; reelected in 1998.
- ❑ **1994** Cease-fire in war with Armenia over Nagorno Karabakh.
- ❑ **1995, 1999** General elections. Noncommunist YAP in power.

SPENDING

GDP/cap. decrease

CONSUMPTION AND SPENDING

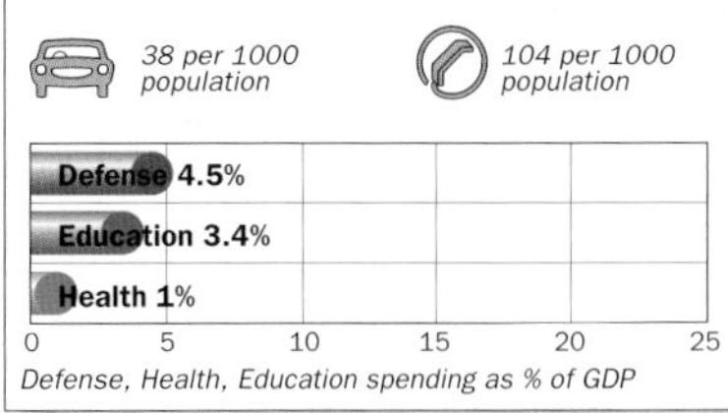
38 per 1000 population
104 per 1000 population
Defense 4.5%
Education 3.4%
Health 1%
0
5
10
15
20
25
Defense, Health, Education spending as % of GDP

New oil revenues are threatening to create a nouveau riche elite without reaching the 60% of Azerbaijan's population currently living in poverty.

WORLD RANKING

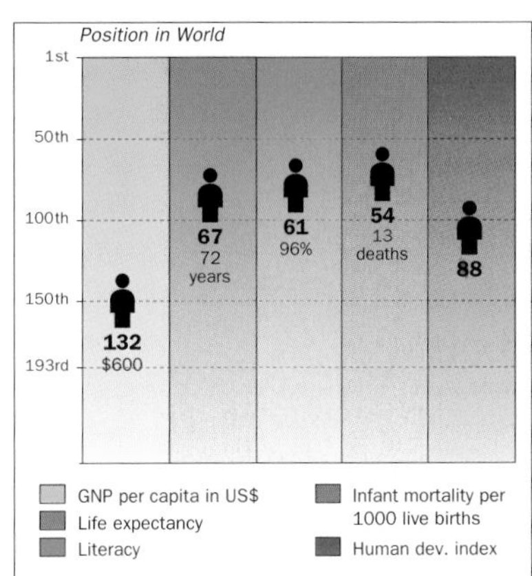
Position in World
1st
50th
100th
150th
193rd
132
$600
67
72 years
61
96%
54
13 deaths
88
GNP per capita in US$
Life expectancy
Literacy
Infant mortality per 1000 live births
Human dev. index

B

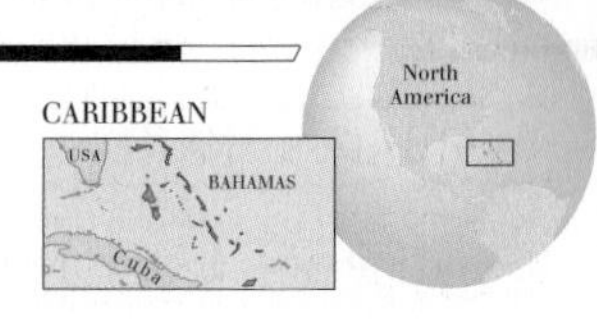

CARIBBEAN
North America
USA
BAHAMAS
Cuba

BAHAMAS

OFFICIAL NAME: Commonwealth of the Bahamas **CAPITAL:** Nassau
POPULATION: 308,000 **CURRENCY:** Bahamian dollar **OFFICIAL LANGUAGE:** English

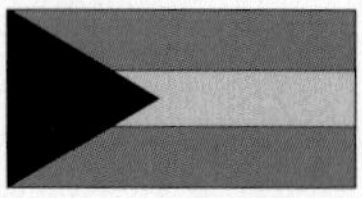

1973 | 1973 | July 10 | BS | -5 | +1242 | .bs

THE BAHAMAS, ALTHOUGH OFTEN bracketed with Caribbean countries, actually lies northeast of Cuba in the western Atlantic. The archipelago has 700 islands and 2400 cays: just 30 are inhabited. Long established as a tourist resort, the Bahamas today is also a major offshore financial center. It has one of the world's largest open-registry fleets; only a tiny fraction is owned by Bahamian nationals.

CLIMATE

Tropical oceanic

WEATHER CHART FOR NASSAU

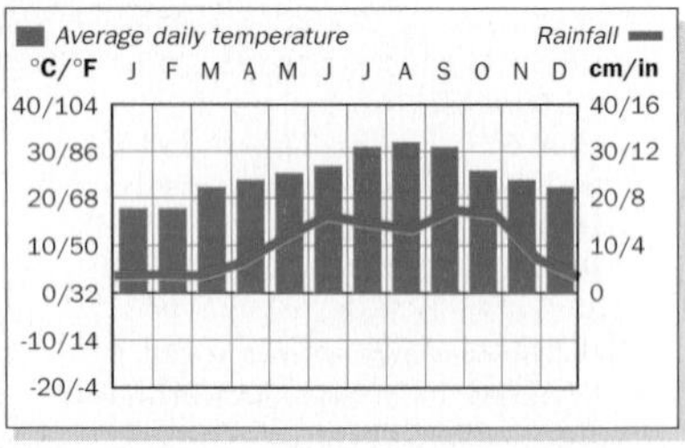

Average daily temperature / Rainfall

°C/°F: 40/104, 30/86, 20/68, 10/50, 0/32, -10/14, -20/-4
J F M A M J J A S O N D
cm/in: 40/16, 30/12, 20/8, 10/4, 0

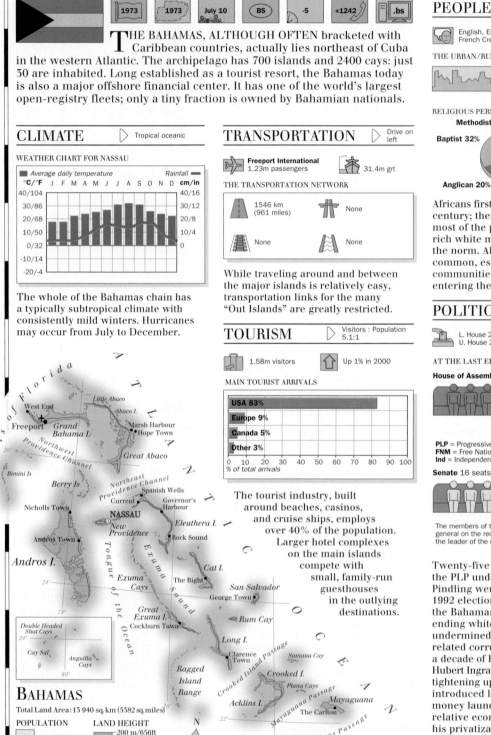

The whole of the Bahamas chain has a typically subtropical climate with consistently mild winters. Hurricanes may occur from July to December.

TRANSPORTATION

Drive on left

Freeport International 1.23m passengers

31.4m grt

THE TRANSPORTATION NETWORK

1546 km (961 miles)

None

None

None

While traveling around and between the major islands is relatively easy, transportation links for the many "Out Islands" are greatly restricted.

TOURISM

Visitors : Population 5.1:1

1.58m visitors

Up 1% in 2000

MAIN TOURIST ARRIVALS

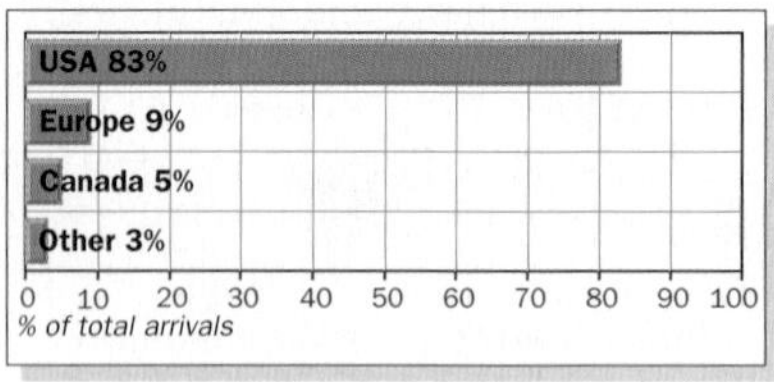

	% of total arrivals
USA	83%
Europe	9%
Canada	5%
Other	3%

0 10 20 30 40 50 60 70 80 90 100
% of total arrivals

The tourist industry, built around beaches, casinos, and cruise ships, employs over 40% of the population. Larger hotel complexes on the main islands compete with small, family-run guesthouses in the outlying destinations.

PEOPLE

Pop. density low

English, English Creole, French Creole

31/km² (80/mi²)

THE URBAN/RURAL POPULATION SPLIT

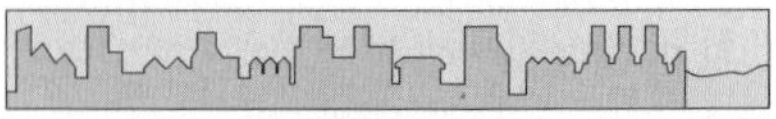

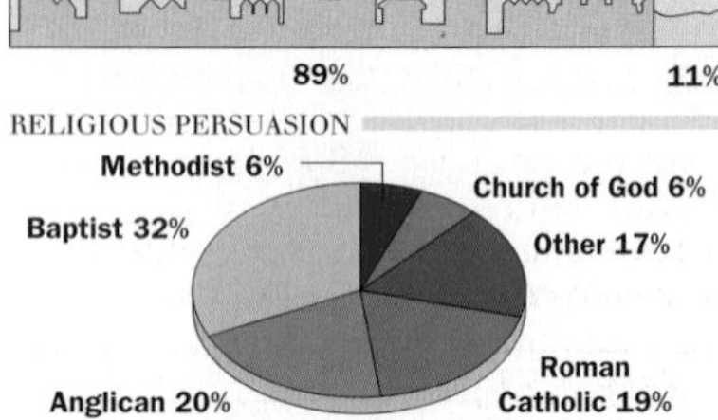

89%
11%
RELIGIOUS PERSUASION
Methodist 6%
Church of God 6%
Baptist 32%
Other 17%
Roman Catholic 19%
Anglican 20%

Africans first arrived as slaves in the 16th century; their descendants constitute most of the population, alongside a rich white minority. Small families are the norm. Absentee fathers are fairly common, especially in outlying fishing communities. More women are now entering the professions.

POLITICS

Multiparty elections

L. House 2002/2007
U. House 2002/2007

H.M. Queen Elizabeth II

AT THE LAST ELECTION

House of Assembly 40 seats

73% PLP
17% FNM
10% Ind

PLP = Progressive Liberal Party
FNM = Free National Movement
Ind = Independents

Senate 16 seats

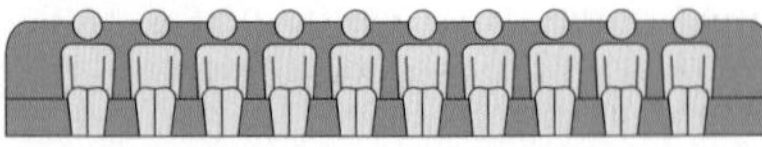

The members of the Senate are appointed by the governor-general on the recommendation of the prime minister and the leader of the opposition

Twenty-five years of unbroken rule by the PLP under Prime Minister Lynden Pindling were brought to an end by the 1992 elections. His legacy of steering the Bahamas to independence and ending white political domination was undermined by allegations of narcotics-related corruption. There followed a decade of FNM government under Hubert Ingraham, who emphasized tightening up ministerial accountability, introduced legislation to counter money laundering, and achieved relative economic success. However, his privatization drive proved deeply unpopular and the FNM was roundly defeated in 2002 by a resurgent PLP now led by Perry Christie.

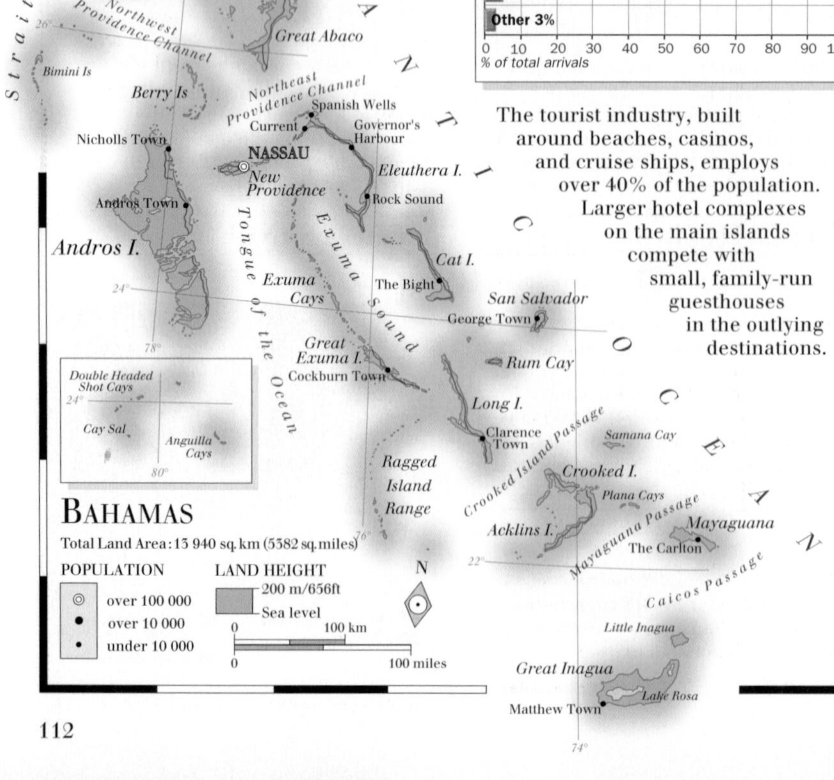

Straits of Florida
ATLANTIC OCEAN
West End
Freeport
Grand Bahama I.
Little Abaco
Abaco I.
Marsh Harbour
Hope Town
Northwest Providence Channel
Great Abaco
Bimini Is
Berry Is
Northeast Providence Channel
Spanish Wells
Current
Governor's Harbour
Nicholls Town
NASSAU
New Providence
Eleuthera I.
Rock Sound
Andros Town
Andros I.
Tongue of the Ocean
Exuma Sound
Cat I.
The Bight
Exuma Cays
San Salvador
George Town
Great Exuma I.
Cockburn Town
Rum Cay
Long I.
Clarence Town
Samana Cay
Crooked Island Passage
Crooked I.
Plana Cays
Ragged Island Range
Acklins I.
Mayaguana Passage
Mayaguana
The Carlton
Caicos Passage
Little Inagua
Great Inagua
Lake Rosa
Matthew Town
Double Headed Shot Cays
Cay Sal
Anguilla Cays

BAHAMAS

Total Land Area: 13 940 sq. km (5382 sq. miles)

POPULATION
over 100 000
over 10 000
under 10 000

LAND HEIGHT
200 m/656ft
Sea level

0 100 km
0 100 miles

WORLD AFFAIRS

Joined UN in 1973

ACS

Caricom

Comm

NAM

OAS

The Bahamas is seen as a center for transshipping narcotics and a money-laundering risk, for which G7 blacklisted the country for a year in 2000. The return of unauthorized immigrants to Haiti and Cuba dominates regional relations.

AID

Recipient

US$6m (receipts)

Down 50% in 2000

Aid is modest. The IDB and the US provide soft development loans. China in 1998 loaned Nassau US$17 million for a convention and theater complex.

DEFENSE

No compulsory military service

US$25m

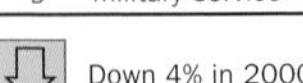
Down 4% in 2000

The UK is the main trainer of and supplier for the small naval defense force. The interception of narcotics and illegal immigrants is the force's main activity. There is no land army.

ECONOMICS

Inflation 2.7% p.a. (1990–2000)

US$4.53bn

1 Bahamian dollar

SCORE CARD

- ❑ WORLD GNP RANKING........................115th
- ❑ GNP PER CAPITA ....US$14,960
- ❑ BALANCE OF PAYMENTS................–US$438m
- ❑ INFLATION 1.6%
- ❑ UNEMPLOYMENT 9%

STRENGTHS

Major international financial services sector, including banking, insurance, and business trade center. Major tourism and cruise ship destination. Growing container port. International ship registration.

WEAKNESSES

Growing competition in financial services and tourism from the countries of the Caribbean.

EXPORTS

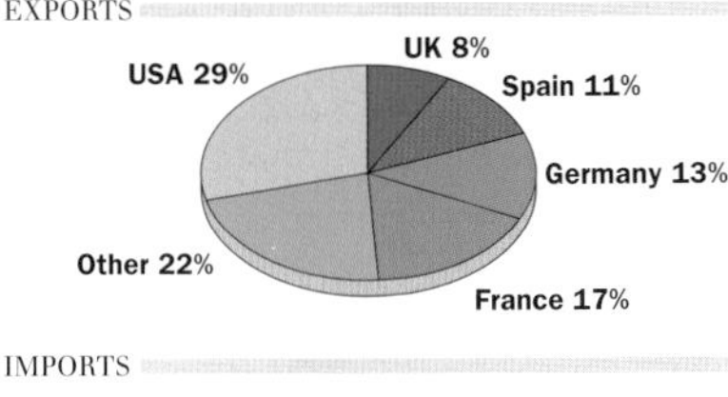
USA 29%
UK 8%
Spain 11%
Germany 13%
Other 22%
France 17%

IMPORTS

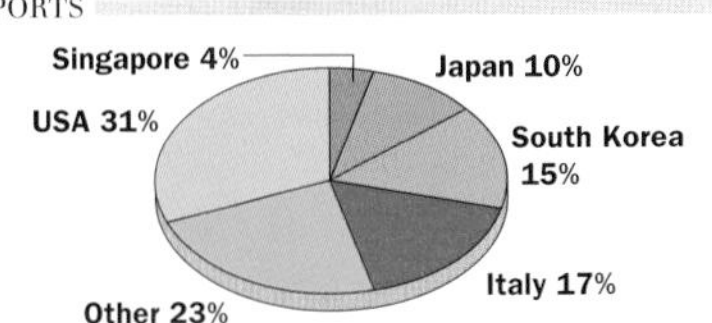
Singapore 4%
Japan 10%
USA 31%
South Korea 15%
Italy 17%
Other 23%

Archetypal island paradise. *Its natural beauty attracts more than five tourists per inhabitant to the Bahamas every year.*

RESOURCES

Electric power 401,000 kw

10,474 tonnes

Not an oil producer

13,852 goats, 6418 sheep, 4873 pigs, 4.5m chickens

Salt, aragonite

The Bahamas has no strategic resources. A 13.5 MW electricity generating plant was opened in 1998.

ENVIRONMENT

Not available

9%

6.1 tonnes per capita

As on many Caribbean islands, hotel overdevelopment is a major cause for concern. Environmental groups have also pointed out the potential for accidents posed by the Bahamas' enormous oil storage depots.

MEDIA

TV ownership medium

Daily newspaper circulation 99 per 1000 people

PUBLISHING AND BROADCAST MEDIA

There are 4 daily newspapers, the *Nassau Guardian*, the *Tribune*, the *Bahama Journal*, and the *Freeport News*

1 state-owned service

5 services: 1 state-owned, 4 independent

The state-owned TV channel faces very stiff competition from Florida-based US broadcasters.

CRIME

Death penalty in use

1401 prisoners

Down 29% in 1999

The death penalty remains in force. Violent crime, ranging from narcotics-related murders to serious vandalism, is on the increase. Tourists can be targets for petty thefts. Illegal weapons are readily available.

EDUCATION

School leaving age: 15

96%

5305 students

Schooling follows the former British selective system. Tertiary students attend the University of the West Indies or colleges in the US.

CHRONOLOGY

Once an English pirate base, the Bahamas, which gained its first parliament in 1729, formally became a British colony in 1783.

- ❑ **1920–1933** US prohibition laws turn Bahamas into prosperous bootlegging center.
- ❑ **1959–1962** Introduction of male suffrage; women gain the vote.
- ❑ **1973** Independence.
- ❑ **1983** Narcotics-smuggling scandals involving the government.
- ❑ **1992** FNM wins elections, ending 25 years of PLP rule.
- ❑ **2002** PLP returned to power.

HEALTH

Welfare state health benefits

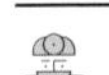
1 per 658 people

Obstetric causes, heart diseases, cancers, murders, accidents

The Bahamian health service combines state and private systems. In the outlying islands access to care relies on the Flying Doctor Service and around 50 local health centers. There are two private clinics on New Providence.

SPENDING

GDP/cap. increase

CONSUMPTION AND SPENDING

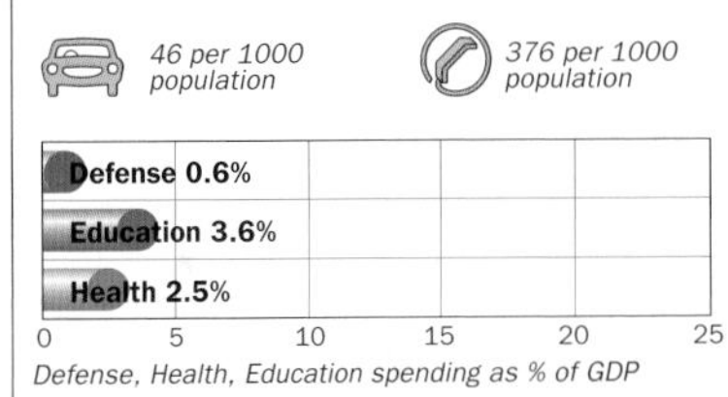
46 per 1000 population
376 per 1000 population
Defense 0.6%
Education 3.6%
Health 2.5%
0
5
10
15
20
25
Defense, Health, Education spending as % of GDP

There are marked wealth disparities: urban professionals who work in the financial sector are at one end of the scale, and the poor fishermen from the outlying islands are near the other. Cuban and Haitian refugees, who have no legal status, are the poorest group of all.

WORLD RANKING

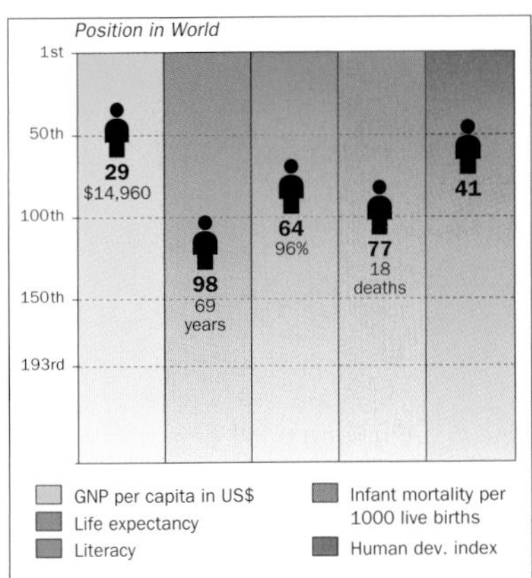
Position in World
1st
50th
100th
150th
193rd
29
$14,960
98
69 years
64
96%
77
18 deaths
41
GNP per capita in US$
Life expectancy
Literacy
Infant mortality per 1000 live births
Human dev. index

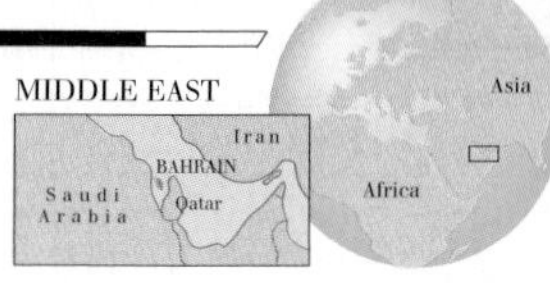
MIDDLE EAST
Iran
BAHRAIN
Saudi Arabia
Qatar
Asia
Africa

BAHRAIN

OFFICIAL NAME: State of Bahrain **CAPITAL:** Manama
POPULATION: 652,000 **CURRENCY:** Bahraini dinar **OFFICIAL LANGUAGE:** Arabic

1971

1971

Dec 16
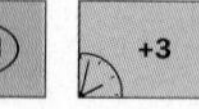
BRN

+3

+973
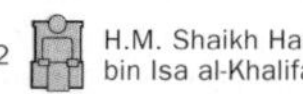
.bh

BAHRAIN IS AN ARCHIPELAGO state situated between the Qatar peninsula and the Saudi Arabian mainland. Only three of its islands are inhabited. Bahrain Island is connected to Saudi Arabia's Eastern Province by a causeway opened in 1986. Bahrain was the first Gulf emirate to export oil; its reserves are now almost depleted. Services such as offshore banking, insurance, and tourism are major employment sectors for skilled Bahrainis.

CLIMATE

Hot desert

WEATHER CHART FOR MANAMA

Average daily temperature — Rainfall
°C/°F J F M A M J J A S O N D cm/in
40/104 40/16
30/86 30/12
20/68 20/8
10/50 10/4
0/32 0
-10/14
-20/-4

Temperatures soar to 40°C (104°F) in June–September. In December–March the weather is pleasantly warm.

TRANSPORTATION

Drive on right

Bahrain International, Manama
3.93m passengers

121 ships
256,235 grt

THE TRANSPORTATION NETWORK

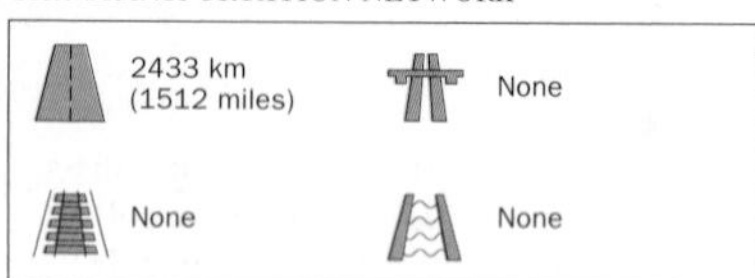
2433 km
(1512 miles)
None
None
None

The King Fahd Causeway connects Bahrain to Saudi Arabia, and ferries travel regularly to Iran. Buses are the main form of public transportation.

TOURISM

Visitors : Population 3.1:1

1.99m visitors

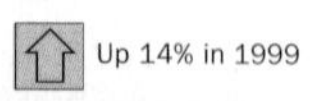
Up 14% in 1999

MAIN TOURIST ARRIVALS

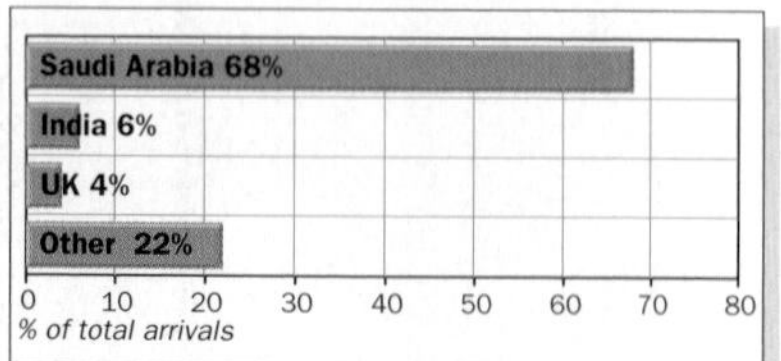
Saudi Arabia 68%
India 6%
UK 4%
Other 22%
0 10 20 30 40 50 60 70 80
% of total arrivals

Bahrain's relatively liberal lifestyle has made it something of a magnet for visitors from neighboring Gulf states. Bahrain has a modern airport and is a center for business conventions.

PEOPLE

Pop. density high

Arabic

924/km² (2388/mi²)

THE URBAN/RURAL POPULATION SPLIT

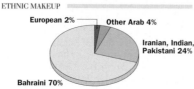

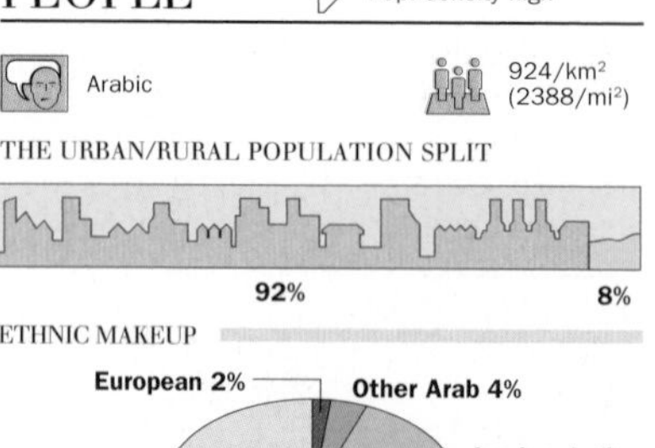
92%
8%
ETHNIC MAKEUP
European 2%
Other Arab 4%
Iranian, Indian, Pakistani 24%
Bahraini 70%

Bahrain is the smallest and most densely populated Arab state. The key division is between Sunni and Shi'a Muslims, about 30% and 70% of the population respectively. Sunnis hold the best jobs in business and government. Shi'a Muslims tend to do menial work and have a lower standard of living. The most impoverished Shi'a Muslims tend to be of Iranian descent.

Bahrain has a smaller expatriate population than many other Arab countries. The ruling al-Khalifa family has responded to declining oil reserves by diversifying the economy to provide service industry jobs for Bahrainis.

Bahrain is the most liberal of the Gulf states. Alcohol is freely available. Women have access to education and are not obliged to wear the veil. Since 2000 they have been entitled to participate in the Consultative Council.

The Grand Mosque, Manama. It is the largest building in Bahrain and can accommodate 7000 people.

POLITICS

In transition

Not applicable/2002

H.M. Shaikh Hamad bin Isa al-Khalifa

LEGISLATIVE OR ADVISORY BODIES

Chamber of Deputies 40 seats (to be elected)

A bicameral legislative body is to be formed. The lower Chamber of Deputies is to be elected in October 2002, and the Consultative Council, first appointed in 1993 and elected in 2001, is to form the appointed upper house.

Consultative Council 40 seats

The al-Khalifa family has dominated politics since 1783 by means of an effectively autocratic system, although the amir has been advised since 1993 by a Consultative Council. A charter of political reforms, approved by the council in 2000, was backed by a popular referendum in 2001. Under the charter, Bahrain was declared a constitutional monarchy in February 2002. H.M. Shaikh Hamad, amir since 1999, supports the economic liberalization begun by his father. The repeal of the State Security Law in 2001 promised an end to the detention of political dissidents. Many were Shi'a opponents of the regime, who are traditionally backed by Iran.

WORLD AFFAIRS

Joined UN in 1971

AL

Damasc

GCC

OIC

OAPEC

Bahrain has good relations with the UK and the US, but is keen to restore relations with Iraq. There is residual tension with Iran. A dispute with Qatar over the Hawar islands was resolved in Bahrain's favor in 2001.

AID

Recipient

$49m (receipts)

Up sharply in 2000

Bahrain receives low levels of aid, but takes the lion's share from the offshore oil field shared with Saudi Arabia, effectively a subsidy from the latter.

DEFENSE

No compulsory military service

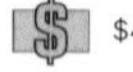
$435m

Down 1% in 2000

The strong defense force includes a small but well-equipped air force. US air bases on Bahrain have been used in recent conflicts, including the "war on terrorism" in Afghanistan in 2001, and the US 5th Fleet is also based there.

B

ECONOMICS

Inflation –0.1% p.a. (1990–1999)

$4.9bn

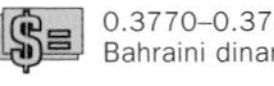
0.3770–0.3771 Bahraini dinars

SCORE CARD

- WORLD GNP RANKING........................110th
- GNP PER CAPITA............................$7640
- BALANCE OF PAYMENTS.....................$113m
- INFLATION......................................–0.4%
- UNEMPLOYMENT.................................15%

STRENGTHS

Oil and gas. Arab world's major offshore banking sector since 1980s. Inward investment. Tourism. Aluminum production.

WEAKNESSES

Depleted oil reserves and insufficient diversification. High unemployment. High levels of government borrowing.

EXPORTS

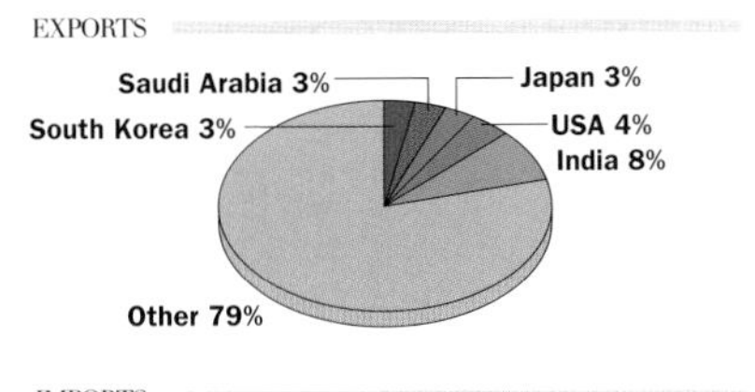
Saudi Arabia 3%
Japan 3%
South Korea 3%
USA 4%
India 8%
Other 79%

IMPORTS

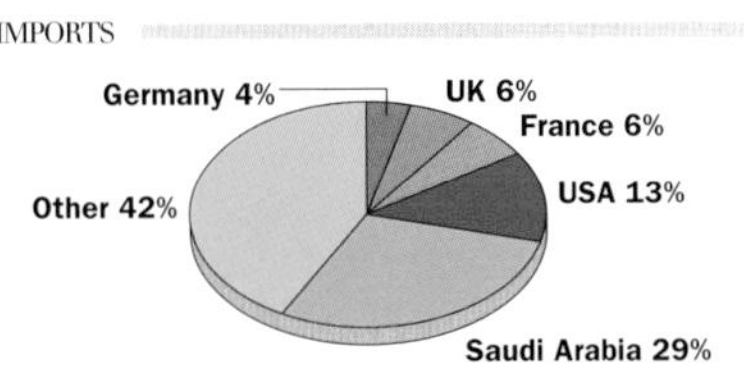
Germany 4%
UK 6%
France 6%
USA 13%
Other 42%
Saudi Arabia 29%

BAHRAIN

Total Land Area : 620 sq. km (239 sq. miles)

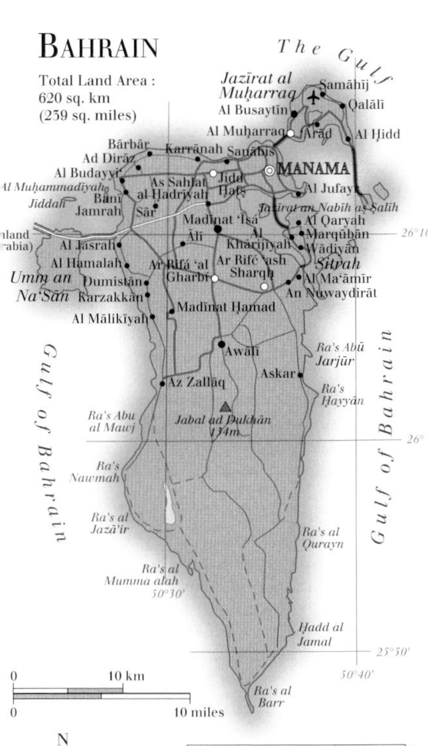
The Gulf
Jazīrat al Muharraq
Samāhij
Qalālī
Al Busaytīn
Al Muharraq
'Arād
Al Hidd
Bārbār
Ad Dirāz
Karrānah
Sanābis
MANAMA
Al Budayyi'
Al Muhammadiyah
Jiddah
Jidd Hafs
As Sahlah
al Hadriyah
Bani Jamrah
Sār
Al Jufayr
Jazīrat an Nabīh as Sālih
Madīnat 'Īsā
'Ālī
Al Qaryah
Marqūbān
Al Jasrah
Al Hamalah
Khārijīyah
Wādiyān
Ar Rifā' al Gharbī
Ar Rifā' ash Sharqī
Sitrah
Umm an Na'sān
Dumistān
Karzakkān
Al Ma'āmir
An Nuwaydirāt
Al Mālikīyah
Madīnat Hamad
Gulf of Bahrain
'Awālī
Ra's Abū Jarjūr
Askar
Az Zallāq
Ra's Hayyān
Ra's Abu al Mawj
Jabal ad Dukhān 134m
Ra's Nawmah
Ra's al Jazā'ir
Ra's al Qurayn
Ra's al Mumma alah
Hadd al Jamal
Ra's al Barr
0 10 km
0 10 miles
N

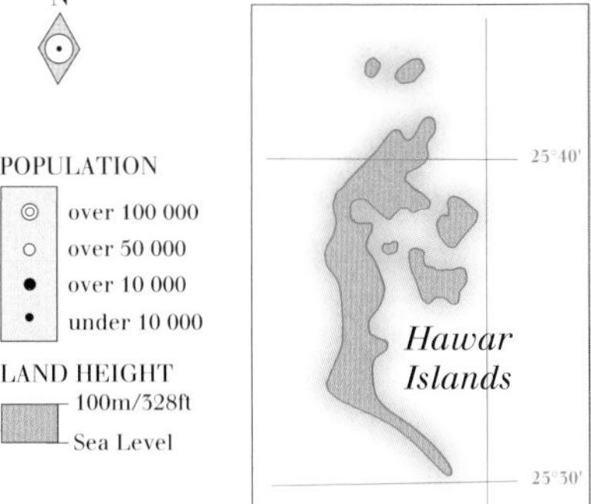
Hawar Islands

POPULATION

- over 100 000
- over 50 000
- over 10 000
- under 10 000

LAND HEIGHT

100m/328ft

Sea Level

RESOURCES

Electric power 1.1m kw

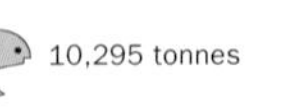
10,295 tonnes

37,473 b/d (reserves 219m barrels)

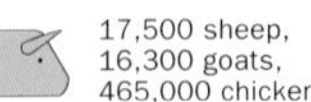
17,500 sheep, 16,300 goats, 465,000 chickens

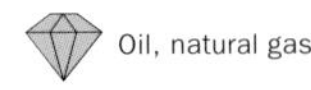
Oil, natural gas

Bahrain remains dependent on its oil and gas industry. Production of crude oil has declined sharply since the 1970s, and there are fears that reserves may run out by 2010. As oil has declined, so gas has assumed greater importance. Most is used to supply local industries, particularly the aluminum plant, which was established in 1972.

ENVIRONMENT

Not available

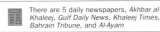

None

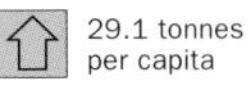
29.1 tonnes per capita

Local marine life, particularly the dugong, is vulnerable to upstream oil pollution from the Gulf. Bahrain and Abu Dhabi (UAE) signed an agreement in 2000 on environmental concerns.

MEDIA

TV ownership high

Daily newspaper circulation 117 per 1000 people

PUBLISHING AND BROADCAST MEDIA

There are 5 daily newspapers, *Akhbar al-Khaleej*, *Gulf Daily News*, *Khaleej Times*, *Bahrain Tribune*, and *Al-Ayam*

1 state-owned service

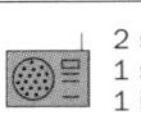
2 services: 1 state-owned, 1 independent

Bahrain has a less authoritarian media regime than most of the Gulf, although government critics have been prosecuted. CNN and BBC satellite TV are freely available.

CRIME

Death penalty in use

911 prisoners

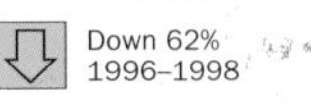
Down 62% 1996–1998

Crime is minimal, and theft and muggings are rare. Suspected political dissidents are monitored by the police.

CHRONOLOGY

Bahrain has been ruled since 1783 by the al-Khalifa family.

- **1971** Independence from Britain.
- **1981** Founder member of GCC.
- **1990–1991** Bahrain supports US-led action expelling Iraq from Kuwait.
- **1994–1996** Shi'a unrest.
- **1999** Accession to throne of Shaikh Hamad bin Isa al-Khalifa.
- **2001** Referendum approves transition to democracy.

EDUCATION

School leaving age: 15

88%

7676 students

Female literacy rates are above the Gulf average. Lack of funding has held up plans for a university.

HEALTH

Welfare state health benefits

1 per 1000 people

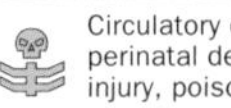
Circulatory diseases, perinatal deaths, injury, poisonings

The high-quality health service is free to Bahraini nationals. Some go abroad for advanced care. The Muharraq Health Center was upgraded in 2001.

SPENDING

GDP/cap. increase

CONSUMPTION AND SPENDING

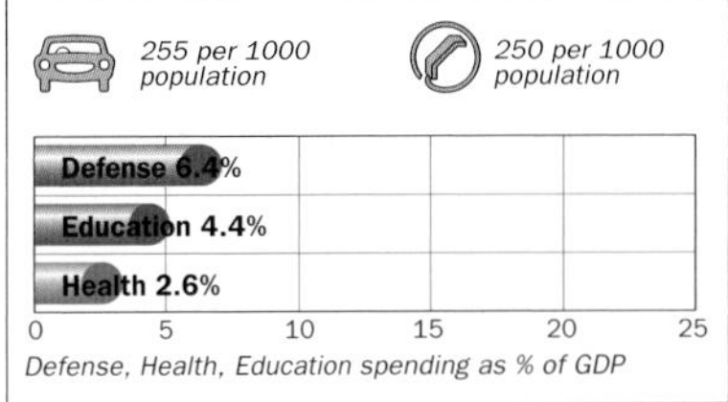
255 per 1000 population
250 per 1000 population
Defense 6.4%
Education 4.4%
Health 2.6%
0
5
10
15
20
25

Defense, Health, Education spending as % of GDP

Beneficiaries of the amir's extensive patronage form the wealthiest group in society. Bahrain's largest religious community, the Shi'a Muslims, is also the poorest.

WORLD RANKING

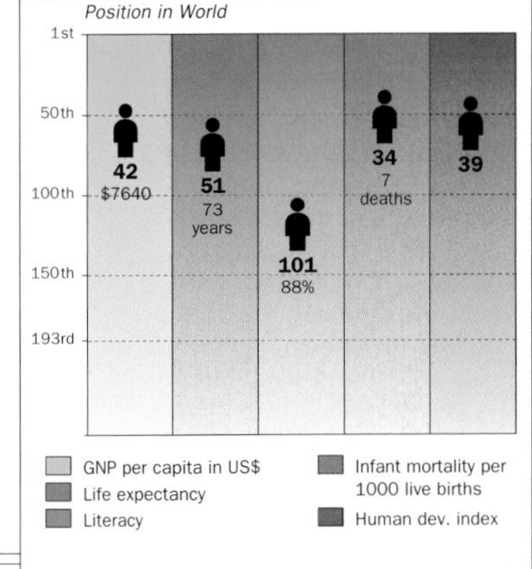
Position in World
1st
50th
100th
150th
193rd
42
$7640
51
73 years
101
88%
34
7 deaths
39
GNP per capita in US$
Life expectancy
Literacy
Infant mortality per 1000 live births
Human dev. index

B

BANGLADESH

SOUTH ASIA

OFFICIAL NAME: People's Republic of Bangladesh **CAPITAL:** Dhaka
POPULATION: 140.4 million **CURRENCY:** Taka **OFFICIAL LANGUAGE:** Bengali

1971

1971

March 26

BD

+6

+880

.bd

LOCATED AROUND the confluence of the mighty Ganges and Jamuna Rivers, Bangladesh is the eastern half of historic Bengal. Most of the country is composed of fertile alluvial plains; the north and northeast are mountainous, as is the Chittagong region in the southeast. After seceding from Pakistan in 1971, Bangladesh had a troubled history of political instability, with periods of emergency rule. Effective democracy was restored in 1991. Bangladesh's major economic sectors are jute production, textiles, and agriculture. Its climate can wreak havoc – in 1991 a massive cyclone killed more than 140,000 people.

CLIMATE

Tropical/subtropical

WEATHER CHART

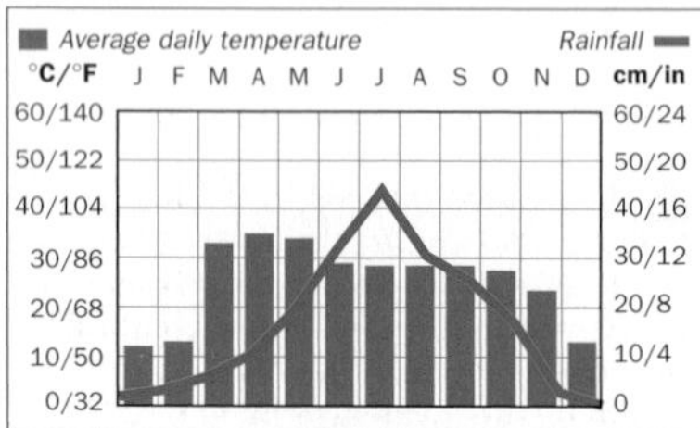
Average daily temperature
Rainfall
°C/°F
J F M A M J J A S O N D
cm/in
60/140
50/122
40/104
30/86
20/68
10/50
0/32
60/24
50/20
40/16
30/12
20/8
10/4
0

During the monsoon, the water level generally rises 6 m (20 ft) above normal, flooding up to two-thirds of the country. The floods are made much worse when the Ganges, Jamuna, and Meghna rivers, which converge in a huge delta in Bangladesh, are swollen by the melting of the Himalayan snows and heavy rain in India. Cyclones build up regularly in the Bay of Bengal, with sometimes devastating effects on the flat coastal region.

TRANSPORTATION

Drive on left

Zia International, Dhaka
2.52m passengers

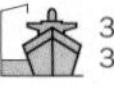

310 ships
370,100 grt

THE TRANSPORTATION NETWORK

19,112 km
(11,876 miles)

None

2705 km
(1681 miles)

8433 km
(5240 miles)

Most transportation in Bangladesh is by water, although government policy is now concentrating on developing road and rail links, including the reopening in mid-2000 of a passenger rail service into India. The Bangabandhu bridge across the Jamuna River, which bisects Bangladesh from north to south, was inaugurated in June 1998, after numerous delays. Bangladesh's two major ports, Mungla and Chittagong, are being upgraded to take advanced container ships.

Begum Khaleda Zia, reelected as prime minister in 2001.

Sheikh Hasina Wajed, AL leader and former prime minister.

TOURISM

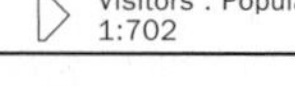

Visitors : Population 1:702

200,000 visitors

Up 16% in 2000

MAIN TOURIST ARRIVALS

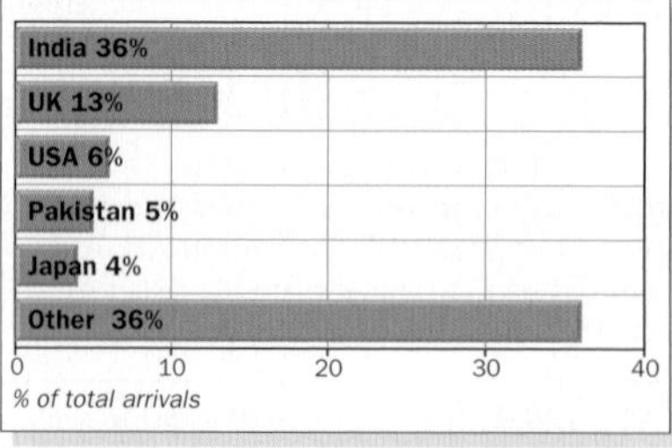
India 36%
UK 13%
USA 6%
Pakistan 5%
Japan 4%
Other 36%
0
10
20
30
40
% of total arrivals

The Mughal architecture in Dhaka and the Pala dynasty (7th–10th centuries) city of Sonargaon, just to the southeast, are major attractions, but tourists may be deterred by social and political unrest. Most visitors are Indian businessmen or Bangladeshis living overseas who return to visit relatives.

Traders on the Meghna River. Life is governed by the vast network of rivers. The floodplains are among the most fertile in the world.

PEOPLE

Pop. density high

Bengali, Urdu, Chakma, Marma (Magh), Garo, Khasi, Santhali, Tripuri, Mro

1048/km² (2716/mi²)

THE URBAN/RURAL POPULATION SPLIT

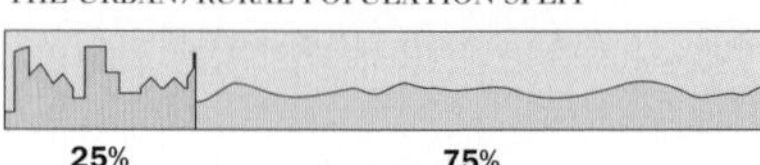
25%
75%

RELIGIOUS PERSUASION

Other 1%
Hindu 12%
Muslim (mainly Sunni) 87%

ETHNIC MAKEUP

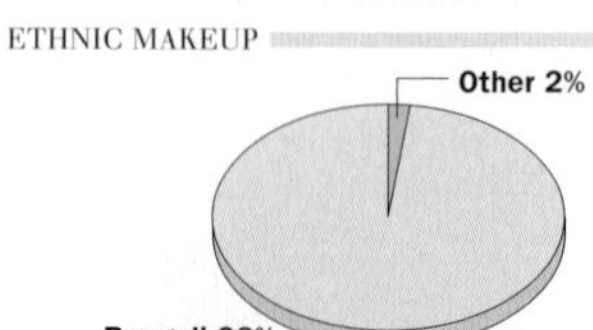
Other 2%
Bengali 98%

Bangladesh is one of the most densely populated countries in the world, despite the fact that three-quarters of the population is rural. As in India, there is considerable Muslim–Hindu tension; in 2001 thousands of Hindus and other religious minorities claimed persecution from the new nationalist government.

Although more than 50% of Bangladeshis, rural and urban, still live below the poverty line, there has been an improvement in living standards over the past decade.

By providing them with an independent income, the textile trade has been a factor in the growing emancipation of Bangladeshi women. They are now included in official employment statistics and are the main customers of the most successful rural bank. Women have led both the government and the opposition. However, Bangladesh was criticized by Amnesty International in 2000 for insufficiently protecting women's rights, and a UN report later that year revealed that almost 50% of Bangladeshi women are victims of domestic violence.

POPULATION AGE BREAKDOWN

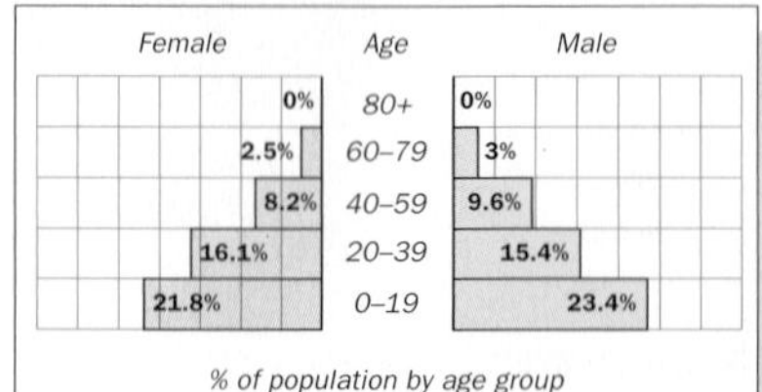

Female	Age	Male
0%	80+	0%
2.5%	60–79	3%
8.2%	40–59	9.6%
16.1%	20–39	15.4%
21.8%	0–19	23.4%

% of population by age group

B

POLITICS

Multiparty elections

2001/2006

Acting President Jamiruddin Sircar

AT THE LAST ELECTION

Parliament 300 seats

66% BNP	20% AL	6% JI	2% Ind	5% JD	1% Others

BNP = Bangladesh Nationalist Party and allies
AL = Awami League **JI** = Jamaat-e-Islami
JD = Jatiya Dal (Ershad) **Ind** = Independents

Bangladesh returned to multiparty democracy in 1991, following a period of military rule.

PROFILE

Between 1975 and 1990 the military was in power in Bangladesh. The overthrow of President Ershad in 1990 saw a return to multiparty politics; the army remains poised, however, to intervene in the event of a breakdown in the rule of law. Bangladesh's first woman prime minister, Begum Khaleda Zia, head of the BNP, was elected in 1991. A change from a presidential to a prime-ministerial system of government followed.

The AL, which had steered Bangladesh to independence in 1971, mounted a sustained campaign against Khaleda Zia's regime, forcing a rerun of elections in 1996. It won the largest number of seats and its leader, Sheikh Hasina Wajed, became the first prime minister to complete a full term. The revived BNP returned to power in the October 2001 poll, amid much electoral violence, and formed a coalition with the hard-line Islamic JI.

MAIN POLITICAL ISSUES

The state sector
Bangladesh is coming under mounting pressure from multilateral lending institutions, which account for the vast majority of the country's capital inflows, to cut costs in the state sector. Simultaneously, state-sector workers are demanding wage increases in line with inflation.

Autonomy for Chittagong Hill Tracts
Buddhist Mongol groups – the Chakma – continue to voice demands for greater autonomy, although the low-level guerrilla war they have waged since 1974 has been contained. Many Chakmas fear persecution by Bengali Muslim settlers, despite a peace treaty signed in 1997 which provides for local autonomy, amnesty, and the return of refugees from India.

BANGLADESH

Total Land Area : 144 000 sq. km (55 598 sq. miles)

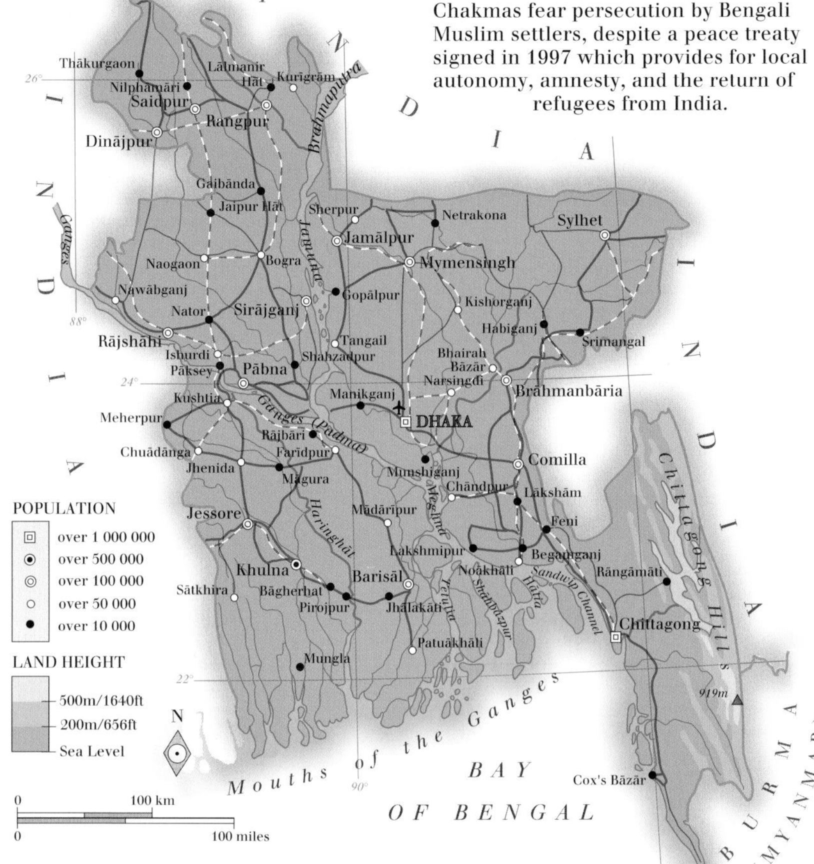
INDIA
Thākurgaon
Nilphāmāri
Saidpur
Lālmanir Hāt
Kurigrām
Rangpur
Dinājpur
Brahmaputra
Gaibānda
Jaipur Hāt
Sherpur
Jamālpur
Netrakona
Sylhet
Mymensingh
Naogaon
Bogra
Nawābganj
Gopālpur
Jamuna
Natore
Sirājganj
Kishorganj
Habiganj
Srimangal
Rājshāhi
Ishurdi
Pāksey
Tangail
Shahzadpur
Bhairab Bāzār
Narsingdi
Pābna
Brāhmanbāria
Kushtia
Manikganj
Meherpur
DHAKA
Ganges (Padma)
Rājbāri
Chuādānga
Farīdpur
Jhenida
Comilla
Munshiganj
Chāndpur
Māgura
Lāksham
Jessore
Mādāripur
Meghna
Feni
Harinighāta
Lakshmipur
Noākhāli
Begamganj
Rāngāmāti
Khulna
Sandwip Channel
Sātkhira
Bāgherhat
Barisāl
Pirojpur
Jhālakāti
Chittagong
Chittagong Hills
Patuākhāli
Mungla
919m
Mouths of the Ganges
BAY OF BENGAL
Cox's Bāzār
BURMA (MYANMAR)
POPULATION
over 1 000 000
over 500 000
over 100 000
over 50 000
over 10 000
LAND HEIGHT
500m/1640ft
200m/656ft
Sea Level
N
0 100 km
0 100 miles

WORLD AFFAIRS

Joined UN in 1974

Comm

NAM

OIC

SAARC

WTO

Good relations with the West, the main source of essential aid, are a priority. Relations with Pakistan have slowly improved since Pakistan's agreement in 1991 to accept the 250,000 pro-Pakistan Bihari Muslims in Bangladeshi refugee camps since 1971. Relations with India are improving. The damaging effects of the construction of the Farakka Dam on the Ganges, which deprived Bangladesh of irrigation water, have been alleviated by a 30-year agreement signed in 1996 guaranteeing the right of both parties to share the Ganges water. Bilateral relations eased further with the 1997 Chittagong Hill Tracts treaty; India had been accused of fomenting unrest.

AID

Recipient

$1.17bn (receipts)

Down 4% in 2000

Aid disbursements to Bangladesh each year are substantially greater than the annual value of foreign investment in the country. Aid also finances the bulk of state capital spending. The Bangladesh Development Aid Consortium meets annually to discuss aid spending under the auspices of the World Bank. One result of the level of aid is that Bangladesh has fallen into one of the traps of an aid-dependent economy: the large middle class has a vested interest in perpetuating a system which provides its members with lucrative contracts and access to external resources.

CHRONOLOGY

Bengal was the first part of the Indian subcontinent to come under British rule when the East India Company was made the *Diwani* (tax collector) by the Mughal emperor in 1765.

- **1905** Muslims persuade British rulers to partition state of Bengal, to create a Muslim-dominated East Bengal.
- **1906** Muslim League established in Dhaka.
- **1912** Partition of 1905 reversed.
- **1947** British withdrawal from India. Partition plans establish a largely Muslim state of East (present-day Bangladesh) and West Pakistan, separated by 1600 km (1000 miles) of Indian, and largely Hindu, territory.
- **1949** AL founded to campaign for autonomy from West Pakistan.
- **1968** Gen. Yahya Khan heads government in Islamabad.

B

CHRONOLOGY *continued*

- **1970** Elections give AL, under Sheikh Mujibur Rahman, clear majority. Rioting and guerrilla warfare following Yahya Khan's refusal to convene assembly. Year ends with worst recorded storms in Bangladesh's history – between 200,000 and 500,000 dead.
- **1971** Civil war, as Sheikh Mujib and AL declare unilateral independence. Ten million Bangladeshis flee to India. Pakistani troops defeated in 12 days by Mukhti Bahini – the Bengal Liberation Army.
- **1972** Sheikh Mujib elected prime minister. Nationalization of key industries, including jute and textile. Bangladesh achieves international recognition and joins Commonwealth. Pakistan withdraws in protest.
- **1974** Severe floods damage rice crop.
- **1975** Sheikh Mujib assassinated. Military coups end with General Zia ur-Rahman taking power. Institution of single-party state.
- **1976** Banning of trade union federations.
- **1977** Gen. Zia assumes presidency. Islam adopted as first principle of the constitution.
- **1981** Gen. Zia assassinated.
- **1982** Gen. Ershad takes over.
- **1983** Democratic elections restored by Ershad. Ershad assumes presidency.
- **1986** Elections. AL and BNP fail to unseat Ershad.
- **1987** Ershad announces state of emergency.
- **1988** Islam becomes constitutional state religion.
- **1990** Ershad resigns following demonstrations.
- **1991** Elections won by BNP. Khaleda Zia becomes prime minister. Ershad imprisoned. Role of the president reduced to ceremonial functions. Floods kill 140,000 people.
- **1994** Author Taslima Nasreen, who is accused of blasphemy, escapes to Sweden.
- **1996** Election returns BNP to power. Results are rejected by opposition parties, which force fresh elections. Sheikh Hasina Wajed of AL then takes power.
- **2001** Supreme Court declares issuing of religious decrees (*fatwas*) to be a criminal offense. BNP returned to power following violence-marred elections.

DEFENSE

No compulsory military service

 $670m

 No change in 2000

The military, which dominated politics between 1975 and 1990, still wields considerable influence, despite the restoration of civilian government. Although there is greater emphasis on poverty alleviation programs, spending on defense is disproportionately high. Plans to improve security by constructing a 4000-km (2500-mile) road along the border with India were announced in mid-2000.

BANGLADESHI ARMED FORCES

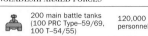

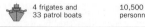

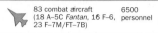

	200 main battle tanks (100 PRC Type–59/69, 100 T–54/55)	120,000 personnel
	4 frigates and 33 patrol boats	10,500 personnel
	83 combat aircraft (18 A–5C *Fantan*, 16 F–6, 23 F–7M/FT–7B)	6500 personnel
	None	

ECONOMICS

Inflation 4% p.a. (1990–2000)

 $47.9bn

 54.1–56.95 taka

SCORE CARD

World GNP Ranking	51st
GNP per Capita	$370
Balance of Payments	$-306m
Inflation	2.3%
Unemployment	3%

EXPORTS

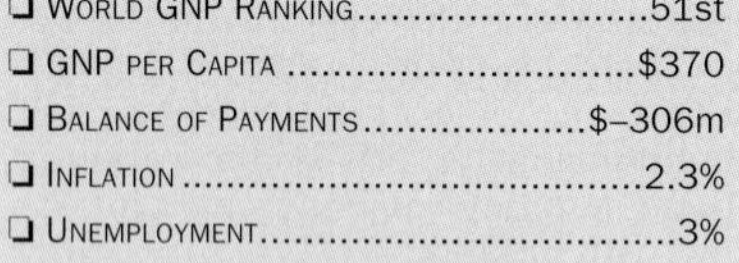
Netherlands 4%
France 5%
UK 8%
Germany 11%
Other 40%
USA 32%

IMPORTS

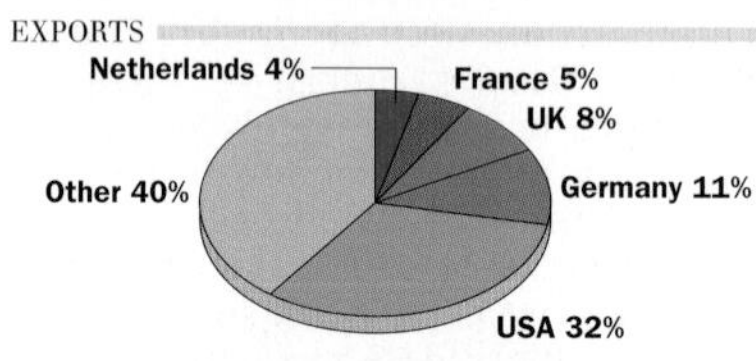
Hong Kong 5%
China 7%
Singapore 8%
Japan 9%
India 11%
Other 60%

Strengths

80% of the world's jute fiber exports come from Bangladesh. Low wages ensure a competitive and expanding textile industry, which provides over three-quarters of manufacturing export earnings.

Weaknesses

The agricultural sector, employing the majority of Bangladeshis, is vulnerable to the violent and unpredictable climate.

Profile

Government ministers like to portray Bangladesh as an emerging NIC, but its economy is still overwhelmingly dependent on agriculture and large aid inflows. Agriculture, which provides jute and tobacco, is productive; Bangladesh's soils, fed by the Ganges, Jamuna, and Meghna rivers, are highly fertile. However, the effects of the weather can be devastating, frequently destroying a whole year's crop. Agricultural wages are among the lowest in the world.

ECONOMIC PERFORMANCE INDICATOR

Consumer Price Index
GDP
160
140
120
100
80
1996
1997
1998
1999
2000
Consumer price index 1995=100
GDP 1996=100

The state sector, which owns large, inefficient and massively loss-making companies, is in difficulty. The World Bank, the source of most aid, wishes to see loss-making concerns cut their workforces or close down. A privatization program was announced in 2002.

Textiles and garments are currently the healthiest sectors. Economic zones (export processing zones) with special concessions have attracted foreign investment, as well as helped to promote a small indigenous electronics industry. Bangladesh receives generous textile import quotas from the EU and NAFTA.

BANGLADESH : MAJOR BUSINESSES

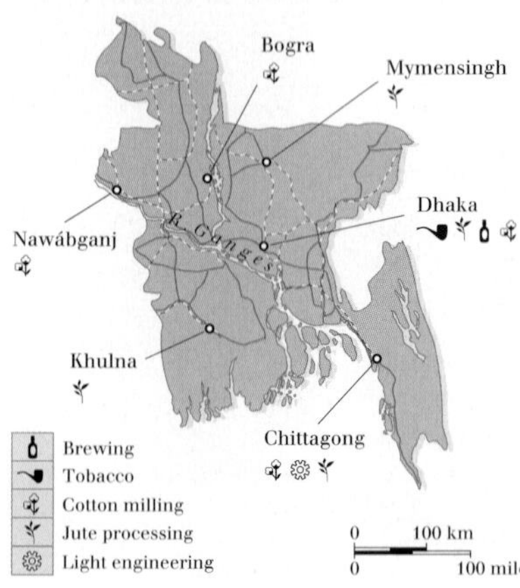
Bogra
Mymensingh
Dhaka
Nawábganj
R. Ganges
Khulna
Chittagong
Brewing
Tobacco
Cotton milling
Jute processing
Light engineering
0 100 km
0 100 miles

B

RESOURCES

Electric power 3.5m kw

 1.54m tonnes

 20 b/d (reserves 70,000 barrels)

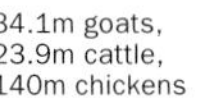 34.1m goats, 23.9m cattle, 140m chickens

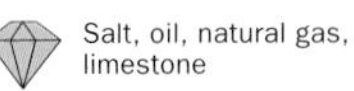 Salt, oil, natural gas, limestone

ELECTRICITY GENERATION

Hydro 6% (0.9bn kwh)
Combustion 94% (13bn kwh)
Nuclear 0%
Other 0%

0 20 40 60 80 100

% of total generation by type

BANGLADESH : LAND USE

R. Jamuna
R. Ganges
Mouths of the Ganges

Cropland
Wetlands
Forest
Rice
Jute - cash crop

0 100 km
0 100 miles

Bangladesh is the world's major jute producer, accounting for 80% of world jute fiber exports and about 50% of world jute manufactures exports.

Bangladesh holds world-class gas reserves, estimated to last as much as 200 years at the present extraction rate. Natural gas from the Bay of Bengal, exploited by the state-owned Bangladesh Oil, Gas, and Minerals Corporation, came on stream in 1988.

ENVIRONMENT

Sustainability rank: 86th

1% (0.7% partially protected)

0.2 tonnes per capita

ENVIRONMENTAL TREATIES

Yes	Yes	Yes
Yes	Yes	Yes

Bangladesh's climate gives rise to devastating floods and cyclones, with consequent huge death tolls and substantial damage to crops. The country is too poor to finance environmental initiatives.

MEDIA

TV ownership low

Daily newspaper circulation 9 per 1000 people

PUBLISHING AND BROADCAST MEDIA

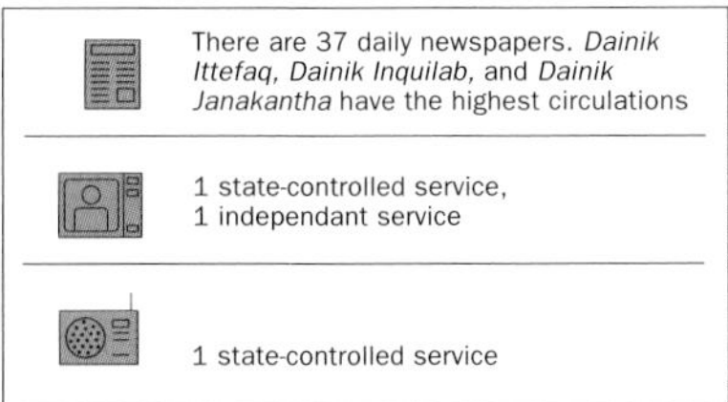

There are 37 daily newspapers. *Dainik Ittefaq*, *Dainik Inquilab*, and *Dainik Janakantha* have the highest circulations

1 state-controlled service, 1 independant service

1 state-controlled service

Press freedom, which emerged after the fall of President Ershad in 1990, has subsequently been gradually eroded under successive civilian governments. Of the daily newspapers, the ten English-language titles appeal mainly to the urban elite. Among political weeklies, the most prominent is *Holiday*. The vast majority – over 70% – of TV programs are produced locally by the state-run services. Foreign satellite channels are increasingly available, but are subject to government bans.

CRIME

Death penalty used

 70,000 prisoners

 Up 17% 1996–1998

CRIME RATES

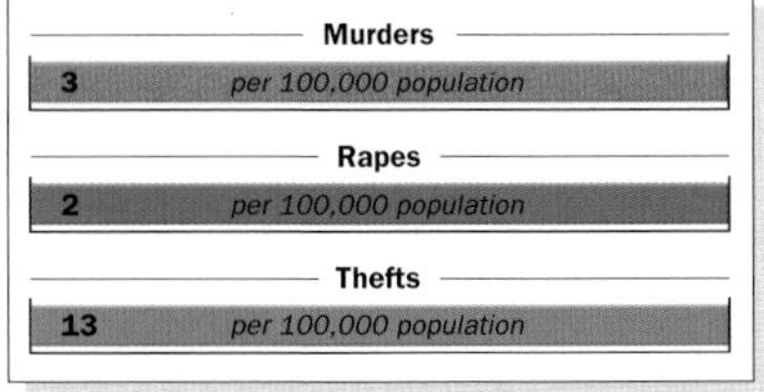
Murders
3 per 100,000 population
Rapes
2 per 100,000 population
Thefts
13 per 100,000 population

Rising levels of political and religious violence have led to the enforcement of antiterrorism legislation, containing provisions for summary justice and heavy penalties, including death. There has been a recent sharp rise in crimes against women, including murder, rape, abduction, and acid attacks. Deaths in Bangladeshi prisons are common, and in addition the human rights record of the security forces, especially the paramilitary Bangladesh Rifles, has been attacked by Amnesty International.

EDUCATION

School leaving age: 10

41%

434,309 students

THE EDUCATION SYSTEM

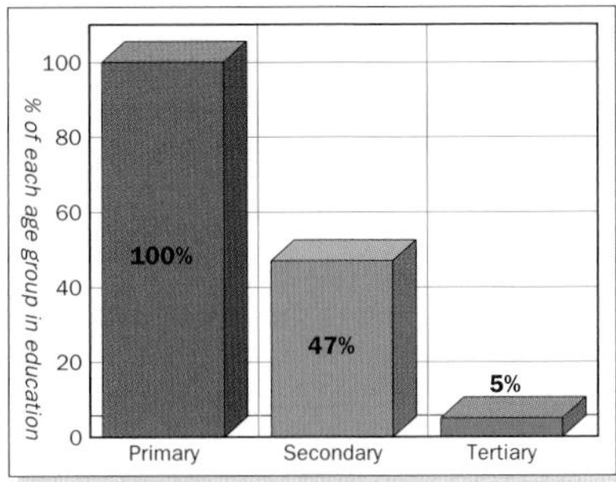
100
80
60
40
20
0
% of each age group in education
100%
47%
5%
Primary
Secondary
Tertiary

Education issues in Bangladeshi society have been poorly addressed, although successive governments have promised to improve literacy levels by increasing spending. A dramatic reduction in the 1990s in the number of child workers meant an accompanying rise in school attendance. Exam cheating is a serious problem. The seven universities are frequently beset by political violence.

HEALTH

Welfare state health benefits

  1 per 5000 people

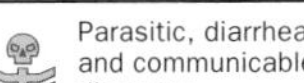 Parasitic, diarrheal, and communicable diseases

More resources are needed to boost health care in rural areas. Although primary health care in these areas improved in the 1990s, Bangladesh's health problems remain severe and are exacerbated by a shortage of medical staff and facilities. Priority for birth control programs has helped reduce the population growth rate by more than 20% over the last 15 years. Half the population is exposed to high levels of arsenic in drinking water.

SPENDING

GDP/cap. increase

CONSUMPTION AND SPENDING

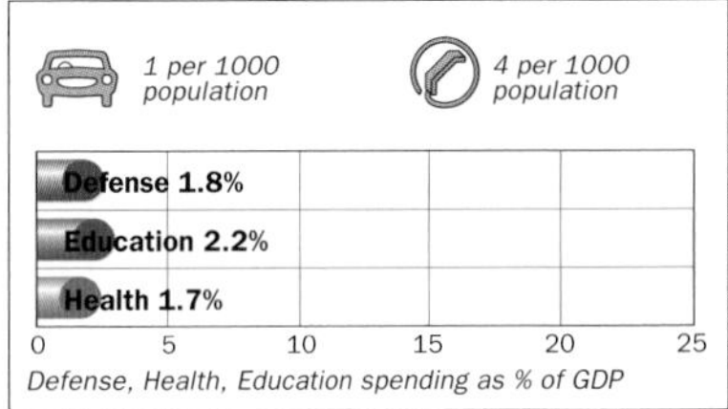
1 per 1000 population
4 per 1000 population
Defense 1.8%
Education 2.2%
Health 1.7%
0 5 10 15 20 25
Defense, Health, Education spending as % of GDP

Average incomes in Bangladesh remain very low, but wealth disparities are not quite as marked as in India or Pakistan. State officials tend to be among the better-off members of society.

WORLD RANKING

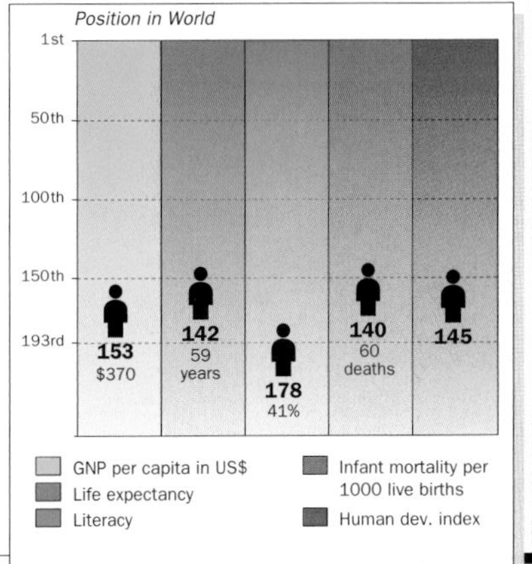
Position in World
1st
50th
100th
150th
193rd
153
$370
142
59 years
178
41%
140
60 deaths
145
GNP per capita in US$
Life expectancy
Literacy
Infant mortality per 1000 live births
Human dev. index

B

BARBADOS

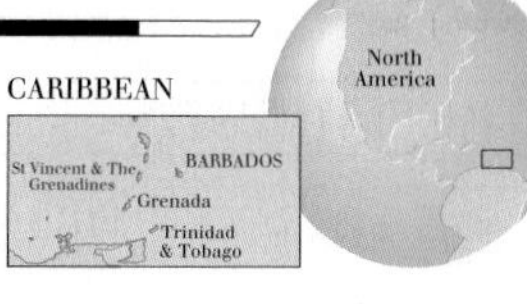
North America
CARIBBEAN
St Vincent & The Grenadines
BARBADOS
Grenada
Trinidad & Tobago

OFFICIAL NAME: Barbados **CAPITAL:** Bridgetown
POPULATION: 268,000 **CURRENCY:** Barbados dollar **OFFICIAL LANGUAGE:** English

1966

1966

Nov 30

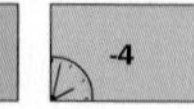
BDS
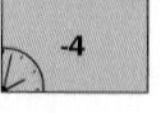

-4

+1246

.bb

SITUATED TO THE NORTHEAST of Trinidad, Barbados is the most easterly of the West Indian Windward Islands. In the 16th century, the Portuguese were the first Europeans to reach the island, then inhabited by Arawak Indians. However, Barbados was not colonized until the 1620s, when British settlers arrived. Popularly referred to by its neighbors as "little England," Barbados now seeks to forge a new national identity for itself.

CLIMATE

Tropical oceanic

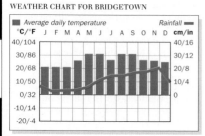

WEATHER CHART FOR BRIDGETOWN

Average daily temperature
Rainfall
°C/°F J F M A M J J A S O N D cm/in
40/104 40/16
30/86 30/12
20/68 20/8
10/50 10/4
0/32 0
-10/14
-20/-4

Barbados has a moderate tropical climate and is sunnier and drier than its more mountainous Caribbean neighbors. Hurricanes may occur in the rainy season.

TRANSPORTATION

Drive on left

Grantley Adams International, Bridgetown
1.21m passengers

79 ships
733,319 grt

THE TRANSPORTATION NETWORK

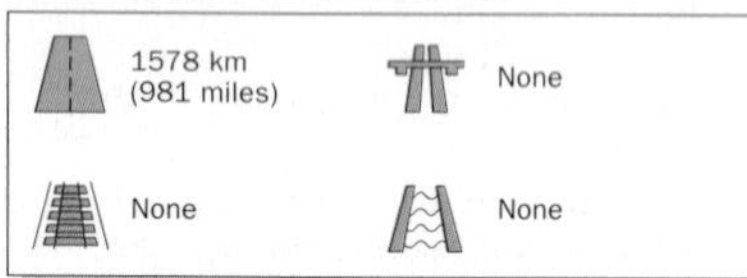
1578 km
(981 miles)
None
None
None

A multimillion dollar expansion program has upgraded facilities at the international airport. Piers at Bridgetown's port have been improved with foreign aid, as have the island's paved roads. There are bus routes over most of the island.

House of Assembly, Trafalgar Square, Bridgetown. *Barbados's parliament, the third oldest in the Commonwealth, dates from 1639.*

TOURISM

Visitors : Population 2.1:1

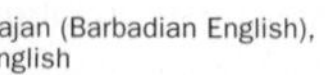
556,000 visitors

Up 8% in 2000

MAIN TOURIST ARRIVALS

UK 39%
USA 20%
Canada 11%
Other 30%
0 10 20 30 40
% of total arrivals

Tourists, essential to the Barbadian economy, are attracted by the white sandy beaches and sporting activities, as well as the post-colonial ambience and the generally laid-back culture.

PEOPLE

Pop. density high

Bajan (Barbadian English), English

623/km² (1614/mi²)

THE URBAN/RURAL POPULATION SPLIT

50% 50%

RELIGIOUS PERSUASION

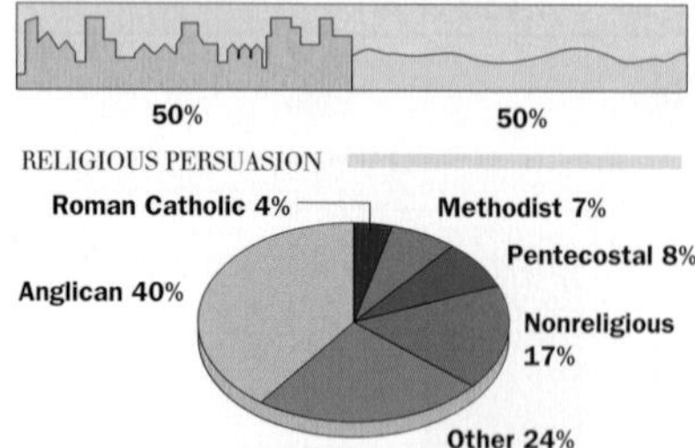
Roman Catholic 4%
Methodist 7%
Pentecostal 8%
Anglican 40%
Nonreligious 17%
Other 24%

Most Bajans are the descendants of Africans brought to the island between the 16th and 19th centuries; there are also small groups of south Asians and Europeans, mainly expatriates from the UK, many of whom take up residence on retirement. There is some latent tension between the white community, which controls most of the economy, and the majority black population, although this rarely spills over into violence. Increasing social mobility has allowed many black Bajans to move into the professions and the civil service. Barbados enjoys a higher standard of living than most Caribbean countries.

POLITICS

Multiparty elections

L. House 1999/2004
U. House 1999/2004

H.M. Queen Elizabeth II

AT THE LAST ELECTION

House of Assembly 28 seats

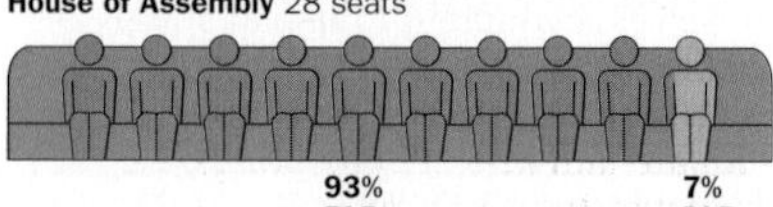
93% BLP
7% DLP

BLP = Barbados Labour Party
DLP = Democratic Labour Party

Senate 21 seats

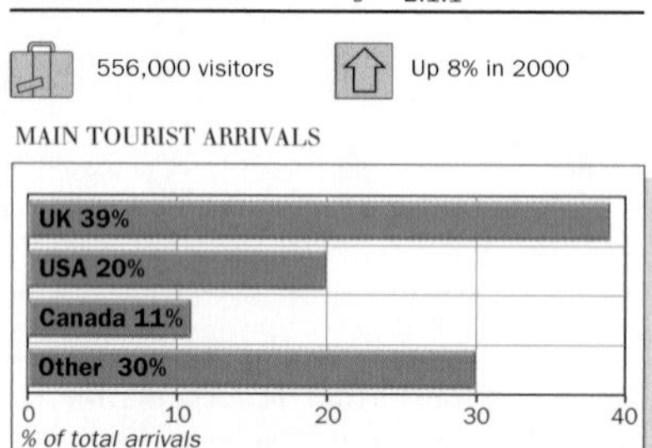

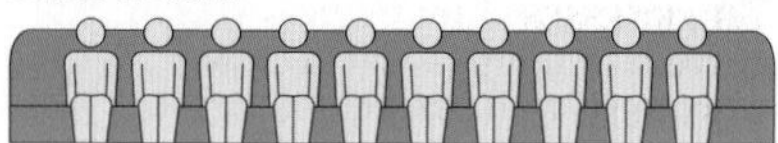

The members of the Senate are appointed. Twelve are chosen by the prime minister, 2 by the leader of the opposition, and 7 independents by the governor-general

Barbados is a multiparty democracy. A primarily European, affluent elite finances the parties and has an indirect influence on government policy. The BLP swept to power in 1994 and won a further landslide in 1999. Owen Arthur, BLP leader and prime minister, prioritizes economic growth and international competitiveness. He has pledged to transform Barbados into a republic, while remaining a member of the Commonwealth.

WORLD AFFAIRS

Joined UN in 1966

ACS
Comm
Caricom
NAM
OAS

Considered to be an international tax haven, Barbados has been under pressure to implement reforms.

AID

Recipient

No net receipts

Down sharply in 1999–2000

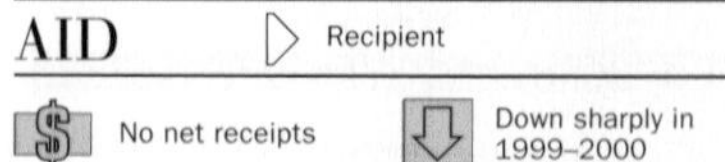

Most aid comes from the EU, the UN and the UK, mainly in the form of development project loans and balance-of-payments support.

DEFENSE

No compulsory military service

US$13m

Up 8% in 2000

The small Barbadian army and the constabulary benefit from financial support and training from the US and UK governments, which also supply equipment. Barbados is the headquarters of the Regional Security System, established in 1982 by the Windward and Leeward Islands, a body which acts as a multinational security force for its members.

B

ECONOMICS

Inflation 3.1% p.a. (1990–2000)

 US$2.47bn

 1.99 Barbados dollars

SCORE CARD

❑ WORLD GNP RANKING	133rd
❑ GNP PER CAPITA	US$9250
❑ BALANCE OF PAYMENTS	–US$126m
❑ INFLATION	2.4%
❑ UNEMPLOYMENT	10%

STRENGTHS

Well-developed tourism based on climate and accessibility. Sugar industry. Information processing and financial services are important new growth sectors.

WEAKNESSES

Narrow economic base, vulnerable to downturns in tourism, failures of sugar harvest, and the sector's dependency on loans and secure markets. Relatively high manufacturing costs.

EXPORTS

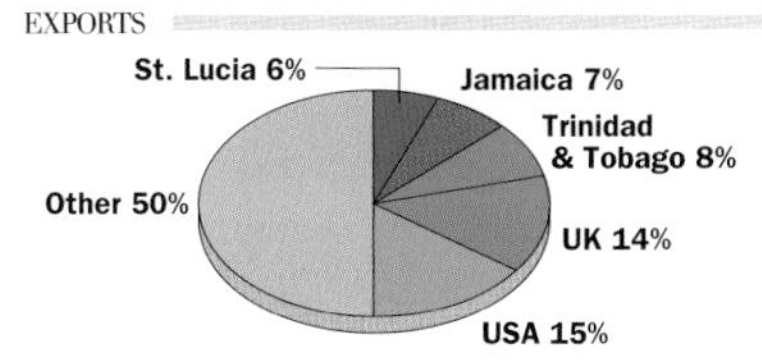
St. Lucia 6%
Jamaica 7%
Trinidad & Tobago 8%
Other 50%
UK 14%
USA 15%

IMPORTS

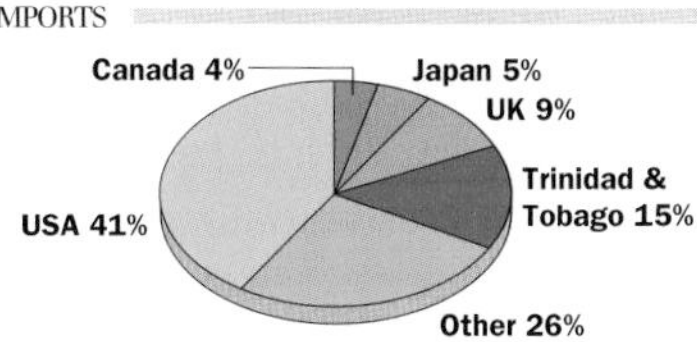
Canada 4%
Japan 5%
UK 9%
Trinidad & Tobago 15%
USA 41%
Other 26%

BARBADOS

Total Land Area : 430 sq. km (166 sq. miles)

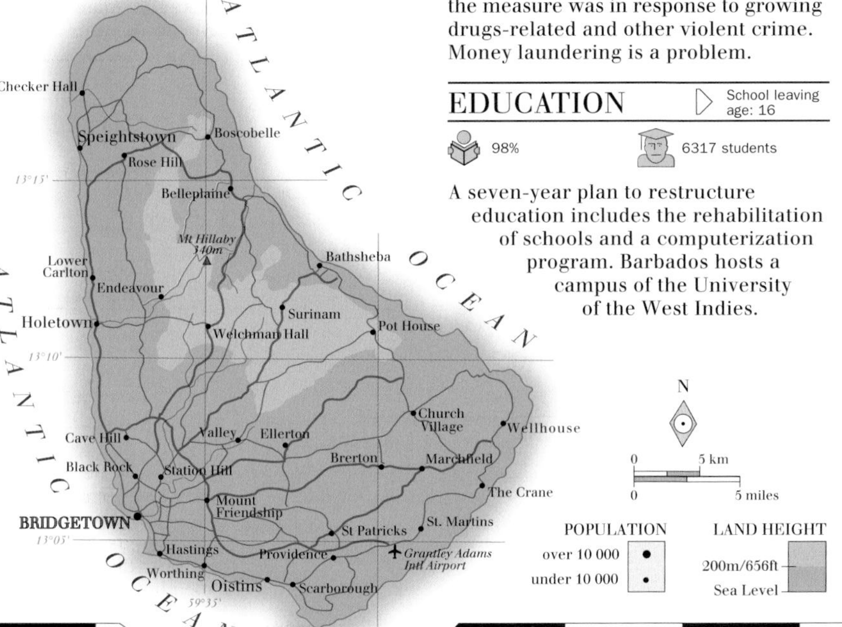
ATLANTIC
ATLANTIC
OCEAN
OCEAN
Checker Hall
Speightstown
Boscobelle
Rose Hill
Belleplaine
Mt Hillaby 340m
Bathsheba
Lower Carlton
Endeavour
Surinam
Holetown
Welchman Hall
Pot House
Church Village
Wellhouse
Cave Hill
Valley
Ellerton
Black Rock
Station Hill
Brerton
Marchfield
The Crane
Mount Friendship
BRIDGETOWN
St Patricks
St. Martins
Hastings
Providence
Grantley Adams Intl Airport
Worthing
Oistins
Scarborough
N
0 5 km
0 5 miles
POPULATION
over 10 000
under 10 000
LAND HEIGHT
200m/656ft
Sea Level

RESOURCES

Electric power 156,000 kw

 3206 tonnes

 1604 b/d (reserves 7.6m barrels)

 41,000 sheep, 33,000 pigs, 3.6m chickens

 Oil, natural gas

Barbados has few strategic resources. The domestic petroleum industry provides about one-third of the country's energy requirements.

ENVIRONMENT

Not available

1%

 3.4 tonnes per capita

Oil slicks created by waste dumped from passing ships are polluting the encircling reef and adversely affecting the life cycle of the flying fish, Barbados's main fish stock.

MEDIA

TV ownership high

 Daily newspaper circulation 199 per 1000 people

PUBLISHING AND BROADCAST MEDIA

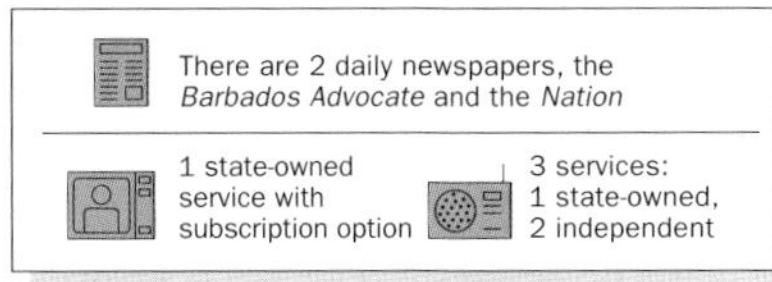

There are 2 daily newspapers, the *Barbados Advocate* and the *Nation*

1 state-owned service with subscription option

3 services: 1 state-owned, 2 independent

There is no political interference in the media. The two daily newspapers are privately owned. Multichannel TV is available on subscription.

CRIME

Death penalty in use

 772 prisoners

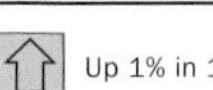 Up 1% in 1999

A firearms amnesty announced in 1998 proposed heavy fines and prison sentences for illegal possession of guns; the measure was in response to growing drugs-related and other violent crime. Money laundering is a problem.

EDUCATION

School leaving age: 16

98%

6317 students

A seven-year plan to restructure education includes the rehabilitation of schools and a computerization program. Barbados hosts a campus of the University of the West Indies.

CHRONOLOGY

Colonized by the British in 1627, Barbados grew rich in the 18th century from sugar produced using slave labor.

- ❑ **1951** Universal adult suffrage introduced.
- ❑ **1961** Full internal self-government.
- ❑ **1966** Independence from the UK.
- ❑ **1983** Barbados supports and provides a base for the US invasion of Grenada.
- ❑ **1994, 1999** The BLP wins two successive general elections.

HEALTH

Welfare state health benefits

 1 per 800 people

 Heart and cerebrovascular diseases, cancers

The health system is based on subsidized government-run clinics and hospitals, supplemented by more expensive private clinics and private doctors. Facilities are within easy reach of all Bajans.

SPENDING

GDP/cap. increase

CONSUMPTION AND SPENDING

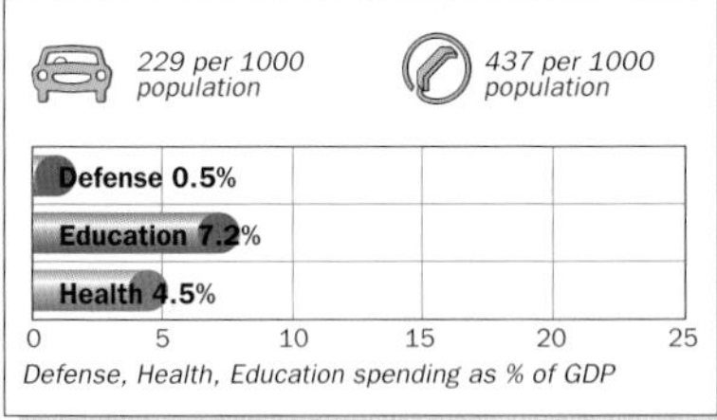
229 per 1000 population
437 per 1000 population
Defense 0.5%
Education 7.2%
Health 4.5%
0
5
10
15
20
25
Defense, Health, Education spending as % of GDP

A significant disparity exists between most Bajans and a small affluent group, its members usually of European origin, which owns and controls business and industry, and parades status symbols such as yachts. Prime Minister Arthur stated in 1998 that "abject poverty" existed in the country.

WORLD RANKING

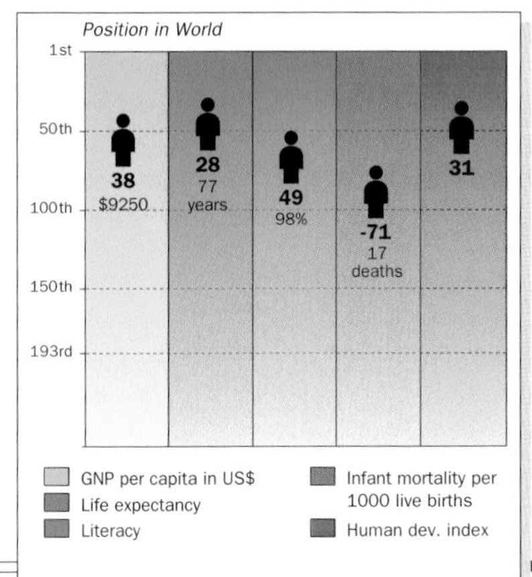
Position in World
1st
50th
100th
150th
193rd
38
$9250
28
77 years
49
98%
-71
17 deaths
31
GNP per capita in US$
Life expectancy
Literacy
Infant mortality per 1000 live births
Human dev. index

B

BELARUS

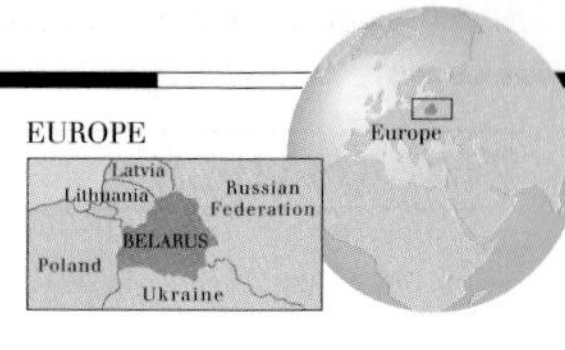
EUROPE
Europe
Latvia
Lithuania
Russian Federation
BELARUS
Poland
Ukraine

OFFICIAL NAME: Republic of Belarus **CAPITAL:** Minsk **POPULATION:** 10.1 million
CURRENCY: Belarussian rouble **OFFICIAL LANGUAGES:** Belarussian and Russian

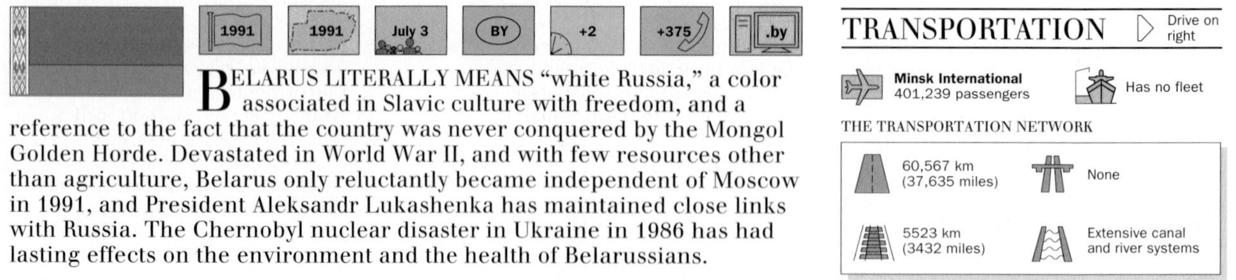
1991
1991
July 3
BY
+2
+375
.by

BELARUS LITERALLY MEANS "white Russia," a color associated in Slavic culture with freedom, and a reference to the fact that the country was never conquered by the Mongol Golden Horde. Devastated in World War II, and with few resources other than agriculture, Belarus only reluctantly became independent of Moscow in 1991, and President Aleksandr Lukashenka has maintained close links with Russia. The Chernobyl nuclear disaster in Ukraine in 1986 has had lasting effects on the environment and the health of Belarussians.

TRANSPORTATION

Drive on right

Minsk International 401,239 passengers

Has no fleet

THE TRANSPORTATION NETWORK

60,567 km (37,635 miles)

None

5523 km (3432 miles)

Extensive canal and river systems

Belarus has no direct access to the sea, but is close to the Baltic ports. Railroad communications are good.

CLIMATE

Continental

WEATHER CHART FOR MINSK

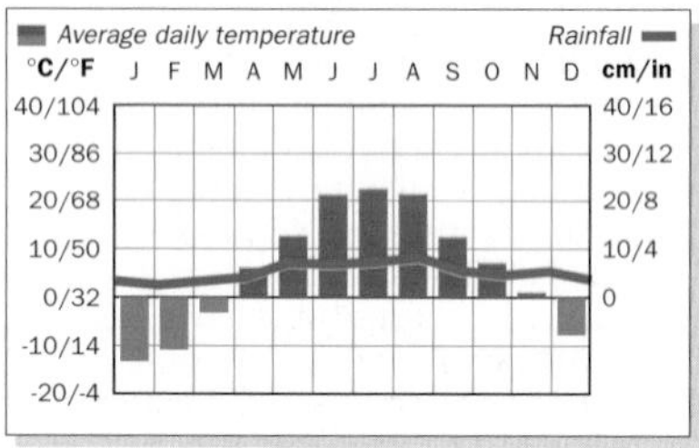

Average daily temperature | Rainfall

°C/°F J F M A M J J A S O N D cm/in

40/104, 30/86, 20/68, 10/50, 0/32, -10/14, -20/4

40/16, 30/12, 20/8, 10/4, 0

Belarus has a continental climate somewhat moderated by the influence of the nearby Baltic Sea. Temperatures in winter drop well below freezing, however, while summers can be hot and humid. Summer is also the main season for rainfall.

Much of southern Belarus is marshy and sparsely populated. It includes the vast Pripet Marshes and the Dnieper lowlands.

TOURISM

Visitors : Population 1:28

355,000 visitors

Up 42% in 1998

MAIN OVERSEAS ARRIVALS

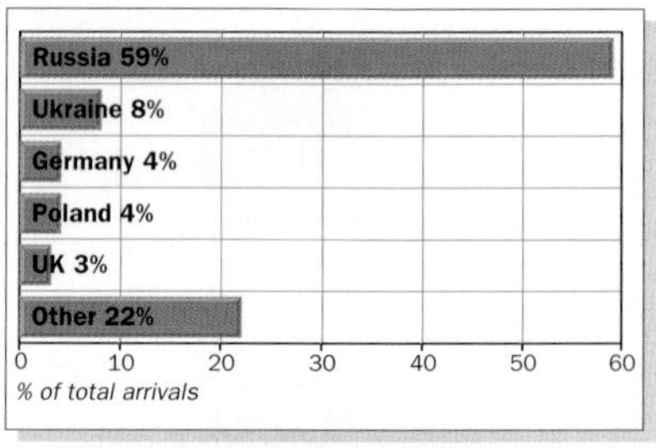

	% of total arrivals
Russia	59%
Ukraine	8%
Germany	4%
Poland	4%
UK	3%
Other	22%

Belarus has fewer tourists than its neighbors. Many of its historic buildings were destroyed during World War II. Minsk was totally flattened, and is now characterized by Stalinist and other high-rise buildings. There is little of mass appeal on which to build a tourist industry.

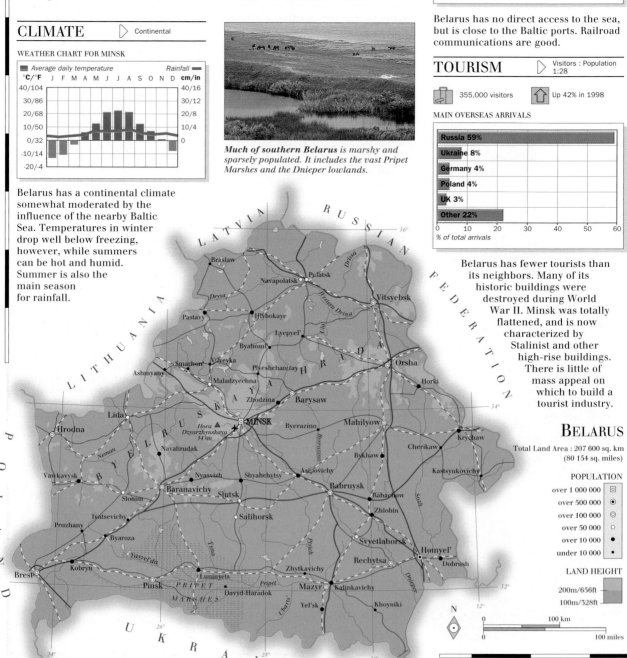

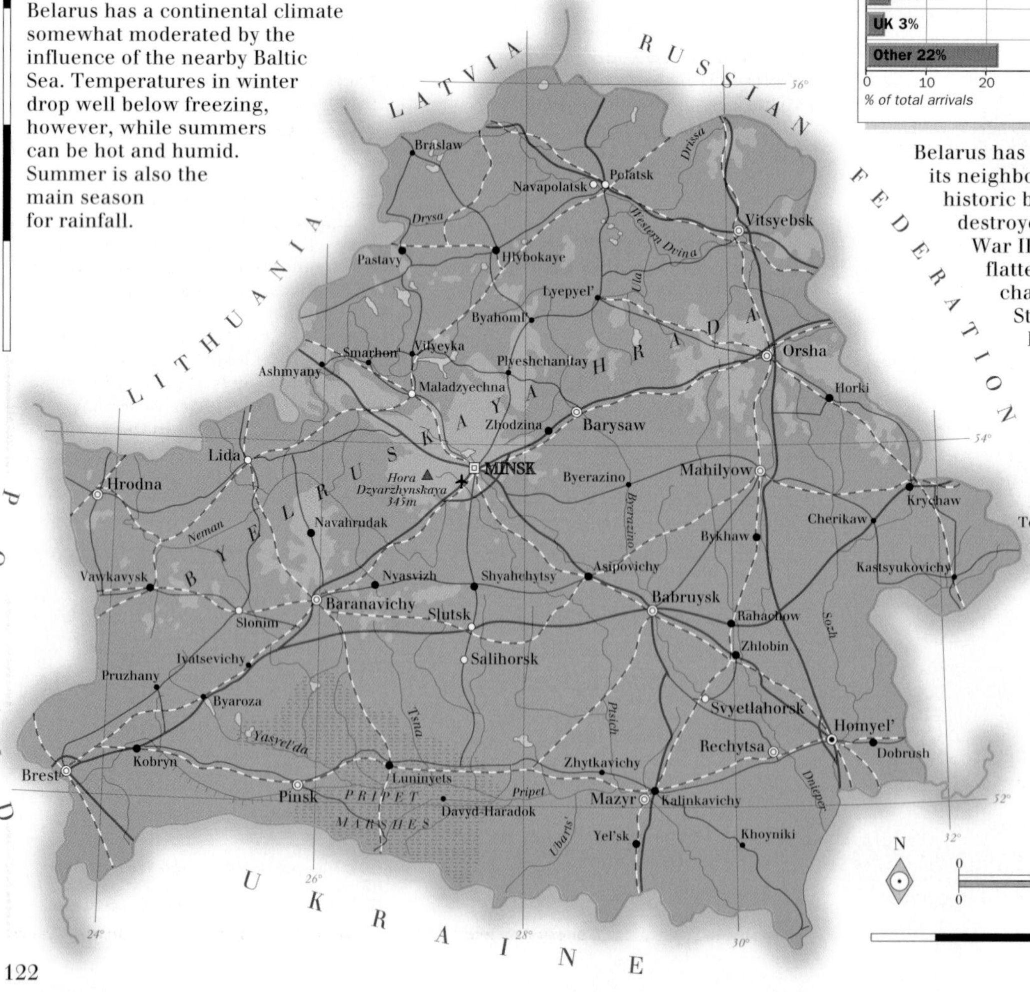

BELARUS

Total Land Area : 207 600 sq. km (80 154 sq. miles)

POPULATION

over 1 000 000
over 500 000
over 100 000
over 50 000
over 10 000
under 10 000

LAND HEIGHT

200m/656ft
100m/328ft

0 100 km
0 100 miles

B

PEOPLE

Pop. density low

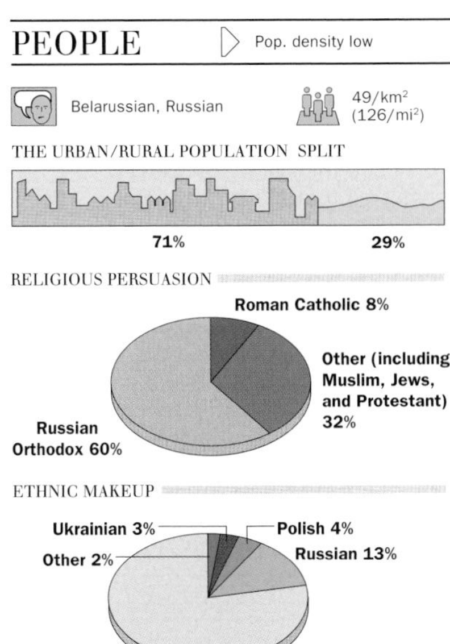
Belarussian, Russian
49/km²
(126/mi²)
THE URBAN/RURAL POPULATION SPLIT
71%
29%
RELIGIOUS PERSUASION
Roman Catholic 8%
Other (including Muslim, Jews, and Protestant) 32%
Russian Orthodox 60%
ETHNIC MAKEUP
Ukrainian 3%
Polish 4%
Other 2%
Russian 13%
Belarussian 78%

Only 2% of the population is non-Slav and there is little ethnic tension. Under a law passed in 1992, the entire population has an automatic right to Belarussian citizenship. Most people speak Russian, and only 11% of the population are fluent in Belarussian, which is used mainly in rural areas. A 1995 referendum and 1998 legislation declared that both languages have equal status.

POPULATION AGE BREAKDOWN

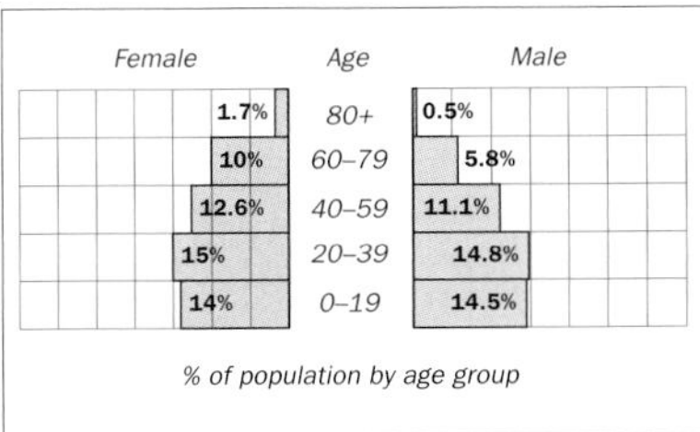
Female
Age
Male
1.7% 80+ 0.5%
10% 60–79 5.8%
12.6% 40–59 11.1%
15% 20–39 14.8%
14% 0–19 14.5%
% of population by age group

POLITICS

Multiparty elections

L. House 2000/2004
U. House 2000/2004
President Aleksandr Lukashenka

AT THE LAST ELECTION

House of Representatives 110 seats

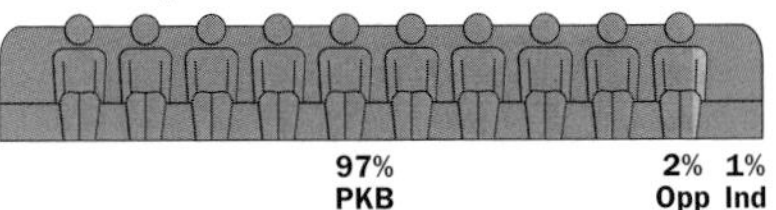
97% PKB
2% Opp
1% Ind

PKB = Party of Communists of Belarus and government supporters **Opp** = Minor opposition parties **Ind** = Independents

Council of the Republic 64 seats

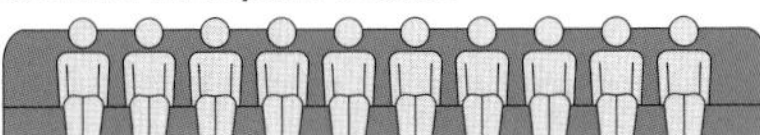

The Council of the Republic is indirectly elected

Under the 1994 constitution, amended in 1996, Belarus has a directly elected president and a bicameral parliament.

PROFILE

Belarus, by far the slowest of the former Soviet states to implement political reform, has struggled to find an identity since 1991. A post-Soviet constitution was not adopted until 1994, and only in 1995 was the first fully fledged post-Soviet parliament elected, dominated by the PKB and its Agrarian Party ally. There is no strong pluralist culture to check authoritarian moves by the president.

Aleksandr Lukashenka was unexpectedly elected as Belarus's first president in 1994. In 1996 a new constitution significantly strengthened his powers. Pro-Lukashenka deputies then voted to replace the Supreme Council with a new House of Representatives. Lukashenka has also extended his own term of office. The opposition contests the legitimacy of this, as does the EU. A clampdown on political opponents effectively invalidated parliamentary elections in late 2000/early 2001, and Lukashenka's reelection in late 2001 was immediately condemned by observers.

MAIN POLITICAL ISSUES

Relationship with Russia

In 1994 an accord (reinforced in 1999 and 2000) was signed on future monetary union with Russia. Lukashenka has sought ever closer relations leading toward a joint presidency. A union treaty was signed in late 1999, but is more symbolic than practical. With Russian reticence prevailing over Belarus's enthusiasm for the union, the treaty has produced no significant developments toward joint state institutions or economic programs.

The environment

The 1986 Chernobyl nuclear disaster continues to cast a shadow. The cleanup operation, slow and laborious, will take decades. It is a major drain on state finances, accounting for 25% of spending. A "Chernobyl tax" is levied on businesses.

President Aleksandr Lukashenka seeks closer ties with the Russian Federation.

Vladimir Goncharik, Lukashenka's main rival in presidential elections in 2001.

WORLD AFFAIRS

Joined UN in 1945

EAPC
CIS
IAEA
CEI
OSCE

Relations with Russia are paramount. Numerous bilateral agreements were signed after independence in 1991. Ties have been strengthened further by the pro-Russian president Aleksandr Lukashenka, although many in Russia fear that closer links will drain Moscow's resources for little strategic gain.

In mid-1998 diplomats were evicted from many official residences. The US, the EU, and others temporarily withdrew their ambassadors in protest. The March 2001 clampdown on political opponents ended moves to relax EU sanctions, introduced four years previously over the authoritarian changes in the constitution.

AID

Recipient

$40m (receipts)

Up 3% in 2000

Although both the World Bank and the IMF provided loans for Belarus in the early 1990s, the lack of structural reforms since Lukashenka's administration came to power in 1994 has meant that further aid has been stalled. Some US bilateral aid continued, but the EU in particular has made it clear that support will depend on human rights improvements and the reversal of authoritarian threats to democracy.

Both the US and the EU extended credits to Belarus to assist in the conversion of the defense industry to nonmilitary production. Belarus also still requires aid to combat the effects of radiation pollution in the wake of the Chernobyl nuclear accident of 1986.

CHRONOLOGY

After forming part of medieval Kievan Rus, Belarus experienced rule by three of its neighbors – Lithuania, Poland, and Russia – before incorporation into the USSR.

- **1918** Belarussian Bolsheviks stage coup. Independence as Belorussian Soviet Socialist Republic (BSSR).
- **1919** Invaded by Poland.
- **1920** Minsk retaken by Red Army. Eastern Belorussia reestablished as Soviet Socialist Republic.
- **1921** Treaty of Riga – Western Belorussia incorporated into Poland.
- **1922** BSSR merges with Soviet Russia and Ukraine to form USSR.
- **1929** Stalin implements collectivization of agriculture.
- **1939** Western Belorussia reincorporated into USSR when Soviet Red Army invades Poland.

B

CHRONOLOGY *continued*

- **1941–1944** Occupied by Germany during World War II.
- **1945** Founding member of UN.
- **1965** K. T. Mazurau, Communist Party of Belorussia (PKB) leader, becomes first deputy chair of Soviet government.
- **1986** Radioactive fallout after Chernobyl accident affects 70% of country.
- **1988** Evidence revealed of mass executions (over 300,000) by Soviet military between 1937 and 1941 near Minsk. Popular outrage fuels formation of nationalist Belorussian Popular Front (BPF), with Zyanon Paznyak as president. PKB authorities crush demonstration.
- **1989** Belarussian adopted as republic's official language.
- **1990** PKB prevents BPF from participating in elections to Supreme Soviet. BPF members join other opposition groups in Belorussian Democratic Bloc (BDB). BDB wins 25% of seats. PKB bows to opposition pressure and issues Declaration of the State Sovereignty of BSSR.
- **1991** March, 83% vote in referendum to preserve union with USSR. April, strikes against PKB and its economic policies. August, independence declared. Republic of Belarus adopted as official name. Stanislau Shushkevich elected chair of Supreme Soviet. December, Belarus, Russia, and Ukraine establish CIS.
- **1992** Supreme Soviet announces that Soviet nuclear weapons must be cleared from Belarus by 1999. Help promised from US.
- **1993** Belarussian parliament ratifies START-I and nuclear nonproliferation treaties.
- **1994** New presidential constitution approved; Aleksandr Lukashenka defeats conservative prime minister Vyacheslav Kebich in elections. Monetary union (reentry into rouble zone) agreed with Russia.
- **1995** First fully fledged post-Soviet parliament elected.
- **1996** Referendum approves changes to constitution strengthening Lukashenka's powers.
- **1997** Belarus and Russia ratify union treaty and Charter.
- **1998** Eviction from embassies sparks withdrawal of Western ambassadors.
- **1999** Union treaty with Russia.
- **2000–2001** Disputed parliamentary elections; clampdown on PKB's political opponents.
- **2001** Lukashenka reelected. Observers label election seriously flawed.

DEFENSE

Compulsory military service

 $366m

 Down 21% in 2000

BELARUSSIAN ARMED FORCES

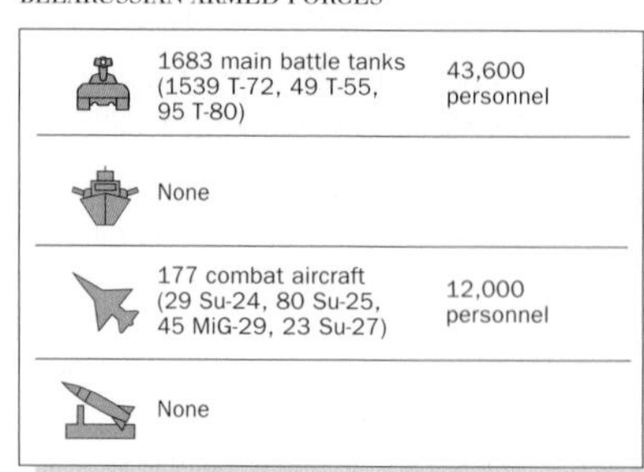
1683 main battle tanks (1539 T-72, 49 T-55, 95 T-80)
43,600 personnel
None
177 combat aircraft (29 Su-24, 80 Su-25, 45 MiG-29, 23 Su-27)
12,000 personnel
None

After the breakup of the Soviet Union in 1991, Belarus briefly adopted a policy of neutrality. It also committed itself to disposing of its inherited nuclear capability. Tactical nuclear weapons were removed by 1993 and strategic nuclear weapons by 1996.

Despite joining the CIS collective security agreement in 1993, Belarus joined NATO's Partnerships for Peace program in 1995. Lukashenka has not developed NATO ties further, preferring to establish stronger military links with Moscow, which include a 1997 Treaty on Military Cooperation and a 1999 agreement on joint procedures for arms exports. Belarus bears some of the costs of Russian troops stationed on its territory.

ECONOMICS

Inflation 355% p.a. (1990–2000)

$28.7bn

1218–1603 Belarussian roubles

SCORE CARD

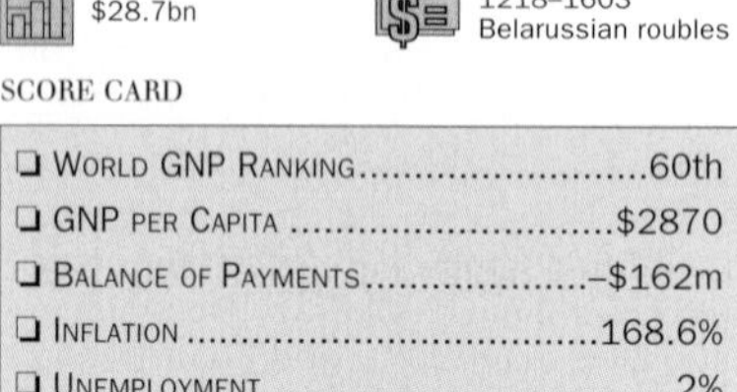

- World GNP Ranking.........................60th
- GNP per Capita $2870
- Balance of Payments....................–$162m
- Inflation 168.6%
- Unemployment 2%

EXPORTS

Lithuania 5%
Latvia 6%
Germany 3%
Ukraine 8%
Russia 51%
Other 27%

IMPORTS

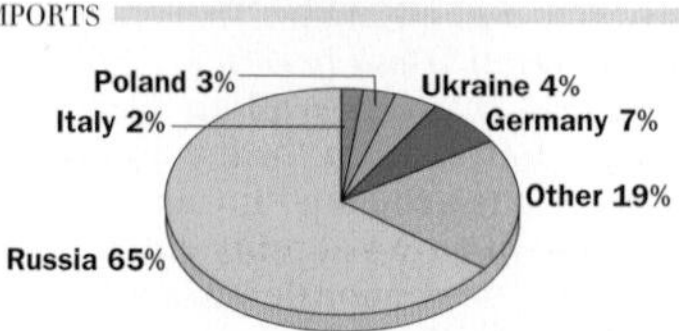
Poland 3%
Ukraine 4%
Italy 2%
Germany 7%
Other 19%
Russia 65%

ECONOMIC PERFORMANCE INDICATOR

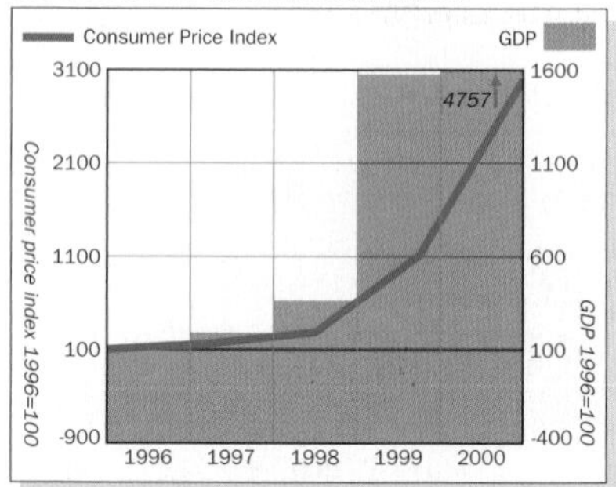
Consumer Price Index
GDP
3100
2100
1100
100
-900
1600
1100
600
100
-400
4757
Consumer price index 1996=100
GDP 1996=100
1996
1997
1998
1999
2000

Strengths

Low unemployment combined with relative social stability. Potential of forestry and agriculture.

Weaknesses

Lack of economic restructuring; support for outmoded businesses. Few natural resources. Dependence on Russia for energy and raw materials. Cleanup costs of Chernobyl.

Profile

After 1991, Belarus adopted a slower pace of economic reform than did other former Soviet states, since attempts to move more quickly to a market economy were thwarted by the largely conservative parliament. Upon election in 1994, Lukashenka suspended privatization moves, resuming them only halfheartedly in 1995, under a policy of "market socialism." Traditional industries continued to receive big subsidies as the government printed money to increase production. A currency crisis in 1998, along with rampant inflation, combined with two successive bad harvests in 1998–1999 to create a major threat to the economy. Inflation receded in 2000–2001, but the rate of economic growth also slowed.

BELARUS : MAJOR BUSINESSES

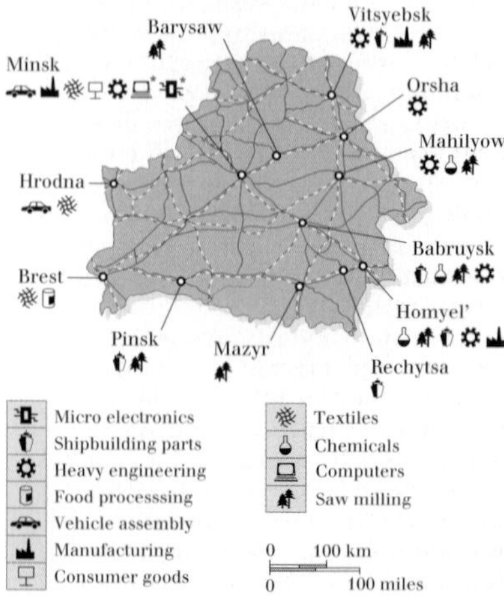
Vitsyebsk
Barysaw
Minsk
Orsha
Mahilyow
Hrodna
Babruysk
Brest
Homyel'
Pinsk
Mazyr
Rechytsa
Micro electronics
Shipbuilding parts
Heavy engineering
Food processsing
Vehicle assembly
Manufacturing
Consumer goods
Textiles
Chemicals
Computers
Saw milling
0 100 km
0 100 miles
* significant multinational ownership

B

RESOURCES

Electric power 7.5m kw

5809 tonnes

36,692 b/d (reserves 201m barrels)

4.22m cattle, 3.43m pigs, 32m chickens

Oil, natural gas, coal, rock salt

ELECTRICITY GENERATION

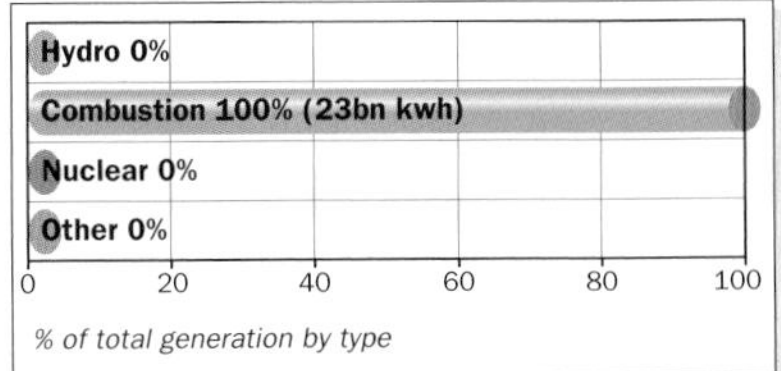
Hydro 0%
Combustion 100% (23bn kwh)
Nuclear 0%
Other 0%
0
20
40
60
80
100
% of total generation by type

Belarus has no significant strategic resources and is heavily dependent on the Russian Federation for fuel and energy supplies. Small quantities of oil and natural gas exist close to the Polish border.

BELARUS : LAND USE

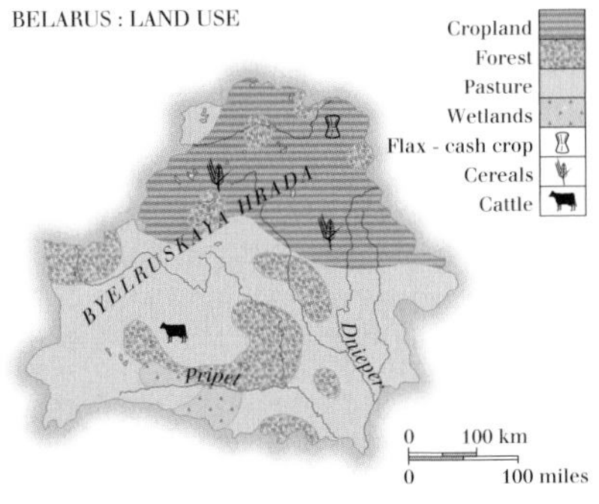
Cropland
Forest
Pasture
Wetlands
Flax - cash crop
Cereals
Cattle
BYELRUSKAYA HRADA
Pripet
Dnieper
0
100 km
0
100 miles

ENVIRONMENT

Sustainability rank: 49th

6%

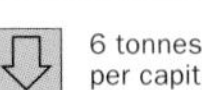

6 tonnes per capita

ENVIRONMENTAL TREATIES

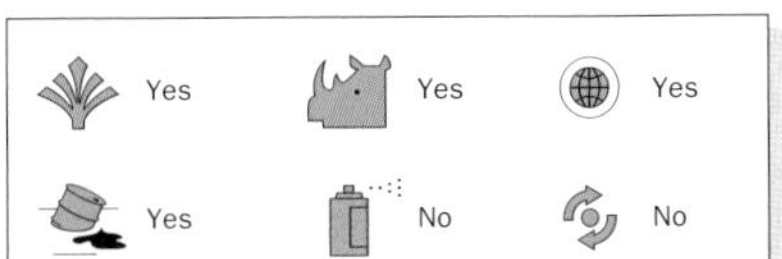
Yes
Yes
Yes
Yes
No
No

The massive leak from Ukraine's Chernobyl nuclear reactor in 1986 released a huge cloud of radiation. Some 70% of the fallout fell on Belarus, including Minsk; 2.3 million people were immediately affected, and cases of leukemia and cancer continue to emerge. Farmland, forests, and water were all contaminated, including underwater streams feeding rivers in eastern Poland. Some areas in the fallout zone are still being farmed. The cleanup program swallows 25% of government finances, despite substantial Western aid.

The Belovezha primeval forest, on the border with Poland, is Europe's largest nature reserve. It is a habitat of the European bison or wisent.

MEDIA

TV ownership high

Daily newspaper circulation 173 per 1000 people

PUBLISHING AND BROADCAST MEDIA

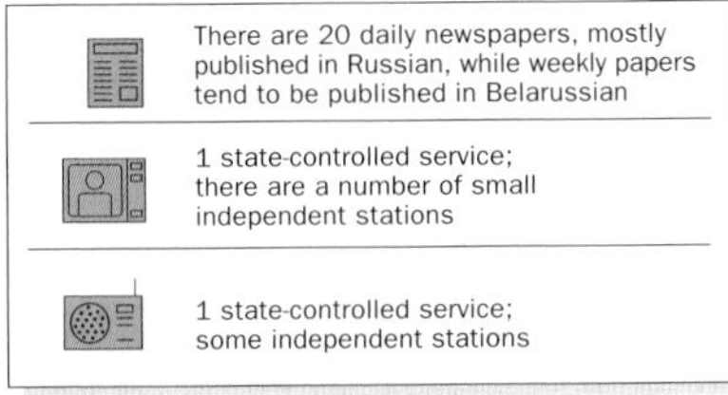
There are 20 daily newspapers, mostly published in Russian, while weekly papers tend to be published in Belarussian
1 state-controlled service; there are a number of small independent stations
1 state-controlled service; some independent stations

There are some independent media outlets, but government critics face harassment. Press freedom is curbed; state-backed publications predominate.

CRIME

Death penalty in use

56,000 prisoners

Up 6% in 1999

CRIME RATES

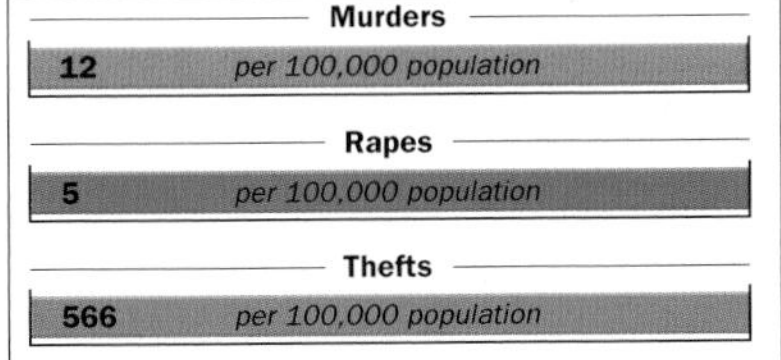
Murders
12
per 100,000 population
Rapes
5
per 100,000 population
Thefts
566
per 100,000 population

As elsewhere in the former Soviet Union, economic hardship and a general breakdown in law and order have resulted in a significant rise in crime. The prison population exceeds the intended capacity of 40,000. Belarus has become a transshipment point for illegal narcotics destined for western Europe, while locally produced opium supplies the internal market.

EDUCATION

School leaving age: 15

99%

301,753 students

THE EDUCATION SYSTEM

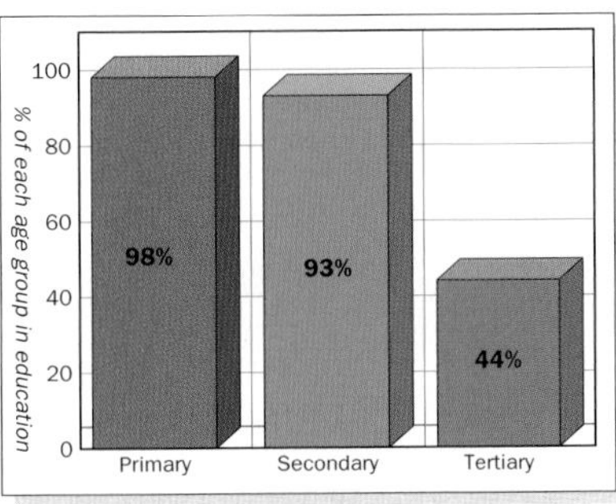
100
80
60
40
20
0
% of each age group in education
98%
93%
44%
Primary
Secondary
Tertiary

Education is officially compulsory for nine years, and teaching is mainly in Russian. Activists complain that because of political bias there is inadequate provision for the teaching of Belarussian. University education – taught in Russian – is of a fairly high standard.

HEALTH

Welfare state health benefits

1 per 227 people

Heart attacks, cancers, accidents, violence

The Chernobyl nuclear disaster placed Belarus's health service, which had hitherto been adequate, under enormous strain. A Chernobyl tax funds assistance for victims of the accident. The number of cancer and leukemia cases has soared, and extra wards and specialist units have had to be built. Many Belarussian doctors are being trained in the latest bone-marrow techniques in Europe and the US.

HIV/AIDS is a growing problem; as at end-2001, 0.3% of the adult population were estimated to be affected.

SPENDING

GDP/cap. decrease

CONSUMPTION AND SPENDING

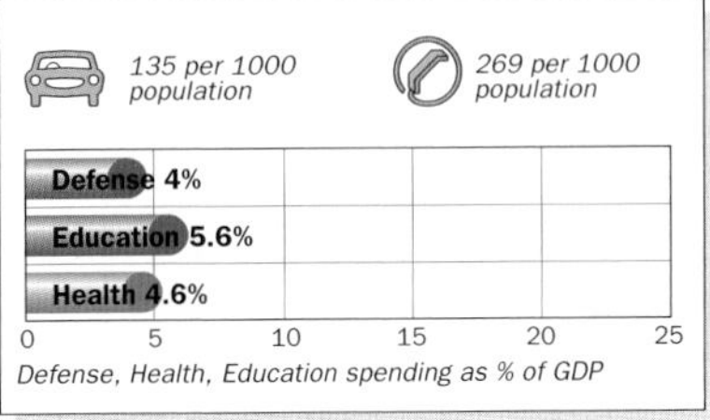
135 per 1000 population
269 per 1000 population
Defense 4%
Education 5.6%
Health 4.6%
0
5
10
15
20
25
Defense, Health, Education spending as % of GDP

The deteriorating economic situation has resulted in an overall drop in living standards. Wealth is concentrated among a small, communist elite which is opposed to market mechanisms. Now that this elite has the upper hand, its members have strengthened their grip on the state's resources. Thus far Belarus has not seen the expansion of entrepreneurial activity found in other former Soviet bloc countries such as Poland or Russia.

Wage increases in 2001 – an election year – brought salaries to levels which were unaffordable, so that enterprise profitability and investment were severely affected.

WORLD RANKING

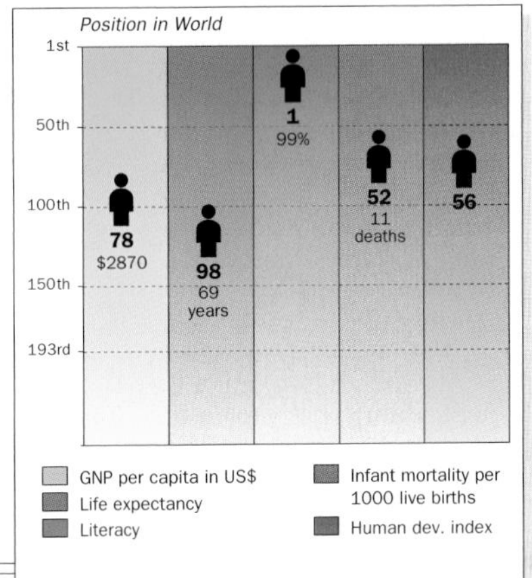
Position in World
1st
50th
100th
150th
193rd
78
$2870
98
69
years
1
99%
52
11
deaths
56
GNP per capita in US$
Life expectancy
Literacy
Infant mortality per 1000 live births
Human dev. index

BELGIUM

EUROPE

OFFICIAL NAME: Kingdom of Belgium **CAPITAL:** Brussels **POPULATION:** 10.3 million
CURRENCY: Euro (Belgian franc until 2002) **OFFICIAL LANGUAGES:** Dutch, French, and German

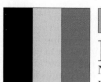

1830

1919
July 21
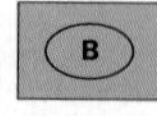
B
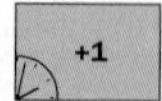
+1

+32

.be

LOCATED BETWEEN GERMANY, France, and the Netherlands, Belgium has a short coastline on the North Sea. The south includes the forested Ardennes region, while the north is crisscrossed by canals. Belgium has been fought over many times in its history; it was occupied by Germany in both world wars. Tensions have existed between the Dutch-speaking Flemings and French-speaking Walloons since the 1830s. These have been somewhat defused by Belgium's move to a federal political structure and the national consensus on the benefits of EU membership.

CLIMATE

Maritime

WEATHER CHART FOR BRUSSELS

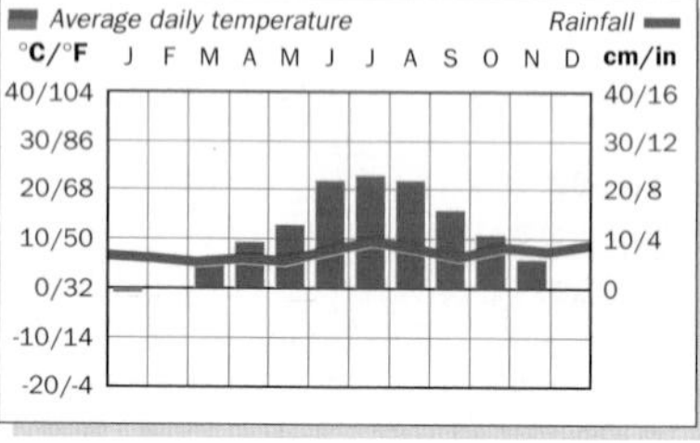
Average daily temperature
Rainfall
°C/°F J F M A M J J A S O N D cm/in
40/104 40/16
30/86 30/12
20/68 20/8
10/50 10/4
0/32 0
-10/14
-20/-4

Belgium has a typical maritime climate and is influenced by the Gulf Stream. Temperatures are mild, accompanied by heavy cloud cover and much rain. Widely fluctuating weather conditions, caused by cyclonic disturbances, can disrupt the climate on the coast. Summers tend to be short.

TRANSPORTATION

Drive on right

Brussels International
21.6m passengers

182 ships
143,900 grt

THE TRANSPORTATION NETWORK

117,701 km
(73,136 miles)

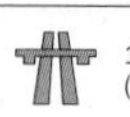
1702 km
(1058 miles)

3472 km
(2158 miles)

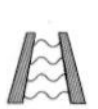
2043 km
(1269 miles)

Belgium can be crossed within four hours by car or train. The expressway network is extensive, and although the railroad system has been reduced since 1970, it still constitutes one of the world's densest networks. Using high-speed TGV lines, Paris is just 80 minutes from Brussels, and London via the Channel Tunnel is 2 hours 40 minutes.

Antwerp is Europe's second-largest port. National carrier Sabena collapsed dramatically in November 2001, amid a slump in the aviation industry. A section of Sabena subsequently formed the basis for a new private airline, SN Brussels Airlines.

TOURISM

Visitors : Population
1:1.6

6.46m visitors

Up 1% in 2000

MAIN TOURIST ARRIVALS

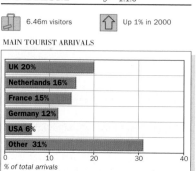

	% of total arrivals
UK	20%
Netherlands	16%
France	15%
Germany	12%
USA	6%
Other	31%

Belgium's main attractions are its historic cities and the museums of Flemish art. Bruges, the capital of West Flanders, is often referred to as the "Venice of the North." With Gothic and Renaissance architecture and a complex canal system, it has become a favored destination for British weekend trippers and Japanese honeymooners. In Brussels, the famous "Grande Place," a cluster of Gothic, Renaissance, and Baroque buildings in a cobbled square, survived bombing during World War II. Much of the rest of the old city center, however, was destroyed. Belgium has 15 resorts on its 62-km (38-mile) coastline, with a single tramline running its entire length. Forests in the Ardennes to the south attract hikers.

The Ardennes, in the southeast, are famous for their forests, cuisine, and lakes. Rivers, such as the Meuse and Semois, dissect the region.

PEOPLE

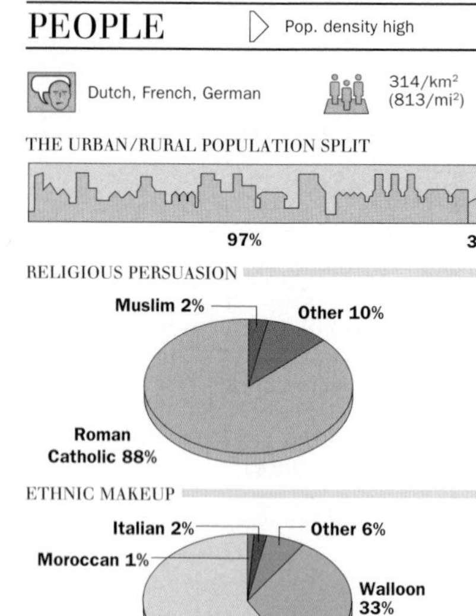
Pop. density high
Dutch, French, German
314/km²
(813/mi²)
THE URBAN/RURAL POPULATION SPLIT
97%
3%
RELIGIOUS PERSUASION
Muslim 2%
Other 10%
Roman Catholic 88%
ETHNIC MAKEUP
Italian 2%
Other 6%
Moroccan 1%
Walloon 33%
Fleming 58%

Belgium has been marked by the divisions between its Flemish and Walloon communities. The majority Dutch-speaking Flemings are concentrated in Flanders. Wallonia is French-speaking and Brussels is 85% francophone. French-speakers were in the ascendancy for many years, their greater economic wealth reinforced by a constitution giving them political control; tensions between Walloons and Flemings occasionally erupted into violence. In the past three decades, however, the situation has been reversed: Wallonia's industries have declined and Flanders is now the wealthier region. To defuse tensions, Belgium began in 1980 to change from being the most centralist to the most federal state in Europe; each community now controls most of its affairs and has its own government. A small German-speaking community in the east has extensive autonomy in educational and cultural matters.

Belgium has a sizable immigrant population. Women account for 40% of the workforce and 19% of administrators and managers.

POPULATION AGE BREAKDOWN

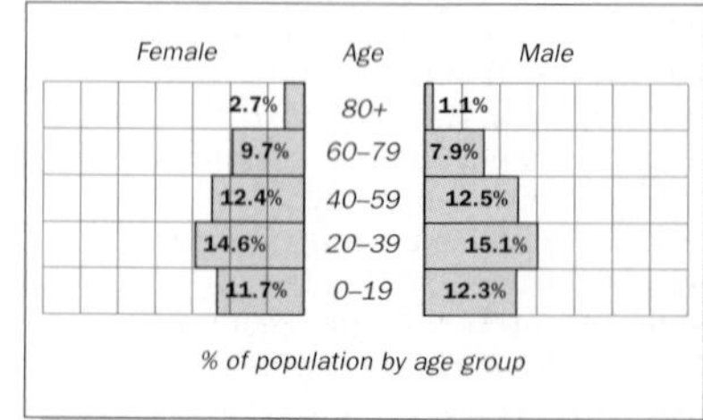
Female
Age
Male
2.7% 80+ 1.1%
9.7% 60-79 7.9%
12.4% 40-59 12.5%
14.6% 20-39 15.1%
11.7% 0-19 12.3%
% of population by age group

BELGIUM

Total Land Area : 30 510 sq. km (11 780 sq. miles)

POPULATION

- over 1 000 000
- over 100 000
- over 50 000
- over 10 000

LAND HEIGHT

- 500m/1640ft
- 200m/656ft
- Sea Level

0 — 40 km

0 — 40 miles

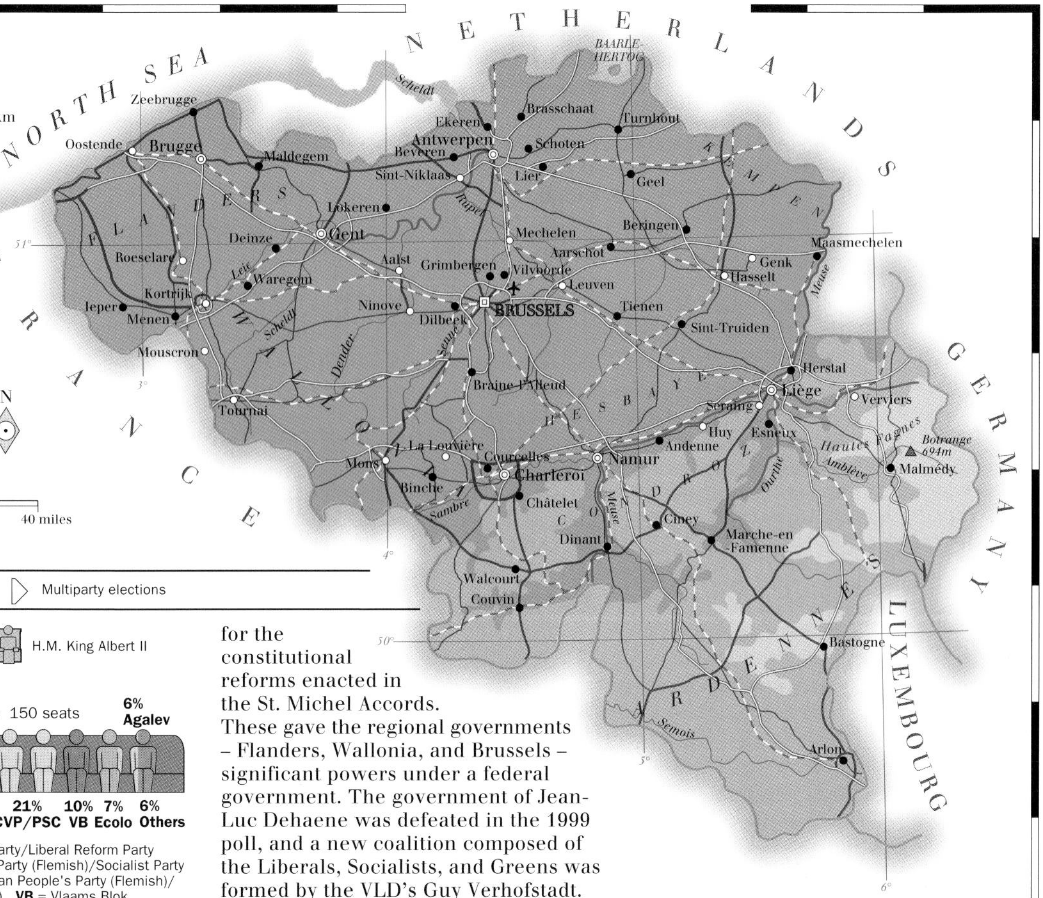

POLITICS

Multiparty elections

L. House 1999/2003
U. House 1999/2003

H.M. King Albert II

AT THE LAST ELECTION

Chamber of Representatives 150 seats

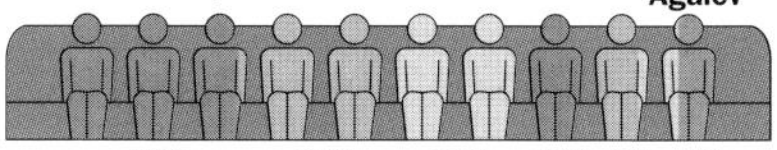

28%	22%	21%	10%	7%	6%	6%
VLD/PRL	**SP/PS**	**CVP/PSC**	**VB**	**Ecolo**	**Others**	**Agalev**

VLD/PRL = Flemish Liberals Party/Liberal Reform Party (Walloon) **SP/PS** = Socialist Party (Flemish)/Socialist Party (Walloon) **CVP/PSC** = Christian People's Party (Flemish)/ Christian Social Party (Walloon) **VB** = Vlaams Blok **Ecolo** = French Greens **Agalev** = Flemish Ecologists **Co-op** = Co-opted members

Senate 71 seats

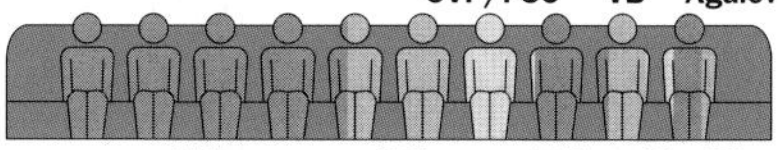

44%	15%	13%	11%	6%	4%	4%	3%
Co-op	**VLD/PRL**	**CVP/PSC**	**SP/PS**	**VB**	**Agalev**	**Ecolo**	**Others**

The Senate has 40 directly elected members and 31 co-opted members.

Until 1970, Belgium was a unitary state. Tensions between language groups led to four waves of federalist reforms from 1980, which culminated in the St. Michel Accords of 1993, confirming the state as a federal monarchy.

PROFILE

Belgian politics are defined along lines of language. Apart from this, a high degree of consensus exists over the benefits of membership of the EU and monetary union. In recent years, support has increased for the racist VB, which objects to Belgium's Turkish, Moroccan, and other African minorities. VB won 28% of the vote in local elections in Antwerp in 1994 and 33% in 2000.

The ruling centrist coalition of the Socialist and Christian Democrat parties of the Flemish and Walloon communities had a parliamentary majority, but had difficulty in securing the necessary majority for the constitutional reforms enacted in the St. Michel Accords. These gave the regional governments – Flanders, Wallonia, and Brussels – significant powers under a federal government. The government of Jean-Luc Dehaene was defeated in the 1999 poll, and a new coalition composed of the Liberals, Socialists, and Greens was formed by the VLD's Guy Verhofstadt.

MAIN POLITICAL ISSUES

Language

Tensions between the two language groups are receding. However, the divisions remain strong. Each community has a right–liberal party (the VLD in Flanders, the PRL in Wallonia), a socialist party (SP/PS), a Christian democratic party (CVP/PSC), and a green party (Agalev/Ecolo).

Police handling of pedophile case

Apparent police incompetence, cover-ups, and corruption in combating the activities of pedophile rings has provoked public anger and protest, focusing in particular on failure to save children from pedophile killer Marc Dutroux. Arrested in 1996, he escaped briefly in 1998, causing two ministers to resign.

King Albert II, succeeded his brother King Baudouin who died in 1993.

Guy Verhofstadt, youthful leader of the VLD/PRL and prime minister since 1999.

WORLD AFFAIRS

Joined UN in 1945

Benelux
CE
EU
OECD
NATO

Belgium's key concern is its role in the EU. It is a keen supporter of economic and monetary union. As a frequent victim of wars between France and Germany, Belgium sees the EU as a guarantor of western European peace. It is also perceived as an important foundation for Belgium's own federalist structure, without which many fear that Belgium could split into two.

Belgium has little in the way of an independent foreign policy, but frequently contributes troops to the UN's operations. Belgian soldiers have served in the Democratic Republic of the Congo and the Middle East in recent years, and a number were killed in Rwanda in 1994.

AID

Donor

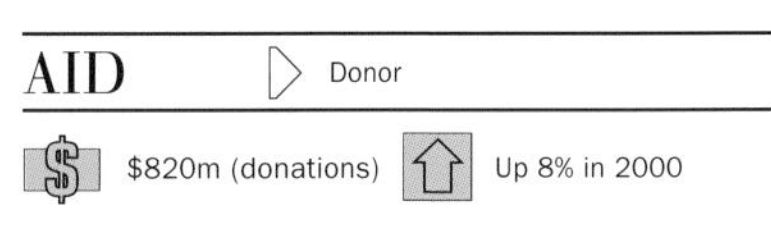
$820m (donations)
Up 8% in 2000

Some 0.33% of GNP goes in overseas development aid. Belgian aid focuses on education and agricultural projects in Africa. The major beneficiaries are the former Belgian colonies of Burundi, Rwanda, and the DRC.

B

CHRONOLOGY

Formerly ruled by the French dukes of Burgundy, Belgium became a Habsburg possession in 1477. It passed to the Austrian Habsburgs in 1713. Belgium was incorporated into France in 1797.

- ❑ **1814–1815** Congress of Vienna; European powers decide to merge Belgium with the Netherlands under King William I of Orange.
- ❑ **1830** Revolt against Dutch; declaration of independence.
- ❑ **1831** European powers place Leopold Saxe Coburg as king.
- ❑ **1865** Leopold II crowned king.
- ❑ **1885** Berlin Conference gives Congo basin to Leopold as colony.
- ❑ **1914** German armies invade. Belgium occupied until 1918.
- ❑ **1921** Belgo-Luxembourg Economic Union formed. Belgian and Luxembourg currencies locked.
- ❑ **1932** Dutch language accorded equal official status with French.
- ❑ **1936** Belgium declares neutrality.
- ❑ **1940** Leopold III capitulates to Hitler. Belgium occupied till 1944.
- ❑ **1948** Customs union with Netherlands and Luxembourg (Benelux) formed.
- ❑ **1950** King abdicates in favor of son, Baudouin.
- ❑ **1957** Becomes one of six original signatories of Treaty of Rome, the principal foundation of what develops into the EU.
- ❑ **1992** Christian Democrat and Socialist government led by Jean-Luc Dehaene takes over federal government.
- ❑ **1993** Culmination of reforms creating federal state. Greater powers for regions and city governments. Death of Baudouin. Succeeded by Albert II.
- ❑ **1995** Allegations of corruption and murder involving French-speaking PS force resignations of Walloon premier, federal deputy premier, and Willy Claes, NATO secretary-general.
- ❑ **1996** Murder and disappearance of young girls arouse fears of international kidnapping and pedophile ring.
- ❑ **1999** February, Claes found guilty of bribery in connection with defense contract to buy helicopters. June, VLD/PRL wins general election. New coalition formed, including Greens for first time.
- ❑ **2001** November, collapse of national airline Sabena.
- ❑ **2002** January, euro fully adopted, with introduction of coins and notes, and withdrawal of Belgian franc from circulation. May, legalization of euthanasia.

DEFENSE

No compulsory military service

$3.34bn

Down 3% in 2000

BELGIAN ARMED FORCES

132 main battle tanks (*Leopard* 1A5)	26,400 personnel
3 frigates	2560 personnel
135 combat aircraft (F-16A, F-16B)	8600 personnel
None	

Belgium spends less on defense than the NATO average of 2.2% of GDP.

In 1994, as part of Belgium's program to reduce government debt, all three military services were targeted for cuts. The government abolished conscription and undertook to cut troop levels. The defense budget was frozen for five years.

However, spending on paratroopers and transport planes has increased. The aim is to allow Belgian forces to fulfill their role in NATO's new rapid reaction forces. It will also make Belgian forces more useful to the UN's worldwide operations. In 1996, the Belgian and Netherlands' navies were brought together under a joint operational command based at Den Helder, the Netherlands.

ECONOMICS

Inflation 1.9% p.a. (1990–2000)

$252bn

1.0651–1.1231 euros

SCORE CARD

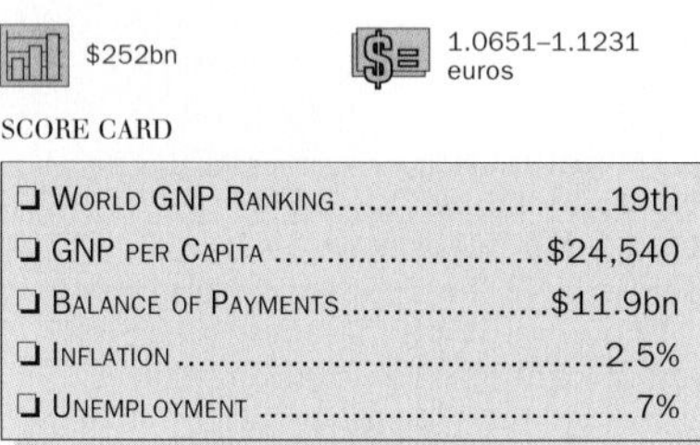

❑ World GNP Ranking	19th
❑ GNP per Capita	$24,540
❑ Balance of Payments	$11.9bn
❑ Inflation	2.5%
❑ Unemployment	7%

EXPORTS

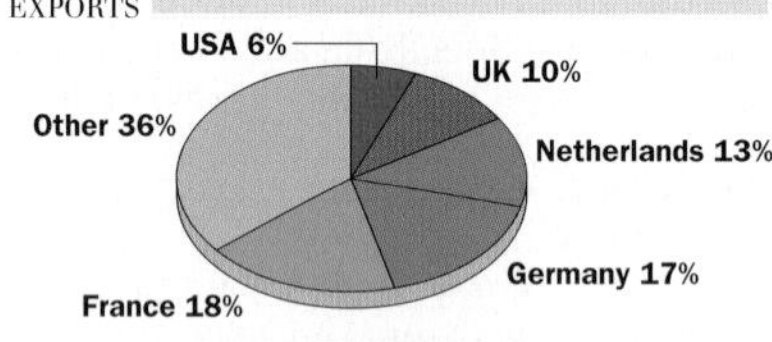
USA 6%
UK 10%
Other 36%
Netherlands 13%
Germany 17%
France 18%

IMPORTS

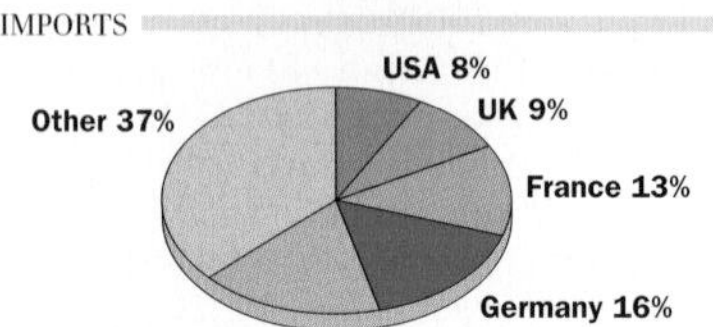
USA 8%
UK 9%
Other 37%
France 13%
Germany 16%
Netherlands 17%

Strengths

One of world's most efficient producers of metal products and textiles. Flanders is a world leader in new high-tech industries. Successful chemicals industry. Highly educated and motivated multilingual workforce: estimates suggest productivity is 20% above that of Germany. Location attractive for US multinationals. Good sea outlets and access to Rhine inland waterway from Antwerp and Ghent.

Weaknesses

Public debt of around 100% of GDP, well over EU target of 60%. High long-term and low-skill joblessness with sharp local variations. Early retirement of large numbers of workers results in high state pension bill. Bureaucracy larger than European average.

ECONOMIC PERFORMANCE INDICATOR

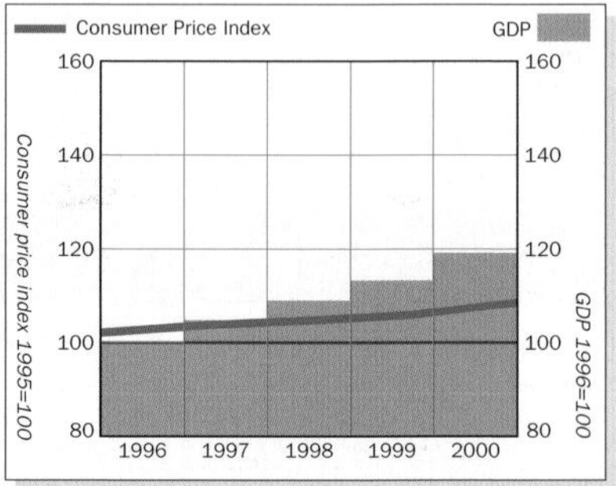
Consumer Price Index
GDP
160
140
120
100
80
Consumer price index 1995=100
GDP 1996=100
1996
1997
1998
1999
2000

Profile

Recession and rising unemployment in the early 1990s prompted the introduction of work-sharing schemes and benefit reforms. Unemployment and the massive public debt are declining gradually. Against a background of a downturn in the world economy, the government is committed to both greater fiscal stringency, in pursuit of a budget surplus, and reduction in taxation over 2002–2005.

BELGIUM : MAJOR BUSINESSES

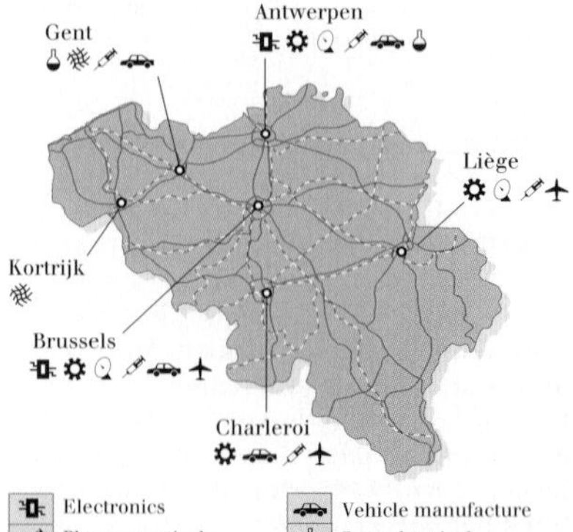
Gent
Antwerpen
Liège
Kortrijk
Brussels
Charleroi
Electronics
Pharmaceuticals
Aerospace industry
Heavy engineering
Telecommunications
Vehicle manufacture
Petrochemicals
Textiles
0 50 km
0 50 miles

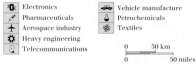

B

RESOURCES

Electric power 15m kw

30,722 tonnes

Not an oil producer; refines 607,000 b/d

7.35m pigs, 3.25m cattle, 38m chickens

Coal, natural gas, shale, marble, sandstone, dolomite

ELECTRICITY GENERATION

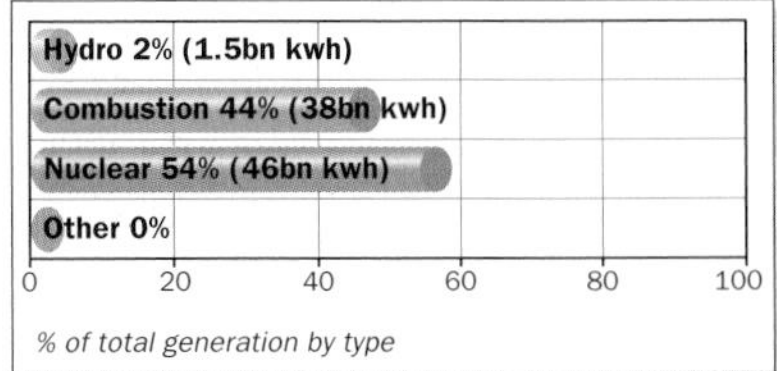
Hydro 2% (1.5bn kwh)
Combustion 44% (38bn kwh)
Nuclear 54% (46bn kwh)
Other 0%
0
20
40
60
80
100
% of total generation by type

Belgium has few natural resources and depends largely on the export of goods and services. The once-rich coal mines of Wallonia are almost depleted. There is some deciduous and conifer forestry in the Ardennes region.

BELGIUM : LAND USE

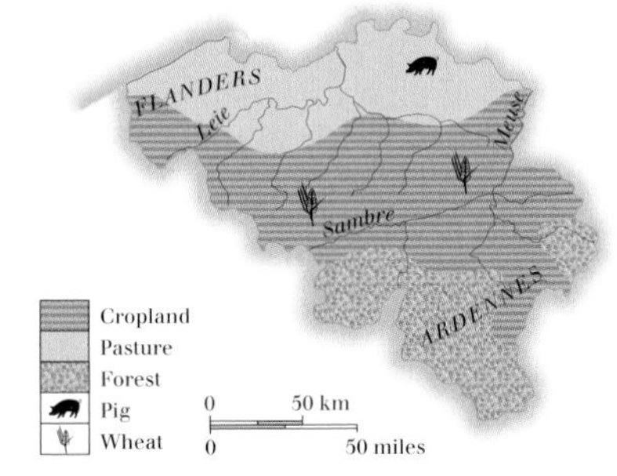
FLANDERS
Leie
Meuse
Sambre
ARDENNES
Cropland
Pasture
Forest
Pig
Wheat
0
50 km
0
50 miles

ENVIRONMENT

Sustainability rank: 125th

3%

9.9 tonnes per capita

ENVIRONMENTAL TREATIES

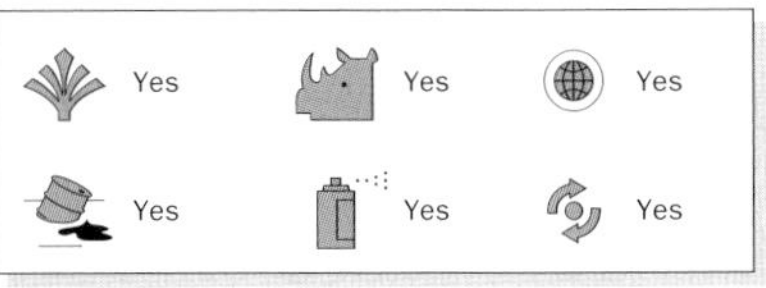
Yes
Yes
Yes
Yes
Yes
Yes

Flanders is concerned about the pollution of groundwater supplies through acid rain, heavy metals, fertilizers, and pesticides. Its government operates an environmental management plan to raise standards. Wallonia has strict laws against illegal tipping of waste, and regulations on air quality and emissions. Awareness of environmental issues is reflected in the rise of the two green parties, which entered government for the first time in the coalition formed in 1999.

MEDIA

TV ownership high

Daily newspaper circulation 161 per 1000 people

PUBLISHING AND BROADCAST MEDIA

There are 30 daily newspapers, published in Dutch, French, and German, including *Het Nieuwsblad*, *Le Soir*, and *De Standaard*
3 state-owned services, broadcasting in Dutch, French, and German, and 5 independent commercial services
3 state-owned services, broadcasting in Dutch, French, and German, and numerous private stations

Newspapers tend to be regional and divided by language. Circulation is low: the most widely read paper has a circulation of only 370,000. Control of broadcasting is divided along linguistic lines between two major corporations. Over 95% of Belgians have cable TV, with access to channels from all over Europe. Commercial TV only began in 1989, with the Flemish station VTM.

CRIME

No death penalty

8671 prisoners

Up 19% 1996–1998

CRIME RATES

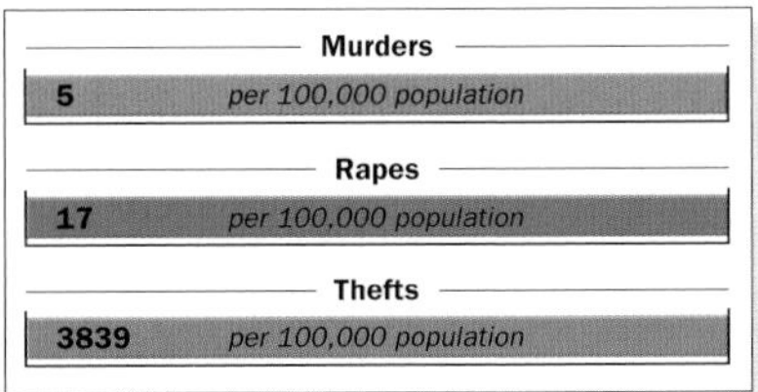
Murders
5 per 100,000 population
Rapes
17 per 100,000 population
Thefts
3839 per 100,000 population

Brussels has one of the lowest murder rates for any capital city. In urban areas pickpocketing is rising and car theft is a problem. Penalties for illegal use of narcotics are strict.

EDUCATION

School leaving age: 18

99%

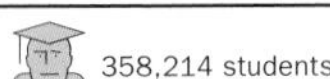
358,214 students

THE EDUCATION SYSTEM

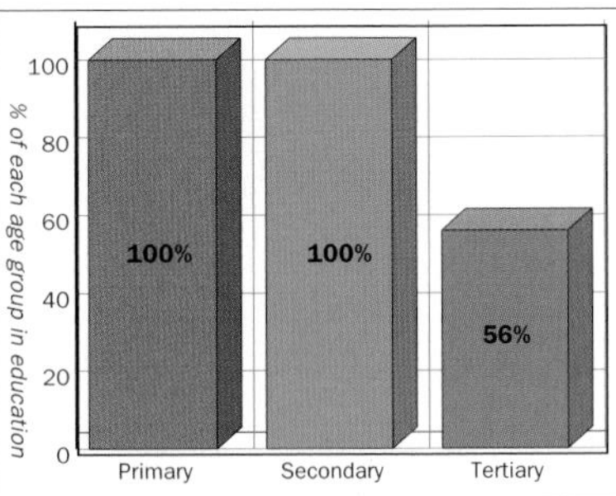
100
80
60
40
20
0
% of each age group in education
100%
100%
56%
Primary
Secondary
Tertiary

In Belgium, parents can choose between schooling provided by the two main language communities, by public authorities, or by private interests. Roman Catholic schools constitute the greatest number of "free" (privately organized) establishments. Since 1989 the system has been administered by the governments of the two main language groups. All universities are split by language.

HEALTH

Welfare state health benefits

1 per 263 people

Heart and respiratory diseases, cancers, car accidents

The quality of health care is among the best in the world, and government spending is high. Belgium is a world leader in fertility treatment and heart and lung transplants. Treatment is not free, but Belgians are able to claim up to 75% of their costs.

Only heart disease and cancer rank before car accidents as a cause of death. There were an estimated 8500 people living with HIV/AIDS at end-2001. In May 2002 Belgium became the second country (after the Netherlands) to legalize euthanasia.

SPENDING

GDP/cap. increase

CONSUMPTION AND SPENDING

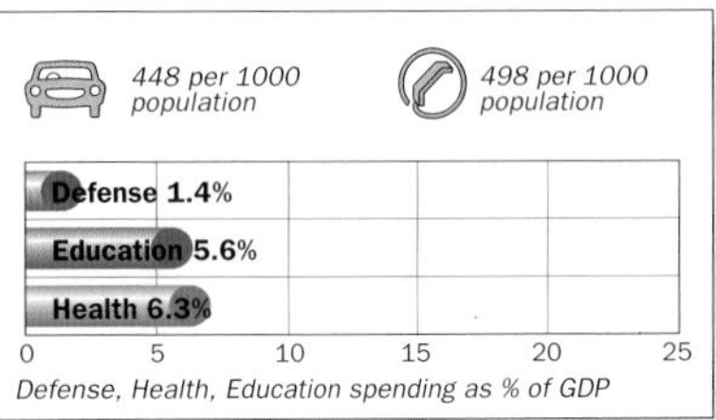
448 per 1000 population
498 per 1000 population
Defense 1.4%
Education 5.6%
Health 6.3%
0
5
10
15
20
25
Defense, Health, Education spending as % of GDP

Despite high levels of state debt and failing traditional industries, Belgium is one of Europe's richest countries. GDP per capita is lower than for Germany but higher than for Italy or the UK.

This statistic masks considerable regional differences. In Flanders, with its many high-tech businesses, the level of unemployment is only half that in Wallonia.

The presence of highly paid EU officials and international company employees and bankers has made Brussels a distinctly wealthy, and expensive, city. The recession of the early 1990s prompted Belgians to save a higher proportion of their income, but the level of savings has fallen since then as consumer confidence has recovered.

WORLD RANKING

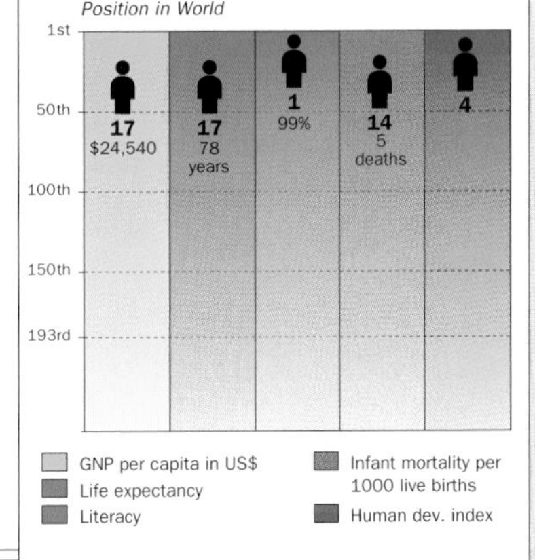
Position in World
1st
50th
100th
150th
193rd
17
$24,540
17
78
years
1
99%
14
5
deaths
4
GNP per capita in US$
Life expectancy
Literacy
Infant mortality per 1000 live births
Human dev. index

BELIZE

CENTRAL AMERICA
Mexico
BELIZE
Guatemala
South America

OFFICIAL NAME: Belize **CAPITAL:** Belmopan
POPULATION: 200,000 **CURRENCY:** Belizean dollar **OFFICIAL LANGUAGE:** English

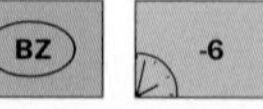

1981
1981
Sept 21
BZ
-6
+501
.bz

FORMERLY BRITISH HONDURAS, Belize was the last Central American country to gain its independence, in 1981. It lies on the southeastern shore of the Yucatan peninsula and shares a border with Mexico along the River Hondo. Belize is Central America's least populous country, and almost half of its land area is still forested. Its swampy coastal plains are protected from flooding by the world's second-largest barrier reef.

Small fishing village near Belize City. About 300 tonnes of Caribbean spiny lobster, the main inshore species, are caught every year.

CLIMATE

Tropical equatorial

WEATHER CHART FOR BELMOPAN

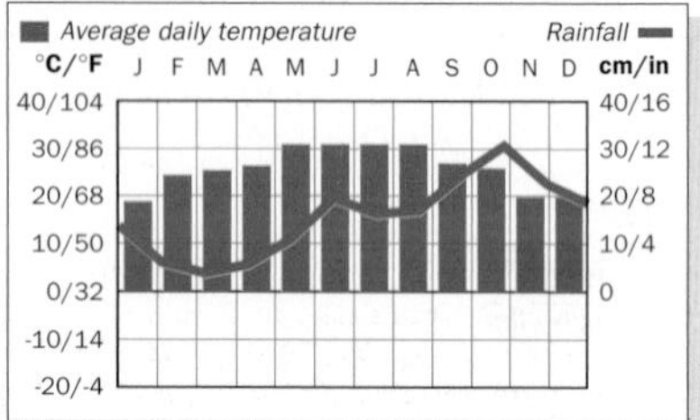
Average daily temperature
Rainfall
°C/°F
J F M A M J J A S O N D
cm/in
40/104
30/86
20/68
10/50
0/32
-10/14
-20/-4
40/16
30/12
20/8
10/4
0

Conditions are hot and humid. Coastal regions are affected by hurricanes, notably Hurricane Iris in late 2001.

TRANSPORTATION

Drive on right

Phillip S. W. Goldson, Belize City
272,000 passengers

1660 ships
2.25m grt

THE TRANSPORTATION NETWORK

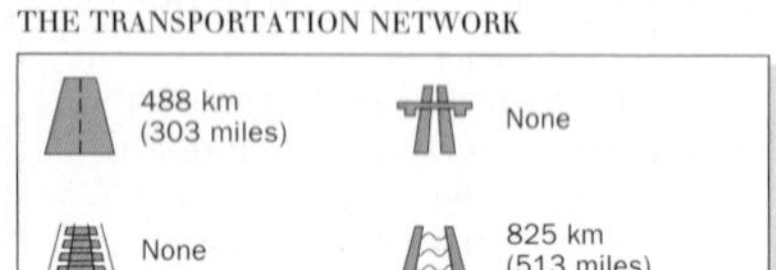
488 km
(303 miles)
None
None
825 km
(513 miles)

A US$16 million IDB loan in 1998 helped improve the country's road network and its feeder roads. A terminal and runway extension have been completed at the international airport near Belize City.

TOURISM

Visitors : Population 1:1.1

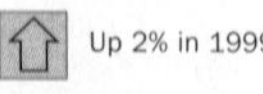
181,000 visitors
Up 2% in 1999

MAIN TOURIST ARRIVALS

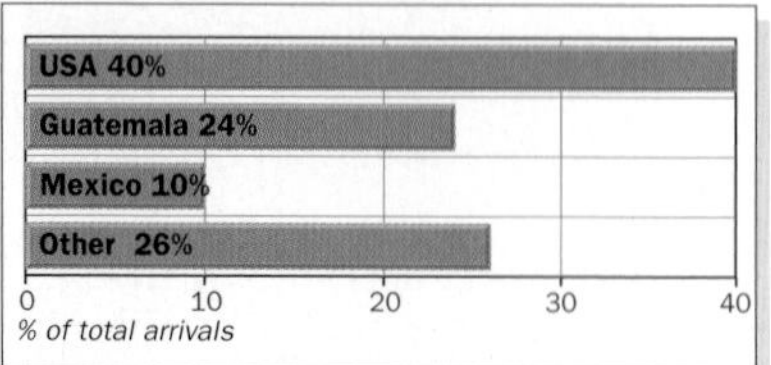
USA 40%
Guatemala 24%
Mexico 10%
Other 26%
0
10
20
30
40
% of total arrivals

The barrier reef, good beaches, and Mayan ruins draw visitors. "Eco" attractions need conservation.

PEOPLE

Pop. density low

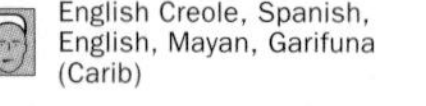
English Creole, Spanish, English, Mayan, Garifuna (Carib)

9/km²
(23/mi²)

THE URBAN/RURAL POPULATION SPLIT

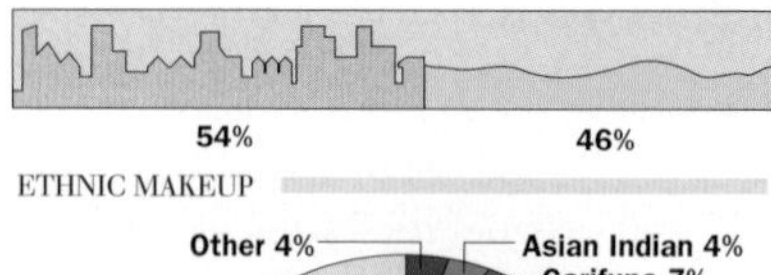
54%
46%

ETHNIC MAKEUP

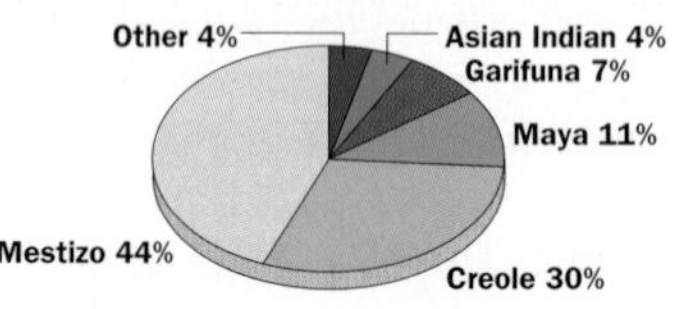
Other 4%
Asian Indian 4%
Garifuna 7%
Maya 11%
Mestizo 44%
Creole 30%

Over 80% of Belizeans are of mixed descent involving African, Amerindian, and European strands. Along with the *mestizo* and Creole populations there are the Afro-Carib *garifuna*. Christianity is dominant, and the Roman Catholic, Anglican, and Methodist Churches run most of the schools.

POLITICS

Multiparty elections

L. House 1998/2003
U. House 1998/2003

H.M. Queen Elizabeth II

AT THE LAST ELECTION

House of Representatives 29 seats

90% PUP
10% UDP

PUP = People's United Party
UDP = United Democratic Party

Senate 8 seats

The members of the Senate are appointed by the governor-general

The desire for independence dominated politics until the 1980s. The PUP, under George Price, negotiated this with the British in 1981. During the 1984–1989 UDP administration, the main concerns were to maintain a pro-US line and the fear of communism in the region. In the absence of any major ideological or policy distinctions, the UDP lost power to the PUP in 1989, winning it back in 1993; the pendulum swung back to the PUP again in 1998. Growth, job creation, "economic citizenship" for foreigners, and border tension with Guatemala are key issues.

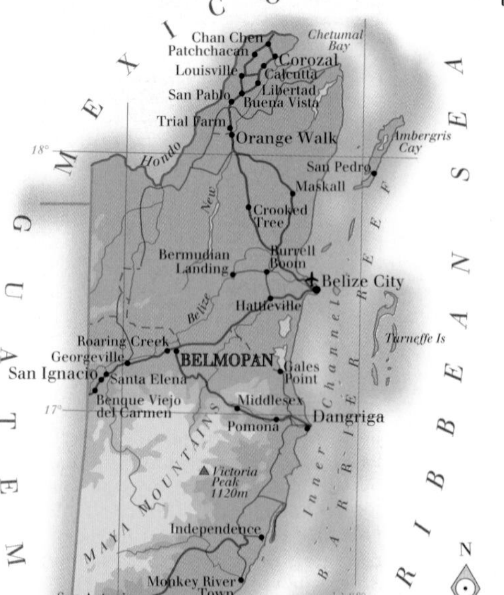
MEXICO
Chan Chen
Patchchacan
Chetumal Bay
Corozal
Louisville
Calcutta
Libertad
San Pablo
Buena Vista
Trial Farm
Orange Walk
Hondo
Ambergris Cay
San Pedro
Maskall
New
Crooked Tree
Bermudian Landing
Burrell Boom
Belize City
Hattieville
Belize
Turneffe Is
Roaring Creek
Georgeville
BELMOPAN
San Ignacio
Santa Elena
Gales Point
Benque Viejo del Carmen
Middlesex
Pomona
Dangriga
Inner Channel
Victoria Peak 1120m
MAYA MOUNTAINS
Independence
BARRIER REEF
Monkey River Town
San Antonio
Toledo Settlement
San Lucas
Punta Gorda
Gulf of Honduras
Sarstoon
GUATEMALA
CARIBBEAN SEA
18°
17°
16°
88°
89°

BELIZE

Total Land Area : 22 966 sq. km (8867 sq. miles)

POPULATION
over 10 000
under 10 000

LAND HEIGHT
1000m/3281ft
500m/1640ft
200m/656ft
Sea Level

N
0 50 km
0 50 miles

WORLD AFFAIRS

Joined UN in 1981

ACS
Comm
Caricom
NAM
OAS

The major concern is Guatemala's historic claim to over half of Belize. Suspended in 1986, the claim was revived in early 2000 as border tension rose.

AID

Recipient

US$15m (receipts)

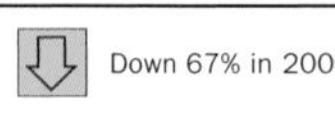
Down 67% in 2000

In 1999 the IDB, the Commonwealth Development Corporation, the European Investment Bank, and the CDB invested in citrus farms. Belize is one of the highest per capita recipients of US aid.

DEFENSE

No compulsory military service

US$17m

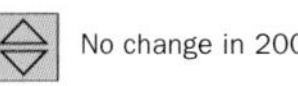
No change in 2000

The small Belize Defense Force took over full responsibility from the UK in 1994 for the country's defense. The UK withdrew its garrison in the same year, but continues to maintain a jungle training school.

ECONOMICS

Inflation 2.7% p.a. (1990–2000)

US$746m

1.97 Belizean dollars

SCORE CARD

- ❑ WORLD GNP RANKING........................160th
- ❑ GNP PER CAPITA.........................US$3110
- ❑ BALANCE OF PAYMENTS................–US$139m
- ❑ INFLATION..0.6%
- ❑ UNEMPLOYMENT.................................13%

STRENGTHS

Sugar, textile manufacture, citrus fruits, bananas, shellfish, forestry, and considerable tourist potential. Sustainable public debt; fair access to concessionary foreign finance.

WEAKNESSES

Narrow export base dependent on preferential market access; reliance on imports of processed foods. Poor fiscal management in late 1990s.

EXPORTS

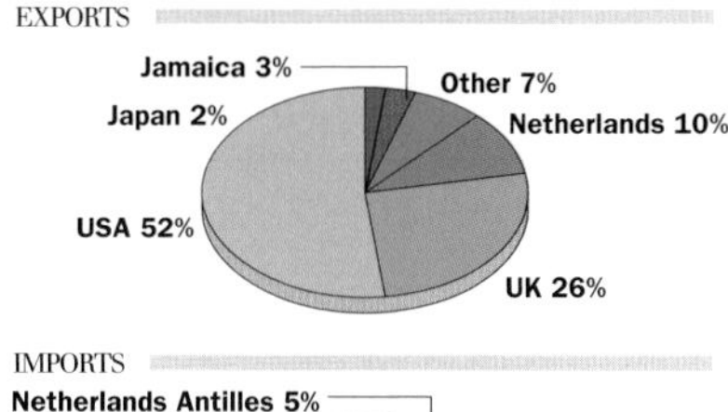
Jamaica 3%
Other 7%
Japan 2%
Netherlands 10%
USA 52%
UK 26%

IMPORTS

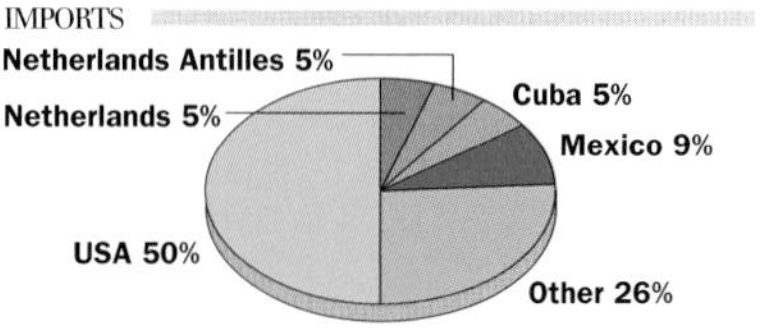
Netherlands Antilles 5%
Cuba 5%
Netherlands 5%
Mexico 9%
USA 50%
Other 26%

RESOURCES

Electric power 43,000 kw

43,103 tonnes

Not an oil producer

45,000 cattle, 24,000 pigs, 1.35m chickens

None

Hopes of finding significant oil and gas deposits in the north of the country have so far proved fruitless.

ENVIRONMENT

Not available

5% partially protected

1.8 tonnes per capita

Tourist developments and logging have depleted the dense tropical forests. Mahogany is endangered so all exports and transshipments now require a certificate of origin. Global warming poses a major threat to the corals of the barrier reef.

MEDIA

TV ownership medium

There are no daily newspapers

PUBLISHING AND BROADCAST MEDIA

There are no daily newspapers. The leading papers are the weekly *Belize Times*, *Amandala*, and *Reporter*

9 services: 1 state-owned, 8 independent

6 independent services

Belize has not suffered the degree of press interference experienced in neighboring states, but successive governments have remained sensitive to even minor criticisms. The two radio stations of the public Broadcasting Corporation of Belize were sold in 1998 to two local stations, but the government has retained ownership of the transmitters. Two official newspapers compete with party-political and independent publications.

CRIME

Death penalty in use

1097 prisoners

Increase in gun-related crime

Belize is a major transit point to the US for cocaine, despite being decertified in 1997 for its antinarcotics efforts. Drugs-related crime is high. Armed robberies by criminal gangs based in neighboring Guatemala are also a major concern. A government ombudsman was appointed in 2000 to investigate police brutality and corruption.

EDUCATION

School leaving age: 14

93%

2853 students

Although most schools are run by the different Churches, a handful are funded by the government, particularly those catering for special needs. The University College of Belize provides for higher education.

CHRONOLOGY

The Mayan heartland included what is now Belize. Between 1798 and 1981 it was effectively a British colony.

- ❑ **1919** Demands for more political rights by black Belizeans returning from World War I.
- ❑ **1936** New constitution.
- ❑ **1950** PUP formed. Voting age limit for women reduced from 30 to 21.
- ❑ **1954** Full adult suffrage.
- ❑ **1972** Guatemala threatens invasion. Britain sends troops.
- ❑ **1981** Full independence.
- ❑ **1998** PUP wins crushing general election victory.
- ❑ **2000** Guatemala revives claim to half of Belize.

HEALTH

Welfare state health benefits

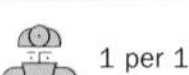
1 per 1818 people

Respiratory, heart, and cerebrovascular diseases

The health service provided by the government includes seven hospitals, more than 30 regional health centers, and numerous mobile clinics. Water supplies and sanitation have been improved; most homes in Belmopan now have both.

SPENDING

GDP/cap. increase

CONSUMPTION AND SPENDING

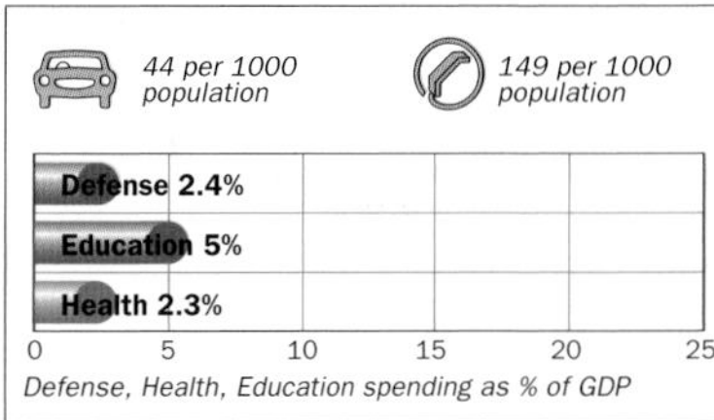
44 per 1000 population
149 per 1000 population
Defense 2.4%
Education 5%
Health 2.3%
0
5
10
15
20
25

Defense, Health, Education spending as % of GDP

The European Development Fund in 1999 granted 3.5 million Belizean dollars toward the reduction of rural poverty. Narcotics trading remains a source of wealth.

WORLD RANKING

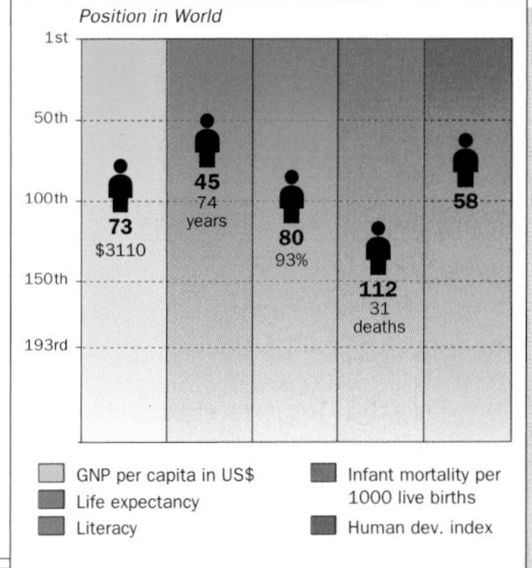
Position in World
1st
50th
100th
150th
193rd
73
$3110
45
74
years
80
93%
112
31
deaths
58
GNP per capita in US$
Life expectancy
Literacy
Infant mortality per 1000 live births
Human dev. index

BENIN

WEST AFRICA
Burkina
Togo
Nigeria
BENIN
Africa

OFFICIAL NAME: Republic of Benin **CAPITAL:** Porto-Novo
POPULATION: 6.4 million **CURRENCY:** CFA franc **OFFICIAL LANGUAGE:** French

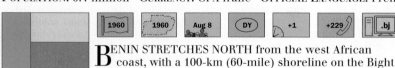

1960 | 1960 | Aug 8 | DY | +1 | +229 | .bj

BENIN STRETCHES NORTH from the west African coast, with a 100-km (60-mile) shoreline on the Bight of Benin. Formerly the kingdom of Dahomey, Benin was under French colonial rule, becoming part of French West Africa, until independence in 1960. In 1990 Benin was a pioneer of multipartyism in Africa, ending 17 years of one-party Marxist–Leninist rule. Benin's economy is based on well-diversified agriculture.

CLIMATE

Tropical wet & dry

WEATHER CHART FOR PORTO-NOVO

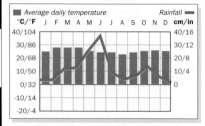

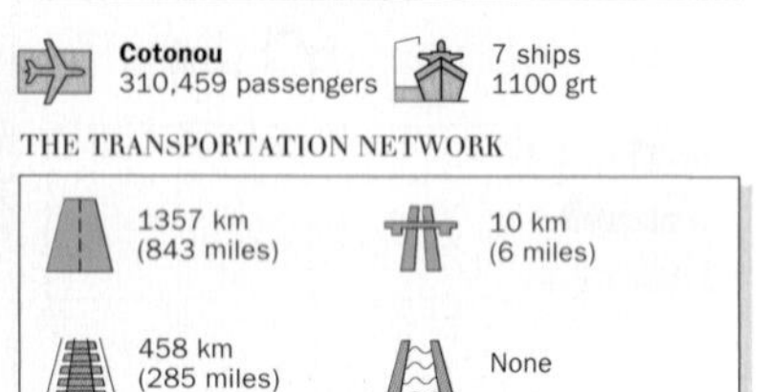
Average daily temperature
Rainfall
°C/°F J F M A M J J A S O N D cm/in
40/104
30/86
20/68
10/50
0/32
-10/14
-20/-4
40/16
30/12
20/8
10/4
0

There are two rainy seasons. The hot, dusty *harmattan* wind characterizes the December to February dry season.

TRANSPORTATION

Drive on right

Cotonou 310,459 passengers

7 ships 1100 grt

THE TRANSPORTATION NETWORK

1357 km (843 miles)

10 km (6 miles)

458 km (285 miles)

None

The joint Benin–Niger railroad runs only as far as Parakou. The Cotonou–Porto-Novo line reopened in 1999.

TOURISM

Visitors : Population 1:42

152,000 visitors

Up 1% in 1998

MAIN TOURIST ARRIVALS

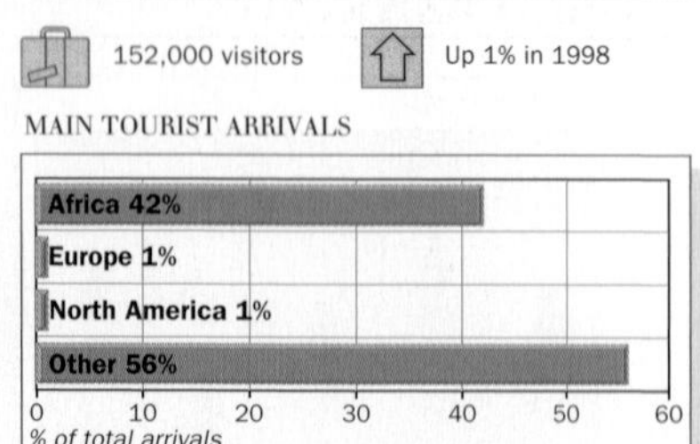
Africa 42%
Europe 1%
North America 1%
Other 56%
0 10 20 30 40 50 60
% of total arrivals

The industry is not well developed; there are plans to increase package tourism. There is safari tourism in the north, in the Atakora Mountains in particular. Benin is popular for weekend breaks for tourists vacationing in Nigeria.

PEOPLE

Pop. density medium

Fon, Bariba, Yoruba, Adja, Houeda, Somba, French

58/km² (150/mi²)

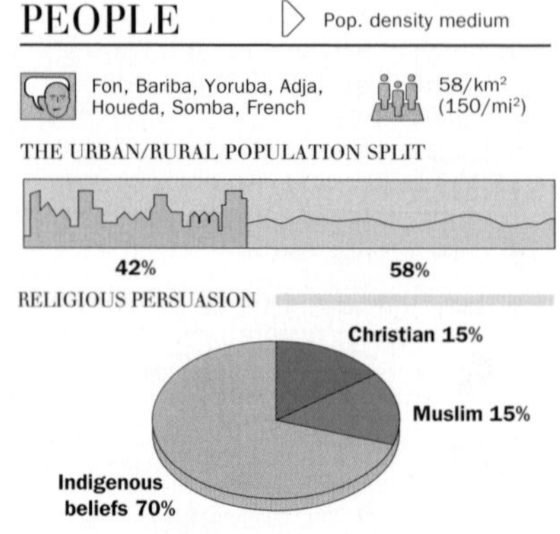
THE URBAN/RURAL POPULATION SPLIT
42%
58%
RELIGIOUS PERSUASION
Christian 15%
Muslim 15%
Indigenous beliefs 70%

Benin is politically dominated by the southern Fon people. There is some north–south tension, partly because the south is more developed, and partly reflecting a Muslim–Christian divide. Women tend to wield power and influence in the retail trade.

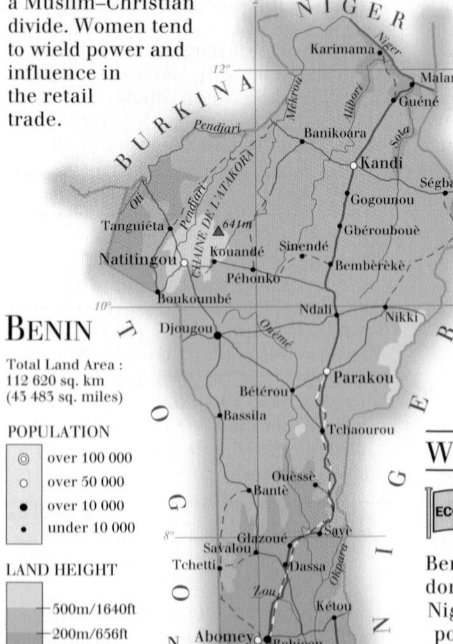
BENIN
Total Land Area : 112 620 sq. km (43 483 sq. miles)
POPULATION
over 100 000
over 50 000
over 10 000
under 10 000
LAND HEIGHT
500m/1640ft
200m/656ft
Sea Level
0 100 km
0 100 miles
NIGER
BURKINA
TOGO
NIGERIA
Karimama
Malanville
Guéné
Banikoara
Kandi
Ségbana
Gogounou
Gbéroubouè
Tanguiéta
Natitingou
Kouandé
Sinendé
Bembèrèkè
Péhonko
Boukoumbé
Ndali
Nikki
Djougou
Parakou
Bétérou
Bassila
Tchaourou
Ouèssè
Banté
Glazoué
Savalou
Savè
Tchetti
Dassa
Kétou
Abomey
Bohicon
Pobè
Lokossa
Allada
Sakété
Ouidah
PORTO-NOVO
Grand-Popo
Cotonou
Bight of Benin
ATLANTIC OCEAN
CHAÎNE DE L'ATAKORA
641m
Pendjari
Mékrou
Alibori
Sota
Oti
Ouémé
Zou
Okpara
Mono
Couffo
Lac Nokoué

POLITICS

Multiparty elections

1999/2003

President Mathieu Kérékou

AT THE LAST ELECTION

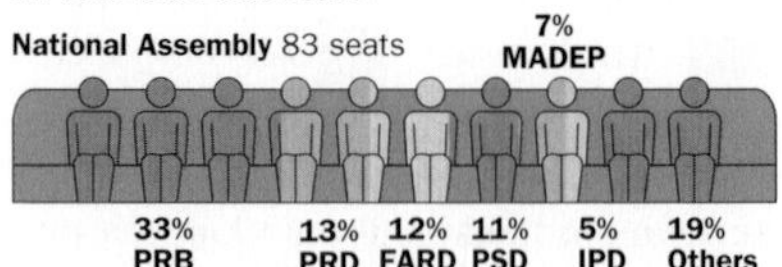
National Assembly 83 seats
7% MADEP
33% PRB
13% PRD
12% FARD
11% PSD
5% IPD
19% Others

PRB = Benin Renaissance Party **PRD** = Party of Democratic Renewal **FARD** = Action Front for Renewal and Development **PSD** = Social Democrat Party **MADEP** = African Movement for Democracy and Progress **IPD** = Impetus for Progress and Democracy

Benin's image as a leader in African democratization was tarnished by allegations of fraud over the 2001 presidential election. Democratization had begun at the National Conference of 1990, when Mathieu Kérékou agreed to hold multiparty elections after years of military one-party rule. The main political parties in Benin tend to be regionally based and depend on the leadership of individuals influential in local communities. Alliances change constantly. Kérékou became the first of the African one-party leaders to hand over power peacefully, to Nicéphore Soglo, a former World Bank official, after elections in 1991. Soglo did not have an automatic majority in the National Assembly, and was forced to include members of the opposition parties in his government. The main political issue became his World Bank-style deregulation of the economy. He was defeated in a controversial election in 1996 which brought Kérékou back to power as president. Kérékou dismissed claims of vote rigging in the presidential election in 2001, saying that democracy was "alive and kicking." He easily won reelection following Soglo's withdrawal from the race.

WORLD AFFAIRS

Joined UN in 1960

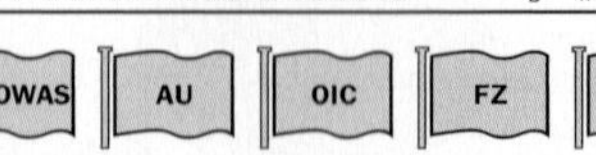
ECOWAS
AU
OIC
FZ
UEMOA

Benin's foreign relations are largely dominated by its giant neighbor, Nigeria, which is by far the most powerful state in the region. The continuation of good relations with France, which is currently the main source of financial aid, is considered to be critical.

AID

Recipient

 $239m (receipts)

 Up 13% in 2000

Benin's poverty is such that the maintenance of aid is at the top of the political agenda. France, the main protector of Benin since independence in 1960, is the major aid donor. Other donors include the World Bank, the US, Germany, the EU, Denmark, Japan, and Italy. Almost all development finance comes from aid, and some has been used to finance debt servicing. There is the usual problem of finding suitable projects, although Benin has a large, well-educated (if top-heavy) civil service, making implementation easier than in many parts of Africa.

DEFENSE

Compulsory military service

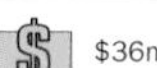 $36m

 Up 6% in 2000

The 4500-strong army is actively involved in the attempt to curb smuggling on the Nigerian border. In 1989 the army was employed internally against rioters.

ECONOMICS

Inflation 8.3% p.a. (1990–2000)

 $2.35bn

 698.7–736.7 CFA francs

SCORE CARD

❑ WORLD GNP RANKING	135th
❑ GNP PER CAPITA	$370
❑ BALANCE OF PAYMENTS	–$168m
❑ INFLATION	4.2%
❑ UNEMPLOYMENT	2%

STRENGTHS

Agriculture-based economy, with good product diversification. Long-overdue devaluation of the CFA franc in January 1994 made exports more competitive.

WEAKNESSES

Large-scale smuggling. Power failures caused by drought brought major economic problems in 1998, and resultant slowdown in GDP growth. Top-heavy civil service.

EXPORTS

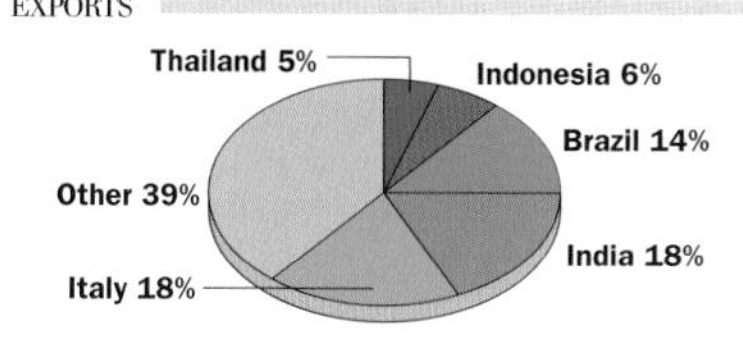
Thailand 5%
Indonesia 6%
Brazil 14%
Other 39%
India 18%
Italy 18%

IMPORTS

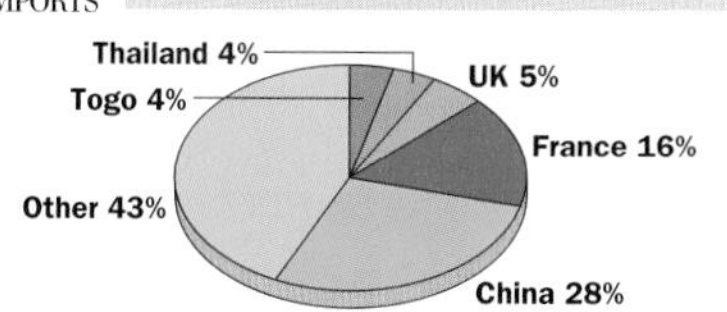
Thailand 4%
Togo 4%
UK 5%
France 16%
Other 43%
China 28%

***Flat landscape near Cotonou,** characteristic of Benin's coastal region. Numerous lagoons lie behind its short coastline.*

RESOURCES

Electric power 15,000 kw

 38,542 tonnes

1243 b/d (reserves 30m barrels)

 1.5m cattle, 1.18m goats, 23m chickens

Oil, limestone, marble, gold

Since 1988 most electricity – which previously had to be imported from Ghana – has been generated by the Nangbeto Dam on the River Mono.

ENVIRONMENT

Sustainability rank: 95th

  7%

0.1 tonnes per capita

Desertification in the north is the major problem. Benin has been used in the past as a dumping ground for toxic waste.

MEDIA

TV ownership low

 Daily newspaper circulation 2 per 1000 people

PUBLISHING AND BROADCAST MEDIA

There are 18 daily newspapers, including *Le Matinal* and *La Nation*

 5 services: 1 state-owned, 4 independent

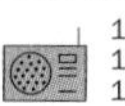 19 services: 1 state-owned, 18 independent

Benin publishes over 50 newspapers and periodicals. The press has considerable freedom, and a media code of practice was introduced in 1999.

CRIME

Death penalty in use

 4961 prisoners

 Up 219% 1996–1998

Armed crime has risen sharply since 1995, despite the reintroduction of the death penalty. Smuggling, including child trafficking, is a major problem.

EDUCATION

School leaving age: 12

 40%

 16,284 students

More is spent on education than on defense, and this is reinforced by Benin's active intellectual community, the "Latin Quarter of Africa." The university at Abomey-Calavi is rated highly in medicine and law.

CHRONOLOGY

In 1625 the Fon, indigenous slave traders, founded the kingdom of Dahomey. Dahomey in turn conquered the neighboring kingdoms of Dan, Allada, and the coast around Porto-Novo.

- ❑ **1857** French establish trading post at Grand-Popo.
- ❑ **1889** French defeat King Behanzin.
- ❑ **1892** French protectorate.
- ❑ **1904** Part of French West Africa.
- ❑ **1960** Full independence.
- ❑ **1975** Renamed Benin.
- ❑ **1989** Marxism–Leninism abandoned as official ideology.
- ❑ **1996** Former ruler Kérékou defeats Soglo in controversial election.
- ❑ **2001** Kérékou reelected to presidency amid claims of electoral fraud.

HEALTH

No welfare state health benefits

1 per 10,000 people

 Communicable and diarrheal diseases, malaria

Outside the major towns, health services and doctors are scarce. It is forecast that by 2030, one million people will have died from AIDS.

SPENDING

GDP/cap. increase

CONSUMPTION AND SPENDING

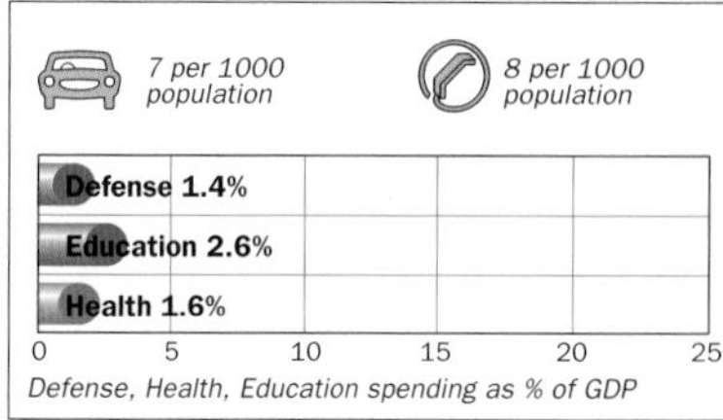
7 per 1000 population
8 per 1000 population
Defense 1.4%
Education 2.6%
Health 1.6%
0
5
10
15
20
25
Defense, Health, Education spending as % of GDP

Substantial differences in wealth reflect the strongly hierarchical nature of society, especially in the south. French cars are considered to be status symbols.

WORLD RANKING

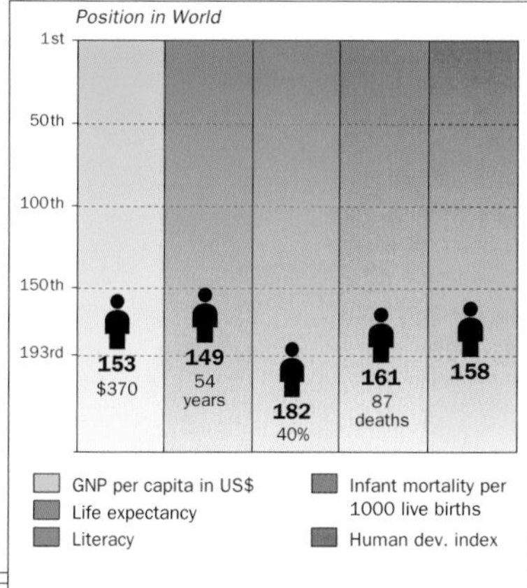
Position in World
1st
50th
100th
150th
193rd
153
$370
149
54 years
182
40%
161
87 deaths
158
GNP per capita in US$
Life expectancy
Literacy
Infant mortality per 1000 live births
Human dev. index

BHUTAN

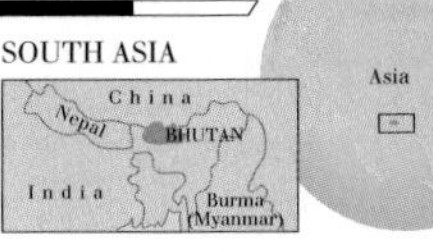
SOUTH ASIA
China
Nepal
BHUTAN
India
Burma (Myanmar)
Asia

B

OFFICIAL NAME: Kingdom of Bhutan **CAPITAL:** Thimphu
POPULATION: 2.1 million **CURRENCY:** Ngultrum **OFFICIAL LANGUAGE:** Dzongkha

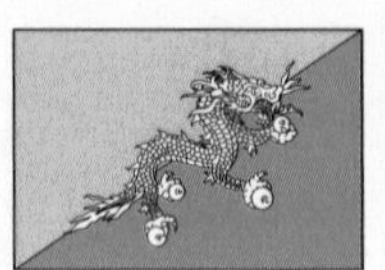

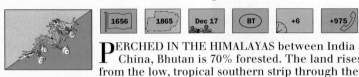

1656

1865
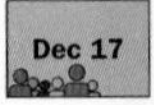
Dec 17
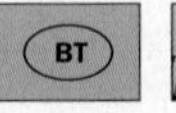
BT

+6

+975

.bt

PERCHED IN THE HIMALAYAS between India and China, Bhutan is 70% forested. The land rises from the low, tropical southern strip through the fertile central valleys to the high Himalayas, inhabited by seminomadic yak herders. Formally a Buddhist state where power is shared by the king and the government, Bhutan began modernizing in the 1960s, but has chosen to do so gradually, and remains largely closed to the outside world.

CLIMATE

Mountain

WEATHER CHART FOR THIMPHU

Average daily temperature | Rainfall
°C/°F J F M A M J J A S O N D cm/in
40/104 | 40/16
30/86 | 30/12
20/68 | 20/8
10/50 | 10/4
0/32 | 0
-10/14
-20/-4

The south is tropical, the north alpine, cold, and harsh. The central valleys are warmer in the east than in the west. The summer monsoon affects all parts.

TRANSPORTATION

Drive on left

Paro International 19,939 passengers

Has no fleet

THE TRANSPORTATION NETWORK

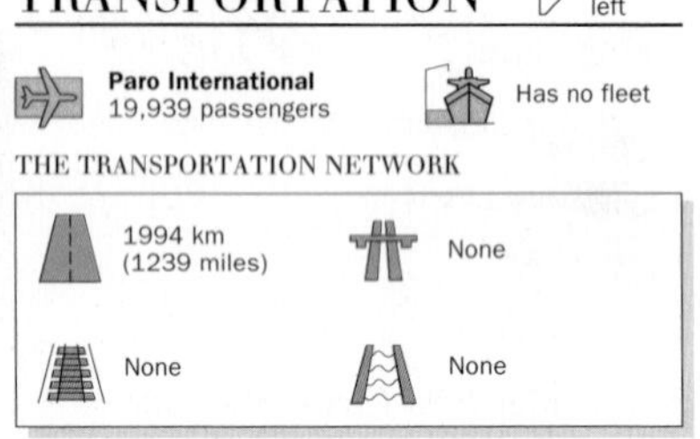
1994 km
(1239 miles)
None
None
None

The main surfaced road runs east–west across central Bhutan. Two others run south into India. Only the national airline, Druk Air, flies into Bhutan.

TOURISM

Visitors : Population 1:339

6200 visitors

Up 4% in 1999

MAIN TOURIST ARRIVALS

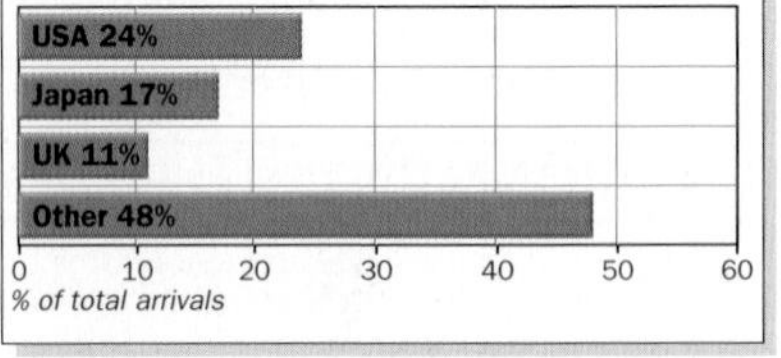
USA 24%
Japan 17%
UK 11%
Other 48%
0 10 20 30 40 50 60
% of total arrivals

Tourism is restricted to protect Bhutan's culture and natural environment; entry has been easier since the industry was privatized in 1991. Most monasteries are closed to tourists. In 1998, fire damaged the famous Taktsang monastery.

Less than 10% of Bhutan is arable, but its fertility allows almost any crop to grow. The diversity of wild plant species inspired its old name: Southern Valleys of the Medicinal Herbs.

PEOPLE

Pop. density low

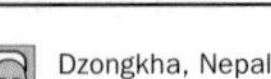
Dzongkha, Nepali, Assamese

45/km²
(116/mi²)

THE URBAN/RURAL POPULATION SPLIT

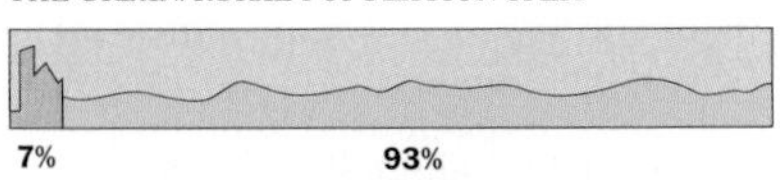
7%
93%

RELIGIOUS PERSUASION

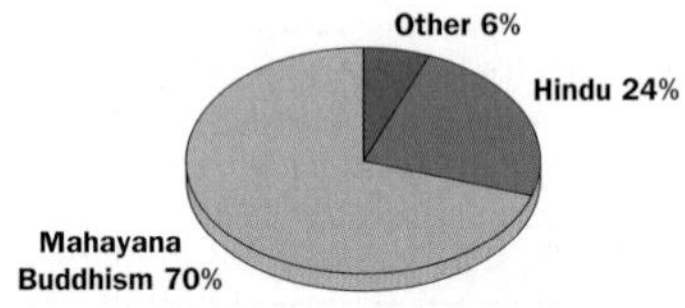
Other 6%
Hindu 24%
Mahayana Buddhism 70%

The majority of the population, the Drukpa peoples, originated from Tibet and are devoutly Buddhist. The Hindu minority is made up of Nepalese who settled in the south from 1910 to 1950. Bhutan has 20 languages. Dzongkha, the language of western Bhutan, native to just 16% of people, was made the official language in 1988. The Nepalese community has reacted strongly, regarding this act as "cultural imperialism."

POLITICS

No multiparty elections

Not applicable

H.M. Druk Gyalpo (Dragon King) Jigme Singye Wangchuk

LEGISLATIVE OR ADVISORY BODIES

National Assembly 150 seats

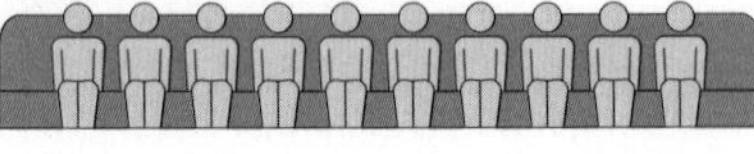

There are no legal political parties; members are elected individually to the National Assembly, to advise the king, who rules as an absolute monarch

The modernization of Bhutan's absolute monarchy began in 1961. Under further changes proposed in 1998, the king relinquished his right to appoint the government in favor of a cabinet elected by the National Assembly. The National Assembly was also empowered to pass a vote of no confidence against the king. These proposals came as a response to a prodemocracy movement fueled by ethnic Nepalese opposed to the Drukpa-dominated political system.

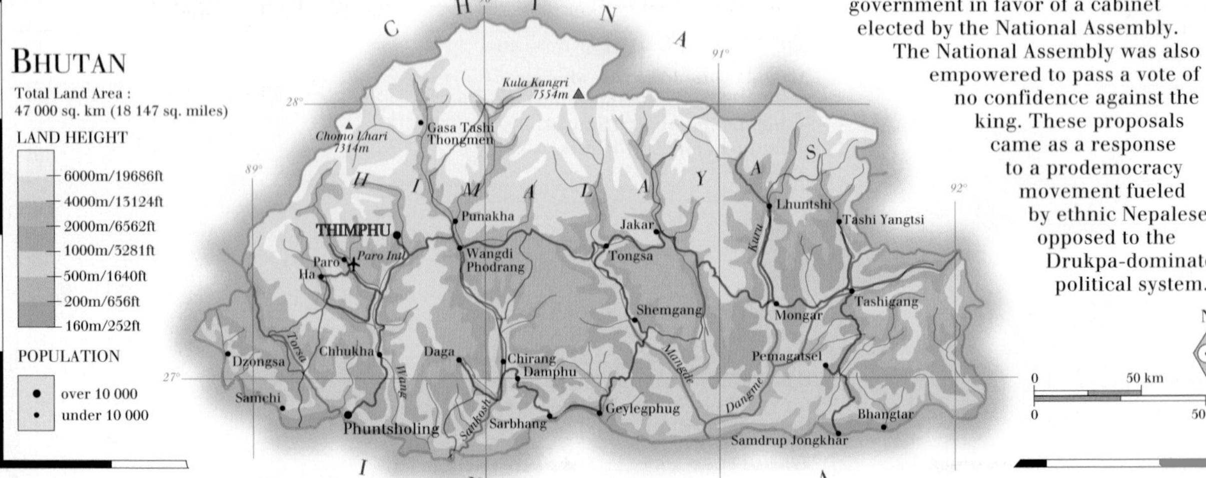
BHUTAN
Total Land Area :
47 000 sq. km (18 147 sq. miles)
LAND HEIGHT
6000m/19686ft
4000m/13124ft
2000m/6562ft
1000m/3281ft
500m/1640ft
200m/656ft
160m/252ft
POPULATION
over 10 000
under 10 000
CHINA
INDIA
HIMALAYAS
Kula Kangri 7554m
Chomo Lhari 7314m
Gasa Tashi Thongmen
Punakha
THIMPHU
Paro
Paro Intl
Ha
Wangdi Phodrang
Jakar
Tongsa
Lhuntshi
Tashi Yangtsi
Shemgang
Mongar
Tashigang
Pemagatsel
Dzongsa
Chhukha
Daga
Chirang
Damphu
Samchi
Phuntsholing
Sarbhang
Geylegphug
Samdrup Jongkhar
Bhangtar
Torsa
Wang
Sankosh
Manas
Kuru
Dangme
0 50 km
0 50 miles
N

B

WORLD AFFAIRS

Joined UN in 1971

CP

IBRD

NAM
SAARC ADB

Bhutan's closest links are with India. Relations with China are cordial, and negotiations to settle the China–Bhutan border have progressed smoothly since 1984. There is tension with Nepal over Bhutan's treatment of its ethnic Nepalese minority and the influx of Bhutanese refugees into Nepal.

AID

Recipient

 $53m (receipts)

 Down 21% in 2000

Bhutan relies on foreign aid for about half of its annual budget. The largest single donor is Japan.

DEFENSE

No compulsory military service

 $20m

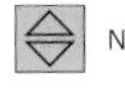 No change in 2000

The army is under the command of the king, and is trained by Indian military instructors. India provides de facto military protection and is obliged to defend Bhutan against attack.

ECONOMICS

Inflation 9.5% p.a. (1990–2000)

 $479m

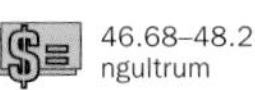 46.68–48.22 ngultrum

SCORE CARD

- ❑ WORLD GNP RANKING........171st
- ❑ GNP PER CAPITA........$590
- ❑ BALANCE OF PAYMENTS........–$87m
- ❑ INFLATION........6.8%
- ❑ UNEMPLOYMENT........Low rate

STRENGTHS

New development of cash crops for Asian markets (cardamoms, apples, oranges, apricots). Hardwoods in south, especially teak, but exploitation currently tightly controlled. Large hydroelectric potential.

WEAKNESSES

Dependence on Indian workers for many public-sector jobs from road building to teaching. The majority of the population are dependent on agriculture. Cultivated land is extremely restricted because of steep mountain slopes. Very little industry. Few mineral resources.

EXPORTS

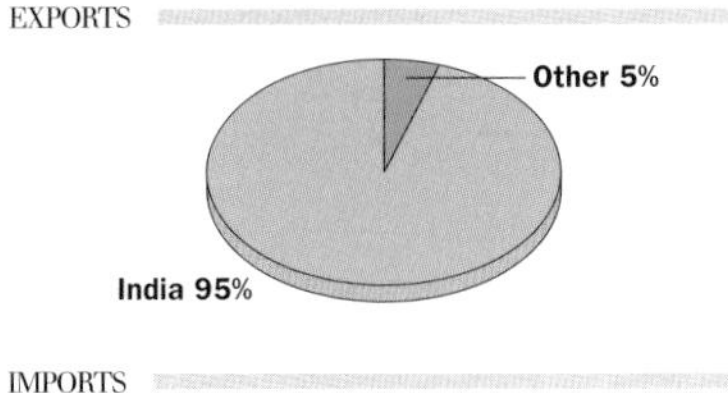
Other 5%
India 95%

IMPORTS

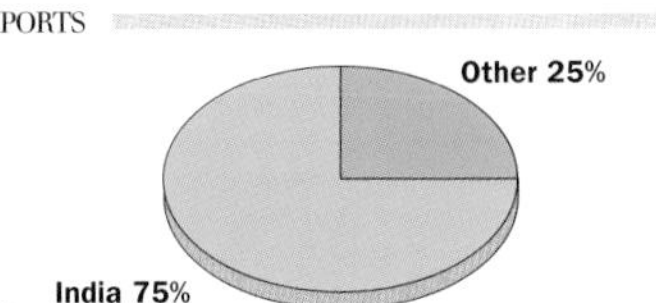
Other 25%
India 75%

RESOURCES

Electric power 356,000 kw

 330 tonnes

 Not an oil producer and has no refineries

 435,000 cattle, 74,900 pigs, 310,000 chickens

 Talc, gypsum, coal, limestone, slate, dolomite

Bhutan's forests remain largely intact, and logging is very strictly controlled. Hydroelectric potential is considerable, but few dams have been built. Power is sold to India from the Chhukha Dam, bringing in substantial foreign earnings.

ENVIRONMENT

Sustainability rank: 30th

 19% (18% partially protected)

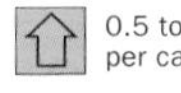 0.5 tonnes per capita

Bhutan's forests stabilize the steep mountainsides and supply the bulk of its fuel needs. Road building, begun in the 1960s, is the biggest cause of deforestation, which has led to topsoil erosion. The high northern pastures are at risk from overgrazing by yaks. Traditional Buddhist values instilling respect for nature and forbidding the killing of animals are still observed.

MEDIA

TV ownership low

 There are no daily newspapers

PUBLISHING AND BROADCAST MEDIA

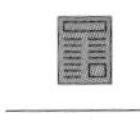 There are no daily newspapers. *Kuensel* is published weekly by the government in Dzongkha, English, and Nepali

 1 state-owned service

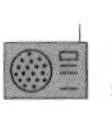 1 state-owned service

Until 1999 TV was banned, in order to protect cultural values. An Internet café opened in Thimphu in 2000.

CRIME

Death penalty not used in practice

 Bhutan does not publish prison figures

 Little variation from year to year

Levels of violent crime and theft are low. In 1991, *Driglam namzha*, an ancient code of conduct including the requirement to wear traditional dress, was revived, with imprisonment or fines for those not in compliance.

EDUCATION

Schooling is not compulsory

 47%

 2683 students

Education is free. A small minority of children attend secondary school. Teaching is in English and Dzongkha. There are no universities.

CHRONOLOGY

The Drukpa, originally from Tibet, united Bhutan in 1656. In 1865 the Drukpa lost the Duars Strip to British India.

- ❑ **1907** Monarchy established.
- ❑ **1949** Independence.
- ❑ **1953** National Assembly inaugurated.
- ❑ **1968** King forms first cabinet.
- ❑ **1971** Bhutan joins UN.
- ❑ **1990** Ethnic Nepalese launch campaign for minority rights.
- ❑ **1998** King proposes to reform government.
- ❑ **1999** First TV service inaugurated.

HEALTH

Welfare state health benefits

 1 per 6510 people

 Diarrheal, respiratory diseases, tuberculosis, malaria, infant deaths

Free clinics, along with Thimphu's hospital, provide basic health care. Progress is being made in child immunization, and monks have been persuaded to teach hygiene. Infant mortality is high. Bhutanese, Tibetan, and Chinese traditional medicines are widely practiced.

SPENDING

GDP/cap. increase

CONSUMPTION AND SPENDING

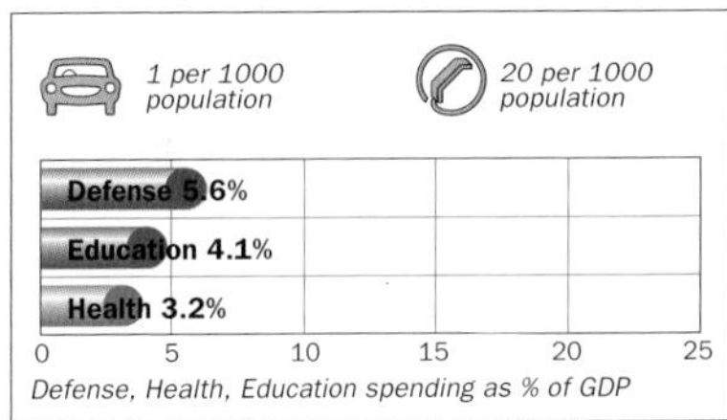
1 per 1000 population
20 per 1000 population
Defense 5.6%
Education 4.1%
Health 3.2%
0
5
10
15
20
25

Defense, Health, Education spending as % of GDP

Most of Bhutan's people are chronically poor, although starvation is virtually unknown. There is a small middle class, consisting of public employees and storekeepers.

WORLD RANKING

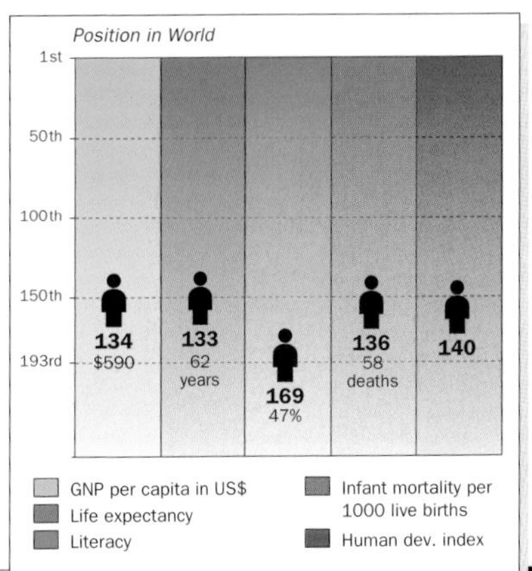
Position in World
1st
50th
100th
150th
193rd
134
$590
133
62
years
169
47%
136
58
deaths
140
GNP per capita in US$
Life expectancy
Literacy
Infant mortality per 1000 live births
Human dev. index

B

BOLIVIA

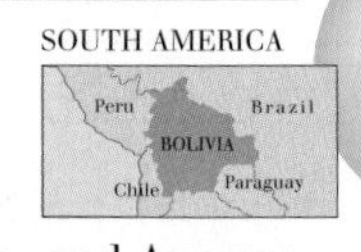
SOUTH AMERICA
Peru
Brazil
BOLIVIA
Chile
Paraguay
South America

OFFICIAL NAME: Republic of Bolivia **CAPITALS:** La Paz (administrative); Sucre (judicial)
POPULATION: 8.5 million **CURRENCY:** Boliviano **OFFICIAL LANGUAGES:** Spanish, Quechua, and Aymara

1825

1938

Aug 6
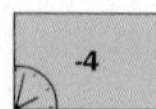
BOL

-4

+591

.bo

BOLIVIA LIES LANDLOCKED high in central South America, and is one of the continent's poorest nations. Over half of the population lives on the *altiplano*, the windswept plateau between two ranges of the Andes, 3500 m (11,500 ft) above sea level. La Paz, the highest capital in the world, has spawned a neighboring large twin, El Alto. Bolivia has the world's highest golf course, ski run, and soccer stadium. The eastern lowland regions are tropical and underdeveloped but are rapidly being colonized.

CLIMATE

Tropical/mountain

WEATHER CHART FOR LA PAZ

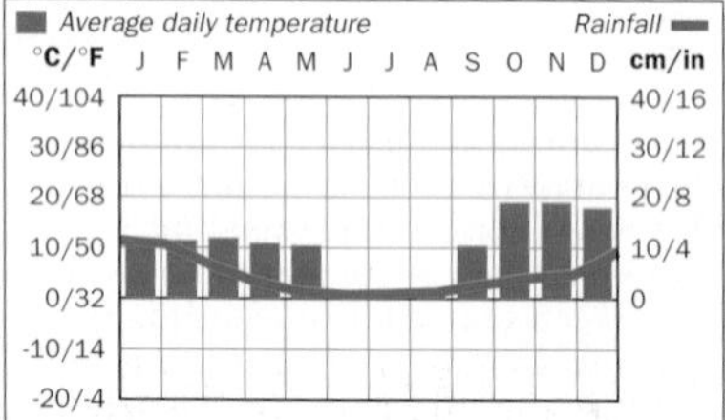
Average daily temperature
Rainfall
°C/°F
J F M A M J J A S O N D
cm/in
40/104
30/86
20/68
10/50
0/32
-10/14
-20/4
40/16
30/12
20/8
10/4
0

The Andean *altiplano* has an extreme tropical highland climate with winter night frosts. Annual rainfall in the west is only 25 cm (10 in). The hot eastern lowlands receive most rain in summer.

TRANSPORTATION

Drive on right

El Alto, La Paz

1 ship
15,800 dwt

THE TRANSPORTATION NETWORK

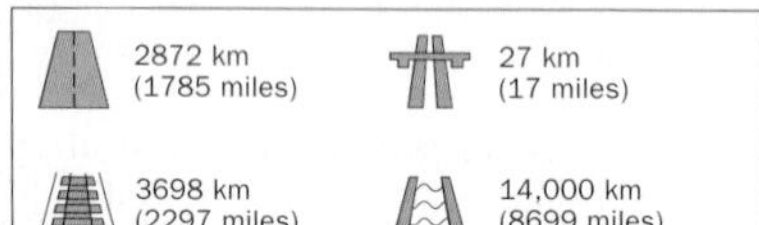
2872 km
(1785 miles)
27 km
(17 miles)
3698 km
(2297 miles)
14,000 km
(8699 miles)

Obtaining port access to the Pacific coast for landlocked Bolivia is important. Only 4% of roads are paved. The national railroad was privatized in 1996. Domestic airlines are generally reliable.

***Potato harvest** on the* altiplano. *Migration to the more fertile lands in the east has been encouraged.*

***Copacabana** on the shores of Lake Titicaca. It lies on a large headland owned by Bolivia on the Peruvian side of the lake.*

TOURISM

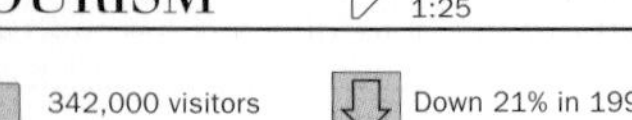
Visitors : Population
1:25

342,000 visitors

Down 21% in 1999

MAIN TOURIST ARRIVALS

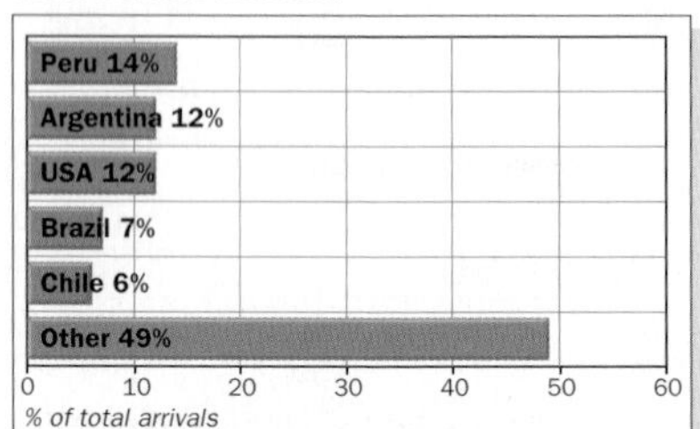
Peru 14%
Argentina 12%
USA 12%
Brazil 7%
Chile 6%
Other 49%
0
10
20
30
40
50
60
% of total arrivals

Foreign tourists are drawn by the traditional festivals, especially carnivals in February or March, the variety of Bolivia's scenery, and its Spanish colonial architecture. Major attractions include the Silver Mountain at Potosí, and Lake Titicaca, the highest navigable lake in the world, covering an area of 8970 sq. km (3463 sq. miles). Recent political stability encouraged some growth in tourism in the 1990s, but potential is limited by Bolivia's isolation, the rugged, inaccessible terrain, and the limited infrastructure.

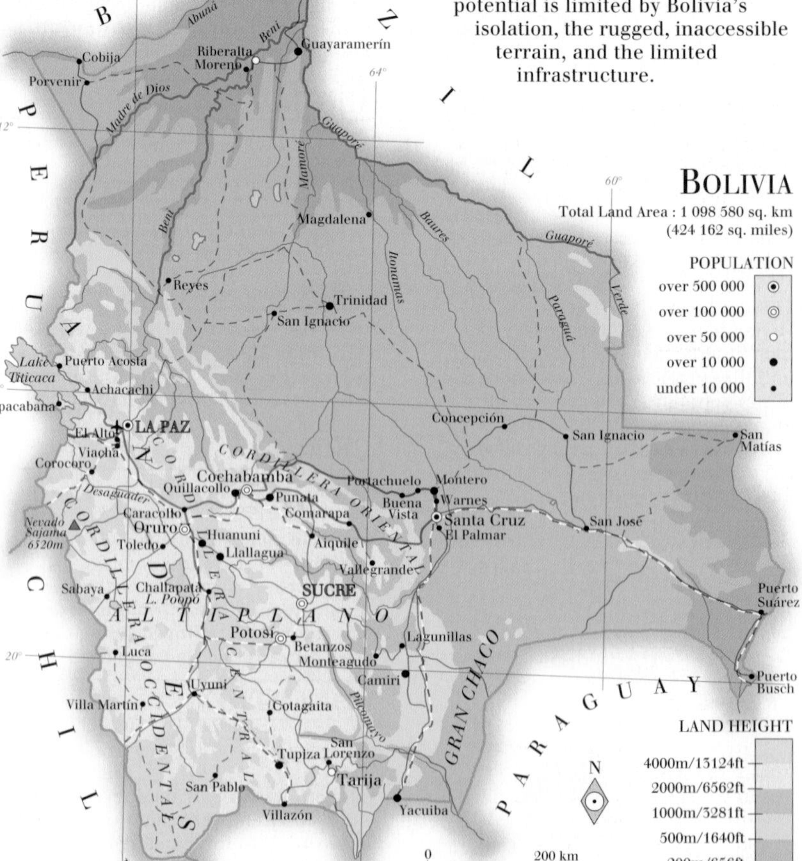
BOLIVIA
Total Land Area : 1 098 580 sq. km
(424 162 sq. miles)
POPULATION
over 500 000
over 100 000
over 50 000
over 10 000
under 10 000
LAND HEIGHT
4000m/13124ft
2000m/6562ft
1000m/3281ft
500m/1640ft
200m/656ft
Sea Level
0
200 km
0
200 miles
N
BRAZIL
PERU
CHILE
ARGENTINA
PARAGUAY
GRAN CHACO
CORDILLERA ORIENTAL
CORDILLERA OCCIDENTAL
ALTIPLANO
Fortaleza
Guayaramerín
Riberalta
Moreno
Cobija
Porvenir
Magdalena
Reyes
Trinidad
San Ignacio
Puerto Acosta
Lake Titicaca
Achacachi
Copacabana
El Alto
LA PAZ
Viacha
Corocoro
Cochabamba
Quillacollo
Punata
Comarapa
Portachuelo
Montero
Warnes
Buena Vista
Santa Cruz
El Palmar
Concepción
San Ignacio
San Matías
San José
Nevado Sajama 6520m
Caracollo
Oruro
Toledo
Huanuni
Llallagua
Aiquile
Vallegrande
Sabaya
Challapata
L. Poopó
SUCRE
Potosí
Betanzos
Monteagudo
Lagunillas
Camiri
Luca
Uyuni
Villa Martín
Cotagaita
Tupiza
San Lorenzo
Tarija
San Pablo
Villazón
Yacuiba
Puerto Suárez
Puerto Busch
Abuná
Beni
Madre de Dios
Guaporé
Mamoré
Baures
Itonamas
Paraguá
Verde
Desaguadero
Pilcomayo

PEOPLE

Pop. density low

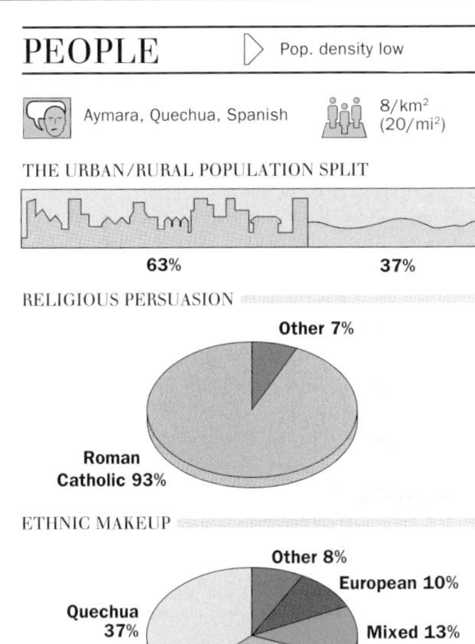
Aymara, Quechua, Spanish
8/km²
(20/mi²)
THE URBAN/RURAL POPULATION SPLIT
63%
37%
RELIGIOUS PERSUASION
Other 7%
Roman Catholic 93%
ETHNIC MAKEUP
Other 8%
European 10%
Quechua 37%
Mixed 13%
Aymara 32%

Two-thirds of Bolivians are Quechua and Aymara Amerindians who historically have been marginalized. In recent years, however, they have played a more active role in politics by supporting new populist parties.

Wealthy city elites, dating back to Spanish colonial rule, retain great influence, but new entrepreneurs with political ambitions have appeared. Most Bolivians are subsistence farmers, miners, small traders, or artisans earning low incomes. Government schemes, spontaneous colonization, and the collapse of tin mining have led to large-scale migration from the Andes to lowland eastern regions in the last few decades.

Family life tends to be close-knit; Amerindians practice Roman Catholicism mixed with their own traditions and culture. Women have low status. There are some 130,000 lowland Amerindians in western regions.

POPULATION AGE BREAKDOWN

Female
Age
Male
1.2% 80+ 1%
2.1% 60–79 1.8%
7.4% 40–59 6.9%
14.6% 20–39 14.3%
24.9% 0–19 25.8%
% of population by age group

WORLD AFFAIRS

Joined UN in 1945

AP | AmCC | NAM | OAS | RG

Gaining improved access to the Pacific via Peru and Chile is Bolivia's historic foreign policy concern. The prospect of a Bolivia–Peru gas pipeline is a long-term project. Current gas export projects, coupled with the opening of the telecoms sector, and the extension of the Apta preferential trade act with the US should strengthen economic growth. Current US aid is conditional on the Bolivian government taking measures to destroy the cocaine-producing and -trafficking industry, involving military and police attacks on impoverished coca growers.

A pointer to the future is that Bolivia is an associate member, along with Chile, of Mercosur. This southern grouping, under the leadership of Brazil, will act as a strong negotiating bloc in upcoming talks for a Free Trade Area of the Americas (FTAA), due in 2005.

Hugo Banzer Suárez*, former dictator, who was elected president (1997–2001).*

Gonzalo Sanchez de Lozada*, president for 1993–1997, was reelected in 2002.*

POLITICS

Multiparty elections

L. House 2002/2007
U. House 2002/2007

President Gonzalo Sanchez de Lozada

AT THE LAST ELECTION

Chamber of Deputies 130 seats

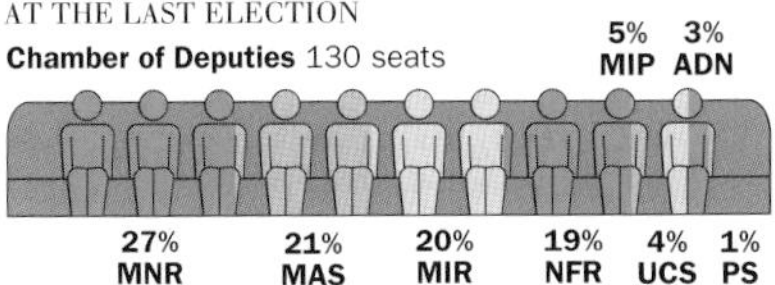
5% MIP
3% ADN
27% MNR
21% MAS
20% MIR
19% NFR
4% UCS
1% PS

MNR = Nationalist Revolutionary Movement **MAS** = Movement for Socialism **MIR** = Movement of the Revolutionary Left **NFR** = New Republican Force **MIP** = Pachakuti Indigenous Movement **UCS** = Union for Civic Solidarity **ADN** = Nationalist Democratic Action **PS** = Socialist Party

Chamber of Senators 27 seats

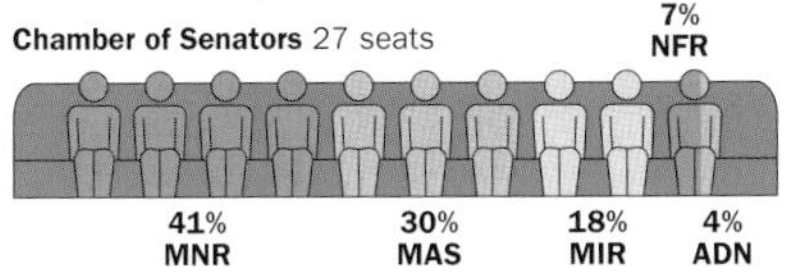
7% NFR
41% MNR
30% MAS
18% MIR
4% ADN

Bolivia is a multiparty democracy.

Profile

From 1825 to the early 1980s, Bolivia experienced, on average, more than one armed coup a year, punctuated by a national revolution in 1952 which delivered important reforms. The fragmented and drug-tainted military finally stepped down in 1982, but full elections were delayed until 1985.

New populist parties have emerged to challenge traditional politics, and even the drift to free-market economics. Coalitions are unstable, nepotism remains rife, and the narcotics trade, the profits of which underpin the economy, is frequently implicated in political corruption scandals. The main trade union federation COB was the traditional focus of opposition, but coca growers and other popular groups have now assumed this role. Peasant leader Evo Morales even gathered enough support to come a close second in presidential elections in 2002.

The austerity policies of the MNR government were continued after its defeat in 1997 by former dictator Hugo Banzer. Ill health forced his resignation in 2001, and the MNR was returned to power in 2002 with Gonzalo Sanchez resuming the presidency.

Main Political Issues

Disillusionment with political system

Public opinion in the run-up to the 2002 presidential elections pointed to a lack of faith in promised electoral reform. The country's veteran politicians are widely viewed as a self-serving, self-perpetuating class tainted by corruption.

Coca growers

Poor farmers oppose the government's forced eradication of coca crops in an anticocaine drive to ensure more US aid. Serious clashes continued in 2001–2002 as troops operating in eastern coca-growing areas removed farmers' road blocks. The *cocaleros* complain that modest government subsidies to switch crops take no account of low prices for bananas, pineapples, and palm hearts.

CHRONOLOGY

The Aymara civilization was conquered by the Incas in the late 1400s. Fifty years later, the Incas were defeated by the *conquistadores* and Upper Peru, as it became, was governed by Spain from Lima.

- **1545** Cerro Rico, the Silver Mountain, discovered at Potosí. Provides Spain with vast wealth.
- **1776** Upper Peru becomes part of Viceroyalty of Río de la Plata centered on Buenos Aires.
- **1809** Simón Bolívar inspires first revolutionary uprisings in Latin America at Chuquisaca (Sucre), La Paz, and Cochabamba, but they fail.
- **1824** Spaniards suffer final defeat by Bolívar's general, José de Sucre.
- **1825** Independence.
- **1836–1839** Union with Peru fails. Internal disorder.
- **1864–1871** Ruthless rule of Mariano Melgarejo. Three Amerindian revolts over seizure of ancestral lands. ➪

B

CHRONOLOGY *continued*

- **1879–1883** War of the Pacific, won by Chile. Bolivia left landlocked.
- **1880–1930** Period of stable governments. Exports from revived mining industry bring prosperity.
- **1903** Acre province ceded to Brazil.
- **1914** Republican Party founded.
- **1920** Amerindian rebellion.
- **1923** Miners bloodily suppressed.
- **1932–1935** Chaco War with Paraguay. Bolivia loses three-quarters of Chaco. Rise of radicalism and labor movement.
- **1951** Víctor Paz Estenssoro of MNR elected president. Military coup.
- **1952** Revolution. Paz Estenssoro and MNR brought back. Land reforms improve Amerindians' status. Education reforms, universal suffrage, tin mines nationalized.
- **1964** Military takes over in coup.
- **1967** Che Guevara killed while trying to mobilize Bolivian workers.
- **1969–1979** Military regimes rule with increasing severity. 1979 coup fails. Interim civilian rule.
- **1980** Military takes over again.
- **1982** President-elect Siles Zuazo finally heads leftist civilian MIR government. Inflation 24,000%.
- **1985** MNR wins elections. Austerity measures bring down inflation.
- **1986** Tin market collapses. 21,000 miners sacked.
- **1989** MIR takes power after close-run elections. President Paz Zamora makes pact with 1970s dictator Gen. Hugo Banzer, leader of ADN.
- **1990** 1.6 million hectares (4 million acres) of rainforest recognized as Amerindian territory.
- **1993** MNR back in power. Gonzalo Sanchez de Lozada president.
- **1997** Banzer wins presidency.
- **1999** Opposition demands inquiry into Banzer's role in regional military repression in 1970s.
- **2001** Banzer resigns due to ill health.
- **2001–2002** Government program of coca eradication and bans on coca trading and marketing provoke clashes with peasant farmers.
- **2002** MNR wins elections. Sanchez returns as president.

AID

Recipient

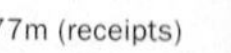

$477m (receipts)

Down 16% in 2000

Most aid comes from the US and depends on progress in coca crop eradication. Smaller amounts come from western European countries. Poor rural areas get project aid from Western religious organizations, NGOs, and charities. The IDB provided emergency aid after flash floods destroyed infrastructure and badly damaged the historic center of La Paz in 2002.

DEFENSE

Compulsory military service

$128m

Down 14% in 2000

BOLIVIAN ARMED FORCES

36 light tanks (36 SK-105 *Kuerassier*)
25,000 personnel
60 riverine craft
3500 personnel
37 combat aircraft (18 AT-33AN)
3000 personnel
None

Although the military has not actively interfered in politics for more than two decades, it is frequently used to quell internal dissent. The army is the main focus of defense spending, with weaponry bought almost entirely from the US. The Bolivian navy consists mainly of gunboats on Lake Titicaca, which borders Peru, and on the Pilcomayo River. The army has worked with US forces against the cocaine business, although its integrity is questioned due to its past associations with narcotics trafficking. The main ambition of the military, apart from protecting its own interests and privileges, is the unrealizable aim of recapturing territory that would allow Bolivia access to the Pacific.

Military service lasts for one year.

ECONOMICS

Inflation 8.5% p.a. (1990–2000)

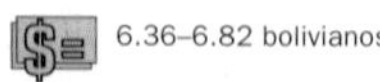

$8.21bn

6.36–6.82 bolivianos

SCORE CARD

- WORLD GNP RANKING.........................93rd
- GNP PER CAPITA $990
- BALANCE OF PAYMENTS...................–$464m
- INFLATION 4.6%
- UNEMPLOYMENT 4%

EXPORTS

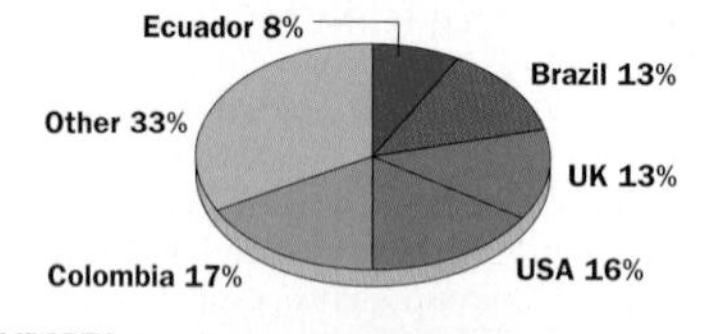
Ecuador 8%
Brazil 13%
Other 33%
UK 13%
Colombia 17%
USA 16%

IMPORTS

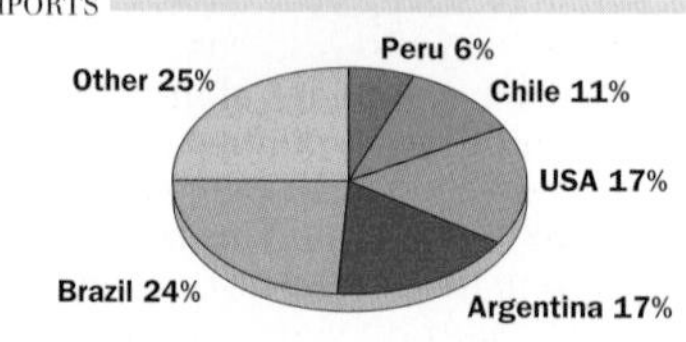
Peru 6%
Other 25%
Chile 11%
USA 17%
Brazil 24%
Argentina 17%

ECONOMIC PERFORMANCE INDICATOR

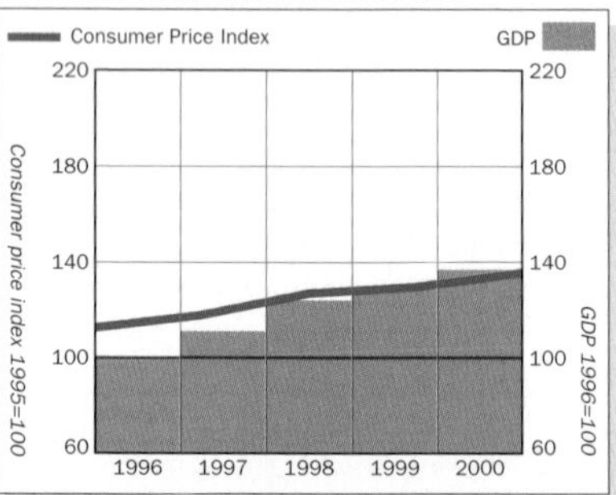
Consumer Price Index
GDP
220
180
140
100
60
Consumer price index 1995=100
GDP 1996=100
1996
1997
1998
1999
2000

STRENGTHS

Mineral riches: gold, silver, zinc, lead, tin. Newly discovered oil and natural gas deposits attracting foreign investment.

WEAKNESSES

Raw materials vulnerable to fluctuating world prices. Lack of processed or manufactured exports with higher added value. Lack of integration between economic sectors and regions. Poor infrastructure.

PROFILE

Traditionally, the state used earnings from the publicly owned state mining sector to control the economy. Years of deep recession in the 1980s, accompanied by accelerating inflation and a collapsing currency, saw the introduction of severe, IMF-approved, austerity policies. These, along with the introduction of a new currency and tax reform, succeeded in curbing inflation, reducing public spending, and restoring international loans, but at the price of great social unrest. Growth was restored in the 1990s and a controversial "capitalization" program was launched which allowed for the 50/50 sell-off of shares in all six state companies on attractive terms to investors and employees. Narcotics revenues remain important for the economy.

BOLIVIA : MAJOR BUSINESSES

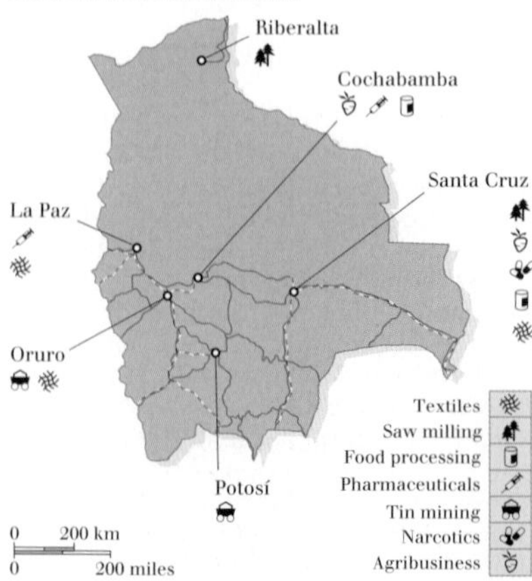
Riberalta
Cochabamba
Santa Cruz
La Paz
Oruro
Potosí
Textiles
Saw milling
Food processing
Pharmaceuticals
Tin mining
Narcotics
Agribusiness
0 200 km
0 200 miles

B

RESOURCES

Electric power 1m kw

6450 tonnes

34,846 b/d (reserves 114m barrels)

8.75m sheep, 6.73m cattle, 2.8m pigs, 74m chickens

Tin, natural gas, oil, zinc, tungsten, gold, antimony, silver, lead

ELECTRICITY GENERATION

Hydro 44% (1.6bn kwh)
Combustion 56% (2.1bn kwh)
Nuclear 0%
Other 0%

0 20 40 60 80 100

% of total generation by type

Bolivia is the world's fifth-largest tin producer. The government is allowing foreign companies to prospect for more oil, and to increase sales of natural gas to Brazil and Argentina.

BOLIVIA : LAND USE

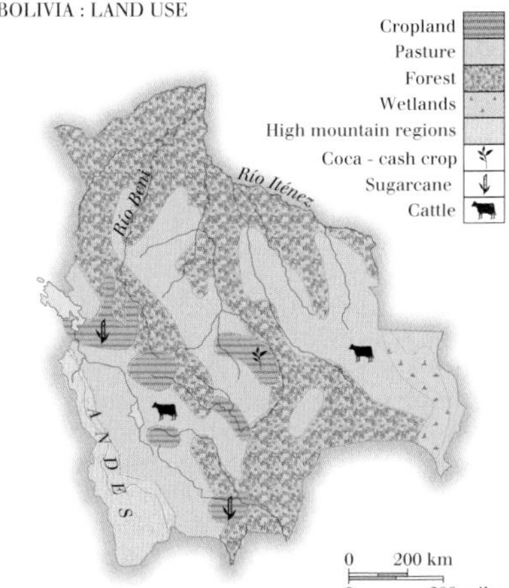
Cropland
Pasture
Forest
Wetlands
High mountain regions
Coca - cash crop
Sugarcane
Cattle
Río Beni
Río Iténez
ANDES
0 200 km
0 200 miles

ENVIRONMENT

Sustainability rank: 21st

14% (5% partially protected)

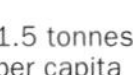

1.5 tonnes per capita

ENVIRONMENTAL TREATIES

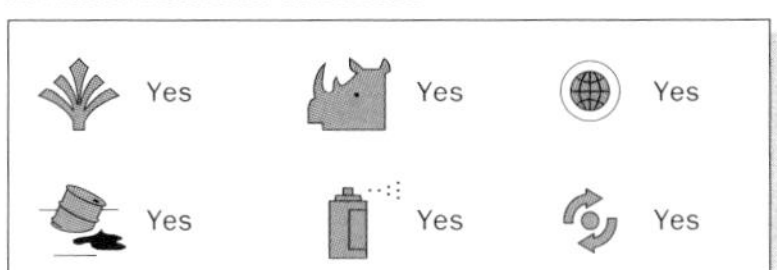
Yes
Yes
Yes
Yes
Yes
Yes

Deforestation is Bolivia's major ecological problem, as throughout the Amazon region. Land clearances are running at 164,000 hectares (398,000 acres) a year. Much of the cleared land is turned over to cattle ranching or the growing of coca. Pesticide and fertilizer overuse in the coca business is a concern. The industry is effectively uncontrolled and rivers in Amazonia have high pollution levels.

Pollution problems are compounded by waste chemicals used in minerals industries. Mercury, used in the extraction of silver, has been found in dangerous quantities in river systems.

MEDIA

TV ownership medium

Daily newspaper circulation 55 per 1000 people

PUBLISHING AND BROADCAST MEDIA

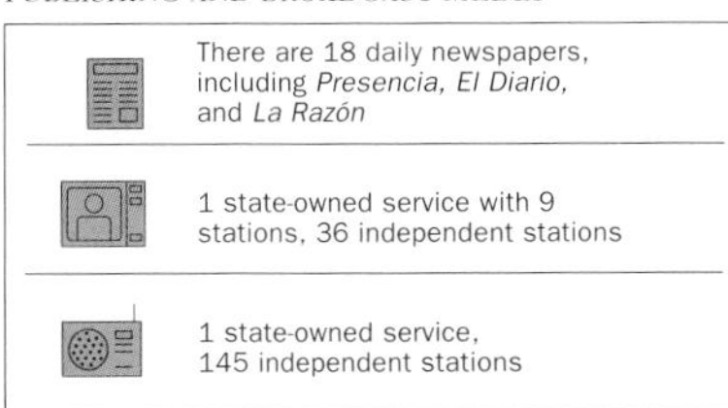
There are 18 daily newspapers, including *Presencia, El Diario,* and *La Razón*
1 state-owned service with 9 stations, 36 independent stations
1 state-owned service, 145 independent stations

Bolivia has strict defamation laws and considerable self-censorship. One of the TV stations is university-run, broadcasting mainly educational programs.

CRIME

Death penalty not used in practice

8315 prisoners

Crime is rising in narcotics-trafficking centers

CRIME RATES

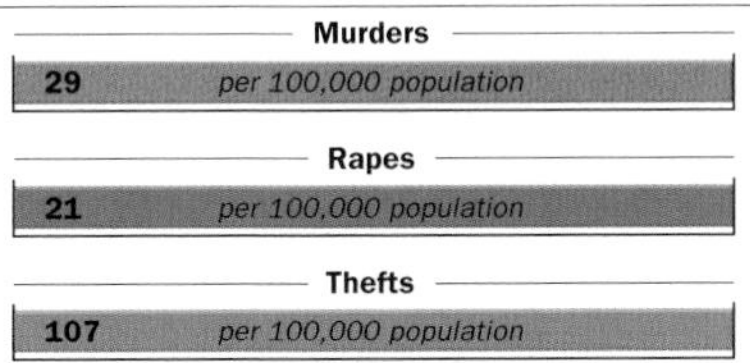
Murders
29 per 100,000 population
Rapes
21 per 100,000 population
Thefts
107 per 100,000 population

Violent crime is centered on narcotics-trafficking towns in the eastern lowlands, particularly Santa Cruz. Main cities are much safer for tourists, and have lower crime rates than cities in neighboring Peru. The police and army have a history of mistreating poor farmers and miners.

EDUCATION

School leaving age: 14

86%

220,000 students

THE EDUCATION SYSTEM

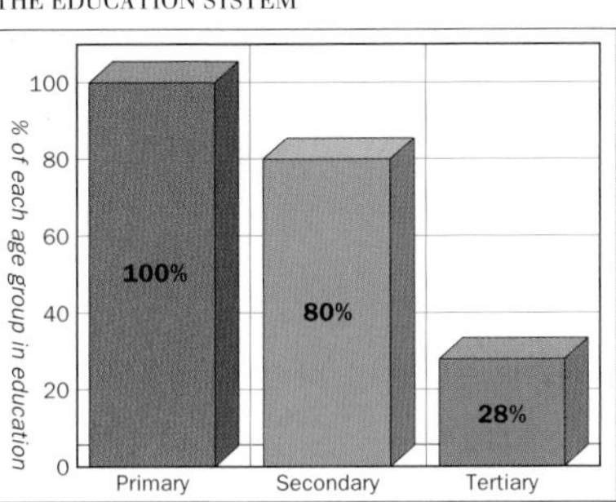
% of each age group in education
100
80
60
40
20
0
100%
80%
28%
Primary
Secondary
Tertiary

IMF targets for increased school attendance are being met, but education, based on a combination of the French and US systems, is seriously underfunded.

Although the majority of people speak indigenous languages, most teaching is in Spanish. Bolivia has one of the lowest literacy rates in South America. Reform and multilateral aid have led to some improvements.

HEALTH

No welfare state health benefits

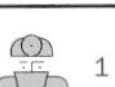

1 per 769 people

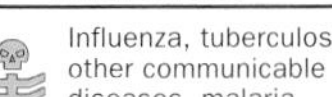

Influenza, tuberculosis, other communicable diseases, malaria

Bolivia has one of the highest child mortality rates in the western hemisphere. Nearly 40% of children up to three years of age suffer from chronic malnutrition, and fewer than half the children under one year are immunized; diseases preventable by vaccination are a major cause of death.

Formal health services are costly, so that over half the population does not use them. Some 60% of births take place at home with no trained assistant present. High maternal mortality was targeted in a UNICEF-backed initiative in the 1990s.

Official figures showed that as at the end of 2001, 4600 people were living with HIV/AIDS.

SPENDING

GDP/cap. increase

CONSUMPTION AND SPENDING

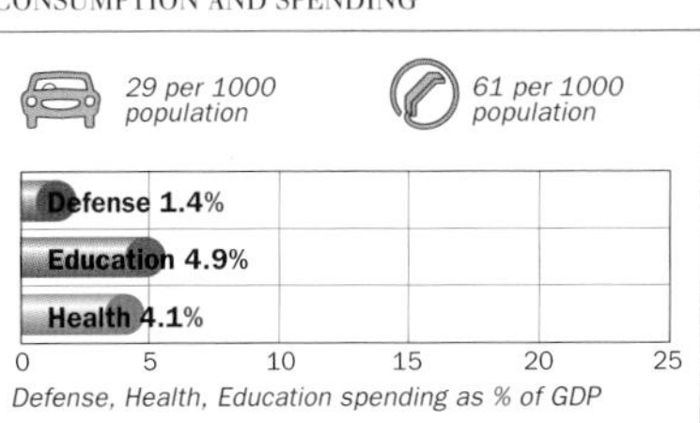
29 per 1000 population
61 per 1000 population
Defense 1.4%
Education 4.9%
Health 4.1%
0 5 10 15 20 25
Defense, Health, Education spending as % of GDP

Havoc created by economic reforms has widened the already huge gap between rich and poor. Generally, the indigenous population who form the rural poor are the worst off. The Andean highlands suffer from grinding poverty that has hardly changed in generations. Migrants to more prosperous eastern regions have faired better, but skewed land ownership remains a big problem. Poor housing, and lack of utilities and regular income are common to urban poverty. Only some 5% of people have bank accounts.

WORLD RANKING

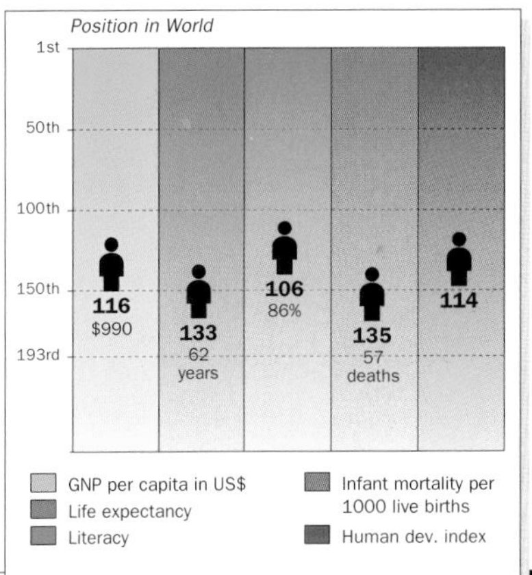
Position in World
1st
50th
100th
150th
193rd
116
$990
133
62 years
106
86%
135
57 deaths
114
GNP per capita in US$
Life expectancy
Literacy
Infant mortality per 1000 live births
Human dev. index

Bosnia & Herzegovina

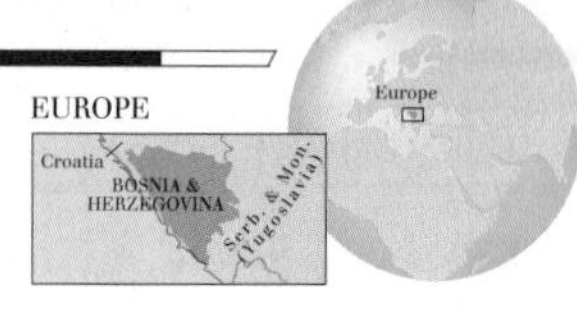
EUROPE
Europe
Croatia
BOSNIA & HERZKGOVINA
Serb. & Mon. (Yugoslavia)

B

OFFICIAL NAME: Bosnia and Herzegovina **CAPITAL:** Sarajevo
POPULATION: 4.1 million **CURRENCY:** Marka **OFFICIAL LANGUAGE:** Serbo-Croat

1992

1992

March 1

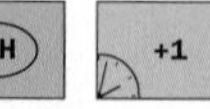
BIH

+1

+387

.ba

BOSNIA AND HERZEGOVINA is a mountainous country in southeastern Europe. A narrow corridor south of Mostar provides access to the Adriatic Sea. The post-1945 Yugoslav regime promoted coexistence between Bosniak Muslims, Croats, and Serbs, but, with Yugoslavia collapsing after 1990, ethnic rivalries led to violent conflict. Around 250,000 people died and more than two million were displaced before the 1995 Dayton peace accord ended three years of war.

CLIMATE

Continental

WEATHER CHART FOR SARAJEVO

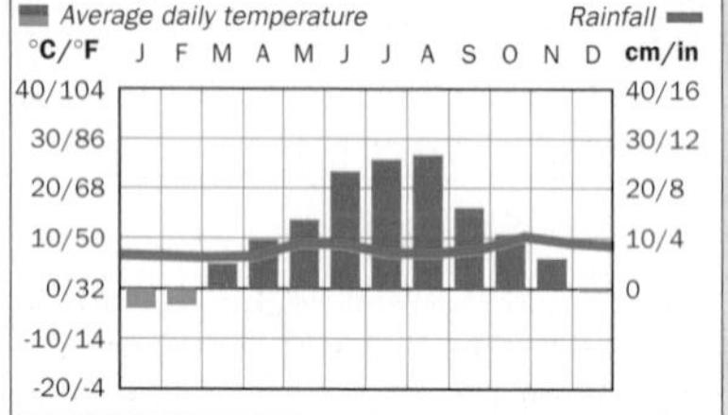
Average daily temperature
Rainfall
°C/°F
J F M A M J J A S O N D
cm/in
40/104
30/86
20/68
10/50
0/32
-10/14
-20/-4
40/16
30/12
20/8
10/4
0

Bosnia has a continental climate with warm summers and bitterly cold winters, often with snow.

TRANSPORTATION

Drive on right

Sarajevo International
354,996 passengers

Has no fleet

THE TRANSPORTATION NETWORK

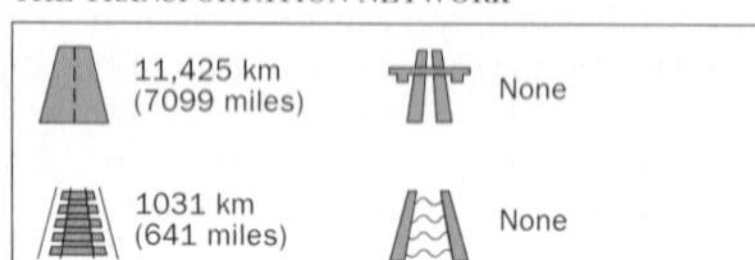
11,425 km
(7099 miles)
None
1031 km
(641 miles)
None

War severely damaged the transportation network, resulting in wrecked bridges, roads, and railroads. De-mining and reconstruction have reopened most main routes. Sarajevo remains the hub of the communications network.

TOURISM

Visitors : Population 1:37

110,000 visitors

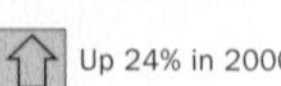
Up 24% in 2000

MAIN TOURIST ARRIVALS

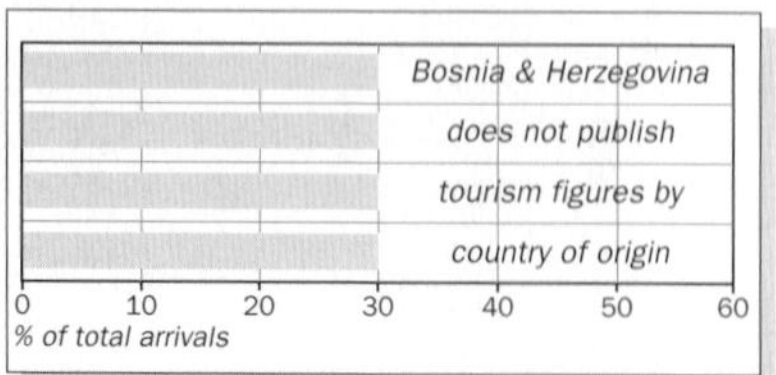
Bosnia & Herzegovina
does not publish
tourism figures by
country of origin
0
10
20
30
40
50
60
% of total arrivals

Even before the war, Bosnia did not have much tourism infrastructure; only the adventurous have begun to visit.

PEOPLE

Pop. density medium

Serbo-Croat

80/km²
(208/mi²)

THE URBAN/RURAL POPULATION SPLIT

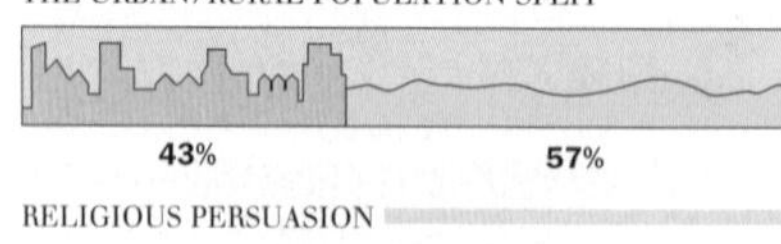
43%
57%

RELIGIOUS PERSUASION

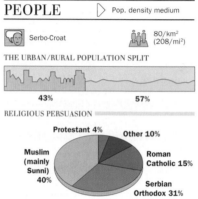

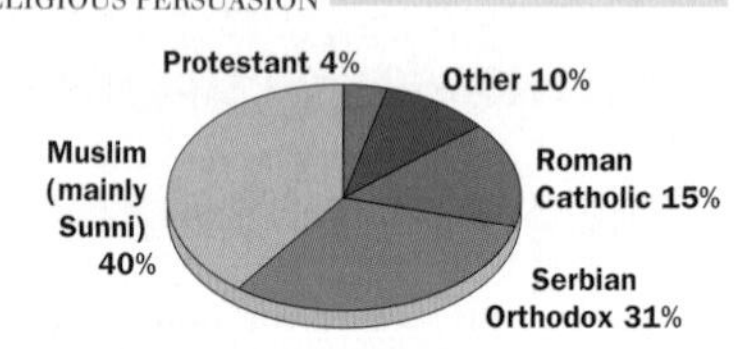
Protestant 4%
Other 10%
Muslim (mainly Sunni) 40%
Roman Catholic 15%
Serbian Orthodox 31%

In 1990, the population was 44% ethnic Bosniak (mostly Muslim), 31% Serb, and 17% Croat, while 8% described themselves as Yugoslav or of other ethnicity. Intermarriage was common and ethnic violence rare. Society was largely secular and materialistic. However, civil war and "ethnic cleansing" displaced some 60% of the population. Bosniak, Serb, and Croat areas became more distinct and separate.

POLITICS

Multiparty elections

L. House 2000/2002
U. House 2000/2002

Chair of the Presidency
Beriz Belkic

AT THE LAST ELECTION

House of Representatives 42 seats

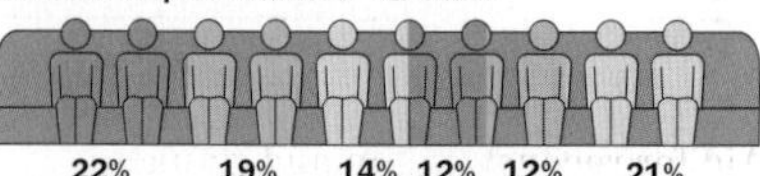
22% SDP
19% SDA
14% SDS
12% SBiH
12% HDZ
21% Others

SDP = Social Democratic Party **SDA** = Party of Democratic Action **SDS** = Serb Democratic Party **SBiH** = Party for Bosnia & Herzegovina **HDZ** = Croatian Democratic Union

The House of Representatives comprises the legislatures of the Federation of Bosnia-Herzegovina and Republika Srpska

House of Peoples 15 seats

10 members are appointed from the Federation of Bosnia-Herzegovina and 5 from the Republika Srpska

Bosnia is subdivided into the Muslim–Croat Federation, the Serbian Republika Srpska (RS), and the multiethnic district of Brčko. As such, there are two levels of government: the overall Republic – with a rotating three-member collective presidency – and the constituent republics, with their own presidents, governments, and parliaments. A UN High Representative retains ultimate control. Nationalist parties have substantial support, although moderate policies are favored in government. Bosniaks have led the way in supporting non-ethnic parties. Croats in the Federation have agitated for greater autonomy.

BOSNIA & HERZEGOVINA
Total Land Area : 51 129 sq. km
(19 741 sq. miles)
POPULATION
over 100 000
over 50 000
over 10 000
under 10 000
LAND HEIGHT
2000m/6562ft
1000m/3281ft
500m/1640ft
200m/656ft
Sea Level
N
0
50 km
0
50 miles
CROATIA
SERBIA & MONTENEGRO (YUGO...
ADRIATIC SEA
Bosanska Gradiška
Bosanski Brod
Bosanski Šamac
Modriča
Derventa
Gradačac
Brčko
Sava
Bijeljina
Cazin
Prijedor
Kozara
Vrbas
Bihać
Banja Luka
Doboj
Gračanica
Tuzla
REPUBLIKA SRPSKA
Grmeč
Sana
Una
Ključ
Teslić
Bosna
Maglaj
DINARA
Zvornik
Jajce
Travnik
Zenica
Vlasenica
Srebrenica
Bugojno
FEDERACIJA BOSNA I HERCEGOVINA
Vranica
Livno
SARAJEVO
Treskavica
Konjic
Foča
Drina
Neretva
Mostar
Maglić 2387m
Trebinje
44°
16°
18°

B

WORLD AFFAIRS

Joined UN in 1992

CEI EBRD OSCE IBRD IAEA

An international High Representative administers Bosnia under the 1995 Dayton agreement. The Stabilization Force (SFOR) is NATO-led; its 35 participant countries include Russia.

AID

Recipient

 $737m (receipts)

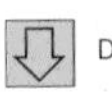 Down 31% in 2000

Aid for reconstruction and refugee return, totaling several billion dollars, has come mainly from the EU and the US. Major objectives have included developing a market economy and job creation, housing, and infrastructure.

DEFENSE

Compulsory military service

 $183m

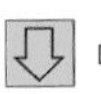 Down 50% in 2000

The international Stabilization Force (SFOR), in place since 1996, has been progressively downsized. The merger of Bosnian and Croat forces under the Dayton accord was completed in 1997.

ECONOMICS

Inflation 2.3% p.a. (1992–2000)

 $4.9bn

 2.083–2.197 maraka

SCORE CARD

- ❑ WORLD GNP RANKING........................111th
- ❑ GNP PER CAPITA $1230
- ❑ BALANCE OF PAYMENTS....................–$565m
- ❑ INFLATION 8%
- ❑ UNEMPLOYMENT.................................40%

STRENGTHS

Potential to revive prewar industries and become a thriving market economy, with solid manufacturing base. Growth in 2000 of 10%. End of cross-border tariffs with Slovenia from 2002.

WEAKNESSES

War damage of $20–$40 billion. Lowest foreign investment in entire region. Corruption. High unemployment.

EXPORTS

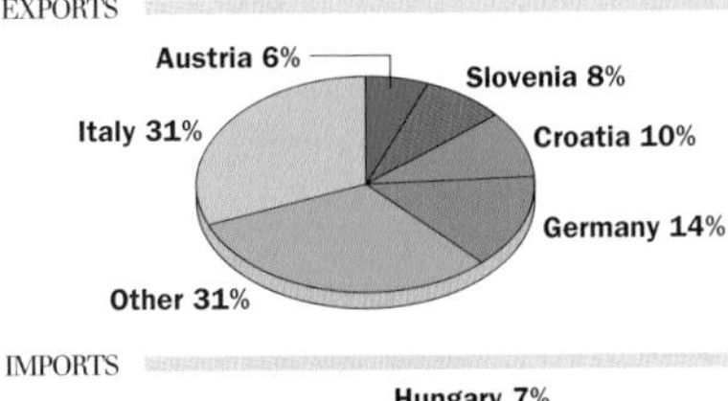
Austria 6%
Slovenia 8%
Italy 31%
Croatia 10%
Germany 14%
Other 31%

IMPORTS

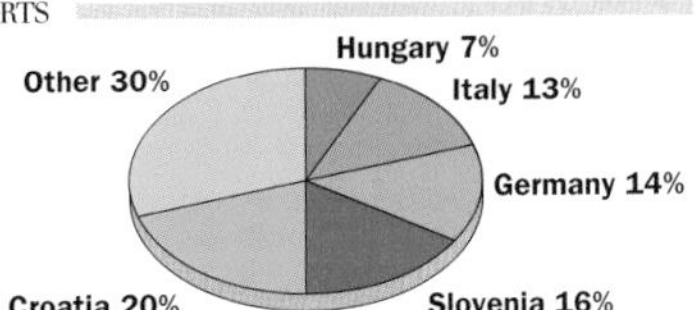
Hungary 7%
Other 30%
Italy 13%
Germany 14%
Croatia 20%
Slovenia 16%

RESOURCES

Electric power 2.7m kw

 2500 tonnes

 640,000 sheep, 440,000 cattle, 4.74m chickens

 Not an oil producer

 Coal, lignite, iron, bauxite, cement

Bosnia's land is not well suited to agriculture, but has mineral deposits, forests, and hydroelectric potential.

ENVIRONMENT

Sustainability rank: 57th

1%

1.2 tonnes per capita

Apart from war damage, Bosnia faces the effects of industrial pollution incurred during the communist regime.

MEDIA

TV ownership medium

 Daily newspaper circulation 152 per 1000 people

PUBLISHING AND BROADCAST MEDIA

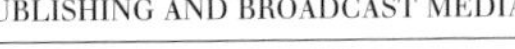

There are 6 daily newspapers. *Oslobodjenje (Liberation)* was published daily throughout the war

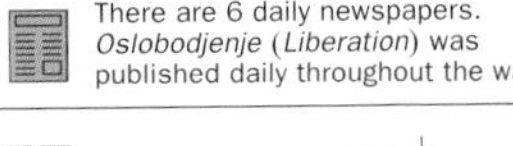

8 independent services

 10 independent services

Violence against journalists is common. Hundreds of print and broadcast outlets, mainly controlled by government or parties, are highly partisan on ethnic lines.

CRIME

Death penalty not used in practice

 1852 prisoners

 Crime is rising

All sides in the war, but especially the Serbs, have been accused of war crimes. The International Criminal Tribunal for the former Yugoslavia (ICTY) in The Hague has indicted over 60 suspects relating to crimes committed in Bosnia. Serb leader Radovan Karadjic and over 20 others have evaded arrest. The RS finally passed laws on cooperating with the tribunal in October 2001.

EDUCATION

School leaving age: 15

 93%

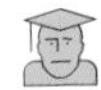 40,000 students

Ethnic bias and educational segregation are being combated by the development of a new coordinated curriculum.

The Muslim town of Mostar. *Its 16th-century bridge at a strategic river crossing, and much of the old town, were destroyed in the war.*

CHRONOLOGY

In 1945, Bosnia and Herzegovina became one of Yugoslavia's six constituent republics.

- ❑ **1990** Nationalists defeat communists in multiparty elections.
- ❑ **1991** Parliament announces republican sovereignty.
- ❑ **1992** EU and USA recognize Bosnia. Serbs declare "Serbian Republic." Civil war begins.
- ❑ **1995** NATO air strikes on Serbs; US-brokered Dayton peace accord.
- ❑ **1996** First international war crimes trial since 1945 opens in The Hague. Elections held under Dayton accord.
- ❑ **1998–2000** Elections: dwindling support for nationalist parties.
- ❑ **2001** Ethnic Croats briefly establish autonomy in Herzegovina (in south).

HEALTH

No welfare state health benefits

1 per 714 people

 Cholera and diphtheria epidemics, violence, deaths from war-stress

War strained health services severely and many died for lack of basic care. Reconstruction includes reform of the primary health care system.

SPENDING

GDP/cap. increase

CONSUMPTION AND SPENDING

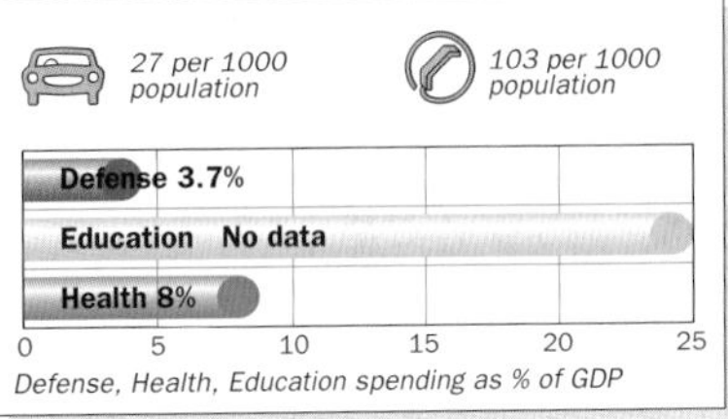
27 per 1000 population
103 per 1000 population
Defense 3.7%
Education No data
Health 8%
0
5
10
15
20
25
Defense, Health, Education spending as % of GDP

With some 60% of the prewar population displaced, housing, jobs, and reintegration are key challenges, while a lack of significant investment means that postwar poverty remains.

WORLD RANKING

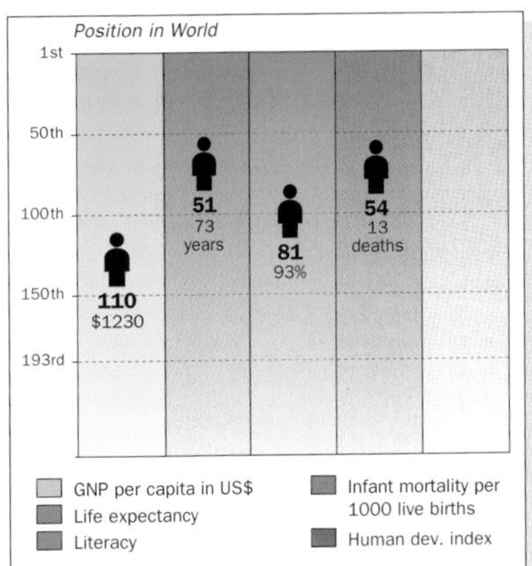
Position in World
1st
50th
100th
150th
193rd
110
$1230
51
73 years
81
93%
54
13 deaths
GNP per capita in US$
Life expectancy
Literacy
Infant mortality per 1000 live births
Human dev. index

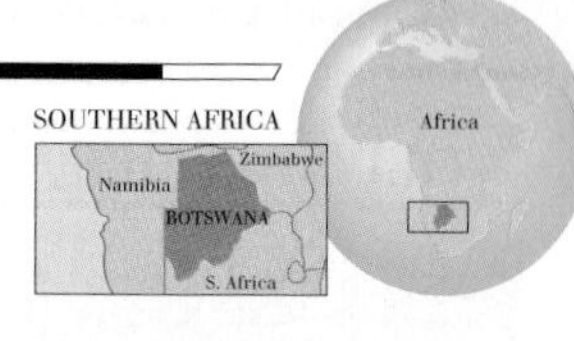
SOUTHERN AFRICA
Africa
Namibia
Zimbabwe
BOTSWANA
S. Africa

BOTSWANA

OFFICIAL NAME: Republic of Botswana **CAPITAL:** Gaborone
POPULATION: 1.6 million **CURRENCY:** Pula **OFFICIAL LANGUAGE:** English

1966
1966
Sept 30
RB
+2
+267
.bw

ARID AND LANDLOCKED, Botswana's central plateau separates the populous eastern grasslands from the Kalahari Desert and the swamps of the Okavango Delta in the west. Diamonds provide Botswana with a prosperous economy, but rain is an even more precious resource, honored in the name of the currency, the pula. Botswana has the world's highest rate of HIV infection among adults: a staggering 38.8% at end-2001.

The Okavango Delta, "jewel of the Kalahari," is the largest inland river delta in the world and home to a rich variety of wildlife.

CLIMATE

Steppe/hot desert

WEATHER CHART FOR GABORONE

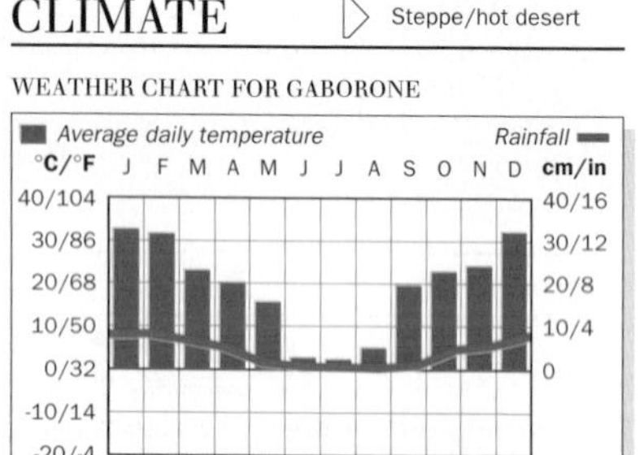
Average daily temperature
Rainfall
°C/°F J F M A M J J A S O N D cm/in
40/104
30/86
20/68
10/50
0/32
-10/14
-20/-4
40/16
30/12
20/8
10/4
0

The subtropical climate is dry and prone to drought. Rainfall declines from 64 cm (25 in) in the north to under 10 cm (4 in) in the Kalahari Desert in the west.

TRANSPORTATION

Drive on left

Sir Seretse Khama International, Gaborone 240,526 passengers

Has no fleet

THE TRANSPORTATION NETWORK

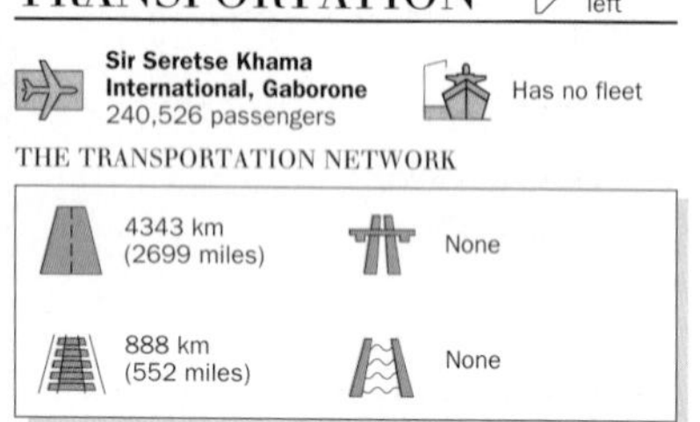
4343 km
(2699 miles)
None
888 km
(552 miles)
None

The opening of the trans-Kalahari road to Namibia in 1998 has reduced Botswana's dependence on South African ports. Upgrading existing rail and road networks is a priority.

TOURISM

Visitors : Population 1:2.1

750,000 visitors

Up 3% in 1998

MAIN TOURIST ARRIVALS

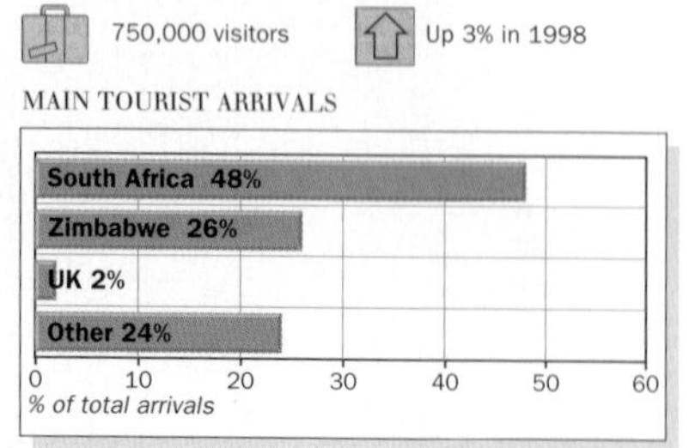
South Africa 48%
Zimbabwe 26%
UK 2%
Other 24%
0 10 20 30 40 50 60
% of total arrivals

Tourism is aimed at wealthy wildlife enthusiasts and focuses on safaris, especially to the Okavango Delta.

PEOPLE

Pop. density low

Tswana, English, Shona, San, Khoikhoi, Ndebele

3/km² (7/mi)²

THE URBAN/RURAL POPULATION SPLIT

50% 50%

ETHNIC MAKEUP

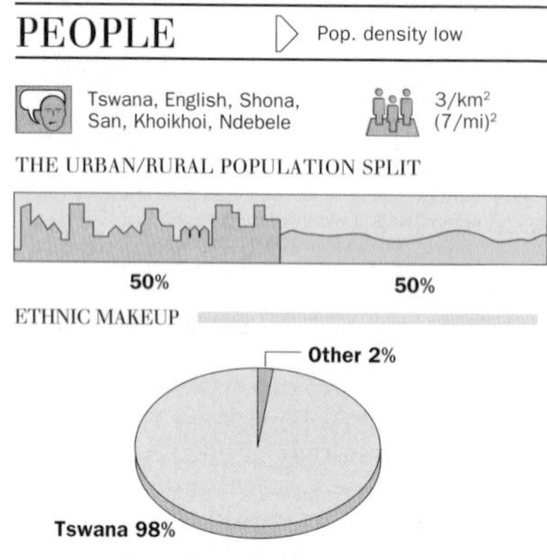
Other 2%
Tswana 98%

Botswana's stability reflects its ethnic homogeneity and the power of traditional authorities, notably the village *kgotla*, or parliament. Almost the whole population is Tswana, with the Bamangwato forming the largest Tswana group. The indigenous San (or Bushmen) of the Kalahari were ordered to abandon their nomadic lifestyle in 2002. Whites continue to dominate the professions.

POLITICS

Multiparty elections

1999/2004

President Festus Mogae

AT THE LAST ELECTION

National Assembly 46 seats

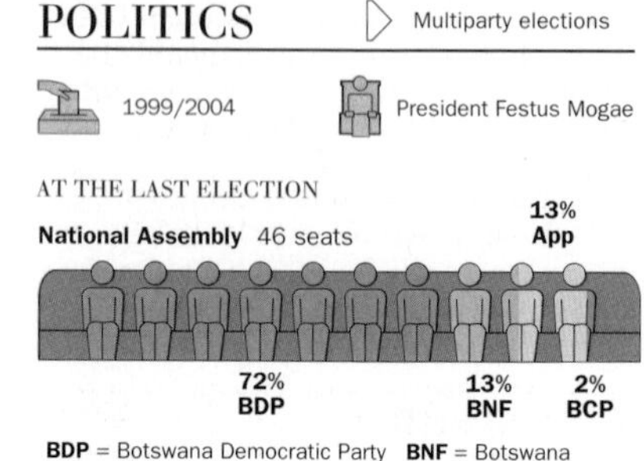
13% App
72% BDP
13% BNF
2% BCP

BDP = Botswana Democratic Party **BNF** = Botswana National Front **App** = Appointed **BCP** = Botswana Congress Party

In addition to 40 elected members, 4 are co-opted and the president and the attorney general are ex-officio members

Although Botswana is formally a multiparty democracy, it has been ruled by the BDP since independence. In 1994, however, economic problems, corruption scandals, and increasing urbanization led to the mainly town-based BNF gaining seats at the expense of the BDP, which nevertheless retained its absolute parliamentary majority. Power transferred smoothly from President Masire to Festus Mogae in 1998. The opposition BNF split in two, and the 1999 elections confirmed the BDP's hold on power.

BOTSWANA

Total Land Area : 600 370 sq. km (231 803 sq. miles)

POPULATION
over 500 000
over 50 000
over 10 000
under 10 000

LAND HEIGHT
1000m/3281ft
500m/1640ft

0 200 km
0 200 miles

WORLD AFFAIRS

Joined UN in 1966

Comm
NAM
AU
SADC
WTO

Botswana has strongly backed a politically and economically stable postapartheid South Africa, and in 1994 appointed its first ambassador to Pretoria since 1966. Potential South African domination of the SADC is a concern. Traditionally pro-Western in orientation, Botswana cherishes its relations with the UK and the US.

AID

Recipient

 $31m (receipts)

 Down 49% in 2000

Botswana's political and economic record has made it a favored aid recipient, notably from the EU, Japan, Germany, and the UK. Some 90% of EU aid goes to projects which try to balance wildlife needs with rural development. Aid also targets transportation projects.

DEFENSE

No compulsory military service

 $245m

 Down 5% in 2000

Reforms of the armed forces in mid-2000 aimed at improving morale included the raising of the compulsory retirement age and the enlistment of women.

ECONOMICS

Inflation 9.7% p.a. (1990–2000)

 $5.28bn

 5.37–7.02 pula

SCORE CARD

❑ WORLD GNP RANKING	109th
❑ GNP PER CAPITA	$3300
❑ BALANCE OF PAYMENTS	$517m
❑ INFLATION	8.6%
❑ UNEMPLOYMENT	40%

STRENGTHS

Diamonds: one of the world's top producers. Economic growth among the highest in the world, averaging 8.5% between 1980 and 1998. Prudent management, large financial reserves, and exchange control liberalization. Lucrative exports of copper, nickel, assembly-produced vehicles, beef.

WEAKNESSES

Spread of AIDS. Overdependence on diamonds. Agriculture and industry weak. Small population, water shortages, and drought. Impact of beef industry on environment. High transportation costs to coast. Widespread unemployment.

EXPORTS

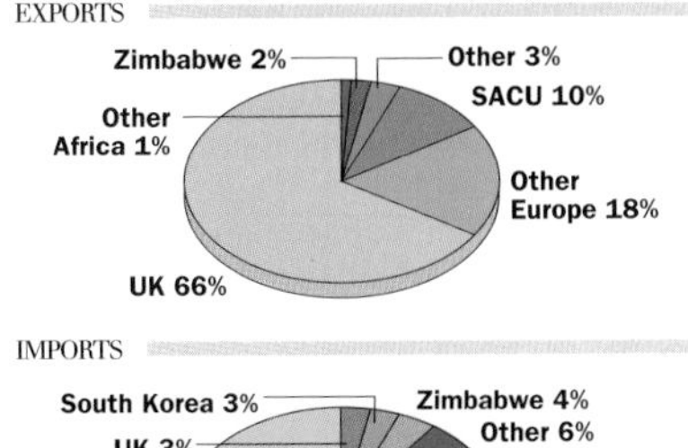
Zimbabwe 2%
Other 3%
Other Africa 1%
SACU 10%
Other Europe 18%
UK 66%

IMPORTS

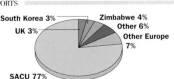

South Korea 3%
Zimbabwe 4%
UK 3%
Other 6%
Other Europe 7%
SACU 77%

RESOURCES

Electric power: Included in South African total

 2000 tonnes

 Not an oil producer and has no refineries

2.4m cattle, 2.25m goats, 370,000 sheep, 4m chickens

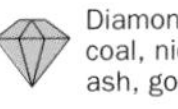 Diamonds, copper, coal, nickel, soda ash, gold

The Orapa 2000 diamond mining project began in May 2000. Large coal deposits are the basis of power grid expansion. Water is Botswana's scarcest resource.

ENVIRONMENT

Sustainability rank: 13th

 19% (2% partially protected)

 2.4 tonnes per capita

Botswana is trying to help communities to earn a living from wildlife protection. A campaign has been launched to curb the use of agrochemicals.

MEDIA

TV ownership low

 Daily newspaper circulation 27 per 1000 people

PUBLISHING AND BROADCAST MEDIA

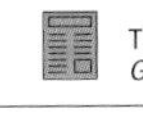 There is 1 daily newspaper, *Dikgang tsa Gompieno*, published by the government

 1 state-owned service

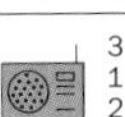 3 services: 1 state-owned, 2 independent

A government-funded TV service began broadcasting in 2000. The government bias of radio and the one daily paper is offset in the many journals.

CRIME

Death penalty in use

 6339 prisoners

 Down 11% 1992–1996

President Mogae warned of a "crime wave" in 1999. Official corruption and diamond smuggling stay major concerns. Human rights are generally respected.

EDUCATION

School leaving age: 15

77%

9595 students

Revenues from the diamond industry have helped to fund educational programs which have improved the country's literacy rate.

CHRONOLOGY

From 1600, Tswana migrations slowly displaced San people. In 1895, at local request, the UK set up the Bechuanaland Protectorate to preempt annexation by South Africa.

- ❑ **1965** BDP, led by Sir Seretse Khama, wins first general election and all subsequent general elections.
- ❑ **1966** Independence declared.
- ❑ **1980** Vice President Quett (later Ketumile) Masire succeeds the late Sir Seretse as president.
- ❑ **1985–1986** South African raids.
- ❑ **1992–1993** Strikes and corruption scandals prompt resignations of senior BDP figures.
- ❑ **1994** BDP support eroded in general election.
- ❑ **1998** Vice President Festus Mogae succeeds Masire as president.

HEALTH

Welfare state health benefits

 1 per 5000 people

 AIDS, tuberculosis, heart diseases, pneumonia

With the world's highest rate of HIV-positive adults (38.8%), Botswana in 2001 became the first sub-Saharan country to provide antiretroviral drugs through its public health service.

SPENDING

GDP/cap. increase

CONSUMPTION AND SPENDING

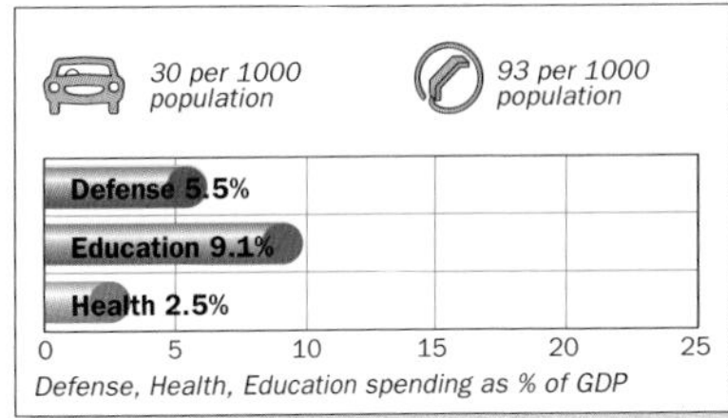
30 per 1000 population
93 per 1000 population
Defense 5.5%
Education 9.1%
Health 2.5%
0
5
10
15
20
25
Defense, Health, Education spending as % of GDP

GNP per capita is among Africa's highest, but about half the population live below the poverty line. Economic growth has exacerbated wealth inequalities.

WORLD RANKING

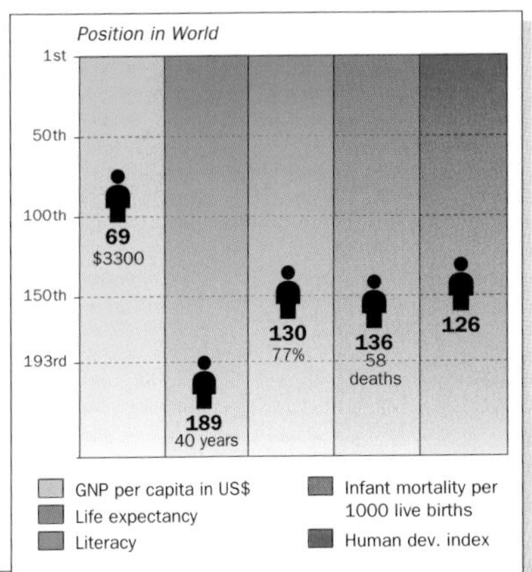
Position in World
1st
50th
100th
150th
193rd
69
$3300
189
40 years
130
77%
136
58 deaths
126
GNP per capita in US$
Life expectancy
Literacy
Infant mortality per 1000 live births
Human dev. index

B

BRAZIL

OFFICIAL NAME: Federative Republic of Brazil **CAPITAL:** Brasília
POPULATION: 172.6 million **CURRENCY:** Real **OFFICIAL LANGUAGE:** Portuguese

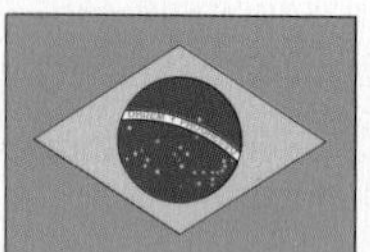

1822

1828
Sept 7

BR

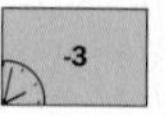
-3

+55

.br

THE LARGEST COUNTRY in South America, Brazil became independent of Portugal in 1822. Today, it is renowned as the site of the world's largest tropical rainforest, the threat to which led to the UN's first international environment conference, held in Rio de Janeiro in 1992. Covering one-third of Brazil's total land area, the rainforest grows around the massive Amazon River and its delta. Apart from the basin of the River Plate in the south, the rest of the country consists of highlands. The mountainous northeast is part forested and part desert. Brazil is the world's leading coffee producer and also has rich reserves of gold, diamonds, oil, and iron ore. Cattle ranching is an expanding industry. The city of São Paulo is the world's fourth-biggest conurbation, with some 18 million inhabitants.

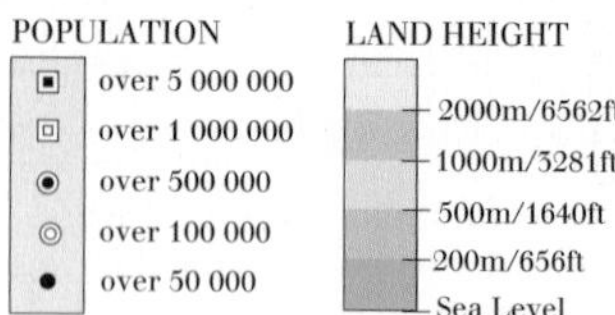
SOUTH AMERICA
Venezuela
Colombia
BRAZIL
Peru
Bolivia
Paraguay
South America

BRAZIL

Total Land Area : 8 511 965 sq. km (3 286 470 sq. miles)

POPULATION	LAND HEIGHT
over 5 000 000	2000m/6562ft
over 1 000 000	1000m/3281ft
over 500 000	500m/1640ft
over 100 000	200m/656ft
over 50 000	Sea Level

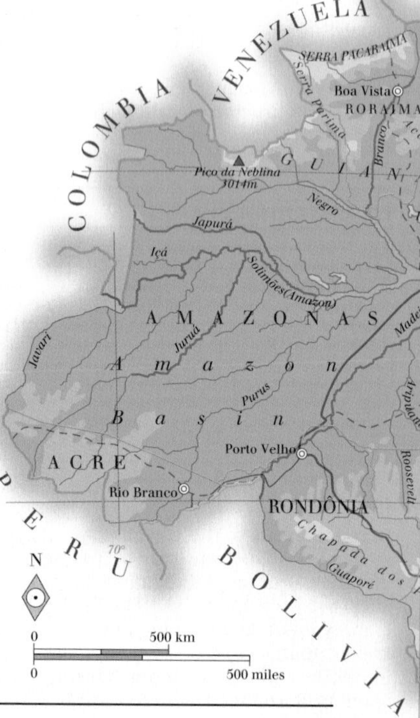
VENEZUELA
COLOMBIA
SERRA PACARAIMA
Serra Parima
Boa Vista
RORAIMA
Branco
Pico da Neblina 3014m
Negro
Japurá
Içá
Solimões (Amazon)
AMAZONAS
Amazon Basin
Javari
Juruá
Purus
Madeira
Aripuanã
Porto Velho
ACRE
Rio Branco
RONDÔNIA
Roosevelt
Chapada dos Pa
Guaporé
PERU
BOLIVIA
70°
60°
N
0
500 km
500 miles

CLIMATE

Tropical equatorial/hot & dry/subtropical/steppe

WEATHER CHART FOR BRASÍLIA

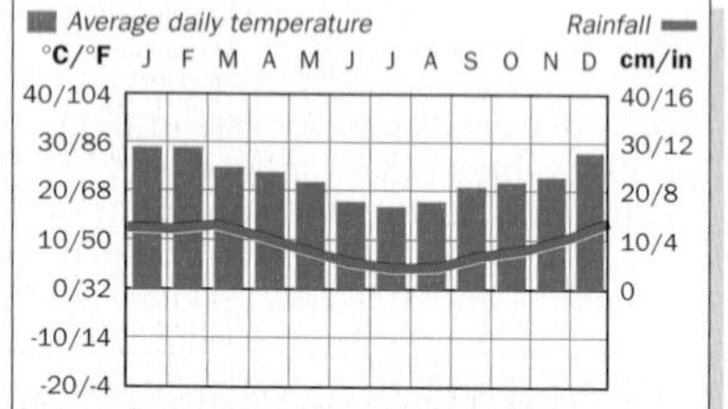
Average daily temperature
Rainfall
°C/°F J F M A M J J A S O N D cm/in
40/104
30/86
20/68
10/50
0/32
-10/14
-20/-4
40/16
30/12
20/8
10/4
0

Brazil's share of the Amazon basin, occupying half of the country, has a model equatorial climate. Its 150–200 cm (59–79 in) of rain are spread throughout the year. Temperatures are high, with almost no seasonal variation, but scarcely ever rise above 38°C (100°F).

The Brazilian plateau, occupying most of the rest of the country, has far greater temperature ranges. Rain falls mainly between October and April. The northeast, the least productive region of Brazil, is very dry and in recent years has been prone to severe drought. However, torrential rains hit Pernambuco and Alagoas in mid-2000, causing the worst flooding for 25 years. The southern states have hot summers and cool winters, when frost may occur.

TRANSPORTATION

Drive on right

Guarulhos International, São Paulo
14.4m passengers

505 ships
3.81m grt

THE TRANSPORTATION NETWORK

184,140 km (114,419 miles)	5000 km (3107 miles)
20,500 km (12,739 miles)	50,000 km (31,069 miles)

Air travel is expensive, while roads are clogged in cities and poor outside.

Proposed cross-border construction projects include a railroad, an intercontinental highway, and a 3442-km (2140-mile) waterway.

Parati, in Rio state, was one of Brazil's major gold-exporting ports in the 17th century. Its colonial architecture is well preserved.

TOURISM

Visitors : Population 1:32

5.31m visitors

Up 4% in 2000

MAIN TOURIST ARRIVALS

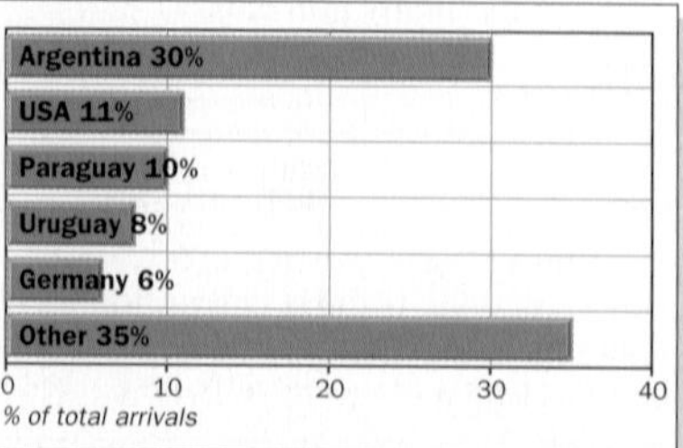
Argentina 30%
USA 11%
Paraguay 10%
Uruguay 8%
Germany 6%
Other 35%
0
10
20
30
40
% of total arrivals

Brazil is underperforming in tourism, with revenues equivalent to 2.5% of GDP in 1996, compared with a world average of around 10%. The situation is slowly improving.

Such attractions as Atlantic beaches stretching 2000 km (1250 miles), the Amazon River basin, the Pantanal – the vast wetland region in the west – and Brazil's world-famous carnivals are offset by the limited availability of medium- to low-cost travel and budget hotels, a factor which deters both domestic and foreign travelers.

In the virtual absence of low-cost charter flights domestic air travel is expensive. This is blamed on high airport charges and inertia in Brazil's aviation department, which is controlled by the air force.

Average overnight hotel rates are higher than in Europe and the US, and the quality of service is generally poor. Basic infrastructure, such as sanitation and water supply, is also deficient.

***Rio de Janeiro** and Sugar Loaf Mountain seen from Corcovado (Hunchback) Peak. With a population of over ten million, the Rio conurbation is Brazil's largest after São Paulo.*

PEOPLE

Pop. density low

Portuguese, German, Italian, Spanish, Polish, Japanese, Amerindian languages

20/km² (53/mi²)

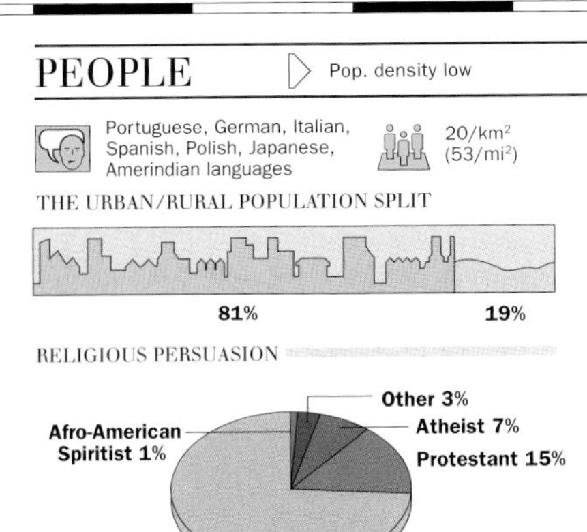
THE URBAN/RURAL POPULATION SPLIT
81%
19%
RELIGIOUS PERSUASION
Other 3%
Atheist 7%
Afro-American Spiritist 1%
Protestant 15%
Roman Catholic 74%

ETHNIC MAKEUP

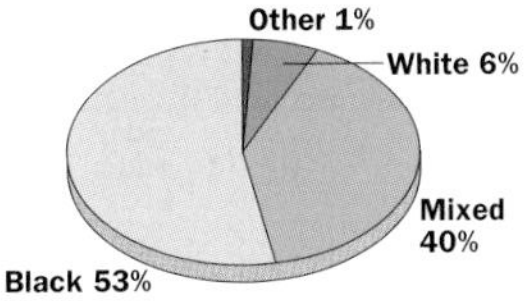

Other 1%
White 6%
Mixed 40%
Black 53%

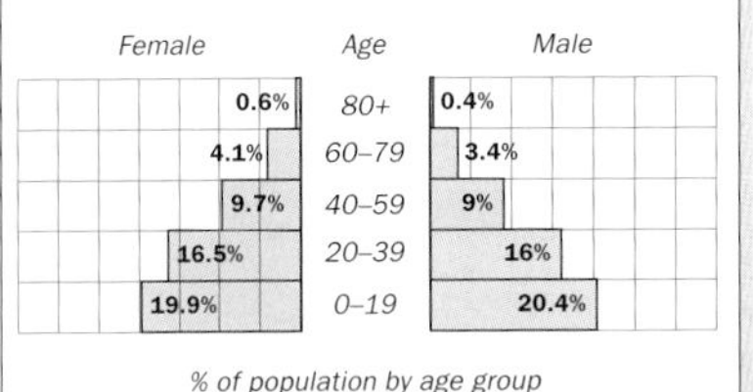
POPULATION AGE BREAKDOWN
Female
Age
Male
0.6%
80+
0.4%
4.1%
60–79
3.4%
9.7%
40–59
9%
16.5%
20–39
16%
19.9%
0–19
20.4%
% of population by age group

Brazil's population is highly diverse. It includes indigenous Amerindian groups, as well as the descendants of its Portuguese colonizers and the Africans brought to work the sugar plantations in the 17th century. More recent immigrant groups include both Italians and Japanese. Extremes of social disadvantage challenge the widely held notion that Brazil is a "racial democracy." Poor, mixed-race migrants are treated as outcasts in cities. Afro-Brazilians suffer higher infant mortality and poverty and more racial and job discrimination than other groups – a program of "positive discrimination" was launched in 2002. Amerindians also experience widespread prejudice, and of the estimated 1000 groups present at the time of the arrival of the first Portuguese, about 200 remain – a total population of about 220,000, who struggle to secure land rights.

Brazil is strongly Catholic, but other religions also flourish. The traditional emphasis on the family is under pressure in urban areas: migrants in search of work often have to leave their families behind.

Women gained the vote in 1934, but are still discriminated against in jobs and politics.

CHRONOLOGY

The first Portuguese, Pedro Alvares Cabral, arrived in Brazil in 1500. By the time Portugal took control of the region, in 1580, it was a thriving colony drawing its wealth from sugar plantations in the northeast, worked by imported Africans, or Amerindians captured from farther inland.

- **1637–1654** Dutch control sugar-growing areas.
- **1763** Rio becomes capital.
- **1788** Inconfidência rebellion, led by Tiradentes, fails.
- **1807** French invade Portugal. King João VI flees to Brazil with British naval escort. In return, Brazil's ports opened to foreign trade.
- **1821** King returns to Portugal; his son Pedro made regent of Brazil.
- **1822** Pedro I declares independence and is made emperor of Brazil.
- **1828** Brazil loses Uruguay.
- **1831** Military revolt after war with Argentina (1825–1828). Emperor abdicates; his five-year-old son succeeds him as Pedro II.
- **1835–1845** Rio Grande secedes.
- **1865–1870** Brazil wins war of Triple Alliance with Argentina and Uruguay against Paraguay.
- **1888** Pedro II abolishes slavery; landowners and military turn against him.
- **1889** First Republic established. Emperor goes into exile in Paris. Increasing prosperity as result of international demand for coffee.
- **1891** Federal constitution established.
- **1914–1918** World War I causes coffee exports to slump.
- **1920s** Working-class and intellectual movements call for end to oligarchic rule.
- **1930** Coffee prices collapse. Revolt led by Dr. Getúlio Vargas, the "Father of the Poor," who becomes president. Fast industrial growth.
- **1937** Vargas's position as benevolent dictator formalized in "New State," based on fascist model.
- **1942** Declares war on Germany. ⇨

B

CHRONOLOGY *continued*

- ❑ **1945** Vargas forced out by military.
- ❑ **1950** Vargas reelected president.
- ❑ **1954** US opposes Vargas's socialist policies. The right, backed by the military, demand his resignation. Commits suicide.
- ❑ **1956–1960** President Juscelino Kubitschek, backed by Brazilian Labor Party (PTB), attracts foreign investment for new industries, especially from US.
- ❑ **1960–1961** Conservative Jânio da Silva Quadros president. Tries to break dependence on US trade.
- ❑ **1961** Brasília, built in three years, becomes new capital. PTB leader, João Goulart, elected president.
- ❑ **1961–1964** President's powers briefly curtailed as right wing reacts to presidential policies.
- ❑ **1964** Bloodless military coup under army chief Gen. Castelo Branco.
- ❑ **1965** Branco assumes dictatorship; bans existing political parties, but creates two official new ones. He is followed by a succession of military rulers. Fast-track economic development, the Brazilian Miracle, is counterbalanced by ruthless suppression of left-wing activists.
- ❑ **1974** World oil crisis marks end of economic boom. Brazil's foreign debt now largest in world.
- ❑ **1979** More political parties allowed.
- ❑ **1980** Huge migrations into Rondônia state begin.
- ❑ **1985** Civilian senator Tancredo Neves wins presidential elections as candidate of new liberal alliance, but dies before taking office. Illiterate adults granted the vote.
- ❑ **1987** Gold found on Yanomami lands in Roraima state; illegal diggers rush in by the thousand.
- ❑ **1988** New constitution promises massive social spending but fails to address land reform. Chico Mendes, rubber-tappers' union leader and environmentalist, murdered.
- ❑ **1989** First environmental protection plan. Yearly inflation reaches 1000%. Fernando Collor de Mello wins first presidential election held under completely free conditions.
- ❑ **1992** Earth Summit in Rio. Collor de Mello resigns and is impeached for corruption.
- ❑ **1994–1995** Plan Real ends hyperinflation. Congress resists constitutional reforms, but passes key privatizations of state monopolies.
- ❑ **1998–1999** Fernando Henrique Cardoso, in power since 1995, reelected president. Real devalued in economic crisis.
- ❑ **2001–2002** Recovery of economy threatened by crisis in Argentina.

POLITICS

Multiparty elections

L. House 1998/2002
U. House 1998/2002

President Fernando Henrique Cardoso

AT THE LAST ELECTION

Chamber of Deputies 513 seats

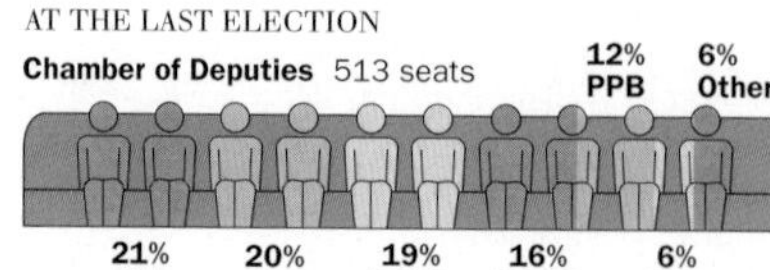

12% PPB
6% Others
21% PFL
20% PT
19% PSDB
16% PMDB
6% PTB

PFL = Liberal Front Party **PT** = PT–leftist coalition (includes the Workers' Party – **PT**, the Democratic Labor Party – **PDT**, and the Brazilian Socialist Party – **PSB**)
PSDB = Brazilian Social Democratic Party **PMDB** = Brazilian Democratic Movement Party **PPB** = Brazilian Progressive Party **PTB** = Brazilian Labor Party
Others include the Popular Socialist Party (**PPS**)

Federal Senate 81 seats

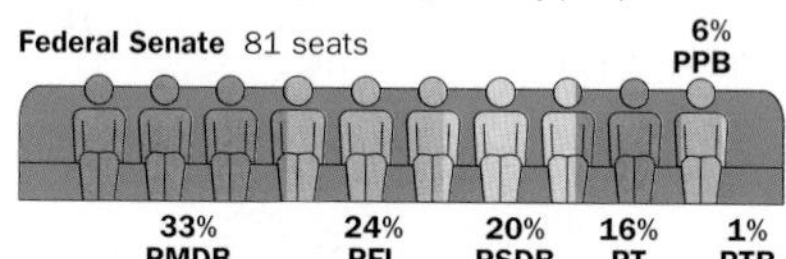

6% PPB
33% PMDB
24% PFL
20% PSDB
16% PT
1% PTB

Brazil is a democratic federal republic with 27 regional parliaments and a national Congress. In 1993, Brazilians voted to retain the direct election of their president.

Profile

Military rule between 1964 and 1985 led to gross human rights abuses, against Amazon Amerindians in particular, and to economic mismanagement, which left Brazil with a legacy of huge debts and inefficient state industries.

Brazil's young democracy is characterized by a weak party system, centered around personalities. Parties do not have set ideological programs, but form shaky coalitions and engage in horsetrading to get legislation through the Congress. A preponderance of small parties and corruption adds to the problems. Former President Collor de Mello was impeached in 1992 on fraud charges.

Dissatisfaction with the center-right provided a boost for the left, led by the influential Luís da Silva, who came second in the 1989 presidential elections. However, his failure to beat Fernando Henrique Cardoso in 1994 or 1998 revealed a lack of fresh ideas and direction in his Workers' Party. Cardoso, the father of the successful anti-inflation plan for the real, held his shaky coalition together until 2002 and emergency fiscal adjustments saved Brazil from a return to persistent economic crisis. Hard-won stability permitted Brazil to avoid the worst effects of the economic crisis in Argentina in 2001–2002.

Main Political Issues

Political stability

Cardoso, who was constitutionally barred from seeking a third term in the 2002 presidential elections, has left his mark on politics, since few parties are prepared to deviate far from his largely successful policy mix which has provided stability and boosted consumer and investor confidence. A ruling by the Supreme Electoral Tribunal, instructing parties forming electoral alliances to retain them from the federal level down to the state, promises cleaner campaigns.

President Fernando Cardoso, *who took office in January 1995.*

Luís Ignacio da Silva, "Lula," *former leader of the left-wing Workers' Party.*

Getúlio Vargas, *the socialist president (1930–1945 and 1950–1954) who was known as the "Father of the Poor."*

Economic management

Rapid recovery from the 1999 currency crisis relied on a tight monetary and fiscal regime backed by a sound macroeconomic policy. This permitted modest growth in 2001 and 2002 despite the global downturn and energy rationing at home. Plaudits from the IMF separated Brazil in investors' minds from the economic turmoil in Argentina. Tax reform and improved energy resources are priorities for the incoming government.

Diplomatic role

Brazil is a leader in South America in the promotion of greater geopolitical unity. Talks with the US on the proposed Free Trade Area of the Americas (FTAA), and a similar trade area with the EU and Mercosur, are among priorities.

Coffee plantation, *São Paulo state. Coffee was introduced into Brazil in the early 18th century. It is declining in importance and now accounts for less than 4% of export revenues.*

WORLD AFFAIRS

Joined UN in 1945

AmCC

G24

Mercosr

OAS

RG

Brazil's regional influence recently helped to abort coups in Paraguay and end one in Ecuador, and settle a border dispute between Peru and Ecuador. Brazil has ambitions to act as main broker in future talks with the US on the FTAA, which is due to be signed in 2005. To bolster its position, it has revived negotiations on a bilateral trade agreement with Mexico, its chief rival for the crown of regional leader. Free trade talks between the EU and Mercosur have raised the prospect of movement by Europe on agriculture, long the main bone of contention between the two regions. Brazil has also strengthened ties with the Andean Community and in 2002 hosted a conference attended by 34 countries invited to join the FTAA.

AID

Recipient

$322m (receipts) Up 72% in 2000

Aid, mainly from Japan and EU states, funds environmental, basic sanitation, road-building, and antipoverty projects. As a consequence of the Argentine economic crisis, the IMF loaned Brazil $10 billion in 2002, and promised $30 billion more in the following years.

DEFENSE

Compulsory military service

$17.5bn

Up 10% in 2000

BRAZILIAN ARMED FORCES

Army	178 main battle tanks (87 *Leopard* 1, 91 M-60A3)	189,000 personnel
Navy	4 submarines, 1 carrier, 14 frigates, 4 corvettes, 50 patrol boats	48,600 personnel
Air force	281 combat aircraft (53 AT-26, 47 F-5E/B/F, 50 AMX, 18 *Mirage* F-103E/D)	50,000 personnel
Missiles	None	

Brazil's dictatorship ended in 1985, but the military still has an important internal security role, for instance controlling vast regions of the north. The arms industry is large, but Brazil states that it has no intention of using its nuclear energy for military purposes: the Comprehensive Test Ban and Nuclear Non-Proliferation treaties were signed in 1998.

Membership of the Mercosur trade bloc has led to increased military cooperation with Argentina, Paraguay, and Uruguay. Brazilian troops have participated in UN peacekeeping, most recently in East Timor.

ECONOMICS

Inflation 208% p.a. (1990–2000)

$610bn

1.950–2.311 reals

SCORE CARD

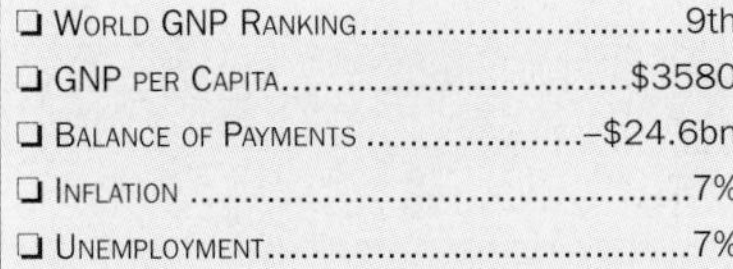

❏ WORLD GNP RANKING	9th
❏ GNP PER CAPITA	$3580
❏ BALANCE OF PAYMENTS	–$24.6bn
❏ INFLATION	7%
❏ UNEMPLOYMENT	7%

EXPORTS

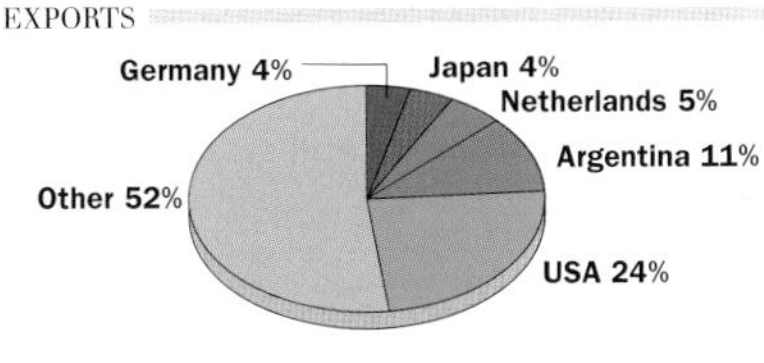
Germany 4%
Japan 4%
Netherlands 5%
Argentina 11%
Other 52%
USA 24%

IMPORTS

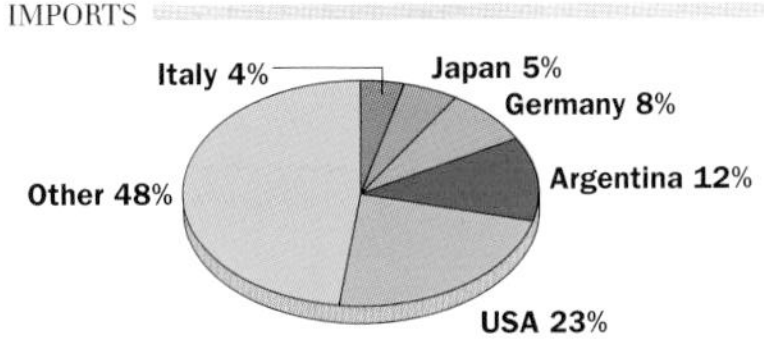
Italy 4%
Japan 5%
Germany 8%
Argentina 12%
Other 48%
USA 23%

STRENGTHS

Dominant economy in region. Strong foreign direct investment flows. Huge growth potential: immense natural resources; major producer of coffee, soy, sugar, oranges; large deposits of gold, silver, and iron; major steel producer; developing oil industry. Development aided by cross-border infrastructural projects and modernization of telecoms.

WEAKNESSES

Expensive domestic borrowing. Weak local capital markets. Vulnerability to external shocks and commodity price fluctuations. Modest productivity. Heavy debt burden. Great social inequalities threaten social unrest. Electricity blackouts caused by badly maintained transmission network or failure of hydropower supply due to drought.

PROFILE

Brazil is the world's ninth-largest economy. Average growth from the start of the 20th century to the early 1970s was over 5%, second only to Japan over a comparable period. Diversification and industrialization transformed Brazil into a producer of cars, computers, and aircraft, but profligate spending produced heavy debts in the 1980s. International lenders demanded belt tightening in return for rescheduling and a steep recession followed in 1990–1992.

The launching of the new currency, the real, in 1994, was the fifth attempt at monetary stabilization since 1986; it contributed to a dramatic fall in inflation. Economic growth in 1994 boosted regional confidence and facilitated the launch of Mercosur. In 1995 a fractious Congress blocked reforms of the tax and social security systems, but finally agreed to end state monopolies in such sectors as telecommunications and oil, thus reviving the privatization program.

The economy grew strongly through 1996 and 1997, but was seriously threatened in 1998 by an international financial crisis. A $41.5 billion rescue package was arranged by the IMF, but foreign currency reserves were heavily depleted in a bid to support the real, which was devalued in 1999 due to speculative pressures. A deep recession was avoided, however, by the successful application over 18 months of tight fiscal and monetary policies, restoring domestic and international confidence.

Modest growth was posted in 2001, but the economic meltdown in Argentina and the effects of global downturn brought the value of the real down in 2002. An end to energy rationing promised to revive industrial production and activity, but deregulation of the labor market, reforms of the social security and tax systems, and high debt-servicing costs remained concerns.

ECONOMIC PERFORMANCE INDICATOR

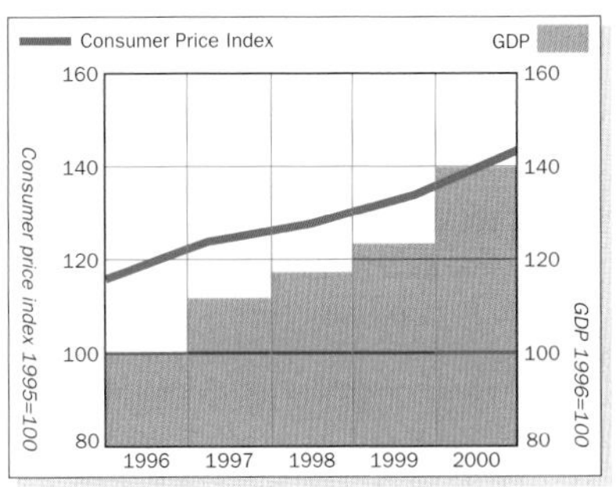
Consumer Price Index
GDP
160
140
120
100
80
Consumer price index 1995=100
GDP 1996=100
1996
1997
1998
1999
2000

BRAZIL : MAJOR BUSINESSES

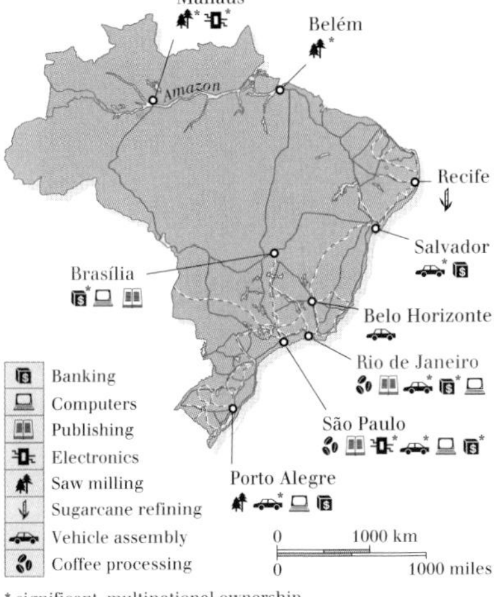
Manaus
Belém
Amazon
Recife
Salvador
Brasília
Belo Horizonte
Rio de Janeiro
São Paulo
Porto Alegre
Banking
Computers
Publishing
Electronics
Saw milling
Sugarcane refining
Vehicle assembly
Coffee processing
0
1000 km
1000 miles
* significant multinational ownership

B

RESOURCES

Electric power 65.2m kw

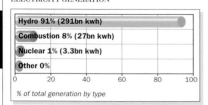

774,750 tonnes

1.34m b/d (reserves 8.5bn barrels)

172m cattle, 29.4m pigs, 15m sheep, 1006m chickens

Iron, manganese, coal, bauxite, nickel, oil, tin, silver, diamonds, gold

ELECTRICITY GENERATION

Hydro 91% (291bn kwh)

Combustion 8% (27bn kwh)

Nuclear 1% (3.3bn kwh)

Other 0%

0 20 40 60 80 100

% of total generation by type

Brazil pumps in gas from Argentina and Bolivia, and has similar plans with Venezuela and Uruguay. Nuclear power has been dogged by controversy and high costs. Hydropower, which already accounts for over 90% of electricity generation, is being expanded further, although output is vulnerable to drought. Producing ethanol from sugar represents an attempt to reduce gasoline imports, and the welcoming of foreign companies in areas of exploration and production is set to increase known oil and natural gas reserves. Exploration of the Amazon's biodiversity was brought under government control in mid-2000, with all new ventures involving any living thing requiring official approval.

Amazon
AMAZON BASIN
BRAZIL : LAND USE
Cropland
Forest
Pasture
Cattle
Coffee – cash crop
Oranges
0 1000 km
0 1000 miles

Equatorial vegetation near Manaus *in the center of Amazonas state. The brown waters of the Rio Solimões and the black waters of the Rio Negro meet near Manaus.*

ENVIRONMENT

Sustainability rank: 20th

4% (1% partially protected)

1.8 tonnes per capita

ENVIRONMENTAL TREATIES

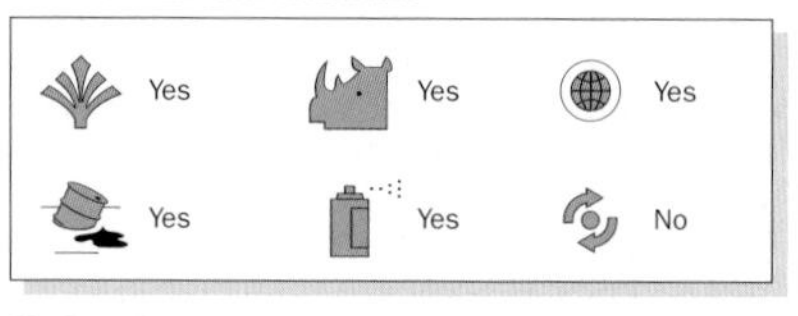

Yes
Yes
Yes
Yes
Yes
No

Federal agencies charged with protecting the Amazon rainforest are underfunded, understaffed, and accused of corruption.

The forest contains an estimated 90% of all the world's plant and animal species. However, 38 million animals a year are smuggled out of the region, and the forest's destruction, usually for cattle pasture and logging, proceeds at a rate of up to 23,000 sq. km (9000 sq. miles) a year; vital genetic diversity is being lost. The government's latest and biggest campaign to stop illegal logging and burning was announced in 2000.

Opencast bauxite mines pollute rivers and threaten indigenous Amerindians, while in 2000 the worst oil spill in 25 years devastated the Iguaçu River. Urban industrial pollution and untreated sewage are major problems.

MEDIA

TV ownership high

Daily newspaper circulation 46 per 1000 people

PUBLISHING AND BROADCAST MEDIA

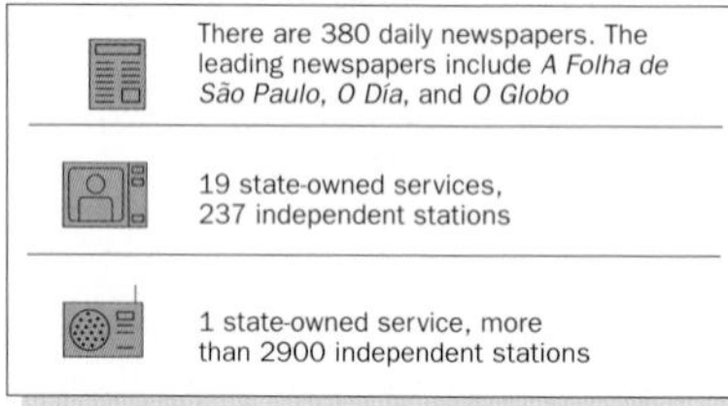

There are 380 daily newspapers. The leading newspapers include *A Folha de São Paulo, O Dia*, and *O Globo*
19 state-owned services, 237 independent stations
1 state-owned service, more than 2900 independent stations

TV and radio licenses are notoriously awarded as political favors. The huge Globo group dominates the home market, with radio, press, and online interests: its TV network is the fifth-largest in the world. However, it is being challenged by the Internet and the growth of new media companies, involving foreign multinationals. A constitutional amendment before the Congress allows foreign groups to take 30% stakes in TV, radio, and the press.

CRIME

Death penalty not used in practice

194,074 prisoners

Rising in 2000–2002

CRIME RATES

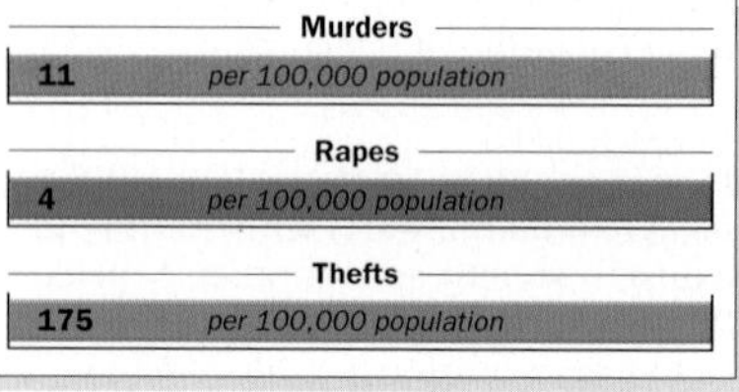

Murders
11 per 100,000 population
Rapes
4 per 100,000 population
Thefts
175 per 100,000 population

Crime levels are among the world's highest, especially in cities, with armed robbery and narcotics-related crime uppermost. Badly paid police are frequently accused of extortion, violence and the murder of citizens. Death squads, thought to be linked to the police, have targeted street children in major cities. A combination of atrocious conditions and overcrowding mean that violent disturbances in prisons are common. An explosion in muggings and kidnappings was a key issue in the 2002 elections.

In the countryside, landless squatters and indigenous peoples have been wounded and murdered in the process of being driven off land by gunmen funded by large landowners. In Roraima state, the discovery of large gold deposits has led to the homelands of Brazil's largest tribe, the Yanomami, being invaded by thousands of gun-toting prospectors, *garimpeiros*.

EDUCATION

School leaving age: 14

85%

1.95m students

THE EDUCATION SYSTEM

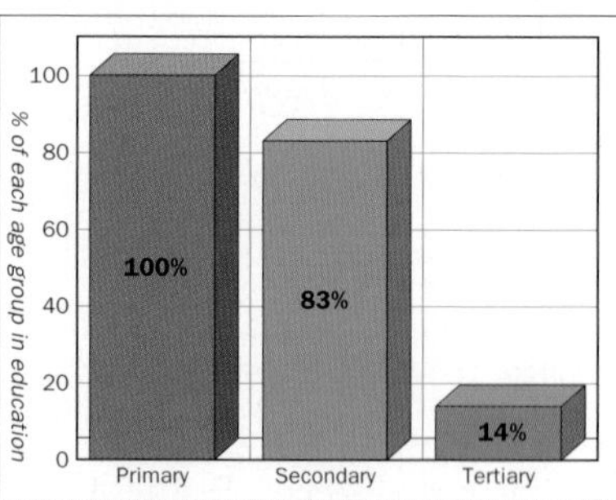

100
80
60
40
20
0
% of each age group in education
100%
83%
14%
Primary
Secondary
Tertiary

The average time spent at school is less than that in other Latin American countries. The portion of GDP spent on education is comparable to that of European countries, but it is misapplied, so that basic primary education remains weak, while many children of wealthy families receive excellent tuition at free public universities. Of Brazil's 95 universities, 55 are administered by the state. Nearly three million children hardly receive any schooling, especially those living in the northeast and in Amazonia, and the urban poor, including street children. However, the number of seven- to 14-year-olds in school has increased during the last decade. Despite an anti-illiteracy campaign, begun in 1971, the adult illiteracy rate is 15%.

B

EXPLOITATION VERSUS PROTECTION IN AMAZONIA

THE VAST AMAZON RIVER basin, extending across northern Brazil, contains one of the world's last great remaining virgin rainforests. Major road-building projects mean that it no longer enjoys the protection of inaccessibility. Since 1970, when it was still 99% intact, 600,000 sq. km (over 230,000 sq. miles) of Amazonia, more than 15% of the total, have been deforested. The rate of destruction did stop accelerating and actually slow down slightly in 2001, but even that year the official record showed a loss of nearly 16,000 sq. km (over 6000 sq. miles). US and Brazilian researchers warned in January 2001 that the government's massive "Advance

Brazil" road- and dam-building project would by 2020 leave at best 28% of the forest untouched. Some of the damage is done by mining, particularly for gold, and the associated pollution of soil and water. Land clearance for ranching is a major cause of forest loss, occurring wherever the building of access roads takes place.

Logging, too, poses a continuing threat. Amazonian tropical hardwoods are highly attractive to the timber industry. A score of foreign-owned multinational companies dominate the timber export business, and some 2500 logging companies and sawmills operate in the Brazilian Amazon. Brazilian mahogany is now protected by a moratorium on exploitation, but illegal logging is widespread; the government estimates that it accounts for 80% of all timber being extracted from the Amazon today.

ENVIRONMENTAL CONSEQUENCES

The rainforest typically has poor and shallow soil. Its ecology depends on the recycling of minerals in the leaf litter on the forest floor. In a damp environment, the leaf litter is broken down rapidly by soil organisms. Once exposed by the loss of tree cover, however, it dries out rapidly and can easily be washed away by the next rain. The forest cannot regenerate growth where substantial areas have been cleared. The deforestation has far-reaching implications. Forests act as "carbon sinks," fixing carbon dioxide produced by the burning of fuel, and thus helping to counteract the buildup of the "greenhouse gases" linked with global warming.

Amazonia is also immensely rich in native plant and animal species. Its resource value, taking account of the potential significance of its genetic pool, for example in medicine or agriculture, was estimated in 2001 by a government research project as $2000 billion. This biodiversity, including much as yet undiscovered by the international scientific community, is dependent on proper protection of rainforest habitat.

INDIGENOUS PEOPLES

Colonization has brought violence and devastating epidemics to its indigenous peoples, along with the loss of their lands, forced removal, confinement to reservations, and the destruction of their lifestyle and culture. Many of them suffered extinction in the last half century. There remain about 200 distinct known forest-dwelling indigenous peoples. One-fifth of the Amazon is Amerindian land and under the 1988 constitution, forest peoples have the right to inhabit their ancestral lands, but they do not have legal title to them. Out of 580 identified Amerindian territories, over a third remain as yet undemarcated. The National Indian Foundation (FUNAI), a government agency responsible for demarcation, has been heavily criticized internationally for failing to provide adequate protection. When a territory is demarcated, its inhabitants have legal rights to protect it from logging, mining, and other industrial activities.

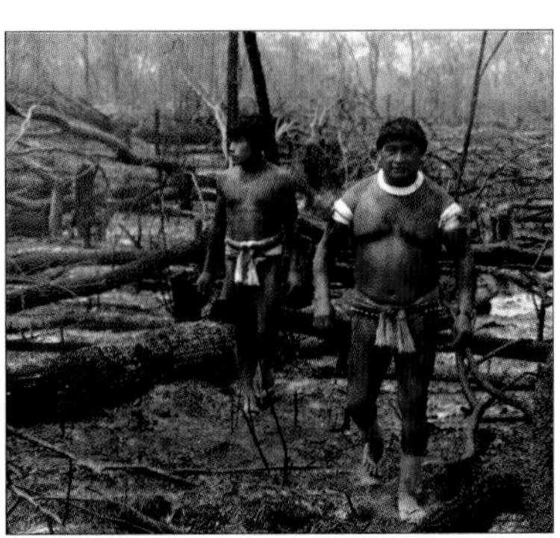

HEALTH

Welfare state health benefits

1 per 769 people

Heart diseases, cancers, accidents, violence

Federal health is underfunded. Fewer than 20% of hospitals are state-run, and they need modernization. Private care is beyond the means of the majority. On average only 15% of the health budget goes to child health, immunization, and other preventive programs. However, infant mortality, at 95 per 1000 children in 1970, had dropped to one-third of that level by 2000; access to potable water increased from 74% of the population in 1992 to 87% in 2000. In 2001 international drug companies and the US dropped patent infringement claims against Brazil for distributing anti-AIDS drugs free to more than 100,000 HIV patients.

SPENDING

GDP/cap. increase

CONSUMPTION AND SPENDING

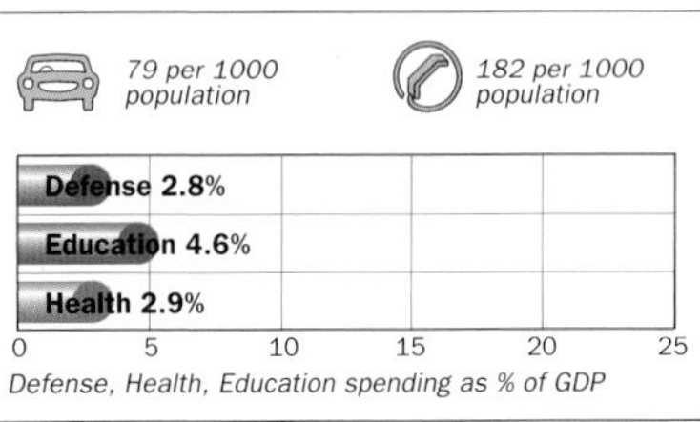
79 per 1000 population
182 per 1000 population
Defense 2.8%
Education 4.6%
Health 2.9%
0
5
10
15
20
25
Defense, Health, Education spending as % of GDP

Brazil's income distribution is among the most skewed in the world. IDB figures in 1999 showed that the richest 10% of the population take 50% of the income and the poorest 50% only 10%. Governments have failed to tackle the problem of homelessness and street children in large cities. An estimated one to five million families remain landless, while nearly 80% of farmland is owned by 10% of landowners. In 2000 the government announced that vast tracts of land of unproven ownership would be redistributed, but widespread invasions by landless workers have continued.

WORLD RANKING

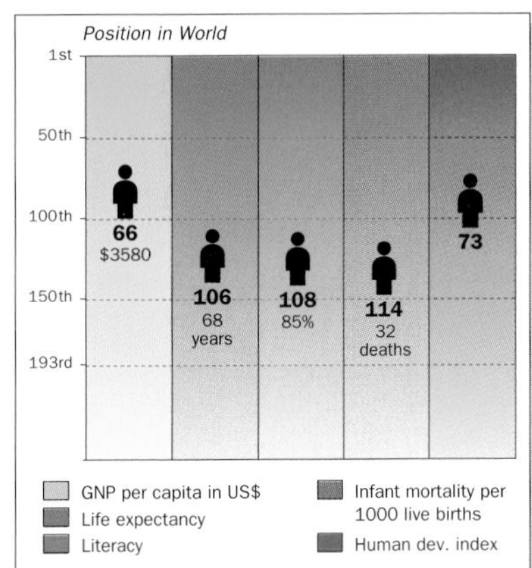
Position in World
1st
50th
100th
150th
193rd
66
$3580
106
68 years
108
85%
114
32 deaths
73
GNP per capita in US$
Life expectancy
Literacy
Infant mortality per 1000 live births
Human dev. index

B

BRUNEI

OFFICIAL NAME: Sultanate of Brunei **CAPITAL:** Bandar Seri Begawan
POPULATION: 335,000 **CURRENCY:** Brunei dollar **OFFICIAL LANGUAGE:** Malay

SOUTHEAST ASIA

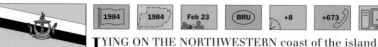

Asia · Malaysia · BRUNEI · Indonesia

1984

1984

Feb 23

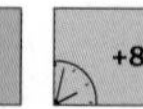
BRU

+8

+673

.bn

LYING ON THE NORTHWESTERN coast of the island of Borneo, Brunei is divided in two by a strip of the surrounding Malaysian state of Sarawak. The interior is mostly rainforest. Independent from the UK since 1984, Brunei is ruled by decree of the sultan. It is undergoing increasing Islamicization. Oil and gas reserves have brought one of the world's highest standards of living.

CLIMATE

Tropical equatorial

WEATHER CHART FOR BANDAR SERI BEGAWAN

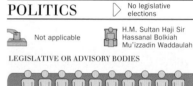

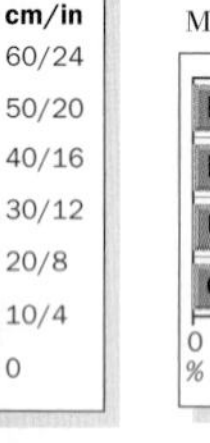
Average daily temperature
Rainfall
°C/°F
J F M A M J J A S O N D
cm/in
60/140
50/122
40/104
30/86
20/68
10/50
0/32
60/24
50/20
40/16
30/12
20/8
10/4
0

Just 480 km (300 miles) north of the equator, Brunei has a long rainy season with extremely high humidity.

TRANSPORTATION

Drive on left

Brunei International, Bandar Seri Begawan 982,041 passengers

97 ships 361,700 grt

THE TRANSPORTATION NETWORK

399 km (248 miles)

None

19 km (12 miles)

209 km (130 miles)

Interest-free loans for civil servants, subsidized gasoline, and limited public transportation account for the high rates of car ownership.

TOURISM

Visitors : Population 2.9:1

964,000 visitors

Up 13% in 1998

MAIN TOURIST ARRIVALS

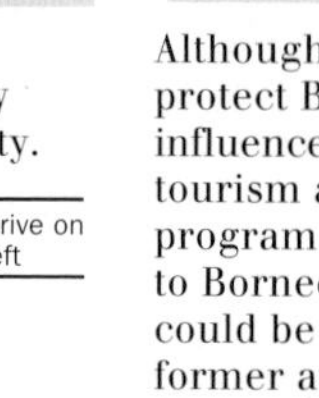
Malaysia 80%
Philippines 3%
UK 3%
Other 14%
0 10 20 30 40 50 60 70 80
% of total arrivals

Although the goverment is keen to protect Bruneians from Western influence, it wants to develop quality tourism as part of its diversification program. Promoted as the "Gateway to Borneo," Brunei's rainforests could be developed for tourism. A former attraction was the Churchill Museum, founded by the late sultan. This has now been superseded by the Museum of Royal Regalia.

PEOPLE

Pop. density medium

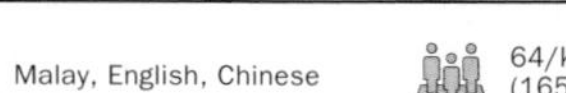
Malay, English, Chinese
64/km² (165/mi²)

THE URBAN/RURAL POPULATION SPLIT

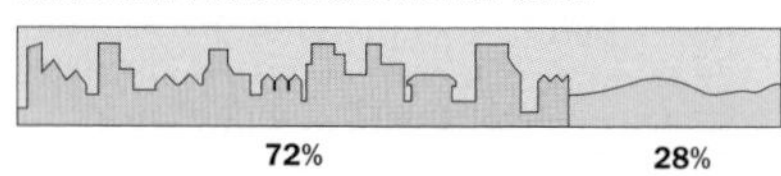
72%
28%

ETHNIC MAKEUP

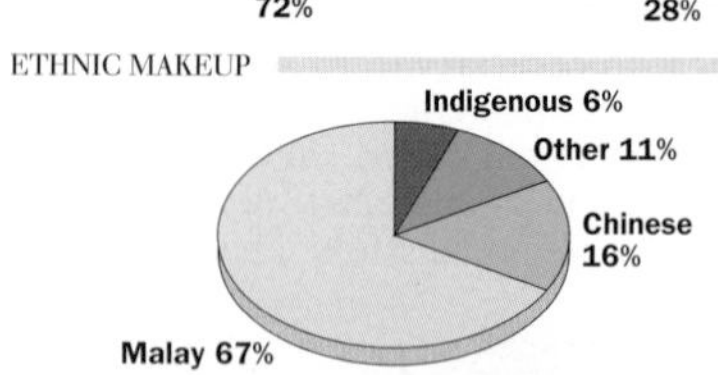
Indigenous 6%
Other 11%
Chinese 16%
Malay 67%

Malays are the beneficiaries of positive discrimination; many in the Chinese community are either stateless or hold British protected person passports. Among indigenous groups, the Murut and Dusuns are favored over the Ibans. Women, less restricted than in some Muslim states, are obliged to wear headscarves but not the veil. Many hold influential posts in the civil service.

POLITICS

No legislative elections

Not applicable

H.M. Sultan Haji Sir Hassanal Bolkiah Mu'izzadin Waddaulah

LEGISLATIVE OR ADVISORY BODIES

Brunei is an absolute monarchy; the sultan consults four advisory councils: Religious Council, Privy Council, Council of Cabinet Ministers, and Council of Succession, which he appoints. Political parties have been banned since 1988.

Since a failed rebellion in 1962, a state of emergency has been in force and the sultan has ruled by decree. Hopes for democracy were dashed when political parties were banned in 1988. In 1990, "Malay Muslim Monarchy" was introduced, promoting Islamic values as the state ideology. This further alienated the large Chinese and expatriate communities. Power is closely tied to the royal family. One of the sultan's brothers holds the foreign affairs portfolio and the sultan himself looks after defense and finance.

WORLD AFFAIRS

Joined UN in 1984

APEC

ASEAN

Comm

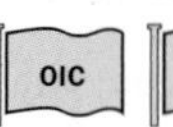
OIC

WTO

Brunei leads calls for a regional free trade area. Political exiles opposed to the government and based in Malaysia are a main concern. Relations with the UK, the ex-colonial power, are good.

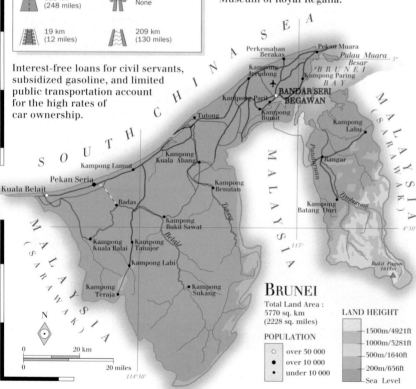

SOUTH CHINA SEA · Perkemahan Berakas · Pekan Muara · Pulau Muara Besar · BRUNEI BAY · Kampong Jerudong · Kampong Paring · BANDAR SERI BEGAWAN · Kampong Parit · Kampong Bunut · Tutong · Kampong Labu · Kampong Kuala Abang · Kampong Lumut · Pekan Seria · Kuala Belait · Kampong Benutan · Bangar · Badas · Kampong Batang Duri · Kampong Bukit Sawat · Kampong Kuala Balai · Kampong Tanajor · Kampong Labi · Kampong Teraja · Kampong Sukang · Bukit Pagon 1618m · MALAYSIA (SARAWAK) · MALAYSIA

BRUNEI

Total Land Area : 5770 sq. km (2228 sq. miles)

POPULATION

- over 50 000
- over 10 000
- under 10 000

LAND HEIGHT

1500m/4921ft · 1000m/3281ft · 500m/1640ft · 200m/656ft · Sea Level

0 – 20 km · 0 – 20 miles

***The magnificent Omar Ali Saifuddin Mosque** is surrounded by an artificial lagoon.*

AID

▷ Donor

Ad hoc handouts from the sultan

Not applicable

Aid spending is largely ad hoc. It has included donations to the Contras in Nicaragua, the Bosnian Muslims, and the homeless of New York.

DEFENSE

▷ No compulsory military service

 US$348m

 Down 13% in 2000

As well as being head of the 5900-strong armed forces, the sultan has a personal bodyguard of 2000 UK-trained Gurkhas. The UK and Singapore are close defense allies.

ECONOMICS

▷ Inflation 1.1% p.a. (1990–1999)

 US$7.8bn

 1.734–1.847 Brunei dollars

SCORE CARD

- ❑ WORLD GNP RANKING........................95th
- ❑ GNP PER CAPITA ....US$24,762
- ❑ BALANCE OF PAYMENTS ....US$2.09bn
- ❑ INFLATION 1%
- ❑ UNEMPLOYMENT 5%

STRENGTHS

Nearly 20 years of known oil reserves; over 30 years of gas. Earnings from massive overseas investments, mainly in the US and Europe, now exceed oil and gas revenues.

WEAKNESSES

Single-product economy. Failure of diversification programs could lead to problems in the future.

EXPORTS

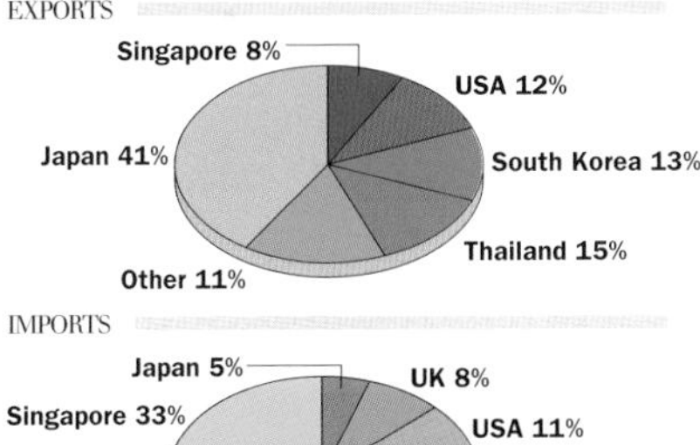
Singapore 8%
USA 12%
Japan 41%
South Korea 13%
Thailand 15%
Other 11%
IMPORTS
Japan 5%
UK 8%
Singapore 33%
USA 11%
Malaysia 19%
Other 24%

RESOURCES

▷ Electric power 479,000 kw

 3308 tonnes

 195,000 b/d (reserves 1.4bn barrels)

40,000 ducks, 8000 cattle, 6000 pigs, 4.7m chickens

 Oil, natural gas

Oil and gas are the major resources. Energy policy is now focused on regulating output in order to conserve stocks, since reserves are of limited duration. Almost all food is imported.

ENVIRONMENT

▷ Not available

14% (5% partially protected)

 17.1 tonnes per capita

The Forestry Strategic Plan aims to protect Brunei's forests (which take up 80% of its land area). It has allocated 64% of their area for protection and recreation and the prevention of soil erosion. However, Brunei's mangrove swamps, the largest on the island of Borneo, remain unprotected.

MEDIA

▷ TV ownership medium

Daily newspaper circulation 69 per 1000 people

PUBLISHING AND BROADCAST MEDIA

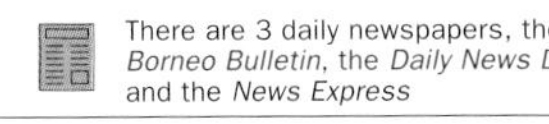
There are 3 daily newspapers, the *Borneo Bulletin*, the *Daily News Digest*, and the *News Express*

 1 state-owned service

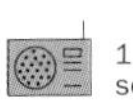 1 state-owned service

The state effectively controls all media. Brunei's television service now has a heavy religious content.

CRIME

▷ Death penalty not used in practice

 333 prisoners

 Up 10% in 1999

Crime levels are low. Most crime involves petty theft or is linked to alcohol and narcotics (both banned). A stolen car often makes TV news headlines. The state of emergency enables the government to detain without charge or trial for indefinitely renewable two-year periods.

EDUCATION

▷ School leaving age: 12

 92%

1878 students

Free schooling is available to the entire population, with the exception of the stateless Chinese, who do not qualify. The University of Brunei Darussalam was opened in 1985.

HEALTH

▷ Welfare state health benefits

 1 per 939 people

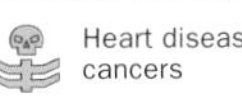
Heart diseases, cancers

The health service is free, although if major surgery is required Bruneians tend to travel to Singapore.

CHRONOLOGY

Under British control since 1841, Brunei became a formal British Protectorate in 1888.

- ❑ **1929** Oil extraction begins.
- ❑ **1959** First constitution enshrines Islam as state religion. Internal self-government.
- ❑ **1962** Prodemocracy rebellion. State of emergency; sultan rules by decree.
- ❑ **1984** Independence from Britain. Brunei joins ASEAN.
- ❑ **1990** Ideology of "Malay Muslim Monarchy" introduced.
- ❑ **1991** Imports of alcohol banned.
- ❑ **1992** Joins Non-Aligned Movement.
- ❑ **1998** Sultan's son, Prince Al-Muhtadee Billah, made crown prince.

SPENDING

▷ GDP/cap. increase

CONSUMPTION AND SPENDING

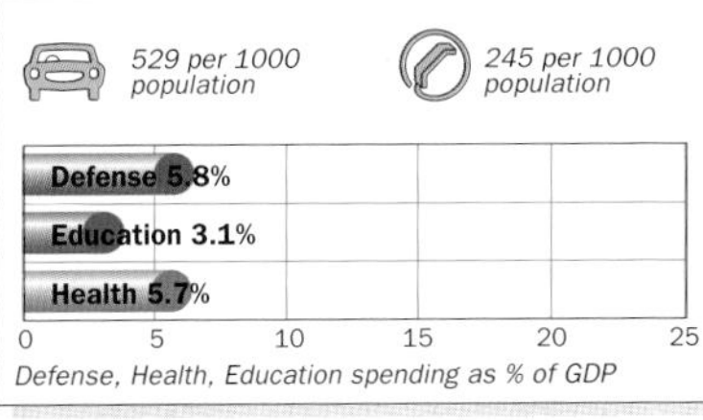
529 per 1000 population
245 per 1000 population
Defense 5.8%
Education 3.1%
Health 5.7%
0
5
10
15
20
25
Defense, Health, Education spending as % of GDP

The wealthiest people in Brunei are those close to the sultan, one of the world's richest men. A generally high standard of living, along with a degree of social mobility among Malays, keeps discontent to a minimum. Bruneians are major consumers of high-tech hi-fi and video equipment, designer-label watches, and Western designer clothes. The sultan's younger brother, Prince Jefri, in 2001 auctioned his possessions, ranging from fire engines to marble baths, after the failure of his business left him with debts of US$3 billion.

WORLD RANKING

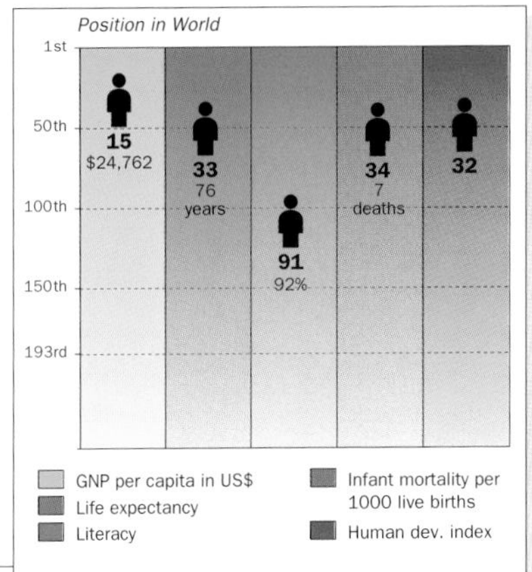
Position in World
1st
50th
100th
150th
193rd
15
$24,762
33
76 years
91
92%
34
7 deaths
32
GNP per capita in US$
Life expectancy
Literacy
Infant mortality per 1000 live births
Human dev. index

B

BULGARIA

EUROPE
Europe
Romania
Serb & Mon.
BULGARIA
Maced.
Greece
Turkey

OFFICIAL NAME: Republic of Bulgaria **CAPITAL:** Sofia
POPULATION: 7.9 million **CURRENCY:** Lev **OFFICIAL LANGUAGE:** Bulgarian

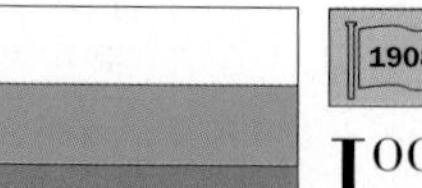

1908
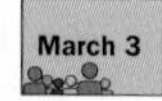
1947

March 3
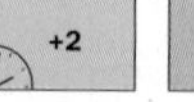
BG

+2

+359

.bg

LOCATED IN SOUTHEASTERN EUROPE, Bulgaria is a mainly mountainous country. The River Danube forms the northern border, while the popular resorts of the Black Sea lie to the east. The most populated areas are around Sofia in the west, Plovdiv in the south, and along the Danube plain. Bulgaria was directly ruled by the Turks from 1396 until 1878. In 1908, it became an independent kingdom and was under communist rule from 1947, with Todor Zhivkov in power from 1954 to 1989. The 1990s brought political instability as Bulgaria adjusted to democracy and economic reconstruction.

***Rila Monastery** in the Rila Mountains. It is famous for its 1200 National Revival Period frescoes dating from the mid-19th century.*

CLIMATE

Mediterranean/ continental

WEATHER CHART FOR SOFIA

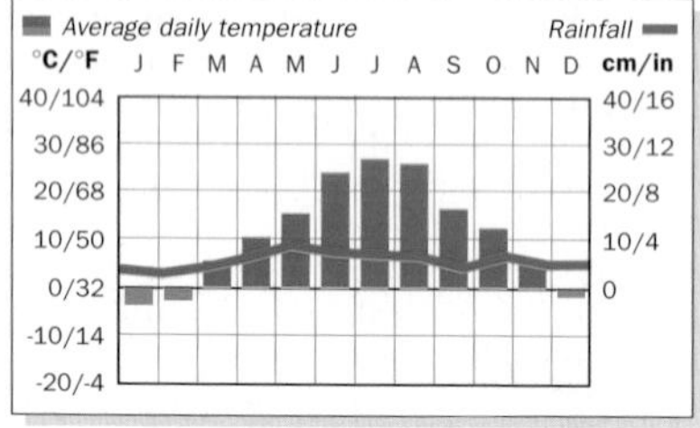
Average daily temperature
Rainfall
°C/°F
J F M A M J J A S O N D
cm/in
40/104
30/86
20/68
10/50
0/32
-10/14
-20/-4
40/16
30/12
20/8
10/4
0

The central valley and the lowlands have warm summers and cold, snowy winters, but hot or cold winds from Russia can bring spells of more extreme weather. The hotter summers on the Black Sea coast have encouraged the growth of tourist resorts. Snow may lie on the high mountain peaks until June.

TRANSPORTATION

Drive on right

Sofia International 1.13m passengers

164 ships 989,600 grt

THE TRANSPORTATION NETWORK

33,818 km (21,013 miles)

319 km (198 miles)

4290 km (2666 miles)

470 km (292 miles)

At the crossroads between Europe and Asia, Bulgarian railroads and expressways were underfunded under Zhivkov (when north–south routes were left undeveloped) and in the economically uncertain 1990s. Funding for modernizing key routes is now in place. Ferries are used for most cross-Danube traffic. In 2000 agreement was reached with Romania on building a second bridge across the river, scheduled to open in 2005.

TOURISM

Visitors : Population 1:2.8

2.8m visitors

Up 13% in 2000

MAIN TOURIST ARRIVALS

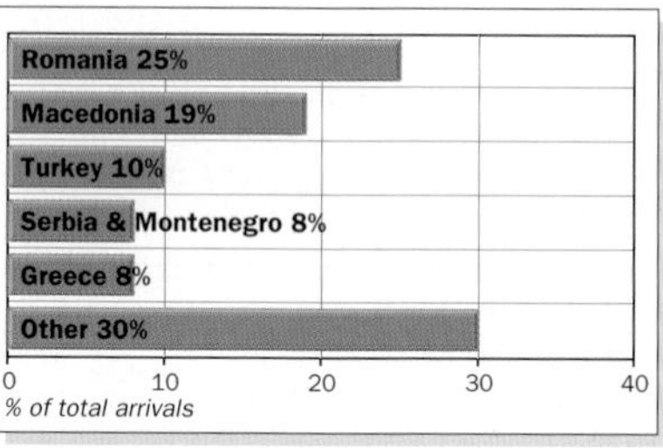
Romania 25%
Macedonia 19%
Turkey 10%
Serbia & Montenegro 8%
Greece 8%
Other 30%
0
10
20
30
40
% of total arrivals

The tourist industry formerly catered for the east European mass-market. Western tourists are attracted by low prices for skiing and beach vacations. Bulgaria is now privatizing the industry and seeks to move it upmarket by stressing its heritage. Since the mid-1990s a slump in earnings has been reversed: Russians are returning in larger numbers and there is a growth in tours from western Europe, especially from Germany.

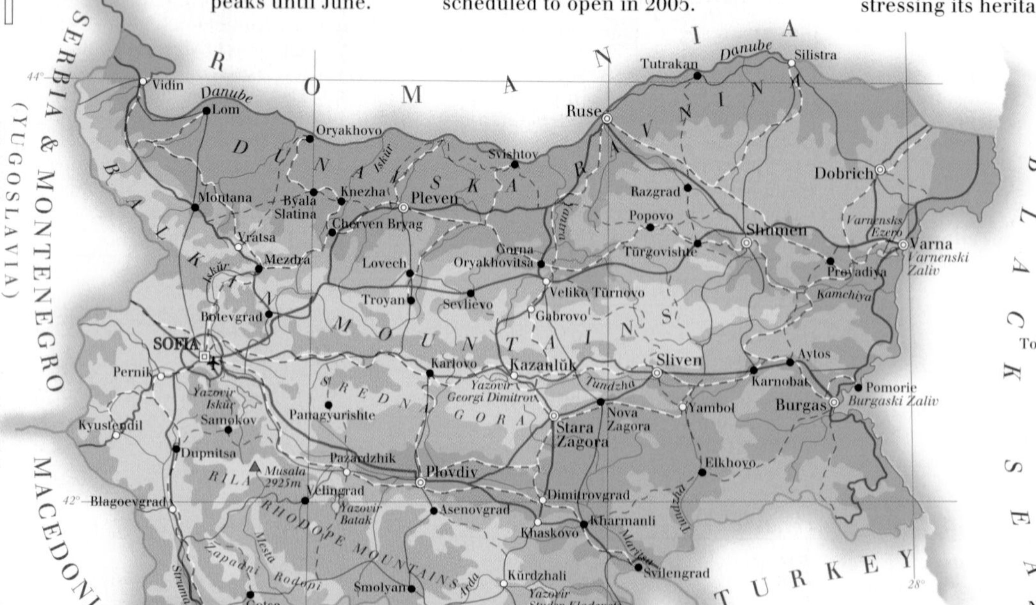
R O M A N I A
SERBIA & MONTENEGRO (YUGOSLAVIA)
MACEDONIA
G R E E C E
T U R K E Y
B L A C K S E A
Danube
Vidin
Lom
Oryakhovo
Svishtov
Ruse
Tutrakan
Silistra
Dobrich
Montana
Byala Slatina
Knezha
Pleven
Cherven Bryag
Vratsa
Mezdra
Lovech
Troyan
Botevgrad
SOFIA
Pernik
Kyustendil
Samokov
Dupnitsa
Blagoevgrad
Petrich
Gotse Delchev
Razgrad
Popovo
Shumen
Türgovishte
Varna
Provadiya
Gorna Oryakhovitsa
Veliko Türnovo
Sevlievo
Gabrovo
Karlovo
Kazanlŭk
Sliven
Aytos
Karnobat
Burgas
Pomorie
Burgaski Zaliv
Yambol
Stara Zagora
Nova Zagora
Panagyurishte
Pazardzhik
Plovdiv
Velingrad
Asenovgrad
Dimitrovgrad
Kharmanli
Khaskovo
Elkhovo
Svilengrad
Kŭrdzhali
Smolyan
Madan
Musala 2925m
D U N A V S K A R A V N I N A
B A L K A N M O U N T A I N S
S R E D N A G O R A
R I L A
R H O D O P E M O U N T A I N S
Zapadni Rodopi
Iztochni Rodopi
Yazovir Georgi Dimitrov
Yazovir Iskŭr
Yazovir Batak
Yazovir Studen Kladenets
Varnenski Zaliv
Kamchiya
Tundzha
Maritsa
Mesta
Struma
Iskŭr
Yantra
44°
42°
24°
26°
28°
0
50 km
50 miles
N

BULGARIA

Total Land Area : 110 910 sq. km (42 822 sq. miles)

POPULATION

over 1 000 000
over 100 000
over 50 000
over 10 000

LAND HEIGHT

2000m/6562ft
1000m/3281ft
500m/1640ft
200m/656ft
Sea Level

B

PEOPLE

Pop. density medium

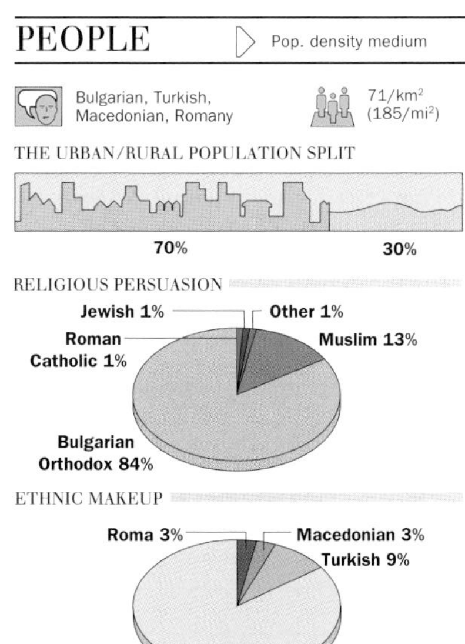
Bulgarian, Turkish, Macedonian, Romany
71/km² (185/mi²)
THE URBAN/RURAL POPULATION SPLIT
70%
30%
RELIGIOUS PERSUASION
Jewish 1%
Other 1%
Roman Catholic 1%
Muslim 13%
Bulgarian Orthodox 84%
ETHNIC MAKEUP
Roma 3%
Macedonian 3%
Turkish 9%
Bulgarian 85%

The communist era was marked by assimilationist policies suppressing minority cultural identities. In the 1970s, Bulgarian Muslims, or Pomaks, were forced to change Muslim names to Bulgarian ones. Bulgarian Turks were particularly targeted in the 1980s.

Linguistic and religious freedom was granted in 1989, but 300,000 ethnic Turks, or 40%, still left for Turkey. Their farming skills have traditionally been important to agriculture. Recent privatizations have left many Turks landless and have provoked new emigration. The MRF, supported mainly by ethnic Turks, is a minority party whose influence can be strong, a fact borne out by its inclusion in the governing coalition formed in 2001.

Macedonians and Roma each account for 3% of the population. The latter minority suffers discrimination at all levels. Women have equal rights in theory, but society remains patriarchal, especially among Turks.

POPULATION AGE BREAKDOWN

Female	Age	Male
1.4%	80+	0.9%
10.5%	60–79	8.6%
13.5%	40–59	12.9%
13.7%	20–39	14%
11.9%	0–19	12.6%

% of population by age group

WORLD AFFAIRS

Joined UN in 1955

BSEC

CE

EAPC

CEFTA
OSCE

Bulgaria is among six "second wave" candidates for EU membership which began negotiations in 2000. It hopes to join NATO.

Bulgaria conscientiously adhered to UN sanctions against Yugoslavia, despite the costs of lost trade. Relations with Russia are no longer close, but are maintained carefully because of heavy Bulgarian dependence on Russia for oil and gas. Relations with Turkey have greatly improved since the tensions of the final years of communist rule.

AID

Recipient

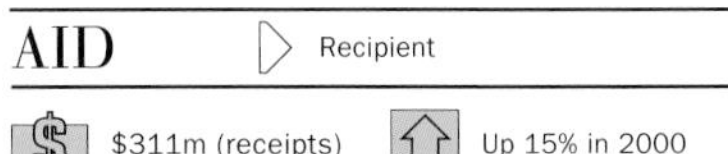
$311m (receipts)
Up 15% in 2000

IMF, World Bank, and EBRD loans are mainly intended for infrastructure improvements. Large-scale EU assistance toward reforms, in preparation for Bulgaria's eventual EU membership, is estimated at 2% of GDP. Humanitarian aid focuses mainly on medical provision and children's homes.

POLITICS

Multiparty elections

2001/2005

President Georgi Purvanov

AT THE LAST ELECTION

National Assembly 240 seats

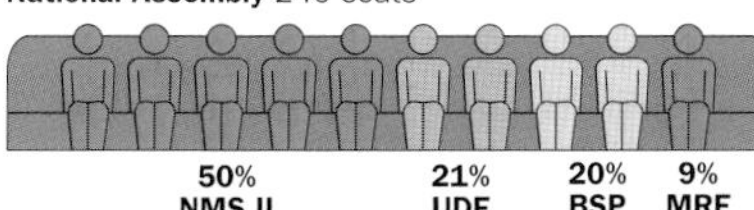
50% NMS II
21% UDF
20% BSP
9% MRF

NMS II = National Movement Simeon II
UDF = Union of Democratic Forces
BSP = Bulgarian Socialist Party
MRF = Movement for Rights and Freedoms

Bulgaria is a multiparty democracy.

PROFILE

Having moved falteringly to a pluralist democratic system after the fall of the communist Zhivkov regime in 1989, Bulgaria suffered during the 1990s from successive weak governments, each brought down by no-confidence votes. The UDF, a broad anticommunist alliance, fell from office in 1992, and by the time of the 1994 general election the former communist BSP appeared to be firmly in the ascendant, winning an overall majority. The BSP government resisted political and economic change. The result was one of the slowest privatization programs in eastern Europe, with the old communist web of patronage still intact. A new UDF government in 1997 launched free-market reforms backed up by the IMF. Its considerable success, and reorientation of policy toward the goals of EU and NATO membership, allowed the UDF to approach the June 2001 elections with some confidence, despite a surge in support for a monarchist party launched by ex-king Simeon II (who had left Bulgaria as a small child in 1946). The poll, however, left the UDF with less than a quarter of the National Assembly seats, exactly half of which went to the NMS II. Uncertainty over whether the ex-king himself would take on the role of prime minister ended when he formed a coalition with the MRF (representing mainly the ethnic Turkish minority) and was sworn in in July. He took the oath to the republic (although he had not formally given up his claim to the throne) as Simeon Saxecoburggotski, promising "spiritual and economic revival," clean government, tax cuts, and more privatization, with integration with the EU and NATO his top priorities.

Ex-king Simeon II *(Saxecoburggotski) returned as prime minister in 2001.*

Georgi Purvanov, *of the BSP, elected president against expectations in 2001.*

CHRONOLOGY

Bulgaria was part of the Ottoman Empire for five centuries until its independence in 1908. Under King Ferdinand, it sided with Germany during World War I, and subsequently lost valuable territory to Greece and Serbia. Under King Boris, Bulgaria once again sided with Germany in World War II.

- **1943** Child king Simeon II accedes.
- **1944** Allies firebomb Sofia. Soviet army invades. Antifascist Fatherland Front coalition, including Agrarian Party and Bulgarian Communist Party (BCP), takes power in bloodless coup. Kimon Georgiev prime minister.
- **1946** September, referendum abolishes monarchy. Republic proclaimed. October, general election results in BCP majority.
- **1947** Prime Minister Georgi Dmitrov discredits Agrarian Party leader Nikola Petkov. Petkov arrested and sentenced to death. International recognition of Dmitrov government. Soviet-style constitution adopted; one-party state established. Country renamed People's Republic of Bulgaria. Nationalization of economy begins.
- **1949** Dmitrov dies, succeeded as prime minister by Vasil Kolarov.
- **1950** Kolarov dies. "Little Stalin" Vulko Chervenkov replaces him ➪

B

CHRONOLOGY *continued*

and begins BCP purge and collectivization.

- ❑ **1953** Stalin dies; Chervenkov's power begins to wane.
- ❑ **1954** Chervenkov yields power to Todor Zhivkov. Zhivkov sets out to make Bulgaria an inseparable part of the Soviet system.
- ❑ **1955–1960** Zhivkov exonerates victims of Chervenkov's purges.
- ❑ **1965** Plot to overthrow Zhivkov discovered by Soviet agents.
- ❑ **1968** Bulgarian troops aid Soviet army in invasion of Czechoslovakia.
- ❑ **1971** New constitution. Zhivkov becomes president of State Council and resigns as premier.
- ❑ **1978** Purge of BCP: 30,000 members expelled.
- ❑ **1984** Turkish minority forced to take Slavic names.
- ❑ **1989** June–August, exodus of 300,000 Bulgarian Turks. November, Zhivkov ousted as BCP leader and head of state. Replaced by Petur Mladenov. Mass protest in Sofia for democratic reform. December, Union of Democratic Forces (UDF) formed.
- ❑ **1990** Economic collapse. Zhivkov arrested. BCP loses constitutional role as leading political party, changes name to Bulgarian Socialist Party (BSP). Elections: BSP victory. Parliament chooses Zhelyu Zhelev, UDF leader, as president. Country renamed Republic of Bulgaria; communist symbols removed from national flag.
- ❑ **1991** February, price controls abolished; steep price rises. July, new constitution adopted. October, UDF wins elections.
- ❑ **1992** Continued political and social unrest. October, UDF resigns after losing vote of confidence. December, Movement for Rights and Freedoms (MRF) forms government. Zhivkov convicted of corruption and human rights abuses.
- ❑ **1993** Ambitious privatization program begins.
- ❑ **1994** General elections return BSP to power.
- ❑ **1995** BSP leader, Zhan Videnov, heads coalition government.
- ❑ **1996** Financial crisis and collapse of lev. Presidential elections won by opposition UDF candidate, Peter Stoyanov.
- ❑ **1997** General election won by UDF, whose leader Ivan Kostov becomes prime minister.
- ❑ **2001** Despite economic upturn, voters turn to new party headed by ex-king, who, as Simeon Saxecoburggotski, becomes prime minister. November, BSP leader Georgi Purvanov elected president.

DEFENSE

Compulsory military service

 $347m

 Down 11% in 2000

BULGARIAN ARMED FORCES

1475 main battle tanks (1042 T–55, 433 T–72)
42,400 personnel
1 submarine, 1 frigate, and 23 patrol boats
5260 personnel
181 combat aircraft (Su–25, MiG–21/23/29)
18,300 personnel
None

Defense spending fell from 14% of GDP in 1985 to 2.8% in 2000. Plans to join NATO, which were announced in 1997, involved a major reorientation in defense thinking.

In late 1999 the government adopted "Plan 2004," which embodies a radical acceleration of its previous plans to restructure the armed forces. The new plan envisages downsizing to only 45,000 personnel by 2004, and moving to an emphasis on rapid reaction capabilities. It was considered that this smaller but combat-ready force would be less costly to maintain in the long run than a larger force.

In 1999 the crisis in Kosovo prompted Bulgaria to make its airspace available to NATO.

ECONOMICS

Inflation 103% p.a. (1990–2000)

$12.4bn

2.084–2.187 leva

SCORE CARD

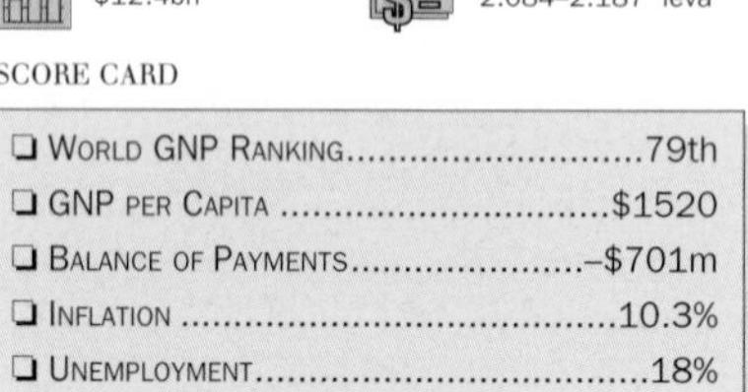

❑ WORLD GNP RANKING	79th
❑ GNP PER CAPITA	$1520
❑ BALANCE OF PAYMENTS	–$701m
❑ INFLATION	10.3%
❑ UNEMPLOYMENT	18%

EXPORTS

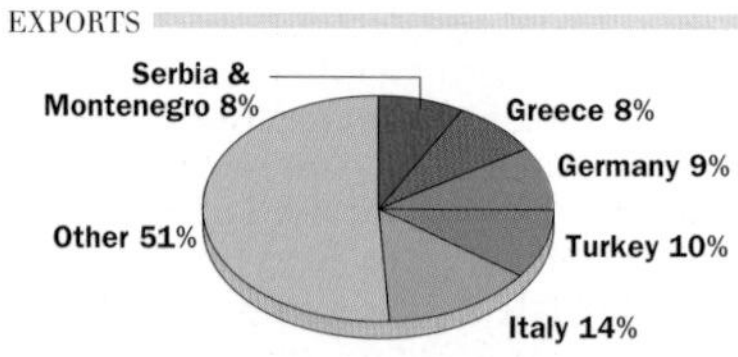
Serbia & Montenegro 8%
Greece 8%
Germany 9%
Turkey 10%
Italy 14%
Other 51%

IMPORTS

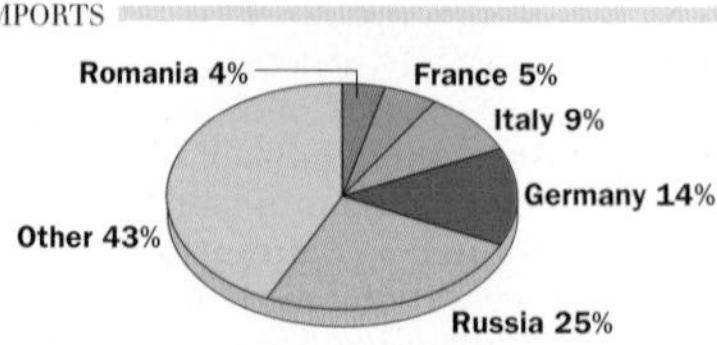
Romania 4%
France 5%
Italy 9%
Germany 14%
Russia 25%
Other 43%

STRENGTHS

Coal and natural gas. Good agricultural production, especially grapes for well-developed wine industry, and tobacco. Strong expertise in computer software.

WEAKNESSES

Outdated infrastructure and equipment, and outstanding debt throughout industry. Slow privatization and structural reform until 1998.

PROFILE

Restructuring the economy is linked to privatization – a process delayed for political and technical reasons until the late 1990s. A financial crisis in 1996 triggered the collapse of the national currency, the lev. Foreign investment is still low, despite laws that since 1992 have allowed foreign firms to own companies outright. Trade has shifted toward the EU, while that with the former Soviet Union has fallen sharply. The UDF government, returned in 1997, followed IMF advice, with free-market reforms, backed by foreign loans, successfully bringing inflation under control. After a setback in 1999, attributed to the regional impact of the crisis over Kosovo, forecasts of economic growth suggested Bulgaria's continuing recovery.

ECONOMIC PERFORMANCE INDICATOR

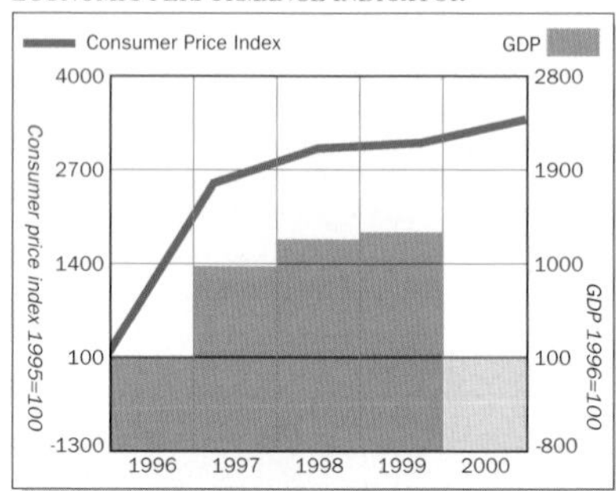
Consumer Price Index
GDP
4000
2700
1400
100
-1300
2800
1900
1000
100
-800
Consumer price index 1995=100
GDP 1996=100
1996
1997
1998
1999
2000

BULGARIA : MAJOR BUSINESSES

Pleven
Ruse
Shumen
Pernik
Varna
Burgas
Stara Zagora
Plovdiv
Sofia
Wine
Steel
Textiles
Shipbuilding
Leather tanning
Food processing
Metal processing
Vehicle assembly
Heavy engineering
Tobacco
Computers
Oil refining
0
200 km
0
200 miles

B

RESOURCES

Electric power 12.1m kw

18,336 tonnes

662 b/d (reserves 15m barrels)

2.29m sheep, 1.14m pigs, 15m chickens

Coal, iron, copper, lead, zinc, natural gas, oil, manganese

ELECTRICITY GENERATION

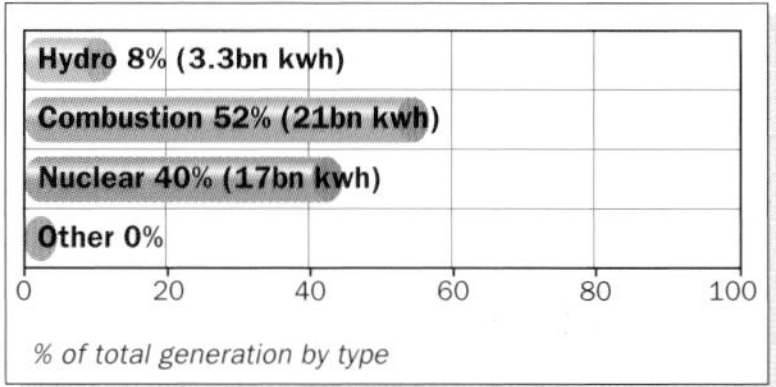

Hydro 8% (3.3bn kwh)
Combustion 52% (21bn kwh)
Nuclear 40% (17bn kwh)
Other 0%
0
20
40
60
80
100
% of total generation by type

Bulgaria has modest oil reserves and rather larger ones of coal and natural gas, but still has to import about 70% of its primary energy needs, much of it from the CIS. Unreliable supplies in the past led to frequent winter power cuts. These have largely disappeared since the mid-1990s, as reduced production in heavy industry and improved domestic supply have lowered import demand. Bulgaria is partly reliant on nuclear power. The EU is providing aid to upgrade two reactors at Kozloduy in return for the promised closure of other reactors there which pose particular safety risks. The government also plans to extend the controversial Chaira Dam, which began generating hydropower in 1993. Bulgaria has the northern hemisphere's largest manganese mine.

BULGARIA : LAND USE

Danube
BALKAN MOUNTAINS
Maritsa
RHODOPE MOUNTAINS
Cropland
Pasture
Forest
Sheep
Cereals
Tobacco
0
200 km
0
200 miles

ENVIRONMENT

Sustainability rank: 71st

5%

5.7 tonnes per capita

ENVIRONMENTAL TREATIES

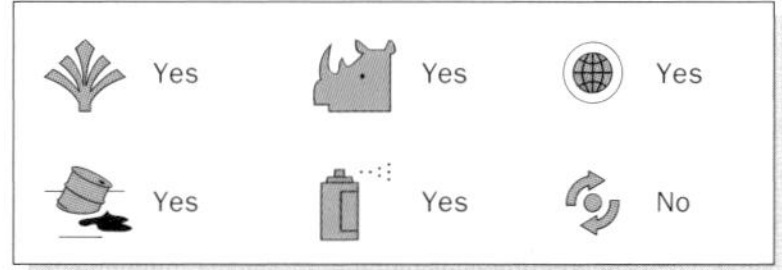

Yes
Yes
Yes
Yes
Yes
No

Environmental degradation led to the foundation in 1989 of the *Ecoglasnost* party. It circulated information on pollution and nuclear waste dump locations, and brought polluters to court. The Kozloduy nuclear complex, east of Lom, restarted in 1995 despite safety concerns, is under pressure to close its oldest reactors. Air pollution has diminished, but problems remain. NATO bombing of Serbian chemical and oil refineries on the Danube in 1999 led to downriver pollution in Bulgaria.

MEDIA

TV ownership high

Daily newspaper circulation 254 per 1000 people

PUBLISHING AND BROADCAST MEDIA

There are 21 daily newspapers, including *Duma*, *Zemya*, and *Trud*
3 services: 2 state-owned, 1 independent
10 services: 1 state-owned and 9 independent

Although it was liberalized in 1989, media freedom remains relative, and under new electronic media laws in 1998 state-run broadcasters retain an effective news monopoly. One group dominates the newspaper market, and of the others most are party-owned. Internet providers are regulated. Journalists also complain of the tough libel laws.

CRIME

No death penalty

9424 prisoners

Down 11% in 1999

CRIME RATES

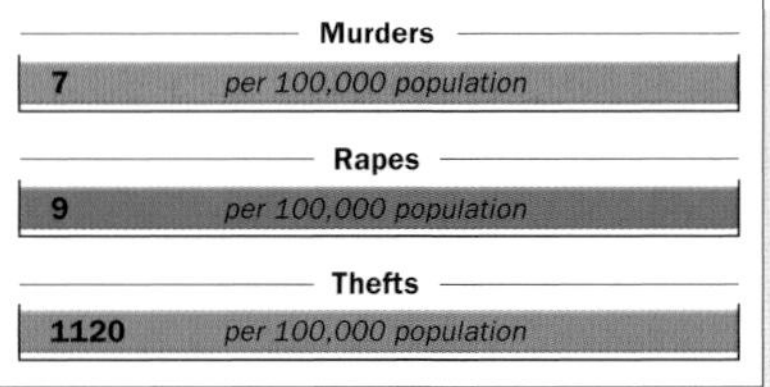

Murders
7 per 100,000 population
Rapes
9 per 100,000 population
Thefts
1120 per 100,000 population

In the 1990s Bulgaria became a key narcotics trafficking route to western Europe. Former security agents, party officials, and prestigious ex-athletes have moved into protection rackets, counterfeiting, and similar activities. Violations of minority rights are a sensitive political issue.

EDUCATION

School leaving age: 16

98%

261,321 students

THE EDUCATION SYSTEM

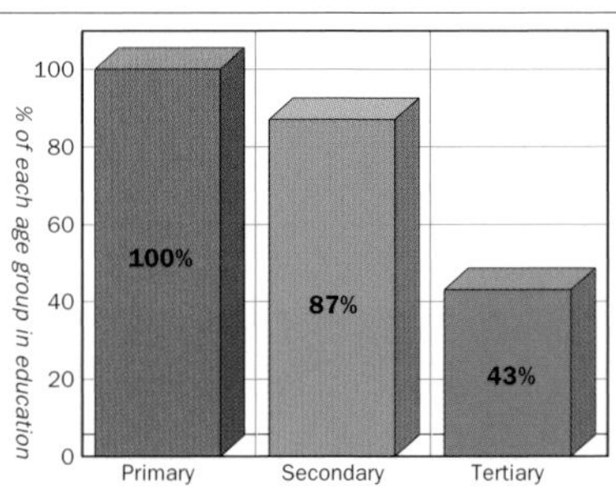

100
80
60
40
20
0
% of each age group in education
100%
87%
43%
Primary
Secondary
Tertiary

Education is free and compulsory between the ages of seven and 16. The system has been changed from a Soviet-inspired to a west European-style model. Government spending on education fell sharply until the 1999 budget. Teaching standards continue to be lowest in the rural and Turkish communities.

HEALTH

Welfare state health benefits

1 per 286 people

Heart and cerebrovascular diseases, cancers

Hospital facilities have kept pace with population growth, but the 1997 economic crisis brought the health service to the brink of collapse. A new health policy was formulated in 1999. The plan of action emphasizes primary care. The Bulgarian Red Cross assists in health administration.

SPENDING

GDP/cap. decrease

CONSUMPTION AND SPENDING

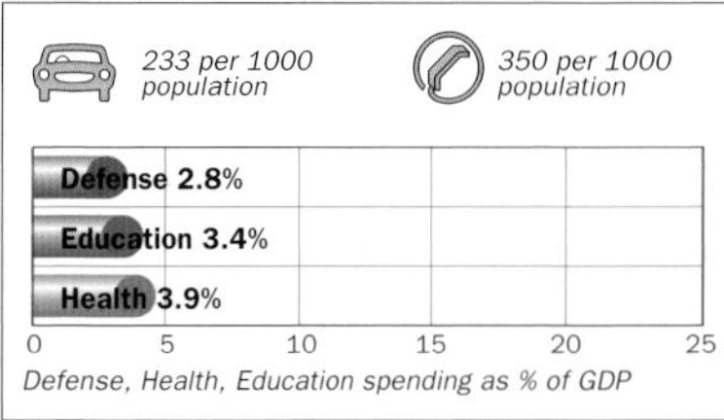

233 per 1000 population
350 per 1000 population
Defense 2.8%
Education 3.4%
Health 3.9%
0
5
10
15
20
25
Defense, Health, Education spending as % of GDP

Conquering the hyperinflation of 1995–1997 has remedied the most acute crisis, and associated hardship has been reduced. Turks and Roma remain the poorest people.

WORLD RANKING

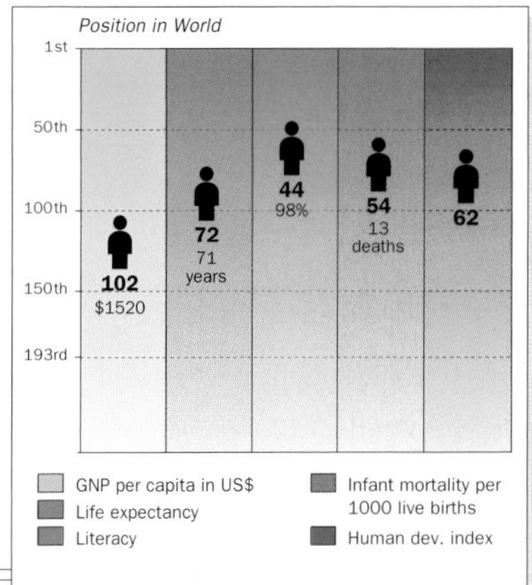

Position in World
1st
50th
100th
150th
193rd
102
$1520
72
71 years
44
98%
54
13 deaths
62
GNP per capita in US$
Life expectancy
Literacy
Infant mortality per 1000 live births
Human dev. index

BURKINA

B

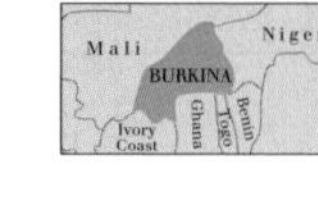
WEST AFRICA
Mali
Niger
BURKINA
Ivory Coast
Ghana
Togo
Benin

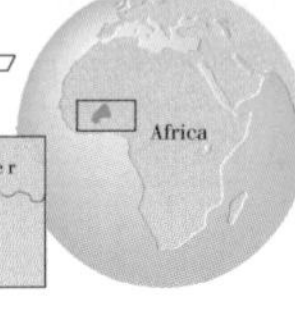
Africa

OFFICIAL NAME: Burkina Faso **CAPITAL:** Ouagadougou
POPULATION: 11.9 million **CURRENCY:** CFA franc **OFFICIAL LANGUAGE:** French

1960

1960

Dec 11

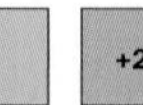
BF

0

+226

.bf

LANDLOCKED IN WEST AFRICA, Burkina (formerly Upper Volta) gained independence from France in 1960. The majority of Burkina lies in the arid fringe of the Sahara known as the Sahel. Ruled by military dictators for much of its postindependence history, Burkina became a multiparty state in 1991. However, much power still rests with President Blaise Compaoré. Burkina's economy remains largely based on agriculture.

CLIMATE

Tropical/steppe

WEATHER CHART FOR OUAGADOUGOU

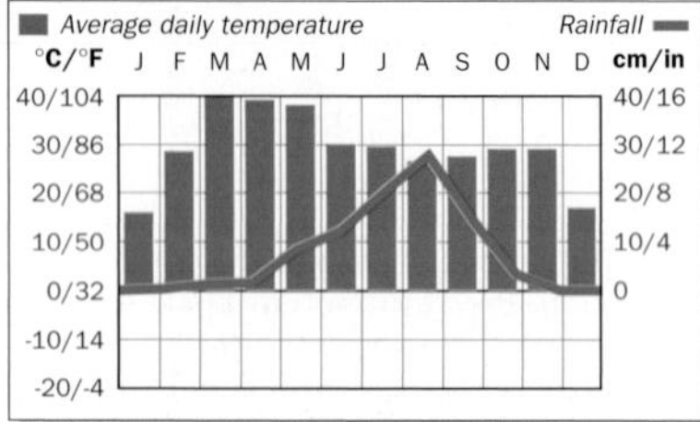
Average daily temperature
Rainfall
°C/°F
J F M A M J J A S O N D
cm/in
40/104
30/86
20/68
10/50
0/32
-10/14
-20/-4
40/16
30/12
20/8
10/4
0

The tropical climate comprises two seasons – unreliable rains from June to October, and a long dry season.

TRANSPORTATION

Drive on right

Ouagadougou International
232,312 passengers

Has no fleet

THE TRANSPORTATION NETWORK

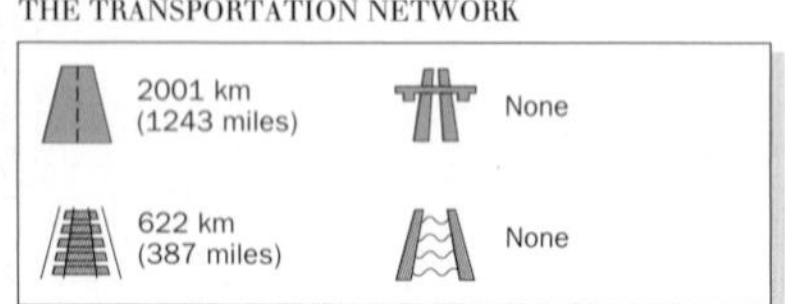
2001 km
(1243 miles)
None
622 km
(387 miles)
None

The railroad to the port of Abidjan in Ivory Coast provides the main commercial route to the sea. Roads through Benin, Togo, and Ghana provide alternative access.

TOURISM

Visitors : Population 1:55

218,000 visitors

Up 36% in 1999

MAIN TOURIST ARRIVALS

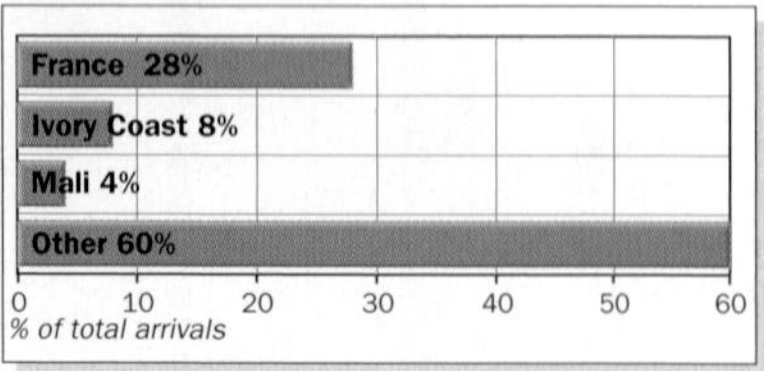
France 28%
Ivory Coast 8%
Mali 4%
Other 60%
0
10
20
30
40
50
60
% of total arrivals

Some potential exists for safari tourism, and the cities offer an attractive mix of colonial and African architecture. Big game hunting is allowed in some areas.

PEOPLE

Pop. density low

Mossi, Fulani, French, Tuareg, Dyula, Songhai

43/km² (113/mi²)

THE URBAN/RURAL POPULATION SPLIT

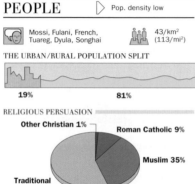

19% **81%**

RELIGIOUS PERSUASION

Other Christian 1%
Roman Catholic 9%
Muslim 35%
Traditional beliefs 55%

No ethnic group is dominant in Burkina, although the Mossi people who live in the area of their old empire around Ouagadougou are the most numerous and have always played an important role in government. Burkina's first president, Maurice Yameogo, and Blaise Compaoré, leader since 1987, are both Mossi. The people from the west are much more ethnically mixed.

The extended family is important, both in rural areas and in the towns and cities. Extreme poverty has led to a strong sense of egalitarianism within society.

The absence of women in public life belies their real power and influence, particularly within the traditional framework of the extended family. However, most women are still denied access to education and senior professional positions.

Camel plowing. Burkina's poor soils and frequent droughts lead many young men to emigrate seasonally in search of work.

POLITICS

Multiparty elections

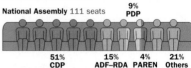

L. House 2002/2007
U. House varies

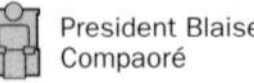
President Blaise Compaoré

AT THE LAST ELECTION

National Assembly 111 seats

9% PDP

51% CDP **15% ADF–RDA** **4% PAREN** **21% Others**

CDP = Congress for Democracy and Progress
ADF–RDA = Alliance for Democracy and Federation–African Democratic Rally **PDP** = Party for Democracy and Progress
PAREN = Party for National Renewal

House of Representatives 178 seats

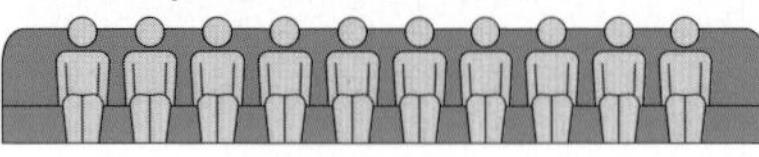

Members of the House of Representatives are appointed or indirectly elected on a nonparty basis by provincial councils and various communities

A multiparty democracy in theory, Burkina is still dominated in practice by former military dictator Blaise Compaoré, and the army remains influential behind the scenes. Compaoré has been in power since the assassination in 1987 of Capt. Thomas Sankara, his former superior. Several of Compaoré's close military colleagues have been murdered. His grip on power in Burkina appears to be solid, and he was reelected president in 1998 with almost 90% of the vote. Most opposition leaders are still living in exile, but opposition parties made small gains in the 2002 polls.

The CDP and the government came under unexpected pressure in 1998 and 1999 after the assassination of a popular newspaper editor, Norbert Zongo, in which leading establishment figures were implicated.

WORLD AFFAIRS

Joined UN in 1960

CILSS

ECOWAS

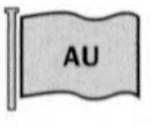
AU

OIC

FZ

Burkina's landlocked position means that good relations with countries to the south are a major concern. The US used the threat of sanctions in 2000 to stem the illegal trade in Burkina of diamonds from Sierra Leone.

AID

Recipient

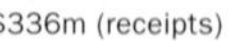
$336m (receipts)

Down 16% in 2000

External aid, mostly from the World Bank and France, is important to the economy. The large number of NGOs has caused organizational problems; there is often difficulty in finding suitable projects for all the prospective donors.

B

BURKINA

Total Land Area : 274 200 sq. km (105 869 sq. miles)

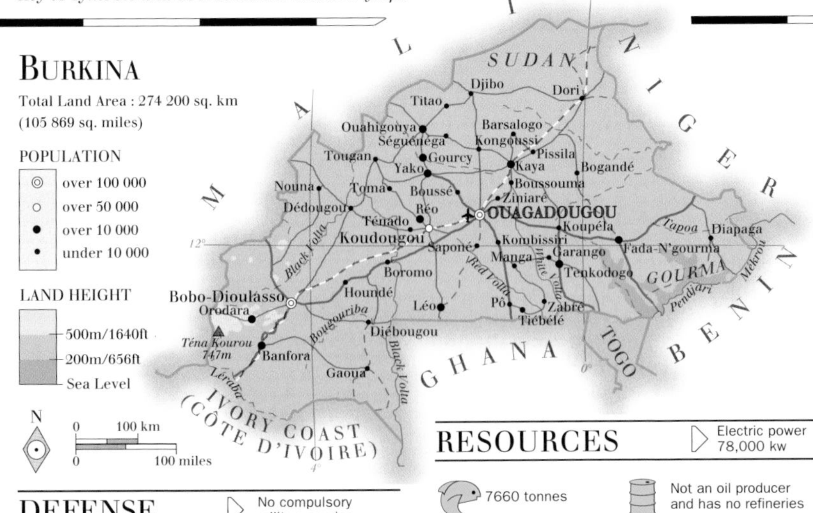
POPULATION
over 100 000
over 50 000
over 10 000
under 10 000
LAND HEIGHT
500m/1640ft
200m/656ft
Sea Level
N
0 100 km
0 100 miles
MALI
NIGER
BENIN
TOGO
GHANA
IVORY COAST (CÔTE D'IVOIRE)
SUDAN
GOURMA
Djibo
Dori
Titao
Ouahigouya
Séguénéga
Barsalogo
Kongoussi
Pissila
Tougan
Gourcy
Yako
Kaya
Bogandé
Nouna
Toma
Boussé
Boussouma
Ziniaré
Dédougou
Réo
OUAGADOUGOU
Ténado
Koupéla
Tapoa
Diapaga
Koudougou
Saponé
Kombissiri
Manga
Garango
Fada-N'gourma
Boromo
Tenkodogo
Pendjari
Mékrou
Houndé
Bobo-Dioulasso
Orodara
Léo
Pô
Zabré
Tiébélé
Diébougou
Ténа Kourou 747m
Banfora
Gaoua
Black Volta
White Volta
Red Volta
Bougouriba
Léraba
12°
0°
4°

DEFENSE

No compulsory military service

$68m

Down 9% in 2000

The main role of the 5600-strong army has been maintaining internal security. Burkina is reliant on France for most equipment and training.

ECONOMICS

Inflation 3.8% p.a. (1990–2000)

$2.42bn

698.7–736.7 CFA francs

SCORE CARD

❑ World GNP Ranking	134th
❑ GNP per Capita	$210
❑ Balance of Payments	–$65m
❑ Inflation	–0.3%
❑ Unemployment	1%

Strengths

Remittances from plantation workers in Ghana and the Ivory Coast. Strongly improved economic management. Low debt burden. Ability to attract foreign aid. Cotton.

Weaknesses

Landlocked. Natural resources not in the main economically viable. Donors' fears over political instability. Food crop fluctuations. Prone to drought. Migrants' remittances have halved to about $80 million since 1988.

EXPORTS

IMPORTS

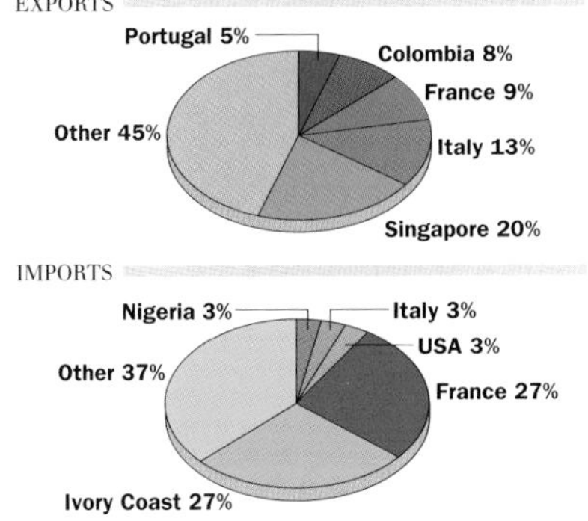
Portugal 5%
Colombia 8%
France 9%
Italy 13%
Singapore 20%
Other 45%
Nigeria 3%
Italy 3%
USA 3%
France 27%
Ivory Coast 27%
Other 37%

RESOURCES

Electric power 78,000 kw

7660 tonnes

Not an oil producer and has no refineries

8.65m goats, 6.78m sheep, 22.4m chickens

Gold, antimony, marble, manganese, silver, zinc

Burkina's considerable mineral wealth includes large manganese and silver deposits. Falling world prices led to the closure in 1999 of the largest gold mine. Three hydroelectric dams will reduce dependence on thermal energy.

ENVIRONMENT

Sustainability rank: 101st

10% (8% partially protected)

0.1 tonnes per capita

Like other countries on the southern rim of the Sahara, desertification is the major ecological issue. The rate of tree cutting for fuel is on the increase.

MEDIA

TV ownership low

Daily newspaper circulation 1 per 1000 people

PUBLISHING AND BROADCAST MEDIA

There are 6 daily newspapers, including *Sidwaya, Le Pays, Le Journal de Soir*, and *L'Observateur Paalga*

4 services: 1 state-owned, 3 independent

1 state-owned service, 46 independent stations

There are a number of small independent newspapers funded by opposition groups. A code of practice was introduced in 1999.

CRIME

Death penalty not used in practice

6100 prisoners

Down 19% 1996–1998

Crime levels have traditionally been low. However, the urbanization of society and the increase in political violence have seen levels increase.

EDUCATION

School leaving age: 16

24%

8911 students

Education is based on the French system. Recently, practical subjects have received more emphasis.

CHRONOLOGY

Ruled by Mossi kings from the 16th century, Upper Volta, a province of French West Africa in the late 19th century, gained independence in 1960.

- ❑ **1980** Ousting of military ruler; Col. Saye Zerbo becomes president.
- ❑ **1982** Capt. Thomas Sankara takes power. People's Salvation Council (PSC) begins radical reforms.
- ❑ **1984** Renamed Burkina.
- ❑ **1987** Sankara assassinated, Capt. Blaise Compaoré takes power.
- ❑ **1991** New constitution. Compaoré elected president.
- ❑ **1997** CDP landslide election victory.
- ❑ **1999** Biggest gold mine closed. General strike.
- ❑ **2001** Meningitis epidemic. HIV infection rate second highest in West Africa.

HEALTH

No welfare state health benefits

1 per 20,000 people

Malaria, diarrheal and respiratory diseases

Health spending focuses on primary health care and vaccination. More than 6% of adults were HIV positive in 2001.

SPENDING

GDP/cap. increase

CONSUMPTION AND SPENDING

4 per 1000 population

4 per 1000 population

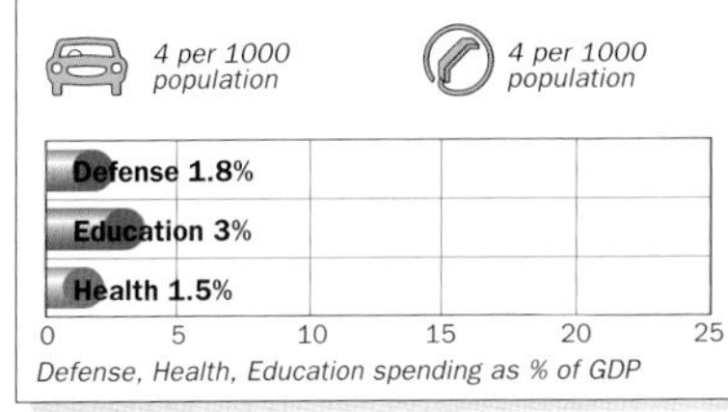
Defense 1.8%
Education 3%
Health 1.5%
0
5
10
15
20
25
Defense, Health, Education spending as % of GDP

Burkina is a country of extreme, almost universal, poverty. Displays of wealth are rare and ownership of high-tech items is limited to a small elite.

WORLD RANKING

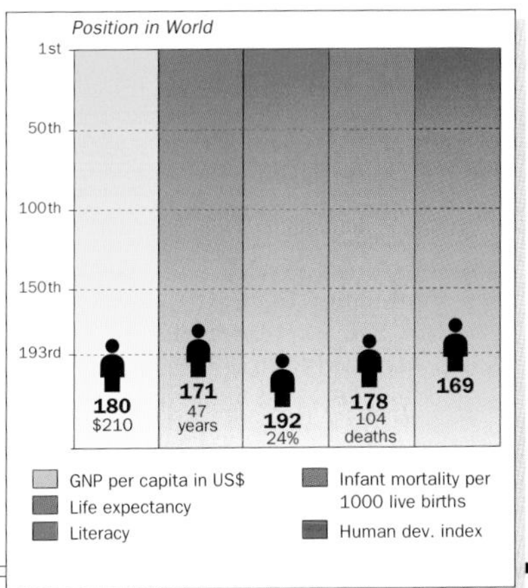
Position in World
1st
50th
100th
150th
193rd
180
$210
171
47 years
192
24%
178
104 deaths
169
GNP per capita in US$
Life expectancy
Literacy
Infant mortality per 1000 live births
Human dev. index

B

BURMA (MYANMAR)

SOUTHEAST ASIA
Bangladesh
China
India
BURMA (MYANMAR)
Laos
Thailand
Asia

OFFICIAL NAME: Union of Myanmar **CAPITAL:** Rangoon (Yangon)
POPULATION: 48.4 million **CURRENCY:** Kyat **OFFICIAL LANGUAGE:** Burmese (Myanmar)

1948 | 1948 | Jan 4 | 
BUR
| +6.5 | +95 | .mm

A PREDOMINANTLY BUDDHIST country on the northeastern shores of the Indian Ocean, Burma is mountainous in the north and east, while the fertile Irrawaddy basin occupies most of the country. Rocked by ethnic conflict ever since gaining independence from the UK in 1948, Burma has been ruled by repressive military regimes since 1962. The National League for Democracy (NLD) gained a majority in free elections in 1990, but has been prevented from taking power by the military. Rich in natural resources, which include fisheries and teak forests, Burma remains a mostly agricultural economy.

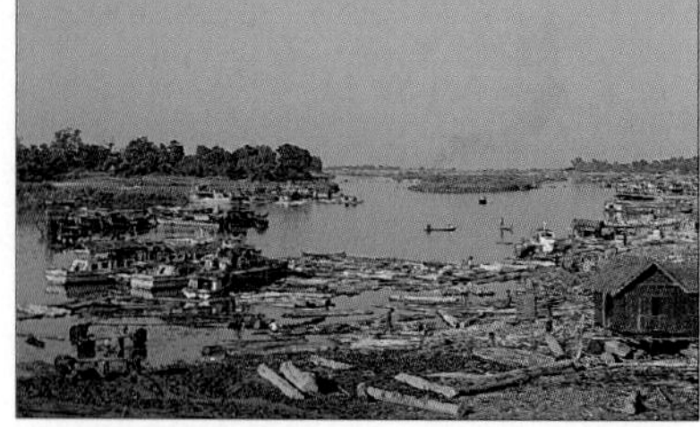

***Transporting timber** on the Irrawaddy River near Mandalay. Burma once had the world's largest reserves of teak.*

CLIMATE

Tropical/mountain

WEATHER CHART FOR RANGOON

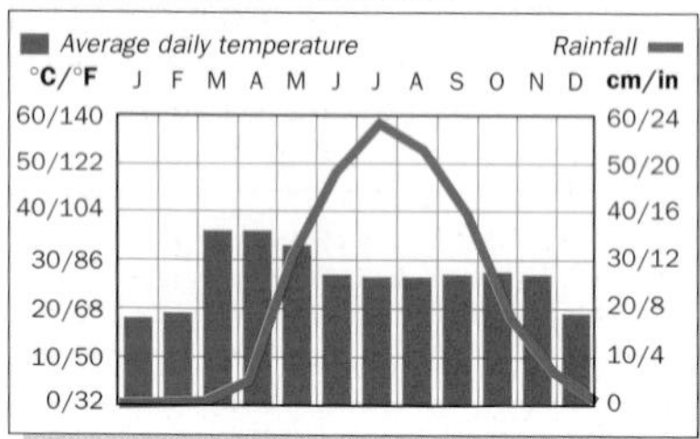
Average daily temperature
Rainfall
°C/°F
J F M A M J J A S O N D
cm/in
60/140
50/122
40/104
30/86
20/68
10/50
0/32
60/24
50/20
40/16
30/12
20/8
10/4
0

There are three seasons: the wet season, when rainfall in the far south Tenasserim region and Irrawaddy delta can reach 500 cm (197 in); summer, when northern Burma experiences 50°C (122°F) and 100% humidity; and winter, when it is rarely cooler than 15°C (59°F) except in the northern mountains.

TRANSPORTATION

Drive on right

Mingaladon, Rangoon
580,000 passengers

124 ships
492,306 grt

THE TRANSPORTATION NETWORK

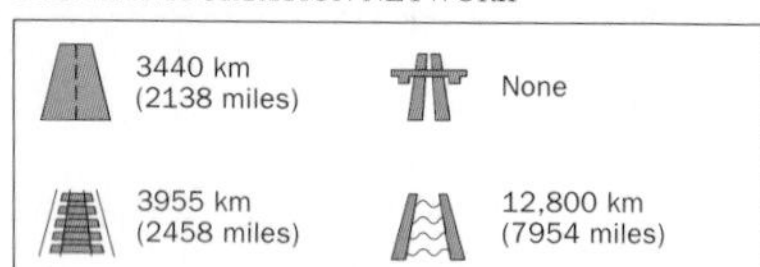
3440 km
(2138 miles)
None
3955 km
(2458 miles)
12,800 km
(7954 miles)

Burma's main transportation corridors run north–south. Most traffic is concentrated between Rangoon and Mandalay. A daily express train runs between these two cities, although visitors are urged to take internal flights on the state-owned carriers. The Irrawaddy River and its tributaries also provide an important artery for travel.

Beyond the Irrawaddy basin transportation is limited and hazardous. The vast majority of roads remain unpaved, and access to neighboring countries is restricted. At certain points on the Chinese border, access is one-way only: out of China.

TOURISM

Visitors : Population 1:233

208,000 visitors

Up 5% in 2000

MAIN TOURIST ARRIVALS

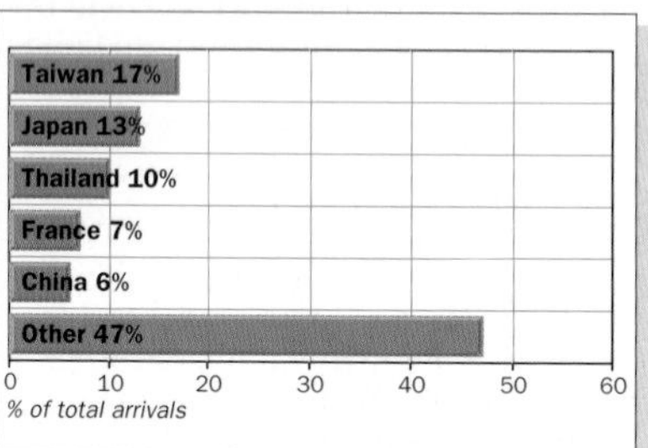
Taiwan 17%
Japan 13%
Thailand 10%
France 7%
China 6%
Other 47%
0
10
20
30
40
50
60
% of total arrivals

Since 1988 the military authorities have courted tourists for their economic value. The previous one-week restriction for visitors arriving to marvel at Burma's stunning Buddhist heritage was increased to 28 days. The state-run tourist agency is heavily promoted while independent travelers are required to spend a minimum of $200 during their stay. However, whole provinces remain entirely off-limits, and elsewhere official guides are often required. Opposition groups wish to discourage people from traveling to Burma at all.

PEOPLE

Pop. density medium

Burmese, Karen, Shan, Chin, Kachin, Mon, Palaung, Wa

74/km² (191/mi²)

THE URBAN/RURAL POPULATION SPLIT

28% 72%

RELIGIOUS PERSUASION

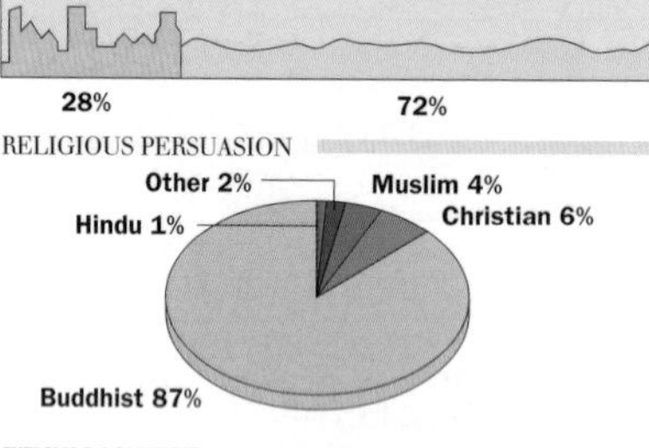
Other 2%
Muslim 4%
Hindu 1%
Christian 6%
Buddhist 87%

ETHNIC MAKEUP

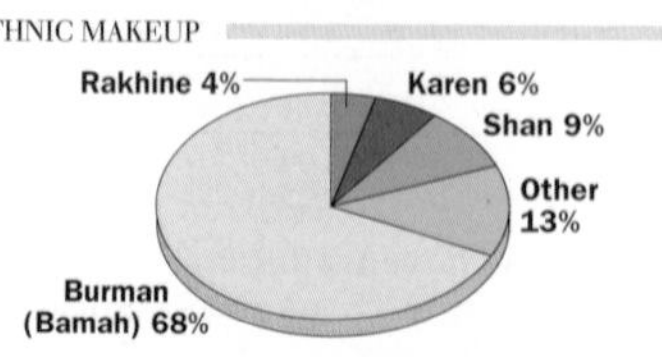
Rakhine 4%
Karen 6%
Shan 9%
Other 13%
Burman (Bamah) 68%

A savage history of ethnic repression at the hands of the majority Burman still plays a large part in the mistrust felt by the smaller minority communities. Each group maintains a distinct cultural identity. At independence the Chin, Kachin, Karen, Karenni, Mon, and Shan all unsuccessfully demanded their own state within a federation. Despite uniting against the military dictatorship in 1988, most factions had by 1996 signed peace agreements. Only the Shan and Karen remain militarily active to any significant degree, the former agreeing to a short-lived cease-fire in 2000. While the Burman claim racial purity, many of them are in fact of mixed blood or ethnically Chinese.

Accusations of forced labor lie at the heart of international criticism of the military regime, with ethnic minorities apparently at highest risk.

Domestic life in Burma is still based around the extended family. Women have a prominent role, and access to education. Many run or own businesses in their own right. However, top jobs in government are still held almost exclusively by men.

POPULATION AGE BREAKDOWN

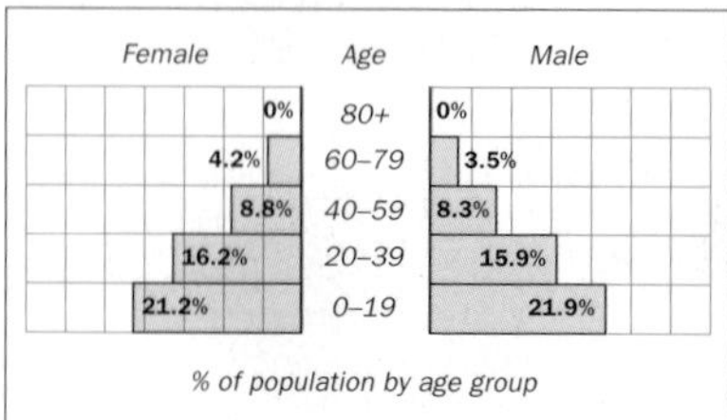
Female
Age
Male
0%
80+
0%
4.2%
60–79
3.5%
8.8%
40–59
8.3%
16.2%
20–39
15.9%
21.2%
0–19
21.9%
% of population by age group

POLITICS

No legislative elections

1990/suspended

Chairman Than Shwe

AT THE LAST ELECTION

Constituent Assembly 485 seats

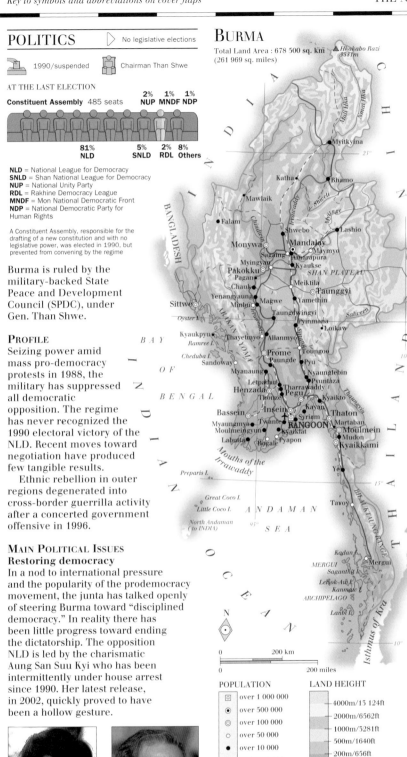

81%	5%	2%	8%	2%	1%	1%
NLD	SNLD	RDL	Others	NUP	MNDF	NDP

NLD = National League for Democracy
SNLD = Shan National League for Democracy
NUP = National Unity Party
RDL = Rakhine Democracy League
MNDF = Mon National Democratic Front
NDP = National Democratic Party for Human Rights

A Constituent Assembly, responsible for the drafting of a new constitution and with no legislative power, was elected in 1990, but prevented from convening by the regime

Burma is ruled by the military-backed State Peace and Development Council (SPDC), under Gen. Than Shwe.

PROFILE

Seizing power amid mass pro-democracy protests in 1988, the military has suppressed all democratic opposition. The regime has never recognized the 1990 electoral victory of the NLD. Recent moves toward negotiation have produced few tangible results.

Ethnic rebellion in outer regions degenerated into cross-border guerrilla activity after a concerted government offensive in 1996.

MAIN POLITICAL ISSUES

Restoring democracy

In a nod to international pressure and the popularity of the prodemocracy movement, the junta has talked openly of steering Burma toward "disciplined democracy." In reality there has been little progress toward ending the dictatorship. The opposition NLD is led by the charismatic Aung San Suu Kyi who has been intermittently under house arrest since 1990. Her latest release, in 2002, quickly proved to have been a hollow gesture.

Aung San Suu Kyi, figurehead of the pro-democracy movement.

Gen. Than Shwe, leader of the military junta since 1992.

BURMA

Total Land Area : 678 500 sq. km (261 969 sq. miles)

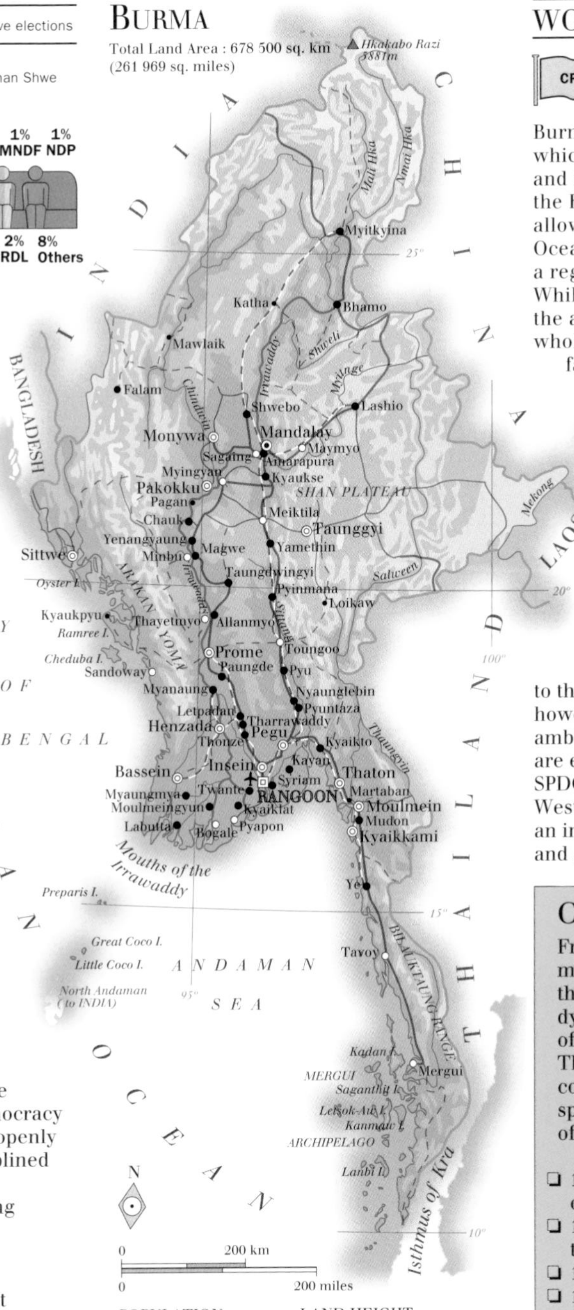

POPULATION: over 1 000 000; over 500 000; over 100 000; over 50 000; over 10 000; under 10 000

LAND HEIGHT: 4000m/13 124ft; 2000m/6562ft; 1000m/3281ft; 500m/1640ft; 200m/656ft; Sea Level

Forced labor

The practice of using slave labor in rural areas was officially banned in 2000. However, many groups, including the International Labor Organization, insist that the military authorities still regularly make use of "slave" laborers.

WORLD AFFAIRS

Joined UN in 1948

CP

IAEA

ASEAN

NAM

WTO

Burma's key relationship is with China, which backs the SPDC military regime and is a major supplier of weapons to the Burmese army. The relationship allows China access to the Indian Ocean and gives it influence over a regime dependent on its support. While Burma's neighbors fear that the arrangement could destabilize the whole of the Asia–Pacific region, many favor a policy of "constructive engagement" with the SPDC. In 1997, Burma was admitted to ASEAN, despite continuing concerns about its human rights record.

The EU and Western members of the UN have strongly condemned the human rights violations in Burma and threatened to impose economic sanctions in response to the regime's policies. In practice, however, the West maintains an ambivalent position. Economic ties are expanding, particularly between SPDC-owned state enterprises and Western multinationals which have an interest in Burmese offshore oil and gas drilling sectors.

CHRONOLOGY

From the 11th century, Burma's many ethnic groups came under the rule of three Tibeto-Burman dynasties, interspersed with periods of rule by the Mongols and the Mon. The Third Dynasty came into conflict with the British in India, sparking the Anglo-Burmese wars of 1824, 1852, and 1885.

- **1886** Burma becomes a province of British India.
- **1930–1931** Economic depression triggers unrest.
- **1937** Separation from India.
- **1942** Japan invades.
- **1945** Antifascist People's Freedom League (AFPFL), led by Aung San, helps Allies reoccupy country.
- **1947** UK agrees to Burmese independence. Aung San wins elections, but is assassinated.
- **1948** Independence under new prime minister, U Nu, who initiates socialist policies. Revolts by ethnic separatists, notably Karen liberation struggle.
- **1958** Ruling AFPFL splits into two. Shan liberation struggle begins.
- **1960** U Nu's faction wins elections.
- **1961** Kachin rebellion begins.

B

CHRONOLOGY *continued*

- **1962** Gen. Ne Win stages military coup. "New Order" policy of "Buddhist Socialism" deepens international isolation. Mining and other industries nationalized. Free trade prohibited.
- **1964** Socialist Program Party declared sole legal party.
- **1976** Social unrest. Attempted military coup. Ethnic liberation groups gain control of 40% of country.
- **1982** Nonindigenous people barred from public office.
- **1988** Thousands die in student riots. Ne Win resigns. Martial law. Aung San Suu Kyi, daughter of Aung San, and others form NLD. Gen. Saw Maung leads military coup. State Law and Order Restoration Council (SLORC) takes power. Ethnic resistance groups form Democratic Alliance of Burma.
- **1989** Army arrests NLD leaders and steps up antirebel activity. Officially renamed Union of Myanmar.
- **1990** Elections permitted. NLD wins landslide. SLORC remains in power, however. More NLD leaders arrested.
- **1991** Aung San Suu Kyi awarded Nobel Peace Prize.
- **1992** Gen. Than Shwe takes over as SLORC leader.
- **1996** Demonstrations against approval of Burma's membership of ASEAN.
- **1997** Ruling SLORC renamed State Peace and Development Council (SPDC). US imposes sanctions and bans further investment in Burma.
- **1998** NLD sets deadline for convening parliament; junta refuses.
- **1999** Aung San Suu Kyi rejects conditions set by SPDC for visiting the UK to see her husband, Michael Aris, who dies of cancer.
- **2000** Negotiations between junta and NLD begin.
- **2002** Aung San Suu Kyi released from house arrest.

AID

Recipient

 $107m (receipts)

 Up 32% in 2000

Moves in late 2000 to begin negotiations with the democratic opposition were welcomed by the UN, but were not enough to lift the sanctions on aid imposed in 1988. There is still some humanitarian assistance provided through agencies such as WHO. Japan has led the way in rewarding the regime's conciliatory gestures and has increased its contributions, making it the largest single donor.

DEFENSE

No compulsory military service

 $2.06bn

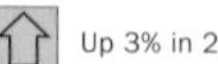 Up 3% in 2000

BURMESE ARMED FORCES

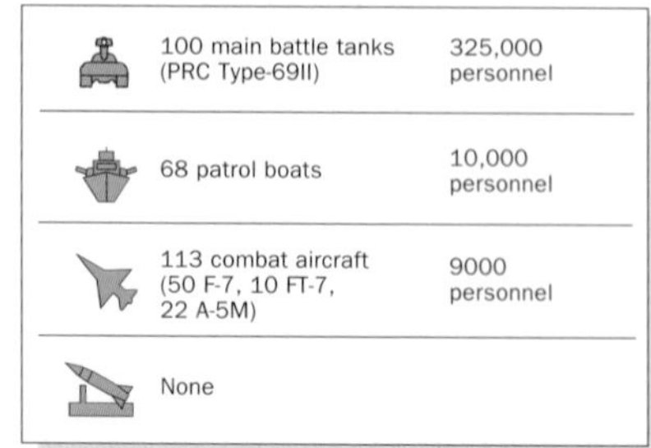

Army	100 main battle tanks (PRC Type-69II)	325,000 personnel
Navy	68 patrol boats	10,000 personnel
Air force	113 combat aircraft (50 F-7, 10 FT-7, 22 A-5M)	9000 personnel
Missiles	None	

The military authorities have steadily increased the country's military power, doubling the size of the army and obtaining modern weapons and military technology from around the world, primarily from China, which since 1989 has delivered arms worth over $1 billion to Burma, including tanks and jet fighters.

Burma's growing military capability is used mainly to control internal dissent, and the army has suppressed most ethnic insurgent campaigns by utilizing its military superiority and cutting numerous deals with rebel leaders. The remaining militant groups are now combated in cooperation with neighboring states. The army is accused of human rights abuses in rural areas.

ECONOMICS

Inflation 26% p.a. (1990–1999)

 $73.9bn

6.526–6.771 kyats

SCORE CARD

- World GNP Ranking..........43rd
- GNP per Capita..........$1500
- Balance of Payments..........–$293m
- Inflation..........–0.1%
- Unemployment..........7%

EXPORTS

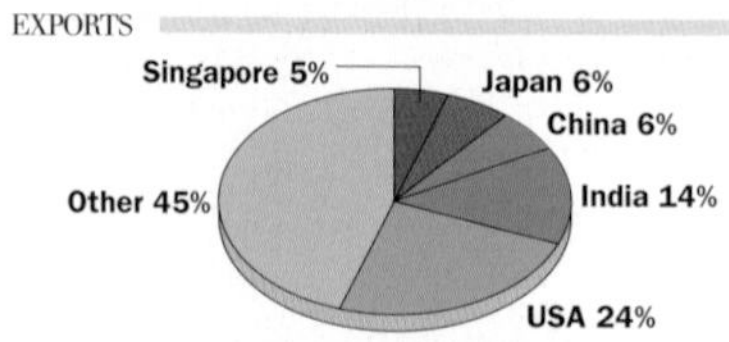

Singapore 5%
Japan 6%
China 6%
India 14%
USA 24%
Other 45%

IMPORTS

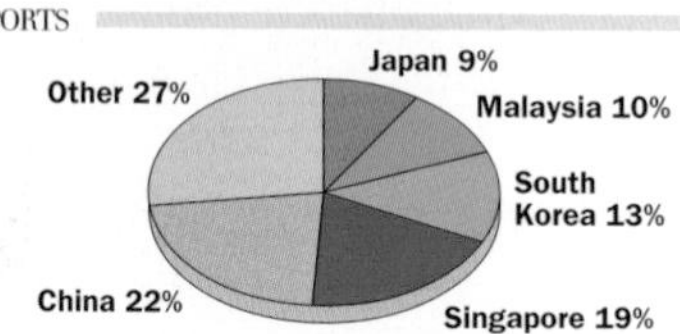

Japan 9%
Malaysia 10%
South Korea 13%
Singapore 19%
China 22%
Other 27%

Strengths

Very rich in natural resources: fertile soil, rich fisheries, timber including diminishing teak reserves, gems, offshore natural gas, and oil.

Weaknesses

Shortage of skilled labor, managers, and technicians. Rudimentary financial systems and institutions. Nationwide black market. Huge external debt. Dependence on imported manufactures.

Profile

Burma's economy is agriculture-based and functions mainly on a cash and barter system. Its key industries are controlled by military-run state enterprises. Every aspect of economic life is permeated by a black market, where prices are rocketing – a reaction to official price controls.

ECONOMIC PERFORMANCE INDICATOR

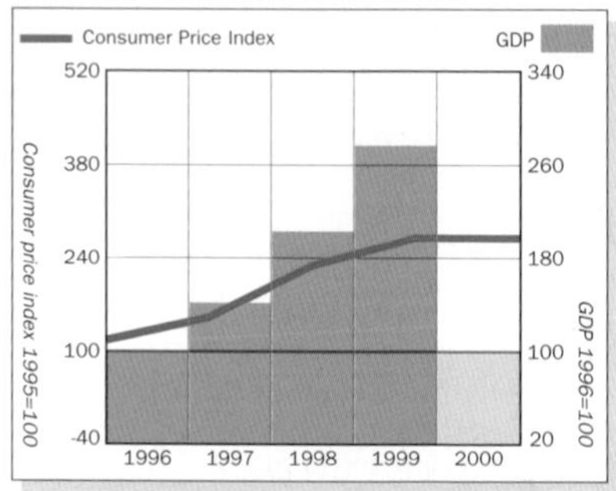

Consumer Price Index
GDP
520
380
240
100
-40
340
260
180
100
20
Consumer price index 1995=100
GDP 1996=100
1996
1997
1998
1999
2000

Since 1989, the SPDC's open-door market-economy policy has brought a flood of foreign investment in oil and gas (by Western companies), and in forestry, tourism, and mining (by Asian companies). The resulting boom in trade with China has turned less developed Upper Burma into a thriving business center. A narcotics-eradication program has been initiated in the northeastern border states, which account for about 60% of the world's heroin, by encouraging farmers to grow food crops instead of poppies. Few plans exist for the manufacturing sector, however, and dependence on imports continues.

BURMA : MAJOR BUSINESSES

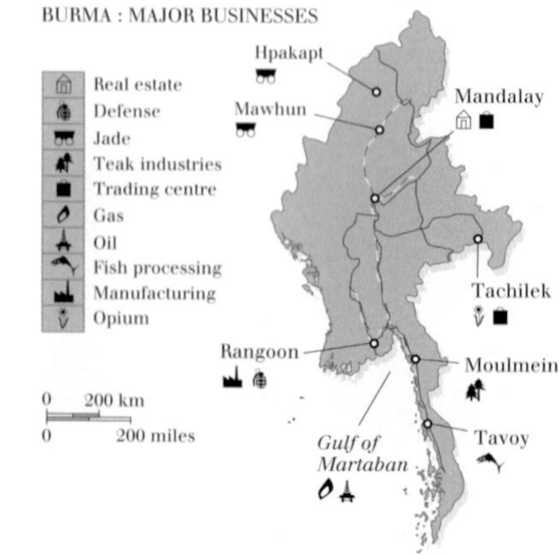

Real estate
Defense
Jade
Teak industries
Trading centre
Gas
Oil
Fish processing
Manufacturing
Opium
Hpakapt
Mawhun
Mandalay
Tachilek
Rangoon
Moulmein
Tavoy
Gulf of Martaban
0 200 km
0 200 miles

B

RESOURCES

Electric power 1.4m kw

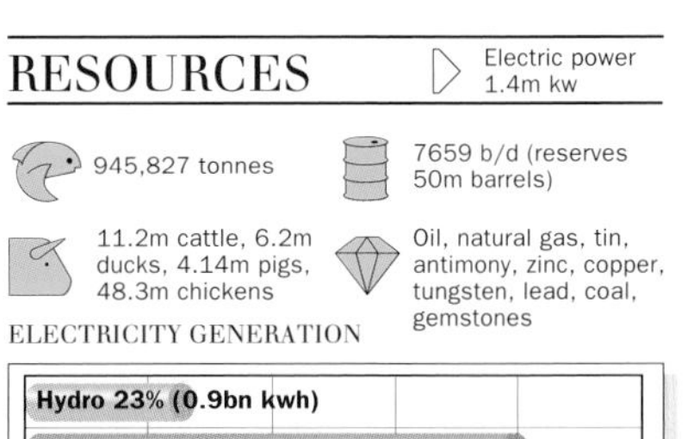
945,827 tonnes
7659 b/d (reserves 50m barrels)
11.2m cattle, 6.2m ducks, 4.14m pigs, 48.3m chickens
Oil, natural gas, tin, antimony, zinc, copper, tungsten, lead, coal, gemstones

ELECTRICITY GENERATION

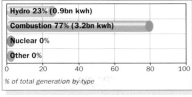

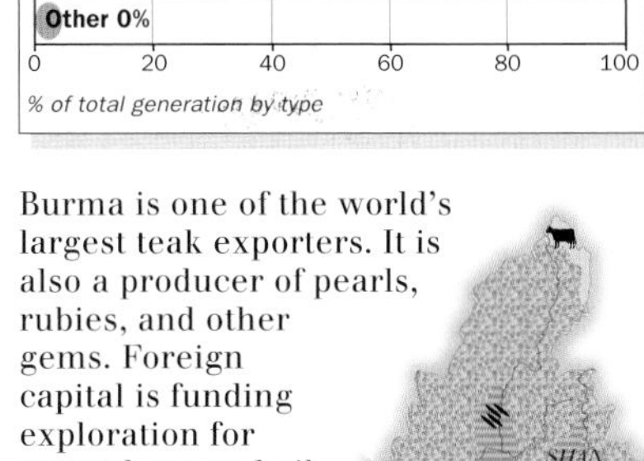
Hydro 23% (0.9bn kwh)
Combustion 77% (3.2bn kwh)
Nuclear 0%
Other 0%
0
20
40
60
80
100
% of total generation by type

Burma is one of the world's largest teak exporters. It is also a producer of pearls, rubies, and other gems. Foreign capital is funding exploration for natural gas and oil in the Tenasserim strip, while offshore fields are already in full production.

SHAN PLATEAU

BURMA : LAND USE

Cropland
Pasture
Forest
Rice
Cattle

0 200 km
0 200 miles

ENVIRONMENT

Sustainability rank: 90th

0.3%

0.2 tonnes per capita

ENVIRONMENTAL TREATIES

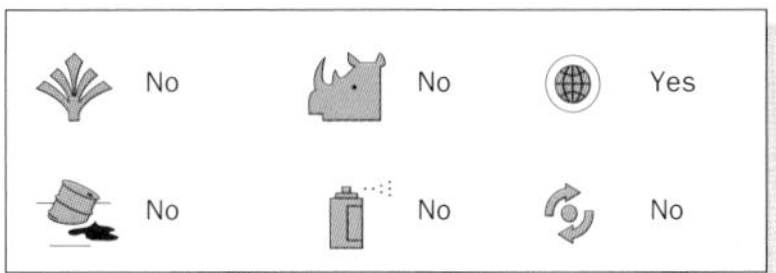
No
No
Yes
No
No
No

Deforestation is a major problem, and it has increased since the 1988 coup. Chinese companies have been given unrestricted logging concessions.

MEDIA

TV ownership low

Daily newspaper circulation 10 per 1000 people

PUBLISHING AND BROADCAST MEDIA

There are 5 daily newspapers, including *Myanma Alin* and *New Light of Myanmar*

1 state-controlled service

1 state-controlled service

Political dissent of any kind is a criminal offense. An underground prodemocracy press produces antigovernment material.

CRIME

Death penalty in use

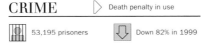

53,195 prisoners

Down 82% in 1999

CRIME RATES

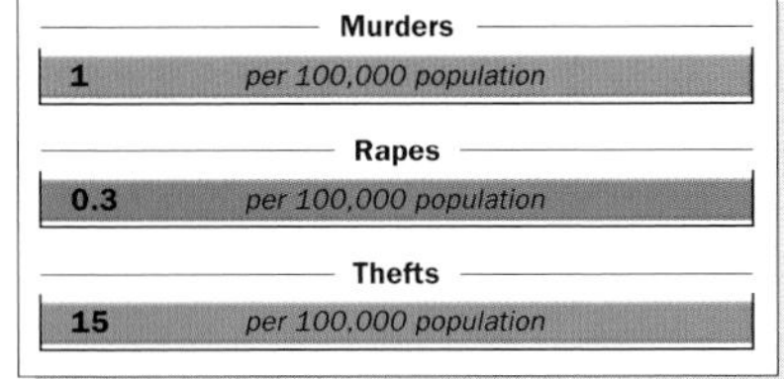
Murders
1
per 100,000 population
Rapes
0.3
per 100,000 population
Thefts
15
per 100,000 population

Compared with similar totalitarian regimes, levels of bribery, corruption, embezzlement, and black marketeering are high. The state is guilty of illegal activity. The UN reports regularly on human rights abuses against civilians, and the murder of innocent civilians including children, women, Buddhist monks, students, minorities, and political dissidents.

There is a nominal civilian judicial system in Burma, but in practice all judges and lawyers are appointed by the junta and all legal functions are executed by the SPDC. The most common charge is that of sedition against the state or the army under the 1975 "Law to Protect the State from Destructionists." Among the SPDC's frequent arbitrary "notices" is Order 2/88, prohibiting assemblies of more than five persons. Most detainees have no legal rights of representation and are either jailed, used as forced labor, or put under house arrest without public trial. Amnesty International is banned.

EDUCATION

School leaving age: 10

85%

385,300 students

THE EDUCATION SYSTEM

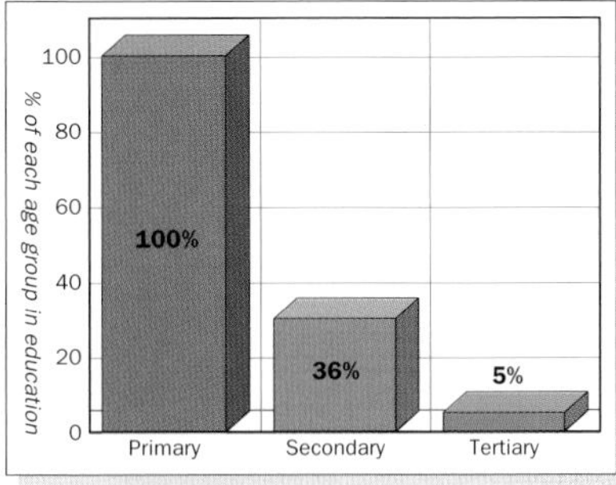
100
80
60
40
20
0
% of each age group in education
100%
36%
5%
Primary
Secondary
Tertiary

The education system offers ten years of schooling. Ethnic-language schools are discouraged. A shortage of teachers, many of whom have left or are in jail, has disrupted education. All but two universities were closed in the late 1990s by the regime, but quietly reopened in 2000. The NLD has criticized the shortened and "sanitized" courses on offer.

HEALTH

Welfare state health benefits

1 per 3333 people

Malaria, fevers, heart and diarrheal diseases

Leprosy, although it affects relatively few people compared with other diseases, has a higher prevalence in Burma than in the rest of Asia. In the last few years Burma has seen an increase in the incidence of malaria.

The growing number of AIDS cases is largely due to migrant prostitution across the Thai–Burmese border. They put an additional strain on health facilities, which are well developed but not comprehensive.

SPENDING

GDP/cap. increase

CONSUMPTION AND SPENDING

1 per 1000 population

6 per 1000 population

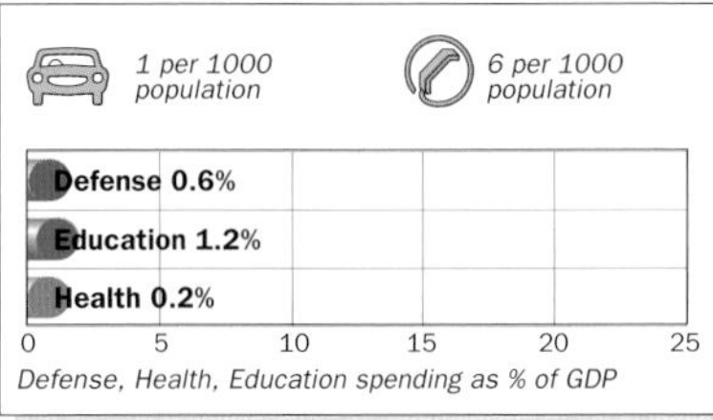
Defense 0.6%
Education 1.2%
Health 0.2%
0
5
10
15
20
25
Defense, Health, Education spending as % of GDP

The state monopoly of the production and distribution of goods by rationing under Gen. Ne Win's administration led to an increase in corruption and the rise of a nationwide black market, with huge disparities between official and unofficial prices. Only the military elite and their supporters could afford to live well. The situation has not changed significantly since 1988. Giant military enterprises grouped under a Defense Services holding company, whose capital amounts to 10% of GDP, now reap wealth and distribute privileges for a minority. Nevertheless, traditional social and economic mobility still exists. Climbing the socioeconomic ladder is mainly a matter of loyalty to the military. Dissidents forced out of their jobs and hill tribes form the poorest groups.

WORLD RANKING

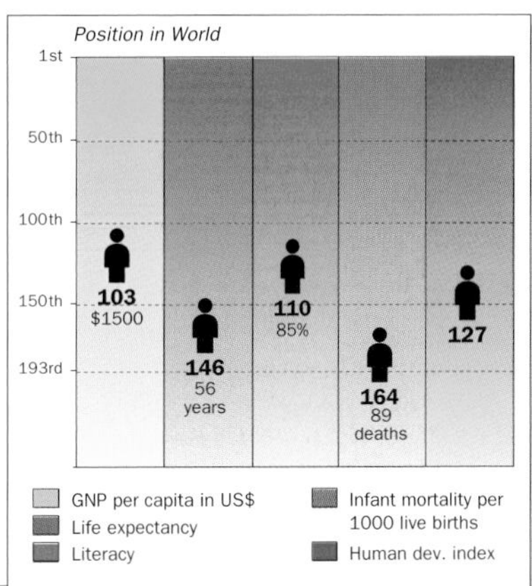
Position in World
1st
50th
100th
150th
193rd
103
$1500
146
56 years
110
85%
164
89 deaths
127
GNP per capita in US$
Life expectancy
Literacy
Infant mortality per 1000 live births
Human dev. index

BURUNDI

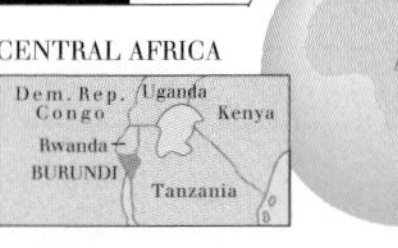
CENTRAL AFRICA
Dem. Rep. Congo
Uganda
Kenya
Rwanda
BURUNDI
Tanzania
Africa

B

OFFICIAL NAME: Republic of Burundi **CAPITAL:** Bujumbura
POPULATION: 6.5 million **CURRENCY:** Burundi franc **OFFICIAL LANGUAGES:** French and Kirundi

1962

1962

July 1
RU
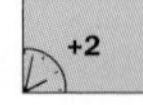
+2

+257

.bi

LANDLOCKED BURUNDI lies just south of the equator on the Nile–Congo watershed. Lake Tanganyika forms part of its border with the Democratic Republic of the Congo (DRC). Tension between the Hutu majority and the dominant Tutsi minority remains the main factor in politics. The current political unrest dates from October 1993, when the assassination of the first Hutu president in a coup by the Tutsi-dominated army sparked terrible violence.

***Pig farming and fish ponds.** The majority of Burundi's population depends on subsistence farming.*

CLIMATE

Tropical wet & dry

WEATHER CHART FOR BUJUMBURA

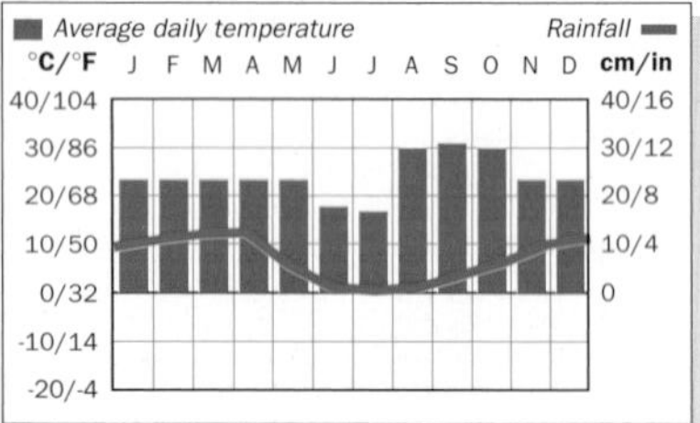
Average daily temperature
Rainfall
°C/°F J F M A M J J A S O N D cm/in
40/104
30/86
20/68
10/50
0/32
-10/14
-20/-4
40/16
30/12
20/8
10/4
0

Burundi is temperate with high humidity, much cloud, and frequent heavy rain. The highlands have frost.

TRANSPORTATION

Drive on right

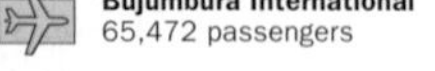
Bujumbura International 65,472 passengers

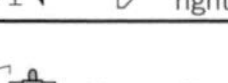
Has no fleet

THE TRANSPORTATION NETWORK

1028 km (639 miles)

None

None

Lake Tanganyika is navigable

The dense road network has been rehabilitated. There are plans for a railroad to link Burundi with Rwanda, Uganda, and Tanzania.

TOURISM

Visitors : Population 1:217

30,000 visitors

Up 15% in 2000

MAIN TOURIST ARRIVALS

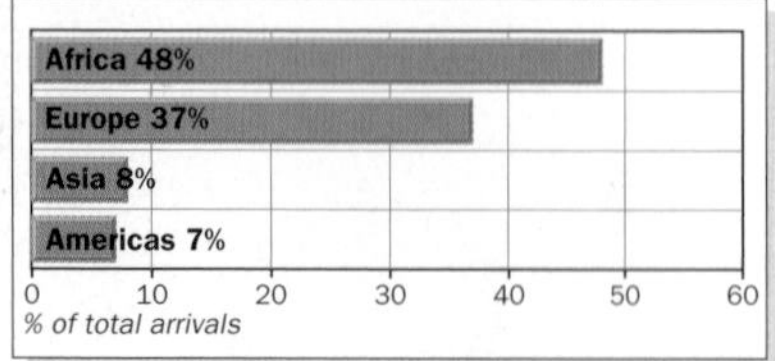
Africa 48%
Europe 37%
Asia 8%
Americas 7%
0 10 20 30 40 50 60
% of total arrivals

A lack of basic infrastructure and violent political strife have deterred tourists. The industry has limited potential, since Burundi lacks its neighbors' spectacular scenery and game parks.

PEOPLE

Pop. density high

Kirundi, French, Kiswahili

253/km² (656/mi²)

THE URBAN/RURAL POPULATION SPLIT

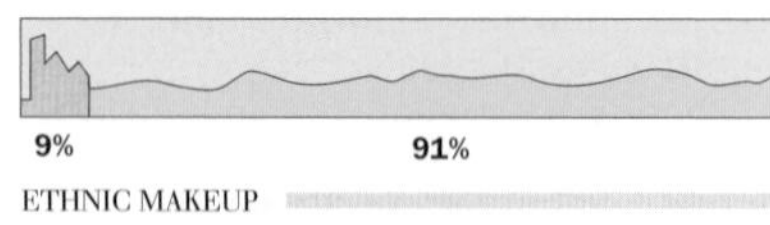
9%
91%

ETHNIC MAKEUP

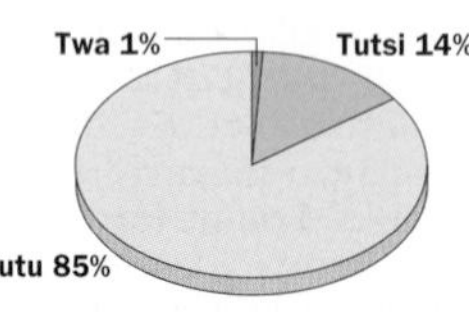
Twa 1%
Tutsi 14%
Hutu 85%

Burundi's history has been marked by violent conflict between the majority Hutu and the Tutsi, formerly the political elite, who still control the army. Large-scale massacres have occurred repeatedly over the past two decades. Hundreds of thousands of people, mostly Hutu, have been killed in political and ethnic conflict since 1993. The Twa pygmy minority has not been greatly affected.

Most Burundians are subsistence farmers, and the vast majority are Roman Catholic.

POLITICS

In transition

Both houses 2002/Transitional

President Pierre Buyoya

AT THE LAST ELECTION

Transitional National Assembly 170 seats

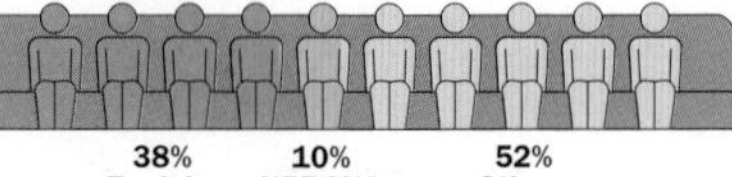
38% Frodebu
10% UPRONA
52% Others

Frodebu = Front for Democracy in Burundi
UPRONA = Union for National Progress

The Transitional National Assembly consists of members of all parties that signed the Arusha peace accord, and members of "civil society."

Transitional Senate 51 seats

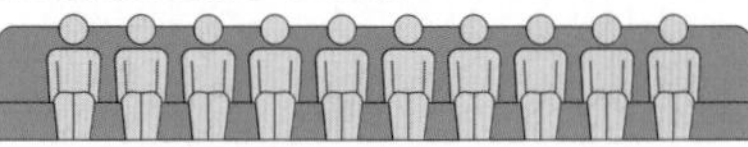

The indirectly elected Transitional Senate comprises 24 Hutus, 24 Tutsis, and 3 Twas, as formulated under the 2000 Arusha peace accord.

From 1966, Tutsi dominated all areas of life. An attempt by Tutsi president Buyoya to promote integration backfired in a bloody coup in 1993, when the first Hutu president, Melchior Ndadaye of Frodebu, was assassinated soon after his election and hundreds of thousands of Hutu were killed or fled; 1994 saw vicious civil war begin. Buyoya was returned to power in 1996 by a military coup, but despite his efforts to regain international acceptance, a regional economic boycott caused serious damage. A peace accord was forged in 2000, and in 2001 a joint Hutu–Tutsi transitional government was installed, but fighting continued, with the Hutu National Liberation Front refusing to sign a cease-fire or join the government.

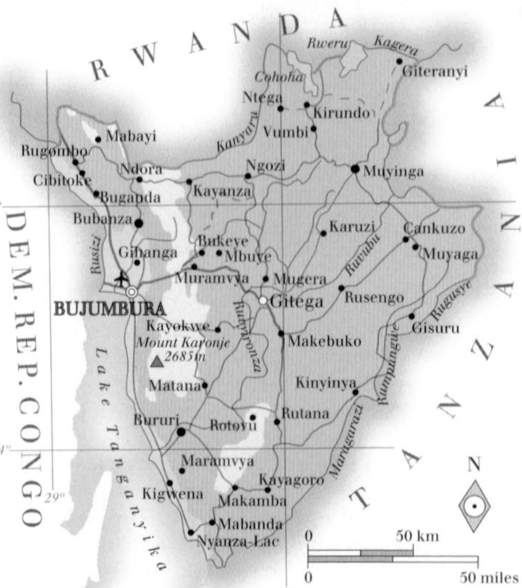
RWANDA
DEM. REP. CONGO
TANZANIA
Lake Tanganyika
Rweru
Kagera
Giteranyi
Cohoha
Ntega
Kirundo
Vumbi
Mabayi
Rugombo
Cibitoke
Ndora
Ngozi
Muyinga
Kayanza
Bugarama
Bubanza
Karuzi
Cankuzo
Muyaga
Gihanga
Bukeye
Mbuye
Muramvya
Mugera
Rusengo
BUJUMBURA
Gitega
Kayokwe
Mount Karonje 2685m
Makebuko
Gisuru
Matana
Kinyinya
Bururi
Rotovu
Rutana
Maramvya
Kayagoro
Kigwena
Makamba
Mabanda
Nyanza-Lac
Ruvubu
Rumpungwe
Malagarazi
0 50 km
0 50 miles
29°
30°
3°
4°

BURUNDI

Total Land Area : 27 830 sq. km (10 745 sq. miles)

LAND HEIGHT
2000m/6562ft
1000m/3281ft
500m/1640ft

POPULATION
over 100 000
over 50 000
over 10 000
under 10 000

WORLD AFFAIRS

Joined UN in 1962

ACP

CEPGL

COMESA

OIF

AU

Since 1995 Burundi has resisted proposals for international intervention to prevent further bloodshed.

AID

Recipient

 $93m (receipts)

 Up 26% in 2000

The flight of hundreds of thousands of people since 1993 has disrupted agriculture, and many people remain dependent on UN food aid.

DEFENSE

No compulsory military service

 $65m

 Down 6% in 2000

The 40,000-strong army is run by Tutsi. The attempt to bring Hutu into officer ranks was a cause of the 1993 coup. A state of virtual civil war now exists between it and rebel Hutu militias. A UNICEF program to demobilize over 14,000 child soldiers from these militias was announced in 2001.

ECONOMICS

Inflation 12% p.a. (1990–2000)

 $732m

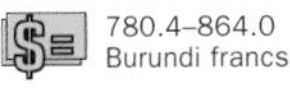 780.4–864.0 Burundi francs

SCORE CARD

- World GNP Ranking........161st
- GNP per Capita........$110
- Balance of Payments........–$49m
- Inflation........24.3%
- Unemployment.......Widespread underemployment

Strengths

Small quantities of gold and tungsten. Potential of massive nickel reserves and oil in Lake Tanganyika.

Weaknesses

Harsh regional sanctions since 1996 coup. Overwhelmingly agricultural economy under pressure from high birthrate and war damage. Little prospect of lasting political stability.

EXPORTS

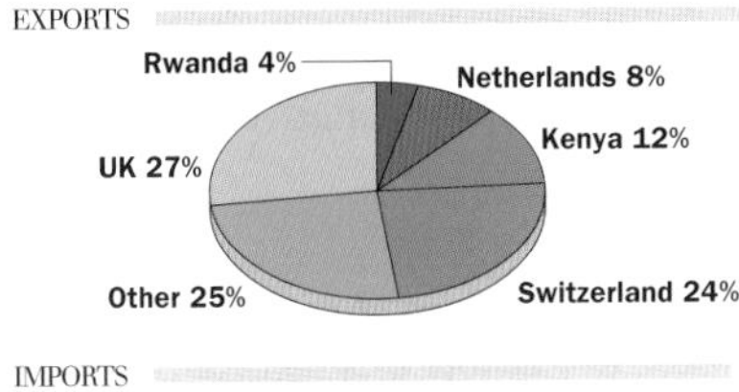
Rwanda 4%
Netherlands 8%
Kenya 12%
UK 27%
Switzerland 24%
Other 25%

IMPORTS

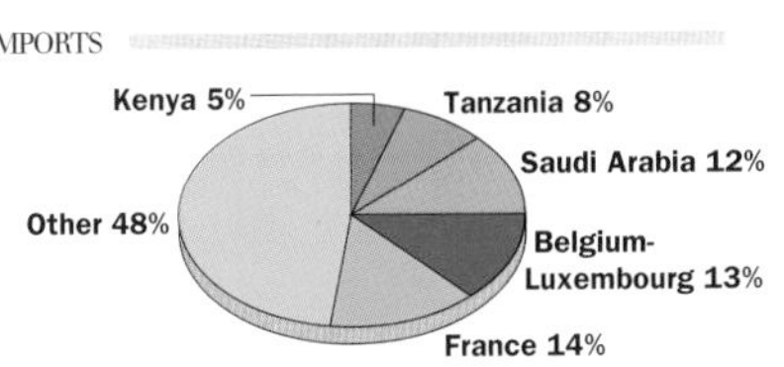
Kenya 5%
Tanzania 8%
Saudi Arabia 12%
Other 48%
Belgium-Luxembourg 13%
France 14%

RESOURCES

Electric power 43,000 kw

 9254 tonnes

 Not an oil producer and has no refineries

 600,000 goats, 315,000 cattle, 4.7m chickens

 Gold, tungsten, nickel, vanadium, uranium, oil

Burundi has around 5% of the world's nickel reserves. Extraction, however, is not economically viable. There are also deposits of gold and vanadium. Surveys in the 1980s detected oil reserves below Lake Tanganyika, but production has yet to begin. Burundi imports gasoline from Iran and electricity from the DRC. New HEP plants at Mugera and Rwegura, in the far north, are intended to meet most domestic electricity requirements.

ENVIRONMENT

Sustainability rank: 115th

6%

 0.03 tonnes per capita

Only 2% of Burundi is forest and even this is now under pressure from one of Africa's highest birthrates. Burundi suffers from the problems associated with deforestation, particularly soil erosion. Some soils are also being exhausted from overuse. Several tree-planting programs have been introduced. UNESCO is also running ecological education initiatives at village level, aimed at women farmers.

MEDIA

TV ownership low

Daily newspaper circulation 3 per 1000 people

PUBLISHING AND BROADCAST MEDIA

 *Le Renouveau du Burundi* is published three times a week by the government

1 state-controlled service

1 state-controlled service, some independent stations

Pro-Hutu/anti-Tutsi radio stations have been broadcasting since 1994. Radio Umwizero, an EU-funded station promoting peace, was launched in 1996.

CRIME

Death penalty in use

 9411 prisoners

 Up 78% 1990–1998

Burundi has an appalling human rights record. There have been frequent massacres of Hutu by the army. The worst pogroms occurred in 1972, 1988, 1993, and 1994.

EDUCATION

School leaving age: 12

 48%

 5037 students

Elementary schooling begins at seven, and is compulsory, though further schooling is not. There are around 70 elementary schoolchildren per teacher. There is one university.

CHRONOLOGY

From the 16th century, Burundi (formerly Urundi) was ruled by the minority Tutsi with the majority Hutu as their serfs. Merged with Rwanda, Burundi was controlled by Germany from 1884 and by Belgium from 1919.

- **1946** UN trust territory.
- **1959** Split from Rwanda.
- **1962** Independence.
- **1966** Army overthrows monarchy.
- **1972** 150,000 Hutu massacred.
- **1993** Ndadaye wins first free elections; killed four months later.
- **1996** Pierre Buyoya retakes power.
- **1999** Talks between warring groups.
- **2000** Renewed violence. Arusha peace accord signed by most groups.
- **2001** Multiethnic transitional government appointed.

B

HEALTH

No welfare state health benefits

1 per 10,000 people

 Communicable infections, parasitic diseases

Over half of the population of Burundi are underfed. Over 30% of the people do not have access to health services, and, as at end-2001, 390,000 people were living with HIV/AIDS.

SPENDING

GDP/cap. decrease

CONSUMPTION AND SPENDING

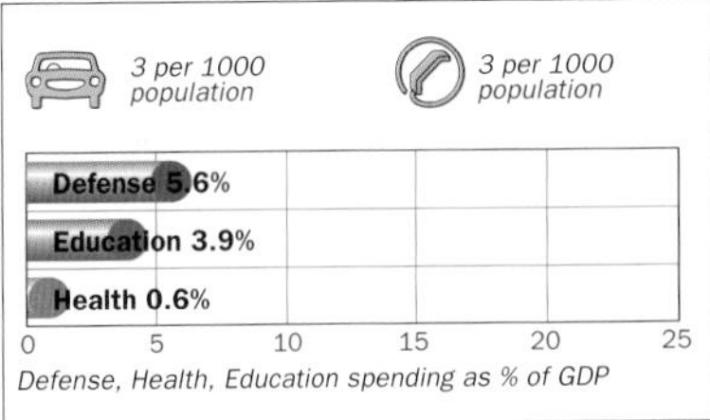
3 per 1000 population
3 per 1000 population
Defense 5.6%
Education 3.9%
Health 0.6%
0
5
10
15
20
25
Defense, Health, Education spending as % of GDP

Wealth is concentrated within the Tutsi political and business elite. Most of Burundi's people live at the level of subsistence farming.

WORLD RANKING

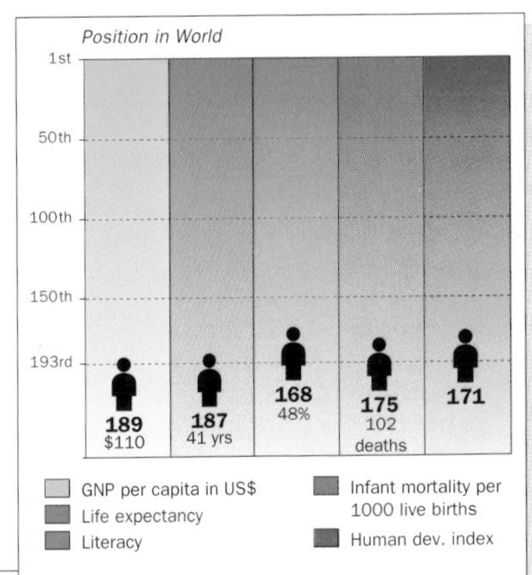
Position in World
1st
50th
100th
150th
193rd
189
$110
187
41 yrs
168
48%
175
102 deaths
171
GNP per capita in US$
Life expectancy
Literacy
Infant mortality per 1000 live births
Human dev. index

CAMBODIA

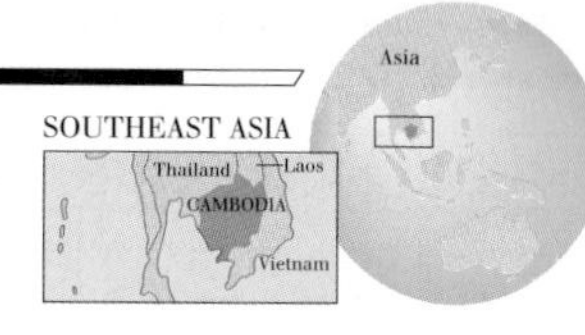
Asia
SOUTHEAST ASIA
Thailand
Laos
CAMBODIA
Vietnam

OFFICIAL NAME: Kingdom of Cambodia **CAPITAL:** Phnom Penh
POPULATION: 13.4 million **CURRENCY:** Riel **OFFICIAL LANGUAGE:** Khmer

1953

1953

Nov 11

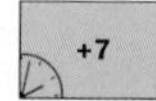
K

+7

+855

.kh

THE ANCIENT KINGDOM of Cambodia emerged from French colonial rule in 1953, only to be plunged into violent civil conflict. Under the extremist Khmer Rouge, headed by the infamous Pol Pot, the country endured one of the world's most brutal totalitarian regimes. Since the withdrawal of Vietnamese troops in 1989 the country has gradually returned to relative stability. The dominating geographic feature is the Tônlé Sap, or Great Lake, which drains into the Mekong River. Over three-quarters of Cambodia is forested, with mangroves lining the coast.

CLIMATE

Tropical monsoon

WEATHER CHART FOR PHNOM PENH

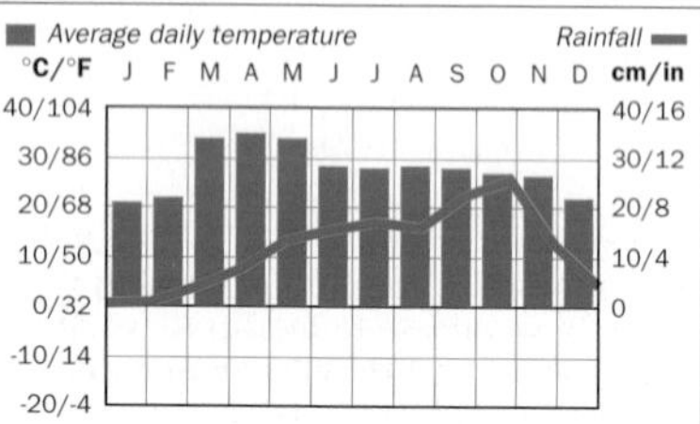
Average daily temperature
Rainfall
°C/°F
J F M A M J J A S O N D
cm/in
40/104
30/86
20/68
10/50
0/32
-10/14
-20/-4
40/16
30/12
20/8
10/4
0

Cambodia has a varied climate. Low-lying regions have moderate rainfall and the most consistent year-round temperatures. The dry season from December to April is characterized by high temperatures and an average of eight hours of sunshine a day. From May to September, winds are southeasterly, while from October to April they are north or northeasterly. During the rainy season, Cambodia is sultry and humid. The monsoons in 2000 caused severe flooding of the Mekong River, which inundated Phnom Penh.

TRANSPORTATION

Drive on right

Pochentong, Phnom Penh
860,613 passengers

405 ships
1.45m grt

THE TRANSPORTATION NETWORK

4165 km (2588 miles)

None

650 km (404 miles)

3700 km (2299 miles)

Years of war led to a near-collapse of Cambodia's rail and road systems. Although international aid has helped fund key projects, the state of the country's roads remains appalling. The government has pledged to make reconstruction a priority. The Mekong River is vital for accessing the interior. Taxi-mopeds, bicycles, and rickshaws dominate urban transportation.

Angkor Wat stands in the ruins of the ancient city of Angkor, once the capital of the Khmer empire. It is now one of Cambodia's leading tourist attractions.

TOURISM

Visitors : Population 1:51

265,000 visitors

Up 1% in 2000

MAIN TOURIST ARRIVALS

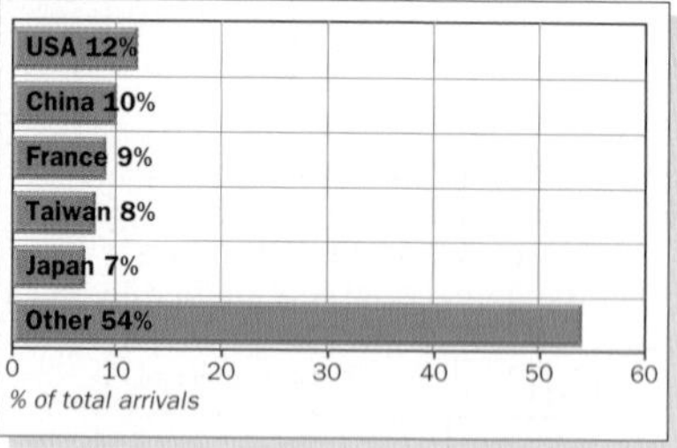
USA 12%
China 10%
France 9%
Taiwan 8%
Japan 7%
Other 54%
0
10
20
30
40
50
60
% of total arrivals

Cambodia, the center of the Khmer empire between 800 and 1400 C.E., has some of the most impressive temples in southeast Asia. The most famous is Angkor Wat, near Siem Reap (Siemreab), which is now largely safe for tourists after the Khmer Rouge relinquished control of the area in 1998. Kidnappings and murders of tourists by the Khmer Rouge kept Cambodia off the backpacker circuit in the mid-1990s. Once the political situation is fully stabilized and land mines have been cleared, there is considerable potential, not just for adventurous independent travelers.

PEOPLE

Pop. density medium

Khmer, French, Chinese, Vietnamese, Cham

76/km² (197/mi²)

THE URBAN/RURAL POPULATION SPLIT

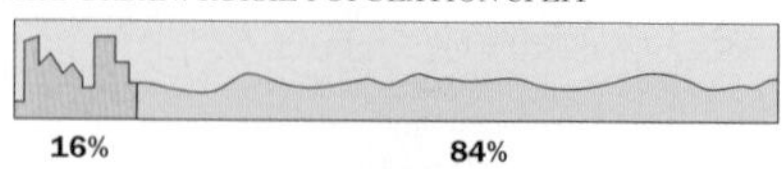
16%
84%

RELIGIOUS PERSUASION

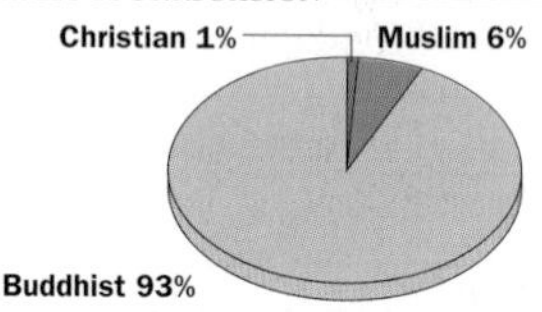
Christian 1%
Muslim 6%
Buddhist 93%

ETHNIC MAKEUP

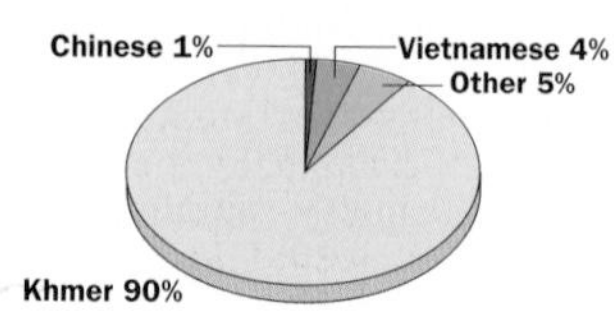
Chinese 1%
Vietnamese 4%
Other 5%
Khmer 90%

Cambodia underwent one of the 20th century's most horrific experiments in social transformation between 1975 and 1979 under Pol Pot's Khmer Rouge regime. Warfare, starvation, exhaustion, or execution killed one in eight of the population. Half a million more fled to Thailand. The Pol Pot regime's extreme radical beliefs led to the scrapping of money, possessions, and hierarchy. "Bourgeois" learning was despised, whereas peasants, soldiers of the revolution, and some industrial workers were officially given higher status. Boys and girls of 13 and 14 were taken from their homes, indoctrinated in the tenets of revolution, and allowed to kill those held guilty of bourgeois crimes. Violence at all levels was sanctioned in the name of revolution. The legacies of the regime are both the emigration of surviving professionals, and one of the world's highest rates of orphans and widows.

Religious and ethnic tensions are minimal, although there is a traditional hostility in Khmer culture toward ethnic Vietnamese.

POPULATION AGE BREAKDOWN

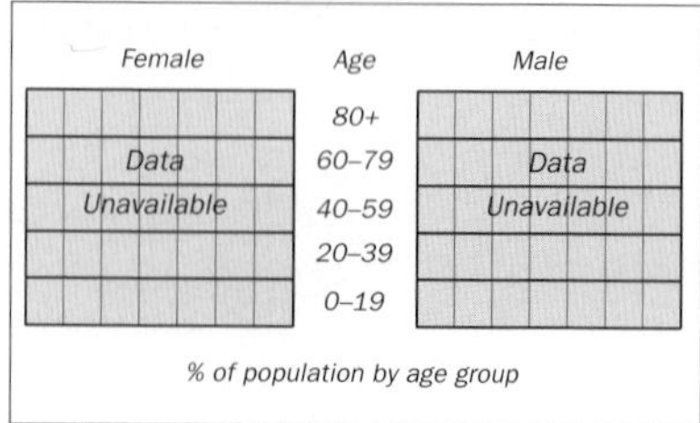
Female
Age
Male
80+
60–79
40–59
20–39
0–19
Data Unavailable
Data Unavailable
% of population by age group

C

POLITICS

Multiparty elections

L. House 1998/2003
U. House 1999/2004

H.M. King Norodom Sihanouk

AT THE LAST ELECTION

National Assembly 122 seats

53% CPP | 35% Funcinpec | 12% SRP

CPP = Cambodian People's Party
Funcinpec = United National Front for an Independent Neutral Peaceful and Cooperative Cambodia
SRP = Sam Rainsy Party

Senate 61 seats

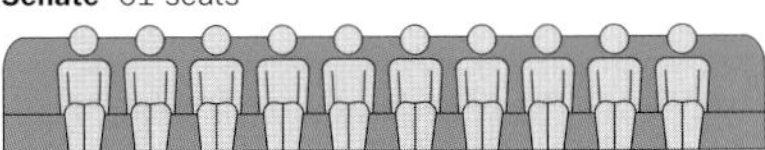

The membership of the Senate, first established in March 1999, is determined in proportion to the results of the 1998 legislative elections

Cambodia is a constitutional monarchy.

PROFILE

In 1975, a US-installed government was overthrown by the Maoist Khmer Rouge under Pol Pot. That extremist and murderous regime was ousted in 1979 thanks to a Vietnamese invasion. The Khmer Rouge then joined a Western-backed anti-Vietnamese exile coalition with the supporters of the then Prince Sihanouk and the Khmer People's National Liberation Front (KPNLF), gaining UN recognition against the Vietnam-backed regime in Phnom Penh. In 1989 Vietnam withdrew its forces, paving the way for UN-supervised elections in 1993. The royalist Funcinpec emerged as the main winners and King Sihanouk formed a coalition government of national reconciliation, but the Khmer Rouge remained outside this coalition and resumed armed resistance until its surrender in 1998. The strife-torn coalition meanwhile had degenerated into open hostility in 1997, when Hun Sen of the communist CPP ousted his co-prime minister Prince Ranariddh. A year later the CPP failed to win elections outright, forcing it to seek a new coalition agreement with Funcinpec.

MAIN POLITICAL ISSUES

Settling accounts with Khmer Rouge

The Khmer Rouge, which had resumed its armed struggle in 1993, surrendered in 1998. Mass defections and the death of Pol Pot earlier that year had weakened the group. The legislature in 2001 approved plans for a tribunal to try the surviving leaders for crimes against humanity.

Royalist–CPP rivalry

Power struggles between the Funcinpec and the CPP came to a head with CPP leader Hun Sen's mid-1997 coup. However, inconclusive elections in July 1998 led the two parties to re-form their uneasy coalition in November. Against expectations, that compromise proved a durable one.

***Prime Minister Hun Sen,** who ousted his co-prime minister in 1997.*

***King Norodom Sihanouk,** the pivotal figure in Cambodian society and politics.*

WORLD AFFAIRS

Joined UN in 1955

ASEAN | IAEA

OIF

Mekong River

NAM

During the civil war that followed the Vietnamese invasion of 1979, the Phnom Penh government (along with Vietnam) was reduced to an international pariah. It was recognized by few countries outside the Soviet bloc, and its seat at the UN was allotted to the exiled resistance coalition, despite one of its components being the Khmer Rouge, which had inflicted appalling violence and suffering on Cambodians.

Although the 1993 constitution aims to make Cambodia a nonaligned "island of peace," a neutral foreign policy was difficult to pursue so long as China continued to advocate the political rehabilitation of the Khmer Rouge. Cambodia's relations with Vietnam also remained problematic, fueled in part by the historic animosity between the two countries. The situation has improved since the late 1990s, and Cambodia's membership of ASEAN, which had been on hold pending the consolidation of full democratic government, was confirmed in 1999.

AID

Recipient

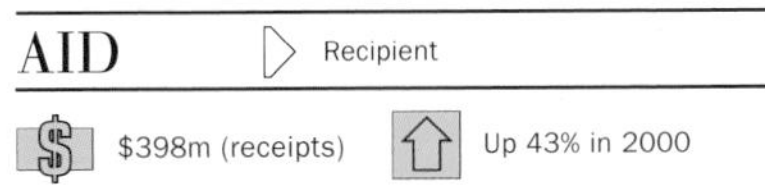
$398m (receipts)
Up 43% in 2000

Aid is crucial to Cambodia's economy, providing the bulk of government revenues. Widespread corruption and political instability prompted some countries to withhold assistance in the late 1990s, but NGOs continued working in the country, and Western donors made fresh pledges in 2000.

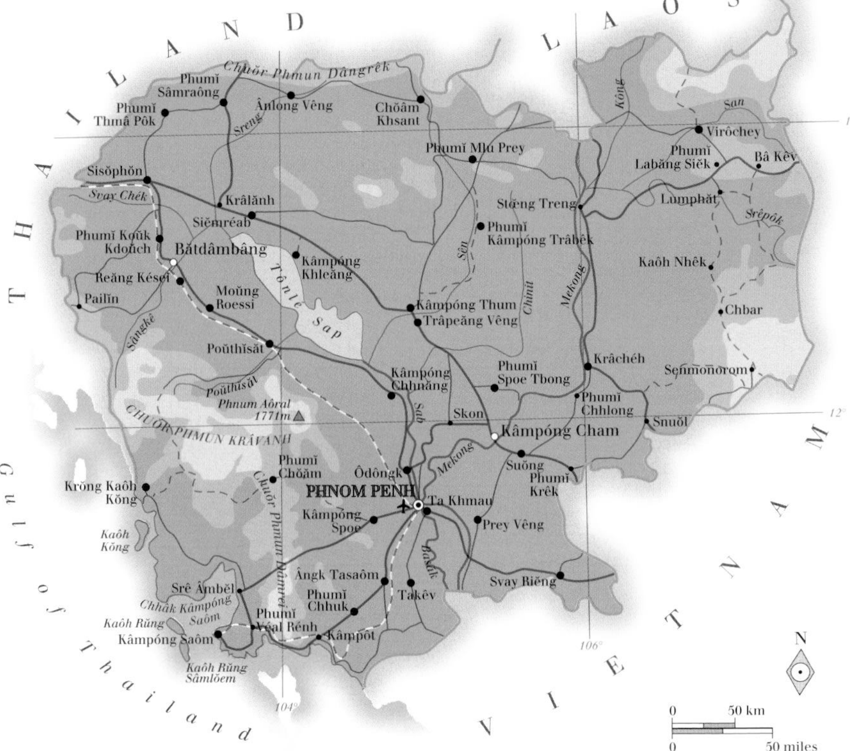
THAILAND
LAOS
VIETNAM
Gulf of Thailand
Chuŏr Phnum Dângrêk
Phumĭ Sâmraông
Phumĭ Thmâ Pŏk
Ânlong Vêng
Choăm Khsant
Sisŏphŏn
Svay Chék
Krâlănh
Siĕmréab
Phumĭ Koŭk Kdoŭch
Bătdâmbâng
Reăng Kései
Pailĭn
Moŭng Roessei
Kâmpóng Khleăng
Tônlé Sap
Poŭthĭsăt
Phnum Aôral 1771m
CHUŎR PHNUM KRÂVANH
Krŏng Kaôh Kŏng
Kaôh Kŏng
Phumĭ Choăm
Chuŏr Phnum Dâmrei
PHNOM PENH
Ŏdôngk
Kâmpóng Spoe
Srê Âmbĕl
Chhăk Kâmpóng Saôm
Kaôh Rŭng
Kâmpóng Saôm
Kaôh Rŭng Sâmlŏem
Phumĭ Veal Rénh
Kâmpôt
Phumĭ Chhuk
Ângk Tasaôm
Takêv
Ta Khmau
Prey Vêng
Svay Riĕng
Suŏng
Phumĭ Krêk
Kâmpóng Cham
Skon
Snuŏl
Phumĭ Chhlong
Krâchéh
Phumĭ Spoe Tbong
Kâmpóng Chhnăng
Kâmpóng Thum
Trâpeăng Vêng
Sên
Chinit
Mékong
Phumĭ Kâmpóng Trâbêk
Stœ̆ng Trêng
Phumĭ Mlu Prey
Kông
San
Virôchey
Phumĭ Labăng Siĕk
Bâ Kêv
Lumphăt
Srêpŏk
Kaôh Nhêk
Chbar
Senmonorom
14°
12°
104°
106°
N
0 50 km
0 50 miles

CAMBODIA

Total Land Area : 181 040 sq. km (69 900 sq. miles)

POPULATION

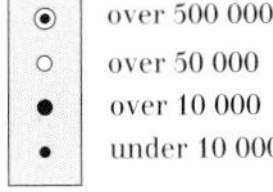
over 500 000
over 50 000
over 10 000
under 10 000

LAND HEIGHT

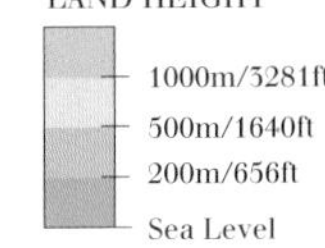
1000m/3281ft
500m/1640ft
200m/656ft
Sea Level

C

CHRONOLOGY

A former French protectorate, Cambodia gained independence in 1953 as a constitutional monarchy with Norodom Sihanouk as king.

- ❑ **1955** Sihanouk abdicates to pursue political career; takes title "prince."
- ❑ **1970** Right-wing coup led by Prime Minister Lon Nol deposes Sihanouk. Exiled Sihanouk forms Royal Government of National Union of Cambodia (GRUNC), backed by communist Khmer Rouge. Lon Nol proclaims Khmer Republic.
- ❑ **1975** GRUNC troops capture Phnom Penh. Prince Sihanouk head of state, Khmer Rouge assumes power. Huge numbers die under radical extremist regime.
- ❑ **1976** Country renamed Democratic Kampuchea. Elections. Sihanouk resigns; GRUNC dissolved. Khieu Samphan head of state; Pol Pot prime minister.
- ❑ **1978** December, Vietnam invades, supported by Cambodian communists opposed to Pol Pot.
- ❑ **1979** Vietnamese capture Phnom Penh. Khmer Rouge ousted by Kampuchean People's Revolutionary Party (KPRP), led by Pen Sovan. Khmer Rouge starts guerrilla war. Pol Pot held responsible for genocide and sentenced to death in absentia.
- ❑ **1982** Government-in-exile including Khmer Rouge and Khmer People's National Liberation Front, headed by Prince Sihanouk, is recognized by UN.
- ❑ **1989** Withdrawal of Vietnamese troops.
- ❑ **1990** UN Security Council approves plan for UN-monitored cease-fire and elections.
- ❑ **1991** Signing of Paris peace accords. Sihanouk reinstated as head of state of Cambodia.
- ❑ **1993** UN-supervised elections won by royalist Funcinpec. Sihanouk takes title of "king."
- ❑ **1994** Khmer Rouge refuses to join peace process.
- ❑ **1995** Former finance minister Sam Rainsy forms opposition party.
- ❑ **1996** Leading Khmer Rouge member Ieng Sary defects.
- ❑ **1997** Joint prime minister Hun Sen mounts coup against royalist co-premier Prince Ranariddh.
- ❑ **1998** April, death of Pol Pot; June, Khmer Rouge surrender; July, parliamentary elections; November, Hun Sen heads coalition government including Funcinpec.
- ❑ **1999** Cambodia admitted to ASEAN.
- ❑ **2001** Law approved on trials of Khmer Rouge leaders for atrocities committed by regime.

DEFENSE

No compulsory military service

 $192m

 Up 9% in 2000

CAMBODIAN ARMED FORCES

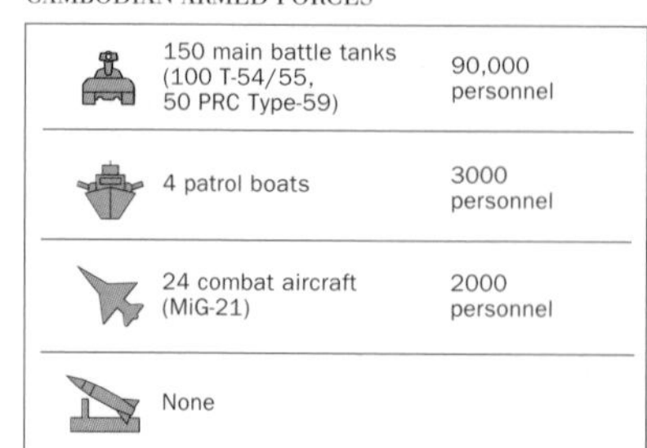

150 main battle tanks (100 T-54/55, 50 PRC Type-59)	90,000 personnel
4 patrol boats	3000 personnel
24 combat aircraft (MiG-21)	2000 personnel
None	

The coalition government's initial defense priority was to unify the command structures of armies under the control of diverse parties. The surrender of Khmer Rouge forces in mid-1998 and the disintegration of remaining pockets of Khmer resistance later that year improved the prospects for a unified national army. Plans for the demobilization of 30,000 soldiers over a three-year period began in May 2000.

Under the nominal overall structure of the Royal Cambodian Armed Forces, there remain in existence three main armies – the CPP's Cambodian People's Armed Forces, Funcinpec's Armée Nationale Sihanoukiste, and the KPNLF's Khmer People's National Liberation Armed Forces. The rivalries between them remain intense, with the two first-named in open conflict as recently as 1997–1998. Although well equipped, their soldiers are poorly paid.

A system of conscription for five years between the ages of 18 and 35 has not been implemented since 1993.

ECONOMICS

Inflation 25% p.a. (1990–2000)

$3.15bn

 3835 riels

SCORE CARD

- ❑ World GNP Ranking........................130th
- ❑ GNP per Capita $260
- ❑ Balance of Payments......................–$19m
- ❑ Inflation ....–0.8%
- ❑ Unemployment 3%

EXPORTS

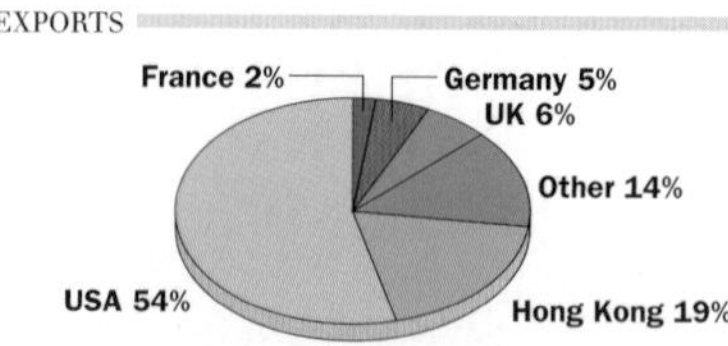

France 2%
Germany 5%
UK 6%
Other 14%
Hong Kong 19%
USA 54%

IMPORTS

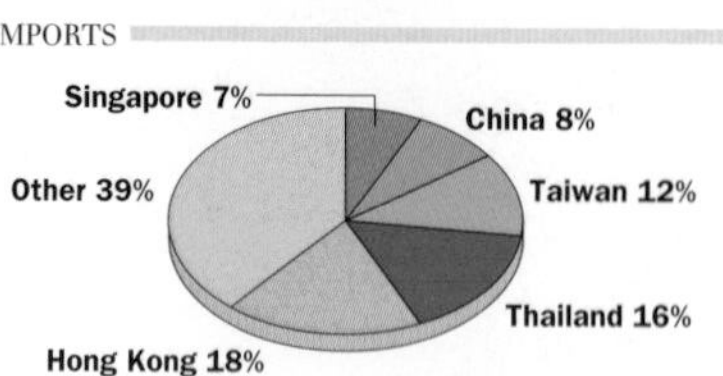

Singapore 7%
China 8%
Taiwan 12%
Thailand 16%
Hong Kong 18%
Other 39%

Strengths

Currently very few, as economy still recovering from long-running conflicts. Considerable future potential. Growth in tourism. Relatively unbureaucratic mentality. Self-sufficiency in rice achieved by 1999. Gems, especially sapphires. Possible offshore oil wealth. Export-oriented garment industry.

Weaknesses

Tiny tax base makes economic reform hard to implement. Dependence on overseas aid; corruption at most levels of government limits its effectiveness. Disputes over land ownership rights.

ECONOMIC PERFORMANCE INDICATOR

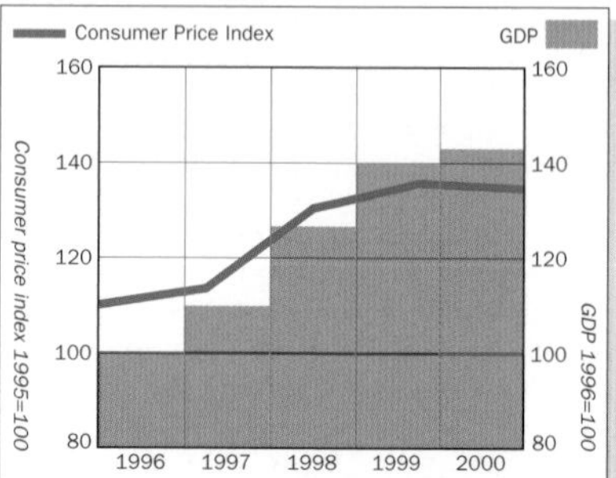

Consumer Price Index
GDP
Consumer price index 1995=100
GDP 1996=100
160
140
120
100
80
1996
1997
1998
1999
2000

Profile

The economy was devastated during the Pol Pot years. The Vietnamese attempted some reconstruction based on central planning, then switched to encouraging the private sector. Investment in the 1990s was heavily aid-dependent. The Asian financial crisis and internal turmoil affected funding after 1997.

CAMBODIA : MAJOR BUSINESSES

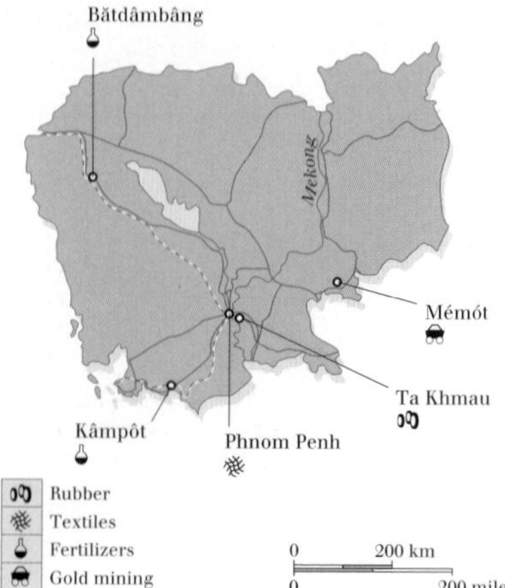

Bătdâmbâng
Mekong
Mémót
Ta Khmau
Kâmpôt
Phnom Penh
Rubber
Textiles
Fertilizers
Gold mining
0 200 km
0 200 miles

RESOURCES

Electric power 35,000 kw

284,100 tonnes

Not an oil producer

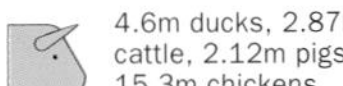
4.6m ducks, 2.87m cattle, 2.12m pigs, 15.3m chickens

Salt, phosphates, gemstones

Few resources are currently exploited, apart from tropical rainforest timber, particularly teak and rosewood, much of which is felled illegally.

ELECTRICITY GENERATION

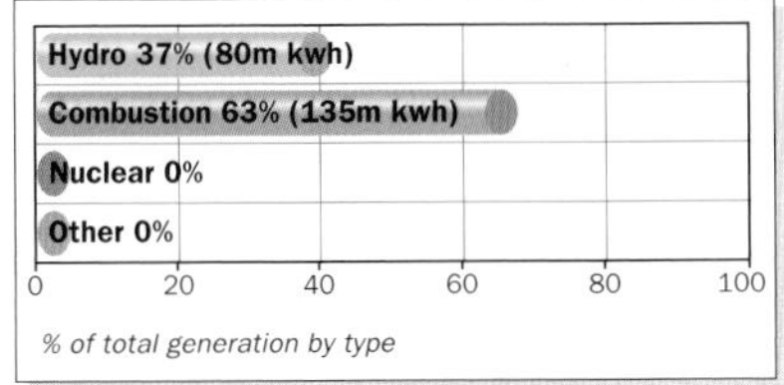

Hydro 37% (80m kwh)
Combustion 63% (135m kwh)
Nuclear 0%
Other 0%
0
20
40
60
80
100
% of total generation by type

CAMBODIA : LAND USE

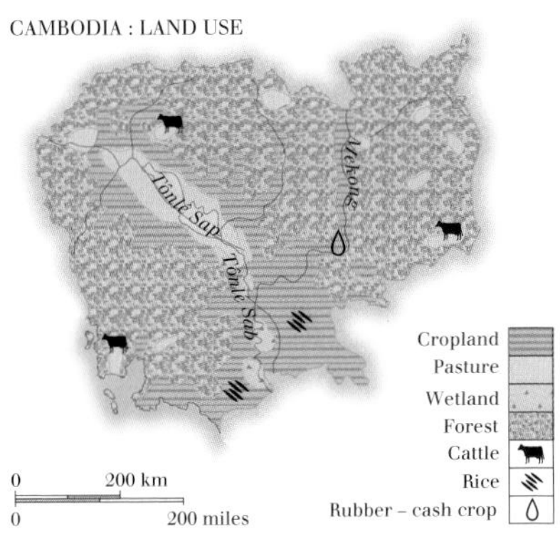

Tonle Sap
Mekong
Cropland
Pasture
Wetland
Forest
Cattle
Rice
Rubber – cash crop
0 200 km
0 200 miles

ENVIRONMENT

Sustainability rank: 97th

16%

0.1 tonnes per capita

ENVIRONMENTAL TREATIES

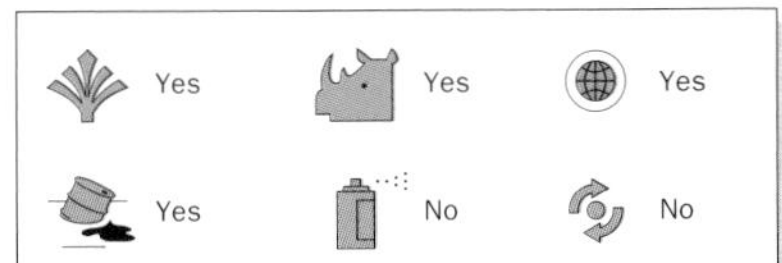

Yes
Yes
Yes
Yes
No
No

Deforestation is one of the most serious problems facing Cambodia. Illegal logging is the main culprit. Timber, one of the country's most valuable assets, was sold in huge quantities by all Cambodian factions to finance their war efforts. A moratorium on logging was declared at the end of 1992, but was largely ignored. Despite international pressure and efforts from 2000 to tighten controls, in many parts of the country logging is impossible to police. Tropical hardwoods extracted illegally from Cambodia find lucrative outlets through Thailand in particular. The environmental consequences – topsoil erosion and increased risk of flooding – are enormous and will hold back Cambodia's reconstruction.

MEDIA

TV ownership low

Daily newspaper circulation 2 per 1000 people

Phnom Penh has several independent TV stations in addition to the national network. However, many are reliant on party political support, which compromises their independence. The government has used a 1995 press law to prosecute numerous newspapers for defamation and disinformation.

PUBLISHING AND BROADCAST MEDIA

There are 2 daily newspapers; only 10 newspapers and magazines publish regularly
6 services: 1 state-run, 5 independent
10 services; 2 state-run, 8 independent

CRIME

No death penalty

5502 prisoners

Civilian crime rates are now fairly stable

CRIME RATES

Statistics for murders, rapes, and thefts are unavailable

There are allegations that Cambodia is becoming Asia's new "narco-state." It is claimed that there has been a proliferation of narcotics trading, money laundering, and illegal banking operations. Mob killings go largely unremarked. Corruption in business is a major issue. Phnom Penh witnessed an increase in violent crime in the aftermath of the 1997 Hun Sen coup, owing to the spread of illegally owned firearms. Until the surrender of the Khmer Rouge in 1998, areas under its command, especially in the west around Pailin and Battambang, were particularly dangerous. Banditry remains rife and policing virtually nonexistent.

EDUCATION

School leaving age: 12

37%

8901 students

THE EDUCATION SYSTEM

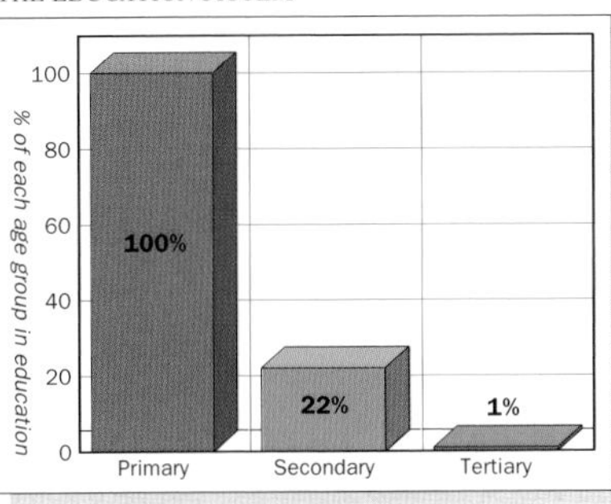

100
80
60
40
20
0
% of each age group in education
100%
22%
1%
Primary
Secondary
Tertiary

The government aims to put in place a nine-year period of education. Currently primary education is compulsory, and lasts for six years between the ages of six and 12. Only 5000 of Cambodia's 20,000 teachers survived the Pol Pot period; the Vietnamese-installed government trained or retrained about 40,000.

HEALTH

No welfare state health benefits

1 per 3333 people

Circulatory and infectious diseases, cancers

The Cambodian health system was effectively destroyed in the Pol Pot period; only 50 doctors survived, and Cambodia's health indicators were among the worst in the world.

Conditions have since improved, but AIDS is widespread, affecting even children in rural areas. Infant mortality remains high, and malaria and cholera are endemic. In 2000, UNICEF helped mount an immunization campaign against tetanus, a major cause of neonatal mortality.

SPENDING

GDP/cap. increase

CONSUMPTION AND SPENDING

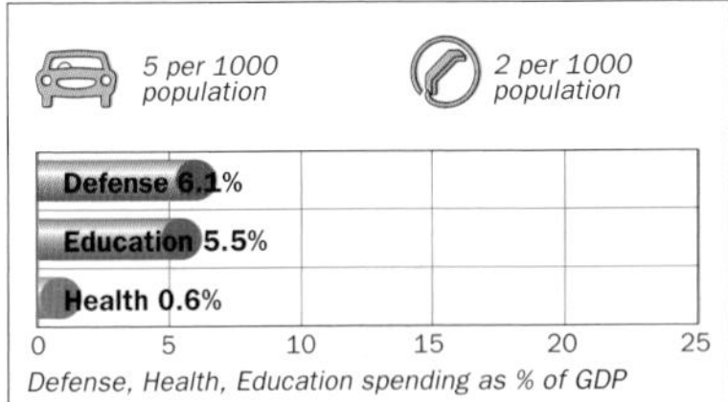

5 per 1000 population
2 per 1000 population
Defense 6.1%
Education 5.5%
Health 0.6%
0
5
10
15
20
25
Defense, Health, Education spending as % of GDP

New industries such as textiles, in which female garment workers may earn $40 a month in vast workshops, help to attract migrants to the towns, although they risk unemployment and homelessness. Cambodians in rural areas face more severe poverty, exacerbated by land shortage.

WORLD RANKING

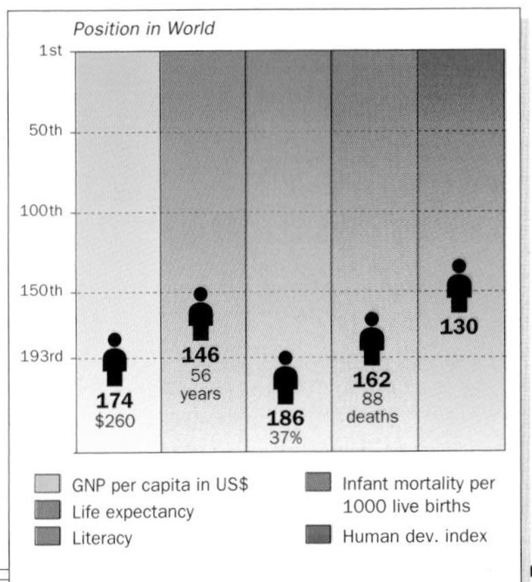

Position in World
1st
50th
100th
150th
193rd
174
$260
146
56 years
186
37%
162
88 deaths
130
GNP per capita in US$
Life expectancy
Literacy
Infant mortality per 1000 live births
Human dev. index

CAMEROON

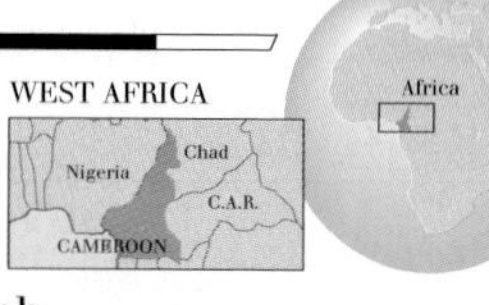
WEST AFRICA
Africa
Nigeria
Chad
C.A.R.
CAMEROON

OFFICIAL NAME: Republic of Cameroon **CAPITAL:** Yaoundé
POPULATION: 15.2 million **CURRENCY:** CFA franc **OFFICIAL LANGUAGES:** French and English

C

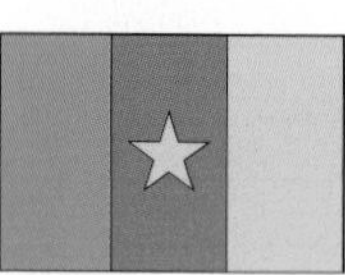

1960

1961

May 20
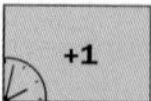
CAM

+1

+237
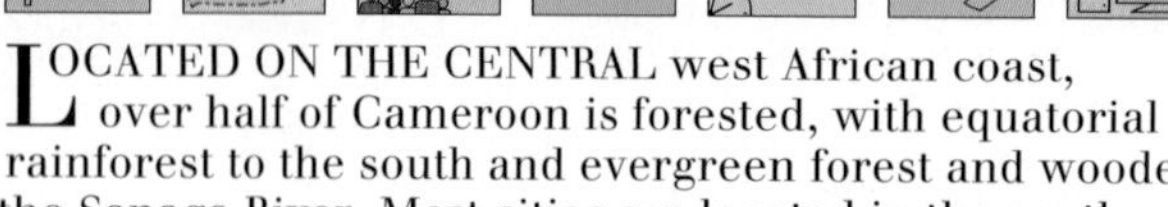
.cm

LOCATED ON THE CENTRAL west African coast, over half of Cameroon is forested, with equatorial rainforest to the south and evergreen forest and wooded savanna north of the Sanaga River. Most cities are located in the south, although there are densely populated areas around Mount Cameroon, a dormant volcano. For 30 years Cameroon was effectively a one-party state. Democratic elections in 1992 returned the former ruling party to power.

Savanna landscape below Mindif Pic in Cameroon's far north. From here, the land slopes down to the hot, arid Lake Chad basin.

CLIMATE

Tropical equatorial

WEATHER CHART

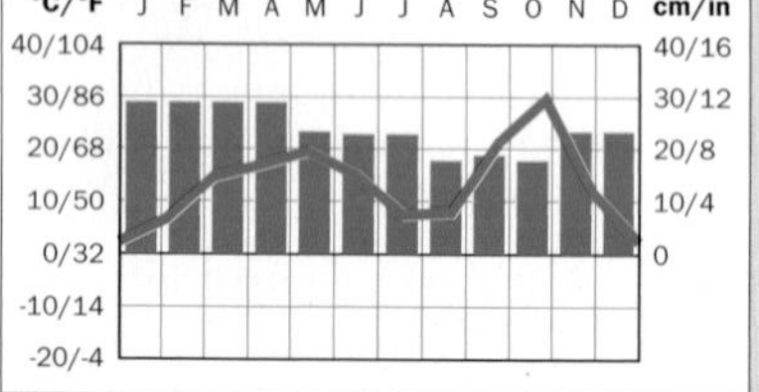
Average daily temperature
Rainfall
°C/°F J F M A M J J A S O N D cm/in
40/104
30/86
20/68
10/50
0/32
-10/14
-20/-4
40/16
30/12
20/8
10/4
0

Climate varies from the equatorial south, with 500 cm (200 in) of rain a year, to the drought-beset Sahelian north.

PEOPLE

Pop. density low

Bamileke, Fang, Fulani, French, English

33/km² (85/mi²)

THE URBAN/RURAL POPULATION SPLIT

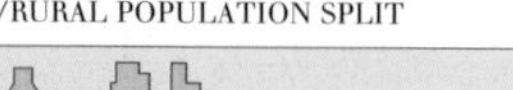

49% 51%

RELIGIOUS PERSUASION

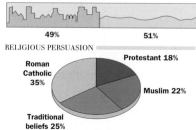

Protestant 18%
Muslim 22%
Traditional beliefs 25%
Roman Catholic 35%

Cameroon is ethnically diverse – there are 230 groups, no single group being dominant. The largest is the Bamileke of the central southwest region, but this group has never held political power. When President Ahidjo, a northern Fulani, retired, he was replaced by Paul Biya of the southeastern Bulu-Beti group. The north–south enmity which affects many other west African states is also present in Cameroon, albeit diminished by the great diversity of peoples. There are tensions between the French- and English-speaking communities, with sections of the latter demanding independence.

POLITICS

Multiparty elections

2002/2007

President Paul Biya

AT THE LAST ELECTION

National Assembly 180 seats

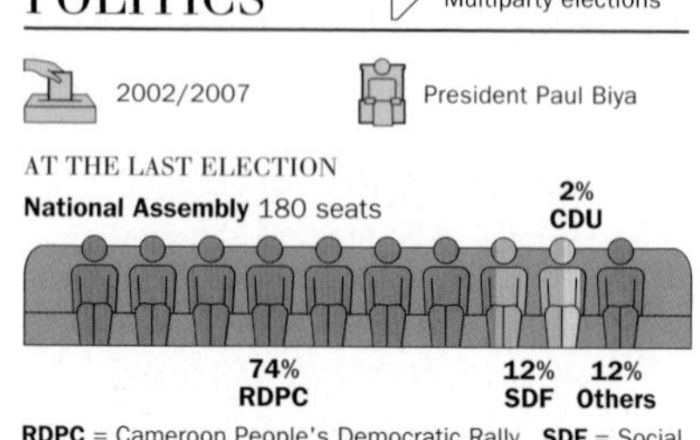

74% RDPC
12% SDF
12% Others
2% CDU

RDPC = Cameroon People's Democratic Rally **SDF** = Social Democratic Front **CDU** = Cameroon Democratic Union

A Senate is to be created under the 1995 constitution.

Incumbent president Paul Biya's RDPC narrowly won control of the new parliament in multiparty elections in 1992, which were boycotted by the main opposition SDF. It has held on to power in the face of SDF claims of corruption, steadily increasing its majority in 1997 and again in 2002. Similarly, Biya's own reelections in 1992 and 1997 were condemned as the products of fraud and intimidation by SDF candidate John Fru Ndi.

TRANSPORTATION

Drive on right

Douala International 439,279 passengers

61 ships 13,600 grt

THE TRANSPORTATION NETWORK

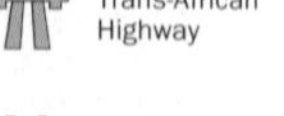

4288 km (2664 miles)	Trans-African Highway
1016 km (631 miles)	2090 km (1299 miles)

Major projects are the east–west Trans-African Highway and the realigning of the Douala–Nkongsamba railroad.

TOURISM

Visitors : Population 1:113

135,000 visitors

Up 32% in 1998

MAIN TOURIST ARRIVALS

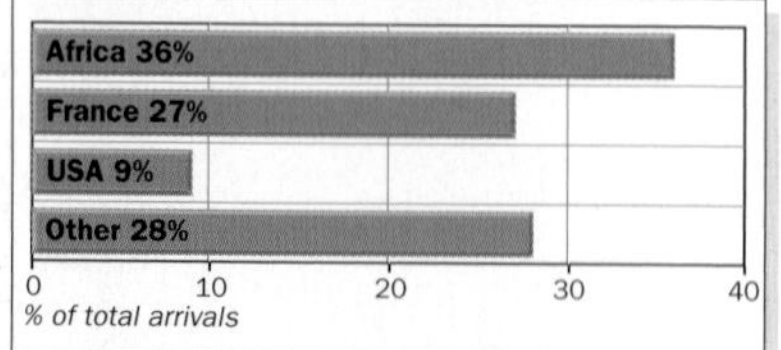
Africa 36%
France 27%
USA 9%
Other 28%
0 10 20 30 40
% of total arrivals

The government and Commonwealth aim to boost visitor numbers, with a yearly target of 500,000. A new airport near Yaoundé will replace the present one. There are beach hotels near Kribi and package tours to northern game parks.

Lake Chad
Kousséri
Logone
Maroua
Yagoua
Guider
Kaélé
Garoua
Benue
Lac de Lagdo
NIGERIA
CHAD
Faro
Mbé
Touboro
Ngaoundéré
ADAMAOUA
Banyo
Tibati
Lac de Mbakaou
Meiganga
Wum
Kumbo
Bamenda
Foumban
Mamfe
Dschang
Bafoussam
Nkongsamba
Bangangté
Bafang
Bafia
Kumba
Loum
Yabassi
Buea
Limbe
Tiko
Douala
Edéa
YAOUNDÉ
Nanga Eboko
Bélabo
Bertoua
Batouri
Abong Mbang
Akonolinga
Mbalmayo
Eséka
Kribi
Ebolowa
Sangmélima
Ambam
Moloundou
CENTRAL AFRICAN REPUBLIC
EQUATORIAL GUINEA
GABON
CONGO
ATLANTIC OCEAN
Gulf of Guinea
Bioko
Fako 4095m
0 200 km
0 200 miles

CAMEROON

Total Land Area : 475 400 sq. km (183 567 sq. miles)

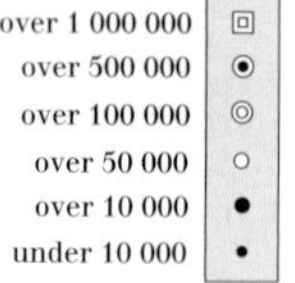
POPULATION
over 1 000 000
over 500 000
over 100 000
over 50 000
over 10 000
under 10 000

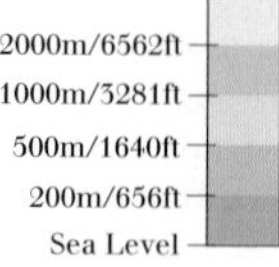
LAND HEIGHT
2000m/6562ft
1000m/3281ft
500m/1640ft
200m/656ft
Sea Level

WORLD AFFAIRS

▷ Joined UN in 1960

BDEAC

Comm

OIC

LCBC
FZ

Cameroon's most important relationship continues to be with France, although the country has attempted to diversify its international links, joining the Commonwealth in 1995. A territorial dispute with Nigeria of long duration concerns sovereignty over the oil-rich Bakassi peninsula, where there were clashes in 1996 and 1998.

AID

▷ Recipient

 $380m (receipts)

 Down 12% in 2000

France is by far the most important donor, even having twice paid up Cameroon's back debts to the IMF to prevent its being blacklisted. Lack of funding has forced many development projects to be abandoned. Despite poor economic performance, relations with the IMF are improving.

DEFENSE

▷ No compulsory military service

 $154m

 No change in 2000

The 11,500-strong army has been active in supporting the regime and maintaining order in the face of prodemocratic protests since before independence. Military equipment and training comes mainly from France. There is also a 9000-strong paramilitary gendarmerie.

ECONOMICS

▷ Inflation 5.1% p.a. (1990–2000)

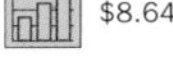 $8.64bn

 698.7–736.7 CFA francs

SCORE CARD

❑ WORLD GNP RANKING	90th
❑ GNP PER CAPITA	$580
❑ BALANCE OF PAYMENTS	$-153m
❑ INFLATION	5.3%
❑ UNEMPLOYMENT	30%

STRENGTHS

French and US companies exploit moderate oil reserves. Very diversified agriculture includes timber, cocoa, bananas, and coffee. Self-sufficiency in food. Strong informal sector. Private sector in relatively good state. Electricity is 95% HEP.

WEAKNESSES

Massive fuel smuggling from Nigeria affects refinery profits. Inflated civil service. Widespread corruption.

EXPORTS

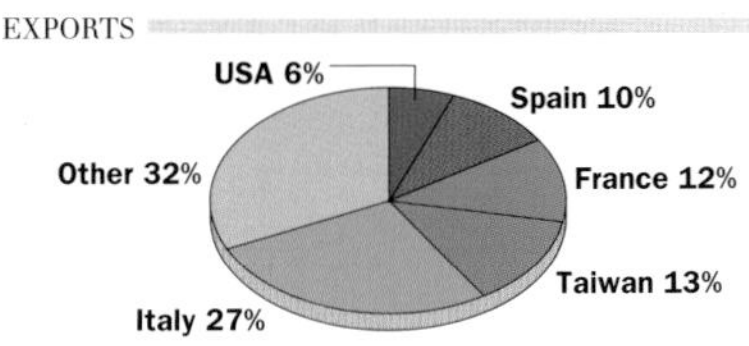
USA 6%
Spain 10%
France 12%
Taiwan 13%
Italy 27%
Other 32%

IMPORTS

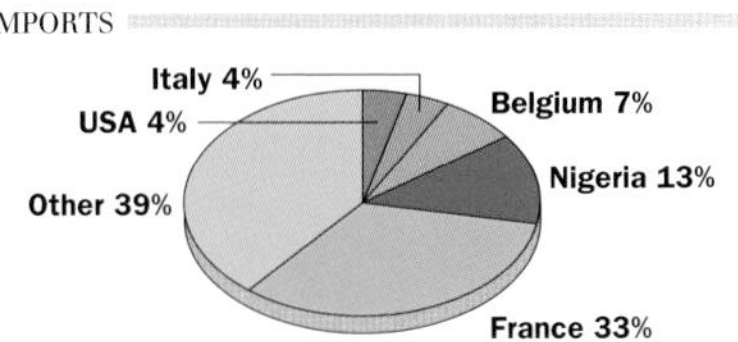
Italy 4%
USA 4%
Belgium 7%
Nigeria 13%
France 33%
Other 39%

RESOURCES

▷ Electric power 627,000 kw

95,067 tonnes

 80,000 b/d (reserves 400m barrels)

5.9m cattle, 4.4m goats, 3.8m sheep, 30m chickens

Oil, coal, tin, natural gas, bauxite, iron, uranium, gold

New oil discoveries may be able to bolster declining extraction rates. In spite of large bauxite deposits, much is imported for the Edea smelter, which takes 50% of national electricity output.

ENVIRONMENT

▷ Sustainability rank: 93rd

 5% (2% partially protected)

 0.1 tonnes per capita

The rate of commercial logging and the planned Cameroon–Chad oil pipeline constitute major threats to Cameroon's environment.

MEDIA

▷ TV ownership low

 Daily newspaper circulation 0.5 per 1000 people

PUBLISHING AND BROADCAST MEDIA

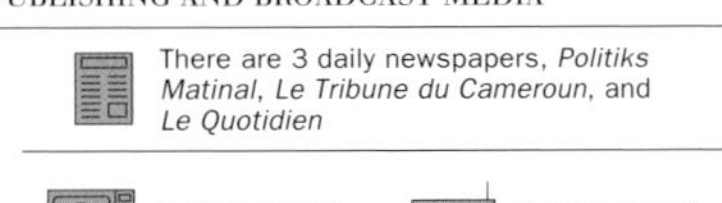

There are 3 daily newspapers, *Politiks Matinal, Le Tribune du Cameroun*, and *Le Quotidien*

1 state-owned service

1 state-owned service

There are frequent allegations of censorship and violence against journalists. English-language media are generally more outspoken.

CRIME

▷ Death penalty used

 15,903 prisoners

 Up 238% 1996–1998

Armed robbery and burglary in Douala and Yaoundé are rising fast. The police are known to use torture.

EDUCATION

▷ School leaving age: 12

 76%

 66,902 students

The French-speaking majority has failed in its attempt to take over the bilingual system. Cameroon has a high literacy rate compared with much of the rest of Africa.

CHRONOLOGY

One of the great trading emporia of west Africa, Cameroon was divided between the French and British in 1919, after 30 years of German rule.

- ❑ **1955** Revolt; French kill 10,000.
- ❑ **1960** French sector independent.
- ❑ **1961** British south joins Cameroon (north joins Nigeria).
- ❑ **1982** Ahidjo dies; Biya president.
- ❑ **1983–1984** Coup attempts. Heavy casualties; 50 plotters executed.
- ❑ **1990** Demonstrations and strikes; declaration of multiparty state.
- ❑ **1992** Multiparty elections.
- ❑ **1997** President and ruling RDPC returned in disputed elections.
- ❑ **2000** World Bank funds for pipeline project, despite environmental fears.
- ❑ **2001** Over 80% of indigenous forests allocated for logging.
- ❑ **2002** RDPC increases its majority.

C

HEALTH

▷ No welfare state health benefits

1 per 10,000 people

 Malaria, diarrheal and respiratory diseases

A sharp fall in government health provision means that more people are using the private health sector or traditional practitioners.

SPENDING

▷ GDP/cap. increase

CONSUMPTION AND SPENDING

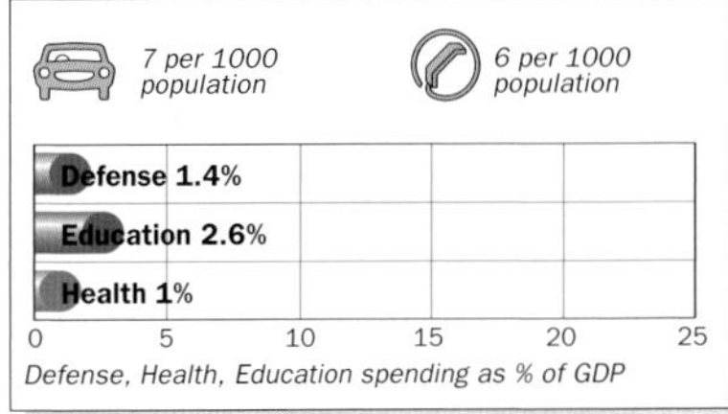
7 per 1000 population
6 per 1000 population
Defense 1.4%
Education 2.6%
Health 1%
0
5
10
15
20
25
Defense, Health, Education spending as % of GDP

Wealth is unevenly distributed and has been declining since the end of the oil boom. There is still a very wealthy, albeit small, sector of the population.

WORLD RANKING

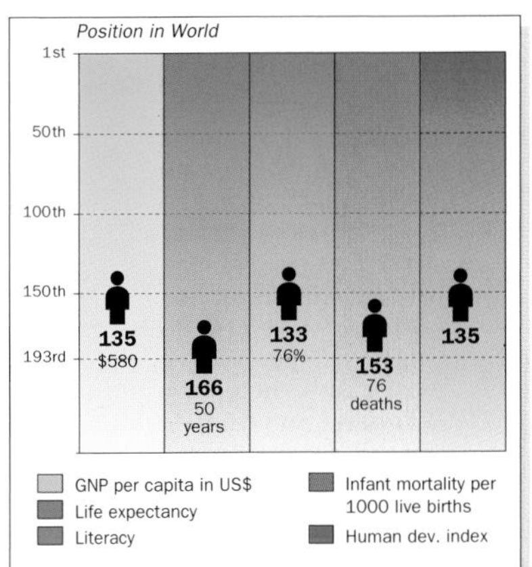
Position in World
1st
50th
100th
150th
193rd
135
$580
166
50 years
133
76%
153
76 deaths
135
GNP per capita in US$
Life expectancy
Literacy
Infant mortality per 1000 live births
Human dev. index

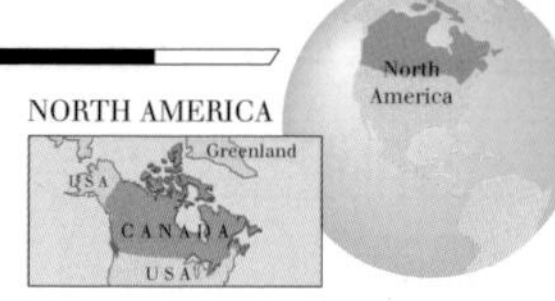
NORTH AMERICA
North America
Greenland
USA
CANADA
USA

CANADA

OFFICIAL NAME: Canada **CAPITAL:** Ottawa **POPULATION:** 31 million
CURRENCY: Canadian dollar **OFFICIAL LANGUAGES:** English and French

1867

1949

July 1

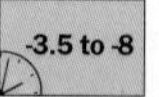
CDN

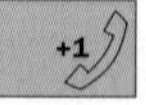
-3.5 to -8

+1

.ca

CANADA IS THE WORLD'S THIRD-LARGEST country, stretching north to Cape Columbia on Ellesmere Island, south to Lake Erie, and across six time zones from Newfoundland to the Pacific seaboard. The interior lowlands around Hudson Bay make up 80% of Canada's land area and include the vast Canadian Shield, with the plains of Saskatchewan and Manitoba and the Rocky Mountains to the west. The St. Lawrence, Yukon, Mackenzie, and Fraser Rivers are among the world's 40 largest. The Great Lakes–St. Lawrence River lowlands are the most populous areas. An Inuit homeland, Nunavut, formerly the eastern part of the Northwest Territories, was created in 1999, covering nearly a quarter of Canada's land area. French-speaking Québec's relationship with the rest of the country causes recurring constitutional arguments.

CANADA

Total Land Area : 9 976 140 sq. km (3 851 788 sq. miles)

POPULATION

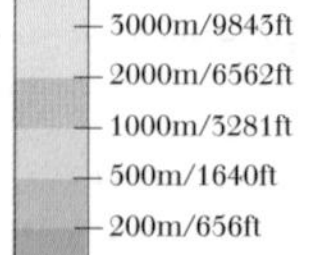

over 1 000 000
over 500 000
over 100 000
over 50 000
over 10 000
under 10 000

LAND HEIGHT

3000m/9843ft
2000m/6562ft
1000m/3281ft
500m/1640ft
200m/656ft
Sea Level

CLIMATE

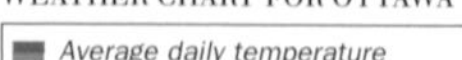
Continental/subarctic/ mountain

WEATHER CHART FOR OTTAWA

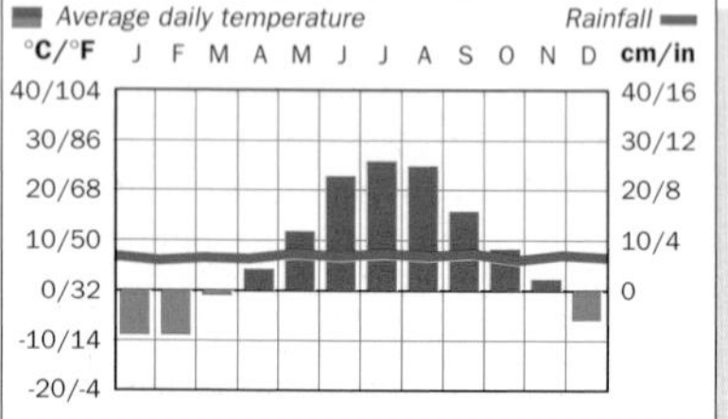
Average daily temperature
Rainfall
°C/°F
J F M A M J J A S O N D
cm/in
40/104
30/86
20/68
10/50
0/32
-10/14
-20/4
40/16
30/12
20/8
10/4
0

Canada's climate ranges from polar and subpolar in the north, to cool in the south. Summers in the interior are hotter, and winters colder and longer than on the coast, with temperatures well below freezing and deep snow. The Pacific coast around Vancouver has the warmest winters, where temperatures rarely fall below zero.

TRANSPORTATION

Drive on right

Lester B. Pearson International, Toronto
28.9m passengers

861 ships
2.66m grt

THE TRANSPORTATION NETWORK

318,371 km
(197,826 miles)

16,571 km
(10,297 miles)

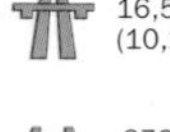
14,000 km
(8700 miles)

3769 km
(2342 miles)

The emergence of a national economy depended on the development of an efficient system of transportation. The Trans-Canada Highway and two transcontinental rail systems are the east–west backbones of the road and rail networks, which also reach into the far north. The Great Lakes–St. Lawrence Seaway system's cheap transportation helped Ontario and Québec dominate the economy for most of the 20th century. Air services were hit by the industry's crisis after 11 September 2001.

TOURISM

Visitors : Population 1:1.6

19.7m visitors

Little change in 2001

MAIN TOURIST ARRIVALS

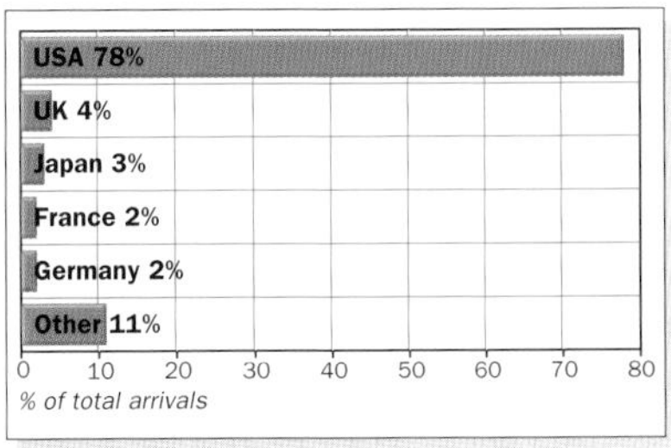

	% of total arrivals
USA	78%
UK	4%
Japan	3%
France	2%
Germany	2%
Other	11%

0 10 20 30 40 50 60 70 80

% of total arrivals

Most tourist visitors come from the US, often on short tours. Efforts to attract European visitors are based around campaigns emphasizing Canada's unpolluted natural beauty. Bizarrely, the fictional home of the eponymous heroine of *Anne of Green Gables* on Prince Edward Island is a magnet for tourists from Japan, where the novels about her by E. M. Montgomery enjoy enormous popularity.

PEOPLE

Pop. density low

English, French, Chinese, Italian, German, Ukrainian, Portuguese, Inuktitut, Cree

3/km²
9/mi²

THE URBAN/RURAL POPULATION SPLIT

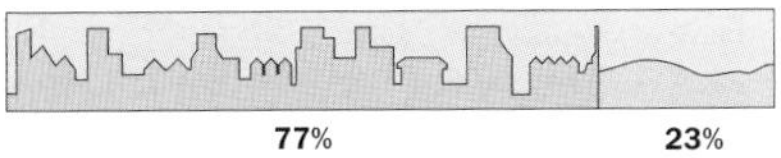

77% 23%

RELIGIOUS PERSUASION

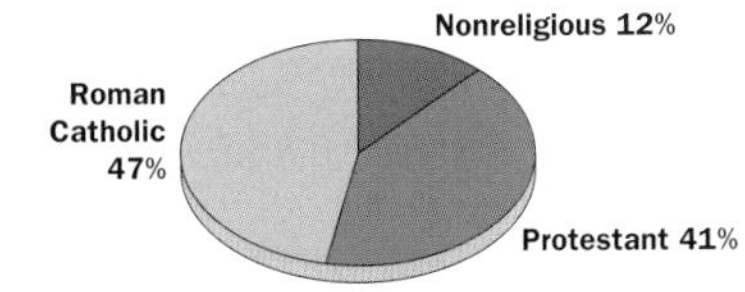

Nonreligious 12%
Roman Catholic 47%
Protestant 41%

ETHNIC MAKEUP

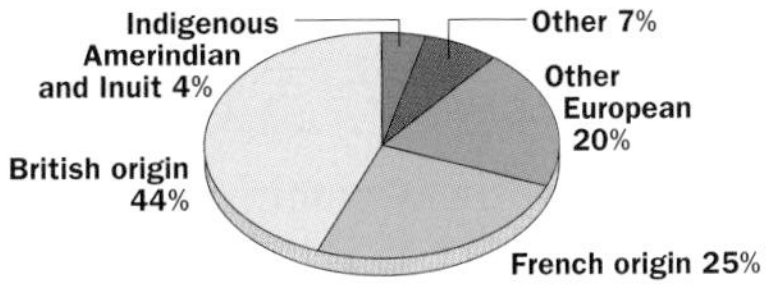

Indigenous Amerindian and Inuit 4%
Other 7%
Other European 20%
British origin 44%
French origin 25%

Relations between French-speaking Québécois and the English-speaking majority in Canada have been the dominant ethnic issue of the past 40 years. Support for separatist parties increased mainly because of the failure of Canada's other provinces to deal with Québec's demand to be recognized as a "distinct society," with powers to preserve its culture and language from further anglicization.

Québec's still controversial 1977 language law made French the province's official language. Two-thirds of Canada's population live in the 5% of its land area taken up by the Great Lakes–St. Lawrence lowlands. However, Canada's ethnic mix has changed significantly since the 1970s, due to a move from a restrictive immigration policy to one which welcomes those with money or skills. Significant numbers of Asians have moved to Canada. The government promotes a policy which encourages each group to maintain its own culture. Canada is now officially a "community of communities."

The largest element of the indigenous population is the 800,000 people of native Amerindian descent, known in Canada as First Nations. There are also 213,000 Métis (French–Amerindians) and an Inuit population of some 50,000 in the north. In 1992 the Inuit successfully settled their long-standing land claim, and in 1999 the Nunavut area, with only 25,000 mainly Inuit inhabitants, gained the status of a territory, the first part of Canada to be governed by indigenous Canadians in modern history. A Supreme Court land rights ruling in 1997, establishing the principle of "aboriginal title," opened the way for the return of ancestral lands claimed by native Amerindian nations, and in 1998 the federal government formally apologized for their past mistreatment.

Canada has a long tradition of state welfare more akin to Scandinavia than the US. Unemployment provision and health care, supported by high taxes, are still generous, despite recent cutbacks. The government has sought to end inequalities. Measures include the "pay-equity" laws, which aim to specify pay rates for jobs done mainly by women – such as receptionists – equivalent to similar skill jobs for men. Women are well represented at most levels of business and government.

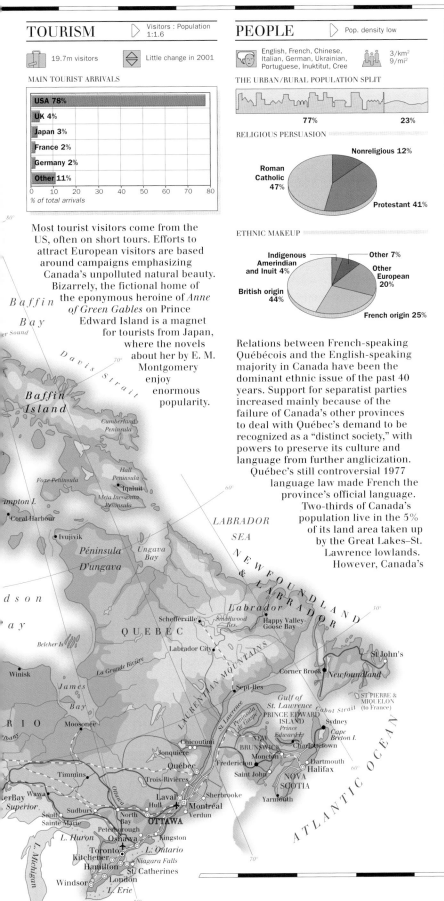

A dude ranch in British Columbia. *Many tourists are attracted by Canada's wide choice of outdoor pursuits.*

POPULATION AGE BREAKDOWN

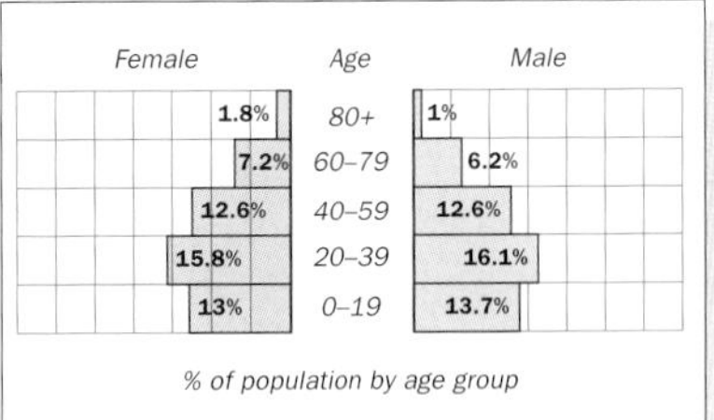

Female	Age	Male
1.8%	80+	1%
7.2%	60–79	6.2%
12.6%	40–59	12.6%
15.8%	20–39	16.1%
13%	0–19	13.7%

% of population by age group

C

CHRONOLOGY

Peopled for centuries by indigenous Inuits and Amerindians, Canada began to experience extensive European settlement following the landing of the English expedition led by John Cabot in 1497 and the French landing of Jacques Cartier in 1534.

- **1754** British fight French and Indian War. France forced to relinquish St. Lawrence and Québec settlements to Britain.
- **1774** Act of Québec recognizes Roman Catholicism, French language, culture, and traditions.
- **1775–1783** American War of Independence. Canada becomes refuge for loyalists to British crown.
- **1867** Federation of Canada created under British North America Act.
- **1885** Transcontinental railroad completed.
- **1897** Klondike gold rush begins.
- **1914–1918, 1939–1945** Canada supports Allies in both world wars.
- **1931** Autonomy within Commonwealth.
- **1949** Founder member of NATO. Newfoundland joins Federation.
- **1968** Liberal Party under Pierre Trudeau in power. Separatist Parti Québécois (PQ) formed.
- **1970s** Québec secessionist movement grows, accompanied by terrorist attacks.
- **1976** In Québec, PQ wins elections.
- **1977** French made official language.
- **1980** Separation of Québec rejected at referendum. Trudeau prime minister again.
- **1982** UK transfers all powers relating to Canada in British law.
- **1984** Trudeau resigns. Elections won by PCP. Brian Mulroney prime minister until 1993.
- **1987** Meech Lake Accord.
- **1989** Canadian–US Free Trade Agreement.
- **1992** Charlottetown Agreement on provincial–federal issues rejected at referendum. Canada, Mexico, and US finalize terms for NAFTA.
- **1993** Crushing election defeat of PCP, rise of regional parties.
- **1994** PQ regains power in Québec. NAFTA takes effect.
- **1995** Narrow "no" vote in second Québec sovereignty referendum.
- **1995** Fishing dispute with EU.
- **1997** Regionalism dominates federal election; Liberals retain power based on support in Ontario.
- **1998** PQ only narrowly holds power in Québec.
- **2000** November, early elections. Liberals retain power.
- **2001** December, Smart Border Declaration and antiterrorist legislation.

POLITICS

Multiparty elections

L. House 2000/2005

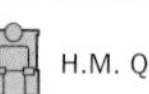
H.M. Queen Elizabeth II

AT THE LAST ELECTION

House of Commons 301 seats

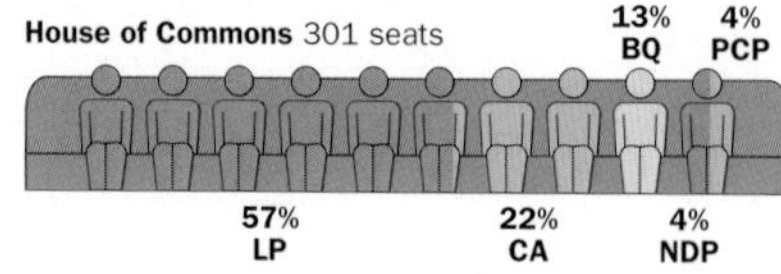
13% BQ
4% PCP
57% LP
22% CA
4% NDP

LP = Liberal Party **CA** = Canadian Reform Conservative Alliance **BQ** = Bloc Québécois **NDP** = New Democratic Party **PCP** = Progressive Conservative Party **Ind** = Independents

Senate 105 seats

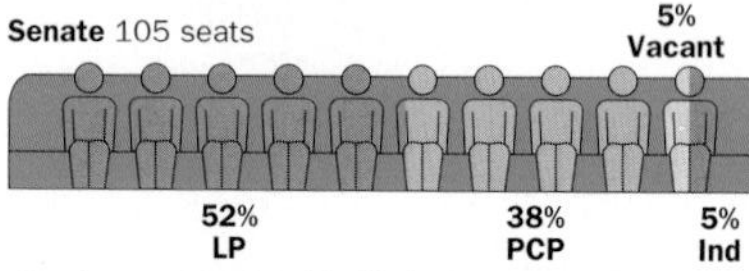
5% Vacant
52% LP
38% PCP
5% Ind

Senators are appointed for life by the governor-general in Council, to a usual maximum of 105; the prime minister may appoint an extra 8 senators.

Canada is a federal multiparty democracy.

Profile

Canadian politics was traditionally dominated by two main parties, the PCP and LP, which had few major ideological differences. The third party, the NDP (which has never held federal power), advocated greater government intervention than the other two.

The 1990s brought a major political shift – the eclipse of the PCP, three successive victories for Jean Chrétien and the LP (1993, 1997, and 2000), and an eventual realignment on the right, based on the emergence of the populist Reform Party (RP) in the western provinces. The trend away from Canada-wide politics, toward parties representing strong regional interests, left the LP with few seats outside its strongholds in the east. The Bloc Québécois (BQ), espousing the separatist cause at federal level, was the second-largest party in the federal parliament between 1994 and 1997. Since then the role of official opposition has been held by the RP and then by its successor, the conservative Canadian Alliance, which was riven in 2001–2002 by a fractious leadership dispute.

Main Political Issues

The unity of the state

Opposition to federal government is not confined to Québec – the 1997 and 2000 federal elections confirmed support for greater autonomy for Canada's western provinces – but Canada has agonized over separatist tendencies in francophone Québec almost since the foundation of the state. Québec did not take part in the 1997 Calgary conference, where a Canadian unity framework was agreed by the other provinces, together with recognition of Québec's "unique character." A series of earlier proposals, to recognize Québec as a distinct society and strengthen the powers of all the federal provinces, had failed to gain ratification or had been rejected by the electorate. The Parti Québécois (PQ), back in power at provincial level since 1994, advocates another referendum on separatism, despite losing those held in 1980 and 1995. The Supreme Court has ruled, however, that secession would require federal approval and the agreement of at least seven of the ten provinces. The April 2000 Clarity Act set strict criteria for any pro-secession referendum result to be valid.

***The Niagara Falls** are situated between Lakes Erie and Ontario on the Canada–US border. Horseshoe Falls, in Canada, are 49 m (160 ft.) high and 790 m (2591 ft.) across.*

North American integration

The North American Free Trade Agreement (NAFTA), a hotly debated issue in Canada when it was being negotiated in the early 1990s, has produced a trade boom, especially for Ontario. However, Canadians have problems competing for foreign investment with Mexico, where labor costs, social welfare, and environmental standards are lower. Most Canadians oppose such ideas as a currency union and ever closer integration with the US.

***Bernard Landry,** the outspoken separatist premier of Québec.*

***Stephen Harper** was elected leader of the CA in 2002.*

***Jean Chrétien,** prime minister since 1993.*

C

WORLD AFFAIRS

Joined UN in 1945

Comm

G8

OECD

NAFTA

NATO

Canada's most important relationship is with the USA, its main trading partner. Canada backed the US-led "war on terrorism" after the attacks on US targets in September 2001; security along the immense US–Canadian border was tightened in an agreement three months later. There are tensions, such as over Cuba – Canada opposes US sanctions – and transboundary pollution. Canadians greatly fear damage from oil transported between Alaska and the rest of the US. Canada has protested to the WTO over US duties on its softwood timber, but a fisheries dispute relating to Pacific salmon was settled in 1999.

In the forefront on debt relief for the poorest countries, Canada also led the world campaign against antipersonnel mines, hosting the conference which agreed the 1997 Ottawa treaty.

AID

Donor

US$1.74bn (donations)

Up 2% in 2000

Canada's aid budget, earmarked for cuts in the 1990s, was given extra funding again at the end of the decade. NGOs supported by the Canadian International Development Agency (CIDA) are prominent on global development issues, and most Canadians support aid.

Aid now aims to provide know-how skills, rather than funding for large-scale development projects. CIDA has pioneered a theme-based approach, stressing human rights, basic needs, gender issues, and good governance. The regional focus of aid has gradually shifted, with less emphasis on Africa. Programs in the 1990s supported recovery and reform in former communist countries. More aid now goes toward the development needs of the Indian subcontinent.

DEFENSE

No compulsory military service

US$7.46bn

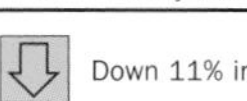

Down 11% in 2000

CANADIAN ARMED FORCES

114 main battle tanks (*Leopard* C-1/C-2)	18,600 personnel
1 submarine, 12 frigates, 4 destroyers, and 14 patrol boats	9000 personnel
140 combat aircraft (122 CF-18)	13,500 personnel
None	

Canada cooperates closely with the US on North American defense and security issues.

Defense spending was cut after the end of the Cold War, and many Canadians continue to advocate further reductions. Canada withdrew its forces stationed in Europe in 1992. The focus of defense planning is now the creation of rapid reaction forces. Canadian troops have served in many UN peacekeeping operations, most recently in Kosovo, East Timor, and Sierra Leone. Their involvement in Somalia, however, which ended in 1993, was tarnished by a scandal over racism, torture, and murder which shocked Canadian society.

ECONOMICS

Inflation 1.4% p.a. (1990–2000)

US$650bn

1.502–1.596 Canadian dollars

SCORE CARD

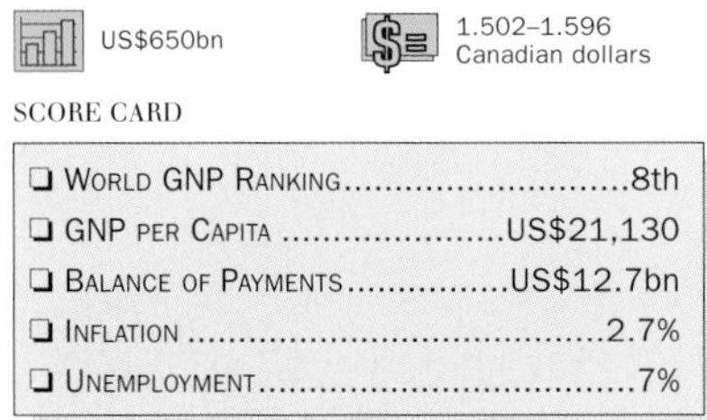

❑ WORLD GNP RANKING	8th
❑ GNP PER CAPITA	US$21,130
❑ BALANCE OF PAYMENTS	US$12.7bn
❑ INFLATION	2.7%
❑ UNEMPLOYMENT	7%

EXPORTS

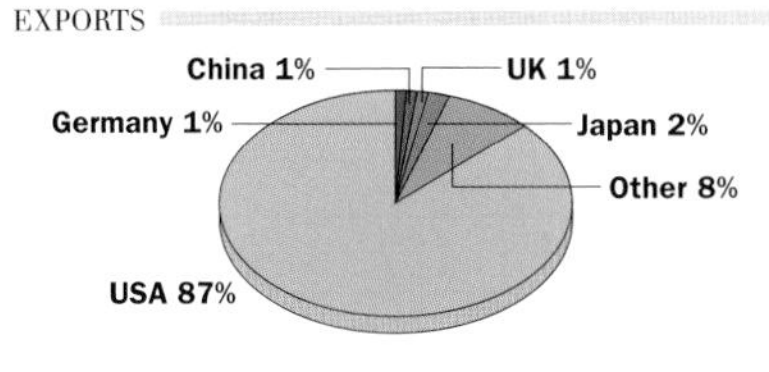
China 1%
UK 1%
Germany 1%
Japan 2%
Other 8%
USA 87%

IMPORTS

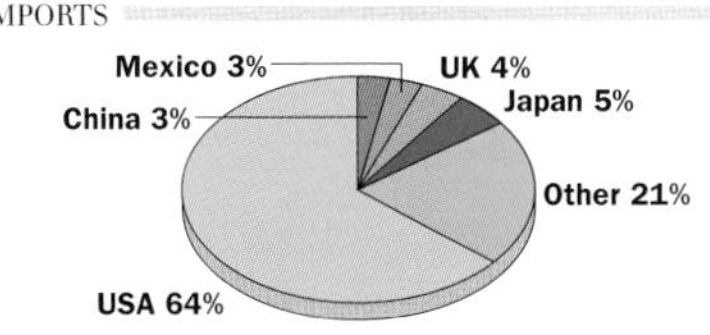
Mexico 3%
UK 4%
Japan 5%
China 3%
Other 21%
USA 64%

STRENGTHS

A broad and rich resource base. Provides exports, raw materials for manufacturing sector, and massive cheap energy, notably HEP; also large oil and gas reserves. Agriculture and forestry contribute 3% of GDP, mining 4%. Successful manufacturing sector, contributes 17% of GDP, especially forestry products, transportation equipment, and chemicals. Strong recovery and growth from mid-1990s to 2001. Access to huge US and Mexican markets through NAFTA. Low inflation.

WEAKNESSES

Problems of competitiveness; higher taxes, more regulations, lower productivity relative to NAFTA; other threats from globalization. Vulnerable to price fluctuations for raw material exports. Federal and provincial budget deficits are still high.

CANADA : MAJOR BUSINESSES

Aerospace industry
Vehicle manufacture
Timber industries
Pulp & paper
Food processing
Fish processing
Electronics
Engineering
Chemicals
Metallurgy
Oil & gas

ECONOMIC PERFORMANCE INDICATOR

Consumer Price Index | GDP

Consumer price index 1995=100 | GDP 1996=100

160, 140, 120, 100, 80

1996 1997 1998 1999 2000

PROFILE

Canada has an enormous resource base, and the UN rates it as having one of the highest standards of living in the world. Since the mid-1980s, however, its manufactured exports have faced increasing competition, while prices for its primary exports fluctuate. Real growth averaged 3.5% a year for most of the 1980s, but then stagnated for five years, while budget deficits rose, forcing restructuring at both federal and provincial levels. Many welfare programs were cut back, while the defense budget was sharply reduced. Growth resumed after 1993 and strengthened before the global economic downturn in 2001. Within NAFTA, Canadian firms have had to become more competitive to maintain exports. Most have been successful, with better productivity and a high-tech shift. Unemployment, at almost 10% in the mid-1990s, was down to just over 6% by 2000. By December 2001, it was back to 8%, but then fell again.

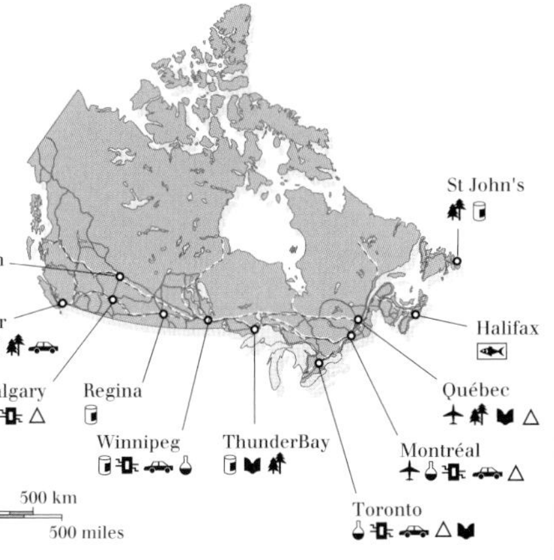
St John's
Edmonton
Vancouver
Halifax
Calgary
Regina
Québec
Winnipeg
ThunderBay
Montréal
Toronto
0 500 km
0 500 miles

C

RESOURCES

Electric power 115m kw

1.14m tonnes

2.76m b/d (reserves 6.6bn barrels)

13m cattle, 12.6m pigs, 5m turkeys, 158m chickens

Coal, oil, gas, gold, zinc, uranium, nickel, potash, asbestos, gypsum

ELECTRICITY GENERATION

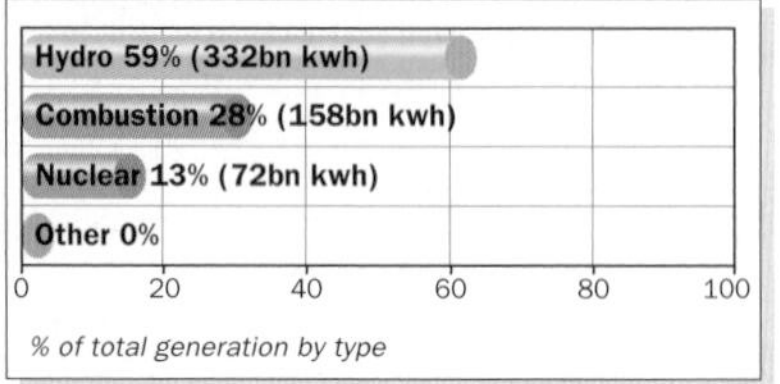

Hydro 59% (332bn kwh)
Combustion 28% (158bn kwh)
Nuclear 13% (72bn kwh)
Other 0%
0
20
40
60
80
100
% of total generation by type

Canada is a country of enormous natural resources. It is the world's largest exporter of forest products and a top exporter of fish, furs, and wheat. Minerals have played a key role in Canada's transformation into an urban–industrial economy. Alberta, British Columbia, Québec, and Saskatchewan are the principal mining regions. Ontario and the Northwest (NWT) and Yukon Territories are also significant producers. Canada is the world's largest producer of uranium, the second-largest of nickel, asbestos, potash, and gypsum, and the third-largest of zinc. Oil and gas are exploited in Alberta, off the Atlantic coast, and in the northwest – huge additional reserves are thought to exist in the high Arctic. Most exports go to the US. Canada is also one of the world's top hydroelectricity producers.

CANADA : LAND USE

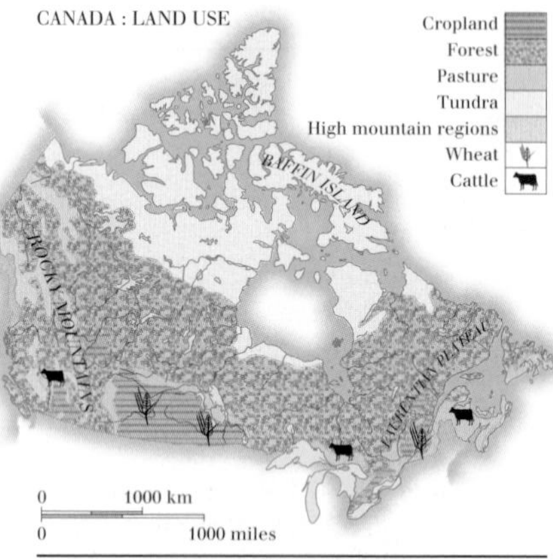

Cropland
Forest
Pasture
Tundra
High mountain regions
Wheat
Cattle
BAFFIN ISLAND
ROCKY MOUNTAINS
LAURENTIAN PLATEAU
0 1000 km
0 1000 miles

ENVIRONMENT

Sustainability rank: 4th

10% (2% partially protected)

15.4 tonnes per capita

ENVIRONMENTAL TREATIES

Yes
Yes
Yes
Yes
Yes
No

With a population of only some 31 million living in the world's third-largest country, Canada is justly renowned for vast tracts of wilderness untroubled by pollution either from industry or from intensive farming methods. A major conservation issue is the battle to stop the logging of virgin forest in northern Ontario and on the west coast. Notable successes were achieved in the late 1990s, pressuring timber companies to adopt more sustainable policies, and a landmark agreement in early 2001 promised protection for British Columbia's coastal Great Bear Rainforest.

Canadians have tighter pollution controls than the neighboring US. Ontario, the most polluted province, has imposed stricter limits on oil refineries and (from 2001) on electricity-generating plants. Carbon dioxide emissions (mainly from cars) are among the highest in the world per capita. Canada has accepted a target of a 6% cut by 2010. Production of hazardous waste is also higher than the European average.

MEDIA

TV ownership high

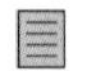
Daily newspaper circulation 158 per 1000 people

PUBLISHING AND BROADCAST MEDIA

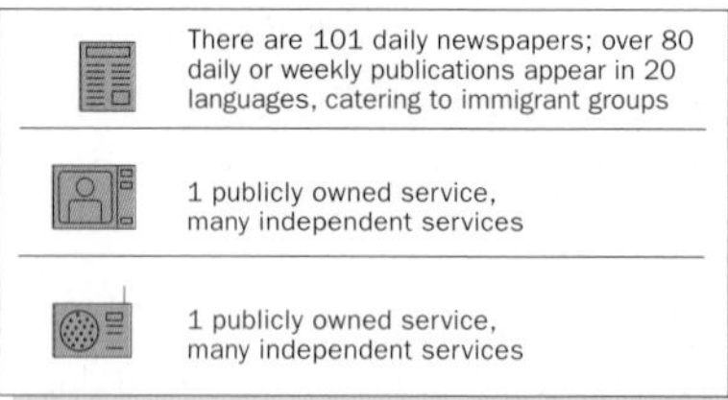

There are 101 daily newspapers; over 80 daily or weekly publications appear in 20 languages, catering to immigrant groups
1 publicly owned service, many independent services
1 publicly owned service, many independent services

The public Canadian Broadcasting Corporation (CBC) runs two national TV channels, in English and French; the Canadian Alliance wants it partially privatized. Local cable services often include multilingual or ethnic channels. Canadian TV is renowned for its news and sports coverage. *La Presse* is the leading French-language daily and the *Globe and Mail* the leading serious newspaper in English.

***Autumn in the tundra** in northern Canada. Trees such as the black spruce are subject to the effects of acid rain originating in the USA's northern industrial regions.*

CRIME

No death penalty

31,600 prisoners

Down 4% in 1999

CRIME RATES

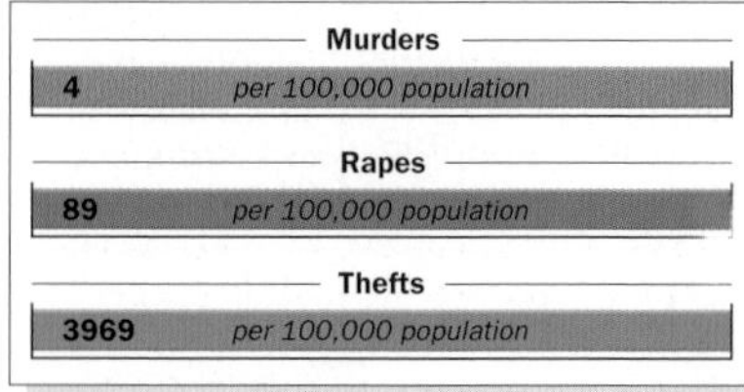

Murders
4 per 100,000 population
Rapes
89 per 100,000 population
Thefts
3969 per 100,000 population

Rates for serious crime are lower in Canada than in the US. Canadians ascribe this to their far stricter gun control laws, which were further tightened in the 1990s.

Newfoundland police began carrying guns routinely only in 1998, the last force in North America to do so. There have been careful efforts to maintain the inner cities as crime-free zones.

To address the narcotics problem, innovative "drug treatment courts" in Toronto and Vancouver link criminal justice with treatment and social care.

EDUCATION

School leaving age: 16

99%

1.56m students

THE EDUCATION SYSTEM

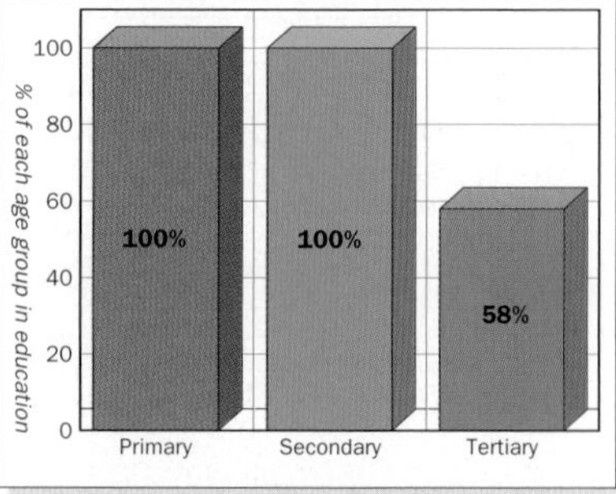

100
80
60
40
20
0
% of each age group in education
100%
100%
58%
Primary
Secondary
Tertiary

Education is a responsibility of the individual provinces, rather than the federal government, and is accorded a very high priority. The period of free compulsory school attendance varies, but is a minimum of nine years.

The prime medium of instruction is English in all provinces except francophone Québec. In several other provinces, French-speaking students are entitled to be taught in French. Multicultural education also helps maintain the cultural identity of immigrant groups.

Canada has 76 universities and some 200 other higher education institutions. Nearly all high school graduates go on to some form of tertiary or further education. This represents the highest proportion in the industrialized world.

QUÉBEC'S DISTINCT SOCIETY AND SEPARATISM

QUÉBEC IS CANADA'S largest province, with an area of 1,667,926 sq. km (594,860 sq. miles). Its population, 7.3 million at the 1994 census, includes more than five million of Canada's 6.1 million French Canadians, most of whom are Catholics. Its capital is Québec City. The leading commercial center, Montréal, has suffered a decline in prestige compared with Toronto, in neighboring Ontario, which is overwhelmingly English-speaking. Québec has massive hydroelectric power resources and vast areas of forest, and its principal industries include timber, pulp, and paper, and mining, particularly for iron ore. Consequently, environmental protection legislation is relatively lax, and there is conflict with those seeking to protect First Nation lands from devastation by logging and massive dam schemes.

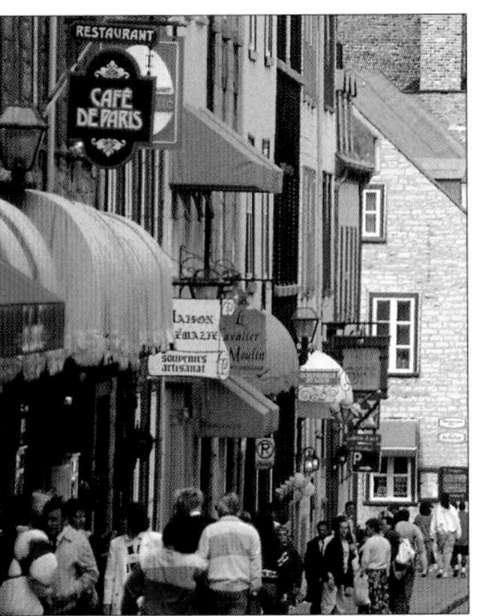
RESTAURANT
CAFÉ DE PARIS

French shop signs *above the streets of Québec City.*

Conquered by the British in the 18th century, the Québécois retained the French civil code under the 1774 Québec Act, but French Canadians only gradually recovered minority language rights suppressed after an unsuccessful rebellion in the 1830s. A "quiet revolution" began in the 1960s, based initially on militant trade unionism. This brought far-reaching changes in the social, economic, and political balance in Québec. The wage gap closed, and francophones now slightly out-earn anglophones, partly because many of the best-educated anglophones tend to leave the province. Francophone-owned businesses were built up through the "Québec Inc." project. Anglophone dominance in government and the civil service was reversed, and higher education opportunities expanded for francophones. Francophone militancy was also channeled into party politics, fueled by a harsh security clampdown in the early 1970s against the guerrilla Québec Liberation Front. The Parti Québécois (PQ) unexpectedly gained control of the provincial assembly in 1976, and a French language charter, Bill 101, was enacted almost immediately. French became not only the official language but compulsory in government and on public signs. Non-francophones protested, and the Supreme Court ruled in 1988 that part of the language charter violated their human rights, but the provincial government managed to enforce these rules on a "temporary" basis.

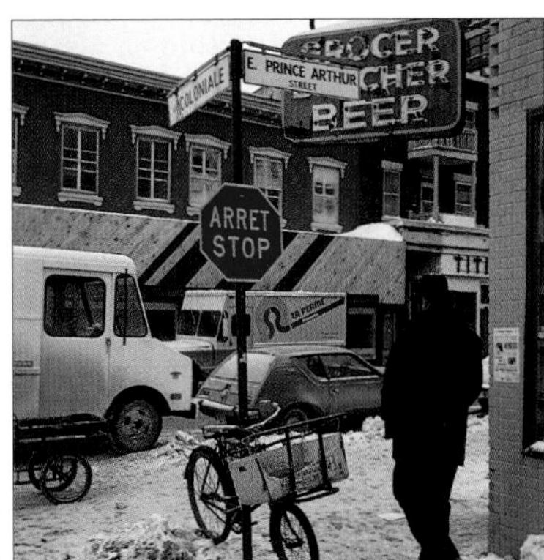
E. PRINCE ARTHUR STREET
GROCER
BEER
ARRET
STOP

Bilingual street signs *in Québec have now largely been replaced by French-only signs.*

The PQ's other initiative was sovereignty for Québec. After its defeat in a referendum in 1980, the idea was revived in the 1990s amid long-running disputes about the Canadian constitution and provincial–federal relationships. The PQ, promising a second referendum (and backed by its newly formed counterpart at federal level, the BQ), returned to power in Québec in 1994. The following year, 60% of Québec's francophone majority voted in favor of sovereignty in association with Canada, a high-water mark for separatism; the proposition was defeated by the narrowest of margins by the non-francophone vote. Although the PQ said that it would seek a third referendum, federal legislation limits the conditions under which it can do so, and the party's commitment to separatism has lost much of its impetus, with disappointing showings in the 1998 provincial and 2000 federal elections. Even francophones, still fiercely protective of their language and culture, appear increasingly disposed to settle for a federal compromise acknowledging Québec's "unique character."

HEALTH

Welfare state health benefits

1 per 476 people

Heart and respiratory diseases, cancers, accidents

The comprehensive state health service is funded from national insurance.

Rising costs are the result of an aging population and the spread of more sophisticated and expensive treatments. Health care was the main issue of the 2000 election campaign. Popular backing for retaining the present publicly funded system has encouraged the LP government to restore spending to earlier levels, after a period of cuts made in an effort to reduce the budget deficit. About 25% of Canadians use private health facilities.

SPENDING

GDP/cap. increase

CONSUMPTION AND SPENDING

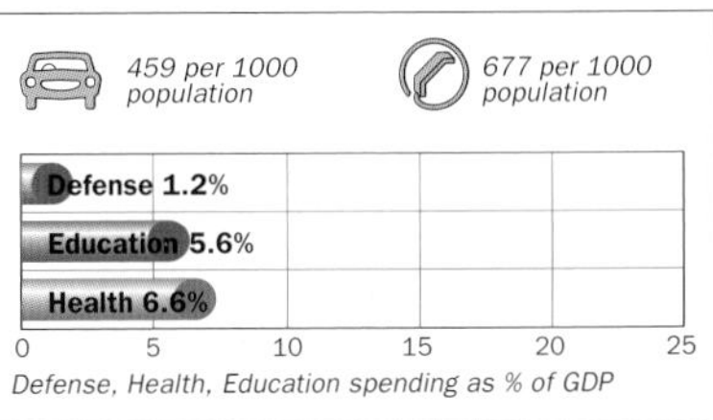
459 per 1000 population
677 per 1000 population
Defense 1.2%
Education 5.6%
Health 6.6%
0
5
10
15
20
25
Defense, Health, Education spending as % of GDP

Despite strains caused by recession during the early 1990s – including a rise in unemployment to over 10% – life for most Canadians remains very good.

The UN ranks Canada as one of the best countries in the world in which to live. In its overall assessment of human development indicators such as income, education, and life expectancy, Canada consistently comes out near the top, and ahead of the USA.

However, disadvantaged groups do exist, in particular among indigenous Canadians. Unemployment, poor housing, and mortality rates for Amerindians and Inuits are well above those for other Canadians; the Inuit suicide rate is three times higher. Those Amerindians who live on reserves are the poorest group.

WORLD RANKING

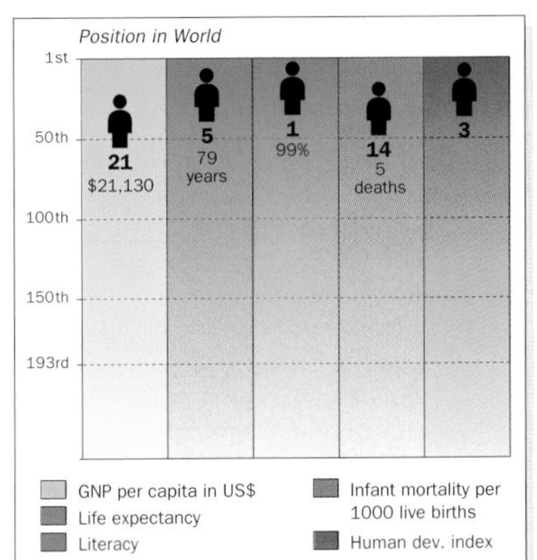
Position in World
1st
50th
100th
150th
193rd
21
$21,130
5
79 years
1
99%
14
5 deaths
3
GNP per capita in US$
Life expectancy
Literacy
Infant mortality per 1000 live births
Human dev. index

C

CAPE VERDE

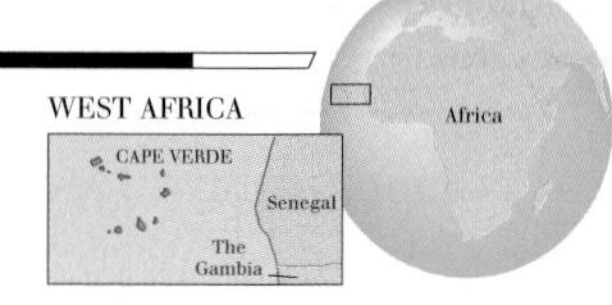
WEST AFRICA
CAPE VERDE
Senegal
The Gambia
Africa

OFFICIAL NAME: Republic of Cape Verde **CAPITAL:** Praia
POPULATION: 437,000 **CURRENCY:** Cape Verde escudo **OFFICIAL LANGUAGE:** Portuguese

1975

1975
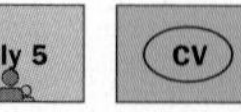
July 5
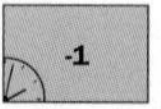
CV

-1

+238
.cv

The Cape Verde archipelago off the west coast of Africa became independent of Portugal in 1975. Most of the islands are mountainous and volcanic; the low-lying islands of Sal, Boa Vista, and Maio have agricultural potential, though they are prone to debilitating droughts. Around 50% of the population live on São Tiago. Following a period of single-party socialist rule, Cape Verde held its first multiparty elections in 1991.

CLIMATE

Tropical oceanic

WEATHER CHART FOR PRAIA

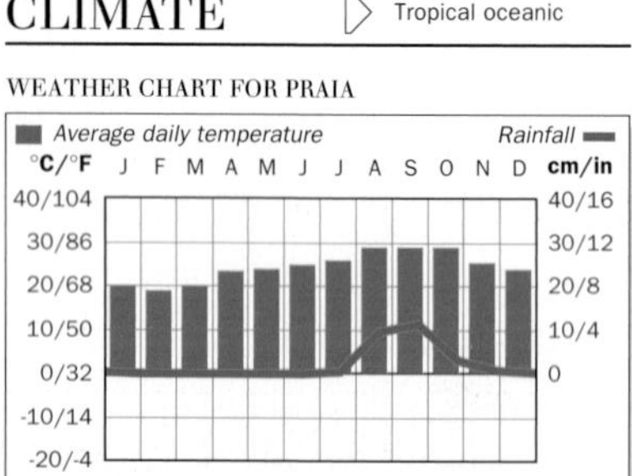
Average daily temperature
Rainfall
°C/°F J F M A M J J A S O N D cm/in
40/104
30/86
20/68
10/50
0/32
-10/14
-20/4
40/16
30/12
20/8
10/4
0

Cape Verde has a very dry climate, subject to droughts that sometimes last for years at a time.

TRANSPORTATION

Drive on right

Amílcar Cabral, Sal Island 635,563 passengers

42 ships 20,800 grt

THE TRANSPORTATION NETWORK

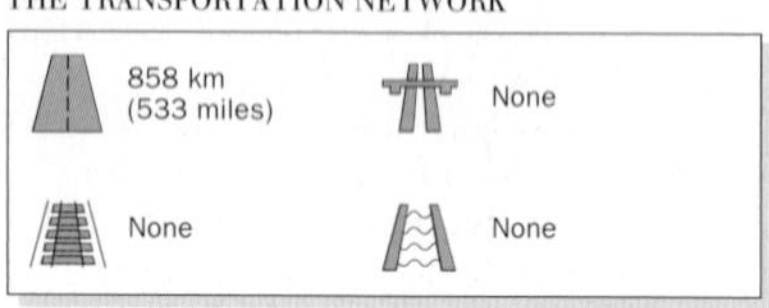
858 km
(533 miles)
None
None
None

Cape Verde has a strategic position on international sea and air routes, which it is beginning to exploit.

TOURISM

Visitors : Population 1:5.9

74,000 visitors

Up 68% in 2000

MAIN TOURIST ARRIVALS

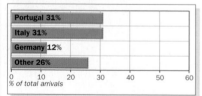

	% of total arrivals
Portugal	31%
Italy	31%
Germany	12%
Other	26%

Tourism has not been a government priority and remains on a modest scale. The islands of São Tiago, Santo Antão, Fogo, and Brava have tourist potential, offering a combination of mountain scenery and extensive beaches.

PEOPLE

Pop. density medium

Portuguese Creole, Portuguese

108/m² (281/mi²)

THE URBAN/RURAL POPULATION SPLIT

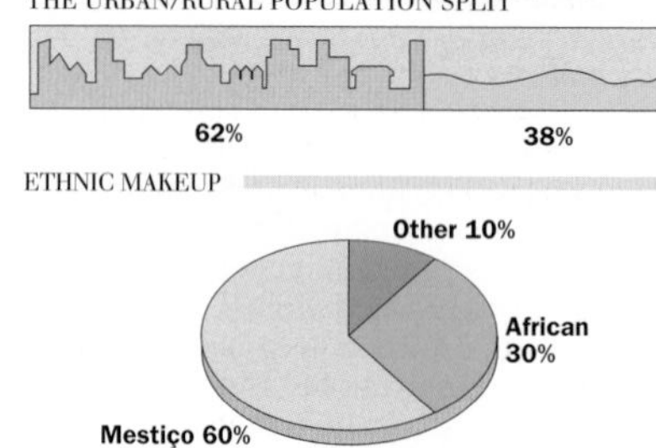
62%
38%
ETHNIC MAKEUP
Other 10%
African 30%
Mestiço 60%

The majority of the population is Portuguese–African *mestiço*; the remainder is largely African, descended either from slaves or from more recent immigrants from the mainland. The Creolization of the culture has led to a relative lack of ethnic tension, though there is some bad feeling between islands. African traditions of the extended family, as well as the Roman Catholic Church, have helped to ensure the vitality of family life. Women outnumber men in Cape Verde, with many single mothers acting as head of the household and main breadwinner.

POLITICS

Multiparty elections

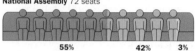

2001/2006

President Pedro Pires

AT THE LAST ELECTION

National Assembly 72 seats

55% PAICV — 42% MPD — 3% ADM

PAICV = African Party for the Independence of Cape Verde
MPD = Movement for Democracy
ADM = Democratic Alliance for Change

Cape Verde experienced a peaceful transition to multipartyism in 1991, when elections brought the MPD to power. Although there had previously been a decade of single-party rule under the PAICV, it had in fact operated a liberal system in which opposition and dissent were tolerated. The large Cape Verdean diaspora had an important influence in effecting the transition to multiparty politics.

The MPD was defeated in legislative elections in January 2001, when the PAICV was returned to power with an absolute majority. Pedro Pires of the PAICV was elected president the following month, beating his MPD rival by just 17 votes. The main issue for the current government is that of economic survival, particularly during periods of drought.

WORLD AFFAIRS

Joined UN in 1975

CPLP ECOWAS OIF NAM AU

Cape Verde wishes to diversify its international contacts in order to secure aid, while maintaining good relations with Portugal, the former colonial power. Within the region, Cape Verde seeks to restore normal relations with Guinea-Bissau, having withdrawn from a proposed union in 1980, and to improve contacts with other mainland states, such as Senegal.

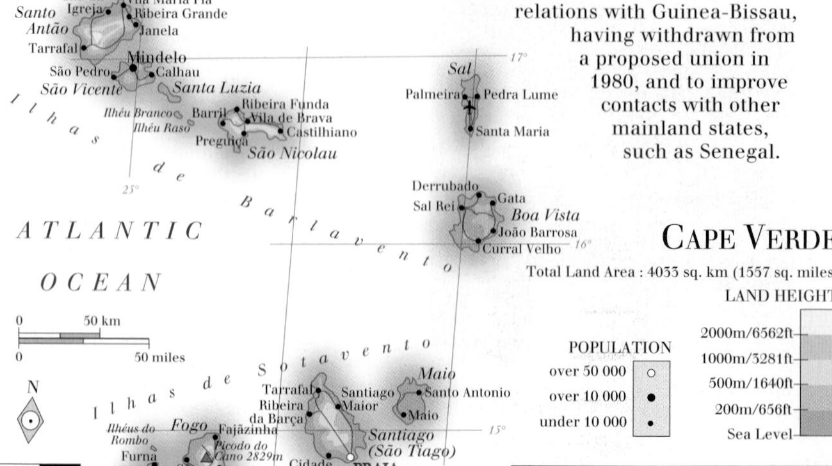
Chã da Igreja
Vila Maria Pia
Ribeira Grande
Janela
Santo Antão
Tarrafal
Mindelo
Calhau
São Pedro
São Vicente
Santa Luzia
Ilhéu Branco
Ilhéu Raso
Barril
Ribeira Funda
Vila de Brava
Castilhiano
Preguiça
São Nicolau
Ilhas de Barlavento
Sal
Palmeira
Pedra Lume
Santa Maria
Derrubado
Gata
Sal Rei
Boa Vista
João Barrosa
Curral Velho
ATLANTIC OCEAN
Ilhas de Sotavento
Maio
Santo Antonio
Tarrafal
Santiago Maior
Ribeira da Barça
Fogo
Fajãzinha
Ilhéus do Rombo
Furna
Fajã
Brava
São Filipe
Pico do Cano 2829m
Cova Figueira
Cidade Velha
Santiago (São Tiago)
PRAIA
50 km
50 miles
N
CAPE VERDE
Total Land Area : 4033 sq. km (1557 sq. miles)
LAND HEIGHT
2000m/6562ft
1000m/3281ft
500m/1640ft
200m/656ft
Sea Level
POPULATION
over 50 000
over 10 000
under 10 000

AID

 Recipient

 $94m (receipts)

 Down 31% in 2000

Portugal is the main source of aid. Other major donors include the World Bank, Germany, Japan, and the Netherlands. An important donor is the EU, which has provided substantial food aid in the wake of recent droughts, as well as funding aid programs. Aid finances almost all development in Cape Verde, which is one of the least industrialized countries in the world.

DEFENSE

Compulsory military service

$7m

No change in 2000

After independence, small armed forces were established, now consisting of a 1000-strong army, a small air force, and a naval coastguard. They have never been called upon to play a political role; their main duties are to protect territorial waters against illegal fishing and to curb smuggling.

ECONOMICS

Inflation 4.9% p.a. (1990–2000)

 $588m

 118.2–119.8 Cape Verde escudos

SCORE CARD

❑ WORLD GNP RANKING	167th
❑ GNP PER CAPITA	$1330
❑ BALANCE OF PAYMENTS	–$67m
❑ INFLATION	4.4%
❑ UNEMPLOYMENT	24%

STRENGTHS

Strategic location, off the westernmost tip of Africa, close to the mid-Atlantic where Africa is nearest to Latin America. This has military and economic advantages, including shipping maintenance and air travel. Low debt-servicing costs.

WEAKNESSES

Permanent threat of drought and water supply problems, despite desalination plants. Lack of agricultural land and dependency on food aid. Difficulties of communications between islands.

EXPORTS

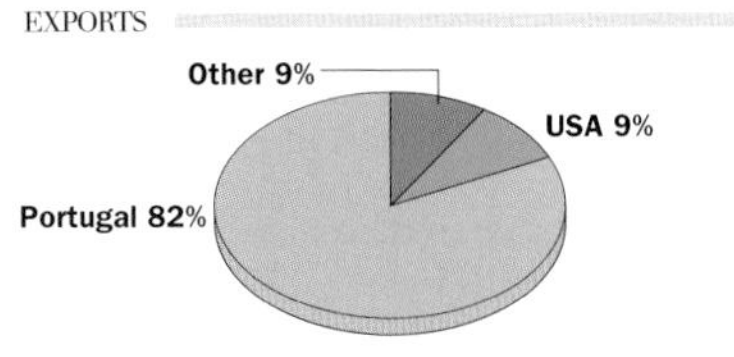
Other 9%
USA 9%
Portugal 82%

IMPORTS

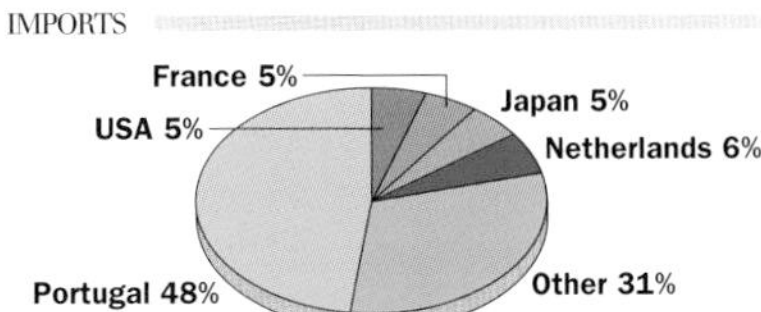
France 5%
Japan 5%
USA 5%
Netherlands 6%
Portugal 48%
Other 31%

***Portuguese colonial-style architecture** on Fogo, one of the larger islands. The volcano in its center is the highest point in Cape Verde.*

RESOURCES

Electric power 7000 kw

10,371 tonnes

Not an oil producer

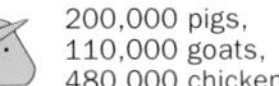 200,000 pigs, 110,000 goats, 480,000 chickens

 Salt, pozzolana

Cape Verde has no known strategic resources. With no oil or gas and no possibility of hydroelectric power, it depends on imported petroleum for energy. However, experimental projects have been carried out to investigate the potential of wave power, windmills, and biogas.

ENVIRONMENT

Not available

 None

 0.3 tonnes per capita

Cape Verde has recently suffered several years of persistent drought, which have affected food production and reduced livestock herds. It is a very active member of CILSS, which struggles against drought in the Sahel region. Environmental initiatives include reforestation, soil conservation, and a water resources program.

MEDIA

TV ownership low

 There are no daily newspapers

PUBLISHING AND BROADCAST MEDIA

There are no daily newspapers. Independent publications suffer from financial pressures

1 state-controlled service

1 state-controlled service

The government publishes three weeklies, but there are no daily newspapers. Press freedom is guaranteed by law. TV and radio broadcasting are in Portuguese and Creole, with the cooperation of the Portuguese service RTPI.

CRIME

No death penalty

 775 prisoners

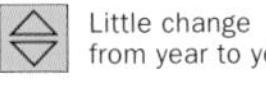 Little change from year to year

Crime is not a serious problem, even in urban centers, though smuggling is fairly widespread.

CHRONOLOGY

Cape Verde was a Portuguese colony from 1462 until 1975, and was ruled jointly with Guinea-Bissau.

- ❑ **1961** Joint struggle for independence of Cape Verde and Guinea-Bissau begins.
- ❑ **1974** Guinea-Bissau independent.
- ❑ **1975** Cape Verde independent.
- ❑ **1981** Final split from Guinea-Bissau.
- ❑ **1991** MPD wins first multiparty poll.
- ❑ **2001** General election returns PAICV to power.

C

EDUCATION

School leaving age: 13

 74%

 Not available

At independence, education became a priority; 80% of children now attend elementary school, and more than 60% go on to secondary education.

HEALTH

No welfare state health benefits

1 per 5882 people

Heart disease, tuberculosis, typhoid, and accidents

Health care has improved since the colonial period, and polio has been virtually eradicated.

SPENDING

GDP/cap. increase

CONSUMPTION AND SPENDING

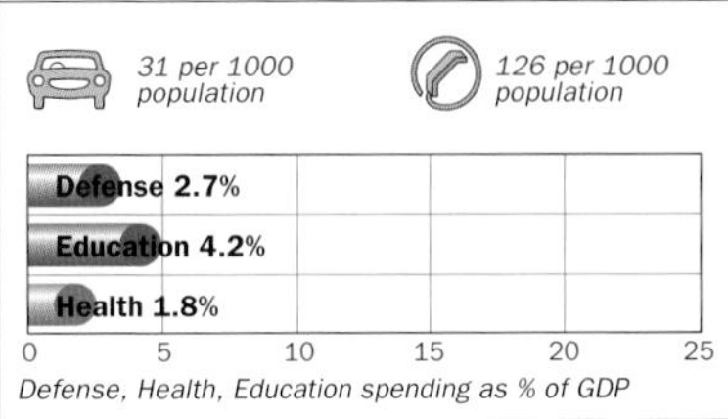
31 per 1000 population
126 per 1000 population
Defense 2.7%
Education 4.2%
Health 1.8%
0
5
10
15
20
25
Defense, Health, Education spending as % of GDP

In comparison with the 90% of the population engaged in primary production, the small business class in Praia is well off.

WORLD RANKING

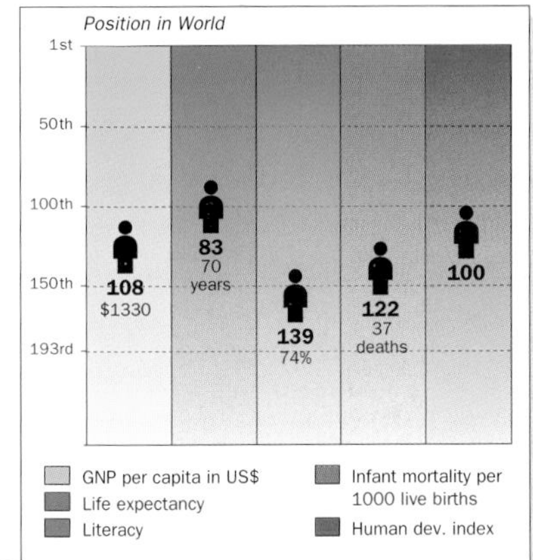
Position in World
1st
50th
100th
150th
193rd
108
$1330
83
70
years
139
74%
122
37
deaths
100
GNP per capita in US$
Life expectancy
Literacy
Infant mortality per 1000 live births
Human dev. index

CENTRAL AFRICAN REPUBLIC

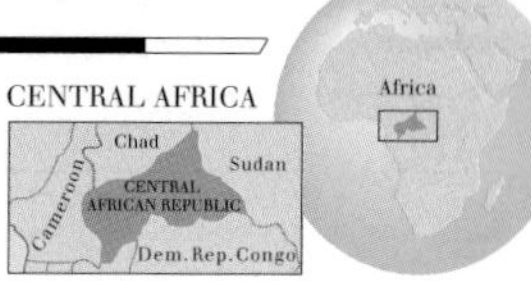
CENTRAL AFRICA
Africa
Chad
Sudan
Cameroon
CENTRAL AFRICAN REPUBLIC
Dem. Rep. Congo

OFFICIAL NAME: Central African Republic **CAPITAL:** Bangui
POPULATION: 3.8 million **CURRENCY:** CFA franc **OFFICIAL LANGUAGE:** French

C

1960
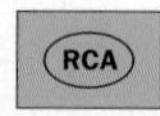
1960
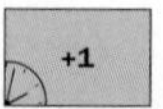
Dec 1

RCA

+1

+236 .cf

LANDLOCKED AT THE WESTERN end of the Sahel, the Central African Republic (CAR) is a low plateau stretching north from one of Africa's great rivers, the Ubangi, which forms its border with the Democratic Republic of the Congo. Almost all the population lives in the equatorial, rainforested south. "Emperor" Bokassa's eccentric rule from 1965 to 1979 was followed by military dictatorship. Democracy was restored in 1993.

CLIMATE

Tropical equatorial

WEATHER CHART FOR BANGUI

Average daily temperature / Rainfall
°C/°F J F M A M J J A S O N D cm/in
40/104, 30/86, 20/68, 10/50, 0/32, -10/14, -20/-4
40/16, 30/12, 20/8, 10/4, 0

The south is equatorial, the north has a savanna-type climate, and the far north lies within the Sahel.

TRANSPORTATION

Drive on right

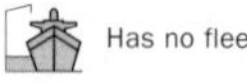
Mpoko, Bangui
76,755 passengers
Has no fleet

THE TRANSPORTATION NETWORK

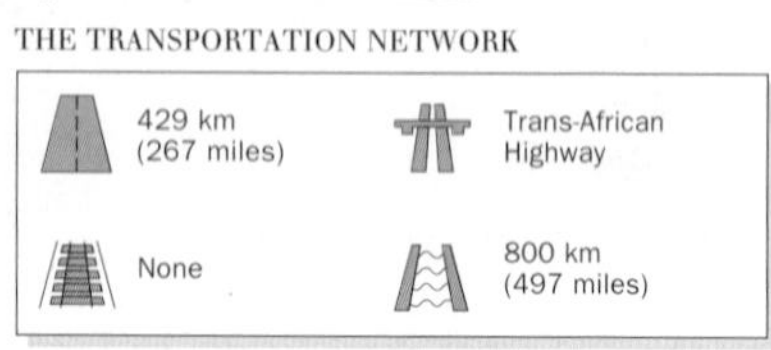
429 km
(267 miles)
Trans-African Highway
None
800 km
(497 miles)

The CAR has a limited transportation system, depending on the river link to Brazzaville, Congo, and rail from there to Pointe-Noire and the Congo River ports.

TOURISM

Visitors : Population 1:380

10,000 visitors

Up 43% in 1999

MAIN TOURIST ARRIVALS

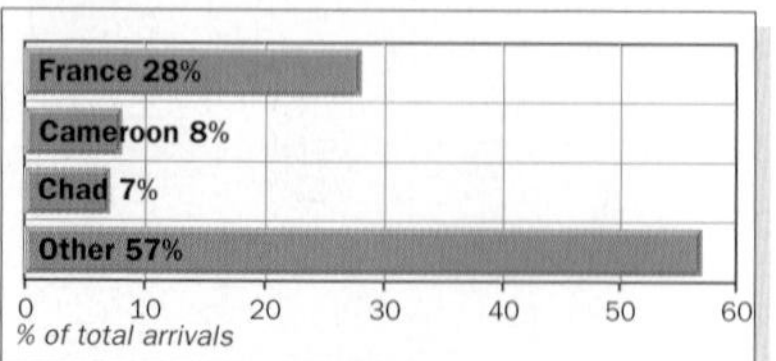
France 28%
Cameroon 8%
Chad 7%
Other 57%
0
10
20
30
40
50
60
% of total arrivals

Tourist promotion is small-scale, but since 1979 there has been a modest increase in national park safaris. A new runway in Bangui will permit air charters, chiefly from France.

PEOPLE

Pop. density low

Sango, Banda, Gbaya, French

6/km² (16/mi²)

THE URBAN/RURAL POPULATION SPLIT

41% 59%

ETHNIC MAKEUP

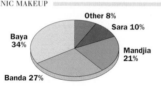

Other 8%
Sara 10%
Mandjia 21%
Banda 27%
Baya 34%

Although the Baya and the Banda are the largest ethnic groups, the lingua franca is Sango. This is spoken by the southern riverine minorities, who provided the political leaders from independence until 1993 (Presidents Dacko and Kolingba and "Emperor" Bokassa). President Patasse is from the interior. Resentment against the river peoples occasionally flares up, as happened after the coup attempt in 2001. As in other non-Muslim African countries, women have considerable power. Elizabeth Domitien was prime minister from 1975 to 1976 and Ruth Rolland ran for president in 1993.

WORLD AFFAIRS

Joined UN in 1960

BDEAC CEMAC FZ LCBC AU

An improving image of political stability since 1996 was shattered by 2001. Libya emerged as the major ally of President Patasse, becoming directly involved in a series of coups and armed rebellions. Relations with Cameroon, which provides key access to the sea, were also threatened in heightened border tension in 2000. Financial help from France is relied upon and will be needed for some time.

AID

Recipient

$76m (receipts)

Down 35% in 2000

Almost all development projects are funded from external aid. France, the former colonial power, provides one-third of the total. The EU and Japan are other major donors. The CAR also receives assistance from the World Bank and the IMF, which has provided funds to support the country's economic program. The Paris Club of creditor nations rescheduled and reduced the CAR's debt in 1998.

DEFENSE

Compulsory military service

$43m

Down 4% in 2000

The well-equipped 3000-strong army is a heavy drain on the budget. Military service is selective. French officers fill senior army posts. In 1996, 1400 resident French troops intervened to quell army rebellions; they began to withdraw in 1997; UNMICAR forces were deployed from 1997 to 2000. Gen. Francois Bozize led the military rebellion in 2001.

POLITICS

Multiparty elections

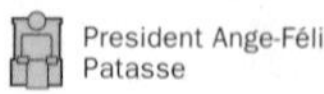
1998/2003

President Ange-Félix Patasse

AT THE LAST ELECTION

National Assembly 109 seats

6% Ind, 6% FPP, 3% PUN, 4% Others

43% MLPC, 18% RDC, 7% MDD, 6% PSD, 5% ADP, 2% PLD

MLPC = Central African People's Liberation Movement **RDC** = Central African Democratic Rally **MDD** = Movement for Democracy and Development **Ind** = Independents **PSD** = Social Democratic Party **FPP** = Patriotic Front for Progress **ADP** = Alliance for Democracy and Progress **PUN** = National Unity Party **PLD** = Liberal Democratic Party

Democratic elections in 1993 ended four years of Gen. André Kolingba's single-party rule, bringing in Ange-Félix Patasse as president. Patasse, Bokassa's prime minister during the 1970s, had been jailed for dissent and subsequently went into exile. Army mutinies in 1996 prompted the formation of a government of national unity, but fighting continued until in February 1997 a French-led multinational force intervened to keep the peace. It was replaced by a UN force (UNMICAR) in April 1998. Patasse's party, the MLPC, remained the most important in the new parliament after the 1998 elections but needs the support of opposition parties to maintain a workable coalition. In 2001, with assistance from Libyan troops, the government weathered a coup attempt and a military insurrection.

CENTRAL AFRICAN REPUBLIC

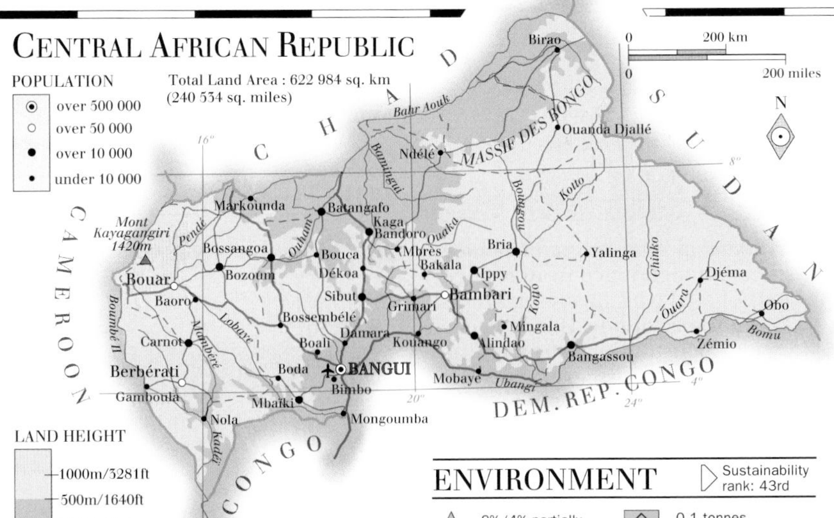
POPULATION
over 500 000
over 50 000
over 10 000
under 10 000
Total Land Area : 622 984 sq. km (240 534 sq. miles)
0 200 km
0 200 miles
N
CHAD
SUDAN
CAMEROON
CONGO
DEM. REP. CONGO
Birao
Bahr Aouk
MASSIF DES BONGO
Ouanda Djallé
Ndélé
Bamingui
Markounda
Batangafo
Kaga Bandoro
Mont Kayagangiri 1420m
Bossangoa
Bouca
Mbrès
Bakala
Bria
Yalinga
Ippy
Bouar
Bozoum
Dékoa
Sibut
Bambari
Djéma
Baoro
Grimari
Bossembélé
Mingala
Obo
Carnot
Boali
Damara
Kouango
Alindao
Zémio
Bangassou
Berbérati
Boda
BANGUI
Bimbo
Mobaye
Gamboula
Mbaïki
Nola
Mongoumba
Ubangi
Ouaka
Kotto
Chinko
Ouara
Bomu
Lobaye
Mambéré
Kadéï
Pende
Ouham
LAND HEIGHT
1000m/3281ft
500m/1640ft
200m/656ft

ECONOMICS

Inflation 4.6% p.a. (1990–2000)

$1.03bn

698.7–736.7 CFA francs

SCORE CARD

- World GNP Ranking 152nd
- GNP per Capita $280
- Balance of Payments.....................Zero
- Inflation ....–1.5%
- Unemployment 6%

STRENGTHS

Self-sufficiency in food. Some diversity of export earnings (diamonds, cotton, timber, iron, coffee). Status as transit zone in central Africa. Trans-African Highway and waterways.

WEAKNESSES

Landlocked. Poor infrastructure. Not enough trained people to run economy.

EXPORTS

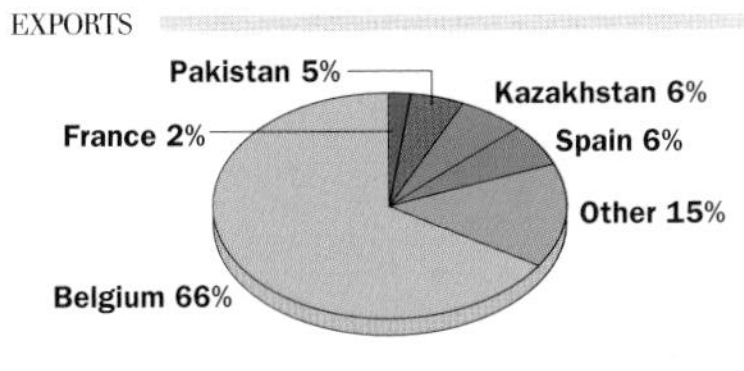
Pakistan 5%
Kazakhstan 6%
France 2%
Spain 6%
Other 15%
Belgium 66%

IMPORTS

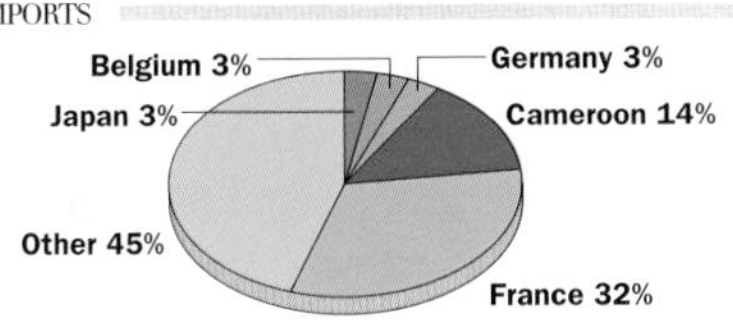
Belgium 3%
Germany 3%
Japan 3%
Cameroon 14%
Other 45%
France 32%

RESOURCES

Electric power 43,000 kw

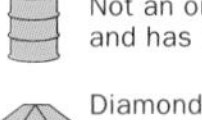

15,117 tonnes

Not an oil producer and has no refineries

3.1m cattle, 2.6m goats, 680,000 pigs, 4.2m chickens

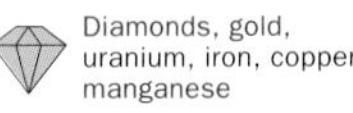

Diamonds, gold, uranium, iron, copper, manganese

Cotton and diamonds are the CAR's major exports. Other minerals are potentially important.

ENVIRONMENT

Sustainability rank: 43rd

8% (4% partially protected)

0.1 tonnes per capita

There has been an attempt to impose a conservationist forest policy. Hunting of elephants was banned in 1985, as numbers had fallen from 80,000 in the mid-1970s to just over 13,000.

MEDIA

TV ownership low

Daily newspaper circulation 2 per 1000 people

PUBLISHING AND BROADCAST MEDIA

There are 3 daily newspapers, *E Le Songo*, *Le Citoyen*, and *Le Novateur*

1 state-owned service

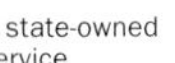

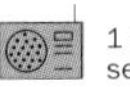

1 state-owned service

The three weeklies and three daily newspapers have only limited circulation. A small opposition press has developed with multipartyism, but is inhibited by lack of resources.

CRIME

Death penalty not used in practice

The CAR does not publish prison figures

Crime is rising

Human rights abuses have been reduced dramatically since the excesses of the Bokassa years. The level of criminality is usually low. The major criminal problem appears to be the increase in urban robbery resulting from continued political instability from 1996.

***Baskets of cotton,** Meme village. Cotton is one of the Central African Republic's most significant export crops.*

CHRONOLOGY

The French established the colony of Ubangi-Chari in 1905 and gave it autonomy as the CAR in 1958.

- **1960** Independence under David Dacko; one-party state.
- **1965** Coup by Jean-Bédel Bokassa.
- **1977** Bokassa crowned "emperor."
- **1979** French help reinstate Dacko.
- **1981** Gen. Kolingba ousts Dacko.
- **1993** Democracy restored.
- **1996** Government of national unity formed following army rebellion.
- **2001** Coup attempt. Army rebellion.

C

EDUCATION

School leaving age: 14

47%

6229 students

Schooling, on the French model, is compulsory, but in practice is only received by 68% of 6–14 year olds.

HEALTH

No welfare state health benefits

1 per 20,000 people

Communicable and parasitic diseases, malnutrition

Colonial neglect and postcolonial maladministration have resulted in a poorly developed health system.

SPENDING

GDP/cap. increase

CONSUMPTION AND SPENDING

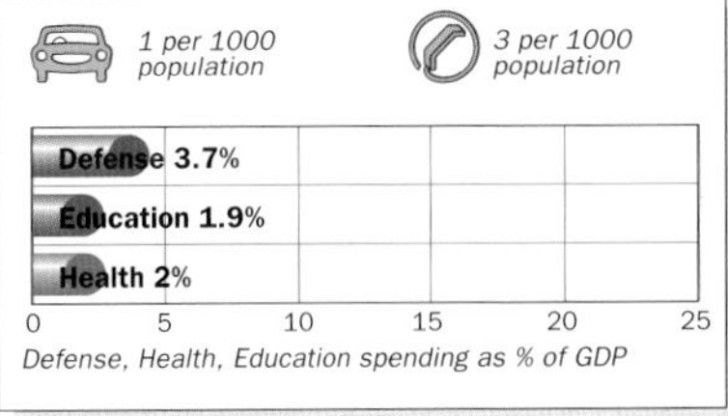
1 per 1000 population
3 per 1000 population
Defense 3.7%
Education 1.9%
Health 2%
0
5
10
15
20
25
Defense, Health, Education spending as % of GDP

There is a small political–military elite in the CAR, which came into being only in postcolonial days. For its members, Paris is the chosen destination and style leader.

WORLD RANKING

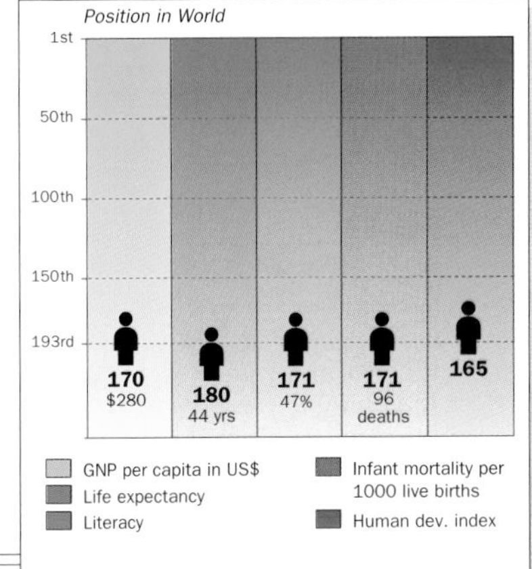
Position in World
1st
50th
100th
150th
193rd
170
$280
180
44 yrs
171
47%
171
96 deaths
165
GNP per capita in US$
Life expectancy
Literacy
Infant mortality per 1000 live births
Human dev. index

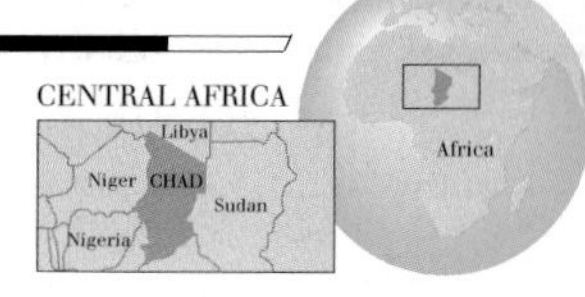
CENTRAL AFRICA
Libya
Niger
CHAD
Sudan
Nigeria
Africa

Chad

Official Name: Republic of Chad **Capital:** N'Djamena **Population:** 8.1 million
Currency: CFA franc **Official Languages:** Arabic and French

C

1960

1960

Aug 11
TCH

+1

+235
.td

LANDLOCKED IN NORTH central Africa, Chad has had a turbulent history since independence from France in 1960. Intermittent periods of civil war, involving French and Libyan troops, followed a coup in 1975. Another coup in 1990 preceded a transition to multipartyism, enshrined in a new constitution. The tropical, cotton-producing south is the most populous region. The discovery of large oil reserves could eventually have a dramatic impact on the economy.

CLIMATE

Hot desert/steppe

WEATHER CHART FOR N'DJAMENA

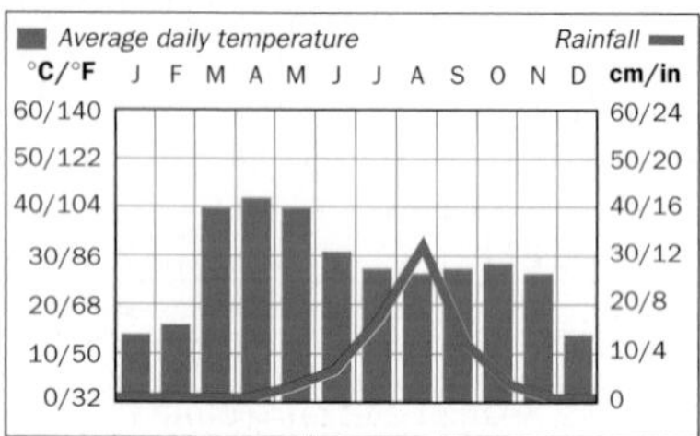
Average daily temperature
Rainfall
°C/°F
J F M A M J J A S O N D
cm/in
60/140
50/122
40/104
30/86
20/68
10/50
0/32
60/24
50/20
40/16
30/12
20/8
10/4
0

There are three distinct zones: the tropical south, the central semiarid Sahelian belt, and the desert north.

TRANSPORTATION

Drive on right

N'Djamena International 16,861 passengers

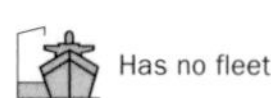
Has no fleet

THE TRANSPORTATION NETWORK

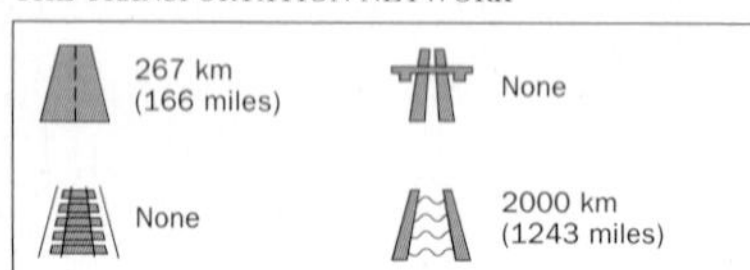
267 km
(166 miles)
None
None
2000 km
(1243 miles)

Chad has a limited transportation infrastructure. The nearest rail links are in Nigeria and Cameroon.

TOURISM

Visitors : Population 1:203

40,000 visitors

Down 15% in 2000

MAIN TOURIST ARRIVALS

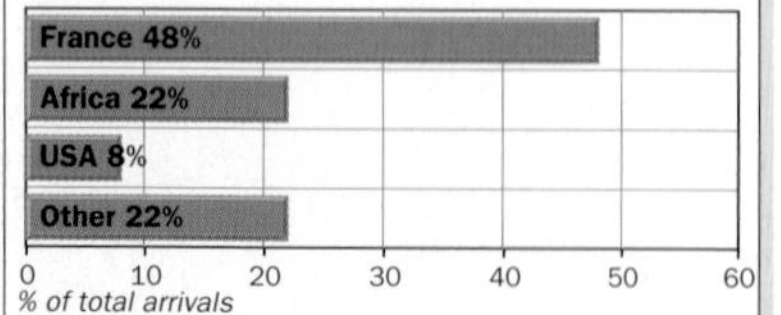
France 48%
Africa 22%
USA 8%
Other 22%
0
10
20
30
40
50
60
% of total arrivals

Tourism is virtually nonexistent. The national parks and game reserves are the main potential attractions. The prehistoric rock painting of the Tibesti plateau and the Muslim cities of central Chad attract the adventurous.

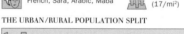

***Watering hole at Oum Hadjer**, a village on the Batha watercourse in central Chad, 145 km (90 miles) east of Ati.*

PEOPLE

Pop. density low

French, Sara, Arabic, Maba

6/km²
(17/mi²)

THE URBAN/RURAL POPULATION SPLIT

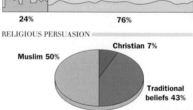

24% 76%

RELIGIOUS PERSUASION

Muslim 50%
Christian 7%
Traditional beliefs 43%

About half the population, mainly the Sara-speaking and related peoples, is concentrated in the south in one-fifth of the national territory. Most of the rest are located in the central sultanates. The northern third of Chad has a population of only 100,000 people, mainly nomadic Muslim Toubou.

POLITICS

Multiparty elections

2002/2006

President Idriss Déby

AT THE LAST ELECTION

National Assembly 155 seats

72% MPS; 7% RDP; 15% Others; 6% FAR

MPS = Patriotic Salvation Movement
RDP = Rally for Democracy and Progress
FAR = Front of Action Forces for the Republic

Idriss Déby overthrew President Hissène Habré in 1990 after an armed invasion from Sudan. He promised multipartyism, and in 1992 – for the first time since the early 1960s – political parties were legalized. After many delays, the transitional process led to a successful referendum in 1996 on a new constitution based on the French model. President Déby was confirmed in office in elections in 1996 and again in 2001. His ruling MPS just achieved an overall majority in the 1997 elections, and expanded its share of seats further in 2002. A shaky Libyan-brokered peace deal in 2002 failed to end a rebellion in the north which had begun in 1999.

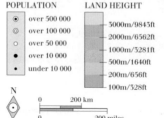

Chad

Total Land Area : 1 284 000 sq. km (495 752 sq. miles)

POPULATION
- over 500 000
- over 100 000
- over 50 000
- over 10 000
- under 10 000

LAND HEIGHT
- 3000m/9843ft
- 2000m/6562ft
- 1000m/3281ft
- 500m/1640ft
- 200m/656ft
- 100m/328ft

N
0 200 km
0 200 miles

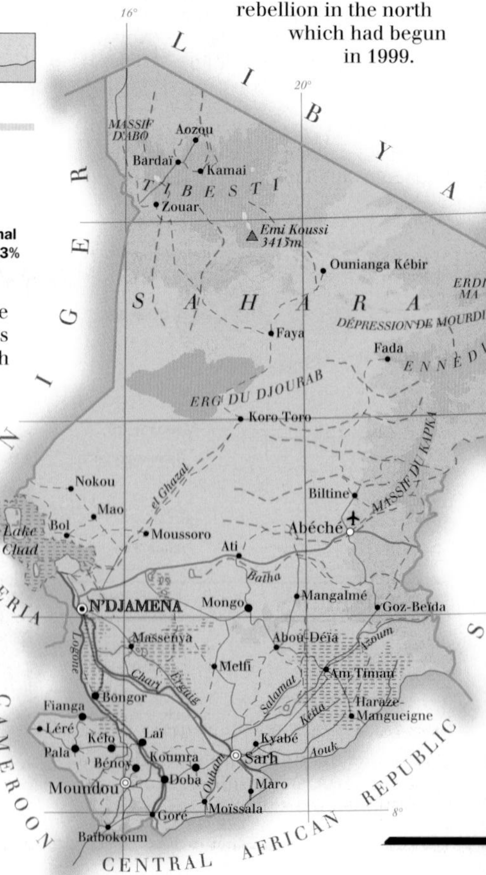
LIBYA
NIGER
NIGERIA
CAMEROON
CENTRAL AFRICAN REPUBLIC
MASSIF D'ABO
Aozou
Bardaï
Kamai
TIBESTI
Zouar
Emi Koussi 3415m
Ounianga Kébir
SAHARA
ERDI MA
DÉPRESSION DE MOURDI
Faya
Fada
ENNEDI
ERG DU DJOURAB
Koro Toro
MASSIF DU KAPKA
Nokou
Mao
Biltine
Lake Chad
Bol
Moussoro
Abéché
Ati
Batha
N'DJAMENA
Mongo
Mangalmé
Goz-Beïda
Massenya
Abou-Déïa
Melfi
Am Timan
Bongor
Fianga
Léré
Pala
Kélo
Laï
Béré
Kyabé
Haraze-Mangueigne
Koumra
Sarh
Moundou
Doba
Maro
Goré
Moïssala
Baïbokoum
16°
20°
8°

Key to symbols and abbreviations on cover flaps

WORLD AFFAIRS

Joined UN in 1960

CILSS

FZ

LCBC

AU

OIC

Chad's most important relationship is with France. Libya occupied the uranium-rich Aozou strip in 1973–1994.

AID

Recipient

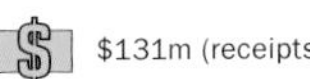

$131m (receipts)

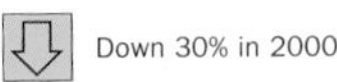

Down 30% in 2000

France is by far the major donor. Other sources include Libya, the EU, the World Bank, Germany, and the Arab members of OPEC. Without assistance to cover civil servants' pay over recent years, the administration would have collapsed.

DEFENSE

Compulsory military service

$47m

No change in 2000

On seizing power, Déby swelled the existing army with irregulars. This policy has now been reversed and the army reduced to 25,000, including former rebels. France provides military aid and personnel.

ECONOMICS

Inflation 7.1% p.a. (1990–2000)

$1.54bn

698.7–736.7 CFA francs

SCORE CARD

- ❏ WORLD GNP RANKING........................143rd
- ❏ GNP PER CAPITA $200
- ❏ BALANCE OF PAYMENTS...................–$158m
- ❏ INFLATION 3.8%
- ❏ UNEMPLOYMENT ....Widespread underemployment

STRENGTHS

Discovery of large oil deposits and Chad–Cameroon pipeline could transform economy. Cotton industry; potential for other agriculture in south. Strategic trading location in heart of Africa. Natron and uranium deposits.

WEAKNESSES

Underdevelopment and poverty. Lack of transportation infrastructure. Political instability. Frequent droughts.

EXPORTS

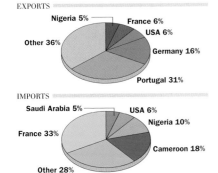

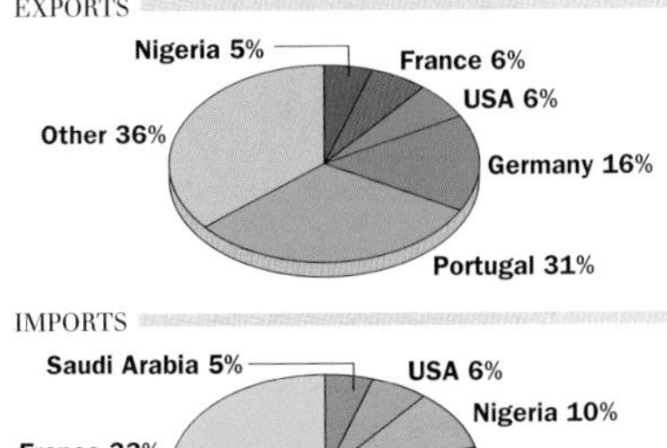
Nigeria 5%
France 6%
USA 6%
Other 36%
Germany 16%
Portugal 31%

IMPORTS

Saudi Arabia 5%
USA 6%
Nigeria 10%
France 33%
Cameroon 18%
Other 28%

RESOURCES

Electric power 29,000 kw

84,000 tonnes

Reserves currently unexploited

5.9m cattle, 5.25m goats, 2.4m sheep, 5m chickens

Natron, uranium, oil, kaolin, soda, rock salt

A consortium of Exxon, Shell, and Elf has discovered large oil reserves in the south, mostly near Doba, which could make Chad a major African producer. Natron, found north of Lake Chad, is the only mineral currently exploited. There is uranium in the Aozou strip.

ENVIRONMENT

Sustainability rank: 96th

9%

0.02 tonnes per capita

President Déby's government has made protection of the environment a priority, with antidesertification measures such as tree-planting campaigns and aid-funded irrigation schemes. There is concern about potential environmental damage which may result from the planned Chad–Cameroon oil pipeline.

MEDIA

TV ownership low

Daily newspaper circulation 0.2 per 1000 people

PUBLISHING AND BROADCAST MEDIA

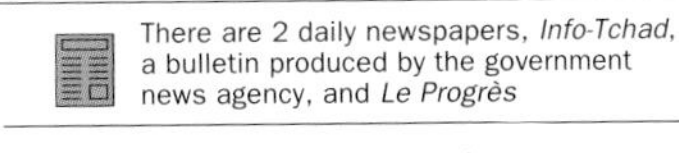
There are 2 daily newspapers, *Info-Tchad*, a bulletin produced by the government news agency, and *Le Progrès*
1 state-controlled service
8 services: 1 state-controlled, 7 independent

Broadcasting is controlled by the government, which sometimes allows the airing of opposition views. There are a few independent publications, of which the best known is the weekly *N'Djamena-Hebdo*.

CRIME

Death penalty in use

2521 prisoners

Crime is rising

The easy availability of weapons in the region in the past two decades has meant that local disputes, usually over water or grazing, often now lead to gun battles. Armed robbery, smuggling, and vandalism are widespread. In several areas, the activities of disaffected former rebels threaten security.

EDUCATION

School leaving age: 12

43%

5901 students

Education is based on the French model, although there are Koranic schools in the north. Primary schooling is officially compulsory, but enrollment is only 46%. Recently, World Bank aid has been directed at elementary schooling.

CHRONOLOGY

France extended its domination of the area now known as Chad after ousting the last Arab ruler in 1900.

- ❏ **1960** Independence. One-party state.
- ❏ **1973** Libyans seize Aozou strip.
- ❏ **1975** Coup by Gen. Félix Malloum.
- ❏ **1979–1982** North–south civil war.
- ❏ **1980** Goukouni Oueddei in power.
- ❏ **1982** Hissène Habré (northerner) defeats Oueddei.
- ❏ **1990** Idriss Déby overthrows Habré, who flees to Senegal.
- ❏ **1994** Libya relinquishes Aozou strip.
- ❏ **1996** National cease-fire; new constitution.
- ❏ **1997** Déby's MPS wins elections.
- ❏ **1999** Rebellion in north.
- ❏ **2001** Déby reelected.
- ❏ **2002** MPS increases its majority.

C

HEALTH

Welfare state health benefits

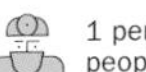

1 per 20,000 people

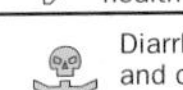

Diarrheal, parasitic, and communicable diseases

There are few city hospitals and fewer than 300 smaller health centers. One in five children die before the age of five.

SPENDING

GDP/cap. increase

CONSUMPTION AND SPENDING

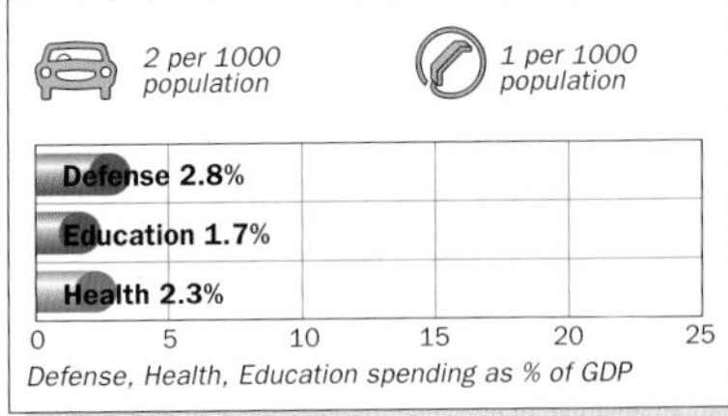
2 per 1000 population
1 per 1000 population
Defense 2.8%
Education 1.7%
Health 2.3%
0
5
10
15
20
25
Defense, Health, Education spending as % of GDP

Poverty is almost universal in Chad; the middle class is very small. There are few wealthy individuals. Habré looted the treasury when he was overthrown in 1990.

WORLD RANKING

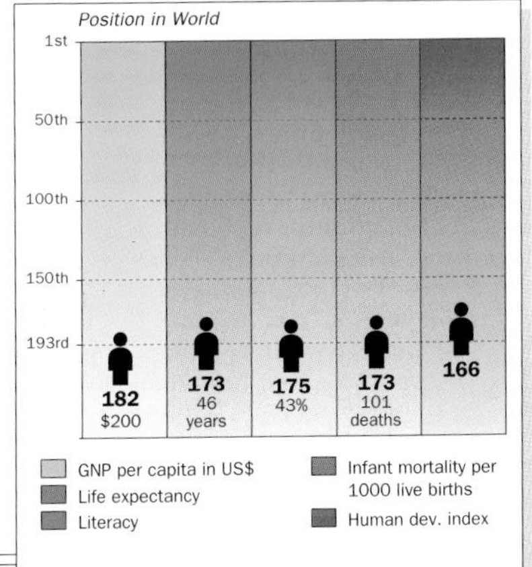
Position in World
1st
50th
100th
150th
193rd
182
$200
173
46 years
175
43%
173
101 deaths
166
GNP per capita in US$
Life expectancy
Literacy
Infant mortality per 1000 live births
Human dev. index

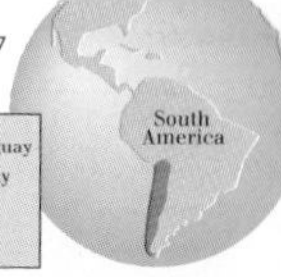
South America

SOUTH AMERICA

Bolivia
Paraguay
Uruguay
CHILE
Argentina

CHILE

OFFICIAL NAME: Republic of Chile **CAPITAL:** Santiago
POPULATION: 15.4 million **CURRENCY:** Chilean peso **OFFICIAL LANGUAGE:** Spanish

C

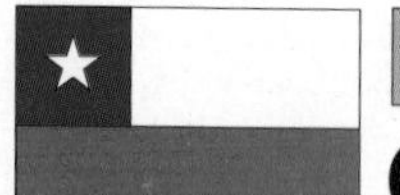

1818

1883

Sept 18

RCH

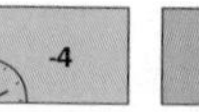
-4

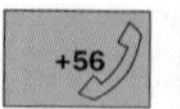
+56

.cl

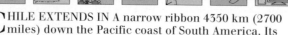

CHILE EXTENDS IN A narrow ribbon 4350 km (2700 miles) down the Pacific coast of South America. Its extraordinary shape means that its physical geography ranges from the deserts of the High Andes in the north to fertile valleys in the center, while in the south are the fjords, lakes, and deep sea channels of the Southern Andes. In 1989, Chile returned to elected civilian rule, following a popular rejection of the Pinochet dictatorship. A collapse in copper prices, coupled with weaker export markets, has interrupted the high growth seen in the 1990s.

General Pinochet, *a dictatorial president rejected by popular referendum in 1989.*

Ricardo Lagos *was narrowly elected president in 2000.*

CLIMATE

Desert/mountain/ maritime

WEATHER CHART FOR SANTIAGO

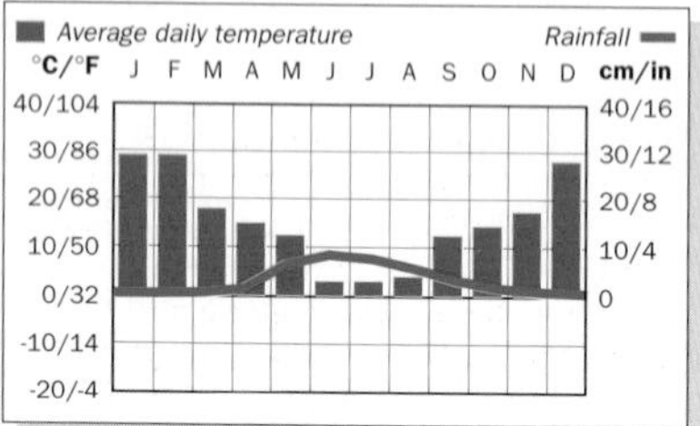
Average daily temperature
Rainfall
°C/°F
J F M A M J J A S O N D
cm/in
40/104
30/86
20/68
10/50
0/32
-10/14
-20/-4
40/16
30/12
20/8
10/4
0

Chile has an immensely varied climate. The north, which includes the world's driest desert, the Atacama, is frequently cloudy and cool for its latitude. The central regions have an almost Mediterranean climate, with changeable winters and hot, dry summers. The higher reaches of the Andes have a typically alpine climate, with glaciers and year-round snow. The south is the wettest region.

TRANSPORTATION

Drive on right

Comodoro Arturo Merino Benítez, Santiago
5.78m passengers

471 ships
842,300 grt

THE TRANSPORTATION NETWORK

11,012 km (6843 miles)

3455 km (2147 miles)

2084 km (1295 miles)

725 km (450 miles)

The state railroad is being upgraded with private capital assistance: suburban services into Santiago are being improved, and line repairs and better rolling stock will shorten journeys from Santiago to Temuco. Sections of the Pan-American Highway, the sole arterial road running from the Peruvian border to Puerto Montt, are being upgraded. The open skies agreement of the US reached with Chile in 1999, the first with a South American country, gives Chile strong regional and international links.

TOURISM

Visitors : Population 1:8.8

1.74m visitors

Up 7% in 2000

MAIN TOURIST ARRIVALS

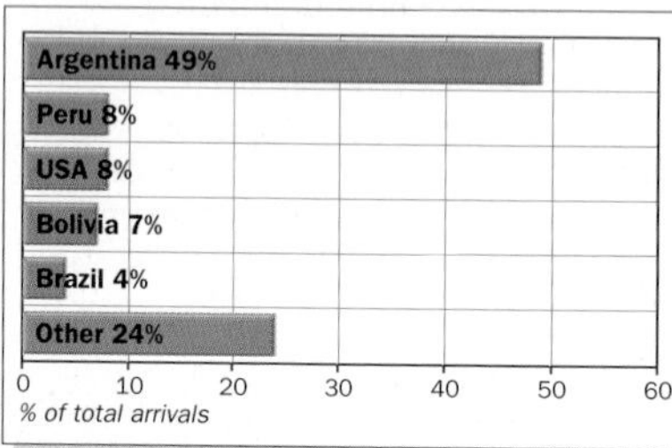
Argentina 49%
Peru 8%
USA 8%
Bolivia 7%
Brazil 4%
Other 24%
0
10
20
30
40
50
60
% of total arrivals

The Pinochet years saw a dramatic decline in tourists from the US and Europe, although the numbers from neighboring countries held up. Since the fall of the regime in 1989, visitors have returned, but over half of them still come from Argentina and Peru.

Investment in the sector totaled $2.17 billion in the first half of 2000, up 3.8% on the same period in 1999. Accordingly, Chile is making more of its stunning Andean scenery, its immensely long coastline, and a number of exceptional sites, including Chuquicamata, the world's largest copper mine, the Elqui Valley wine-growing region, and the spectacular glaciers and fjords of southern Chile. Easter Island in the Pacific is another major attraction.

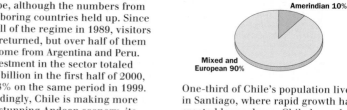

Peaks in the Paine range, southern Chile. *Fjords, glaciers, and myriad islands typify Chile's very wet, wild, and stormy south.*

PEOPLE

Pop. density low

Spanish, Amerindian languages

21/km² (53/mi²)

THE URBAN/RURAL POPULATION SPLIT

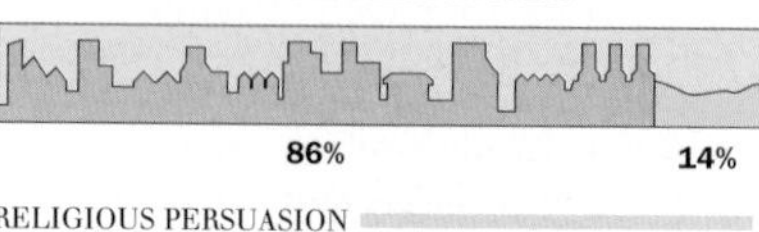
86%
14%

RELIGIOUS PERSUASION

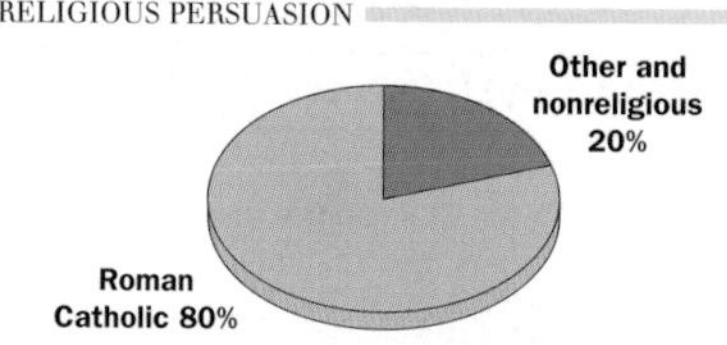
Other and nonreligious 20%
Roman Catholic 80%

ETHNIC MAKEUP

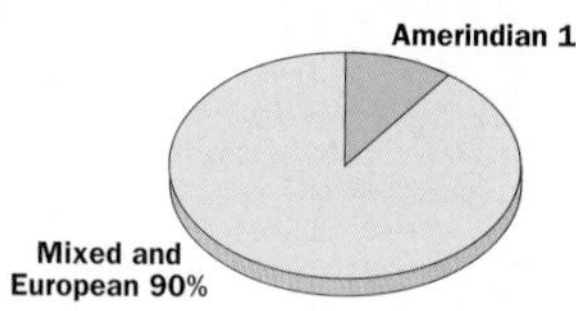
Amerindian 10%
Mixed and European 90%

One-third of Chile's population lives in Santiago, where rapid growth has created large slums. Chile has relatively few immigrants; most people are of mixed Spanish–Amerindian descent. A Commission for Historic Truth was set up in 2000 to address the problems of Amerindians. There are some 80,000 Mapuche Amerindians around Temuco in the south, 20,000 Aymara in the northern Chilean Andes, and 2000 Rapa Nui on Easter Island.

Over 25% of working women are employed in domestic service.

POPULATION AGE BREAKDOWN

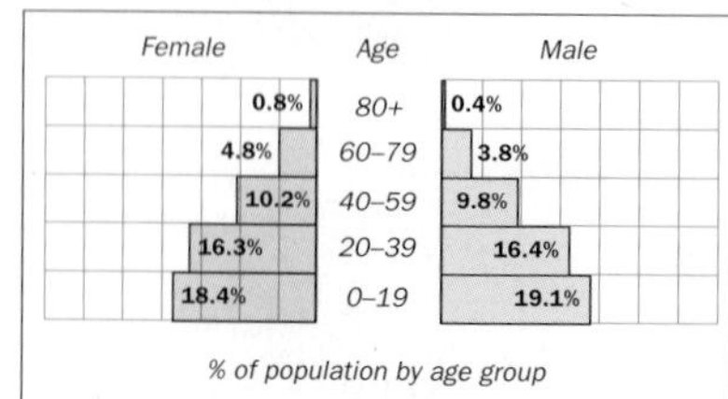
Female
Age
Male
0.8%
80+
0.4%
4.8%
60–79
3.8%
10.2%
40–59
9.8%
16.3%
20–39
16.4%
18.4%
0–19
19.1%
% of population by age group

Juan Fernández Is
I. Alejandro Selkirk
San Juan Bautista
I. Robinson Crusoe
0 100 km
0 100 miles
Easter I.
Terevaka
Hanga Roa
0 10 km
0 10 miles
PERU
BOLIVIA
ARGENTINA
PACIFIC OCEAN
Arica
Iquique
Atacama Desert
Chuquicamata
Calama
Antofagasta
Chañaral
Nevado Ojos del Salado 6880m
Copiapó
Vallenar
La Serena
Coquimbo
Ovalle
Illapel
Los Andes
Viña del Mar
Valparaíso
Quilpué
SANTIAGO
Llolleo
San Bernardo
Puente Alto
Rancagua
Curicó
Rengo
Talca
San Javier de Loncomilla
Linares
Parral
San Carlos
Talcahuano
Concepción
Chillán
Los Ángeles
Curanilahue
Mulchén
Traiguén
Victoria
Temuco
Lautaro
Valdivia
La Unión
Puerto Varas
Puerto Montt
Isla de Chiloé
Castro
Puerto Aisén
Coihaique
Lago General Carrera
Chile Chico
Isla Wellington
Cerro Paine
Puerto Natales
Punta Arenas
Porvenir
Strait of Magellan
Beagle Channel
I. Picton
I. Nueva
I. Lennox
Cape Horn
Drake Passage
CHILE
Total Land Area : 756 950 sq. km (292 258 sq. miles)
POPULATION
over 1 000 000
over 100 000
over 50 000
over 10 000
under 10 000
LAND HEIGHT
4000m/13124ft
2000m/6562ft
1000m/3281ft
200m/656ft
Sea Level
N
0 300 km
0 300 miles

POLITICS

Multiparty elections

L. House 2001/2005
U. House 2001/2005

President Ricardo Lagos Escobar

AT THE LAST ELECTION

Chamber of Deputies 118 seats

52% CPD	47% APC	1% Ind

CPD = Concertación – Coalition of Parties for Democracy (Christian Democratic Party–**PDC**, Party for Democracy–**PPD**, and Socialist Party of Chile–**PS**) **APC** = Alliance for Chile (Independent Democratic Union–**UDI** and National Renewal Party–**RN**) **Ind** = Independents **App** = Appointed **SL** = Senator-for-Life

Senate 48 seats

13% UDI	6% PPD

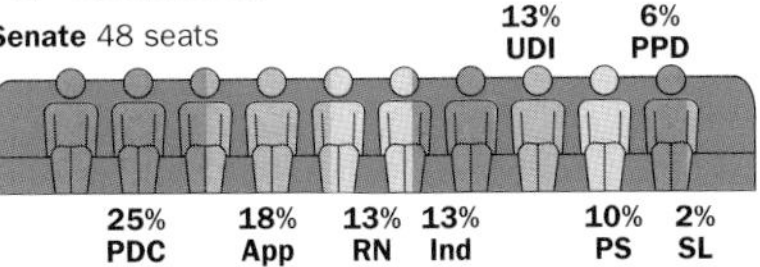

25% PDC	18% App	13% RN	13% Ind	10% PS	2% SL

There are 38 elected members and 9 appointed senators. Former president Frei is a senator-for-life

After 16 years of military rule under Gen. Augusto Pinochet, Chile returned to multiparty democracy in 1989.

PROFILE

Chilean politics is still strongly affected by the legacy of the military dictatorship of 1973–1989, which began when Pinochet's coup overthrew the elected Marxist government of Salvador Allende.

The CIA backed the Pinochet coup, anxious to halt Allende's program of nationalization of the largely US-owned copper mines. As a result thousands of Chileans were killed by the military or "disappeared," and a further 80,000 were taken as political prisoners.

Pinochet's nationalist politics drew on the example of Franco's Spain, while his economic policy was one of the first experiments in the free-market Chicago School of monetarism. Chile's business and middle classes prospered, while opposition, which was brutally suppressed by the DINA secret police, came most visibly from the Church and the urban poor.

In 1988 Pinochet, attempting to secure a popular mandate for continuing his regime, was surprised when the population emphatically voted for democracy. Pinochet stepped down, but remained head of the army. Patricio Aylwin won presidential elections held in 1989, heading Concertación, a center-left coalition.

Under Aylwin, politics became more stable, partly as a result of a cross-party consensus on economic policy. Continued growth and some progressive social measures attracted the support of the trade unions. These policies were continued under Eduardo Frei of the PS party, who was elected president in 1993.

When Pinochet retired as army chief in 1998, heated disagreements over his entry to the Senate as a senator-for-life split Concertación along broadly left–right lines. Disagreements over Pinochet's subsequent arrest and detention in Europe on human rights charges further complicated the picture. Ricardo Lagos of the PS emerged as front runner for the 1999 presidential elections, in which both the PS and the PDC presented themselves as the guarantors of peace and democracy. Right-wing opposition parties, however, also began to downplay their past links with Pinochet to broaden their electoral appeal. Lagos narrowly won a run-off poll in 2000. Facing a better organized opposition than hitherto, the ruling center-left Concertación performed reasonably well in the 2001 elections retaining its majority in the Chamber of Deputies but losing it in the Senate.

MAIN POLITICAL ISSUES

Political consolidation

President Lagos, still Chile's most charismatic and trusted politician, reshuffled what was seen as a weak cabinet after the 2001 legislative elections to reinvigorate his government for his remaining four years in office. Top priority was to guarantee a stable environment for economic growth and job creation. The strong electoral performance of the UDI, senior partner in the right-wing Alliance for Chile opposition coalition, boosted the presidential hopes of its leader, Joaquín Lavín, currently the mayor of Santiago.

Improved relations with military

Relations between government and military are improving. Lagos appointed Chile's first woman defense minister and installed as new army commander a moderate who had accepted that Pinochet should respect the authority of the courts, before the human rights charges against the former dictator were dropped on grounds of dementia.

CHRONOLOGY

The Spanish first attempted the conquest of Chile against the fierce indigenous Araucanian people in 1535. Santiago was founded in 1541. Chile was subject to Spanish rule until independence in 1818.

- **1817–1818** Bernardo O'Higgins leads republican Army of the Andes in victories against royalist forces.
- **1879–1883** War of the Pacific with Bolivia and Peru. Chile gains valuable nitrate regions.
- **1891–1924** Parliamentary republic ends with growing political chaos.

C

CHRONOLOGY *continued*

- ❑ **1936–1946** Communist, Radical, and Socialist parties form influential Popular Front coalition.
- ❑ **1943** Chile backs US in World War II.
- ❑ **1946–1964** Right-wing Chilean presidents follow US McCarthy policy and marginalize the left.
- ❑ **1970** Salvador Allende elected. Reforms provoke strong reaction from the right.
- ❑ **1973** Allende dies in army coup. Brutal dictatorship of Gen. Pinochet begins.
- ❑ **1988** Referendum votes "no" to Pinochet staying in power.
- ❑ **1989** Democracy peacefully restored; Pinochet steps down after Aylwin election victory.
- ❑ **1998** Pinochet detained in UK pending extradition to Spain on human rights charges.
- ❑ **2000** Ricardo Lagos (PS) sworn in as president. Pinochet, deemed unfit to face trial, returns to Chile, where charges are suspended in 2001.

WORLD AFFAIRS

Joined UN in 1945

APEC
G15
NAM
OAS
RG

Chile hopes to consolidate the key relationship with the US, its main trading partner, through a bilateral free trade agreement. During the Allende period in the early 1970s, the US had actively worked against the government, fearing that the spread of socialism would jeopardize its investments in Chile and the rest of Latin America. However, the human rights record of the Pinochet regime eventually became an embarrassment.

In 1999, in response to Pinochet's detention in the UK and planned extradition to Spain to stand trial on human rights charges, Chile argued that this amounted to an infringement of its sovereignty and the jurisdiction of its courts. Relations with Spain and the UK improved in 2000 following Pinochet's return to Chile.

An associate member of Mercosur, Chile also jealously guards its right to pursue its own bilateral economic agreements; a free trade agreement between Chile and the EU was initialled in June 2002. Border disputes with Bolivia and Peru are ongoing.

AID

Recipient

$49m (receipts)

Down 30% in 2000

The majority of aid is in the form of debts rescheduled by the World Bank at the instigation of the US.

DEFENSE

Compulsory military service

$2.89bn

Up 12% in 2000

CHILEAN ARMED FORCES

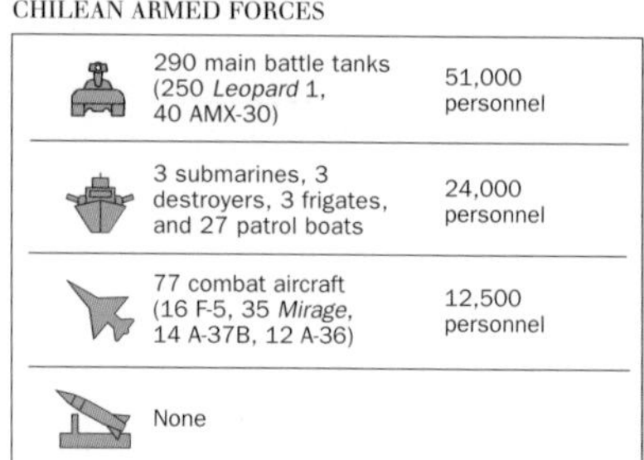

290 main battle tanks (250 *Leopard* 1, 40 AMX-30)	51,000 personnel
3 submarines, 3 destroyers, 3 frigates, and 27 patrol boats	24,000 personnel
77 combat aircraft (16 F-5, 35 *Mirage*, 14 A-37B, 12 A-36)	12,500 personnel
None	

Most of the officers linked to human rights abuses under the dictatorship (1973–1989) are now retired, but dozens still face investigation and possible trial. The army has a moderate commander, and for the first time a woman, also a member of the PS, has been appointed defense minister. Despite domestic concerns on pressing social needs, President Lagos signed up to reequip the air force with expensive, US-made F-16 fighters, still to be part-funded by 10% of copper exports under a controversial law inherited from the years of military rule. To dampen fears of arms escalation, Chile forewarned Argentina, Brazil, Peru, and Bolivia about the deal. Modernization of the navy, which was lobbying for four new frigates, is on hold.

ECONOMICS

Inflation 7.3% p.a. (1990–2000)

$69.9bn

573.75–661.15 Chilean pesos

SCORE CARD

- ❑ World GNP Ranking....44th
- ❑ GNP per Capita....$4590
- ❑ Balance of Payments....–$991m
- ❑ Inflation....3.8%
- ❑ Unemployment....9%

EXPORTS

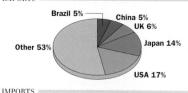

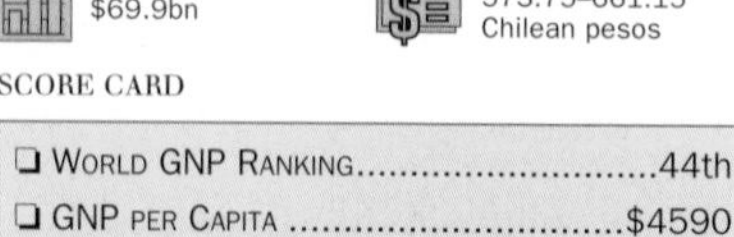
Brazil 5%
China 5%
UK 6%
Japan 14%
USA 17%
Other 53%

IMPORTS

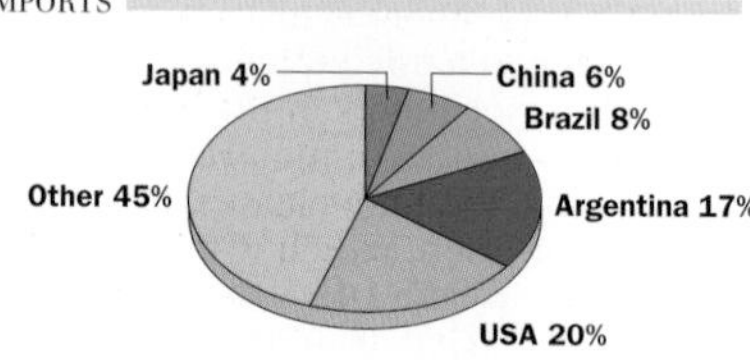
Japan 4%
China 6%
Brazil 8%
Argentina 17%
USA 20%
Other 45%

ECONOMIC PERFORMANCE INDICATOR

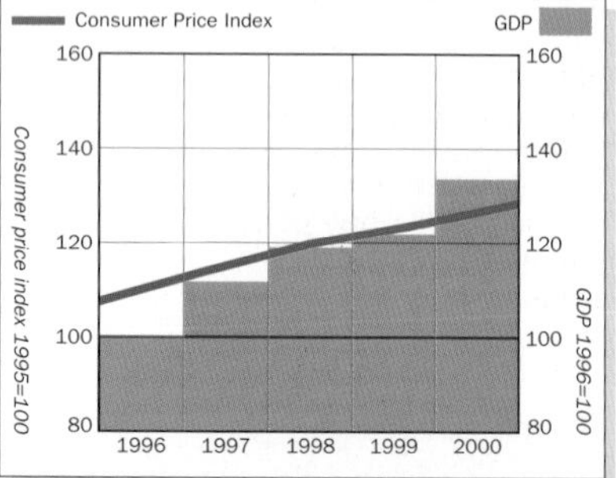
Consumer Price Index
GDP
160
140
120
100
80
Consumer price index 1995=100
GDP 1996=100
1996
1997
1998
1999
2000

Strengths

World's largest copper producer. Fresh fruit exports. Strong investment inflows allowing rapid economic growth. Highest credit rating due to fiscal and monetary stability and highly liquid financial system. Development of nontraditional industries such as wine and fresh and prepared fish.

Weaknesses

Dependence on US as single largest trading partner. Prolonged weakness in Asian export markets. Vulnerability of copper revenues, representing 40% of exports, to low world market prices.

Profile

Competing ideologies have battled over Chile's economy. Allende's socialism brought huge corporations into the state sector. The Pinochet dictatorship introduced radical monetarist policies. Drastic cutting of the state sector and the selling-off of state enterprises at below market value led to large profits for investors and speculators. Tough economic measures, irrespective of the social consequences, brought Chile's inflation rate down from 400%.

The Aylwin and Frei governments continued with neoliberal policies, including privatizing the pension system. However, some 30 companies, including the large Codelco copper company, remain in the state sector.

Though it avoided the worst of the Argentine crisis, the economy in 2002 suffered from sluggish international conditions and modest copper prices.

CHILE : MAJOR BUSINESSES

Oil
Oil refining
Copper mining
Manufacturing
Pharmaceuticals
Heavy engineering
Fish processing
Agribusiness
Iquique
Chuquicam
Vina del Mar
Santiago
Teniente
Talcahuano
Concepción
Punta Arenas
Straits of Magellan
0 300 km
0 300 miles

C

RESOURCES

Electric power 8.4m kw

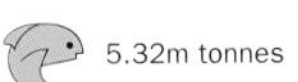
5.32m tonnes

4852 b/d (reserves 285m barrels)

4.2m sheep, 4.15m cattle, 2.5m pigs, 78m chickens

Coal, copper, gold, silver, iron, lithium, molybdenum, iodine, natural gas, oil

ELECTRICITY GENERATION

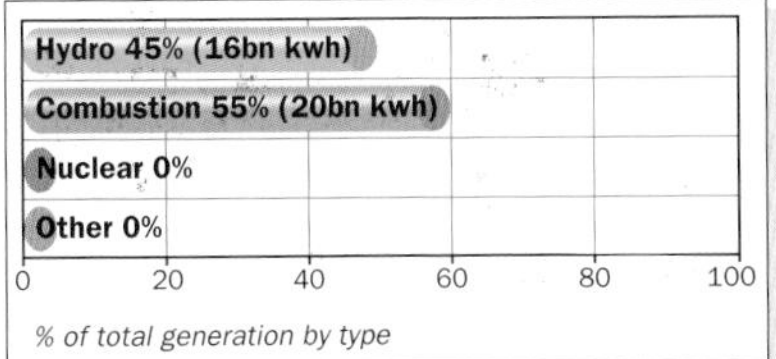
Hydro 45% (16bn kwh)
Combustion 55% (20bn kwh)
Nuclear 0%
Other 0%
0
20
40
60
80
100
% of total generation by type

Chile is the world's largest producer of copper, which accounts for a major proportion of its export revenues. There are important deposits of lithium, molybdenum, and especially of gold. Chile also has reserves of natural gas, oil, and coal, and plenty of hydroelectric potential. In addition, it is a leading producer of fishmeal, and has a flourishing wine industry.

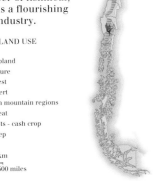

CHILE : LAND USE

ATACAMA DESERT

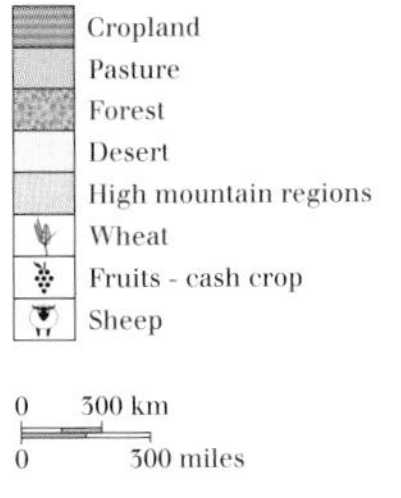
Cropland
Pasture
Forest
Desert
High mountain regions
Wheat
Fruits - cash crop
Sheep

0 500 km
0 500 miles

ENVIRONMENT

Sustainability rank: 35th

19% (7% partially protected)

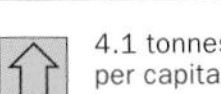
4.1 tonnes per capita

ENVIRONMENTAL TREATIES

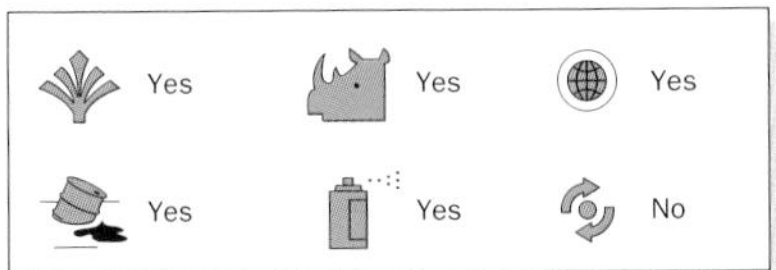
Yes
Yes
Yes
Yes
Yes
No

Environmental concerns do not rank highly on the political agenda. Severe smogs still cover Santiago, due in part to diesel fumes from the city's 14,500 buses. The chief concern is logging in the south by Japanese and other foreign companies. The huge growth of the salmon industry, which fences off sea lakes, is resulting in dolphins losing their natural habitats. Overfishing of swordfish has led to friction with the EU, particularly Spain.

MEDIA

TV ownership medium

Daily newspaper circulation 98 per 1000 people

PUBLISHING AND BROADCAST MEDIA

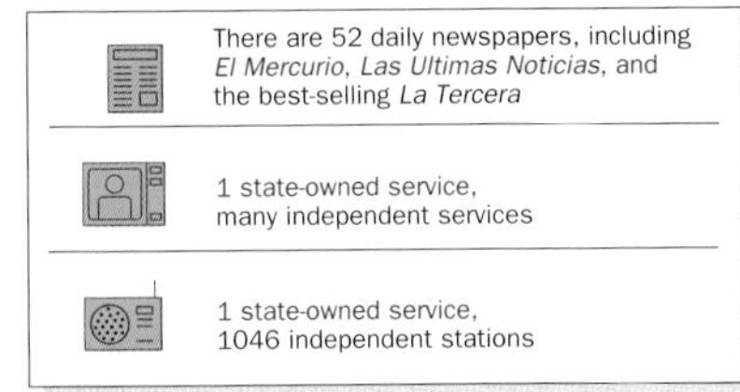
There are 52 daily newspapers, including *El Mercurio*, *Las Ultimas Noticias*, and the best-selling *La Tercera*
1 state-owned service, many independent services
1 state-owned service, 1046 independent stations

A long-delayed liberalized press law was finally introduced in 2001. Military courts will no longer be able to try journalists, and political authorities will be denied the special procedures used to sue reporters for slander.

CRIME

Death penalty not used in practice

30,852 prisoners

Up 1% in 1999

CRIME RATES

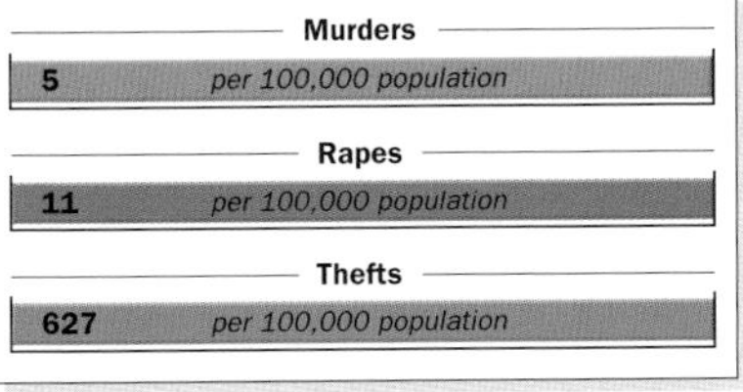
Murders
5 per 100,000 population
Rapes
11 per 100,000 population
Thefts
627 per 100,000 population

The judiciary has been slow to pursue human rights cases from the Pinochet regime, despite the discoveries during the 1990s of mass graves of victims of the DINA (secret police). Mapuche leaders were among the "disappeared." Levels of child abuse are exceptionally high, although now starting to fall.

EDUCATION

School leaving age: 13

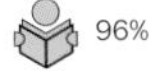
96%

424,672 students

THE EDUCATION SYSTEM

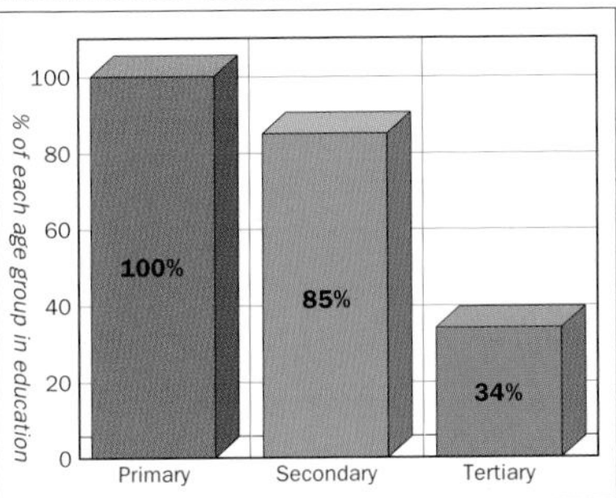
100
80
60
40
20
0
% of each age group in education
100%
85%
34%
Primary
Secondary
Tertiary

Economic growth has permitted public spending on education to increase substantially, but the sector suffers from budgetary cuts. Free primary education is officially compulsory for eight years. Human rights issues now appear in school curricula.

HEALTH

Welfare state health benefits

1 per 909 people

Heart diseases, cancers, murders

Recent growth has meant increased public spending on health; however, 2002 saw cuts. The public health service covers 80% of people, but is mostly found in urban areas. There is private care for the rich. Infant mortality has fallen to one-third of the 1980 level.

SPENDING

GDP/cap. increase

CONSUMPTION AND SPENDING

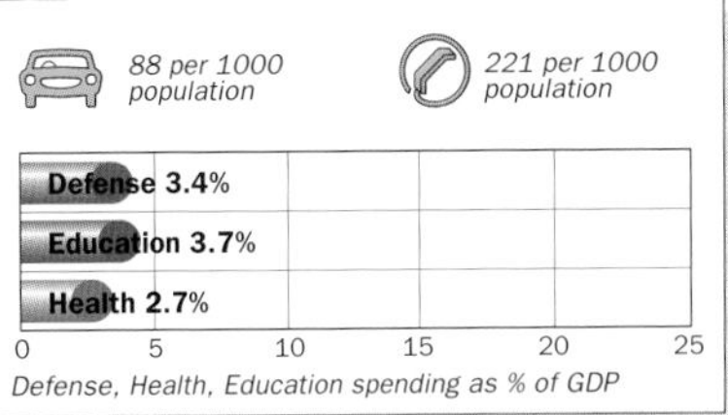
88 per 1000 population
221 per 1000 population
Defense 3.4%
Education 3.7%
Health 2.7%
0
5
10
15
20
25
Defense, Health, Education spending as % of GDP

Chile's traditionally large middle class did well under Pinochet and the economic policies of the Chicago School. The wealthiest sections benefited considerably from the sale of state assets at 40%–50% of their true market value. Five years into the regime, wealth had become highly concentrated, with just nine economic conglomerates controlling the assets of the top 250 businesses, 82% of banking, and 64% of all financial loans. The regime's artificially high domestic interest rates enabled those with access to international finance to earn an estimated $800 million between 1977 and 1980, simply by borrowing abroad and lending at home. These groups have retained their position.

The poor, by contrast, are over 15% worse off than in 1970, with an estimated four million people living just above the UN poverty line and one million below it.

WORLD RANKING

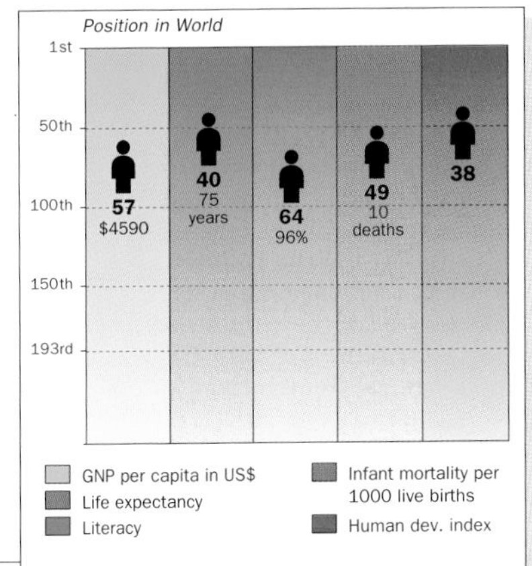
Position in World
1st
50th
100th
150th
193rd
57
$4590
40
75
years
64
96%
49
10
deaths
38
GNP per capita in US$
Life expectancy
Literacy
Infant mortality per 1000 live births
Human dev. index

CHINA

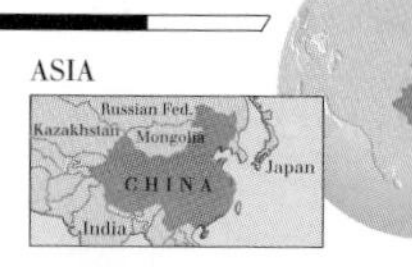
ASIA
Russian Fed.
Kazakhstan
Mongolia
CHINA
Japan
India
Asia

OFFICIAL NAME: People's Republic of China **CAPITAL:** Beijing
POPULATION: 1.29 billion **CURRENCY:** Renminbi (known as yuan) **OFFICIAL LANGUAGE:** Mandarin

960

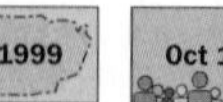
1999

Oct 1

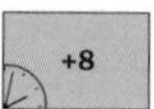
VRC

+8

+86

.cn

COVERING A VAST AREA of eastern Asia, China is bordered by 14 countries; to the east it has a long Pacific coastline. Two-thirds of China is uplands: the southwestern mountains include the Tibetan Plateau; in the northwest, the Tien Shan Mountains separate the Tarim and Dzungarian basins. The low-lying east is home to two-thirds of the population. China was dominated by Mao Zedong from the founding of the Communist People's Republic in 1949 until his death in 1976. Despite the major disasters of the 1950s Great Leap Forward and the 1960s Cultural Revolution, it became an industrial and nuclear power. Today, China is rapidly developing a market-oriented economy. The current leadership remains set on accomplishing this without political liberalization, instead enforcing single-party rule as was advocated by "elder statesman" Deng Xiaoping, who died in 1997.

Li River (Xi Jiang), Guangxi, China's most beautiful region. Its spectacular scenery has encouraged large-scale tourist development.

CLIMATE

Mountain/tropical/continental/steppe

WEATHER CHART FOR BEIJING

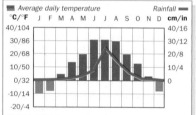

Average daily temperature — Rainfall
°C/°F J F M A M J J A S O N D cm/in
40/104 40/16
30/86 30/12
20/68 20/8
10/50 10/4
0/32 0
-10/14
-20/-4

China is divided into two main climatic regions. The north and west are semiarid or arid, with extreme temperature variations. The south and east are warmer and more humid, with year-round rainfall.

Winter temperatures vary with latitude and are warmest on the subtropical southeast coast, where they average about 16°C (60°F). Summer temperatures are more uniform, rising above 21°C (70°F) throughout China; on the southeast coast, the July average is about 30°C (86°F). In the north and west, temperate summers contrast with harsh winters. In northern Manchuria, rivers freeze for five months and temperatures can fall to –25°C (–13°F). In the deserts of Xinjiang province, temperatures range from –11°C (12°F) in winter to 33°C (91°F) in summer.

Summer and autumn are China's wettest seasons. Winds from the Pacific during the summer monsoon bring rains to most of the country. The south and east also have wet winters, but elsewhere the winter monsoon brings cold, dry air from Siberia.

Floods are frequent and sometimes catastrophic, as in 2002. Droughts can be even more devastating; that of 1959–1962 contributed to a famine which killed millions.

RUSSIAN FEDERATION
KAZAKHSTAN
Altay
Karamay
Yining
Ürümqi
KYRGYZSTAN
TIEN SHAN
Hami
Tarim He
Korla
Kashi
XINJIANG UYGHUR ZIZHIQU
Lop Nur
TAJIKISTAN
Ruoqiang
AFGHANISTAN
TAKLIMAKAN SHAMO
PAKISTAN
Hotan
ALTUN SHAN
QILIAN
K2 8611m
KUNLUN SHAN
Golmud
Aksai Chin
INDIA
QINGH
QING-ZANG GAOYUAN
XIZANG ZIZHIQU (TIBET)
TANGGULA SHAN
Nagqu
Salween
NYAINQENTANGLHA SHAN
Lhasa
Baxoi
Brahmaputra
Mount Everest 8848m
NEPAL
HIMALAYAS
INDIA
BHUTAN
INDIA
80°
90°
BURMA (MYANMAR)
Baoshan
Yumen

CHINA

Total Land Area : 9 596 960 sq. km (3 705 386 sq. miles)

POPULATION
- over 5 000 000
- over 1 000 000
- over 500 000
- over 100 000
- over 50 000
- over 10 000
- Great Wall of China

LAND HEIGHT
- 6000m/19686ft
- 4000m/13124ft
- 3000m/9843ft
- 2000m/6562ft
- 1000m/3281ft
- 500m/1640ft
- 200m/656ft
- Sea Level

N

0 400 km
0 400 miles

TOURISM

Visitors : Population 1:39

33.2m visitors

Up 6% in 2001

MAIN TOURIST ARRIVALS

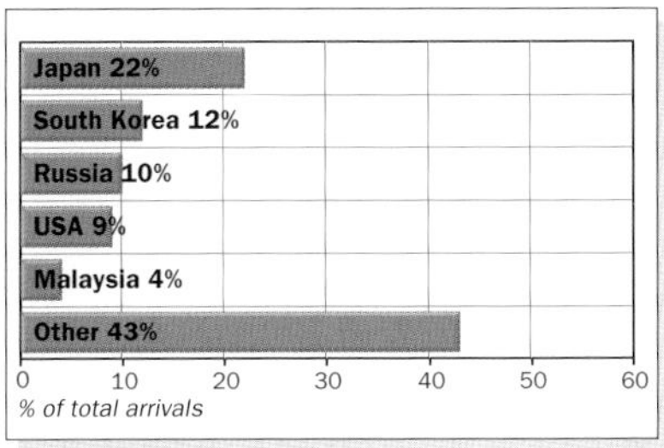

	% of total arrivals
Japan	22%
South Korea	12%
Russia	10%
USA	9%
Malaysia	4%
Other	43%

The easing of restrictions since the 1980s has led to the rapid growth of all kinds of tourism, from luxury tours to budget packages and backpacking. Hong Kong is a major entry point. Most of China is now open to visitors, and there have been some moves to allow tourists into Tibet, although access to Xinjiang in western China, and other areas, is sometimes impossible. The Great Wall, the Forbidden City in Beijing, and the terra-cotta warriors at Xi'an remain among the top attractions. The number of Chinese able to travel abroad now exceeds ten million.

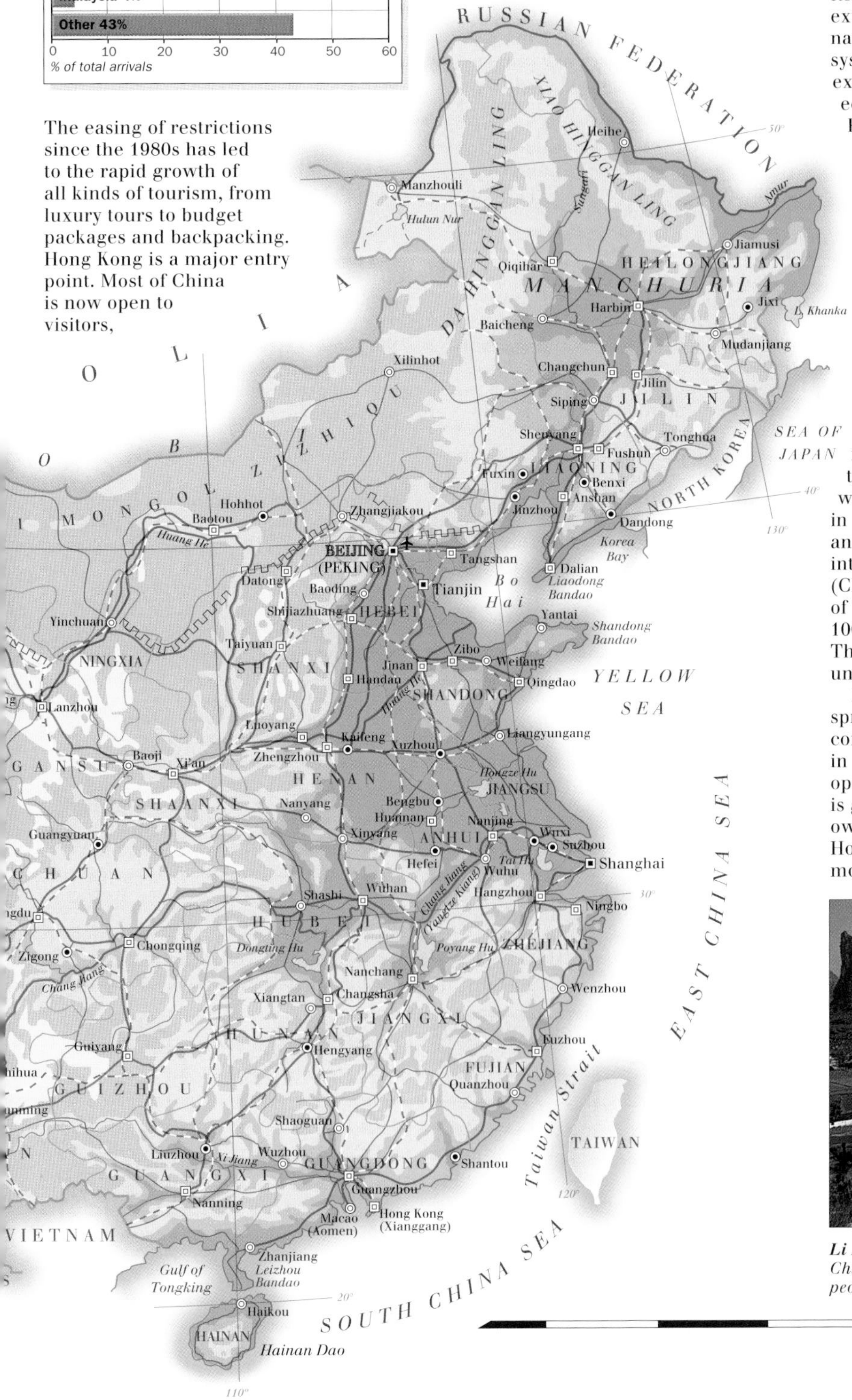

TRANSPORTATION

Drive on right

Chek Lap Kok, Hong Kong 32.8m passengers

3319 ships 16.5m grt

THE TRANSPORTATION NETWORK

271,300 km (168,578 miles)

24,474 km (15,207 miles)

67,394 km (41,879 miles)

110,263 km (68,514 miles)

Roads and railroads have been extended since 1949 to provide a basic national network. The transportation system is now being modernized and expanded to support the push for economic growth. The Ninth Five-Year Plan (1996–2000) provided for 8100 km (5000 miles) of new railroad. A high-speed rail link between Beijing and Shanghai is in prospect under the Tenth Plan. In 2001, plans were announced to connect the Tibetan capital, Lhasa, to the railroad system by extending the track from Golmud. If and when it is completed, this will present one of the world's highest railroads.

Container shipping is growing fast. Shanghai handled one-third of all Chinese container traffic before the reversion of Hong Kong (which is the world's biggest container port) to Chinese rule in 1997. The inland waterway system, which was hitherto in a state of disrepair, is being upgraded and now handles one-third of all internal freight. The Yangtze River (Chang Jiang) is navigable by ships of over 1000 tonnes for more than 1000 km (620 miles) from the coast. This capacity is planned to increase under the Three Gorges Dam project.

Nine small airlines which had sprung up since 1988 were consolidated into three carriers in 2002. Hong Kong's new airport opened in 1999. Air transportation is growing rapidly, like private car ownership, as wealth increases. However, bicycles still form the main mode of personal transportation.

Li River (Xi Jiang) valley. *Irrigation helps Chinese farmers to feed 20% of the world's people, using only 7% of the world's farmland.*

C

PEOPLE

Pop. density medium

Mandarin, Wu, Cantonese, Hsiang, Min, Hakka, Kan

138/km² (357/mi²)

THE URBAN/RURAL POPULATION SPLIT

32% 68%

RELIGIOUS PERSUASION

Muslim 2%, Buddhist 6%, Other 13%, Traditional beliefs 20%, Nonreligious 59%

ETHNIC MAKEUP

Zhaung 1%, Hui 1%, Other 5%, Han 93%

The vast majority of China's population is Han Chinese. The rest belong to one of 55 minority nationalities, or recognized ethnic groups. The minorities have disproportionate political significance because many, like the Mongolians, Tibetans, or Muslim Uyghurs in Xinjiang, live in strategic border areas.

The deeply resented policy of resettling Han Chinese in remote regions has led to ruthlessly suppressed uprisings in Xinjiang and Tibet. Han Chinese are now a majority in Xinjiang and Nei Mongol Zizhiqu (Inner Mongolia). Tibetan calls for greater political and cultural autonomy get much more international attention.

A one-child policy was adopted in 1979. Most Han Chinese still face strict family-planning controls, though these are widely flouted. Cases of female infanticide at birth have produced a demographic imbalance, and rules were relaxed for minorities after some small groups came near to extinction.

Chinese society is patriarchal in practice, and generations tend to live together. However, economic change is putting pressure on family life, breaking down the social controls of the Mao era. Divorce and unemployment are rising; materialism has replaced the puritanism of the past. The Falun Gong spiritual movement, perceived as a rival to CCP authority, was banned in 1999.

POPULATION AGE BREAKDOWN

Female	Age	Male
0.6%	80+	0.3%
5%	60–79	4.7%
10.6%	40–59	10.9%
17.4%	20–39	17.5%
15.7%	0–19	17.3%

% of population by age group

POLITICS

No multiparty elections

1998/2003

President Jiang Zemin

AT THE LAST ELECTION

National People's Congress 2979 seats

The Communist Party of China (CCP) is the only permitted party

China is a single-party state, dominated by the CCP, the world's largest political party. The National People's Congress, indirectly elected every five years, is theoretically the supreme organ of state power. It appoints the president and executive State Council, headed by the prime minister. The real focus of power, however, is the 22-member Politburo of the CCP and, in particular, its Standing Committee of seven.

Profile

The death in 1997 of Deng Xiaoping marked in effect the passing of the dominance of the "Immortals" – those who took part with Mao Zedong in the 1934–1935 Long March. Deng, the architect of China's economic reforms, had worked hard behind the scenes forming alliances to promote his reformist ideas and followers. Jiang Zemin consolidated his position as president and CCP general secretary after Deng's death, but a "fourth generation" leadership stands ready to emerge at the 2002 party congress, with Vice President Hu Jintao its most prominent figure.

Main Political Issues

Economic change and CCP authority

After the death of Mao Zedong in 1976, China embarked on economic reform, while seeking to secure the dominance of the CCP and avoid political upheaval. The "great helmsman" of this process for two decades was Deng Xiaoping, China's paramount leader, even after he had relinquished all official posts. Advocating a fast-track move to a "socialist market economy," Deng and his followers looked to South Korea and Taiwan as achieving high growth without political reform. At the 1997 party congress the reformers, led by

Nanjing Donglu (Nanking Road), in central Shanghai, is one of China's most famous shopping streets. A magnet for foreign investment, Shanghai is China's largest city.

Jiang Zemin, took their opportunity to realign formal party policy with their desire to privatize large areas of state-run industry. The transfer of much of the huge state economic system into private ownership has been a challenge to the CCP's ability to monopolize power. The 22 provinces, particularly those in the southeast, are acting increasingly independently of Beijing. At a popular level, there is growing rural discontent over widening wealth differentials. However, the party has allowed no political opposition to surface.

Shifting balances in the top leadership

The prodemocracy protests of 1989, culminating in the Tiananmen Square massacre, enabled conservatives within the party, under the then premier Li Peng, to gain the upper hand until Deng moved to restore the balance. His longevity shifted the advantage toward his heir apparent, President Jiang Zemin, who subsequently strengthened his own power base and international stature, while Premier Zhu Rongji spearheaded the economic reform and anticorruption campaigns. Li Peng, since 1998 president of the National People's Congress, still commands support, but all three are scheduled to stand down from the Politburo at the 2002 party congress. A major overhaul of party ideology in 2001, under Jiang's "Three Represents" doctrine, encouraged the equal promotion of business, culture, and the rural masses.

***Deng Xiaoping** was China's paramount leader until his death in 1997.*

***Jiang Zemin,** CCP leader and China's president since Deng resigned the post.*

***Li Peng,** president of the National People's Congress.*

***Vice President Hu Jintao,** prominent among "fourth generation" leaders.*

C

WORLD AFFAIRS

Joined UN in 1945

APEC
IAEA
WTO
ADB
SCO

The push for economic modernization, and concerns about regional stability, dominate Chinese foreign policy. Investment, technology, and trade considerations outweigh ideology.

Despite lingering concerns over human rights, relations with the West have rebounded from the low of the 1989 Tiananmen Square massacre. China was awarded Most Favored Nation trading status by the US in 2000 and entered the WTO in 2001. Despite initial tensions in relations with the new US administration in 2001, the two countries had reached a "common understanding" by the end of the year on the issue of terrorism, which had shot to the top of the global agenda.

Relations with Russia have improved steadily. The two states condemned the UN bombing campaign in Yugoslavia in 1999, and are against the proposed US missile defense shield. Regionally, ties have been normalized with Vietnam and strengthened with South Korea.

Taiwan remains the dark spot of Chinese external relations. Beijing strongly rejects any moves to recognize even de facto independence for the island, and cross-straits relations are frequently damaged by displays of military might and repetitive threats. Despite such alarming flare-ups, however, closer links have been forged since 2000, with an increase in tourist traffic and indirect trade.

AID

Recipient

$1.74bn (receipts) | Down 27% in 2000

In the 1970s aid was an important part of Chinese diplomacy, going mostly to Africa, but other communist and southeast Asian states were also recipients. Outward aid flows almost ceased in the late 1970s, as the economic reform process turned China itself into a major aid recipient. Japan is the biggest bilateral donor to China, but the potential of the Chinese market means that most developed states provide aid. A significant portion of funding is linked to a donor country's interest in opportunities created by China's huge infrastructure problems, and used to finance high-tech imports.

DEFENSE

Compulsory military service

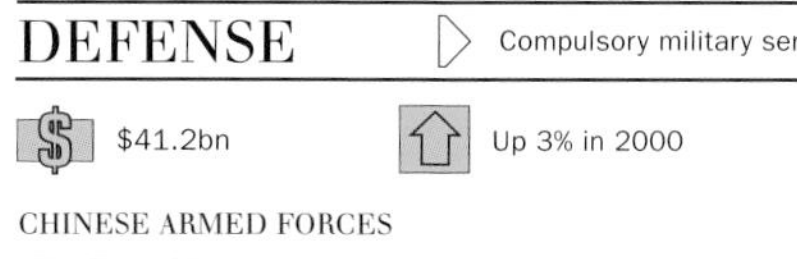
$41.2bn
Up 3% in 2000

CHINESE ARMED FORCES

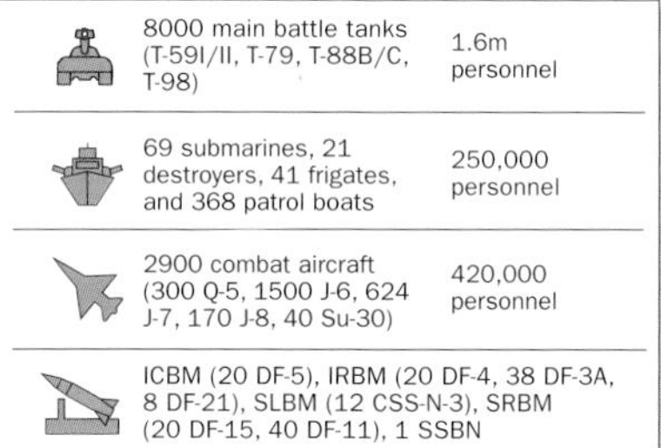

8000 main battle tanks (T-59I/II, T-79, T-88B/C, T-98)	1.6m personnel
69 submarines, 21 destroyers, 41 frigates, and 368 patrol boats	250,000 personnel
2900 combat aircraft (300 Q-5, 1500 J-6, 624 J-7, 170 J-8, 40 Su-30)	420,000 personnel
ICBM (20 DF-5), IRBM (20 DF-4, 38 DF-3A, 8 DF-21), SLBM (12 CSS-N-3), SRBM (20 DF-15, 40 DF-11), 1 SSBN	

The People's Liberation Army (PLA) is being cut back as part of a modernization process, reducing both its numbers and its involvement in the economy through army-run industries. Closely linked with the ruling CCP, it had appeared unassailable until the mid-1990s, when it numbered as many as three million personnel. A key instrument of ensuring the party's dominance, it was used in 1967 to restore order after the chaos of the Cultural Revolution, and in 1989 to suppress prodemocracy protests in Tiananmen Square, as well as to stamp out dissent in Tibet.

China has a large weapons industry, and has extended its nuclear weapons capability to include the neutron bomb. It is a significant arms exporter.

ECONOMICS

Inflation 7.1% p.a. (1990–2000)

$1063bn | 8.2774–8.2766 yuan

SCORE CARD

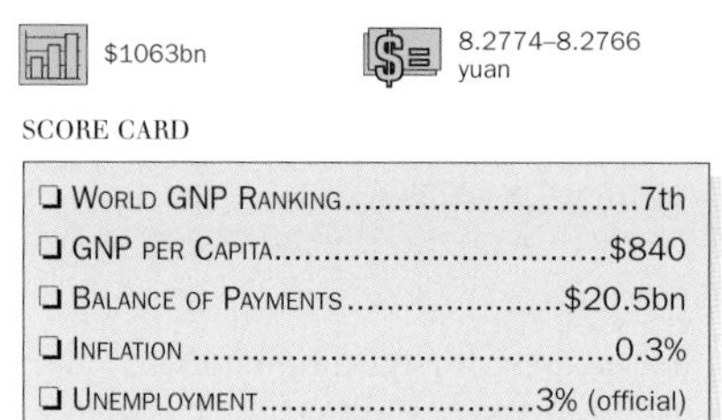

- ❑ WORLD GNP RANKING..............7th
- ❑ GNP PER CAPITA..............$840
- ❑ BALANCE OF PAYMENTS..............$20.5bn
- ❑ INFLATION..............0.3%
- ❑ UNEMPLOYMENT..............3% (official)

STRENGTHS

Huge domestic market. Food self-sufficiency. Mineral reserves. Diversified industrial sector. Low wage costs. Rapid sustained growth. Growing export sector. Hong Kong as financial center. Heavy investment in communications and IT.

ECONOMIC PERFORMANCE INDICATOR

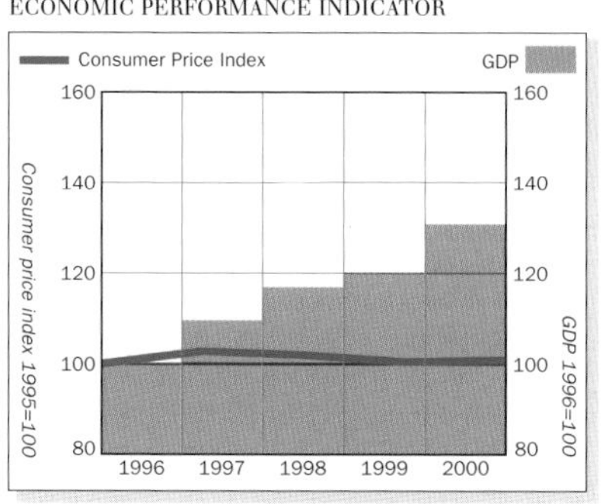
Consumer Price Index
GDP
160
140
120
100
80
1996
1997
1998
1999
2000
Consumer price index 1995=100
GDP 1996=100

EXPORTS

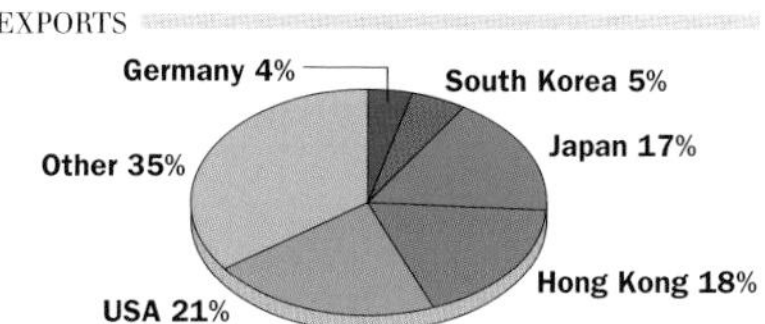
Germany 4%
South Korea 5%
Japan 17%
Hong Kong 18%
USA 21%
Other 35%

IMPORTS

Germany 5%
South Korea 10%
USA 10%
Taiwan 11%
Japan 18%
Other 46%

WEAKNESSES

Poor transportation. Debt-ridden state sector. Massive underemployment, rising unemployment (132 million admitted in 2002). Unevenly distributed resources.

CHINA : MAJOR BUSINESSES

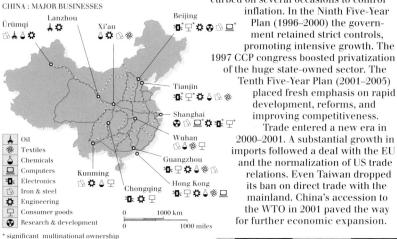

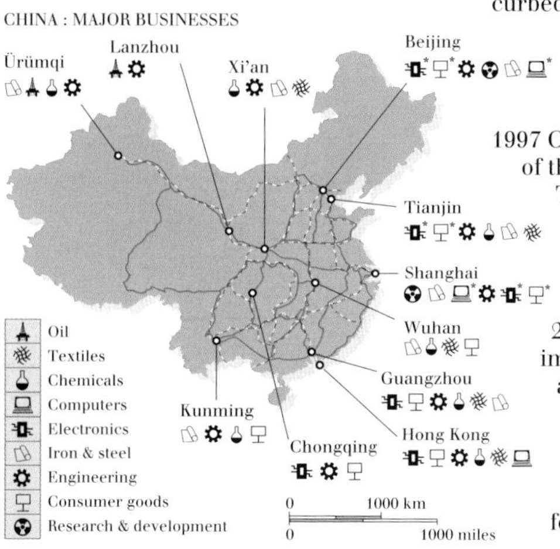
Ürümqi
Lanzhou
Xi'an
Beijing
Tianjin
Shanghai
Wuhan
Guangzhou
Hong Kong
Chongqing
Kunming
Oil
Textiles
Chemicals
Computers
Electronics
Iron & steel
Engineering
Consumer goods
Research & development
0
1000 km
0
1000 miles
* significant multinational ownership

PROFILE

China has demonstrably shifted from a centrally planned to a market-oriented economy since the 1980s; liberalization is greatest in the south. Growth had to be curbed on several occasions to control inflation. In the Ninth Five-Year Plan (1996–2000) the government retained strict controls, promoting intensive growth. The 1997 CCP congress boosted privatization of the huge state-owned sector. The Tenth Five-Year Plan (2001–2005) placed fresh emphasis on rapid development, reforms, and improving competitiveness.

Trade entered a new era in 2000–2001. A substantial growth in imports followed a deal with the EU and the normalization of US trade relations. Even Taiwan dropped its ban on direct trade with the mainland. China's accession to the WTO in 2001 paved the way for further economic expansion.

MINORITIES IN THE AUTONOMOUS REGIONS

AUTONOMOUS REGIONS OF CHINA
XINJIANG UYGHUR ZIZHIQU
INNER MONGOLIA (NEI MONGOL ZIZHIQU)
NINGXIA
TIBET (XIZANG ZIZHIQU)
GUANGXI

CHINA'S CONSTITUTION states that "regional autonomy is practiced in areas where people of minority nationalities live in compact communities." The Manchu, Miao, Yi, and many others among the 55 recognized ethnic minorities are not regarded as meeting this criterion, but there are five areas which do officially have the status of autonomous region or *zizhiqu*. The two most westerly such regions, sparsely populated Tibet and the home of the Uyghur people in Xinjiang, are both particularly sensitive border areas, and opposition to Chinese rule there has been suppressed.

TIBET

The vast mountainous region of Tibet has China's lowest population density, at only two inhabitants per sq. km (five per sq. mile), spread across an area of 1,221,600 sq. km (472,000 sq. miles). Lhasa is its capital. Tibetans were officially recorded as making up 95% of its total population of 2.2 million at the 1990 census. Han Chinese, whose immigration has aroused strong local resentment, were recorded as under 4%, though another census in 1995 gave a total population figure of nearly three million.

In 1950, Chinese forces invaded Tibet, which had been part of the Chinese empire from the 18th century until 1911, and ruthlessly crushed the 1959 independence uprising. The Dalai Lama, spiritual head of Tibetan Buddhism, fled to India amid an exodus of refugees and established a government in exile, but he has recently come close to accepting Chinese rule in an effort to win greater real self-government and cultural freedom for the Tibetan people. In 1965 Tibet was made a region of China, known as Xizang Zizhiqu. Opponents of Chinese rule were imprisoned or executed and many Buddhist monasteries were destroyed. Clashes between nationalists and Chinese troops in 1987 led to a renewed clampdown. In 1988, in an apparently more conciliatory spirit, the Chinese government accepted Tibetan as a "major official language."

Since 1995, the Chinese authorities have detained a young boy recognized by Tibetan Buddhists as the latest incarnation of their second-ranking spiritual leader, the Panchen Lama. The Chinese government named its own candidate to this office. Tibetans see this as unwarranted interference in their religious life. They are also alarmed by Chinese promotion of Tibet as a tourist destination, and the start of major restoration projects on the Potala and Norbulinka palaces in Lhasa.

XINJIANG UYGHUR

Consisting largely of desert, and bordering on Kyrgyzstan and Kazakhstan, the Xinjiang Uyghur Zizhiqu has a total area of 1,646,900 sq. km (636,000 sq. miles), or nearly 18% of China's total land mass. Han Chinese are narrowly in the majority in its population of 15.4 million, the total number of (mainly Muslim) Uyghurs being 7.2 million. Access is restricted and dissent is firmly suppressed, with a major clampdown on "terrorists" in 2002. The regional capital, Urumqi, is an industrial city noted for iron and steel, oil, and chemicals.

The Kumbum dagoba at the Palkhor Tschöde monastery, Gyangze (Chiang-tzu), Tibet. At the tip of this 15th-century structure is a Buddhist chapel.

INNER MONGOLIA

Inner Mongolia, or Nei Mongol Zizhiqu, occupies most of the area of northern China along the long land border with Mongolia (i.e. Outer Mongolia, as seen from the Chinese perspective). This totals 1,177,500 sq. km (455,000 sq. miles). Han Chinese are the majority in a total population of 21.1 million, of whom 4.8 million, or less than a quarter, are ethnic Mongolian.

NINGXIA HUI

The smallest of all the autonomous regions, Ningxia Hui Zizhiqu occupies an area of 66,000 sq. km (25,000 sq. miles), just south of Nei Mongol Zizhiqu. Its capital is Yinchuan. The population was 4.7 million at the 1990 census. Many of the Hui people, who number some 8.6 million in all, live outside the region.

GUANGXI ZHUANG

This autonomous region lies in the south, close to China's booming economic heartland. Relatively small in terms of area (220,000 sq. km – 85,000 sq. miles), it is by far the largest in terms of population, with a density of 192 per sq. km (497 per sq. mile) and 42.5 million inhabitants at the 1990 census. The capital is Nanning. The Zhuang people, although China's largest ethnic minority, number only 15.5 million, and thus are substantially in the minority in the region.

XINJIANG
KUNLUN SHAN
Golmud
Qinghai Hu
QINGHAI
K2 8611m
TIBET (XIZANG ZIZHIQU)
INDIA
Toling
Mangnang
HIMALAYAS
Brahmaputra
NEPAL
TANGGULA SHAN
Drepung
Nyethang
Nagqu
Xigazê
Sa'gya
Dinggyê
Riwoqê
Salween
Qamdo
Lhari
Lhasa
Bowo
Markam
Buchuser
Baxoi
Mt. Everest 8848m
Samyai
Gyangzê
Dagla Gompo
Thel Densathil
Mekong
HIMALAYAS
BHUTAN
INDIA
400 km
400 miles
Monasteries

C

SPECIAL ECONOMIC ZONES, OPEN CITIES, AND SPECIAL ADMINISTRATIVE REGIONS

ECONOMIC REFORMS FIRST INSTITUTED by Deng Xiaoping in 1978 began opening China to foreign business. This has spread from a small number of zones operating under special tax regimes to other cities all along the coastal belt and inland, and has fueled an extraordinary urban investment boom.

Nanchang
Changsha
HUNAN
JIANGXI
Hengyang
Fuzhou
Ganzhou
FUJIAN
Quanzhou
Shaoguan
Xiamen
GUANGXI
GUANGDONG
TAIWAN
Wuzhou
Xi Jiang
Guangzhou
Shantou
Shenzen
Zhuhai
SOUTH CHINA SEA
Macao
Hong Kong
Zhanjiang
Haikou
HAINAN
Special Economic Zones (SEZs)
Special Administrative Regions (SARs)

SPECIAL ECONOMIC ZONES

The creation of five special economic zones (SEZs) on the south coast was a major early milestone. Shenzhen, the trailblazer for the concept, was established in 1980, adjoining Kowloon in Hong Kong. Also in Guangdong province are Zhuhai, adjoining Macao, and Shantou further east, while Xiamen SEZ is in Fujian province, facing Taiwan. These locations reflect the aim of attracting investment from the 30 million Chinese overseas, especially in Hong Kong and Taiwan, whose ancestral homes were in Guangdong and Fujian. The fifth and least dynamic of the SEZs comprises the southern island province of Hainan.

FOREIGN INVESTMENT

Foreign direct investment reached over $45 billion in 1998, more than for any other country except the US. The inflow then began falling, for the first time for a decade. Most investment goes into the Open Coastal Belt, based on 14 cities, from Dalian in the north to Zhanjiang and Beihai in the south, which were picked in the mid-1980s for the dual role of "windows," opening to the outside world, and "radiators," spreading the development of an export-oriented economy.

Among the most successful development areas is the great port of Shanghai, where the stock exchange reopened in 1990. Investors in the city's Pudong New Zone, on the east bank of the Huangpu River, enjoy more preferential conditions than do the SEZs, including the right to sell goods and financial services. Pudong has attracted major foreign companies keen to establish a foothold in the potentially massive Chinese market, including General Motors, NEC, Sharp, Hitachi, Siemens, Unilever, BASF, and Pilkington. Pudong also forms the "dragon head" for a chain of open cities, extending up the Yangtze River (Chang Jiang) valley, where foreign investment has been encouraged since 1990. China has also designated a new set of open cities since 1992, this time in areas near its land borders, adjoining Russia, Mongolia, Kazakhstan, Burma, and Vietnam, to develop infrastructure and promote trade and the growth of export-oriented industries.

INDUSTRIAL GROWTH IN GUANGDONG

Guangdong province, with Guangzhou and Shenzhen at its heart, promises to become Asia's largest industrial region. One in ten of the world's top 500 multinational companies has invested in the Guangzhou Economic and Technical Development District (GET). Shenzhen, meanwhile, was the first Chinese city to start selling state-owned apartments freehold. It is being revamped as the science and technology city of the future, complete with parks, pedestrian precincts, civic amenities, and a sophisticated communications infrastructure. The process of reform was extended to politics in 2002 in an experiment to tackle rampant corruption.

HONG KONG AND MACAO SARS

Hong Kong, recession-hit after its reversion to Chinese sovereignty in mid-1997, recovered to record impressively high growth in 2000, but then suffered the knock-on effects of the global economic slowdown. As a British colony, Hong Kong had flourished through its textile industry, subsequently expanding into electronics, but it was above all a trade and financial services center. This role is preserved under its status as a special administrative region (SAR). Hong Kong's population is estimated at 6.69 million, in an area of only 1098 sq. km (424 sq. miles). Its Basic Law guarantees a high level of autonomy for 50 years under the "one country, two systems" formula. Its Chinese-appointed chief executive, Tung Chee-hwa, has been unpopular, however – increasingly at odds with members of his administration (several of whom resigned in 2000–2001), and with the region's Legislative Council. The number of directly elected seats on the Council was increased somewhat (to 24 out of 60) for the September 2000 elections, half of which were won by the opposition Democratic Party, but public apathy produced a low turnout. English and Chinese are both official languages, and the freely convertible Hong Kong dollar remains as the SAR's currency, while Hong Kong has its own separate membership of key international organizations, notably the WTO.

China's other SAR, Macao, ceased to be a Portuguese colony at midnight on 19–20 December 1999. Macao too has free port status, its own currency, the pataca, two official languages (Portuguese and Chinese), a partly elective Legislative Assembly, and a guarantee under its Basic Law that the "one country, two systems" formula will apply for 50 years. Macao's area of only 25.4 sq. km (9.8 sq. miles) was due to increase by 20% on completion of the Nam Van lakes project. The population of 414,128 at the 1996 census included more than 100,000 people with Portuguese nationality. The economy, based on tourism and gambling, has recently been in decline.

PHILIPS

Hong Kong's return to Chinese sovereignty after 157 years of British rule, took place at midnight on 30 June 1997.

C

CHRONOLOGY

China's recorded history begins 4000 years ago with the Shang dynasty, which was founded in the north in 1766 BCE. Succeeding dynasties expanded China's boundaries; it reached its greatest extent under the Manchu (Qing) dynasty in the 18th century. Chinese isolationism frustrated Europe's attempts to expand trade into the empire until the 19th century, when China had fallen behind the industrializing West. For the previous 3000 years, it had been one of the world's most advanced nations.

- ❑ **1839–1860** Opium Wars with Britain. China defeated; forced to open ports to foreigners.
- ❑ **1850–1873** Internal rebellions against Manchu Empire.
- ❑ **1895** Defeat by Japan in war over Korean peninsula.
- ❑ **1900** Boxer Rebellion to expel all foreigners suppressed.
- ❑ **1911** Manchu Empire overthrown by nationalists led by Sun Yat-sen. Republic of China declared.
- ❑ **1912** Sun Yat-sen forms National People's Party (Guomindang).
- ❑ **1916** Nationalists factionalize. Sun Yat-sen sets up government in Guangdong. Rest of China under control of rival warlords.
- ❑ **1921** CCP founded in Shanghai.
- ❑ **1923** CCP joins Soviet-backed Guomindang to fight warlords.
- ❑ **1925** Chiang Kai-shek becomes Guomindang leader on death of Sun Yat-sen.
- ❑ **1927** Chiang turns on CCP. CCP leaders escape to rural south.
- ❑ **1930–1934** Mao Zedong formulates strategy of peasant-led revolution.
- ❑ **1931** Japan invades Manchuria.
- ❑ **1934** Chiang forces CCP out of its southern bases. Start of 12,000-km (7450-mile) Long March.
- ❑ **1935** Long March ends in Yanan, Shaanxi province. Mao becomes CCP leader.
- ❑ **1937–1945** War against Japan: CCP Red Army in north, Guomindang in south. Japan defeated.
- ❑ **1945–1949** War between Red Army and Guomindang. US-backed Guomindang retreats to Taiwan.
- ❑ **1949** 1st October, Mao proclaims People's Republic of China.
- ❑ **1950** Invasion of Tibet. Mutual assistance treaty with USSR.
- ❑ **1950–1958** Land reform; culminates in setting up of communes. First Five-Year Plan (1953–1958) fails.
- ❑ **1958** "Great Leap Forward" to boost production fails; contributes to millions of deaths during 1959–1961 famine. Mao resigns ➪

RESOURCES

Electric power 231m kw

41.5m tonnes

3.31m b/d (reserves 24bn barrels)

636m ducks, 454m pigs, 208m geese, 3.77bn chickens

Coal, oil, natural gas, salt, iron, molybdenum, titanium, tungsten

ELECTRICITY GENERATION

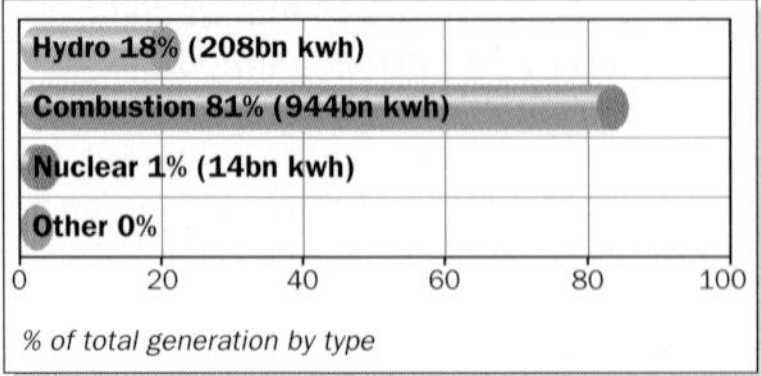
Hydro 18% (208bn kwh)
Combustion 81% (944bn kwh)
Nuclear 1% (14bn kwh)
Other 0%
0
20
40
60
80
100
% of total generation by type

China dominates the world market in molybdenum, titanium, and tungsten; it has the world's largest deposits of more than a dozen minerals, and commercial deposits of most others.

China is the world's second-largest coal producer, with reserves of about 800 billion tonnes, primarily in the Shaanxi and Sichuan basins. Annual output (over one billion tonnes) considerably exceeds demand; this and appalling safety records are forcing the closure of many mines.

Power generation, previously well behind demand, expanded rapidly to create overcapacity by the late 1990s. Nuclear power capacity in 2002 was just over 2100 MW from three reactors, with major expansion to 50,000 MW planned by 2020. The world's largest hydropower plant, the "Three Gorges" scheme on the Yangtze River, is due for completion in 2009, but faces sustained controversy over its proposed benefits, costs, and environmental consequences.

Crude oil production has risen only slightly since reaching 160 million tonnes in 1997. Eastern oil fields are depleted, and hopes now center on enormous reserves in the Tarim basin in the far west. A 4000-km (2500-mile) pipeline is due to come on line in 2004.

CHINA : LAND USE

TIEN SHAN
HIMALAYAS
Mekong

Cropland
Forest
Pasture
Wetlands
Desert
High mountain regions
Sheep
Tea
Rice

0 200 km
0 200 miles

ENVIRONMENT

Sustainability rank: 129th

6%

2.5 tonnes per capita

ENVIRONMENTAL TREATIES

Yes
Yes
Yes
Yes
No
No

Climate and geology mean that natural disasters are quite frequent in China, and their impact is often made worse by human actions. The economic policies of the 1950s turned drought into a devastating famine, while poor building standards helped push the death toll in the 1976 Tangshan earthquake to over 500,000.

Economic growth is the priority of China's leaders. Industrial pollution and environmental degradation, already widespread, are increasing. However, the environment is a growing concern among educated Chinese. The campaign, unsuccessful in 1992, to stop the Three Gorges hydroelectric scheme was revived with growing open criticism in the late 1990s. The government is becoming less suspicious of Western pressure for environmental controls, and is taking steps to respond to acute problems of urban air pollution, deforestation, and water quality in particular.

MEDIA

TV ownership high

Daily newspaper circulation 40 per 1000 people

PUBLISHING AND BROADCAST MEDIA

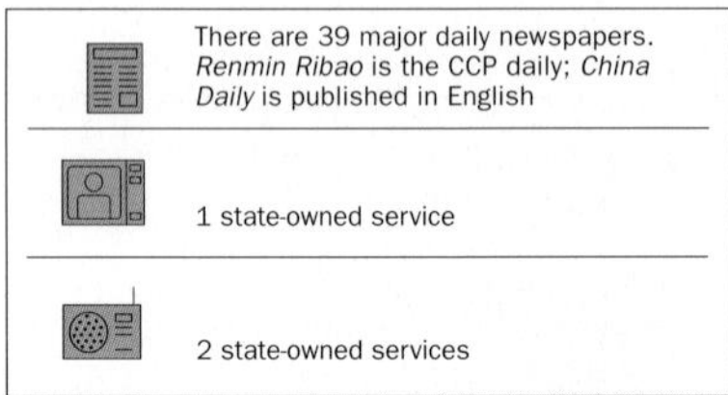
There are 39 major daily newspapers. *Renmin Ribao* is the CCP daily; *China Daily* is published in English
1 state-owned service
2 state-owned services

China's more open, market-oriented economy has created access to nonofficial sources of information. TV ownership is rising and satellite-dish owners can choose what to view, while growing Internet usage makes central control even more difficult. Since 2000, Internet sites have had to obtain official approval and have been held responsible for their content. Many Internet cafés were closed in 2001, and new rules in 2002 restricted the media's use of Internet sources.

The ideological influence of the huge-circulation party newspaper *Renmin Ribao* (*People's Daily*) and the trade union *Gongren Ribao* (*Workers' Daily*) is much diminished, but "undesirable" papers have their licenses removed in periodic cleanups. Beijing and the provinces have their own dailies.

C

CHRONOLOGY *continued*

as CCP chairman; succeeded by Liu Shaoqi.

- ❑ **1960** Sino-Soviet split.
- ❑ **1961–1965** More pragmatic economic approach led by Liu and Deng Xiaoping.
- ❑ **1966** Cultural Revolution initiated by Mao to restore his supreme power. Youthful Red Guards attack all authority. Mao rules, with Military Commission under Lin Biao and State Council under Zhou Enlai.
- ❑ **1967** Army intervenes to restore order amid countrywide chaos. Liu and Deng purged from party.
- ❑ **1969** Mao regains chair of CCP. Lin Biao designated his successor, but quickly attacked by Mao.
- ❑ **1971** Lin dies in plane crash.
- ❑ **1972** US president Nixon visits. More open foreign policy initiated.
- ❑ **1973** Mao's wife Jiang Qing, Zhang Chunquio, and other "Gang of Four" members elected to CCP Politburo. Deng Xiaoping rehabilitated.
- ❑ **1976** Death of Zhou Enlai. Mao strips Deng of posts. September, Mao dies. October, Gang of Four arrested.
- ❑ **1977** Deng regains party posts, begins to extend power base.
- ❑ **1978** Decade of economic modernization launched. Open door policy to foreign investment; farmers allowed to farm for profit.
- ❑ **1980** Deng emerges as China's paramount leader. Economic reform gathers pace, but hopes for political change suppressed.
- ❑ **1983–1984** Elderly conservative leaders attempt to slow reform.
- ❑ **1984** Industrial reforms announced.
- ❑ **1989** Prodemocracy demonstrations in Tiananmen Square. Crushed by army; 1000–5000 dead.
- ❑ **1992–1995** Trials of prodemocracy activists continue. Plans for market economy accelerated.
- ❑ **1993** Jiang Zemin president.
- ❑ **1997** February, Deng Xiaoping dies at 92. July, UK hands back Hong Kong. September, party congress confirms reformist policies.
- ❑ **1999** China develops neutron bomb. Portugal hands back Macao. Friction over Taiwanese claim of statehood. Clampdown on Falun Gong sect.
- ❑ **2000** Taiwanese presidential election causes tension. USA normalizes trade relations.
- ❑ **2001** Major diplomatic incident with USA when Chinese pilot is killed and US spy-plane is forced down on Hainan Island. December, accession to WTO.
- ❑ **2002** Crackdown on Uyghur separatism. Severe flooding: 750 killed.

CRIME

Death penalty in use

1.43m prisoners

Down 33% 1990–1996

CRIME RATES

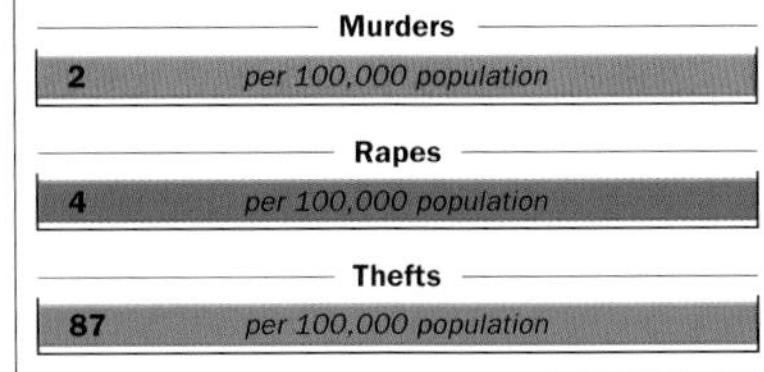
Murders
2 per 100,000 population
Rapes
4 per 100,000 population
Thefts
87 per 100,000 population

China's legal system is a mix of custom and statute; judges have only been required to hold law degrees since 2002.

Economic reform and social changes have been paralleled by a rise in violent crime and corruption. Many party officials were condemned in China's largest anticorruption trial in 2000. A crackdown on people trafficking followed the breaking of rings smuggling Chinese into Europe. The death penalty is used extensively – China carried out over 80% of the world's executions in 2001. Many detainees have been released since the clampdown on dissent after the 1989 Tiananmen Square massacre, but many more remain incarcerated.

EDUCATION

School leaving age: 16

84%

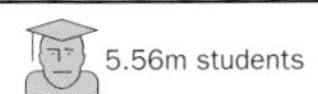
5.56m students

THE EDUCATION SYSTEM

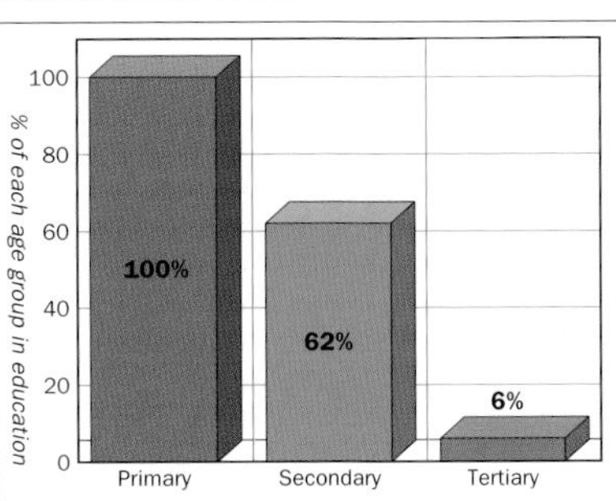
% of each age group in education
100
80
60
40
20
0
100%
62%
6%
Primary
Secondary
Tertiary

Despite the expansion of education since 1949, illiteracy and semiliteracy are still widespread. School attendance fell when fees at all levels were introduced in the 1980s, but now most children of secondary school age are in school – including those catching up with primary education.

Selection for higher education is now based on academic rather than political criteria. Internet-based distance learning degrees, first allowed in 1998, are especially popular in English, computing, and business studies. The government legalized private schools (tacitly permitted since the early 1980s) in 2001, in an effort to regulate and profit from them. About seven million pupils are thought to attend them.

HEALTH

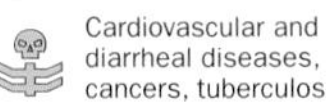
Welfare state health benefits

1 per 588 people

Cardiovascular and diarrheal diseases, cancers, tuberculosis

China's network of primary health care, combining traditional and Western medicine, used to extend to the remotest regions, and a free health system meant that the Chinese enjoyed a life expectancy on a par with many richer countries. The change to a market-oriented economy, however, has produced a two-tier system. A gaping divide exists between city and rural provision, fees for treatment are rising, and fewer people are covered by free health care as a benefit of state employment. The UN estimated in 2002 that there would be over ten million HIV/AIDS sufferers in China by 2010.

SPENDING

GDP/cap. increase

CONSUMPTION AND SPENDING

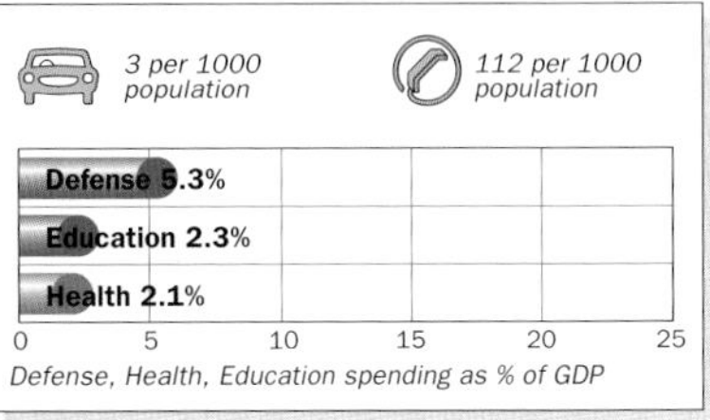
3 per 1000 population
112 per 1000 population
Defense 5.3%
Education 2.3%
Health 2.1%
0
5
10
15
20
25
Defense, Health, Education spending as % of GDP

The majority of Chinese are still farmers, whose living standards are threatened by rising production costs. Economic change has led to widening wealth disparities. The burgeoning small-business class and employees of companies with foreign investment have benefited most. They mainly live in the east, especially the southeast, where there are a number of dollar millionaires. The main losers are the 150 million "surplus" agricultural workers, many of whom have migrated to the cities in search of jobs. By July 2001, there were more mobile phones in China than in the US, 120.6 million in total, about one for every ten Chinese.

WORLD RANKING

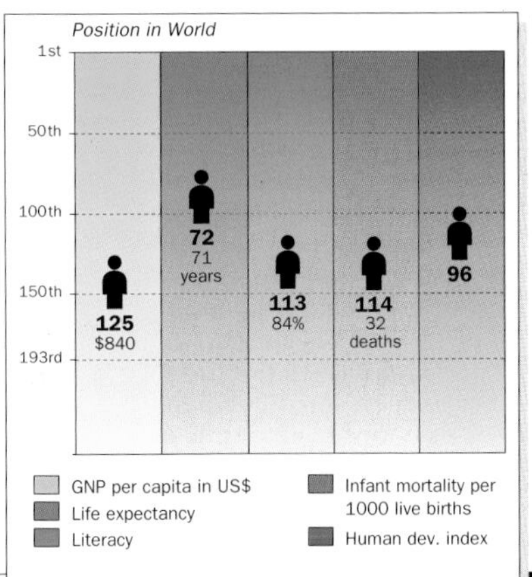
Position in World
1st
50th
100th
150th
193rd
125
$840
72
71 years
113
84%
114
32 deaths
96
GNP per capita in US$
Life expectancy
Literacy
Infant mortality per 1000 live births
Human dev. index

C

COLOMBIA

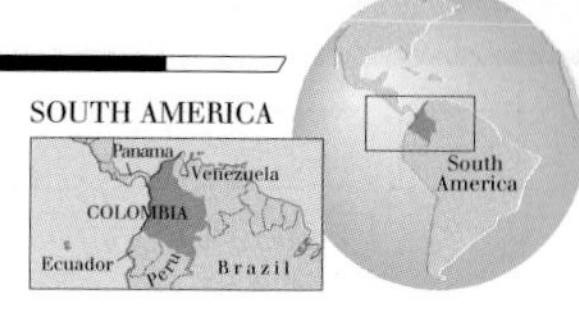
SOUTH AMERICA
Panama
Venezuela
COLOMBIA
Ecuador
Peru
Brazil
South America

OFFICIAL NAME: Republic of Colombia **CAPITAL:** Bogotá
POPULATION: 42.8 million **CURRENCY:** Colombian peso **OFFICIAL LANGUAGE:** Spanish

1819

1903

July 20

CO

-5
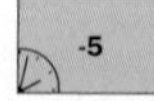
+57
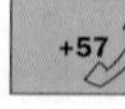

.co

LYING IN NORTHWEST South America, Colombia has coastlines on both the Caribbean and the Pacific. The east is densely forested and sparsely populated, and separated from the western coastal plains by the Andes mountains. The Andes divide into three ranges (cordilleras) in Colombia. The eastern range is divided from the two western ranges by the densely populated Magdalena River valley. The Colombian lowlands are very wet, hot, and fertile, supporting two harvests and allowing many crops to be planted at any time of year. A state plagued by instability and violence, Colombia is noted for its coffee, emeralds, gold, and narcotics trafficking.

CLIMATE

Tropical/mountain

WEATHER CHART FOR BOGOTÁ

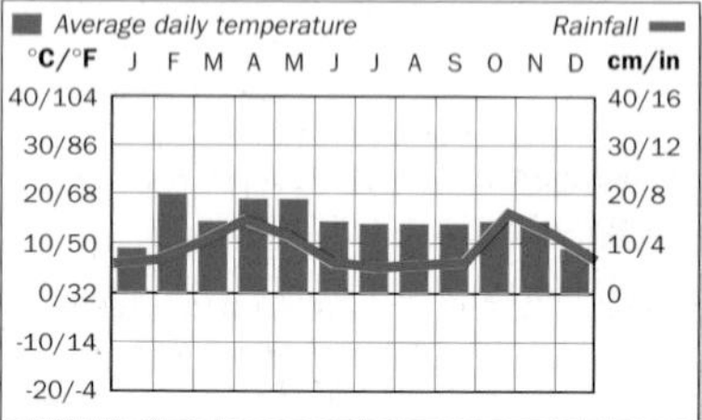
Average daily temperature
Rainfall
°C/°F J F M A M J J A S O N D cm/in
40/104 40/16
30/86 30/12
20/68 20/8
10/50 10/4
0/32 0
-10/14
-20/-4

Most of Colombia is wet, and the hot Pacific coastal areas receive up to 500 cm (200 in) of rain a year. The Caribbean coast is a little drier. The Andes have three climatic regions: the *tierra caliente* (hot lowlands), *tierra templada* (temperate uplands), and *tierra fría* (cold highlands); the last has year-round springlike conditions such as those found in Bogotá. The equatorial east has two wet seasons.

TRANSPORTATION

Drive on right

El Dorado, Bogotá
4.66m passengers

111 ships
81,400 grt

THE TRANSPORTATION NETWORK

13,868 km
(8617 miles)

Caribbean Trunk Highway

2113 km
(1313 miles)

18,140 km
(11,272 miles)

Roads in the north are in reasonable condition. Those in the south and east tend to be rutted and badly affected by the frequent rains. Continuing instability means that roads are frequently blocked by the guerrillas and the military. Most of the railroad is closed.

Rivers are an important means of transportation; the Magdalena, Orinoco, Atrato, and Amazon river systems are all extensively navigable. Plans exist to connect Colombia to the Pan-American Highway.

TOURISM

Visitors : Population 1:81

530,000 visitors

Down 3% in 2000

MAIN TOURIST ARRIVALS

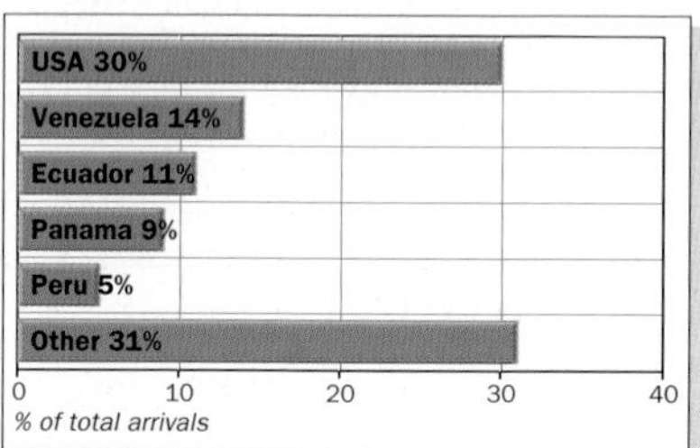
USA 30%
Venezuela 14%
Ecuador 11%
Panama 9%
Peru 5%
Other 31%
0 10 20 30 40
% of total arrivals

Tourism in Colombia is largely limited to the beaches of the Caribbean coast. Cartagena, Barranquilla, and Santa Marta are the main resorts. Cartagena has also been developed as a major Latin American conference center.

The expansion of tourism has been limited by Colombia's political instability and the prevalence of narcotics-related crime. The well-publicized activities of drugs cartels in Medellín and Cali, and instances of kidnappings in Bogotá, are major deterrents for travelers.

Limited infrastructure makes many regions of the country, particularly Amazonia to the east of the Andes, almost inaccessible. The Caribbean coast is also barely exploited.

Simón Bolívar and Cristóbal Colón, twin peaks with a height of 5775 m (18,947 ft) are the highest in the Colombian Andes.

PEOPLE

Pop. density low

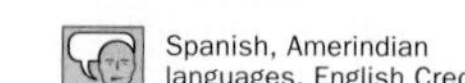
Spanish, Amerindian languages, English Creole

41/km² (107/mi²)

THE URBAN/RURAL POPULATION SPLIT

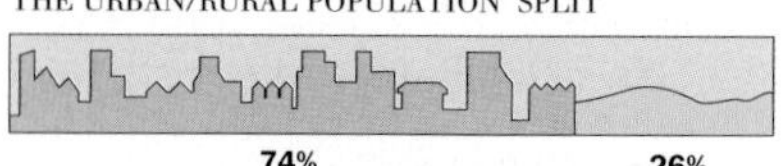
74%
26%

RELIGIOUS PERSUASION

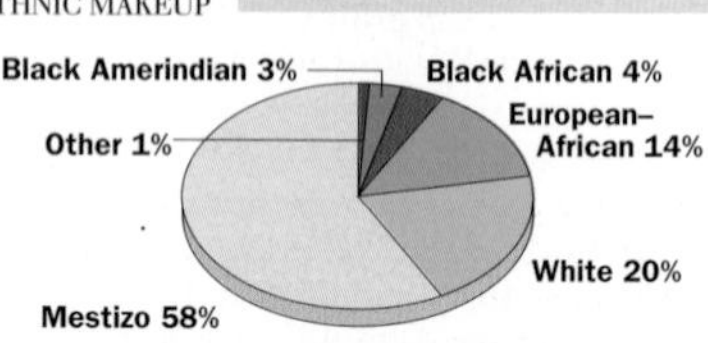
Other 5%
Roman Catholic 95%

ETHNIC MAKEUP

Black Amerindian 3%
Black African 4%
European–African 14%
Other 1%
White 20%
Mestizo 58%

The majority of Colombians are people of mixed blood. An estimated 450,000 indigenous Amerindians are largely concentrated in the southwest and Amazonia, although some communities are scattered throughout the country. A small black population lives along both coasts, particularly in Chocó, Colombia's poorest region. Blacks are the most unrepresented group.

Some progress has been made in giving Amerindians a greater political voice. In 1991, constitutional reforms reserved two seats in the Senate for indigenous representatives, and Amerindian pressure groups are increasingly active. Harassment by landowners and narcotics traffickers continues in Amazonia, and very few investigations into suspected human rights violations against Amerindians have led to prosecutions.

Women in Colombia have a higher profile than in much of the rest of Latin America. Many are prominent in the professions, though few reach the top in politics. The traditional extended Catholic family is still the norm. Regional identity is strong.

POPULATION AGE BREAKDOWN

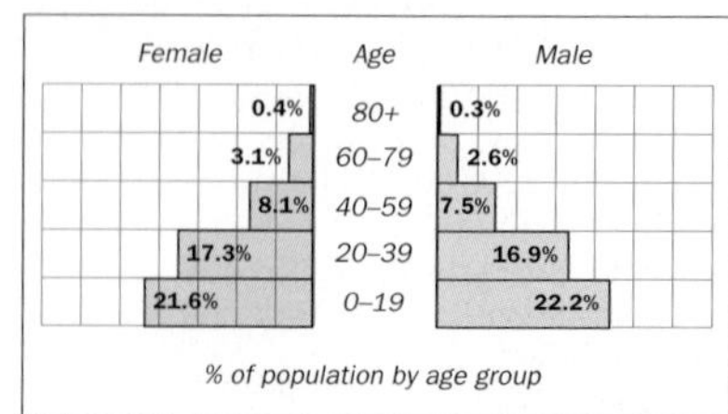
Female
Age
Male
0.4% 80+ 0.3%
3.1% 60–79 2.6%
8.1% 40–59 7.5%
17.3% 20–39 16.9%
21.6% 0–19 22.2%
% of population by age group

C

POLITICS

Multiparty elections

L. House 2002/2006
U. House 2002/2006

President Alvaro Uribe Velez

Colombia is a presidential democracy, with a bicameral Congress. Presidents may not serve two consecutive terms.

PROFILE

The two-party system which had held sway from the late 1950s appeared to have fractured by the 21st century. The dominance of the PSC and the PL, with few ideological differences, was undermined by electoral breakthroughs by new, smaller parties, culminating in the presidential victory of right-wing independent Alvaro Uribe Velez in 2002.

Pervasive official corruption and the violence associated with drugs cartels, guerrillas, paramilitaries, and the military have seriously weakened confidence in the state and the government, and have deterred foreign investors.

MAIN POLITICAL ISSUES

Elusive peace

Against a background of decades of civil conflict, former president Andres Pastrana Arango (1998–2002) took political risks in pursuit of a lasting peace, granting "safe havens" in remote areas to the Armed Revolutionary Forces of Colombia (FARC) and National Liberation Army (ELN). Mutual suspicions undermined these peace efforts, which were abandoned in early 2002. A renewed military offensive focused mainly on the FARC.

Violence

At least 2856 people were kidnapped in 2001, including 35 foreigners, with guerrilla groups and right-wing paramilitaries blamed for 85% of cases. Unions said that 152 labor leaders were killed. The murder in 2002 of the Archbishop of Cali rocked the nation.

AT THE LAST ELECTION

House of Representatives 166 seats

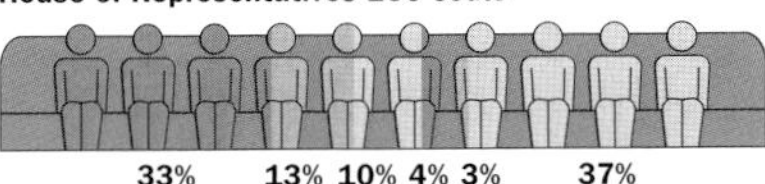

33% PL
13% PCC
10% C
4% CR
3% AL
37% Others

PL = Liberal Party PCC = Colombian Conservative Party C = Coalition CR = Radical Change AL = Liberal Opening MN = National Movement

Senate 102 seats

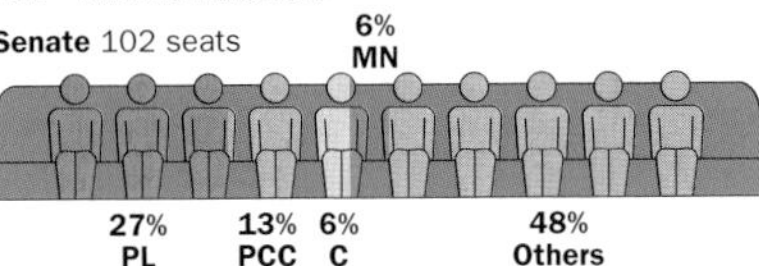

6% MN
27% PL
13% PCC
6% C
48% Others

Two special representatives of the Amerindian communities are appointed to the Senate, and one is elected.

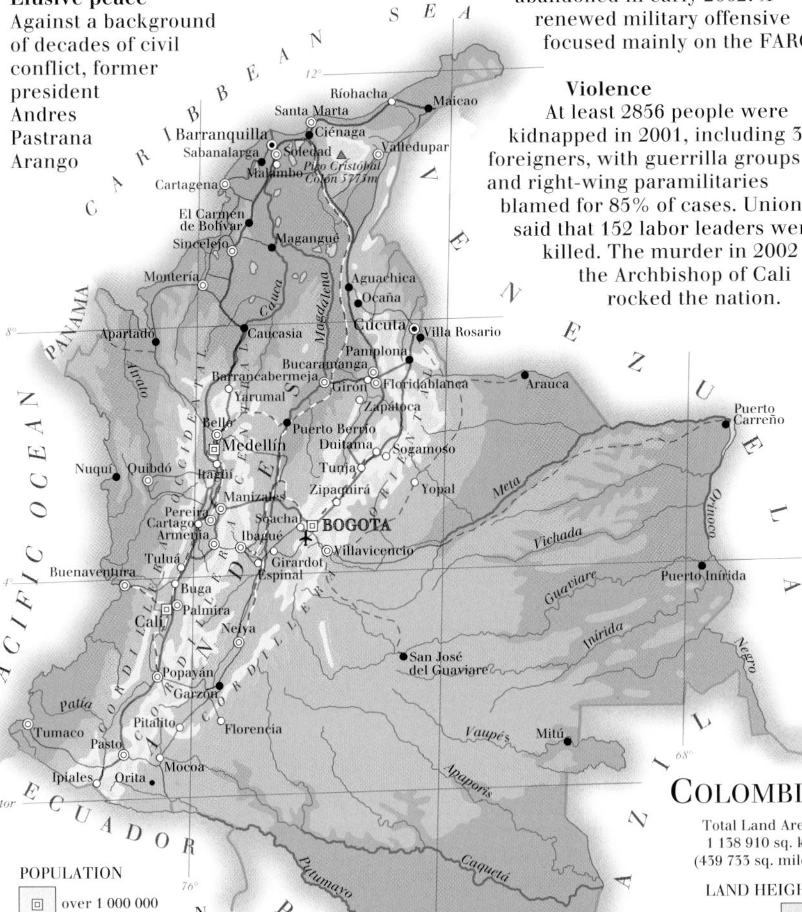

CARIBBEAN SEA
PACIFIC OCEAN
PANAMA
VENEZUELA
ECUADOR
PERU
BRAZIL
Ríohacha
Maicao
Santa Marta
Barranquilla
Ciénaga
Sabanalarga
Soledad
Valledupar
Malambo
Pico Cristóbal Colón 5775m
Cartagena
El Carmen de Bolívar
Magangué
Sincelejo
Montería
Aguachica
Ocaña
Cúcuta
Villa Rosario
Apartadó
Caucasia
Pamplona
Bucaramanga
Barrancabermeja
Girón
Floridablanca
Arauca
Yarumal
Zapatoca
Puerto Carreño
Bello
Puerto Berrío
Medellín
Duitama
Sogamoso
Nuquí
Quibdó
Itagüí
Tunja
Zipaquirá
Yopal
Manizales
Pereira
Cartago
Armenia
Soacha
BOGOTA
Ibagué
Villavicencio
Tuluá
Girardot
Espinal
Buenaventura
Buga
Palmira
Cali
Neiva
Puerto Inírida
San José del Guaviare
Popayán
Garzón
Pitalito
Florencia
Tumaco
Pasto
Mocoa
Mitú
Ipiales
Orita
Leticia
Equator
Magdalena
Cauca
Atrato
Meta
Vichada
Guaviare
Inírida
Orinoco
Negro
Vaupés
Apaporis
Caquetá
Putumayo
Amazon
Patía
COLOMBIA
Total Land Area : 1 138 910 sq. km (439 733 sq. miles)
POPULATION
over 1 000 000
over 500 000
over 100 000
over 50 000
over 10 000
under 10 000
LAND HEIGHT
3000m/9843ft
2000m/6562ft
1000m/3281ft
500m/1640ft
Sea Level
0 200 km
0 200 miles
N

Alvaro Uribe Velez, the right-leaning independent elected president in 2002.

***Pablo Escobar,** ex-leader of Medellín drugs cartel, gunned down in 1993.*

WORLD AFFAIRS

Joined UN in 1945

ACS

AP

AmCC

OAS

RG

Good relations with the US are conditional on tough measures to fight the narcotics trade. A $1.3 billion military aid package, approved in 2000, provided training, intelligence, and hardware, ostensibly to assist the "Plan Colombia" antidrugs program. The boundary between this aim and the targeting of guerrilla groups was always blurred.

The army's invasion in 2002 of the FARC's demilitarized haven reportedly was assisted by US military advisers. However, the US administration fought shy of describing events as a chapter in the US-led "war on terrorism," insisting that it was operating within legal limits set down by the Congress. Colombia's neighbors fear a "spillover" of violence and refugees.

AID

Recipient

$187m (receipts) Down 38% in 2000

US military "antinarcotics aid" forms some 75% of the US total. The World Bank approved a loan of $400 million in 2002 to aid fiscal reform and improve delivery of public health services.

CHRONOLOGY

In 1525, Spain began the conquest of Colombia, which became its chief source of gold.

- ❑ **1819** Simón Bolívar defeats the Spanish at Boyacá. Republic of Gran Colombia formed with Venezuela, Ecuador, and Panama.
- ❑ **1830** Venezuela and Ecuador split away during revolts and civil wars.
- ❑ **1849** The Centralist Conservative and federalist Liberal parties are established.
- ❑ **1861–1886** Liberals hold monopoly on power.
- ❑ **1886–1930** Conservative rule.
- ❑ **1899–1903** Liberal "War of 1000 Days" revolt fails; 120,000 die.

C

CHRONOLOGY *continued*

- **1903** Panama secedes, but is not recognized by Colombia until 1921.
- **1930** Liberal President Olaya Herrera elected by coalition in first peaceful change of power.
- **1946** Conservatives take over.
- **1948** Shooting of Liberal mayor of Bogotá and riot known as *El Bogotazo* spark civil war, *La Violencia*, to 1957; 300,000 killed.
- **1953–1957** Military dictatorship of Rojas Pinilla.
- **1958** Conservatives and Liberals agree to alternate government in a National Front until 1974. Other parties banned.
- **1965** Left-wing guerrilla National Liberation Army and Maoist Popular Liberation Army founded.
- **1966** Pro-Soviet FARC guerrilla group formed.
- **1968** Constitutional reform allows new parties, but two-party parity continues. Guerrilla groups proliferate from now on.
- **1984** Minister of justice assassinated for attempting to enforce antinarcotics campaign.
- **1985** M-19 guerrillas blast their way into ministry of justice; 11 judges and 90 others killed. Patriotic Union (UP) party formed.
- **1986** Liberal Virgilio Barco Vargas wins presidential elections, ending power-sharing. UP wins ten seats in parliament. Right-wing paramilitary start murder campaign against UP politicians. Violence by both left-wing groups and death squads run by narcotics cartels continues.
- **1989** M-19 reaches peace agreement with government, including the granting of a full pardon. Becomes legal party.
- **1990** Presidential candidates of UP and PL murdered during campaign. Liberal César Gaviria elected on antidrugs platform.
- **1991** New constitution legalizes divorce, prohibits extradition of Colombian nationals. Indigenous peoples' democratic rights guaranteed, but territorial claims are not addressed.
- **1992–1993** Medellín drugs cartel leader, Pablo Escobar, captured, escapes, and shot dead by police.
- **1995–1996** President Samper cleared of charges of receiving Cali cartel drug funds for elections.
- **1999** Earthquake kills thousands.
- **2001** US-backed spraying of coca plantations and destruction of food crops by herbicides provokes resentment.
- **2002** Peace talks abandoned. Renewed military offensive. Independent candidate Alvaro Uribe Velez elected president.

DEFENSE

Compulsory military service

$1.96bn

Down 10% in 2000

COLOMBIAN ARMED FORCES

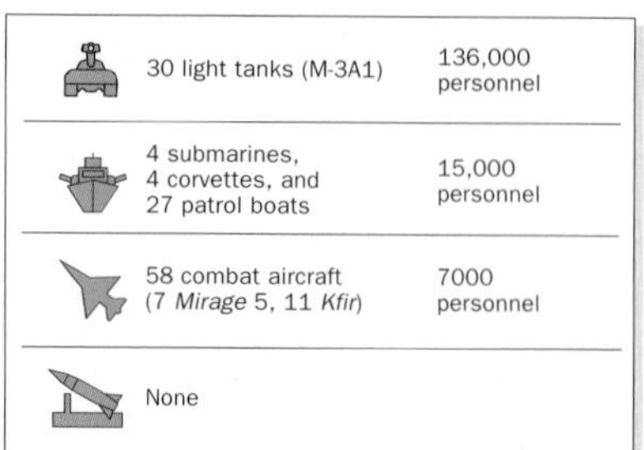
30 light tanks (M-3A1)
136,000 personnel
4 submarines, 4 corvettes, and 27 patrol boats
15,000 personnel
58 combat aircraft (7 Mirage 5, 11 Kfir)
7000 personnel
None

The military is powerful, but rarely intervenes directly in politics. Human rights groups accuse the armed forces and their paramilitary allies of gross and systematic abuses, involving torture and murder, in their fight against guerrilla groups and the production of narcotics. Despite its restructuring in 1998, the army high command remained suspicious of peace negotiations and supported a tougher stance against the rebels, successfully exploiting tensions to expand substantially under President Pastrana, with numbers set to rise still more. Orders were given in February 2002 to retake the FARC "safe haven."

Colombia participates in the joint Latin American Defense Force. The US supplies most arms and training, especially through "Plan Colombia."

ECONOMICS

Inflation 21% p.a. (1990–2000)

$85.3bn

2236–2278 Colombian pesos

SCORE CARD

World GNP Ranking	40th
GNP per Capita	$2020
Balance of Payments	$41m
Inflation	9.5%
Unemployment	20%

EXPORTS

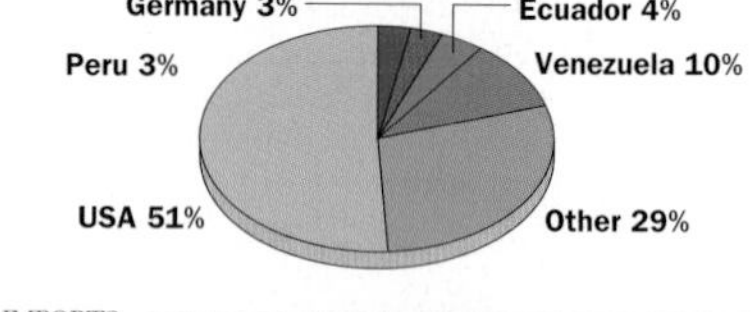
Germany 3%
Ecuador 4%
Peru 3%
Venezuela 10%
USA 51%
Other 29%

IMPORTS

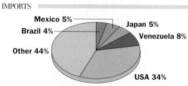

Mexico 5%
Japan 5%
Brazil 4%
Venezuela 8%
Other 44%
USA 34%

Strengths

Substantial oil and coal deposits plus well-developed hydroelectric power makes Colombia almost self-sufficient in energy. Diversified export sector – especially coffee and coal. Light manufactures. Worldwide market for cocaine.

Weaknesses

Narcotics-related violence, corruption, and political instability discourage foreign investors. Domestic industry uncompetitive owing to protection. High unemployment. Coffee and oil subject to world price fluctuations.

Profile

Of all the Latin American economies, Colombia's is probably the closest to the US model. The state has traditionally played a relatively minor role, and

ECONOMIC PERFORMANCE INDICATOR

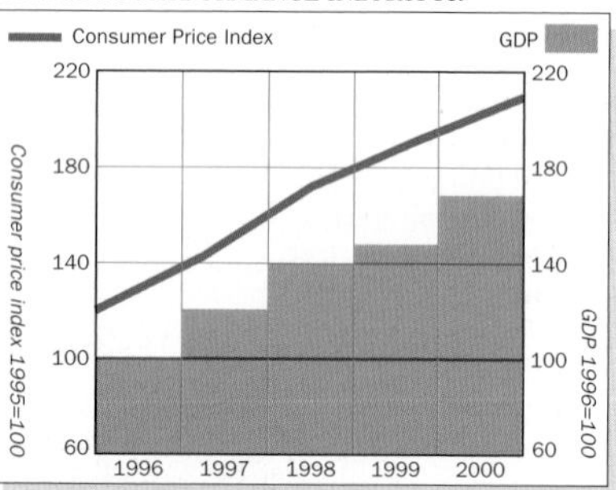
Consumer Price Index
GDP
220
180
140
100
60
Consumer price index 1995=100
GDP 1996=100
1996
1997
1998
1999
2000

Colombia has a successful private export sector. A program of privatization and macroeconomic stabilization was rewarded with IMF assistance of $1.5 billion in 2002.

Regional disparities remain marked. Most wealth is found in the Bogotá, Medellín, and Cali regions. Rural areas are largely underdeveloped. The main obstacle to growth is the instability caused by the narcotics business and protracted conflict. Given stability and investment, Colombia's potential for growth is considerable.

COLOMBIA : MAJOR BUSINESSES

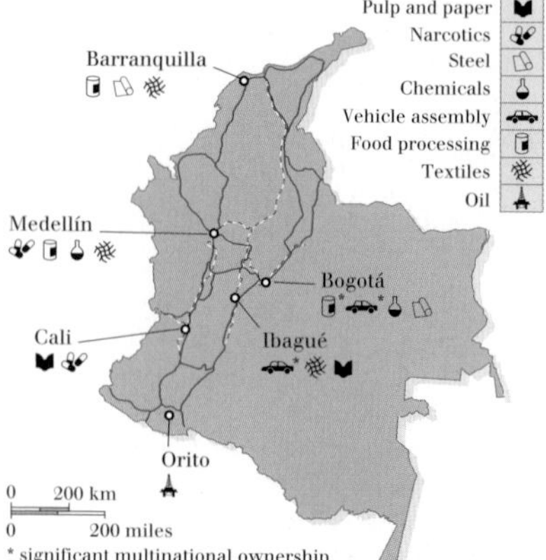
Pulp and paper
Narcotics
Steel
Chemicals
Vehicle assembly
Food processing
Textiles
Oil
Barranquilla
Medellín
Bogotá
Cali
Ibagué
Orito
0 200 km
0 200 miles
* significant multinational ownership

RESOURCES

Electric power 11.7m kw

170,896 tonnes

627,000 b/d (reserves 1.8bn barrels)

28.3m cattle, 2.75m pigs, 2.6m horses, 110m chickens

Oil, coal, natural gas, silver, emeralds, gold, platinum

ELECTRICITY GENERATION

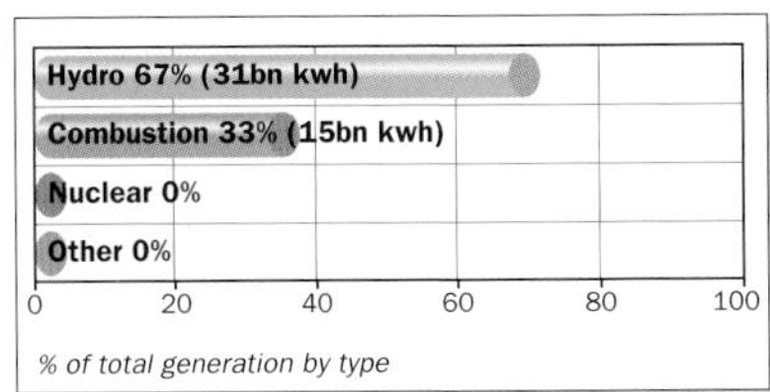

Hydro 67% (31bn kwh)
Combustion 33% (15bn kwh)
Nuclear 0%
Other 0%
0
20
40
60
80
100
% of total generation by type

Colombia has substantial oil reserves but needs increasing investment to maintain production. Coal and gas are important, and it is a major producer of gold, platinum, silver, and emeralds.

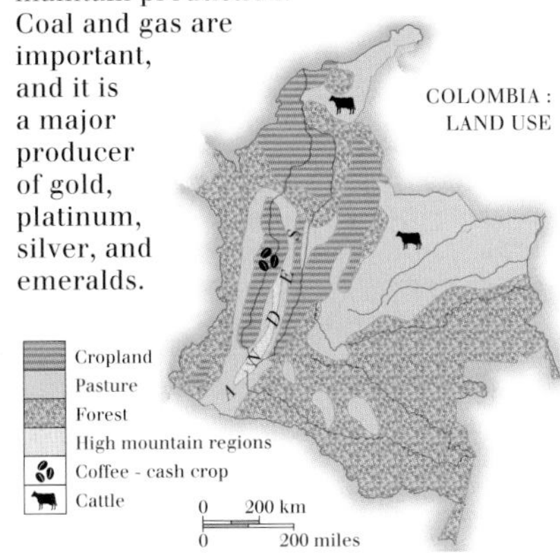

COLOMBIA : LAND USE
Cropland
Pasture
Forest
High mountain regions
Coffee - cash crop
Cattle
0 200 km
0 200 miles

ENVIRONMENT

Sustainability rank: 22nd

9%

1.7 tonnes per capita

ENVIRONMENTAL TREATIES

Yes Yes Yes

Yes Yes Yes

Cattle ranching, logging, coca growing, and spraying cause soil degradation and loss of natural habitats.

MEDIA

TV ownership high

Daily newspaper circulation 26 per 1000 people

PUBLISHING AND BROADCAST MEDIA

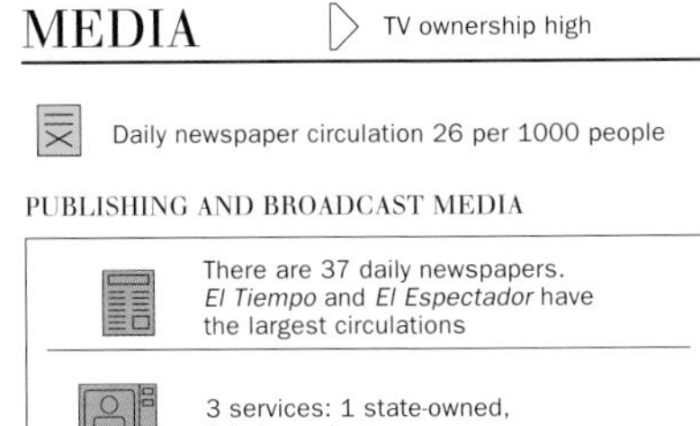

There are 37 daily newspapers. *El Tiempo* and *El Espectador* have the largest circulations

3 services: 1 state-owned, 2 independent

589 stations: 31 state-owned, 558 independent

The independent press is small. Journalists have been murdered by paramilitaries and held by guerrillas.

CRIME

No death penalty

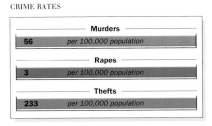

57,068 prisoners

Up 35% 1996–1998

CRIME RATES

Murders	56 per 100,000 population
Rapes	3 per 100,000 population
Thefts	233 per 100,000 population

Colombia is one of the most violent countries in the world. Armed groups assassinated 20 mayoral candidates and 20 mayors, and kidnapped 200 other candidates, in regional and local elections in 2000. In late 2001, the local monitoring group Fundación País Libre reported 2856 cases of kidnapping that year, mostly by paramilitaries and guerrillas, the rest blamed on other criminal groups. Anonymous gunmen murdered the Archbishop of Cali in 2002. NGOs estimate that two million people have been displaced over the last 14 years. Homicide is the main cause of death among young men in cities; overall, it rates third in the mortality stakes. Much of the violence is narcotics-related, and Bogotá, Cali, Barrancabermeja, and Medellín can all be dangerous; the army, the police, paramilitaries, and guerrillas are thought to be involved. Frequent armed robberies and kidnappings make wealthy residents extremely security conscious.

A relatively new phenomenon is that of "social cleansing" – the murder of street children and beggars by armed gangs, some in Bogotá funded by businesses.

EDUCATION

School leaving age: 15

92%

644,188 students

THE EDUCATION SYSTEM

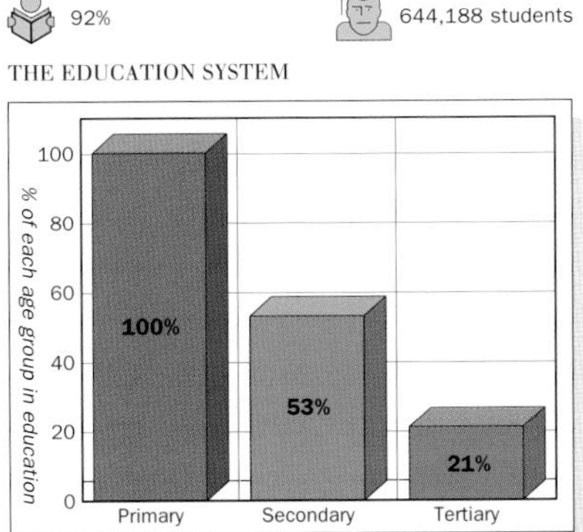

100
80
60
40
20
0
% of each age group in education
100%
53%
21%
Primary
Secondary
Tertiary

Education in Colombia is free and compulsory, and is a mix of French and US models, with a *baccalauréat* examination taken at the end of secondary school. Provision in rural areas is poor and absenteeism high. Where provided, public and university education is generally of a high standard, but the resources available to public education have decreased due to budget cuts. The rich send their children to private schools and universities in the US.

HEALTH

Welfare state health benefits

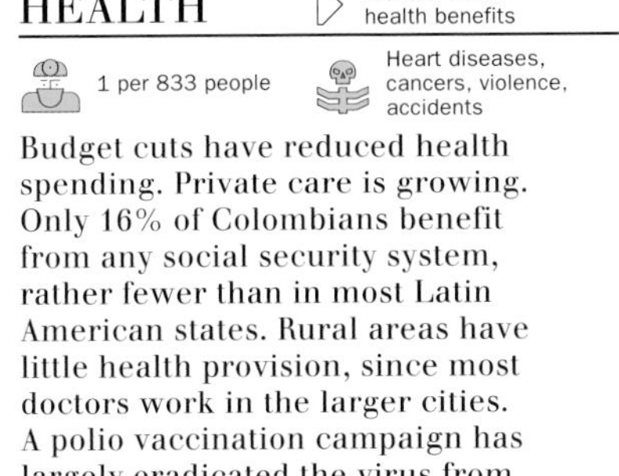

1 per 833 people

Heart diseases, cancers, violence, accidents

Budget cuts have reduced health spending. Private care is growing. Only 16% of Colombians benefit from any social security system, rather fewer than in most Latin American states. Rural areas have little health provision, since most doctors work in the larger cities. A polio vaccination campaign has largely eradicated the virus from Colombia, except in coastal regions.

SPENDING

GDP/cap. increase

CONSUMPTION AND SPENDING

43 per 1000 population

169 per 1000 population

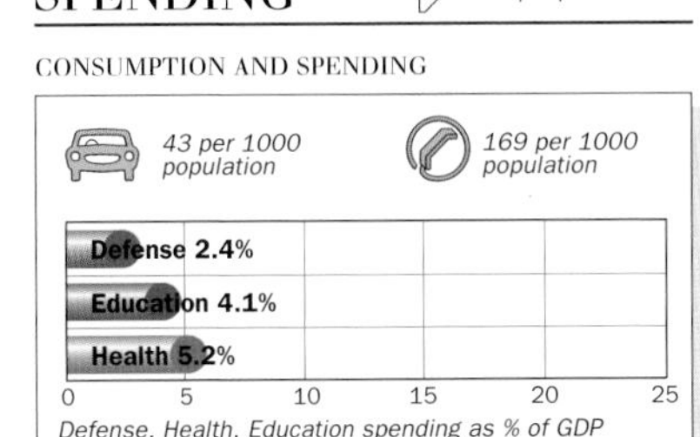

Defense 2.4%
Education 4.1%
Health 5.2%
0
5
10
15
20
25
Defense, Health, Education spending as % of GDP

There is little social mobility; the historically wealthy Spanish families are still dominant in political and business life, but the entry of narcotics-related money has created new layers of rich in cities and among landowners. Drug money also finances the import of consumer goods such as TV sets, computers, and perfume. The wealthy go to the US for medical treatment and educate their children overseas. The rural poor are mostly landless. The inhabitants of shanty towns in Barranquilla, Buenaventura, Cali, and Cartagena form the poorest groups.

WORLD RANKING

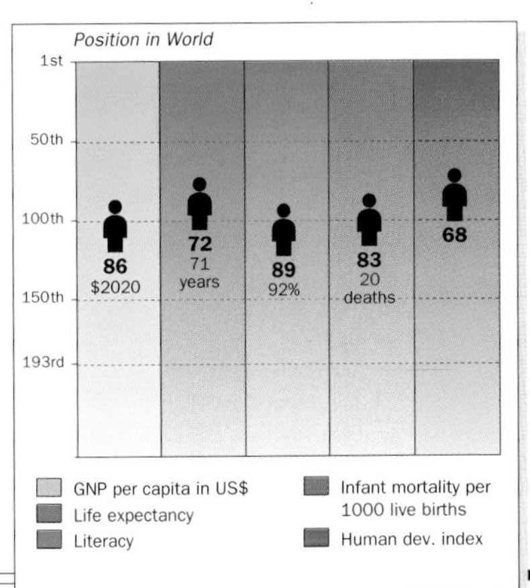

Position in World
1st
50th
100th
150th
193rd
86
$2020
72
71
years
89
92%
83
20
deaths
68
GNP per capita in US$
Life expectancy
Literacy
Infant mortality per 1000 live births
Human dev. index

COMOROS

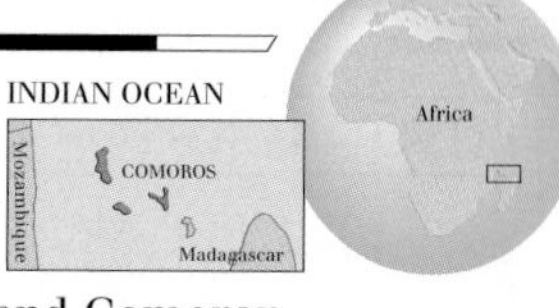
INDIAN OCEAN
Mozambique
COMOROS
Madagascar
Africa

OFFICIAL NAME: Federal Islamic Republic of the Comoros **CAPITAL:** Moroni
POPULATION: 727,000 **CURRENCY:** Comoros franc **OFFICIAL LANGUAGES:** Arabic, French, and Comoran

C

1975

1975
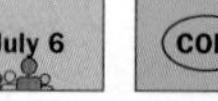
July 6
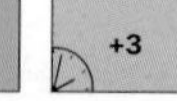
COM

+3

+269
.km

THE ARCHIPELAGO republic of the Comoros lies off the east African coast, between Mozambique and Madagascar. It consists of three main islands and a number of islets. Most of the population are subsistence farmers. In 1975, the Comoros islands, except for Mayotte, became independent of France. Since then instability has plagued this poor region, with countless coups and countercoups, and repeated attempts at secession by smaller islands.

Moroni, the capital, on Grande Comore. *The Comoros islands are fertile and heavily forested. Many are ringed by coral reefs.*

CLIMATE

Tropical oceanic

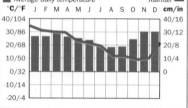

WEATHER CHART FOR MORONI

Average daily temperature / Rainfall

°C/°F: 40/104, 30/86, 20/68, 10/50, 0/32, -10/14, -20/-4
J F M A M J J A S O N D
cm/in: 40/16, 30/12, 20/8, 10/4, 0

The islands are tropical; it is hot and humid on the coasts and cooler higher up, notably on Mount Kartala.

TRANSPORTATION

Drive on right

Moroni–Hahaya, Grande Comore
118,565 passengers

3 ships
2959 grt

THE TRANSPORTATION NETWORK

673 km (418 miles)

None

None

None

Recent projects have included the development of the port at Moroni and upgrading the international airport.

TOURISM

Visitors : Population 1:30

24,000 visitors

Down 11% in 1999

MAIN TOURIST ARRIVALS

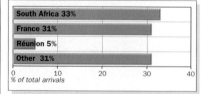

	% of total arrivals
South Africa	33%
France	31%
Réunion	5%
Other	31%

In 1988, Sun International of South Africa joined a major project to build four hotels designed to attract 12,000 visitors a year. However, political instability has wrecked the islands' tourism prospects.

PEOPLE

Pop. density high

Arabic, Comoran, French

326/km² (844/mi²)

THE URBAN/RURAL POPULATION SPLIT

33% / 67%

RELIGIOUS PERSUASION

Other 1%
Roman Catholic 1%
Muslim (mainly Sunni) 98%

The Comoros has absorbed Polynesians, Africans, Indonesians, Persians, and Arabs over time, as well as immigrants from Portugal, Holland, France, and India. Some communities retain their individual character; for instance, Mwali (Mohéli) is still primarily African. Ethnic tension is rare, partly owing to the unifying force of Islam, the predominant religion. A more potent divisive factor, especially on Anjouan, is regionalism.

POLITICS

In transition

1996/2000 (postponed)

President Assoumani Azzali

AT THE LAST ELECTION

Federal Assembly 43 seats

91% RND
5% FNJ
2% VA
2% Ind

RND = National Rally for Development
FNJ = National Front for Justice
Ind = Independent **VA**= Votes annulled

The Senate was abolished under the 1996 constitution

Attempts to introduce democracy have been consistently undermined by repeated coups. Anjouan declared independence in 1997, since when rival militias have favored either cohabitation with Grande Comore or reattachment to France. Renewed violence in April 1999 provided Col. Assoumani Azzali with the pretext to install himself as Comoran president. In a new "Union of the Comoros" a loose confederation gives each island its own president and legislature, with a federal president having overall authority.

In 2002 Azzali was returned to power unopposed, although his victory was undermined by low voter turnout.

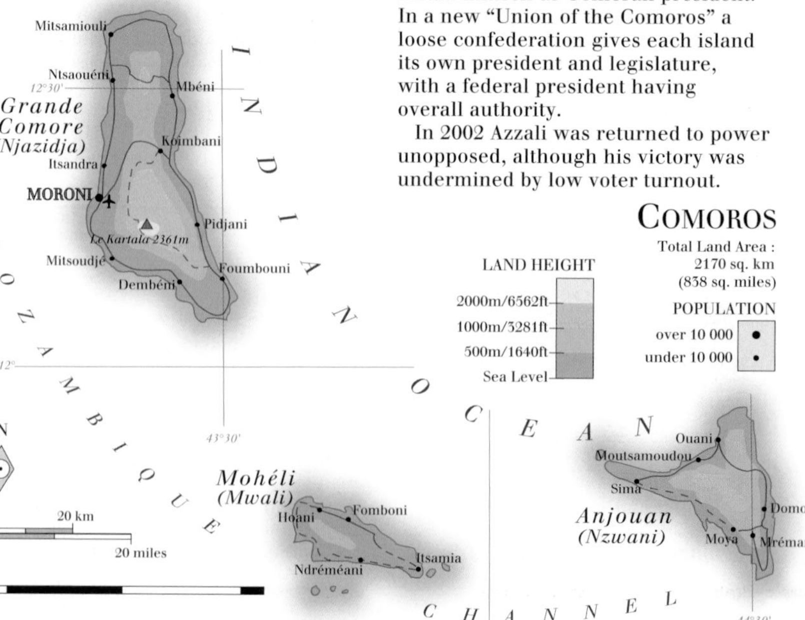
COMOROS
Total Land Area : 2170 sq. km (838 sq. miles)
LAND HEIGHT
2000m/6562ft
1000m/3281ft
500m/1640ft
Sea Level
POPULATION
over 10 000
under 10 000
Mitsamiouli
Ntsaouéni
Mbéni
Grande Comore (Njazidja)
Koimbani
Itsandra
MORONI
Le Kartala 2361m
Pidjani
Mitsoudjé
Foumbouni
Dembéni
INDIAN OCEAN
MOZAMBIQUE CHANNEL
Mohéli (Mwali)
Hoani
Fomboni
Itsamia
Ndrémeani
Anjouan (Nzwani)
Ouani
Moutsamoudou
Sima
Domoni
Moya
Mrémani
N
0 20 km
0 20 miles
12°30'
12°
43°30'
44°
44°30'

WORLD AFFAIRS

Joined UN in 1975

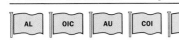

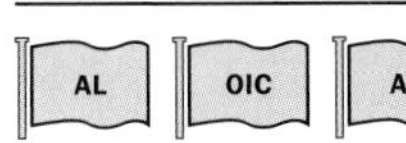
AL
OIC
AU
COI
FZ

France remains the main aid donor, although economic ties with South Africa are also strong. The turbulent situation on the war-torn islands necessitated the evacuation in 1999 of a visiting OAU assessment team from Anjouan. An army of European mercenaries attempted to take over Mohéli in December 2001.

AID

Recipient

$19m (receipts)

Down 10% in 2000

Foreign aid, mainly from France, the EU, the UN, and the World Bank, accounts for over 40% of GDP. Because of its Islamic links, the Comoros also gets aid from Arab states and OPEC. In 1998, major donors attacked the government for spending more than 70% on "political superstructure."

DEFENSE

No compulsory military service

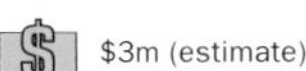
$3m (estimate)

No significant change

France and South Africa finance the small presidential guard, the principal security force. Mauritian aid was also sought after clashes on Anjouan.

ECONOMICS

Inflation 4.9% p.a. (1990–2000)

$212m

528–558 Comoros francs

SCORE CARD

- ❑ WORLD GNP RANKING........................183rd
- ❑ GNP PER CAPITA $380
- ❑ BALANCE OF PAYMENTS.......................–$1m
- ❑ INFLATION 3.5%
- ❑ UNEMPLOYMENT.................................20%

STRENGTHS

Vanilla, ylang-ylang, and cloves are the main cash crops.

WEAKNESSES

Subsistence-level farming. Over 50% of food requirements are imported. Lack of basic infrastructure, especially electricity and transportation. Open to allegations of financial mismanagement. Persistent political instability is hindering growth of tourism.

EXPORTS

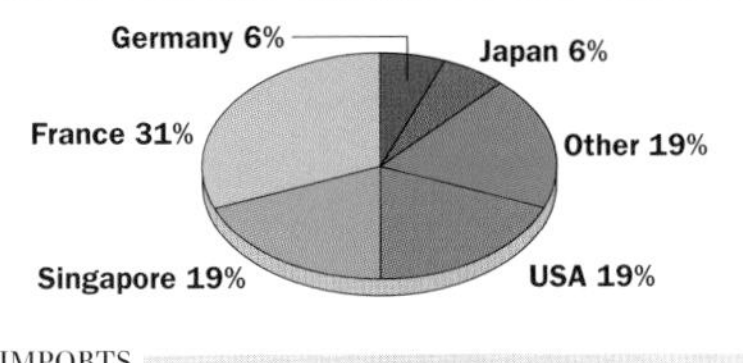
Germany 6%
Japan 6%
France 31%
Other 19%
Singapore 19%
USA 19%

IMPORTS

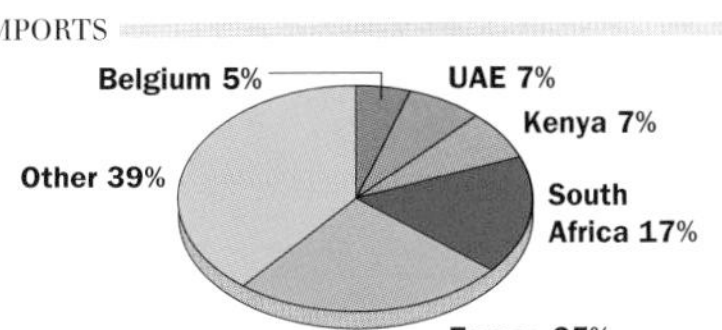
Belgium 5%
UAE 7%
Kenya 7%
Other 39%
South Africa 17%
France 25%

RESOURCES

Electric power 5000 kw

12,200 tonnes

Not an oil producer

172,000 goats, 52,000 cattle, 490,000 chickens

None

There are few strategic resources. An HEP plant is under construction on Anjouan, but most fuel for energy is still imported. Fishing remains a neglected source of future growth.

ENVIRONMENT

Not available

None

0.1 tonnes per capita

The environment is not a major priority in the Comoros; natural disasters, such as the volcanic eruption in 1977 which left 20,000 people homeless, are of more immediate concern. The government is promoting tourism and recognizes the long-term commercial value of imposing environmental controls on new developments.

MEDIA

TV ownership low

There are no daily newspapers

PUBLISHING AND BROADCAST MEDIA

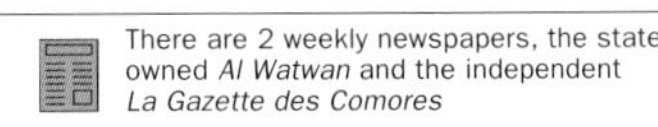
There are 2 weekly newspapers, the state-owned *Al Watwan* and the independent *La Gazette des Comores*

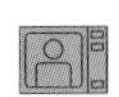
No TV service

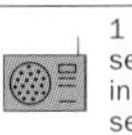
1 state-controlled service, some independent services

Radio is strictly controlled, and there is no single national newspaper. Although there is criticism of the government, self-censorship is reportedly common.

CRIME

Death penalty in use

200 prisoners

Crime is rising

A climate of lawlessness has been created by the continuing power struggles between rival militias – particularly on the island of Anjouan since 1997.

EDUCATION

School leaving age: 16

60%

649 students

There is a very limited education system beyond secondary level. Schools are equipped to teach only basic literacy, hygiene, and agricultural techniques. Pupil–teacher ratios are high.

CHRONOLOGY

The Comoros was ruled by matrilineally inherited sultanates until shortly before becoming a French protectorate in 1886.

- ❑ **1961** Internal self-government.
- ❑ **1975** Independence.
- ❑ **1978** Mercenaries restore Ahmed Abdallah to power.
- ❑ **1989** Abdallah assassinated.
- ❑ **1992** Chaotic first multiparty polls.
- ❑ **1996** Mohammed Taki Abdoulkarim elected president.
- ❑ **1997** Anjouan separatists beat off government troops.
- ❑ **1999** Col. Azzali seizes power. Anjouan militias clash.
- ❑ **2000** Fomboni declaration signed with Anjouan.
- ❑ **2001** Successive coups on Anjouan.
- ❑ **2002** New constitution. Azzali returned to office.

HEALTH

No welfare state health benefits

1 per 14,000 people

Malaria, infectious intestinal and bacterial diseases

Health care is rudimentary, except that there are two maternity clinics and 30 health centers.

SPENDING

GDP/cap. decrease

CONSUMPTION AND SPENDING

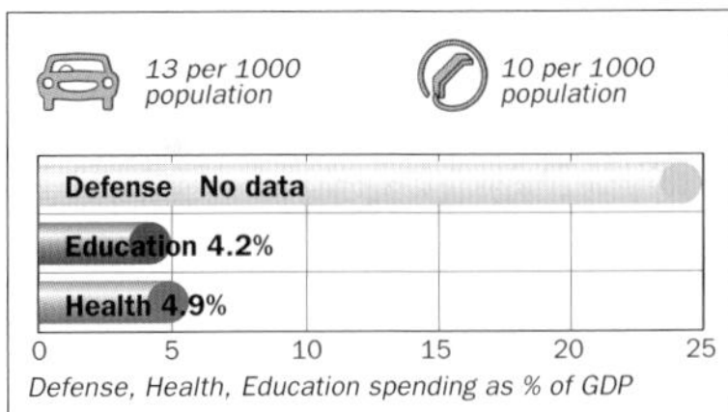
13 per 1000 population
10 per 1000 population
Defense No data
Education 4.2%
Health 4.9%
0
5
10
15
20
25
Defense, Health, Education spending as % of GDP

A political and business elite controls most of the wealth. Bridegrooms win social status according to the size of their weddings. Government workers often suffer from wage arrears.

WORLD RANKING

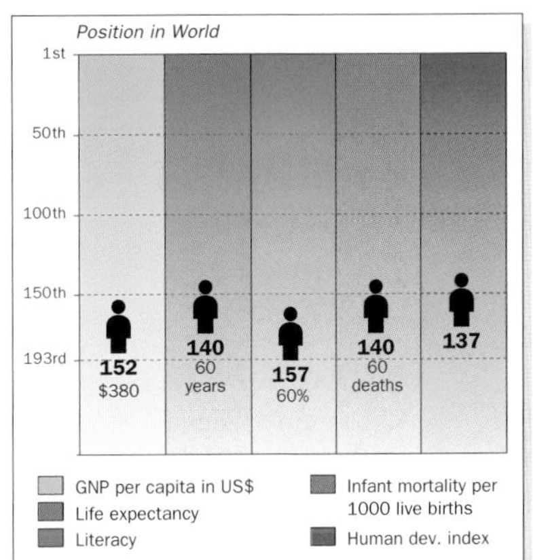
Position in World
1st
50th
100th
150th
193rd
152
$380
140
60 years
157
60%
140
60 deaths
137
GNP per capita in US$
Life expectancy
Literacy
Infant mortality per 1000 live births
Human dev. index

C

CONGO

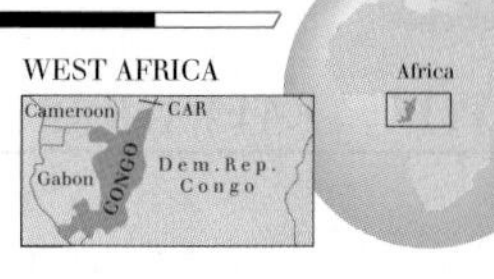
WEST AFRICA
Africa
Cameroon
CAR
Gabon
CONGO
Dem. Rep. Congo

OFFICIAL NAME: Republic of the Congo **CAPITAL:** Brazzaville
POPULATION: 3.1 million **CURRENCY:** CFA franc **OFFICIAL LANGUAGE:** French

C

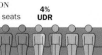

1960

1960

Aug 15
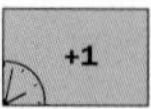
RCB
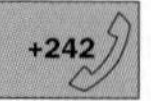
+1

+242

.cg

STRADDLING THE EQUATOR in west central Africa, Congo achieved independence from France in 1960, soon falling under a Marxist–Leninist form of government which discouraged much foreign investment. Multiparty democracy was achieved in 1991, but was soon overshadowed by years of violence.

CLIMATE

Tropical equatorial

WEATHER CHART FOR BRAZZAVILLE

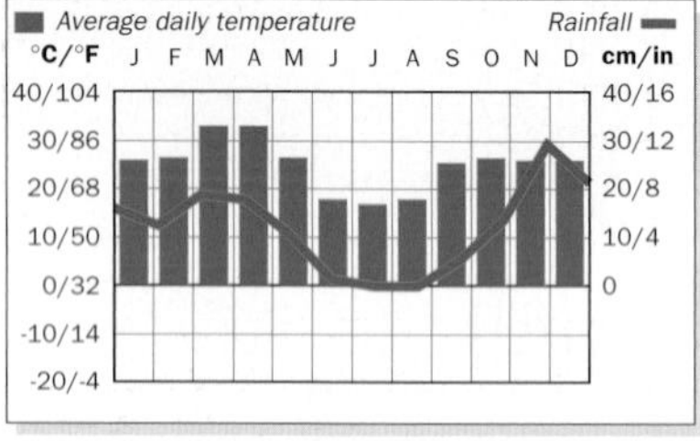
Average daily temperature
Rainfall
°C/°F J F M A M J J A S O N D cm/in
40/104
30/86
20/68
10/50
0/32
-10/14
-20/4
40/16
30/12
20/8
10/4
0

In most years there are two wet seasons and two dry seasons in Congo. The rainfall is heaviest in the coastal regions.

TRANSPORTATION

Drive on right

Brazzaville International
411,678 passengers

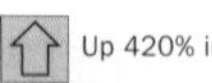

18 ships
3400 grt

THE TRANSPORTATION NETWORK

1242 km (772 miles)

None

895 km (556 miles)

4385 km (2725 miles)

Pointe-Noire is a major port, used by the Central African Republic, Chad, and Cameroon. The Congo Ocean Railroad (to Brazzaville) reopened in 2000. There are plans for a second, and larger, international airport near Ewo.

TOURISM

Visitors : Population 1:119

26,000 visitors

Up 420% in 2000

MAIN TOURIST ARRIVALS

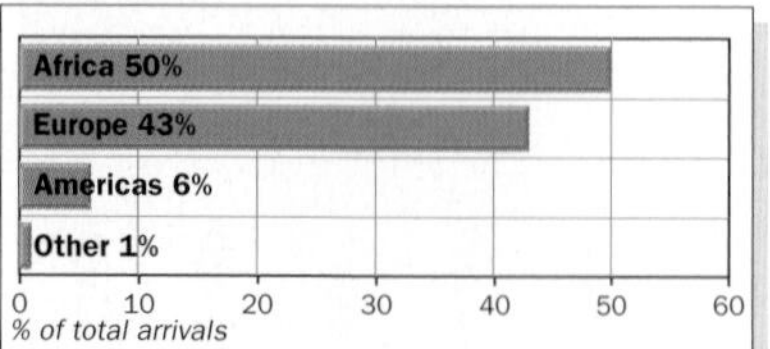
Africa 50%
Europe 43%
Americas 6%
Other 1%
0 10 20 30 40 50 60
% of total arrivals

The Marxist–Leninist regime did not seek to develop tourism, and visitors, mostly on safaris and business-related trips, are still rare, though increasing.

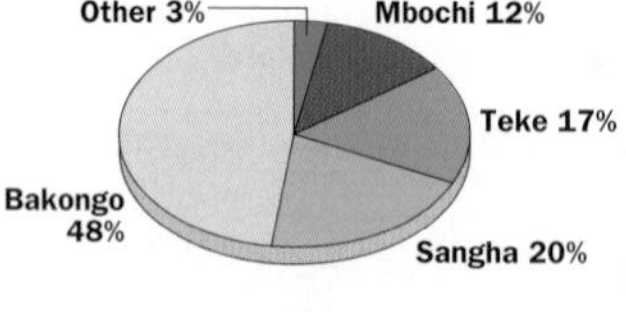

The Loufoulakari Falls, near Brazzaville. The Congo River is a key transportation artery for the region.

PEOPLE

Pop. density low

Kongo, Teke, Lingala, French

9/km² (24/mi²)

THE URBAN/RURAL POPULATION SPLIT

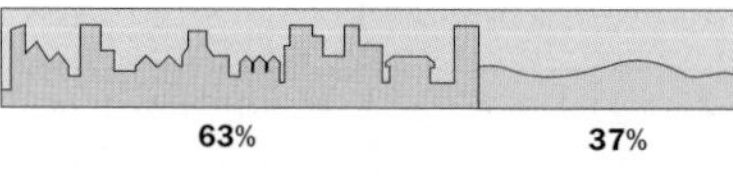
63%
37%

ETHNIC MAKEUP

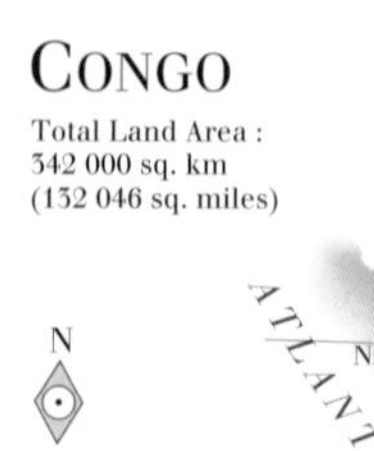
Other 3%
Mbochi 12%
Teke 17%
Bakongo 48%
Sangha 20%

Congo is one of the most tribally conscious countries in Africa. The main tensions are between the Bakongo, who live in the north, and the Mbochi, who are concentrated in the more prosperous south. Since the 1950s, women have achieved considerable freedom.

POLITICS

Multiparty elections

L. House 2002/2007
U. House 2002/2008

President Denis Sassou-Nguesso

AT THE LAST ELECTION

National Assembly 137 seats

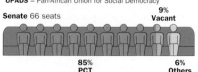

61% PCT, **4% UDR**, **6% Vacant**, **3% UPADS**, **26% Others**

PCT = Congolese Labor Party and allies
UDR = Union for Democracy and the Republic
UPADS = Pan-African Union for Social Democracy

Senate 66 seats

85% PCT, **9% Vacant**, **6% Others**

Former Marxist dictator Denis Sassou-Nguesso seized power in 1997, amid intense fighting which left thousands dead. Relative peace was secured in 1999. A new constitution giving greater power to the presidency was approved in 2002, and Sassou-Nguesso was easily elected. Fighting broke out again later that year, but legislative elections were eventually held, and were won by Sassou-Nguesso's supporters.

CONGO

Total Land Area : 342 000 sq. km (132 046 sq. miles)

CENTRAL AFRICAN REPUBLIC, CAMEROON, GABON, DEM. REP. CONGO, ATLANTIC OCEAN, CABINDA (Angola), Bétou, Souanké, Sembé, Ouésso, Bomassa, Dongou, Impfondo, Epéna, Mbomo, Pikounda, Etoumbi, Kellé, Makoua, Owando, Liranga, Loukoléla, Ewo, Oyo, Okoyo, Gamboma, Mayoko, Lékana, Ngo, Bambama, Djambala, Mossendjo, Komono, Ngabé, Sibiti, Kindamba, Loubomo, Loudima, Madingou, Nzambi, Madingo-Kayes, Nkayi, Mfouati, Mindouli, Kinkala, BRAZZAVILLE, Boko, Pointe-Noire, Stanley Pool, Loufoulakari falls, Monts de la Lékéti 1040m, PLATEAUX BATÉKÉ, Lac Tchibenda, Equator

POPULATION: over 500 000, over 50 000, over 100 000, over 10 000, under 10 000

LAND HEIGHT: 500m/1640ft, 200m/656ft, Sea Level

0 100 km
0 100 miles

WORLD AFFAIRS

Joined UN in 1960

BDEAC
CEMAC
FZ
NAM
ACP

Carefully balancing relations with France and the US is a priority, since both seek to extend their stake in the oil industry; nevertheless relations with old Eastern bloc allies remain strong. Congo has been susceptible in recent years to political instability and wars in the neighboring Democratic Republic of the Congo and Angola.

AID

Recipient

$33m (receipts) | Down 77% in 2000

Until 1990, the USSR, Cuba, and China were major donors. Most aid now comes from Italy and France. The IMF has begun moves to clear Congo's high level of debt built up during the 1970s and 1980s.

DEFENSE

No compulsory military service

$72m | Down 1% in 2000

The militias of the various political forces were being integrated into the 8000-strong army, until fighting broke out again between them in mid-1997. The air force numbers 1200, and is equipped with 12 MiG-21s.

ECONOMICS

Inflation 11% p.a. (1990–2000)

$1.74bn

698.7–736.7 CFA francs

SCORE CARD

- World GNP Ranking: 141st
- GNP per Capita: $570
- Balance of Payments: –$252m
- Inflation: –0.9%
- Unemployment: Widespread underemployment

STRENGTHS

Increase in importance of oil, now providing 95% of export revenues. Significant timber supplies. Skilled and well-trained workforce helps sustain substantial industrial base in the capital and Pointe-Noire.

WEAKNESSES

Massive debt burden. Top-heavy bureaucracy. Overdependence on oil. Political instability. Large refugee population.

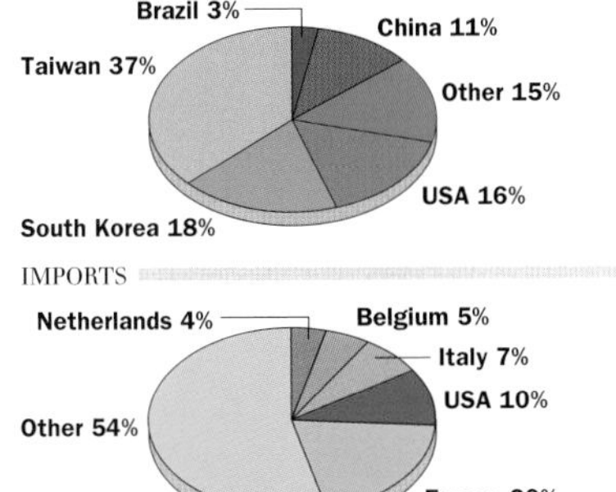
EXPORTS
Brazil 3%
China 11%
Taiwan 37%
Other 15%
USA 16%
South Korea 18%
IMPORTS
Netherlands 4%
Belgium 5%
Italy 7%
USA 10%
Other 54%
France 20%

RESOURCES

Electric power 118,000 kw

43,886 tonnes

271,000 b/d (reserves 1.5bn barrels)

280,000 goats, 96,000 sheep, 1.9m chickens

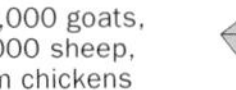

Oil, natural gas, zinc, gold, copper, potash, diamonds

Oil is by far the Congo's most important resource. Natural gas reserves have yet to be exploited. There are some deposits of diamonds and potash. Bauxite and iron ore reserves are not large enough to be profitably mined and phosphate production was abandoned in 1977. Chinese aid has helped build two hydroelectric dams, on the Bouenza and Djoué rivers. A third is currently being built on the Léfini at Imboulou.

ENVIRONMENT

Sustainability rank: 40th

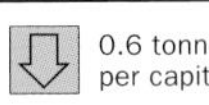

5% | 0.6 tonnes per capita

The 1999 Yaoundé Declaration should help control exploitation of tropical timber. Congo has been used in the past as a dumping ground for dangerous toxic waste from the West.

MEDIA

TV ownership low

Daily newspaper circulation 8 per 1000 people

PUBLISHING AND BROADCAST MEDIA

There are 64 daily newspapers, including *Mweti* and *Aujourd'hui*

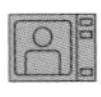
1 state-controlled service

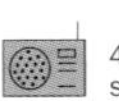
4 state-controlled services

With only moderate media control, the press often takes an antigovernment stance. During World War II, Radio Brazzaville, still the official state radio, was vital to de Gaulle's French forces.

CRIME

Death penalty not used in practice

918 prisoners

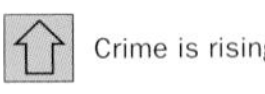
Crime is rising

Armed robbery and smuggling are widespread. Years of conflict and instability in neighboring countries mean that guns are easily available.

EDUCATION

School leaving age: 11

81%

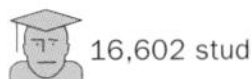
16,602 students

Originally pioneered by French Roman Catholic missions, schools are still subject to inspection from Paris.

CHRONOLOGY

The kingdoms of Teke and Loango were incorporated as the Middle Congo (part of French Equatorial Africa) between 1880 and 1883.

- **1960** Independence.
- **1964** Marxist–Leninist National Revolution Movement (MNR) sole legal party.
- **1977** Yhompi-Opango head of state after President Ngoumbi's murder.
- **1979** Col. Denis Sassou-Nguesso president.
- **1991** Multiparty democracy.
- **1992** Pascal Lissouba president.
- **1993** Elections: Lissouba's UPADS party gains majority.
- **1997** Sassou-Nguesso ousts Lissouba.
- **1999** Cease-fire signed.
- **2001** IMF starts to clear debt.
- **2002** New constitution approved. Sassou-Nguesso wins elections.

HEALTH

Welfare state health benefits

1 per 3333 people | Diarrheal, parasitic and respiratory diseases, malaria

The health service, set up by French military doctors at the start of the 20th century, has been devastated by civil war.

SPENDING

GDP/cap. decrease

CONSUMPTION AND SPENDING

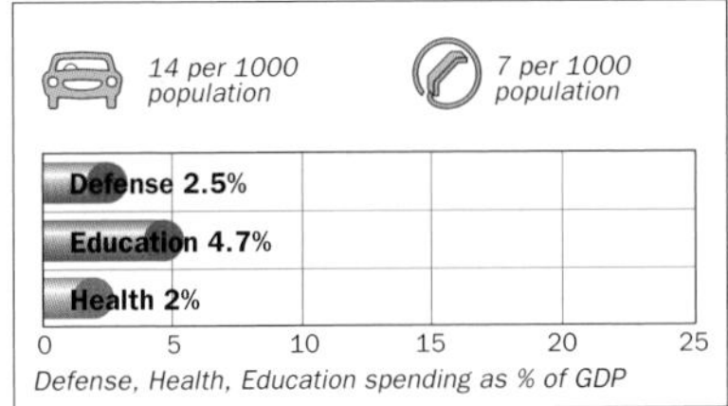
14 per 1000 population
7 per 1000 population
Defense 2.5%
Education 4.7%
Health 2%
0
5
10
15
20
25
Defense, Health, Education spending as % of GDP

Wealth generated from oil extraction has sustained an active and confident middle class. French-label products are considered to be status symbols.

WORLD RANKING

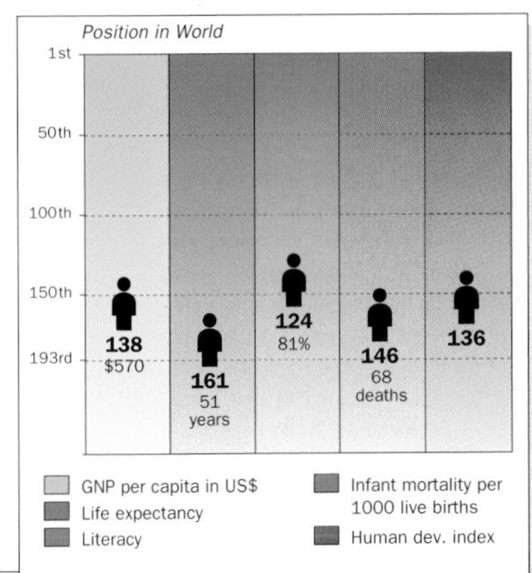
Position in World
1st
50th
100th
150th
193rd
138
$570
161
51 years
124
81%
146
68 deaths
136
GNP per capita in US$
Life expectancy
Literacy
Infant mortality per 1000 live births
Human dev. index

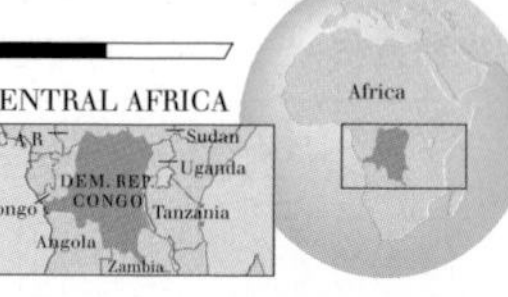
CENTRAL AFRICA
Africa
C.A.R.
Sudan
Uganda
DEM. REP. CONGO
Congo
Tanzania
Angola
Zambia

CONGO (DEMOCRATIC REPUBLIC)

OFFICIAL NAME: Democratic Republic of the Congo **CAPITAL:** Kinshasa
POPULATION: 52.5 million **CURRENCY:** Congolese franc **OFFICIAL LANGUAGE:** French

C

1960

1960

June 6

CGO

+1

+243
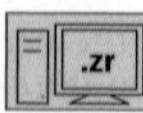
.zr

LYING IN EAST CENTRAL AFRICA, the Democratic Republic of the Congo (DRC), known as Zaire from 1971 to 1997, is one of Africa's largest countries. The rainforested basin of the Congo River occupies 60% of the land area. On independence in 1960, civil war broke out. The notoriously corrupt Gen. Mobutu ruled from 1965 until his overthrow in 1997 by rebel forces under Laurent-Désiré Kabila. A rebellion launched in 1998 plunged the country into renewed chaos, and spiraled into regional conflict. Peace initiatives were revitalized by Joseph Kabila's succession in January 2001.

The Congo River is navigable for 1357 km (848 miles) and provides one of the most convenient ways of traveling in the country.

CLIMATE

Tropical equatorial/ wet & dry

WEATHER CHART FOR KINSHASA

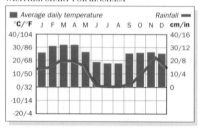

Average daily temperature / Rainfall
°C/°F J F M A M J J A S O N D cm/in
40/104 30/86 20/68 10/50 0/32 -10/14 -20/4
40/16 30/12 20/8 10/4 0

The climate is tropical and humid. Temperatures average 25°C (77°F) and vary little through the year. Annual rainfall is around 150–200 cm (59–79 in); mountainous areas are wetter. The equator passes through the north of the country, causing marked regional variations. To its south, well-differentiated wet and dry seasons are October–May and June–September respectively. North of the equator, a short dry season lasts from December to February; the rest of the year is wet.

TRANSPORTATION

Drive on right

N'Djili, Kinshasa
273,563 passengers

20 ships
12,900 grt

THE TRANSPORTATION NETWORK

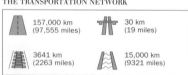

157,000 km (97,555 miles)

30 km (19 miles)

3641 km (2263 miles)

15,000 km (9321 miles)

The Congo River and its many tributaries provide the main means of communication. The size of the country and the fact that most of it is covered by dense rainforest have severely limited the development of road and rail networks. Many forest settlements are inaccessible except by air. Road maintenance, always poor, has virtually ceased outside the main towns since 1990, isolating even more settlements away from the main rivers.

TOURISM

Visitors : Population 1:991

53,000 visitors

Down in 1997–2001

MAIN TOURIST ARRIVALS

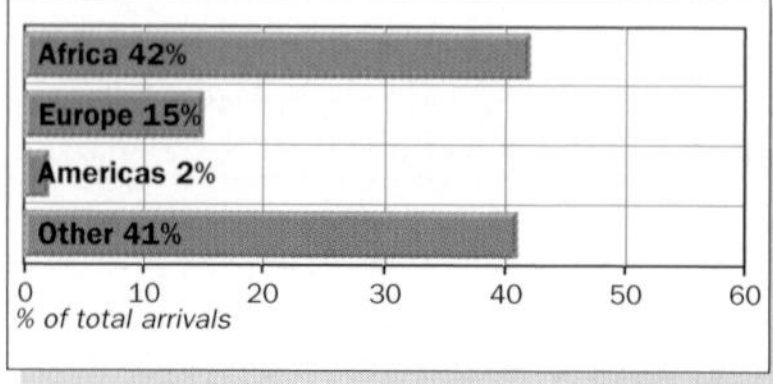
Africa 42%
Europe 15%
Americas 2%
Other 41%
0
10
20
30
40
50
60
% of total arrivals

Political turmoil and widespread anarchy since early 1997 ensure that the country remains off the itinerary for most travelers.

Potential tourist attractions consist mainly of scenery – mountains and lakes – and wildlife, but there are few facilities for tourists even in the capital. The Congo, 16 km (10 miles) wide in places, is Africa's second-longest river after the Nile. Visitors were formerly also attracted by the vibrant music of Kinshasa's many bands.

The once-large number of visitors on business has also collapsed as a consequence of the chronic instability of the 1990s.

PEOPLE

Pop. density low

Kiswahili, Tshiluba, Kikongo, Lingala, French

23/km² (60/mi²)

THE URBAN/RURAL POPULATION SPLIT

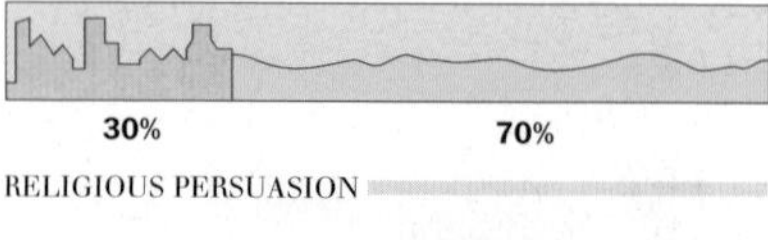
30%
70%

RELIGIOUS PERSUASION

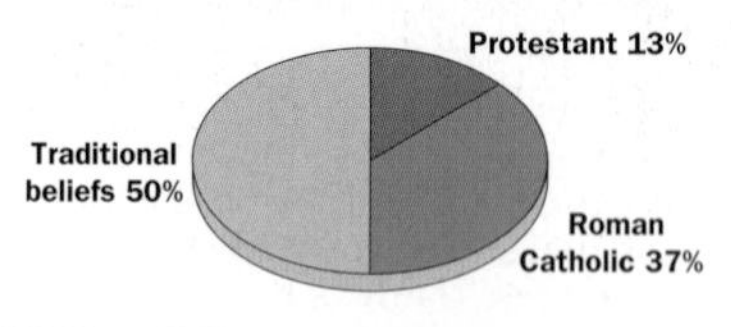
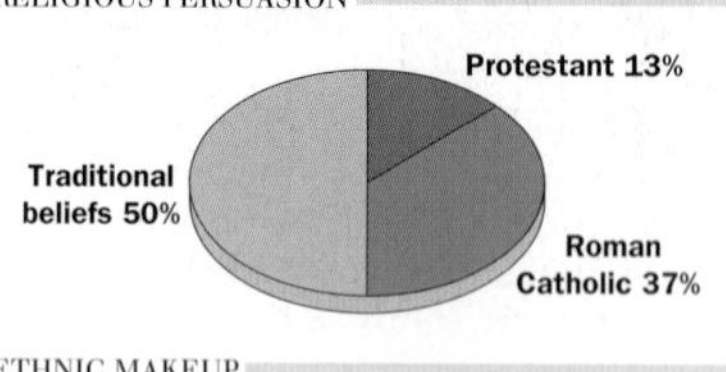
Protestant 13%
Traditional beliefs 50%
Roman Catholic 37%

ETHNIC MAKEUP

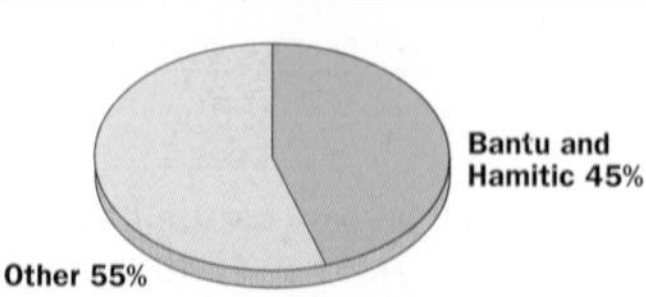
Bantu and Hamitic 45%
Other 55%

The Shaba mining area and major urban centers are densely populated, while the rainforests have a density of fewer than three people per sq. km, mostly subsisting on the margins of the cash economy. There is great ethnic diversity, with more than 12 main groups and around 190 smaller ones. The majority are of Bantu origin, but there are also large Hamitic and Nilotic populations, mainly in the north and northeast. The original inhabitants, the forest pygmies, today form a tiny and marginalized group.

Ethnic tensions inherited from the colonial period were contained under Mobutu until the 1990s. Ethnic violence in southeastern provinces in 1993 cost many thousands of lives, and a Hutu refugee influx from Rwanda the following year provoked serious tension among Tutsis in eastern areas; revenge killings on a large scale soon became commonplace. Regarded by Mobutu as foreigners, Tutsis provided the backbone of the 1996–1997 insurgency that overthrew him, and subsequently turned against Laurent Kabila also.

POPULATION AGE BREAKDOWN

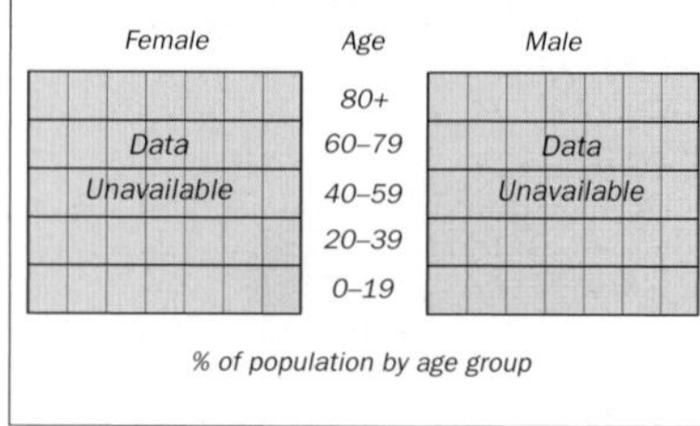
Female
Age
Male
80+
60–79
40–59
20–39
0–19
Data Unavailable
Data Unavailable
% of population by age group

CONGO, DEMOCRATIC REPUBLIC

Total Land Area : 2 345 410 sq. km (905 563 sq. miles)

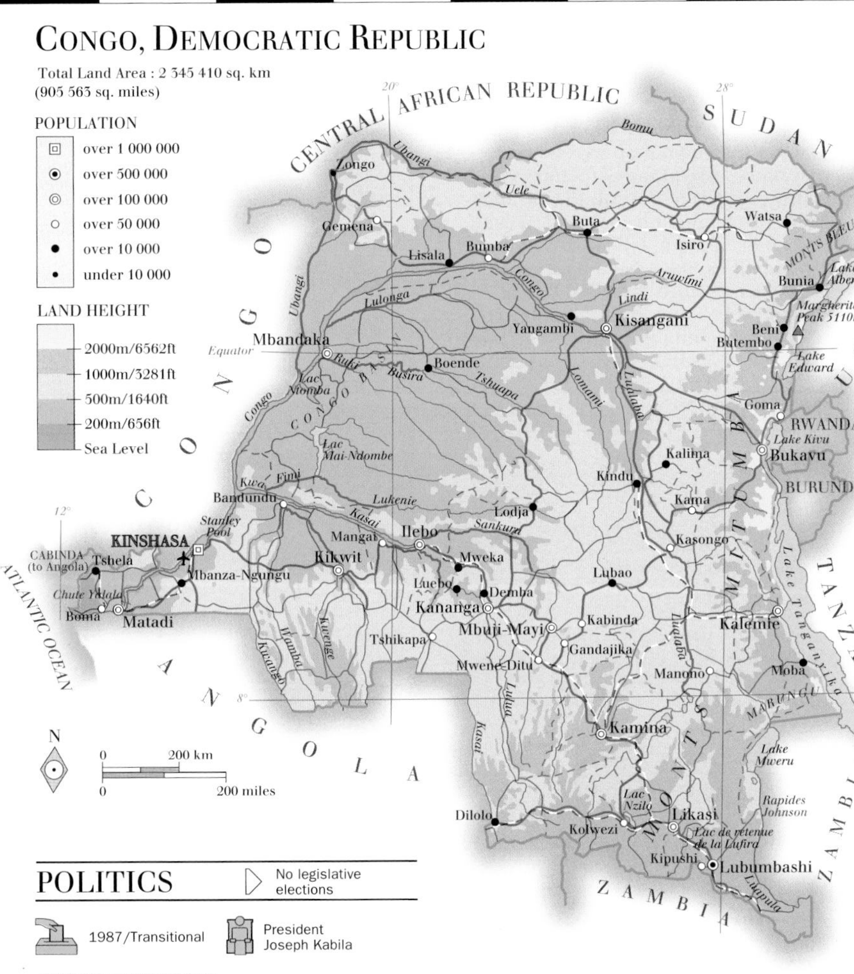
POPULATION
over 1 000 000
over 500 000
over 100 000
over 50 000
over 10 000
under 10 000
LAND HEIGHT
2000m/6562ft
1000m/3281ft
500m/1640ft
200m/656ft
Sea Level
CENTRAL AFRICAN REPUBLIC
SUDAN
UGANDA
RWANDA
BURUNDI
TANZANIA
ZAMBIA
ANGOLA
CONGO
CABINDA (to Angola)
ATLANTIC OCEAN
KINSHASA
Zongo
Gemena
Lisala
Bumba
Buta
Isiro
Watsa
Bunia
Beni
Butembo
Kisangani
Yangambi
Mbandaka
Boende
Goma
Bukavu
Kalima
Kindu
Kama
Kasongo
Bandundu
Ilebo
Lodja
Mangai
Kikwit
Mweka
Tshela
Mbanza-Ngungu
Boma
Matadi
Luebo
Demba
Kananga
Mbuji-Mayi
Kabinda
Lubao
Tshikapa
Gandajika
Mwene-Ditu
Manono
Kalemie
Moba
Kamina
Dilolo
Kolwezi
Likasi
Kipushi
Lubumbashi
Ubangi
Uele
Bomu
Congo
Lulonga
Aruwimi
Lindi
Ruki
Busira
Tshuapa
Lomami
Lualaba
Kwa
Fimi
Kasai
Lukenie
Sankuru
Kwango
Kwilu
Lulua
Luvua
Lukuga
Luapula
Lac Mai-Ndombe
Lac Tumba
Lake Albert
Lake Edward
Lake Kivu
Lake Tanganyika
Lake Mweru
Lac Nzilo
Lac de retenue de la Lufira
Stanley Pool
Chute Yelala
Margherita Peak 5110m
MONTS BLEUS
MONTS MITUMBA
MARUNGU
Rapides Johnson
Equator
N
200 km
200 miles

POLITICS

No legislative elections

1987/Transitional

President Joseph Kabila

AT THE LAST ELECTION

Constituent and Legislative Assembly 300 seats

The last legislative poll was in 1987 when members were chosen from Mobutu's Popular Revolutionary Movement (**MPR**). An interim legislature was appointed in July/August 2000 with 300 members chosen by Laurent Kabila.

The authoritarian regime of Laurent-Désiré Kabila continues under his son.

PROFILE

Kabila overthrew Mobutu's 32-year dictatorial regime in 1997, but his commitment to pluralist democracy soon came under question as he dissolved parliament and scrapped the constitution. A new Constituent and Legislative Assembly did not convene until August 2000. Meanwhile Kabila's ethnic Tutsi supporters rose against him in 1998, with backing from Uganda and Rwanda. As neighboring countries joined the fighting, the war became a regional crisis. The arrival of a UN peacekeeping mission was consistently delayed by Kabila in 2000.

Kabila's murder in January 2001 left a vacuum that exposed the extent of his grip on power. His son Joseph, head of the armed forces, was rapidly appointed as his successor and has promised to promote political pluralism and economic liberalization.

***Joseph Kabila,** who succeeded his father as president in 2001.*

***Mobutu,** the ousted dictator, held power from 1965–1997.*

MAIN POLITICAL ISSUE

Securing peace and democracy

Positive change after Mobutu's overthrow proved elusive. The 1999 elections did not take place as fighting spiraled, with the government in true control of only half of the country. Kabila's death in 2001 brought all factions together for talks but fighting continued. Talks in South Africa in 2002 again failed to make any real progress, although a deal was signed with Rwanda to bring security to the east.

WORLD AFFAIRS

Joined UN in 1960

CEPGL | COMESA | OIF | G24 | AU

The protracted civil war which has devastated this huge country has irrevocably muddied the DRC's international relations. While the West has grown frustrated with the slow pace of peace, neighboring countries have been drawn into one of the continent's largest conflicts.

The collapse of the autocratic Mobutu regime in 1997 did not bring the long-hoped for stability, and instead prompted a savage rebellion by the erstwhile Rwandan and Ugandan allies of new leader Laurent Kabila. The vast potential wealth of the Congolese forests and mineral deposits prompted a quick spiral into regional war. Angola, Burundi, Chad, Namibia, Sudan, and Zimbabwe all sent troops, leaving the DRC divided into government- and rebel-controlled areas.

Fighting continued despite a 1999 cease-fire signed in Lusaka. Laurent Kabila attracted intense international criticism in 2000 for suspending the accords, for taking an autocratic approach to a new transitional assembly, and for obstructing the arrival of the UN peacekeeping mission, MONUC. Talks after Kabila's death led to the withdrawal of some Ugandan, Rwandan, and Burundian troops, but fighting continued, and in early 2002 both Rwanda and Uganda reinforced troops. However, in July Rwanda agreed to recall its forces in return for Congolese assurances of security.

The involvement of foreign firms in the running of the long-neglected mines in the south of the country has drawn sharp internal criticism.

AID

Recipient

$184m (receipts)

Up 39% in 2000

The regime's importance to the West during the Cold War brought in aid revenues on a large scale. Between 1970 and 1989, it received $8.3 billion in economic aid as well as large-scale military assistance.

By 1990, changing political priorities led the US to act on long-deferred problems of human rights abuses and misappropriation of aid. It suspended all but humanitarian aid; most other donors quickly followed suit, and the IMF declared the government "noncooperative" over its $10 billion foreign debt. Joseph Kabila's accession in early 2001 improved the country's international standing; aid was resumed and debt cancellations followed.

CHRONOLOGY

The modern Congo was the site of the Kongo and other powerful African kingdoms, and a focus of the slave trade. Belgium's King Leopold II claimed most of the Congo basin after 1876 as his personal possession.

- ❑ **1885** Brutal colonization of Congo Free State (CFS) as King Leopold's private fief.
- ❑ **1908** Belgium takes over CFS after international outcry.
- ❑ **1960** Independence of Republic of Congo. Katanga (Shaba) province secedes. UN intervenes.
- ❑ **1963** Katanga secession collapses.
- ❑ **1965** Gen. Joseph-Désiré Mobutu seizes power.
- ❑ **1970** Mobutu elected president; his MPR becomes sole legal party.
- ❑ **1971** Country renamed Zaire.
- ❑ **1977–1978** Two invasions by former Katanga separatists repulsed with Western help.
- ❑ **1982** Opposition parties set up Union for Democracy and Social Progress (UDPS).
- ❑ **1986–1990** Civil unrest and foreign criticism of human rights abuses.
- ❑ **1990** Belgium suspends aid after security forces kill prodemocracy demonstrators. Mobutu announces transition to multiparty rule.
- ❑ **1991** Opposition leader Etienne Tshisekedi heads short-lived "crisis government" formed by Mobutu.
- ❑ **1992–1993** Rival governments claim legitimacy.
- ❑ **1994** Combined High Council of the Republic–Transitional Parliament established.
- ❑ **1995** Regime demands international assistance to support a million Rwandan Hutu refugees.
- ❑ **1996** Major insurgency launched in east by Alliance of Democratic Forces for the Liberation of the Congo (AFDL) including Laurent Kabila's Popular Revolutionary Party (PRP), with disaffected ethnic Tutsi Banyamulunge.
- ❑ **1997** Forces led by Kabila sweep south and west. Kabila takes power. Country renamed DRC. Mobutu dies in exile.
- ❑ **1998** Banyamulunge join Kabila's opponents and launch rebellion in the east, backed by Rwanda and Uganda. Southern African states, give military backing to Kabila.
- ❑ **2000** UN approves peacekeeping mission; arrival stalled by Kabila.
- ❑ **2001** Laurent Kabila assassinated; succeeded by son Joseph. Peace talks start, with troop withdrawals from front line.
- ❑ **2002** April, peace talks inconclusive. May, 180 massacred in Kisangani by Rwandan-backed forces.

DEFENSE

▷ No compulsory military service

$392m

Down 5% in 2000

Although the military strongly backed Mobutu's regime, it offered no real resistance when Laurent Kabila's insurgents swept the country in 1996–1997. Government troops, poorly paid and undisciplined, but supported by foreign allies, have been fighting in the civil war since 1998, led by Kabila's son and successor Joseph Kabila. The UN peacekeeping force, MONUC, had established a small presence by 2001.

CONGOLESE ARMED FORCES

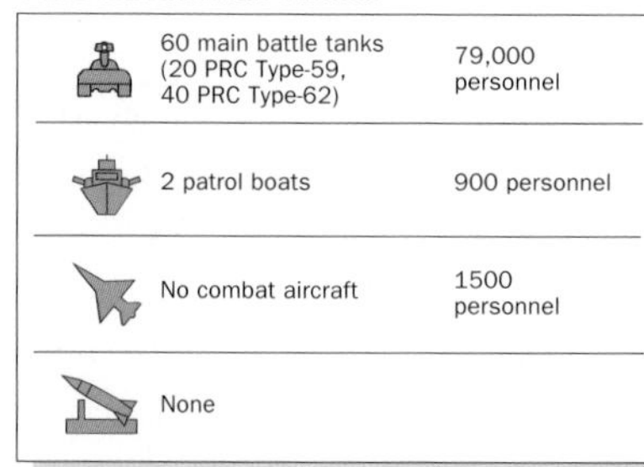
60 main battle tanks (20 PRC Type-59, 40 PRC Type-62)
79,000 personnel
2 patrol boats
900 personnel
No combat aircraft
1500 personnel
None

ECONOMICS

▷ Inflation 1423% p.a. (1990–1999)

$5.4bn

4.5–313 Congolese francs

SCORE CARD

- ❑ World GNP Ranking........................108th
- ❑ GNP per Capita..............................$110
- ❑ Balance of Payments....................–$798m
- ❑ Inflation......................................540%
- ❑ Unemployment.............................Very high

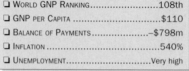

EXPORTS

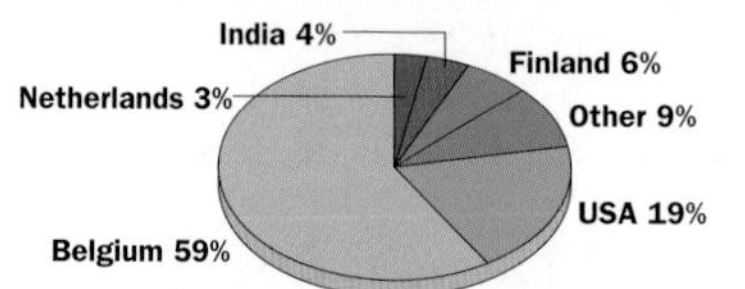
India 4%
Finland 6%
Netherlands 3%
Other 9%
USA 19%
Belgium 59%

IMPORTS

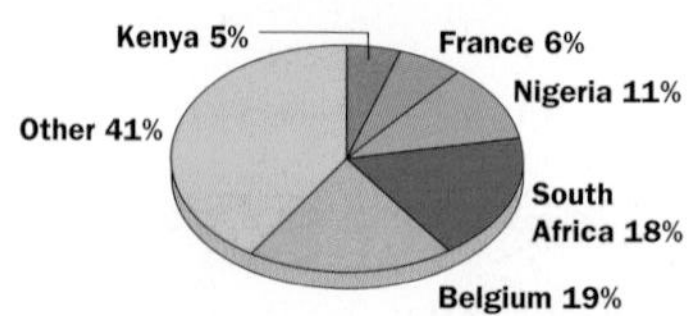
Kenya 5%
France 6%
Nigeria 11%
Other 41%
South Africa 18%
Belgium 19%

ECONOMIC PERFORMANCE INDICATOR

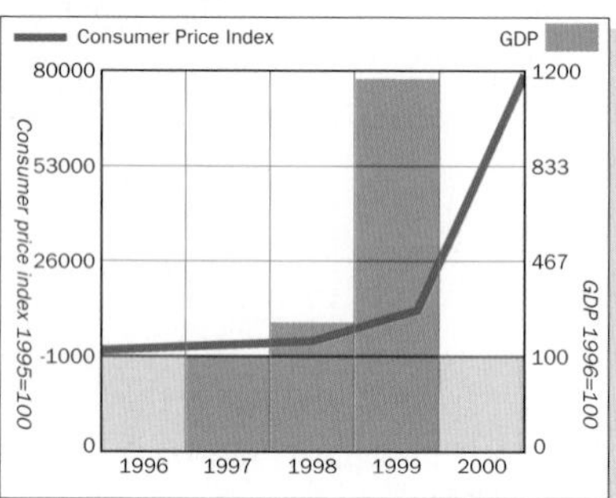
Consumer Price Index
GDP
80000
53000
26000
-1000
0
1200
833
467
100
0
Consumer price index 1995=100
GDP 1996=100
1996
1997
1998
1999
2000

Strengths

Rich resource base. Minerals – notably copper, cobalt, diamonds – provide 85% of export earnings. Energy: oil; possibly Africa's largest hydropower potential. Rich soil; much unutilized arable land.

Weaknesses

Decades of mismanagement and corruption: $12 billion foreign debt; lack of food self-sufficiency; inadequate, disintegrating infrastructure. Political instability. Hyperinflation. Loss of export income. Mineral resources plundered by foreign troops.

Profile

Political instability, systematic corruption, long-term mismanagement, and outright civil war have brought what is potentially a leading African economy to a state of collapse.

By the mid-1990s real GDP was falling by 10% or more each year. The government budget ran record deficits, and inflation ran virtually out of control from 1994 onward. Lack of spares and power cuts have since closed many mines and halted most other industry. Strikes and riots over plummeting living standards hastened the flight of foreign capital. Subsistence farming and petty trade keep most people going.

The Kabila regime, despite its originally Marxist background, claims to want an effective free-market economy. Resumption of essential large-scale aid and debt relief will depend on difficult reforms and paying off arrears to the IMF and other creditors. Restructuring of state-owned enterprises in 2001 aimed to attract foreign investment back.

CONGO, DEM. REP. : MAJOR BUSINESSES

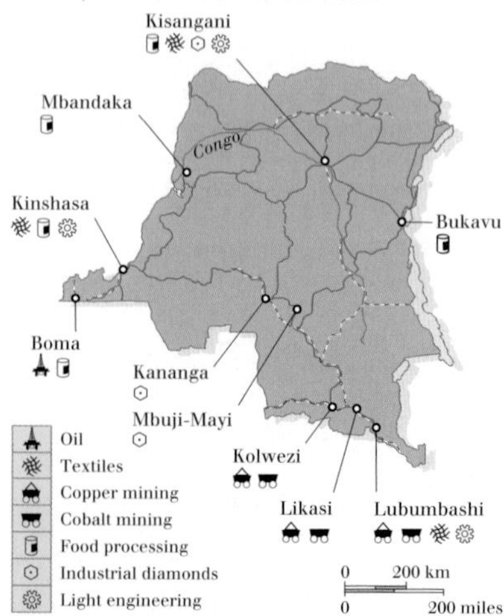
Kisangani
Mbandaka
Congo
Kinshasa
Bukavu
Boma
Kananga
Mbuji-Mayi
Kolwezi
Likasi
Lubumbashi
Oil
Textiles
Copper mining
Cobalt mining
Food processing
Industrial diamonds
Light engineering
0 200 km
0 200 miles

RESOURCES

Electric power 3.2m kw

208,862 tonnes

23,158 b/d (reserves 194m barrels)

4.07m goats, 999,748 pigs, 20.6m chickens

Copper, diamonds, oil, coltan, cobalt, zinc, uranium, manganese

ELECTRICITY GENERATION

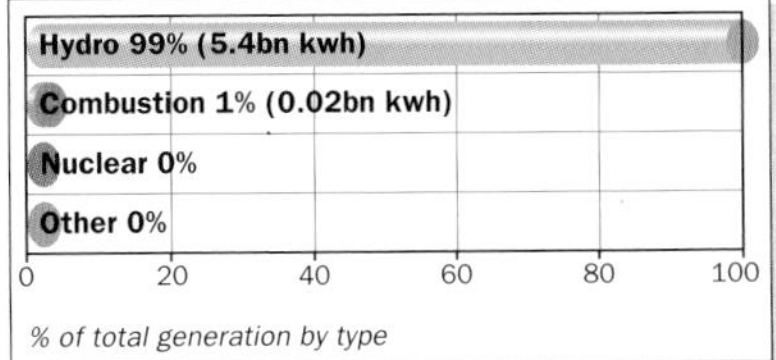
Hydro 99% (5.4bn kwh)
Combustion 1% (0.02bn kwh)
Nuclear 0%
Other 0%
0
20
40
60
80
100
% of total generation by type

What should be a prosperous country, with its rich resources, is instead one of the world's poorest states, exploited and mismanaged by its rulers for decades and plundered further by occupying forces during the civil war. In the 1980s, the country was the world's largest cobalt exporter and second-largest industrial diamond exporter. Since 1990, copper and cobalt output have collapsed and diamond smuggling is booming. There are oil reserves and hydroelectric installations with sufficient potential capacity to export power, but instead lack of maintenance has shut down many turbines and most urban areas face power cuts. Despite rich soils and the fact that 60% of people are involved in farming, the DRC is not even self-sufficient in food.

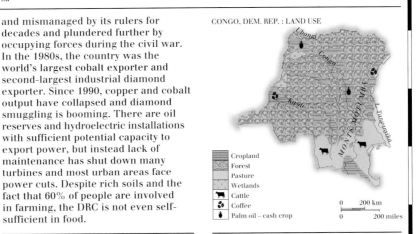

CONGO, DEM. REP. : LAND USE

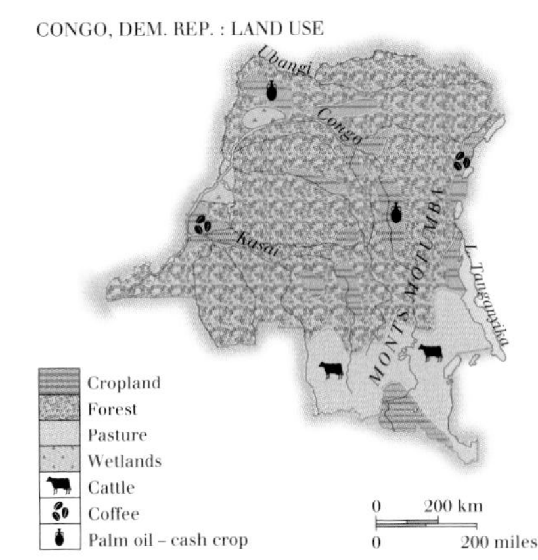
Ubangi
Congo
Kasai
MONTS MITUMBA
L. Tanganyika
Cropland
Forest
Pasture
Wetlands
Cattle
Coffee
Palm oil – cash crop
0 200 km
0 200 miles

ENVIRONMENT

Sustainability rank: 109th

5%

0.1 tonnes per capita

ENVIRONMENTAL TREATIES

Yes	Yes	Yes
Yes	Yes	No

Rainforests cover over 60% of the country, representing almost 6% of the world's and 50% of Africa's remaining woodlands. They are home to several endangered species. The poor transportation network has so far prevented large-scale commercial exploitation of timber, but clearance for fuelwood is a problem. The collapse of many urban refuse and sewage disposal systems has led to major health and pollution problems. Environmental damage caused by the civil war is estimated at $320 million.

CRIME

Death penalty in use

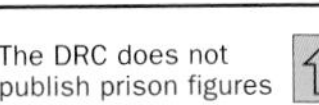
The DRC does not publish prison figures

Violence and crime have risen rapidly since 1990

CRIME RATES

All types of crime are on the increase in the DRC

Political crisis and economic collapse exacerbate long-standing problems of corruption and human rights abuses. Violence and crime of all kinds, including extortion, robbery, rape, and murder, are widespread and on the increase, and war zones have seen the abandonment of the rule of law. Ethnic violence, suppressed after 1965, resurfaced in the 1990s in the south and between the Hema and Lendu tribes in the northeast.

HEALTH

No welfare state health benefits

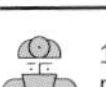
1 per 10,000 people

Malaria, respiratory and diarrheal diseases

State services have now virtually collapsed. Disease and death rates are rising, especially in rural areas. A new health insurance plan was announced in 2001, designed to enable greater access to health care. As at end-2001, 1.3 million people were estimated to be HIV/AIDS infected.

SPENDING

GDP/cap. decrease

CONSUMPTION AND SPENDING

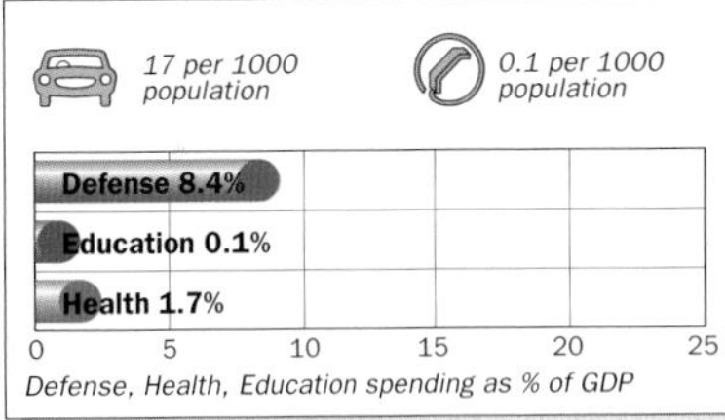
17 per 1000 population
0.1 per 1000 population
Defense 8.4%
Education 0.1%
Health 1.7%
0
5
10
15
20
25
Defense, Health, Education spending as % of GDP

Before his death in exile in 1997, ex-dictator Mobutu was one of the world's richest men, worth an estimated $4 billion. Most of his former subjects live in poverty, exacerbated by civil war.

MEDIA

TV ownership low

Daily newspaper circulation 3 per 1000 people

PUBLISHING AND BROADCAST MEDIA

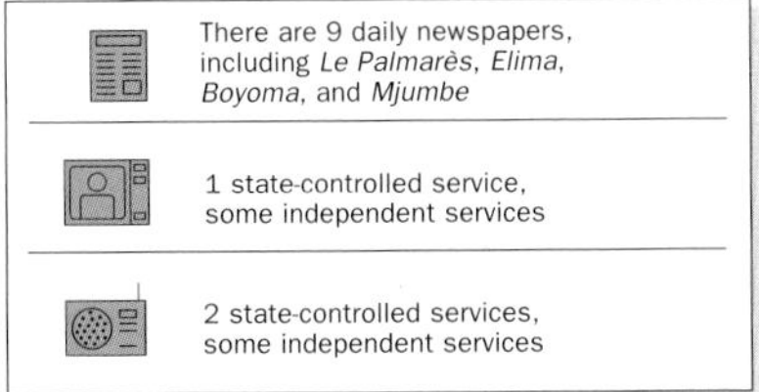
There are 9 daily newspapers, including *Le Palmarès*, *Elima*, *Boyoma*, and *Mjumbe*
1 state-controlled service, some independent services
2 state-controlled services, some independent services

Fighting makes services less accessible outside the capital. Unlike the broadcast media, press outlets are privately owned, and many newspapers openly criticize the authorities, although self-preservation requires a degree of self-censorship. A ban on foreign radio broadcasts was lifted in 2001. MONUC operates the only countrywide radio station, Radio Okapi.

EDUCATION

School leaving age: 12

61%

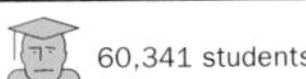
60,341 students

THE EDUCATION SYSTEM

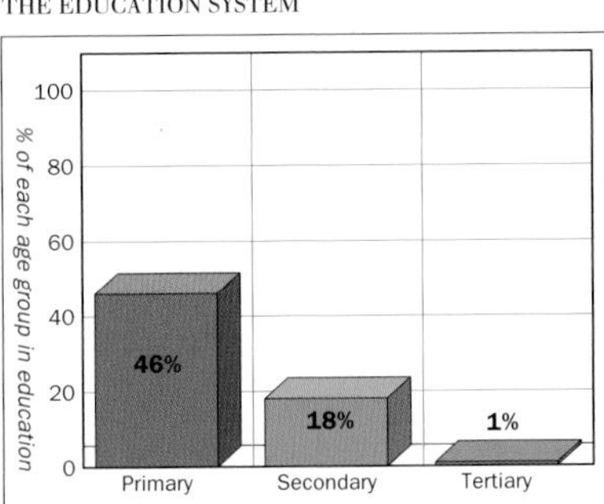
100
80
60
40
20
0
% of each age group in education
46%
18%
1%
Primary
Secondary
Tertiary

In 1997 just over 37% of secondary age children were attending classes, but this figure has dropped sharply during the civil war. State provision, as with health care, is patchy and has faced sharp budget cuts since 1980. About 70% of schooling is now provided by the Roman Catholic Church.

WORLD RANKING

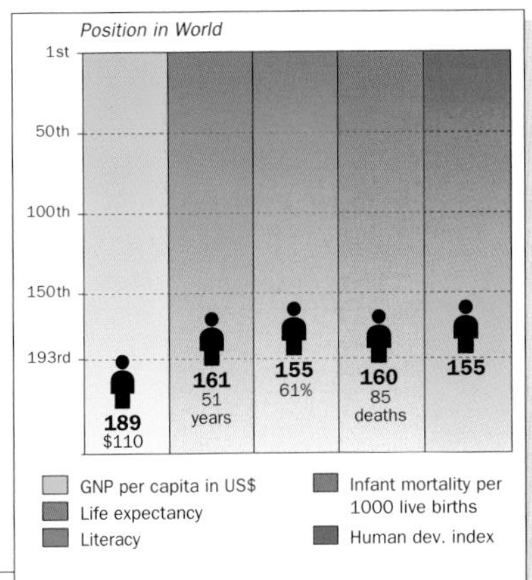
Position in World
1st
50th
100th
150th
193rd
189
$110
161
51 years
155
61%
160
85 deaths
155
GNP per capita in US$
Life expectancy
Literacy
Infant mortality per 1000 live births
Human dev. index

COSTA RICA

CENTRAL AMERICA

OFFICIAL NAME: Republic of Costa Rica **CAPITAL:** San José
POPULATION: 4.1 million **CURRENCY:** Costa Rican colón **OFFICIAL LANGUAGE:** Spanish

C

1838

1838

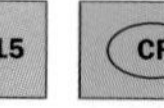
Sept 15

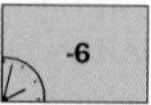
CR

-6

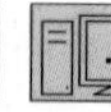
+506

.cr

SPANNING THE CENTRAL AMERICAN isthmus and wedged between Nicaragua and Panama, Costa Rica was under Spanish rule until 1821 and gained full independence in 1838. From 1948 until the end of the 1980s, it had the most developed welfare state in Central America. Costa Rica is nominally a multiparty democracy, but two parties dominate. Its army was abolished in 1948; the 1949 constitution then forbade national armies.

CLIMATE

Tropical wet & dry

WEATHER CHART FOR SAN JOSÉ

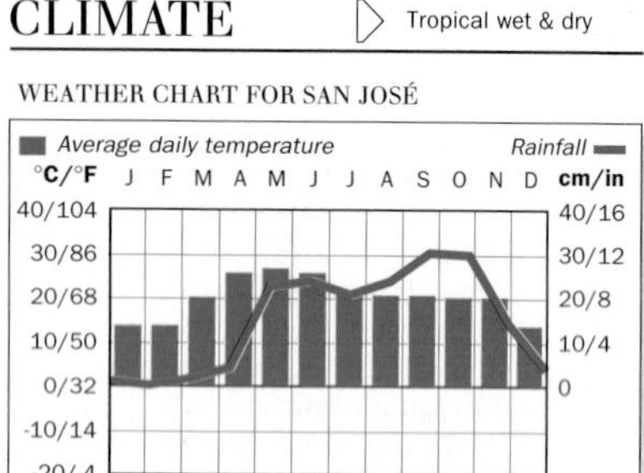
Average daily temperature
Rainfall
°C/°F J F M A M J J A S O N D cm/in
40/104
30/86
20/68
10/50
0/32
-10/14
-20/4
40/16
30/12
20/8
10/4
0

The Caribbean coast has heavy rainfall, while the Pacific coast is much drier. The central uplands are temperate.

TRANSPORTATION

Drive on right

Juan Santamaría, San José
988,000 passengers

15 ships
5700 grt

THE TRANSPORTATION NETWORK

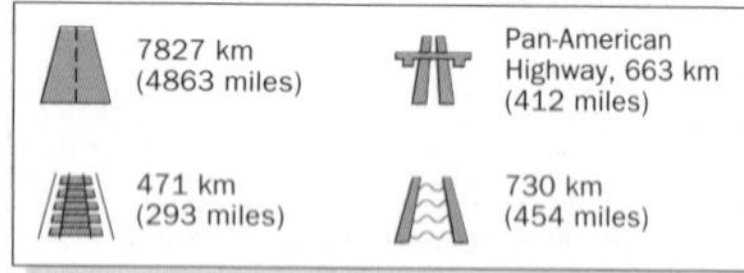
7827 km (4863 miles)
Pan-American Highway, 663 km (412 miles)
471 km (293 miles)
730 km (454 miles)

San José is the hub of a well-used bus network. Railroads, badly earthquake-damaged in 1991, are being reconstructed.

TOURISM

Visitors : Population 1:3.8

1.09m visitors

Up 5% in 2000

MAIN TOURIST ARRIVALS

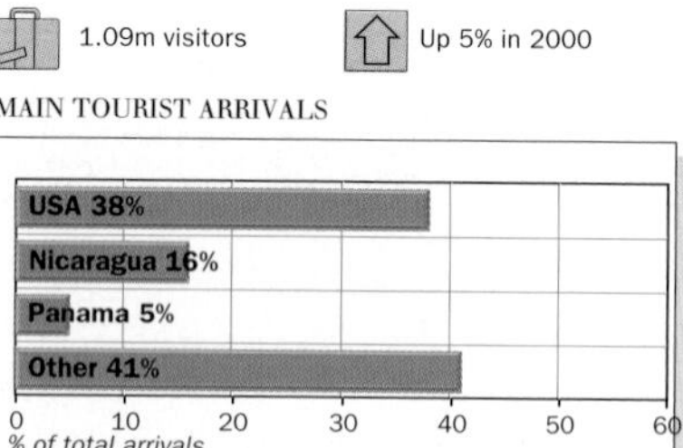
USA 38%
Nicaragua 16%
Panama 5%
Other 41%
0 10 20 30 40 50 60
% of total arrivals

Tourism brought in $1 billion in 1999 and is expanding, with the help of both domestic and foreign investment. A new 154-room hotel in San José and a comparable one on the Pacific coast attract wealthy visitors.

PEOPLE

Pop. density medium

Spanish, English Creole, Bribri, Cabecar

80/km²
(208/mi²)

THE URBAN/RURAL POPULATION SPLIT

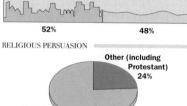

52% 48%

RELIGIOUS PERSUASION

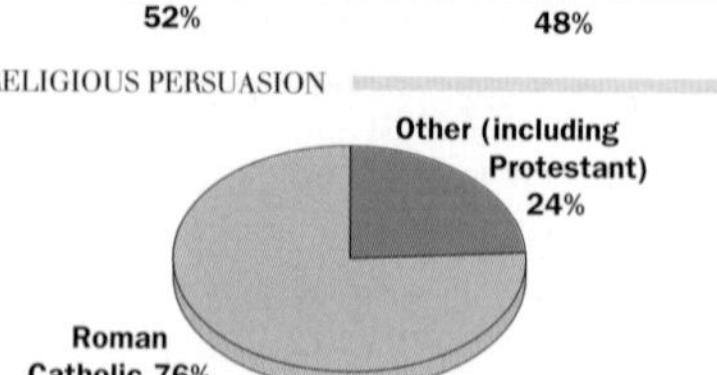
Other (including Protestant) 24%
Roman Catholic 76%

The majority of the population is *mestizo*, of partly Spanish origin. One-third of people in the Limón area are black and often English-speaking. There are only about 5000 indigenous Amerindians.

POLITICS

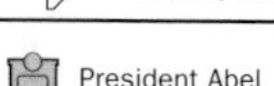
Multiparty elections

2002/2006

President Abel Pacheco de la Espriella

AT THE LAST ELECTION

Legislative Assembly 57 seats

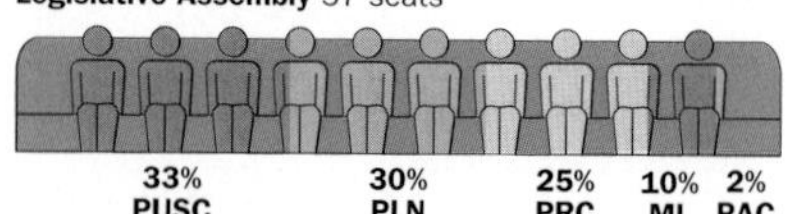
33% PUSC
30% PLN
25% PRC
10% ML
2% PAC

PUSC = Social Christian Unity Party **PLN** = National Liberation Party **PRC** = Costa Rican Renewal Party **ML** = Liberty Movement **PAC** = Citizens' Action Party

Politics has long been dominated by the PUSC and PLN, both of which have close ties to major banana- and coffee-growing families. Historically the US has exercised a very powerful influence on politics.

The PLN in 1994 promised reforms to its previous austerity policies, but soon came under pressure from international financial organizations to reduce the budget deficit. Harsh structural adjustment measures proved highly unpopular.

In 1998 the PUSC regained power. President Miguel Angel Rodríguez launched a three-year plan to reduce inflation and poverty, create thousands of jobs, and stimulate foreign investment in state companies. His chosen successor, Abel Pacheco, needed an unprecedented second round to clinch the presidency in 2002, when voter turnout hit an all-time low.

WORLD AFFAIRS

Joined UN in 1945

ACS

Geplac

RG

OAS

San José

Trade ties with the US and protection of prices for coffee and bananas are priorities. Trade ties have been agreed with Canada and Chile. Tensions with Nicaragua over their mutual border were resolved in 2000, but illegal immigrants remain an issue.

Pineapple plantation near Buenos Aires, crossed by the Pan-American Highway which runs for 663 km (412 miles) through Costa Rica.

COSTA RICA

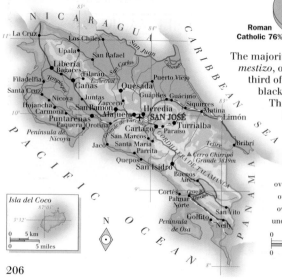

Total Land Area : 51 100 sq. km (19 730 sq. miles)

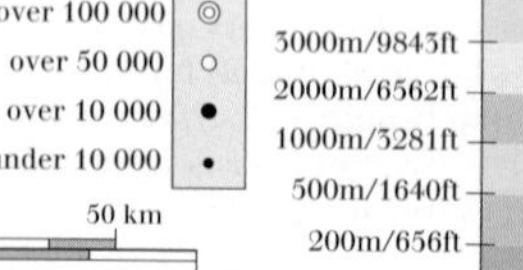
POPULATION
over 100 000
over 50 000
over 10 000
under 10 000
LAND HEIGHT
3000m/9843ft
2000m/6562ft
1000m/3281ft
500m/1640ft
200m/656ft
Sea level
0 50 km
0 50 miles

AID

Recipient

 $12m (receipts)

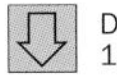 Down 60% in 1999–2000

During the 1980s Costa Rica was a large recipient of US aid designed to inoculate it against left-wing insurgencies such as those in El Salvador, Guatemala, and neighboring Nicaragua. Peace in the region has led to a sharp decline in such aid, especially given the country's relatively high per capita income. World Bank aid will help modernize Juan Santamaría international airport.

DEFENSE

No compulsory military service

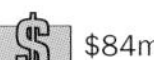 $84m

 Up 22% in 2000

Costa Rica emerged from the 1948 civil war as a neutral, demilitarized modern state. A 4400-strong Civil Guard is complemented by a largely military-trained police force. Spending on security has long been the lowest in the region. Lack of a common command structure hinders the influence of the security forces but also renders them less open to public control. Right-wing paramilitary groups are known to exist.

ECONOMICS

Inflation 17% p.a. (1990–2000)

 $14.5bn

 317.5–341.3 Costa Rican colones

SCORE CARD

- ❑ WORLD GNP RANKING........76th
- ❑ GNP PER CAPITA........$3810
- ❑ BALANCE OF PAYMENTS........–$650m
- ❑ INFLATION........11%
- ❑ UNEMPLOYMENT........5%

EXPORTS

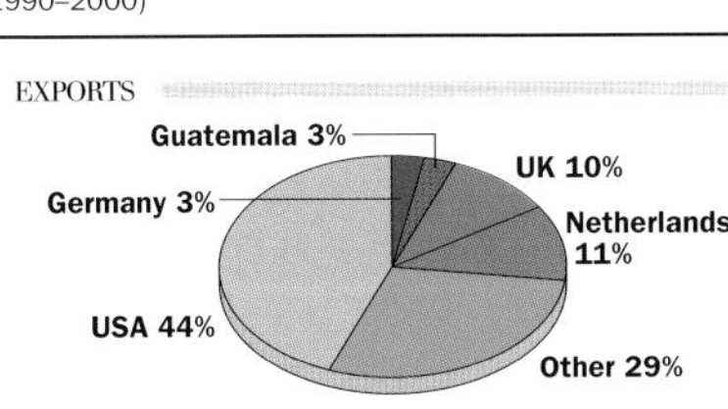
Guatemala 3%
UK 10%
Germany 3%
Netherlands 11%
USA 44%
Other 29%

IMPORTS

Japan 4%
Venezuela 5%
Mexico 5%
Guatemala 3%
Other 42%
USA 41%

STRENGTHS

Major coffee, beef, and banana exports. Expanding tourism also fueling construction. Strong inward investment. Favorable WTO ruling on banana access to EU market.

WEAKNESSES

Coffee, beef, and bananas all vulnerable to falling prices. History of high inflation. Dependence on imported oil. Large domestic debt. Competitiveness hindered by insufficient investment in infrastructure. State monopolies have deterred investment in energy, telecommunications, and insurance sectors. Inefficient management.

RESOURCES

Electric power 1.5m kw

 35,003 tonnes

 Not an oil producer

1.72m cattle, 430,000 pigs, 17m chickens

 Bauxite, gold, silver, manganese, mercury

Costa Rica has large bauxite deposits in the south – aluminum smelting is an important industry. Small quantities of gold, silver, manganese, and mercury are also mined. Self-sufficiency in energy is being pursued through the development of hydroelectric power. Forests cover 31% of the country.

ENVIRONMENT

Sustainability rank: 9th

 14% (3% partially protected)

 1.4 tonnes per capita

Despite good environmental regulation, reckless economic development has contributed to extensive deforestation. Pasture land now covers some 46% of the territory and pesticide abuse by agribusiness has poisoned rivers and threatened species. Urban sprawl has degraded the fertile central valley.

MEDIA

TV ownership medium

 Daily newspaper circulation 72 per 1000 people

PUBLISHING AND BROADCAST MEDIA

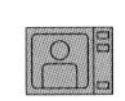 There are 8 daily newspapers, including *La Nación*, *La República*, *La Prensa Libre*, and *Diario Extra*

8 stations: 1 state-owned, 7 independent

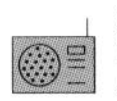 State-owned and independent stations

The media are free but dominated by conservative opinion. Entry into journalism is strictly licensed.

CRIME

No death penalty

 8526 prisoners

 Up 71% 1989–1994

Costa Rica is the least violent Central American country. Attacks on and kidnappings of tourists are rare but have dented its image as a safe haven. Drug cartels use the country to transfer cocaine to the US and Europe. Police show hostility toward immigrants from neighboring countries.

CHRONOLOGY

Costa Rica, ruled since the 16th century by Spain, became an independent state in 1838.

- ❑ **1948** Disputed elections lead to civil war; ended by Social Democratic Party (later the PLN) forming provisional government under José Ferrer. Army abolished.
- ❑ **1949** New constitution promulgated.
- ❑ **1987** Central American Peace Plan initiated by President Arias.
- ❑ **1998** PUSC returns to power.

C

EDUCATION

School leaving age: 14

 96%

 59,947 students

Costa Rica has the highest literacy rate in the isthmus, and is home to the University of Central America.

HEALTH

Welfare state health benefits

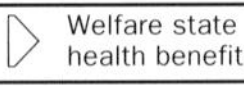 1 per 1111 people

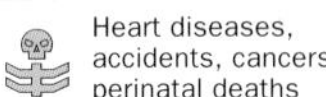 Heart diseases, accidents, cancers, perinatal deaths

The public health system is one of the most developed in Latin America. Health was allocated 29.3% of total public spending in 1999.

SPENDING

GDP/cap. increase

CONSUMPTION AND SPENDING

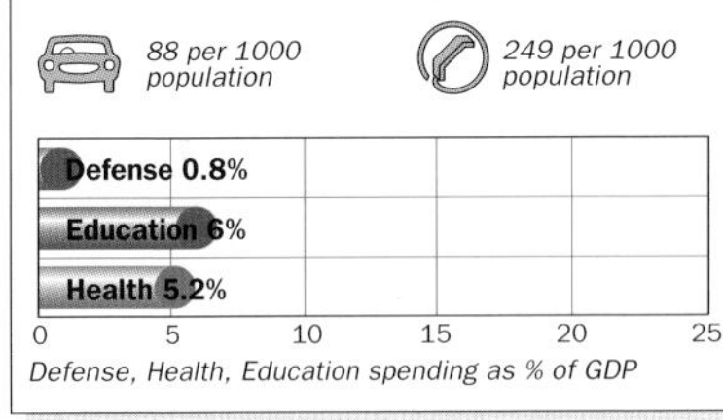
88 per 1000 population
249 per 1000 population
Defense 0.8%
Education 6%
Health 5.2%
0
5
10
15
20
25
Defense, Health, Education spending as % of GDP

The plantation-owning families are the wealthiest group; official figures claim that one-fifth of Costa Rica's population is living in poverty.

WORLD RANKING

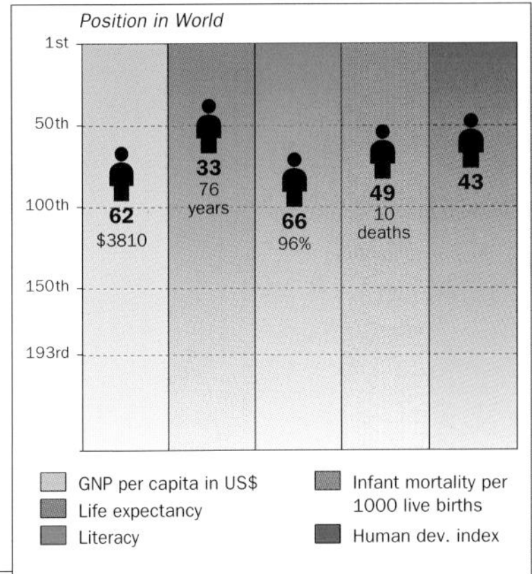
Position in World
1st
50th
100th
150th
193rd
62
$3810
33
76 years
66
96%
49
10 deaths
43
GNP per capita in US$
Life expectancy
Literacy
Infant mortality per 1000 live births
Human dev. index

C

CROATIA

EUROPE

Europe

Hungary
Slovenia
CROATIA
Bosnia & Herzegovina
Serb. & Mont.

OFFICIAL NAME: Republic of Croatia **CAPITAL:** Zagreb
POPULATION: 4.7 million **CURRENCY:** Kuna **OFFICIAL LANGUAGE:** Croatian

1991

1991

May 30
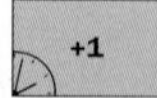
HR

+1

+385

.hr

ALTHOUGH IT WAS CONTROLLED by Hungary from medieval times and was a part of the Yugoslav state for much of the 20th century, Croatia still has a strong national identity. It includes the historic provinces of Slavonia, Istria, and Dalmatia (Dinara). Actively involved in the conflicts which broke up Yugoslavia in the early 1990s, Croatia only regained full control of Serb-occupied Eastern Slavonia, around Vukovar, in 1998.

Dubrovnik, Dalmatia. This historic city on the Adriatic coast was shelled and besieged by the Yugoslav federal army in 1991.

CLIMATE

Mediterranean/ continental

WEATHER CHART FOR ZAGREB

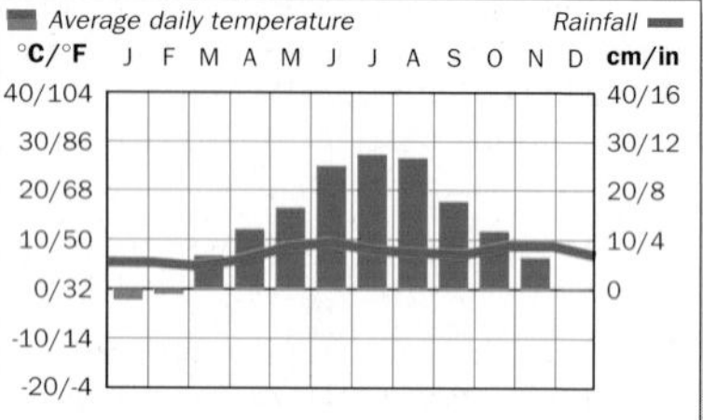
Average daily temperature
Rainfall
°C/°F
J F M A M J J A S O N D
cm/in
40/104
30/86
20/68
10/50
0/32
-10/14
-20/-4
40/16
30/12
20/8
10/4
0

Northern Croatia has a temperate continental climate. Its Adriatic coast has a Mediterranean climate.

TRANSPORTATION

Drive on right

Pleso International, Zagreb
1.15m passengers

246 ships
734,300 grt

THE TRANSPORTATION NETWORK

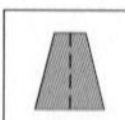
23,497 km (14,600 miles)

330 km (205 miles)

2726 km (1694 miles)

785 km (488 miles)

Zagreb has recovered from the effects of war, and of sanctions against Yugoslavia, and is once again an important regional road and rail hub. The Adriatic Highway affords fantastic views along the Dalmatian coast.

TOURISM

Visitors : Population 1.2:1

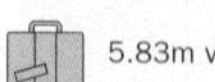
5.83m visitors

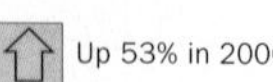
Up 53% in 2000

MAIN TOURIST ARRIVALS

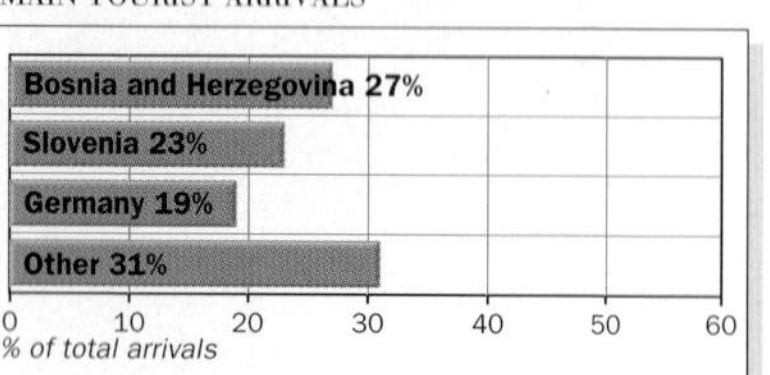
Bosnia and Herzegovina 27%
Slovenia 23%
Germany 19%
Other 31%
0 10 20 30 40 50 60
% of total arrivals

The Adriatic coast is regaining its popularity as a tourist destination. There are also many historical sites.

PEOPLE

Pop. density medium

Croatian

83/km² (215/mi²)

THE URBAN/RURAL POPULATION SPLIT

58% 42%

RELIGIOUS PERSUASION

Muslim 1%
Orthodox 11%
Other 12%
Roman Catholic 76%

Before the breakup of Yugoslavia, Croats made up nearly 80% of the population, and Serbs 12%. In 1991, the Serbs, alienated by Croatian nationalism, proclaimed the Republic of Serbian Krajina in areas where they formed a majority. Croatian forces retook most Serb-held territory in offensives in 1995, precipitating a huge exodus of Serb refugees. The last rebel Serb enclave, Eastern Slavonia, was reintegrated into Croatia in 1998. Many of the 250,000 Bosnian refugees who fled to Croatia have now settled there. By 2001 Serbs comprised only 4% of the population.

WORLD AFFAIRS

Joined UN in 1992

CE

WTO

PfP

CEI

OSCE

Tudjman's nationalist stance after independence worked against closer relations with the West. Croatia's military successes in 1995, hastening the end of the Bosnian war, were reflected in the Bosnian peace accord. Tudjman's successors seek closer cooperation with the EU. In 2000 Croatia joined NATO's Partnerships for Peace program.

AID

Recipient

$66m (receipts)

Up 38% in 2000

EU states have spent over $1 billion on reconstruction in Croatia since 1991.

DEFENSE

Compulsory military service

$509m

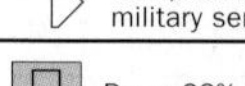
Down 23% in 2000

The army proved its effectiveness when in 1995 it recaptured territory held by Serb forces. Under 2002 proposals its size will be halved by 2005.

POLITICS

Multiparty elections

2000/2004

President Stipe Mesic

AT THE LAST ELECTION

House of Representatives 151 seats

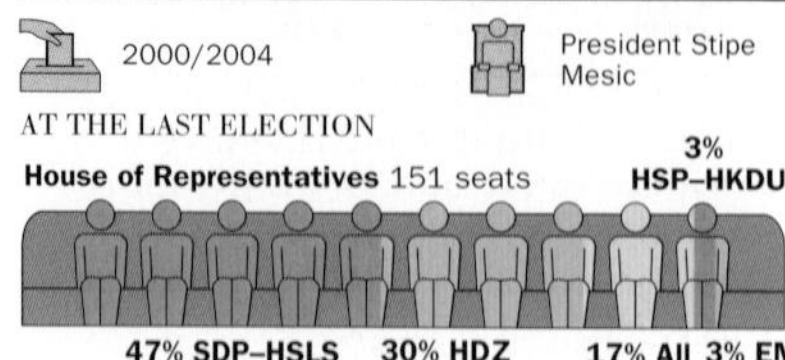
3% HSP–HKDU
47% SDP–HSLS
30% HDZ
17% All
3% EM

SDP–HSLS = *SDP*–Social Democratic Party, *HSLS*–Croatian Social Liberal Party **HDZ** = Croatian Democratic Union **All** = Alliance (*HSS*–Croatian Peasant Party, *IDS*–Istrian Democratic Assembly, *HNS*–Croatian People's Party, *LS*–Liberal Party, *ASH*–Croatian Social Democrats' Action) **HSP–HKDU** = *HSP*–Croatian Party of Rights, *HKDU*–Croatian Christian Democratic Union **EM** = Ethnic minorities

Six seats are reserved for representatives of Croats living abroad (all are currently held by the HDZ). Five seats are reserved for ethnic minorities (one is currently held by the HSS)

Croatia left the former Yugoslavia in 1991 under the direction of the right-wing HDZ led by Franjo Tudjman. However, the country suffered growing international isolation because of its nationalist stance and ambiguous support for the 1995 Bosnian peace accord. The HDZ was left rudderless after Tudjman's death in December 1999 and, mired in corruption and spying scandals, was swept from power in elections in January 2000 by left-leaning parties. The eclipse of the HDZ was confirmed in presidential elections the following month, won by one-time Yugoslav federal president Stipe Mesic. The new government's priorities were to rehabilitate Croatia internationally and to reduce the authoritarian powers of the presidency.

ECONOMICS

Inflation 86% p.a. (1990–2000)

$20.2bn

8.087–8.254 kuna

SCORE CARD

- ❑ World GNP Ranking.........................61st
- ❑ GNP per Capita $4620
- ❑ Balance of Payments....................–$399m
- ❑ Inflation 5.4%
- ❑ Unemployment..................................20%

Strengths

Moderate growth as economy made recovery from 1998–1999 recession. Progress in reducing government overspending, backed by IMF.

Weaknesses

War damage estimated at $50 billion. Slow privatization until 2001. High and persistent unemployment.

EXPORTS

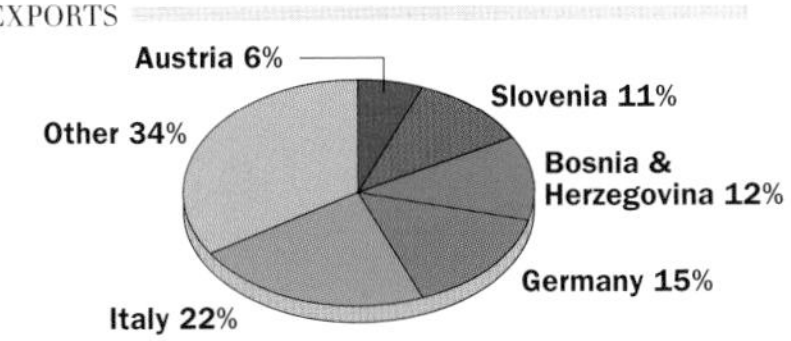

Austria 6%
Slovenia 11%
Other 34%
Bosnia & Herzegovina 12%
Germany 15%
Italy 22%

IMPORTS

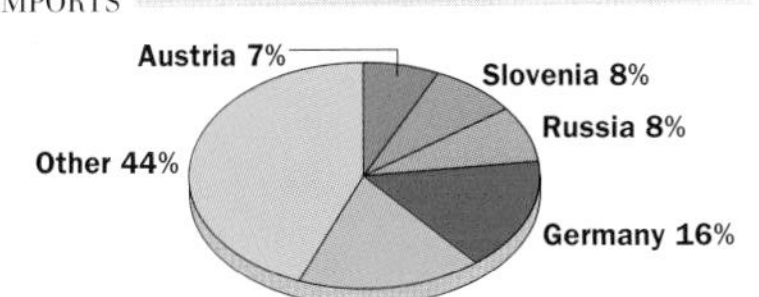
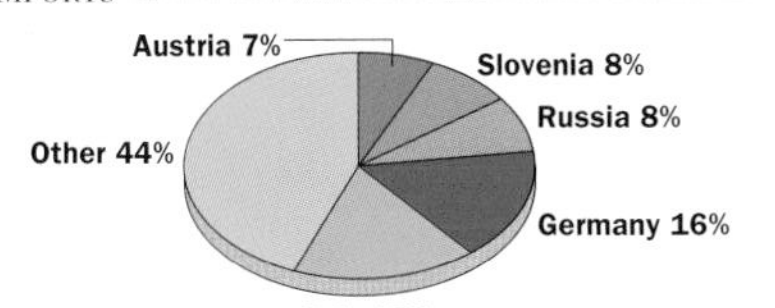

Austria 7%
Slovenia 8%
Russia 8%
Other 44%
Germany 16%
Italy 17%

RESOURCES

Electric power 3.8m kw

25,519 tonnes

31,859 b/d (reserves 47m barrels)

1.23m pigs, 710,000 turkeys, 10.8m chickens

Coal, bauxite, iron, oil, china clay, natural gas

Croatia generates 40% of its energy needs from combustion and 60% from hydroelectric sources. It has very few minerals, although it does have oil and gas fields. The rich fishing grounds of the Adriatic are a major resource.

ENVIRONMENT

Sustainability rank: 12th

8%

4.5 tonnes per capita

Croatia was the first Yugoslav republic to create reserves in order to protect endangered and unique wetlands.

MEDIA

TV ownership high

Daily newspaper circulation 115 per 1000 people

PUBLISHING AND BROADCAST MEDIA

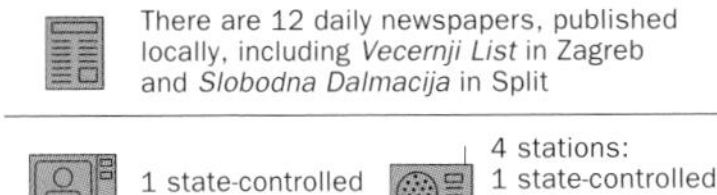
There are 12 daily newspapers, published locally, including *Vecernji List* in Zagreb and *Slobodna Dalmacija* in Split

1 state-controlled service

4 stations: 1 state-controlled, 3 independent

The three TV channels are state-owned. Media freedoms, eroded under Tudjman, have gradually improved.

CRIME

No death penalty

2617 prisoners

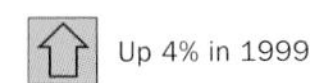
Up 4% in 1999

Under Tudjman, former Croat HOS militiamen escaped prosecution for "ethnic cleansing" in Bosnia, but the post-Tudjman government ordered a number of arrests.

HUNGARY
SLOVENIA
SERBIA & MONTENEGRO (YUGOSLAVIA)
BOSNIA & HERZEGOVINA
ADRIATIC SEA
ZAGREB
Čakovec
Varaždin
Koprivnica
Križevci
Samobor
Sesvete
Velika Gorica
Bjelovar
Virovitica
Beli Manastir
Slatina
Osijek
Borovo
Vukovar
Vinkovci
Županja
Karlovac
Sisak
Petrinja
Glina
Kutina
Slavonska Požega
Nova Gradiška
Daković
Slavonski Brod
Rijeka
Pazin
Rovinj
Pula
Crikvenica
Senj
Ogulin
Gospić
Zadar
Knin
Šibenik
Trogir
Split
Solin
Sinj
Makarska
Metković
Dubrovnik
Dinara 1831m
CROATIA
Total Land Area : 56 542 sq. km (21 851 sq. miles)
LAND HEIGHT
1000m/3281ft
500m/1640ft
200m/656ft
Sea Level
POPULATION
over 500 000
over 100 000
over 50 000
over 10 000
under 10 000
0 50 km
0 50 miles

CHRONOLOGY

Between 1945 and 1991 Croatia was a republic of the Yugoslav federation.

- ❑ **1991** Independence. Rebel Croatian Serb republic proclaimed.
- ❑ **1992** Tudjman president. Involvement in Bosnian civil war.
- ❑ **1995** Krajina and Western Slavonia recaptured. Dayton agreement ends fighting.
- ❑ **1998** Eastern Slavonia reintegrated.
- ❑ **1999** Death of Tudjman.
- ❑ **2000** Center-left wins elections.

EDUCATION

School leaving age: 15

98%

100,297 students

The education system is well developed. There are four universities, at Zagreb, Rijeka, Osijek, and Split.

HEALTH

Welfare state health benefits

1 per 435 people

Cerebrovascular and heart diseases, cancers

Most Croats are covered by a health insurance scheme. However, an extra strain on already scarce funds is created by the demands of refugees and disabled war veterans.

SPENDING

GDP/cap. increase

CONSUMPTION AND SPENDING

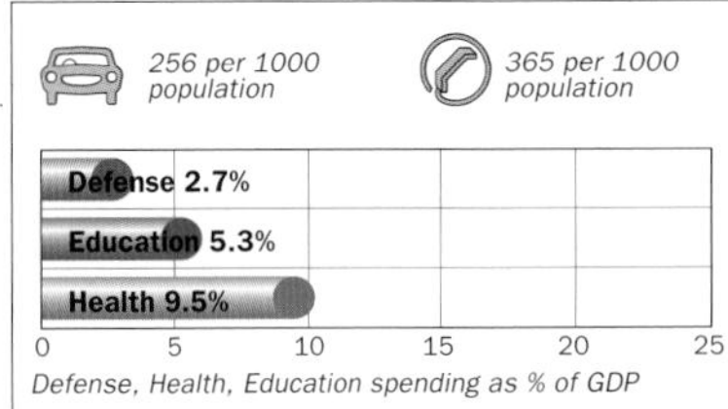

256 per 1000 population
365 per 1000 population
Defense 2.7%
Education 5.3%
Health 9.5%
0
5
10
15
20
25
Defense, Health, Education spending as % of GDP

Wage rises in the mid-1990s and again in 1999 led to spending booms. Consumers' high expectations were reined in by tighter wage policies.

WORLD RANKING

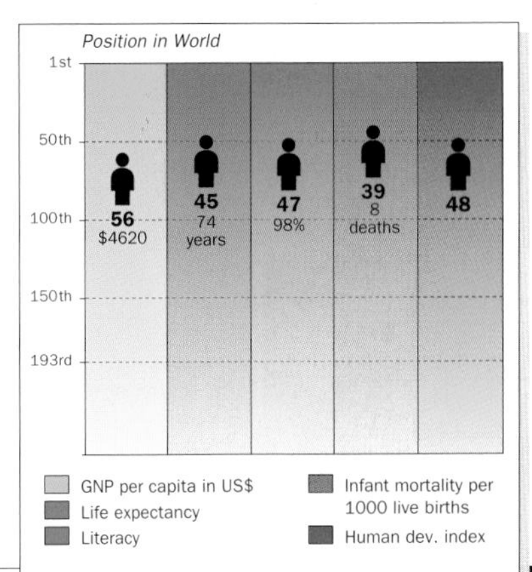

Position in World
1st
50th
100th
150th
193rd
56
$4620
45
74 years
47
98%
39
8 deaths
48
GNP per capita in US$
Life expectancy
Literacy
Infant mortality per 1000 live births
Human dev. index

CUBA

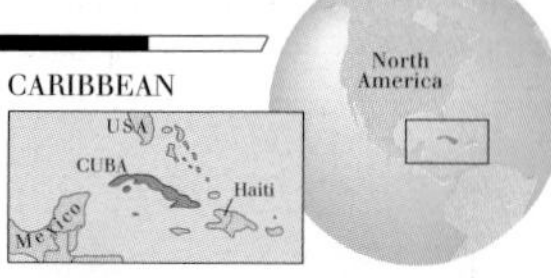
CARIBBEAN
North America
USA
CUBA
Haiti
Mexico

OFFICIAL NAME: Republic of Cuba **CAPITAL:** Havana
POPULATION: 11.2 million **CURRENCY:** Cuban peso **OFFICIAL LANGUAGE:** Spanish

C

1902

1902
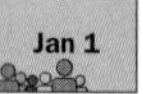
Jan 1

C

-5

+53

.cu

THE CARIBBEAN'S LARGEST ISLAND, Cuba has widely cultivated lowlands which fall between three mountainous areas. The fertile soil of the lowlands supports the sugarcane, rice, and coffee plantations. Sugar, the country's major export, suffers from underinvestment, low yields, and fluctuating world prices. A former Spanish colony, Cuba in 1959 became the only communist state in the Americas. In 1962, the deployment of Soviet nuclear missiles on the island shocked the US and brought the two superpowers close to war. Veteran president Fidel Castro is still very much in control, but, since the collapse of the USSR, the US sees Cuba as less of a threat.

Valle de Viñales, Pinar del Río province. Cuba's undulating countryside is ideal for growing the main export crop, sugar.

CLIMATE

Tropical oceanic

WEATHER CHART FOR HAVANA

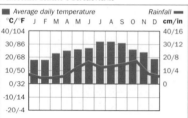

Average daily temperature — Rainfall
°C/°F J F M A M J J A S O N D cm/in
40/104, 30/86, 20/68, 10/50, 0/32, -10/14, -20/-4
40/16, 30/12, 20/8, 10/4, 0

Cuba's subtropical climate is hot all year round and very hot in the summer. Rainfall is heaviest in the mountains, which receive up to 250 cm (98 in) a year. Generally, the north is wetter than the south; the Guantánamo area receives only 20 cm (8 in) of rainfall annually. In winter, the west is affected sometimes by cold air from the US, but only for a day or two at a time.

TRANSPORTATION

Drive on right

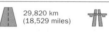

José Martí, Havana 2.5m passengers

99 ships 120,000 grt

THE TRANSPORTATION NETWORK

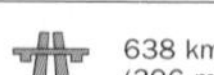

29,820 km (18,529 miles)

638 km (396 miles)

14,331 km (8905 miles)

240 km (149 miles)

Public transportation in Cuba has been extremely cheap, although fuel shortages have made it increasingly erratic and unreliable. Cubans rely mostly on traditional black bicycles, imported by the thousand from China. Havana owes much of its charm to the number of 50-year-old Chevrolets and Oldsmobiles still being driven around. This is another result of sanctions, but keeps the many inventive local spare-parts workshops in business.

TOURISM

Visitors : Population 1:6.4

1.74m visitors

No change in 2001

MAIN TOURIST ARRIVALS

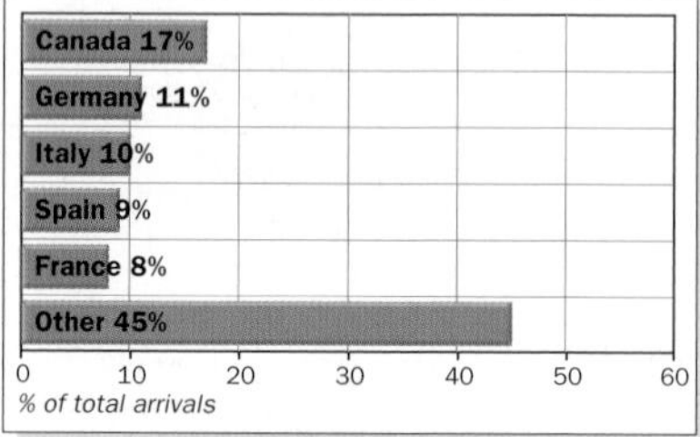
Canada 17%
Germany 11%
Italy 10%
Spain 9%
France 8%
Other 45%
0 10 20 30 40 50 60
% of total arrivals

Tourism began to develop after 1977 (when the US relaxed some travel restrictions), and Cuba is now among the Caribbean's most popular tourist destinations. Tourism has supplanted sugar as the most important motor of the economy and largest generator of foreign exchange. Official estimates are that the number of arrivals will rise to more than five million by 2010. The government seeks to promote family tourism by cracking down on prostitutes who target Havana's main hotels.

Guanabo, 25 km east of Havana, is a low-key holiday resort favored by Cubans. The most modern cars in Cuba are imported, along with computers, in exchange for sugar in a special trading deal with Japan.

CUBA

Total Land Area : 110 860 sq. km (42 803 sq. miles)

POPULATION
- over 1 000 000
- over 500 000
- over 100 000
- over 50 000
- over 10 000
- under 10 000

LAND HEIGHT
- 1000m/3281ft
- 500m/1640ft
- 200m/656ft
- Sea Level

0 50 km
0 50 miles

C

PEOPLE

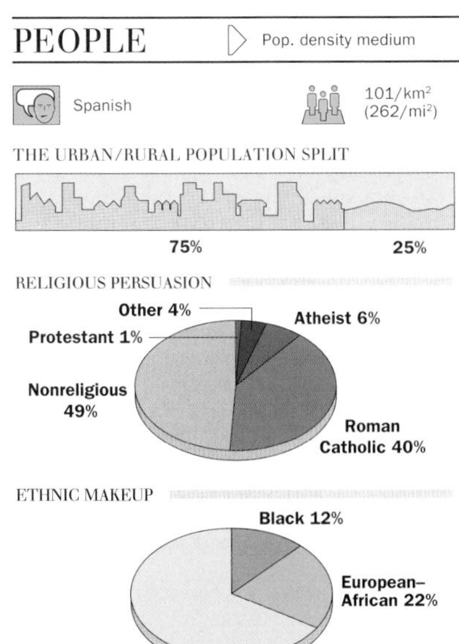
Pop. density medium
Spanish
101/km² (262/mi²)
THE URBAN/RURAL POPULATION SPLIT
75%
25%
RELIGIOUS PERSUASION
Other 4%
Protestant 1%
Atheist 6%
Nonreligious 49%
Roman Catholic 40%
ETHNIC MAKEUP
Black 12%
European–African 22%
White 66%

Ethnic tension in Cuba is minimal. About 70% of Cubans are of Spanish descent, mainly from the settlers, but also from the more recent influx of exiles from Franco's Spain. The black population is descended from the slaves and migrants from neighboring states, in particular Jamaica.

Living standards in Cuba fell dramatically in the early 1990s after the collapse of the east European communist bloc, previously its main trading partner, and rationing for most basic foodstuffs was subsequently introduced. The "dollarization" of the economy in recent years has led to great divisions between those who survive on pesos and the more than 50% of the population who have access to dollars. Since the early 1990s the number of those trying to leave, legally or otherwise, has risen markedly.

An increasing number of women are playing a prominent role in politics, the professions, and the armed forces. Child care facilities are widespread.

POPULATION AGE BREAKDOWN

Female	Age	Male
0.5%	80+	0.4%
5.9%	60–79	5.7%
11.1%	40–59	10.9%
18%	20–39	18.2%
14.3%	0–19	15%

% of population by age group

WORLD AFFAIRS

Joined UN in 1945

ACP    
IAEA
SELA
NAM
ACS

Since the 1959 revolution, and particularly after the 1962 stand-off over Soviet missiles, the US has considered Cuba a danger. The US trade blockade, first imposed in 1961, has left Cuba economically isolated despite regular votes in the UN condemning sanctions. The end of Soviet aid in the 1990s was another serious blow to Castro's embattled regime, but it has forced Cuba to soften its anti-Western stance in the search for alternative aid. Partly as a result, the US embargo has been progressively loosened since 1999 and now allows for more flights, direct mail, essential medicines, and more food imports.

Ties to Russia have been diluted further but relations with the US remain fraught, despite the arrival in 2001 of the first direct trade between the two countries' governments, in the form of emergency aid. In 2002 the US included Cuba in its "axis of evil" terrorist-sponsoring states. Relations with Europe are now improving, after being dented over the treatment of dissidents.

POLITICS

No multiparty elections

1998/2003

President Fidel Castro Ruz

AT THE LAST ELECTION

National Assembly of the People's Power 601 seats

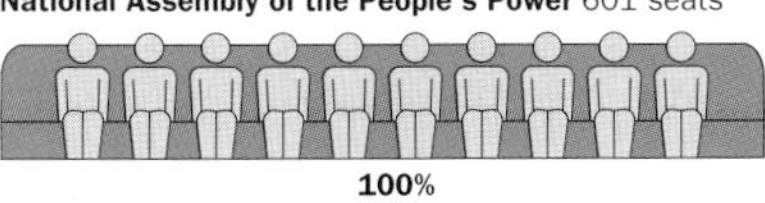

100% PCC

PCC = Cuban Communist Party

Fidel Castro, who has led Cuba since 1959, founded its one-party communist system set out in the 1976 constitution.

Profile

The 1959 popular revolution, led by Castro, toppled the corrupt Batista dictatorship and launched a far-reaching program of social, economic, and political reforms.

In the 1990s the revolution seemed under siege in the wake of the collapse of the Soviet Union and tightened trade sanctions by the US. Supporters continue to see Cuba as living proof of the triumph of socialist development over adversity, but critics offer the view that the Castro administration is an intolerant dictatorship.

Main Political Issues

The succession

The aging Castro remains firmly in place, but his successor is a major issue. Some observers predict that a younger, reform-minded leadership would normalize relations with the US and steer Cuba toward Western-style democracy. Others warn that Castro's departure will lead to a power vacuum, vulnerable to social unrest and internal divisions between reformers and communist hard-liners.

Democracy

The 1999 trial of four moderate dissidents was part of a tough clampdown on internal opposition. This has reduced hopes for the type of opening seen in the economic sphere and threatens diplomatic and economic ties. It offers the US justification for its trade embargo.

***Raúl Castro,** brother of Fidel and the minister of defense.*

***Fidel Castro,** Cuba's leader since 1959. The US is keen to oust his regime.*

N
Moa
Baracoa
El Salvador
Guantánamo
GUANTANAMO BAY (to US)
Windward Passage

AID

Recipient

$44m (receipts)

Down 25% in 2000

Spain, France, and UNICEF have given aid, and China loaned $400 million in 2001. A cash purchase of food and supplies from the US was made after Hurricane Michelle in late 2001.

CHRONOLOGY

Originally inhabited by the Arawak people, Cuba was claimed for Spain by Columbus in 1492. Development of the sugar industry from the 18th century, using imported slave labor, made Cuba the world's third-largest producer by 1860.

- ❑ **1868** End of the slave trade.
- ❑ **1868–1878** Ten Years' War for independence from Spain.
- ❑ **1895** Second war of independence. Thousands die in Spanish concentration camps.
- ❑ **1898** In support of Cuban rebels US declares war on Spain to protect strong American financial interests in Cuba.
- ❑ **1899** US takes Cuba and installs military interim government.
- ❑ **1901** US is granted intervention rights and military bases, including Guantánamo Bay naval base.

⇨

C

CHRONOLOGY *continued*

- **1902** Tomás Estrada Palma takes over as first Cuban president. US leaves Cuba, but intervenes in 1906–1909 and 1919–1924.
- **1909** Liberal presidency of José Miguel Goméz. Economy prospers; US investment in tourism, gambling, and sugar.
- **1925–1933** Dictatorship of President Gerardo Machado.
- **1933** Years of guerrilla activity end in revolution. Sgt. Fulgencio Batista takes over; military dictatorship.
- **1955** Fidel Castro exiled after two years' imprisonment for subversion.
- **1956–1958** Castro returns to lead a guerrilla war in the Sierra Maestra.
- **1959** Batista flees. Castro takes over. Wholesale nationalizations; Cuba reorganized on Soviet model.
- **1960** US breaks off relations.
- **1961** US-backed invasion of Bay of Pigs by anti-Castro Cubans fails. Cuba declares itself Marxist–Leninist. US economic and political blockade.
- **1962** Missile crisis: Soviet deployment of nuclear weapons in Cuba leads to extreme Soviet–US tension; war averted by Khrushchev ordering withdrawal of weapons.
- **1965** One-party state formalized.
- **1972** Cuba joins COMECON (communist economic bloc).
- **1976** New socialist constitution. Cuban troops in Angola until 1991.
- **1977** Sends troops to Ethiopia.
- **1980** 125,000 Cubans, including "undesirables," flee to US.
- **1982** US tightens sanctions and bans flights and tourism to Cuba.
- **1983** US invasion of Grenada. Cuba involved in clashes with US forces.
- **1984** Agreement with US on Cuban emigration and repatriation of "undesirables" is short-lived.
- **1986** Soviet-style *glasnost* rejected.
- **1988** UN's second veto of US attempt to accuse Cuba of human rights violations. Diplomatic relations established with EC.
- **1989** Senior military executed for arms and narcotics smuggling.
- **1991** Preferential trade agreement with USSR ends. Severe rationing.
- **1992–1993** US tightens blockade. All former Soviet military leave.
- **1994–1995** Economic reforms to boost foreign trade and investment.
- **1996** US Helms-Burton Act tightens sanctions.
- **1998** Visit of Pope John Paul II.
- **1999** Leading moderate dissidents put on trial.
- **2001** Devastation by Hurricane Michelle.
- **2002** Guantánamo Bay used as high-security prison for captives from US "war on terrorism."

DEFENSE

Compulsory military service

 $735m

  Down 2% in 2000

CUBAN ARMED FORCES

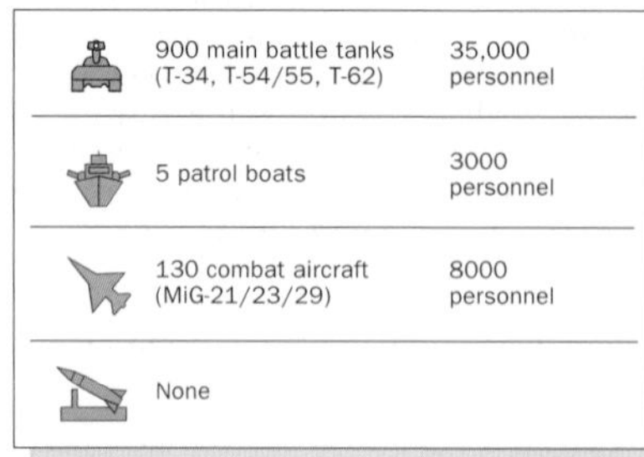

900 main battle tanks (T-34, T-54/55, T-62)	35,000 personnel
5 patrol boats	3000 personnel
130 combat aircraft (MiG-21/23/29)	8000 personnel
None	

From 1959 to the 1980s, Cuba's efficient military, well represented in the Council of Ministers and the Politburo, was one of the achievements of the revolution. Under Castro's brother Raúl, it succeeded in repelling the US-sponsored Bay of Pigs invasion in 1961, and saw effective action in Africa in the 1970s, preventing South Africa from taking control of Angola, and Somalia from occupying the Ogaden region in Ethiopia.

Today, with communist regimes collapsed around the world, it has lost much of its prestige. Russia is still the main source of arms. A siege mentality associated with the US economic embargo keeps the military on the alert for perceived internal and external threats.

ECONOMICS

Inflation 1.1% p.a. (1990–1999)

$18.3bn

21.00 Cuban pesos

SCORE CARD

- WORLD GNP RANKING..........................70th
- GNP PER CAPITA $1650
- BALANCE OF PAYMENTS ....In deficit
- INFLATION 0.3%
- UNEMPLOYMENT 6%

EXPORTS

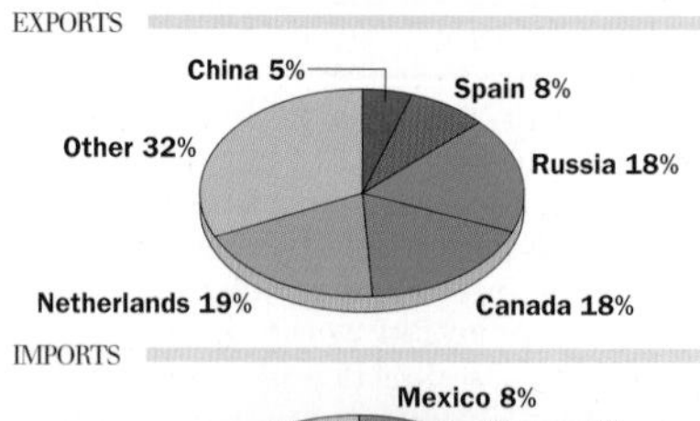

China 5%
Spain 8%
Other 32%
Russia 18%
Netherlands 19%
Canada 18%

IMPORTS

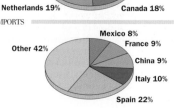

Mexico 8%
France 9%
Other 42%
China 9%
Italy 10%
Spain 22%

ECONOMIC PERFORMANCE INDICATOR

Consumer Price Index
GDP
160
140
120
100
80
Consumer price index unavailable
GDP 1994=100
1994
1995
1996
1997
1998

STRENGTHS
Buoyant tourism attracts strong foreign investment. Major exporter of sugar and nickel. Premium Cuban cigars. Strengthening banking sector.

WEAKNESSES
Denied a major market and investment capital by US trade embargo. Acute shortage of hard currency. Vulnerability of sugar and nickel to world price fluctuations. Difficult terms of trade and weak legal framework deter investment. Deficient infrastructure. Shortages of fuel, fertilizers, spare parts, and other inputs. Severe hurricane damage in 2001.

PROFILE
After a brief flirtation with the open market, total state control was reimposed in 1986. The collapse of the USSR meant the loss of some $5 billion in annual aid and led to a deep recession in the early 1990s. A cautious adoption of some capitalist-style reforms in the mid-1990s, including the free use of the US dollar, stimulated the growth of a dollarized sector centered on tourism, which has attracted strong foreign investment. Tourism rather than sugar now dominates the economy, and benefits some 160,000 self-employed and small businesses. Foreign companies are also involved in joint ventures in banking and the oil and gas sectors. There is a very large informal sector. The government lost significant revenue from 2001 when Russia terminated the lease of its information-gathering center.

CUBA : MAJOR BUSINESSES

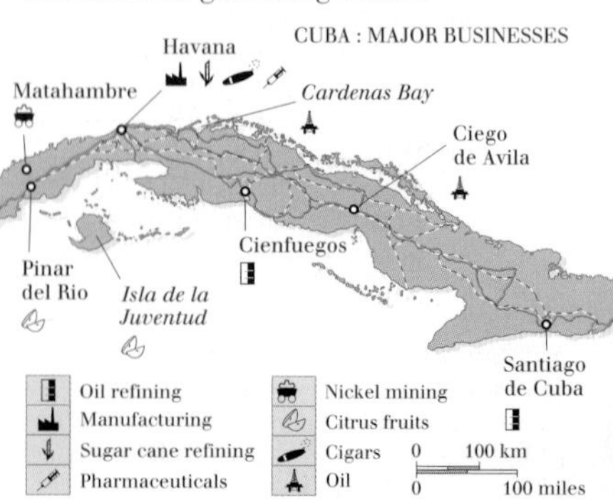

Havana
Matahambre
Cardenas Bay
Ciego de Avila
Cienfuegos
Pinar del Rio
Isla de la Juventud
Santiago de Cuba
Oil refining
Manufacturing
Sugar cane refining
Pharmaceuticals
Nickel mining
Citrus fruits
Cigars
Oil
0
100 km
0
100 miles

RESOURCES

Electric power 4m kw

122,425 tonnes

37,473 b/d (reserves 123m barrels)

4.4m cattle, 2.7m pigs, 13.3m chickens

Iron, nickel, cobalt, chromite, gold, manganese, oil

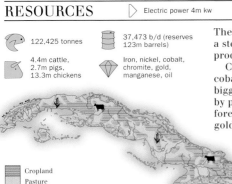

Cropland
Pasture
Forest
Wetlands
Sugarcane – cash crop
Cattle

CUBA : LAND USE
0 100 km
0 100 miles

ELECTRICITY GENERATION

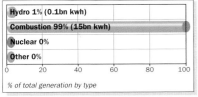

Hydro 1% (0.1bn kwh)
Combustion 99% (15bn kwh)
Nuclear 0%
Other 0%

0 20 40 60 80 100

% of total generation by type

The collapse of the USSR precipitated a steep decline in demand for sugar; production hit a 50-year low in 1998.

Cuba seeks to expand nickel and cobalt production, traditionally its biggest merchandise exports, assisted by private mining ventures. Several foreign companies are prospecting for gold, silver, and other metals, and for heavy crude oil and gas, through concessions. Work at Juraguá on a Russian-built nuclear reactor was abandoned in December 2000.

ENVIRONMENT

Sustainability rank: 58th

17% (2% partially protected)

2.2 tonnes per capita

ENVIRONMENTAL TREATIES

Yes
Yes
Yes
Yes
Yes
Yes

At the time of the revolution in 1959, only 14% of the country's forest cover remained, but a strong drive to replant has raised the tree cover level to over 20%. The intensive use of irrigation without adequate drainage has caused salinization and waterlogging. There is regional concern about the never-completed nuclear reactor at Juraguá.

MEDIA

TV ownership medium

Daily newspaper circulation 118 per 1000 people

PUBLISHING AND BROADCAST MEDIA

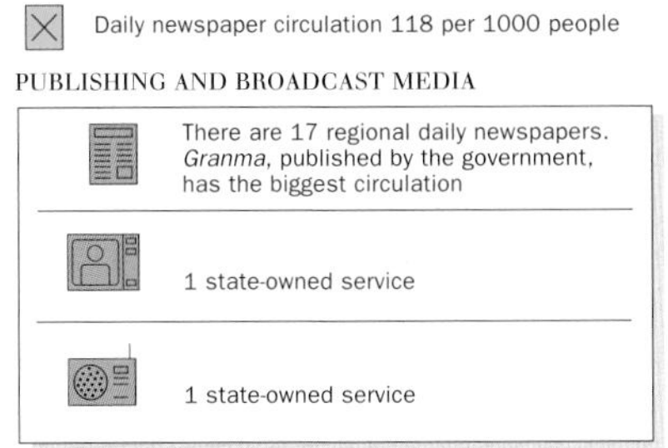
There are 17 regional daily newspapers. *Granma*, published by the government, has the biggest circulation
1 state-owned service
1 state-owned service

A catch-all anticrime law restricts and penalizes investigative reporting by independent journalists which is judged to be assisting the US foreign policy against Cuba.

CRIME

Death penalty in use

33,000 prisoners

Crime is rising

CRIME RATES

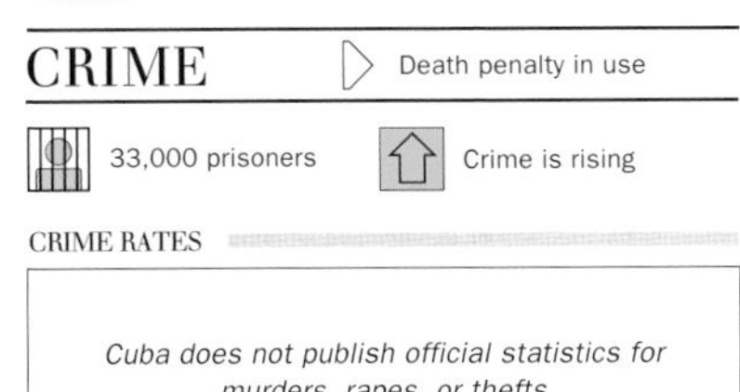
Cuba does not publish official statistics for murders, rapes, or thefts

Violent crime is officially viewed as a threat to national stability. In 1999 the penal code was amended to extend the death penalty to robbery involving firearms, certain narcotics offenses, attacks on security officers, and sexual corruption of minors. Widening inequality has led to growing crime, especially theft from state factories. The US base at Guantánamo Bay houses a prison for suspected members of the al-Qaida terrorist network.

EDUCATION

School leaving age: 15

97%

116,700 students

THE EDUCATION SYSTEM

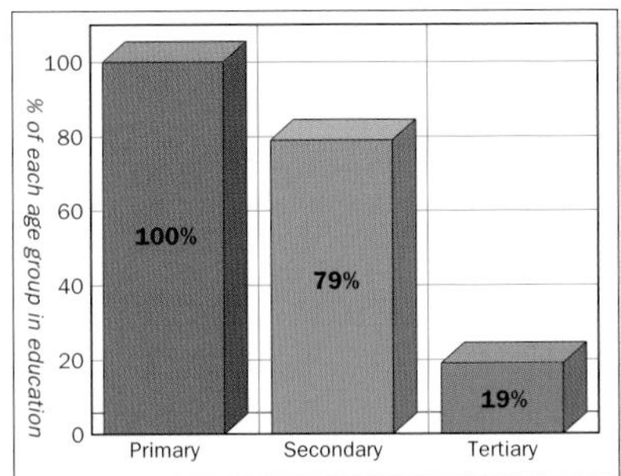
100
80
60
40
20
0
% of each age group in education
100%
79%
19%
Primary
Secondary
Tertiary

Education, which is universal and free at all levels, combines academic with manual work, in line with Marxist–Leninist principles. The importance given to education under Castro, which is reflected in the high literacy rate, is now being promoted to attract foreign investment in high-tech industries, particularly biotechnology. Spending on education was increased in the late 1990s to around 11.5% of the budget.

HEALTH

Welfare state health benefits

1 per 189 people

Heart & cerebrovascular diseases, cancers, nutritional disorders

Spending on health accounted for around 10% of government expenditure in the late 1990s. Average life expectancy in Cuba is among the highest in Latin America, which is a reflection of its efficient, countrywide health service. The US trade embargo has led to shortages of hospital equipment and raw materials for drugs. The latter are normally supplied by Havana's sizable pharmaceuticals industry. Cuba's advanced surgery techniques attract patients from overseas.

SPENDING

GDP/cap. increase

CONSUMPTION AND SPENDING

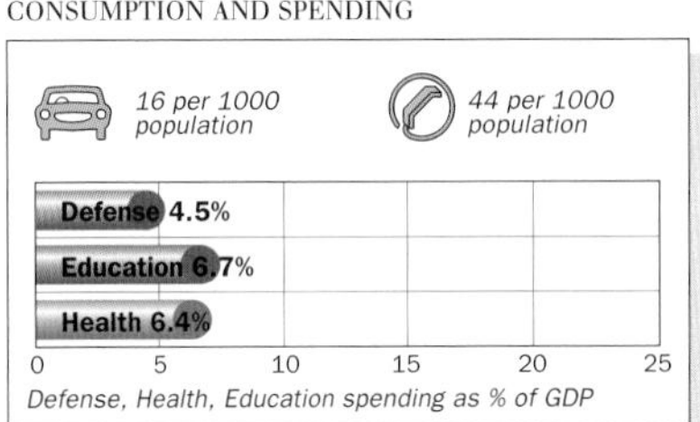
16 per 1000 population
44 per 1000 population
Defense 4.5%
Education 6.7%
Health 6.4%
0
5
10
15
20
25
Defense, Health, Education spending as % of GDP

Under Batista there were huge wealth disparities, and Cuba was a playground for the rich. The 1959 revolution succeeded in reducing the disparities, partly by taking over all businesses, from oil companies to barbers' shops, and partly by prescribing not only minimum but also maximum wages. Economic regulations have varied since then; for a brief period in 1985, different wage rates were allowed in an attempt to provide incentives for hard workers, but this decision was reversed in 1986. Economic liberalization in the mid-1990s has created a large gulf between that half of the population with access to US dollars and those left in the peso economy who have to subsist on lower salaries.

WORLD RANKING

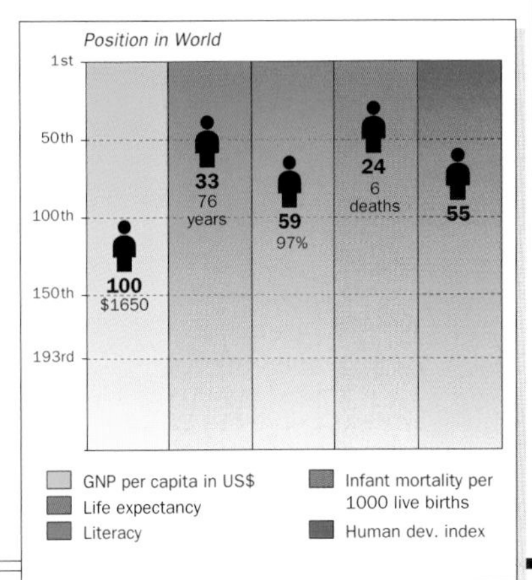
Position in World
1st
50th
100th
150th
193rd
100
$1650
33
76 years
59
97%
24
6 deaths
55
GNP per capita in US$
Life expectancy
Literacy
Infant mortality per 1000 live births
Human dev. index

C

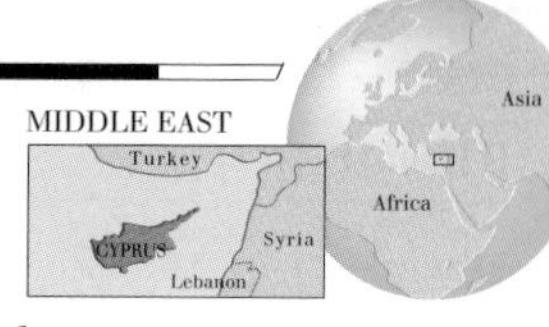
MIDDLE EAST
Turkey
CYPRUS
Syria
Lebanon
Asia
Africa

CYPRUS

OFFICIAL NAME: Republic of Cyprus **CAPITAL:** Nicosia **POPULATION:** 790,000
CURRENCY: Cyprus pound (Turkish lira in TRNC) **OFFICIAL LANGUAGES:** Greek and Turkish

C

1960

1960
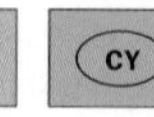
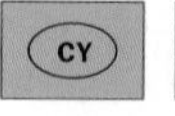
Oct 1
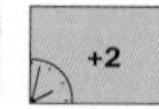
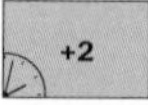
CY

+2

+357

.cy

THE ISLAND OF Cyprus, which rises from a central plateau to a high point at Mount Olympus, lies south of Turkey in the eastern Mediterranean. It was partitioned in 1974, following an invasion by Turkish troops. The south of the island is the Greek Cypriot Republic of Cyprus (Cyprus); the self-proclaimed Turkish Republic of Northern Cyprus (TRNC) is recognized only by Turkey.

CLIMATE

Mediterranean

WEATHER CHART FOR NICOSIA

Average daily temperature / Rainfall

°C/°F J F M A M J J A S O N D cm/in

40/104, 30/86, 20/68, 10/50, 0/32, -10/14, -20/4 — 40/16, 30/12, 20/8, 10/4, 0

The climate is typically Mediterranean: summers are hot and dry and winters mild, though there is mountain snow.

TRANSPORTATION

Drive on left

Larnaca 4.88m passengers

1475 ships 23.2m grt

THE TRANSPORTATION NETWORK

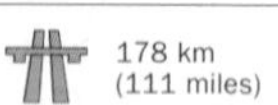
6249 km (3883 miles)

178 km (111 miles)

None

None

Travel between the two zones is impeded. The south regards the airport at Ercan as an illegal point of entry.

TOURISM

Visitors : Population 3.4:1

2.7m visitors

Up 10% in 2000

MAIN TOURIST ARRIVALS

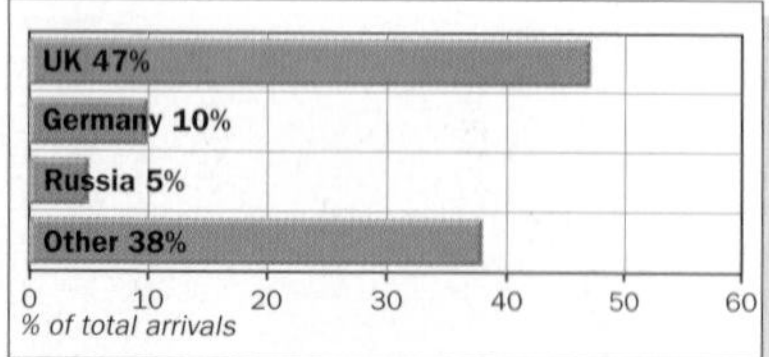
UK 47%
Germany 10%
Russia 5%
Other 38%
0 10 20 30 40 50 60
% of total arrivals

Tourism expanded rapidly in the 1980s in the south, and more recently in the north. A ten-year plan aims to double the number of visitors by 2010. Tourists come for beaches, archeology, or the abundant wildlife, notably on the Akamas peninsula and the Troodos Mountains.

PEOPLE

Pop. density medium

Greek, Turkish

85/km² (221/mi²)

THE URBAN/RURAL POPULATION SPLIT

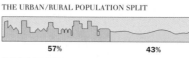

57% 43%

ETHNIC MAKEUP

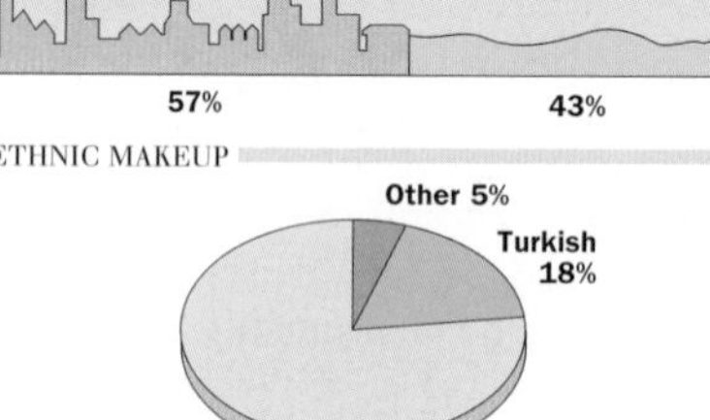
Other 5%
Turkish 18%
Greek 77%

Cyprus's Greek majority are Orthodox Christian. The Turkish minority are Muslim. Most Turkish Cypriots are the descendants of Turks who settled on the island from the 16th century, while it was under the rule of the Ottoman empire. Both Cypriot communities have suffered great upheavals: in 1974 the island was partitioned following the Turkish invasion, and 200,000 Greek Cypriots were forced to flee to the south, while 65,000 Turkish Cypriots fled in the other direction. Northern Cyprus is officially recognized as a political entity only by Turkey. Thousands of mainland Turks have settled there.

East European contract labor is brought in to staff hotels in the south, where wage levels are on average three times higher than in the north. Unemployment levels in the north, meanwhile, are rising.

***The 2nd-century theater** at Curium, 14 km (19 miles) west of Limassol. Curium was the site of a flourishing Mycenaean colony before 1100 BCE.*

POLITICS

Multiparty elections

2001/2006 Cyprus
1998/2003 TRNC

President Glafcos Clerides (Cyprus)
President Rauf Denktash (TRNC)

AT THE LAST ELECTION

House of Representatives 80 seats

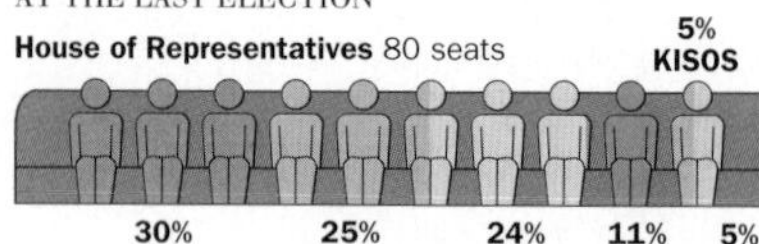
5% KISOS
30% TC
25% AKEL
24% DISY
11% DIKO
5% Others

TC = Reserved for Turkish Cypriots **AKEL** = Progressive Party of the Working People **DISY** = Democratic Rally **DIKO** = Democratic Party **KISOS** = Movement of Social Democrats

The 24 seats reserved for Turkish Cypriots have not been occupied since December 1963

Legislative Assembly (TRNC) 50 seats

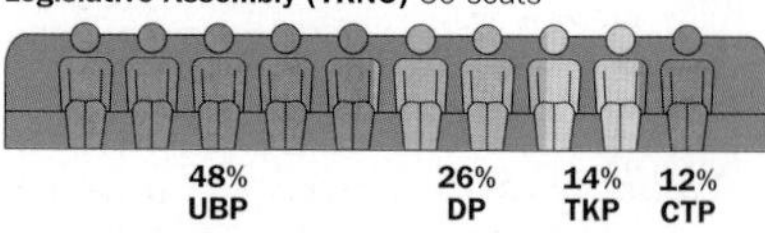
48% UBP
26% DP
14% TKP
12% CTP

UBP = National Unity Party **DP** = Democrat Party
TKP = Social Welfare Party **CTP** = Republican Turkish Party

The UN-backed proposal of a two-zone federation for Cyprus would give each community its own territory while sharing a number of government functions and ministries. TRNC president Rauf Denktash, mindful of the Greek Cypriots' repression of the Turks prior to 1974, is unwilling to accept a plan that does not ensure full sovereignty and political equality for Turks. Greek Cypriots, in turn, fear the plan would give too much influence over their affairs to the small Turkish minority, who would be able to veto all government decisions. Fresh impetus was given to talks in 2002 by the imminence of EU membership for Cyprus.

WORLD AFFAIRS

Joined UN in 1960

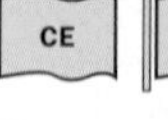
CE

Comm

IBRD

NAM

OSCE
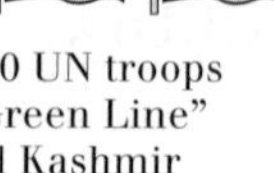

The presence of over 1000 UN troops since 1974 staffing the "Green Line" (only the Middle East and Kashmir have longer-standing UN peacekeeping forces) costs $43 million a year. Only Turkey recognizes the TRNC.

Cyprus applied to join the EU in 1990, and formal negotiations opened in 1998.

AID

Recipient

$54m (receipts) (Cyprus)

Up 2% in 2000 (Cyprus)

Cyprus receives aid from international agencies, as well as from the EU and individual countries such as the UK. The TRNC is dependent on aid from Turkey of more than $60 million a year.

CYPRUS

Total Land Area : 9250 sq. km (3571 sq. miles)

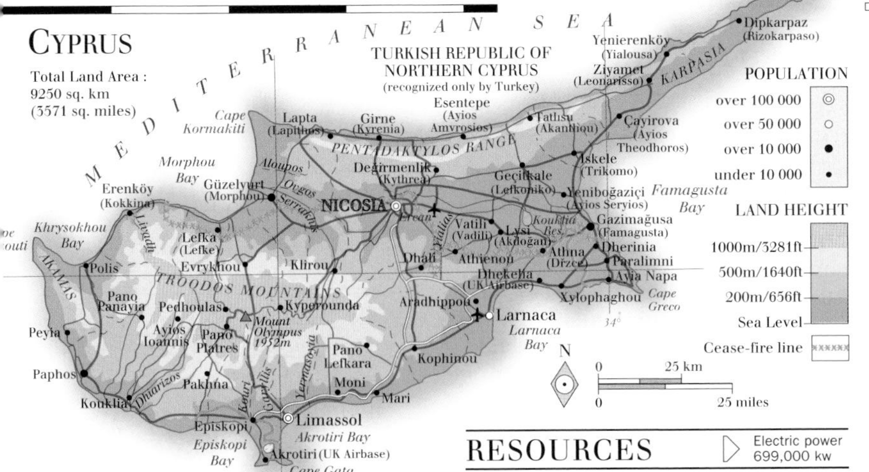
MEDITERRANEAN SEA
TURKISH REPUBLIC OF NORTHERN CYPRUS
(recognized only by Turkey)
Cape Apostolos Andreas
Dipkarpaz (Rizokarpaso)
Yenierenköy (Yialousa)
Ziyamet (Leonarisso)
KARPASIA
Esentepe (Ayios Amvrosios)
Tatlısu (Akanthou)
Çayirova (Ayios Theodhoros)
Cape Kormakiti
Lapta (Lapithos)
Girne (Kyrenia)
PENTADAKTYLOS RANGE
Değirmenlik (Kythrea)
İskele (Trikomo)
Geçitkale (Lefkoniko)
Yeniboğaziçi (Ayios Seryios)
Famagusta Bay
Morphou Bay
Erenköy (Kokkina)
Güzelyurt (Morphou)
NICOSIA
Gazimağusa (Famagusta)
Vatili (Vadili)
Lysi (Akdogan)
Athna (Dizce)
Dherinia
Paralimni
Ayia Napa
Cape Greco
Khrysokhou Bay
Lefka (Lefke)
Evrykhou
Polis
Klirou
Dhali
Athienou
Dhekelia (UK Airbase)
Xylophaghou
AKAMAS
TROODOS MOUNTAINS
Pano Panayia
Pedhoulas
Kyperounda
Mount Olympus 1952m
Aradhippou
Larnaca
Larnaca Bay
Peyia
Ayios Ioannis
Pano Platres
Pano Lefkara
Kophinou
Paphos
Pakhna
Moni
Mari
Kouklia
Episkopi
Episkopi Bay
Limassol
Akrotiri Bay
Akrotiri (UK Airbase)
Cape Gata
POPULATION
over 100 000
over 50 000
over 10 000
under 10 000
LAND HEIGHT
1000m/3281ft
500m/1640ft
200m/656ft
Sea Level
Cease-fire line
0 25 km
0 25 miles

CHRONOLOGY

Cyprus was dominated, in turn, by Egypt, Greece, the Byzantines, the Ottomans, and the UK, before independence in 1960.

- ❑ **1963** Turkish Cypriots abandon parliament.
- ❑ **1974** President Makarios deposed by Greek military junta. Turkey invades. Partition.
- ❑ **1983** Self-proclamation of TRNC.
- ❑ **1998** Talks on EU membership start, following 1990 application to join.
- ❑ **2001** President Clerides visits TRNC.

DEFENSE

Compulsory military service

$453m (Cyprus)

Up 28% in 2000 (Cyprus)

In addition to two sovereign British bases and UN forces, there are 36,000 Turkish troops in northern Cyprus and 1250 Greek troops in the buffer zone. The 10,000-strong Greek Cypriot army and the 5000-strong Turkish Cypriot army both rely heavily on conscripts.

ECONOMICS

Inflation 3.5% p.a. (1990–2000)

$9.36bn

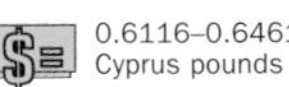

0.6116–0.6461 Cyprus pounds

SCORE CARD

- ❑ WORLD GNP RANKING 86th
- ❑ GNP PER CAPITA $12,370
- ❑ BALANCE OF PAYMENTS .... –$456m
- ❑ INFLATION 4.1%
- ❑ UNEMPLOYMENT 4%

STRENGTHS

Booming tourism industry, accounting for over 20% of GDP. Manufacturing sector and provision of services to Middle Eastern countries.

WEAKNESSES

Pressure for tighter supervision of offshore finance and crackdown on tax evasion. Limited liberalization. TRNC starved of foreign investment.

EXPORTS

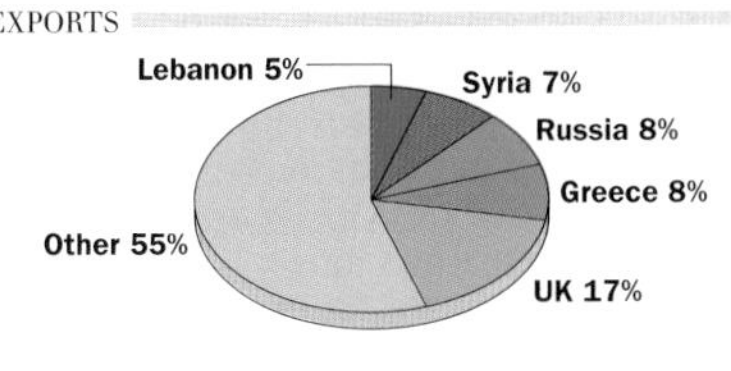
Lebanon 5%
Syria 7%
Russia 8%
Greece 8%
UK 17%
Other 55%

IMPORTS

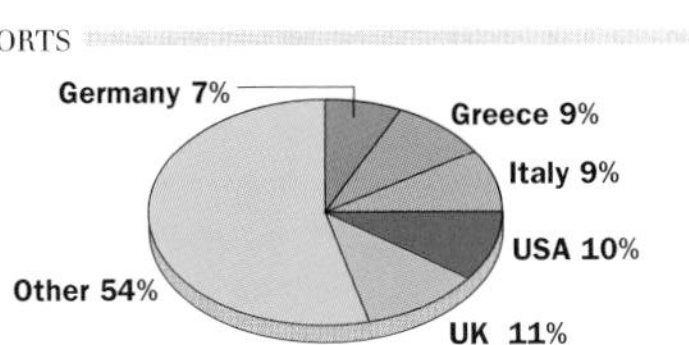
Germany 7%
Greece 9%
Italy 9%
USA 10%
UK 11%
Other 54%

RESOURCES

Electric power 699,000 kw

6695 tonnes

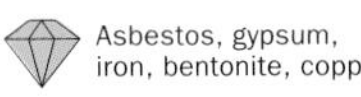

Not an oil producer

418,500 pigs, 378,600 goats, 3.2m chickens

Asbestos, gypsum, iron, bentonite, copper

Cyprus continues to supply electricity to TRNC, but has not been paid for this. The possibility of offshore oil and gas to the south has attracted interest. Water is precious but new desalinization plants have reduced shortages.

ENVIRONMENT

Not available

0.2% partially protected

7.9 tonnes per capita

Campaigners demand that the 155 sq. km (60 sq. miles) of the Akamas peninsula be fully protected from the threat of being sold for tourist development. Akamas is home to an unusual variety of plant and bird life, and contains breeding sites of the rare green turtle.

MEDIA

TV ownership high

Daily newspaper circulation 69 per 1000 people

PUBLISHING AND BROADCAST MEDIA

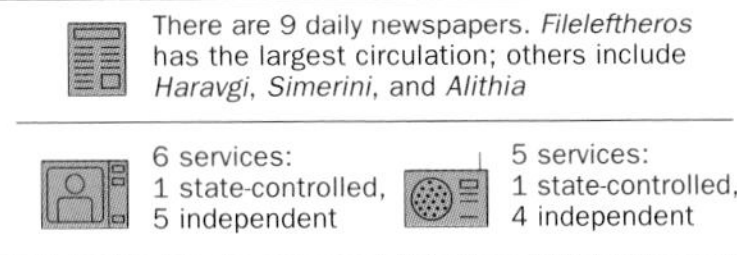

There are 9 daily newspapers. *Fileleftheros* has the largest circulation; others include *Haravgi*, *Simerini*, and *Alithia*

6 services: 1 state-controlled, 5 independent

5 services: 1 state-controlled, 4 independent

Cyprus's press is lively and tends to be highly politicized. The radio and TV services for British troops based in Cyprus are also popular.

CRIME

Death penalty not used in practice

247 prisoners

Down 6% 1997–1998

Crime rates are low and violence is rare. The unruly and sometimes violent behavior of foreign forces has on occasions led Cypriots to object to their presence. The arrest of a Cypriot politician during protests at a UK military base in 2001 provoked riots.

EDUCATION

School leaving age: 15

97%

25,163 students

Education is free and compulsory up to the age of 15. Many Greek Cypriots go abroad to university.

HEALTH

Welfare state health benefits

1 per 392 people

Heart diseases, accidents, cancers

Health care is more advanced in the south; sophisticated surgery is carried out at Lefkosia General Hospital.

SPENDING

GDP/cap. increase

CONSUMPTION AND SPENDING

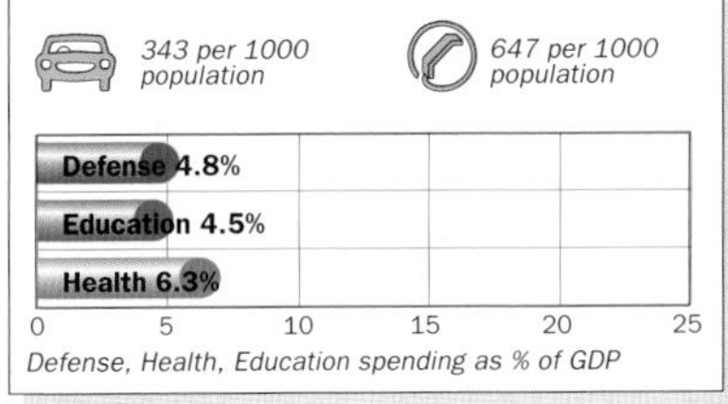
343 per 1000 population
647 per 1000 population
Defense 4.8%
Education 4.5%
Health 6.3%
0
5
10
15
20
25
Defense, Health, Education spending as % of GDP

The average income per capita in the southern part of Cyprus is higher than in Greece and Portugal, but slightly lower than that in Spain.

WORLD RANKING

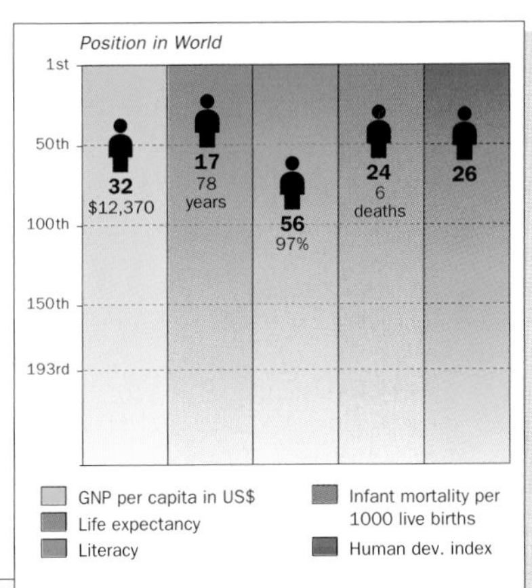
Position in World
1st
50th
100th
150th
193rd
32 $12,370
17 78 years
56 97%
24 6 deaths
26
GNP per capita in US$
Life expectancy
Literacy
Infant mortality per 1000 live births
Human dev. index

Czech Republic

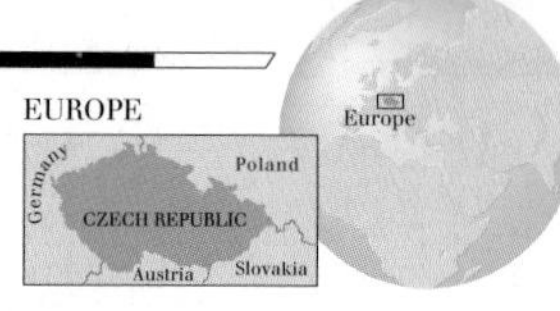
EUROPE
Europe
Germany
Poland
CZECH REPUBLIC
Austria
Slovakia

Official Name: Czech Republic **Capital:** Prague
Population: 10.3 million **Currency:** Czech koruna **Official Language:** Czech

C

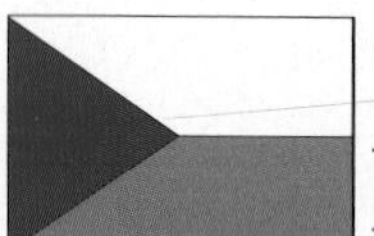

1993

1993
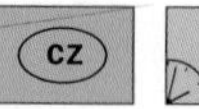
Oct 28
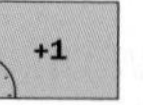
CZ

+1

+420

.cz

LANDLOCKED IN CENTRAL Europe, the Czech Republic comprises the territories of Bohemia and Moravia, and for most of the 20th century it was part of Czechoslovakia. In 1989, the "Velvet Revolution" ended four decades of communist rule, and free elections followed in 1990. In 1993 the Czech Republic and Slovakia peacefully dissolved their federal union to become two independent states.

CLIMATE

Continental

WEATHER CHART

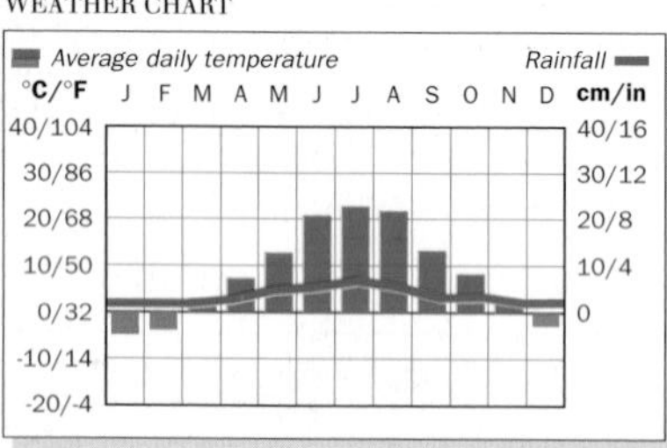
Average daily temperature
Rainfall
°C/°F
J F M A M J J A S O N D
cm/in
40/104
30/86
20/68
10/50
0/32
-10/14
-20/-4
40/16
30/12
20/8
10/4
0

The Czech climate is more moderate than that of Slovakia, though easterly winds bring low temperatures in winter.

TRANSPORTATION

Drive on right

Ruzyné, Prague
5.55m passengers

18 ships
228,000 grt

THE TRANSPORTATION NETWORK

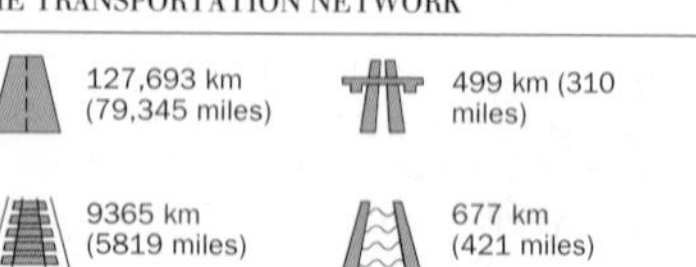
127,693 km (79,345 miles)
499 km (310 miles)
9365 km (5819 miles)
677 km (421 miles)

There are new expressways and rail links to Germany. Prague is a busy regional center for passenger air traffic.

TOURISM

Visitors : Population 1:2.2

4.67m visitors

Down 17% in 2000

MAIN TOURIST ARRIVALS

	% of total arrivals
Germany	41%
Poland	19%
Slovakia	11%
Other	29%

Revenues from tourism amount to nearly $4 billion a year and are an invaluable source of foreign earnings for the Czech economy. Germans are the most numerous among the millions of visiting tourists, mainly from Europe. Prague, which rivals Paris as the most beautiful capital in Europe, is still the main destination for visitors, although a growing proportion now seek other attractions such as spa towns and skiing in the Carpathian Mountains.

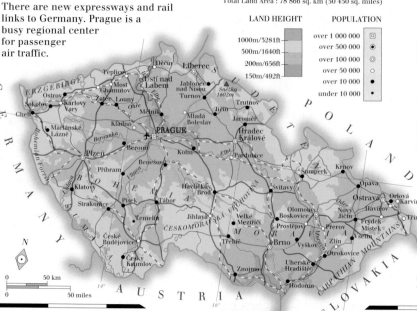

PEOPLE

Pop. density medium

Czech, Slovak, Hungarian

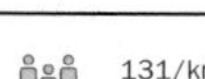
131/km² (338/mi²)

THE URBAN/RURAL POPULATION SPLIT

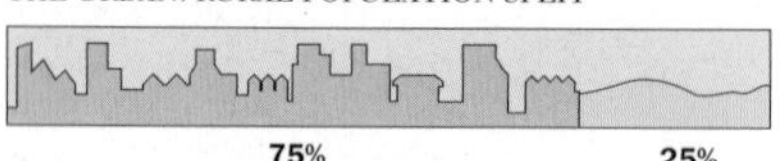
75%
25%

RELIGIOUS PERSUASION

- Protestant 3%
- Hussites 2%
- Other 18%
- Roman Catholic 39%
- Atheist 38%

Czechs make up over 80% of the population; Moravians are the next largest group. Some 300,000 Slovaks were left in the country after partition, and dual citizenship is now permitted. Ethnic tensions are few, except that the Roma community faces discrimination. Divorce rates are high.

POLITICS

Multiparty elections

L. House 2002/2006
U. House 2000/2002

President Vaclav Havel

AT THE LAST ELECTION

Chamber of Deputies 200 seats

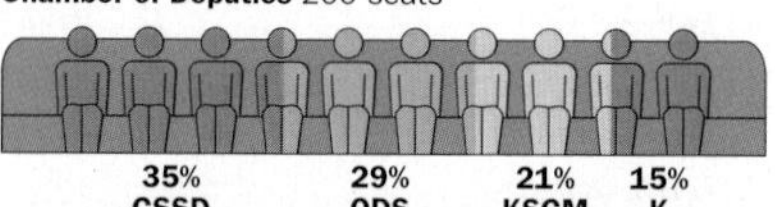
35% CSSD
29% ODS
21% KSCM
15% K

CSSD = Czech Social Democratic Party **ODS** = Civic Democratic Party **KSCM** = Communist Party of Bohemia and Moravia **K** = Coalition of the Christian Democratic Union–Czech People's Party (**KDU–CSL**) and the Freedom Union (**US**) **QC** = Quad Coalition, comprising the KDU–CSL, the US, the Civic Democratic Alliance (ODA), and the Democratic Union (DEU) **Ind** = Independents

Senate 81 seats

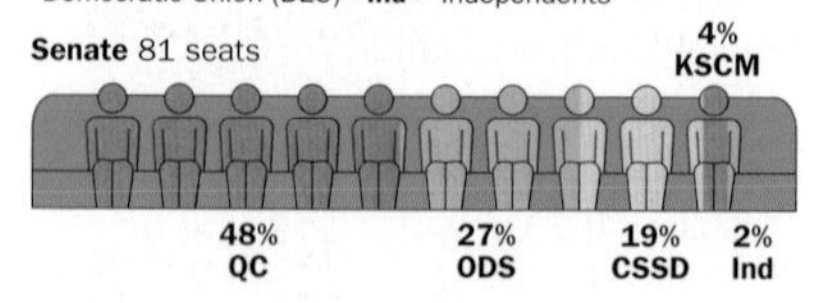
4% KSCM
48% QC
27% ODS
19% CSSD
2% Ind

The pro-democratic solidarity of 1989–1990, which saw the election of the Civic Forum and dissident playwright Vaclav Havel as president, soon gave way to a two-party system. The right-of-center ODS government pursued market economics, and oversaw the split with the Slovak Republic in 1993, but it lost its overall majority in 1996. It then gave tacit support to the social-democratic CSSD's minority government on the understanding that constitutional reforms would favor the two larger parties. The CSSD under Vladimir Spidla was able to form a majority government, without ODS backing, in 2002.

Czech Republic

Total Land Area : 78 866 sq. km (30 450 sq. miles)

LAND HEIGHT: 1000m/3281ft, 500m/1640ft, 200m/656ft, 150m/492ft

POPULATION: over 1 000 000, over 500 000, over 100 000, over 50 000, over 10 000, under 10 000

C

WORLD AFFAIRS

Joined UN in 1993

CE

CEFTA

NATO

OECD

OSCE

The Czech Republic began formal negotiations on EU membership in 1998. It joined NATO in 1999.

Relations with Germany are a priority. The issue of the forced expulsion of Germans in 1945 remains a stumbling block. Austria and Germany strongly opposed the opening of the Temelín nuclear plant in 2000.

AID

Recipient

 $438m (receipts)

 Up 35% in 2000

Aid for economic restructuring has been crucial for modernizing infrastructure such as telecommunications.

DEFENSE

Compulsory military service

 $1.13bn

Down 2% in 2000

The split with Slovakia left an oversized, expensive army. In 1994, plans to cut the military by 20,000 were approved. Professional soldiers with a communist past were the first to go. The Czech armaments industry has a long tradition based on precision engineering, and the Czech Republic is among the world's 20 largest arms exporters.

ECONOMICS

Inflation 12% p.a. (1990–2000)

$53.9bn

 37.63-35.56 Czech koruny

SCORE CARD

❑ World GNP Ranking	46th
❑ GNP per Capita	$5250
❑ Balance of Payments	–$2.34bn
❑ Inflation	3.9%
❑ Unemployment	9%

Strengths

Skilled industrial labor force. Good industrial base. Speed of privatization of state industries. Attractive to German investors. Draw of Prague for tourists.

Weaknesses

Lack of diversification in sectors usually attractive to overseas investors. Limited restructuring, banking sector problems. Pressure to cut government expenditure to reduce serious budget deficit.

EXPORTS

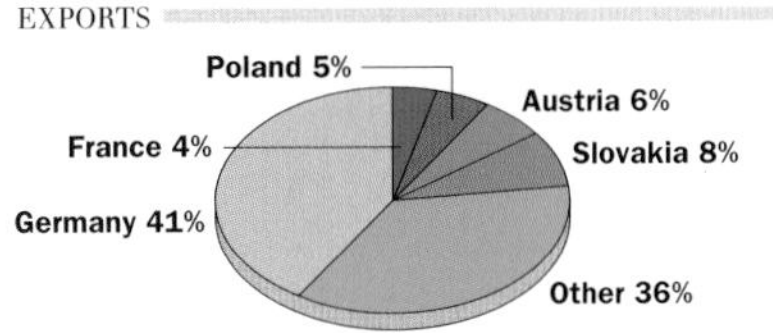
Poland 5%
Austria 6%
France 4%
Slovakia 8%
Germany 41%
Other 36%

IMPORTS

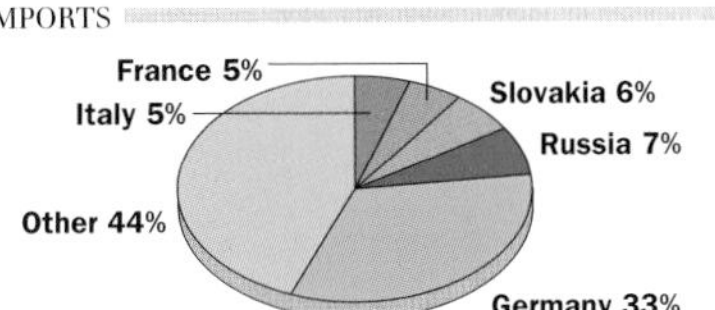
France 5%
Slovakia 6%
Italy 5%
Russia 7%
Other 44%
Germany 33%

RESOURCES

Electric power 13.9m kw

 22,965 tonnes

 3589 b/d (reserves 88m barrels)

 3.59m pigs, 1.58m cattle, 14.7m chickens

Oil, natural gas, copper, lead, zinc, coal, uranium

Copper, lead, zinc, and coal are the Czech Republic's chief resources. The government is aiming to phase out the worst-polluting coal-fired power plants. Opposition to a planned 2000 MW Soviet-designed nuclear power station at Temelín delayed its completion until late 2000.

ENVIRONMENT

Sustainability rank: 64th

 16%

 11.5 tonnes per capita

Pollution from the power, chemical, and cement industries and the new Temelín nuclear plant are key concerns.

MEDIA

TV ownership high

 Daily newspaper circulation 254 per 1000 people

PUBLISHING AND BROADCAST MEDIA

There are 21 daily newspapers. *Mladá Fronta Dnes* has the largest circulation

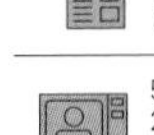 5 services: 2 state-owned, 3 independent

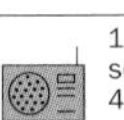 1 state-owned service, over 44 independent services

Government-influenced appointments of senior media officials provoked mass protests and a change in the law in 2001.

CRIME

No death penalty

 21,538 prisoners

 Little change in 1999

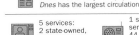

Prostitution is becoming a growing problem, especially in regions bordering Austria and Germany.

EDUCATION

School leaving age: 15

 99%

 233,865 students

Schooling has reverted to the pre-1945 system. Charles University in Prague was founded in the 13th century.

HEALTH

Welfare state health benefits

 1 per 333 people

Cancers, heart and cerebrovascular diseases, accidents

Health care expenditure as a share of GDP has increased by 30% since 1990. Wealthy Czechs travel to Germany for complex surgery.

The Vltava River in Prague. Millions of tourists, mainly from Europe and the US, visit Prague every year.

CHRONOLOGY

Formerly part of the Austro-Hungarian Empire, the Republic of Czechoslovakia was established in 1918. It was invaded by Hitler in 1939.

- ❑ **1968** "Prague Spring." Invasion by Warsaw Pact countries.
- ❑ **1989** "Velvet Revolution."
- ❑ **1990** Free elections won by Civic Forum; Vaclav Havel president.
- ❑ **1993** Split with Slovakia.
- ❑ **1998** Start of EU membership negotiations. Elections: CSSD forms minority government.
- ❑ **1999** Joins NATO.
- ❑ **2002** CSSD reelected.

SPENDING

GDP/cap. increase

CONSUMPTION AND SPENDING

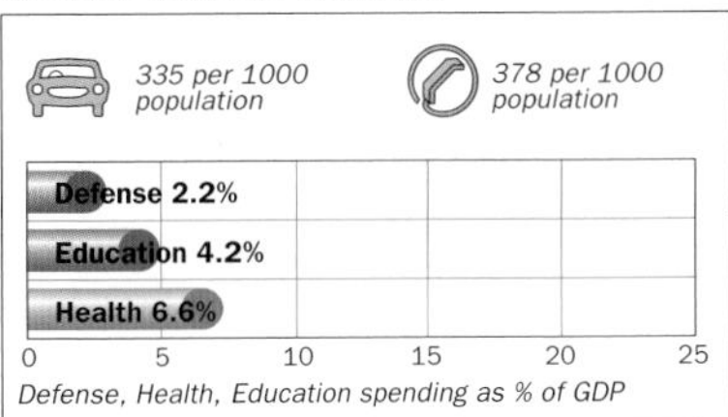
335 per 1000 population
378 per 1000 population
Defense 2.2%
Education 4.2%
Health 6.6%
0
5
10
15
20
25
Defense, Health, Education spending as % of GDP

A new entrepreneurial business class has emerged since the "Velvet Revolution" in 1989. Many Czechs have shares in privatized enterprises.

WORLD RANKING

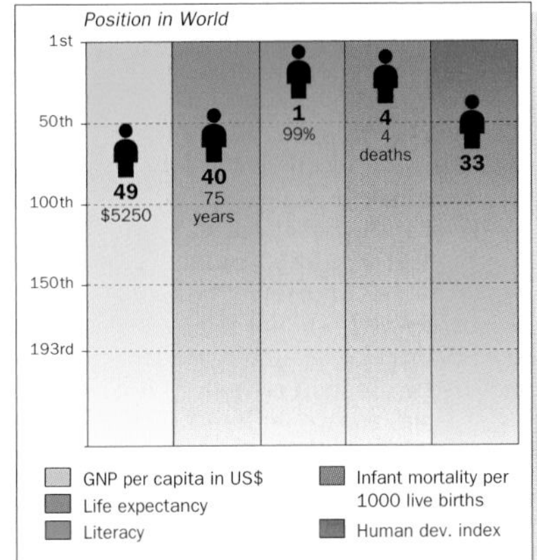
Position in World
1st
50th
100th
150th
193rd
49
$5250
40
75 years
1
99%
4
4 deaths
33
GNP per capita in US$
Life expectancy
Literacy
Infant mortality per 1000 live births
Human dev. index

DENMARK

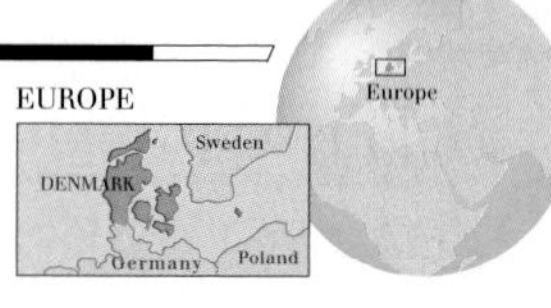
EUROPE
Europe
Sweden
DENMARK
Germany
Poland

OFFICIAL NAME: Kingdom of Denmark **CAPITAL:** Copenhagen
POPULATION: 5.3 million **CURRENCY:** Danish krone **OFFICIAL LANGUAGE:** Danish

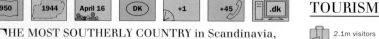

950
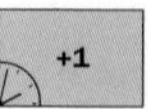
1944

April 16

DK

+1 +45 .dk

THE MOST SOUTHERLY COUNTRY in Scandinavia, Denmark occupies the Jutland (Jylland) peninsula, the islands of Sjælland, Fyn, Lolland, and Falster, and more than 400 smaller islands. Its terrain is among the flattest in the world. The Faeroe Islands and Greenland in the North Atlantic are self-governing associated territories. Politically, Denmark is stable, despite a preponderance of minority governments since 1945. It possesses a long liberal tradition and was one of the first countries to establish a welfare system, in the 1930s.

TOURISM

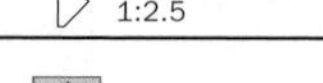
Visitors : Population
1:2.5

2.1m visitors

Up 3% in 2000

MAIN TOURIST ARRIVALS

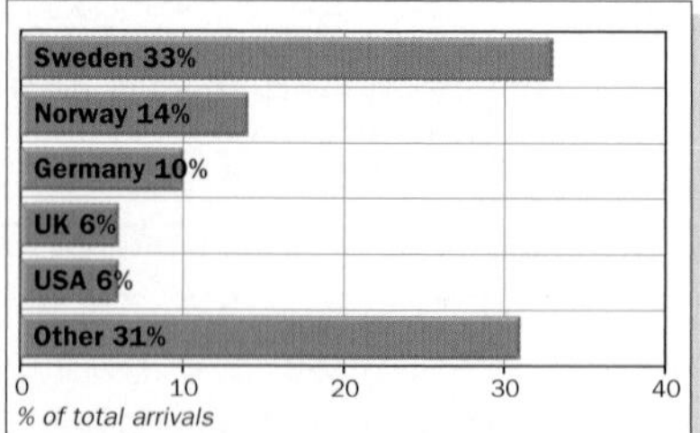
Sweden 33%
Norway 14%
Germany 10%
UK 6%
USA 6%
Other 31%
0
10
20
30
40
% of total arrivals

Principal attractions for tourists are Copenhagen (with its Tivoli Gardens and 18th-century architecture), Legoland, the countryside, and seaside resorts. Since 1959, only Danes have been allowed to own vacation homes.

CLIMATE

Maritime

WEATHER CHART FOR COPENHAGEN

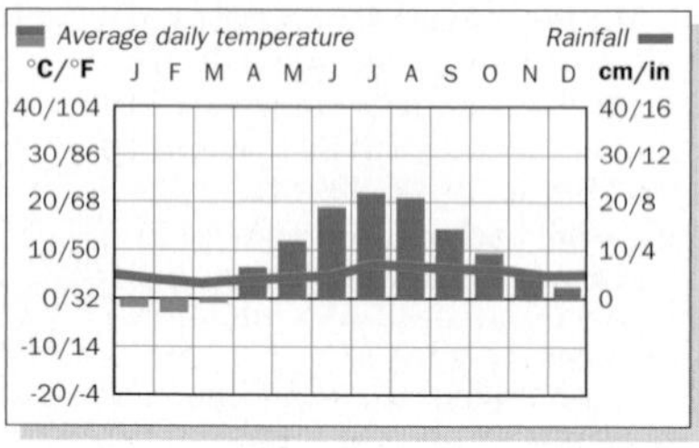

Average daily temperature / Rainfall
°C/°F J F M A M J J A S O N D cm/in
40/104 40/16
30/86 30/12
20/68 20/8
10/50 10/4
0/32 0
-10/14
-20/-4

The island of Fyn, like the rest of Denmark, is flat and depends on coastal defenses to prevent flooding by the sea.

Denmark's temperate, damp climate is one of the keys to its agricultural success. The Faeroes are windy, foggy, and cool. Greenland's climate ranges north–south from arctic to subarctic.

TRANSPORTATION

Drive on right

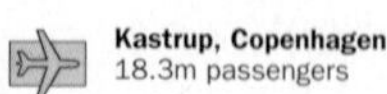
Kastrup, Copenhagen
18.3m passengers

942 ships
6.82m grt

THE TRANSPORTATION NETWORK

71,681 km (44,540 miles)	953 km (592 miles)
2324 km (1444 miles)	417 km (259 miles)

There is an extensive, well-integrated transportation network of bus, rail, and ferry services. State-owned companies predominate, although privatization of some ferry and rail services has been mooted. Denmark wishes to reduce significant state transportation subsidies. A few private companies operate in the Faeroes and Greenland with state support.

Major new construction projects focus on bridge and tunnel links, such as the Storebælt project connecting the islands of Fyn and Sjælland. A 16-km (10-mile) Øresund road and rail link by bridge and tunnel, connecting Copenhagen with Malmö in Sweden, opened in July 2000. Copenhagen's new Metro light rail system has now been completed.

DENMARK
Total Land Area : 43 094 sq. km (16 639 sq. miles)

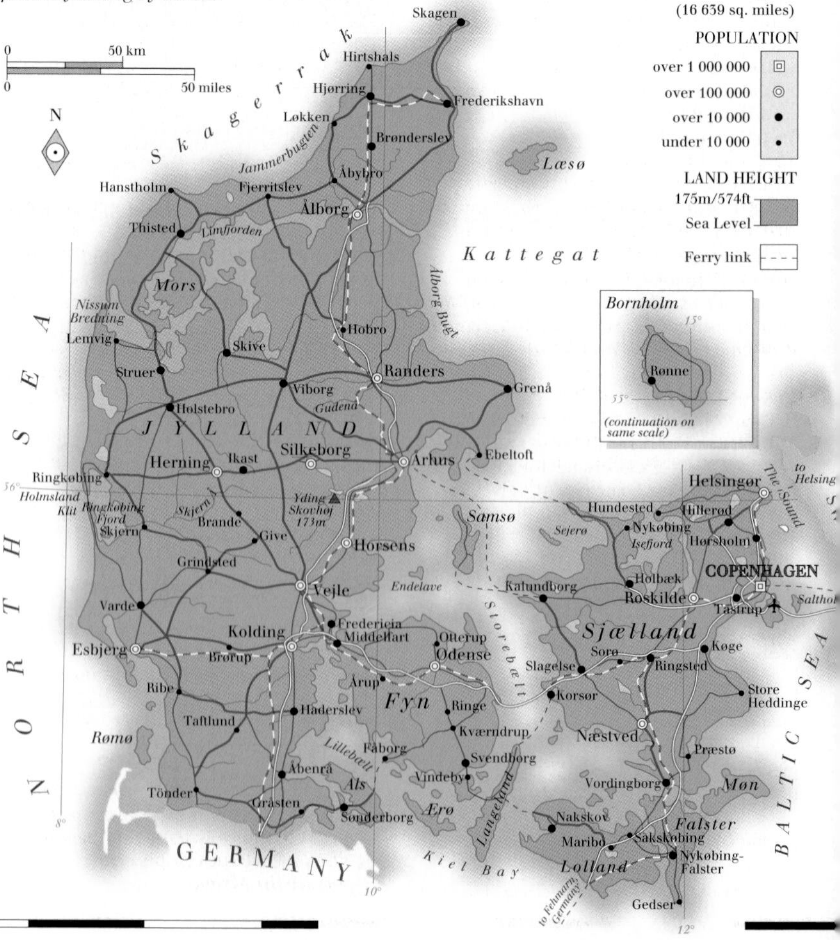

POPULATION
over 1 000 000
over 100 000
over 10 000
under 10 000

LAND HEIGHT
175m/574ft
Sea Level
Ferry link

0 50 km
0 50 miles

Skagerrak
Kattegat
NORTH SEA
BALTIC SEA
GERMANY
JYLLAND
Fyn
Sjælland
COPENHAGEN
Bornholm
Rønne
(continuation on same scale)

PEOPLE

Pop. density medium

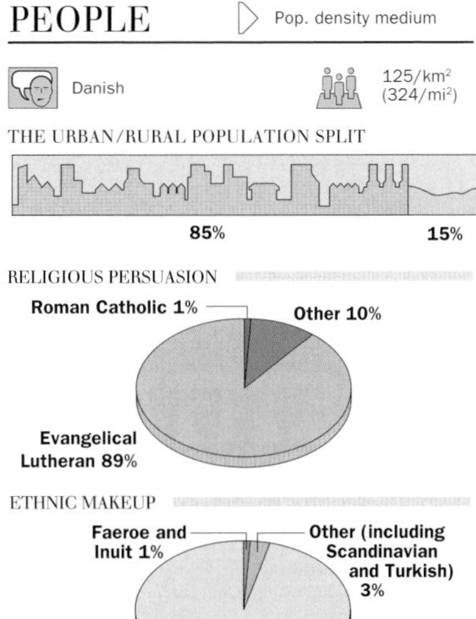
Danish
125/km² (324/mi²)
THE URBAN/RURAL POPULATION SPLIT
85%
15%
RELIGIOUS PERSUASION
Roman Catholic 1%
Other 10%
Evangelical Lutheran 89%

ETHNIC MAKEUP

Faeroe and Inuit 1%
Other (including Scandinavian and Turkish) 3%
Danish 96%

Danish society is homogeneous, but the small population of foreign citizens doubled over 1984–1999, and the current right-leaning government has pledged to curb immigration. The most visible minority groups are the Inuit, Greenland's indigenous inhabitants, and the Turkish community. Rising unemployment has engendered some ethnic tension, although racially motivated attacks are still rare.

Denmark has undergone profound social changes over the last 20 years, and the role of women has been transformed. Helped by extensive social and educational provision, three-quarters of women now work in part-time or full-time jobs. Denmark provides the best state child support in Europe: almost half of children under two, and two-thirds of three- to six-year-olds, are in day nurseries.

Less than half the population lives in a nuclear family, in part reflecting the high divorce rate. Marriage is becoming less common; almost 40% of children are brought up by unmarried couples or single parents. Cohabiting couples now have the same legal rights as those who are married. In 1990, Denmark became the first country to allow registered partnerships between homosexual couples, effectively granting them the same legal status as heterosexual couples.

POPULATION AGE BREAKDOWN

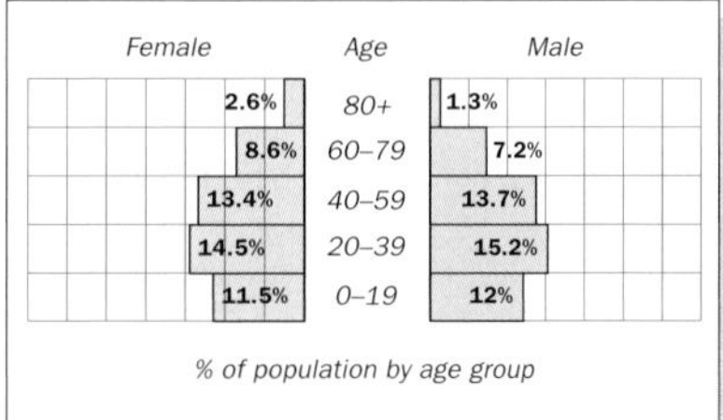
Female
Age
Male
2.6% 80+ 1.3%
8.6% 60–79 7.2%
13.4% 40–59 13.7%
14.5% 20–39 15.2%
11.5% 0–19 12%
% of population by age group

POLITICS

Multiparty elections

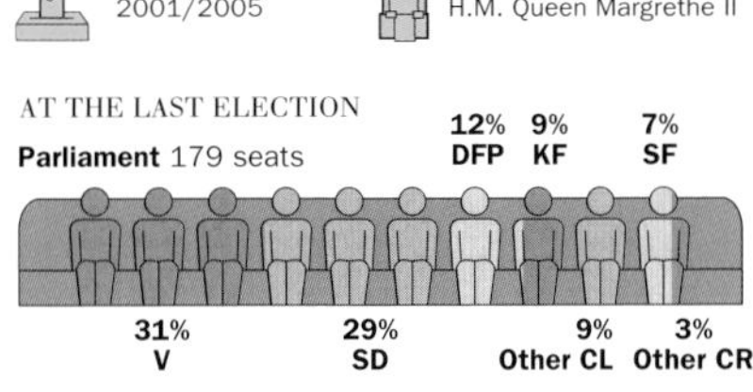
2001/2005
H.M. Queen Margrethe II
AT THE LAST ELECTION
Parliament 179 seats
12% DFP
9% KF
7% SF
31% V
29% SD
9% Other CL
3% Other CR

V = Liberal Party (Venstre) **SD** = Social Democrats **DFP** = Danish People's Party **KF** = Conservative People's Party **Other CL** = Other Center Left **SF** = Socialist People's Party **Other CR** = Other Center Right

Greenland and the Faeroe Islands send 2 members each to the Parliament

Denmark is a constitutional monarchy and a multiparty democracy. The associated territories of Greenland and the Faeroe Islands have home rule. The latter are divided over the issue of independence.

PROFILE

The intricate proportional electoral system ensures that Parliament truly reflects voters' wishes, but also tends to lead to minority governments. After a decade of Conservative–Liberal rule, the SD regained power in 1993, at the head of a center-left coalition under Poul Nyrup Rasmussen. The pendulum swung back with a victory in 2001 for the Liberals and Anders Fogh Rasmussen. Tax and immigration are important issues, but major policy differences between the two main political groups are few.

MAIN POLITICAL ISSUES

Relations with the EU

In recent years, Denmark's left-of-center parties have been suspicious of further EU integration. In 1992, ratification of the Maastricht Treaty, approved by Parliament, was rejected in a referendum, voters objecting to monetary union, a common defense force, and local election voting rights for European citizens living in Denmark. Later that year an EU summit allowed Denmark to opt out of monetary union, defense, and European citizenship. A referendum in 1993 finally approved the Maastricht Treaty, and voters approved the Amsterdam Treaty, its successor, in 1998, but a referendum in 2000 on joining the euro (which Denmark had decided not to introduce from its outset in 1999) produced a highly significant, and unexpected, "no" result.

Immigration

Despite Danish liberal traditions, the situation of immigrants and refugees – under 5% of the population – is a controversial issue. The election of a right-wing government in 2001 signaled a change to an overtly restrictive policy and a greater emphasis on integration for those immigrants already there. The ruling coalition relies for support on the far-right, anti-immigrant Danish People's Party, which had almost doubled its representation. Legislation passed in June 2002 – on the eve of Denmark's taking over the EU presidency – prevents the foreign (non-EU) spouse of a Danish citizen living in or emigrating to the country.

Anders Fogh Rasmussen, heads a right-wing, minority coalition.

Queen Margrethe II, who succeeded to the throne in 1972.

WORLD AFFAIRS

Joined UN in 1945

CE | EU | NATO | OECD | OSCE

Relations with the rest of Europe are the major foreign policy concern, notably the issues of a common defense policy and monetary union. Denmark has decided against introducing the euro, but the krone is pegged to it and economic policies follow those of the participating states. Promoting economic ties with Norway, Sweden, and Finland is a priority, as are improving links with former Eastern bloc states, especially those on the Baltic – not least to assist pollution reduction, a serious concern. Denmark is also a strong supporter of Third World development, especially in Africa.

CHRONOLOGY

Founded in the 10th century, Denmark's monarchy is Europe's oldest. It was the dominant Baltic power until the 17th century, when it was eclipsed by Sweden.

- **1815** Denmark forced to cede Norway to Swedish rule.
- **1849** Creation of first democratic constitution.
- **1864** Denmark forced to cede provinces of Schleswig and Holstein after losing war with Prussia.
- **1914–1918** Denmark neutral in World War I.
- **1915** Universal adult suffrage introduced. Rise of SD.
- **1920** Northern Schleswig votes to return to Danish rule.

D

CHRONOLOGY *continued*

- **1929** First full SD government takes power under Prime Minister Thorvald Stauning.
- **1930s** Implementation of advanced social welfare legislation and other liberal reforms under SD.
- **1939** Outbreak of World War II; Denmark reaffirms neutrality.
- **1940** Nazi occupation. National coalition government formed.
- **1943** Danish Resistance successes lead Nazis to take full control.
- **1944** Iceland declares independence from Denmark.
- **1945** Denmark recognizes Icelandic independence. After defeat of Nazi Germany, SD leads postwar coalition governments.
- **1948** Faeroes granted home rule.
- **1952** Founder member of the Nordic Council.
- **1953** Constitution reformed; single-chamber, proportionally elected parliament created.
- **1959** Denmark joins EFTA.
- **1973** Denmark joins European Communities.
- **1979** Greenland granted home rule.
- **1975–1982** SD's Anker Jorgensen heads series of coalitions; elections in 1977, 1979, and 1981. Final coalition collapses over economic policy differences.
- **1982** Poul Schlüter first Conservative prime minister since 1894.
- **1992** Referendum rejects Maastricht Treaty on European Union.
- **1993** Schlüter resigns over "Tamilgate" scandal. Center-left government led by Poul Nyrup Rasmussen. Danish voters ratify revised Maastricht Treaty.
- **1994–1998** Elections: Rasmussen heads SD-led minority coalition.
- **2000** Referendum rejects joining eurozone.
- **2001** Elections: Liberals return to power. Anders Fogh Rasmussen appointed prime minister.

AID

Donor

$1.66bn (donations)

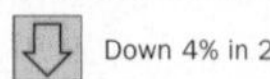

Down 4% in 2000

During the 1990s, Denmark was the world's leading aid donor in GNP terms, contributing an average 1% of national income. It supports both economic and social development projects and policy reforms. The Liberal government elected in 2001 is committed to reducing the foreign aid budget.

Denmark provides aid to Asia and Latin America, but its closest ties are with Africa. Tanzania is the largest single aid recipient. Denmark has also provided considerable support to the other SADC states.

DEFENSE

Compulsory military service

$2.4bn

Down 10% in 2000

Denmark was neutral until 1945. Apart from NATO commitments, defense has a low priority; spending is less than 2% of GDP (well below the NATO average). Denmark provides troops for the NATO-led forces in former Yugoslavia and observers for other UN peacekeeping operations. One-quarter of its armed forces are conscripts, and its reserves include a Home Guard. Denmark has observer status at the WEU.

DANISH ARMED FORCES

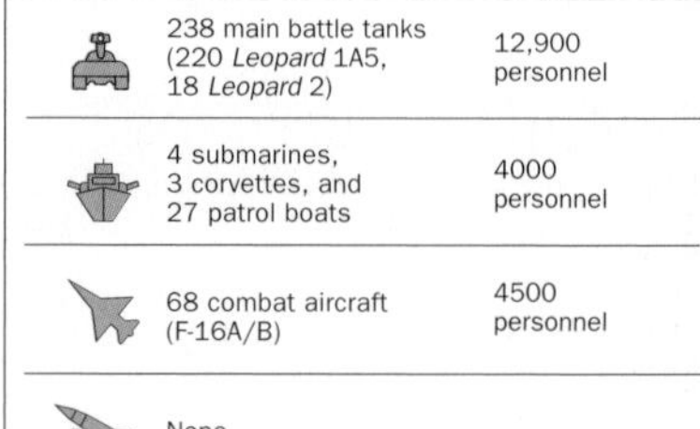
238 main battle tanks (220 Leopard 1A5, 18 Leopard 2)
12,900 personnel
4 submarines, 3 corvettes, and 27 patrol boats
4000 personnel
68 combat aircraft (F-16A/B)
4500 personnel
None

ECONOMICS

Inflation 2.2% p.a. (1990–2000)

$172bn

7.9499–8.3504 Danish kroner

SCORE CARD

World GNP Ranking	24th
GNP per Capita	$32,280
Balance of Payments	$3.35bn
Inflation	2.9%
Unemployment	5%

EXPORTS

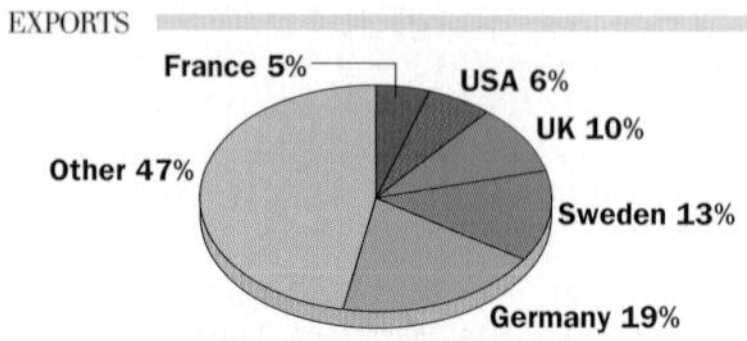
France 5%
USA 6%
UK 10%
Other 47%
Sweden 13%
Germany 19%

IMPORTS

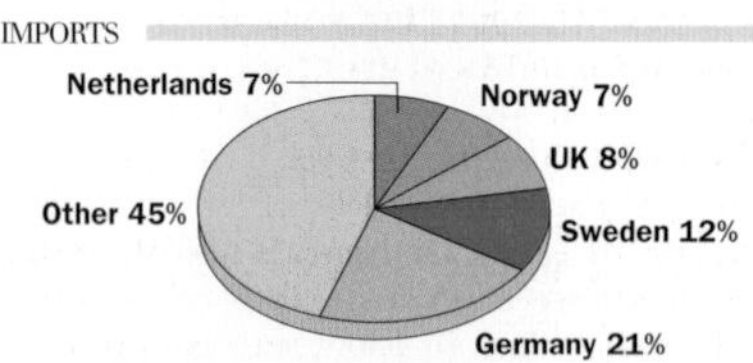
Netherlands 7%
Norway 7%
UK 8%
Other 45%
Sweden 12%
Germany 21%

ECONOMIC PERFORMANCE INDICATOR

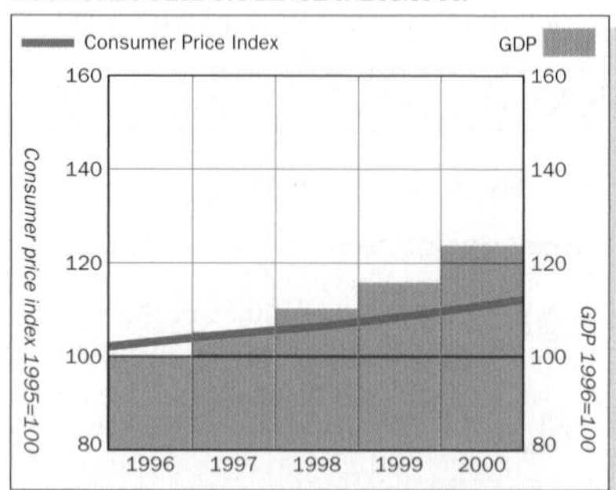
Consumer Price Index
GDP
160
140
120
100
80
Consumer price index 1995=100
GDP 1996=100
1996
1997
1998
1999
2000

Strengths

Low inflation and unemployment. Substantial balance-of-payments surplus. Gas and oil reserves. Strong high-tech, high-profit manufacturing sector. Skilled workforce.

Weaknesses

Heavy tax burden. Labor costs and historically strong currency affect competitiveness.

Profile

Denmark's mix of a large state sector and a private sector has been successful. GDP per capita is one of the highest among the OECD countries. However, total taxation, at about 50%, is among the world's highest. In November 2001 the Liberal government promised to fund increased welfare spending and lower taxation by extending privatization and reducing development aid.

In the 1980s the government stabilized the exchange rate and tightened budget controls in order to reduce inflation and reverse the balance of payments deficit. Growth slowed in the early 1990s, but an economic upturn between 1993 and 2000 was first led by private consumption and then buoyed by exports and business investment. Growth slowed significantly in 2001, but showed signs of regaining momentum by mid-2002.

Voters have refused to accept EU monetary union, most recently in a "no" vote in 2000 in a referendum on the euro. However, Denmark meets the EU's convergence criteria for monetary union, the krone is pegged to the euro, and economic policies follow those of the participating states.

DENMARK : MAJOR BUSINESSES

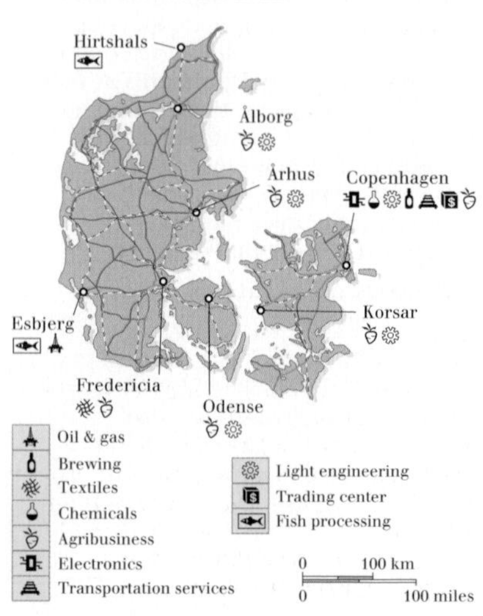
Hirtshals
Ålborg
Århus
Copenhagen
Korsar
Esbjerg
Fredericia
Odense
Oil & gas
Brewing
Textiles
Chemicals
Agribusiness
Electronics
Transportation services
Light engineering
Trading center
Fish processing
0 100 km
0 100 miles

RESOURCES

Electric power 12.1m kw

1.45m tonnes

342,000 b/d (reserves 1.1bn barrels)

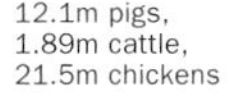
12.1m pigs, 1.89m cattle, 21.5m chickens

Natural gas, oil

ELECTRICITY GENERATION

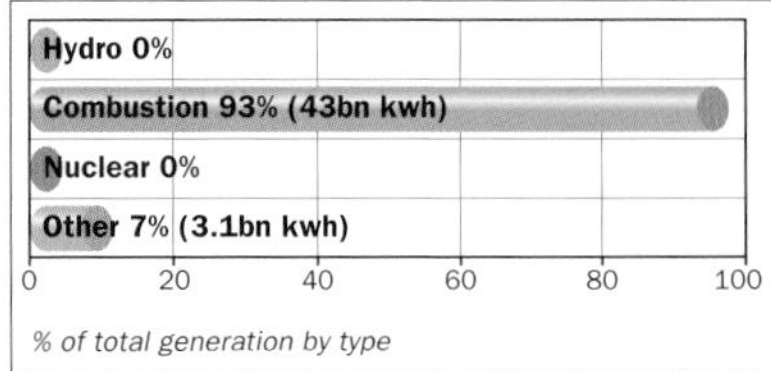

Hydro 0%
Combustion 93% (43bn kwh)
Nuclear 0%
Other 7% (3.1bn kwh)
0
20
40
60
80
100
% of total generation by type

Despite expansion of North Sea oil and gas output, Denmark is still an overall importer of energy. Agriculture is highly efficient, and Denmark is the world's biggest exporter of pork. The use of wind power is expanding rapidly. Denmark is a world leader in this technology.

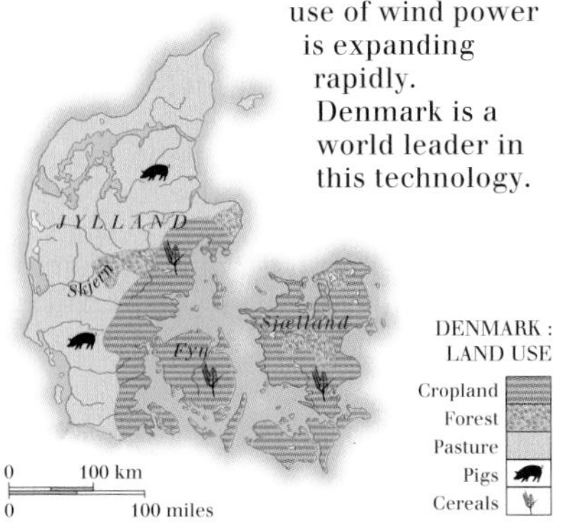

JYLLAND
Skjern
Fyn
Sjælland
0
100 km
0
100 miles
DENMARK : LAND USE
Cropland
Forest
Pasture
Pigs
Cereals

ENVIRONMENT

Sustainability rank: 31st

33%

10.1 tonnes per capita

ENVIRONMENTAL TREATIES

Yes
Yes
Yes
Yes
Yes
Yes

Denmark has some of the strictest regulations in Europe, including those aimed at reducing ozone-destroying emissions and water pollution, and met its 2000 target – recycling 54% of all waste – a year early. There was a marked change in policy under the incoming Liberal government of 2001. The bans on house-building in state forests and on the sale of beer in cans were lifted, three planned windpower plants were shelved, and the environment ministry's budget was cut by one-third.

MEDIA

TV ownership high

Daily newspaper circulation 311 per 1000 people

PUBLISHING AND BROADCAST MEDIA

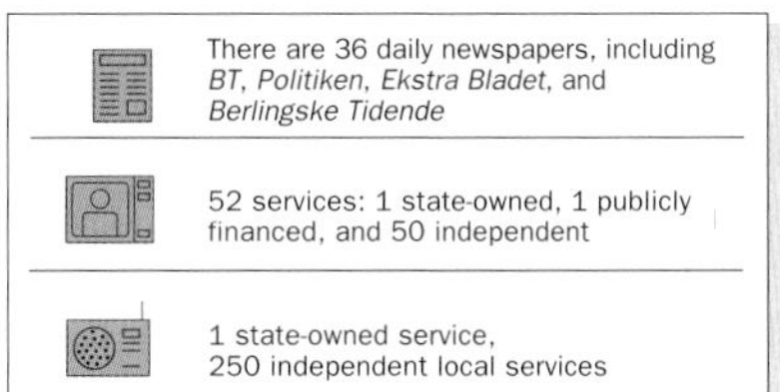

There are 36 daily newspapers, including *BT, Politiken, Ekstra Bladet,* and *Berlingske Tidende*
52 services: 1 state-owned, 1 publicly financed, and 50 independent
1 state-owned service, 250 independent local services

The media have a long history of political independence, and objectivity is prized. The tone of both TV and the press is serious; there is no scandal-mongering tabloid press as found in the US, the UK, and Germany. Invasion of privacy laws are strict. Legislation proposed in May 2002 would extend competition in broadcasting and prepare for the privatization of TV2.

CRIME

No death penalty

3279 prisoners

Down 1% in 1999

CRIME RATES

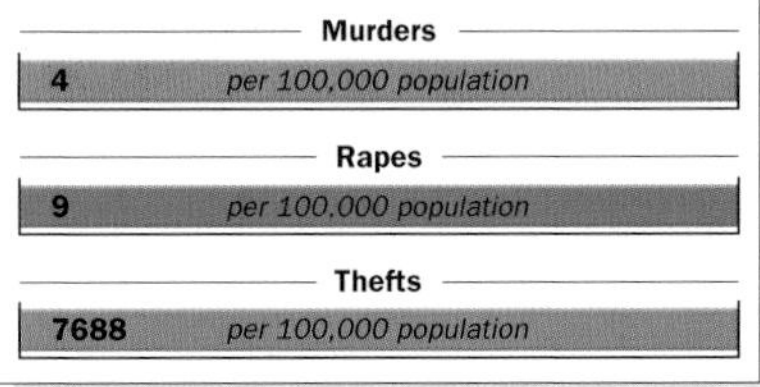

Murders
4 per 100,000 population
Rapes
9 per 100,000 population
Thefts
7688 per 100,000 population

Problems are the potential importing of Mafia-style organized crime from eastern Europe, computer hacking, and drug trafficking. In 2001, stricter penalties for violence and rape were promised.

EDUCATION

School leaving age: 16

99%

170,169 students

THE EDUCATION SYSTEM

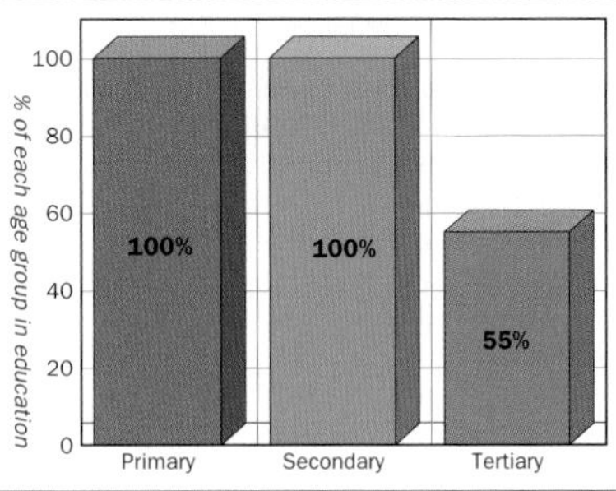

100
80
60
40
20
0
% of each age group in education
100%
100%
55%
Primary
Secondary
Tertiary

The educational level is generally high, in part reflecting the need for a skilled workforce. Formal schooling begins at age seven and is mandatory for nine years. However, most children receive preschool education, and over 90% of pupils go on at the age of 16 to further academic or vocational training. Traditional folk high schools offer a wide range of further education options, but no professional qualifications.

HEALTH

Welfare state health benefits

1 per 294 people

Heart diseases, cancers, accidents

Denmark was one of the first countries to introduce a state social welfare system. The national health service, which still provides free treatment for almost everything, is the main reason for Denmark's high taxes – in 2000 almost a quarter of government spending was allocated to social services. Any attempts to reduce expenditure will meet with strong opposition. Repeated surveys show that most Danes prefer their system to those based on private health insurance. In the early 1980s, Denmark had the highest incidence of AIDS in Europe, but after peaking in 1993, it has dropped markedly, due to the free availability of drug therapy.

SPENDING

GDP/cap. increase

CONSUMPTION AND SPENDING

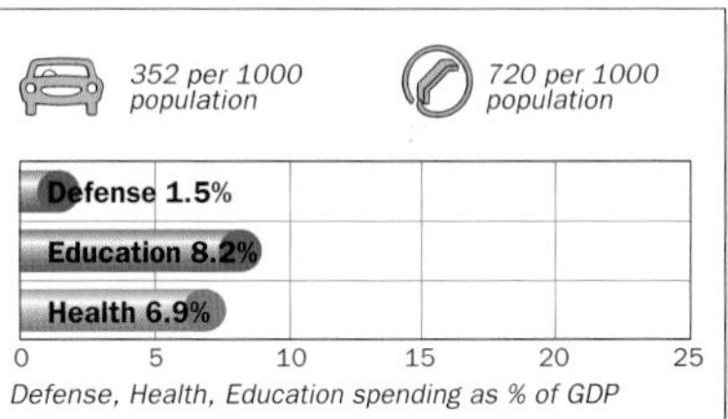

352 per 1000 population
720 per 1000 population
Defense 1.5%
Education 8.2%
Health 6.9%
0
5
10
15
20
25
Defense, Health, Education spending as % of GDP

Most Danes are comfortably off. Income distribution is more even than in many Western countries and social mobility is high. Free higher education means that access to the professions is more a question of ability than wealth or connections. Denmark forms one of the world's most egalitarian societies. The generous social security system means that Danes suffer little from social deprivation. The SD government of 1994–2001 has created more kindergarten places and increased time off for those with young children. Refugees and recent immigrants tend to be the most disadvantaged members of Danish society.

WORLD RANKING

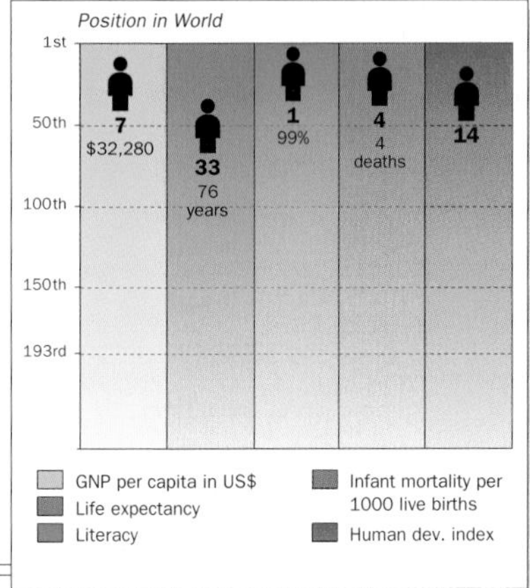

Position in World
1st
50th
100th
150th
193rd
7
$32,280
33
76 years
1
99%
4
4 deaths
14
GNP per capita in US$
Life expectancy
Literacy
Infant mortality per 1000 live births
Human dev. index

See also OVERSEAS TERRITORIES *p.640*

DJIBOUTI

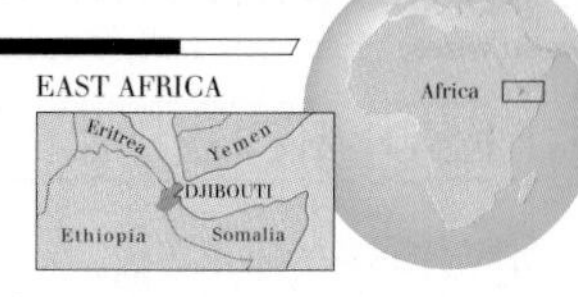
EAST AFRICA
Africa
Eritrea
Yemen
DJIBOUTI
Ethiopia
Somalia

OFFICIAL NAME: Republic of Djibouti **CAPITAL:** Djibouti
POPULATION: 644,000 **CURRENCY:** Djibouti franc **OFFICIAL LANGUAGES:** Arabic and French

1977

1977
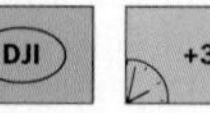
June 27

DJI

+3
+253 .dj

A CITY WITH A DESERT HINTERLAND, Djibouti lies in northeast Africa on the strait linking the Red Sea and the Indian Ocean. Known from 1967 as the French Territory of the Afars and Issas, Djibouti became independent in 1977. Its economy relies on the main port, the railroad to Addis Ababa, and French aid. A guerrilla war which erupted in 1991 as a result of tension between the Issas in the south and the Afars in the north has largely been resolved.

CLIMATE

Hot desert

WEATHER CHART FOR DJIBOUTI

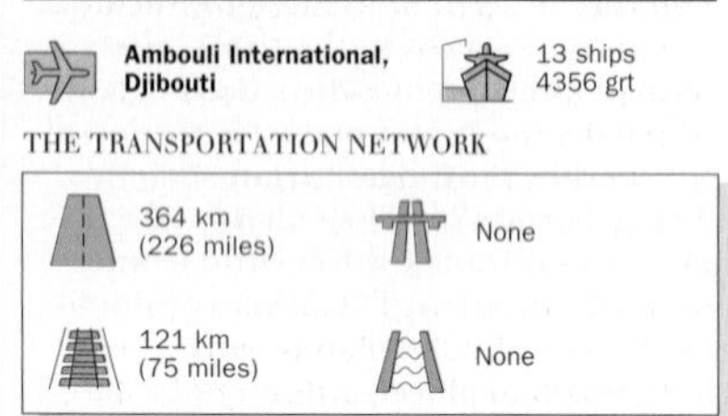
Average daily temperature
Rainfall
°C/°F J F M A M J J A S O N D cm/in
50/122
40/104
30/86
20/68
10/50
0/32
-10/14
50/20
40/16
30/12
20/8
10/4
0

Despite extremely low rainfall, the monsoon season is characterized by very humid conditions. Even locals find the heat in June–August hard to bear.

TRANSPORTATION

Drive on right

Ambouli International, Djibouti 13 ships 4356 grt

THE TRANSPORTATION NETWORK

364 km (226 miles) None
121 km (75 miles) None

Djibouti's port, created by the French in the 19th century and now a modern container facility, is its key asset. Landlocked Ethiopia's vital link to the sea is the Addis Ababa–Djibouti railroad.

TOURISM

Visitors : Population 1:31

21,000 visitors Little change 1995–1998

MAIN TOURIST ARRIVALS

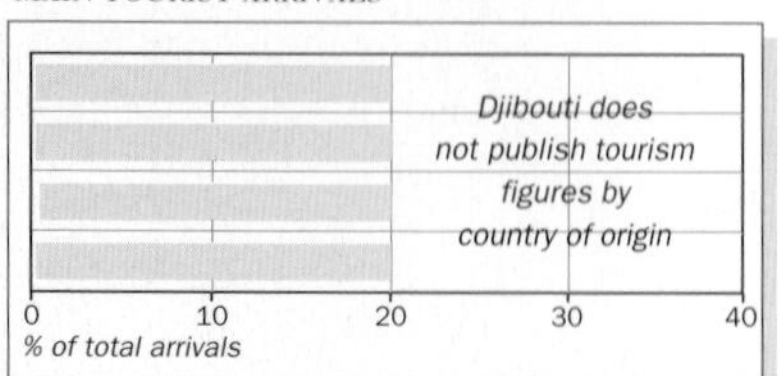
Djibouti does not publish tourism figures by country of origin
0
10
20
30
40
% of total arrivals

Most visitors are passing through on their way to Ethiopia, or coming to see relatives working in Djibouti port.

Nomadic Djiboutian village, *close to Balho near the Ethiopian border.*

PEOPLE

Pop. density low

Somali, Afar, French, Arabic
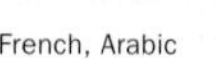
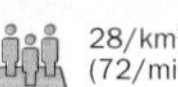
28/km²
(72/mi²)

THE URBAN/RURAL POPULATION SPLIT

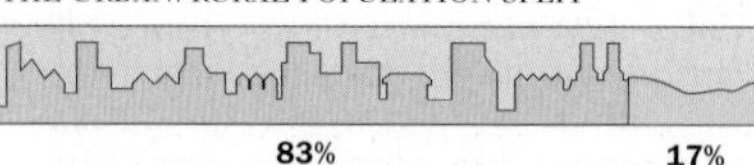
83%
17%

ETHNIC MAKEUP

Other 5%
Afar 35%
Issa 60%

The main ethnic groups are the Afars and Issas; tension between these groups developed into a guerrilla war in 1991. The population was swelled in 1992 by 20,000 Somali refugees. The rural people are mostly nomadic.

POLITICS

Multiparty elections

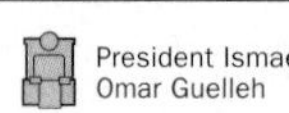
President Ismael Omar Guelleh
1997/2002

AT THE LAST ELECTION

National Assembly 65 seats

100% RPP–FRUD

RPP–FRUD = Alliance of the Popular Rally for Progress (RPP) and the Front for the Restoration of Unity and Democracy (FRUD)

President Hassan Gouled Aptidon, an Issa, backed by France, dominated politics from independence in 1977 until his retirement in 1999. Afar fears of Issa domination erupted in 1991, and the Afar guerrilla group FRUD took control of much of the country. The French intervened militarily to keep Gouled in power, but forced him to hold elections in 1992, which were won by the RPP. The FRUD became a legal political party following a 1996 peace agreement. An alliance of the RPP and FRUD made a clean sweep of elections in 1997. Presidential elections in 1999 were won by Ismael Omar Guelleh, a former close aide of Gouled, amid opposition claims of electoral fraud.

DJIBOUTI

Total Land Area : 22 000 sq. km (8494 sq. miles)

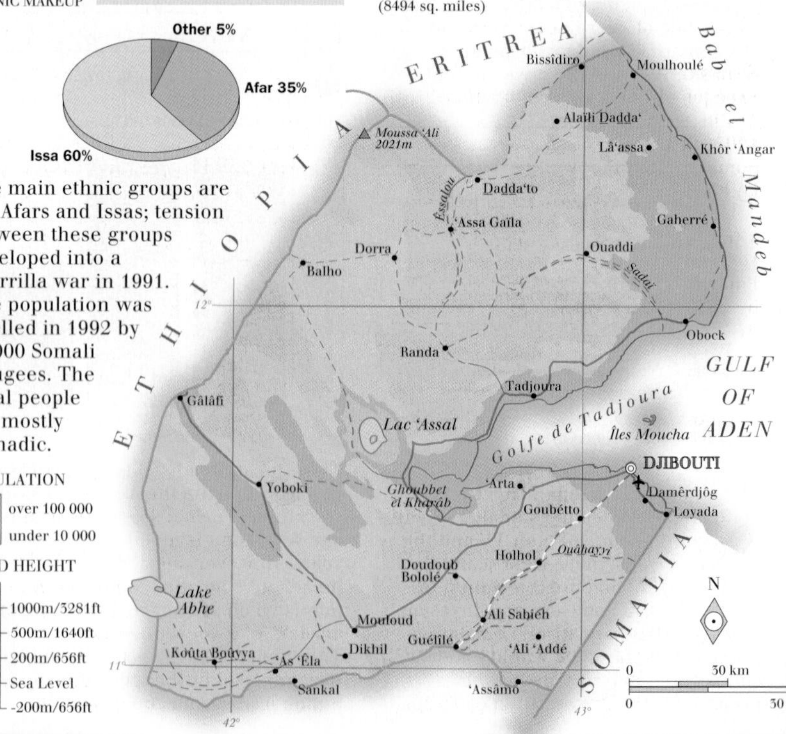
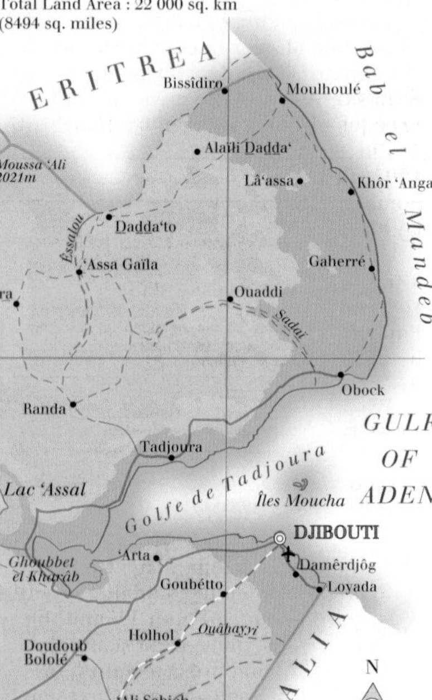
ERITREA
ETHIOPIA
SOMALIA
Bab el Mandeb
GULF OF ADEN
Golfe de Tadjoura
Bissidiro
Moulhoulé
Alaïli Dadda'
Moussa 'Ali 2021m
Lâ'assa
Khôr 'Angar
Dadda'to
Assa Gaïla
Gaherré
Dorra
Ouaddi
Balho
Obock
Randa
Tadjoura
Gâlâfi
Lac 'Assal
Îles Moucha
DJIBOUTI
'Arta
Damêrdjôg
Yoboki
Ghoubbet el Kharâb
Goubétto
Loyada
Holhol
Doudoub Bololé
Lake Abhe
Ali Sabîeh
Mouloud
Guélilé
Dikhil
'Ali 'Addé
Koûta Boûyya
As 'Ela
Sankal
'Assâmo
POPULATION
over 100 000
under 10 000
LAND HEIGHT
1000m/3281ft
500m/1640ft
200m/656ft
Sea Level
-200m/656ft
0 30 km
0 30 mi

WORLD AFFAIRS

Joined UN in 1977

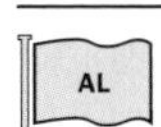
AL

OIF

IGAD

AU

OIC

France, with a key military presence, is pressing for greater democratization. Djibouti, Ethiopia, and Eritrea all seek to contain Afar secessionism. In 2000 the southern town of Arta hosted the Somali reconciliation conference.

AID

Recipient

$71m (receipts)

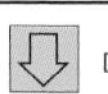
Down 5% in 2000

France is the major donor, effectively financing one-third of government expenditure. Djibouti has also received aid from Saudi Arabia and Kuwait.

DEFENSE

No compulsory military service

$23m

Up 5% in 2000

The size of the armed forces is a state secret, but is estimated at 9600 personnel; former FRUD guerrillas were integrated into the army. There is a 3200-strong French garrison.

ECONOMICS

Inflation 3.7% p.a. (1990–2000)

$553m

174.8–170.0 Djibouti francs

SCORE CARD

- ❑ World GNP Ranking........169th
- ❑ GNP per Capita........$880
- ❑ Balance of Payments........–$14m
- ❑ Inflation........2%
- ❑ Unemployment........50%

Strengths

Free port in key Red Sea location; large profits from 1991 Gulf War and from 1992 US and UN intervention in Somalia. Continuing development of Djibouti and Tadjoura port facilities.

Weaknesses

Dependent on French aid and garrison. Planned Saudi investment held back by civil war in 1990s. Other ports on Red Sea now providing stiff competition.

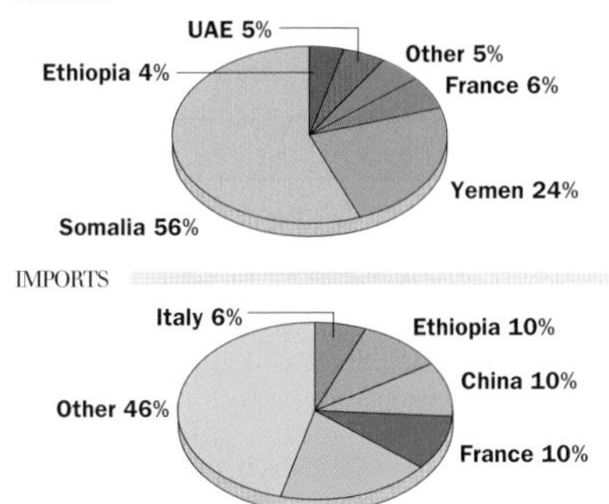
EXPORTS
UAE 5%
Other 5%
Ethiopia 4%
France 6%
Yemen 24%
Somalia 56%
IMPORTS
Italy 6%
Ethiopia 10%
China 10%
Other 46%
France 10%
Saudi Arabia 18%

RESOURCES

Electric power 85,000 kw

350 tonnes

Not an oil producer

513,000 goats, 465,000 sheep, 70,000 camels

Gypsum, mica, amethyst, sulfur, natural gas

The few mineral resources are scarcely exploited. Geothermal energy is being developed and natural gas has recently been found. The guerrilla war delayed attempts to develop underground water supplies for agriculture.

ENVIRONMENT

Not available

0.4%

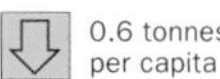
0.6 tonnes per capita

The concentration of business around Djibouti port means that inland desert areas are largely untouched. Ecological issues are not a national concern.

MEDIA

TV ownership low

There are no daily newspapers

PUBLISHING AND BROADCAST MEDIA

There are no daily newspapers.
The weekly *La Nation de Djibouti* is published by the government
1 state-controlled service
1 state-controlled service

Djibouti is a member of the Arab Satellite Communications Organization. The media are largely state-controlled, but there is one opposition newspaper.

CRIME

No death penalty

384 prisoners

Up 74% 1996–1998

The government accused the FRUD of atrocities, but the state's own human rights record has since been criticized by Amnesty International. While drug smuggling and prostitution are rife, petty crime, rather than violence, is the norm.

EDUCATION

School leaving age: 12

65%

478 students

Schooling is mostly in French, although there has been a growing emphasis on Islamic teaching, particularly as Saudi Arabia has declared an interest in providing aid for education. Djibouti has no university.

HEALTH

Welfare state health benefits

1 per 7143 people

Respiratory and heart diseases

AIDS is a growing problem in Djibouti port, with its large prostitute population. UN estimates for 1999 suggested that there were some 37,000 HIV/AIDS sufferers. Small French-financed hospitals cater for the urban elite.

CHRONOLOGY

The French set up a coaling station at Djibouti in the 1880s, to balance the British presence in Aden.

- ❑ **1917** Railroad from Addis Ababa reaches Djibouti port.
- ❑ **1977** Independence.
- ❑ **1981–1992** One-party state.
- ❑ **1989** Eruption of violence between Afars and Issas.
- ❑ **1991** FRUD launches armed insurrection.
- ❑ **1994** Peace agreement with FRUD.
- ❑ **1999** Ismael Omar Guelleh becomes president.
- ❑ **2000** Coup attempt by police officers fails.

D

SPENDING

GDP/cap. decrease

CONSUMPTION AND SPENDING

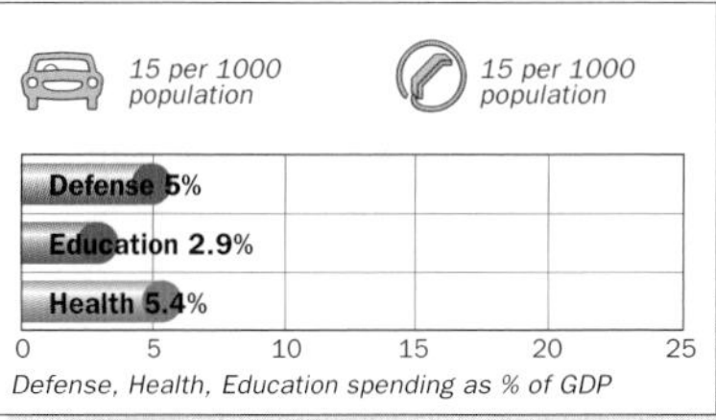
15 per 1000 population
15 per 1000 population
Defense 5%
Education 2.9%
Health 5.4%
0
5
10
15
20
25
Defense, Health, Education spending as % of GDP

As happens in many African states, the wealth in Djibouti tends to be concentrated among those closest to government. Djiboutians working in the ports also do well, although much port labor is expatriate. The guerrilla war has had little effect on port life, since it is almost completely isolated from the rest of the country. The nomads of the interior are the poorest group.

Trade in the mild narcotic *qat*, or "green gold," which is grown in Ethiopia and shipped through Djibouti, is highly lucrative, to the extent that the state is now taking its share of the profits. In Djibouti, *qat* chewing is an age-old social ritual.

WORLD RANKING

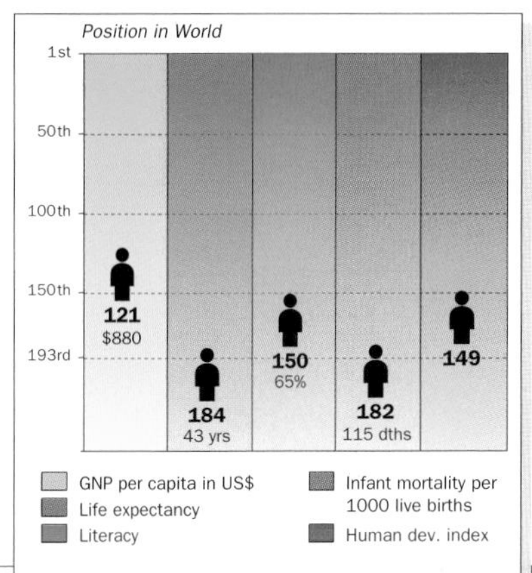
Position in World
1st
50th
100th
150th
193rd
121
$880
184
43 yrs
150
65%
182
115 dths
149
GNP per capita in US$
Life expectancy
Literacy
Infant mortality per 1000 live births
Human dev. index

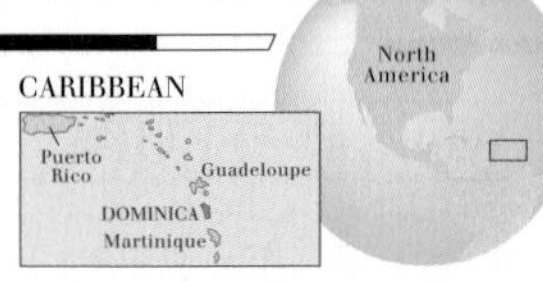
CARIBBEAN
North America
Puerto Rico
Guadeloupe
DOMINICA
Martinique

DOMINICA

OFFICIAL NAME: Commonwealth of Dominica **CAPITAL:** Roseau

POPULATION: 73,000 **CURRENCY:** Eastern Caribbean dollar **OFFICIAL LANGUAGE:** English

1978

1978
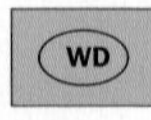
Nov 3
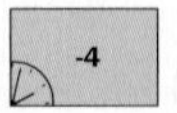
WD

-4

+1767

.dm

DOMINICA IS RENOWNED as the Caribbean island that resisted European colonization until the 18th century, when it came under French control, passing to the UK in 1759. It is known as the "Nature Island" because of its spectacular, lush, and abundant flora and fauna, protected by extensive national parks. The most mountainous of the Lesser Antilles, Dominica is located between Guadeloupe and Martinique in the West Indian Windward Islands group. Its volcanic origin has given it very fertile soils and the second-largest boiling lake in the world.

CLIMATE

Tropical oceanic

WEATHER CHART FOR ROSEAU

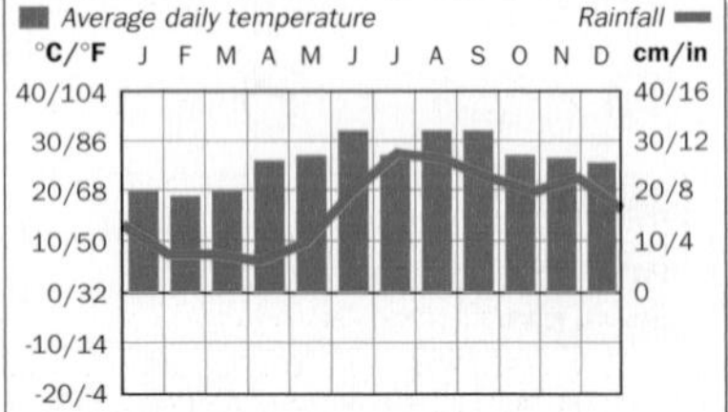
Average daily temperature
Rainfall
°C/°F J F M A M J J A S O N D cm/in
40/104 40/16
30/86 30/12
20/68 20/8
10/50 10/4
0/32 0
-10/14
-20/4

Like the other Windward Islands in the eastern Caribbean, Dominica is subject to constant trade winds. The rainy season is in the summer, and tropical depressions and hurricanes are likely between June and November. Short, thundery showers in the late afternoon and evening are common throughout the year.

TRANSPORTATION

Drive on left

Canefield, Roseau 108,179 passengers

7 ships 2233 grt

THE TRANSPORTATION NETWORK

393 km (244 miles)

None

None

None

Both airports take only small propeller aircraft. Roads are well maintained. There is no speed limit in rural areas.

TOURISM

Visitors : Population 1:1

74,000 visitors

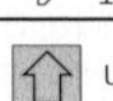

Up 12% in 1999

MAIN TOURIST ARRIVALS

	% of total arrivals
Caribbean	58%
USA	21%
UK	9%
Other	12%

The national parks, with their rare indigenous birds, hot springs, and sulfur pools, are a major attraction for tourists. However, the lack of an airport able to take commercial jetliners (visitors use connecting flights from Barbados or Antigua) has made Dominica less accessible to mass-market tourism than its neighbors.

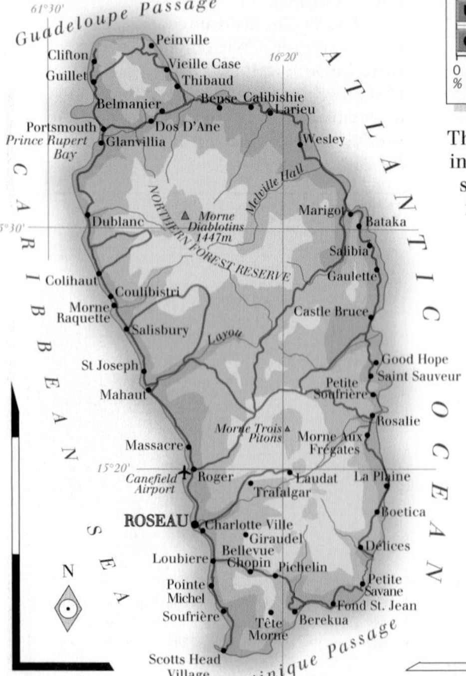
Guadeloupe Passage
61°30'
Peinville
Clifton
Vieille Case
Guillet
Thibaud
16°20'
Belmanier
Bense
Calibishie
Larieu
Portsmouth
Dos D'Ane
Prince Rupert Bay
Glanvillia
Wesley
Melville Hall
NORTHERN FOREST RESERVE
Morne Diablotins 1447m
Dublanc
Marigot
Bataka
15°30'
Salibia
Colihaut
Gaulette
Coulibistri
Morne Raquette
Castle Bruce
Salisbury
Layou
St Joseph
Good Hope
Saint Sauveur
Mahaut
Petite Soufrière
Rosalie
Massacre
Morne Trois Pitons
Morne Aux Frégates
15°20'
Canefield Airport
Roger
Laudat
La Plaine
Trafalgar
ROSEAU
Charlotte Ville
Boetica
Giraudel
Loubiere
Bellevue Chopin
Délices
Pichelin
Pointe Michel
Petite Savane
Fond St. Jean
Soufrière
Tête Morne
Berekua
Scotts Head Village
15°10'
Martinique Passage
ATLANTIC OCEAN
CARIBBEAN SEA
N

DOMINICA

Total Land Area : 754 sq. km (291 sq. miles)

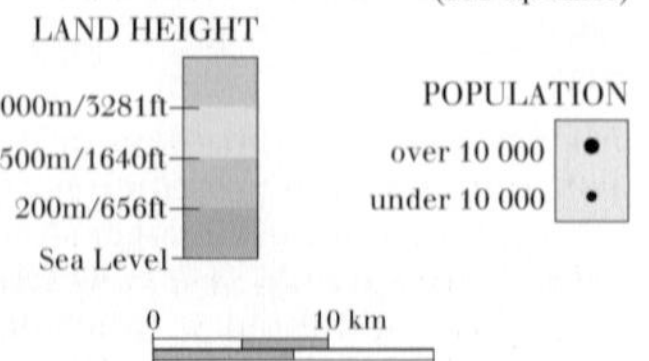
LAND HEIGHT
1000m/3281ft
500m/1640ft
200m/656ft
Sea Level
POPULATION
over 10 000
under 10 000
0
10 km
0
10 miles

PEOPLE

Pop. density medium

French Creole, English

97/km² (252/mi²)

THE URBAN/RURAL POPULATION SPLIT

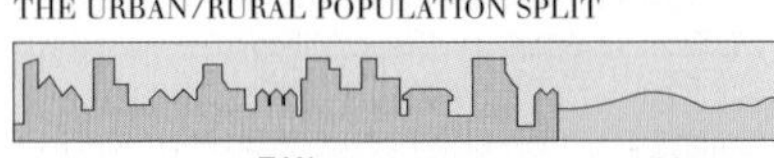
71%
29%

RELIGIOUS PERSUASION

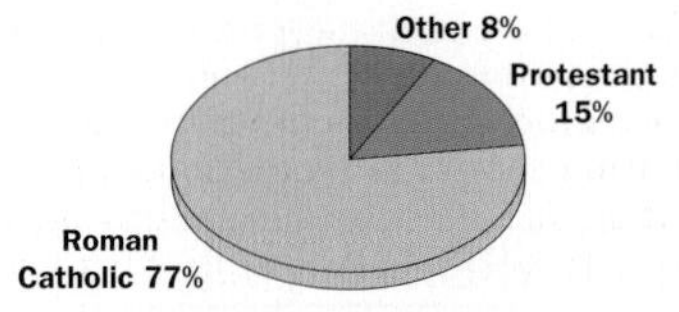
Other 8%
Protestant 15%
Roman Catholic 77%

The majority of Dominicans are descendants of Africans brought over to work the banana plantations. The Carib Territory on the northeast of the island contains the only surviving Carib population in the Caribbean.

POLITICS

Multiparty elections

2000/2005

President Vernon Shaw

AT THE LAST ELECTION

House of Assembly 30 seats

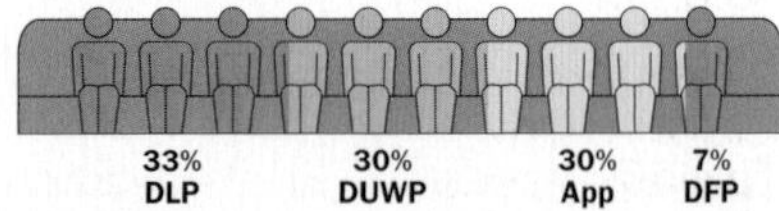
33% DLP
30% DUWP
30% App
7% DFP

DLP = Dominica Labour Party **DUWP** = Dominica United Workers' Party **App** = Appointed **DFP** = Dominica Freedom Party

9 senators are appointed to the House of Assembly by the head of state

Politicians tend to come from the professional classes – usually young lawyers and doctors. Occasionally the larger farmers, who provide most party funding, stand for election. The center-left DUWP narrowly won the 1995 elections, ending 15 consecutive years of rule by the right-wing DFP. A further swing to the left produced a DLP victory in January 2000, and its leader Rosie Douglas became prime minister, Pierre Charles succeeding him on his sudden death in October. The main political issue is how to cope with the ending of preferential banana exports to the EU.

WORLD AFFAIRS

Joined UN in 1978

ACS
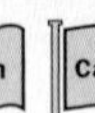
Comm

Caricom

OAS

OECS

Dominica has close links with France and the US. Preferential access to the EU for Caribbean bananas, crucial for Dominica's economy, was lost after a successful protest to the WTO in 1999 by the US.

***Inshore fishing boats,** which mostly supply the domestic market, on a typical Dominican beach.*

AID

▷ Recipient

 US$15m (receipts)  Up 50% in 2000

The European Development Fund allocated money in 1998 for sustainable tourism projects; Japan and Taiwan gave aid for firefighting and education.

DEFENSE

▷ No compulsory military service

Defense forces were officially disbanded in 1981 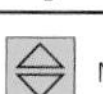 Not applicable

Dominica has no armed forces, but it does participate in the US-sponsored Regional Security System.

ECONOMICS

▷ Inflation 3% p.a. (1990–2000)

 US$238m  2.7 Eastern Caribbean dollars

SCORE CARD

❑ WORLD GNP RANKING	181st
❑ GNP PER CAPITA	US$3260
❑ BALANCE OF PAYMENTS	–US69m
❑ INFLATION	0.8%
❑ UNEMPLOYMENT	20%

STRENGTHS

Bananas, though this sector has declined since the loss of EU preferential access. Offshore business center and "economic citizenship" scheme. Growing services sector.

WEAKNESSES

Dependence on US and EU markets for its banana crop, threatened by WTO ruling. Low productivity in public sector. Poor infrastructure.

EXPORTS

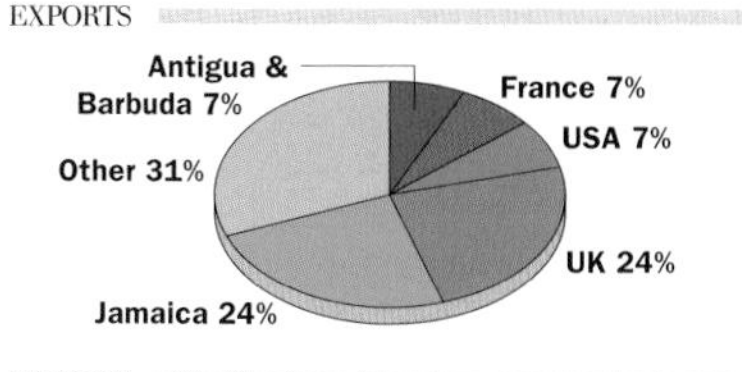
Antigua & Barbuda 7%
France 7%
USA 7%
Other 31%
UK 24%
Jamaica 24%

IMPORTS

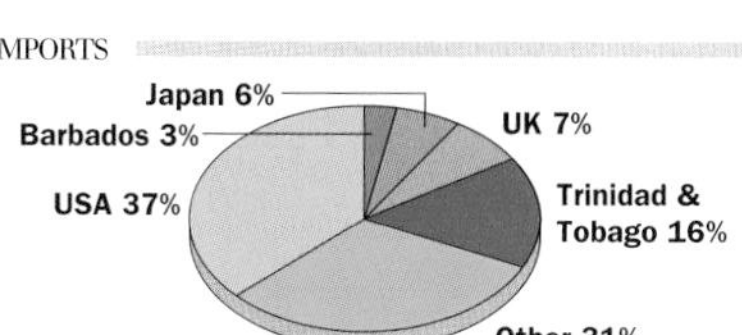
Japan 6%
Barbados 3%
UK 7%
USA 37%
Trinidad & Tobago 16%
Other 31%

RESOURCES

▷ Electric power 8000 kw

 1205 tonnes

 Not an oil producer

13,400 cattle, 9700 goats, 7600 sheep, 190,000 chickens

None

Dominica has no natural resources. A hydroelectric power plant in the Morne Trois Pitons national park provides most of the island's power.

ENVIRONMENT

▷ Not available

9%  1.2 tonnes per capita

Increased agriculture and timber harvesting is threatening Dominica's rainforest; already there is more land under cultivation than planned by the government. The current promotion of the rainforest as a tourist attraction poses a threat, as does a possible expansion in HEP generators. Two species of parrot – the imperial, or sisserou, and the red-necked – are threatened, despite conservation orders. Endangered hawksbill turtles, living on coral reefs off the island, are traditionally hunted.

MEDIA

▷ TV ownership medium

 There are no daily newspapers

PUBLISHING AND BROADCAST MEDIA

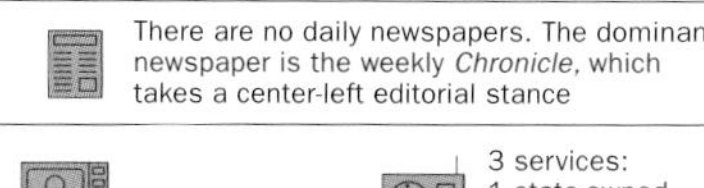

There are no daily newspapers. The dominant newspaper is the weekly *Chronicle*, which takes a center-left editorial stance

No TV service 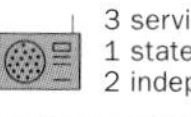 3 services: 1 state-owned, 2 independent

Local franchises, offering cable TV with selected US networks, serve one-third of the island. Broadcasts from other Caribbean states can also be received. There are five weekly newspapers.

CRIME

▷ Death penalty in use

 298 prisoners  Up 8% in 1999

Dominica has a lower crime rate than most of its Caribbean neighbors. Burglary and armed robbery are the major concerns; murders are rare. Justice is based on British common law and administered by the Eastern Caribbean Supreme Court, which is based on the island of St. Lucia.

EDUCATION

▷ School leaving age: 16

 94%  461 students

Dominican education is based on the British model, and retains the selective 11-plus exam for entrance into high school. Students go on to the University of the West Indies or, increasingly, to colleges in the US and the UK.

CHRONOLOGY

Colonized first by the French, Dominica came under British control in 1759.

- ❑ **1975** Morne Trois Pitons national park established.
- ❑ **1978** Independence from UK. Patrick John first prime minister.
- ❑ **1980** Eugenia Charles becomes Caribbean's first woman prime minister.
- ❑ **1981** Two coup attempts, backed by Patrick John, foiled.
- ❑ **1995** Opposition DUWP defeats DFP. Dame Eugenia Charles retires after 27 years in politics.
- ❑ **1999** WTO ruling on preferential access for bananas to EU market.
- ❑ **2000** DLP wins elections.

HEALTH

▷ Welfare state health benefits

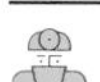 1 per 2041 people  Heart and respiratory diseases, cancers

There are numerous health centers. Difficult communications hamper emergency hospital access for people living in the interior.

SPENDING

▷ GDP/cap. increase

CONSUMPTION AND SPENDING

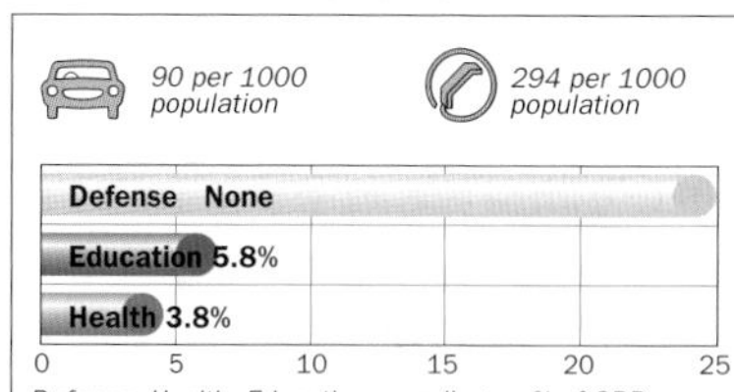
90 per 1000 population
294 per 1000 population
Defense None
Education 5.8%
Health 3.8%
0
5
10
15
20
25
Defense, Health, Education spending as % of GDP

Wealth disparities are not as marked in Dominica as they are on the larger Caribbean islands, but the alleviation of poverty has become a major plank of government policy. Measures taken include increased benefits and help for the country's pensioners.

WORLD RANKING

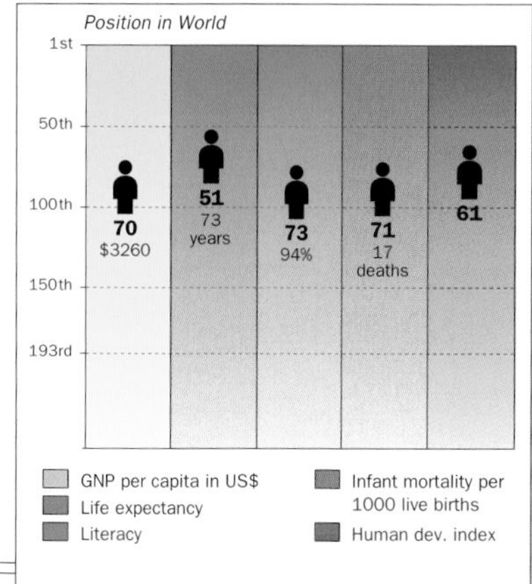
Position in World
1st
50th
100th
150th
193rd
70
$3260
51
73 years
73
94%
71
17 deaths
61
GNP per capita in US$
Life expectancy
Literacy
Infant mortality per 1000 live births
Human dev. index

Dominican Republic

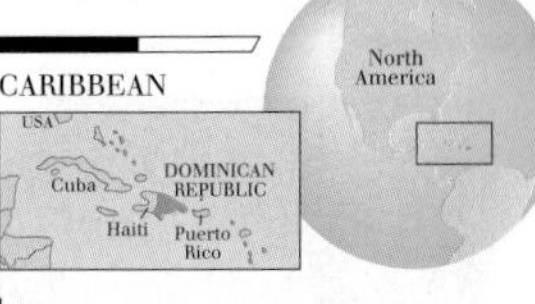
CARIBBEAN
USA
Cuba
DOMINICAN REPUBLIC
Haiti
Puerto Rico
North America

Official Name: Dominican Republic **Capital:** Santo Domingo
Population: 8.5 million **Currency:** Dominican Republic peso **Official Language:** Spanish

1865

1865
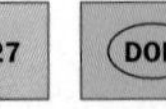
Feb 27
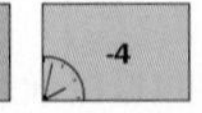
DOM

-4

+1809

.do

THE MOST POPULAR tourist destination in the Caribbean, the Dominican Republic lies 970 km (600 miles) southeast of Florida. Once ruled by Spain, it occupies the eastern two-thirds of the island of Hispaniola and boasts both the region's highest point (Pico Duarte, 3088 m – 10,131 ft) and its lowest (Lake Enriquillo, 44 m – 144 ft – below sea level). Spanish-speaking, it seeks closer ties with the anglophone West Indies.

View south from Pico Duarte along the fertile banks of the Río Yaque del Norte.

CLIMATE

Tropical equatorial/ oceanic

WEATHER CHART FOR SANTO DOMINGO

Average daily temperature — Rainfall
°C/°F J F M A M J J A S O N D cm/in
40/104 · 30/86 · 20/68 · 10/50 · 0/32 · -10/14 · -20/-4
40/16 · 30/12 · 20/8 · 10/4 · 0

The trade winds blow all year round, providing relief from the tropical heat and humidity. The hurricane season runs from June until November.

TRANSPORTATION

Drive on right

Aeropuerto International de las Américas, Santo Domingo 4.65m passengers

24 ships 10,400 grt

THE TRANSPORTATION NETWORK

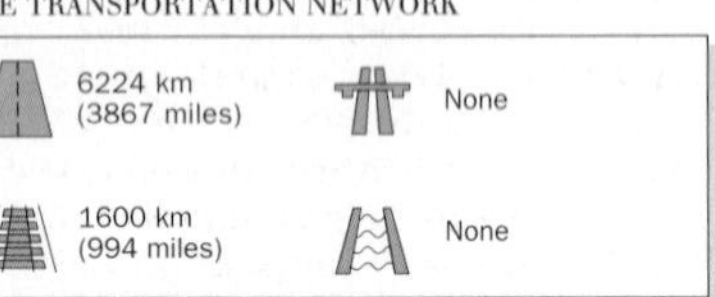
6224 km
(3867 miles)
None
1600 km
(994 miles)
None

Urban and rural transportation is poor; railroads are mainly for transporting sugarcane and ores. An international consortium in 1999 won a 30-year concession to operate four airports.

TOURISM

Visitors : Population 1:2.9

2.97m visitors

Up 12% in 2000

MAIN TOURIST ARRIVALS

	% of total arrivals
USA	18%
Germany	17%
Canada	7%
Other	58%

0 10 20 30 40 50 60 — % of total arrivals

Ample accommodation and excellent beaches attract many tourists each year, mainly from Europe and North America.

PEOPLE

Pop. density medium

Spanish, French Creole

176/km² (455/mi²)

THE URBAN/RURAL POPULATION SPLIT

65% 35%

RELIGIOUS PERSUASION

Other and nonreligious 8%
Roman Catholic 92%

The white population, primarily the descendants of Spanish settlers, still owns most of the land. The mixed race majority – about 73% – controls much of the republic's commerce, and forms the bulk of the professional middle classes. Blacks, the descendants of Africans, are mainly small-scale farmers and often the victims of latent racism, especially in business. Women in the black community work the farms; in the white and mixed race communities women are starting to make professional careers.

POLITICS

Multiparty elections

L. House 2002/2006
U. House 2002/2006

President Hipolito Mejia

AT THE LAST ELECTION

Chamber of Deputies 150 seats

49% PRD · 27% PLD · 24% PRSC

PRD = Dominican Revolutionary Party **PLD** = Dominican Liberation Party **PRSC** = Christian Social Reform Party

Senate 32 seats

91% PRD · 6% PRSC · 3% PLD

Joaquín Balaguer of the PRSC, a political patriarch since the 1960s and representative of the white elite and the military, achieved a bogus victory in a 1994 poll. He was forced to agree to fresh elections in 1996 which were narrowly won by Leonel Fernández of the more moderate PLD. The PRSC and PLD later joined forces against the opposition center-left PRD, which won control of the Congress in 1998. The PRD called for a fair presidential election in 2000; it was won by the PRD candidate, Hipolito Mejia.

Dominican Republic

Total Land Area : 48 380 sq. km (18 679 sq. miles)

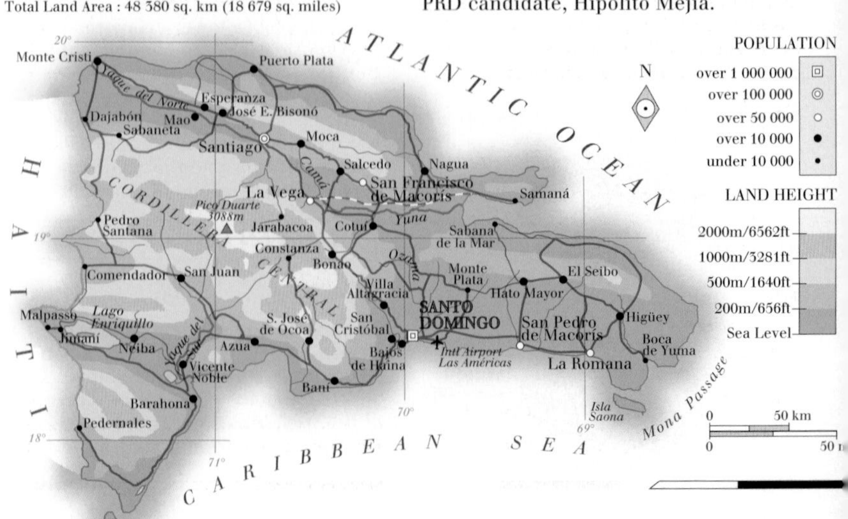
ATLANTIC OCEAN
CARIBBEAN SEA
HAITI
CORDILLERA CENTRAL
Monte Cristi
Puerto Plata
Yaque del Norte
Esperanza
José E. Bisonó
Dajabón
Mao
Sabaneta
Santiago
Moca
Salcedo
Nagua
San Francisco de Macorís
Samaná
La Vega
Camú
Pico Duarte 3088m
Jarabacoa
Cotuí
Yuna
Pedro Santana
Constanza
Bonao
Sabana de la Mar
Comendador
San Juan
Villa Altagracia
Ozama
Monte Plata
Hato Mayor
El Seibo
Malpasso
Lago Enriquillo
S. José de Ocoa
San Cristóbal
SANTO DOMINGO
San Pedro de Macorís
Higüey
Jimaní
Neiba
Azua
Bajos de Haina
Intl Airport Las Américas
La Romana
Boca de Yuma
Vicente Noble
Baní
Barahona
Isla Saona
Mona Passage
Pedernales
Isla Beata
N
POPULATION
over 1 000 000
over 100 000
over 50 000
over 10 000
under 10 000
LAND HEIGHT
2000m/6562ft
1000m/3281ft
500m/1640ft
200m/656ft
Sea Level
0 50 km

D

WORLD AFFAIRS

Joined UN in 1945

ACS
Geplac
IBRD
OAS
SELA

Relations with Haiti, with which it shares the island of Hispaniola, are important. The Dominican Republic favors a "strategic alliance" between the Caribbean and Central America.

AID

Recipient

$62m (receipts)

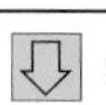
Down 68% in 2000

Multilateral and bilateral aid of some $235 million was granted in 1998 to repair severe hurricane damage.

DEFENSE

No compulsory military service

$112m

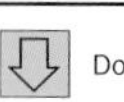
Down 2% in 2000

The military has economic and political interests, but no longer holds the defense portfolio. It focuses on illegal immigration from Haiti. The main arms supplier is the US.

ECONOMICS

Inflation 9.4% p.a. (1990–2000)

$17.8bn

16.12–16.40 Dominican Republic pesos

SCORE CARD

- ❑ World GNP Ranking..........71st
- ❑ GNP per Capita..........$2130
- ❑ Balance of Payments..........–$1.03bn
- ❑ Inflation..........6.5%
- ❑ Unemployment..........14%

Strengths
Sustained tourism growth. Mining – mainly of nickel and gold – and sugar major sectors. Hand-made cigars are biggest sellers in US. Large hidden economy based on transshipment of narcotics to US.

Weaknesses
Major sectors severely affected by fluctuating world prices and cutbacks in US import quotas. Poor creditworthiness. Failure to diversify. Stalled privatizations.

EXPORTS

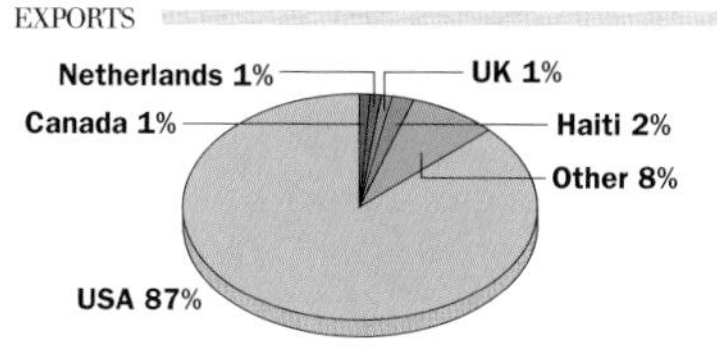
Netherlands 1%
UK 1%
Canada 1%
Haiti 2%
Other 8%
USA 87%

IMPORTS

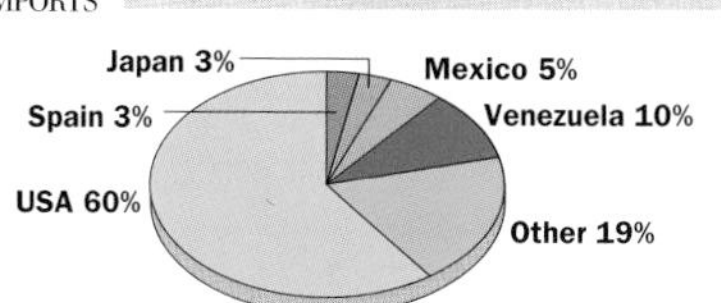
Japan 3%
Mexico 5%
Spain 3%
Venezuela 10%
USA 60%
Other 19%

RESOURCES

Electric power 2.2m kw

9269 tonnes

Not an oil producer

2.11m cattle, 565,529 pigs, 47.4m chickens

Ferro-nickel, bauxite, copper, gold, silver

The Dominican Republic is a net energy importer: hydroelectric generators are the only domestic source of power, and electricity blackouts can be a major problem. Oil prospecting has been unsuccessful, and oil is imported from Mexico and Venezuela on preferential terms under the San José Agreement. The Dominican Republic's quota from Venezuela was increased under the 2000 Caracas Accord.

ENVIRONMENT

Sustainability rank: 79th

32% (10% partially protected)

2.5 tonnes per capita

Forests are threatened by destructive agricultural practices and also by the use of wood as fuel by rural communities. Deforestation has accelerated soil erosion.

MEDIA

TV ownership medium

Daily newspaper circulation 52 per 1000 people

PUBLISHING AND BROADCAST MEDIA

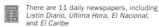

There are 11 daily newspapers, including *Listín Diario, Ultima Hora, El Nacional,* and *El Caribe*

7 services: 1 state-owned, 6 independent

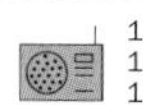
131 services: 1 state-owned, 130 independent

Television broadcasts from both Mexico and the US can easily be received in the Dominican Republic.

CRIME

No death penalty

15,340 prisoners

Thefts decreased 1996–1998

The Dominican Republic is increasingly used by narcotics cartels as a transit point to the US. Narcotics trafficking and arms smuggling are linked to the high levels of violent crime.

EDUCATION

School leaving age: 17

84%

176,995 students

State schools are badly underfunded. The state university of Santo Domingo suffered a financial crisis in 1999. The rich send their children to study in the US and Spain.

HEALTH

Welfare state health benefits

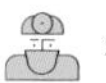
1 per 455 people

Heart attacks, infectious and parasitic diseases

Wealthy Dominicans fly to Cuba and the US for treatment. The poor rely on a basic public service, inadequately provided by some 20 state hospitals.

CHRONOLOGY

The 1697 Franco-Spanish partition of Hispaniola left Spain with the eastern two-thirds of the island, now the Dominican Republic.

- ❑ **1865** Independence from Spain.
- ❑ **1930–1961** Gen. Molina dictator.
- ❑ **1965** Civil war. US intervention.
- ❑ **1966** Joaquín Balaguer begins first of seven presidential terms over next 30 years.
- ❑ **1996** Center-left PRD candidate Fernández succeeds Balaguer.
- ❑ **1998** Major hurricane damage.
- ❑ **2000** Hipolito Mejia of PRD wins presidency.

SPENDING

GDP/cap. increase

CONSUMPTION AND SPENDING

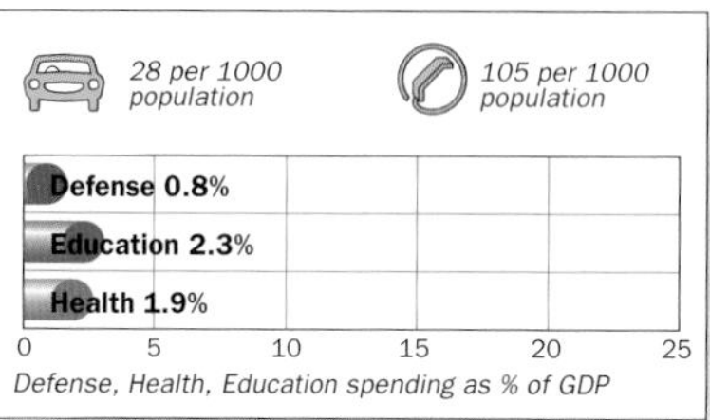
28 per 1000 population
105 per 1000 population
Defense 0.8%
Education 2.3%
Health 1.9%
0
5
10
15
20
25
Defense, Health, Education spending as % of GDP

Great disparities exist between rich and poor. The government in 1998 announced a seven-year plan to relieve poverty and reduce the level of malnutrition affecting well over two million people. Black Dominicans remain at the bottom of the economic and social ladder, accounting for the major proportion of small farmers and unemployed. Haitian immigrants are poorly paid, badly treated, and liable to be deported at short notice. Mixed races have shown most upward mobility in recent years, but, nevertheless, the old Spanish families still form the wealthiest section of society and retain their grip on valuable estates.

WORLD RANKING

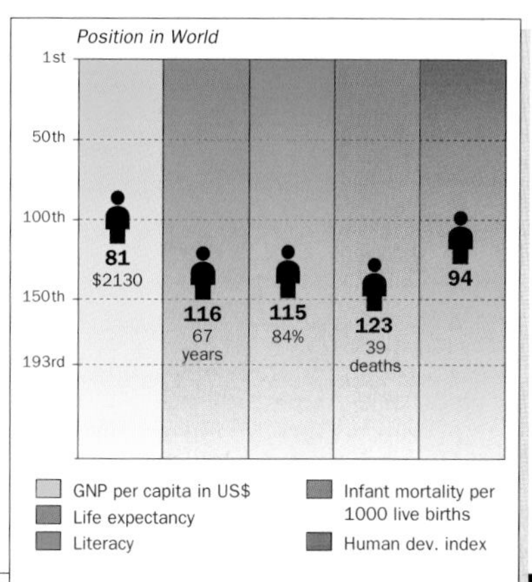
Position in World
1st
50th
100th
150th
193rd
81
$2130
116
67 years
115
84%
123
39 deaths
94
GNP per capita in US$
Life expectancy
Literacy
Infant mortality per 1000 live births
Human dev. index

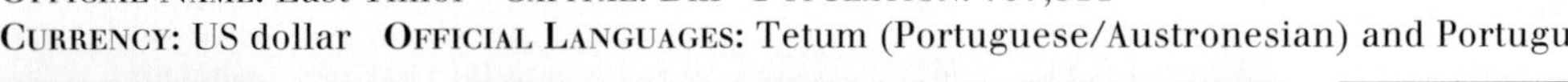

East Timor

Official Name: East Timor **Capital:** Dili **Population:** 737,811
Currency: US dollar **Official Languages:** Tetum (Portuguese/Austronesian) and Portuguese

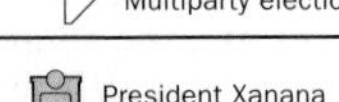
SOUTHEAST ASIA
Asia
Malaysia
Papua New Guinea
Indonesia
EAST TIMOR
Australia

2002

2002

May 20

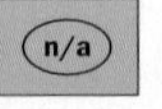
n/a

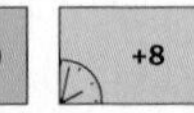
+8

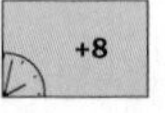
+670

.tp

Lying north of Australia across the Timor Sea, the island of Timor has a narrow coastal plain giving way to forested highlands. Its mountainous backbone rises to 2963 m (9715 ft). The eastern half was colonized for over 400 years by the Portuguese, then occupied from 1975 by Indonesia, whose forces hunted down all resistance. A referendum in 1999 launched a turbulent transition to independence in May 2002.

CLIMATE

Tropical equatorial

WEATHER CHART FOR DILI

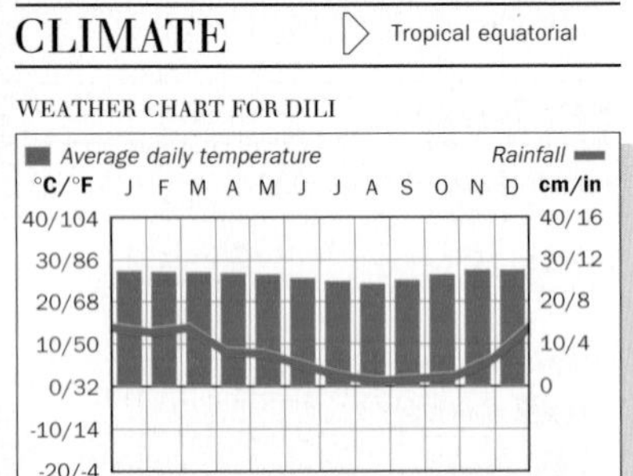
Average daily temperature
Rainfall
°C/°F
J F M A M J J A S O N D
cm/in
40/104
30/86
20/68
10/50
0/32
-10/14
-20/-4
40/16
30/12
20/8
10/4
0

The climate is tropical, with heavy rain from December to March, then dry and increasingly hot weather for the rest of the year, especially in the north.

TRANSPORTATION

Not available

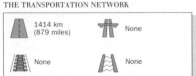

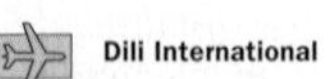
Dili International

Not available

THE TRANSPORTATION NETWORK

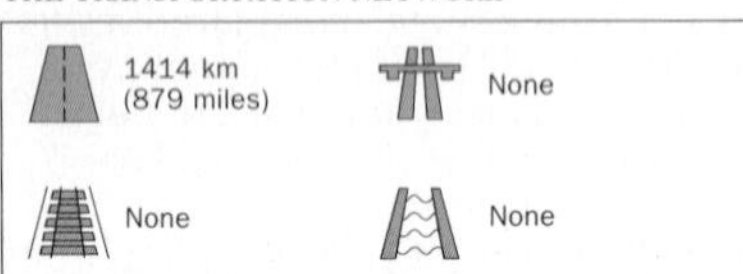
1414 km
(879 miles)
None
None
None

Roads are of poor quality and public transportation beyond Dili is unreliable and sparse.

TOURISM

Visitors : Population 1:246

3000 visitors

Down 70% in 1998

MAIN TOURIST ARRIVALS

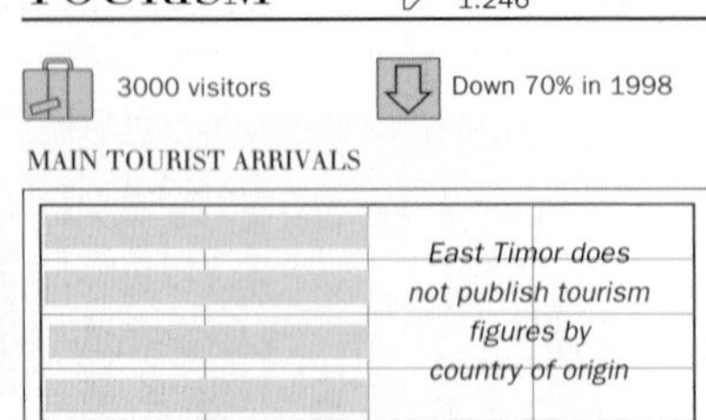
East Timor does not publish tourism figures by country of origin
0
10
20
30
40

The number of tourists fell dramatically after the preindependence violence in the region. Visitors are now generally discouraged by the high levels of crime and the almost complete lack of a tourist infrastructure.

PEOPLE

Pop. density medium

Tetum (Portuguese/Austronesian), Bahasa Indonesia, and Portuguese

51/km² (131/mi²)

THE URBAN/RURAL POPULATION SPLIT

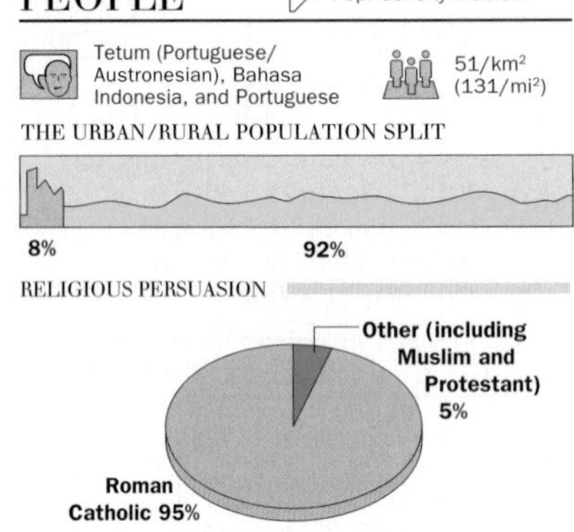
8%
92%
RELIGIOUS PERSUASION
Other (including Muslim and Protestant) 5%
Roman Catholic 95%

East Timor is almost entirely Roman Catholic. Preindependence violence was politically rather than ethnically motivated. The Timorese are a mix of Malay and Papuan peoples. Many indigenous Papuan tribes survive and there is an urban Chinese minority. Ethnic Indonesian settlers became numerous after annexation, and consisted 20% of the population by 1999.

Women do not enjoy a high profile in public life. The incidence of domestic violence is notably high.

POLITICS

Multiparty elections

2001/Transitional

President Xanana Gusmão

AT THE LAST ELECTION

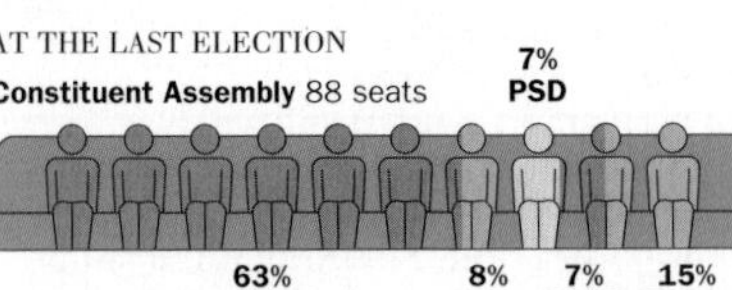
Constituent Assembly 88 seats
7% PSD
63% Fretilin
8% PD
7% ASDT
15% Others

Fretilin = Revolutionary Front of an Independent East Timor
PD = Democratic Party **PSD** = Social-Democratic Party
ASDT = Timorese Social-Democratic Association

East Timor is an emerging multiparty democracy. The Fretilin movement was the leading voice in the long struggle for independence. Turnout at recent elections was high.

After years of unrest and human rights abuses committed by the Indonesian army, the Indonesian government in 1999 conceded a referendum on East Timor's future. Pro-Indonesian militias went on the rampage, murdering hundreds in indiscriminate attacks, and forcing thousands of people into the Indonesian-controlled western half of the island. An international peacekeeping force, led by Australia, eventually secured relative calm and organized the promised vote on 30 August 1999. An overwhelming 80% of voters endorsed independence.

The UN Mission in East Timor was given full power over the territory in October 1999. Fretilin emerged as the outright victor in elections in late 2001 to the new Constituent Assembly. Its popular leader, Xanana Gusmão, reversed his decision to retire from politics and was duly elected president in 2002. Independence took effect on 20 May that year.

The new government, headed by Prime Minister Mari Alkatiri, has pledged to concentrate spending on health and education.

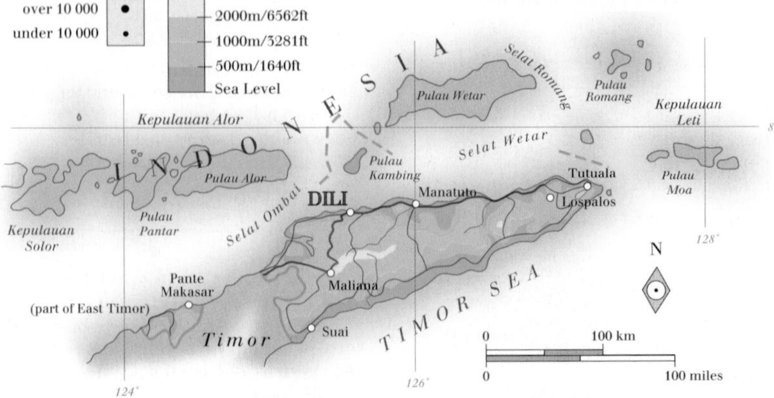
East Timor
Total Land Area : 14,874 sq. km (5756 sq. miles)
POPULATION
LAND HEIGHT
over 10 000
under 10 000
2000m/6562ft
1000m/3281ft
500m/1640ft
Sea Level
INDONESIA
Selat Romang
Pulau Romang
Pulau Wetar
Kepulauan Leti
Kepulauan Alor
Selat Wetar
Pulau Kambing
Pulau Alor
Tutuala
Pulau Moa
Manatuto
Lospalos
DILI
Selat Ombai
Pulau Pantar
Kepulauan Solor
Pante Makasar
(part of East Timor)
Maliana
Timor
Suai
TIMOR SEA
N
0
100 km
100 miles
124°
126°
128°
8°

WORLD AFFAIRS

Joined UN in 2002

IMF

Relations with Indonesia remain strained over the issue of justice for past human rights abuses. Portugal, the former colonial power, had led international opposition to Indonesia's annexation of the territory. Australia, one of the few Western countries to recognize that annexation, later swung vital support behind the cause of independence. It is also a key player in providing future aid and assistance for reconstruction.

East Timor applied to join the UN and ASEAN shortly after it achieved independence in 2002.

AID

Recipient

 $233m (receipts)  Up 52% in 2000

International development aid provides the backbone of East Timor's GDP. Australia and Portugal are by far the biggest donors.

DEFENSE

No compulsory military service

 Not available  Not available

The East Timorese Defense Force was established in 2001 from the remains of pro-independence militia. Its role is largely as an extension of the police force, as external defense is still provided by UN troops.

ECONOMICS

Not available

 $403m 

SCORE CARD

- ❑ World GNP Ranking 172nd
- ❑ GNP per Capita $458
- ❑ Balance of Payments $8m
- ❑ Inflation 0%
- ❑ Unemployment 4%

STRENGTHS

Potential from oil and natural gas reserves in Timor Sea. Traditional agricultural base: coffee the main export.

WEAKNESSES

Infrastructure devastated by 1999 violence. Insecurity deterred investment. Undeveloped industrial sector.

EXPORTS/IMPORTS

Export and import figures are not available for East Timor

Despite its young age, East Timor has a strong national identity, based largely on the domination of Roman Catholicism.

RESOURCES

Electric power 14,400 kw

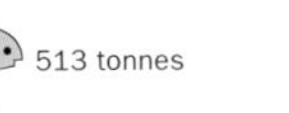  Oil figures not available

378,000 pigs, 187,000 cattle, 1.02m chickens  Oil, natural gas, gold, manganese, marble

The country has few natural resources. In 2001 East Timor successfully negotiated with Australia for a greater share of the oil under the Timor Sea.

ENVIRONMENT

Not available

 Not available  Not available

Unrestricted logging under Indonesian rule has greatly diminished important species and contributed to erosion of the country's poor quality soil.

MEDIA

TV ownership low

 Daily newspaper circulation figures are not available

PUBLISHING AND BROADCAST MEDIA

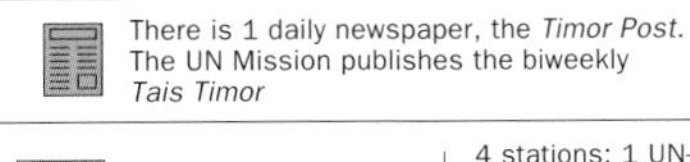
There is 1 daily newspaper, the *Timor Post*. The UN Mission publishes the biweekly *Tais Timor*

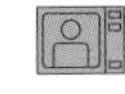 1 service, run by the UN  4 stations: 1 UN-run, 1 run by the Catholic Church

Official newsletters have the highest circulation. A number of private papers were established in 2000, including the *Timor Post*, which was formed from the remains of the main preindependence daily, *Suara Timor Timur*.

CRIME

No death penalty

 East Timor does not publish prison figures  Crime levels are low but rising

Petty and violent crimes are common. Most of the perpetrators of violence in 1999 have yet to be brought to justice.

EDUCATION

Not available

 41%  7500 students

During Indonesian domination, classes were taught in Bahasa Indonesia. The number of students attending school and university has already begun to recover from the sharp decline in 1999.

CHRONOLOGY

The Portuguese arrived in Timor in the 1520s. It was formally divided by Portugal and the Netherlands in 1859.

- ❑ **1949** Dutch west Timor becomes part of Indonesia.
- ❑ **1975** Fretilin declares East Timor independent; Indonesia invades.
- ❑ **1991** Massacre of pro-independence demonstrators in Dili.
- ❑ **1996** Timorese leaders receive Nobel Peace Prize, raising global awareness.
- ❑ **1999** Indonesian government agrees to hold referendum; resulting violence quelled by UN force.
- ❑ **2001** Elections to new Constituent Assembly; Fretilin wins majority.
- ❑ **2002** Xanana Gusmão elected president. Independence.

E

HEALTH

Not available

 1 per 40,000 people  Not available

Life expectancy in East Timor has traditionally been lower than in the rest of the Indonesian archipelago. Most doctors left the province in 1999.

SPENDING

Not available

CONSUMPTION AND SPENDING

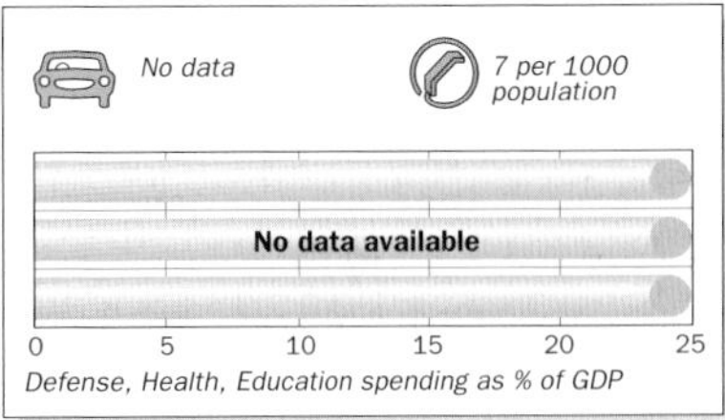
No data
7 per 1000 population
No data available
0
5
10
15
20
25
Defense, Health, Education spending as % of GDP

Living standards, already relatively low for the region, were made worse by the events of 1999. Thousands were left homeless. Well-paid UN staff enjoy a sharply contrasting lifestyle.

WORLD RANKING

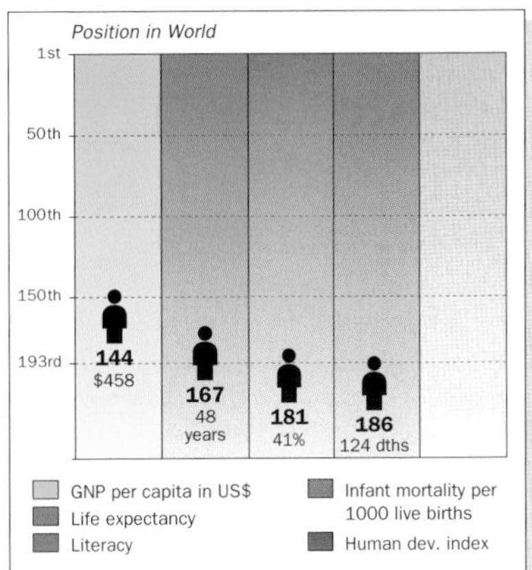
Position in World
1st
50th
100th
150th
193rd
144
$458
167
48 years
181
41%
186
124 dths
GNP per capita in US$
Life expectancy
Literacy
Infant mortality per 1000 live births
Human dev. index

ECUADOR

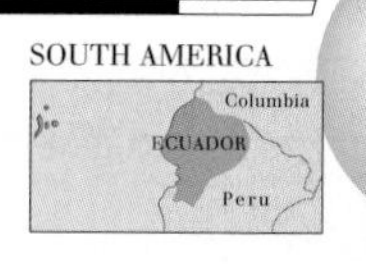
SOUTH AMERICA
Columbia
ECUADOR
Peru
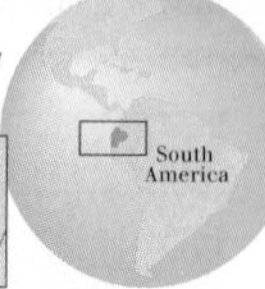
South America

OFFICIAL NAME: Republic of Ecuador **CAPITAL:** Quito
POPULATION: 12.9 million **CURRENCY:** US dollar **OFFICIAL LANGUAGE:** Spanish

1830

1941

Aug 10
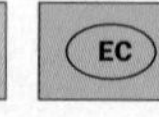
EC

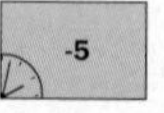
-5

+593

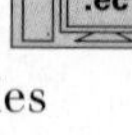
.ec

ONCE PART OF THE INCA heartland, Ecuador lies on the western coast of South America. It was ruled by Spain from 1533, when the last Inca emperor was executed, until independence in 1830. Most Ecuadorians live either in the lowland coastal region or in the Andean Sierra. The Amazonian Amerindians are now pressing for their land rights to be recognized. Massive depreciation of the sucre forced the government to dollarize the currency in 2000.

CLIMATE

Tropical/mountain

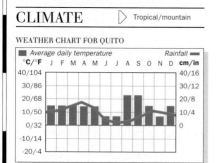

WEATHER CHART FOR QUITO

Average daily temperature — Rainfall

°C/°F J F M A M J J A S O N D cm/in

40/104, 30/86, 20/68, 10/50, 0/32, -10/14, -20/-4 — 40/16, 30/12, 20/8, 10/4, 0

Climate varies from hot equatorial in the Amazon forests, to dry heat in the south and "perpetual spring" in Quito.

TRANSPORTATION

Drive on right

Mariscal Sucre, Quito 2.04m passengers

171 ships 300,900 grt

THE TRANSPORTATION NETWORK

8165 km (5073 miles)

Pan-American Highway

956 km (594 miles)

1500 km (932 miles)

The road network and railroad are grossly underfunded. Serious political unrest in 2000–2001 blocked roads.

TOURISM

Visitors : Population 1:21

615,000 visitors

Up 21% in 2000

MAIN TOURIST ARRIVALS

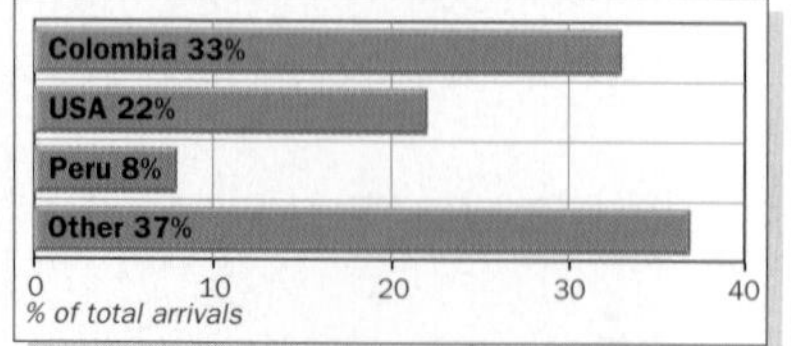
Colombia 33%
USA 22%
Peru 8%
Other 37%
0
10
20
30
40
% of total arrivals

Tourism is growing. Quito, once the capital of the Inca empire, has restored many of its Spanish imperial buildings, including 86 churches. Access to the Galapagos Islands is restricted to 60,000 visitors a year.

PEOPLE

Pop. density low

Spanish, Quechua, other Amerindian languages

47/km² (121/mi²)

THE URBAN/RURAL POPULATION SPLIT

65% 35%

RELIGIOUS PERSUASION

Protestant, Jewish and Other 7%

Roman Catholic 93%

Over half of the population is of Amerindian–Spanish extraction (*mestizo*). Black communities exist on the coast. The Amerindians, who make up about one-quarter of the population, are pressing for Ecuador to be described as a plurinational state, within which the different indigenous communities are recognized as distinct nationalities. The strong and largely unified Amerindian movement is at the forefront of social protests.

POLITICS

Multiparty elections

1998/2002

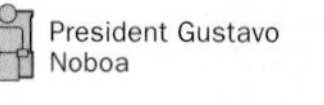

President Gustavo Noboa

AT THE LAST ELECTION

National Congress 121 seats

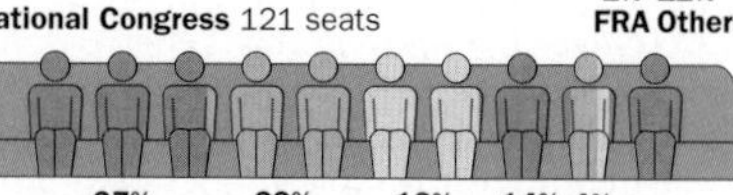
2% FRA
12% Others
27% DP
23% PSC
18% PRE
14% ID
4% NMN–PP

DP = Popular Democracy **PSC** = Social Christian Party
PRE = Ecuadorian Roldosist Party **ID** = Democratic Left
NMN–PP = New Country–Pachakutik Movement
FRA = Alfarist Radical Front

101 provincial members are elected for two-year terms, and 20 national members are directly elected for four-year terms

After the instability and corruption of the late 1990s, a new broad-based alliance was formed in 1999. Severe austerity reforms provoked widespread protests; the army intervened and Vice President Gustavo Noboa took over the presidency in 2000. The dollarization of the economy in that year led to popular protest, which resurfaced more widely and violently in response to IMF-backed austerity in 2001.

WORLD AFFAIRS

Joined UN in 1945

AP AmCC NAM OAS RG

Access to US and EU markets for bananas and oil prices are major concerns. There are serious security problems on the border with Colombia.

ECUADOR
Total Land Area : 283 560 sq. km (109 483 sq. miles)
POPULATION
over 1 000 000
over 500 000
over 100 000
over 50 000
over 10 000
under 10 000
PACIFIC OCEAN
COLOMBIA
PERU
ANDES
QUITO
San Lorenzo
Esmeraldas
Tulcán
Ibarra
Otavalo
Cayambe
Nueva Loja
Puerto El Carmen de Putumayo
Santo Domingo de los Colorados
Equator
Caráquez
Chone
Manta
Calceta
Portoviejo
Velasco Ibarra
Latacunga
Cotopaxi
Tena
Jipijapa
Balzar
Ambato
Baños
Chimborazo 6310 m
Guaranda
Riobamba
Puyo
Vinces
Babahoyo
Daule
Milagro
Salinas
La Libertad
Guayaquil
Eloy Alfaro
Alausí
Macas
Gulf of Guayaquil
Isla Puná
Cuenca
Azogues
Sigsig
Machala
Pasaje
Santa Rosa
Loja
Zamora
Macará
Nuevo Rocafuerte
San Miguel
Aguarico
Napo
Curaray
Pastaza
Galapagos Islands
I. Darwin
I. Wolf
I. Pinta
I. Marchena
I. Genovesa
I. Fernandina
I. Santiago
I. Santa Cruz
I. Isabela
I. San Cristóbal
Puerto Baquerizo Moreno
I. Santa María
I. Española
100 km
100 miles
LAND HEIGHT
4000m/13124ft
2000m/6562ft
500m/1640ft
Sea Level
N

Quito is the highest capital in the world after La Paz in Bolivia. It lies in an Andean valley, lined with 30 volcanoes.

AID

 Recipient

 $147m (receipts)

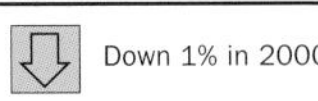 Down 1% in 2000

Aid from the US, Japan, Spain, and the IDB alleviates the heavy foreign debt burden. The Galapagos Islands receive generous grants from UNESCO.

DEFENSE

Compulsory military service

 $314m

 Down 7% in 2000

The army kept out of politics from the mid-1970s until its intervention in 2000. Moves to reduce the military's 50% "royalty" share of oil revenues, its prime source of funding, have raised tensions.

ECONOMICS

Inflation 37% p.a. (1990–2000)

 $15.3bn

 Currency is US dollar

SCORE CARD

- ❑ WORLD GNP RANKING..........74th
- ❑ GNP PER CAPITA..........$1210
- ❑ BALANCE OF PAYMENTS..........$928m
- ❑ INFLATION..........96.1%
- ❑ UNEMPLOYMENT..........13%

STRENGTHS

Net oil exporter. World's second-biggest banana producer. Fishing industry.

WEAKNESSES

Poor infrastructure and land productivity. Energy crises. High inflation. Financial instability. Confusion over adoption of US dollar as official currency in 2000.

EXPORTS

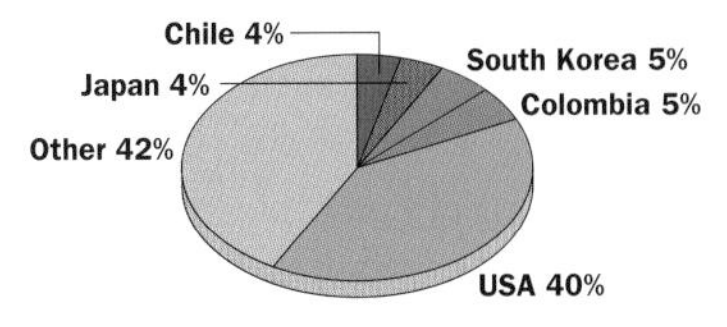
Chile 4%
Japan 4%
South Korea 5%
Colombia 5%
Other 42%
USA 40%

IMPORTS

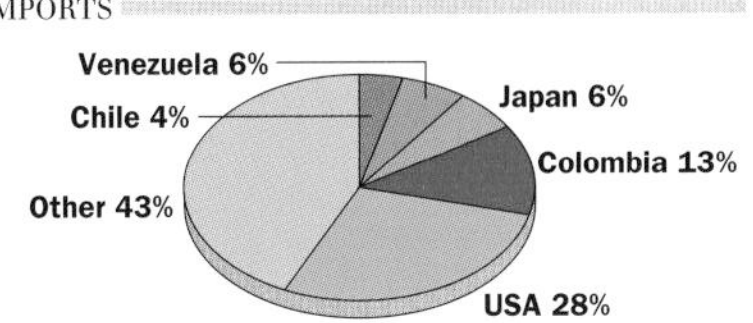
Venezuela 6%
Chile 4%
Japan 6%
Colombia 13%
Other 43%
USA 28%

RESOURCES

Electric power 3.9m kw

 625,247 tonnes

 416,000 b/d (reserves 2.1bn barrels)

 5.57m cattle, 2.39m pigs, 1.98m sheep, 138m chickens

 Oil, natural gas, gold, silver, copper, zinc

The government is encouraging faster oil exploration and higher output. Ecuador left OPEC in 1992. Overfishing is threatening mackerel and squid stocks.

ENVIRONMENT

Sustainability rank: 41st

44%

 2.2 tonnes per capita

Oil drilling in new areas of Amazonia threatens indigenous tribes. Tourism, some of it illegal, has upset the delicate ecosystems of the Galapagos Islands; the land iguana is endangered, and black coral is stolen in quantity for souvenirs. The breaching of the *Jessica* oil tanker just offshore in 2001 raised concerns about shipping oil through ecologically sensitive areas.

MEDIA

TV ownership medium

 Daily newspaper circulation 70 per 1000 people

PUBLISHING AND BROADCAST MEDIA

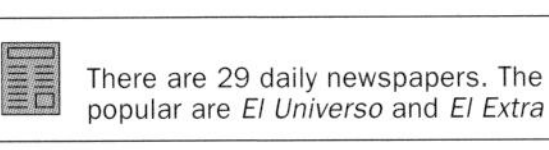 There are 29 daily newspapers. The most popular are *El Universo* and *El Extra*

 67 independent services

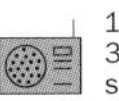 1 state-owned, 320 independent stations

The press is largely independent. It is highly regionalized, based either in the Quito region or around Guayaquil on the coast. The latter is also a center for commercial radio stations. There are ten cultural and ten religious radio stations.

CRIME

No death penalty

 8520 prisoners

 Up 1% in 1999

Right-wing paramilitaries were blamed for the murders of a trade union leader in 1998 and a left-wing congressman in 1999. The paramilitaries are rumored to be supported by Colombians. Left-wing urban guerrillas are also reported. Unprecedented numbers of citizens are applying for arms permits, while the illegal arms trade is thriving.

EDUCATION

School leaving age: 15

 91%

 206,541 students

Some 20% of Ecuadorians in the relevant age group receive higher education at 16 universities.

Programs have been launched to combat high levels of adult illiteracy in rural areas. Secondary schools are badly underfunded.

CHRONOLOGY

Alternating republican and military governments ruled Ecuador from independence in 1830 to 1978.

- ❑ **1941–1942** Loss of mineral-rich El Oro region to Peru.
- ❑ **1948–1960** Prosperity from bananas.
- ❑ **1972** Oil production starts.
- ❑ **1979** Return to democracy.
- ❑ **1992** Amerindians win land in Amazonia.
- ❑ **1996–1997** Abdalá Bucarám Ortíz removed from presidency on grounds of mental incapacity.
- ❑ **1998–1999** Jamil Mahuad of DP wins elections; forms new majority alliance. Economic crisis.
- ❑ **2000** Army sides with Amerindian protestors. Vice President Gustavo Noboa replaces Mahuad.

HEALTH

Welfare state health benefits

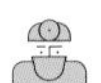 1 per 588 people

 Malnutrition, intestinal infectious diseases, pneumonia, accidents

Health care is seriously underfunded. Some services exist in poor urban districts but are still unavailable in many rural areas. Severe budget cuts mean that any improvement will depend on more outside aid.

SPENDING

GDP/cap. increase

CONSUMPTION AND SPENDING

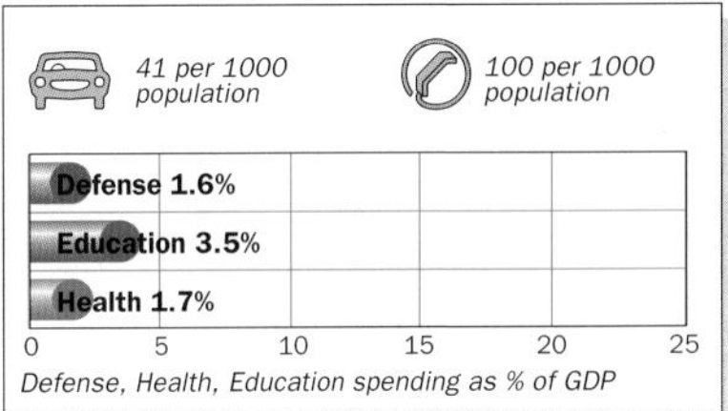
41 per 1000 population
100 per 1000 population
Defense 1.6%
Education 3.5%
Health 1.7%
0
5
10
15
20
25
Defense, Health, Education spending as % of GDP

An estimated 60% of the population live in poverty, but disparity between rich and poor is not as great as in other South American countries.

WORLD RANKING

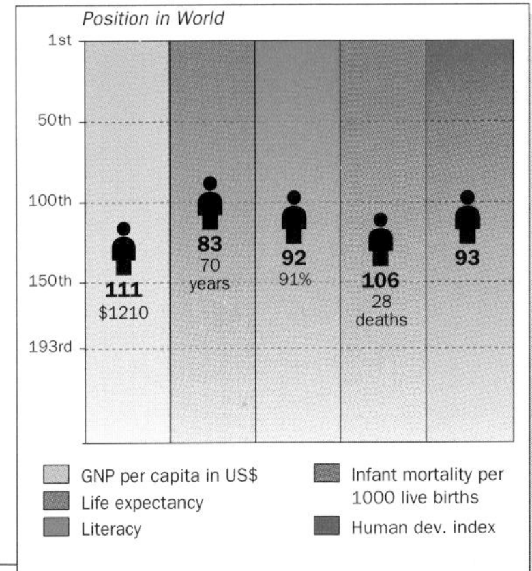
Position in World
1st
50th
100th
150th
193rd
111
$1210
83
70 years
92
91%
106
28 deaths
93
GNP per capita in US$
Life expectancy
Literacy
Infant mortality per 1000 live births
Human dev. index

E

EGYPT

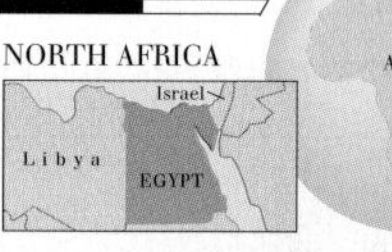
NORTH AFRICA
Israel
Libya
EGYPT
Africa

OFFICIAL NAME: Arab Republic of Egypt **CAPITAL:** Cairo
POPULATION: 69.1 million **CURRENCY:** Egyptian pound **OFFICIAL LANGUAGE:** Arabic

1936

1982

July 23

ET
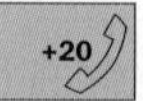
+2

+20

.eg

OCCUPYING THE NORTHEAST corner of Africa, Egypt is bisected by the highly fertile Nile valley separating the arid western desert from the smaller semiarid eastern desert. Egypt's 1979 peace treaty with Israel brought security, the return of the Sinai, and large injections of US aid. Its essentially pro-Western military-backed regime is now being challenged by an increasingly influential Islamic fundamentalist movement.

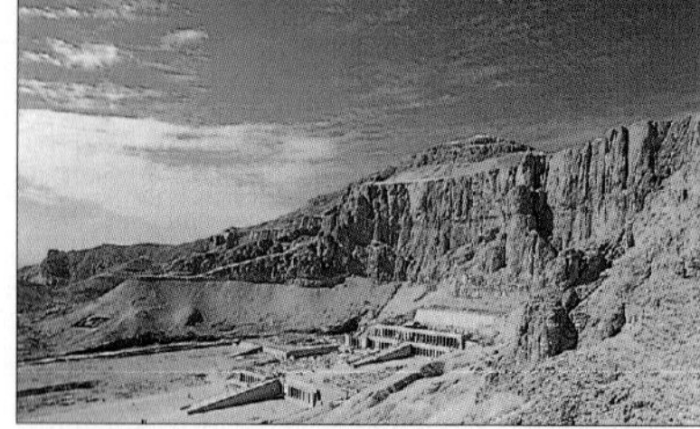

18th-Dynasty Temple of Queen Hatshepsut *dating from the Middle Kingdom, c.1480 BCE. It is at Deir el-Bahri on the west bank of the Nile opposite Thebes, Egypt's capital at the time.*

CLIMATE

Hot desert/ Mediterranean

WEATHER CHART FOR CAIRO

Average daily temperature | Rainfall
°C/°F J F M A M J J A S O N D cm/in
40/104 | 40/16
30/86 | 30/12
20/68 | 20/8
10/50 | 10/4
0/32 | 0
-10/14
-20/-4

Summers are very hot, especially in the south, but winters are cooler. The only significant rain falls in winter along the Mediterranean coast.

TRANSPORTATION

Drive on right

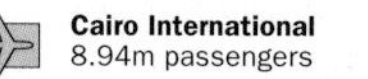
Cairo International
8.94m passengers

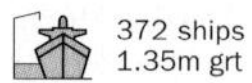
372 ships
1.35m grt

THE TRANSPORTATION NETWORK

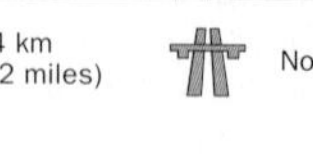
49,684 km
(30,872 miles)

None

5024 km
(3122 miles)

3500 km
(2175 miles)

Cities are linked by adequate roads, but railroads are the main transportation arteries, and trains are frequently overcrowded. The Suez Canal is a vital international shipping lane.

TOURISM

Visitors : Population 1:16

4.32m visitors

Down 16% in 2001

MAIN TOURIST ARRIVALS

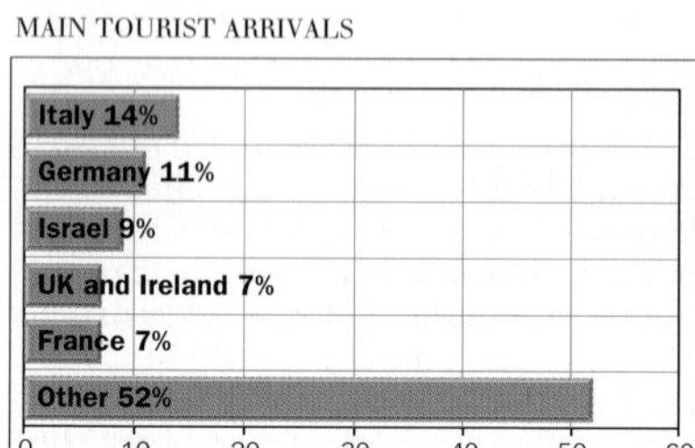
Italy 14%
Germany 11%
Israel 9%
UK and Ireland 7%
France 7%
Other 52%
0
10
20
30
40
50
60
% of total arrivals

Egypt's wealth of antiquities from its ancient civilizations have made it a key tourist destination since the 1880s. Today, it also offers Nile cruises and some of the world's best scuba diving, notably on the coral reefs near Hurghada on the Red Sea.

The industry went into sharp decline, however, when militant Islamists began targeting Western tourists in the mid-1990s; in an attack in Luxor in November 1997, 58 tourists were killed. Heightened security measures allowed for a brief recovery, but the situation worsened following the September 2001 attacks in the US; the resultant contraction in the global tourism industry affected Islamic countries in particular.

MEDITERRANEAN SEA
Nile Delta
Gulf of Sollum
Matrûh
Kafr el Dauwâr
Alexandria
Rashîd
Disûq
El Mahalla el Kubra
Dumyât
Port Said
El Mansûra
El 'Arîsh
GAZA STRIP
ISRAEL
Damanhûr
Kafr el Sheikh
Tanta
Zagazig
Ismâ'iliya
Shibîn el Kôm
Bilbeis
Suez Canal
Bitter Lakes
LOWER EGYPT
Benha
Shubrâ el Kheima
CAIRO
El Giza
Suez
Helwân
LIBYAN PLATEAU
Monkhafad el Qattâra
JORDAN
SINAI
Sinnûris
Birket Qârûn
El Faiyûm
Beni Suef
Gulf of Suez
Gebel Katherina 2637m
Gulf of Aqaba
SAUDI ARABIA
GREAT SAND SEA
Nile
El Minya
MAAZA PLATEAU
28°
Mallawi
Wadi Qena
Hurghada
Asyût
SAHARA EL GHARQÎYA
SAHARA EL SHARQÎYA
Akhmîm
Sohâg
UPPER EGYPT
Qena
Qus
Deir el-Bahri
Luxor
El Khârga
Isna
RED SEA
Idfu
LIBYAN DESERT
El Wâhât el Khârga
Wadi el-Kharit
Aswân Dam
Aswân
24°
LIBYA
Buheiret Nâsir
Wadi el 'Allâqi
GILF KEBIR PLATEAU
Administrative border
Political border
SUDAN
32°
36°

EGYPT

Total Land Area : 1 001 450 sq. km (386 660 sq. miles)

POPULATION

over 5 000 000
over 1 000 000
over 500 000
over 100 000
over 50 000
over 10 000
under 10 000

LAND HEIGHT

2000m/6562ft
1000m/3281ft
500m/1640ft
200m/656ft
Sea Level
-200m/-656ft

0 200 km
0 200 miles

E

PEOPLE

 Pop. density medium

Arabic, French, English, Berber

69/km² (180/mi²)

THE URBAN/RURAL POPULATION SPLIT

45% 55%

RELIGIOUS PERSUASION

Coptic Christian and Other 6%

Muslim (mainly Sunni) 94%

ETHNIC MAKEUP

Other (Nubian, Armenian, Greek) 10%

Eastern Hamitic 90%

Egypt has a long tradition of ethnic and religious tolerance, though the rise in Islamic fundamentalism has sparked sectarian clashes between Muslims and Coptic Christians. Most Egyptians speak Arabic, though many also have French or English as a second language. There are Berber-speaking communities in the western oases. Small colonies of Greeks and Armenians live in the larger towns. Islam is the dominant religion, followed by Coptic Christianity. Although many Jews left Egypt for Israel after 1948, a small community remains in Cairo.

Cairo is the second most populous city in Africa, and a key social question in Egypt is the high birthrate. In 1985 the government set up the National Population Council, which made birth control readily available. Since then, the birthrate has dropped from 39 to fewer than 25 per 1000 people, but population growth is still high. The population is predicted to reach almost 100 million by 2025. The growing influence of Islamic fundamentalists, who oppose contraception, could see the rate accelerate once more.

Egyptian women have been among the most liberated in the Arab world, and under a law passed in 2000 they now have the right to initiate divorce proceedings. The steady rise of Islamic fundamentalism, however, threatens their position, particularly in rural areas.

POPULATION AGE BREAKDOWN

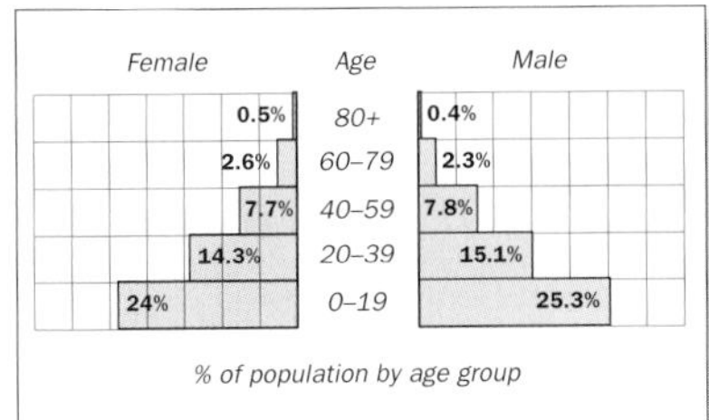
Female
Age
Male
0.5%
80+
0.4%
2.6%
60–79
2.3%
7.7%
40–59
7.8%
14.3%
20–39
15.1%
24%
0–19
25.3%
% of population by age group

POLITICS

 Multiparty elections

2000/2005

President Mohammed Hosni Mubarak

AT THE LAST ELECTION

People's Assembly 454 seats

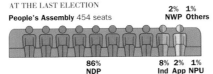

86% NDP; 8% Ind; 2% App; 1% NPU; 2% NWP; 1% Others

NDP = National Democratic Party **Ind** = Independents **App** = Appointed **NWP** = New Wafd Party **NPU** = National Progressive Unionist Party

444 members of the Assembly are elected and 10 are appointed by the head of state

Egypt is a multiparty system in theory. In practice, the ruling NDP, backed by the military, runs a one-party state.

PROFILE

Egypt has been politically stable since World War II, with just three leaders since 1954 when Nasser came to power. Anwar Sadat was assassinated in 1981, but was immediately replaced by Hosni Mubarak, a man in the same mold. The NDP retains its grip on the political process by means of the state of emergency, and has close links with the military. Elections in 2000 were more transparent than before and the Islamic opposition fared slightly better, but many candidates elected as independents then joined the NDP.

While Nasser promoted Arab socialism, influenced by the Soviet model, Sadat and Mubarak encouraged private enterprise and a liberalized economy. However, there has been no parallel liberalization in politics – one reason for the growing success of Islamist militants.

MAIN POLITICAL ISSUES

Islamic fundamentalism

The NDP government is engaged in a struggle against Islamist terrorist groups seeking to turn Egypt into a Muslim theocracy along Iranian lines. Extremists have been responsible for numerous attacks on police and tourists. The fundamentalist message, with promises of improved conditions, has proved attractive to both urban and rural poor. Mosques are often the main providers of education and health services that parallel the state's. Although the government uses draconian measures to counter the terrorist threat, and banned the only legal Islamic party, the Labor Party, in May 2000, it continues to allow religious organizations to pursue their social programs.

Hosni Mubarak, president since the assassination of Anwar Sadat in 1981.

Gemal Abd al-Nasser, pan-Arab nationalist, president from 1954 to 1970.

The state of emergency

The ruling NDP have repeatedly extended the national state of emergency, in force since the assassination of President Sadat by Islamic terrorists in 1981. The most recent extension was in February 2002 for a further three years. Emergency laws have been invoked to justify the ban on religious parties, especially the Muslim Brotherhood. It is claimed by human rights groups that emergency powers are routinely applied to silence the NDP's political opponents.

WORLD AFFAIRS

Joined UN in 1945

AL

Damasc

OAPEC

AU

OIC

Egypt has close relations with the West, particularly the US. Its support for the alliance against Iraq in the Gulf War was crucial, and Egypt received a massive economic reward from Saudi Arabia. It is also one of only two Arab countries to be technically at peace with Israel, for which it faces criticism from hard-line Islamic states. However, the Israeli military action in 2002 has weakened Egypt's support, and it continues to lobby for the creation of an autonomous Palestinian state.

Relations with Iran are particularly tense: Iran actively supports the Islamist groups operating against the NDP government, and characterizes Egypt as a corrupt state under US influence.

Egypt has expressed concern over international treatment of Iraq; President Mubarak calls for a diplomatic solution. He has opposed recent US-led air strikes against Iraq and urged restraint in the "war on terrorism."

Egypt's diplomatic service is the Arab world's largest, and many Egyptians, such as former UN Secretary General Boutros Boutros Ghali, have served on international bodies.

AID

Recipient

 $1.33bn (receipts)

 Down 16% in 2000

Egypt has received massive levels of US military aid since the late 1970s, and currently heads the list of US aid recipients, having overtaken Israel. International aid was pledged in 2002 to make up for lost tourism revenue.

E

CHRONOLOGY

Egypt's centuries-long Ottoman occupation ended in 1914, when it came under direct British rule. It became fully independent in 1936. Army officers led by Lt. Col. Gemal Abd al-Nasser seized power in 1952.

- ❑ **1953** Political parties dissolved, monarchy abolished. Republic proclaimed with Gen. Mohammed Neguib as president.
- ❑ **1954** Nasser deposes Neguib to become president.
- ❑ **1956** Suez Crisis over nationalization of Suez Canal. Israeli, British, and French forces invade, but withdraw after pressure from UN and US.
- ❑ **1957** Suez Canal reopens after UN salvage fleet clears blockade.
- ❑ **1958** Egypt merges with Syria as United Arab Republic.
- ❑ **1960–1970** Aswan High Dam built.
- ❑ **1961** Syria breaks away from union with Egypt.
- ❑ **1967** Six-Day War with Israel; loss of Sinai.
- ❑ **1970** Nasser dies; succeeded by Anwar Sadat.
- ❑ **1971** Readopts the name Egypt. Islam becomes state religion.
- ❑ **1972** Soviet military advisers dismissed from Egypt.
- ❑ **1974–1975** US brokers partial Israeli withdrawal from Sinai.
- ❑ **1977** Sadat visits Jerusalem: first-ever meeting between Egyptian president and Israeli prime minister.
- ❑ **1978** Camp David accords, brokered by US, signed by Egypt and Israel.
- ❑ **1979** Egypt and Israel sign peace treaty, alienating most Arab states.
- ❑ **1981** Sadat assassinated; succeeded by Hosni Mubarak.
- ❑ **1982** Last Israeli troops leave Sinai.
- ❑ **1986** President Mubarak meets Israeli prime minister Shimon Peres to discuss Middle East peace.
- ❑ **1989** After 12-year rift, Egypt and Syria resume diplomatic relations.
- ❑ **1990–1991** Egypt participates in UN operation to liberate Kuwait.
- ❑ **1991** Damascus Declaration provides for a defense pact among Egypt, Syria, and GCC countries against Iraq.
- ❑ **1994–1998** Islamist extremists begin campaign of terrorism, killing civilians and tourists. Government steps up countermeasures.
- ❑ **1999** Banned Gamaat Islamiya ends campaign to overthrow government.
- ❑ **2000** Egypt recalls ambassador to Israel because of escalating Israeli aggression against Palestinians.
- ❑ **2001** Heavy decline in tourist numbers following September 2001 attack on US.

DEFENSE

▷ Compulsory military service

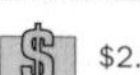 $2.82bn

  Down 6% in 2000

EGYPTIAN ARMED FORCES

Army	3860 main battle tanks (T-54/55, M1A1 *Abrams*, *Ramses* II, T-62, M-60)	320,000 personnel
Navy	4 submarines, 1 destroyer, 10 frigates, and 38 patrol boats	19,000 personnel
Air force	580 combat aircraft (*Alpha Jet*, PRC J-6, F-4E, *Mirage* 5E2)	29,000 personnel
Missiles	None	

Egypt's armed forces, the largest in the Arab world, are battle-hardened from successive wars with Israel and from participation in Operation Desert Storm to liberate Kuwait in 1991. More than 500,000 reservists augment the regular troops.

After the 1978 Camp David framework agreements were reached with Israel, Egypt stopped buying Soviet weapons and aircraft, and turned instead to Western suppliers. Cooperation with the US has reaped dividends in the form of access to more sophisticated defense equipment and improved training. Egypt has a small arms industry and sells light weapons, notably its version of the Soviet-developed AK-47 assault rifle, to other developing countries.

ECONOMICS

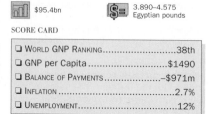

▷ Inflation 8.2% p.a. (1990–2000)

$95.4bn

3.890–4.575 Egyptian pounds

SCORE CARD

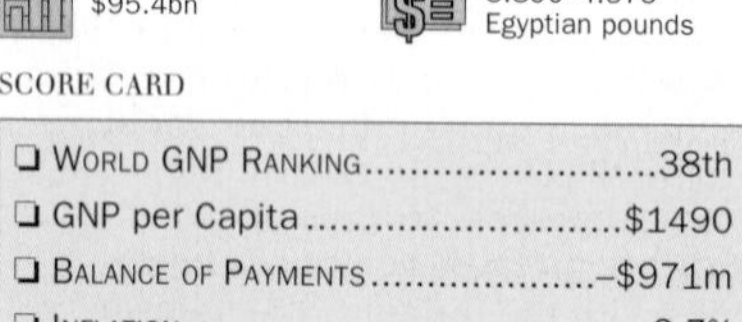

❑ World GNP Ranking	38th
❑ GNP per Capita	$1490
❑ Balance of Payments	–$971m
❑ Inflation	2.7%
❑ Unemployment	12%

EXPORTS

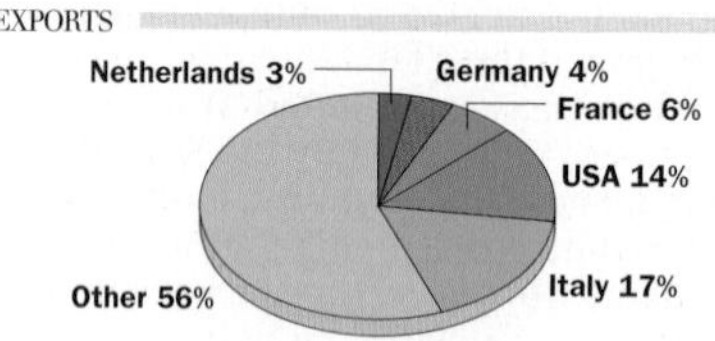
Netherlands 3%
Germany 4%
France 6%
USA 14%
Italy 17%
Other 56%

IMPORTS

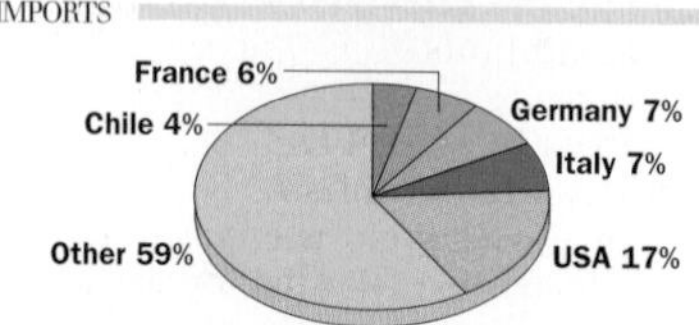
France 6%
Chile 4%
Germany 7%
Italy 7%
USA 17%
Other 59%

ECONOMIC PERFORMANCE INDICATOR

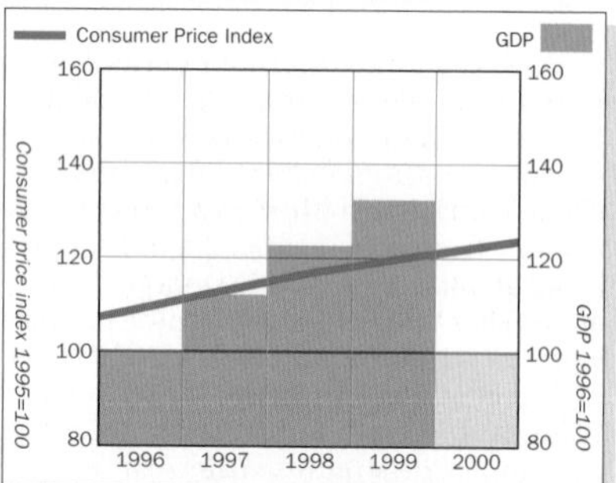
Consumer Price Index
GDP
160
140
120
100
80
Consumer price index 1995=100
GDP 1996=100
1996
1997
1998
1999
2000

Strengths

Oil and gas revenues. Well-developed tourist infrastructure. Remittances from Egyptians working throughout the region. Suez Canal tolls. Agricultural produce, especially cotton. Light industry and manufacturing.

Weaknesses

Tourism revenues hit by "war on terrorism." Dependence on imported technology. High birthrate.

Profile

Under President Nasser, Egypt followed an economic policy inspired by the Soviet model. Rigid and highly centralized, it gave Egypt one of the largest public sectors of all developing countries. Economic restrictions were first relaxed in 1974. President Sadat's open-door policy allowed joint ventures with foreign partners for the first time, although the business classes were the only ones to profit. Most Egyptians suffered from new austerity measures.

Under President Mubarak, economic reform has quickened and there is more awareness of poverty and the high levels of unemployment. Priorities now are to encourage manufacturing, sustain economic growth, and reduce the gap between rich and poor.

EGYPT : MAJOR BUSINESSES

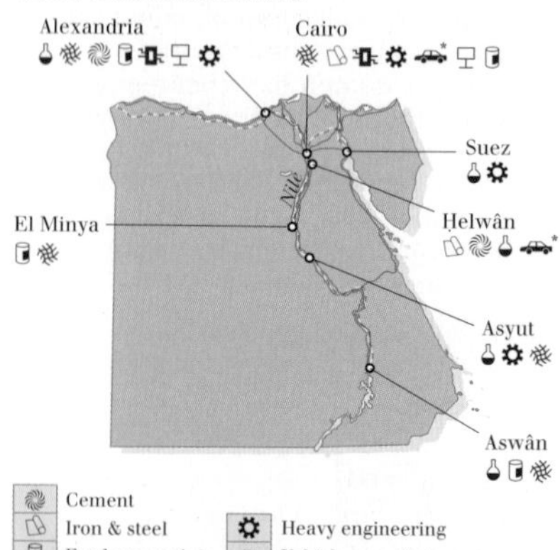
Alexandria
Cairo
Suez
Helwân
El Minya
Asyut
Aswân
Nile
Cement
Iron & steel
Food processing
Consumer goods
Electronics
Heavy engineering
Vehicle manufacture
Chemicals
Textiles
* significant multinational ownership
0 200 km
0 200 miles

RESOURCES

Electric power 17.5m kw

606,780 tonnes

758,000 b/d (reserves 2.9bn barrels)

9.2m ducks, 9.1m geese, 4.6m sheep, 88m chickens

Natural gas, oil, phosphates, manganese, uranium

ELECTRICITY GENERATION

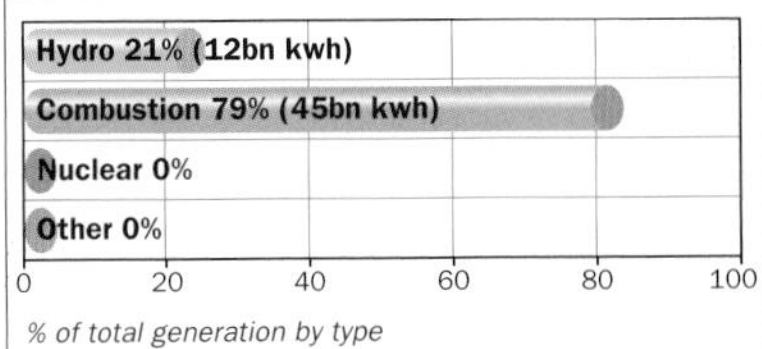
Hydro 21% (12bn kwh)
Combustion 79% (45bn kwh)
Nuclear 0%
Other 0%
0
20
40
60
80
100
% of total generation by type

Oil and gas are Egypt's most valuable resources. Oil multinationals are involved in new explorations, but more competitive oil-rich countries, such as Algeria and Yemen, are more profitable; 55% of Egypt's oil production is consumed locally.

Most electricity is derived from coal and hydroelectric power. The Aswan High Dam, built between 1960 and 1970 and with a maximum output of ten billion kWh, provides the bulk of hydroelectricity. Within four years, revenue from it had covered its construction costs. The US gave aid in 2000 to upgrade the power plant, with work going ahead in 2001.

EGYPT : LAND USE

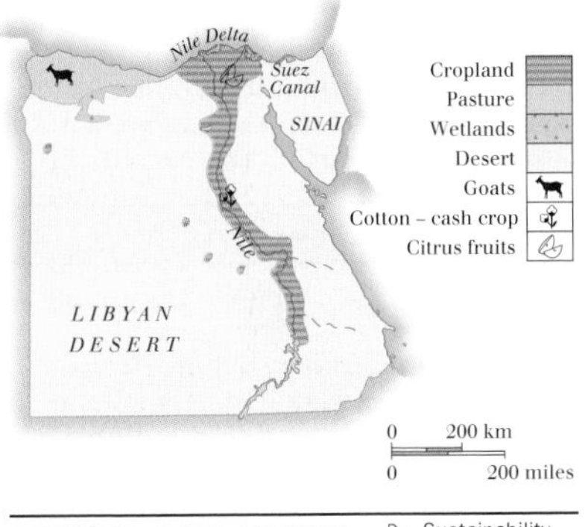
Nile Delta
Suez Canal
SINAI
Nile
LIBYAN DESERT
Cropland
Pasture
Wetlands
Desert
Goats
Cotton – cash crop
Citrus fruits
0 200 km
0 200 miles

ENVIRONMENT

Sustainability rank: 74th

0.8% (0.7% partially protected)

1.7 tonnes per capita

ENVIRONMENTAL TREATIES

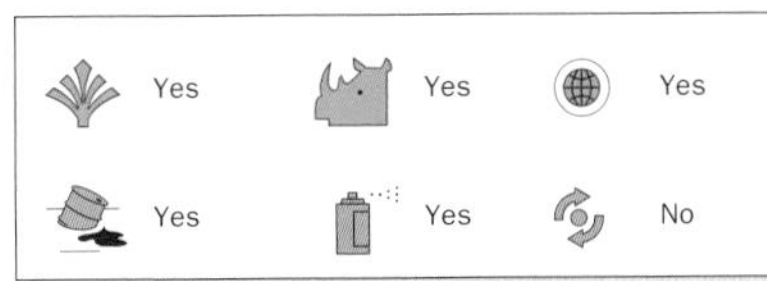
Yes
Yes
Yes
Yes
Yes
No

Egypt suffers from a chronic lack of water. The Nile, the only perennial source, is increasingly saline because of its much-reduced flow, due to irrigation use and the Aswan High Dam. The main cities suffer heavy industrial pollution, and environmental controls are few. In Cairo a sewerage system has improved sanitary conditions.

MEDIA

TV ownership medium

Daily newspaper circulation 38 per 1000 people

Pressure from Islamists has resulted in more airtime for Islamic sermons. Severe restrictions were imposed in 1998, after criticism of the government's security clampdown. Egypt was the first Arab state to have its own satellite, Nilesat 101, and is now the center of a flourishing satellite TV industry.

PUBLISHING AND BROADCAST MEDIA

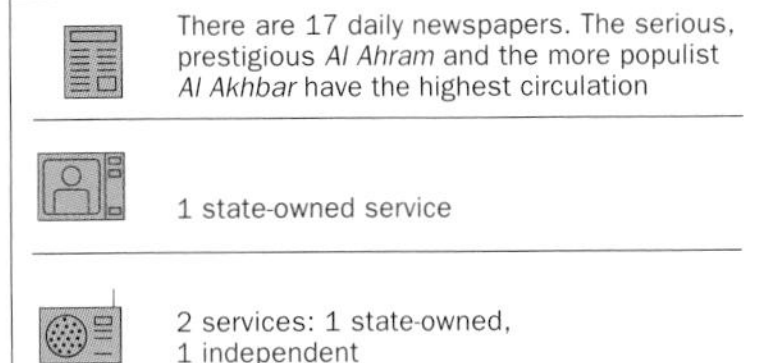
There are 17 daily newspapers. The serious, prestigious *Al Ahram* and the more populist *Al Akhbar* have the highest circulation
1 state-owned service
2 services: 1 state-owned, 1 independent

CRIME

Death penalty in use

80,000 prisoners

Crime is rising

CRIME RATES

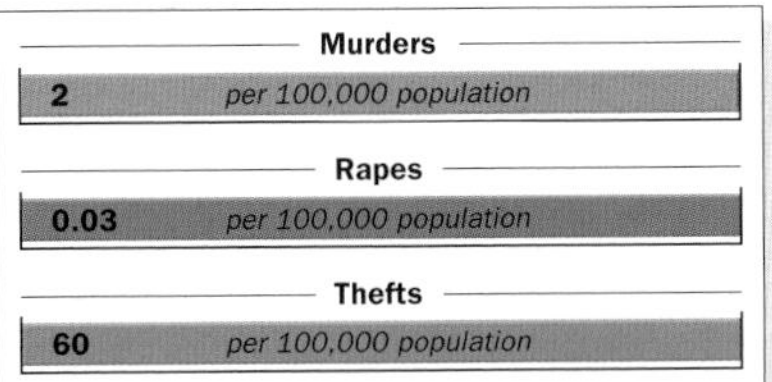
Murders
2 per 100,000 population
Rapes
0.03 per 100,000 population
Thefts
60 per 100,000 population

Terrorist attacks have tarnished Egypt's reputation as a law-abiding country; street crime and muggings were previously rare.

Intercommunity violence – in particular between Muslims and Christians – has become more common, as have attacks on Western tourists by Islamic extremists. Human rights groups have criticized the police for their abuse of current emergency laws, which results in the routine torture and death in police custody of scores of political prisoners.

EDUCATION

School leaving age: 14

55%

850,051 students

THE EDUCATION SYSTEM

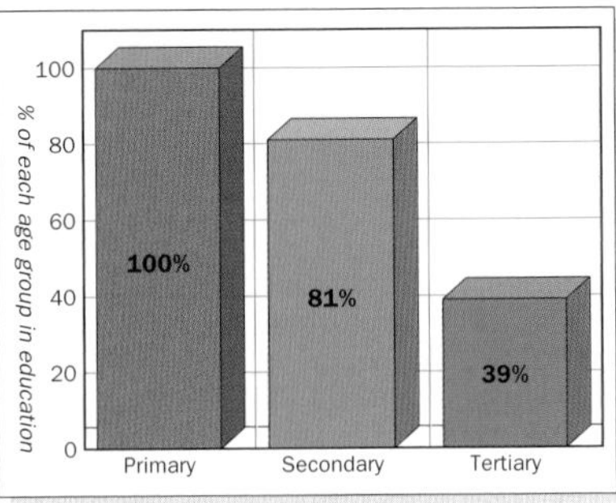
100
80
60
40
20
0
% of each age group in education
100%
81%
39%
Primary
Secondary
Tertiary

Most Egyptians attend elementary school until the age of 11, but not many complete secondary education. A small majority of men, but a minority of women, are literate. A government initiative to improve girls' primary education was launched in 2000. The quality of the education given by Egyptian universities is widely respected in the Arab world.

HEALTH

Welfare state health benefits

1 per 625 people

Digestive, respiratory, and heart diseases, perinatal deaths

Health care, although improved, remains basic – there is only one hospital bed for every 500 people. Islamic medical centers based on the mosque organization are spreading, and are replacing the state system. In 1996 the government banned female genital mutilation, a move upheld in 1997 by the Supreme Consitutional Court after being overturned by a lower court.

SPENDING

GDP/cap. increase

CONSUMPTION AND SPENDING

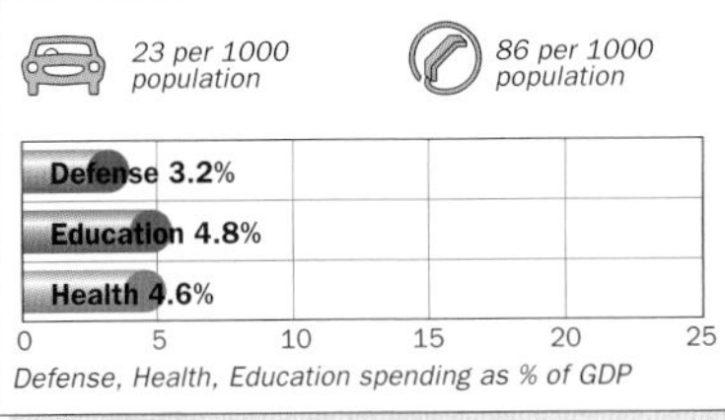
23 per 1000 population
86 per 1000 population
Defense 3.2%
Education 4.8%
Health 4.6%
0
5
10
15
20
25
Defense, Health, Education spending as % of GDP

Wealth disparities are highly marked in Egypt. The largely urban Coptic Christian community is the group with the country's highest standard of living. Most Egyptians remain subsistence farmers wih low incomes. The return of many unemployed workers from the Gulf states to their mainly rural homes has further depressed conditions in the countryside.

WORLD RANKING

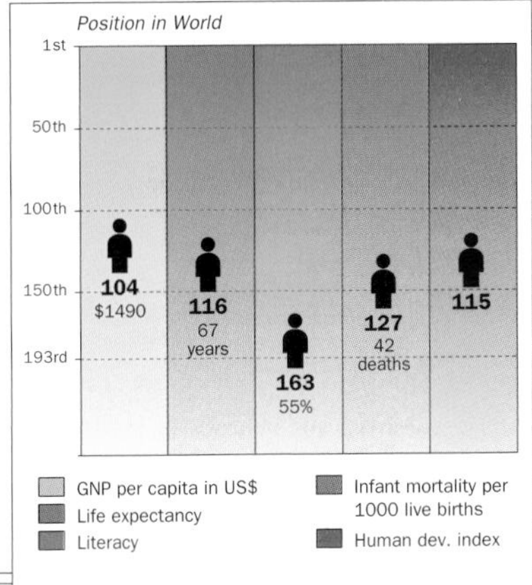
Position in World
1st
50th
100th
150th
193rd
104
$1490
116
67 years
163
55%
127
42 deaths
115
GNP per capita in US$
Life expectancy
Literacy
Infant mortality per 1000 live births
Human dev. index

E

El Salvador

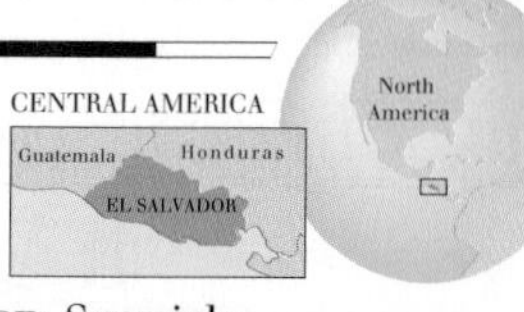
CENTRAL AMERICA
Guatemala
Honduras
EL SALVADOR
North America

Official Name: Republic of El Salvador **Capital:** San Salvador
Population: 6.4 million **Currencies:** Salvadorean colón & US dollar **Official Language:** Spanish

1841

1841
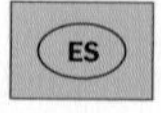
Sept 15
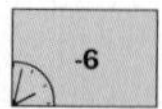
ES

-6

+503
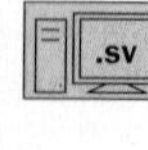
.sv

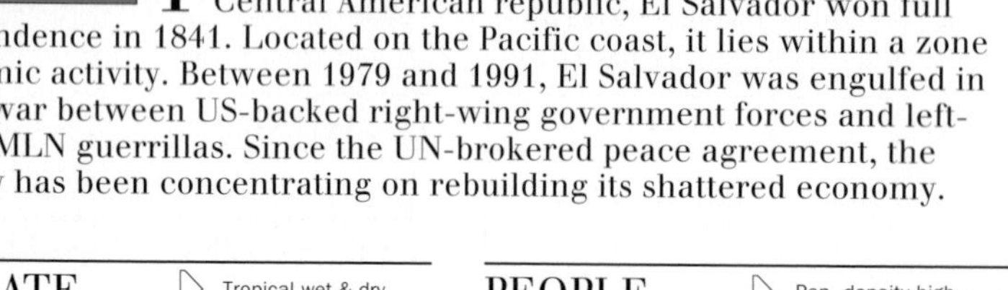

The smallest and most densely populated Central American republic, El Salvador won full independence in 1841. Located on the Pacific coast, it lies within a zone of seismic activity. Between 1979 and 1991, El Salvador was engulfed in a civil war between US-backed right-wing government forces and left-wing FMLN guerrillas. Since the UN-brokered peace agreement, the country has been concentrating on rebuilding its shattered economy.

View over the capital, San Salvador. It lies in a depression in the southern and higher of El Salvador's two mountain ranges, which is punctuated by more than 20 volcanoes.

CLIMATE

Tropical wet & dry

WEATHER CHART FOR SAN SALVADOR

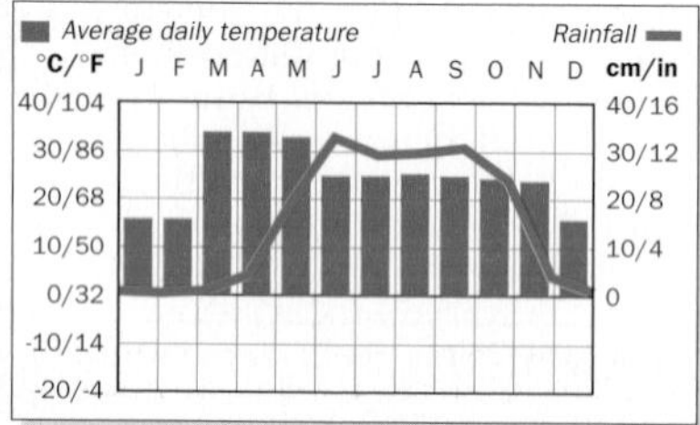
Average daily temperature
Rainfall
°C/°F J F M A M J J A S O N D cm/in
40/104
30/86
20/68
10/50
0/32
-10/14
-20/-4
40/16
30/12
20/8
10/4
0

The tropical coastal *tierra caliente* is very hot, with seasonal rains. The low hills are cooler at night; the higher *tierra templada* is drier and also cooler.

TRANSPORTATION

Drive on right

Cuscatlan, San Salvador 1.34m passengers

13 ships 1600 grt

THE TRANSPORTATION NETWORK

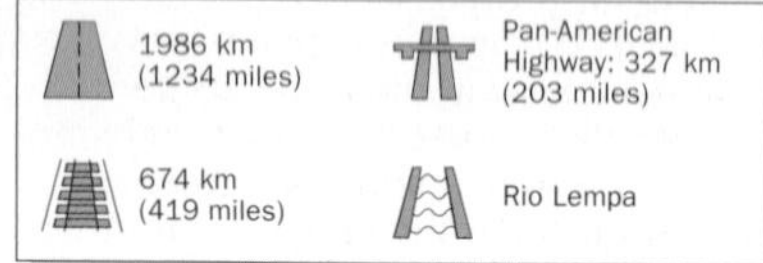
1986 km
(1234 miles)
Pan-American Highway: 327 km
(203 miles)
674 km
(419 miles)
Rio Lempa

Earthquakes in 2001 further damaged the already war-ravaged road and rail networks. Reconstruction will take many years.

TOURISM

Visitors : Population 1:8.1

795,000 visitors

Up 21% in 2000

MAIN TOURIST ARRIVALS

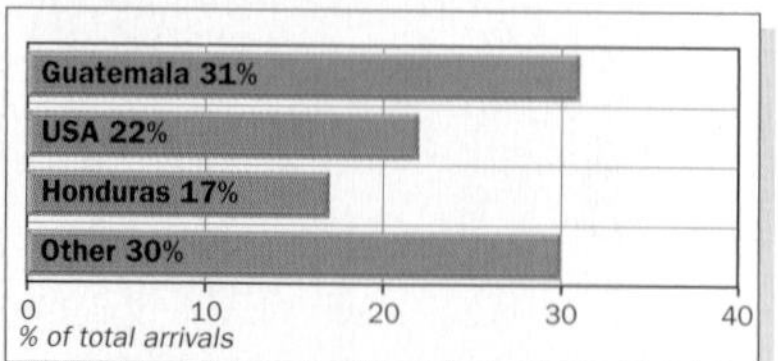
Guatemala 31%
USA 22%
Honduras 17%
Other 30%
0 10 20 30 40
% of total arrivals

Peace has brought visitors back to the unspoiled beach resorts, but crime, earthquakes, and high prices for rooms and air travel hinder tourist expansion.

PEOPLE

Pop. density high

Spanish

309/km² (800/mi²)

THE URBAN/RURAL POPULATION SPLIT

47% 53%

RELIGIOUS PERSUASION

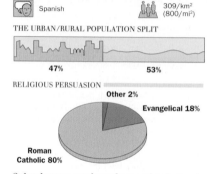

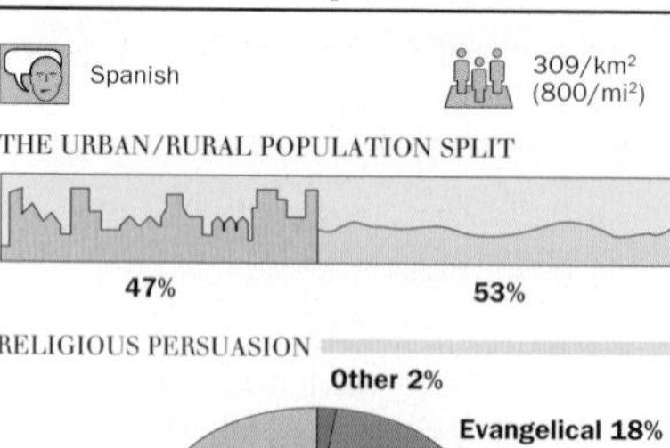
Other 2%
Evangelical 18%
Roman Catholic 80%

Salvadorans are largely *mestizo* (mixed race); there are few ethnic tensions. The civil war was fought over gross economic disparities, which still exist.

POLITICS

Multiparty elections

2000/2003

President Francisco Flores

AT THE LAST ELECTION

Legislative Assembly 84 seats

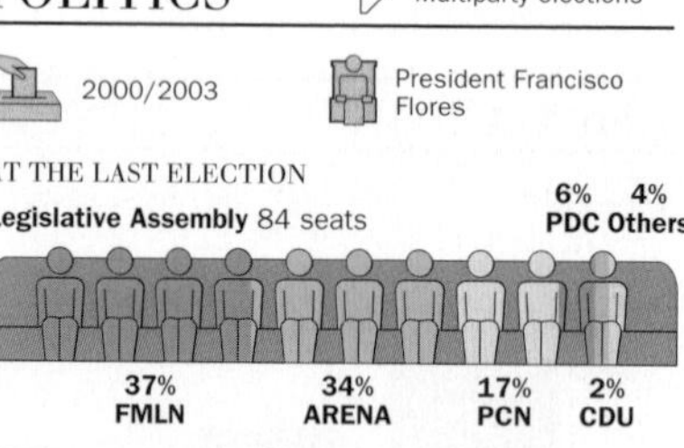
37% FMLN
34% ARENA
17% PCN
2% CDU
6% PDC
4% Others

FMLN = Farabundo Martí National Liberation Front **ARENA** = Nationalist Republican Alliance **PCN** = National Conciliation Party **PDC** = Christian Democratic Party **CDU** = United Democratic Center

El Salvador had been dominated by the centrist PDC and right-wing ARENA. The latter, however, now faces greater opposition from the FMLN, leftist former guerrillas, who in 1997 won the mayorship of San Salvador and half the state capitals.

In the 1999 presidential election the FMLN, split between center-left pragmatists and hard-liners, came a poor second to ARENA's Francisco Flores, who promised reduced poverty and income redistribution. By March 2000, however, with the economy in difficulties, voters punished ARENA by returning the FMLN as the largest party in parliament.

WORLD AFFAIRS

Joined UN in 1945

ACS

Geplac
IBRD

OAS

San José

El Salvador was an international pariah in the 1980s because of the human rights abuses committed by military death squads. Today it cooperates with its neighbors in pressing the US on key issues such as trade and immigration. It relied heavily on US aid in 2001 after three devastating earthquakes. In 2000 it signed a free trade treaty with Mexico, Guatemala, and Honduras. A territorial dispute with Honduras was finally settled in 1998, when those affected were permitted to choose between Honduran and Salvadoran citizenship.

AID

Recipient

$180m (receipts)

Down 2% in 2000

Post-civil war aid focused on efforts to secure peace and achieve national reconciliation by funding rebuilding and refugee resettlement programs. The current emphasis is on a shift toward supporting growth.

The UN received a slow international response in 2001 to its appeal for $34.8 million in emergency housing, medicine, and disaster prevention programs after El Salvador's devastating earthquakes.

DEFENSE

Compulsory military service

$168m

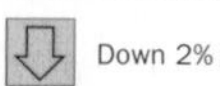
Down 2% in 2000

Between 1979 and 1991, the role of the US-backed military was to fight an unrestricted war against the FMLN. Human rights were in effect suspended and governments that opposed the military were overthrown. Under the peace accords the military agreed to withdraw from politics and internal security matters, but it remains a potent force capable of intervention.

ECONOMICS

Inflation 7.4% p.a. (1990–2000)

$12.6bn

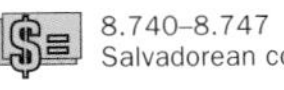
8.740–8.747 Salvadorean colones

SCORE CARD

- ❑ WORLD GNP RANKING........................78th
- ❑ GNP PER CAPITA $2000
- ❑ BALANCE OF PAYMENTS.....................–$48m
- ❑ INFLATION 2.3%
- ❑ UNEMPLOYMENT.................................10%

STRENGTHS
Coffee. Foreign investment in *maquila* assembly plants. Sizable family remittances from US.

WEAKNESSES
Exports uncompetitive. High tax evasion and unemployment. Low savings. Vast reconstruction needed after earthquakes in 2001.

EXPORTS

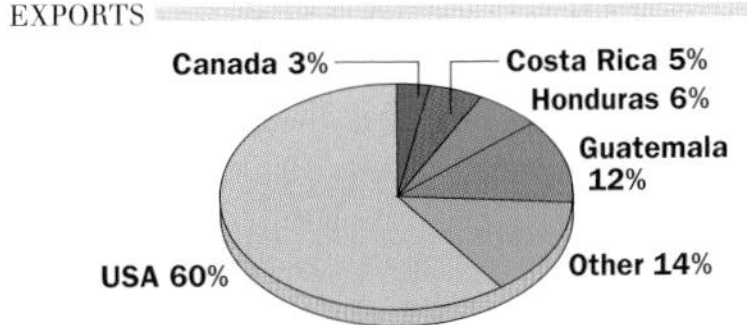
Canada 3%
Costa Rica 5%
Honduras 6%
Guatemala 12%
Other 14%
USA 60%

IMPORTS

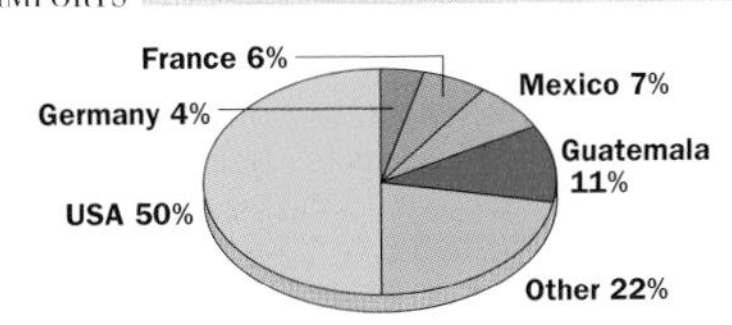
France 6%
Germany 4%
Mexico 7%
Guatemala 11%
USA 50%
Other 22%

RESOURCES

Electric power 996,000 kw

15,467 tonnes

1.22m cattle, 150,000 pigs, 8.1m chickens

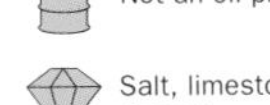
Not an oil producer

Salt, limestone, gypsum

No significant resources. Several volcanoes facilitate abundant and relatively cheap geothermal energy.

ENVIRONMENT

Sustainability rank: 75th

0.3% (0.2% partially protected)

1 tonne per capita

Deforestation has led to erosion and desertification – worsening landslides during the earthquakes of 2001. Overuse of pesticides is a major problem.

MEDIA

TV ownership medium

Daily newspaper circulation 48 per 1000 people

PUBLISHING AND BROADCAST MEDIA

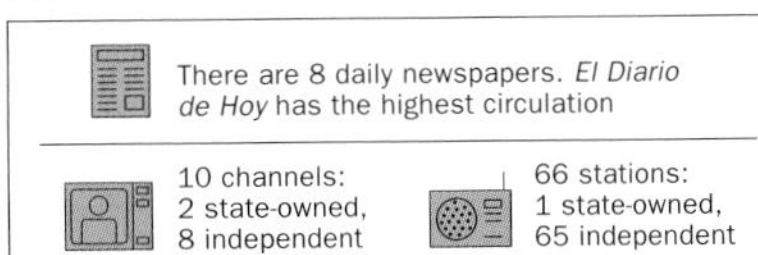
There are 8 daily newspapers. *El Diario de Hoy* has the highest circulation

10 channels: 2 state-owned, 8 independent

66 stations: 1 state-owned, 65 independent

The media is owned by powerful groups, such as the Dutriz family. Intimidation and self-censorship exist.

CRIME

Death penalty not used in practice

6914 prisoners

Falling, but still high by regional standards

A corrupt judiciary and police force have failed to stem a postwar crime wave fueled by readily available arms; armed robberies, kidnappings, and murders deter investment and tourism. Uncompleted elements of the peace accords, particularly land transfers, often lead to violence.

EDUCATION

School leaving age: 15

79%

114,675 students

Education is based on the US system and is limited in rural areas. During the civil war, state universities were closed by the military and replaced by private universities which continue to thrive despite their low standards. A 1995 reform bill tried to address the negative impact of deregulation.

CHRONOLOGY

El Salvador was a Spanish colony until 1821. Part of the United Provinces of Central America in 1823–1839, it became fully independent in 1841.

- ❑ **1932** Army crushes popular insurrection led by Farabundo Martí.
- ❑ **1944–1979** Army rules through PCN.
- ❑ **1979** Reformist officers overthrow PCN government.
- ❑ **1981** Left-wing FMLN launches civil war.
- ❑ **1991** UN-brokered peace. FMLN recognized as a political party.
- ❑ **1997** Leftist wins San Salvador mayoralty.
- ❑ **2001** Devastating earthquakes kill hundreds; dollarization of economy.

E

HEALTH

Welfare state health benefits

1 per 909 people

Accidents, violence, circulatory diseases, infections

Health spending, almost halved during the civil war, has been slow to recover. The wealthy go to the US for surgery.

SPENDING

GDP/cap. increase

CONSUMPTION AND SPENDING

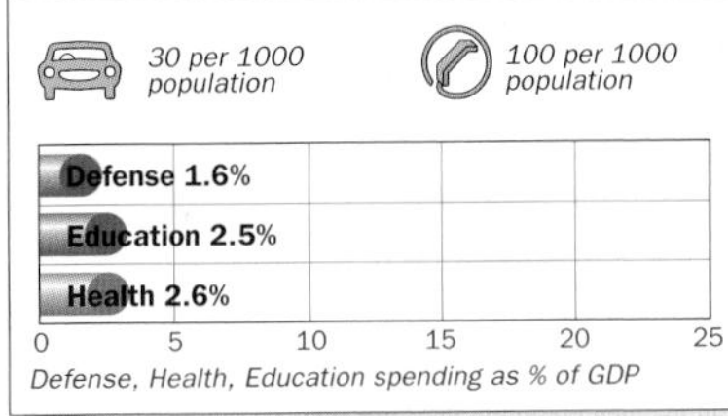
30 per 1000 population
100 per 1000 population
Defense 1.6%
Education 2.5%
Health 2.6%
0
5
10
15
20
25
Defense, Health, Education spending as % of GDP

Gross wealth disparities see 20% of the population owning 70% of national wealth. Land distribution remains highly skewed, and some three million – nearly half the population – live in poverty.

WORLD RANKING

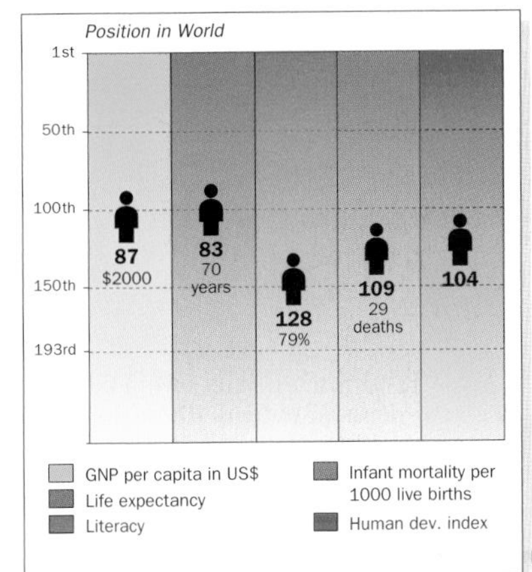
Position in World
1st
50th
100th
150th
193rd
87
$2000
83
70 years
128
79%
109
29 deaths
104
GNP per capita in US$
Life expectancy
Literacy
Infant mortality per 1000 live births
Human dev. index

EL SALVADOR

Total Land Area : 21 040 sq. km (8124 sq. miles)

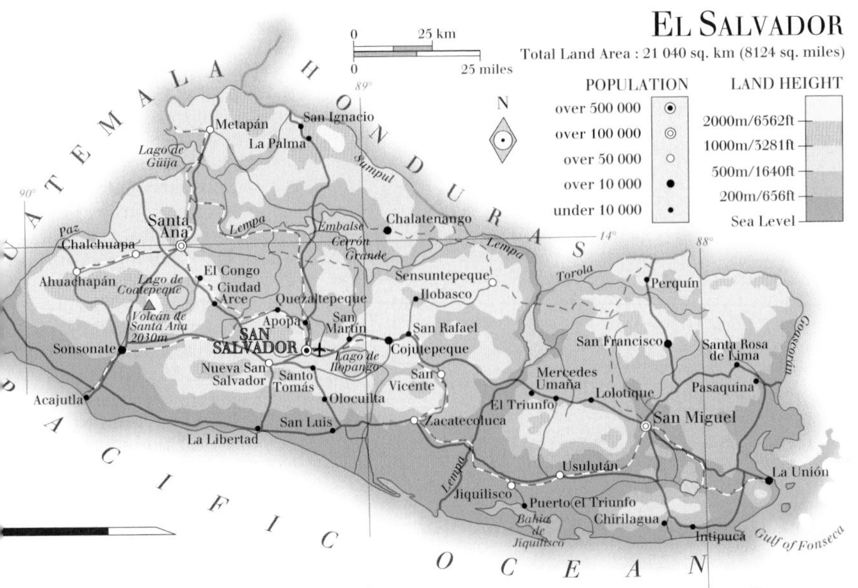
0
25 km
0
25 miles
POPULATION
over 500 000
over 100 000
over 50 000
over 10 000
under 10 000
LAND HEIGHT
2000m/6562ft
1000m/3281ft
500m/1640ft
200m/656ft
Sea Level
N
GUATEMALA
HONDURAS
PACIFIC OCEAN
Metapán
San Ignacio
La Palma
Lago de Güija
Sumpul
Santa Ana
Chalchuapa
Chalatenango
Lempa
Embalse Cerrón Grande
Ahuachapán
Lago de Coatepeque
El Congo
Ciudad Arce
Quezaltepeque
Volcán de Santa Ana 2030m
Sonsonate
Apopa
San Martín
SAN SALVADOR
Nueva San Salvador
Lago de Ilopango
Cojutepeque
Sensuntepeque
Ilobasco
San Rafael
Santo Tomás
Olocuilta
San Vicente
Acajutla
San Luis
La Libertad
Zacatecoluca
Torola
Perquín
San Francisco
Santa Rosa de Lima
Mercedes Umaña
Lolotique
Pasaquina
El Triunfo
San Miguel
Usulután
La Unión
Jiquilisco
Puerto el Triunfo
Bahía de Jiquilisco
Chirilagua
Intipucá
Gulf of Fonseca
Goascorán

E

EQUATORIAL GUINEA

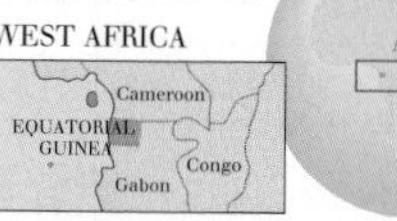
WEST AFRICA
Cameroon
EQUATORIAL GUINEA
Gabon
Congo
Africa

OFFICIAL NAME: Republic of Equatorial Guinea **CAPITAL:** Malabo
POPULATION: 470,000 **CURRENCY:** CFA franc **OFFICIAL LANGUAGES:** Spanish and French

1968

1968

Oct 12
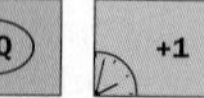
GQ

+1

+240

.gq

COMPRISING FIVE ISLANDS and the territory of Río Muni on the west coast of Africa, Equatorial Guinea lies just north of the equator. Mangrove swamps border the mainland coast. The republic gained its independence in 1968 after 190 years of Spanish rule. Multipartyism was accepted in 1991, but the fairness of subsequent general elections has been questioned.

Bioko, formerly Fernando Po. Although the volcanic land is very fertile, cocoa production fell by 90% during the Macías years.

CLIMATE

Tropical equatorial

WEATHER CHART FOR MALABO

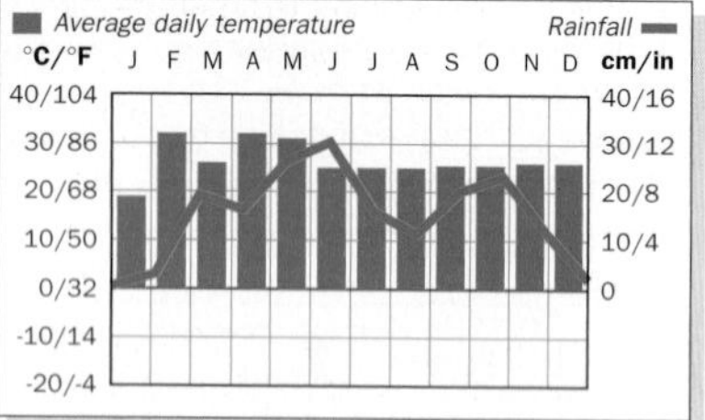
Average daily temperature
Rainfall
°C/°F
J F M A M J J A S O N D
cm/in
40/104
30/86
20/68
10/50
0/32
-10/14
-20/-4
40/16
30/12
20/8
10/4
0

The island of Bioko is extremely wet and humid, with an annual rainfall of 200 cm (79 in), while the mainland is only marginally drier and cooler.

TRANSPORTATION

Drive on right

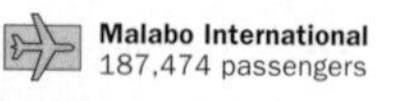
Malabo International
187,474 passengers

64 ships
45,838 grt

THE TRANSPORTATION NETWORK

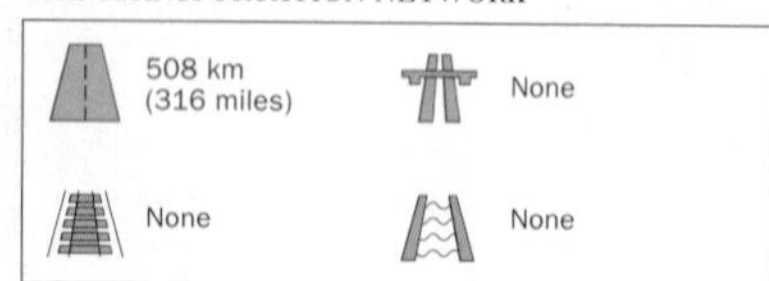
508 km
(316 miles)
None
None
None

Apart from once- or twice-weekly flights to Madrid by the Spanish airline Iberia, all air links are through neighboring countries. There are six flights a week between Malabo and Bata.

TOURISM

Not available

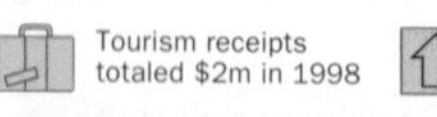
Tourism receipts totaled $2m in 1998

Numbers are increasing slowly

MAIN TOURIST ARRIVALS

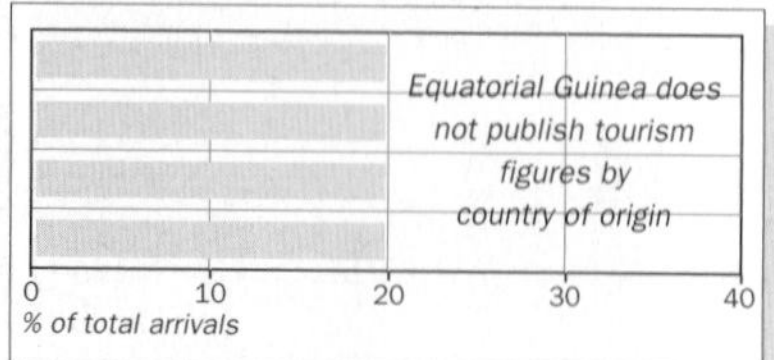
Equatorial Guinea does not publish tourism figures by country of origin
0
10
20
30
40
% of total arrivals

Equatorial Guinea is only of interest to the adventurous, independent tourist, despite the potential attraction of its beaches and the island of Bioko's spectacular mountain scenery.

PEOPLE

Pop. density low

Spanish, Fang, Bubi

17/km²
(43/mi²)

THE URBAN/RURAL POPULATION SPLIT

48% 52%

RELIGIOUS PERSUASION

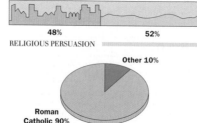

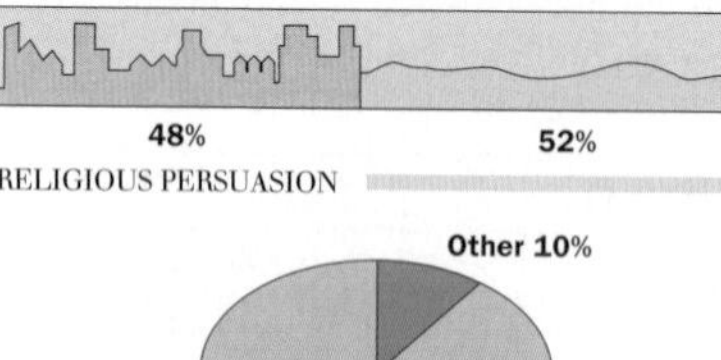
Other 10%
Roman Catholic 90%

The mainland has a majority of Fang, a people who also inhabit Cameroon and north Gabon. Bioko is populated by a majority of Bubi and a minority of Creoles, known as Fernandinos. The Macías dictatorship consolidated the power of the Fang, especially the Mongomo clan, from which both Macías and his successor Obiang come. The extended family has maintained its solidarity, despite disruptive social pressure during the Macías dictatorship.

POLITICS

Multiparty elections

1999/2004

President Teodoro Obiang Nguema Mbasogo

AT THE LAST ELECTION

House of Representatives of the People 80 seats

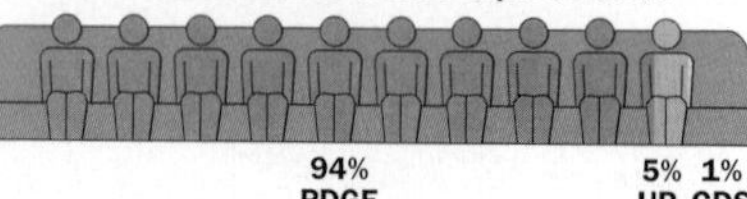
94% PDGE
5% UP
1% CDS

PDGE = Equatorial Guinea Democratic Party **UP** = Popular Union **CDS** = Convergence for Social Democracy

Despite officially being a multiparty state since 1991, some of the several exiled political parties have not yet found it safe to return. The ruling PDGE was set up in 1987 by Teodoro Obiang Nguema Mbasogo, nephew of the dictator Francisco Macías Nguema, whom he overthrew in 1979. It replaced Macías' National Workers' Party (PUNT). The PDGE benefits from heavy government patronage, receiving 3% of all salaries.

The continuing movement toward multipartyism – which was initiated in 1988 following the first elections for 20 years – has been marked by instability. The 1993 parliamentary elections were boycotted by the main opposition parties, while the presidential poll in 1996, in which Obiang was the only candidate, was declared farcical by foreign observers. The 1999 legislative elections were won easily by the PDGE, but denounced by the opposition. In 2002, 68 opposition leaders were jailed in connection with a 1997 plot to overthrow the government.

EQUATORIAL GUINEA

Total Land Area : 28 051 sq. km (10 830 sq. miles)

POPULATION
over 10 000
under 10 000

LAND HEIGHT
2000m/6562ft
1000m/3281ft
500m/1640ft
200m/656ft
Sea Level

0 50 km
0 50 miles

Bioko MALABO Baney Pico de Basilé 3008m Luba Riaba Bococo Ureca Punta Santiago Bight of Biafra

Gulf of Guinea ATLANTIC OCEAN CAMEROON GABON Río Muni Ayamiken Ncue Micomeseng Ebebiyin Bata Niefang Añisoc Nsoc-Nsomo Cabo Dos Puntas Senye Mbini Nkumekie Mongomo Evinayong Etembue Aconibe Cabo San Juan Cogo Nsoc Acurnam Isla de Corisco

WORLD AFFAIRS

Joined UN in 1968

BDEAC

ACP

FZ

NAM

AU

After a period of extreme isolation at the time of the Macías dictatorship, Equatorial Guinea sought to rebuild links, especially with Spain, the former colonial power and traditionally a haven for political dissenters. A maritime border dispute with Nigeria was finally settled in 2000, paving the way for the exploitation of large oil reserves in the Gulf of Guinea. Wealth from these reserves is likely to affect regional relations greatly in the future.

AID

Recipient

$21m (receipts)

Up 5% in 2000

Equatorial Guinea is poorly developed and therefore heavily dependent on aid. Inefficiency, corruption, and a shortage of skilled people hinder the planning and implementation of projects, and the government's political record threatens funding. Spain and France are the main donors while aid from UN agencies is also important. An IMF program was suspended in 1997 after the government failed to implement reforms.

DEFENSE

No compulsory military service

$12m

Up 20% in 2000

The main concern for the military and paramilitary force is internal security. Cuba and North Korea provided Macías with a presidential guard, while Obiang has been protected by Moroccan troops. Nigeria, Cameroon, and Gabon have interests in maintaining the autonomy of the Malabo and Río Muni regions.

RESOURCES

Electric power 5000 kw

7001 tonnes

181,000 b/d (reserves 3.6m barrels)

37,600 sheep, 30,000 ducks, 320,000 chickens

Oil, natural gas, gold

Oil production levels doubled between 1998 and the end of 2001. President Obiang has pledged to use income from oil to promote development. The region around Bata is served by a 3.2 MW hydropower station.

ENVIRONMENT

Not available

None

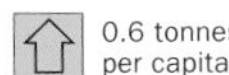
0.6 tonnes per capita

The government has failed to take any serious measures to stop timber companies depleting the rainforest.

MEDIA

TV ownership low

Daily newspaper circulation 5 per 1000 people

PUBLISHING AND BROADCAST MEDIA

There is no regular daily press. The formerly daily newspaper *Poto Poto* now appears irregularly

1 state-owned service

3 services: 1 state-owned, 2 independent

The press remains tightly controlled, with little liberalization, despite the state's adoption of multipartyism. The press association was closed down by the police in February 2001.

CRIME

Death penalty in use

Equatorial Guinea does not publish prison figures

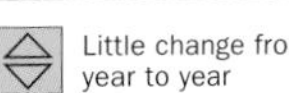
Little change from year to year

The level of recorded crime is relatively low, although many offenses do not get reported. Many human rights abuses still occur.

CHRONOLOGY

Equatorial Guinea remained a backwater of Spanish colonialism until development began after 1939.

- **1968** Independence. President Macías begins reign of terror.
- **1979** Coup puts nephew in power.
- **1991** Multiparty constitution.
- **1999** Ruling party wins majority in election condemned as fraudulent.
- **2001** Government ministers resign over allegations of mismanagement and corruption.

E

EDUCATION

School leaving age: 11

83%

578 students

Education declined in the Macías years, when attendance rates fell from 90% to 55%. Although education is declared the state's first priority, funding is poor.

HEALTH

No welfare state health benefits

1 per 4000 people

Diarrheal and respiratory diseases, malaria

Life expectancy – just 37 years in 1960 – had risen substantially by 2000. There are 25 doctors to every 100,000 people.

SPENDING

GDP/cap. increase

CONSUMPTION AND SPENDING

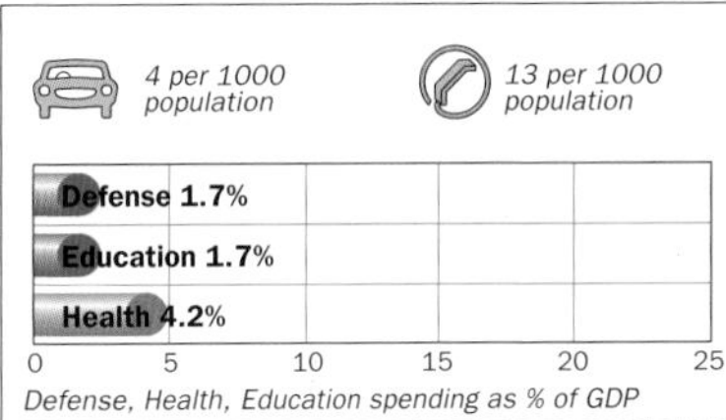

4 per 1000 population
13 per 1000 population
Defense 1.7%
Education 1.7%
Health 4.2%
0
5
10
15
20
25
Defense, Health, Education spending as % of GDP

What wealth there is in Equatorial Guinea tends to be concentrated in the ruling clan. There is also a remnant of the former Spanish plutocracy.

ECONOMICS

Inflation 15% p.a. (1990–2000)

$363m

698.7–736.7 CFA francs

SCORE CARD

- World GNP Ranking 175th
- GNP per Capita $800
- Balance of Payments ....–$344m
- Inflation 6%
- Unemployment 30%

Strengths

Fertile soils. Timber. Cocoa and coffee. Extensive territorial waters, with potential for fisheries. The economy is strengthening as oil and gas reserves are exploited.

Weaknesses

Lasting effects of economic regression under Macías dictatorship. Maladministration and ideological attacks on the educated have restricted growth; under Macías, cocoa production slumped by 90%. Deterioration of rural economy under successive brutal regimes. Undeveloped natural resources.

EXPORTS

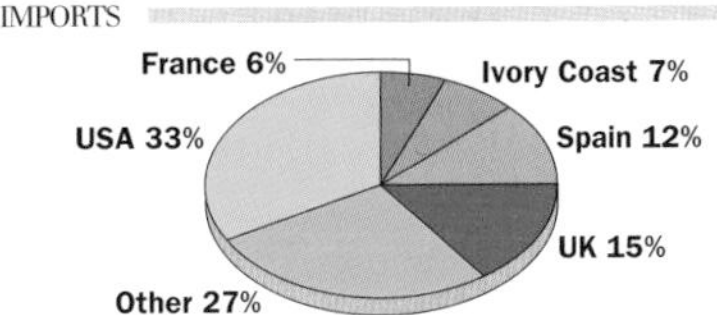

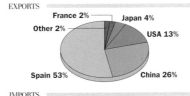

France 2%
Japan 4%
Other 2%
USA 13%
Spain 53%
China 26%

IMPORTS

France 6%
Ivory Coast 7%
USA 33%
Spain 12%
UK 15%
Other 27%

WORLD RANKING

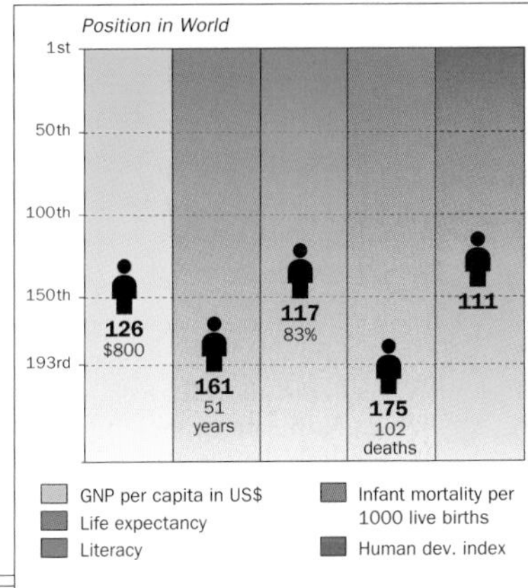

Position in World
1st
50th
100th
150th
193rd
126
$800
161
51 years
117
83%
175
102 deaths
111
GNP per capita in US$
Life expectancy
Literacy
Infant mortality per 1000 live births
Human dev. index

ERITREA

EAST AFRICA
Africa
Saudi Arabia
Sudan
ERITREA
Yemen
Ethiopia

OFFICIAL NAME: State of Eritrea **CAPITAL:** Asmara
POPULATION: 3.8 million **CURRENCY:** Nakfa **OFFICIAL LANGUAGE:** Tigrinya

1993

2002

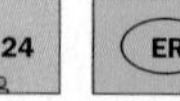
May 24

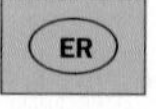
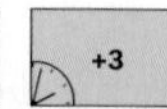
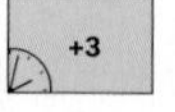
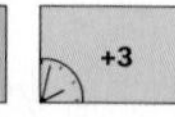
ER

+3

+291

.er

LYING ON THE SHORES of the Red Sea, Eritrea's landscape is one of rugged mountains, bush, and desert. A former Italian colony later annexed by Ethiopia, Eritrea fought a long war to win independence in 1993. Like its southern neighbor, Eritrea is prone to recurring droughts and the threat of famine. War with Ethiopia in 1998–2000 brought heavy losses on both sides, until the signing of a comprehensive peace agreement in December 2000.

CLIMATE

Hot desert/mountain

WEATHER CHART FOR ASMARA

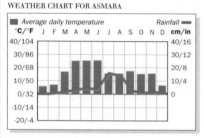

■ *Average daily temperature* — *Rainfall*

°C/°F J F M A M J J A S O N D cm/in
40/104 40/16
30/86 30/12
20/68 20/8
10/50 10/4
0/32 0
-10/14
-20/-4

Eritrea's harvest is dependent on mid-year rainfall in the highlands. Lowland temperatures may exceed 50°C (122°F).

TRANSPORTATION

Drive on right

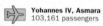

Yohannes IV, Asmara
103,161 passengers

10 ships
15,913 grt

THE TRANSPORTATION NETWORK

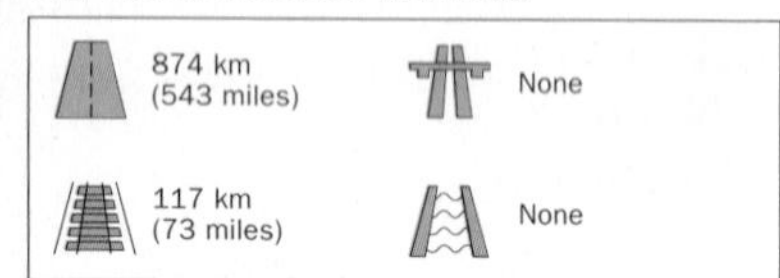

874 km
(543 miles)
None
117 km
(73 miles)
None

All transportation infrastructure requires massive investment. Ports have potential as transit points for Ethiopia.

TOURISM

Visitors : Population 1:54

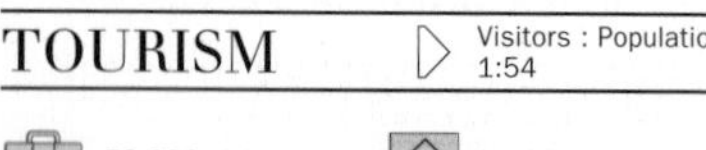

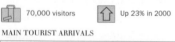

70,000 visitors
Up 23% in 2000

MAIN TOURIST ARRIVALS

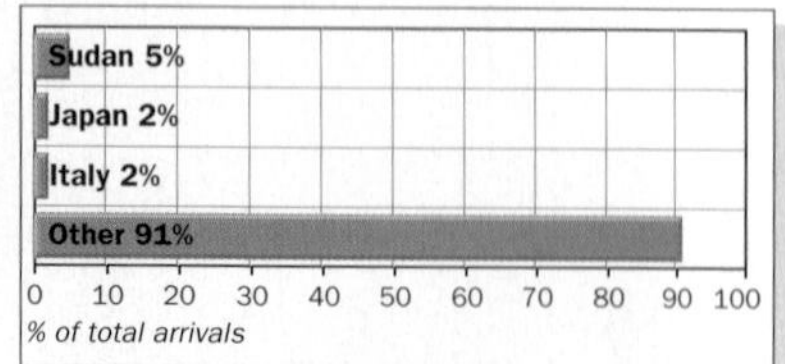

Sudan 5%
Japan 2%
Italy 2%
Other 91%
0 10 20 30 40 50 60 70 80 90 100
% of total arrivals

There is currently very little tourism, but Eritrea has considerable long-term potential, especially along the Red Sea coast, with its underwater attractions, and in the spectacular Danakil depression. Guides are essential.

PEOPLE

Pop. density low

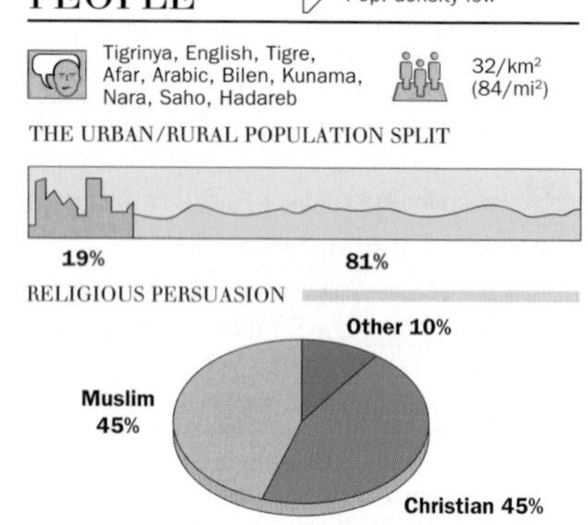

Tigrinya, English, Tigre, Afar, Arabic, Bilen, Kunama, Nara, Saho, Hadareb
32/km²
(84/mi²)
THE URBAN/RURAL POPULATION SPLIT
19%
81%
RELIGIOUS PERSUASION
Other 10%
Muslim 45%
Christian 45%

Tigrinya-speakers, mainly Orthodox Christians, form the largest of Eritrea's nine main ethnic groups. A strong sense of nationhood has been forged by the 30-year struggle for independence. Women played an important role in the war; from 1973, 30,000 fought alongside men, some in positions of command. The nomadic peoples of the Danakil desert remain fiercely independent. Subsistence farmers account for 80% of the population.

ERITREA

Total Land Area : 121 320 sq. km (46 842 sq. miles)

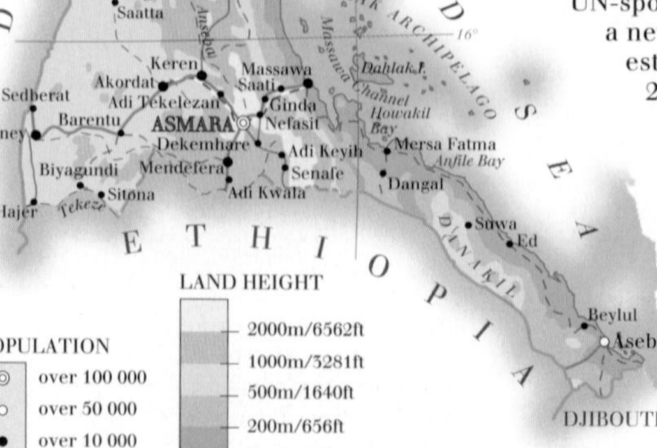

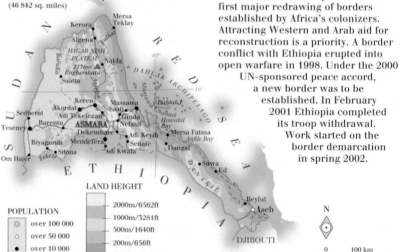

Mersa Teklay
Kerora
Algena
HAGAR NISH PLATEAU
2576m
Enghershatu
Nakfa
Saatta
Barka
Anseba
DAHLAK ARCHIPELAGO
Keren
Massawa
Dahlak I.
Akordat
Adi Tekelezan
Saati
Ginda
Sedberat
Barentu
ASMARA
Nefasit
Teseney
Dekemhare
Adi Keyih
Mersa Fatma
Anfile Bay
Biyagundi
Mendefera
Senafe
Sitona
Adi Kwala
Dangal
Om Hajer
Tekeze
Suwa
Ed
Beylul
Aseb
DJIBOUTI
SUDAN
RED SEA
ETHIOPIA
DANAKIL
LAND HEIGHT
2000m/6562ft
1000m/3281ft
500m/1640ft
200m/656ft
Sea Level
-200m/-656ft
POPULATION
over 100 000
over 50 000
over 10 000
under 10 000

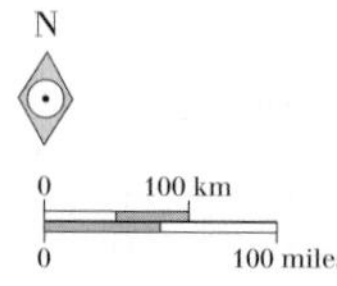

N
0 100 km
0 100 miles

POLITICS

In transition

Elections not yet held

President Issaias Afewerki

AT THE LAST ELECTION

National Assembly 150 seats

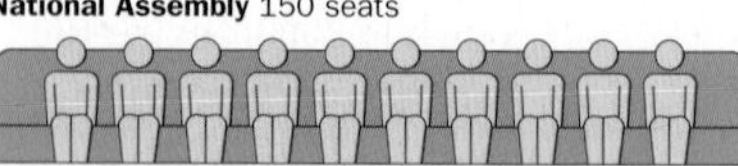

The National Assembly comprises 75 People's Front for Democracy and Justice (**PFDJ**) central committee members and 75 directly elected members, including 11 seats reserved for women. Elections expected in 1997 under the new constitution have not yet taken place

A former Italian colony, Eritrea was dominated by Ethiopia in a federation set up in 1952. Ethiopia annexed the country in 1961, prompting a long secessionist struggle. The Eritrean People's Liberation Front (EPLF) and its Tigrean allies helped defeat the Ethiopian regime in 1991. In 1993 a referendum gave overwhelming support to independence.

Pending multiparty elections, the country is run by a core leadership from the EPLF (now the PFDJ), with the National Assembly only convening sporadically. The 1997 constitution forbids parties based on religious or ethnic affiliations. Issaias Afewerki, a Christian, has been careful to include Muslims in his transitional cabinet. He lashed out at political opponents in 2001 and faced unprecedented criticism, even from within the PFDJ.

WORLD AFFAIRS

Joined UN in 1993

COMESA | IBRD | IGAD | NAM | AU

Eritrea's secession represented the first major redrawing of borders established by Africa's colonizers. Attracting Western and Arab aid for reconstruction is a priority. A border conflict with Ethiopia erupted into open warfare in 1998. Under the 2000 UN-sponsored peace accord, a new border was to be established. In February 2001 Ethiopia completed its troop withdrawal. Work started on the border demarcation in spring 2002.

AID

 Recipient

 $176m (receipts) Up 18% in 2000

The economy is highly aid-dependent, and 75% of the population survive on food aid. This is an obvious and pressing need, given the country's vulnerability to famine, but Western donors have been less generous with aid for the $2 billion reconstruction costs. Emergency UN aid was requested in mid-2000 to assist over a million people displaced by the Ethiopian incursion. WHO supplied emergency medical aid.

DEFENSE

Compulsory military service

  $206m Down 33% in 2000

Defense expenditure is massive. The 50,000-strong permanent army (of whom about a third are women) is swelled by vast numbers of conscripts. Troops were being reintegrated into the economy on "food for work" schemes until the latest war with Ethiopia, which inflicted heavy losses. Mass demobilization was restarted in 2002.

ECONOMICS

Inflation 9.4% p.a. (1993–2000)

 $696m  10.2–13.55 nakfa

SCORE CARD

❏ World GNP Ranking	162nd
❏ GNP per Capita	$170
❏ Balance of Payments	–$208m
❏ Inflation	14%
❏ Unemployment	Widespread underemployment

Strengths
Resourceful, hard-working population. Strategic position on Red Sea – tourism and transportation. Potential for mining and oil industry. Government committed to cutting dependence on food aid.

Weaknesses
Destruction of infrastructure and equipment; port of Massawa heavily bombed. Dependent on aid. Most of population living at subsistence level. Susceptibility to drought and famine. Return of some 750,000 refugees.

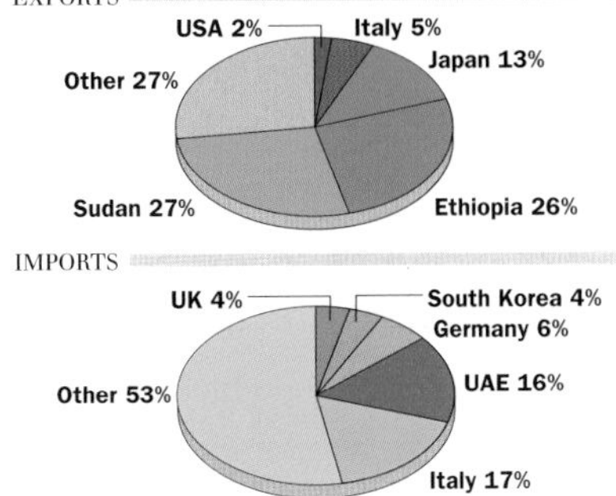
EXPORTS
USA 2%
Italy 5%
Japan 13%
Other 27%
Sudan 27%
Ethiopia 26%
IMPORTS
UK 4%
South Korea 4%
Germany 6%
UAE 16%
Other 53%
Italy 17%

RESOURCES

  Electric power: Not available

7042 tonnes

Not an oil producer; oil refinery at Assab

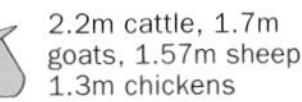 2.2m cattle, 1.7m goats, 1.57m sheep, 1.3m chickens

 Copper, potash, gold, iron, silver, zinc, oil, silica, granite, marble

Eritrea has substantial copper reserves, and lesser ones of silver, zinc, and gold. High-quality silica, granite, and marble deposits could be exploited. Onshore and offshore oil deposits are believed to exist, but exploration work is at an early stage. There is potential for power generation from geothermal sources.

ENVIRONMENT

 Not available

5%

0.03 tonnes per capita

Deforestation and soil erosion are major problems. The Ethiopian army uprooted trees to destroy the cover they provided for Eritrean soldiers. Since 1991, 22 million seedlings have been grown in a replanting scheme. The Red Sea coast is a conservation priority.

MEDIA

TV ownership low

 There are no daily newspapers

PUBLISHING AND BROADCAST MEDIA

 *New Eritrea*, owned by the PFDJ, is published every 3 days in English, Tigrinya, and Arabic

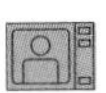 1 state-controlled service  1 state-controlled service

The media are largely controlled by the PFDJ, which runs both the radio and TV services. Independent newspapers are not encouraged.

CRIME

Death penalty in use

 Eritrea does not publish prison figures  Crime levels remain low

Crime has not been a major problem since independence. The judiciary and police answer to the PFDJ. There are a number of political prisoners.

EDUCATION

School leaving age: 13

 56%  3994 students

Very few schools functioned during the war. There is one university. In an attempt to reduce potential ethnic tension, all children above the age of 11 are being taught in English.

HEALTH

No welfare state health benefits

 1 per 20,000 people  Malaria, potential risk of famine

The risk of famine overrides normal health concerns. Eritreans built their own hospitals during the independence struggle. Health provision is basic.

Seasonal river beds *carry rain from the Ethiopian highlands into Eritrea, providing essential irrigation for agriculture.*

E

CHRONOLOGY

British military rule replaced Italian colonial authority in 1941.

- ❏ **1952** Ethiopia absorbs Eritrea.
- ❏ **1961** Beginning of armed struggle.
- ❏ **1987** EPLF refuses offer of autonomy; fighting intensifies.
- ❏ **1991** EPLF takes Asmara.
- ❏ **1993** Formal independence.
- ❏ **1998** Border war with Ethiopia.
- ❏ **2000** OAU peace treaty signed.
- ❏ **2001** Ethiopia completes troop withdrawal.
- ❏ **2002** Border demarcation begins.

SPENDING

GDP/cap. increase

CONSUMPTION AND SPENDING

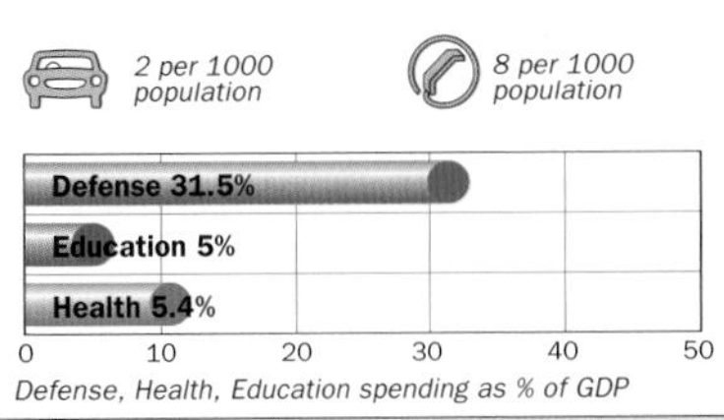
2 per 1000 population
8 per 1000 population
Defense 31.5%
Education 5%
Health 5.4%
0
10
20
30
40
50
Defense, Health, Education spending as % of GDP

Over 80% of Eritrea's population are subsistence farmers. A few of the 150,000 refugees who fled to Arab and Western countries have built up some personal savings.

WORLD RANKING

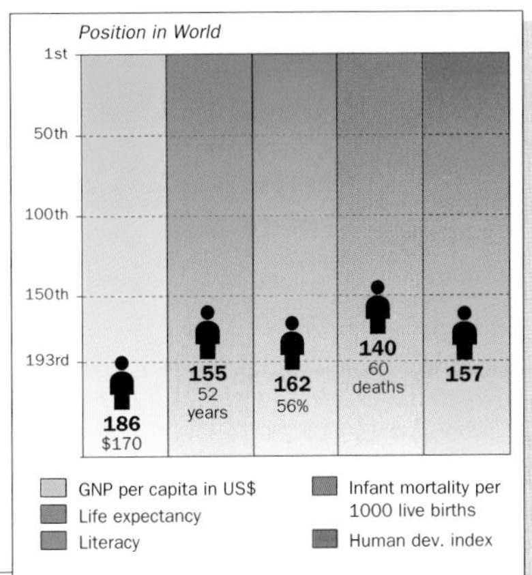
Position in World
1st
50th
100th
150th
193rd
186
$170
155
52 years
162
56%
140
60 deaths
157
GNP per capita in US$
Life expectancy
Literacy
Infant mortality per 1000 live births
Human dev. index

E

ESTONIA

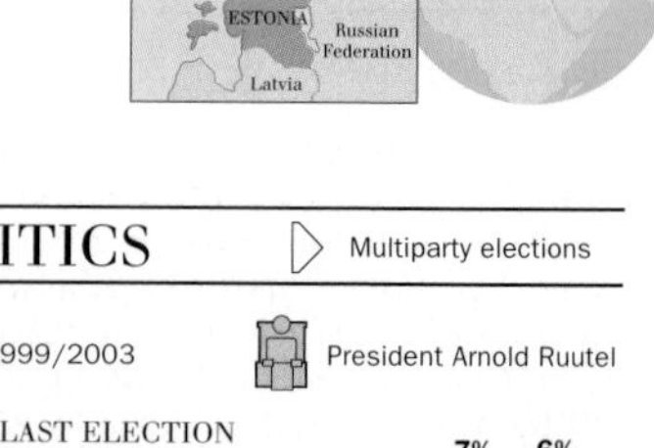
EUROPE
Europe
Finland
ESTONIA
Russian Federation
Latvia

OFFICIAL NAME: Republic of Estonia **CAPITAL:** Tallinn
POPULATION: 1.4 million **CURRENCY:** Kroon **OFFICIAL LANGUAGE:** Estonian

1991

1991

Feb 24
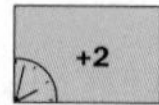
EST

+2

+372

.ee

TRADITIONALLY THE MOST Western-oriented of the Baltic states, Estonia is bordered by Latvia and the Russian Federation. Its terrain is flat, boggy, and partly wooded, and includes more than 1500 islands. Estonia formally regained its independence as a multiparty democracy in 1991. In contrast to the peoples of Latvia and Lithuania, Estonians are Finno-Ugric, speaking a language related to Finnish.

CLIMATE

Continental

WEATHER CHART FOR TALLINN

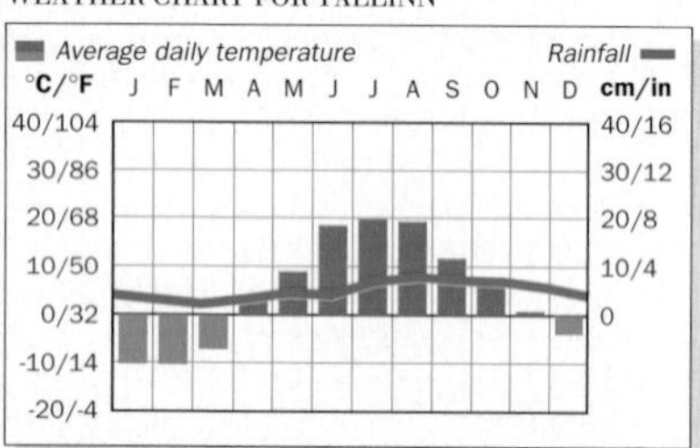
Average daily temperature
Rainfall
°C/°F
J F M A M J J A S O N D
cm/in
40/104
30/86
20/68
10/50
0/32
-10/14
-20/4
40/16
30/12
20/8
10/4
0

Estonia's coastal location gives it cool summers, and cold winters when the Baltic Sea freezes.

TRANSPORTATION

Drive on right

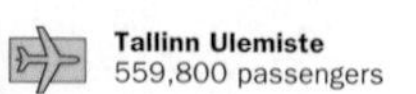
Tallinn Ulemiste
559,800 passengers

209 ships
379,110 grt

THE TRANSPORTATION NETWORK

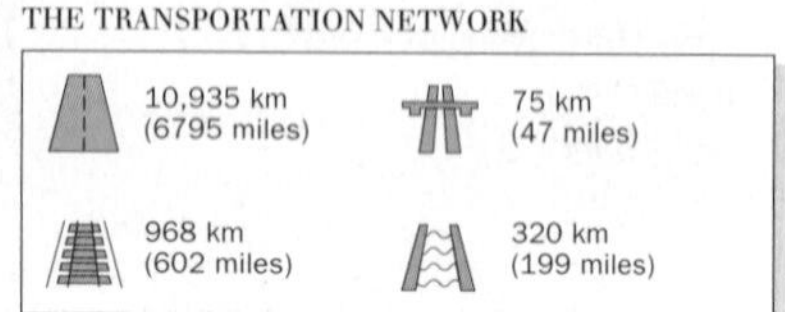
10,935 km
(6795 miles)
75 km
(47 miles)
968 km
(602 miles)
320 km
(199 miles)

Railroads have improved and buses are reliable. Baltic ferries link Tallinn with Finland, Sweden, and Germany.

TOURISM

Visitors : Population 1:1.3

1.24m visitors

Up 31% in 2000

MAIN TOURIST ARRIVALS

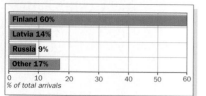

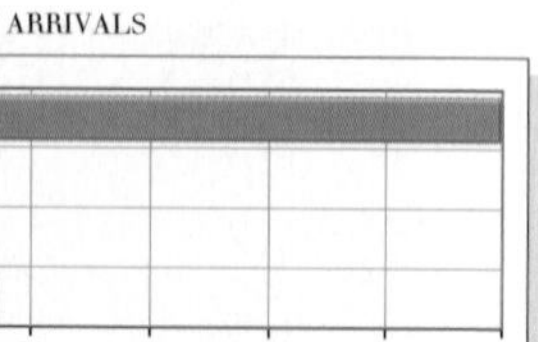
Finland 60%
Latvia 14%
Russia 9%
Other 17%
0
10
20
30
40
50
60
% of total arrivals

Estonia is particularly popular with Finns. Water sports, winter sports, folk and architectural heritage, and nature tours, are the main attractions. Tallinn's medieval center is a major draw for tourists on short visits.

PEOPLE

Pop. density low

Estonian, Russian

31/km² (80/mi²)

THE URBAN/RURAL POPULATION SPLIT

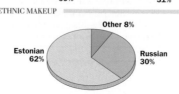

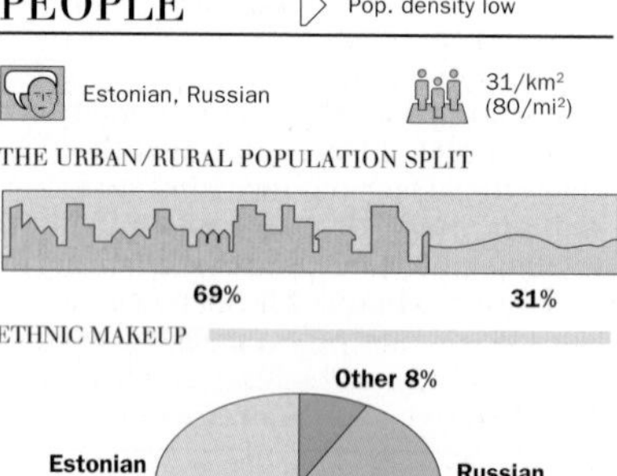
69%
31%
ETHNIC MAKEUP
Other 8%
Estonian 62%
Russian 30%

After the decades of Soviet rule, the relationship between Estonians and the Russian minority was problematic. Rules on citizenship, introduced in 1992 and 1995, excluded many ethnic Russians who could not meet the Estonian language and minimum residency requirements. Over 100,000 took Russian rather than Estonian citizenship. A 2000 language law met international demands for an end to discrimination against the Russian-speaking minority. Estonians are predominantly Protestant. Families are small; divorce rates are high.

POLITICS

Multiparty elections

1999/2003

President Arnold Ruutel

AT THE LAST ELECTION

Parliament 101 seats

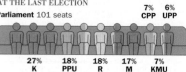

7% **CPP** 6% **UPP**

27% **K** 18% **PPU** 18% **R** 17% **M** 7% **KMU**

K = Center Party **PPU** = Pro Patria Union **R** = Reform Party
M = Moderates **CPP** = Rural People's Party
KMU = Coalition Party **UPP** = United People's Party

Coalition government has been the norm since the end of communist rule. In elections in 1999 the Center Party won the most seats, on a platform of graduated income tax to help overcome wealth disparities. However, Mart Laar, who had led the government in power from 1992 to 1995, became prime minister, leading a center-right coalition of the PPU, the Reform Party, and the Moderates. Although hindered by a slim majority, Laar's efforts to pursue free-market reforms became easier as the economy grew strongly from the beginning of 2000. However, cracks in the coalition led to his resignation in January 2002. The new government led by Siim Kallas included the Center Party.

ESTONIA

Total Land Area : 45 226 sq. km (17 462 sq. miles)

LAND HEIGHT
200m/565ft
Sea Level

POPULATION
over 500 000
over 100 000
over 50 000
over 10 000
under 10 000

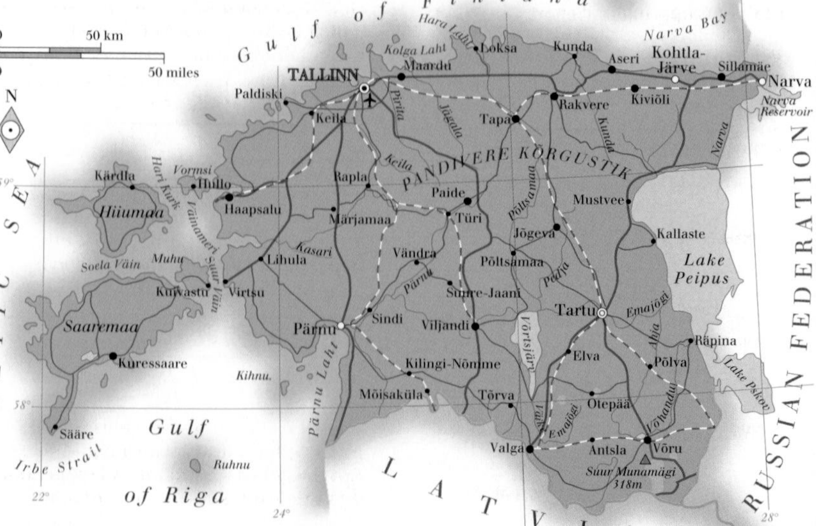
Gulf of Finland
0 50 km
0 50 miles
TALLINN
Paldiski
Keila
Maardu
Loksa
Kunda
Aseri
Kohtla-Järve
Sillamäe
Narva
Narva Bay
Narva Reservoir
Kiviõli
Rakvere
Tapa
PANDIVERE KÕRGUSTIK
Rapla
Paide
Türi
Märjamaa
Haapsalu
Kärdla
Hiiumaa
Vormsi
Lihula
Virtsu
Kuivastu
Muhu
Saaremaa
Kuressaare
Sääre
Irbe Strait
Gulf of Riga
Ruhnu
Kihnu
Pärnu
Pärnu Laht
Sindi
Vändra
Suure-Jaani
Viljandi
Põltsamaa
Jõgeva
Mustvee
Kallaste
Lake Peipus
Tartu
Elva
Räpina
Põlva
Lake Pskov
Kilingi-Nõmme
Mõisaküla
Tõrva
Otepää
Valga
Antsla
Võru
Suur Munamägi 318m
Võrtsjärv
Emajõgi
BALTIC SEA
LATVIA
RUSSIAN FEDERATION

WORLD AFFAIRS

Joined UN in 1991

CBSS
CE
EAPC
OSCE
PfP

Estonia's trade with the West has been growing, and "fast track" negotiations on full EU membership began in 1998. Ties with other Baltic countries and with Scandinavia have been particularly emphasized. Integration with the EU and NATO is the top priority.

Estonia has now accepted the de facto border with Russia, having effectively ceded a portion of its territory during the Soviet period.

AID

Recipient

 $64m (receipts)

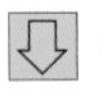 Down 24% in 2000

Although an aid recipient, since 1997 Estonia has also been an aid donor, mainly through technical assistance.

DEFENSE

Compulsory military service

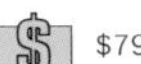 $79m

 Up 11% in 2000

The government agreed in July 2000 to shorten compulsory military service from 12 to eight months. Initial US opposition to full membership of NATO has now been changed to support for the Baltic states' entry into the organization.

ECONOMICS

Inflation 53% p.a. (1990–2000)

$4.89bn

 16.675–17.577 krooni

SCORE CARD

- ❑ WORLD GNP RANKING........................112th
- ❑ GNP PER CAPITA..............................$3580
- ❑ BALANCE OF PAYMENTS....................–$315m
- ❑ INFLATION..4%
- ❑ UNEMPLOYMENT..................................13%

STRENGTHS

Improved productivity and stable currency are pegged to the euro. Simplicity of tax regime. More advantage is being taken of natural resources, including timber and oil shale. Transportation infrastructure has been upgraded. Exports growing.

WEAKNESSES

Poor raw materials base. Dependence on imported energy supplies.

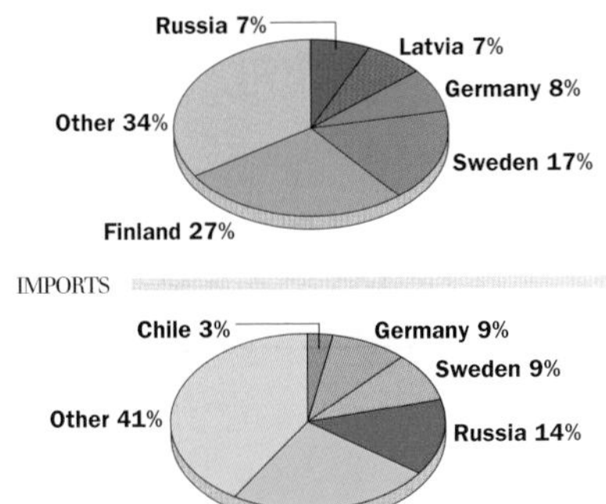
EXPORTS
Russia 7%
Latvia 7%
Germany 8%
Sweden 17%
Finland 27%
Other 34%
IMPORTS
Chile 3%
Germany 9%
Sweden 9%
Russia 14%
Finland 24%
Other 41%

RESOURCES

Electric power 2.6m kw

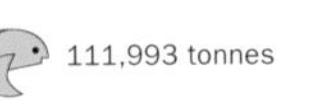 111,993 tonnes

 Oil figures not published

 300,200 pigs, 252,800 cattle, 2.37m chickens

 Oil shale, coal, peat, phosphorite

The chief energy resource is oil shale. Phosphorite mining has been stopped. Timber is processed to make paper.

ENVIRONMENT

Sustainability rank: 18th

 12%

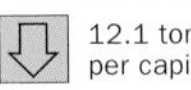 12.1 tonnes per capita

Industrial pollution comes especially from power stations burning oil shale. Danger of radioactive leaks from former Soviet bases remains. Water supply and sewage treatment have improved.

MEDIA

TV ownership high

 Daily newspaper circulation 174 per 1000 people

PUBLISHING AND BROADCAST MEDIA

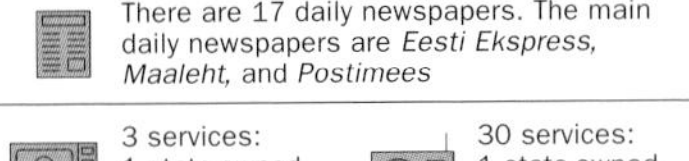

There are 17 daily newspapers. The main daily newspapers are *Eesti Ekspress*, *Maaleht*, and *Postimees*

3 services: 1 state-owned, 2 independent

30 services: 1 state-owned, 29 independent

The media are mostly pro-government. The number of Russian-language programs is declining. Estonians have been able to receive Finnish satellite TV for some years.

CRIME

No death penalty

 4742 prisoners

 Up 13% in 1999

Robbery and narcotics are the main crime problems. Generally, however, crime levels are still relatively low.

EDUCATION

School leaving age: 16

 99%

50,814 students

Education is becoming increasingly Westernized. There are two main universities, in Tallinn and Tartu.

HEALTH

Welfare state health benefits

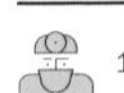 1 per 333 people

Heart diseases, cancers, accidents, violence

The health system, improved since the collapse of communism, is better than that of most former Soviet republics.

***The Russian Orthodox convent** of Pühtitsa at Kuremäe in Estonia's marshy north. Most of the population is Evangelical Lutheran.*

E

CHRONOLOGY

After Swedish and then Russian rule, Estonia briefly enjoyed independence from 1921 until its incorporation into the Soviet Union in 1940.

- ❑ **1990** Unilateral declaration of independence; achieved in 1991.
- ❑ **1992** First multiparty elections: election of center-right government.
- ❑ **1996** President Lennart Meri wins second term of office.
- ❑ **1998** EU opens negotiations.
- ❑ **1999** Elections: new center-right government.
- ❑ **2001** Communist-era leader Arnold Ruutel elected president.

SPENDING

GDP/cap. decrease

CONSUMPTION AND SPENDING

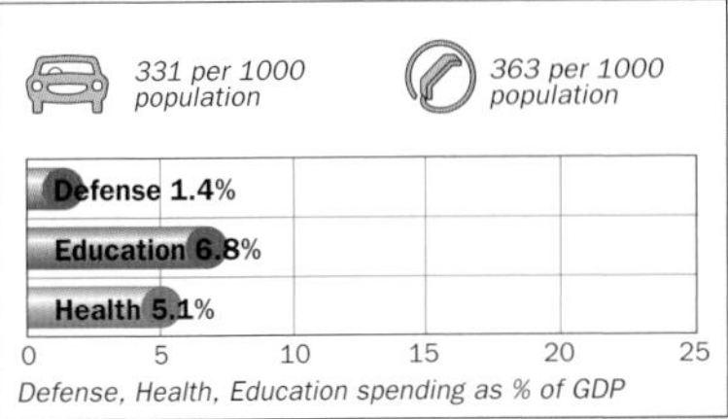
331 per 1000 population
363 per 1000 population
Defense 1.4%
Education 6.8%
Health 5.1%
0
5
10
15
20
25
Defense, Health, Education spending as % of GDP

Market reforms have led to increased prosperity. A few have become very rich. Average wages are higher than in other Baltic states.

WORLD RANKING

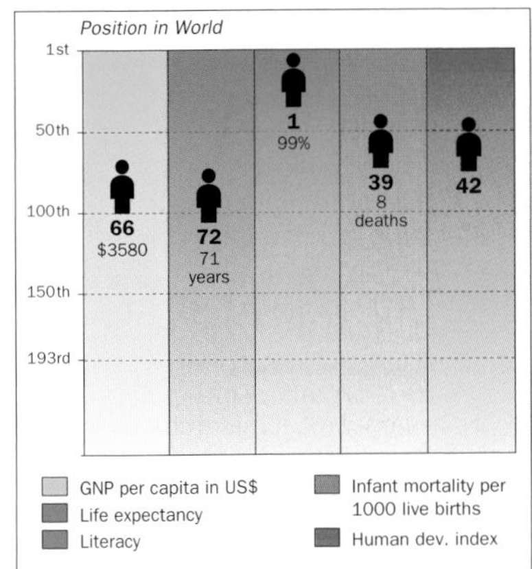
Position in World
1st
50th
100th
150th
193rd
66
$3580
72
71
years
1
99%
39
8
deaths
42
GNP per capita in US$
Life expectancy
Literacy
Infant mortality per 1000 live births
Human dev. index

E

ETHIOPIA

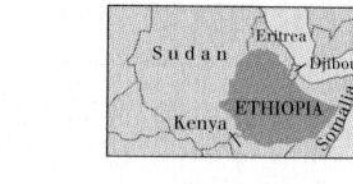

EAST AFRICA
Africa
Sudan
Eritrea
Djibouti
ETHIOPIA
Kenya
Somalia

OFFICIAL NAME: Federal Democratic Republic of Ethiopia **CAPITAL:** Addis Ababa
POPULATION: 64.5 million **CURRENCY:** Ethiopian birr **OFFICIAL LANGUAGE:** Amharic

1896

2002

May 28

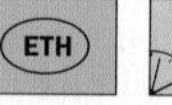
ETH

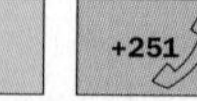
+3

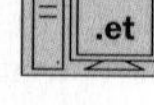
+251

.et

LOCATED IN NORTHEAST Africa, the former empire of Ethiopia is the cradle of an ancient civilization which adopted Orthodox Christianity in the 4th century. It has been landlocked since 1993, when Eritrea, on the Red Sea, seceded. Ethiopia is mountainous except for desert lowlands in the northeast and southeast, and is prone to devastating drought and famine. A long civil war ended in 1991 with the defeat of the Stalinist military dictatorship that had ruled since 1974. A free-market, multiparty democratic system now provides substantial regional autonomy. War with Eritrea in 1998–2000 brought heavy losses on both sides before a peace agreement was signed in December 2000. Arbitrators began redefining the border in spring 2002.

CLIMATE

Mountain/steppe

WEATHER CHART FOR ADDIS ABABA

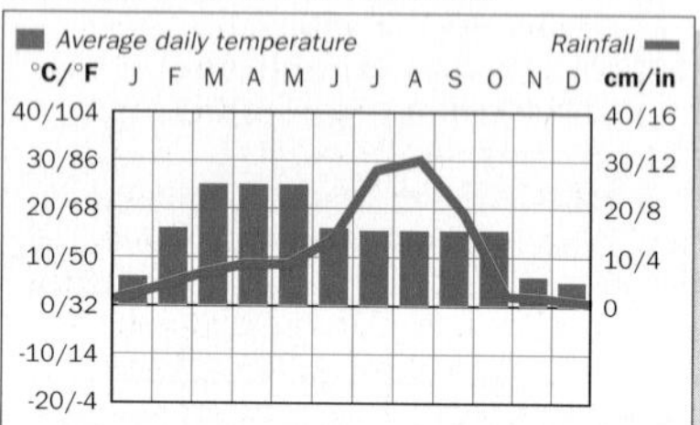

Average daily temperature
Rainfall
°C/°F
J F M A M J J A S O N D
cm/in
40/104
30/86
20/68
10/50
0/32
-10/14
-20/-4
40/16
30/12
20/8
10/4
0

In general, the climate is moderate, except in the lowlands of the Danakil and the Ogaden deserts, which are hot all year round and can suffer severe drought. The highlands are temperate, with night frost in the mountains. The single rainy season in the west brings twice as much rain as do the two wet seasons in the east. During these cloudy periods, thunderstorms occur almost daily.

TRANSPORTATION

Drive on right

Bole International, Addis Ababa
1.04m passengers

12 ships
92,434 grt

THE TRANSPORTATION NETWORK

4275 km (2656 miles)

Trans-East Africa Highway

681 km (423 miles)

None

The single railroad linking Addis Ababa with Djibouti has grown in strategic importance due to the conflict with Eritrea. Ethiopia's main access to the sea by road has been through the Red Sea ports of Assab and Massawa, now part of an independent Eritrea. Inland, pack mules and donkeys are widely used. Ethiopian Airlines has good services to much of Africa, and to European and US cities.

TOURISM

Visitors : Population 1:516

125,000 visitors

Up 36% in 2000

MAIN TOURIST ARRIVALS

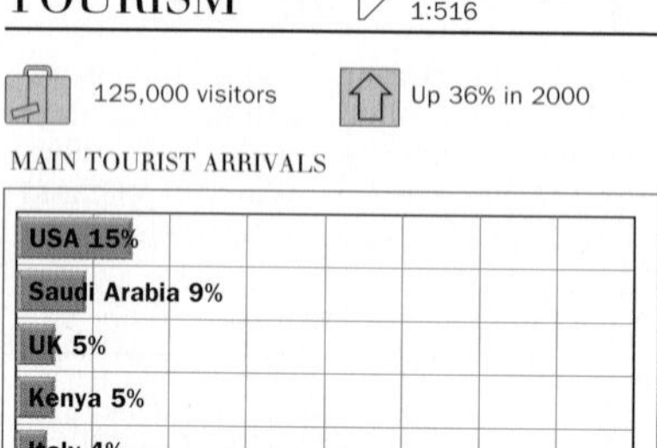

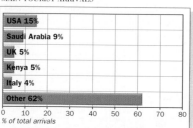

USA 15%
Saudi Arabia 9%
UK 5%
Kenya 5%
Italy 4%
Other 62%
0 10 20 30 40 50 60 70 80
% of total arrivals

Despite Ethiopia's unique attractions, tourism is on a small scale, although since 1991 there has been a sizable increase in the number of visitors, mostly on organized tours. Several new hotels are being built. The Rift Valley lakes, Lake Tana, the Gonder castles, and the Blue Nile gorge, with its spectacular scenery, are popular tourism destinations, but guides are essential. Ancient churches and cities such as Aksum, the royal capital of the first Ethiopian kingdom, are now accessible. Visitors are also attracted to the five national parks.

***Lalibela** lies 120 km (75 miles) northwest of Desē in Ethiopia's plateau region, and is famous for the rock-hewn churches created by King Lalibela of the Zagwe dynasty.*

PEOPLE

Pop. density medium

Amharic, Tigrinya, Galla, Sidamo, Somali, English, Arabic

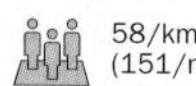

58/km² (151/mi²)

THE URBAN/RURAL POPULATION SPLIT

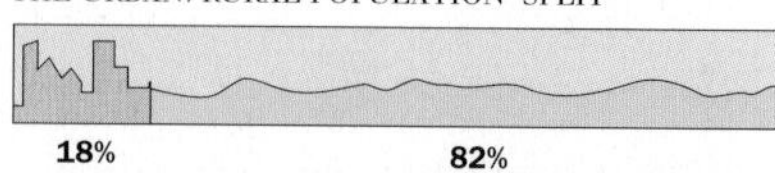

18%
82%

RELIGIOUS PERSUASION

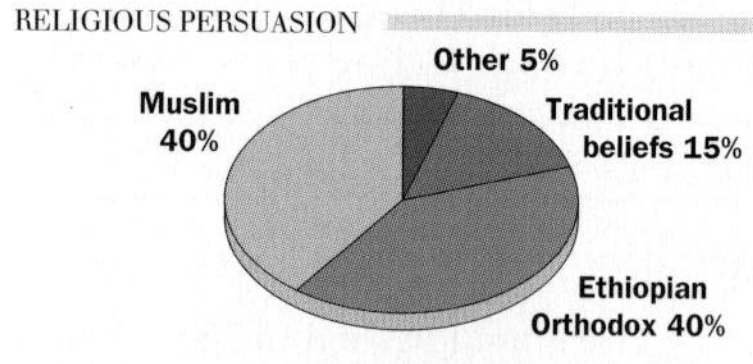

Other 5%
Muslim 40%
Traditional beliefs 15%
Ethiopian Orthodox 40%

ETHNIC MAKEUP

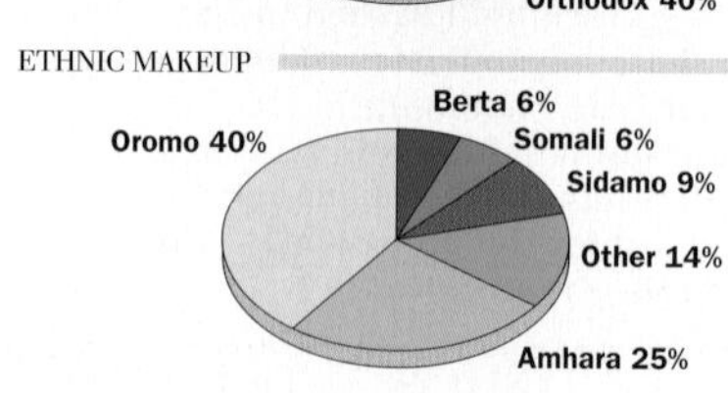

Berta 6%
Oromo 40%
Somali 6%
Sidamo 9%
Other 14%
Amhara 25%

There are 76 ethnic nations in Ethiopia, speaking 286 languages. Oromos (or Gallas) form the largest group, whereas less than 5% of the population are Tigreans.

Civil war was sparked by fighting between different ethnic groups, but they later united in opposition to the Mengistu regime. Ethnic tensions are still near the surface, in spite of the new federal structure, and there have been reports of boundary disputes in several regions. The Oromos withdrew from the Tigrean-dominated government in 1992. Hostility to the government has also been voiced by disaffected Amharas, who had been dominant for several centuries, and by the Orthodox Church. The aspirations of ethnic Somalis in the southeast are another source of tension.

Most of the small Jewish community, which has lived in Ethiopia for 2000 years, was evacuated to Israel in 1991, but more than 20,000 remain, waiting for Israel to offer them citizenship.

The participation of women in rural organizations is increasing, reflecting the key role women played in the war.

POPULATION AGE BREAKDOWN

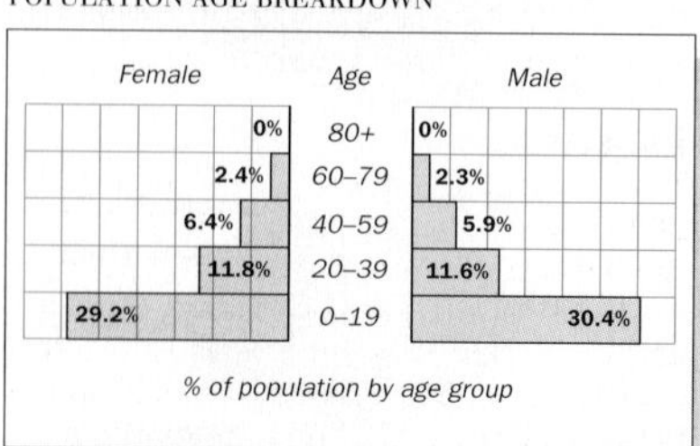

Female
Age
Male
0%
80+
0%
2.4%
60–79
2.3%
6.4%
40–59
5.9%
11.8%
20–39
11.6%
29.2%
0–19
30.4%
% of population by age group

Key to symbols and abbreviations on cover flaps

ETHIOPIA

Total Land Area : 1 127 127 sq. km (435 184 sq. miles)

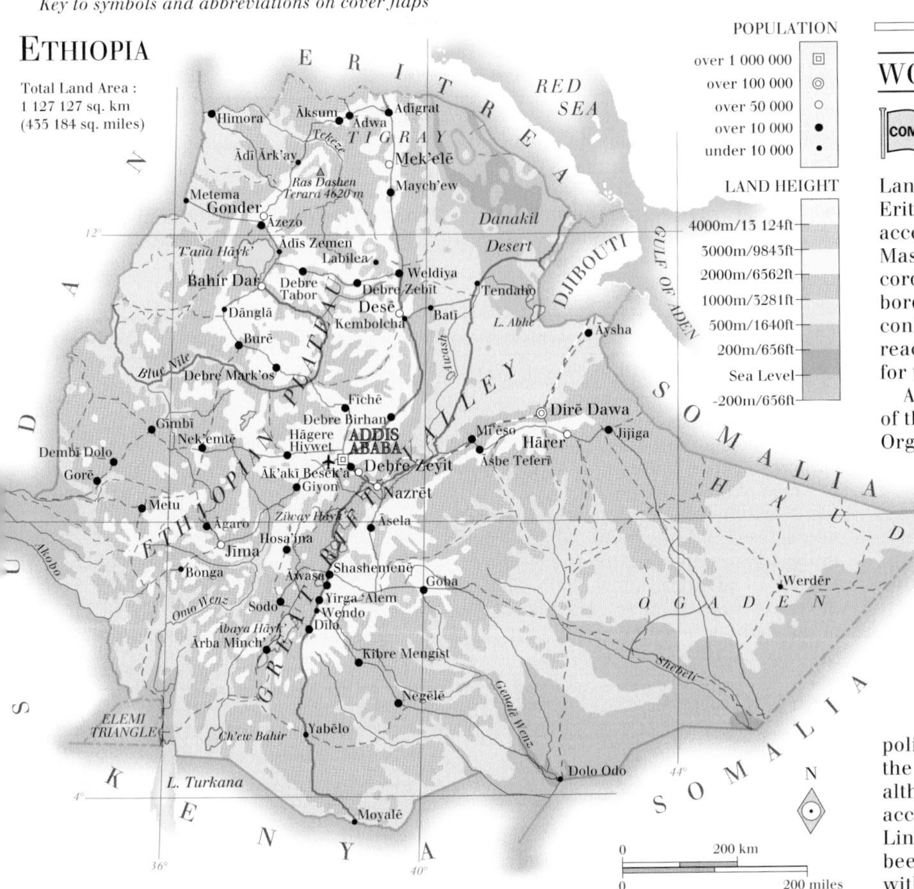

POPULATION
over 1 000 000
over 100 000
over 50 000
over 10 000
under 10 000
LAND HEIGHT
4000m/13 124ft
3000m/9843ft
2000m/6562ft
1000m/3281ft
500m/1640ft
200m/656ft
Sea Level
-200m/656ft
ERITREA
RED SEA
DJIBOUTI
GULF OF ADEN
SOMALIA
SUDAN
KENYA
TIGRAY
Ḥimora
Āksum
Ādwa
Ādīgrat
Mek'elē
Maych'ew
Ādī Ārk'ay
Ras Dashen Terara 4620 m
Metema
Gonder
Āzezo
Danakil Desert
T'ana Hāyk'
Ādīs Zemen
Labilea
Bahir Dar
Debre Tabor
Debre Zebit
Weldiya
Tendaho
Dese
Kembolcha
Bati
L. Abhe
Āysha
Dānglā
Burē
Blue Nile
Debre Mark'os
Fichē
Debre Birhan
Dirē Dawa
Gimbī
Nek'emtē
Hāgere Hiywet
ADDIS ABABA
Mī'ēso
Hārer
Jijiga
Dembī Dolo
Gorē
Āk'akī Besek'a
Debre Zeyit
Āsbe Teferi
Giyon
Nazrēt
Metu
Āgaro
Zīway Hāyk'
Āsela
Jīma
Hosa'ina
ETHIOPIAN PLATEAU
GREAT RIFT VALLEY
Bonga
Āwasa
Shashemenē
Goba
Omo Wenz
Sodo
Yirga 'Alem
Wendo
Dila
Werdēr
OGADEN
HAUD
Abaya Hāyk'
Ārba Minch'
Kibre Mengist
Shebeli
Negēlē
Genale Wenz
ELEMI TRIANGLE
Ch'ew Bahir
Yabēlo
Dolo Odo
L. Turkana
Moyalē
Akobo
N
0 200 km
0 200 miles

WORLD AFFAIRS

Joined UN in 1945

COMESA | G24 | IGAD | NAM | AU

Landlocked since the secession of Eritrea, Ethiopia needs continued access to the Red Sea ports of Massawa and Assab, and maintained cordial relations with Eritrea until a border dispute escalated into armed conflict in 1998. A peace accord reached two years later allowed for the demarcation of a new border.

Addis Ababa is the headquarters of the AU (successor to the Organization of African Unity) and of the UN Economic Commission for Africa. Ethiopia is active in regional diplomacy, including numerous attempts at brokering peace in Somalia, although tension with Somali factions during 1998 and 1999 escalated into armed intervention by Ethiopia.

The government's official policy is one of noninterference in the affairs of neighboring countries, although Sudan and Somalia have accused Ethiopia of supporting rebels. Links with other African states have been strengthened, as have those with the US, the EU, and Israel.

POLITICS

Multiparty elections

L. House 2000/2005
U. House 2000/2005

President Girma Wolde Giorgis

AT THE LAST ELECTION

House of People's Representatives 550 seats

32% OPDO	24% ANDM	7% TPLF	27% Others	8% Vacant	2% Ind

OPDO = Oromo People's Democratic Organization
ANDM = Amhara National Democratic Movement
TPLF = Tigre People's Liberation Front **Ind** = Independents

The Ethiopian People's Revolutionary Democratic Front (**EPRDF**) includes all the main parties and controls over 90% of the lower house

Federal Council 108 seats

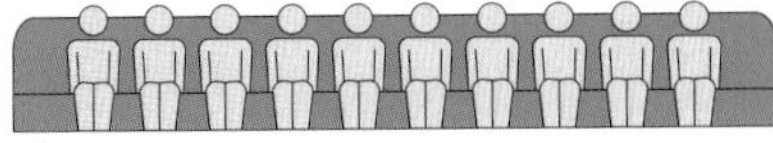

The Federal Council is elected indirectly on a nonparty basis

The transitional period which followed the collapse of the Mengistu military dictatorship in 1991 ended in 1995 with multiparty elections.

PROFILE

The current government, reelected in 2000, succeeded one set up in 1991 by the EPRDF, the strongest of the groups that fought Mengistu's Marxist regime and chiefly responsible for winning the civil war. Prime Minister Meles Zenawi is the leader of the Tigrean People's Liberation Front, the largest group within the EPRDF. There is growing opposition from the Oromos and Amharas to the dominance of Tigreans. The nine states are largely governed by elected governments dominated by local liberation movements.

MAIN POLITICAL ISSUE
Ethnic representation

The 1994 constitution establishing a nine-state federation grants the states considerable autonomy, including the right to secede, as Eritrea did in 1993. The EPRDF government believes this to be the best way to prevent secessionist conflict and maintain national unity. The ruling broad-based coalition is ideologically dominated by Tigrean politicians.

***Prime Minister Meles Zenawi**, leader of the EPRDF, which ousted the Mengistu regime.*

***Mengistu Haile Mariam**, who ran Ethiopia on Soviet lines from 1977–1991.*

CHRONOLOGY

After repelling a devastating Muslim invasion in 1523, Ethiopia developed as an isolated empire until Egyptian and Sudanese incursions in the 1850s led to its renewed political power under Emperor Teodros. His successor, Menelik II, doubled the empire southward and eastward.

- ❑ **1896** Italian invasion of Tigre defeated. Europeans recognize Ethiopia's independence.
- ❑ **1913** Menelik II dies.
- ❑ **1916** His son, Lij Iyasu, is deposed for his conversion to Islam and a proposed alliance with Turkey. Menelik's daughter, Zauditu, becomes empress with Ras (Prince) Tafari as regent.
- ❑ **1923** Joins League of Nations.
- ❑ **1930** Zauditu dies. Ras Tafari crowned Emperor Haile Selassie.
- ❑ **1936** Italians occupy Ethiopia. League of Nations fails to react.
- ❑ **1941** British oust Italians and restore Haile Selassie, who sets up a constitution, parliament, and cabinet, but retains personal power and the feudal system.
- ❑ **1952** Eritrea, ruled by Italy until 1941, then under British mandate, federated with Ethiopia.

➪

E

CHRONOLOGY *continued*

- **1962** Unitary state created; Eritrea loses its autonomy despite demands of secessionists.
- **1972–1974** Famine kills 200,000.
- **1974** Strikes and army mutinies at Haile Selassie's autocratic rule and country's economic decline. Dergue (Military Committee) stages coup.
- **1975** Becomes socialist state: nationalizations, worker cooperatives, and health reforms.
- **1977** Col. Mengistu Haile Mariam takes over. Somali invasion of Ogaden defeated with Soviet and Cuban help.
- **1978–1979** Thousands of political opponents killed or imprisoned.
- **1984** Workers' Party of Ethiopia (WPE) set up on Soviet model. One million die in famine after drought and years of war. Live Aid concert raises funds for relief.
- **1986** Eritrean rebels now control the whole northeastern coast.
- **1987** Serious drought again threatens famine.
- **1988** Eritrean and Tigrean People's Liberation Fronts (EPLF and TPLF) begin new offensives. Mengistu's budget is for "Everything to the War Front." Diplomatic relations with Somalia restored.
- **1989** Military coup attempt fails. TPLF in control of most of Tigre. TPLF and Ethiopian People's Revolutionary Movement form alliance – EPRDF.
- **1990** Military gains by opponents of Mengistu regime. Moves toward market economy and restructuring of ruling party to include non-Marxists. Distribution of food aid for victims of new famine is hampered by government and rebel forces.
- **1991** Mengistu accepts military defeat and flees country. EPRDF enters Addis Ababa, sets up provisional government, promising representation for all ethnic groups. Outbreaks of fighting continue, between mainly Tigrean EPRDF and opposing groups.
- **1993** Eritrean independence recognized following referendum.
- **1995** Transitional rule ends. EPRDF wins landslide in multiparty elections, sets up first democratic government. New nine-state federation is formed.
- **1998–2000** Border war with Eritrea.
- **2000** OAU peace treaty signed. Haile Selassie's remains buried in Trinity Cathedral, Addis Ababa.
- **2001** Ethiopia completes troop withdrawal from Eritrea.
- **2002** Over 120 human rights demonstrators killed by police.

AID

Recipient

 $693m (receipts)  Up 8% in 2000

The World Food Program and the EU are the largest sources of assistance, while the US has taken over from Italy and the former Soviet Union as the major bilateral donor. Aid per capita is low by regional standards. However, long-term development assistance and balance-of-payments support look set to continue their recent growth. Aid is now playing an increasingly important part in the economy. The emphasis was shifting from food aid toward credit for infrastructure development – until the 2000 drought disaster.

DEFENSE

No compulsory military service

 $448m Up 1% in 2000

Ethiopia is one of the most heavily militarized states in Africa. Its sizable standing army is boosted by conscription at times of crisis. Heavy losses have been sustained in fighting with Eritrea, and a 12-month international arms embargo was imposed following the Ethiopian advance in May 2000. The government is trying to gain control of the many ethnic and clan-based militias throughout the country.

ETHIOPIAN ARMED FORCES

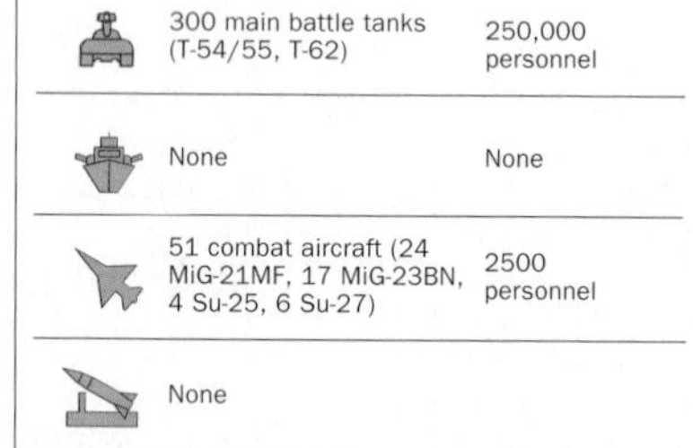
300 main battle tanks (T-54/55, T-62)
250,000 personnel
None
None
51 combat aircraft (24 MiG-21MF, 17 MiG-23BN, 4 Su-25, 6 Su-27)
2500 personnel
None

ECONOMICS

Inflation 7% p.a. (1990–2000)

 $6.74bn  8.225–8.450 Ethiopian birr

SCORE CARD

- World GNP Ranking 99th
- GNP per Capita $100
- Balance of Payments $16m
- Inflation 5.9%
- Unemployment 63%

EXPORTS

Italy 7%, Saudi Arabia 8%, Djibouti 11%, Japan 11%, Germany 18%, Other 45%

IMPORTS

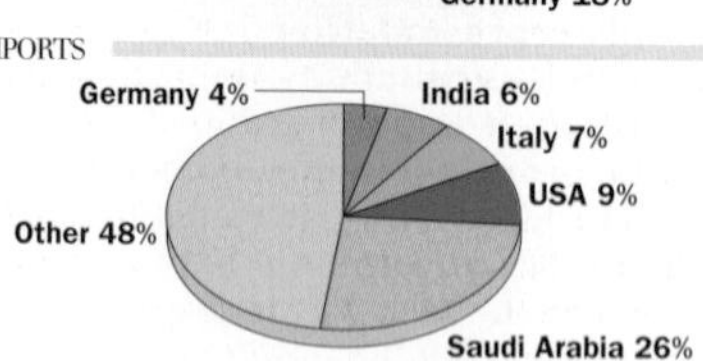
Germany 4%
India 6%
Italy 7%
USA 9%
Other 48%
Saudi Arabia 26%

Strengths

Increased economic aid in 1990s. End of total state control. Coffee production.

Weaknesses

Overwhelming dependence on agriculture. Periodic serious droughts. War-damaged infrastructure. Massive displacement of population by war and drought. Small industrial base. Lack of skilled workers. Legacy of Mengistu regime's centrally planned economy.

Profile

After the end of the civil war in 1991, Ethiopia began moving toward a market economy by encouraging foreign investment and reforming land tenure. Economic decline was reversed in 1993 as agricultural and industrial output grew, with foreign aid used to fund the purchase of parts and raw materials for manufacturing. These gains were undermined by war with Eritrea in 1998–2000 and the renewed danger of severe drought-related famine in 2000.

ECONOMIC PERFORMANCE INDICATOR

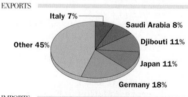

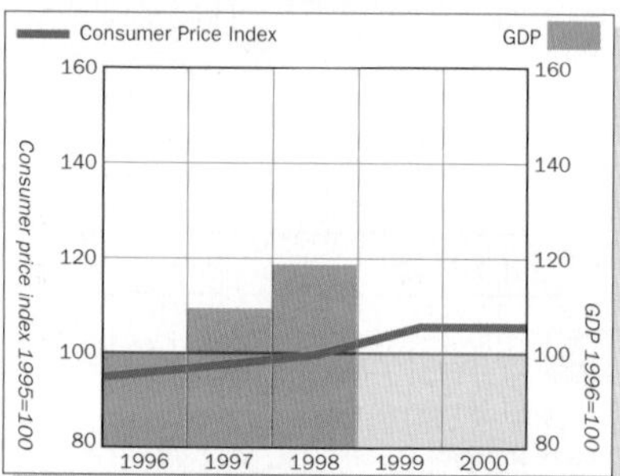
Consumer Price Index
GDP
160
140
120
100
80
Consumer price index 1995=100
GDP 1996=100
1996
1997
1998
1999
2000

ETHIOPIA : MAJOR BUSINESSES

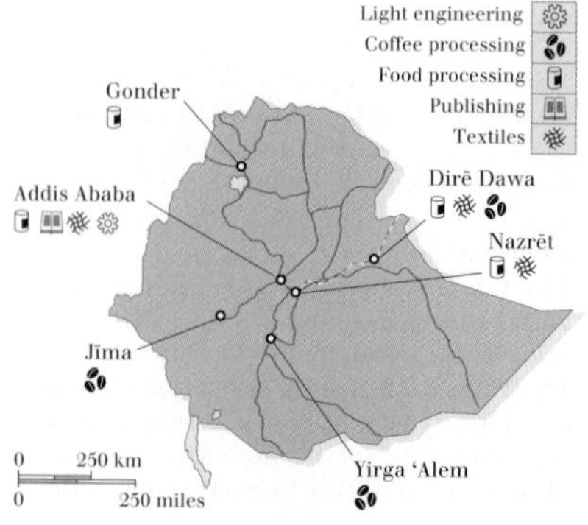
Light engineering
Coffee processing
Food processing
Publishing
Textiles
Gonder
Addis Ababa
Dirē Dawa
Nazrēt
Jīma
Yirga 'Alem
0 250 km
0 250 miles

RESOURCES

Electric power 456,000 kw

15,858 tonnes

Oil reserves currently unexploited

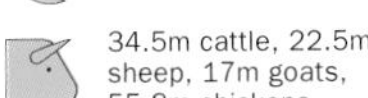
34.5m cattle, 22.5m sheep, 17m goats, 55.8m chickens

Oil, gold, platinum, copper, potash, iron, natural gas

ELECTRICITY GENERATION

Hydro 94% (1.6bn kwh)
Combustion 6% (0.1bn kwh)
Nuclear 0%
Other 0%
0 20 40 60 80 100
% of total generation by type

Manpower and financial constraints have prevented a systematic survey of mineral resources. At present, mining contributes less than 1% of GDP. Ethiopia has great potential for hydroelectric power which, in the long run, could offset a domestic reliance on fuelwood and slow massive deforestation and soil erosion. Current exploration for oil and gas has revealed reserves in the Ogaden, but exploitation has not begun. When Eritrea seceded in 1993, Ethiopia lost other substantial oil reserves and many oil concessions.

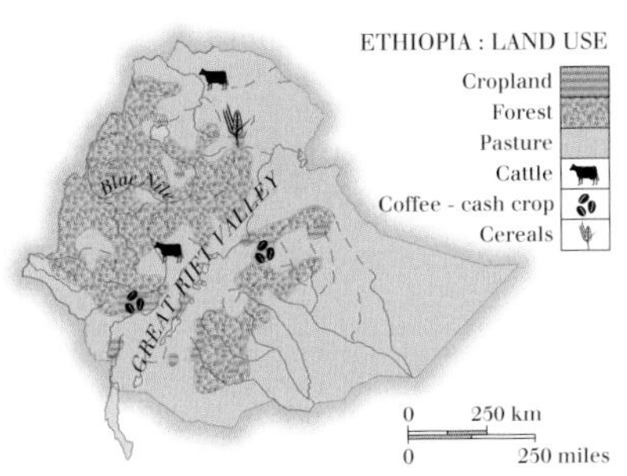
ETHIOPIA : LAND USE
Cropland
Forest
Pasture
Cattle
Coffee - cash crop
Cereals
Blue Nile
GREAT RIFT VALLEY
0 250 km
0 250 miles

ENVIRONMENT

Sustainability rank: 113th

6%

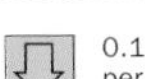
0.1 tonnes per capita

ENVIRONMENTAL TREATIES

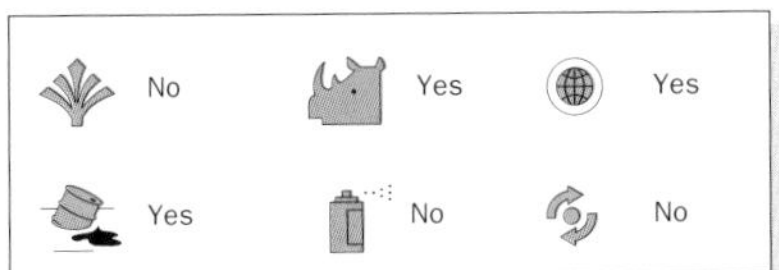
No
Yes
Yes
Yes
No
No

Deforestation for fuelwood and the resultant rapid soil erosion, particularly in the highlands, are serious problems. Forest cover has fallen from 40% in 1900 to only 2% today. Dung is being used for fuel, instead of as a fertilizer. The amount used – with a fertilizing value of $123 million a year – would increase annual grain harvests by up to 1.5 million tonnes. Local projects include terracing hillsides to prevent soil and water run-off – 36,000 km (22,370 miles) of terraces were built in Tigre in 1992.

MEDIA

TV ownership low

Daily newspaper circulation 2 per 1000 people

The government remains uneasy about the post-Mengistu independent press, which has become prolific and critical, although circulation is small. Legal action has been taken to silence several publications. All main newspapers and the TV broadcasting station are government-owned and operated.

PUBLISHING AND BROADCAST MEDIA

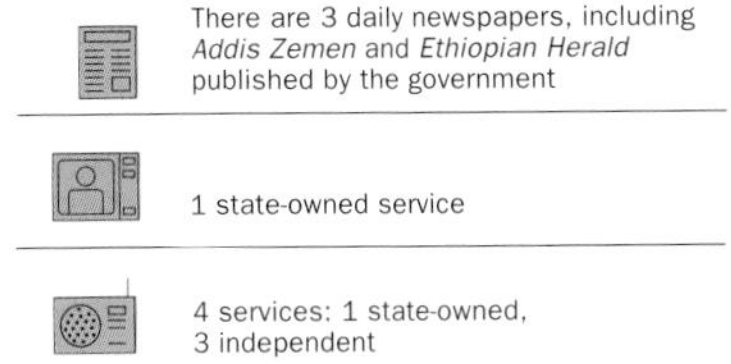
There are 3 daily newspapers, including *Addis Zemen* and *Ethiopian Herald* published by the government
1 state-owned service
4 services: 1 state-owned, 3 independent

CRIME

Death penalty in use

13,585 prisoners

Down 19% in 1999

CRIME RATES

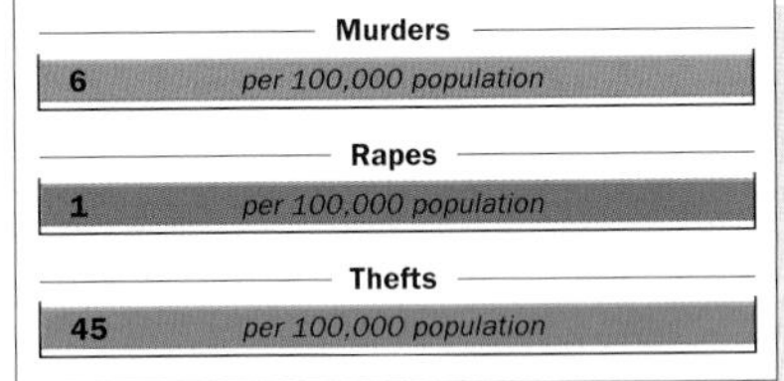
Murders
6 per 100,000 population
Rapes
1 per 100,000 population
Thefts
45 per 100,000 population

A number of human rights abuses by the transitional government have been documented by the independent Ethiopian Human Rights Council. These include detention without trial, "disappearances," and extrajudicial killings. There is some concern over indiscipline among EPRDF forces, who provide a de facto police force in many regions. In many rural areas, the state system has yet to replace traditional forms of justice.

EDUCATION

School leaving age: 13

38%

67,682 students

THE EDUCATION SYSTEM

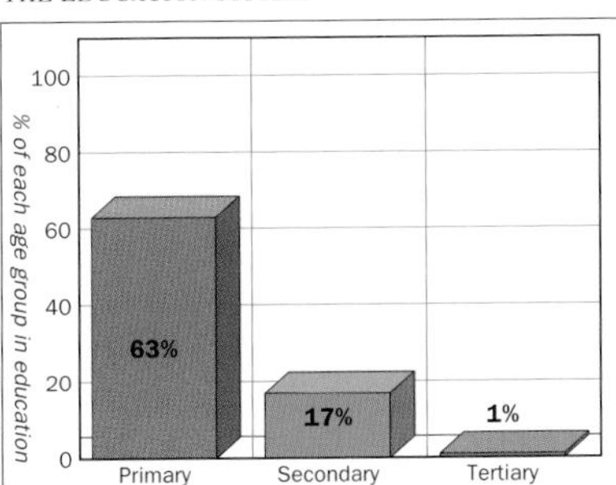
% of each age group in education
100
80
60
40
20
0
63%
17%
1%
Primary
Secondary
Tertiary

Secondary education is in English and Amharic. Schools are basic and classes crowded; education was severely disrupted during the civil war. Addis Ababa University, a center of political activity (usually anti-EPRDF), suffers periodic closures and the dismissal of leading academics.

HEALTH

No welfare state health benefits

1 per 20,000 people

Diarrheal and respiratory diseases, tuberculosis, malaria

Only about half of the population lives within 12 km (8 miles) of a health unit. Hospital building, distribution of resources to rural areas, outpatient visits, and referrals are all very slow. Skin and eye diseases are common. Mission hospitals are of a reasonably high standard, although the use of traditional remedies is widespread. The incidence of HIV/AIDS is falling.

SPENDING

GDP/cap. increase

CONSUMPTION AND SPENDING

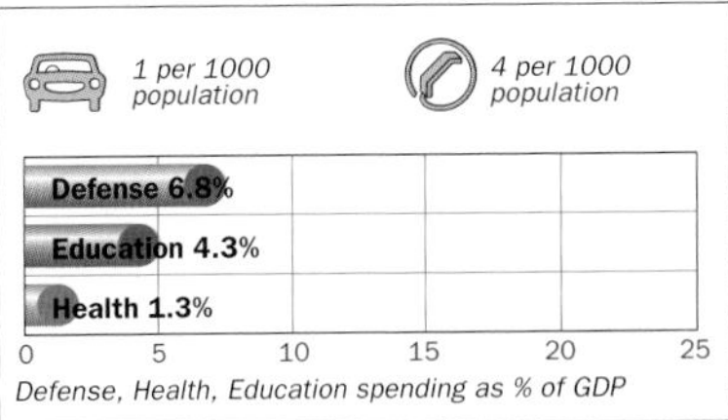
1 per 1000 population
4 per 1000 population
Defense 6.8%
Education 4.3%
Health 1.3%
0 5 10 15 20 25
Defense, Health, Education spending as % of GDP

Most Ethiopians are extremely poor, many of the country's wealthier families having fled into exile in recent years. Ethiopian Christian culture places more value on maintaining traditional social structures than on realising individual ambition. Living at subsistence level and a reliance on traditional agriculture remain the general expectation.

WORLD RANKING

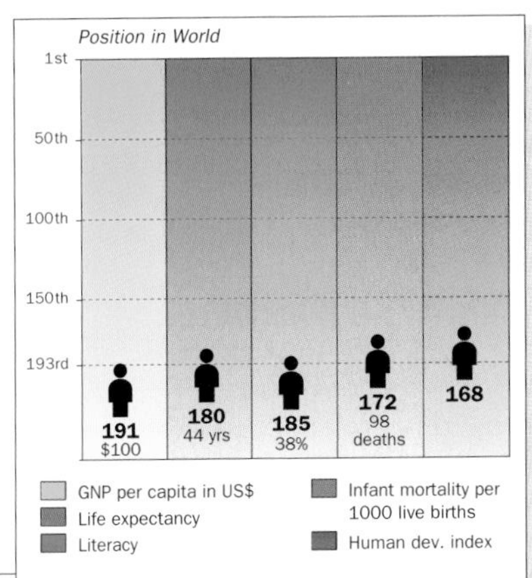
Position in World
1st
50th
100th
150th
193rd
191 $100
180 44 yrs
185 38%
172 98 deaths
168
GNP per capita in US$
Life expectancy
Literacy
Infant mortality per 1000 live births
Human dev. index

E

FIJI

SOUTH PACIFIC

Tuvalu, Wallis & Futuna, Vanuatu, FIJI, Tonga, New Caledonia, Pacific Ocean

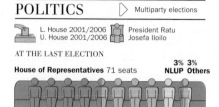

OFFICIAL NAME: Republic of the Fiji Islands **CAPITAL:** Suva
POPULATION: 823,000 **CURRENCY:** Fiji dollar **OFFICIAL LANGUAGE:** English

1970

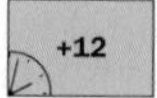
1970

Oct 10 | FJI | +12 | +679 | .fj

FIJI IS A VOLCANIC archipelago in the southern Pacific Ocean, comprising two main islands and nearly 900 smaller islands and islets. The Melanesian Fijian population was outnumbered in the post-1945 period by ethnic Indians, descended from workers brought over by the British in 1879–1916. Coups led by Fijian supremacists between 1987 and 2000 led to a mass exodus of Indo-Fijians, reversing the ethnic balance and seriously damaging the economy.

POLITICS

Multiparty elections

L. House 2001/2006
U. House 2001/2006

President Ratu Josefa Iloilo

AT THE LAST ELECTION

House of Representatives 71 seats

45% SDL | 38% FLP | 8% MV | 3% Ind | 3% NLUP | 3% Others

SDL = Fijian People's Party **FLP** = Fiji Labor Party **MV** = Conservative Alliance (Matanitu Vanua) **NLUP** = New Labor Unity Party **Ind** = Independents

Senate 32 seats

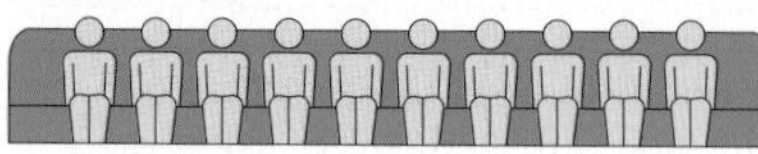

The Senate is appointed by the president

The issue of political representation for Fiji's large Indian population has dominated politics since the 1980s. The 1987 coup, justified as defending ethnic Fijian land rights, was condemned internationally. The traditional Great Council of Chiefs emerged as the power broker after the 2000 coup, despite its lack of constitutional power. It ensured that the Indian-dominated FLP government, elected in 1999, was barred from returning to power and that the multiethnic 1997 constitution remained buried.

CLIMATE

Tropical oceanic

WEATHER CHART FOR SUVA

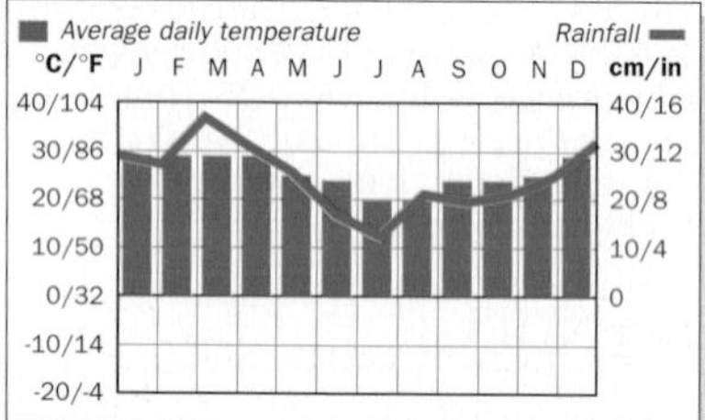
Average daily temperature
Rainfall
°C/°F
J F M A M J J A S O N D
cm/in
40/104
30/86
20/68
10/50
0/32
-10/14
-20/-4
40/16
30/12
20/8
10/4
0

The eastern sides of the main islands are wettest, having more than twice the annual rainfall of the western flanks. Fiji lies in a cyclone path.

TRANSPORTATION

Drive on left

Nadi International 926,677 passengers

53 ships 29,500 grt

THE TRANSPORTATION NETWORK

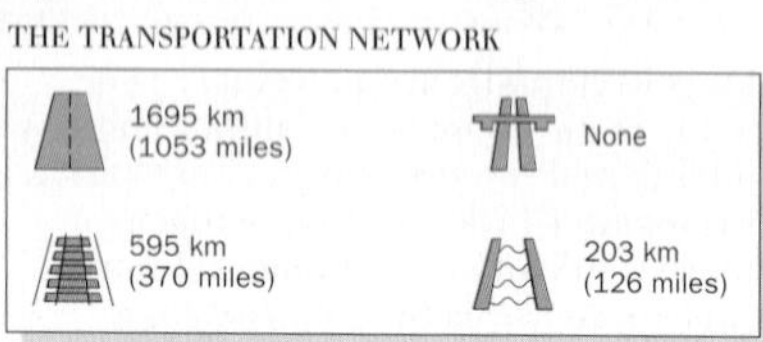
1695 km
(1053 miles)
None
595 km
(370 miles)
203 km
(126 miles)

On the axis of Australian–US west coast air routes, Fiji is well served by international flights. An international airport is proposed for Vanua Levu.

TOURISM

Visitors : Population 1:2.8

294,000 visitors

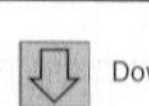

Down 28% in 2000

MAIN TOURIST ARRIVALS

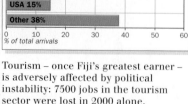

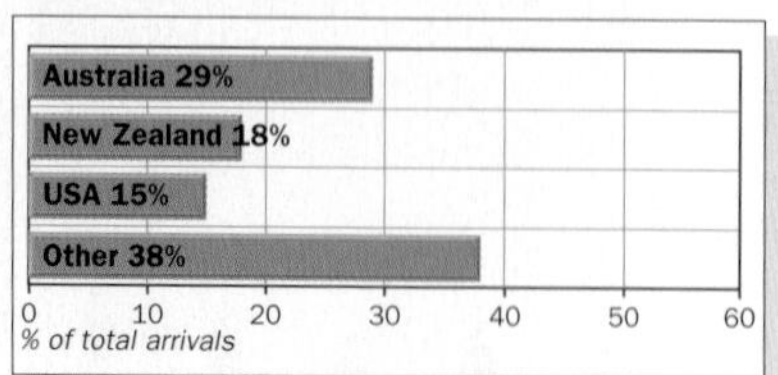
Australia 29%
New Zealand 18%
USA 15%
Other 38%
0
10
20
30
40
50
60
% of total arrivals

Tourism – once Fiji's greatest earner – is adversely affected by political instability: 7500 jobs in the tourism sector were lost in 2000 alone.

PEOPLE

Pop. density low

Fijian, English, Hindi, Urdu, Tamil, Telugu

45/km² (117/mi²)

THE URBAN/RURAL POPULATION SPLIT

49% | 51%

RELIGIOUS PERSUASION

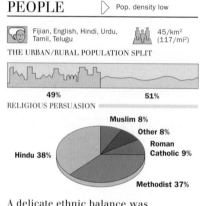

Hindu 38% | Methodist 37% | Roman Catholic 9% | Other 8% | Muslim 8%

A delicate ethnic balance was shattered by the exodus of Indo-Fijians in 1987–1989 and again in 2000–2001. The lawlessness accompanying the recent upheavals exaggerated ethnic tensions and brought racist rhetoric back to the political mainstream. A substantial population of Polynesians live on Rotuma and have traditionally pressed for autonomy. Women are lobbying for more rights.

FIJI

Total Land Area : 18 270 sq. km (7054 sq. miles)

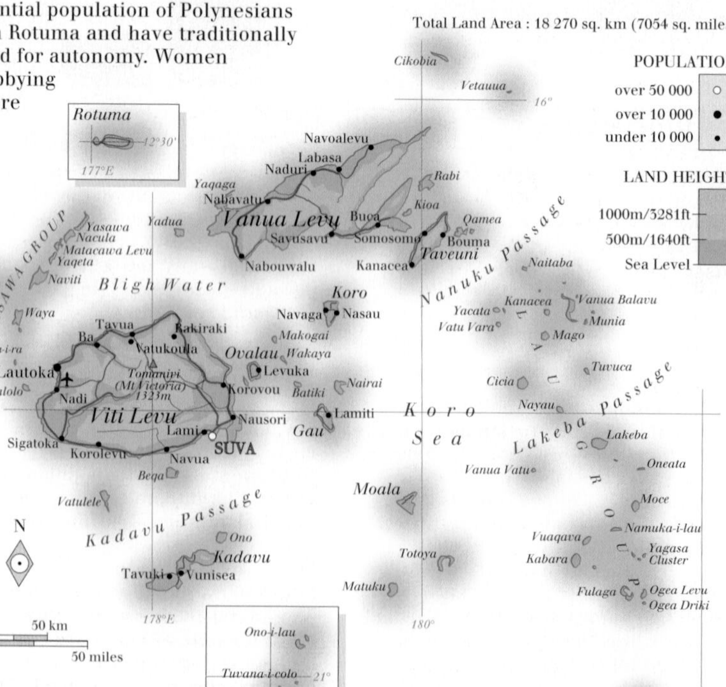
POPULATION
over 50 000
over 10 000
under 10 000
LAND HEIGHT
1000m/3281ft
500m/1640ft
Sea Level
Rotuma
Cikobia
Vetauua
Navoalevu
Labasa
Naduri
Nabavatu
Vanua Levu
Savusavu
Buca
Somosomo
Taveuni
Bouma
Qamea
Rabi
Kioa
Kanacea
Nabouwalu
Yadua
Yasawa
Nacula
Matacawa Levu
Yaqeta
Naviti
Waya
Viwa
YASAWA GROUP
Bligh Water
Tavua
Rakiraki
Vatukoula
Ba
Lautoka
Nadi
Malolo
Mamanuca-i-ra Group
Tomanivi (Mt Victoria) 1323m
Viti Levu
Sigatoka
Korolevu
Navua
Lami
SUVA
Nausori
Korovou
Beqa
Vatulele
Koro
Navaga
Nasau
Makogai
Ovalau
Wakaya
Levuka
Batiki
Nairai
Gau
Lamiti
Koro Sea
Moala
Totoya
Matuku
Kadavu Passage
Ono
Kadavu
Tavuki
Vunisea
Nanuku Passage
Naitaba
Vanua Balavu
Munia
Yacata
Vatu Vara
Mago
Tuvuca
Cicia
Nayau
Lakeba Passage
Lakeba
LAU GROUP
Vanua Vatu
Oneata
Moce
Namuka-i-lau
Vuaqava
Kabara
Yagasa Cluster
Fulaga
Ogea Levu
Ogea Driki
Vatoa
Ono-i-lau
Tuvana-i-colo
Tuvana-i-ra
N
0 50 km
0 50 miles
16°
18°
20°
21°
12°30'
177°E
178°E
179°E
180°
178°W

WORLD AFFAIRS

Joined UN in 1970

ACP CP Comm PC PIF

Fiji's international reputation has been severely damaged by its discrimination against Indo-Fijians and the recent coups. Fiji has been intermittently suspended from the Commonwealth.

AID

Recipient

 US$29m (receipts)

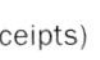
 Down 17% in 2000

Fiji traditionally received a lot of overseas aid, but international reaction to the 2000 coup prompted drastic cuts.

DEFENSE

No compulsory military service

 US$32m

 Down 9% in 2000

Of the almost entirely ethnic Fijian military, significant numbers – around 20% – are assigned to UN duties and have served in Lebanon and Egypt.

ECONOMICS

Inflation 3.3% p.a. (1990–2000)

 US$1.48bn

 2.176–2.298 Fiji dollars

SCORE CARD

- World GNP Ranking 145th
- GNP per Capita .... US$1820
- Balance of Payments .... US$13m
- Inflation 1.1%
- Unemployment 5%

Strengths

Relatively well-diversified economy, with strong tourist infrastructure. Location on Pacific air routes. Many regional and international organizations located in Suva.

Weaknesses

2000 coup caused dramatic contraction in economy – 12.5%. Migration of many Indo-Fijian professionals. Sugar crops vulnerable to drought. Major exports – sugar, copra, and gold – subject to large fluctuations in world prices.

EXPORTS

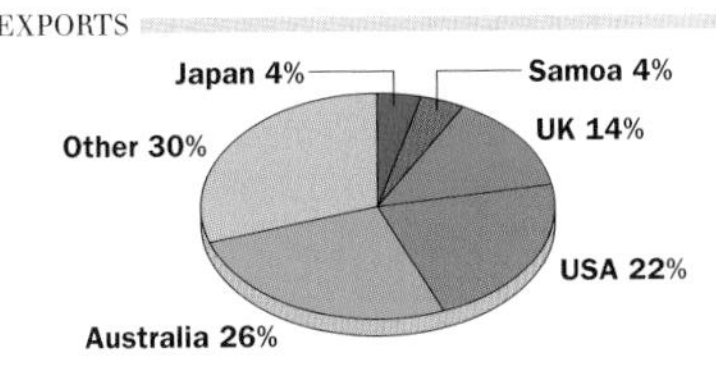
Japan 4%
Samoa 4%
UK 14%
Other 30%
USA 22%
Australia 26%

IMPORTS

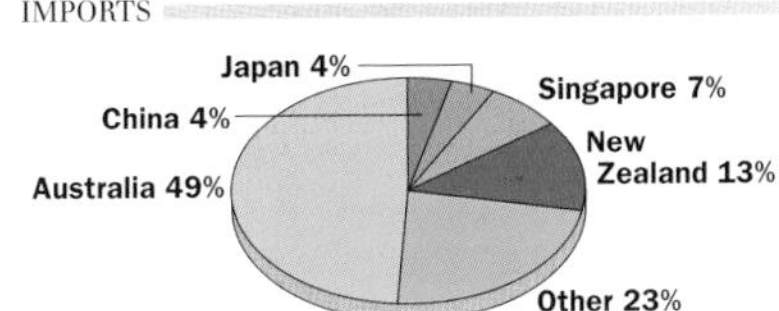
Japan 4%
China 4%
Singapore 7%
New Zealand 13%
Australia 49%
Other 23%

***Cane field** on the west side of Viti Levu, between Nadi and Lautoka. Sugar accounts for about one-third of Fiji's exports.*

RESOURCES

Electric power 200,000 kw

 38,471 tonnes

 Not an oil producer

 245,749 goats, 340,000 cattle, 3.7m chickens

 Gold, silver

The varied terrain allows diversified agriculture. Gold and minerals are mined. A hydroelectric plant at Monasavu provides 95% of electricity.

ENVIRONMENT

Not available

0.3%

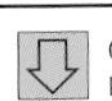 0.9 tonnes per capita

Governments are environmentally aware. Tourism is damaging coral reefs. Fiji was downwind of French Pacific nuclear tests. Fertilizers are overused.

MEDIA

TV ownership low

 Daily newspaper circulation 51 per 1000 people

PUBLISHING AND BROADCAST MEDIA

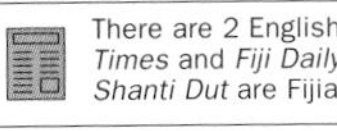 There are 2 English-language dailies, *Fiji Times* and *Fiji Daily Post*. *Nai Lalakai* and *Shanti Dut* are Fijian and Indian weeklies

2 services: 1 state-owned, 1 independent

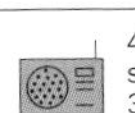 4 services: 1 state-controlled, 3 independent

Freedom of the press is championed by the government, and cases of corruption are often reported in the media. However, the police have blocked politically sensitive broadcasts.

CRIME

Death penalty not used in practice

 1185 prisoners

 Up in 2000

Usually theft and drink-related violence top the crime list. The 2000 coup spurred Fijians to settle old scores by force.

EDUCATION

Schooling is not compusory

 93%

 9208 students

Education, originally modeled on the British system, is now mostly run by local committees and is increasingly racially segregated. The use of the birch in schools was banned in 2002.

CHRONOLOGY

The British decision to import Indian sugar workers in 1879–1916 dramatically changed Fijian society.

- **1970** Independence from Britain.
- **1987** Election win for Indo-Fijian coalition. Sitiveni Rabuka's coups secure minority ethnic Fijian rule. Ejected from Commonwealth.
- **1989** Mass Indo-Fijian emigration.
- **1990** Constitution discriminating against Indo-Fijians introduced.
- **1992** Rabuka wins legislative polls.
- **1997** Census shows ethnic Fijians outnumber Indo-Fijians. Fiji rejoins Commonwealth. New constitution.
- **1999** General election won by FLP. First Indo-Fijian prime minister.
- **2000** Civilian-led coup; new ethnic Fijian government.
- **2001** Nationalists win elections.

HEALTH

Welfare state health benefits

 1 per 1922 people

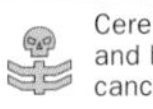 Cerebrovascular and heart diseases, cancers, accidents

Medical treatment is provided for all at a nominal charge. Fiji is free of almost all tropical diseases. Suicide is increasing, particularly among ethnic Indians.

SPENDING

GDP/cap. increase

CONSUMPTION AND SPENDING

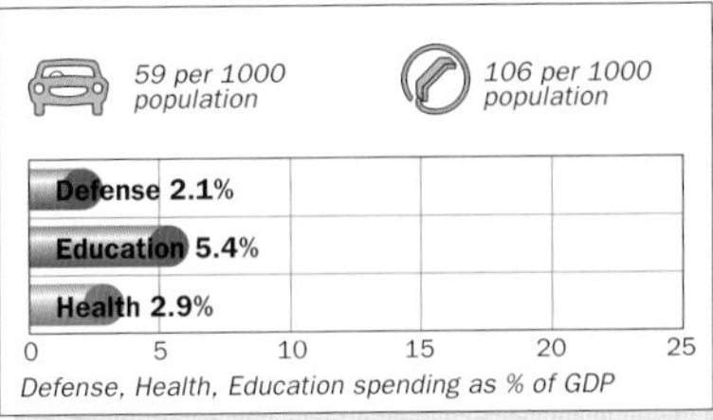
59 per 1000 population
106 per 1000 population
Defense 2.1%
Education 5.4%
Health 2.9%
0
5
10
15
20
25
Defense, Health, Education spending as % of GDP

Ostentatious displays of wealth are rare in Fiji; prestige derives from family and landholdings. The professional middle class, traditionally dominated by Indo-Fijians, is becoming more mixed.

WORLD RANKING

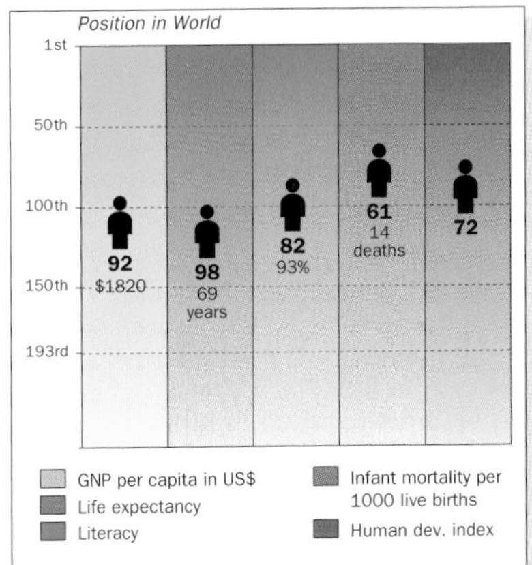
Position in World
1st
50th
100th
150th
193rd
92
$1820
98
69 years
82
93%
61
14 deaths
72
GNP per capita in US$
Life expectancy
Literacy
Infant mortality per 1000 live births
Human dev. index

F

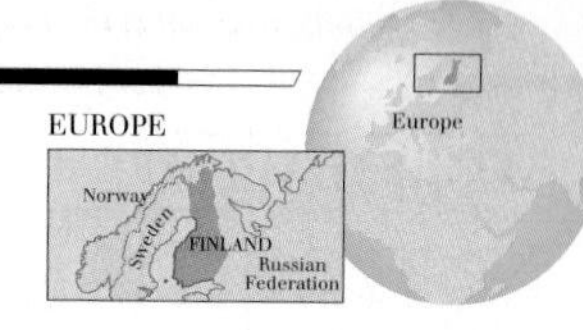

EUROPE
Europe
Norway
Sweden
FINLAND
Russian Federation

FINLAND

OFFICIAL NAME: Republic of Finland **CAPITAL:** Helsinki **POPULATION:** 5.2 million
CURRENCY: Euro (markka until 2002) **OFFICIAL LANGUAGES:** Finnish and Swedish

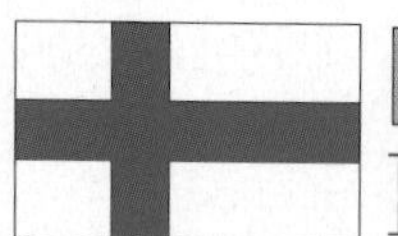

1917

1947

Dec 6
FIN
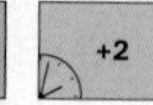
+2

+358

.fi

BORDERED TO THE north and west by Norway and Sweden, and to the east by Russia, Finland is a low-lying country of forests and 187,888 lakes. Politics is based on consensus, and the country has been stable despite successive short-lived coalitions. Russia annexed Finland in 1809, ruling it until 1917, and subsequently Finland accepted a close relationship with the USSR as the price of maintaining its independence. It joined the EU in 1995 and, despite popular suspicion of Brussels bureaucracy, Finland was among the 12 EU states to adopt the euro from 2002.

F

CLIMATE

Subarctic/continental

WEATHER CHART FOR HELSINKI

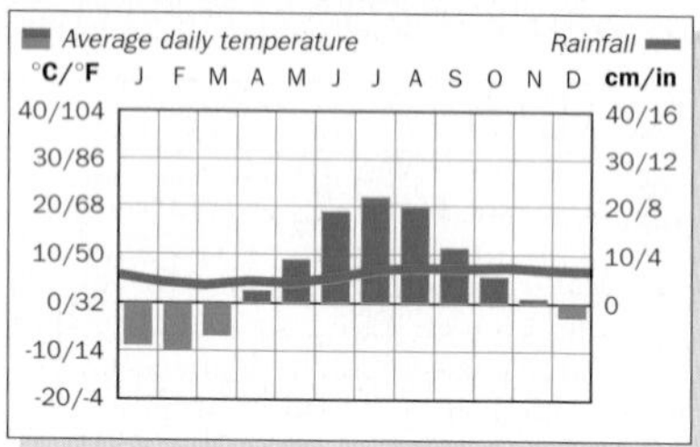

Average daily temperature
Rainfall
°C/°F
J F M A M J J A S O N D
cm/in
40/104
30/86
20/68
10/50
0/32
-10/14
-20/-4
40/16
30/12
20/8
10/4
0

North of the Arctic Circle the climate is extreme. Temperatures fall to –30°C (–22°F) in the six-month winter and rise to 27°C (81°F) during the 73 days of summer midnight sun. In the south, summers are mild and short, winters are cold.

TRANSPORTATION

Drive on right

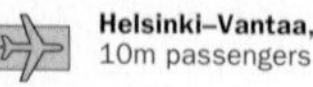
Helsinki–Vantaa, 10m passengers

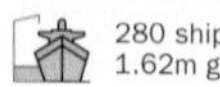
280 ships 1.62m grt

THE TRANSPORTATION NETWORK

49,853 km (30,977 miles)

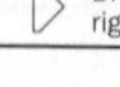
512 km (318 miles)

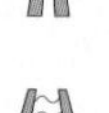
5836 km (3626 miles)

6715 km (4172 miles)

The transportation system is well integrated. The railroad connects with the Swedish and Russian networks. There are frequent air services to most neighboring states, and links with Baltic states are being expanded. With one of the densest domestic networks in Europe, internal air travel is important, particularly north of the Arctic Circle.

With 187,888 lakes and a major river network, Finland has Europe's largest inland waterway system. It still carries freight, but is now used mainly for recreation. Finland's international ports handle around 70 million tonnes a year. Kotka is the chief export port. Helsinki, with five specialized harbors, handles most imports.

TOURISM

Visitors : Population 1:1.9

2.71m visitors

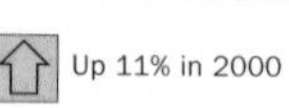
Up 11% in 2000

MAIN TOURIST ARRIVALS

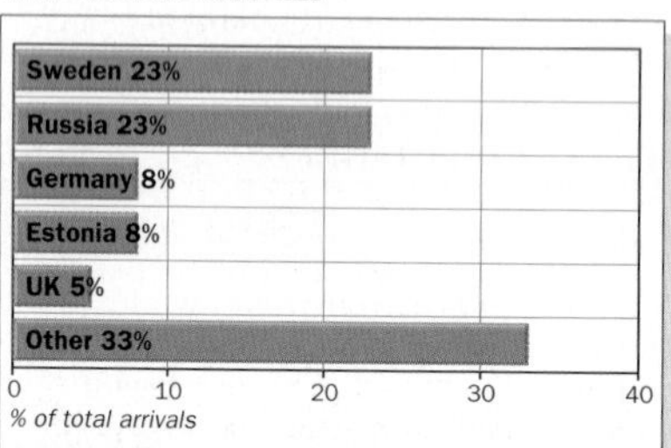

Sweden 23%
Russia 23%
Germany 8%
Estonia 8%
UK 5%
Other 33%
0
10
20
30
40
% of total arrivals

The scenery of the southern lakes and the vast forests of its Arctic north are Finland's main attractions. Helsinki is an important cultural center and hosts an annual arts festival. There are many first-class restaurants and its opera house has an international reputation. Most tourists try a sauna, a Finnish invention, and the local vodka, which is reputedly among the world's finest.

Visitors come largely from Sweden, Russia, the Baltic states, and Germany. By 2001 Finland emerged as the seventh most popular European destination for employers offering vacations as an incentive to their workforce.

A summer's night at Kilpisjärvi, "The Way of the Four Winds," which lies at the point where Finland, Sweden, and Norway meet.

PEOPLE

Pop. density low

Finnish, Swedish, Sami

17/km² (44/mi²)

THE URBAN/RURAL POPULATION SPLIT

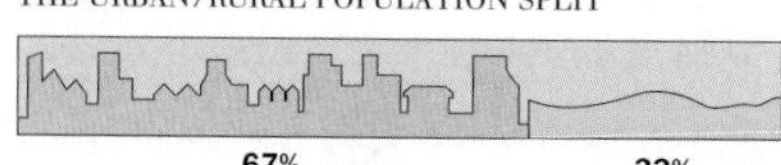

67% 33%

RELIGIOUS PERSUASION

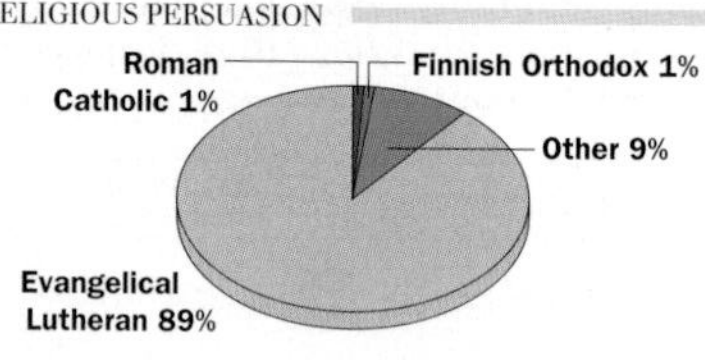

Roman Catholic 1%
Finnish Orthodox 1%
Other 9%
Evangelical Lutheran 89%

ETHNIC MAKEUP

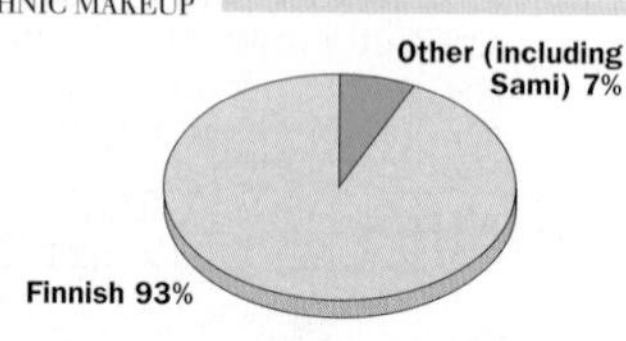

Other (including Sami) 7%
Finnish 93%

Most Finns are of Scandinavian–Baltic extraction. Finnish belongs to the small Finno-Ugric linguistic group and is a legacy of the country's early Asian invaders. These tribes integrated with local and surrounding European peoples, but preserved their distinct language. Sami, also a Finno-Ugric language, is spoken by the small Sami population, who live above the Arctic Circle. Around 6% of the population speak Swedish, most of whom live in the southwestern coastal regions and on the Åland Islands.

More than 50% of Finns live in the five southernmost districts around Helsinki. Families tend to be close-knit, although divorce rates are high. The sauna is an integral part of everyday life; there are 1.5 million saunas among 5.2 million Finns.

Finnish women have a long tradition of political and economic participation. They were the first in Europe to get the vote, in 1906, and the first in the world able to stand for parliament. Almost 50% of women now work outside the home, and the president and one-third of the cabinet are female.

POPULATION AGE BREAKDOWN

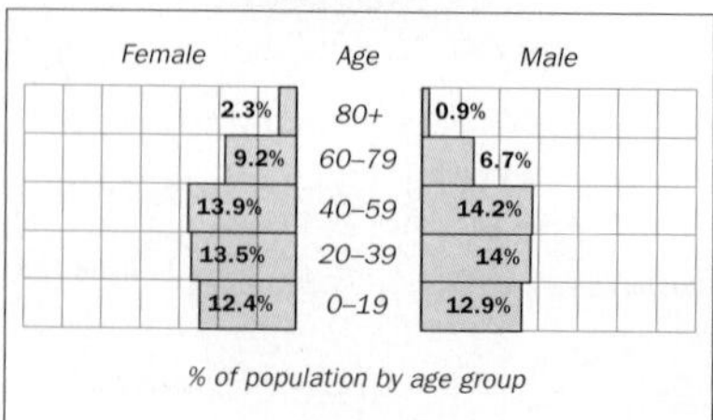

Female
Age
Male
2.3%
80+
0.9%
9.2%
60–79
6.7%
13.9%
40–59
14.2%
13.5%
20–39
14%
12.4%
0–19
12.9%
% of population by age group

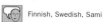

POLITICS

Multiparty elections

1999/2003

President Tarja Halonen

Finland's constitution combines parliamentary government with a strong presidency. The external territory of the Åland Islands, inhabited by Swedish-speakers, has internal self-government.

PROFILE

Proportional representation has led to government by coalition, usually dominated by the SDP or KESK. The emphasis on consensus has favored stability but resulted in slow decision-making. The current "rainbow" coalition comprises parties from across the political spectrum.

AT THE LAST ELECTION

Parliament 200 seats

25% SDP, 24% KESK, 23% KOK, 10% VL, 6% SFP, 6% G, 6% Others

SDP = Social Democratic Party **KESK** = Center Party
KOK = National Coalition Party **VL** = Left-wing Alliance
SFP = Swedish People's Party **G** = Greens

MAIN POLITICAL ISSUES

EU membership

Finland joined the EU in 1995, many Finns supporting entry as a way of identifying with western Europe. The small but influential farming community was hostile to membership. Others feared that welfare cuts would be more far-reaching if the economy were liberalized in line with EU expectations. In the event, after a dose of austerity, EU membership became associated with greater prosperity from the late 1990s. Finland's decision to join the eurozone in January 1999 contrasted with the nonmembership of its closest EU neighbor, Sweden.

Unemployment

The victory of SDP candidate Martti Ahtisaari in the 1994 presidential election was a sign of discontent with the conservative coalition then in power, whose handling of the recession resulted in record unemployment levels and welfare cuts. The 1995 general election led to the return of an SDP-led coalition which continued many of the previous government's austerity policies. Unemployment has since been substantially reduced.

FINLAND

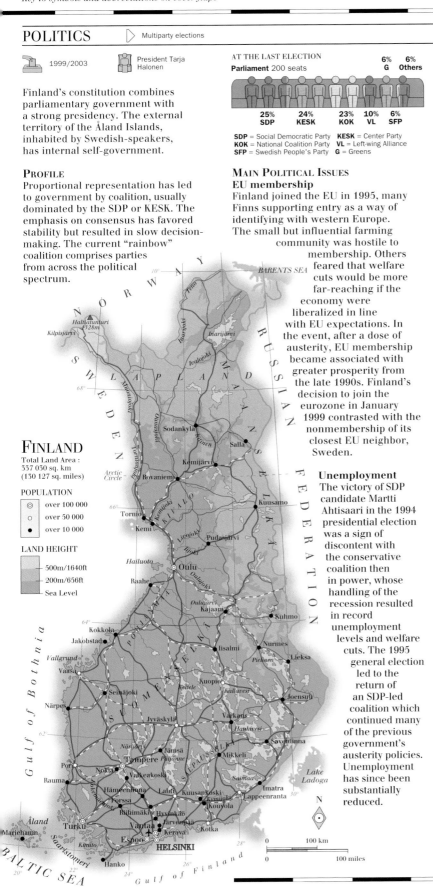

Total Land Area : 337 030 sq. km (130 127 sq. miles)

POPULATION
- over 100 000
- over 50 000
- over 10 000

LAND HEIGHT
- 500m/1640ft
- 200m/656ft
- Sea Level

0 100 km
0 100 miles

Tarja Halonen, who became Finland's first female president in 2000.

Prime Minister Paavo Lipponen, heads the SDP-led coalition.

WORLD AFFAIRS

Joined UN in 1955

CE

EU

OECD

OSCE
PfP

After carefully balancing its relations with the USSR and the West during the Cold War, Finland has now decided that its national interest lies with western Europe. In addition to joining the EU, it has observer status at the WEU. However, acknowledging historical and geographic realities, the government is also keen to maintain a special relationship with Russia.

AID

Donor

$371m (donations)

Down 11% in 2000

Finland's aid budget is still well below the UN target of 0.7% of GNP, despite vigorous campaigning. The main recipients are China, Bosnia and Herzegovina, and southern Africa.

CHRONOLOGY

Finland's history has been closely linked with the competing interests of Sweden and Russia.

- **1323** Treaty of Pähkinäsaari. Finland part of Swedish Kingdom.
- **1809** Treaty of Fredrikshamn, Sweden cedes Finland to Russia. Finland becomes a Grand Duchy enjoying considerable autonomy.
- **1812** Helsinki becomes capital.
- **1863** Finnish becomes an official language alongside Swedish.
- **1865** Grand Duchy acquires its own monetary system.
- **1879** Conscription law lays the foundation for a Finnish army.
- **1899** Czar Nicholas II begins process of Russification. Labor Party founded.
- **1900** Gradual imposition of Russian as the official language begins.
- **1901** Finnish army disbanded, Finns ordered into Russian units. Disobedience campaign prevents men being drafted into the army. ➪

F

F

CHRONOLOGY *continued*

- **1903** Labor Party becomes SDP.
- **1905** National strike forces restoration of 1899 status quo.
- **1906** Parliamentary reform. Universal suffrage introduced.
- **1910** Responsibility for important legislation passed to Russian Duma.
- **1917** Russian revolution allows Finland to declare independence.
- **1918** Civil war between Bolsheviks and right-wing government. Gen. Gustav Mannerheim leads government to victory at Battle of Tampere.
- **1919** Finland becomes republic. Kaarlo Ståhlberg elected president with wide political powers.
- **1920** Treaty of Tartu: USSR recognizes Finland's borders.
- **1921** London Convention. Åland Islands become part of Finland.
- **1939** August, Hitler–Stalin nonaggression pact gives USSR a free hand in Finland. November, Soviet invasion; strong Finnish resistance in ensuing Winter War.
- **1940** Treaty of Moscow. Finland cedes a tenth of national territory.
- **1941** Finnish troops join Germany in its invasion of USSR.
- **1944** June, Red Army invades. August, President Risto Ryti resigns. September, Finland, led by Marshal Mannerheim, signs armistice.
- **1946** President Mannerheim resigns, Juho Paasikivi president.
- **1948** Signs friendship treaty with USSR. Agrees to resist any attack on USSR made through Finland by Germany or its allies.
- **1952** Payment of $570 million in war reparations completed.
- **1956** Uhro Kekkonen, leader of the Agrarian Party, becomes president.
- **1956–1991** A series of coalition governments involving SDP and Agrarians, renamed KESK in 1965, hold power.
- **1981** President Kekkonen resigns.
- **1982** Mauno Koivisto president.
- **1989** USSR recognizes Finnish neutrality for first time.
- **1991** Non-SDP government elected. Austerity measures.
- **1992** January, signs ten-year agreement with Russia which, for first time since World War II, involves no military agreement.
- **1994** SDP candidate Martti Ahtisaari elected president.
- **1995** Finland joins EU. General election returns SDP-led coalition under Paavo Lipponen.
- **1999** General election returns Lipponen's coalition to power.
- **2000** Tarja Halonen elected as first woman president.
- **2002** Euro fully adopted. Fifth nuclear power plant approved.

DEFENSE

Compulsory military service

 $1.52bn

 Down 7% in 2000

Finland is a neutral country. Its armed forces, the majority of whom are conscripts, are backed up by 500,000 active reservists and 3400 border guards. Russia's relative instability has reinforced concern about border security, the main defense issue. Finland participates in NATO's Partnerships for Peace program and has WEU observer status. Military service lasts for up to 12 months.

FINNISH ARMED FORCES

230 main battle tanks (70 T-55, 160 T-72)	24,550 personnel
10 patrol boats	5000 personnel
64 combat aircraft (57 F/A-18C)	2700 personnel
None	

ECONOMICS

Inflation 1.9% p.a. (1990–2000)

$130bn

1.0651–1.1231 euros

SCORE CARD

World GNP Ranking	28th
GNP per Capita	$25,130
Balance of Payments	$8.89bn
Inflation	3.4%
Unemployment	10%

EXPORTS

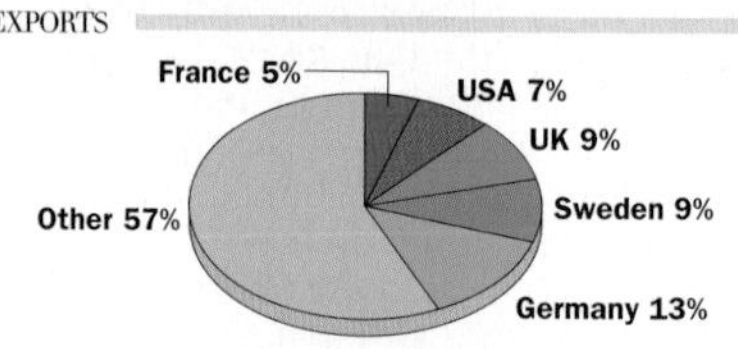
France 5%
USA 7%
UK 9%
Sweden 9%
Germany 13%
Other 57%

IMPORTS

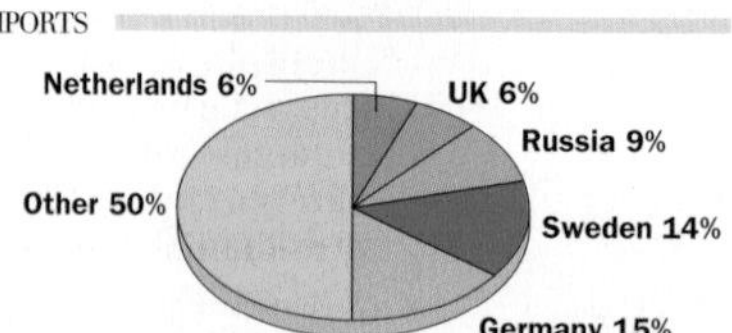
Netherlands 6%
UK 6%
Russia 9%
Sweden 14%
Germany 15%
Other 50%

Strengths

Industry export- and quality-oriented. Large high-tech sector, especially Nokia mobile phones and Internet services. World leader in pulp and paper. Exports quick to recover from recession. Low inflation, averaging about 2% a year. Improved foreign investment incentives. Gateway to Russian and Baltic economies. Membership of eurozone. Strong recent economic growth. Finland named as world's most competitive economy by World Economic Forum in 2001.

Weaknesses

Severe recession in 1991–1993 when real GDP declined 15%. Rapidly aging population and low retirement age. High level of public and foreign debt. High unemployment. Small domestic market. Peripheral position in Europe.

Profile

Finland is a wealthy market economy. In the early 1990s it experienced the worst recession in 60 years, chiefly as a result of the collapse of the former Soviet Union. Russia took only 4% of Finland's exports in 1999, compared with over 25% to the Soviet Union before 1990.

A rapid rise in unemployment and business failures after 1990 pushed up government spending. The floating of the markka in 1992 and austerity measures improved competitiveness, overturned substantial fiscal deficits and allowed tax cuts in 2002.

Unemployment has fallen, but is still around 10%. Rapid growth in 2000 was followed by a slowdown in 2001. Finland was one of the 12 EU countries to adopt the euro in January 2002.

ECONOMIC PERFORMANCE INDICATOR

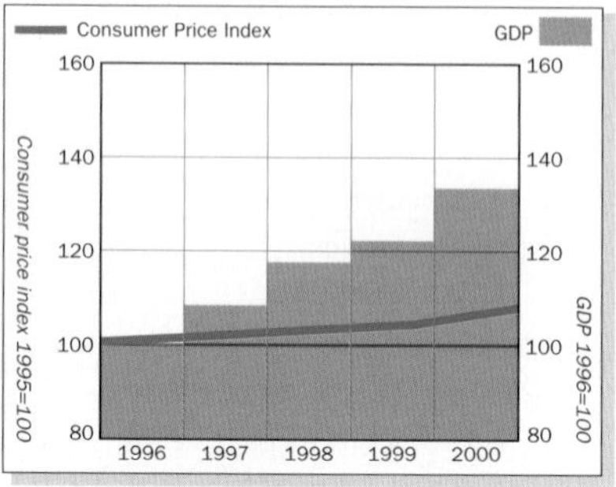
Consumer Price Index
GDP
160
140
120
100
80
Consumer price index 1995=100
GDP 1996=100
1996
1997
1998
1999
2000

FINLAND : MAJOR BUSINESSES

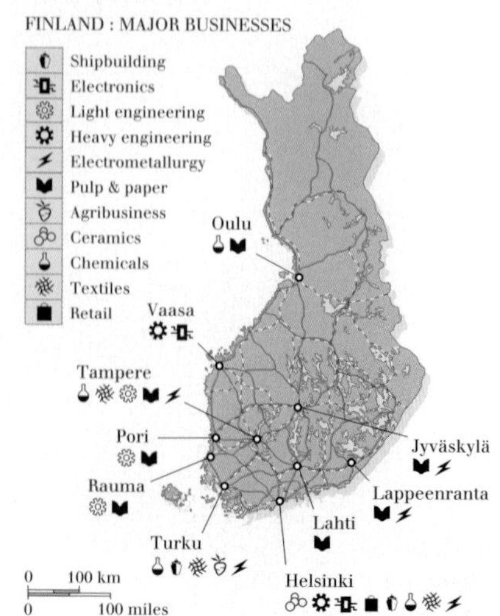
Shipbuilding
Electronics
Light engineering
Heavy engineering
Electrometallurgy
Pulp & paper
Agribusiness
Ceramics
Chemicals
Textiles
Retail
Oulu
Vaasa
Tampere
Pori
Rauma
Turku
Helsinki
Lahti
Jyväskylä
Lappeenranta
0 100 km
0 100 miles

RESOURCES

Electric power 14.9m kw

176,018 tonnes

Not an oil producer; refines 200,000 b/d

1.3m pigs, 1.09m cattle, 100,000 sheep, 6m chickens

Gold, copper, zinc, iron, lead, silver

ELECTRICITY GENERATION

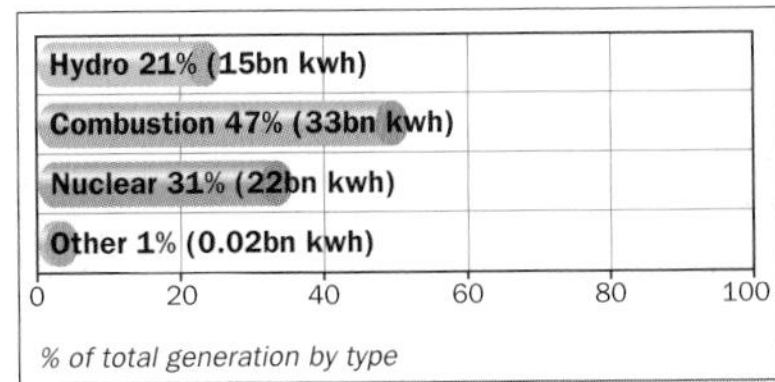
Hydro 21% (15bn kwh)
Combustion 47% (33bn kwh)
Nuclear 31% (22bn kwh)
Other 1% (0.02bn kwh)
0
20
40
60
80
100
% of total generation by type

Finland's trees are its prime natural resource. Commercial forests cover 65% of the land, and wood products account for 30% of exports. Finland has no oil, but has significant hydroelectric resources. The high energy demands of industry are met chiefly by combustion and nuclear power. A fifth nuclear power plant was approved in 2002. Oil import costs have risen since 1990, when the collapse of the USSR ended a 42-year agreement on the exchange of Finnish manufactures for Soviet oil.

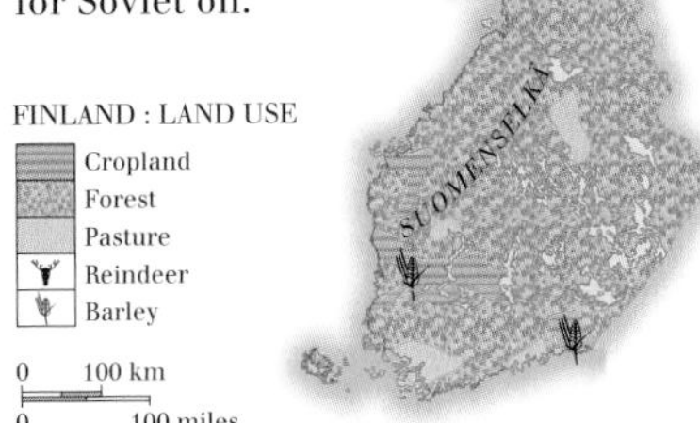
LAPLAND
MAANSELKÄ
SUOMENSELKÄ
FINLAND : LAND USE
Cropland
Forest
Pasture
Reindeer
Barley
0 100 km
0 100 miles

ENVIRONMENT

Sustainability rank: 1st

6%

10.3 tonnes per capita

ENVIRONMENTAL TREATIES

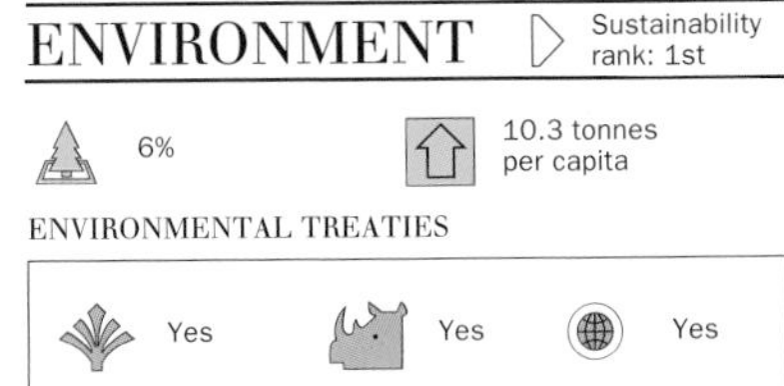
Yes
Yes
Yes
Yes
Yes
Yes

Finland has strict laws on industrial emissions. Energy efficiency is a priority; nearly half of all homes are connected to district heating systems. Although there is opposition to the fifth nuclear plant, proponents argue that it will help the country meet its target emissions for greenhouse gases. The government is funding nuclear safety programs in Russia. Rising levels of pollution in the Baltic have given rise to concern.

MEDIA

TV ownership high

Daily newspaper circulation 452 per 1000 people

PUBLISHING AND BROADCAST MEDIA

There are 200 daily newspapers. The most important are *Helsingin Sanomat, Aamulehti, Ilta-Sanomat, Turun Sanomat,* and *Kaleva*

3 services: 1 state-owned, 2 independent

5 services: 1 state-owned, 4 independent

Nine out of ten adult Finns read a daily paper, the world's third-highest per capita ratio. Regional papers dominate; the only national is the independent *Helsingin Sanomat*. There is no censorship, but the press shows restraint in criticizing the government.

CRIME

No death penalty

2703 prisoners

Up 7% in 1999

CRIME RATES

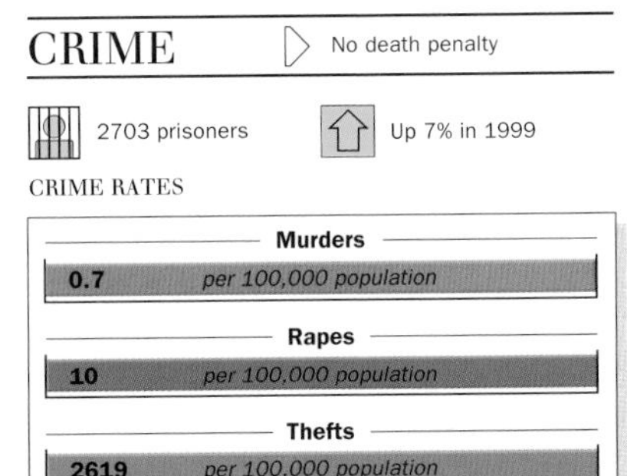
Murders
0.7 per 100,000 population
Rapes
10 per 100,000 population
Thefts
2619 per 100,000 population

The jump in unemployment in the early 1990s was seen as one of the causes of rising crime. There is concern about links with organized crime in Russia.

EDUCATION

School leaving age: 16

99%

271,215 students

THE EDUCATION SYSTEM

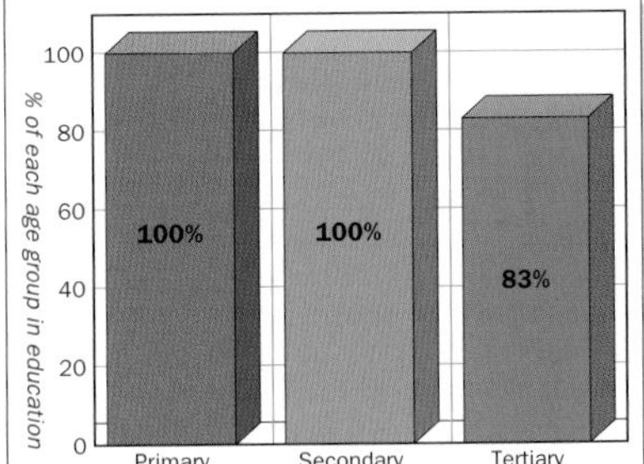
100
80
60
40
20
0
% of each age group in education
100%
100%
83%
Primary
Secondary
Tertiary

Compulsory education, introduced in 1921, lasts from seven to 16 years of age. By 1977/1978 the whole country had transferred to a new comprehensive system. Almost all children receive preschool education and also go on to three years of upper secondary education. Tough examinations mean that only 35% of entrants qualify to attend one of the 20 universities.

HEALTH

Welfare state health benefits

1 per 323 people

Cerebrovascular and heart diseases, cancers, suicides

Of total government expenditure, 14% is spent on Finland's well-developed health system. Every Finn is legally guaranteed access to a local health center which is staffed by up to four doctors, as well as nurses and a midwife. Most non-hospital medical costs are covered by national health insurance; hospital fees are moderate. Diabetes and osteoporosis are increasing, and obesity is a growing health problem.

SPENDING

GDP/cap. increase

CONSUMPTION AND SPENDING

403 per 1000 population

550 per 1000 population

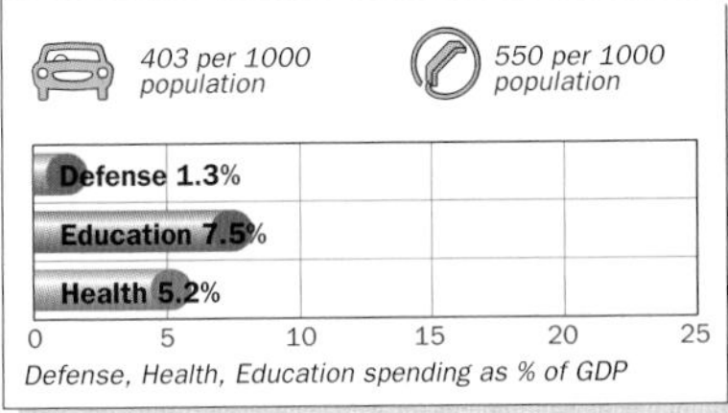
Defense 1.3%
Education 7.5%
Health 5.2%
0
5
10
15
20
25
Defense, Health, Education spending as % of GDP

Income disparities are more marked in Finland than in other Scandinavian countries. However, the economic boom and labor shortages of the 1980s led to a sharp rise in all living standards. Personal consumption reached Swedish levels, and many families were able to take two vacations a year. Social security benefits were extended.

During the deep recession which began in 1990, this improvement was reversed. Wealth disparities widened and expenditure cuts led to lower social security benefits for the jobless. Those in work had to accept lower pay rises and higher taxes. Average real disposable incomes dropped sharply. The situation started to improve in 2000, with a temporary downturn in 2001.

Estonian immigrants form the poorest group in Finnish society.

WORLD RANKING

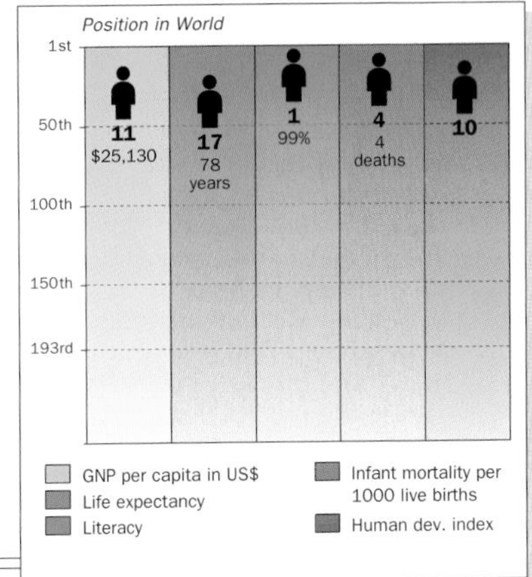
Position in World
1st
50th
100th
150th
193rd
11
$25,130
17
78
years
1
99%
4
4
deaths
10
GNP per capita in US$
Life expectancy
Literacy
Infant mortality per 1000 live births
Human dev. index

F

F

FRANCE

OFFICIAL NAME: French Republic **CAPITAL:** Paris **POPULATION:** 59.5 million
CURRENCY: Euro (French franc until 2002) **OFFICIAL LANGUAGE:** French

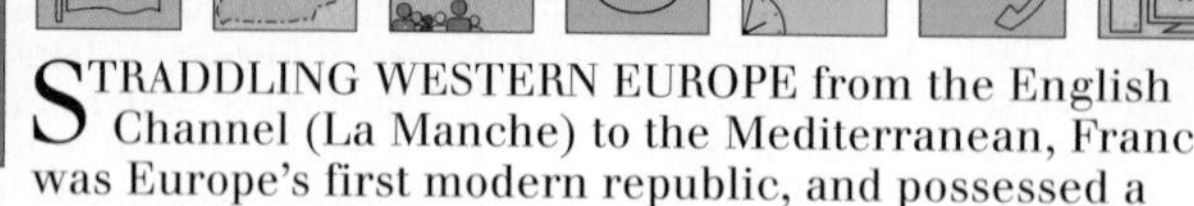
987
1919
July 14
F
+1
+33
.fr

STRADDLING WESTERN EUROPE from the English Channel (La Manche) to the Mediterranean, France was Europe's first modern republic, and possessed a colonial empire second only to that of the UK. Today, it is one of the world's major industrial powers and its fourth-largest exporter. Industry is the leading economic sector, but the agricultural lobby remains powerful – French farmers will mount the barricades in defense of their interests. France's focus is very much on Europe. Together with Germany it was a founder member of the European Economic Community (EEC), and has supported successive steps to build a more closely integrated European Union. Paris, the French capital, is generally considered to be one of the world's most beautiful cities. Some of the most influential artists, writers, and filmmakers of the modern era have lived there.

Le Plessis-Bourré, Loire Valley. *The region is famous for its many chateaux, which attract thousands of visitors every year.*

CLIMATE

Maritime/Mediterranean/ mountain/continental

WEATHER CHART FOR PARIS

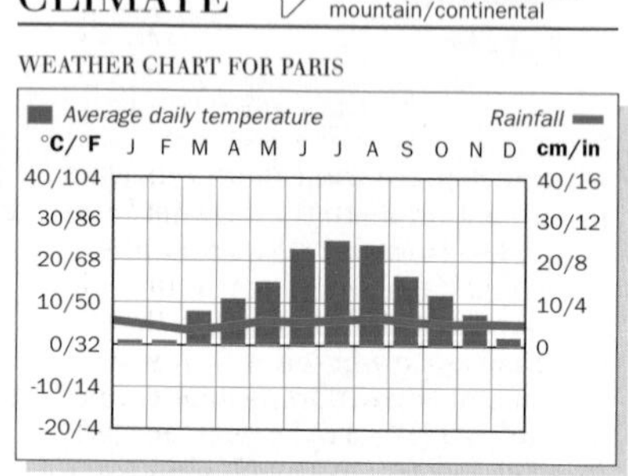
Average daily temperature
Rainfall
°C/°F
J F M A M J J A S O N D
cm/in
40/104
30/86
20/68
10/50
0/32
-10/14
-20/-4
40/16
30/12
20/8
10/4
0

France's climate differs significantly from region to region. The northwest, in particular Brittany, is mild but damp. The east has hot summers and stormy winters, whereas summers in the south are dry and hot, and forest fires are a frequent occurrence.

TRANSPORTATION

Drive on right

Charles de Gaulle, Paris 48.2m passengers

700 ships 1.5m grt

THE TRANSPORTATION NETWORK

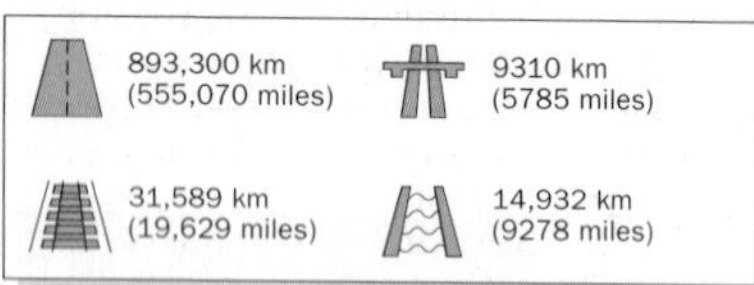
893,300 km
(555,070 miles)
9310 km
(5785 miles)
31,589 km
(19,629 miles)
14,932 km
(9278 miles)

France led the world in high-speed train technology in 1983 with the TGV (*train à grande vitesse*) from Paris to Lyon. TGV lines have since been extended, and also link up with Belgium, Italy, Spain, the Channel Tunnel, and the Mediterranean; a Paris–Strasbourg line is planned for 2005. Traffic through Paris's two airports is set to double to 140 million by 2020, but plans for a third airport at Chaulnes, 130 km (80 miles) to the north, have become a political football.

TOURISM

Visitors : Population 1.3:1

76.5m visitors

Up 1% in 2001

MAIN TOURIST ARRIVALS

	% of total arrivals
Germany	21%
UK & Ireland	16%
Netherlands	16%
Belgium & Luxembourg	12%
Italy	9%
Other	26%

France is the world's leading tourist destination, with over 76 million visitors a year. It ranks high as a destination for visitors from neighboring countries, particularly Germany and the UK. Most French people also prefer to take vacations in their own country, although many do visit Spain and Italy.

Paris is the most visited city in Europe. Its attractions include the Eiffel Tower, Nôtre Dame cathedral, Eurodisney, the Pompidou Center, and the Louvre, the world's most popular art museum.

The Côte d'Azur in the southeast became a byword for fashionable tourism when royalty and other notables flocked to resorts such as Nice at the end of the 19th century. Today Cannes hosts the world's leading film festival, and has a growing business convention trade. Other destinations throughout the country attract tourists for a variety of reasons such as wine production, historic and archaeological sites, and good beaches. There are resorts for skiing and hiking in the Alps and Pyrenees, and sailing off the varied coastline is also popular.

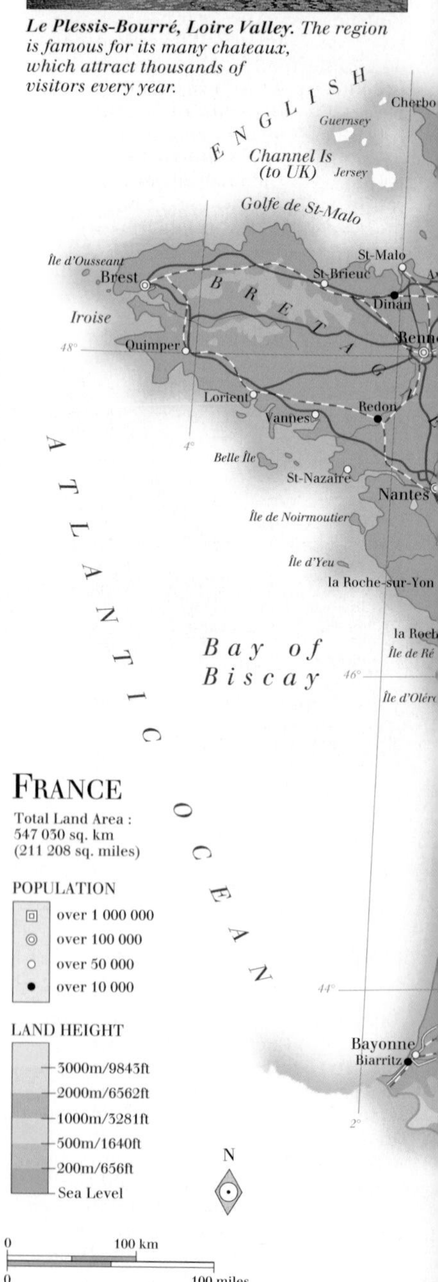
ENGLISH
Channel Is (to UK)
Guernsey
Jersey
Golfe de St-Malo
Île d'Ousseant
Brest
Iroise
Quimper
BRETAGNE
St-Malo
St-Brieuc
Dinan
Lorient
Vannes
Redon
Belle Île
St-Nazaire
Nantes
Île de Noirmoutier
Île d'Yeu
la Roche-sur-Yon
ATLANTIC OCEAN
Bay of Biscay
Bayonne
Biarritz
FRANCE
Total Land Area : 547 030 sq. km (211 208 sq. miles)
POPULATION
over 1 000 000
over 100 000
over 50 000
over 10 000
LAND HEIGHT
3000m/9843ft
2000m/6562ft
1000m/3281ft
500m/1640ft
200m/656ft
Sea Level
N
0
100 km
100 miles

F

PEOPLE

Pop. density medium

French, Provençal, German, Breton, Catalan, Basque

108/km² (280/mi²)

There is a strong sense of national identity, and compulsory use of French has traditionally been promoted as a unifying force. The cultural traditions of Bretons, Flemings, Alsatians, Basques, Occitans, Catalans, and Corsicans are now respected, but the Constitutional Court has struck down legislation on the use of regional language in government.

France encouraged immigration until the early 1970s, but laws have since become more restrictive. The 1999 census recorded 3.26 million resident foreigners (5.6% of the population) and a similar number of foreign-born immigrants have been naturalized. Racist National Front (FN) propaganda has periodically whipped up anti-immigrant feeling, but there is also a strong current of youth solidarity among "black, blanc, beur" ("black, white, Arab"), spectacularly boosted in 1998 by the World Cup success of the multiracial national soccer team.

The Roman Catholic Church is still dominant, but there are sizable Muslim, Protestant, Buddhist, and Jewish minorities. Abortion and birth control were both legalized in the 1970s, despite strong Catholic opposition, and couples now commonly live together before marriage. Some two million unmarried couples of two or more years' standing, including gay couples, gained legal status with social and tax rights under 1998 legislation recognizing the civil solidarity pact.

Women did not get the vote until 1944. Although there was a woman prime minister, Edith Cresson, in 1991–1992, and women took five senior cabinet posts in the incoming Socialist-led government in 1997, the proportion of women in parliament was the second lowest in the EU until the introduction of a "parity" law in 2000, which requires an equal number of male and female candidates on party electoral lists.

THE URBAN/RURAL POPULATION SPLIT

76% 24%

POPULATION AGE BREAKDOWN

Female	Age	Male
2.7%	80+	1.2%
8.7%	60–79	7%
11.8%	40–59	11.8%
15%	20–39	15%
13.1%	0–19	13.7%

% of population by age group

RELIGIOUS PERSUASION

Buddhist 1%
Protestant 2%
Muslim 8%
Jewish 1%
Roman Catholic 88%

ETHNIC MAKEUP

Other (including Corsicans) 1%
German (Alsace) 2%
Breton 1%
North African (mainly Algerian) 6%
French 90%

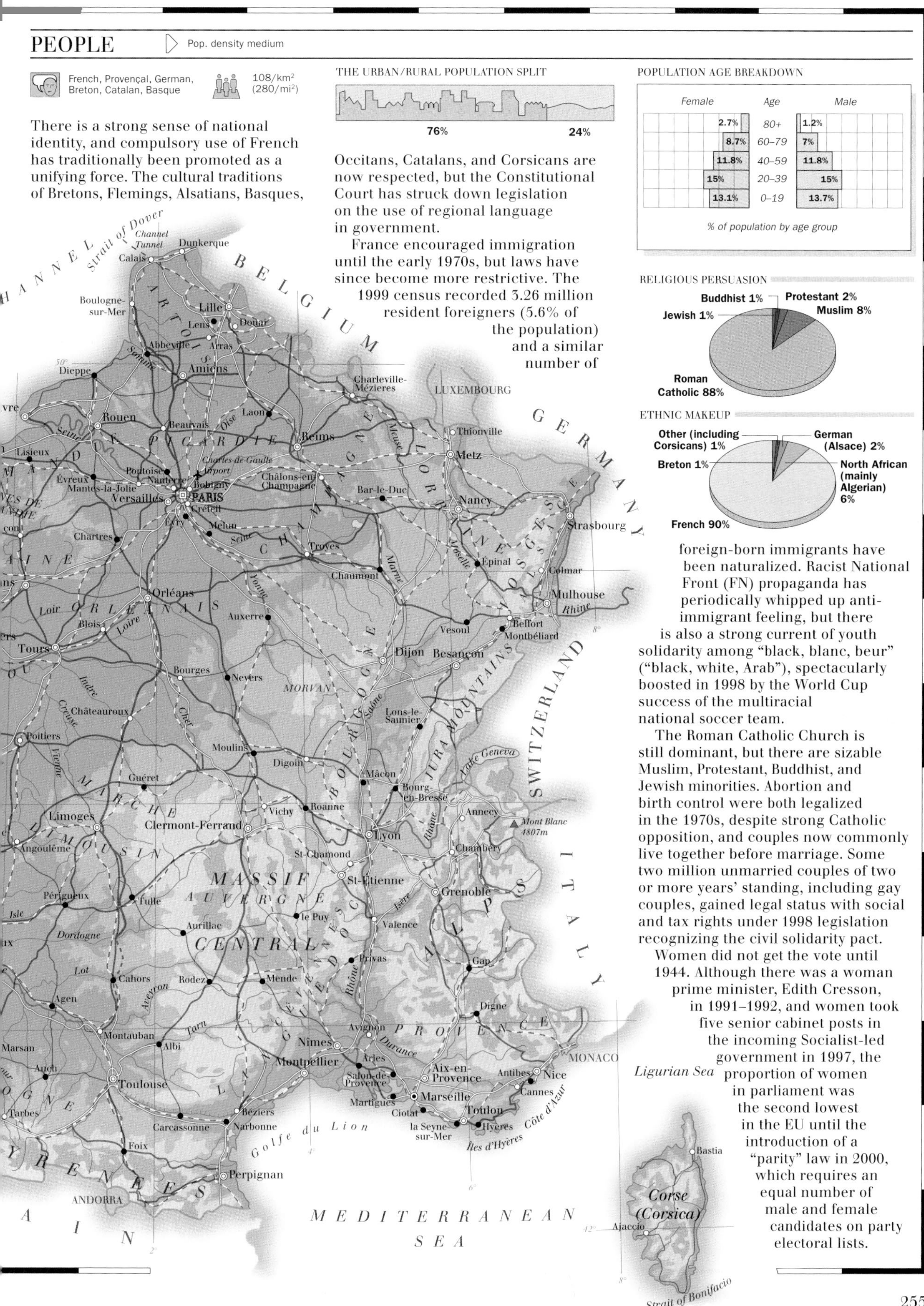
Strait of Dover
Channel Tunnel
Calais
Dunkerque
BELGIUM
Boulogne-sur-Mer
Lille
Lens
Douai
Arras
ARTOIS
Abbeville
Somme
Amiens
Dieppe
Charleville-Mézières
LUXEMBOURG
GERMANY
Rouen
Seine
Beauvais
Oise
Laon
PICARDIE
Reims
Thionville
Metz
Lisieux
Évreux
Pontoise
Nanterre
Charles-de-Gaulle Airport
Bobigny
Mantes-la-Jolie
Versailles
PARIS
Créteil
Évry
Melun
Châlons-en-Champagne
Bar-le-Duc
Nancy
CHAMPAGNE
LORRAINE
Meuse
Moselle
Strasbourg
VOSGES
ALSACE
Chartres
Troyes
Épinal
Marne
Chaumont
Colmar
Orléans
Mulhouse
Rhine
ORLÉANAIS
Loir
Blois
Loire
Auxerre
Yonne
Vesoul
Belfort
Montbéliard
Tours
Dijon
Besançon
Bourges
Nevers
MORVAN
Indre
Creuse
Châteauroux
Cher
Saône
Lons-le-Saunier
JURA MOUNTAINS
SWITZERLAND
Poitiers
Vienne
Moulins
Digoin
Lake Geneva
BOURGOGNE
Mâcon
Guéret
Bourg-en-Bresse
MARCHE
Vichy
Roanne
Annecy
Limoges
Clermont-Ferrand
Mont Blanc 4807m
LIMOUSIN
Lyon
Rhône
Angoulême
St-Chamond
Chambéry
MASSIF
AUVERGNE
St-Étienne
Grenoble
ITALY
Périgueux
Tulle
Isère
Isle
le Puy
Valence
Aurillac
Dordogne
CENTRAL
ALPS
Privas
Gap
Lot
Cahors
Rodez
Mende
Aveyron
Agen
Digne
Montauban
Tarn
Albi
PROVENCE
Avignon
Durance
Nîmes
Arles
Montpellier
MONACO
Ligurian Sea
Auch
Toulouse
Salon-de-Provence
Aix-en-Provence
Antibes
Nice
Cannes
Martigues
Marseille
Côte d'Azur
Toulon
Ciotat
Tarbes
Béziers
Narbonne
Carcassonne
Golfe du Lion
la Seyne-sur-Mer
Hyères
Îles d'Hyères
Foix
Bastia
Perpignan
PYRENEES
Corse (Corsica)
ANDORRA
MEDITERRANEAN SEA
Ajaccio
Strait of Bonifacio

F

CHRONOLOGY

The French Revolution of 1789–1794 overthrew a monarchy that had lasted for more than 800 years. It ushered in a period of alternating republicanism, Napoleonic imperialism, and monarchism. In 1870 the founding of the Third Republic established France firmly in the republican tradition.

- **1914–1918** 1.4 million Frenchmen killed in World War I.
- **1918–1939** Economic recession and political instability; 20 prime ministers and 44 governments.
- **1940** Capitulation to Germany. Puppet Vichy regime. Gen. de Gaulle leads "Free French" abroad.
- **1944** Liberation of France.
- **1946–1958** Fourth Republic. Political instability: 26 governments. Nationalizations. France takes leading role in EEC formation.
- **1958** Fifth Republic. De Gaulle president with strong powers.
- **1960** Most French colonies gain independence.
- **1962** Algerian independence after bitter war with France.
- **1966** France withdraws from NATO military command.
- **1968** General strike and riots over education policy and low wages. Assemblée Nationale dissolved; Gaullist victory in June elections.
- **1969** De Gaulle resigns after defeat in referendum on regional reform; replaced by Georges Pompidou.
- **1974** Valéry Giscard d'Estaing president. Center-right coalition.
- **1981** Left wins elections; François Mitterrand president.
- **1983–1986** Government U-turn on economic policy.
- **1986** Cohabitation between socialist president and new right-wing government led by Jacques Chirac. Privatization program introduced.
- **1988** Mitterrand wins second term. PS-led coalition returns.
- **1991** Edith Cresson becomes first woman prime minister.
- **1993** Center-right wins elections. Second period of cohabitation.
- **1995** Jacques Chirac president.
- **1995–1996** Controversial series of Pacific nuclear tests.
- **1996** Unpopular austerity measures to prepare economy for European monetary union.
- **1997** PS-led government takes office in reversed cohabitation.
- **1999** France introduces euro.
- **2000** 35-hour week becomes law.
- **2002** January, euro fully adopted. April–June, center-right victory in presidential and legislative elections. July, Chirac unharmed in assassination attempt.

POLITICS

Multiparty elections

L. House 2002/2007
U. House 2001/2004

President Jacques Chirac

AT THE LAST ELECTION

National Assembly 577 seats

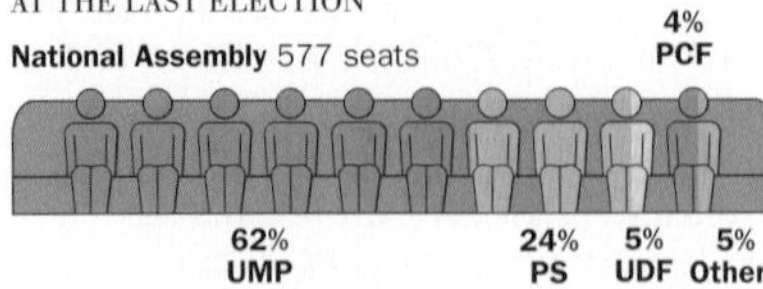
4% PCF
62% UMP
24% PS
5% UDF
5% Others

UMP = Union for a Presidential Majority (Rally for the Republic – **RPR** and Liberal Democracy – **DL**)
PS = Socialist Party **UDF** = Union for French Democracy
PCF = Communist Party of France **UC** = Centrist Union
Rep = Republicans and Independents **RCC** = Republicans, Communists, and Citizens **RDSE** = European Democratic and Social Rally **Ind** = Independents **G** = Greens

Senate 321 seats

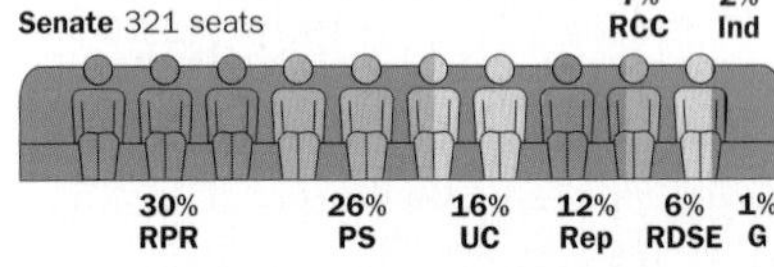
7% RCC
2% Ind
30% RPR
26% PS
16% UC
12% Rep
6% RDSE
1% G

France is a multiparty democracy. The constitution of the Fifth Republic, framed by Charles de Gaulle in 1958, ensures that the president has strong executive powers, but rules in tandem with a government and prime minister chosen by the Assemblée Nationale. Under changes agreed in 2000, the president and parliament are no longer elected according to separate timetables; the 2002 elections chose both for five-year terms. Traditionally the president attends to foreign policy and defense issues, while the government focuses on domestic and economic policy.

Profile

Apart from the two-term presidency of François Mitterrand of the PS (1981–1995), French presidents of the Fifth Republic have all been right-of-center. From 1986 onwards, however, both Mitterrand and his successor Jacques Chirac had to coexist for much of the time with a government and parliament dominated by political opponents. Thus Chirac, a conservative Gaullist, appointed Lionel Jospin of the PS as prime minister in 1997, heading a government coalition with the Communists, Greens, and Radical Socialists. In 2002, however, after Chirac won reelection as president in May, the sweeping victory of his UMP in legislative elections the following month marked a decisive rejection of so-called "cohabitation."

The far left has declined since 1945, when the PCF had 25% of the vote. The Greens suffered a serious setback in the 2002 elections, winning just three Assembly seats. The racist National Front (FN) failed to win any seats, but had created a major shock in the first round of the presidential poll when its leader Jean-Marie Le Pen finished second. Although the "democratic" parties united to ensure that he was trounced by Chirac in the run-off, his score of almost 18% was still a high-water mark for the FN.

Main Political Issues

Liberal economic reform

The center-right program which brought electoral success in 2002 was built around promises of immediate income tax cuts and pro-business measures.

Racism and "exclusion"

Exploiting concerns about crime, urban violence, and unemployment, the racist right has several times turned such fears to its political advantage by blaming immigrants. Legislation on immigration has been tightened and the new government in June 2002 promised tough new measures on maintaining order. Inner-city deprivation and "exclusion" of the unemployed and homeless, although widely recognized as divisive, were issues on which the PS-led government of 1997–2002 failed to find effective policies.

European integration and globalization

Opposition to European integration has grown since the early 1990s, fueled by fears of losing French sovereignty, while opponents of "globalization" tap into similar concerns that French jobs and culture are under threat. With their country now part of the eurozone, the French have been encouraged to look for benefits in terms of stability and economic growth, but many still see giving up the franc as losing a symbolic part of their national identity. The most strident opponent of EU integration is the FN, but the center-right has also promised a stout defense of French interests, notably over agricultural reform.

Jacques Chirac, *president of France since 1995.*

Jean-Marie Le Pen, *leader of the far-right FN and presidential runner-up in 2002.*

Jean-Pierre Raffarin *was appointed prime minister by Chirac in 2002.*

WORLD AFFAIRS

Joined UN in 1945

EU
G8
NATO
OECD
OSCE

French foreign policy has followed two, apparently contradictory, strands since World War II – maintenance of a strongly independent line and furtherance of French interests within a united Europe. France's leading role within the EU is seen as a way of combining the two strands. Along with Germany, France called for greater European integration, and fully supports the EU's eastward expansion.

Relations with the UK have been troubled from time-to-time, notably over French health-related restrictions on the entry of UK farm products, and because illegal immigrants passing through France persistently use the Channel Tunnel to enter the UK.

France seeks to offset US dominance in both foreign affairs and culture. It left NATO's military command in 1966 and maintained an independent nuclear deterrent (which it insisted on testing in the Pacific in 1995–1996 despite a wave of international criticism). France has latterly not supported the US and the UK in a hard-line attitude to Saddam Hussein's regime in Iraq, but it has backed other recent US-led military campaigns, including the "war on terrorism." It also seeks to maintain its influence over its former empire and francophone regions in general.

AID

Donor

$4.11bn (donations)

Down 27% in 2000

France is one of the world's major aid donors. Its motives are not simply commercial; it also wishes to maintain the influence of the French language, particularly in west Africa, the main aid recipient. Médecins sans Frontières is part of a tradition of active involvement through NGO aid agencies.

ECONOMICS

Inflation 1.5% p.a. (1990–2000)

$1438bn

1.0651–1.1231 euros

SCORE CARD

❑ WORLD GNP RANKING	5th
❑ GNP PER CAPITA	$24,090
❑ BALANCE OF PAYMENTS	$20.4bn
❑ INFLATION	1.7%
❑ UNEMPLOYMENT	10%

EXPORTS

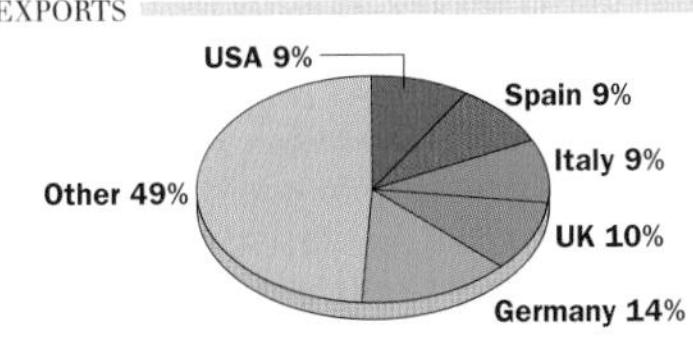
USA 9%
Spain 9%
Italy 9%
UK 10%
Germany 14%
Other 49%

IMPORTS

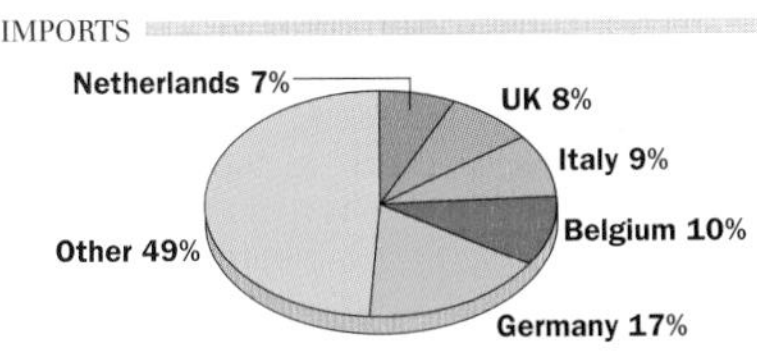
Netherlands 7%
UK 8%
Italy 9%
Belgium 10%
Germany 17%
Other 49%

STRENGTHS

Engineering, reflected in the TGV and nuclear industries. Specializations such as cars (Citroën, Peugeot, and Renault) and telecommunications (Alcatel). Defense sector a major exporter, as are pharmaceutical and chemical industries. Success in attracting inward investment. Strong technocratic traditions: unlike in USA or UK, top graduates are attracted into engineering. Luxury goods, cosmetics, perfumes, and quality wines. Most agriculture well modernized; France is Europe's leading agricultural producer.

WEAKNESSES

High taxes, social charges, and labor costs. France is losing its positions in traditional industries such as iron and steel, metallurgy, and textiles. Some major high-tech industries, such as telecommunications, run partly to further national pride, rather than on a strictly commercial basis.

ECONOMIC PERFORMANCE INDICATOR

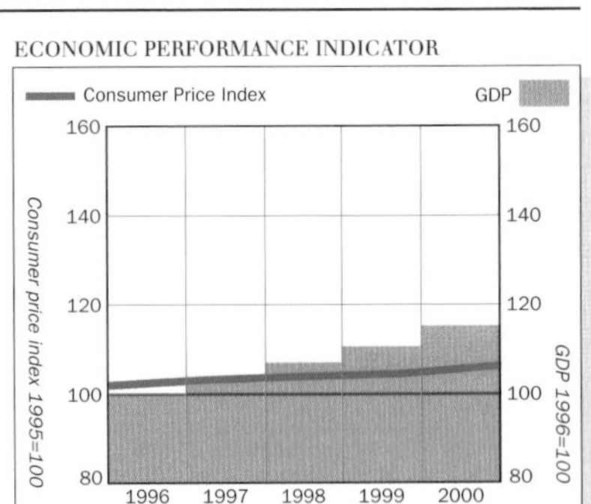
Consumer Price Index
GDP
160
140
120
100
80
Consumer price index 1995=100
GDP 1996=100
1996
1997
1998
1999
2000

PROFILE

At first slow to industrialize, protectionist France started competing in world markets and modernizing its industry in the 1950s and 1960s. Integration in western Europe, starting with the coal and steel industry in the 1950s, placed France at the heart of the EU. It was one of the 12 EU countries fully to adopt the euro in 2002. One of the world's top exporters, France's foreign trade balance runs a healthy surplus. It has a long tradition of state involvement in running the economy. Nationalization of key industries began in the late 1930s, with a fresh burst in 1981–1983, but since then both right-of-center and socialist governments have pursued privatization with vigor, reaching into the defense industry, aviation, banking, telecommunications, and insurance. Regional hubs are of growing economic significance. France is the EU's largest agricultural producer, and its farmers form a powerful political lobby. Active trade unions succeeded in getting a maximum 35-hour week introduced in 2000.

FRANCE : MAJOR BUSINESSES

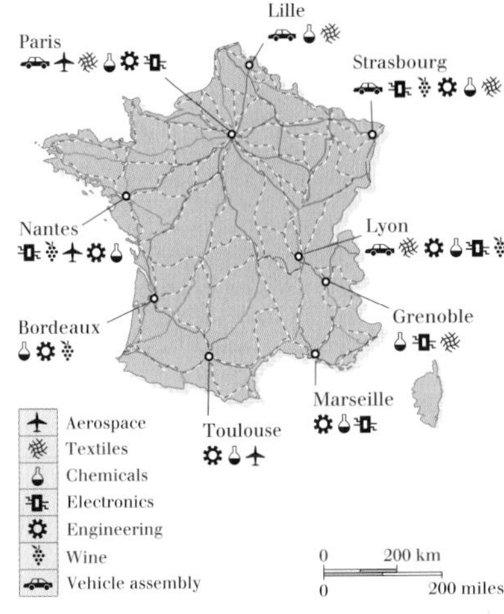
Lille
Paris
Strasbourg
Nantes
Lyon
Bordeaux
Grenoble
Marseille
Toulouse
Aerospace
Textiles
Chemicals
Electronics
Engineering
Wine
Vehicle assembly
0
200 km
0
200 miles

DEFENSE

No compulsory military service

$34.3bn

Down 9% in 2000

FRENCH ARMED FORCES

809 main battle tanks (498 AMX-30B2, 311 *Leclerc*)	150,000 personnel
1 carrier, 10 submarines, 1 cruiser, 3 destroyers, 30 frigates, & 39 patrol boats	45,600 personnel
532 combat aircraft (352 *Mirage* F-1B/1C/1CR, and 120 *Jaguar*	63,000 personnel
64 SLBM in 4 SSBN	

France was a founder member of NATO, but left its military command in 1966 in opposition to US domination. It maintained an independent nuclear deterrent through the Cold War, but went though a rapprochement with NATO in the 1990s. Joint participation between France and Germany in European army units is partly symbolic of reconciliation, as well as an expression of the need for an EU defense structure.

The influence of the army, which was once very strong, is now much diminished. Compulsory military service ended in 2001.

France has one of the world's largest and most export-oriented defense industries, producing its own tanks, jet fighter aircraft, and missiles.

F

RESOURCES

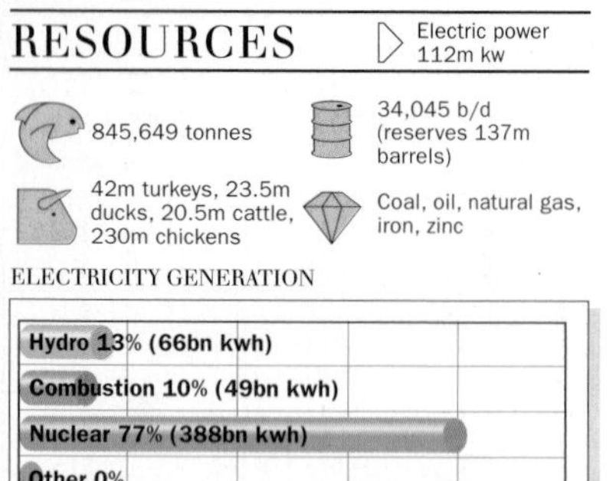
Electric power 112m kw
845,649 tonnes
34,045 b/d (reserves 137m barrels)
42m turkeys, 23.5m ducks, 20.5m cattle, 230m chickens
Coal, oil, natural gas, iron, zinc
ELECTRICITY GENERATION
Hydro 13% (66bn kwh)
Combustion 10% (49bn kwh)
Nuclear 77% (388bn kwh)
Other 0%
0
20
40
60
80
100
% of total generation by type

France is the world's most committed user of nuclear energy, which provides over three-quarters of its electricity requirements. The policy reflects a desire for national energy self-sufficiency. Coal, once plentiful in the north and Lorraine, is now mostly exhausted, as are the gas fields off the southwest coast.

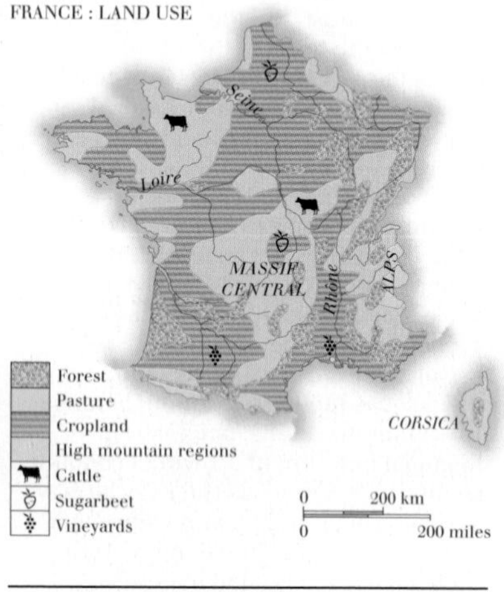
FRANCE : LAND USE
Seine
Loire
MASSIF CENTRAL
Rhône
ALPS
CORSICA
Forest
Pasture
Cropland
High mountain regions
Cattle
Sugarbeet
Vineyards
0
200 km
0
200 miles

ENVIRONMENT

Sustainability rank: 33rd

14% (9% partially protected)

6.3 tonnes per capita

ENVIRONMENTAL TREATIES

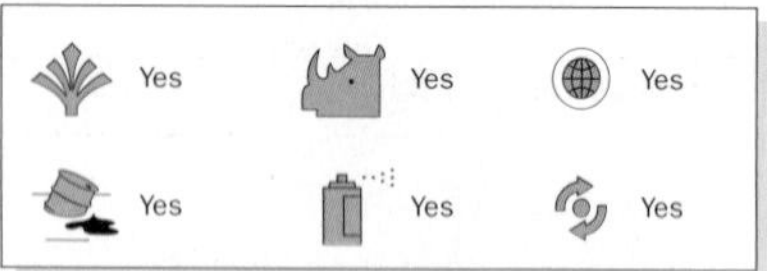
Yes
Yes
Yes
Yes
Yes
Yes

Awareness of "green" issues has risen with a series of campaigns against major infrastructure projects. Nuclear power's importance, however, puts the environmentalist lobby in perspective. Transportation of oil by sea poses the threat of pollution of the Atlantic coast. Brittany's beaches and fisheries were severely affected by the wreck of the *Erika* in 1999. Severe storms devastated woodlands in 2000, focusing attention on the dangers of global warming.

THE IMPACT OF MODERN TRANSPORTATION

While the French regions retain distinctive characteristics, the country's modern transportation and telecommunications network, with Paris at its hub, has brought the main provincial cities within easy reach of the capital. It also helps France to play a central role in European business.

The autoroute network

The toll-charging *autoroutes*, or freeways, with a speed limit of 130 km/h (81 mph), now reach almost every corner of the map except the Cherbourg peninsula and Brittany in the northwest. They have relieved the previous generation of town-to-town highways of much of their former traffic congestion, but have also encouraged the upsurge in trans-European heavy freight vehicle movement, carrying through traffic to and from Spain, Italy, Switzerland, Germany, the Benelux countries, and the Channel ports. There has been much controversy about the environmental impact of several as yet uncompleted cross-country links, notably the E11 route south from Clermont-Ferrand to join the Mediterranean network.

French leadership in high-speed trains

French pride in leading-edge engineering, and a capacity for ambitious centralized planning and state-backed investment, have been apparent in its high speed train system, the *train à grande vitesse* (TGV). Rivaled only by the Japanese "bullet train" as the fastest in passenger service, the TGV can run at sustained speeds of over 300 km/h (186 mph). It has hit domestic airline traffic hard on some of the prime routes, by offering a combination of comfort and shorter door-to-door journey times in many instances. During the 1990s, some double-decker trains were introduced, to help meet high demand for seats, and a new generation of TGV trains was unveiled by the engineers Alstom in 1998–1999.

The first element was the TGV southeast. Construction began in 1975 and the Paris–Lyon section came into service in 1983, with subsequent extensions southward, culminating in the high-profile opening in 2001 of the direct service from Paris to Marseille. By the end of the 1980s a western route from Paris, initially to

Calais
Lille
Le Havre
Rouen
PARIS
Metz
Brest
Strasbourg
Le Mans
Orléans
Nantes
Dijon
Tours
Poitiers
Limoges
Lyon
Bordeaux
Montpellier
Nice
Toulouse
Marseille
Autoroute
TGV route
Other rail

FRANCE : ROAD AND RAIL

Le Mans, was also in service. The northern Paris–Lille route opened in 1993 with extensions to Brussels, Calais, and the Channel Tunnel.

Extending the train network

The existing TGV lines, which use mainly modernized but also some purpose-built track, are a highly profitable part of the French rail network. Seeing the likely economic benefits, many towns have lobbied hard to be included. The next planned line, however, from Paris east to Strasbourg, has faced more opposition as well as escalating cost estimates. Construction began in early 1999, aiming for completion in six to seven years. To make money, this line will need to provoke a big shift in existing traveling habits. The French government's commitment is also in part a strategic political decision, designed to reinforce Strasbourg's role "at the heart of European integration."

A link via Tours to Bordeaux is intended to be part of the next stage, along with further international links – from Lyon to Turin, from Marseille and Montpellier to Barcelona, and via Strasbourg to Stuttgart and Frankfurt and thus linking into the German high-speed ICE train network.

Modern office blocks *in Montpellier, a city whose new dynamic image owes much to its transportation links and investment in communications technology.*

MEDIA

TV ownership high

Daily newspaper circulation 218 per 1000 people

PUBLISHING AND BROADCAST MEDIA

There are 117 daily newspapers, including *Le Monde*, *Libération*, and *Le Figaro*. *Ouest-France* has the highest circulation

10 services: 5 state-controlled, 5 independent

7 services: 3 state-controlled, 4 independent

TV and radio were freed from direct state influence in the 1980s. Two of the main TV channels are still state-owned, but TF1 was privatized in 1987. Canal Plus mixes pay-per-view and advertising-backed services. A Breton-language station started up in 2000. Commercial channels have multiplied with the growth of satellite and cable.

The once innovative Minitel electronic communications system is now overshadowed by the Internet. Circulation of prestigious national newspapers has dwindled and regional papers too have suffered from a gradual shift to electronic media.

CRIME

No death penalty

46,376 prisoners

Little change in 1999

CRIME RATES

	Murders
3	per 100,000 population

	Rapes
14	per 100,000 population

	Thefts
3849	per 100,000 population

The Code Napoléon, enacted in 1804 by Napoléon I, still forms the basis of French law. Criminal justice is based on inquisitorial rather than adversarial principles. The *juge d'instruction* has considerable powers to examine witnesses and assess evidence. The press are not restricted by *sub judice* rules in reporting trials. Political corruption cases, reaching into government, attract much attention.

Public concern about rising petty crime and violence has encouraged successive governments to promote tough policing. An "antiterrorism" bill was introduced in October 2001, giving police extra powers.

HEALTH

Welfare state health benefits

1 per 333 people

Heart and cerebro-vascular diseases, cancers, accidents

The French consume more medicines per capita than any other nation, and a significant number take medically approved, and prescribed, cures at health spas. French health care was rated the most efficient in the world by WHO in 2000. Under the national health system patients pay for treatment, and then get the majority of the cost reimbursed by an insurance company paid by the social services. Although health awareness has risen in recent years, a 1992 law banning smoking in public places is widely ignored and alcoholism remains a problem, with cirrhosis of the liver not uncommon as a cause of death.

F

SPENDING

GDP/cap. increase

CONSUMPTION AND SPENDING

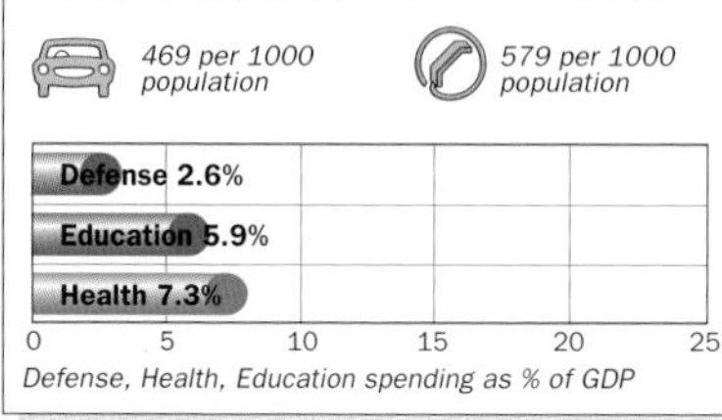
469 per 1000 population
579 per 1000 population
Defense 2.6%
Education 5.9%
Health 7.3%
0
5
10
15
20
25
Defense, Health, Education spending as % of GDP

Wealth and income disparities in France are higher than in most OECD states. The Socialists narrowed the gap a little in the 1980s with the introduction of the legal minimum wage (*le SMIC*). Most tax is indirect – a result of a long French tradition of income-tax evasion. Major tax cuts announced in 2000 aimed to redress the imbalance of income tax on the rich and poor. The wealthy take exotic vacations to the Himalayas, the Andes, and Polynesia. The French lag behind their European neighbors in using the Internet; fewer than one-fifth of the population had done so by 2001.

EDUCATION

School leaving age: 16

99%

2.09m students

THE EDUCATION SYSTEM

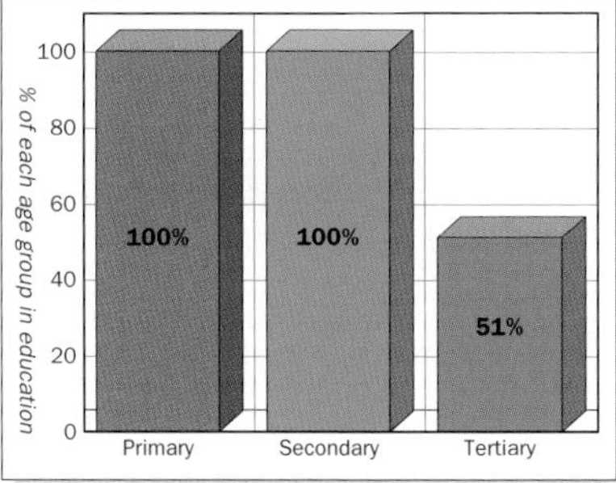
100
80
60
40
20
0
% of each age group in education
100%
100%
51%
Primary
Secondary
Tertiary

Education is highly centralized, a situation which is slowly generating a desire for greater flexibility. The education ministry organizes the curriculum, sets examinations, including the *baccalauréat*, and decides staffing issues. Roman Catholic schools, which take most of the 17% of privately educated children (but are not fee paying and receive large state subsidies), are the exception. However, they are still obliged to follow the national curriculum.

The focus in the classroom remains the acquisition of a broad range of knowledge. Pupils' academic records are impressive, despite frequent staff strikes.

France has more than 70 universities – 13 in Paris – and higher education bodies with 1.2 million students. Entry is not competitive, but based on passing the secondary-level exam, the *baccalauréat*. Most students attend the university nearest to home. The universities have not been given the funds or staff to cope with the huge increase in student numbers in recent years. The 150 *Grandes Ecoles*, the most influential tertiary institutions, are outside the university system, and each takes just a few hundred carefully selected students. They groom the future governing elite, opening the way for their successful graduates to gain the top civil service and professional jobs.

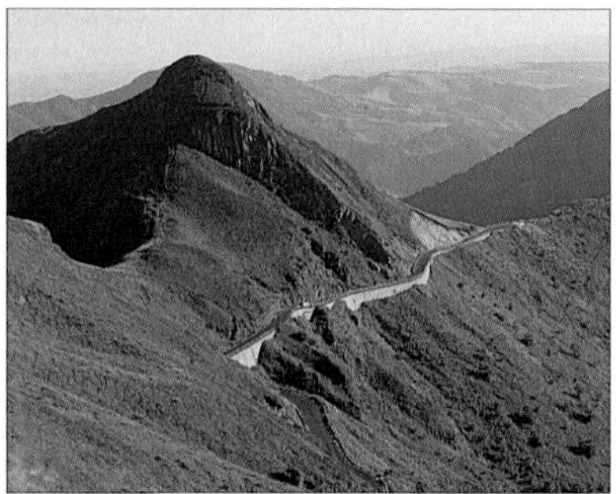

Massif Central, Auvergne. *The Massif's lonely granite plateaus and extinct volcanoes are France's oldest rock formations.*

WORLD RANKING

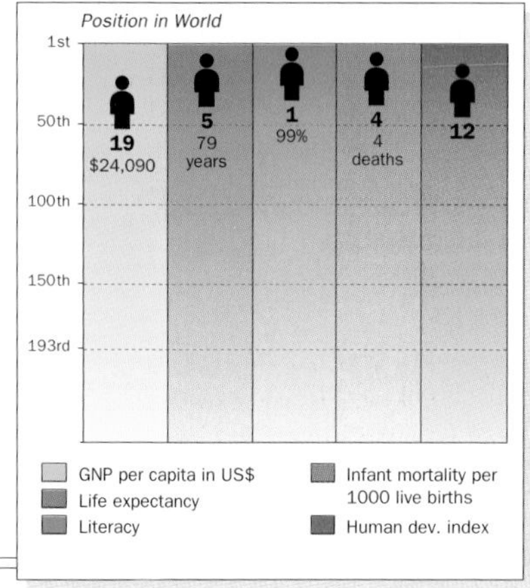
Position in World
1st
50th
100th
150th
193rd
19
$24,090
5
79 years
1
99%
4
4 deaths
12
GNP per capita in US$
Life expectancy
Literacy
Infant mortality per 1000 live births
Human dev. index

See also OVERSEAS TERRITORIES *p.640*

GABON

OFFICIAL NAME: Gabonese Republic **CAPITAL:** Libreville
POPULATION: 1.3 million **CURRENCY:** CFA franc **OFFICIAL LANGUAGE:** French

WEST AFRICA
Cameroon
Equatorial Guinea
GABON
Congo
Africa

  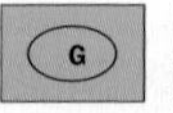 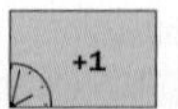  
1960
1960
Aug 17
G
+1
+241
.ga

AN EQUATORIAL COUNTRY on the west coast of Africa, Gabon's major economic activity is the production of oil. Only a small area of Gabon is cultivated, and more than two-thirds of it constitutes one of the world's finest virgin rainforests. Gabon became independent of France in 1960. A single-party state from 1968, it returned to multiparty democracy in 1990. Gabon's population is small, and the government is encouraging its increase.

CLIMATE

Tropical equatorial

WEATHER CHART FOR LIBREVILLE

Average daily temperature / Rainfall
°C/°F J F M A M J J A S O N D cm/in
40/104 – 40/16; 30/86 – 30/12; 20/68 – 20/8; 10/50 – 10/4; 0/32 – 0; -10/14; -20/-4

The climate is heavily equatorial – hot all year round with a long rainy season from October to May. The cold Benguela current lowers coastal temperatures.

TRANSPORTATION

Drive on right

Léon M'Ba, Libreville
753,557 passengers

26,532 grt

THE TRANSPORTATION NETWORK

629 km (391 miles)

30 km (19 miles)

814 km (506 miles)

1600 km (994 miles)

The Trans-Gabon Railroad from Owendo port near Libreville to Massoukou is the key transportation link.

Air transportation is well developed, and most big companies have airstrips.

TOURISM

Visitors : Population 1:8.4

155,000 visitors

Down 11% in 2000

MAIN TOURIST ARRIVALS

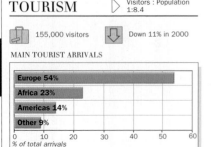

	% of total arrivals
Europe	54%
Africa	23%
Americas	14%
Other	9%

0 10 20 30 40 50 60
% of total arrivals

Despite Libreville's many hotels, Gabon has little tourism, in part a reflection of its lack of good beaches.

PEOPLE

Pop. density low

Fang, French, Punu, Sira, Nzebi, Mpongwe

5/km² (13/mi²)

THE URBAN/RURAL POPULATION SPLIT

81% 19%

ETHNIC MAKEUP

French 2%
European and other African 9%
Eshira 25%
Other Bantu 29%
Fang 35%

The largest ethnic group in Gabon is the Fang, who live mainly in the north. President Omar Bongo, from a subgroup of the Bateke in the southeast, has artfully united the common interests of other ethnic groups to keep the Fang from government. The Myene group around Port-Gentil consider themselves as the aristocrats of Gabonese society owing to their long-standing ex-colonial contacts. Oil wealth has led to the growth of a distinct bourgeoisie. Gabon is one of Africa's most urbanized countries.

POLITICS

Multiparty elections

L. House 2001/2006
U. House 1997/2003

President Omar Bongo

AT THE LAST ELECTION

National Assembly 120 seats

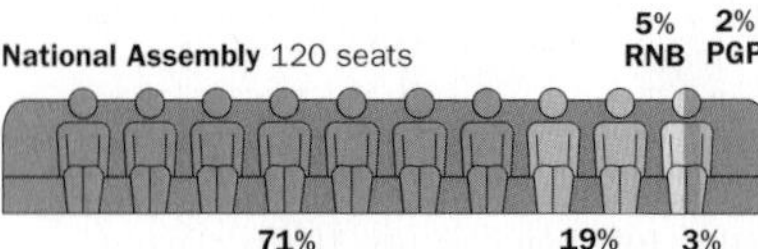

71% PDG **19% Vacant** **3% Others** **5% RNB** **2% PGP**

PDG = Gabonese Democratic Party **RNB** = National Rally of Woodcutters **PGP** = Gabonese Progress Party

Senate 91 seats

The Senate is indirectly elected by regional councils

Gabon has had a constitution enshrining a multiparty system since 1990, when elections confirmed in power the former sole ruling party – Omar Bongo's PDG. Bongo, in power since 1967, won the first multiparty presidential poll in 1993, but its fairness was widely disputed. He was reelected president in 1998, this time for a seven-year term. In parliamentary elections in December 2001, the PDG won a majority in the National Assembly, as they had in 1996.

WORLD AFFAIRS

Joined UN in 1960

FZ    ACP
G24
AU
OIC

Gabon maintains close links with France, although US companies are also making inroads into Gabon's oil-rich economy. Gabon remains influential regionally (the Gulf of Guinea Commission was launched in Libreville in 2001), and relations further afield, particularly with OPEC, are also important.

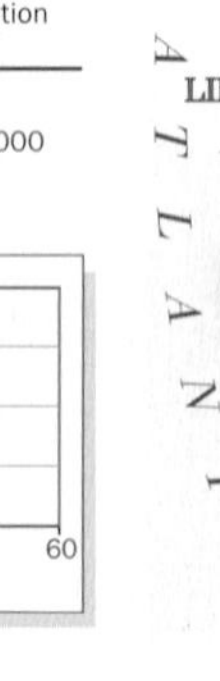

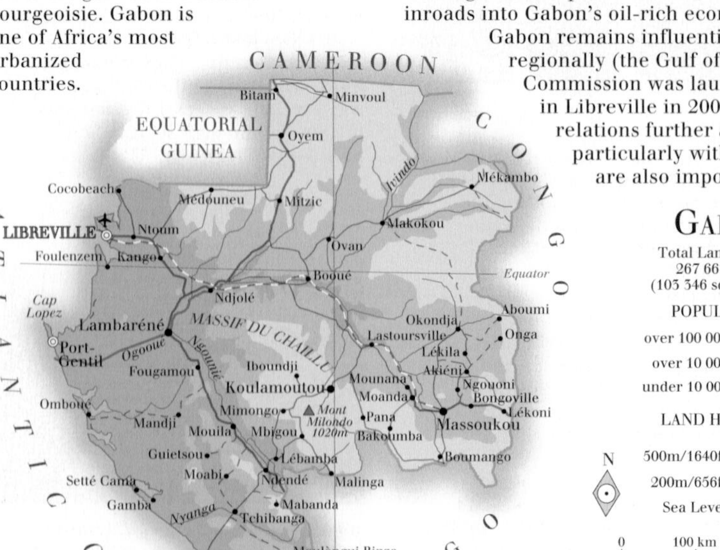
CAMEROON
EQUATORIAL GUINEA
CONGO
ATLANTIC OCEAN
Bitam
Minvoul
Oyem
Cocobeach
Médouneu
Mitzic
Mékambo
Makokou
LIBREVILLE
Ntoum
Foulenzem
Kango
Ovan
Booué
Equator
Ndjolé
Cap Lopez
Lambaréné
Port-Gentil
MASSIF DU CHAILLU
Okondja
Aboumi
Onga
Lastoursville
Lékila
Akiéni
Fougamou
Iboundji
Ngouoni
Bongoville
Lékoni
Omboué
Koulamoutou
Mounana
Moanda
Massoukou
Mimongo
Mont Milondo 1020m
Pana
Bakoumba
Mandji
Mouila
Mbigou
Guietsou
Lébamba
Boumango
Moabi
Ndendé
Malinga
Setté Cama
Mabanda
Gamba
Tchibanga
Mayumba
Moulèngui Binza
Ndindi
Ogooué
Ngounié
Nyanga
Ivindo
4°
12°

GABON
Total Land Area : 267 667 sq. km (103 346 sq. miles)

POPULATION
over 100 000
over 10 000
under 10 000

LAND HEIGHT
500m/1640ft
200m/656ft
Sea Level

0 100 km
0 100 miles

AID

Recipient

 $12m (receipts)

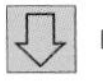 Down 75% in 2000

France is by far the major aid donor, providing two-thirds of total receipts. For a middle-income country with one of the highest GNPs per capita in the developing world, Gabon has benefited from considerable aid. Its indebtedness is the result of excessive borrowing, which was encouraged by Western banks in the 1970s. Much aid goes to servicing this debt.

DEFENSE

No compulsory military service

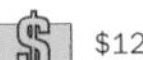  $123m

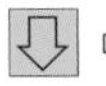 Down 9% in 2000

President Bongo's background in the military is reflected in Gabon's large defense budget and prestige weaponry, which includes French *Mirage* jets. France guarantees Gabon's security and keeps a 750-strong garrison in Libreville. A recruitment drive was launched in April 2001 to recruit 1500 18- to 25-year-olds by 2006.

ECONOMICS

Inflation 6.2% p.a. (1990–2000)

 $3.93bn

 698.7–736.7 CFA francs

SCORE CARD

- ❏ World GNP Ranking........................118th
- ❏ GNP per Capita $3190
- ❏ Balance of Payments $385m
- ❏ Inflation 1.5%
- ❏ Unemployment.................................21%

Strengths

Oil and a relatively small population give Gabon a high per capita GNP. The country's other abundant resources – including some of the world's best tropical hardwoods – are just beginning to be tapped.

Weaknesses

Large debt burden incurred in the 1970s. Continuing dependence on French technical assistance.

EXPORTS

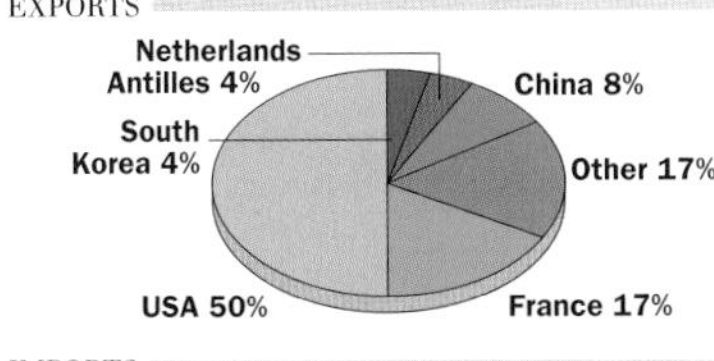
Netherlands Antilles 4%
China 8%
South Korea 4%
Other 17%
USA 50%
France 17%

IMPORTS

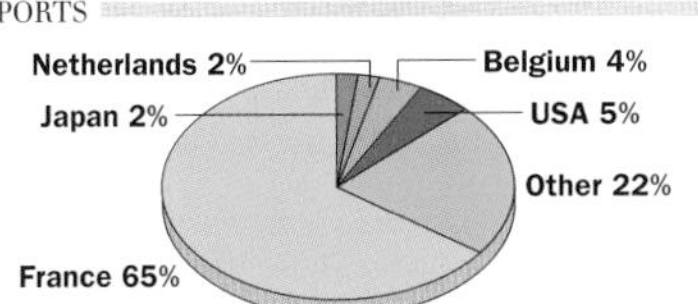
Netherlands 2%
Belgium 4%
Japan 2%
USA 5%
Other 22%
France 65%

RESOURCES

Electric power 379,000 kw

 53,440 tonnes

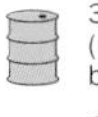 301,000 b/d (reserves 2.5bn barrels)

 213,000 pigs, 198,000 sheep, 3.2m chickens

Oil, manganese, uranium, gold, iron, natural gas

Oil is the major export earner. Gabon also has large deposits of uranium and over 100 years' reserves of manganese. The unexploited iron ore deposits at Bélinga are the world's largest.

ENVIRONMENT

Sustainability rank: 36th

 3%

 2.4 tonnes per capita

The Trans-Gabon Railroad has sliced through one of the world's finest virgin rainforests and has opened the interior to indiscriminate exploitation of rare woods such as oleoirme. Gabon abandoned plans for nuclear power following the 1986 Chernobyl disaster.

MEDIA

TV ownership high

 Daily newspaper circulation 30 per 1000 people

PUBLISHING AND BROADCAST MEDIA

 There are 2 daily newspapers, *L'Union* and *Gabon-Matin*

 3 services: 1 state-owned, 2 independent

 7 services: 2 state-controlled, 5 independent

The media are mostly government-controlled. There was a crackdown in 1998 on independent media – which in the 1990s had become quite diverse – raising concerns about freedom of expression.

CRIME

Death penalty in use

 Gabon does not publish prison figures

 Recorded crime rose sharply from 1992–1996

Urban crime rates are rising. A campaign to combat child trafficking was launched in 2001.

Gabon's human rights record has improved in recent years.

***Albert Schweitzer Hospital,** Lambaréné, on the lower Ogooué River. Schweitzer won a Nobel Prize for his pioneering work in Africa.*

CHRONOLOGY

Gabon became a French colony in 1886, administered as part of French Equatorial Africa.

- ❏ **1960** Independence. Léon M'ba president.
- ❏ **1967** Albert-Bernard (later Omar) Bongo president.
- ❏ **1968** Single-party state instituted.
- ❏ **1990** Multiparty democracy.
- ❏ **1998** Bongo reelected president.
- ❏ **2001** Elections: ruling PDG retains majority.

G

EDUCATION

School leaving age: 16

 71%

 7473 students

Education follows the French system. Université Omar Bongo in Libreville, founded in the 1970s, now has more than 4000 students.

HEALTH

No welfare state health benefits

1 per 5000 people

Heart and diarrheal diseases, pneumonia, accidents

Oil revenues have allowed substantial investment in the health service, which is now among the best in Africa.

SPENDING

GDP/cap. increase

CONSUMPTION AND SPENDING

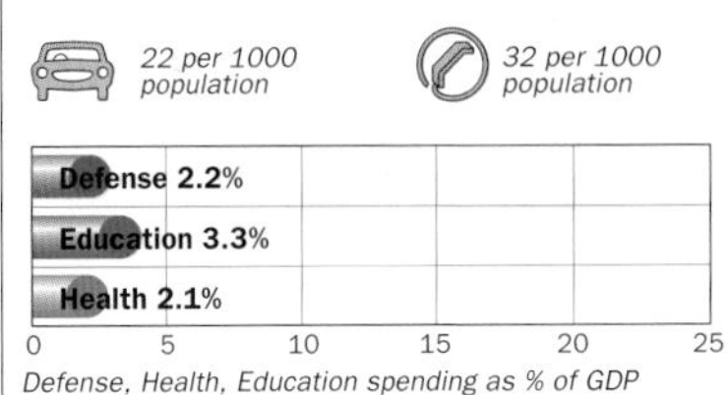
22 per 1000 population
32 per 1000 population
Defense 2.2%
Education 3.3%
Health 2.1%
0
5
10
15
20
25
Defense, Health, Education spending as % of GDP

Oil wealth has led to the growth of an affluent bourgeoisie. Menial and low-income jobs are done by immigrant workers.

WORLD RANKING

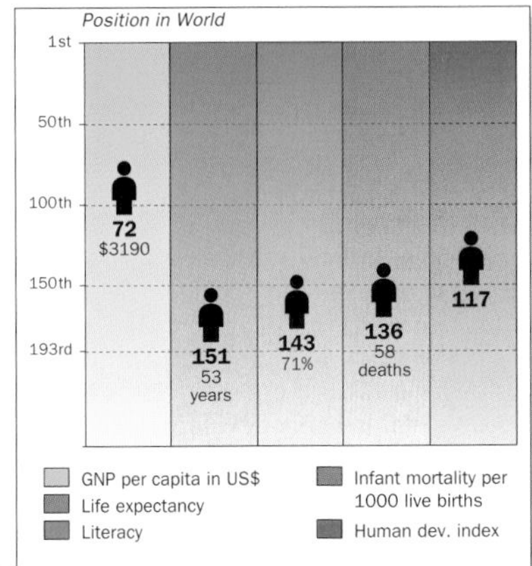
Position in World
1st
50th
100th
150th
193rd
72
$3190
151
53 years
143
71%
136
58 deaths
117
GNP per capita in US$
Life expectancy
Literacy
Infant mortality per 1000 live births
Human dev. index

WEST AFRICA
Africa
Senegal
GAMBIA
Mali
Guinea Bissau
Guinea

GAMBIA

OFFICIAL NAME: Republic of the Gambia **CAPITAL:** Banjul
POPULATION: 1.34 million **CURRENCY:** Dalasi **OFFICIAL LANGUAGE:** English

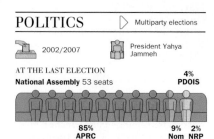

1965

1965

Feb 18

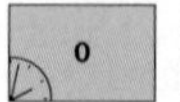
WAG

0

+220

.gm

A NARROW COUNTRY on the western coast of Africa, the Gambia was renowned as a stable democracy until an army coup in 1994. Agriculture accounts for 65% of GDP, yet many Gambians are leaving rural areas for the towns, where average incomes are four times higher. Its position as an enclave within Senegal seems likely to endure, following the failure of an experiment in federation in the 1980s.

CLIMATE

Tropical wet & dry

WEATHER CHART FOR BANJUL

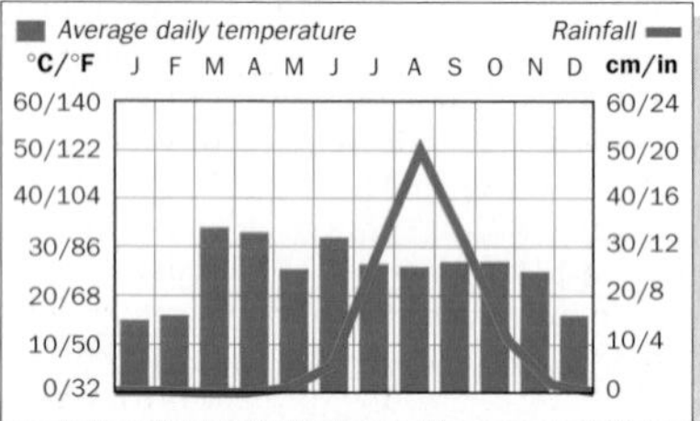
Average daily temperature
Rainfall
°C/°F
J F M A M J J A S O N D
cm/in
60/140
50/122
40/104
30/86
20/68
10/50
0/32
60/24
50/20
40/16
30/12
20/8
10/4
0

The subtropical and sunny dry season is punctuated by intermittent hot *harmattan* winds.

TRANSPORTATION

Drive on right

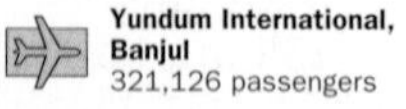
Yundum International, Banjul
321,126 passengers

8 ships
1884 grt

THE TRANSPORTATION NETWORK

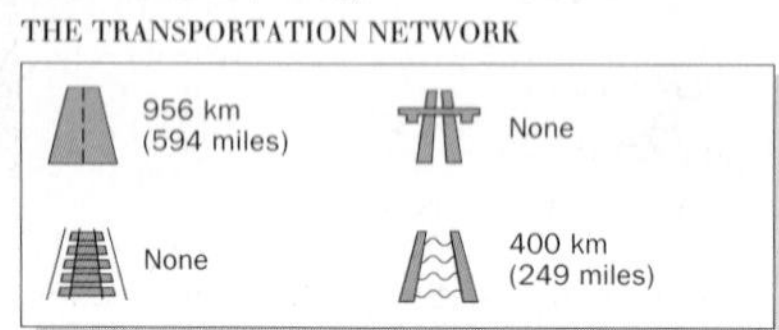
956 km
(594 miles)
None
None
400 km
(249 miles)

The Gambia River carries more traffic than the roads – ships of up to 3000 tonnes can reach Georgetown. Yundum airport was upgraded by NASA in 1989 for US space shuttle emergency landings.

TOURISM

Visitors : Population 1:14

96,000 visitors

Up 5% in 1999

MAIN TOURIST ARRIVALS

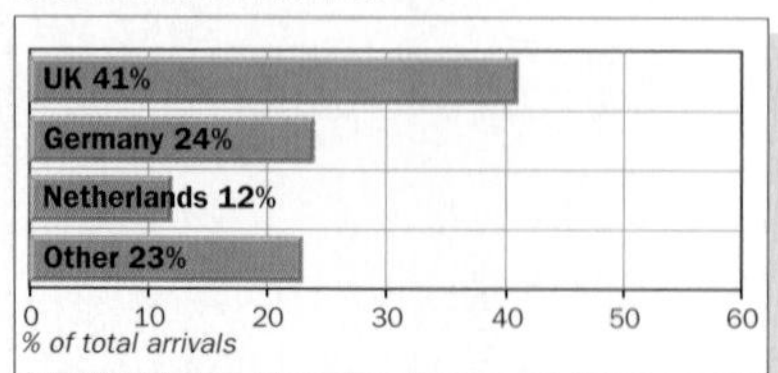
UK 41%
Germany 24%
Netherlands 12%
Other 23%
0 10 20 30 40 50 60
% of total arrivals

Ecotourism is being developed, but most of those enjoying the beaches and resort hotel life are Europeans escaping winter, including many single women.

PEOPLE

Pop. density medium

Mandinka, Fulani, Wolof, Jola, Soninke, English

134/km²
(346/mi²)

THE URBAN/RURAL POPULATION SPLIT

33% 67%

ETHNIC MAKEUP

Other 5%, Serahuli 9%, Jola 10%, Wolof 16%, Fulani 18%, Mandinka 42%

Until the 1994 coup, led by Lt. Col. Yahya Jammeh, the 1962–1994 government of President Sir Dawda Jawara had tried to offset minority resentment of the Mandinka's domination of politics, distributing political offices fairly according to ethnic origins. Jammeh, a fervent Muslim, is from the minority Jola (or Diola) community, numerous across the border in Senegal, where they are active in a local rebellion. About 85% of Gambians follow Islam, although there is no official state religion. There is a yearly influx of migrants, who come from Senegal, Guinea, and Mali to trade in groundnuts. The Gambia is still a very poor country, with 80% of the labor force engaged in agriculture. Women are active as traders in an otherwise male-dominated society.

Fishing village. *Overfishing in the waters off the Gambia and Senegal, mainly by distant nations, is a growing problem.*

POLITICS

Multiparty elections

2002/2007

President Yahya Jammeh

AT THE LAST ELECTION

National Assembly 53 seats

4%
PDOIS
85%
APRC
9%
Nom
2%
NRP

APRC = Alliance for Patriotic Reorientation and Construction
Nom = Nominated **PDOIS** = People's Democratic Organization for Independence and Socialism
NRP = National Reconciliation Party

The People's Progressive Party (PPP) was in government from 1962 until 1994, for most of which time the Gambia was one of Africa's few democracies. A ban, imposed after the 1994 coup on the PPP and the three main parties which had been in opposition, was lifted in 2001 in time for elections.

During the army's coup, Sir Dawda Jawara took refuge aboard a visiting US warship, and he then went into exile in Britain. The coup's leaders claimed that it had been initiated in a bid to end corruption and pledged to preserve democracy. In the new government several portfolios went to civil servants who had served in the Jawara administration. Military leader Yahya Jammeh was elected president in controversial elections in September 1996, and the following January his APRC won the majority of seats in a parliamentary election. Jammeh and the APRC were reelected in 2001–2002; the PPP won no seats.

WORLD AFFAIRS

Joined UN in 1965

CILSS

Comm

ECOWAS

AU

OIC

Commonwealth and Western criticism of the 1994 coup which brought President Jammeh to power has softened over recent years, but an apparent crackdown on opponents following Jammeh's reelection in 2001 reawoke international concern. Relations with Senegal have been troublesome since the end of federation in 1989.

AID

Recipient

$49m (receipts)

Up 48% in 2000

Western aid flows, suspended after the 1994 coup, have largely resumed. The World Bank, the IMF, the AfDB, the UK, the US, Japan, Libya, Germany, the Gulf states, the UN, and the EU are all significant donors.

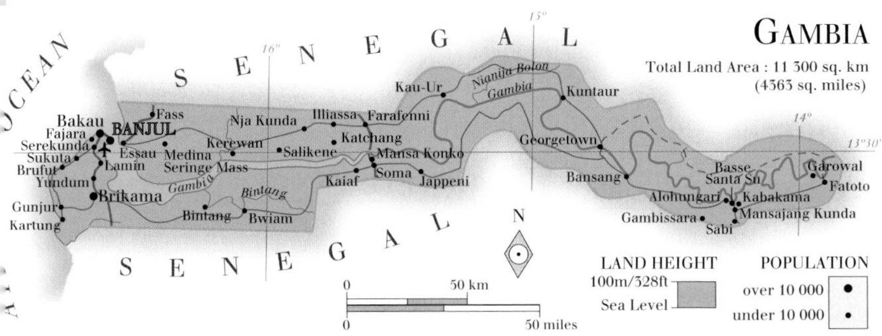
GAMBIA
Total Land Area : 11 300 sq. km (4363 sq. miles)
S E N E G A L
S E N E G A L
OCEAN
Bakau
Fajara
Serekunda
Sukuta
Brufut
Yundum
BANJUL
Fass
Essau
Lamin
Medina
Seringe Mass
Brikama
Gunjur
Kartung
Nja Kunda
Kerewan
Illiassa
Farafenni
Salikene
Katchang
Mansa Konko
Soma
Kaiaf
Jappeni
Bintang
Bwiam
Kau-Ur
Kuntaur
Georgetown
Bansang
Basse Santa Su
Garowal
Fatoto
Alohungari
Kabakama
Mansajang Kunda
Gambissara
Sabi
N
LAND HEIGHT
100m/328ft
Sea Level
POPULATION
over 10 000
under 10 000
0
50 km
0
50 miles

DEFENSE

No compulsory military service

 $15m

The Gambia National Army, with two infantry battalions, takes about half of the defense budget; the rest finances the 600-strong gendarmerie. Most arms are bought from the UK, although supplies are now increasingly coming from Nigeria too. A defense pact with Senegal collapsed along with the federation in 1989.

ECONOMICS

Inflation 4.1% p.a. (1990–2000)

SCORE CARD

- ❑ World GNP Ranking 172nd
- ❑ GNP per Capita $340
- ❑ Balance of Payments ......–$48m
- ❑ Inflation 0.8%
- ❑ Unemployment ......Widespread underemployment

Strengths
Low tariffs make the Gambia a focus of regional trade. Natural deepwater harbor at Banjul, one of the finest on the west African coast. Well-managed economy, favorably viewed by donors.

Weaknesses
Small size of market can inhibit investment. Smuggling: deprives government of significant revenues. Lack of resources, little agricultural diversification; consequent over-reliance on groundnuts, the main crop.

EXPORTS

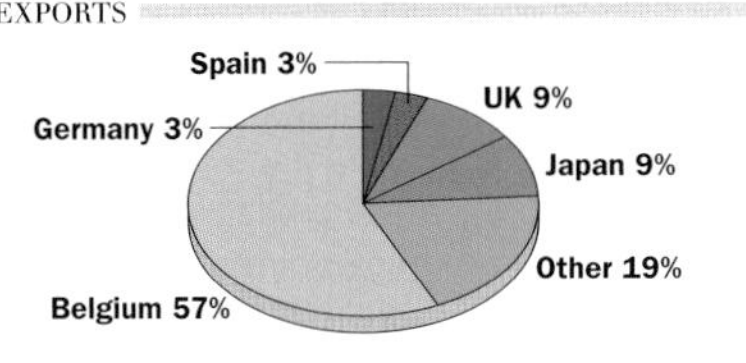
Spain 3%
UK 9%
Germany 3%
Japan 9%
Other 19%
Belgium 57%

IMPORTS

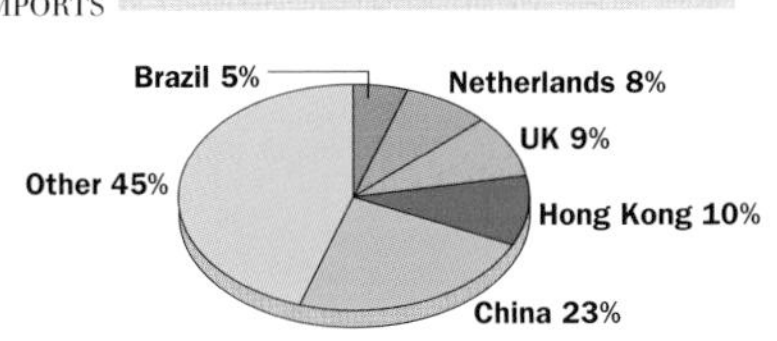
Brazil 5%
Netherlands 8%
UK 9%
Other 45%
Hong Kong 10%
China 23%

RESOURCES

Electric power 29,000 kw

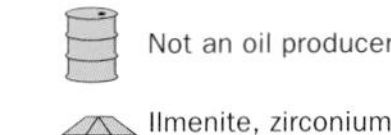

30,004 tonnes — Not an oil producer

365,000 cattle, 145,000 goats, 780,000 chickens — Ilmenite, zirconium, rutile, kaolin, tin, oil

The Gambia River is one of Africa's few good waterways, but it is underused owing to its separation from its natural hinterland by the Gambia–Senegal border. Irrigation is at present provided by a single dam; plans for further dams for power generation have met with opposition. Oil deposits are believed to exist offshore.

ENVIRONMENT

Sustainability rank: 103rd

The impact of tourism and of overfishing in Gambian waters are major concerns, as are desertification and deforestation.

MEDIA

TV ownership low

PUBLISHING AND BROADCAST MEDIA

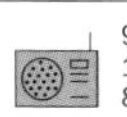

The independent media are severely restricted. Radio news broadcasts are dominated by state-controlled Radio Gambia, and the government runs the only national television station.

CRIME

Death penalty not used in practice

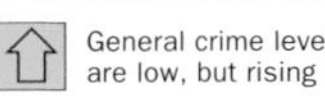

Crime levels are relatively low in what is a peaceful society compared with many other states in the region.

EDUCATION

Schooling is not compulsory

The aims are to increase enrollment to 75% in primary and 20% in secondary schools, and to improve teacher quality. A university was established in 1998.

CHRONOLOGY

Mandinka traders brought Islam in the 13th century and were the main influence until the 18th century. The 1700s and 1800s saw colonial rivalry between Britain and France.

- ❑ **1888** British possession.
- ❑ **1959** Dawda Jawara founds PPP.
- ❑ **1965** Independence from Britain.
- ❑ **1970** Republic; Jawara president.
- ❑ **1982–1989** Federation with Senegal.
- ❑ **1994** Jawara ousted in army coup.
- ❑ **1996** Yahya Jammeh wins presidential election.
- ❑ **2000** Military coup foiled.
- ❑ **2001** $2 million antipoverty program launched by government.
- ❑ **2002** Jammeh's party sweeps parliamentary elections.

HEALTH

No welfare state health benefits

1 per 20,000 people — Malaria, tuberculosis, parasitic diseases

Most people have access to basic medicines, but these are no longer free. Advanced medical care in the public sector is limited. An HIV/AIDS awareness campaign was launched in 2002.

SPENDING

GDP/cap. increase

CONSUMPTION AND SPENDING

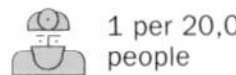

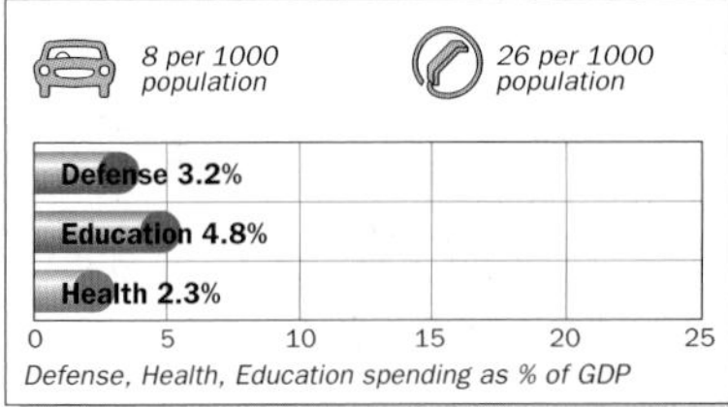
8 per 1000 population
26 per 1000 population
Defense 3.2%
Education 4.8%
Health 2.3%
0
5
10
15
20
25
Defense, Health, Education spending as % of GDP

Public service and the professions have created wealth and some people are comfortably off, but great wealth is not a feature of Gambian life. Unemployed young men in Banjul are regarded as the underclass.

WORLD RANKING

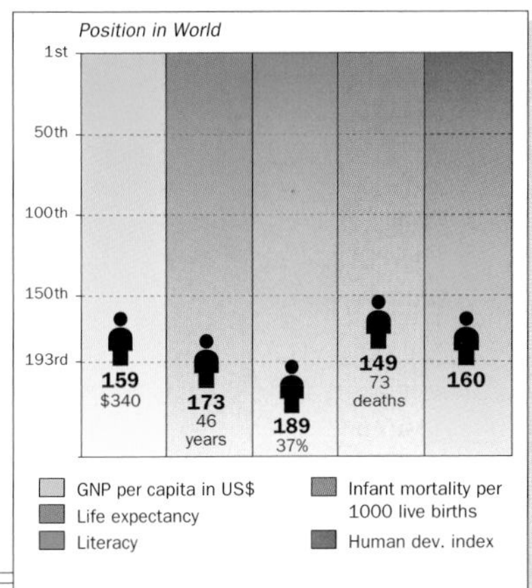
Position in World
1st
50th
100th
150th
193rd
159
$340
173
46 years
189
37%
149
73 deaths
160
GNP per capita in US$
Life expectancy
Literacy
Infant mortality per 1000 live births
Human dev. index

G

GEORGIA

OFFICIAL NAME: Georgia **CAPITAL:** Tbilisi
POPULATION: 5.2 million **CURRENCY:** Lari **OFFICIAL LANGUAGE:** Georgian; Abkhazian (in Abkhazia)

EUROPE

Europe

Russian Federation
GEORGIA
Turkey
Armenia

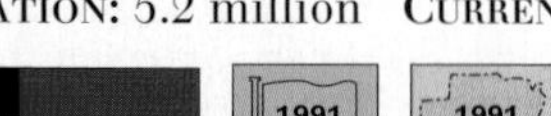
1991

1991

May 26

GE
+4

+995
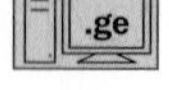
.ge

SANDWICHED BETWEEN the Greater and Lesser Caucasus, Georgia is a mountainous country, with a Black Sea coastline running north–south from Abkhazia to Ajaria. Georgia was one of the first republics to demand independence from the Soviet Union, but has been plagued over recent years by civil war and ethnic disputes in Abkhazia and South Ossetia. Georgia is a primarily agricultural country, and is noted for its wine.

CLIMATE

Mountain/subtropical

WEATHER CHART FOR TBILISI

Average daily temperature | Rainfall
°C/°F J F M A M J J A S O N D cm/in
40/104, 30/86, 20/68, 10/50, 0/32, -10/14, -20/-4
40/16, 30/12, 20/8, 10/4, 0

Georgia's climate is continental inland and subtropical along the coast, where grapes, citrus fruit, and tea are grown.

Tbilisi, Georgia's capital since the 5th century. Its buildings rise in steep terraces from both banks of the Kura River.

TRANSPORTATION

Drive on right

Novo Alexeyevka, Tbilisi 270,011 passengers

118 ships 118,600 grt

THE TRANSPORTATION NETWORK

19,354 km (12,026 miles)

None

1545 km (960 miles)

None

Civil war has seriously disrupted transportation. A new rail route and oil pipeline from Baku to the Black Sea ports of Poti and Supsa was opened in 1999.

TOURISM

Visitors : Population 1:14

384,000 visitors

Up 21% in 1999

MAIN TOURIST ARRIVALS

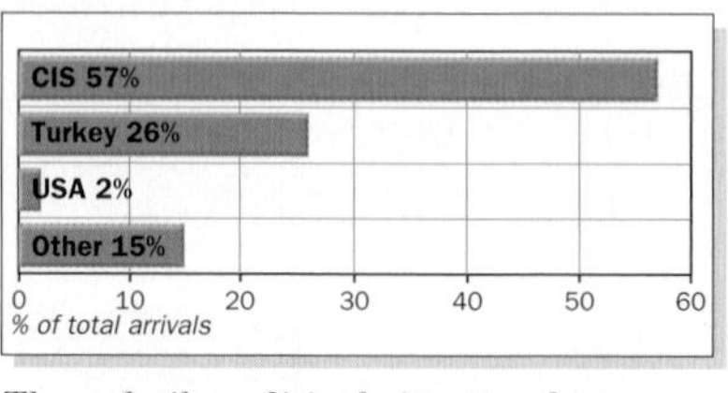
CIS 57%
Turkey 26%
USA 2%
Other 15%
0 10 20 30 40 50 60
% of total arrivals

The volatile political situation has discouraged tourism, but numbers are rising again. Most tourists still come from former Soviet states.

GEORGIA
Total Land Area :
69 700 sq. km
(26 911 sq. miles)
POPULATION
over 1 000 000
over 100 000
over 50 000
over 10 000
under 10 000
LAND HEIGHT
3000m/9843ft
2000m/6562ft
1000m/3281ft
500m/1640ft
200m/656ft
Sea Level
0 50 km
0 50 miles
RUSSIAN FEDERATION
BLACK SEA
CAUCASUS
LESSER CAUCASUS
TURKEY
ARMENIA
AZERBAIJAN
ABKHAZIA
AJARIA
SOUTH OSSETIA
Gagra
Sokhumi
Tqvarch'eli
Mestia
Och'amch'ire
Ch'ikhareshi
Oni
Kazbek 5047m
Qazbegi
Zugdidi
K'ut'aisi
Tqibuli
Chiatura
Tskhinvali
P'ot'i
Samtredia
Gori
Kaspi
Ozurget'i
Khashuri
T'elavi
Qvareli
Bat'umi
Gurjaani
TBILISI
Rust'avi
Marneuli
Akhalts'ikhe
Ninotsminda
K'vemo K'edi
Dedoplistsqaro
Enguri
Rioni
Kura
Aragvi
Alazani
Iori

PEOPLE

Pop. density medium

Georgian, Russian, Abkhazian
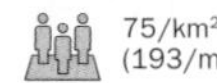
75/km² (193/mi²)

THE URBAN/RURAL POPULATION SPLIT

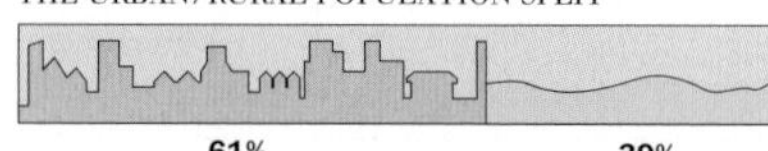
61%
39%

ETHNIC MAKEUP

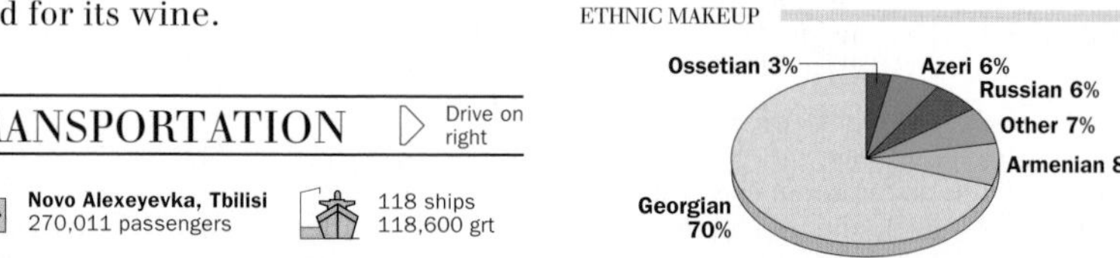
Ossetian 3%
Azeri 6%
Russian 6%
Other 7%
Armenian 8%
Georgian 70%

Georgia is a paternalistic society, with strong family and cultural traditions. Approximately 70% of the population is Georgian, with Armenian, Russian, Azeri, Ossetian, Greek, and Abkhaz minorities. More than 300,000 people were displaced by the violent internal conflicts of the 1990s.

POLITICS

Multiparty elections

1999/2003

President Eduard Shevardnadze

AT THE LAST ELECTION

Parliament 235 seats

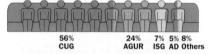

56% CUG | 24% AGUR | 7% ISG | 5% AD | 8% Others

CUG = Citizens' Union of Georgia **AGUR** = All Georgian Union of Revival **ISG** = Industry Will Save Georgia **AD** = Abkhazian Deputies

10 Abkhazian seats were not contested in 1999.

The political situation remains volatile. Eduard Shevardnadze, reelected as president in 2000, has been the target of several assassination attempts. He came to power in 1992 amid civil war with the "Zviadists," supporters of the ex-president Zviad Gamsakhurdia, who committed suicide while under fire in late 1993.

Fighting raged simultaneously in Abkhazia, where ethnic Abkhaz attempted to secede; ethnic Georgians were expelled. Fighting still flares up sporadically. A UN-brokered peace process begun in mid-2000 soon stalled over the basic issue of the future status of Abkhazia.

WORLD AFFAIRS

Joined UN in 1992

BSEC CIS CE OSCE PfP

Relations with Russia are strained over regional security and pipelines. Georgia is a focus of the US "war on terrorism."

AID

Recipient

 $170m (receipts)

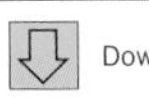 Down 31% in 2000

As well as aid for infrastructure projects, Georgia receives Western support for institutional and financial sector reform.

DEFENSE

Compulsory military service

 $116m

 Up 5% in 2000

The army's main focus remains Abkhazia. However, US concerns over the presence of Islamic terrorists prompted the arrival in 2002 of US forces to train the regular Georgian army; the security service is already CIA-trained. Russian troops, once numbering over 9000, have been steadily withdrawn since 2000.

ECONOMICS

Inflation 389% p.a. (1990–2000)

$3.18bn

 1.960–2.055 lari

SCORE CARD

- ❑ WORLD GNP RANKING........129th
- ❑ GNP PER CAPITA........$630
- ❑ BALANCE OF PAYMENT........–$162m
- ❑ INFLATION........19.1%
- ❑ UNEMPLOYMENT........15%

STRENGTHS

Gateway to West for Azeri oil through pipelines across Georgia to Black Sea and Mediterranean ports. Hyperinflation brought under control in mid-1990s.

WEAKNESSES

War damage and severance of links with other ex-Soviet republics. Large black economy and influential Mafia. Drought and currency crisis in 1998. Serious budget deficit problems. Negative trade balance.

EXPORTS

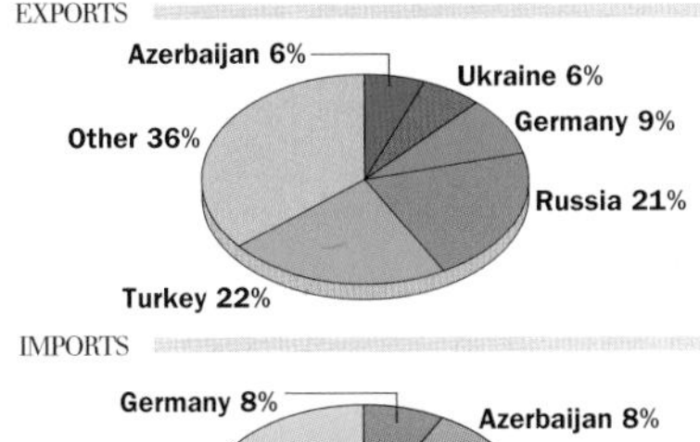
Azerbaijan 6%
Ukraine 6%
Germany 9%
Other 36%
Russia 21%
Turkey 22%
IMPORTS
Germany 8%
Azerbaijan 8%
USA 10%
Other 46%
Russia 13%
Turkey 15%

RESOURCES

Electric power 4.6m kw

 1600 tonnes

 2386 b/d (reserves 37m barrels)

 1.18m cattle, 545,000 sheep, 7.83m chickens

 Manganese, coal, oil, natural gas, zinc, cobalt, vanadium

Known oil reserves are as yet barely developed. Georgia is dependent on Russia for much of its energy supply, although a new US–Georgian oil refinery was opened in eastern Georgia in 1998. Georgia is a predominantly agricultural country, and food processing and wine production continue to be the major industries. Manganese and small quantities of zinc, cobalt, and vanadium are mined.

ENVIRONMENT

Not available

 3%

 1 tonne per capita

Radiation from materials left by departing Russian soldiers is a growing problem, as is Black Sea pollution and the protection of upland pastures.

MEDIA

TV ownership high

 Daily newspaper circulation figures not available

PUBLISHING AND BROADCAST MEDIA

 There are 3 daily newspapers, *Rezonansi*, published in Georgian, *Georgian Messenger* and *Georgian Times* published in English

 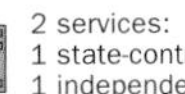 2 services: 1 state-controlled, 1 independent

 1 state-controlled service

Government newspapers are subsidized. The independent TV channel, Rustavi–2, was permitted to reopen in 1998, but has faced harassment.

CRIME

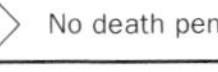 No death penalty

 10,158 prisoners

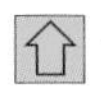 Up 7% 1996–1998

Organized crime under the control of Mafia-style groups has flourished since independence in 1991. The judicial system currently favors Shevardnadze and his supporters.

EDUCATION

School leaving age: 14

 99%

 128,372 students

Since independence, education has stressed Georgian language and history. All levels of education are seriously underfunded. Tbilisi University was formerly of a high standard.

HEALTH

Welfare state health benefits

 1 per 227 people

 Circulatory and respiratory diseases, cancers, accidents

The health system was limited under the control of the Soviet Union. Internal strife and a lack of resources have prevented any recent investment.

CHRONOLOGY

A Russian protectorate from 1763, Georgia was absorbed into the Russian Empire in 1801. It was established as an independent state under a Menshevik socialist government in 1918.

- ❑ **1920** Recognized by Soviet Russia as an independent state.
- ❑ **1921** Soviet Red Army invades. Effectively part of USSR.
- ❑ **1922–1936** Incorporated into Transcaucasian Soviet Federative Socialist Republic (TSFSR).
- ❑ **1989** Pro-independence riots in Tbilisi put down by Soviet troops.
- ❑ **1990** Declares sovereignty.
- ❑ **1991** Independence. Zviad Gamsakhurdia elected president.
- ❑ **1992** Gamsakhurdia flees Tbilisi. Shevardnadze elected chair of Supreme Soviet and State Council.
- ❑ **1992–1993** Abkhazia conflict.
- ❑ **1995** Shevardnadze narrowly survives assassination attempt, subsequently elected president.
- ❑ **1999** Opening of pipeline from Caspian to Black Sea.
- ❑ **2000** Shevardnadze reelected. Russian troop withdrawal begins.

G

SPENDING

GDP/cap. decrease

CONSUMPTION AND SPENDING

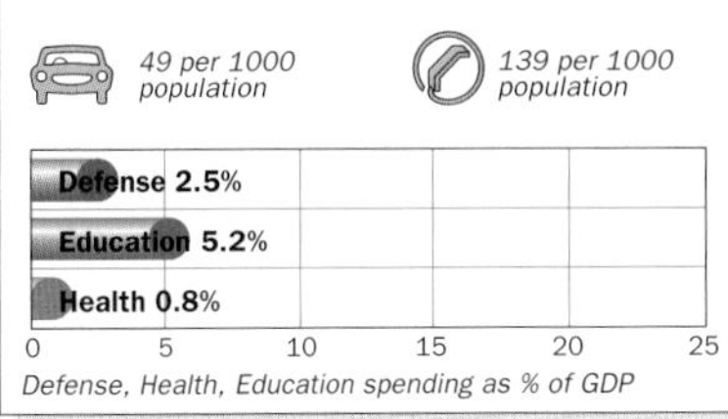
49 per 1000 population
139 per 1000 population
Defense 2.5%
Education 5.2%
Health 0.8%
0
5
10
15
20
25
Defense, Health, Education spending as % of GDP

There is a small wealthy and extravagant elite, but most Georgians live in poverty. Wages and welfare are often in arrears.

WORLD RANKING

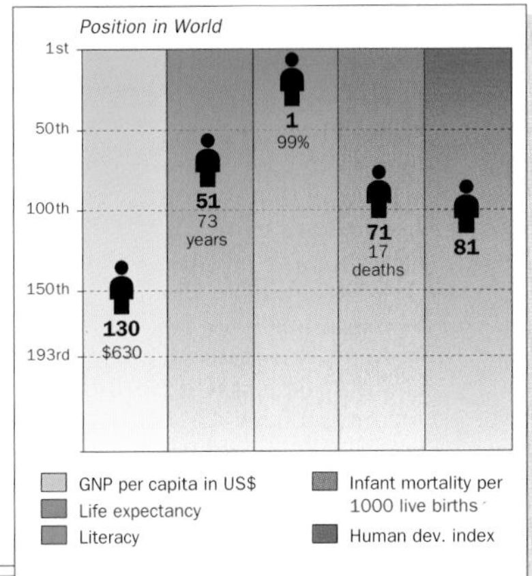
Position in World
1st
50th
100th
150th
193rd
130
$630
51
73
years
1
99%
71
17
deaths
81
GNP per capita in US$
Life expectancy
Literacy
Infant mortality per 1000 live births
Human dev. index

GERMANY

OFFICIAL NAME: Federal Republic of Germany **CAPITAL:** Berlin **POPULATION:** 82 million
CURRENCY: Euro (deutsche mark until 2002) **OFFICIAL LANGUAGE:** German

1871

1990

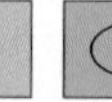
Oct 3

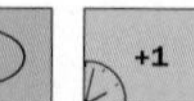
D

+1

+49
.de

WITH COASTLINES on both the Baltic and North Seas, Germany is bordered by nine countries. Plains and rolling hills in the north give way to more mountainous terrain in the south. Europe's foremost industrial power, and its most populous country apart from Russia, Germany is the world's second-biggest exporter. Unified in the 1870s, it was divided after the defeat of the Nazi regime in 1945. The communist-ruled east was part of the Soviet bloc until the collapse of the East German regime in 1989, which paved the way for reunification in 1990. Tensions created by wealth differences between east and west were then exacerbated by record levels of unemployment. The government committed itself to European union and adopted the single currency, the euro, even though the stable deutsche mark had been a symbol of German pride.

CLIMATE

Continental/maritime

WEATHER CHART FOR BERLIN

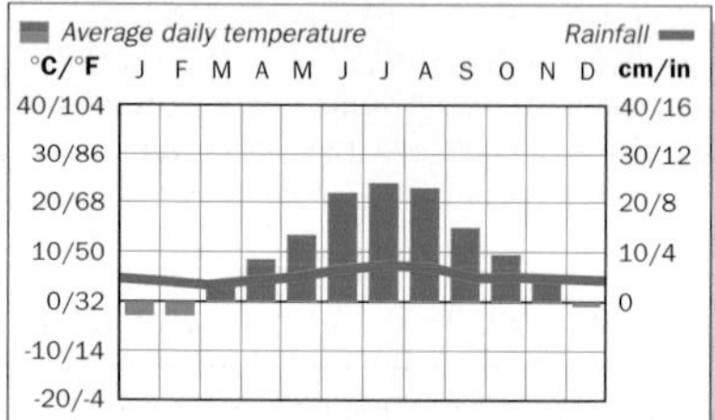
Average daily temperature
Rainfall
°C/°F J F M A M J J A S O N D cm/in
40/104
30/86
20/68
10/50
0/32
-10/14
-20/-4
40/16
30/12
20/8
10/4
0

Germany has a broad climatic range. The upper Rhine valley is very mild and suitable for wine making. The Bavarian Alps, the Harz Mountains, and the Black Forest are by contrast cold, with heavy falls of snow in winter.

TRANSPORTATION

Drive on right

Frankfurt/Main International
49.4m passengers

994 ships
6.55m grt

THE TRANSPORTATION NETWORK

650,891 km
(404,444 miles)

11,515 km
(7155 miles)

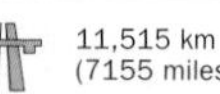

37,559 km
(23,339 miles)

7467 km
(4640 miles)

Germany virtually invented the modern highway with the first *Autobahnen* in the 1930s. These have since become Europe's most elaborate highway network; there are generally no tolls and few speed limits, despite protests from environmentalists. The efficient railroad system has been restructured as a first step toward privatization. Germany's high-speed ICE railroad opened its main north–south routes in 1991 and has expanded greatly since then. Urban transportation systems are highly efficient.

TOURISM

Visitors : Population
1:4.6

17.9m visitors

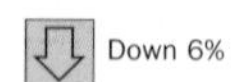

Down 6% in 2001

MAIN TOURIST ARRIVALS

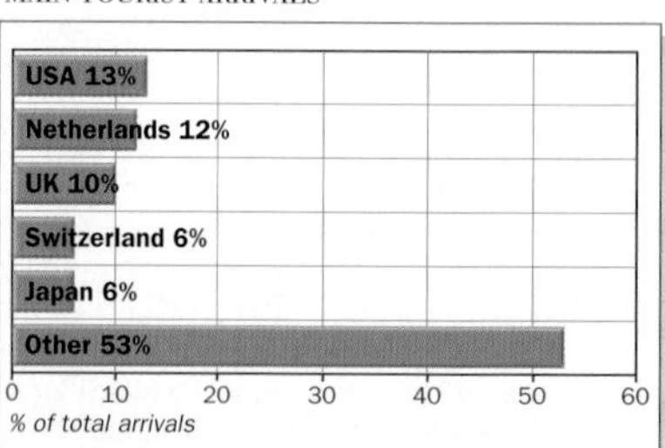
USA 13%
Netherlands 12%
UK 10%
Switzerland 6%
Japan 6%
Other 53%
0 10 20 30 40 50 60
% of total arrivals

Northerly beaches and a colder climate make Germany less of a tourist destination than France or Italy. Skiing in the Bavarian Alps, the historic castles of the Rhine valley, the Black Forest, and Germany's excellent beer are all major attractions. Berlin, even before 1989, drew tourists with its rich cultural life and its Wall separating capitalist West and communist East. Now capital of the reunified Germany, and with a dynamic and vibrant atmosphere, it has undergone massive reconstruction.

***The Stillach Valley, Allgäu Alps,** Bavaria (Bayern). Alarm at acid rain damage to Germany's forests sparked off the rise of the Greens, who joined the government in 1998.*

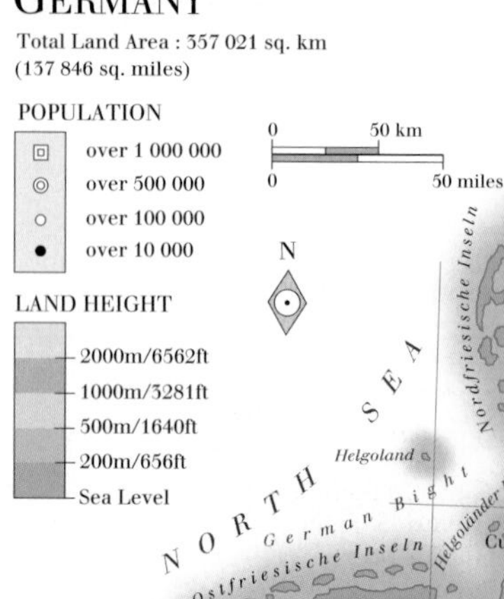
EUROPE
Europe
Netherlands
Denmark
Belgium
GERMANY
Poland
Luxembourg
Czech Rep.
France
Switz.
Austria
GERMANY
Total Land Area : 357 021 sq. km
(137 846 sq. miles)
POPULATION
over 1 000 000
over 500 000
over 100 000
over 10 000
LAND HEIGHT
2000m/6562ft
1000m/3281ft
500m/1640ft
200m/656ft
Sea Level
0 50 km
0 50 miles
N
NORTH SEA
Helgoland
Nordfriesische Inseln
German Bight
Ostfriesische Inseln
Cuxhaven
Wilhelmshaven
Emden
Bremer

NETHERLANDS
Oldenburg
Weser
Ems
NORDDEUTSCHES
NIEDERSA
Osnabrück
Münster
Bielefel
NORDRHEIN-WESTFALEN
Recklinghausen
Herne
Hamm
Gelsenkirchen
Bochum
Oberhausen
Dortmund
Duisburg
Essen
RUHR
Hagen
Mülheim
Wuppertal
Mönchengladbach
Düsseldorf
Solingen
VALLEY
Leverkusen
Köln
Siegen
Aachen
Bonn
Rhine

BELGIUM
LUXEMBOURG
Koblenz
RHEINISCHES
EIFEL
Wiesbaden
Mainz
Frankfurt am Main
Offenbach
Darmstadt
Mosel
RHEINLAND-PFALZ
Trier
Kaiserslautern
SAARLAND
Ludwigshafen
Mannheim
Neckar
Heidelberg
Saarbrücken

Heilbronn
Karlsruhe
Baden-Baden
FRANCE
Offenburg
BLACK FOREST
Freiburg im Breisgau
SWITZERLAND
Friedrichsh

G

PEOPLE

Pop. density high

German, Turkish

235/km² (608/mi²)

THE URBAN/RURAL POPULATION SPLIT

88% 12%

RELIGIOUS PERSUASION

Muslim 3%
Other 30%
Protestant 34%
Roman Catholic 33%

ETHNIC MAKEUP

Turkish 2%
Other European 3%
Other 3%
German 92%

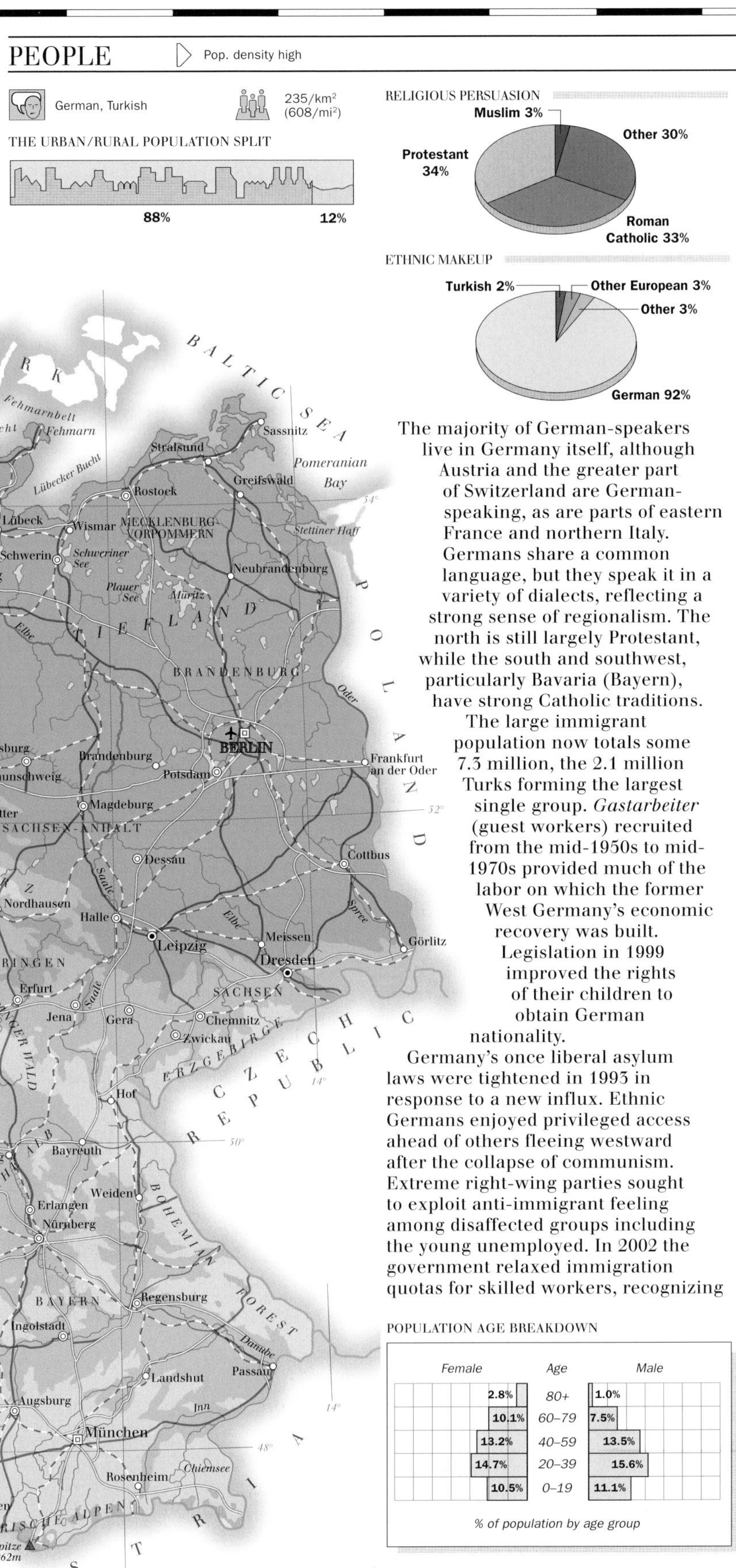
BALTIC SEA
Fehmarnbelt
Fehmarn
Bucht
Lübecker Bucht
Sassnitz
Stralsund
Pomeranian Bay
Greifswald
Rostock
Lübeck
Wismar
MECKLENBURG-VORPOMMERN
Stettiner Haff
Schwerin
Schweriner See
Neubrandenburg
Plauer See
Müritz
TIEFLAND
Elbe
BRANDENBURG
Oder
POLAND
BERLIN
Wolfsburg
Brandenburg
Frankfurt an der Oder
Braunschweig
Potsdam
Salzgitter
Magdeburg
SACHSEN-ANHALT
Dessau
Cottbus
Nordhausen
Saale
Spree
Halle
Elbe
Meissen
Görlitz
Leipzig
Dresden
THÜRINGEN
SACHSEN
Erfurt
Jena
Gera
Chemnitz
Zwickau
ERZGEBIRGE
CZECH REPUBLIC
Hof
Bayreuth
Weiden
Erlangen
Nürnberg
BOHEMIAN FOREST
Regensburg
BAYERN
Ingolstadt
Danube
Passau
Landshut
Augsburg
Inn
München
Rosenheim
Chiemsee
BAYERISCHE ALPEN
Zugspitze 2962m
AUSTRIA

The majority of German-speakers live in Germany itself, although Austria and the greater part of Switzerland are German-speaking, as are parts of eastern France and northern Italy. Germans share a common language, but they speak it in a variety of dialects, reflecting a strong sense of regionalism. The north is still largely Protestant, while the south and southwest, particularly Bavaria (Bayern), have strong Catholic traditions.

The large immigrant population now totals some 7.3 million, the 2.1 million Turks forming the largest single group. *Gastarbeiter* (guest workers) recruited from the mid-1950s to mid-1970s provided much of the labor on which the former West Germany's economic recovery was built. Legislation in 1999 improved the rights of their children to obtain German nationality.

Germany's once liberal asylum laws were tightened in 1993 in response to a new influx. Ethnic Germans enjoyed privileged access ahead of others fleeing westward after the collapse of communism. Extreme right-wing parties sought to exploit anti-immigrant feeling among disaffected groups including the young unemployed. In 2002 the government relaxed immigration quotas for skilled workers, recognizing the need for specific talents, but concurrently toughened the asylum procedure.

Family ties in Germany are little different from those in the US or the UK. Millions of couples live together in common-law arrangements, although this is frowned on by the Roman Catholic Church. In rural districts, notably in Bavaria (Bayern), more traditional habits are still observed. The birthrate is one of Europe's lowest, and the population would have fallen were it not for the influx of immigrants since the 1950s.

Germany has a tradition of strong feminism. Women have full rights under the law and play a bigger role in politics than in most other European countries. Over 30% of Bundestag (Federal Assembly) members elected in 2002 are women, and women ministers occupy several top cabinet posts. From 2001, women were permitted to take on combat roles in the armed forces. However, they are less well represented in top jobs in business and industry. Abortion remains a charged issue. Women in former East Germany once had the right to abortion on demand, but the Constitutional Court, after strong Catholic lobbying, overruled a relatively liberal 1992 compromise law for the whole country. The current regulations, dating from mid-1995, allow abortions (but only after counseling) within three months of conception.

POPULATION AGE BREAKDOWN

Female	Age	Male
2.8%	80+	1.0%
10.1%	60–79	7.5%
13.2%	40–59	13.5%
14.7%	20–39	15.6%
10.5%	0–19	11.1%

% of population by age group

G

CHRONOLOGY

German unification in the 19th century brought together a mosaic of states with a common linguistic, but varied cultural, heritage.

- **1815** German Confederation under nominal Austrian leadership.
- **1834** Zollverein Customs Union of 18 states, including Prussia.
- **1862** Otto von Bismarck appointed Prussian chancellor.
- **1864–1870** Prussia defeats Austrians, Danes, and French; north German states under Prussian control.
- **1871** Southern states join Prussian-led unified German Empire under Wilhelm I.
- **1870s** Rapid industrialization.
- **1890** Kaiser Wilhelm II accedes, with aspirations for German world role. Bismarck sacked.
- **1914–1918** World War I.
- **1918** Germany signs armistice; Weimar Republic created.
- **1919** Treaty of Versailles: colonies lost and reparations paid. Rhineland demilitarized.

⇨

G

CHRONOLOGY *continued*

- **1923** France occupies Ruhr; financial collapse, hyperinflation.
- **1933** Hitler chancellor after Nazis become largest party in elections. One-party rule; rearmament.
- **1935** Nuremberg Laws; official persecution of Jews begins.
- **1936** German entry into Rhineland. Axis alliance with Italy.
- **1938** Annexation of Austria and Sudetenland.
- **1939** Invasion of Poland starts World War II.
- **1940** France, Belgium, the Netherlands, and Norway invaded.
- **1941** USSR invaded.
- **1942–1943** Germans defeated by Red Army at Stalingrad.
- **1945** German surrender; Allies control four occupation zones.
- **1949** Germany divided: communist East led by Walter Ulbricht 1951–1971, Erich Honecker 1971–1989; liberal democratic West led by CDU's Konrad Adenauer, first Chancellor, 1949–1963.
- **1955** West Germany joins NATO.
- **1961** Berlin Wall built.
- **1966–1969** West German "grand coalition" of CDU and SPD.
- **1969–1982** SPD-led West German governments under Willy Brandt (1969–1974) and Helmut Schmidt (1974–1982).
- **1973** Both Germanies join UN.
- **1982** Helmut Kohl West German chancellor, CDU–FDP coalition.
- **1989** Fall of Berlin Wall.
- **1990** Reunification of Germany. First all-German elections since 1933; Kohl heads government.
- **1998** Gerhard Schröder heads coalition of SPD and Greens.
- **2000** Disgrace of Kohl in party funding scandal.
- **2001** Historic Berlin city government coalition of SPD and former communist PDS.
- **2002** Euro fully adopted. SPD–Green coalition reelected.

***The Messeturm**, Frankfurt, is the tallest office building in Europe. Frankfurt is Germany's financial services center, and many of its leading companies are located there.*

POLITICS

Multiparty elections

L. House 2002/2006
U. House varying

President
Johannes Rau

AT THE LAST ELECTION

Federal Assembly 603 seats

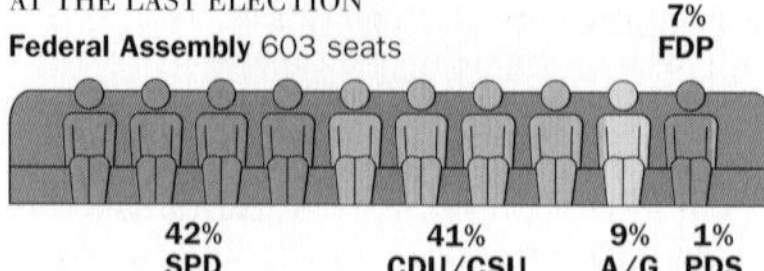

7% **FDP**

42% **SPD** 41% **CDU/CSU** 9% **A/G** 1% **PDS**

SPD = Social Democratic Party of Germany **CDU/CSU** = Christian Democratic Union/Christian Social Union **A/G** = Alliance 90/Greens **FDP** = Free Democratic Party **PDS** = Party of Democratic Socialism

Federal Council 69 seats

Each of the 16 states (*Länder*) is represented by between 3 and 6 members in the Federal Council (Bundesrat), who are appointed after the elections in each *Land*.

Germany is a federal democratic republic of 16 *Länder* (states). The government is led by the federal chancellor, elected by the Bundestag (Federal Assembly). The president's role is largely ceremonial. The "Basic Law" of West Germany, drawn up in 1948, became the 1990 federal constitution of reunified Germany.

Profile

Germany's politics are now strongly democratic, with a long tradition of federative association. Before 1871, Germany was a mass of separate principalities, kingdoms, and city-states, a situation in many ways respected by Bismarck's unification constitution. The 1933–1945 Nazi period, during which the federal system was abolished, was very much a hiatus. The Allies reestablished the federal system in West Germany in 1949; in the east, the *Länder* were restored after reunification in 1990. In many ways, the *Länder* are at the heart of German political life, each with its own elected parliament and largely controlling its finances. By general consensus the system delivers efficient and commercially astute government. There have been few major differences on domestic policy between the postwar ruling coalitions. All parties support the social market economy on which prosperity was built.

Germany has enjoyed stable governments, with coalitions of the center-left and center-right each holding sway since the "grand coalition" of 1966–1969. In 1998 the electorate chose moderate SPD leader Gerhard Schröder in a vote for change, ousting long-serving CDU chancellor Helmut Kohl. In opposition the CDU was beset by a party-funding scandal. Kohl was disgraced. Angela Merkel succeeded him, the first person from the former East Germany to become CDU leader. She accepted that Edmund Stoiber, leader of the CDU's Bavarian-based sister party, the CSU, was the stronger candidate to contest the chancellorship in September 2002. In the poll, Stoiber's challenge fell just short of toppling Schröder's "red–green" coalition.

Main Political Issues

The economy

Recent recession has come as a shock to Germany, used to constant growth since the 1950s. Public spending has been reined in since the mid-1990s to meet targets for European monetary union, despite unrest as unemployment peaked at over four million. A brief return to growth enabled the SPD to start tackling pension reform, but the economy still dominates the political agenda.

East and west

Most Germans supported reunification after the fall of the Berlin Wall in 1989, but feelings soured as the true costs became clear. Over DM 1000 billion have been spent on reconstruction in the east, financed partly by a "solidarity surcharge" on income tax, but the tenth anniversary of reunification in November 2000 was less a celebration than a reflection on past mistakes. The east remains poorer, and people moving west, still seen as "*ossis*," may find it hard to fit in. Support for the former communist PDS is confined almost entirely to the east.

Far-right violence

Unemployment and resentment of "foreigners" led to a rise in support for far-right parties. Foreign workers, particularly Turks, and asylum seekers have been subject to shocking attacks. The problem of racism, even if no worse than in many other European states, is particularly sensitive, given Germany's history.

***Helmut Kohl**, long-serving chancellor until 1998.*

***Gerhard Schröder**, chancellor from 1998, reelected in 2002.*

***Joschka Fischer**, leading Green politician and a popular foreign minister from 1998.*

WORLD AFFAIRS

Joined UN in 1973

EU
G8
NATO
OECD
OSCE

During the Cold War, a divided Germany was forced to play a subservient role in international affairs, the West closely adhering to US interests and the East taking its orders directly from the USSR. Since reunification in 1990, the emphasis has changed, and Germany has begun to voice a foreign policy which reflects its position as the most powerful country in Europe. Still firmly Western-oriented, Germany has championed European integration and has considerable economic power over the former Soviet bloc countries through large-scale investment there.

2001 proved to be a seminal year for German foreign policy. Germany was given command of the NATO peacekeeping mission in Macedonia and German troops were involved in the "war on terrorism" in Afghanistan.

AID

Donor

$5.03bn (donations)

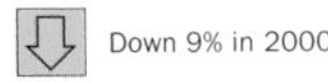
Down 9% in 2000

Unlike the US, the UK, and France, Germany's aid programs are not directly motivated by its desire for political influence in the world's poorer regions. Most are multilateral, although there is also a strong tradition of direct aid. Much comes from Church organizations such as the Protestant Brot für die Welt. Many German volunteers and missionaries work overseas on aid programs.

DEFENSE

Compulsory military service

$28.2bn

Down 9% in 2000

The large German army, designed for national defense, is being slimmed down to focus more on mobility and providing support to allied states. The Constitutional Court ruled in 1994 that army units could take part in collective defense activities abroad, opening the way for German participation in NATO operations; the 1999 action against Serbia was a landmark.

GERMAN ARMED FORCES

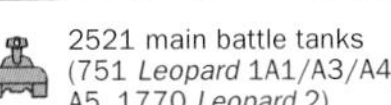
2521 main battle tanks (751 *Leopard* 1A1/A3/A4/A5, 1770 *Leopard* 2) — 211,800 personnel

14 submarines, 2 destroyers, 12 frigates, and 28 patrol boats — 26,050 personnel

434 combat aircraft (131 F-4, 267 *Tornado*, 23 MiG-29) — 70,550 personnel

None

ECONOMICS

Inflation 2% p.a. (1990–2000)

$2064bn

1.0651–1.1231 euros

SCORE CARD

❑ World GNP Ranking	3rd
❑ GNP per Capita	$25,120
❑ Balance of Payments	–$18.7bn
❑ Inflation	1.9%
❑ Unemployment	11%

EXPORTS

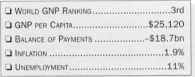

Netherlands 6%, Italy 8%, UK 8%, USA 10%, France 11%, Other 57%

IMPORTS

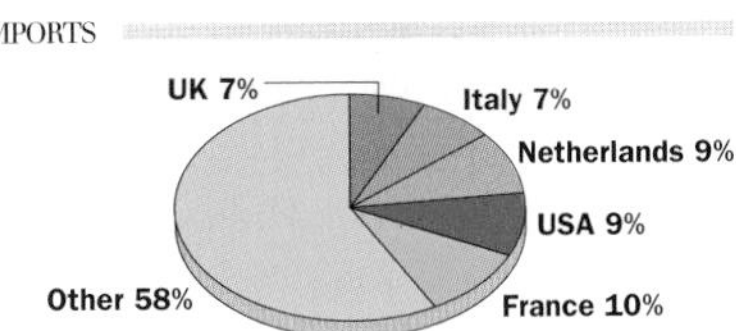
UK 7%
Italy 7%
Netherlands 9%
USA 9%
France 10%
Other 58%

Strengths

Europe's major industrial power. Efficient industry benefits from low inflation environment. Workers and managers live up to reputation for thoroughness and hard work. Strong sectors are cars, heavy engineering, electronics, and chemicals.

Weaknesses

Underestimation of costs of updating inefficient eastern German economy. High welfare costs; pension obligations (despite reforms in 2001) with an aging population. Rising unemployment. Competition from efficient Asian economies. Relatively few small firms, a short working week in terms of hours, poorly developed service sector.

ECONOMIC PERFORMANCE INDICATOR

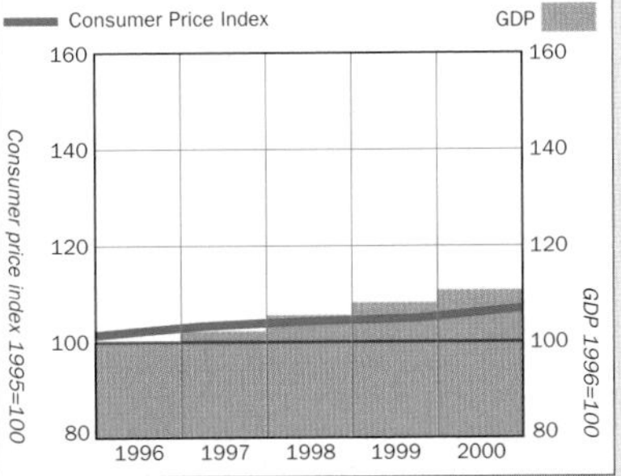
Consumer Price Index
GDP
160
140
120
100
80
1996
1997
1998
1999
2000
Consumer price index 1995=100
GDP 1996=100

Profile

West Germany's remarkable postwar recovery, to become the world's third-strongest economy, was based on the concept of a social market economy, under which the state provided welfare and ensured workers' rights, while the economy was largely in private hands. Major banks and businesses are privately owned, except for the partly state-owned Volkswagen. After reunification in 1990, there was massive investment in the former East Germany, where state concerns were sold off.

Germany was one of the 12 EU states fully to adopt the euro in 2002. The SPD-led government elected in 1998 promised to tackle unemployment and to maintain growth, which reached a ten-year high of 3.1% in 2000. The $46 billion sale of "third generation" mobile phone operators' licenses reduced the budget deficit in 2001. Major tax reforms aimed to balance the budget by 2006. However, global economic slowdown brought Germany into recession at the end of 2001, with unemployment above 10% (meaning over four million out of work) and rising once again.

GERMANY : MAJOR BUSINESSES

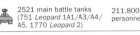

Kiel
Hamburg
Berlin
Ruhr Valley
Dresden
Nürnberg
Frankfurt am Main
München
Stuttgart
Optics
Shipbuilding
Chemicals
Electronics
Engineering
Research & development
Vehicle assembly
0 200 km
0 200 miles

Potsdamer Platz, Berlin, was rapidly reconstructed after reunification and is once again the commercial center of the capital.

G

RESOURCES

Electric power 115m kw

 312,492 tonnes

 58,827 b/d (reserves 215m barrels)

 25.8m pigs, 14.6m cattle, 2.14m sheep, 108m chickens

 Coal, oil, natural gas, copper, salt, potash, tin, nickel

ELECTRICITY GENERATION

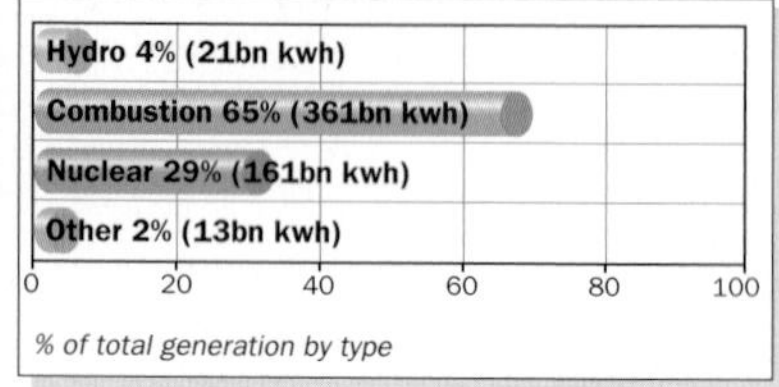
Hydro 4% (21bn kwh)
Combustion 65% (361bn kwh)
Nuclear 29% (161bn kwh)
Other 2% (13bn kwh)
0
20
40
60
80
100
% of total generation by type

With relatively few natural resources, Germany imports over 50% of its energy needs, mainly oil and gas. Coal, the basis of industrialization, now accounts for under a quarter of energy consumption. West Germany invested less heavily than France in nuclear power, and Soviet-built plants in the east have been shut down. The "red–green" coalition government decided in mid-2000 to phase out nuclear power (which provides over 25% of electricity). Renewable resources account for 2% of primary energy consumption, but 6% of electricity production (with a target of 50% by 2050); Germany is the world's leading user of wind power.

GERMANY : LAND USE

Cropland
Forest
Pasture
Vineyards
Pigs
Cattle

0 200 km
0 200 miles

NORDDEUTSCHES TIEFLAND
Elbe
Rhine
BLACK FOREST

ENVIRONMENT

Sustainability rank: 50th

27%

10.1 tonnes per capita

ENVIRONMENTAL TREATIES

Yes
Yes
Yes
Yes
Yes
Yes

Germans are among the world's most environmentally conscious people. Environmental campaigns led by the Green Party, which emerged as a powerful political force in the 1980s, have influenced the policies of all major parties. At national level, the Greens are a significant force in the Bundestag; they joined the SPD-led federal government coalition in 1998, and are strongly represented in *Land* parliaments and local councils.

Germany has some of the strictest pollution controls in the world, with ambitious targets for reducing carbon dioxide emissions, compelling businesses to become more energy-efficient. Germans recycle half of their waste paper, and three-quarters of their used tires and glass.

The nuclear debate has been vigorously fought and won by the Greens; a gradual program of closing existing nuclear power plants was approved in 2001, although waste disposal is still an issue. Fears in the 1980s that up to 50% of trees were sick or dying because of car fumes and industrial pollution led to Germany becoming the first European country to insist that new cars be fitted with catalytic converters. The east had the highest per capita rate of sulfur emissions in the world, but these have been reduced by the closure of industrial plants and the elimination of the noxious Trabant cars.

MEDIA

TV ownership high

 Daily newspaper circulation 300 per 1000 people

PUBLISHING AND BROADCAST MEDIA

 There are 375 daily newspapers, including the *Frankfurter Allgemeine Zeitung*, the *Süddeutsche Zeitung*, and *Die Welt*

 3 public service and several independent commercial channels

 13 public service and several independent networks

TV is supervised by the political parties to ensure a balance of views. Satellite and cable TV have taken much of the audience once shared between the main public service channels, ARD and ZDF. Media conglomerates such as Kirch and Bertelsmann have expanded abroad. Newspapers are mostly regional and serious. An exception is *Bild Zeitung*, a right-wing, sensationalist tabloid, which sells 4.5 million copies daily.

***Neuschwanstein Castle**, Bavaria (Bayern), one of Germany's major tourist attractions. It was built for the eccentric King Ludwig II.*

EDUCATION

 School leaving age: 18

 99%

 1.77m students

THE EDUCATION SYSTEM

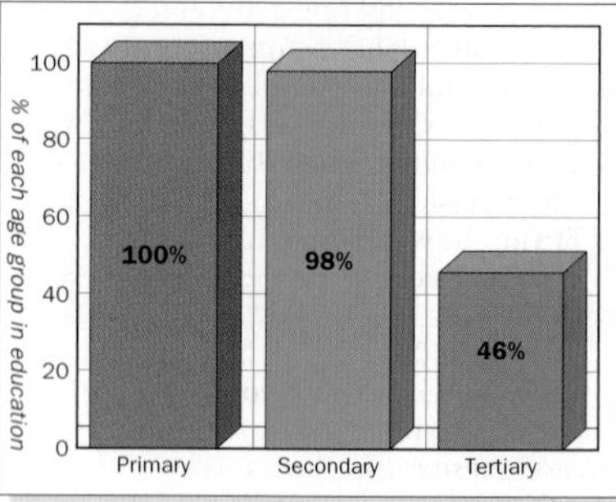
100
80
60
40
20
0
% of each age group in education
100%
98%
46%
Primary
Secondary
Tertiary

Nearly one-tenth of total government expenditure goes on education, which is run by the *Länder*. They coordinate teaching policies, but have autonomy within their borders. The German approach to education stresses academic and vocational achievement. Sporting or cultural activities tend to be organized informally. Nearly all schools have Internet access.

Young people wanting to leave school must continue studying at least part-time until 18. Those who wish to go to university attend the upper-secondary *Gymnasien* to prepare for the *Abitur* exam. Students were taking an average of seven years to complete degrees, until new legislation added shorter bachelor's and master's degrees as in other countries. Research is done as much by major companies as by the universities.

CRIME

No death penalty

 79,348 prisoners

 Down 2% in 1999

CRIME RATES

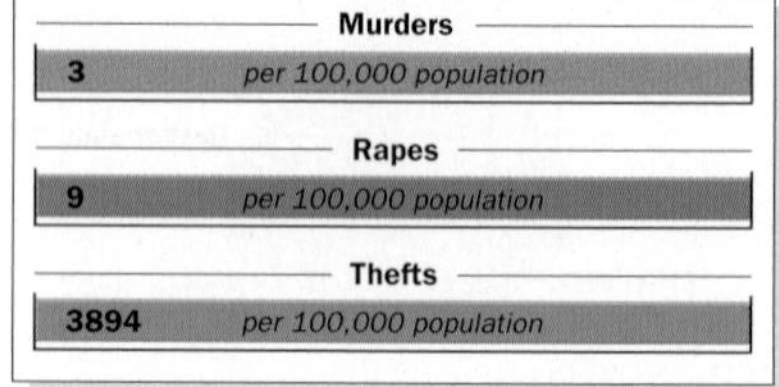
Murders
3 per 100,000 population
Rapes
9 per 100,000 population
Thefts
3894 per 100,000 population

Crime rates in Germany are lower than in most other European countries. This is largely the result of a genuine respect for the law, coupled with a strong police force. Recently, however, higher unemployment has led to an increase in petty theft and a wave of violence, notably against immigrants.

German politicians, once with an enviably clean reputation, have suffered several corruption scandals. Civil service corruption remains rare. People convicted under environmental laws can face ten-year jail sentences.

G

BERLIN, REUNIFICATION, AND CENTRAL EUROPE

THE SO-CALLED "BONN REPUBLIC," created in West Germany in the postwar period, ended symbolically in September 1999. Bonn could justifiably claim to have been home for some 50 years to Germany's only enduringly successful parliamentary democracy. The decision to move the capital to Berlin, however, symbolizing German unity, was enshrined in the 1990 unification treaty only months after the fall of the Berlin Wall in November 1989. It took time to confirm that both government and parliament would be transferred – the lower house of parliament, the Bundestag, only voted in 1994 to move – but five years after that vote, in the same month that Federal Chancellor Schröder moved into his office in Berlin, the Bundestag held its first full session there.

***The dramatic glass dome** of the new Reichstag building in Berlin.*

PROSPECTS FOR BERLIN

The parliament's new home is the former Reichstag building, impressively redesigned by English architect Norman Foster. Topped with a transparent dome and lit up at night from within, it epitomizes the emphasis on architectural and engineering achievements in modern Berlin. New government, commercial, and tourist facilities have brought a lengthy construction boom in Berlin, which with a population of some 3.5 million is Germany's largest city. The proliferation of cultural and artistic activity has also contributed to a strong sense of excitement, boosting the city's image with the international media and the public. Berlin nevertheless faces many problems, including an unemployment rate well above the national average and the need to revitalize its declining industrial base. The outward movement of business and population, from central urban areas to surrounding regions, has been an established trend for decades in major western German cities, but is now happening much more rapidly in Berlin, whose western half was for years an enclave within East Germany. The city also has a long way to go in attracting major companies to make their headquarters there; only 12 of the largest German companies are located in Berlin, compared with over 40 each in Hamburg, Munich, and Frankfurt.

COSTLY REUNIFICATION

The costs of reconstruction of the former East Germany far outweighed initial expectations. Achievements stand out in telecommunications and rail transportation, but most infrastructure in the east is still well below the standard of the west, despite subsidies amounting to the transfer to the east of some 7% of GDP of western *Länder* per year for a decade. Unemployment is higher, labor productivity lower, and living conditions less attractive to the majority of Germans.

GERMANY'S NEW CENTER OF GRAVITY

The transfer of the capital reinforces the shift in Germany's center of gravity brought about by reunification. Coinciding with the collapse of communism across the whole former Soviet bloc came a revival of interest in Germany's role in central Europe or *Mitteleuropa*. In the former communist countries German economic influence is now particularly strong. West Germany was firmly anchored in western Europe, in economic terms by the European Economic Community (now the EU), and in a political–military sense by its membership of the NATO alliance. The imminent eastward expansion of the EU, however, suggests the emergence of France–Germany–Poland as a new and powerful axis in the Europe of the 21st century.

Arbeit!

***Unemployed** Germans demonstrate for the right to work.*

HEALTH

Welfare state health benefits

1 per 286 people

Heart and cerebrovascular diseases, cancers, accidents

The German social security system, pioneered by Bismarck, is one of the most comprehensive in the world. Health insurance is compulsory, and employer and employee contributions are high. Although most hospitals are run by the *Länder*, some are still owned by Germany's wealthy churches. Almost one-third of health spending is now private.

Germans are increasingly health-conscious, paying great attention to diet. Nearly a million people go on cures every year to the country's 200-plus spas. In the east there is a higher incidence of lung diseases, the legacy of industrial pollution.

SPENDING

GDP/cap. increase

CONSUMPTION AND SPENDING

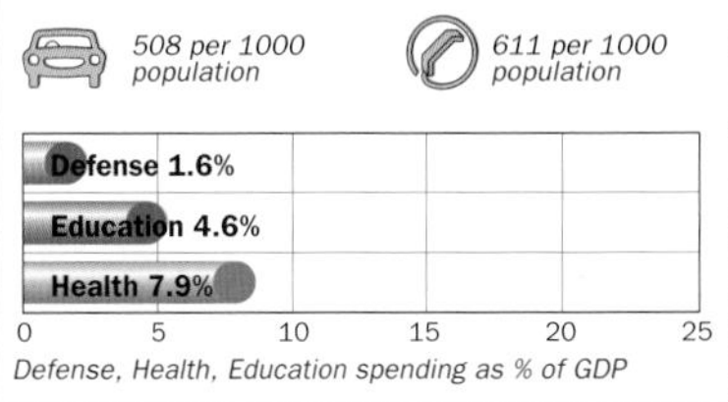
508 per 1000 population
611 per 1000 population
Defense 1.6%
Education 4.6%
Health 7.9%
0
5
10
15
20
25
Defense, Health, Education spending as % of GDP

The effects of the Nazi period, which discredited many of the ruling class, and the destruction of the property of millions of families in the war, explain the relatively classless nature of society. Status is now more closely linked to wealth than to birth. In the west, there are fewer disparities than in most of Europe; workers are generally well paid and social security is generous. Wages in the east, however, are 10% below western rates, and a disproportionate number of unemployed live on welfare benefit. Most Germans own a mobile phone, and almost a third of them had accessed the Internet by 2001.

WORLD RANKING

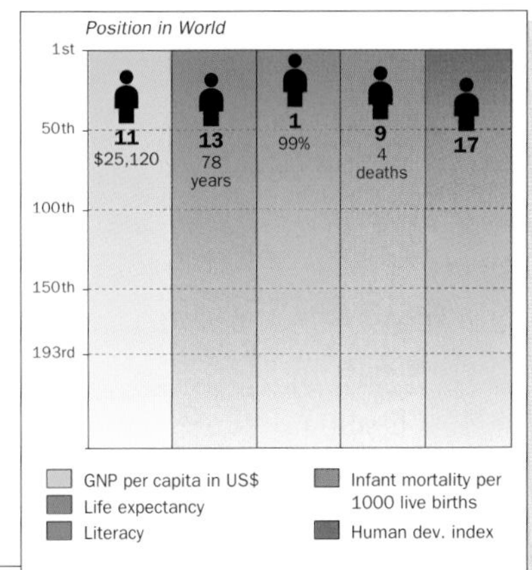
Position in World
1st
50th
100th
150th
193rd
11
$25,120
13
78 years
1
99%
9
4 deaths
17
GNP per capita in US$
Life expectancy
Literacy
Infant mortality per 1000 live births
Human dev. index

G

GHANA

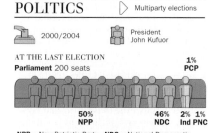

OFFICIAL NAME: Republic of Ghana **CAPITAL:** Accra
POPULATION: 19.7 million **CURRENCY:** Cedi **OFFICIAL LANGUAGE:** English

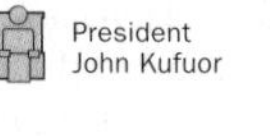
WEST AFRICA
Burkina
Togo
Ivory Coast
Africa

1957

1957

March 6
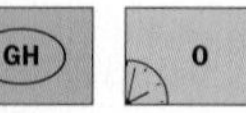
GH
0
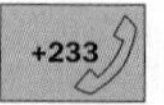
+233

.gh

THE HEARTLAND OF THE ancient Ashanti kingdom, modern Ghana is a union of the former British colony of the Gold Coast and the British-administered part of the UN Trust Territory of Togoland. Ghana gained independence in 1957, the first British colony to do so. Multiparty democracy was embraced in 1992, and the handover of power to the main opposition party in 2000 confirmed the shift away from Ghana's recent history of intermittent military rule.

CLIMATE

Tropical wet & dry/ equatorial

WEATHER CHART FOR ACCRA

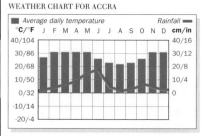

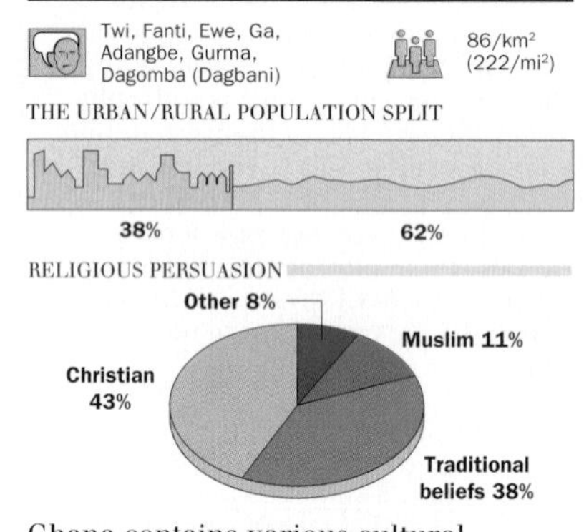
Average daily temperature
Rainfall
°C/°F J F M A M J J A S O N D cm/in
40/104 40/16
30/86 30/12
20/68 20/8
10/50 10/4
0/32 0
-10/14
-20/-4

Southern Ghana has two rainy seasons: from April to July and September to November. The drier north has just one, from April to September.

TRANSPORTATION

Drive on right

Kotoka International, Accra 673,504 passengers

210 ships 118,700 grt

THE TRANSPORTATION NETWORK

11,653 km (7241 miles)

30 km (19 miles)

1300 km (808 miles)

168 km (104 miles)

In 1983, work began to restore Ghana's roads, which had fallen into disrepair in the 1960s and 1970s; the network is now improving.

TOURISM

Visitors : Population 1:53

373,000 visitors

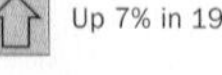
Up 7% in 1999

MAIN TOURIST ARRIVALS

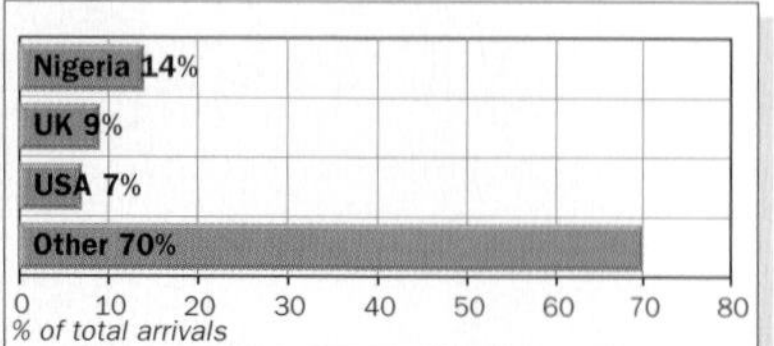
Nigeria 14%
UK 9%
USA 7%
Other 70%
0 10 20 30 40 50 60 70 80
% of total arrivals

Tourism is still small-scale; most visitors come from the rest of Africa, the UK, and the US. Good beaches and old coastal forts are major attractions.

PEOPLE

Pop. density medium

Twi, Fanti, Ewe, Ga, Adangbe, Gurma, Dagomba (Dagbani)

86/km² (222/mi²)

THE URBAN/RURAL POPULATION SPLIT

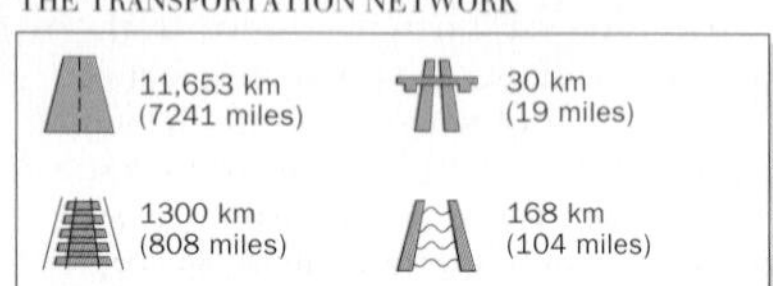
38%
62%

RELIGIOUS PERSUASION

Other 8%
Muslim 11%
Christian 43%
Traditional beliefs 38%

Ghana contains various cultural-linguistic groups. The largest is the Akan, who include the Ashanti and Fanti peoples. Other important groups are the Mole-Dagbani in the north, Ga-Adangbe around Accra, and Ewe in the southeast. Tribal tensions are rare, but can flare up, most recently in 2002.

POLITICS

Multiparty elections

2000/2004

President John Kufuor

AT THE LAST ELECTION

Parliament 200 seats

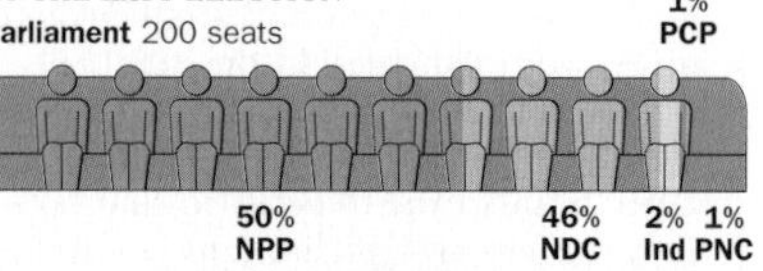
1% PCP
50% NPP
46% NDC
2% Ind
1% PNC

NPP = New Patriotic Party **NDC** = National Democratic Congress **Ind** = Independents **PNC** = People's National Convention **PCP** = People's Convention Party

Ghana's return to multiparty rule in 1992 marked the legitimization of the military government of Jerry Rawlings. An air force flight-lieutenant of Ewe-Scottish descent and one of the great survivors of African politics, Rawlings staged coups in 1979 and 1981, and led the 1981–1992 Provisional National Defense Council (PNDC) military government. As the NDC candidate, Rawlings won 58% of the vote in the 1992 presidential election. Opposition parties boycotted the following parliamentary elections, which the NDC won easily. Elections in 1996 gave Rawlings a further and final term of office. In December 2000 the opposition NPP gained a historic victory when it stripped the NDC of its parliamentary majority and NPP candidate John Kufuor won the presidency.

WORLD AFFAIRS

Joined UN in 1957

Comm | ECOWAS | G24 | IAEA | AU

Good relations with the West, which provides the bulk of Ghana's military and development aid, are a priority. Ghana has played a significant part in UN peacekeeping operations. It was also the main contributor, after Nigeria, to the ECOWAS forces (ECOMOG) in war-torn Liberia from 1990 to 1997. Ghana maintains good relations with its French-speaking neighbors, despite periods of strain with Togo.

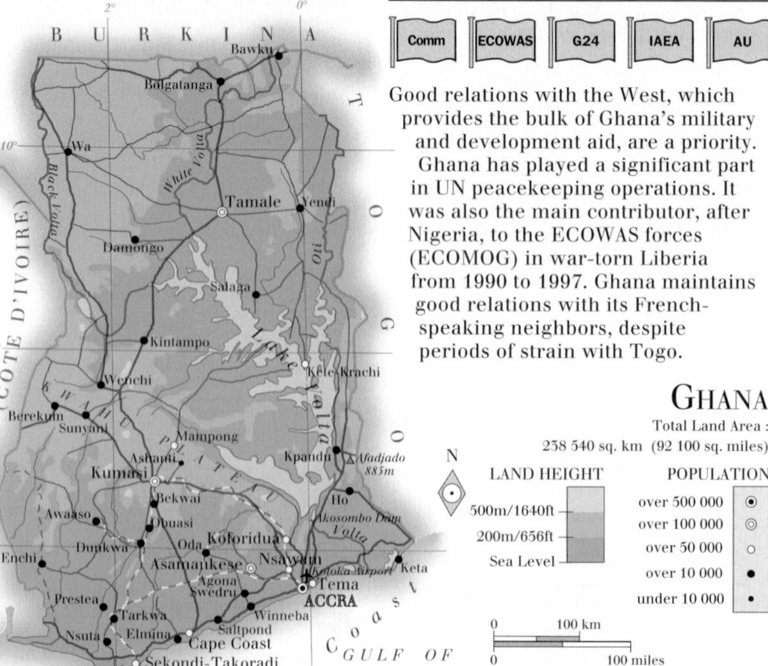
BURKINA
TOGO
IVORY COAST (CÔTE D'IVOIRE)
Gold Coast
GULF OF GUINEA
ATLANTIC OCEAN
Bawku
Bolgatanga
Wa
White Volta
Black Volta
Tamale
Yendi
Damongo
Oti
Salaga
Kintampo
Lake Volta
Kete-Krachi
Wenchi
Berekum
Sunyani
Mampong
Ashanti
Kumasi
Kpandu
Afadjado 885m
Ho
Akosombo Dam
Volta
Bekwai
Awaso
Obuasi
Dunkwa
Oda
Koforidua
Enchi
Asamankese
Nsawam
Agona Swedru
Kotoka Airport
Keta
Tema
ACCRA
Prestea
Tarkwa
Winneba
Nsuta
Elmina
Saltpond
Cape Coast
Sekondi-Takoradi
Axim
Dixcove
N
GHANA
Total Land Area : 238 540 sq. km (92 100 sq. miles)
LAND HEIGHT
500m/1640ft
200m/656ft
Sea Level
POPULATION
over 500 000
over 100 000
over 50 000
over 10 000
under 10 000
0 100 km
0 100 miles

AID

 Recipient

 $609m (receipts)  Little change in 2000

Ghana is the third-largest recipient of World Bank aid (after India and China), supporting a largely successful economic recovery program which began in 1983. The most recent World Bank/IMF debt relief program started in April 2001.

DEFENSE

No compulsory military service

$95m — Down 21% in 2000

In 1966, 1972, 1979, and 1981, the military mounted successful coups, and there have also been several unsuccessful coups. Outside Ghana, the 5000-strong army has been deployed mainly in UN and ECOWAS operations. Ghana's navy is small, with four patrol boats. The air force has 19 combat aircraft.

ECONOMICS

Inflation 27% p.a. (1990–2000)

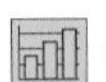 $6.59bn  7275–7400 cedis

SCORE CARD

- ❏ World GNP Ranking 101st
- ❏ GNP per Capita $340
- ❏ Balance of Payments .... –$413m
- ❏ Inflation 25.2%
- ❏ Unemployment 20%

Strengths

GNP rose by 5% a year throughout the 1990s, following economic recovery policies begun in 1983. Second-largest gold producer in Africa. Since 1996, the Ashanti Goldfields Co. Ltd. has expanded into a multinational active in 12 African countries. Cocoa production accounts for 13% of world total.

Weaknesses

High budget deficits and debt repayments; the cedi was devalued in 1983 and has since tended to float downward. Foreign investors generally invest solely in gold mining. High inflation levels.

EXPORTS

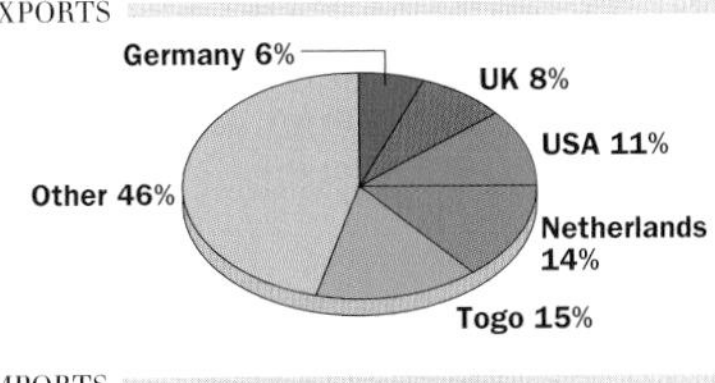
Germany 6%
UK 8%
USA 11%
Netherlands 14%
Togo 15%
Other 46%

IMPORTS

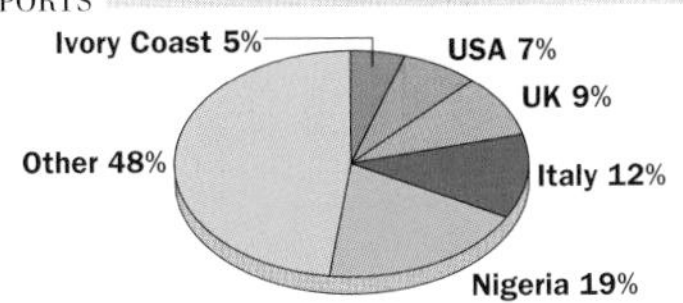
Ivory Coast 5%
USA 7%
UK 9%
Italy 12%
Nigeria 19%
Other 48%

Dixcove harbor, close to Ghana's most southerly cape. The majority of Ghanaians lead a traditional subsistence existence.

RESOURCES

Electric power 1.2m kw

 493,206 tonnes  Reserves of 500,000 barrels

 3.08m goats, 2.74m sheep, 1.3m cattle, 20.5m chickens  Gold, diamonds, bauxite, manganese

Gold production has expanded strongly since the mid-1980s; by 1993, gold had overtaken cocoa as the major export. Diamonds, bauxite, and manganese are also exported. Hydropower from the Volta Dam is exported to Togo and Benin, but is hit by periodic droughts.

ENVIRONMENT

Sustainability rank: 65th

 5%  0.2 tonnes per capita

Cutting of wood for fuel, timber, and farming has destroyed 70% of forests. Devastation caused by mining is now being tackled under a World Bank project.

MEDIA

TV ownership medium

 Daily newspaper circulation 14 per 1000 people

PUBLISHING AND BROADCAST MEDIA

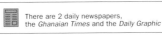

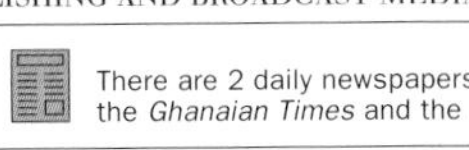 There are 2 daily newspapers, the *Ghanaian Times* and the *Daily Graphic*

 1 state-controlled service  1 state-controlled service

New independent weeklies reflect the increase in private press ownership. Radio and TV tend to follow government reporting guidelines.

CRIME

Death penalty in use

 10,444 prisoners  Down 20% 1997–1999

The judiciary has little independence and the government often resorts to ad hoc "people's tribunals." Corruption is now less of a problem.

EDUCATION

School leaving age: 15

 72%  38,305 students

All sectors of the education system are oversubscribed. There are a few high-quality boarding schools and four universities.

CHRONOLOGY

In 1874 Kumasi, capital of the Ashanti kingdom, was sacked by a British force to create the Gold Coast colony.

- ❏ **1957** Independence under Kwame Nkrumah.
- ❏ **1964** Single-party state.
- ❏ **1966** Army coup.
- ❏ **1972–1979** "Kleptocracy" of Gen. Acheampong. Executed 1979.
- ❏ **1979** Flt. Lt. Jerry Rawlings leads coup. Civilian Hilla Limann wins elections.
- ❏ **1981** Rawlings takes power again.
- ❏ **1992, 1996** Rawlings and NDC win multiparty elections.
- ❏ **2000** Opposition NPP wins elections; John Kufuor wins presidency.

G

HEALTH

No welfare state health benefits

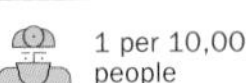 1 per 10,000 people  Malaria, diarrheal diseases, tuberculosis

The health of most of the population has benefited more from improvements in public hygiene than in medical care. Private health care is available.

SPENDING

GDP/cap. increase

CONSUMPTION AND SPENDING

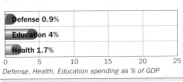

5 per 1000 population — *12 per 1000 population*

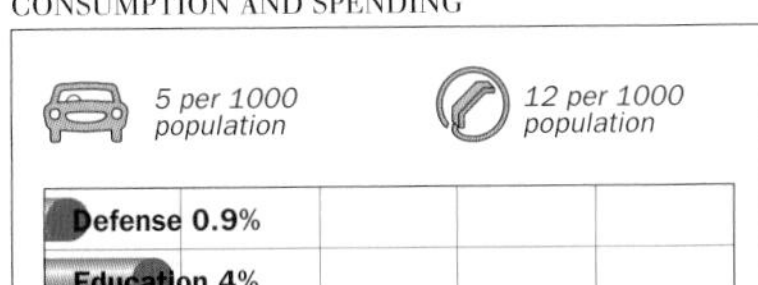
Defense 0.9%
Education 4%
Health 1.7%
0
5
10
15
20
25

Defense, Health, Education spending as % of GDP

Political uncertainty brought few opportunities for advancement, and many Ghanaians emigrated, but the situation is now improving. The main economic disparity is still between the poorer rural north and the richer, more urban, south.

WORLD RANKING

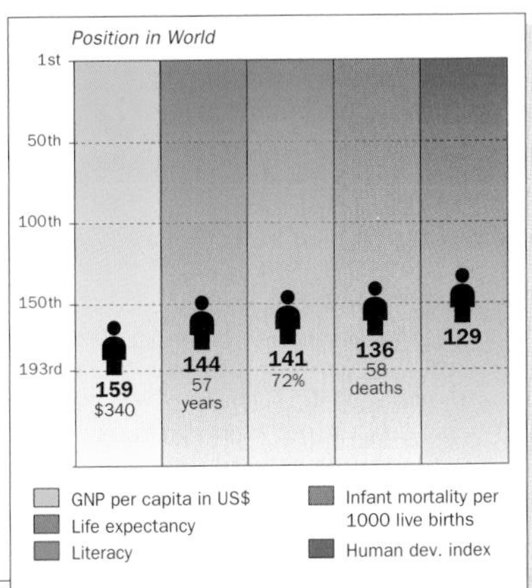
Position in World
1st
50th
100th
150th
193rd
159
$340
144
57 years
141
72%
136
58 deaths
129
GNP per capita in US$
Life expectancy
Literacy
Infant mortality per 1000 live births
Human dev. index

GREECE

EUROPE

OFFICIAL NAME: Hellenic Republic **CAPITAL:** Athens **POPULATION:** 10.6 million
CURRENCY: Euro (drachma until 2002) **OFFICIAL LANGUAGE:** Greek

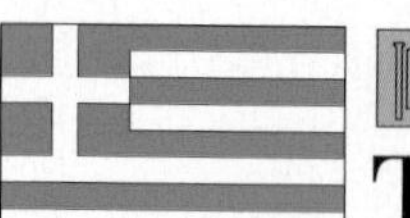

1829

1947

March 25

GR

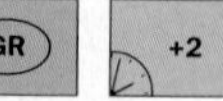
+2

+30

.gr

THE SOUTHERNMOST COUNTRY of the Balkans, Greece is embraced by the Aegean, Ionian, and Cretan Seas. Its mainly mountainous territory includes more than 2000 islands. Only one-third of the land is cultivated. There is a strong seafaring tradition, and some of the world's biggest shipowners are Greek. Greece is rich in minerals – including chromium, whose occurrence is rare. Relations with Turkey, marked by conflict and territorial disputes, have improved in recent years. To the north, however, upheavals in Albania and the conflicts in former Yugoslavia have made for greater instability.

CLIMATE

Mediterranean

WEATHER CHART FOR ATHENS

Average daily temperature / Rainfall

The climate varies from region to region. The northwest is alpine, while parts of Crete border on the subtropical. The large central plain experiences high summer temperatures. Water is a problem, particularly on the many barren Aegean islands.

TRANSPORTATION

Drive on right

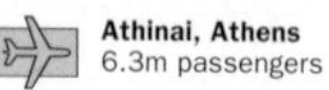
Athinai, Athens
6.3m passengers

1529 ships
26.4m grt

THE TRANSPORTATION NETWORK

107,406 km (66,739 miles)

470 km (292 miles)

2299 km
(1429 miles)

80 km
(50 miles)

The easiest and cheapest method of transportation between the islands and the mainland is by boat or hovercraft. A major ferry disaster in 2000 prompted government moves to improve standards. Greece has 444 ports, of which Piraeus is the main one, and 123 are large enough to handle passenger or freight traffic. A new airport at Spata, 30 km east of Athens, opened in 2001. Greece has a good, if increasingly congested, road network. Two expressway routes are nearing completion, with the help of EU funds, as is the upgrading of the Athens metro. An interurban bus system and a fleet of air-conditioned tourist Pullmans offer an extensive service.

TOURISM

Visitors : Population 1.2:1

12.5m visitors

Up 3% in 2000

MAIN TOURIST ARRIVALS

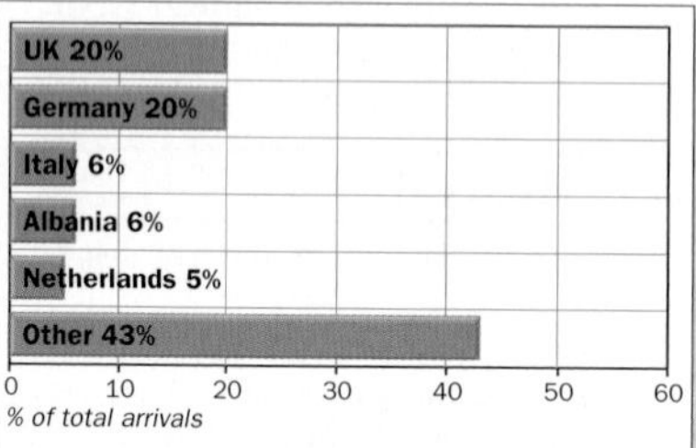
UK 20%
Germany 20%
Italy 6%
Albania 6%
Netherlands 5%
Other 43%
0 10 20 30 40 50 60
% of total arrivals

Tourism is a mainstay of the Greek economy, with an annual turnover of some $10 billion, and is a major source of foreign exchange. Until recently, the state gave grants for hotel development and many third-grade hotels were built, especially on Crete and Rhodes. Smaller islands often lack sufficient water supplies or sandy beaches. To offset falling visitor numbers in the mid-1990s the industry has been encouraged to move upmarket, and is also promoting year-round activity vacations and conference tourism. The 2004 Olympics, to be held in Athens, are a stimulus to upgrade the city's facilities. A museum is planned to house the Parthenon Marbles, currently held in the British Museum (UK).

The theater at Dodona. *Classical sites, such as this amphitheater in northwestern Greece, have helped to make tourism one of Greece's most important industries.*

PEOPLE

Pop. density medium

Greek, Turkish, Macedonian, Albanian

81/km² (210/mi²)

THE URBAN/RURAL POPULATION SPLIT

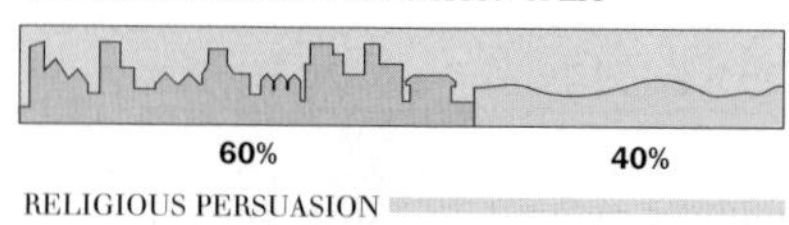
60%
40%

RELIGIOUS PERSUASION

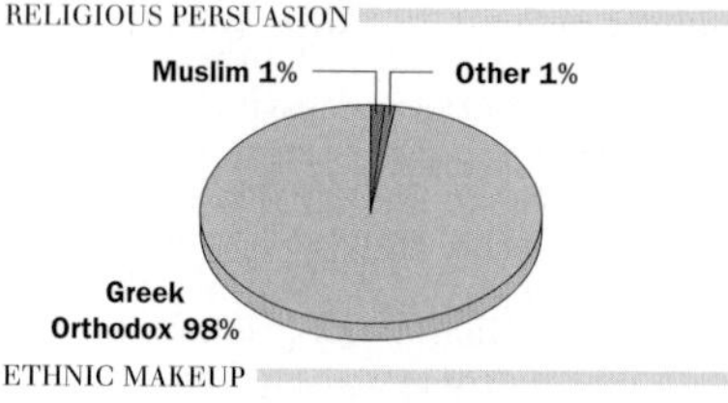
Muslim 1%
Other 1%
Greek Orthodox 98%

ETHNIC MAKEUP

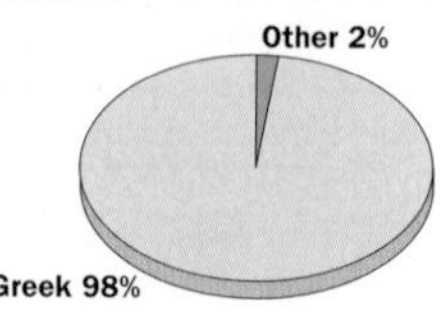
Other 2%
Greek 98%

The Greeks were for many centuries a largely agrarian and seafaring nation. The German occupation during World War II, and the civil war that followed, destroyed much of the fabric of rural life and there was rapid urbanization after the 1950s. There was also extensive emigration in the 1950s and 1960s to northern Europe, Australia, the US, Canada, and southern Africa. However, many people returned to Greece in the 1980s, putting pressure on the labor market. The socialist PASOK governments of 1981–1989 spent large sums, mostly from EU sources, on developing the infrastructure and business life of the rural regions with a view to halting emigration to the cities. The policy was mostly successful, but a majority still lives in or near the capital, Athens, and Thessaloníki in the north.

Some 98% of the population belong to the Greek Orthodox Church. Civil marriage and divorce only became legal in 1982. There are minorities of Muslims, Roman Catholics, and Jews, and a recent influx of illegal immigrants, mainly from Albania.

POPULATION AGE BREAKDOWN

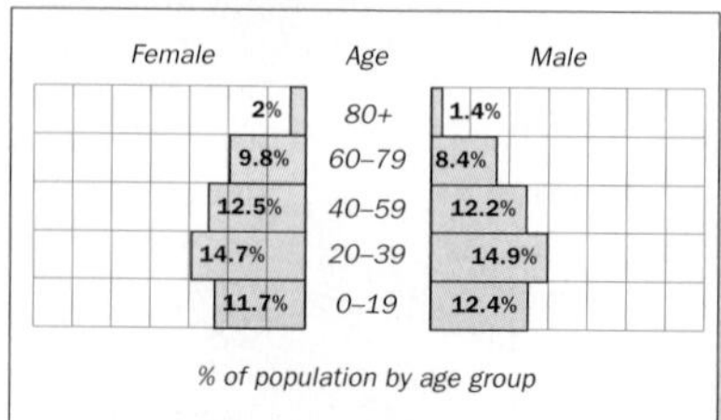
Female
Age
Male
2% 80+ 1.4%
9.8% 60–79 8.4%
12.5% 40–59 12.2%
14.7% 20–39 14.9%
11.7% 0–19 12.4%
% of population by age group

POLITICS

Multiparty elections

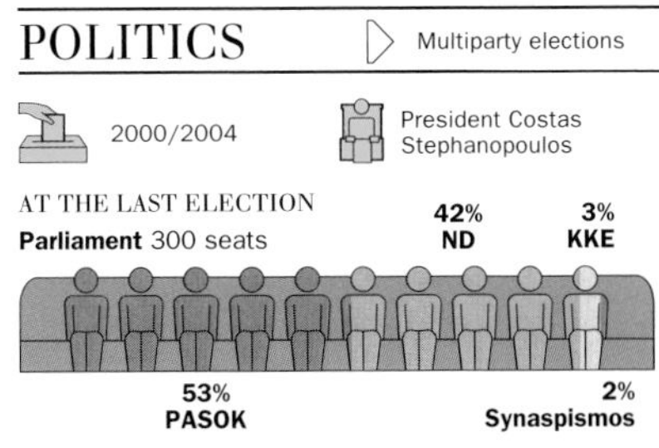
2000/2004
President Costas Stephanopoulos
AT THE LAST ELECTION
Parliament 300 seats
42% ND
3% KKE
53% PASOK
2% Synaspismos

PASOK = Pan-Hellenic Socialist Movement
ND = New Democracy **KKE** = Communist Party of Greece
Synaspismos = Left Coalition

Greece is a multiparty democracy. A military regime held power in 1967–1974.

PROFILE

On PASOK's election victory in 1981 its founder Andreas Papandreou headed the first socialist government in Greek history, which remained in power until 1989. Since 1993 PASOK has again held power continuously, but its economic policies have differed little from the previous conservative government; it only narrowly won the 2000 election. Kostas Simitis has led the party since Papandreou's resignation and death in 1996.

MAIN POLITICAL ISSUES

Closer European union
Greece joined the eurozone in 2001, but only after stringent austerity policies, which evoked widespread protests. Greece favors EU enlargement, and backs EU membership for Cyprus, to the annoyance of Turkey. Currently the poorest EU country, Greece could lose funding to new members, especially if Turkey joins.

Relations with FYRM
In 1995, Greece, with its northern province of Macedonia, finally agreed to recognize the sovereignty of the Former Yugoslav Republic of Macedonia. The crisis in that country in 2001 over ethnic Albanian separatism increased Greek regional security concerns.

Albanian refugees
Thousands of Albanians of Greek descent entered Greece illegally after 1990. Willing to work for very low wages, they swelled the thriving black economy.

Kostas Simitis, *prime minister since January 1996.*

Andreas Papandreou, *founded PASOK, prime minister 1981–1989, 1993–1996.*

A 1998 legalization program resulted in the registration of 375,000 Albanians.

WORLD AFFAIRS

Joined UN in 1945

EU
NATO
OECD
OSCE
CE

Although closely allied to the West, Greece has sympathies with Russians and Serbs, who share its Orthodox heritage. Greece withdrew from NATO's military command in 1974, in protest at its failure to prevent the Turkish invasion of Cyprus, but has since rejoined. The island remains a sore point in Greece's otherwise improving ties with Turkey. Friction with FYRM over the use of the name "Macedonia" has eased, but relations remain tense. The current priority is Greece's role within an expanding EU.

AID

Donor

$226m (donations) Up 16% in 2000

Greece's contribution to overseas development assistance is the lowest in western Europe. However, Greek companies have invested elsewhere in the southern Balkans, and increasingly in Turkey since 1999.

Greece receives regional development assistance from the EU, especially from the EU's structural and cohesion funds, its share of which was estimated to amount to around $3.5 billion over the period 1994–1999. Some of the money has been used to reverse the decline of northeast Greece – the EU's least developed region. Emergency humanitarian aid was given to Turkey following the severe earthquake there in 1999.

GREECE

Total Land Area : 131 940 sq. km (50 942 sq. miles)

POPULATION
- over 1 000 000
- over 500 000
- over 100 000
- over 50 000
- over 10 000

LAND HEIGHT
- 2000m/6562ft
- 1000m/3281ft
- 500m/1640ft
- 200m/656ft
- Sea Level

0 – 100 km
0 – 100 miles

CHRONOLOGY

Greece was occupied by Nazi Germany between 1941 and 1944. After liberation by the Allies, communists and royalists fought a five-year civil war. This ended with communist defeat, and King Paul became the constitutional monarch.

- ❑ **1964** King Constantine succeeds his father, King Paul.
- ❑ **1967** Military coup. King in exile. Col. Giorgios Papadopoulos premier.
- ❑ **1973** Greece declared a republic, with Papadopoulos as president. Papadopoulos overthrown in military coup. Lt. Gen. Ghizikis becomes president, Adamantios Androutsopoulos prime minister.
- ❑ **1974** Greece leaves NATO in protest at Turkish occupation of northern Cyprus. "Colonels' regime" falls. Constantinos Karamanlis becomes premier and his ND party wins subsequent elections.
- ❑ **1975** Konstantinos Tsatsou becomes president.
- ❑ **1977** Elections: ND reelected.
- ❑ **1980** Karamanlis president. Georgios Rallis prime minister. Greece rejoins NATO.
- ❑ **1981** PASOK wins elections. Andreas Papandreou first-ever socialist premier. Greece joins European Communities.
- ❑ **1985** Proposals to limit power of president. Karamanlis resigns. Christos Sartzetakis president. Greece and Albania reopen borders, closed since 1940.
- ❑ **1985–1989** Civil unrest caused by economic austerity program.
- ❑ **1988** Cabinet implicated in financial scandal. Leading members resign.
- ❑ **1989** Defense agreement with US. Two inconclusive elections lead to formation of all-party coalition.
- ❑ **1990** Coalition government collapses. ND wins elections. Konstantinos Mitsotakis prime minister, Karamanlis president.
- ❑ **1990–1992** Strikes against economic reform.
- ❑ **1992** EU persuaded not to recognize independent Macedonia.
- ❑ **1993** PASOK wins election, Andreas Papandreou premier.
- ❑ **1995** Costas Stephanopoulos elected president; recognition of sovereignty of Former Yugoslav Republic of Macedonia (FYRM).
- ❑ **1996** Andreas Papandreou resigns as prime minister; succeeded by Kostas Simitis.
- ❑ **1999** Earthquakes in Greece and Turkey. Sympathetic response shows improvement in relations.
- ❑ **2000** PASOK wins general election.
- ❑ **2001** Armed conflict in FYRM.
- ❑ **2002** Euro fully adopted.

DEFENSE

▷ Compulsory military service

$5.46bn

Up 5% in 2000

Greece spends a higher percentage of GDP on defense than any other NATO country except Turkey, whose perceived threat is its main concern, although tensions with that country are now less acute. In 1998 a law was passed on the conscription of women (for four days a year) for the defense of border regions. Greece has committed 3500 troops to the planned European rapid response force.

GREEK ARMED FORCES

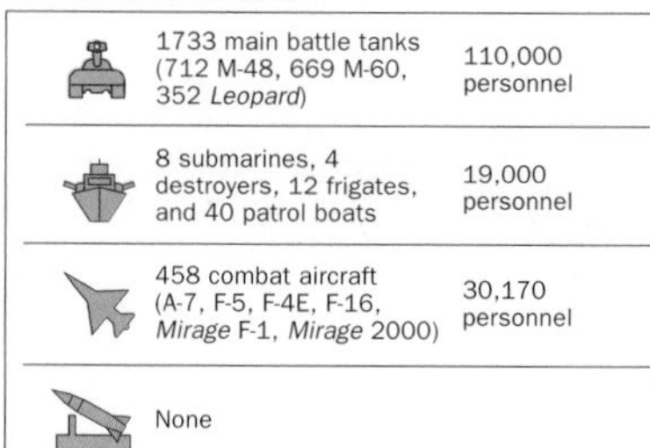

1733 main battle tanks (712 M-48, 669 M-60, 352 *Leopard*)	110,000 personnel
8 submarines, 4 destroyers, 12 frigates, and 40 patrol boats	19,000 personnel
458 combat aircraft (A-7, F-5, F-4E, F-16, *Mirage* F-1, *Mirage* 2000)	30,170 personnel
None	

ECONOMICS

▷ Inflation 9.2% p.a. (1990–2000)

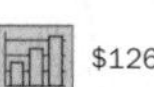
$126bn

1.0651–1.1231 euros

SCORE CARD

❑ World GNP Ranking	30th
❑ GNP per Capita	$11,960
❑ Balance of Payments	–$9.82bn
❑ Inflation	3.2%
❑ Unemployment	11%

EXPORTS

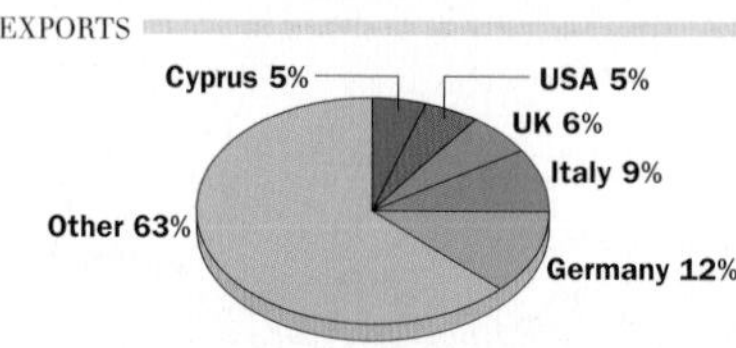
Cyprus 5%
USA 5%
UK 6%
Italy 9%
Other 63%
Germany 12%

IMPORTS

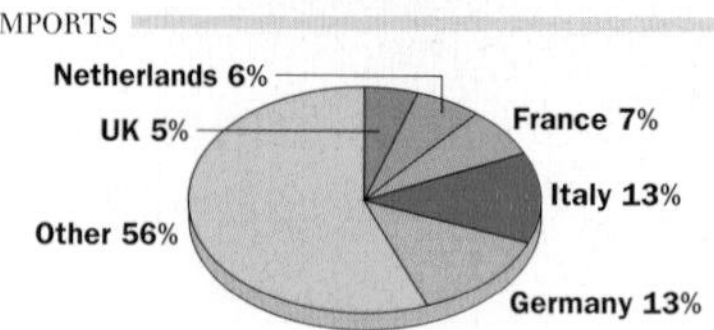
Netherlands 6%
UK 5%
France 7%
Italy 13%
Other 56%
Germany 13%

Strengths

One of the major tourist destinations in Europe. Efficient agricultural exporter. Shipping: the world's largest beneficially owned fleet.

Weaknesses

High levels of public debt. Until recently, interest rates and bureaucratic banking system discouraged private initiative. State-owned sector, often poorly managed, remains large, as does the black economy.

Profile

Greece took longer than most other countries to recover from World War II. It was not until the 1960s that any substantial investment occurred. The Colonels' dictatorship curbed inflationary pressures with a wage freeze. When civilian government was restored in 1974, a spate of high wage settlements and the oil price shocks of 1973 and 1979 drove inflation over 20%. Greece's largest companies made substantial losses, until the socialists' controversial austerity program of 1986–1987 reined in labor costs.

Greece failed in 1999 to meet the economic convergence criteria for introducing the euro. It then tackled the problems with determination, balancing the budget and bringing inflation under control, although public-sector debt remains high. In January 2001 Greece became the 12th member of the eurozone, and fully adopted the currency in 2002. Unemployment is still high and GDP per capita is the lowest in the EU.

ECONOMIC PERFORMANCE INDICATOR

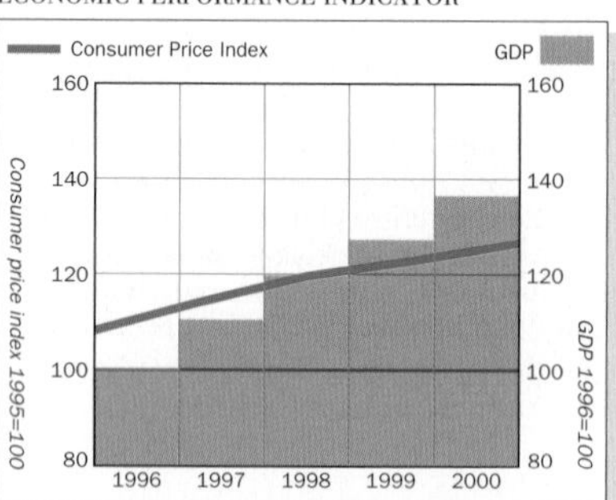
Consumer Price Index
GDP
160
140
120
100
80
Consumer price index 1995=100
GDP 1996=100
1996
1997
1998
1999
2000

GREECE : MAJOR BUSINESSES

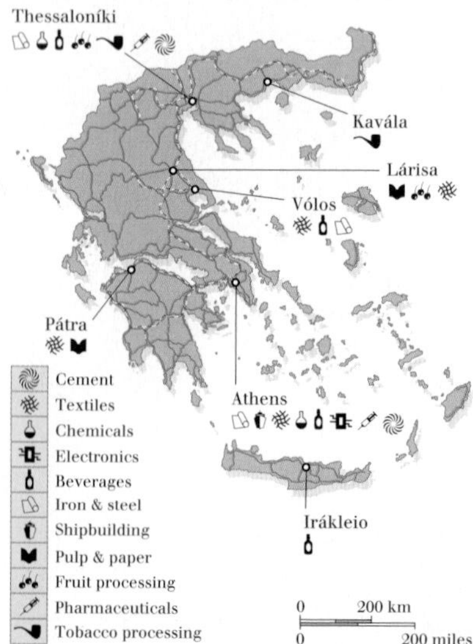
Thessaloníki
Kavála
Lárisa
Vólos
Pátra
Athens
Irákleio
Cement
Textiles
Chemicals
Electronics
Beverages
Iron & steel
Shipbuilding
Pulp & paper
Fruit processing
Pharmaceuticals
Tobacco processing
0
200 km
0
200 miles

RESOURCES

Electric power 10.1m kw

215,964 tonnes

5875 b/d (reserves 15m barrels)

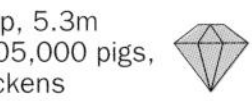
9m sheep, 5.3m goats, 905,000 pigs, 28m chickens

Oil, gas, coal, iron, bauxite, marble, nickel, magnesite, chromium

ELECTRICITY GENERATION

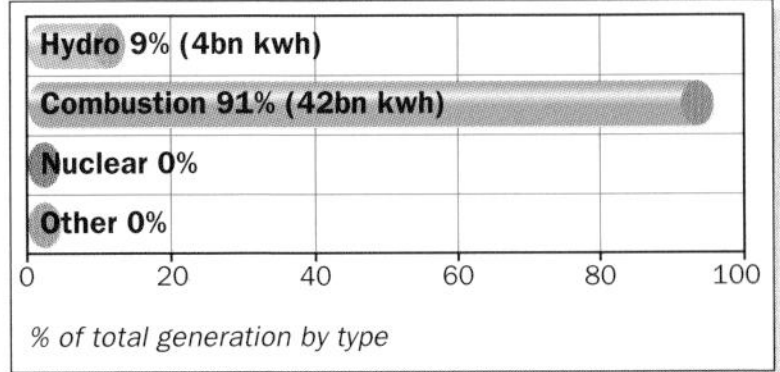
Hydro 9% (4bn kwh)
Combustion 91% (42bn kwh)
Nuclear 0%
Other 0%
0
20
40
60
80
100
% of total generation by type

Greece has an oil and gas field off the coast of Thasos island. There may also be exploitable reserves in eastern waters, ownership of which is contested by Turkey. Coal, iron, and other mining contributes less than 2% to GDP. Greece is a leading producer of marble.

ENVIRONMENT

Sustainability rank: 60th

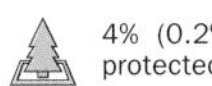
4% (0.2% partially protected)

8.1 tonnes per capita

ENVIRONMENTAL TREATIES

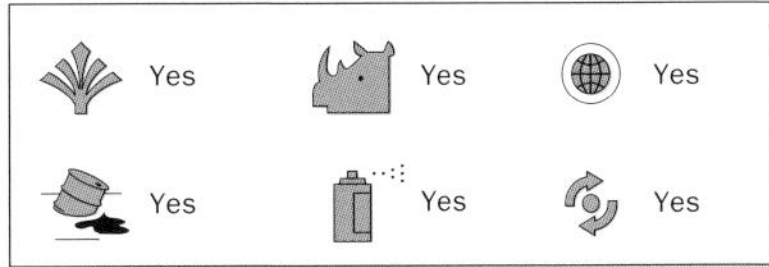
Yes
Yes
Yes
Yes
Yes
Yes

Local fishing interests have formed a successful antipollution organization, HELMEPA. Smog in Athens is irritating to the eyes and throat and highly damaging to ancient monuments: the Parthenon in Athens has suffered more erosion in the last two decades than in the previous 2000 years. Forest fires regularly cause havoc, damaging flora and fauna.

MEDIA

TV ownership high

Daily newspaper circulation 64 per 1000 people

PUBLISHING AND BROADCAST MEDIA

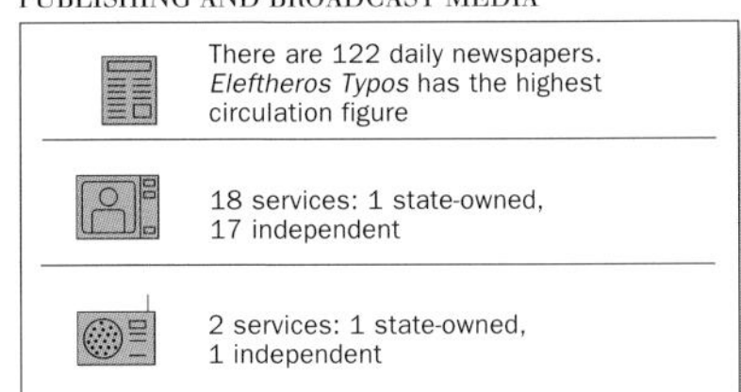
There are 122 daily newspapers. *Eleftheros Typos* has the highest circulation figure
18 services: 1 state-owned, 17 independent
2 services: 1 state-owned, 1 independent

After the state broadcasting monopoly ended in 1990, many private TV and radio networks emerged; the two national public TV channels account for only abut 10% of viewing. Commercial broadcasting has made politicians more answerable to the public, and has also had a cultural impact, with the import of more foreign, particularly US, programs.

GREECE : LAND USE

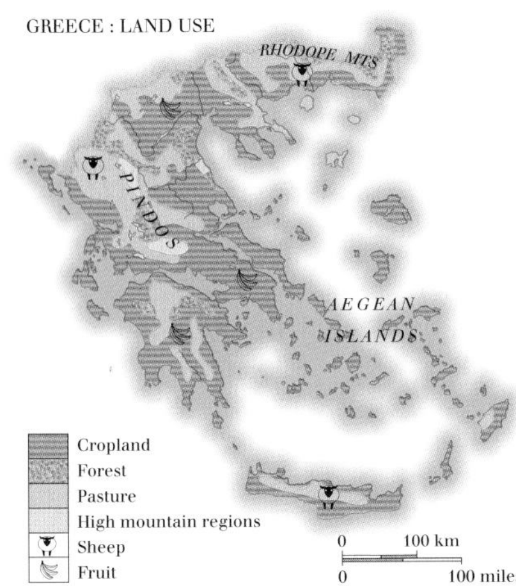
RHODOPE MTS
PINDOS
AEGEAN ISLANDS
Cropland
Forest
Pasture
High mountain regions
Sheep
Fruit
0
100 km
0
100 miles

CRIME

Death penalty not used in practice

8343 prisoners

Down 3% in 1999

CRIME RATES

Murders	3 per 100,000 population
Rapes	2 per 100,000 population
Thefts	741 per 100,000 population

An influx of migrants is blamed for an increase in violent crime. The terrorist group November 17, which had carried out high-profile assassinations, was finally tracked down in 2002.

EDUCATION

School leaving age: 15

97%

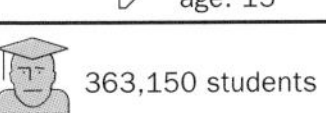
363,150 students

THE EDUCATION SYSTEM

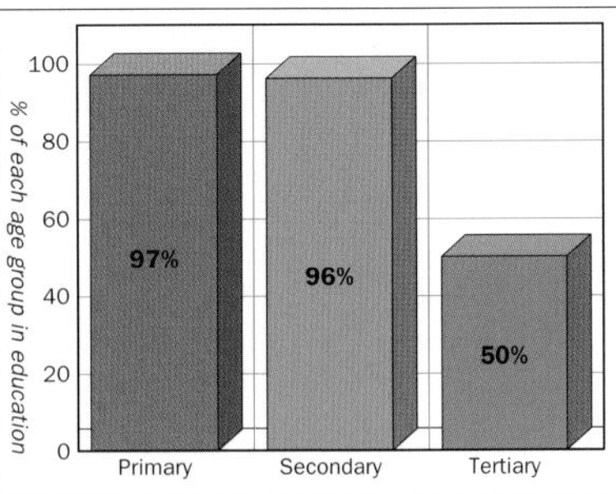
100
80
60
40
20
0
% of each age group in education
97%
96%
50%
Primary
Secondary
Tertiary

Some 9% of total government spending is on education, which is free and officially compulsory for nine years. Teachers are poorly paid and qualifications are low. University places are limited and many students go abroad for tertiary education. Technical courses, funded by the EU, have increased since the 1990s.

HEALTH

Welfare state health benefits

1 per 244 people

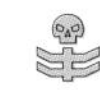
Heart and cerebrovascular diseases, cancers, accidents

The socialists (PASOK) introduced a national health service and a national pharmaceuticals industry. Some 12% of government expenditure goes on health, and every Greek is entitled to sickness benefit. Greece now has the second-highest number of doctors per head of population in the EU; however, primary care is poor, as is that in state hospitals. In the early 1990s the ND attempted to upgrade private medicine and to incorporate its activities with those in state hospitals. Many Greeks needing major surgery go abroad for treatment.

SPENDING

GDP/cap. increase

CONSUMPTION AND SPENDING

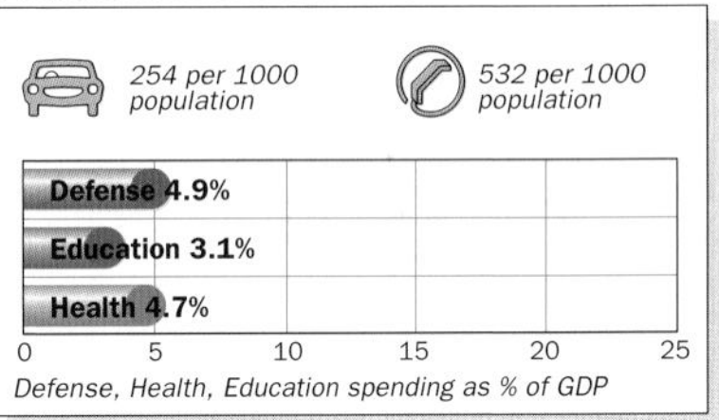
254 per 1000 population
532 per 1000 population
Defense 4.9%
Education 3.1%
Health 4.7%
0
5
10
15
20
25
Defense, Health, Education spending as % of GDP

Greek society changed dramatically in the postwar period. Formerly a largely agricultural society living in isolated communities, it was rapidly urbanized in the 1950s. Former agricultural workers made fortunes, many by grabbing opportunities presented by the shipping industry. Among these were the now prominent Niarchos and Onassis families.

The advent of the republic in 1975 reflected the social changes which had occurred since the war. New wealth and success became more admired than aristocratic birth or prestige. Greece is now a socially mobile society. Living standards have improved universally since the 1950s.

WORLD RANKING

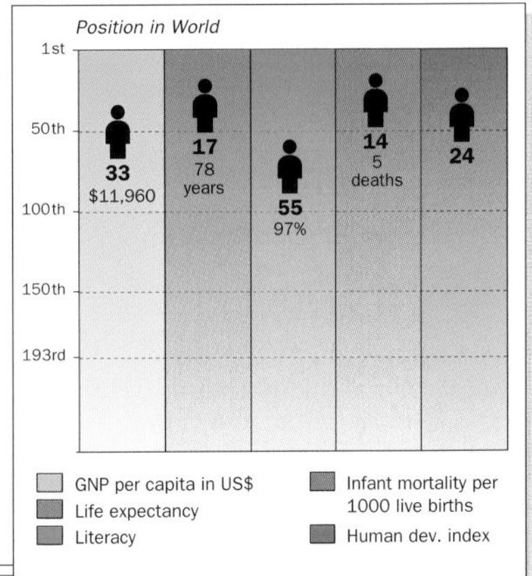
Position in World
1st
50th
100th
150th
193rd
33
$11,960
17
78
years
55
97%
14
5
deaths
24
GNP per capita in US$
Life expectancy
Literacy
Infant mortality per 1000 live births
Human dev. index

G

GRENADA

CARIBBEAN

OFFICIAL NAME: Grenada **CAPITAL:** St. George's
POPULATION: 98,000 **CURRENCY:** Eastern Caribbean dollar **OFFICIAL LANGUAGE:** English

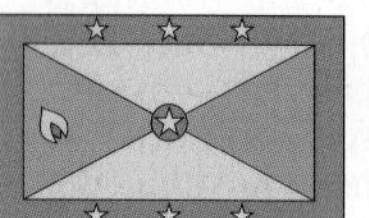

1974

1974

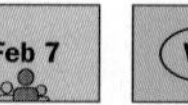
Feb 7

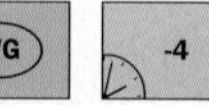
WG

-4

+1473
.gd

THE MOST SOUTHERLY of the Windward Islands, Grenada also includes the southern Grenadine islands of Carriacou and Petite Martinique. It is the world's second-largest nutmeg producer. One of the seven members of the OECS, Grenada became a focus of international attention in 1983 when the US, with token backing from several Caribbean states, mounted an invasion to sever its growing links with Castro's Cuba.

CLIMATE

Tropical oceanic

WEATHER CHART FOR ST. GEORGE'S

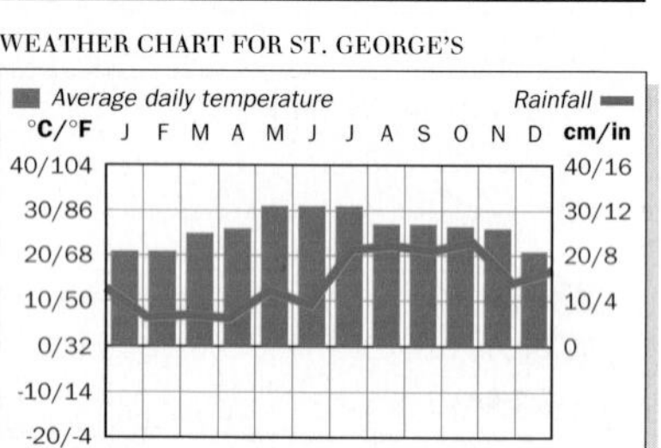
Average daily temperature
Rainfall
°C/°F J F M A M J J A S O N D cm/in
40/104 40/16
30/86 30/12
20/68 20/8
10/50 10/4
0/32 0
-10/14
-20/-4

Rainfall totals 150 cm (60 in) on the coast, and twice that in the mountains. Hurricanes occur in the rainy season.

TRANSPORTATION

Drive on left

Point Salines, St. George's
438,987 passengers

6 ships
1009 grt

THE TRANSPORTATION NETWORK

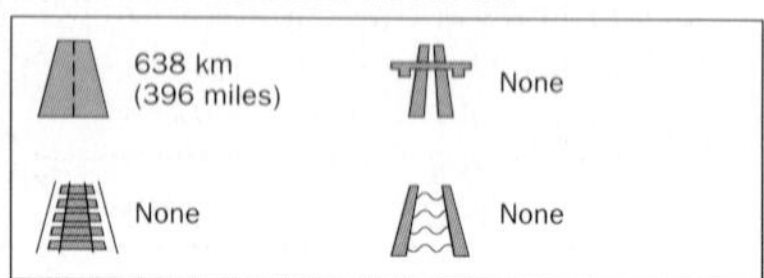
638 km
(396 miles)
None
None
None

Roads in the interior are poor. Catamarans provide the fastest link between Grenada and Carriacou.

TOURISM

Visitors : Population 1.3:1

129,000 visitors

Up 3% in 2000

MAIN TOURIST ARRIVALS

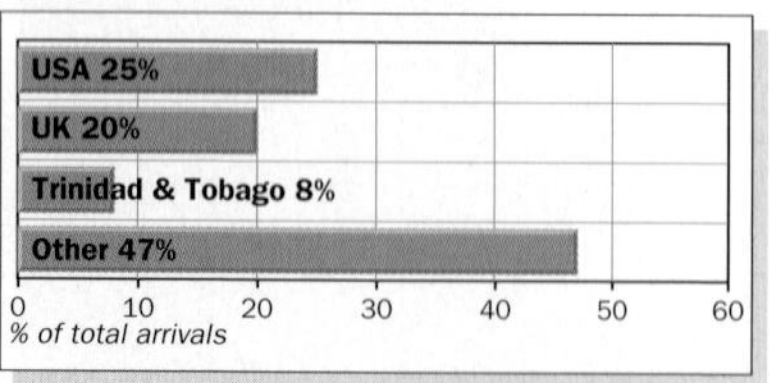
USA 25%
UK 20%
Trinidad & Tobago 8%
Other 47%
0 10 20 30 40 50 60
% of total arrivals

Tourism has developed since the 1984 completion of the international airport, though there has been a concurrent decline in the arrival of cruise ships; the Crown Dynasty company folded in 1999, with a major impact on tourist numbers.

PEOPLE

Pop. density high

English, English Creole

288/km² (748/mi²)

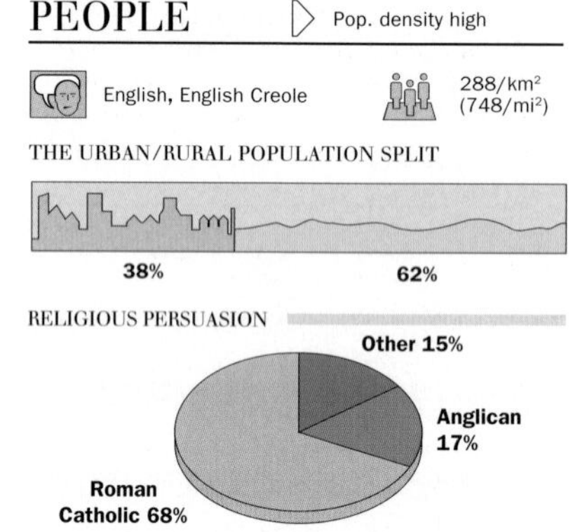
THE URBAN/RURAL POPULATION SPLIT
38%
62%
RELIGIOUS PERSUASION
Other 15%
Anglican 17%
Roman Catholic 68%

Most Grenadians are descendants of Africans brought over to work sugar plantations in the 16th to 19th centuries. Intermarriage between this group and the small numbers of Europeans and indigenous Amerindians has meant that there is little racial tension. As in other Caribbean states, extended families with absentee fathers are not uncommon.

GRENADA

Total Land Area :
340 sq. km (131 sq. miles)

POPULATION
• over 10 000
• under 10 000

LAND HEIGHT
500m/1640ft
200m/656ft
Sea Level

0 8 km
0 8 miles

POLITICS

Multiparty elections

L. House 1999/2004
U. House 1999/2004

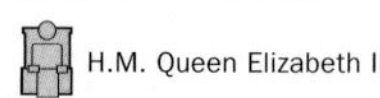
H.M. Queen Elizabeth II

AT THE LAST ELECTION

House of Representatives 15 seats

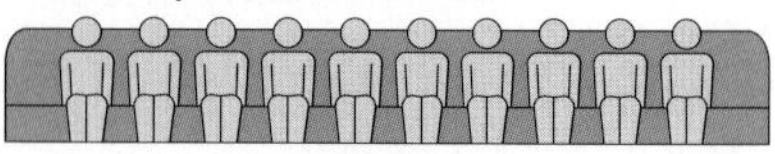

100% NNP

NNP = New National Party

Senate 13 seats

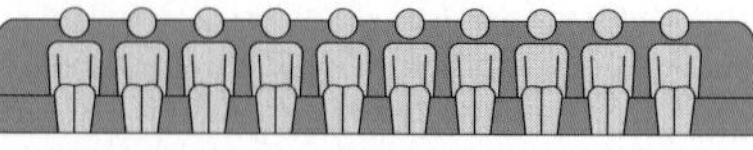

The members of the Senate are appointed by the prime minister and the leader of the opposition

The past 25 years have seen Grenada move from a position of political isolation toward integration with the rest of the region. The late Sir Eric Gairy, when prime minister, was as well known for his eccentric requests to the UN Security Council – he once asked it to investigate UFOs on the island – as for his intimidation of political opponents by means of organized gangs. Gairy was overthrown in 1979 by armed militants of the New Jewel Movement led by Maurice Bishop, a charismatic socialist who in turn was deposed, and executed by former allies, in 1983. This coup was the pretext for the US invasion, the primary motive of which was to end the perceived Cuban influence in Grenada. A new government was elected in 1984, and the US provided large amounts of aid. Politics has since been center-right, and there is little to choose ideologically between the parties. The NNP, led by Keith Mitchell, gained power in 1995, and went on to achieve an unparalleled victory over a divided opposition in the early general election held in 1999, taking all 15 seats. Mitchell promised "explosive" growth and lower unemployment.

WORLD AFFAIRS

Joined UN in 1974

ACS

Caricom

Comm

OAS

OECS

Priorities are the relations with the rest of the Windward Islands group, access to the EU for bananas, and strategies with Indonesia aimed at steadying world nutmeg prices. Since 1983, it has supported US policy in the Caribbean.

AID

Recipient

US$17m (receipts)

Up 70% in 2000

Main aid sources are Arab donors, Japan, the Caribbean Development Bank, and the EU. Cuba helped build the pre-1983 airport at Point Salines.

DEFENSE

No compulsory military service

Minimal expenditure

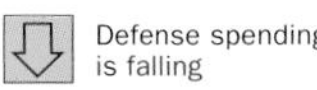
Defense spending is falling

The People's Revolutionary Army, created by Maurice Bishop in the wake of his 1979 coup, was replaced in 1983 by a paramilitary defense unit trained by the US and the UK.

ECONOMICS

Inflation 2.5% p.a. (1990–2000)

US$370m

2.7 Eastern Caribbean dollars

SCORE CARD

❑ WORLD GNP RANKING	174th
❑ GNP PER CAPITA	US$3770
❑ BALANCE OF PAYMENTS	–US$79m
❑ INFLATION	0.2%
❑ UNEMPLOYMENT	15%

STRENGTHS

Second-largest producer of nutmeg after Indonesia. Important sectors are tourism, cocoa, bananas, construction, and financial services.

WEAKNESSES

Weak tax base, lack of diversification. Poor infrastructure. Low productivity. Large avoidance of customs duties. Smuggling.

EXPORTS

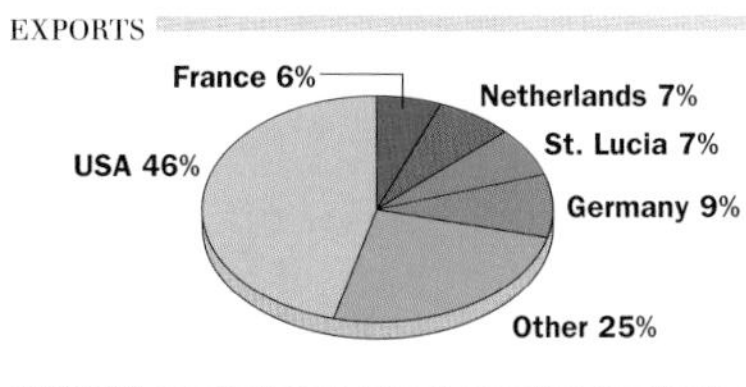
France 6%
Netherlands 7%
USA 46%
St. Lucia 7%
Germany 9%
Other 25%

IMPORTS

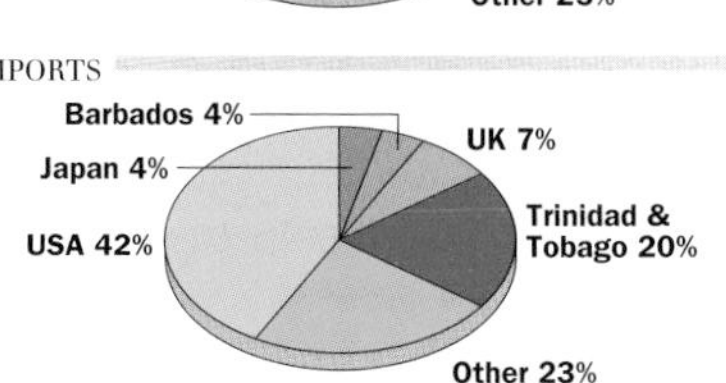
Barbados 4%
Japan 4%
UK 7%
USA 42%
Trinidad & Tobago 20%
Other 23%

RESOURCES

Electric power 15,000 kw

1632 tonnes

Not an oil producer

13,100 sheep, 7100 goats, 220,000 chickens

None

Grenada has no strategic resources and has to import most of its energy. The major asset is Grenadian nutmeg, which is highly prized for its quality, but production levels fluctuate.

ENVIRONMENT

Not available

None

1.9 tonnes per capita

Tourism threatens some key environmental sites, including a remnant of rainforest. Resort projects have caused serious beach erosion, in turn requiring costly coastal defenses. An environmental levy on visitors is opposed by cruise companies.

MEDIA

TV ownership high

There are no daily newspapers

PUBLISHING AND BROADCAST MEDIA

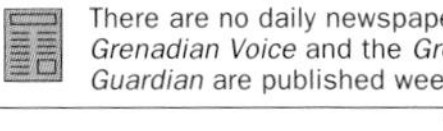

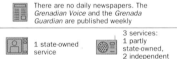

There are no daily newspapers. The *Grenadian Voice* and the *Grenada Guardian* are published weekly

1 state-owned service

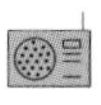

3 services: 1 partly state-owned, 2 independent

The government sold a 60% share in the then Grenada Broadcasting Corporation (GBC) in 1999. There is an independent private press.

CRIME

Death penalty not used n practice

327 prisoners

Up 345% 1989–1996

The doubling of poverty during the 1990s and high unemployment have contributed to a rising crime rate. Narcotics trafficking is also a growing problem. However, while there is street crime, the level of violence is low.

EDUCATION

School leaving age: 16

96%

651 students

Education follows the former British selective 11-plus system. Many students go on to the University of the West Indies, or to college in the US.

HEALTH

Welfare state health benefits

1 per 2000 people

Heart diseases, cancers, nutritional disorders

After 1979, Cuban physicians provided a basic health care system. There are free weekly clinics in each district, and treatment in subsidized state hospitals now matches the Caribbean average. In 1999 Cuba began its promised expansion of the general hospital.

St. George's harbor. *The newest hotel developments are on the beaches to the south of the capital.*

CHRONOLOGY

A French colony from 1650, Grenada was captured by the British in 1762.

- ❑ **1951** Universal suffrage introduced.
- ❑ **1967–1974** Internal self-government. Full independence from the UK.
- ❑ **1979** Coup. Maurice Bishop prime minister. Growing links with Cuba.
- ❑ **1983** US invasion establishes pro-US administration.
- ❑ **1999** NNP reelected, taking all 15 seats in House of Representatives.

SPENDING

GDP/cap. increase

CONSUMPTION AND SPENDING

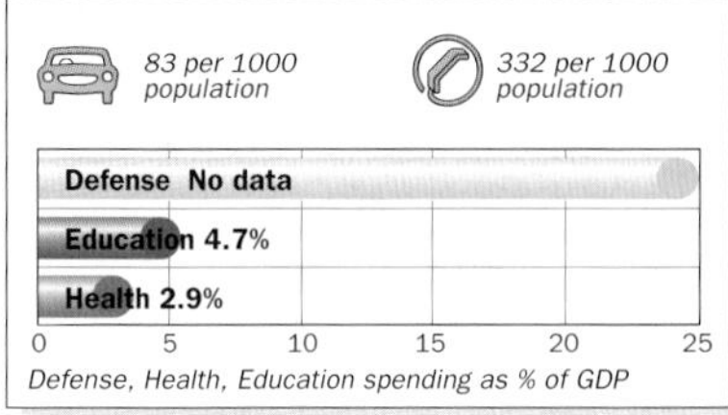
83 per 1000 population
332 per 1000 population
Defense No data
Education 4.7%
Health 2.9%
0
5
10
15
20
25
Defense, Health, Education spending as % of GDP

Wealth disparities in Grenada are less marked than in most Caribbean states, but poverty is growing. The wealthiest groups are those in control of the nutmeg trade.

WORLD RANKING

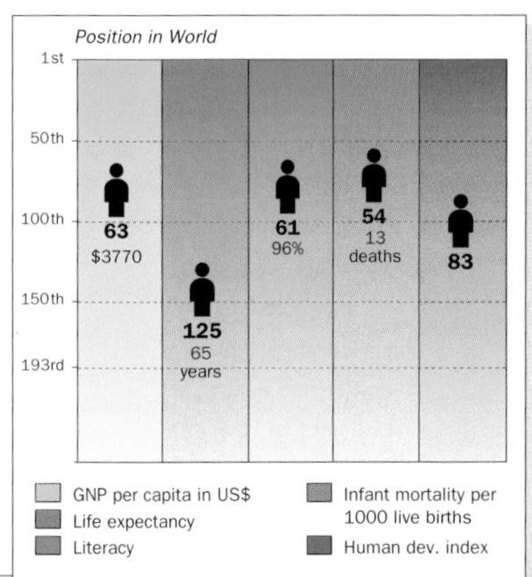
Position in World
1st
50th
100th
150th
193rd
63
$3770
125
65
years
61
96%
54
13
deaths
83
GNP per capita in US$
Life expectancy
Literacy
Infant mortality per 1000 live births
Human dev. index

G

GUATEMALA

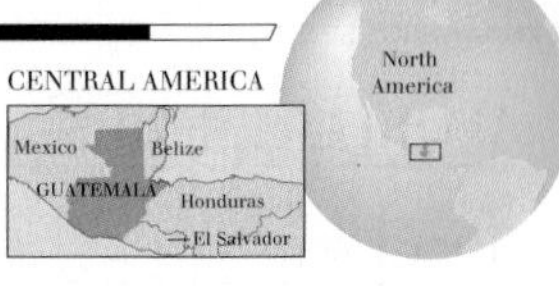
CENTRAL AMERICA
North America
Mexico
Belize
GUATEMALA
Honduras
El Salvador

OFFICIAL NAME: Republic of Guatemala **CAPITAL:** Guatemala City
POPULATION: 11.7 million **CURRENCY:** Quetzal **OFFICIAL LANGUAGE:** Spanish

1838 | 1838 | Sept 15 | GCA | -6 | +502 | .gt

THE LARGEST AND MOST POPULOUS of the states of the Central American isthmus, Guatemala was home to the ancient Mayan civilization. Its fertile Pacific and Caribbean coastal lowlands give way to the highlands which dominate the country. Independent since 1838, Guatemala fell under military rule in 1954. Civilian rule was not restored for over three decades, in 1986. Some 60% of people live below the poverty line.

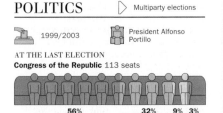

POLITICS

Multiparty elections

1999/2003

President Alfonso Portillo

AT THE LAST ELECTION

Congress of the Republic 113 seats

56% FRG | 32% PAN | 9% NNA | 3% Others

FRG = Guatemalan Republican Front **PAN** = National Advancement Party **NNA** = New Nation Alliance

A military government which came to power in 1954 with US backing brutally suppressed opposition and persecuted the highland Amerindians until the return of democracy in 1986. Civil war effectively continued until President Arzú of the PAN concluded a peace agreement with the Guatemalan National Revolutionary Unity (URNG) guerrillas in 1996. The 36-year war had claimed 200,000 lives, mostly innocent civilians. In presidential elections in 1999 Alfonso Portillo of the right-wing FRG, running on a strong law and order platform, defeated the PAN candidate.

Efraín Ríos Montt, military ruler in 1982–1986, remains a force in politics and in 2001 was elected president of the Congress, despite being investigated by a Spanish court on charges of genocide. Constant tension exists between the legislative and executive branches of government.

CLIMATE

Tropical equatorial/ wet & dry

WEATHER CHART FOR GUATEMALA CITY

Average daily temperature — Rainfall
°C/°F J F M A M J J A S O N D cm/in
40/104, 30/86, 20/68, 10/50, 0/32, -10/14, -20/-4 — 40/16, 30/12, 20/8, 10/4, 0

The climate varies with altitude: daytime temperatures average 28°C (82°F) in tropical coast areas and 20°C (68°F) in the more temperate central highlands.

TRANSPORTATION

Drive on right

La Aurora, Guatemala City 939,000 passengers

9 ships 4700 grt

THE TRANSPORTATION NETWORK

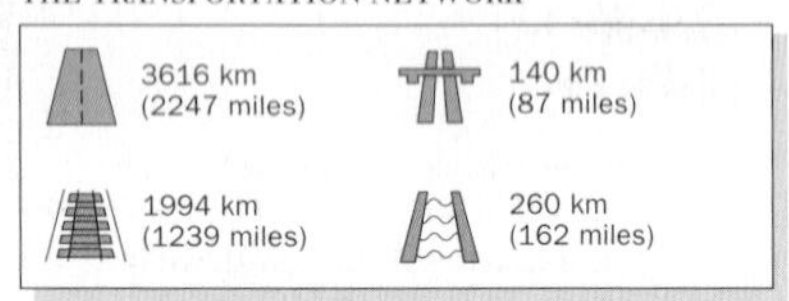
3616 km (2247 miles)
140 km (87 miles)
1994 km (1239 miles)
260 km (162 miles)

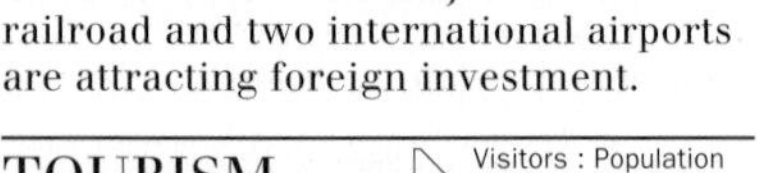

Good roads link the major towns. The railroad and two international airports are attracting foreign investment.

TOURISM

Visitors : Population 1:14

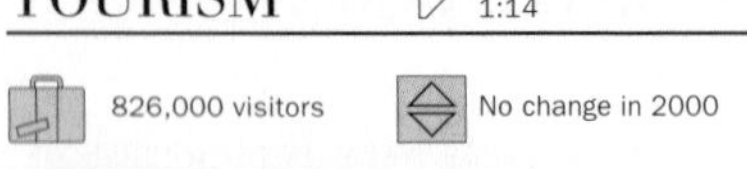
826,000 visitors
No change in 2000

MAIN TOURIST ARRIVALS

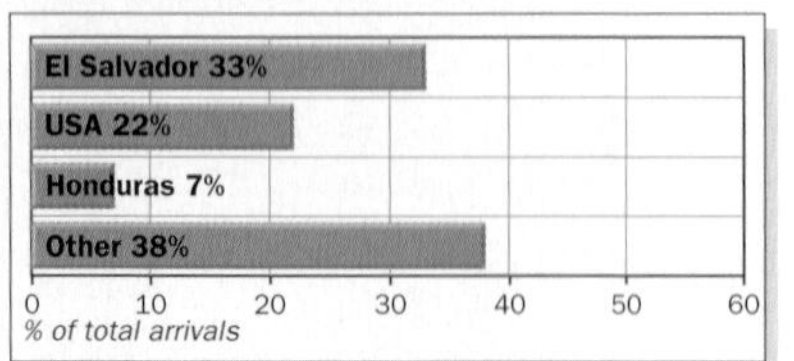
El Salvador 33%
USA 22%
Honduras 7%
Other 38%
0 10 20 30 40 50 60
% of total arrivals

Tourism rapidly revived after the military excesses in the 1980s, but postwar crime, including an increase in mob violence, deters visitors. Mayan ruins are the top attractions.

PEOPLE

Pop. density medium

Quiché, Mam, Cakchiquel, Kekchí, Spanish

108/km² (279/mi²)

THE URBAN/RURAL POPULATION SPLIT

40% | 60%

ETHNIC MAKEUP

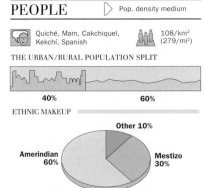

Amerindian 60% | Mestizo 30% | Other 10%

More than half of Guatemalans are Amerindians, descendants of the original Mayans. Culture and language distinguish them from *ladino* groups. *Ladinos* include a white elite, a large mixed race group, and also Amerindians rejecting traditional dress and language to avoid discrimination. Political power and 65% of land are held by a few *ladino* families. Amerindians mainly live in the highlands, by subsistence farming. In a 1999 plebiscite, *ladinos* rejected proposed reforms recognizing 23 Amerindian languages and the right of Amerindians to have judicial hearings in their own languages.

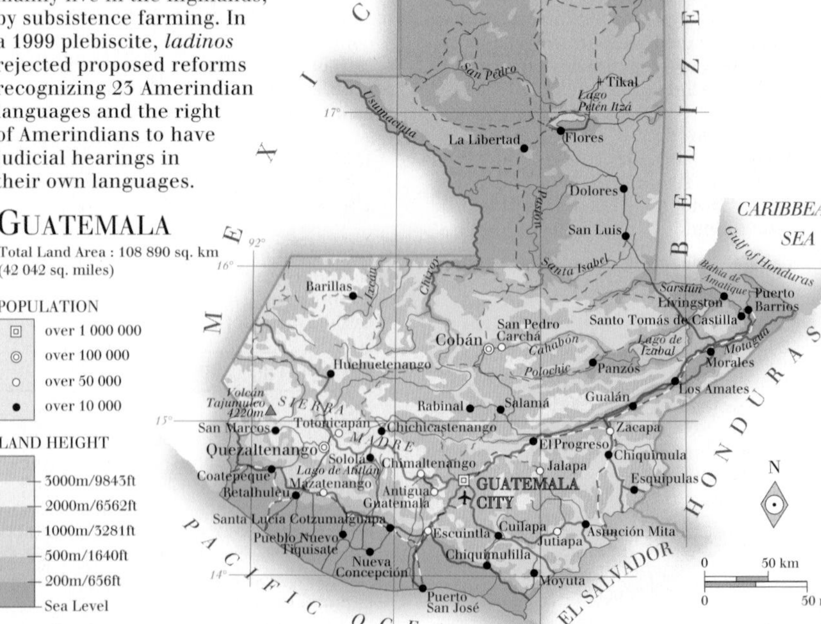
GUATEMALA
Total Land Area : 108 890 sq. km (42 042 sq. miles)
POPULATION
over 1 000 000
over 100 000
over 50 000
over 10 000
LAND HEIGHT
3000m/9843ft
2000m/6562ft
1000m/3281ft
500m/1640ft
200m/656ft
Sea Level
MEXICO
BELIZE
HONDURAS
EL SALVADOR
PACIFIC OCEAN
CARIBBEAN SEA
Gulf of Honduras
Tikal
Lago Petén Itzá
Flores
La Libertad
Dolores
San Luis
Barillas
Cobán
San Pedro Carchá
Huehuetenango
Puerto Barrios
Santo Tomás de Castilla
Livingston
Lago de Izabal
Morales
Los Amates
Panzós
Gualán
Rabinal
Salamá
Zacapa
El Progreso
Chiquimula
Jalapa
Esquipulas
San Marcos
Quetzaltenango
Totonicapán
Chichicastenango
Sololá
Chimaltenango
GUATEMALA CITY
Antigua Guatemala
Coatepeque
Retalhuleu
Mazatenango
Santa Lucía Cotzumalguapa
Pueblo Nuevo Tiquisate
Escuintla
Cuilapa
Jutiapa
Asunción Mita
Chiquimulilla
Nueva Concepción
Puerto San José
Moyuta
Volcán Tajumulco 4220m
SIERRA MADRE
Usumacinta
San Pedro
Pasión
Santa Isabel
Chixoy
Motagua
Polochic
Cahabón
N
0 50 km
0 50 mi

WORLD AFFAIRS

Joined UN in 1945

ACS
Geplac
NAM
OAS
San José

Economic relations with the US and neighboring states are priorities. A UN mission was deployed for six months to oversee the 1996 peace accord. A claim to half of Belize was revived in early 2000.

AID

Recipient

 $264m (receipts)

 Down 10% in 2000

In 1998 the government agreed the disbursement of the second half of $1.9 billion pledged by the international donors for postwar reconstruction. In the same year $1.4 million were also loaned as hurricane relief.

DEFENSE

Compulsory military service

 $115m

 Down 23% in 2000

A damning "truth commission" report in 1999 found the armed forces and their allies guilty of 93% of human rights violations during the civil war. The army remains largely unreformed and a potent sociopolitical force.

ECONOMICS

Inflation 10% p.a. (1990–2000)

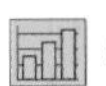 $19.2bn

 7.782–7.910 quetzales

SCORE CARD

❑ WORLD GNP RANKING	67th
❑ GNP PER CAPITA	$1680
❑ BALANCE OF PAYMENTS	–$1.05bn
❑ INFLATION	6%
❑ UNEMPLOYMENT	8%

STRENGTHS

Main exports are coffee, sugar, beef, bananas, and cardamom. Privatizations boost foreign investor confidence.

WEAKNESSES

Traditional exports vulnerable to world price changes. Shaky financial system. Inequalities in land and wealth limit domestic market. Tax evasion.

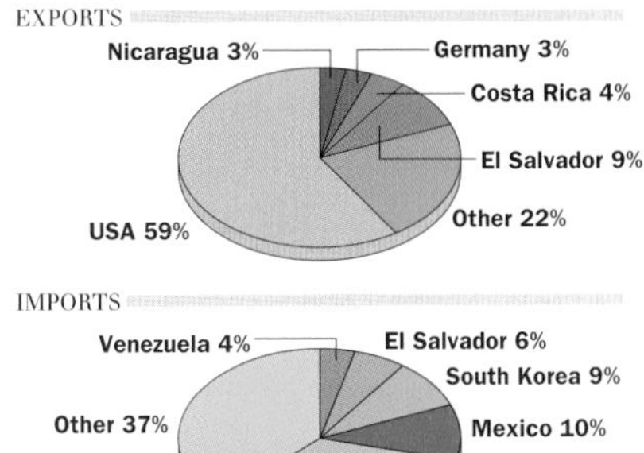
EXPORTS
Nicaragua 3%
Germany 3%
Costa Rica 4%
El Salvador 9%
Other 22%
USA 59%
IMPORTS
Venezuela 4%
El Salvador 6%
South Korea 9%
Other 37%
Mexico 10%
USA 34%

***North Acropolis, Tikal, Petén.** One of the largest lowland Maya cities, Tikal was virtually abandoned by about 900 CE.*

RESOURCES

Electric power 1.5m kw

 15,878 tonnes

 25,503 b/d (reserves 196m barrels)

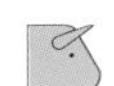 2.5m cattle, 1.45m pigs, 35m chickens

 Oil, antimony, lead, tungsten, nickel, copper

Agriculture provides over 20% of GDP and about 70% of export earnings. Guatemala is the world's largest producer of cardamom. The civil war hindered the exploitation of oil reserves and hydroelectric potential.

ENVIRONMENT

Sustainability rank: 67th

 17% (0.5% partially protected)

 0.9 tonnes per capita

Forest cover has been halved to 35% since 1954 due to intensive agriculture. The excessive use of pesticides, many banned in the US, threatens health. Pollution and waste are serious problems, especially in the capital, Guatemala City.

MEDIA

TV ownership medium

 Daily newspaper circulation 33 per 1000 people

PUBLISHING AND BROADCAST MEDIA

 There are 7 daily newspapers, including *Prensa Libre, Siglo Veintiuno, El Periódico,* and the state *Diario de Centroamérica*

 5 services: 1 state-owned, 4 independent

 85 stations: 5 state-owned, 80 independent

Powerful groups own the media, but newspapers can be hard-hitting. The four independent TV stations are controlled by a single owner.

CRIME

Moratorium on death penalty

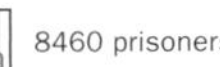 8460 prisoners

 Crime is rising

Violence is high. In 2000 the interior ministry purged the police, which had been corrupted by drug trafficking.

EDUCATION

School leaving age: 14

 69%

 146,291 students

Education is only for the privileged. A literacy rate of 69% makes Guatemala one of the worst educated countries in Latin America.

CHRONOLOGY

The site of the Mayan civilization, Guatemala declared independence from Spain in 1821. It became a fully independent nation in 1838.

- ❑ **1954** US-backed coup topples reformist government.
- ❑ **1966–1984** Counterinsurgency war; highlands "pacification."
- ❑ **1986–1993** Return of civilian rule; President Serrano elected. Flees country after abortive "self-coup."
- ❑ **1996** President Arzú elected; peace deal with URNG guerrillas, ending 36 years of civil war.
- ❑ **1998** Bishop Juan Gerardi, human rights campaigner, murdered.
- ❑ **1999** "Truth Commission" blames army for most human rights abuses. Portillo and FRG win elections.

G

HEALTH

Welfare state health benefits

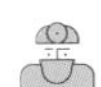 1 per 1111 people

 Gastrointestinal infections, tuberculosis, heart disease, violence

Health spending is a budget priority as a result of pressure from the UN and multilateral lenders. Gastrointestinal and other infections directly linked to poverty remain the main causes of death.

SPENDING

GDP/cap. increase

CONSUMPTION AND SPENDING

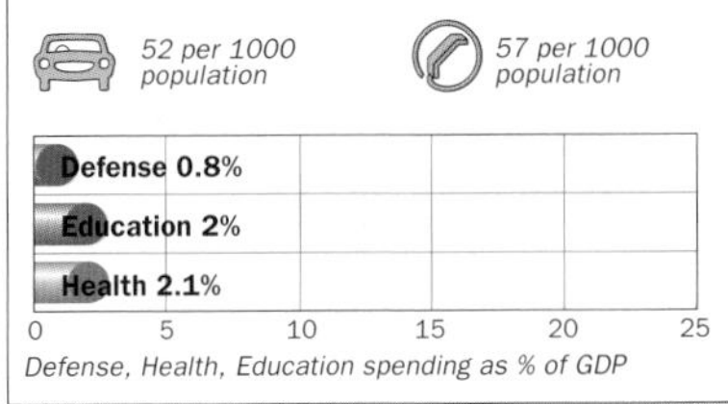
52 per 1000 population
57 per 1000 population
Defense 0.8%
Education 2%
Health 2.1%
0
5
10
15
20
25
Defense, Health, Education spending as % of GDP

Poverty has risen since 1980: 60% of the population now live below the poverty line. The richest 10% control an estimated 46% of the national wealth.

WORLD RANKING

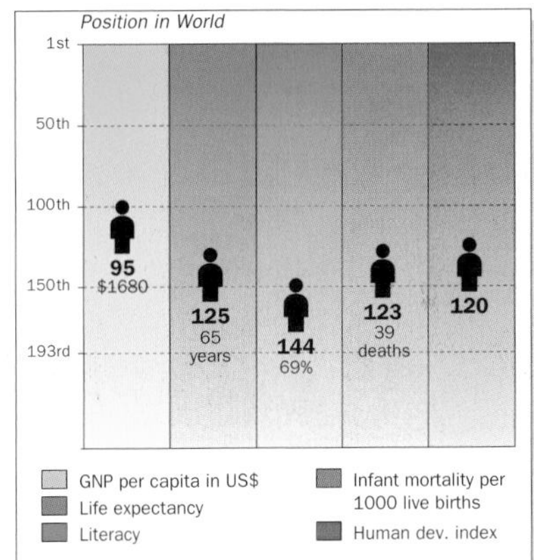
Position in World
1st
50th
100th
150th
193rd
95
$1680
125
65 years
144
69%
123
39 deaths
120
GNP per capita in US$
Life expectancy
Literacy
Infant mortality per 1000 live births
Human dev. index

GUINEA

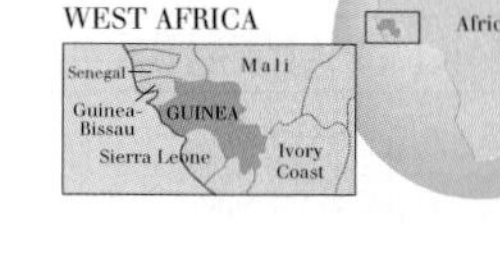
WEST AFRICA
Africa
Senegal
Mali
Guinea-Bissau
GUINEA
Sierra Leone
Ivory Coast

OFFICIAL NAME: Republic of Guinea **CAPITAL:** Conakry
POPULATION: 8.3 million **CURRENCY:** Guinea franc **OFFICIAL LANGUAGE:** French

1958

1958

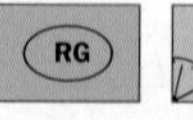
Apr 3

RG

0

+224 .gn

GUINEA LIES ON the western coast of Africa. Central highlands, either densely forested or savanna covered, slope down to coastal plains and swamps; the north is semidesert. Military rule from 1984 ended with disputed elections in 1995. Neighboring civil wars have spilled over into domestic conflict in Guinea.

CLIMATE

Tropical monsoon

WEATHER CHART FOR CONAKRY

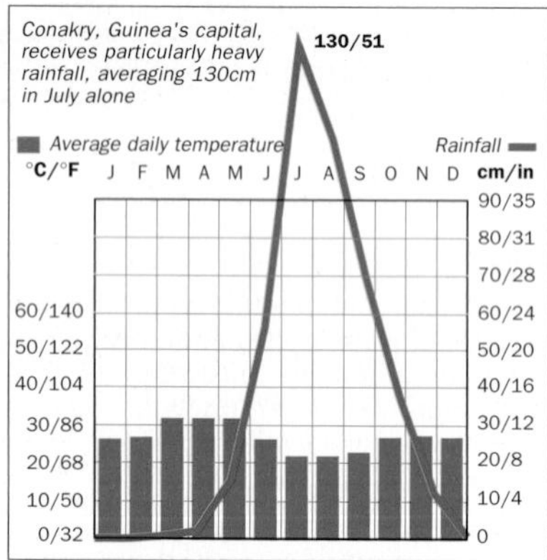
Conakry, Guinea's capital, receives particularly heavy rainfall, averaging 130cm in July alone
130/51
Average daily temperature
Rainfall
°C/°F
J F M A M J J A S O N D
cm/in
90/35
80/31
70/28
60/140
60/24
50/122
50/20
40/104
40/16
30/86
30/12
20/68
20/8
10/50
10/4
0/32
0

Guinea's climate is similar to Sierra Leone's. Conakry's six-month rainy season peaks dramatically in July.

TRANSPORTATION

Drive on right

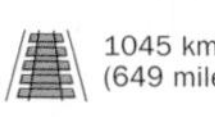
Conakry–Gbessia
293,484 passengers

33 ships
11,300 grt

THE TRANSPORTATION NETWORK

5033 km (3127 miles)	None
1045 km (649 miles)	1295 km (805 miles)

Major roads and rail lines are being rebuilt with World Bank and French aid. Much of the rail network is exclusively for the use of the bauxite industry.

A small mosque in Conakry. Muslims make up 85% of the population; 8% are Christian. The remainder follow traditional beliefs.

TOURISM

Visitors : Population 1:252

33,000 visitors

Up 22% in 2000

MAIN TOURIST ARRIVALS

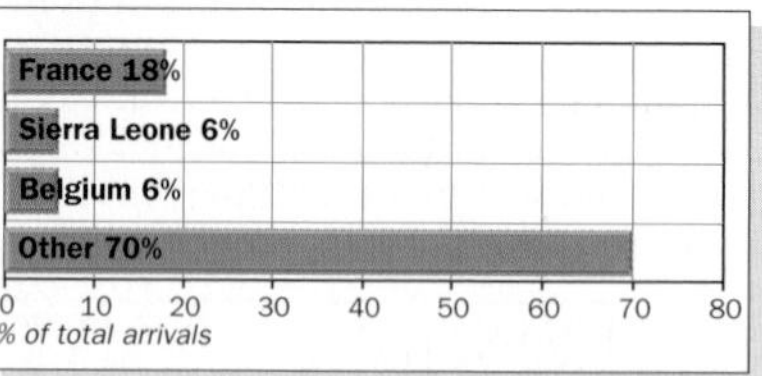
France 18%
Sierra Leone 6%
Belgium 6%
Other 70%
0 10 20 30 40 50 60 70 80
% of total arrivals

Limited infrastructure means that Guinea cannot exploit the tourist potential of its beaches, scenery, and rich culture.

PEOPLE

Pop. density low

Fulani, Malinke, Soussou, French

34/km²
(87/mi²)

THE URBAN/RURAL POPULATION SPLIT

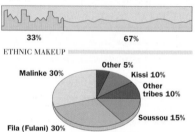

33% 67%

ETHNIC MAKEUP

Malinke 30%, Other 5%, Kissi 10%, Other tribes 10%, Soussou 15%, Fila (Fulani) 30%

Since the death of Marxist dictator Sekou Touré in 1984, traditional rivalries have reemerged between ethnic groups. The two largest groups are the Fulani, based in the highland region of Fouta Djallon, and the Malinke, who lost the power they had held under Touré, and have suffered reprisals. Today, the coastal peoples, including the Soussou, are dominant, benefiting from renewed rivalry between the Malinke and Fulani.

The extended family system survived the climate of suspicion generated by paid informers under Sekou Touré. Women acquired influence within his Marxist party, but a Muslim revival since 1984 has reversed this trend.

Hundreds of thousands of refugees, fleeing from conflicts in neighboring countries, are now caught up in fighting in the southern border region.

POLITICS

Multiparty elections

2002/2007

President Lansana Conté

AT THE LAST ELECTION

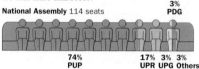

National Assembly 114 seats

74% PUP; 17% UPR; 3% UPG; 3% Others; 3% PDG

PUP = Party of Unity and Progress **UPR** = Union for Progress and Renewal **UPG** = Union for the Progress of Guinea **PDG** = Democratic Party of Guinea

The death in 1984 of Sekou Touré, head of the Marxist single-party PDG regime from 1958, opened the way for the military to intervene, with promises of multiparty elections. A 1990 referendum overwhelmingly approved democratic changes, but the military appointed a Transitional Committee, delaying elections until 1993. Incumbent leader Gen. Lansana Conté heavily defeated the Malinke leader, Alpha Condé, who had been in exile until 1992. Amid claims that the election was rigged, serious violence erupted. A disputed victory for Conté's PUP in the 1995 legislative elections was followed by his own reelection in 1998. Fighting escalated into civil war in 2000, with cross-border incursions from rebels based in Sierra Leone and Liberia. A controversial constitutional referendum in 2001 allowed Conté to stand for a possible third term, but the legislative polls set for that December were postponed. When they were eventually held in July 2002, the PUP increased its majority.

WORLD AFFAIRS

Joined UN in 1958

ECOWAS

OIF

AU

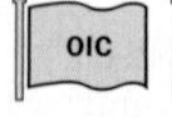
OIC

OMVG

Balancing the interests of the major aid donors, France and the US, is a growing concern. Relations with neighboring states, badly affected by conflicts in border regions, were patched up symbolically in February 2002.

AID

Recipient

$153m (receipts)

Down 36% in 2000

In 1969, the World Bank funded the Boké bauxite project, then one of its most ambitious projects. Since 1986, Western aid has grown to finance over 85% of all development projects. The 1997–2000 World Bank/IMF structural reform program foresees an annual growth rate of 5%.

GUINEA

Total Land Area : 245 857 sq. km (94 925 sq. miles)

POPULATION
- over 500 000
- over 50 000
- over 10 000
- under 10 000

LAND HEIGHT
- 1000m/3281ft
- 500m/1640ft
- 200m/656ft
- Sea Level

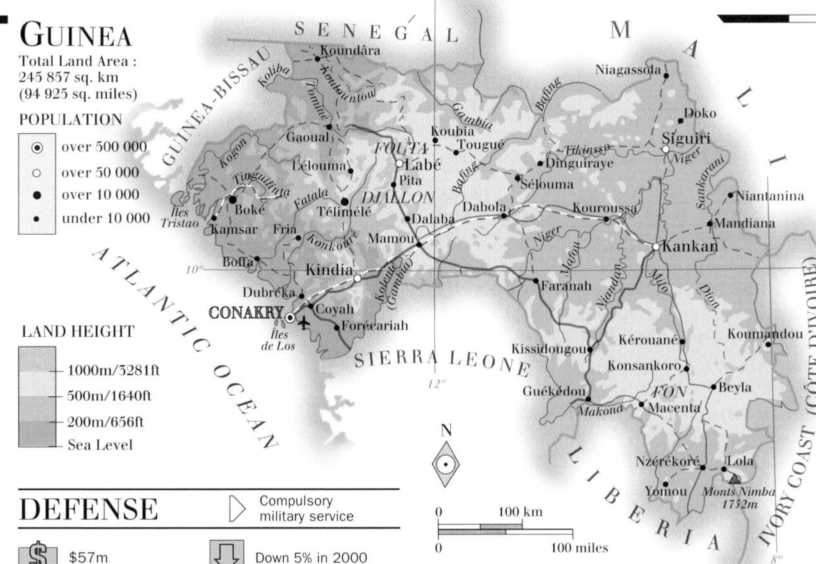
SENEGAL
MALI
GUINEA-BISSAU
ATLANTIC OCEAN
SIERRA LEONE
LIBERIA
IVORY COAST (CÔTE D'IVOIRE)
Koundâra
Niagassola
Doko
Siguiri
Koubia
Tougué
Gaoual
Labé
Pita
FOUTA DJALLON
Dinguiraye
Sélouma
Niantanina
Mandiana
Kankan
Kouroussa
Dabola
Dalaba
Mamou
Télimélé
Boké
Kamsar
Fria
Boffa
Kindia
Dubréka
CONAKRY
Coyah
Forécariah
Îles de Los
Îles Tristao
Faranah
Kissidougou
Guékédou
Kérouané
Konsankoro
Koumandou
Beyla
Macenta
FON
Makona
Nzérékoré
Lola
Yomou
Monts Nimba 1752m
Niger
Milo
Tinkisso
Konkouré
Kolente
Gambia
Bafing
Kogon
Koliba
Tominé
Sankarani
Dion
Mafou
Niandan
N
0 100 km
0 100 miles

DEFENSE

Compulsory military service

 $57m

Down 5% in 2000

Defense forces consist of an army and a militia – partly merged since the 1984 coup – a gendarmerie, and a tiny navy and air force. China, North Korea, and the Soviet bloc used to be the main arms procurement markets. Weaponry is now supplied by France and the US.

ECONOMICS

Inflation 5.1% p.a. (1990–2000)

$3.3bn

1880–1970 Guinea francs

SCORE CARD

- ❑ WORLD GNP RANKING........................128th
- ❑ GNP PER CAPITA $450
- ❑ BALANCE OF PAYMENTS....................–$165m
- ❑ INFLATION 6%
- ❑ UNEMPLOYMENT..........Widespread underemployment

STRENGTHS

Natural resources including bauxite, gold, and diamonds. Major iron ore deposits at Mount Nimba. Good soil and climate give high cash-crop yields. Relatively low inflation.

WEAKNESSES

Legacy of maladministration from Touré years. Poor infrastructure. 1990–1997 Liberian civil war set back major joint projects. Current refugee situation a drain on resources.

EXPORTS

IMPORTS

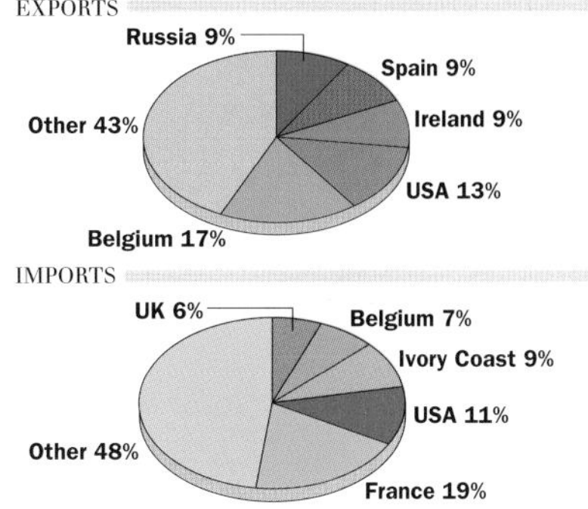
Russia 9%
Spain 9%
Ireland 9%
USA 13%
Belgium 17%
Other 43%
UK 6%
Belgium 7%
Ivory Coast 9%
USA 11%
France 19%
Other 48%

RESOURCES

Electric power 186,000 kw

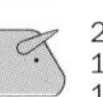 87,314 tonnes

 Not an oil producer

2.68m cattle, 1.01m goats, 11.9m chickens

Bauxite, diamonds, gold, iron

Bauxite accounts for over 90% of export earnings. Guinea, with 30% of known world reserves, is the world's largest producer after Australia. Demand for electricity for bauxite processing is high. Gold production has grown steadily since 1995.

ENVIRONMENT

Sustainability rank: 98th

 0.7%

 0.2 tonnes per capita

Uncontrolled deforestation, particularly of large areas of rainforest, is the major long-term problem.

MEDIA

TV ownership low

 Daily newspaper circulation 2 per 1000 people

PUBLISHING AND BROADCAST MEDIA

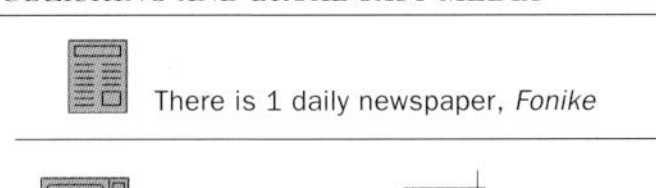
There is 1 daily newspaper, *Fonike*
1 state-owned service
1 state-owned service

Guinea's limited broadcast media are state-owned. The main newspaper, *Horoya*, is a weekly. There has been a slight relaxation in censorship.

CRIME

 Death penalty in use

 4014 prisoners

 Up 20% in 1992

The death penalty was reintroduced in 2001 in an attempt to crack down on spiraling crime. Cross-border diamond smuggling is at the root of anarchy in the south.

CHRONOLOGY

France colonized Guinea in 1890, strongly opposed by the Fulani Muslim empire of Fouta Djallon.

- ❑ **1958** Full independence under Sekou Touré.
- ❑ **1984** Touré dies. Army coup.
- ❑ **1993–1995** Disputed elections.
- ❑ **1998** Conté reelected president.
- ❑ **2000** Cross-border rebel attacks place Guinea in a state of civil war.
- ❑ **2002** PUP wins delayed elections.

EDUCATION

School leaving age: 13

 41%

 8151 students

French was readopted as the main teaching language in 1984, after Sekou Touré's Marxist-inspired experiments.

HEALTH

No welfare state health benefits

1 per 10,000 people

 Malaria, diarrheal and respiratory diseases, tuberculosis

Health provision is very poor, reflected in Guinea's high infant mortality rate and low average life expectancy. Private health care was legalized in 1984.

SPENDING

GDP/cap. increase

CONSUMPTION AND SPENDING

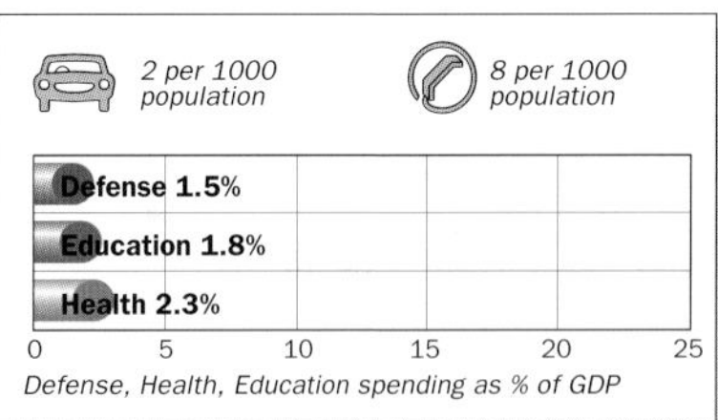
2 per 1000 population
8 per 1000 population
Defense 1.5%
Education 1.8%
Health 2.3%
0
5
10
15
20
25
Defense, Health, Education spending as % of GDP

Private enterprise has brought with it a new business class and Guinea now has some wealthy exiles, but much of the country remains poor and underdeveloped; GNP is below $500 per capita.

WORLD RANKING

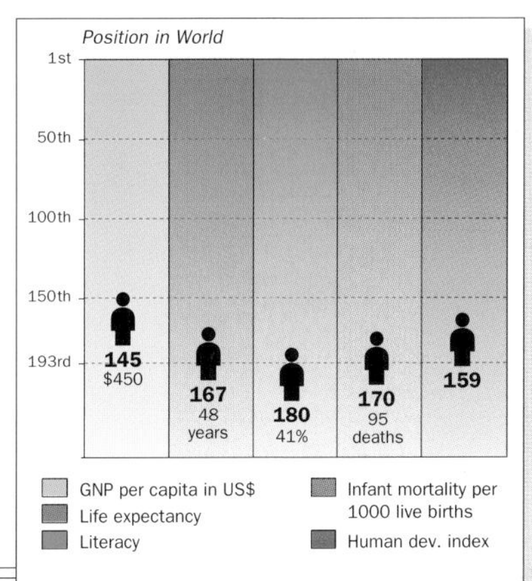
Position in World
1st
50th
100th
150th
193rd
145
$450
167
48 years
180
41%
170
95 deaths
159
GNP per capita in US$
Life expectancy
Literacy
Infant mortality per 1000 live births
Human dev. index

G

GUINEA-BISSAU

WEST AFRICA
Senegal
Mali
GUINEA-BISSAU
Guinea
Africa

OFFICIAL NAME: Republic of Guinea-Bissau **CAPITAL:** Bissau
POPULATION: 1.2 million **CURRENCY:** CFA franc **OFFICIAL LANGUAGE:** Portuguese

1974

1974

Sept 24

GNB

0

+245 .gw

LYING ON AFRICA'S west coast, impoverished Guinea-Bissau, a former Portuguese territory, is bordered by Senegal to the north and Guinea to the south and east. Apart from savanna highlands in the northeast, the country is low-lying. The PAIGC initiated a process of change to multiparty democracy in 1990, and elections were held in 1994. A military coup in 1999 followed army rebellion the previous year, but legislative and presidential elections have since been held.

CLIMATE

Tropical monsoon

WEATHER CHART FOR BISSAU

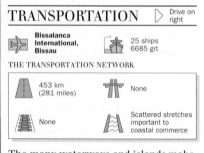

Average daily temperature — Rainfall
°C/°F J F M A M J J A S O N D cm/in
60/140, 50/122, 40/104, 30/86, 20/68, 10/50, 0/32
90/35, 80/31, 70/28, 60/24, 50/20, 40/16, 30/12, 20/8, 10/4, 0

The climate is tropical. The north is affected by the Sahel, the wetter south by the Atlantic. Droughts can occur.

TRANSPORTATION

Drive on right

Bissalanca International, Bissau

25 ships 6685 grt

THE TRANSPORTATION NETWORK

453 km (281 miles)

None

None

Scattered stretches important to coastal commerce

The many waterways and islands make water transportation as vital as the roads. Both are being improved.

TOURISM

Not available

A small number of visitors

No significant change from year to year

MAIN TOURIST ARRIVALS

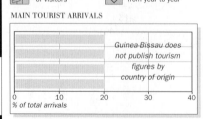

Guinea-Bissau does not publish tourism figures by country of origin

0 10 20 30 40
% of total arrivals

Lack of tourist facilities means that the country remains a destination for only the most independent of travelers.

PEOPLE

Pop. density low

Portuguese Creole, Balante, Fulani, Malinke, Portuguese

43/km² (111/mi²)

THE URBAN/RURAL POPULATION SPLIT

24% 76%

RELIGIOUS PERSUASION

Christian 8%
Muslim 40%
Indigenous beliefs 52%

About 98% of Guinea-Bissau's people come from indigenous ethnic groups. The largest is the southern Balante, who form almost one-third of the population. Mixed-race *mestiço* and European minorities make up just 2% of the population. Although small in number, the *mestiços* – many of whom derive from Cape Verde, Portugal's other former west African colony – still dominate the bureaucracy. Resentment at this, especially among the Balante, who provided most of the PAIGC troops in the independence war, was one cause of the 1980 coup. The majority of the population live and work on small family farms, grouped in self-contained villages. The bulk of the urban population live in the capital, Bissau.

POLITICS

Multiparty elections

1999/2003

President Kumba Yalla

AT THE LAST ELECTION

National People's Assembly 102 seats

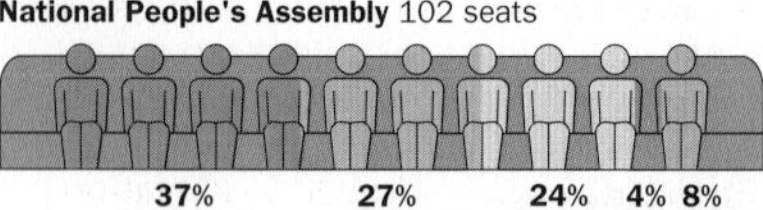
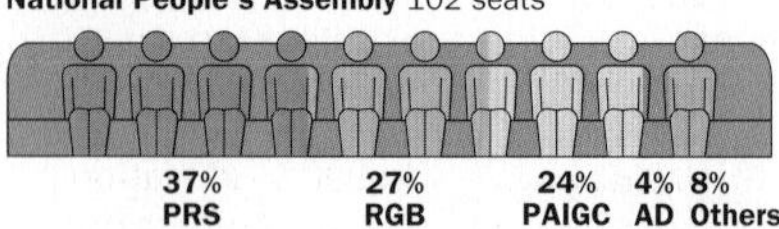

37% **PRS** **27%** **RGB** **24%** **PAIGC** **4%** **AD** **8%** **Others**

PRS = Party for Social Renewal **RGB** = Guinea-Bissau Resistance **PAIGC** = African Party for the Independence of Guinea and Cape Verde **AD** = Alliance for Democracy

Twenty years of one-party rule ended in 1994 with the holding of multiparty elections. However, opposition groups disputed the ruling PAIGC's victory in these polls.

A period of instability led to an army rebellion in 1998 and eight months of fighting between loyalists of President João Bernardo Vieira and the army chief, Gen. Ansumane Mane, which displaced about half the population. ECOWAS troops intervened and a national unity government was formed, only to be overthrown in May 1999.

Mane, the coup leader, convened fresh elections which were won by the PRS, whose candidate Kumba Yalla was elected president in January 2000. Mane was killed in another, failed, coup attempt, but Yalla now faces resistance from the Assembly itself over his increasingly erratic style of government.

WORLD AFFAIRS

Joined UN in 1974

ECOWAS

CPLP

OIF

AU

OIC

Relations with neighboring states are extremely tense due to the activities of various rebel militias on the borders. Senegalese and Guinean troops intervened in mid-1998 to suppress the army rebellion.

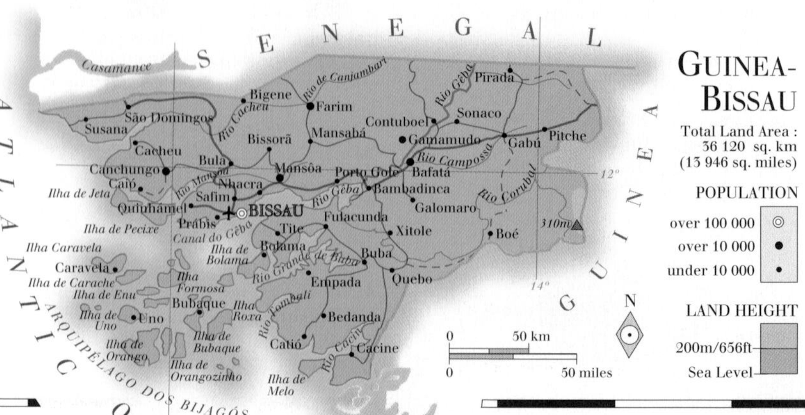
SENEGAL
GUINEA
ATLANTIC OCEAN
ARQUIPÉLAGO DOS BIJAGÓS
Casamance
São Domingos
Susana
Cacheu
Bigene
Farim
Rio Cacheu
Rio de Canjambari
Bissorã
Mansabá
Contuboel
Pirada
Sonaco
Rio Gêba
Gamamudo
Gabú
Pitche
Rio Campossa
Bula
Canchungo
Mansôa
Porto Gole
Bafatá
Bambadinca
Galomaro
Rio Corubal
Caió
Ilha de Jeta
Nhacra
Safim
Quinhamel
BISSAU
Prábis
Ilha de Pecixe
Canal do Gêba
Fulacunda
Xitole
Boé
310m
Tite
Ilha Caravela
Caravela
Ilha de Bolama
Bolama
Rio Grande de Buba
Buba
Ilha de Carache
Ilha de Enu
Ilha Formosa
Empada
Quebo
Bubaque
Ilha Roxa
Ilha de Uno
Uno
Rio Tombali
Bedanda
Ilha de Orango
Ilha de Bubaque
Catió
Rio Cacine
Cacine
Ilha de Orangozinho
Ilha de Melo
12°
14°
16°
0 50 km
0 50 miles
N

GUINEA-BISSAU

Total Land Area : 36 120 sq. km (13 946 sq. miles)

POPULATION

over 100 000
over 10 000
under 10 000

LAND HEIGHT

200m/656ft
Sea Level

AID

 Recipient

$80m (receipts) — Up 54% in 2000

Portugal is Guinea-Bissau's most important single aid donor. Balance-of-payments support is critical to the economy. Export earnings are small, compared with the costs of imports and debt servicing. Donor support was frozen in 1991 beause of the country's World Bank arrears, but the government pushed ahead with economic reforms begun in the mid-1980s, and the World Bank and the IMF agreed a $790 million debt relief package in 2001. Education, the infrastructure, and health care are the main targets of project aid.

DEFENSE

Compulsory military service

$6m — No change in 2000

There are around 9000 troops. The army led coups in 1980 and 1999, and suffered internal rebellions in 1998 and 2000, continuing a history of military interference in politics. ECOWAS soldiers have intervened to restore order on a number of occasions.

ECONOMICS

Inflation 33% p.a. (1990–2000)

 $217m

 698.7–736.7 CFA francs

SCORE CARD

- ❑ WORLD GNP RANKING.......183rd
- ❑ GNP PER CAPITA.......$180
- ❑ BALANCE OF PAYMENTS.......–$27m
- ❑ INFLATION.......8.6%
- ❑ UNEMPLOYMENT.......Widespread underemployment

STRENGTHS

Minimal at present, but good potential in fisheries and timber. Offshore oil potential.

WEAKNESSES

Lack of sufficiency in rice staple. Fish stocks depleted by poaching. Few exports, mainly cashew nuts, groundnuts. Minimal industry. High illiteracy. Poor state economic management.

EXPORTS

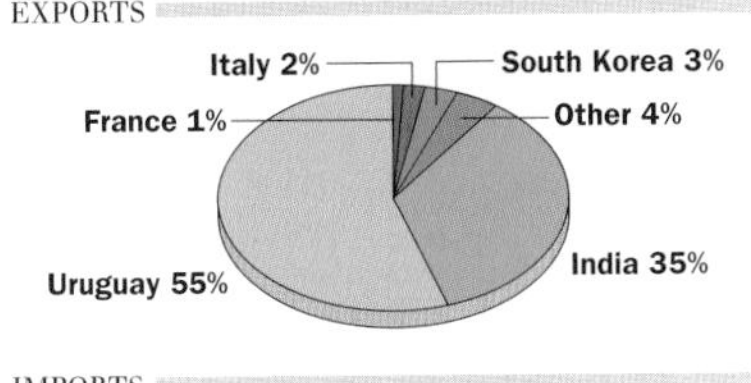
Italy 2%
South Korea 3%
France 1%
Other 4%
India 35%
Uruguay 55%

IMPORTS

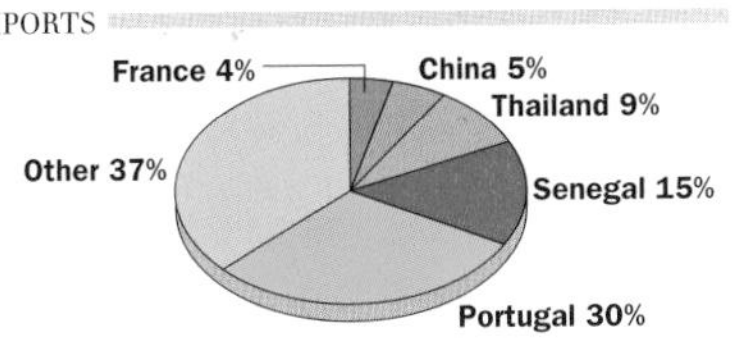
France 4%
China 5%
Thailand 9%
Other 37%
Senegal 15%
Portugal 30%

Bafatá, the chief town in central Guinea-Bissau. It lies on the Gêba River and is also an important inland port.

RESOURCES

Electric power 11,000 kw

5000 tonnes

 Not an oil producer

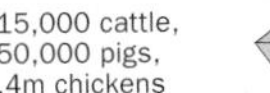 515,000 cattle, 350,000 pigs, 1.4m chickens

 Bauxite, phosphates

Fish and timber are the main natural resources, but local exploitation is only a tiny proportion of the sustainable levels. There is considerable potential for developing hydropower and for offshore oil production.

ENVIRONMENT

Sustainability rank: 127th

None

 0.2 tonnes per capita

Drought and locust plagues are serious natural hazards. A small population and minimal industry mean that there are few serious environmental problems.

MEDIA

TV ownership low

 Daily newspaper circulation 5 per 1000 people

PUBLISHING AND BROADCAST MEDIA

There is 1 daily newspaper, *Nô Printcha*, published by the government

 1 state-owned service

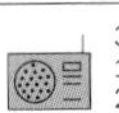 3 services: 1 state-owned, 2 independent

There are three independently owned newspapers that publish intermittently due to the high costs of paper and distribution. Portugal helps to fund the TV service, started in 1989.

CRIME

No death penalty

Guinea-Bissau does not publish prison figures

 Up 66% in 1992

The death penalty was abolished in 1993. Reform of the legal system is in progress to make it more independent. Human rights abuses increased during the instability in the late 1990s.

EDUCATION

School leaving age: 13

 39%

 463 students

Around 65% of children receive rudimentary education. Guinea-Bissau has no university.

CHRONOLOGY

Explored by the Portuguese in the 15th century, Portuguese Guinea was established in 1879. A war for independence began in the 1960s.

- ❑ **1974** Independence. PAIGC takes power.
- ❑ **1980** Military coup.
- ❑ **1990** Multiparty politics accepted.
- ❑ **1994** Multiparty elections.
- ❑ **1998** Army rebellion led by Gen. Mane. ECOWAS intervention.
- ❑ **1999** Transitional government. May, army seizes power. November, PRS defeats PAIGC in elections.
- ❑ **2000** Kumba Yalla president. Mane killed in failed coup attempt.

G

HEALTH

No welfare state health benefits

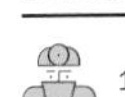 1 per 5000 people

 Parasitic, diarrheal, and communicable diseases, malaria

Guinea-Bissau's health statistics are among the world's worst, due partly to the minimal medical facilities. Average life expectancy is just 45 years; infant mortality is 126 per 1000 live births; the maternal death rate is high. In mid-2000 the AfDB provided $500,000 in funding for an emergency health program.

SPENDING

GDP/cap. increase

CONSUMPTION AND SPENDING

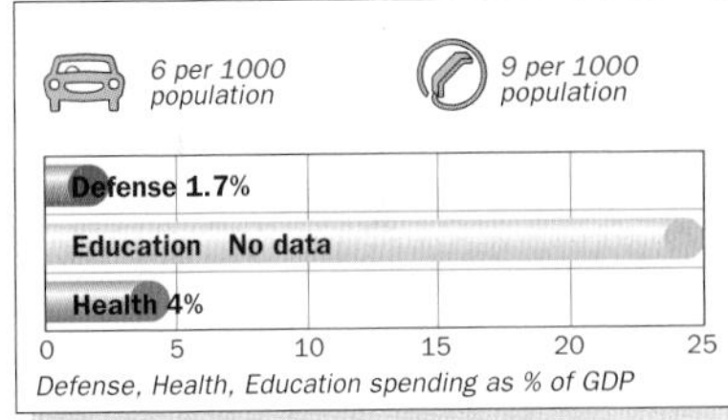
6 per 1000 population
9 per 1000 population
Defense 1.7%
Education No data
Health 4%
0
5
10
15
20
25
Defense, Health, Education spending as % of GDP

Living conditions for the majority of Guinea-Bissau's people are extremely poor; over 70% of the population are unable to meet their basic needs. The tiny elite is mainly *mestiço*.

WORLD RANKING

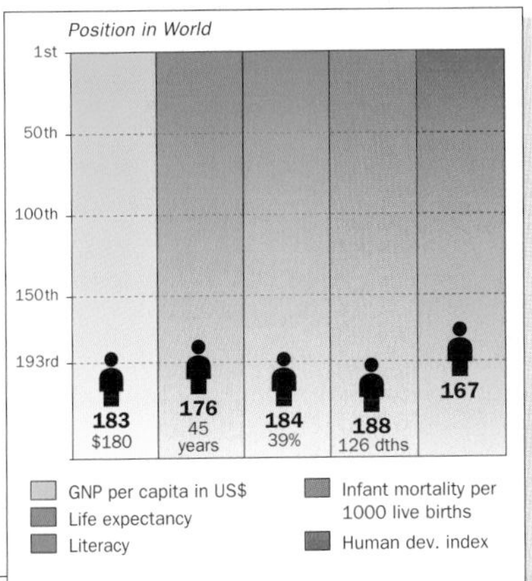
Position in World
1st
50th
100th
150th
193rd
183
$180
176
45 years
184
39%
188
126 dths
167
GNP per capita in US$
Life expectancy
Literacy
Infant mortality per 1000 live births
Human dev. index

GUYANA

SOUTH AMERICA
Venezuela
GUYANA
Suriname
Brazil
South America

OFFICIAL NAME: Cooperative Republic of Guyana **CAPITAL:** Georgetown
POPULATION: 763,000 **CURRENCY:** Guyana dollar **OFFICIAL LANGUAGE:** English

1966

1966

Feb 23
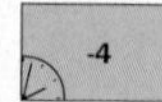
GUY

-4

+592
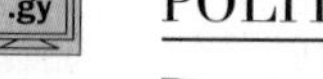
.gy

LYING ON THE NORTHERN EDGE of South America, Guyana stretches 600 km (375 miles) from dense tropical rainforests, through broad savanna and mountains dotted with waterfalls, to the narrow Atlantic coastal plain where most of the population lives. A British colony from 1814 until independence in 1966, Guyana has closer ties with the mostly anglophone Caribbean than with its Spanish-, Portuguese-, and Dutch-speaking neighbors.

CLIMATE

Tropical equatorial

WEATHER CHART FOR GEORGETOWN

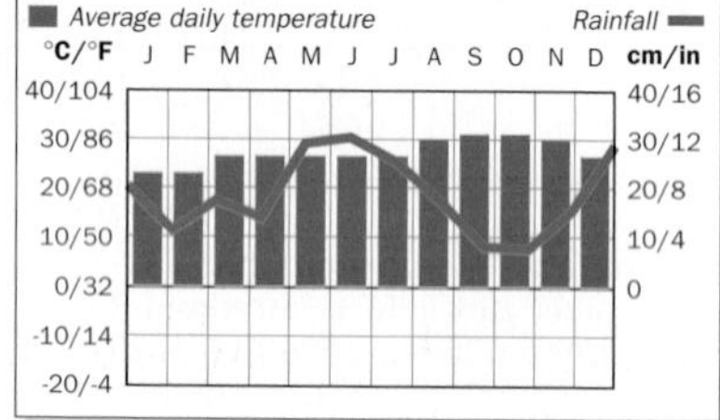
Average daily temperature
Rainfall
°C/°F
J F M A M J J A S O N D
cm/in
40/104
30/86
20/68
10/50
0/32
-10/14
-20/-4
40/16
30/12
20/8
10/4
0

The lowlands are very humid with a constant temperature. The highlands are a little cooler, especially at night.

TRANSPORTATION

Drive on left

Timehri International, Georgetown
270,500 passengers

61 ships
15,669 grt

THE TRANSPORTATION NETWORK

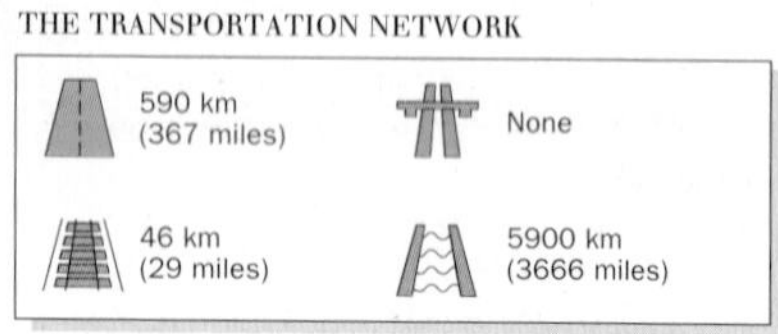
590 km
(367 miles)
None
46 km
(29 miles)
5900 km
(3666 miles)

Reliable travel to the interior is by air or river; most paved roads are coastal. The only international airport is Timehri.

TOURISM

Visitors : Population 1:10

75,000 visitors
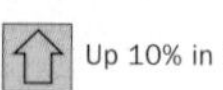
Up 10% in 1999

MAIN TOURIST ARRIVALS

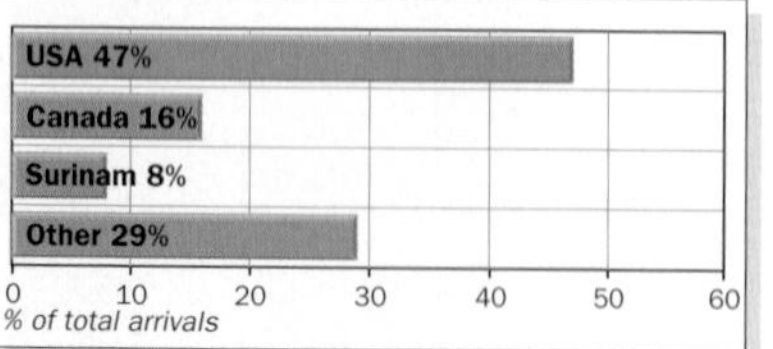
USA 47%
Canada 16%
Surinam 8%
Other 29%
0
10
20
30
40
50
60
% of total arrivals

The government promotes tourism, but the number of tourists is modest. Guyana means Land of Many Waters; the Kaieteur Falls are among the world's most impressive. Old Dutch wooden architecture characterizes Georgetown.

Modest homes, Georgetown. Most buildings are made of wood. The cathedral is one of the world's tallest freestanding wooden buildings.

PEOPLE

Pop. density low

English Creole, Hindi, Tamil, Amerindian languages, English

4/km² (10/mi²)

THE URBAN/RURAL POPULATION SPLIT

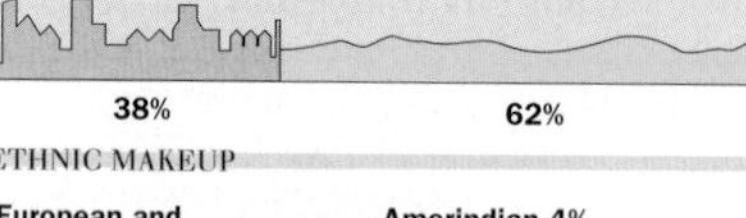
38%
62%
ETHNIC MAKEUP
European and Chinese 2%
Amerindian 4%
Other 4%
East Indian 52%
Black African 38%

Tension exists between the Afro-Guyanese, descended from Africans brought over in the 17th to 19th centuries, and the Indo-Guyanese, descendants of south Asian laborers brought from India in the 19th century. This is currently displayed in the hostility existing between the opposition PNC, representing Afro-Guyanese, and the ruling PPP, traditionally representing the Indo-Guyanese.

POLITICS

Multiparty elections

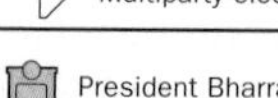
2001/2006

President Bharrat Jagdeo

AT THE LAST ELECTION

National Assembly 65 seats
3% GAP–WPA
2% TUF
52% PPP–CIVIC
41% PNC
2% ROAR

PPP–CIVIC = People's Progressive Party–CIVIC
PNC = People's National Congress
GAP–WPA = Guyana Action Party–Working People's Alliance
ROAR = Rise, Organize, and Rebuild **TUF** = The United Force

The success of the PPP in the 1992 elections, widely seen as the first fair poll since independence, ended the dominance of the pro-Afro-Guyanese PNC. Veteran PPP leader Cheddi Jagan, a Marxist before adopting free-market policies, died in office in 1997. The PNC violently contested the succession of his wife Janet. A Caricom-brokered peace restored calm, though tension flared in early 2001 when the High Court condemned the 1997 poll. Violence erupted once again when elections in 2001 confirmed Janet Jagan's successor, Bharrat Jagdeo, as president.

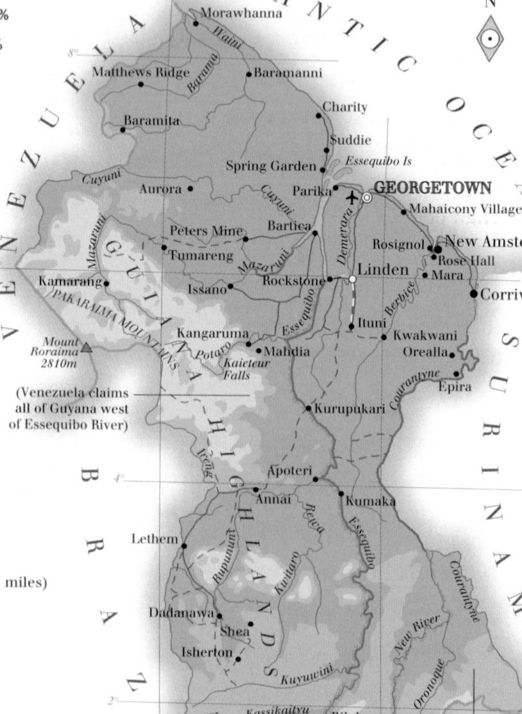
ATLANTIC OCEAN
VENEZUELA
BRAZIL
SURINAME
GEORGETOWN
Morawhanna
Matthews Ridge
Baramanni
Charity
Baramita
Suddie
Spring Garden
Essequibo Is
Aurora
Parika
Mahaicony Village
Peters Mine
Bartica
Tumareng
Rosignol
New Amsterdam
Rose Hall
Linden
Mara
Kamarang
Issano
Rockstone
Corriver
PAKARAIMA MOUNTAINS
GUIANA HIGHLANDS
Mount Roraima 2810m
Ituni
Kangaruma
Mahdia
Kwakwani
Orealla
Kaieteur Falls
Epira
(Venezuela claims all of Guyana west of Essequibo River)
Kurupukari
Apoteri
Annai
Kumaka
Lethem
Dadanawa
Shea
Isherton
Biloku
ACARAI MOUNTAINS
(claimed by Suriname)
Cuyuni
Mazaruni
Demerara
Berbice
Essequibo
Potaro
Courantyne
Rupununi
Kuyuwini
Kassikaityu
New River
0
100 km
100 miles
N

GUYANA

Total Land Area : 214 970 sq. km (83 000 sq. miles)

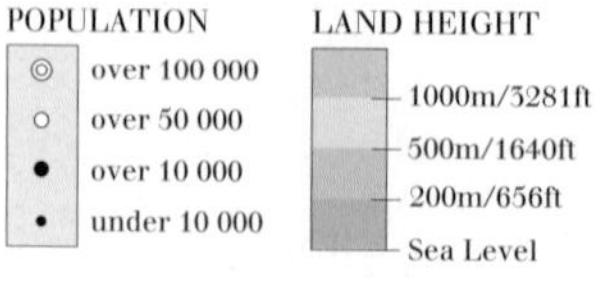
POPULATION
over 100 000
over 50 000
over 10 000
under 10 000
LAND HEIGHT
1000m/3281ft
500m/1640ft
200m/656ft
Sea Level

WORLD AFFAIRS

Joined UN in 1966

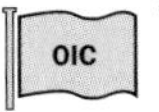

ACS | OIC | Caricom | Comm | OAS

Rescheduling debt with Western creditor states is paramount. Also important are the long-standing border dispute with Venezuela and closer integration with the Caribbean.

AID

Recipient

 US$108m (receipts)

 Up 35% in 2000

Most aid comes from the US, the EU, and the UK. Recent grants covered public health projects, business development, and protection of the rainforest.

DEFENSE

No compulsory military service

 US$7m

 No change in 2000

The security forces, which include a small land army, benefit from financial support and training provided by the US and UK governments.

ECONOMICS

Inflation 13% p.a. (1990–2000)

 US$652m

 180.5 Guyana dollars

SCORE CARD

- World GNP Ranking 163rd
- GNP per Capita .... US$860
- Balance of Payments .... –US$117m
- Inflation 6.1%
- Unemployment 12%

STRENGTHS

Gold, rice, sugar, diamonds, bauxite, and timber production. Good tourism potential. Debt reduction plan agreed with multilateral agencies.

WEAKNESSES

High per capita foreign debt. Political instability dents investor confidence. Currency vulnerable to exchange rate pressure. Main exports vulnerable to fluctuations in international commodity prices. Weak manufacturing base.

EXPORTS

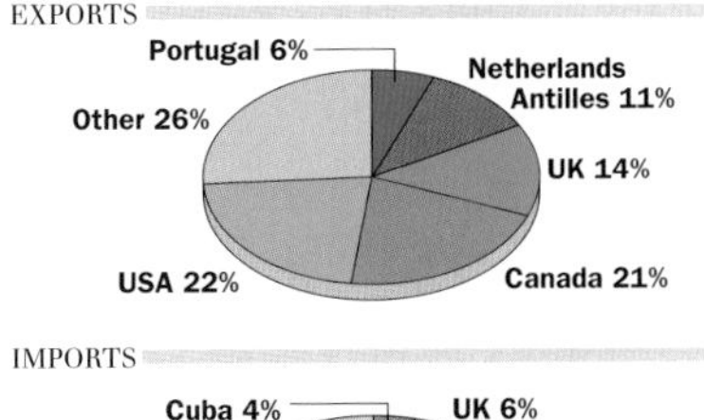
Portugal 6%
Netherlands Antilles 11%
Other 26%
UK 14%
USA 22%
Canada 21%

IMPORTS

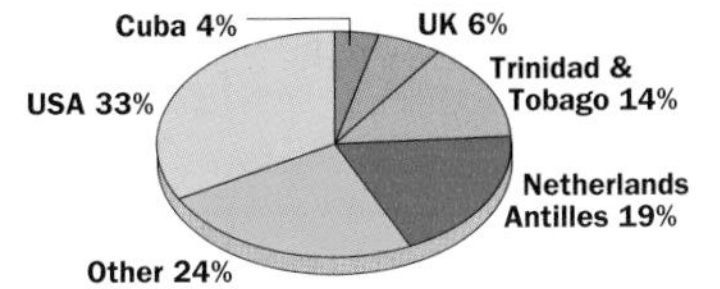
Cuba 4%
UK 6%
USA 33%
Trinidad & Tobago 14%
Netherlands Antilles 19%
Other 24%

RESOURCES

Electric power 202,000 kw

54,450 tonnes

Minimal oil production

220,000 cattle, 130,000 sheep, 12.5m chickens

 Gold, diamonds, bauxite, gemstones, oil, manganese, uranium

Gold, diamonds, bauxite, and timber are major resources. Offshore and onshore prospecting for oil has not reduced the need for petroleum imports for electricity generation. More hydroelectric power plants are being constructed.

ENVIRONMENT

Not available

0.05%

2.2 tonnes per capita

The state of disrepair of the 18th-century sea defense system endangers the urbanized coastline that lies below sea level. Commercial logging threatens to deplete the rainforest. The pollution of rivers due to mining activities is now a serious problem.

MEDIA

TV ownership medium

 Daily newspaper circulation 50 per 1000 people

PUBLISHING AND BROADCAST MEDIA

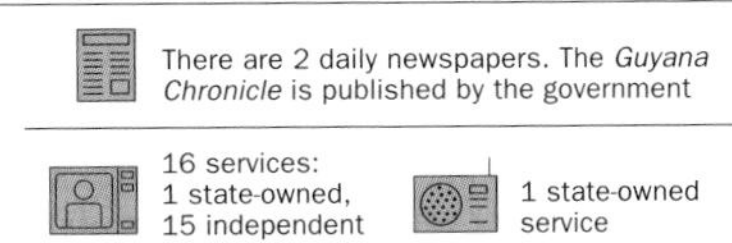

There are 2 daily newspapers. The *Guyana Chronicle* is published by the government

16 services: 1 state-owned, 15 independent

1 state-owned service

The PPP and the PNC both publish periodicals. The government owns one TV service and the sole radio service as well as publishing a daily newspaper.

CRIME

Death penalty in use

 1507 prisoners

 Down 6% in 1999

The police are strongly criticized for corruption and ineffectiveness in the face of rising urban crime. Serious violence between PNC and PPP–CIVIC supporters erupted in 1998, 1999, and 2001.

EDUCATION

School leaving age: 15

 98%

 8965 students

Education is based on the former British system. Entry to high schools is by 11-plus examination. There is a state-financed university, though many students go to the US or the UK.

HEALTH

Welfare state health benefits

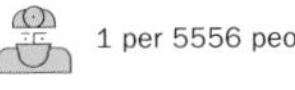 1 per 5556 people

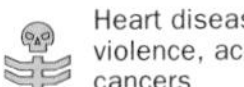 Heart diseases, violence, accidents, cancers

Nearly all of Guyana's population have access to the country's mainly state-run health service. The referral system is relatively good.

CHRONOLOGY

During the 17th and 18th centuries, the Dutch founded three colonies, Essequibo, Demerara, and Berbice, in the region. In 1814, these came under British control, and were later combined to form British Guiana.

- **1953** First universal elections won by PPP under Cheddi Jagan; parliament later suspended by UK.
- **1966** Independence from UK.
- **1973** PPP boycotts parliament, accusing PNC of electoral fraud.
- **1992** Fair elections won by PPP. Jagan president.
- **1997–1998** Jagan dies in office; PNC rejects his widow's election victory. Political crisis.
- **1999** Caricom-brokered peace deal. Janet Jagan resigns; Bharrat Jagdeo takes over as president.
- **2001** Jagdeo and PPP reelected. Political violence flares again.

SPENDING

GDP/cap. increase

CONSUMPTION AND SPENDING

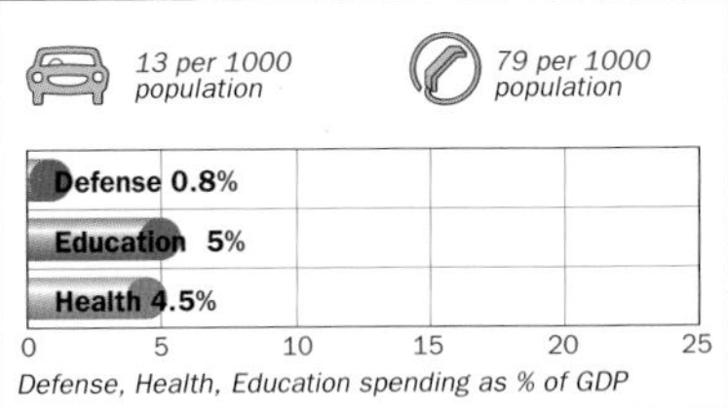
13 per 1000 population
79 per 1000 population
Defense 0.8%
Education 5%
Health 4.5%
0
5
10
15
20
25
Defense, Health, Education spending as % of GDP

Significant urban and rural poverty in Guyana has forced the government to make provision in the budget for poverty alleviation. Redundancies in the public sector exacerbate the problem. The poorest group in society are Amerindian subsistence farmers. There are a few very affluent urban families who derive their wealth not only from business but also from rural farming interests.

WORLD RANKING

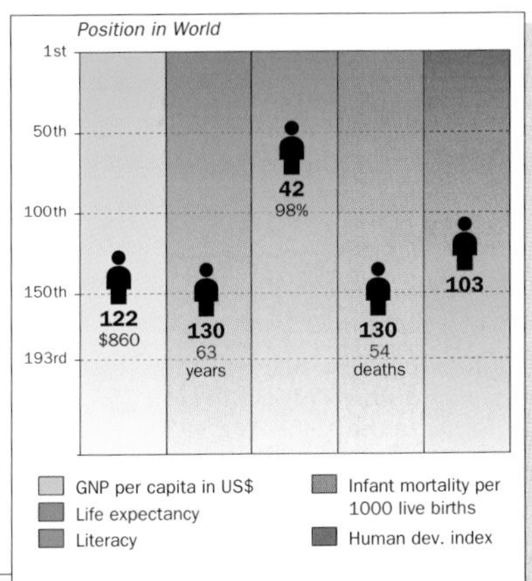
Position in World
1st
50th
100th
150th
193rd
122
$860
130
63 years
42
98%
130
54 deaths
103
GNP per capita in US$
Life expectancy
Literacy
Infant mortality per 1000 live births
Human dev. index

G

HAITI

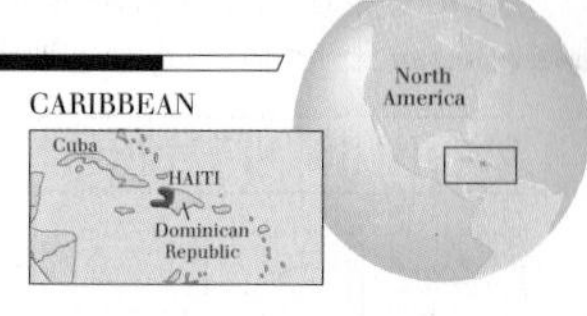
North America
CARIBBEAN
Cuba
HAITI
Dominican Republic

OFFICIAL NAME: Republic of Haiti **CAPITAL:** Port-au-Prince **POPULATION:** 8.3 million
CURRENCY: Gourde **OFFICIAL LANGUAGES:** French and French Creole

1804

1844
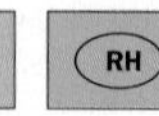
Jan 1
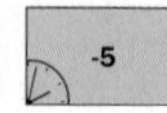
RH

-5

+509

.ht

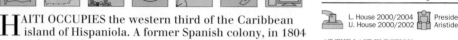

HAITI OCCUPIES the western third of the Caribbean island of Hispaniola. A former Spanish colony, in 1804 it was the first Caribbean state to become independent, and has been in a state of political chaos virtually ever since. Democracy did not materialize with the exile of the dictator Jean-Claude Duvalier in 1986. Elections were held in 1990, but by 1991 the military were back in power and were ousted in 1994 only through US intervention.

CLIMATE

Tropical equatorial/ oceanic

WEATHER CHART FOR PORT-AU-PRINCE

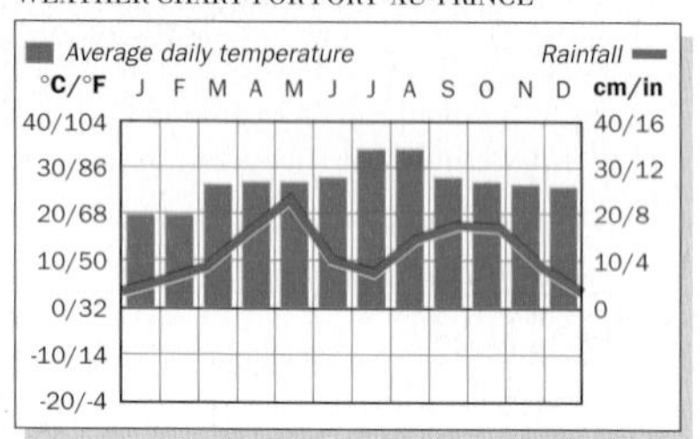
Average daily temperature
Rainfall
°C/°F J F M A M J J A S O N D cm/in
40/104
30/86
20/68
10/50
0/32
-10/14
-20/-4
40/16
30/12
20/8
10/4
0

Haiti lies mostly in the rain shadow of the central mountains, so humidity is lower than the Caribbean average.

TRANSPORTATION

Drive on right

Port-au-Prince
962,599 passengers

4 ships
1200 grt

THE TRANSPORTATION NETWORK

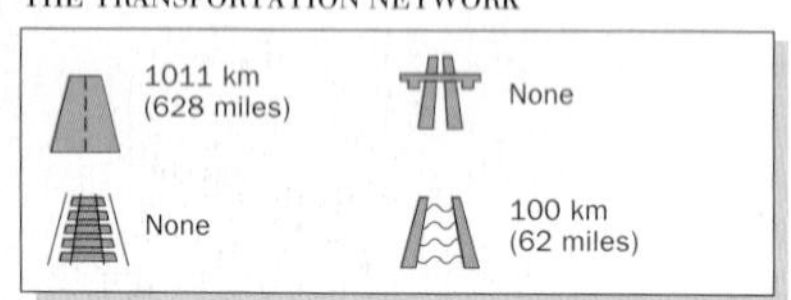
1011 km
(628 miles)
None
None
100 km
(62 miles)

Roads are poor, especially in the interior. Ferries provide the main transportation to the southern peninsula.

TOURISM

Visitors : Population 1:58

143,000 visitors

Down 3% in 1999

MAIN TOURIST ARRIVALS

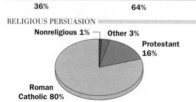

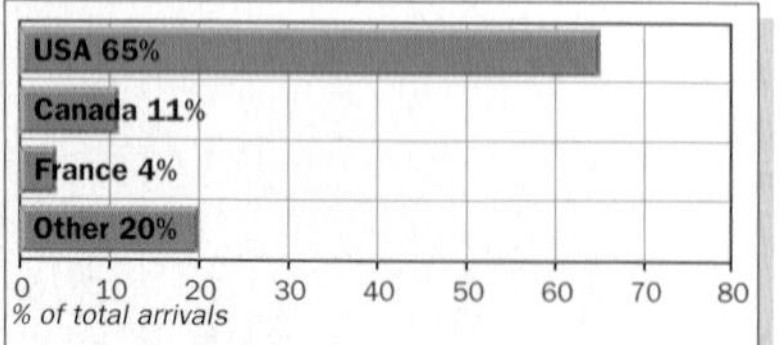
USA 65%
Canada 11%
France 4%
Other 20%
0 10 20 30 40 50 60 70 80
% of total arrivals

Haiti's location, history, and culture provided much of its attraction for tourists in the 1960s and 1970s. Political instability and violence in the 1980s, however, led to the industry's near collapse and it has yet to recover.

PEOPLE

Pop. density high

French Creole, French

301/km² (780/mi²)

THE URBAN/RURAL POPULATION SPLIT

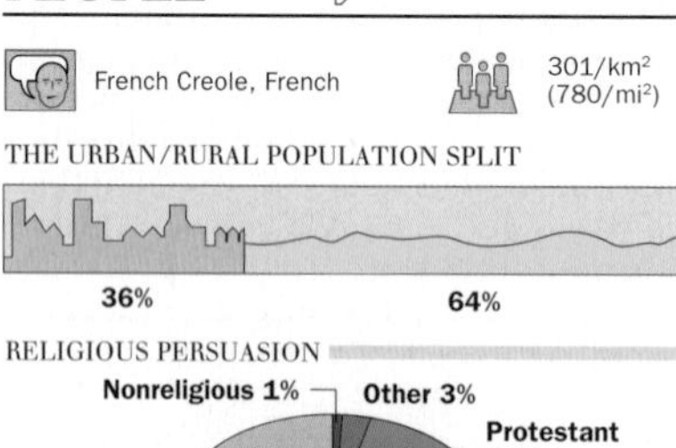
36%
64%
RELIGIOUS PERSUASION
Nonreligious 1%
Other 3%
Protestant 16%
Roman Catholic 80%

Most Haitians are the descendants of Africans; a few have European roots, primarily French. The majority of the population lives in extreme poverty: Haiti is the poorest country in the Americas, and Port-au-Prince has the worst slums in the Caribbean. Social tensions run high, and focus on class rather than race. In recent years, the combination of political repression and a collapsing economy led many to emigrate illegally to the US, or across the border to the neighboring Dominican Republic.

POLITICS

Multiparty elections

L. House 2000/2004
U. House 2000/2002

President Jean-Bertrand Aristide

AT THE LAST ELECTION

Chamber of Deputies 83 seats

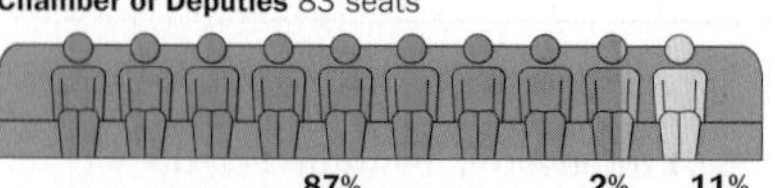

87% Lavalas coalition | 2% Vacant | 11% Others

Senate 27 seats

96% Lavalas coalition | 4% Others

A wealthy elite, backed by the military, supported the Duvalier dictatorships and regularly financed coups after "Baby Doc" Duvalier's overthrow in 1986. UN sanctions followed the 1991 coup and US forces restored the elected president, Jean-Bertrand Aristide, in 1994. His left-wing Lavalas party won legislative elections in 1995 and René Préval was installed as the new president, with the party's backing, in 1996. A backlash against austerity policies insisted on by the US forced the prime minister to step down in 1997 and soured relations between the presidency and the legislature. New elections, finally held in May 2000, resulted in a strongly disputed but decisive victory for the Lavalas coalition. A similarly controversial presidential poll in November was won by Aristide. The opposition rejected his victory, leading to effective political stalemate.

HAITI

Total Land Area : 27 750 sq. km (10 714 sq. miles)

POPULATION
- over 1 000 000
- over 500 000
- over 10 000
- under 10 000

LAND HEIGHT
- 1000m/3281ft
- 500m/1640ft
- 200m/656ft
- Sea Level

0 50 km
0 50 miles

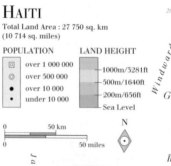

WORLD AFFAIRS

Joined UN in 1945

ACS | ACP | Geplac | OAS | WTO

Continued political instability has cut off international aid and relations with major donors. Major issues are illegal immigration to the US and relations with the Dominican Republic.

AID

Recipient

$208m (receipts)

Down 21% in 2000

The IMF granted $21 million in emergency aid in 1998 for hurricane damage. The IDB approved loans for water and health problems, and Taiwan granted $60.4 million in aid.

DEFENSE

No compulsory military service

$48m

Down 4% in 2000

In 1994, the military were ousted and democracy was restored. The armed forces and police were disbanded and an interim public security force was created. A 5300-strong new national police force has now been formed, funded and trained by the US.

ECONOMICS

Inflation 20% p.a. (1990–2000)

$4.06bn

21.00-25.74 gourdes

SCORE CARD

- ❑ WORLD GNP RANKING........................117th
- ❑ GNP PER CAPITA..............................$510
- ❑ BALANCE OF PAYMENTS.....................–$38m
- ❑ INFLATION..13.7%
- ❑ UNEMPLOYMENT.................................70%

STRENGTHS

Coffee exports. Remittances by Haitians living abroad. US demand for goods assembled in Haiti. Large profits from transshipment of narcotics to US.

WEAKNESSES

Huge tax avoidance. Foreign investment and promised aid deterred by political instability.

EXPORTS

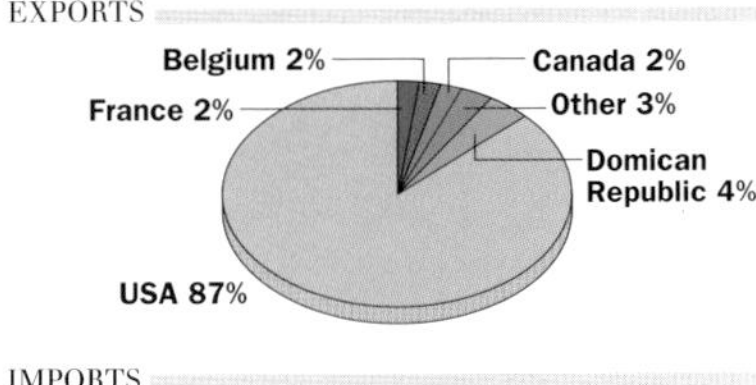
Belgium 2%
Canada 2%
France 2%
Other 3%
Domican Republic 4%
USA 87%

IMPORTS

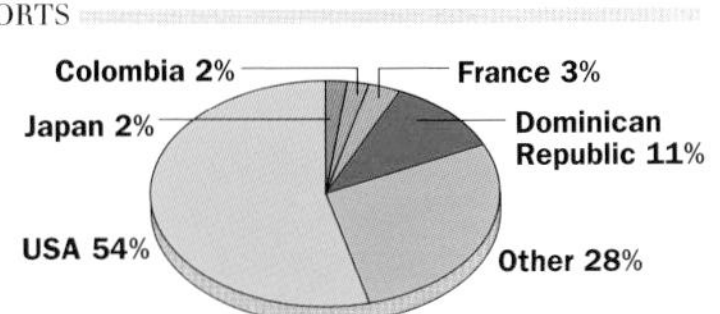
Colombia 2%
France 3%
Japan 2%
Dominican Republic 11%
USA 54%
Other 28%

Haiti: the poorest country in the Americas. *In remote villages, most houses are made of earth and do not have glass in their windows.*

RESOURCES

Electric power 264,000 kw

5000 tonnes

Not an oil producer

1.94m goats, 1.44m cattle, 1m pigs, 5.5m chickens

Marble, limestone, clay, silver, gold, natural asphalt

Haiti has no strategic resources. Under prolonged economic sanctions, it had to find unofficial sources of oil; much was imported from Europe.

ENVIRONMENT

Sustainability rank: 137th

0.4%

0.2 tonnes per capita

One-third of soil is seriously eroded and forest cover is now only 1.5% of total land area. The removal of 4000 tonnes of toxic waste, illegally dumped in 1988 near Gonaïves, finally began in 1998.

MEDIA

TV ownership low

Daily newspaper circulation 3 per 1000 people

PUBLISHING AND BROADCAST MEDIA

There are 2 daily newspapers, *Le Nouvelliste* and *Le Matin*

5 services: 1 state-owned, 4 independent

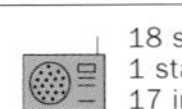
18 services: 1 state-owned, 17 independent

Under military rule, the media were largely controlled through intimidation. The transition to multiparty democracy has produced a more open press.

CRIME

No death penalty

4152 prisoners

Crime is rising

Extrajudicial killings, torture, and brutality continue, despite the ending of military dictatorship. Narcotics trafficking is highly organized. Police are inexperienced and the judicial system is slow and open to corruption.

EDUCATION

School leaving age: 15

50%

6288 students

The run-down state system is based on the French model; the *baccalauréat* pass rate is consistently under 10%. The wealthy educate their children abroad.

CHRONOLOGY

In 1697, Spain ceded the west of Hispaniola to France. Ex-slave Toussaint Louverture's rebellion in 1791 led to independence in 1804.

- ❑ **1915–1934** US occupation.
- ❑ **1957–1971** François "Papa Doc" Duvalier's brutal dictatorship.
- ❑ **1971–1986** His son Jean-Claude, "Baby Doc," rules; eventually flees.
- ❑ **1986–1988** Military rule.
- ❑ **1990** Jean-Bertrand Aristide elected; exiled in 1991 coup.
- ❑ **1994–1995** US forces oust military. Aristide reinstated; elections.
- ❑ **1997–1999** Political deadlock.
- ❑ **2000** Lavalas coalition and Aristide reelected.

H

HEALTH

No welfare state health benefits

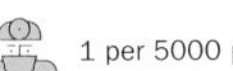
1 per 5000 people

Malaria, other parasitic diseases, tuberculosis

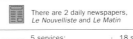

Most Haitians cannot afford health care. In rural areas, help is often sought from voodoo priests.

SPENDING

GDP/cap. decrease

CONSUMPTION AND SPENDING

4 per 1000 population

9 per 1000 population

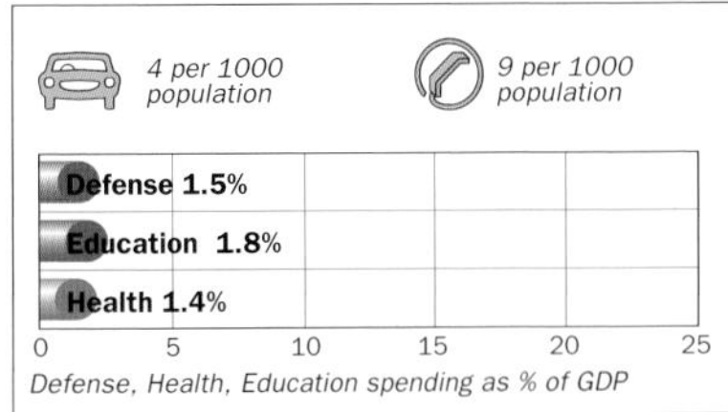
Defense 1.5%
Education 1.8%
Health 1.4%
0
5
10
15
20
25

Defense, Health, Education spending as % of GDP

Haiti's rigid class structure maintains extreme disparities of wealth between a few affluent families and the mass of the population, who live in slums without running water or proper sanitation. According to the UN, around 80% of Haitians cannot meet their basic daily needs.

WORLD RANKING

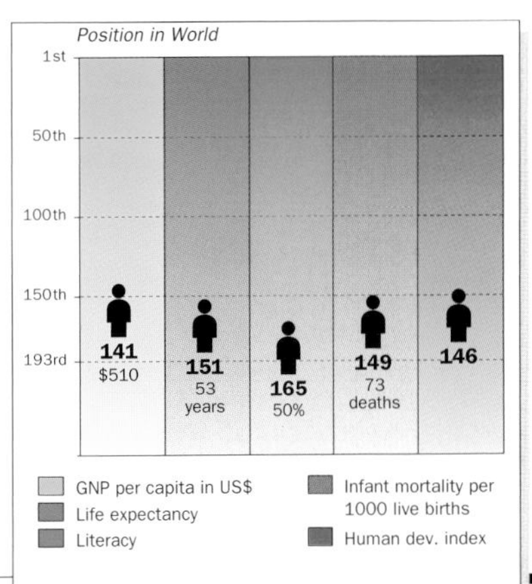
Position in World
1st
50th
100th
150th
193rd
141
$510
151
53 years
165
50%
149
73 deaths
146
GNP per capita in US$
Life expectancy
Literacy
Infant mortality per 1000 live births
Human dev. index

HONDURAS

OFFICIAL NAME: Republic of Honduras **CAPITAL:** Tegucigalpa
POPULATION: 6.6 million **CURRENCY:** Lempira **OFFICIAL LANGUAGE:** Spanish

CENTRAL AMERICA

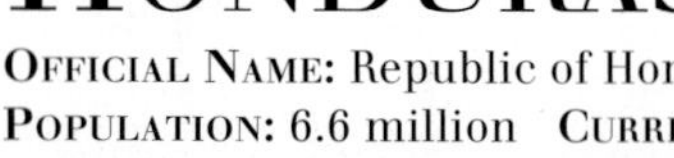

1838

1838

Sept 15

HN

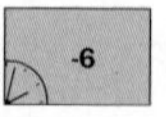
-6

+504

.hn

MOST OF HONDURAS is mountainous terrain, with a small sheltered Pacific coast to the south and a broad Caribbean shoreline to the north, including part of the virtually uninhabited Mosquito Coast. After a succession of military governments it returned to full civilian rule in 1984. In 1998 Honduras was devastated by Hurricane Mitch, which resulted in the death of at least 5600 people and damage estimated at some $3 billion.

CLIMATE

Tropical equatorial

WEATHER CHART FOR TEGUCIGALPA

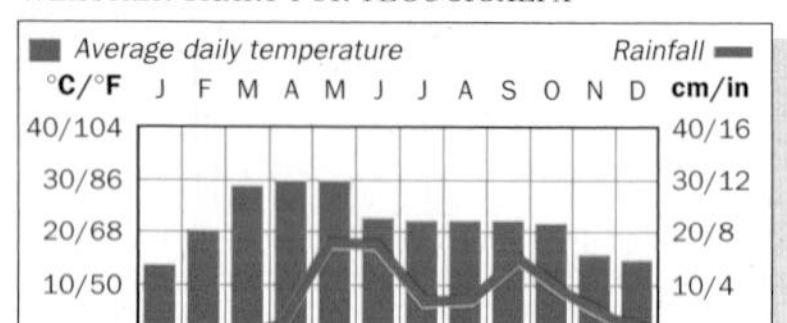
Average daily temperature
Rainfall
°C/°F J F M A M J J A S O N D cm/in
40/104 30/86 20/68 10/50 0/32 -10/14 -20/-4
40/16 30/12 20/8 10/4 0

Honduras' Caribbean coastline is generally extremely hot. The rest of the country is much cooler.

TRANSPORTATION

Drive on right

La Mesa, San Pedro Sula 635,859 passengers

1407 ships 1.11m grt

THE TRANSPORTATION NETWORK

3126 km (1942 miles)

None

996 km (619 miles)

465 km (289 miles)

In 1998 Hurricane Mitch destroyed roads and bridges across the country; reconstruction will take many years.

TOURISM

Visitors : Population 1:16

408,000 visitors

Up 10% in 2000

MAIN TOURIST ARRIVALS

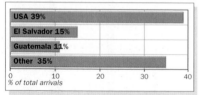

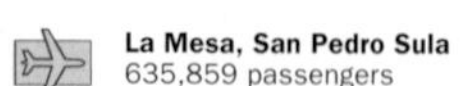
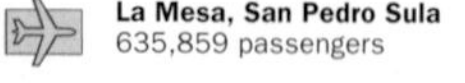
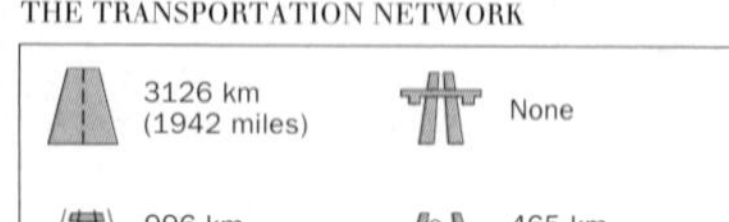
USA 39%
El Salvador 15%
Guatemala 11%
Other 35%
0 10 20 30 40
% of total arrivals

Caribbean coast resorts and the Bay Islands are popular, while exploring the remote region inland from the Mosquito Coast and jungle rafting appeal to the adventurous. The ruined Maya temples of Copán are a major draw.

PEOPLE

Pop. density medium

Spanish, Black Carib, English Creole

59/km² (153/mi²)

THE URBAN/RURAL POPULATION SPLIT

53% 47%

ETHNIC MAKEUP

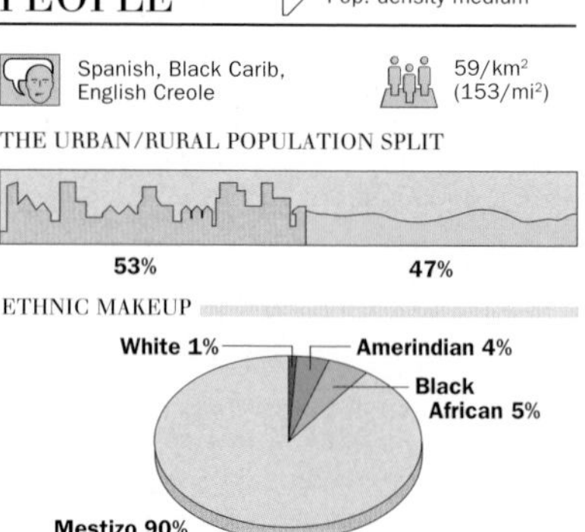
White 1%
Amerindian 4%
Black African 5%
Mestizo 90%

As in most of Central America, very few pure indigenous groups remain. The estimated 45,000 Miskito Amerindians, and an English-speaking *garífuna* (black) population on the Caribbean coast united in 1999 to oppose a constitutional amendment allowing foreigners to buy land in coastal areas, traditionally their communal lands. Poverty is at the root of social tension; whites still have the best opportunities.

Rural poverty and strong Roman Catholicism (97% are Roman Catholic) mean that the family is a powerful unifying force. The status of women is low; many work in domestic service.

POLITICS

Multiparty elections

2001/2005

President Ricardo Maduro

AT THE LAST ELECTION

National Congress 128 seats

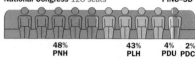

48% PNH · 43% PLH · 4% PDU · 2% PDCH · 3% PINU–SD

PNH = National Party of Honduras **PLH** = Liberal Party of Honduras **PDU** = Party of Democratic Unification **PINU–SD** = Innovation and Unity Party–Social Democracy **PDCH** = Honduran Christian Democratic Party

The traditional power brokers have been the military, the US embassy, and the United Fruit Company (now called Chiquita), the country's biggest banana producer.

The military held power intermittently from the mid-1950s, until pressure from the US government forced it to restore civilian rule in 1984. During the 1980s, US President Reagan effectively converted the country into a US "aircraft carrier" to counter a perceived communist threat from El Salvador and Nicaragua. Peace in the region then saw a cut in US aid.

The PNH and PLH have few real ideological differences. Presidents, able to serve only one four-year term, have tended to be weak. The PLH introduced unpopular austerity measures in 1994, but also began reducing the autonomy of the military by abolishing conscription. PLH president Carlos Flores, elected in 1997, continued this "demilitarization" process by naming a civilian defense minister in 1999. The presidency was won back by the PNH when Ricardo Maduro was elected in 2001.

Reconstruction after the devastation of Hurricane Mitch in 1998 will be a long-term undertaking.

HONDURAS

Total Land Area : 112 090 sq. km (43 278 sq. miles)

LAND HEIGHT: 2000m/6562ft · 1000m/3281ft · 500m/1640ft · 200m/656ft · Sea Level

POPULATION: over 500 000 · over 100 000 · over 50 000 · over 10 000 · under 10 000

WORLD AFFAIRS

Joined UN in 1945

ACS · Geplac · NAM · OAS · San José

Hurricane aid, and trade with and immigration to the US are key issues. In 2001 free trade was agreed with El Salvador, Guatemala, and Mexico.

AID

Recipient

 $449m (receipts)

 Down 45% in 2000

Aid from the IMF and the World Bank on favorable terms followed the 1998 earthquake. Western countries agreed debt relief of $1.2 billion.

DEFENSE

No compulsory military service

 $93m

 Down 2% in 2000

Until 1994 the military operated with virtual impunity. The first civilian defense minister was appointed in 1999, completing the "demilitarization" process begun with the return to civilian rule in 1984.

ECONOMICS

Inflation 19% p.a. (1990–2000)

$5.52bn

15.10–15.88 lempiras

SCORE CARD

- ❑ WORLD GNP RANKING 107th
- ❑ GNP PER CAPITA $860
- ❑ BALANCE OF PAYMENTS .... –$204m
- ❑ INFLATION 11.1%
- ❑ UNEMPLOYMENT 28%

STRENGTHS

Coffee, flowers, fruit. Economic boost due to hurricane reconstruction. Barely exploited mineral deposits. Hardwoods.

WEAKNESSES

Servicing of foreign debt. Vulnerability of coffee exports. Banana industry yet to recover from hurricane damage. Slow rate of privatizations. Corruption. Lack of land reform. High unemployment and underemployment. Weak industrial base. Overdependence on hydroelectric power.

EXPORTS

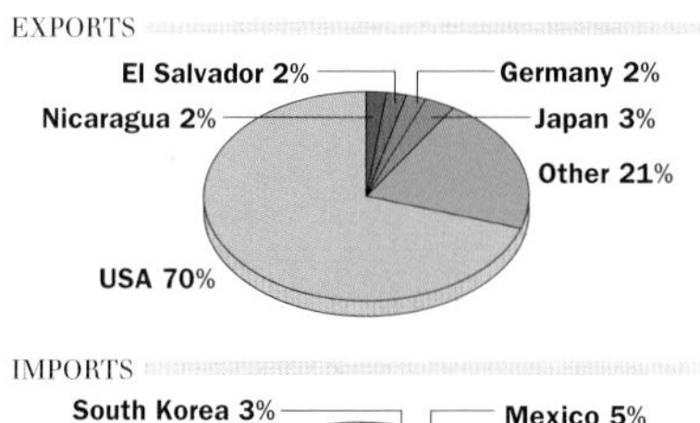
El Salvador 2%
Germany 2%
Nicaragua 2%
Japan 3%
Other 21%
USA 70%

IMPORTS

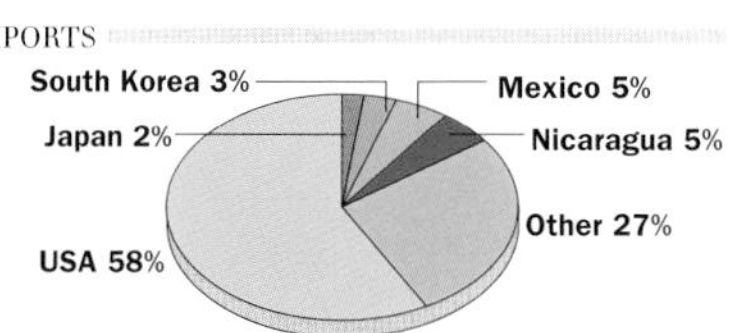
South Korea 3%
Mexico 5%
Japan 2%
Nicaragua 5%
Other 27%
USA 58%

Tobacco field. As an export earner tobacco is overshadowed by coffee, shrimps, and melons.

RESOURCES

Electric power 518,000 kw

15,395 tonnes

Not an oil producer

1.72m cattle, 480,000 pigs, 18m chickens

Lead, zinc, silver, gold, copper, iron, tin, coal

Coffee exports dropped dramatically in 2001, due to low world prices. Oil and mineral deposits are being explored. Hydroelectric supply is erratic.

ENVIRONMENT

Sustainability rank: 47th

 6%

 0.8 tonnes per capita

The unregulated timber, cotton, and cattle industries, land colonization, and pesticides have led to ecological crisis.

MEDIA

TV ownership medium

 Daily newspaper circulation 55 per 1000 people

PUBLISHING AND BROADCAST MEDIA

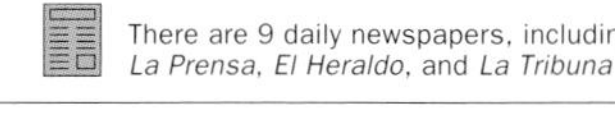 There are 9 daily newspapers, including *La Prensa*, *El Heraldo*, and *La Tribuna*

 6 independent services

 5 services: 1 state-owned, 4 independent

Self-censorship, dependence on US sources, corruption, and intimidation guarantee a largely compliant media.

CRIME

No death penalty

 10,869 prisoners

 Violent crime is rising

Overzealous security forces are frequently accused of murdering child members of violent street gangs.

EDUCATION

School leaving age: 13

 75%

 76,573 students

State-run education follows the US system. The drop-out rate from secondary schools is high.

HEALTH

Welfare state health benefits

 1 per 1250 people

 Circulatory, infectious, and parasitic diseases, malaria

Only 66% of people have easy access to health services. Plans were announced in 2000 to privatize the health service.

CHRONOLOGY

Honduras was a Spanish possession until 1821. In 1823, it formed the United Provinces of Central America with four neighboring nations.

- ❑ **1838** Declares full independence.
- ❑ **1890s** US banana plantations set up.
- ❑ **1932–1949** Dictatorship of General Tiburcio Carías Andino of PNH.
- ❑ **1954–1957** Elected PLH president Villeda Morales deposed, reelected.
- ❑ **1963** Military coup.
- ❑ **1969** 13-day Soccer War with El Salvador sparked by World Cup.
- ❑ **1980–1983** PLH wins elections but General Gustavo Alvarez holds real power. Military maneuvers with US. Trades unionists arrested; death squads operate.
- ❑ **1984** Return to democracy.
- ❑ **1988** 12,000 Contra rebels forced out of Nicaragua into Honduras.
- ❑ **1995** Military defies human rights charges.
- ❑ **1998** Hurricane Mitch wreaks havoc.
- ❑ **1999** Appointment of first civilian defense minister.

H

SPENDING

GDP/cap. increase

CONSUMPTION AND SPENDING

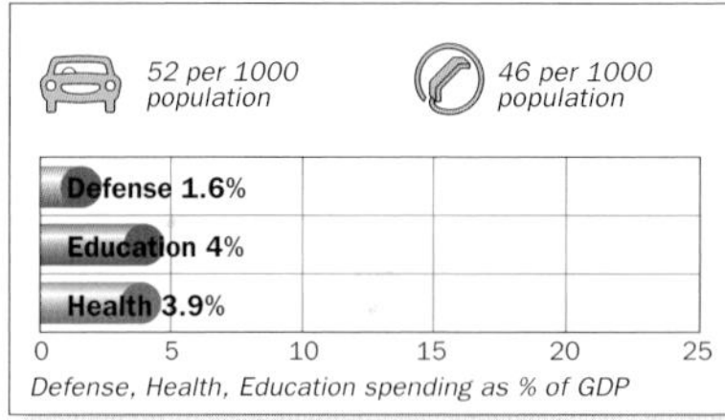
52 per 1000 population
46 per 1000 population
Defense 1.6%
Education 4%
Health 3.9%
0
5
10
15
20
25
Defense, Health, Education spending as % of GDP

The social structure of Honduras is characterized by great inequalities: 4% of people own 60% of the land. Relief agencies estimate that 85% of people now live below the poverty line, compared with 80% before the devastation of Hurricane Mitch.

WORLD RANKING

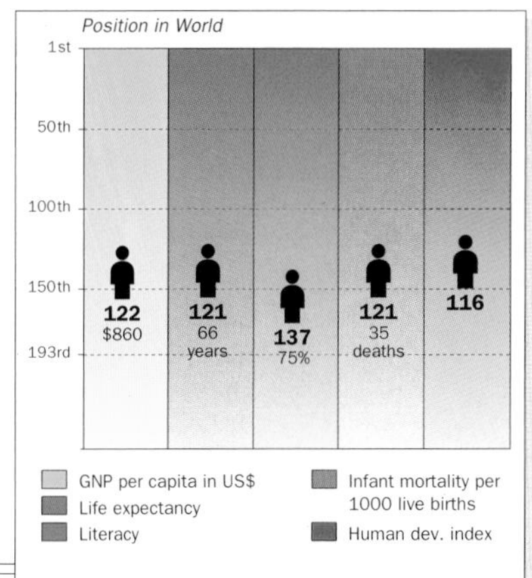
Position in World
1st
50th
100th
150th
193rd
122
$860
121
66 years
137
75%
121
35 deaths
116
GNP per capita in US$
Life expectancy
Literacy
Infant mortality per 1000 live births
Human dev. index

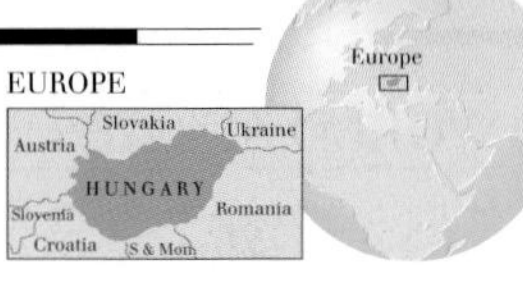
Europe
EUROPE
Slovakia
Ukraine
Austria
HUNGARY
Romania
Slovenia
Croatia
S & Mon

HUNGARY

OFFICIAL NAME: Republic of Hungary **CAPITAL:** Budapest
POPULATION: 9.9 million **CURRENCY:** Forint **OFFICIAL LANGUAGE:** Hungarian

1918 | 1947 | Aug 20 | H |

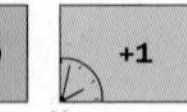
+1

+36

.hu

LYING AT THE HEART of central Europe, Hungary is landlocked and has borders with seven states. Historically, Hungary has been a cosmopolitan cultural center, and during its years of market socialism was more prosperous than the other Eastern bloc countries. Economic and political reforms have brought it closer to the EU, which it expects to join in the first "wave" of eastward enlargement; Hungary has also become a member of the NATO alliance. In foreign policy it is particularly sensitive about the treatment of Hungarian minorities in neighboring states.

CLIMATE

Continental

WEATHER CHART FOR BUDAPEST

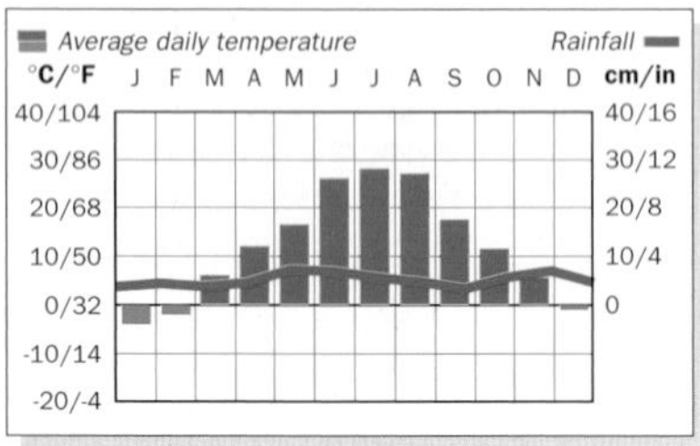

Average daily temperature / Rainfall
°C/°F J F M A M J J A S O N D cm/in
40/104 40/16
30/86 30/12
20/68 20/8
10/50 10/4
0/32 0
-10/14
-20/-4

Hungary has a continental climate, with wet springs, late summers, and cold, cloudy winters. There are no great differences of weather and climate within the country. Conditions in summer and winter may, however, differ from one year to the next. The transition between seasons tends to be sudden.

TRANSPORTATION

Drive on right

Budapest Ferihegy 4.68m passengers

Has no fleet

THE TRANSPORTATION NETWORK

80,680 km (50,132 miles)	448 km (278 miles)
7988 km (4964 miles)	1373 km (853 miles)

Freight travels mainly via the rail link from Budapest to the Austrian border. Most foreign investment is located along this corridor. A direct link to Slovenia opened in mid-2001. The Budapest–Vienna expressway was the first of four big EU-backed road projects to be completed.

TOURISM

Visitors : Population 1.5:1

15.3m visitors

Down 2% in 2001

MAIN TOURIST ARRIVALS

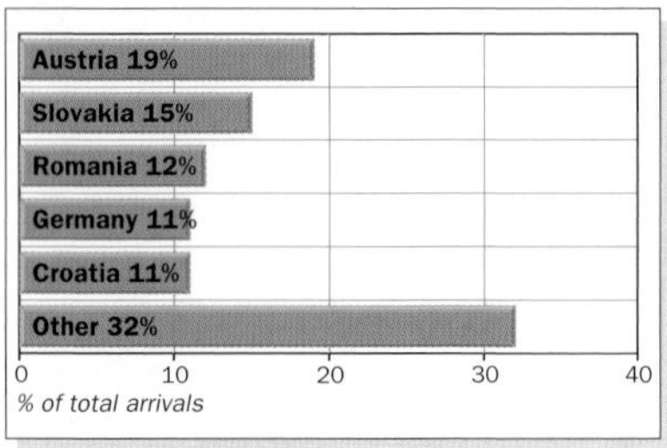

	% of total arrivals
Austria	19%
Slovakia	15%
Romania	12%
Germany	11%
Croatia	11%
Other	32%

0 10 20 30 40
% of total arrivals

Lake Balaton, the traditional destination for summer vacations, was a magnet for east European visitors during the communist period. Since then, Hungary has invested heavily in its tourist facilities, and the number of travel agents and hotels has risen dramatically. Austrians and Slovaks are most numerous among the new clientele the country now attracts. Budapest's baths, some of which date from the Ottoman period, are a distinctive feature, and the capital also promotes itself as an international business convention center.

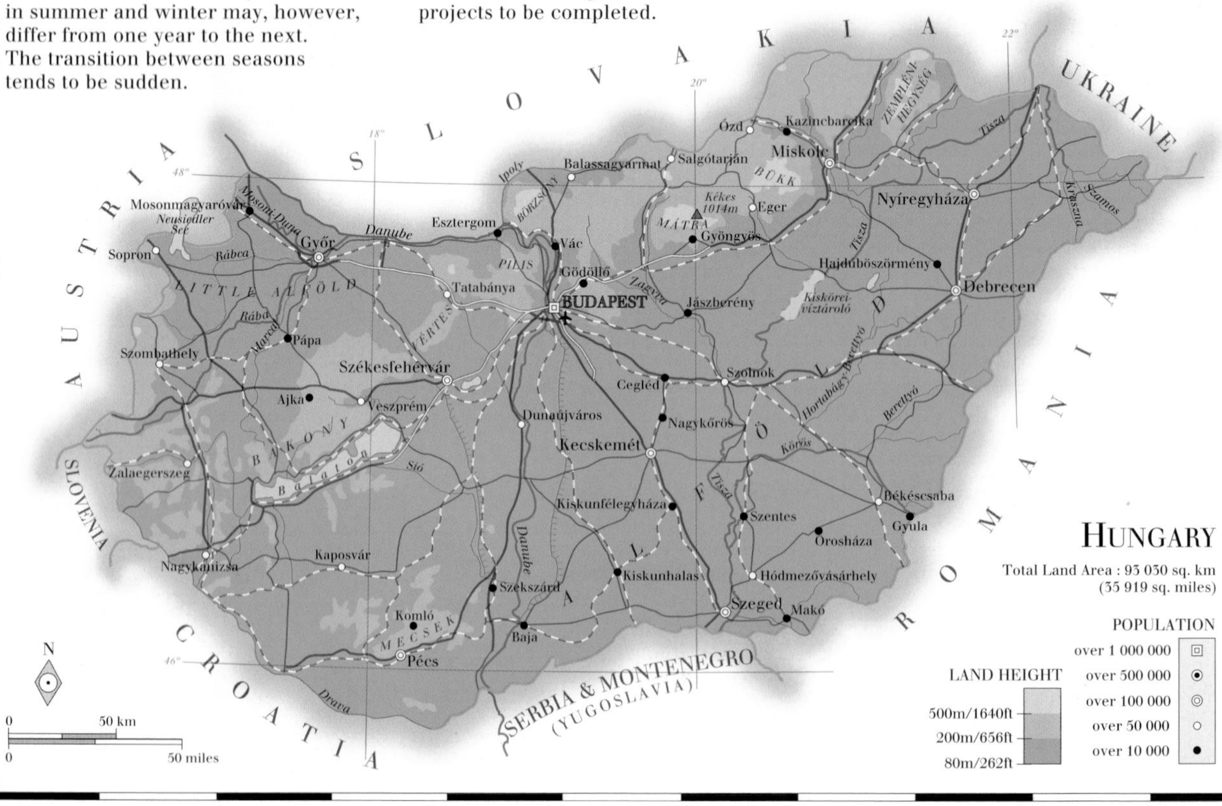

HUNGARY

Total Land Area : 93 030 sq. km (35 919 sq. miles)

POPULATION
over 1 000 000
over 500 000
over 100 000
over 50 000
over 10 000

LAND HEIGHT
500m/1640ft
200m/656ft
80m/262ft

0 50 km
0 50 miles

PEOPLE

Pop. density medium

Hungarian

107/km² (278/mi²)

THE URBAN/RURAL POPULATION SPLIT

64% 36%

RELIGIOUS PERSUASION

Greek Orthodox 3%, Lutheran 4%, Other 2%, Nonreligious 7%, Calvinist 20%, Roman Catholic 64%

ETHNIC MAKEUP

Roma 1%, German 2%, Slovak 1%, Other 6%, Magyar 90%

Hungary is ethnically homogeneous, although there are small minorities of Germans, Slovaks, Roma, Serbs, Croats, Romanians, and Hungarian Jews. There is little ethnic tension at home, but considerable concern about the treatment of Hungarian minorities in Romanian Transylvania, Serbian Vojvodina, and Slovakia. New legislation in June 2001 gave them special status in Hungary, including the right to work there for three months a year. The 100,000-strong Jewish community in Hungary is the largest in the region, although a fraction of the pre-Holocaust numbers, and is the target of occasional antisemitic outbursts. Prejudice against Roma is widespread and there have been major discrimination rows.

A new bourgeoisie is emerging, but for the unskilled and unemployed life is tougher than under communism. Hungary's suicide rate was the highest in the world throughout almost the whole of the 20th century.

POPULATION AGE BREAKDOWN

Female	Age	Male
1.9%	80+	0.8%
9.9%	60–79	6.8%
14.2%	40–59	13%
13.7%	20–39	14.1%
12.5%	0–19	13.1%

% of population by age group

POLITICS

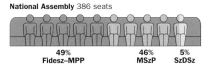

Multiparty elections

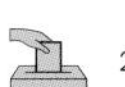

2002/2006

President Ferenc Mádl

AT THE LAST ELECTION

National Assembly 386 seats

49% **Fidesz–MPP** 46% **MSzP** 5% **SzDSz**

Fidesz–MPP = Young Democrats–Hungarian Civic Party
MSzP = Hungarian Socialist Party
SzDSz = Alliance of Free Democrats

Hungary has been a multiparty democracy since 1990.

PROFILE

Hungary's governments since the fall of communism have been relatively stable coalitions. The electoral pendulum has swung at four-yearly intervals between right and left.

József Antall, leader of the Christian democratic nationalist Hungarian Democratic Forum (MDF) and prime minister from 1990, was the dominant figure in Hungarian democratic politics until his death in 1993. However, party disintegrations and disappointing economic results increased apathy and disillusionment among voters, who returned the former communists to power in 1994. The victorious MSzP under Gyula Horn nevertheless preferred to work in coalition in order to ease the passage of economic and social reforms through parliament.

The right-of-center Fidesz–MPP coalition led by Viktor Orbán, which took office after the 1998 election, drove the country toward completing its transition to a market economy ready for EU membership. However, Orbán was narrowly defeated by a coalition of the MSzP and the SzDSz in the 2002 elections. Peter Medgyessy was appointed prime minister.

MAIN POLITICAL ISSUE
Social welfare versus free-market economics

Reforms to assist transition to a market economy have led to strong economic recovery in the Budapest area and the western part of the country. Widening income differentials between young, skilled workers and those in education, health, and other state sectors have provoked protests and strikes, as the new prosperity eludes others.

Ferenc Mádl, *known as "Mr. Professor," was elected president in 2000.*

Peter Medgyessy, *leader of the center-left MSzP and prime minister since 2002.*

The majestic Danube River *divides the modern city of Budapest between the ancient towns of Buda (foreground) and Pest.*

WORLD AFFAIRS

Joined UN in 1955

CE

CEFTA

NATO

OECD

OSCE

Hungary gained WEU associate status in 1994. In a 1997 referendum 85% of voters endorsed joining NATO, and in 1999 Hungary became a full NATO member. Joining the EU is a slower process. With an association agreement since early 1994, in 1998 Hungary was one of six applicant countries to open formal membership negotiations.

Hungary has a cooperation and friendship treaty with Russia, but relations have been strained by Hungary's open courting of the West. Difficult relations with Slovakia and Romania were eased by friendship treaties concluded in the mid-1990s, but were troubled again by the controversial status law in regard to Hungarians resident abroad.

CHRONOLOGY

The region today occupied by Hungary was first settled by the Finno-Ugrian Magyar peoples from the 8th century. In the 16th and 17th centuries, it came under Austrian domination, lasting until 1867, when Austria-Hungary was formed.

- **1918** Hungarian Republic created as successor state to Austria-Hungary.
- **1919** Béla Kún leads a short-lived communist government. Romania intervenes militarily and hands power to Adm. Horthy.
- **1938–1941** Hungary gains territory from Czechoslovakia, Yugoslavia, and Romania in return for supporting Nazi Germany.
- **1941** Hungary drawn into World War II on Axis side when Hitler attacks Soviet Union.
- **1944** Nazi Germany preempts Soviet advance on Hungary by invading. Deportation of Hungarian Jews and Roma to extermination camps begins. Soviet Red Army enters in October. Horthy forced to resign.
- **1945** Liberated by Red Army. Soviet-formed provisional government

H

CHRONOLOGY *continued*

installed. Imre Nagy introduces land reform.

- ❑ **1947** Communists emerge as largest party in second postwar election.
- ❑ **1948** Forcible merger of Social Democrats with communists; known as Hungarian Socialist Workers' Party (HSWP) from 1956.
- ❑ **1949** New constitution; formally becomes People's Republic.
- ❑ **1950–1951** First Secretary Mátyás Rákosi uses authoritarian powers to collectivize agriculture and industrialize the economy.
- ❑ **1953** Nagy, Rákosi's rival, becomes premier and reduces political terror.
- ❑ **1955** Nagy deposed by Rákosi.
- ❑ **1956** Rákosi out. Student demonstrations, demanding withdrawal of Soviet troops and Nagy's return, become popular uprising. Nagy appointed premier and János Kádár First Secretary. Nagy announces Hungary will leave Warsaw Pact. Three days later, Soviet forces suppress protests. About 25,000 killed. Kádár becomes premier.
- ❑ **1958** Nagy executed.
- ❑ **1968** Kádár introduces New Economic Mechanism to bring market elements to socialism.
- ❑ **1986** Police suppress commemoration of 1956 uprising. Democratic opposition demands Kádár resign.
- ❑ **1987** Party reformers establish MDF as a political movement.
- ❑ **1988** Kádár ousted. Protests force suspension of plans for Nagymaros Dam on the Danube.
- ❑ **1989** Parliament votes to allow independent parties. Posthumous rehabilitation of Nagy, who is given state funeral. Round table talks between HSWP and opposition.
- ❑ **1990** József Antall's MDF wins multiparty elections decisively. Speed of economic reform hotly debated. Árpád Göncz president.
- ❑ **1991** Warsaw Pact dissolved. Last Soviet troops leave.
- ❑ **1994** Hungary joins NATO's Partnerships for Peace program. Former communist MSzP wins general election. Austerity program prompts protests.
- ❑ **1998** EU entry negotiations open. Elections: Viktor Orbán (Fidesz–MPP) forms right-of-center coalition.
- ❑ **1999** Joins NATO. Airspace used in NATO bombing of Yugoslavia.
- ❑ **2000** Ferenc Mádl succeeds Göncz as president.
- ❑ **2002** Elections won by socialist and free democrat alliance. Peter Medgyessy prime minister.

H

AID

Recipient

 $252m (receipts)  Up 1% in 2000

Hungary received substantial Western aid in 1990–1996, but by the end of the decade was considered able to attract investment mainly on commercial terms. EU and World Bank assistance moved to focus on targeting disadvantaged social groups, raising environmental standards, and strengthening market institutions.

DEFENSE

Compulsory military service

 $777m  Up 1% in 2000

Troop numbers were more than halved, and conventional arms and the military hierarchy were updated in advance of NATO membership in 1999. The emphasis has switched toward more flexibility and rapid response. Almost immediately on joining NATO, Hungary permitted it to use its airspace to bomb Yugoslavia. Military service has been shortened to six months with effect from 2002.

HUNGARIAN ARMED FORCES

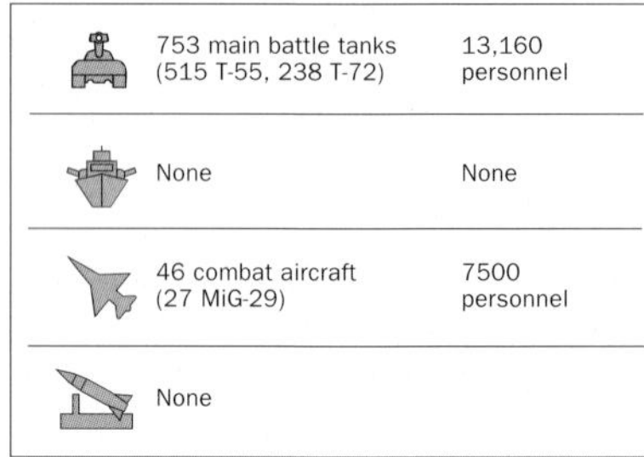

753 main battle tanks (515 T-55, 238 T-72)	13,160 personnel
None	None
46 combat aircraft (27 MiG-29)	7500 personnel
None	

ECONOMICS

Inflation 19% p.a. (1990–2000)

 $47.2bn  282.3–274.8 forint

SCORE CARD

- ❑ World GNP Ranking 52nd
- ❑ GNP per Capita $4710
- ❑ Balance of Payments .... –$1.49bn
- ❑ Inflation 9.8%
- ❑ Unemployment 6%

EXPORTS

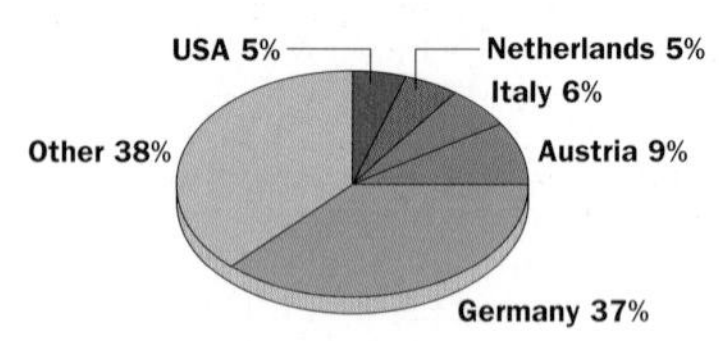

USA 5%
Netherlands 5%
Italy 6%
Austria 9%
Other 38%
Germany 37%

IMPORTS

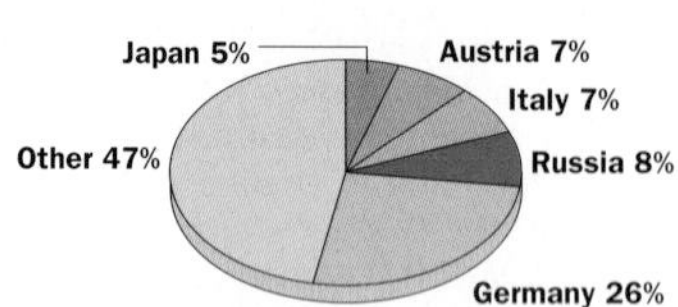

Japan 5%
Austria 7%
Italy 7%
Russia 8%
Other 47%
Germany 26%

Strengths

Openness to foreign direct investment, especially since 1998. Favorable tax regime, streamlined bureaucracy. Strong export-led growth since late 1990s. High industrial production, especially at new, state-of-the-art factories. Currency fully convertible from mid-2001. Inflation dropping.

Weaknesses

Low energy efficiency. East–west split as development bypasses rural eastern areas. Widening income differentials. Money laundering a challenge to finance industry regulators; Hungary has featured on OECD blacklists.

ECONOMIC PERFORMANCE INDICATOR

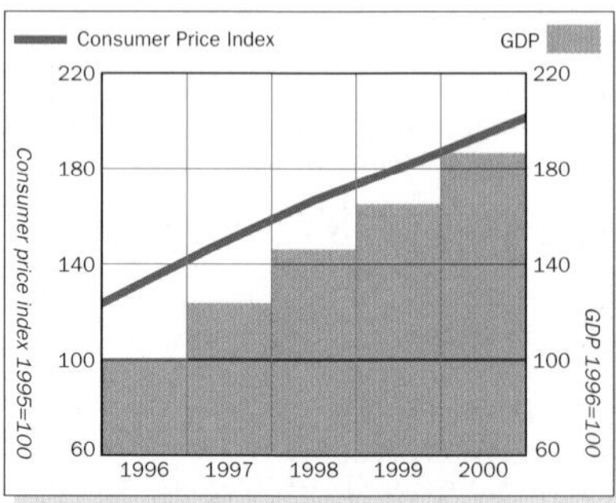

Consumer Price Index
GDP
220
180
140
100
60
Consumer price index 1995=100
GDP 1995=100
1996
1997
1998
1999
2000

Profile

The collapse of COMECON (communist economic bloc) caused a reorientation of trade toward western Europe. Exports increased rapidly and competitiveness has improved. However, the economy did not recover to its pre-1989 level until 1999. Privatization has reduced the state-owned share of the economy, from 85% to 15%, and has helped cut external debt.

HUNGARY : MAJOR BUSINESSES

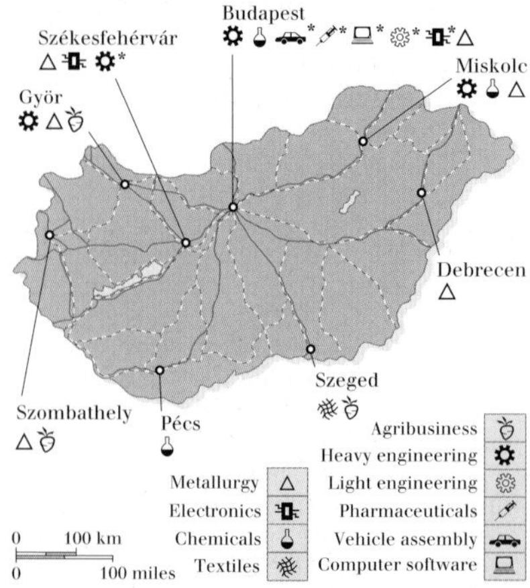

Budapest
Székesfehérvár
Miskolc
Györ
Debrecen
Szeged
Szombathely
Pécs
Agribusiness
Heavy engineering
Light engineering
Pharmaceuticals
Vehicle assembly
Computer software
Metallurgy
Electronics
Chemicals
Textiles
0 100 km
0 100 miles
* significant multinational ownership

RESOURCES

Electric power 7m kw

19,461 tonnes

25,263 b/d (reserves 55m barrels)

4.83m pigs, 3.3m turkeys, 30.7m chickens

Bauxite, coal, oil, natural gas, lignite

ELECTRICITY GENERATION

Hydro 0%
Combustion 62% (23bn kwh)
Nuclear 38% (14bn kwh)
Other 0%
0 20 40 60 80 100
% of total generation by type

HUNGARY : LAND USE

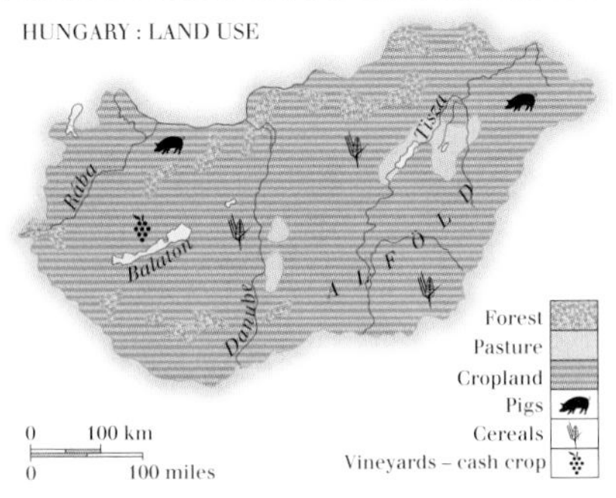

Raba
Tisza
Balaton
Duna
ALFOLD
Forest
Pasture
Cropland
Pigs
Cereals
Vineyards – cash crop
0 100 km
0 100 miles

Hungary has bauxite, brown coal, lignite, and natural gas reserves. It depends for about 40% of its electricity on nuclear energy from the Paks complex, north of Baja. Fertile farmlands provide grains, sugar beet, and potatoes. Wine production is also important.

ENVIRONMENT

Sustainability rank: 11th

7% (5% partially protected)

5.8 tonnes per capita

ENVIRONMENTAL TREATIES

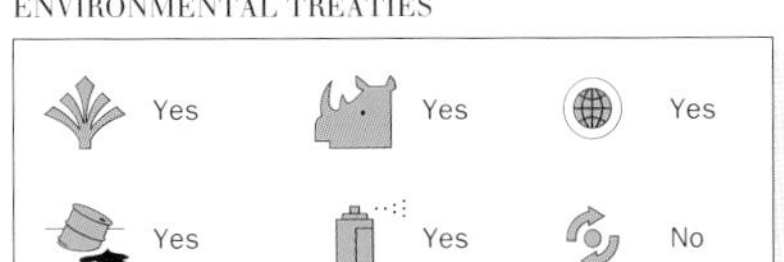

Yes
Yes
Yes
Yes
Yes
No

A high sulfur content in Hungary's fossil fuels exacerbates the serious air pollution in industrial zones. The energy industry must meet new standards by 2005. A "green card" system has been introduced to favor the use of cars with catalytic convertors and reduce serious pollution from older vehicles.

Industrial pollution of the Sajo, Tisza, and Danube river systems, sometimes originating in neighboring countries, is a major problem.

MEDIA

TV ownership high

Daily newspaper circulation 186 per 1000 people

PUBLISHING AND BROADCAST MEDIA

There are 40 daily newspapers, including *Népszabadság* and *Népszava*
2 services: 1 state-owned, 1 independent
4 services: 1 state-owned, 3 independent

Newspapers and magazines are fiercely independent and critical of government policy. In 1994, the Constitutional Court declared that state interference in the media was unlawful, but allegations of interference persist. The boards controlling state TV and radio must have equal representation from government and opposition under a 1996 media law, but the Orbán government was accused of bending or ignoring the rules.

CRIME

No death penalty

15,588 prisoners

Down 15% in 1999

CRIME RATES

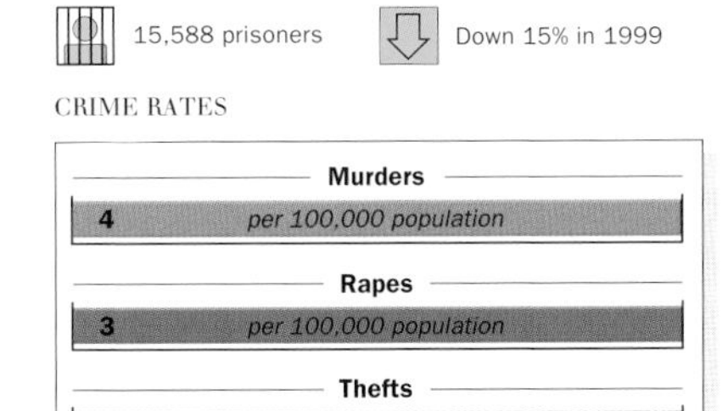

Murders
4 per 100,000 population
Rapes
3 per 100,000 population
Thefts
2591 per 100,000 population

An alarming trend in the late 1990s was the increase in murders of elderly people for financial gain. Organized crime, money laundering, and smuggling of illegal immigrants are rising.

EDUCATION

School leaving age: 16

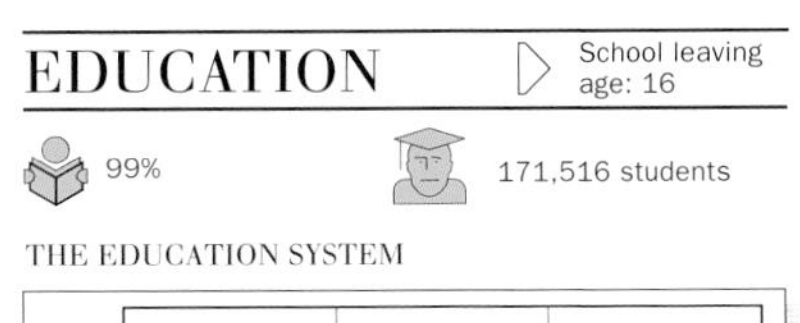

99%

171,516 students

THE EDUCATION SYSTEM

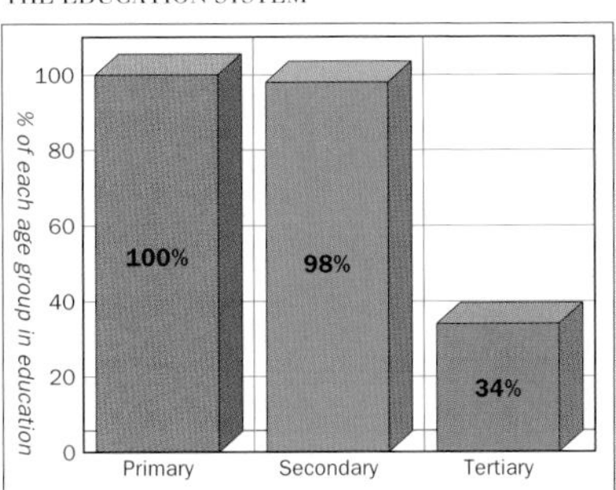

% of each age group in education
100
80
60
40
20
0
100%
98%
34%
Primary
Secondary
Tertiary

Education is free and compulsory from the age of six to 16. Bilingual schools have been established in southern Hungary to promote the languages of the national minorities. In 1999–2000 a major transformation of the education system took place, as a result of which, with effect from 2000, there are 30 universities and colleges run by the state, 26 run by the Church, and six colleges run by various foundations.

HEALTH

Welfare state health benefits

1 per 313 people

Cerebrovascular and heart diseases, cancers, accidents

Medical treatment has traditionally been free to all, although there is a contribution to prescription costs. State sickness benefits remain relatively generous. Spending on the health service has fallen in recent years in real terms; at $600 per capita, it is only one-third the OECD average, and there is concern that Hungary's health care sector is among the least developed of OECD countries. The ratio of doctors to patients is high, but there is a shortage of nurses. Family physician services are being privatized rapidly under a law passed in 2000.

SPENDING

GDP/cap. increase

CONSUMPTION AND SPENDING

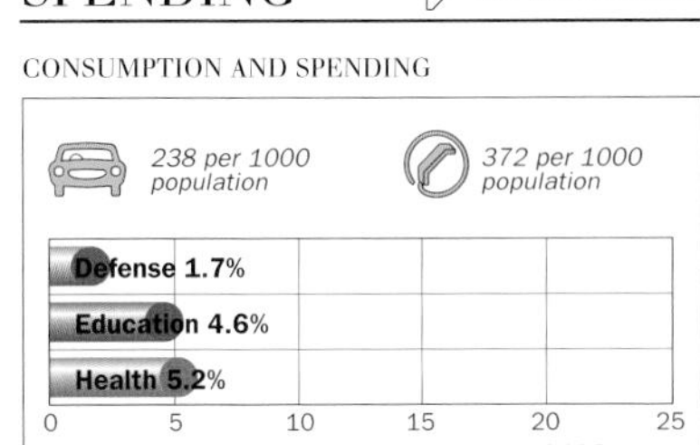

238 per 1000 population
372 per 1000 population
Defense 1.7%
Education 4.6%
Health 5.2%
0 5 10 15 20 25
Defense, Health, Education spending as % of GDP

Hungary enjoys a higher standard of living than other former communist countries except Slovenia, and demand for luxury goods is rising. Access to mobile phones and the Internet is relatively high. Real wages, which fell by 15% in the mid-1990s, had regained most of this ground by 2000. Hungarians still have to work longer hours to pay for basic consumer goods than workers in western Europe. Salaries in the public services have not kept pace with the rising cost of living, and as a result there is a growing disparity between those working in the state and private sectors. The Roma minority suffers particularly over access to housing, which is in short supply.

WORLD RANKING

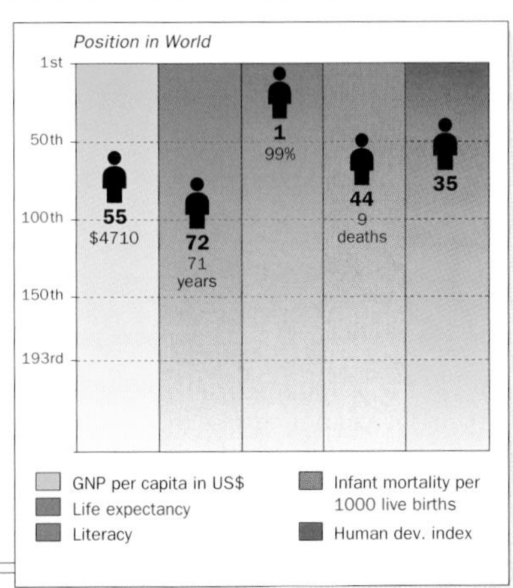

Position in World
1st
50th
100th
150th
193rd
55 $4710
72 71 years
1 99%
44 9 deaths
35
GNP per capita in US$
Life expectancy
Literacy
Infant mortality per 1000 live births
Human dev. index

H

ICELAND

OFFICIAL NAME: Republic of Iceland **CAPITAL:** Reykjavík
POPULATION: 281,000 **CURRENCY:** Icelandic króna **OFFICIAL LANGUAGE:** Icelandic

EUROPE
Greenland
ICELAND
Europe

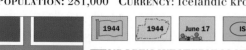

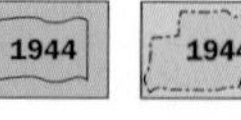
1944

1944

June 17

IS

0

+354

.is

EUROPE'S WESTERNMOST country, Iceland has a strategic location in the North Atlantic, just south of the Arctic Circle. Its position, on the rift where the North American and European continental plates are pulling apart, accounts for its 200 volcanoes and its numerous geysers and solfataras. Previously a Danish possession, Iceland became fully independent in 1944. Most settlements are along the coast, where ports remain ice-free in winter.

CLIMATE

Subarctic

WEATHER CHART FOR REYKJAVÍK

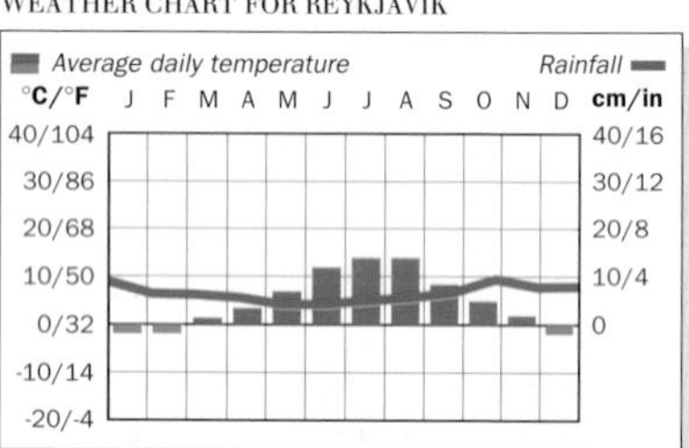

Average daily temperature
Rainfall
°C/°F
J F M A M J J A S O N D
cm/in
40/104
30/86
20/68
10/50
0/32
-10/14
-20/-4
40/16
30/12
20/8
10/4
0

Iceland sits in the Gulf Stream. Winters are consequently mild. Summers are cool, with fine, long sunny days.

TRANSPORTATION

Drive on right

Keflavík International, Reykjavík
1.46m passengers

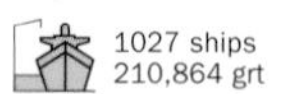

1027 ships
210,864 grt

THE TRANSPORTATION NETWORK

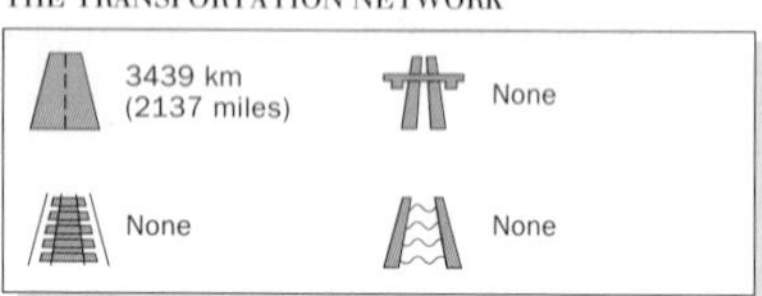

3439 km
(2137 miles)
None
None
None

Icelanders rely entirely on cars and internal airplane and helicopter flights. Most freight moves by sea. The only main road circles the island.

TOURISM

Visitors : Population
1.1:1

303,000 visitors

Up 15% in 2000

MAIN TOURIST ARRIVALS

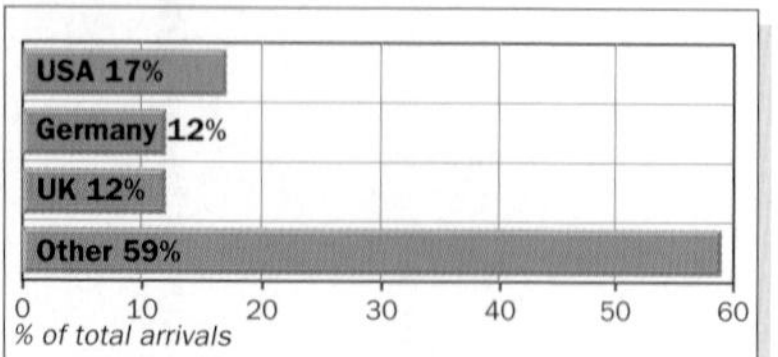

USA 17%
Germany 12%
UK 12%
Other 59%
0 10 20 30 40 50 60
% of total arrivals

Iceland is promoting itself, especially in Japan, as an upmarket destination for tourists who are attracted by its spectacular scenery, glaciers, green valleys, fjords, and hot springs.

PEOPLE

Pop. density low

Icelandic

3/km²
(7/mi²)

THE URBAN/RURAL POPULATION SPLIT

93% 7%

RELIGIOUS PERSUASION

Other Christian 1%
Nonreligious 6%
Evangelical Lutheran 93%

Descended from Norwegians and Celts, Icelanders form an ethnically homogeneous society. Almost all follow the Evangelical Lutheran Church. More than half the population live in or near Reykjavík. Living standards are high, and there are few social tensions.

POLITICS

Multiparty elections

1999/2003
President Olafur Ragnar Grimsson

AT THE LAST ELECTION

Parliament 63 seats

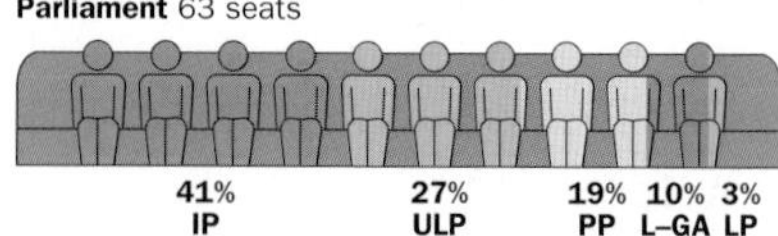

41% IP
27% ULP
19% PP
10% L–GA
3% LP

IP = Independence Party **ULP** = United Left Party
PP = Progressive Party **L–GA** = Left–Green Alliance
LP = Liberal Party

From independence Iceland has been ruled by coalitions, but in the 1980s the traditional four-party system began to splinter. After the 1991 election, a new IP–Social Democratic coalition promoted market-led reforms. Arguments over whether or not to join the EU were defused in 1992 with the successful negotiation of the EEA, giving Iceland access to the key EU market.

The coalition collapsed after the 1995 general election, when both parties had lost support, and was replaced by a center-right government led by the IP, with David Oddsson as prime minister. He has successfully built on an economic recovery under way since 1994, and strengthened his position in the 1999 general election.

ICELAND

Total Land Area : 103 000 sq. km
(39 768 sq. miles)

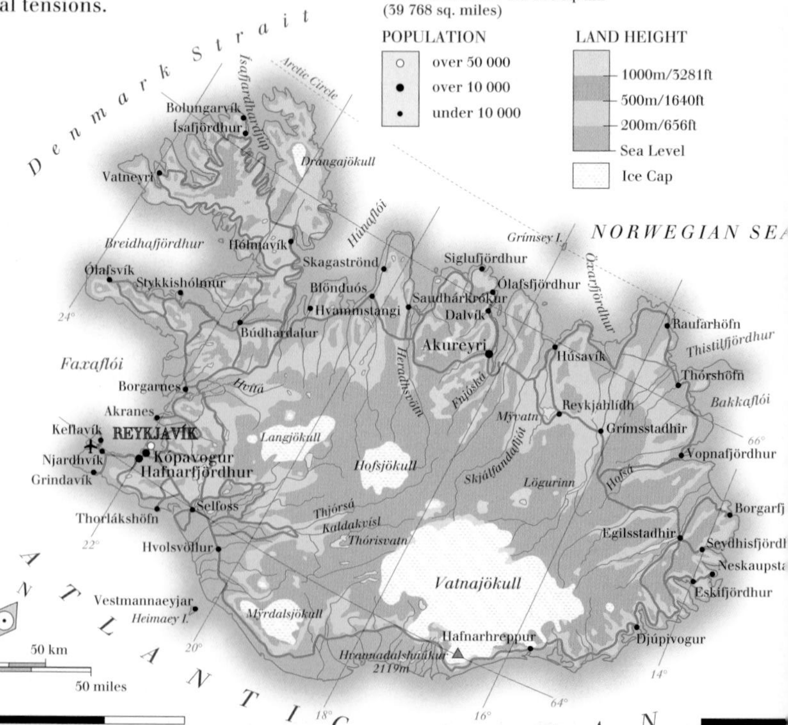

POPULATION
over 50 000
over 10 000
under 10 000
LAND HEIGHT
1000m/3281ft
500m/1640ft
200m/656ft
Sea Level
Ice Cap
Denmark Strait
Arctic Circle
NORWEGIAN SEA
ATLANTIC OCEAN
Bolungarvík
Ísafjördhur
Ísafjardhardjúp
Drangajökull
Vatneyri
Húnaflói
Breidhafjördhur
Hólmavík
Grímsey I.
Skagaströnd
Siglufjördhur
Ólafsvík
Stykkishólmur
Ólafsfjördhur
Blönduós
Hvammstangi
Saudhárkrókur
Dalvík
Eyjafjördhur
Raufarhöfn
Búdhardalur
Thistilfjördhur
Akureyri
Húsavík
Faxaflói
Héradsvötn
Thórshöfn
Borgarnes
Hvítá
Fnjóská
Bakkaflói
Akranes
Mývatn
Reykjahlídh
Grímsstadhir
Keflavík
REYKJAVÍK
Langjökull
Vopnafjördhur
Njardhvík
Kópavogur
Hafnarfjördhur
Skjálfandafljót
Hofsjökull
Lögurinn
Grindavík
Selfoss
Thjórsá
Kaldakvísl
Thórisvatn
Borgarfj
Thorlákshöfn
Egilsstadhir
Hvolsvöllur
Seydhisfjördh
Neskaupst
Vatnajökull
Eskifjördhur
Vestmannaeyjar
Heimaey I.
Mýrdalsjökull
Hafnarhreppur
Djúpivogur
Hvannadalshnúkur 2119m
0 50 km
0 50 miles
24°
22°
20°
18°
16°
14°
66°
64°

WORLD AFFAIRS

Joined UN in 1946

CE | NATO | OECD | OSCE | EEA

Although a member of NATO and EFTA, Iceland has traditionally maintained arm's length relations with the EU and the US. Major disputes have concerned fishing rights. Iceland left the IWC in 1992 when a commercial whaling ban was extended, but rejoined in 2001. Links with other Nordic states are strong.

AID

 Donor

 $9m (donations)

 Up in 2000

Aid donations are modest, and form a smaller proportion of the budget than in other Scandinavian states.

DEFENSE

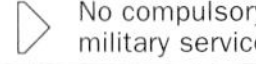

No compulsory military service

 Coastguard is only military force

 Not applicable

A NATO member and an associate WEU member, Iceland has no armed forces. US troops are based at Keflavík.

ECONOMICS

Inflation 3.2% p.a. (1990-2000)

$8.54bn

84.7–102.9 Icelandic krónur

SCORE CARD

- ❑ WORLD GNP RANKING 91st
- ❑ GNP PER CAPITA $30,390
- ❑ BALANCE OF PAYMENTS ....–$851m
- ❑ INFLATION 5.2%
- ❑ UNEMPLOYMENT 1%

STRENGTHS

High-tech fishing industry with exclusive access to prime fishing grounds. Strong economic recovery in late 1990s; low inflation and unemployment. Very cheap geothermal power.

WEAKNESSES

Over 70% of export earnings derived from single source: fish. State-owned banking sector restricts flexibility.

EXPORTS

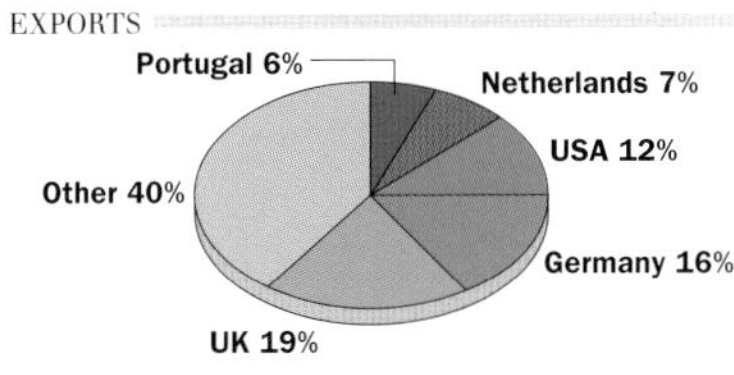
Portugal 6%
Netherlands 7%
USA 12%
Other 40%
Germany 16%
UK 19%

IMPORTS

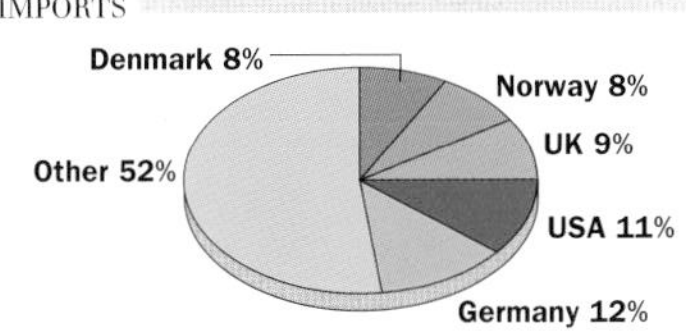
Denmark 8%
Norway 8%
UK 9%
Other 52%
USA 11%
Germany 12%

***Lava towers,** near Lake Mývatn in northern Iceland – an area of grassy lowlands. Iceland's center consists of lava desert and glaciers.*

RESOURCES

Electric power 1.2m kw

1.74m tonnes

 Not an oil producer

465,000 sheep, 78,000 horses, 180,000 chickens

Diatomite

Iceland has virtually no minerals. All energy needs are met by geothermal and hydroelectric sources. It has implemented measures to try to restore once abundant fish stocks.

ENVIRONMENT

Sustainability rank: 8th

9% (7% partially protected)

 7.6 tonnes per capita

Iceland has no nuclear or coal-fired power plants. Pollution levels are low. Believing that minke whales eat valuable cod stocks, Iceland resumed whale hunting in 1992. The 1996 eruption of the Loki volcano under Vatna glacier caused extensive flooding and damage.

MEDIA

TV ownership high

 Daily newspaper circulation 535 per 1000 people

PUBLISHING AND BROADCAST MEDIA

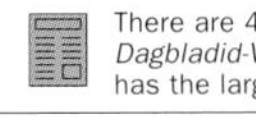

There are 4 daily newspapers, including *Dagbladid-Visir* and *Morgunbladid*, which has the largest circulation

 11 services: 1 state-owned, 10 independent

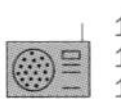 17 services: 1 state-owned, 16 independent

Iceland is renowned for having one of the highest per capita newspaper circulations in the world.

CRIME

No death penalty

 82 prisoners

 Crime rates are rising

Crime rates are comparatively low. The rate of alcohol-related murders is higher than the European average.

EDUCATION

School leaving age: 16

 99%

 10,543 students

Icelanders buy more books per capita than any other nation. Education is state-run; some 40% of school students go on to university at Reykjavík or Akureyri, or to colleges in the US.

CHRONOLOGY

Settled in the 9th century by Norwegians, Iceland was ruled by Denmark from 1380 to 1944, becoming fully self-governing in 1918.

- ❑ **1940–1945** Occupied by UK and US.
- ❑ **1944** Independence as republic.
- ❑ **1949** Founder member of NATO.
- ❑ **1951** US air base built at Keflavík despite strong local opposition.
- ❑ **1972–1976** Extends fishing limits to 50 miles; two "cod wars" with UK.
- ❑ **1975** Sets 200-mile fishing limit.
- ❑ **1980** Vigdís Finnbogadóttir world's first elected woman head of state.
- ❑ **1985** Declares nuclear-free status.
- ❑ **1995** Formation of enter-right coalition under David Oddsson after general election; reelected in 1999.

I

HEALTH

Welfare state health benefits

 1 per 307 people

 Heart disease, cancers, accidents

The state health system is free to all Icelanders. Iceland has one of the lowest infant mortality rates and one of the highest longevity rates in the world.

SPENDING

GDP/cap. increase

CONSUMPTION AND SPENDING

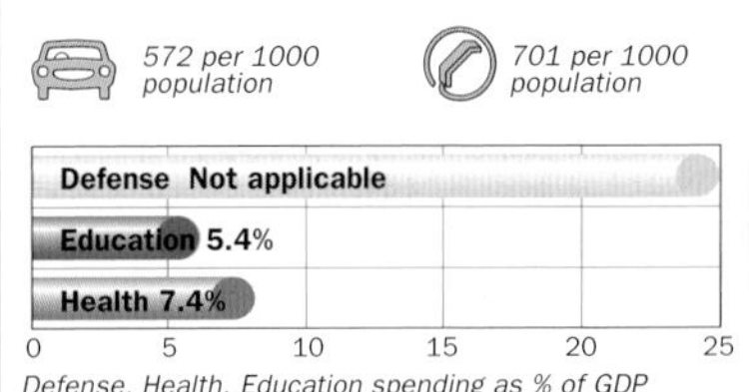
572 per 1000 population
701 per 1000 population
Defense Not applicable
Education 5.4%
Health 7.4%
0
5
10
15
20
25
Defense, Health, Education spending as % of GDP

Wealth distribution in Iceland is comparatively even and social mobility is high. Domestic heating, extracted from geothermal sources, is provided at almost no cost.

WORLD RANKING

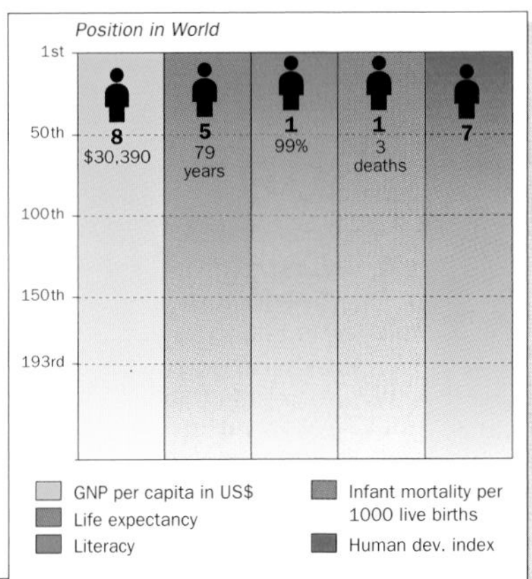
Position in World
1st
50th
100th
150th
193rd
8
$30,390
5
79 years
1
99%
1
3 deaths
7
GNP per capita in US$
Life expectancy
Literacy
Infant mortality per 1000 live births
Human dev. index

INDIA

OFFICIAL NAME: Republic of India **CAPITAL:** New Delhi
POPULATION: 1.03 billion **CURRENCY:** Indian rupee **OFFICIAL LANGUAGES:** Hindi and English

SOUTH ASIA

Pakistan Nepal China Bhutan Bangladesh INDIA Asia

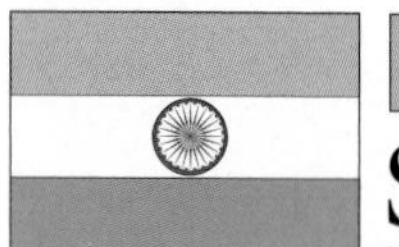

1947

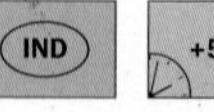
1947

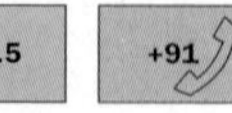
Jan 26

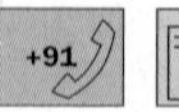
IND

+5.5

+91 .in

SEPARATED FROM the rest of Asia by the Himalaya mountain range, India forms the bulk of a subcontinent. As well as the Himalayas, there are two other main geographic regions, the Indo-Gangetic plain, which lies between the foothills of the Himalayas and the Vindhya Mountains, and the central-southern plateau. India is the world's largest democracy and second most populous country after China. The birthrate has recently been falling, but even at its current level India's population will probably overtake China's by 2030. After years of protectionism, India is opening up its economy to the outside world. The hope is that the free market will go some way to alleviating one of the country's major problems, poverty.

CLIMATE

Tropical/subtropical/desert/mountain/monsoon

WEATHER CHART FOR NEW DELHI

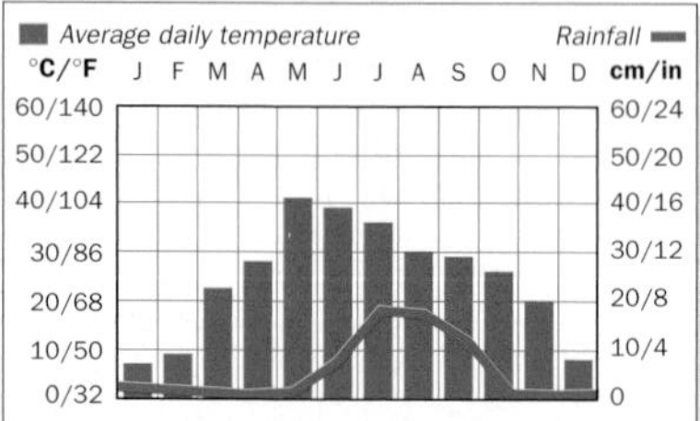
Average daily temperature
Rainfall
°C/°F
J F M A M J J A S O N D
cm/in
60/140
50/122
40/104
30/86
20/68
10/50
0/32
60/24
50/20
40/16
30/12
20/8
10/4
0

During the hot season, temperatures in the north can reach 40°C (105°F). The monsoon breaks in June and peters out in September or October. In the cool season, average temperatures are 5°C–15°C (40°F–60°F) in the north and the weather is mainly dry. The south has a less variable climate. Chennai (Madras) is always hot: average temperatures range from 24°C (75°F) in January to 32°C (90°F) in May and June.

TRANSPORTATION

Drive on left

Mumbai International
12.1m passengers

987 ships
6.66m grt

THE TRANSPORTATION NETWORK

1.52m km
(942,666 miles)

33,500 km
(20,816 miles)

62,810 km
(39,030 miles)

16,180 km (10,054 miles)

India's state-owned railroad system, the largest in Asia, spans all the major cities. Rail carries 40% of passenger traffic and 65% of freight. Strict controls on diesel emissions from cars and buses to check pollution were enforced in 2001. Cycle and scooter rickshaws are common in urban centers. Kolkata (Calcutta) still has rickshaws pulled by hand.

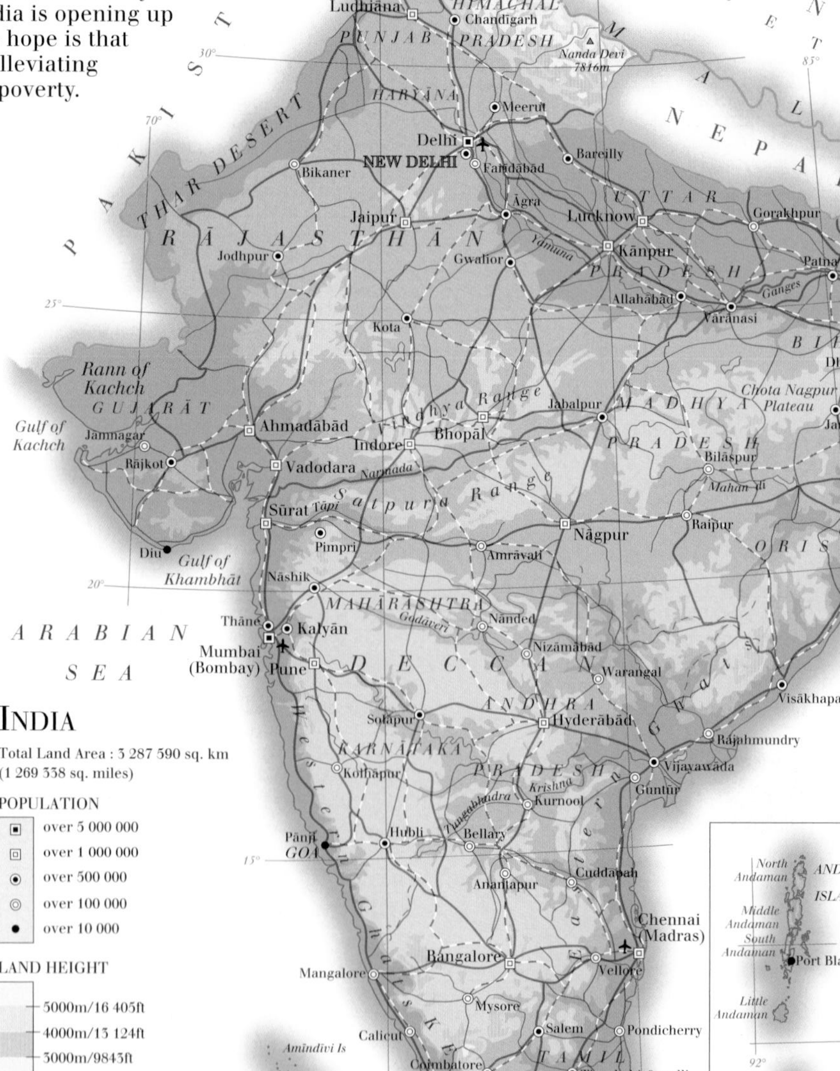
Aksai Chin
Line of control
KARAKORAM RANGE
Indus
Srinagar
JAMMU & KASHMIR
Jammu
CHINA
TIBET
PAKISTAN
Amritsar
Jalandhar
Ludhiana
Chandigarh
HIMACHAL PRADESH
PUNJAB
Nanda Devi 7816m
HARYANA
Meerut
NEPAL
THAR DESERT
Delhi
NEW DELHI
Faridabad
Bikaner
Bareilly
Agra
UTTAR PRADESH
Lucknow
Gorakhpur
Jaipur
RAJASTHAN
Jodhpur
Gwalior
Kanpur
Yamuna
Patna
Allahabad
Ganges
Varanasi
Kota
BIHAR
Chota Nagpur Plateau
Rann of Kachch
GUJARAT
Gulf of Kachch
Jamnagar
Rajkot
Ahmadabad
Vindhya Range
Bhopal
Jabalpur
MADHYA PRADESH
Indore
Vadodara
Narmada
Bilaspur
Mahanadi
Satpura Range
Surat
Tapi
Raipur
Nagpur
Pimpri
Amravati
ORISSA
Diu
Gulf of Khambhat
Nashik
MAHARASHTRA
Godavari
Nanded
ARABIAN SEA
Thane
Kalyan
Mumbai (Bombay)
Pune
Nizamabad
DECCAN
Warangal
Visakhapatnam
Solapur
ANDHRA PRADESH
Hyderabad
Rajahmundry
KARNATAKA
Kolhapur
Vijayawada
Krishna
Kurnool
Guntur
Tungabhadra
Panji
GOA
Hubli
Bellary
Western Ghats
Eastern Ghats
Cuddapah
Anantapur
Chennai (Madras)
Bangalore
Mangalore
Vellore
Mysore
Calicut
KERALA
Salem
Pondicherry
Amindivi Is
Coimbatore
TAMIL NADU
Tiruchchirappalli
LAKSHADWEEP (Laccadive Is)
Kavaratti I.
Kalpeni I.
Trichur
Dindigal
Palk Strait
Madurai
Cochin
Trivandrum
Tuticorin
SRI LANKA
Minicoy I.
Gulf of Mannar
Nagercoil
INDIAN OCEAN
North Andaman
Middle Andaman
South Andaman
Port Blair
Little Andaman
NICOBAR ISLANDS
Great Nicobar

INDIA

Total Land Area : 3 287 590 sq. km (1 269 338 sq. miles)

POPULATION

- over 5 000 000
- over 1 000 000
- over 500 000
- over 100 000
- over 10 000

LAND HEIGHT

- 5000m/16 405ft
- 4000m/13 124ft
- 3000m/9843ft
- 2000m/6562ft
- 1000m/3281ft
- 500m/1640ft
- 200m/656ft
- Sea Level

0 200 km

0 200 miles

***A religious festival.** Such festivals are a frequent occurrence and form an important part of Hindu culture.*

ARUNACHAL PRADESH
BHUTAN
Brahmaputra
Guwahati
ASSAM
Imphal
BANGLADESH
BURMA (MYANMAR)
Calcutta (Kolkata)
Mouths of the Ganges
BAY OF BENGAL

TOURISM

Visitors : Population 1:388

2.64m visitors

Up 6% in 2000

MAIN TOURIST ARRIVALS

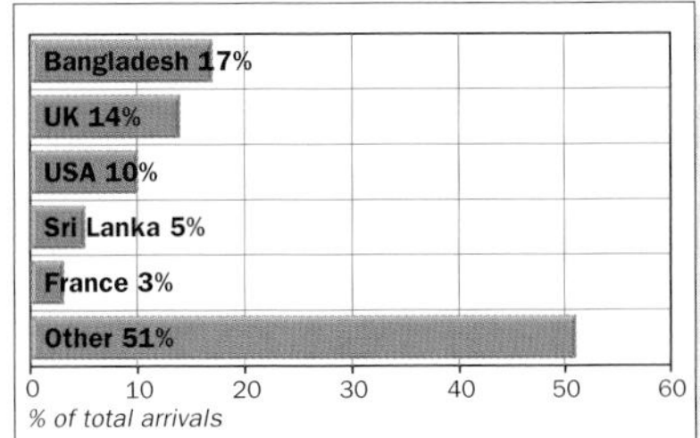
Bangladesh 17%
UK 14%
USA 10%
Sri Lanka 5%
France 3%
Other 51%
0 10 20 30 40 50 60
% of total arrivals

Tourism is India's sixth-largest foreign exchange earner. More luxury hotels are being built, and wildlife and adventure tourism are being promoted. However, India has only a small share of the world tourism market, and has suffered recently from security worries over the repercussions of the US-led "war on terrorism" and acute tensions with Pakistan, particularly over Kashmir.

PEOPLE

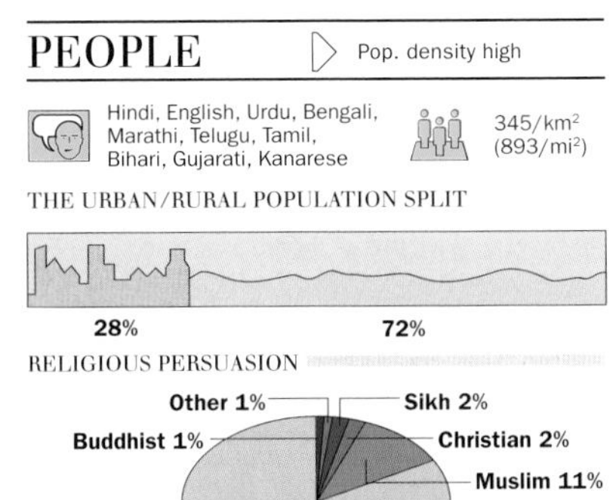
Pop. density high
Hindi, English, Urdu, Bengali, Marathi, Telugu, Tamil, Bihari, Gujarati, Kanarese
345/km² (893/mi²)
THE URBAN/RURAL POPULATION SPLIT
28%
72%
RELIGIOUS PERSUASION
Other 1%
Sikh 2%
Buddhist 1%
Christian 2%
Muslim 11%
Hindu 83%

ETHNIC MAKEUP

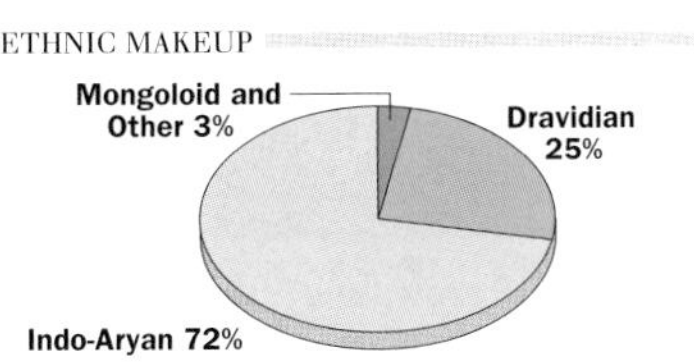
Mongoloid and Other 3%
Dravidian 25%
Indo-Aryan 72%

India is the world's second most populous country after China, officially passing the one billion mark in 2000. Despite a major birth control program, the decrease in population growth has been marginal. Nationwide awareness campaigns aim to promote the idea of smaller families. India's planners consider the rise in the population the most significant brake on development. Cultural and religious pressures encourage large families, however, and the extended family is seen as essential security for old age.

The fertile rice-growing areas of the Gangetic plain and delta are very densely populated. The northern state of Uttar Pradesh has the largest population, followed by neighboring Bihar and the western state of Maharashtra. Maharashtra is also the most urbanized state, with more than half of its people living in towns or cities. Elsewhere, most Indians live in rural areas, although poverty continues to drive many to the swelling cities.

The overwhelming majority of the population are Hindus, who belong to thousands of castes and subcastes, which largely determine status, occupation, and whom they marry. Tension between Hindus and Muslims has grown in recent years and escalated sharply in 2002 following violent clashes in Gujarat.

POPULATION AGE BREAKDOWN

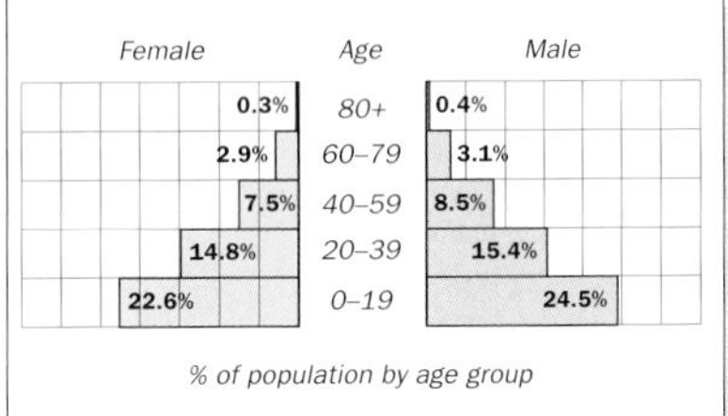
Female
Age
Male
0.3% 80+ 0.4%
2.9% 60–79 3.1%
7.5% 40–59 8.5%
14.8% 20–39 15.4%
22.6% 0–19 24.5%
% of population by age group

CHRONOLOGY

The origins of an Indus Valley civilization may be traced back to the third millennium BCE. By the 3rd century BCE, the Mauryan kingdom under Ashoka encompassed most of modern India. Following the Battle of Plassey in 1757, British rule – through the East India Company – was consolidated.

- ❑ **1885** Formation of Indian National Congress.
- ❑ **1919** Act of Parliament introduces "responsible government."
- ❑ **1920–1922** Mahatma Gandhi's first civil disobedience campaign.
- ❑ **1935** Government of India Act grants autonomy to provinces.
- ❑ **1936** First elections under new constitution.
- ❑ **1942–1943** "Quit India" movement.
- ❑ **1947** August, independence and partition into India and Pakistan. Jawarhalal Nehru becomes first prime minister.
- ❑ **1948** Assassination of Mahatma Gandhi. War with Pakistan over Kashmir. India becomes a republic.
- ❑ **1951–1952** First general election won by Congress party.
- ❑ **1957** Congress party reelected. First elected communist state government installed in Kerala.
- ❑ **1960** Bombay divided into states of Gujarat and Maharashtra.
- ❑ **1962** Congress party reelected. Border war with China.
- ❑ **1964** Death of Nehru. Lal Bahadur Shastri becomes prime minister.
- ❑ **1965** Second war with Pakistan over Kashmir.
- ❑ **1966** Shastri dies; Indira Gandhi (daughter of Jawarhalal Nehru) becomes prime minister.
- ❑ **1969** Congress party splits into two factions; larger faction led by Indira Gandhi.
- ❑ **1971** Indira Gandhi's Congress party wins elections. Third war with Pakistan, over creation of Bangladesh.
- ❑ **1972** Simla (peace) Agreement signed with Pakistan.
- ❑ **1974** Explosion of first nuclear device in underground test.
- ❑ **1975–1977** Imposition of state of emergency.
- ❑ **1977** Congress loses general election. People's Party (JD) takes power at the center.
- ❑ **1978** New political group, Congress (Indira) – Congress (I) – formally established.
- ❑ **1980** Indira Gandhi's C(I) wins general election.
- ❑ **1984** Indian troops storm Sikh Golden Temple in Amritsar. Assassination of Indira Gandhi ⇨

CHRONOLOGY *continued*

by Sikh bodyguard; her son Rajiv becomes prime minister and C(I) leader. Gas explosion at US-owned Union Carbide Corporation plant in Bhopal kills 2000 people in India's worst industrial disaster.

- **1985** Peace accords with militant separatists in Assam and Punjab.
- **1987** Deployment of Indian peacekeeping force in Sri Lanka to combat Tamil Tigers.
- **1989** General election; National Front forms minority government with BJP support. C(I) implicated in Bofors scandal.
- **1990** Withdrawal of troops from Sri Lanka.
- **1991** Rajiv Gandhi assassinated. Narasimha Rao becomes prime minister of a C(I) minority government and initiates economic liberalization.
- **1992** Demolition of the Babri Masjid mosque at Ayodhya by Hindu extremists triggers widespread violence; 1200 people die.
- **1993** Resurgence of Hindu–Muslim riots. Bomb explosions in Bombay (Mumbai). Border troop agreement with China.
- **1994** Rupee made fully convertible. C(I) routed in key state elections amid increasing allegations of corruption in ruling party. Outbreak of pneumonic plague.
- **1995** Punjab chief minister assassinated by Sikh extremists.
- **1996** Corruption scandal triggers political crisis. C(I) suffers its worst electoral defeat. Leftist United Front coalition government takes office.
- **1997** Successive governments fall as C(I) withdraws support.
- **1998** General election; BJP led by Atal Bihari Vajpayee forms coalition government. Sonia Gandhi, widow of Rajiv Gandhi, becomes president of C(I). India tests nuclear bomb.
- **1999** Vajpayee travels to Pakistan to inaugurate bus service between India and Pakistan. India and Pakistan test nuclear missiles, and engage in violent confrontation in Kashmir. BJP returned to power after elections triggered by vote of no confidence.
- **2001** Earthquake kills more than 25,000 in Gujarat. BJP government implicated in major bribery scandal.
- **2001–2002** Terrorist attacks by Kashmiri separatists precipitate crisis with Pakistan.
- **2002** More than 2000, mainly Muslims, killed in Gujarat following worst intercommunal riots since independence. Heatwave kills more than 600.

I

POLITICS

Multiparty elections

L. House 1999/2003
U. House 2002/2004

President A. P. J. Abdul Kalam

AT THE LAST ELECTION

House of the People (Lok Sabha) 545 seats

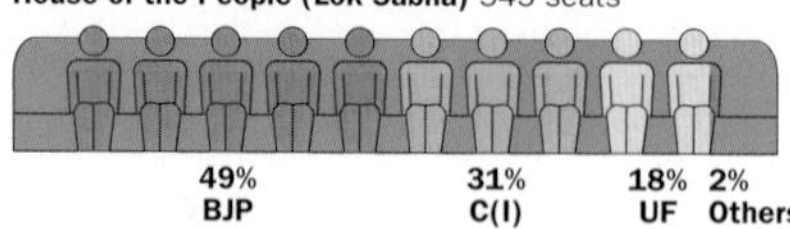

49% BJP **31%** C(I) **18%** UF **2%** Others

BJP = Bharatiya Janata Party and Allies, including the All-India Anna Dravida Munnetra Kazhagam
C(I) = Congress (I) and Allies
UF = United Front and Allies, including the Communist Party of India (Marxist) – **CPI(M)**
Nom = Nominated

Others include the Rashtriya Janata Dal, independents, and two appointed seats reserved for Anglo-Indians

Council of States (Rajya Sabha) 245 seats

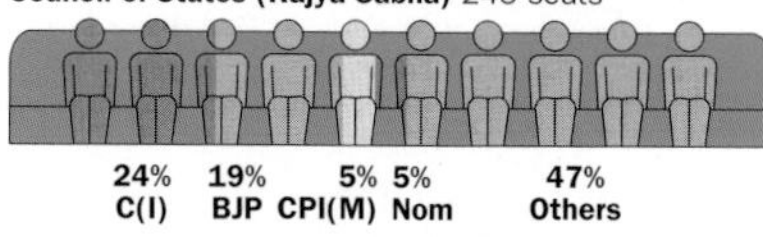

24% C(I) **19%** BJP **5%** CPI(M) **5%** Nom **47%** Others

233 members are elected to the Rajya Sabha by State Legislative Assemblies, and 12 "distinguished citizens" are nominated by the head of state

India is a multiparty democracy. The Lok Sabha (lower house) is directly elected by universal adult suffrage, while the Rajya Sabha (upper house) is indirectly elected by the state assemblies. There are 28 self-governing states. Of the seven union territories, Delhi and Pondicherry have their own assemblies.

Profile

The ascendancy of the Hindu nationalist BJP, in government since 1998, has relegated the C(I) – founded in 1978 as successor to the historic Congress Party which led India to independence in 1947 – to an unprecedented period of opposition at union level. Under P. V. Narasimha Rao's leadership in 1991–1996, a bold program of economic liberalization broke with the C(I)'s traditionally left-of-center policies. However, allegations of corruption undermined the party, resulting in heavy electoral defeats in 1996 and 1998. After 1998 hopes for C(I)'s political revival rested on its president Sonia Gandhi, who, as the widow of the assassinated former prime minister Rajiv Gandhi, restored the influence of the Nehru family over the party.

In 1996 regional parties formed the core of a center-left United Front coalition government in power until 1997. However, the 1998 election established the BJP with a strong enough mandate to lead a coalition government headed by its leader, A. B. Vajpayee. Elections held in 1999 returned the coalition to power with an overall majority.

Main Political Issues

Hindu militancy

The right-wing Hindu BJP has emerged as a credible alternative to the C(I). In 1996 it won most parliamentary seats, despite being tainted by corruption allegations. Although it failed on that occasion to form a viable administration, the BJP was returned to power in 1998 and 1999. Its rise has encouraged the spread of Hindu nationalism and raised fears about the future of India's secular constitution. In 2002 the BJP state government in Gujarat was accused by its critics of sanctioning mob attacks against the minority Muslim community in Ahmadabad during communal riots in the state.

Political corruption

Allegations of political corruption have dominated Indian politics. In 1989, the C(I) prime minister Rajiv Gandhi was accused of accepting bribes from a Swedish arms company. The C(I) was also implicated in a financial scandal in 1992. In 1996 corruption forced the resignations of several C(I) government ministers and the leader of the opposition BJP. The issue resurfaced in 2001 when the BJP government was implicated in a bribery scandal over arms sales which led to the resignation of the defense minister.

The free market

The introduction of the free-market economy has been vigorously resisted. Critics contend that free trade would undermine local production and participation by foreign companies damage the national economy. In the mid-1990s governments increased spending on rural development programs in order to soften the impact of economic liberalization. The BJP-led coalition, although more sympathetic to urban interests, is generally opposed to competition from foreign businesses.

A. B. Vajpayee, BJP leader and prime minister since 1998.

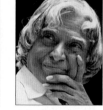

***A. P. J. Abdul Kalam,** a Muslim nuclear scientist, elected president in 2002.*

***Sonia Gandhi** has revived the fortunes of the opposition Congress (I).*

WORLD AFFAIRS

Joined UN in 1945

Comm | G15 | G24 | NAM | SAARC

The overriding preoccupation in foreign policy is the dispute with Pakistan over Kashmir, which has sparked two wars, in 1948 and 1965, and intensified the nuclear race between the two countries. High-level talks occasionally raise hopes of better relations, but border clashes stoke tensions. In 2002 a series of attacks across the Line of Control in Kashmir by Muslim militants, allegedly sponsored by Pakistan, pushed the two countries to the brink of war.

In 1998 India carried out nuclear weapons tests, prompting Pakistan to follow suit and sparking international condemnation. The US imposed sanctions on India for ignoring the ban on nuclear weapons testing, but India has still not signed the Comprehensive Test Ban Treaty. US sanctions were lifted in 2001 following India's support for the "war on terrorism."

AID

Recipient

$1.49bn (receipts) | Little change in 2000

India does not depend on aid. The US suspended aid following nuclear tests in 1998, but restored payments in late 2001. The World Bank has withdrawn funding for the Narmada Dam project. International relief aid helped victims of the 2001 Gujarat earthquake.

DEFENSE

No compulsory military service

$14.5bn | Up 4% in 2000

INDIAN ARMED FORCES

3414 main battle tanks (700 T-55, 1500 T-72M1, 1200 *Vijayanta*, 14 *Arjun*)	1.1m personnel
16 submarines, 1 carrier, 8 destroyers, 11 frigates, 7 corvettes, and 39 patrol boats	53,000 personnel
738 combat aircraft (84 *Jaguar* S(I), MiG-21/23/27/29)	110,000 personnel
Capability undisclosed; weapons tested in 1998	

India considers the possession of a nuclear deterrent to be vital, and tested its own weapons in 1998. It has the world's fourth-largest army, and produces its own *Arjun* battle tank. The *Agni*-II intermediate-range missile, capable of carrying a nuclear warhead to all parts of Pakistan, was tested in 2001. Also that year a much-delayed jet fighter program reached test flight stages. However, the replacement of India's outdated foreign weaponry was disrupted by US sanctions between 1998 and 2001. Russia signed a $10 billion air defense deal in 2001.

ECONOMICS

Inflation 8.5% p.a. (1990–2000)

$455bn | 46.68–48.22 Indian rupees

SCORE CARD

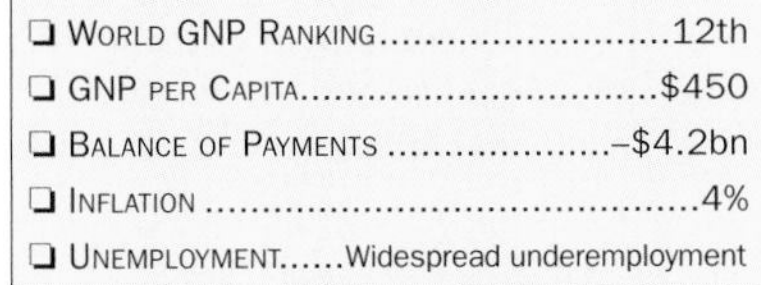

❑ WORLD GNP RANKING	12th
❑ GNP PER CAPITA	$450
❑ BALANCE OF PAYMENTS	–$4.2bn
❑ INFLATION	4%
❑ UNEMPLOYMENT	Widespread underemployment

EXPORTS

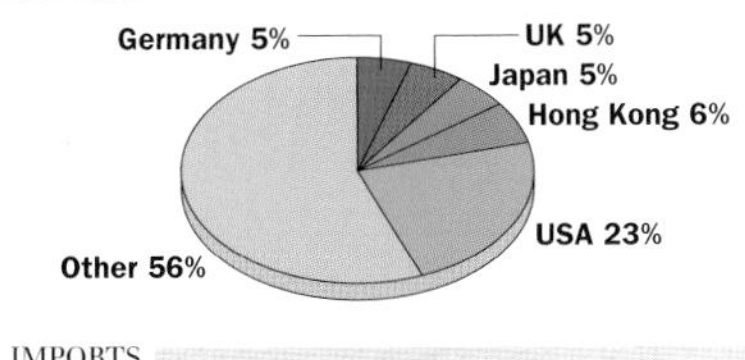
Germany 5%
UK 5%
Japan 5%
Hong Kong 6%
USA 23%
Other 56%

IMPORTS

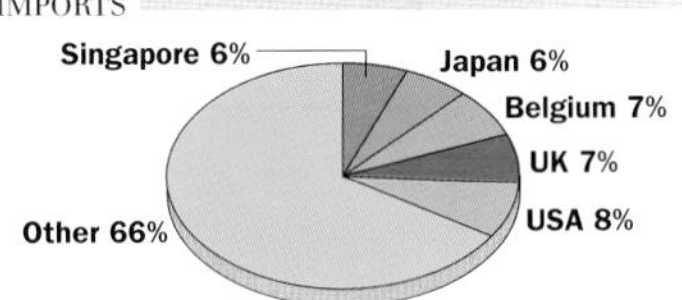
Singapore 6%
Japan 6%
Belgium 7%
UK 7%
USA 8%
Other 66%

STRENGTHS

Massive home market of over one billion people. Cheap labor. Some of the workforce possess skills for new high-tech industries such as software programming, a major boom area of the economy. Vibrant film industry: "Bollywood." Highly efficient textile sector and garment manufacturers. Growing competitiveness in world market, reflected in strong export growth. Competition is encouraging firms to manufacture to international standards. Despite strong objections from opposition parties, India ratified the GATT world trade agreement in 1995.

Since the economy was opened up to foreign competition in 1991, foreign direct investment has risen massively. Much of this went into the power sector. Large multinationals, such as Coca-Cola and IBM, are expanding, despite protests from some sections of the ruling BJP hostile to the growing presence of foreign businesses.

WEAKNESSES

Large budget deficit dogs economy. Governments have found it politically difficult to move away from the old system of widespread subsidies. Value of rupee has declined sharply. Inflation compounded by rise in wholesale food prices. Poor communications systems and power shortages hinder growth.

PROFILE

India has the fastest growing economy in Asia after China. From a highly protectionist mixed economy, which succeeded in building the basis of a modern industrial state, India has to a large extent converted to a free-market economy and is entering the global marketplace. Wide-ranging reforms, from lowering trade barriers to attracting foreign investment, have been put in place. The United Front and BJP-led governments in power since 1996 have not undone these reforms, though they have been criticized for not being wholehearted about driving them forward. Meanwhile, in the rural economy millions of people grapple with the problems of subsistence farming.

ECONOMIC PERFORMANCE INDICATOR

Consumer Price Index
GDP
160
140
120
100
80
Consumer price index 1995=100
GDP 1996=100
1996
1997
1998
1999
2000

INDIA : MAJOR BUSINESSES

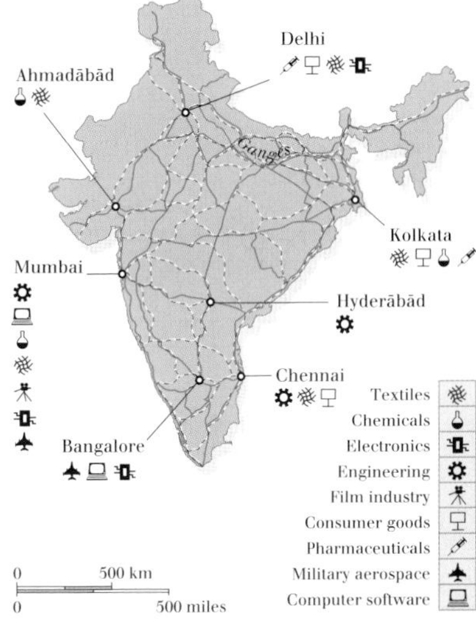
Delhi
Ahmadābād
Ganges
Kolkata
Mumbai
Hyderābād
Chennai
Bangalore
Textiles
Chemicals
Electronics
Engineering
Film industry
Consumer goods
Pharmaceuticals
Military aerospace
Computer software
0
500 km
0
500 miles

Hillside monastery in Ladakh, *Kashmir, northern India. The Ladakhi Buddhists maintain their traditional farming existence and are known for their friendliness.*

I

RESOURCES

Electric power 107m kw

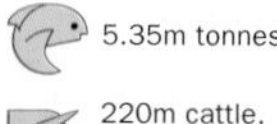

5.35m tonnes

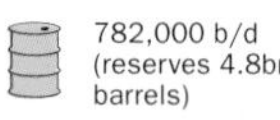

782,000 b/d (reserves 4.8bn barrels)

220m cattle, 124m goats, 413m chickens

Iron, diamonds, coal, limestone, zinc, lead

ELECTRICITY GENERATION

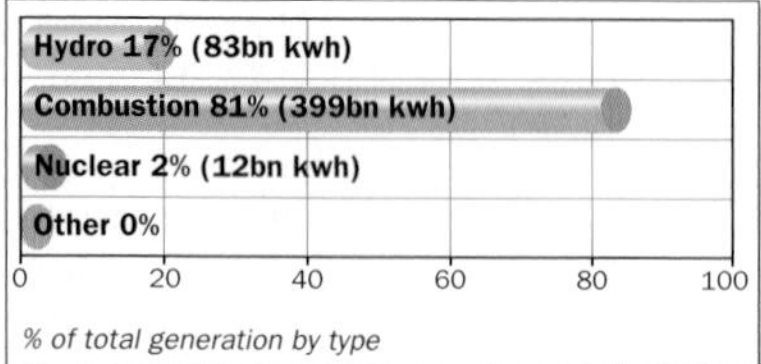

Hydro 17% (83bn kwh)
Combustion 81% (399bn kwh)
Nuclear 2% (12bn kwh)
Other 0%
0
20
40
60
80
100
% of total generation by type

India's most significant mineral exports are iron ore and cut diamonds. In addition, there are large coal reserves; India is the world's fourth-largest coal producer. The steel industry has been privatized. Steel imports are now subject to lower duties, but the industry has so far withstood external competition, and exports have increased. However, steel production, which consumes up to twice as much energy as the method used by some foreign competitors, is inefficient by international standards.

The state is currently unable to meet the country's demand for electricity. Petroleum and coal are the main sources of energy generation, although these are imported. A scheme in Maharashtra to increase output by attracting investment from a US-led consortium was mired in controversy in the mid-1990s when the BJP state government temporarily suspended negotiations in response to nationalist groups opposed to foreign businesses.

ENVIRONMENT

Sustainability rank: 116th

5% (3% partially protected)

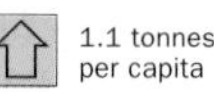

1.1 tonnes per capita

ENVIRONMENTAL TREATIES

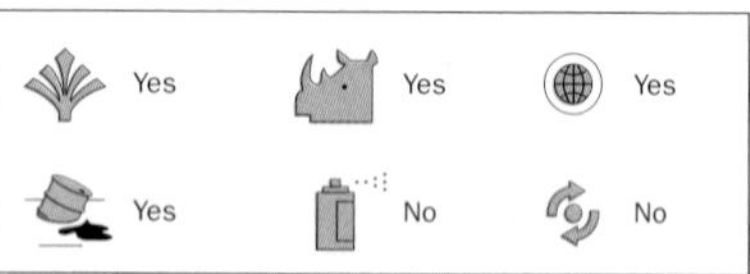

Yes
Yes
Yes
Yes
No
No

Deforestation is one of India's most pressing environmental problems. Industrial and agricultural pressures have felled almost 90% of original forest cover. This results in major soil erosion, the silting up of dams, and landslides. Unusually serious flooding in eastern states in 2000 was largely attributed to deforestation. On the other hand, dealing with water scarcities, such as the drought which affected much of the northwest in 2000, has become a major public policy issue.

INDIA - LAND USE

HIMALAYAS
GREAT INDIAN DESERT
Ganges
Narmada

Cropland
Forest
Pasture
Wetlands
Desert
High mountain regions
Cattle
Tea – cash crop
Cotton – cash crop
Rice
0 500 km
0 500 miles

MEDIA

TV ownership medium

Daily newspaper circulation 28 per 1000 people

PUBLISHING AND BROADCAST MEDIA

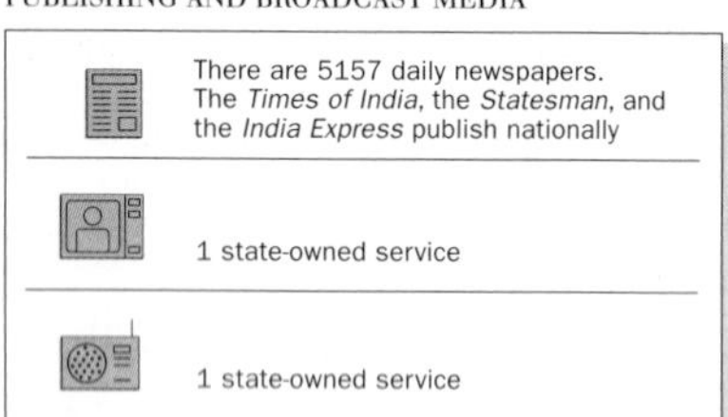

There are 5157 daily newspapers. The *Times of India*, the *Statesman*, and the *India Express* publish nationally
1 state-owned service
1 state-owned service

Satellite TV is increasingly popular in India. Services range from the BBC World Service to CNN, Hindi-language Zee TV to MTV, and one state-run channel. More than seven million households are estimated to have acquired dishes. State-run terrestrial TV has suffered as a result. Critics fear a Western onslaught on Indian values. Recent newspaper launches include the *Asian Age*, which is simultaneously published in London by satellite and claims to be India's first truly international paper.

CRIME

Death penalty in use

381,147 prisoners

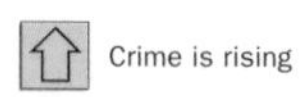

Crime is rising

CRIME RATES

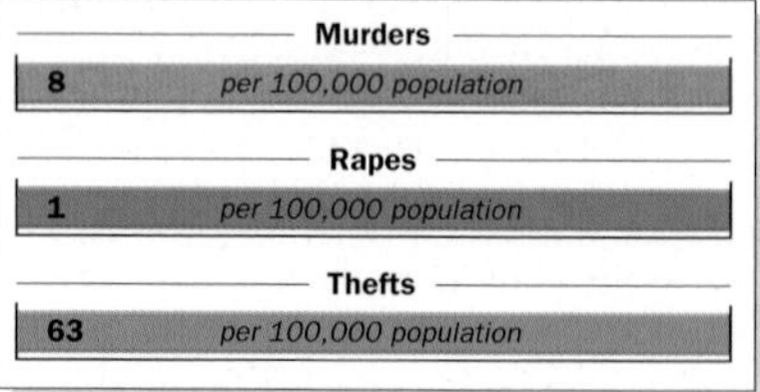

Murders
8 per 100,000 population
Rapes
1 per 100,000 population
Thefts
63 per 100,000 population

Interreligious violence is sporadic but serious. Attacks on Christians drew particular attention, but were overshadowed by Hindu–Muslim violence in Gujarat in 2002. Security forces gained increased powers that year under the Prevention of Terrorism Ordinance, passed in response to Kashmiri separatist attacks in Delhi and other cities.

Violent crime is increasing, especially in the big cities. Gangs have made vast profits from smuggling, prostitution, narcotics, protection, and extortion rackets. Theft has risen sharply as consumer spending increases, and Internet crime is a problem.

Dacoits still operate in large areas of central India. Modeled on the *thugee* gangs of the 19th century, they are outlaws who live by highway robbery and terrorizing small rural communities.

EDUCATION

School leaving age: 12

57%

6.06m students

THE EDUCATION SYSTEM

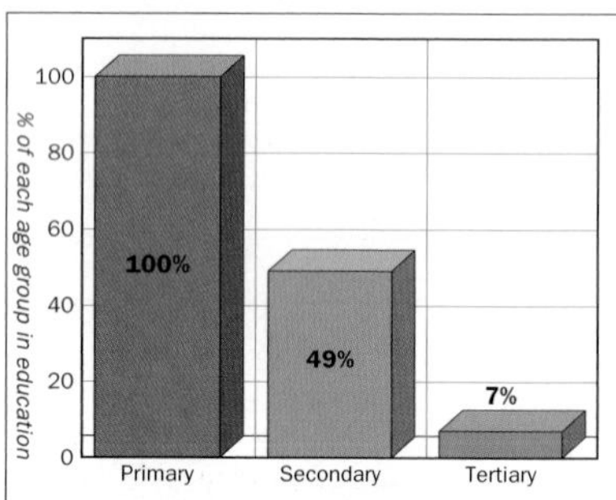

100
80
60
40
20
0
% of each age group in education
100%
49%
7%
Primary
Secondary
Tertiary

Education is primarily the responsibility of the individual state governments. There is now a primary school in every village across the subcontinent, but many children drop out of school to provide supplementary income for their families. There are more than 50 million students at secondary level, and an estimated 10 million graduates from more than 200 universities. Women make up almost 10% of those enrolled in higher education; this is a good percentage for a low-income economy. Although the high level of illiteracy is a brake on development, India has one of the largest pools of science graduates anywhere in the world.

Terraced fields *in central India. In addition to rice, wheat, sorghum, maize, millet, and barley are also important cereal crops.*

THE SANGH MOVEMENT FOR A HINDU NATIONAL IDENTITY

THE IDEA OF *HINDUTVA*, a specifically Hindu-based national identity, has since preindependence days existed in Indian politics. For decades, however, its supporters were no more than a minor strand of the opposition to the secularist Congress party, which ruled almost uninterruptedly for over 40 years. In the late 1980s, however, the rise of the BJP made *Hindutva* a national issue. When the party came to power, heading its first coalition government in 1998, Prime Minister A. B. Vajpayee kept its ideology in the background, but it is pressed by the nationalist Sangh movement to deliver on its pro-Hindu aims. The so-called "saffron brigade" calls in particular for a Hindu temple to be built in place of a Muslim mosque at Ayodhya, Uttar Pradesh, the supposed birthplace of the Hindu god Rama. In 1992 Hindu activists destroyed the mosque, setting off communal rioting in which 1200 people died. Another major flare-up came in 2002, sparked by plans to begin construction of the proposed temple. When Hindu activists returning from Ayodhya were massacred on a train, revenge attacks set off intercommunal violence in Gujarat in which some 2000 people died.

THE SANGH MOVEMENT

The BJP's ideological origins lay in the Sangh movement built around the Rashtriya Swayamsevak Sangh (RSS). The RSS, whose name translates as the National Union of Selfless Servants, was first formed in Nagpur in 1925 to recruit men and women as *swayemsevaks*, volunteers dedicated to the creation of a Hindu nation (Hindu *rashtra*). Banned briefly in 1948 over its association with the Hindu extremist assassin of Mahatma Gandhi, the RSS has no clear organizational structure or membership. It has educational and trade union-based arms, and a religious wing in the militant World Hindu Council (VHP). The Sangh's influence is still strong in the BJP, although in 2000 the BJP-led government was compelled by the strength of pro-secular parties to reaffirm a ban on RSS members in the civil service. The BJP also tried to distance itself from the Sangh leader K. S. Sudarshan's increasingly vocal campaign for the expulsion of Christian missionaries and the replacement of "foreign churches" with a Hindu national church. The movement is linked with fringe organizations blamed for attacks on Christians.

Multifaith demonstrators *condemned the intercommunal violence which left two thousand dead in Gujarat in early 2002.*

REGIONAL APPEAL

The appeal of *Hindutva* is strongest in the Hindi-speaking heartland, the "cow belt," centered on Uttar Pradesh and Madhya Pradesh, with Bihar and Orissa to the east and in the west Maharashtra and Gujarat. While seeking to be inclusive and to appeal to all Hindus, the movement is rooted in Brahmin culture, and faces rivalry in some states from parties combating the oppression of the lowest castes (*dalits*).

In Maharashtra, the RSS inspires a more extreme local rival Hindu chauvinist party, Shiv Sena (the "army of Shivaji," a 17th-century warrior king). Shiv Sena leader Bal Thackeray, an open admirer of Adolf Hitler, is the main force in local politics. Thackeray also led the campaign for Bombay's change of name to Mumbai in 1995.

The growth of regional parties has made coalition building a feature of government, and has been a major restraint on the BJP's more militant ideologues at national level. The party under Vajpayee's leadership at first downplayed controversial policies, but he provoked a storm in 2000 by describing the temple-building movement as embodying national sentiment and part of an unfinished agenda.

Young Hindu militants *on a pilgrimage to Ayodhya in Uttar Pradesh, northern India.*

HEALTH

Welfare state health benefits

1 per 2500 people

Respiratory, nutritional, and diarrheal diseases, malaria

Malnutrition is extremely common, increasing infant mortality, much of which is due to preventable diseases. Air pollution from the domestic use of solid fuel kills half a million children every year. AIDS has accelerated and HIV infection rates are rising. State governments are responsible for most health programs, but there are various national health projects, including a massive polio eradication program. The government announced plans in 2000 to subsidize health insurance for the poorest of the poor.

SPENDING

GDP/cap. increase

CONSUMPTION AND SPENDING

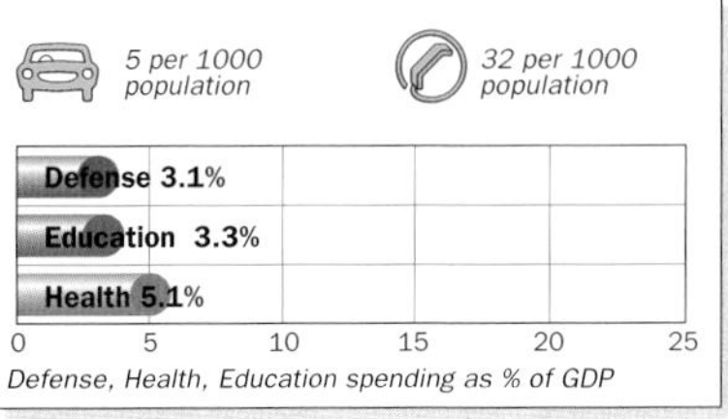
5 per 1000 population
32 per 1000 population
Defense 3.1%
Education 3.3%
Health 5.1%
0
5
10
15
20
25
Defense, Health, Education spending as % of GDP

According to the government, 240 million people (25% of the population at that time), mostly in rural areas, were living below the poverty line in the late 1980s. Recent studies dispute whether this figure is rising or falling. Extremes of wealth, particularly with the opening up of the economy, are frequently seen alongside extremes of poverty. The middle classes, who number some 150–200 million, have an exceedingly comfortable lifestyle, with servants and plush housing. Many of the slums in cities such as Mumbai (Bombay) and Kolkata (Calcutta) have five to nine people living in one room; few slum houses have sanitation. In Mumbai alone, over 100,000 people live on the sidewalks.

WORLD RANKING

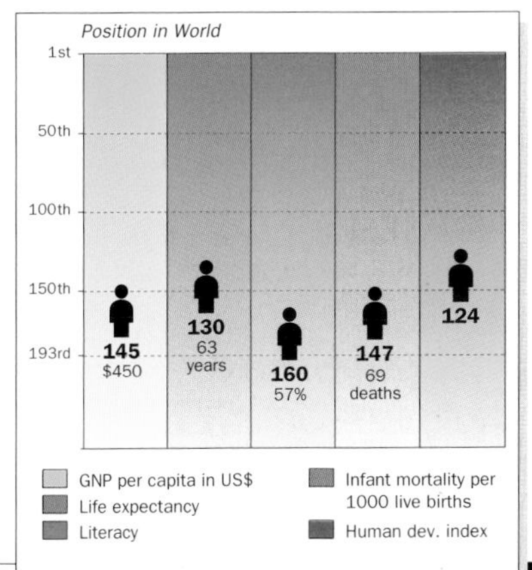
Position in World
1st
50th
100th
150th
193rd
145
$450
130
63 years
160
57%
147
69 deaths
124
GNP per capita in US$
Life expectancy
Literacy
Infant mortality per 1000 live births
Human dev. index

Indonesia

Asia
SOUTHEAST ASIA
Malaysia
Papua New Guinea
INDONESIA
East Timor
Australia

Official Name: Republic of Indonesia **Capital:** Jakarta
Population: 214 million **Currency:** Rupiah **Official Language:** Bahasa Indonesia

1949

1999

Aug 17
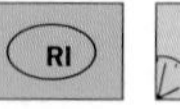
RI
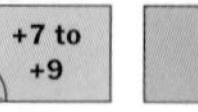
+7 to +9

+62

.id

INDONESIA IS THE WORLD'S LARGEST archipelago. Its 13,677 islands stretch 5000 km (3100 miles) from the Indian Ocean to New Guinea. Java, Kalimantan, Papua, Sulawesi, and Sumatra are mountainous, volcanic, and densely forested. Politics after independence was dominated by the military for over three decades, until the fall of the Suharto regime in 1998, when a partial "civilianization" began. In outlying regions, the forcibly suppressed demands for greater autonomy have flared up, bringing renewed violence. East Timor, which Indonesia invaded in 1975 and then annexed, voted for independence in 1999 and became a fully sovereign state in 2002.

***Rice terraces on Bali,** one of Indonesia's 13,677 islands and its most popular tourist destination. Rice is the staple food crop.*

CLIMATE

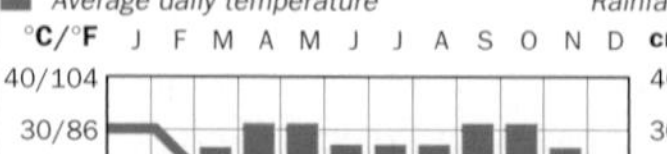
Tropical equatorial/ monsoon

WEATHER CHART FOR JAKARTA

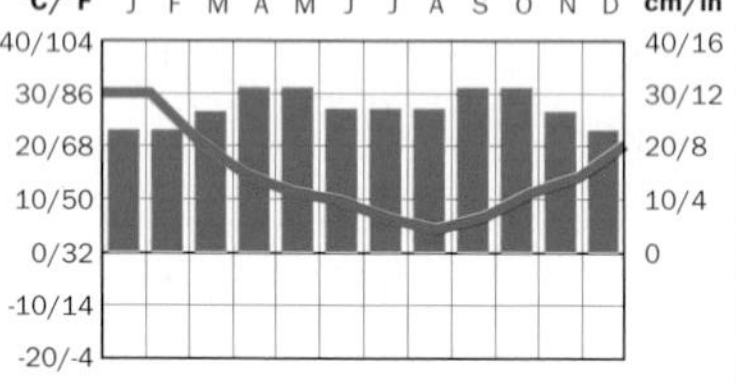
Average daily temperature
Rainfall
°C/°F
J F M A M J J A S O N D
cm/in
40/104
30/86
20/68
10/50
0/32
-10/14
-20/-4
40/16
30/12
20/8
10/4
0

Indonesia's climate is predominantly tropical. Variations relate mainly to differences in latitude, but hilly areas are cooler overall. Rain falls throughout the year, often in thunderstorms, but there is a relatively dry season from June to September. December to March is the wettest period, except in the Moluccas, which receive the bulk of their rain between June and September.

TRANSPORTATION

Drive on left

Sukarno-Hatta, Jakarta 10.7m passengers

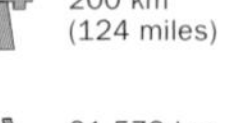

2480 ships 3.38m grt

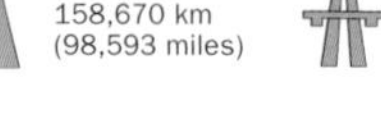

THE TRANSPORTATION NETWORK

	158,670 km (98,593 miles)		200 km (124 miles)
	6485 km (4030 miles)		21,579 km (13,409 miles)

For a multi-island state spread across three time zones, communications are an obvious government priority. Indonesia was an early entrant into satellite communications, providing an international satellite-based telephone system as early as 1976.

However, provisions vary greatly in the different provinces. Road surfaces in Java and Sumatra are excellent. Rail services are restricted to these two islands. In contrast, roads in Kalimantan and Papua are poor, and most travel is by air or river.

TOURISM

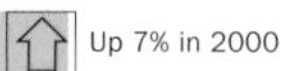
Visitors : Population 1:42

5.06m visitors

Up 7% in 2000

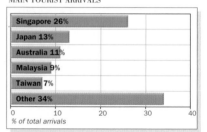

MAIN TOURIST ARRIVALS

	% of total arrivals
Singapore	26%
Japan	13%
Australia	11%
Malaysia	9%
Taiwan	7%
Other	34%

Tourism took off during the 1980s. In 1997, the number of tourists exceeded five million, although political unrest in 1998 and 1999 discouraged many visitors. Bali, Java, and Sumatra are the most popular destinations. Expansion has been encouraged by a major investment in hotels and the opening of Bali to airlines other than the national carrier, Garuda Indonesia.

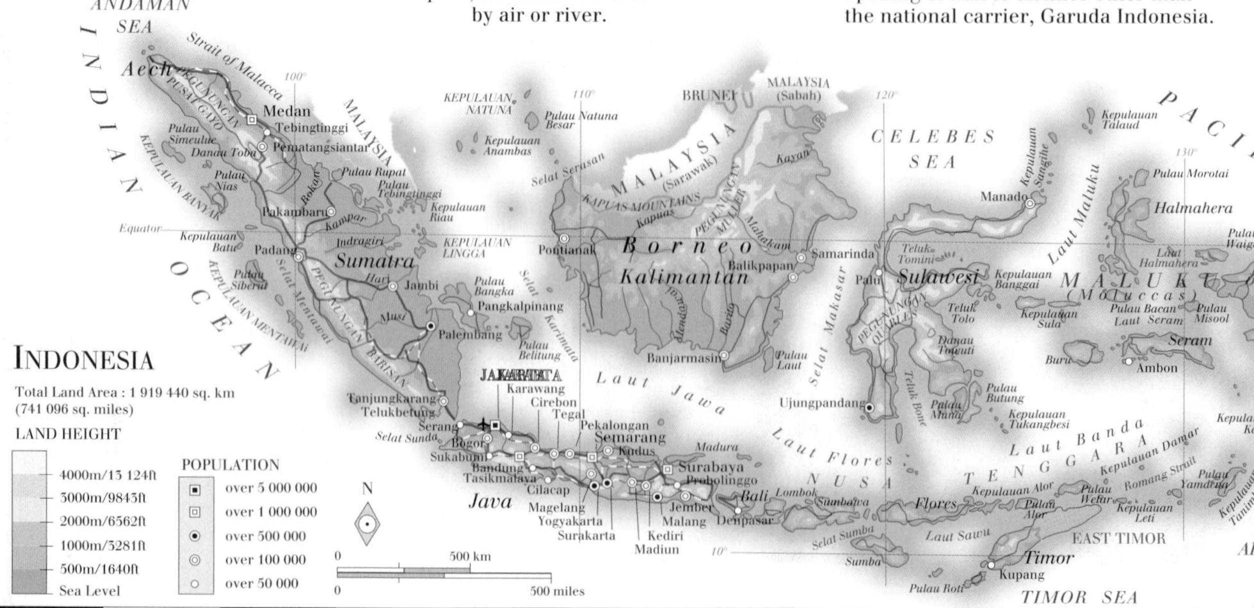
INDONESIA
Total Land Area : 1 919 440 sq. km (741 096 sq. miles)
LAND HEIGHT
4000m/13 124ft
3000m/9843ft
2000m/6562ft
1000m/3281ft
500m/1640ft
Sea Level
POPULATION
over 5 000 000
over 1 000 000
over 500 000
over 100 000
over 50 000
N
0 500 km
0 500 miles
ANDAMAN SEA
INDIAN OCEAN
Strait of Malacca
Aceh
Medan
Tebingtinggi
Pematangsiantar
Pulau Simeulue
Danau Toba
Pulau Nias
Pakanbaru
Padang
Sumatra
Jambi
Palembang
Tanjungkarang-Telukbetung
Pulau Siberut
Selat Mentawai
Pulau Bangka
Pangkalpinang
Pulau Belitung
Kepulauan Riau
Kepulauan Natuna
Pulau Natuna Besar
Kepulauan Anambas
BRUNEI
MALAYSIA (Sabah)
MALAYSIA (Sarawak)
Pontianak
Borneo
Kalimantan
Samarinda
Balikpapan
Banjarmasin
JAKARTA
Karawang
Cirebon
Tegal
Pekalongan
Semarang
Kudus
Serang
Bogor
Sukabumi
Bandung
Tasikmalaya
Cilacap
Java
Magelang
Yogyakarta
Surakarta
Madiun
Kediri
Malang
Jember
Probolinggo
Surabaya
Madura
Bali
Denpasar
Lombok
Sumbawa
Flores
Sumba
Timor
Kupang
EAST TIMOR
TIMOR SEA
Laut Jawa
Laut Flores
Laut Sawu
Laut Banda
CELEBES SEA
Manado
Sulawesi
Palu
Ujungpandang
Selat Makasar
Kepulauan Banggai
Kepulauan Sula
MALUKU (Moluccas)
Halmahera
Pulau Morotai
Seram
Ambon
Pulau Buru
Kepulauan Talaud
PACIFIC

PEOPLE

Pop. density medium

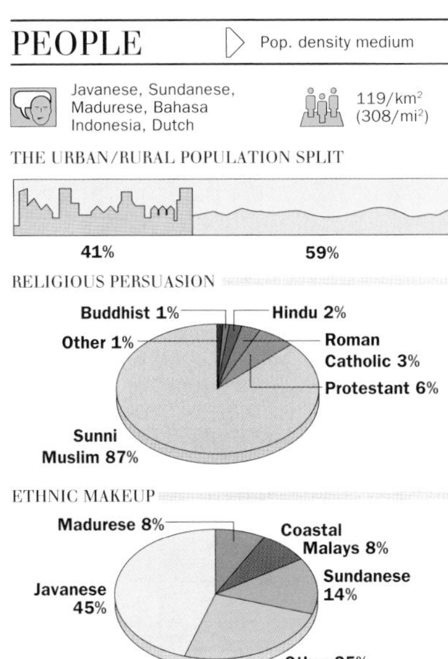
Javanese, Sundanese, Madurese, Bahasa Indonesia, Dutch
119/km² (308/mi²)
THE URBAN/RURAL POPULATION SPLIT
41%
59%
RELIGIOUS PERSUASION
Buddhist 1%
Hindu 2%
Other 1%
Roman Catholic 3%
Protestant 6%
Sunni Muslim 87%
ETHNIC MAKEUP
Madurese 8%
Coastal Malays 8%
Sundanese 14%
Javanese 45%
Other 25%

The basic Melanesian–Malay ethnic division disguises a diverse society. The national language, Bahasa Indonesia, coexists with at least 250 other spoken languages or dialects. Attempts by the Javanese political elite to suppress local cultures have been vigorously opposed, especially by the East Timorese, the Aceh of northern Sumatra, and the Papuans of New Guinea.

Religious and interethnic hostility is increasing. Since 1998 there have been violent clashes between Muslims and Christians on Sulawesi, and in the Moluccas. Similar clashes occurred in Kalimantan in 1999 and 2001 between indigenous Dayaks and ethnic Madurese immigrants. Aceh introduced *sharia* (Islamic law) in 2000.

Discrimination against ethnic Chinese has encouraged vicious attacks on their businesses, as in Jakarta in 1998. Gender equality is enshrined in law, and women are active in public life.

POPULATION AGE BREAKDOWN

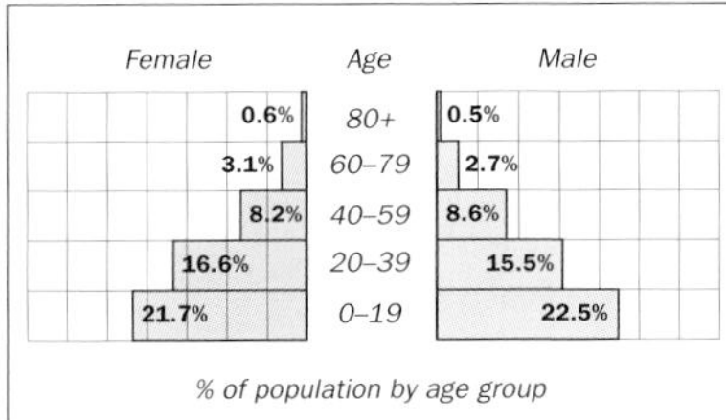
Female
Age
Male
0.6% 80+ 0.5%
3.1% 60–79 2.7%
8.2% 40–59 8.6%
16.6% 20–39 15.5%
21.7% 0–19 22.5%
% of population by age group

WORLD AFFAIRS

Joined UN in 1950

APEC | ASEAN | G15 | OIC | OPEC

Indonesia pursues a largely pro-Western foreign policy, although the government is under pressure to improve its human rights record. The scale and nature of the East Timor massacres in 1999 severely damaged its standing. Internal security in the far-flung provinces tests regional relations.

China remains a major foreign policy concern, despite the restoration of diplomatic ties in 1990. Indonesia and Australia signed a groundbreaking security cooperation agreement in late 1995 but the movement of illegal immigrants through Indonesia creates tensions.

AID

Recipient

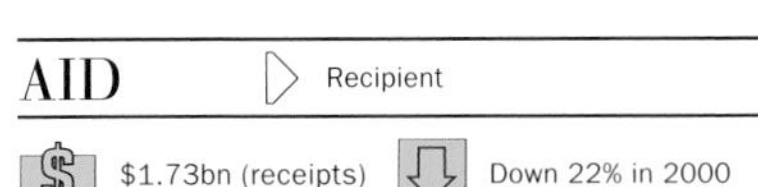
$1.73bn (receipts)
Down 22% in 2000

Japan accounts for the bulk of bilateral aid. Multilateral aid comes above all from the World Bank, although it called for urgent reform in late 2001 before loaning more money. Aid has been notoriously subject to "leakage."

POLITICS

Multiparty elections

1999/2004

President Megawati Sukarnoputri

AT THE LAST ELECTION

House of Representatives 500 seats

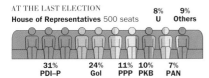

31% PDI–P	24% Gol	11% PPP	10% PKB	7% PAN	8% U	9% Others

PDI–P = Indonesian Democratic Party of Struggle
Gol = Golkar **PPP** = United Development Party
PKB = National Awaking Party **U** = Unelected (army) seats
PAN = National Mandate Party

Indonesia is a multiparty democracy. The president is indirectly elected by a 700-member People's Consultative Assembly.

PROFILE

Under Gen. Suharto, the autocratic ruler from 1968, Golkar – an amalgam of interest groups – and the army were dominant. Suharto was forced to resign in 1998 amid widespread protest over corruption, economic mismanagement, and denial of democratic rights. After his fall, the mainly Muslim PDI–P, led by Megawati Sukarnoputri, daughter of the first president, won elections, but Abdurrahman Wahid of the PKB was elected president; Megawati became vice president. Disenchantment with Wahid pressured him into devolving many of his duties to her in 2000, and increasingly vehement opposition in parliament culminated in his removal in July 2001, whereupon Megawati became president herself.

MAIN POLITICAL ISSUES

The army

The army's involvement in politics remains a serious problem. In early 2000, the influential Gen. Wiranto left the government, but the army refused to relinquish its 38-seat entitlement in the legislature any earlier than 2009.

Separatist movements

Separatism – the legacy of Suharto's centralist policies – and violence between Muslims and religious minorities threaten national unity. The movements in Aceh (northern Sumatra) and Papua continue to gather momentum; both were granted greater autonomy in 2001.

***Abdurrahman Wahid**, dismissed as president in 2001.*

***President Megawati Sukarnoputri**, Wahid's successor.*

CHRONOLOGY

Hindu, Buddhist, and Islamic interest preceded 16th-century European rivalries over valuable spices. The Dutch had won control by the 17th century, when colonization began on Java. By 1910, the Dutch East Indies encompassed present-day Indonesia.

- **1912** Sarekat Islam party formed.
- **1920** Indonesian Communist Party (PKI) formed; leads revolt in West Java (1926) and Sumatra (1927).
- **1927** Indonesian National Party formed under Dr. Sukarno.
- **1930s** Dutch repress nationalists.
- **1942–1945** Japanese occupation. Autonomy within "Greater East Asia" state promised. Sukarno works with Japanese while promoting independence.
- **1945** Sukarno declares Indonesian independence. Dutch forces attempt to reassert control.
- **1945–1949** Nationalist guerrilla war.
- **1949** Dutch grant independence to federal-style United States of Indonesia under Sukarno as president.
- **1950** Union dissolved, Sukarno grows increasingly authoritarian.
- **1950s** Moluccas declare independence from Indonesia, fight unsuccessful separatist war.
- **1957–1959** Sukarno introduces authoritarian Guided Democracy.

CHRONOLOGY *continued*

- **1959** Sukarno extends presidential powers. Civilian legislature replaced by military. Extreme nationalist and pro-Chinese policies.
- **1962** Dutch relinquish Western New Guinea.
- **1965** Communist PKI alliance with military ends. Army led by Gen. Suharto crushes abortive coup and acts to eliminate banned PKI; up to one million killed.
- **1966** Sukarno hands over power to Gen. Suharto. Temporary handover becomes permanent in following year.
- **1968** Suharto becomes president: declares "New order"; introduces pro-Western liberal economic policies while transferring real power from cabinet to small group of officers.
- **1971** First elections for 16 years. Government-sponsored Golkar wins landslide. Opposition parties now passive partners of government.
- **1975** Invasion of East Timor; its incorporation as 27th province in 1976 is not recognized by UN.
- **1984** Muslim protesters clash with troops in Jakarta. Start of resurgence of Islamic protest.
- **1985** Fretilin movement declares East Timor independent.
- **1989** Growing discontent with authoritarian government; student protests, unrest in Java and Sumbawa. Demands for Suharto to retire. Low-key official response.
- **1991** Indonesian troops massacre pro-independence demonstrators in East Timor. New organizations, including Democratic Forum and League for Restoration of Democracy, allowed to form in response to growing demands for "openness."
- **1993** Suharto wins sixth term in office.
- **1996** Antigovernment demonstrations in Jakarta.
- **1997** Economic recession. Smog across region from forest fires.
- **1998** Suharto resigns amid unrest.
- **1999** Election victory for opposition led by Megawati Sukarnoputri. East Timor referendum backing independence triggers violent backlash. Abdurrahman Wahid elected president, Megawati named vice president.
- **2000** Aceh becomes first province to introduce Islamic *sharia* law. Violence erupts again in Moluccas.
- **2001** Wahid removed, replaced by Megawati.
- **2002** January, autonomy officially granted to Papua. May, East Timor independent.

I

DEFENSE

Compulsory military service

 $1.49bn

 Down 1% in 2000

INDONESIAN ARMED FORCES

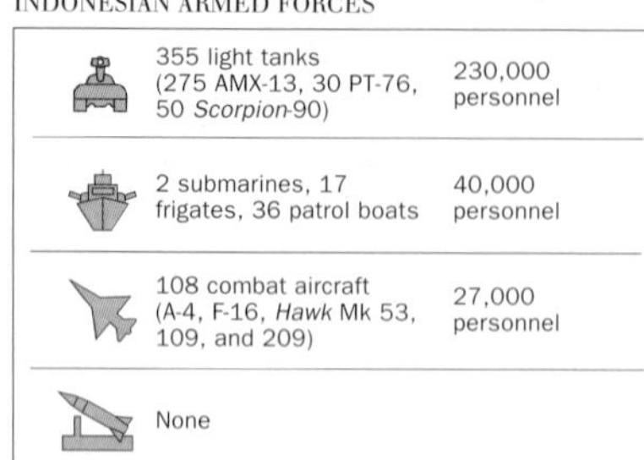

355 light tanks (275 AMX-13, 30 PT-76, 50 *Scorpion*-90)	230,000 personnel
2 submarines, 17 frigates, 36 patrol boats	40,000 personnel
108 combat aircraft (A-4, F-16, *Hawk* Mk 53, 109, and 209)	27,000 personnel
None	

Defense spending is low by regional standards, despite the military's high profile. The constitution enshrines the military's political role, and even now it remains the key influence in Indonesia. The recent "civilianization" of political parties, the bureaucracy, and state companies has reduced the presence of the military in these areas, if not their influence. The government is under pressure from the IMF to audit "off-budget funds," which often supplement official defense allocations. Western arms sales are increasingly being made dependent on the improvement of Indonesia's human rights record. The main defense issues are internal security and the perceived Chinese threat.

Military service is selective and lasts for two years.

ECONOMICS

Inflation 16% p.a. (1990–2000)

 $120bn

9675–10,400 rupiahs

SCORE CARD

- World GNP Ranking 32nd
- GNP per Capita $570
- Balance of Payments $7.99bn
- Inflation 3.7%
- Unemployment 5%

EXPORTS

China 4%
South Korea 7%
Singapore 11%
USA 14%
Japan 23%
Other 41%

IMPORTS

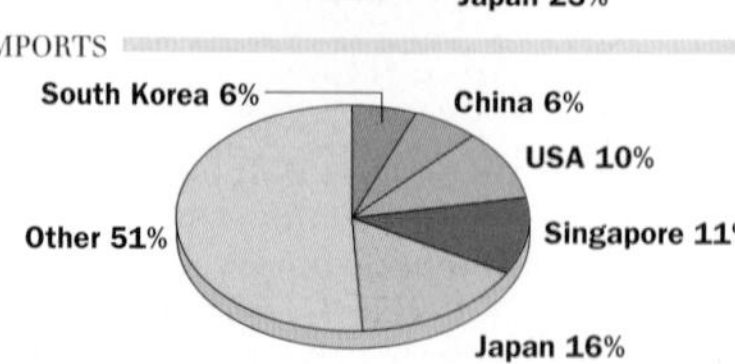
South Korea 6%
China 6%
USA 10%
Singapore 11%
Japan 16%
Other 51%

Strengths

Varied resources, especially oil. Signs of return to high growth. Debt successfully rescheduled. International credit rating improved.

Weaknesses

High level of bureaucracy. Endemic corruption. Huge wealth disparities. Regional insecurity deters investment. High underemployment.

ECONOMIC PERFORMANCE INDICATOR

Consumer Price Index
GDP
400
300
200
100
0
Consumer price index 1995=100
GDP 1996=100
1996
1997
1998
1999
2000

Profile

Under Suharto the economy grew rapidly, fueled largely by oil, until its collapse in 1997–1998. State-owned corporations played a significant role, protected from foreign competition. Non-oil exports were diversified, but the debt burden used up a third of export earnings. Reform was delayed by conflict between "technologists" who favored industrialization over profit for state concerns and advocates of deregulation. Further reforms have been stymied by the government's poor representation in parliament. Corruption remains rife, embroiling then president Wahid in 2001. Persistent weaknesses are likely to continue to hold back real growth. However, relations with the IMF and other donors were sufficiently improved by 2002 for Indonesia's international credit rating to be upgraded.

INDONESIA : MAJOR BUSINESSES

Rubber
Heavy engineering
Gas
Chemicals
Timber industries
Oil
Oil refining
Electronics
Vehicle assembly
Aerospace industry
* significant multinational ownership
0 500 km
0 500 miles
Medan
Banjarmasin
Balikpapan
Kendari
Sorong
Palembang
Jakarta
Bandung
Surabaya
Ujung Pandang

RESOURCES

Electric power 22.9m kw

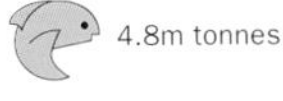
4.8m tonnes

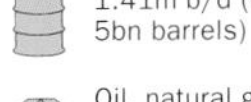
1.41m b/d (reserves 5bn barrels)

28.1m ducks, 12.5m goats, 751m chickens

Oil, natural gas, coal, bauxite, nickel, copper, gold, tin

ELECTRICITY GENERATION

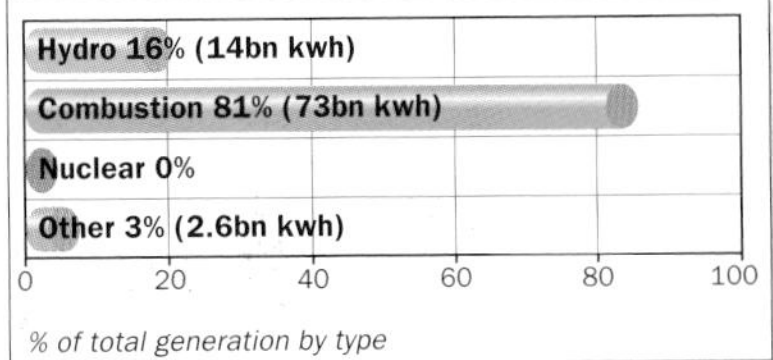
Hydro 16% (14bn kwh)
Combustion 81% (73bn kwh)
Nuclear 0%
Other 3% (2.6bn kwh)
0
20
40
60
80
100
% of total generation by type

Indonesia is rich in energy sources. The main export earners are oil and liquefied natural gas (LNG) – Indonesia is the world's largest LNG exporter. However, oil output has been falling. Combined with rapid growth in domestic energy demand, this could turn the country into an oil importer in the next decade. The government is therefore encouraging oil exploration in remote regions. It is also considering developing geothermal and nuclear energy sources. Indonesia's other main resources are coal, bauxite, and nickel, along with agricultural products such as rubber and palm oil. Processed wood products are a significant export commodity; the rapid depletion of the rainforests has given rise to attempts at control over log exports.

INDONESIA : LAND USE

Cropland
Forest
Pasture
Wetlands
Rice
Nutmeg - cash crop
Cattle
0 500 km
0 500 miles
Sumatra
Kalimantan
Sulawesi
Maluku
Irian Jaya
Laut Jawa
Laut Banda
Java
Timor
PACIFIC OCEAN

ENVIRONMENT

Sustainability rank: 100th

11% (3% partially protected)

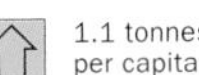
1.1 tonnes per capita

ENVIRONMENTAL TREATIES

Yes
Yes
Yes
Yes
Yes
No

Environment legislation is poorly enforced, which means that the rich tropical forests are threatened by excessive logging and that rare flora and fauna are disappearing. In 2002, the government asked Greenpeace to help monitor illegal logging in remote areas. Smog from forest fires is a serious problem, causing a regional health alert in 1997. Frequent oil spillages in the Malacca Strait are a hazard.

MEDIA

TV ownership medium

Daily newspaper circulation 23 per 1000 people

PUBLISHING AND BROADCAST MEDIA

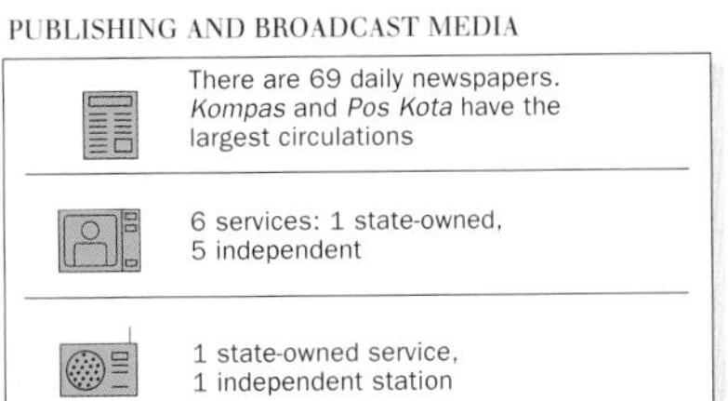
There are 69 daily newspapers. *Kompas* and *Pos Kota* have the largest circulations
6 services: 1 state-owned, 5 independent
1 state-owned service, 1 independent station

The 1999 press law prohibits censorship, but journalists can still be fined for violating "religious and moral norms." Independent press and broadcasting have flourished since 1998.

CRIME

Death penalty in use

53,399 prisoners

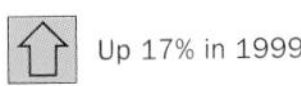
Up 17% in 1999

CRIME RATES

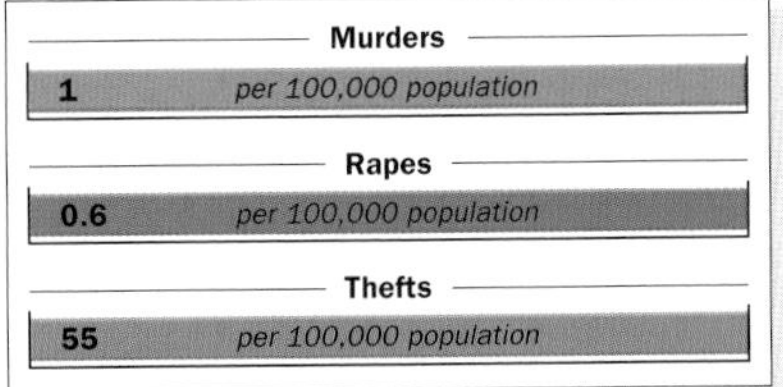
Murders
1 per 100,000 population
Rapes
0.6 per 100,000 population
Thefts
55 per 100,000 population

Suppression of secessionists is harsh. There is brutal ethnic violence in the provinces.

EDUCATION

School leaving age: 12

87%

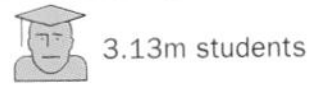
3.13m students

THE EDUCATION SYSTEM

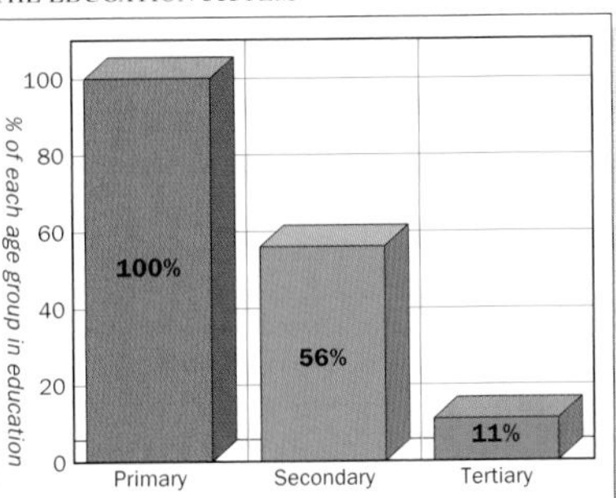
100
80
60
40
20
0
% of each age group in education
100%
56%
11%
Primary
Secondary
Tertiary

Primary education is subsidized by the state, compulsory, and often provided by Islamic schools. In contrast, good secondary education is hard to find in rural areas. University students come predominantly from the richer elites.

HEALTH

Welfare state health benefits

1 per 5000 people

Lower respiratory and diarrheal diseases

An extensive network of clinics, down to village level, means that access to health care is reasonable. As a result, health indicators have improved significantly. The death rate declined from 2% in 1965 to 0.7% in 2000, thus helping to increase life expectancy, while infant mortality has more than halved over this period. However, malnutrition, particularly among children, continues to remain a real problem, although chronic levels of vitamin A deficiency have been largely eradicated. HIV infections among intravenous drug users, particularly in prisons, are rising.

SPENDING

GDP/cap. increase

CONSUMPTION AND SPENDING

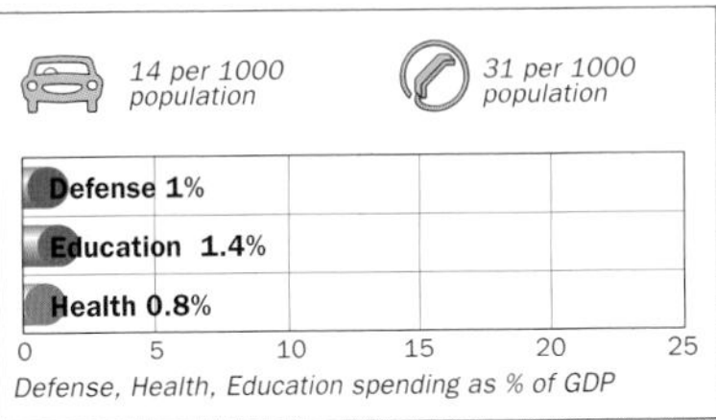
14 per 1000 population
31 per 1000 population
Defense 1%
Education 1.4%
Health 0.8%
0
5
10
15
20
25
Defense, Health, Education spending as % of GDP

Many Indonesians live in relative poverty and those on the peripheral islands in real poverty; large wealth disparities exist between the Javanese middle classes and the subsistence farmers and tribesmen of Papua and Kalimantan. This reflects both a concentration of wealth in the hands of a limited number of key political and business figures, and a concentration of development and investment on the main islands, particularly Java. Since 1998 attempts have been made in the courts to tackle the issues of corruption and concentration of wealth in the hands of close associates and relatives of former President Suharto and his political successors.

WORLD RANKING

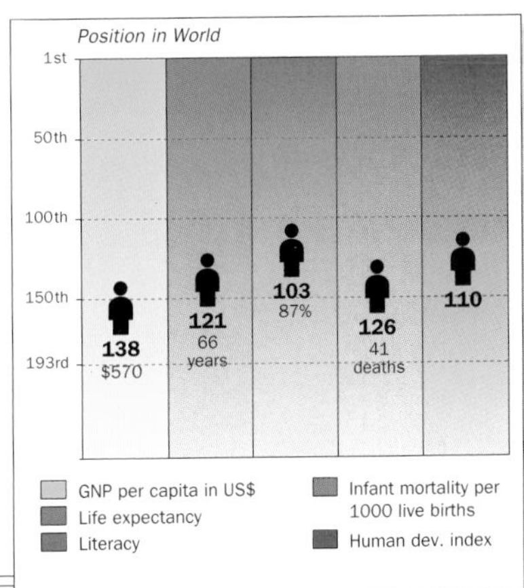
Position in World
1st
50th
100th
150th
193rd
138
$570
121
66 years
103
87%
126
41 deaths
110
GNP per capita in US$
Life expectancy
Literacy
Infant mortality per 1000 live births
Human dev. index

I

IRAN

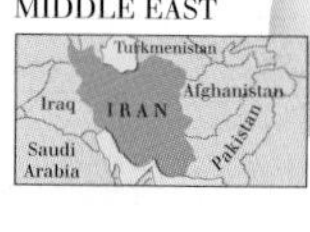
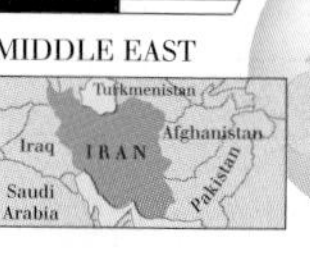
MIDDLE EAST
Turkmenistan
Iraq
IRAN
Afghanistan
Saudi Arabia
Pakistan
Asia
Africa

OFFICIAL NAME: Islamic Republic of Iran **CAPITAL:** Tehran
POPULATION: 71.4 million **CURRENCY:** Iranian rial **OFFICIAL LANGUAGE:** Farsi

1502

1990

Feb 11
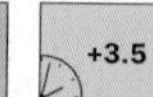
IR

+3.5

+98

.ir

IRAN IS SURROUNDED by turbulent neighbors, with republics of the former Soviet Union to the north, Afghanistan and Pakistan to the east, and Iraq and Turkey to the west. The south faces the Persian Gulf and the Gulf of Oman. Since 1979, when a revolution led by Ayatollah Khomeini deposed the shah, Iran has become the world's largest theocracy and the leading center for militant Shi'a Islam. Iran's active support for Islamic fundamentalist movements has led to strained relations with central Asian, Middle Eastern, and north African states, as well as the US and Europe.

The Reshteh-ye Kuhhā-ye Alborz (Elburz Mountains). Their Caspian Sea slopes are rainy and forested; the southern slopes are dry.

CLIMATE

Mountain/cold desert

WEATHER CHART FOR TEHRAN

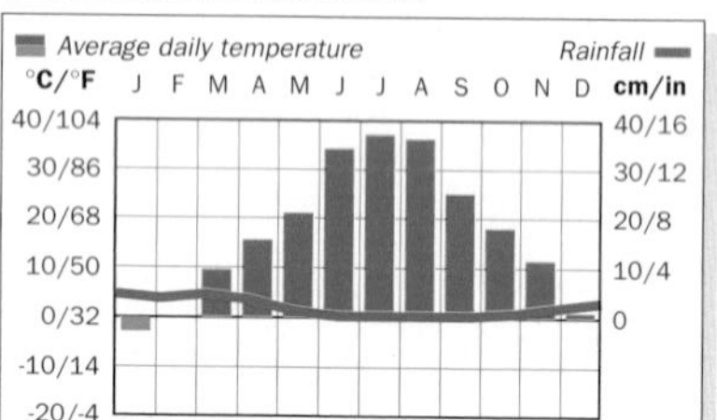
Average daily temperature
Rainfall
°C/°F
J F M A M J J A S O N D
cm/in
40/104
30/86
20/68
10/50
0/32
-10/14
-20/4
40/16
30/12
20/8
10/4
0

The area bordering the Caspian Sea is Iran's most temperate region. Most of the country has a desert climate.

TRANSPORTATION

Drive on right

Mehrabad International, Tehran
8.47m passengers

395 ships
4.23m grt

THE TRANSPORTATION NETWORK

49,440 km (30,721 miles)

470 km (292 miles)

6398 km (3976 miles)

904 km (562 miles)

Adequate roads link main towns, but rural areas are less well served. Most freight travels by rail. A ferry runs from Bandar-e Abbas to the UAE.

PEOPLE

Pop. density low

Farsi, Azerbaijani, Gilaki, Mazanderani, Kurdish, Baluchi, Arabic, Turkmen

44/km² (113/mi²)

THE URBAN/RURAL POPULATION SPLIT

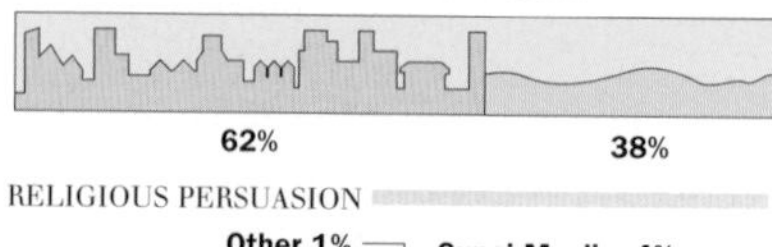
62%
38%

RELIGIOUS PERSUASION

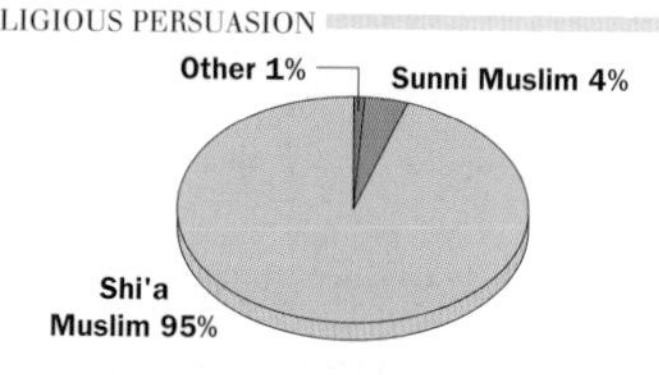
Other 1%
Sunni Muslim 4%
Shi'a Muslim 95%

ETHNIC MAKEUP

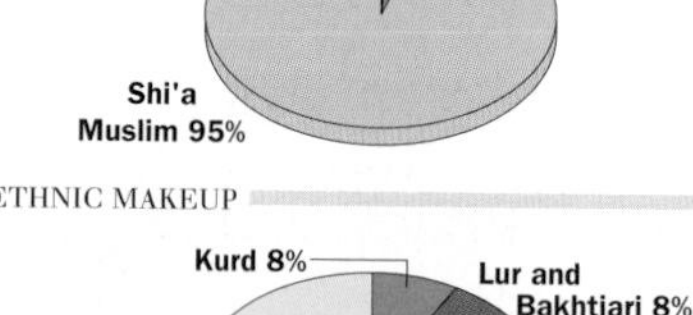
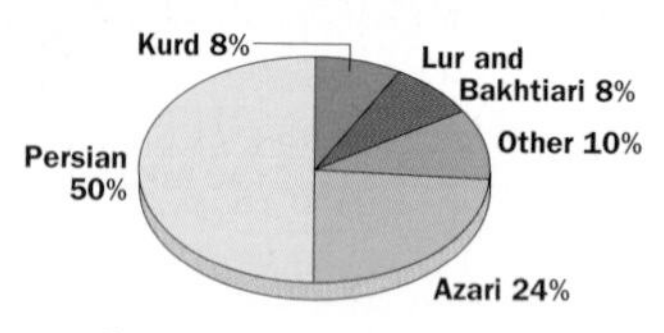
Kurd 8%
Lur and Bakhtiari 8%
Other 10%
Persian 50%
Azari 24%

The population comprises several ethnic groups. The people of the north and center – about half of all Iranians – speak Farsi (Persian), while about a quarter speak related languages, including Kurdish in the west and Baluchi in the southeast. Another quarter of the population speaks Turkic languages, primarily the Azaris and the Turkmen in the northwest. Smaller groups, such as the Circassians and Georgians, are found in the northern provinces.

Until the 16th century, much of Iran followed the Sunni interpretation of Islam, but since then the Shi'a sect has been dominant. Religious minorities, accounting for just 1% of the population, include followers of the Bahai faith, who suffer discrimination, Zoroastrians, Christians, and Jews. The regime has a remarkably liberal attitude to refugees of the Muslim faith. Nearly three million Afghan refugees were received during the height of the Afghan civil war in the 1980s–1990s, although many have since returned. In Khorosan province in the east, refugees account for nearly a quarter of the population; near the Turkish border they constitute half the total population. Many are young, resulting in intense competition with Iranians for jobs and consequent ethnic tensions.

One of the main consequences of the 1979 Islamic revolution was to reverse the policy of female emancipation. The revolution restricted the public role of women and enforced a strict dress code, obliging women to cover themselves from head to foot in the *chadoor* or veil. More liberal attitudes have gradually emerged, and reform of the divorce laws has been proposed, under which it would become possible for proceedings to be initiated by the wife.

POPULATION AGE BREAKDOWN

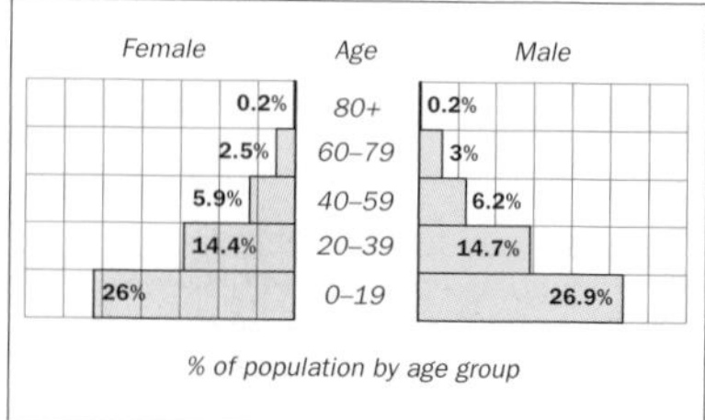
Female
Age
Male
0.2%
80+
0.2%
2.5%
60–79
3%
5.9%
40–59
6.2%
14.4%
20–39
14.7%
26%
0–19
26.9%
% of population by age group

TOURISM

Visitors : Population 1:42

1.7m visitors

Up 29% in 2000

MAIN TOURIST ARRIVALS

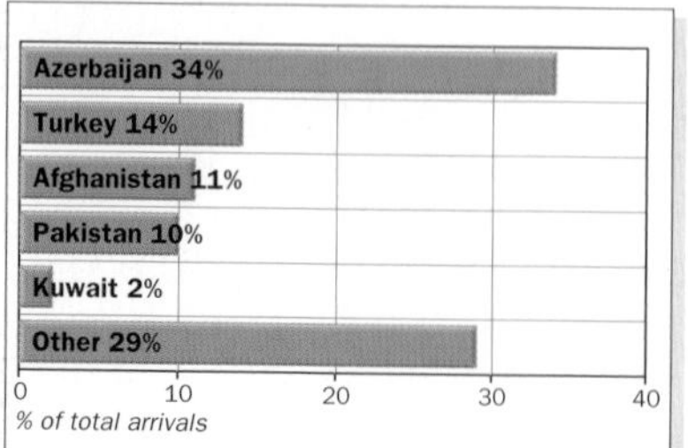
Azerbaijan 34%
Turkey 14%
Afghanistan 11%
Pakistan 10%
Kuwait 2%
Other 29%
0
10
20
30
40
% of total arrivals

Iran's historical heritage, mosques, and bazaars formerly attracted sizable numbers of tourists. This flow was cut off by the 1979 revolution, which deterred visitors, especially from the West. In the 1990s, however, there was a rise in the number of business people visiting Iran. Procedures at Tehran's Mehrabad airport have been simplified and the capital's hotels refurbished. In late 1998 President Khatami's more liberal regime welcomed a delegation of US tourists, despite opposition from conservative groups.

POLITICS

Multiparty elections

2000/2004

President Mohammad Khatami

AT THE LAST ELECTION

Consultative Council (Majlis) 290 seats

65% **IIPF** · **17%** **C** · **11%** **Ind** · **5%** **Vacant** · **2%** **Rel**

IIPF = Islamic Iran Participation Front (reformists)
C = Coalition of Followers of the Line of Imam (conservatives)
Ind = Independents **Rel** = Religious Minorities

Iran is a theocracy. Tension exists between the conservative mullahs and the reformist government.

Ayatollah Khamenei, who became spiritual leader after the death of Ayatollah Khomeini.

Mohammad Khatami, reformist president, in office since 1997.

PROFILE

Iran's religious revolution of 1979 was fueled by popular outrage at the corruption, repression, and inequalities of the shah's regime. Since the time of Ayatollah Khomeini, successive Iranian governments have maintained that the clergy have a religious duty to establish a just social system. Accordingly, the legislature, the executive, and the judiciary may, in theory, be overruled by the religious leadership; former president Hashemi Rafsanjani's moderate policies were questioned by radical clergymen advocating "permanent revolution." However, the mullahs' failure to address Iran's economic problems has eroded their political standing. Reformists were encouraged by the election of President Mohammad Khatami in 1997, but the clergy remains powerful. Huge student demonstrations in 1999 and 2000 in favor of reform were offset by crackdowns on reformist politicians and newspapers. The mullahs, so far unbowed by reformist election victories, have continued to confront the modernizers.

MAIN POLITICAL ISSUE

Mosque versus secular state

A power struggle between the clergy and the secular state has arisen from the ill-defined division of power between the two. The conservative faction in parliament lost its overall majority in 1996 and has been steadily displaced by reformists, who in 2000 made sweeping gains in parliamentary elections. In 2001 reformist president Mohammad Khatami was reelected, this time with an overwhelming 77% of the vote.

Khatami is committed to modernizing the economy, but is strongly opposed by the mullahs, for whom adherence to religious values is more important than material welfare. Student proreform protests were heavily suppressed by hard-liners, who remain a force, despite the 2000 elections, through the powerful Council of Guardians.

I

WORLD AFFAIRS

Joined UN in 1945

ECO
G24

NAM

OIC

OPEC

Following the 1979 revolution, Iran assumed international significance as the voice of militant Shi'a Islam. Iran is accused of backing terrorist activity by Muslim extremists and of fostering unrest throughout the Middle East and central Asia.

Under President Khatami, Iran has tried to convey a less confrontational image. Improved relations with Saudi Arabia, troubled since Iran's seizure of the islands of Abu Musa and the Tunbs in 1970, resulted in the signing of a pact in 2001. Ties with the West have also improved, although relations with the US remain tense and suffered a sharp setback in early 2002 after US President George W. Bush cast Iran (with Iraq and North Korea) as part of an international "axis of evil."

Iran's most pressing security preoccupation is with Iraq, which allows *mujahideen* guerrillas to mount attacks on Iran from its territory.

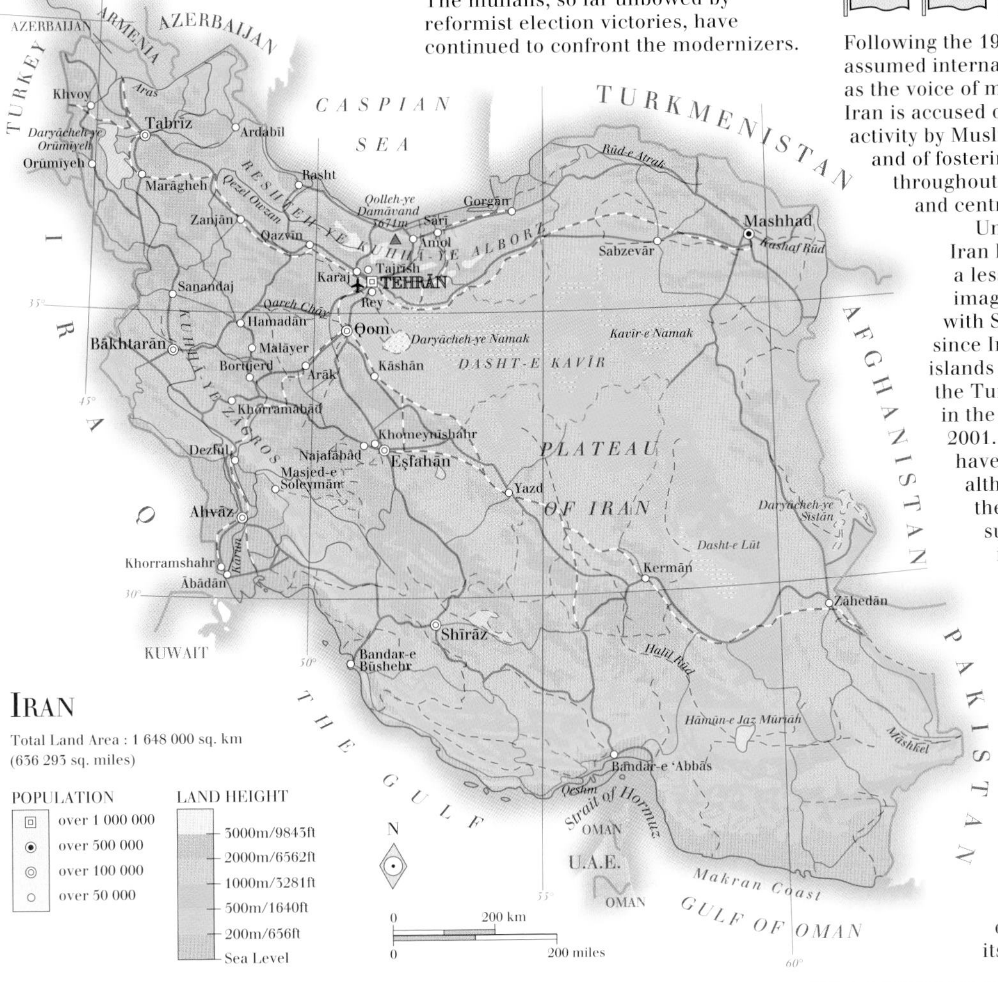

AZERBAIJAN ARMENIA AZERBAIJAN TURKEY CASPIAN SEA TURKMENISTAN IRAQ AFGHANISTAN PAKISTAN KUWAIT THE GULF OMAN U.A.E. GULF OF OMAN
Khvoy Tabrīz Ardabīl Daryācheh-ye Orūmīyeh Orūmīyeh Marāgheh Rasht Zanjān Qazvīn Gorgān Sārī Āmol Qolleh-ye Damāvand 5671m Mashhad Sabzevār Karaj Tajrīsh TEHRĀN Rey Sanandaj Hamadān Qom Bākhtarān Malāyer Borūjerd Arāk Kāshān Khorramābād Dezfūl Najafābād Khomeynīshahr Eşfahān Masjed-e Soleymān Yazd Ahvāz Khorramshahr Ābādān Kermān Zāhedān Shīrāz Bandar-e Būshehr Bandar-e 'Abbās Qeshm
Aras Qezel Owzan Qareh Chāy Kārūn Rūd-e Atrak Kashaf Rūd Halīl Rūd Māshkel Hāmūn-e Jaz Mūrīān Daryācheh-ye Namak Kavīr-e Namak Daryācheh-ye Sīstān Dasht-e Lūt
RESHTEH-YE KŪHHĀ-YE ALBORZ KŪHHĀ-YE ZAGROS DASHT-E KAVĪR PLATEAU OF IRAN Strait of Hormuz Makran Coast

IRAN

Total Land Area: 1 648 000 sq. km (636 293 sq. miles)

POPULATION
over 1 000 000
over 500 000
over 100 000
over 50 000

LAND HEIGHT
3000m/9843ft
2000m/6562ft
1000m/3281ft
500m/1640ft
200m/656ft
Sea Level

N
0 200 km
0 200 miles

CHRONOLOGY

Persia was ruled by the shahs as an absolute monarchy until 1906, when the first constitution was approved. The Pahlavis took power in 1925 and changed the country's name to Iran in 1935.

- ❑ **1957** SAVAK, shah's secret police, established to control opposition.
- ❑ **1964** Ayatollah Khomeini is exiled to Iraq for criticizing secular state.
- ❑ **1971** Shah celebrates 2500th anniversary of Persian monarchy.
- ❑ **1975** Agreement with Iraq over Shatt al Arab waterway.
- ❑ **1977** Khomeini's son dies. Anti-shah demonstrations during mourning.
- ❑ **1978** Riots and strikes. Khomeini settles in Paris.
- ❑ **1979** Shah goes into exile. Ayatollah Khomeini returns, declares an Islamic republic. Students seize 63 hostages at US embassy in Tehran.
- ❑ **1980** Shah dies in exile. Start of eight-year Iran–Iraq war.
- ❑ **1981** US hostages released. Hojatoleslam Ali Khamenei elected president.
- ❑ **1985** Khamenei reelected.
- ❑ **1987** Around 275 Iranian pilgrims killed in riots in Mecca.
- ❑ **1988** USS *Vincennes* shoots down Iranian airliner; 290 killed. End of Iran–Iraq war.
- ❑ **1989** Khomeini issues *fatwa* condemning UK author Salman Rushdie to death for blasphemy. Khomeini dies. President Ali Khamenei appointed Supreme Religious Leader. Hashemi Rafsanjani elected president.
- ❑ **1990** Earthquake in northern Iran kills 45,000 people.
- ❑ **1992** Majlis elections.
- ❑ **1993** Rafsanjani reelected president.
- ❑ **1995** Imposition of US sanctions.
- ❑ **1996** Majlis elections. Society for Combatant Clergy loses ground to more liberal Servants of Iran's Construction.
- ❑ **1997** Earthquake south of Mashhad kills 1500 people. Mohammad Khatami elected president.
- ❑ **1998** Khatami government dissociates itself from *fatwa* against Salman Rushdie.
- ❑ **1999** First nationwide local elections since 1979. President Khatami visits Italy: first Iranian leader to be welcomed by a Western government since 1979.
- ❑ **2000** Sweeping election victory for reformists. Crackdown on reformist newspapers.
- ❑ **2001** Khatami reelected, winning 77% of vote.

AID

▷ Recipient

 $130m (receipts)  Down 20% in 2000

As an oil exporter, Iran does not qualify for much aid, and hard-liners are opposed to money from the West. However, Iran receives some UN aid for its millions of refugees from Afghanistan and Iraq.

Concern that Iran supports Muslim terrorism has affected aid programs. In 1994, the World Bank suspended loans, and US sanctions imposed in 1995 ended bilateral assistance, although European oil companies have more recently announced new deals.

DEFENSE

▷ Compulsory military service

 $7.33bn ⇧ Up 28% in 2000

Iran has more than 500,000 men under arms, including the 125,000-strong Revolutionary Guard Corps (*Pasdaran Inqilab*), and is regarded by neighboring states as a serious military threat. The testing of medium-range cruise and ballistic missiles in 1998 heightened concern over Iran's possible military objectives.

Before the 1979 revolution Iran was part of a pro-Western alliance structure. The long war with Iraq in the 1980s diminished the military power of the revolutionary regime.

IRANIAN ARMED FORCES

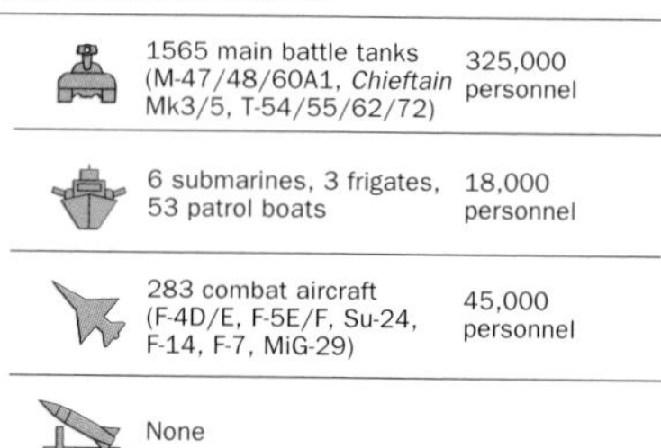
1565 main battle tanks (M-47/48/60A1, *Chieftain* Mk3/5, T-54/55/62/72)
325,000 personnel
6 submarines, 3 frigates, 53 patrol boats
18,000 personnel
283 combat aircraft (F-4D/E, F-5E/F, Su-24, F-14, F-7, MiG-29)
45,000 personnel
None

A new defense agreement with Russia allowing for the sale of arms to Iran was reached in 2000.

ECONOMICS

▷ Inflation 26% p.a. (1990–2000)

 $107bn $≡ 1747.5–1750.0 Iranian rials

SCORE CARD

❑ World GNP Ranking	34th
❑ GNP per Capita	$1680
❑ Balance of Payments	$12.6bn
❑ Inflation	14.5%
❑ Unemployment	25%

EXPORTS

China 6%
Saudi Arabia 7%
Italy 7%
South Korea 8%
Japan 18%
Other 54%

IMPORTS

France 5%
Italy 5%
UAE 6%
South Korea 9%
Germany 10%
Other 65%

Strengths

OPEC's second-biggest oil producer; soaring world oil prices in 2000. Potential for related industries and increased production of traditional exports: carpets, pistachio nuts, and caviar.

Weaknesses

Theocratic authorities restrict contact with West and access to technology. High unemployment and inflation. Excessive foreign debts.

ECONOMIC PERFORMANCE INDICATOR

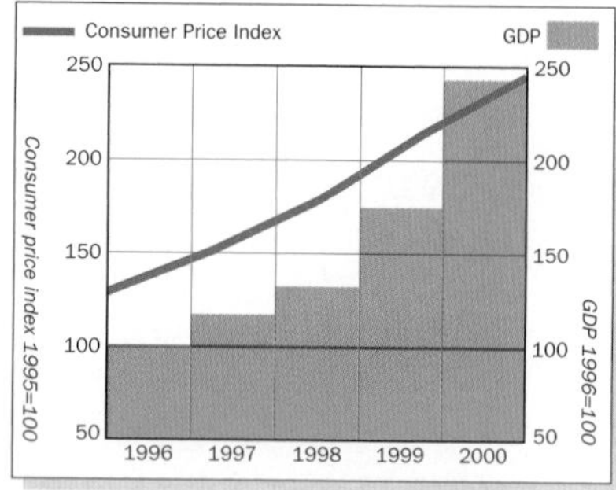
Consumer Price Index
GDP
250
200
150
100
50
Consumer price index 1995=100
GDP 1996=100
1996
1997
1998
1999
2000

Profile

With few industries other than oil, US sanctions and fluctuations in oil prices made foreign earnings volatile; higher prices in 2000 held out the prospect of being able to invest in diversification.

IRAN : MAJOR BUSINESSES

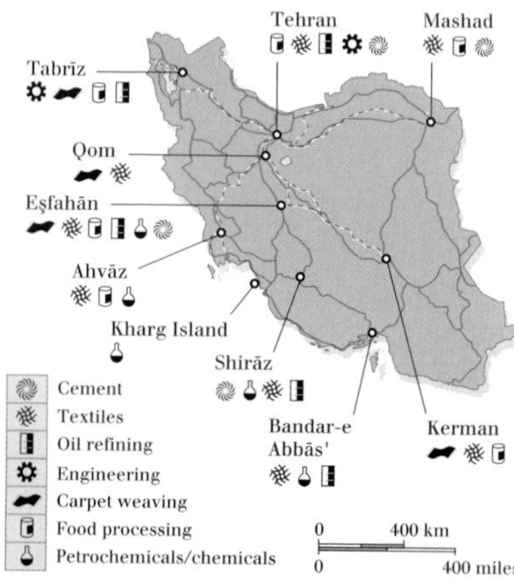
Tehran
Mashad
Tabrīz
Qom
Eşfahān
Ahvāz
Kharg Island
Shirāz
Bandar-e Abbās'
Kerman
Cement
Textiles
Oil refining
Engineering
Carpet weaving
Food processing
Petrochemicals/chemicals
0
400 km
0
400 miles

RESOURCES

Electric power 30.6m kw

419,000 tonnes

3.69m b/d (reserves 89.7bn barrels)

53m sheep, 25.2m goats, 7m cattle, 260m chickens

Iron, copper, lead, oil, zinc, chromite, coal, manganese, gypsum

ELECTRICITY GENERATION

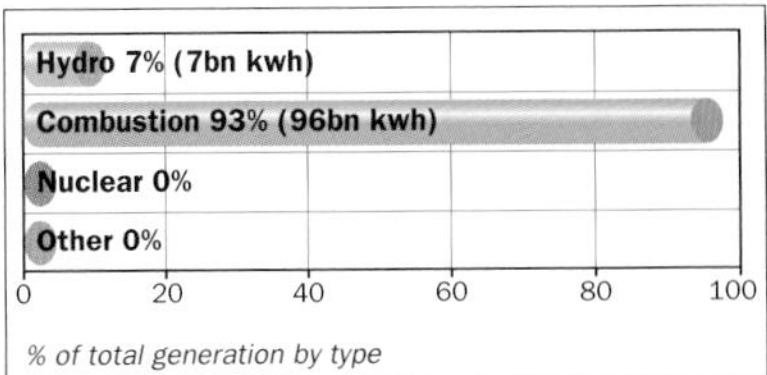
Hydro 7% (7bn kwh)
Combustion 93% (96bn kwh)
Nuclear 0%
Other 0%
0
20
40
60
80
100
% of total generation by type

Iran has substantial oil reserves. It also has metal, coal, and salt deposits, but these are relatively undeveloped. The agricultural sector is an important part of Iran's economy. Principal crops are wheat, barley, rice, sugar beet, tobacco, and pistachio nuts.

Iran was once an opium exporter, but its cultivation and use have since been banned. The vodka industry has also been closed down. Enough wool is produced to supply the carpet weaving industry. Iran has insufficient livestock to supply the domestic meat market and has to import large quantities. The Caspian Sea fisheries are controlled by the state, which sells caviar for export.

IRAN : LAND USE

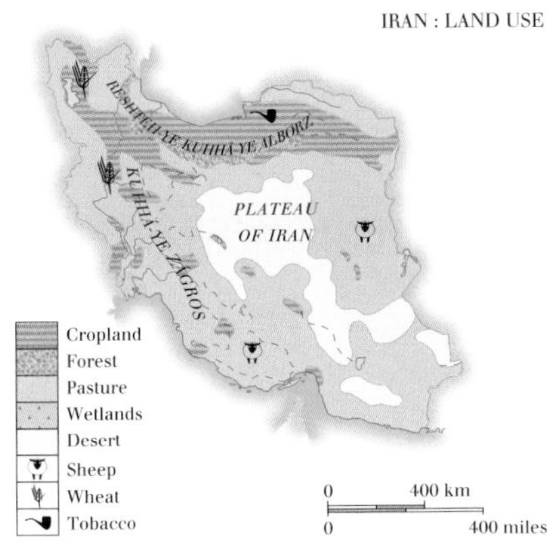
RESHTEH-YE KUHHA-YE ALBORZ
KUHHA-YE ZAGROS
PLATEAU OF IRAN
Cropland
Forest
Pasture
Wetlands
Desert
Sheep
Wheat
Tobacco
0
400 km
0
400 miles

ENVIRONMENT

Sustainability rank: 104th

5% (3% partially protected)

4.7 tonnes per capita

ENVIRONMENTAL TREATIES

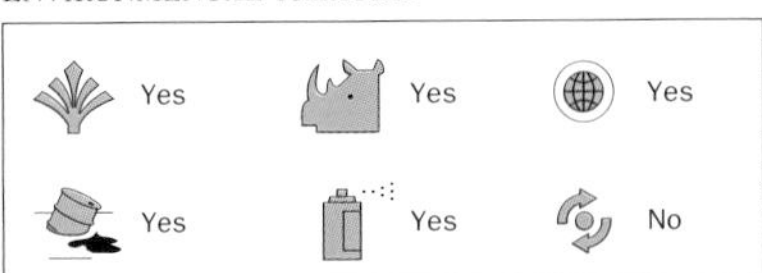
Yes
Yes
Yes
Yes
Yes
No

War damage to southern Iran, especially at Bandar Khomeini, the tanker terminal at Kharg Island, and the refinery at Abadan, has caused significant environmental damage. Environmental issues are not of concern to the religious leadership.

MEDIA

TV ownership medium

Daily newspaper circulation 26 per 1000 people

PUBLISHING AND BROADCAST MEDIA

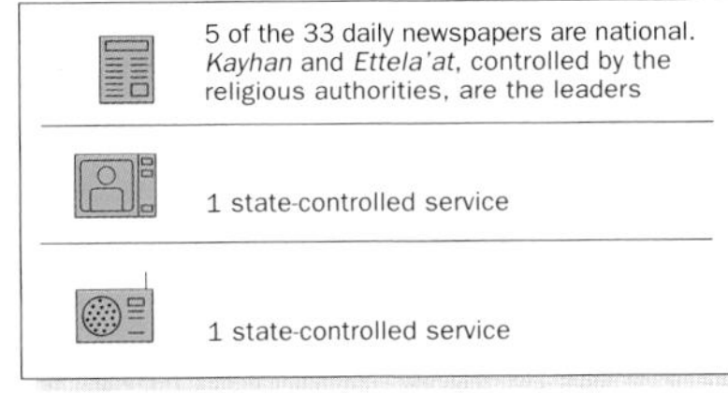
5 of the 33 daily newspapers are national. Kayhan and Ettela'at, controlled by the religious authorities, are the leaders
1 state-controlled service
1 state-controlled service

Radio and TV are state-controlled. Satellite dishes are banned. Closures of reformist newspapers by the conservative Council of Guardians, and prosecutions of their editors, continue.

CRIME

Death penalty in use

158,000 prisoners

Little change from year to year

CRIME RATES

Iran does not publish crime statistics. However, general crime rates are relatively low.

Revolutionary guards enforce the law. More than 100 offenses carry the death sentence. Executions, of both men and women, are common for political "crimes." Western governments accuse Iran of supporting international terrorism carried out by Muslim extremists abroad. In 2000 the government finally admitted that narcotics addiction, prostitution, and the violent abuse of women were rife.

EDUCATION

School leaving age: 14

77%

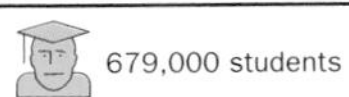
679,000 students

THE EDUCATION SYSTEM

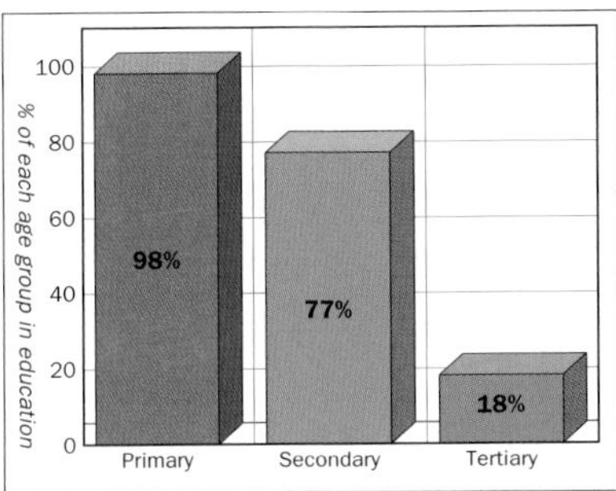
100
80
60
40
20
0
% of each age group in education
98%
77%
18%
Primary
Secondary
Tertiary

Most of the population is literate. Primary education, which lasts for five years from the age of six, is free, as are universities, but a small fee is charged for secondary education. Most schools are single-sex. There are 36 universities. Students are strong supporters of liberalization and the reformist agenda of President Khatami.

HEALTH

Welfare state health benefits

1 per 1111 people

Heart and respiratory diseases, injuries, neonatal deaths

Although an adequate system of primary health care exists in the cities, conditions in rural areas are basic. Under Khomeini, having children became a political and religious duty, but the high birthrate has now forced the introduction of birth control programs, and sterilization and contraception are now officially promoted. Growing drug addiction has resulted in rehabilitation programs and antidrugs propaganda. AIDS is spreading; some 20,000 adults were estimated to be living with HIV/AIDS in 2001.

SPENDING

GDP/cap. increase

CONSUMPTION AND SPENDING

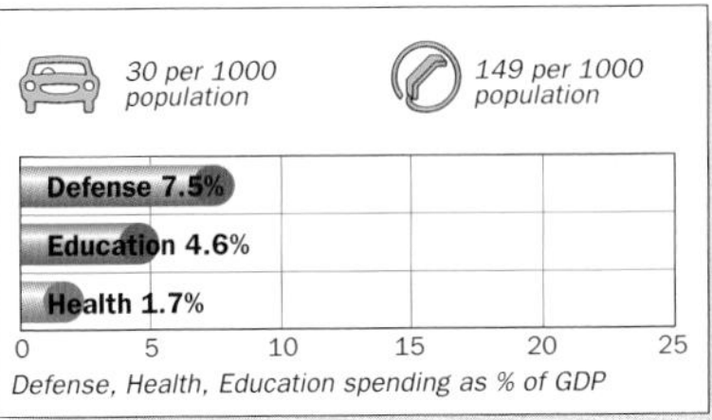
30 per 1000 population
149 per 1000 population
Defense 7.5%
Education 4.6%
Health 1.7%
0
5
10
15
20
25
Defense, Health, Education spending as % of GDP

After the 1979 revolution, living standards in Iran declined markedly. A shortage of foreign exchange has stifled imports of consumer goods. Rationing, brought in during the war with Iraq, is still partly in force, and smuggling from the Arab Gulf states is rife. Unemployment is high, and few Iranians are able to gain access to modern technology such as telephones. Official figures for income per capita do not relate to conditions on the ground. In reality, oil wealth fails to reach the economically deprived. Private businesses have gradually emerged in Iran since the launch in 1994 of the country's first private savings and loans associations.

WORLD RANKING

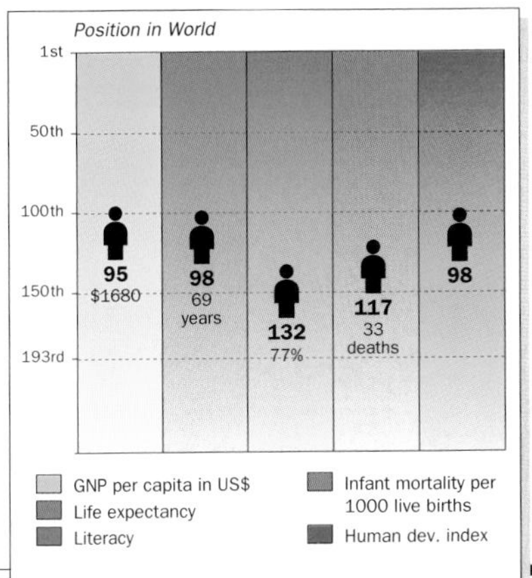
Position in World
1st
50th
100th
150th
193rd
95
$1680
98
69 years
132
77%
117
33 deaths
98
GNP per capita in US$
Life expectancy
Literacy
Infant mortality per 1000 live births
Human dev. index

I

MIDDLE EAST
Turkey
Syria
IRAQ
Iran
Jordan
Saudi Arabia
Kuwait
Asia
Africa

IRAQ

OFFICIAL NAME: Republic of Iraq **CAPITAL:** Baghdad
POPULATION: 23.6 million **CURRENCY:** Iraqi dinar **OFFICIAL LANGUAGE:** Arabic

1932

1990

July 17

IRQ

+3

+964
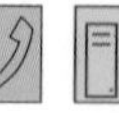
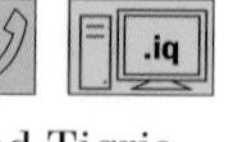
.iq

OIL-RICH IRAQ, divided by the Euphrates and Tigris Rivers, shares borders with Iran, Turkey, Syria, Jordan, Saudi Arabia, and Kuwait. The Euphrates valley is fertile, but most of the country is desert or mountains. Iraq was the site of the ancient civilization of Babylon. Today, it encompasses Shi'a Muslim holy shrines. After the removal of the monarchy in 1958, it experienced domestic political turmoil. Despite Iraq's defeat in the 1991 Gulf War, the current regime (in place since 1968) retains power – through repression.

Golden Mosque at Sāmarrā' on the Tigris. Among the extensive remains of its ancient city are those of the Great Mosque built in 847 CE.

CLIMATE

Hot desert/steppe

WEATHER CHART FOR BAGHDAD

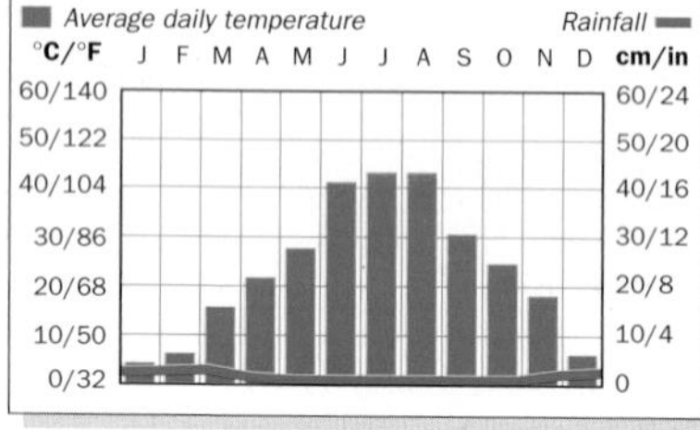

Average daily temperature
Rainfall
°C/°F
J F M A M J J A S O N D
cm/in
60/140
50/122
40/104
30/86
20/68
10/50
0/32
60/24
50/20
40/16
30/12
20/8
10/4
0

The weather is dry and rainfall is low and unreliable, except in the north. Iraq experiences a wide range of temperatures. The south has a desert climate, with hot, dry summers and mild winters. In mountainous Iranian and Turkish border regions winters can be harsh, with frost and heavy falls of snow. In the Mesopotamian plain huge dust storms are a regular feature of the summer.

TRANSPORTATION

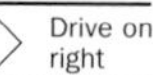
Drive on right

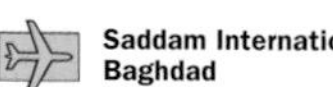
Saddam International, Baghdad

99 ships 511,000 grt

THE TRANSPORTATION NETWORK

38,400 km (23,861 miles)

1264 km (785 miles)

2339 km (1453 miles)

1015 km (631 miles)

Some international air links have been reopened, but the main road access remains through Ramadi to Jordan.

TOURISM

Visitors : Population 1:303

78,000 visitors

Up 53% 1999–2000

MAIN TOURIST ARRIVALS

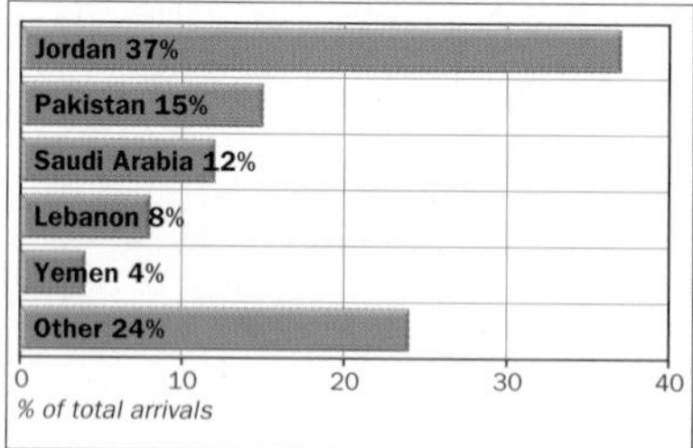

Jordan 37%
Pakistan 15%
Saudi Arabia 12%
Lebanon 8%
Yemen 4%
Other 24%
0
10
20
30
40
% of total arrivals

The Shi'a holy shrines attract thousands of pilgrims each year, but Iraq is effectively closed to the tourists who once visited its many archaeological sites. In particular, the ruins of Babylon, and its fabled hanging gardens, were once a major tourist attraction. Tourists also used to visit the marshlands close to the Shatt al Arab waterway. However, this area of great ecological importance continues to be drained as part of the effort to suppress the mainly Shi'a Marsh Arabs.

IRAQ

Total Land Area : 437 072 sq. km (168 753 sq. miles)

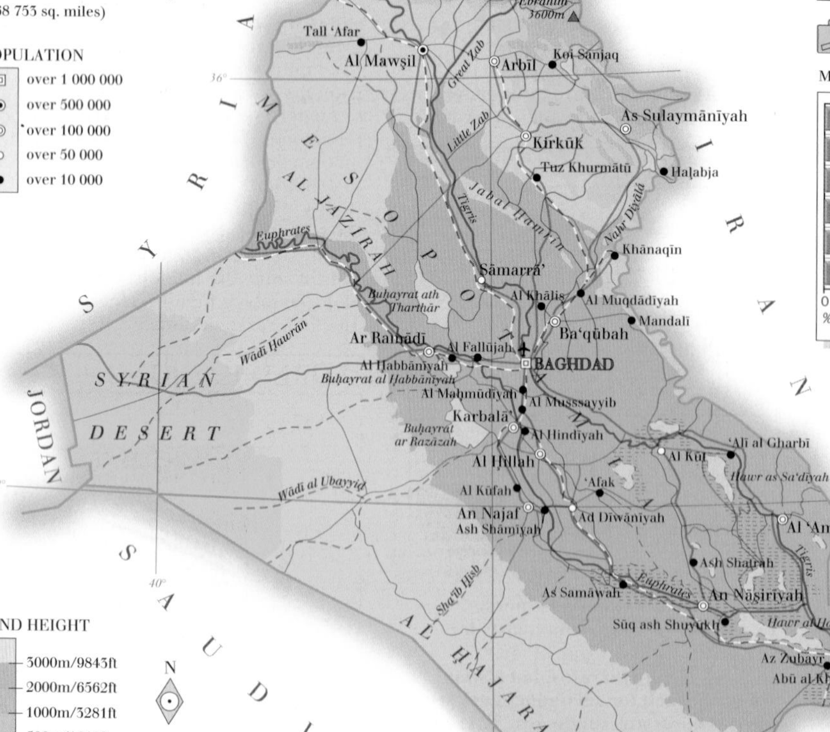

POPULATION
over 1 000 000
over 500 000
over 100 000
over 50 000
over 10 000
LAND HEIGHT
3000m/9843ft
2000m/6562ft
1000m/3281ft
500m/1640ft
200m/656ft
Sea Level
0
100 km
100 miles
TURKEY
SYRIA
IRAN
JORDAN
SAUDI ARABIA
KUWAIT
The Gulf
Zākhō
Dahūk
Kūh-e Hājī Ebrāhīm 3600m
Tall 'Afar
Al Mawşil
Arbīl
Koi Sanjaq
Great Zab
Little Zab
As Sulaymānīyah
Kirkūk
Tuz Khurmātū
Halabja
Jabal Hamrīn
Tigris
Nahr Diyālá
Khānaqīn
MESOPOTAMIA
AL JAZĪRAH
Euphrates
Sāmarrā'
Buhayrat ath Tharthār
Al Khāliş
Al Muqdādīyah
Mandalī
Ba'qūbah
Ar Ramādī
Al Fallūjah
BAGHDAD
Al Habbānīyah
Buhayrat al Habbānīyah
Wādī Hawrān
SYRIAN DESERT
Al Mahmūdīyah
Al Musayyib
Karbalā'
Buhayrat ar Razāzah
Al Hindīyah
Al Hillah
Al Kūt
'Alī al Gharbī
Hawr as Sa'dīyah
Wādī al Ubayyid
Al Kūfah
'Afak
An Najaf
Ash Shāmīyah
Ad Dīwānīyah
Al 'Amārah
Ash Shatrah
Sha'īb Hisb
As Samāwah
An Nāşirīyah
Sūq ash Shuyūkh
Hawr al Hammār
Al Başrah
Az Zubayr
Abū al Khaşīb
Shatt al 'Arab
AL HAJARAH

PEOPLE

Pop. density medium

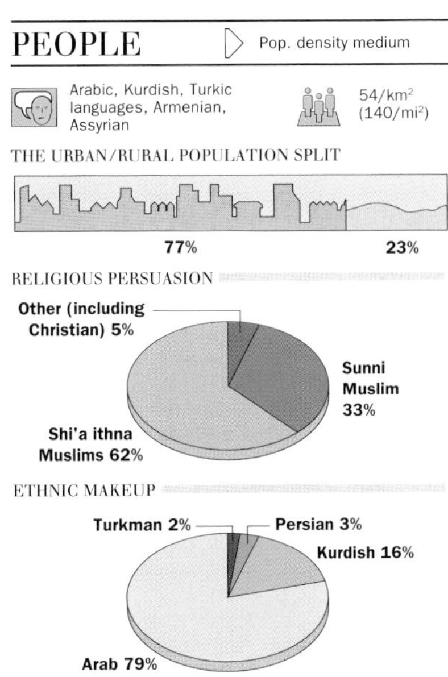
Arabic, Kurdish, Turkic languages, Armenian, Assyrian
54/km² (140/mi²)
THE URBAN/RURAL POPULATION SPLIT
77%
23%
RELIGIOUS PERSUASION
Other (including Christian) 5%
Sunni Muslim 33%
Shi'a ithna Muslims 62%
ETHNIC MAKEUP
Turkman 2%
Persian 3%
Kurdish 16%
Arab 79%

In addition to the Arab and Kurdish populations, Iraq has a small number of minority groups, such as Turkmen and Persians. Over 90% of the population are Muslim, while the rest comprise a variety of Christian sects. Since the creation of Israel, most Iraqi Jews have emigrated. The Arab Muslims are divided into Sunni and Shi'a sects. The Shi'a form the largest single religious group; however, Shi'a divines do not have as intimate a connection with the people as they do in Iran and their influence on government is limited.

Since the mid-1970s, many Iraqis have moved, or been forced to move, to the cities, where some three-quarters of the population now live.

In the marshes of the southeast, communities of mainly Shi'a Marsh Arabs survive. After the 1991 Gulf War, some of these attempted a rebellion against the state, which continues to drain the marshes in order to destroy both the people and their culture.

POPULATION AGE BREAKDOWN

Female
Age
Male
0.3% 80+ 0.3%
2.3% 60–79 2.2%
5.4% 40–59 5.7%
13.4% 20–39 14.4%
27.2% 0–19 28.8%
% of population by age group

WORLD AFFAIRS

Joined UN in 1945

AL

NAM

OAPEC
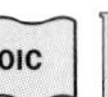
OIC

OPEC

The legacy of the Iran–Iraq war in the 1980s, the invasion of Kuwait in 1990, and the Gulf War in 1991 still dominate foreign relations.

A US-led military force, assisted by several Arab states, inflicted a crushing defeat on Iraq in 1991. Since then the UN has imposed a severe sanctions regime. Continued tensions still lead to punitive air strikes by US and UK forces. Russia, France, and several Arab states led the way in late 2000 in undermining sanctions on humanitarian grounds. Although some limited economic links have been resumed, no major Western states have yet restored diplomatic relations as the regime continues to deny access to UN weapons inspectors. By early 2002 Iraq was high on the list of possible targets in the US-led "war on terrorism," despite the government's pledge never to invade Kuwait again.

Though there have been some signs of improvement, relations with Iran remain tense. Iranian opposition groups continue to use Iraq as a base for their operations.

I

POLITICS

No multiparty elections

2000/2004

President Saddam Hussein

AT THE LAST ELECTION

National Assembly 250 seats

The National Assembly is composed of Ba'athists and their allies. Thirty seats representing the Kurdish region were not elected.

President Saddam Hussein has dominated Iraqi politics since overthrowing his predecessor in 1979. In theory, the highest state authority rests with the nine-member Revolutionary Command Council, of which Saddam Hussein is the chairman.

Profile

Iraq's repressive regime is controlled by Saddam Hussein and trusted members of his Takriti tribe. The defection to Jordan in 1995 of Saddam Hussein's relative, senior government minister Gen. Hussein Kamil, suggested dissent within the ruling circle, but his assassination in 1996 effectively neutralized a potential threat.

Saddam Hussein has promoted a massive personality cult. His regime relies on terror and a ruthless intelligence network. Iraq has the highest rate of "disappearances" in the world. Legislative elections allow voters only to choose between approved candidates, and produce results such as the 99.99% vote for Saddam's son Uday. The formation of a new Communist Party was announced in 2000, but it forms a section of the Ba'ath Party.

Main Political Issues

Sanctions

Iraq's invasion of Kuwait in 1990 and its defeat in the 1991 Gulf War resulted in the imposition of UN sanctions. These remain in force pending Iraq's full compliance with UN Gulf War resolutions. In 1994, Iraq recognized Kuwait but continued to defy UN weapons inspection programs. Campaigns for the lifting of sanctions focus on the impact they have on innocent civilians, particularly children. Since 1996 Iraq has been authorized by the UN to sell limited quantities of oil to purchase humanitarian supplies. Despite the sanctions, Iraq pursues a program of reconstruction. From 2000 limited oil revenue was permitted to be reinvested in the oil industry. In 2002 the list of sanctioned items was restricted to items of potential military use.

Threats to the regime

There is little unity among the opposition groups. The most significant are the Tehran-based Supreme Council for the Islamic Revolution in Iraq and the Iraqi National Congress operating from London. The murder in 1996 of Gen. Hussein Kamil was a blow to the opposition and a further demonstration of the regime's ruthlessness.

Tariq Aziz, deputy prime minister and mediator between Iraq and the UN.

Saddam Hussein, *Iraq's dictatorial leader since he seized power in 1979.*

The separatist Kurdish minority in the north suffers factional disputes which triggered a civil war and the intervention of Iraqi troops in 1996. Iraq's enemies continue to be reluctant to endorse Kurdish plans for Iraq's territorial dismemberment.

AID

Recipient

$101m (receipts)

Up 33% in 2000

Before it invaded Kuwait, Iraq received economic aid from neighboring Gulf states. Under UN sanctions, Iraq has been entitled only to humanitarian aid, but there is mounting evidence of covert trade, especially through Jordan and Turkey.

CHRONOLOGY

Iraq became independent in 1932. In 1958, the Hashemite dynasty was overthrown when King Faisal died in a coup led by the military under Brig. Kassem. He was initially supported by the Iraqi Ba'ath Party.

- ❑ **1961** Start of Kurdish rebellion. Iraq claims sovereignty over Kuwait on the eve of Kuwait's independence.
- ❑ **1963** Kassem overthrown. Col. Abd as-Salem Muhammad Aref takes power. Kuwait's sovereignty recognized.
- ❑ **1964** Ayatollah Khomeini, future leader of Iran, takes refuge at Najaf in Iraq.
- ❑ **1966** Aref is succeeded by his brother, Abd ar-Rahman.
- ❑ **1968** Ba'athists under Ahmad Hassan al-Bakr take power.
- ❑ **1970** Revolutionary Command Council agrees manifesto on Kurdish autonomy.
- ❑ **1972** Nationalization of Western-controlled Iraq Petroleum Company.
- ❑ **1978** Iraq and Syria form economic and political union.
- ❑ **1979** Saddam Hussein replaces al-Bakr as president.
- ❑ **1980** Outbreak of Iraq–Iran war.
- ❑ **1982** Shi'a leader Mohammed Baqir al-Hakim, exiled in Tehran, forms Supreme Council of the Islamic Revolution in Iraq.
- ❑ **1988** Iraq and Iran agree cease-fire. Iraqi chemical weapons attack on Kurdish village of Halabja.
- ❑ **1990** British journalist Farzad Bazoft hanged for spying. Iraq and Iran restore diplomatic relations. Iraq invades Kuwait. UN imposes trade sanctions.
- ❑ **1991** Gulf War. US-led military coalition defeats Iraq and liberates Kuwait. Iraqi regime suppresses Shi'a rebellion. Northern no-fly zone enforced.
- ❑ **1992** Western powers proclaim air exclusion zone over south.
- ❑ **1994** Outbreak of Kurdish civil war. Iraq recognizes Kuwaiti sovereignty.
- ❑ **1995** Government minister Gen. Hussein Kamil defects to Jordan, and is murdered on his return to Iraq in January 1996.
- ❑ **1996** First legislative elections since 1989 won by ruling Ba'ath Party. UN supervises limited sales of Iraqi oil to purchase humanitarian supplies.
- ❑ **1998–1999** UN weapons inspection teams refused reentry into Iraq; US and UK mount punitive air strikes.
- ❑ **2002** Threat of US military attack to oust Saddam Hussein.

DEFENSE

Compulsory military service

$1.47bn

Down 2% in 2000

IRAQI ARMED FORCES

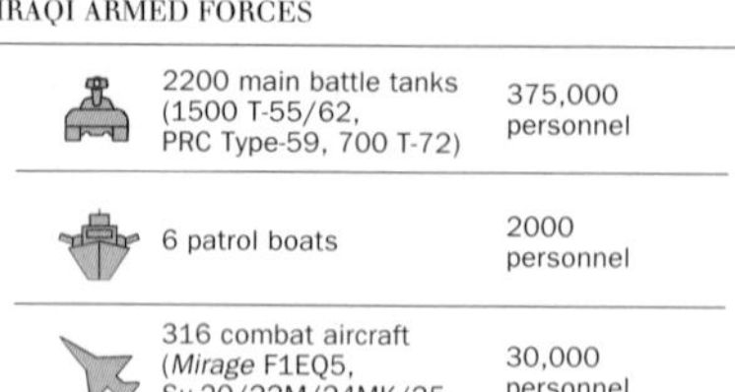

2200 main battle tanks (1500 T-55/62, PRC Type-59, 700 T-72)	375,000 personnel
6 patrol boats	2000 personnel
316 combat aircraft (*Mirage* F1EQ5, Su-20/22M/24MK/25, MiG-21/23/25/29)	30,000 personnel
None	

Iraq's military defeat by the US-led coalition in 1991 led to the destruction of much of its arsenal. Since then UN Security Council resolutions have required the elimination of the bulk of Iraq's weapons of mass destruction, and inspection teams have sought with some effect to enforce this. There is a shortage in Iraq of high-tech weaponry that could match the kind acquired by Kuwait and Saudi Arabia from US and other Western suppliers since the Gulf War. The army is large, but poorly trained and equipped; military service, lasting between 18 months and two years, is compulsory for all men at the age of 18. The military relies on tanks and aircraft from the former Soviet Union and China. The air force has some French *Mirage* fighters and US helicopters.

ECONOMICS

Not available

$20bn

0.3124–0.3110 Iraqi dinars

SCORE CARD

❑ World GNP Ranking	64th
❑ GNP per Capita	$950
❑ Balance of Payments	Not available
❑ Inflation	100%
❑ Unemployment	Not available

EXPORTS

Jordan 4%
France 8%
Spain 8%
Italy 11%
Other 28%
USA 41%

IMPORTS

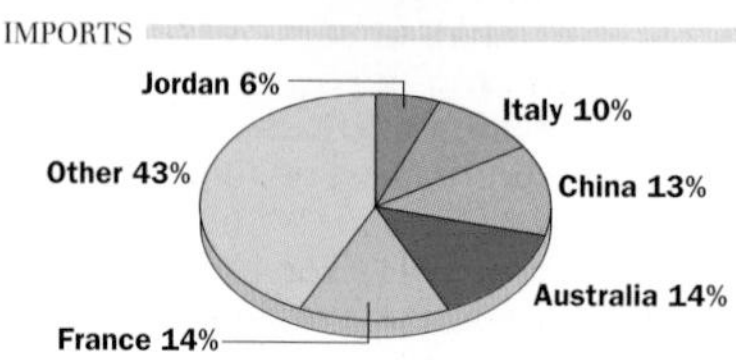

Jordan 6%
Italy 10%
China 13%
Australia 14%
France 14%
Other 43%

ECONOMIC PERFORMANCE INDICATOR

Consumer Price Index
GDP
Consumer price index unavailable
GDP 1991=100
12797
12000
8000
4000
0
-50
1991
1992
1993
1994
1995

Strengths

Second-largest crude oil and natural gas reserves in OPEC. Large labor force.

Weaknesses

Inability to sell oil on the international market; Iraq's GNP halved by UN sanctions. Once-thriving agricultural sector devastated by war.

Profile

Before 1990, Iraq was the world's third-largest oil supplier. Under sanctions, oil was produced only for domestic consumption. Limited oil exports under strict UN supervision were resumed for the first time in 1996, and in 2000 the UN Security Council approved a resolution permitting Iraq to buy parts and equipment for the oil industry.

The denial of Western assistance following the 1991 Gulf War has stifled Iraq's economy, although the resumption of some informal economic links and the revision of UN sanctions in 2002 may lead to some improvement. The once thriving agricultural sector was badly affected by the war. The manufacturing industry is at a standstill. The introduction of draconian penalties, including the death sentence, have failed to curb the black market or halt the sharp depreciation in the value of the dinar.

IRAQ : MAJOR BUSINESSES

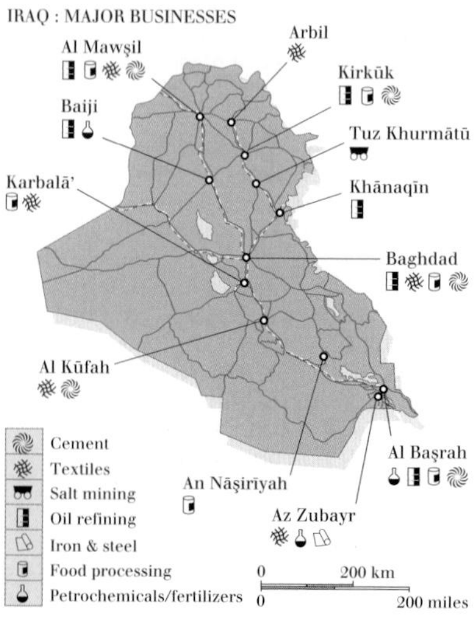

Al Mawşil
Arbil
Kirkūk
Baiji
Tuz Khurmātū
Karbalā'
Khānaqīn
Baghdad
Al Kūfah
Al Başrah
An Nāşirīyah
Az Zubayr
Cement
Textiles
Salt mining
Oil refining
Iron & steel
Food processing
Petrochemicals/fertilizers
0
200 km
0
200 miles

RESOURCES

Electric power 9.5m kw

26,789 tonnes

2.41m b/d (reserves 112.5bn barrels)

6.78m sheep, 1.6m goats, 1.35m cattle, 23m chickens

Oil, natural gas, phosphates, sulfur, gypsum

ELECTRICITY GENERATION

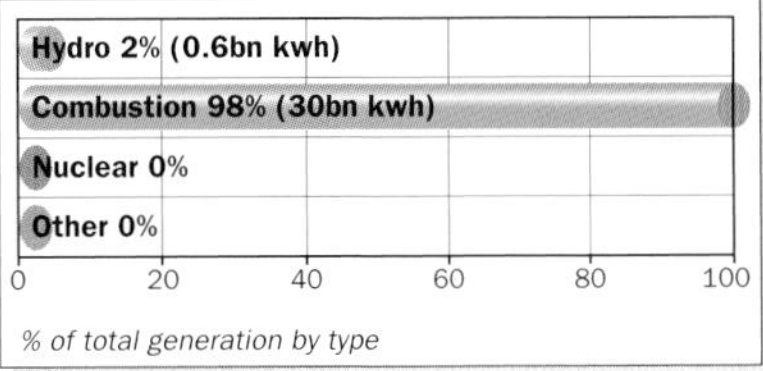
Hydro 2% (0.6bn kwh)
Combustion 98% (30bn kwh)
Nuclear 0%
Other 0%
0
20
40
60
80
100
% of total generation by type

Iraq has huge reserves of oil and gas. The oil industry is controlled by the Iraqi National Oil Company. Total gas reserves, many of which are associated with oil, are proven to be 3.11 trillion cu. m (110 trillion cu. ft), with estimates of a further 4.25 trillion cu. m (150 trillion cu. ft). Most electricity is generated from oil, although hydroelectric power also makes a small contribution. Reserves of phosphates, sulfur, gypsum, and salt are also exploited.

Before the invasion of Kuwait and subsequent war, Iraq supplied 80% of the world's trade in dates. Production is now sharply down; food is now produced simply for domestic consumption. Iraq has, however, achieved a degree of self-sufficiency in such crops as wheat, rice, and sugarcane.

IRAQ : LAND USE

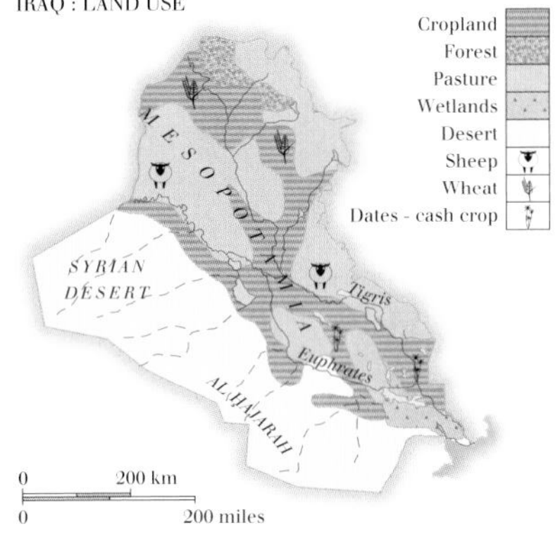
Cropland
Forest
Pasture
Wetlands
Desert
Sheep
Wheat
Dates - cash crop
MESOPOTAMIA
SYRIAN DESERT
Tigris
Euphrates
AL HIJARAH
0
200 km
0
200 miles

ENVIRONMENT

Sustainability rank: 139th

None

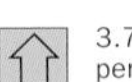
3.7 tonnes per capita

ENVIRONMENTAL TREATIES

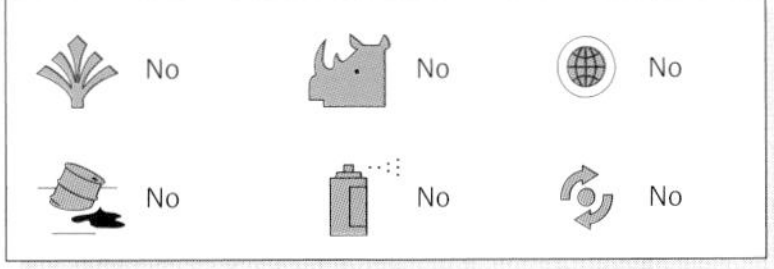
No
No
No
No
No
No

Wars with Iran and with the US-led alliance over the Kuwait occupation led to massive environmental damage. Hundreds of thousands of land mines remain in the Kuwait border regions, posing lethal hazards to farmers, livestock, and wild animals. The north has been affected by chemical weapons, used by the regime against the Kurds. In the southeast, an entire wetland ecosystem is being destroyed by a program to drain the marshes for political reasons.

MEDIA

TV ownership medium

Daily newspaper circulation 20 per 1000 people

PUBLISHING AND BROADCAST MEDIA

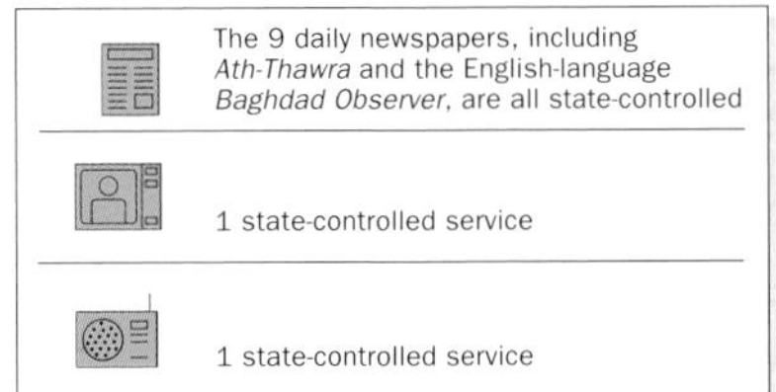
The 9 daily newspapers, including *Ath-Thawra* and the English-language *Baghdad Observer*, are all state-controlled
1 state-controlled service
1 state-controlled service

The media are strictly controlled, although rebel groups circulate clandestine newspapers. Saddam Hussein's son Uday controls two newspapers, one of which is the influential *Babil*, which opposes UN Gulf War resolutions. All foreign journalists are vetted. Opposition groups began satellite broadcasts with US assistance in 2001.

CRIME

Death penalty in use

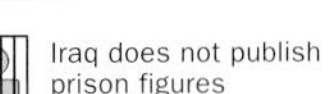
Iraq does not publish prison figures

Crime is rising

CRIME RATES

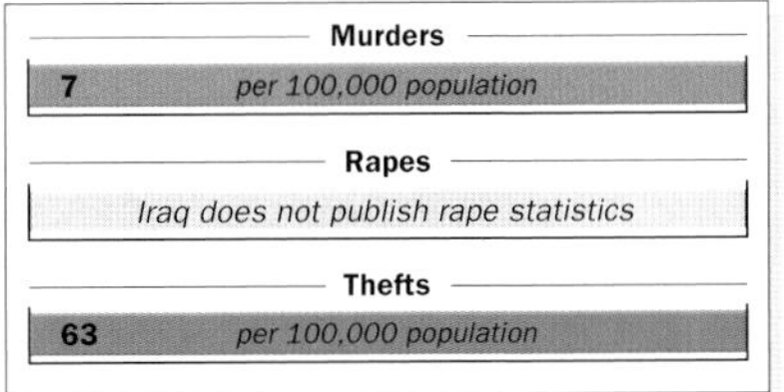
Murders
7 per 100,000 population
Rapes
Iraq does not publish rape statistics
Thefts
63 per 100,000 population

Economic collapse has sent crime rates soaring, especially in cities. Theft has been made a capital offense, encouraging thieves to murder their victims in order to escape detection.

EDUCATION

School leaving age: 11

56%

209,818 students

THE EDUCATION SYSTEM

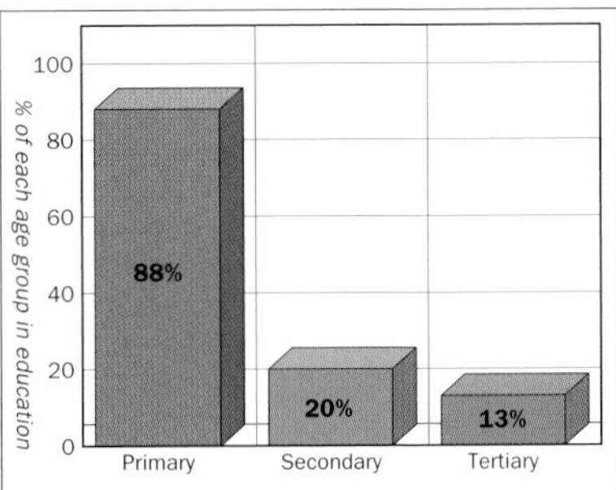
% of each age group in education
100
80
60
40
20
0
88%
20%
13%
Primary
Secondary
Tertiary

Education is free and universal, except in remote rural areas. Primary education has been made compulsory in an effort to reduce illiteracy. There are six universities. Academics authorized the organized plunder of antiquities and university equipment from Kuwait during the 1990 occupation. University scientists work closely with the regime on weapons research programs.

HEALTH

Welfare state health benefits

1 per 2000 people

Pneumonia, influenza, cancers, heart diseases

An effect of UN sanctions has been to aggravate the shortage of medical supplies and equipment, and deaths among children and the elderly have spiraled sharply. The increase in the number of children born with birth defects since 1991 is attributed to the Allies' use of depleted uranium shells during the Gulf War.

SPENDING

GDP/cap. decrease

CONSUMPTION AND SPENDING

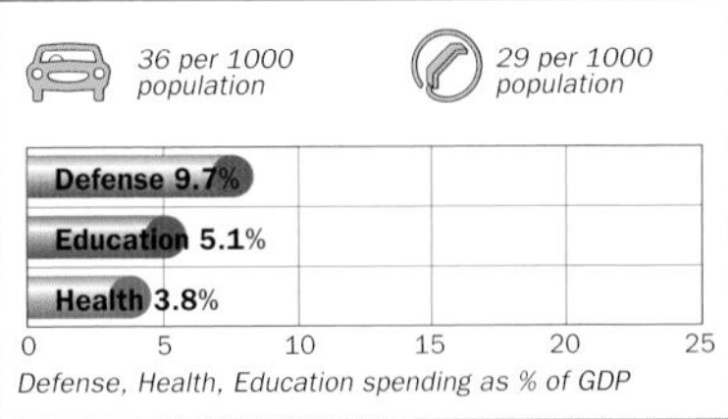
36 per 1000 population
29 per 1000 population
Defense 9.7%
Education 5.1%
Health 3.8%
0
5
10
15
20
25
Defense, Health, Education spending as % of GDP

Vulnerable sections of society have been particularly affected by UN sanctions. Middle-class citizens and traders are able to benefit from Iraq's open border with Jordan.

WORLD RANKING

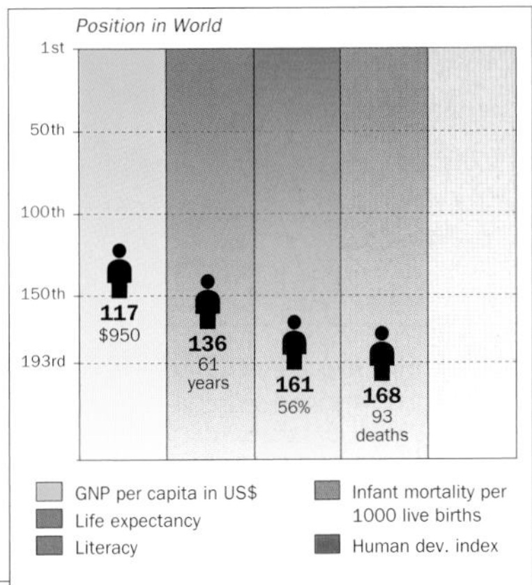
Position in World
1st
50th
100th
150th
193rd
117
$950
136
61 years
161
56%
168
93 deaths
GNP per capita in US$
Life expectancy
Literacy
Infant mortality per 1000 live births
Human dev. index

I

EUROPE

IRELAND

OFFICIAL NAME: Ireland **CAPITAL:** Dublin **POPULATION:** 3.8 million
CURRENCY: Euro (punt until 2002) **OFFICIAL LANGUAGES:** Irish and English

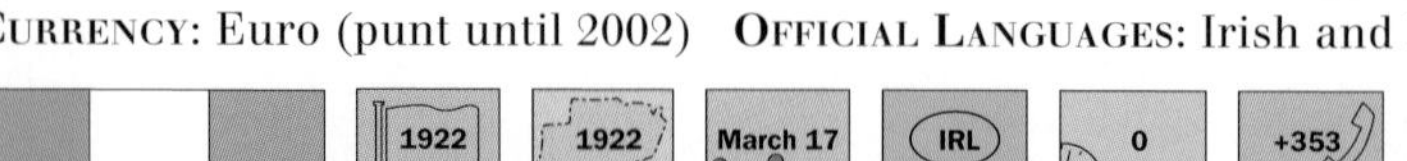

1922
1922
March 17
IRL
0
+353
.ie

LYING IN THE Atlantic Ocean, off the west coast of Great Britain, the Irish republic occupies about 85% of the island of Ireland. Low coastal ranges surround a central basin with lakes, hills, and peat bogs. Centuries of struggle against English domination led in 1922 to the formation of the Irish Free State and in 1937 full sovereignty. Efforts to resolve the Northern Ireland conflict center on the 1998 Good Friday accord, under which Ireland gave up its territorial claim.

CLIMATE

Maritime

WEATHER CHART FOR DUBLIN

Average daily temperature · Rainfall

°C/°F J F M A M J J A S O N D cm/in

40/104 · 30/86 · 20/68 · 10/50 · 0/32 · -10/14 · -20/-4 — 40/16 · 30/12 · 20/8 · 10/4 · 0

Moderated by the Gulf Stream, the Irish climate is mild, equable, and wet. The mean annual temperature is 12°C (54°F).

TRANSPORTATION

Drive on left

Dublin International 13.8m passengers

172 ships 248,236 grt

THE TRANSPORTATION NETWORK

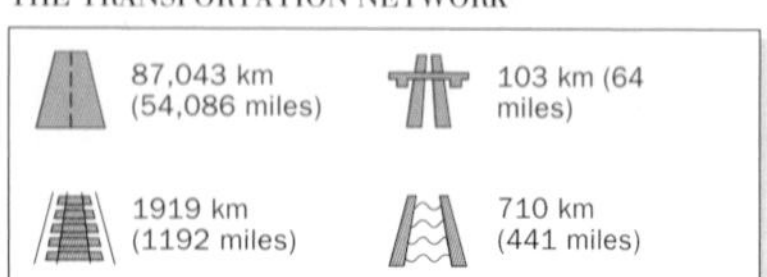

87,043 km
(54,086 miles)
103 km (64
miles)
1919 km
(1192 miles)
710 km
(441 miles)

EU funds have improved road networks, especially around Dublin, which still, however, suffers from congestion.

TOURISM

Visitors : Population 1.8:1

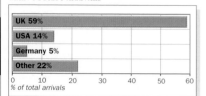

6.75m visitors

Up 5% in 2000

MAIN TOURIST ARRIVALS

	% of total arrivals
UK	59%
USA	14%
Germany	5%
Other	22%

0 10 20 30 40 50 60
% of total arrivals

Tourist numbers have increased steadily in recent years and exceed six million a year. Vibrant Dublin attracts many on "city breaks." Other draws are scenery, Ireland's "clean" environmental image, and the relaxed lifestyle.

PEOPLE

Pop. density medium

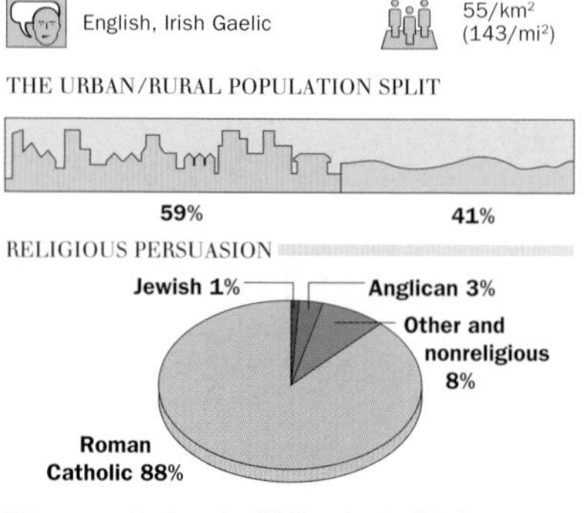

English, Irish Gaelic
55/km²
(143/mi²)
THE URBAN/RURAL POPULATION SPLIT
59%
41%
RELIGIOUS PERSUASION
Jewish 1%
Anglican 3%
Other and nonreligious 8%
Roman Catholic 88%

The population is 95% ethnic Irish. The influence of the Roman Catholic Church is declining. Ireland is now a country of net immigration, against the trend of the past 150 years.

POLITICS

Multiparty elections

L. House 2002/2007
U. House 2002/2007
President Mary McAleese

AT THE LAST ELECTION

House of Representatives 166 seats

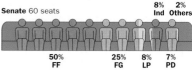

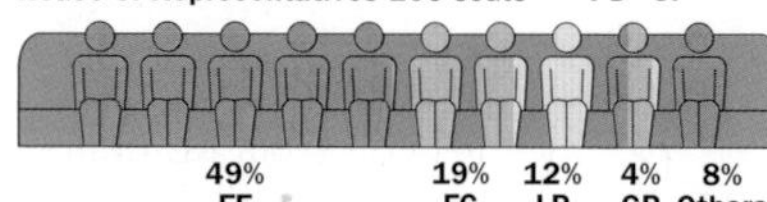

5% PD
3% SF

49% FF · 19% FG · 12% LP · 4% GP · 8% Others

FF = Fianna Fail **FG** = Fine Gael **LP** = Labour Party
PD = Progressive Democrats **GP** = Green Party
SF = Sinn Fein **Ind** = Independents

Senate 60 seats

8% Ind · 2% Others

50% FF · 25% FG · 8% LP · 7% PD

In 1973, an FG–LP coalition took power, marking the end of FF as the traditional party of government – a position which it had held since 1932. Government leadership then alternated between FG and FF. The latter dominated an FF–PD coalition under Bertie Ahern as *taoiseach* (prime minister) from 1997, but although it improved its position in the April 2002 elections, it still failed to win an outright majority of seats in the legislature.

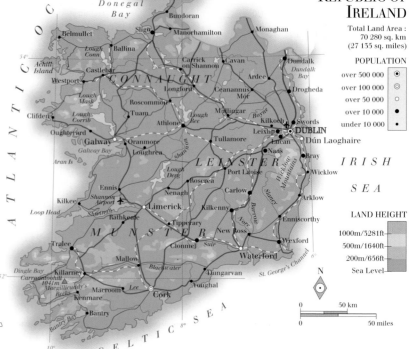

REPUBLIC OF IRELAND

Total Land Area : 70 280 sq. km (27 135 sq. miles)

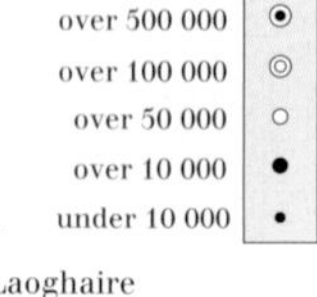

POPULATION
over 500 000
over 100 000
over 50 000
over 10 000
under 10 000

LAND HEIGHT

1000m/3281ft · 500m/1640ft · 200m/656ft · Sea Level

0 50 km · 0 50 miles

WORLD AFFAIRS

Joined UN in 1955

CE
ESA
EU
OECD
OSCE

Ireland, the UK, and the US are involved in the Northern Ireland peace process. A 2001 referendum rejected the EU's Nice treaty, embarrassing the government.

AID

Donor

$235m (donations)

Down 4% in 2000

Africa is the main recipient of Irish aid. Ireland itself benefits from EU funding.

DEFENSE

No compulsory military service

$684m

Down 8% in 2000

Ireland is determined to maintain its neutrality, despite EU moves to establish a common European defense policy. It has observer status at the WEU.

ECONOMICS

Inflation 3.5% p.a. (1990–2000)

$86bn

1.0651–1.1231 euros

SCORE CARD

❑ WORLD GNP RANKING	39th
❑ GNP PER CAPITA	$22,660
❑ BALANCE OF PAYMENTS	–$593m
❑ INFLATION	5.6%
❑ UNEMPLOYMENT	4%

STRENGTHS

The "Celtic tiger" – one of Europe's fastest-growing economies: real GDP growth of 9% a year in the latter half of the 1990s. Trade surplus. Low inflation. Efficient agriculture and food processing. Expanding high-tech sector: electronics account for 25% of exports. Large recipient of EU infrastructure aid. Highly educated workforce.

WEAKNESSES

Many key sectors owned by overseas multinationals. Danger of economy overheating. Housing shortage. Rapid growth is now straining the country's infrastructure.

EXPORTS

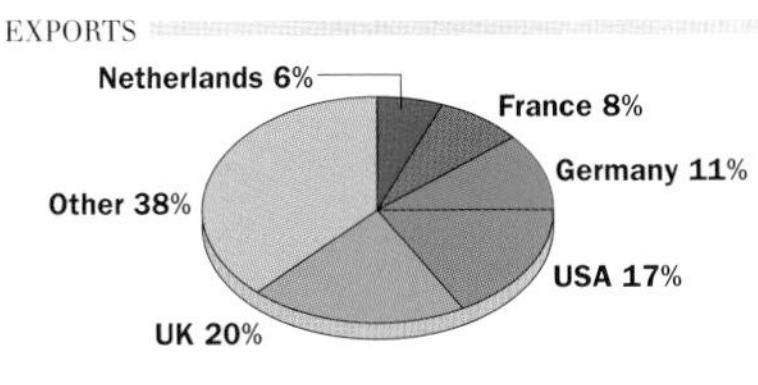
Netherlands 6%
France 8%
Germany 11%
Other 38%
USA 17%
UK 20%

IMPORTS

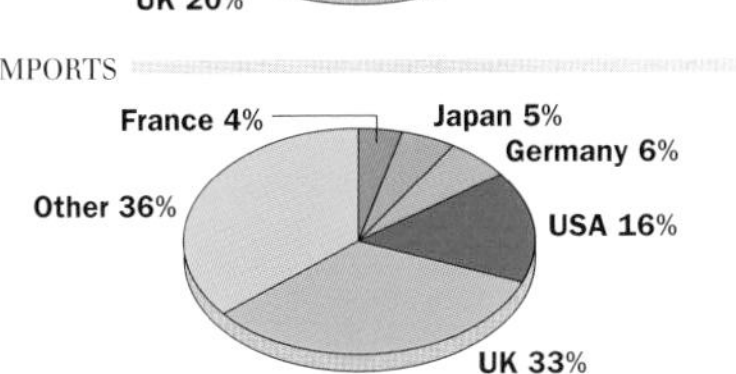
France 4%
Japan 5%
Germany 6%
Other 36%
USA 16%
UK 33%

RESOURCES

Electric power 4.4m kw

329,777 tonnes

Not an oil producer; refines 56,000 b/d

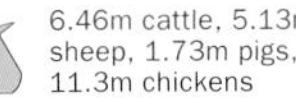
6.46m cattle, 5.13m sheep, 1.73m pigs, 11.3m chickens

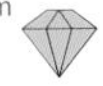
Lead, zinc, natural gas, silver, coal, oil

Oil has been found off the south coast. Studies suggest that this may be in commercially exploitable quantities.

ENVIRONMENT

Sustainability rank: 37th

0.9%

10.3 tonnes per capita

The main environmental concerns are overexploitation of the country's peat bogs for fuel and the recent expansion of conifer plantations. While Ireland's levels of forest cover will increase in the next few years, most new planting is of conifers. In 1994 stringent new laws increased pollution controls. A small windpower industry is expanding.

MEDIA

TV ownership high

Daily newspaper circulation 149 per 1000 people

PUBLISHING AND BROADCAST MEDIA

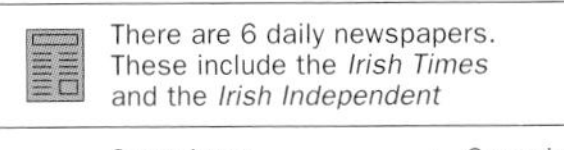
There are 6 daily newspapers. These include the *Irish Times* and the *Irish Independent*

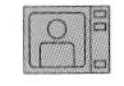
2 services: 1 state-owned, 1 independent

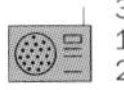
3 services: 1 state-owned, 2 independent

The national Radio Telefís Éireann (RTE) dominates broadcasting. Radió na Gaeltachta and TG4 broadcast in Irish. Access to UK media is widespread.

CRIME

No death penalty

2948 prisoners

Down 26% in 1999

Rural Ireland has one of the EU's lowest crime rates. Urban crime and narcotics are a problem in Dublin and Cork.

EDUCATION

School leaving age: 16

99%

113,338 students

The Roman Catholic Church runs many schools. Increased education spending has resulted in a skilled workforce.

***Clew Bay in County Mayo,** on the western coast of Connaught, viewed from the slopes of neighboring Croagh Patrick.*

CHRONOLOGY

English colonization, begun in 1167, was reinforced after 1558 by anti-Catholic legislation and settlement of Scottish Protestants in the north.

- ❑ **1845–1855** Famine. One million die, 1.5 million emigrate.
- ❑ **1919–1921** Anglo-Irish war after Sinn Féin proclaims independence.
- ❑ **1922** Irish Free State established.
- ❑ **1973** FG/LP alliance wins elections, ending FF's ascendancy since 1932.
- ❑ **1990** Mary Robinson elected first woman president.
- ❑ **1995** Referendum favors divorce.
- ❑ **1998** Good Friday accord on Northern Ireland.
- ❑ **2001** Voters reject EU treaty on expansion.
- ❑ **2002** Euro fully adopted.

I

HEALTH

Welfare state health benefits

1 per 435 people

Heart diseases, cancers, accidents

Free care is means tested; about one-third of people qualify. Others pay to visit their doctor and for prescriptions, and there is a modest charge for hospital care.

SPENDING

GDP/cap. increase

CONSUMPTION AND SPENDING

272 per 1000 population

420 per 1000 population

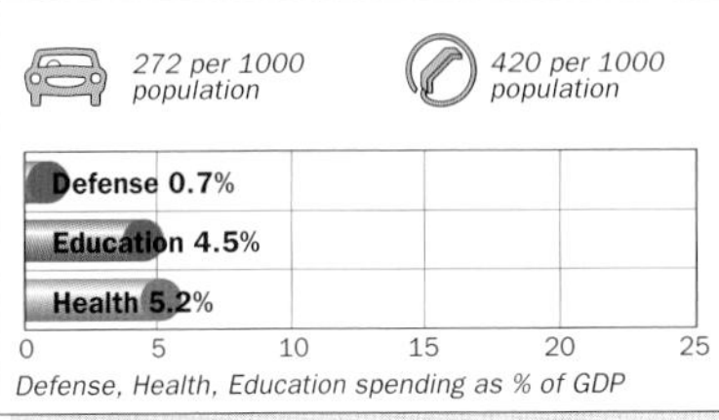
Defense 0.7%
Education 4.5%
Health 5.2%
0
5
10
15
20
25
Defense, Health, Education spending as % of GDP

Living standards for those with jobs are rising steadily. Welfare for those not in work is low by OECD standards.

WORLD RANKING

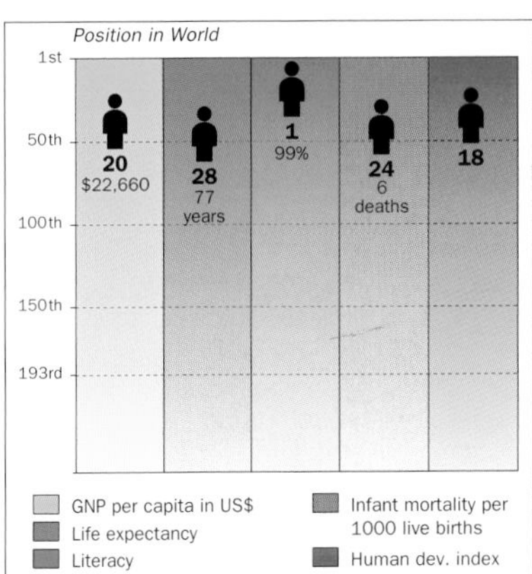
Position in World
1st
50th
100th
150th
193rd
20
$22,660
28
77
years
1
99%
24
6
deaths
18
GNP per capita in US$
Life expectancy
Literacy
Infant mortality per 1000 live births
Human dev. index

ISRAEL

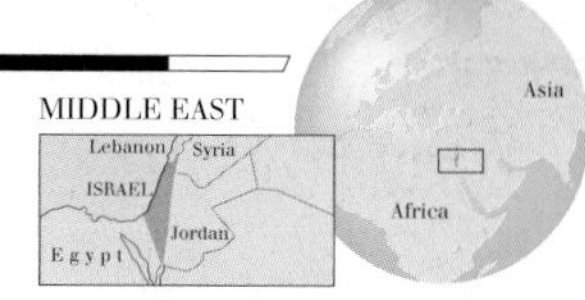
MIDDLE EAST
Lebanon
Syria
ISRAEL
Jordan
Egypt
Asia
Africa

OFFICIAL NAME: State of Israel **CAPITAL:** Jerusalem (not internationally recognized)
POPULATION: 6.2 million **CURRENCY:** Shekel **OFFICIAL LANGUAGES:** Hebrew and Arabic

1948

1994
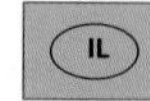
May 12
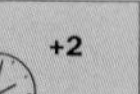
IL

+2

+972
.il

THE CREATION OF ISRAEL in 1948 in the then UN-mandated territory of Palestine fulfilled the Zionist ambition for a Jewish state. Subsequent military victories over its Arab neighbors enabled Israel to annex or occupy additional territory, some of which it returned to Egypt under the 1978 Camp David agreement. Hopes in the 1990s for a "land for peace" deal to end the Israeli–Palestinian conflict were mired in a cycle of violence by 2002.

CLIMATE

Hot desert/ Mediterranean

WEATHER CHART FOR JERUSALEM

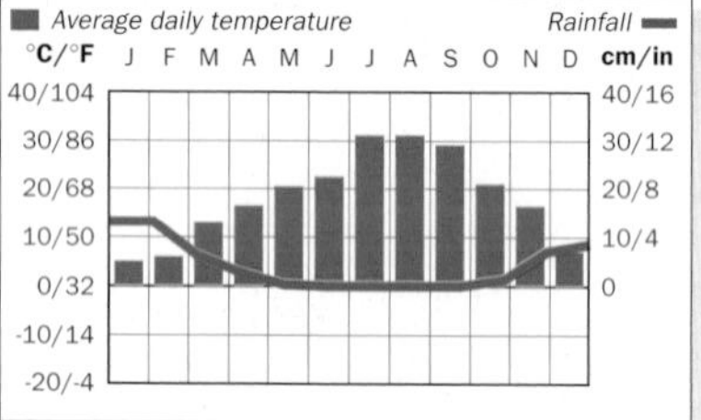
Average daily temperature
Rainfall
°C/°F
J F M A M J J A S O N D
cm/in
40/104
30/86
20/68
10/50
0/32
-10/14
-20/-4
40/16
30/12
20/8
10/4
0

Summers are hot and dry. The wet season is between November and March, when the weather is mild.

TRANSPORTATION

Drive on right

Ben-Gurion International, Tel Aviv–Yafo
9.88m passengers

49 ships
611,730 grt

THE TRANSPORTATION NETWORK

15,965 km (9920 miles)

56 km (35 miles)

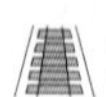
663 km (412 miles)

None

Railroads are being extended. There are three commercial ports. Ben-Gurion international airport is being expanded.

TOURISM

Visitors : Population 1:2.6

2.42m visitors

Up 5% in 2000

MAIN TOURIST ARRIVALS

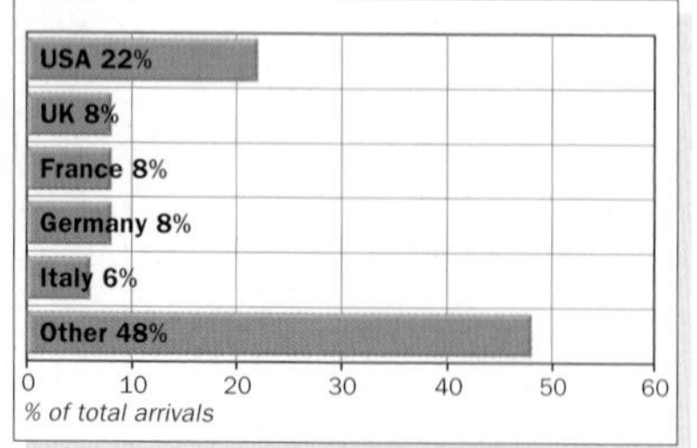
USA 22%
UK 8%
France 8%
Germany 8%
Italy 6%
Other 48%
0
10
20
30
40
50
60
% of total arrivals

Tourism has been damaged by the sustained violence of the Israeli–Palestinian conflict.

PEOPLE

Pop. density high

Hebrew, Arabic, Yiddish, German, Russian, Polish, Romanian, Persian

305/km² (790/mi²)

THE URBAN/RURAL POPULATION SPLIT

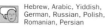

91% 9%

RELIGIOUS PERSUASION

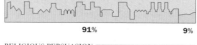

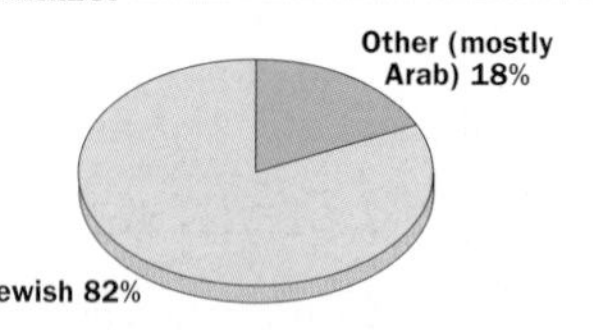
Christian 2%
Druze and Other 2%
Muslim (mainly Sunni) 14%
Jewish 82%

ETHNIC MAKEUP

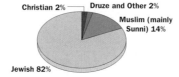

Other (mostly Arab) 18%
Jewish 82%

Large numbers of Jewish immigrants settled in Palestine before Israel was founded. Sephardi Jews from the Middle East and Mediterranean are now probably in the majority, but Ashkenazi Jews, most of central European origin, still dominate society. Hundreds of thousands of Russian Jews have arrived since 1989. Israel's non-Jewish population – mostly Arab and predominantly Muslim (with Christian and Druze minorities) – totals more than one million, but remains sidelined in Israeli life. There are tensions between secular and Orthodox Jews and between left and right over the pursuit of peace with the Palestinians.

POPULATION AGE BREAKDOWN

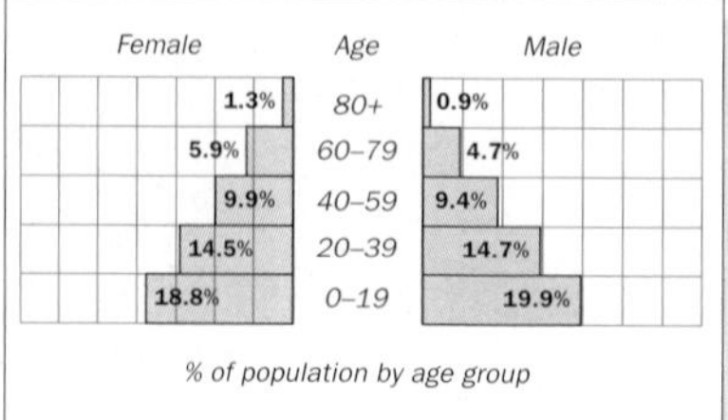
Female
Age
Male
1.3%
80+
0.9%
5.9%
60–79
4.7%
9.9%
40–59
9.4%
14.5%
20–39
14.7%
18.8%
0–19
19.9%
% of population by age group

POLITICS

Multiparty elections

1999/2003

President Moshe Katzav

AT THE LAST ELECTION

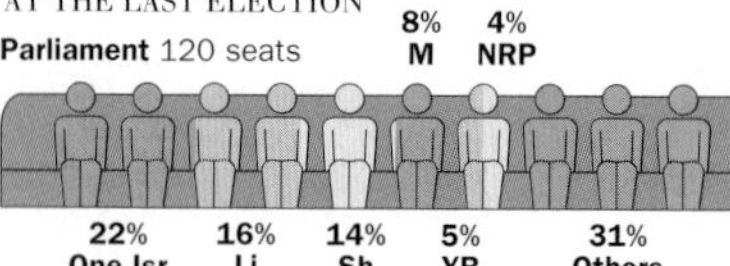
Parliament 120 seats
8% M
4% NRP
22% One Isr
16% Li
14% Sh
5% YB
31% Others

One Isr = One Israel (comprising Labor, Gesher, and Meymad) **Li** = Likud **Sh** = Shas
M = Meretz–Democratic Israel **YB** = Yisrael Ba'aliya
NRP = National Religious Party

Israel is a multiparty democracy.

PROFILE

Conflict with the Palestinians dominates the agenda. Peace initiatives of the 1990s have foundered. Labor prime minister Ehud Barak was replaced by hard-liner Ariel Sharon, elected in 2001. Fractious coalitions are endemic and Sharon's broad "government of national unity" is plagued by interparty disputes.

MAIN POLITICAL ISSUES

Peace with the Palestinians

Whether Israel's security should be based on strength, or on agreements with Palestinians and neighboring countries, is a central issue. Small but influential religious parties support Jewish settlement of the occupied territories, the main obstacle to any "land for peace" deal. The Oslo Peace Process, begun in 1993, seemingly heralded the gradual end of occupation, but Palestinian terror attacks and, latterly, suicide bombings of civilian targets, reinforced Israeli skepticism about prospects for peace.

Political unity

The national unity government merely papers over major differences. Sharon's Likud vehemently backs a hard-line, aggressive stance against the Palestinians and their Arab supporters. The centrist Labor party favors negotiation, which most religious groups reject.

Ariel Sharon, Israel's hard-line prime minister elected in 2001.

Yasser Arafat, the militant-turned-moderate leader of the PLO.

I

WORLD AFFAIRS

Joined UN in 1949

EBRD
IAEA
IBRD
WTO
IMF

Israel is technically at war with all Arab states except Egypt and Jordan. In 2000 it withdrew from southern Lebanon. Israel maintains close ties with the USA, but its harsh response to Palestinian militancy makes it increasingly isolated.

AID

Recipient

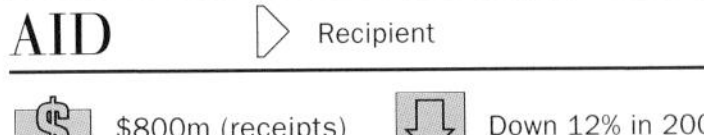
$800m (receipts)
Down 12% in 2000

Israel receives massive military and economic aid from the US. Large ad hoc donations are also received from Jewish NGOs.

ISRAEL

Total Land Area : 20 770 sq. km (8019 sq. miles)

POPULATION
- over 100 000
- over 50 000
- over 10 000

LAND HEIGHT
- 1000m/3281ft
- 500m/1640ft
- 200m/656ft
- Sea Level
- -200m/-656ft

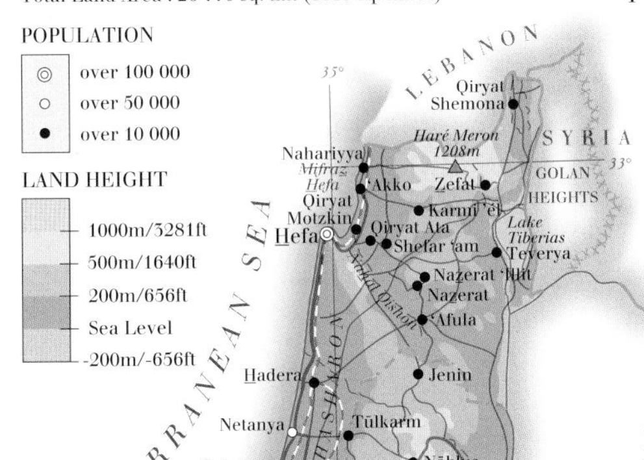
LEBANON
SYRIA
Qiryat Shemona
Haré Meron 1208m
Nahariyya
'Akko
Zefat
Karmi'el
Qiryat Motzkin
Hefa
Qiryat Ata
Shefar'am
Lake Tiberias
Teverya
Nazerat 'Illit
Nazerat
'Afula
GOLAN HEIGHTS
MEDITERRANEAN SEA
Hadera
Jenin
Netanya
Tulkarm
Nablus
Ra'anana
Herzliyya
Kefar Sava
WEST BANK
Tel Aviv-Yafo
Petah Tiqwa
Bat Yam
Ramat Gan
Holon
Rishon Le Ziyyon
Lod
Rehovot
Ramla
Ramallah
Jericho
Ashdod
JERUSALEM
Jordan

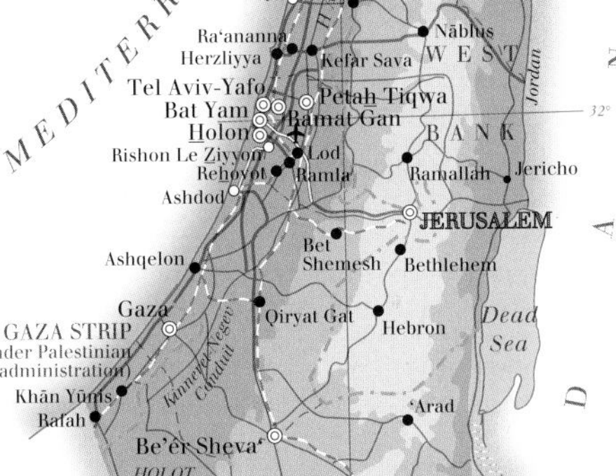
Ashqelon
Bet Shemesh
Bethlehem
Gaza
Qiryat Gat
Hebron
Dead Sea
GAZA STRIP (under Palestinian administration)
Khan Yunis
Rafah
'Arad
Be'er Sheva'
HOLOT
HALUZA
Dimona
HANEGEV
EGYPT
JORDAN
N
0 50 km
0 50 miles
Elat
Gulf of Aqaba

DEFENSE

Compulsory military service

$9.37bn
Up 6% in 2000

Israel, the only known nuclear-armed power in the Middle East, has a small regular defense force, which can be boosted by nearly 600,000 reservists. Equipped with some of the latest US technology, it is vastly superior in firepower (and training) to the armies of its Arab neighbors. To counter the Palestinian *intifada* it uses punitive strikes and counterinsurgency methods rather than conventional tactics.

ISRAELI ARMED FORCES

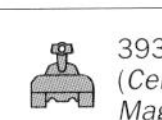

3930 main battle tanks (*Centurion*, M-48A5/60A3, *Magach* 7, *Merkava* I/II/III) — 120,000 personnel

3 submarines and 47 patrol boats — 6500 personnel

696 combat aircraft (50 F-4E-2000, 73 F-15, 237 F-16) — 37,000 personnel

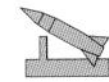

Widely believed that Israel has a nuclear capacity with up to 100 warheads. Delivery via *Jericho* 1 and *Jericho* 2 missiles

ECONOMICS

Inflation 10% p.a. (1990–2000)

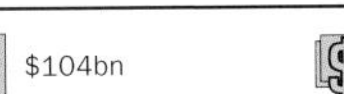

$104bn
4.044–4.412 sheqalim

STRENGTHS

Modern infrastructure; educated population. Huge potential of agriculture, manufacturing, and high-tech industry. Important banking sector.

WEAKNESSES

Violence and regional instability. Large defense budget. Financial burdens of absorbing immigrants and subsidizing settlement. Little trade with Arab neighbors. Corruption. Rising unemployment.

PROFILE

The government seeks ways to reduce massive state spending. The state owns most of the land and controls over 20% of all industries and services. Public companies are being privatized and there are plans to end restrictive labor practices. Agriculture, highly specialized and profitable, has been eclipsed by high-tech industries. The state aims to boost the service sector.

Israel's economy expanded through the 1990s, benefiting from mass immigration of Jews, many highly educated, from the former Soviet Union. Although unemployment rose as a result, new skills and contacts also helped the Israeli economy toward sustained export-led growth. Pockets of poverty remain in "development towns."

The Palestinian uprising has since 2000 seriously injured the economy, hitting trade, tourism, and investment. The Israeli military response has inflicted over $300 million of damage to Palestinian infrastructure.

ECONOMIC PERFORMANCE INDICATOR

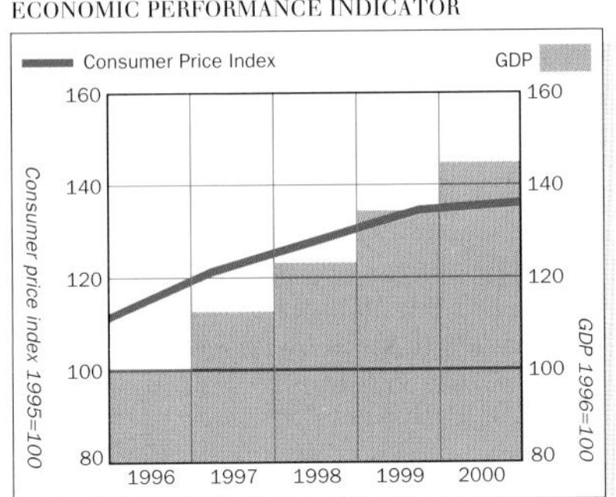
Consumer Price Index
GDP
160
140
120
100
80
Consumer price index 1995=100
GDP 1996=100
1996
1997
1998
1999
2000

SCORE CARD

- ❑ WORLD GNP RANKING........35th
- ❑ GNP PER CAPITA........$16,710
- ❑ BALANCE OF PAYMENTS........–$1.42bn
- ❑ INFLATION........1.1%
- ❑ UNEMPLOYMENT........9%

EXPORTS

UK 4%
Germany 5%
Hong Kong 4%
Belgium 6%
Other 44%
USA 37%

IMPORTS

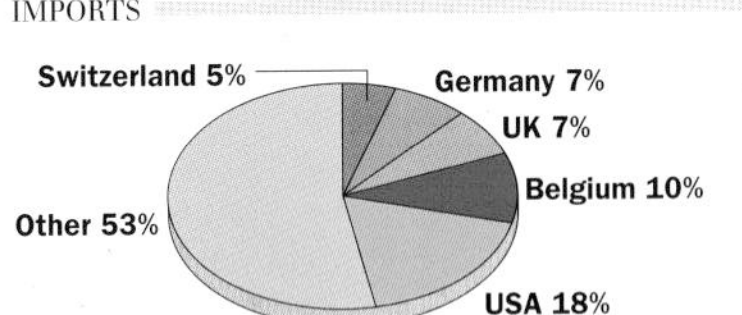
Switzerland 5%
Germany 7%
UK 7%
Belgium 10%
Other 53%
USA 18%

ISRAEL : MAJOR BUSINESSES

Hefa
Nazerat
Tel Aviv-Yafo
Ashqelon
Jerusalem
Be'ér Sheva'
Textiles
Chemicals
Computers
Food processing
Pharmaceuticals
Consumer goods
0 50 km
0 50 miles

I

ISRAEL AND THE PALESTINIANS

The conflict between Israel and the Palestinians is crucial in Middle East politics. The creation of Israel in 1948 effectively turned the almost one million Arabs living in the former British Mandate of Palestine into refugees overnight. Some 300,000 more left territories later occupied by Israel in 1967. Not until the 1993 accords were the Palestinians granted limited self-rule in the West Bank and Gaza – inhabited by two million Palestinians and 300,000 Jewish settlers – under the jurisdiction of the Palestinian National Authority (PNA), headed by Palestine Liberation Organization (PLO) leader Yasser Arafat. From 2000, hopes of a peaceful resolution of "permanent status" issues, and accession to full statehood, were lost from sight amid a return to violent conflict.

Background

In 1947 a UN plan partitioned Palestine into separate Jewish and Arab states. When in May 1948 the British Mandate ended, Arab states (which had all rejected the plan) invaded Palestine but were pushed back well beyond the UN partition lines by Israeli forces. The 1949 armistices left only East Jerusalem, the West Bank (5900 sq. km – 2300 sq. miles), and the Egyptian-administered Gaza Strip (1000 sq. km – 400 sq. miles) outside Israel. Jordan declared East Jerusalem and the West Bank to be part of its territory, only renouncing this claim formally in 1988. In the 1967 Six-Day War, Israel annexed the Palestinian territories. Jewish settlers began moving in, regarding the new borders as encompassing the biblical-era "Land of Israel."

Palestinian autonomy, part of the 1978 Camp David agreement, did not happen; there was diplomatic deadlock. The PLO did not recognize Israel's right to exist until 1988, while Israel refused to "negotiate with terrorists."

***Poor slum housing** in Gaza.*

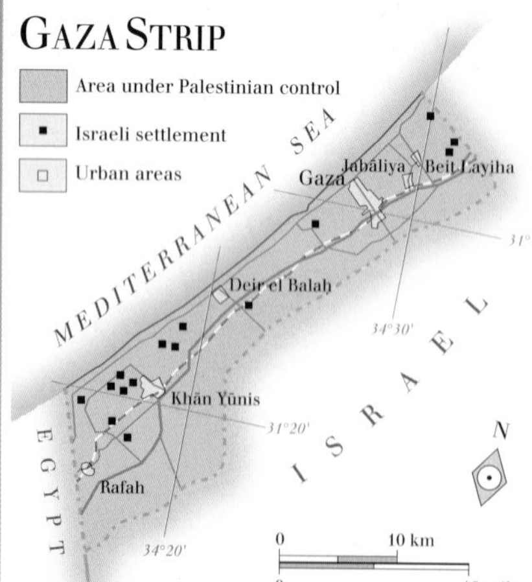
GAZA STRIP
Area under Palestinian control
Israeli settlement
Urban areas
MEDITERRANEAN SEA
Gaza
Jabāliya
Beit Lāyiha
Deir el Balah
Khān Yūnis
Rafah
EGYPT
ISRAEL
31°30'
34°30'
31°20'
34°20'
N
0
10 km
10 miles

Palestinian autonomy

Turning away from armed struggle, the PLO concluded a historic "land for peace" deal with Israel in 1993, known as the "Oslo Accords." The two sides formally recognized one another, and Arafat and Israeli Prime Minister Yitzhak Rabin signed a Declaration of Principles in Washington D.C. A five-year timetable for "permanent status" negotiations would tackle the future of Jewish settlements and the Palestinian demand for East Jerusalem as their capital. Palestinians were to get interim self-rule, initially in Gaza and Jericho (achieved in 1994), and gradually in the whole West Bank. This meant Palestinian police taking over from the Israeli military, who had been struggling since 1987 to end an uprising (*intifada*) led by the radical Islamic organization Hamas. The PNA was established, based in Gaza, and Arafat made a triumphal return in July 1994. He was elected president of the newly created 88-member Palestinian Legislative Council in January 1996.

The derailing of the peace process

The "Oslo B" accord in 1995 extended PNA rule to six more West Bank towns. After repeated delays Israel also gave up control of Hebron (but not of rural areas). Mutual mistrust and violence, however, threatened to derail the peace process before it reached the stage of "final status" talks and Arafat risked losing credibility among radical Palestinians. Attacks by rogue Hamas guerrillas, including suicide bombings, dominated mid-1996 elections in Israel. A new right-wing government took a tougher stance and did little to restrain Jewish settlers, fierce opponents of "land for peace."

A Labor-led government, in power from 2000, laid out an ambitious timetable for a permanent agreement with the PNA. Intensive US diplomacy helped keep this plan alive until the eleventh hour, but heavy-handed Israeli retaliation to violent incidents in September–October provoked Palestinian rage and a return to the tactics of *intifada*. In a bitter climate of atrocity, retaliation, and counterstrike, a new Israeli government was formed by Ariel Sharon, elected prime minister in 2001. Denouncing Palestinian militants who mounted suicide bomb attacks on civilian targets, Sharon blamed Arafat for failing to control them, while instituting a brutal policy of military reprisals and "targeted assassinations." Determined to sideline Arafat completely, the Sharon government gained US support in mid-2002 for the idea that only a Palestinian leadership change could bring Israel back to the negotiating table.

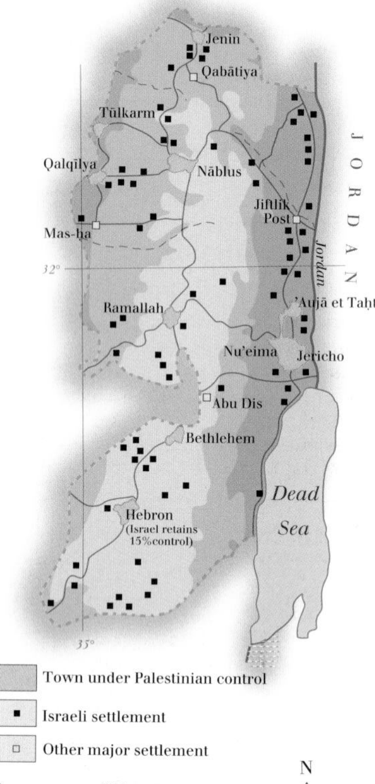
WEST BANK
Jenin
Qabātiya
Tūlkarm
Qalqīlya
Nāblus
Jiftlik Post
Mas-ha
JORDAN
Jordan
32°
Ramallah
'Aujā et Taḥtā
Nu'eima
Jericho
Abu Dis
Bethlehem
Hebron
(Israel retains 15% control)
Dead Sea
35°
Town under Palestinian control
Israeli settlement
Other major settlement
N
0
25 km
25 miles

***Bethlehem,** situated in the troubled West Bank.*

RESOURCES

Electric power 7.9m kw

24,661 tonnes

Low levels of oil production

5m turkeys, 1.4m geese, 30m chickens

Natural gas, oil, salt, potash, copper, gold, magnesium, bromine

ELECTRICITY GENERATION

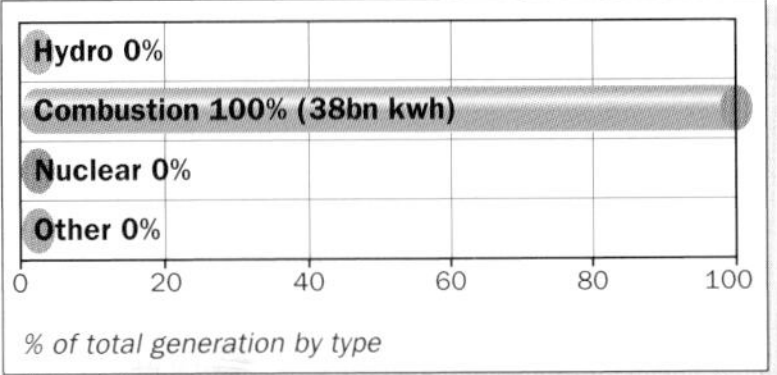

Hydro 0%
Combustion 100% (38bn kwh)
Nuclear 0%
Other 0%
0
20
40
60
80
100
% of total generation by type

The country's most valuable deposits of minerals are potash, bromine (of which Israel is the world's largest exporter), and other salts mined near the Dead Sea. Reserves of copper ore and gold were discovered in 1988. In the coastal plain, mixed farming, vineyards, and citrus groves are plentiful. Former desert areas now have extensive irrigation systems supporting specialized agriculture.

Israel's most critical resource is water. The 1994 treaty with Jordan provided for sharing the water of the Jordan River.

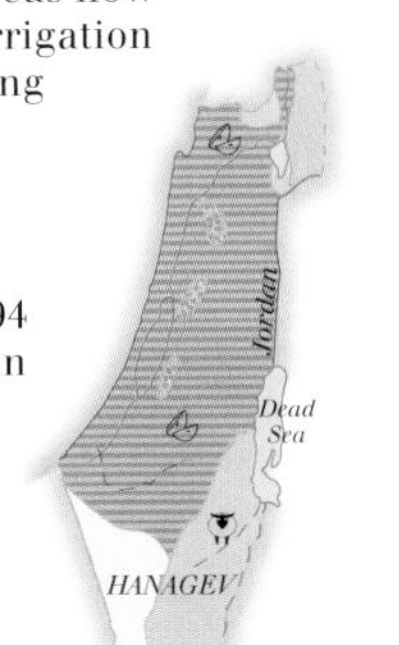

Jordan
Dead Sea
HANAGEV

ISRAEL : LAND USE

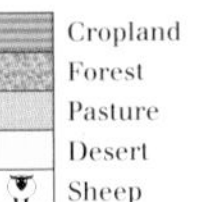

Cropland
Forest
Pasture
Desert
Sheep
Citrus fruit – cash crop

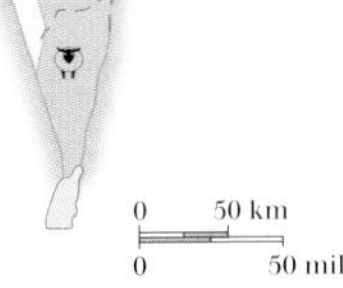

0
50 km
0
50 miles

ENVIRONMENT

Sustainability rank: 63rd

16%

10.1 tonnes per capita

Since 1993–1994, designated Environment Year, the government has aimed to promote recycling schemes, the cleanup of rivers, and a healthier urban environment.

ENVIRONMENTAL TREATIES

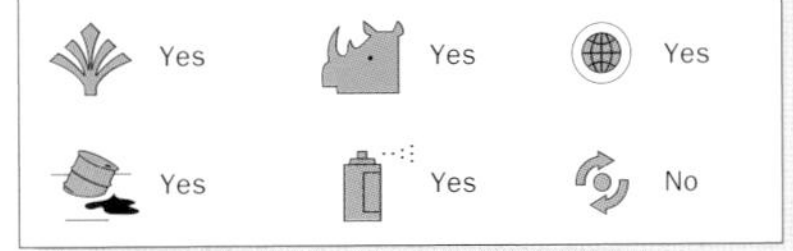

Yes
Yes
Yes
Yes
Yes
No

MEDIA

TV ownership high

Daily newspaper circulation 288 per 1000 people

PUBLISHING AND BROADCAST MEDIA

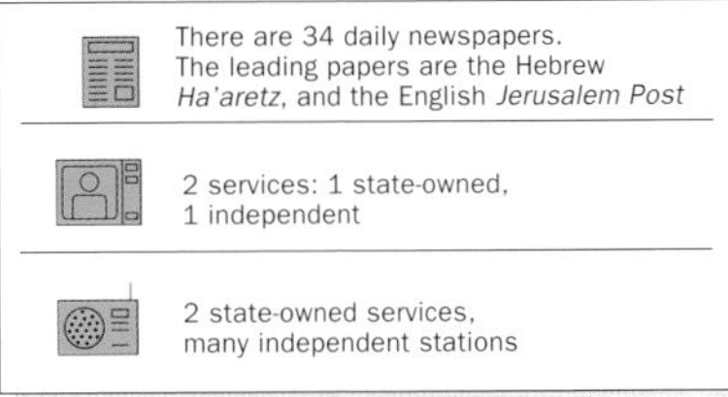

There are 34 daily newspapers. The leading papers are the Hebrew *Ha'aretz*, and the English *Jerusalem Post*
2 services: 1 state-owned, 1 independent
2 state-owned services, many independent stations

The left-wing press favors the peace process. The number of private radio stations, many right-wing, is rising.

CRIME

Death penalty not used in practice

9421 prisoners

Up 25% in 1999

CRIME RATES

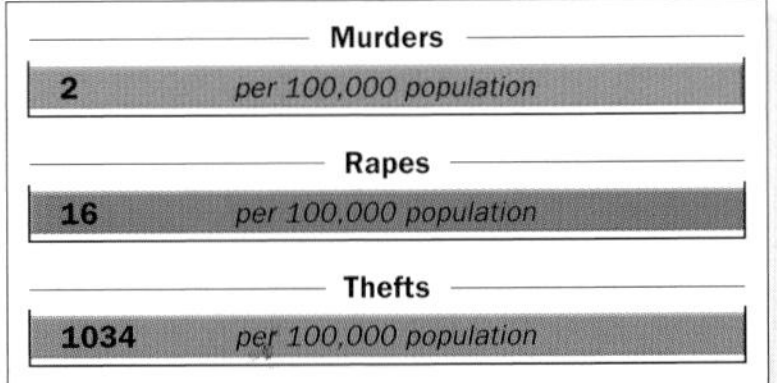

Murders
2 per 100,000 population
Rapes
16 per 100,000 population
Thefts
1034 per 100,000 population

Car theft is a rising concern. The vast majority of violent attacks are due to the Israeli–Palestinian conflict.

EDUCATION

School leaving age: 16

96%

198,766 students

THE EDUCATION SYSTEM

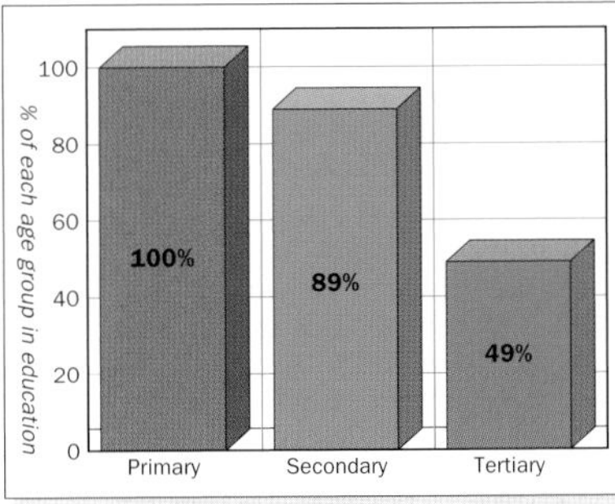

% of each age group in education
100
80
60
40
20
0
100%
89%
49%
Primary
Secondary
Tertiary

A highly educated population has been the engine of Israel's economic growth. State schools have religious (Jewish), secular, and Arab streams. Ultraorthodox and Sephardi Jews increasingly run their own private establishments.

HEALTH

Welfare state health benefits

1 per 256 people

Heart and cerebrovascular diseases, cancers

The ratio of doctors to the total population in Israel is one of the highest in the world. Primary health care reaches all communities. Israel's hospitals have pioneered many innovative treatments.

CHRONOLOGY

War with the neighboring Arab states followed immediately upon the creation of the state of Israel in 1948.

- ❑ **1967** Israeli victory in Six-Day War.
- ❑ **1973** Egypt and Syria attack Israel.
- ❑ **1978** Camp David accords with Egypt.
- ❑ **1979** Formal peace treaty, Sinai returned to Egypt.
- ❑ **1982** Israel invades Lebanon.
- ❑ **1987** Palestinians launch *intifada*.
- ❑ **1993** Oslo Accords.
- ❑ **1994** Palestinian autonomy begins in Gaza and Jericho.
- ❑ **1995** Prime Minister Yitzhak Rabin assassinated.
- ❑ **1996** Palestinian elections.
- ❑ **1998** Government stalls on US-backed plan to revive peace process.
- ❑ **1999** Ehud Barak (Labor) prime minister. Renewed peace process with Palestinians and Syria.
- ❑ **2000** Israeli withdrawal from southern Lebanon. Large-scale Israeli–Palestinian clashes.
- ❑ **2001** Ariel Sharon (Likud) elected prime minister; forms unity government. Conflict worsens.
- ❑ **2002** Arafat briefly besieged in Ramallah. Suicide attacks and Israeli reprisals intensify.

I

SPENDING

GDP/cap. increase

CONSUMPTION AND SPENDING

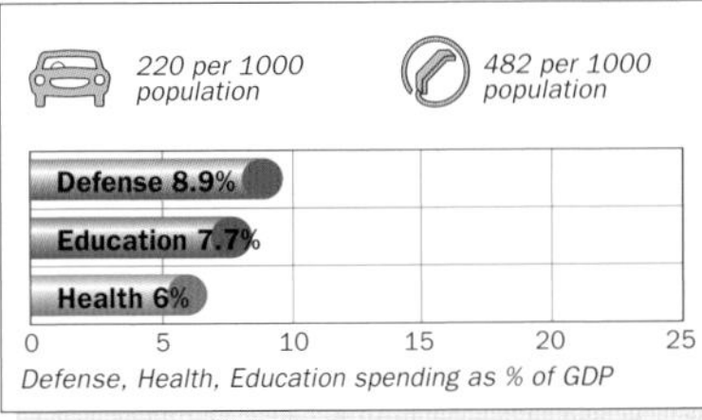

220 per 1000 population
482 per 1000 population
Defense 8.9%
Education 7.7%
Health 6%
0
5
10
15
20
25
Defense, Health, Education spending as % of GDP

Income per head is high, but taxation is heavy. In theory, those living in communes (*kibbutzim*) eschew personal material wealth.

WORLD RANKING

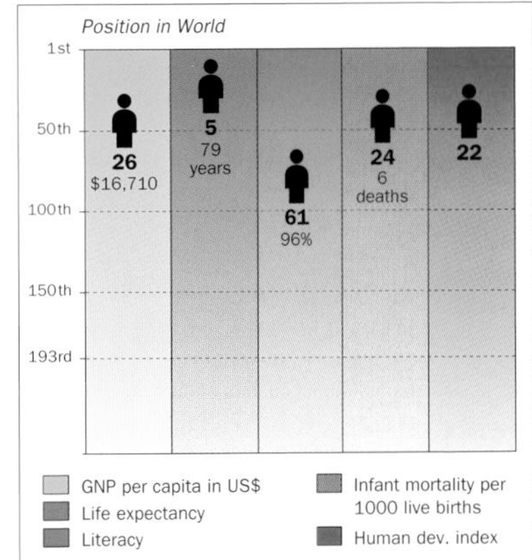

Position in World
1st
50th
100th
150th
193rd
26
$16,710
5
79 years
61
96%
24
6 deaths
22
GNP per capita in US$
Life expectancy
Literacy
Infant mortality per 1000 live births
Human dev. index

I

ITALY

EUROPE

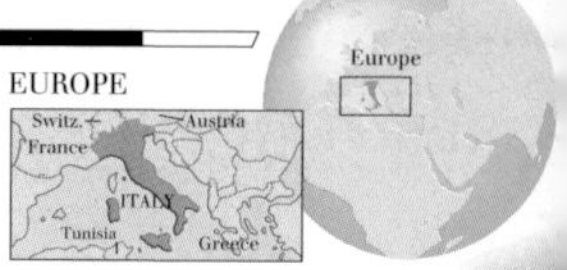

OFFICIAL NAME: Italian Republic **CAPITAL:** Rome **POPULATION:** 57.5 million
CURRENCY: Euro (Italian lira until 2002) **OFFICIAL LANGUAGE:** Italian

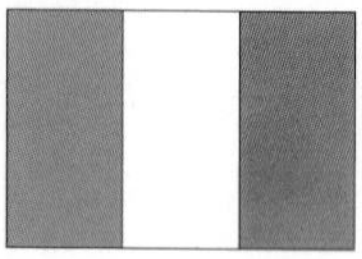

1861

1947

June 2

I

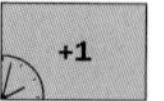
+1

+39

.it

THE BOOT-SHAPED Italian peninsula stretches 800 km (500 miles) southward into the Mediterranean, while the Alps form a natural boundary to the north. Italy also includes Sicily, Sardinia, and several smaller islands. The south is an area of seismic activity, with two famous volcanoes, Vesuvius and Etna. Rival city-states flourished in Renaissance Italy, a unified country only in Roman times and since 1861. Fascist rule under Mussolini from 1922 ended with Italy's defeat in World War II. The Christian Democrats (DC) then dominated Italy's notoriously short-lived governments for decades, until in the 1990s the established parties and patronage systems were shaken up by corruption investigations. New groupings emerged, power alternating between a right-wing coalition and a broad center-left Olive Tree alliance.

CLIMATE

Mediterranean/mountain

WEATHER CHART FOR ROME

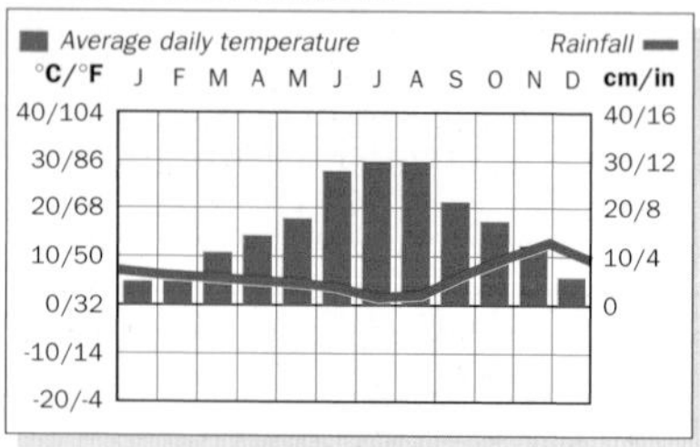

Average daily temperature — Rainfall

°C/°F J F M A M J J A S O N D cm/in

40/104, 30/86, 20/68, 10/50, 0/32, -10/14, -20/-4 — 40/16, 30/12, 20/8, 10/4, 0

A Mediterranean climate in the south contrasts with more temperate conditions in the north. Summers are hot and dry, especially in the south; Sardinia and Sicily have highs of more than 30°C (86°F). The Adriatic coast suffers from cold winds such as the *bora*. Southern winters are mild; northern ones are cooler and wetter, with heavy snow in the mountains.

TRANSPORTATION

Drive on right

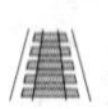

Leonardo da Vinci (Fiumicino), Rome
26.3m passengers

1457 ships
9.05m grt

THE TRANSPORTATION NETWORK

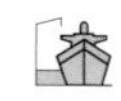

654,676 km (406,796 miles)

6957 km (4323 miles)

16,108 km (10,010 miles)

2400 km (1491 miles)

Italy's roads, which carry most its trade via Switzerland and Austria, are badly congested. The *autostrada* (expressway) network lacks key links, and serious bottlenecks affect the main north–south artery. Rail services are extensive and cheap. Luxurious Pendolino trains link Rome and Milan, but the high-speed Naples–Turin TAV project is behind schedule and over budget.

TOURISM

Visitors : Population 1:1.5

39m visitors

Down 5% in 2001

MAIN TOURIST ARRIVALS

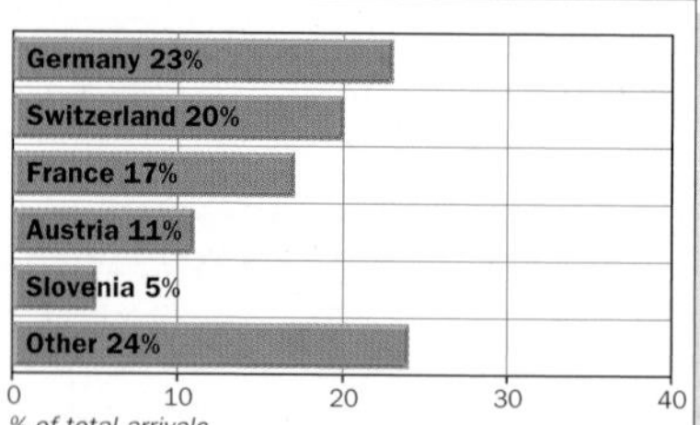
Germany 23%
Switzerland 20%
France 17%
Austria 11%
Slovenia 5%
Other 24%
0 10 20 30 40
% of total arrivals

Italy has been a tourist destination since the 16th century. Roman popes consciously aimed to make their city the most beautiful in the world to attract travelers. In the 18th century, Italy was the focus of any Grand Tour. Today, its many unspoiled centers of Renaissance and ancient culture continue to make Italy one of the world's major tourism destinations. The industry accounts for 3% of Italy's GDP, and employs a million people – 5% of the workforce.

Most visitors travel to the northern half of Italy, to cities such as Venice and Florence, and to Rome. Tourists are also drawn to the northern lakes, while beach resorts such as Rimini attract a large, youthful crowd in summer. Italy is also growing in popularity as a skiing destination. In the south the breathtaking ruins of Roman Pompeii are a particular magnet for visitors.

Fears have been expressed about the detrimental impact of tourism on Italy's environment. The pressure of visitors to Venice is such that in summer one-way systems for pedestrians have to be introduced and day trippers are often turned away.

Tuscan landscape. *Chianti wine is produced in this region, where many northern Europeans own holiday homes.*

ITALY

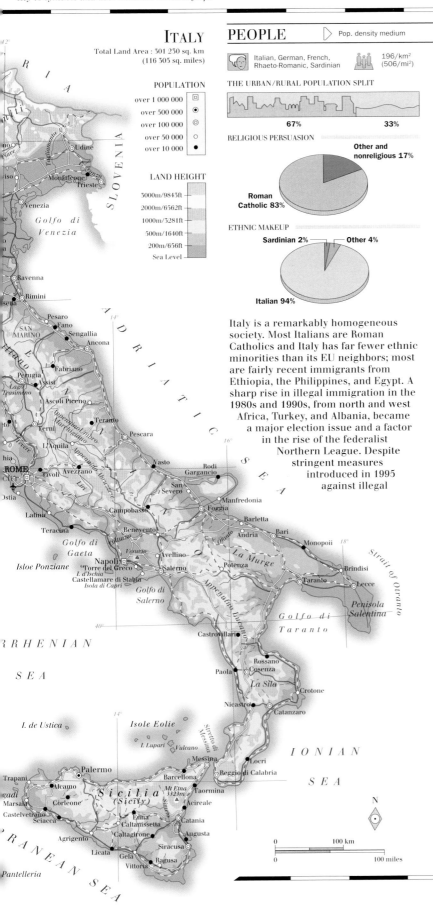

Total Land Area : 301 230 sq. km (116 305 sq. miles)

POPULATION
- over 1 000 000
- over 500 000
- over 100 000
- over 50 000
- over 10 000

LAND HEIGHT
- 3000m/9843ft
- 2000m/6562ft
- 1000m/3281ft
- 500m/1640ft
- 200m/656ft
- Sea Level

0 — 100 km
0 — 100 miles

PEOPLE

Pop. density medium

Italian, German, French, Rhaeto-Romanic, Sardinian

196/km² (506/mi²)

THE URBAN/RURAL POPULATION SPLIT

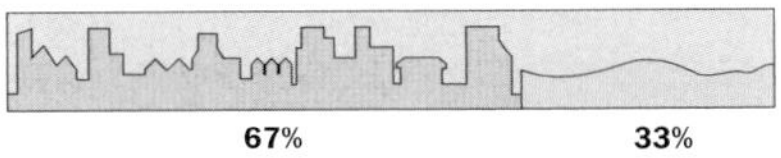

67% | 33%

RELIGIOUS PERSUASION

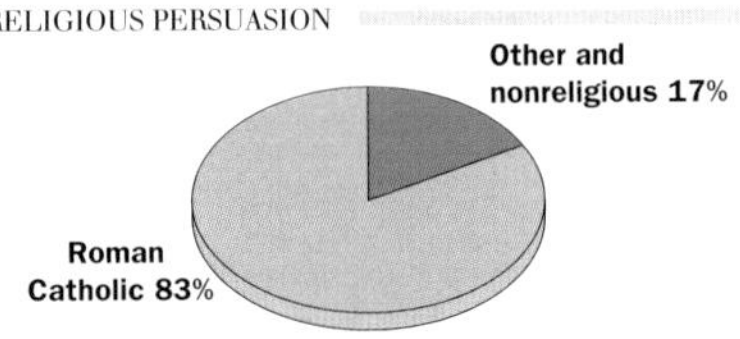

Roman Catholic 83%
Other and nonreligious 17%

ETHNIC MAKEUP

Sardinian 2%
Other 4%
Italian 94%

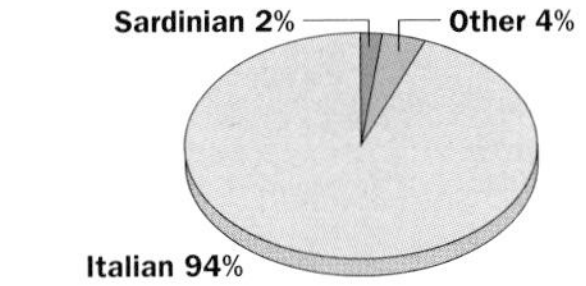

Italy is a remarkably homogeneous society. Most Italians are Roman Catholics and Italy has far fewer ethnic minorities than its EU neighbors; most are fairly recent immigrants from Ethiopia, the Philippines, and Egypt. A sharp rise in illegal immigration in the 1980s and 1990s, from north and west Africa, Turkey, and Albania, became a major election issue and a factor in the rise of the federalist Northern League. Despite stringent measures introduced in 1995 against illegal immigrants, the number of new arrivals continues to increase.

Difficult economic conditions caused many Italians to emigrate in the 1950s and 1960s. There are now five million Italians living abroad. About half live in other EU countries, the rest mainly in the USA, South America, and Australia. Most migrants then, as now, are from the poorer south – the Mezzogiorno. Within Italy, prejudice still exists in the north against southern Italians.

Sport – especially soccer – has an unusual ability to bring out a strong sense of national identity among Italians. In other spheres, with state institutions viewed as inefficient and corrupt, most people feel a stronger allegiance to the region, or the community, and above all to the family. The extended family remains Italy's key social and economic support system. Most Italians live at home before marriage. Marriage rates are among the highest in Europe and divorce rates the lowest. Catholicism, however, has not stopped Italy having the lowest birthrate and one of the highest abortion rates in the EU.

Italians tend to dress well. Their preoccupation with style reflects the traditional importance of *bella figura* – image, cutting a dash – in Italian life as much as the high living standards which most now enjoy.

POPULATION AGE BREAKDOWN

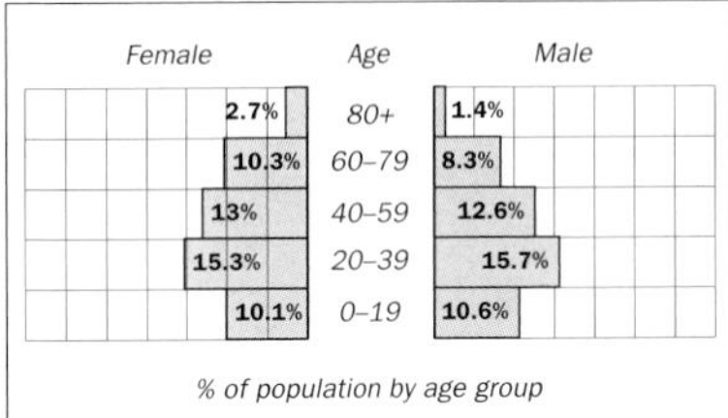
Female
Age
Male
2.7%
80+
1.4%
10.3%
60–79
8.3%
13%
40–59
12.6%
15.3%
20–39
15.7%
10.1%
0–19
10.6%
% of population by age group

CHRONOLOGY

Previously a collection of independent city states, dukedoms, and monarchies, Italy became a unified state in 1861.

- ❑ **1922** Mussolini asked to form government by king.
- ❑ **1928** One-party rule by Fascists.
- ❑ **1929** Lateran Treaties with Vatican recognize sovereignty of Holy See.
- ❑ **1936–1937** Axis formed with Nazi Germany. Abyssinia (Ethiopia) conquered.
- ❑ **1939** Albania annexed.
- ❑ **1940** Italy enters World War II on German side.
- ❑ **1943** Invaded by Allies. Mussolini imprisoned by Victor Emmanuel III. Armistice with Allies. Italy declares war on Germany.

⇨

I

I

CHRONOLOGY *continued*

- **1945** Mussolini released; establishes puppet regime in north; executed by Italian partisans.
- **1946** Referendum votes in favor of Italy becoming a republic.
- **1947** Italy signs peace treaty, ceding border areas to France and Yugoslavia, Dodecanese to Greece, and giving up colonies.
- **1948** Elections: DC under De Gaspieri heads coalition.
- **1949** Founder member of NATO.
- **1950** Agreement reached on US bases in Italy.
- **1951** Joins European Coal and Steel Community.
- **1957** Founder member of European Economic Community. Aided by funds from that organization and by Marshall Aid, industrial growth accelerates.
- **1964** DC government under Aldo Moro forms coalition with Socialist Party (PSI).
- **1969** Red Brigades, extreme left terrorist group, formed.
- **1972** Support for extreme right reaches postwar peak (9%). Rise in urban terrorism by both extreme left and right.
- **1976** Communist Party (PCI) support reaches a peak of 34% under Enrico Berlinguer's Eurocommunist philosophy.
- **1978** Aldo Moro abducted and murdered by Red Brigades.
- **1980** Extreme right bombing of Bologna station kills 84, wounds 200.
- **1983–1987** Center-left coalition formed under Bettino Craxi.
- **1990** LN attacks immigration policies and subsidies for the south.
- **1992** Corruption scandal, involving bribes for public contracts, uncovered in Milan. Government members accused.
- **1994** General election: DC support collapses; coalition government formed between Silvio Berlusconi's Forza Italia, LN, and neo-fascists.
- **1995–1996** Technocrat government tackles budget, pensions, media, and regional issues.
- **1996** Center-left Olive Tree alliance wins general election; Romano Prodi prime minister.
- **1998** May, Italy qualifies to join euro currency from January 1999. October, Prodi government falls, Massimo D'Alema prime minister.
- **1999** Carlo Ciampi president.
- **2000** D'Alema replaced by Giuliano Amato.
- **2001** May, Berlusconi victory in general election. June, right-wing government includes "post-fascist" National Alliance.
- **2002** Euro fully adopted.

POLITICS

Multiparty elections

L. House 2001/2006
U. House 2001/2006

President Carlo Azeglio Ciampi

AT THE LAST ELECTION

Chamber of Deputies 630 seats

Party	Share
PdL	58%
U	38%
PRC	2%
Others	1%
U–SVP	1%

PdL = Freedom Alliance (includes Forza Italia, National Alliance – **AN**, and Northern League – **LN**)
U = Olive Tree alliance (includes Democrats of the Left – **DS** and Party of Italian Communists – **PdCI**)
PRC = Communist Refoundation Party
U–SVP = Olive Tree–South Tyrolese People's Party
Nom = Nominated

Senate 324 seats

Party	Share
PdL	55%
U	38%
Nom	3%
PRC	1%
U–SVP	1%
Others	2%

The Senate comprises 315 elected members and several life senators

Italy is a multiparty democracy.

Profile

A coalition of Silvio Berlusconi's Forza Italia, Gianfranco Fini's "post-fascist" AN, and Umberto Bossi's xenophobic, secessionist LN won elections in 1994, but had collapsed by the end of that year. A technocratic government took over until 1996, when fresh elections resulted in a historic victory for the center-left Olive Tree alliance headed by Romano Prodi.

Bossi failed to rouse mass support for his declaration in September 1996 of an independent northern state of "Padania."

Success in qualifying for membership of the single European currency crowned the Prodi government's achievements in the economic sphere. After two years in office, it fell in October 1998 when the communists challenged its budget. The premiership passed to Massimo D'Alema of the DS (one of the successor groups to the Communist Party – PCI), then to Giuliano Amato, retaining a broadly similar coalition formula. Aware of the strength of the challenge from Berlusconi and the right, Amato then stood aside so that the popular mayor of Rome, Francesco Rutelli, could lead the center-left campaign in the May 2001 elections. The right-wing victory in that poll was essentially Berlusconi's personal triumph, although his new government repeated his 1994 formula by including the LN and the AN. It faced mass public protest in 2002 against proposed labor law reforms.

Romano Prodi, ex-premier (1996–1998) who heads European Commission.

Umberto Bossi, NL leader and minister in the Berlusconi government.

Silvio Berlusconi, conservative prime minister, elected for a second time in 2001.

The church of Santa Maria della Salute marks the entrance to Venice. The city-state managed to retain its independence until Napoleon Bonaparte's invasion of Italy.

Main Political Issues

Corruption

The 1990s *mani pulite* (clean hands) investigations, initially concerned with financial scandals in Milan, went on to reveal a nationwide network of corruption permeating political and business life. The issue destroyed the old political order, and many public figures were disgraced. Berlusconi, who emerged as a leading figure in the country's new political makeup, was himself dogged by bribery-related charges. Although he was acquitted on several counts during 2000, his return to power the following year left the corruption issue still simmering, as he pushed through reforms which meant that outstanding charges against him could be quashed.

Institutional Reform

The old proportional representation (PR) electoral system, blamed for a lack of strong government, was much modified in the early 1990s, but the process lost impetus after the 1994 elections. Twice, in 1999 and in 2000, referendums failed to abolish the 25% of seats still elected by PR, not because voters opposed this, but merely due to inadequate turnout. Berlusconi favors a system with greater presidential powers, akin to that in France (and has made explicit his desire to go on to be head of state under such a system).

WORLD AFFAIRS

Joined UN in 1955

EU | G8 | NATO | OECD | OSCE

Italy was one of the founders of the EU, but the return of a right-wing government in 2001 encouraged the opponents of further integration. Italy's strategic position in the Mediterranean makes it a central member of NATO, whose South European Command is based in Naples. Despite a pro-Western orientation, Italy often contributes to mediation in eastern Europe and the Middle East. Dependence on Libya for energy supplies made it especially keen to see UN sanctions lifted.

Major concerns in the 1990s were upheaval in Albania and conflict in the former Yugoslavia. NATO used Italian bases for air strikes against Yugoslavia in 1999, and Italy backed the US-led "war on terrorism" in Afghanistan in 2001.

AID

Donor

$1.38bn (donations) | Down 24% in 2000

A relatively small aid program is nearly all carried out through international organizations. Albania and the countries of the former Yugoslavia have received funding to ward off a feared influx of "economic" migrants, the Italian military running special relief operations.

DEFENSE

Phasing out conscription

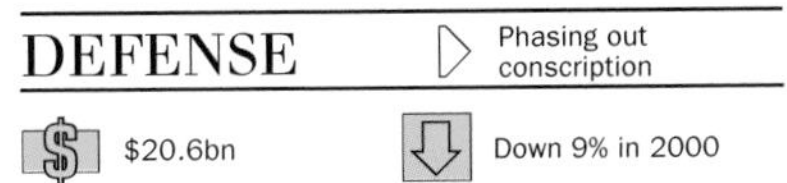
$20.6bn
Down 9% in 2000

ITALIAN ARMED FORCES

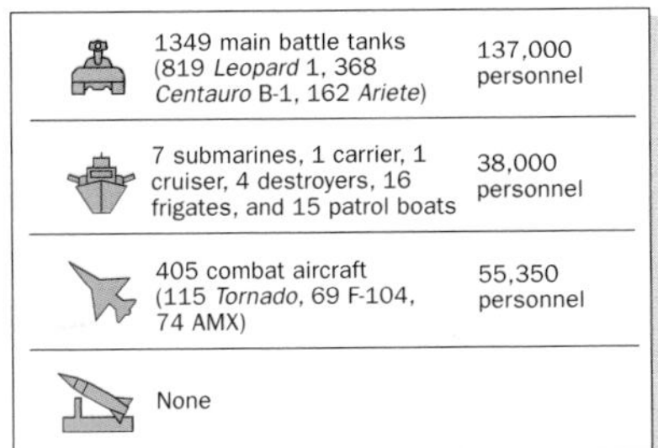

1349 main battle tanks (819 *Leopard* 1, 368 *Centauro* B-1, 162 *Ariete*)	137,000 personnel
7 submarines, 1 carrier, 1 cruiser, 4 destroyers, 16 frigates, and 15 patrol boats	38,000 personnel
405 combat aircraft (115 *Tornado*, 69 F-104, 74 AMX)	55,350 personnel
None	

Since the ending of the Cold War, conflicts in former Yugoslavia have helped refocus defense priorities. A "New Model Defense" was announced in 1992, and in 2000 the legislature approved a law ending conscription by 2005 and also introducing women soldiers. The envisaged professional army is to play a rapid-intervention role on NATO's southern flank, while the navy fulfills Mediterranean coastal functions rather than retaining ocean-going capabilities. Defense spending remains low, despite pressures to modernize weapons systems. In 2001, 2700 troops were deployed in the "war on terrorism."

ECONOMICS

Inflation 3.8% p.a. (1990–2000)

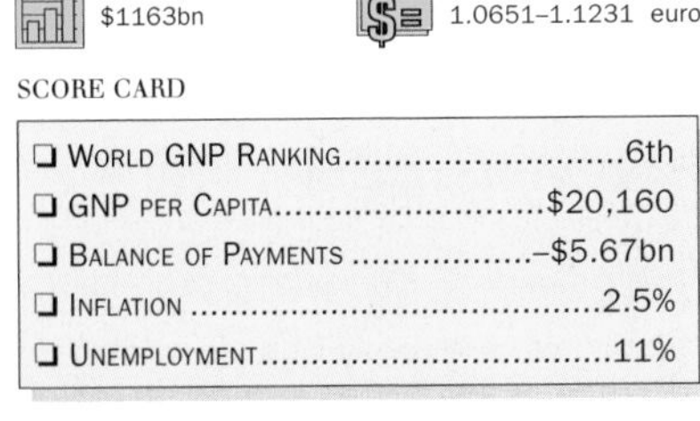

$1163bn | 1.0651–1.1231 euros

SCORE CARD

❑ WORLD GNP RANKING	6th
❑ GNP PER CAPITA	$20,160
❑ BALANCE OF PAYMENTS	–$5.67bn
❑ INFLATION	2.5%
❑ UNEMPLOYMENT	11%

EXPORTS

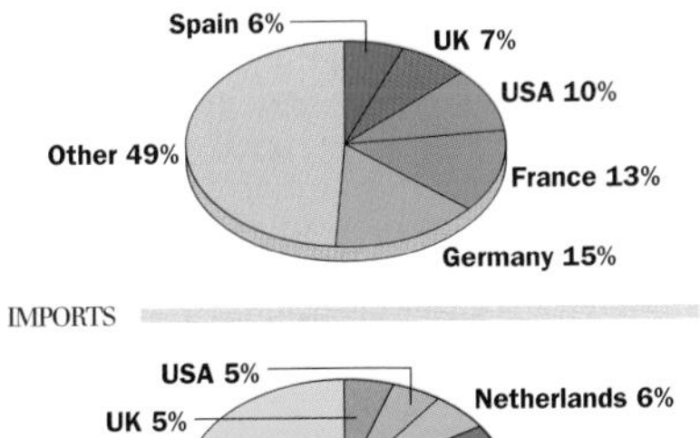
Spain 6%
UK 7%
USA 10%
Other 49%
France 13%
Germany 15%

IMPORTS

USA 5%
Netherlands 6%
UK 5%
France 11%
Other 56%
Germany 17%

STRENGTHS

Highly competitive, innovative small- to medium-size business sector. World leader in industrial and product design, textiles, and household appliances. Several highly innovative firms include Fiat (cars), Montedison (plastics), Olivetti (communications), and Benetton (clothes). Strong tourism and agriculture sectors, prestigious fashion houses.

WEAKNESSES

Public deficit and government debt remain high. Slow growth. Inefficient public sector undergoing major privatizations. Uneven wealth distribution: northern Italy far richer than the south, which suffers three times more unemployment. Poor record on tax collection, although now much improved. Relatively small companies facing foreign competition. Heavy dependence on imported energy.

PROFILE

Since World War II, Italy has developed from a mainly agricultural society into a world industrial power. The economy is characterized by a large state sector, a mass of family-owned businesses, relatively high levels of protectionism, and strong regional differences. Italy also has relatively few multinationals compared with other G7 economies.

The Institute for Industrial Reconstruction (IRI), a state-owned holding company dating from the Fascist era, progressively privatized its electronics, steel, engineering, shipbuilding, telecommunications, transportation, and aerospace companies, until closing down itself in 2000. The National Hydrocarbons Group (ENI), one of the world's top players in the energy and chemicals sectors, has been privatized, as has Telecom Italia, and the electricity corporation Enel. City and regional authorities own utilities, banks, and other businesses.

Family-owned businesses, which are the backbone of the private sector, include Fiat, whose interests include aero engines, telecommunications, and bioengineering, as well as cars. Similar businesses tend to congregate, encouraging local competition which has translated into national success.

The Mezzogiorno remains an exception. State attempts to attract new investment have met with success in areas immediately south of Rome, but elsewhere organized crime has deterred investors and siphoned off state funds. Anger at the misuse of state funds in the south was a powerful factor in the growth of the LN, with its demands for autonomy. One-third of Italian tax revenue is generated in Italy's industrial heartland of Milan.

ECONOMIC PERFORMANCE INDICATOR

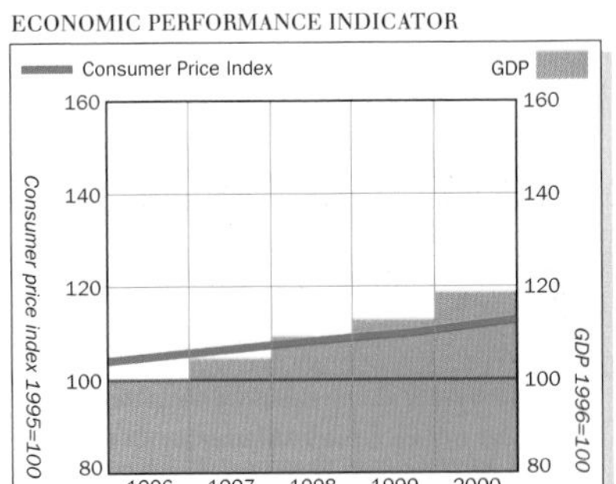
Consumer Price Index
GDP
160
140
120
100
80
Consumer price index 1995=100
GDP 1996=100
1996
1997
1998
1999
2000

ITALY : MAJOR BUSINESSES

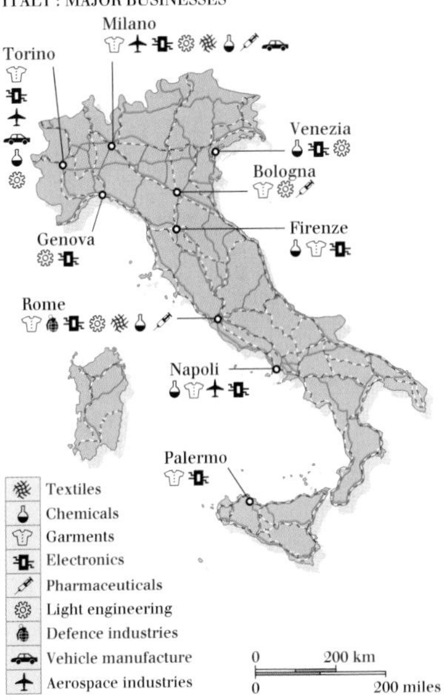
Milano
Torino
Venezia
Bologna
Firenze
Genova
Rome
Napoli
Palermo
Textiles
Chemicals
Garments
Electronics
Pharmaceuticals
Light engineering
Defence industries
Vehicle manufacture
Aerospace industries
0
200 km
0
200 miles

I

***Remains of the Greek theater** at Taormina, eastern Sicily. It was rebuilt by the Romans in the 2nd century CE. Today, the theater is the venue for an annual arts festival.*

RESOURCES

Electric power 68.4m kw

540,523 tonnes

79,000 b/d (reserves 600m barrels)

25m turkeys, 11.1m sheep, 8.33m pigs, 100m chickens

Coal, oil, lignite, pyrites, fluorite, barytes, bauxite

ELECTRICITY GENERATION

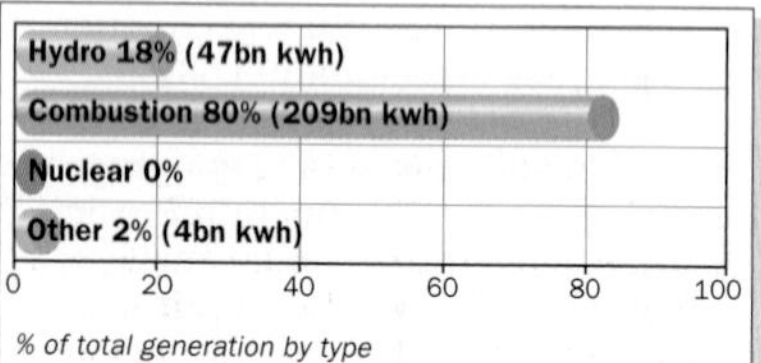
Hydro 18% (47bn kwh)
Combustion 80% (209bn kwh)
Nuclear 0%
Other 2% (4bn kwh)
0
20
40
60
80
100
% of total generation by type

Italy has very few natural resources. It produces just 1% of its oil needs and is highly vulnerable to both fluctuations in world prices and political instability in its north African suppliers. It has reduced its exposure since 1973, when oil accounted for 71% of its needs. Even so, oil still accounts for over 50% of energy consumption. Some power is generated from hydro and geothermal sources. Nuclear power was rejected in a 1987 referendum, and development has effectively been abandoned. Italy's mineral assets are small and the sector contributes little to national wealth.

ITALY : LAND USE

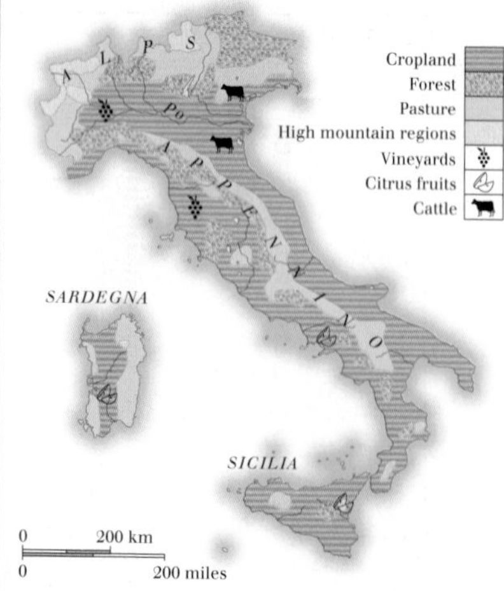
Cropland
Forest
Pasture
High mountain regions
Vineyards
Citrus fruits
Cattle
SARDEGNA
SICILIA
Po
APPENNINO
0 200 km
0 200 miles

ENVIRONMENT

Sustainability rank: 84th

8%

7.2 tonnes per capita

ENVIRONMENTAL TREATIES

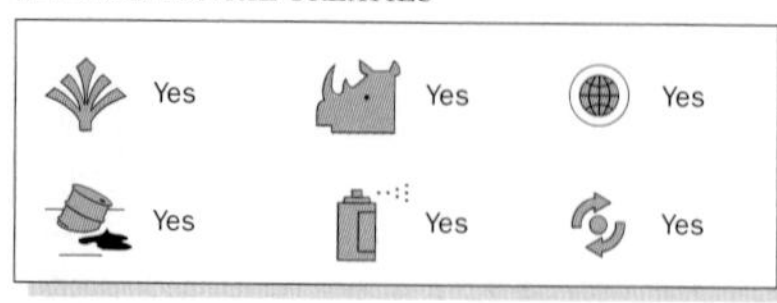
Yes
Yes
Yes
Yes
Yes
Yes

Italy has extensive environmental legislation, but has faced problems in enforcing directives. Wildlife successes include the return of the endangered lynx and brown bear, and growing numbers of wolves in the Appenines. The hunting of migrant birds, a popular sport in Italy, attracts international criticism. The use of drift nets, prone to catching dolphins and turtles as well as fish, has been made illegal under EU law. The right-wing government of the mid-1990s, returned to office in 2001, is suspicious of energy taxes and laws on waste recycling, not wanting to restrict business competitiveness. Green Party members in government in the Olive Tree alliance from 1996 to 2001 had insisted on a more active environmental stance.

Pollution in cities such as Naples and Rome is a major concern. Bans on traffic for up to seven hours during windless days are not uncommon. Acid rain has also damaged forests; 10% of trees are affected.

MEDIA

TV ownership high

Daily newspaper circulation 104 per 1000 people

PUBLISHING AND BROADCAST MEDIA

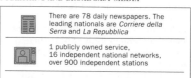

There are 78 daily newspapers. The leading nationals are *Corriere della Serra* and *La Repubblica*

1 publicly owned service, 16 independent national networks, over 900 independent stations

1 publicly owned service, over 2100 independent stations

Italy's media are dominated by a few conglomerates. Mediaset, owned by Prime Minister Silvio Berlusconi, is the main commercial operator. The three state TV RAI channels have nearly 50% of the market. Until reforms in the 1990s, they were highly politicized, in that RAI 1 was apportioned to the Christian Democrats, RAI 2 to the Socialists, and RAI 3 to the Communists. News Corporation has brought out Vivendi to gain a near-monopoly on pay-TV. All the media reflect the Italian love of sport, especially soccer. *La Gazzetta dello Sport* has one of the largest circulations of the national dailies. The press is highly regionalized.

CRIME

No death penalty

53,481 prisoners

Down 1% 1996–1998

CRIME RATES

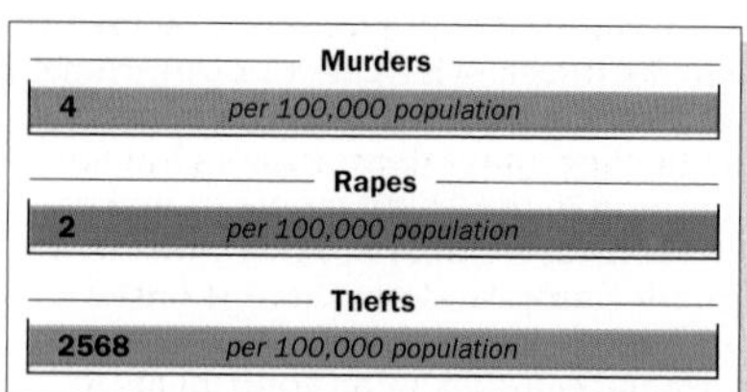
Murders
4 per 100,000 population
Rapes
2 per 100,000 population
Thefts
2568 per 100,000 population

Over 25% of prisoners are foreigners, many held for narcotics offenses. There is a huge backlog of cases. Organized crime has been weakened by the anticorruption drive and a cleaned-up bureaucracy. The Sicilian Mafia, the Cosa Nostra, in particular, was hit hard by arrests and trials in which former members provided key evidence. However, the Cosa Nostra and its counterparts in Naples and Calabria – Camorra and 'ndrangheta – still control wholesale agricultural markets and much of the narcotics trade, bleed businesses of protection money, and manipulate public works contracts.

EDUCATION

School leaving age: 16

98%

1.89m students

THE EDUCATION SYSTEM

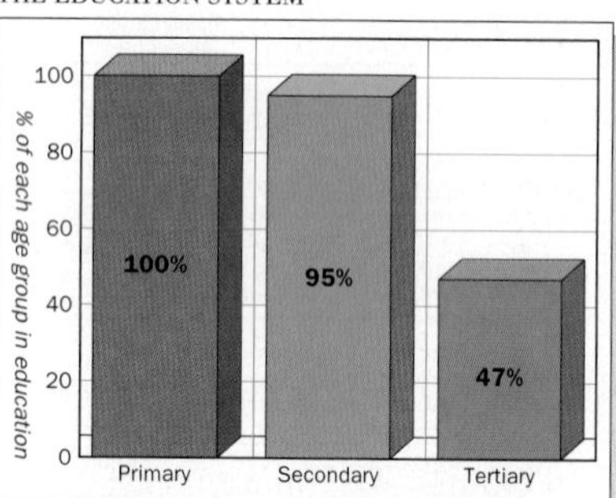
100
80
60
40
20
0
% of each age group in education
100%
95%
47%
Primary
Secondary
Tertiary

Schooling is state-run, apart from a few religious and elite private institutions. The pupil–teacher ratio in Italian schools is one of the best in Europe. In 1993, the minimum school leaving age was raised from 14 to 16 years, bringing Italy into line with most of Europe. However, the dropout rate in schools – as high as 50% in Sicily – remains a problem. An educational credit system aims to tackle shortcomings in information technology training.

Universities in Italy are oversubscribed. Rome has 180,000 students at any one time, only 30% of whom gain a degree. Many Italian educationalists wish to restrict entry. Another concern is the fact that Italy devotes only 1.4% of its GNP to research, compared with a European average of 2.5%.

REGIONALISM AND NORTHERN SEPARATISM

The economic division of Italy into the more dynamic and prosperous north and the impoverished south or Mezzogiorno is a source of continuing tension. It helps to explain why, in a country as homogeneous as Italy in terms of ethnicity, language, and religion, there is a strong demand for more devolution of power to regional level. In the mid-1990s, a political party calling for the breakup of the country was able to gain a significant following. The Northern League (LN), created in 1991 and built around the earlier Lombard League, won over 10% of the vote in the 1996 elections. Although it has lost momentum since then, regional feeling remained sufficient for the LN to be drafted into the right-wing government formed after the 2001 elections, and for a referendum in October to secure the first step toward limited devolution.

Run-down housing *in the poor, rural area of Calabria in southern Italy.*

Regions with more dynamic economies

GDP per capita is almost three times higher in the richest northern areas, such as Bologna in Emilia-Romagna and Milan in Lombardy, than in southern areas such as Reggio di Calabria. After the 1996 recession, the north achieved more rapid growth, creating more new jobs faster. Lombardy and Piedmont attract the lion's share of inward investment. Turin, the capital of Piedmont and home of Fiat, is boosting its industrial image by completing high-speed rail links and staging the Winter Olympics in 2006. Some southern areas try to attract investment by emphasizing that labor is cheaper there, and getting unions to accept lower wages locally in so-called "territorial pacts," but poor infrastructure remains a major obstacle. A road–rail bridge project linking Sicily to the Italian mainland was only revived in 2002.

Padania

The LN championed the idea of creating a "Republic of Padania" to secede from the rest of Italy (Rome,

Aosta
Trento
Milano
Venezia
Trieste
Torino
Genova
Bologna
PADANIA
Firenze
Ancona
Perugia

its surrounding Lazio region, and the south). The LN leader Umberto Bossi declared Padania's "independence" in a ceremony in 1996 after a pilgrimage the length of the river Po, the symbolic artery of "Padania," from Piedmont in northwest Italy to the Adriatic south of Venice. The LN set up a self-styled government and parliament, a national guard, and a flag, while allowing a year to "negotiate" terms of separation with the Italian government. Mantua was described as the Padanian capital, but it remained unclear precisely how far its intended territory extended. Usually understood to refer to the nine most northerly regions, as far south as Florence and the rest of Tuscany, it was sometimes defined by the LN as also including the Umbria and Marche regions.

The Italian government refused to take the Padania idea seriously, dismissing it as a publicity stunt. LN leaders were subsequently persuaded, by a series of bad local election results including losing the mayorships of Mantua and Milan, that the time was not right to press forward with their project. Some of its impetus – the desire of northern businesses to be at the center of European Union integration, not held back by the south – also disappeared when Italy qualified to join the European single currency, the euro, in 1999.

Shopping center *in Milan, Italy's center of fashion and commerce.*

HEALTH

Welfare state health benefits

1 per 169 people

Heart and cerebrovascular diseases. cancers, accidents

Italy's health care system is rated by WHO as the second most efficient in the world.

Standards of health care vary across the country, as the services are run by the regions. The state-run health system introduced in the 1970s initially provided services free at the point of use, but charges have been levied since 1988 for some dental and prescription costs; patients also have to pay a daily hospital charge and a yearly health fee. AIDS patients are exempt.

SPENDING

GDP/cap. increase

CONSUMPTION AND SPENDING

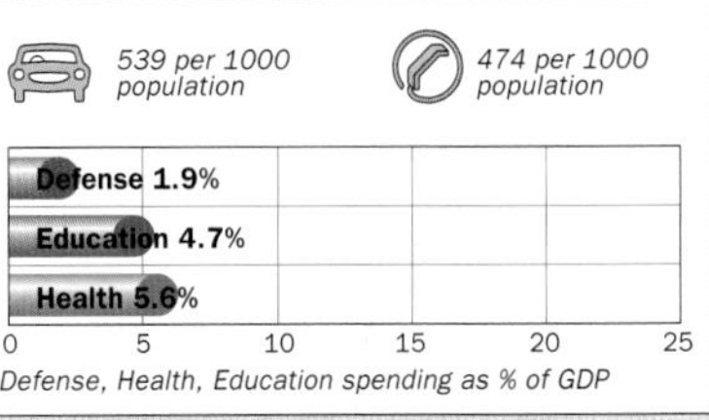
539 per 1000 population
474 per 1000 population
Defense 1.9%
Education 4.7%
Health 5.6%
0
5
10
15
20
25
Defense, Health, Education spending as % of GDP

Italians, particularly in the north, are today among the world's wealthiest people in terms of disposable income. This is a result not only of economic growth, but also of the structure of Italian society.

Many Italians have more than one job. The extended families in which most people still live often have access to more than one income. Few people have mortgages, and savings and tax avoidance levels are high.

The main exceptions are in parts of the south. Although inward investment has been attracted to the Bari area, many people still live in poverty in other places, such as Naples and the Calabria region, where investment has been lowest, unemployment is highest, and even tourism is underdeveloped.

WORLD RANKING

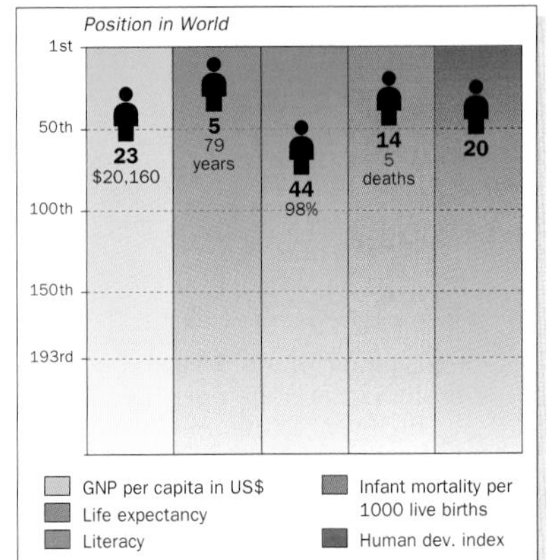
Position in World
1st
50th
100th
150th
193rd
23
$20,160
5
79 years
44
98%
14
5 deaths
20
GNP per capita in US$
Life expectancy
Literacy
Infant mortality per 1000 live births
Human dev. index

I

Ivory Coast

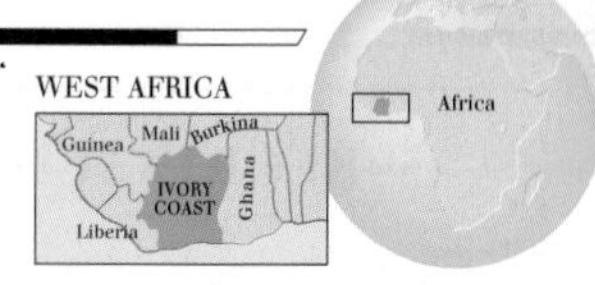
WEST AFRICA
Guinea
Mali
Burkina
IVORY COAST
Ghana
Liberia
Africa

Official Name: Republic of Côte d'Ivoire **Capital:** Yamoussoukro
Population: 16.3 million **Currency:** CFA franc **Official Language:** French

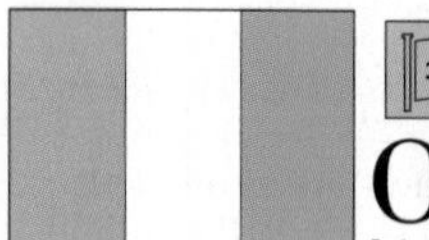

1960

1960
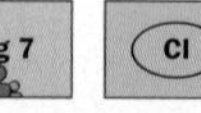
Aug 7
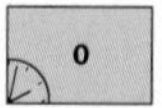
CI

0

+225

.ci

One of the larger west African coastal countries, Ivory Coast – officially Côte d'Ivoire – is the world's biggest cocoa producer. The forested interior, apart from the capital, is more sparsely populated than the sandy coastal strip. Pro-Western President Houphouët-Boigny ruled from independence in 1960 until 1993. A period of turbulence followed his death. Military rule in 1999 gave way in 2000 to an elected president after a chaotic poll.

CLIMATE

Tropical wet & dry

WEATHER CHART FOR YAMOUSSOUKRO

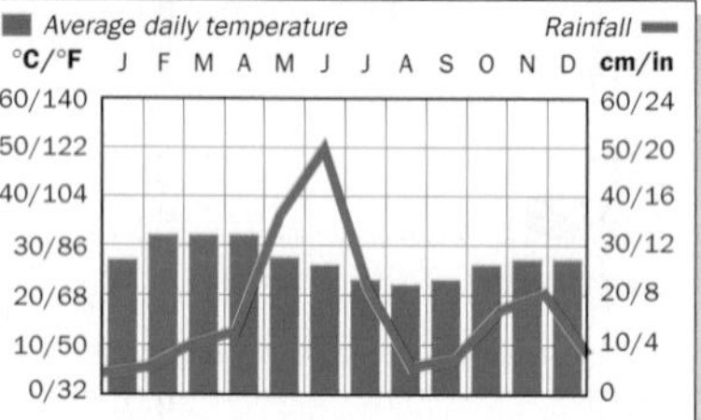
Average daily temperature
Rainfall
°C/°F
J F M A M J J A S O N D
cm/in
60/140
50/122
40/104
30/86
20/68
10/50
0/32
60/24
50/20
40/16
30/12
20/8
10/4
0

The south's four seasons – two rainy and two dry – merge in the north into a single wet season with lower rainfall.

TRANSPORTATION

Drive on right

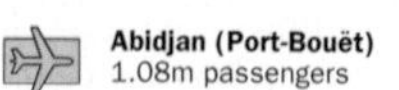
Abidjan (Port-Bouët)
1.08m passengers

33 ships
9200 grt

THE TRANSPORTATION NETWORK

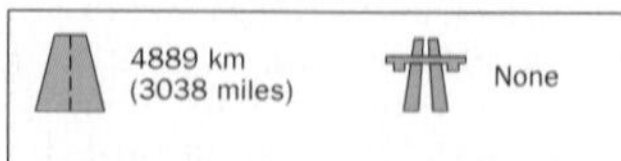
4889 km
(3038 miles)
None

1955 km (1215 miles)

980 km (609 miles)

The relatively good transportation system focuses on Abidjan, the premier port of francophone west Africa.

TOURISM

Visitors : Population 1:54

301,000 visitors
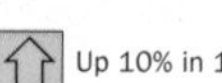
Up 10% in 1998

MAIN TOURIST ARRIVALS

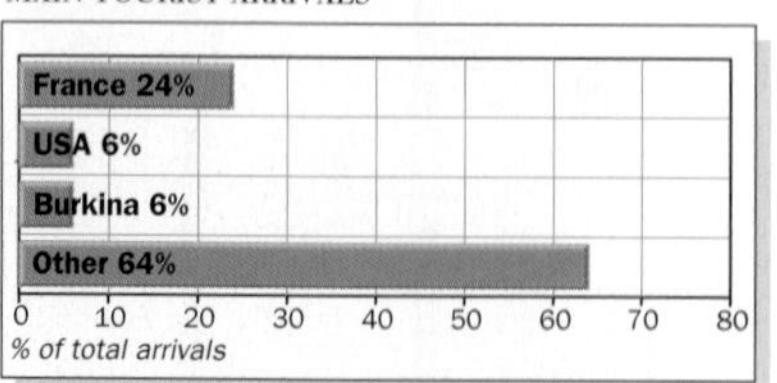
France 24%
USA 6%
Burkina 6%
Other 64%
0 10 20 30 40 50 60 70 80
% of total arrivals

Ambitious plans for an "African Riviera" east of Abidjan and the opening of a hotel by the French Club Méditerranée have helped increase tourism. The giant Roman Catholic basilica at Yamoussoukro is an attraction.

PEOPLE

Pop. density medium

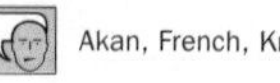
Akan, French, Kru, Voltaic

51/km² (133/mi²)

THE URBAN/RURAL POPULATION SPLIT

46% 54%

RELIGIOUS PERSUASION

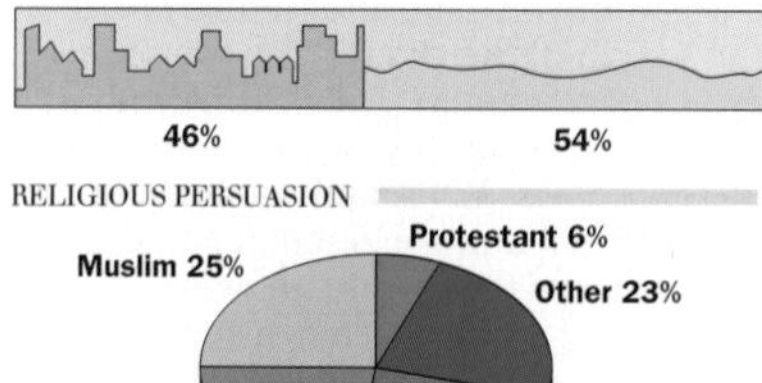
Protestant 6%
Muslim 25%
Other 23%
Roman Catholic 23%
Traditional beliefs 23%

There are more than 60 tribes in Ivory Coast. Larger groups among them can be recognized on the basis of cultural identity, the key ones being the Baoulé in the center, the Agri in the east, the Senufo in the north, the Dioula in the northwest and west, the Bété in the center-west, and the Dan-Yacouba in the west. President Houphouët-Boigny promoted his own group, the Baoulé, who account for 23% of the population.

The extended family is an important force, particularly in the shanty towns of Abidjan. Migrants from other west African countries account for up to 40% of the population.

POLITICS

Multiparty elections

2000/2005

President Laurent Gbagbo

AT THE LAST ELECTION

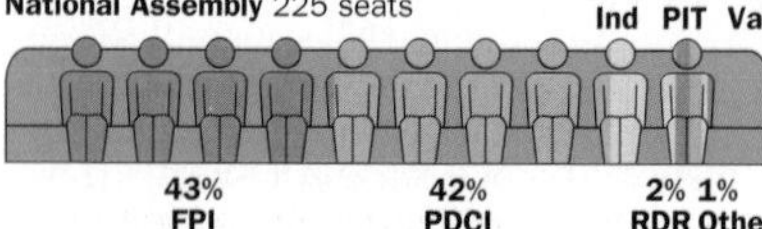
National Assembly 225 seats
10% Ind
1% PIT
1% Vac
43% FPI
42% PDCI
2% RDR
1% Others

FPI = Ivorian Popular Front **PDCI** = Democratic Party of Ivory Coast **Ind** = Independents **RDR** = Rally of the Republicans **PIT** = Ivorian Labor Party **Vac** = Vacant

A power struggle to replace President Houphouët-Boigny in 1993 was won by arch-conservative, and French-backed, National Assembly president Henri Konan Bédié. His main rival, the Muslim northerner Alassane Ouattara of the RDR, returned in 1999 to contest the presidency. Konan Bédié's attempts to prevent him from standing provoked a coup led by Gen. Robert Guei. Guei stood as a candidate in the 2000 presidential elections, while barring both Konan Bédié and Ouattara from the poll. A popular revolt was incited by Guei's attempts to falsify the results and he fled the country. Supporters of veteran socialist Laurent Gbagbo, who had led in the poll, then clashed violently with Ouattara's followers. Legislative elections in December were won by Gbagbo's FPI. A national reconciliation forum convened in October–December 2001 successfully promoted a dialog among the main political parties.

Ivory Coast

Total Land Area : 322 460 sq. km (124 502 sq. miles)

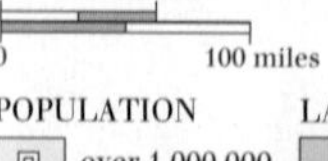
0 100 km
0 100 miles

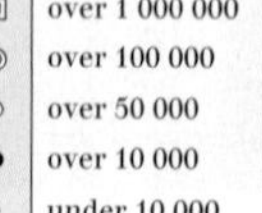
POPULATION
over 1 000 000
over 100 000
over 50 000
over 10 000
under 10 000

LAND HEIGHT
1000m/3281ft
500m/1640ft
200m/656ft
Sea Level

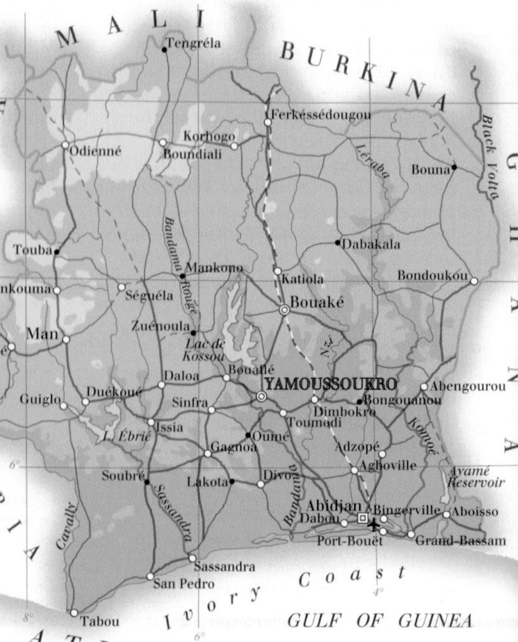
MALI
BURKINA
GUINEA
LIBERIA
GHANA
Tengréla
Ferkéssédougou
Korhogo
Boundiali
Odienné
Bouna
Black Volta
Léraba
Touba
Dabakala
Mankono
Katiola
Bondoukou
Biankouma
Séguéla
Bouaké
Mount Nimba 1752m
Man
Danané
Zuénoula
Lac de Kossou
Bouaflé
YAMOUSSOUKRO
Abengourou
Guiglo
Duékoué
Daloa
Sinfra
Bongouanou
Dimbokro
Toumodi
Issia
Oumé
L. Ébrié
Gagnoa
Adzopé
Agboville
Soubré
Lakota
Divo
Ayamé Réservoir
Abidjan
Dabou
Bingerville
Aboisso
Port-Bouët
Grand-Bassam
Sassandra
San Pedro
Tabou
Cavally
Bandama
Ivory Coast
GULF OF GUINEA
ATLANTIC OCEAN

WORLD AFFAIRS

Joined UN in 1960

Instability in other west African countries, such as neighboring Liberia, is a major concern. Relations with the main creditors, the World Bank group, and France, the chief source of private investment and bilateral loans, also form a major issue. Ivory Coast plays an important role in international cocoa and coffee organizations.

AID

Recipient

 $352m (receipts)

 Down 21% in 2000

France is the largest source of bilateral aid. Structural adjustment loans from the World Bank were particularly important in easing the acute burden of a debt accumulated on the strength of overinflated oil hopes.

DEFENSE

Compulsory military service

 $132m

 Up 2% in 2000

France is the main supplier of equipment and trainer of officers for the 6800-strong army. It also maintains a garrison near Abidjan. Ivory Coast has a large paramilitary gendarmerie.

ECONOMICS

Inflation 7.5% p.a. (1990–2000)

 $9.59bn

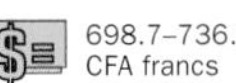 698.7–736.7 CFA francs

SCORE CARD

- World GNP Ranking..........85th
- GNP per Capita..........$600
- Balance of Payments..........–$13m
- Inflation..........2.5%
- Unemployment..........13%

Strengths

Well-developed agriculture: major cocoa and coffee producer. Relatively good infrastructure. Expanding oil and gas industries. Healthy foreign investment levels. Successful debt rescheduling.

Weaknesses

Failure to invest adequately in education and professional training. Overdependence on cocoa and coffee. Slave labor on plantations.

EXPORTS

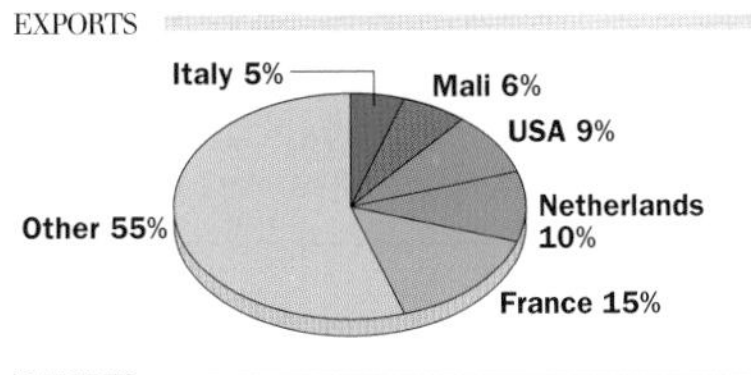
Italy 5%
Mali 6%
USA 9%
Netherlands 10%
France 15%
Other 55%

IMPORTS

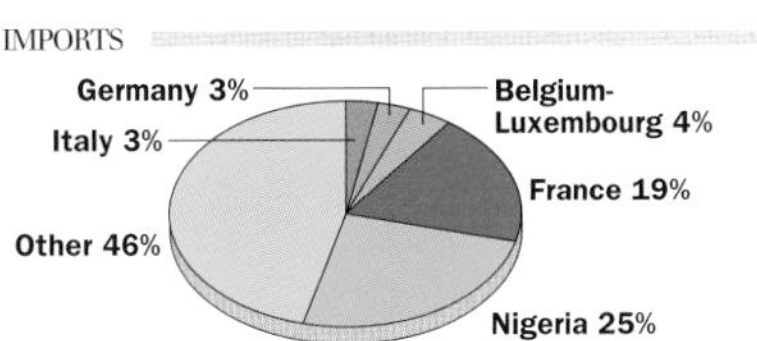
Germany 3%
Italy 3%
Belgium-Luxembourg 4%
France 19%
Nigeria 25%
Other 46%

RESOURCES

Electric power 1.2m kw

 77,000 tonnes

 25,704 b/d (reserves 103m barrels)

 1.45m sheep, 1.41m cattle, 29.4m chickens

Oil, diamonds, cobalt, gold, iron, manganese, nickel

Close to self-sufficiency in energy. Substantial offshore oil and gas reserves. Forest resources badly depleted.

ENVIRONMENT

Sustainability rank: 108th

6% (0.3% partially protected)

 0.9 tonnes per capita

The government imposed a ban on unprocessed timber exports in 1995 to protect Ivorian forests.

MEDIA

TV ownership medium

 Daily newspaper circulation 17 per 1000 people

PUBLISHING AND BROADCAST MEDIA

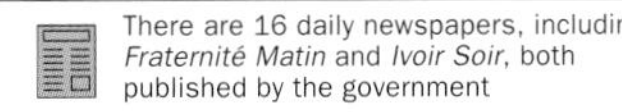 There are 16 daily newspapers, including *Fraternité Matin* and *Ivoir Soir*, both published by the government

 2 services: 1 state-owned, 1 independent

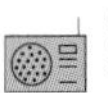 8 services: 1 state-owned, 7 independent

Although heavy censorship in Ivory Coast has eased since the early 1990s, there are still cases of official harassment of the media.

EDUCATION

School leaving age: 13

 47%

 96,681 students

Baccalauréat pass rates are low. Cuts in spending have triggered student protests. Primary education fees ended in 2001.

CRIME

No death penalty

 13,670 prisoners

 Up 32% 1992–1996

Foreign immigrants are often blamed for the widespread crime in Abidjan. Human rights abuses are common.

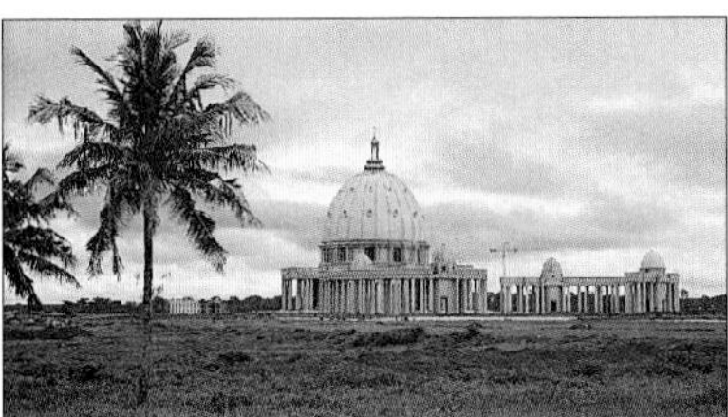

The basilica, Yamoussoukro. *Built in the new capital, Houphouët-Boigny's birthplace, it is modeled on St. Peter's, Rome.*

CHRONOLOGY

One of the great trading emporia of west Africa, the Ivory Coast was made a French colony in 1893. By 1918, the French had defeated the Malinke empire and the forest peoples of the interior.

- **1903–1935** Plantations developed.
- **1960** Félix Houphouët-Boigny declares independence.
- **1990** First contested polls: Houphouët-Boigny and PDCI win.
- **1993** Houphouët-Boigny dies.
- **1998** Power of president increased, Ouattara apparently barred from standing in presidential elections.
- **1999** Military coup by Gen. Guei.
- **2000** Guei ousted after false election victory claim. Gbagbo president.
- **2001** National Reconciliation Forum set up.

HEALTH

No welfare state health benefits

 1 per 10,000 people

 Malaria, communicable diseases, neonatal deaths

The incidence of HIV/AIDS is high, affecting 9.7% of adults as at end-2001. In 2001 drugs companies agreed to cut the prices of treatments by 80–90%.

SPENDING

GDP/cap. increase

CONSUMPTION AND SPENDING

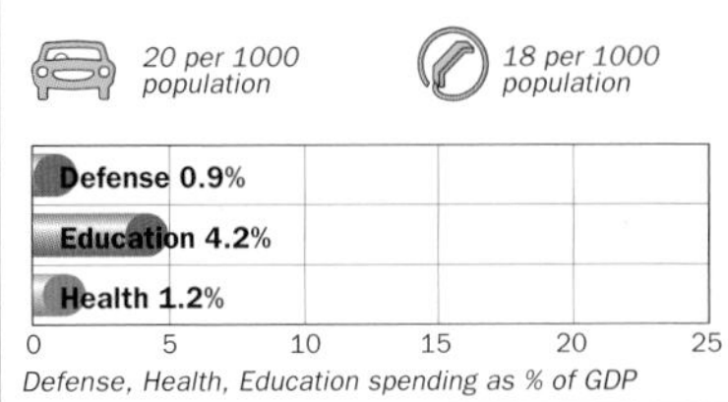
20 per 1000 population
18 per 1000 population
Defense 0.9%
Education 4.2%
Health 1.2%
0
5
10
15
20
25
Defense, Health, Education spending as % of GDP

A large bourgeoisie grew rich in the boom years. Urban living standards in Ivory Coast are better than in many African countries.

WORLD RANKING

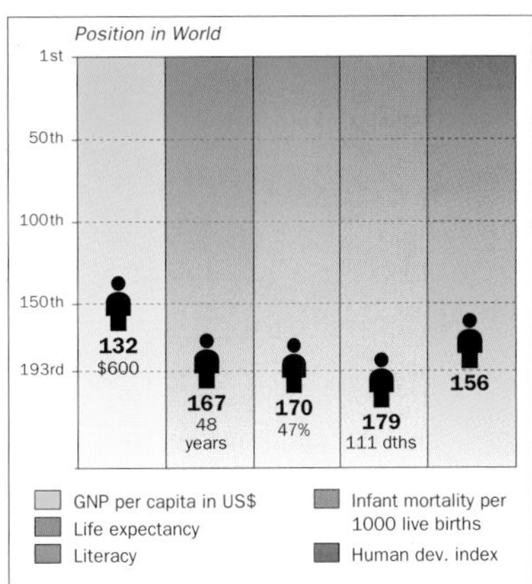
Position in World
1st
50th
100th
150th
193rd
132
$600
167
48 years
170
47%
179
111 dths
156
GNP per capita in US$
Life expectancy
Literacy
Infant mortality per 1000 live births
Human dev. index

I

JAMAICA

CARIBBEAN
Cuba
Haiti
Dominican Republic
JAMAICA
North America

OFFICIAL NAME: Jamaica **CAPITAL:** Kingston
POPULATION: 2.6 million **CURRENCY:** Jamaican dollar **OFFICIAL LANGUAGE:** English

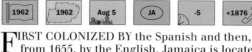

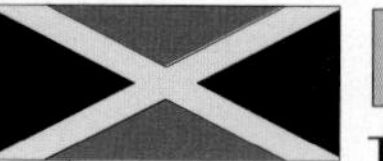

1962

1962

Aug 5
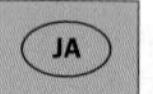
JA
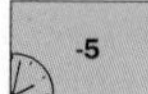
-5

+1876

.jm

FIRST COLONIZED BY the Spanish and then, from 1655, by the English, Jamaica is located in the Caribbean, 145 km (90 miles) south of Cuba. It was the first of the Caribbean island countries to become independent in the postwar years, and remains an active force in Caribbean politics. Jamaica is also influential on the world music scene; *reggae* and *ragga* (or dancehall) developed in the tough conditions of Kingston's poor districts.

CLIMATE

Tropical oceanic

WEATHER CHART FOR KINGSTON

Average daily temperature
Rainfall
°C/°F J F M A M J J A S O N D cm/in
40/104 30/86 20/68 10/50 0/32 -10/14 -20/-4
40/16 30/12 20/8 10/4 0

Tropical and humid conditions at sea level give way to temperate weather in mountain areas. Rainfall is seasonal, with marked regional variations.

TRANSPORTATION

Drive on left

Donald Sangster International, Montego Bay
2.99m passengers

9 ships
3600 grt

THE TRANSPORTATION NETWORK

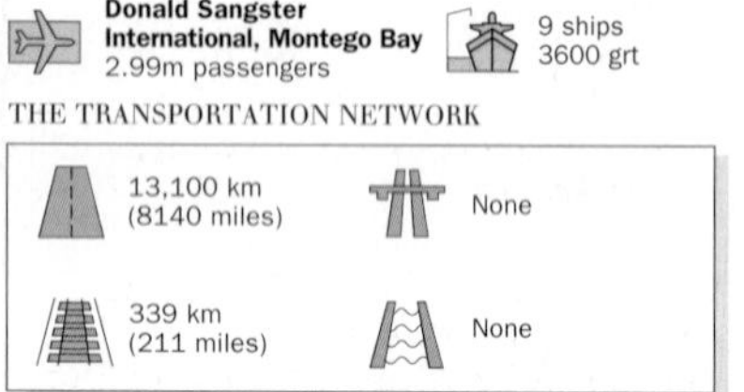
13,100 km
(8140 miles)
None
339 km
(211 miles)
None

Kingston's harbor has been expanded and its airport improved. The road network is extensive. Private buses provide public transportation.

TOURISM

Visitors : Population 1:2

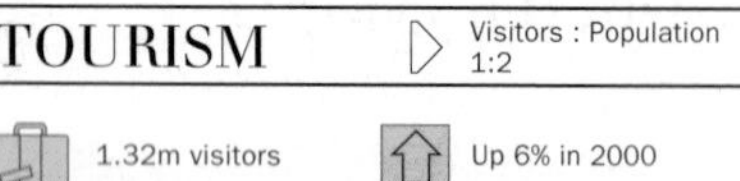
1.32m visitors
Up 6% in 2000

MAIN TOURIST ARRIVALS

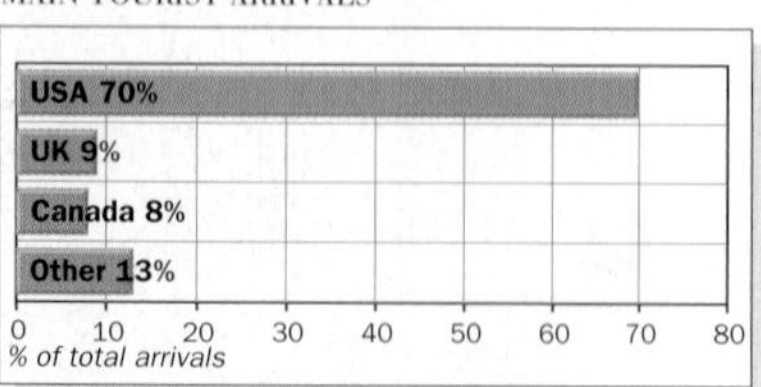
USA 70%
UK 9%
Canada 8%
Other 13%
0 10 20 30 40 50 60 70 80
% of total arrivals

Tourism is the major earner of foreign exchange. Most tourists stay in large, enclosed beach resorts. Recent bouts of social unrest have damaged the sector.

PEOPLE

Pop. density high

English Creole, English

240/km²
(622/mi²)

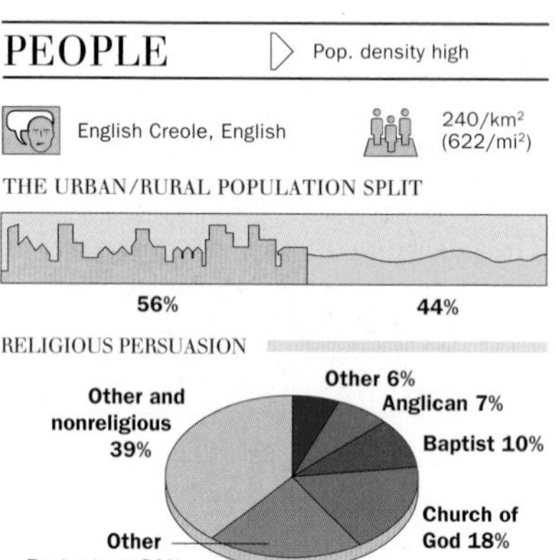
THE URBAN/RURAL POPULATION SPLIT
56%
44%
RELIGIOUS PERSUASION
Other 6%
Anglican 7%
Baptist 10%
Church of God 18%
Other Protestant 20%
Other and nonreligious 39%

Jamaicans are a broad ethnic mix. Most are the descendants of Africans brought to the island between the 16th and 19th centuries, but there are minorities of Europeans, Indians, Chinese, and Arabs. Jamaica is also home to the Rastafarians, followers of Haile Selassie, the former emperor of Ethiopia.

Most social tension is the result of the marked disparities in wealth. The Caribbean women's rights movement originated in Jamaica, and today many Jamaican women hold senior positions in economic and political life.

Although life revolves around the family, absentee fathers are common. Many career women are single parents by choice. Life in the ghettos of Kingston is often violent and based largely on gun law. Kingston slums have their own patois.

Bauxite mine and terminal, *Runaway Bay. Bauxite – from which aluminum is extracted – is the main source of foreign income.*

POLITICS

Multiparty elections

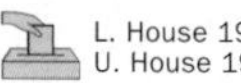
L. House 1997/2002
U. House 1998/2003
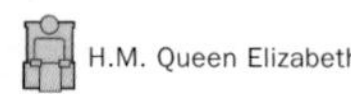
H.M. Queen Elizabeth II

AT THE LAST ELECTION

House of Representatives 60 seats

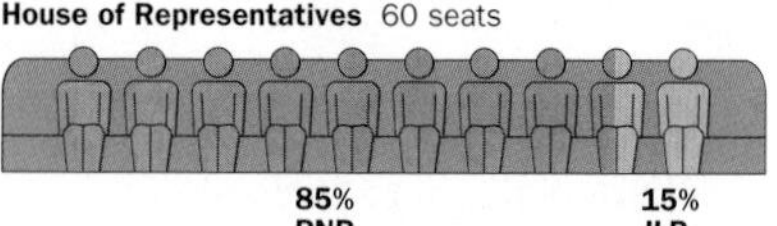
85% PNP
15% JLP

PNP = People's National Party **JLP** = Jamaica Labour Party

Senate 21 seats

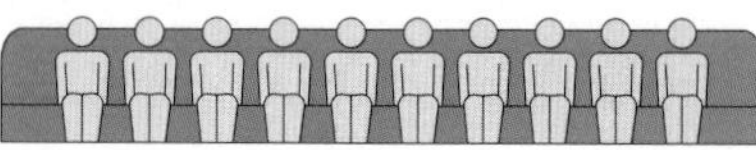

The members of the Senate are appointed. Thirteen members are chosen by the prime minister and 8 by the leader of the opposition

The country's political complexion changed markedly in the late 1980s, as the ideologies of the once socialist PNP and the conservative JLP converged toward a moderate free-market economic approach. A general election in 1997 gave the PNP government a third consecutive term in office. The party subsequently swept the local elections in 1998.

Violent disturbances in 1998 and 1999 were in response to the government's attempts to deal with economic recession and a large fiscal deficit. The unrest, which led to several deaths, gave new life to the internally troubled JLP, as it then identified itself with opposition to fuel tax increases.

WORLD AFFAIRS

Joined UN in 1962

ACS

Caricom

Geplac

Comm
OAS

Antidrugs cooperation with the US and future relations within Caricom and the Commonwealth predominate.

AID

Recipient

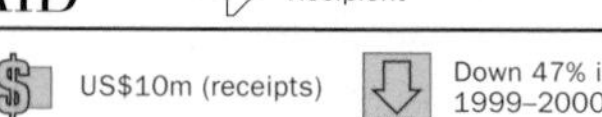
US$10m (receipts)
Down 47% in 1999–2000

Most aid comes from the US, the EU, and Japan. It includes both project loans and balance-of-payments support.

DEFENSE

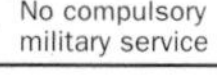
No compulsory military service

US$49m

Down 4% in 2000

Jamaica's defense force buys its arms from the US, but is trained by the UK. Today, the defense force is used against narcotics smugglers and to assist the police to break up unrest, as in 1999.

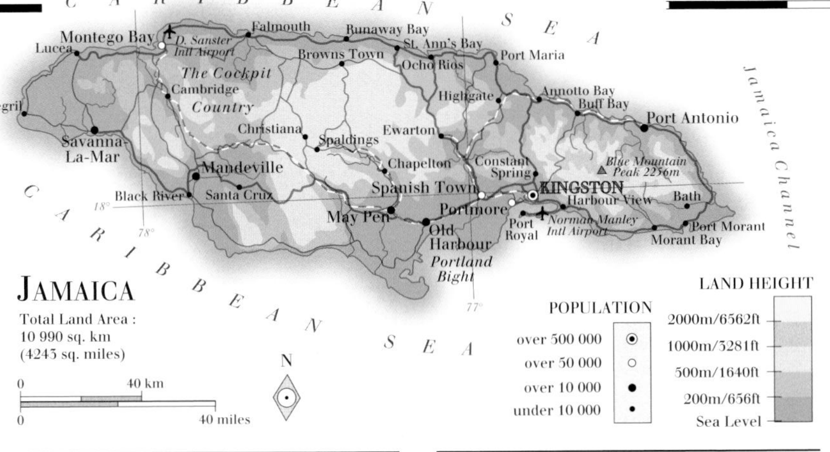
CARIBBEAN SEA
Montego Bay
Lucea
D. Sangster Intl Airport
Falmouth
Runaway Bay
St. Ann's Bay
Port Maria
Browns Town
Ocho Rios
The Cockpit Country
Cambridge
Highgate
Annotto Bay
Buff Bay
Port Antonio
Negril
Christiana
Ewarton
Savanna-La-Mar
Spaldings
Mandeville
Chapelton
Constant Spring
Blue Mountain Peak 2256m
Black River
Santa Cruz
Spanish Town
KINGSTON
Harbour View
Bath
May Pen
Portmore
Port Royal
Norman Manley Intl Airport
Port Morant
Morant Bay
Old Harbour
Portland Bight
Jamaica Channel
JAMAICA
Total Land Area : 10 990 sq. km (4243 sq. miles)
0 40 km
0 40 miles
N
POPULATION
over 500 000
over 50 000
over 10 000
under 10 000
LAND HEIGHT
2000m/6562ft
1000m/3281ft
500m/1640ft
200m/656ft
Sea Level

ECONOMICS

Inflation 24% p.a. (1990–2000)

 US$6.88bn

 45.10–47.05 Jamaican dollars

SCORE CARD

- ❑ WORLD GNP RANKING..........................98th
- ❑ GNP PER CAPITA ....US$2610
- ❑ BALANCE OF PAYMENTS................–US$275m
- ❑ INFLATION 8.2%
- ❑ UNEMPLOYMENT.................................16%

STRENGTHS

Relatively diversified economy. Tourism. Mining and refining of bauxite for aluminum. Agriculture, including sugar, bananas, rum, and coffee. Light manufacturing and data processing for US companies are growing sectors.

WEAKNESSES

Banking and insurance sectors. Financing of sugar production. Stagnant growth. High debt burden.

EXPORTS

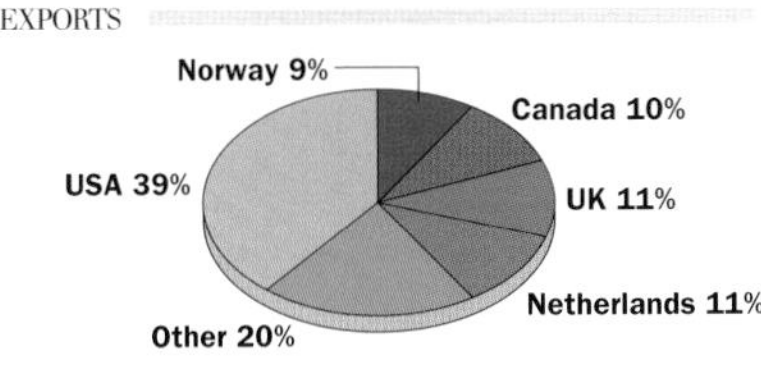
Norway 9%
Canada 10%
USA 39%
UK 11%
Netherlands 11%
Other 20%

IMPORTS

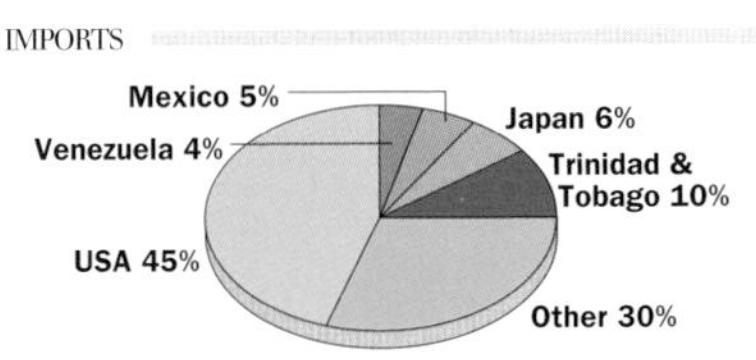
Mexico 5%
Venezuela 4%
Japan 6%
Trinidad & Tobago 10%
USA 45%
Other 30%

RESOURCES

Electric power 1.2m kw

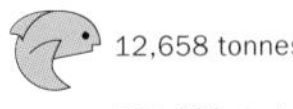 12,658 tonnes

 Not an oil producer; refines 32,000 b/d

440,000 goats, 400,000 cattle, 11m chickens

 Bauxite, marble, gypsum, silica, clay

Jamaica is the world's third-largest producer of bauxite. Sugar and bananas are major exports.

ENVIRONMENT

Sustainability rank: 122nd

0.1%

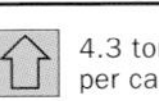 4.3 tonnes per capita

Acidic dust from bauxite processing is a major problem, as is urban pollution in Kingston and its bay. Broad-leaved tropical forests have largely disappeared.

MEDIA

TV ownership medium

Daily newspaper circulation 63 per 1000 people

PUBLISHING AND BROADCAST MEDIA

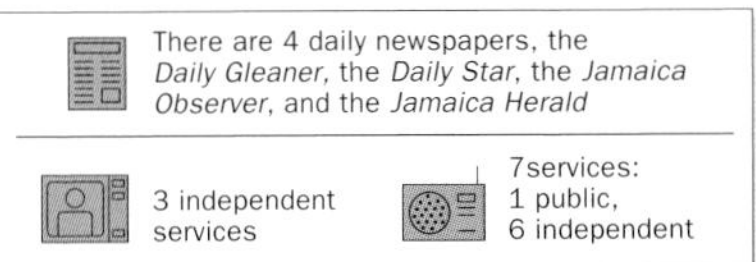
There are 4 daily newspapers, the *Daily Gleaner*, the *Daily Star*, the *Jamaica Observer*, and the *Jamaica Herald*
3 independent services
7services: 1 public, 6 independent

The government has loosened its hold on broadcasting. The Jamaican press is one of the most influential in the Caribbean.

CRIME

Death penalty in use

 4288 prisoners

 Down 18% 1996–1998

Armed crime is a major problem. Many murders are the result of armed robberies linked to narcotics gangs competing for territory. Much of the world crack trade is still controlled from Jamaica. Large areas of Kingston are ruled by violent gang leaders. The armed police are also frequently accused of the arbitrary shooting of suspects. An agreement to create a new Caribbean Court of Justice increased the likelihood of executions being carried out. The last hangings were in 1988 but sentences are still being imposed.

EDUCATION

School leaving age: 12

 87%

 21,217 students

Education is based on the former British 11-plus selection system. Jamaica hosts the largest of the three campuses of the University of the West Indies.

CHRONOLOGY

Spain occupied the island in 1510, wiping out the indigenous Arawak population. Britain seized it in 1655.

- ❑ **1958–1961** West Indies Federation.
- ❑ **1962** Independence under JLP.
- ❑ **1972** PNP elected. Reforms fail; street violence begins.
- ❑ **1980** Unpopular IMF austerity measures lead to JLP election win.
- ❑ **1989, 1993, 1997** PNP returned and austerity continues.
- ❑ **1999** Violent protests over fuel tax increases.

HEALTH

Welfare state health benefits

 1 per 714 people

Cerebrovascular and heart diseases, cancers, diabetes

The once-efficient state health service is now seriously underfunded. There are fewer doctors and nurses than in the 1980s. Hospitals generally have a shortage of drugs and there is only rudimentary medical equipment.

SPENDING

GDP/cap. increase

CONSUMPTION AND SPENDING

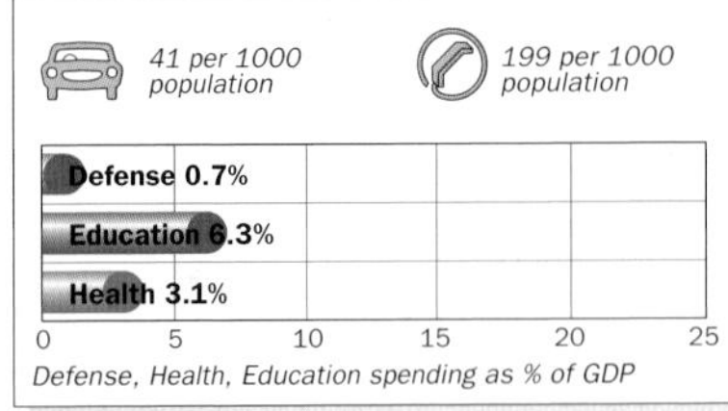
41 per 1000 population
199 per 1000 population
Defense 0.7%
Education 6.3%
Health 3.1%
0
5
10
15
20
25
Defense, Health, Education spending as % of GDP

Wealth disparities are highly marked in Jamaica, although better education has seen an increase in the number of Afro-Jamaicans taking more lucrative, white-collar jobs. The poorest in Jamaica, mostly migrants from rural areas, live in the slums of Kingston.

WORLD RANKING

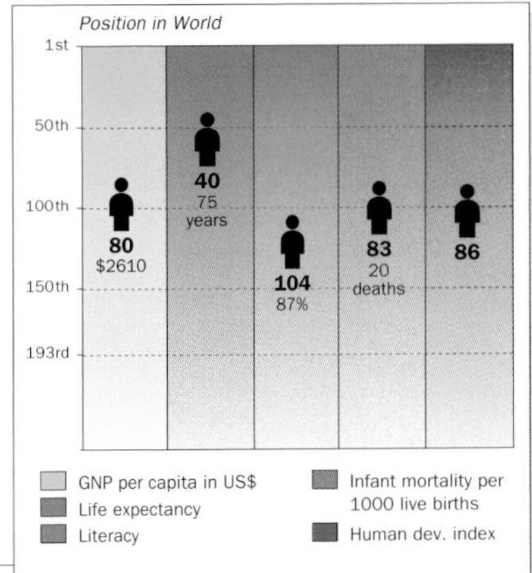
Position in World
1st
50th
100th
150th
193rd
80
$2610
40
75 years
104
87%
83
20 deaths
86
GNP per capita in US$
Life expectancy
Literacy
Infant mortality per 1000 live births
Human dev. index

J

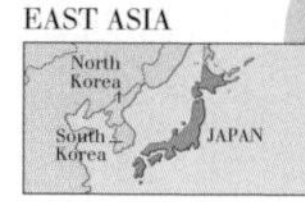
EAST ASIA
North Korea
South Korea
JAPAN

JAPAN

OFFICIAL NAME: Japan **CAPITAL:** Tokyo
POPULATION: 127.3 million **CURRENCY:** Yen **OFFICIAL LANGUAGE:** Japanese

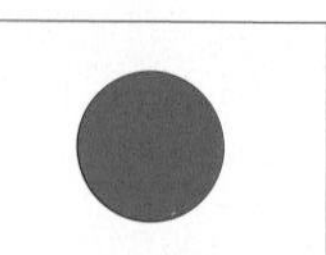

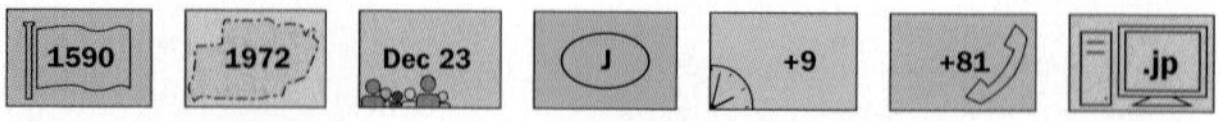

1590

1972

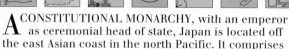

Dec 23

J

+9

+81

.jp

A CONSTITUTIONAL MONARCHY, with an emperor as ceremonial head of state, Japan is located off the east Asian coast in the north Pacific. It comprises four principal islands and more than 3000 smaller islands. Sovereignty over the most southerly islands in the Kurile chain is disputed with the Russian Federation. The terrain is mostly mountainous, with fertile coastal plains; over two-thirds is woodland. The Pacific coast is vulnerable to tsunamis – tidal waves triggered by submarine earthquakes. Most cities are located by the sea; Tokyo, Kawasaki, and Yokohama together constitute the most populous and heavily industrialized area. Hokkaido is the most rural of the main islands. Japan's power in the global economy, with annual trade surpluses exceeding $100 billion and massive overseas investments, has been shaken since the early 1990s by a series of bad debt crises, bankruptcies in the financial sector, and two recessions.

The Shinkansen "bullet train" *is the second fastest train in the world. Its speed is matched by its punctuality.*

CLIMATE

Continental/subtropical

WEATHER CHART FOR TOKYO

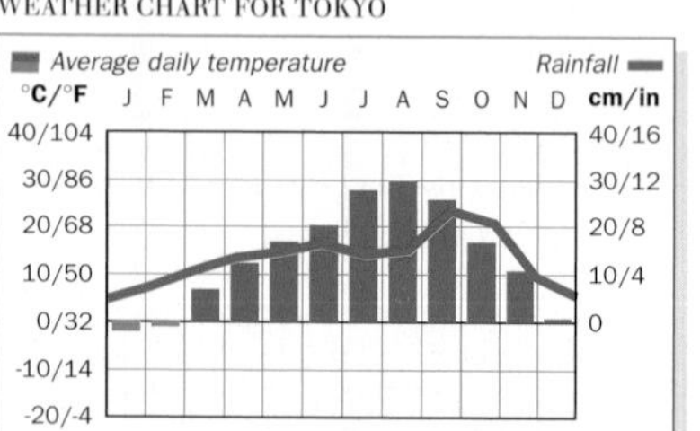
Average daily temperature
Rainfall
°C/°F
J F M A M J J A S O N D
cm/in
40/104
30/86
20/68
10/50
0/32
-10/14
-20/4
40/16
30/12
20/8
10/4
0

The Sea of Japan has a moderating influence on the climate. Winters are less cold than on the Asian mainland, and rainfall is much higher. Spring is perhaps the most pleasant season, with warm, sunny days without the sultry, oppressive heat and rainfall of the summer. Recent freak storms and heavy floods have raised concern over the implications of global climate change.

TRANSPORTATION

Drive on left

Haneda, Tokyo
56.4m passengers

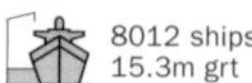
8012 ships
15.3m grt

THE TRANSPORTATION NETWORK

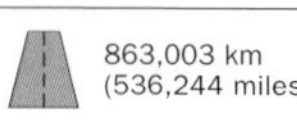
863,003 km
(536,244 miles)

6617 km
(4112 miles)

20,165 km
(12,531 miles)

1770 km
(1100 miles)

Railroads are the most important means of transportation in Japan. The Shinkansen, known in the West as the bullet train, is the second-fastest in the world. It is renowned as much for its reliability – timed to the second – as for its speed. The Tokyo–Sapporo air route is the busiest in the world.

JAPAN

Total Land Area : 377 835 sq. km
(145 882 sq. miles)

POPULATION

- over 5 000 000
- over 1 000 000
- over 500 000
- over 100 000
- over 50 000
- over 10 000

LAND HEIGHT

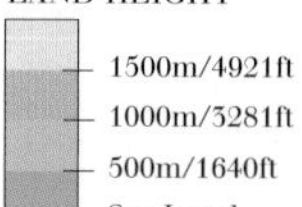
1500m/4921ft
1000m/3281ft
500m/1640ft
Sea Level

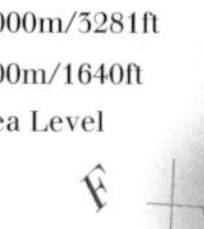

0 100 km
0 100 miles

Iturup
Kuril'sk
PACIFIC OCEAN
0 50 km
0 50 miles
HOKKAIDŌ
Kunashir
KURILE IS
Asahikawa
Kitami
Kussharo-ko
Asahi-dake 2290m
HABOMAI IS
Obihiro
Kushiro
Horoshiri-dake 2051m

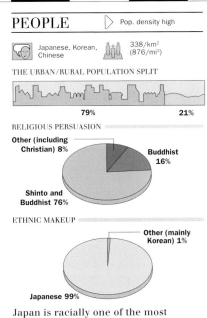

The Kurile islands are administered by the Russian Federation, but claimed by Japan

P A C I F I C O C E A N

TOURISM

Visitors : Population 1:27

4.76m visitors

Up 7% in 2000

MAIN TOURIST ARRIVALS

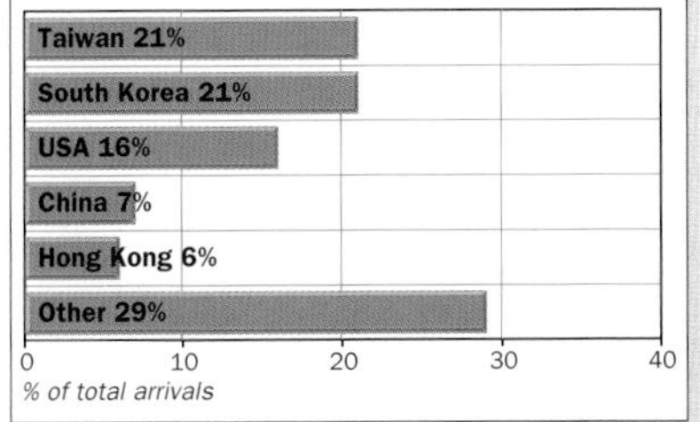
Taiwan 21%
South Korea 21%
USA 16%
China 7%
Hong Kong 6%
Other 29%
0
10
20
30
40
% of total arrivals

Japan is expensive for foreign tourists, despite reductions in the yen exchange rate. Most come from Taiwan, South Korea, the US, and, increasingly, China. The ancient imperial capital, Kyoto, and the temples and gardens of Nara are popular tourist destinations. Other attractions include the extraordinary variety of energetic high-tech urban living in Tokyo and Osaka. Traditional agricultural life can be found in rural areas such as Tohoku in northern Honshu. Wilderness areas of Hokkaido attract mainly Japanese climbers and hikers.

High Street, Ginza District, Tokyo. *Japan's well-policed cities are among the safest in the world.*

PEOPLE

Pop. density high

Japanese, Korean, Chinese

338/km² (876/mi²)

THE URBAN/RURAL POPULATION SPLIT

79% 21%

RELIGIOUS PERSUASION

Other (including Christian) 8%
Buddhist 16%
Shinto and Buddhist 76%

ETHNIC MAKEUP

Other (mainly Korean) 1%
Japanese 99%

Japan is racially one of the most homogeneous societies in the world. Its sense of order is reflected in the tradition of the lifetime employer. Many Japanese men define themselves by the company they work for rather than their job. An employer's influence stretches into employees' social time, and even to encouraging and approving marriages.

Traditionally women run the home and supervise the all-important education of their children. Many pursue careers until marriage, then continue to work part-time. However, trends are changing and some women are beginning to take on long-term careers, particularly in the medical and legal professions. Takako Doi became the country's first female political party leader in 1991.

Social form remains extremely important in Japanese society. Respect for elders and for social and business superiors is strongly ingrained. There is little tradition of generation rebellion, but the youth market is powerful and current fashions are geared toward teenagers. Many may still follow their parents' lifestyles, but established attitudes are under challenge. Working for the same company for life, and giving up evenings and weekends to entertain company clients, have become harder to justify since the economically turbulent 1990s.

POPULATION AGE BREAKDOWN

Female	Age	Male
2.2%	80+	1.1%
9.8%	60–79	8.1%
14.5%	40–59	14.3%
13.7%	20–39	14.1%
10.8%	0–19	11.4%

% of population by age group

CHRONOLOGY

Japan's tendency to limit its contacts with the outside world ended in 1853, when a US naval squadron coerced trading concessions from the last of the Tokugawa shoguns.

- **1868** Meiji Restoration; overthrow of Tokugawa regime and restoration of imperial power.
- **1872** Modernization along Western lines. Japan's strong military tradition becomes state-directed.
- **1889** Constitution modeled on Bismarck's Germany adopted.
- **1894–1895** War with China; ends in Japanese victory.
- **1904–1905** War with Russia; ends in Japanese victory. Formosa (Taiwan) and Korea later annexed.
- **1914** Joins World War I on Allied side. Sees limited naval action.
- **1919** Versailles peace conference gives Japan limited territorial gains in the Pacific.
- **1923** Yokohama earthquake kills 140,000.
- **1927** Japan enters period of radical nationalism, and introduces the notion of a "coprosperity sphere" in southeast Asia under Japanese control. Interpreted in the US as a threat to its Pacific interests.
- **1931** Chinese Manchuria invaded and renamed Manchukuo.
- **1937** Japan launches full-scale invasion of China proper.
- **1938** All political parties placed under one common banner; Japan effectively ruled by militarists.
- **1939** Undeclared border war with Soviet Union; Japan defeated.
- **1940** Fall of France in Europe; Japan occupies French Indo-China.
- **1941** US imposes total trade embargo, including oil, on Japan thereby threatening to stifle its military machine. Japan responds in December by launching attack on US fleet at Pearl Harbor and invading US, British, and Dutch possessions in the Pacific.
- **1942** Japan loses decisive naval battle of Midway.
- **1945** Huge US bombing campaign culminates in atomic bombing of Hiroshima and Nagasaki. Soviet Union declares war on Japan. Emperor Hirohito surrenders, gives up divine status. Japan placed under US military government with Gen. MacArthur installed in Japan as supreme commander of Allied Powers.
- **1947** New Japanese constitution: modeled on that of the US, but retains emperor in ceremonial role.
- **1950** Korean War. US army contracts lead to quick expansion of Japanese economy.

J

CHRONOLOGY *continued*

- **1952** Treaty of San Francisco. Japan regains independence. Industrial production recovers to 15% above 1936 levels.
- **1955** Formation of LDP, which governs for next 38 years.
- **1964** Tokyo Olympics. Bullet train (Shinkansen) inaugurated. Japan admitted to OECD.
- **1973** Oil crisis. Economic growth cut. Government-led economic reassessment decides to concentrate on high-tech industries.
- **1976** LDP shaken by Lockheed bribery scandal; in subsequent election it remains in power but loses outright majority for first time.
- **1979** Second oil crisis. Growth continues at 6% per year.
- **1980** Elections: restoration of LDP overall majority.
- **1982** Honda establishes first car factory in US.
- **1988** Japan becomes world's largest aid donor and overseas investor.
- **1989** Death of Emperor Hirohito. Recruit–Cosmos bribery scandal leads to resignation of Prime Minister Noburo Takeshita; replaced by Sosuke Uno, who is in turn forced to resign over sexual scandal.
- **1990** Tokyo stock market crash.
- **1991–1992** LDP torn by factional disputes, further financial scandals, and the issue of electoral reform.
- **1993** Reformists split from LDP and create new parties. Elections; LDP loses power. Morihiro Hosokawa becomes prime minister at head of seven-party coalition.
- **1994** Hosokawa resigns. Withdrawal of SDPJ causes collapse of coalition. New three-party coalition includes LDP and SDPJ. Opposition parties unified by creation of Shinshinto. Implementation of far-reaching political and electoral reforms designed to eradicate "money politics."
- **1995** Kobe earthquake kills more than 5000 people.
- **1996** Elections: LDP minority government. Copper trader Yasuo Yamanaka sentenced to eight years in prison for incurring losses of $2.6 billion while acting for the Sumitomo Corporation.
- **1997** Severe economic recession.
- **1998** Crisis over reform of banking and financial system.
- **2000** April, Prime Minister Keizo Obuchi falls into coma, replaced by Yoshiro Mori. June, LDP loses overall majority in general election.
- **2001** April, LDP turns to populist right-winger Junichiro Koizumi as prime minister; five women appointed to cabinet.

J

POLITICS

Multiparty elections

L. House 2000/2004
U. House 2001/2004

Emperor Tsegu no Miya Akihito

AT THE LAST ELECTION

House of Representatives 480 seats

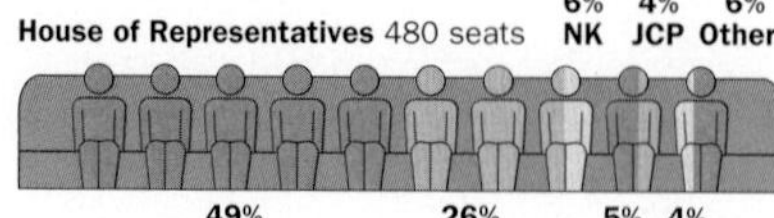
49% LDP
26% DPJ
5% LP
4% SDPJ
6% NK
4% JCP
6% Others

LDP = Liberal Democratic Party **DPJ** = Democratic Party of Japan **NK** = New Komeito **LP** = Liberal Party **JCP** = Japan Communist Party **SDPJ** = Social Democratic Party of Japan

House of Councillors 247 seats

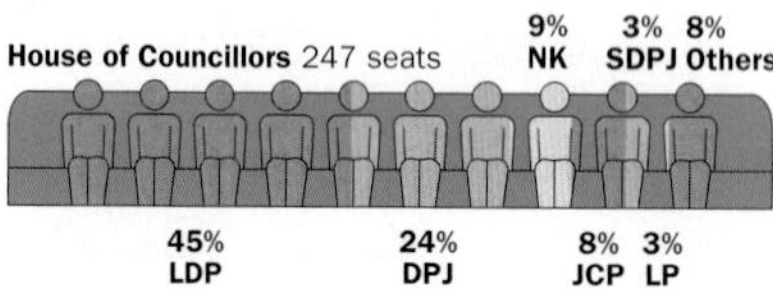
45% LDP
24% DPJ
8% JCP
3% LP
9% NK
3% SDPJ
8% Others

Japan is a multiparty democracy. The emperor has a purely ceremonial role.

Profile

Morihiro Hosokawa, who became the first non-LDP prime minister in 1993, was helped by not being associated with the tainted world of Tokyo politics. Four LDP prime ministers had resigned because they were implicated in scandals, or had failed to stamp out corruption. However, after becoming prime minister Hosokawa was also accused of financial irregularities, and resigned in 1994. His government, a fragile coalition, nevertheless laid the basis for electoral reform, apologized for Japan's war crimes, and began the process of institutional deregulation.

In 1996, the LDP returned to power in a coalition including the SDPJ, with Ryutaro Hashimoto as prime minister. Elections later that year were mainly a contest for the center-right vote. The LDP emerged as the largest party, ahead of Shinshinto, formed two years earlier by Ichiro Ozawa as a merger of opposition groups. Hashimoto formed a minority LDP government, but was replaced as LDP leader in mid-1998 by Keizo Obuchi. His coalition was shattered in April 2000 when Ozawa withdrew his repeatedly sidelined Liberal Party. Obuchi himself fell into a coma after suffering a stroke and died. His successor as prime minister and LDP leader, Yoshiro Mori, led a lack-luster campaign in a general election in 2000, from which the LDP emerged with a reduced representation but still as the largest party and the main force in government. Mori's unpopularity finally led to his replacement as LDP leader (and prime minister) in April 2001 by charismatic newcomer Junichiro Koizumi.

Junichiro Koizumi, *populist premier who has promised economic reform.*

Yoshiro Mori, *prime minister (2000–2001), but not liked by the voters.*

Emperor Akihito. *He acceded in 1989 on the death of his father, Hirohito.*

Traditional paddy field *in Hokkaido. Rice farming is among the most protected sectors of the Japanese economy.*

Main Political Issues

LDP strength and opposition weakness

Japanese postwar politics was until 1993 dominated by a system of patronage linking big business, the bureaucracy, and the ruling LDP. Numerous scandals and public disaffection with the LDP then saw the party briefly ousted from government. Far-reaching electoral reforms adopted in 1994 abolished multimember constituencies.

This weakened the system of LDP factions (although these later returned), striking at the "money politics" by which they were financed.

A shortlived unification of opposition groups, creating a Shinshinto alliance, then promised a realignment toward a two-party system. Adroit maneuvering, however, restored the LDP's position, which was bolstered by its hold on the centralized bureaucracy.

The military issue

The Japanese constitution enshrines pacifism. This prohibition of the use of force is a matter of hot debate within the country; critics argue that its economic power should be better reflected in foreign policy and the defense arena. Japan's role in the US-led "war on terrorism" from 2001 brought it a step closer to this aim.

WORLD AFFAIRS

Joined UN in 1956

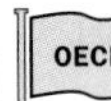
APEC
G8
IAEA
WTO
OECD

After years of limiting its role on the world stage, Japan began to make its influence felt in the 1990s. Its eventual aim is a seat on the UN's Security Council, which would be commensurate with its economic influence. Tentative moves were made in 1993, when Japanese forces joined UN peacekeepers in Cambodia, and in 2001, when Japan assisted in the US-led "war on terrorism." However, the lobby fearing a resurgence of militarism in Japan is still strong. Relations with the West are seriously strained over Japan's continuing to carry out "scientific" whale hunts. In Asia, Japan remains burdened by the legacy of its wartime aggression, exacerbated by revision of school history texts downplaying its crimes.

AID

Donor

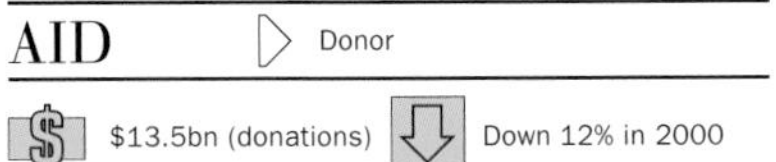
$13.5bn (donations)
Down 12% in 2000

Japan's aid donations are the largest of any single country in the world. Most aid goes to Asia and the Pacific, particularly China. Polynesian islands are heavily dependent on Japanese aid in support of their main livelihood, fishing. In 2001, Japan admitted to "buying" support for whaling.

DEFENSE

No compulsory military service

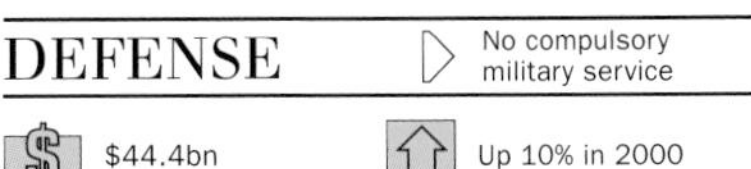
$44.4bn
Up 10% in 2000

JAPANESE ARMED FORCES

1050 main battle tanks (840 Type-74, 210 Type-90)	148,700 personnel
16 submarines, 12 frigates, 42 destroyers, 3 patrol boats	44,200 personnel
297 combat aircraft (F-1, F-4EJ, F-15J/DJ)	45,400 personnel
None	

Article 9 of the Japanese constitution renounces war as a means of settling international disputes, and in the postwar period any military activity in Japan has aroused fierce debate. Involvement even in UN peacekeeping duties, which Japan first undertook in Cambodia in 1993, is hotly contested by pacifists. Japan's Self-Defense Forces, however, have grown quite large. Since 1999 force has been used to deter intrusions by North Korean vessels, sparking fresh debate. A bill was passed in 2001 allowing noncombat assistance specifically to the "war on terrorism."

ECONOMICS

Inflation 0.1% p.a. (1990–2000)

$4519bn

114.20–131.06 yen

ECONOMIC PERFORMANCE INDICATOR

Consumer Price Index
GDP
Consumer price index 1995=100
GDP 1996=100
160
140
120
100
80
1996
1997
1998
1999
2000

STRENGTHS

Established market leader in high-tech products and cars. Commitment to long-term research. Talent for capitalizing on imported ideas. Manufacturing plants already established in the West. Domestic economy heavily protected from outside competition. Weak yen has promoted exports.

WEAKNESSES

Recent recession. Dependence on oil imports. Secretive and debt-ridden financial system. Falling industrial production, high-profile bankruptcies, and record unemployment levels. Trade surplus damages international relations.

PROFILE

Once among the world's strongest performing economies, Japan's strengths have been overshadowed for the last decade by its growing weaknesses. The country entered a brief recession for the second time in five years in 2001.

The 1990 crash of the Tokyo stock market marked the end of a period of remarkable growth. The government managed to spend its way out of disaster, effectively delaying the full impact of the downturn.

In an attempt to appease Western discontent over Japan's huge trade surplus, the government encouraged a move away from Japan's dependence on export revenues through stimulation of the domestic economy. However, the contracting trade balance remains above $100 billion a year. Elsewhere the Koizumi government has attempted to promote greater flows of bilateral trade by steering Japan into its first free trade agreement: with Singapore from April 2002.

However, the domestic situation remains critical. The financial sector is in desperate need of reform. The prominent corporate collapses of 1997 were repeated in 2001, with record losses reported across the high-tech industries in particular. Koizumi has promised radical change, rejecting the standard increase in government spending in favor of basic structural reform. Banks' bad loans have been cleared and the system of privileged "special public institutions" has been overhauled.

Nonetheless, critics have slated Koizumi's plans as overcautious, and have pointed to the increasing pressure from traditional economic power brokers which has put a brake on the pace of change.

SCORE CARD

❑ WORLD GNP RANKING	2nd
❑ GNP PER CAPITA	$35,620
❑ BALANCE OF PAYMENTS	$117bn
❑ INFLATION	–0.6%
❑ UNEMPLOYMENT	5%

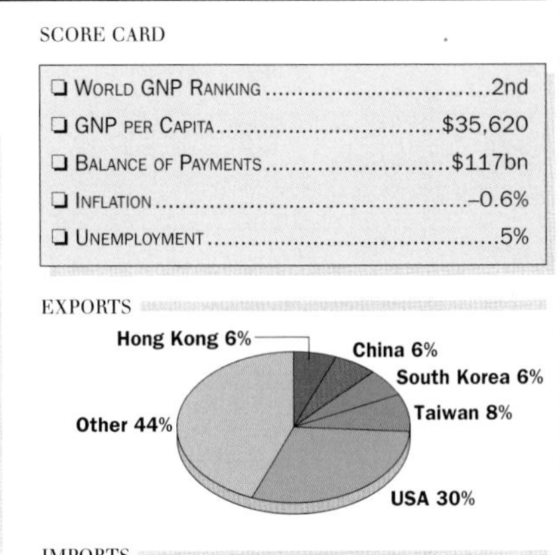
EXPORTS
Hong Kong 6%
China 6%
South Korea 6%
Taiwan 8%
Other 44%
USA 30%
IMPORTS
South Korea 5%
Taiwan 5%
Indonesia 4%
China 15%
Other 52%
USA 19%

J

JAPAN : MAJOR BUSINESSES

Research & development
Vehicle manufacture
Heavy engineering
Consumer goods
Shipbuilding
Iron & steel
Electronics
Chemicals
Brewing
Textiles
Computers
Banking & Finance
Sapporo
Toyama
Kóbe
Hitachi
Hiroshima
Tokyo
Kitakyúshu
Yokohama
Nagoya
Nagasaki
Kyóto
Osaka
0 300 km
0 300 miles

RESOURCES

Electric power 245m kw

5.94m tonnes

9023 b/d (reserves 53m barrels)

9.79m pigs, 4.53m cattle, 297m chickens

Limestone, sulfur, coal

ELECTRICITY GENERATION

Hydro	10% (103bn kwh)
Combustion	58% (608bn kwh)
Nuclear	32% (332bn kwh)
Other	0%

% of total generation by type

Japan has few exploitable resources. High production costs have made it the world's largest coal importer. In an attempt to reduce dependence on imported fuels, Japan has developed alternative energy sources. It is now the world's third-biggest generator of nuclear power. However, environmentalists strongly oppose any expansion of this sector. Nuclear safety became a priority issue after a serious accident at the Tokaimura plant in 1999.

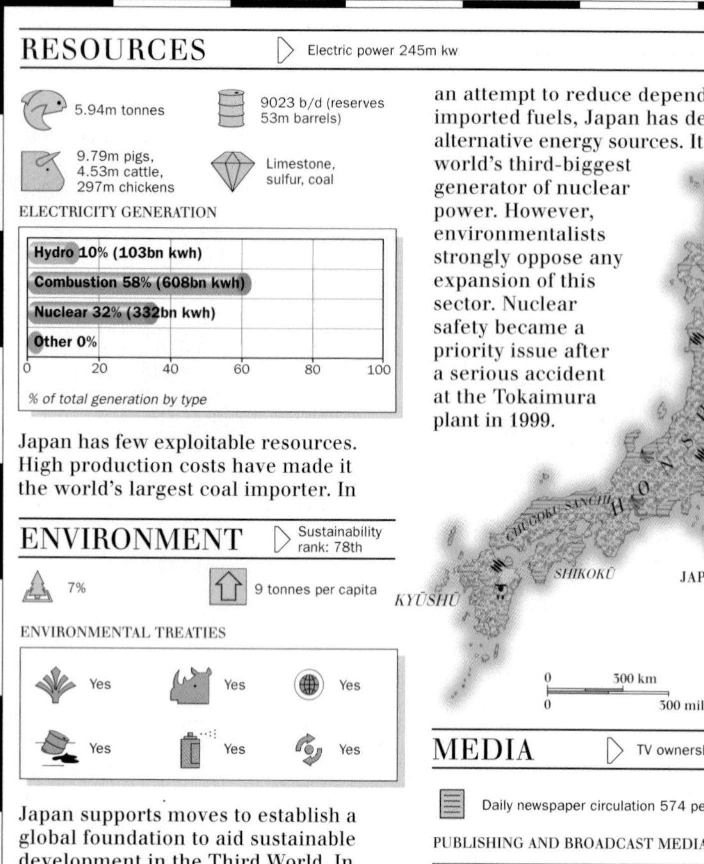
HOKKAIDŌ
HONSHU
CHUGOKU-SANCHI
SHIKOKU
KYŪSHŪ
JAPAN : LAND USE
Cropland
Forest
Pasture
Sheep
Fruits
Rice
0 300 km
0 300 miles

ENVIRONMENT

Sustainability rank: 78th

7%

9 tonnes per capita

ENVIRONMENTAL TREATIES

Yes · Yes · Yes

Yes · Yes · Yes

Japan supports moves to establish a global foundation to aid sustainable development in the Third World. In 1997 it played host to the Kyoto climate conference, although it only agreed to a modest cut in its "greenhouse gas" emissions. It faces strong criticism for its consumption of tropical timber, overfishing, and continuing to catch whale species under the umbrella of "scientific research."

Traditional Japanese respect for nature has spawned a vigorous grassroots ecological movement which prevented a second runway at Tokyo's Narita airport, and opposes nuclear power expansion and waste processing. The most serious environmental disasters have been a nuclear accident at Tokaimura in 1999 and the breakup in early 1997 of a Russian oil tanker along Japan's western shoreline.

***Datsetsusan National Park**, Hokkaido. Japan's northerly island is the least populous of the main group.*

MEDIA

TV ownership high

Daily newspaper circulation 574 per 1000 people

PUBLISHING AND BROADCAST MEDIA

There are 122 daily newspapers. *Asahi Shimbun*, *Mainichi Shimbun*, and *Yomiuri Shimbun* are among the most popular

128 services: 1 publicly owned, 127 commercial

100 services: 1 publicly owned, 99 commercial

The Japanese are among the world's most avid newspaper readers. Major papers are issued in simultaneous editions in the main urban centers. Most dailies are owned by large media groups who also have TV and cable interests. Weekly newspapers carry more tabloid journalism. *Manga,* Japanese comics, are hugely popular, with their characteristic artwork influencing design and art across Japanese culture. They now account for 40% of all published material in Japan: the most popular title, *Jump*, sells over six million copies a week.

Japanese technology has defined the world's media. Along with the personal stereo, Japanese companies effectively created the huge international computer games market. Nintendo, a leading games company, is among the most profitable in Japan. Ironically, the Internet was slow to take off, although by 2001 Japan had the third-highest number of people "online." That year also saw Japan launch the world's first "third-generation" mobile phone service.

EDUCATION

School leaving age: 15

99%

3.13m students

THE EDUCATION SYSTEM

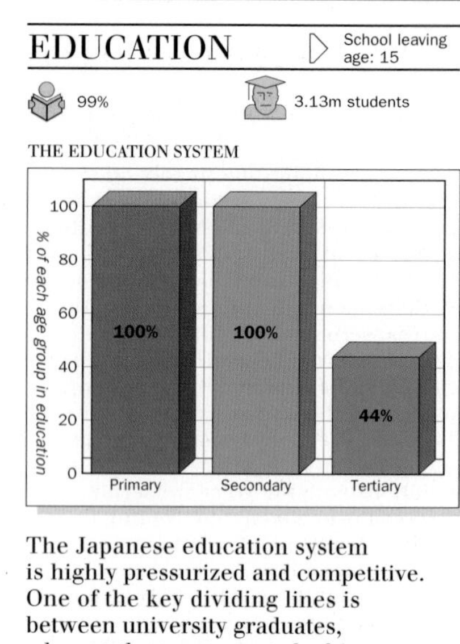
100
80
60
40
20
0
% of each age group in education
100%
100%
44%
Primary
Secondary
Tertiary

The Japanese education system is highly pressurized and competitive. One of the key dividing lines is between university graduates, who get the most coveted white-collar jobs, and nongraduates, who have difficulty reaching management level.

Competition for university places is intense, and starts with the choice of kindergarten, which the Japanese attend from the age of four. Academic pressure diminishes once at university. Graduates from Tokyo, Kyoto, Waseda, and Keio, which are the most prestigious universities, have access to top civil service and business jobs. The system succeeds in producing a uniformly well-educated workforce. However, it has also been criticized for not fostering individual responsibility, flexibility, or entrepreneurship.

CRIME

Death penalty in use

56,133 prisoners

Up 6% in 1999

CRIME RATE

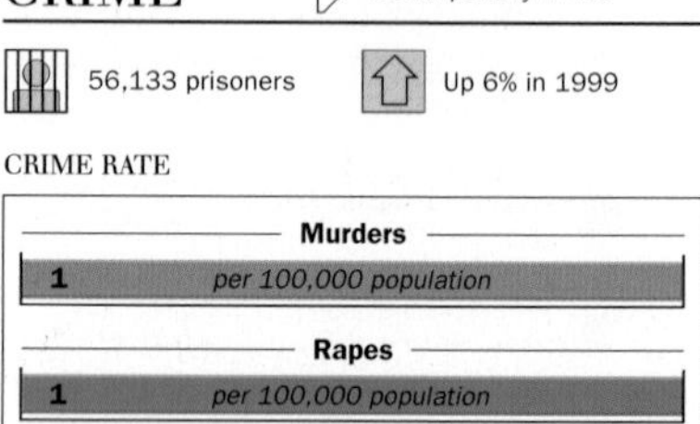
Murders
1 per 100,000 population
Rapes
1 per 100,000 population
Thefts
1508 per 100,000 population

Japan has one of the Western world's lowest crime rates, despite petty crime levels being at a 50-year high. Cities are safe, with police kiosks at frequent intervals on street corners. However, crime is involving more young people, and narcotics abuse is increasing.

The major crime problem is fraud and the activities of the kumi, organized Mafia-style syndicates. The authorities have been reluctant to challenge these groups, seeking to contain rather than halt their activities. Kumi are suspected of having connections with the political extreme right.

J

THE 2002 WORLD CUP: A GAME OF TWO HALVES

SEEKING BOTH AN economic boost, and a way of improving rocky relations across the Korea Strait, Japan successfully bid to cohost the world's premier soccer tournament, the FIFA World Cup, with South Korea in 2002. It was the first World Cup to be held in Asia and the first to be staged by two countries. Japan's bid ignored the relative infancy of soccer in the country, where it only recently became a professional game. The Japanese Professional Football League itself (popularly known as the J League) has been in existence only since 1993, prompted by the return in 1986 of Japan's first professional player, Yasuhiko Okudera, who had been playing in the West German Bundesliga.

***The Sapporo Dome** features a "floating" pitch, which is kept outside when not in use.*

CAUGHT OFFSIDE

The principle behind sharing the monthlong competition was to spread both the cost and the expected economic rewards. However, the 2002 World Cup was by far the most costly ever staged in the tournament's 72-year history and the much-vaunted financial benefits failed to materialize. Under FIFA rules at least eight stadiums must be supplied for each tournament. Japan and its neighbor went well beyond this minimum requirement. Each built eight brand new arenas from scratch, refurbished a further two, and upgraded local facilities and infrastructure. The cost for Japan alone was estimated at $4.5 billion. To offset this outlay the authorities predicted rewards totaling somewhere near $30 billion from ticket sales, advertising, and "soccer tourists." In the event, although companies selling televisions saw a 38% increase in trade, and imports of foreign beer tripled, actual revenues were far less spectacular. An embarrassing mix-up with ticketing left each game being played before a significant number of empty seats. Advertisers declined to risk the early morning/midnight exposure which the time difference imposed on key Western audiences, and "soccer tourists" came in far fewer numbers than the extravagant 800,000 predicted. Those who did come spent little, and left quickly. Of the stadiums constructed in Japan, few are likely to play host to professional soccer again, leaving remote communities with annual bills of over $2 million in maintenance for impressive but now empty arenas.

A "Corean" fan at the World Cup in 2002.

THE FAILED FRIENDLY

The World Cup had also been expected to contribute to a warming of relations between its ambitious cohosts. However, in the year leading up to the opening match the intensified press attention seemed only to put greater strain on the two neighbors. Junichiro Koizumi's new populist administration in Japan stirred Korean ire by staging visits to the Yasakuni war shrine (which includes Japanese war criminals among those it commemorates) and by approving "revisionist" school textbooks which, in Korean eyes, glossed over the Imperial army's war crimes in occupied east Asia during World War II. Korean fans took the opportunity presented by the Cup to publicize the popular campaign to reinstate their country's English-language spelling as "Corea." The "K" spelling, they claim, came about through a conspiracy by their imperial overlords after 1910, designed to give Japan precedence in alphabetical ranking.

Although the ailing Japanese economy saw little overall benefit, and relations with Korea (or Corea) failed to transcend traditional enmities, the tournament itself was celebrated around the world. Expected underdogs, including Japan and South Korea, did well enough to challenge the supposed supremacy of European and South American teams. In the end, however, the competition had its most fitting finale as the world's most celebrated team, Brazil, won through to take the title for a record fifth time.

HEALTH

Welfare state health benefits

1 per 526 people

Heart and circulatory diseases, cancers, accidents

Japan's health care system, which is ranked by WHO as the best in the world, delivers some of the highest longevity and lowest infant mortality rates. The poorest in society receive free treatment; expensive high-tech hospital facilities can also offer the latest techniques. Contributory national health insurance is based on earnings-related premiums, and the cost of medical care for the elderly and the self-employed is subsidized, though the rapidly aging population presents a major future funding challenge.

SPENDING

GDP/cap. increase

CONSUMPTION AND SPENDING

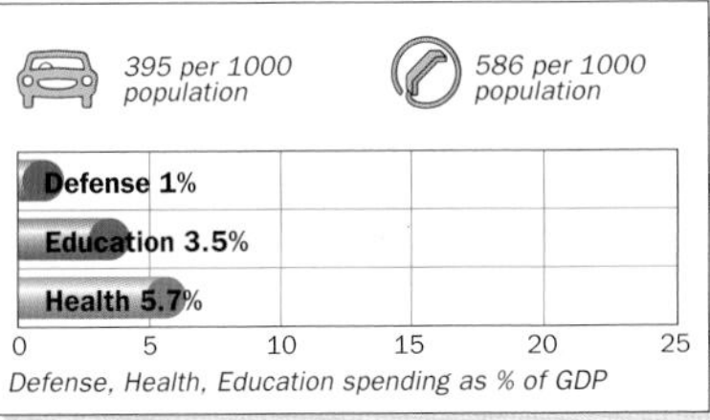

395 per 1000 population
586 per 1000 population
Defense 1%
Education 3.5%
Health 5.7%
0
5
10
15
20
25
Defense, Health, Education spending as % of GDP

Measured in consumer goods, the Japanese are wealthy; car ownership is only low because city parking is so restricted. Most households have substantial savings, enabling them to withstand economic recession.

The country's wealthiest men remain extremely wealthy; the fortunes of the top ten averaged $7 billion in 2000. The richest, software king Masayoshi Son, tripled his fortune in one year, to $19.4 billion. Tokyo living costs are high and most who work there live outside the city center, facing a long, cramped commuter journey. Girls and young women still living in their parents' homes are one group with a high disposable income.

WORLD RANKING

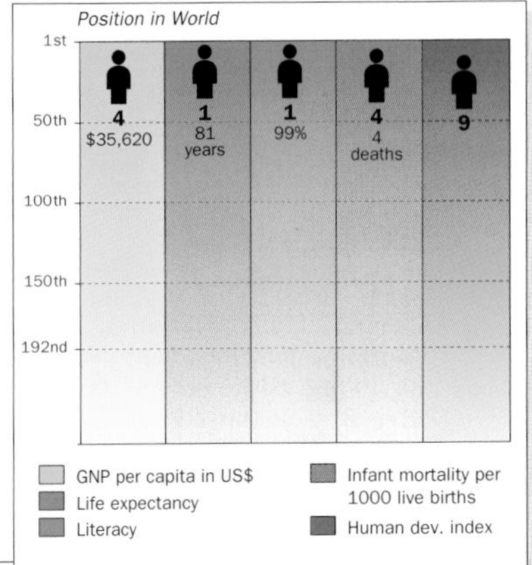

Position in World
1st
50th
100th
150th
192nd
4
$35,620
1
81 years
1
99%
4
4 deaths
9
GNP per capita in US$
Life expectancy
Literacy
Infant mortality per 1000 live births
Human dev. index

JORDAN

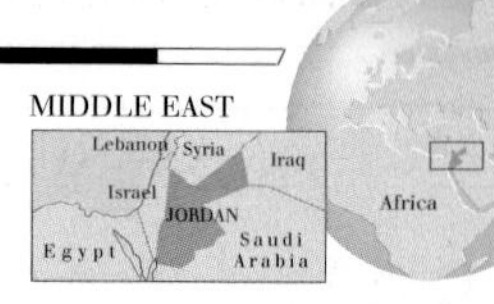
MIDDLE EAST
Lebanon
Syria
Iraq
Israel
JORDAN
Egypt
Saudi Arabia
Africa
Asia

OFFICIAL NAME: Hashemite Kingdom of Jordan **CAPITAL:** Amman
POPULATION: 5.1 million **CURRENCY:** Jordanian dinar **OFFICIAL LANGUAGE:** Arabic

1946

1967

May 25

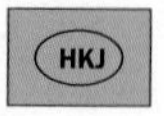
HKJ
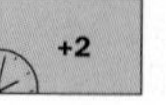
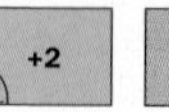
+2

+962

.jo

SURROUNDED ON ALL SIDES by the deserts of the Middle East, Jordan has just 26 km (16 miles) of maritime coastline on the Gulf of Aqaba. The vast majority of the population lives in the northwest, on the east bank of the Jordan River. Jordan ceded its claim to the West Bank of the river to the aspiring Palestinian state in 1988. Tourism, associated with important historical sites such as Petra, and phosphates are the mainstays of the economy.

CLIMATE

Hot desert/steppe/ Mediterranean

WEATHER CHART FOR AMMAN

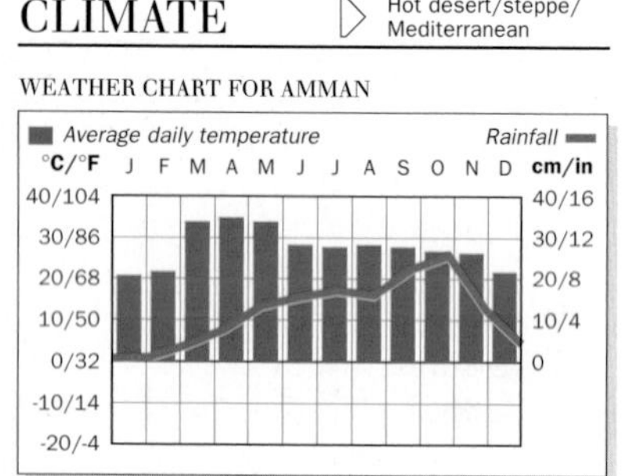
Average daily temperature
Rainfall
°C/°F J F M A M J J A S O N D cm/in
40/104
30/86
20/68
10/50
0/32
-10/14
-20/-4
40/16
30/12
20/8
10/4
0

Summers are hot and dry, winters cool and wet. Areas below sea level are very hot in summer and warm in winter.

TRANSPORTATION

Drive on right

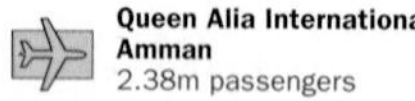
Queen Alia International, Amman
2.38m passengers

9 ships
41,800 grt

THE TRANSPORTATION NETWORK

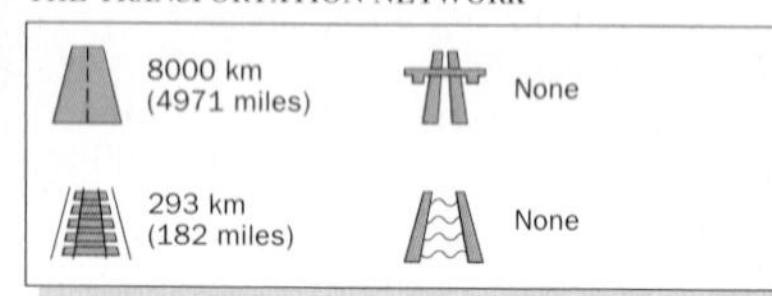
8000 km
(4971 miles)
None
293 km
(182 miles)
None

Adequate roads link main cities. A railroad links the port of Aqaba with the Syrian capital, Damascus.

TOURISM

Visitors : Population 1:3.5

1.48m visitors

Up 4% in 2001

MAIN TOURIST ARRIVALS

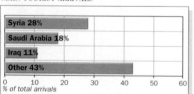

	% of total arrivals
Syria	28%
Saudi Arabia	18%
Iraq	11%
Other	43%

0 10 20 30 40 50 60
% of total arrivals

Aqaba offers fine beaches, water sports, and scuba diving, while the ancient city of Petra attracts those visitors interested in Nabatean remains. Amman is developing as a center for Arab culture and the arts.

PEOPLE

Pop. density medium

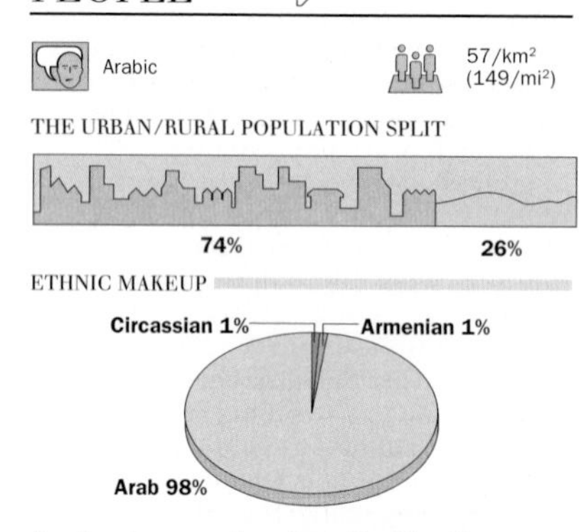
Arabic
57/km²
(149/mi²)
THE URBAN/RURAL POPULATION SPLIT
74%
26%
ETHNIC MAKEUP
Circassian 1%
Armenian 1%
Arab 98%

Jordan is a predominantly Muslim country drawn from Bedouin roots, with a Christian minority. About half the population are Palestinian in origin. The monarchy's power base lies among the rural tribes, which also provide the backbone of the military. National identity is strong.

POLITICS

Multiparty elections

L. House 1997/2002
U. House 2001/2005

H.M. King Abdullah II

AT THE LAST ELECTION

House of Deputies 80 seats

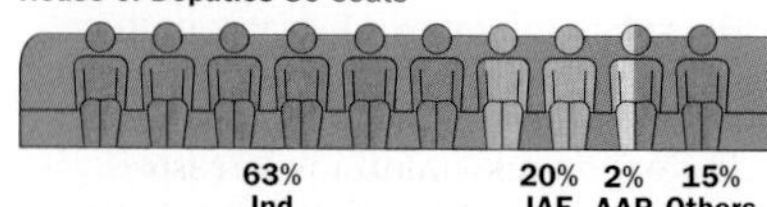

63% Ind 20% IAF 2% AAP 15% Others

Ind = Independents **IAF** = Islamic Action Front
AAP = Al-Ahd Party

Senate 40 seats

The members of the Senate (Majlis Al-Aayan) are appointed by the king

King Abdullah II acceded to the throne in February 1999 upon the death of his father, King Hussein. Although lacking in political experience, he is respected by the army and enjoys the support of Jordan's tribal leaders. Multiparty elections, initiated in 1993, have benefited pro-government parties, despite a strong Islamist opposition lobby. The appointment in 2000 of Prime Minister Ali Abu al-Ragheb marked a shift toward a modernizing and pro-business government.

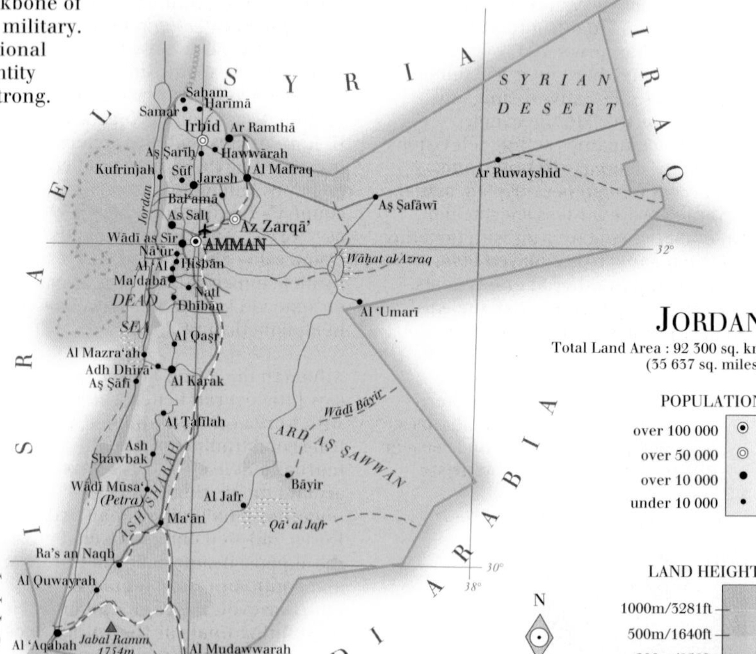
SYRIA
IRAQ
SAUDI ARABIA
ISRAEL
EGYPT
SYRIAN DESERT
Saham
Samar
Harimā
Irbid
Ar Ramthā
Aş Şarīh
Hawwārah
Kufrinjah
Şūf
Jarash
Al Mafraq
Ar Ruwayshid
Bal'amā
Aş Şafāwī
As Salt
Az Zarqā'
Wādī as Sīr
Nā'ūr
AMMAN
Al 'Āl
Hisbān
Wāhat al Azraq
Ma'dabā
Nāil
DEAD SEA
Dhībān
Al 'Umarī
Al Qaşr
Al Mazra'ah
Adh Dhirā'
Aş Şāfī
Al Karak
Wādī Bāyir
At Tafīlah
ARD AŞ ŞAWWĀN
Ash Shawbak
Bāyir
Wādī Mūsa'
(Petra)
ASH SHARAH
Al Jafr
Ma'ān
Qā' al Jafr
Ra's an Naqb
Al Quwayrah
Jabal Ramm
1754m
Al 'Aqabah
Gulf of Aqaba
Al Mudawwarah
Jordan
32°
30°
36°
38°
N
0 50 km
0 50 miles

JORDAN

Total Land Area : 92 300 sq. km (35 637 sq. miles)

POPULATION
over 100 000
over 50 000
over 10 000
under 10 000

LAND HEIGHT
1000m/3281ft
500m/1640ft
200m/656ft
Sea Level
-200m/-656ft

WORLD AFFAIRS

Joined UN in 1955

AL
AMF
WTO
NAM
OIC

Jordan's position as a key player in Middle East politics is under question. Policy toward the emerging Palestinian state is uncertain since Jordan's relations with Israel remain much less hostile than those of other Arab countries. The US signed a ten-year free trade agreement with Jordan in 2000.

Jordan has recently been at the forefront of the rehabilitation of Iraq, having refused to join the anti-Iraq coalition formed by the Gulf states in 1991.

AID

Recipient

 $552m (receipts)

 Up 28% in 2000

The Gulf states undertook to restore aid to Jordan when King Hussein moved to distance himself from Iraq in 1995.

DEFENSE

No compulsory military service

 $510m

 Down 13% in 2000

The armed forces are loyal to the monarchy. They have a reputation for thorough training and professionalism. The forces are dependent on Western support for credit in purchasing advanced arms and equipment, but Jordanian forces played no part in the 1991 Gulf War.

ECONOMICS

Inflation 3.2% p.a. (1990–2000)

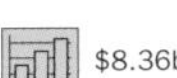 $8.36bn

 0.7100–0.7088 Jordanian dinars

SCORE CARD

❑ WORLD GNP RANKING	92nd
❑ GNP PER CAPITA	$1710
❑ BALANCE OF PAYMENTS	$59m
❑ INFLATION	0.7%
❑ UNEMPLOYMENT	15%

STRENGTHS

Major exporter of phosphates. Skilled workforce. Recovery of tourist industry after 1991 Gulf War. The port of Aqaba is a special economic zone.

WEAKNESSES

Reliant on imports of energy. Poor export to import ratio. Unemployment, exacerbated by influx of refugees from Kuwait after Gulf crisis. Little arable land.

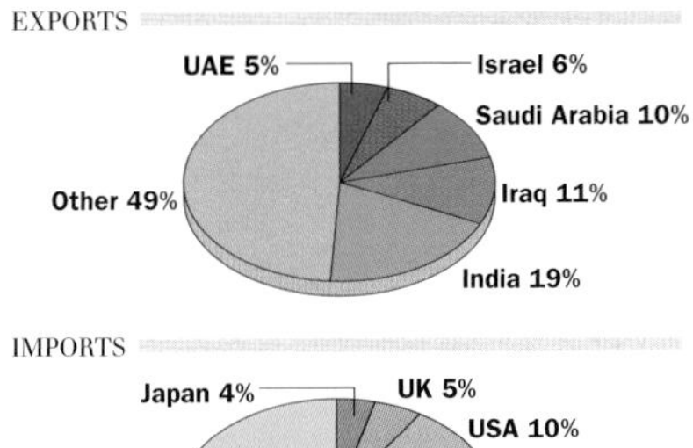
EXPORTS
UAE 5%
Israel 6%
Saudi Arabia 10%
Iraq 11%
India 19%
Other 49%
IMPORTS
Japan 4%
UK 5%
USA 10%
Germany 12%
Iraq 15%
Other 54%

RESOURCES

Electric power 1.3m kw

 1025 tonnes

 40 b/d

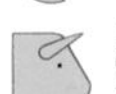 1.85m sheep, 640,000 goats, 23.7m chickens

Oil, phosphates, potash

Oil deposits have been discovered. Phosphates, livestock, and crops such as tomatoes, wheat, olives, and vegetables are the main resources.

ENVIRONMENT

Sustainability rank: 53rd

 3%

 3 tonnes per capita

Conservation is a government priority. Rare animals are protected and species that became extinct in the wild in the 1950s are being reintroduced into controlled environments.

MEDIA

TV ownership medium

 Daily newspaper circulation 42 per 1000 people

PUBLISHING AND BROADCAST MEDIA

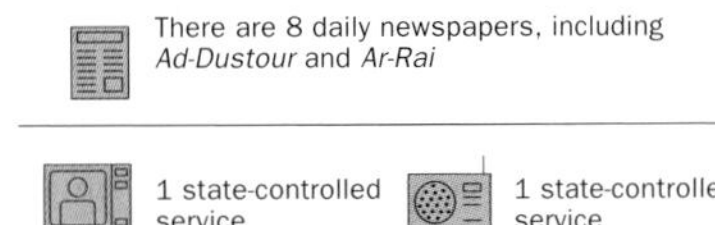
There are 8 daily newspapers, including *Ad-Dustour* and *Ar-Rai*
1 state-controlled service
1 state-controlled service

A restrictive press and publications law was enacted in 1998. Radio and TV are controlled by the state.

CRIME

Death penalty in use

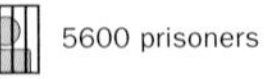 5600 prisoners

 Up 7% in 1999

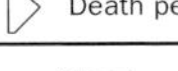

Jordan is largely peaceful. Crime levels are generally low, although theft in urban areas is rising.

EDUCATION

School leaving age: 16

90%

142,190 students

Men and women receive the same education. Jordanian teachers work all over the Middle East.

HEALTH

Welfare state health benefits

 1 per 588 people

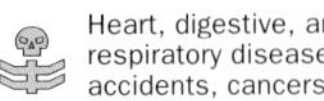 Heart, digestive, and respiratory diseases, accidents, cancers

Health care is subsidized by the government. Hospitals are well distributed throughout the country.

CHRONOLOGY

Jordan, previously the British-mandated territory of Transjordan, became independent in 1946.

- ❑ **1953** Hussein becomes king.
- ❑ **1967** Israel seizes West Bank territories.
- ❑ **1970** Massive crackdown on Palestine Liberation Organization.
- ❑ **1988** Jordan cedes claims to West Bank to PLO.
- ❑ **1994** Peace treaty with Israel.
- ❑ **1999** Death of King Hussein; succession of King Abdullah II.

The King's Highway, seen from the castle at Al Karak, a strategic fortress built by Crusader knights in the 12th century.

SPENDING

GDP/cap. increase

CONSUMPTION AND SPENDING

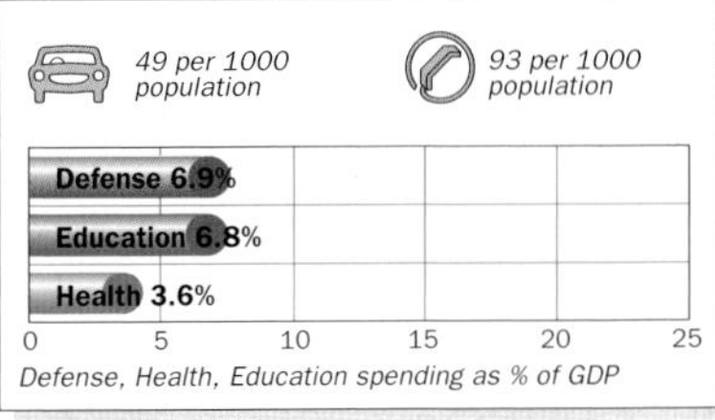
49 per 1000 population
93 per 1000 population
Defense 6.9%
Education 6.8%
Health 3.6%
0
5
10
15
20
25
Defense, Health, Education spending as % of GDP

Poverty is relatively rare, though refugee camps still exist and 25% unemployment damaged many family incomes in the late 1990s.

WORLD RANKING

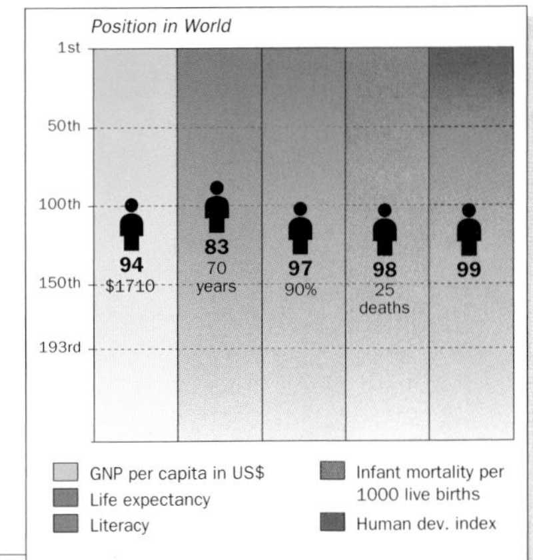
Position in World
1st
50th
100th
150th
193rd
94
$1710
83
70 years
97
90%
98
25 deaths
99
GNP per capita in US$
Life expectancy
Literacy
Infant mortality per 1000 live births
Human dev. index

J

Kazakhstan

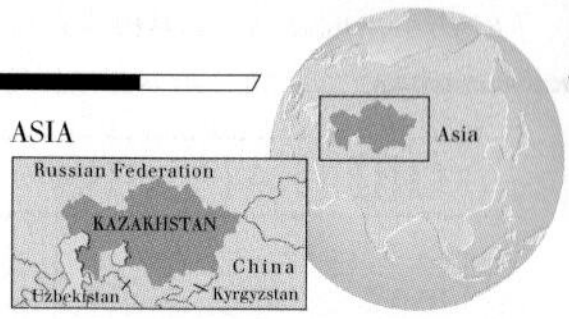

ASIA

Russian Federation

KAZAKHSTAN

China

Uzbekistan

Kyrgyzstan

Asia

Official Name: Republic of Kazakhstan **Capital:** Astana
Population: 16.1 million **Currency:** Tenge **Official Language:** Kazakh

1991

1991

Oct 25

KZ

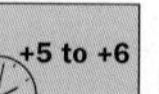
+5 to +6

+7

.kz

THE SECOND-LARGEST of the former Soviet republics, Kazakhstan extends almost 3000 km (1900 miles) from the Caspian Sea in the west to the Altai Mountains in the east and 1600 km (1000 miles) north to south. It borders Russia to the north and China to the east. Kazakhstan was the last Soviet republic to declare its independence, in 1991. In 1999, elections confirmed the former communist Nursultan Nazarbayev and his supporters in power. Kazakhstan has considerable economic potential, and many Western companies seek to exploit its mineral resources.

***The Altai Mountains**, eastern Kazakhstan. Subject to harsh continental winters, the Altai range is a cold, inhospitable place. Rivers carry meltwater down onto the vast steppe.*

CLIMATE

Cold desert/steppe

WEATHER CHART FOR ASTANA

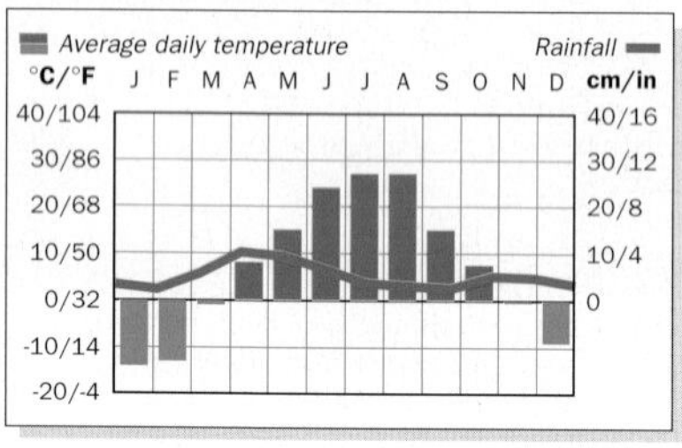

Average daily temperature — Rainfall

°C/°F J F M A M J J A S O N D cm/in

40/104, 30/86, 20/68, 10/50, 0/32, -10/14, -20/-4

40/16, 30/12, 20/8, 10/4, 0

Kazakhstan has a continental climate with large temperature variations: average January temperatures range from –18°C (0°F) on the northern Kazakh steppe to –3°C (27°F) in the deserts 1600 km (1000 miles) to the south; July temperatures average 19°C (66°F) and 30°C (86°F) respectively. As the Caspian Sea never freezes, winters are mildest on Kazakhstan's southwestern coast.

TRANSPORTATION

Drive on right

Astana
242,936 passengers

18 ships
9253 grt

THE TRANSPORTATION NETWORK

103,272 km (64,170 miles)

None

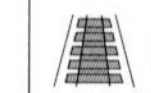

13,601 km (8452 miles)

3900 km (2423 miles)

Transportation networks focus on the north and east as the key economic areas. Most of the roads in Kazakhstan are in urgent need of repair. In 1998, measures to restructure and privatize the railroad were announced. The railroad system links into that of the Russian Federation. Helicopters are frequently used to reach remote destinations.

TOURISM

Not available

Visitors are mainly on business

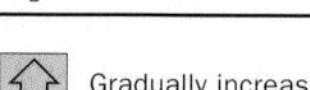

Gradually increasing

MAIN TOURIST ARRIVALS

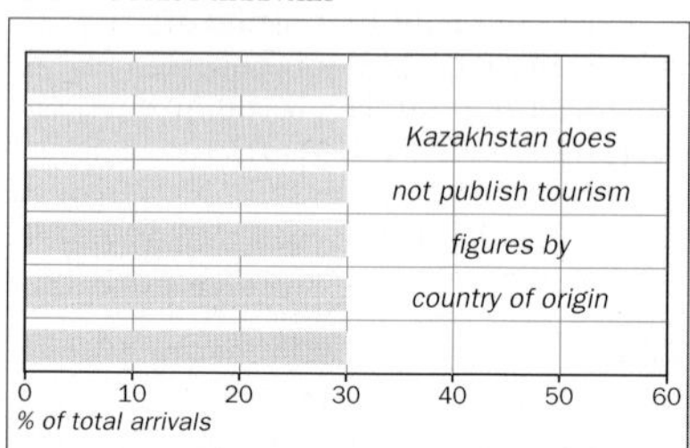
Kazakhstan does not publish tourism figures by country of origin
0
10
20
30
40
50
60
% of total arrivals

The number of visitors to Kazakhstan is increasing, but very few come solely as tourists. The majority are business travelers, and a dense web of contacts with foreign companies has evolved. Of the central Asian states, Kazakhstan has cultivated the closest links with the West. Most foreign businesses are concentrated in Almaty.

Kazakhstan

Total Land Area : 2 717 300 sq. km (1 049 150 sq. miles)

RUSSIAN FEDERATION
Petropavlovsk
Kostanay
Rudnyy
Kokshetau
Shchuchinsk
Pavlodar
Yermak
Ekibastuz
Kushmurun
Zhitikara
Ural'sk
Aksay
Aktobe
Alga
Uil
Temir
KIRGHIZ STEPPE
Emba
Atyrau
Turgay
Arkalyk
Atbasar
Ishim
ASTANA
Ozero Tengiz
Temirtau
Karaganda
Saran
Shakhtinsk
Semipalatinsk
Leninogorsk
Gora Belukha 4506m
Zyryanovsk
Shar
Ust'-Kamenogorsk
Ozero Zaysan
Karkaralinsk
MELKOSOPOCHNIK
KAZAKHSKIY
Ayaguz
Chelkar
Satpayev
Zhezkazgan
Balkhash
Ozero Balkhash
Ozero Alakol'
Aral'sk
Ayteke Bi
Aral Sea
Syr Darya
Dzhusaly
Beyneu
Fort-Shevchenko
Caspian Sea
Aktau
USTYURT PLATEAU
Zhanaozen
PESKI MUYUNKUM
Kzylorda
Ili
Taldykorgan
Tekeli
Panfilov
Shu
Almaty
Turkestan
Kentau
Karatau
Taraz
Arys'
Shymkent
UZBEKISTAN
KYRGYZSTAN
CHINA
TURKMENISTAN
POPULATION
over 500 000
over 100 000
over 50 000
over 10 000
under 10 000
LAND HEIGHT
3000m/9843ft
2000m/6562ft
1000m/3281ft
500m/1640ft
200m/656ft
Sea Level
-200m/-656ft
0 200 km
0 200 miles

PEOPLE

Pop. density low

Kazakh, Russian, Uighur, Korean, German

6/km² (15/mi²)

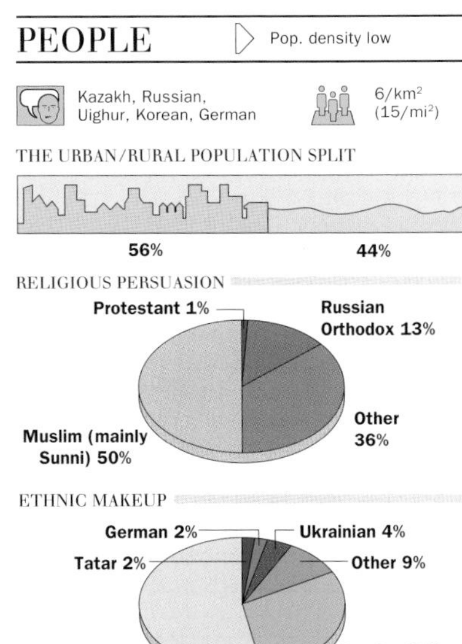
THE URBAN/RURAL POPULATION SPLIT
56%
44%
RELIGIOUS PERSUASION
Protestant 1%
Russian Orthodox 13%
Other 36%
Muslim (mainly Sunni) 50%
ETHNIC MAKEUP
German 2%
Ukrainian 4%
Tatar 2%
Other 9%
Kazakh 53%
Russian 30%

POPULATION AGE BREAKDOWN

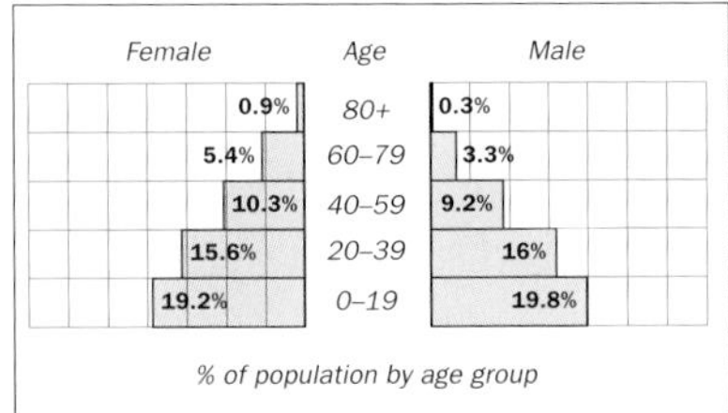
Female
Age
Male
0.9%
80+
0.3%
5.4%
60–79
3.3%
10.3%
40–59
9.2%
15.6%
20–39
16%
19.2%
0–19
19.8%
% of population by age group

Kazakhstan's ethnic diversity arose mainly from forced settlement of Tatars, Germans, and Russians during the Soviet era. By 1959, Kazakhs were outnumbered by ethnic Russians. This balance has been redressed by the immigration of ethnic Kazakhs from neighboring states and the departure in the 1990s of some 1.5 million ethnic Russians. In addition, a majority of ethnic Germans have opted to live in Germany, although in 2000 the government announced a campaign to try to lure some of them back.

In 1995, ethnic Russians criticized the country's new constitution for preventing dual citizenship with Russia and refusing to recognize Russian as an official language. Central control over ethnic Russians has been reinforced by shifting the capital to Astana (formerly Akmola) in the north, where the majority of ethnic Russians reside.

Few Kazakhs retain their traditional nomadic life. Commitment to Islam and loyalty to the clan remain strong.

POLITICS

Multiparty elections

L. House 1999/2004
U. House 1999/2002

President Nursultan Nazarbayev

AT THE LAST ELECTION

Majlis 77 seats

4% AP | 47% Others | 31% Otan | 14% CPK | 4% CP

Otan = Fatherland Republican Party of Kazakhstan
CPK = Civil Party of Kazakhstan **AP** = Agrarian Party
CP = Communist Party of Kazakhstan

Senate 39 seats

Two members are elected by each of 16 districts and 7 are nominated by the president

Legislative authority is vested in the bicameral Parliament. The president, who must be fluent in Kazakh, has supreme executive power.

Profile

Despite a democratic government, the president enjoys political dominance, and the patronage of the Kazakh clans is still important. Since coming to power in 1989, President Nazarbayev has concentrated on market reforms. There have been allegations of electoral fraud, and domestic and international criticism of Nazarbayev's attempts to expand the scope of presidential powers. In 2000 the Majlis granted him special powers to advise future presidents after his term expires in 2006, and in 2002 opposition parties were effectively neutered by reform of the party registration process.

Main Political Issue

Presidential powers

The increased powers of Nazarbayev are the focus of political controversy. Critics accuse him of developing a personality cult. The 1995 constitution strengthened presidential powers, conferring a veto over the decisions of the Constitutional Council. A referendum extended Nazarbayev's term of office until 2000, but in 1998 the legislature approved constitutional amendments forcing him to hold a presidential election in 1999. Although reelected, Nazarbayev was tarnished by allegations of voting irregularities. In April 2002 Nazarbayev's credibility was again shaken by the revelation that he had moved $1 billion of state oil revenues into a secret overseas bank account, without parliament's knowledge.

***President Nursultan Nazarbayev**, who steered Kazakhstan to independence.*

WORLD AFFAIRS

Joined UN in 1992

CIS | SCO | EAPC | OIC | OSCE

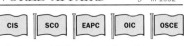

Relations with Russia, although strained at times by Moscow's concern over Kazakhstan's ethnic Russians, have been cemented by a 25-year cooperation treaty. However, Kazakhstan's earlier enthusiasm for greater integration of the former Soviet states has cooled.

Kazakhstan's location, bridging Europe and Asia, has led to close relations with a variety of partners, and its rich mineral resources have attracted investors from Europe, the US, and Asia. Relations with China have improved, with agreements in 1998 and 1999 on border issues, and it joins China, Russia, and three other central Asian republics in the Shanghai Cooperation Organization. However, it competes with Uzbekistan, with which its has a border dispute, to be the acknowledged central Asian regional power.

AID

Recipient

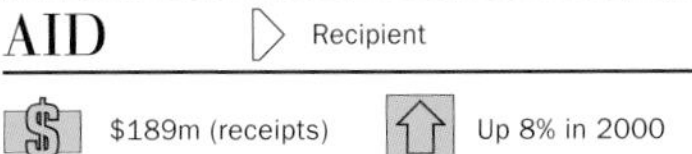
$189m (receipts)
Up 8% in 2000

Kazakhstan joined the IMF and the World Bank in 1992, and is also a member of the EBRD. Most multilateral and bilateral aid is aimed at supporting economic reform and improving health care, transportation, and communications.

CHRONOLOGY

Once part of the Mongol empire, Kazakhstan was absorbed by the Russian Empire in the 19th century. Ethnic Russians began to settle on land used by nomadic Kazakhs. Russian settlement intensified after the 1917 Revolution and Kazakhstan was subjected to intensive industrial and agricultural development.

- **1916** Rebellion against Russian rule brutally suppressed.
- **1917** Russian Revolution inspires civil war in Kazakhstan between Bolsheviks, anti-Bolsheviks, and Kazakh nationalists.
- **1918** Kazakh nationalists set up autonomous republic.
- **1920** Bolsheviks take control. Kirghiz Autonomous Soviet Socialist Republic (ASSR) set up within Russian Soviet Federative Socialist Republic.
- **1925** Kirghiz ASSR renamed Kazakh ASSR.
- **1936** Kazakhstan becomes full union republic of the USSR as Kazakh SSR.

⇨

CHRONOLOGY *continued*

- **1930s** Stalin's collectivization program leads to increase in Russian settlement and the deaths of an estimated one million Kazakhs. Large penal settlements established for victims of Stalinist purges.
- **1941–1945** Large-scale deportations from Russia of Germans, Jews, Crimean Tatars, and others to Kazakhstan.
- **1949–1989** Nuclear test site at Semipalatinsk carries out nearly 500 nuclear explosions.
- **1954–1960** Khrushchev's policy to plow "Virgin Lands" for grain most vigorously followed in Kazakhstan. Russian settlement reaches peak.
- **1986** Riots in Almaty after ethnic Russian Gennadi Kolbin appointed head of Kazakhstan Communist Party (CPK) to replace Kazakh Dinmukhamed Kunyev.
- **1989** Kolbin replaced by Nursultan Nazarbayev, ethnic Kazakh and chair of Council of Ministers. Reform of political and administrative system.
- **1990** CPK wins elections to Supreme Soviet by overwhelming majority. Nazarbayev appointed first president of Kazakhstan. Kazakhstan declares sovereignty.
- **1991** Kazakhstan votes to preserve USSR as union of sovereign states. USSR authorities hands over control of enterprises in Kazakhstan to Kazakh government. CPK ordered to cease activities in official bodies following abortive August coup in Moscow. CPK restructures itself as Socialist Party of Kazakhstan (SPK). Independence of Republic of Kazakhstan declared; joins CIS. Announcement of closure of Semipalatinsk nuclear test site.
- **1992** Opposition demonstrations against dominance of reformed communists in Supreme Soviet, now Supreme Kenges. Nationalist groups form Republican Party, Azat.
- **1993** Adoption of new constitution. Introduction of new currency, the tenge.
- **1994** Legislative elections annulled after proof of widespread voting irregularities.
- **1995** Adoption of new constitution broadening presidential powers; referendum extends Nazarbayev's term until 2000; legislative elections.
- **1998** Legislature approves constitutional amendments, including the holding of early presidential election.
- **1999** Nazarbayev reelected president for seven more years.

K

DEFENSE

Phasing out conscription

$357m

Up 23% in 2000

KAZAKH ARMED FORCES

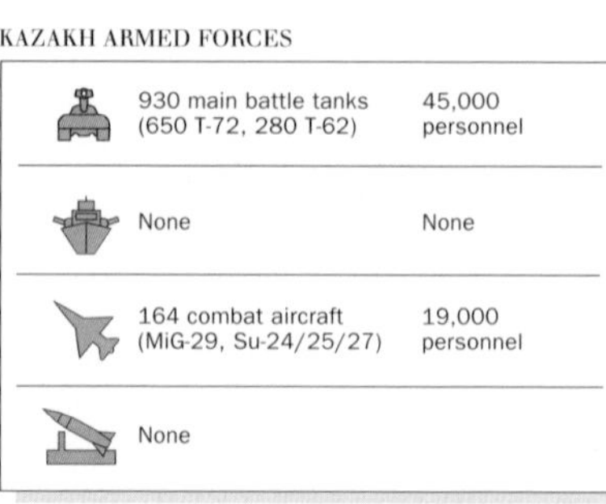
930 main battle tanks (650 T-72, 280 T-62)
45,000 personnel
None
None
164 combat aircraft (MiG-29, Su-24/25/27)
19,000 personnel
None

As a former nuclear power and the largest of the five former Soviet central Asian republics, Kazakhstan is a potential guarantor of regional peace and stability. Kazakhstan ratified the START-I nuclear reduction treaty in 1992 and the NPT in 1993. The US agreed in 1993 to grant Kazakhstan $84 million to enable it to dismantle its nuclear weapons armory. In 1995, Kazakhstan announced that all its nuclear weapons had been transferred to Russia or else had been destroyed.

A restructuring of Kazakhstan's armed forces was launched in 2001. This included moves to end conscription, whose duration had been halved, from two years to one in 1998. It is expected that the restructuring will take at least three years.

ECONOMICS

Inflation 205% p.a. (1990–2000)

$18.8bn

145.56-150.90 tenge

SCORE CARD

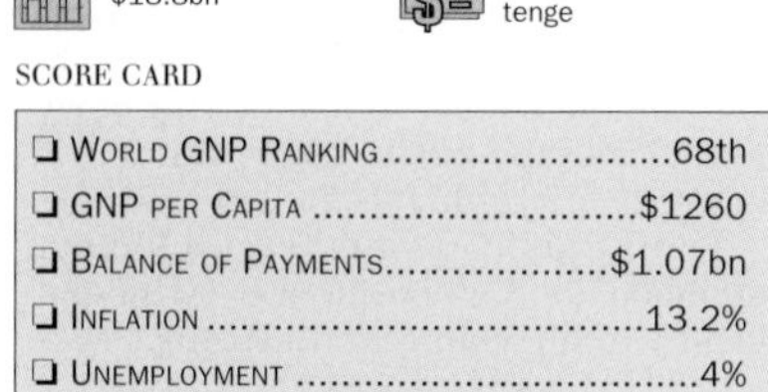

World GNP Ranking	68th
GNP per Capita	$1260
Balance of Payments	$1.07bn
Inflation	13.2%
Unemployment	4%

EXPORTS

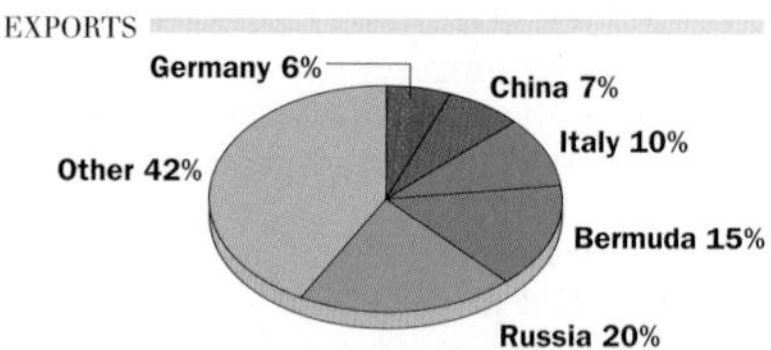
Germany 6%
China 7%
Italy 10%
Bermuda 15%
Russia 20%
Other 42%

IMPORTS

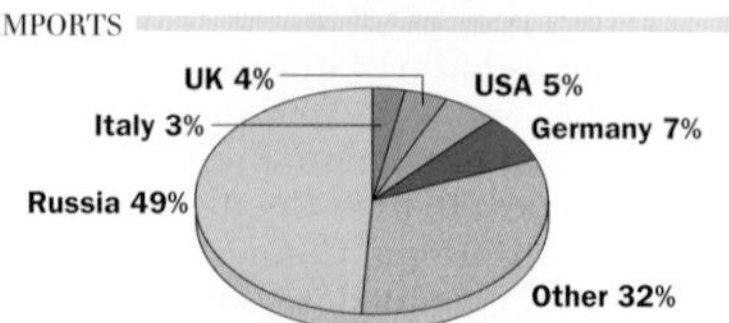
UK 4%
USA 5%
Italy 3%
Germany 7%
Russia 49%
Other 32%

ECONOMIC PERFORMANCE INDICATOR

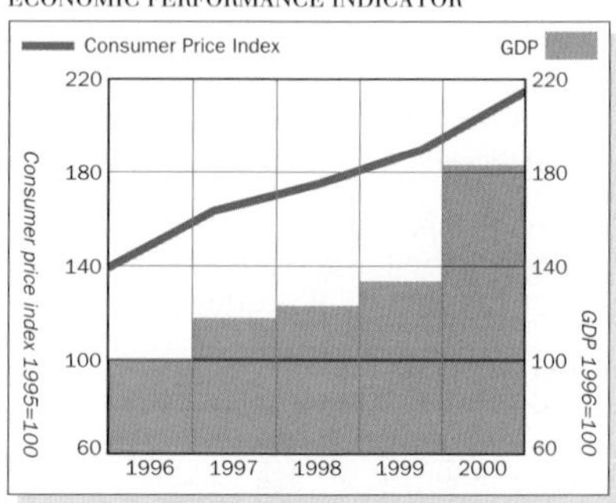
Consumer Price Index
GDP
Consumer price index 1995=100
GDP 1996=100
220
180
140
100
60
1996
1997
1998
1999
2000

Strengths

Mineral resources, notably oil and gas, and bismuth and cadmium, used in electronics industry. Joint oil and gas ventures with Western companies. Mass privatization program launched in 1994.

Weaknesses

Collapse of former Soviet economic and trading system. Reliance on imported consumer goods. Rapid introduction of the tenge in 1993 increased instability and fueled sharp price rises. Inefficient industrial plants.

Profile

Kazakhstan has moved faster than other former Soviet republics to establish a market economy. Prices have been freed, foreign trade deregulated, and the tax system reformed. Petroleum revenues have been the main engine of growth, and inflation, at one time rising sharply, has been brought under control.

Foreign direct investment is mainly in the energy sector. Outdated equipment and inadequate distribution networks mean that energy has to be imported, although Kazakhstan exports fossil fuel.

An oil price boom set the tone for the country's first "five-year plan" in 2000, part of Nazarbayev's "Kazakhstan 2030" program. It promised long-term land leases, although not full private ownership, from 2001.

KAZAKHSTAN : MAJOR BUSINESSES

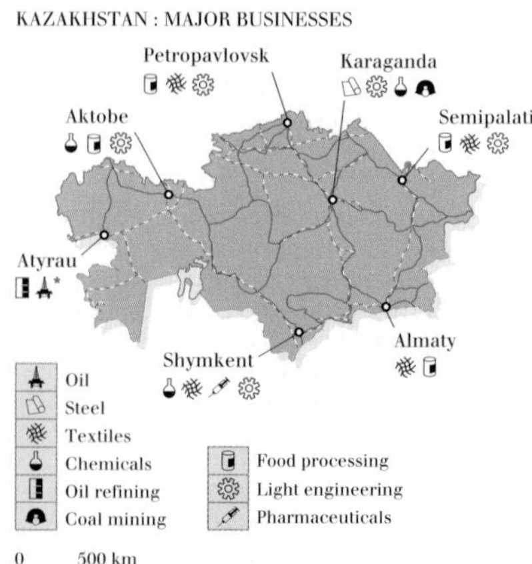
Petropavlovsk
Karaganda
Aktobe
Semipalatinsk
Atyrau
Shymkent
Almaty
Oil
Steel
Textiles
Chemicals
Oil refining
Coal mining
Food processing
Light engineering
Pharmaceuticals
0 500 km
0 500 miles
* significant multinational ownership

RESOURCES

Electric power 19m kw

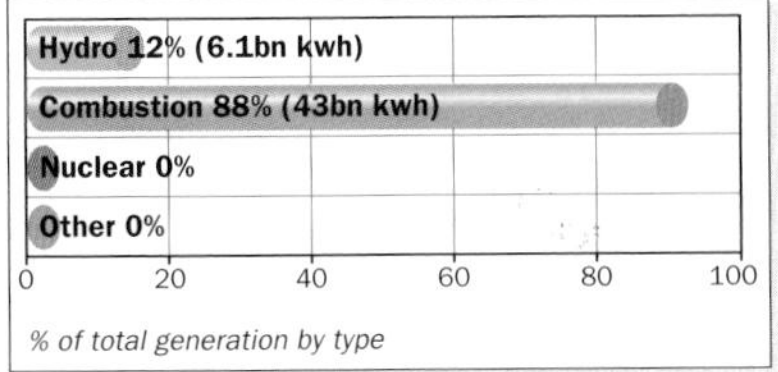
26,951 tonnes
828,000 b/d
(reserves 8bn barrels)
8.94m sheep,
4.11m cattle,
19.7m chickens
Oil, gas, manganese, gold, silver, coal, iron, tungsten, chromite, bismuth, cadmium

ELECTRICITY GENERATION

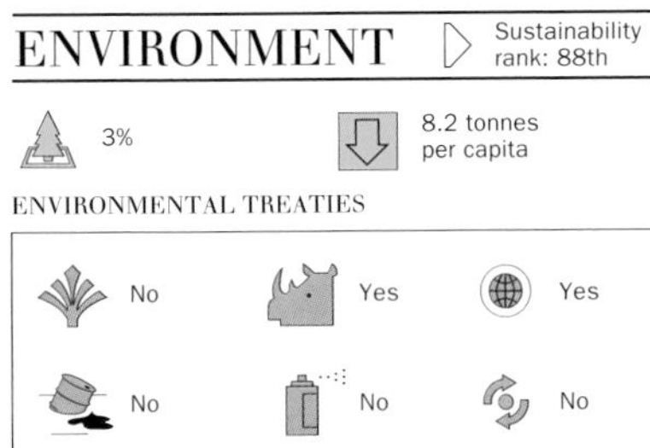
Hydro 12% (6.1bn kwh)
Combustion 88% (43bn kwh)
Nuclear 0%
Other 0%
0
20
40
60
80
100
% of total generation by type

Mining is the single most important industry in Kazakhstan. The US company Chevron has been developing the huge Tengiz oil field since 1993. Joint ventures to exploit substantial oil and gas reserves in the Caspian Sea were agreed with Russia in 1995, and the US and Japan in 1998. Kazakhstan also possesses vast iron ore and gold reserves.

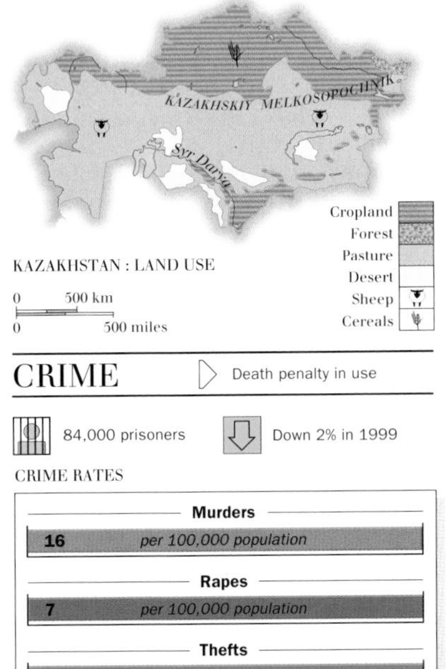
KAZAKHSKIY MELKOSOPOCHNIK
Syr Darya
Cropland
Forest
Pasture
Desert
Sheep
Cereals
0 500 km
0 500 miles

KAZAKHSTAN : LAND USE

ENVIRONMENT

Sustainability rank: 88th

3%

8.2 tonnes per capita

ENVIRONMENTAL TREATIES

No	Yes	Yes
No	No	No

Environmental damage caused by intensive industrial and agricultural development is a major concern. The eastern cities are heavily polluted and farmlands are being eroded. The Aral Sea, polluted by overuse of fertilizers, has shrunk by 50% owing to the Soviet-era diversion of rivers for irrigation.

In 1991, environmental pressure groups succeeded in ending more than 40 years of nuclear testing at Semipalatinsk. The green lobby now presses for tighter pollution controls.

MEDIA

TV ownership medium

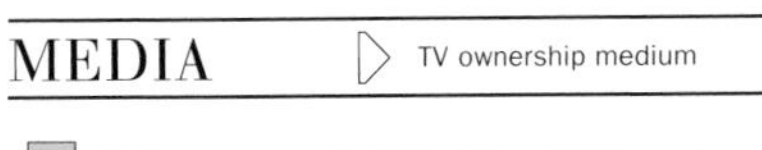

Daily newspaper circulation 30 per 1000 people

PUBLISHING AND BROADCAST MEDIA

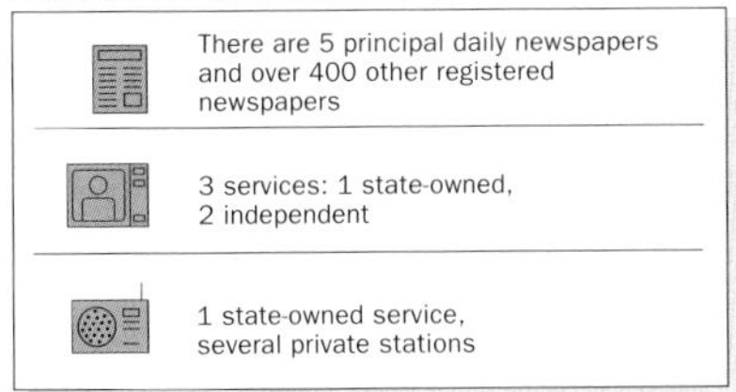
There are 5 principal daily newspapers and over 400 other registered newspapers
3 services: 1 state-owned, 2 independent
1 state-owned service, several private stations

The state-owned media operate alongside independent publications and privately owned radio and television stations. Amendments in 2001 to the Media Law strengthened state control over small broadcast outlets and Internet sites, and created more grounds for libel charges against editors and proprietors. Such state pressure has produced a drastic decline in the number of media outlets.

CRIME

Death penalty in use

84,000 prisoners

Down 2% in 1999

CRIME RATES

Murders	16 per 100,000 population
Rapes	7 per 100,000 population
Thefts	334 per 100,000 population

Narcotics smuggling is increasing. Corruption is rife. Kazakhstan retains and uses the death penalty, but does not publish statistics on its use.

EDUCATION

School leaving age: 15

99%

419,460 students

THE EDUCATION SYSTEM

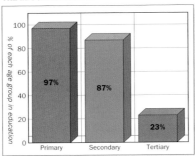

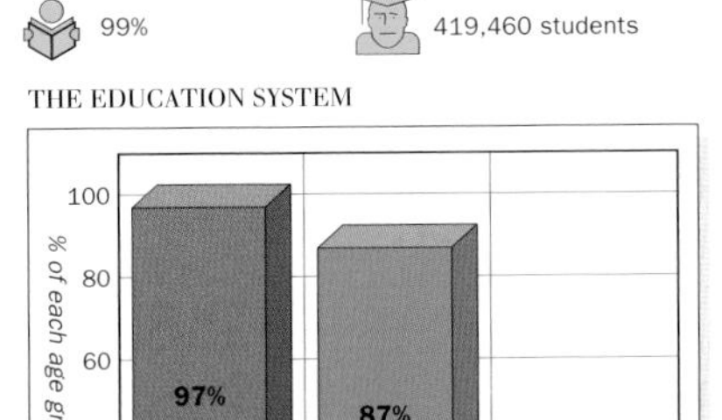
% of each age group in education
100
80
60
40
20
0
97%
87%
23%
Primary
Secondary
Tertiary

Education remains based on the Soviet model. Since its adoption as the state language in 1995, Kazakh is gradually replacing Russian as the main instruction medium in schools, but there is a shortage of Kazakh textbooks and Kazakh-speaking teachers. There are a large number of higher-education institutions and medical schools.

HEALTH

Welfare state health benefits

1 per 286 people

Heart attacks, cancers, accidents, violence, tuberculosis

Kazakhstan's ill-equipped and poorly funded health system has produced the lowest average life expectancy in central Asia.

The health system is limited in terms of both facilities and coverage. Rural people have minimal access to clinics. The country's size means that extending coverage and improving the quality of care will be costly. Attempts are therefore being made to attract foreign investment into the health sector. Many doctors have emigrated to Russia.

SPENDING

GDP/cap. decrease

CONSUMPTION AND SPENDING

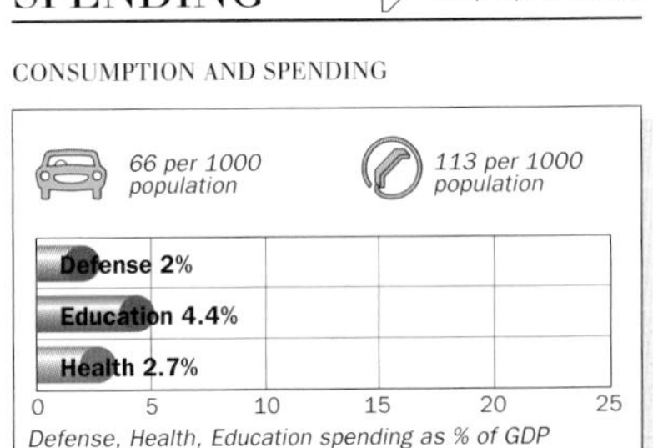
66 per 1000 population
113 per 1000 population
Defense 2%
Education 4.4%
Health 2.7%
0
5
10
15
20
25
Defense, Health, Education spending as % of GDP

Life for the majority of Kazakhs has always been hard, and has grown even more difficult since 1989. Unemployment has risen and living standards have deteriorated as a result of the market-oriented reforms within Kazakhstan. In addition, the liberalization of the economy has had the effect of fueling sharp price rises for essential commodities.

The rural population, the poorest group in Kazakhstan, has been badly affected. The small wealthy elite is made up mainly of former communist officials, many of whom have benefited from privatization, or belong to President Nazarbayev's clan.

WORLD RANKING

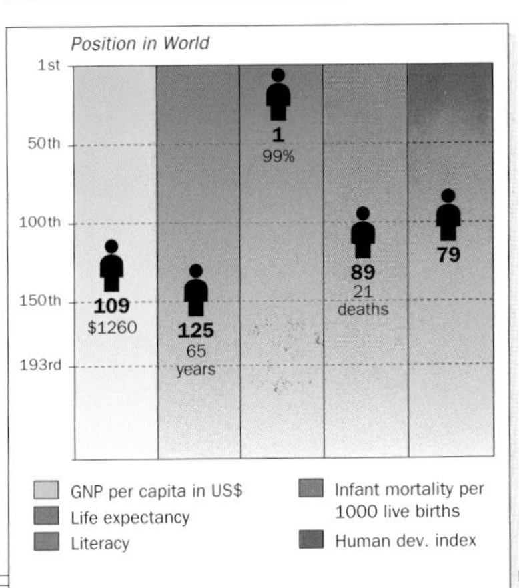
Position in World
1st
50th
100th
150th
193rd
109
$1260
125
65
years
1
99%
89
21
deaths
79
GNP per capita in US$
Life expectancy
Literacy
Infant mortality per 1000 live births
Human dev. index

K

KENYA

EAST AFRICA

OFFICIAL NAME: Republic of Kenya **CAPITAL:** Nairobi
POPULATION: 31.3 million **CURRENCY:** Kenya shilling **OFFICIAL LANGUAGES:** Kiswahili and English

1963

1963

Dec 12

EAK

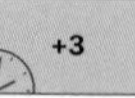
+3

+254

.ke

KENYA STRADDLES the equator on Africa's east coast, its central plateau bisected by the Great Rift Valley. The land to the north is desert, while to the east lies a fertile coastal belt. After independence from the UK in 1963, politics was dominated by Jomo Kenyatta. He was succeeded as president in 1978 by Daniel arap Moi, whose divide-and-rule policies have drawn accusations of favoritism and of fomenting ethnic hatreds. His KANU won elections easily in 1992 and 1997, amid accusations of fraud. Economic mainstays are tourism and agriculture. High population growth is a major problem.

Kenyatta Conference Center, Nairobi. The modern skyline of the business center contrasts sharply with the slums on the city's outskirts.

CLIMATE

Steppe/mountain/ tropical

WEATHER CHART FOR NAIROBI

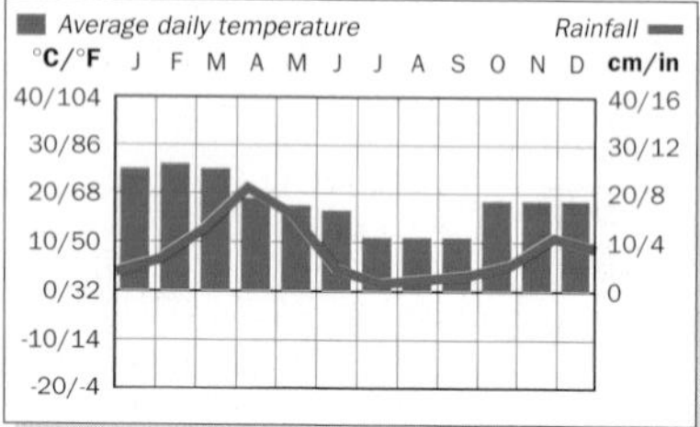
Average daily temperature
Rainfall
°C/°F
J F M A M J J A S O N D
cm/in
40/104
30/86
20/68
10/50
0/32
-10/14
-20/-4
40/16
30/12
20/8
10/4
0

The coast and Great Rift Valley are hot and humid, the plateau interior is temperate, and the northeastern desert hot and dry. Rain generally falls from April to May and October to November.

TOURISM

Visitors : Population 1:33

943,000 visitors

Up 10% in 1999

Tourism is vital to the economy and a key foreign exchange earner. After a boom in package safaris and beach vacations during the 1980s, Kenya saw a decline in visitor numbers in the 1990s due to world recession, reports of instability, and the much-publicized murder of several tourists. A slight upturn in 1999–2000 was set back again by the global decline in the industry in 2001.

MAIN TOURIST ARRIVALS

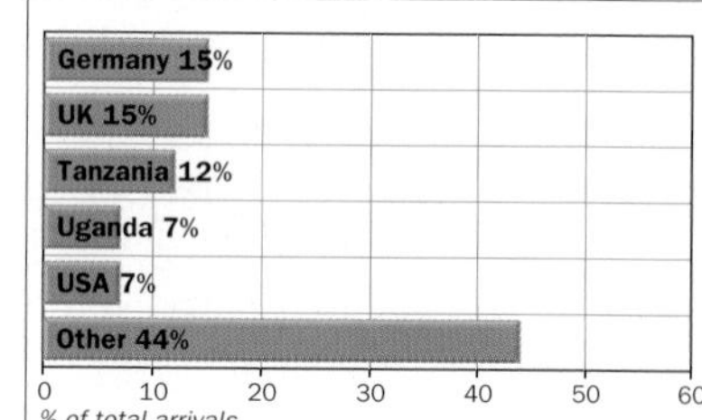
Germany 15%
UK 15%
Tanzania 12%
Uganda 7%
USA 7%
Other 44%
0
10
20
30
40
50
60
% of total arrivals

TRANSPORTATION

Drive on left

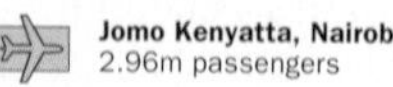

Jomo Kenyatta, Nairobi 2.96m passengers

39 ships 20,900 grt

THE TRANSPORTATION NETWORK

8868 km (5510 miles)

None

2634 km (1637 miles)

Lake Victoria is navigable

Kenya's railroads, ports, and main airport are currently being upgraded, while in addition a five-year road improvement program was begun in 2000.

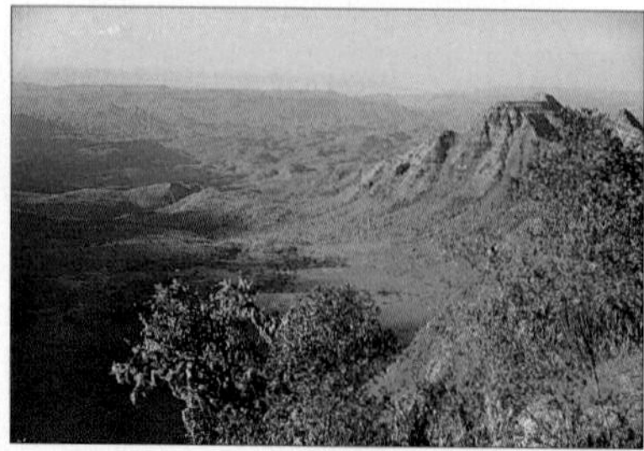

***Great Rift Valley**, Kenya. This huge crack in the Earth's crust runs from the River Jordan right through Africa to the Zambezi River.*

KENYA

Total Land Area : 582 650 sq. km (224 961 sq. miles)

POPULATION

- over 1 000 000
- over 500 000
- over 100 000
- over 50 000
- over 10 000
- under 10 000

LAND HEIGHT

- 5000m/9843ft
- 2000m/6562ft
- 1000m/3281ft
- 500m/1640ft
- 200m/656ft
- Sea Level

0 100 km
0 100 miles

PEOPLE

Pop. density medium

Kiswahili, English, Kikuyu, Luo, Kamba

55/km² (143/mi²)

THE URBAN/RURAL POPULATION SPLIT

33% 67%

RELIGIOUS PERSUASION

Muslim 6%
Other 9%
Traditional beliefs 25%
Christian 60%

ETHNIC MAKEUP

Kamba 11%
Other 30%
Kalenjin 11%
Luo 13%
Kikuyu 21%
Luhya 14%

Kenya's ethnic diversity, with about 70 different groups, reflects its past as a focus of population movements. Asians, Europeans, and Arabs form 1% of the population. The rural majority retains strong clan and extended family links, although these are being weakened by urban migration. Poverty, severe drought, and a high population growth rate are the root causes of a land hunger which has fueled a surge in ethnic violence. There is much violence in western Kenya, where Kikuyu, the formerly dominant tribe, are the main targets of attacks by other groups. Several hundred thousand Kikuyu are believed to have been displaced from their villages.

Female genital mutilation, criminalized in 2001, remains widespread.

POPULATION AGE BREAKDOWN

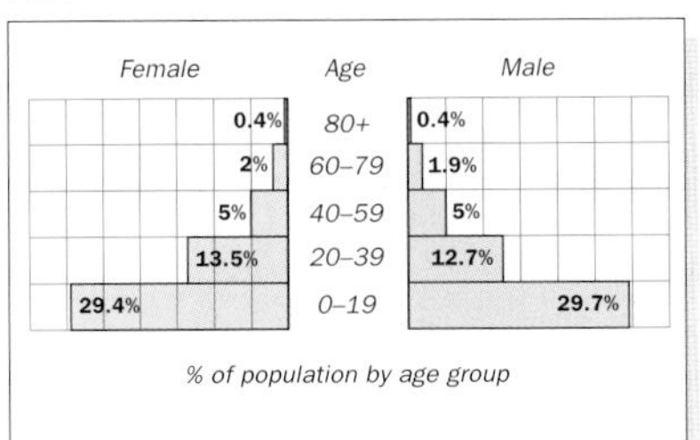
Female
Age
Male
0.4%
80+
0.4%
2%
60–79
1.9%
5%
40–59
5%
13.5%
20–39
12.7%
29.4%
0–19
29.7%
% of population by age group

POLITICS

Multiparty elections

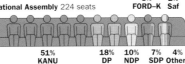

1997/2002

President Daniel arap Moi

AT THE LAST ELECTION

National Assembly 224 seats

8% FORD–K
2% Saf
51% KANU
18% DP
10% NDP
7% SDP
4% Others

KANU = Kenya African National Union
DP = Democratic Party **NDP** = National Development Party
FORD–K = Forum for the Restoration of Democracy–Kenya
SDP = Social Democratic Party **Saf** = Safina
Others include FORD–People, Kenya Social Congress, Shirikisho, FORD–Asili, and two ex-officio members.

The National Assembly comprises 210 elected members, 12 nominated by the president, and two ex-officio members

Kenya has been led by President Daniel arap Moi since 1978, and became a multiparty democracy in 1992.

PROFILE

After 1982, President Moi's efforts to entrench KANU in power stirred up demands at home for the introduction of multiparty politics and provoked condemnation abroad for the human rights abuses committed. Forced in 1992 to concede free elections, Moi helped ensure KANU's victory by curtailing the campaign period. His reelection in 1997 was marred by widespread allegations of intimidation and electoral fraud.

Moi has attempted to break down barriers with some opponents. In 1999 he appointed paleontologist Richard Leakey, a former political opponent, to head a drive against official corruption, which was completed in 2001. In 2002, KANU absorbed the "opposition" NDP, which had joined the government the previous year.

MAIN POLITICAL ISSUE

Ethnic violence

The ethnic polarization of political parties in Kenya and rising poverty are fueling ethnic violence. Determined to ensure KANU dominance, President Moi, a Kalenjin, is turning the party into an alliance of smaller ethnic groups opposed to the once dominant Kikuyu. The latter are the largest ethnic group, the main victims of violence, and the leading supporters of the opposition.

Daniel arap Moi, succeeded Kenyatta as president in 1978.

Mwai Kibaki, opposition leader.

AID

Recipient

$512m (receipts)

Up 65% in 2000

Kenya has been a major recipient of aid from donors including the UK, Japan, the EU, the World Bank, and the IMF. Little, however, has trickled down to the majority of the population. This is partly because of the high proportion of aid tied to construction projects and donor-country firms, and partly because of mismanagement and official corruption. In 1996 Western creditor governments decided to link aid disbursements to improvements in human rights, and in 2001 the IMF and the World Bank withheld all aid pending anticorruption reforms.

WORLD AFFAIRS

Joined UN in 1963

Comm
COMESA
EAC
IGAD
AU

Relations with neighboring states and with key Western donors are Kenya's priorities. The US embassy closed temporarily in 1998 after a terrorist bombing caused carnage there. In 1991, human rights concerns were partly responsible for a two-year suspension of aid. Talks were held in spring 2001 in an effort to end Kenya's border dispute with Sudan. Relations with Rwanda improved in 2000, when it was agreed to reopen the Kenyan embassy in Kigali and to establish a joint commission for bilateral relations. Kenya is greatly concerned by the situation in Burundi and the DRC, as well as by tensions in Sudan, Somalia, and Ethiopia.

CHRONOLOGY

From the 10th century, Arab coastal settlers mixed with indigenous peoples in the region. Britain's need for a route to landlocked Uganda led to the formation in 1895 of the British East African Protectorate in the coastal region.

- **1900–1918** White settlement.
- **1920** Interior becomes British colony.
- **1930** Jomo Kenyatta goes to UK; stays 14 years.
- **1944** Kenyan African Union (KAU) formed; Kenyatta returns to lead it.
- **1952–1956** *Mau Mau*, Kikuyu-led violent campaign to restore African lands. State of emergency; 13,000 people killed.
- **1953** KAU banned. Kenyatta jailed.
- **1960** State of emergency ends. Tom Mboya and Oginga Odinga form KANU.
- **1961** Kenyatta freed; takes up presidency of KANU.
- **1963** KANU wins elections. Kenyatta prime minister. Full independence declared.
- **1964** Republic of Kenya formed with Kenyatta as president and Odinga as vice president.
- **1966** Odinga defects to form Kenya People's Union (KPU).

K

CHRONOLOGY *continued*

- 1969 KANU sole party to contest elections (also 1974). Tom Mboya of KANU assassinated. Unrest. KPU banned and Odinga arrested.
- 1978 Kenyatta dies. Vice President Daniel arap Moi succeeds him.
- 1982 Kenya declared a one-party state. Opposition to Moi. Abortive air force coup. Odinga rearrested.
- 1986 Open "queue-voting" replaces secret ballot in first stage of general elections. Other measures to extend Moi's powers incite opposition.
- 1988 Moi wins third term and extends his control over judiciary.
- 1990 Government implicated in deaths of Foreign Minister Robert Ouko and Anglican archbishop. Riots. Odinga and others form FORD, outlawed by government.
- 1991 Arrest of FORD leaders and attempts to stop prodemocracy demonstrations. Donors suspend aid. Moi agrees to introduce multiparty system. Ethnic violence increases.
- 1992 FORD splits into factions led by ex-minister Kenneth Matiba and Odinga. Opposition weakness helps Moi win December elections.
- 1994 Odinga dies.
- 1997 December, Moi wins further term in widely criticized elections.
- 1998 Bomb at US embassy kills 230.
- 1999 Moi appoints paleontologist Richard Leakey to lead government drive against corruption.
- 2000 Worst drought since 1947.
- 2001 Leakey resigns. Drought threatens starvation for millions.
- 2002 Thousands displaced by recurrence of major floods.

DEFENSE

No compulsory military service

$307m

Down 6% in 2000

KENYAN ARMED FORCES

78 main battle tanks (Vickers Mk 3)
20,000 personnel
4 patrol boats
1400 personnel
29 combat aircraft (9 F-5E/F)
3000 personnel
None

Destabilization of the northeastern border by insecurity in Somalia is the main defense issue. The army has recently been deployed to suppress tribal fighting in the Rift Valley. Military assistance is given by the UK and the US.

ECONOMICS

Inflation 14% p.a. (1990–2000)

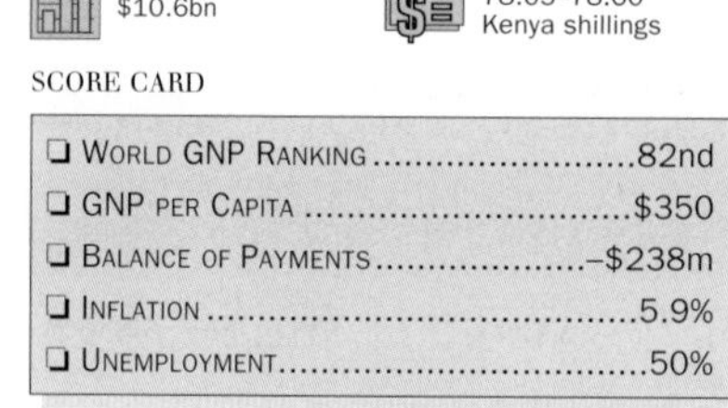

$10.6bn

78.05–78.60 Kenya shillings

SCORE CARD

- World GNP Ranking 82nd
- GNP per Capita $350
- Balance of Payments .... –$238m
- Inflation 5.9%
- Unemployment 50%

EXPORTS

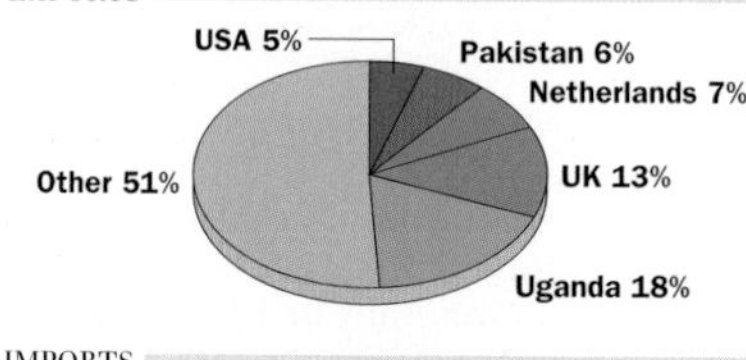

USA 5%
Pakistan 6%
Netherlands 7%
UK 13%
Uganda 18%
Other 51%

IMPORTS

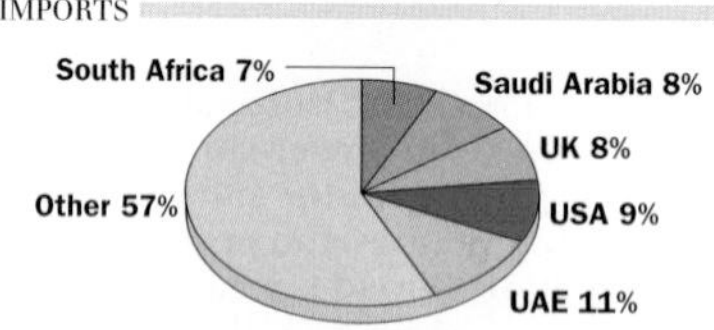

South Africa 7%
Saudi Arabia 8%
UK 8%
USA 9%
UAE 11%
Other 57%

Strengths

Tourism – largest foreign exchange earner. Broad agricultural base, especially cash crops such as coffee and tea. East Africa's largest, most diversified manufacturing sector.

Weaknesses

Fluctuating world prices for coffee and tea. Corruption. Poor recent GDP growth. High population growth. Land shortage means uneconomic small units. Country's image problem affects tourism.

Profile

Kenya has been hailed as an example to the rest of Africa of the benefits of a mainly free-market economy. Government involvement has been relatively limited, and recently further reduced by privatization. Foreign investment has been encouraged, with some success. Tourism has become the leading foreign exchange earner over the past 20 years, despite suffering serious setbacks since the 1990s. Manufacturing now accounts for 21% of GDP, and is the most diversified sector in east Africa, but needs to expand rapidly so as to create more jobs.

Economic growth was good by African standards during the 1980s, averaging over 4% a year. However, it was barely sufficient to compensate for one of the world's highest population growth rates, approximately 3.5%, although the UN estimates that this rate will have almost halved by 2015. For the majority of Kenyans, farming ever-smaller landholdings or earning a living in the informal sector, life has

ECONOMIC PERFORMANCE INDICATOR

Consumer Price Index — GDP — Consumer price index 1995=100 — GDP 1996=100 — 1996 1997 1998 1999 2000

become harsher. The situation was exacerbated in 2001 by a ban on all trade with Somalia, which affects exports of the mild narcotic, *qat*, widely grown in Kenya. Severe drought since 2000 has crippled the agricultural sector and put 20 million people at risk of starvation.

Other problems, including inflation, a heavy debt burden, and growing dependence on balance-of-payments support had come to a head in the early 1990s, when economic growth gave way to recession. Real GDP growth fell to 0.4% in 1992 and has remained low, falling to –0.2% in 2000. The rise in poverty-linked violence and political unrest hit tourism; earnings fell by 15% in the early 1990s, and the industry has yet to recover fully, a global downturn causing further problems in 2001.

Partly as a response to pressure from donors, the government has implemented some economic liberalization measures. These include floating the Kenya shilling, raising interest rates, and giving exporters direct access to their hard currency earnings. However, sustained growth is likely to remain elusive until Kenya overcomes the official corruption which drains vital resources, and the poor image affecting its tourist industry.

KENYA : MAJOR BUSINESSES

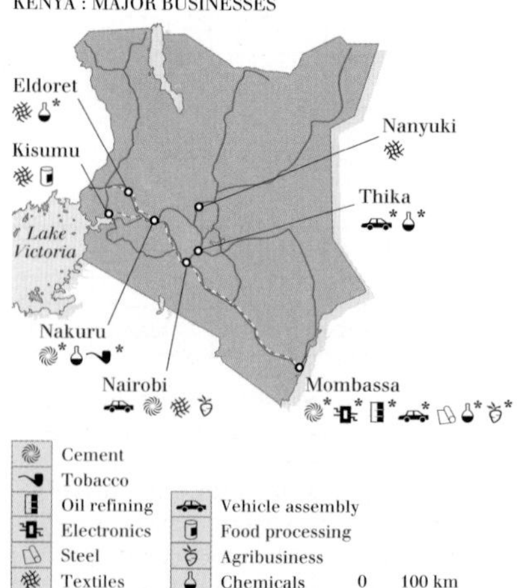

Eldoret
Kisumu
Lake Victoria
Nakuru
Nairobi
Nanyuki
Thika
Mombassa
Cement
Tobacco
Oil refining
Electronics
Steel
Textiles
Vehicle assembly
Food processing
Agribusiness
Chemicals
0 100 km
0 100 miles
* significant multinational ownership

RESOURCES

Electric power 889,000 kw

205,587 tonnes

Oil reserves not currently exploited

12.5m cattle, 9m goats, 6.5m sheep, 32m chickens

Soda ash, fluorite, limestone, rubies, gold, vermiculite, oil

ELECTRICITY GENERATION

Hydro	74% (3.3bn kwh)
Combustion	14% (0.61bn kwh)
Nuclear	0%
Other	12% (0.55bn kwh)

0 20 40 60 80 100

% of total generation by type

Agriculture is still the largest sector of the economy. Kenya's varied topography means that tropical, subtropical, and temperate crops may be grown. Coffee and tea, the main export crops, have been affected by falling world prices. Efforts to reduce dependence on these have led to the growth of a successful export-oriented horticultural industry.

Kenya has few mineral resources, though oil exploration has revealed deposits in Turkana District. Hydroelectric and geothermal sources are being developed to reduce energy imports – currently 70% of total requirements. However, droughts caused power shortages in 2000 and 2001, leading to the imposition of daily power cuts.

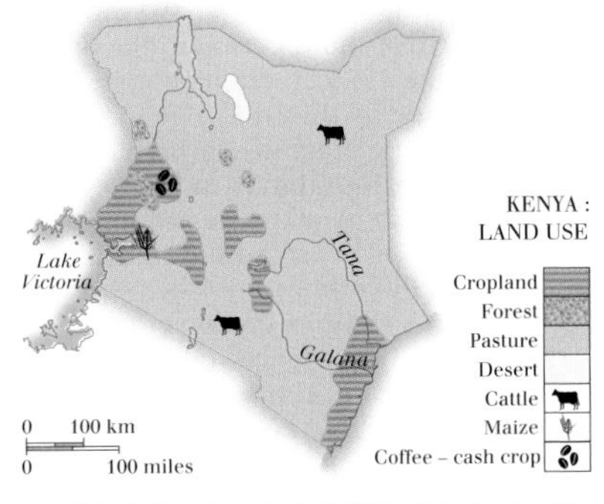
KENYA : LAND USE
Lake Victoria
Tana
Galana
Cropland
Forest
Pasture
Desert
Cattle
Maize
Coffee – cash crop
0 100 km
0 100 miles

ENVIRONMENT

Sustainability rank: 89th

6%

0.3 tonnes per capita

ENVIRONMENTAL TREATIES

Yes	Yes	Yes
Yes	Yes	No

The importance to tourism of wildlife conservation is recognized, and recent elephant protection schemes have been a success, but proposed national reserves compete with agriculture for land. Opposition to government plans to reallocate some national park land to squatters is growing.

MEDIA

TV ownership low

Daily newspaper circulation 8 per 1000 people

PUBLISHING AND BROADCAST MEDIA

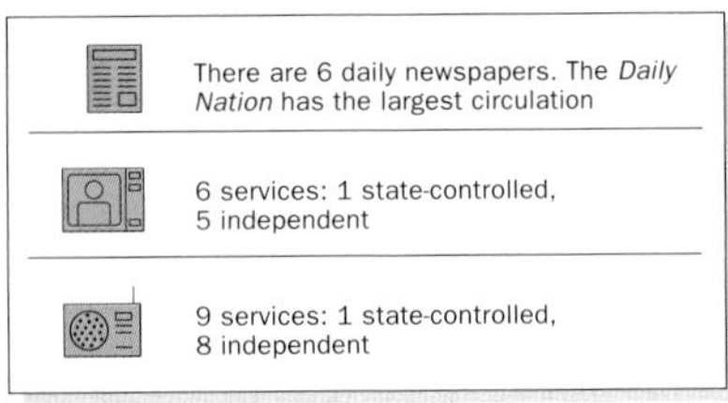
There are 6 daily newspapers. The *Daily Nation* has the largest circulation
6 services: 1 state-controlled, 5 independent
9 services: 1 state-controlled, 8 independent

Government intolerance of criticism is long-standing and includes plays and novels as well as the media. Ngugi wa Thiongo, Kenya's most famous novelist, was exiled for his criticism of KANU.

CRIME

Death penalty in use

35,340 prisoners

Crime is rising

CRIME RATES

Murders	6 per 100,000 population
Rapes	2 per 100,000 population
Thefts	84 per 100,000 population

Nairobi's high crime levels are spreading countrywide, as a result of worsening poverty, ethnic violence, and rising banditry in the northeast. An increase in the use of guns underlies the rapid increase in violent crime.

EDUCATION

School leaving age: 14

82%

44,411 students

THE EDUCATION SYSTEM

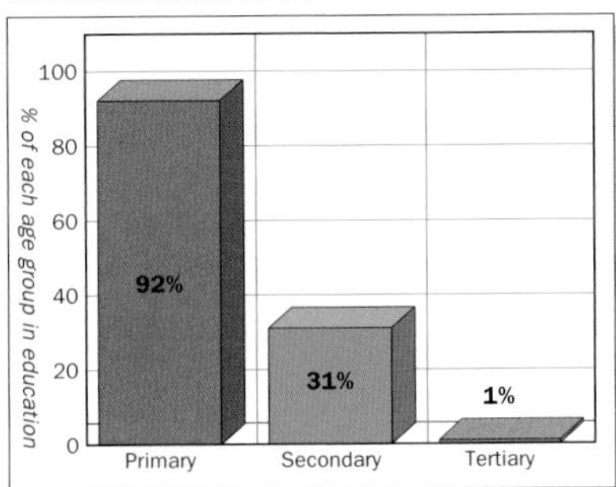
100
80
60
40
20
0
% of each age group in education
92%
31%
1%
Primary
Secondary
Tertiary

The education system is loosely based on the British model. Primary education, from six to 14 years of age, is free and compulsory; some 92% of children attend. The drop-out rate at secondary level is high, with only about 31% attendance. In higher education, the emphasis is on vocational training. There are five state universities.

HEALTH

Welfare state health benefits

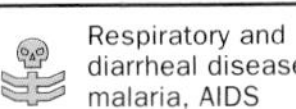
1 per 10,000 people

Respiratory and diarrheal diseases, malaria, AIDS

The health system comprises a mixture of state and private facilities, the latter mainly run by charities and missions. The state system has been badly hit by recession, worsening the already limited access of the rural majority. Poverty-related illnesses, particularly among women and children, are increasing. HIV/AIDS reached epidemic proportions in some areas in the 1990s, and the level of infection with HIV/AIDS was estimated at 15% of the adult population as at end-2001. Kenya has ten doctors and 23 qualified nurses for every 100,000 people.

SPENDING

GDP/cap. increase

CONSUMPTION AND SPENDING

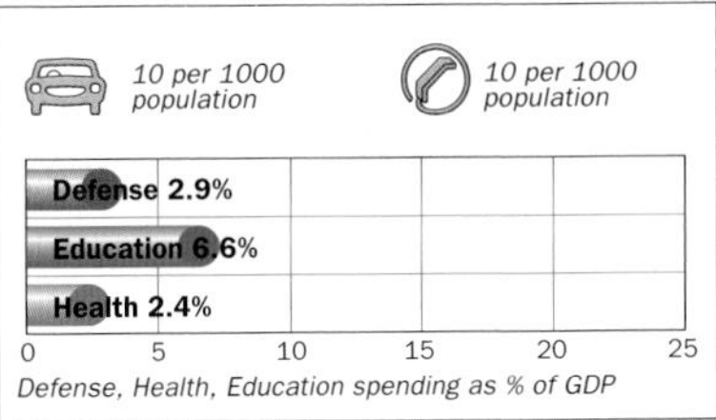
10 per 1000 population
10 per 1000 population
Defense 2.9%
Education 6.6%
Health 2.4%
0 5 10 15 20 25
Defense, Health, Education spending as % of GDP

Wealth disparities in Kenya are large and growing, exacerbated by land hunger and migration to the cities, where jobs are few and existence depends on the informal economy. More than half of all town dwellers live in slums, and the slum dwellers of Nairobi's Amarthi Valley are among Africa's poorest, worst-nourished people. Their lives contrast sharply with those of the country's elite – top government officials with access to patronage; white Kenyans, who derive their wealth largely from agricultural estates; and the largely Asian business community. Wealthy Kenyans often send their children abroad for higher education.

WORLD RANKING

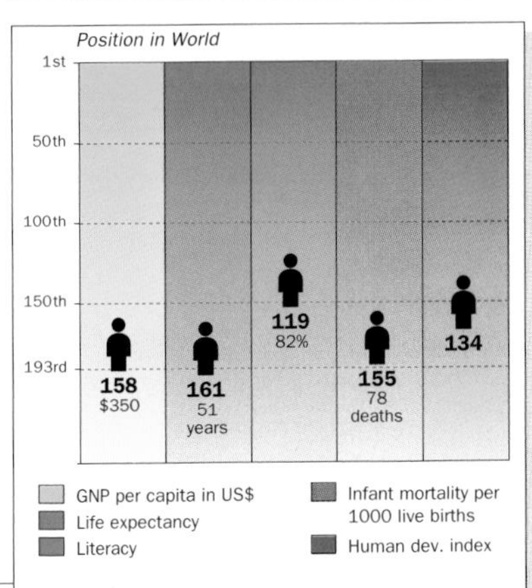
Position in World
1st
50th
100th
150th
193rd
158
$350
161
51 years
119
82%
155
78 deaths
134
GNP per capita in US$
Life expectancy
Literacy
Infant mortality per 1000 live births
Human dev. index

K

KIRIBATI

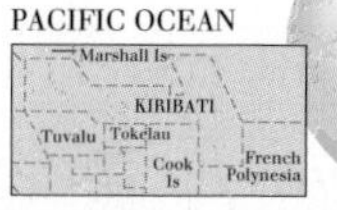
PACIFIC OCEAN
Marshall Is
KIRIBATI
Tuvalu
Tokelau
Cook Is
French Polynesia
Pacific Ocean

OFFICIAL NAME: Republic of Kiribati **CAPITAL:** Bairiki (Tarawa Atoll)
POPULATION: 92,000 **CURRENCY:** Australian dollar **OFFICIAL LANGUAGE:** English

1979

1979

July 12

KIR

+12

+686

.ki

FORMERLY PART OF THE colony of the Gilbert and Ellice Islands, the Gilberts became independent from Britain in 1979 and took the name Kiribati (pronounced "Kir-ee-bahs"). British interest in the Gilbert Islands rested solely on the exploitation of the phosphate deposits on Banaba; these ran out in 1980. In 1981, Kiribati won damages (but not the costs of litigation) from the British for decades of phosphate exploitation.

Banreaba Island, Tarawa Atoll. None of the atolls is more than 8 m (26 ft) high except Banaba, once the source of phosphates.

CLIMATE

Tropical oceanic

WEATHER CHART FOR BAIRIKI

Average daily temperature — Rainfall
°C/°F J F M A M J J A S O N D cm/in
40/104, 30/86, 20/68, 10/50, 0/32, -10/14, -20/-4
40/16, 30/12, 20/8, 10/4, 0

Kiribati's small land area in the vast Pacific means that some atolls can often go for months without rain. In 1999, a nationwide drought emergency was declared.

TRANSPORTATION

Drive on right

Bonriki International, Tarawa 51,000 passengers

4200 grt

THE TRANSPORTATION NETWORK

483 km (300 miles)

None

None

5 km (3 miles)

Kiribati has a limited air link with Fiji. Transportation around and between the atolls is provided mostly by small canoes.

TOURISM

Visitors : Population 1:92

1000 visitors

No change in 2000

MAIN TOURIST ARRIVALS

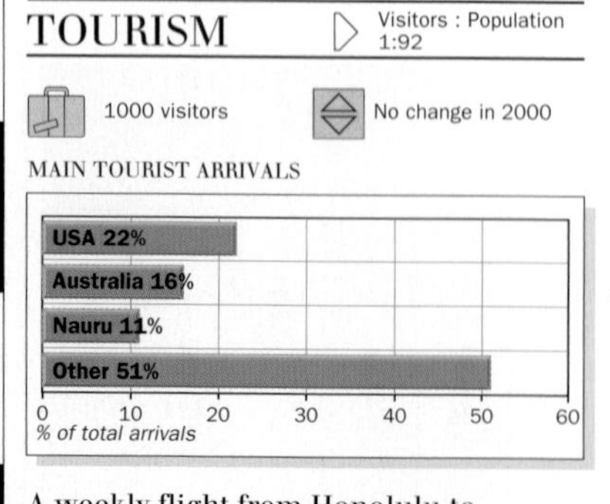
USA 22%
Australia 16%
Nauru 11%
Other 51%
0 10 20 30 40 50 60
% of total arrivals

A weekly flight from Honolulu to Kiritimati (Christmas Island) ensures a small but steady stream of visitors.

PEOPLE

Pop. density medium

English, Micronesian dialect

130/km² (336/mi²)

THE URBAN/RURAL POPULATION SPLIT

36% 64%

RELIGIOUS PERSUASION

Other 8%
Kiribati Protestant Church 39%
Roman Catholic 53%

Locals still refer to themselves as Gilbertese. Almost all Gilbertese are Micronesian, although the inhabitants of Banaba employed anthropologists to establish their racial distinctness. Tension with the Banabans is intense, mostly fueled by the historic value of Banaba's phosphate deposits. Most Gilbertese are poor. Many go to Nauru as guest workers, living in barrack-room conditions, or work as merchant shipping crew. Those who stay at home go through a circular migration from the outlying islands to Tarawa, returning to see relatives. Women play a prominent role, especially on outlying islands, where they run most of the farms.

POLITICS

Nonparty elections

1998/2002

President Teburoro Tito

AT THE LAST ELECTION
House of Assembly 42 seats

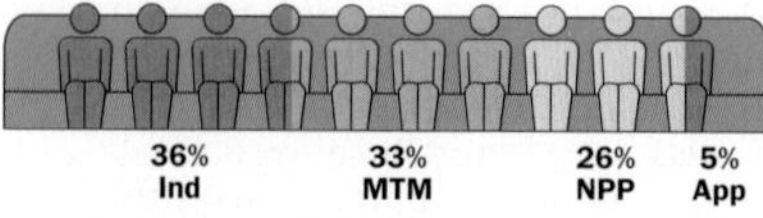

36% Ind **33% MTM** **26% NPP** **5% App**

Ind = Independents **MTM** = Maneaban Te Mauri
NPP = National Progressive Party **App** = Appointed

The House of Assembly has one appointed member and one ex-officio member

The traditional chiefs still effectively rule Kiribati, through a party system on the British model. Victory for the MTM in the 1994 elections ended 15 years of rule by the NPP. The main concern is the economy, which is extremely vulnerable to any fluctuations in world demand for coconuts. The overpopulation of Tarawa is the other major issue. Possible restrictions on travel to the island have been discussed. In part, the problem of migration is caused by the poverty and lack of opportunity on the outer islands. A resettlement program, aiming to move people out of Tarawa, began in 1998.

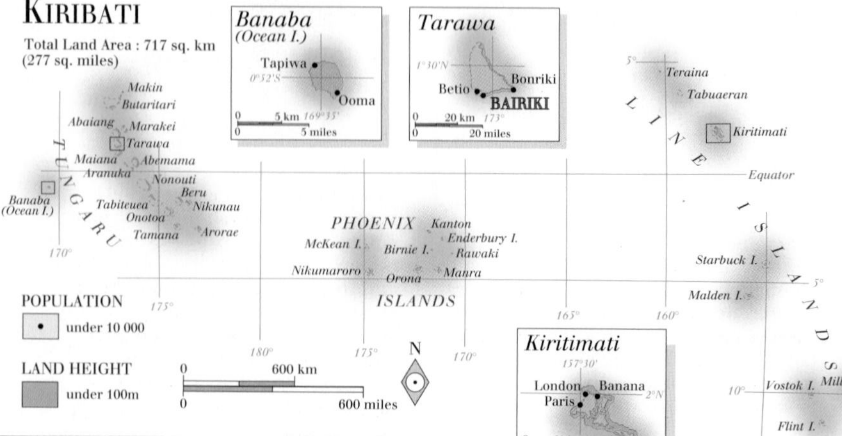
KIRIBATI
Total Land Area : 717 sq. km (277 sq. miles)
Banaba (Ocean I.)
Tapiwa
Ooma
5 km
5 miles
Tarawa
Betio
Bonriki
BAIRIKI
20 km
20 miles
Makin
Butaritari
Abaiang
Marakei
Tarawa
Maiana
Abemama
Aranuka
Nonouti
Beru
Nikunau
Tabiteuea
Onotoa
Tamana
Arorae
Banaba (Ocean I.)
TUNGARU
PHOENIX ISLANDS
Kanton
Enderbury I.
McKean I.
Birnie I.
Rawaki
Nikumaroro
Orona
Manra
LINE ISLANDS
Teraina
Tabuaeran
Kiritimati
Equator
Starbuck I.
Malden I.
Vostok I.
Millennium
Flint I.
POPULATION
under 10 000
LAND HEIGHT
under 100m
600 km
600 miles
N
Kiritimati
London
Banana
Paris
20 km
20 miles

WORLD AFFAIRS

Joined UN in 1999

ACP
Comm
PC
PIF
ADB

Kiribati has little international significance because of its tiny size and remote location, but is able to make its voice heard regionally through the Pacific Islands Forum. In 1986, Kiribati was a signatory to a deal between the US and a number of Pacific Island states that resulted in the US paying $60 million in return for access to Pacific fishing grounds. In the Cold War era Kiribati played the USSR off against the US, extracting a high price for fishing leases, which allowed boats to spy on US nuclear testing on the neighboring Kwajalein Atoll in the Marshall Islands.

AID

Recipient

US$18m (receipts)

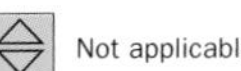
Down 14% in 2000

Aid is mostly offered for small projects to improve infrastructure. New Zealand is the most generous donor, granting US$1.7 million in 2000.

DEFENSE

No compulsory military service

Kiribati has no defense budget

Not applicable

Australia and New Zealand provide de facto protection, with regular antisubmarine patrols.

ECONOMICS

Inflation 3.8% p.a. (1990–2000)

US$86m

1.7997–1.9535 Australian dollars

SCORE CARD

❑ WORLD GNP RANKING	188th
❑ GNP PER CAPITA	US$950
❑ BALANCE OF PAYMENTS	US$1m
❑ INFLATION	2%
❑ UNEMPLOYMENT	2%

STRENGTHS

Subsistence economy; only Tarawa imports food. Coconuts provide some export income: the EU is the biggest market. Fisheries have limited potential. Upgraded port facilities at Betio.

WEAKNESSES

Lack of resources. High levels of poverty. Isolation, and large distances between islands. Heavy dependence on international aid. Almost no economic potential.

EXPORTS

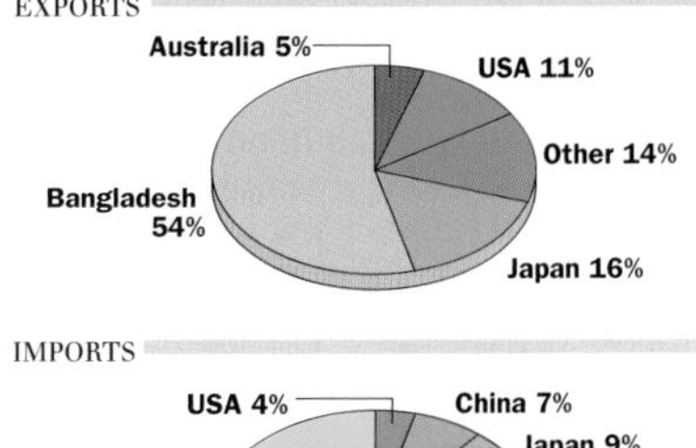
Australia 5%
USA 11%
Other 14%
Bangladesh 54%
Japan 16%

IMPORTS

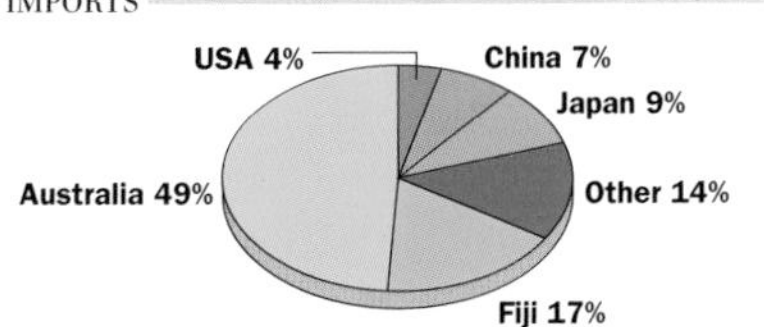
USA 4%
China 7%
Japan 9%
Other 14%
Australia 49%
Fiji 17%

RESOURCES

Electric power 2000 kw

48,218 tonnes

Not an oil producer

13,000 pigs, 400,000 chickens

None

Phosphate deposits on Banaba ran out in 1980. All energy supplies have to be imported. An underwater agriculture scheme is under development.

ENVIRONMENT

Not available

39% (including marine and semi-protected areas)

0.3 tonnes per capita

Rising sea levels cause coastal erosion and ultimately threaten Kiribati's existence. Global warming is a critical issue both for this reason, and because of its damaging effects on the coral reef which protects Tarawa from the sea and holds important inshore fish stocks in the lagoon. The coral has also suffered from pollution by untreated effluent.

MEDIA

TV ownership low

There are no daily newspapers

PUBLISHING AND BROADCAST MEDIA

There are no daily newspapers. The weekly newspapers are *Butim'aea Manin te Euangkerio, Kiribati Newstar*, and *Te Uekera*

1 independent service

1 independent service

The independent *Newstar* competes with the state-owned *Te Uekera* and the Protestant Church's paper. Churches also produce monthly newsletters.

CRIME

No death penalty

57 prisoners

Crime is minimal

Crime, apart from brawls resulting from drunkenness, is minimal. The islands' judicial system is based on the British model.

EDUCATION

School leaving age: 15

98%

568 students

Education is British-inspired and compulsory from six to 15. The best students go on to university in Fiji.

CHRONOLOGY

The British established the phosphate-producing colony of the Gilbert and Ellice Islands in 1892.

- ❑ **1957** British nuclear tests take place near Kiritimati.
- ❑ **1979** Independence as two states, Kiribati and Tuvalu. NPP in power.
- ❑ **1981** Kiribati wins damages for phosphate mining from UK.
- ❑ **1986** Kiribati–US fishing deal.
- ❑ **1994** NPP loses election to MTM. Teburoro Tito elected president.
- ❑ **1999** National drought emergency.

HEALTH

Welfare state health benefits

1 per 7600 people

Heart diseases, diabetes

Most Gilbertese are healthy despite the high levels of poverty. Free medical care is provided to all. Nutrition is becoming a problem on Tarawa where overpopulation limits agriculture.

SPENDING

GDP/cap. increase

CONSUMPTION AND SPENDING

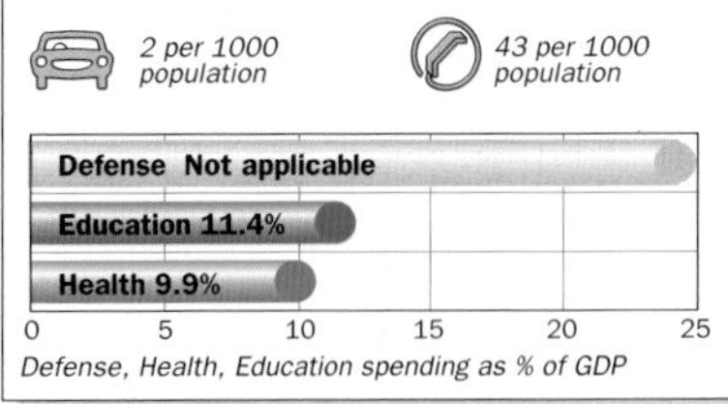
2 per 1000 population
43 per 1000 population
Defense Not applicable
Education 11.4%
Health 9.9%
0
5
10
15
20
25
Defense, Health, Education spending as % of GDP

Life in Kiribati is modest. Most Gilbertese live by subsistence farming and fishing. Civil servants in Bairiki form the wealthiest group. The cost of living in Tarawa is higher than in the outlying islands due to the need to import food, although fish is abundant and cheap everywhere.

WORLD RANKING

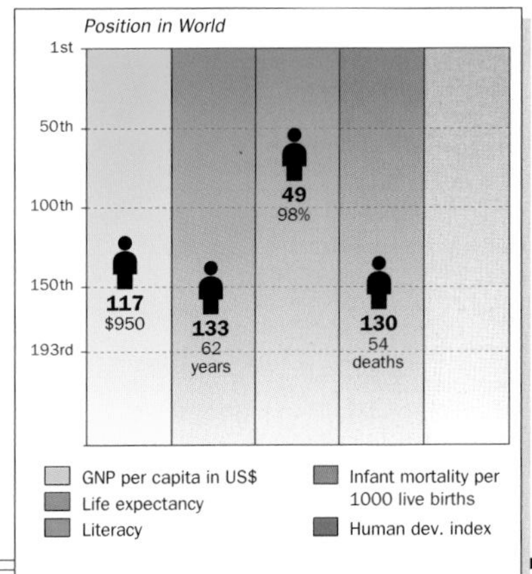
Position in World
1st
50th
100th
150th
193rd
117
$950
133
62
years
49
98%
130
54
deaths
GNP per capita in US$
Life expectancy
Literacy
Infant mortality per 1000 live births
Human dev. index

K

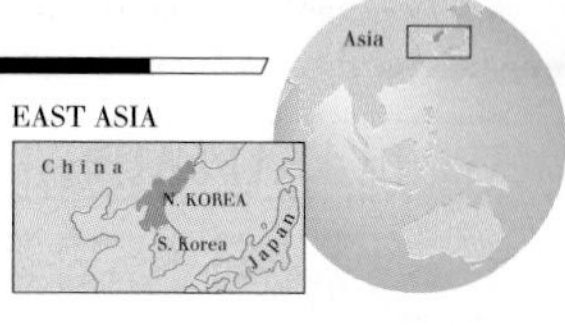
Asia

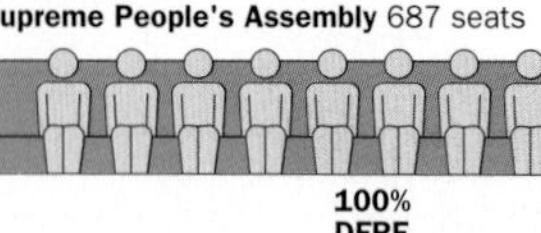
EAST ASIA
China
N. KOREA
S. Korea
Japan

North Korea

Official Name: Democratic People's Republic of Korea **Capital:** Pyongyang
Population: 22.4 million **Currency:** North Korean won **Official Language:** Korean

1948

1953

Sept 9

DRK

+9

+850

.kp

COMPRISING THE NORTHERN half of the Korean peninsula, North Korea is separated from the US-dominated South by an armistice line straddling the 38th parallel. Much of the country is mountainous; the Chaeryong and Pyongyang plains in the southwest are the most fertile regions. An independent communist republic from 1948, it remains largely isolated. With its economy starved of capital, it now faces a food crisis requiring large-scale international assistance.

CLIMATE

Continental

WEATHER CHART FOR PYONGYANG

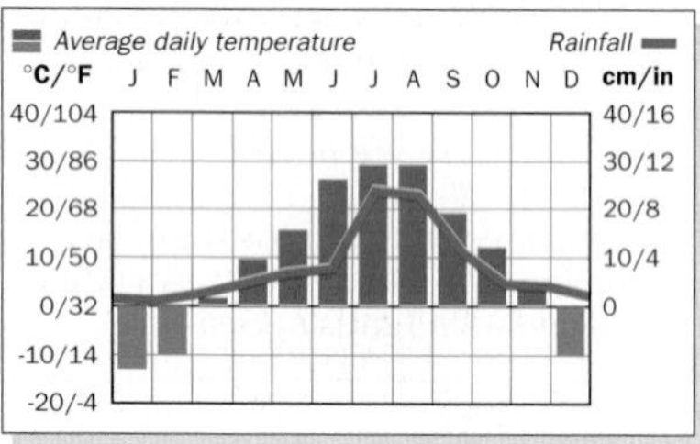
Average daily temperature
Rainfall
°C/°F
J F M A M J J A S O N D
cm/in
40/104
30/86
20/68
10/50
0/32
-10/14
-20/4
40/16
30/12
20/8
10/4
0

North Korea has a typically continental climate. Winters in the north can be extreme.

TRANSPORTATION

Drive on right

Sunan, Pyongyang

182 ships 631,000 grt

THE TRANSPORTATION NETWORK

1997 km (1241 miles)

524 km (326 miles)

5214 km (3240 miles)

2253 km (1400 miles)

The railroad network built by the occupying Japanese is heavily relied on. Highways are open only to very limited, officially approved traffic. Improving relations with South Korea in 2000 led to plans to construct cross-border links.

TOURISM

Visitors : Population 1:172

130,000 visitors

Up 2% 1995–1998

MAIN TOURIST ARRIVALS

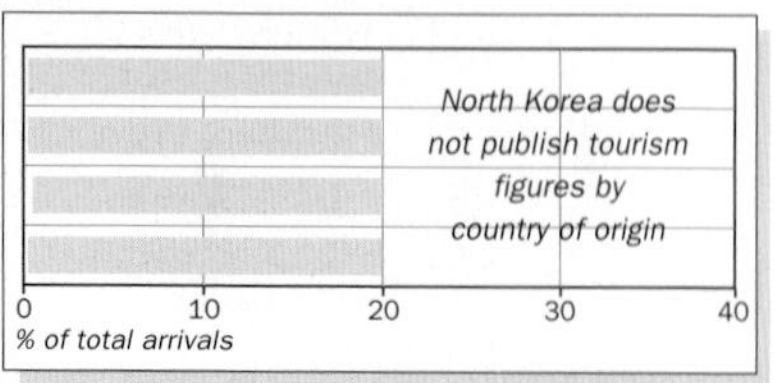
North Korea does not publish tourism figures by country of origin
0
10
20
30
40
% of total arrivals

Economic need has forced limited tourism. South Korean firms have developed resorts such as Mt. Kumgang.

Rice paddy field. The hot, wet summers are ideal for rice growing. Most farms are run as cooperatives.

PEOPLE

Pop. density medium

Korean, Chinese

186/km² (482/mi²)

THE URBAN/RURAL POPULATION SPLIT

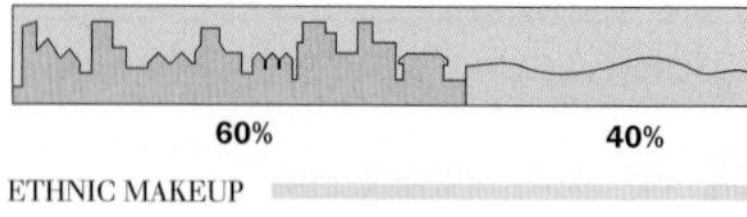
60%
40%

ETHNIC MAKEUP

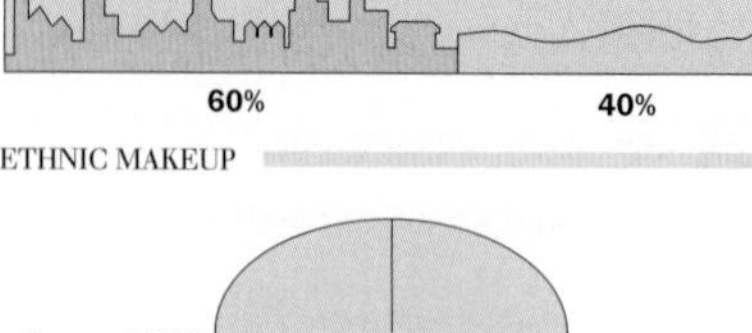
Korean 100%

The Korean peninsula is unusual in having been inhabited by a single ethnic group for the last 2000 years. There is a tiny Chinese minority in North Korea.

The religions practiced under strict state control are Buddhism, Christianity, and Chondogyo, a combination of Confucianism, Buddhism, and Christianity which is peculiar to Korea.

North Koreans live highly regulated lives. Divorce is nonexistent and extramarital sex highly frowned upon. Women form more than 50% of the workforce, but are also expected to run the home; it is not uncommon for them to rise at 4 a.m., and end their working day at 7 p.m. From an early age, children are looked after by an extensive system of state-run crèches. The privileged lifestyle of the political elite – some 200,000 in number – is rumored to be a source of popular resentment.

POLITICS

No multiparty elections

1998/2003

Eternal President Kim Il Sung

AT THE LAST ELECTION

Supreme People's Assembly 687 seats

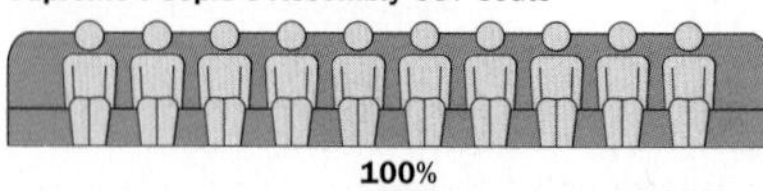
100%
DFRF

DFRF = Democratic Front for the Reunification of the Fatherland
The Democratic Front for the Reunification of the Fatherland (led by the Korean Workers' Party – **KWP**) was the sole party permitted to take part in the 1998 elections

The three million-strong KWP is the only legal party; membership is essential for individual advancement. Kim Il Sung, the subject of a lavish personality cult, died in 1994 after almost 50 years as leader. Since then the key question has been how Kim Jong Il, his son and chosen successor, would handle the leadership. He lacks his father's authority, and has yet to seal the succession; in 1998 Kim Il Sung, by then four years dead, was declared "Eternal President."

WORLD AFFAIRS

Joined UN in 1991

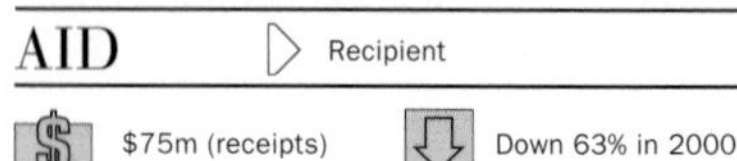
NAM

The collapse of Soviet communism in 1991 destroyed North Korea's framework of foreign relations, leaving China as its closest ally. A breakthrough in North–South relations came in 2000, with their leaders' first direct encounter since the 1953 armistice. In 2001 Kim Jong Il paid a landmark visit to Russia. Despite North Korea's nuclear missile program, several countries have established diplomatic links, and many US trade sanctions were lifted; however, the US designated North Korea part of an "axis of evil" in 2002.

AID

Recipient

$75m (receipts)

Down 63% in 2000

Rice harvests were badly damaged by alternate drought and flood in the mid-1990s, precipitating a famine. Efforts to stave off starvation using international aid are threatened by lack of funds.

DEFENSE

Compulsory military service

$2.05bn

Down 2% in 2000

North Korea is thought to have manufactured nuclear weapons prior to the 1994 freeze on its nuclear program. It developed, and exports, missiles.

North Korea

Total Land Area : 120 540 sq. km (46 540 sq. miles)

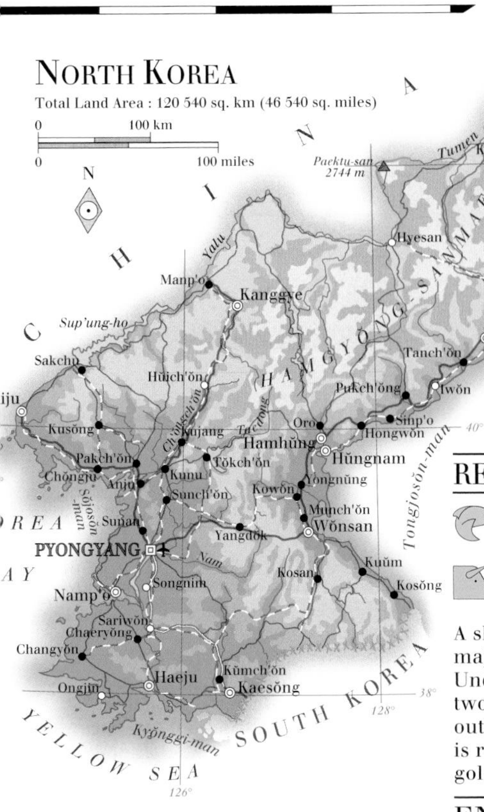
0 100 km
0 100 miles
N
CHINA
Onsŏng
Kyŏnghŭng
Hoeryŏng
RUSSIAN FEDERATION
Musan
Komusan
Najin
Tumen
Paektu-san 2744 m
Ch'ŏngjin
Hyesan
Yalu
Manp'o
Kanggye
Myŏngch'ŏn
Kilchu
Sup'ung-ho
Sakchu
Hŭich'ŏn
HAMGYŎNG-SANMAEK
Tanch'ŏn
Kimch'aek
Pukch'ŏng
Iwŏn
Sinp'o
Hongwŏn
Oro
Kusŏng
Kujang
Hamhŭng
Hŭngnam
Pakch'ŏn
Chŏngju
Anju
Kunu
Tŏkch'ŏn
Sunch'ŏn
Kowŏn
Yŏngnŭng
Munch'ŏn
Wŏnsan
Sunan
Yangdŏk
PYONGYANG
Kŭmŭm
Kosan
Kosŏng
Namp'o
Songnim
Sariwŏn
Chaeryŏng
Changyŏn
Haeju
Kŭmch'ŏn
Kaesŏng
Ongjin
SOUTH KOREA
YELLOW SEA
Kyŏnggi-man
Tongjosŏn-man

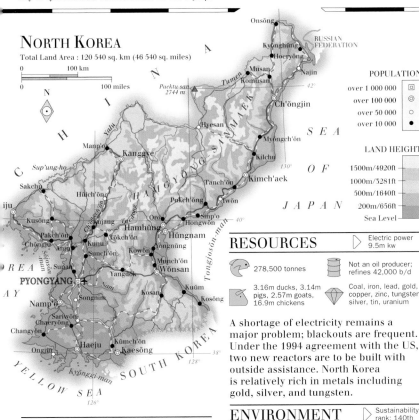

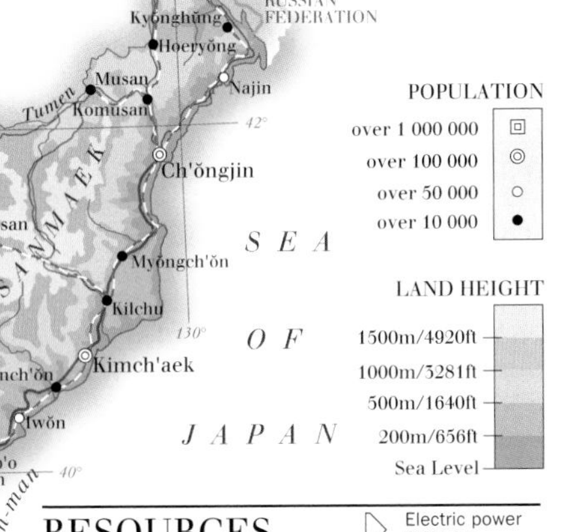
SEA OF JAPAN
POPULATION
over 1 000 000
over 100 000
over 50 000
over 10 000
LAND HEIGHT
1500m/4920ft
1000m/3281ft
500m/1640ft
200m/656ft
Sea Level

ECONOMICS

Not available

$12.6bn

2.2 North Korean won

SCORE CARD

- ❑ World GNP Ranking 77th
- ❑ GNP per Capita $573
- ❑ Balance of Payments .... *Closed economy;*
- ❑ Inflation .... *does not publish*
- ❑ Unemployment .... *any figures*

Strengths

Other than minerals, North Korea's economy has few strengths.

Weaknesses

GNP has declined steadily since 1990. The acute shortage of foreign capital and technology has been catastrophic.

EXPORTS

IMPORTS

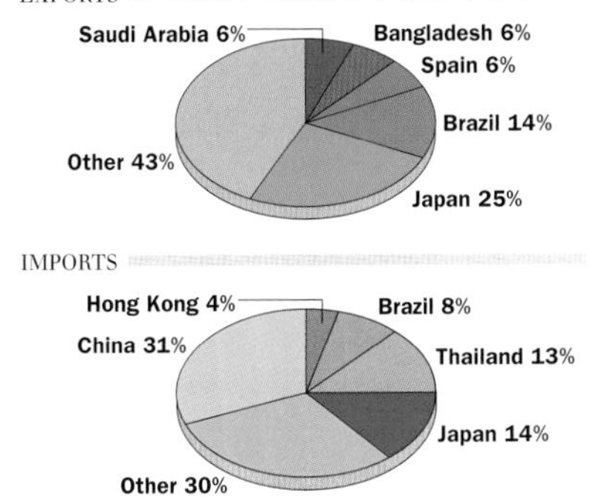
Saudi Arabia 6%
Bangladesh 6%
Spain 6%
Brazil 14%
Other 43%
Japan 25%
Hong Kong 4%
Brazil 8%
China 31%
Thailand 13%
Japan 14%
Other 30%

RESOURCES

Electric power 9.5m kw

278,500 tonnes

Not an oil producer; refines 42,000 b/d

3.16m ducks, 3.14m pigs, 2.57m goats, 16.9m chickens

Coal, iron, lead, gold, copper, zinc, tungsten, silver, tin, uranium

A shortage of electricity remains a major problem; blackouts are frequent. Under the 1994 agreement with the US, two new reactors are to be built with outside assistance. North Korea is relatively rich in metals including gold, silver, and tungsten.

ENVIRONMENT

Sustainability rank: 140th

3% (0.1% partially protected)

10.3 tonnes per capita

Excessive use of fertilizers and unchecked pollution from heavy industry are the major problems.

MEDIA

TV ownership medium

Daily newspaper circulation 199 per 1000 people

PUBLISHING AND BROADCAST MEDIA

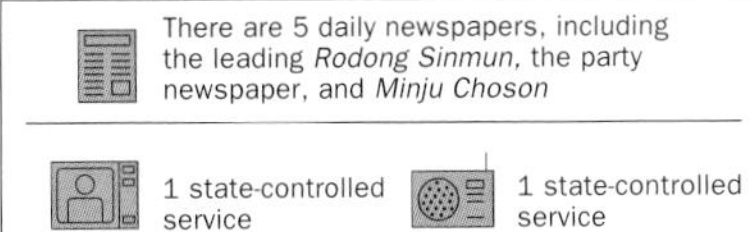

There are 5 daily newspapers, including the leading *Rodong Sinmun*, the party newspaper, and *Minju Choson*

1 state-controlled service

1 state-controlled service

TV consists mostly of musical shows praising Kim Il Sung and Kim Jong Il, and anti-American tirades directed against the Korean War.

CRIME

Death penalty in use

North Korea does not publish prison figures

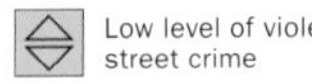

Low level of violent street crime

At an individual level, crime is officially said hardly to exist. The criminal code is weighted to protect the state against "subversion," rather than the rights of the individual. North Korea has a very poor human rights record and there is a *gulag* of more than 100,000 "subversives," where whole families are sent along with those accused, and where torture is routine.

CHRONOLOGY

Annexed by Japan in 1910, the peninsula was divided in 1945 at the 38th parallel; North Korea was made an independent state in 1948.

- ❑ **1950–1953** Korean War.
- ❑ **1994** Withdrawal from IAEA. Kim Il Sung dies; declared "Eternal President" four years later.
- ❑ **1997** Threat of famine worsens. Kim Jong Il becomes party leader.
- ❑ **2000** Historic North–South summit.

EDUCATION

School leaving age: 16

95%

390,000 students

English is compulsory as a second language at the age of 14. Kim Il Sung, Pyongyang, is the only university.

HEALTH

Welfare state health benefits

1 per 333 people

Heart disease, cancers, digestive diseases

Health care is free. Reasonable life expectancy is now threatened by malnutrition and outright starvation.

SPENDING

GDP/cap. decrease

CONSUMPTION AND SPENDING

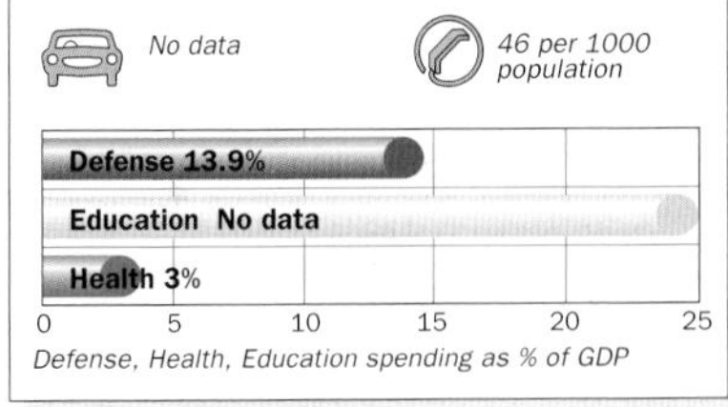
No data
46 per 1000 population
Defense 13.9%
Education No data
Health 3%
0
5
10
15
20
25
Defense, Health, Education spending as % of GDP

An elite within the KWP lives well, with access to specialist shops and consumer goods such as VCRs. Ownership of telephones, private cars, and, in many areas, bicycles, is forbidden.

WORLD RANKING

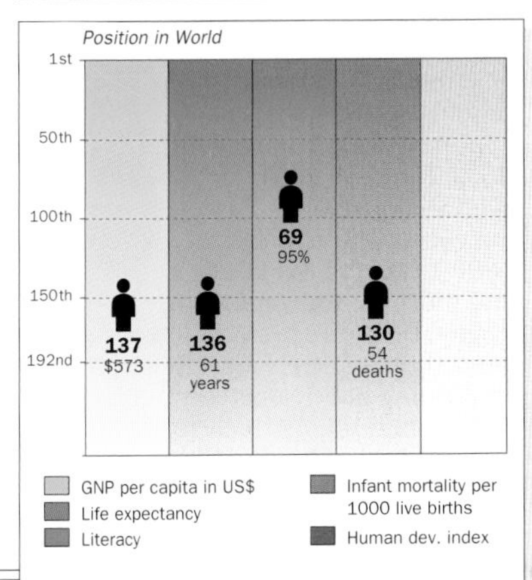
Position in World
1st
50th
100th
150th
192nd
137
$573
136
61
years
69
95%
130
54
deaths
GNP per capita in US$
Life expectancy
Literacy
Infant mortality per 1000 live births
Human dev. index

K

SOUTH KOREA

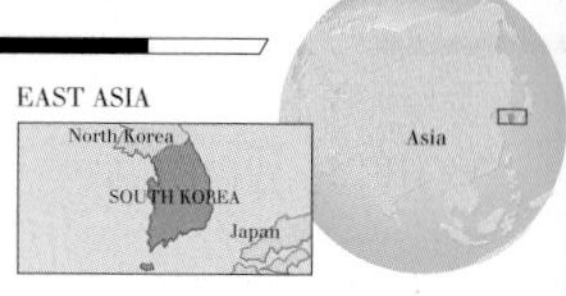
EAST ASIA
North Korea
SOUTH KOREA
Japan
Asia

OFFICIAL NAME: Republic of Korea **CAPITAL:** Seoul
POPULATION: 47.1 million **CURRENCY:** South Korean won **OFFICIAL LANGUAGE:** Korean

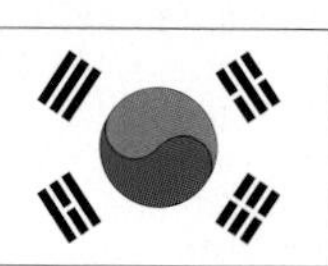

1948

1953
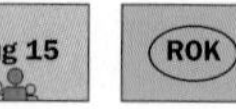
Aug 15
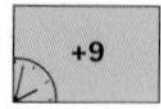
ROK

+9

+82

.kr

OCCUPYING THE SOUTHERN half of the Korean peninsula in East Asia, over 80% of South Korea is mountainous and two-thirds is forested. Rice is the major agricultural product, grown by over 85% of South Korea's three million farmers. The whole peninsula was annexed by Japan from 1910 to 1945, and the split between South Korea and the communist North originated with the arrival of rival US and Soviet armies in 1945. Although the two states have discussed reunification, the legacy of hostility arising from the 1950–1953 Korean War remains a major obstacle.

TOURISM

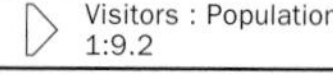
Visitors : Population 1:9.2

5.15m visitors

Down 3% in 2001

MAIN TOURIST ARRIVALS

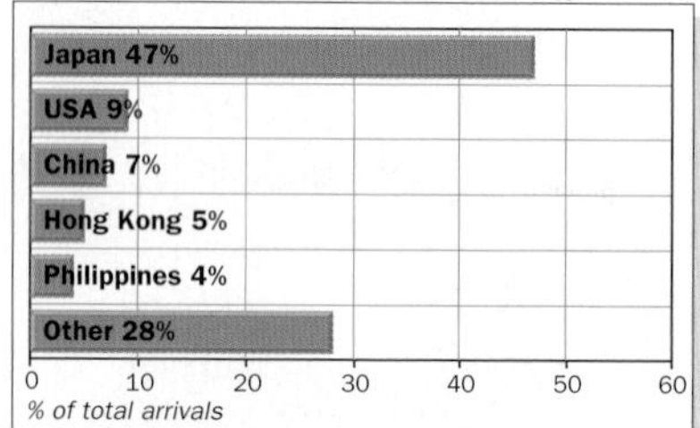
Japan 47%
USA 9%
China 7%
Hong Kong 5%
Philippines 4%
Other 28%
0
10
20
30
40
50
60
% of total arrivals

Overseas tourism to South Korea has increased ten-fold since 1969. Most visitors are Japanese, who come for the golf and Seoul's nightlife; Jeju-do (Cheju-do) is a favored honeymoon destination. Whereas visiting relations of US army personnel once made up 13% of all tourists, today Los Angeles-based Korean–Americans make up the greatest proportion of US visitors. However, despite the publicity generated by the 1988 Seoul Olympics and the 2002 soccer World Cup, South Korea is still not seen in the West as a prime tourist destination.

CLIMATE

Continental

WEATHER CHART FOR SEOUL

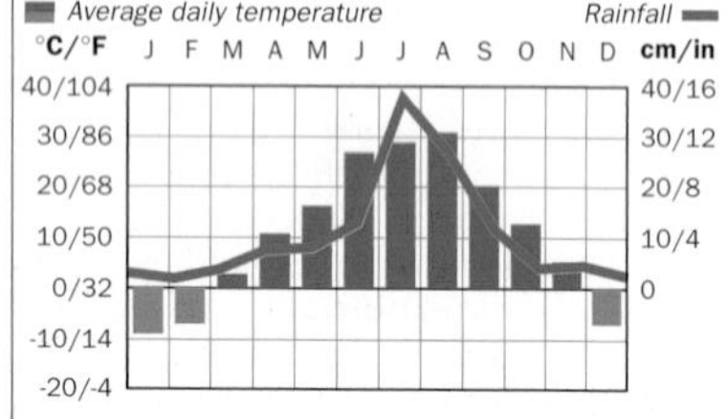
Average daily temperature
Rainfall
°C/°F
J F M A M J J A S O N D
cm/in
40/104
30/86
20/68
10/50
0/32
-10/14
-20/-4
40/16
30/12
20/8
10/4
0

South Korea has four distinct seasons. Winters are dry and can be bitterly cold. Summers are hot and humid, especially during July and August.

TRANSPORTATION

Drive on right

Kimpo International, Seoul 36.7m passengers

2381 ships 5.69m grt

THE TRANSPORTATION NETWORK

64,808 km (40,270 miles)

2477 km (1539 miles)

3098 km (1925 miles)

1609 km (1000 miles)

South Korea has an efficient and highly integrated public transportation system. A toll-based nationwide motor expressway network joins most major urban centers. Air travel has expanded rapidly as a convenient way to traverse the mountainous interior. Korean Air competes with Asiana. Buses, trains, boats, and planes are integrated in one timetable, and have a reputation for punctuality. A high-speed rail link will connect Seoul and Busan (Pusan).

Improving relations with North Korea in 2000 produced proposals for the construction of cross-border rail and road links and north–south connecting flights. Concrete projects have been slow to materialize, due to continuing political uncertainty, but the first flights since 1953 took place in mid-2002, albeit limited to teams of construction workers in the nuclear industry.

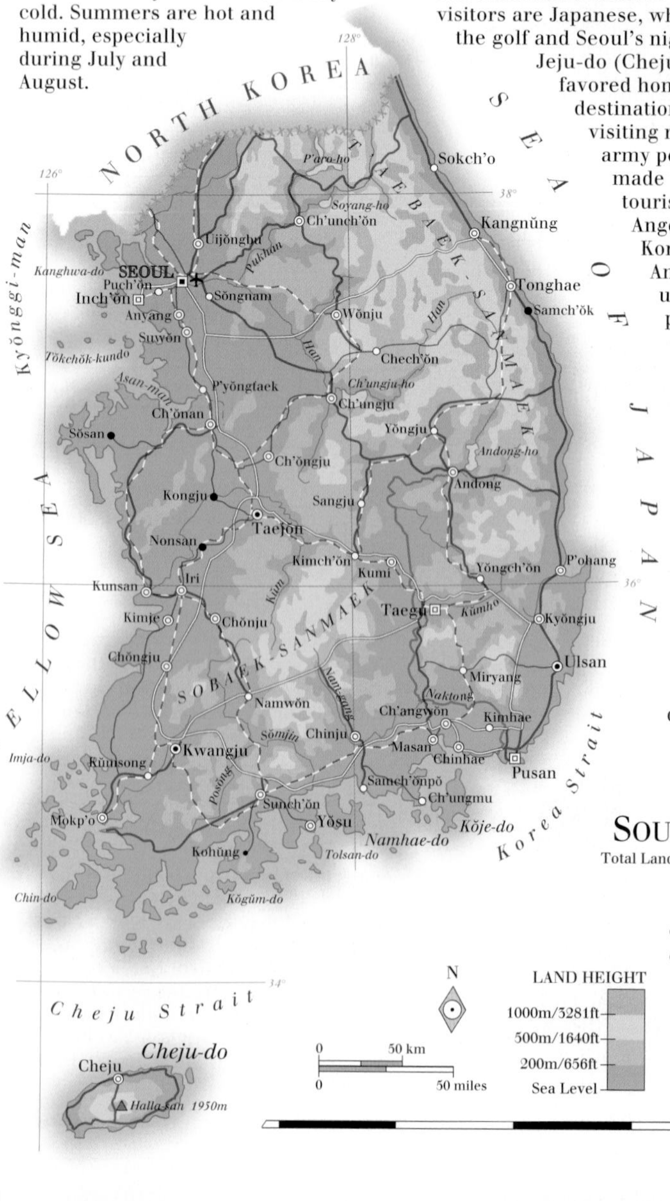

SOUTH KOREA

Total Land Area : 98 480 sq. km (38 023 sq. miles)

POPULATION: over 5 000 000; over 1 000 000; over 500 000; over 100 000; over 50 000; over 10 000; under 10 000

LAND HEIGHT: 1000m/3281ft; 500m/1640ft; 200m/656ft; Sea Level

0 – 50 km; 0 – 50 miles

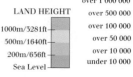

PEOPLE

Population density high

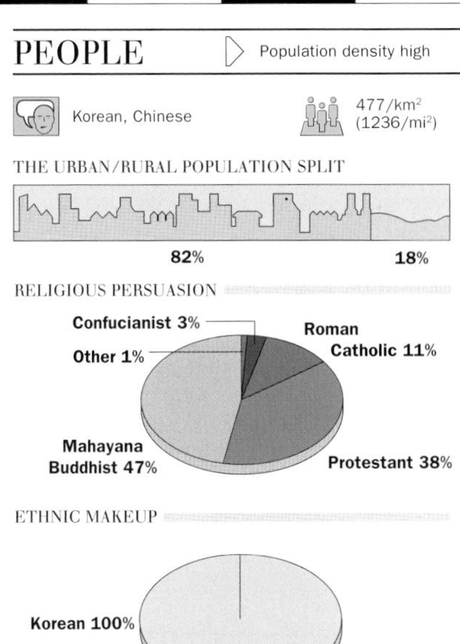
Korean, Chinese
477/km² (1236/mi²)
THE URBAN/RURAL POPULATION SPLIT
82%
18%
RELIGIOUS PERSUASION
Confucianist 3%
Other 1%
Roman Catholic 11%
Mahayana Buddhist 47%
Protestant 38%
ETHNIC MAKEUP
Korean 100%

The Korean peninsula is unusual in having been inhabited by one ethnic group for the last 2000 years. Family life is a central and clearly defined part of society, although the nuclear family model is becoming the norm rather than the old-style household of the extended family. Most Koreans can trace their ancestry back thousands of years. Regional origin is important in determining blood heritage, since there are only 270 Korean surnames and half the population is named Kim, Lee, Park, or Choi. Chondogyo, combining elements of Shamanism, Buddhism, and Christianity, is peculiar to the Koreas but has only a tiny following. Traditional values condition attitudes to women, and it is still not respectable for those who are married to have a job.

Economic growth has attracted illegal immigrants from the poorer Asian countries, who take menial jobs that South Koreans now refuse to do.

POPULATION AGE BREAKDOWN

Female	Age	Male
0.6%	80+	0.2%
5%	60–79	3.5%
10.6%	40–59	10.8%
18.5%	20–39	19.2%
15.1%	0–19	16.5%

% of population by age group

POLITICS

Multiparty elections

2000/2004 President Kim Dae Jung

AT THE LAST ELECTION

National Assembly 273 seats

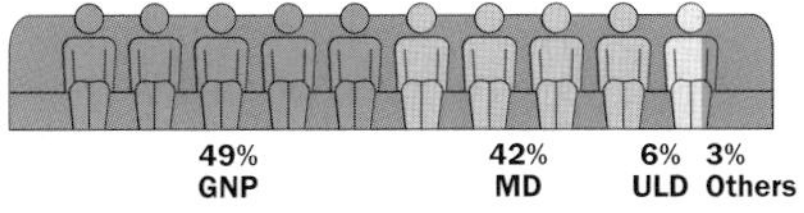
49% GNP
42% MD
6% ULD
3% Others

GNP = Grand National Party **MD** = Millennium Democratic Party **ULD** = United Liberal Democrats

Officially a democracy since its inception, South Korea was in practice ruled by military dictators until 1987.

PROFILE

South Korea's politics changed radically in 1987 with the introduction of direct presidential elections and a parliament with enhanced powers.

In 1993, Kim Young Sam became president, the first nonmilitary leader in 30 years. He launched a popular anticorruption campaign, targeting former presidents. His New Korea Party (NKP) (formerly the Democratic Liberal Party – DLP), returned to power with a reduced majority in 1996, was brought down amid a steel scandal in 1997. Veteran opposition leader Kim Dae Jung was elected president later that year in the first peaceful transfer of power to an opposition politician in South Korea's history. Supporters of the new president gained ground in legislative elections in 2000, but political instability continued, with frequent changes of prime minister and opposition byelection victories. Chang Sang was appointed the first female prime minister in 2002, but was rejected by parliament.

MAIN POLITICAL ISSUES

The economy

South Korea's economy, once one of the most impressive in the world, was severely shaken by the regional crisis of late 1997. Severe financial austerity and retrenchment failed to prevent large-scale bankruptcies in 2000.

Relations with North Korea

The sudden flowering of North–South relations in 2000 soon lost popularity. Expensive cross-border projects, and promises of aid for the impoverished North, raised fears of the mounting cost for the embattled southern economy.

President Kim Dae Jung. *A veteran activist, he has won the Nobel Peace Prize.*

Chang Sang, *first woman appointed as premier, but rejected by parliament in 2002.*

WORLD AFFAIRS

Joined UN in 1991

APEC

CP

IAEA

OECD

WTO

Since the division of Korea, relations with the North have dominated foreign policy. A historic summit meeting in 2000 in the North Korean capital, Pyongyang, opened a new phase, and cross-border diplomatic and economic cooperation briefly flourished. Reunification remains the ultimate goal of both Koreas, but with doubts about its social and economic costs. Military tensions persist, and a large detachment of US troops remains on the border – the most heavily defended in the world. Relations with China, the closest ally of North Korea, have improved. Japan is also a major trading partner, although South Koreans continue to harbor resentment over the 1910–1945 Japanese annexation.

AID

Donor

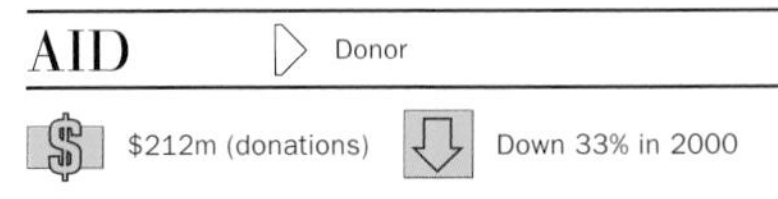
$212m (donations)
Down 33% in 2000

Once a massive recipient of US aid, and then of Japanese war reparations, South Korea emerged in the 1970s and 1980s as a major aid donor. However, the economic crisis which hit South Korea in 1997–1998 forced it to seek international financial assistance to salvage key sectors of its threatened economy.

CHRONOLOGY

The Yi dynasty, founded in Seoul in 1392, ruled the kingdom of Korea until 1910. Korea became a vassal state of China in 1644.

- **1860** Korea reacts to French and British occupation of Peking by preventing Western influence: becomes the "Hermit Kingdom."
- **1904–1905** Russo-Japanese War. Japan conquers Korea.
- **1910** Japan annexes Korea.
- **1919** Independence protests violently suppressed.
- **1945** US and Soviet armies arrive. Korea split at 38°N. South comes under de facto US rule.
- **1948** Republic of South Korea created; Syngman Rhee becomes president at head of an increasingly authoritarian regime.
- **1950** Hostilities between North and South, each aspiring to rule a united Korea. North invades, sparking Korean War. US, with UN backing, enters on South's side; China unofficially assists North. In 1951 fighting stabilizes near 38th parallel.
- **1953** Armistice; de facto border at cease-fire line, close to the 38th parallel.

K

CHRONOLOGY *continued*

- ❑ **1960** Syngman Rhee resigns in face of popular revolt.
- ❑ **1961** Military coup leads to authoritarian junta led by Park Chung Hee.
- ❑ **1963** Pressure for civilian government. Park reelected as president (also in 1967 and 1971). Strong manufacturing base and exports drive massive economic development program.
- ❑ **1965** Links restored with Japan.
- ❑ **1966** 45,000 troops engaged in South Vietnam.
- ❑ **1972** Martial law stifles political opposition. New constitution with greater presidential powers.
- ❑ **1979** Park assassinated. Gen. Chun Doo Hwan, intelligence chief, leads coup. Kim Young Sam, opposition leader, expelled from parliament.
- ❑ **1980** Chun chosen as president. Kim Dae Jung and other opposition leaders arrested.
- ❑ **1986** Car exports start.
- ❑ **1987** Emergence of prodemocracy movement. Roh Tae Woo, Chun's chosen successor, elected president.
- ❑ **1988** Inauguration of Sixth Republic which includes genuine multiparty democracy. Restrictions on foreign travel lifted.
- ❑ **1990** Government party and two opposition parties, including Kim Young Sam's, merge to form DLP.
- ❑ **1991** South Korea joins UN.
- ❑ **1992** Diplomatic links with China established. December, Kim Young Sam elected president.
- ❑ **1996** Chun sentenced to death on charges of organizing 1979–1980 overthrow of civilian government; Roh given a lengthy prison term. Both sentences were rescinded.
- ❑ **1997** Violent protests against new labor laws. Steel scandal brings down government. Economic crisis.
- ❑ **1998** Kim Dae Jung president.
- ❑ **2000** Historic North–South summit in Pyongyang.
- ❑ **2002** Short-lived appointment of Chang Sang as first female premier.

Seoul lit up at night. *The city is home to more than ten million people – one-quarter of South Korea's population. Seoul means "capital."*

K

DEFENSE

▷ Compulsory military service

 $12.5bn

 Up 3% in 2000

SOUTH KOREAN ARMED FORCES

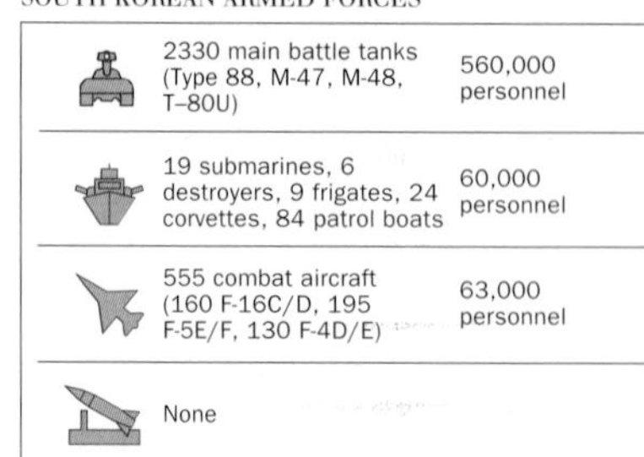

2330 main battle tanks (Type 88, M-47, M-48, T-80U)	560,000 personnel
19 submarines, 6 destroyers, 9 frigates, 24 corvettes, 84 patrol boats	60,000 personnel
555 combat aircraft (160 F-16C/D, 195 F-5E/F, 130 F-4D/E)	63,000 personnel
None	

The main defense concern is North Korea. South Korea has fewer troops, tanks, artillery, and aircraft, but it claims parity through the permanent presence of 35,000 US troops on its territory and its superior technology. The manufacture of missiles capable of striking any target in North Korea was legalized in 2001. However, US computer simulations question South Korea's ability to resist an invasion by the North, since Seoul is only 55 km (35 miles) from the demilitarized zone.

In the mid-1990s, the army's role and standing in national politics was sharply downgraded after a vigorous government campaign to investigate corruption in the armed forces, especially pertaining to arms procurement and past military involvement in politics.

ECONOMICS

▷ Inflation 5% p.a (1990–2000)

$421bn

1265–1314 South Korean won

SCORE CARD

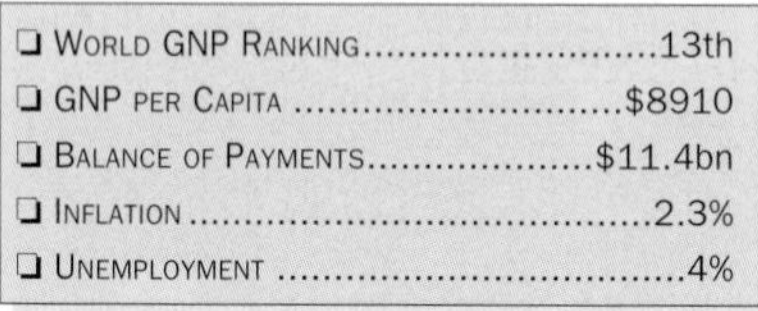

❑ World GNP Ranking	13th
❑ GNP per Capita	$8910
❑ Balance of Payments	$11.4bn
❑ Inflation	2.3%
❑ Unemployment	4%

EXPORTS

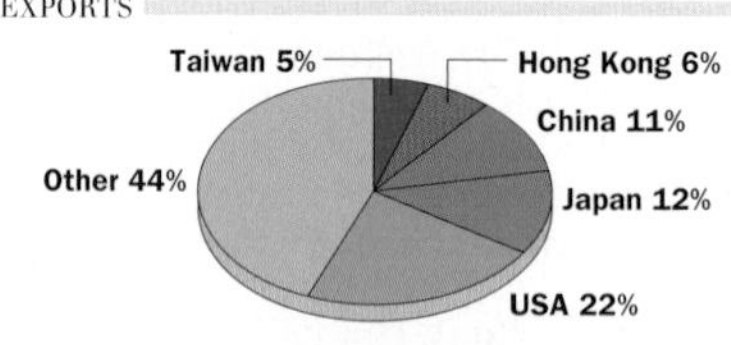
Taiwan 5%
Hong Kong 6%
China 11%
Japan 12%
USA 22%
Other 44%

IMPORTS

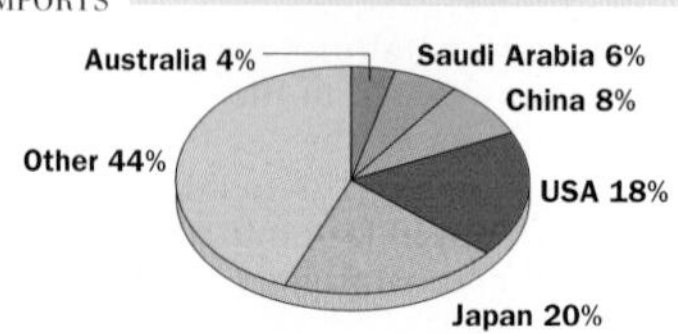
Australia 4%
Saudi Arabia 6%
China 8%
USA 18%
Japan 20%
Other 44%

Strengths

World's most successful shipbuilder, with 45% of the market. Strong yen makes South Korean exports more competitive than Japan's. Demand from China, particularly for cars.

Weaknesses

High level of indebtedness and vulnerability to international capital movements. Increasingly militant workforce since 1997. State sector a burden on the economy. Strong competition from Japan.

Profile

South Korea's economic miracle began with centralized planning. *Chaebol* (conglomerates) such as Samsung achieved impressive growth rates in strategic industries such as car making, shipbuilding, and semiconductors.

ECONOMIC PERFORMANCE INDICATOR

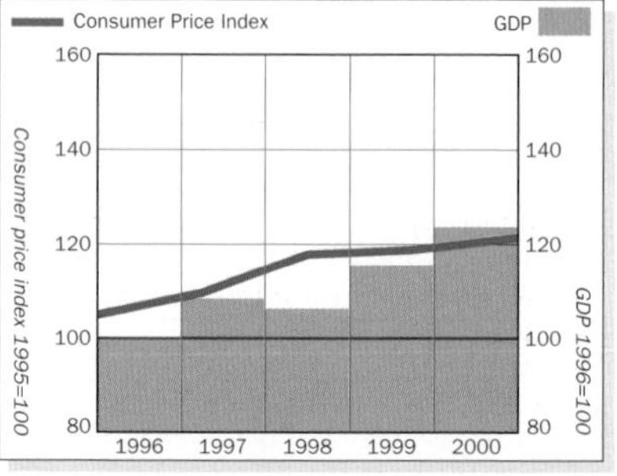
Consumer Price Index
GDP
160
140
120
100
80
Consumer price index 1995=100
GDP 1996=100
1996
1997
1998
1999
2000

A well-educated workforce and cheap state credit gave Korea a competitive edge. The government then encouraged foreign investment and an emphasis on smaller industries to maintain growth. In 1996 South Korea joined the OECD. The following year, however, a major financial crisis and the threat of a debt implosion forced the government to turn to the IMF for a huge credit agreement. By 2001 the banking system showed signs of recovery, although global slowdown continued to affect the economy.

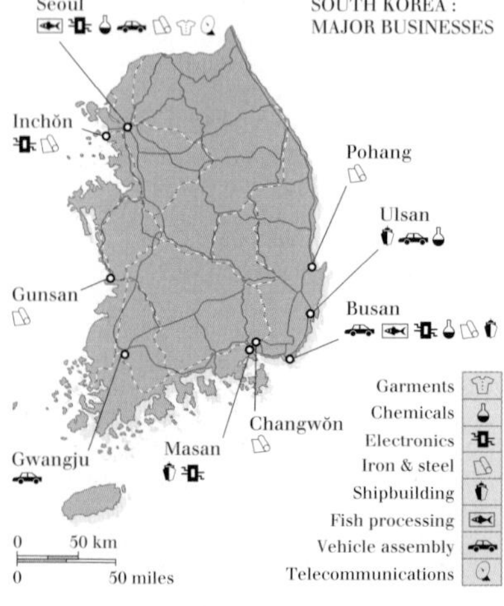
SOUTH KOREA : MAJOR BUSINESSES
Seoul
Inchŏn
Pohang
Ulsan
Gunsan
Busan
Changwŏn
Masan
Gwangju
Garments
Chemicals
Electronics
Iron & steel
Shipbuilding
Fish processing
Vehicle assembly
Telecommunications
0 50 km
0 50 miles

RESOURCES

Electric power 48m kw

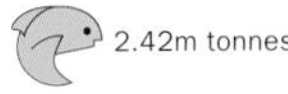
2.42m tonnes

Not an oil producer; refines 1.15m b/d

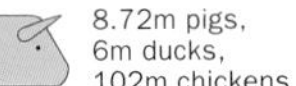
8.72m pigs, 6m ducks, 102m chickens

Coal, iron, lead, zinc, tungsten, gold, graphite, fluorite

ELECTRICITY GENERATION

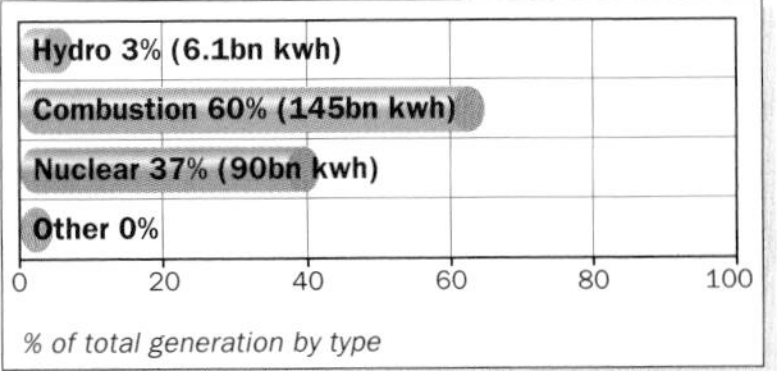
Hydro 3% (6.1bn kwh)
Combustion 60% (145bn kwh)
Nuclear 37% (90bn kwh)
Other 0%
0
20
40
60
80
100
% of total generation by type

South Korea has few natural resources. It has to import all of its oil and has built a series of nuclear reactors for generating electricity. Under the terms of the 1994 agreement between North Korea and the US, South Korea is constructing two reactors in the North which, in the event of reunification, will be connected to the national grid.

Agriculture remains a highly protected sector of the economy. Plans to open up the rice market have in the past provoked massive demonstrations in Seoul.

SOUTH KOREA : LAND USE

Cropland
Pasture
Forest
Poultry
Rice
Cereals

Han-gang
TAEBAEK-SANMAEK
Naktong-gang
Cheju-do

0 50 km
0 50 miles

ENVIRONMENT

Sustainability rank: 135th

7%

7.8 tonnes per capita

ENVIRONMENTAL TREATIES

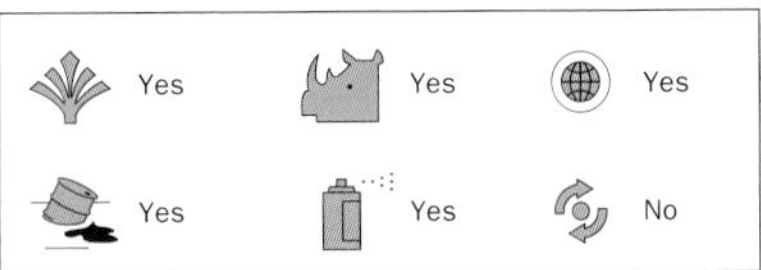
Yes
Yes
Yes
Yes
Yes
No

Environmental groups remain hostile to South Korea's nuclear power program. Rapid industrialization has resulted in environmental problems. The government has signed but not yet ratified the Kyoto Protocol, which aims to reduce carbon dioxide emissions. More recently, steps have been taken to address the severe problem of air pollution in urban areas, particularly in Seoul. Rivers in rural areas have been polluted by fertilizers and chemicals.

MEDIA

TV ownership high

Daily newspaper circulation 394 per 1000 people

PUBLISHING AND BROADCAST MEDIA

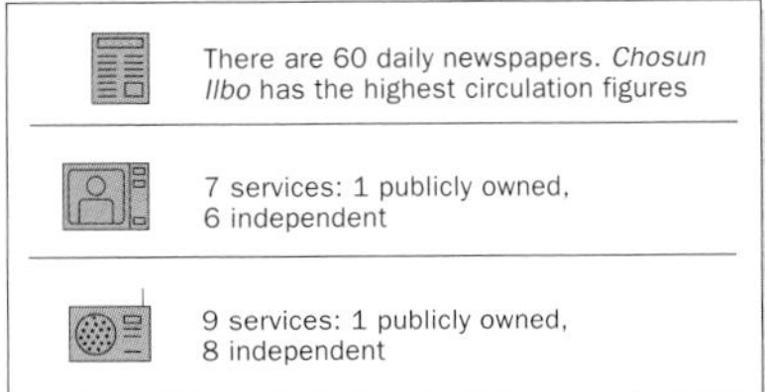
There are 60 daily newspapers. *Chosun Ilbo* has the highest circulation figures

7 services: 1 publicly owned, 6 independent

9 services: 1 publicly owned, 8 independent

South Korea's media have been freed of most restrictions since the advent of full multiparty democracy. However, criticisms of the armed forces are still frowned upon, and journalists tend to avoid altogether the subject of the role of the military in society. Caution also has to be exercised in reporting facts about North Korea. In the past, South Korean journalists who have made favorable mention of the North Korean communist regime have suffered harassment and intimidation.

CRIME

Death penalty in use

64,038 prisoners

Up 159% in 1999

CRIME RATES

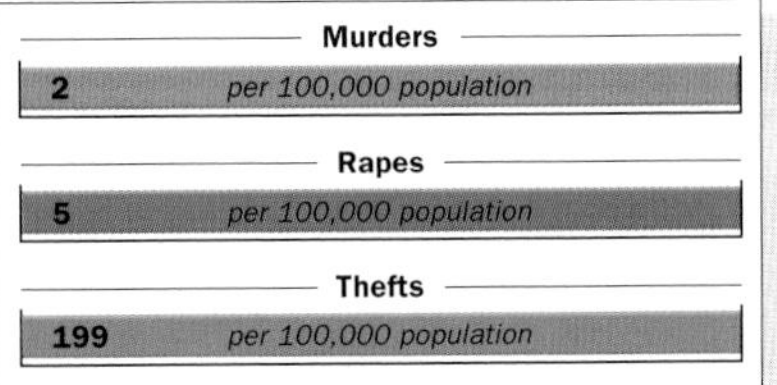
Murders
2 per 100,000 population
Rapes
5 per 100,000 population
Thefts
199 per 100,000 population

Crime rates are low but rising; violent crime is uncommon. The government has begun to treat corruption as a crime. Since 1987, the internal security forces' operations have been restricted, although left-wing activists are still harassed. Striking workers and student demonstrators encounter – and prepare for – forceful and confrontational crowd control.

EDUCATION

School leaving age: 15

98%

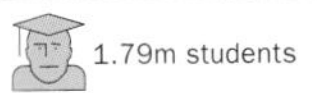
1.79m students

THE EDUCATION SYSTEM

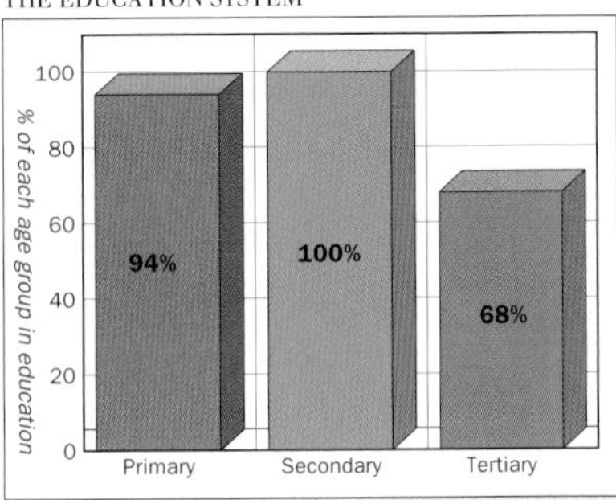
% of each age group in education
100
80
60
40
20
0
94%
100%
68%
Primary
Secondary
Tertiary

South Korea began a concentrated education program in the 1950s, and a well-educated workforce has been the foundation of impressive economic growth. Secondary education, which begins at 12 years of age, comprises two three-year cycles, of which one is compulsory. Tertiary enrollment is nearly 70% – one of the highest rates in the world.

HEALTH

Welfare state health benefits

1 per 769 people

Cancers, accidents, heart and cerebro-vascular diseases, TB

The health service has improved in line with economic growth. Most hospitals are equipped with modern facilities, and many offer advanced treatments comparable with those in the US and western Europe. Health indicators such as infant mortality and longevity have likewise improved.

SPENDING

GDP/cap. increase

CONSUMPTION AND SPENDING

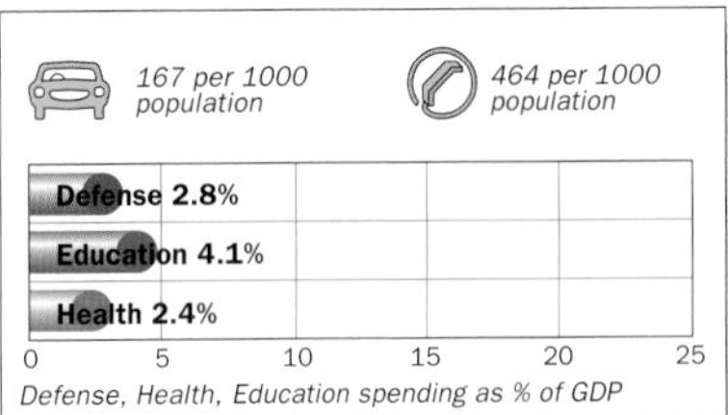
167 per 1000 population
464 per 1000 population
Defense 2.8%
Education 4.1%
Health 2.4%
0
5
10
15
20
25
Defense, Health, Education spending as % of GDP

Most South Koreans have felt the effects of economic growth – and recession. The Jolla (Cholla) region in the southwest remains the poorest.

WORLD RANKING

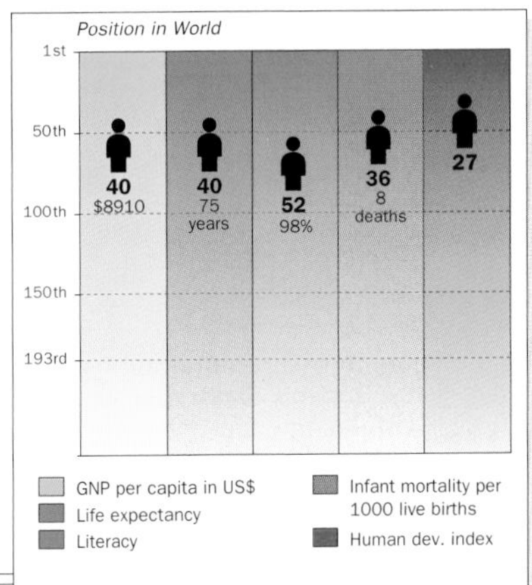
Position in World
1st
50th
100th
150th
193rd
40 $8910
40 75 years
52 98%
36 8 deaths
27
GNP per capita in US$
Life expectancy
Literacy
Infant mortality per 1000 live births
Human dev. index

K

KUWAIT

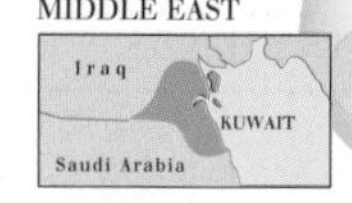
MIDDLE EAST
Iraq
KUWAIT
Saudi Arabia
Asia
Africa

OFFICIAL NAME: State of Kuwait **CAPITAL:** Kuwait City
POPULATION: 2 million **CURRENCY:** Kuwaiti dinar **OFFICIAL LANGUAGE:** Arabic

1961 | 1961 | Feb 25 | KWT | +3 | +965 | .kw

AT THE NORTHWEST EXTREME of the Gulf, Kuwait is dwarfed by its neighbors. The flat, almost featureless landscape conceals huge oil and gas reserves which put Kuwait among the world's first oil-rich states. In 1990 Iraq invaded, claiming it as its 19th province. A US-led alliance, under the aegis of the UN, expelled Iraqi forces following a short war in 1991 and restored the rule of the al-Sabah dynasty. Kuwait has now built a wall separating its territory from Iraq.

Saffar Towers in the business center of Kuwait City. Rebuilding Kuwait's postwar economy is estimated to have cost $25 billion.

CLIMATE

Hot desert

WEATHER CHART FOR KUWAIT CITY

Average daily temperature | Rainfall
°C/°F J F M A M J J A S O N D cm/in
40/104, 30/86, 20/68, 10/50, 0/32, -10/14, -20/-4 | 40/16, 30/12, 20/8, 10/4, 0

Summer temperatures can soar to over 40°C (104°F), but winters can be cold, with frost at night.

TRANSPORTATION

Drive on right

Kuwait International, Kuwait City
3.87m passengers

202 ships
2.46m grt

THE TRANSPORTATION NETWORK

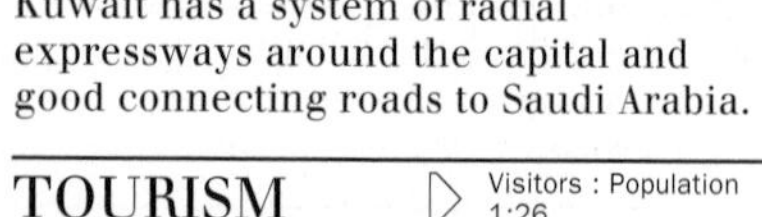

3590 km (2231 miles)

280 km (174 miles)

None

None

Kuwait has a system of radial expressways around the capital and good connecting roads to Saudi Arabia.

TOURISM

Visitors : Population 1:26

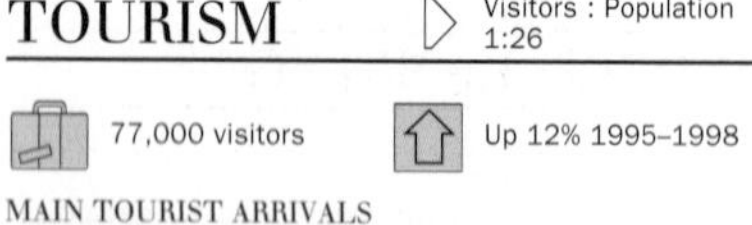

77,000 visitors

Up 12% 1995–1998

MAIN TOURIST ARRIVALS

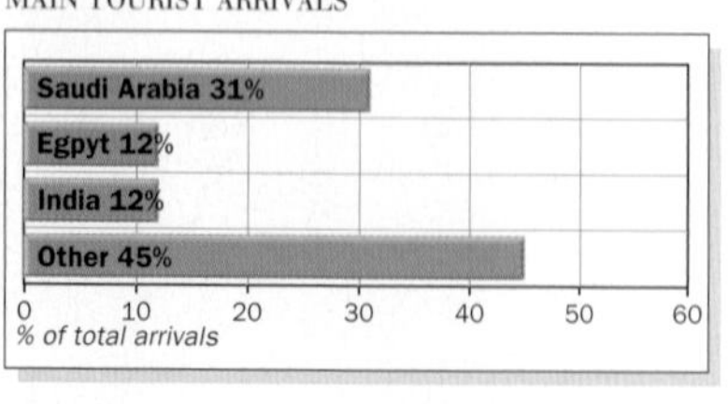
Saudi Arabia 31%
Egypt 12%
India 12%
Other 45%
0 10 20 30 40 50 60
% of total arrivals

The limited tourism from neighboring Arab states, notably Saudi Arabia, has not recovered since the 1990–1991 Gulf War. Most Western visitors to Kuwait go specifically to see relatives working in the oil industry.

PEOPLE

Pop. density medium

Arabic, English

112/km² (291/mi²)

THE URBAN/RURAL POPULATION SPLIT

98% | 2%

ETHNIC MAKEUP

Kuwaiti 45%
Other Arab 35%
South Asian 9%
Other 7%
Iranian 4%

Kuwait is a conservative Sunni Muslim society (27% of the population are Shi'a). Women have considerable freedom, although the amir's decree providing for female enfranchisement has been repeatedly rejected by the National Assembly.

Kuwait's oil wealth has drawn in thousands of workers from other Arab countries and south Asia. The Palestine Liberation Organization's support for the Iraqi invasion led to most Palestinians, hitherto more numerous in Kuwait than elsewhere in the Arabian peninsula, being driven out. Native Kuwaitis are outnumbered by resident foreign nationals.

POLITICS

No multiparty elections

1999/2003

Amir Shaikh Jabir al-Ahmad al-Jabir al-Sabah

AT THE LAST ELECTION

National Assembly 50 seats

The electorate comprises civilian men over 21 years of age whose families have been resident in Kuwait since before 1921. Elections on 3 July 1999 were contested by independents. The 50 seats were split evenly between Islamists, liberals, and government supporters.

In 1992 Amir Shaikh Jabir restored the National Assembly. There was then a government of "national unity" until 1999, when elections strengthened the amir's Islamist and liberal opponents. The Council of Ministers resigned in 2001 in the face of Assembly criticism. In the new cabinet the ruling al-Sabah family still held the top posts.

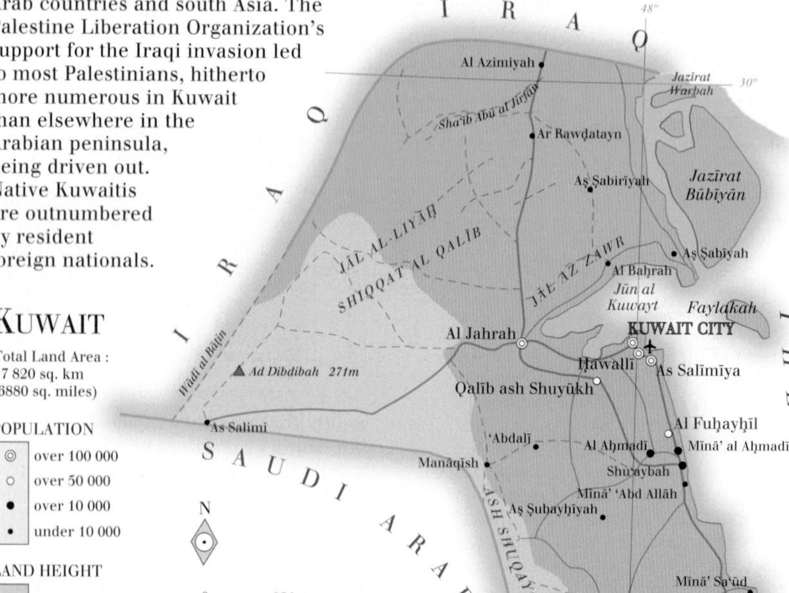
IRAQ
IRAQ
SAUDI ARABIA
THE GULF
Al Azimiyah
Jazirat Warbah
Sha'ib Abu al Jirfan
Ar Rawḑatayn
Aş Şabiriyah
Jazirat Būbiyān
Aş Şabiyah
JAL AL-LIYAH
SHIQQAT AL QALIB
JAL AZ ZAWR
Al Bahrah
Jūn al Kuwayt
Faylakah
Al Jahrah
KUWAIT CITY
Ḩawallī
As Salīmīya
Qalib ash Shuyūkh
Wadi al Batin
Ad Dibdibah 271m
As Salimi
'Abdalī
Al Aḩmadi
Al Fuḩayḩil
Mīnā' al Aḩmadī
Manāqīsh
Shu'aybah
Mīnā' 'Abd Allāh
Aş Şubayḩiyah
ASH SHUQAYQ
Mīnā' Sa'ūd
Al Khīrān
Al Wafrā'
48°
30°
N
0 25 km
0 25 miles

KUWAIT

Total Land Area : 17 820 sq. km (6880 sq. miles)

POPULATION
- over 100 000
- over 50 000
- over 10 000
- under 10 000

LAND HEIGHT
- 200m/656ft
- Sea Level

WORLD AFFAIRS

Joined UN in 1963

AL
Damasc
GCC
OIC
OPEC

Strategically important as a major exporter of crude oil and natural gas, Kuwait has always maintained very close links with the West, which have strengthened since the war with Iraq, and on which it depends for its future security. In 2002 Iraq promised to respect the common border, but has yet to return POWs and detained Kuwaiti nationals.

AID

Donor

 $165m (donations)  Up 12% in 2000

The Kuwait Fund for Arab Economic Development continued to give aid even during the invasion crisis.

DEFENSE

Compulsory military service

 $3.21bn  Down 2% in 2000

In August 1990 Kuwait's 11,000-strong, partly volunteer army was easily overrun by vastly superior Iraqi forces. Since its liberation, defense pacts have been signed with the US, the UK, France, and Russia. Kuwait rearmed fast, with weapons purchased from major Western suppliers.

ECONOMICS

Inflation 3% p.a. (1990–1999)

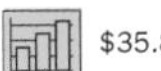 $35.8bn  0.3055–0.3071 Kuwaiti dinars

SCORE CARD

❑ World GNP Ranking	54th
❑ GNP per Capita	$18,030
❑ Balance of Payments	$14.9bn
❑ Inflation	1.8%
❑ Unemployment	2%

STRENGTHS

Recovery of oil and gas production. Large overseas investments. Stable banking system.

WEAKNESSES

Overreliance of economy on oil and gas. Adverse consequences of Iraqi invasion. Strategic vulnerability deters investment. Reliance on imported skilled labor, food, and raw materials. Delays in enacting privatization package.

EXPORTS

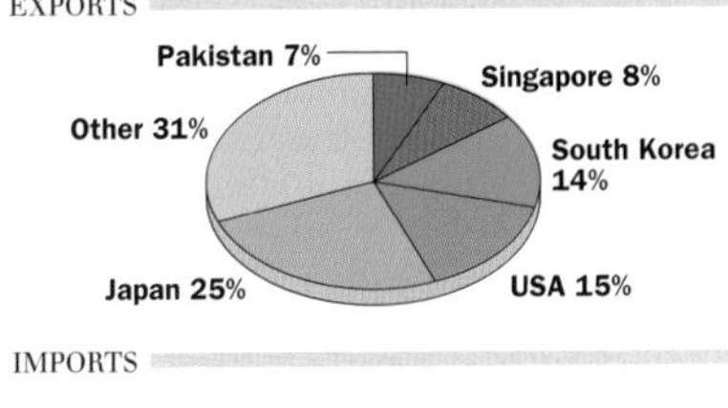
Pakistan 7%
Singapore 8%
Other 31%
South Korea 14%
Japan 25%
USA 15%

IMPORTS

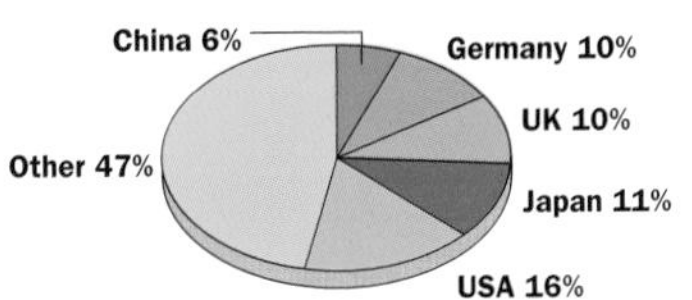
China 6%
Germany 10%
UK 10%
Other 47%
Japan 11%
USA 16%

RESOURCES

Electric power 7m kw

 6535 tonnes  2.14m b/d (reserves 96.5bn barrels)

 630,000 sheep, 130,000 goats, 32.5m chickens — Oil, natural gas, salt

The oil industry is Kuwait's most profitable sector, accounting for over 80% of export earnings. Although badly hit by the Gulf War, when a number of wells were deliberately fired, it was quickly rehabilitated. Kuwait also possesses valuable reserves of natural gas. Other resources are dates, fish, ammonia, and chemicals.

ENVIRONMENT

Sustainability rank: 142nd

 2%  26.3 tonnes per capita

The Iraqi invasion and the subsequent war caused an ecological disaster. Although the effects of this did not prove as grave as some observers first feared, marine life was damaged and many thousands of hectares of cultivated land were obliterated. Millions of land mines still litter border areas. Water is scarce.

MEDIA

TV ownership high

 Daily newspaper circulation 377 per 1000 people

PUBLISHING AND BROADCAST MEDIA

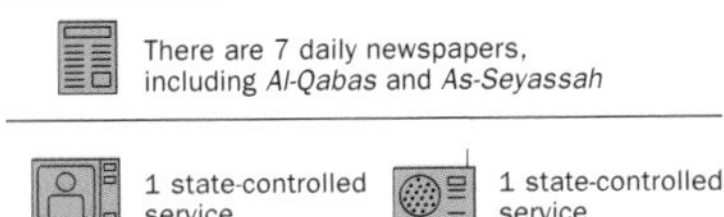
There are 7 daily newspapers, including *Al-Qabas* and *As-Seyassah*

1 state-controlled service — 1 state-controlled service

Radio and TV are state-controlled, but satellite TV is freely available. Press freedom exists in theory.

CRIME

Death penalty in use

1735 prisoners  Up 8% 1996–1998

Isolated acts of terrorism occur. There have been complaints of human rights abuses.

EDUCATION

School leaving age: 14

 83%  29,509 students

Kuwaiti citizens receive free education from nursery to university. Since the liberation, more emphasis has been placed on technology in the curriculum.

CHRONOLOGY

Kuwait traces its independence to 1710, but was under British rule from the late 18th century until 1961. The government denies any historical link with Iraq.

- ❑ **1961** Independence from UK. Iraq claims Kuwait.
- ❑ **1976** Amir suspends National Assembly.
- ❑ **1990** Iraqi invasion.
- ❑ **1991** Liberation following Gulf War.
- ❑ **1992** National Assembly elections.
- ❑ **1999** Elections; Islamists and liberals win most seats.

HEALTH

Welfare state health benefits

 1 per 526 people  Heart diseases, accidents, cancers, perinatal deaths

Despite theft of equipment during the Iraqi invasion, Kuwait has restored its Western-standard health care service. Nationals receive free treatment.

SPENDING

GDP/cap. increase

CONSUMPTION AND SPENDING

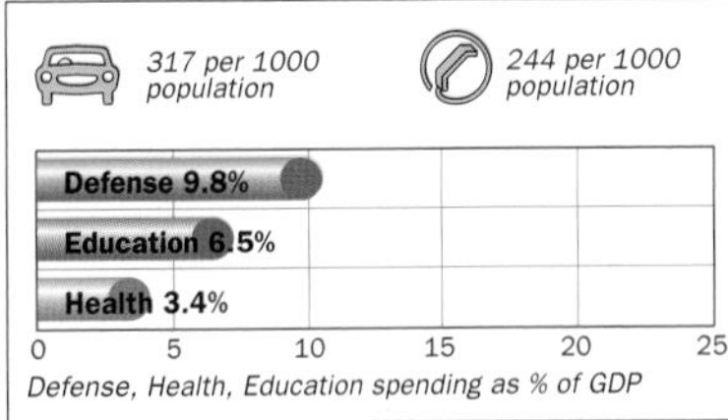
317 per 1000 population
244 per 1000 population
Defense 9.8%
Education 6.5%
Health 3.4%
0
5
10
15
20
25
Defense, Health, Education spending as % of GDP

As well as the oil-rich elite, most Kuwaitis enjoy high incomes, and the government has repeatedly rescued citizens who have suffered stock market or other financial losses. School and university leavers are guaranteed jobs. Capital is easily transferred abroad and there are effectively no exchange controls.

WORLD RANKING

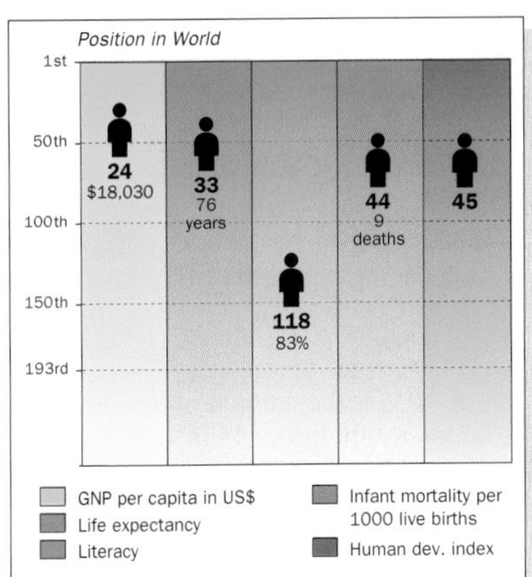
Position in World
1st
50th
100th
150th
193rd
24
$18,030
33
76 years
118
83%
44
9 deaths
45
GNP per capita in US$
Life expectancy
Literacy
Infant mortality per 1000 live births
Human dev. index

K

KYRGYZSTAN

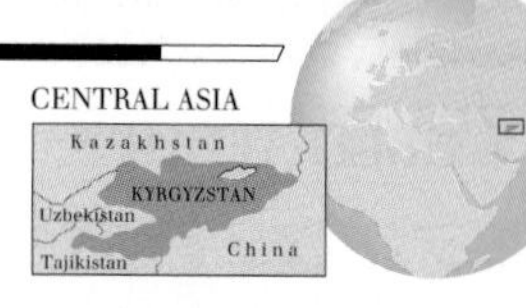
CENTRAL ASIA
Kazakhstan
KYRGYZSTAN
Uzbekistan
Tajikistan
China
Asia

OFFICIAL NAME: Kyrgyz Republic **CAPITAL:** Bishkek
POPULATION: 5 million **CURRENCY:** Som **OFFICIAL LANGUAGES:** Kyrgyz and Russian

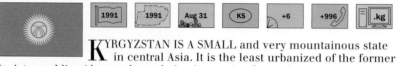

1991

1991

Aug 31
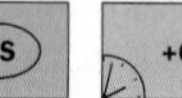
KS
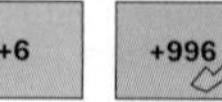
+6

+996

.kg

KYRGYZSTAN IS A SMALL and very mountainous state in central Asia. It is the least urbanized of the former Soviet republics (the rural population is growing faster than that in the towns) and was among the last to develop its own cultural nationalism. Its increasingly autocratic government tries to steer between Kyrgyz nationalist pressures and ensuring that the minority Russians are not alienated, since they tend to possess the skills necessary to run a market-based economy.

CLIMATE

Mountain

WEATHER CHART FOR BISHKEK

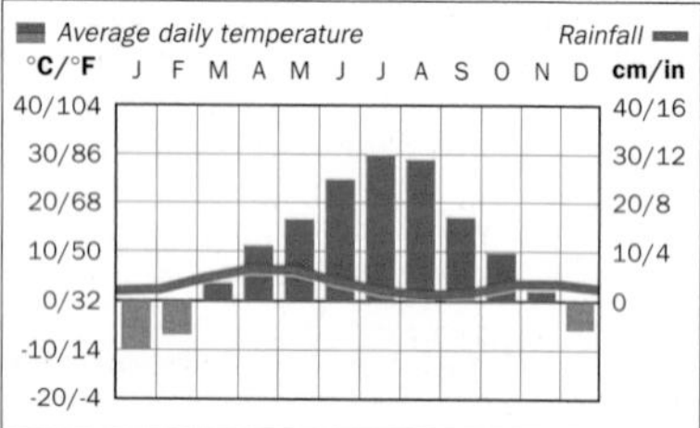
Average daily temperature
Rainfall
°C/°F
J F M A M J J A S O N D
cm/in
40/104
30/86
20/68
10/50
0/32
-10/14
-20/-4
40/16
30/12
20/8
10/4
0

Conditions vary from permanent snow and cold deserts at altitude to hot deserts in low regions. Intermediate slopes and valleys receive some rain.

TRANSPORTATION

Drive on right

Bishkek International

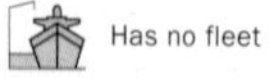
Has no fleet

THE TRANSPORTATION NETWORK

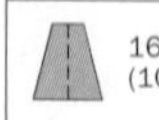
16,854 km (10,473 miles)

140 km (87 miles)

417 km (259 miles)

600 km (373 miles)

Kyrgyzstan does not have the financial budget to improve on its poor mountain road network.

TOURISM

Visitors : Population 1:72

69,000 visitors

Up 17% in 1999

MAIN TOURIST ARRIVALS

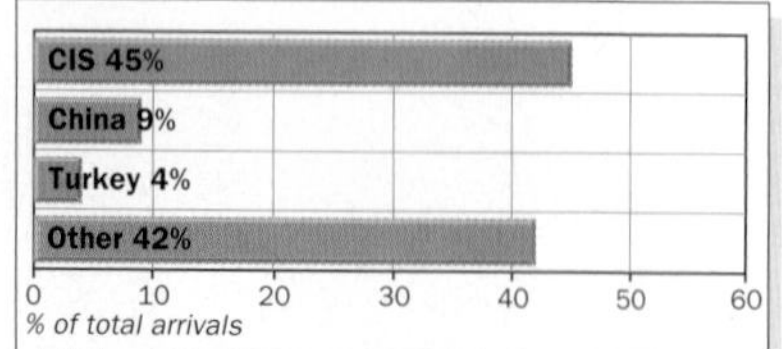
CIS 45%
China 9%
Turkey 4%
Other 42%
0
10
20
30
40
50
60
% of total arrivals

Tourism is undeveloped; most visitors are on business or working on multilateral aid projects. Tourism promotion centers on Kyrgyzstan's position on the Silk Road.

PEOPLE

Pop. density low

Kyrgyz, Russian

25/km² (65/mi²)

THE URBAN/RURAL POPULATION SPLIT

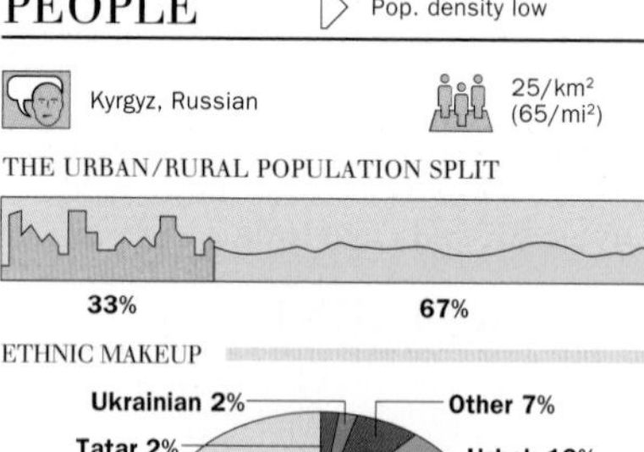
33%
67%
ETHNIC MAKEUP
Ukrainian 2%
Other 7%
Tatar 2%
Uzbek 13%
Russian 19%
Kyrgyz 57%

Like other former Soviet republics, Kyrgyzstan has witnessed the rise of militant nationalism. Relations are most strained with the large Uzbek minority. The preference given to Kyrgyz in the political system and in particular in the land laws, which exclude all others from full title, has aggravated ethnic tensions. The trend in politics is toward greater Islamization, which is linking religion and race issues more closely and adding social pressure on "foreigners," particularly ethnic Russians, to leave.

Since 1989 a high birthrate has enabled the Kyrgyz to resume their position as the main ethnic group, replacing the Russian community which until recently controlled the economy. However, the government moved to stem the tide of Russian emigration by declaring Russian an official language with full equal status with effect from 2000.

Loess landscape, Naryn valley. *Kyrgyzstan is dominated by the ice-capped Tien Shan mountains, but valleys are green and fertile.*

POLITICS

Multiparty elections

L. House 2000/2005
U. House 2000/2005

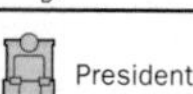
President Askar Akayev

AT THE LAST ELECTION

Legislative Assembly 60 seats

Election results by party were announced only for the 15 national list seats. The other 45 seats are elected on a constituency basis. The Party of Communists of Kyrgyzstan is the largest party and the Union of Democratic Forces the second-largest.

Assembly of People's Representatives 45 seats

The members represent Kyrgyzstan's different regional and ethnic communities

President Akayev's administration has become increasingly autocratic, ending Kyrgyzstan's reputation as one of the region's most liberal societies. Already damaged democratic credentials were destroyed by accusations of fraud during the 2000 legislative and presidential elecions, and the main opposition leader, Felix Kulov, was imprisoned only weeks before the presidential poll. The temporary imprisonment in March 2002 of his successor, Azimbek Beknazarov, prompted mass protests, when five demonstrators were shot dead by police. Spiraling protests forced the entire cabinet to resign in May.

Akayev himself has been accused of fostering a personality cult. Already in power under the Soviet regime before independence, he was reelected in 1995 and in 2000, after objections to a third term were overruled by the Constitutional Court.

WORLD AFFAIRS

Joined UN in 1992

CIS

SCO
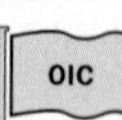
OIC

OSCE
EAPC

Kyrgyzstan is working to reduce its dependence on Russia. Turkey is developing close links based on ethnic similarities and aimed at restraining Iranian influence. Relations with Uzbekistan, which allegedly supports some of the antigovernment forces in Kyrgyzstan, are tense, although in 2000 both countries joined with Tajikistan to combat Islamist militants in the region.

AID

Recipient

$215m (receipts)

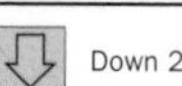
Down 24% in 2000

The US and Japan are the main aid donors. The World Bank is also an important source of financial assistance.

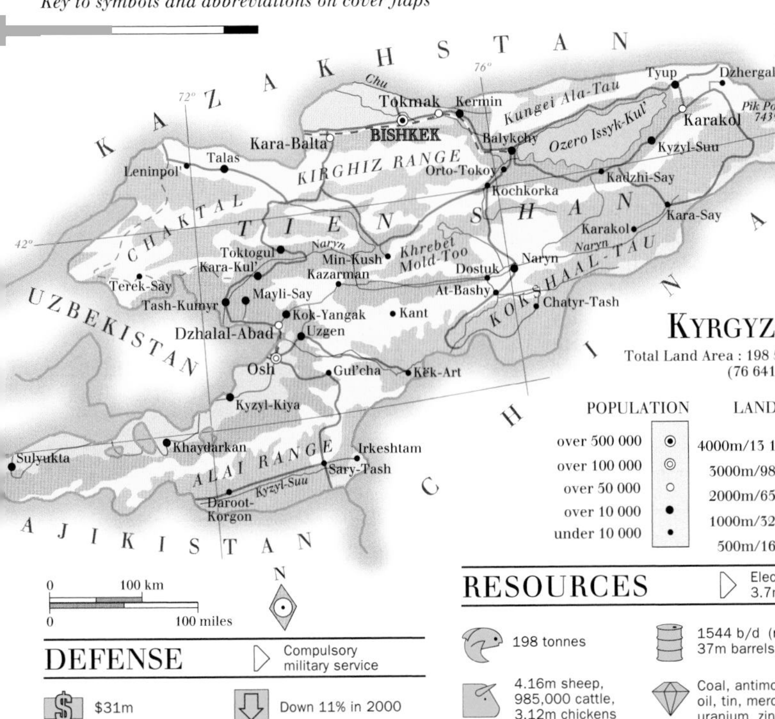
KAZAKHSTAN
UZBEKISTAN
TAJIKISTAN
CHINA
BISHKEK
Tokmak
Kermin
Kara-Balta
Talas
Leninpol'
KIRGHIZ RANGE
Orto-Tokoy
Balykchy
Kungei Ala-Tau
Ozero Issyk-Kul'
Tyup
Dzhergalan
Karakol
Pik Pobedy 7439m
Kyzyl-Suu
Kadzhi-Say
Kochkorka
TIEN SHAN
CHAKTAL
Kara-Say
Karakol
Naryn
Toktogul
Kara-Kul'
Min-Kush
Khrebet Mold-Too
Dostuk
Kazarman
Terek-Say
Tash-Kumyr
Mayli-Say
At-Bashy
KOKSHAAL-TAU
Chatyr-Tash
Kok-Yangak
Kant
Dzhalal-Abad
Uzgen
Osh
Gul'cha
Kok-Art
Kyzyl-Kiya
Khaydarkan
Sulyukta
ALAI RANGE
Irkeshtam
Sary-Tash
Kyzyl-Suu
Daroot-Korgon
Chu
Naryn
72°
76°
80°
42°

0 100 km
0 100 miles

N

KYRGYZSTAN

Total Land Area : 198 500 sq. km (76 641 sq. miles)

POPULATION	
over 500 000	⦿
over 100 000	◎
over 50 000	○
over 10 000	●
under 10 000	•

LAND HEIGHT

4000m/13 124ft
3000m/9843ft
2000m/6562ft
1000m/3281ft
500m/1640ft

CHRONOLOGY

The Kyrgyz first developed a recognizable ethnic consciousness in the late 18th century.

- ❑ **1860s** Expansion of Russian Empire into Kyrgyz lands.
- ❑ **1924–1991** Incorporated in USSR.
- ❑ **1995** New constitution adopted.
- ❑ **2000** Legislative and presidential elections; Akayev reelected.
- ❑ **2002** Government resigns after police shoot demonstrators.

DEFENSE

▷ Compulsory military service

$31m

Down 11% in 2000

In 1992, a national army was set up and a defense treaty was signed with five other CIS states. The army is weak and not influential in politics.

ECONOMICS

▷ Inflation 110% p.a. (1990–2000)

$1.35bn

48.24–47.72 soms

SCORE CARD

- ❑ WORLD GNP RANKING........................148th
- ❑ GNP PER CAPITA...............................$270
- ❑ BALANCE OF PAYMENTS......................–$77m
- ❑ INFLATION..18.7%
- ❑ UNEMPLOYMENT..................................6%

STRENGTHS

Agricultural self-sufficiency. Private land ownership since 2000. Gold and mercury exports. Hydropower potential.

WEAKNESSES

Dominant state and collective farming mentality. Sharp economic decline since breakup of USSR, on which it depended for trade and supplies. Chronic inflation.

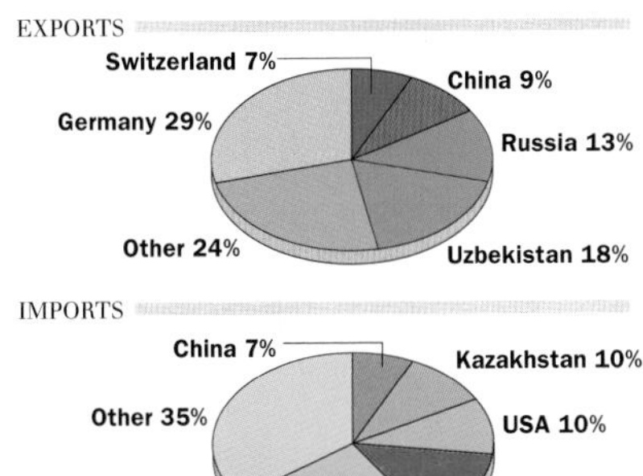
EXPORTS
Switzerland 7%
China 9%
Germany 29%
Russia 13%
Other 24%
Uzbekistan 18%
IMPORTS
China 7%
Kazakhstan 10%
Other 35%
USA 10%
Uzbekistan 14%
Russia 24%

RESOURCES

▷ Electric power 3.7m kw

198 tonnes

1544 b/d (reserves 37m barrels)

4.16m sheep, 985,000 cattle, 3.12m chickens

Coal, antimony, gas, oil, tin, mercury, iron, uranium, zinc, gold

Kyrgyzstan has small quantities of commercially exploitable coal, oil, and gas, and great hydroelectric power potential. Energy policy, which relies on Western aid and technology, is primarily aimed at developing these further in order to reduce dependence on supplies from Russia, and eventually to achieve self-sufficiency in energy.

ENVIRONMENT

▷ Sustainability rank: 56th

4%

1.3 tonnes per capita

The major problem is the salination of the soil caused by excessive irrigation of cotton crops. Kyrgyzstan has a poor record in limiting industrial pollution.

MEDIA

▷ TV ownership low

Daily newspaper circulation 9 per 1000 people

PUBLISHING AND BROADCAST MEDIA

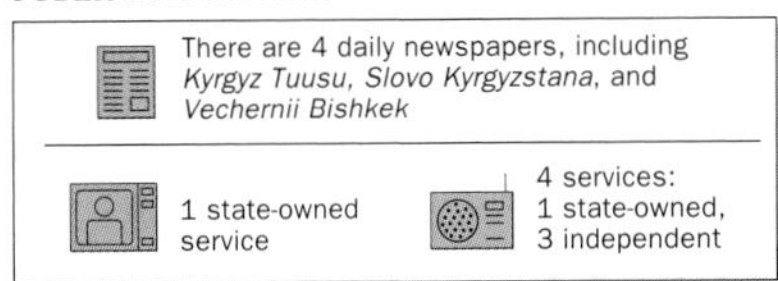
There are 4 daily newspapers, including *Kyrgyz Tuusu, Slovo Kyrgyzstana,* and *Vechernii Bishkek*
1 state-owned service
4 services: 1 state-owned, 3 independent

The Kyrgyz press is increasingly under pressure from the government's efforts to silence opposition.

CRIME

▷ Moratorium on death penalty

20,000 prisoners

The crime rate is rising

Ethnic tension fuels violence. The narcotics trade flourishes. A 1998 moratorium on the death penalty was extended in 2002 for a further year.

EDUCATION

▷ School leaving age: 15

97%

49,744 students

Replacing Russian as the main teaching language is proving an enormous task. Russian is likely to survive at tertiary level, as the Kyrgyz language lacks key technical and scientific terms.

HEALTH

▷ Welfare state health benefits

1 per 333 people

Heart diseases, cancers, accidents, violence, tuberculosis

Kyrgyzstan has one of the least developed public health systems in central Asia. Infant mortality is high.

SPENDING

▷ GDP/cap. decrease

CONSUMPTION AND SPENDING

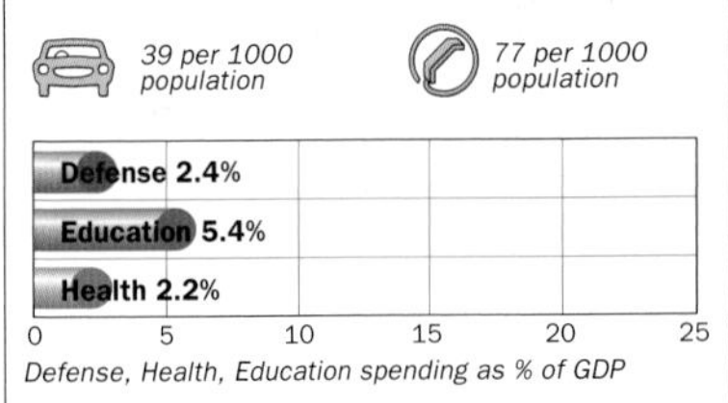
39 per 1000 population
77 per 1000 population
Defense 2.4%
Education 5.4%
Health 2.2%
0
5
10
15
20
25
Defense, Health, Education spending as % of GDP

In 2000, almost 90% of the population were estimated as living in poverty. The old Communist Party *nomenklatura* are the main beneficiaries of privatization.

WORLD RANKING

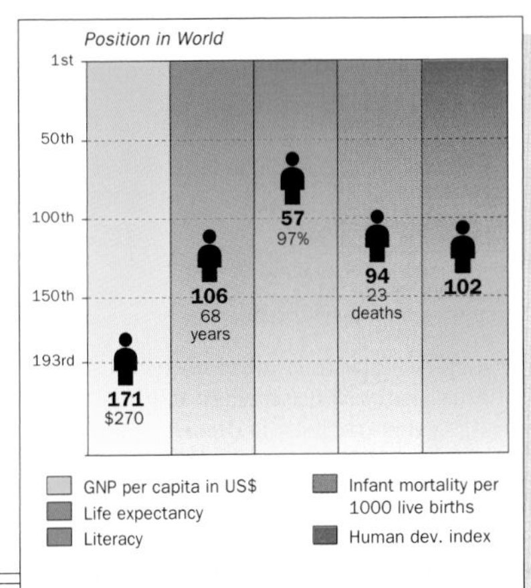
Position in World
1st
50th
100th
150th
193rd
171
$270
106
68 years
57
97%
94
23 deaths
102
GNP per capita in US$
Life expectancy
Literacy
Infant mortality per 1000 live births
Human dev. index

K

LAOS

SOUTHEAST ASIA
Burma (Myanmar) China LAOS Thailand Vietnam Cambodia Asia

OFFICIAL NAME: Lao People's Democratic Republic **CAPITAL:** Vientiane
POPULATION: 5.4 million **CURRENCY:** New kip **OFFICIAL LANGUAGE:** Lao

1953

1953

Dec 2

LAO

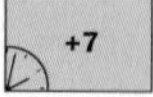
+7

+856

.la

THE MEKONG RIVER forms Laos's main thoroughfare and feeds the fertile lowlands of the Mekong valley. Two decades of civil war followed independence from France in 1953, and Laos was bombed heavily during the Vietnam War. The communist LPRP has held power since 1975. Market-oriented reforms began to be introduced in 1986. A transfer of power to a younger generation within the LPRP took place during the 1990s.

CLIMATE

Tropical monsoon

WEATHER CHART FOR VIENTIANE

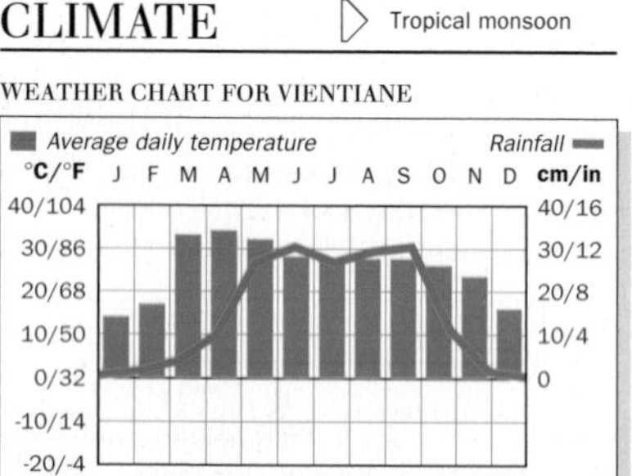
Average daily temperature
Rainfall
°C/°F J F M A M J J A S O N D cm/in
40/104
30/86
20/68
10/50
0/32
-10/14
-20/-4
40/16
30/12
20/8
10/4
0

The tropical southerly monsoon brings heavy rains from May to September.

TRANSPORTATION

Drive on right

Wattay, Vientiane
165,000 passengers

1 ship
2400 grt

THE TRANSPORTATION NETWORK

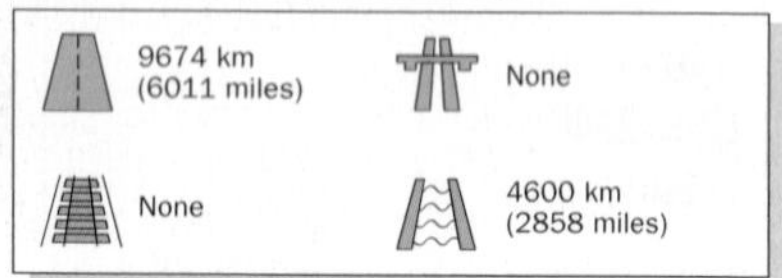
9674 km
(6011 miles)
None
None
4600 km
(2858 miles)

A major new Thailand–Vietnam road is planned via Savannakhét. Freight goes mainly by river; there is no railroad and roads are poor and few.

TOURISM

Visitors : Population 1:18

300,000 visitors
Up 16% in 2000

MAIN TOURIST ARRIVALS

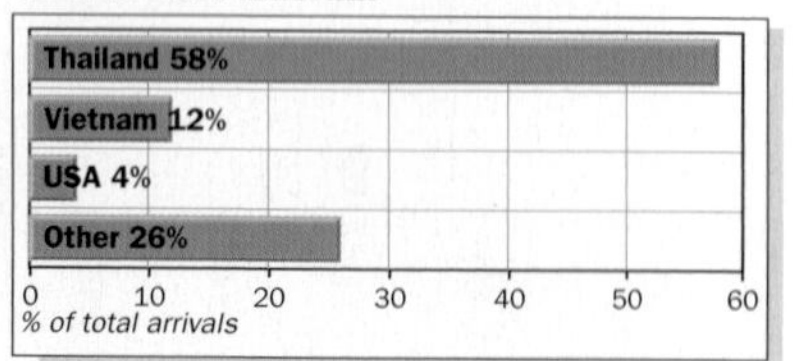
Thailand 58%
Vietnam 12%
USA 4%
Other 26%
0 10 20 30 40 50 60
% of total arrivals

Tourists were first allowed into Laos in 1989; numbers have risen rapidly since then. Mass tourism is discouraged and preference given to small package tours. Hotels are few, and travel outside Vientiane is difficult.

PEOPLE

Pop. density low

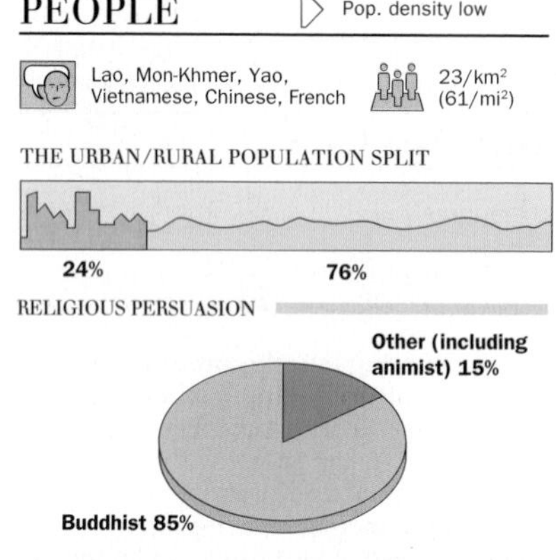
Lao, Mon-Khmer, Yao, Vietnamese, Chinese, French
23/km² (61/mi²)
THE URBAN/RURAL POPULATION SPLIT
24%
76%
RELIGIOUS PERSUASION
Other (including animist) 15%
Buddhist 85%

There are more than 60 ethnic groups in Laos and this considerable diversity has hindered national integration. Society is broadly divided by geography, but by altitude rather than by region. The lowland Laotians (*Lao Loum*) make up some 60% of the population. The upland Laotians (*Lao Theung*), who live in the hills above the valleys, account for 30%.

The small minority of mountain-top Laotians (*Lao Soung*), who include the Hmong, Yao, and Man groups, have resisted government efforts to introduce substitutes for traditional cash crops such as opium. The government continues to face small pockets of Hmong resistance.

Two-thirds of the population speak Lao, and a large number of tribal dialects are also spoken. Buddhism is the main religion, but there are some Christians and animists.

POLITICS

No multiparty elections

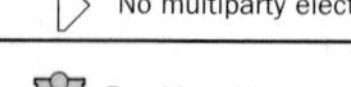
2002/2007
President Khamtay Siphandone

AT THE LAST ELECTION

National Assembly 109 seats

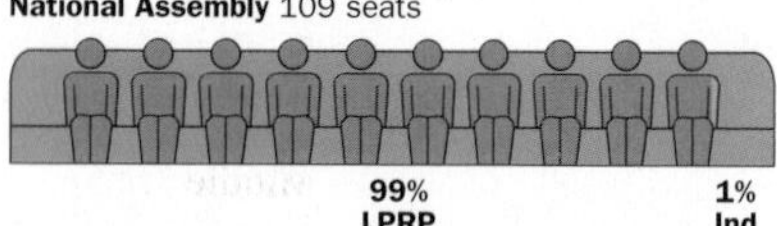
99% LPRP
1% Ind

LPRP = Lao People's Revolutionary Party (the sole legal political party) **Ind** = Independent

All candidates were approved by the LPRP

Laos has been ruled by the same circle of communist revolutionaries, one of the world's most durable and closely knit hierarchies, since 1975. The military, the LPRP, and the executive branch remain closely intertwined, with party chairman Gen. Khamtay Siphandone becoming the country's president in February 1998. Despite limited moves toward political reform, the LPRP, which is modeled on the Communist Party of Vietnam, continues to dominate political life at every level. The long-standing problem of corruption, sometimes at high levels, has become a matter of concern as Laos has opened to foreign investors. Economic reform has not been accompanied by political liberalization. Tensions continue to be felt between the government and the rural areas, where there is particular resistance to attempts to alter traditional farming methods.

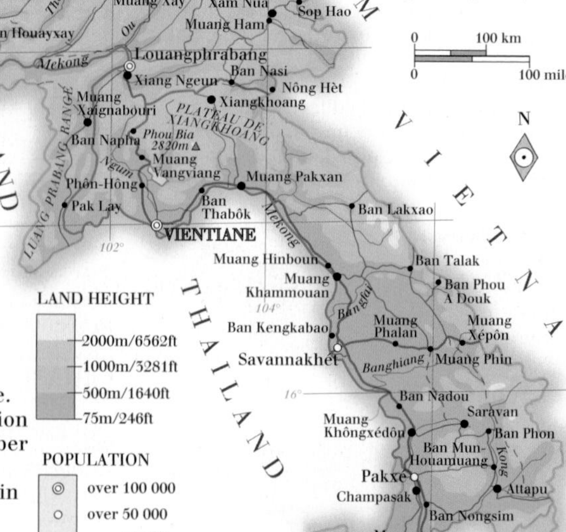
LAOS
Total Land Area : 236 800 sq. km
(91 428 sq. miles)
0 100 km
0 100 miles
N
CHINA
VIETNAM
BURMA (MYANMAR)
THAILAND
CAMBODIA
Phôngsali
Louang Namtha
Muang Sing
Muang Namo
Ban Na Môn
Sôp Xai
Viangphoukha
Muang Xay
Xam Nua
Sop Hao
Muang Ham
Ban Houayxay
Louangphrabang
Ban Nasi
Xiang Ngeun
Nông Hèt
Muang Xaignabouri
Xiangkhoang
PLATEAU DE XIANGKHOANG
Ban Napha
Phou Bia 2820m
Muang Vangviang
Phôn-Hông
Muang Pakxan
Pak Lay
Ban Thabôk
VIENTIANE
Ban Lakxao
LUANG PRABANG RANGE
Mekong
Muang Hinboun
Ban Talak
Muang Khammouan
Ban Phou A Douk
Ban Kengkabao
Muang Phalan
Muang Xépôn
Savannakhét
Muang Phin
Ban Nadou
Saravan
Muang Khôngxédôn
Ban Phon
Ban Mun-Houamuang
Pakxe
Champasak
Attapu
Ban Nongsim
Muang Không
Ban Xot
LAND HEIGHT
2000m/6562ft
1000m/3281ft
500m/1640ft
75m/246ft
POPULATION
over 100 000
over 50 000
over 10 000
under 10 000

WORLD AFFAIRS

Joined UN in 1955

ASEAN

CP

OIF

Mekong River

NAM

Vietnam was Laos's most important ally from 1975 until the late 1980s, when the LPRP began to seek improved relations with Thailand and the West. The change was mainly due to the need for foreign aid. In 1992, Laos acceded to the Treaty of Amity and Concord of ASEAN, marking the beginning of a new relationship with former adversaries. Laos was admitted to full membership of ASEAN in 1997.

AID

Recipient

 $281m (receipts)

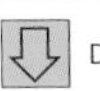 Down 5% in 2000

Laos has one of the highest per capita aid inflows in the developing world, Japan being a major donor.

DEFENSE

Compulsory military service

  $19m

 Down 14% in 2000

The armed forces are estimated by the West to number around 30,000 personnel. This total is further swelled by a paramilitary militia. Eighteen months' military service is compulsory for all Laotian men.

ECONOMICS

Inflation 27% p.a. (1990–2000)

$1.52bn

7600 new kips

SCORE CARD

❑ WORLD GNP RANKING	144th
❑ GNP PER CAPITA	$290
❑ BALANCE OF PAYMENTS	–$121m
❑ INFLATION	25.1%
❑ UNEMPLOYMENT	6%

EXPORTS

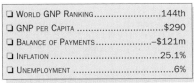

- Belgium 3%
- Germany 5%
- France 6%
- Thailand 16%
- Other 37%
- Vietnam 33%

IMPORTS

- Singapore 5%
- China 5%
- France 4%
- Other 13%
- Thailand 59%
- Vietnam 14%

STRENGTHS
Rising levels of investment from overseas. Potential of garment manufacturing, timber plantations, mining, wood processing, tourism, banking, and aviation. Minerals and possible oil and gas deposits.

WEAKNESSES
One of the world's least developed countries. Lack of technical expertise. Imbalance in sources of foreign investment – most is Thai. Problems in targeting aid efficiently.

RESOURCES

Electric power 256,000 kw

 60,403 tonnes

 Not an oil producer

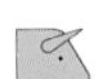 1.7m ducks, 1.5m pigs, 14m chickens

 Tin, gypsum, iron, coal, copper, potash, lead, limestone, antimony

Laos's most important agricultural resources are timber and coffee. It has awarded foreign countries concessions to mine for gold and precious stones.

ENVIRONMENT

Sustainability rank: 32nd

None

0.1 tonnes per capita

Bombing and the use of defoliants in the Vietnam War did serious ecological damage. Slash and burn farming and illegal logging are destroying forests.

MEDIA

TV ownership low

 Daily newspaper circulation 4 per 1000 people

PUBLISHING AND BROADCAST MEDIA

 There are 3 daily newspapers, including the government-published *Vientiane Mai*

 2 state-owned services

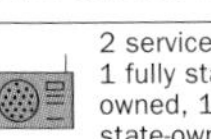
2 services:
1 fully state-owned, 1 part state-owned

Newspapers are owned and controlled by the LPRP; one is published by the Lao People's Army. Revelations of corruption by state officials are not uncommon, but criticism of the party and its leaders remains taboo.

CRIME

Death penalty in use

 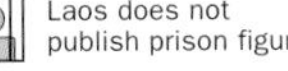 Laos does not publish prison figures

 Rising overall, particularly corruption

Laos is the world's third-largest opium producer. The US has provided funds to replace poppies with alternative cash crops in the northeast provinces.

Farm in northeastern Laos. *The only lowlands are along the Mekong River. Three-quarters of Laotians are subsistence farmers.*

CHRONOLOGY

In 1899, the three small Lao kingdoms were unified under the French.

- ❑ **1953** Independence.
- ❑ **1963** Left-wing armed struggle, overshadowed by Vietnam War.
- ❑ **1975** LPRP seizes power.
- ❑ **1986** Market-oriented reforms.
- ❑ **1997** Accession to ASEAN.
- ❑ **1999** Protests demanding greater political freedom.
- ❑ **2001** Prime minister resigns over economic mismanagement.

EDUCATION

School leaving age: 11

 49%

 12,732 students

Adult education is being expanded. Since 1990, private schools have been allowed, in order to help meet demand.

HEALTH

Welfare state health benefits

1 per 5000 people

Diarrheal, respiratory, and parasitic diseases, malaria, influenza

Since 1975, public health care has developed steadily. Malaria and hemorrhagic fever are on the increase.

L

SPENDING

GDP/cap. increase

CONSUMPTION AND SPENDING

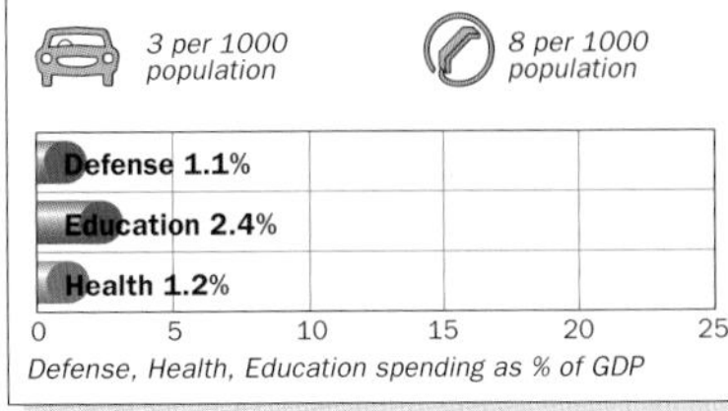
3 per 1000 population
8 per 1000 population
Defense 1.1%
Education 2.4%
Health 1.2%
0
5
10
15
20
25
Defense, Health, Education spending as % of GDP

While a rapidly expanding group of entrepreneurs profits from the gradual liberalization of the country's economy, many in highland and mountainous regions lead a subsistence existence.

WORLD RANKING

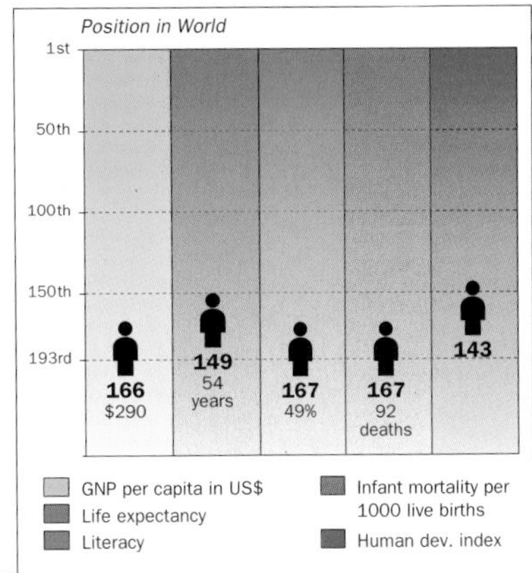
Position in World
1st
50th
100th
150th
193rd
166
$290
149
54 years
167
49%
167
92 deaths
143
GNP per capita in US$
Life expectancy
Literacy
Infant mortality per 1000 live births
Human dev. index

LATVIA

EUROPE

Estonia
Russian Federation
LATVIA
Lithuania
Belarus
Europe

OFFICIAL NAME: Republic of Latvia **CAPITAL:** Riga
POPULATION: 2.4 million **CURRENCY:** Lats **OFFICIAL LANGUAGE:** Latvian

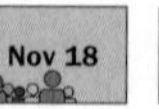
1991

1991
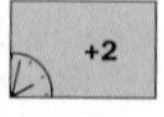
Nov 18

LV

+2

+371
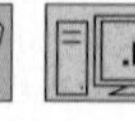
.lv

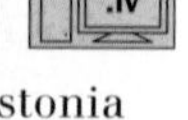

LATVIA IS ONE of the three Baltic states (with Estonia to the north and Lithuania to the south) which regained independence from Soviet rule in 1991. It lies on a low plain, which nowhere rises above 300 m (975 ft). Almost one-third of the population lives in the capital, Riga. Defense-related industries and agriculture play an important role in the economy. Only just over half of the population are ethnic Latvians.

CLIMATE

Continental

WEATHER CHART FOR RIGA

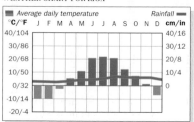

Average daily temperature — Rainfall
°C/°F J F M A M J J A S O N D cm/in
40/104 — 40/16
30/86 — 30/12
20/68 — 20/8
10/50 — 10/4
0/32 — 0
-10/14
-20/-4

Latvia's coastal position moderates its continental-type climate, but winters are cold and summers are cool.

TRANSPORTATION

Drive on right

Riga International 574,870 passengers

186 ships 118,000 grt

THE TRANSPORTATION NETWORK

22,843 km (14,194 miles)

None

2413 km (1499 miles)

300 km (186 miles)

Riga has the busiest container port in the Baltic. The EU-backed Via Baltica highway, linking Poland and Finland, runs north–south through Latvia. An east–west link of similar standard is a priority.

The Russian Orthodox Cathedral in Riga. Used as a planetarium during the Soviet era, its interior has now been restored.

TOURISM

Visitors : Population 1:4.9

490,000 visitors

Down 14% in 1999

MAIN TOURIST ARRIVALS

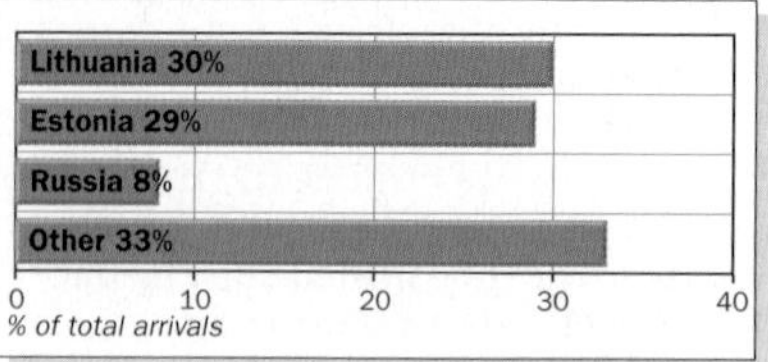
Lithuania 30%
Estonia 29%
Russia 8%
Other 33%
0
10
20
30
40
% of total arrivals

Riga is the main tourist destination, with many hotels and restaurants. Its medieval center is being restored.

PEOPLE

Pop. density low

Latvian, Russian

37/km² (96/mi²)

THE URBAN/RURAL POPULATION SPLIT

69% 31%

ETHNIC MAKEUP

Polish 2%
Ukrainian 3%
Other 2%
Belarussian 4%
Latvian 57%
Russian 32%

Latvians make up just over half the population, but are a minority in Riga. Naturalization procedures were simplified in 1998, easing tension with the country's large minority population of ethnic Russians, but a restrictive language law reopened controversy. In mid-2000 Latvian was proclaimed the only official language for the public and private sectors. The divorce rate is high.

POLITICS

Multiparty elections

1998/2002

President Vaira Vike-Freiberga

AT THE LAST ELECTION

Parliament 100 seats

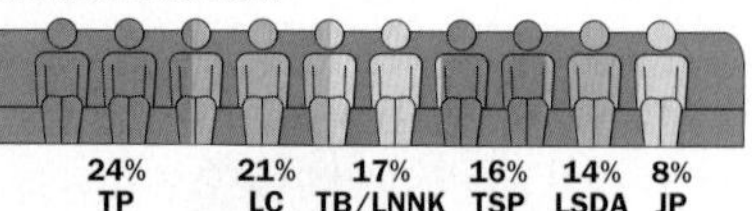
24% TP
21% LC
17% TB/LNNK
16% TSP
14% LSDA
8% JP

TP = People's Party **LC** = Latvia's Way **TB/LNNK** = Fatherland and Freedom **TSP** = National Harmony Party **LSDA** = Social Democratic Party **JP** = New Party

The 1998 general election boosted center-right parties, all in favor of EU membership and continuing market reforms. The largest single party was the recently formed TP, led by former premier Andris Skele, whose 1995–1997 government prepared the way for economic recovery. The TP was excluded from the new government, but this coalition was short-lived, the populist Skele taking over again in 1999. In May 2000, however, Riga's high-profile mayor, Andris Berzins, was appointed prime minister as Skele's coalition disintegrated.

LATVIA

Total Land Area : 64 589 sq. km (24 938 sq. miles)

POPULATION
over 500 000
over 100 000
over 50 000
over 10 000
under 10 000

LAND HEIGHT
200m/656ft
Sea Level

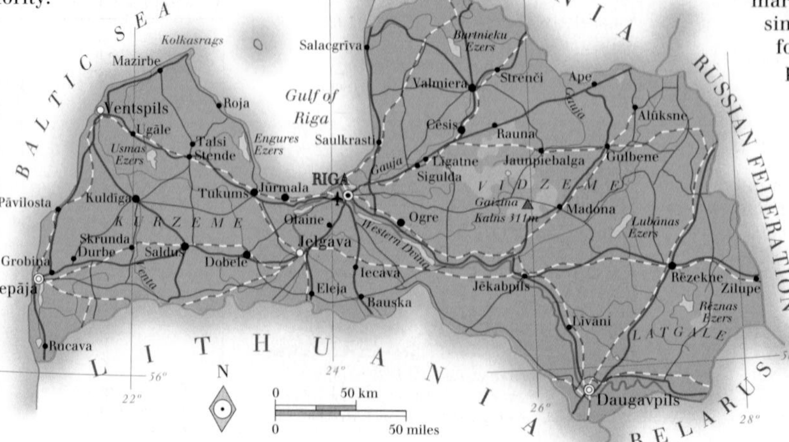
ESTONIA
RUSSIAN FEDERATION
BELARUS
LITHUANIA
BALTIC SEA
Gulf of Riga
Kolkasrags
Mazirbe
Salacgrīva
Burtnieku Ezers
Valmiera
Strenči
Ape
Gauja
Alūksne
Ventspils
Ugāle
Roja
Talsi
Stende
Usmas Ezers
Engures Ezers
Saulkrasti
Cēsis
Rauna
Līgatne
Sigulda
Jaunpiebalga
Gulbene
RIGA
Jūrmala
Tukums
Kuldīga
Pāvilosta
KURZEME
VIDZEME
Gaiziņa Kalns 311m
Madona
Lubānas Ezers
Olaine
Ogre
Skrunda
Durbe
Saldus
Jelgava
Western Dvina
Grobiņa
Liepāja
Dobele
Iecava
Rēzekne
Zilupe
Eleja
Bauska
Jēkabpils
Rāznas Ezers
Līvāni
LATGALE
Rucava
Daugavpils
N
0 50 km
0 50 miles
58°
56°
22°
24°
26°
28°

WORLD AFFAIRS

Joined UN in 1991

CE
CBSS
OSCE
EAPC
PfP

Latvia is one of six "second wave" candidates for admission to the EU which began membership negotiations in 2000. It also has US backing for entry to NATO. Discrimination against Russian-speakers, exemplified by laws passed in 2000 making Latvian the only official language, have strained relations with Russia.

AID

Recipient

$91m (receipts) | Down 9% in 2000

Aid to Latvia comes mainly from the World Bank, the IMF, and the EU. Most goes toward improving the country's infrastructure.

DEFENSE

Compulsory military service

$70m | Up 21% in 2000

Building up the military is a priority, and there is now US backing for entry into NATO. Latvia has been participating in NATO's Partnership for Peace program. In February 2000, Russian forces finished dismantling their last military installation in Latvia, the Skrunda radar station.

ECONOMICS

Inflation 49% p.a. (1990–2000)

$6.93bn | 0.6178–0.6292 lats

SCORE CARD

- ❑ WORLD GNP RANKING........................97th
- ❑ GNP PER CAPITA $2920
- ❑ BALANCE OF PAYMENTS...................–$485m
- ❑ INFLATION 2.7%
- ❑ UNEMPLOYMENT 8%

STRENGTHS

Industrial production improving after slump. Service sector now providing more than half of GDP. Inflation under control. Foreign investment rising.

WEAKNESSES

Dependence on imported oil and natural gas for energy. Lack of raw materials. Farming is technically backward after the dismantling of the collective farms.

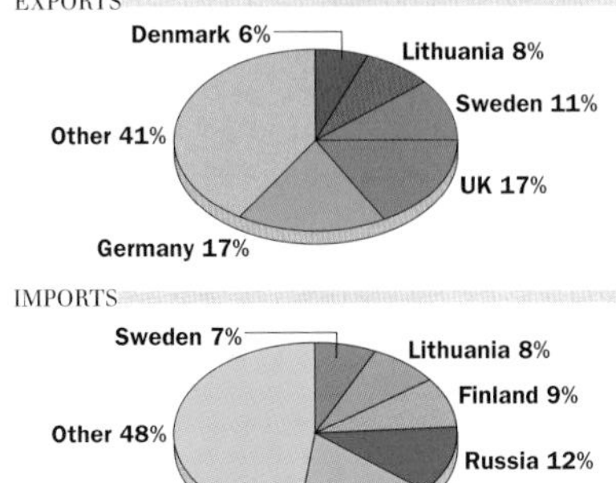
EXPORTS
Denmark 6%
Lithuania 8%
Sweden 11%
Other 41%
UK 17%
Germany 17%
IMPORTS
Sweden 7%
Lithuania 8%
Finland 9%
Other 48%
Russia 12%
Germany 16%

RESOURCES

Electric power 2.1m kw

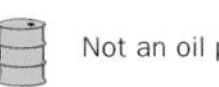

125,857 tonnes | Not an oil producer

437,000 turkeys, 393,500 pigs, 3.11m chickens | Amber, dolomite, gravel, gypsum, limestone, peat, sand

Latvia has very limited natural resources, and is dependent on imports (mainly from Russia) to meet its energy needs. Peat is still burned for energy. Electricity comes chiefly from hydroelectric power and imports from Lithuania and Estonia. Offshore oil exploration is planned. Ventspils deepwater port is being promoted as an oil terminal.

ENVIRONMENT

Sustainability rank: 10th

13% | 3.2 tonnes per capita

Peat extraction has damaged valuable bog habitat. Pollution of the Baltic Sea and air and water quality in industrial centers are also of concern.

Environmental issues featured in pre-independence protests and have retained a high profile.

MEDIA

TV ownership high

Daily newspaper circulation 247 per 1000 people

PUBLISHING AND BROADCAST MEDIA

There are 23 daily newspapers, including *Diena* and *Neatkariga Rita Avize*

2 services: 1 state-owned, 1 independent | 16 services: 1 state-owned, 15 independent

The press is now relatively free from state interference. Previously, the media were predominantly in Russian. Since 1991 the state, aiming to broaden the use of the official language, has actively promoted Latvian publications.

CRIME

Death penalty not used in practice

8603 prisoners | Up 41% in 1999

Organized crime is a growing problem, but general crime levels are lower than in other former Soviet states.

EDUCATION

School leaving age: 16

99% | 101,278 students

Nearly three-quarters of pupils receive teaching solely in Latvian. Schools have opened for many minority ethnic groups.

CHRONOLOGY

Latvia was dominated by Germany and, briefly, Sweden before Russia completed its conquest in 1795.

- ❑ **1917** Opposes Russian Bolshevik revolution. Declares independence.
- ❑ **1918–1920** Invaded by Bolsheviks and Germany.
- ❑ **1920** Gains independence.
- ❑ **1944** Incorporated into USSR.
- ❑ **1989** Popular Front wins elections; declares independence.
- ❑ **1991** Independence recognized.
- ❑ **1995** TP-led coalition formed.
- ❑ **1998** Elections; LC-led coalition. Naturalization procedure eased.
- ❑ **1999** First woman president. Andris Skele of TP returns as premier.
- ❑ **2000** Skele resigns; replaced by Andris Berzins.

HEALTH

Welfare state health benefits

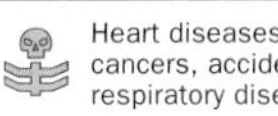

1 per 357 people | Heart diseases, cancers, accidents, respiratory diseases

The state-run system suffers shortages of medicines and equipment. Some improvements have been made, but it is still seriously underfunded.

L

SPENDING

GDP/cap. decrease

CONSUMPTION AND SPENDING

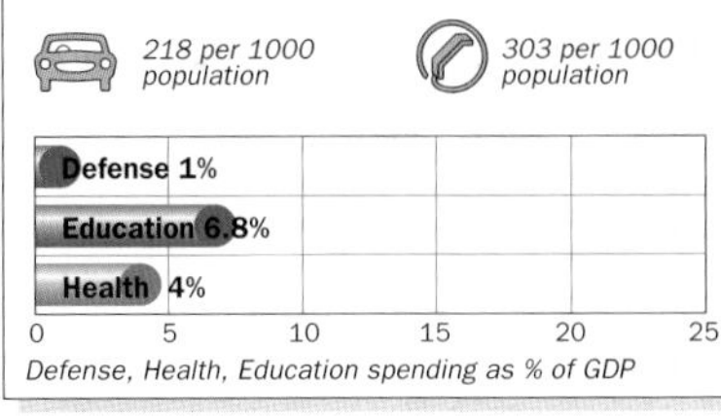
218 per 1000 population
303 per 1000 population
Defense 1%
Education 6.8%
Health 4%
0
5
10
15
20
25

Defense, Health, Education spending as % of GDP

The old bureaucracy has retained its privileged status and contacts, and remains the wealthiest group. Farmers are among the poorest.

WORLD RANKING

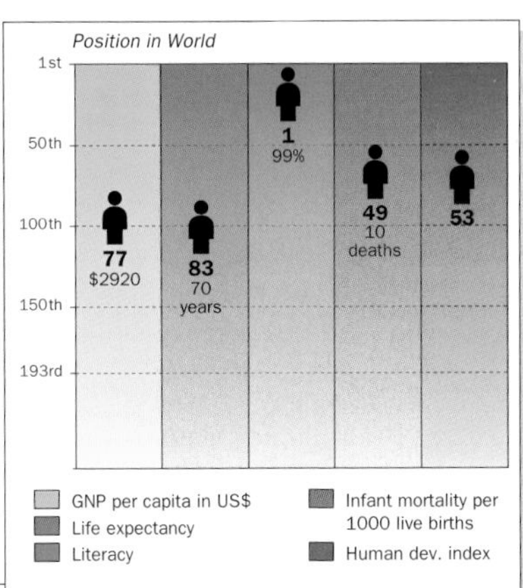
Position in World
1st
50th
100th
150th
193rd
77
$2920
83
70
years
1
99%
49
10
deaths
53
GNP per capita in US$
Life expectancy
Literacy
Infant mortality per 1000 live births
Human dev. index

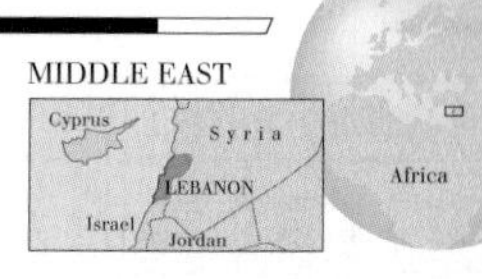
MIDDLE EAST
Cyprus
Syria
LEBANON
Israel
Jordan
Asia
Africa

Lebanon

Official Name: Republic of Lebanon **Capital:** Beirut
Population: 3.6 million **Currency:** Lebanese pound **Official Language:** Arabic

1941

1941

Nov 22

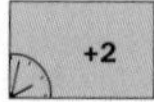
RL
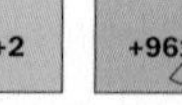

+2
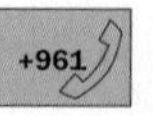

+961
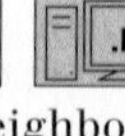
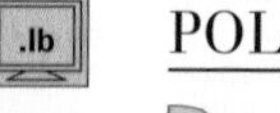
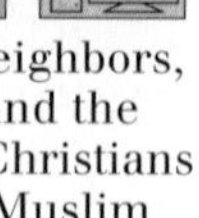
.lb

LEBANON LIVES in the shadow of its powerful neighbors, Syria and Israel. The coastal strip is fertile and the hinterland mountainous. The minority Maronite Christians have traditionally dominated the government. Civil war between Muslim and Christian factions from 1975, complicated by an Israeli invasion in 1982, threatened a breakup of the state, until Saudi Arabia brokered a peace deal in 1989. Greater political stability and reconstruction have ensued.

CLIMATE

Mediterranean/ mountain

WEATHER CHART FOR BEIRUT

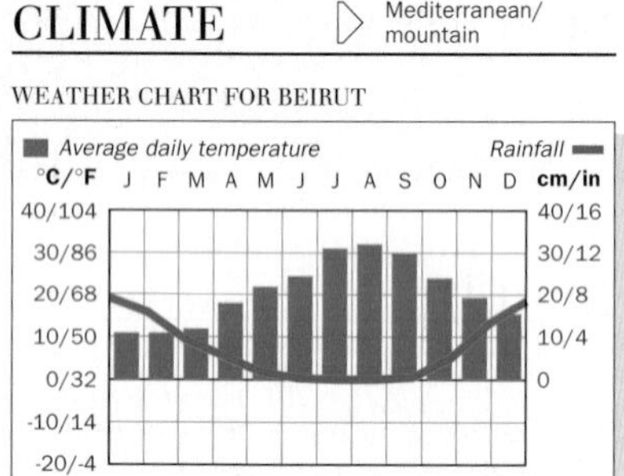
Average daily temperature
Rainfall
°C/°F J F M A M J J A S O N D cm/in
40/104
30/86
20/68
10/50
0/32
-10/14
-20/-4
40/16
30/12
20/8
10/4
0

Winters are mild and summers hot, with high humidity on the coast. Snow falls on high ground in the winter.

TRANSPORTATION

Drive on right

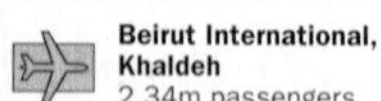

Beirut International, Khaldeh
2.34m passengers

106 ships
263,500 grt

THE TRANSPORTATION NETWORK

6200 km (3852 miles)

None

40 km (25 miles)

None

The redevelopment of Beirut could see it regain its position as one of the Middle East's major entrepôts.

TOURISM

Visitors : Population 1:4.9

742,000 visitors

Up 10% in 2000

MAIN TOURIST ARRIVALS

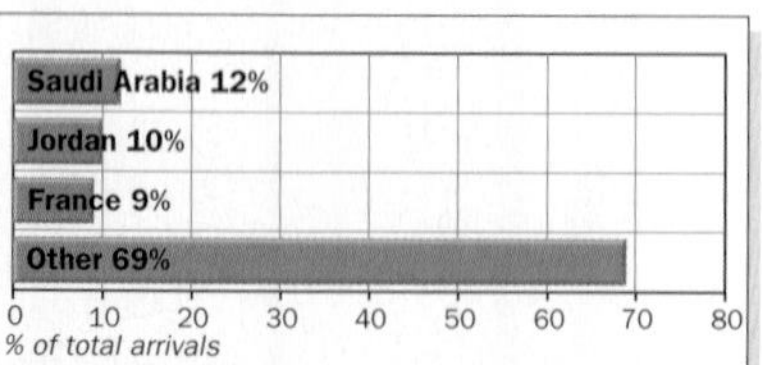
Saudi Arabia 12%
Jordan 10%
France 9%
Other 69%
0 10 20 30 40 50 60 70 80
% of total arrivals

Tourists have gradually returned since the devastation of the civil war. Beirut remains the main destination, its battle scars even adding to its attraction. In 1998 the US lifted its restrictions on travel by US citizens to Lebanon.

PEOPLE

Pop. density high

Arabic, French, Armenian, Assyrian

352/km² (911/mi²)

THE URBAN/RURAL POPULATION SPLIT

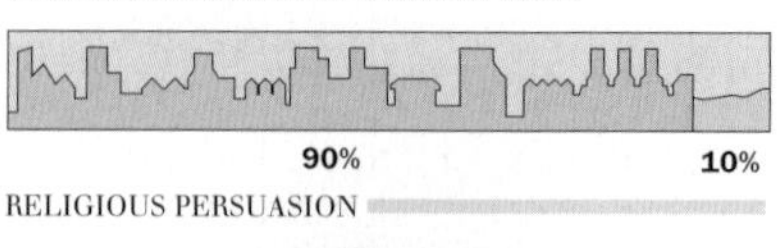
90%
10%

RELIGIOUS PERSUASION

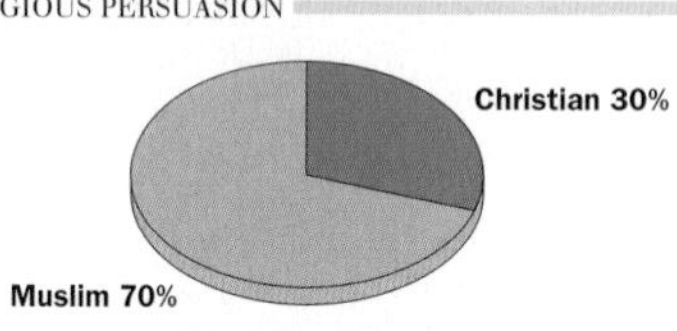
Christian 30%
Muslim 70%

The Lebanese population is fragmented in religious terms into subsects of Christians and Muslims, but retains a strong sense of national identity. There has been a large Palestinian refugee population in the country since 1948. Islamic fundamentalism is influential among poorer Shi'a Muslims, who constitute the largest single group.

POLITICS

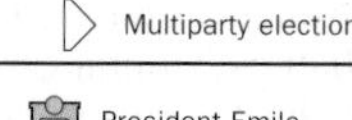

Multiparty elections

2000/2004

President Emile Lahoud

AT THE LAST ELECTION

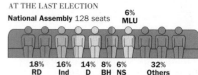

National Assembly 128 seats

18% RD, **16%** Ind, **14%** D, **8%** BH, **6%** NS, **6%** MLU, **32%** Others

RD = Resistance and Development List **Ind** = Independents **D** = Dignity **BH** = Baalbek–Hermel List **NS** = National Struggle List **MLU** = Mount Lebanon Unity

The Arab-brokered 1989 Taif peace agreement ending the civil war redressed the constitutional balance between Christians and Muslims and guaranteed power-sharing. Relative stability has been maintained under Rafiq al-Hariri, who won the first postwar legislative elections in 1992. After a brief term as premier by Salim al-Hoss, Hariri was elected to a third term by the National Assembly in 2000. Gen. Emile Lahoud was elected president in 1998. Syria remains the main power broker in Lebanon, especially following the withdrawal of Israeli troops in 2000.

LEBANON
Total Land Area : 10 400 sq. km (4015 sq. miles)
0 20 km
0 20 miles
N
SYRIA
MEDITERRANEAN SEA
ISRAEL
El Mina
Tripoli
Zgharta
Qoubaïyât
Hermel
Amioun
3087m
QORNET ES SAOUDA
Batroûn
Bcharré
Orontes
Jbaïl
Qartaba
LIBAN
ANTI-LEBANON MTS
Joûnié
Baalbek
34°
BEIRUT
Baabda
Zahlé
Aaley
Bhamdoun
Rayak
Damoûr
JEBEL
BEQAA
Aanjar
Barouk
Saïda
Lac de Qaraoun
Jezzine
Rachaïya
36°
Nabatîyé
Litani
Marjayoun
El Khiyam
Soûr
Tibnine
GOLAN HEIGHTS
En Nâqoûra
Bent Jbaïl
33°
LAND HEIGHT
3000m/9843ft
2000m/6562ft
1000m/3281ft
500m/1640ft
200m/656ft
Sea Level
POPULATION
over 1 000 000
over 100 000
over 10 000
under 10 000

WORLD AFFAIRS

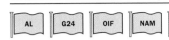

Joined UN in 1945

AL
G24
OIF
NAM
OIC

The 1989 Taif Agreement ending the civil war has left Syria with enormous influence in domestic politics. Anti-Israeli rhetoric remains the political lingua franca. The Hezbollah militia frequently skirmished with Israeli occupying forces (and their proxy militias) until the Israeli withdrawal in 2000, and both sides continue to exchange fire across the border, despite the presence of a UN patrol force.

AID

Recipient

 $197m (receipts)

 Up 2% in 2000

The government has sought billions of dollars for reconstruction. The World Bank prefers working through NGOs.

DEFENSE

Compulsory military service

$553m

Down 2% in 2000

The army has over 70,000 troops. Hezbollah guerrillas rapidly regained control of southern Lebanon after the Israeli withdrawal in 2000. There is a UN peacekeeping force on the Israeli border. Syrian forces, whose dominant security role had been formalized in 1991, withdrew from Beirut in 2001.

ECONOMICS

Inflation 17% p.a. (1990–2000)

$17.4bn

 1507.0–1513.8 Lebanese pounds

SCORE CARD

- ❑ World GNP Ranking 72nd
- ❑ GNP per Capita $4010
- ❑ Balance of Payments ....–$3.07bn
- ❑ Inflation 0%
- ❑ Unemployment 18%

Strengths

Peace will allow Lebanon to regain its position as an Arab center for banking and services. Potentially a major producer of wine and fruit. Tight fiscal policy has kept inflation down.

Weaknesses

Dependent on imported oil and gas. Agriculture still at 40% of prewar levels. High public debt. Alleged Syrian "dumping" of cheap products.

EXPORTS

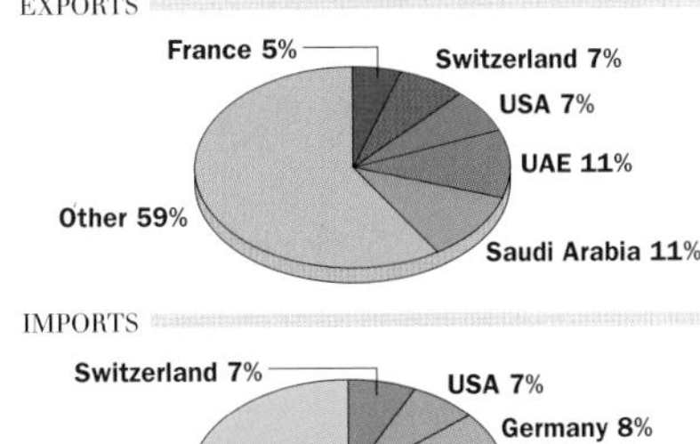
France 5%
Switzerland 7%
USA 7%
UAE 11%
Other 59%
Saudi Arabia 11%

IMPORTS

Switzerland 7%
USA 7%
Germany 8%
France 8%
Other 59%
Italy 11%

RESOURCES

Electric power 1.4m kw

 3860 tonnes

 Not an oil producer; refines 37,500 b/d

 445,000 goats, 380,000 sheep, 32m chickens

Lignite, iron ore

Wine, cotton, fruit, and vegetables are the main crops. Power plants are fueled by imported petroleum.

ENVIRONMENT

Sustainability rank: 106th

 0.5%

3.9 tonnes per capita

Lack of central authority during the civil war allowed unregulated building, logging, and quarrying to flourish.

MEDIA

TV ownership high

 Daily newspaper circulation 74 per 1000 people

PUBLISHING AND BROADCAST MEDIA

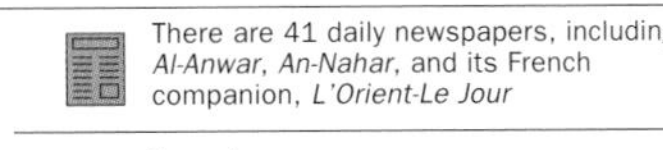

There are 41 daily newspapers, including *Al-Anwar*, *An-Nahar*, and its French companion, *L'Orient-Le Jour*

5 services: 1 state-controlled, 4 independent

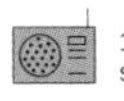 1 state-owned service

In the late 1990s the government banned news and political programs on private satellite television channels.

CRIME

Death penalty in use

 7296 prisoners

 Up 13% in 1999

The kidnapping of hostages and the breakdown of law during the civil war made Beirut a dangerous city for visitors.

Politically motivated violence has recently declined, though the risk of urban terrorism remains. Rural areas untouched by the conflict have low levels of crime.

***The Corniche, Beirut,** was rebuilt after the civil war by US consultant engineers and architects in a privately financed scheme.*

CHRONOLOGY

Under French mandate from 1920, Lebanon declared independence in 1941, achieving full autonomy in 1946.

- ❑ **1975** Civil war erupts.
- ❑ **1982** Israeli invasion.
- ❑ **1989** Taif Agreement ends civil war.
- ❑ **1992** First election in 20 years. Rafiq al-Hariri prime minister.
- ❑ **1996** Israeli attack kills over 100 civilians at UN base in Qana.
- ❑ **1998** Emile Lahoud president.
- ❑ **2000** Israeli forces withdraw. Hariri reelected by a landslide.

EDUCATION

School leaving age: 11

 86%

 81,588 students

Lebanon has one of the highest literacy rates in the Arab world. Education was severely disrupted by the war.

HEALTH

Welfare state health benefits

1 per 476 people

Heart disease, infectious and parasitic diseases

An adequate system of primary health care exists. Hospital staffing is returning to prewar levels.

SPENDING

GDP/cap. increase

CONSUMPTION AND SPENDING

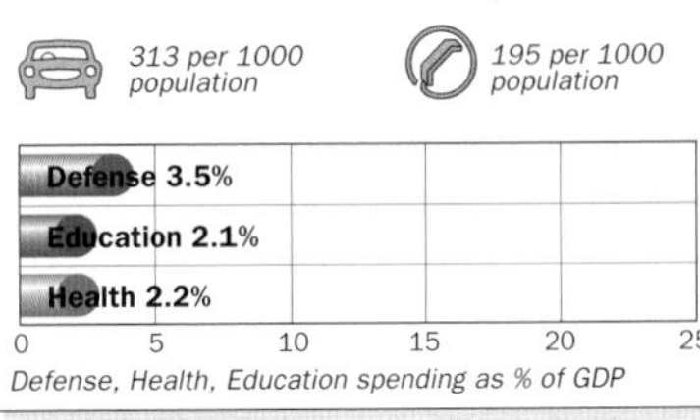
313 per 1000 population
195 per 1000 population
Defense 3.5%
Education 2.1%
Health 2.2%
0
5
10
15
20
25
Defense, Health, Education spending as % of GDP

Average income per capita statistics conceal the fact that a huge gulf exists between the poor and a small, massively rich elite.

WORLD RANKING

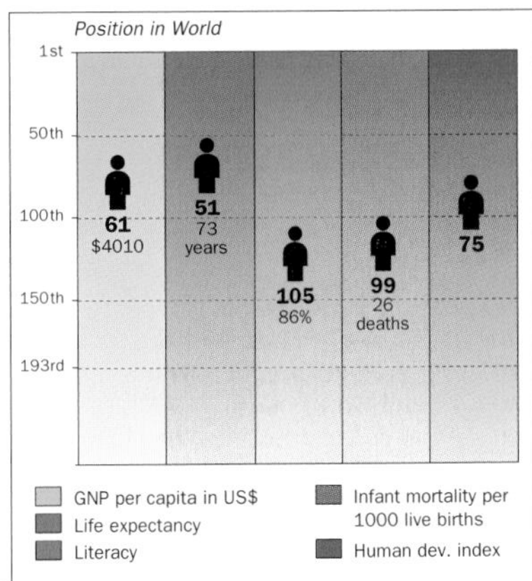
Position in World
1st
50th
100th
150th
193rd
61
$4010
51
73 years
105
86%
99
26 deaths
75
GNP per capita in US$
Life expectancy
Literacy
Infant mortality per 1000 live births
Human dev. index

L

LESOTHO

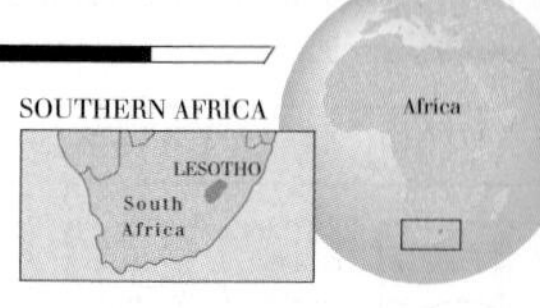

SOUTHERN AFRICA
LESOTHO
South Africa
Africa

OFFICIAL NAME: Kingdom of Lesotho **CAPITAL:** Maseru
POPULATION: 2.1 million **CURRENCY:** Loti **OFFICIAL LANGUAGES:** English and Sesotho

1966

1966

Oct 4

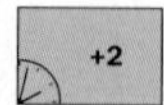
LS

+2

+266

.ls

A MOUNTAINOUS AND landlocked country entirely surrounded by South Africa, Lesotho is economically dependent on its larger neighbor. However, Lesotho is beginning to benefit from the export of energy from the recently completed Highlands Water Scheme. Elections in 1993 ended a period of military rule, but South Africa had to send in its troops when serious political unrest erupted in 1998.

CLIMATE

Mountain

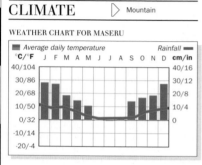

WEATHER CHART FOR MASERU

Average daily temperature | Rainfall
°C/°F J F M A M J J A S O N D cm/in
40/104 30/86 20/68 10/50 0/32 -10/14 -20/-4
40/16 30/12 20/8 10/4 0

Drought is often followed by torrential rain storms. Snow is frequent in winter in the mountains.

TRANSPORTATION

Drive on left

Moshoeshoe International, Maseru 43,000 passengers

Has no fleet

THE TRANSPORTATION NETWORK

887 km (551 miles)

None

3 km (2 miles)

None

Lesotho has to rely on South African road and rail outlets. New roads have been constructed to service the Highlands Water Scheme.

TOURISM

Visitors : Population 1:11

186,000 visitors

Up 24% in 1999

MAIN TOURIST ARRIVALS

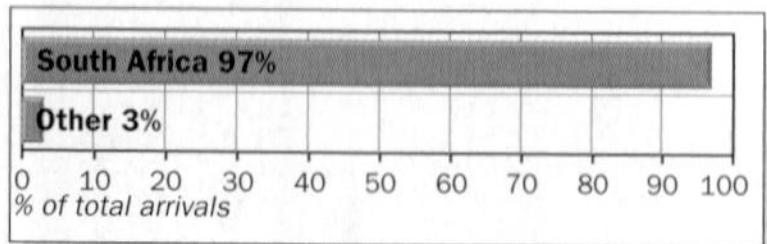

South Africa 97%
Other 3%
0 10 20 30 40 50 60 70 80 90 100
% of total arrivals

Tourists, mainly South African, are attracted to the dramatic mountain scenery and watersports on artificially created lakes. Another draw is Thaba-Bosiu, King Moshoeshoe the Great's mountain stronghold. Political violence in 1998 deterred many visitors.

PEOPLE

Pop. density medium

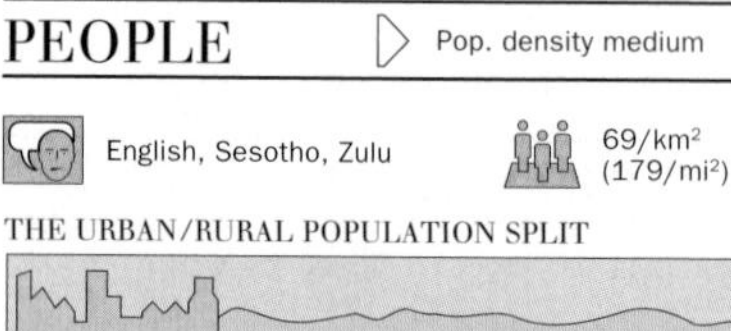

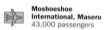

English, Sesotho, Zulu
69/km² (179/mi²)
THE URBAN/RURAL POPULATION SPLIT
28%
72%

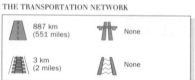

ETHNIC MAKEUP

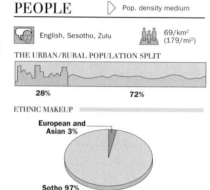

European and Asian 3%

Sotho 97%

The overwhelming majority of the population are Sotho, though there is a small community of European origin, as well as south Asian and Chinese minorities, which are active in the retailing business. Ethnic homogeneity and a strong sense of national identity have tended to minimize ethnic tension.

The export of male contract labor to South African mines means that women head 72% of households; they also run farming, regarded by Lesotho men as "women's work."

POLITICS

Multiparty elections

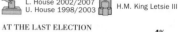

L. House 2002/2007
U. House 1998/2003

H.M. King Letsie III

AT THE LAST ELECTION

National Assembly 120 seats

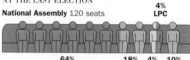

4% LPC
64% LCD
18% BNP
4% NIP
10% Others

LCD = Lesotho Congress for Democracy Party
BNP = Basotho National Party **NIP** = National Independent Party **LPC** = Lesotho People's Congress

Senate 33 seats

The Senate comprises 22 principal chiefs and 11 other members named by the king

The armed forces have been key political players in Lesotho since a bloodless coup in 1986. Although direct military rule ended in 1993, and a free and peaceful general election resulted in a sweeping victory for the Basotho Congress Party (BCP), the army maintained its powers over national security. Tensions between the army and the civilian administration escalated in 1994, when mutinous troops killed the deputy prime minister. King Moshoeshoe II was restored to the throne, and was succeeded by his son Letsie III in 1996.

Accusations of vote rigging and mass protests greeted a general election win in 1998 by the LCD. After an attempted coup in September, the South African military intervened to restore democracy, brokering an agreement between the king and Lesotho's 12 parties. When elections were finally held in 2002, the LCD retained its majority.

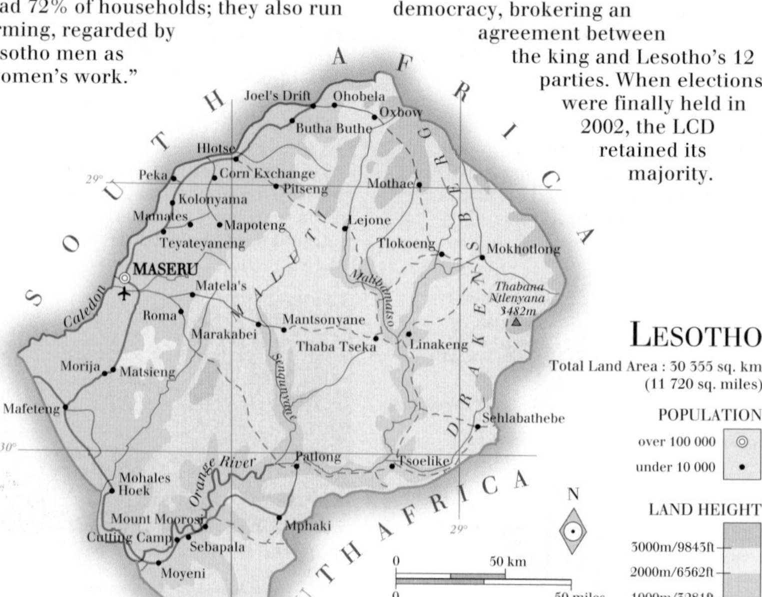

SOUTH AFRICA
Joel's Drift
Ohobela
Oxbow
Butha Buthe
Hlotse
Peka
Corn Exchange
Pitseng
Mothae
Kolonyama
Mamates
Mapoteng
Lejone
Teyateyaneng
Tlokoeng
Mokhotlong
MASERU
Matela's
Thabana Ntlenyana 3482m
Caledon
Roma
Mantsonyane
Marakabei
Thaba Tseka
Linakeng
Morija
Matsieng
Mafeteng
Sehlabathebe
Orange River
Patlong
Tsoelike
Mohales Hoek
Mount Moorosi
Mphaki
Cutting Camp
Sebapala
Moyeni
LESOTHO
Total Land Area : 30 355 sq. km (11 720 sq. miles)
POPULATION
over 100 000
under 10 000
LAND HEIGHT
3000m/9843ft
2000m/6562ft
1000m/3281ft
0 50 km
0 50 miles
N

WORLD AFFAIRS

Joined UN in 1966

Comm
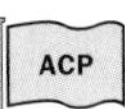
ACP

NAM

AU

SADC

Foreign policy is dominated by the nature of Lesotho's relationship with South Africa. Lesotho currently has duty-free access to the EU for most manufactured goods, and also has preferential access to US and Scandinavian markets.

AID

Recipient

 $41m (receipts)

 Up 32% in 2000

Aid, mostly devoted to agricultural development, accounts for 26% of Lesotho's GNP; about half comes from SACU. A "national famine" was declared in 2002 to prompt donations of emergency assistance.

DEFENSE

No compulsory military service

$29m

 Down 15% in 2000

Lesotho's 2000-strong army relied on South African assistance to quell political violence in 1998.

ECONOMICS

Inflation 9.9% p.a. (1990–2000)

$1.18bn

7.570–11.995 maloti

SCORE CARD

- ❑ World GNP Ranking........150th
- ❑ GNP per Capita........$580
- ❑ Balance of Payments........–$151m
- ❑ Inflation........6.1%
- ❑ Unemployment........45%

Strengths

Potential of educated workforce. Boom in textiles and other manufacturing. Membership of SACU. Future revenues from water sales.

Weaknesses

Dependent on South Africa. Loss of workers to mining in South Africa. Weak agricultural sector. Retail sector affected by sporadic unrest. High unemployment.

EXPORTS

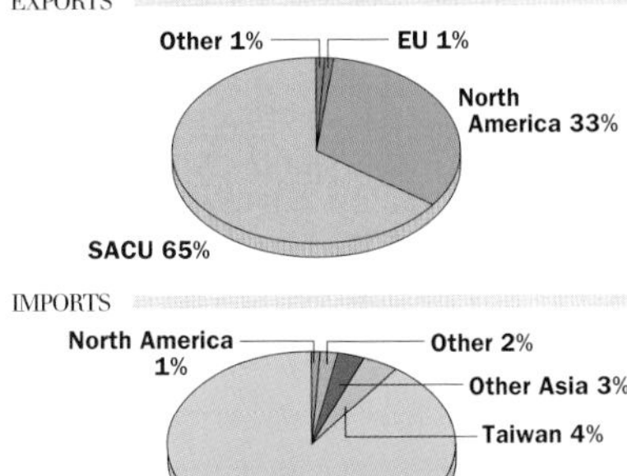
Other 1%
EU 1%
North America 33%
SACU 65%

IMPORTS

North America 1%

Other 2%

Other Asia 3%

Taiwan 4%

SACU 90%

Landscape near ***Mohales Hoek*** *in Lesotho's lowest lands – over 1300 m (4260 ft) above sea level.*

RESOURCES

Electric power: Included in South African total

 34 tonnes

 Not an oil producer

 730,000 sheep, 570,000 goats, 1.7m chickens

Diamonds

The Highlands Water hydroelectric scheme has the capacity to supply all of Lesotho's energy requirements, as well as 62 cu. m (2200 cu. ft) of water per second for South African use. Diamonds are mined in the northeast.

ENVIRONMENT

Not available

 0.2%

 Negligible emissions per capita

Climate and overgrazing have seriously eroded the land. The Highlands Water Scheme has flooded acres of peasant farmland. Supporters of this massive dam project stress encouragement for wildlife in reservoirs and on bird-friendly pylons.

MEDIA

TV ownership low

 Daily newspaper circulation 8 per 1000 people

PUBLISHING AND BROADCAST MEDIA

There are 2 daily newspapers. *Leselinyana la Lesotho* is a popular religious periodical

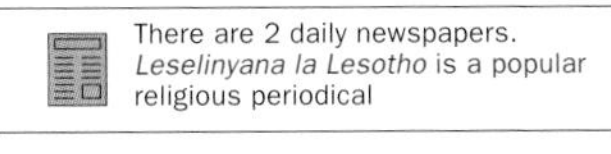

1 state-owned service

1 state-owned service

State control of the media was used for political ends in 1998. The *Mirror* is the only independent paper.

CRIME

Death penalty in use

 2552 prisoners

 Up 25% 1993–1999

The 1998 political crisis increased crime levels, which previously were much lower than in South Africa.

EDUCATION

School leaving age: 13

 83%

 4046 students

Lesotho's schools have very high enrollment levels and one of the highest literacy rates in Africa.

CHRONOLOGY

As Basutoland, Lesotho became a British Crown colony in 1884.

- ❑ **1966** Independent kingdom.
- ❑ **1986** Military coup.
- ❑ **1990** King Moshoeshoe II exiled. Son installed as Letsie III.
- ❑ **1993** Free elections.
- ❑ **1994** Return of Moshoeshoe II.
- ❑ **1996** Letsie III succeeds to throne.
- ❑ **1998** New LCD wins polls. South Africa intervenes after coup attempt, and reconciles king and parties.
- ❑ **2002** Food emergency follows successive poor harvests. LCD wins long-postponed elections.

HEALTH

Welfare state health benefits

 1 per 10,000 people

 Tuberculosis, AIDS, parasitic diseases, nutritional disorders

Private health organizations and NGOs account for half of all health services. A government-operated flying doctor service covers the highlands. An estimated 25% of adults are HIV positive and life expectancy is declining. Food shortages in 2002 exacerbated the country's health problems.

SPENDING

GDP/cap. increase

CONSUMPTION AND SPENDING

6 per 1000 population

10 per 1000 population

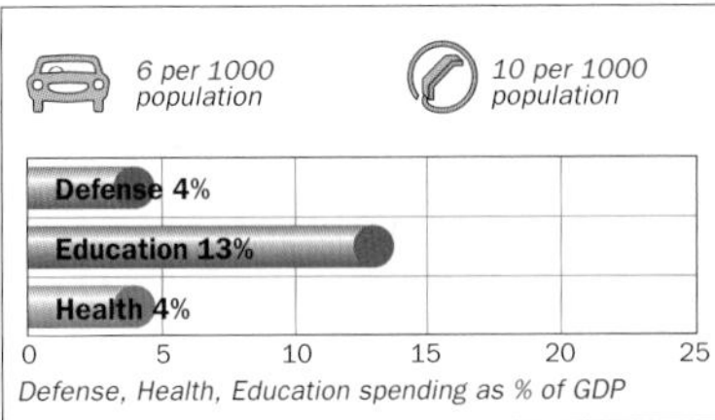
Defense 4%
Education 13%
Health 4%
0
5
10
15
20
25
Defense, Health, Education spending as % of GDP

Social mobility is limited in Lesotho; the ruling elite keeps a tight control on power and wealth. Over 90% of the population live below the poverty line and many are migrant laborers.

WORLD RANKING

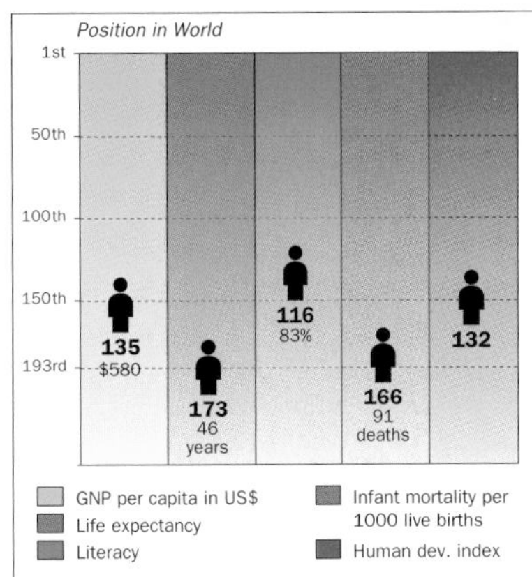
Position in World
1st
50th
100th
150th
193rd
135
$580
173
46
years
116
83%
166
91
deaths
132
GNP per capita in US$
Life expectancy
Literacy
Infant mortality per 1000 live births
Human dev. index

L

LIBERIA

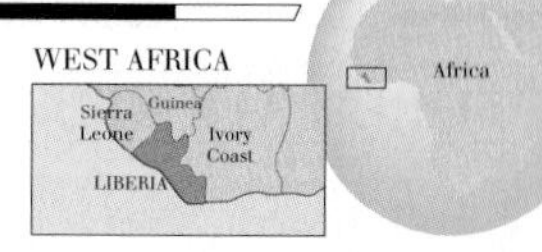
WEST AFRICA
Sierra Leone
Guinea
Ivory Coast
LIBERIA
Africa

OFFICIAL NAME: Republic of Liberia **CAPITAL:** Monrovia
POPULATION: 3.1 million **CURRENCY:** Liberian dollar **OFFICIAL LANGUAGE:** English

1847

1847

July 26
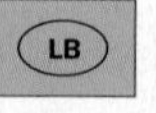
LB
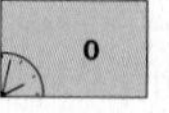
0
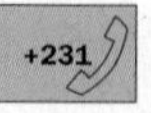
+231

.lr
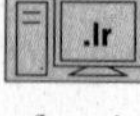

FOUNDED IN 1847 by freed US slaves, Liberia today is struggling to recover from a civil war which reduced it to anarchy in the 1990s but new rebel factions have relaunched the fighting. Facing the Atlantic in equatorial west Africa, most of its coastline is characterized by lagoons and mangrove swamps. Inland, a grassland plateau supports the limited agriculture (just 1% of land is arable). Liberia has the world's largest flag of convenience merchant fleet.

CLIMATE

Tropical equatorial

WEATHER CHART FOR MONROVIA

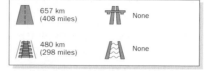

Average daily temperature / Rainfall

°C/°F: 60/140, 50/122, 40/104, 30/86, 20/68, 10/50, 0/32 — J F M A M J J A S O N D — cm/in: 90/35, 80/31, 70/28, 60/24, 50/20, 40/16, 30/12, 20/8, 10/4, 0

There is one long rainy season from May to October, with a brief interlude in most of the country of about two weeks in August.

Temperatures are consistently high. During the dry season, when the dust-laden *harmattan* wind blows, they rise even higher inland.

TRANSPORTATION

Drive on right

Roberts Field International, Monrovia

1717 ships 60.5m grt

THE TRANSPORTATION NETWORK

657 km (408 miles)

None

480 km (298 miles)

None

Most roads in Liberia are unpaved. The 480-km (298-mile) railroad was built to transport iron ore and carries little other traffic. Roberts Field airport was built by the US during World War II.

TOURISM

Not available

Tourists deterred by civil war

Little change from year to year

MAIN TOURIST ARRIVALS

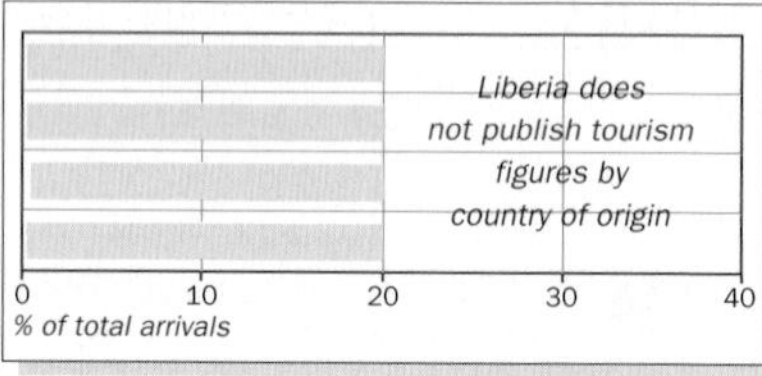
Liberia does not publish tourism figures by country of origin
0
10
20
30
40
% of total arrivals

As a result of the civil war and continuing insecurity, tourism, never a significant activity in Liberia, is now nonexistent.

PEOPLE

Pop. density low

Kpelle, Vai, Bassa, Kru, Grebo, Kissi, Gola, Loma, English

33/km² (83/mi²)

THE URBAN/RURAL POPULATION SPLIT

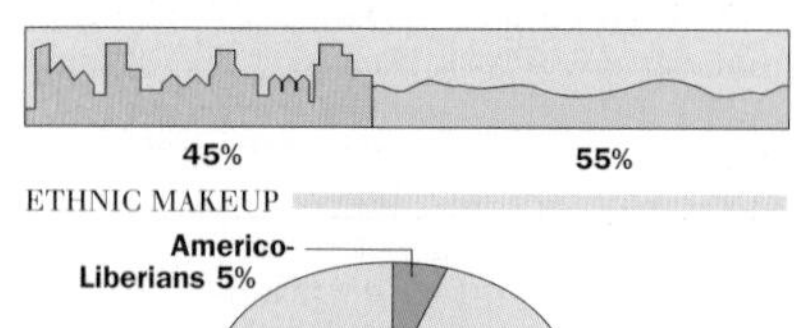
45%
55%

ETHNIC MAKEUP

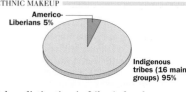

Americo-Liberians 5%

Indigenous tribes (16 main groups) 95%

A key distinction in Liberia has been between Americo-Liberians, the descendants of those freed from slavery (known as "civilized persons"), and the majority indigenous "tribals." The latter were long held in contempt by the Americos, but intermarriage and political assimilation since 1944 have softened attitudes. Intertribal tension in Liberia is now a much more serious problem, and was the main cause of the civil war which erupted in 1990.

POLITICS

Multiparty elections

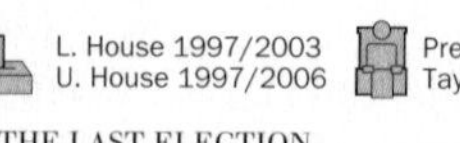
L. House 1997/2003
U. House 1997/2006
President Charles Taylor

AT THE LAST ELECTION

House of Representatives 64 seats

76% NPP · 5% ALCP · 11% UP · 8% Others

NPP = National Patriotic Party **UP** = Unity Party
ALCP = All Liberia Coalition Party

Senate 26 seats

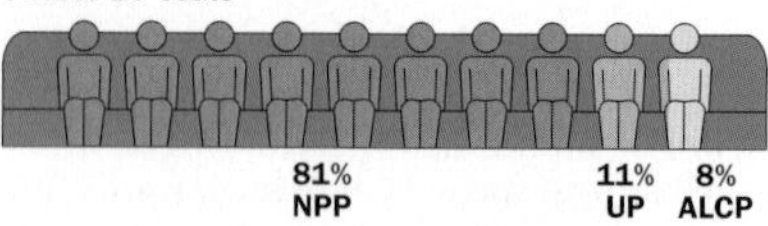
81%
NPP
11%
UP
8%
ALCP

Liberian politics collapsed after 1990 into a chaotic, bloody, and many-sided conflict. A peace agreement finally signed in 1996 provided for presidential and legislative elections in 1997, won by Charles Taylor and his NPP, formerly the National Patriotic Front of Liberia (NPFL), the predominant armed faction. Some 700,000 refugees began returning, but instability continued, and intense fighting broke out in the north in mid-2000 following the formation of a new rebel faction, Liberians United for Reconciliation and Democracy. Thousands of civilians fled the area in 2001 and 2002 as fighting escalated.

GUINEA · SIERRA LEONE · IVORY COAST (CÔTE D'IVOIRE) · ATLANTIC OCEAN · Grain Coast · Sielo · Voinjama · Kolahun · Zelimai · Zorzor · Yekepa · Mount Nimba 1752m · Sanniquellie · Kahnple · Ganta · Bendurna · Bopolu · Tubmanburg · Gbarnga · Robertsport · Nyenyen · Zienzu · Salala · Totota · Bomi · Kakata · Brewerville · Bensonville · Mount Barclay · MONROVIA · Harbel · Tapeta · Dolobli · Marshall · Barclayville · Poabil · Hartford · Edina · Buchanan · Zwedru · Zia Town · River Cess · Juarzon · Dodwekon · Kahnwia · Greenville · Nyaake · Sasstown · Grand Cess · Plibo · Harper · Lofa · St. Paul · St. John · Cestos · Cavalla · 50 km · 50 miles

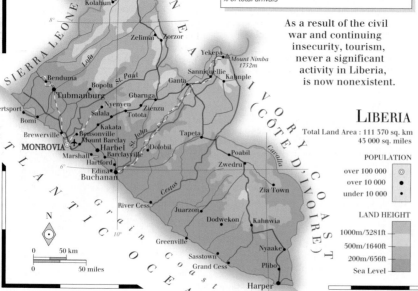

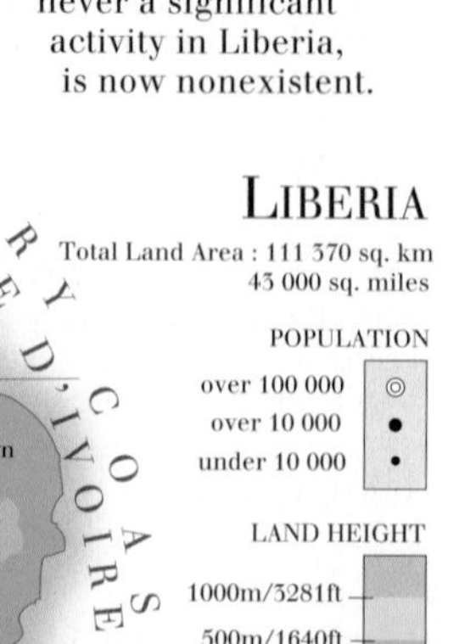
LIBERIA
Total Land Area : 111 370 sq. km
43 000 sq. miles
POPULATION
over 100 000
over 10 000
under 10 000
LAND HEIGHT
1000m/3281ft
500m/1640ft
200m/656ft
Sea Level

WORLD AFFAIRS

Joined UN in 1945

ACP
ECOWAS
IAEA
NAM
AU

The ECOWAS peacekeeping force ECOMOG, backed chiefly by Nigeria and Ghana, arrived in 1990, and it was ECOWAS that kept the UN-led peace process on track in the mid-1990s. The UN imposed sanctions in 2001, accusing Liberia of fomenting war in west Africa. Relations with neighboring states, badly affected by these conflicts, were patched up symbolically in February 2002.

AID

Recipient

 US$68m (receipts)
 Down 28% in 2000

Liberia remains heavily dependent on foreign aid. In mid-2000 the EU cut an aid package, accusing the regime of selling arms to rebels in Sierra Leone.

DEFENSE

No compulsory military service

 US$25m
 No change in 2000

Peace agreements in 1995 and 1996 provided for the demobilization of the various warring factions and the formation of a single national army.

ECONOMICS

Not available

 US$1.03bn
 1 Liberian dollar

SCORE CARD

- ❑ World GNP Ranking 153rd
- ❑ GNP per Capita .... US$330
- ❑ Balance of Payments .... –US$145m
- ❑ Inflation 5%
- ❑ Unemployment 70%

Strengths

Potential for reviving the Firestone rubber plantation and huge LAMCO iron ore mine. Tropical timber, but reserves are declining.

Weaknesses

Little commercial activity, and low confidence. Political instability. UN sanctions on diamond trade from 2001.

EXPORTS

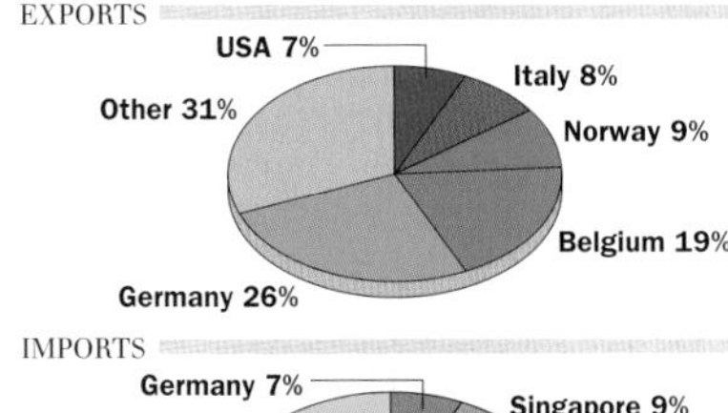
USA 7%
Italy 8%
Norway 9%
Belgium 19%
Germany 26%
Other 31%

IMPORTS

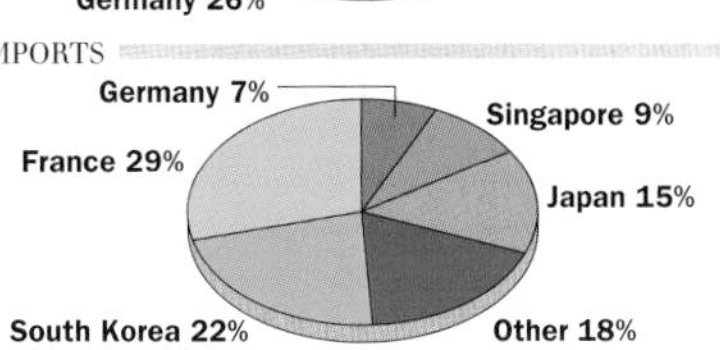
Germany 7%
Singapore 9%
Japan 15%
Other 18%
South Korea 22%
France 29%

Village near Gbarnga. *The Kpelle, the largest of Liberia's 16 indigenous ethnic groups, are concentrated in this part of Liberia.*

RESOURCES

Electric power 334,000 kw

 15,472 tonnes

Not an oil producer

220,000 goats, 210,000 sheep, 4m chickens

Iron ore, diamonds, gold, barytes, kyanite, columbite, manganese

Liberia has an estimated billion tonnes of iron ore reserves at Mount Nimba, but even when peaceful conditions return, the current state of world demand would barely justify exploitation.

ENVIRONMENT

Sustainability rank: 130th

 1%
 0.1 tonnes per capita

The NPFL and other armed groups cut down tropical forests to help finance their armies.

MEDIA

TV ownership low

 Daily newspaper circulation 16 per 1000 people

PUBLISHING AND BROADCAST MEDIA

There are 6 daily newspapers, including the independent *Daily Observer*

 1 partly state-owned service
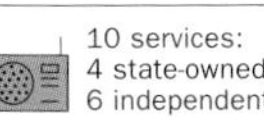 10 services: 4 state-owned, 6 independent

The Liberian Communication Network, owned by President Charles Taylor and his party, effectively controls the print and broadcast media.

CRIME

Death penalty in use

 Liberia does not publish prison figures
 Crime is rampant. There are no enforcing agencies

In the 1990s warring factions regularly massacred civilians, press-ganged armies, and displaced thousands into seeking refuge in neighboring states. As fighting erupted again in 2001, it was government forces that were accused of human rights abuses.

EDUCATION

School leaving age: 16

 54%
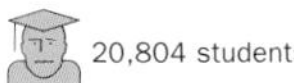 20,804 students

Originally based on the US model, the education system effectively collapsed during the civil war.

CHRONOLOGY

Between 1816 and 1892, 22,000 liberated slaves, most from the US, settled in Liberia, established as a republic in 1847.

- ❑ **1980** Coup. President assassinated by Samuel Doe.
- ❑ **1990** Outbreak of civil war.
- ❑ **1991** Doe assassinated.
- ❑ **1996** Second peace agreement.
- ❑ **1997** Charles Taylor president.
- ❑ **2001** Borders with Guinea and Sierra Leone closed. Conflict with rebels escalates.
- ❑ **2002** State of emergency declared.

HEALTH

No welfare state health benefits

 1 per 20,000 people
 Communicable, diarrheal, parasitic, and heart diseases

Very few people have access to basic health care. Liberia's infant mortality rate remains among the highest in the world.

SPENDING

GDP/cap. increase

CONSUMPTION AND SPENDING

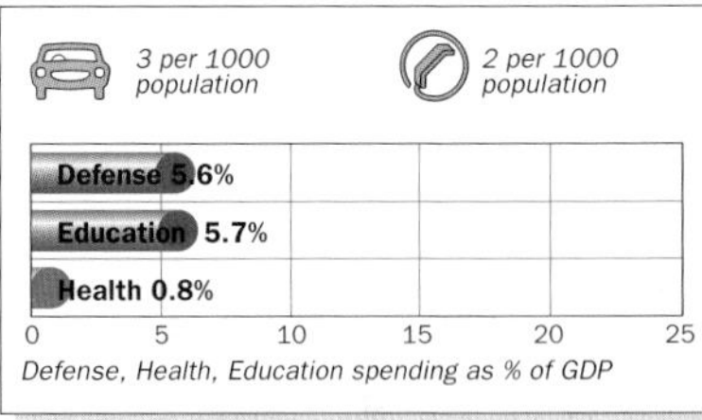
3 per 1000 population
2 per 1000 population
Defense 5.6%
Education 5.7%
Health 0.8%
0
5
10
15
20
25
Defense, Health, Education spending as % of GDP

Power and wealth have a very direct connection in Liberia. Both the Americo-Liberian regimes and the Doe regime which replaced them, saw the state as a source of plunder in the form of well-paid jobs and kickbacks from contracts. The factions in the 1990–1996 civil war sought similar power. Most ordinary Liberians live in rural poverty.

WORLD RANKING

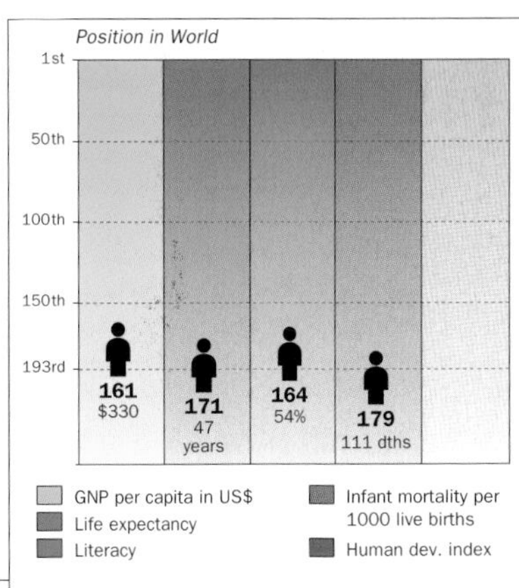
Position in World
1st
50th
100th
150th
193rd
161
$330
171
47 years
164
54%
179
111 dths
GNP per capita in US$
Life expectancy
Literacy
Infant mortality per 1000 live births
Human dev. index

L

LIBYA

NORTH AFRICA

Africa

Algeria LIBYA Egypt Niger Chad Sudan

OFFICIAL NAME: Great Socialist People's Libyan Arab Jamahiriyah
CAPITAL: Tripoli **POPULATION:** 5.4 million **CURRENCY:** Libyan dinar **OFFICIAL LANGUAGE:** Arabic

1951

1951
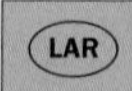
Sept 1
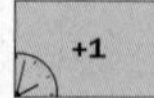
LAR

+1

+218

.ly

LIBYA IS SITUATED between Egypt and Algeria on the Mediterranean coast of north Africa, with Chad and Niger on its southern borders. Apart from the coastal strip and the mountains in the south, it is desert or semidesert. Libya's strategic position in north Africa and its abundant oil and gas resources made it an important trading partner for European states. It has for many years been politically marginalized by the West for its links with terrorist groups, but UN sanctions were suspended in 1999, when it handed over the two men suspected of the 1988 Lockerbie bombing.

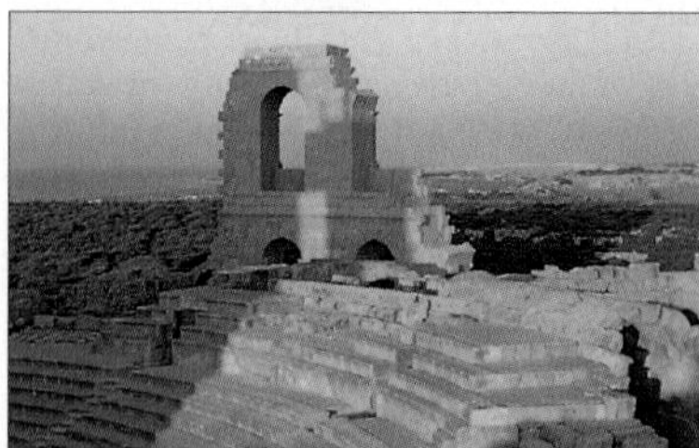

Roman amphitheater, Sabrata. *Libya's impressive classical heritage testifies to its importance in ancient times.*

CLIMATE

Hot desert

WEATHER CHART FOR TRIPOLI

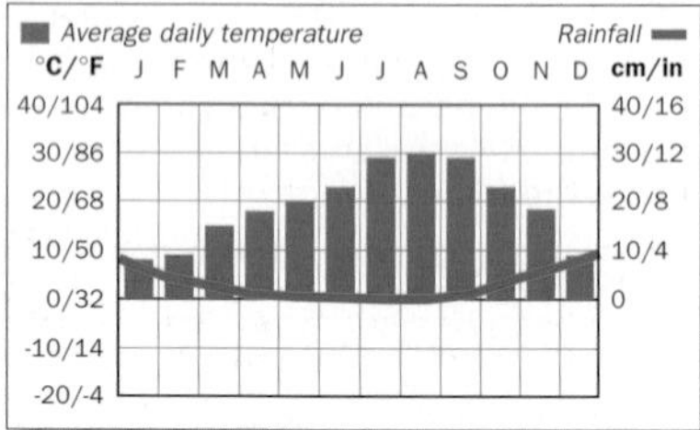

Average daily temperature
Rainfall
°C/°F J F M A M J J A S O N D cm/in
40/104
30/86
20/68
10/50
0/32
-10/14
-20/-4
40/16
30/12
20/8
10/4
0

The coastal region has a warm, temperate climate, with mild, wet winters and hot, dry summers.

TRANSPORTATION

Drive on right

Tripoli International

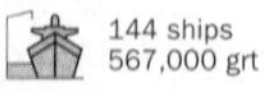

144 ships
567,000 grt

THE TRANSPORTATION NETWORK

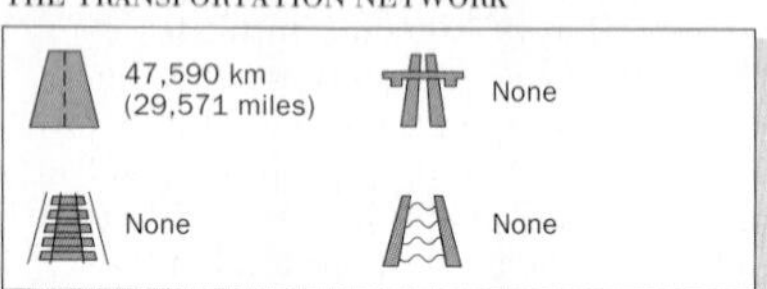

47,590 km
(29,571 miles)
None
None
None

The National Coast Road runs 1825 km (1135 miles) between the Tunisian and Egyptian borders, linking the principal urban centers. There are no railroads, but some are planned. Since sanctions were lifted in 1999, international airlines have resumed flights to Libya.

Al Kufrah Oasis. *As 90% of Libya is arid rock and sand, oases provide essential agricultural land, besides being tourist attractions.*

TOURISM

Visitors : Population 1:135

40,000 visitors

Up 25% in 1999

Libya possesses a rich Roman and Greek heritage, centered on the ancient Roman coastal towns of Labdah (Leptis Magna) and Sabrata near Tripoli, and Shahhat (Cyrene) further east. There are fine beaches at Tripoli. A $2–3 billion investment program launched in 2000 aims to attract thousands of visitors. Western tourists have begun to return since sanctions were lifted in 1999.

MAIN TOURIST ARRIVALS

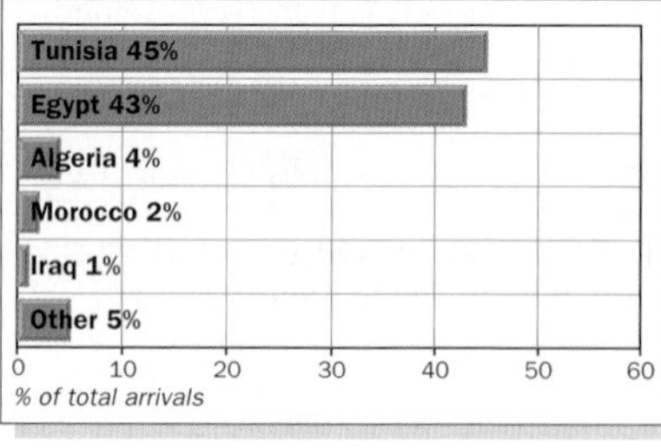

Tunisia 45%
Egypt 43%
Algeria 4%
Morocco 2%
Iraq 1%
Other 5%
0 10 20 30 40 50 60
% of total arrivals

PEOPLE

Pop. density low

Arabic, Tuareg

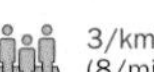

3/km² (8/mi²)

THE URBAN/RURAL POPULATION SPLIT

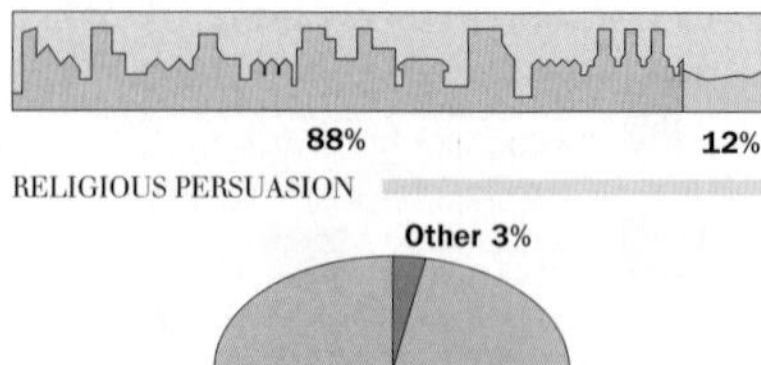

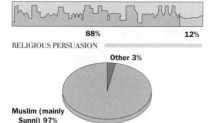

88%
12%
RELIGIOUS PERSUASION
Other 3%
Muslim (mainly Sunni) 97%

ETHNIC MAKEUP

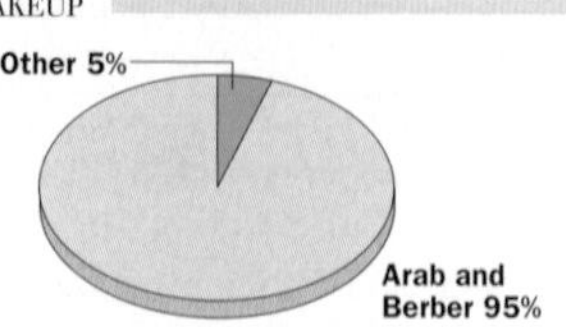

Other 5%
Arab and Berber 95%

POPULATION AGE BREAKDOWN

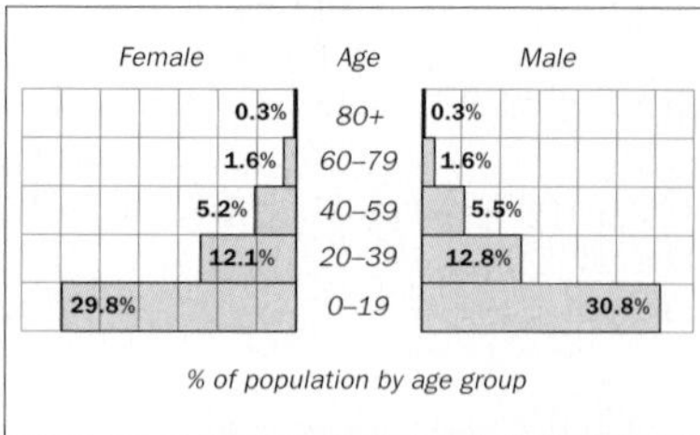

Female
Age
Male
0.3% 80+ 0.3%
1.6% 60–79 1.6%
5.2% 40–59 5.5%
12.1% 20–39 12.8%
29.8% 0–19 30.8%
% of population by age group

Arabs and Berbers, split into many tribal groupings, form 95% of the population. They were artificially brought together when Libya was created in 1951 by the unification of three historic Ottoman provinces. The pro-Western monarchy then perpetuated the dominance of Cyrenaican tribes and the Sanusi religious order.

The 1969 revolution brought to the fore Arab nationalist Col. Muammar al-Gaddafi, who embodied the character and aspirations of the rural Sirtica tribes from Fazzan: fierce independence, deep Islamic convictions, belief in a communal lifestyle, and hatred for the urban rich. His revolution wiped out private enterprise and the middle class, banished European settlers and Jews, undermined the religious Muslim establishment, and imposed a form of popular democracy through the *jamahiriyah* (state of the masses). However, resentment of the regime grew as it became clear that power now lay mainly with the Sirtica tribes, especially Gaddafi's own clan, the Qadhadhfa.

Since the revolution, Libya has become a society where most are city dwellers. Jews have been invited to return as investors, and immigrants from sub-Saharan Africa have been drawn in to provide low-cost labor. However, clashes in 2000, in which 100 died, highlighted unresolved social issues.

POLITICS

No multiparty elections

Not applicable

Leader of the Revolution Col. Muammar al-Gaddafi

LEGISLATIVE OR ADVISORY BODIES

General People's Congress 750 seats

The constitution makes no provision for direct elections. Last renewal May 2000

Executive power is exercised by the General People's Committee. The General People's Congress elects the head of state, the Leader of the Revolution.

PROFILE

In 1977, a new form of direct democracy was promulgated, through which some 2000 People's Congresses sought to involve every adult in policy-making. In theory, their wishes are carried out by popular committees. In practice, ultimate control rests with Col. Gaddafi and his collaborators, many of whom date from the 1969 revolution. In recent years some are thought to have been alienated from Gaddafi, including his deputy, Maj. Abdessalem Jalloud, who in 1994 was reportedly marginalized after expressing differences with him. In 1995, another of Gaddafi's close associates, Khoueldi Hamidi, a defense commander, was also said to have become disillusioned with Gaddafi. In 2000, Gaddafi embraced African unity – an unpopular concept among most Libyans, increasing Gaddafi's alienation from his fomer associates. He is now believed to rely on members of his own clan, particularly his five sons.

MAIN POLITICAL ISSUES

Repression

Political dissidents, including Islamist militants, have been violently suppressed. Libyan dissidents have been murdered abroad, allegedly by government agents. Political parties were banned in 1971, but opposition groups are active in Egypt and Sudan.

The regime's public image

In the past few years, the regime has made an effort to improve its image. Measures have included freeing political prisoners, welcoming back exiles, permitting foreign travel, and accepting responsibility for past acts of terrorism.

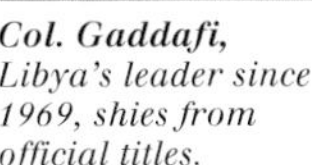

Col. Gaddafi, *Libya's leader since 1969, shies from official titles.*

Ex-king Idris *was deposed by Col. Gaddafi in 1969.*

WORLD AFFAIRS

Joined UN in 1955

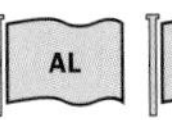
AL

AMU

NAM

OIC

OPEC

Gaddafi has attempted to style himself as the champion of African integration and regional stability. He was a chief architect of the African Union, and has hosted various peace negotiations in the trans-Saharan area. This transformation has also involved a less confrontational stance toward the West, dropping key policies which had left Libya internationally isolated in the past – support for various terrorist groups and strong opposition to Israel – and pursuing a crackdown on internal corruption. UN sanctions imposed in 1992 were eased, and relations with the UK were resumed in 1999. Gaddafi tacitly approved the US-led "war on terrorism" in 2001. However, the US still added Libya to the "axis of evil" in 2002.

L

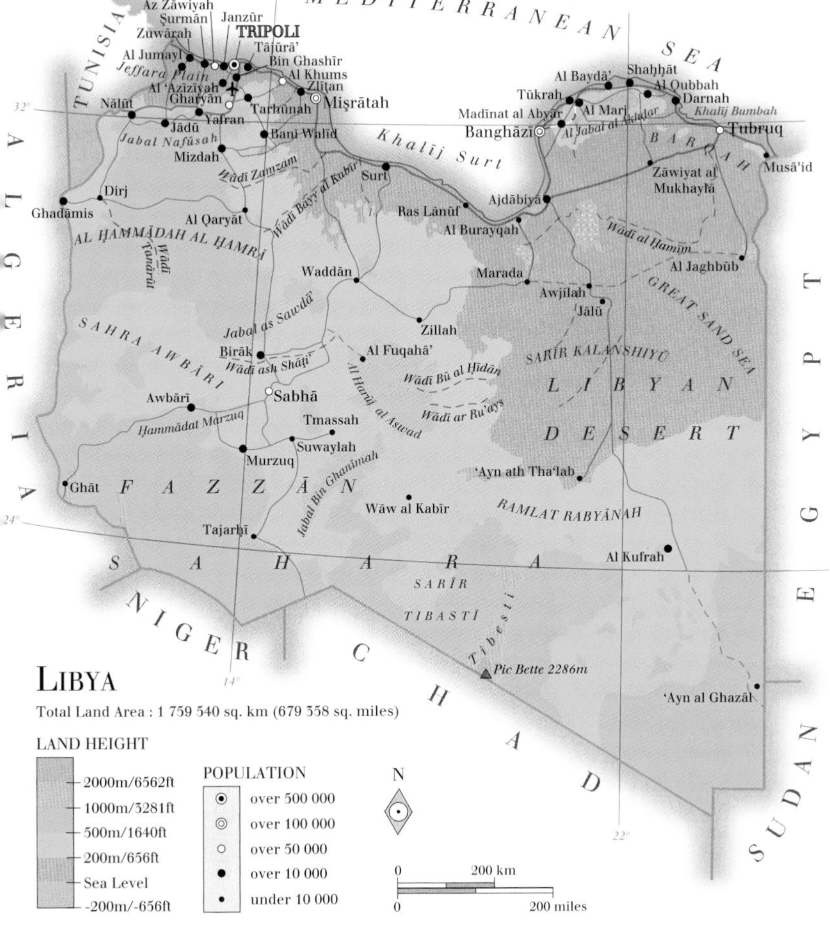
MEDITERRANEAN SEA
TUNISIA
ALGERIA
EGYPT
SUDAN
CHAD
NIGER
Az Zāwiyah
Şurmān
Janzūr
Zuwārah
TRIPOLI
Tājūrā'
Al Jumayl
Bin Ghashīr
Al Khums
Jeffara Plain
Al 'Azīzīyah
Zlīţan
Gharyān
Miṣrātah
Nālūt
Tarhūnah
Yafran
Jādū
Banī Walīd
Jabal Nafūsah
Mizdah
Khalīj Surt
Wādī Zamzam
Surt
Dirj
Ghadāmis
Al Qaryāt
Wādī Bayy al Kabīr
AL ḨAMMĀDAH AL ḨAMRĀ'
Wādī Tanārūt
Waddān
Ras Lānūf
Al Burayqah
Ajdābiyā
Al Bayḑā'
Shaḩḩāt
Al Qubbah
Darnah
Khalīj Bumbah
Tūkrah
Madīnat al Abyār
Al Marj
Banghāzī
Al Jabal al Akhḑar
Ţubruq
BARQAH
Zāwiyat al Mukhaylā
Mūsā'id
Wādī al Ḩamīm
Al Jaghbūb
Marada
Awjilah
Jālū
GREAT SAND SEA
Zillah
Jabal as Sawdā'
SAḨRĀ AWBĀRĪ
Birāk
Wādī ash Shāţi'
Al Fuqahā'
Wādī Bū al Ḩidān
SARĪR KALANSHIYŪ
LIBYAN DESERT
Awbārī
Sabhā
Al Harūj al Aswad
Wādī ar Ru'ays
Ḩammādat Marzūq
Tmassah
Suwaylah
Murzuq
Jabal Bin Ghanīmah
'Ayn ath Tha'lab
Ghāt
FAZZĀN
Wāw al Kabīr
RAMLAT RABYĀNAH
Tajarhī
Al Kufrah
SAHARA
SARĪR TIBASTĪ
Tibesti
Pic Bette 2286m
'Ayn al Ghazāl
32°
24°
14°
22°
LIBYA
Total Land Area : 1 759 540 sq. km (679 358 sq. miles)
LAND HEIGHT
2000m/6562ft
1000m/3281ft
500m/1640ft
200m/656ft
Sea Level
-200m/-656ft
POPULATION
over 500 000
over 100 000
over 50 000
over 10 000
under 10 000
N
0 200 km
0 200 miles

CHRONOLOGY

Italy occupied Libya and expelled the Turks in 1911. Britain and France agreed to a UN plan for an independent monarchy in 1951.

- **1969** King Idris deposed in coup by Revolutionary Command Council led by Col. Gaddafi. Tripoli Charter sets up revolutionary alliance with Egypt and Sudan.
- **1970** UK and US military ordered out. Property belonging to Italians and Jews confiscated. Western oil company assets nationalized, a process completed in 1973.
- **1973** Libya forms abortive union with Egypt. Gaddafi launches Cultural Revolution. Libya occupies Aozou Strip in Chad.
- **1974** Libya forms union of Libya and Tunisia.
- **1977** Official name changed to the Great Socialist People's Libyan Arab Jamahiriyah.
- **1979** Members of Revolution Command Council replaced by elected officials. Gaddafi remains Leader of the Revolution.

CHRONOLOGY *continued*

- ❑ **1981** US shoots down two Libyan aircraft over Gulf of Sirte.
- ❑ **1984** Gunman at Libyan embassy in UK kills British policewoman; UK severs diplomatic relations (until 1999). Oudja Accord signed with Morocco for Arab Africa Federation.
- ❑ **1985** Libya expels 30,000 foreign workers. Tunisia cuts diplomatic links.
- ❑ **1986** US aircraft bomb Libya, killing 101 people and destroying Gaddafi's residence.
- ❑ **1988** Pan-Am airliner explodes over Lockerbie, Scotland; allegations of Libyan complicity.
- ❑ **1988–1989** Army restructured.
- ❑ **1989** Arab Maghreb Union established with Algeria, Morocco, Mauritania, and Tunisia. Cease-fire in Aozou Strip.
- ❑ **1990** Libya expels Palestinian splinter group led by Abu Abbas.
- ❑ **1991** Opening of first branch of Great Man-Made River project.
- ❑ **1992–1993** UN sanctions imposed as Libya fails to hand over Lockerbie suspects; sanctions made stricter.
- ❑ **1994** Religious leaders obtain right to issue religious decrees (*fatwas*) for first time since 1969. Return of Aozou strip to Chad.
- ❑ **1996** US legislation imposes penalties on foreign companies investing in Libya's energy sector.
- ❑ **1999** Lockerbie suspects handed over for trial in the Netherlands under Scottish law; UN sanctions eased.
- ❑ **2001** Lockerbie trial verdict: one suspect convicted. Sanctions eased further.
- ❑ **2002** US–Libya talks aim to mend relations. Lockerbie bomber begins life sentence.

L

AID

Recipient

$15m (receipts)

Up 114% in 2000

As an oil-exporting state, Libya fails to qualify for much international aid, despite being a developing country. During the 1970s, Col. Gaddafi aided several African liberation movements, such as the ANC in South Africa. He backed Hissène Habré's forces in Chad, and helped dissidents by training them in his Pan-African legion. He has also financed or supplied arms to the PLO in the Middle East, Irish republicans in Northern Ireland, the Moros in the southern Philippines, and the Basques, Corsicans, and other separatist causes in Europe. In 1993, Libya granted aid totaling $27 million, despite UN sanctions and a lack of surplus resources.

DEFENSE

Compulsory military service

$1.18bn

Down 10% in 2000

LIBYAN ARMED FORCES

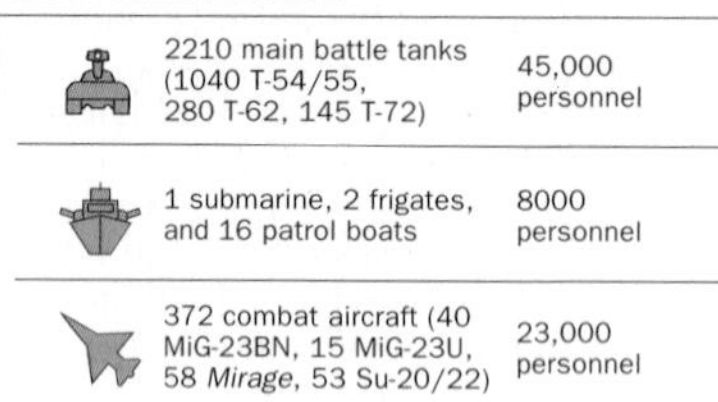

Army	2210 main battle tanks (1040 T-54/55, 280 T-62, 145 T-72)	45,000 personnel
Navy	1 submarine, 2 frigates, and 16 patrol boats	8000 personnel
Air force	372 combat aircraft (40 MiG-23BN, 15 MiG-23U, 58 *Mirage*, 53 Su-20/22)	23,000 personnel
Missiles	None	

The armed forces suffered a blow in 1987 with the loss of thousands of men and equipment worth $1.4 billion when Libya became embroiled in the Chad civil war; in 1994 it agreed to hand back to Chad the Aozou Strip, which it had occupied in 1973. In 1988–1989 the armed forces were replaced by "the Armed People." Conscription is selective, and can last up to two years. In addition, there is a People's Militia of 40,000.

Attempts to depoliticize the army received a setback following confirmation of an abortive military coup in 1993. UN sanctions resulted in military hardware becoming outdated. Despite the suspension of the sanctions in 1999, fresh arms contracts would still be too controversial for most potential suppliers.

ECONOMICS

Not available

$29.2bn

0.5423–0.6416 Libyan dinars

SCORE CARD

- ❑ World GNP Ranking.........................59th
- ❑ GNP per capita $5220
- ❑ Balance of Payments...................$1.48bn
- ❑ Inflation 18.5%
- ❑ Unemployment.................................30%

EXPORTS

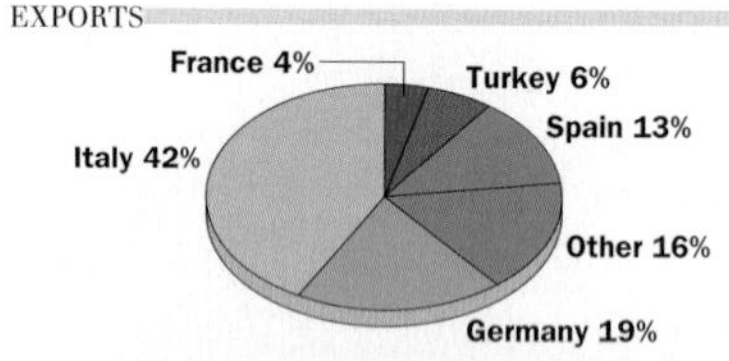
France 4%
Turkey 6%
Spain 13%
Italy 42%
Other 16%
Germany 19%

IMPORTS

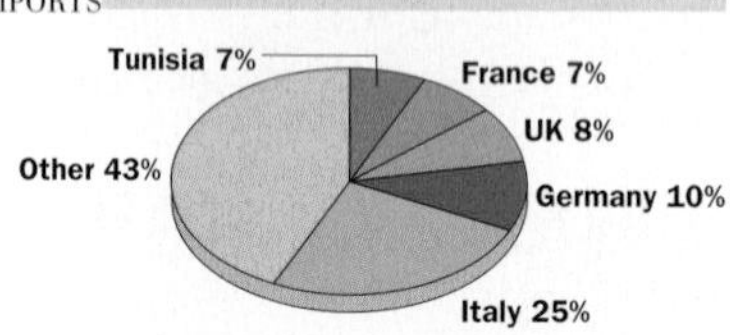
Tunisia 7%
France 7%
UK 8%
Other 43%
Germany 10%
Italy 25%

Strengths

Oil and gas production. High investment in downstream industries – petrochemicals, refineries, fertilizers, and aluminum smelting.

Weaknesses

Single-resource economy subject to oil-market fluctuations. Most food is imported. Reliance on foreign labor. Lack of water for agriculture. History of international unreliability.

Profile

Western oil companies had close business ties with Libya until the imposition in 1992 of UN sanctions over the Lockerbie affair. In 1993, Gaddafi called for the program of privatization, authorized by the General People's Congress in 1992, to be revived, but there have been few tangible results.

ECONOMIC PERFORMANCE INDICATOR

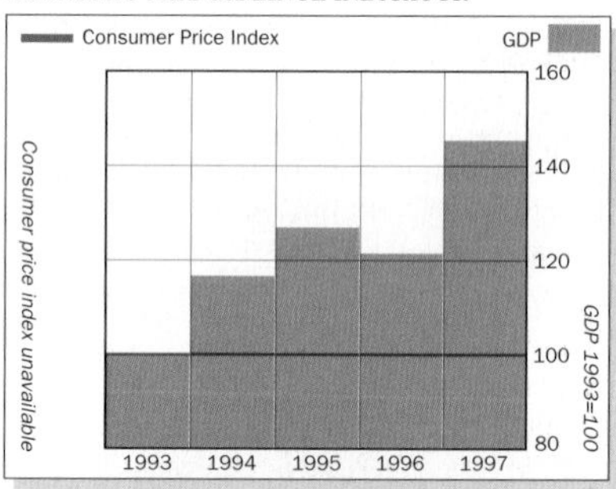
Consumer Price Index
GDP
Consumer price index unavailable
GDP 1993=100
160
140
120
100
80
1993
1994
1995
1996
1997

In the 1970s, an ambitious program of industrialization was launched. Gaddafi's most controversial economic project has been the Great Man-Made River. Started in 1984 and engineered by European and Korean companies, this scheme was designed to bring underground water from the Sahara to the coast, but the pipes are already corroding, with water leaking into the sand.

LIBYA : MAJOR BUSINESSES

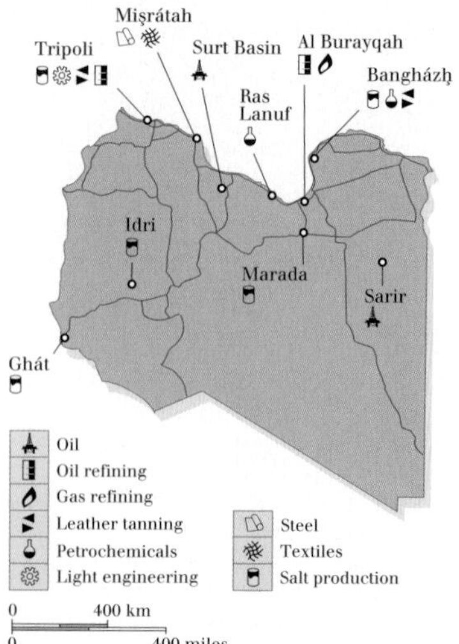
Tripoli
Mişrátah
Surt Basin
Al Burayqah
Bangházh
Ras Lanuf
Idri
Marada
Sarir
Ghát
Oil
Oil refining
Gas refining
Leather tanning
Petrochemicals
Light engineering
Steel
Textiles
Salt production
0
400 km
0
400 miles

RESOURCES

Electric power 4.6m kw

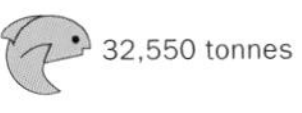

32,550 tonnes

1.43m b/d (reserves 29.5bn barrels)

5.1m sheep, 1.95m goats, 25m chickens

Oil, natural gas, iron, potassium, sulfur, magnesium, gypsum

ELECTRICITY GENERATION

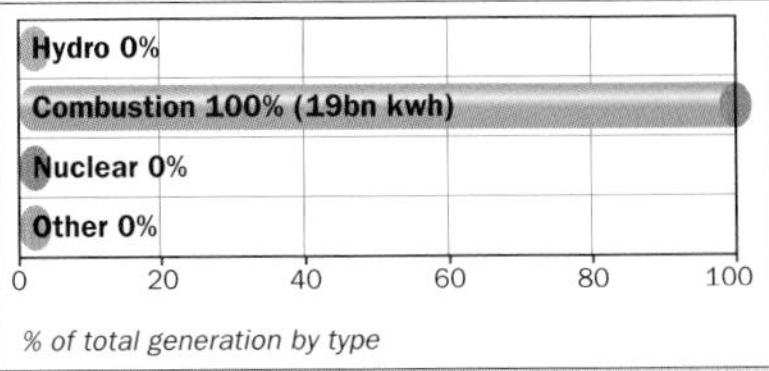
Hydro 0%
Combustion 100% (19bn kwh)
Nuclear 0%
Other 0%
0
20
40
60
80
100
% of total generation by type

Libya's economy depends almost entirely on petroleum and natural gas resources. It has considerable crude oil reserves and is likely to remain an oil-exporting country well into the century. Natural gas potential is more limited but, provided links are developed with other north African states, the future is assured. Libya also has reserves of iron ore, potassium, sulfur, magnesium, and gypsum. The Great Man-Made River project means that the area of irrigated land has grown, but 90% of Libya is desert. Animal husbandry is the basis of farming, but some cereal crops are grown, as well as dates, olives, and citrus fruits. Cement production is sufficient to meet national demand; raw materials are local, but most other manufacturing inputs are imported.

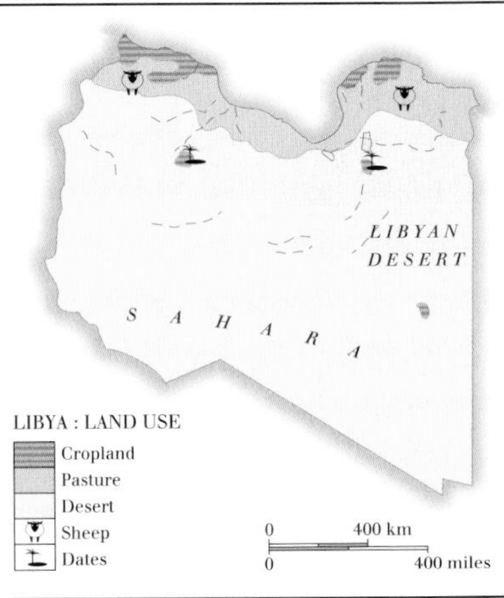
LIBYAN DESERT
S A H A R A
LIBYA : LAND USE
Cropland
Pasture
Desert
Sheep
Dates
0
400 km
0
400 miles

ENVIRONMENT

Sustainability rank: 124th

0.1%

7.2 tonnes per capita

ENVIRONMENTAL TREATIES

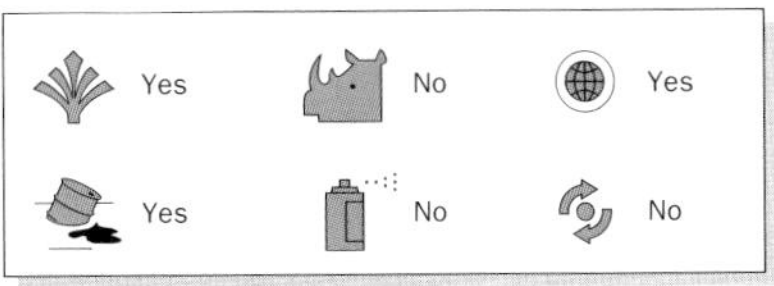
Yes
No
Yes
Yes
No
No

The UN Development Program has described Libya as more than 90% "wasteland." Both nature and man have conspired against the environment. Apart from two coastal strips – the Jeffara Plain and the Jabal al Akhdar in Cyrenaica – together with the Fazzan Oasis, most of Libya is desert. Much of the irrigated area is saline because of unwise use of naturally occurring water from artesian wells. Near Tripoli, seawater has penetrated the water table as far as 20 km (12 miles) inland.

MEDIA

TV ownership medium

Daily newspaper circulation 14 per 1000 people

PUBLISHING AND BROADCAST MEDIA

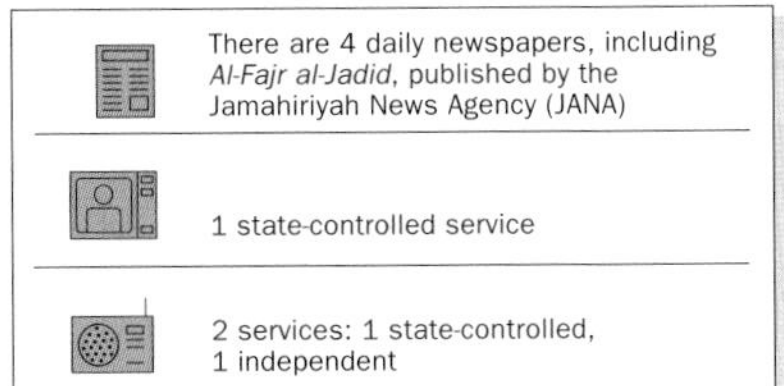
There are 4 daily newspapers, including *Al-Fajr al-Jadid*, published by the Jamahiriyah News Agency (JANA)
1 state-controlled service
2 services: 1 state-controlled, 1 independent

Libya's press and TV are a mouthpiece for the leadership. Satellite TV and the Internet are widely available, but heavily censored. The main daily newspaper is published in Arabic and has a circulation of 40,000 readers. The TV station broadcasts mostly in Arabic. Radio broadcasts in Kiswahili, Hausa, Fulani, and Amharic were due to begin in 1999, but were postponed.

CRIME

Death penalty in use

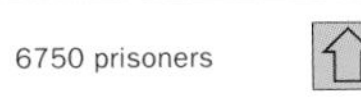

6750 prisoners

Up 10% in 1999

CRIME RATES

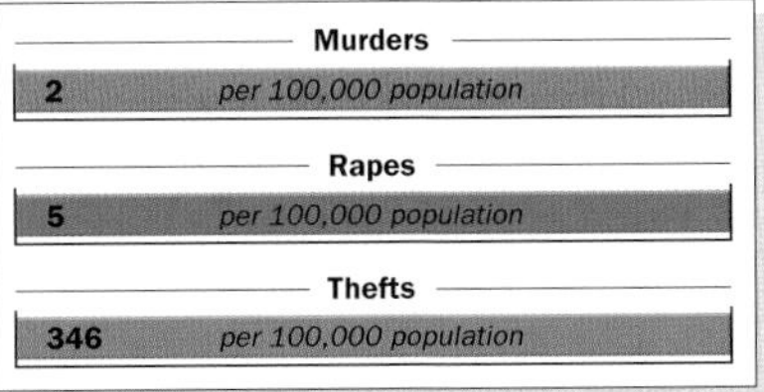
Murders
2 per 100,000 population
Rapes
5 per 100,000 population
Thefts
346 per 100,000 population

Policing is often in the hands of gangs appointed by Gaddafi's lieutenants to root out student protestors and other dissidents. Hit squads allegedly operate abroad against Libyan exiles.

EDUCATION

School leaving age: 15

80%

126,348 students

THE EDUCATION SYSTEM

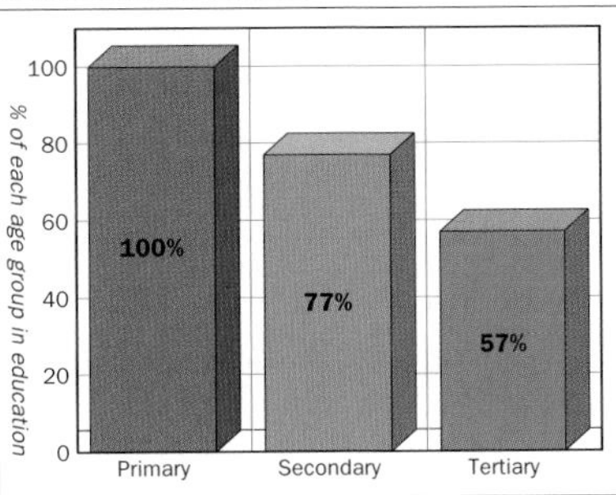
100
80
60
40
20
0
% of each age group in education
100%
77%
57%
Primary
Secondary
Tertiary

Some one million Libyans are in formal education. It is compulsory between the ages of six and 15, and rates of attendance are very high, but it varies in quality and can be rudimentary in rural areas. Secondary education, from the age of 15, lasts for three years. There are 13 universities, and institutes for vocational training. The literacy rate has more than doubled from a level of 39% in 1970.

HEALTH

Welfare state health benefits

1 per 769 people

Pneumonia, diarrheal diseases, accidents, cancers

An adequate system of free primary health care exists except in remote areas, and there are two big hospitals, in Benghazi and Tripoli. However, hospitals lack equipment, and there is a shortage of medical supplies.

L

SPENDING

GDP/cap. decrease

CONSUMPTION AND SPENDING

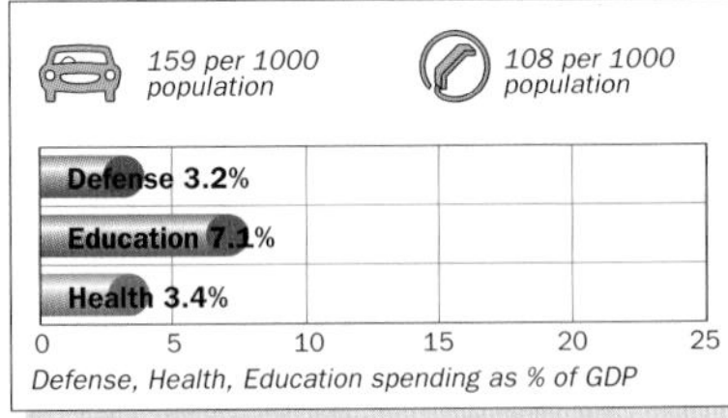
159 per 1000 population
108 per 1000 population
Defense 3.2%
Education 7.1%
Health 3.4%
0
5
10
15
20
25
Defense, Health, Education spending as % of GDP

There is widespread poverty after years of import constraints; UN sanctions worsened the situation. Gaddafi refuses to use oil revenues for basic expenses, such as salaries – teachers earn about $1200 a year.

WORLD RANKING

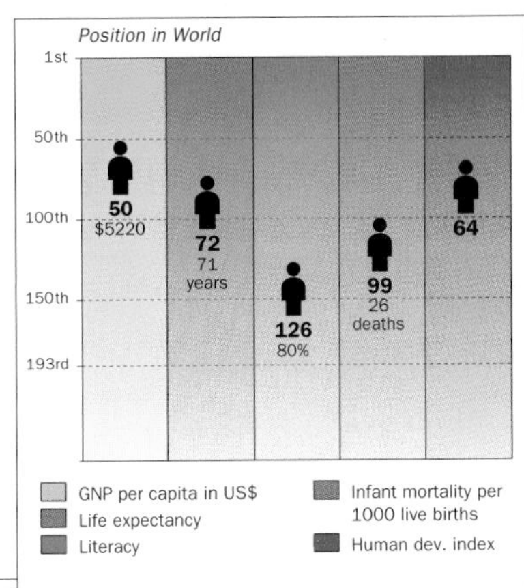
Position in World
1st
50th
100th
150th
193rd
50
$5220
72
71
years
126
80%
99
26
deaths
64
GNP per capita in US$
Life expectancy
Literacy
Infant mortality per 1000 live births
Human dev. index

LIECHTENSTEIN

OFFICIAL NAME: Principality of Liechtenstein **CAPITAL:** Vaduz
POPULATION: 32,200 **CURRENCY:** Swiss franc **OFFICIAL LANGUAGE:** German

1719

1719

Aug 15
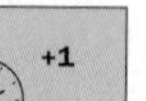
FL

+1

+423

.li

PERCHED IN THE ALPS, Liechtenstein is rare among small states in having both a thriving banking sector and a well-diversified manufacturing economy. It is closely allied to Switzerland, which handles its foreign relations and defense. Life in Liechtenstein is stable and conservative. The traditional secrecy surrounding the country's financial industry, and low taxes, means that many overseas trusts, banks, and investment companies are located there.

CLIMATE

Mountain

WEATHER CHART FOR VADUZ

Average daily temperature; Rainfall
°C/°F J F M A M J J A S O N D cm/in

Climate varies with altitude. Excellent skiing conditions are the result of heavy settling snow from December to March. Summers are warm and wet.

TRANSPORTATION

Drive on right

None — Has no fleet

THE TRANSPORTATION NETWORK

250 km (155 miles) — None — 19 km (12 miles) — 26 km (16 miles)

Public transportation in Liechtenstein is mostly by the postal bus network. The single-track railroad has few stops. Zürich, a two-hour drive away, is the nearest airport.

TOURISM

Visitors : Population 1.9:1

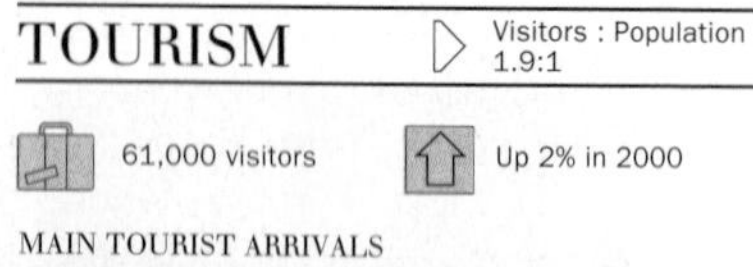
61,000 visitors
Up 2% in 2000

MAIN TOURIST ARRIVALS

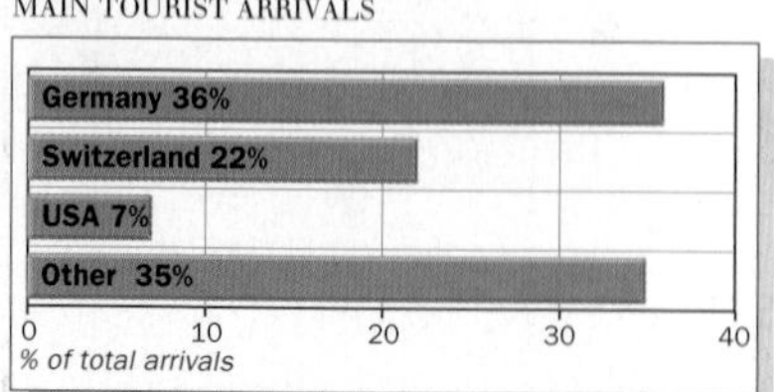
Germany 36%
Switzerland 22%
USA 7%
Other 35%
0
10
20
30
40
% of total arrivals

Liechtenstein's alpine scenery attracts skiers in the winter, and climbers and hikers in the summer.

PEOPLE

Pop. density high

German, Alemannish dialect, Italian — 201/km² (521/mi²)

THE URBAN/RURAL POPULATION SPLIT

21% 79%

RELIGIOUS PERSUASION

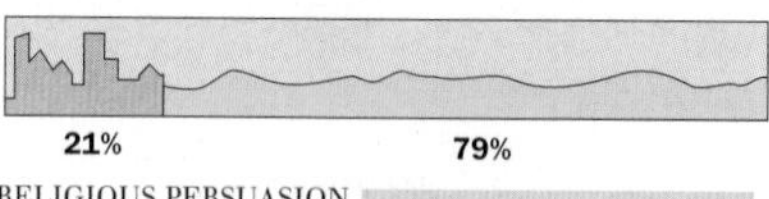
Protestant 7%
Other 12%
Roman Catholic 81%

Liechtenstein's role as a financial center accounts for the many foreign residents (over 35% of the population), of whom half are Swiss and the rest mostly German. The high standard of living results in few ethnic or social tensions. Family life is very traditional; women received the vote only in 1984, after much controversy. A proposal to enshrine equal rights for women in the constitution was rejected in a referendum in 1985 by a large majority, and was only finally passed in 1992.

POLITICS

Multiparty elections

2001/2005
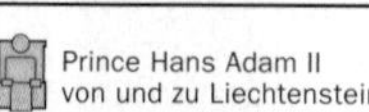
Prince Hans Adam II von und zu Liechtenstein

AT THE LAST ELECTION

Parliament 25 seats

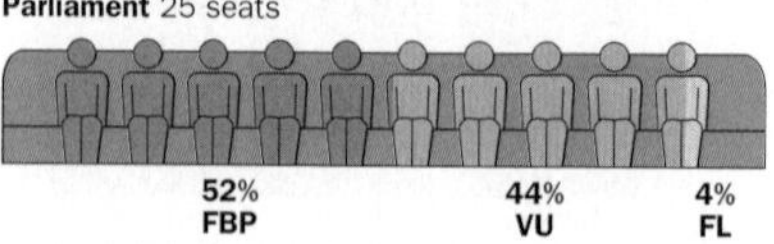
52% FBP
44% VU
4% FL

FBP = Progressive Citizens' Party **VU** = Fatherland Union **FL** = Free List

Between 1938 and 1997 the VU and the FBP alternated as coalition leaders, except briefly in 1993. This partnership ended in 1997 when Mario Frick, at the age of 28 Europe's youngest premier, formed a VU-only government. In elections in 2001 the FBP under Otmar Hasler overtook the VU. Referendums have increasingly been used to decide policy issues.

EUROPE
LIECHTENSTEIN Austria Switzerland Europe

WORLD AFFAIRS

Joined UN in 1990

CE

EEA

IAEA

OSCE

WTO

Liechtenstein effectively gave up control of its external relations in 1924 when it signed a Customs Union Treaty with Switzerland. This agreement requires Swiss approval for any treaty arrangements between Liechtenstein and a third state. Liechtenstein became a member of the UN only in 1990. It joined EFTA and the EBRD in 1991, and has been a participant in the EEA since 1995. However, Swiss rejection of EU membership in 1992 effectively ended any prospect of Liechtenstein joining the EU in the foreseeable future.

AID

Donor

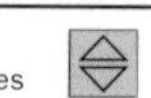
Donor, but does not publish figures

Not available

Although overseas aid donations are small and aid issues have little political importance, Liechtenstein has helped to fund shelter and reconstruction projects in former Yugoslavia and local development projects in Bulgaria.

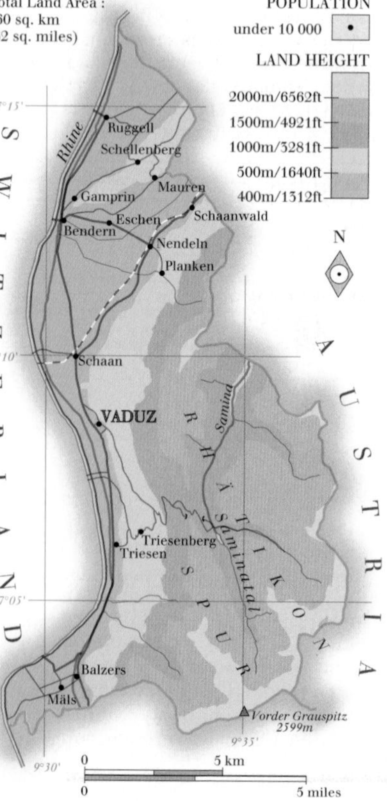
LIECHTENSTEIN
Total Land Area : 160 sq. km (62 sq. miles)
POPULATION
under 10 000
LAND HEIGHT
2000m/6562ft
1500m/4921ft
1000m/3281ft
500m/1640ft
400m/1312ft
SWITZERLAND
AUSTRIA
Rhine
Ruggell
Schellenberg
Mauren
Gamprin
Eschen
Schaanwald
Bendern
Nendeln
Planken
Schaan
VADUZ
Samina
RHÄTIKON
Triesenberg
Triesen
Saminatal
Balzers
Mäls
Vorder Grauspitz 2599m
N
0 5 km
0 5 miles

Alpine scenery near Vaduz. *The state budget includes 2% allocated to restoring mountain vegetation and coordinating land use.*

DEFENSE

No compulsory military service

 No defense force

 Not applicable

There has been no standing army since 1868, and there is only a small police force. De facto protection is provided by Switzerland. In theory, any male under 60 is liable for military service during a national emergency, although this law has never been invoked.

ECONOMICS

Inflation 2.9% p.a. (1985–1996)

 $1.6bn

 1.6205–1.6603 Swiss francs

SCORE CARD

- ❑ WORLD GNP RANKING 142nd
- ❑ GNP PER CAPITA $50,000
- ❑ BALANCE OF PAYMENTS ..... Included in Swiss total
- ❑ INFLATION 0.5%
- ❑ UNEMPLOYMENT 2%

STRENGTHS

Stability and customs union with Switzerland make Liechtenstein a favored tax haven; its lack of EU membership makes the banking sector less vulnerable to future changes in EU banking laws. The economy is well diversified; chemicals, furniture, coatings for the electro-optical industry, construction services, and precision instruments are all thriving sectors.

WEAKNESSES

Very few. Need to balance integration with other countries with safeguarding economic independence.

EXPORTS

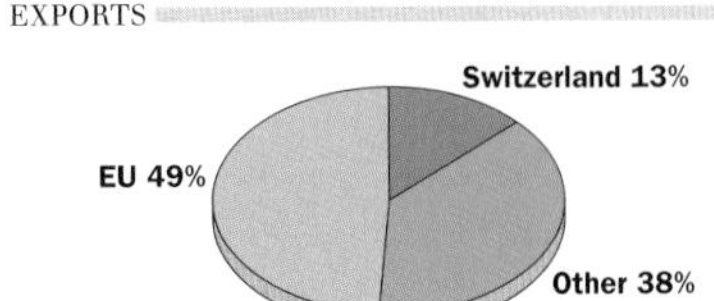
Switzerland 13%
EU 49%
Other 38%

IMPORTS

With a limited domestic market, Liechtenstein's industry is export-oriented. Liechtenstein has a customs union with Switzerland and does not publish separate import figures.

RESOURCES

Electric power: Included in Swiss total

 None

 Not an oil producer

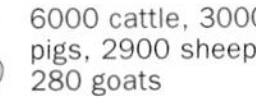 6000 cattle, 3000 pigs, 2900 sheep, 280 goats

 None

Liechtenstein has to import most of its energy. Almost all of its electricity comes from German power plants.

ENVIRONMENT

Not available

 38% partially protected

Not available

Protection of Liechtenstein's alpine scenery is high enough on the political agenda for one of the five councillors, or ministers, to have responsibility for the environment. As in Switzerland, the greatest worry is the effect of high rates of car use and of through traffic. However, a 1988 trial in providing free public bus transportation proved a failure, as Liechtensteiners remained firmly wedded to their automobiles.

MEDIA

TV ownership high

 Daily newspaper circulation 602 per 1000 people

PUBLISHING AND BROADCAST MEDIA

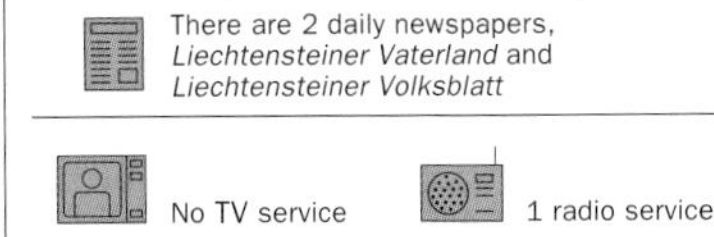

There are 2 daily newspapers, *Liechtensteiner Vaterland* and *Liechtensteiner Volksblatt*

No TV service

1 radio service

The two daily newspapers, although free of formal state control, are both run by political parties: the *Vaterland* by the VU, the *Volksblatt* by the FBP. Both have circulations of around 10,000.

CRIME

No death penalty

 24 prisoners

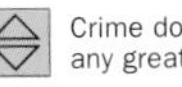 Crime does not pose any great problems

Crime is a minor problem, a result of the relatively even distribution of wealth and high average living standard. Liechtenstein has also taken great care to protect its tax-haven status by careful regulation of its financial sector. It has avoided major scandals, and took steps in 2000 to tighten precautions against the growing problem of money laundering.

EDUCATION

School leaving age: 16

 99%

 Not available

Education, modeled on the German system, includes two types of school at secondary level – the more academic *Gymnasium* and the *Realschule*. Liechtenstein has no university; students go on to colleges in Austria, Switzerland, or Germany, or to business schools in the US.

CHRONOLOGY

In 1719 Liechtenstein became an independent principality of the Holy Roman Empire.

- ❑ **1924** Customs union with Switzerland.
- ❑ **1990** Joins UN.
- ❑ **1995** Joins EEA.
- ❑ **1997** End of VU–FBP coalition dominant since 1938. Mario Frick heads VU government.
- ❑ **2001** FBP wins majority in elections; Otmar Hasler premier.

HEALTH

Welfare state health benefits

 1 per 948 people

Heart and respiratory diseases, cancers

Although clinics and hospitals are few, the health system provides advanced care. Many Liechtensteiners have private health insurance arrangements, so that they have access to Swiss medical expertise and facilities as well as their own.

SPENDING

GDP/cap. increase

CONSUMPTION AND SPENDING

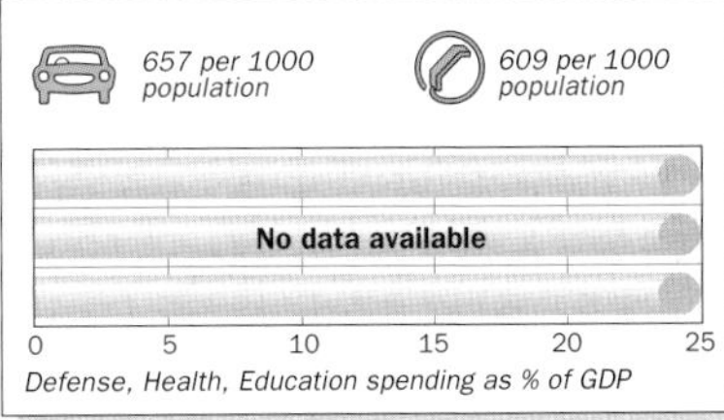
657 per 1000 population
609 per 1000 population
No data available
0
5
10
15
20
25
Defense, Health, Education spending as % of GDP

Unlike other tax havens, Liechtenstein displays a more conservative prosperity. Private deposit accounts are not a key part of its banking business, but an increase in money-laundering activities and the country's appearance on a blacklist of financial centers led to a ban on anonymous accounts in mid-2000.

WORLD RANKING

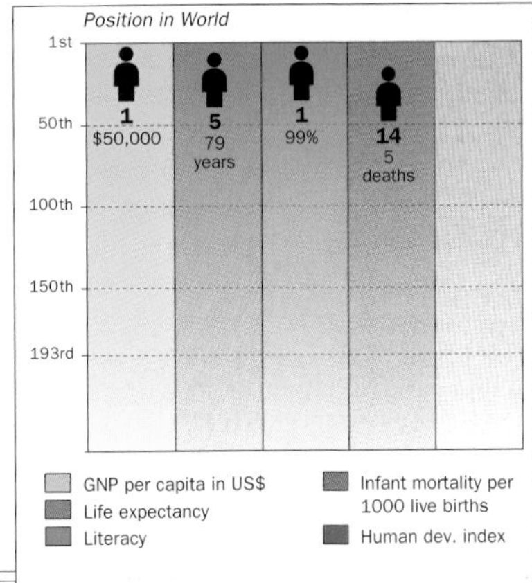
Position in World
1st
50th
100th
150th
193rd
1
$50,000
5
79 years
1
99%
14
5 deaths
GNP per capita in US$
Life expectancy
Literacy
Infant mortality per 1000 live births
Human dev. index

L

LITHUANIA

EUROPE

Latvia Russian Fed. LITHUANIA Belarus Poland

OFFICIAL NAME: Republic of Lithuania **CAPITAL:** Vilnius
POPULATION: 3.7 million **CURRENCY:** Litas **OFFICIAL LANGUAGE:** Lithuanian

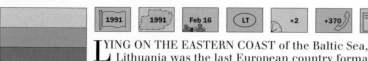

1991

1991

Feb 16

LT

+2

+370

.lt

LYING ON THE EASTERN COAST of the Baltic Sea, Lithuania was the last European country formally to embrace Christianity, around 1400. Its terrain is mostly flat, with many lakes, moors, and bogs. Now a multiparty democracy, Lithuania regained independence from the former USSR in 1991. Industrial production and agriculture are the mainstays of the economy. Russia finally withdrew all its troops from Lithuania in 1993.

CLIMATE

Continental

WEATHER CHART FOR VILNIUS

Average daily temperature — Rainfall
°C/°F J F M A M J J A S O N D cm/in
40/104 30/86 20/68 10/50 0/32 -10/14 -20/4
40/16 30/12 20/8 10/4 0

Lithuania's coastal position moderates an otherwise continental-type climate. Summers are cool.

TRANSPORTATION

Drive on right

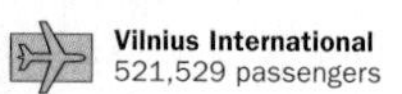
Vilnius International
521,529 passengers

206 ships
481,100 grt

THE TRANSPORTATION NETWORK

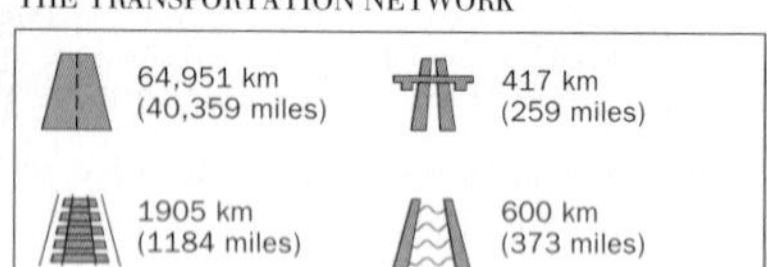
64,951 km
(40,359 miles)
417 km
(259 miles)
1905 km
(1184 miles)
600 km
(373 miles)

Lithuania is crossed by international rail routes, and Klaipeda provides extensive connections to other Baltic ports.

TOURISM

Visitors : Population 1:3.4

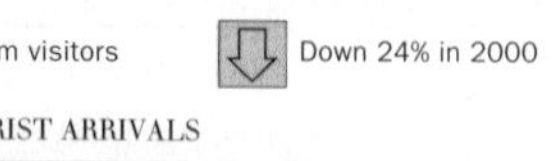
1.08m visitors
Down 24% in 2000

MAIN TOURIST ARRIVALS

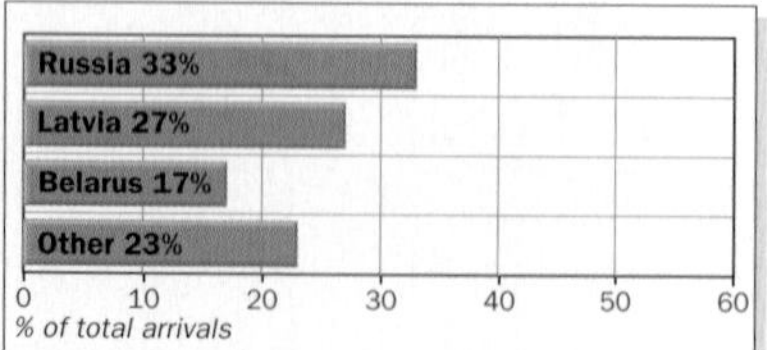
Russia 33%
Latvia 27%
Belarus 17%
Other 23%
0 10 20 30 40 50 60
% of total arrivals

The tourist industry boomed after independence. Vilnius is well preserved; its historic center survived German and Russian occupation. Trakai, the capital of the Grand Duchy in the 16th century, is also popular with visitors.

PEOPLE

Pop. density medium

Lithuanian, Russian

57/km² (147/mi²)

THE URBAN/RURAL POPULATION SPLIT

ETHNIC MAKEUP

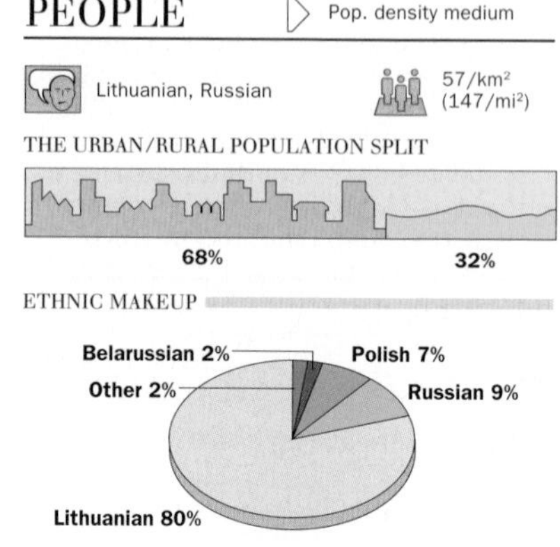
68%
32%
Belarussian 2%
Polish 7%
Other 2%
Russian 9%
Lithuanian 80%

With a mainly Catholic population, Lithuania has strong historical links with Poland, with which it was once united, although there is tension between ethnic Lithuanians and Poles. Relations with the Jewish minority are strained. More than 90% of nonethnic Lithuanians have been granted citizenship. Of all the Baltic states, Lithuania has the best relations with ethnic Russians, but there are fewer of them who live there.

POLITICS

Multiparty elections

2000/2004

President Valdas Adamkus

AT THE LAST ELECTION

Parliament 141 seats

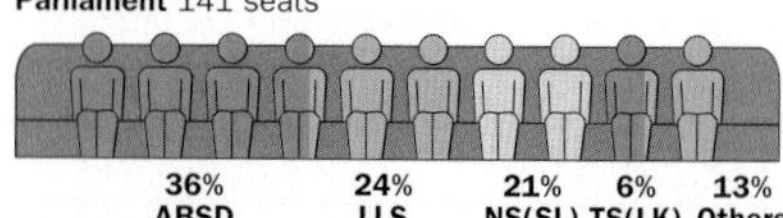
36% ABSD
24% LLS
21% NS(SL)
6% TS(LK)
13% Others

ABSD = A. Brazauskas Social Democratic Coalition
LLS = Lithuanian Liberal Union
NS(SL) = New Union (Social Liberals)
TS(LK) = Homeland Union (Lithuanian Conservatives)

Lithuania was the most politically stable of the three Baltic republics in the 1990s. Strongly nationalist when it reestablished independence from the disintegrating Soviet Union in 1991, it unexpectedly voted the former communists back into office in 1992 and 1993. Former communist leader Algirdas Brazauskas held the presidency for five years until returning emigré businessman Valdas Adamkus was elected in 1998. Meanwhile the conservative TS(LK) had won the 1996 legislative elections in the wake of a banking scandal, but their popularity dwindled as successive prime ministers were forced to resign. The TS(LK) suffered a massive defeat in the 2000 legislative elections. Adamkus controversially bypassed Brazauskas' ABSD, the largest bloc in the new Parliament, in favor of a pro-market coalition. This collapsed in mid-2001, and Brazauskas became prime minister in a government dominated by the ABSD.

LITHUANIA

Total Land Area : 65 200 sq. km (25 174 sq. miles)

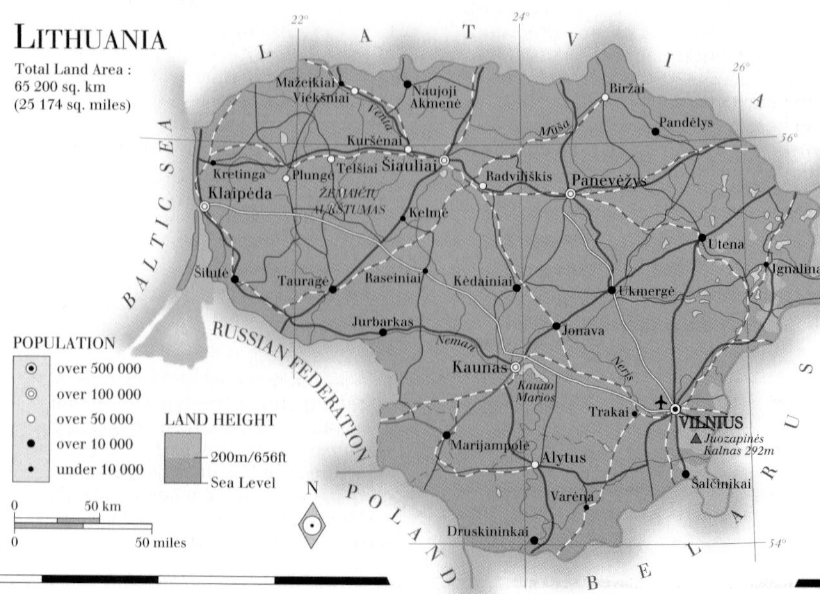
LATVIA
BALTIC SEA
RUSSIAN FEDERATION
POLAND
BELARUS
Mažeikiai
Viekšniai
Naujoji Akmenė
Biržai
Pandėlys
Kuršėnai
Šiauliai
Telšiai
Plungė
Kretinga
Klaipėda
Radviliškis
Panevėžys
Kelmė
Utena
Ignalina
Šilutė
Tauragė
Raseiniai
Kėdainiai
Ukmergė
Jurbarkas
Jonava
Kaunas
Kauno Marios
Trakai
VILNIUS
Juozapinės Kalnas 292m
Marijampolė
Alytus
Šalčininkai
Varėna
Druskininkai
Neman
Neris
POPULATION
over 500 000
over 100 000
over 50 000
over 10 000
under 10 000
LAND HEIGHT
200m/656ft
Sea Level
0 50 km
0 50 miles

WORLD AFFAIRS

Joined UN in 1991

CE

CBSS

WTO

OSCE

PfP

As a "second wave" applicant, Lithuania began EU membership negotiations in 2000. It currently has the best relations with Russia of all the Baltic states, but is keen for all three to join NATO.

AID

Recipient

 $99m (receipts)

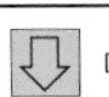 Down 26% in 2000

Aid, mostly from the IMF and the EU, is used for infrastructure projects and to promote private enterprise.

DEFENSE

Compulsory military service

 $195m

 Up 82% in 2000

Lithuania's security is in the hands of its army, the small navy and air force, and a large National Guard formed to patrol its frontiers. Twelve months' military service is compulsory. The US is now supporting Lithuania's entry into NATO.

ECONOMICS

Inflation 75% p.a. (1990–2000)

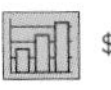 $10.8bn

 3.9990–3.9984 litai

SCORE CARD

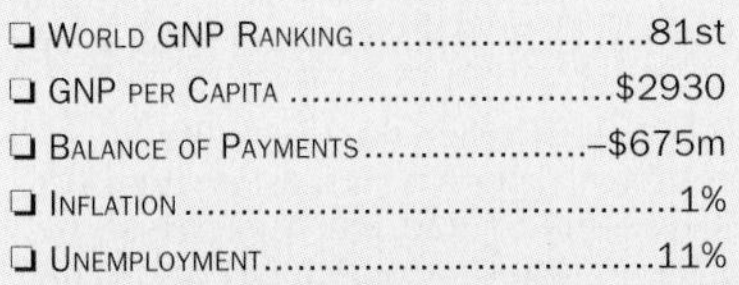

❑ World GNP Ranking	81st
❑ GNP per Capita	$2930
❑ Balance of Payments	–$675m
❑ Inflation	1%
❑ Unemployment	11%

Strengths

Privatization has stimulated economy to some extent. Inflation under control. Successful liberal reforms. Litas pegged to euro.

Weaknesses

Agriculture in the doldrums following decollectivization. Exports dependent on health of Russian economy. Poor raw materials base. Difficulty in attracting significant foreign investment.

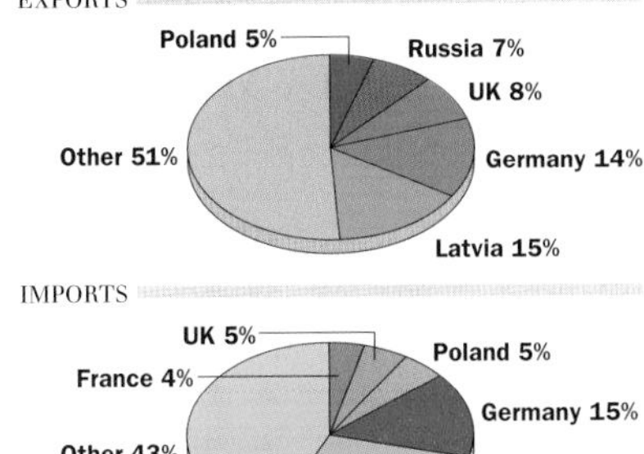
EXPORTS
Poland 5%
Russia 7%
UK 8%
Germany 14%
Latvia 15%
Other 51%
IMPORTS
UK 5%
France 4%
Poland 5%
Germany 15%
Russia 28%
Other 43%

One of Lithuania's 3000 lakes. The entire country is low-lying. Its coast, fringed by sand dunes and pine forests, is famous for amber.

RESOURCES

Electric power 5.8m kw

 35,244 tonnes

 5554 b/d (reserves 14m barrels)

 855,600 pigs, 748,300 cattle, 5.58m chickens

Sand, gravel, clay, limestone, gypsum

Lithuania has significant reserves of peat and materials used in construction industry. The Ignalina nuclear plant provides more than 80% of the country's electricity. Oil is mostly imported from Russia.

ENVIRONMENT

Sustainability rank: 27th

 12%

 4.2 tonnes per capita

Radioactive leaks and the risk of accident at the giant Chernobyl-type nuclear plant at Ignalina cause much concern. In 2002 the government agreed to decommission it by 2009, after pressure from the EU, which will share the $2.6 billion cost.

MEDIA

TV ownership high

 Daily newspaper circulation 93 per 1000 people

PUBLISHING AND BROADCAST MEDIA

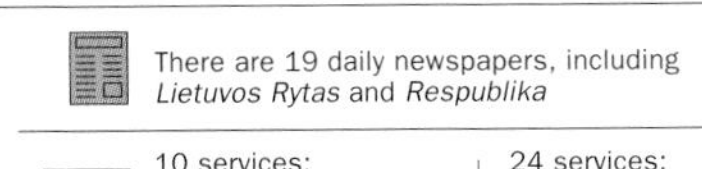

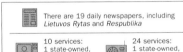

There are 19 daily newspapers, including *Lietuvos Rytas* and *Respublika*

10 services: 1 state-owned, 9 independent

24 services: 1 state-owned, 23 independent

The mainstream media, Russian under communism, now publish and broadcast mainly in Lithuanian.

CRIME

No death penalty

 8960 prisoners

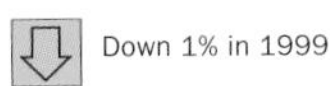 Down 1% in 1999

Levels of crime are low compared with other parts of the former USSR. Robbery is a growing problem.

EDUCATION

School leaving age: 15

 99%

 84,300 students

Teaching at all levels is in Lithuanian, making access to higher education harder for minorities; 8% of the population are graduates.

CHRONOLOGY

Russia annexed Lithuania in 1795. The suppression of rebellions in 1831 and 1863 failed to undermine its nationalist movement.

- ❑ **1915** Occupied by German troops.
- ❑ **1918** Independence declared.
- ❑ **1926** Military coup; one-party rule.
- ❑ **1940** Annexed by Soviet Union.
- ❑ **1941–1944** Nazi occupation.
- ❑ **1945** Incorporated into USSR.
- ❑ **1991** Achieves full independence.
- ❑ **1992** First multiparty elections.
- ❑ **1993** Russian troops withdraw.
- ❑ **1996** Prime minister forced from office by banking scandal. General election; conservative TS(LK) wins.
- ❑ **1998** Valdas Adamkus president.
- ❑ **2000** Brief center-left coalition.
- ❑ **2001** Ex-president Brazauskas becomes prime minister.

HEALTH

Welfare state health benefits

 1 per 250 people

Heart disease, cancers, accidents, tuberculosis

The 1997 reorganization of the health service involves replacing state funding with finance from insurance funds.

SPENDING

GDP/cap. decrease

CONSUMPTION AND SPENDING

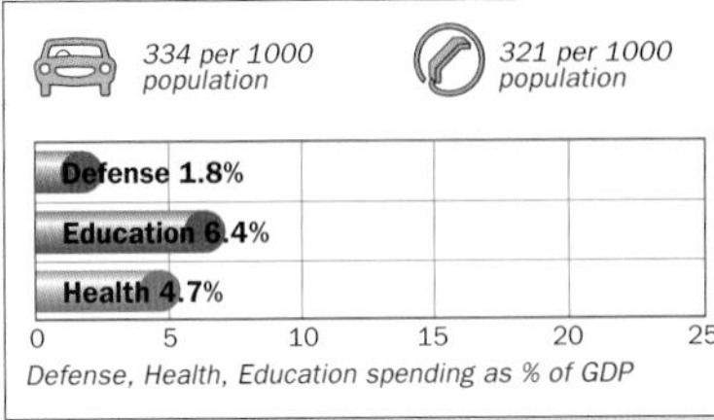
334 per 1000 population
321 per 1000 population
Defense 1.8%
Education 6.4%
Health 4.7%
0
5
10
15
20
25
Defense, Health, Education spending as % of GDP

Lithuanians are on average poorer than their neighbors in the other Baltic states. Since 1991 a large gap has opened between the incomes of rich and poor.

WORLD RANKING

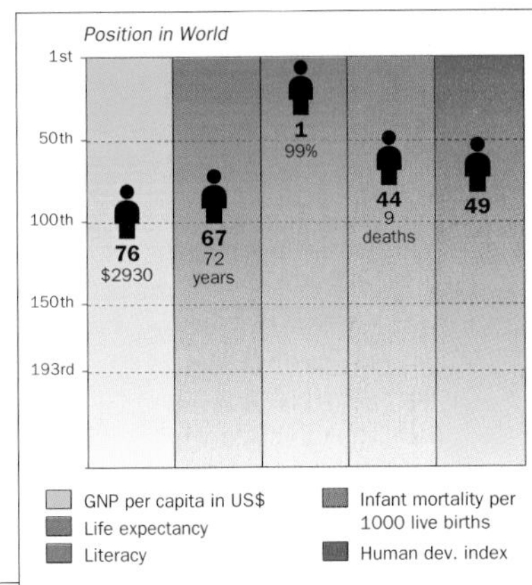
Position in World
1st
50th
100th
150th
193rd
76
$2930
67
72
years
1
99%
44
9
deaths
49
GNP per capita in US$
Life expectancy
Literacy
Infant mortality per 1000 live births
Human dev. index

L

LUXEMBOURG

EUROPE
Europe
Belgium
Germany
LUXEMBOURG
France

OFFICIAL NAME: Grand Duchy of Luxembourg **CAPITAL:** Luxembourg-Ville
POPULATION: 442,000 **CURRENCY:** Euro **OFFICIAL LANGUAGES:** French, German, & Luxemburgish

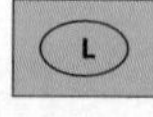
1867
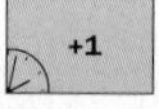
1867

June 23

L

+1 +352 .lu

LUXEMBOURG SHARES BORDERS with the industrial regions of Germany, France, and Belgium, and has the highest per capita income in the EU. Making up part of the plateau of the Ardennes, its countryside is undulating and forested. Its prosperity was once based on steel; before World War II it produced more per capita than the US. Today, it is known as a tax haven and banking center, and as the headquarters of key EU institutions.

CLIMATE

Maritime

WEATHER CHART FOR LUXEMBOURG-VILLE

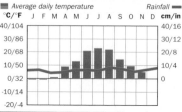

Average daily temperature / Rainfall
°C/°F J F M A M J J A S O N D cm/in
40/104 30/86 20/68 10/50 0/32 -10/14 -20/-4
40/16 30/12 20/8 10/4 0

The south, where vines grow, is the warmest area. Winter is cold and snowy, especially in the Ardennes.

TRANSPORTATION

Drive on right

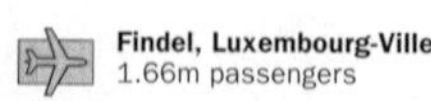
Findel, Luxembourg-Ville
1.66m passengers

Has no fleet

THE TRANSPORTATION NETWORK

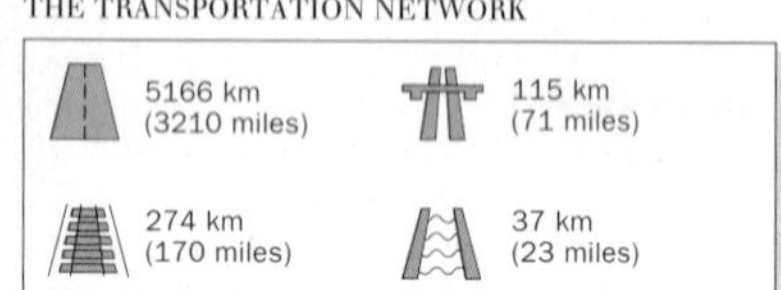
5166 km
(3210 miles)
115 km
(71 miles)
274 km
(170 miles)
37 km
(23 miles)

There is an excellent road network, though congestion is a problem. Rail and bus services are integrated.

TOURISM

Visitors : Population 1.8:1

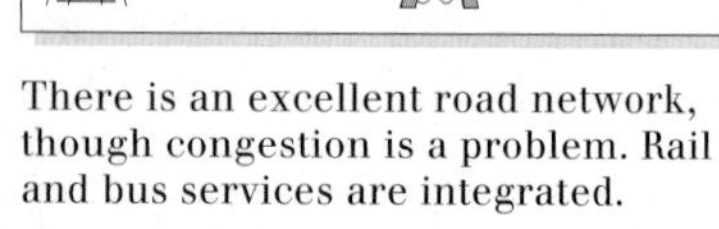
807,000 visitors
Down 3% in 2000
MAIN TOURIST ARRIVALS

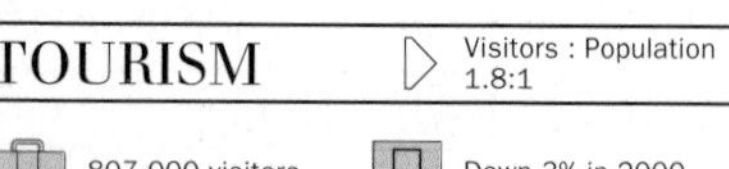
Belgium 28%
Germany 15%
Netherlands 14%

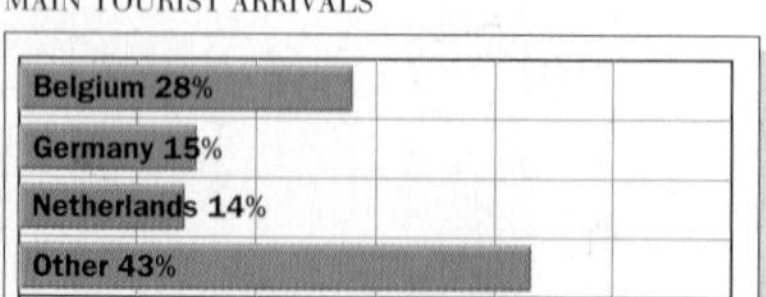
Other 43%
0 10 20 30 40 50 60
% of total arrivals

Key attractions are the mountains and forests, 76 castles, and the Benedictine abbey at Echternach. Foreign hotel workers learn about the history, language, and culture of the duchy under a government initiative.

PEOPLE

Pop. density medium

Luxemburgish, German, French

171/km² (443/mi²)

THE URBAN/RURAL POPULATION SPLIT

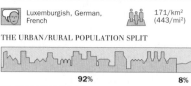

92% 8%

RELIGIOUS PERSUASION

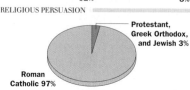

Protestant, Greek Orthodox, and Jewish 3%
Roman Catholic 97%

Nearly a third of its residents and half of Luxembourg's workers are foreigners. Integration has been straightforward; most are fellow western Europeans and Roman Catholics, mainly from Italy and Portugal. Life in Luxembourg is comfortable. Salaries are high, unemployment very low, and social tensions few.

POLITICS

Multiparty elections

L. House 1999/2004

Grand Duke Henri

AT THE LAST ELECTION

Chamber of Deputies 60 seats

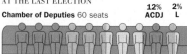

32% CSV/PCS **25% DP/PD** **21% LSAP/POSL** **12% ACDJ** **8% G** **2% L**

CSV/PCS = Christian Social Party **DP/PD** = Democratic Party **LSAP/POSL** = Luxembourg Socialist Workers' Party **ACDJ** = Action Committee for Democracy and Justice **G** = Greens **L** = The Left

Council of State 21 seats

The members of the Council of State are appointed for life by the grand duke

There is remarkable political consensus, and governments are characterized by coalitions and long-serving prime ministers; the grand duke's role is mostly ceremonial. The main issues relate to European integration.

WORLD AFFAIRS

Joined UN in 1945

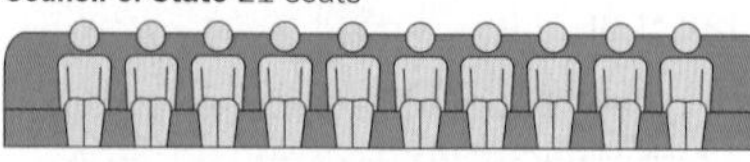
Benelux
EU
NATO
OECD
OSCE

Luxembourg has long been the keenest member of the EU. It was during its EU presidency that the Maastricht agreement for closer European union was brokered; Luxembourg was not only the first member state to meet all the economic, financial, and legal requirements of union under Maastricht, but did so a year early. This commitment to the EU reflects the tremendous benefits Luxembourg has gained from membership. It is home to both the Secretariat of the European Parliament and the Court of Justice. In 1995, Prime Minister Jacques Santer left office to become president of the European Commission, but had to resign in 1999 amid allegations of corruption in the Commission.

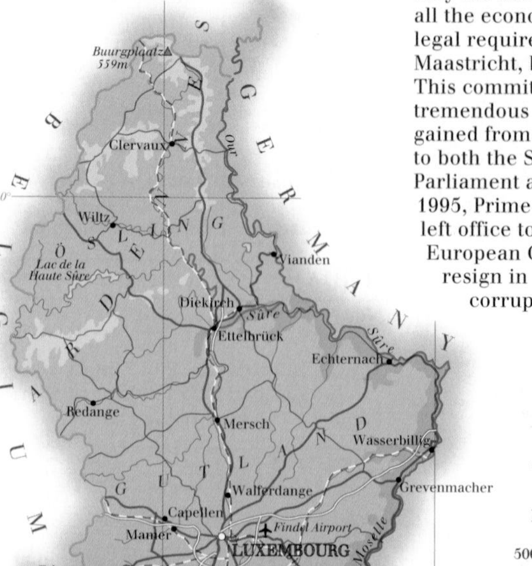
Buurgplaatz 559m
Clervaux
Wiltz
Lac de la Haute Sûre
Vianden
Diekirch
Sûre
Ettelbrück
Echternach
Redange
Mersch
Wasserbillig
Walferdange
Grevenmacher
Capellen
Mamer
Findel Airport
LUXEMBOURG
Pétange
Hesperange
Sanem
Differdange
Schifflange
Bettembourg
Remich
Esch-sur-Alzette
Kayl
Dudelange
Moselle
Our
BELGIUM
GERMANY
FRANCE
GUTLAND
ÖSLING
ARDENNES

LUXEMBOURG
Total Land Area : 2586 sq. km (998 sq. miles)

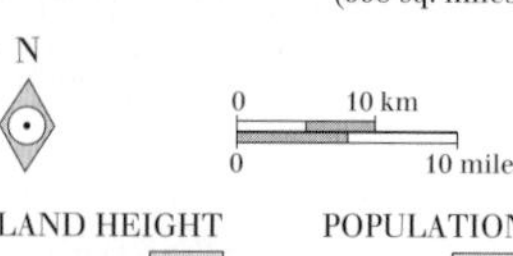
N
0 10 km
0 10 miles
LAND HEIGHT
500m/1640ft
200m/656ft
Sea Level
POPULATION
over 50 000
over 10 000
under 10 000

Charlotte Bridge, Luxembourg. *The modern road system provides excellent links with the rest of Europe.*

AID

Donor

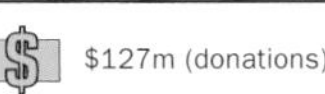 $127m (donations)

 Up 7% in 2000

Aid has been increased substantially and equaled 0.69% of GNP in 2000. Most goes to sub-Saharan Africa.

DEFENSE

No compulsory military service

 $126m

 Down 7% in 2000

The army numbers around 900 full-time soldiers. Spending has fallen slightly in recent years.

ECONOMICS

Inflation 2.2% p.a. (1990–2000)

 $18.4bn

 1.0651–1.1231 euros

SCORE CARD

- ❑ WORLD GNP RANKING..........69th
- ❑ GNP PER CAPITA..........$42,060
- ❑ BALANCE OF PAYMENTS..........$1.59bn
- ❑ INFLATION..........3.1%
- ❑ UNEMPLOYMENT..........3%

STRENGTHS

Location for some EU institutions. Banking secrecy and expertise make the capital home to around 1000 investment funds and over 200 banks – more than in any other city in the world.

WEAKNESSES

International service industries account for 65% of GDP, making Luxembourg vulnerable to changing conditions overseas. Downturn in steel market.

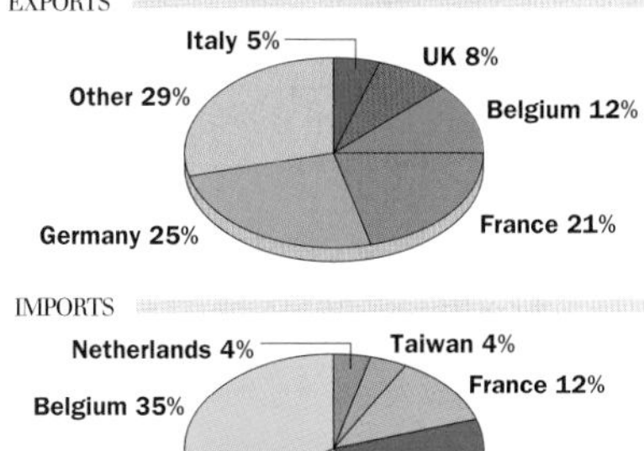
EXPORTS
Italy 5%
UK 8%
Other 29%
Belgium 12%
Germany 25%
France 21%
IMPORTS
Netherlands 4%
Taiwan 4%
France 12%
Belgium 35%
Other 21%
Germany 24%

RESOURCES

Electric power 1.3m kw

 Not available

 Not an oil producer

 207,862 cattle, 85,830 pigs, 8220 sheep, 2818 horses

 Iron

Most energy is imported; Luxembourg produces only a small amount of hydroelectricity. Arbed is the world's third-largest steel producing company.

ENVIRONMENT

Not available

None

 18 tonnes per capita

Acid rain from European industry has affected 20% of Luxembourg's trees and, in the worst cases, 30% of trees in mature stands. Luxembourg is a member of an international committee on reducing pollution of the Rhine.

MEDIA

TV ownership high

 Daily newspaper circulation 328 per 1000 people

PUBLISHING AND BROADCAST MEDIA

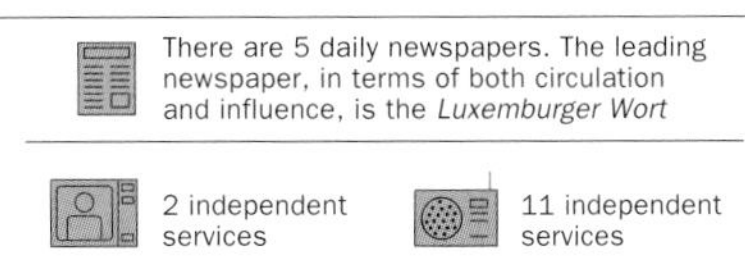

There are 5 daily newspapers. The leading newspaper, in terms of both circulation and influence, is the *Luxemburger Wort*

2 independent services

11 independent services

Broadcasting is dominated by RTL (Radio–Television Luxembourg), one of the largest media groups in Europe, which exports its programs in a variety of languages.

CRIME

No death penalty

 394 prisoners

 Down 2% in 1999

Luxembourg's banking secrecy rules have provoked international criticism, as they can provide a cover for both tax evasion and fraud. Violent crime remains uncommon.

EDUCATION

School leaving age: 15

 99%

 2662 students

Teaching is mainly in German at primary and French at secondary level. Higher education is limited and many students go to universities in other European countries. Training given by Luxembourg banks is reputed to be the best in Europe.

HEALTH

Welfare state health benefits

 1 per 368 people

 Cerebrovascular and heart diseases, cancers, accidents

There are no private commercial hospitals in Luxembourg; they are run either by the state or by nuns. Patients' fees are refunded from the state sickness fund.

CHRONOLOGY

Until 1867, Luxembourg was ruled by a succession of neighboring European powers.

- ❑ **1890** Link with Dutch throne ends.
- ❑ **1921** Economic union with Belgium. End of German ties.
- ❑ **1940–1944** German occupation.
- ❑ **1948** Benelux treaty (1944) creating a customs union comes into effect.
- ❑ **1957** One of six signatories of Treaty of Rome, the principal foundation of what develops into the EU.
- ❑ **1995** Premier Jacques Santer is president of European Commission.
- ❑ **1999** Santer resigns amid corruption allegations. Socialist election losses.
- ❑ **2000** Grand Duke Jean abdicates in favor of his son, Henri.
- ❑ **2002** Euro fully adopted.

SPENDING

GDP/cap. increase

CONSUMPTION AND SPENDING

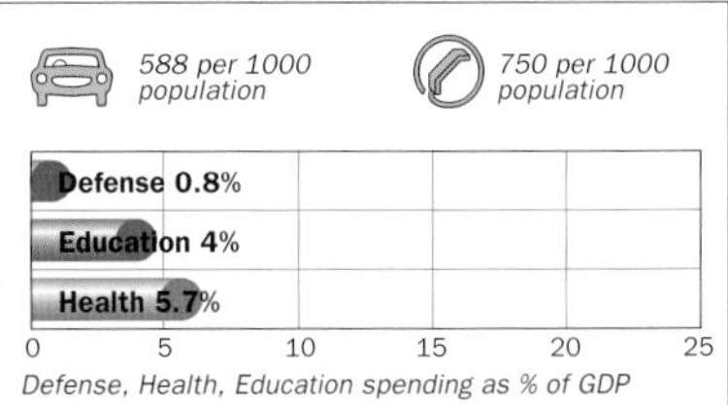
588 per 1000 population
750 per 1000 population
Defense 0.8%
Education 4%
Health 5.7%
0
5
10
15
20
25
Defense, Health, Education spending as % of GDP

With the world's second-highest per capita income, Luxembourgers enjoy a comfortable lifestyle. Recent strong economic performance has allowed them to benefit both from lower taxes and increased social security spending. Low unemployment has led to the recruitment of foreign workers, mainly from neighboring countries or from other EU countries such as Portugal and Italy, to take less well-paid jobs. As elsewhere in western Europe, financing care of the aging population is likely to be a burden in the future.

WORLD RANKING

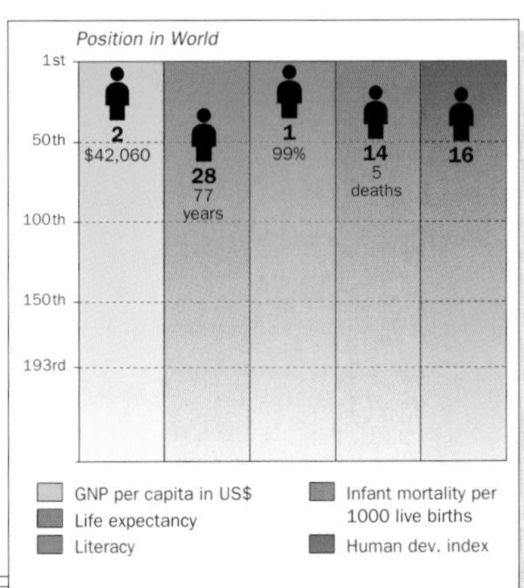
Position in World
1st
50th
100th
150th
193rd
2
$42,060
28
77
years
1
99%
14
5
deaths
16
GNP per capita in US$
Life expectancy
Literacy
Infant mortality per 1000 live births
Human dev. index

L

MACEDONIA

EUROPE
Europe
Serbia & Montenegro
Bulgaria
Albania
MACEDONIA
Greece

OFFICIAL NAME: Former Yugoslav Republic of Macedonia **CAPITAL:** Skopje
POPULATION: 2 million **CURRENCY:** Macedonian denar **OFFICIAL LANGUAGES:** Macedonian and Albanian

1991

1991

Aug 2

MK

+1

+389 .mk

THE FORMER YUGOSLAV REPUBLIC of Macedonia (FYRM) is landlocked in southeastern Europe. Despite the signing of an accord in 1995, Greece remains suspicious that it harbors ambitions about absorbing northern Greece – also called Macedonia – in a "Greater Macedonia." A militant movement among ethnic Albanians erupted into violent conflict in March–September 2001, but a peace agreement was reached after the involvement of a NATO force.

A ***fisherman's hut*** *on Lake Dojran. The lake lies on the border with Greece in southeastern Macedonia and is shared by the two countries.*

CLIMATE

Continental

WEATHER CHART FOR SKOPJE

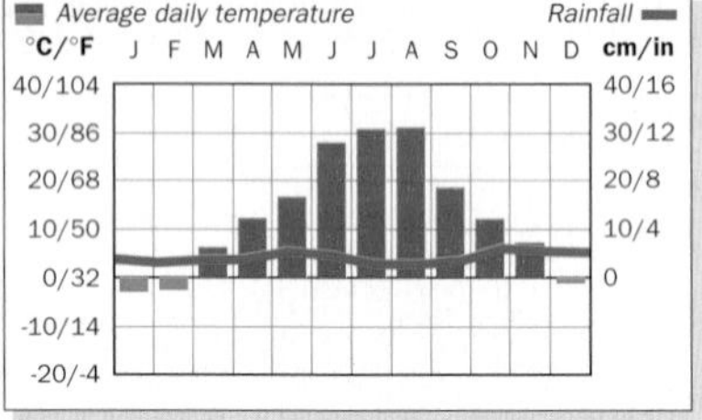
Average daily temperature
Rainfall
°C/°F
J F M A M J J A S O N D
cm/in
40/104
30/86
20/68
10/50
0/32
-10/14
-20/-4
40/16
30/12
20/8
10/4
0

The FYRM has a continental climate. Winter snow supports skiing.

TRANSPORTATION

Drive on right

Skopje International 1.01m passengers

Has no fleet

THE TRANSPORTATION NETWORK

5540 km (3442 miles)

133 km (83 miles)

699 km (434 miles)

None

An east–west road and rail route from Tirana in Albania through Macedonia to Sofia in Bulgaria is planned, to reduce reliance on routes through Serbia.

TOURISM

Visitors : Population 1:8.9

224,000 visitors

Up 24% in 2000

MAIN TOURIST ARRIVALS

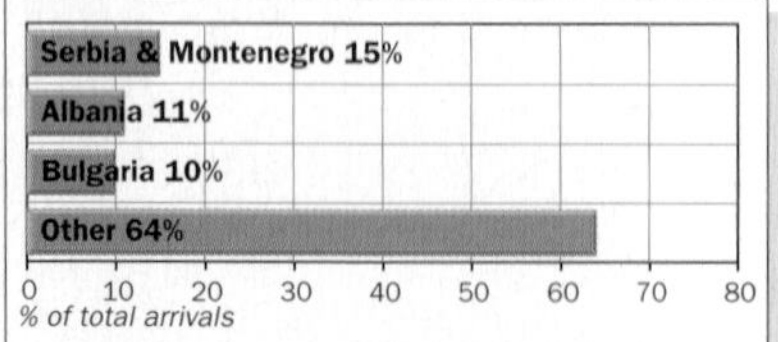
Serbia & Montenegro 15%
Albania 11%
Bulgaria 10%
Other 64%
0 10 20 30 40 50 60 70 80
% of total arrivals

The major attraction is the ecclesiastical center of Ohrid, situated on Europe's deepest lake, with Roman and Byzantine ruins. Other lake resorts and skiing in the Sara mountains in the northwest, have potential once stability is restored.

PEOPLE

Pop. density medium

Macedonian, Albanian, Serbo-Croat

78/km² (201/mi²)

THE URBAN/RURAL POPULATION SPLIT

62% 38%

ETHNIC MAKEUP

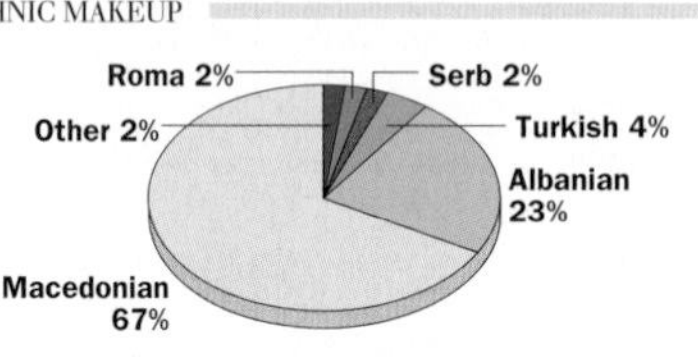
Roma 2%
Serb 2%
Other 2%
Turkish 4%
Albanian 23%
Macedonian 67%

Slav Macedonians, speaking a language akin to Bulgarian, are in the majority. The large Albanian minority, claiming to amount to over one-third of the population, maintains strong links with Albanians in neighboring states. Months of violent conflict sparked by ethnic Albanian insurgents seeking greater rights resulted in a new constitution in 2001 guaranteeing equality.

Macedonians are mostly Orthodox Christians, but there are a substantial number of Slavic Muslims (Pomaks), whose ancestors converted during the Ottoman occupation. Ethnic Albanians are mostly Muslim.

WORLD AFFAIRS

Joined UN in 1993

CE

IAEA

EAPC

PfP OSCE

Macedonia's hosting of NATO troops in the Kosovo conflict in 1999 placed it firmly in the Western fold. Nationalists resented Western pressure in 2001 to compromise with Albanian militants.

AID

Recipient

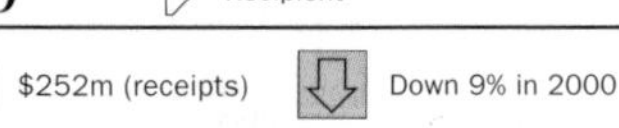

$252m (receipts)

Down 9% in 2000

The World Bank and the EU are the main channels for economic development assistance. The EU and US pledged $515 million for reconstruction in 2002. Regional security fears limit foreign investment.

DEFENSE

Compulsory military service

$76m

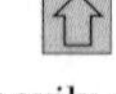

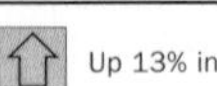

Up 13% in 2000

The army relies heavily on officer training in NATO countries. A major overhaul has been announced, with the aim being to tackle weaknesses revealed by the 2001 conflict.

POLITICS

Multiparty elections

2002/2006

President Boris Trajkovski

AT THE LAST ELECTION

Assembly of the Republic 120 seats

49%	28%	13%	6%	2%	1%	1%
ZMZ	VMRO–DPMNE	BDI	PDS	PPD	PDK	SPM

ZMZ = Together for Macedonia, headed by the Social Democratic Alliance of Macedonia (**SDSM**) **VMRO–DPMNE** = Internal Macedonian Revolutionary Organization–Democratic Party for Macedonian National Unity **BDI** = Democratic Union for Integration **PDS** = Democratic Party of Albanians **PPD** = Party of Democratic Prosperity **PDK** = National Democratic Party **SPM** = Socialist Party of Macedonia

Political parties divide along ethnic lines. The Social Democrats and the right-wing VMRO–DPMNE are predominantly Slav, while Albanians vote mainly for the BDI, the PDS, and the smaller PPD. The September 2002 elections gave the Social Democrat-led ZMZ alliance a slight ascendancy.

Previously, VMRO–DPMNE Prime Minister Ljubco Georgievski had led a panethnic right-wing coalition (1998–2001), then a fragile "national unity" government formed in the face of armed insurrection by Albanian rebels (UCK). A peace deal in August 2001 ushered in a new, more equal constitution and guarantees on minority rights. With peace secured at the end of 2001, the "unity" government fractured, but Georgievski remained in power until the 2002 polls.

M

ECONOMICS

Inflation 79% p.a. (1990–2000)

$3.7bn

66.26–69.17 Macedonian denari

SCORE CARD

❑ World GNP Ranking	124th
❑ GNP per Capita	$1820
❑ Balance of Payments	–$107m
❑ Inflation	–1.3%
❑ Unemployment	32%

EXPORTS

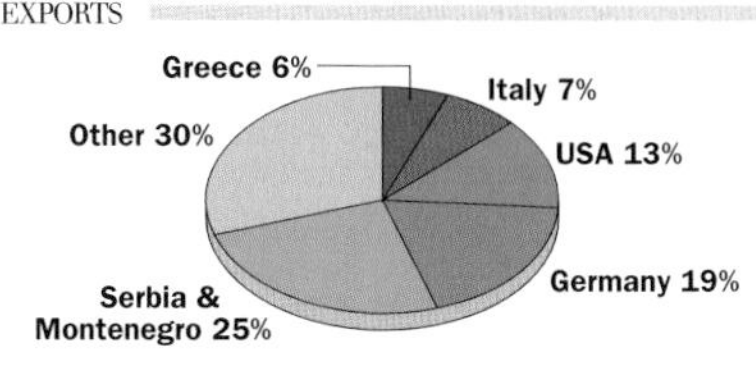
Greece 6%
Italy 7%
Other 30%
USA 13%
Germany 19%
Serbia & Montenegro 25%

IMPORTS

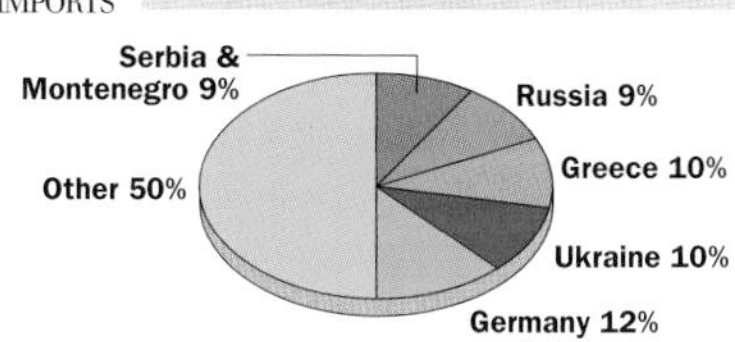
Serbia & Montenegro 9%
Russia 9%
Other 50%
Greece 10%
Ukraine 10%
Germany 12%

Strengths

Growth in private sector and foreign investment. Mineral resources.

Weaknesses

Poorest of former Yugoslav republics. Loss of trade in mid-1990s due to Greek embargo and sanctions on Yugoslavia. Dependence on oil, gas, and machinery imports. Disruption caused by Kosovo conflict and 2001 violence.

RESOURCES

Electric power 1.5m kw

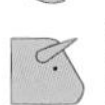
1804 tonnes

Not an oil producer

1.25m sheep, 265,000 cattle, 3.35m chickens

Coal, copper, bauxite, iron, antimony, chromium, lead, zinc

Minerals remain underexploited. South-facing fertile plains produce early fruit and vegetables for EU markets.

ENVIRONMENT

Sustainability rank: 83rd

7%

6.1 tonnes per capita

Industrial pollution affects water quality. The Titov Veles lead and zinc smelter is the worst culprit for toxic waste.

MEDIA

TV ownership high

Daily newspaper circulation 21 per 1000 people

PUBLISHING AND BROADCAST MEDIA

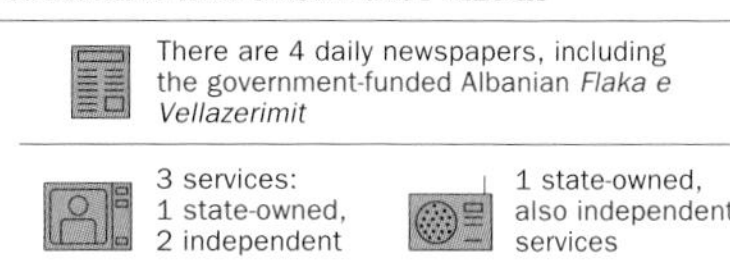

There are 4 daily newspapers, including the government-funded Albanian *Flaka e Vellazerimit*

3 services: 1 state-owned, 2 independent

1 state-owned, also independent services

Newspaper sales have expanded rapidly. In 2001, two independent newspapers "voluntarily" ceased publication.

CRIME

No death penalty

1394 prisoners

Down 11% 1996–1998

Cigarette smuggling is dominated by Albanian gangs also involved in the illegal arms trade and heroin trafficking.

EDUCATION

School leaving age: 15

94%

36,922 students

A new university at Tetovo, offering teaching in Albanian as well as Macedonian, officially opened in 2001.

CHRONOLOGY

The end of Ottoman rule saw historic Macedonia divided between Serbia, Bulgaria, and Greece in 1912–1913. What is now the FYRM was incorporated into Serbia.

- ❑ **1944** Tito establishes republic, stressing Macedonian identity.
- ❑ **1989–1990** Multiparty elections.
- ❑ **1991** Independence declared. EU recognition delayed by Greeks.
- ❑ **1995** Accord with Greece.
- ❑ **1998–1999** Right-wing VMRO–DPMNE coalition wins elections.
- ❑ **1999** Upheaval over Kosovo conflict.
- ❑ **2001** New constitution after NATO involvement to end conflict with ethnic Albanian militants.
- ❑ **2002** Left-of-center ZMZ alliance gains upper hand in elections.

HEALTH

Welfare state health benefits

1 per 455 people

Cerebrovascular and heart diseases, cancers

In theory, the state guarantees universal health care, but effective and speedy treatment is increasingly only available in the private sector.

SPENDING

GDP/cap. decrease

CONSUMPTION AND SPENDING

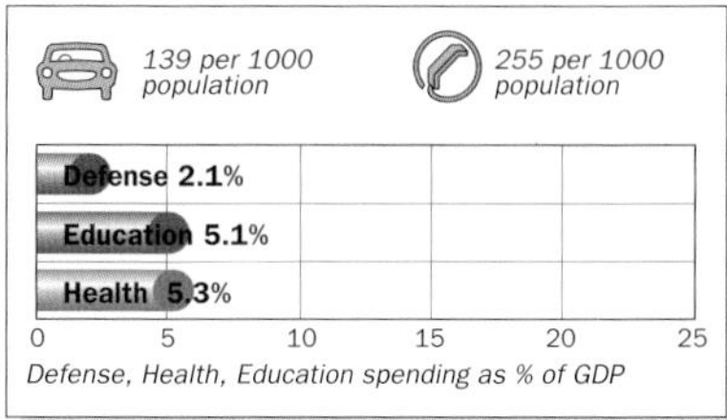
139 per 1000 population
255 per 1000 population
Defense 2.1%
Education 5.1%
Health 5.3%
0
5
10
15
20
25
Defense, Health, Education spending as % of GDP

Incomes have fallen by more than two-thirds since 1990, although smuggling and organized crime have made a few people conspicuously wealthy.

WORLD RANKING

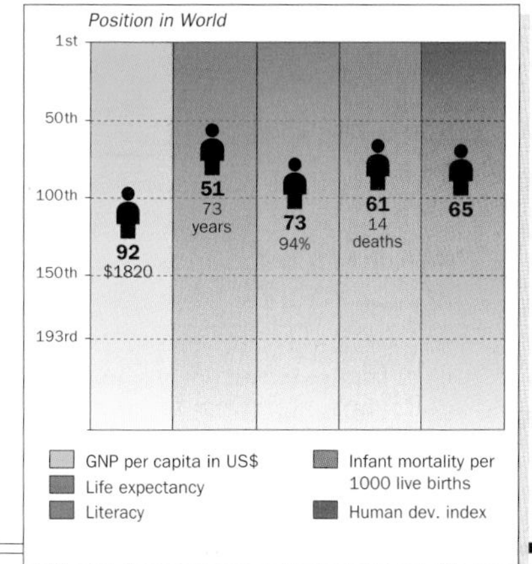
Position in World
1st
50th
100th
150th
193rd
92
$1820
51
73 years
73
94%
61
14 deaths
65
GNP per capita in US$
Life expectancy
Literacy
Infant mortality per 1000 live births
Human dev. index

FORMER YUGOSLAV REPUBLIC OF MACEDONIA

Total Land Area : 25 333 sq. km (9781 sq. miles)

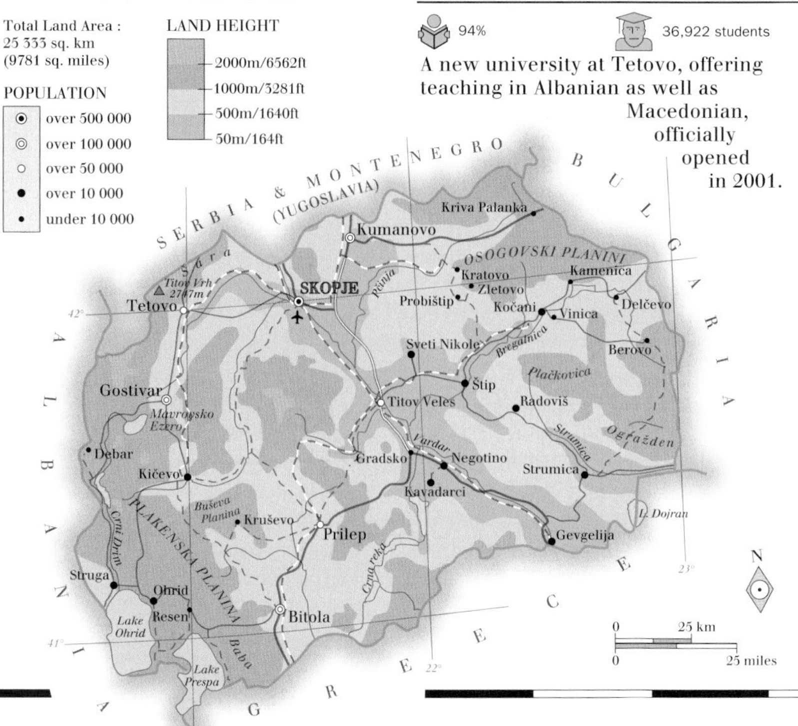
LAND HEIGHT
2000m/6562ft
1000m/3281ft
500m/1640ft
50m/164ft
POPULATION
over 500 000
over 100 000
over 50 000
over 10 000
under 10 000
SERBIA & MONTENEGRO (YUGOSLAVIA)
BULGARIA
ALBANIA
GREECE
Kumanovo
Kriva Palanka
OSOGOVSKI PLANINI
Kratovo
Kamenica
Zletovo
SKOPJE
Probištip
Kočani
Delčevo
Vinica
Tetovo
Titov Vrh 2747m
Šar Planina
Pčinja
Sveti Nikole
Bregalnica
Berovo
Štip
Plačkovica
Gostivar
Mavrovsko Ezero
Titov Veles
Radoviš
Strumica
Ogražden
Debar
Gradsko
Vardar
Negotino
Strumica
Kičevo
Kavadarci
L. Dojran
PLAKENSKA PLANINA
Buševa Planina
Kruševo
Prilep
Crni Drim
Crna reka
Gevgelija
Struga
Ohrid
Resen
Lake Ohrid
Bitola
Baba
Lake Prespa
0
25 km
25 miles
N

M

MADAGASCAR

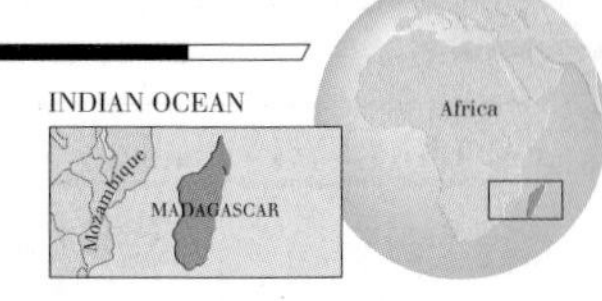
INDIAN OCEAN
Africa
Mozambique
MADAGASCAR

OFFICIAL NAME: Republic of Madagascar **CAPITAL:** Antananarivo

POPULATION: 16.4 million **CURRENCY:** Malagasy franc **OFFICIAL LANGUAGES:** French and Malagasy

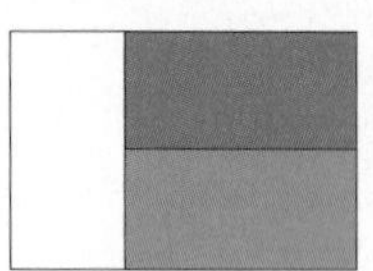

1960

1960 June 26

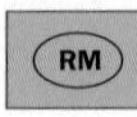
RM

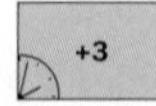
+3

+261

.mg

LYING IN THE INDIAN Ocean, Madagascar is the world's fourth-largest island. Its isolation means that there is a host of unique wildlife and plants. To the east, the large central plateau drops precipitously through forested cliffs to the coast; in the west, gentler gradients give way to fertile plains. It became independent from France in 1960, and after 18 years of radical socialism under Didier Ratsiraka, became a multiparty democracy in 1993.

CLIMATE

Tropical

WEATHER CHART FOR ANTANANARIVO

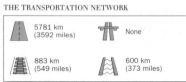

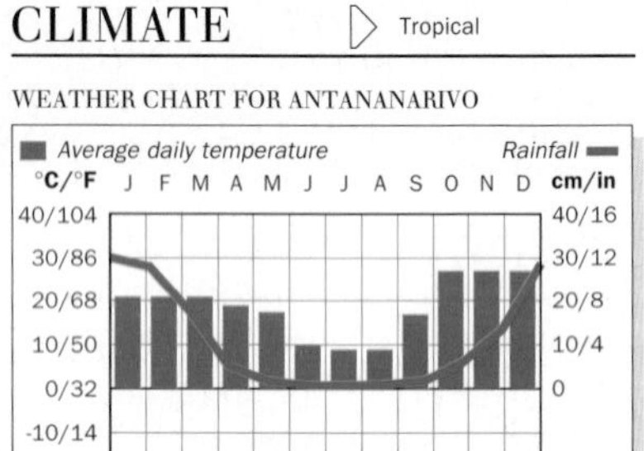
Average daily temperature
Rainfall
°C/°F J F M A M J J A S O N D cm/in
40/104
30/86
20/68
10/50
0/32
-10/14
-20/-4
40/16
30/12
20/8
10/4
0

Tropical Madagascar often has cyclones. The coastal lowlands are humid, but the central plateau is cooler.

TRANSPORTATION

Drive on right

Ivato, Antananarivo 707,304 passengers

101 ships 41,700 grt

THE TRANSPORTATION NETWORK

5781 km (3592 miles)

None

883 km (549 miles)

600 km (373 miles)

An extensive domestic air network – due for privatization – compensates for a very limited rail network and roads that are impassable during the rains.

TOURISM

Visitors : Population 1:103

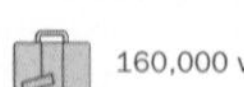

160,000 visitors

Up 16% in 2000

MAIN TOURIST ARRIVALS

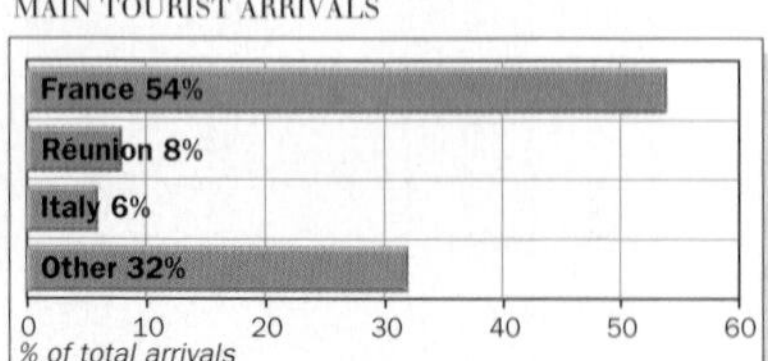
France 54%
Réunion 8%
Italy 6%
Other 32%
0 10 20 30 40 50 60
% of total arrivals

Extensive tropical beaches and unique flora and fauna offer great tourism potential. Political stability after 1993 led to a marked increase in arrivals which was threatened by the chaos in 2002.

PEOPLE

Pop. density low

Malagasy, French

28/km² (73/mi²)

THE URBAN/RURAL POPULATION SPLIT

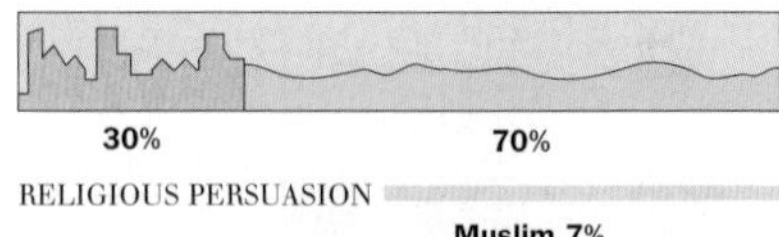
30%
70%

RELIGIOUS PERSUASION

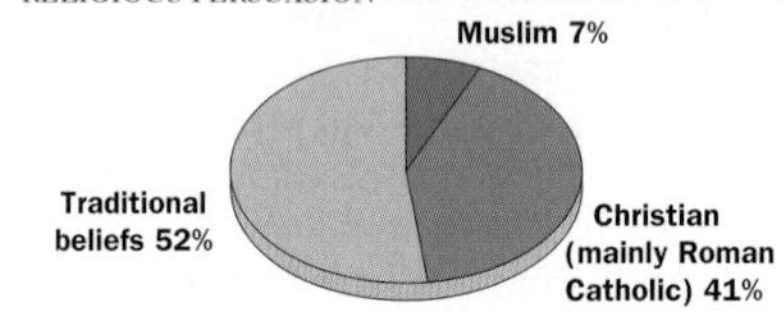
Muslim 7%
Traditional beliefs 52%
Christian (mainly Roman Catholic) 41%

Madagascans are essentially Malay–Indonesian in origin. Their ancestors migrated across the Indian Ocean from the 1st century. Later migrants from the African mainland intermixed. Arab traders provided another ingredient to the mix. The main ethnic division is between the central plateau and *côtier* peoples. Of more pronounced Malay extraction, the plateau Merina were Madagascar's historic rulers. They remain the social elite – to the resentment of the poorer *côtiers*, the group to which long-term ruler Didier Ratsiraka belongs. The extended family remains the focus of social life for the rural majority.

POLITICS

Multiparty elections

L. House 1998/2003
U. House 2001/2007

President Marc Ravalomanana

AT THE LAST ELECTION

National Assembly 150 seats

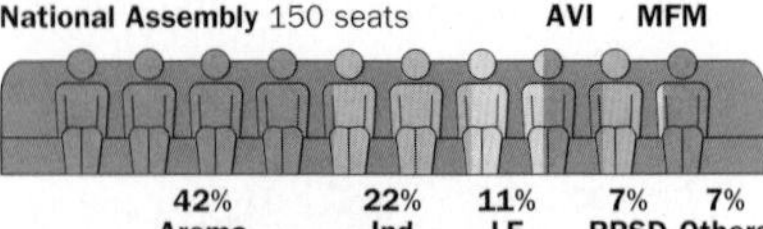
9% AVI
2% MFM
42% Arema
22% Ind
11% LF
7% RPSD
7% Others

Arema = Association for the rebirth of Madagascar
Ind = Independents **LF** = Leader–Torch (Fanilo)
AVI = "People are judged by the work they do"
RPSD = Rally for Socialism and Democracy
MFM = Militant Party for the Development of Madagascar

Senate 90 seats

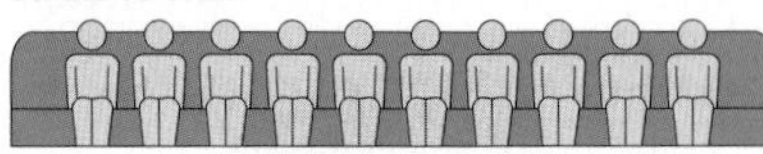

Two-thirds of Senate members are elected by regional governments; the remainder are nominated by the president

From 1975 until 2002 Madagascan politics was dominated by radical socialist president Didier Ratsiraka. A return to multipartyism resulted in his removal in 1993, but in 1997 he was returned to power. Continuing economic gloom saw him defeated in elections in late 2001. Refusing to accept the victory of the liberal opposition leader Marc Ravalomanana, Ratsiraka precipitated an eight-month power struggle which violently divided the island. Eventually support for Ravalomanana proved overwhelming, and he was recognized internationally as president in June 2002.

MADAGASCAR

Total Land Area : 587 040 sq. km (226 656 sq. miles)

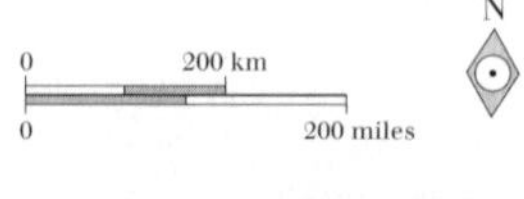
0 200 km
0 200 miles
N

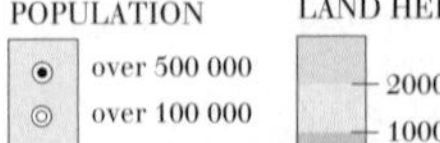
POPULATION
over 500 000
over 100 000
over 50 000
over 10 000
under 10 000

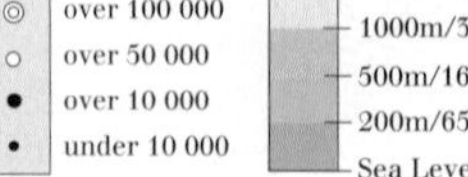
LAND HEIGHT
2000m/6562ft
1000m/3281ft
500m/1640ft
200m/656ft
Sea Level

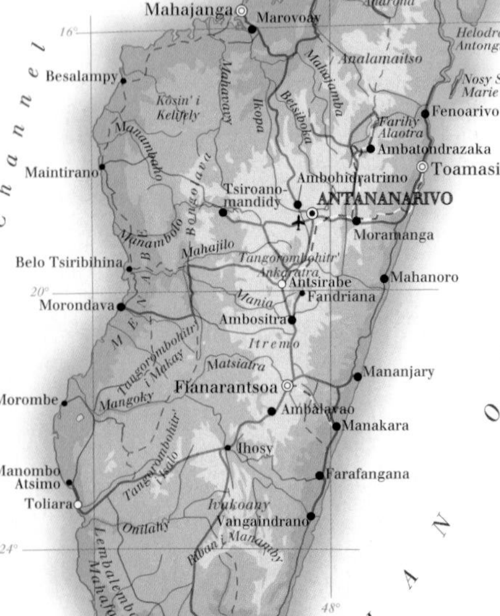
Antsirañana
Nosy Be
Andoany
Ambanja
Ambilobe
Maromokotro 2876m
Sambava
Analalava
Antsohihy
Andapa
Antalaha
Maroantsetra
Mahajanga
Marovoay
Besalampy
Fenoarivo Atsinanana
Ambatondrazaka
Toamasina
Maintirano
Ambohidratrimo
Tsiroanomandidy
ANTANANARIVO
Moramanga
Belo Tsiribihina
Antsirabe
Mahanoro
Fandriana
Morondava
Ambositra
Mananjary
Fianarantsoa
Morombe
Ambalavao
Manakara
Ihosy
Farafangana
Manombo Atsimo
Toliara
Vangaindrano
Amboasary
Tôlañaro (Fort Dauphin)
Mozambique Channel
INDIAN OCEAN

WORLD AFFAIRS

Joined UN in 1960

COMESA

OIF

IAEA

COI

AU

Once-close ties with Russia and North Korea waned as Madagascar cemented relations with its main Western trading partners, especially with France and the US. Since 1997 cooperation with the IMF has improved as well. Regionally, Madagascar has reestablished ties with South Africa and in 1994 joined COMESA.

AID

 Recipient

 $322m (receipts)  Down 10% in 2000

International donors pledged $2.3 billion in July 2002 in a four-year emergency aid package to support the new government and fund reconstruction.

DEFENSE

Compulsory military service

 $41m  Down 5% in 2000

The army was briefly divided during the 2002 turmoil. It intervened against federalist *côtiers* in 1992.

ECONOMICS

Inflation 19% p.a. (1990–2000)

 $3.87bn 
6220–6370
Malagasy francs

SCORE CARD

❑ WORLD GNP RANKING	120th
❑ GNP PER CAPITA	$250
❑ BALANCE OF PAYMENTS	–$260m
❑ INFLATION	12%
❑ UNEMPLOYMENT	Widespread underemployment

STRENGTHS

Varied agricultural base; vanilla, coffee, and clove exports. Offshore oil and gas. Prawns. Tourism. Literate workforce.

WEAKNESSES

Losing out to cheaper vanilla exporters. Vulnerability to drought and cyclone damage, severe in early 2000. Economic reforms yet to bear fruit. Not self-sufficient in rice, the food staple.

EXPORTS

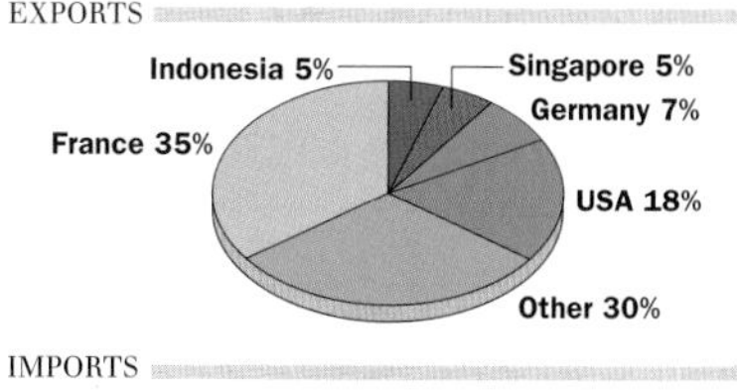
Indonesia 5%
Singapore 5%
Germany 7%
France 35%
USA 18%
Other 30%

IMPORTS

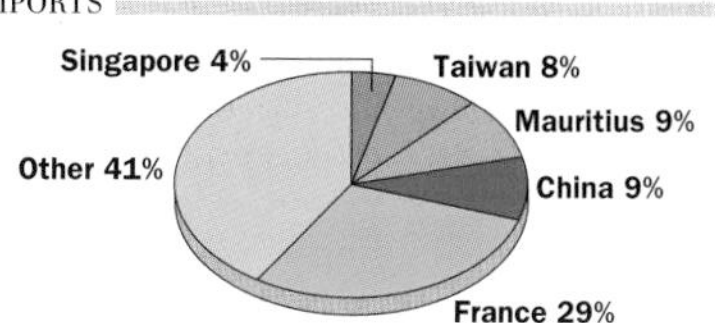
Singapore 4%
Taiwan 8%
Mauritius 9%
Other 41%
China 9%
France 29%

Tôlañaro *(also known as Fort Dauphin), a port on the southeast coast. This was the area first settled by the French in the 16th century.*

RESOURCES

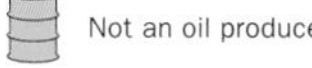
Electric power
220,000 kw

141,057 tonnes | Not an oil producer

 10.3m cattle, 3.8m ducks, 3m geese, 19m chickens | Chromite, graphite, oil, mica, iron, bitumen, gemstones, marble, gas

Electricity is hydrogenerated. There are underexploited mineral reserves and offshore oil and gas. There were large finds of high-quality sapphires in 1998.

ENVIRONMENT

Sustainability rank: 128th

 2% (1% partially protected) 
0.1 tonnes
per capita

Madagascar's environment is a unique resource; 80% of its plant species and many animal species, such as the lemur, are found nowhere else. Aid helps to combat deforestation and soil erosion.

MEDIA

TV ownership low

 Daily newspaper circulation 5 per 1000 people

PUBLISHING AND BROADCAST MEDIA

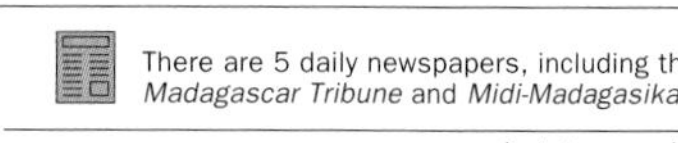
There are 5 daily newspapers, including the *Madagascar Tribune* and *Midi-Madagasikara*

 1 state-owned service 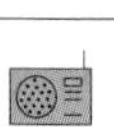 1 state-owned service, many independent stations

Even before the return of multiparty democracy in 1993, there was a flourishing opposition press. There are 127 local radio stations.

CRIME

Death penalty not used in practice

 20,109 prisoners  Crime is rising

Urban crime levels are rising, with theft a particular concern. The army faces accusations of abusing human rights and of shooting federalists in 1993.

EDUCATION

School leaving age: 11

 66% 
31,013 students

Primary education will soon be based on French, not Malagasy. Attendance at secondary level is rising, as is the number of teachers. The OPEC Fund is backing a government literacy drive.

CHRONOLOGY

Increasing European contacts after the 16th century culminated in the 1895 French invasion. Madagascar became a French colony and the Merina monarchy was abolished.

- ❑ **1947–1948** French troops kill thousands in nationalist uprisings.
- ❑ **1960** Independence.
- ❑ **1975** Radical socialist Didier Ratsiraka takes power.
- ❑ **1991** Forces Vives (CFV) coalition set up, led by Albert Zafy. Mass strikes.
- ❑ **1992** Civilian rule restored.
- ❑ **1993** Zafy's CFV defeats Ratsiraka's MFM coalition in free elections.
- ❑ **1996** Zafy impeached.
- ❑ **1997** Ratsiraka elected president.
- ❑ **2002** Country divided after opposition leader Marc Ravalomanana claims victory in 2001 presidential election.

HEALTH

Welfare state health benefits

 1 per 10,000 people  Malaria, enteric, and respiratory diseases

Private health care was legalized in 1993. State care is free but inadequate. Malaria is at epidemic levels. There are outbreaks of bubonic plague.

SPENDING

GDP/cap. increase

CONSUMPTION AND SPENDING

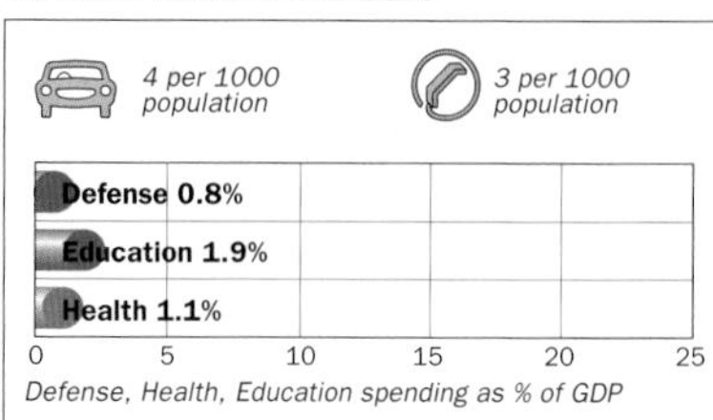
4 per 1000 population
3 per 1000 population
Defense 0.8%
Education 1.9%
Health 1.1%
0
5
10
15
20
25
Defense, Health, Education spending as % of GDP

Most of Madagascar's people are terribly poor, although those who live on the central plateau are richer than the *côtier* farmers and fishermen.

WORLD RANKING

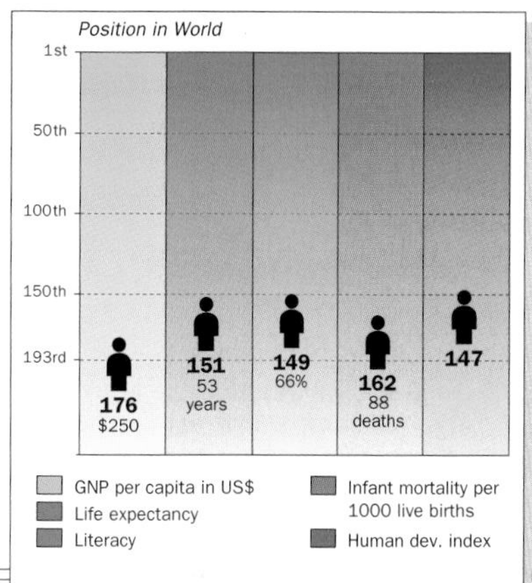
Position in World
1st
50th
100th
150th
193rd
176
$250
151
53 years
149
66%
162
88 deaths
147
GNP per capita in US$
Life expectancy
Literacy
Infant mortality per 1000 live births
Human dev. index

M

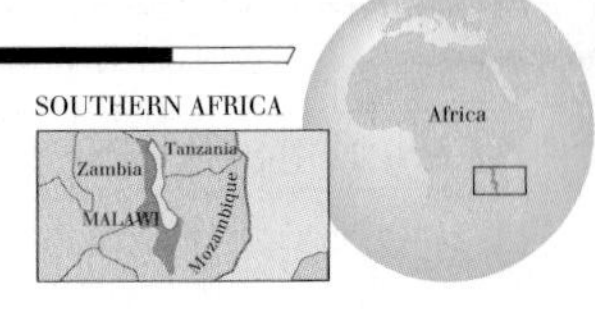
SOUTHERN AFRICA
Tanzania
Zambia
MALAWI
Mozambique
Africa

MALAWI

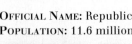

OFFICIAL NAME: Republic of Malawi **CAPITAL:** Lilongwe
POPULATION: 11.6 million **CURRENCY:** Malawi kwacha **OFFICIAL LANGUAGE:** English

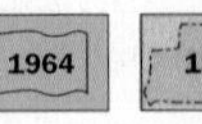
1964

1964

July 6

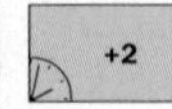
MW

+2

+265

.mw

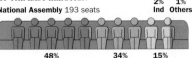

LANDLOCKED IN SOUTHEAST Africa, Malawi lies on the Great Rift Valley and is dominated by Lake Nyasa, Africa's third-largest expanse of water. In the 1980s Malawi hosted large numbers of Mozambican refugees, at some cost to its fragile economy; food shortages are a recurrent and serious threat. A former British colony, Malawi established democracy in 1994 after three decades of one-party rule under Hastings Banda.

CLIMATE

Tropical wet & dry

WEATHER CHART FOR LILONGWE

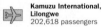

Average daily temperature · Rainfall
°C/°F J F M A M J J A S O N D cm/in
40/104 · 30/86 · 20/68 · 10/50 · 0/32 · -10/14 · -20/-4
40/16 · 30/12 · 20/8 · 10/4 · 0

The south is hot and humid. The rest of Malawi is warm and very sunny in the dry season, but cooler in the highlands.

TRANSPORTATION

Drive on left

Kamuzu International, Lilongwe 202,618 passengers

Has no fleet

THE TRANSPORTATION NETWORK

5254 km (3265 miles)

None

797 km (495 miles)

144 km (89 miles)

The Kamuzu Highway has been upgraded, and the Nacala Rail Corridor, a vital link to the distant sea, has attracted private investment.

TOURISM

Visitors : Population 1:51

228,000 visitors

Down 10% in 2000

MAIN TOURIST ARRIVALS

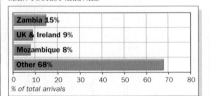

	% of total arrivals
Zambia	15%
UK & Ireland	9%
Mozambique	8%
Other	68%

0 10 20 30 40 50 60 70 80
% of total arrivals

The national parks and Lake Nyasa's fishing and water sports are the main tourist attractions. The opening of international airports at Blantyre and Lilongwe has increased accessibility.

PEOPLE

Pop. density medium

Chewa, Lomwe, Yao, Ngoni, English

123/km² (319/mi²)

THE URBAN/RURAL POPULATION SPLIT

25% · 75%

RELIGIOUS PERSUASION

Traditional beliefs 5%
Muslim 20%
Roman Catholic 20%
Protestant 55%

Ethnicity has not been exploited for political ends as has been the case in neighboring states. Most Malawians share a common Bantu origin. Of the various groups, the Chewa are dominant in the central region, Nyanja in the south, Tumbuka in the north, the mostly Muslim Yao in the southeast, and the Ngoni, a Zulu offshoot, in the lowlands. Other groups include the Chieoka and Tonga. Northerners felt ignored by Banda and his MCP, but the UDF government has largely succeeded in reducing the resulting tensions.

The election of President Muluzi, a member of Malawi's 20% Muslim minority, arguably signals the failure of Banda's plan to enforce Protestant domination in Malawi. Many of Malawi's Muslim Asians work in the retail sector.

***Fruit and vegetable sellers** offering their wares on the Mozambican border. The south of the country is intensively cultivated.*

POLITICS

Multiparty elections

1999/2004

President Bakili Muluzi

AT THE LAST ELECTION

National Assembly 193 seats

48% UDF · 34% MCP · 15% AFORD · 2% Ind · 1% Others

UDF = United Democratic Front **MCP** = Malawi Congress Party **AFORD** = Alliance for Democracy **Ind** = Independents

For 30 years from independence in 1964 Malawi was ruled by the autocratic Hastings Banda. His single-party regime outlawed dissent; torture and imprisonment without trial were common. In 1992 international aid was suspended because of the regime's human rights record. A referendum forced Banda to introduce multiparty politics in 1994, when the mainly southern-based UDF scored a dramatic victory, and its leader Bakili Muluzi won the presidency. He shrewdly recruited several prominent MCP politicians to his team, and vowed to restore personal and religious freedom, liberalize and revive the shattered economy, and improve Malawi's regional standing. Muluzi was narrowly reelected in 1999, but opposition leader Gwanda Chakuamba contested the results, and violence against Muslims and UDF supporters erupted in the north. A new opposition National Democratic Alliance emerged in 2001. Its leader Brown Mpinganjira was arrested for treason later that year.

WORLD AFFAIRS

Joined UN in 1964

Comm

COMESA

NAM

AU

SADC

Malawi aims to protect its restored status as a recipient of Western aid. In 1998 the UK pledged to support President Muluzi's fiscal policies and antipoverty drive. Malawi also wants to preserve ties with South Africa, unbroken since 1967, a unique record in black Africa. One in ten Mozambicans fled to Malawi as refugees in the 1980s.

AID

Recipient

$445m (receipts)

Little change in 2000

Nonhumanitarian aid resumed with the advent of democracy; international donors have pledged a total of $1.2 billion. The World Food Program launched an emergency aid program in April 2001 to help flood victims.

M

DEFENSE

No compulsory military service

$26m

Down 4% in 2000

Muluzi's government has the backing of the 5300-strong army. In the last days of Banda's rule, the military lost confidence in the ruling MCP, forcing the pace of democratization. In 1993, it disarmed the Young Pioneers, a militarized section of the MCP.

ECONOMICS

Inflation 34% p.a. (1990–2000)

$1.74bn

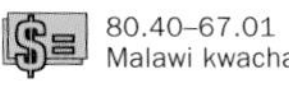
80.40–67.01 Malawi kwacha

SCORE CARD

- ❑ WORLD GNP RANKING........................140th
- ❑ GNP PER CAPITA..............................$170
- ❑ BALANCE OF PAYMENTS...................–$523m
- ❑ INFLATION..29.5%
- ❑ UNEMPLOYMENT....................................1%

STRENGTHS

Tobacco, earning 76% of foreign exchange. Tea and sugar production. Unexploited bauxite, asbestos, and coal reserves. Much tourism potential.

WEAKNESSES

Agriculture vulnerable to drought and price fluctuations. Only 14% of GDP derived from industry. Small domestic market, few skilled workers. Strain of housing Mozambican refugees.

EXPORTS

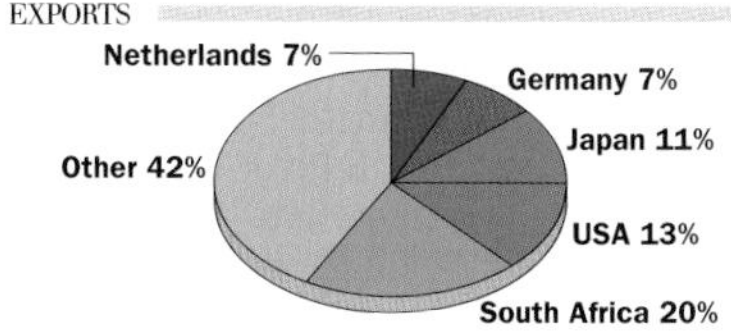
Netherlands 7%
Germany 7%
Japan 11%
Other 42%
USA 13%
South Africa 20%

IMPORTS

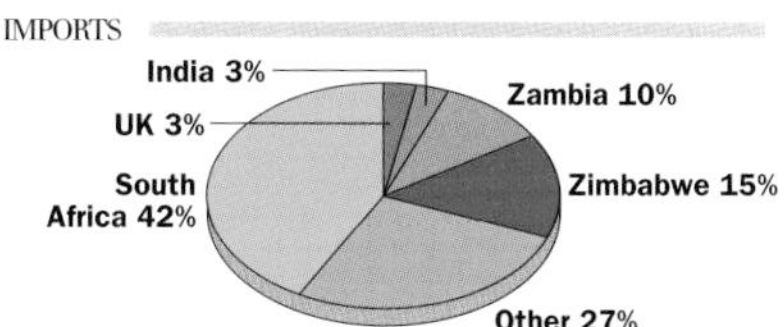
India 3%
UK 3%
Zambia 10%
South Africa 42%
Zimbabwe 15%
Other 27%

RESOURCES

Electric power 185,000 kw

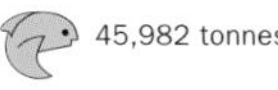
45,982 tonnes

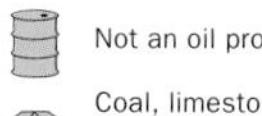
Not an oil producer

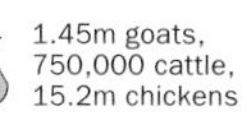
1.45m goats, 750,000 cattle, 15.2m chickens

Coal, limestone, gemstones, bauxite, graphite, uranium

One 215 MW hydropower plant on Shire River accounts for nearly 85% of generating capacity, but only 3% of total energy use. Most rely on fuelwood for their energy needs. Malawi now encourages privatization, crop diversification, improved irrigation, and regional economic integration via the SADC to exploit its naturally limited resources. A deep-seam coal mine currently operates at Rumphi.

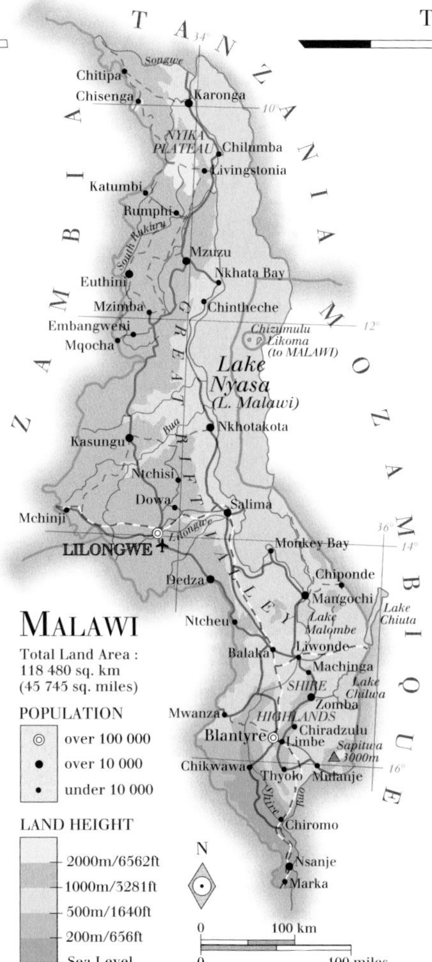
TANZANIA
ZAMBIA
MOZAMBIQUE
Chitipa
Chisenga
Karonga
NYIKA PLATEAU
Chilumba
Livingstonia
Katumbi
Rumphi
Mzuzu
Nkhata Bay
Euthini
Mzimba
Chintheche
Embangweni
Mqocha
Chizumulu
Likoma (to MALAWI)
Lake Nyasa (L. Malawi)
Nkhotakota
Kasungu
GREAT RIFT VALLEY
Nchisi
Dowa
Salima
Mchinji
LILONGWE
Monkey Bay
Dedza
Chiponde
Mangochi
Lake Malombe
Lake Chiuta
Ntcheu
Balaka
Liwonde
Machinga
SHIRE HIGHLANDS
Lake Chilwa
Zomba
Mwanza
Blantyre
Chiradzulu
Limbe
Sapitwa 3000m
Chikwawa
Thyolo
Mulanje
Chiromo
Nsanje
Marka
MALAWI
Total Land Area : 118 480 sq. km (45 745 sq. miles)
POPULATION
over 100 000
over 10 000
under 10 000
LAND HEIGHT
2000m/6562ft
1000m/3281ft
500m/1640ft
200m/656ft
Sea Level
N
0 100 km
0 100 miles

ENVIRONMENT

Sustainability rank: 82nd

11%

0.1 tonnes per capita

Drought, with its devastating effects on agriculture, eclipses all other problems. Ecological husbandry attracts tourism.

MEDIA

TV ownership low

Daily newspaper circulation 3 per 1000 people

PUBLISHING AND BROADCAST MEDIA

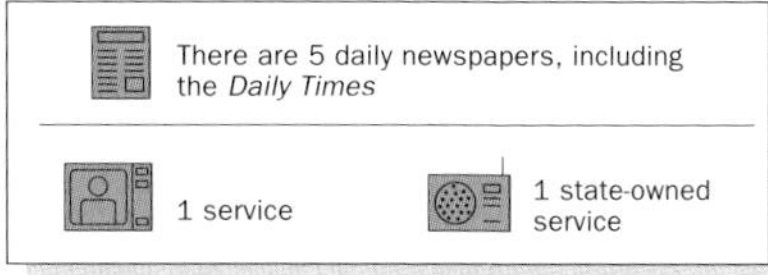
There are 5 daily newspapers, including the *Daily Times*
1 service
1 state-owned service

Violence against the staff of privately owned newspapers causes concern among human rights groups.

CRIME

Death penalty in use

7024 prisoners

Crime is rising

Urban crime is on the increase. The proliferation of weapons, particularly guns, is contributing to a rise in cases of armed robbery.

CHRONOLOGY

After strong Scottish missionary activity, Malawi came under British rule as Nyasaland in 1891.

- ❑ **1964** Independence under Hastings Banda.
- ❑ **1966** One-party state.
- ❑ **1992** Antigovernment riots. Illegal prodemocracy groups unite.
- ❑ **1993** Referendum for multipartyism.
- ❑ **1994** Muluzi's UDF wins elections.
- ❑ **2002** Worst-ever cholera epidemic exacerbated by food shortages.

EDUCATION

School leaving age: 13

60%
3179 students

Primary-level education is widespread, with 73% of boys and 60% of girls attending school regularly.

HEALTH

Welfare state health benefits

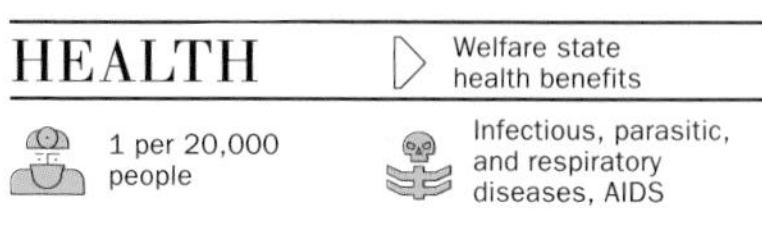
1 per 20,000 people
Infectious, parasitic, and respiratory diseases, AIDS

Life expectancy is around only 40 years, owing to the high incidence of HIV/AIDS.

SPENDING

GDP/cap. increase

CONSUMPTION AND SPENDING

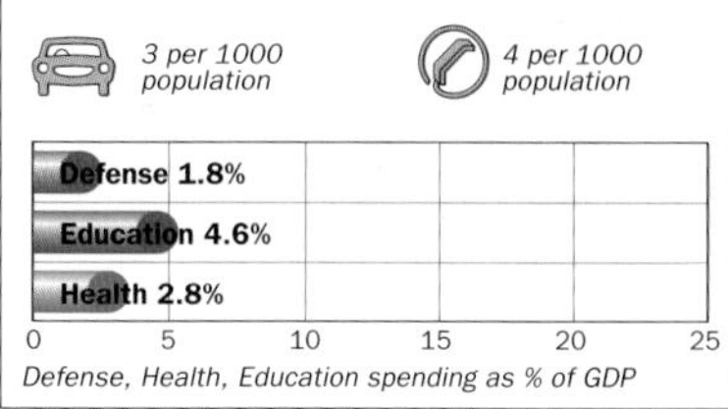
3 per 1000 population
4 per 1000 population
Defense 1.8%
Education 4.6%
Health 2.8%
0 5 10 15 20 25
Defense, Health, Education spending as % of GDP

The ousted MCP elite grew wealthy, allegedly through embezzlement. However, 80% of Malawians remain mired in poverty, and are forced to survive on less than $1 a day.

WORLD RANKING

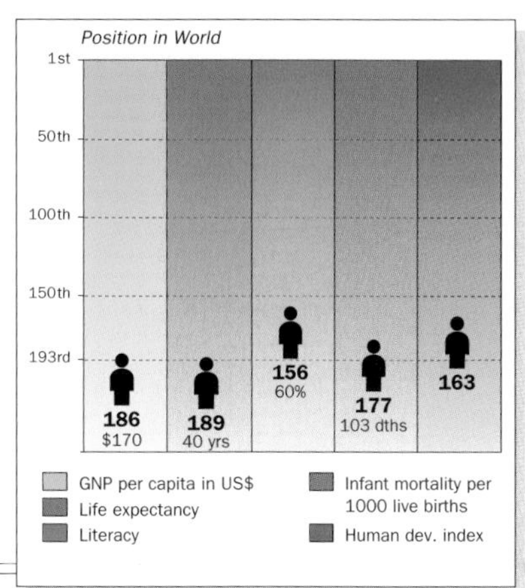
Position in World
1st
50th
100th
150th
193rd
186 $170
189 40 yrs
156 60%
177 103 dths
163
GNP per capita in US$
Life expectancy
Literacy
Infant mortality per 1000 live births
Human dev. index

M

MALAYSIA

OFFICIAL NAME: Federation of Malaysia **POPULATION:** 22.6 million **CURRENCY:** Ringgit
CAPITALS: Kuala Lumpur; Putrajaya (administrative) **OFFICIAL LANGUAGE:** Bahasa Malaysia

1963

1965

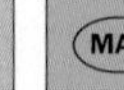
Aug 31

MAL

+8

+60

.my

MALAYSIA COMPRISES the three territories of Peninsular Malaysia, Sarawak, and Sabah, stretching over 2000 km (1240 miles) from the edge of the Indian Ocean to the northeastern end of the island of Borneo. A central mountain chain separates the coastal lowlands of Peninsular Malaysia; Sarawak and Sabah have swampy coastal plains rising to mountains on the border with Indonesia. Putrajaya, just south of Kuala Lumpur, is a high-tech new development intended as the home of government. The United Malays National Organization has dominated politics since independence.

CLIMATE

Tropical equatorial

WEATHER CHART FOR KUALA LUMPUR

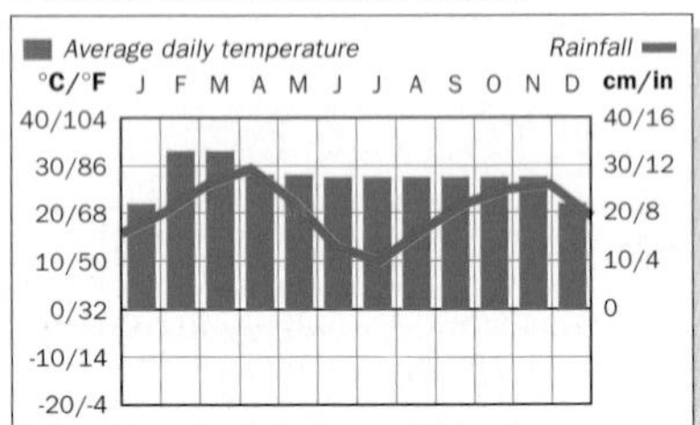
Average daily temperature
Rainfall
°C/°F J F M A M J J A S O N D cm/in
40/104
30/86
20/68
10/50
0/32
-10/14
-20/-4
40/16
30/12
20/8
10/4
0

The whole of Malaysia has an equatorial climate. While there are two distinct rainy seasons, from March to May and from September to November, there is rain throughout the year: almost everywhere has rain on between 150 and 200 days a year. Coastal areas are also subject to monsoon winds, which alternate in direction between the southwest and northeast.

***Tea plantations** and colonial-style houses and gardens make the Cameron Highlands, in Peninsular Malaysia, one of Asia's most popular mountain resorts.*

TRANSPORTATION

Drive on left

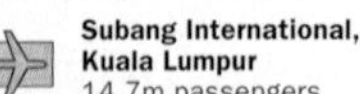

Subang International, Kuala Lumpur
14.7m passengers

828 ships
5.21m grt

THE TRANSPORTATION NETWORK

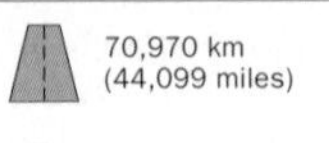

70,970 km (44,099 miles)

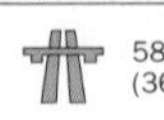

580 km (360 miles)

1622 km (1008 miles)

7296 km (4534 miles)

Transportation in Peninsular Malaysia is well developed. A major north–south highway connects the urban centers of the west coast. Roads in Sabah are also good, with an efficient bus network linking the towns. Travel in Sarawak, on the other hand, is hindered by poorly maintained roads and a lack of public transportation. East Malaysia is most effectively traversed by air.

TOURISM

Visitors : Population 1:2.2

10.2m visitors

Up 29% in 2000

MAIN TOURIST ARRIVALS

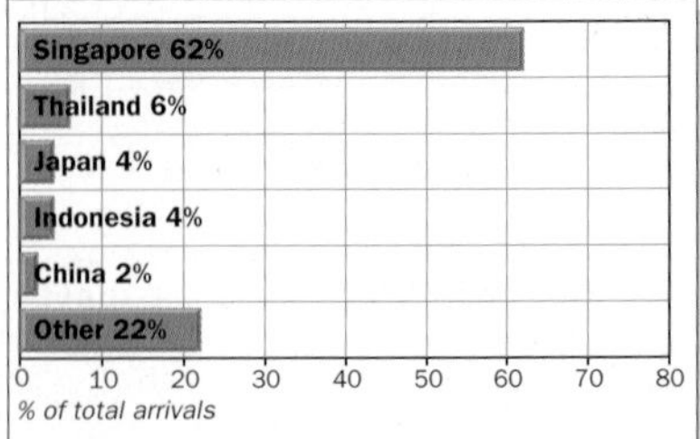
Singapore 62%
Thailand 6%
Japan 4%
Indonesia 4%
China 2%
Other 22%
0 10 20 30 40 50 60 70 80
% of total arrivals

Malaysia is southeast Asia's major tourist destination. Most tourists come for the excellent tropical beaches on the peninsula's east coast, to hike in the Cameron Highlands, or to trek in the world's oldest rainforests in Borneo. There has recently been an increase in the international business convention trade, and hotel capacity has been growing at 10% a year.

By 1990, when the government ran the Visit Malaysia Year campaign, tourism had become Malaysia's third-biggest foreign exchange earner. Two other such campaigns were launched in 1994 and 1998. However, the resurgence since 1999 of pro-Islamic parties, which favor stricter dress codes for women and a ban on alcohol, has deterred some Western tourists. In 2000 Malaysia backed an integrated tourism package with Thailand, Indonesia, and Singapore to enable tourists to visit the four countries under a common program.

SOUTHEAST ASIA
Asia
Thailand
Brunei
MALAYSIA
Indonesia

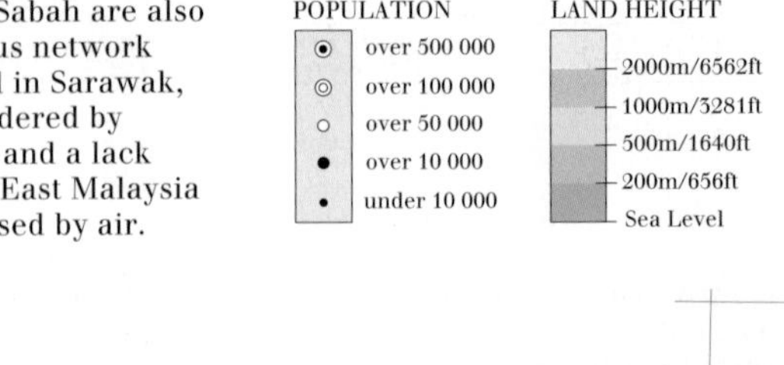
MALAYSIA
Total Land Area : 329 750 sq. km (127 316 sq. miles)
POPULATION
over 500 000
over 100 000
over 50 000
over 10 000
under 10 000
LAND HEIGHT
2000m/6562ft
1000m/3281ft
500m/1640ft
200m/656ft
Sea Level

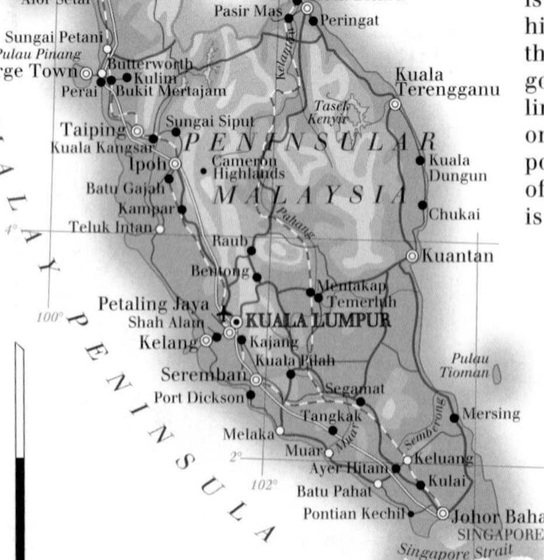
THAILAND
Kangar
Pulau Langkawi
Alor Setar
Sungai Petani
Pulau Pinang
George Town
Butterworth
Kulim
Perai
Bukit Mertajam
Tumpat
Kota Bharu
Pasir Mas
Peringat
Kuala Terengganu
Tasek Kenyir
Taiping
Sungai Siput
Kuala Kangsar
Ipoh
Cameron Highlands
PENINSULAR MALAYSIA
Kuala Dungun
Batu Gajah
Kampar
Teluk Intan
Chukai
Raub
Kuantan
Bentong
Mentakap
Temerloh
Petaling Jaya
Shah Alam
KUALA LUMPUR
Kelang
Kajang
Kuala Pilah
Seremban
Port Dickson
Segamat
Pulau Tioman
Mersing
Tangkak
Melaka
Muar
Ayer Hitam
Keluang
Kulai
Batu Pahat
Pontian Kechil
Johor Baharu
SINGAPORE
Singapore Strait
MALAY PENINSULA
Kepulauan Anambas (INDONESIA)
Kepulauan Natuna (INDONESIA)
SOUTH CHINA SEA
Kepulauan Tambelan (INDONESIA)
Selat Serasan
Teluk Datu
Kuching
Bandar Sri Aman
Pulau Beruit
Sibu
Bintulu
SARAWAK
EAST
Rajang
INDONESIA

PEOPLE

Pop. density medium

Bahasa Malaysia, Malay, Chinese, Tamil, English

69/km² (178/mi²)

THE URBAN/RURAL POPULATION SPLIT

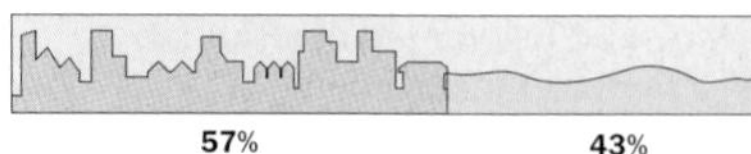
57%
43%

RELIGIOUS PERSUASION

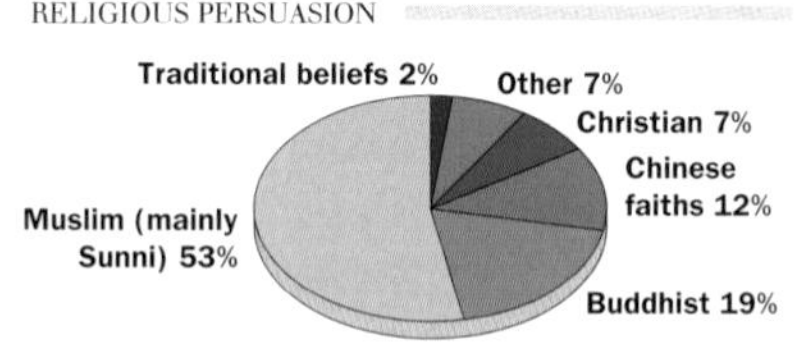
Traditional beliefs 2%
Other 7%
Christian 7%
Chinese faiths 12%
Muslim (mainly Sunni) 53%
Buddhist 19%

ETHNIC MAKEUP

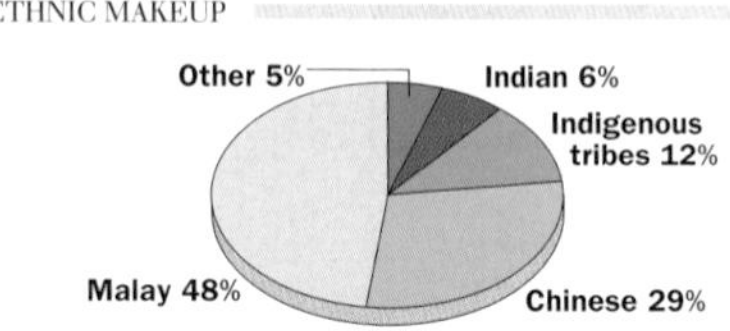
Other 5%
Indian 6%
Indigenous tribes 12%
Malay 48%
Chinese 29%

POPULATION AGE BREAKDOWN

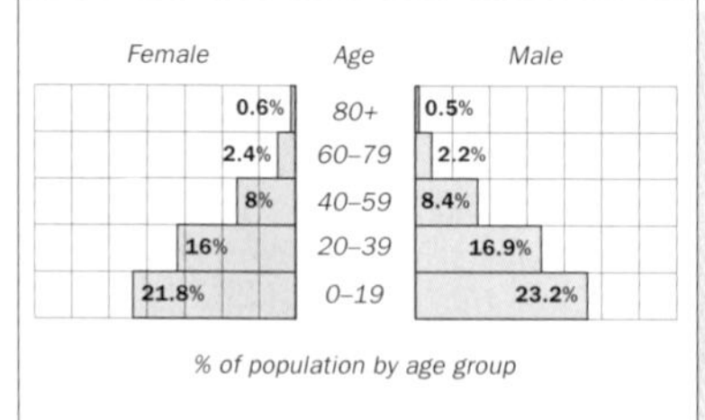
Female
Age
Male
0.6%
80+
0.5%
2.4%
60–79
2.2%
8%
40–59
8.4%
16%
20–39
16.9%
21.8%
0–19
23.2%
% of population by age group

The key distinction in Malaysian society is between the indigenous Malays, termed the Bumiputras ("sons of the soil"), and the Chinese. The Malays form the largest group, accounting for just under half of the population. However, the Chinese have traditionally controlled most business activity. The New Economic Policy (NEP), introduced in the 1970s, was designed to address this imbalance by offering positive opportunities to the Malays through the education system and by making jobs available to them in both the state and private sectors. There are estimated to be more than one million Indonesian and Filipino immigrants in Malaysia, a dearth of employment in their own countries giving additional attraction to Malaysia's need for labor. In addition, nearly 255,000 Vietnamese refugees were offered temporary refuge in Malaysia between 1975 and 1997; most have now been resettled in third countries, but around 6000 remain. Gender discrimination was only outlawed in 2001. Muslim women are encouraged to wear a veil.

POLITICS

Multiparty elections

L. House 1999/2004
U. House varying

Raja Tuanku Syed Sirajuddin ibni al-Marhum Syed Putra Jamalullail

AT THE LAST ELECTION

House of Representatives 193 seats

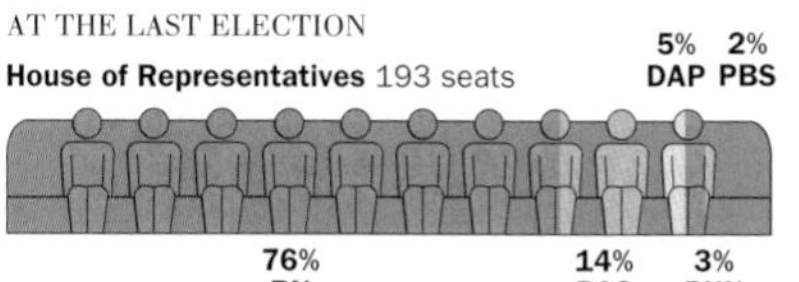
5% DAP
2% PBS
76% BN
14% PAS
3% PKN

BN = National Front (dominated by the United Malays National Organization – **UMNO**) **PAS** = Pan-Malaysian Islamic Party **DAP** = Democratic Action Party **PKN** = National Justice Party (Keadilan) **PBS** = United Sabah Party

The DAP, the PAS, and the PKN form the Alternative Front

Senate 69 seats

The Senate comprises 26 members indirectly elected by the State Legislative Assemblies, and 43 appointed by the head of state

Balabac Strait
Pulau Banggi
Kudat
SULU SEA
Gunung Kinabalu 4094m
Telukan Labuk
Kota Kinabalu
Labuk
Sandakan
Pulau Labuan
Brunei Bay
CROCKER
Kinabatangan
Sibutu Passage
SABAH
Banjaran Brassey
Telukan Lahad Datu
Tawau
116°
118°
4°
CELEBES SEA
N
0 100 km
0 100 miles

Supreme power rests in theory with the monarch, acting on the advice of parliament. In practice, the prime minister wields executive authority. Opposition parties, while legal, are under tight control.

Profile

Malaysia has been dominated by UMNO, part of the ruling BN coalition, since Malay independence in 1957. It controls a huge network of patronage. In 2002, Prime Minister Mahathir Mohamed shocked the country by tearfully announcing that he would retire in 2003, after 22 years at the helm. He had appeared unassailable, although his authority had been shaken by the economic crisis of 1997–1998 and dissent within the ruling coalition. In 1998, Anwar Ibrahim, deputy prime minister and once Mahathir's chosen successor, was dismissed after challenging the government's economic policy and calling for political reform. He was convicted in 1999 on corruption charges; his prison sentence was lengthened to 15 years in 2000 after his conviction for sodomy. The initial verdict sparked riots and gained support for the new opposition PKN headed by Anwar's wife, Wan Azizah. PKN activists have been hounded by the government. In the 1999 elections, the BN coalition retained its large majority but Mahathir's own UMNO lost ground.

***Anwar Ibrahim,** in 1998 controversially dismissed by Mahathir.*

***Mahathir Mohamed,** prime minister since 1981.*

Main Political Issue

Malay dominance of government

Mahathir's administration has declared that it no longer wishes to discriminate positively in favor of Malays, but the Chinese community accuses the government of corruption and uncompetitive practices, declaring that Malays are still favored for the placing of government contracts. The Chinese are further alienated by the more restrictive nature of Islamic society.

WORLD AFFAIRS

Joined UN in 1957

APEC
ASEAN
Comm
G15
OIC

Mahathir sees himself as one of the developing world's leading voices. He maintains a strongly anti-US line in his public speeches and has chastised the West for singling out Islamic countries in its campaign against international terrorism. Mahathir's pro-Malay policies have caused tensions with Singapore, exacerbated by the latter's dependence on Malaysia for water.

AID

Recipient

$45m (receipts)

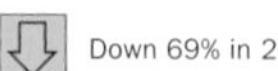
Down 69% in 2000

Most Western aid to Malaysia was used until recently to finance large infrastructure projects. The economic crisis which affected southeast Asia in 1997–1998 forced Malaysia to seek foreign assistance to support an economic recovery program.

M

CHRONOLOGY

The former British protectorate of Malaya gained independence in 1957. The federation of Malaysia, incorporating Singapore, Sarawak, and Sabah, was founded in 1963.

- ❑ **1965** Singapore leaves federation, reducing Malaysian states to 13.
- ❑ **1970** Malay–Chinese ethnic tension forces resignation of Prime Minister Tunku Abdul Rahman. New prime minister, Tun Abdul Razak, creates the BN coalition.
- ❑ **1976-1978** Guerrilla attacks by banned Communist Party of Malaya (CPM), based in southern Thailand.
- ❑ **1976** Death of Tun Abdul Razak.
- ❑ **1977** Unrest in Kelantan following expulsion of its chief minister from Pan-Malaysian Islamic Party (PAS). National emergency declared. PAS expelled from BN.
- ❑ **1978** Elections consolidate BN power. PAS marginalized. Government rejects plans for Chinese university.
- ❑ **1978-1989** Unrestricted asylum given to Vietnamese refugees.
- ❑ **1981** Mahathir Mohamed becomes prime minister.
- ❑ **1982** General election returns BN with increased majority.
- ❑ **1985** BN defeated by PBS in Sabah state elections.
- ❑ **1986** PBS joins BN coalition. Dispute between Mahathir and his deputy, Dakuk Musa, triggers general election, won by BN.
- ❑ **1987** Detention without trial of 106 politicians from all parties suspected of Chinese sympathies. Media censored.
- ❑ **1989** Disaffected UMNO members join PAS. Screening of Vietnamese refugees introduced. CPM signs peace agreement with Malaysian and Thai governments.
- ❑ **1990** General election. BN returned to power with reduced majority.
- ❑ **1993** Sultans lose powers, including legal immunity.
- ❑ **1995** BN wins landslide victory in the country's ninth general election.
- ❑ **1997** Major financial crisis ends decade of spectacular economic growth.
- ❑ **1998–1999** Deputy Prime Minister Anwar Ibrahim dismissed from office. Launches Reformasi (Reform) movement. Found guilty of corruption, later convicted of sodomy and six-year sentence extended to 15 years; his wife, Wan Azizah, forms National Justice Party (PKN). November, UMNO loses ground in general election.
- ❑ **2002** Mahathir announces that he will step down in 2003.

M

DEFENSE

No compulsory military service

$2.71bn

Down 14% in 2000

MALAYSIAN ARMED FORCES

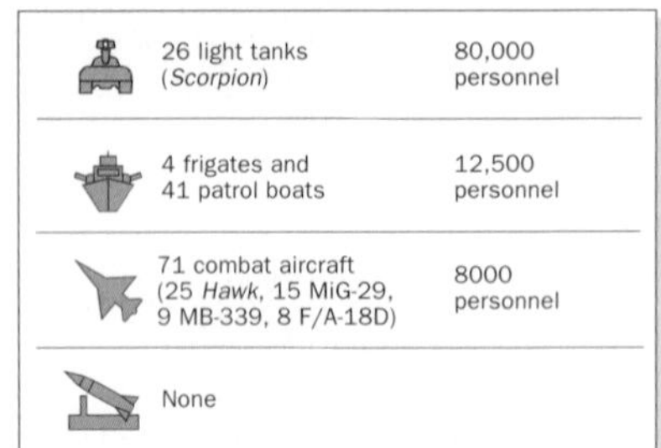
26 light tanks (*Scorpion*)
80,000 personnel
4 frigates and 41 patrol boats
12,500 personnel
71 combat aircraft (25 *Hawk*, 15 MiG-29, 9 MB-339, 8 F/A-18D)
8000 personnel
None

Malaysia's armed forces are predominantly composed of Malays. Main defense concerns are Singapore, with its large and highly mechanized army, and more recently, though to a lesser extent, Indonesia. Also important to Malaysia is the growing Chinese influence in the South China Sea. Patrolling east and west Malaysia is a key function of the navy, which is large by regional standards.

Malaysia is an important market for Western arms suppliers. However, since 1994 Malaysia has bought a number of Russian MiG-29 fighter aircraft.

ECONOMICS

Inflation 3.9% p.a. (1990–2000)

$78.7bn

3.8 ringgits

SCORE CARD

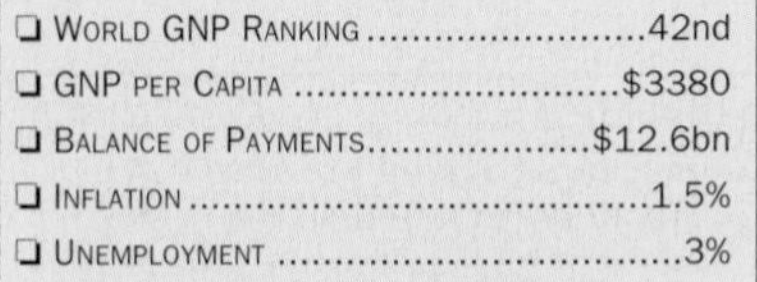

❑ World GNP Ranking	42nd
❑ GNP per Capita	$3380
❑ Balance of Payments	$12.6bn
❑ Inflation	1.5%
❑ Unemployment	3%

EXPORTS

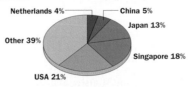

Netherlands 4%
China 5%
Japan 13%
Other 39%
Singapore 18%
USA 21%

IMPORTS

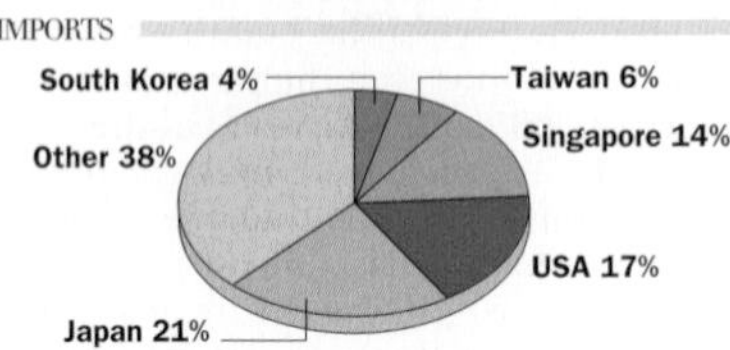
South Korea 4%
Taiwan 6%
Other 38%
Singapore 14%
USA 17%
Japan 21%

ECONOMIC PERFORMANCE INDICATOR

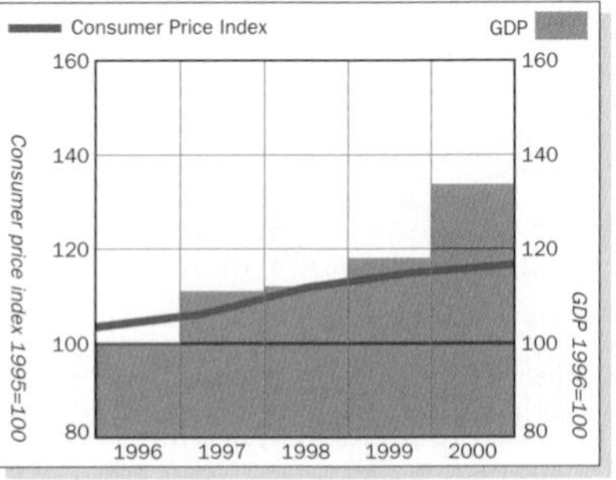
Consumer Price Index
GDP
160
140
120
100
80
Consumer price index 1995=100
GDP 1996=100
1996
1997
1998
1999
2000

STRENGTHS

Electronics, computer hardware, and electrical appliances. Tourism. Heavy industries such as steel. Palm oil. Latex, rubber, chemical products. Success of "national car," the Proton. Return to growth from 2000.

WEAKNESSES

High level of debt. Shortage of skilled labor. High interest rates deter private investors. High government budget spending. Competition from NICs.

PROFILE

From 1987, for almost a decade, Malaysia expanded faster than any other southeast Asian nation, at an average yearly rate of 8%, with much of the growth state-directed. However, plans for full industrialization, named "Vision 2020," were revised after the 1997 financial crisis. The construction of Putrajaya provided a renewed stimulus to growth at the end of the 1990s. A project for a Multimedia Super Corridor (MSC), located south of Kuala Lumpur and aimed at attracting world-class companies, is expected to be completed by 2003.

MALAYSIA : MAJOR BUSINESSES

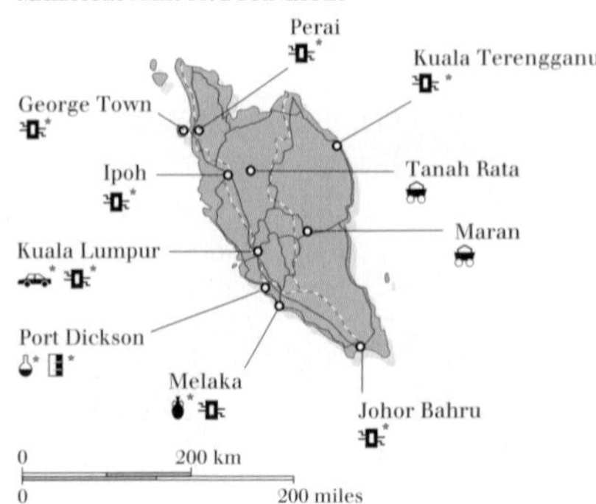
Perai
Kuala Terengganu
George Town
Ipoh
Tanah Rata
Maran
Kuala Lumpur
Port Dickson
Melaka
Johor Bahru
0
200 km
0
200 miles

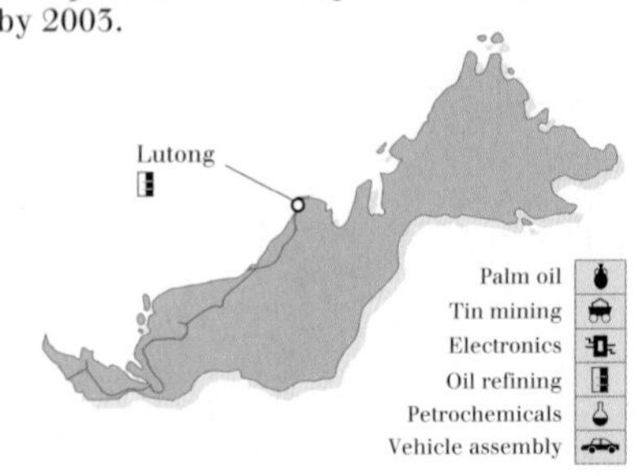
Lutong
Palm oil
Tin mining
Electronics
Oil refining
Petrochemicals
Vehicle assembly

* significant multinational ownership

RESOURCES

Electric power 13.6m kw

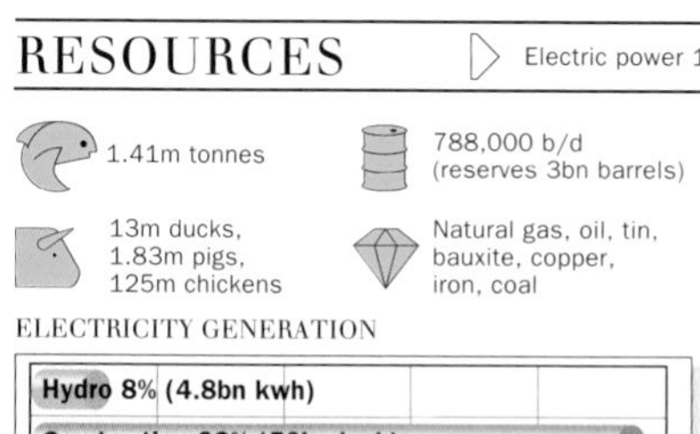
1.41m tonnes
788,000 b/d (reserves 3bn barrels)
13m ducks, 1.83m pigs, 125m chickens
Natural gas, oil, tin, bauxite, copper, iron, coal

ELECTRICITY GENERATION

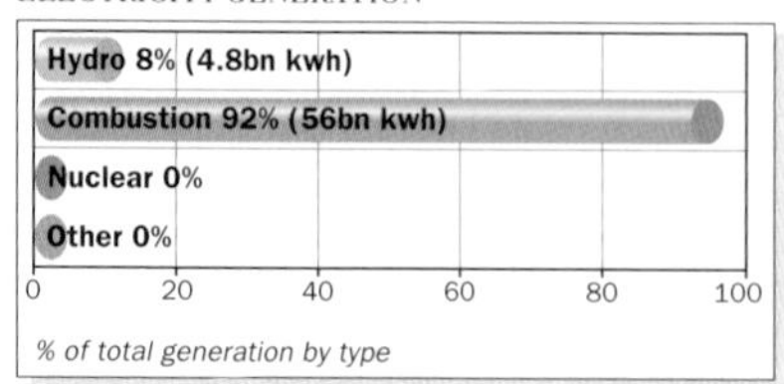
Hydro 8% (4.8bn kwh)
Combustion 92% (56bn kwh)
Nuclear 0%
Other 0%
0
20
40
60
80
100
% of total generation by type

Palm oil, of which Malaysia is the world's largest producer, is now the major export product, whereas rubber production is now overshadowed by that of neighboring Thailand. Malaysia is a significant exporter of petroleum and natural gas. Reserves lie offshore from Sabah and Sarawak. The petroleum is high grade, so that most is exported, while crude imports are refined.

PENINSULAR MALAYSIA

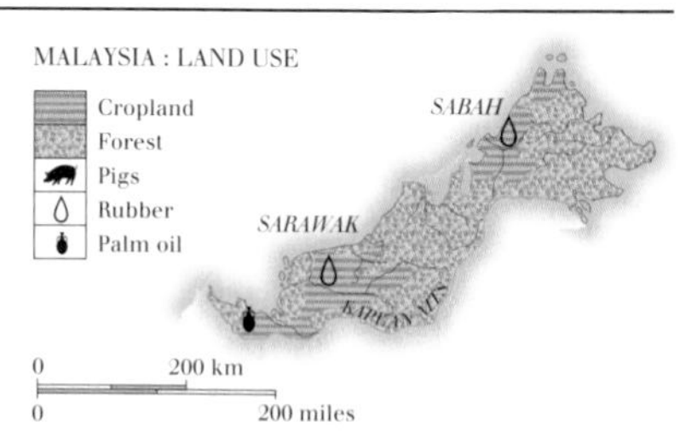
MALAYSIA : LAND USE
Cropland
Forest
Pigs
Rubber
Palm oil
SABAH
SARAWAK
0
200 km
0
200 miles

Malaysia accounts for nearly half of world timber exports, most of which come from Sarawak.

ENVIRONMENT

Sustainability rank: 68th

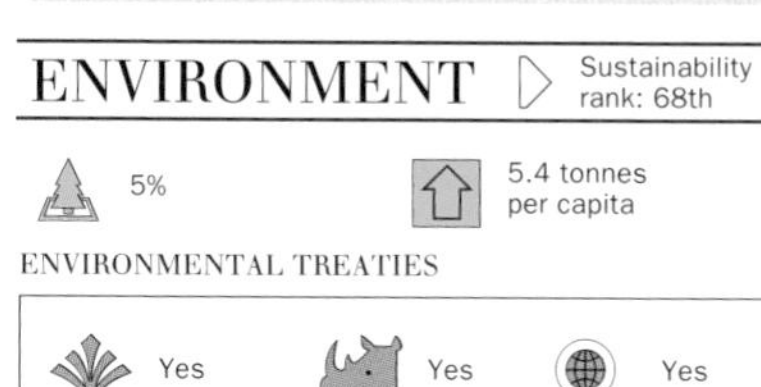
5%
5.4 tonnes per capita

ENVIRONMENTAL TREATIES

Yes	Yes	Yes
Yes	Yes	No

Logging is the overwhelming concern. World Bank estimates suggest that trees are being cut down at four times the sustainable rate. Indigenous forest communities are being destroyed, and some species of wood are near extinction. The government pledged in 1997 to plant 20 million trees across the country by 2000.

Smog caused by burning forests and scrub in Indonesia creates a pollution and health threat to which the whole region is alerted.

Traditional lifestyles are threatened by grandiose modernization schemes, and the Bakun Dam project, shelved in 1997 due to a lack of investment confidence, was restarted in 2000.

MEDIA

TV ownership medium

Daily newspaper circulation 112 per 1000 people

PUBLISHING AND BROADCAST MEDIA

There are 42 daily newspapers. The most influential are the *New Straits Times*, *Utusan Malaysia*, and *Sin Chew Jit Poh*

7 services: 3 state-controlled, 4 independent

5 services: 3 state-controlled, 2 independent

Almost all newspapers, TV stations, and radio broadcasts are strictly controlled by the state and UMNO, which claim a need to protect the country from un-Islamic influences. A series of laws passed in the 1980s regulate content. The Internet provides a medium for free reporting. The leading website is Malaysiakini.

CRIME

Death penalty in use

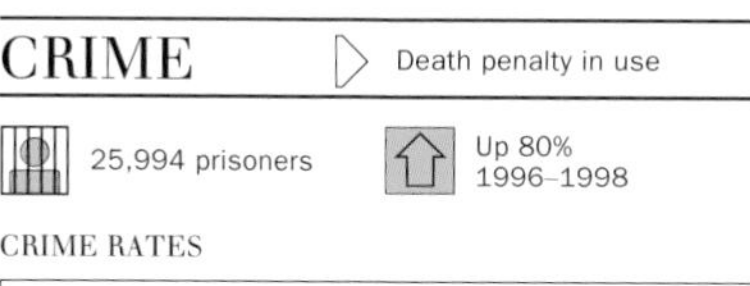
25,994 prisoners
Up 80% 1996–1998

CRIME RATES

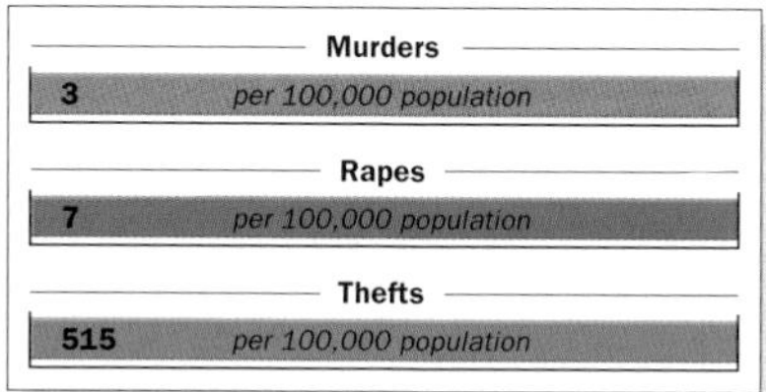
Murders
3 per 100,000 population
Rapes
7 per 100,000 population
Thefts
515 per 100,000 population

The death sentence for possession of narcotics is mandatory. The Internal Security Act allows detention without trial. Some Islamic-dominated states have attempted to implement the harsh punishments of the Islamic penal code, including stoning and amputation.

EDUCATION

School leaving age: 15

88%

230,000 students

THE EDUCATION SYSTEM

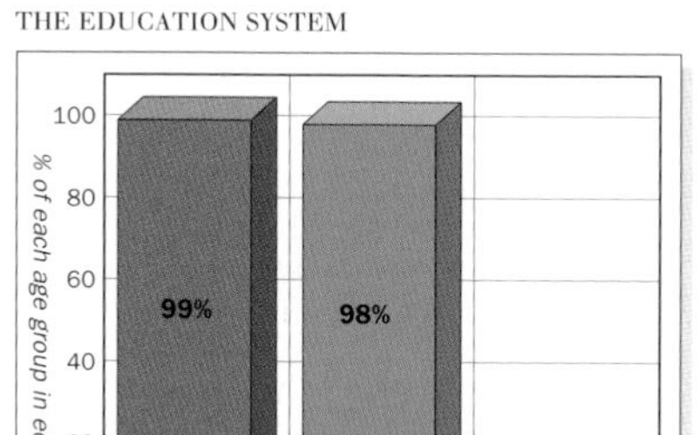
100
80
60
40
20
0
% of each age group in education
99%
98%
12%
Primary
Secondary
Tertiary

Racial integration in multiethnic schools is encouraged, but teaching is to remain in separate languages.

At tertiary level a quota system gives Malays preference for places. The Chinese community has its own schools, but plans for a private Chinese university were vetoed by the government. Many students, particularly the Chinese, complete their studies in the UK or the US. From 2002, university students and staff have had to swear allegiance to the state.

HEALTH

Welfare state health benefits

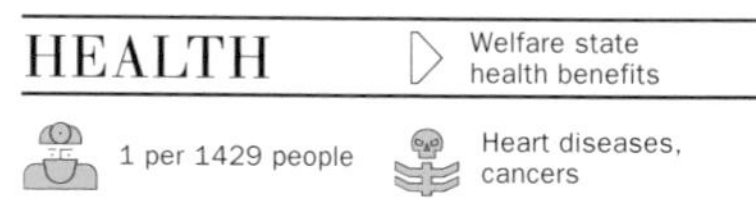
1 per 1429 people
Heart diseases, cancers

There is growing disparity between the modern facilities available in cities and the traditional medicine practiced in rural and outlying areas. Traditional practices such as acupuncture and herbal medicine continue to be used by the Chinese community.

SPENDING

GDP/cap. increase

CONSUMPTION AND SPENDING

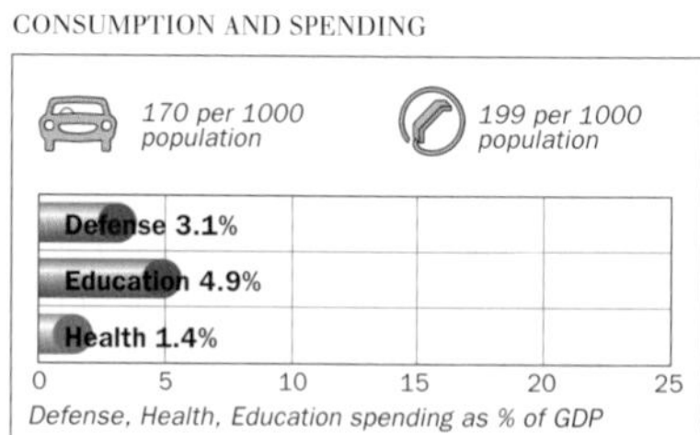
170 per 1000 population
199 per 1000 population
Defense 3.1%
Education 4.9%
Health 1.4%
0
5
10
15
20
25
Defense, Health, Education spending as % of GDP

The Chinese remain the wealthiest community in Malaysia. However, following riots in 1970, the UMNO government embarked on a deliberate program of achieving 30% Malay ownership of the corporate sector. The "extremely rich" were barred from government service in 2001 in an effort to stamp out "money politics."

WORLD RANKING

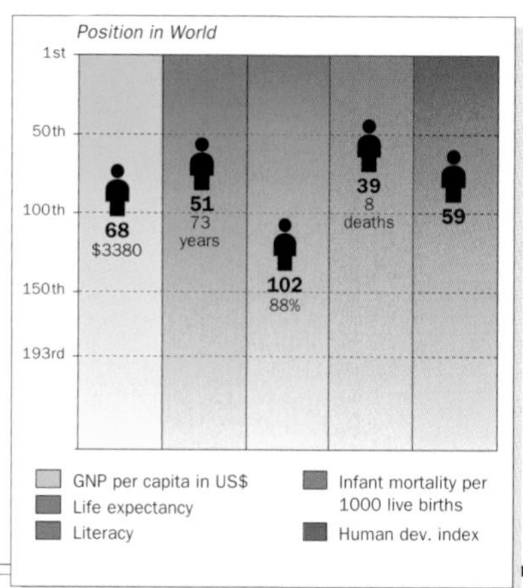
Position in World
1st
50th
100th
150th
193rd
68
$3380
51
73 years
102
88%
39
8 deaths
59
GNP per capita in US$
Life expectancy
Literacy
Infant mortality per 1000 live births
Human dev. index

M

Maldives

Official Name: Republic of Maldives **Capital:** Male'
Population: 300,000 **Currency:** Rufiyaa **Official Language:** Dhivehi

INDIAN OCEAN
India
Asia
MALDIVES
Sri Lanka

1965

1965

July 26

MV

+5

+960

.mv

An Islamic sultanate until 1968, the Maldives is an archipelago of 1190 small coral islands or atolls (a word derived from the local Dhivehi language), set in the Indian Ocean southwest of India. The islands, none of which rise above 1.8 m (6 ft), are protected by encircling reefs or faros. Only 200 are inhabited. Tourism has grown in recent years, though vacation islands are separate from settled islands.

Traditional Maldivian trading yacht. *The 1190 coral islands are grouped in natural atolls, derived from the Dhivehi word "atolu."*

CLIMATE

Tropical oceanic

WEATHER CHART FOR MALE'

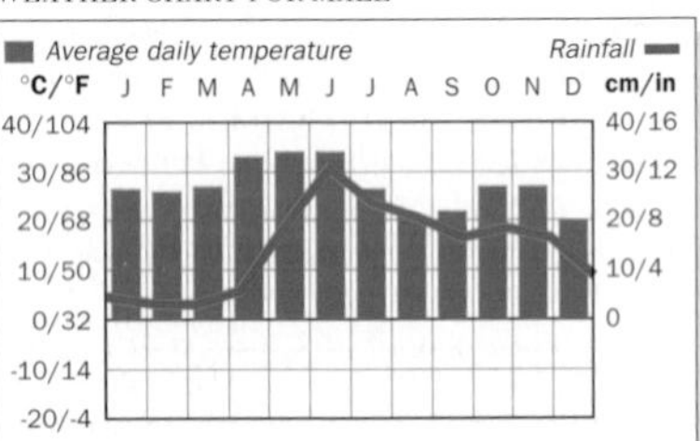
Average daily temperature
Rainfall
°C/°F
J F M A M J J A S O N D
cm/in
40/104
30/86
20/68
10/50
0/32
-10/14
-20/-4
40/16
30/12
20/8
10/4
0

The Maldives has a tropical climate, with abundant rainfall and high temperatures throughout the year. The northern islands are occasionally affected by violent storms caused by tropical cyclones. Most rain falls in the southern islands.

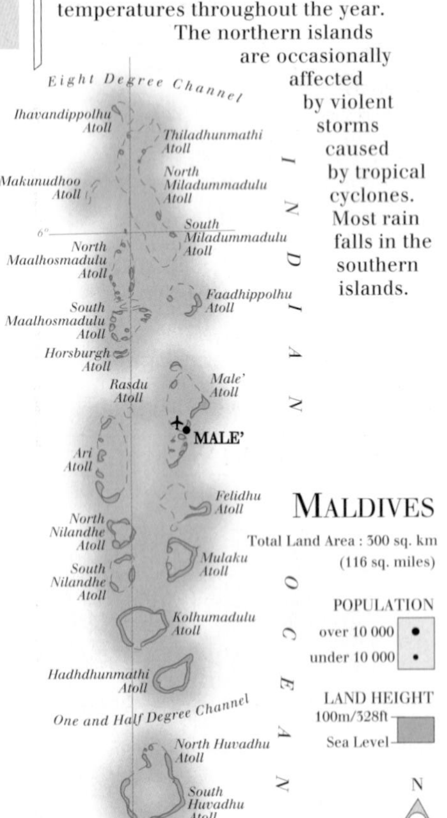
Eight Degree Channel
Ihavandippolhu Atoll
Thiladhunmathi Atoll
Makunudhoo Atoll
North Miladummadulu Atoll
South Miladummadulu Atoll
North Maalhosmadulu Atoll
South Maalhosmadulu Atoll
Faadhippolhu Atoll
Horsburgh Atoll
Rasdu Atoll
Male' Atoll
MALE'
Ari Atoll
Felidhu Atoll
North Nilandhe Atoll
South Nilandhe Atoll
Mulaku Atoll
Kolhumadulu Atoll
Hadhdhunmathi Atoll
One and Half Degree Channel
North Huvadhu Atoll
South Huvadhu Atoll
Equatorial Channel
Equator
Fuammulah
Addu Atoll
Gan
INDIAN OCEAN
MALDIVES
Total Land Area : 300 sq. km (116 sq. miles)
POPULATION
over 10 000
under 10 000
LAND HEIGHT
100m/328ft
Sea Level
N
0 100 km
0 100 miles

TRANSPORTATION

Drive on left

Male' International, Hulule Island
1.71m passengers

68 ships
101,100 grt

THE TRANSPORTATION NETWORK

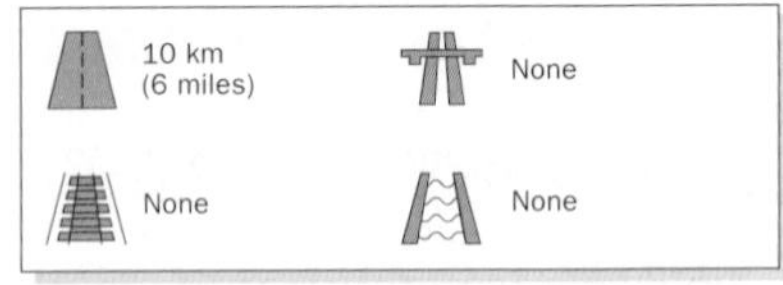
10 km
(6 miles)
None
None
None

It is possible to walk across Male' island in 20 minutes. Interisland travel is mostly by ferry and traditional *dhoni*.

TOURISM

Visitors : Population
1.6:1

467,000 visitors

Up 9% in 2000

MAIN TOURIST ARRIVALS

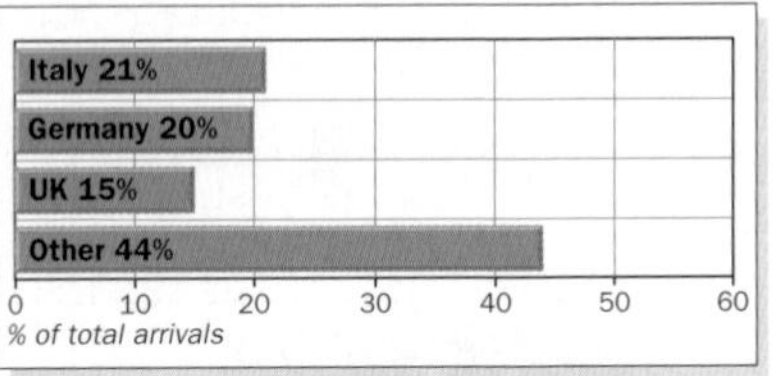
Italy 21%
Germany 20%
UK 15%
Other 44%
0 10 20 30 40 50 60
% of total arrivals

Tourism is the largest source of foreign exchange, accounting for almost 20% of GDP. The first resort was opened in 1972, and hotels financed by local and foreign capital have since been built on the uninhabited islands. There are now nearly half a million visitors a year.

POLITICS

Nonparty elections

1999/2004

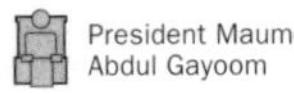
President Maumoon Abdul Gayoom

AT THE LAST ELECTION

Citizens' Assembly 50 seats

There are no political parties. 42 members of the Majlis (Assembly) are elected, and 8 appointed by the president

Politics in the Maldives is the preserve of a small group of influential families. Most were already dominant under the sultanate. Formal parties with ideological objectives are virtually nonexistent, politics being organized around family and clan loyalties.

Former president Ibrahim Nasir abolished the premiership in 1975 and substantially strengthened the presidency. The main figure now is Maumoon Abdul Gayoom, a wealthy businessman who has been president since 1978. His brother-in-law, Ilyas Ibrahim, is regarded as his main rival.

A young Westernized elite has increased the pressure for political reform. Under a constitution effective since 1998, rival candidates may seek to be parliament's presidential nominee; only one name then goes forward for popular endorsement in a referendum.

PEOPLE

Pop. density high

Dhivehi (Maldivian)

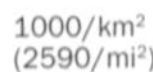
1000/km²
(2590/mi²)

THE URBAN/RURAL POPULATION SPLIT

26% 74%

RELIGIOUS PERSUASION

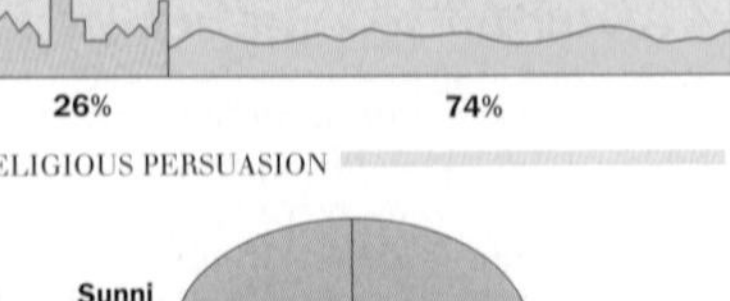
Sunni Muslim 100%

It is believed that the islands were inhabited as early as 1500 BCE. Aryan immigrants arrived around 500 BCE. The islands were then discovered by Arab traders. The people, who are all Sunni Muslims, live on only 200 of the 1190 islands. About 25% of the total population live on the island capital of Male'. It is estimated that 12,000 guest workers from neighboring Sri Lanka and India work in the Maldives. The country's newfound prosperity has seen the emergence of a commercial elite.

WORLD AFFAIRS

Joined UN in 1965

Comm

NAM

OIC

SAARC

WTO

The Maldives is a long-standing member of the NAM. The government continues to support it and rejects the view that the organization does not have a role to play in the post-Cold War world. The Maldives' international standing was enhanced in 1990, when it hosted the fifth SAARC summit meeting, held in Male'.

AID

Recipient

 $19m (receipts)

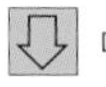 Down 39% in 2000

Aid has helped to finance the development of port and airport facilities. Japan is the most important bilateral aid donor. The Maldives is classed by the UN as a Least Developed Country, which grants it access to special financial programs; it hopes to graduate from this listing.

DEFENSE

No compulsory military service

 $44m

 Up 7% in 2000

The British military presence ended in 1975, when troops were withdrawn from the staging post on Gan, in the Addu atoll. The Maldives follows a policy of nonalignment, but in 1988 called on India for military assistance to help suppress a coup attempt.

ECONOMICS

Inflation 5.8% p.a. (1990–2000)

 $541m

 11.77 rufiyaa

SCORE CARD

❑ WORLD GNP RANKING	170th
❑ GNP PER CAPITA	$1960
❑ BALANCE OF PAYMENTS	–$60m
❑ INFLATION	–1.1%
❑ UNEMPLOYMENT	1%

STRENGTHS

Boom in tourism. Thriving fishing industry, especially tuna. Shipping. Clothing. Coconut production. Economic reforms since 1989 have eased import restrictions and encouraged foreign investment.

WEAKNESSES

Too dependent on fluctuating tourist industry. Growing trade deficit. Skilled labor shortage. Small manufacturing base. Cottage industries employ 25% of workforce; little scope for expansion.

EXPORTS

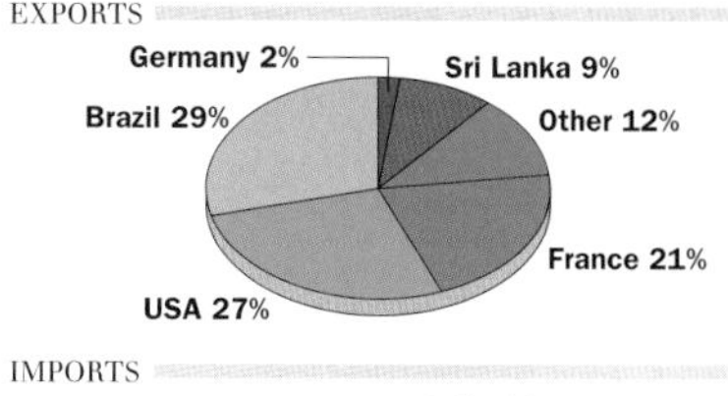
Germany 2%
Sri Lanka 9%
Brazil 29%
Other 12%
France 21%
USA 27%

IMPORTS

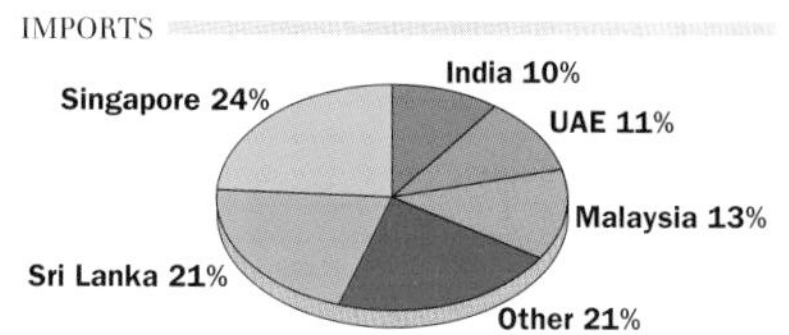
India 10%
Singapore 24%
UAE 11%
Malaysia 13%
Sri Lanka 21%
Other 21%

RESOURCES

Electric power 25,000 kw

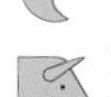 133,547 tonnes

 Not an oil producer

 31,000 cattle, 20,000 goats, 11,000 sheep

 None

Natural resources include abundant stocks of fish, particularly tuna. Fishing, still carried out by the traditional pole and line method to help conserve stocks, employs over 20% of the working population. Coconut production is also important. All oil products and virtually all staple foods are imported.

ENVIRONMENT

Not available

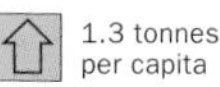 None

1.3 tonnes per capita

Rising sea levels due to global warming and climate change threaten the islands, which have an average height of just 1.5 m (5 ft). A sea wall has been built around the capital island.

Other environmental concerns are sewerage, waste disposal, and the mining of coral for building.

MEDIA

TV ownership low

 Daily newspaper circulation 19 per 1000 people

PUBLISHING AND BROADCAST MEDIA

There are 3 daily newspapers, including *Haveeru Daily* and *Aafathis*, published in Dhivehi and English

  1 state-owned service

 2 services

There is a marked degree of press self-censorship; in the past, journalists have been imprisoned. An Internet café opened in Male' in 1998.

CRIME

Death penalty not used in practice

1098 prisoners

Up 90% 1992–1996

The Maldives is a strict Islamic society. Narcotics crimes are heavily punished. Political prisoners are banished to outer islands. The judiciary and executive are closely linked.

CHRONOLOGY

The Maldives was a British protectorate from 1887 and gained its independence in 1965.

- ❑ **1932** First written constitution.
- ❑ **1968** Sultanate abolished. Declared a republic. Ibrahim Nasir elected as first president.
- ❑ **1978** Gayoom becomes president.
- ❑ **1994** Nonparty legislative elections.
- ❑ **1998** New constitution; Gayoom reelected for fifth five-year term.

EDUCATION

Schooling is not compulsory

 96%

 Not available

Primary education has been improved. Secondary education is less developed in the outer islands; the first school outside Male' was opened in 1992.

HEALTH

Welfare state health benefits

 1 per 1328 people

Infectious and parasitic diseases, tuberculosis, perinatal deaths

There is a lack of general equipment and facilities. Health care is less developed on the outlying islands.

SPENDING

GDP/cap. increase

CONSUMPTION AND SPENDING

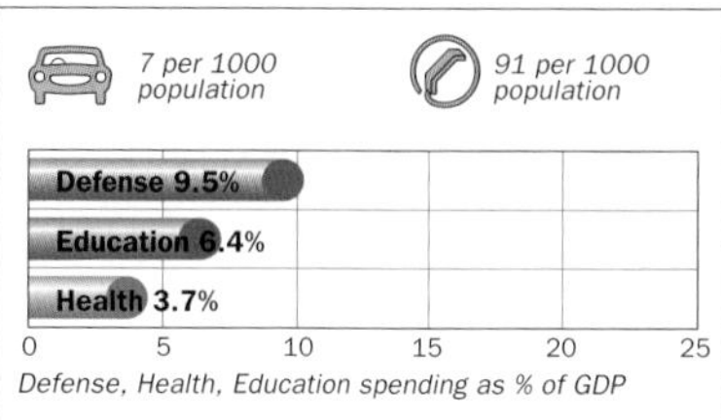
7 per 1000 population
91 per 1000 population
Defense 9.5%
Education 6.4%
Health 3.7%
0
5
10
15
20
25
Defense, Health, Education spending as % of GDP

Great disparities of wealth exist between the people who live in Male' and those who live on the more distant outer islands.

WORLD RANKING

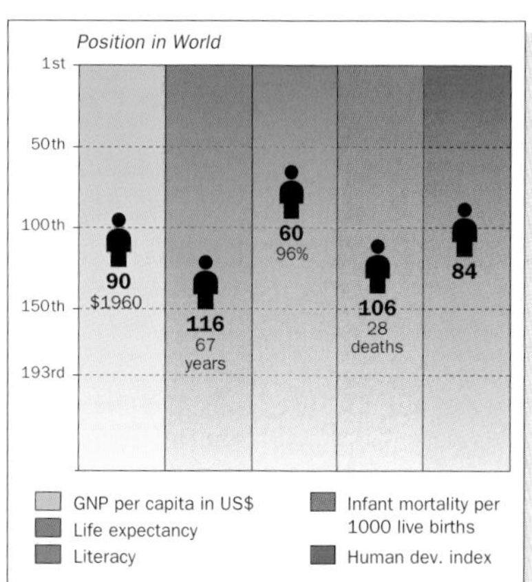
Position in World
1st
50th
100th
150th
193rd
90
$1960
116
67 years
60
96%
106
28 deaths
84
GNP per capita in US$
Life expectancy
Literacy
Infant mortality per 1000 live births
Human dev. index

M

MALI

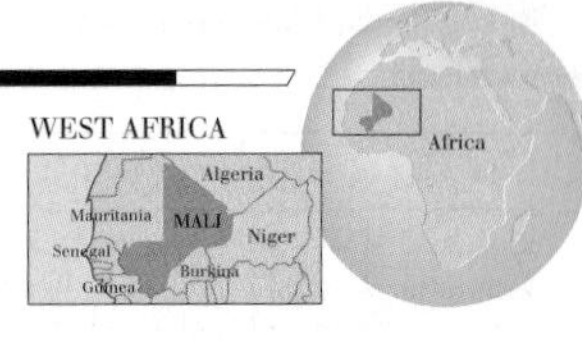
WEST AFRICA
Algeria
Mauritania
MALI
Niger
Senegal
Burkina
Guinea
Africa

OFFICIAL NAME: Republic of Mali **CAPITAL:** Bamako
POPULATION: 11.7 million **CURRENCY:** CFA franc **OFFICIAL LANGUAGE:** French

1960

1960

Sept 22

RMM

0

+223
.ml

MALI IS LANDLOCKED in the heart of west Africa. Its mostly flat terrain comprises virtually uninhabited Saharan plains in the north and more fertile savanna land in the south, where most of the population lives. The Niger River irrigates the central and southwestern regions. Mali achieved independence from France in 1960. Multiparty democratic elections under a new constitution, in 1992 and then in 1997, provoked accusations of severe irregularities.

CLIMATE

Hot desert/steppe

WEATHER CHART FOR BAMAKO

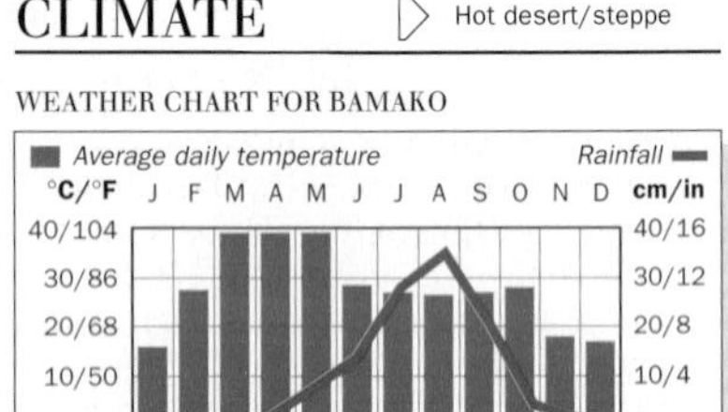
Average daily temperature
Rainfall
°C/°F
J F M A M J J A S O N D
cm/in
40/104
30/86
20/68
10/50
0/32
-10/14
-20/-4
40/16
30/12
20/8
10/4
0

In the south, intensely hot, dry weather precedes the westerly rains. Mali's northern half is almost rainless.

TRANSPORTATION

Drive on right

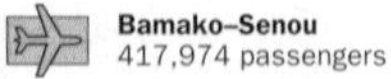
Bamako-Senou
417,974 passengers

Has no fleet

THE TRANSPORTATION NETWORK

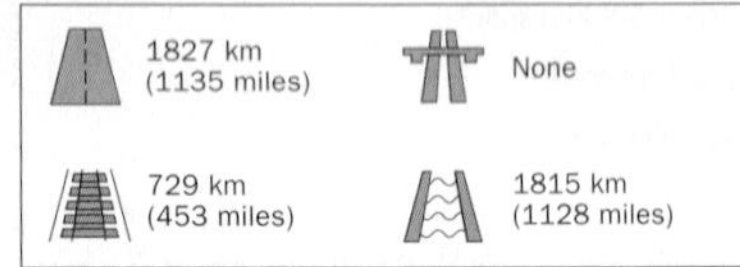
1827 km
(1135 miles)
None
729 km
(453 miles)
1815 km
(1128 miles)

Mali is linked by rail with the port of Dakar in Senegal, and by good roads to the port of Abidjan in Ivory Coast.

TOURISM

Visitors : Population 1:129

91,000 visitors

Up 5% in 2000

MAIN TOURIST ARRIVALS

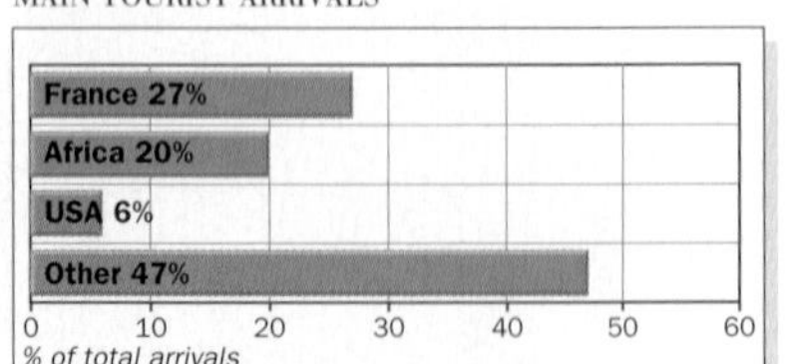
France 27%
Africa 20%
USA 6%
Other 47%
0
10
20
30
40
50
60
% of total arrivals

Tourism is largely safari-oriented, though the historic cities of Djénné, Gao, and Mopti, lying on the banks of the Niger River, also attract visitors. A national domestic airline began operating in 1990.

PEOPLE

Pop. density low

Bambara, Fulani, Senufo, Soninke, French

10/km² (25/mi²)

THE URBAN/RURAL POPULATION SPLIT

RELIGIOUS PERSUASION

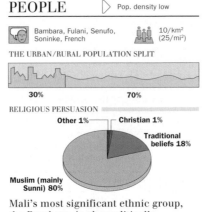

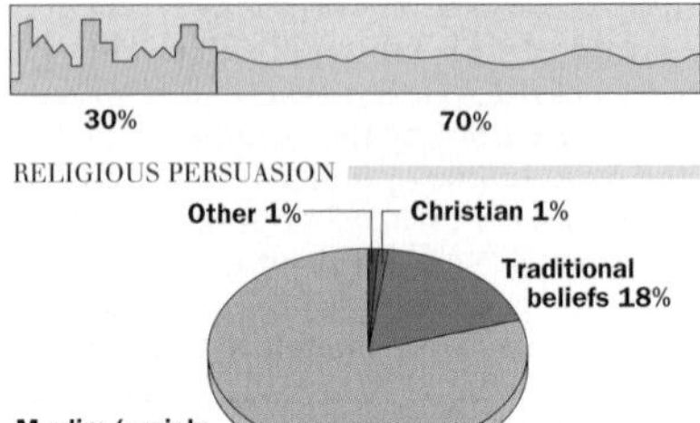
30%
70%
Other 1%
Christian 1%
Traditional beliefs 18%
Muslim (mainly Sunni) 80%

Mali's most significant ethnic group, the Bambara, is also politically dominant. The Bambara speak the *lingua franca* of the Niger River, which is shared with other groups including the Malinke. The relationship between the Bambara–Malinke majority and the Tuareg nomads of the Saharan north is tense and sometimes violent. The extended family is a vital social security system and link between the urban and rural poor. There are a few powerful women in Mali but, in general, women have little status.

POLITICS

Multiparty elections

2002/2007

President Amadou Toumani Touré

AT THE LAST ELECTION

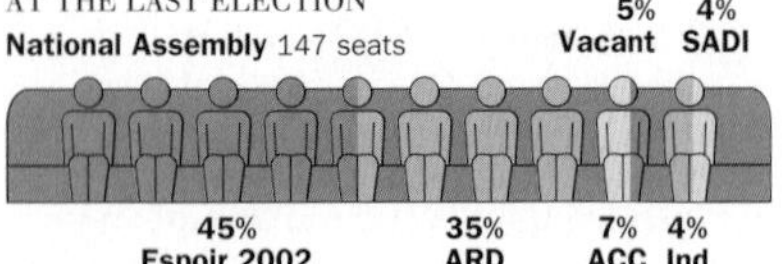
National Assembly 147 seats
5% Vacant
4% SADI
45% Espoir 2002
35% ARD
7% ACC
4% Ind

Espoir 2002 = Hope 2002 (led by the Rally for Mali – **RPM**)
ARD = Alliance for the Republic and Democracy (led by the Alliance for Democracy in Mali – **ADEMA**)
ACC = Convergence for Rotation and Change
Ind = Independents **SADI** = Party for African Solidarity, Democracy, and Integration

The successful transition to multiparty politics in 1992 followed the overthrow the previous year of Moussa Traoré, Mali's dictator for 23 years. The army's role was crucial in leading the coup, while Col. Amadou Toumani Touré, who acted as interim president, was responsible for the swift return to civilian rule in less than a year. For a decade President Alpha Oumar Konaré's ADEMA government attempted to alleviate poverty while placating the opposition. However, its economic austerity measures proved unpopular. Col. Touré was returned to power in the 2002 presidential elections with the support of all main opposition parties, and in the legislative polls ADEMA saw its majority disappear. It sought to form a coalition with smaller parties. Maintaining good relations with the Tuareg remains a key issue.

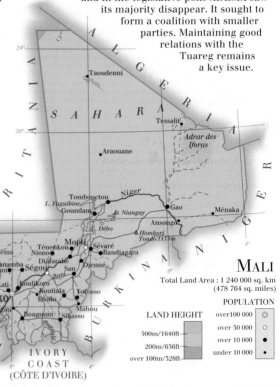

MALI

Total Land Area : 1 240 000 sq. km (478 764 sq. miles)

LAND HEIGHT: 500m/1640ft, 200m/656ft, over 100m/328ft

POPULATION: over 100 000, over 50 000, over 10 000, under 10 000

0 – 200 km; 0 – 200 miles

WORLD AFFAIRS

Joined UN in 1960

ECOWAS

FZ

AU

OIC
OIF

Mali concentrates on maintaining good relations with the ECOWAS countries to its south and northern neighbors such as Algeria. Relations with Libya, which is suspected of fomenting Tuareg revolt, are tense. There are good relations with the US and other Western aid providers.

AID

Recipient

 $360m (receipts)  Up 2% in 2000

Mali is highly dependent on foreign aid, which principally comes from the World Bank, France, the US, Germany, the Netherlands, Japan, and the EU.

DEFENSE

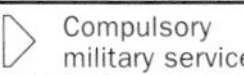
Compulsory military service

 $29m  Down 15% in 2000

Mali's 7800-strong armed forces have stayed out of politics since the overthrow of President Traoré in 1991.

ECONOMICS

Inflation 7.1% p.a. (1990–2000)

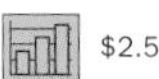 $2.55bn  698.7–736.7 CFA francs

SCORE CARD

❑ WORLD GNP RANKING	132nd
❑ GNP PER CAPITA	$240
❑ BALANCE OF PAYMENTS	–$178m
❑ INFLATION	–0.7%
❑ UNEMPLOYMENT	Widespread underemployment

STRENGTHS
Producer of high-quality cotton. Irrigation potential from the Niger and Senegal rivers. Rapid expansion of gold production now under way.

WEAKNESSES
Serious poverty and underdevelopment. Communications difficulties of vast landlocked country. Drought-prone climate. Cotton exports vulnerable to low world prices.

EXPORTS

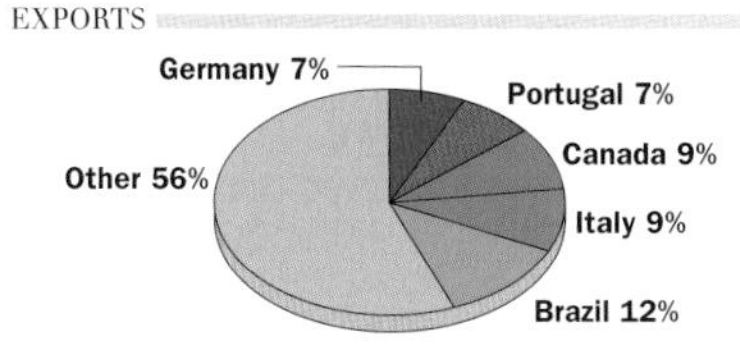
Germany 7%
Portugal 7%
Canada 9%
Italy 9%
Brazil 12%
Other 56%

IMPORTS

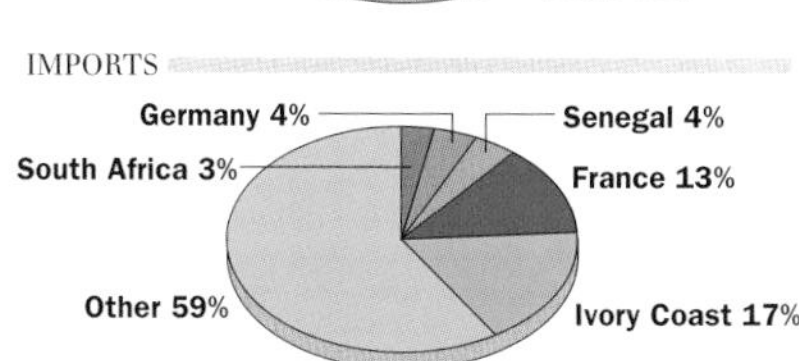
Germany 4%
Senegal 4%
South Africa 3%
France 13%
Ivory Coast 17%
Other 59%

Village near Bandiagara. *The low, broken hills typical of the east and southeast of Mali are the homeland of the Dogon people.*

RESOURCES

Electric power 114,000 kw

 98,766 tonnes 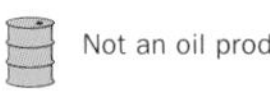 Not an oil producer

9.9m goats, 6.82m cattle, 6.4m sheep, 25m chickens

Gold, salt, marble, phosphates, tungsten, diamonds, oil

Gold deposits are now being mined, and prospecting is under way for tungsten, diamonds, and oil. Exploitation of natural resources is hampered by Mali's poor infrastructure and landlocked situation. Electric power comes from the Selingue Dam on the Niger and the Manantali Dam on the Senegal. The latter produced its first electricity in 2001 – 13 years after it was completed.

ENVIRONMENT

Sustainability rank: 85th

 4% 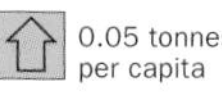 0.05 tonnes per capita

Severe drought in 1983 destroyed herds and accelerated desertification and deforestation. The Selingue Dam seriously affects the levels of the Niger, even in years of good rainfall.

MEDIA

TV ownership low

 Daily newspaper circulation 1 per 1000 people

PUBLISHING AND BROADCAST MEDIA

There are 5 daily newspapers, including the pro-government *L'Essor – La Voix du Peuple*

3 services: 1 state-owned, 2 independent  15 services: 1 state-owned, 14 independent

Even before the 1991 coup, previously rigid controls were being relaxed. The 1992 constitution guarantees the freedom of the press, and Mali's broadcast and print media are now among the freest in Africa.

CRIME

Death penalty not used in practice

 3135 prisoners  Crime is rising slowly

Crime is not particularly prevalent compared with some other countries in the region, owing at least in part to the relative lack of urbanization. In towns, robbery, juvenile delinquency, and smuggling are problems.

CHRONOLOGY

Mali was a major trans-Saharan trading empire. The French colonized the area between 1881 and 1895.

- ❑ **1960** Independence.
- ❑ **1968** Coup by Gen. Moussa Traoré.
- ❑ **1990** Pro-democracy demonstrations.
- ❑ **1991** Traoré arrested.
- ❑ **1992** Free multiparty elections.
- ❑ **1997** President Konaré and ADEMA party reelected in disputed polls.
- ❑ **2002** Elections: Col. Touré president; ADEMA loses majority.

EDUCATION

School leaving age: 13

 41%  18,662 students

Only 25% of children go to primary school and just 7% to secondary school. A ten-year program to raise education levels for girls was launched in 2001.

HEALTH

No welfare state health benefits

1 per 10,000 people

Malaria, pneumonia, parasitic and diarrheal diseases

A health program began in 1998, with the aim of higher immunization rates for children and more health care access.

SPENDING

GDP/cap. increase

CONSUMPTION AND SPENDING

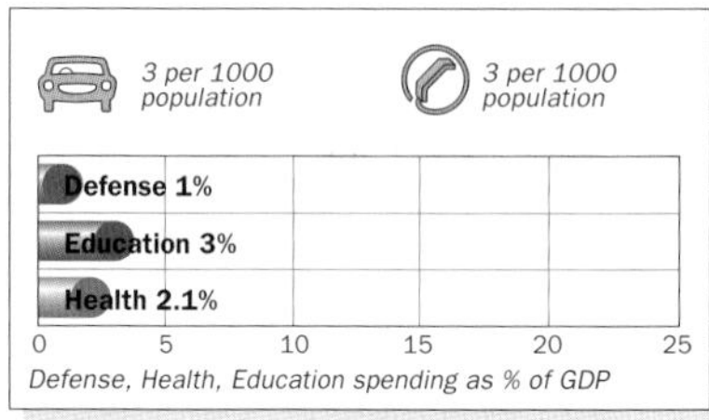
3 per 1000 population
3 per 1000 population
Defense 1%
Education 3%
Health 2.1%
0
5
10
15
20
25
Defense, Health, Education spending as % of GDP

Poverty is widespread, and wealth is limited to a very small group; Malians disapprove of flaunted wealth and public ostentation is rare.

WORLD RANKING

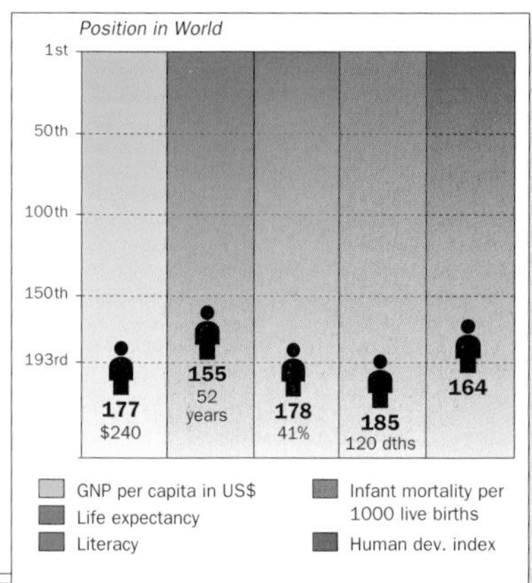
Position in World
1st
50th
100th
150th
193rd
177
$240
155
52 years
178
41%
185
120 dths
164
GNP per capita in US$
Life expectancy
Literacy
Infant mortality per 1000 live births
Human dev. index

M

MALTA

EUROPE

Italy Greece MALTA Tunisia Europe

OFFICIAL NAME: Republic of Malta **CAPITAL:** Valletta

POPULATION: 392,000 **CURRENCY:** Maltese lira **OFFICIAL LANGUAGES:** Maltese and English

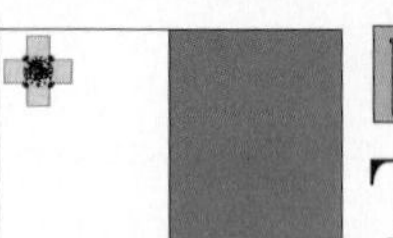

1964

1964
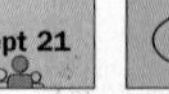
Sept 21

M

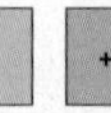
+1

+356

.mt

THE MALTESE ARCHIPELAGO is strategically located, lying between Europe and north Africa. Controlled throughout its history by successive colonial powers, Malta finally gained independence from the UK in 1964. The islands are mainly low-lying, with rocky coastlines; only Malta, Gozo (Ghawdex), and Kemmuna are inhabited. Tourism is Malta's chief source of income, with an influx of tourists each year of over three times the islands' population.

CLIMATE

Mediterranean

WEATHER CHART FOR VALLETTA

Average daily temperature — Rainfall

°C/°F J F M A M J J A S O N D cm/in

40/104, 30/86, 20/68, 10/50, 0/32, -10/14, -20/-4 — 40/16, 30/12, 20/8, 10/4, 0

The climate is typical of the southern Mediterranean – with at least six hours of sunshine a day, even in winter.

TRANSPORTATION

Drive on left

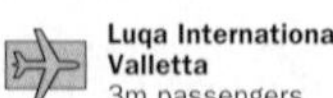

Luqa International, Valletta 3m passengers

3189 ships 28.6m grt

THE TRANSPORTATION NETWORK

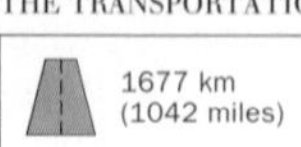

1677 km (1042 miles)

None

None

None

Malta Freeport at Marsaxlokk exploits Malta's strategic shipping location in the Mediterranean. In summer, a five-minute helicopter flight from the international airport links the islands of Malta and Gozo. There is a well-developed public transportation system, with ferry and hovercraft services and buses on both islands.

Traditionally painted **luzzus** *at St. Julian's harbor. The fish caught are now only for domestic and tourist consumption.*

TOURISM

Visitors : Population 3.1:1

1.2m visitors

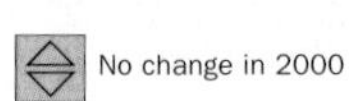

No change in 2000

MAIN TOURIST ARRIVALS

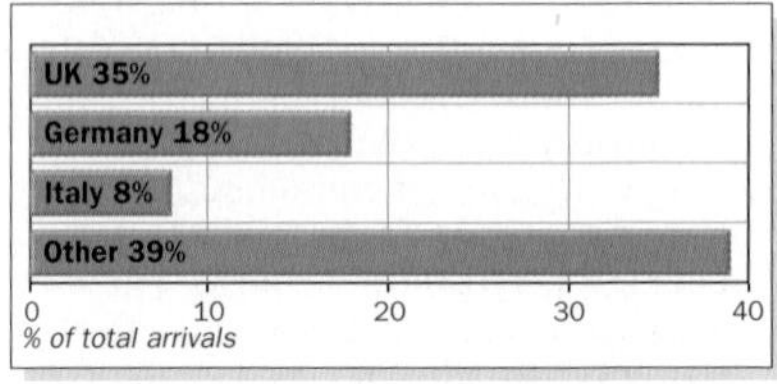
UK 35%
Germany 18%
Italy 8%
Other 39%
0 10 20 30 40
% of total arrivals

Tourism is vital to the economy and accounts for more than 30% of GDP, even though most visitors are budget vacationers. In addition to beaches and scenery, there are the historical attractions of Mdina and Valletta. Development on the quieter island of Gozo is limited to luxury-grade hotels.

PEOPLE

Pop. density high

Maltese, English

1225/km² (3173/mi²)

THE URBAN/RURAL POPULATION SPLIT

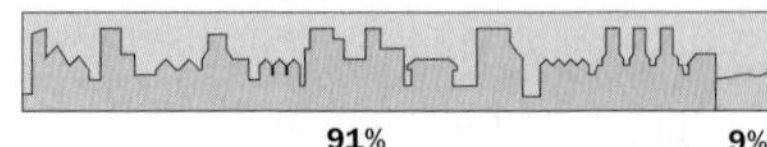

91% 9%

RELIGIOUS PERSUASION

Other and nonreligious 2%

Roman Catholic 98%

Malta's population has been subject over the centuries to diverse Arabic, Sicilian, Norman, Spanish, English, and Italian influences. Today, much of the younger Maltese population goes abroad to find work, especially to the US or Australia; opportunities for them on the islands are few.

The Maltese are staunch Roman Catholics, on a percentage basis more so than virtually any other nation. The remainder are mainly Anglicans, who are included within the diocese of Gibraltar. Divorce is illegal.

POLITICS

Multiparty elections

1998/2003

President Guido de Marco

AT THE LAST ELECTION

House of Representatives 65 seats

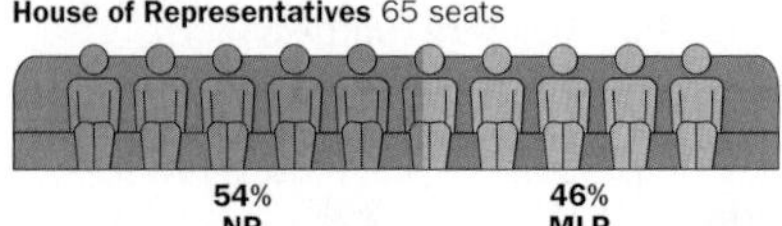
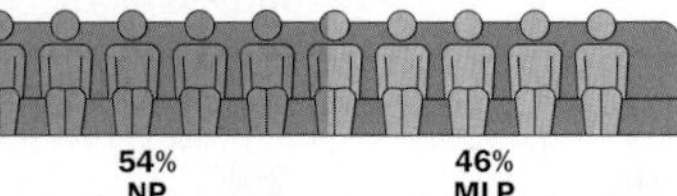

54% NP **46% MLP**

NP = Nationalist Party **MLP** = Malta Labour Party

Maltese politics is strongly adversarial and evenly split between the right-wing NP and the left-wing MLP. The latter dominated the government in the 1970s and 1980s, ensuring state control of industry and pursuing a nonaligned foreign policy.

The 1990s, however, saw a switch in favor of the NP, with Prime Minister Edward Fenech Adami at the helm in 1987–1996 and since 1998. Under Fenech Adami, Malta moved toward ever closer ties with Europe, and favors a free-market approach to its economy. The NP secured reelection in 1992, largely due to a rise in living standards. A modernized MLP ended the NP's nine-year reign in 1996. Under Alfred Sant, a leading writer and Harvard MBA, it diluted traditional links with the unions and "froze" Malta's EU application. However, the MLP's small parliamentary majority undermined the government, and the NP won early elections in 1998. Fenech Adami has now reset Malta on its course for membership of the EU.

WORLD AFFAIRS

Joined UN in 1964

CE

Comm

IBRD

NAM

OSCE

Malta has made the most of its location on the fringe of Europe, with a staunchly nonaligned foreign policy. Ties are traditionally strong with the Arab world and north Africa, and relations with Libya remain good. There are also close commercial links with Russia and China.

However, it is the island's relationship with Europe that has dominated recent policy. Malta's bid for EU membership, launched in 1990, was derailed by the anti-EU MLP government in 1996. With the application frozen, Malta was denied a place in the "first wave" of potential EU members. However, the return to power of the pro-EU NP in 1998 restarted the bid and Malta joined talks in March 2000 as part of the "second wave."

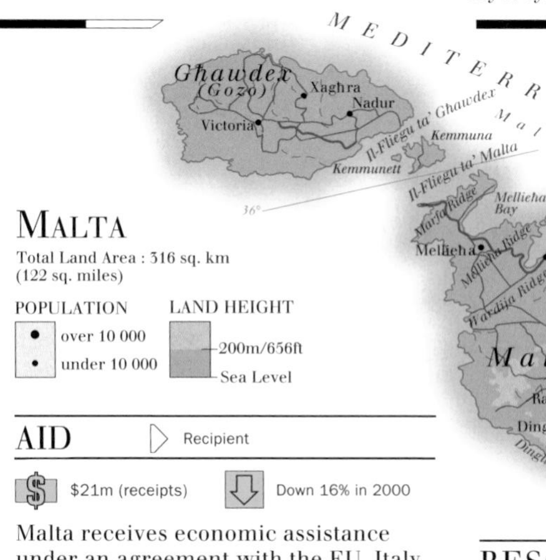
MEDITERRANEAN SEA
Ghawdex (Gozo)
Xaghra
Nadur
Victoria
Il-Fliegu ta' Ghawdex
Kemmuna
Kemmunett
Il-Fliegu ta' Malta
Malta Channel
Marfa Ridge
Mellieha Bay
Mellieha
Mellieha Ridge
Bugibba
San Pawl il Bahar
St Julian's
Sliema
Gzira
Msida
VALLETTA
Grand Harbour
Vittoriosa
Zabbar
Mosta
Birkirkara
Hamrun
Qormi
Cospicua
Paola
Malta
Marsa
Tarxien
Zejtun
Marsaxlokk
Rabat
Mdina
Zebbug
Luqa
Siggiewi
Dingli
Dingli Cliffs
253m
Birzebbuga
Marsaxlokk Bay
Zurrieq
0 5 km
0 5 miles
N

MALTA

Total Land Area : 316 sq. km (122 sq. miles)

POPULATION
- over 10 000
- under 10 000

LAND HEIGHT
- 200m/656ft
- Sea Level

AID

Recipient

$21m (receipts) — Down 16% in 2000

Malta receives economic assistance under an agreement with the EU. Italy is the main bilateral source of aid.

DEFENSE

No compulsory military service

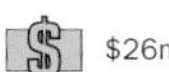

$26m

Down 4% in 2000

The Maltese army, advised by the Libyans in the 1980s, now receives training and equipment from Italy, Germany, and the UK.

ECONOMICS

Inflation 2.8% p.a. (1990–2000)

$3.56bn

0.4357–0.4503 Maltese liri

SCORE CARD

❑ WORLD GNP RANKING	127th
❑ GNP PER CAPITA	$9120
❑ BALANCE OF PAYMENTS	–$515m
❑ INFLATION	2.4%
❑ UNEMPLOYMENT	5%

STRENGTHS

Tourism and naval dockyards. Schemes to attract foreign high-tech industry. Malta Freeport container distribution center. Offshore banking. Strategic position between Europe and Africa, on main Mediterranean shipping lines.

WEAKNESSES

Cut-rate competition from Africa and Asia in traditional textile industry. Need to import almost all requirements.

EXPORTS

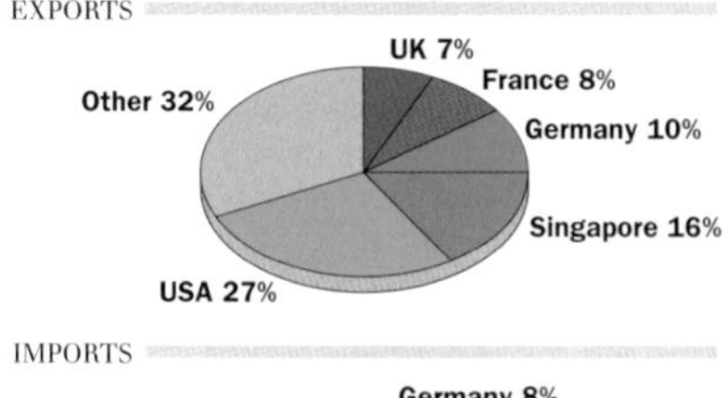
UK 7%
France 8%
Germany 10%
Singapore 16%
USA 27%
Other 32%

IMPORTS

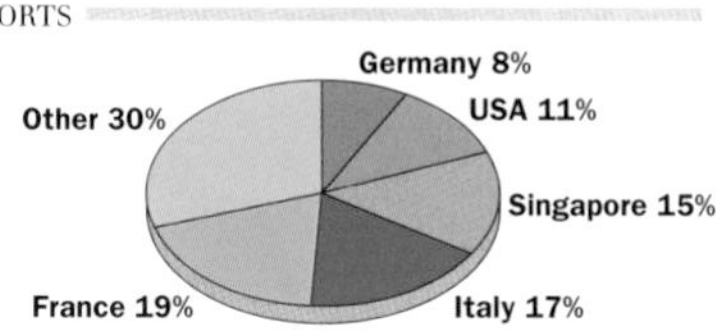
Germany 8%
USA 11%
Singapore 15%
Italy 17%
France 19%
Other 30%

RESOURCES

Electric power 250,000 kw

3035 tonnes

Reserves under exploration

80,074 pigs, 19,200 cattle, 820,000 chickens

Stone, sand, oil

Malta is dependent on desalination plants for most of its water supply. All oil has to be imported, mostly from Libya. However, there are petroleum reserves currently under exploration in Maltese waters.

ENVIRONMENT

Not available

None

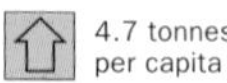

4.7 tonnes per capita

The main environmental concern is linked to the tourist industry. A lack of planning controls in the 1970s was responsible for unsightly beach developments. These are now tightly controlled, particularly on Gozo.

MEDIA

TV ownership high

Daily newspaper circulation 127 per 1000 people

PUBLISHING AND BROADCAST MEDIA

There are 4 daily newspapers, *In-Nazzjon*, *L-Orizzont*, *The Times*, and *The Malta Independent*

7 services: 1 state-owned, 6 independent

12 services: 1 state-owned, 11 independent

The Maltese press is largely party politically oriented. Two of the three main press groups are affiliated to the NP or MLP; one is independent.

CRIME

No death penalty

260 prisoners

Down 19% 1992–1996

Crime rates are low compared with those on the European mainland. There has been an increase in narcotics transshipment and associated crimes.

CHRONOLOGY

Ancient Malta was dominated by Phoenicians, Carthaginians, Greeks, and Romans; later it was ruled by the Arabs, Norman Sicily, Spain, France, and, finally, the UK.

- ❑ **1947** Internal self-government.
- ❑ **1964** Full independence from UK.
- ❑ **1971** Dom Mintoff's MLP in power.
- ❑ **1987–1996** Edward Fenech Adami (NP) is premier.
- ❑ **1998** Early elections bring pro-EU Fenech Adami back to power after brief MLP interlude.

EDUCATION

School leaving age: 16

92%

8260 students

One-third of pupils attend non-state schools, including heavily subsidized church-run institutions. There is a state university in Valletta.

HEALTH

Welfare state health benefits

1 per 383 people

Cerebrovascular and heart diseases, cancers, diabetes

Malta has five state-run and a couple of private hospitals. Diabetes is prevalent, as on other Mediterranean islands.

SPENDING

GDP/cap. increase

CONSUMPTION AND SPENDING

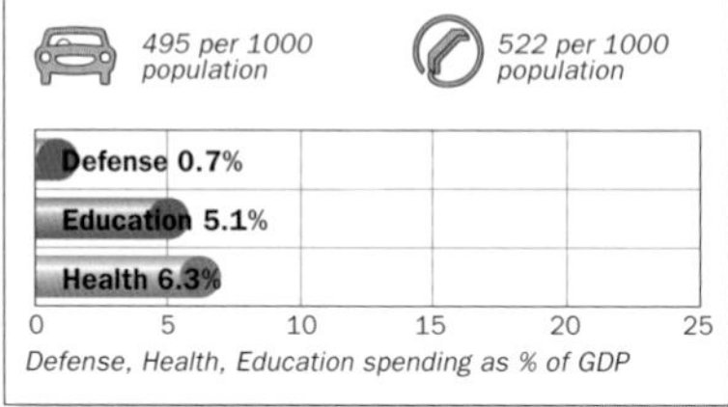
495 per 1000 population
522 per 1000 population
Defense 0.7%
Education 5.1%
Health 6.3%
0
5
10
15
20
25
Defense, Health, Education spending as % of GDP

Remittances from Maltese working abroad are an important source of income for many island families.

WORLD RANKING

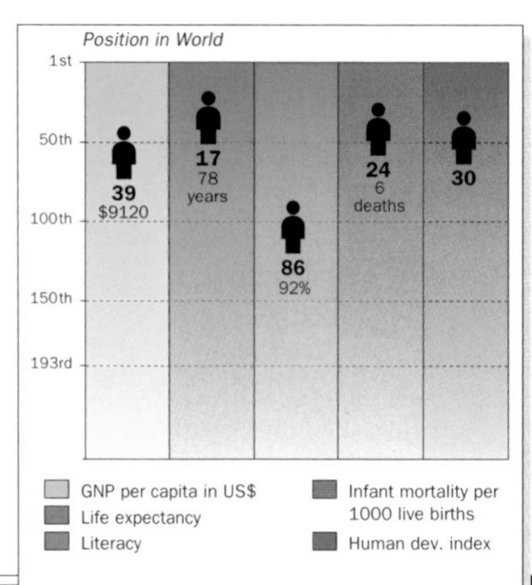
Position in World
1st
50th
100th
150th
193rd
39 $9120
17 78 years
86 92%
24 6 deaths
30
GNP per capita in US$
Life expectancy
Literacy
Infant mortality per 1000 live births
Human dev. index

M

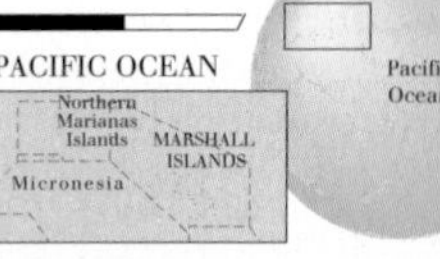
PACIFIC OCEAN
Northern Marianas Islands
MARSHALL ISLANDS
Micronesia
Pacific Ocean

Marshall Islands

Official Name: Republic of the Marshall Islands **Capital:** Majuro
Population: 68,100 **Currency:** US dollar **Official Languages:** English and Marshallese

1986
1986
May 1
MH
+12
+692
.mh

THE MARSHALL ISLANDS comprise a group of 34 widely scattered atolls in the central Pacific Ocean, formerly under US rule as part of the UN Trust Territory of the Pacific Islands. An agreement which granted internal sovereignty in free association with the US became operational in 1986, and the Trust was formally dissolved in 1990. The economy is almost entirely dependent on US aid and rent for the US missile base on Kwajalein Atoll.

Ebeye District on Kwajalein Atoll. *Population pressures on the island have led to the disappearance of most tree and grass cover.*

CLIMATE

Tropical oceanic

WEATHER CHART FOR MAJURO

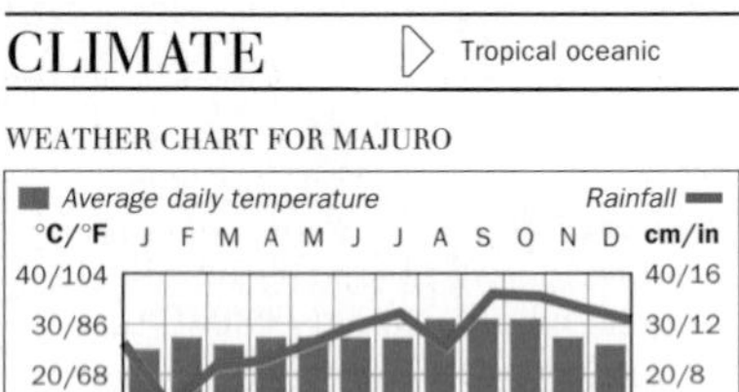
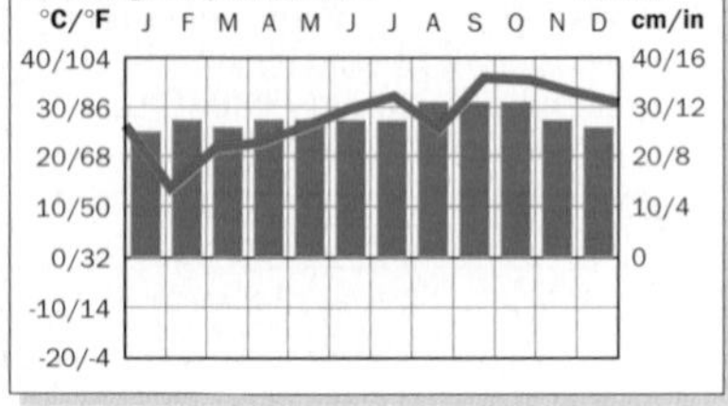
Average daily temperature
Rainfall
°C/°F
J F M A M J J A S O N D
cm/in
40/104
30/86
20/68
10/50
0/32
-10/14
-20/-4
40/16
30/12
20/8
10/4
0

The climate is tropical oceanic with little seasonal variation; temperatures average around 30°C (86°F).

TRANSPORTATION

Drive on right

Majuro International

207 ships
6.44m grt

THE TRANSPORTATION NETWORK

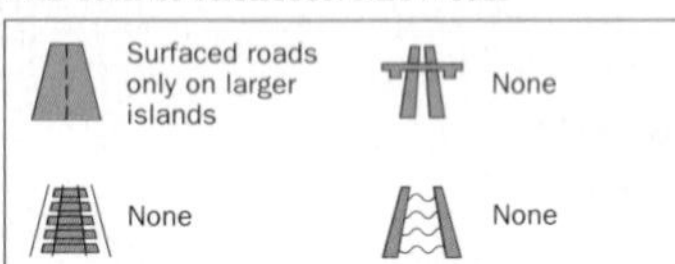
Surfaced roads only on larger islands
None
None
None

The transportation system is limited, although there is some interisland shipping. State carrier Air Marshalls has experienced economic difficulties.

TOURISM

Visitors : Population 1:14

5000 visitors
No change in 2000

MAIN TOURIST ARRIVALS

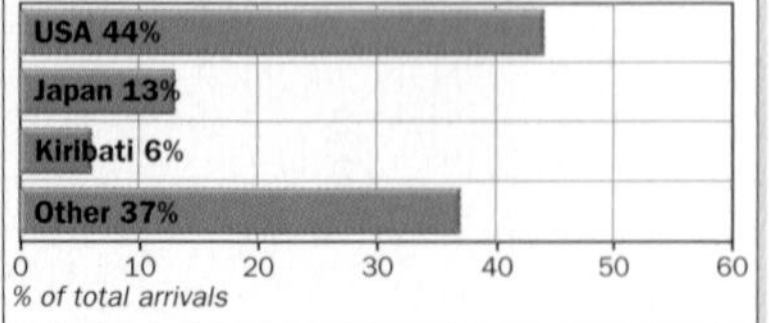
USA 44%
Japan 13%
Kiribati 6%
Other 37%
0
10
20
30
40
50
60
% of total arrivals

In the late 1990s major resort complexes were established on Majuro and on Mili Atoll. Attractions include diving, game fishing, and exploring the sites and relics of World War II battles.

PEOPLE

Pop. density high

Marshallese, English, Japanese, German

376/km² (975/mi²)

THE URBAN/RURAL POPULATION SPLIT

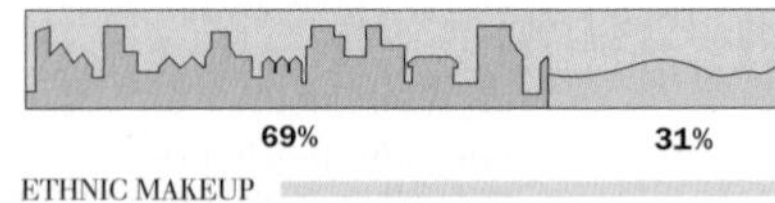
69%
31%

ETHNIC MAKEUP

Other 3%
Micronesian 97%

Of the 34 atolls making up the Marshall Islands, 24 are inhabited. Majuro, the capital and commercial center, is home to almost half of the population, many of whom live in its overcrowded slums. The other main center of population is Ebeye Island in the Kwajalein Atoll, where tensions are high due to poor living conditions. Most of Kwajalein Atoll's inhabitants were forcibly relocated to Ebeye in 1947 to make way for a US missile tracking, testing, and interception base; many still travel daily to work at the base. Life on the outlying islands is still centered on subsistence agriculture and fishing. Society is traditionally matrilineal.

POLITICS

Multiparty elections

1999/2003

President Kessai Note

AT THE LAST ELECTION

Parliament 33 seats

55% UDP **45% K**

UDP = United Democratic Party **K** = Pro-Kabua Grouping
The 33 members are elected from 25 districts

Council of Chiefs 12 seats

All 12 members are high chiefs

Politics is traditionally dominated by chiefs. Amata Kabua, the islands' high chief and first president until his death in 1996, was succeeded in early 1997 by his cousin Imata Kabua. However, the 1999 elections were won by the United Democratic Party, whose presidential candidate, commoner and former parliamentary speaker Kessai Note, was elected in early January 2000. Just over a year later Imata Kabua instigated an unsuccessful vote of no confidence in Note's administration – only the second in the islands' history. The vote was motivated by criticism of the government approach to the crucial renegotiation of the Compact of Free Association with the US. The original treaty, which provided most of the islands' revenue and defense, expired in 2001; a new treaty is under negotiation.

Marshall Islands

Total Land Area : 181 sq. km (70 sq. miles)

LAND HEIGHT
100m/328ft
Sea Level

PACIFIC OCEAN
RATAK CHAIN
RALIK CHAIN
Bokak
Bikar
Enewetak
Bikini
Rongelap
Rongerik
Ailinginae
Utirik
Taka
Ailuk
Mejit
Wotho
Jemo
Likiep
Wotje
Ujelang
Ujae
Lae
Kwajalein
Erikub
Maloelap
Lib
Namu
Aur
Jabat
MAJURO
Ailinglaplap
Arno
Jaluit
Mili
Narikrik
Namorik
Kili
Ebon
164°
170°
10°

Majuro Atoll
Rongrong District
Majuro District
Aeankan District
Jarej District
Arrak District
Woja District
Dalap District
Ajeltake District
Rairok District
171°
7°
0 10 km
0 10 miles

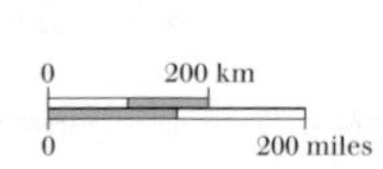
0 200 km
0 200 miles

WORLD AFFAIRS

Joined UN in 1991

IAEA | PIF | ACP | PC | ADB

The Compact of Free Association has made ties to the USA of key importance. From 1986 the US provided $1 billion in return for the use of Kwajalein Atoll as a missile range and has determined the islands' foreign and defense policies. A new Compact is under discussion. Taiwan has become a source of funding for development, provoking controversy over diplomatic recognition.

AID

Recipient

 $57m (receipts)

 Down 10% in 2000

US aid accounts for around two-thirds of the islands' revenue. Australia and Japan also provide some assistance.

DEFENSE

No compulsory military service

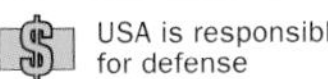

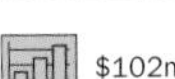 USA is responsible for defense

Not applicable

There is no defense force. All defense is provided by the US under the Compact of Free Association. The US does not have offensive weapons sited in the Marshalls, but its navy patrols regularly.

ECONOMICS

Inflation 5.6% p.a. (1990–2000)

$102m

Currency is US dollar

SCORE CARD

- ❏ WORLD GNP RANKING........................187th
- ❏ GNP PER CAPITA............................$1970
- ❏ BALANCE OF PAYMENTS......................$21m
- ❏ INFLATION...5%
- ❏ UNEMPLOYMENT.................................31%

STRENGTHS

US guarantee against economic collapse to preserve strategic influence. Aid from the US, on which islands almost totally depend. Copra. Huge tourism potential.

WEAKNESSES

High unemployment. Dependence on imports, which are twice as large as exports. All fuel has to be imported. Vulnerability to storm damage. Large state sector employs 75% of workers.

EXPORTS

The Marshall Islands' main export partners are the US, Australia, and Japan.

IMPORTS

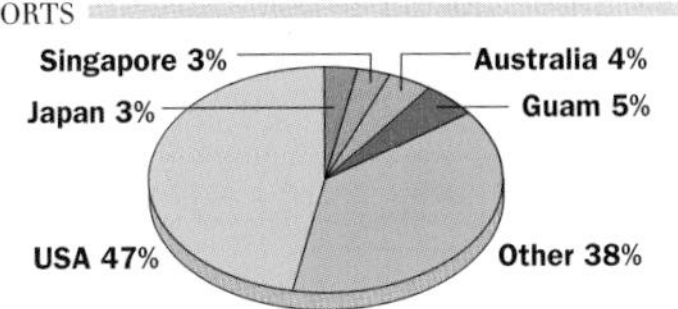
Singapore 3%
Australia 4%
Japan 3%
Guam 5%
USA 47%
Other 38%

RESOURCES

Not available

 400 tonnes

 Not an oil producer

 Not available

 Phosphates

There are few known strategic resources. Exploratory tests have revealed some high-grade phosphate deposits, but not in economically viable quantities. Small diesel generators are used for electricity production.

ENVIRONMENT

Not available

 None

 Not available

Between 1946 and 1958, Bikini, Enewetak, and neighboring atolls were rendered uninhabitable by a series of US nuclear military tests. Enewetak residents were allowed to return in 1980, and Rongelap was declared habitable in 2001. A 1999 tribunal adopted stringent standards for further decontamination. The US has now paid out over $101 million to victims of nuclear testing. Nuclear waste imports were banned in 1999. The effects of rising sea levels are a major concern. Erosion affects beaches and soil is being lost and also contaminated by brackish water.

MEDIA

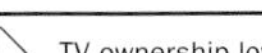

TV ownership low

 There are no daily newspapers

PUBLISHING AND BROADCAST MEDIA

There are no daily newspapers. The one weekly newspaper, the *Marshall Islands Journal*, is privately owned

2 independent services

2 services: 1 state-owned, 1 independent

Radio is the major source of information in the Marshalls. The main TV service is subscription-only. The US personnel stationed on Kwajalein have their own TV and radio stations.

CRIME

No death penalty

 23 prisoners

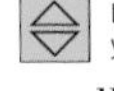 Little change from year to year

Crime levels are generally low; however, the rate is up in Ebeye. Outlying islands are crime-free.

EDUCATION

School leaving age: 14

 91%

 251 students

Education, compulsory between the ages of six and 14 years, is based on the US model. The number of secondary school graduates exceeds the availability of suitable employment in the Marshall Islands. Many go on to university in the US.

CHRONOLOGY

After a period under Spanish rule, the Marshall Islands became a German protectorate in 1885; Japan took possession at the start of World War I. The islands were transferred to US control in 1945.

- ❏ **1946** US nuclear testing begins.
- ❏ **1947** UN Trust Territory of the Pacific Islands established.
- ❏ **1961** Kwajalein becomes US army missile range.
- ❏ **1979** Constitution approved in referendum. Government set up.
- ❏ **1986** Compact of Free Association with US operational.
- ❏ **1990** Trust terminated by UN.
- ❏ **1997** Imata Kabua elected president after death in office of Amata Kabua, his cousin.
- ❏ **2000** Kessai Note president after opposition election victory.

HEALTH

No welfare state health benefits

 1 per 3294 people

 Respiratory, heart, and diarrheal diseases

Medical facilities are rudimentary. Complex operations are performed in Hawaii. Levels of malnutrition and vitamin A deficiency are high.

SPENDING

Not available

CONSUMPTION AND SPENDING

No data

62 per 1000 population

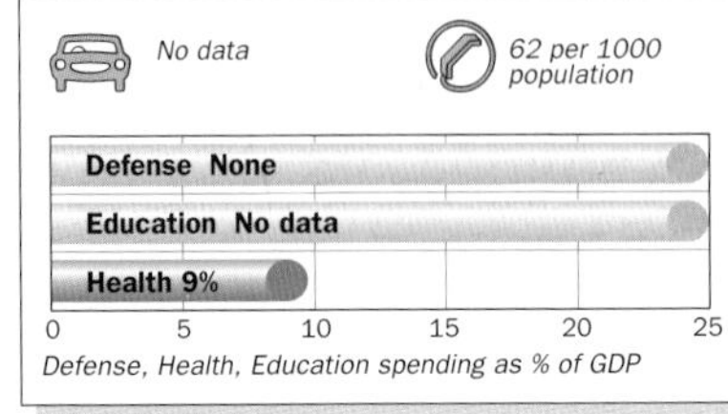
Defense None
Education No data
Health 9%
0
5
10
15
20
25
Defense, Health, Education spending as % of GDP

Wealth disparities are small. Very few citizens can afford luxuries such as air conditioning and cars.

WORLD RANKING

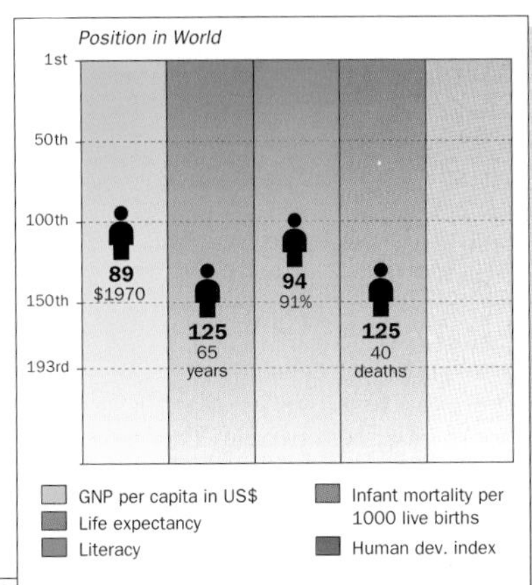
Position in World
1st
50th
100th
150th
193rd
89
$1970
125
65
years
94
91%
125
40
deaths
GNP per capita in US$
Life expectancy
Literacy
Infant mortality per 1000 live births
Human dev. index

M

NORTH AFRICA
Western Sahara
Algeria
MAURITANIA
Mali
Senegal
Africa

Mauritania

Official Name: Islamic Republic of Mauritania **Capital:** Nouakchott
Population: 2.7 million **Currency:** Ouguiya **Official Language:** Arabic

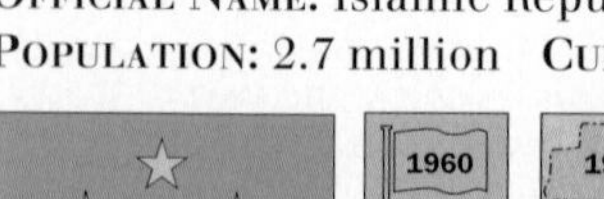

1960

1960

Nov 28

RIM

0

+222

.mr

LOCATED IN NORTHWEST AFRICA, Mauritania is a member of both the AU and the Arab League. Formerly a French colony, the country has taken a strongly Arab direction since 1964; today, it is the Maures who control political life and dominate the minority black population. The Sahara extends across two-thirds of Mauritania's territory; the only productive land is that drained by the Senegal River in the south and southwest.

CLIMATE

Hot desert

WEATHER CHART FOR NOUAKCHOTT

Average daily temperature — Rainfall
°C/°F J F M A M J J A S O N D cm/in
40/104 30/86 20/68 10/50 0/32 -10/14 -20/-4
40/16 30/12 20/8 10/4 0

The dusty Saharan *harmattan* wind often aggravates the very hot, dry conditions. Some rain falls in the south.

TRANSPORTATION

Drive on right

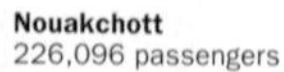
Nouakchott
226,096 passengers

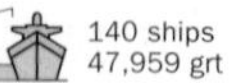
140 ships
47,959 grt

THE TRANSPORTATION NETWORK

866 km (538 miles)

None

740 km (460 miles)

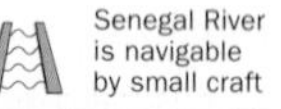
Senegal River is navigable by small craft

The transportation system is limited and unevenly developed. There are two major roads, but shifting sands mean that they require constant maintenance.

TOURISM

Visitors : Population 1:113

24,000 visitors

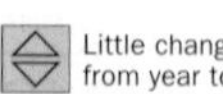
Little change from year to year

MAIN TOURIST ARRIVALS

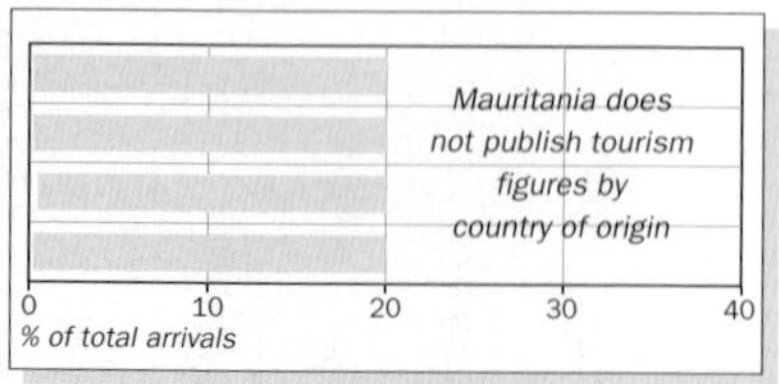
Mauritania does not publish tourism figures by country of origin
0
10
20
30
40
% of total arrivals

There are few tourists apart from desert safari enthusiasts. The more mountainous areas are especially dramatic, but access is difficult. Nouakchott has some hotels.

PEOPLE

Pop. density low

Hassaniyah Arabic, Wolof, French

3/km² (7/mi²)

THE URBAN/RURAL POPULATION SPLIT

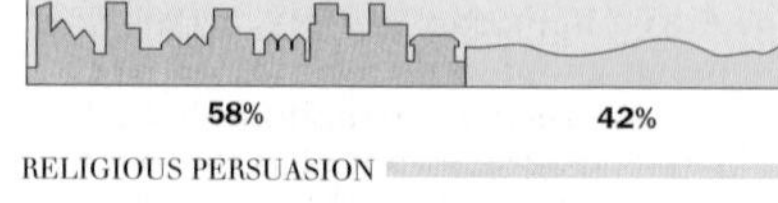
58%
42%

RELIGIOUS PERSUASION

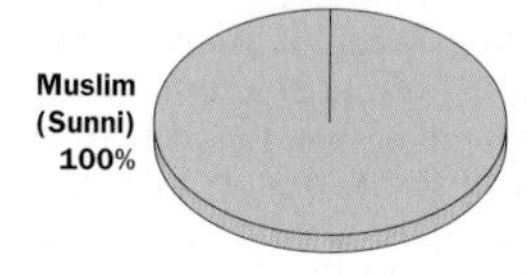
Muslim (Sunni) 100%

The politically dominant Maures make up the majority of the population. The black population is composed of the Havalin, the Senegalese, and the Peulh, Tukolor, and Wolof groups. Ethnic tension centers on the oppression of blacks by Maures. The old black bourgeoisie has now been superseded by a Maurish class; tens of thousands of blacks are estimated to be in slavery. The arrival of 200,000 Maures from Senegal in 1989 caused ethnic tension to come to a head. There were attacks on Senegalese in Mauritania and many fled or were deported to refugee camps along the Senegal River.

Family solidarity among nomads is particularly strong.

POLITICS

Multiparty elections

L. House 2001/2006
U. House 2002/2004

President Maaouya ould Sid Ahmed Taya

AT THE LAST ELECTION

National Assembly 81 seats

79% PRDS
5% AC
16% Others

PRDS = Democratic and Social Republican Party **AC** = Action for Change **Rep** = Representatives of Mauritanians living abroad **Ind** = Independents

Senate 56 seats

93% PRDS
5% Rep
2% Ind

The Senate is indirectly elected

Mauritania officially adopted multiparty democracy in 1991. However, the 1992 and 1997 presidential elections simply returned to power the incumbent military ruler, President Maaouya ould Sid Ahmed Taya, with around 90% of the vote. Opposition parties have accused the government of electoral fraud; initially boycotting legislative elections, they increased their representation in the 2001 poll, but the AC was subsequently banned. The opposition parties are mainly Maure-led. The blacks of the south support exiled parties, such as the Senegal-based African Liberation Forces of Mauritania (FLAM).

Mauritania

Total Land Area : 1 030 700 sq. km (397 953 sq. miles)

POPULATION
over 500 000
over 10 000
under 10 000

LAND HEIGHT
500m/1640ft
200m/656ft
Sea Level

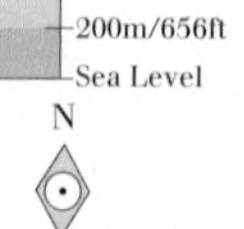
N

0 200 km
0 200 miles

WESTERN SAHARA
ALGERIA
ATLANTIC OCEAN
SAHARA
MALI
SENEGAL
Bîr Mogreïn
'Aïn Ben Tili
Sebkhet Iguetti
Sebkhet Oumm ed Drous Telli
Fdérik
Zouérat
Kediet ej Jill 915m
Nouâdhibou
Atâr
Chinguetti
Ouadane
Et Tidra
Akjoujt
Sebkhet Te-n-Dghâmcha
NOUAKCHOTT
Ouâd Nâga
Moudjéria
Tidjikja
Tichît
Magta' Lahjar
Boutilimit
Keur Massène
Rosso
Lac Rkiz
Aleg
Boûmdeïd
Tâmchekket
'Ayoûn el 'Atroûs
Oualâta
Néma
Bogué
Bababé
Mbagne
Kaédi
Guérou
Kiffa
Mbout
Tintâne
Timbédra
Kankossa
Kobenni
Djiguéni
Amourj
Bassikounou
Ould Yenjé
Sélibabi
Senegal

WORLD AFFAIRS

Joined UN in 1961

AL | CILSS | AU | OIF | OIC

Mauritania seeks to maintain a balance between sub-Saharan Africa and the Arab world, but has had tensions with all its neighbors. It has now effectively withdrawn from the Western Sahara dispute. Relations with Senegal have improved since the conflicts of 1989.

AID

Recipient

$212m (receipts)

Down 3% in 2000

France, Germany, the IMF, OPEC, and Iraq are all donors. Most aid is used for development projects, such as the EU-funded Trans-Mauritanian Highway.

DEFENSE

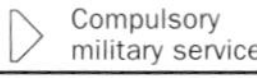

Compulsory military service

$23m

Down 4% in 2000

The 15,000-strong army is a strain on Mauritania's budget. Troops are used increasingly in public works projects. France is the main arms supplier.

ECONOMICS

Inflation 5.9% p.a. (1990–2000)

$978m

250.97–264.01 ouguiyas

SCORE CARD

- ❑ World GNP Ranking........................155th
- ❑ GNP per Capita $370
- ❑ Balance of Payments $90m
- ❑ Inflation 3.3%
- ❑ Unemployment...................................23%

Strengths

Iron from the Cominor mine at Zouérat. Largest gypsum deposits in the world. Copper, yet to be properly exploited. Offshore fishing among the best in West Africa. Significant debt cancellations in 2002.

Weaknesses

Poor land. Drought, locust attacks. Fluctuating commodity prices. Very hot, dry desert climate.

EXPORTS

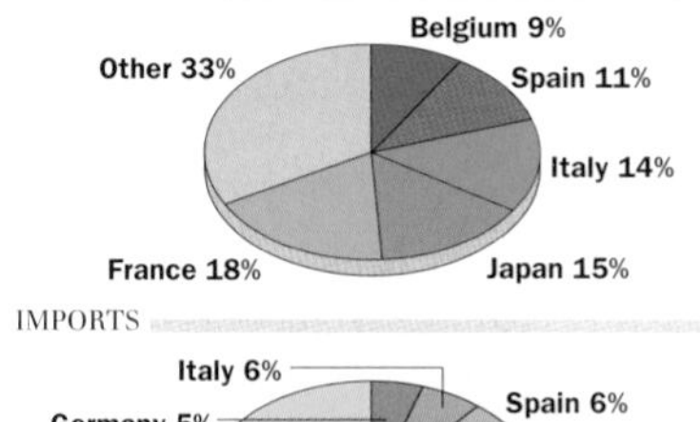
Belgium 9%
Spain 11%
Italy 14%
Japan 15%
France 18%
Other 33%

IMPORTS

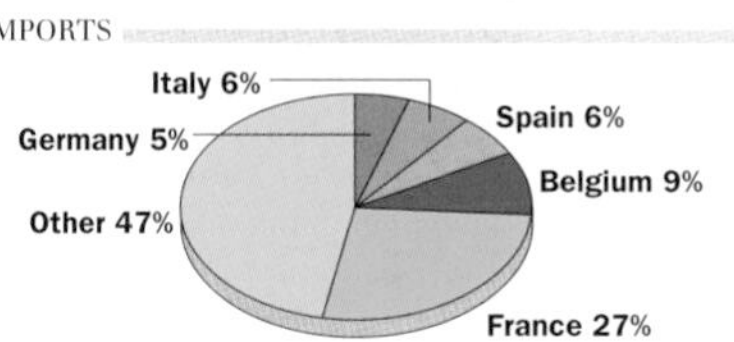
Italy 6%
Spain 6%
Germany 5%
Belgium 9%
Other 47%
France 27%

Mauritania's extreme aridity means that only 1% of the land is arable. Two-thirds of the country are part of the Sahara Desert; sparse vegetation over the rest supports some livestock.

RESOURCES

Electric power 105,000 kw

47,811 tonnes

Not an oil producer

7.6m sheep, 5.1m goats, 1.5m cattle, 4.1m chickens

Iron, gypsum, copper, gold, phosphates, yttrium, diamonds

Iron continues to be exploited, despite low world prices. There are some gold and diamond deposits. Mining and fisheries represent 99.7% of exports. Electricity generation expanded by 40% between 1989 and 1996, and further expansion is expected to come from the Manantali Dam. Phosphates have been found near the Senegal River. Offshore oil exploration started in April 2001.

ENVIRONMENT

Sustainability rank: 126th

2% (0.2% partially protected)

1.2 tonnes per capita

The chief environmental problem in Mauritania is that of the encroaching Sahara Desert, a situation worsened by the droughts of 1973 and 1983, which caused widespread loss of grazing land. The consequent exodus of people away from the land has raised Nouakchott's population from 20,000 in 1960 to almost a million today.

MEDIA

TV ownership medium

Daily newspaper circulation 0.5 per 1000 people

PUBLISHING AND BROADCAST MEDIA

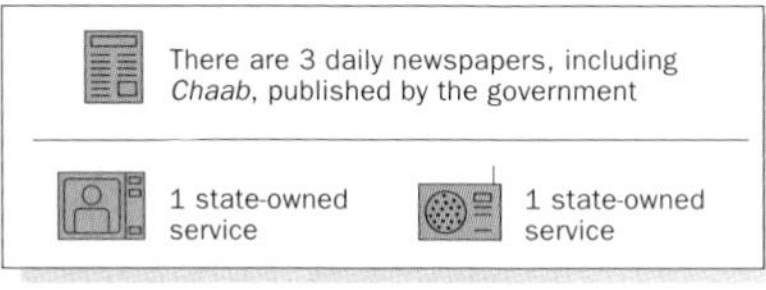

There are 3 daily newspapers, including *Chaab*, published by the government

1 state-owned service | 1 state-owned service

The press is heavily censored, and the broadcast media are state-owned. *Chaab*, the government newspaper, is also published in French (*Horizons*).

CRIME

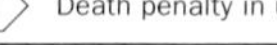

Death penalty in use

1400 prisoners

Down 45% 1997–1999

Key issues are smuggling, robbery, and a growing number of abandoned children in Nouakchott and other towns.

CHRONOLOGY

Once part of the Islamic Almoravid state, Mauritania became a French colony in 1814.

- ❑ **1960** Independence; one-party state.
- ❑ **1972** Peace with Polisario in war waged over Western Sahara.
- ❑ **1984** Col. Maaouya Taya takes power in bloodless coup.
- ❑ **1992** First multiparty elections.
- ❑ **1997** Taya reelected as president.

EDUCATION

School leaving age: 16

42%

12,912 students

Despite improvements in education, over half the population continues to be illiterate. Arabic has been compulsory in all schools since 1988.

HEALTH

No welfare state health benefits

1 per 10,000 people

Diarrheal and respiratory diseases, influenza, tuberculosis

Historic regional inequalities persist and the best facilities are in the capital. The overall level of care is on a par with neighboring states.

SPENDING

GDP/cap. increase

CONSUMPTION AND SPENDING

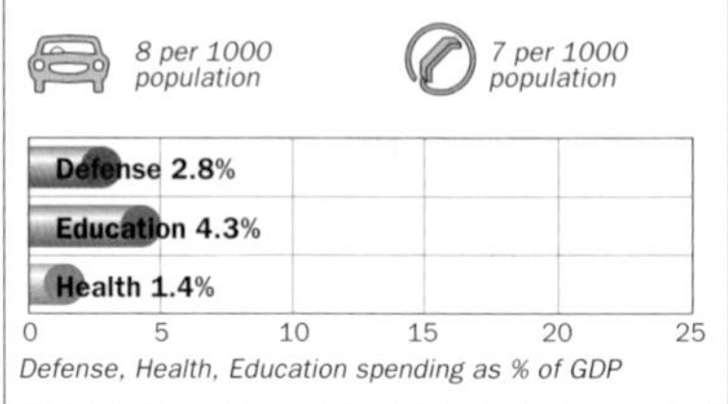
8 per 1000 population
7 per 1000 population
Defense 2.8%
Education 4.3%
Health 1.4%
0
5
10
15
20
25
Defense, Health, Education spending as % of GDP

The small ruling Maurish elite forms the richest sector. Wealthy Maures travel to Mecca, Saudi Arabia, to perform the *haj* (Muslim pilgrimage).

WORLD RANKING

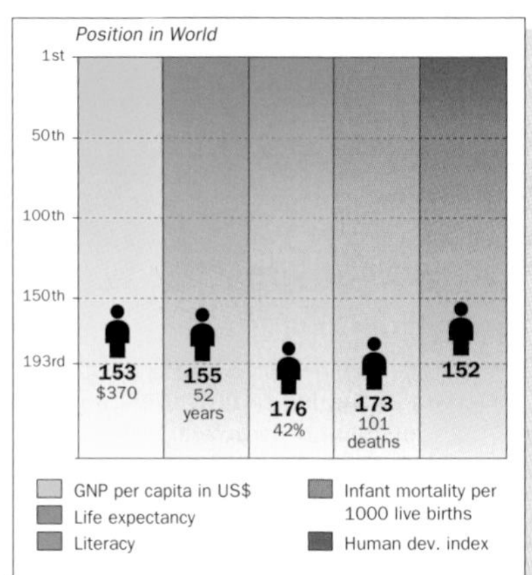
Position in World
1st
50th
100th
150th
193rd
153
$370
155
52 years
176
42%
173
101 deaths
152
GNP per capita in US$
Life expectancy
Literacy
Infant mortality per 1000 live births
Human dev. index

M

MAURITIUS

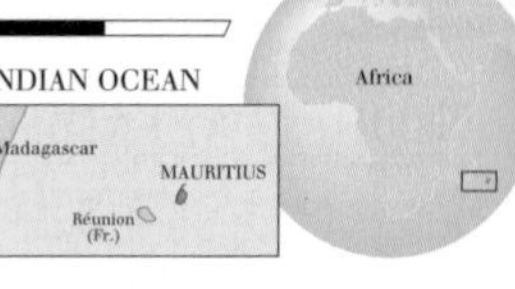
INDIAN OCEAN
Africa
Madagascar
MAURITIUS
Réunion (Fr.)

OFFICIAL NAME: Republic of Mauritius **CAPITAL:** Port Louis
POPULATION: 1.2 million **CURRENCY:** Mauritian rupee **OFFICIAL LANGUAGE:** English

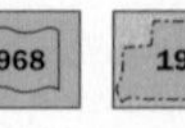
1968

1968

March 12
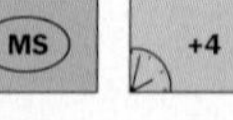
MS

+4

+230

.mu

THE ISLANDS THAT MAKE UP Mauritius lie in the Indian Ocean east of Madagascar. The main island, from which the country takes its name, is of volcanic origin and surrounded by coral reefs. Along with Rodrigues to the east, the country includes the Agalega Islands and the Cargados Carajos Shoals (500 km – 300 miles – to the north). Mauritius has enjoyed considerable economic success following recent industrial diversification and the expansion of tourism.

CLIMATE

Tropical oceanic

WEATHER CHART FOR PORT LOUIS

Average daily temperature | Rainfall

°C/°F J F M A M J J A S O N D cm/in

40/104 30/86 20/68 10/50 0/32 -10/14 -20/-4 | 40/16 30/12 20/8 10/4 0

The climate is subtropical and humid. December to March are the hottest and wettest months. Tropical cyclones are an occasional threat during this time.

TRANSPORTATION

Drive on left

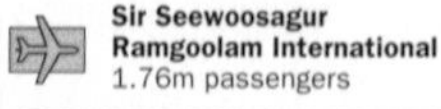
Sir Seewoosagur Ramgoolam International
1.76m passengers

51 ships
206,000 grt

THE TRANSPORTATION NETWORK

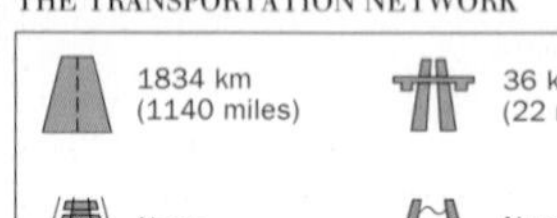
1834 km (1140 miles)
36 km (22 miles)
None
None

Roads are extensive, but often congested. Plans exist for a monorail link between Port Louis and Curepipe.

TOURISM

Visitors : Population 1:1.8

656,000 visitors

Up 14% in 2000

MAIN TOURIST ARRIVALS

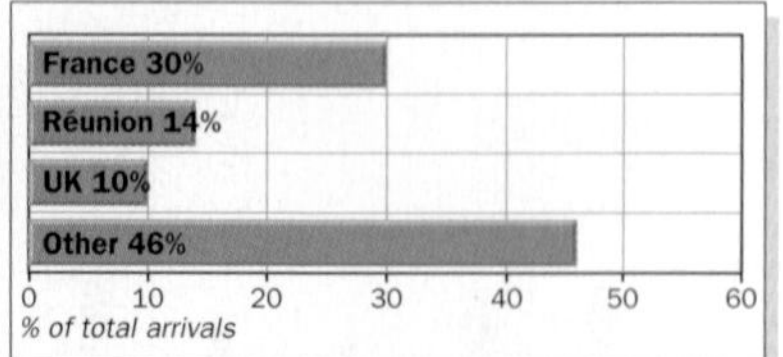
France 30%
Réunion 14%
UK 10%
Other 46%
0 10 20 30 40 50 60
% of total arrivals

Tourism expanded rapidly in the 1990s. Spectacular beaches, water sports, and big game fishing are major attractions. Around 30% of visitors each year come from France.

PEOPLE

Pop. density high

French Creole, Hindi, Urdu, Tamil, Chinese, English, French

645/km² (1671/mi²)

THE URBAN/RURAL POPULATION SPLIT

41% | 59%

RELIGIOUS PERSUASION

Protestant 2% | Other 3% | Muslim 17% | Hindu 52% | Roman Catholic 26%

Mauritius is one of the world's most densely populated countries. The majority of the population descend from indentured Indian laborers brought over in the 19th century. Creoles make up 27% of the population, while 3% are of Chinese origin. Clashes between Hindus, Muslims, and Creoles no longer occur, although Creoles complain of discrimination.

POLITICS

Multiparty elections

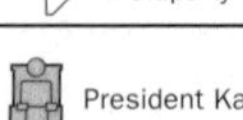
President Karl Offmann

2000/2005

AT THE LAST ELECTION

National Assembly 70 seats

83% MSM/MMM | 11% PTr/PMXD | 3% MR | 3% OPR

MSM/MMM = Mauritian Socialist Movement/Mauritian Militant Movement **PTr/PMXD** = Labour Party/Mauritian Social Democratic Party of Xavier Duval **OPR** = Organization of the People of Rodrigues **MR** = Mouvement Rodriguais

62 members of the National Assembly are directly elected, and up to eight are appointed from the highest losers

Mauritius became a republic in 1992. Navin Ramgoolam of the PTr became prime minister in 1995, promoting regional integration and economic liberalization to attract investment. However, corruption scandals led to early elections in 2000. The PTr was defeated and Sir Anerood Jugnauth, Ramgoolam's predecessor, now heads a new coalition between his MSM and the MMM.

WORLD AFFAIRS

Joined UN in 1968

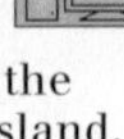

Comm | COMESA | COI | AU | SADC

Mauritius hosted a francophone nations summit in 1995, and the first OAU human rights conference in 1999. Links with South Africa and India are important. Disputes persist over UK-administered Diego Garcia and the French-ruled island of Tromelin.

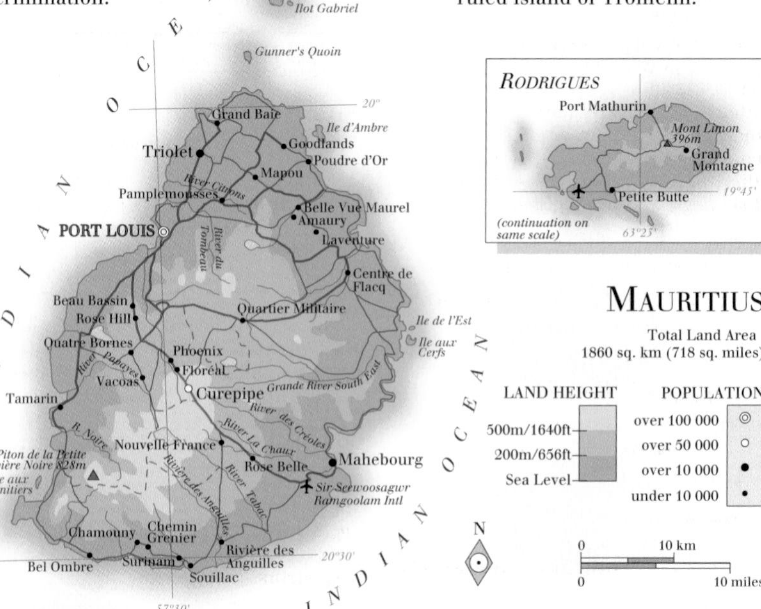
INDIAN OCEAN
Round Island
Flat Island
Ilot Gabriel
Gunner's Quoin
Grand Baie
Ile d'Ambre
Goodlands
Triolet
Poudre d'Or
Mapou
Pamplemousses
Belle Vue Maurel
Amaury
PORT LOUIS
Laventure
Centre de Flacq
Beau Bassin
Rose Hill
Quartier Militaire
Ile de l'Est
Ile aux Cerfs
Quatre Bornes
Phoenix
Floréal
Vacoas
Tamarin
Curepipe
Grande River South East
Nouvelle France
Mahebourg
Rose Belle
Sir Seewoosagur Ramgoolam Intl
Piton de la Petite Rivière Noire 828m
Ile aux Bénitiers
Chamouny
Chemin Grenier
Bel Ombre
Surinam
Souillac
Rivière des Anguilles
RODRIGUES
Port Mathurin
Mont Limon 396m
Grand Montagne
Petite Butte
(continuation on same scale)
MAURITIUS
Total Land Area : 1860 sq. km (718 sq. miles)
LAND HEIGHT
500m/1640ft
200m/656ft
Sea Level
POPULATION
over 100 000
over 50 000
over 10 000
under 10 000
10 km
10 miles

M

AID

 Recipient

$20m (receipts) — Down 52% in 2000

Aid is predominantly bilateral, with the EU and France as the main donors. Mauritius also receives aid from Arab aid agencies and Japan, and from the UN and other international organizations. The World Bank assisted a five-year conservation program, starting in 1990, and promised $53 million toward transforming Port Louis into a free port.

DEFENSE

No compulsory military service

 $87m

 Down 4% in 2000

Mauritius has no standing defense forces. There is, however, a 1100-strong special police mobile unit to ensure internal security. There is also a coastguard numbering 500.

ECONOMICS

Inflation 5.9% p.a. (1990–2000)

 $4.45bn

 27.82–30.25 Mauritian rupees

SCORE CARD

- ❑ WORLD GNP RANKING........116th
- ❑ GNP PER CAPITA........$3750
- ❑ BALANCE OF PAYMENTS........–$33m
- ❑ INFLATION........4.2%
- ❑ UNEMPLOYMENT........6%

STRENGTHS

Strong economic growth. The sugar industry accounts for 30% of export earnings. Export processing zone (EPZ), especially for clothing manufacture. Tourism. Highly educated workforce. Ranked as the most competitive economy in Africa by the World Economic Forum in 1999. Development as offshore financial center.

WEAKNESSES

Vulnerability to fluctuating world sugar price and droughts. 75% of food requirements are imported. Few crops other than sugar can be grown. Lack of strategic resources. Remoteness.

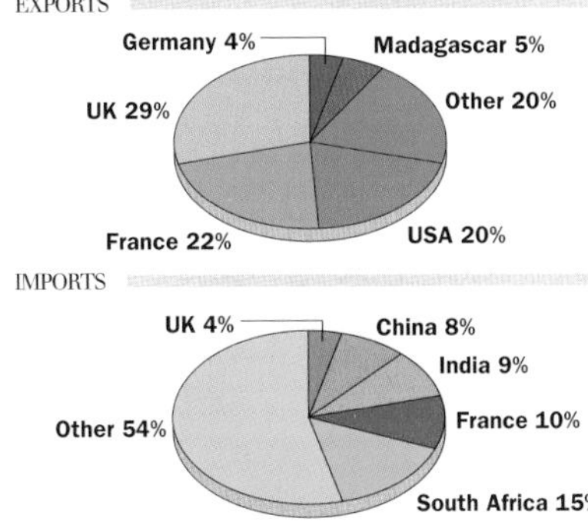
EXPORTS
Germany 4%
Madagascar 5%
UK 29%
Other 20%
France 22%
USA 20%
IMPORTS
UK 4%
China 8%
India 9%
Other 54%
France 10%
South Africa 15%

Villagers at a water source in the center of Mauritius island. Mauritius' main rivers are used for hydropower generation.

RESOURCES

Electric power 364,000 kw

 12,089 tonnes

 Not an oil producer

 95,000 goats, 28,000 cattle, 4.5m chickens

 None

Mauritius has to import oil, so the government has invested heavily in alternative indigenous energy schemes, including HEP generation, and power plants fueled by bagasse (a by-product of the sugar industry). Industrialization and agricultural diversification make up for limited natural resources.

ENVIRONMENT

Not available

 8% (0.1% partially protected)

 1.5 tonnes per capita

Rapid industrialization as well as unchecked hotel building have caused environmental problems. Coral reefs are under threat from both coral sand mining and the discharging of untreated sewage into the sea.

MEDIA

TV ownership high

 Daily newspaper circulation 76 per 1000 people

PUBLISHING AND BROADCAST MEDIA

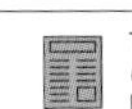 There are 10 daily newspapers. *Le Quotidien, L'Express,* and *Le Mauricien* have the largest circulations

 1 independent service

 1 independent service

Mauritius has an active press, subject to few regulations and with a wide readership. Newspapers are published in English, French, Creole, Hindi, Chinese, and Tamil. Opposition parties complain that TV and radio broadcasts are consistently biased toward the government.

CRIME

No death penalty

 1985 prisoners

 Down 19% in 1999

Crime rates on the main island are fairly low. There has been a small increase in thefts and narcotics smuggling. Outlying islands are virtually crime-free.

CHRONOLOGY

Mauritius was colonized and ruled by the Dutch in the 17th century, the French (1710–1810), and the British.

- ❑ **1959** First full elections.
- ❑ **1968** Independence. Riots between Creoles and Muslims.
- ❑ **1982–1995** Sir Anerood Jugnauth prime minister; forms MSM.
- ❑ **1992** Becomes a republic.
- ❑ **1995** Elections won by PTr–MMM.
- ❑ **2000** Return of Jugnauth.

EDUCATION

School leaving age: 12

  85%

 7559 students

Educational provision is good, and 91% of Mauritians under 30 are literate. The University of Mauritius has about 2000 students.

HEALTH

Welfare state health benefits

 1 per 1111 people

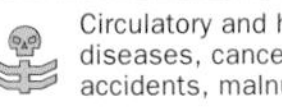 Circulatory and heart diseases, cancers, accidents, malnutrition

In Mauritius free health care is universally available. There are 14 state hospitals and six private clinics.

SPENDING

GDP/cap. increase

CONSUMPTION AND SPENDING

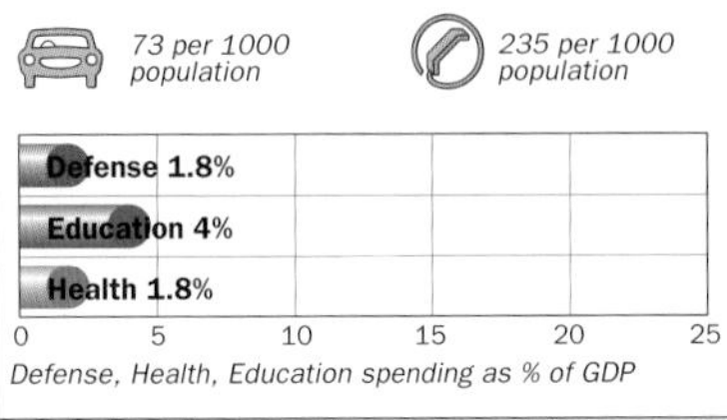
73 per 1000 population
235 per 1000 population
Defense 1.8%
Education 4%
Health 1.8%
0
5
10
15
20
25
Defense, Health, Education spending as % of GDP

French-descended hotel and plantation owners form the country's wealthiest social group. Government employees are well paid.

WORLD RANKING

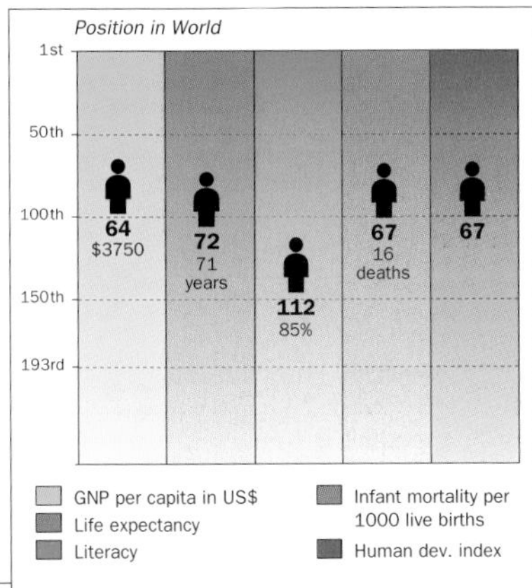
Position in World
1st
50th
100th
150th
193rd
64
$3750
72
71
years
112
85%
67
16
deaths
67
GNP per capita in US$
Life expectancy
Literacy
Infant mortality per 1000 live births
Human dev. index

M

Mexico

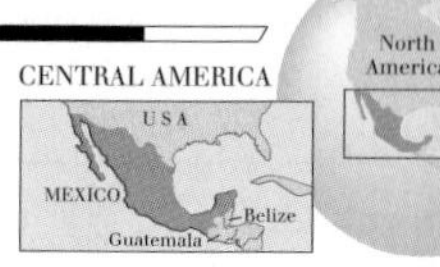
CENTRAL AMERICA
North America
USA
MEXICO
Belize
Guatemala

Official Name: United Mexican States **Capital:** Mexico City
Population: 100.4 million **Currency:** Mexican peso **Official Language:** Spanish

1836

1848

Sept 16
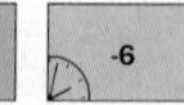
MEX
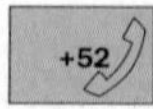
-6

+52

.mx

INCREASINGLY CONSIDERED a part of North rather than Central America, Mexico separates the US from the rest of Latin America. Coastal plains along its Pacific and Caribbean seaboards rise into an arid central plateau, which includes one of the world's biggest conurbations, Mexico City, built on the site of the Aztec capital, Tenochtitlán. Colonized by the Spanish for its silver mines, Mexico achieved independence in 1836. In the "Epic Revolution" of 1910–1920, in which 250,000 died, much of modern Mexico's structure was established. In 1994, Mexico signed the North American Free Trade Agreement (NAFTA).

The cathedral of Santa Prisca at Taxco near Cuernavaca. It was built in Spanish Churriguera style between 1748 and 1758.

CLIMATE

Tropical/mountain/desert

WEATHER CHART FOR MEXICO CITY

Average daily temperature / Rainfall

°C/°F J F M A M J J A S O N D cm/in

40/104, 30/86, 20/68, 10/50, 0/32, -10/14, -20/-4 — 40/16, 30/12, 20/8, 10/4, 0

The plateau and high mountains are warm for much of the year. The Pacific coast has a tropical climate.

TRANSPORTATION

Drive on right

Benito Juárez International, Mexico City 21m passengers

626 ships 1.09m grt

THE TRANSPORTATION NETWORK

96,221 km (59,789 miles)

6335 km (3936 miles)

26,595 km (16,526 miles)

2900 km (1802 miles)

A privately financed $14 billion road network, some 6000 km (3730 miles) of toll roads, is seriously underused and a commercial failure. Regional travel is mainly by bus; the unreliable railroad is largely for freight. Plans to construct a new international airport serving Mexico City were abandoned in 2002 as local farmers staged violent protests against the implied acquisition of their lands.

TOURISM

Visitors : Population 1:5.1

19.8m visitors

Down 4% in 2001

MAIN TOURIST ARRIVALS

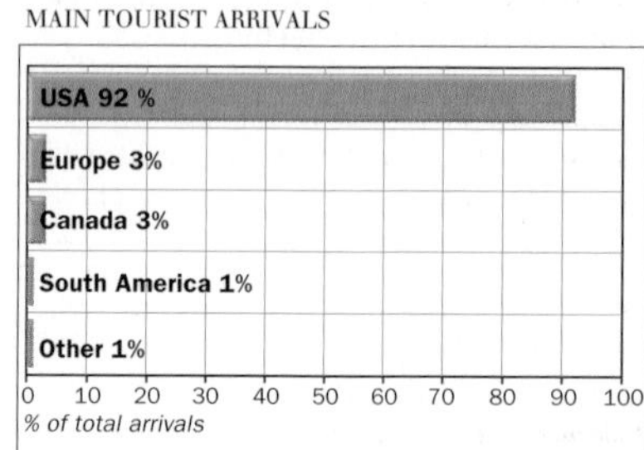
USA 92 %
Europe 3%
Canada 3%
South America 1%
Other 1%
0 10 20 30 40 50 60 70 80 90 100
% of total arrivals

Tourism is one of the largest employment sectors in Mexico, and a major source of foreign exchange. Attractions include excellent beach resorts such as Acapulco on the Pacific coast, and the new resorts of the Peninsula de Yucatán on the Caribbean coast. Impressive coastal scenery, volcanoes, the Sierra Madre, and archaeological remains of Aztec and Mayan civilizations, designated as World Heritage sites, are major draws, as are the many Spanish colonial cities, such as Morelia and Guadalajara, which have remained virtually intact since their construction after the conquest.

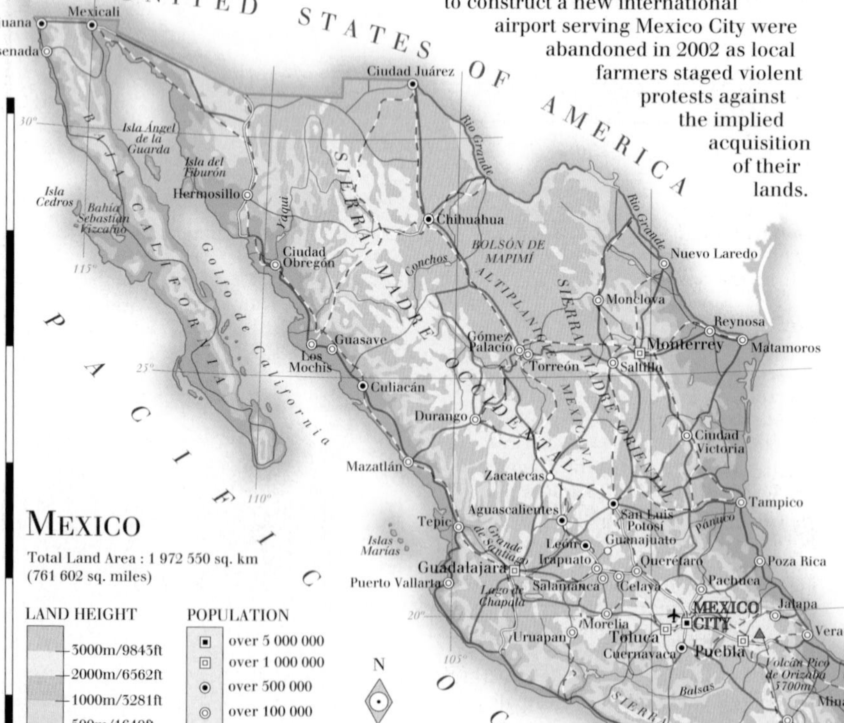
UNITED STATES OF AMERICA
PACIFIC OCEAN
GULF OF MEXICO
CARIBBEAN SEA
Golfo de California
Bahía de Campeche
Golfo de Tehuantepec
Yucatan Channel
Tijuana
Mexicali
Ensenada
Ciudad Juárez
Hermosillo
Chihuahua
Ciudad Obregón
Nuevo Laredo
Monclova
Reynosa
Matamoros
Monterrey
Saltillo
Torreón
Gómez Palacio
Los Mochis
Guasave
Culiacán
Durango
Mazatlán
Zacatecas
Ciudad Victoria
Tampico
San Luis Potosí
Aguascalientes
Tepic
León
Guanajuato
Irapuato
Querétaro
Guadalajara
Puerto Vallarta
Salamanca
Celaya
Pachuca
Poza Rica
Morelia
Uruapan
Toluca
Cuernavaca
MEXICO CITY
Puebla
Jalapa
Veracruz
Volcán Pico de Orizaba 5700m
Coatzacoalcos
Minatitlán
Villahermosa
Oaxaca
Acapulco
Tuxtla Gutiérrez
Tapachula
Mérida
Campeche
PENÍNSULA DE YUCATÁN
Isla de Cozumel
BELIZE
GUATEMALA
SIERRA MADRE OCCIDENTAL
SIERRA MADRE ORIENTAL
SIERRA MADRE DEL SUR
BOLSÓN DE MAPIMÍ
ALTIPLANICIE MEXICANA
ISTMO DE TEHUANTEPEC
BAJA CALIFORNIA
Isla Ángel de la Guarda
Isla del Tiburón
Isla Cedros
Bahía Sebastián Vizcaíno
Islas Marías
Lago de Chapala
Río Grande
Conchos
Yaqui
Pánuco
Balsas
Grijalva
Usumacinta
Hondo
MEXICO
Total Land Area : 1 972 550 sq. km (761 602 sq. miles)
LAND HEIGHT
3000m/9843ft
2000m/6562ft
1000m/3281ft
500m/1640ft
200m/656ft
Sea Level
POPULATION
over 5 000 000
over 1 000 000
over 500 000
over 100 000
over 50 000
N
0 200 km
0 200 miles

PEOPLE

Pop. density medium

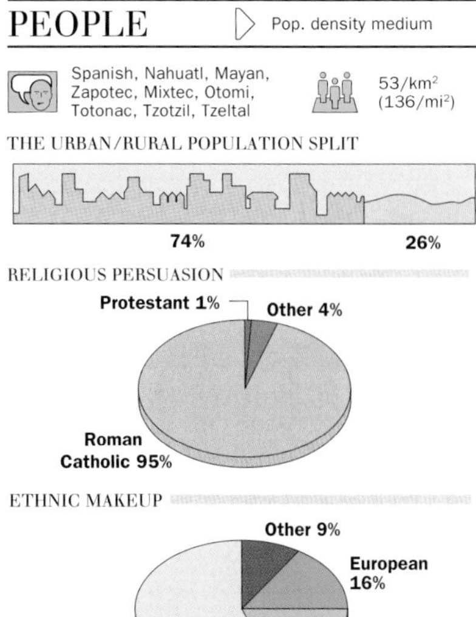
Spanish, Nahuatl, Mayan, Zapotec, Mixtec, Otomi, Totonac, Tzotzil, Tzeltal
53/km² (136/mi²)
THE URBAN/RURAL POPULATION SPLIT
74%
26%
RELIGIOUS PERSUASION
Protestant 1%
Other 4%
Roman Catholic 95%

ETHNIC MAKEUP

Other 9%

European 16%

Indigenous Amerindian 20%

Mestizo 55%

POPULATION AGE BREAKDOWN

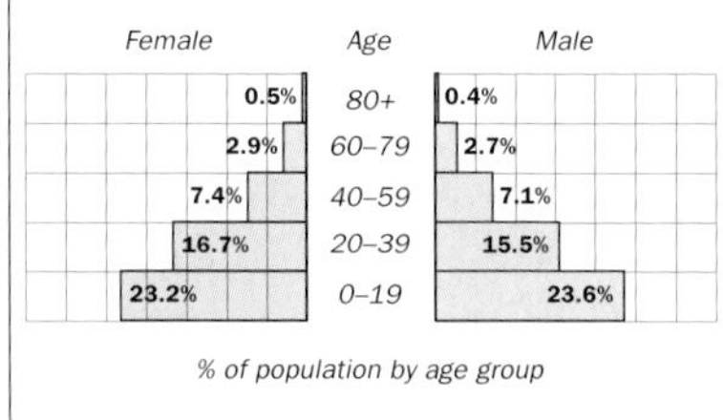
Female
Age
Male
0.5%
80+
0.4%
2.9%
60–79
2.7%
7.4%
40–59
7.1%
16.7%
20–39
15.5%
23.2%
0–19
23.6%
% of population by age group

While most Mexicans are *mestizo* (mixed race), it is Mexico's Amerindian culture which is promoted by the state. This obscures the fact that rural Amerindians are largely segregated from Hispanic society, a situation that dates back to the Spanish colonial period and which has only recently been seriously challenged. The 1994 Chiapas Zapatista (EZLN) guerrilla uprising was on behalf of Amerindian rights, and in protest against the poverty of landless Amerindians. President Vicente Fox promised to act, but the Indigenous Rights and Culture Bill, watered down by a hostile Congress and enacted in 2001, was rejected by the EZLN and all the main indigenous groups.

The small black community, which is concentrated along the eastern coast, is well integrated.

As in much of Latin America, men retain their dominance in business and relatively few women take part in the political process.

POLITICS

Multiparty elections

L. House 2000/2003
U. House 2000/2006

President Vicente Fox

AT THE LAST ELECTION

Chamber of Deputies 500 seats

45% PAN–PVEM	42% PRI	13% PRD–PT

PAN–PVEM = Alliance for Change (National Action Party and Green Party) **PRI** = Institutional Revolutionary Party **PRD–PT** = Alliance for Mexico (Party of the Democratic Revolution and Labor Party)

Senate 128 seats

45% PRI	42% PAN–PVEM	13% PRD–PT

Mexico was a multiparty democracy in name only until electoral reform in 1997.

Profile

The PRI dominated Mexico from 1929. Regional elections in the 1990s showed the strength of opposition parties, and after grudging electoral reform, the PRI lost its monopoly on power in 1997. President Vicente Fox's PAN became the largest party in the Chamber of Deputies in the 2000 elections, but it lacks an overall majority. Fox promised a broadly conceived "government of transition."

Main Political Issues

President Fox's administration

Strong opposition in the Congress has frustrated President Fox's program of structural reform: by relying on advisers and cabinet rule, he neglected to consult the opposition and failed to build a legislative consensus.

Indigenous rights

Fox initially honored some conditions of the Zapatista National Liberation Army (EZLN) for new peace talks, by releasing prisoners and withdrawing troops from Chiapas. Serious revisions by the Congress to the Indigenous Rights and Culture Bill were rejected by the EZLN and indigenous groups, setting the peace process back.

Future of the PRI

Election defeats left the PRI faction-ridden and rudderless. Roberto Madrazo, a hard-liner elected party leader in 2002, has to unite the PRI to win the 2006 presidential elections.

***Vicente Fox**, elected president in 2000, ending 70 years of PRI dominance.*

***Subcomandante Marcos**, leader of the Zapatista National Liberation Army.*

WORLD AFFAIRS

Joined UN in 1945

G15
NAFTA

OECD

OAS

RG

NAFTA has bonded the economies of Mexico and the US. Mexican imports have undercut US jobs, as has the southward relocation of multinationals to benefit from cheap Mexican labor. Heightened US border security post-11 September complicated a bilateral agreement to deal with high flows of illegal immigrants but also hindered powerful narcotics cartels supplying the US market.

Mexico has free trade agreements with 32 countries, as well as with the EU and EFTA, and competes with Brazil to play a leading negotiating role for Latin America in the formation of a Free Trade Area of the Americas (FTAA) due to be signed in 2005.

Mexico's swing vote in the UN in 2002, which triggered a resolution on human rights in Cuba, inflamed relations with the Castro regime and set President Fox against the Senate on the issue of who decides foreign policy.

AID

Recipient

$37m (receipts)

Loan repayments exceeded aid received in 2000

In 2002 the IDB made its biggest loan – $1 billion – to support a six-year antipoverty project in Mexico. European and US NGOs provide assistance.

M

CHRONOLOGY

The Aztec kingdom of Montezuma II was defeated in war by the Spaniard, Hernán Cortés, in 1521. By 1546, the Spaniards had discovered large silver mines at Zacatecas. Mexico, then known as New Spain, became a key part of the Spanish colonial empire.

- **1810** Fr. Miguel Hidalgo leads abortive rising against Spanish.
- **1821** Spanish viceroy forced to leave by Agustín de Iturbide.
- **1822** Federal Republic established.
- **1823** Texas opened to US immigration.
- **1829** Spanish military expedition fails to regain control.
- **1836** US is first country to recognize Mexico's independence. Spain follows suit. Texas declares its independence from Mexico.
- **1846** War breaks out with US.
- **1848** Loses modern-day New Mexico, Arizona, Nevada, Utah, California, and part of Colorado.
- **1858–1861** War of Reform won by anticlerical Liberals.
- **1862** France, Britain, and Spain launch military expedition.

⇨

CHRONOLOGY *continued*

- **1863** French troops capture Mexico City. Maximilian of Austria established as Mexican emperor.
- **1867** Mexico recaptured by Benito Juárez. Maximilian shot.
- **1876** Porfirio Díaz president. Economic growth; rail system built.
- **1901** First year of oil production.
- **1910–1920** Epic Revolution provoked by excessive exploitation by foreign companies and desire for land reform. 250,000 killed.
- **1911** Díaz overthrown by Francisco Madero. Guerrilla war breaks out in north. Emilio Zapata leads peasant revolt in the south.
- **1913** Madero murdered.
- **1917** New constitution limits power of Church. Minerals and subsoil rights reserved for the nation.
- **1926–1929** Cristero rebellion led by militant Catholic priests.
- **1929** National Revolutionary Party (later PRI) formed.
- **1934** Gen. Cárdenas president. Land reform accelerated, cooperative farms established, railroads nationalized, and US and UK oil companies expelled.
- **1940s** US war effort helps Mexican economy to grow.
- **1970** Accelerating population growth reaches 3% a year.
- **1982** Mexico declares it cannot repay its foreign debt of over $800 billion. IMF insists on economic reforms to reschedule the debt.
- **1984** Government contravenes constitution by relaxing laws on foreign investment.
- **1985** Earthquake in Mexico City. Official death toll 7000. Economic cost estimated at $425 million.
- **1988** Carlos Salinas de Gortari, minister of planning during the earthquake, elected president.
- **1990** Privatization program begun.
- **1994–1995** Guerrilla rebellion in southern Chiapas state brutally suppressed by army: 100 dead. Mexico joins NAFTA. PRI presidential candidate Luis Colosio murdered. Ernesto Zedillo replaces him and is elected. Economic crisis.
- **1997–1999** PRI's monopoly on power in Congress ended. Banks bailed out.
- **2000** July, PAN wins presidency and elections, ending 70 years of PRI rule. December, President Vicente Fox takes office.
- **2001** EZLN guerrillas and supporters make 16-day motorcade from Chiapas to Mexico City to push for indigenous rights law.
- **2002** Former governor of Tabasco, Roberto Madrazo, traditionalist hard-liner, elected new PRI leader.

DEFENSE

Compulsory military service

 $5.23bn

 Up 22% in 2000

Mexico has no ambitions beyond its borders, and the army acts to defend internal security. The military has, on the whole, avoided direct interference in politics. Most arms procurement is from the US and France. In 1994, the role of controlling the border with the US was passed to the police.

The Zapatista rebellion in Chiapas in 1994 elicited a brutal response from the army, acting on PRI orders. The increasing militarization of the state over the next six years hindered the peace process and led to a proliferation of paramilitaries, with the tacit blessing of the local PRI, who were blamed by human rights groups for the massacre of Amerindians. The PAN government has withdrawn some forces from key areas of the state.

MEXICAN ARMED FORCES

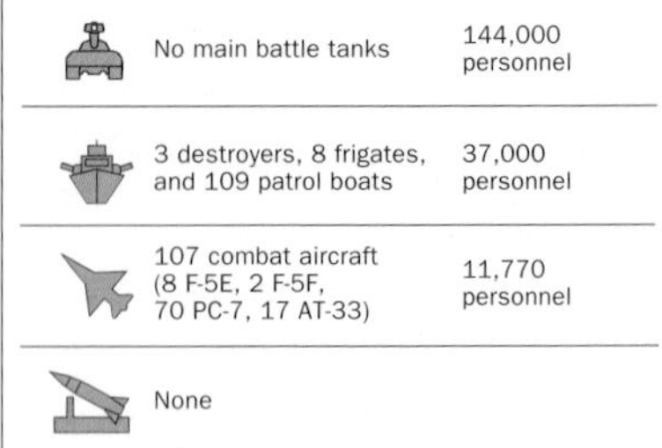

No main battle tanks	144,000 personnel
3 destroyers, 8 frigates, and 109 patrol boats	37,000 personnel
107 combat aircraft (8 F-5E, 2 F-5F, 70 PC-7, 17 AT-33)	11,770 personnel
None	

ECONOMICS

Inflation 19% p.a. (1990–2000)

$497bn

9.609–9.169 Mexican pesos

SCORE CARD

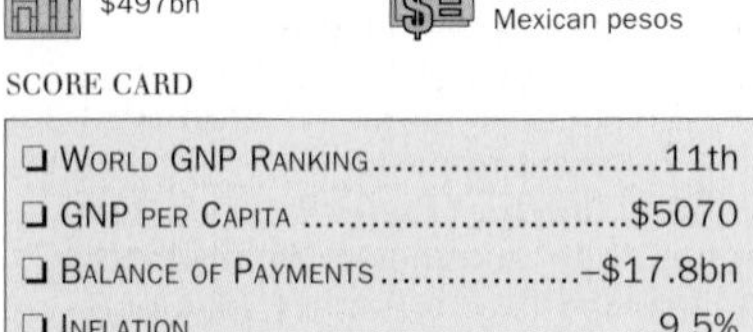

World GNP Ranking	11th
GNP per Capita	$5070
Balance of Payments	–$17.8bn
Inflation	9.5%
Unemployment	2%

EXPORTS

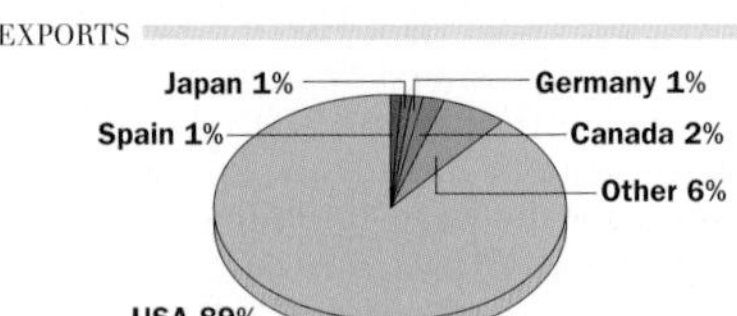
Japan 1%
Germany 1%
Spain 1%
Canada 2%
Other 6%
USA 89%

IMPORTS

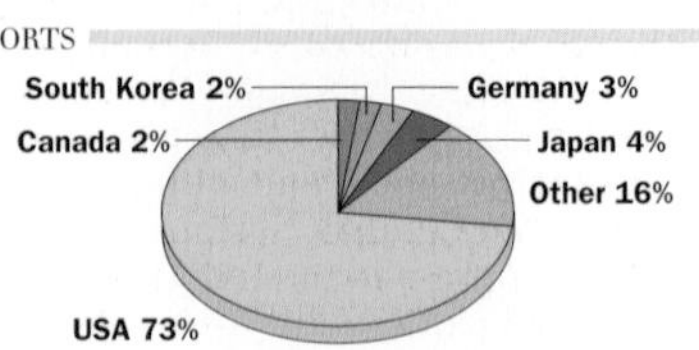
South Korea 2%
Germany 3%
Canada 2%
Japan 4%
Other 16%
USA 73%

Strengths

Global oil producer, with substantial reserves. Extensive mineral resources. Strong foreign direct investment. Diversification of exports. NAFTA membership. Low overheads.

Weaknesses

Debt burden. Vulnerable currency. Corruption. Affected by oil price changes and US slowdown. Weak tax system.

Profile

While in power, the PRI effectively ran the economy. The debt crisis of the 1980s, however, forced privatizations. The 1994 peso crisis needed a US-led $20 billion international bailout and resulted in a severe slump. The Zedillo government launched tough reforms, but a global loss of confidence in emerging markets affected growth.

ECONOMIC PERFORMANCE INDICATOR

Consumer Price Index
GDP
400
300
200
100
0
280
220
160
100
40
Consumer price index 1995=100
GDP 1996=100
1996
1997
1998
1999
2000

Tighter fiscal management was rewarded by 2000, when investor confidence improved. The Fox government's tight fiscal and monetary stance secured the approval of the IMF and major credit agencies, giving the country greater access to cheaper foreign capital. However, pledges on social spending and poverty relief suffered, and promises of large job creation projects also went unfulfilled, as the export-led economy echoed the slowdown in the US in late 2001.

MEXICO : MAJOR BUSINESSES

Food processing
Vehicle assembly
Petrochemicals
Oil refining
Computers
Silver mining
Electronics
Brewing
Textiles
Tijuana
Ciudad Juárez
Monterrey
Reynosa
Tampico
Minatitlán
Durango
Guadalajara
Salamanca
Mexico City
0 400 km
0 400 miles
* significant multinational ownership

RESOURCES

Electric power 45.6m kw

1.25m tonnes

3.56m b/d (reserves 26.9bn barrels)

30.6m cattle, 17.8m pigs, 9.1m goats, 496m chickens

Oil, gas, gold, silver, copper, coal, fluorite, mercury, antimony

ELECTRICITY GENERATION

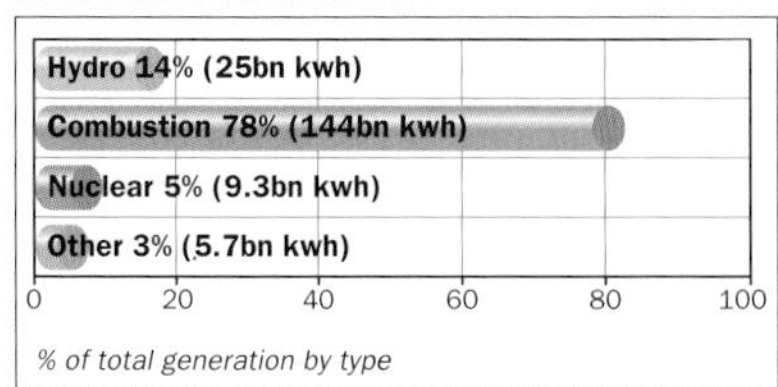
Hydro 14% (25bn kwh)
Combustion 78% (144bn kwh)
Nuclear 5% (9.3bn kwh)
Other 3% (5.7bn kwh)
0
20
40
60
80
100
% of total generation by type

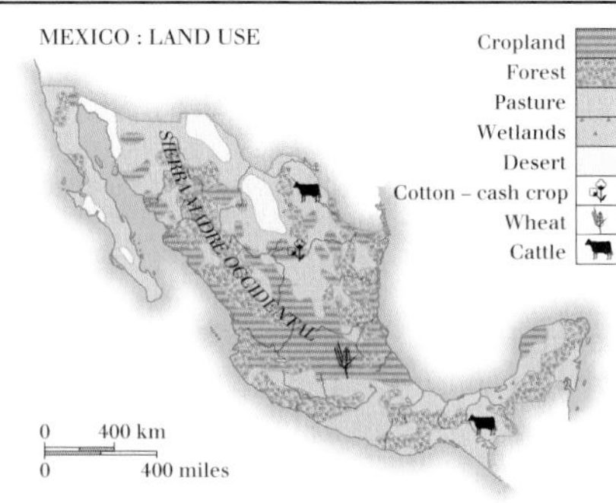
MEXICO : LAND USE
Cropland
Forest
Pasture
Wetlands
Desert
Cotton – cash crop
Wheat
Cattle
SIERRA MADRE OCCIDENTAL
0 400 km
0 400 miles

Mexico is one of the largest oil exporters outside OPEC. Most oil production comes from offshore drilling platforms in the Gulf of Mexico. The industry was state-owned and state-run by PEMEX, the world's fifth-largest oil company, employing 120,000 people. The decision to privatize petrochemical plants has provoked serious social unrest, and further sell-offs and deregulation remain politically highly sensitive. Despite its oil reserves, Mexico has embarked on a nuclear power program and projects to modernize the national electricity grid and boost natural gas production to overcome an energy crisis.

ENVIRONMENT

Sustainability rank: 92nd

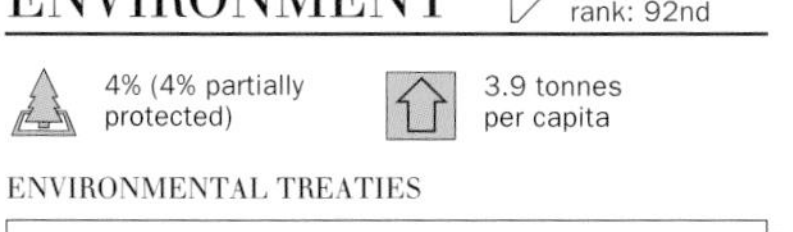
4% (4% partially protected)
3.9 tonnes per capita

ENVIRONMENTAL TREATIES

Yes Yes Yes

Yes Yes Yes

Mexico City, largely unplanned, struggles to accommodate around 20 million inhabitants as the absence of environmental controls contributes to perhaps the world's worst air quality and waste problems. PEMEX (the state petroleum company) stands accused of massive pollution. *Maquiladoras* – assembly plants on the Mexico–US border – have no effective environmental controls and are usually surrounded by slums. Environmentalists oppose the intense development of tourism along the coast and are concerned about high rates of deforestation.

MEDIA

TV ownership high

Daily newspaper circulation 97 per 1000 people

PUBLISHING AND BROADCAST MEDIA

There are 295 daily newspapers. *Excélsior* is a prominent newspaper both within Mexico and the rest of Latin America

Many state-owned and independent services

Many state-owned and independent services

The PRI in particular has historically manipulated the media, being accused of denying electoral opponents airtime.

CRIME

Death penalty not used in practice

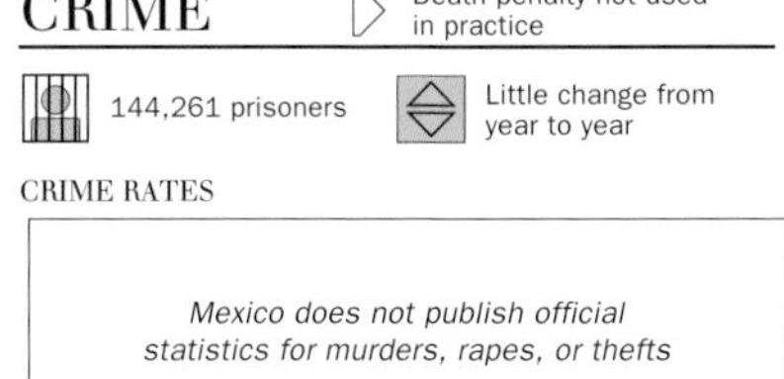
144,261 prisoners
Little change from year to year

CRIME RATES

Mexico does not publish official statistics for murders, rapes, or thefts

Northern Mexico is a major center for narcotics shipments to the US. Antidrugs police are accused by the US of corruption. Guns are rife and minor incidents may end in shootings. The high crime rate in Mexico City is a major political issue. Reforms of the corrupt judiciary and the whole police force are perennial issues.

EDUCATION

School leaving age: 14

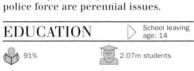

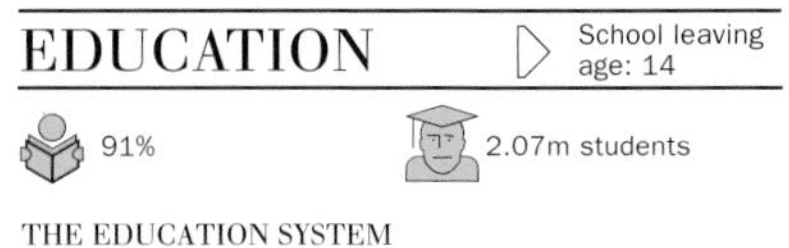
91%
2.07m students

THE EDUCATION SYSTEM

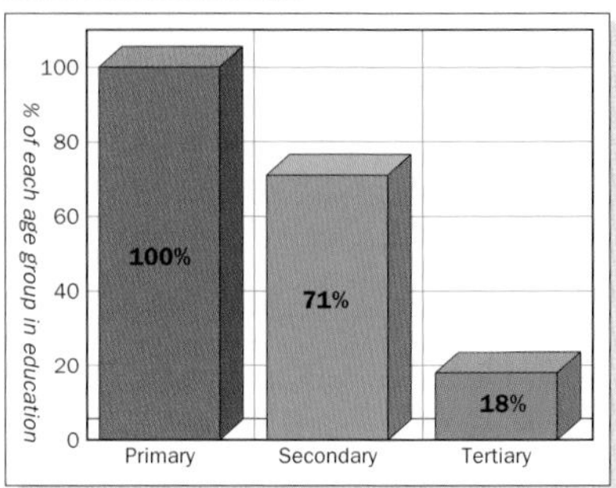
100
80
60
40
20
0
% of each age group in education
100%
71%
18%
Primary
Secondary
Tertiary

Public education, officially compulsory for the first nine years, is underfunded and rural provision is poor. The system is a mixture of the French and US models. There is a well-developed public university system.

HEALTH

Welfare state health benefits

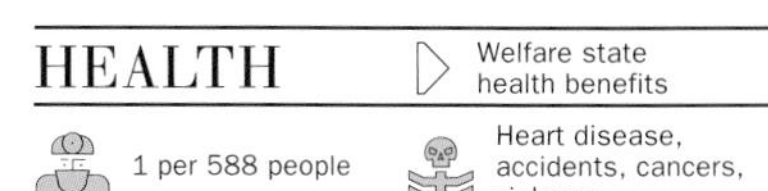
1 per 588 people
Heart disease, accidents, cancers, violence

The national health care system is basic and badly underfunded, although an ambitious scheme was launched in 1996 to improve access to health services in marginalized areas. Mexico has a good reputation for surgery and dentistry, but this is mostly in the private sector. The rich also go to the US for treatment.

SPENDING

GDP/cap. increase

CONSUMPTION AND SPENDING

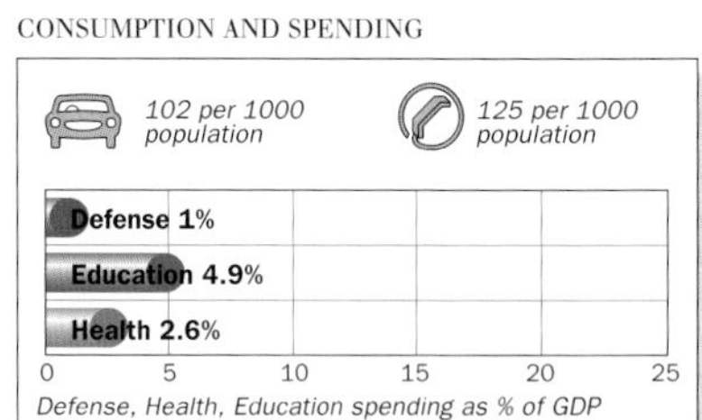
102 per 1000 population
125 per 1000 population
Defense 1%
Education 4.9%
Health 2.6%
0
5
10
15
20
25
Defense, Health, Education spending as % of GDP

Mexico has enormous wealth disparities. Preliminary official figures in 2002 had six million families, or 26 million people, in extreme poverty. There is little social mobility; the old Spanish families retain their hold on institutions. In the past, the wealthy did not generally pay taxes and often benefited from the large state machine. Tax evasion remains a serious problem.

Rural Amerindians are probably the most disadvantaged group. In the last decade, poverty has forced them into city slums to work in factories or *maquiladoras*, where conditions and pay are poor. The 1994 Chiapas rebellion was fed by demands for more land and more assistance in farming it. The flow of poor rural migrants to the US stems largely from the need to subsidize families back home.

WORLD RANKING

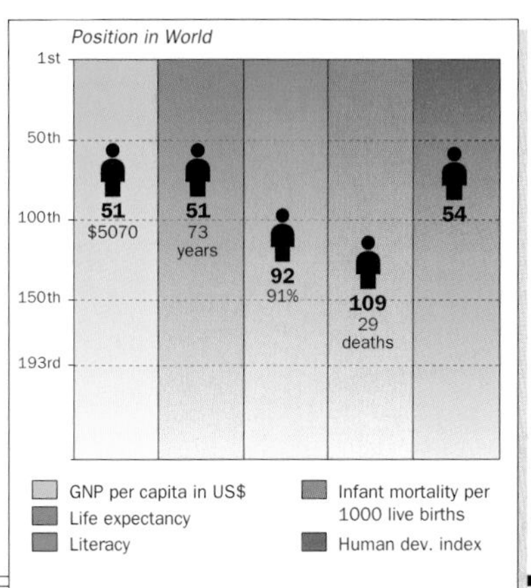
Position in World
1st
50th
100th
150th
193rd
51
$5070
51
73
years
92
91%
109
29
deaths
54
GNP per capita in US$
Life expectancy
Literacy
Infant mortality per 1000 live births
Human dev. index

M

MICRONESIA

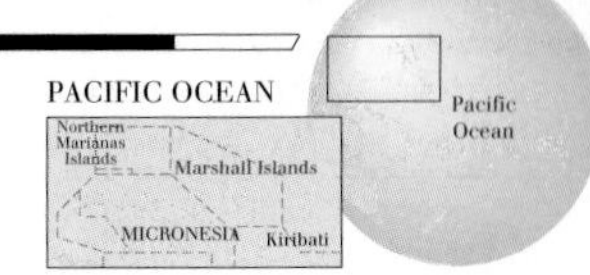
PACIFIC OCEAN
Pacific Ocean
Northern Marianas Islands
Marshall Islands
MICRONESIA
Kiribati

OFFICIAL NAME: Federated States of Micronesia **CAPITAL:** Palikir (Pohnpei Island)
POPULATION: 133,000 **CURRENCY:** US dollar **OFFICIAL LANGUAGE:** English

1986 | 1986 | May 10 | FSM | +10 to +11 | +691 | .fm

THE FEDERATED STATES of Micronesia (FSM), situated in the Pacific Ocean, encompasses all the Caroline Islands except Palau. It is composed of four island cluster states: Pohnpei, Kosrae, Chuuk, and Yap. The FSM was formerly under US rule as part of the UN Trust Territory of the Pacific Islands. An agreement which granted internal sovereignty in free association with the US became operational in 1986, and the Trust was formally dissolved in 1990. The islands continue to receive considerable aid from the US.

CLIMATE

Tropical oceanic

WEATHER CHART FOR PALIKIR

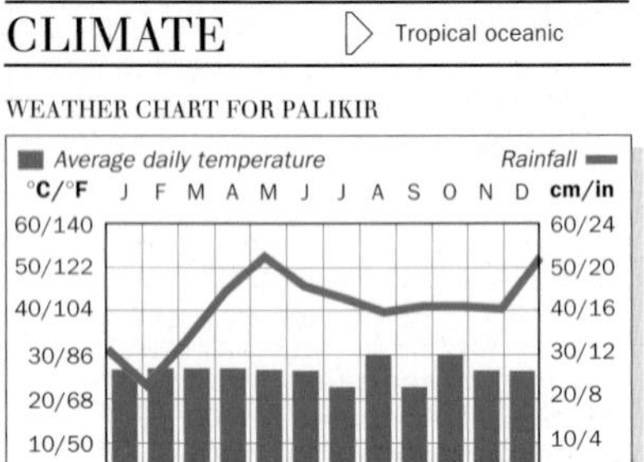
Average daily temperature
Rainfall
°C/°F J F M A M J J A S O N D cm/in
60/140 50/122 40/104 30/86 20/68 10/50 0/32
60/24 50/20 40/16 30/12 20/8 10/4 0

The islands are humid and fairly hot all year round, and the daily temperature range is small. Rainfall is abundant.

TRANSPORTATION

Drive on right

Pohnpei 58,029 passengers

19 ships 10,400 grt

THE TRANSPORTATION NETWORK

42 km (26 miles)

None

None

None

The inauguration in 2000 of flights by a Boeing 737 opened the way for greater air traffic between the islands. Shipping is mainly used for bulk cargoes and copra. Some island roads are surfaced with coral.

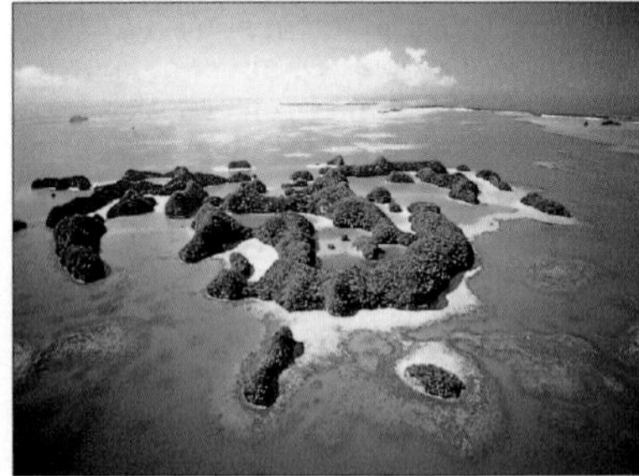

Micronesia, aerial view of rock islands. *Like many Pacific states, Micronesia fears rising sea levels as a result of global warming.*

TOURISM

Visitors : Population 1:4

33,000 visitors

Up 65% 1997–2000

MAIN TOURIST ARRIVALS

Origin	% of total arrivals
Japan	40%
USA	34%
Europe	9%
Other	17%

0 10 20 30 40
% of total arrivals

Outlying islands remain untouched and unspoiled. Chuuk's underwater war wreckage and Kosrae's beaches attract visitors. Lack of infrastructure tends to hinder the growth of tourism.

PEOPLE

Pop. density medium

Trukese, Pohnpeian, Mortlockese, Losrean, English

190/km² (491/mi²)

THE URBAN/RURAL POPULATION SPLIT

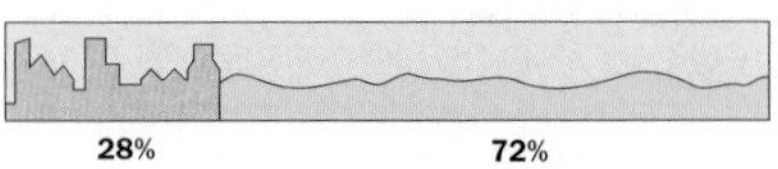
28%
72%

RELIGIOUS PERSUASION

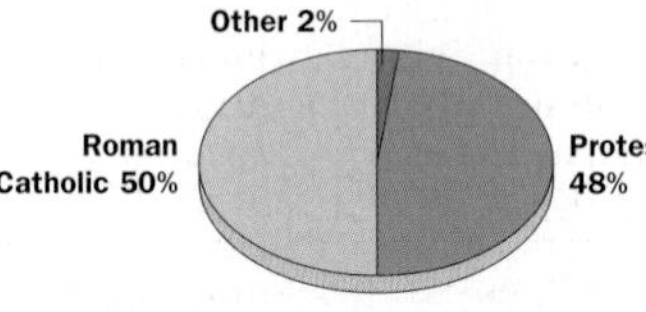
Other 2%
Roman Catholic 50%
Protestant 48%

The Micronesians are physically, linguistically, and culturally diverse. Melanesians live on Yap, and Polynesians occupy southwestern atolls in Pohnpei state. Most islanders live without electricity or running water, and many are effectively recipients of US welfare. Society is traditionally matrilineal.

POLITICS

Nonparty elections

2001/2003

President Leo Falcam

AT THE LAST ELECTION

Congress 14 seats

There are no political parties. Congress has 14 members, 10 senators directly elected for a two-year term and four "at-large" senators (one from each state) who are elected for a four-year term

Under the federal structure, the president and vice president are elected from among the four "at-large" senators (one from each state) by the federal legislature. However, the power of the traditional chiefs in politics remains very strong. Increasing Micronesia's economic independence is the key political issue and has dominated negotiations since 2000 on the outline of the next Compact of Free Association with the US.

MICRONESIA

Total Land Area : 702 sq. km (271 sq. miles)

POPULATION
• under 10 000

LAND HEIGHT
100m/328ft
Sea Level

0 200 km
0 200 miles

PHILIPPINE SEA · Ulithi · Fais · Yap · STATE OF YAP · Ngulu · Sorol · Gaferut · Faraulep · West Fayu · Pikelot · Namonuito · East Fayu I. · Murilo · HALL ISLANDS · Minto Reef · STATE OF CHUUK · Pulap · Nomwin · CHUUK IS. · Weno · Kuop · Losap · Puluwat · Pulusuk · Olimarao · Lamotrek · Elato · Satawal · Woleai · Ifalik · Eauripik · UPPER MORTLOCKS · LOWER MORTLOCKS · Namoluk · Satawan · POHNPEI ISLANDS · Oroluk · Pakin · PALIKIR · Pohnpei · Mokil · Pingelap · STATE OF POHNPEI · Ngatik · Kosrae · KOSRAE STATE · WEST CAROLINE ISLANDS · EAST CAROLINE ISLANDS · Nukuoro · Kapingamarangi · PACIFIC OCEAN · 140° · 145° · 150° · 155° · 160° · 10° · 5°

WORLD AFFAIRS

Joined UN in 1991

ADB

PIF

ACP

PC

IBRD

Micronesia's most important relationship is with the US, which administered the islands from 1947 as part of the UN Trust Territory of the Pacific Islands. Under the Compact of Free Association, the US has exclusive control over the FSM's foreign and defense policies. Alternative funds were sought after 2001 to compensate for the reduction in annual aid likely under a revised Compact. Japan is also important, with the Tokyo government providing aid, and the FSM has recently cultivated strong links with China.

AID

Recipient

$102m (receipts) | Down 6% in 2000

The US is the principal donor of aid, which funds hospitals, schools, food stamps, and construction projects.

DEFENSE

No compulsory military service

US is responsible for defense | Not applicable

Defense is entirely in the hands of the US. Airstrips in the FSM were used by the US in the Vietnam War.

ECONOMICS

Inflation 2.6% p.a. (1990–2000)

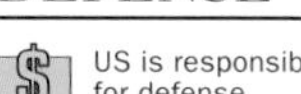

$250m | Currency is US dollar

SCORE CARD

❑ WORLD GNP RANKING	179th
❑ GNP PER CAPITA	$2110
❑ BALANCE OF PAYMENTS	$67m
❑ INFLATION	2.6%
❑ UNEMPLOYMENT	16%

STRENGTHS

Access to US economy, especially for garment manufacture, through preferential trading rights. Construction industry largest private-sector activity. Tourism, fishing, and copra production. US strategic interest in Micronesia, and US budget subsidies.

WEAKNESSES

Dependence on US for imports, especially for fuel. Heavy indebtedness. Acute shortage of water limits development potential. High levels of underemployment.

EXPORTS

IMPORTS

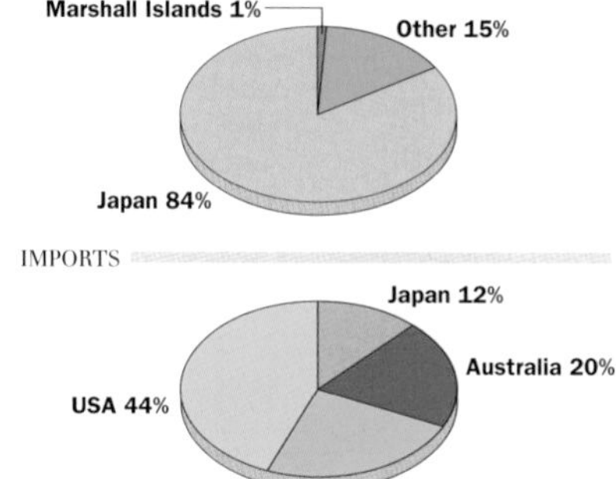
Marshall Islands 1%
Other 15%
Japan 84%
Japan 12%
Australia 20%
USA 44%
Other 24%

RESOURCES

Not available

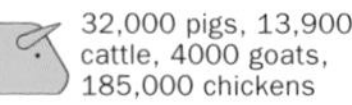

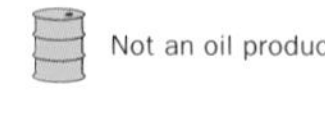

11,886 tonnes | Not an oil producer

32,000 pigs, 13,900 cattle, 4000 goats, 185,000 chickens | None

The FSM is entirely dependent on external sources for its energy supply. Almost all electricity is produced by small diesel generators. The main resources are copra and valuable fish stocks, especially tuna.

ENVIRONMENT

Not available

None | Not available

The FSM does not face pollution on the scale of that in the neighboring Marshall Islands. However, Chuuk suffers serious droughts; occasionally water rationing has had to be introduced for short periods. The growth of marine-based tourism is monitored by the South Pacific Environment Program, which aims to promote sustainable development.

MEDIA

TV ownership medium

There are no daily newspapers

PUBLISHING AND BROADCAST MEDIA

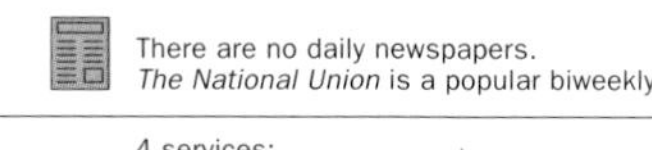

There are no daily newspapers. *The National Union* is a popular biweekly

4 services: 1 state-owned, 3 independent | 1 state-owned service

Press freedoms have not been infringed since the strongly criticized expulsion of a Canadian journalist in 1997.

CRIME

No death penalty

39 prisoners | Little change from year to year

Crime is rare and the outlying islands are crime-free. Some alcohol-related assault occurs on Chuuk.

EDUCATION

School leaving age: 14

89% | 1461 students

Education is compulsory between the ages of six and 14 years. Most university students are supported by US grants and a large number attend US colleges.

CHRONOLOGY

The Caroline Islands were first colonized by the Spanish. Sold to Germany in 1899, the islands were occupied by Japan from 1914 and served as an important base in World War II. US control of the islands began in 1945.

- ❑ **1947** UN Trust Territory of the Pacific Islands established.
- ❑ **1979** Becomes independent.
- ❑ **1986** Compact of Free Association with US operational.
- ❑ **1990** Official termination of trusteeship agreement.
- ❑ **1991** Joins UN.
- ❑ **1999** Leo Falcam elected president.
- ❑ **2001** Submerged remains of USS *Mississinewa* causes oil leak in Yap.

HEALTH

Welfare state health benefits

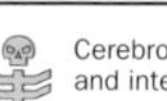

1 per 1640 people | Cerebrovascular, heart, and intestinal diseases

Basic health care is accessible to all. Diabetes and drug abuse are growing problems. An increase in imported food has led to dietary problems.

SPENDING

GDP/cap. decrease

CONSUMPTION AND SPENDING

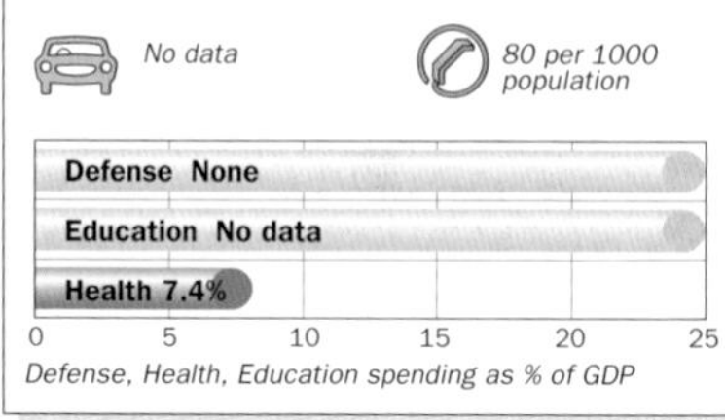
No data
80 per 1000 population
Defense None
Education No data
Health 7.4%
0
5
10
15
20
25
Defense, Health, Education spending as % of GDP

The gap between rich and poor is increasing as Micronesia's businessmen and local officials exploit US aid donations.

WORLD RANKING

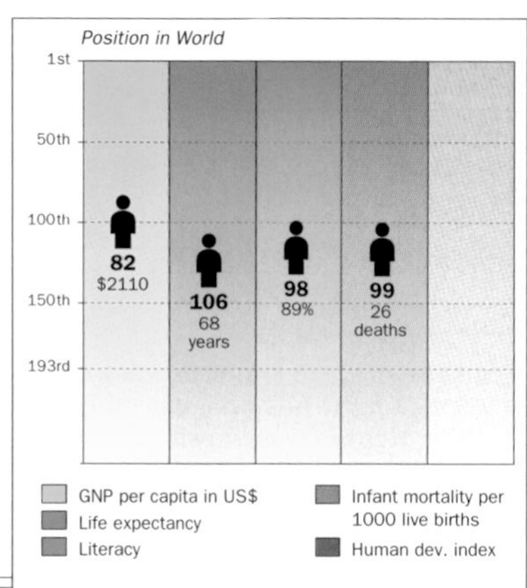
Position in World
1st
50th
100th
150th
193rd
82
$2110
106
68 years
98
89%
99
26 deaths
GNP per capita in US$
Life expectancy
Literacy
Infant mortality per 1000 live births
Human dev. index

M

EUROPE
Europe
Ukraine
MOLDOVA
Romania

MOLDOVA

OFFICIAL NAME: Republic of Moldova **CAPITAL:** Chişinău
POPULATION: 4.3 million **CURRENCY:** Moldovan Leu **OFFICIAL LANGUAGE:** Moldovan

1991 | 1991 | Aug 27 | MD | +2 | +373 | .md

MOSTLY UNDULATING steppe country, Moldova is the most densely populated of the former Soviet republics. Once a part of Romania, it was incorporated into the Soviet Union in 1940. Independence in 1991 brought with it the expectation that Moldova would be reunited with Romania. In a 1994 plebiscite, however, Moldovans voted against the proposal. Most of Moldova's population is engaged in intensive agriculture.

Agricultural landscape. Warm summers and even rainfall are ideal for cereal and fruit farming. Moldova is famous for its wine.

CLIMATE

Continental

WEATHER CHART FOR CHISINAU

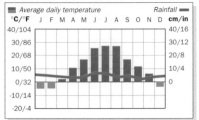

Average daily temperature / Rainfall
°C/°F: 40/104, 30/86, 20/68, 10/50, 0/32, -10/14, -20/-4
J F M A M J J A S O N D
cm/in: 40/16, 30/12, 20/8, 10/4, 0

Warm summers, mild winters, and moderate rainfall give Moldova an ideal climate for cultivation.

TRANSPORTATION

Drive on right

Chisinau International 254,283 passengers

Small Black Sea fleet

THE TRANSPORTATION NETWORK

10,738 km (6672 miles)		None	
1140 km (708 miles)		424 km (263 miles)	

The transportation infrastructure is to be part of a planned "Transport Corridor Europe–Caucasus–Asia" (TRACECA).

TOURISM

Visitors : Population 1:253

17,000 visitors

Up 21% in 2000

MAIN TOURIST ARRIVALS

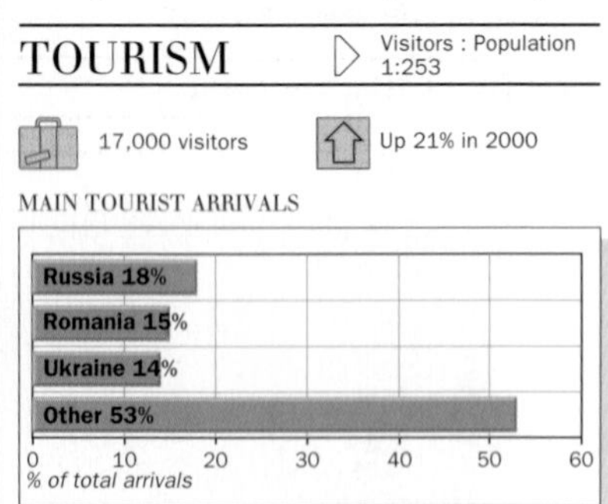
Russia 18%
Romania 15%
Ukraine 14%
Other 53%
0 10 20 30 40 50 60
% of total arrivals

Few tourists go to Moldova, although some visitors to Romania do combine the two. Hopes for expansion of tourism focus on vineyards and underground wine vault "streets" as the main attractions.

PEOPLE

Pop. density medium

Moldovan, Romanian, Russian

128/km² (330/mi²)

THE URBAN/RURAL POPULATION SPLIT

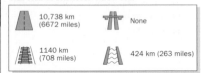

46% 54%

ETHNIC MAKEUP

Gagauz 4%, Other 4%, Russian 13%, Ukrainian 14%, Moldovan 65%

Moldovans are ethnically identical to Romanians. There are 153,000 Gagauz (Orthodox Christian Turks) in the south, and a population of mixed Russian–Moldovan–Ukrainian parentage on the eastern bank of the Dniester.

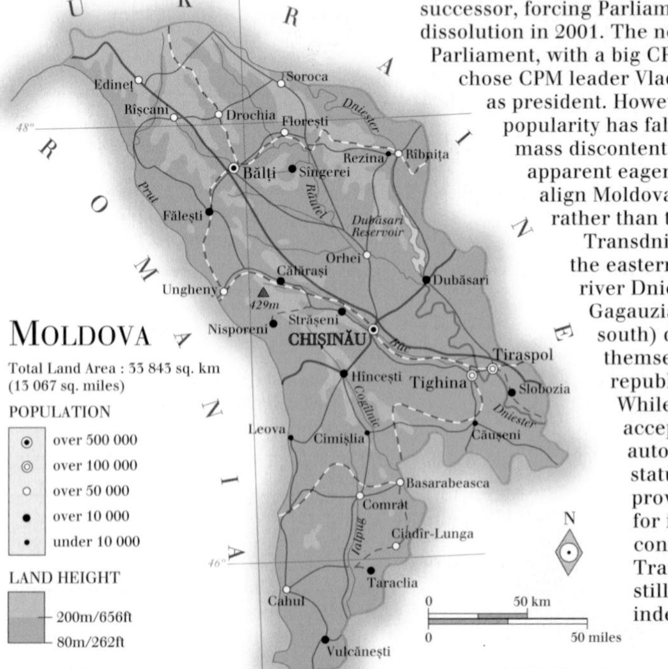
UKRAINE
ROMANIA
Edineţ
Soroca
Rîşcani
Drochia
Floreşti
Dniester
Rezina
Ribniţa
Bălţi
Sîngerei
Prut
Răut
Făleşti
Dubăsari Reservoir
Orhei
Călăraşi
Dubăsari
Ungheny
429m
Nisporeni
Străşeni
CHIŞINĂU
Bîc
Tiraspol
Hînceşti
Tighina
Slobozia
Cogîlnic
Leova
Cimişlia
Căuşeni
Basarabeasca
Comrat
Ialpug
Ciadîr-Lunga
Taraclia
Cahul
Vulcăneşti
48°
46°
28°
MOLDOVA
Total Land Area : 33 843 sq. km (13 067 sq. miles)
POPULATION
over 500 000
over 100 000
over 50 000
over 10 000
under 10 000
LAND HEIGHT
200m/656ft
80m/262ft
0 50 km
0 50 miles
N

POLITICS

Multiparty elections

2001/2005

President Vladimir Voronin

AT THE LAST ELECTION

Parliament 101 seats

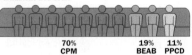

70% CPM, 19% BEAB, 11% PPCD

CPM = Communist Party of Moldova
BEAB = Electoral Bloc Braghis Alliance
PPCD = Christian-Democratic People's Party

Moldova declared its independence in 1991. Reformist Petru Lucinschi was elected president in 1996 but faced stiff opposition from the increasingly powerful left: the revived CPM won most seats in the 1998 elections. Parliament ended direct presidential elections in 2000, but deadlock ensued over the appointment of Lucinschi's successor, forcing Parliament's dissolution in 2001. The new Parliament, with a big CPM majority, chose CPM leader Vladimir Voronin as president. However, the left's popularity has faltered, with mass discontent over its apparent eagerness to align Moldova with Russia rather than the West. Transdniestria (on the eastern bank of the river Dniester) and Gagauzia (in the south) declared themselves as republics in 1990. While Gagauzia accepted autonomous status as provided for in the 1994 constitution, Transdniestria still seeks independence.

M

WORLD AFFAIRS

Joined UN in 1992

BSEC

CE

CIS

EAPC

OSCE

Moldova has not sought NATO membership, and in 2001 showed interest in joining a Union State with Russia and Belarus. Ties with countries in the Black Sea Economic Zone, including Romania and Ukraine, are being developed. The creation of a free economic zone near the mouth of the Danube is under discussion.

AID

Recipient

 $123m (receipts)

 Up 15% in 2000

The World Bank resumed lending in 2002, granting $30 million. The US, the EU states, and the IMF are also important sources of aid.

DEFENSE

Compulsory military service

 $21m

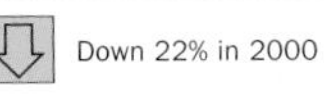
Down 22% in 2000

In 1999 plans were announced to cut army personnel by 30%. Military service has been reduced from 18 months to 12. Russian forces were withdrawn from Transdniestria by the end of 2001.

ECONOMICS

Inflation 120% p.a. (1990–2000)

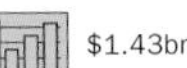
$1.43bn

12.395–13.085 Moldovan lei

SCORE CARD

- ❑ WORLD GNP RANKING........................147th
- ❑ GNP PER CAPITA...............................$400
- ❑ BALANCE OF PAYMENTS....................–$121m
- ❑ INFLATION.......................................31.3%
- ❑ UNEMPLOYMENT.................................11%

STRENGTHS

Agriculture – notably wine, tobacco, and cotton – and food processing. Light manufacturing.

WEAKNESSES

Dependent on Russia as source of raw materials and fuel, and main market for exports. Dramatic shrinking of economy since independence. Isolated location; weak transportation network. Slow pace of reform. Cumbersome bureaucracy. Strong black economy. Foreign debt – over 50% of GDP – costs 20% of export earnings to service.

EXPORTS

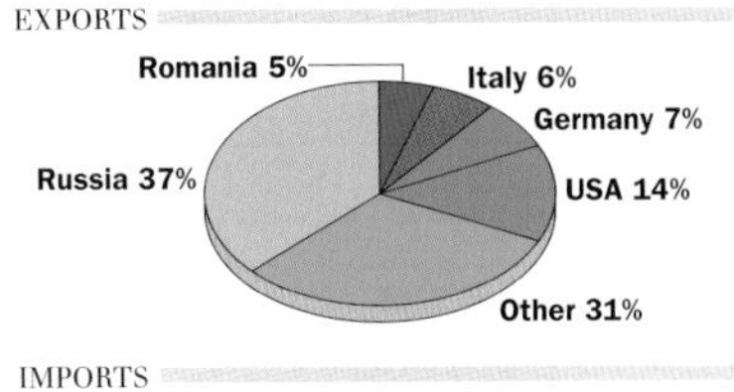
Romania 5%
Italy 6%
Germany 7%
Russia 37%
USA 14%
Other 31%

IMPORTS

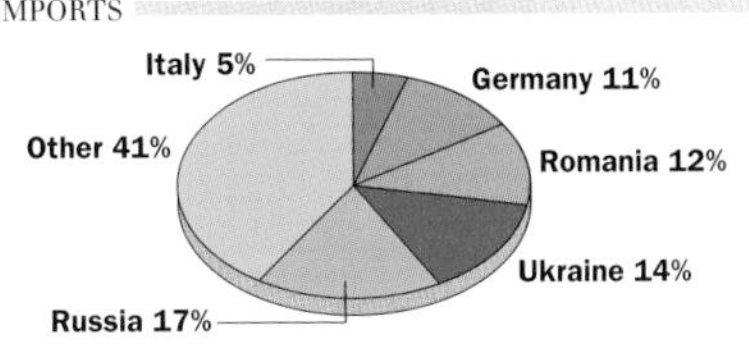
Italy 5%
Germany 11%
Other 41%
Romania 12%
Ukraine 14%
Russia 17%

RESOURCES

Electric power 1m kw

1630 tonnes

866,000 sheep, 543,000 pigs, 13m chickens

Oil and gas reserves not exploited

Lignite, phosphates, gypsum, oil, natural gas

Moldova has few mineral resources. It has to import all its fuel and most of its electricity.

ENVIRONMENT

Sustainability rank: 39th

1%

2.2 tonnes per capita

Overuse of agricultural chemicals and pesticides on tobacco farms is a problem, as is soil erosion. There is little spending on environmental improvement.

MEDIA

TV ownership high

 Daily newspaper circulation 60 per 1000 people

PUBLISHING AND BROADCAST MEDIA

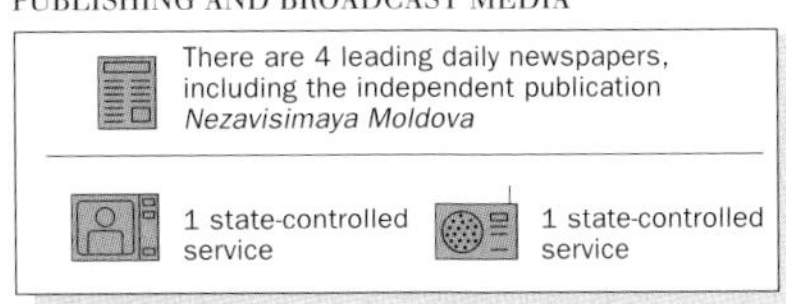
There are 4 leading daily newspapers, including the independent publication *Nezavisimaya Moldova*
1 state-controlled service
1 state-controlled service

The many new publications represent widely differing interest groups. Russian-language broadcasting is restricted.

CRIME

No death penalty

 9837 prisoners

 Up 5% 1996–1998

Economic decline has caused crime to increase. The unstable situation in Transdniestria has encouraged smuggling, particularly of Russian arms. The Council of Europe has accused the police of routinely using torture.

EDUCATION

School leaving age: 15

99%

93,759 students

Education has followed a Romanian (French-inspired) system since 1991. Mass protests met plans in 2002 to make Russian compulsory in schools.

HEALTH

Welfare state health benefits

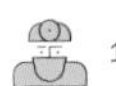 1 per 286 people

 Circulatory diseases, cancers, accidents

The centralized health service is poor by regional standards. There are serious shortages of medical supplies.

CHRONOLOGY

Modern Moldova corresponds roughly to the eastern part of the Romanian principality of Moldavia, which existed for 500 years from 1359. Most of it was annexed by Russia in 1812 as Bessarabia.

- ❑ **1918** Bessarabia joins Romania.
- ❑ **1924** Moldovan Autonomous Soviet Republic formed within USSR.
- ❑ **1940** Romania cedes Bessarabia to Ukrainian and Moldovan SSRs.
- ❑ **1941–1945** Bessarabia again under Romanian control.
- ❑ **1945** Returns to Soviet control.
- ❑ **1990** Declares sovereignty.
- ❑ **1991** Independence.
- ❑ **1993–1994** Pro-unification parties' election defeat; referendum rejects Romanian unification. Rejoins CIS.
- ❑ **1998** Elections: communist revival.
- ❑ **2001** CPM wins big majority. Voronin becomes president.
- ❑ **2002** Mass protests over education plans.

SPENDING

GDP/cap. decrease

CONSUMPTION AND SPENDING

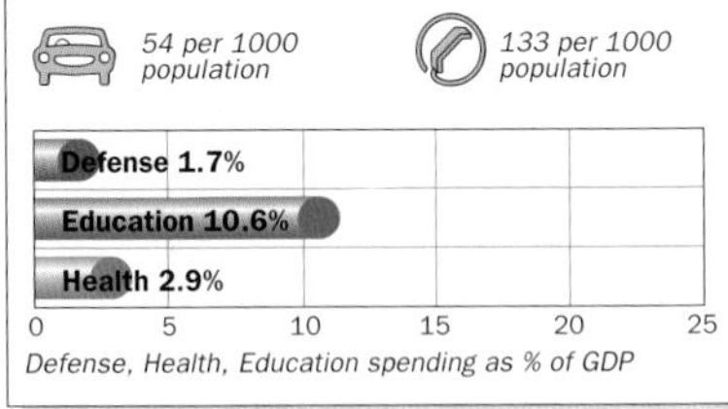
54 per 1000 population
133 per 1000 population
Defense 1.7%
Education 10.6%
Health 2.9%
0
5
10
15
20
25
Defense, Health, Education spending as % of GDP

Former communist officials have been well placed to benefit from the sale of state-owned businesses. Car ownership is low but rising. However, pensions and wages are often months in arrears. In 1998 the benefits for low-income families and veterans were scrapped. Ethnic Gagauz (Orthodox Christian Turks) are the poorest group.

WORLD RANKING

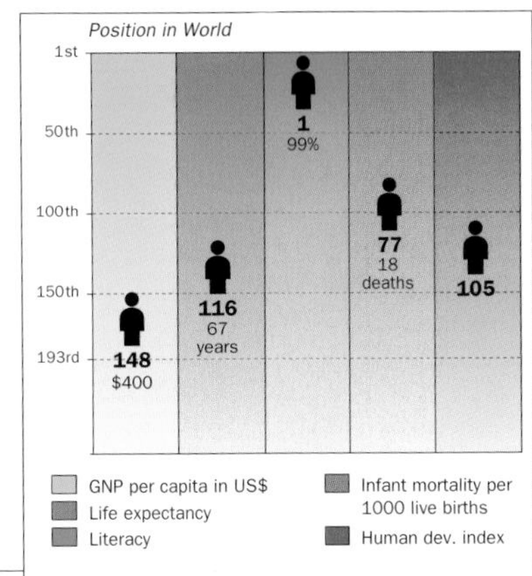
Position in World
1st
50th
100th
150th
193rd
148
$400
116
67 years
1
99%
77
18 deaths
105
GNP per capita in US$
Life expectancy
Literacy
Infant mortality per 1000 live births
Human dev. index

M

MONACO

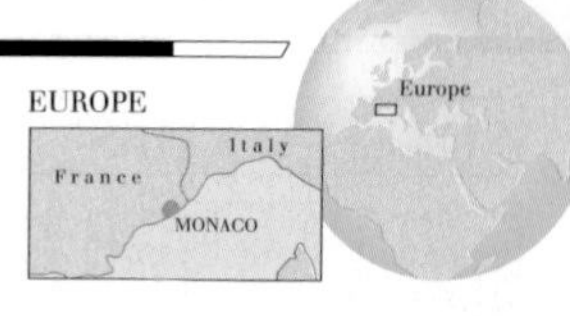
EUROPE
Europe
Italy
France
MONACO

OFFICIAL NAME: Principality of Monaco **CAPITAL:** Monaco **POPULATION:** 31,700
CURRENCY: Euro (French franc until 2002) **OFFICIAL LANGUAGE:** French

1861
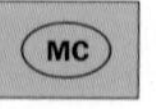
1861
Nov 19

MC

+1
+377
.mc

MONACO IS A TINY ENCLAVE on the French Côte d'Azur. Its destiny changed radically in 1863, when Prince Charles III, after whom Monte Carlo is named, opened the casino. Today, Monaco is a lucrative banking and services center, as well as a tourist destination. Prince Rainier's marriage to film star Grace Kelly, and some astute management of the economy, successfully transformed Monaco into a center for the international jet set. In 1962, the prince's absolute authority was abolished by a new, democratic constitution.

CLIMATE

Mediterranean

WEATHER CHART FOR MONACO

Average daily temperature / Rainfall
°C/°F J F M A M J J A S O N D cm/in
40/104, 30/86, 20/68, 10/50, 0/32, -10/14, -20/-4 | 40/16, 30/12, 20/8, 10/4, 0

Summers are hot and dry; days with 12 hours of sunshine are not uncommon. Winters are mild and sunny.

TRANSPORTATION

Drive on right

Héliport de Monaco, Fontvieille
152,600 passengers

8 ships

THE TRANSPORTATION NETWORK

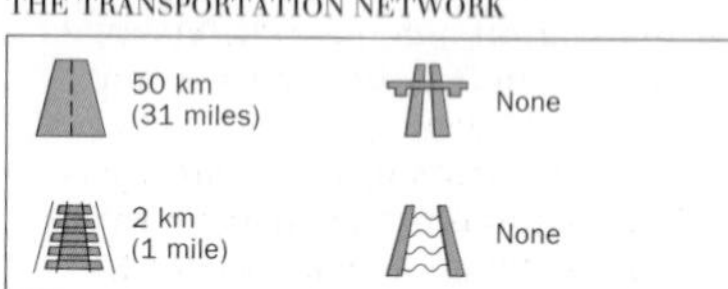
50 km
(31 miles)
None
2 km
(1 mile)
None

An underground railroad system connecting to Cap d'Ail in France and opened in 1999, meant that 2% of Monaco's area could be reclaimed. Access to Monaco is mainly by land, although Air Monaco provides helicopter flights from Nice airport.

TOURISM

Visitors : Population 9.5:1

300,000 visitors

Up 8% in 2000

MAIN TOURIST ARRIVALS

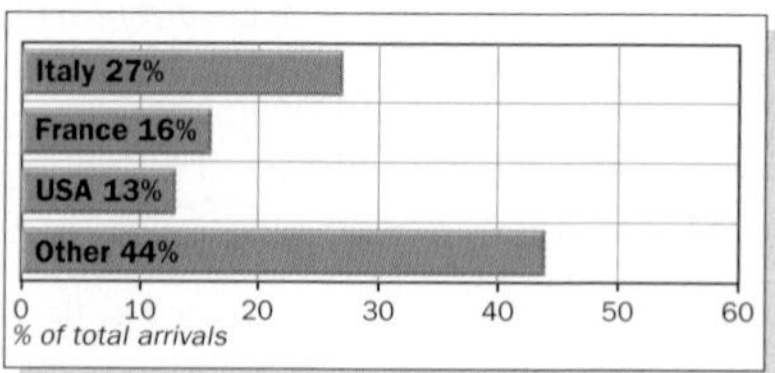
Italy 27%
France 16%
USA 13%
Other 44%
0 10 20 30 40 50 60
% of total arrivals

Huge numbers of tourists, greatly outnumbering the inhabitants, are attracted to Monaco, most coming from Italy and France. Almost all are day trippers drawn by the casinos and Monaco's conspicuous high society. Around 75% of hotel rooms are classed as "four-star deluxe," and the principality is a particular favorite of wealthy Italians. The Grimaldi Forum conference center, which opened in 2000, hopes to attract more business travelers.

A number of social and sporting events draw particularly large crowds each spring, including the Rose Ball (March), the Tennis Open (April), and the Grand Prix (May).

FRANCE
Lycée de l'Annonciade
Musée National
Larvotto
Centre de la Culture et d'Expositions
Monte-Carlo Sporting Club d'Été
Hospitalier Grace
Railway Station
La Condamine
Casino
Centre de Congrès Monte-Carlo
Palais du Prince
Port de Monaco
Stade Louis II
Port de Fontvieille
Palais de Justice
Cathédrale
Ministère d'Etat
MONACO
Fontvieille
Musée Océanographique
Heliport
Côte d'Azur
MONACO
Total Land Area : 1.95 sq. km (0.75 sq. miles)
Places of Interest
Parks and Gardens
Grand Prix Circuit
0 500 m
0 1000 yds

PEOPLE

Pop. density high

French, Italian, Monégasque, English

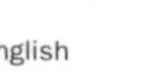
16,253/km² (42,095/mi²)

THE URBAN/RURAL POPULATION SPLIT

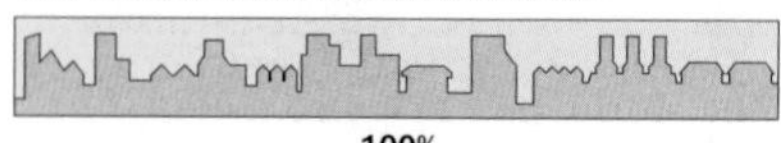
100%

RELIGIOUS PERSUASION

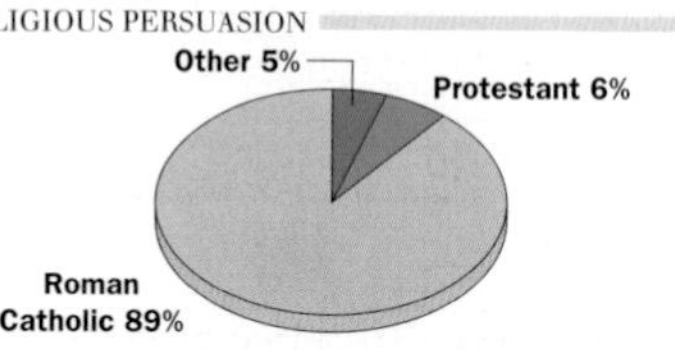
Other 5%
Protestant 6%
Roman Catholic 89%

Less than a fifth of Monaco's residents are Monégasque. Around half are French, the rest Italian, American, British, and Belgian. Monégasques enjoy considerable privileges, including housing subsidies to protect them from Monaco's high property prices, and the right of first refusal before a job can be offered to a foreigner. Women have equal status, but only acquired the vote in the constitutional changes of 1962.

POLITICS

Multiparty elections

1998/2003

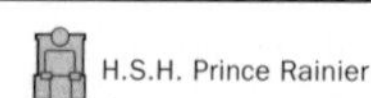
H.S.H. Prince Rainier III

AT THE LAST ELECTION

National Council 18 seats

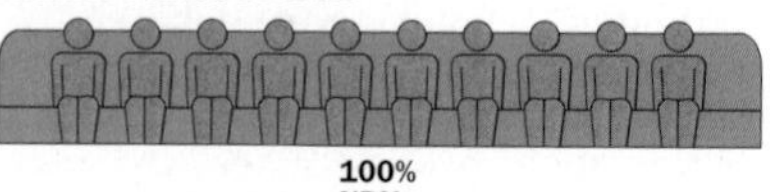
100%
NDU

NDU = National and Democratic Union
There are no formal political parties

The Grimaldi princes have been hereditary rulers of Monaco for more than 700 years. Prince Rainier III renounced absolute rule in 1962 but retains considerable power. The executive minister of state is appointed by the prince from a list of French diplomats. National Council elections – based on personalities rather than parties – were last held in 1998.

WORLD AFFAIRS

Joined UN in 1993

FZ

IAEA

OSCE
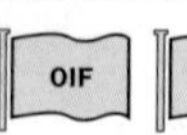
OIF
IWC

A key concern is to protect both banking secrecy and the liberal tax regime from EU regulation, although the principality has adopted the euro. France is particularly critical, and French citizens have been banned from banking in Monaco since 1962.

***Monte Carlo** with its luxury hotels and yacht harbor. The only space for new development is on land reclaimed from the sea.*

AID

Not applicable

 Monaco has no aid receipts or donations

 Not applicable

Monaco neither receives nor gives aid, and the issue is not of concern to Monégasques.

DEFENSE

No compulsory military service

 France responsible for defense

 Not applicable

Monaco has no armed forces and no defense budget. France, as the protecting power, bears responsibility for the defense of the principality.

ECONOMICS

Inflation 2.6% p.a. (1985–1996)

 $800m

 1.0651–1.1231 euros

SCORE CARD

- ❑ WORLD GNP RANKING........................158th
- ❑ GNP PER CAPITA..........................$25,000
- ❑ BALANCE OF PAYMENTS.....Included in French total
- ❑ INFLATION.....................Included in French total
- ❑ UNEMPLOYMENT..................................3%

STRENGTHS

Strict banking confidentiality and low taxes attract billions of dollars of overseas deposits. Strong tourism sector. Assets managed by Monaco banks increased by 18% a year in the late 1990s. No formal debt and reserves of over 2.3 billion euros. Very low unemployment.

WEAKNESSES

Continuing vulnerability to money laundering despite revised banking secrecy laws under the 1994 accord with France obliging banks to furnish details of suspicious accounts. Subject to fluctuations of French and Italian economies. Dependence on VAT for 55% of revenues. Pressure from EU states to end privileged banking and tax laws. Total dependence on imports because of lack of natural resources.

EXPORTS/IMPORTS

Monaco has a full customs union with France

RESOURCES

Electric power: Included within French total

 3 tonnes

 Not an oil producer

 Included within French total

 None

Monaco has no strategic resources and imports all its energy from France. It has no agricultural land.

ENVIRONMENT

Not available

 None

 4.4 tonnes per capita

Monaco has built the most extensive underground car parking facilities in the world to tackle congestion. The quality of the built environment around the harbor occasionally arouses local passions. Important populations of red coral are under threat from land reclamation and pollution.

MEDIA

TV ownership high

 Daily newspaper circulation 251 per 1000 people

PUBLISHING AND BROADCAST MEDIA

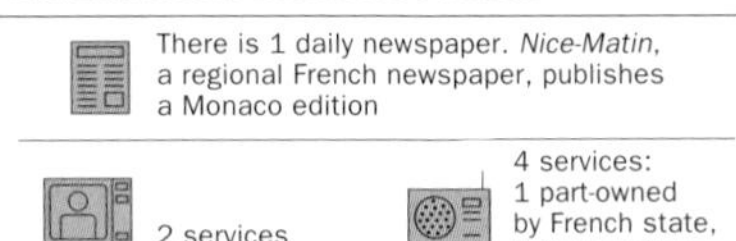

There is 1 daily newspaper. *Nice-Matin*, a regional French newspaper, publishes a Monaco edition

2 services

4 services: 1 part-owned by French state, 3 independent

In addition to its domestic radio and TV, Monaco receives all the mainstream French and Italian channels.

CRIME

No death penalty

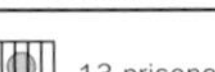 13 prisoners

 Up 3% in 1999

Low crime rates make it safe for the rich to sport their furs and jewelry in public. In late 1998 the appeals court upheld Monaco's first conviction of an individual for money laundering.

EDUCATION

School leaving age: 16

 99%

 Not available

The education system is essentially the same as that of France, with students studying for the *baccalauréat* exam. Most go on to university in France, but then return to claim good jobs in Monaco. The Catholic Church exerts considerable influence and is still responsible for primary schooling.

HEALTH

Welfare state health benefits

 1 per 333 people

 Heart and cerebrovascular diseases, cancers

Most medical care is provided by private health insurance. Doctors train in France. The Princess Grace Hospital can serve 60,000 people, also catering for patients from outside Monaco.

CHRONOLOGY

In 1297, the Grimaldis established themselves as the principality's hereditary rulers.

- ❑ **1861** Independent under French protection.
- ❑ **1949** Rainier III accedes to throne.
- ❑ **1962** Constitution rewritten: end of absolute authority of the prince.
- ❑ **1963** Democratic legislative elections held for first time.
- ❑ **1982** Princess Grace dies following car accident.
- ❑ **2002** Euro introduced.

SPENDING

GDP/cap. increase

CONSUMPTION AND SPENDING

680 per 1000 population

1030 per 1000 population

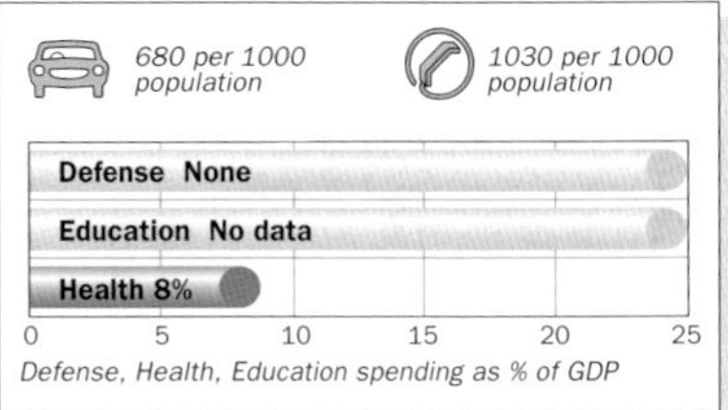
Defense None
Education No data
Health 8%
0
5
10
15
20
25
Defense, Health, Education spending as % of GDP

Monaco's image abroad has changed dramatically since Prince Rainier acceded in 1949. From being considered a simple gambling spot, it is now ranked as one of the world's most glamorous international jet set destinations. In part, this was the result of Prince Rainier's wedding to Grace Kelly, then a leading Hollywood star, which brought Monaco to the attention of US high society. More important was the prince's work in turning Monaco into a major tax haven and an upmarket resort, by making the most of its Mediterranean coastal location. Many tax exiles have taken up residence, among them Luciano Pavarotti and Wall Street investment guru Bob Beckman.

WORLD RANKING

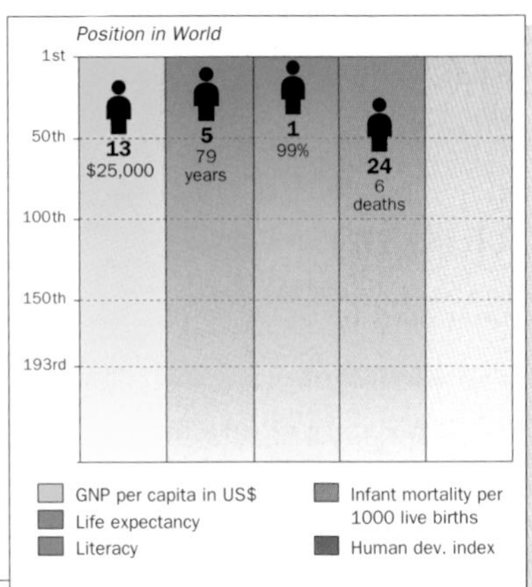
Position in World
1st
50th
100th
150th
193rd
13
$25,000
5
79 years
1
99%
24
6 deaths
GNP per capita in US$
Life expectancy
Literacy
Infant mortality per 1000 live births
Human dev. index

M

MONGOLIA

OFFICIAL NAME: Mongolia **CAPITAL:** Ulan Bator **POPULATION:** 2.6 million
CURRENCY: Tugrik (tögrög) **OFFICIAL LANGUAGE:** Khalkha Mongolian

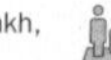

1924 | 1924 | July 11 | MGL | +8 | +976 | .mn

LANDLOCKED BETWEEN Siberia and China's Mongolian provinces, Mongolia rises from the semiarid Gobi Desert to mountainous steppe. The traditionally nomadic Mongols were first unified by Genghis Khan in 1206. "Outer" Mongolia achieved independence from China as a communist state in 1924 and was officially aligned with the USSR from 1936. In 1990, it abandoned communist rule; widespread poverty ensued. Extremely harsh winters in 1999–2001 devastated the rural economy.

CLIMATE

Mountain/cold desert/steppe

WEATHER CHART FOR ULAN BATOR

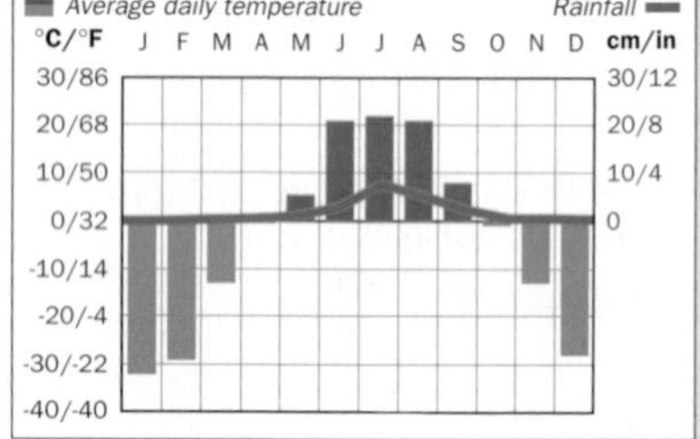

Average daily temperature
Rainfall
°C/°F J F M A M J J A S O N D cm/in
30/86 20/68 10/50 0/32 -10/14 -20/-4 -30/-22 -40/-40
30/12 20/8 10/4 0

Temperature variations are extreme. Dry summers combine with severe winters, known as *zud*, to devastate livestock, as happened in 1999 and 2000.

TRANSPORTATION

Drive on right

Buyant-Ukhaa, Ulan Bator

Has no fleet

THE TRANSPORTATION NETWORK

1674 km (1040 miles)	None
1810 km (1125 miles)	397 km (247 miles)

Lack of investment has left Mongolia's infrastructure to decay, increasing transportation and distribution costs. Links to China and the Pacific are priorities. Gasoline shortages have meant a large increase in the use of animal labor.

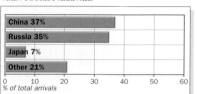

Traditional gers in the Gobi Desert. *Many Mongolians still choose to pursue a nomadic lifestyle, living in felt tents called gers.*

TOURISM

Visitors : Population 1:16

158,000 visitors

Down 1% in 2000

MAIN TOURIST ARRIVALS

	% of total arrivals
China	37%
Russia	35%
Japan	7%
Other	21%

Tourism has expanded overall since the easing of visa restrictions in 1991. Under communism, all travel was arranged through the state agency, Zhuuichin, but private companies are now entering the market.

PEOPLE

Pop. density low

Khalkha Mongolian, Kazakh, Chinese, Russian

2/km² (4/mi²)

THE URBAN/RURAL POPULATION SPLIT

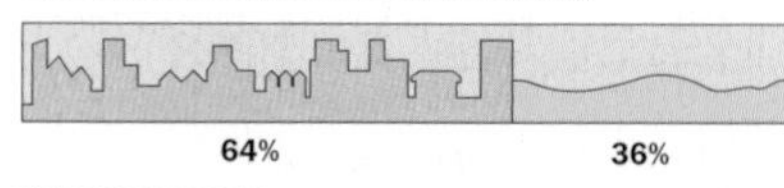

64%
36%

ETHNIC MAKEUP

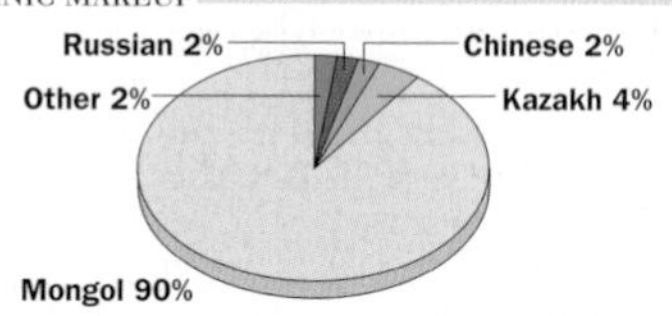

Russian 2%
Chinese 2%
Other 2%
Kazakh 4%
Mongol 90%

Khalkh Mongols, who adhere to Tibetan Buddhism, are the main ethnic group. Although economic pressures keep many people near urban centers, most remain nomadic. One-third live in Ulan Bator. Turkic Kazakhs in the west form the largest minority, but emigration to Kazakhstan since 1990 has reduced their numbers. Tensions exist with the Chinese and Russian minorities.

POLITICS

Multiparty elections

2000/2004

President Natsagyn Bagabandi

AT THE LAST ELECTION

State Great Hural 76 seats

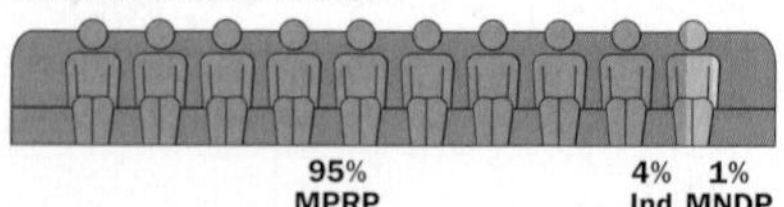

95% MPRP
4% Ind
1% MNDP

MPRP = Mongolian People's Revolutionary Party
Ind = Independents
MNDP = Mongolian National Democratic Party

The end of communism and the advent of democracy in 1990 revolutionized Mongolian politics. The shock of economic reform led many Mongolians to regret the lost certainties of the communist era. In 1992 the democrats lost power to the renamed communists (MPRP), but their failure to revive the economy swung the pendulum back in favor of a democratic coalition in 1996. An uneasy cohabitation between President Natsagyn Bagabandi of the MPRP, who took office in 1997, and an MNDP-led government existed until the sweeping MPRP election victory in 2000.

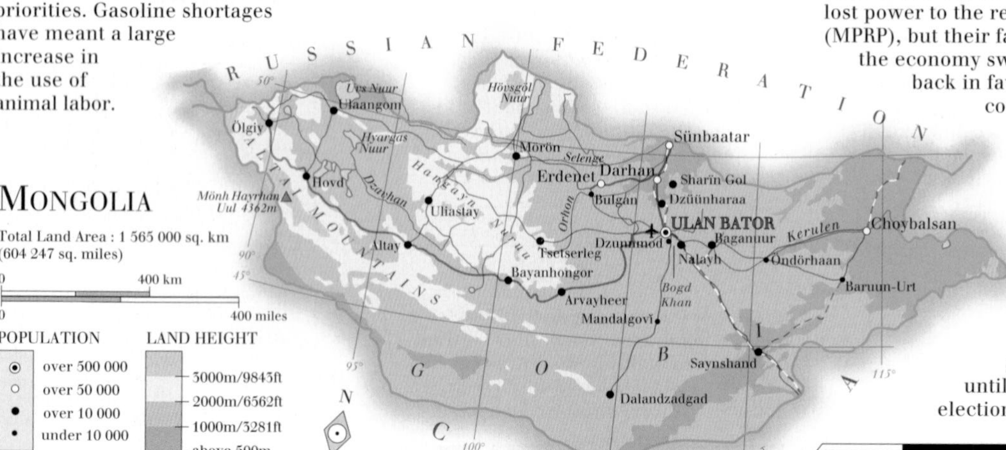

MONGOLIA
Total Land Area : 1 565 000 sq. km (604 247 sq. miles)
POPULATION
over 500 000
over 50 000
over 10 000
under 10 000
LAND HEIGHT
3000m/9843ft
2000m/6562ft
1000m/3281ft
above 500m
RUSSIAN FEDERATION
CHINA
GOBI
ALTAI MOUNTAINS
ULAN BATOR
Ölgiy
Ulaangom
Uvs Nuur
Hövsgöl Nuur
Hyargas Nuur
Mörön
Hovd
Uliastay
Altay
Erdenet
Darhan
Sühbaatar
Sharin Gol
Dzüünharaa
Bulgan
Tsetserleg
Bayanhongor
Arvayheer
Mandalgovi
Dalandzadgad
Saynshand
Baruun-Urt
Öndörhaan
Choybalsan
Baganuur
Nalayh
Dzuunmod
Mönh Hayrhan Uul 4362m
Kerulen
Selenge
Orhon
Dzavhan
Bogd Khan

M

WORLD AFFAIRS

Joined UN in 1961

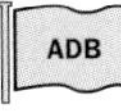

IAEA
IBRD
NAM
ADB
WTO

Closer relations with Japan and other east Asian states have failed to weaken Mongolia's ties with Russia and China. There are residual tensions with China, since the majority of ethnic Mongols actually resides in the adjoining Chinese province of Inner Mongolia, but there is no longer a fear of Chinese designs on Mongolian sovereignty.

AID

Recipient

 $217m (receipts)

 Down 2% in 2000

A large balance-of-payments deficit and severe weather make aid vital. The main donors are Japan and the ADB.

DEFENSE

Compulsory military service

$19m

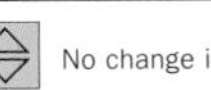 No change in 2000

The last Soviet forces left in 1992. However, ties are still strong, and under agreements reached in 2000 and 2001 Russia is helping to reform the greatly reduced and poorly equipped Mongolian forces.

ECONOMICS

Inflation 58% p.a. (1990–2000)

$947m

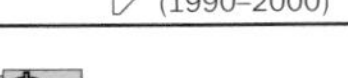 1097–1102 tugriks

SCORE CARD

❑ WORLD GNP RANKING	156th
❑ GNP PER CAPITA	$390
❑ BALANCE OF PAYMENTS	–$112m
❑ INFLATION	7.6%
❑ UNEMPLOYMENT	5%

STRENGTHS

Copper and cashmere. Largely untapped coal and oil reserves. Traditional and efficient rural economy.

WEAKNESSES

Harsh winters ravaged livestock between 1999 and 2001. Decaying infrastructure. Rising poverty.

EXPORTS

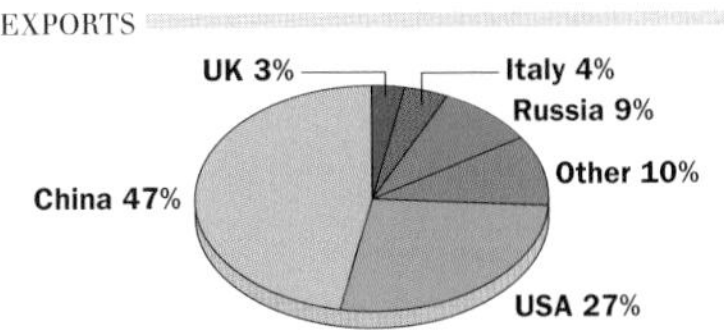
UK 3%
Italy 4%
Russia 9%
Other 10%
China 47%
USA 27%

IMPORTS

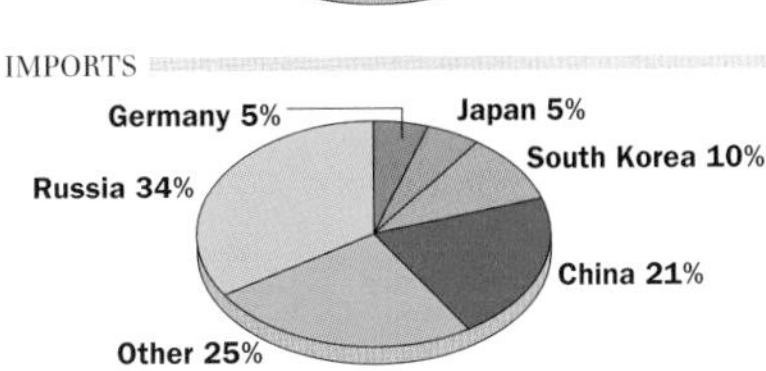
Germany 5%
Japan 5%
South Korea 10%
Russia 34%
China 21%
Other 25%

RESOURCES

 Electric power 901,000 kw

524 tonnes

 Contracts have been signed with oil prospectors

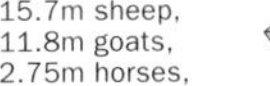 15.7m sheep, 11.8m goats, 2.75m horses,

 Oil, coal, copper, lead, fluorite, tungsten, tin, gold, uranium

Under communism, Mongolia's vast mineral resources were barely exploited, and prospecting has only recently begun. A uranium mining joint venture with Russia has been established. Mongolia is rich in oil, with sufficient reserves to meet future domestic needs. In 1999 an oil extraction agreement was signed with China.

ENVIRONMENT

Sustainability rank: 42nd

 12%

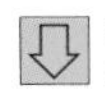 3.3 tonnes per capita

Industrial pollution around Ulan Bator is a health hazard. Lake Hövsgöl is seriously polluted. Renewed efforts were launched in 2001 to preserve the Bogd Khan, the world's oldest protected area, dating from 1788, from illegal logging, hunting, and air pollution.

MEDIA

TV ownership medium

 Daily newspaper circulation 17 per 1000 people

PUBLISHING AND BROADCAST MEDIA

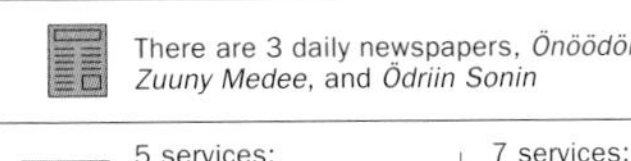

There are 3 daily newspapers, *Önöödör, Zuuny Medee,* and *Ödriin Sonin*

5 services: 1 state-owned, 4 independent

7 services: 1 state-owned, 6 independent

Since 1990, Mongolia's press has enjoyed unlimited freedom; there are no slander or libel laws. Legislation enacted in 1999 eased remaining curbs on the media. However, the shortage of paper and fuel supplies has restricted both the number of publications and their distribution.

CRIME

Death penalty in use

 6414 prisoners

 Up 8% in 1999

Crime rose rapidly in the early 1990s, particularly organized crime and muggings by knife gangs. Ulan Bator is the most dangerous area, especially for foreigners; Russians, Chinese, and dollar-carrying US tourists are the main targets.

EDUCATION

School leaving age: 16

 99%

 84,970 students

Education is modeled on the former Soviet system. The majority of teachers are women on low salaries. Private-sector schools emphasizing Mongol culture are beginning to open.

CHRONOLOGY

In the 17th century, the Manchus took control of Mongolia. It stayed in Chinese hands until 1911.

- ❑ **1919** China reoccupies Mongolia.
- ❑ **1924** Independent communist state.
- ❑ **1989–1990** Prodemocracy protests; communist election defeat.
- ❑ **1992** Former communists, renamed MPRP, returned to power.
- ❑ **1996** Democratic Union coalition wins general election.
- ❑ **1997** MPRP wins presidency.
- ❑ **1999–2001** Severe winters.
- ❑ **2000** Landslide electoral victory for MPRP.

HEALTH

Welfare state health benefits

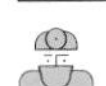 1 per 417 people

 Heart, parasitic, and respiratory diseases

Shortages of drugs and equipment have renewed interest in traditional Mongolian herbal medicine. As well as the state-run system, some Buddhist monasteries provide health care.

SPENDING

GDP/cap. increase

CONSUMPTION AND SPENDING

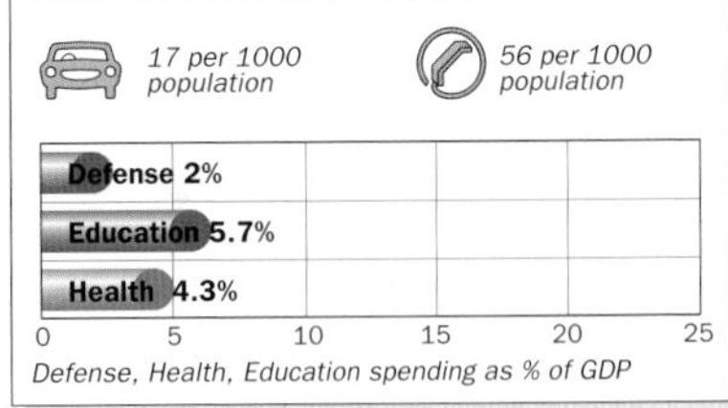
17 per 1000 population
56 per 1000 population
Defense 2%
Education 5.7%
Health 4.3%
0
5
10
15
20
25
Defense, Health, Education spending as % of GDP

Economic liberalization has fueled great disparities in wealth. An estimated 40% of the population now live below the poverty line; the poorest cannot even afford to buy bread. Starvation threatened after the severe winters of 1999–2001.

WORLD RANKING

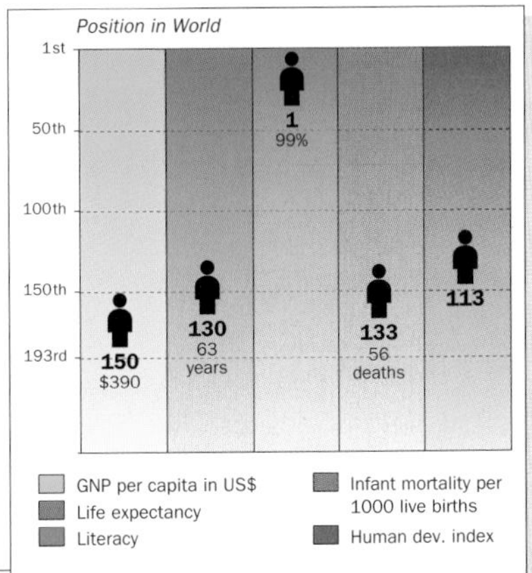
Position in World
1st
50th
100th
150th
193rd
150
$390
130
63 years
1
99%
133
56 deaths
113
GNP per capita in US$
Life expectancy
Literacy
Infant mortality per 1000 live births
Human dev. index

M

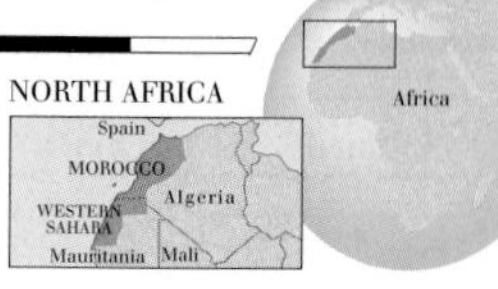
NORTH AFRICA
Spain
MOROCCO
Algeria
WESTERN SAHARA
Mauritania
Mali
Africa

Morocco

Official Name: Kingdom of Morocco **Capital:** Rabat
Population: 30.4 million **Currency:** Moroccan dirham **Official Language:** Arabic

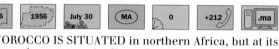

1956

1956
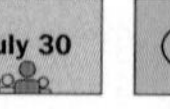
July 30
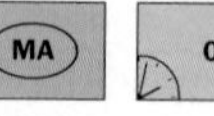
MA
0
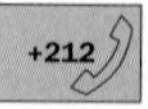
+212

.ma

MOROCCO IS SITUATED in northern Africa, but at its northernmost point lies only 12 km (8 miles) from mainland Europe, across the Strait of Gibraltar. The northern regions have a Mediterranean climate, while the south comprises semiarid desert. The late King Hassan's international prestige gave Morocco status out of proportion to its wealth. The main issues facing the country are the internal threat of Islamic militancy and the stalled political process in Western Sahara, the former Spanish colony occupied by Morocco since 1975. Key economic strengths are tourism, phosphates, and agriculture.

CLIMATE

Hot desert/mountain/ Mediterranean

WEATHER CHART FOR RABAT

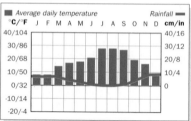

Average daily temperature — Rainfall

°C/°F: 40/104, 30/86, 20/68, 10/50, 0/32, -10/14, -20/-4
J F M A M J J A S O N D
cm/in: 40/16, 30/12, 20/8, 10/4, 0

The climate ranges from warm and temperate in the north to semiarid in the south, but temperatures are cooler in the mountains, especially in the high Atlas. During the summer, the effects of the *sirocco* and *chergui*, hot winds from the Sahara, are felt.

TRANSPORTATION

Drive on right

Mohammed V, Casablanca 3.57m passengers

496 ships 444,000 grt

THE TRANSPORTATION NETWORK

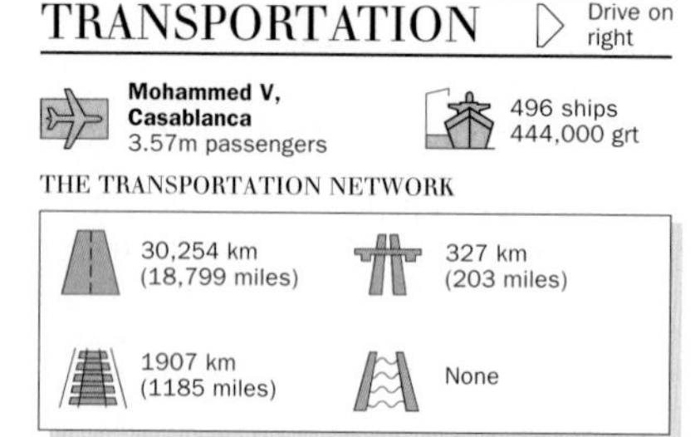
30,254 km (18,799 miles)
327 km (203 miles)
1907 km (1185 miles)
None

Morocco has six international airports. A highway links Rabat and Casablanca, and plans for a new trans-Sahara highway from Tangiers to Lagos, Nigeria, were announced in mid-2000. In rural areas, however, roads tend to peter out.

TOURISM

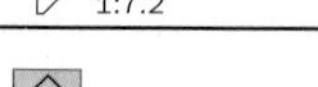
Visitors : Population 1:7.2

4.22m visitors

Up 3% in 2001

MAIN TOURIST ARRIVALS

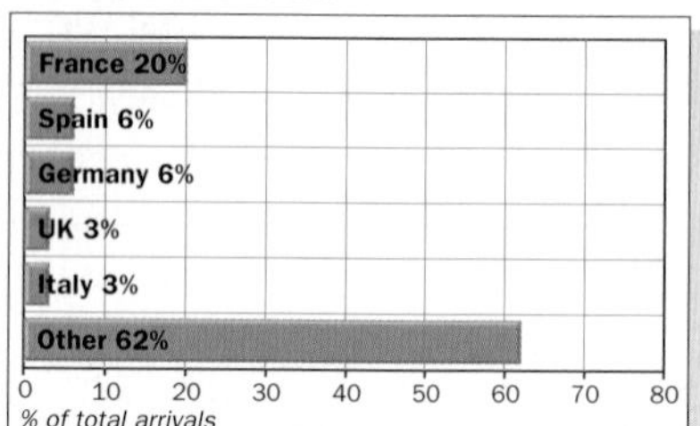
France 20%
Spain 6%
Germany 6%
UK 3%
Italy 3%
Other 62%
0 10 20 30 40 50 60 70 80
% of total arrivals

Tourism is vital to the Moroccan economy. Good beaches abound; Agadir has 300 days of sunshine a year. Fès and Marrakech offer cultural interest, while the Atlas Mountains attract walkers and skiers. Desert safaris are offered in the Sahara. Most Western tourists come from France, Germany, and Spain.

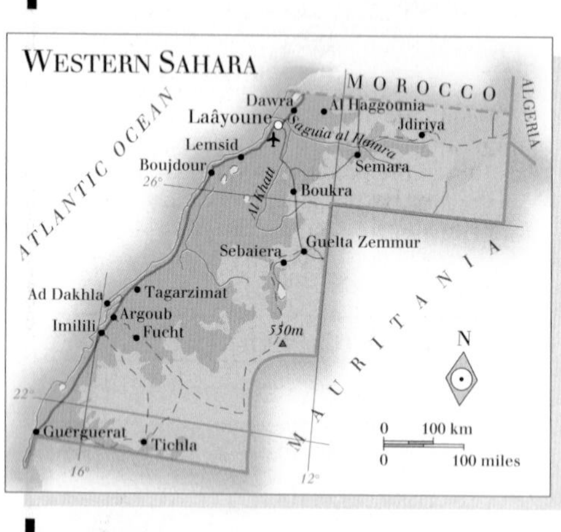
WESTERN SAHARA
MOROCCO
ALGERIA
ATLANTIC OCEAN
MAURITANIA
Dawra
Al Haggounia
Laâyoune
Jdiriya
Lemsid
Saguia al Hamra
Semara
Boujdour
Al Khatt
Boukra
Sebaiera
Guelta Zemmur
Ad Dakhla
Tagarzimat
Argoub
Imlili
Fucht
550m
Guerguerat
Tichla
N
0 100 km
0 100 miles

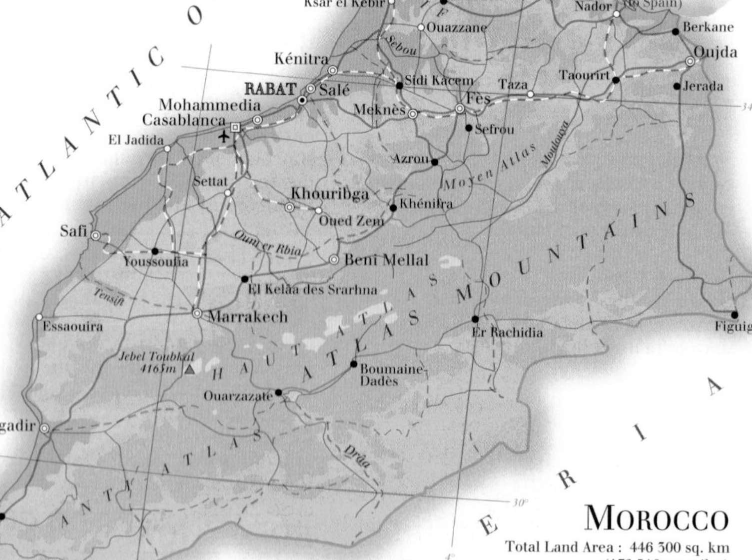
Strait of Gibraltar
MEDITERRANEAN SEA
ATLANTIC OCEAN
Tanger
CEUTA (to Spain)
Tétouan
Larache
Ksar el Kebir
Chaouèn
Al Hoceima
MELILLA (to Spain)
Nador
Berkane
Ouazzane
Oujda
Kénitra
Sidi Kacem
Taza
Taourirt
Jerada
RABAT
Salé
Fès
Mohammedia
Casablanca
Meknès
Sefrou
El Jadida
Azrou
Moyen Atlas
Settat
Khouribga
Khénifra
Oued Zem
Safi
Beni Mellal
Youssoufia
El Kelâa des Srarhna
Marrakech
Essaouira
Jebel Toubkal 4165m
HAUT ATLAS
ATLAS MOUNTAINS
Er Rachidia
Figuig
Boumaine-Dadès
Ouarzazate
Agadir
ANTI ATLAS
Tiznit
Tan-Tan
WESTERN SAHARA
ALGERIA
MOROCCO
Total Land Area : 446 300 sq. km (172 316 sq. miles)
POPULATION
over 1 000 000
over 500 000
over 100 000
over 50 000
over 10 000
under 10 000
LAND HEIGHT
3000m/9843ft
2000m/6562ft
1000m/3281ft
500m/1640ft
200m/656ft
Sea Level
0 100 km
0 100 miles

PEOPLE

Pop. density medium

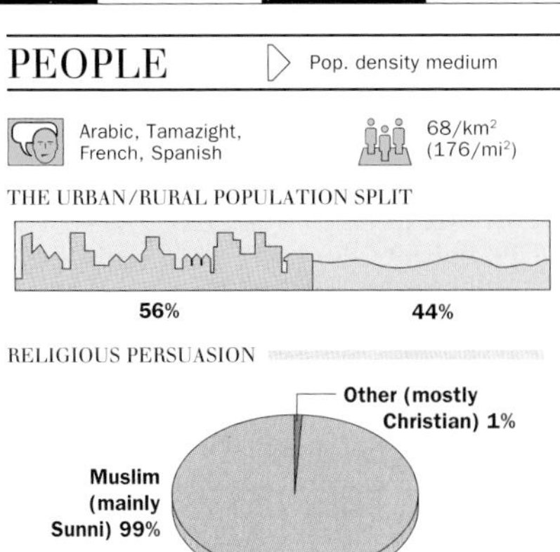
Arabic, Tamazight, French, Spanish
68/km² (176/mi²)
THE URBAN/RURAL POPULATION SPLIT
56%
44%
RELIGIOUS PERSUASION
Other (mostly Christian) 1%
Muslim (mainly Sunni) 99%

ETHNIC MAKEUP

European 1%

Berber 29%

Arab 70%

Morocco, the westernmost of the Maghreb states, is the main refuge for descendants of the original Berber inhabitants of northwest Africa. About 35% of Moroccans speak Berber (Tamazight is the main dialect). They live mainly in mountain villages, while the Arab majority inhabit the lowlands. Before independence from France, 450,000 Europeans lived in Morocco; numbers have since greatly diminished. Some 45,000 Jews enjoy religious freedom and full civil rights – a position in society unique among Arab countries. Most people speak Arabic, and French is also spoken in urban areas.

Sunni Islam is the religion of most of the population. The king is the spiritual leader through his position as Commander of the Faithful.

Female emancipation has been slow to take root in Morocco, but women are starting to take a more prominent role in society.

POPULATION AGE BREAKDOWN

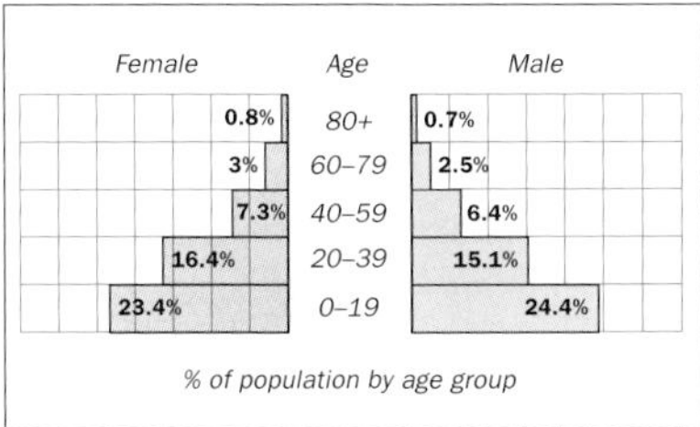
Female
Age
Male
0.8%
80+
0.7%
3%
60–79
2.5%
7.3%
40–59
6.4%
16.4%
20–39
15.1%
23.4%
0–19
24.4%
% of population by age group

***The town of Boumaine-Dadès** lies in the southern foothills of the Atlas Mountains. The region's outstanding scenery makes it one of Morocco's major tourist attractions.*

POLITICS

Multiparty elections

L. House 1997/2002
U. House 2000/2003

H.M. King Mohammed VI

AT THE LAST ELECTION

House of Representatives 325 seats

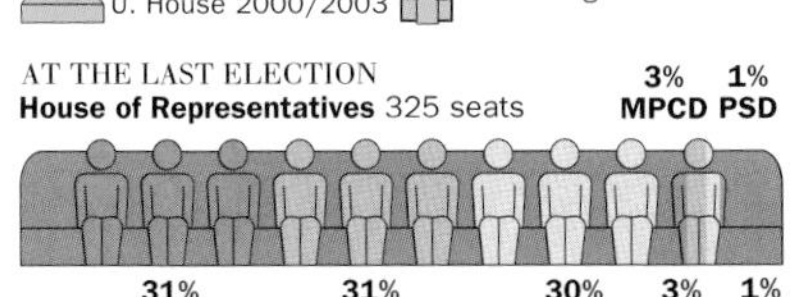
3% MPCD
1% PSD
31% K
31% W
30% C
3% FFD
1% Others

K = Koutla bloc (includes Socialist Union of Popular Forces and Istiqlal) **W** = Wifaq bloc (includes Constitutional Union & Popular Movement) **C** = Center bloc (includes National Rally of Independents and Democratic Popular Movement) **MPCD** = Constitutional & Democratic Popular Movement **FFD** = Democratic Forces Front **PSD** = Democratic Socialist Party

House of Councillors 270 seats

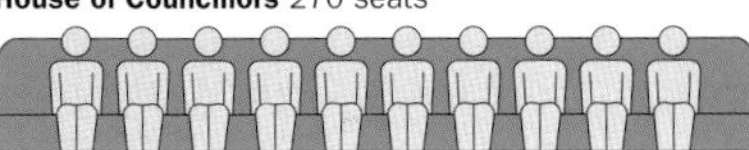

An indirectly elected House of Councillors was formed in December 1997

Morocco is a constitutional monarchy with a bicameral legislature.

Profile

Although the majority parliamentary party now chooses the government, the king reserved the right to appoint or dismiss the prime minister. Legislative elections in 1997 left parliament split three ways, with only nine seats going to Islamists. The eventual formation of a socialist-led government was seen as the clearest expression to date of the increasing role of the party system.

Main Political Issues

The post-Hassan monarchy

King Mohammed VI is seen as a less dominating figure than his late father King Hassan II. His accession encouraged expectations of change. The king's marriage in 2002 to computer engineer Salma Bennani was presented as evidence of a more modern and open outlook. The queen is expected to provide a role model for Moroccan women.

Islamic militancy

Mohammed VI and the government in office since 1998 have ushered in a more liberal stance toward Islamist activists. In 2000, Abdessalam Yassine, spiritual leader of the banned Islamic movement Justice and Good Deeds, was released after ten years' imprisonment without trial. Popular support for Islamic fundamentalism is fueled by the fear that Morocco is losing its Islamic identity. Pro-Islamist rallies have far outnumbered those by supporters of greater rights for women.

***Abderrahmane el Youssoufi**, prime minister since 1998.*

***King Mohammed VI**, who succeeded his father, Hassan II.*

WORLD AFFAIRS

Joined UN in 1956

AL

AMU

IBRD

NAM

OIC

Morocco's important role in the quest for lasting peace in the Middle East was underlined by Israeli prime minister Yitzhak Rabin's visit to Rabat following the signing in Washington D.C. of the 1993 peace accord with the Palestine Liberation Organization. King Hassan's foreign policy was ambiguous, for while he negotiated with Israel he also headed the Jerusalem Committee of the Organization of the Islamic Conference (OIC). Generally more pro-Western than other Arab states, Morocco has also earned respect by protecting its Jewish minority.

International disapproval has focused on Morocco's occupation since 1975 of the former Spanish colony of Western Sahara. Resistance by Polisario Front guerrillas, who are fighting for an independent Western Sahara, commenced in 1983 and has continued, despite a UN-brokered peace plan in 1991. Since then, UN proposals have included a referendum on self-determination and autonomy for Western Sahara under Moroccan sovereignty. The process remained stalled in 2002, with the UN extending its mandate to the end of July.

Relations with the EU were strengthened with the signing of an association agreement in late 1995, envisaging free trade in industrial goods within 12 years.

Sovereignty disputes with Spain over rocky islets in the Mediterranean were at the heart of a high-profile confrontation in 2002.

AID

Recipient

$419m (receipts)

Down 38% in 2000

Saudi Arabia wrote off $2.7 billion of Moroccan debt after the Gulf War. The World Bank has given help to Morocco, but the country receives little aid.

M

CHRONOLOGY

Independence from France in 1956 was only the first step in ending colonial rule for the oldest kingdom in the Arab world, even though the present Alaoui dynasty has been in power for three centuries.

- ❑ **1956** France recognizes Moroccan independence under Sultan Mohammed ibn Yousif. Morocco joins UN. Spain renounces control over most of its territories.
- ❑ **1957** Sultan Mohammed king.
- ❑ **1961** Hassan succeeds his father.
- ❑ **1967** Morocco backs Arab cause in Six-Day War with Israel.
- ❑ **1969** Spain returns Ifni to Morocco.
- ❑ **1972** King Hassan survives assassination attempt.
- ❑ **1975** International Court of Justice grants right of self-determination to Western Saharan people. Moroccan forces seize Saharan capital.
- ❑ **1976** Morocco and Mauritania partition Western Sahara.
- ❑ **1979** Mauritania renounces claim to part of Western Sahara, which is added to Morocco's territory.
- ❑ **1984** King Hassan signs Oujda Treaty with Col. Gaddafi of Libya as first step toward Maghreb union. Morocco leaves OAU after criticism of its role in Western Sahara.
- ❑ **1986** Morocco abrogates Oujda Treaty.
- ❑ **1987** Defensive wall built around Western Sahara.
- ❑ **1989** Arab Maghreb Union (AMU) creates no-tariff zone between Morocco, Algeria, Tunisia, Libya, and Mauritania.
- ❑ **1991** Morocco accepts UN plan for referendum in Western Sahara.
- ❑ **1992** New constitution grants majority party in parliament right to choose the government.
- ❑ **1993** First general election for nine years. After major parties refuse his invitation, king appoints nonparty government.
- ❑ **1994** King Hassan replaces veteran prime minister Karim Lamrani with Abdellatif Filali.
- ❑ **1995** Islamist opposition leader Mohamed Basri returns to Morocco after 28 years of exile.
- ❑ **1998** Socialists enter government with Abderrahmane el Youssoufi as prime minister.
- ❑ **1999** Death of King Hassan. Mohammed VI enthroned. Liberalization program announced.
- ❑ **2001** UN special representative proposes a ten-year trial period for Western Sahara as part of Morocco.
- ❑ **2002** UN extends mandate in Western Sahara as political process stalls.

M

DEFENSE

Compulsory military service

$1.68bn

Down 5% in 2000

MOROCCAN ARMED FORCES

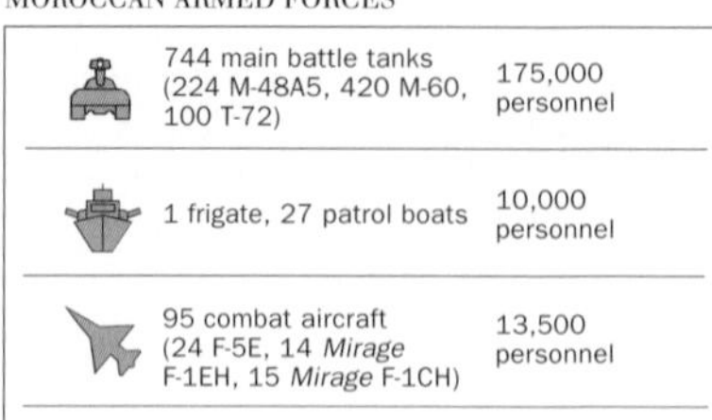

744 main battle tanks (224 M-48A5, 420 M-60, 100 T-72)	175,000 personnel
1 frigate, 27 patrol boats	10,000 personnel
95 combat aircraft (24 F-5E, 14 *Mirage* F-1EH, 15 *Mirage* F-1CH)	13,500 personnel
None	

Morocco's long struggle in Western Sahara against Polisario Front guerrillas has given the kingdom's forces a formidable reputation. Moroccans have also fought as mercenaries in the Gulf. In the 1980s, Moroccan sappers constructed a 2500-km (1550-mile) defensive wall to cordon off Western Sahara in an attempt to prevent incursions from Polisario guerrillas based in Algeria. The Polisario forces themselves number some 3000–6000.

Morocco's pro-Western stance has allowed its forces access to sophisticated weapons and training from the West, particularly the US – unlike neighboring north African states, which are dependent on the former Soviet bloc.

The air force was formed in 1956 and flies US and European aircraft, notably *Mirage* interceptors. The navy uses Western-supplied ships, but is insignificant in regional terms. In addition, there are 42,000 paramilitaries.

Around 5% of national income is spent on defense – a relatively high figure for a developing country. Military service, lasting 18 months, is compulsory.

ECONOMICS

Inflation 2.8% p.a. (1990–2000)

$33.9bn

10.563–11.519 Moroccan dirhams

SCORE CARD

❑ World GNP Ranking	56th
❑ GNP per Capita	$1180
❑ Balance of Payments	–$475m
❑ Inflation	1.9%
❑ Unemployment	23%

EXPORTS

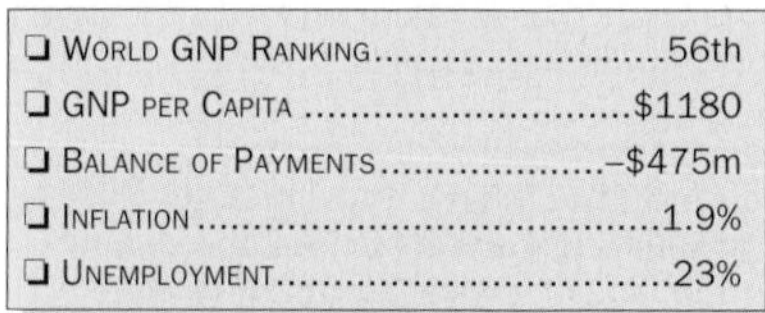
Germany 5%
Italy 6%
UK 8%
Spain 10%
France 26%
Other 45%

IMPORTS

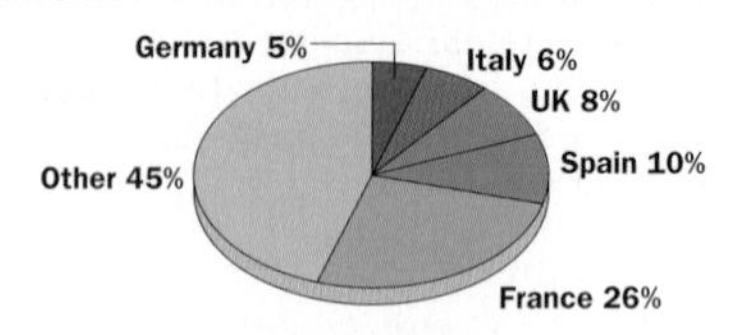
UK 5%
Italy 6%
Germany 6%
Spain 11%
France 25%
Other 47%

Strengths

Probusiness policies and abundant labor attract foreign investment. Low inflation. Great potential for tourist industry (already important), phosphates, and agriculture.

Weaknesses

High unemployment and population growth. Droughts have hit agriculture. Cannabis production (providing Europe's main source of resin) complicates closer EU links.

Profile

The government's large-scale privatization program, which began in

ECONOMIC PERFORMANCE INDICATOR

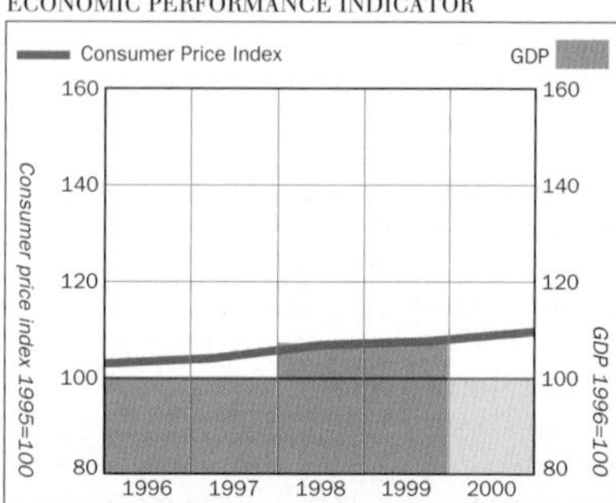
Consumer Price Index
GDP
160
140
120
100
80
Consumer price index 1995=100
GDP 1996=100
1996
1997
1998
1999
2000

1992, was designed to attract investment, particularly from Europe. Severe drought in 1995 made austerity measures necessary. The new socialist-led government has given social policy a higher priority. Expected revenue from oil reserves will be channeled into the development of rural areas.

MOROCCO : MAJOR BUSINESSES

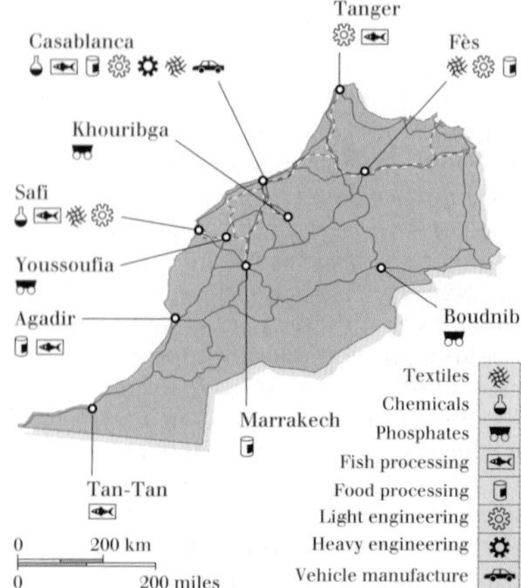
Casablanca
Tanger
Fès
Khouribga
Safi
Youssoufia
Agadir
Boudnib
Marrakech
Tan-Tan
0
200 km
0
200 miles
Textiles
Chemicals
Phosphates
Fish processing
Food processing
Light engineering
Heavy engineering
Vehicle manufacture

RESOURCES

▷ Electric power 4m kw

748,183 tonnes

241 b/d (reserves 659m barrels)

17.3m sheep, 5.2m goats, 2.7m cattle, 137m chickens

Phosphates, oil, gas, coal, iron, barytes, lead, copper, zinc

ELECTRICITY GENERATION

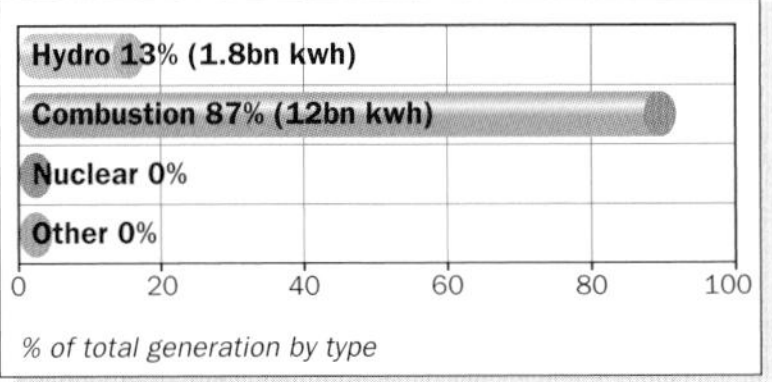
Hydro 13% (1.8bn kwh)
Combustion 87% (12bn kwh)
Nuclear 0%
Other 0%
0
20
40
60
80
100
% of total generation by type

Morocco possesses 75% of the world's phosphate reserves. The discovery of large oil and gas deposits in the northeastern desert in mid-2000 could yield an annual revenue of $400 million.

MOROCCO : LAND USE

Cropland
High mountain regions
Pasture
Forest
Desert
Sheep
Fruit
Wheat

Sebou
Moulouya
ATLAS MOUNTAINS

0 200 km
0 200 miles

ENVIRONMENT

▷ Sustainability rank: 73rd

0.7%

1.2 tonnes per capita

ENVIRONMENTAL TREATIES

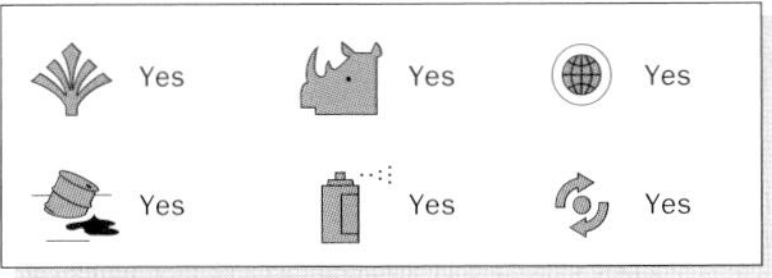
Yes
Yes
Yes
Yes
Yes
Yes

Morocco's wealth of plant and animal life has suffered severely from long periods of drought, most recently in the early 1980s and early 1990s. The unplanned development of tourist resorts is posing a threat to fragile coastal ecosystems.

MEDIA

▷ TV ownership medium

Daily newspaper circulation 27 per 1000 people

PUBLISHING AND BROADCAST MEDIA

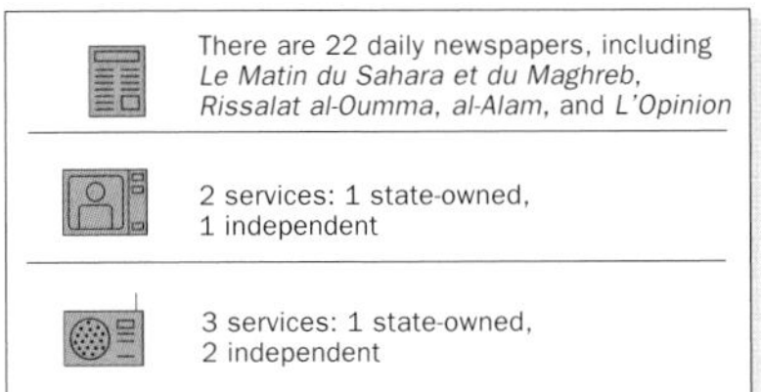

There are 22 daily newspapers, including *Le Matin du Sahara et du Maghreb*, *Rissalat al-Oumma*, *al-Alam*, and *L'Opinion*

2 services: 1 state-owned, 1 independent

3 services: 1 state-owned, 2 independent

The succession of Mohammed VI fueled hopes of a more liberal climate, but the media remain strictly controlled, particularly over reporting of the Western Sahara issue. In 2000, the outspoken French-language weekly *Demain* was banned. The sports pages, especially the soccer reports, are the most dynamic sections of the press. State-owned TV began transmissions in Arabic and French in 1962. Radio broadcasts are in Arabic, Berber, French, Spanish, and English from Rabat and Tangier. Morocco can receive broadcasts from Spanish radio and television stations.

CRIME

▷ Death penalty in use

48,600 prisoners

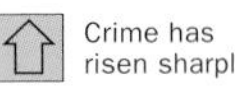

Crime has risen sharply

CRIME RATES

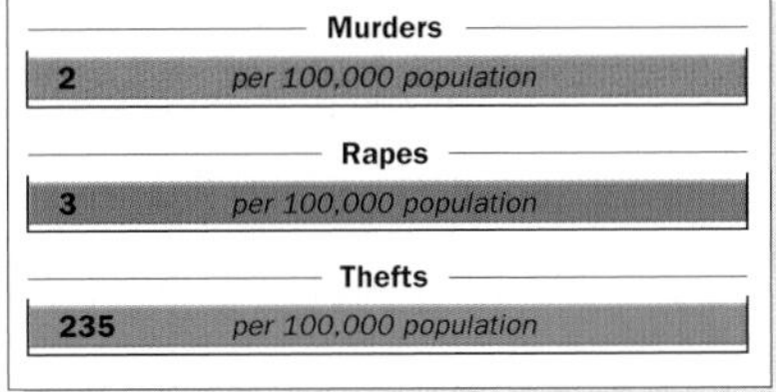
Murders
2 per 100,000 population
Rapes
3 per 100,000 population
Thefts
235 per 100,000 population

Urban crime is increasing, but muggings are rare. Apart from a 1990 strike that led to 40 deaths in Fès, there has been little civil unrest. Prisons are overcrowded, and conditions are poor.

EDUCATION

▷ School leaving age: 14

49%

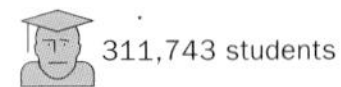

311,743 students

THE EDUCATION SYSTEM

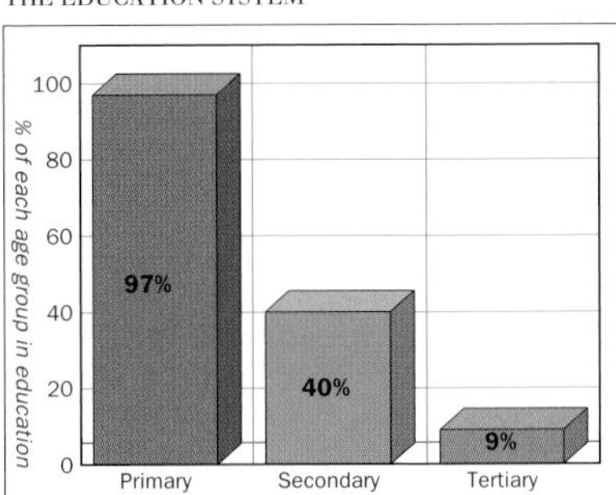
100
80
60
40
20
0
% of each age group in education
97%
40%
9%
Primary
Secondary
Tertiary

Literacy in rural areas – at only 14% – is much lower than in the cities. The literacy level and elementary school enrollment rates are well below average for countries with similar living standards; child labor is widely used. There are both state-controlled and private schools. In 1988 the secondary school graduation examination, the *baccalauréat*, was replaced by a system of continuous assessment.

HEALTH

▷ Welfare state health benefits

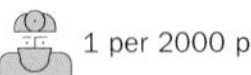

1 per 2000 people

Neonatal causes, cerebrovascular and heart diseases

There is one hospital bed for every 1000 people. Despite recent progress, child mortality and nutritional standards for the poorest Moroccans remain substantially below the average. Outside the cities, primary health care is virtually nonexistent, with the result that people depend on traditional remedies for illnesses. All employees are required to contribute to a social welfare fund, which operates a system of benefits in the event of illness, occupational accidents, and old age.

SPENDING

▷ GDP/cap. increase

CONSUMPTION AND SPENDING

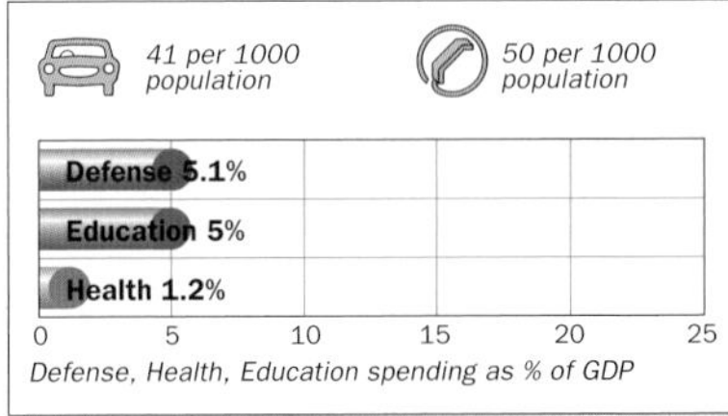
41 per 1000 population
50 per 1000 population
Defense 5.1%
Education 5%
Health 1.2%
0
5
10
15
20
25
Defense, Health, Education spending as % of GDP

Average income per head is much lower than in the neighboring countries of Algeria and Tunisia. Almost one in five Moroccans live below the poverty line, and the rural–urban gap in wealth is considerable; the majority of the population lives in rural areas. A period of drought in the 1990s encouraged urban drift.

Unrest has largely been avoided owing to the fact that Morocco has a thriving informal sector. This provides jobs in food processing, clothes manufacturing, goods transportation, and the hotel and building trades. In addition, there is work to be found in the illegal hashish trade and the smuggling of alcohol and Western goods.

WORLD RANKING

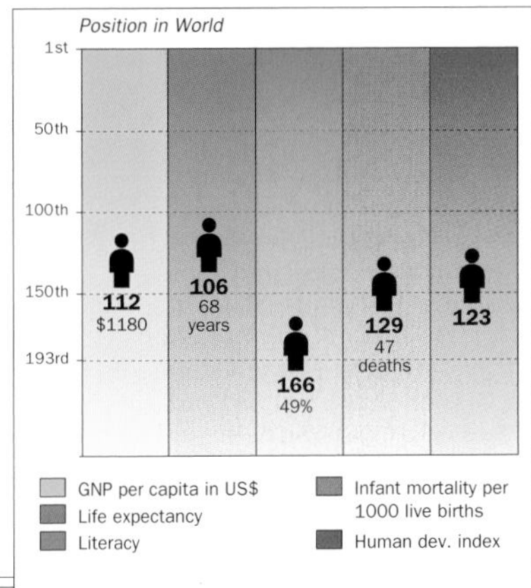
Position in World
1st
50th
100th
150th
193rd
112
$1180
106
68
years
166
49%
129
47
deaths
123
GNP per capita in US$
Life expectancy
Literacy
Infant mortality per 1000 live births
Human dev. index

M

Mozambique

SOUTHERN AFRICA

Official Name: Republic of Mozambique **Capital:** Maputo

Population: 18.6 million **Currency:** Metical **Official Language:** Portuguese

1975

1975

June 25

MOC

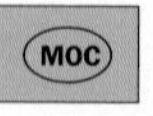

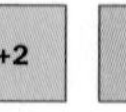
+2

+258

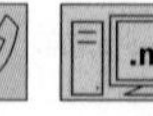

.mz

SITUATED ON THE SOUTHEAST African coast, Mozambique is bisected by the Zambezi River. South of the Zambezi lies a semiarid savanna lowland. The more fertile north-central delta provinces around Tete are home to most of Mozambique's ethnically diverse population. Following independence from Portugal in 1975, Mozambique was torn apart by civil war between the (then Marxist) FRELIMO government and the South African-backed Mozambique National Resistance (RENAMO). The conflict finally ended in 1992 after UN arbitration. Multiparty elections in 1994 returned FRELIMO to power. Devastating floods in 2000 and 2001 created a desperate situation for this impoverished country, with famine a major threat.

CLIMATE

Tropical wet & dry

WEATHER CHART FOR MAPUTO

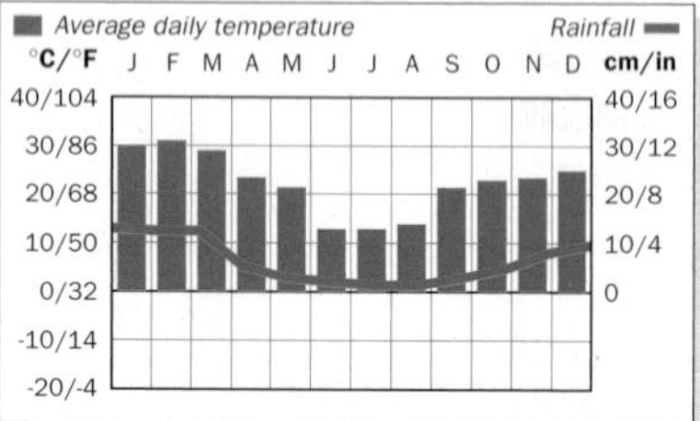
Average daily temperature
Rainfall
°C/°F
J F M A M J J A S O N D
cm/in
40/104
30/86
20/68
10/50
0/32
-10/14
-20/-4
40/16
30/12
20/8
10/4
0

In theory, Mozambique has a rainy and a dry season. The coast at Beira and at Quelimane and the highlands west of Nampula are the wettest areas. The Zambezi valley is the hottest region.

Mozambique is prone to extremes of rainfall. In the 1980s, frequent failure of the rains contributed to two disastrous famines. Devastating floods occurred in 2000 and 2001.

TRANSPORTATION

Drive on left

Mavalane International, Maputo
436,849 passengers

124 ships
35,300 grt

THE TRANSPORTATION NETWORK

5685 km (3532 miles)	None
3114 km (1935 miles)	3750 km (2330 miles)

The billion-dollar Maputo Corridor, launched in 1995, reconnects South African industrial centers with the Mozambican coast, and should also facilitate port modernization. CFM, the state-owned railroad company, is cooperating with other neighboring states. The national airline is returning profits. Even so, millions of land mines still hamper access, damaged bridges have yet to be rebuilt, and remote communities remain isolated.

Tea picking. Other important cash crops are cashew nuts, cotton, sugar, copra, and citrus fruits. Agriculture employs 83% of workers.

TOURISM

Not available

Tourism has still not recovered after war

Little change from year to year

MAIN TOURIST ARRIVALS

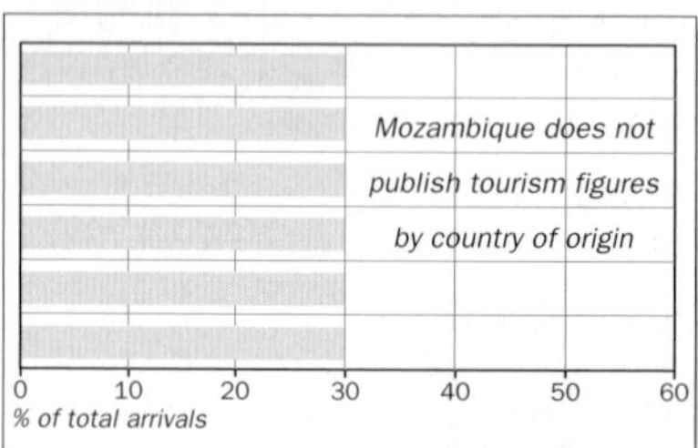
Mozambique does not publish tourism figures by country of origin
0
10
20
30
40
50
60
% of total arrivals

Mozambique used to attract around 300,000 South Africans and Rhodesians a year in the 1970s, but the tourist industry was destroyed by the civil war and is only slowly being rebuilt. Land mines still render travel outside the capital hazardous, while food shortages, poor infrastructure, and costly international flights are added obstacles. Further setbacks followed the floods in 2000 and 2001.

Given political stability, though, Mozambique could yet exploit its excellent beaches and game reserves, which include the Gorongosa Game Park. Some hotel groups are once more targeting Maputo as a luxury tourist and conference venue.

PEOPLE

Pop. density low

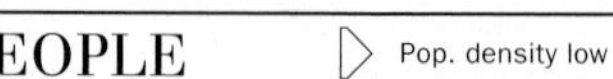

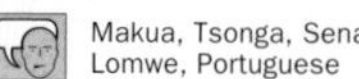

Makua, Tsonga, Sena, Lomwe, Portuguese

24/km² (61/mi²)

THE URBAN/RURAL POPULATION SPLIT

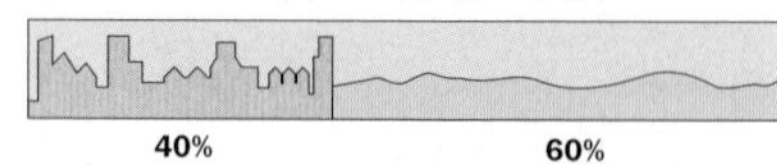
40%
60%

RELIGIOUS PERSUASION

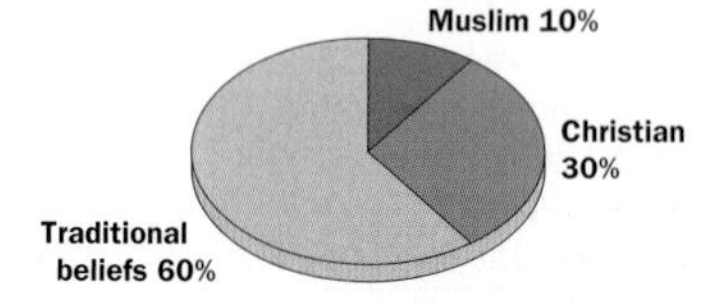
Muslim 10%
Christian 30%
Traditional beliefs 60%

ETHNIC MAKEUP

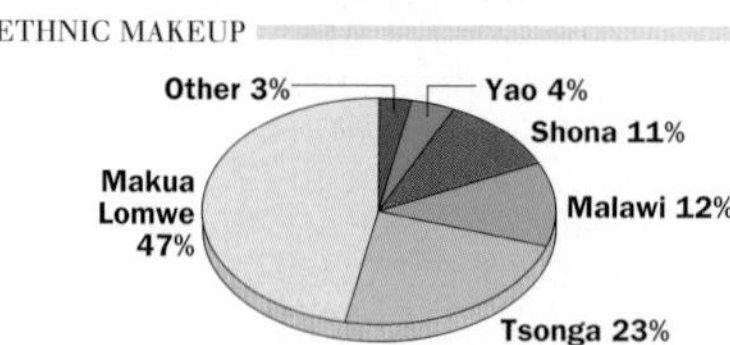
Other 3%
Yao 4%
Shona 11%
Malawi 12%
Tsonga 23%
Makua Lomwe 47%

Mozambique is ethnically diverse, with the overwhelming black African majority divided into numerous groups, and tiny minorities of whites, mixed-race groups, and Asians. However, the predominant social tensions are regional: RENAMO, strong in the north and central regions, accuses the FRELIMO government of consistently favoring the south. Antiwhite feelings are growing too, as "Africanists" claim that whites enjoy excessive political influence – incendiary charges in a country where average life expectancy is well below 50 years and poverty is endemic.

Mozambican society centers on the extended family. In some provinces, notably Zambezia, Cabo Delgado, and Tete, this is matriarchal. Polygamy is fairly widespread among men who can support second wives. FRELIMO pays special attention to women's rights. Many women served in FRELIMO armies, and are now protected by divorce, child-custody, and husband-desertion laws. The Mozambican Women's Organization encourages participation in political life.

POPULATION AGE BREAKDOWN

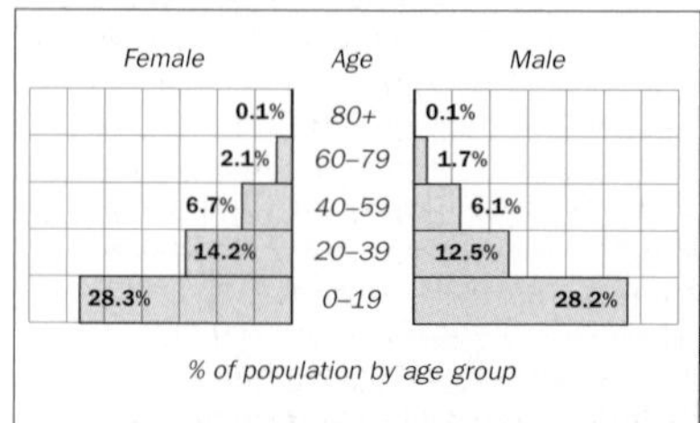
Female
Age
Male
0.1%
80+
0.1%
2.1%
60–79
1.7%
6.7%
40–59
6.1%
14.2%
20–39
12.5%
28.3%
0–19
28.2%
% of population by age group

M

POLITICS

Multiparty elections

1999/2004

President Joaquim Alberto Chissano

AT THE LAST ELECTION

Assembly of the Republic 250 seats

53% FRELIMO	47% RENAMO

FRELIMO = Front for the Liberation of Mozambique
RENAMO = Mozambique National Resistance

Mozambique held its first multiparty elections in 1994.

PROFILE

Between 1977 and 1990, Mozambique was a one-party state run by the Soviet-backed FRELIMO. It had campaigned for independence since the 1960s. White South Africa and Southern Rhodesia (now Zimbabwe) both backed anti-Marxist RENAMO rebels.

Changing international realities persuaded FRELIMO to adopt a democratic constitution in 1990. Meanwhile, RENAMO lost its external sponsors. Today little distinguishes the two ideologically. Although FRELIMO is the biggest party in parliament, RENAMO is clearly popular, and demands recognition for its 15 years of struggle. New groups, such as the antiwhite PALMO, COINMO, and UNAMO, have recently emerged; FRELIMO is pushing ahead with plans to decentralize power. Continuing disputes since 1999 between RENAMO and the government about provincial representation have rekindled fears for the future of multiparty democracy.

Joaquim Chissano, president since 1986, has pushed toward political pluralism.

Afonso Dhlakama, RENAMO leader, has turned from militarism to politics.

MAIN POLITICAL ISSUES

The move to democracy

In 1993, the UN secured, with difficulty, the $260 million and the 7500-strong multinational forces required to demobilize the warring factions and to stage the first democratic elections.

Despite last-minute hitches, elections were held in 1994 which returned FRELIMO to power. However, support for RENAMO was stronger than had been expected. Their leader Afonso Dhlakama polled strongly in the presidential elections in both 1994 and 1999, contesting Joaquim Chissano's claim of victory.

Reconstruction

The government faces a huge task in rebuilding a country where devastating floods came on top of the ravages of civil war. The fighting had left 900,000 dead, one million refugees, and an estimated 90% of the remaining population living below the poverty line.

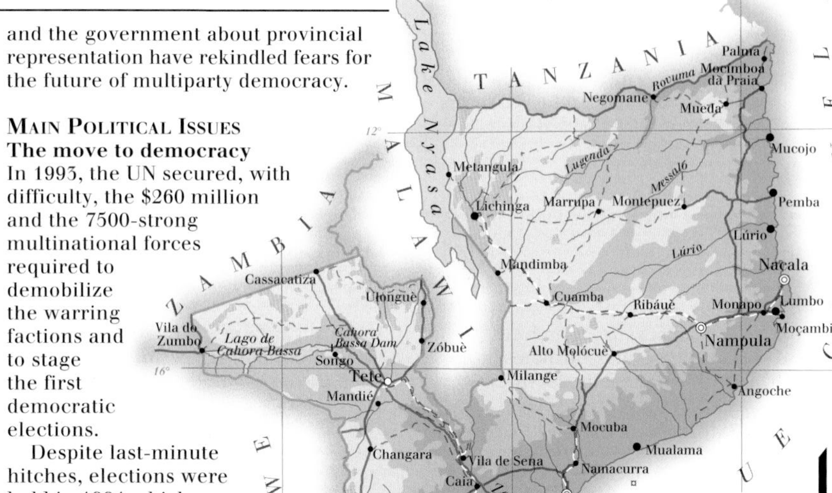

MOZAMBIQUE
Total Land Area : 801 590 sq. km
(309 494 sq. miles)
POPULATION
over 1 000 000
over 100 000
over 50 000
over 10 000
under 10 000
LAND HEIGHT
2000m/6562ft
1000m/3281ft
500m/1640ft
200m/656ft
Sea Level
200 km
200 miles
TANZANIA
MALAWI
ZAMBIA
ZIMBABWE
SOUTH AFRICA
SWAZILAND
MOZAMBIQUE CHANNEL
INDIAN OCEAN
Lake Nyasa
MAPUTO
Beira
Nampula
Quelimane
Tete
Chimoio
Pemba
Lichinga
Inhambane
Xai-Xai
Baía de Maputo
Baía de Sofala
Monte Binga 2436m
Gorongosa National Park
Lago de Cahora Bassa
Cahora Bassa Dam
Zambezi
Limpopo
Save
Lúrio
Rovuma
Lugenda
Messalo
Púngoè
Búzi
Changane

WORLD AFFAIRS

Joined UN in 1975

Comm | CPLP | AU | OIC | SADC

Mozambique was a key Cold War battleground between Soviet-backed Marxism, and capitalism sponsored by the US and South Africa. The resulting civil war devastated the country between 1977 and 1992.

In the early 1980s, however, the FRELIMO government's position began to shift as Soviet aid became erratic. Responding to President Samora Machel's overtures, the US lifted its ban on economic assistance in 1984. Britain agreed to train FRELIMO's forces in 1987. South Africa continued tacitly to support RENAMO until at least 1990. Zimbabwean troops helped Mozambique guard the strategically important Beira and Limpopo corridors, but left in 1993.

In 1995, the UN withdrew its peacekeepers and a democratic Mozambique joined the Commonwealth, despite having no formal links with the old British Empire. President Chissano became deputy head of the SADC, but regional tensions persisted, with Mozambique accusing South Africans of gun-running, and Swaziland claiming Maputo Province as its own.

CHRONOLOGY

The Portuguese tapped the local trade in slaves, gold, and ivory in the 16th century and made Mozambique a colony in 1752. Large areas were run by private companies until 1929.

- **1964** FRELIMO starts war of liberation.
- **1975** Independence. FRELIMO leader Samora Machel is president.
- **1976** Resistance movement RENAMO set up inside Mozambique by Rhodesians.
- **1976–1980** Mozambique closes Rhodesian border and supports Zimbabwean freedom fighters. Reprisals by RENAMO.
- **1977** FRELIMO constitutes itself as Marxist–Leninist party.
- **1980** South Africa takes over backing of RENAMO.

M

CHRONOLOGY *continued*

- **1982** Zimbabwean troops arrive to guard Mutare–Beira corridor.
- **1984** Nkomati Accord: South Africa agrees to stop support for RENAMO, and Mozambique for ANC, but fighting continues.
- **1986** RENAMO declares war on Zimbabwe. Tanzanian troops reinforce FRELIMO. Machel dies in mysterious air crash in South Africa. Joaquim Chissano replaces him.
- **1988** Nkomati Accord reactivated. Mozambicans allowed back to work in South African mines.
- **1989** War and malnutrition said to claim one million lives. FRELIMO drops Marxism–Leninism.
- **1990** Multipartyism and free-market economy in new constitution. RENAMO breaches cease-fire.
- **1992** Chissano signs peace agreement with RENAMO.
- **1994** Democratic elections return FRELIMO to power.
- **1995** Joins Commonwealth. Economic reforms begun.
- **1999** G7 chooses Mozambique as flagship for international debt relief initiative. RENAMO disputes results of December elections.
- **2000–2001** Thousands displaced by devastating floods.

AID

Recipient

 $876m (receipts)  Up 9% in 2000

Mozambique is the world's second most aid-dependent country. Aid accounts for fully 60% of national earnings, and pays for the food needs of some seven million citizens. In 1999, Mozambique became one of only four countries to receive the G7 debt relief scheme for HIPCs, which is worth nearly $3 billion. The main donors are the World Bank, Portugal, the US, the EU and its member states, and the IMF. Debts from earlier Soviet aid have been written off.

DEFENSE

Compulsory military service

 $85m Down 10% in 2000

About 2.5 million men were deemed "fit for military service" in 1998, but since the civil war ended in 1992, the military's once dominant role in society has greatly diminished. Military figures, once prominent in the FRELIMO government, have been largely stripped of political influence.

Mozambique's new post-peace, British-trained permanent army was formally inaugurated in 1994. Truly national in character, and only around 10,000 strong, it contains both former government and RENAMO troops.

However, one by-product of reorganization was the demobilization of some 75,000 battle-hardened soldiers. Their severance pay ended in mid-1996, and it has not been easy to retrain them, or reintegrate them into civilian life. Some have turned to banditry.

The war's end also saw the departure of external forces, such as UN peacekeepers and the Zimbabwean troops who once guarded strategic railroads against RENAMO attack.

MOZAMBICAN ARMED FORCES

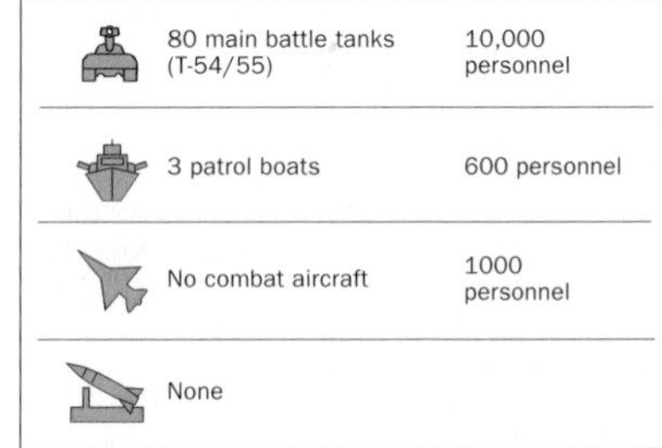
80 main battle tanks (T-54/55)
10,000 personnel
3 patrol boats
600 personnel
No combat aircraft
1000 personnel
None

M

ECONOMICS

Inflation 33% p.a. (1990–2000)

$3.75bn

17,175–22,885 meticais

SCORE CARD

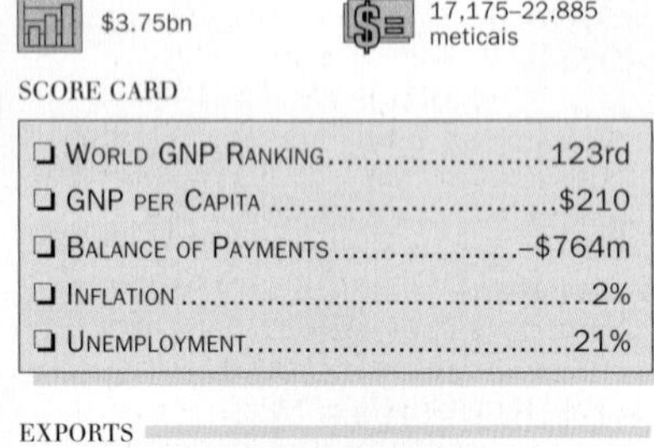

World GNP Ranking	123rd
GNP per Capita	$210
Balance of Payments	–$764m
Inflation	2%
Unemployment	21%

EXPORTS

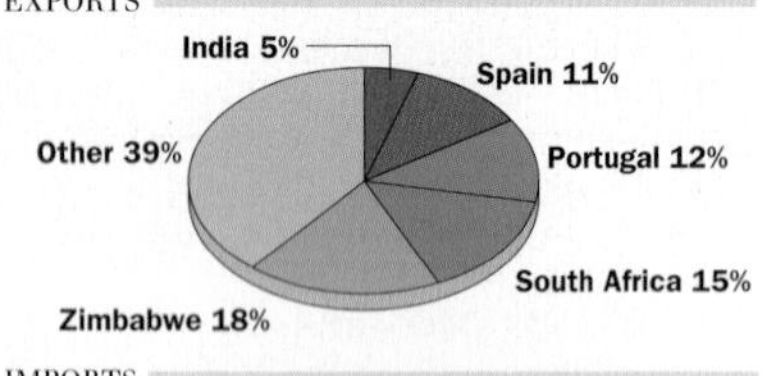
India 5%
Spain 11%
Other 39%
Portugal 12%
South Africa 15%
Zimbabwe 18%

IMPORTS

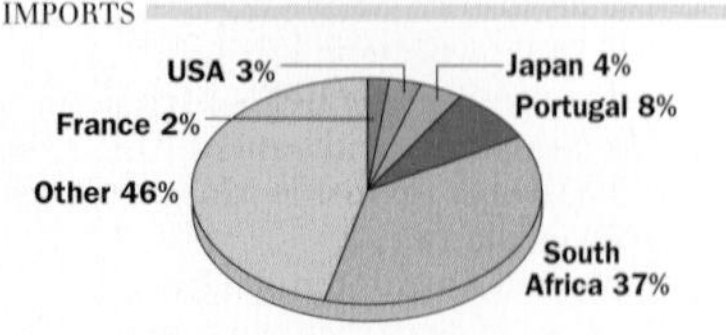
USA 3%
Japan 4%
Portugal 8%
France 2%
Other 46%
South Africa 37%

Strengths

Following IMF advice, the government has adopted privatization, exchange rate reforms, and trade liberalization. This has enabled Mozambique to attract aid and increase exports.

ECONOMIC PERFORMANCE INDICATOR

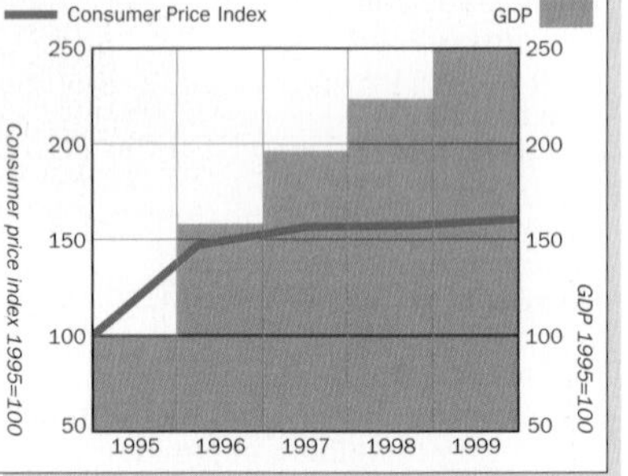
Consumer Price Index
GDP
250
200
150
100
50
Consumer price index 1995=100
GDP 1995=100
1995
1996
1997
1998
1999

Massive rural development programs target agriculture, which employs 85% of the workforce. Fisheries industry has great potential. Improved transportation links with Maputo, Africa's second-largest harbor, will help it to service southern Africa's landlocked regions. Go-ahead for a new port at Dobela, in the south, was given in 2002.

Weaknesses

Overseas aid is essential to prevent at least half the population starving. Overdependence on foreign donors and companies is another long-term concern. The country is susceptible to drought, floods, and cyclones. Skilled workers often choose to work in other countries; their absence has delayed the return to normal economic activity.

Profile

Although Mozambique has enormous problems, the government in 1995 produced an optimistic plan, based on World Bank recommendations, to eradicate poverty and raise annual GDP growth to 8–9% by 2000. Devastation resulting from the floods of 2000 and 2001 has effectively destroyed such hopes, and GDP growth in 2000 was only 2.1%.

MOZAMBIQUE : MAJOR BUSINESSES

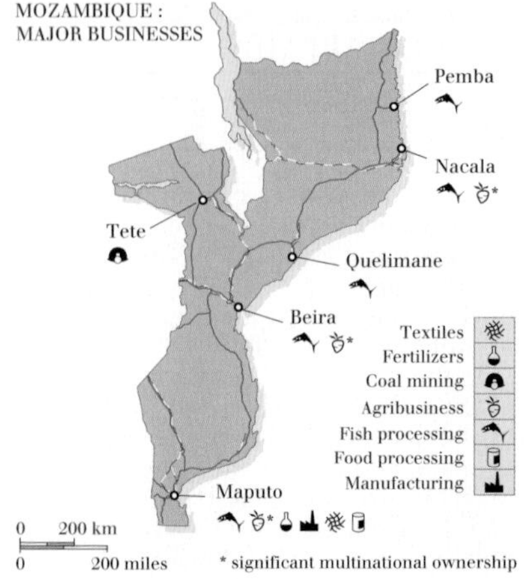
Pemba
Nacala
Tete
Quelimane
Beira
Maputo
Textiles
Fertilizers
Coal mining
Agribusiness
Fish processing
Food processing
Manufacturing
0 200 km
0 200 miles
* significant multinational ownership

RESOURCES

Electric power 2.4m kw

 35,560 tonnes

 Not an oil producer

 1.32m cattle, 670,000 ducks, 28m chickens

Coal, iron, tantalite, uranium, gold, bauxite titanium, copper, gas

ELECTRICITY GENERATION

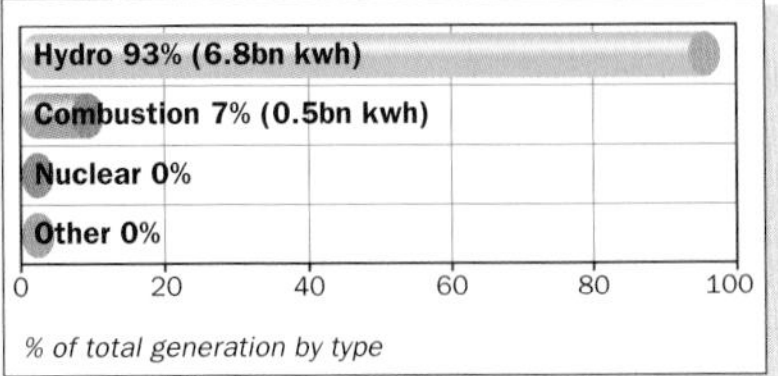
Hydro 93% (6.8bn kwh)
Combustion 7% (0.5bn kwh)
Nuclear 0%
Other 0%
0
20
40
60
80
100
% of total generation by type

Reserves of coal, iron, bauxite, uranium, and gas are underexploited, but a gas pipeline to South Africa is to be completed by 2004. Cotton vies with cashew nuts as the chief crop. Fishing is a vital sector. Electricity supplies are being restored, and the Mozal aluminum smelter opened in 2000.

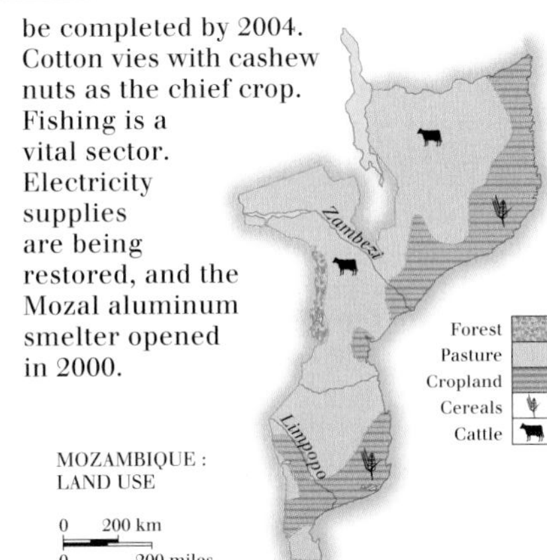
Zambezi
Limpopo
Forest
Pasture
Cropland
Cereals
Cattle
MOZAMBIQUE : LAND USE
0 200 km
0 200 miles

HEALTH

Welfare state health benefits

 1 per 20,000 people

 Tuberculosis, gastroenteric infections, pneumonia, AIDS

Thousands of people lost limbs from land mines, or suffered other appalling injuries and psychological trauma, during Mozambique's savage civil war. Health services have improved since the war's end, and preventive medicines and antenatal care are provided free. Doctors serve a mandatory two years in rural areas. Many private clinics have been established since 1987. However, cholera, a lingering by-product of war, is still a serious issue in the far northern province of Cabo Delgado. An estimated 1.1 million Mozambicans have contracted HIV.

ENVIRONMENT

Sustainability rank: 59th

 6%

 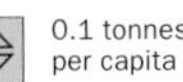 0.1 tonnes per capita

ENVIRONMENTAL TREATIES

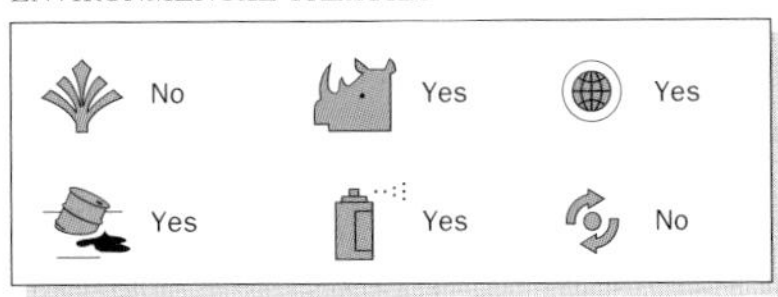
No
Yes
Yes
Yes
Yes
No

Floods followed by droughts are often devastating. The floods of early 2000, resulting from a combination of cyclones and torrential rain, affected an estimated one million people. More than 200,000 were displaced by the floods of 2001. The worst drought in living memory was in 1982–1984; it killed 100,000 and left four million people close to starvation. Civil war pushed rural populations toward the cities and coasts, resulting in overcrowding, disease, pollution, and desertification on abandoned farms. However, ecological concerns are still low on the agenda.

MEDIA

TV ownership low

 Daily newspaper circulation 3 per 1000 people

PUBLISHING AND BROADCAST MEDIA

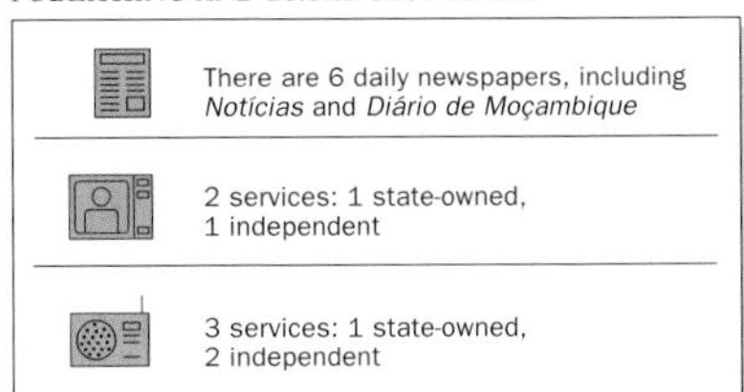
There are 6 daily newspapers, including *Notícias* and *Diário de Moçambique*
2 services: 1 state-owned, 1 independent
3 services: 1 state-owned, 2 independent

The press, hitherto a FRELIMO publicity machine, has enjoyed greater freedom in the 1990s. The killing in 2000 of a popular and outspoken editor, Carlos Cardoso, shocked the country. TV sets are still a rarity. The state-owned radio station broadcasts in Portuguese, English, and vernacular languages.

CRIME

No death penalty

 10,800 prisoners

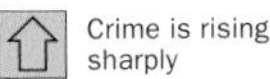 Crime is rising sharply

CRIME RATES

Murders

47 *per 100,000 population*

Rapes

Mozambique does not publish rape statistics

Thefts

1300 *per 100,000 population*

Weapons are easily obtainable. In rural areas there are many bandits, often former soldiers; road travel is unsafe. Senior officials stand accused of misappropriating food aid money.

EDUCATION

School leaving age: 13

  44%

 7143 students

THE EDUCATION SYSTEM

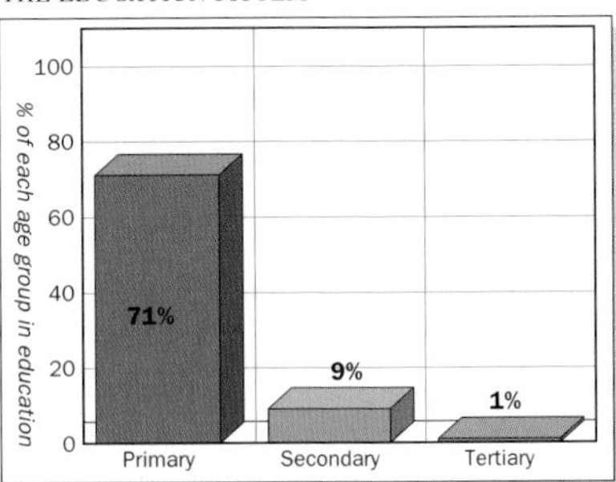
100
80
60
40
20
0
% of each age group in education
71%
9%
1%
Primary
Secondary
Tertiary

At independence, between 85% and 95% of the adult population were illiterate, and school closures during the civil war created a lost generation of uneducated people. A target of educating at least 86% of children was set for 2000, concentrating on rural areas. The government used World Bank/IMF debt relief to strengthen the education budget in 2001.

SPENDING

GDP/cap. increase

CONSUMPTION AND SPENDING

1 per 1000 population

4 per 1000 population

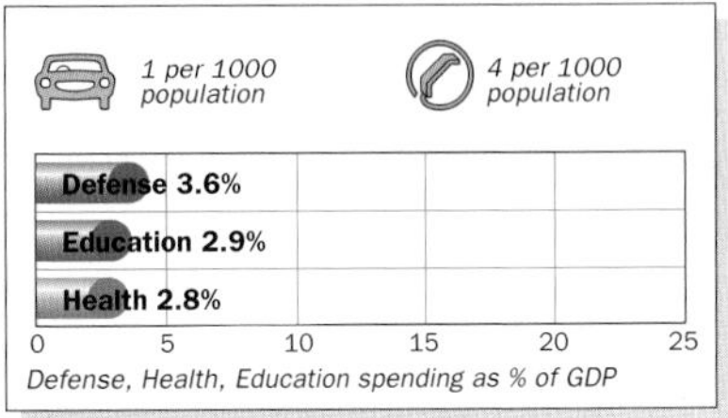
Defense 3.6%
Education 2.9%
Health 2.8%
0
5
10
15
20
25
Defense, Health, Education spending as % of GDP

Mozambique is one of the world's poorest countries, with over 90% of the people living below the breadline even before the floods of 2000 and 2001, which left thousands homeless. Measures adopted in the 1990s to attract Western aid made conditions tougher, raising the price of rice by 600%. The recent export boom has generally bypassed the traditional subsistence farmer. Only the higher echelons of FRELIMO, RENAMO, and other political parties have cars, air-conditioning, and brick-built apartments. Free-market reforms, however, are gradually increasing access to consumer goods.

WORLD RANKING

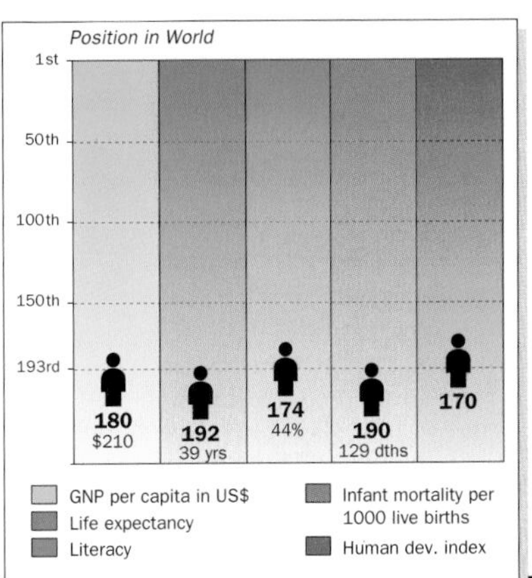
Position in World
1st
50th
100th
150th
193rd
180
$210
192
39 yrs
174
44%
190
129 dths
170
GNP per capita in US$
Life expectancy
Literacy
Infant mortality per 1000 live births
Human dev. index

M

NAMIBIA

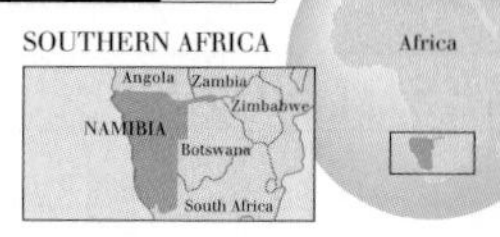
SOUTHERN AFRICA
Angola
Zambia
Zimbabwe
NAMIBIA
Botswana
South Africa
Africa

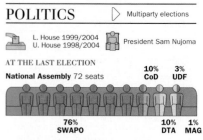

OFFICIAL NAME: Republic of Namibia **CAPITAL:** Windhoek
POPULATION: 1.8 million **CURRENCY:** Namibian dollar **OFFICIAL LANGUAGE:** English

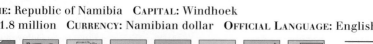

1990

1994

March 21

NAM

+2

+264 .na

LOCATED IN SOUTHWESTERN Africa, Namibia has an arid coastal strip formed by the Namib Desert. After many years of guerrilla warfare, Namibia won independence from South Africa in 1990. Despite the move away from apartheid, Namibia's economy remains reliant on the expertise of the small white population, a legacy of the previously poor education for blacks. Namibia is Africa's fourth-largest minerals producer.

CLIMATE

Hot desert/steppe

WEATHER CHART FOR WINDHOEK

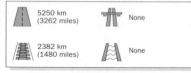

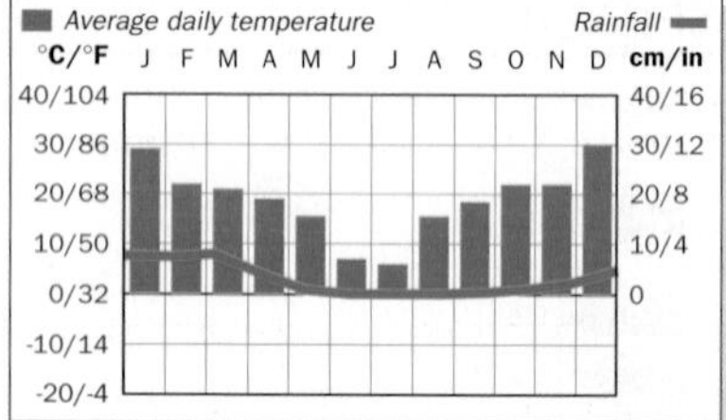
Average daily temperature
Rainfall
°C/°F J F M A M J J A S O N D cm/in
40/104
30/86
20/68
10/50
0/32
-10/14
-20/-4
40/16
30/12
20/8
10/4
0

Namibia is almost rainless. The coast is usually shrouded in thick, cold fog unless the hot, very dry *berg* blows.

TRANSPORTATION

Drive on left

Windhoek International 481,419 passengers

105 ships 54,794 grt

THE TRANSPORTATION NETWORK

5250 km (3262 miles)

None

2382 km (1480 miles)

None

Large-scale industry is well served by road and rail. Plans exist to build a new harbor at Walvis Bay.

TOURISM

Visitors : Population 1:2.9

614,000 visitors

Up 54% 1995–1998

MAIN TOURIST ARRIVALS

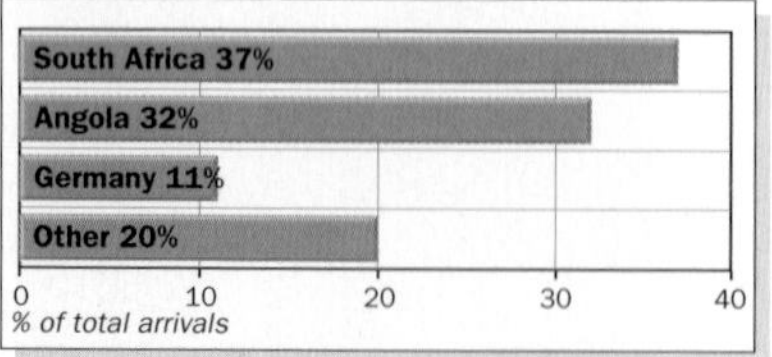
South Africa 37%
Angola 32%
Germany 11%
Other 20%
0
10
20
30
40
% of total arrivals

Tourists, largely from South Africa, make a very limited contribution to GDP. German tourists come to see Windhoek's German sector. There are plans to limit tourists to 300,000 a year to preserve Namibia's fragile desert ecology.

Spitzkoppe, west of Karibib. *Unique scenery such as this is attracting increasing numbers of tourists to Namibia.*

PEOPLE

Pop. density low

Ovambo, Kavango, English, Bergdama, German, Afrikaans

2/km² (6/mi²)

THE URBAN/RURAL POPULATION SPLIT

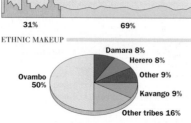

31% 69%

ETHNIC MAKEUP

Ovambo 50%, Damara 8%, Herero 8%, Other 9%, Kavango 9%, Other tribes 16%

The largest ethnic group, the Ovambo, tend to live in the sparsely populated north of the country. Whites – 60% of whom speak Afrikaans – are concentrated in Windhoek. The capital is also home to a wealthy century-old German community. Namibia's original inhabitants, the San and Khoi (once called Bushmen) now constitute a tiny, marginalized minority.

The ethnic strife predicted in 1990 has not materialized. For the most part, black Namibians, predominantly subsistence farmers, have accepted the greater wealth of the white community. Families are large in Namibia; many black women have six or more children. The constitution supports gender equality and discriminates in favor of women; few, however, have official jobs or own property. Homosexuality is not tolerated.

POLITICS

Multiparty elections

L. House 1999/2004
U. House 1998/2004

President Sam Nujoma

AT THE LAST ELECTION

National Assembly 72 seats

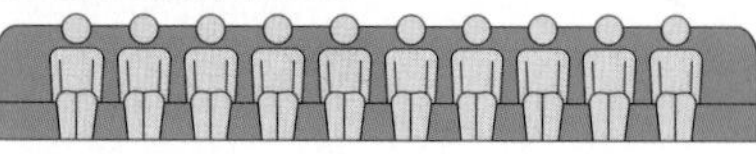
10% CoD
3% UDF
76% SWAPO
10% DTA
1% MAG

SWAPO = South West Africa People's Organization **CoD** = Congress of Democrats **DTA** = Democratic Turnhalle Alliance **UDF** = United Democratic Front **MAG** = Monitor Action Group
Six additional nonvoting members may be appointed to the National Assembly by the president

National Council 26 seats

Two members are elected by each of the 13 Regional Councils to the National Council

SWAPO guerrillas fought for and won independence from South Africa in 1990. Namibia switched from a system of apartheid to a state-wide, multiparty democracy. The center-left SWAPO has dominated politically ever since. In 1998 a constitutional amendment allowed President Nujoma to run for a third term, which he duly won in late 1999. SWAPO's main opposition comes from the center-right DTA, a coalition of 11 parties favoring free-market practices.

WORLD AFFAIRS

Joined UN in 1990

Comm

COMESA

NAM

AU

SADC

In 1992, South Africa settled its border dispute with Namibia, and in 1994 relinquished control of the enclave of Walvis Bay – Namibia's only deepwater port. South Africa has also written off Namibia's earlier debts. Withdrawal of Namibian troops from the war-torn DRC began in 2001.

AID

Recipient

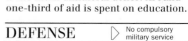

US$152m (receipts)

Down 15% in 2000

The EU provides most aid; Germany is the main unilateral donor. Around one-third of aid is spent on education.

DEFENSE

No compulsory military service

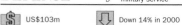

US$103m

Down 14% in 2000

Fishing grounds are patrolled to prevent raids by foreign trawlers. Namibian soldiers performed a peacekeeping role in the DRC.

N

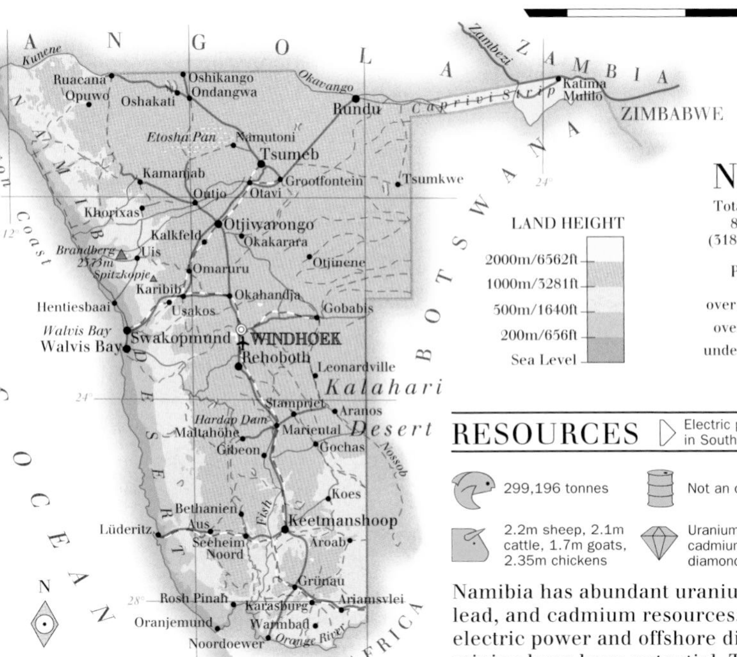
ANGOLA
ZAMBIA
ZIMBABWE
BOTSWANA
SOUTH AFRICA
Kunene
Zambezi
Okavango
Caprivi Strip
Ruacana
Opuwo
Oshakati
Oshikango
Ondangwa
Rundu
Katima Mulilo
Etosha Pan
Namutoni
Tsumeb
Tsumkwe
Kamanjab
Grootfontein
Outjo
Otavi
Khorixas
Otjiwarongo
Okakarara
Kalkfeld
Brandberg 2573m
Uis
Spitzkopje
Omaruru
Otjinene
Karibib
Okahandja
Hentiesbaai
Usakos
Gobabis
Walvis Bay
Swakopmund
WINDHOEK
Rehoboth
Leonardville
Kalahari Desert
Stampriet
Aranos
Hardap Dam
Maltahöhe
Mariental
Gochas
Gibeon
Nossob
Koes
Bethanien
Aus
Lüderitz
Keetmanshoop
Seeheim
Noord
Aroab
Grünau
Rosh Pinah
Karasburg
Ariamsvlei
Oranjemund
Warmbad
Orange River
Noordoewer
Fish
NAMIB DESERT
Skeleton Coast
ATLANTIC OCEAN
200 km
200 miles

NAMIBIA

Total Land Area : 825 418 sq. km (318 694 sq. miles)

LAND HEIGHT

2000m/6562ft
1000m/3281ft
500m/1640ft
200m/656ft
Sea Level

POPULATION

over 100 000
over 10 000
under 10 000

CHRONOLOGY

In 1915, South Africa took over the former German colony as a League of Nations' mandate known as South West Africa.

- ❑ **1966** Apartheid laws imposed. SWAPO begins armed struggle.
- ❑ **1968** Renamed Namibia.
- ❑ **1973** UN recognizes SWAPO.
- ❑ **1990** Independence.
- ❑ **1994** South Africa relinquishes Walvis Bay.
- ❑ **1999** President Sam Nujoma wins third term.

ECONOMICS

Inflation 9.5% p.a. (1990–2000)

US$3.57bn

7.570–11.995 Namibian dollars

SCORE CARD

- ❑ World GNP Ranking........................126th
- ❑ GNP per Capita ....US$2030
- ❑ Balance of Payments ....US$162m
- ❑ Inflation 8.6%
- ❑ Unemployment..................................40%

Strengths

Varied mineral resources. Market conditions attractive to private investors. Rich fishing grounds. Potential of Walvis Bay as conduit for landlocked neighbors. Low external debt.

Weaknesses

Most goods imported. Fluctuations in mineral prices. Recessionary ripple effect of currency pegged to South African rand. Lack of skilled labor; high unemployment. Severe drought in 1996.

EXPORTS

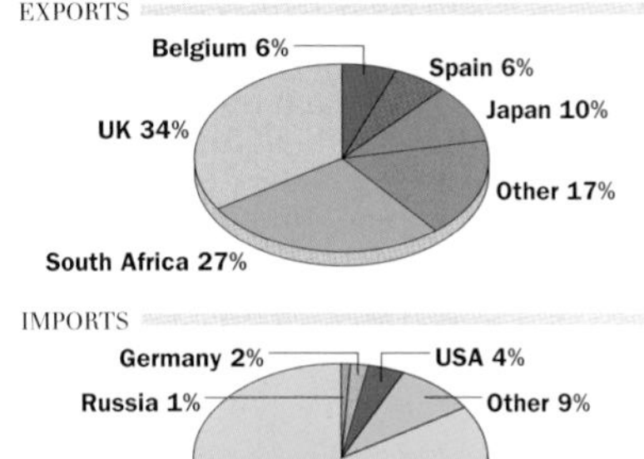
Belgium 6%
Spain 6%
Japan 10%
UK 34%
Other 17%
South Africa 27%

IMPORTS

Germany 2%
USA 4%
Russia 1%
Other 9%
South Africa 84%

RESOURCES

Electric power: Included in South African total

299,196 tonnes

Not an oil producer

2.2m sheep, 2.1m cattle, 1.7m goats, 2.35m chickens

Uranium, lead, gold, cadmium, oil, copper, diamonds, zinc, silver,

Namibia has abundant uranium, lead, and cadmium resources. Hydro-electric power and offshore diamond mining have huge potential. The Okavango river system carries more water than all South Africa's rivers combined. Large oil deposits were discovered in 2000.

ENVIRONMENT

Sustainability rank: 26th

13%

Negligible emissions per capita

Illegal poaching and the presence of anthrax threaten the unique Namibian desert-adapted elephant (fewer than 50 remain) and the black rhino. Vast expanses of the fragile, unspoiled Namib and Kalahari Desert ecosystems are protected. The government is generally sensitive to environmental issues (the annual seal cull to protect fish stocks is an exception) and wishes to attract ecotourists rather than invest in mass-market developments.

MEDIA

TV ownership low

Daily newspaper circulation 19 per 1000 people

PUBLISHING AND BROADCAST MEDIA

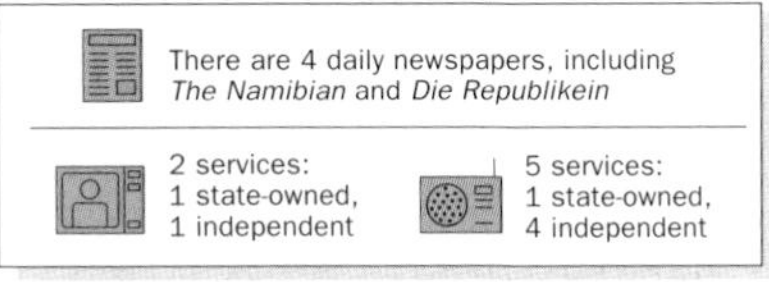
2 services:
1 state-owned,
1 independent
5 services:
1 state-owned,
4 independent

There are 4 daily newspapers, including *The Namibian* and *Die Republikein*

State radio transmits in 11 languages, including German and English. An active press targets corrupt politicians.

CRIME

No death penalty

4397 prisoners

Down 51% in 1999

Burglary and theft are rising, particularly in urban areas. Ostrich smuggling to the US is common.

EDUCATION

School leaving age: 16

82%

11,209 students

Nearly 90% of children attend primary school, but illiteracy among black adults remains a legacy of apartheid.

HEALTH

Welfare state health benefits

1 per 3333 people

AIDS, respiratory, heart, and intestinal diseases

Preventive care and rural health care have top priority. Most areas lack safe water. AIDS is the leading cause of death.

SPENDING

GDP/cap. increase

CONSUMPTION AND SPENDING

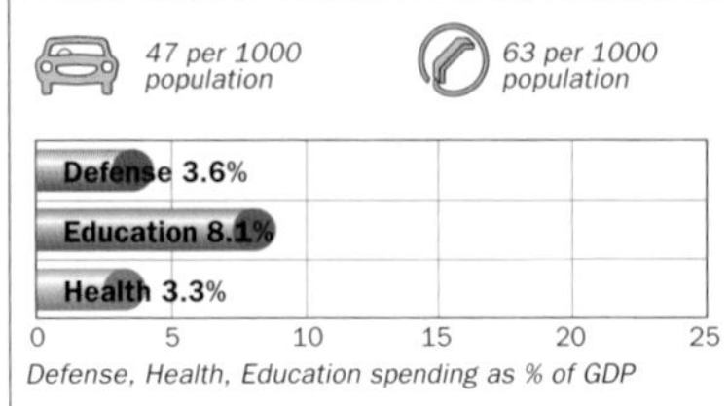
47 per 1000 population
63 per 1000 population
Defense 3.6%
Education 8.1%
Health 3.3%
0
5
10
15
20
25
Defense, Health, Education spending as % of GDP

Gross disparities in wealth persist throughout Namibia: the top 1% of households consumes as much as the poorest 50%.

WORLD RANKING

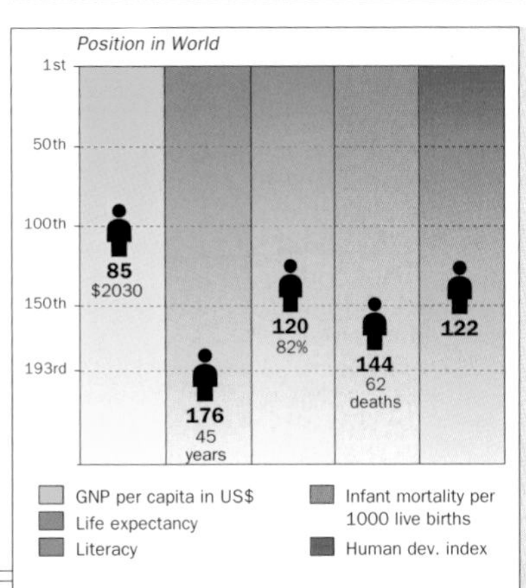
Position in World
1st
50th
100th
150th
193rd
85
$2030
176
45 years
120
82%
144
62 deaths
122
GNP per capita in US$
Life expectancy
Literacy
Infant mortality per 1000 live births
Human dev. index

N

NAURU

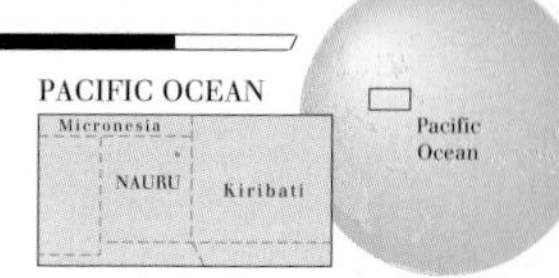
PACIFIC OCEAN
Micronesia
NAURU
Kiribati
Pacific Ocean

OFFICIAL NAME: Republic of Nauru **CAPITAL:** *No official capital*
POPULATION: 11,800 **CURRENCY:** Australian dollar **OFFICIAL LANGUAGE:** Nauruan

1968

1968

Jan 31

NAU
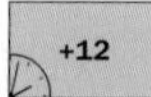
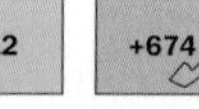
+12

+674

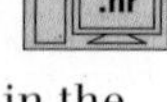
.nr

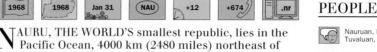

NAURU, THE WORLD'S smallest republic, lies in the Pacific Ocean, 4000 km (2480 miles) northeast of Australia. Once a British colony exploited for its phosphates by the UK, Australia, and New Zealand, it became independent in 1968. The phosphates industry made Nauruans among the wealthiest people in the world, but economic mismanagement and the imminent end of phosphate reserves has left Nauru facing financial ruin, prompting economic reform.

CLIMATE

Tropical oceanic

WEATHER CHART FOR NAURU

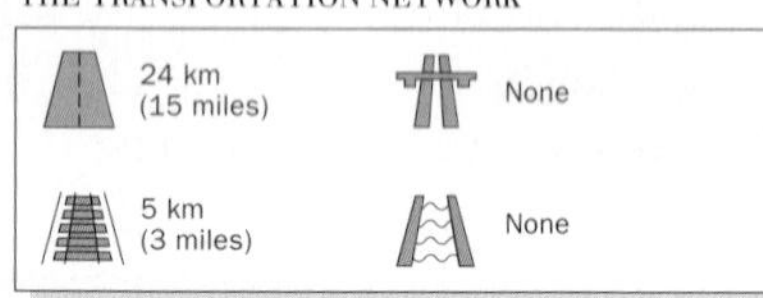
Average daily temperature
Rainfall
°C/°F J F M A M J J A S O N D cm/in
40/104
30/86
20/68
10/50
0/32
-10/14
-20/-4
40/16
30/12
20/8
10/4
0

Nauru's tiny size means that rain clouds often miss the island; years can pass without rain.

TRANSPORTATION

Drive on left

Nauru Island International

3 ships 1000 grt

THE TRANSPORTATION NETWORK

24 km (15 miles) — None

5 km (3 miles) — None

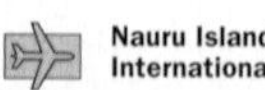

Nauru operates its own airline with a Boeing 737 flown by Australian pilots. The Nauru Pacific Line provides infrequent commercial services to Australia. However, all external travel is very expensive. Nauru has no harbor: to load cargoes of phosphates, ships have to dock, engines still running, with huge concrete caissons floating out at sea. The circular ring road is often littered with abandoned cars, as it has been much cheaper for Nauruans to import new vehicles than to attempt to repair existing ones. The number of car accident fatalities is one of the highest in the South Pacific.

TOURISM

Not available

Minimal tourist arrivals

Little variation from year to year

MAIN TOURIST ARRIVALS

Nauru does not publish tourism figures by country of origin

0 10 20 30 40

% of total arrivals

Even if Nauru had any conventional tourist attractions, the enormous cost of getting there would dissuade most tourists from making the journey. The main feature of interest on the island is the bizarre lunar landscape created by over 80 years of phosphate extraction. There are no beaches and only a few basic hotels.

NAURU

Total Land Area : 21 sq. km (8.1 sq. miles)

LAND HEIGHT
200m/565ft
Sea Level

Urban area

Phosphate mineworks

0 1 km
0 1 mile

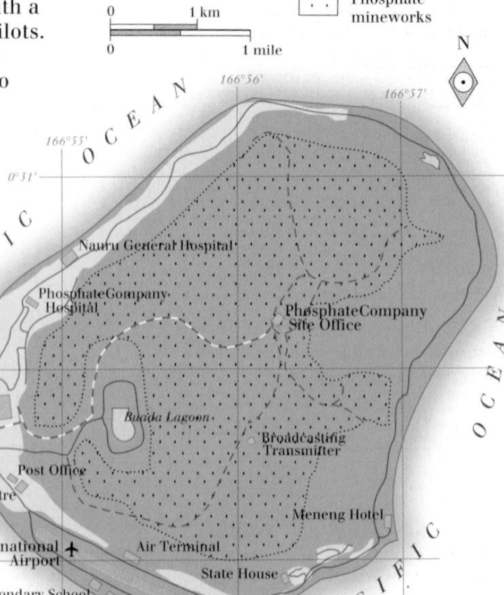
N
166°56'
166°57'
166°55'
0°31'
0°32'
0°33'
PACIFIC OCEAN
OCEAN
PACIFIC
Nauru General Hospital
Phosphate Company Hospital
Phosphate Company Site Office
Phosphate Company Works
Buada Lagoon
Broadcasting Transmitter
YAREN
Post Office
Nauru Civic Centre
Meneng Hotel
Nauru International Airport
Air Terminal
State House
Nauru Secondary School
Police Station

PEOPLE

Pop. density high

Nauruan, Kiribati, Chinese, Tuvaluan, English

564/km² (1461/mi²)

THE URBAN/RURAL POPULATION SPLIT

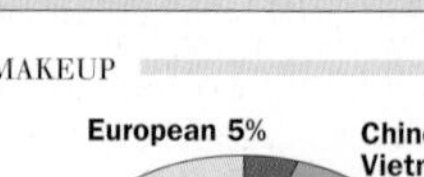
Nauru is 100% semi-urban

ETHNIC MAKEUP

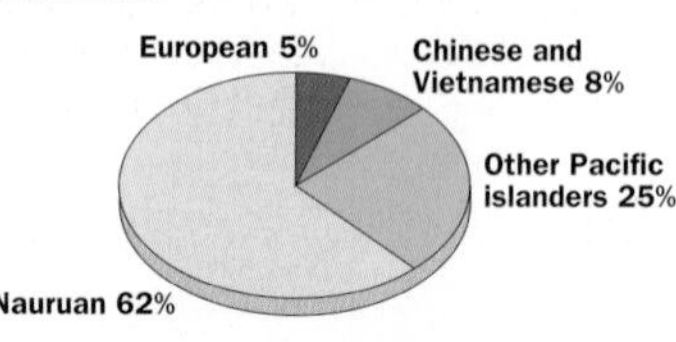
European 5%
Chinese and Vietnamese 8%
Other Pacific islanders 25%
Nauruan 62%
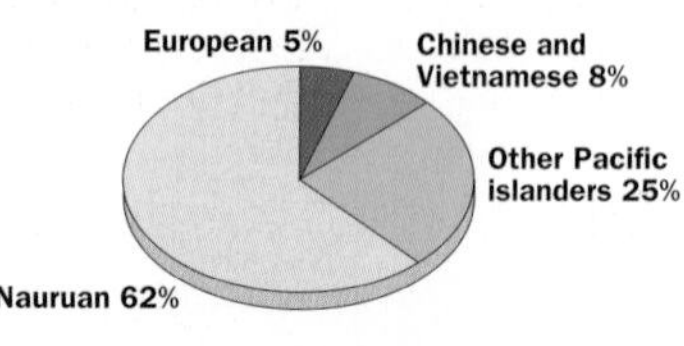

Indigenous Nauruans are a homogeneous blend of Melanesian, Micronesian, and Polynesian strands. They have traditionally held posts in government service, while a large imported workforce – mainly from Kiribati – mines the phosphates.

A society of just over 10,000 people, Nauru is mostly self-regulating. There is some tension between younger Nauruans, who go to Australia to study but have little incentive to do well, and their parents, who fought hard for independence. As the phosphates run out, an increasing feeling of futility is gripping the young. Many see their future in Australia or New Zealand, but fear a drop in living standards and the loss of the luxury of sovereignty. These fears led Nauruans to reject the offer of resettlement on an island off the Queensland coast of Australia.

POLITICS

Nonparty elections

2000/2003

President Rene Harris

AT THE LAST ELECTION

Parliament 18 seats

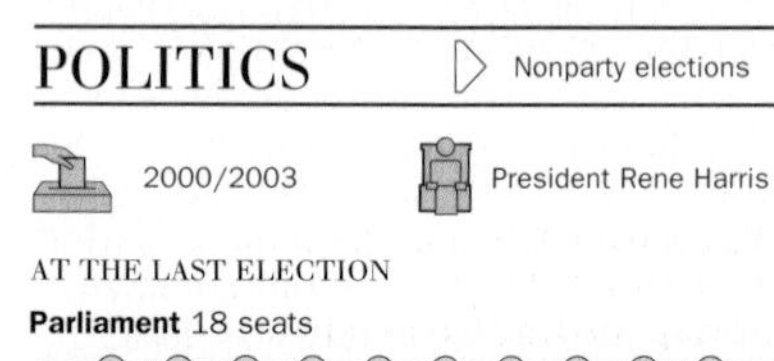

All members are elected as independents

Parliament is based on the British Westminster model, but traditional leaders are the dominant figures. Members of the legislature often switch between temporary groupings based more on personalities than ideologies. Hammer DeRoburt, the island's first post-independence president, dominated the political landscape for two decades until a 1989 vote of no confidence.

Since DeRoburt's downfall, the presidency (elected by parliament) has changed hands frequently in a more volatile political atmosphere.

N

WORLD AFFAIRS

Joined UN in 1999

Comm
PC
PIF
ADB
ACP

The case for compensation for phosphate exploitation brought by Nauru against the UK government was rejected in 1992 after the longest suit in British legal history. However, an Australian settlement that year brought payments eventually totaling US$79 million. Nauru's main concern is participation in the Pacific Islands Forum and the management of trust funds to support Nauruans when phosphate deposits run out. Nauru joined the UN in 1999. It seeks a voice on environmental issues.

AID

Recipient

 US$4m (receipts)

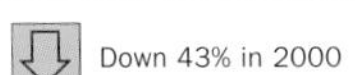 Down 43% in 2000

Nauru receives or donates little aid, except as a member of the Pacific Islands Forum.

DEFENSE

No compulsory military service

 Australia responsible for defense

Not applicable

Nauru has no defense force. Australia, under a de facto arrangement, is responsible for the island's security.

ECONOMICS

Not available

US$80m

 1.7997–1.9535 Australian dollars

SCORE CARD

- ❑ WORLD GNP RANKING........190th
- ❑ GNP PER CAPITA........US$7270
- ❑ BALANCE OF PAYMENTS........Not available
- ❑ INFLATION........–3.6%
- ❑ UNEMPLOYMENT........Minimal unemployment

STRENGTHS

Considerable investments overseas. Tax haven. Offshore banking potential. Trust funds for post-phosphate era. Strong Australian dollar.

WEAKNESSES

Phosphate revenues due to end in 2003. High cost of rehabilitating the 80% of the island where mining occurred. Virtually no other resources. Bad investment decisions in 1990s.

EXPORTS

Nauru's only export commodity is phosphates, in which it trades with Australia and New Zealand.

IMPORTS

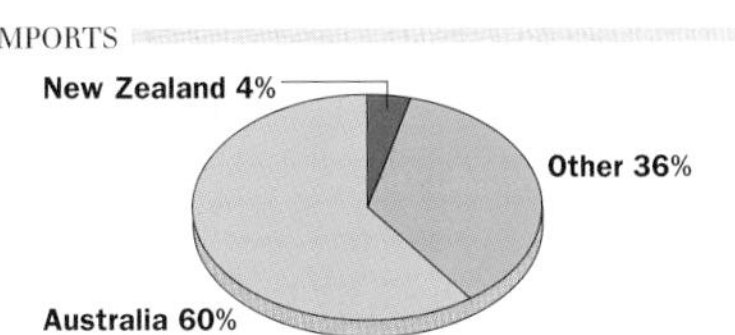
New Zealand 4%
Other 36%
Australia 60%

RESOURCES

Electric power 10,000 kw

 250 tonnes

 Not an oil producer

 2.09m sheep, 2800 pigs, 5000 chickens

 Guano (phosphates)

Since 1888 Nauru has been exploited by the Germans, British, Australians, New Zealanders, and recently by Nauruans themselves, for its valuable phosphate reserves. Extraction has destroyed 80% of the island, and the deposits have been virtually exhausted. Nauru has no other resources. It is entirely dependent on outside energy supplies, and the cost of oil is 50% higher than the Pacific average, since Nauru does not lie on any shipping routes. Most electricity is produced by small diesel generators.

ENVIRONMENT

Not available

 None

 Not available

Nauru is an environmental disaster area. Mining has destroyed 80% of its ecosystem and, like other Pacific islands, it faces the increasing threat of rising sea levels. Also of concern is contamination from the nearby former French nuclear test sites in the Pacific.

MEDIA

TV ownership low

 There are no daily newspapers

PUBLISHING AND BROADCAST MEDIA

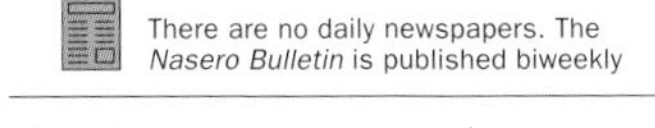
There are no daily newspapers. The *Nasero Bulletin* is published biweekly

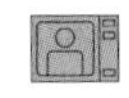 1 state-owned service

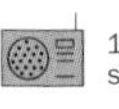 1 state-owned service

Nauru has one national TV broadcasting service and one radio station. Both are state-run.

CRIME

Death penalty not used in practice

 Nauru does not publish prison figures

 Crime levels are rising slightly

Theft is almost nonexistent. Assaults and dangerous driving as a result of drunkenness are the major problems.

Nauru is almost circular with a 16-km (10-mile) ring road. The overcrowded coastal strip is the sole habitable land.

CHRONOLOGY

Colonized by Germany in 1888, from 1919 the island was administered jointly by the UK, Australia, and New Zealand.

- ❑ **1968** Independence.
- ❑ **1970** Gains phosphate control.
- ❑ **1992** Australia agrees compensation for phosphate extraction.

EDUCATION

School leaving age: 16

 99%

 Not available

Many Nauruans attend boarding school in Australia from a young age. Few go on to university.

HEALTH

Welfare state health benefits

 1 per 700 people

 Tuberculosis, vitamin deficiencies, diabetes

A diet of processed imported foods and widespread obesity are the major problems. Over one-third of the population suffers from non-insulin-dependent diabetes. Industrial accidents are treated in Australia.

SPENDING

GDP/cap. increase

CONSUMPTION AND SPENDING

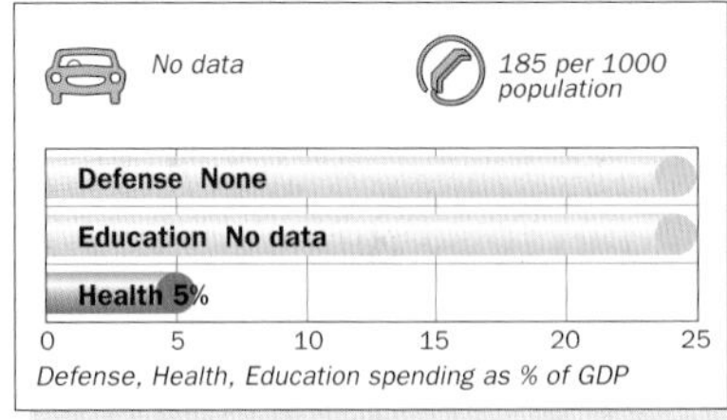
No data
185 per 1000 population
Defense None
Education No data
Health 5%
0
5
10
15
20
25
Defense, Health, Education spending as % of GDP

Nauru is carrying out a major economic adjustment program, to be funded by the ADB. The program is intended to allow it to adjust to the loss of income when its phosphate reserves are exhausted.

WORLD RANKING

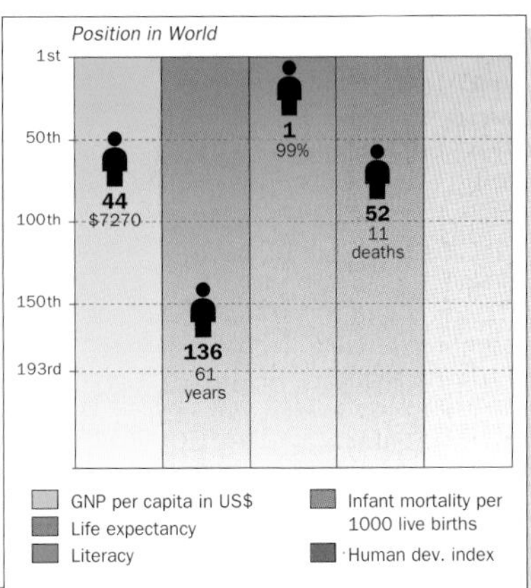
Position in World
1st
50th
100th
150th
193rd
44
$7270
136
61 years
1
99%
52
11 deaths
GNP per capita in US$
Life expectancy
Literacy
Infant mortality per 1000 live births
Human dev. index

N

NEPAL

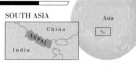

SOUTH ASIA

China · NEPAL · India · Asia

OFFICIAL NAME: Kingdom of Nepal **CAPITAL:** Kathmandu
POPULATION: 23.6 million **CURRENCY:** Nepalese rupee **OFFICIAL LANGUAGE:** Nepali

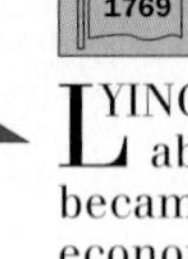

1769 | 1769 | Dec 28 | NEP | +5.75 | +977 | .np

LYING ALONG the southern Himalayas, Nepal was an absolute monarchy until 1990, since when its politics became increasingly turbulent. The mainly agricultural economy depends heavily on the prompt arrival of the monsoon. Hopes for development have been invested in hydropower, despite the adverse impact of large dams. A Maoist insurgency, begun in 1999, threw the country into chaos in 2001.

CLIMATE

Mountain/subtropical

WEATHER CHART FOR KATHMANDU

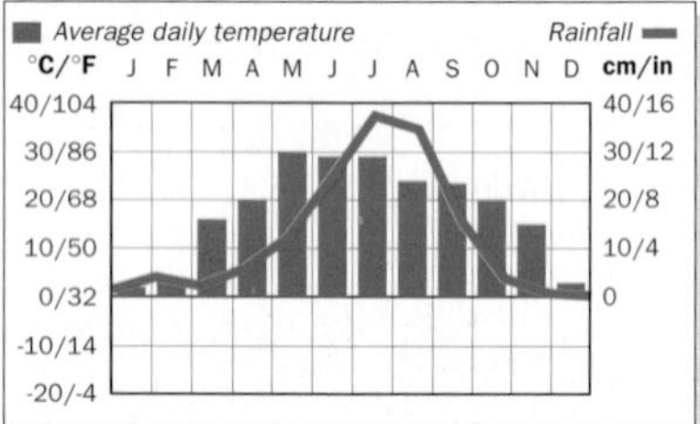

Average daily temperature
Rainfall
°C/°F J F M A M J J A S O N D cm/in
40/104 30/86 20/68 10/50 0/32 -10/14 -20/-4
40/16 30/12 20/8 10/4 0

The warm July to October monsoon affects the whole country, causing flooding in the hot Terai plain, but generally decreases northward and westward. The rest of the year is dry, sunny, and mild, except in the Himalayas, where valley temperatures in winter may average –10°C (14°F).

TRANSPORTATION

Drive on left

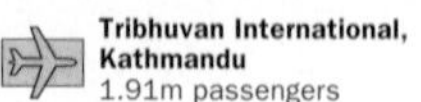

Tribhuvan International, Kathmandu 1.91m passengers

Has no fleet

THE TRANSPORTATION NETWORK

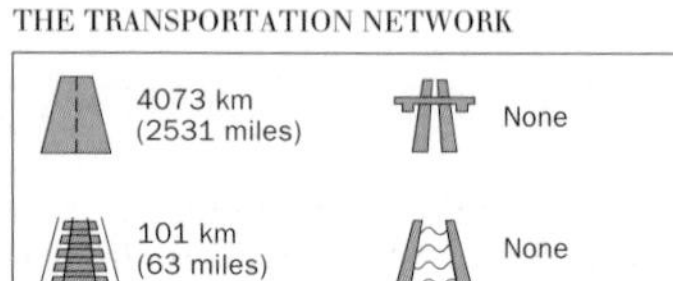

4073 km (2531 miles)
None
101 km (63 miles)
None

Domestic flights link the main towns. There are paved roads in the south and in the Kathmandu valley; only one runs north to China. Two short stretches of railroad cross into India.

Himalayan harvest. *Steep mountainsides and easily eroded soils mean that most fields are terraced. A majority of Nepalese are farmers.*

TOURISM

Visitors : Population 1:51

464,000 visitors

Down 6% in 2000

MAIN TOURIST ARRIVALS

	% of total arrivals
India	29%
USA	8%
Japan	8%
Other	55%

The wish to preserve the environment conflicts with the desire for tourist revenue. Backpackers have been a major source of income, but recent terrorist activity, such as the 1999 hijacking of an Air India aircraft bound for Nepal and the increased violence in 2001 of the Maoist insurgency, threaten to deter visitors.

Child labor was banned in the tourism industry from 2000.

PEOPLE

Pop. density medium

Nepali, Maithilli, Bhojpuri

173/km² (447/mi²)

THE URBAN/RURAL POPULATION SPLIT

12% 88%

RELIGIOUS PERSUASION

Hindu 90%, Buddhist 4%, Muslim 3%, Other 2%, Christian 1%

There are few tensions among diverse ethnic groups such as the Sherpas in the north, "Hill Hindu" Brahmins and Chhettris, Newars, and others in the Kathmandu valley, and Terai in the south. The Sherpa and other Buddhist women are less restricted than Hindus. Polygamy is practiced in the hills. Since 1990 many ethnic Nepalese refugees from Bhutan have settled in Nepal.

POLITICS

Multiparty elections

L. House 1999/2002
U. House 2001/2003

H.M. King Gyanendra Bir Bikram Shah Dev

AT THE LAST ELECTION

House of Representatives 205 seats

54% NCP, 35% CPN–UML, 6% Others, 5% NDP

NCP = Nepali Congress Party **CPN–UML** = Communist Party of Nepal–United Marxist–Leninist **NDP** = National Democratic Party **App** = Appointed by the king

National Council 60 seats

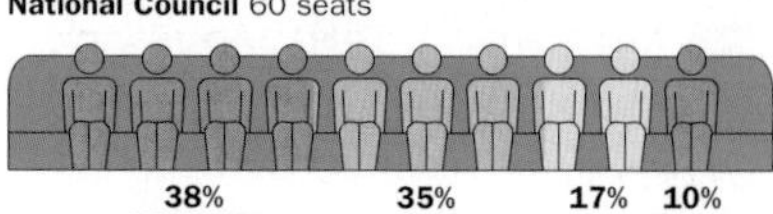

38% CPN–UML, 35% NCP, 17% App, 10% Others

A multiparty system introduced in 1990 produced a short-lived communist government in 1994, then a series of unstable coalitions until the NCP won elections in 1999. Instability returned in 2001: the royal family was murdered and the Maoist rebel insurgency intensified. To tackle the Maoists, Prime Minister Sher Bahadur Deuba imposed a state of emergency, and its unilateral extension in 2002, supported by unpopular new King Gyanendra, sparked a political crisis which saw Deuba expelled from his own NCP.

WORLD AFFAIRS

Joined UN in 1955

CP | IBRD | NAM | SAARC | ADB

The government has revived links with India, and security relations are questioned by the UML, which is more pro-Chinese. Relations with Bhutan are strained over ethnic Nepali Bhutanese refugees in Nepal. The Maoist insurgency has increased regional tension.

AID

Recipient

$390m (receipts)

Up 11% in 2000

Nepal's strategic position has made it a focus for powerful donors, including the US, China, India, Japan, and member states of the CIS.

DEFENSE

No compulsory military service

$49m

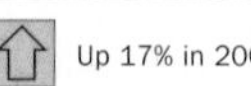

Up 17% in 2000

The army, at 46,000 men, is small and has no tanks or combat aircraft. Weapons come from India and the UK, in whose own army the Nepalese Gurkhas serve.

N

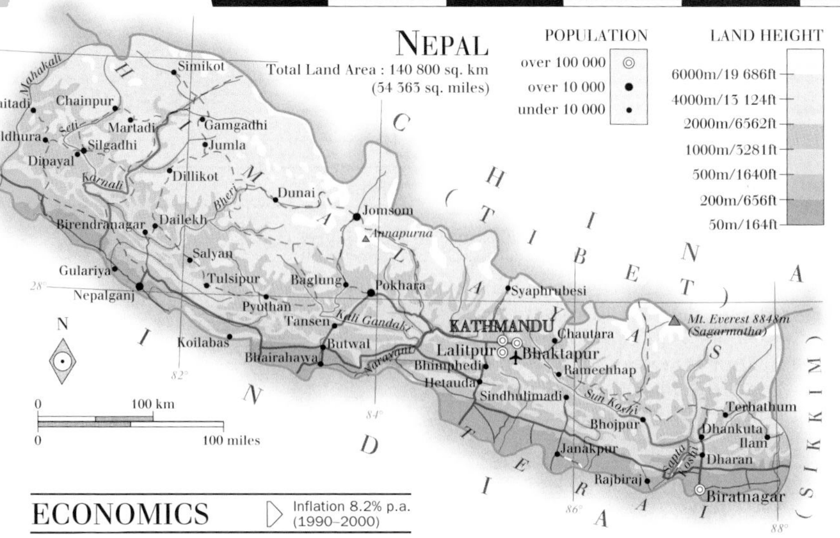
NEPAL
Total Land Area : 140 800 sq. km (54 363 sq. miles)
POPULATION
over 100 000
over 10 000
under 10 000
LAND HEIGHT
6000m/19 686ft
4000m/13 124ft
2000m/6562ft
1000m/3281ft
500m/1640ft
200m/656ft
50m/164ft
Simikot
Chainpur
Baitadi
Martadi
Gamgadhi
Jumla
Silgadhi
Dipayal
Dillikot
Dunai
Jomsom
Annapurna
Birendranagar
Dailekh
Salyan
Gulariya
Nepalganj
Tulsipur
Baglung
Pokhara
Pyuthan
Tansen
Kali Gandaki
Syaphrubesi
KATHMANDU
Chautara
Mt. Everest 8848m (Sagarmatha)
Koilabas
Butwal
Bhairahawa
Lalitpur
Bhaktapur
Bhimphedi
Hetauda
Ramechhap
Sindhulimadi
Sun Kosi
Terhathum
Bhojpur
Dhankuta
Ilam
Janakpur
Dharan
Rajbiraj
Biratnagar
CHINA (TIBET)
INDIA
(SIKKIM)
N
0 100 km
0 100 miles

ECONOMICS

Inflation 8.2% p.a. (1990–2000)

 $5.58bn

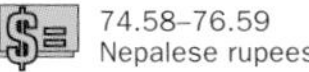 74.58–76.59 Nepalese rupees

SCORE CARD

- ❑ WORLD GNP RANKING........................106th
- ❑ GNP PER CAPITA $240
- ❑ BALANCE OF PAYMENTS...................–$293m
- ❑ INFLATION 1.5%
- ❑ UNEMPLOYMENT 1%

STRENGTHS

Self-sufficiency in grain most years. Economic liberalization under NCP government. Potential for hydroelectric power generation. Low debt level.

WEAKNESSES

Agricultural dependency: only 10% of GDP from manufacturing. Landlocked status. Low savings rate. Absence of active entrepreneurial class.

EXPORTS

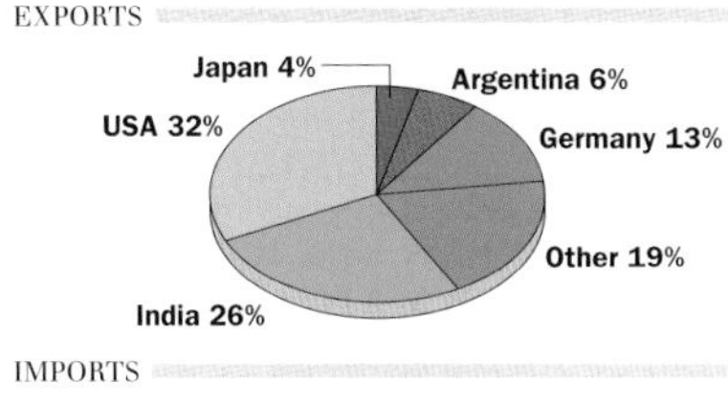
Japan 4%
Argentina 6%
USA 32%
Germany 13%
Other 19%
India 26%

IMPORTS

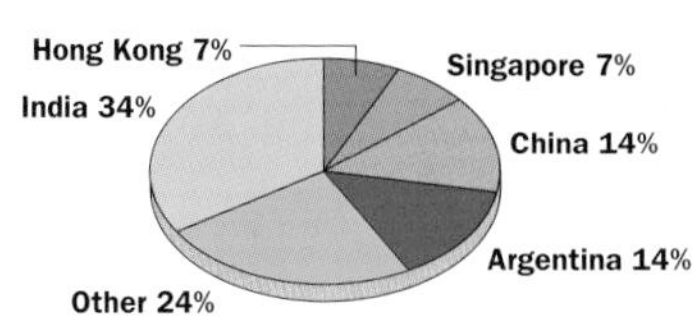
Hong Kong 7%
Singapore 7%
India 34%
China 14%
Argentina 14%
Other 24%

RESOURCES

Electric power 327,000 kw

 25,780 tonnes

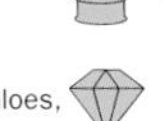 Not an oil producer

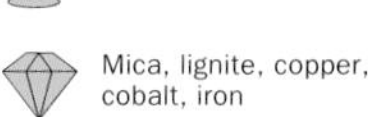 7m cattle, 6.5m goats, 3.6m buffaloes, 19.8m chickens

Mica, lignite, copper, cobalt, iron

The first privately owned power plant, situated near Ramechhap, opened in mid-2000.

ENVIRONMENT

Sustainability rank: 99th

 8% (1% partially protected)

 0.1 tonnes per capita

Kathmandu has chronic traffic and pollution problems. Deforestation and soil erosion are serious. The native tiger is fast disappearing. Approval of the controversial Arun III hydroelectric project was granted in mid-2000.

MEDIA

TV ownership low

 Daily newspaper circulation 11 per 1000 people

PUBLISHING AND BROADCAST MEDIA

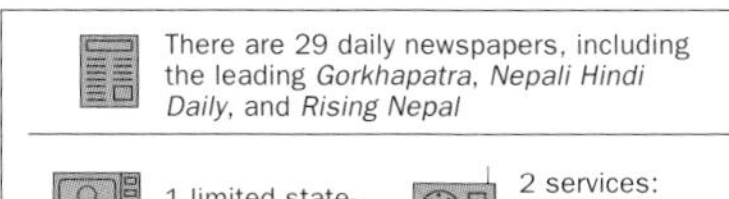
There are 29 daily newspapers, including the leading *Gorkhapatra*, *Nepali Hindi Daily*, and *Rising Nepal*

1 limited state-owned service

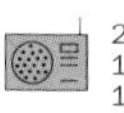 2 services: 1 state-owned, 1 independent

The Nepal TV service began in 1986; under 25% of the population receives it. The press is mainly Kathmandu-based with low circulations. Press watchdogs warned of censorship under the state of emergency imposed in 2001.

CRIME

No death penalty

5878 prisoners

 Up 5% in 1999

Petty theft and smuggling are the main problems. The legal provision for detention without trial is used, and police suppression of demonstrations is often brutal.

EDUCATION

School leaving age: 11

 42%

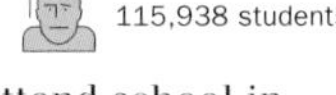 115,938 students

Over 80% of boys attend school in Nepal, but still only a minority of girls. Nepal's literacy rate is among the lowest in the world.

CHRONOLOGY

The foundations of the Nepalese state were laid in 1769, when King Prithvi Narayan Shah conquered the region.

- ❑ **1816–1923** Quasi-British protectorate.
- ❑ **1959** First multiparty constitution.
- ❑ **1960** Constitution suspended.
- ❑ **1962–1990** *Panchayat* nonparty system.
- ❑ **1972** Birendra succeeds to throne.
- ❑ **1991** NCP victory in elections.
- ❑ **1994** First communist government.
- ❑ **1995–1998** Succession of weak coalition governments.
- ❑ **1999** NCP election victory. Maoist insurgency in rural areas.
- ❑ **2001** King and family shot; Gyanendra crowned amid unrest. Upsurge in Maoist violence; state of emergency declared.

HEALTH

Welfare state health benefits

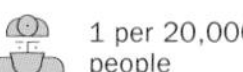 1 per 20,000 people

 Respiratory and diarrheal diseases, maternal deaths

There are about 100 *dharmi-jhankri* (faith healers) for every health worker. Maternal mortality is high, the result of harmful traditional birth practices; a re-education program for midwives has been established.

SPENDING

GDP/cap. increase

CONSUMPTION AND SPENDING

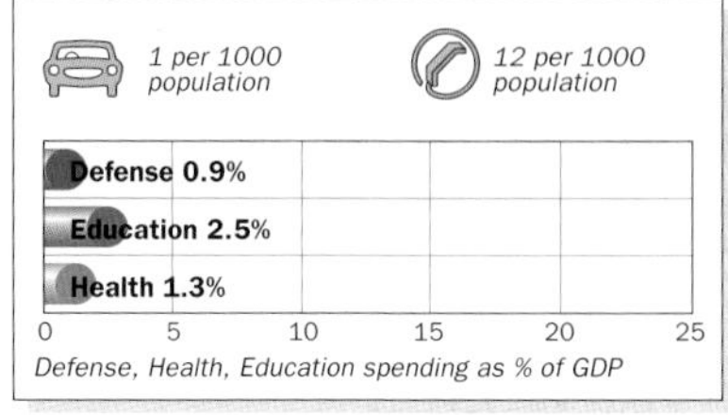
1 per 1000 population
12 per 1000 population
Defense 0.9%
Education 2.5%
Health 1.3%
0
5
10
15
20
25
Defense, Health, Education spending as % of GDP

Nepal is one of the poorest countries in the world. Bonded labor was abolished in mid-2000, releasing 36,000 people.

WORLD RANKING

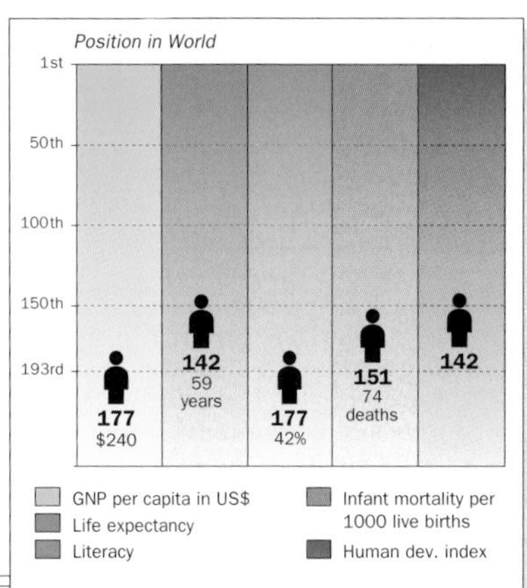
Position in World
1st
50th
100th
150th
193rd
177
$240
142
59 years
177
42%
151
74 deaths
142
GNP per capita in US$
Life expectancy
Literacy
Infant mortality per 1000 live births
Human dev. index

N

NETHERLANDS

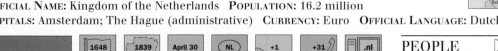

EUROPE
Europe
NETHERLANDS
Germany
Belgium

OFFICIAL NAME: Kingdom of the Netherlands **POPULATION:** 16.2 million
CAPITALS: Amsterdam; The Hague (administrative) **CURRENCY:** Euro **OFFICIAL LANGUAGE:** Dutch

1648
1839

April 30

NL

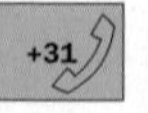
+1

+31

.nl

THE NETHERLANDS IS LOCATED at the delta of four major rivers in northwest Europe. The few hills in the eastern and southern part of the country fall into a flat coastal area, bordered by the North Sea to the north and west. This is protected by a giant infrastructure of dunes, dikes, and canals, as 27% of the coast is below sea level. The Netherlands became one of the world's first confederative republics after Spain recognized its independence in 1648. Its highly successful economy has a long trading tradition, and Rotterdam is the world's largest port.

CLIMATE

Maritime

WEATHER CHART FOR AMSTERDAM

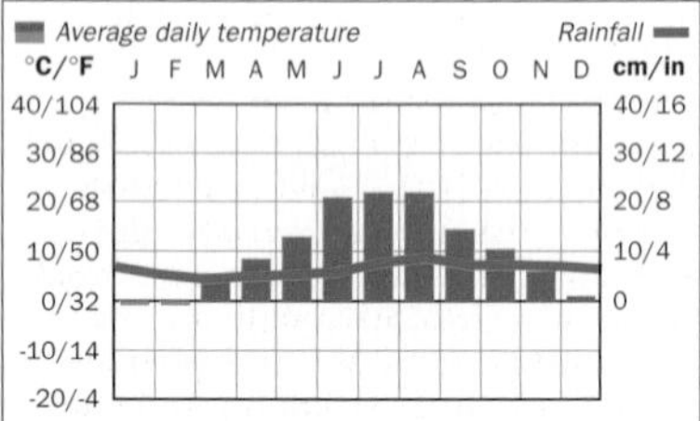
Average daily temperature
Rainfall
°C/°F
J F M A M J J A S O N D
cm/in
40/104
30/86
20/68
10/50
0/32
-10/14
-20/-4
40/16
30/12
20/8
10/4
0

The Netherlands has a temperate climate, with mild winters which rarely fall much below freezing, and cool summers with a mean temperature of 20°C (68°F). The country's coastal areas have the mildest climate, though northerly gales are fairly frequent, particularly in autumn and winter.

TRANSPORTATION

Drive on right

Schiphol, Amsterdam
39.6m passengers
1214 ships
4.26m grt

THE TRANSPORTATION NETWORK

113,018 km (70,226 miles)

2200 km (1367 miles)

2808 km (1745 miles)

5043 km (3134 miles)

Rotterdam, the key transshipment port for northern Europe, is also the world's largest. Schiphol airport is one of the air transportation hubs of Europe. A high-speed passenger rail line is due to link Amsterdam and Rotterdam with Brussels and Paris in 2005, and a high-speed freight line from Rotterdam to Germany should be completed in 2004.

TOURISM

Visitors : Population 1:1.6

10m visitors

Up 1% in 2000

MAIN TOURIST ARRIVALS

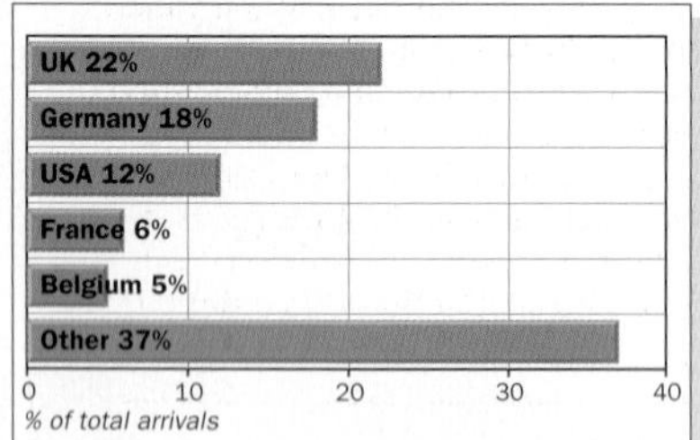
UK 22%
Germany 18%
USA 12%
France 6%
Belgium 5%
Other 37%
0
10
20
30
40
% of total arrivals

Tourism is a major business in the Netherlands. Visitors go mainly to Amsterdam, although cities such as Groningen and Maastricht are growing in popularity. Amsterdam caters for a diverse tourism market. Its world-famous museums include the Rijksmuseum, with its collection of Vermeers and Rembrandts, while its network of canals is popular. Amsterdam is also renowned for its liberal attitude to sex; its red-light district draws millions every year. In the past decade, the city has become a center for the European gay community, with celebrations on 30 April (Queen's Day – the monarch's official birthday) and in August (Amsterdam Pride). A thriving club scene and liberal drug laws draw enthusiasts from neighboring countries. In spring and summer, the tulip fields and North Sea beaches attract large numbers of visitors.

Windmill at Baambrugge, *near Amsterdam. A century ago there were 10,000 in the country compared with today's 1000. A protective ring of 900 mills kept Amsterdam from flooding.*

PEOPLE

Pop. density high

Dutch, Frisian

478/km² (1237/mi²)

THE URBAN/RURAL POPULATION SPLIT

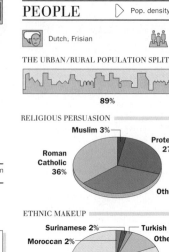

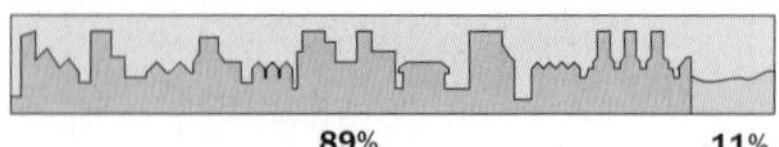
89%
11%

RELIGIOUS PERSUASION

Muslim 3%
Protestant 27%
Roman Catholic 36%
Other 34%

ETHNIC MAKEUP

Surinamese 2%
Turkish 2%
Moroccan 2%
Other 12%
Dutch 82%

The Dutch see their country as the most tolerant in Europe, and it has a long history of welcoming refugees seeking religious and political asylum. In the 20th century, immigrants from former colonies settled in the Netherlands and became fully accepted as citizens. They came first from Indonesia and then from Suriname, the Netherlands Antilles, and Morocco. The small Turkish community, however, does not enjoy full citizenship.

The tradition of tolerance is reflected in liberal attitudes to sexuality. In 2001 same-sex marriages were legalized, giving gay couples full equality, including the right of adoption (after three years of marriage).

The state does not try to impose a particular morality on its citizens. Drug taking is seen as a matter of personal choice, and in 2001 the Netherlands became the first country in the world to legalize euthanasia, albeit under strict conditions.

Women enjoy equal rights and hold 37% of seats in the Second Chamber of the States-General, but are not well represented in boardrooms.

POPULATION AGE BREAKDOWN

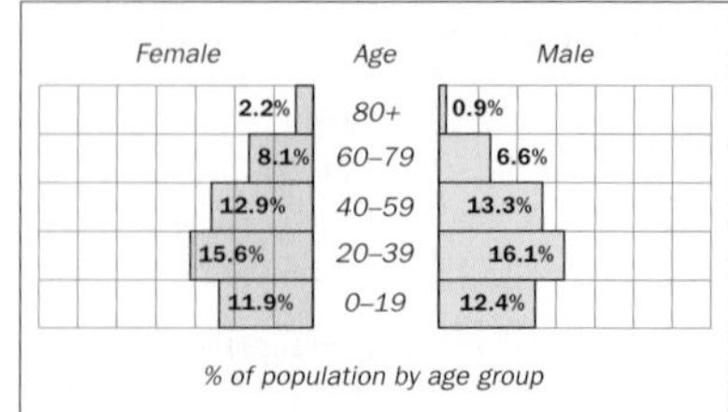
Female
Age
Male
2.2%
80+
0.9%
8.1%
60–79
6.6%
12.9%
40–59
13.3%
15.6%
20–39
16.1%
11.9%
0–19
12.4%
% of population by age group

N

POLITICS

Multiparty elections

L. House 2002/2006
U. House 1999/2003

H.M. Queen Beatrix

AT THE LAST ELECTION

Second Chamber of the States-General 150 seats

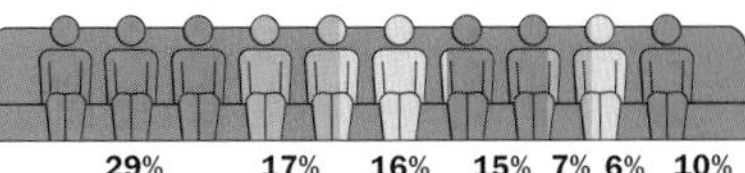

29% CDA	17% LPF	16% VVD	15% PvdA	7% GL	6% SP	10% Others

CDA = Christian Democratic Appeal **LPF** = List Pim Fortuyn
VVD = People's Party for Freedom and Democracy
PvdA = Labor Party **GL** = Green Left **SP** = Socialist Party
D66 = Democrats 66

First Chamber of the States-General 75 seats

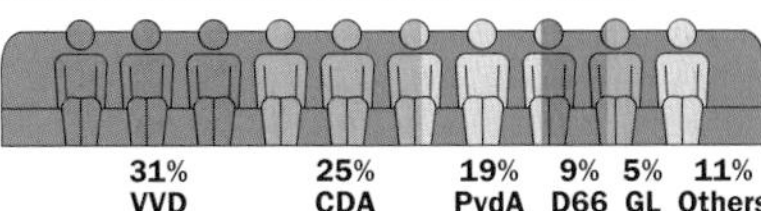

31% VVD	25% CDA	19% PvdA	9% D66	5% GL	11% Others

The First Chamber of the States-General is indirectly elected

The Netherlands is a constitutional monarchy. Legislative power is vested in parliament, and the monarch has only nominal power.

PROFILE

Dutch politics show a high degree of consensus. Since the early 1980s, governments have employed the "polder model," which focuses on pay moderation, job creation, economic deregulation, and generous social protection.

The CDA has traditionally led two-party coalitions, either with the left-of-center PvdA or with the right-wing VVD. However, after the 1994 election the PvdA under Wim Kok led the government, in coalition with the VVD and the left-liberal D66.

Reelected in 1998, this administration resigned in April 2002 after a report criticized Dutch troops serving with the UN in Bosnia in 1995 for failing to stop the Srebrenica massacre. In 2002 the emergence of the ultranationalist Pim Fortuyn challenged current policy on immigration and integration. Fortuyn's influence, articulating a sense of widespread public alienation from a complacent political establishment, was expected to be lasting, despite his assassination only days before the May election. The strong sympathy vote boosted a swing to the right, resulting in victory for the CDA under Jan Peter Balkenende and second place for the LPF.

Queen Beatrix, who acceded in 1980 and rebuilt support for the Dutch monarchy.

CDA leader Jan Peter Balkenende *was appointed prime minister in 2002.*

MAIN POLITICAL ISSUES

The future of social welfare

Despite cutbacks in the 1980s, the Dutch still have one of Europe's most generous welfare systems. Most parties accepted that levels of welfare could not be maintained indefinitely. The debate thus focuses on how much and in which areas cuts should be made.

Refugees and asylum seekers

Responding to rising numbers of asylum seekers in the Netherlands, asylum laws have been tightened since 1994. First- and second-generation immigrants made up 18% of the national population in 2002, but 40% (30% non-European) in Rotterdam, where the LPF won 35% of the vote in the March 2002 local elections. Fortuyn opposed multiculturalism and advocated amending the constitutional ban on discrimination.

NETHERLANDS

Total Land Area :
41 526 sq. km
(16 033 sq. miles)

NORTH SEA
WADDENEILANDEN
GERMANY
BELGIUM
Schiermonnikoog
Ameland
Terschelling
Vlieland
Texel
Waddenzee
Ems
Delfzijl
Groningen
Leeuwarden
Hoogezand
Sappemeer
Harlingen
Assen
Stadskanaal
Den Helder
Heerenveen
Emmen
IJSSELMEER
Enkhuizen
Meppel
Hoogeveen
Alkmaar
Hoorn
Purmerend
Lelystad
Zwolle
Beverwijk
Velsen
Zaanstad
Almelo
Haarlem
Almere
AMSTERDAM
Amstelveen
Huizen
Hilversum
Aalsmeer
Apeldoorn
Deventer
Hengelo
Enschede
Katwijk aan Zee
Leiden
Nieuwkoop
Amersfoort
'S-GRAVENHAGE (THE HAGUE)
Alphen aan de Rijn
Maarssen
Barneveld
Utrecht
Zeist
Ede
Rheden
Zoetermeer
Gouda
Doetinchem
Delft
Nieuwegein
Arnhem
Schiedam
Lek
Neder-Rijn
Vlaardingen
Rotterdam
Tiel
Spijkenisse
Zwijndrecht
Waal
Nijmegen
Goeree
Ridderkerk
Grevelingen
Oss
Overflakkee
Dordrecht
Schouwen
Maas
Oosterhout
's-Hertogenbosch
Oosterschelde
Noord-Beveland
Breda
Tholen
Walcheren
Roosendaal
Tilburg
Helmond
Middelburg
Goes
Vlissingen
Zuid-Beveland
Bergen op Zoom
Eindhoven
Westerschelde
Venlo
Terneuzen
Roermond
Sittard
Geleen
Heerlen
Kerkrade
Maastricht
POPULATION
over 1 000 000
over 500 000
over 100 000
over 50 000
over 10 000
LAND HEIGHT
100m/328ft
Sea Level
-100m/-328ft
0 40 km
0 40 miles
N

WORLD AFFAIRS

Joined UN in 1945

Benelux | EU | NATO | OECD | OSCE

Political and monetary integration within the EU have strong popular support. In 1995, internal border controls were lifted under the Schengen Convention and in 2002 the Netherlands fully adopted the euro. Traditionally the Netherlands favored enlargement of the EU, but the CDA government expressed reservations when it came to power in May 2002. The International Court of Justice, the International Criminal Tribunal for the former Yugoslavia, and the International Criminal Court sit in The Hague.

AID

Donor

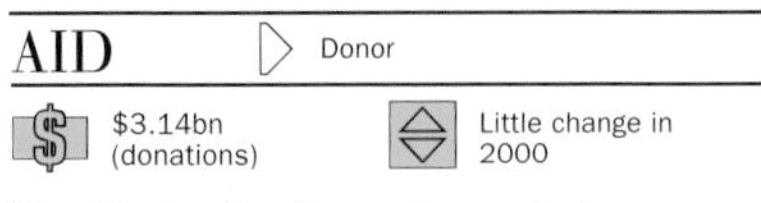

$3.14bn (donations)

Little change in 2000

The Netherlands continues to be one of the few countries which exceeds the UN target of devoting 0.7% of GNP to development aid. The government actively pursues a policy of linking foreign aid and human rights. It also gives priority to projects which link longer-term development goals with efforts to manage and reduce intergroup conflict.

N

CHRONOLOGY

Suppression of Protestantism by the ruling Spanish Habsburgs led to the revolt of the Netherlands and the declaration of independence of the northern provinces as a republic in 1581, recognized by Spain in 1648.

- ❑ **1813** Dutch oust French after 18 years of French rule and choose to become a constitutional monarchy.
- ❑ **1815** United Kingdom of Netherlands formed to include Belgium and Luxembourg.
- ❑ **1839** Recognition of 1830 secession of Catholic southern provinces as Belgium.
- ❑ **1848** New constitution – ministers to be accountable to parliament.
- ❑ **1897–1901** Wide-ranging social legislation enacted. Development of strong trade unions.
- ❑ **1898** Wilhelmina succeeds to throne, ending Luxembourg union, where male hereditary Salic Law is in force.
- ❑ **1914–1918** Dutch neutrality respected in World War I.
- ❑ **1922** Women fully enfranchised.
- ❑ **1940** Dutch assert neutrality in World War II, but Germany invades. Fierce resistance.
- ❑ **1942** Japan invades Dutch East Indies.
- ❑ **1944–1945** "Winter of starvation" in German-occupied western provinces.
- ❑ **1945** Liberation. International Court of Justice set up in The Hague.
- ❑ **1946–1958** PvdA leads center-left coalitions.
- ❑ **1948** Juliana becomes queen.
- ❑ **1949** Joins NATO. Most of East Indies colonies gain independence as Indonesia.
- ❑ **1957** Founder member of EEC.
- ❑ **1960** Economic union with Belgium and Luxembourg comes into effect.
- ❑ **1973** PvdA wins power after 15 years spent mainly in opposition. Center-left coalition until 1977.
- ❑ **1980** CDA alliance of the "confessional" parties forms a single party. Beatrix becomes queen.
- ❑ **1982–1994** CDA-led coalitions under Ruud Lubbers.
- ❑ **1990** 20-year National Environment Policy (NEP) introduced.
- ❑ **1992** Licensed brothels legalized.
- ❑ **1994** Elections: Wim Kok of PvdA heads coalition with VVD and D66.
- ❑ **2001** Euthanasia and gay marriage legalized.
- ❑ **2002** January, euro fully adopted. April, government resigns after publication of report blaming Dutch military in Bosnia for Srebrenica massacre in 1995. May, ultra-nationalist leader Pim Fortuyn assassinated. Elections result in CDA-led coalition under Jan Peter Balkenende.

DEFENSE

No compulsory military service

$6.39bn

Up 3% in 2000

DUTCH ARMED FORCES

320 main battle tanks (*Leopard* 2)	23,100 personnel
4 submarines, 2 destroyers, and 10 frigates	12,130 personnel
157 combat aircraft (F-16A/B)	10,000 personnel
None	

The Dutch military has undergone major restructuring since the end of the Cold War with the aim of making it a rapidly deployable, more flexible military force as befits a NATO member state. Compulsory military service was abolished in 1996 and personnel cut by 44%, with the number of army divisions reduced from three to two. In 1995, a joint Dutch–German army corps numbering 28,000 was inaugurated. The Netherlands sent troops to the international peacekeeping effort in Bosnia; a 2002 report criticized them for failing to prevent the massacre at Srebrenica in 1995.

The Netherlands also has a large defense industry, which specializes in submarines, weapons systems, and aircraft.

ECONOMICS

Inflation 1.9% p.a. (1990–2000)

$398bn

1.0651–1.1231 euros

SCORE CARD

❑ World GNP Ranking	14th
❑ GNP per Capita	$24,970
❑ Balance of Payments	$16.3bn
❑ Inflation	2.5%
❑ Unemployment	3%

EXPORTS

- Italy 6%
- UK 11%
- France 11%
- Belgium 12%
- Germany 27%
- Other 33%

IMPORTS

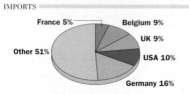

- France 5%
- Belgium 9%
- UK 9%
- USA 10%
- Germany 16%
- Other 51%

Strengths

Highly skilled, educated, multilingual workforce. Sophisticated infrastructure. Many blue-chip multinationals, including Philips and Shell. Strong consensus between employers and employees. Low inflation and unemployment. Tradition of high-tech innovation, including development of music cassette and CD.

Weaknesses

Costly welfare system, resulting in high taxes and social insurance premiums; one-third of national income spent on social security. High labor costs.

Profile

Trade has been central to the success of the economy of the Netherlands since the 16th century. Most goods travel through Rotterdam, the world's biggest port. As well as high-tech industries such as electronics, telecommunications, and chemicals, there is a successful agricultural sector. Dependence on trade makes the economy vulnerable to world economic fluctuations; thus high growth rates in 1997–2000 were followed by a marked downturn in 2001.

ECONOMIC PERFORMANCE INDICATOR

Consumer Price Index
GDP
160
140
120
100
80
Consumer price index 1995=100
GDP 1996=100
1996
1997
1998
1999
2000

NETHERLANDS : MAJOR BUSINESSES

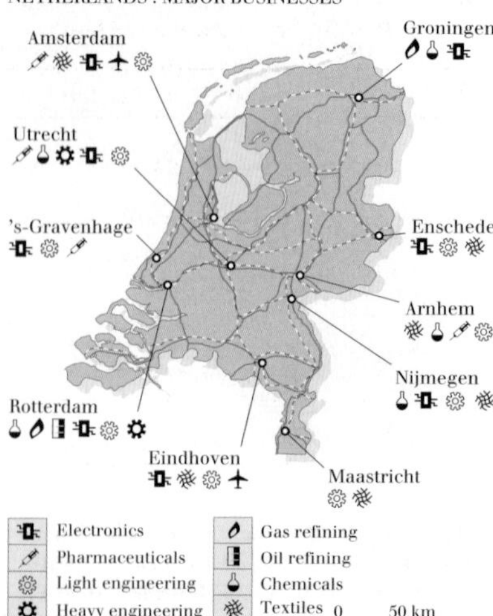
Amsterdam
Groningen
Utrecht
's-Gravenhage
Enschede
Arnhem
Nijmegen
Rotterdam
Eindhoven
Maastricht
Electronics
Pharmaceuticals
Light engineering
Heavy engineering
Aerospace industry
Gas refining
Oil refining
Chemicals
Textiles
0
50 km
0
50 miles

N

RESOURCES

Electric power 20.2m kw

623,396 tonnes

34,366 b/d (reserves 113m barrels)

12.8m pigs, 4.05m cattle, 1.5m turkeys, 107m chickens

Natural gas, oil

ELECTRICITY GENERATION

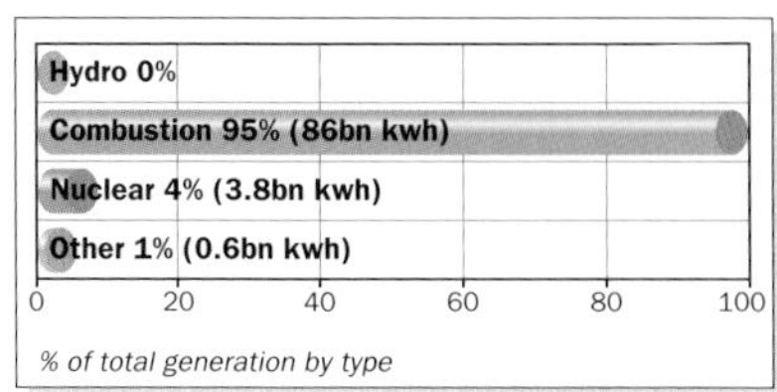
Hydro 0%
Combustion 95% (86bn kwh)
Nuclear 4% (3.8bn kwh)
Other 1% (0.6bn kwh)
0
20
40
60
80
100
% of total generation by type

There are large natural gas reserves in the north. There is some oil production from offshore drilling in the North Sea.

NETHERLANDS : LAND USE

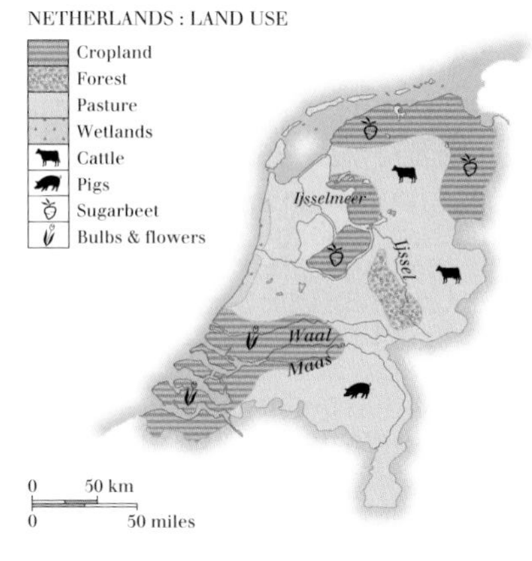
Cropland
Forest
Pasture
Wetlands
Cattle
Pigs
Sugarbeet
Bulbs & flowers
IJsselmeer
IJssel
Waal
Maas
0 50 km
0 50 miles

ENVIRONMENT

Sustainability rank: 34th

7%

10.4 tonnes per capita

ENVIRONMENTAL TREATIES

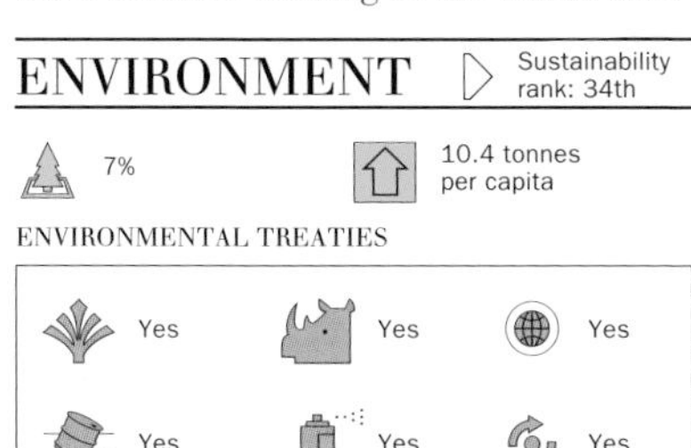
Yes
Yes
Yes
Yes
Yes
Yes

There is a strong environmental tradition, a legacy in part of living in one of the most densely populated states in the world. NGOs such as Greenpeace are well supported and the Green Left party is well represented in parliament.

The Dutch recycle domestic trash, have a good record on energy efficiency, and have developed innovative projects in housing and local transportation. An eco-tax on energy users was introduced in 1996 – the first of its kind in the West – though big businesses are exempt.

Serious flooding of the rivers Maas and Waal (an arm of the Rhine) in 1993 and 1995 raised concern about the state of the country's flood defenses and the use of floodplains for development.

MEDIA

TV ownership high

Daily newspaper circulation 305 per 1000 people

PUBLISHING AND BROADCAST MEDIA

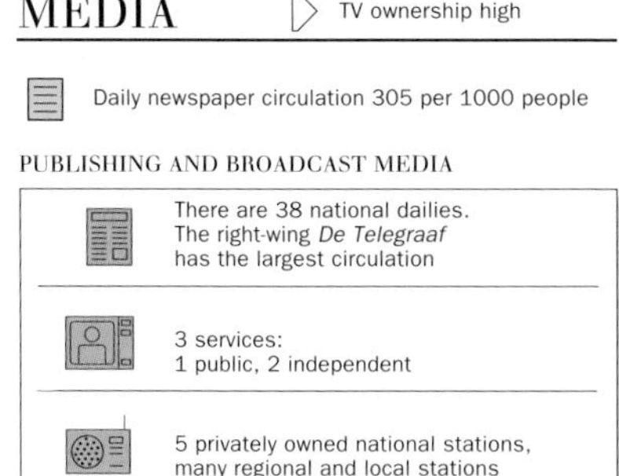
There are 38 national dailies. The right-wing *De Telegraaf* has the largest circulation
3 services: 1 public, 2 independent
5 privately owned national stations, many regional and local stations

Newspaper circulation is high. While editorially independent, broadcasting is strongly regulated. Dutch law does not recognize a right of reply or a right to protect information sources.

CRIME

No death penalty

13,847 prisoners

Down 16% 1996–1998

CRIME RATES

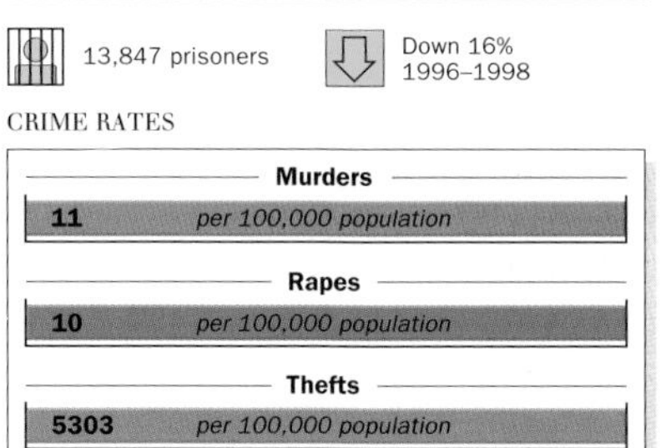
Murders
11 per 100,000 population
Rapes
10 per 100,000 population
Thefts
5303 per 100,000 population

The Netherlands treats the use of hard drugs more as a medical and social than a criminal issue. Other member states of Europe's Schengen Convention, particularly France, fear that this makes Dutch ports a soft point of entry for narcotics. Possessing cannabis for personal use has been decriminalized – stopping short of actual legalization.

EDUCATION

School leaving age: 16

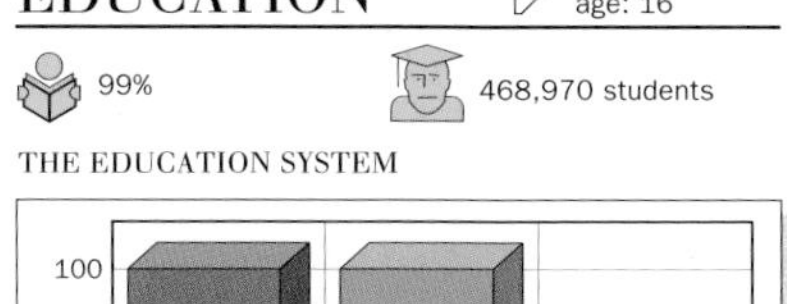
99%
468,970 students

THE EDUCATION SYSTEM

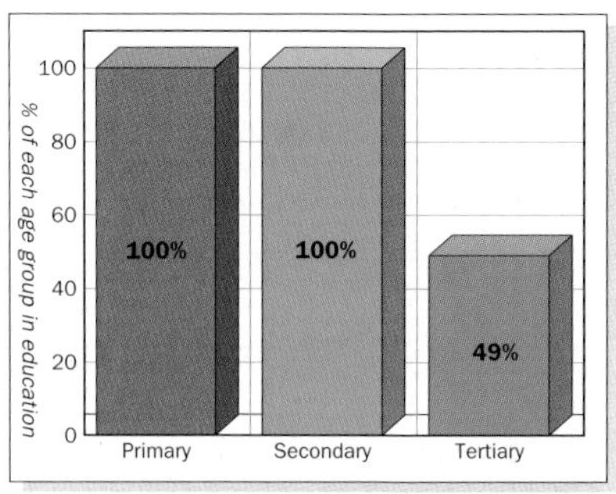
100
80
60
40
20
0
% of each age group in education
100%
100%
49%
Primary
Secondary
Tertiary

Corporate funding plays an important part in university research.

Public, municipally run schools attract 35% of pupils, and private (mostly denominational) schools are attended by 65%. Both types are fully funded by the state.

HEALTH

Welfare state health benefits

1 per 323 people

Heart and respiratory diseases, cancers

Health care is largely funded by the state, though around 25% of funding comes from private sources. High spending ensures that the care that is offered is among the best in the world, but it is threatened by a rapidly aging population. The Netherlands was the first country to legalize abortion, but has the lowest rate of terminations in the world. Major health problems are similar to those in the rest of western Europe. Incidence of AIDS is higher than in Sweden or the UK but lower than in Switzerland, France, or Spain.

SPENDING

GDP/cap. increase

CONSUMPTION AND SPENDING

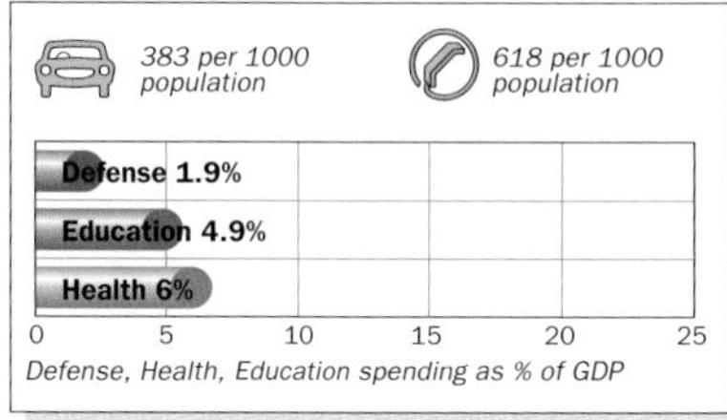
383 per 1000 population
618 per 1000 population
Defense 1.9%
Education 4.9%
Health 6%
0
5
10
15
20
25
Defense, Health, Education spending as % of GDP

The Netherlands is, per capita, one of the richest countries in the world. Oil executives, stock market traders, and businessmen are among the wealthiest sector of the population. A progressive taxation system and extensive social welfare mean that wealth is quite evenly distributed. A small elite have considerable inherited wealth, but extravagant displays of affluence are rare.

Class does not play a big part in Dutch society. Most citizens would consider themselves middle class. Immigrant communities are the exception; they often live on the edges of towns in deprived areas. The poorest group of all are the illegal immigrants.

WORLD RANKING

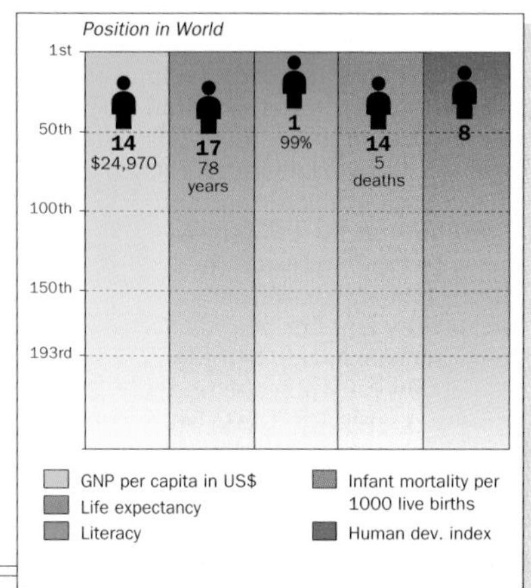
Position in World
1st
50th
100th
150th
193rd
14
$24,970
17
78 years
1
99%
14
5 deaths
8
GNP per capita in US$
Life expectancy
Literacy
Infant mortality per 1000 live births
Human dev. index

N

NEW ZEALAND

AUSTRALASIA
Australia
NEW ZEALAND
Pacific Ocean

OFFICIAL NAME: New Zealand **CAPITAL:** Wellington **POPULATION:** 3.8 million
CURRENCY: New Zealand dollar **OFFICIAL LANGUAGES:** English and Maori

1947 | 1947 | Feb 6 | 
NZ
| 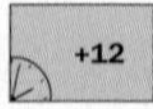
+12
| +64 | 
.nz

LYING IN THE SOUTH PACIFIC, 1600 km (992 miles) southeast of Australia, New Zealand comprises the main North and South Islands, separated by the Cook Strait, and a number of smaller islands. South Island is the more mountainous; North Island contains hot springs and geysers, and the bulk of the population. The political tradition is liberal and egalitarian, and has been dominated by the National and Labour Parties. Radical, and often unpopular, reforms since 1984 have restored economic growth, speeded up economic diversification, and strengthened New Zealand's position within the Pacific Rim countries.

CLIMATE

Maritime/subtropical

WEATHER CHART FOR WELLINGTON

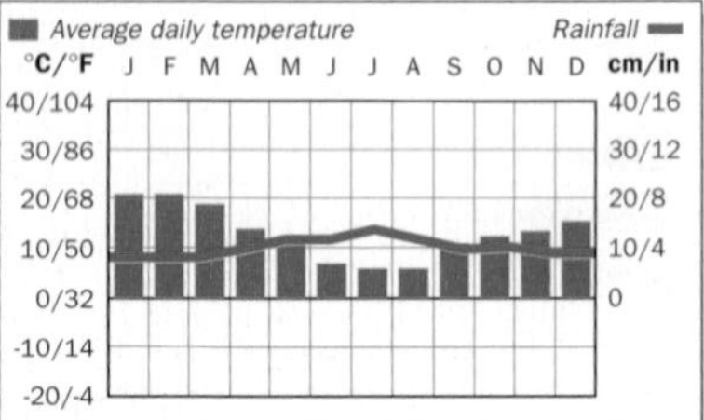
Average daily temperature
Rainfall
°C/°F
J F M A M J J A S O N D
cm/in
40/104
30/86
20/68
10/50
0/32
-10/14
-20/-4
40/16
30/12
20/8
10/4
0

New Zealand's climate is generally temperate and damp, with an average temperature of 12°C (54°F). There are differences between the islands, which extend north–south nearly 2000 km (1240 miles). The extreme north is almost subtropical; southern winters are cold. It is windy: Wellington, in particular, is known for bouts of blustery weather that can last for days.

TRANSPORTATION

Drive on left

Auckland International
7.95m passengers

173 ships
336,278 grt

THE TRANSPORTATION NETWORK

53,568 km (33,286 miles)

226 km (140 miles)

3900 km (2423 miles)

1609 km (1000 miles)

Although both the main islands are well served by transportation services, the more populous North Island's road and rail network is more extensive than the South's. Air and ferry services complement the land networks and provide links between the North and South Islands, as well as with the numerous smaller islands. Cargo ferry services are particularly important for Antarctic bases in the Ross Dependency. Links with New Zealand's other associated territories – the Cook Islands, Niue, and the Tokelau atolls – are underdeveloped.

TOURISM

Visitors : Population 1:2.1

1.8m visitors

Up 11% in 2000

MAIN TOURIST ARRIVALS

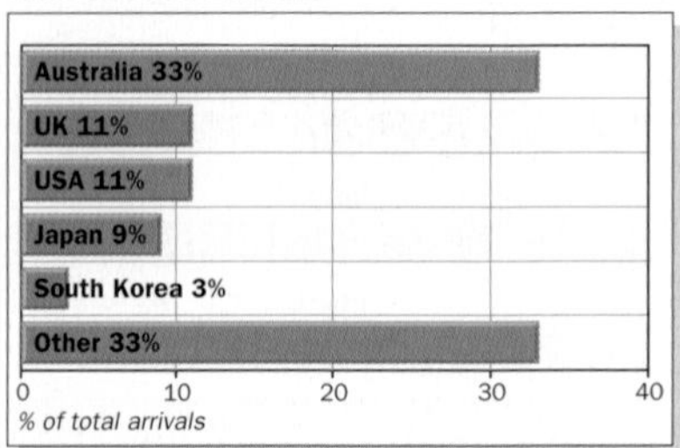
Australia 33%
UK 11%
USA 11%
Japan 9%
South Korea 3%
Other 33%
0
10
20
30
40
% of total arrivals

New Zealand's prime attraction is its scenery. Unspoiled and, relative to the country's size, the most varied in the world, it offers mountains, fjords and lakes, glaciers, rainforests, beaches, boiling mud pools, and geysers. Other attractions are the Maori culture and outdoor activities such as river rafting, fishing, skiing, whale watching, and bungee jumping – a local invention.

Tourists come mainly from Australia, the US, the UK, and Japan. Tourism, the largest single foreign-exchange earner, continues to grow, although the 1997–1998 Asian economic crisis saw the number of Asian tourists drop by 10%. An increase in visitor numbers has followed the huge success of the first *Lord of the Rings* film: the whole series was shot in New Zealand.

Mount Egmont, *an extinct volcano, is one of the numerous popular natural attractions of New Zealand's North Island.*

PEOPLE

Pop. density low

English, Maori

14/km² (37/mi²)

THE URBAN/RURAL POPULATION SPLIT

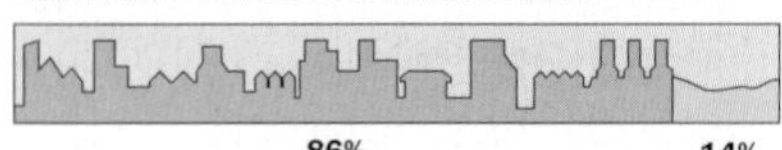
86%
14%

RELIGIOUS PERSUASION

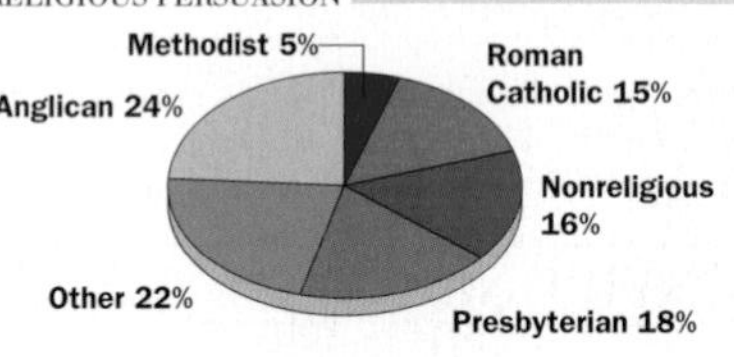
Methodist 5%
Roman Catholic 15%
Anglican 24%
Nonreligious 16%
Other 22%
Presbyterian 18%

ETHNIC MAKEUP

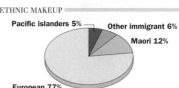

Pacific islanders 5%
Other immigrant 6%
Maori 12%
European 77%

New Zealand is a country of migrants. The first settlers, the Maoris, migrated from Polynesia about 1200 years ago. Today's majority European population is mainly descended from British migrants who settled after 1840. Newer migrants include Asians from Hong Kong and Malaysia, and Polynesians. The government is keen to attract skilled South Americans, Russians, Chinese, and Africans to revitalize the economy.

The living standards and unemployment rates of the Maoris compare adversely with those of the European-descended majority, and relations can be tense. The Waikato Raupatu Claims Settlement Act was signed in 1995 and an official apology to the Maoris was made. In 1998 the Waitangi Tribunal ordered the return of confiscated land.

New Zealand became the first country in the world to give women the vote – in 1893. In 2001, the posts of prime minister, leader of the opposition, and governor-general were all filled by women.

POPULATION AGE BREAKDOWN

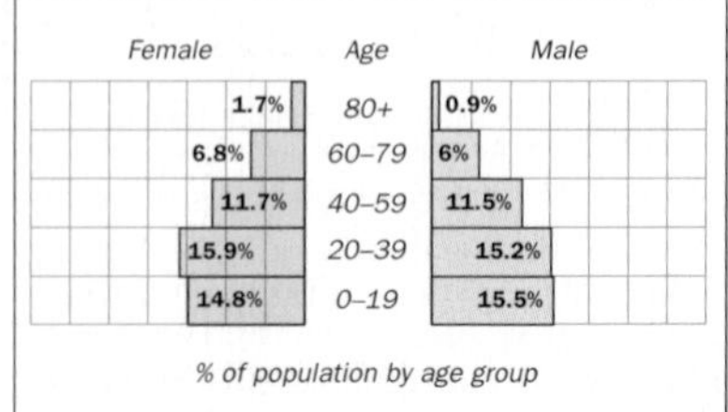
Female
Age
Male
1.7%
80+
0.9%
6.8%
60–79
6%
11.7%
40–59
11.5%
15.9%
20–39
15.2%
14.8%
0–19
15.5%
% of population by age group

N

POLITICS

Multiparty elections

2002/2005 H.M. Queen Elizabeth II

AT THE LAST ELECTION

House of Representatives 120 seats

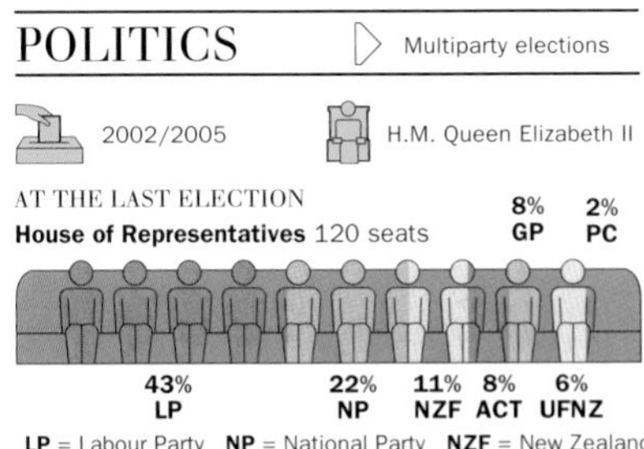
43% LP
22% NP
11% NZF
8% ACT
6% UFNZ
8% GP
2% PC

LP = Labour Party **NP** = National Party **NZF** = New Zealand First Party **ACT** = ACT New Zealand (Association of Consumers and Taxpayers) **GP** = Green Party **UFNZ** = United Future **PC** = Progressive Coalition

Helen Clark, *LP leader and prime minister since 1999.*

Jenny Shipley, *the first woman prime minister (NP, 1997–1999).*

New Zealand is a parliamentary democracy. The Cook Islands and Niue are self-governing territories.

PROFILE

Since 1984 the economy has undergone massive reforms; cuts to the welfare system and privatization of public assets have been unpopular. The NP and the LP dominated politics until proportional representation (PR) was introduced in 1996. Former NP leader Jenny Shipley became the country's first woman prime minister in 1997 at the head of a coalition with the small NZF, and from 1998 led a minority administration when the NZF withdrew. A new minority coalition of the LP and the Alliance assumed power under LP leader Helen Clark in 1999. The LP remained in power after the 2002 elections, forming a minority coalition with the PC.

MAIN POLITICAL ISSUE

Electoral reform

New Zealand shifted from a first-past-the-post electoral system to PR for the 1996 general election. Endorsement of this reform in a referendum in 1993 reflected widespread disillusionment with the NP and LP. The new German-style system strengthened the role of smaller parties. As predicted, the first election to use the system in 1996 produced a coalition government, led by the NP. Forced into opposition in 1998, the NP has, unsuccessfully, called for a review of the PR system.

WORLD AFFAIRS

Joined UN in 1945

Comm | APEC | OECD | PIF | PC

Many New Zealanders are strongly committed to the British monarchy and the Commonwealth, but the UK's EU involvement has forced New Zealand to reorient its trade and foreign policy toward its Pacific Rim neighbors, especially Australia, now New Zealand's single largest trading partner. Their 1983 Closer Economic Relationship (CER) treaty was strengthened in 1996 by the signing of a mutual recognition agreement.

Relations with Asia are growing in importance. The 1997–1998 Asian economic crisis significantly affected trade, particularly tourism.

Relations with the US are improving after a low point when New Zealand's antinuclear stance led to its exclusion from the ANZUS pact. Official ties with France, cut in 1985 after French agents bombed Greenpeace's *Rainbow Warrior* in Auckland harbor, were restored in 1997.

AID

Donor

US$113m (donations) Down 16% in 2000

Over half of New Zealand's overseas aid is bilateral. Particular areas of focus are the Pacific states and Pacific-wide organizations. New Zealand is a major supporter of the Pacific Islands Forum, the University of the South Pacific, and the Pacific Environment Program. It also offers scholarships to overseas students for study or training in New Zealand.

NEW ZEALAND

Total Land Area : 268 680 sq. km (103 737 sq. miles)

LAND HEIGHT: 2000m/6562ft, 1000m/3281ft, 500m/1640ft, 200m/656ft, Sea Level

POPULATION: over 500 000, over 100 000, over 50 000, over 10 000, under 10 000

0 – 100 km; 0 – 100 miles

Chatham Is (continuation on same scale)

CHRONOLOGY

A former British colony, New Zealand became a dominion in 1907, self-governing from 1926, and fully independent in 1947.

- ❑ **1962** Western Samoa (now Samoa) gains independence.
- ❑ **1965** Cook Islands gain autonomy.
- ❑ **1975** Conservative NP wins elections. Economic austerity program introduced.
- ❑ **1976** Immigration cut by over 80%.
- ❑ **1984** LP elected; David Lange prime minister. Auckland harbor headland restored to Maoris.
- ❑ **1985** New Zealand prohibits nuclear vessels from ports and waters. French agents sink Greenpeace ship *Rainbow Warrior* in Auckland harbor.
- ❑ **1986** US suspends military obligations under ANZUS Treaty.
- ❑ **1987** LP wins elections. Introduction of controversial privatization plan. Nuclear ban enshrined in legislation.
- ❑ **1990** LP defeated by NP in elections. James Bolger prime minister.
- ❑ **1991** Widespread protests at spending cuts.
- ❑ **1992** Maoris win South Island fishing rights. Majority vote for electoral reform in referendum.
- ❑ **1993** Docking of first French naval ship for eight years. NP returned with single-seat majority in election. Proportional representation introduced by referendum.
- ❑ **1994** Senior-level US contacts restored; agrees not to send nuclear-armed ships to New Zealand ports. Maoris reject government ten-year land claims settlement of US$660 million.
- ❑ **1995** Waitangi Day celebrations abandoned after Maori protests. Crown apologizes to Maoris and signs Waikato Raupatu Claims Act. UK warship visits resume.
- ❑ **1996** NP forms coalition to preserve overall legislative majority. First general election under new proportional representation system.
- ❑ **1997** NP forms coalition with New Zealand First (NZF) party. Bolger resigns. Jenny Shipley becomes first woman prime minister.
- ❑ **1998** Shipley sacks NZF leader Winston Peters as deputy prime minister. Waitangi Tribunal orders government to return to Maoris US$3.3 million of confiscated land.
- ❑ **1999** LP led by Helen Clark wins general election.
- ❑ **2001** Air New Zealand renationalized.
- ❑ **2002** Combat wing of air force taken out of service. July, elections: LP reelected

N

DEFENSE

▷ No compulsory military service

 US$788m

 Down 4% in 2000

Military cuts announced in May 2001 emphasized the aim to refocus defense policy on small-scale peacekeeping. The move put renewed stress on the 1951 security pact with Australia and the US (ANZUS). The 1984 decision to refuse access to nuclear warships from 1985 damaged defense cooperation with the US and other Western powers for ten years, forcing New Zealand to seek closer links with Australia. Senior-level contacts were resumed in 1994, the US announcing that it would not send nuclear-armed warships to New Zealand ports. Since then, the UK has also resumed naval visits.

NEW ZEALAND ARMED FORCES

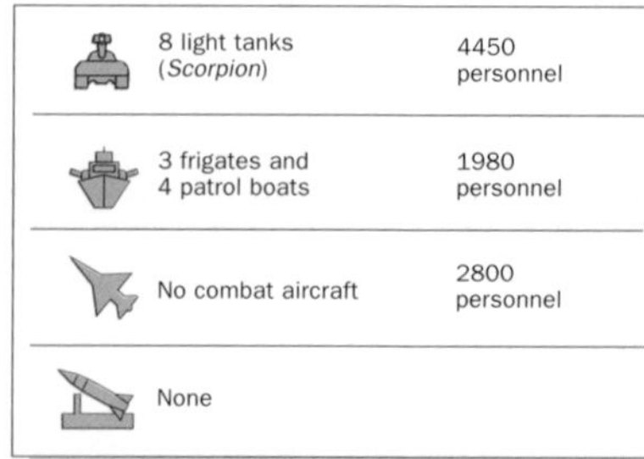
8 light tanks (*Scorpion*)
4450 personnel
3 frigates and 4 patrol boats
1980 personnel
No combat aircraft
2800 personnel
None

New Zealand's defense strategy centers on protection against low-level economic threats, terrorism, regional security, and peacekeeping.

ECONOMICS

▷ Inflation 1.5% p.a. (1990–2000)

US$49.8bn

2.26–2.40 New Zealand dollars

SCORE CARD

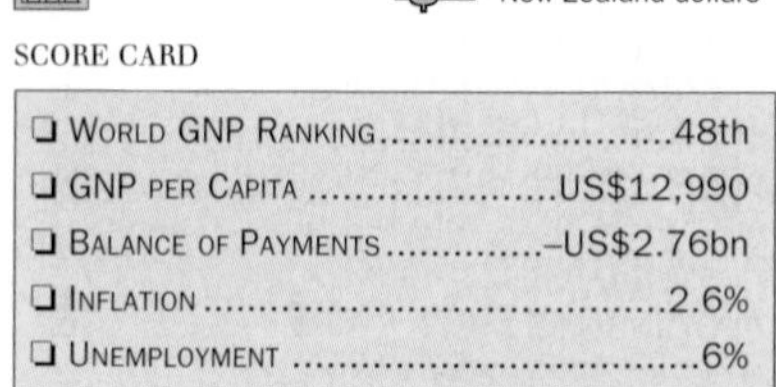

❑ World GNP Ranking	48th
❑ GNP per Capita	US$12,990
❑ Balance of Payments	–US$2.76bn
❑ Inflation	2.6%
❑ Unemployment	6%

EXPORTS

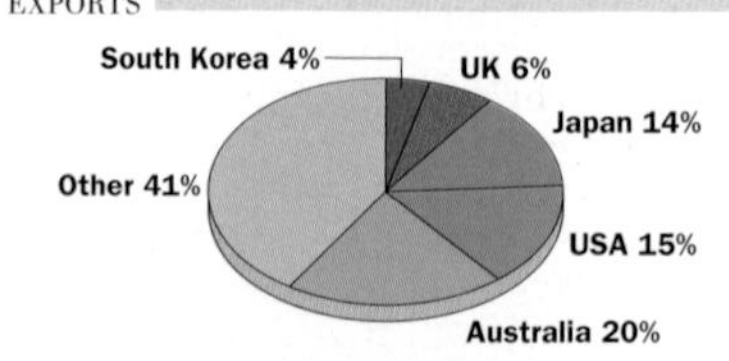
South Korea 4%
UK 6%
Japan 14%
Other 41%
USA 15%
Australia 20%

IMPORTS

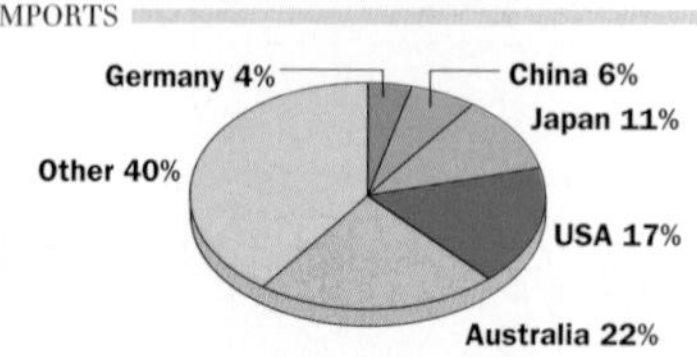
Germany 4%
China 6%
Japan 11%
Other 40%
USA 17%
Australia 22%

ECONOMIC PERFORMANCE INDICATOR

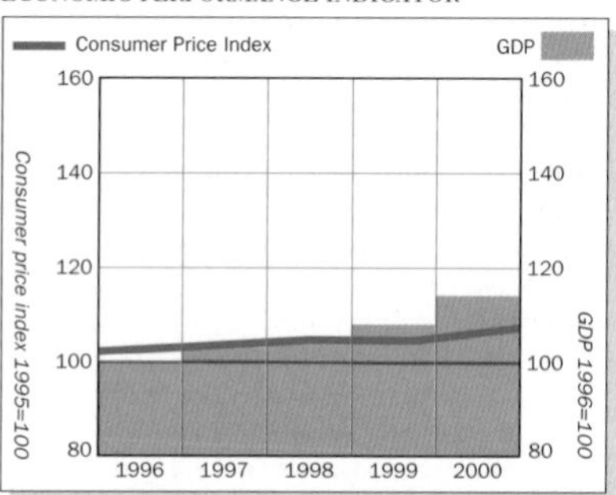
Consumer Price Index
GDP
160
140
120
100
80
Consumer price index 1995=100
GDP 1996=100
1996
1997
1998
1999
2000

Strengths

Modern agricultural sector; world's biggest exporter of butter and (per capita) of wool. Rapidly expanding tourist sector. Manufacturing, with emphasis on high tech. One of world's most open economies. Strong trade links within Pacific Rim.

Weaknesses

One of the highest levels of public debt outside developing world. Continuing reliance on imported manufactured goods and foreign investment.

Profile

Since 1984, New Zealand has changed from being one of the most regulated to one of the most open economies in the world. Radical reforms and drastic cuts in social security and related government spending helped to restore growth and cut inflation to a minimum. Diversification into new markets and products recovered after the 1997–1998 Asian economic crisis. Prime Minister Clark dropped objections in 2001 to the idea of a unified Australia–New Zealand dollar. High public debt and poor levels of private investment remain a problem.

NEW ZEALAND : MAJOR BUSINESSES

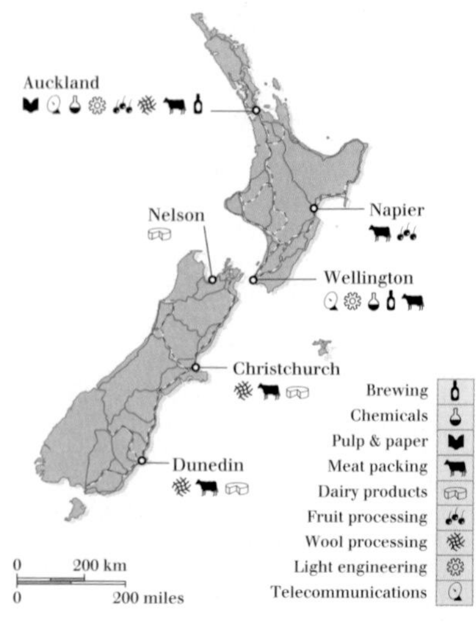
Auckland
Nelson
Napier
Wellington
Christchurch
Dunedin
Brewing
Chemicals
Pulp & paper
Meat packing
Dairy products
Fruit processing
Wool processing
Light engineering
Telecommunications
0 200 km
0 200 miles

RESOURCES

Electric power 8m kw

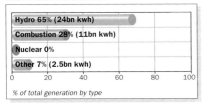

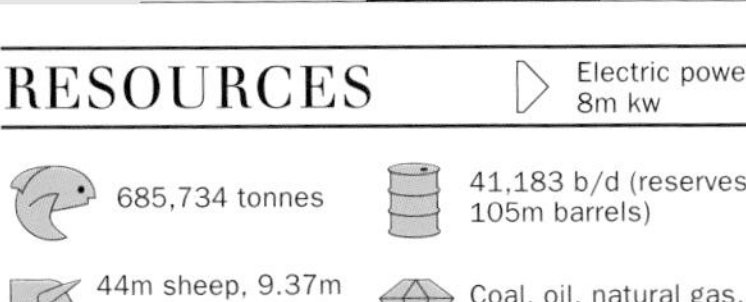
685,734 tonnes
41,183 b/d (reserves 105m barrels)
44m sheep, 9.37m cattle, 354,489 pigs, 13m chickens
Coal, oil, natural gas, iron, gold, silica sand

ELECTRICITY GENERATION

Hydro 65% (24bn kwh)
Combustion 28% (11bn kwh)
Nuclear 0%
Other 7% (2.5bn kwh)
0 20 40 60 80 100
% of total generation by type

New Zealand's rich pastures, a result of even rainfall throughout the year, have traditionally been its key resource. The sheep, wool, and dairy products on which the country's wealth was built are still important. Newer export industries include products such as fruit, vegetables, fish, cork, wood, wine, and textile fibers.

New Zealand is well endowed with energy resources. It has coal, oil, and natural gas reserves, but most energy is generated by hydroelectric plants.

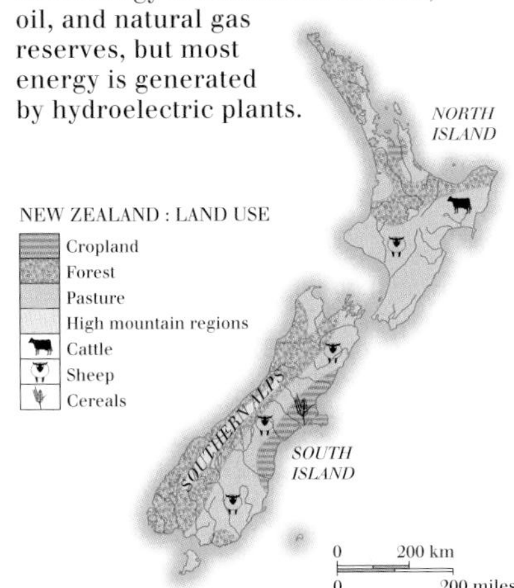
NORTH ISLAND
NEW ZEALAND : LAND USE
Cropland
Forest
Pasture
High mountain regions
Cattle
Sheep
Cereals
SOUTHERN ALPS
SOUTH ISLAND
0 200 km
0 200 miles

ENVIRONMENT

Sustainability rank: 19th

24%

7.9 tonnes per capita

ENVIRONMENTAL TREATIES

Yes
Yes
Yes
Yes
Yes
No

New Zealand's isolation, small population, and limited industry have helped to keep it one of the world's most pollution-free countries. It was a leading opponent of French nuclear testing in the Pacific and has banned nuclear vessels from its ports. Ozone depletion over Antarctica, deforestation, and protection of native flora and fauna are major issues.

MEDIA

TV ownership high

Daily newspaper circulation 216 per 1000 people

PUBLISHING AND BROADCAST MEDIA

There are 29 daily newspapers. The leading newspaper is the *New Zealand Herald*
7 services: 1 state-owned, 6 independent
3 services: 1 state-owned, 2 independent

Deregulated in 1988, New Zealand television is one of the most liberal media in the world. Ruia Mai, the first Maori-language radio station, began broadcasting in 1996.

CRIME

No death penalty

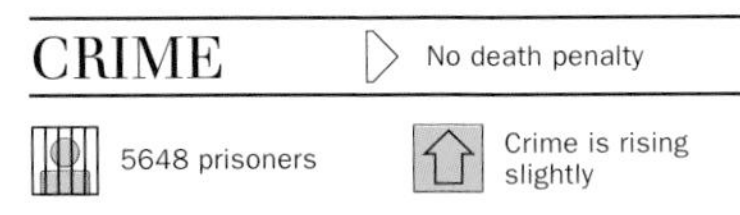
5648 prisoners
Crime is rising slightly

CRIME RATES

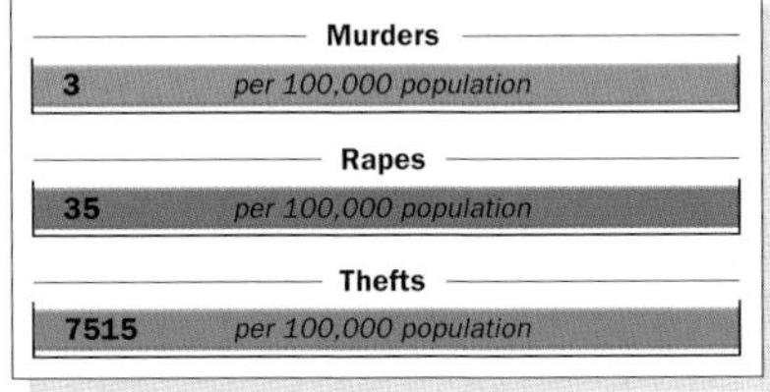
Murders
3 per 100,000 population
Rapes
35 per 100,000 population
Thefts
7515 per 100,000 population

Crime rates in New Zealand's urban areas have increased in recent years. However, overall, the country remains one of the world's safest and most peaceful places in which to live.

EDUCATION

School leaving age: 16

99%

237,781 students

THE EDUCATION SYSTEM

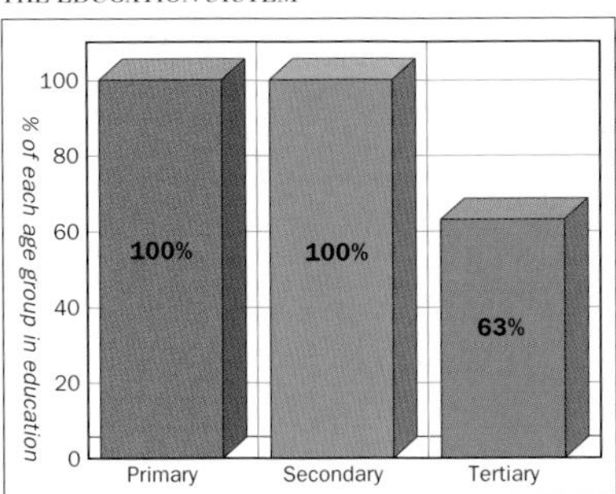
100
80
60
40
20
0
% of each age group in education
100%
100%
63%
Primary
Secondary
Tertiary

Education is free, and compulsory between six and 16. A number of schools are composite, providing both primary and secondary education. New Zealand has one of the highest proportions of the population with tertiary qualifications in the OECD. Nearly all adults are literate, but the level of literacy is not always very high: a government initiative, More than Words, was launched in 2000.

HEALTH

Welfare state health benefits

1 per 435 people

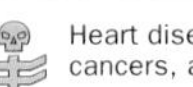

Heart disease, cancers, accidents

In 1936 New Zealand was the first country to introduce a full welfare state. Government efforts since 1991 to impose UK-style market systems on the health service have been very unpopular. While life expectancy continues to improve, the nation's OECD health ranking is falling. In comparison with other OECD countries New Zealand has high mortality rates for heart disease, respiratory disease, breast and bowel cancer, motor vehicle accidents, and suicide.

SPENDING

GDP/cap. increase

CONSUMPTION AND SPENDING

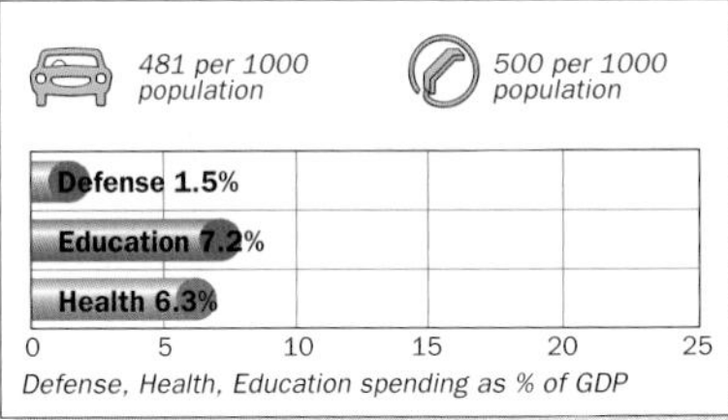
481 per 1000 population
500 per 1000 population
Defense 1.5%
Education 7.2%
Health 6.3%
0 5 10 15 20 25
Defense, Health, Education spending as % of GDP

The years since 1984 have been very difficult for New Zealanders, who are used to affluence within a generous welfare state. A rash of economic and social reforms has held back wages, raised unemployment, and cut welfare benefits. Even so, average living standards are still high, and a strong egalitarian tradition means that wealth remains quite evenly distributed.

The quality of life in New Zealand is among the world's highest, in terms of access to basic necessities, and a pure, healthy, urban and rural environment. Social mobility is fairly high. Wealthier people tend to spend their money on houses close to the water. Yachts are a major status symbol.

WORLD RANKING

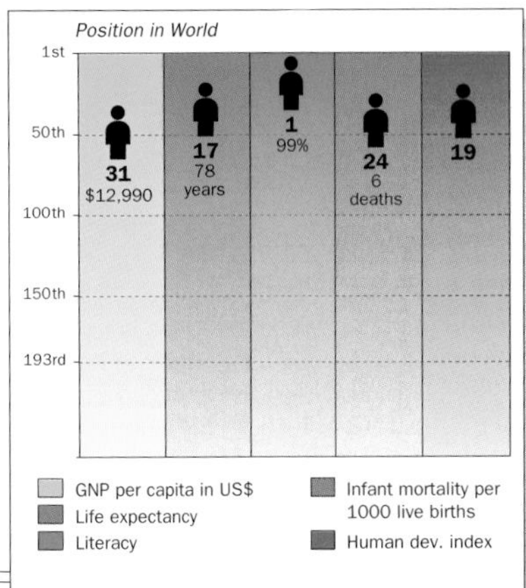
Position in World
1st
50th
100th
150th
193rd
31
$12,990
17
78
years
1
99%
24
6
deaths
19
GNP per capita in US$
Life expectancy
Literacy
Infant mortality per 1000 live births
Human dev. index

N

NICARAGUA

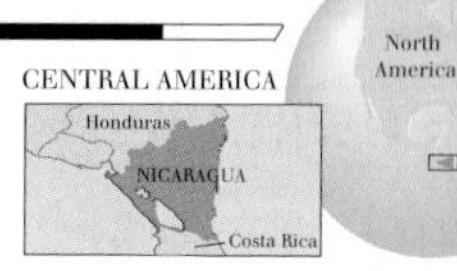
North America
CENTRAL AMERICA
Honduras
NICARAGUA
Costa Rica

OFFICIAL NAME: Republic of Nicaragua **CAPITAL:** Managua
POPULATION: 5.2 million **CURRENCY:** Córdoba oro **OFFICIAL LANGUAGE:** Spanish

1838

1838

Sept 15

NIC

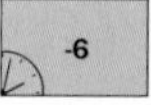
-6

+505

.ni

BOUNDED BY the Pacific Ocean to the west and the Caribbean Sea to the east, Nicaragua lies at the heart of Central America. After more than 40 years of dictatorship, the Sandinista revolution in 1978 led to social reforms, but also to 11 years of civil war, which almost destroyed the economy. Right-wing parties have held power since the Sandinistas unexpectedly lost the 1990 elections. Despite the devastation of Hurricane Mitch in 1998, the economy is slowly strengthening.

Oil refinery at Bluefields, *on the Caribbean coast. Under the Sandinistas, most crude oil came from the Soviet Union, via Cuba.*

CLIMATE

Tropical equatorial/ wet & dry

WEATHER CHART FOR MANAGUA

Average daily temperature | Rainfall
°C/°F J F M A M J J A S O N D cm/in
40/104 40/16
30/86 30/12
20/68 20/8
10/50 10/4
0/32 0
-10/14
-20/-4

The climate is tropical and often violent, as evidenced by seasonal hurricanes, such as Hurricane Mitch in 1998.

TRANSPORTATION

Drive on right

Augusto C. Sandino International, Managua 816,054 passengers

28 ships 4300 grt

THE TRANSPORTATION NETWORK

1818 km (1130 miles)

Pan-American Highway: 384 km (239 miles)

None

2220 km (1379 miles)

Nicaragua lacks a Caribbean deepwater port. Hurricane Mitch damaged major roads and destroyed 35 key bridges.

TOURISM

Visitors : Population 1:11

486,000 visitors

Up 4% in 2000

MAIN TOURIST ARRIVALS

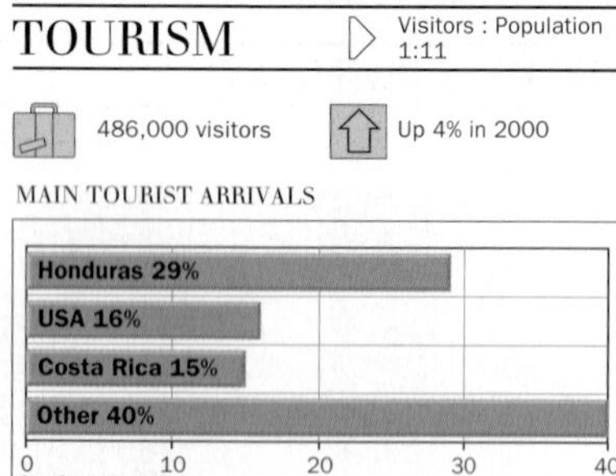
Honduras 29%
USA 16%
Costa Rica 15%
Other 40%
0 10 20 30 40
% of total arrivals

The civil war caused the near-collapse of tourism, and slow recovery has been interrupted by the devastation caused by Hurricane Mitch. Foreign direct investment in the sector grew in 1997–1998.

PEOPLE

Pop. density low

Spanish, English Creole, Miskito

44/km² (113/mi²)

THE URBAN/RURAL POPULATION SPLIT

65% 35%

ETHNIC MAKEUP

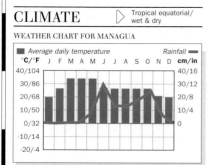

Zambos 4%
Amerindian 5%
Black 8%
White 14%
Mestizo 69%

The Caribbean regions, which in 1987 achieved limited autonomy, are isolated from the more populous Pacific regions. The indigenous Miskito tribes and the descendants of Africans, brought over by Spanish colonists in the 18th century to work the plantations, are concentrated along the Caribbean coast, where English Creole is widely spoken. The Sandinista revolution improved the status of women through changes in the legal system and the incorporation of women into economic and political life. However, poverty and lack of permanent employment have since forced many women into prostitution.

POLITICS

Multiparty elections

2001/2006

President Enrique Bolanos

AT THE LAST ELECTION

National Assembly 92 seats

51% PLC **47% FSLN** **2% PCN**

PLC = Liberal Constitutionalist Party
FSLN = Sandinista National Liberation Front
PCN = Conservative Party of Nicaragua

Defeated in the 1990 and 1996 polls by right-wing parties, the FSLN underwent an internal crisis. A PLC-dominated alliance took office in 1997, promising to unite the country but it quickly became unpopular as a result of its austerity measures and of allegations of corruption. Recent two-party domination, strengthened by a controversial pact between the ruling PLC and FSLN in 2000, has weakened democracy. Enrique Bolanos of the PLC was elected president in 2001.

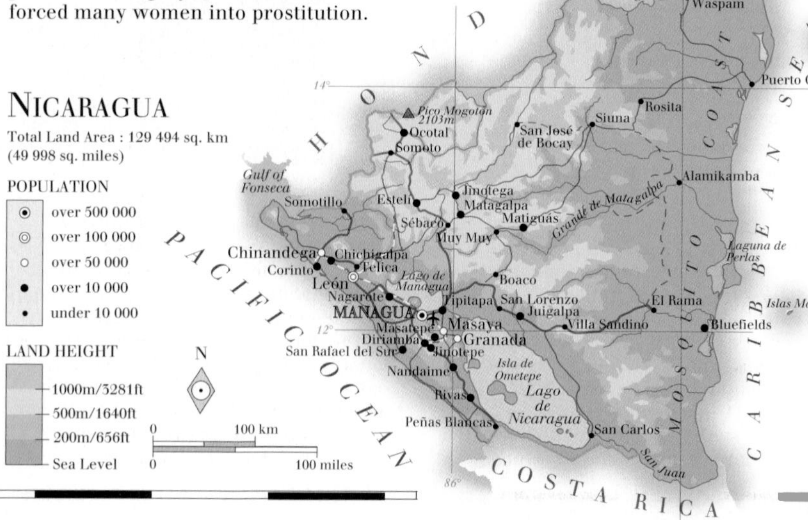
NICARAGUA
Total Land Area : 129 494 sq. km (49 998 sq. miles)
POPULATION
over 500 000
over 100 000
over 50 000
over 10 000
under 10 000
LAND HEIGHT
1000m/3281ft
500m/1640ft
200m/656ft
Sea Level
N
0 100 km
0 100 miles
HONDURAS
PACIFIC OCEAN
CARIBBEAN SEA
COSTA RICA
MOSQUITO COAST
Gulf of Fonseca
Pico Mogotón 2103m
Ocotal
Somoto
Somotillo
Estelí
Jinotega
Matagalpa
Matiguás
Sébaco
Muy Muy
Chinandega
Chichigalpa
Corinto
Telica
León
Lago de Managua
Nagarote
MANAGUA
Tipitapa
San Lorenzo
Juigalpa
Boaco
Masaya
Masatepe
Granada
Diriamba
Jinotepe
San Rafael del Sur
Nandaime
Rivas
Isla de Ometepe
Lago de Nicaragua
Peñas Blancas
San Carlos
San Juan
Villa Sandino
El Rama
Bluefields
Islas Maíz
Laguna de Perlas
Alamikamba
Grande de Matagalpa
San José de Bocay
Siuna
Rosita
Coco
Waspam
Puerto Cabe
14°
12°
86°
84°

WORLD AFFAIRS

Joined UN in 1945

ACS
Geplac
NAM
OAS
San José

Main issues are debt relief in the wake of Hurricane Mitch, cooperation with neighboring countries for increased US trade access, and the treatment of over 300,000 Nicaraguan immigrants in Costa Rica. A free trade agreement with Mexico is important. A dispute over the common border with Costa Rica was resolved in 2000. Ongoing border and navigation rights disputes exist with Honduras and Colombia.

AID

Recipient

 $562m (receipts)

 Down 17% in 2000

Hurricane Mitch damage produced new World Bank and IDB loans. Cuba, France, Finland, and Spain pardoned all or part of Nicaragua's debt. The US and other Western creditors canceled outstanding obligations when the IMF included Nicaragua in the Highly Indebted Poor Countries initiative.

DEFENSE

Compulsory military service

 $26m

 Up 4% in 2000

FSLN forces once formed the basis of the army, which was cut from a civil war peak of 134,000 to 10,000 by 1995. Senior Sandinistas were among officers retired in 1998. The army is to be involved in more community-based roles focused on the defense of natural resources and mine clearance. Russia agreed in 2001 to help upgrade the military.

ECONOMICS

Inflation 34% p.a. (1990–2000)

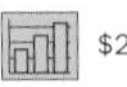 $2.05bn

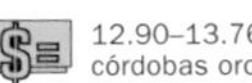 12.90–13.76 córdobas oro

SCORE CARD

- ❑ WORLD GNP RANKING........136th
- ❑ GNP PER CAPITA........$400
- ❑ BALANCE OF PAYMENTS........–$505m
- ❑ INFLATION........11.2%
- ❑ UNEMPLOYMENT........11%

STRENGTHS

Coffee, sugar, and grain exports. Foreign aid and public and private reconstruction work after Hurricane Mitch will benefit tourism, energy, services, and construction.

WEAKNESSES

Heavy debt burden. Main exports subject to commodity price fluctuations. High unemployment. Poor energy supply and infrastructure. Lack of investment and diversification. Weak banks. Delays in privatization. Skewed land ownership and protracted disputes over property. Corruption.

EXPORTS

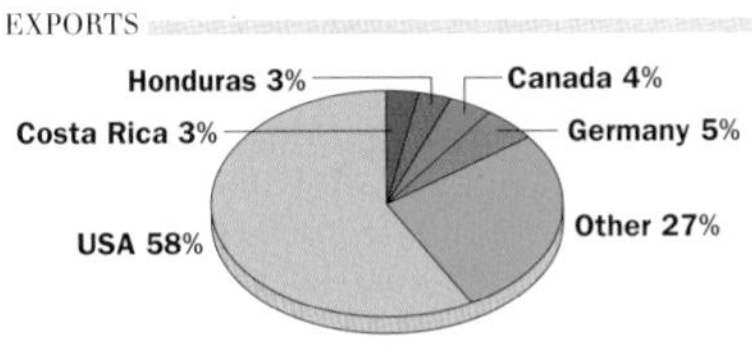
Honduras 3%
Canada 4%
Costa Rica 3%
Germany 5%
USA 58%
Other 27%

IMPORTS

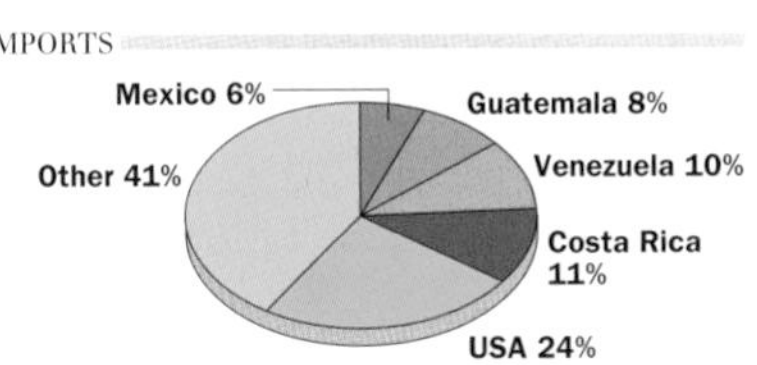
Mexico 6%
Guatemala 8%
Other 41%
Venezuela 10%
Costa Rica 11%
USA 24%

RESOURCES

Electric power 614,000 kw

 24,767 tonnes

 Not an oil producer

 2.28m cattle, 400,000 pigs, 17m chickens

 Gold, silver, lead, zinc, copper, tungsten, salt

Nicaragua has small quantities of gold and silver. New thermal generation projects are planned to overcome energy deficits. There is possible offshore oil.

ENVIRONMENT

Sustainability rank: 52nd

 8%

0.7 tonnes per capita

Deforestation over large areas and the widespread use of pesticides are major problems.

MEDIA

TV ownership medium

Daily newspaper circulation 30 per 1000 people

PUBLISHING AND BROADCAST MEDIA

There are 8 daily newspapers, including *La Prensa* and *El Nuevo Diario*
7 services
62 stations: 1 state-owned, 61 independent

Since the civil war, radio, TV, and newspapers have tended to ally themselves with the government or the opposition; there is little room for political neutrality.

CRIME

No death penalty

 7198 prisoners

 Crime is rising

Former combatants have menaced parts of central and northern regions. Violent crime is rising, as is drug trafficking.

CHRONOLOGY

Nicaragua became an independent state in 1838. Guerrilla forces, led by Gen. Sandino, opposed the US marine presence in the early 1930s.

- ❑ **1978** FSLN, formed in 1961, ends 44-year Somoza dictatorship; ensuing conflict between FSLN and Contras.
- ❑ **1984** Daniel Ortega, FSLN leader, elected president.
- ❑ **1990** End of civil war. Center-right National Opposition Union (UNO) unexectedly wins elections.
- ❑ **1998** Hurricane Mitch causes havoc.
- ❑ **2002** Bolanos elected president.

EDUCATION

School leaving age: 12

 69%

 56,558 students

The Sandinista "Literacy Crusade" achieved dramatic results in the 1980s, but has long since died away.

HEALTH

Welfare state health benefits

 1 per 1111 people

Heart and intestinal infectious diseases, accidents, tuberculosis

Real spending on health fell by 71% between 1988 and 1993, and still has to recover.

SPENDING

GDP/cap. increase

CONSUMPTION AND SPENDING

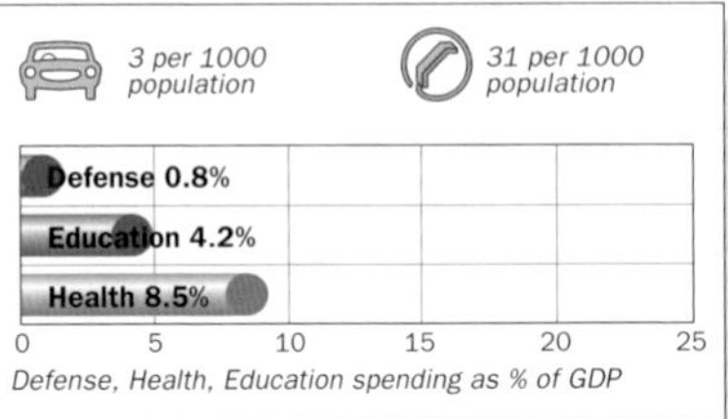
3 per 1000 population
31 per 1000 population
Defense 0.8%
Education 4.2%
Health 8.5%
0
5
10
15
20
25
Defense, Health, Education spending as % of GDP

A 1998 UNDP study revealed that 44% of Nicaragua's population survive on less than $1 a day.

WORLD RANKING

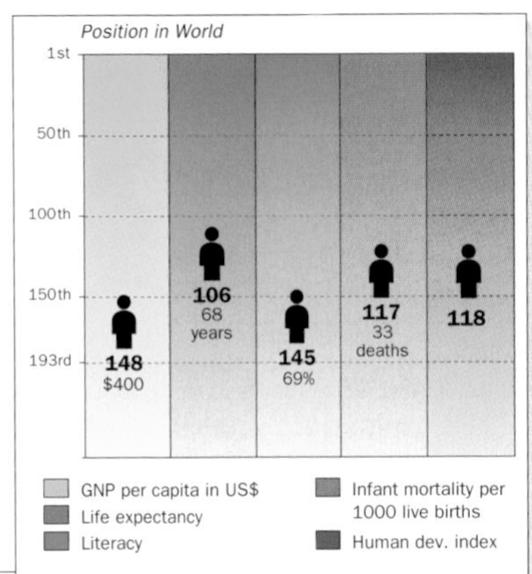
Position in World
1st
50th
100th
150th
193rd
148
$400
106
68 years
145
69%
117
33 deaths
118
GNP per capita in US$
Life expectancy
Literacy
Infant mortality per 1000 live births
Human dev. index

N

NIGER

WEST AFRICA
Algeria Libya Mali NIGER Chad Burkina Nigeria
Africa

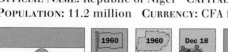

OFFICIAL NAME: Republic of Niger **CAPITAL:** Niamey
POPULATION: 11.2 million **CURRENCY:** CFA franc **OFFICIAL LANGUAGE:** French

1960

1960
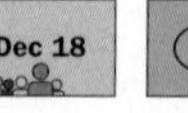
Dec 18

RN
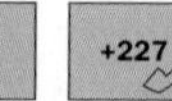
+1
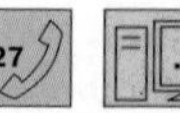
+227

.ne

LANDLOCKED IN THE WEST of Africa, Niger is linked to the sea by the Niger River. Saharan conditions prevail in the northern regions, in the area around the Aïr Mountains, and, particularly, in the vast uninhabited northeast. Niger was ruled by one-party or military regimes until 1992. A much-troubled democratic process was then disrupted by military coups in 1996 and 1999. Niger is one of the poorest countries in Africa.

CLIMATE

Hot desert/steppe

WEATHER CHART FOR NIAMEY

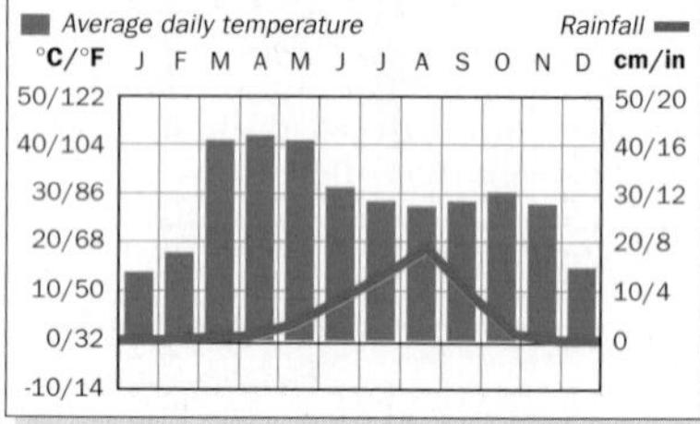
Average daily temperature
Rainfall
°C/°F J F M A M J J A S O N D cm/in
50/122 40/104 30/86 20/68 10/50 0/32 -10/14
50/20 40/16 30/12 20/8 10/4 0

The Saharan north is virtually rainless. The south, in the Sahel belt, has an unreliable rainy season, preceded by a period of extreme daytime heat.

TRANSPORTATION

Drive on right

Niamey International 84,096 passengers

Has no fleet

THE TRANSPORTATION NETWORK

798 km (496 miles)

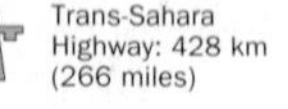

Trans-Sahara Highway: 428 km (266 miles)

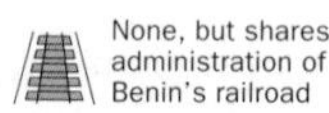

None, but shares administration of Benin's railroad

300 km (186 miles)

A very small proportion of Niger's road network is paved. There are international airports at Niamey and Agadez. There is no railroad.

TOURISM

Visitors : Population 1:224

50,000 visitors

Up 16% in 2000

MAIN TOURIST ARRIVALS

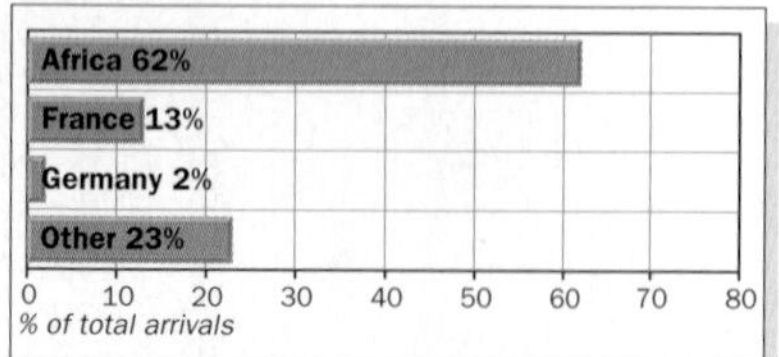
Africa 62%
France 13%
Germany 2%
Other 23%
0 10 20 30 40 50 60 70 80
% of total arrivals

The Aïr Mountains, southern Hausa cities, and Saharan Tuareg culture attract some tourists in spite of Niger's limited infrastructure and its instability.

PEOPLE

Pop. density low

Hausa, Djerma, Fulani, Tuareg, Teda, French

9/km² (23/mi²)

THE URBAN/RURAL POPULATION SPLIT

21% 79%

ETHNIC MAKEUP

Other 6%
Tuareg 9%
Fulani 10%
Djerma and Songhai 21%
Hausa 54%

Considerable tensions exist in Niger between the Tuaregs in the north and the southern groups. The Tuaregs' sense of alienation from mainstream Nigerien politics has increased since the 1973 and 1984 droughts, which disrupted the Tuaregs' nomadic way of life. A five-year rebellion by northern Tuaregs ended in 1995 with a peace agreement. In eastern Niger, Toubou and Arab groups have also been in revolt.

A more subtle antagonism exists between the Djerma and Hausa groups. The Djerma elite from the southwest dominated politics for many years until 1993, when control passed to the Hausa majority.

Niger is an overwhelmingly Islamic society. Women have, on the whole, only limited rights and restricted access to education.

***Testing boating poles in the market** at Ayorou on the Niger River, the country's only major permanent watercourse.*

POLITICS

Multiparty elections

1999/2004

President Mamadou Tandja

AT THE LAST ELECTION

National Assembly 83 seats

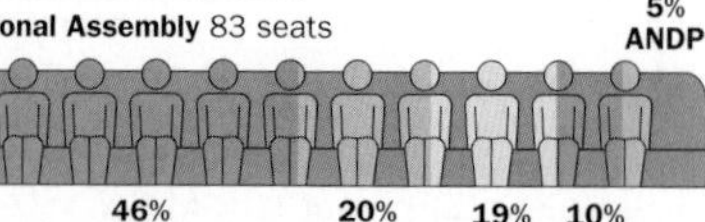
5% ANDP
46% MNSD
20% CDS
19% PNDS
10% RDP

MNSD = National Movement for the Development of Society **CDS** = Democratic and Social Convention **PNDS** = Niger Party for Democracy and Socialism **RDP** = Rally for Democracy and Progress **ANDP** = Niger Alliance for Democracy and Progress

The death in 1987 of the military dictator, President Seyni Kountché, paved the way for prodemocracy demonstrations and eventually led to multiparty elections in 1993. An ensuing power struggle between President Mahamane Ousmane and his political opponents provoked a military coup in 1996. Gen. Ibrahim Barre Mainassara promulgated a new constitution and won a presidential election condemned as fraudulent by the opposition. Mainassara was assassinated by his presidential guard in early 1999. The new military leadership drew up yet another constitution. MNSD leader Mamadou Tandja won the presidential poll later that year. His party, allied with the CDS of former president Ousmane, dominates the new legislature.

WORLD AFFAIRS

Joined UN in 1960

CILSS

ECOWAS

FZ

AU

OIC

Relations with Libya and Algeria have improved since the end of the Tuareg rebellion in 1995. ECOWAS members and the OAU condemned the 1999 coup, as did all key donors, led by France.

AID

Recipient

$211m (receipts)

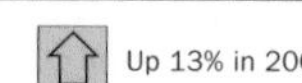

Up 13% in 2000

France is the principal donor, followed by the World Bank and Japan. Most aid was frozen immediately following the 1999 coup, but in late 2000 the IMF approved a three-year loan under its Poverty Reduction and Growth Facility.

DEFENSE

Compulsory military service

US$26m

Down 7% in 2000

Niger's armed forces and paramilitary elements total 10,700. Politics has been dominated by the military since 1974.

N

NIGER

Total Land Area : 1 267 000 sq. km (489 188 sq. miles)

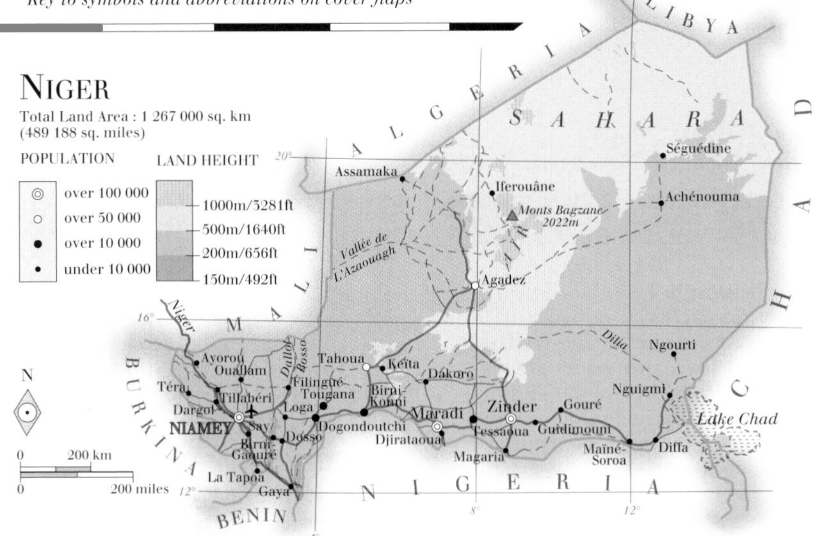
POPULATION
over 100 000
over 50 000
over 10 000
under 10 000
LAND HEIGHT
1000m/3281ft
500m/1640ft
200m/656ft
150m/492ft
0 200 km
0 200 miles
LIBYA
ALGERIA
SAHARA
CHAD
MALI
BURKINA
NIGERIA
BENIN
Séguédine
Achénouma
Assamaka
Iferouâne
Monts Bagzane 2022m
Vallée de l'Azaouagh
Agadez
Niger
Dilia
Ngourti
Ayorou
Ouallam
Tahoua
Keita
Dakoro
Téra
Filingué
Tillabéri
Tougana
Birni-Konni
Nguigmi
Dargol
NIAMEY
Say
Loga
Dosso
Maradi
Zinder
Gouré
Lake Chad
Dogondoutchi
Djirataoua
Tessaoua
Guidimouni
Birni-Gaouré
Magaria
Mainé-Soroa
Diffa
La Tapoa
Gaya

CHRONOLOGY

The powerful Islamic Sokoto empire dissolved as the French took over Niger between 1883 and 1901.

- **1960** Independence.
- **1968** French open uranium mines.
- **1973** Drought; 60% of livestock die.
- **1974** Military coup. Gen. Seyni Kountché bans political parties.
- **1984** New drought; Niger River dries up. Uranium boom ends.
- **1987** Kountché dies. Gen. Ali Saibou eases transition to democracy.
- **1990–1995** Tuareg rebellion.
- **1992** Multiparty constitution.
- **1993** Democratic elections.
- **1996** Military coup. Staged elections.
- **1999** New constitution. Gen. Mainassara assassinated. Multiparty elections won by Mamadou Tandja.
- **2001** Hunting banned in effort to save wildlife.

ECONOMICS

Inflation 6% p.a. (1990–2000)

$1.94bn

698.7–736.7 CFA francs

SCORE CARD

World GNP Ranking	139th
GNP per Capita	$180
Balance of Payments	–$168m
Inflation	2.9%
Unemployment	3%

STRENGTHS

Vast uranium deposits; gold and oil discoveries in late 1990s revived hopes for economic viability.

WEAKNESSES

Aid-dependent. Collapse of uranium prices in 1980s created large debt burden. Only 3% of land cultivable. Weak infrastructure. Frequent droughts. Political instability.

EXPORTS

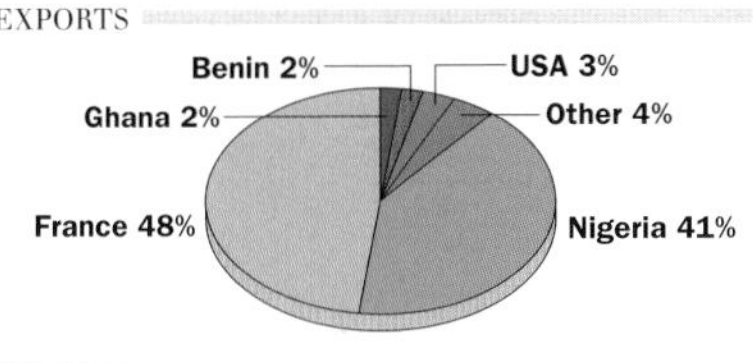
Benin 2%
USA 3%
Ghana 2%
Other 4%
France 48%
Nigeria 41%

IMPORTS

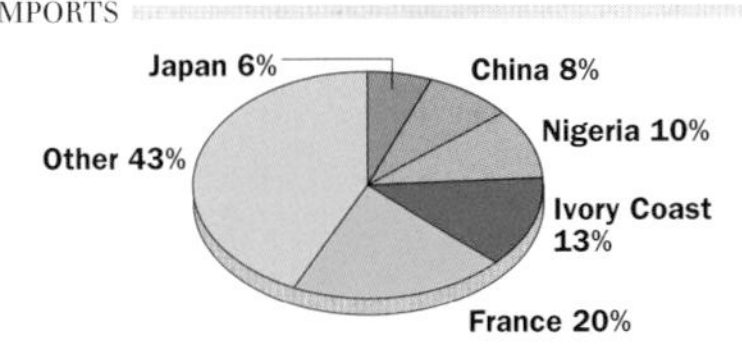
Japan 6%
China 8%
Other 43%
Nigeria 10%
Ivory Coast 13%
France 20%

ENVIRONMENT

Sustainability rank: 123rd

8%

0.1 tonnes per capita

Serious droughts intensify desertification. Hunting was banned in 2001, in an effort to preserve wildlife numbers.

MEDIA

TV ownership low

Daily newspaper circulation 0.2 per 1000 people

PUBLISHING AND BROADCAST MEDIA

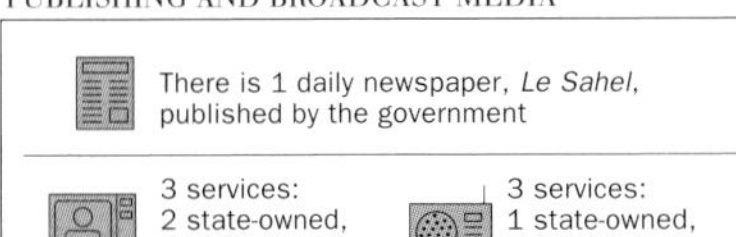
There is 1 daily newspaper, Le Sahel, published by the government
3 services: 2 state-owned, 1 independent
3 services: 1 state-owned, 2 independent

The government controls most broadcasting. The BBC World Service's Hausa programming is influential.

CRIME

Death penalty not used in practice

5263 prisoners

Down 36% 1996–1998

Rural banditry is common, often involving access to grazing and water. Urban crime levels are low, but in border areas smuggling is a way of life.

EDUCATION

School leaving age: 12

16%

4513 students

Local languages are emphasized more strongly than in most francophone states. School attendance is only 30%.

RESOURCES

Electric power 105,000 kw

11,014 tonnes

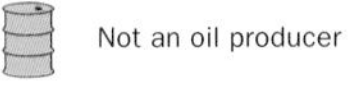
Not an oil producer

6.9m goats, 4.5m sheep, 2.26m cattle, 24m chickens

Uranium, tin, gypsum, coal, salt, tungsten, oil, iron, phosphates, gold

During the 1970s, Niger's uranium mines boomed, but output collapsed in the 1980s when world prices slumped. Other mining is small-scale and oil reserves, discovered in the Lake Chad area, are not yet commercially viable. Salt is a traditionally exploited resource, as are such plant resources as the doum and palmyra palms.

HEALTH

No welfare state health benefits

1 per 20,000 people

Malaria, tuberculosis, meningitis, measles, malnutrition

In spite of progress in rural health care, immunization, malaria control, and child nutrition are still limited.

SPENDING

GDP/cap. increase

CONSUMPTION AND SPENDING

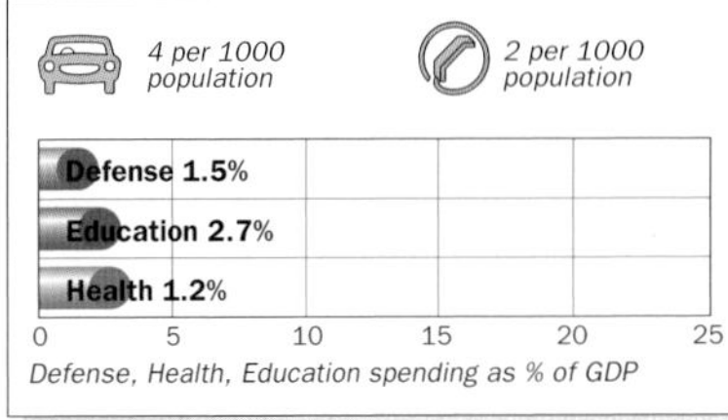
4 per 1000 population
2 per 1000 population
Defense 1.5%
Education 2.7%
Health 1.2%
0
5
10
15
20
25
Defense, Health, Education spending as % of GDP

A small circle of secretive trading families controls much of Niger's wealth. They are successful in evading taxation.

WORLD RANKING

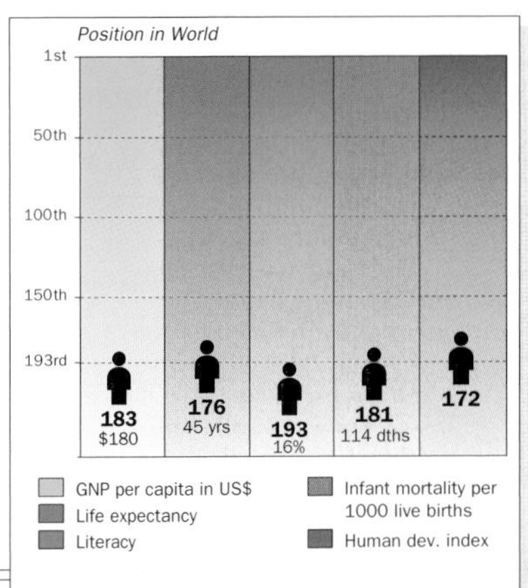
Position in World
1st
50th
100th
150th
193rd
183
$180
176
45 yrs
193
16%
181
114 dths
172
GNP per capita in US$
Life expectancy
Literacy
Infant mortality per 1000 live births
Human dev. index

N

NIGERIA

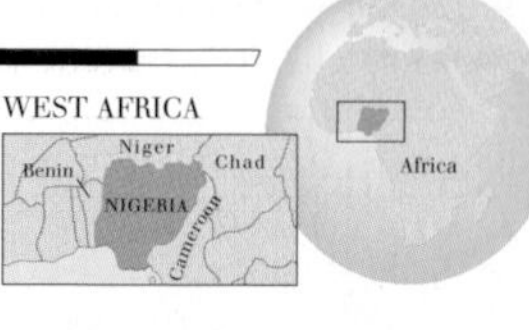
WEST AFRICA
Niger
Chad
Benin
NIGERIA
Cameroon
Africa

OFFICIAL NAME: Federal Republic of Nigeria **CAPITAL:** Abuja
POPULATION: 116.9 million **CURRENCY:** Naira **OFFICIAL LANGUAGE:** English

1960
1961
Oct 1
NGR
+1
+234
.ng

AFRICA'S MOST POPULOUS state, Nigeria gained its independence from Britain in 1960. Bordered by Benin, Niger, Chad, and Cameroon, its terrain varies from tropical rainforest and swamps in the south to savanna in the north. Nigeria has been dominated by military governments since 1966. After many delays, a promised return to civilian rule came about in 1999, with the election as president of Olusegun Obasanjo, a former general who had been head of state from 1976 to 1979. Nigeria is a major OPEC oil producer, but it has experienced a fall in living standards since the 1970s oil boom.

***Village beneath Tengele Peak** in Bauchi State. A large proportion of Nigerians live from subsistence agriculture.*

CLIMATE

Tropical/steppe

WEATHER CHART FOR ABUJA

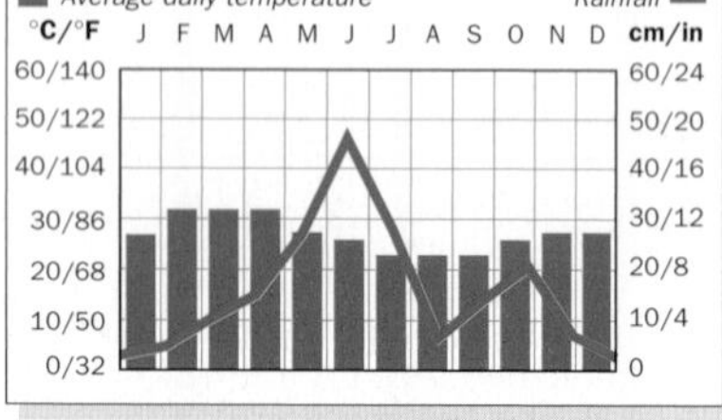
Average daily temperature
Rainfall
°C/°F
J F M A M J J A S O N D
cm/in
60/140
50/122
40/104
30/86
20/68
10/50
0/32
60/24
50/20
40/16
30/12
20/8
10/4
0

The south is hot, rainy, and humid for most of the year. The arid north experiences only one, uncomfortably humid, rainy season from May to September. Its very hot dry season is marked by the *harmattan* wind. The Jos Plateau and the eastern highlands are cooler than the rest of Nigeria. Forcados, in the Niger delta, gets most rain, with 380 cm (150 in.) a year.

TRANSPORTATION

Drive on right

Murtala Muhammad, Lagos
2.7m passengers

293 ships
451,900 grt

THE TRANSPORTATION NETWORK

60,068 km (37,324 miles)

1194 km (742 miles)

3557 km (2210 miles)

8575 km (5328 miles)

Nigeria relies almost entirely on road transportation. During the oil-boom years of the 1970s, new long-distance roads and stretches of freeway were built. The road network is now badly maintained and in urgent need of repair. The road accident rate is among the world's highest and there is severe and chronic traffic congestion in Lagos. In mid-2000 plans for a new trans-Sahara highway from Lagos to Tangiers, Morocco, were announced. Work started in 2001 on a $40 million ports project that will link the five southeastern states.

TOURISM

Visitors : Population 1:144

813,000 visitors

Up 5% in 2000

Nigeria has attempted to build a tourist industry, but numbers remain low. Year-round tropical temperatures and poor infrastructure have limited its growth. The major deterrent to visitors, however, is crime. Travel can be hazardous, and Lagos has one of the world's highest crime rates.

MAIN TOURIST ARRIVALS

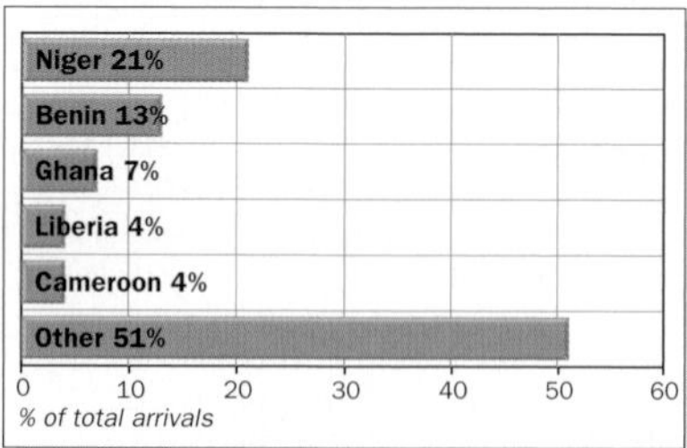
Niger 21%
Benin 13%
Ghana 7%
Liberia 4%
Cameroon 4%
Other 51%
0 10 20 30 40 50 60
% of total arrivals

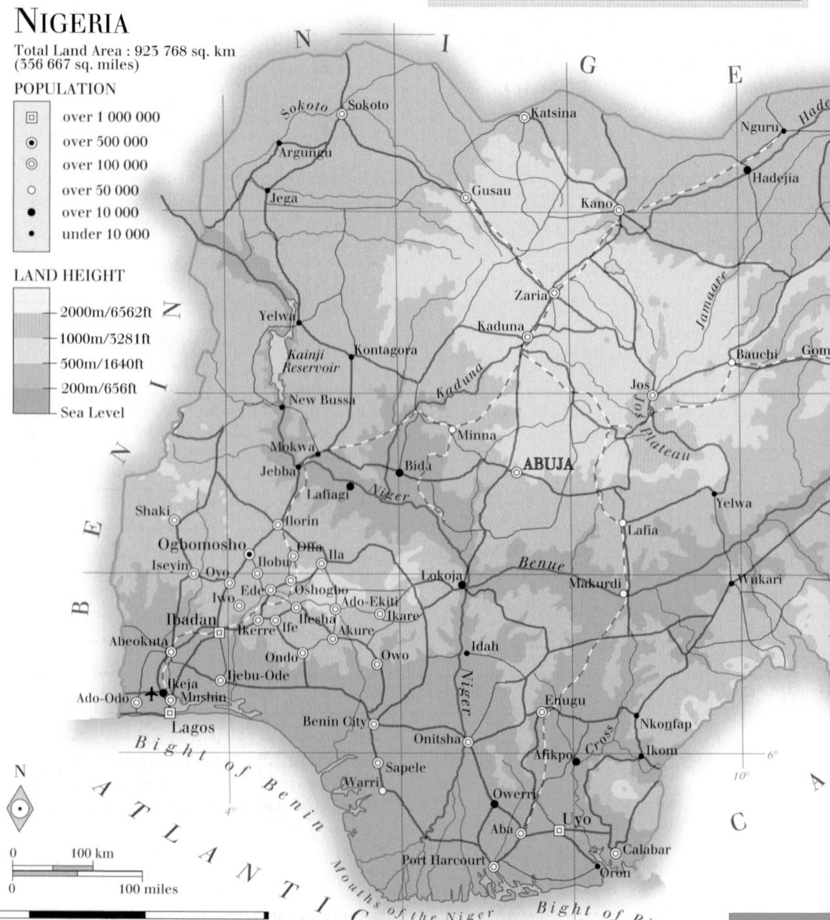
NIGERIA
Total Land Area : 923 768 sq. km (356 667 sq. miles)
POPULATION
over 1 000 000
over 500 000
over 100 000
over 50 000
over 10 000
under 10 000
LAND HEIGHT
2000m/6562ft
1000m/3281ft
500m/1640ft
200m/656ft
Sea Level
0 100 km
0 100 miles
NIGER
BENIN
ATLANTIC OCEAN
Bight of Benin
Bight of Biafra
Mouths of the Niger
Sokoto
Katsina
Nguru
Hadejia
Argungu
Gusau
Kano
Jega
Zaria
Yelwa
Kaduna
Kainji Reservoir
Kontagora
Bauchi
Jos
Jos Plateau
New Bussa
Minna
Mokwa
Jebba
Bida
ABUJA
Lafiagi
Niger
Shaki
Ilorin
Ogbomosho
Offa
Ila
Iseyin
Oyo
Ilobu
Ede
Oshogbo
Iwo
Ado-Ekiti
Ibadan
Ikerre
Ife
Ilesha
Ikare
Akure
Abeokuta
Ondo
Owo
Ijebu-Ode
Ikeja
Mushin
Ado-Odo
Lagos
Lokoja
Idah
Lafia
Benue
Makurdi
Wukari
Yelwa
Enugu
Nkonfap
Benin City
Onitsha
Afikpo
Ikom
Sapele
Warri
Owerri
Aba
Uyo
Calabar
Port Harcourt
Oron
Cross
Jamaare
Kaduna

PEOPLE

Pop. density medium

Hausa, English, Yoruba, Ibo

128/km² (332/mi²)

THE URBAN/RURAL POPULATION SPLIT

44% 56%

RELIGIOUS PERSUASION

Traditional beliefs 10%
Muslim 50%
Christian 40%

ETHNIC MAKEUP

Fulani 11%
Other 29%
Ibo 18%
Yoruba 21%
Hausa 21%

The Hausa, Yoruba, and Ibo are the three main ethnic groups, and there are 245 smaller ones. There is intense ethnic rivalry, and religion is also a source of tension. Outbreaks of violence, particularly in the north, are often caused by clashes between Islamist militants and Christian proselytizers. When in 1999 Zamfara state introduced *sharia* – Islamic law – President Obasanjo, a Christian, would not attend the ceremony. The introduction of *sharia* in other states in the north has sparked violence, leaving over 400 dead in Kaduna in 2000. Religious tensions were exacerbated by the September 2001 attacks on the US, and 500 died in subsequent riots in Jos.

Traditionally, except in the Islamic north, women have had independent economic status. In recent years they have, however, been subjected to some prejudice in professional circles.

POPULATION AGE BREAKDOWN

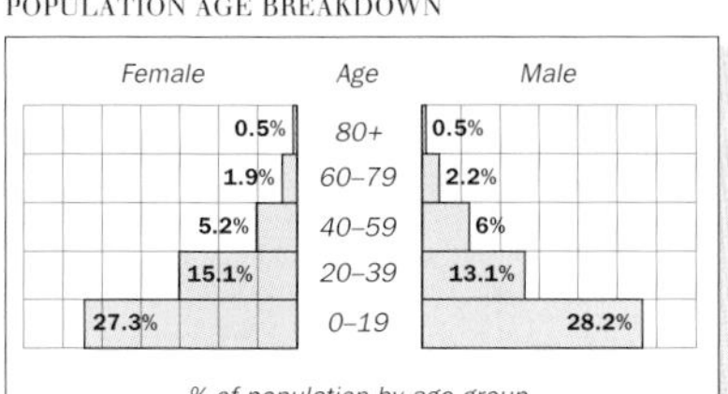
Female
Age
Male
0.5%
80+
0.5%
1.9%
60–79
2.2%
5.2%
40–59
6%
15.1%
20–39
13.1%
27.3%
0–19
28.2%
% of population by age group

POLITICS

Multiparty elections

L. House 1999/2003
U. House 1999/2003

President Olusegun Obasanjo

AT THE LAST ELECTION

House of Representatives 360 seats

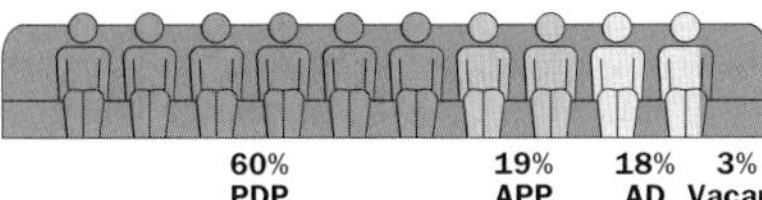

60% PDP	19% APP	18% AD	3% Vacant

PDP = People's Democratic Party **APP** = All People's Party **AD** = Alliance for Democracy

Senate 109 seats

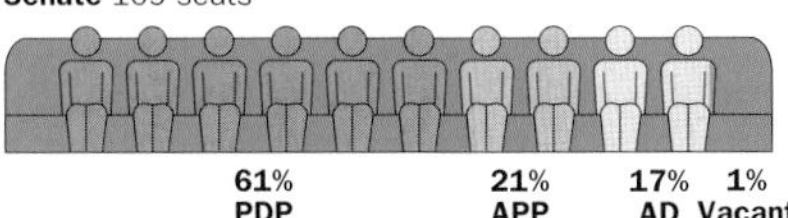

61% PDP	21% APP	17% AD	1% Vacant

Since May 1999 Nigeria has had a civilian constitution, after 16 years of military dictatorships. President Olusegun Obasanjo and his PDP promised national reconciliation.

Profile

The sudden death of military ruler Sani Abacha in 1998, followed by that of the imprisoned Chief Moshood Abiola, the presumed winner of the annulled 1993 presidential elections, left Gen. Abdulsalam Aboubakar to usher in civilian rule. Olusegun Obasanjo, a popular general who was head of state in 1976–1979, won elections held in 1999.

Main Political Issues

Corruption

Corruption has been identified as a major cause of Nigeria's debt levels. Accusations at all levels of government peaked with the impeachment in mid-2000 of the Senate president. President Obasanjo has promised to stamp out corruption.

Ethnic and religious tensions

The Obasanjo regime faces difficulties in reversing the rivalries among the Hausa, Yoruba, and Ibo. The situation deteriorated from early 2000, with ethnic violence involving Yoruba separatists. The adoption of *sharia* law, with strict punishments meted out for crimes such as theft and adultery, by many northern states has provoked rioting between Christians and the majority Muslims. Clashes in Jos in May 2002 left 500 dead.

***Olusegun Obasanjo**, elected president in 1999.*

***Gen. Abacha.** Head of state from 1993 until 1998.*

CHRONOLOGY

Before formal colonization by the British, begun only in 1861, Nigeria was a collection of African states owing their considerable wealth to trans-Saharan and transatlantic trade. During the 18th century the principal commodity was slaves: over 15,000 people a year were exported from the Bight of Benin and another 15,000 from the Bight of Biafra.

- **1885** Royal Niger Company given official responsibility for British sphere of influence along Niger and Benue Rivers. British armed forces coerce local rulers into accepting British rule.
- **1897** West Africa Frontier Force (WAFF) established; subjugation of the north begins.
- **1898** The Royal Niger Company's charter revoked.
- **1900** British Protectorate of Northern Nigeria established.
- **1906** Lagos incorporated into the Protectorate of Southern Nigeria.
- **1914** Protectorates of Northern and Southern Nigeria joined to form colony of Nigeria.
- **1960** Independence. Nigeria established as a federation.
- **1961** Northern part of UK-administered UN Trust Territory of the Cameroons incorporated as part of Nigeria's Northern Region.
- **1966** January, first military coup, led by Maj. Gen. Ironsi. July, countercoup mounted by group of northern army officers. Ironsi murdered. Thousands of Ibo in Northern Region massacred. Gen. Gowon in control of north and west.
- **1967–1970** Civil war. Lt. Col. Ojukwu calls for secession of oil-rich east under the new name Biafra. Over one million Nigerians die before secessionists defeated by federal forces.
- **1970** Gen. Gowon in power.
- **1975** Gowon toppled in bloodless coup. Brig. Mohammed takes power.
- **1976** Mohammed murdered in abortive coup. Succeeded by Gen. Olusegun Obasanjo.
- **1978** Political parties legalized, on condition they represent national, not tribal, interests.
- **1979** Elections won by Alhaji Shehu Shagari and the National Party of Nigeria (NPN), marking return to civilian government.
- **1983** Military coup. Maj. Gen. Mohammed Buhari heads Supreme Military Council.
- **1985** Maj. Gen. Ibrahim Babangida heads bloodless coup, promising a return to democracy.
- **1993** August, elections annulled; Babangida resigns; military sets

N

CHRONOLOGY *continued*

up Interim National Government (ING). November, ING dissolved. Military, headed by Gen. Sani Abacha, takes over.

- **1994** Moshood Abiola arrested, opposition harassed.
- **1995** Ban on parties lifted. Obasanjo and 39 others convicted of plotting coup. Execution of Ken Saro-Wiwa and eight other Ogoni activists: EU sanctions, suspension of Commonwealth membership.
- **1998** Abacha dies; Abiola dies; return to civilian rule timetabled.
- **1999** Elections: presidency won by Obasanjo. Sanctions lifted, Commonwealth membership restored. Zamfara becomes first state to introduce *sharia*.
- **2000** Ethnic violence escalates, threatens national unity.
- **2001** 200 villagers massacred by army in apparent revenge killing.
- **2002** 1000 killed in Lagos munitions dump explosion and ensuing chaos. 500 killed in clashes between Muslims and Christians in Jos.

WORLD AFFAIRS

Joined UN in 1960

Comm

ECOWAS

AU

OIC
OPEC

Liking to be regarded as Africa's leading voice, Nigeria is a keen sponsor of ECOWAS and of the AU, and has been the main contributor to ECOMOG forces. However, the military regime's violations of human rights (most notoriously the execution of Ken Saro-Wiwa and eight other Ogoni activists in 1995) prompted UN condemnation and suspension from the Commonwealth. Membership was restored only after Obasanjo came to power in 1999.

Nigeria was an implacable opponent of apartheid, and relations with South Africa were restored only after democratic elections there in 1994. A maritime border dispute with Equatorial Guinea was settled in 2001, opening up the further exploitation of undersea oil reserves.

AID

Recipient

$185m (receipts)

Up 22% in 2000

The 1981 drop in world oil prices turned Nigeria from an aid donor into a major receiver of World Bank assistance. Aid flows interrupted in late 1995 were resumed in 2000. Aid agreed under the G8 summit in June 2002 was welcomed by President Obasanjo – one of the architects of the New Partnership for Africa's Development (NEPAD) plan.

DEFENSE

No compulsory military service

$2.34bn

Up 5% in 2000

The Nigerian government has contributed a significant number of forces to ECOMOG, the regional peacekeeping force set up in 1990 by ECOWAS, initially to restore order in Liberia. Nigerian forces served there until 1999, and in Sierra Leone in 1993–2000 and from 2002.

Since 2001 the armed forces have also been deployed to restore internal control after outbreaks of religious unrest in northern and central states, in which hundreds of civilians have been killed. However, government troops themselves were reported to have massacred some 200 villagers in the eastern state of Benue in October 2001, in retaliation for the abduction of 19 members of the armed forces who had been deployed to quell unrest.

NIGERIAN ARMED FORCES

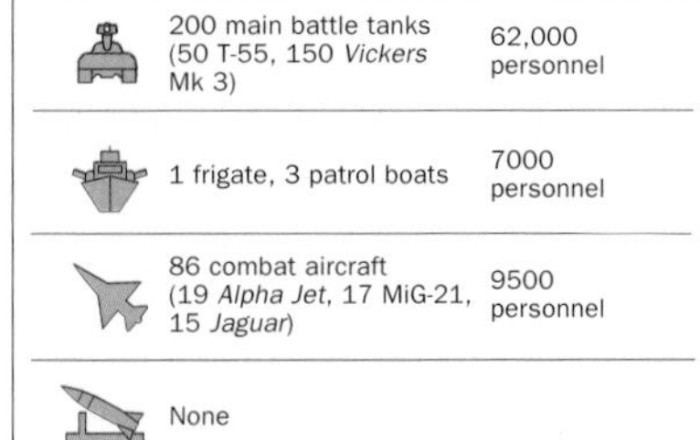
200 main battle tanks (50 T-55, 150 *Vickers* Mk 3)
62,000 personnel
1 frigate, 3 patrol boats
7000 personnel
86 combat aircraft (19 *Alpha Jet*, 17 MiG-21, 15 *Jaguar*)
9500 personnel
None

ECONOMICS

Inflation 29% p.a. (1990–2000)

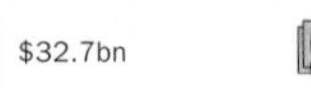
$32.7bn
110–120 naira

SCORE CARD

- World GNP Ranking..........57th
- GNP per Capita..........$260
- Balance of Payments..........$6.98bn
- Inflation..........6.9%
- Unemployment..........28%

EXPORTS

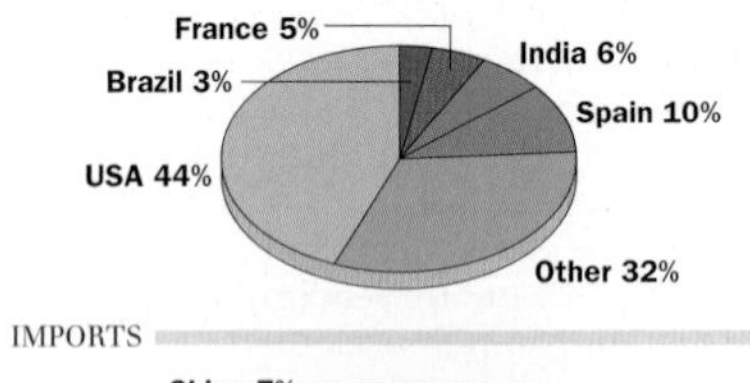
France 5%
India 6%
Brazil 3%
Spain 10%
USA 44%
Other 32%

IMPORTS

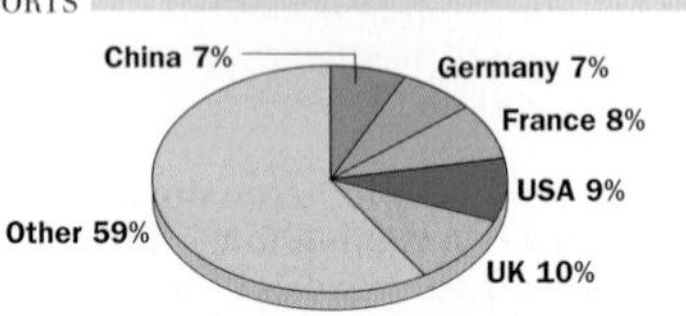
China 7%
Germany 7%
France 8%
USA 9%
Other 59%
UK 10%

Strengths

One of world's top oil producers at 2.15 million b/d. Vast reserves of natural gas, still only partly exploited. Soaring world oil prices in 2000 signified recovery from 1986 collapse. Almost self-sufficient in food. Strong entrepreneurial class. Large domestic market.

Weaknesses

Overdependence since the 1970s on oil, which encourages massive state inefficiency. Advantages of a large domestic market mitigated by low per capita purchasing power and high unit transportation costs. Entrepreneurs focus on trade rather than production. Of Nigeria's traditional agricultural exports only cocoa remains. Notorious corruption and maladministration undermine investors' confidence.

ECONOMIC PERFORMANCE INDICATOR

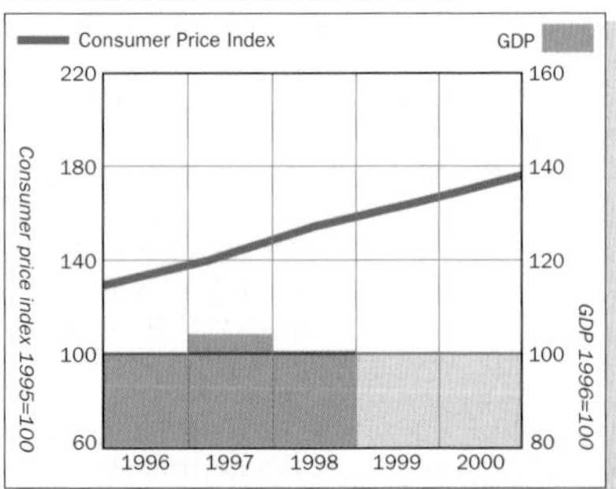
Consumer Price Index
GDP
220
180
140
100
60
160
140
120
100
80
Consumer price index 1995=100
GDP 1996=100
1996
1997
1998
1999
2000

Profile

The economy has been characterized by massive government spending and the running up of debts which could not be serviced after the 1981 oil price fall. Led by the IMF, creditors want major cuts in spending – especially on loss-making public-sector companies – and subsidies. Gasoline subsidies alone are estimated to have cost $2.4 billion a year. Such changes are politically fraught. In May 1999 the government began an anticorruption drive.

NIGERIA : MAJOR BUSINESSES

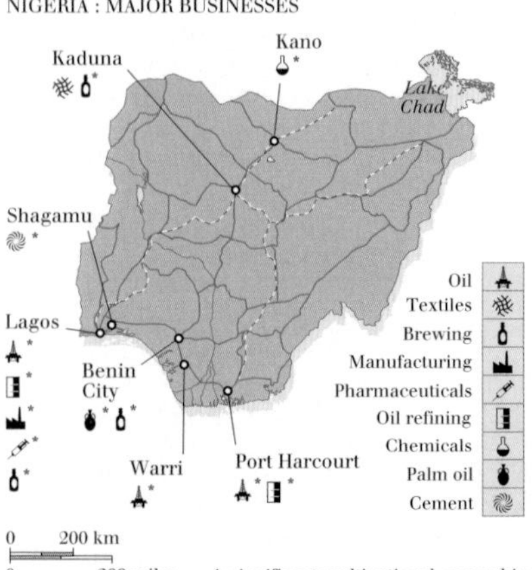
Kano
Kaduna
Lake Chad
Shagamu
Lagos
Benin City
Warri
Port Harcourt
Oil
Textiles
Brewing
Manufacturing
Pharmaceuticals
Oil refining
Chemicals
Palm oil
Cement
0 200 km
0 200 miles
* significant multinational ownership

RESOURCES

Electric power 5.9m kw

477,365 tonnes

2.15m b/d (reserves 24bn barrels)

24.3m goats, 20.5m sheep, 19.8m cattle, 126m chickens

Oil, natural gas, coal, tin, iron, bauxite, columbite, lead

ELECTRICITY GENERATION

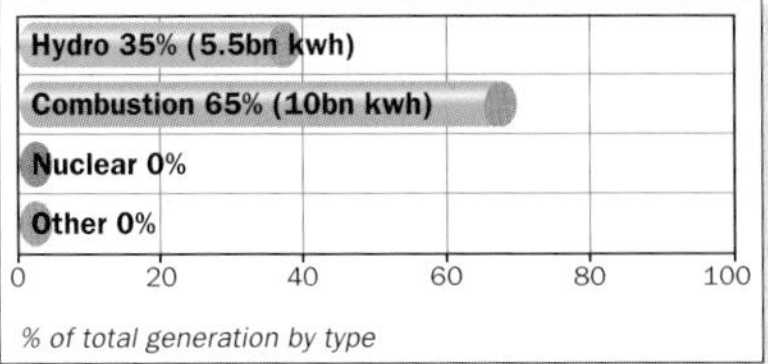
Hydro 35% (5.5bn kwh)
Combustion 65% (10bn kwh)
Nuclear 0%
Other 0%
0
20
40
60
80
100
% of total generation by type

Oil has been Nigeria's main resource since the 1970s. Government policy is to increase output to 3 million b/d by 2003. Domestic demand is 300,000 b/d, much of it smuggled to neighboring countries. Nigeria's vast gas deposits are still underexploited. The state retains 60% control of the oil and gas industry. Shell is the main foreign shareholder, but most oil multinationals are represented.

Nigeria has sizable iron ore deposits. These are not yet utilized in the state-run steel industry; imported ore is used instead. Bauxite deposits are also currently underexploited. There are, however, plans for establishing an aluminum industry. Nigeria also has deposits of coal and tin.

NIGERIA : LAND USE

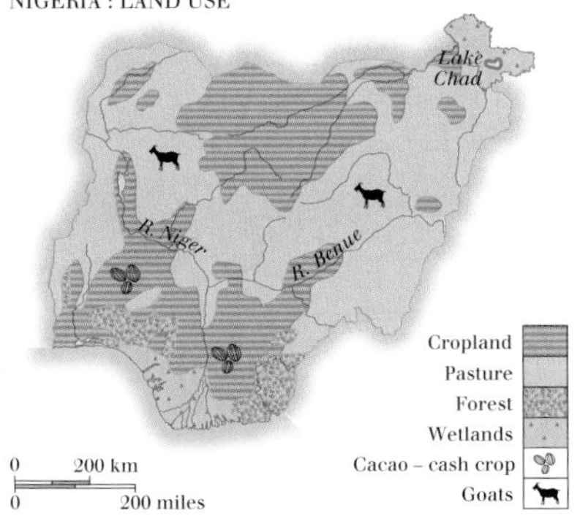
Lake Chad
R. Niger
R. Benue
Cropland
Pasture
Forest
Wetlands
Cacao – cash crop
Goats
0 200 km
0 200 miles

ENVIRONMENT

Sustainability rank: 133rd

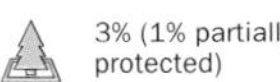
3% (1% partially protected)

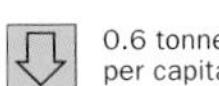
0.6 tonnes per capita

ENVIRONMENTAL TREATIES

Yes
Yes
Yes
Yes
Yes
No

Oil industry pollution in the Niger delta, a major local concern, came to international attention in 1995. Shell has been particularly condemned. Before the discovery of a highly toxic cargo in 1988, Nigeria was a dumping ground for European chemical waste.

MEDIA

TV ownership medium

Daily newspaper circulation 27 per 1000 people

The Nigerian press is traditionally one of Africa's liveliest. Media freedom has improved since the return to civilian government. There are about 90 TV stations; the state-run Nigerian Television Authority runs 32.

PUBLISHING AND BROADCAST MEDIA

There are 25 daily newspapers. The *Daily Times*, published by the government, has the highest circulation
1 state-controlled service, about 60 private stations
5 services: 2 state-controlled, 3 independent

CRIME

Death penalty in use

44,450 prisoners

Crime is rising

CRIME RATES

Murders
94 per 100,000 population

Rapes
Nigeria does not publish rape statistics

Thefts
1256 per 100,000 population

Nigeria has one of the highest crime rates in the world. Murder often accompanies even minor burglaries. Rich Nigerians live in high-security compounds. A special armed police unit was formed to restore order in Lagos in 2002. It is claimed that brutal *sharia* punishments have reduced crime in northern states. Vigilante gangs are used to fight crime in the southeastern states.

EDUCATION

School leaving age: 15

64%

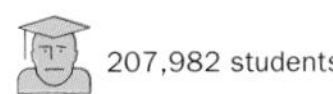
207,982 students

THE EDUCATION SYSTEM

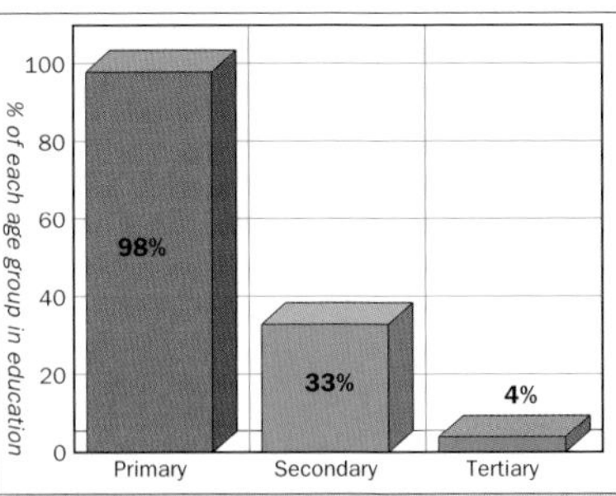
100
80
60
40
20
0
% of each age group in education
98%
33%
4%
Primary
Secondary
Tertiary

Responsibility for education is shared between the federal and the state governments. Education has suffered from the government's massive debt repayment burden. During the oil-boom years, Nigeria concentrated on creating 31 universities with prestigious medical and scientific schools. However, standards in primary education, which has not received the same level of investment, have fallen since the 1970s. Almost all children attend primary school, but only about 30% receive secondary education.

HEALTH

Welfare state health benefits

1 per 5000 people

Yellow fever, malaria, trachoma, yaws

The health service functions mainly in urban areas and has suffered from the crisis in government revenues; a 1999 government report stated that only 49% of the population had access to water and health services. More than 3.5 million Nigerians were living with HIV/AIDS as at end-2001. Free generic versions of AIDS treatments became available in 2001 and the government supports a successful AIDS awareness strategy.

SPENDING

GDP/cap. increase

CONSUMPTION AND SPENDING

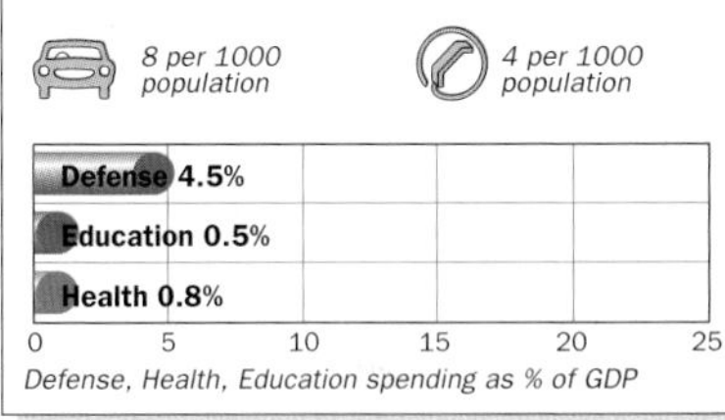
8 per 1000 population
4 per 1000 population
Defense 4.5%
Education 0.5%
Health 0.8%
0
5
10
15
20
25
Defense, Health, Education spending as % of GDP

Nigerians with access to the rich pickings of political office spent on a massive scale during the country's oil boom – on expensive cars and on overseas education for their children. Much was financed by government loans. Habits have not changed with the fall in oil revenues: borrowing has simply grown.

WORLD RANKING

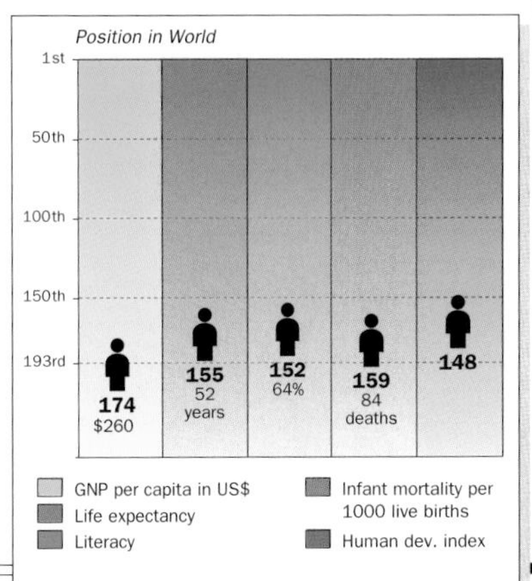
Position in World
1st
50th
100th
150th
193rd
174
$260
155
52 years
152
64%
159
84 deaths
148
GNP per capita in US$
Life expectancy
Literacy
Infant mortality per 1000 live births
Human dev. index

N

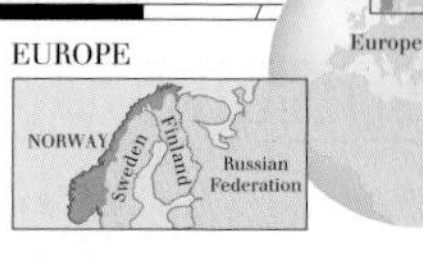
EUROPE
Europe
NORWAY
Sweden
Finland
Russian Federation

NORWAY

OFFICIAL NAME: Kingdom of Norway **CAPITAL:** Oslo
POPULATION: 4.5 million **CURRENCY:** Norwegian krone **OFFICIAL LANGUAGE:** Norwegian

1905

1905

May 17
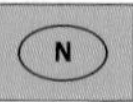
N
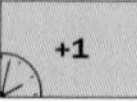
+1

+47

.no

OCCUPYING THE WESTERN PART of Scandinavia, Norway's western coastline is characterized by numerous fjords and islands. Large oil and gas revenues have brought prosperity. Gro Harlem Brundtland, Norway's first woman prime minister, went on to take top UN posts. Despite the Europe-wide recession in the early 1990s, Norway was able to contain rising unemployment, which peaked at 6% in 1993. A constitutional requirement is that government creates conditions that enable every person to find work.

***The village of Reine** on Moskenesøya, deep inside the Arctic Circle in the Lofoten Islands. It is a popular destination for summer visitors.*

CLIMATE

Maritime/subarctic

WEATHER CHART FOR OSLO

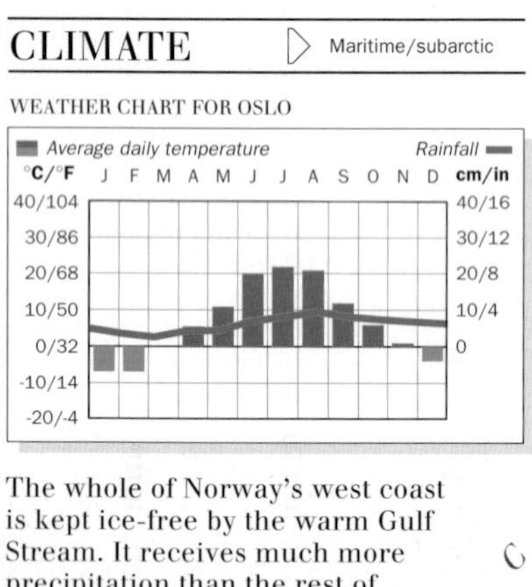
Average daily temperature
Rainfall
°C/°F
J F M A M J J A S O N D
cm/in
40/104
30/86
20/68
10/50
0/32
-10/14
-20/-4
40/16
30/12
20/8
10/4
0

The whole of Norway's west coast is kept ice-free by the warm Gulf Stream. It receives much more precipitation than the rest of the country; Bergen has a yearly average of 225 cm (89 in). Norway enjoys the highest mean temperatures in Scandinavia, but in winter the temperature in Oslo can drop to –25°C (–13°F).

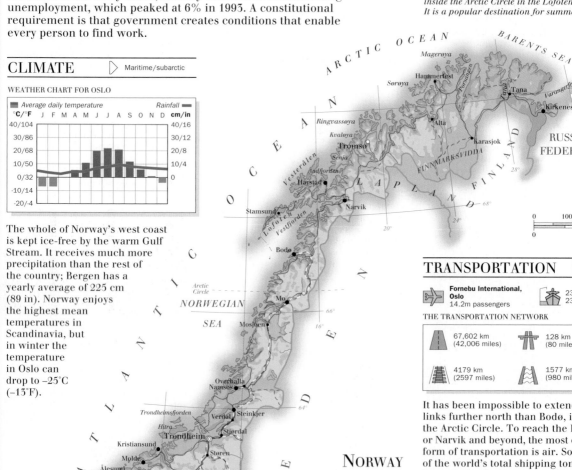

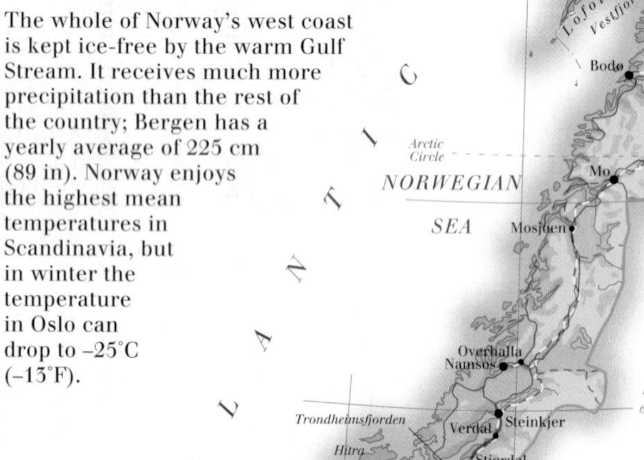

ARCTIC OCEAN
BARENTS SEA
ATLANTIC OCEAN
NORWEGIAN SEA
NORTH SEA
Skagerrak
SWEDEN
FINLAND
LAPLAND
FINNMARKSVIDDA
RUSSIAN FEDERATION
Magerøya
Hammerfest
Sørøya
Porsangen
Tana
Varangerfjorden
Kirkenes
Ringvassøya
Alta
Kvaløya
Tromsø
Karasjok
Senja
Vesterålen
Andfjorden
Harstad
Narvik
Stamsund
Lofoten
Vestfjorden
Bodø
Arctic Circle
Mo
Mosjøen
Overhalla
Namsos
Trondheimsfjorden
Verdal
Steinkjer
Hitra
Trondheim
Stjørdal
Kristiansund
Molde
Støren
Ålesund
Røros
Oppdal
Ørsta
DOVREFJELL
Glomma
Glittertind 2470m
JOTUNHEIMEN
Lågen
Hermansverk
Sognefjorden
Lillehammer
Gjøvik
Elverum
Voss
Hamar
Gol
Stange
Mjøsa
Bergen
Begna
Eidsvoll
Hardangerfjorden
Hønefoss
Kongsvinger
OSLO
Haukeligrend
Kongsberg
Lillestrøm
Drammen
Åkrehamn
Haugesund
Notodden
Horten
Moss
Boknafjorden
Oslofjorden
Skien
Sarpsborg
Stavanger
Porsgrunn
Halden
Sola
Sandnes
Otra
Larvik
Fredrikstad
Sandefjord
Tønsberg
Egersund
Arendal
Vennesla
Mandal
Kristiansand
70°
68°
66°
64°
62°
60°
58°
32°
28°
24°
20°
16°
12°
8°
N
0 100 km
0 100 miles

TRANSPORTATION

Drive on right

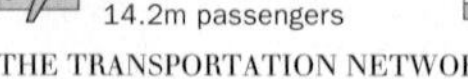

Fornebu International, Oslo
14.2m passengers

2325 ships
23.1m grt

THE TRANSPORTATION NETWORK

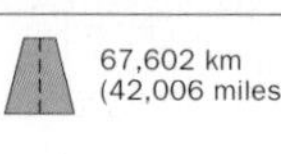
67,602 km (42,006 miles)

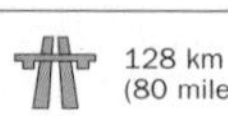
128 km (80 miles)

4179 km (2597 miles)

1577 km (980 miles)

It has been impossible to extend rail links further north than Bodø, inside the Arctic Circle. To reach the Lofotens or Narvik and beyond, the most common form of transportation is air. Some 5% of the world's total shipping tonnage is controlled by Norway, making it the sixth-largest shipping nation.

NORWAY

Total Land Area : 324 220 sq. km (125 181 sq. miles)

LAND HEIGHT

2000m/6562ft
1000m/3281ft
500m/1640ft
200m/656ft
Sea Level

POPULATION

over 100 000
over 50 000
over 10 000
under 10 000

***The royal palace, Oslo.** This is situated near the national theater, at one end of the Karl Johanisgate, the city's main thoroughfare.*

TOURISM

Visitors : Population 1:1

4.35m visitors

Down 3% in 2000

Norway is a popular destination, receiving visitors from Germany, Sweden, the UK, Denmark, and the US. Its winter tourism industry is based on skiing and was boosted by the 1994 Winter Olympics taking place in Lillehammer. Cruising along the fjords is popular with summer visitors. Areas within the Arctic Circle are a particular attraction in June, when tourists go in search of the midnight sun. Oslo has a reputation for good classical music and jazz. However, the strength of the krone and the high cost of living make Norway expensive.

MAIN TOURIST ARRIVALS

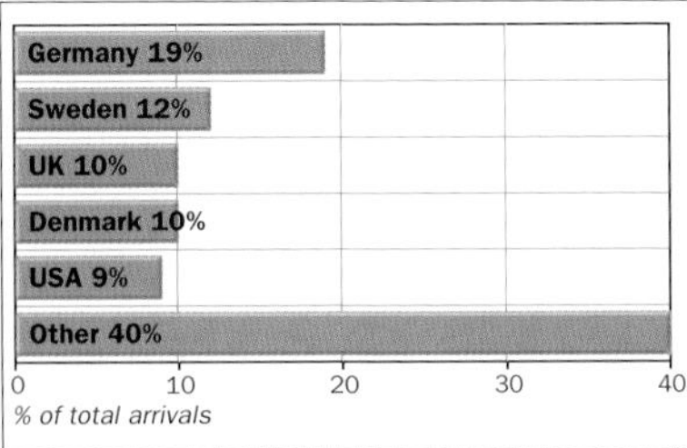
Germany 19%
Sweden 12%
UK 10%
Denmark 10%
USA 9%
Other 40%
0
10
20
30
40
% of total arrivals

PEOPLE

Pop. density low

Norwegian (*Bokmål* "book language" and *Nynorsk* "new Norsk"), Sami

15/km² (38/mi²)

THE URBAN/RURAL POPULATION SPLIT

76% 24%

RELIGIOUS PERSUASION

Roman Catholic 1%

Other and nonreligious 10%

Evangelical Lutheran 89%

ETHNIC MAKEUP

Sami 1%

Other 6%

Norwegian 93%

Norway has a small but growing immigrant population, forming 5.6% of the population in 2001. In the early 1990s the number of asylum seekers increased, especially as a result of the Bosnian conflict. Some refugees have been attacked by right-wing groups.

The family is traditionally close and nuclear. Men are expected to share responsibility for raising children, who frequently attend day schools from under the age of two years. Women enjoy considerable power and freedom, and comprise at least 40% of members of the government. Over half of marriages end in divorce.

POPULATION AGE BREAKDOWN

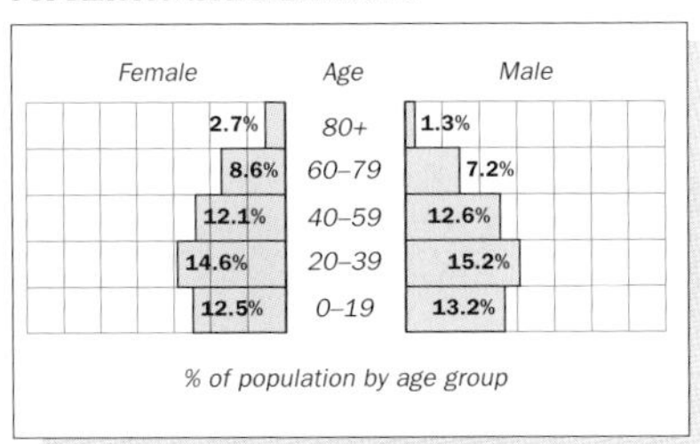
Female
Age
Male
2.7% 80+ 1.3%
8.6% 60–79 7.2%
12.1% 40–59 12.6%
14.6% 20–39 15.2%
12.5% 0–19 13.2%
% of population by age group

POLITICS

Multiparty elections

2001/2005

H.M. King Harald V

AT THE LAST ELECTION

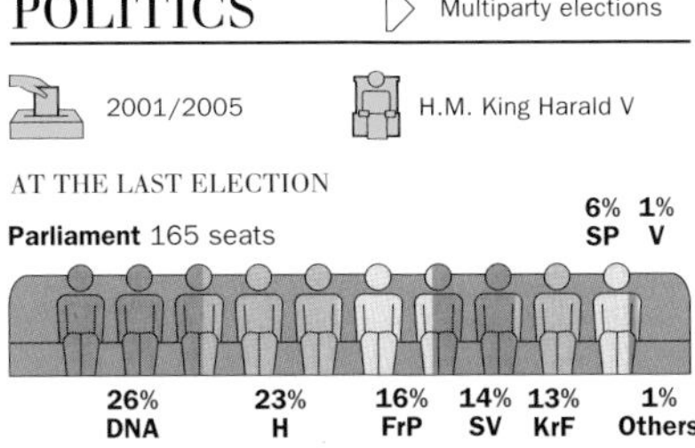
Parliament 165 seats
6% SP
1% V
26% DNA
23% H
16% FrP
14% SV
13% KrF
1% Others

DNA = Norwegian Labor Party **H** = Hoeyre (Conservative Party) **FrP** = Progress Party **SV** = Socialist Left Party **KrF** = Christian Democratic Party **SP** = Center Party **V** = Venstre (Liberal Party)

The Parliament (Storting) is elected as one body but divides itself for most legislative purposes into an upper chamber (Lagting, with 42 members) and a lower chamber (Odelsting, with 123 members)

Norway is a constitutional monarchy, with a king as head of state and an elected parliament.

PROFILE

Political decisions are based on consensus building between the government, parliament, and the strong trade unions. In 1993 opposition to EU membership boosted support for the SP but the DNA remained the largest party.

In 1997 the DNA lost ground to the center-right, who formed a coalition led by Kjell Magne Bondevik of the KrF. The DNA returned to office in 2000, but received heavy losses in the 2001 polls. Bondevik, who had campaigned on a platform of using oil revenues to fund tax cuts and public service improvements, returned as prime minister. A notable feature of the election was the capture of 26 seats by the anti-immigration FrP, which had emerged during the 1990s.

MAIN POLITICAL ISSUE

EU membership

In a 1994 referendum, 52% voted against EU membership, previously rejected in 1972. Terms for accession, supported by government and industry, had been agreed, but opponents argued successfully that it would lead to a loss of control of national resources, notably fisheries and offshore oil. Membership was not an electoral issue in 2001.

WORLD AFFAIRS

Joined UN in 1945

CE
NATO
OECD
OSCE
EEA

A founder member of NATO, Norway continues to offer it strong support. Norway also became an associate member of the WEU in 1992.

The 1994 referendum decision to reject EU membership means that the European Economic Area (EEA) offers Norway its chief access to the European single market. As a Nordic Council member it is also associated with the Schengen Convention. In January 2002 Bondevik stated that there would be no referendum on joining the EU. However, a concurrent opinion poll suggested that a majority now favored membership.

Norway has played peacemaker in a number of major international conflicts, notably the Israeli–Palestinian conflict during the mid-1990s, and in Sri Lanka in 2002.

The government has been unable to control the ecological effects of acid rain, which is destroying its forests. Representatives of 25 European countries and Canada met in Oslo in 1994 and signed a UN protocol on reducing sulfur emissions.

***King Harald V**, who succeeded his father King Olaf V in 1991.*

***Prime Minister Kjell Magne Bondevik**, reelected in 2002.*

CHRONOLOGY

Norway gained independence from the Swedish crown in 1905 and elected its own king, Håkon VII.

- **1935** DNA forms government.
- **1940–1945** Nazi occupation. Puppet regime led by Vidkun Quisling.
- **1945** DNA resumes power.
- **1949** Founder member of NATO.
- **1957** King Håkon dies. Succeeded by son, Olaf V.
- **1960** Becomes member of EFTA.
- **1962** Unsuccessfully applies to join the European Communities (EC).
- **1965** DNA electoral defeat by SP coalition led by Per Borten.
- **1967** Second bid for EC membership.
- **1971** Prime Minister Per Borten resigns following disclosure of secret negotiations to join EC. DNA government, led by Trygve Bratteli.

N

CHRONOLOGY *continued*

- ❑ **1972** EC membership rejected in popular referendum by 5% majority. Bratteli resigns. Center coalition government takes power. Lars Korvald prime minister.
- ❑ **1973** Elections. Bratteli returns to power as prime minister.
- ❑ **1976** Bratteli succeeded by Odvar Nordli.
- ❑ **1981** Nordli resigns owing to ill health. Gro Harlem Brundtland becomes first woman prime minister. Elections bring to power Norway's first Conservative Party (H) government for 53 years. Kare Willoch prime minister.
- ❑ **1983** Conservatives form coalition with SP and KrF.
- ❑ **1985** Election. Willoch's H–SP–KrF coalition returned. Norway agrees to suspend commercial whaling.
- ❑ **1986** 100,000 demonstrate for better working conditions. Brundtland forms minority DNA government. Currency devalued by 12%.
- ❑ **1989** Brundtland resigns. H–KrF coalition in power. Soviet Union agrees exchange of information after fires on Soviet nuclear submarines off Norwegian coast.
- ❑ **1990** H–KrF coalition breaks up over closer ties with EU (formerly EC). Brundtland and DNA in power.
- ❑ **1991** Olaf V dies; succeeded by son, King Harald V.
- ❑ **1994** EEA comes into effect. Referendum rejects EU membership.
- ❑ **1996** Brundtland resigns; replaced by Thorbjørn Jagland (also DNA).
- ❑ **1997** Kjell Magne Bondevik forms center-right coalition.
- ❑ **2000** Jens Stoltenberg (DNA) heads three-party coalition.
- ❑ **2001** Right-wing victory in elections. Bondevik heads coalition government.

AID

Donor

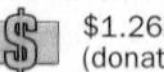
$1.26bn (donations)

Down 8% in 2000

Norway has been granting more than the UN development target of 0.7% of GNP in aid every year since 1975. Although Norway's ratio of aid to GNP has declined somewhat, to 0.92% in 1999, it remains one of the highest in the world. The vast majority of Norway's bilateral aid goes to the least developed countries of southeast Africa, south Asia, and Central America, though Palestine and Bosnia are also important recipients. The Norwegian government allocates funds to various multilateral assistance programs, as well. The 1999 budget included a debt relief program to help reduce developing country indebtedness.

DEFENSE

Compulsory military service

$2.86bn

Down 12% in 2000

Plans have been announced to almost halve Norway's conscript army, which traditionally has absorbed most of the defense budget. Norway joined NATO in 1949. The overriding defense issue is the stability of Russia and the security of their common border. Five Russian diplomats were expelled in 1998 after a double agent revealed that Russia had extensive information on Norwegian defenses and oil industry.

NORWEGIAN ARMED FORCES

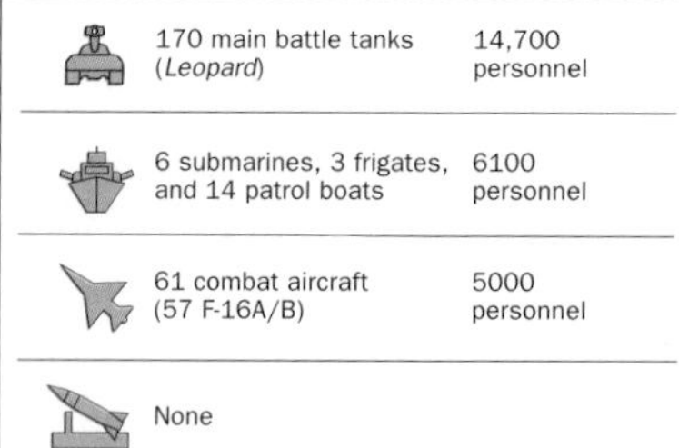

170 main battle tanks (*Leopard*)	14,700 personnel
6 submarines, 3 frigates, and 14 patrol boats	6100 personnel
61 combat aircraft (57 F-16A/B)	5000 personnel
None	

ECONOMICS

Inflation 2.8% p.a. (1990–2000)

$155bn

8.819–8.969 Norwegian kroner

SCORE CARD

❑ World GNP Ranking	26th
❑ GNP per Capita	$34,530
❑ Balance of Payments	$23bn
❑ Inflation	3.1%
❑ Unemployment	3%

EXPORTS

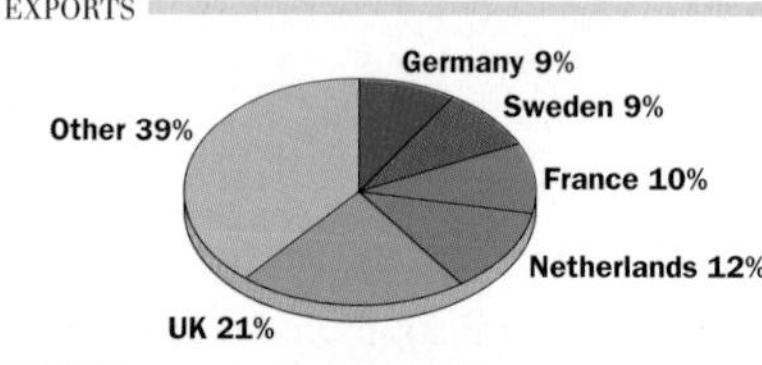
Germany 9%
Sweden 9%
France 10%
Netherlands 12%
UK 21%
Other 39%

IMPORTS

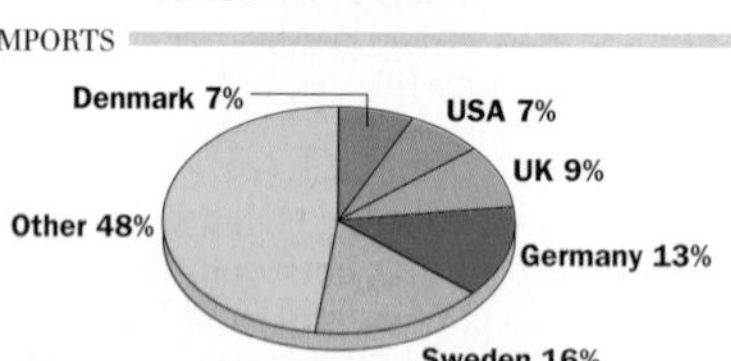
Denmark 7%
USA 7%
UK 9%
Germany 13%
Sweden 16%
Other 48%

Strengths

Western Europe's biggest producer and exporter of oil and natural gas. Mineral reserves. Hydroelectric power satisfies much of country's energy demands, allowing most oil to be exported. Soaring world oil prices in 2000. Petroleum fund for current profits to provide for future generations. Large merchant shipping fleet. Low inflation and unemployment compared with rest of Europe.

Weaknesses

Overdependence on oil revenue. Small home market and relatively remote location. Shortage of skilled labor. Harsh climate limits agriculture.

Profile

The state is interventionist by nature. In 1991, it stepped in to rescue most of the main commercial banks, which had been hit by bad loans. It began returning them to the private sector in 1994. The state also manages the distribution of offshore oil and gas licenses, and owns 45% of the Norsk Hydro conglomerate. In 2000 the state sold a 20% share in its oil and telecommunications companies.

Norway's immediate future prosperity is guaranteed by its offshore sector. There is a shortage of skilled labor, partly eased by the arrival of workers from other Scandinavian countries, but creating upward pressure on wages. Continuing the strong regional policy of redirecting resources from the more prosperous south to the isolated north is likely to remain a priority, both for social and strategic reasons.

ECONOMIC PERFORMANCE INDICATOR

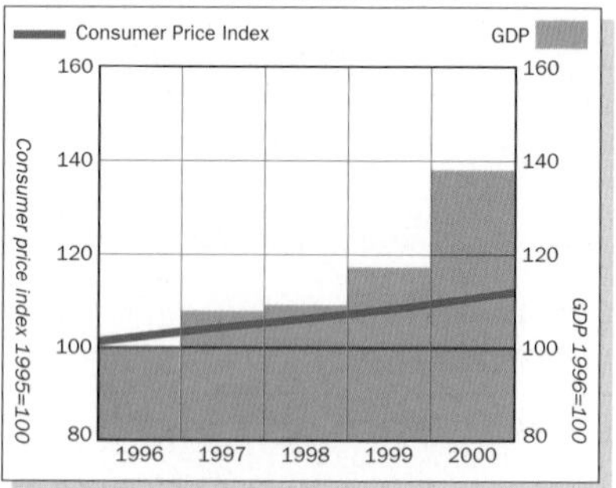
Consumer Price Index
GDP
160
140
120
100
80
Consumer price index 1995=100
GDP 1996=100
1996
1997
1998
1999
2000

NORWAY : MAJOR BUSINESSES

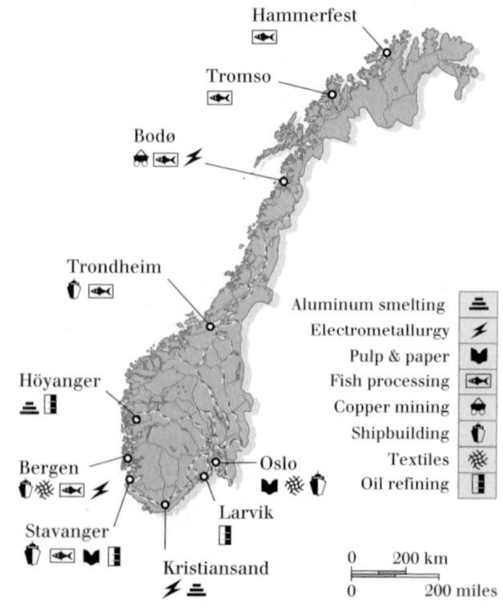
Hammerfest
Tromso
Bodø
Trondheim
Höyanger
Bergen
Stavanger
Kristiansand
Larvik
Oslo
Aluminum smelting
Electrometallurgy
Pulp & paper
Fish processing
Copper mining
Shipbuilding
Textiles
Oil refining
0 200 km
0 200 miles

N

RESOURCES

Electric power 29.3m kw

3.09m tonnes

3.41m b/d (reserves 9.4bn barrels)

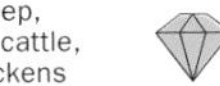
2.4m sheep, 979,800 cattle, 3.2m chickens

Oil, natural gas, iron, coal, copper, lead, zinc

ELECTRICITY GENERATION

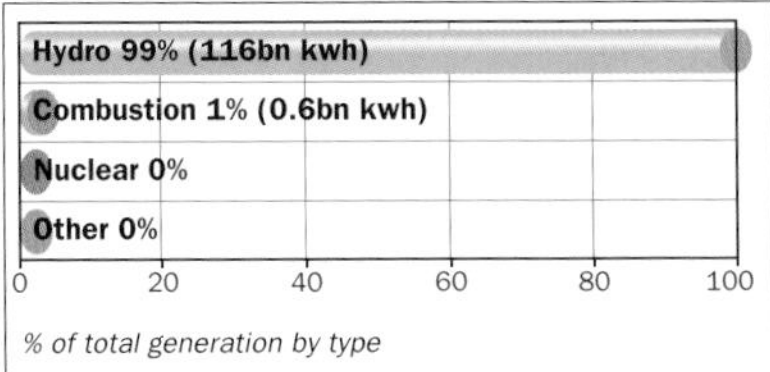
Hydro 99% (116bn kwh)
Combustion 1% (0.6bn kwh)
Nuclear 0%
Other 0%
0
20
40
60
80
100
% of total generation by type

Norway is Europe's largest oil producer; it also has sizable gas reserves. Most of Norway's electricity is produced by hydropower. In summer, the HEP surplus is exported. Fish and forestry are traditionally significant sectors – salmon farming, managed with particular efficiency, has grown rapidly – although with agriculture, they account for only 2.5% of GDP and 5% of the workforce.

A March 2002 report warned that cod stocks in the North Sea were under threat from oil-drilling waste products.

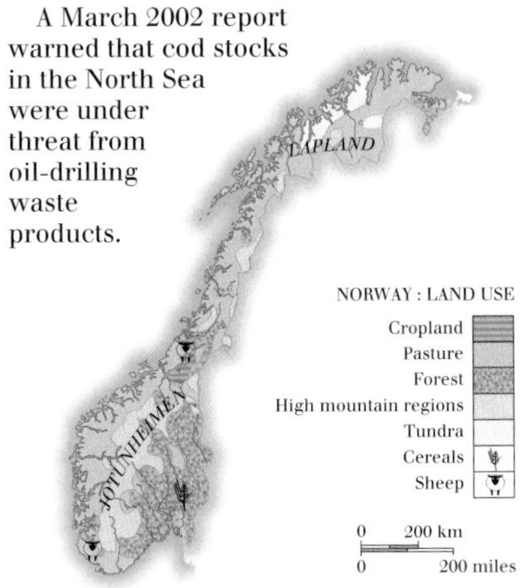
LAPLAND
JOTUNHEIMEN
NORWAY : LAND USE
Cropland
Pasture
Forest
High mountain regions
Tundra
Cereals
Sheep
0 200 km
0 200 miles

ENVIRONMENT

Sustainability rank: 2nd

31%

7.6 tonnes per capita

ENVIRONMENTAL TREATIES

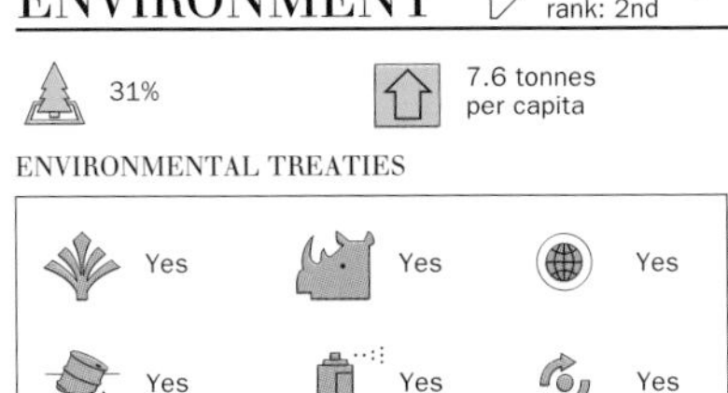
Yes
Yes
Yes
Yes
Yes
Yes

In 1986 northern Norway suffered radioactive contamination after the Chernobyl nuclear disaster. Norway has a tax on carbon dioxide emissions and was instrumental in securing agreement on the 1997 Kyoto Protocol on greenhouse gas emissions. In 1993, it lifted a ban on fishing minke whales, and in 2001 it allowed the export of whale products. It was also criticized in 2001 for a cull of (endangered) gray wolves and for plans to develop a coalfield on Svalbard.

MEDIA

TV ownership high

Daily newspaper circulation 583 per 1000 people

PUBLISHING AND BROADCAST MEDIA

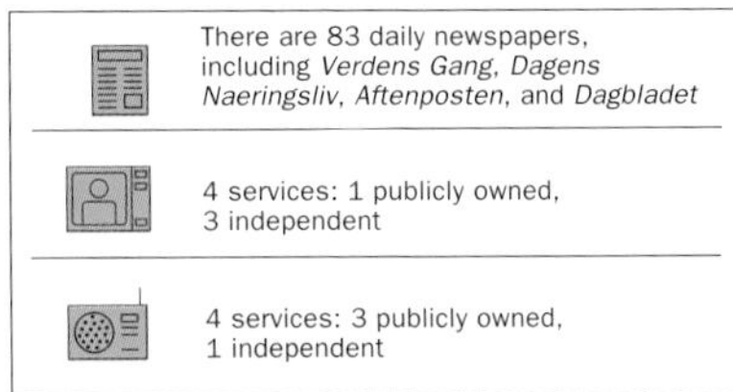
There are 83 daily newspapers, including *Verdens Gang, Dagens Naeringsliv, Aftenposten,* and *Dagbladet*
4 services: 1 publicly owned, 3 independent
4 services: 3 publicly owned, 1 independent

The state broadcaster NRK enjoys comparatively generous funding and has a 38% share of viewers. Newspapers have the second-highest proportion of readers in the world.

CRIME

No death penalty

2643 prisoners

Down 3% in 1999

CRIME RATES

Murders	
2	per 100,000 population
Rapes	
12	per 100,000 population
Thefts	
4577	per 100,000 population

Norway has low levels of crime, even by Scandinavian standards. Violent crime barely exists – the murder rate is one-fifth of that of Sweden, and there are considerably fewer assaults and robberies.

EDUCATION

School leaving age: 16

99%

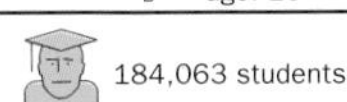
184,063 students

THE EDUCATION SYSTEM

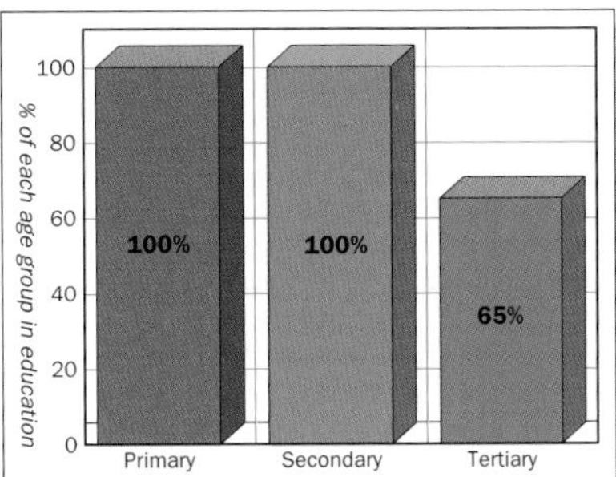
100
80
60
40
20
0
% of each age group in education
100%
100%
65%
Primary
Secondary
Tertiary

The period of compulsory schooling was increased from nine to ten years, with effect from the school year 1997/1998. Most schools are run by municipalities. There are four universities; specialized colleges include the Nordic College of Fisheries. Promotion of continuing education kept youth unemployment down during the early 1990s recession.

HEALTH

Welfare state health benefits

1 per 357 people

Heart and cerebrovascular diseases, cancers, accidents

WHO ranks Norway as the world's third-best provider of health care. The country's infant mortality rate is among the world's lowest, and life expectancy at birth is one of the highest. Spending is among the highest of OECD states.

A new health plan announced in 2002 will place the high-spending hospitals under central control.

Telemedicine (online remote audio and image diagnosis) allows remote northern hospitals to obtain specialist consultations without having to send patients to the regional hospital.

SPENDING

GDP/cap. increase

CONSUMPTION AND SPENDING

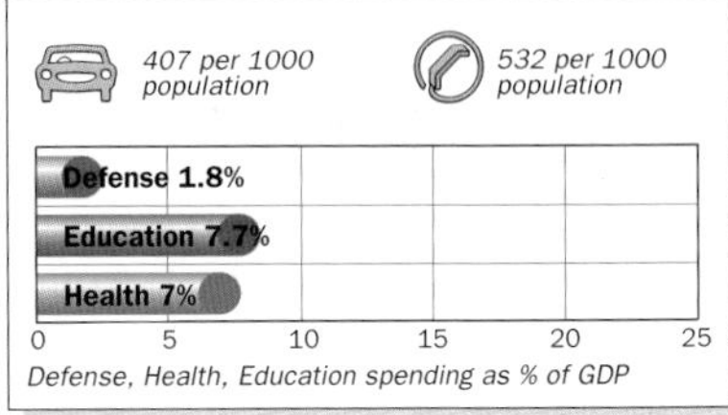
407 per 1000 population
532 per 1000 population
Defense 1.8%
Education 7.7%
Health 7%
0
5
10
15
20
25
Defense, Health, Education spending as % of GDP

In terms of income distribution, the Scandinavian countries are the most egalitarian in the world, and the richest 10% of Norway's population owns much less of the country's wealth than is the case in other developed countries. Homelessness and social deprivation are very rare. Recent refugees from the Bosnian conflict are the most disadvantaged group.

The discrepancy between men's and women's pay is greater than in either Sweden or Finland, although still well below the European average. Social provision has been maintained even through economic recession. Benefits are generous.

WORLD RANKING

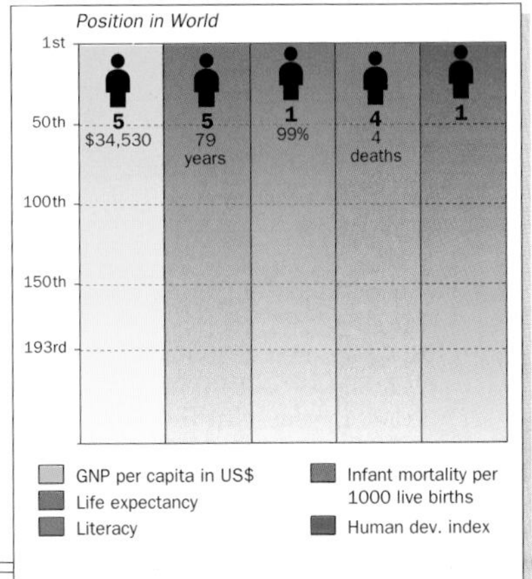
Position in World
1st
50th
100th
150th
193rd
5
$34,530
5
79 years
1
99%
4
4 deaths
1
GNP per capita in US$
Life expectancy
Literacy
Infant mortality per 1000 live births
Human dev. index

N

OMAN

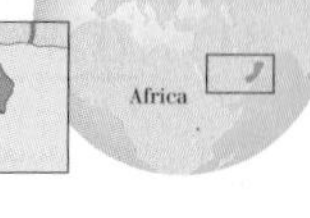
MIDDLE EAST
Saudi Arabia
UAE
Yemen
Asia
Africa

OFFICIAL NAME: Sultanate of Oman **CAPITAL:** Muscat
POPULATION: 2.6 million **CURRENCY:** Omani rial **OFFICIAL LANGUAGE:** Arabic

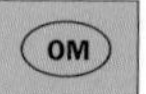
1951
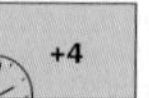
1951

Nov 18

OM

+4 +968 .om

OCCUPYING THE EASTERN corner of the Arabian peninsula, Oman commands a strategic position at the entrance to the Persian Gulf. It is the least developed of the Gulf states. The most densely populated areas are the northern coast and the southern Salalah plain. Oil exports have given Oman modest prosperity under a paternalistic sultan. A Marxist-led insurgency supported by southern Dhofaris was defeated in the 1970s.

CLIMATE

Hot desert

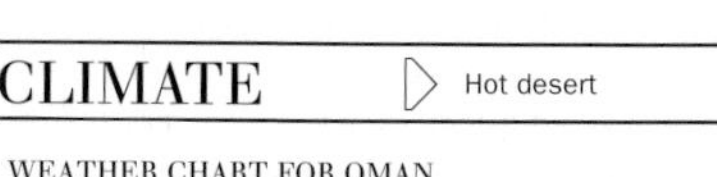

WEATHER CHART FOR OMAN

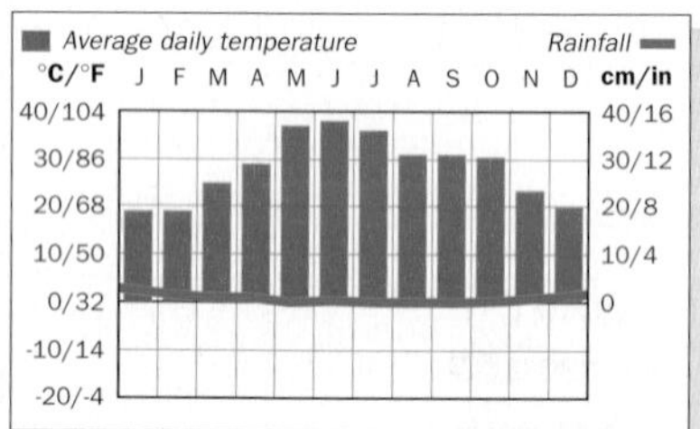
Average daily temperature
Rainfall
°C/°F J F M A M J J A S O N D cm/in
40/104
30/86
20/68
10/50
0/32
-10/14
-20/-4
40/16
30/12
20/8
10/4
0

In the north temperatures often climb above 45°C (113°F) in summer. The south has a monsoon climate.

TRANSPORTATION

Drive on right

Seeb International, Muscat 2.72m passengers

20 ships 15,000 grt

THE TRANSPORTATION NETWORK

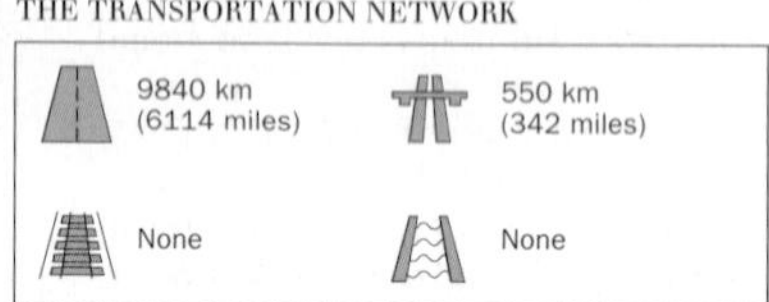
9840 km (6114 miles)
550 km (342 miles)
None
None

Northern cities are well served by good roads, but some places, in the south particularly, are best reached by air.

TOURISM

Visitors : Population 1:5.2

502,000 visitors

Down 19% in 1999

MAIN TOURIST ARRIVALS

	% of total arrivals
Europe	34%
Asia	26%
Middle East	23%
Other	17%

0 10 20 30 40
% of total arrivals

Until the late 1980s, Oman was closed to all but business or official visitors. The sultanate's rich cultural heritage, fine beaches, and luxury hotels are now enjoyed by thousands of Western visitors each year.

PEOPLE

Pop. density low

Arabic, Baluchi

12/km² (32/mi²)

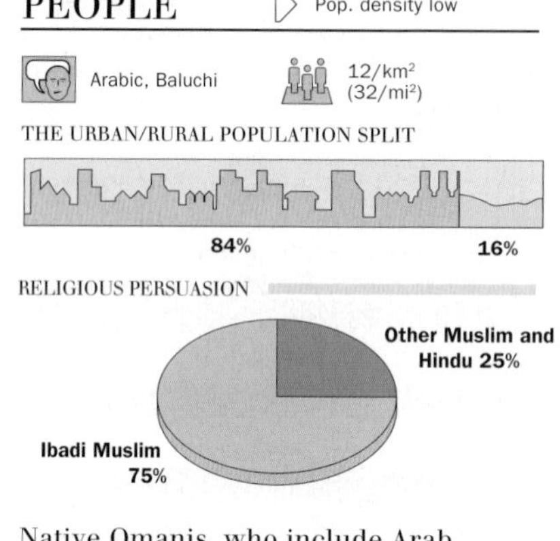
THE URBAN/RURAL POPULATION SPLIT
84%
16%
RELIGIOUS PERSUASION
Other Muslim and Hindu 25%
Ibadi Muslim 75%

Native Omanis, who include Arab refugees who fled Zanzibar in the 1960s, make up three-quarters of the population. Baluchis are the largest foreign grouping. Expatriates pose no threat to the regime and Westerners enjoy considerable freedom. Urban drift has taken place, and most Omanis now live in cities. Oman has a number of distinct minorities; the most numerous are the Jebalis in Dhofar – nomadic herdsmen who speak a language resembling Ethiopian. Many Dhofaris supported the Marxist-led insurgents in the 1970s, but they are now considered loyal. Most Omanis are Ibadi Muslims who follow an appointed leader, the imam. Ibadism does not oppose freedom for women, and a few enjoy positions of authority; two were elected to the Consultative Council in September 2000.

POLITICS

No legislative elections

2000/2003

Sultan Qaboos bin Said

LEGISLATIVE OR ADVISORY BODIES

Consultative Council 82 seats

There are no political parties. The members of the Consultative Council (Majlis ash-shoura) were directly elected for the first time in 2000 by electoral committees in each province, and included two women.

Sultan Qaboos is an authoritarian but paternalistic monarch, whose dynasty traces its roots to the 18th century. As well as being head of state, the sultan is prime minister and minister for foreign affairs, defense, and finance. Family members hold other key positions. The regime faces no serious challenge, although Qaboos keeps a careful eye on the religious right wing. In 1991, he created the Consultative Council (Majlis ash-shoura), which gives a semblance of democracy. From 2000 its members were directly elected by provincial committees, rather than being appointed. Major political issues include the planned privatization of medium-sized government projects, and improving Oman's self-defense capability.

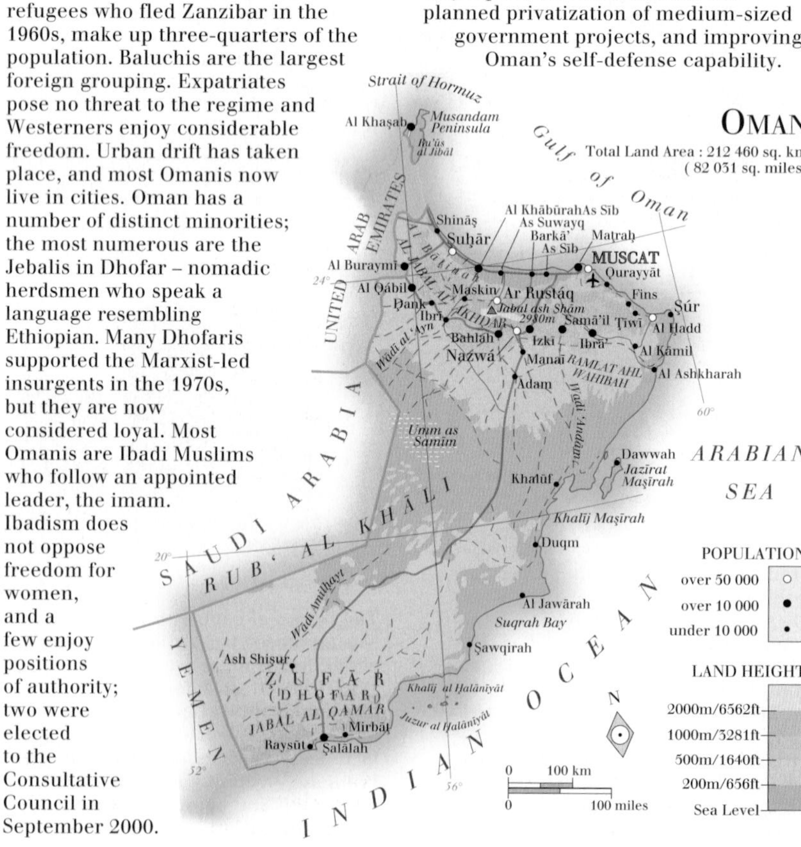
OMAN
Total Land Area : 212 460 sq. km (82 031 sq. miles)
Strait of Hormuz
Al Khaşab
Musandam Peninsula
Gulf of Oman
UNITED ARAB EMIRATES
Shināş
Şuḩār
Al Khābūrah
As Suwayq
Barkā'
As Sīb
Maţraḩ
MUSCAT
Qurayyāt
Al Buraymī
Al Qābil
Ḑank
Ibrī
Maskin
Ar Rustāq
Jabal ash Sham 2980m
Samā'il
Ţīwī
Şūr
Al Hadd
Bahlāh
Nazwá
Izkī
Ibrā'
Al Kāmil
Manaḩ
Al Ashkharah
Ramlat Ahl Wahibah
Ādam
Umm as Samīm
SAUDI ARABIA
RUB' AL KHALI
Dawwah
Jazīrat Maşīrah
ARABIAN SEA
Khalūf
Khalīj Maşīrah
Duqm
Al Jawārah
Suqrah Bay
Şawqirah
YEMEN
Ash Shişur
ZUFĀR (DHOFAR)
JABAL AL QAMAR
Mirbāţ
Raysūt
Şalālah
Khalīj al Ḩalānīyāt
Juzur al Ḩalānīyāt
INDIAN OCEAN
0 100 km
0 100 miles
POPULATION
over 50 000
over 10 000
under 10 000
LAND HEIGHT
2000m/6562ft
1000m/3281ft
500m/1640ft
200m/656ft
Sea Level

WORLD AFFAIRS

Joined UN in 1971

AL

AMF

Damasc

GCC

OIC

Relations with Israel were cut off after the renewal of the Palestinian *intifada* in 2000. Although still basically pro-Western, and supportive of the US-led "war on terrorism," Oman has ties with Iran and calls for an easing of sanctions against Iraq.

A watchtower above an oasis. *Most of Oman is gravelly desert. The only large area of cultivation is the 20-km-wide Al Batinah plain.*

CHRONOLOGY

The present Albusaidi dynasty has ruled in Oman since 1749.

- **1932** Sultan bin Taimur in power.
- **1951** Sovereignty recognized by UK.
- **1970** Sultan Qaboos bin Said seizes power from his father.
- **1975** Suppression of Dhofar revolt.
- **1991** Consultative Council set up.
- **2000** Consultative Council members elected for first time.

AID

Recipient

 $46m (receipts)

 Up 15% in 2000

Aid used to come mainly from the West, and particularly the US, but Arab donors are now of greater significance. Japan is also a major contributor. Oman makes occasional donations of its own to Arab and Muslim causes.

DEFENSE

No compulsory military service

$1.73bn

Up 6% in 2000

The UK is the main supplier of equipment. In the 1991 Gulf War, Oman provided communications and services to US and UK forces. Oman's Defense Council, established in 1996, has replenished tanks, ships, and aircraft in recent years. Baluchi mercenaries supplement army strength.

ECONOMICS

Inflation 0.4% p.a. (1991–2000)

$11bn

0.3851 Omani rials

SCORE CARD

World GNP Ranking	80th
GNP per Capita	$4778
Balance of Payments	$3.35bn
Inflation	–1.1%
Unemployment	5%

STRENGTHS

Oil industry, led by Royal Dutch/Shell. Has benefited from staying out of OPEC and selling oil at spot prices without quotas. Soaring world oil prices in 2000 signified recovery from 1986 collapse. Rich Indian Ocean coastal waters have potential for sizable fishing industry.

WEAKNESSES

Overdependence on oil (90% of GNP), with less than 20 years' known reserves. Services sector could be better developed. Foreign workers needed in all economic sectors.

EXPORTS

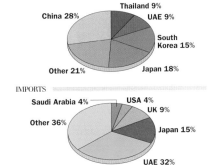

China 28%
Thailand 9%
UAE 9%
South Korea 15%
Japan 18%
Other 21%

IMPORTS

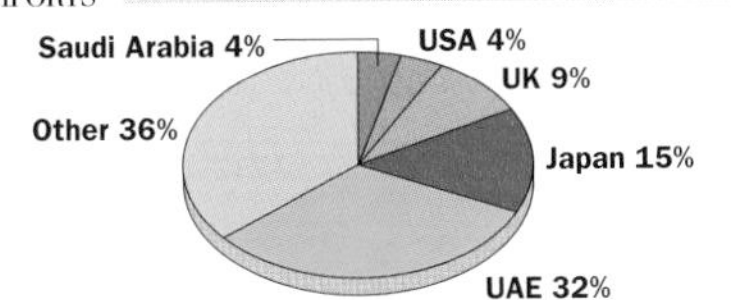
Saudi Arabia 4%
USA 4%
UK 9%
Other 36%
Japan 15%
UAE 32%

RESOURCES

Electric power 2.1m kw

 108,819 tonnes

 959,000 b/d (reserves 5.5bn barrels)

980,000 goats, 290,000 cattle, 3.4m chickens

 Oil, natural gas, copper, chromite, marble, gypsum

Limiting oil production to conserve resources ended in 1993 following a number of exploration successes, but reserve levels are now falling again.

ENVIRONMENT

Sustainability rank: 120th

 16%

 8.8 tonnes per capita

Overpumping of groundwater causes seawater to seep into traditional irrigation areas. Nature reserves and antihunting laws protect rich wildlife.

MEDIA

TV ownership high

 Daily newspaper circulation 28 per 1000 people

PUBLISHING AND BROADCAST MEDIA

There are 5 daily newspapers, including *Al-Watan, Oman Daily Newspaper*, and the English-language *Oman Daily Observer*

1 state-controlled service

2 state-controlled services

Nothing critical of the government may be published in Oman, despite a 1984 law allowing for "freedom of opinion."

CRIME

Death penalty in use

Oman does not publish prison figures

 Up 18% in 1999

Reckless driving by young Omani males is a problem. A "flying court" serves remote communities.

EDUCATION

Schooling is not compulsory

 72%

 16,032 students

Education has improved, but rural illiteracy is still high. Between 1996 and 2000, over 200 new schools were built.

HEALTH

Welfare state health benefits

 1 per 769 people

 Heart and cerebrovascular diseases, accidents

There is a policy of replacing expatriate medical staff with Omani nationals. Rural areas are served by clinics.

SPENDING

GDP/cap. increase

CONSUMPTION AND SPENDING

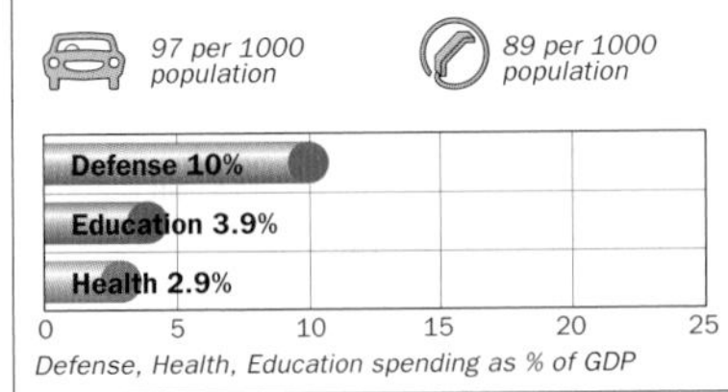
97 per 1000 population
89 per 1000 population
Defense 10%
Education 3.9%
Health 2.9%
0
5
10
15
20
25
Defense, Health, Education spending as % of GDP

Omanis in urban areas enjoy the same high living standards that are to be found in other Gulf states. Hunting trips to Pakistan are popular among the rich Omani elite, and a *khanjar*, a curved dagger, is seen as a status symbol.

WORLD RANKING

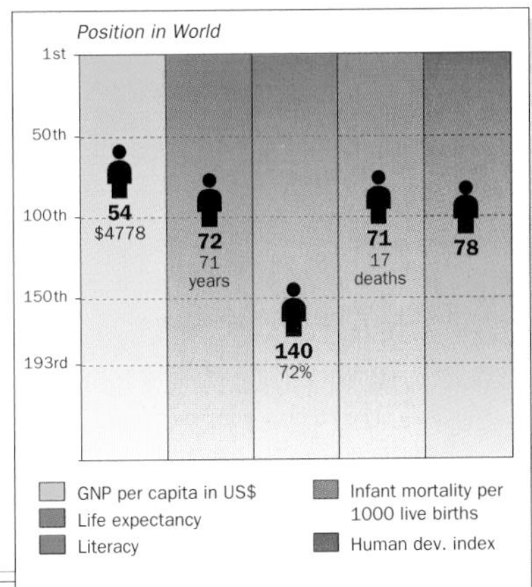
Position in World
1st
50th
100th
150th
193rd
54
$4778
72
71 years
140
72%
71
17 deaths
78
GNP per capita in US$
Life expectancy
Literacy
Infant mortality per 1000 live births
Human dev. index

O

PAKISTAN

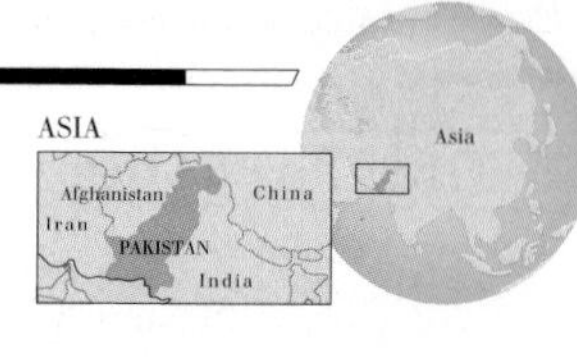
ASIA
Asia
Afghanistan
Iran
China
PAKISTAN
India

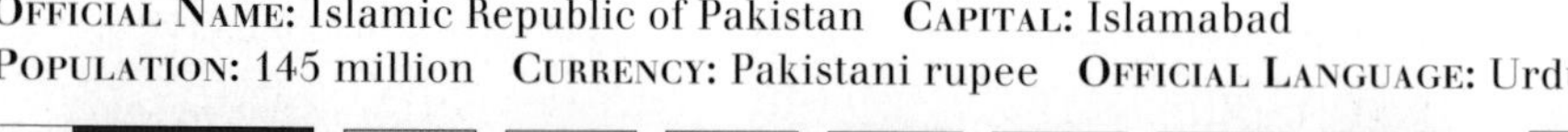

OFFICIAL NAME: Islamic Republic of Pakistan **CAPITAL:** Islamabad
POPULATION: 145 million **CURRENCY:** Pakistani rupee **OFFICIAL LANGUAGE:** Urdu

1947
1971
March 23
PK
+5
+92
.pk

ONCE A PART OF BRITISH INDIA, Pakistan was created in 1947 in response to the demand for an independent and predominantly Muslim Indian state. Initially the new nation included East Pakistan, present-day Bangladesh, which seceded from Pakistan in 1971. Eastern and southern Pakistan, the flood plain of the Indus River, is highly fertile and produces cotton, the basis of the large textile industry.

Barren landscape in Kachhi, Baluchistan. *This area of Pakistan has some of the highest May-to-September temperatures in the world.*

CLIMATE

Mountain/steppe/ hot desert

WEATHER CHART FOR ISLAMABAD

Average daily temperature — Rainfall
°C/°F J F M A M J J A S O N D cm/in
40/104, 30/86, 20/68, 10/50, 0/32, -10/14, -20/-4 — 40/16, 30/12, 20/8, 10/4, 0

Temperatures can soar to 50°C (122°F) in Sindh and Baluchistan and fall to –20°C (–4°F) in the northern mountains.

TRANSPORTATION

Drive on left

Karachi International
4.94m passengers
57 ships
401,200 grt

THE TRANSPORTATION NETWORK

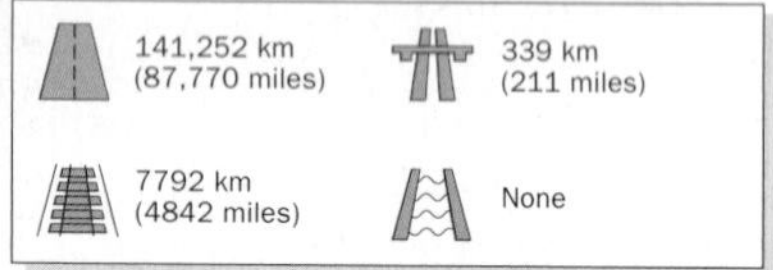
141,252 km
(87,770 miles)
339 km
(211 miles)
7792 km
(4842 miles)
None

Most roads are poorly maintained. A modern highway linking Islamabad and Lahore was inaugurated in 1997.

TOURISM

Visitors : Population 1:267

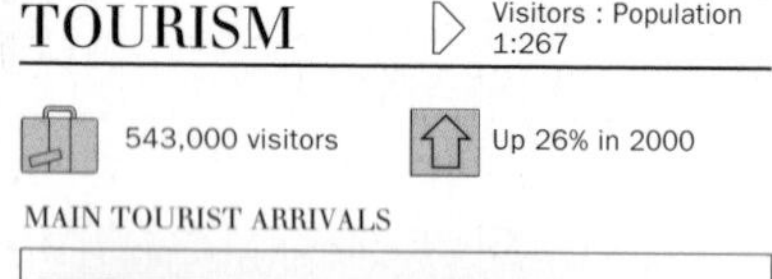
543,000 visitors
Up 26% in 2000
MAIN TOURIST ARRIVALS

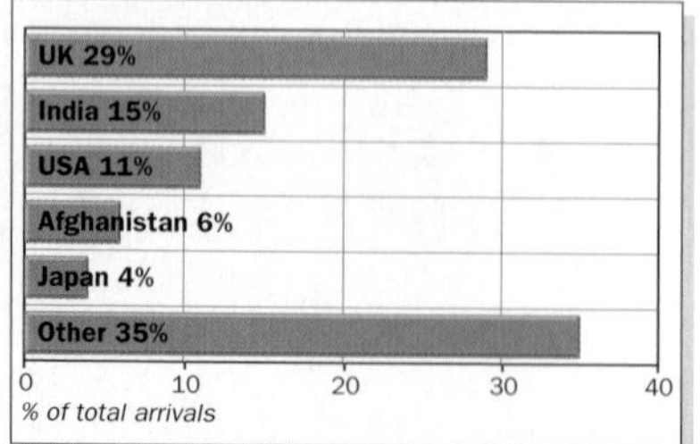
UK 29%
India 15%
USA 11%
Afghanistan 6%
Japan 4%
Other 35%
0
10
20
30
40
% of total arrivals

Relatively few tourists visit Pakistan, despite its rich cultural heritage and unspoiled natural beauty.

PEOPLE

Pop. density medium

Punjabi, Sindhi, Pashtu, Urdu, Baluchi, Brahui

188/km² (487/mi²)

THE URBAN/RURAL POPULATION SPLIT

37% 63%

RELIGIOUS PERSUASION

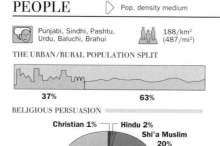

Christian 1%, Hindu 2%, Shi'a Muslim 20%, Sunni Muslim 77%

ETHNIC MAKEUP

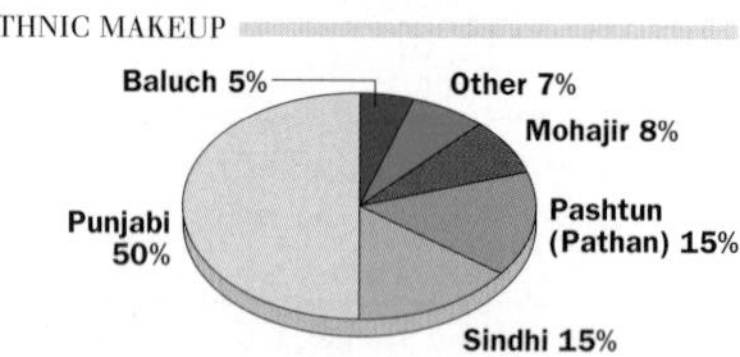
Baluch 5%
Other 7%
Mohajir 8%
Pashtun (Pathan) 15%
Punjabi 50%
Sindhi 15%

Punjabis account for 50% of the population, while Sindhis, Pathans (Pashtun), and Baluch are also prominent. *Mohajirs* – Urdu-speaking immigrants from pre-partition India – predominate in Karachi and Hyderabad. Punjabi political and military dominance of the centralized state has spawned many separatist and autonomy movements. Pathans have frequently threatened to establish a homeland with ethnic kin in Afghanistan. Tensions between the Baluch and Pashtun refugees from Afghanistan sporadically erupt into violence, as do those between native Sindhis and immigrant *mohajirs.*

The gap between rich and poor, for example between the feudal landowning class and their serfs, is considerable. There is an expanding middle class of small-scale traders and manufacturers.

Recent years have witnessed a marked increase in Islamist militancy, accompanied by growing discrimination against religious minorities. After the 1999 coup, the Musharraf regime trod a fine line in trying to avoid conflict

POPULATION AGE BREAKDOWN

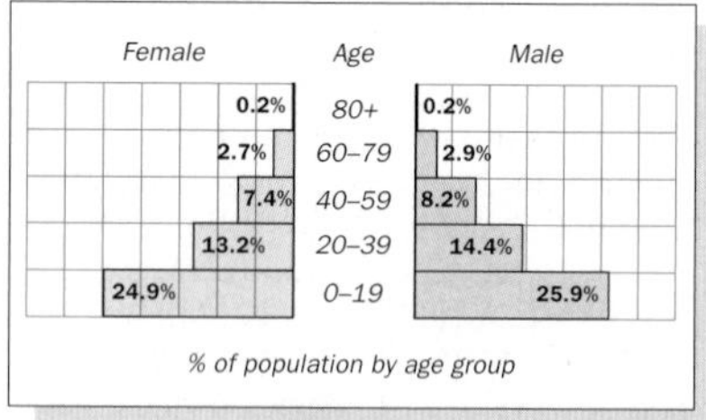
Female
Age
Male
0.2%
80+
0.2%
2.7%
60–79
2.9%
7.4%
40–59
8.2%
13.2%
20–39
14.4%
24.9%
0–19
25.9%
% of population by age group

with Islamic militants, both over issues such as the strict application of Islamic *sharia*, and over foreign policy.

The extended family is an enduring institution, and ties between its members are strong, reflected in the dynastic and nepotistic nature of the political system. Although some women hold prominent positions, and Benazir Bhutto has twice been prime minister, relatively few are allowed to work by their religiously conservative menfolk. Pakistan has one of the world's lowest ratios of females to males, implying widespread neglect and some female infanticide. Amnesty International criticized Pakistan in 2000 for its failure to give women's rights sufficient protection. Women's rights groups are mainly based in cities, and have made little overall impact.

CHĀGAI HILLS
IRAN
Turbat
Dasht Kaur
Gwādar
ARABI

P

POLITICS

No legislative elections

L. House 1997/2002
U. House 1997/2002

President Pervez Musharraf

AT THE LAST ELECTION

National Assembly (dissolved)
237 seats

5% MQM(A) · 4% N-M · 13% Others

58% PML · 8% W · 8% PPP · 4% ANP

PML = Pakistan Muslim League **W** = Women
PPP = Pakistan People's Party **MQM(A)** = Mohajir Qaumi Movement (A) Haq Parast Group **ANP** = Awami National Party **N-M** = Non-Muslim minorities
Others include Baluchistan National Party, Jamiat-Ulema-e-Pakistan, Pakistan People's Party Jamhoori Watan Party (Republican Nation Party), National People's Party, and independents

Senate (dissolved) 87 seats

The Senate is indirectly elected

The military suspended multiparty democracy in 1999, and the National Assembly was dissolved in June 2001.

PROFILE

Throughout the 1990s fragile coalitions had to rule in cooperation with the president and the army, and were hampered by a large bureaucracy. The current military regime, which removed Nawaz Sharif in 1999, remains committed to restoring democracy by 2003, and faces pressure from all opposition parties to accelerate the transition. In June 2001 Gen. Pervez Musharraf appointed himself president, and in May 2002 he was endorsed for a further five years following a controversial referendum.

The National Security Council (NSC) acts as a cabinet under Musharraf. The popularity of the 1999 coup indicated the loss of respect for corrupt politicians and the country's much-abused democratic institutions. The ousted prime minister Nawaz Sharif was found guilty of treason in 2000.

MAIN POLITICAL ISSUE
Militant Islam

Under pressure to curb the ambitions of radical Islamists in Kashmir, as in Afghanistan, the Musharraf regime treads a tightrope in domestic politics, where militant Islam has a powerful grip on the urban poor. There was a fresh upsurge in 2002 in Sunni extremist violence against the Shi'a minority.

***Benazir Bhutto,** PPP leader and former prime minister.*

***Pervez Musharraf,** military ruler and self-styled president.*

WORLD AFFAIRS

Joined UN in 1947

IAEA
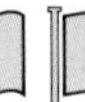
ECO

NAM

OIC

SAARC

In 1998, Pakistan carried out a series of nuclear tests in response to similar tests by India, provoking international condemnation and three years of US sanctions. The last of three wars between India and Pakistan was three decades ago, but their dispute over Kashmir has repeatedly stoked tensions. In late 2001–early 2002, a series of attacks by Muslim Kashmiri separatists, allegedly supported by Pakistan, pushed the two countries again to the brink of war.

One of only three countries to recognize the Afghan *taliban* regime, Pakistan in September 2001 urged the *taliban* to hand over terrorism suspect Osama bin Laden, and then backed US-led action against the regime.

Pakistan's membership of the Commonwealth was suspended after the 1999 coup.

PAKISTAN

Total Land Area : 803 940 sq. km (310 401 sq. miles)

LAND HEIGHT

6000m/19 686ft
4000m/13 124ft
3000m/9843ft
2000m/6562ft
1000m/3281ft
500m/1640ft
200m/656ft
Sea Level

POPULATION

over 5 000 000
over 1 000 000
over 500 000
over 100 000
over 50 000
over 10 000

0 — 200 km
0 — 200 miles

***Rice paddy fields,** with monsoon rains threatening from the Himalaya mountains. Rice is the second most valuable agricultural export after cotton.*

CHRONOLOGY

From the 8th to the 16th centuries, Islamic rule extended to northwest and northeast India. Punjab and Sindh, annexed by the British East India Company in the 1850s, were ceded to the British Raj in 1857.

- **1906** Muslim League founded as organ of Indian Muslim separatism.
- **1947** Partition of India. Pakistan divided by 1600 km (994 miles) of Indian territory into East and West Pakistan. Millions displaced by large-scale migration. Muhammad Ali Jinnah first governor-general.
- **1948** First India-Pakistan war over Kashmir.
- **1949** New Awami League (AL) demands East Pakistan's autonomy.
- **1956** Constitution establishes Pakistan as an Islamic republic.
- **1958** Martial law. Gen. Muhammad Ayub Khan takes over; elected president two years later.

P

CHRONOLOGY *continued*

- **1965** Second India–Pakistan war over Kashmir.
- **1970** Ayub Khan resigns. Gen. Agha Yahya Khan takes over. First direct elections won by AL; West Pakistani parties reject results. War with India over East Pakistan.
- **1971** East Pakistan secedes as Bangladesh. PPP leader Zulfikar Ali Bhutto president.
- **1972** Simla (peace) Agreement with India.
- **1973** Bhutto, now prime minister, initiates Islamic socialism.
- **1977** General election. Riots over allegations of vote rigging. Gen. Zia ul-Haq stages military coup.
- **1979** Bhutto executed.
- **1986** Bhutto's daughter Benazir returns from exile to lead PPP.
- **1988** Zia killed in air crash. PPP wins general election.
- **1990** Ethnic violence in Sindh. President dismisses Benazir Bhutto. Nawaz Sharif becomes premier.
- **1991** Muslim *sharia* incorporated in legal code.
- **1992** Violence between Sindhis and *mohajirs* escalates in Sindh.
- **1993** President Khan and Prime Minister Sharif resign. Elections; Bhutto returns to power.
- **1996** President dismisses Bhutto.
- **1997** PML wins landslide election victory; Sharif prime minister.
- **1998** Nuclear tests.
- **1999–2000** Military coup. Sharif found guilty of treason.
- **2001** Parliament suspended, Gen. Musharraf appoints himself president. Pakistan key ally in US-led "war on terrorism."
- **2002** Controversial referendum endorses Musharraf as president. US and French nationals killed in terrorist attacks in Islamabad and Karachi. Threat of war with India over Kashmir.

AID

Recipient

 $703m (receipts)  Down 4% in 2000

Pakistan is heavily dependent on aid, although the government has a long history of misdirecting aid payments. Aid intended for major projects has regularly been used to fund the current-account deficit. In mid-1998 the IMF agreed to help Pakistan meet its international debt obligations after the US and other Western aid donors cut off aid in protest against Pakistan's nuclear tests. The US subsequently agreed to resume aid, and boosted its support in late 2001 to reward the regime for supporting the "war on terrorism." Japan and Germany are among other main bilateral donors.

DEFENSE

No compulsory military service

 $3.58bn  Up 2% in 2000

Pakistan has emerged as a significant regional arms trader. It established itself as a nuclear power by conducting a number of successful nuclear tests in 1998. Defense spending has a high priority, accounting for about a quarter of all government expenditure. The US, once the most important arms supplier, in 2001 lifted the sanctions imposed in 1990 and 1998. Pakistan's other main defense procurements are from France, the UK, and China. The army has been highly significant in politics throughout the period since independence, even when not actually in power in a military regime. Two years prior to the 1999 coup, it had already assumed a formal role in civilian decision-making by its inclusion in a National Security Council.

PAKISTANI ARMED FORCES

2300 main battle tanks (PRC Type-59/69/85, M-47/48A5, T-54/55/80)	555,000 personnel
10 submarines, 8 frigates, and 9 patrol boats	25,000 personnel
353 combat aircraft (52 *Mirage* 5, 42 Q-5)	45,000 personnel
Capability undisclosed; weapons tested in 1998	

ECONOMICS

Inflation 10% p.a. (1990–2000)

$61bn   57.6–59.9 Pakistani rupees

SCORE CARD

WORLD GNP RANKING	45th
GNP PER CAPITA	$440
BALANCE OF PAYMENTS	–$2.21bn
INFLATION	4.4%
UNEMPLOYMENT	6%

EXPORTS

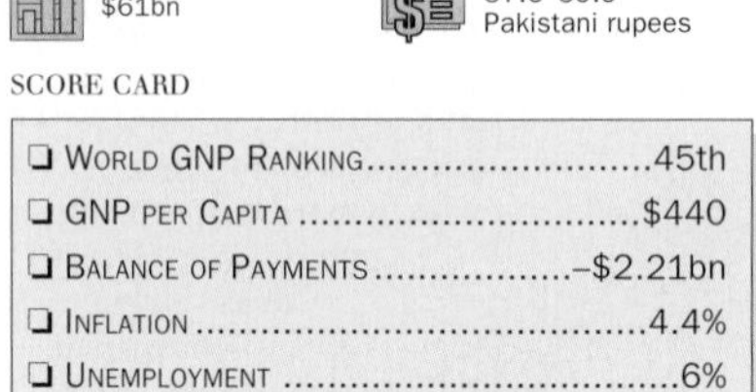

UAE 6%
Germany 6%
Hong Kong 6%
UK 7%
Other 50%
USA 25%

IMPORTS

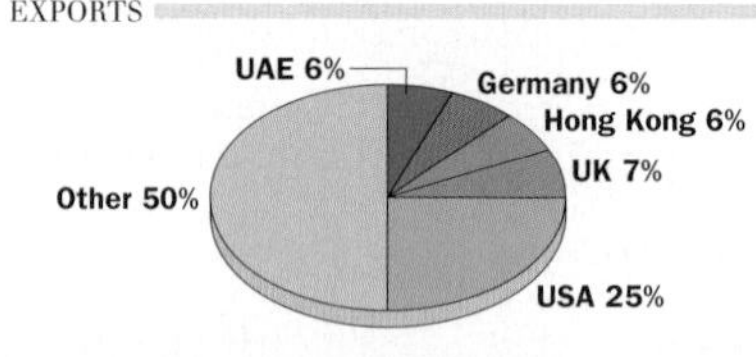

USA 6%
Japan 6%
UAE 11%
Saudi Arabia 11%
Other 54%
Kuwait 12%

ECONOMIC PERFORMANCE INDICATOR

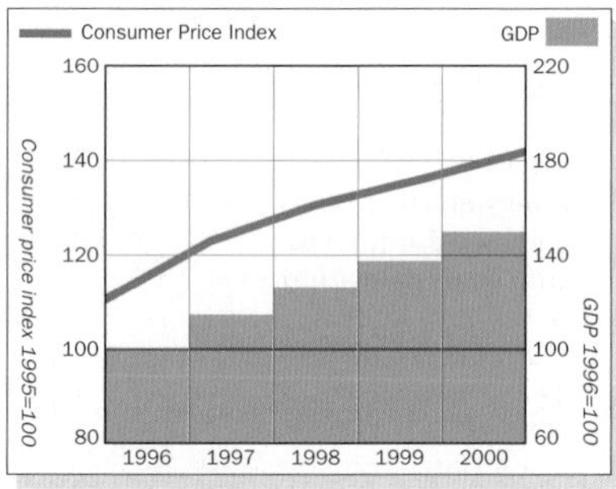

Consumer Price Index
GDP
Consumer price index 1995=100
GDP 1996=100
160
140
120
100
80
220
180
140
100
60
1996
1997
1998
1999
2000

STRENGTHS

Gas, water, coal, oil. Substantial untapped natural resources. Low labor costs. Potentially huge market. One of the world's leading producers of cotton and a major exporter of rice.

WEAKNESSES

Production and sales of cotton and rice vulnerable to weather conditions. History of inefficient and haphazard government economic policies. Weak and overstretched infrastructure.

PROFILE

Pakistan has yet to show progress in tackling its considerable economic problems. Although successive governments have reversed the nationalization policies instituted in the 1970s, private enterprise has been stifled by the rules of a massive bureaucracy. There is some foreign investment in previously state-only sectors such as banking, and water and other utilities. However, corruption at all levels of government undermined economic confidence throughout the 1990s, and it was particularly acute under the administration of Benazir Bhutto. Efforts by the military government to tackle corruption and poverty were praised by the World Bank in 2001. Defense spending remains high.

PAKISTAN : MAJOR BUSINESSES

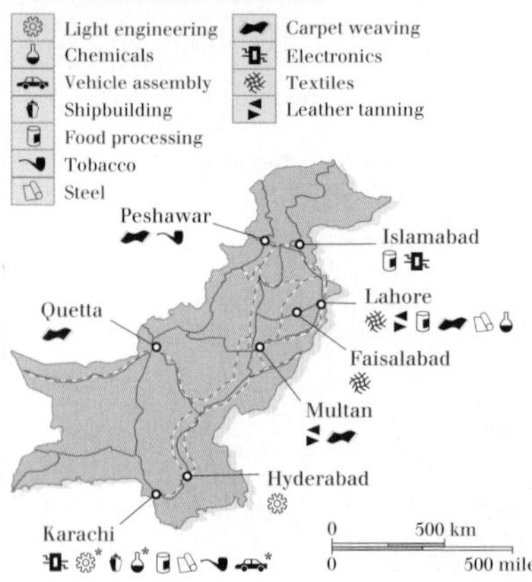

Light engineering
Chemicals
Vehicle assembly
Shipbuilding
Food processing
Tobacco
Steel
Carpet weaving
Electronics
Textiles
Leather tanning
Peshawar
Islamabad
Lahore
Quetta
Faisalabad
Multan
Hyderabad
Karachi
0
500 km
0
500 miles

* significant multinational ownership

RESOURCES

Electric power 15.7m kw

674,606 tonnes

55,258 b/d (reserves 222m barrels)

49.1m goats, 24.2m sheep, 155m chickens

Oil, natural gas, coal, limestone, salt, gypsum, silica sand

ELECTRICITY GENERATION

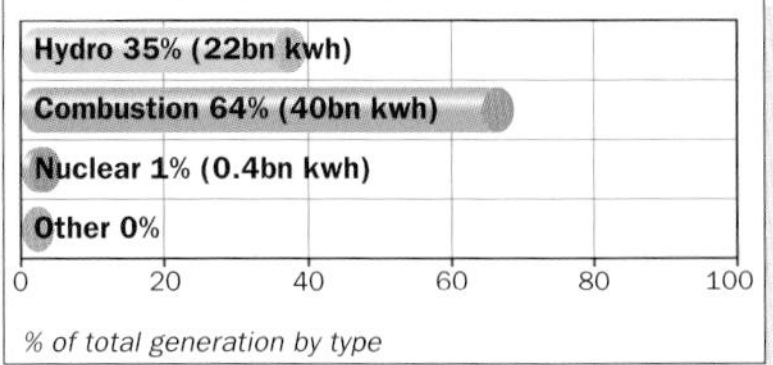
Hydro 35% (22bn kwh)
Combustion 64% (40bn kwh)
Nuclear 1% (0.4bn kwh)
Other 0%
0
20
40
60
80
100
% of total generation by type

Apart from cotton and rice, Pakistan's major resources are oil, coal, gas, and water. The state hopes that the privatization of the utilities industries will reduce energy imports and shortages – peak electricity demand, for example, exceeds supply by 20%. Refining capacity is also well below present demand. Steps are being taken to attract more foreign investment in oil and gas exploration, extraction, and distribution, but privatization plans went on hold in October 2001 because of the region's acute instability.

PAKISTAN : LAND USE

0 500 km
0 500 miles

Cropland
Pasture
Forest
Desert
Wetlands
High mountain regions
Sugarcane
Wheat
Cattle

Hindu Kush
Indus
Thar Desert

ENVIRONMENT

Sustainability rank: 112th

5% (3% partially protected)

0.7 tonnes per capita

ENVIRONMENTAL TREATIES

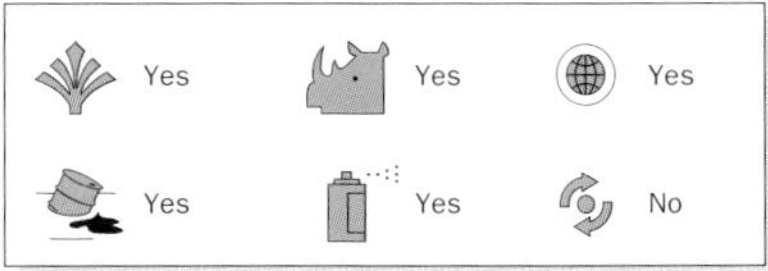
Yes
Yes
Yes
Yes
Yes
No

Tough measures are in force to curb illegal logging. Urban pollution affects many cities. Local groups increasingly voice environmental concerns.

MEDIA

TV ownership medium

Daily newspaper circulation 21 per 1000 people

PUBLISHING AND BROADCAST MEDIA

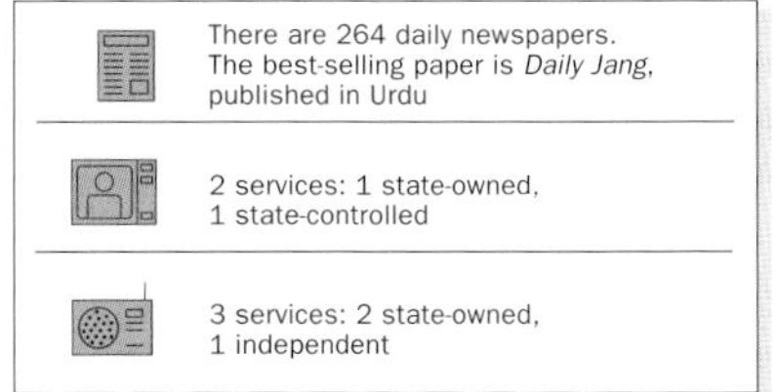
There are 264 daily newspapers. The best-selling paper is *Daily Jang*, published in Urdu
2 services: 1 state-owned, 1 state-controlled
3 services: 2 state-owned, 1 independent

State-run services dominate the mass media. Journalists who challenge official views are systematically harassed.

EDUCATION

Schooling is not compulsory

46%

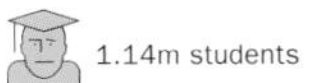
1.14m students

THE EDUCATION SYSTEM

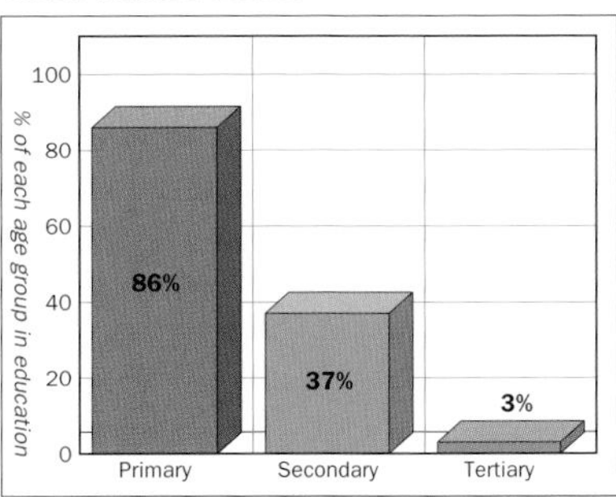
100
80
60
40
20
0
% of each age group in education
86%
37%
3%
Primary
Secondary
Tertiary

Although universal free primary education is a constitutional right, it is not compulsory. Literacy rates are among the lowest in the world. The education system is heavily Islamized, and weighted toward educating males; a large majority of children enrolled in primary schools are boys.

There are 23 universities, 675 arts and sciences colleges, and 99 professional colleges, all of which have a heavy preponderance of arts students. Wealthy parents frequently choose to send their children abroad for higher education, mainly to colleges in the UK or the US.

HEALTH

No welfare state health benefits

1 per 1667 people

Malaria, tuberculosis, diarrheal diseases

Availability of doctors and hospital beds is low, and there is a shortage of equipment and medicines. Uncontrolled counterfeit drugs are common. A specialized cancer hospital in Lahore, opened in 1995, offers modern facilities and advanced treatment. Pakistan has a high incidence of heroin addicts, due largely to its proximity to Afghanistan.

SPENDING

GDP/cap. increase

CONSUMPTION AND SPENDING

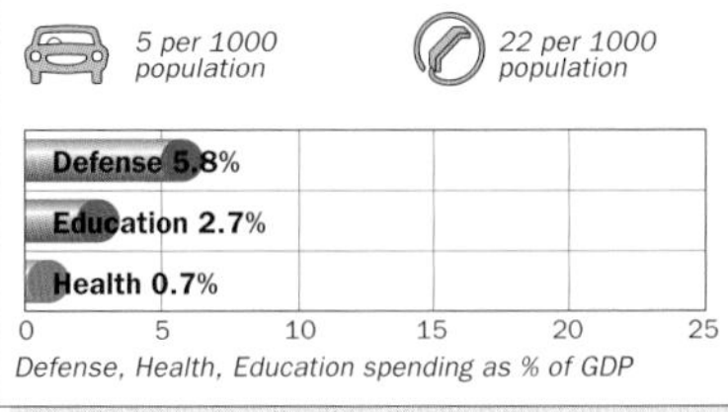
5 per 1000 population
22 per 1000 population
Defense 5.8%
Education 2.7%
Health 0.7%
0
5
10
15
20
25
Defense, Health, Education spending as % of GDP

Members of the bureaucratic and political elite tend to be extremely rich, as are some of the top military. Despite Pakistan's considerable economic potential, many people live below the poverty line.

CRIME

Death penalty in use

78,938 prisoners

Down 4% in 1999

CRIME RATES

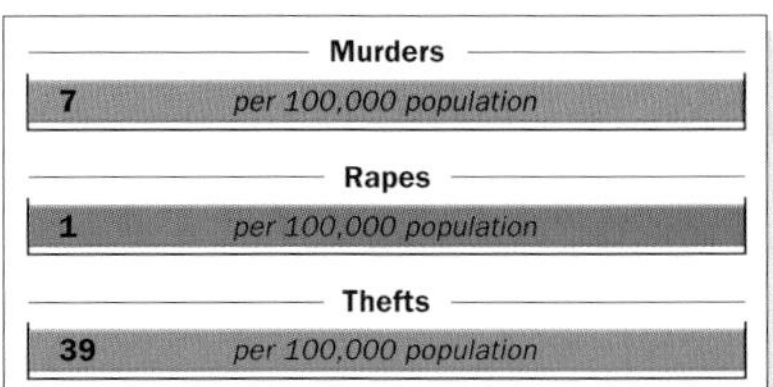
Murders
7 per 100,000 population
Rapes
1 per 100,000 population
Thefts
39 per 100,000 population

Compared with similar Islamic states, rates of murder, kidnapping, narcotics trafficking, rape, and robbery are high. Corruption and the abuse of women are major causes for concern; reports of deaths or death threats for refusing to accept arranged marriages are rising. Torture and rape of prisoners and deaths in custody are frequent. The most dangerous area is Sindh: Karachi is terrorized by severe factional violence. Militant sectarian groups are also blamed for a recent rise in crime in Punjab. Pressure from Islamic parties has forced the government of North West Frontier Province to replace British-based civil law with the rulings of *sharia* courts. Special part-military courts were established in 2002 to combat terrorism and other "serious" crimes.

WORLD RANKING

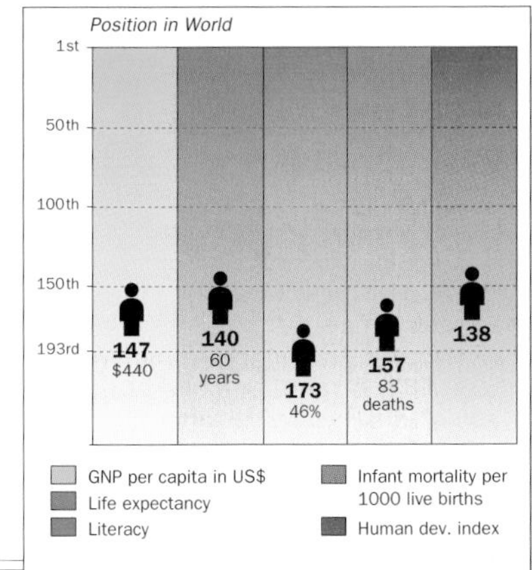
Position in World
1st
50th
100th
150th
193rd
147
$440
140
60
years
173
46%
157
83
deaths
138
GNP per capita in US$
Life expectancy
Literacy
Infant mortality per 1000 live births
Human dev. index

P

PALAU

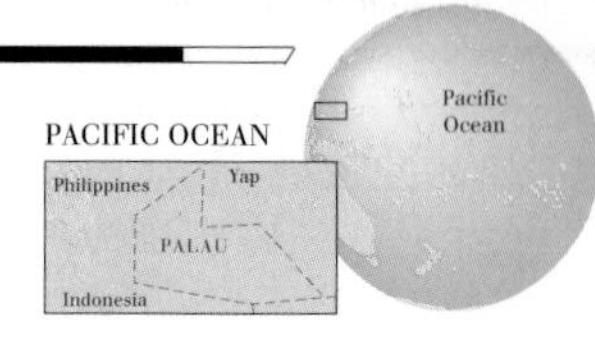
PACIFIC OCEAN
Pacific Ocean
Philippines
Yap
PALAU
Indonesia

OFFICIAL NAME: Republic of Palau **CAPITAL:** Koror
POPULATION: 19,100 **CURRENCY:** US dollar **OFFICIAL LANGUAGES:** Palauan and English

1994

1994

Oct 1
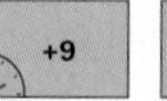
PAL

+9

+680

.pw

THE REPUBLIC OF PALAU (locally known as Belau) is situated in the western Pacific and comprises more than 300 islands in the Caroline Islands archipelago, only nine of which are inhabited. Formerly a part of the US-administered Trust Territory of the Pacific Islands, Palau became independent in association with the US in 1994, but continues to be heavily dependent on US aid.

CLIMATE

Tropical oceanic

WEATHER CHART FOR KOROR

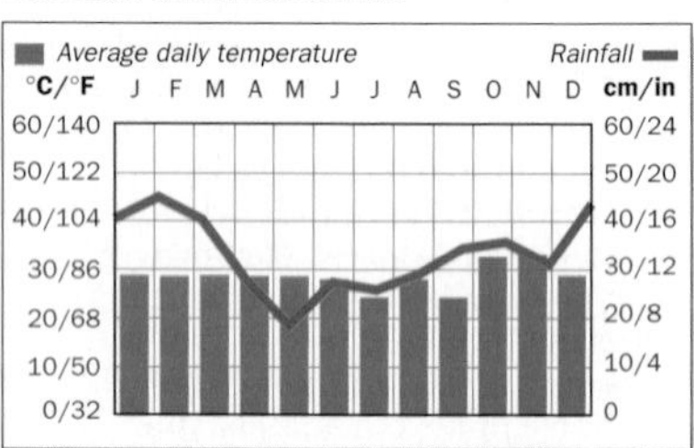
Average daily temperature
Rainfall
°C/°F
J F M A M J J A S O N D
cm/in
60/140
50/122
40/104
30/86
20/68
10/50
0/32
60/24
50/20
40/16
30/12
20/8
10/4
0

The islands are humid, with fairly constant temperatures and heavy rainfall all year round. The mean temperature is 27°C (81°F).

TRANSPORTATION

Drive on right

Palau International, Koror
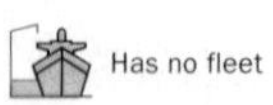
Has no fleet

THE TRANSPORTATION NETWORK

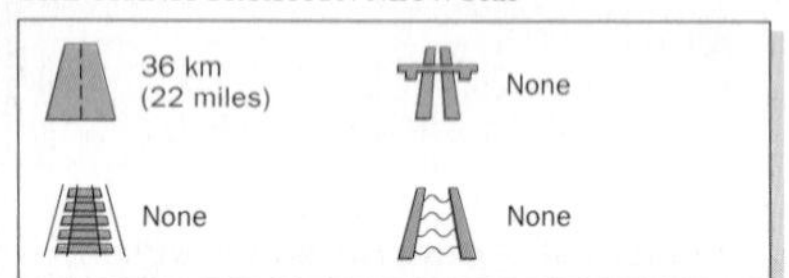
36 km
(22 miles)
None
None
None

There are limited air and sea links between islands. A new bridge, opened in 2002, connects Koror to Babelthuap.

TOURISM

Visitors : Population 2.9:1

55,000 visitors

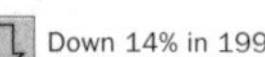
Down 14% in 1999

MAIN TOURIST ARRIVALS

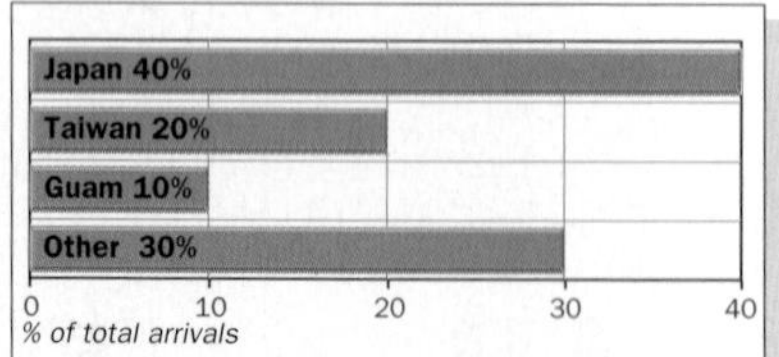
Japan 40%
Taiwan 20%
Guam 10%
Other 30%
0
10
20
30
40
% of total arrivals

Tourism is becoming more important. Improved transportation access is planned, amid concerns about impact on traditional culture. Several islands have battle sites from the Pacific War. The outlying islands remain unspoiled.

PEOPLE

Pop. density low

Palauan, English, Japanese, Angaur, Tobi, Sonsorolese

38/km²
(98/mi²)

THE URBAN/RURAL POPULATION SPLIT

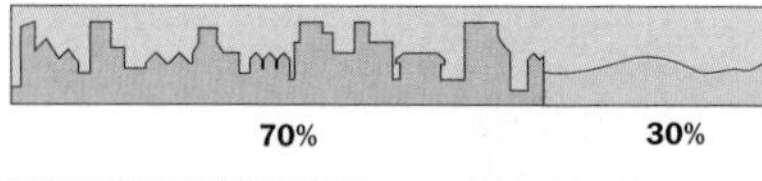
70%
30%

RELIGIOUS PERSUASION

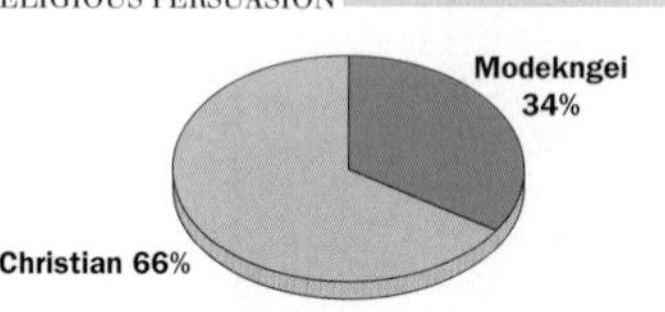
Modekngei 34%
Christian 66%

Palauans can claim a diverse genetic heritage. Palau was first colonized by southeast Asian peoples some 3000 years ago. There has also been mixing with other Melanesian and Polynesian peoples. More recently, low-skilled Filipinos have added to the population. Immigration from south Asia was banned in 2001 due to rising tensions.

Some 70% of Palauans live on the island city of Koror (a new capital is being constructed on neighboring Babelthuap island). The thinly scattered remaining population are linguistically diverse, with distinct separate languages in the most southern islands. Cultural influence from the US and Japan has been strong, though in the more remote islands a more traditional way of life has been maintained. Society remains largely matrilineal. The indigenous Modekngei religion is a unique blend of traditional beliefs and Christianity.

Palau's islands have many idyllic beaches, but a lack of resources means that tourism remains underdeveloped.

POLITICS

Nonparty elections

L. House 2000/2004
U. House 2000/2004

President Tommy Remengesau

AT THE LAST ELECTION

House of Delegates 16 seats

One member is elected to the House of Delegates to represent each of the 16 states

Senate 14 seats

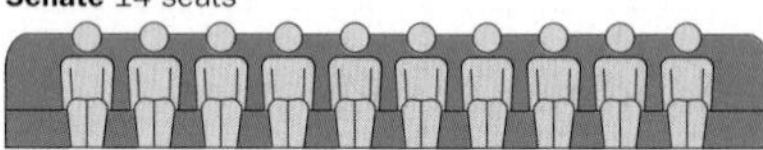

The 14 senators represent geographical districts, according to population

The Compact of Free Association with the US formed the basis of Palau's independence. One clause, requiring the transit and storage of nuclear materials, was the subject of several referendums until a simple majority in 1993 brought sovereignty the following year. Vice President Tommy Remengesau was elected president in November 2000.

WORLD AFFAIRS

Joined UN in 1994

PC

PIF

IBRD

ACP

IWC

The US has exclusive control over Palau's foreign affairs and defense policies under the 1994 conditions of the Compact of Free Association. Cordial relations with the Pacific Islands Forum were restored in 1999 after tensions caused by Palau's bid to give Japan the right to veto the establishment of a whale sanctuary in the South Pacific.

AID

Recipient

$39m (receipts)

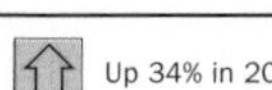
Up 34% in 2000

Palau is heavily dependent on aid, which accounts for around 30% of GDP. Under the 15-year Compact of Free Association, signed with the US in 1994, Palau is set to receive up to $700 million in return for the use of military facilities. However, in recent years Japan has overtaken the US as the country's major donor.

DEFENSE

No compulsory military service

There are no armed forces

Not applicable

Under the 1994 Compact of Free Association, the US is responsible for Palau's defense.

P

ECONOMICS

Inflation 4% p.a. (1994–2000)

 $82m

Currency is US dollar

SCORE CARD

- ❑ WORLD GNP RANKING........................189th
- ❑ GNP PER CAPITA $5000
- ❑ BALANCE OF PAYMENTS....................$17.2bn
- ❑ INFLATION 3%
- ❑ UNEMPLOYMENT 2%

STRENGTHS

Relationship with aid donors – US and Japan – provides economic stability and access to lucrative markets. Improved transportation infrastructure. Tourism industry growing. Fishing and copra production important. Increasing regional trade. Trust funds established from Compact money.

EXPORTS

Palau does not publish export figures by country

IMPORTS

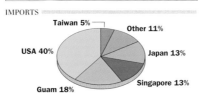

Taiwan 5%
Other 11%
Japan 13%
Singapore 13%
Guam 18%
USA 40%

WEAKNESSES

Heavy dependence on aid. Remote location. Underemployment. Poor transportation links between islands and to other countries. Few resources.

RESOURCES

Electric power 62,000 kw

 1801 tonnes

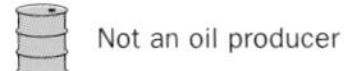 Not an oil producer

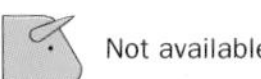 Not available

 Gold

On some islands the soil is highly fertile, although the terrain of the larger islands makes farming difficult. Some islands are densely forested. Palau has copra and some gold deposits. There is also the possibility of exploitation of reserves of minerals on the seabed. Palau has a small fishing industry with the potential for development.

ENVIRONMENT

Not available

None

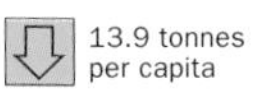 13.9 tonnes per capita

Palau suffers from inadequate facilities for the disposal of solid waste. Sand and coral dredging and illegal fishing practices pose a significant threat to the marine ecosystem. Palau and its surrounding waters are Micronesia's richest habitat; however, there is concern about the commercial export of fruit bats to neighboring islands as a delicacy. Typhoons sometimes cause severe damage to infrastructure.

MEDIA

TV ownership high

 There are no daily newspapers

PUBLISHING AND BROADCAST MEDIA

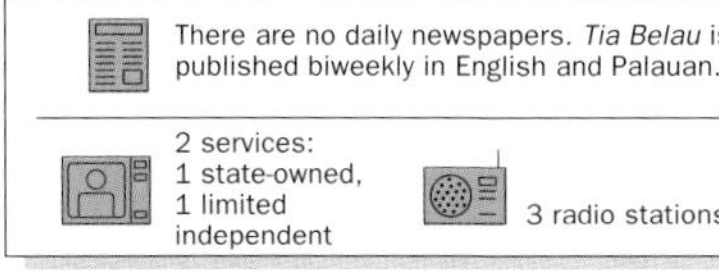
There are no daily newspapers. *Tia Belau* is published biweekly in English and Palauan.
2 services: 1 state-owned, 1 limited independent
3 radio stations

The country's TV and radio stations tend to deal in material which is largely derived from the US.

CRIME

No death penalty

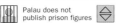

 Palau does not publish prison figures

 Little change from year to year

There is a little alcohol-related crime, but much of the country, particularly the outlying islands, is crime-free.

EDUCATION

School leaving age: 14

 92%

 305 students

Elementary education is compulsory between the ages of six and 14. The Micronesian Occupational College, based in Palau, provides two-year training programs.

PALAU

Total Land Area : 458 sq. km (177 sq. miles)

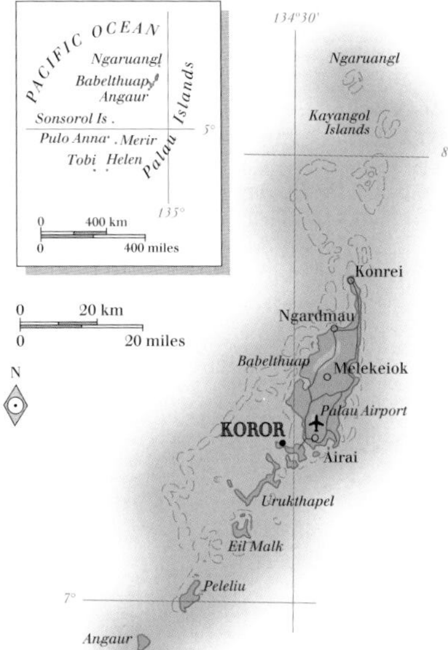
PACIFIC OCEAN
Ngaruangl
Babelthuap
Angaur
Sonsorol Is.
Pulo Anna
Merir
Tobi
Helen
Palau Islands
0 400 km
0 400 miles
134°30'
Ngaruangl
Kayangol Islands
Konrei
Ngardmau
Babelthuap
Melekeiok
Palau Airport
KOROR
Airai
Urukthapel
Eil Malk
Peleliu
Angaur
0 20 km
0 20 miles
N

CHRONOLOGY

The Caroline Islands were colonized in turn by Spain, Germany, and Japan, coming under US control in 1945.

- ❑ **1947** UN Trust Territory of the Pacific Islands established.
- ❑ **1982** Compact of Free Association with US signed.
- ❑ **1993** Compact approved.
- ❑ **1994** Independence in free association with US.
- ❑ **2001** Tommy Remengesau president.

HEALTH

Welfare state health benefits

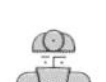 1 per 83 people

 Heart, cerebrovascular, and intestinal diseases

Basic health care is available. Many outlying islands do not have easy access to qualified doctors and therefore often rely on nurses or traditional health remedies. An epidemic of mosquito-borne dengue fever hit Palau in 2000.

SPENDING

Not available

CONSUMPTION AND SPENDING

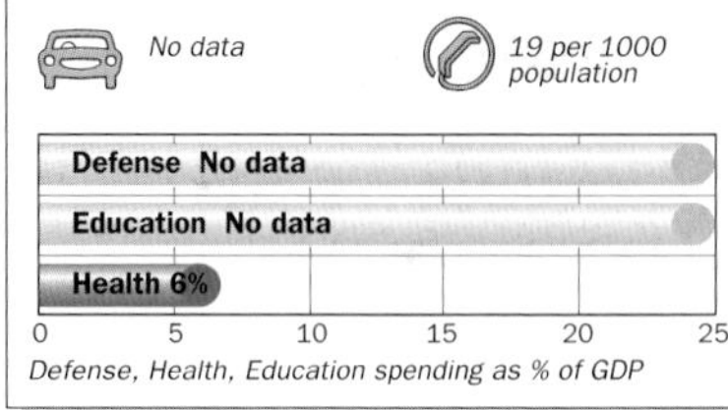
No data
19 per 1000 population
Defense No data
Education No data
Health 6%
0 5 10 15 20 25
Defense, Health, Education spending as % of GDP

The gap between rich and poor is growing steadily, as entrepreneurs and government officials exploit aid and develop the tourist industry. In 2001, a program providing cheap rental housing for low-income families received a US grant of $200,000.

WORLD RANKING

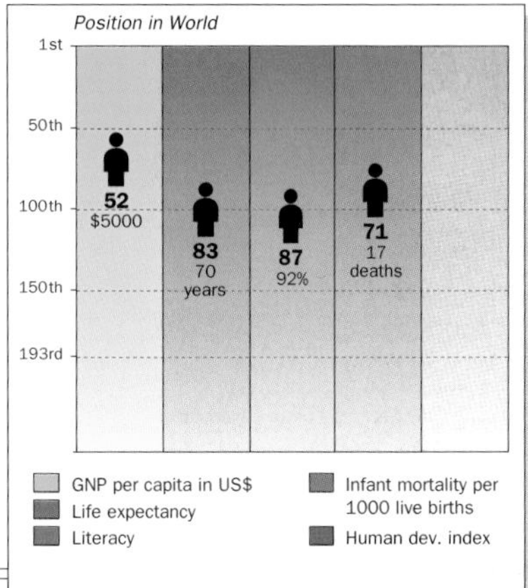
Position in World
1st
50th
100th
150th
193rd
52 $5000
83 70 years
87 92%
71 17 deaths
GNP per capita in US$
Life expectancy
Literacy
Infant mortality per 1000 live births
Human dev. index

P

PANAMA

CENTRAL AMERICA
Costa Rica
PANAMA
Colombia
North America

OFFICIAL NAME: Republic of Panama **CAPITAL:** Panama City
POPULATION: 2.9 million **CURRENCY:** Balboa **OFFICIAL LANGUAGE:** Spanish

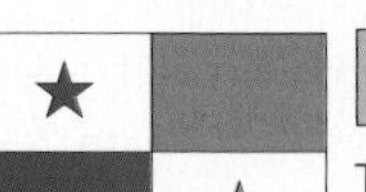

1903

1903
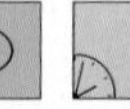
Nov 3

PA
-5

+507

.pa

PANAMA IS THE SOUTHERNMOST of the seven countries occupying the isthmus that joins North and South America. The rainforests of the southeastern Darien Peninsula are some of the wildest areas left in the Americas. Elected governments have held power since the US invasion of 1989. Panama's traditional economic strength is its banking sector. The US returned control of the Panama Canal Zone to Panama on 31 December 1999.

***Cruise liner** on the Panama Canal. The canal shortens the sea route between the east coast of the US and Japan by 4800 km (3000 miles).*

CLIMATE

Tropical wet & dry

WEATHER CHART FOR PANAMA CITY

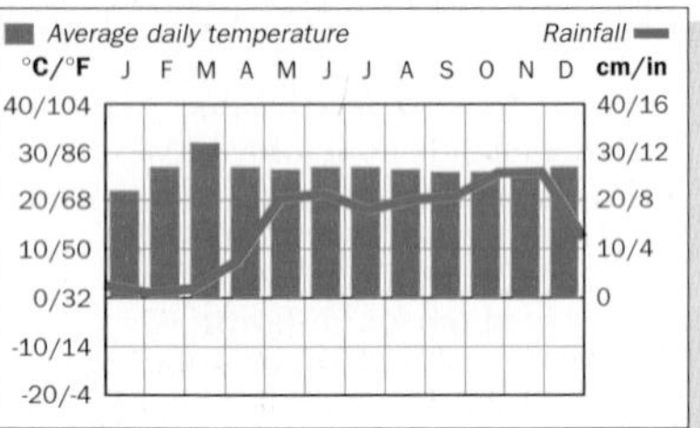
Average daily temperature
Rainfall
°C/°F J F M A M J J A S O N D cm/in
40/104 30/86 20/68 10/50 0/32 -10/14 -20/-4
40/16 30/12 20/8 10/4 0

Panama has a humid tropical climate; rainfall is twice as heavy on the Caribbean coast as on the Pacific coast.

TRANSPORTATION

Drive on right

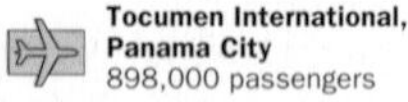
Tocumen International, Panama City
898,000 passengers
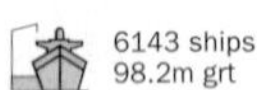
6143 ships
98.2m grt

THE TRANSPORTATION NETWORK

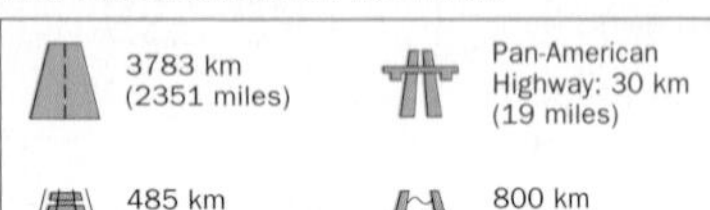
3783 km (2351 miles)
Pan-American Highway: 30 km (19 miles)
485 km (301 miles)
800 km (497 miles)

The 80-km (50-mile) Panama Canal cuts around two weeks off the voyage between the Atlantic and Pacific Oceans. Boats form the most convenient means of transportation for the coastal regions. Many roads are in disrepair.

TOURISM

Visitors : Population 1:6.1

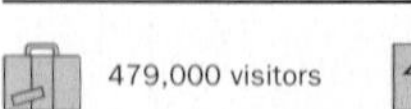
479,000 visitors
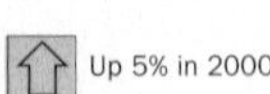
Up 5% in 2000

MAIN TOURIST ARRIVALS

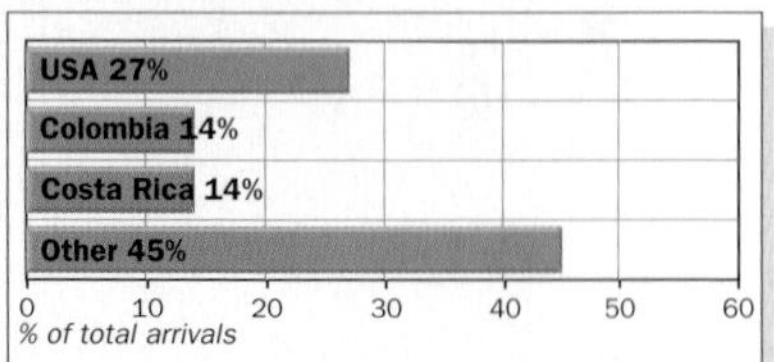
USA 27%
Colombia 14%
Costa Rica 14%
Other 45%
0 10 20 30 40 50 60
% of total arrivals

Portobelo and Panama City have old Spanish colonial buildings. In 2000, new cruise-ship facilities opened in Colón.

PEOPLE

Pop. density low

English Creole, Spanish, Amerindian languages, Chibchan

38/km² (99/mi²)

THE URBAN/RURAL POPULATION SPLIT

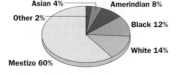

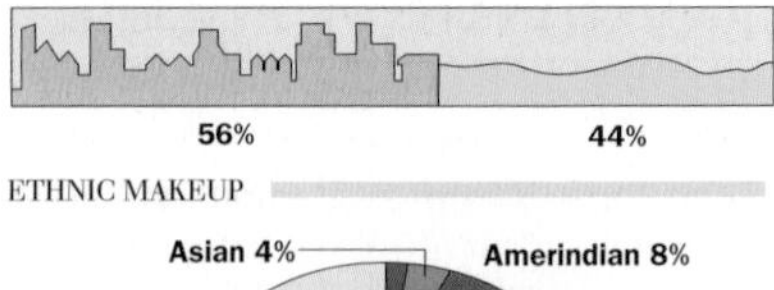
56%
44%

ETHNIC MAKEUP

Asian 4%
Amerindian 8%
Other 2%
Black 12%
White 14%
Mestizo 60%

The northwest coast has a large black community, mostly descended from African immigrants who worked the plantations. The majority speak English Creole rather than Spanish. About 8% of the population are Amerindians mainly from the Guaymies, Chocoes, Kunas, and Ngobe-Buglé tribes. Roman Catholicism and the extended family remain strong, although the canal and the former US military bases have given society a cosmopolitan outlook.

POLITICS

Multiparty elections

1999/2004

President Mireya Moscoso

AT THE LAST ELECTION

Legislative Assembly 71 seats

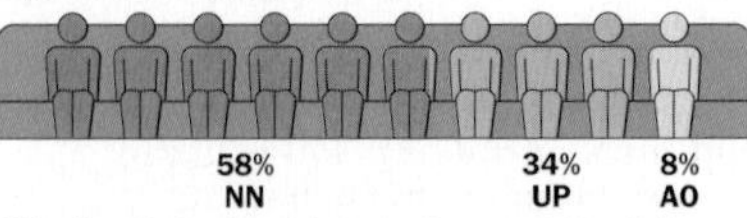
58% NN
34% UP
8% AO

NN = New Nation (comprises the Democratic Revolutionary Party (**PRD**), the Solidarity Party, the National Liberal Party, and the Papa Egoro Movement)
UP = Union for Panama (comprises the Arnulfisto Party (**PA**), MOLIRENA, the Party for Democratic Change, and MORENA)
AO = Action for the Opposition (comprises the Christian Democratic Party, the Liberal Party, the Popular Nationalist Party, and the Civil Renovation Party)

In 1989, the US invaded Panama and arrested its ruler, Gen. Manuel Noriega, for narcotics smuggling. US forces installed the compliant Endara government, criticized for corruption. The 1994 elections were won by Ernesto Pérez Balladares and the PRD, Noriega's old party, but the new government was largely pro-US, and its economic reforms attracted widespread discontent. A 1998 referendum denied Pérez Balladares a second consecutive term. In 1999 opposition leader Mireya Moscoso of the PA was elected as Panama's first woman president. She promised social justice for the poor, but has followed an IMF-backed program and seen her popularity decline.

PANAMA

Total Land Area : 78 200 sq. km (30 193 sq. miles)

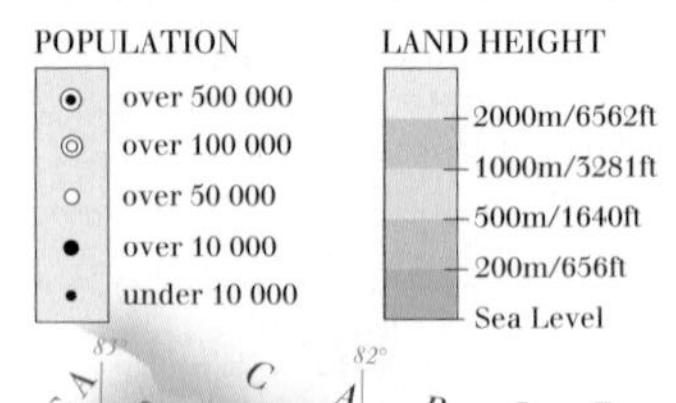
POPULATION
over 500 000
over 100 000
over 50 000
over 10 000
under 10 000
LAND HEIGHT
2000m/6562ft
1000m/3281ft
500m/1640ft
200m/656ft
Sea Level

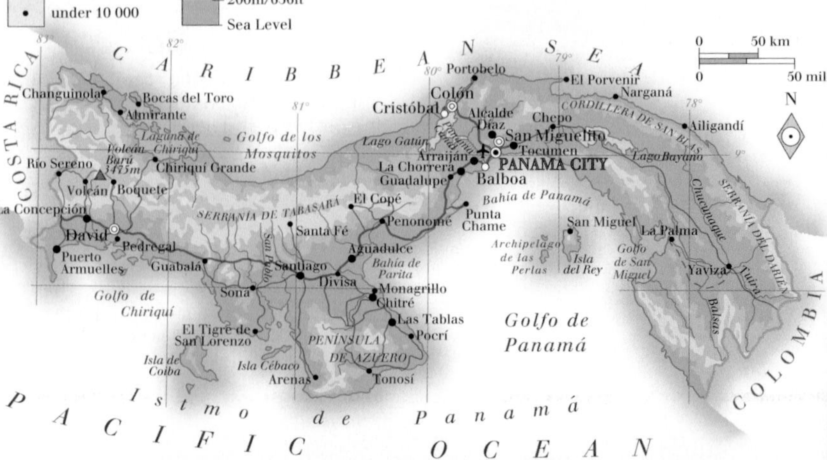
CARIBBEAN SEA
COSTA RICA
COLOMBIA
PACIFIC OCEAN
Istmo de Panamá
0 50 km
0 50 miles
Changuinola
Bocas del Toro
Almirante
Laguna de Chiriquí
Golfo de los Mosquitos
Volcán Barú 3475m
Chiriquí Grande
Río Sereno
Volcán
Boquete
La Concepción
David
Pedregal
Puerto Armuelles
Guabalá
Golfo de Chiriquí
Serranía de Tabasará
Santa Fé
Santiago
Soná
El Tigre de San Lorenzo
Isla de Coiba
Isla Cébaco
Península de Azuero
Arenas
Tonosí
Divisa
Aguadulce
Bahía de Parita
Monagrillo
Chitré
Las Tablas
Pocrí
Penonomé
El Copé
Portobelo
Colón
Cristóbal
Lago Gatún
La Chorrera
Guadalupe
Arraiján
Alcalde Díaz
San Miguelito
Tocumen
PANAMA CITY
Balboa
Bahía de Panamá
Punta Chame
Chepo
El Porvenir
Narganá
Cordillera de San Blas
Ailigandí
Lago Bayano
Archipiélago de las Perlas
Isla del Rey
San Miguel
Golfo de San Miguel
La Palma
Yaviza
Serranía del Darién
Golfo de Panamá

WORLD AFFAIRS

Joined UN in 1945

RG
Geplac
NAM
OAS
San José

The Canal Zone reverted to Panama on December 31, 1999. After failing to reach agreement to allow some troops to stay on in Panama, the US, which had maintained 14 military bases there, has re-based its antinarcotics activities in Florida.

Panama has had observer status with the Andean Community since 1995.

AID

Recipient

 $17m (receipts)

 Up 13% in 2000

The IDB pledged $3.3 million in 2000 to rehabilitate a former US base; Japan has loaned $1 million for the project.

DEFENSE

No compulsory military service

 $127m

 Down 1% in 2000

The National Guard and defense forces were disbanded in 1990 following the 1989 US invasion. They were replaced by the Panamanian Public Force, numbering some 11,800 and comprising the National Police, the National Air Service, and the National Maritime Service.

ECONOMICS

Inflation 1.9% p.a. (1990–2000)

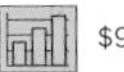 $9.31bn

 1 balboa

SCORE CARD

❑ World GNP Ranking	87th
❑ GNP per Capita	$3260
❑ Balance of Payments	–$927m
❑ Inflation	1.4%
❑ Unemployment	13%

Strengths

Colón Free Trade Zone second-largest in world. Strong banking, financial, insurance, and other allied services. Banana, shrimp exports. Merchant shipping payments for sailing under the Panamanian flag.

Weaknesses

History of political instability and corruption. Large foreign debt. High unemployment, underemployment. Poor infrastructure.

EXPORTS

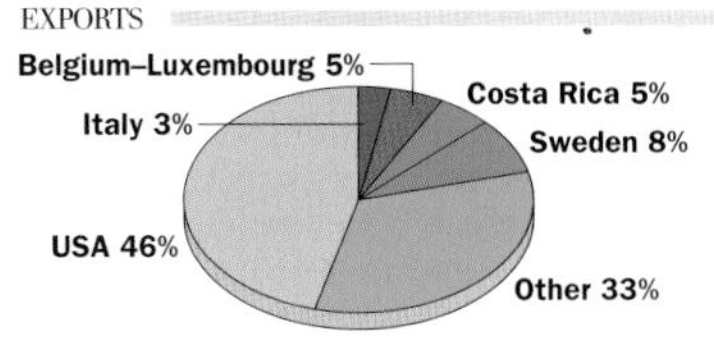
Belgium–Luxembourg 5%
Italy 3%
Costa Rica 5%
Sweden 8%
USA 46%
Other 33%

IMPORTS

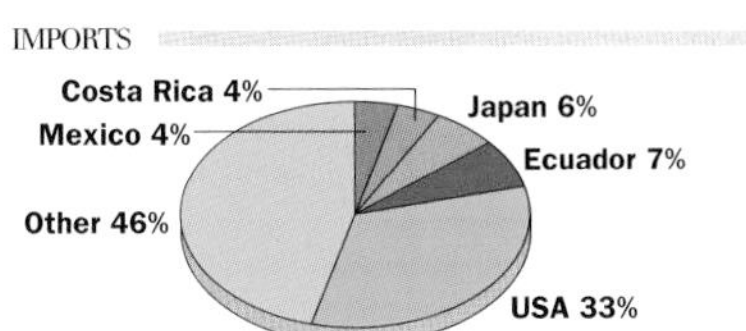
Costa Rica 4%
Mexico 4%
Japan 6%
Ecuador 7%
Other 46%
USA 33%

RESOURCES

Electric power 1m kw

 123,734 tonnes

Not an oil producer; refines 100,000 b/d

 1.34m cattle, 277,900 pigs, 14.5m chickens

Copper, coal, gold, silver, manganese, salt, clay

The Petaquilla area, west of the canal, has great copper and gold potential. To reduce the country's dependence on oil imports, the government has stepped up hydroelectric production; four state energy plants were privatized in 1999. Tropical hardwoods are being cut down at an alarming rate.

ENVIRONMENT

Sustainability rank: 17th

19%

 2.1 tonnes per capita

The destruction of rainforests is proceeding at an increasingly rapid rate, resulting in widespread soil erosion. Large numbers of rare bird and animal species are threatened. Sewage from Panama City and Colón is discharged directly into coastal waters, canals, and ditches. Stretches of mangrove swamps are cut down for urban development, shrimp farms, and resorts.

MEDIA

TV ownership medium

 Daily newspaper circulation 62 per 1000 people

PUBLISHING AND BROADCAST MEDIA

There are 8 daily newspapers, including *La Prensa* and *La Estrella de Panamá*

 5 independent services

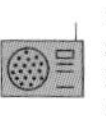 1 state-owned service, over 200 independent stations

A more independent press has flourished since Noriega's overthrow. Radio reaches the greatest number.

CRIME

No death penalty

 8517 prisoners

Down 44% 1996–1998

Panama City and Colón have high crime levels. Money laundering, narcotics trafficking, and corruption are rife.

EDUCATION

School leaving age: 15

 92%

 84,452 students

Schooling is based on the US model. Provision for the urban poor, blacks, and indigenous people is limited.

CHRONOLOGY

On independence from Spain in 1821, Panama was incorporated into Gran Colombia. Panama gained independence from Colombia with US support in 1903.

- ❑ **1903** US buys concession for Panama Canal.
- ❑ **1914–1939** Canal opens to traffic. US protectorate status ended.
- ❑ **1968–1981** Rule of Col. Torrijos Herrera.
- ❑ **1989** Indicted narcotics trafficker Gen. Noriega annuls elections to retain power. US invasion.
- ❑ **1994** PRD wins presidency and is largest party in parliament.
- ❑ **1999** PA's Mireya Moscoso elected first woman president. December, control of Canal Zone reverts from US to Panama.

HEALTH

Welfare state health benefits

 1 per 688 people

Heart disease, cancers, violence, accidents, tuberculosis

Primary health care is accessible to some two-thirds of the rural population. The isolation of many villages hinders efforts to improve the system.

SPENDING

GDP/cap. increase

CONSUMPTION AND SPENDING

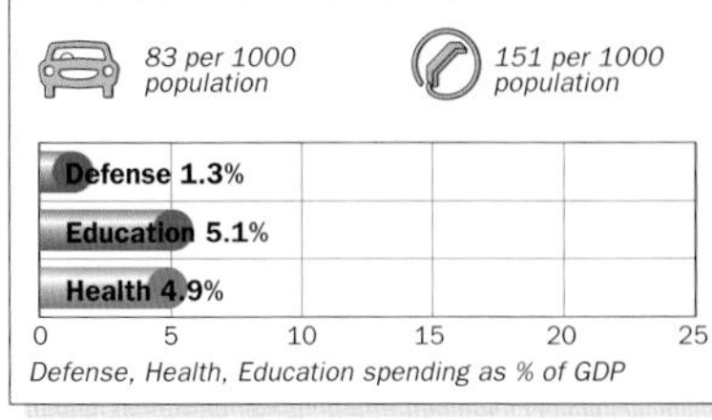
83 per 1000 population
151 per 1000 population
Defense 1.3%
Education 5.1%
Health 4.9%
0
5
10
15
20
25
Defense, Health, Education spending as % of GDP

Wealth disparities are large. Almost 40% of the population are estimated to live below the poverty line – clustered in the cities rather than in rural areas.

WORLD RANKING

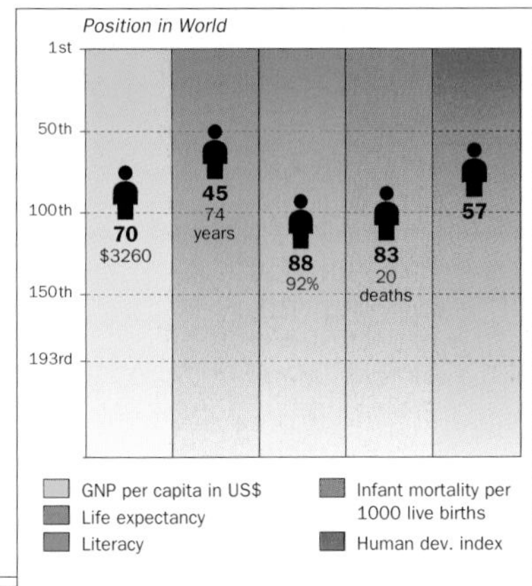
Position in World
1st
50th
100th
150th
193rd
70
$3260
45
74 years
88
92%
83
20 deaths
57
GNP per capita in US$
Life expectancy
Literacy
Infant mortality per 1000 live births
Human dev. index

P

Papua New Guinea

Asia
SOUTHEAST ASIA
Indonesia
Solomon Islands
PAPUA NEW GUINEA

Official name: Independent State of Papua New Guinea **Capital:** Port Moresby
Population: 5.2 million **Currency:** Kina **Official language:** English

1975

1975
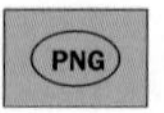

Sept 16
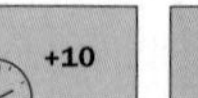
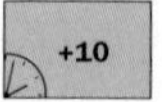
PNG

+10

+675

.pg

THE MOST LINGUISTICALLY diverse country in the world, with approximately 750 languages, Papua New Guinea (PNG) achieved independence from Australia in 1975. The country occupies the eastern end of New Guinea, the world's third-largest island, and several other groups of islands. Much of the country is still isolated and much of the rural population experiences basic living conditions.

CLIMATE

Tropical equatorial/monsoon

WEATHER CHART FOR PORT MORESBY

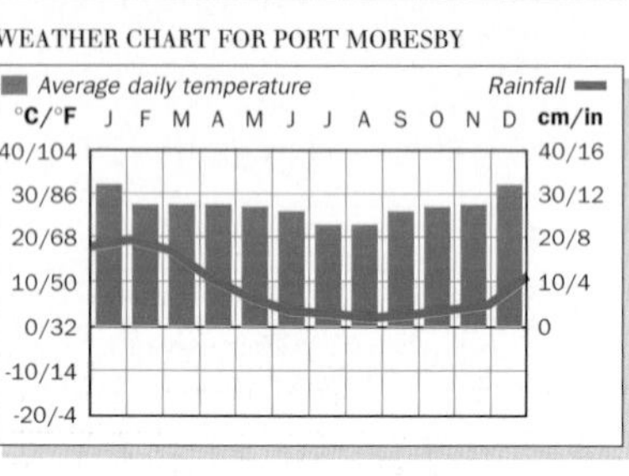
Average daily temperature
Rainfall
°C/°F J F M A M J J A S O N D cm/in
40/104 40/16
30/86 30/12
20/68 20/8
10/50 10/4
0/32 0
-10/14
-20/4

Unvaryingly hot lowlands contrast with snow on Mount Victoria. Severe weather followed El Niño of 1997–1998.

TRANSPORTATION

Drive on left

Jacksons, Port Moresby 745,000 passengers

102 ships 61,000 grt

THE TRANSPORTATION NETWORK

686 km (426 miles)

None

None

10,940 km (6798 miles)

Infrastructure is improving with the construction and upgrading of major link roads, airports, and port facilities.

Papua New Guinea's 600 or so outer islands are mainly mountainous and volcanic, with lush vegetation and fringing coral reefs.

TOURISM

Visitors : Population 1:90

58,000 visitors

Down 13% in 2000

MAIN TOURIST ARRIVALS

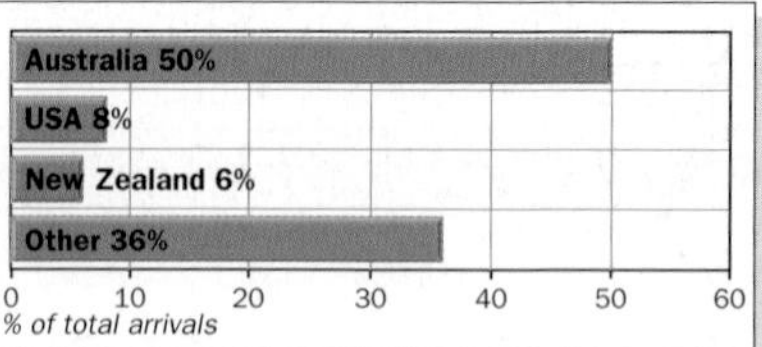
Australia 50%
USA 8%
New Zealand 6%
Other 36%
0 10 20 30 40 50 60
% of total arrivals

Tourism has great potential. However, it is hampered by the high rates of poverty-related violent crime, particularly in urban centers.

PEOPLE

Pop. density low

Pidgin English, Papuan, English, Motu, 750 (est) native languages

11/km² (30/mi²)

THE URBAN/RURAL POPULATION SPLIT

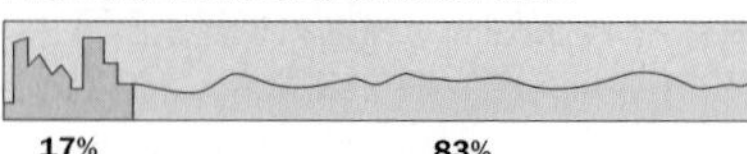
17%
83%

RELIGIOUS PERSUASION

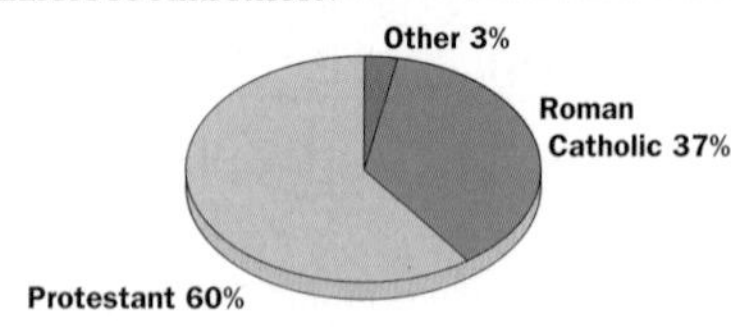
Other 3%
Roman Catholic 37%
Protestant 60%

PNG has an extraordinary diversity of peoples, with around 750 different languages and even more tribes. The key distinction is between the lowlanders, who have frequent contact with the outside world, and the very isolated highlanders. Highland tribes see all strangers as potentially hostile. Vendettas can last for generations and tribal battles are not infrequent. A majority of people are nominally Christian, but indigenous beliefs and practices are widespread.

POLITICS

Multiparty elections

2002/2007

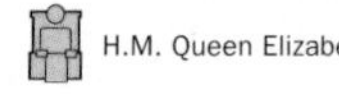
H.M. Queen Elizabeth II

AT THE LAST ELECTION

National Parliament 109 seats

5% PAP
17% NAP
16% Ind
11% PDM
7% PPP
6% V
6% PP
32% Others

NAP = National Alliance Party **Ind** = Independents
PDM = People's Democratic Movement
PPP = People's Progress Party **V** = Vacant
PP = Pangu Pati **PAP** = People's Action Party

The many parties lack clear ideological foundations, creating long-term political instability. The patronage required to maintain coalitions bred corruption. A PDM-led government worked to overhaul the system, and claimed the first simple majority in 2001. However, the party was overshadowed in the 2002 elections by the NAP. Veteran NAP leader Sir Michael Somare was appointed prime minister. A ten-year insurgency by separatists on Bougainville ended with the 1998 cease-fire. Autonomy has been promised, along with a future referendum on independence. Elsewhere, strong local traditions and communications problems have made centralization difficult.

Papua New Guinea

Total Land Area : 462 840 sq. km (178 703 sq. miles)

POPULATION
over 100 000
over 50 000
over 10 000
under 10 000

LAND HEIGHT
3000m/9843ft
2000m/6562ft
1000m/3281ft
500m/1640ft
200m/656ft
Sea Level

NORTH PACIFIC OCEAN
Hermit Is
Ninigo Group
Manus I.
Lorengau
Mussau I.
ADMIRALTY ISLANDS
Ysabel Channel
BISMARCK ARCHIPELAGO
Kavieng
Vanimo
Aitape
Lumi
Wewak
Angoram
INDONESIA (PAPUA)
BISMARCK SEA
New Ireland
Namatanai
Rabaul
Sepik
New Guinea
Karkar I.
CENTRAL RANGE
Madang
New Britain
Ok Tedi
Sologomik
Porgera
Wabag
Mt Wilhelm 4509m
Arop I.
Umboi
Vitiaz Strait
GREEN IS
Buka I.
Pomio
Kiunga
Mendi
Mount Hagen
Kundiawa
Kimbe
Bougainville Island
SOLOMON IS
Arawa
Kieta
Panguna
Buin
Goroka
Lae
Kandrian
Lake Murray
Bulolo
Wau
Huon Gulf
SOLOMON SEA
Balimo
Kerema
Morehead
Gulf of Papua
Mt Victoria
Popondetta
KIRIWINA ISLANDS
Daru
PORT MORESBY
OWEN STANLEY RANGE
Goodenough I.
Muruа Island
Fergusson I.
D'ENTRECASTEAUX IS
Torres Strait
Esa'ala
Normanby I.
Great Barrier Reef
Kupiano
Alotau
Samarai
LOUISIADE ARCHIPELAGO
Tagula I.
CORAL SEA
N
0 200 km
0 200 miles

WORLD AFFAIRS

Joined UN in 1975

APEC

Comm

NAM

PC

PIF

Accusations of PNG support for separatists in Papua, the neighboring Indonesian province, have strained relations. A pro-Taiwan policy, in return for aid funding, provoked Chinese anger in 1999.

AID

Recipient

$275m (receipts)

Up 27% in 2000

Australia is the major aid donor. The World Bank in 2000 endorsed government reforms with a $90 million loan.

DEFENSE

No compulsory military service

$55m

Up 20% in 2000

The army has expressed its doubts over political and economic reforms, most recently in a mutiny in March 2001.

ECONOMICS

Inflation 7.9% p.a. (1990–2000)

$3.61bn

3.035–3.767 kina

SCORE CARD

❑ WORLD GNP RANKING	125th
❑ GNP PER CAPITA	$700
BALANCE OF PAYMENTS	–$8m
❑ INFLATION	15.6%
❑ UNEMPLOYMENT	8%

STRENGTHS

Significant mineral reserves of copper, gold, nickel, cobalt, oil, and natural gas. A proposed gas pipeline between the highlands and Australia is expected to net $219 million a year. Agriculture sustains the population.

WEAKNESSES

Agricultural production and mining were significantly disrupted by severe drought caused by El Niño of 1997–1998. Poor transportation and banking infrastructures. Political instability. Foreign exploitation of resources.

EXPORTS

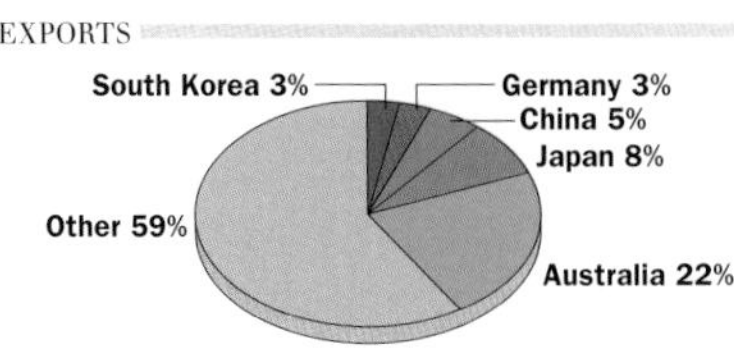
South Korea 3%
Germany 3%
China 5%
Japan 8%
Other 59%
Australia 22%

IMPORTS

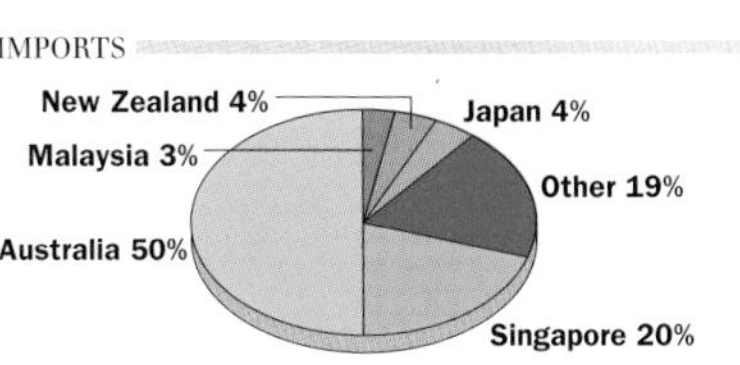
New Zealand 4%
Japan 4%
Malaysia 3%
Other 19%
Australia 50%
Singapore 20%

RESOURCES

Electric power 490,000 kw

53,763 tonnes

57,000 b/d (reserves 200m barrels)

1.6m pigs, 88,000 cattle, 9000 ducks, 3.7m chickens

Copper, gold, silver, gas, oil, nickel, chromite, cobalt

PNG is rich in minerals. The Ok Tedi gold/copper mine in the Star Mountains is the most productive in the country; the Porgera gold mine is one of the world's largest. The severity of the 1998 drought significantly affected production at both. Prospecting has revealed extensive oil and natural gas reserves.

ENVIRONMENT

Sustainability rank: 51st

None

0.5 tonnes per capita

Deforestation and heavy-metal pollution are major issues. Cyanide poisoning from an Australian-owned mine in 2000 caused serious water pollution. Subduction of continental plates has forced the relocation of thousands of people from the low-lying islands.

MEDIA

TV ownership low

Daily newspaper circulation 15 per 1000 people

PUBLISHING AND BROADCAST MEDIA

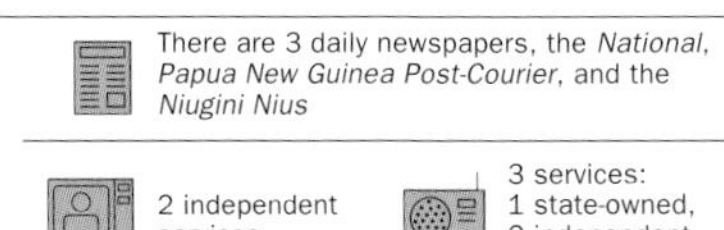
There are 3 daily newspapers, the *National, Papua New Guinea Post-Courier,* and the *Niugini Nius*
2 independent services
3 services: 1 state-owned, 2 independent

The dismissal of the head of the National Broadcasting Commission in 2001 raised concerns over the freedom of the media.

CRIME

Death penalty not used in practice

3566 prisoners

Up 22% in 1992

Violent crime by gangs of "Rascals" is very common. A cultural tradition of vendettas persists in rural communities.

EDUCATION

School leaving age: 16

64%

14,755 students

Education is not compulsory. Equipment charges and fees have been introduced. Universities are suffering funding cuts.

HEALTH

Welfare state health benefits

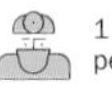
1 per 10,000 people

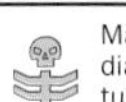
Malaria, pneumonia, diarrheal diseases, tuberculosis

The health system has suffered from recent cuts. HIV and tuberculosis co-infections are at crisis level. Life expectancy rates are among the lowest in the Pacific. Access to clean water and sanitation are major issues.

CHRONOLOGY

The British annexed the southeast and the Germans the northeast of the island of New Guinea in 1884.

- ❑ **1904** Australia takes over British sector; renamed Papua in 1906.
- ❑ **1914** German sector occupied by Australia.
- ❑ **1942–1945** Japanese occupation.
- ❑ **1964** National Parliament created.
- ❑ **1971** Renamed Papua New Guinea.
- ❑ **1975** Independence under Michael Somare, leader since 1972.
- ❑ **1988** Bougainville Revolutionary Army begins guerrilla campaign.
- ❑ **1997** El Niño causes severe drought and tsunamis. Prime Minister Julius Chan resigns over use of Western-led mercenaries in Bougainville.
- ❑ **2000** Loloata Understanding promises autonomy for Bougainville.
- ❑ **2001** PDM claims parliamentary majority. Final peace accord with Bougainville ratified after three-year cease-fire.
- ❑ **2002** NAP wins elections. Somare returns as prime minister.

SPENDING

GDP/cap. increase

CONSUMPTION AND SPENDING

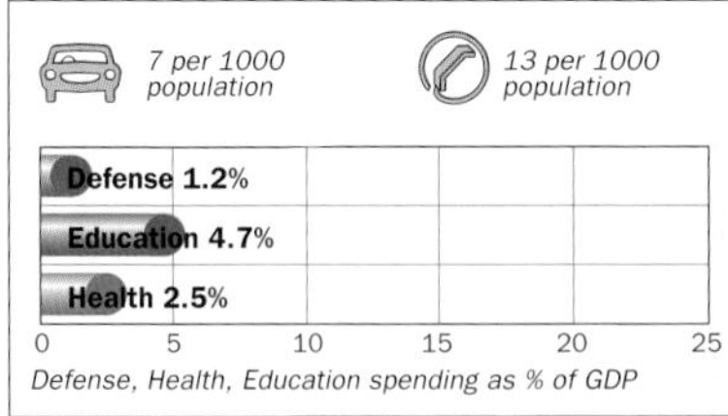
7 per 1000 population
13 per 1000 population
Defense 1.2%
Education 4.7%
Health 2.5%
0
5
10
15
20
25
Defense, Health, Education spending as % of GDP

There is a growing gap between rich and poor, particularly in urban areas. Spending on education and health was cut in 1998, when the kina dropped in value, but it was increased again in 2000, and the 2002 budget proposed the reintroduction of free schooling.

WORLD RANKING

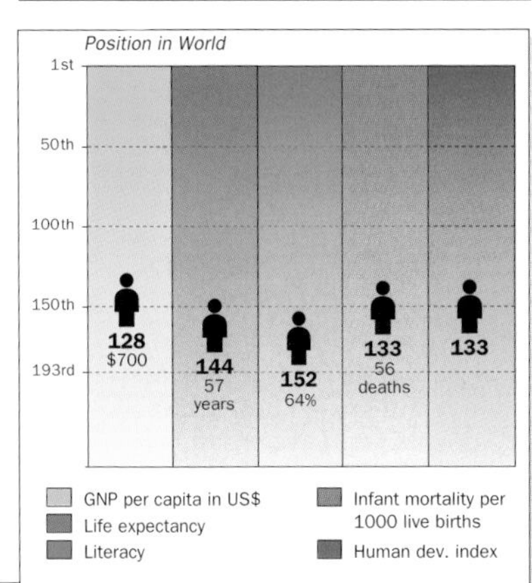
Position in World
1st
50th
100th
150th
193rd
128
$700
144
57 years
152
64%
133
56 deaths
133
GNP per capita in US$
Life expectancy
Literacy
Infant mortality per 1000 live births
Human dev. index

P

PARAGUAY

OFFICIAL NAME: Republic of Paraguay **CAPITAL:** Asunción
POPULATION: 5.6 million **CURRENCY:** Guaraní **OFFICIAL LANGUAGE:** Spanish

1811

1938
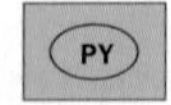
May 14

PY

-4

+595

.py

SOUTH AMERICA

LANDLOCKED IN SOUTH America and a Spanish possession until 1811, Paraguay gained large tracts of land from Bolivia in 1938. From then until the overthrow in 1989 of Gen. Alfredo Stroessner, South America's longest-surviving dictator, it experienced periods of anarchy and military rule. The Paraguay River divides the eastern hills and fertile plains, where 90% of people live, from the almost uninhabited Chaco in the west. Paraguay's economy is largely agricultural.

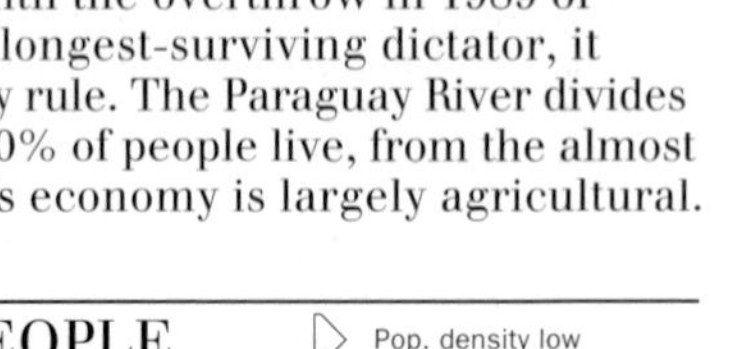

CLIMATE

Tropical/subtropical

WEATHER CHART FOR ASUNCIÓN

Average daily temperature — Rainfall
°C/°F J F M A M J J A S O N D cm/in
40/104, 30/86, 20/68, 10/50, 0/32, -10/14, -20/-4 — 40/16, 30/12, 20/8, 10/4, 0

Paraguay is subtropical, with all parts experiencing floods and droughts, but the Chaco is generally drier and hotter.

TRANSPORTATION

Drive on right

Silvio Pettirossi International, Asunción
465,664 passengers

47 ships
44,900 grt

THE TRANSPORTATION NETWORK

15,000 km (9321 miles)

Pan-American Highway: 700 km (435 miles)

370 km (230 miles)

3100 km (1926 miles)

Roads badly need upgrading. The government wants to part-privatize the near-paralyzed FCCAL (railroad).

TOURISM

Visitors : Population 1:25

221,000 visitors

Down 18% in 2000

MAIN TOURIST ARRIVALS

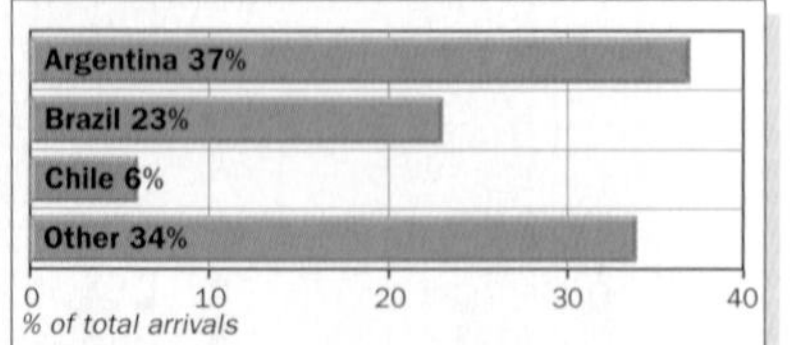
Argentina 37%
Brazil 23%
Chile 6%
Other 34%
0
10
20
30
40
% of total arrivals

Tourist numbers are small. Most visitors are cross-border day trippers from Brazil and Argentina, who flock to Ciudad del Este to buy cheap, mainly Far Eastern, electrical goods. The Chaco attracts tourists for safaris.

PEOPLE

Pop. density low

Guaraní, Spanish

14/km² (37/mi²)

THE URBAN/RURAL POPULATION SPLIT

56% 44%

ETHNIC MAKEUP

Amerindian 2% Other 8% Mestizo 90%

Most Paraguayans are of combined Spanish and native Guaraní origin. The majority are bilingual, although outside the large cities Guaraní is spoken almost exclusively. Rural Guaraní, deprived of ancestral lands, have been forced into marginal labor and prostitution. While most immigrants historically have been European, Japanese, Koreans, and South Africans are recent arrivals.

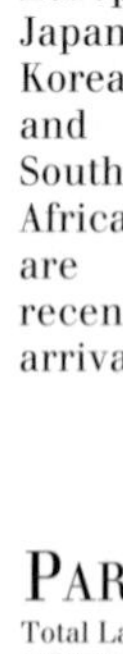

POLITICS

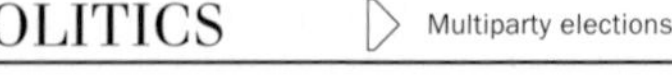

Multiparty elections

L. House 1998/2003
U. House 1998/2003

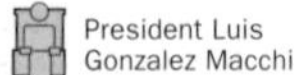

President Luis Gonzalez Macchi

AT THE LAST ELECTION

Chamber of Deputies 80 seats

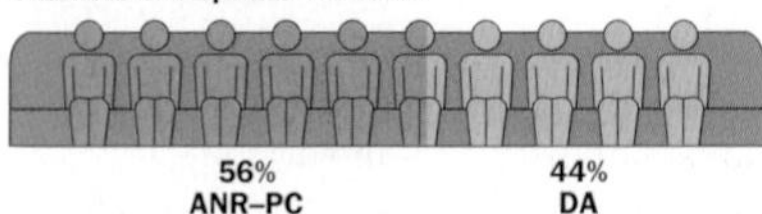

56% ANR–PC — 44% DA

ANR–PC = National Republican Association–Colorado Party
DA = Democratic Alliance (led by the Authentic Radical Liberal Party – **PLRA**)

Senate 45 seats

53% ANR–PC — 45% DA — 2% Other

A 1989 coup ended Gen. Stroessner's 34-year dictatorship. In 1993, his PC won the first free elections in 60 years, but with continued reliance on the military. The PC again won in 1998, despite the last-minute annulment of former army chief Gen. Lino Oviedo's presidential candidacy. Unrest in 1999 after the assassination of Vice President Luis Argaña, Oviedo's main opponent in the PC, forced the resignation of President Raúl Cubas, a supporter of Oviedo. A fragile coalition government survived a coup in May 2000 led by Oviedo – his third such attempt since 1993.

Vice-presidential elections in August 2000 led to an unprecedented cohabitation between a PC president and his PLRA deputy.

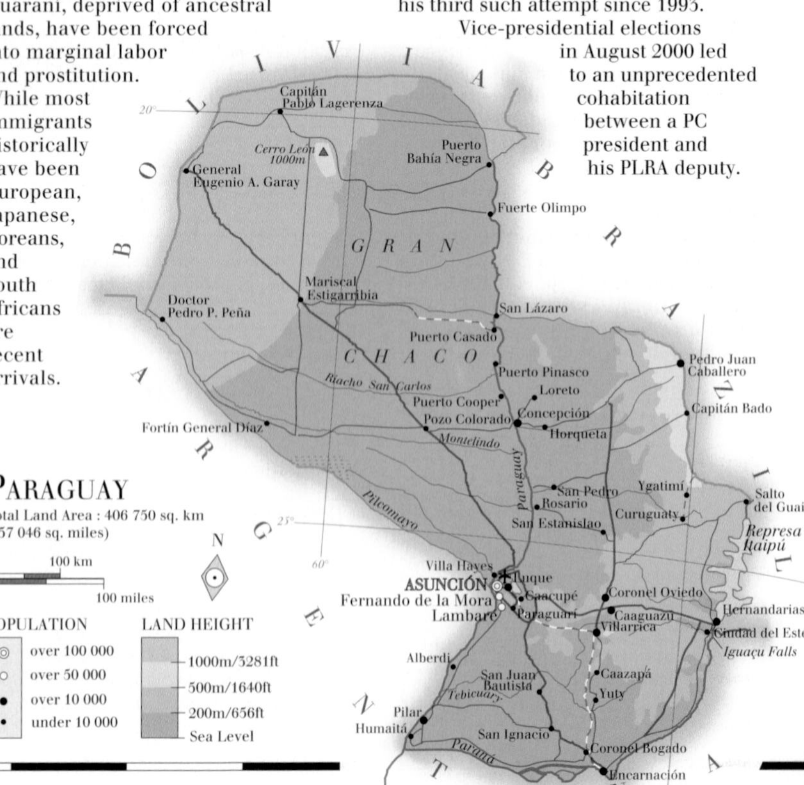
BOLIVIA
BRAZIL
ARGENTINA
Capitán Pablo Lagerenza
Cerro León 1000m
Puerto Bahía Negra
General Eugenio A. Garay
Fuerte Olimpo
GRAN CHACO
Mariscal Estigarribia
Doctor Pedro P. Peña
San Lázaro
Puerto Casado
Puerto Pinasco
Pedro Juan Caballero
Riacho San Carlos
Loreto
Puerto Cooper
Pozo Colorado
Concepción
Capitán Bado
Fortín General Díaz
Montelindo
Horqueta
Paraguay
San Pedro
Rosario
Ygatimí
Salto del Guairá
Pilcomayo
Curuguaty
San Estanislao
Represa Itaipú
Villa Hayes
ASUNCIÓN
Luque
Fernando de la Mora
Caacupé
Coronel Oviedo
Lambaré
Paraguarí
Caaguazú
Villarrica
Hernandarias
Ciudad del Este
Iguaçu Falls
Alberdi
San Juan Bautista
Caazapá
Tebicuary
Yuty
Pilar
Humaitá
San Ignacio
Coronel Bogado
Paraná
Encarnación
PARAGUAY
Total Land Area : 406 750 sq. km (157 046 sq. miles)
0 100 km
0 100 miles
POPULATION
over 100 000
over 50 000
over 10 000
under 10 000
LAND HEIGHT
1000m/3281ft
500m/1640ft
200m/656ft
Sea Level

P

WORLD AFFAIRS

Joined UN in 1945

IBRD
IAEA
Geplac
Mercosr
RG

The main aims are fairer integration in the Mercosur common market and good relations with the US.

AID

Recipient

 $82m (receipts)  Up 5% in 2000

The World Bank offers development aid, the IMF conditional loans. NGO charities run small programs in rural areas.

DEFENSE

Compulsory military service

 $121m 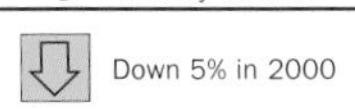 Down 5% in 2000

Under Stroessner, the military controlled political and economic life. In 1994–1995, Congress tried to limit its powers, but President Juan Carlos Wasmosy endorsed its political and institutional role. The pact between the military and the PC, in power since 1947, has been weakened by factionalism.

ECONOMICS

Inflation 13% p.a. (1990–2000)

 $7.93bn  3545–4635 guaranies

SCORE CARD

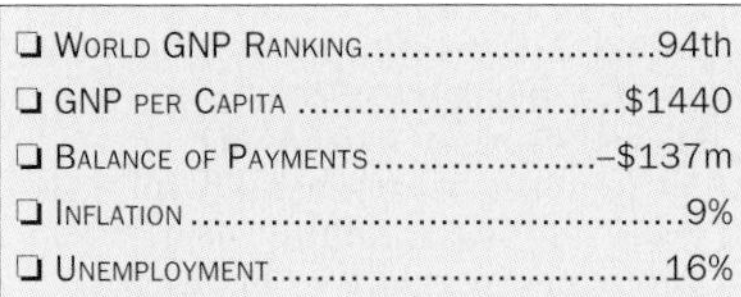

❑ WORLD GNP RANKING	94th
❑ GNP PER CAPITA	$1440
❑ BALANCE OF PAYMENTS	–$137m
❑ INFLATION	9%
❑ UNEMPLOYMENT	16%

STRENGTHS

Electricity exporter – earnings obtain foreign exchange. Self-sufficiency in wheat and other staple foodstuffs. Cotton, oilseeds, notably soybeans.

WEAKNESSES

Reliance on agriculture and shaky Brazilian and Argentine markets. No hydrocarbons produced. Weak banking and financial sectors. High unemployment. Political instability deters foreign investment.

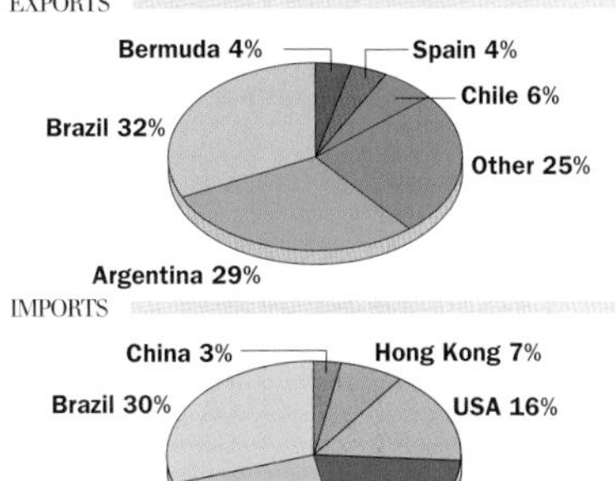
EXPORTS
Bermuda 4%
Spain 4%
Chile 6%
Brazil 32%
Other 25%
Argentina 29%
IMPORTS
China 3%
Hong Kong 7%
Brazil 30%
USA 16%
Other 21%
Argentina 23%

***The Iguaçu Falls,** on the border with Brazil and Argentina, are composed of over 20 cataracts, separated by rocks and tree-covered islands.*

RESOURCES

Electric power 7.1m kw

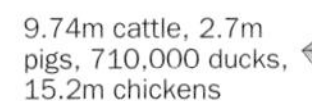 25,095 tonnes  Not an oil producer

9.74m cattle, 2.7m pigs, 710,000 ducks, 15.2m chickens  Gypsum, marble, clay, kaolin, iron, manganese, uranium

The joint Paraguay–Brazil Itaipú hydroelectric dam is the world's largest. The massive Yacyretá Dam is operated with Argentina.

ENVIRONMENT

Sustainability rank: 25th

4%  0.9 tonnes per capita

Apart from the destruction of forests for farming and for dams, a major ecological worry is the smuggling abroad of endangered species.

MEDIA

TV ownership medium

Daily newspaper circulation 43 per 1000 people

PUBLISHING AND BROADCAST MEDIA

 There are 7 daily newspapers, including *ABC Color, Noticias, Popular,* and *Ultima Hora*

 4 independent services 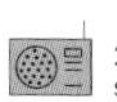 21 independent services

The media, historically sponsored by political parties, flourished after the fall of Stroessner, publishing details of corruption and abuses of human rights. The constitution nominally protects the rights of columnists to air their views.

CRIME

No death penalty

 4088 prisoners  Down 8% in 1999

Paraguay is the contraband capital of Latin America, with trade in everything from cars to cocaine. Jungle airstrips near Brazil provide a route for narcotics.

EDUCATION

School leaving age: 12

 93%  42,302 students

Education is compulsory only to the age of 12, and fewer than half of children go on to secondary school. Provision is limited in remote rural areas.

CHRONOLOGY

Paraguay was controlled by Spain from 1536 until 1811.

- ❑ **1864–1870** Loses War of the Triple Alliance against Argentina, Brazil, and Uruguay.
- ❑ **1928–1935** Two Chaco Wars against Bolivia over disputed territory.
- ❑ **1938** Boundary with Bolivia fixed; Paraguay awarded large tracts.
- ❑ **1954–1989** Rule of Gen. Stroessner; repressive military regime.
- ❑ **1993** First democratic elections.
- ❑ **1996** Coup attempt by Gen. Lino Oviedo.
- ❑ **1998–1999** Raúl Cubas elected president; resigns after assassination of vice president; Cubas and Oviedo leave country.
- ❑ **2000** Further failed coup attempt by Oviedo.

HEALTH

Welfare state health benefits

1 per 909 people; Heart disease, cancers, obstetric causes, tuberculosis

Hepatitis, typhoid, dysentery, and tuberculosis are endemic and leprosy is common. Medical care is expensive.

SPENDING

GDP/cap. increase

CONSUMPTION AND SPENDING

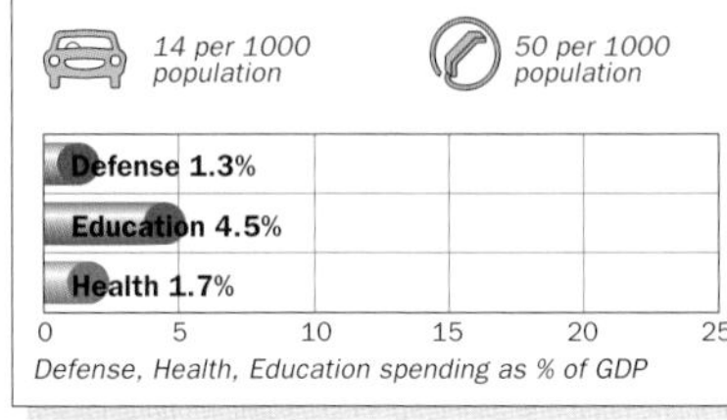
14 per 1000 population
50 per 1000 population
Defense 1.3%
Education 4.5%
Health 1.7%
0
5
10
15
20
25
Defense, Health, Education spending as % of GDP

Income inequality is great and rural poverty serious. Top military ranks, business leaders, and landed elites control wealth.

WORLD RANKING

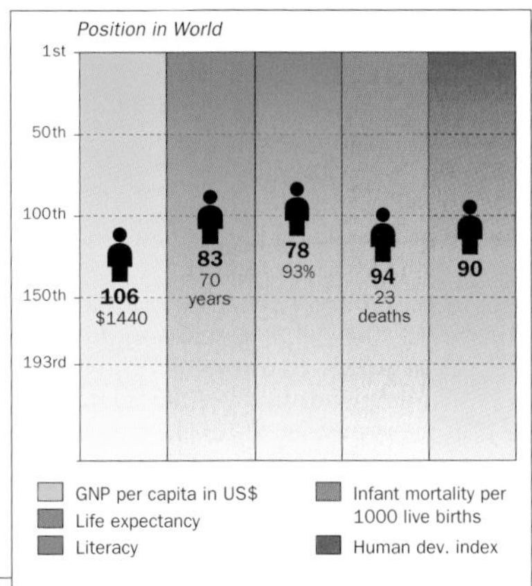
Position in World
1st
50th
100th
150th
193rd
106
$1440
83
70 years
78
93%
94
23 deaths
90
GNP per capita in US$
Life expectancy
Literacy
Infant mortality per 1000 live births
Human dev. index

P

PERU

SOUTH AMERICA

Ecuador Columbia Brazil PERU Bolivia Chile

South America

OFFICIAL NAME: Republic of Peru **CAPITAL:** Lima **POPULATION:** 26.1 million
CURRENCY: Nuevo sol **OFFICIAL LANGUAGES:** Spanish and Quechua

1824 | 1941 | July 28 | PE | -5 | +51 | .pe

LYING JUST SOUTH of the equator, on the Pacific coast of South America, Peru became independent of Spain in 1824. It rises from an arid coastal strip to the Andes, dominated in the south by volcanoes; about half of Peru's population live in mountain regions. Peru's border with Bolivia to the south runs through Lake Titicaca, the highest navigable lake in the world. In 1995, Peru was involved in a brief border war with Ecuador, its northern neighbor, and the issue was finally settled in 1998.

CLIMATE

Tropical/mountain/desert

WEATHER CHART FOR LIMA

Average daily temperature
Rainfall
°C/°F J F M A M J J A S O N D cm/in
40/104 40/16
30/86 30/12
20/68 20/8
10/50 10/4
0/32 0
-10/14
-20/-4

Peru has several distinct climatic regions. The arid or desert coastal region experiences the *garúa*, persistent low cloud and fog, giving Lima cool "winters" even though it is close to the equator. The temperate slopes of the Andes have large daily temperature ranges and one rainy season, while the tropical Amazon Basin receives year-round rains.

TRANSPORTATION

Drive on right

Jorge Chávez International, Lima 4.51m passengers

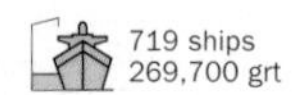
719 ships
269,700 grt

THE TRANSPORTATION NETWORK

8700 km (5406 miles)

Pan-American Highway: 2495 km (1550 miles)

1639 km (1018 miles)

8600 km (5344 miles)

The World Bank in 2001 estimated that Peru had a public infrastructure deficit of $7 billion. Most roads remain unpaved. Work on a transcontinental highway from Ilo, a free port on the Pacific, via Puerto Suárez in Bolivia, to the port of Portos in Brazil is ongoing. The two rail networks, the Central and Southern, are as yet unconnected. The La Oroya–Huancayo line is the world's highest stretch of standard-gauge railroad. River transportation provides major access to Iquitos in Amazonia. As well as four important international airports, there are more than 130 airstrips scattered throughout the country.

Spanish colonial church near Urubamba. *The Urubamba River with its deep gorges was known as the Sacred Valley to the Incas.*

TOURISM

Visitors : Population 1:25

1.03m visitors

Up 9% in 2000

MAIN TOURIST ARRIVALS

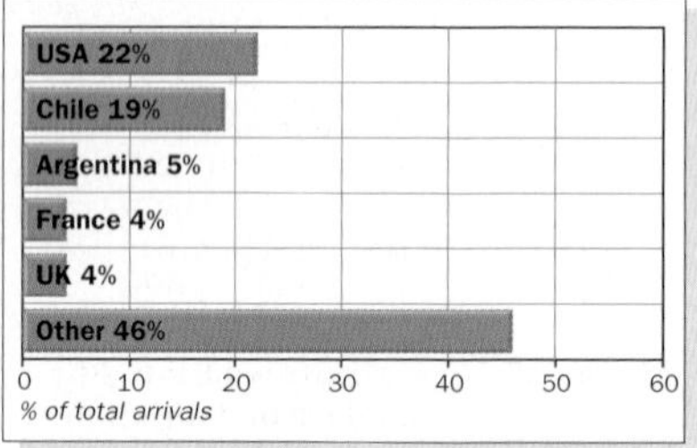
USA 22%
Chile 19%
Argentina 5%
France 4%
UK 4%
Other 46%
0 10 20 30 40 50 60
% of total arrivals

Tourism is gradually recovering, after being plunged into crisis in the early 1990s by guerrilla activity, crime, and cholera fears. The heavily indebted industry has been unable to take full advantage of new investment opportunities, but privatization programs have seen the sale of state hotels. Visitors face poor infrastructure and accommodation to see incomparable sites such as the Inca ruins at Machu Picchu in the Andes. Tourism to the Amazon is also growing, but environmentalists are concerned about the impact on indigenous people. The patterns in the desert made by the Nazca people (known as the Nazca lines), dating from the 2nd century BCE, are also a major attraction. Lake Titicaca and the Spanish colonial architecture of Lima are other draws.

PEOPLE

Pop. density low

Spanish, Quechua, Aymara
20/km²
(53/mi²)

THE URBAN/RURAL POPULATION SPLIT

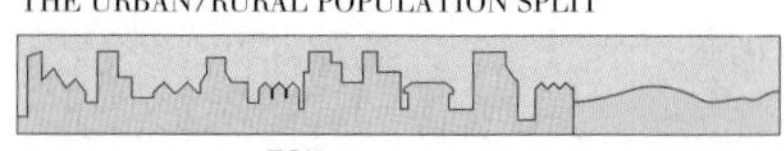
73%
27%

RELIGIOUS PERSUASION

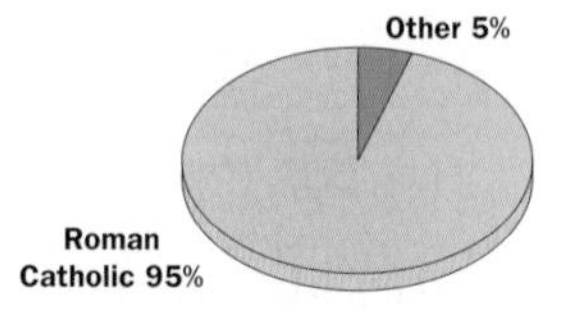
Other 5%
Roman Catholic 95%

ETHNIC MAKEUP

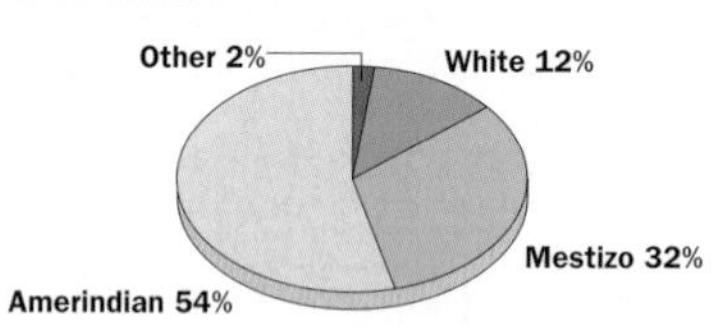
Other 2%
White 12%
Mestizo 32%
Amerindian 54%

Most Peruvians are Amerindian or *mestizo* (mixed race). The small elite of Spanish descendants retain a strong hold on the economy, power, and social standing. A few Chinese and Japanese live in the northern cities.

Previously remote Andean Amerindians are increasingly informed of events in Lima and the coastal strip by radio and by relatives in cities. This has compensated for problems associated with the marginalization of their native Quechua and Aymara languages in a Spanish-speaking culture. A further 250,000 Amazonian Amerindians live in the eastern lowlands. Together with the small community of Africans (descendants of plantation workers), they tend to suffer the worst discrimination in towns.

The extended family remains strong. A part of traditional native Amerindian traditions, its role as a social bond was strengthened by Roman Catholicism. In recent years, economic difficulties have raised its profile as the key social support system for most Peruvians.

POPULATION AGE BREAKDOWN

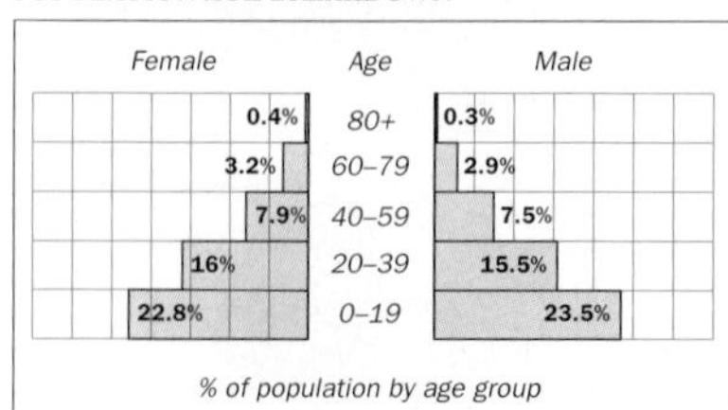
Female Age Male
0.4% 80+ 0.3%
3.2% 60–79 2.9%
7.9% 40–59 7.5%
16% 20–39 15.5%
22.8% 0–19 23.5%
% of population by age group

POLITICS

Multiparty elections

2001/2006

President Alejandro Toledo

Alberto Fujimori, president 1990–2000, ran an increasingly autocratic regime.

***President Alejandro Toledo,** Peru's first Amerindian head of state, elected in 2001.*

AT THE LAST ELECTION

Congress of the Republic 120 seats

Party	Share
PP	38%
APRA	23%
NU	14%
FIM	9%
UPP	5%
SP	3%
Others	8%

PP = Peru Posible **APRA** = American Popular Revolutionary Alliance **NU** = National Unity **FIM** = Independent Moralizing Front **UPP** = Union for Peru **SP** = Somos Peru

Peru is a multiparty democracy with an executive presidency.

PROFILE

The long tradition of large parties dominating politics ended with President Alberto Fujimori's election in 1990. His "self-coup" created a compliant legislature and judiciary, and the approval of a new constitution permitted his reelection in 1995. His popularity was greatly boosted by successes against hyperinflation and the Sendero Luminoso (Shining Path) guerrillas. It faded in the late 1990s as he tightened his personal control of the government, and appeared increasingly reliant on the army. Few checks on the executive remained, and Fujimori was able to obtain his third term in 2000. However, blatant electoral fraud and the corrupt use of power were exposed when his security service chief Vladimiro Montesinos was videoed bribing opposition legislators. Fujimori's position was irrevocably damaged, and he resigned in November, having fled to Japan. Fresh presidential and legislative elections in April 2001 were won by populist Alejandro Toledo and his Peru Posible party.

MAIN POLITICAL ISSUES

Disenchantment with Toledo

President Toledo promised a new start for Peru, but his low ratings in opinion polls testified to early impatience with his administration. Respondents expressed disappointment that he had failed to fulfill election promises, especially on job creation, as the government committed itself to IMF goals on structural reform and fiscal restraint. Polls also pointed to public unease at the continuation of nepotism in high places, a much resented aspect of the Fujimori years. The dismissal of three cabinet ministers revealed discontent in the ruling PP congressional bloc with the defection of the highly popular Cecilia Tait, a former sporting star.

Corruption

The public endorses investigations into widespread corruption during the Fujimori era, but is growing fatigued by endless rounds of legislative commissions and probes. Many would prefer politicians to refocus on sorting out the economy.

Reform of military

Senior officers loyal to Fujimori and Montesinos have been sacked, but the military, and the militarized police, need reforming to make them professional forces. The likelihood of this happening soon is slim.

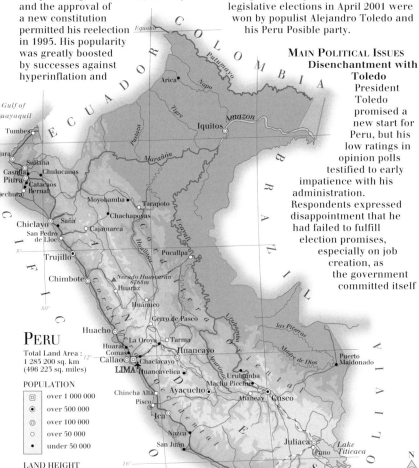

PERU

Total Land Area : 1 285 200 sq. km (496 225 sq. miles)

POPULATION
- over 1 000 000
- over 500 000
- over 100 000
- over 50 000
- under 50 000

LAND HEIGHT
- 4000m/13124ft
- 2000m/6562ft
- 500m/1640ft
- Sea Level

0 200 km
0 200 miles

CHRONOLOGY

Francisco Pízarro's arrival in 1532 during a war of succession between two Inca rulers marked the start of the Spanish colonization of Peru, and the end of the Inca empire.

- ❑ **1821** Independence proclaimed in Lima after its capture by Argentine liberator, José de San Martín, who had just freed Chile.
- ❑ **1824** Spain suffers final defeats at battles of Junín and Ayacucho by Simón Bolívar and Gen. Sucre, liberators of Venezuela and Colombia.
- ❑ **1836–1839** Peru and Bolivia joined in short-lived confederation.
- ❑ **1866** Peruvian–Spanish War.
- ❑ **1879–1884** War of the Pacific. Chile defeats Peru and Bolivia. Peru loses territory in south.
- ❑ **1908** Augusto Leguía y Salcedo's dictatorial rule begins.
- ❑ **1924** Dr. Víctor Raúl Haya de la Torre founds nationalist APRA in exile in Mexico.
- ❑ **1930** Leguía ousted. APRA moves to Peru as first political party.
- ❑ **1931–1945** APRA banned.
- ❑ **1939–1945** Moderate, pro-US civilian government.
- ❑ **1948** Gen. Manuel Odría takes power. APRA banned again.
- ❑ **1956** Civilian government restored.
- ❑ **1962–1963** Two military coups.
- ❑ **1963** Election of Fernando Belaúnde Terry. Land reform, but military used to suppress communist-inspired insurgency.

P

CHRONOLOGY *continued*

- **1968** Military junta takes over. Attempts to alleviate poverty. Large-scale nationalizations.
- **1975–1978** New right-wing junta.
- **1980** Belaúnde reelected. Maoist Sendero Luminoso (Shining Path) begins armed struggle.
- **1981–1998** Border war with Ecuador over Cordillera del Cóndor, given to Peru by a 1942 protocol. Ecuador wants access to Amazon.
- **1982** Deaths and "disappearances" start to escalate as army cracks down on guerrillas and narcotics.
- **1985** Electoral win for left-wing APRA under Alán García Pérez.
- **1987** Peru bankrupt. Plans to nationalize banks blocked by new Libertad movement led by writer Mario Vargas Llosa.
- **1990** Over 3000 political murders. Alberto Fujimori, an independent, elected president on anticorruption platform. Severe austerity program.
- **1992–95** Fujimori "self-coup." New constitution. Fujimori reelected.
- **1996–1997** Left-wing Tupac Amarú guerrillas seize hundreds of hostages at Japanese ambassador's residence in four-month siege.
- **2000** November, Fujimori seeks refuge in Japan and resigns amid corruption scandal despite having won controversial third term in May.
- **2001** Fresh presidential elections, won by Alejandro Toledo in run-off against García.

P

WORLD AFFAIRS

Joined UN in 1945

AP | AmCC | NAM | OAS | RG

Cooperation with the US, which is the main source of aid, extends to the war on cocaine, although Peru remains one of the world's largest coca producers.

Reported incursions by guerrillas, paramilitaries, and narcotics traffickers highlight the security of the border with Colombia as a key problem.

Peru competes with Chile for an alliance with Bolivia to process and pipe natural gas.

AID

Recipient

$401m (receipts) | Down 11% in 2000

Aid from the US is mostly directed at antinarcotics activity. Loans worth some $1.3 billion in recent years from the IDB, the World Bank, and Japan were conditional on Peru's meeting specific health and educational targets and on making progress on privatizations.

DEFENSE

Compulsory military service

 $861m

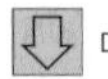 Down 3% in 2000

The military, in power from 1968 to 1980, supported President Fujimori's 1992 presidential coup. A quarter of national territory remained under states of emergency until early 2000, despite the apparent defeat of the Sendero Luminoso guerrillas. Fujimori's control over promotions and the National Intelligence Service (SIN) guaranteed a loyal armed forces leadership. After his resignation and the dissolution of the SIN in late 2000, the interim government moved to cut back military influence. Bomb attacks in Lima prior to US president Bush's 2002 visit were widely blamed on guerrillas, but an option studied by the Interior Ministry was that elements within the military, intelligence, or police still sympathetic to Montesinos were responsible.

PERUVIAN ARMED FORCES

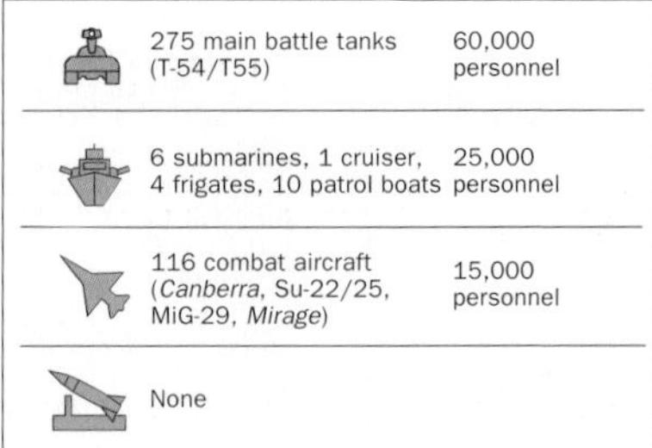

275 main battle tanks (T-54/T55)	60,000 personnel
6 submarines, 1 cruiser, 4 frigates, 10 patrol boats	25,000 personnel
116 combat aircraft (*Canberra*, Su-22/25, MiG-29, *Mirage*)	15,000 personnel
None	

ECONOMICS

Inflation 27% p.a. (1990–2000)

$53.4bn | 3.527–3.444 new soles

SCORE CARD

World GNP Ranking	47th
GNP per Capita	$2080
Balance of Payments	–$1.63bn
Inflation	3.8%
Unemployment	8%

EXPORTS

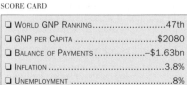

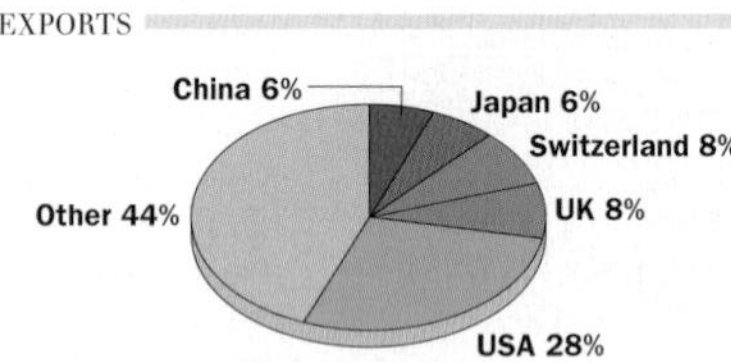
China 6%
Japan 6%
Switzerland 8%
UK 8%
USA 28%
Other 44%

IMPORTS

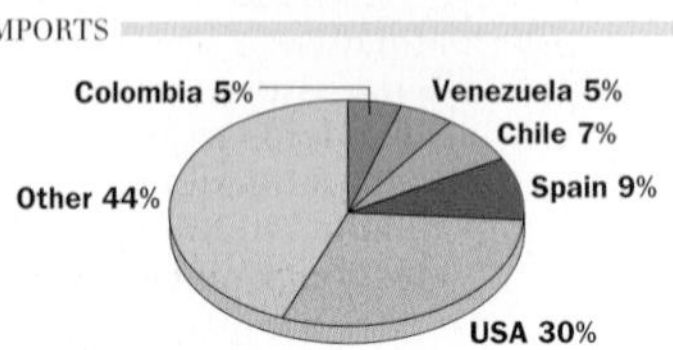
Colombia 5%
Venezuela 5%
Chile 7%
Spain 9%
USA 30%
Other 44%

Strengths

Abundant mineral resources, including oil. Rich Pacific fish stocks. Wide climatic variation allows diverse and productive agriculture; cotton and coffee are important. Well-developed textile industry.

Weaknesses

Overdependence on metals and commodities whose fluctuating prices undermine trade and investment. Stalled privatization. Corruption and poor infrastructure deterring investment. Weak banks.

Profile

Economic activity is largely confined to the cities of the coastal plain. The inhabitants of the Andean uplands are subsistence farmers or coca producers. Peru's strict fiscal and monetary policy continued under Fujimori. Growth was hit hard in 1998 by the disruption of fishing by El Niño-generated storms, the Asian economic crises, Russian-provoked turmoil in emerging markets, and depressed world commodity prices. In 1999 the IMF granted a three-year loan package, extended in 2002, to support comprehensive structural reform and fiscal restraint, with the aim of gradually reducing the state-sector deficit. Meanwhile public belief in Toledo's promises on jobs and poverty waned.

ECONOMIC PERFORMANCE INDICATOR

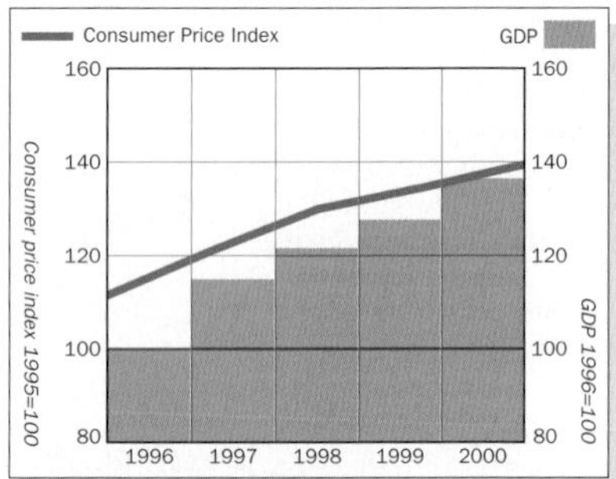
Consumer Price Index
GDP
160
140
120
100
80
Consumer price index 1995=100
GDP 1996=100
1996
1997
1998
1999
2000

PERU : MAJOR BUSINESSES

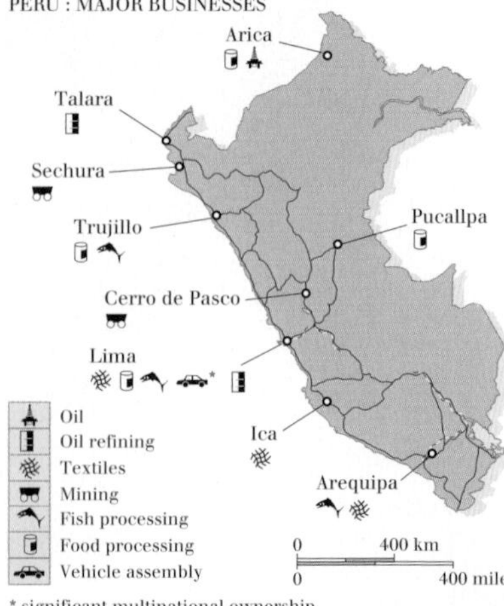
Arica
Talara
Sechura
Trujillo
Pucallpa
Cerro de Pasco
Lima
Ica
Arequipa
Oil
Oil refining
Textiles
Mining
Fish processing
Food processing
Vehicle assembly
0
400 km
0
400 miles
* significant multinational ownership

RESOURCES

Electric power 5.2m kw

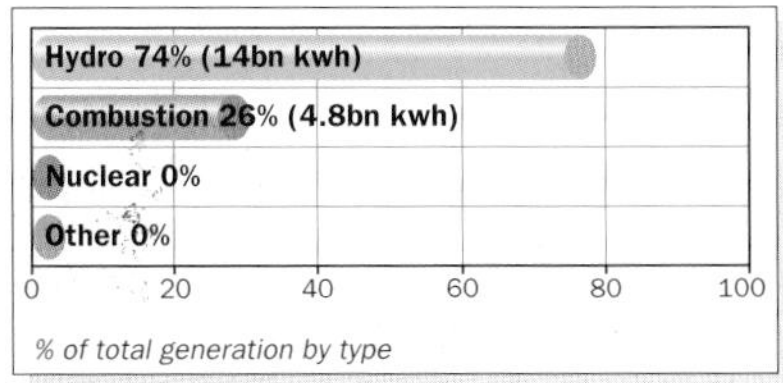
8.44m tonnes
102,000 b/d (reserves 300m barrels)
14.5m sheep, 4.93m cattle, 2.8m pigs, 90m chickens
Oil, coal, lead, zinc, silver, iron, gold, copper

ELECTRICITY GENERATION

Hydro 74% (14bn kwh)
Combustion 26% (4.8bn kwh)
Nuclear 0%
Other 0%

0 20 40 60 80 100

% of total generation by type

Peru is an important exporter of copper and lead. Development of the huge Antamina copper and zinc deposit is under way. Development of the $3 billion Camisea hydrocarbon project is ongoing. The full extent of the large oil reserves still needs to be explored. A priority is the further development of hydroelectric power.

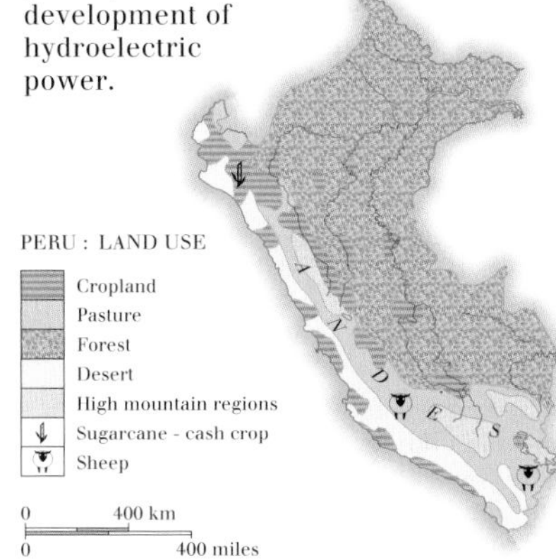
PERU : LAND USE
Cropland
Pasture
Forest
Desert
High mountain regions
Sugarcane - cash crop
Sheep
A N D E S
0 400 km
0 400 miles

ENVIRONMENT

Sustainability rank: 29th

3%

1.1 tonnes per capita

ENVIRONMENTAL TREATIES

Yes Yes Yes
Yes Yes No

Environmentalists have long been concerned about coastal industrial pollution and the activities of the fishing industry. Overfishing of anchovies almost resulted in their extinction in the 1970s. Today, attention has switched to the rising number of dolphins being caught in drift nets. Unchecked urban and industrial pollution, especially in Lima, is a major problem.

Environmentalists fear that Peru's policy of using powerful air-sprayed herbicides to destroy coca crops is adding to river pollution in the Andes, where mining also causes severe environmental problems.

MEDIA

TV ownership medium

Daily newspaper circulation 4 per 1000 people

PUBLISHING AND BROADCAST MEDIA

There are 74 daily newspapers. These include the conservative *El Comercio* and *Expreso*, and the left-wing *La República*
11 services: 1 state-owned, 10 independent
3 state-owned services, many independent services

Main cities have their own newspapers; Lima's papers cater to diverse political readerships. State TV is not popular.

CRIME

Death penalty not used in practice

27,452 prisoners

Down 81% 1992–1998

CRIME RATES

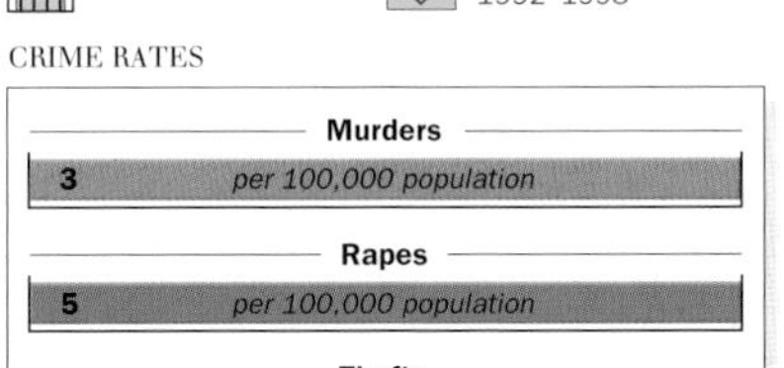
Murders
3 per 100,000 population
Rapes
5 per 100,000 population
Thefts
67 per 100,000 population

Kidnappings, murders, armed robberies, and drugs-related crime remain serious problems, especially in Lima. Corruption is deep-seated in the police and security forces. Despite the near-destruction of the left-wing Sendero Luminoso guerrillas, main cities frequently have curfews, and those who can afford it protect themselves with high-security homes and armed guards.

EDUCATION

School leaving age: 16

90%

657,586 students

THE EDUCATION SYSTEM

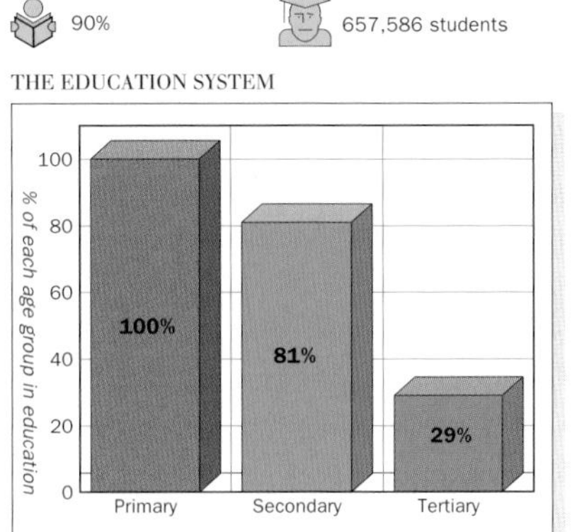
% of each age group in education
100
80
60
40
20
0
100%
81%
29%
Primary
Secondary
Tertiary

Education is based on the US system; spending has been declining. The provision of state education, especially for the poor, remains a major challenge. State and private universities are accessible to a small minority.

HEALTH

Welfare state health benefits

1 per 1111 people

Respiratory, heart, infectious, and parasitic diseases

The poor public health system almost collapsed in the 1980s. In many areas primary care is nonexistent. Advanced treatment is available only to private patients in city clinics. Goiter, a thyroid abnormality, is widespread, especially in mountain areas. Infant mortality is rising due to social deprivation, diarrheal diseases, and tuberculosis. Malaria is again widespread, and cholera reached epidemic proportions in 1994. Thousands of poor women were forcibly sterilized in the late 1990s as part of a government program to lower the birthrate. Social welfare is compulsory, and benefits cover sickness, disability, and old age.

SPENDING

GDP/cap. increase

CONSUMPTION AND SPENDING

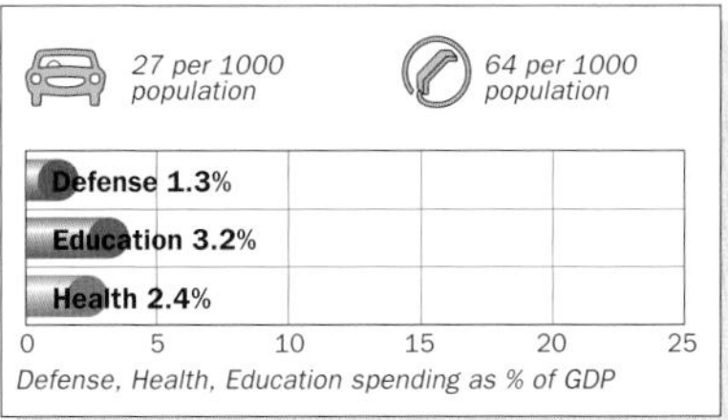
27 per 1000 population
64 per 1000 population
Defense 1.3%
Education 3.2%
Health 2.4%
0 5 10 15 20 25
Defense, Health, Education spending as % of GDP

Most wealth and power in Peru is still retained by old Spanish families. Indigenous peoples remain excluded from both. The rich live in a state of siege; a key status symbol is the number of armed guards and security cameras protecting family property. Overpopulation and rural migration accentuate poverty in Lima, where some 2.7 million people live in shanty towns, many of them lacking such basic utilities as running water and electricity. The UN estimates that over 30% of Peruvians live below the poverty line.

WORLD RANKING

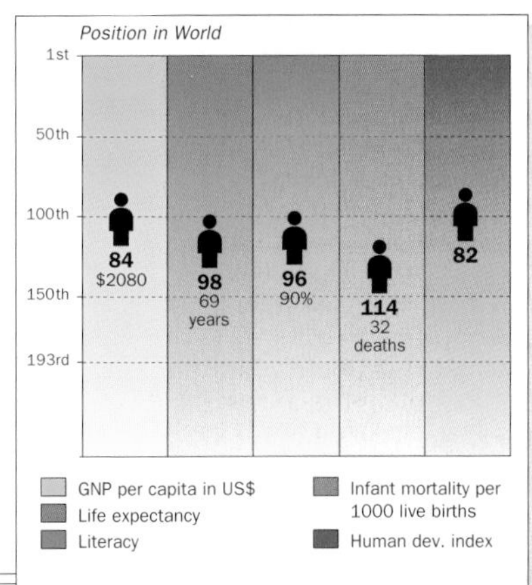
Position in World
1st
50th
100th
150th
193rd
84 $2080
98 69 years
96 90%
114 32 deaths
82
GNP per capita in US$
Life expectancy
Literacy
Infant mortality per 1000 live births
Human dev. index

P

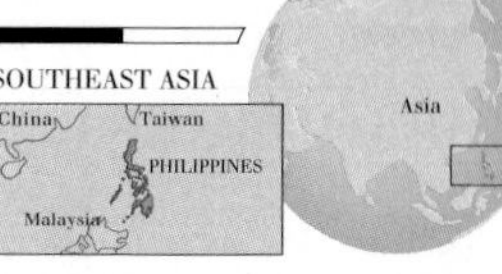
SOUTHEAST ASIA
China
Taiwan
PHILIPPINES
Malaysia
Asia

PHILIPPINES

OFFICIAL NAME: Republic of the Philippines **CAPITAL:** Manila
POPULATION: 77.1 million **CURRENCY:** Philippine peso **OFFICIAL LANGUAGES:** English and Filipino

1946

1946

June 12

RP

+8

+63

.ph

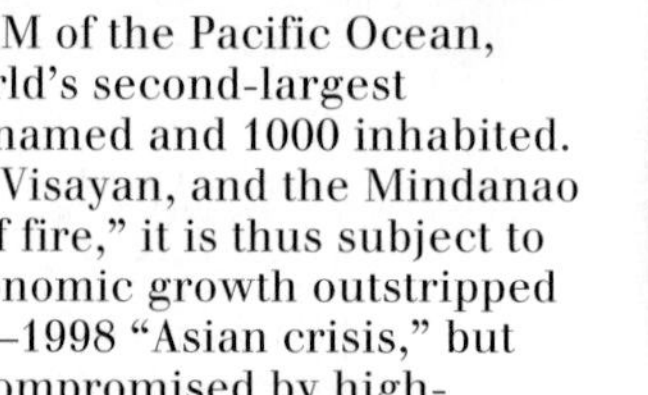

LYING ON THE WESTERN RIM of the Pacific Ocean, the Philippines is the world's second-largest archipelago-state. Of its 7107 islands, 4600 are named and 1000 inhabited. There are three main island groupings: Luzon, Visayan, and the Mindanao and Sulu islands. Located on the Pacific "ring of fire," it is thus subject to frequent earthquakes and volcanic activity. Economic growth outstripped population increase in the 1990s, until the 1997–1998 "Asian crisis," but efforts to build a stable democracy have been compromised by high-level corruption, leading to the ouster of President Estrada in 2001.

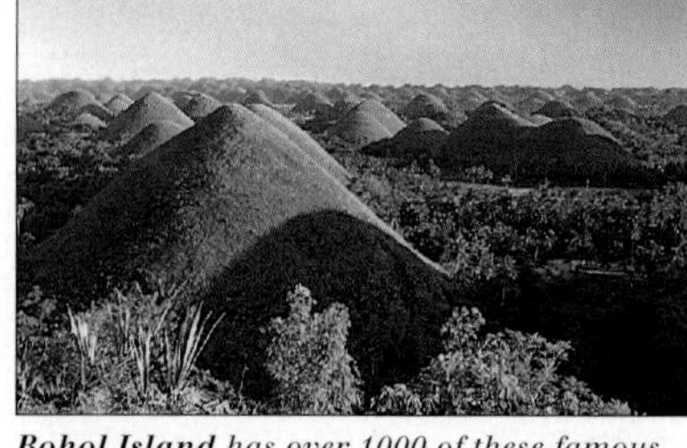

***Bohol Island** has over 1000 of these famous mounds, known as "the chocolate hills."*

CLIMATE

Tropical monsoon/equatorial

WEATHER CHART FOR MANILA

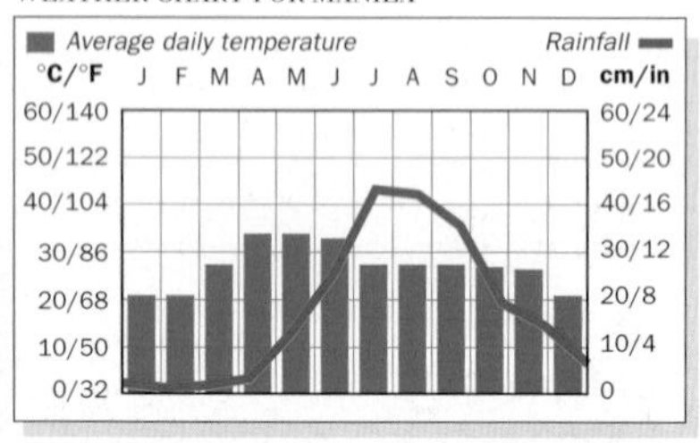
Average daily temperature
Rainfall
°C/°F
J F M A M J J A S O N D
cm/in
60/140
50/122
40/104
30/86
20/68
10/50
0/32
60/24
50/20
40/16
30/12
20/8
10/4
0

The Philippines is warm and humid all year. The rainy season lasts from June to October. Humidity falls from 85% in September to 71% in March.

TRANSPORTATION

Drive on right

Ninoy Aquino International, Manila
12.8m passengers

1726 ships
8.51m grt

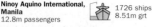

THE TRANSPORTATION NETWORK

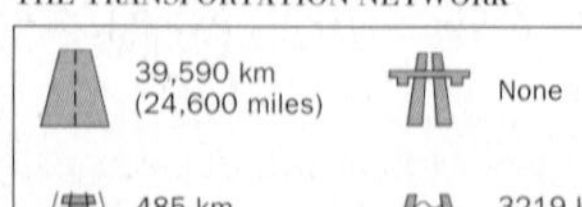
39,590 km
(24,600 miles)
None
485 km
(301 miles)
3219 km
(2000 miles)

Basic infrastructure lacks investment and many main roads are in desperate need of repair. Chronic traffic congestion in Manila holds back economic growth.

Air travel is the only means of getting around the islands quickly. Philippines Airlines, privatized in 1992, has invested heavily in new aircraft and in expanding its regional route network. Work began in 2000 on a spectacular new terminal at Manila's international airport, nearly trebling its capacity.

Subic Bay, a massive US naval base until 1992, is now being exploited as a commercial asset, thanks to its prime location. Opening on to the South China Sea, its deep natural harbor has been developed as a free port and enterprise zone. The Taiwanese are the biggest investors in this project.

TOURISM

Visitors : Population 1:36

2.17m visitors

Down 8% in 2000

MAIN TOURIST ARRIVALS

	% of total arrivals
USA	21%
Japan	18%
Hong Kong	7%
Taiwan	7%
South Korea	6%
Other	41%

There is less tourism in the Philippines than in other regional NICs. Dubious images conveyed by sex-tourism have become a liability. International pressure to end this abuse has intensified.

Muslim secessionists have seized tourists on neighboring Malaysian islands as hostages.

The tiny island of Boracay, off Panay, is a popular resort, and Palawan retains most of its tropical rainforest and coral lagoons, although coral reefs elsewhere are badly damaged. The rice terraces of northern Luzon are another attraction.

PHILIPPINES

Total Land Area : 300 000 sq. km (115 830 sq. miles)

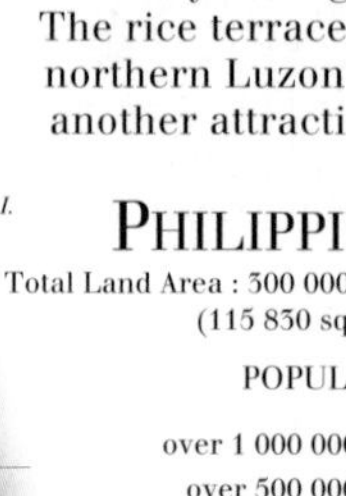
POPULATION
over 1 000 000
over 500 000
over 100 000
over 50 000
LAND HEIGHT
2000m/6562ft
1000m/3281ft
500m/1640ft
200m/656ft
Sea Level
0
200 km
200 miles

PEOPLE

Pop. density high

Filipino, Tagalog, Cebuano, Hiligaynon, Samaran, Ilocano, Bikol, English

259/km² (670/mi²)

THE URBAN/RURAL POPULATION SPLIT

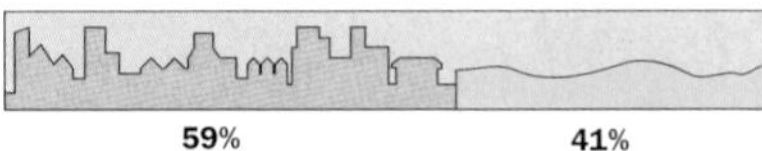

59% 41%

RELIGIOUS PERSUASION

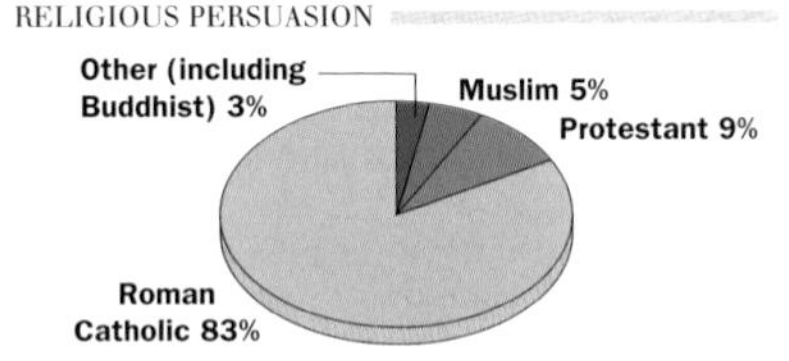
Other (including Buddhist) 3%
Muslim 5%
Protestant 9%
Roman Catholic 83%

ETHNIC MAKEUP

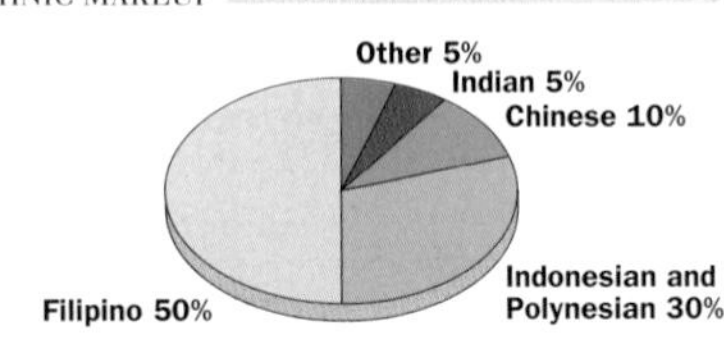
Other 5%
Indian 5%
Chinese 10%
Indonesian and Polynesian 30%
Filipino 50%

POPULATION AGE BREAKDOWN

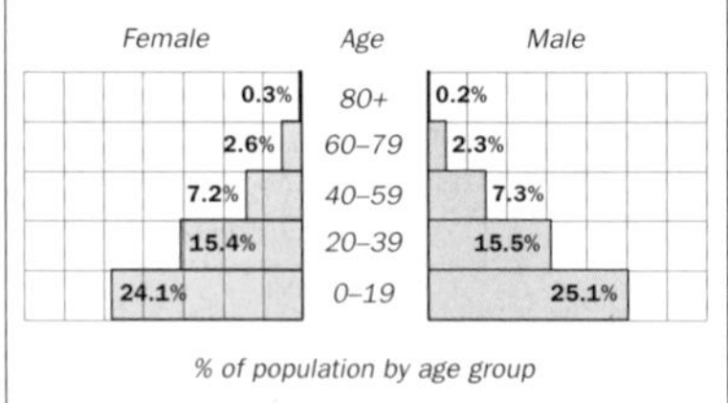
Female
Age
Male
0.3%
80+
0.2%
2.6%
60–79
2.3%
7.2%
40–59
7.3%
15.4%
20–39
15.5%
24.1%
0–19
25.1%
% of population by age group

The Philippines encompasses more than 100 distinct ethnic groups, many of Malay origin. The national language, Filipino, is based on Tagalog, spoken by the largest of the various groups. Other groups include Cebuano, Ilocan, Longgo, Bicolano, Waray, Pampangan, and Pangasinan. They are concentrated on the main island, Luzon, and are also a majority on Mindanao. Most Muslims live on Mindanao, but many are also found in the Sulu archipelago. The Chinese minority, which was well established by 1603, has remained significant in business and trade. More than 120 Chinese schools have ensured that it has retained a distinct identity.

There are also a number of cultural minorities who practice animist religions. They include the Ifugaos, Bontocks, Kalingas, and Ibalois on Luzon, the Manobo and Bukidnon on Mindanao, and the Mangyans on Palawan. Many of these groups speak Malayo-Polynesian dialects. Limited intermarriage with other peoples has meant that groups in the more remote regions have managed to retain their traditional ways of life.

The Philippines is one of only two Christian states in Asia. Over 80% of Filipinos are Roman Catholics, and the Church is the dominant cultural force. It opposes state-sponsored preventive family planning programs aimed at curbing accelerating population growth. Abortion is illegal but widespread.

Women have traditionally played a prominent part in Philippine public and professional life. Inheritance laws give them equal rights to men. Many go into politics, banking, and business, and in several professional sectors they form a majority.

POLITICS

Multiparty elections

L. House 2001/2004
U. House 2001/2004

President Gloria Macapagal Arroyo

AT THE LAST ELECTION

House of Representatives 216 seats

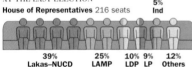

39% Lakas–NUCD	25% LAMP	10% LDP	9% LP	12% Others	5% Ind

Lakas–NUCD = Lakas – National Union of Christian Democrats **LAMP** = Party of the Filipino Masses **LDP** = Fight of Democratic Filipinos **LP** = Liberal Party **Ind** = Independents **PP** = People's Power (coalition led by Lakas–NUCD and including the LP) **PnM** = Strength of the Masses (coalition of LAMP and LDP)

Senate 24 seats

50% PP	46% PnM	4% Ind

The Philippines is a multiparty democracy.

PROFILE

The 21-year dictatorship of Ferdinand Marcos was shown by 1986 to have no popular legitimacy as "people power" massed in support of his opponent Corazon Aquino, who was declared the true winner of the presidential elections. Losing the backing of the US, Marcos was forced into exile. Aquino was able to hand over power to Fidel Ramos through fair elections in 1992. His dependence on loose coalition arrangements in Congress slowed the economic liberalization program. His successor in 1998, the populist flamboyant former film star Joseph Estrada of LAMP, was accused of links to a gambling syndicate, and in late 2000 impeachment proceedings began. A repeat of the mass "people power" demonstrations of 1986 toppled him in 2001. Vice President Gloria Macapagal Arroyo of the NUCD, who led the united political opposition to Estrada, was appointed president on his downfall.

MAIN POLITICAL ISSUES

Political stability

The downfall of the Marcos regime through popular demonstration ensured a return to democracy. The subsequent two presidents passed on power smoothly to an elected successor, but Estrada was quickly mired in corruption scandals. Moves for his impeachment were overtaken by mass rallies in Manila. Out of office he retained widespread support among the rural poor, questioning the legitimacy of Vice President Arroyo's assumption of the presidency.

Insurgency and separatism

Communist and Muslim separatists have been fighting Manila-based governments for over 30 years, with more than 10,000 armed confrontations with rebels recorded by the army. Much of the support for secession has been fueled by the failure of successive governments to curb poverty.

Since 1992, the government has been seeking peace with all armed groups. The communist New People's Army (NPA), once regarded as a heroic army of the oppressed and as an alternative to traditional politics, has declined in

***Joseph Estrada,** film-star-turned-president, ousted in 2001.*

***Gloria Arroyo,** succeeded Estrada in a popular uprising.*

significance, but launched new offensives in 2000. The Moro National Liberation Front (MNLF), representing secessionist Muslim rebels on Mindanao, signed a peace agreement in 1996. A militant breakaway Muslim Islamic Liberation Front (MILF) continued fighting, forcing the government to abandon a massive irrigation project. Fighting peaked in 2000, and the MILF joined the peace process in mid-2001. The smaller but internationally backed Abu Sayyaf continues low-level fighting in the Sulu region.

WORLD AFFAIRS

Joined UN in 1945

APEC

ASEAN G24 NAM WTO

Regionally, there are sporadic frictions with Malaysia. Manila's claim to the Spratly Islands conflicts notably with China's. The resumption of US naval visits since 1999 indicate improved security cooperation. US troops provided logistical support against Muslim separatists from 2001.

P

CHRONOLOGY

Ceded to the US by Spain in 1898, the Philippines became self-governing in 1935, and an independent republic in 1946.

- ❑ **1965** Ferdinand Marcos president.
- ❑ **1972** Marcos declares martial law. Opposition leaders arrested, parliament suspended, press censored.
- ❑ **1977** Ex-Liberal Party leader Benigno Aquino sentenced to death. Criticism forces Marcos to delay execution.
- ❑ **1978** Elections won by Marcos's New Society (KBL). He is named president and prime minister.
- ❑ **1980** Aquino allowed to travel to US for medical treatment.
- ❑ **1981** Martial law ends. Marcos reelected president by referendum.
- ❑ **1983** Aquino shot dead on return from US. Inquiry blames military conspiracy.
- ❑ **1986** US compels presidential election. Result disputed. Army rebels led by Gen. Fidel Ramos, and public demonstrations, bring Aquino's widow, Corazon, to power. Marcos exiled to US.
- ❑ **1987** New constitution. Aquino-led coalition wins Congress elections.
- ❑ **1988** Marcos and wife Imelda indicted for massive racketeering.
- ❑ **1989** Marcos dies in US.
- ❑ **1990** Imelda Marcos acquitted of fraud charges in US. Earthquake in Baguio City leaves 1600 dead.
- ❑ **1991** Mt. Pinatubo erupts. US leaves Clark Air Base.
- ❑ **1992** Ramos wins presidential election. US withdrawal from Subic Bay base.
- ❑ **1996** Peace agreement with Muslim MNLF secessionists.
- ❑ **1998** Joseph Estrada president.
- ❑ **1999** First execution in 22 years.
- ❑ **2000** Tourists kidnapped by Islamic extremists.
- ❑ **2001** Estrada overthrown by popular protest. Gloria Macapagal Arroyo assumes presidency. August, Muslim MILF joins peace process.
- ❑ **2002** Local elections, described as "peaceful" despite 86 deaths.

AID

 Recipient

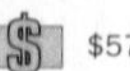 $578m (receipts)  Down 17% in 2000

The Philippines' main bilateral aid donors are Japan and the US. Many NGOs operate in the outlying islands. Large remittances are also received from Filipinos working overseas. Funds received from these hundreds of thousands of emigrant workers exceeded $6 billion in 2000.

DEFENSE

No compulsory military service

 $1.5bn  Down 8% in 2000

The military retains political influence. The armed forces are undergoing modernization with US assistance.

The historic tie to the US has been translated in modern times into a close strategic relationship. President Arroyo quickly pledged her country's support for the US-led "war on terrorism" in 2001, and US forces were deployed to help the government combat Muslim rebels in the south.

PHILIPPINE ARMED FORCES

40 light tanks (*Scorpion*)
67,000 personnel
1 frigate and 58 patrol boats
24,000 personnel
44 combat aircraft (8 F-5A/B)
16,000 personnel
None

ECONOMICS

Inflation 8.4% p.a. (1990–2000)

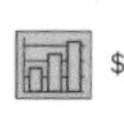 $78.8bn  50.0–51.6 Philippine pesos

SCORE CARD

- ❑ World GNP Ranking.........................41st
- ❑ GNP per Capita $1040
- ❑ Balance of Payments...................$9.08bn
- ❑ Inflation 4.4%
- ❑ Unemployment................................10%

EXPORTS

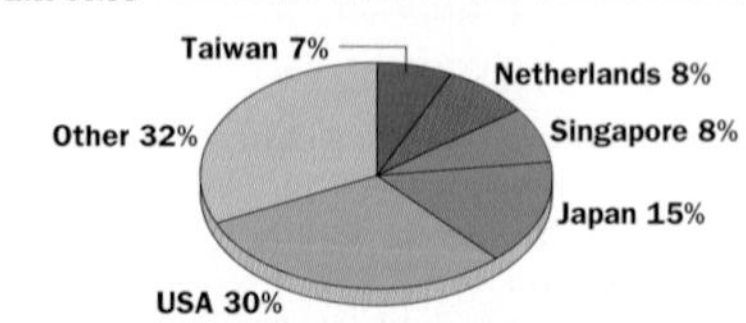
Taiwan 7%
Netherlands 8%
Singapore 8%
Japan 15%
USA 30%
Other 32%

IMPORTS

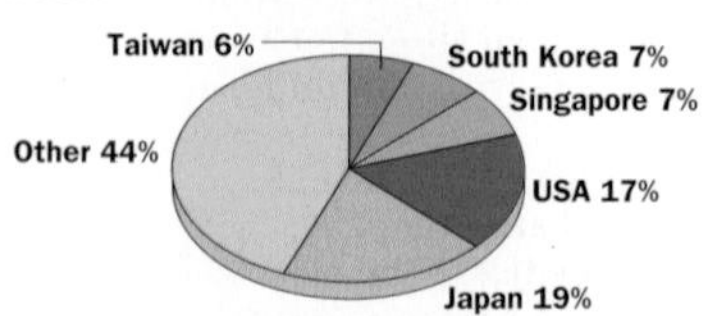
Taiwan 6%
South Korea 7%
Singapore 7%
USA 17%
Japan 19%
Other 44%

Strengths

Now fully open to outside investment. Agricultural productivity rising. Strong pineapple and banana export industries. Substantial remittances from Filipinos working overseas.

Weaknesses

Power failures limit scope for expansion. Rudimentary infrastructure. Low domestic savings rates mean reliance on foreign finance. Low productivity in subsistence agriculture.

Profile

Once one of Asia's strongest economies, the Philippines has now fallen behind once much poorer countries such as Thailand, Malaysia, and South Korea. Around 50% of the rural population live on the poverty line, fueling many of the secessionist movements undermining the stability of successive governments.

The financial successes of the early 1990s – opening up to foreign investment, cutting back private monopolies – were effectively nullified by the crippling 1997–1998 Asian economic and financial crises. The government of Joseph Estrada failed to shore up the economy and ran up a crippling budget deficit, and investor confidence vanished amid political scandals. His successor, President Arroyo, has done much to regain the approval of the IMF and the international community, mostly through further deregulation and privatization. Although growth is back on track, despite the global slowdown in 2001, many public works projects have simply been put on hold.

ECONOMIC PERFORMANCE INDICATOR

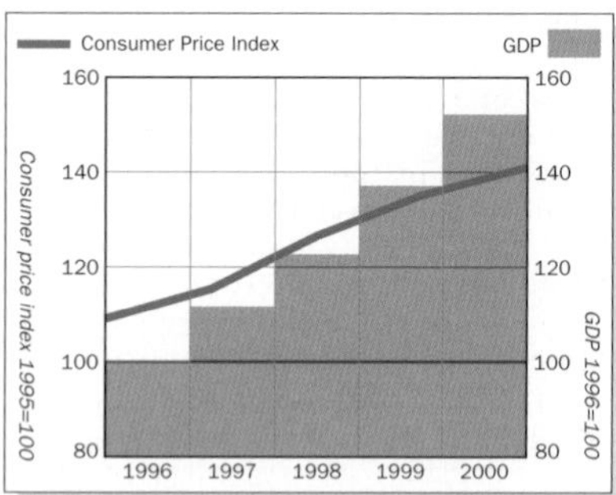
Consumer Price Index
GDP
160
140
120
100
80
1996
1997
1998
1999
2000
Consumer price index 1995=100
GDP 1996=100

PHILIPPINES : MAJOR BUSINESSES

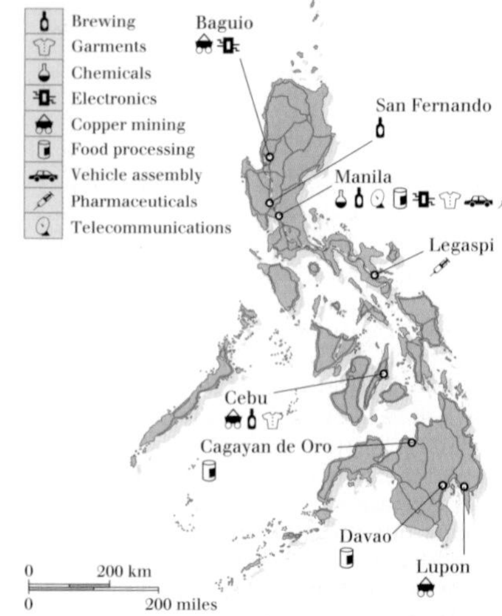
Brewing
Garments
Chemicals
Electronics
Copper mining
Food processing
Vehicle assembly
Pharmaceuticals
Telecommunications
Baguio
San Fernando
Manila
Legaspi
Cebu
Cagayan de Oro
Davao
Lupon
0 200 km
0 200 miles

RESOURCES

Electric power 11.6m kw

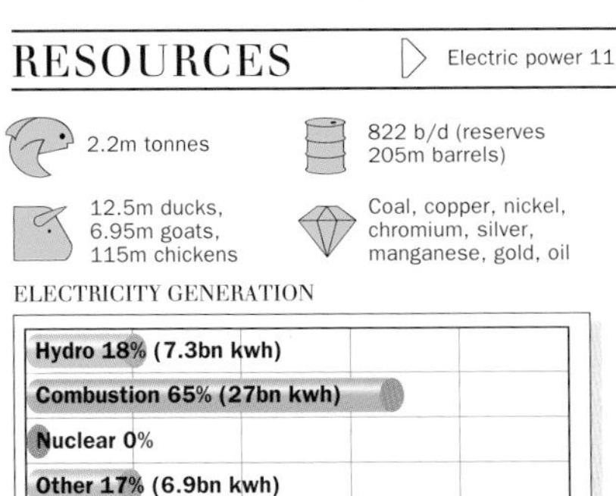
2.2m tonnes
822 b/d (reserves 205m barrels)
12.5m ducks, 6.95m goats, 115m chickens
Coal, copper, nickel, chromium, silver, manganese, gold, oil

ELECTRICITY GENERATION

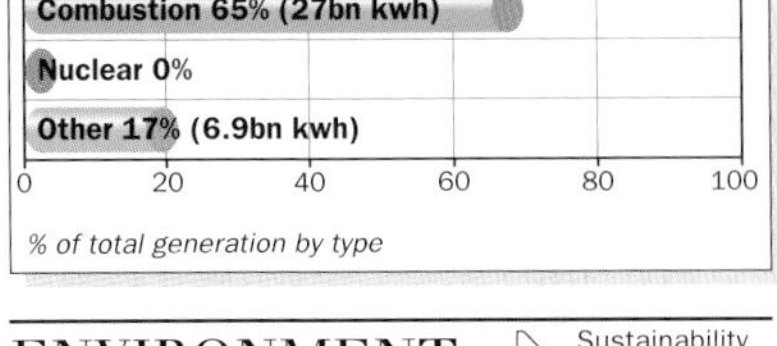
Hydro 18% (7.3bn kwh)
Combustion 65% (27bn kwh)
Nuclear 0%
Other 17% (6.9bn kwh)
0
20
40
60
80
100
% of total generation by type

The Philippines is the world's biggest supplier of refractory chrome. Copper is also a significant export. Substantial gold reserves have been mined since 1996. However, more than 90% of mineral potential remains undeveloped. Oil production off Palawan began in 1979. The Philippines is the world's second-biggest user of geothermal power after the US. Almost 25% of electricity on Luzon is provided by this method. Although timber exports were halted in 1989, illegal logging continues to cause deforestation.

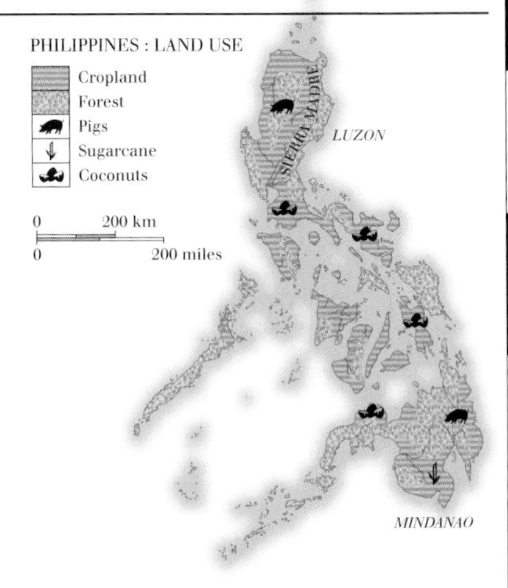
PHILIPPINES : LAND USE
Cropland
Forest
Pigs
Sugarcane
Coconuts
0
200 km
0
200 miles
SIERRA MADRE
LUZON
MINDANAO

ENVIRONMENT

Sustainability rank: 117th

5% (1% partially protected)

1 tonne per capita

ENVIRONMENTAL TREATIES

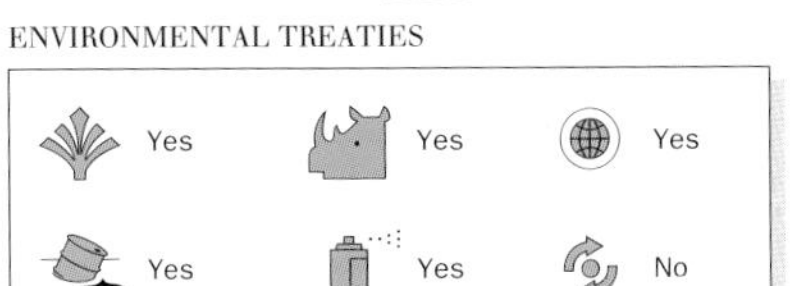
Yes
Yes
Yes
Yes
Yes
No

The environment has become a major issue. Most of the tropical rainforest has been destroyed, except for pockets such as the island of Palawan. Unique coral habitats have been dynamited, and fishermen continue to use cyanide and muro-ami (reef-hunting) techniques to increase the size of their catches.

The government recognizes the costs of environmental damage, as soil run-off silts rivers and reduces the power generated by hydroelectric dams, and fast-depleting coral habitats reduce the attraction of the Philippines for tourists.

Enforcement of a ban on logging is difficult; many loggers have their own private armies. In addition, continued use of slash and burn farming has contributed to deforestation.

MEDIA

TV ownership medium

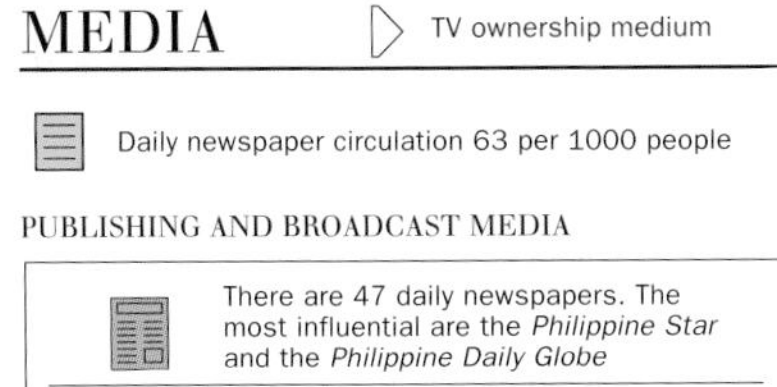
Daily newspaper circulation 63 per 1000 people

PUBLISHING AND BROADCAST MEDIA

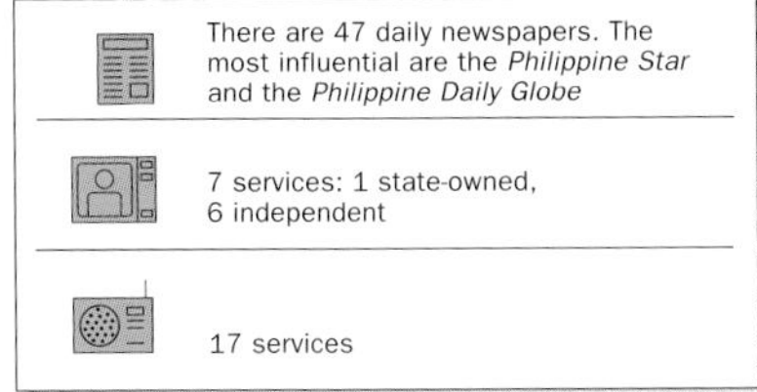
There are 47 daily newspapers. The most influential are the Philippine Star and the Philippine Daily Globe
7 services: 1 state-owned, 6 independent
17 services

The lifting of censorship following Corazon Aquino's election in 1986 led to a burgeoning of the media. As well as the national press, there are more than 250 regional newspapers in local dialects. State TV and radio broadcast in English and Filipino. Four independent television stations serve Metro Manila.

CRIME

Moratorium on death penalty

70,383 prisoners

Crime is rising

CRIME RATES

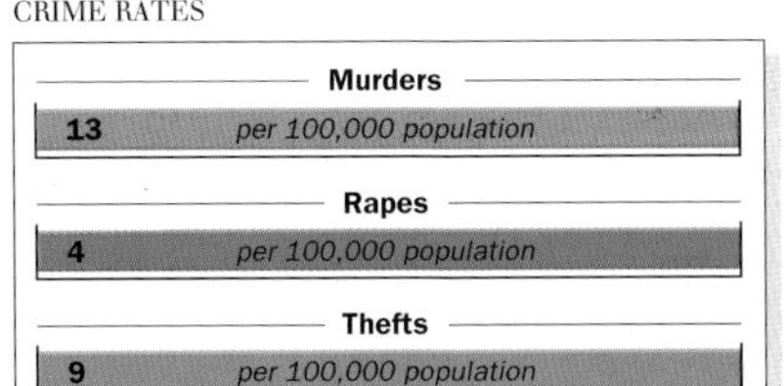
Murders
13 per 100,000 population
Rapes
4 per 100,000 population
Thefts
9 per 100,000 population

The death penalty, reinstated in 1993, was suspended again in 2000, but President Arroyo made an exception for 95 kidnappers in 2001.

EDUCATION

School leaving age: 12

95%

2.02m students

THE EDUCATION SYSTEM

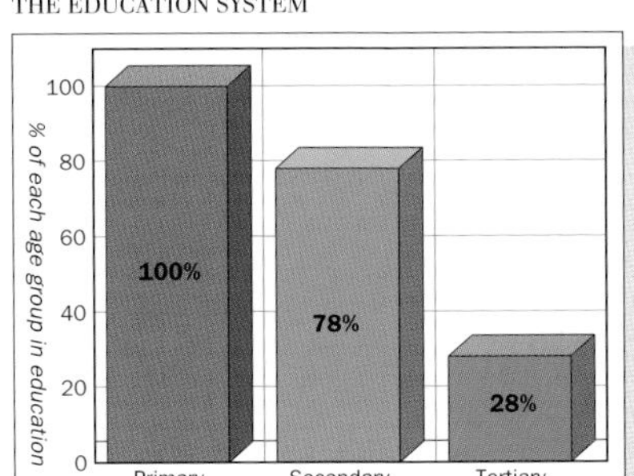
100
80
60
40
20
0
% of each age group in education
100%
78%
28%
Primary
Secondary
Tertiary

The Philippines has one of the highest literacy rates among developing countries. The education system is based on the US model, but with a higher proportion of private schools. The main teaching languages are English and Filipino/Tagalog.

Although there is a national curriculum up to age 15, sectarianism is common; the Chinese community has its own schools. Most colleges and universities are also run privately. The universities of San Carlos in Cebu and Santo Tomas in Manila are Spanish colonial foundations, dating from 1595 and 1611 respectively.

HEALTH

No welfare state health benefits

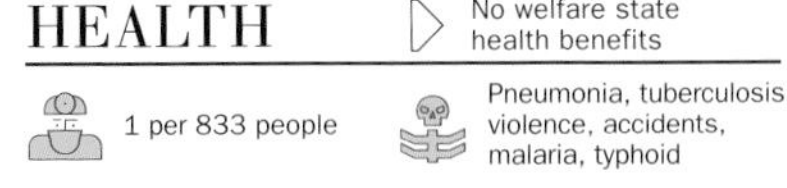
1 per 833 people
Pneumonia, tuberculosis, violence, accidents, malaria, typhoid

Most general hospitals are privately run. Malaria, which was once a major problem, has been eradicated in all but remote areas. Poor sanitation and disease are common in the sprawling slums around Manila.

SPENDING

GDP/cap. increase

CONSUMPTION AND SPENDING

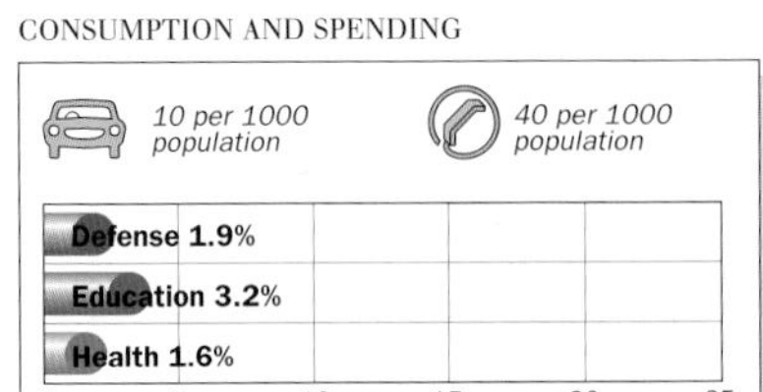
10 per 1000 population
40 per 1000 population
Defense 1.9%
Education 3.2%
Health 1.6%
0
5
10
15
20
25
Defense, Health, Education spending as % of GDP

The contrast between extremes of wealth and poverty is particularly marked. Wealth remains highly concentrated in a few select business families which are based in Manila.

WORLD RANKING

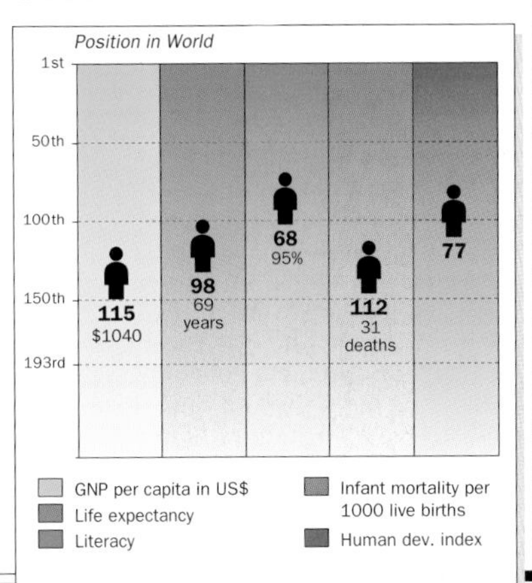
Position in World
1st
50th
100th
150th
193rd
115
$1040
98
69 years
68
95%
112
31 deaths
77
GNP per capita in US$
Life expectancy
Literacy
Infant mortality per 1000 live births
Human dev. index

P

POLAND

EUROPE

Russian Fed. Lithuania Belarus Germany POLAND Ukraine Czech Republic Slovakia

Europe

OFFICIAL NAME: Republic of Poland **CAPITAL:** Warsaw
POPULATION: 38.6 million **CURRENCY:** Zloty **OFFICIAL LANGUAGE:** Polish

1918

1945

May 3
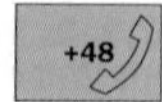
PL

+1

+48
.pl

LOCATED IN THE HEART of Europe, Poland's low-lying plains extend from the Baltic shore in the north to the Tatra Mountains on its southern border with Slovakia. Since the collapse of communism, Poland has undergone massive social, economic, and political change. Opting for a radical form of economic "shock therapy" in the early 1990s to kick-start the switch to a market economy, it has experienced rapid growth, is one of the front-runners in negotiations to join the EU, and has already been accepted as a member of NATO.

CLIMATE

Continental

WEATHER CHART FOR WARSAW

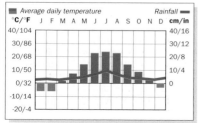

Average daily temperature / Rainfall

°C/°F: 40/104, 30/86, 20/68, 10/50, 0/32, -10/14, -20/-4

J F M A M J J A S O N D

cm/in: 40/16, 30/12, 20/8, 10/4, 0

Most of the country experiences a similar climate. Summers are hot, with heavy rainfall often accompanied by thunder. Winters are severe, with snow covering the ground on the southern mountains, for as much as 60–70 days in the east.

TRANSPORTATION

Drive on right

Okecie International, Warsaw
4.33m passengers

447 ships
1.42m grt

THE TRANSPORTATION NETWORK

249,966 km (155,321 miles)

358 km (222 miles)

22,891 km (14,224 miles)

3812 km (2369 miles)

The national airline LOT has increased its charter business as more middle-class Poles vacation abroad. Russian aircraft have all been replaced with Western models. A 15-year roads expansion program was begun in 1997. "Fast tram" systems for cities and long-distance high-speed rail links need major investment.

Telecommunications have been affected by the advent of mobile phones, and by the privatization of Telekomunikacja Polska and the end of its monopoly on long-distance calls. The government plans to slim down the rail workforce and to commercialize and part-privatize Polish State Railways (PKP) by 2003.

The medieval administrative center of Lublin lies in Poland's southeastern agricultural heartland.

TOURISM

Visitors : Population 1:2.6

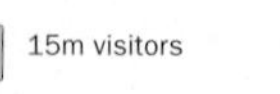
15m visitors
Down 14% in 2001

MAIN TOURIST ARRIVALS

	% of total arrivals
Germany	60%
Czech Republic	15%
Ukraine	6%
Slovakia	5%
Belarus	5%
Other	9%

0 10 20 30 40 50 60
% of total arrivals

Despite environmental problems, Poland is renowned for its skiing and hiking, especially in the Tatra Mountains. Kraków's medieval core has been preserved, while Toruń has restored its historic German Hanseatic buildings.

Warsaw's historic center has been reconstructed following the destruction of 80% of it by the German army in 1944. More hotels and restaurants are being opened.

Poznań has exploited its location between Warsaw and Berlin to create an international exhibition and business convention industry.

Airlines have increased their flights from the West to take advantage of the country's tourist potential.

PEOPLE

Pop. density medium

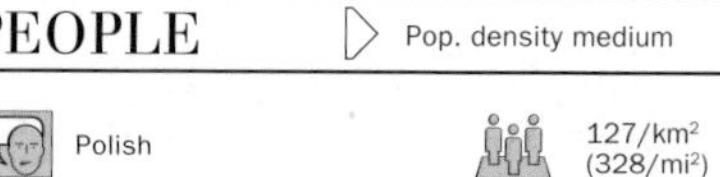
Polish
127/km² (328/mi²)

THE URBAN/RURAL POPULATION SPLIT

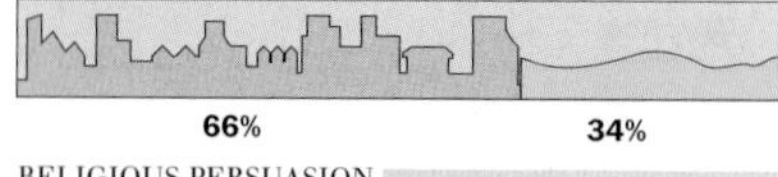
66%
34%

RELIGIOUS PERSUASION

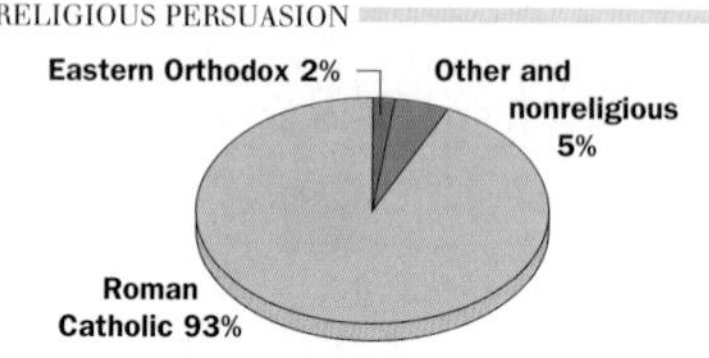
Eastern Orthodox 2%
Other and nonreligious 5%
Roman Catholic 93%

ETHNIC MAKEUP

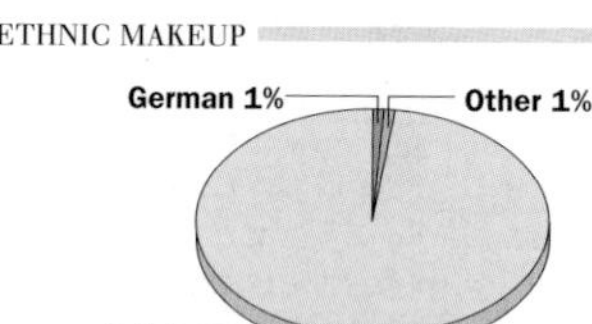
German 1%
Other 1%
Polish 98%

Poland has a strongly Roman Catholic population, and in addition there is little ethnic diversity. The Church believes that stronger links with the West, especially through joining the EU, will weaken its influence. Abortion is still a major issue, and attempts to liberalize the law in 1996 were overturned by the Constitutional Tribunal.

Some small ethnic groups have opened schools and cultural and religious centers. Others, particularly the Germans in Silesia, are becoming more assertive. Jews are still resentful of past discrimination, and there is some evidence of residual antisemitism at a high level. Disputes over the special significance of the site of the World War II Auschwitz concentration camp, near Kraków, have caused conflict between Jews and Catholics.

Wealth disparities are small, although the growing wealth of the entrepreneurial class is causing tension. The major political parties on left and right agree on continuing economic reform.

POPULATION AGE BREAKDOWN

Female	Age	Male
1.4%	80+	0.6%
8.1%	60–79	5.8%
12.7%	40–59	12.1%
14.2%	20–39	14.6%
14.9%	0–19	15.6%

% of population by age group

P

POLITICS

Multiparty elections

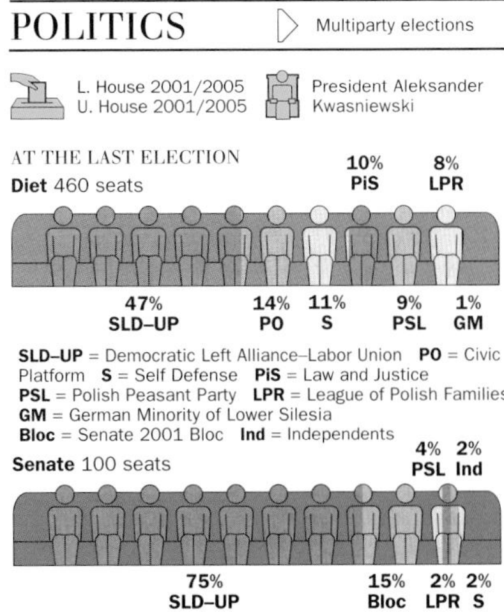
L. House 2001/2005
U. House 2001/2005
President Aleksander Kwasniewski
AT THE LAST ELECTION
Diet 460 seats
10% PiS
8% LPR
47% SLD–UP
14% PO
11% S
9% PSL
1% GM
SLD–UP = Democratic Left Alliance–Labor Union PO = Civic Platform S = Self Defense PiS = Law and Justice
PSL = Polish Peasant Party LPR = League of Polish Families
GM = German Minority of Lower Silesia
Bloc = Senate 2001 Bloc Ind = Independents
Senate 100 seats
4% PSL
2% Ind
75% SLD–UP
15% Bloc
2% LPR
2% S

Since 1989, Poland has been a multiparty parliamentary democracy.

Profile

From 1993 until 1997, successive governments were formed by the reformed communists of the SLD and the PSL, which pursued a policy of market reforms. Aleksander Kwasniewski, leader of the SLD, was elected president in 1995, and reelected in 2000. More right-wing groups held sway for the 1997– 2001 parliamentary term, with Jerzy Buzek, a member of Solidarity since its formation in 1980, as prime minister. His Solidarity Electoral Action (AWS) alliance, a right-wing grouping with vocal Catholic and nationalist elements, formed a coalition with the liberal Freedom Union (UW). A threatened split in early 1999 over controversial health service reforms was averted, but further contentious issues led to the withdrawal of the UW in May 2000. The AWS remained in office thereafter as a minority government. Buzek, who resisted pressure for early elections, came under attack from within the AWS itself.

In May 2001 the Solidarity trade union wing voted to withdraw from politics, and AWS lost all its seats in the legislative elections that September. A new left-of-center coalition was brought together under former communist Leszek Miller.

Main Political Issues

Coalition rule

Poland's emerging party system has been hindered by a superfluity of political factions, and sustaining coalitions has proved difficult. Parties are required to have at least 5% of the vote in order to gain a seat and 8% to be eligible to join a coalition government.

Former president Lech Walesa. He was awarded the Nobel Peace Prize in 1983.

President Aleksander Kwasniewski, leader of the SLD, who was elected in 1995.

Church–state relations

The Roman Catholic Church has been outspoken in its views on social and political policy. Debates over abortion, worship in schools, and values in the media have fueled a heated dialogue over the proper role of the Church. Abortion laws were eased in 1996 until the Constitutional Tribunal ruled that this move was unconstitutional.

POLAND

Total Land Area : 312 685 sq. km (120 728 sq. miles)

POPULATION: over 1 000 000; over 500 000; over 100 000; over 50 000

LAND HEIGHT: 1000m/3281ft; 500m/1640ft; 200m/656ft; Sea Level

0 100 km; 0 100 miles

WORLD AFFAIRS

Joined UN in 1945

CE | CEFTA | OECD | NATO | OSCE

Poland has good relations with the Baltic states and other countries of central Europe. It was admitted to NATO in 1999, and is among the region's "first wave" of applicants to join the EU, hoping for accession in 2004. A treaty has been signed with Germany recognizing the postwar border. The EU wants tighter controls along the eastern border with Belarus and Ukraine.

CHRONOLOGY

Poland has Europe's second-oldest written constitution. In 1795, it was partitioned between Austria-Hungary, Prussia, and Russia.

- ❑ **1918** Polish state recreated.
- ❑ **1921** Democratic constitution.
- ❑ **1926–1935** Marshal Jozef Pilsudski heads military coup. Nine years of authoritarian rule.
- ❑ **1939** Germany invades and divides Poland with Russia.
- ❑ **1941** First concentration camps built on Polish soil.
- ❑ **1944** Warsaw Uprising.
- ❑ **1945** Potsdam and Yalta Conferences set present borders and determine political allegiance to Soviet Union.
- ❑ **1947** Communists manipulate elections to gain power.

P

CHRONOLOGY *continued*

- **1956** More than 50 killed in rioting in Poznań.
- **1970** Food price increases lead to strikes and riots in the Baltic port cities. Hundreds are killed.
- **1979** Cardinal Karol Wojtyla of Kraków is elected pope and takes the name of John Paul II.
- **1980** Strikes force the government to negotiate with the Solidarity union. Resulting Gdańsk Accords grant the right to strike and to form free trade unions.
- **1981** Gen. Wojciech Jaruzelski becomes prime minister.
- **1981–1983** Martial law. Solidarity forced into underground existence. Many of its leaders, including Lech Walesa, are interned.
- **1983** Walesa awarded Nobel Peace Prize.
- **1986** Amnesty for political prisoners.
- **1987** Referendum rejects government austerity program.
- **1988** Renewed industrial unrest.
- **1989** Ruling party holds talks with Solidarity, which is relegalized. Partially free elections are held. First postwar noncommunist government formed.
- **1990** Launch of market reforms. Walesa elected president.
- **1991** Free elections lead to fragmented parliament.
- **1992** Last Russian troops leave.
- **1993** Elections: reformed communists head coalition government.
- **1994** Launch of mass privatization.
- **1995** Leader of reformed communists Aleksander Kwasniewski is elected president.
- **1996** Historic Gdańsk shipyard declared bankrupt and closed down.
- **1997** April, parliament finally adopts new postcommunist constitution. September, legislative elections end former communist majority with big swing to AWS coalition. December, EU agrees to open membership negotiations.
- **1999** Joins NATO.
- **2001** Elections: AWS routed, left-of-center coalition formed under Leszek Miller.

AID

Recipient

 $1.4bn (receipts)

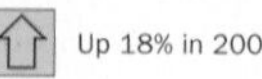 Up 18% in 2000

Large-scale aid for economic transformation was a phenomenon of the early 1990s. The IMF, the EBRD, and the EU all supported Poland's stabilization and reform program. EU aid now focuses on helping Poland to prepare to meet the environmental and other standards required of its members.

DEFENSE

Compulsory military service

 $3.19bn

Down 1% in 2000

Poland joined NATO in March 1999. Its standing army is among the largest in Europe, and there are also large paramilitary units, including border guards. A 15-year program to modernize the armed forces was introduced by the government in 1997. A civilian alternative to military service was first offered in 1998, and military service was reduced from 18 months to a year from 1999.

POLISH ARMED FORCES

	1677 main battle tanks (786 T-55, 685 T-72, 206 PT-91)	120,300 personnel
	3 submarines, 2 frigates, 1 destroyer, and 23 patrol boats	16,760 personnel
	212 combat aircraft (91 MiG-21, 22 MiG-29, 99 Su-22)	43,735 personnel
	None	

ECONOMICS

Inflation 23.4% p.a. (1990–2000)

 $162bn

4.133–3.952 zlotys

SCORE CARD

World GNP Ranking	25th
GNP per Capita	$4190
Balance of Payments	–$10bn
Inflation	10.1%
Unemployment	16%

EXPORTS

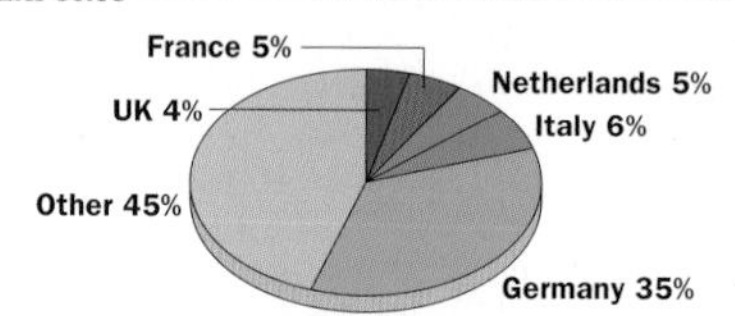
France 5%
Netherlands 5%
UK 4%
Italy 6%
Other 45%
Germany 35%

IMPORTS

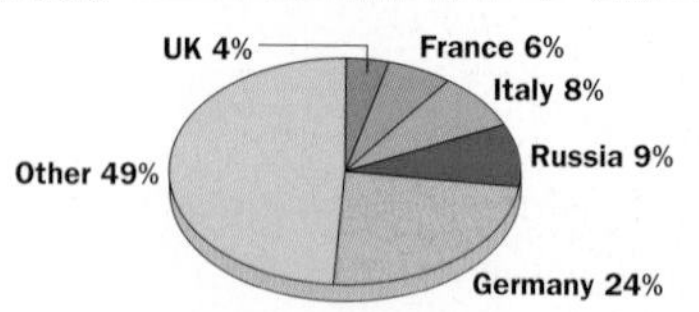
UK 4%
France 6%
Italy 8%
Other 49%
Russia 9%
Germany 24%

Strengths

Restructuring of loss-making coal industry began in 1998. Successful privatizations accelerated again in late 1990s. High rates of foreign investment reflect status as largest market in central Europe. Modernization of electricity supply and oil refining. Booming construction industry.

Weaknesses

Agriculture suffers from overmanning, tiny farms, and lack of investment. Compensation for communist-era property expropriations unresolved. Heavy industries not competitive.

Profile

After a decade of economic crisis, the postcommunist government in 1990 drove through the most determined plan in the whole region to make the transition to a market economy. Most prices were freed, trade was opened and the zloty was made convertible. Economic growth and foreign investment soared, especially after

ECONOMIC PERFORMANCE INDICATOR

Consumer Price Index
GDP
220
180
140
100
60
Consumer price index 1995=100
GDP 1996=100
1996
1997
1998
1999
2000

Western creditors agreed to cancel half of the country's foreign debt in 1994. Poland now attracts the most foreign capital in central and eastern Europe.

There are still large-scale heavy industrial plants left over from the communist era, but some have been converted or reorganized successfully. Many state farms have been liquidated, but agricultural efficiency is improving only slowly. Some 26% of the workforce is employed in farming.

Economic growth slowed at the end of the 1990s, but was back to 4% in 2000. Inflation has remained high and unemployment hit a postcommunist record level in 2002.

POLAND : MAJOR BUSINESSES

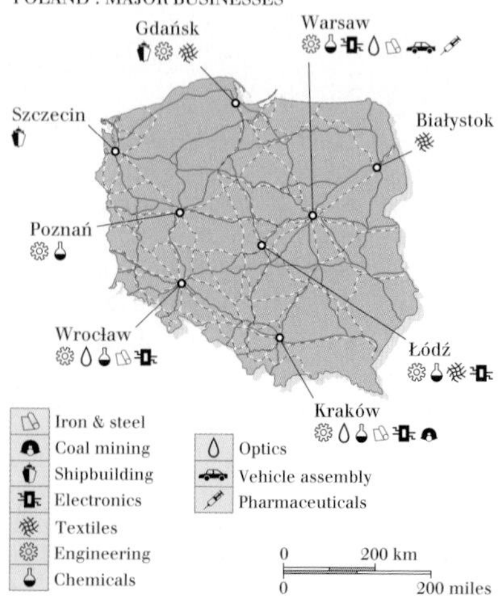
Gdańsk
Warsaw
Szczecin
Białystok
Poznań
Wrocław
Łódź
Kraków
Iron & steel
Coal mining
Shipbuilding
Electronics
Textiles
Engineering
Chemicals
Optics
Vehicle assembly
Pharmaceuticals
0
200 km
0
200 miles

RESOURCES

Electric power 29.5m kw

268,822 tonnes

7218 b/d (reserves 37m barrels)

17m pigs, 5.72m cattle, 3.57m ducks, 48.3m chickens

Coal, copper, silver, sulfur, natural gas, lead, salt, iron

ELECTRICITY GENERATION

Hydro	3% (4.3bn kwh)
Combustion	97% (138bn kwh)
Nuclear	0%
Other	0%

% of total generation by type

Poland has significant quantities of coal, copper, silver, sulfur, natural gas, lead, and salt.

The government's aim is to achieve self-sufficiency in energy resources and eventually to be able to export them; plans are in place to privatize both the fuel and energy industries. Coal supplies two-thirds of the country's electricity generation.

POLAND : LAND USE

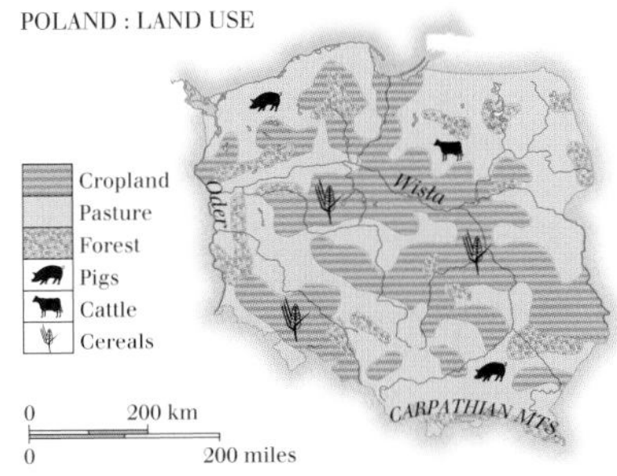
Cropland
Pasture
Forest
Pigs
Cattle
Cereals
Oder
Wisla
CARPATHIAN MTS.
0 200 km
0 200 miles

ENVIRONMENT

Sustainability rank: 87th

10%

8.3 tonnes per capita

ENVIRONMENTAL TREATIES

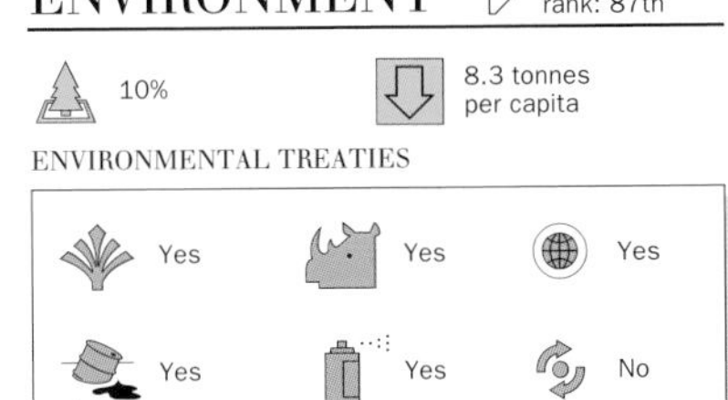
Yes
Yes
Yes
Yes
Yes
No

Pollution problems are serious, but improving. Upper Silesia and the Kraków area are still badly affected, but industry there only emits a third of the pollutants it emitted in 1990. Now that much heavy industry has been cleaned up or closed down, there is more concern about small factories, domestic coal fires, and the increased use of private cars.

Water pollution is a major problem, mainly from untreated sewage and industrial discharges. Rivers flowing into the Baltic are badly affected by nitrates and phosphates used in farming. Polish standards, themselves widely disregarded, need to be raised to meet EU minimum requirements.

MEDIA

TV ownership high

Daily newspaper circulation 37 per 1000 people

The constitution guarantees media freedom. *Gazeta Wyborcza*, set up by Solidarity in 1989, is still the leading daily, and its owners are expanding into other media. The head of election coverage at Polish Public Television (TVP) resigned in 2000, accused of favoring President Kwasniewski.

PUBLISHING AND BROADCAST MEDIA

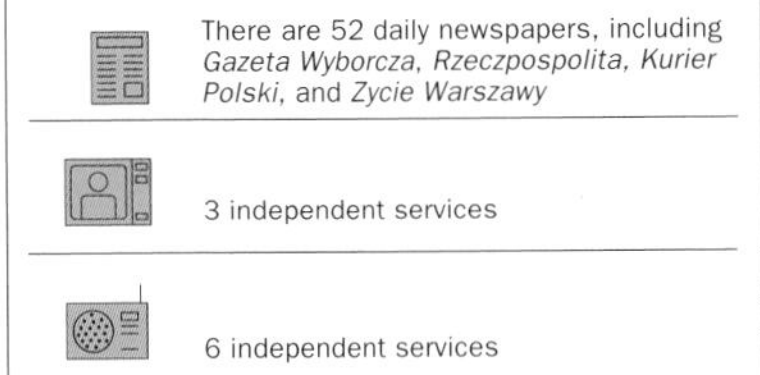
There are 52 daily newspapers, including *Gazeta Wyborcza, Rzeczpospolita, Kurier Polski*, and *Zycie Warszawy*
3 independent services
6 independent services

CRIME

No death penalty

70,544 prisoners

Up 5% in 1999

CRIME RATES

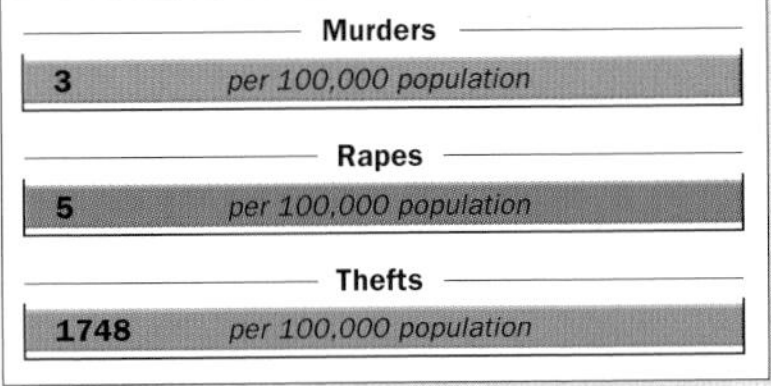
Murders
3 per 100,000 population
Rapes
5 per 100,000 population
Thefts
1748 per 100,000 population

Smuggling is the most significant problem, and Warsaw is a main center for this. Narcotics are transferred westward to Germany and expensive cars eastward to Russia. A National Remembrance Institute was set up in mid-2000 to investigate and prosecute the Nazi- and communist-era crimes of 1939–1989.

EDUCATION

School leaving age: 18

99%

1.43m students

THE EDUCATION SYSTEM

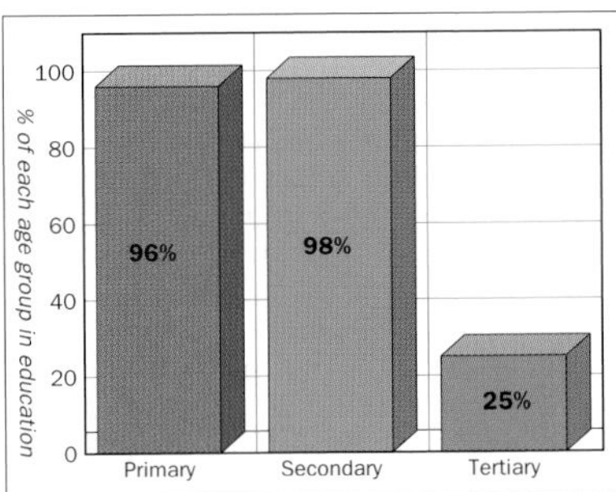
% of each age group in education
96%
98%
25%
Primary
Secondary
Tertiary

Primary education lasts from the age of seven to 13; lower secondary level follows until 16. At upper secondary level, exam-based selection separates the academic, technical, and vocational schools. A standard curriculum is followed in all schools. Despite the high official literacy figures, a relatively large proportion of school-leavers still lack basic skills. Public spending on education fell in real terms in the 1990s. Since 1989 the Roman Catholic Church has been allowed to operate schools. Most of the higher education institutions offer business-related courses.

HEALTH

Welfare state health benefits

1 per 435 people

Arteriosclerosis, heart disease, cancers, accidents

Fundamental reforms introduced in 1999 created a "market" health system, giving patients the right to choose where to go for treatment. Intended to be decentralized and less bureaucratic, the new system was confusing for some patients, with hospitals and doctors competing for business. Medical care is free for most people, but there are now a number of private health clinics.

SPENDING

GDP/cap. increase

CONSUMPTION AND SPENDING

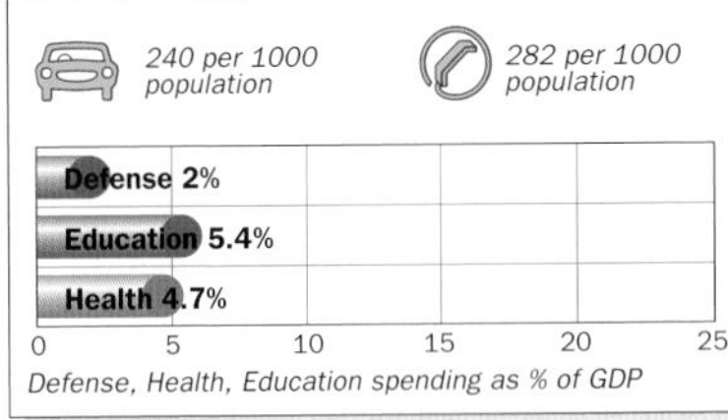
240 per 1000 population
282 per 1000 population
Defense 2%
Education 5.4%
Health 4.7%
Defense, Health, Education spending as % of GDP

Market reforms have led to some structural unemployment, and the inevitable hardship that this represents. More restructuring of heavy industry is planned. Pensioners have enjoyed benefits amounting to a higher percentage of GDP than in most countries, but state cutbacks are making private pensions more necessary.

WORLD RANKING

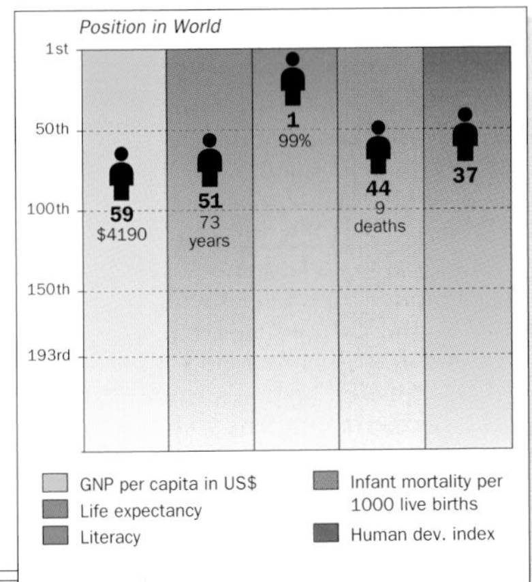
Position in World
59 $4190
51 73 years
1 99%
44 9 deaths
37
GNP per capita in US$
Life expectancy
Literacy
Infant mortality per 1000 live births
Human dev. index

P

EUROPE
France
PORTUGAL
Spain

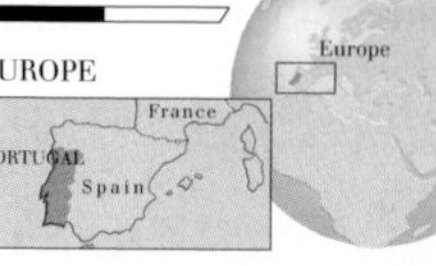
Europe

PORTUGAL

OFFICIAL NAME: Republic of Portugal **CAPITAL:** Lisbon **POPULATION:** 10 million
CURRENCY: Euro (escudo until 2002) **OFFICIAL LANGUAGE:** Portuguese

1139

1640

June 10

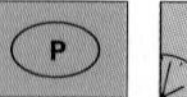
P

0

+351

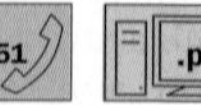
.pt

PORTUGAL, WITH ITS long Atlantic coast, lies on the western side of the Iberian peninsula. The river Tagus divides the more mountainous north from the lower, undulating terrain to the south. In 1974, a bloodless military coup overthrew a long-standing conservative dictatorship. A constituent assembly was elected in 1975 and the armed forces withdrew from politics thereafter. Portugal then began a substantial program of economic modernization and accompanying social change. Membership of the EU has helped underpin this process.

Santa Marta de Penaguiao, a small village in the heart of Portugal's wine-producing region, which is centered on the Douro valley.

CLIMATE

Mediterranean/ maritime

WEATHER CHART FOR LISBON

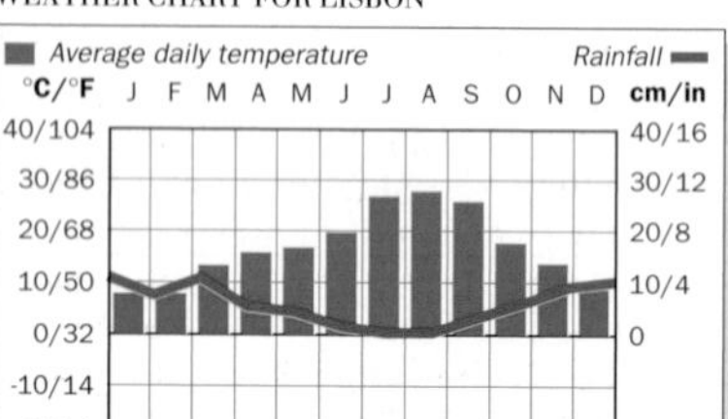
Average daily temperature
Rainfall
°C/°F
J F M A M J J A S O N D
cm/in
40/104
30/86
20/68
10/50
0/32
-10/14
-20/-4
40/16
30/12
20/8
10/4
0

Portugal has a mild, Mediterranean climate, which is moderated by the influence of the Atlantic. Summers can be hot and sultry, while winters are relatively mild. Inland areas have more variable weather than coastal regions. Rainfall is generally higher in the mountainous north, while the central areas are more temperate. The southern Algarve region is predominantly dry and sunny.

TRANSPORTATION

Drive on right

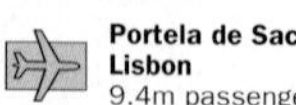

Portela de Sacavém, Lisbon
9.4m passengers

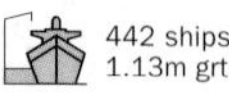

442 ships
1.13m grt

THE TRANSPORTATION NETWORK

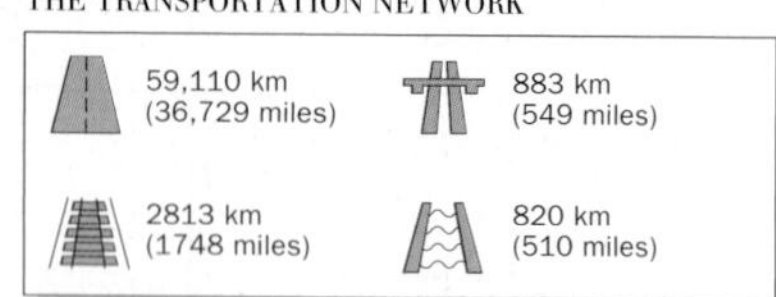
59,110 km
(36,729 miles)
883 km
(549 miles)
2813 km
(1748 miles)
820 km
(510 miles)

Road links with Spain remain limited, despite modernization schemes and the new southern Guadiana bridge. The Lisbon–Madrid expressway was finally completed in 1999, the year after the 18-km Vasco da Gama bridge in Lisbon opened. Poor road construction, heavy traffic, and dangerous driving mean that Portugal has Europe's highest rate of road deaths. Lisbon's small, efficient metro complements its trams, but Porto's metro remains uncompleted. A high speed rail link with Spain is due for completion in 2008.

TOURISM

Visitors : Population
1.2:1

12m visitors

Up 4% in 2000

MAIN TOURIST ARRIVALS

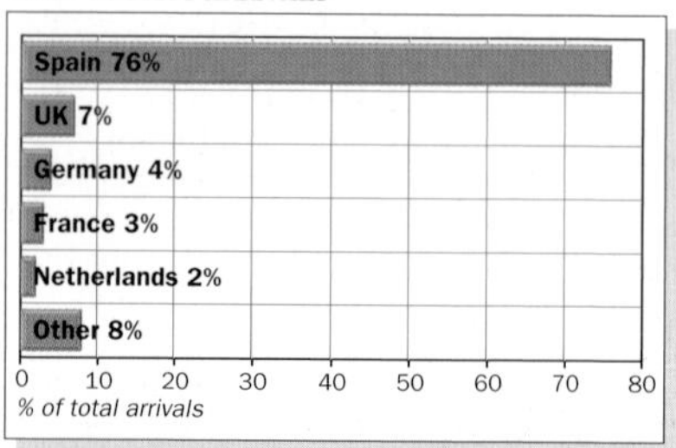
Spain 76%
UK 7%
Germany 4%
France 3%
Netherlands 2%
Other 8%
0 10 20 30 40 50 60 70 80
% of total arrivals

From the 1960s, Portugal's popularity as a tourist destination has been linked in part to qualities which reflected its relatively poor economic development, such as low prices and little crime. Thus some of the consequences of its substantial economic growth may have eroded part of Portugal's appeal, but since it now has around 12 million visitors a year, tourism remains a major income-earner. The most popular destination is the Algarve, the southernmost province, followed by the western resorts of Figueira da Foz and the Tróia Peninsula. Visitors are also attracted by Portugal's architecture, notably that dating from the Manueline period (1490–1520), and by its handicrafts, such as ceramics, lace, and tapestries. In addition, Portugal is noted for being the location of some of Europe's finest golf courses.

PORTUGAL

Total Land Area : 92 391 sq. km (35 672 sq. miles)

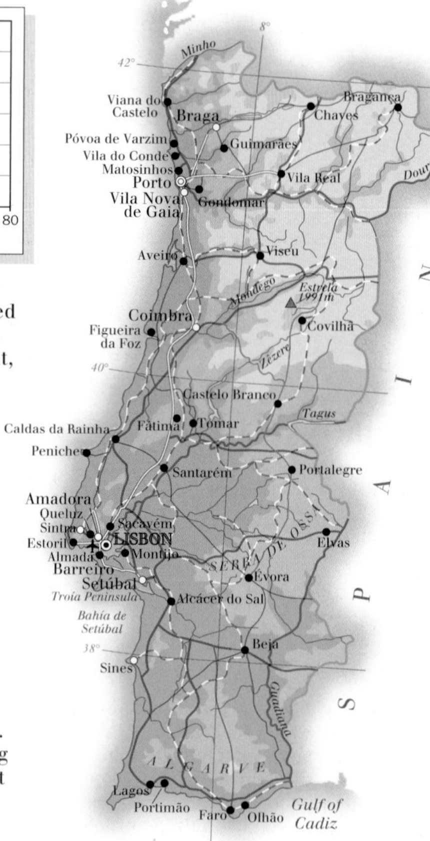
Minho
Viana do Castelo
Braga
Bragança
Chaves
Póvoa de Varzim
Vila do Conde
Guimarães
Matosinhos
Porto
Vila Real
Douro
Vila Nova de Gaia
Gondomar
Aveiro
Viseu
Mondego
Estrela 1991m
Coimbra
Figueira da Foz
Covilhã
Zêzere
Castelo Branco
Caldas da Rainha
Fátima
Tomar
Tagus
Peniche
Santarém
Portalegre
Amadora
Queluz
Sintra
Sacavém
Estoril
LISBON
Montijo
Almada
Barreiro
Elvas
SERRA DE OSSA
Setúbal
Évora
Tróia Peninsula
Alcácer do Sal
Bahia de Setúbal
Beja
Sines
Guadiana
ALGARVE
Lagos
Portimão
Faro
Olhão
Gulf of Cadiz
SPAIN

Azores
Corvo
Flores
São Jorge
Graciosa
Terceira
Faial
Pico
São Miguel
Ponta Delgada
Santa Maria
0 200 km
0 200 miles

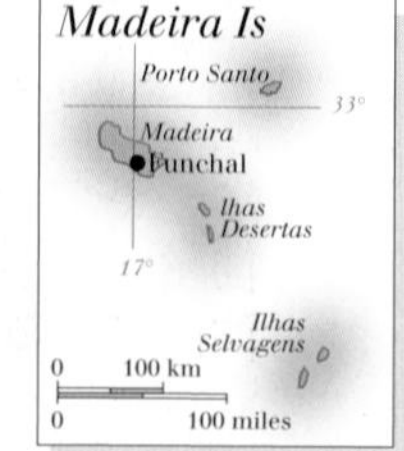
Madeira Is
Porto Santo
Madeira
Funchal
Ilhas Desertas
Ilhas Selvagens
0 100 km
0 100 miles

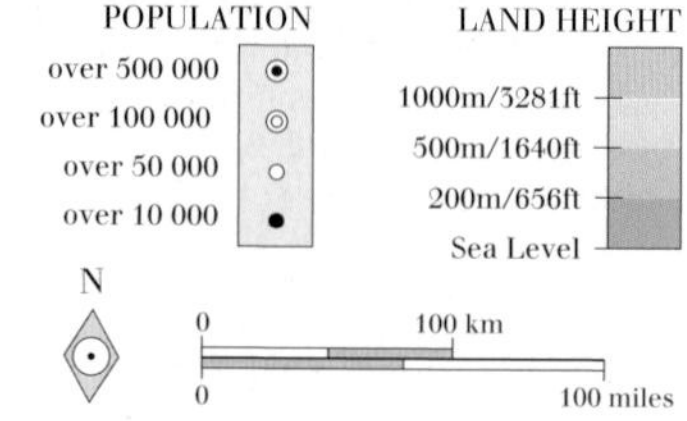
POPULATION
over 500 000
over 100 000
over 50 000
over 10 000
LAND HEIGHT
1000m/3281ft
500m/1640ft
200m/656ft
Sea Level
N
0 100 km
0 100 miles

P

PEOPLE

Population density medium

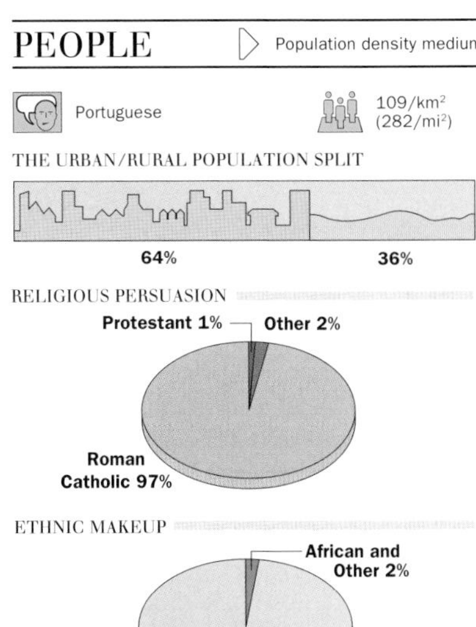
Portuguese
109/km²
(282/mi²)
THE URBAN/RURAL POPULATION SPLIT
64%
36%
RELIGIOUS PERSUASION
Protestant 1%
Other 2%
Roman Catholic 97%

ETHNIC MAKEUP

African and Other 2%

Portuguese 98%

Portuguese society, once regarded as rather inward-looking, has become much more egalitarian since the 1974 revolution. It is increasingly integrated into the rest of western Europe.

The Roman Catholic Church has lost some of its social influence, as shown by falling birthrates and more liberal attitudes to abortion, divorce, and unmarried mothers (almost one in five children are born outside marriage). Apart from urban areas, the north is still devoutly Catholic.

Ethnic and religious tensions are limited. Immigration increased after 1974, and foreigners now constitute 2% of the population. Early arrivals came from the former African colonies, but recently there has been an influx of east European workers.

Family ties remain all-important. Women got the vote only in 1976; now 60% of university students are women, and 63% of women of working age have jobs.

POPULATION AGE BREAKDOWN

Female	Age	Male
1.9%	80+	1%
9.7%	60–79	7.6%
12.6%	40–59	11.5%
15.4%	20–39	15.2%
12.3%	0–19	12.8%

% of population by age group

WORLD AFFAIRS

Joined UN in 1955

EU
CE
NATO
OECD
OSCE

Since 1986, Portugal's foreign policy has dealt mainly with the consequences of EU membership. It is a committed NATO member, though its relative strategic importance declined after Spain joined. Relations with its former African colonies, occasionally turbulent, are a high priority, as are those with Brazil. Portugal backed East Timor's struggle to reverse its annexation by Indonesia. Relations with China were cordial enough to ensure the smooth return of Macao to the latter at the end of 1999.

AID

Donor

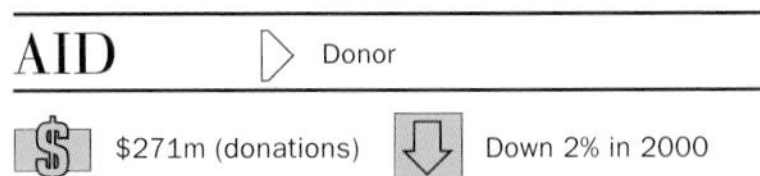
$271m (donations)
Down 2% in 2000

Portugal is a major beneficiary of EU aid from the so-called structural funds. It currently earmarks 0.25% of its GDP for aid to developing countries. More than 60% goes to former colonies in Africa, especially Mozambique, where Portuguese funding helped rebuild the massive war-damaged Cahora Bassa Dam and power plant.

POLITICS

Multiparty elections

2002/2006

President Jorge Sampão

AT THE LAST ELECTION

Assembly of the Republic 230 seats

46% PSD | 42% PS | 6% PP | 1% BE | 5% CDU

PSD = Social Democratic Party
PS = Socialist Party **PP** = People's Party
CDU = United Democratic Coalition **BE** = Left Bloc

Portugal is a multiparty democracy.

PROFILE

A decade of center-right government ended in the 1995 elections, when the PSD lost to the PS. Having campaigned on a platform of social reform, the PS, under António Guterres, gave priority to fiscal control. Despite a privatization program, however, the size of the public sector actually increased in 1995–1999. The PS was reelected in 1999, but soon lost popularity, and was ousted by the PSD in early elections in March 2002. PSD leader José Manuel Durão Barroso was appointed prime minister. Lacking an overall majority, the PSD allied with the right-wing PP. The new government pledged to cut public spending and accelerate privatization.

MAIN POLITICAL ISSUES

Transformation

Portugal has been transformed in recent decades. The consolidation of democracy since the 1974 "carnation revolution" and EU membership since 1986 have brought Portugal into the European mainstream. There are high expectations about completing the "catching-up" process. Expo '98 in Lisbon showed off the new confidence. In 2002 Portugal was among the first 12 EU countries fully to adopt the euro.

Presidency and parliament

For ten years up to 1995, the presidency and the government were controlled by opposing parties, a situation which encouraged conflict and obstruction. The relationship then began to function more smoothly, with PS presidents and minority governments. Former PS leader Jorge Sampãio, who succeeded Mário Soares as president in 1996, was reelected comfortably in early 2001, only to see a PSD government reclaim power a year later.

***President Jorge Sampãio**, socialist president since 1996.*

***José Manuel Durão Barroso**, PSD prime minister since 2002.*

CHRONOLOGY

Portugal has existed as a nation state since 1139, although it was frequently challenged by Spain. It reached its zenith in the 16th century, before being annexed by Spain in 1580.

- **1640** Independence from Spain
- **1755** Earthquake destroys Lisbon.
- **1793** Joins coalition against revolutionary France.
- **1807** France invades; royal family flees to Brazil.
- **1808** British troops arrive under Wellington. Start of Peninsular War.
- **1810** French leave Portugal.
- **1820** Liberal revolution.
- **1822** King John VI returns and accepts first Portuguese constitution. His son Dom Pedro declares independence of Brazil.
- **1834** Dom Pedro returns to Portugal to end civil war and installs his daughter as Queen Mary II.
- **1875–1876** Republican and Socialist parties founded.
- **1891** Republican uprising in Porto.
- **1908** Assassination of King Carlos I and heir to the throne.
- **1910** Abdication of Manuel II and proclamation of the Republic. Church and state separated.
- **1916** Portugal joins Allied side in World War I.
- **1917–1918** New Republic led by Sidónio Pais.

P

CHRONOLOGY *continued*

- ❑ **1926** Army overturns republic.
- ❑ **1928** António Salazar joins government as finance minister. Economy improves significantly.
- ❑ **1932** Salazar prime minister.
- ❑ **1933** Promulgation of the constitution of the "New State," instituting right-wing dictatorship.
- ❑ **1936–1939** Salazar assists Franco in Spanish Civil War.
- ❑ **1939–1945** Portugal neutral during World War II, but lets UK use air bases in Azores.
- ❑ **1949** Founder member of NATO.
- ❑ **1955** Joins UN.
- ❑ **1958** Américo Thómas appointed president, following fraudulent defeat of Gen. Humberto Delgado.
- ❑ **1961** India annexes Goa. Guerrilla warfare breaks out in Angola, Mozambique, and Guinea.
- ❑ **1970** Death of Salazar, incapacitated since 1968; succeeded by Marcelo Caetano.
- ❑ **1971** Caetano attempts liberalization.
- ❑ **1974** Carnation Revolution – left-wing Armed Forces Movement overthrows Caetano.
- ❑ **1974–1975** Portuguese possessions in Africa attain independence. Some 750,000 Portuguese expatriates return to Portugal.
- ❑ **1975** Communist takeover foiled by moderates and Mário Soares' PS.
- ❑ **1975–1976** Indonesia seizes former Portuguese East Timor unopposed.
- ❑ **1976** Gen. António Eanes elected president. New constitution. Soares appointed prime minister.
- ❑ **1978** Period of nonparty technocratic government instituted.
- ❑ **1980** Center-right wins elections. Gen. Eanes reelected.
- ❑ **1982** Full civilian government formally restored.
- ❑ **1983** Soares becomes caretaker prime minister; PS is majority party.
- ❑ **1985** Anibal Cavaco Silva becomes prime minister. Minority PSD government.
- ❑ **1986** Soares elected president. Portugal joins EU, which funds major infrastructure and construction projects.
- ❑ **1987** Cavaco Silva wins absolute majority in parliament.
- ❑ **1991** Soares reelected president.
- ❑ **1995** PS wins elections; António Guterres becomes prime minister.
- ❑ **1996** Former PS leader Jorge Sampãio elected president.
- ❑ **1999** PS strengthens its position in general election. December, Macao returned to China.
- ❑ **2001** Sampãio reelected.
- ❑ **2002** Euro fully adopted. PSD wins early elections, forms coalition with PP. José Manuel Durão Barroso becomes prime minister.

DEFENSE

▷ Phasing out conscription

 $2.2bn

  Down 5% in 2000

Portugal has been a member of NATO since 1949. It has a small but relatively modern navy. The army and air force are less efficient. The government announced in 1999 that it planned to abolish compulsory miltary service (currently of four months, with a seven-month civilian alternative). The US, which is the major arms supplier, has a strategic air base in the Azores.

PORTUGUESE ARMED FORCES

187 main battle tanks (86 M-48A5, 101 M-60)	25,400 personnel
2 submarines, 6 frigates, and 31 patrol boats	10,800 personnel
66 combat aircraft (40 *Alpha Jet*, 20 F-16A/B)	7400 personnel
None	

ECONOMICS

▷ Inflation 5.3% p.a. (1990–2000)

 $111bn

 1.0651–1.1231 euros

SCORE CARD

❑ World GNP Ranking	33rd
❑ GNP per Capita	$11,120
❑ Balance of Payments	–$10.6bn
❑ Inflation	2.9%
❑ Unemployment	4%

EXPORTS

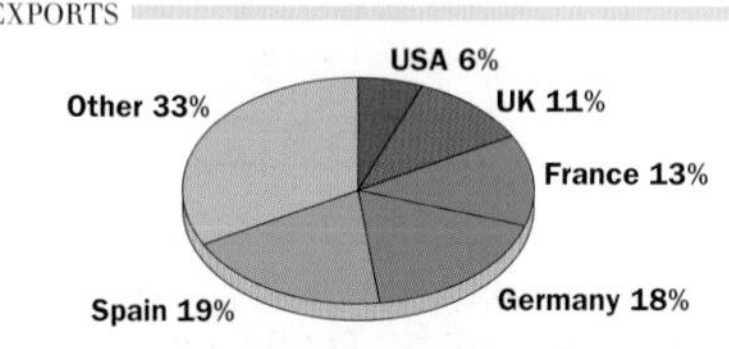
USA 6%
UK 11%
France 13%
Germany 18%
Spain 19%
Other 33%

IMPORTS

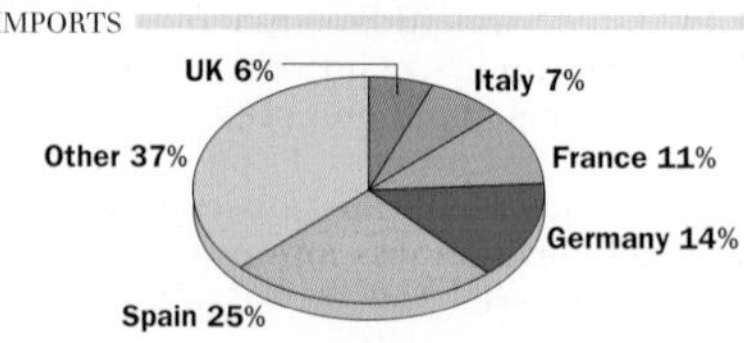
UK 6%
Italy 7%
France 11%
Germany 14%
Spain 25%
Other 37%

Strengths

Relatively low, if rapidly rising, labor costs. Flexible labor market. High domestic and direct foreign investment. Strong banking sector. Potential for growth of tourism, now earning 6% of GDP – highest ratio in EU. Clothing and shoe manufacturing now joined by cars (notably Volkswagens) and machinery as major exports. Fast-track improvement of transportation infrastructure. Good deepwater port at Lisbon. Wine, especially port. Tomatoes, citrus fruit, cork, sardines.

Weaknesses

High dependence on imported oil. Reliance on public works to drive economic growth. Large agricultural sector (4% of GDP, 13% of workforce) most inefficient in EU. Outdated farming methods, small landholdings, low crop yields; product prices undercut by Spain.

Profile

EU membership in 1986 brought a sharp increase in foreign investment to largely rural Portugal. Exports rose dramatically. The economy recorded growth of over 3% a year in the second half of the 1990s. Maintaining growth thereafter required a big emphasis on raising labor productivity.

At one time, participation in EU monetary union seemed a distant prospect. In the event the socialist government of António Guterres (1995–2001) was able to meet the required economic criteria with relative ease, and in January 2002 Portugal was among the 12 EU countries which fully adopted the euro. While Portuguese wages are about 70% of the EU average, the unemployment rate is among the lowest in the EU.

ECONOMIC PERFORMANCE INDICATOR

Consumer Price Index
GDP
160
140
120
100
80
1996
1997
1998
1999
2000
Consumer price index 1995=100
GDP 1996=100

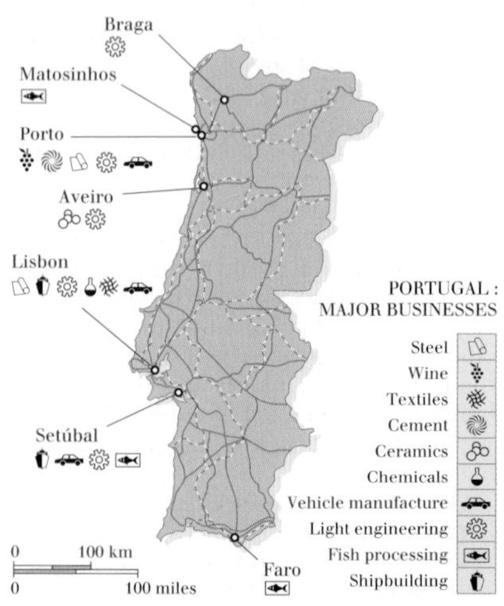
Braga
Matosinhos
Porto
Aveiro
Lisbon
Setúbal
Faro
0
100 km
100 miles
PORTUGAL : MAJOR BUSINESSES
Steel
Wine
Textiles
Cement
Ceramics
Chemicals
Vehicle manufacture
Light engineering
Fish processing
Shipbuilding

P

RESOURCES

Electric power 9.8m kw

215,230 tonnes

Not an oil producer; refines 294,000 b/d

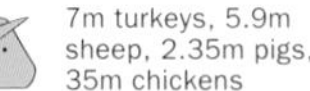

7m turkeys, 5.9m sheep, 2.35m pigs, 35m chickens

Coal, limestone, granite, marble, tin, copper, tungsten

ELECTRICITY GENERATION

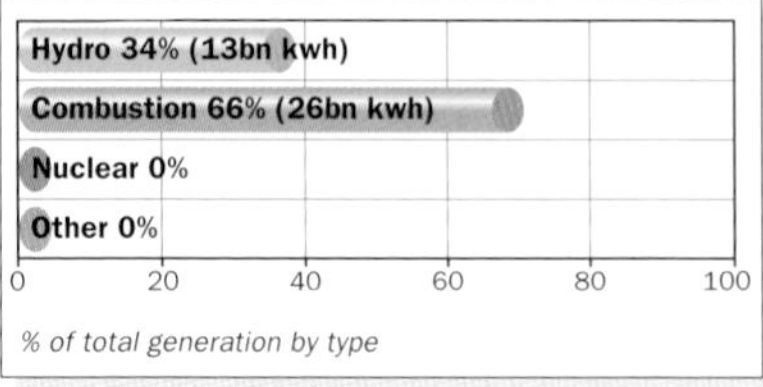
Hydro 34% (13bn kwh)
Combustion 66% (26bn kwh)
Nuclear 0%
Other 0%
0
20
40
60
80
100
% of total generation by type

Portugal is disadvantaged by a lack of natural resources, including water. Mining has historically been important, notably for tungsten, copper, and tin. The last coal mine closed in the mid-1990s. The fish catch, once central to the economy, has been declining in recent years.

PORTUGAL : LAND USE

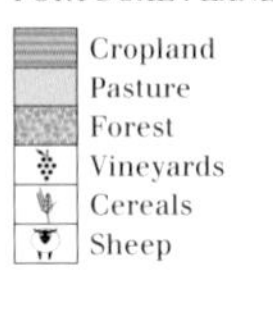
Cropland
Pasture
Forest
Vineyards
Cereals
Sheep
0
100 km
0
100 miles

Rio Douro
SERRA DE ESTRELA
Tagus
SERRA DE OSSA

ENVIRONMENT

Sustainability rank: 28th

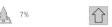

7%

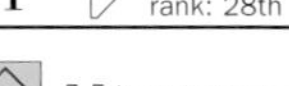

5.5 tonnes per capita

ENVIRONMENTAL TREATIES

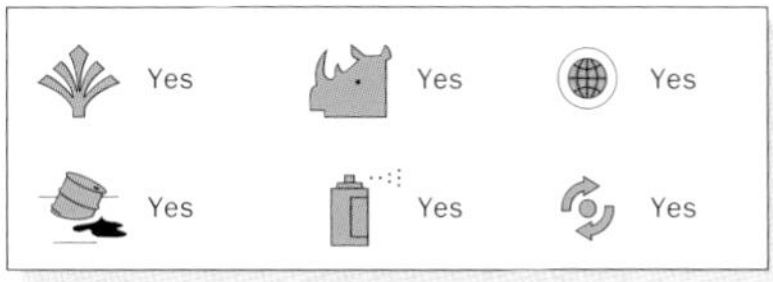
Yes
Yes
Yes
Yes
Yes
Yes

The unrestricted development of tourist resorts in the Algarve and major infrastructure projects are having detrimental effects on natural habitats. EU agricultural grants for projects such as draining meadows, and monoculture afforestation, notably of eucalyptus and pine, are degrading biodiversity. Much toxic waste is dumped on any available land, as few official controls or infill sites exist. New waste management regulations are being introduced.

MEDIA

TV ownership high

Daily newspaper circulation 73 per 1000 people

PUBLISHING AND BROADCAST MEDIA

There are 31 daily newspapers. The most prestigious is the *Diário de Notícias*, but the weekly *Expresso* has a wider circulation
1 state-owned service, several independent services
1 state-owned service, several independent national and regional services

Most newspapers have only regional distribution. TV is the dominant medium. In 1992–1993 two independent TV stations were launched. The Catholic TVI was sold in 1998 to Media Capital, one of four groups that control most of the press and broadcasting. The government increased its control over state-owned RTP in 2002, raising fears over its political independence.

CRIME

No death penalty

13,106 prisoners

Up 1% in 1999

CRIME RATES

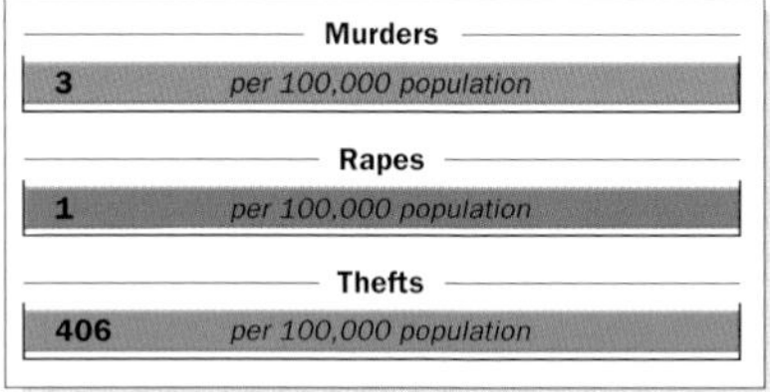
Murders
3 per 100,000 population
Rapes
1 per 100,000 population
Thefts
406 per 100,000 population

Portugal enjoys a low crime rate. Consumption and possession of small quantities of narcotics were decriminalized in mid-2000 and legalized a year later.

EDUCATION

School leaving age: 15

92%

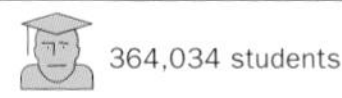

364,034 students

THE EDUCATION SYSTEM

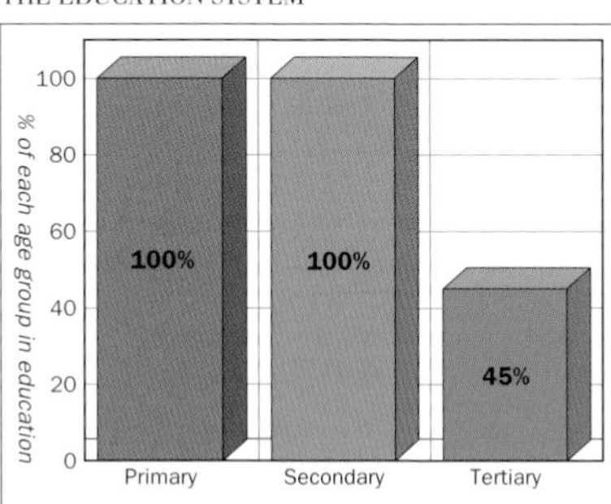
100
80
60
40
20
0
% of each age group in education
100%
100%
45%
Primary
Secondary
Tertiary

Portuguese is the sixth most widely spoken language in the world.

Free state education is available to all pupils between the ages of three and 15. Nursery provision has been greatly expanded, although the preschool stage is not compulsory. Middle-class parents rely heavily on the private sector. State universities have been expanded to ease the pressure on places. There are several prestigious private universities.

HEALTH

Welfare state health benefits

1 per 313 people

Heart and cerebro-vascular diseases, cancers, accidents

Of total government expenditure nearly 10% is spent on health.

Portugal has had a publicly funded, free national health service since 1979; Spending on health has increased markedly in recent years, but care remains below the EU average. There are strong regional differences in facilities. Larger urban hospitals are modern and well equipped. Private health care schemes, which are allowed to coexist, are both affordable and good value for money; over 40% of the population use the private system.

SPENDING

GDP/cap. increase

CONSUMPTION AND SPENDING

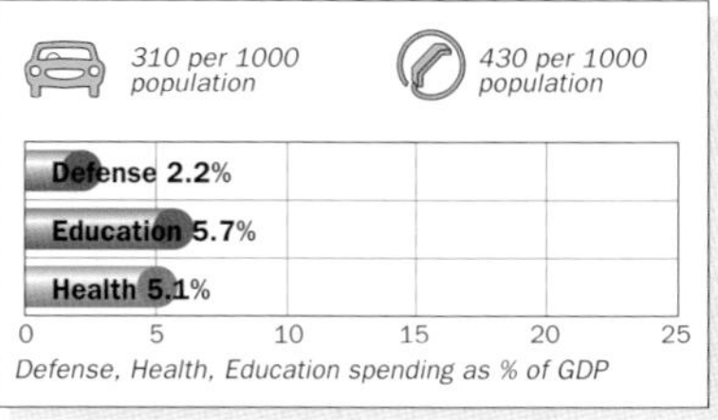
310 per 1000 population
430 per 1000 population
Defense 2.2%
Education 5.7%
Health 5.1%
0
5
10
15
20
25
Defense, Health, Education spending as % of GDP

Wealth differentials in Portugal are smaller than in most EU countries. The 1976 constitution committed Portugal to making the transition to socialism, and since then governments have introduced limited wealth redistribution measures.

Internal investment is directed chiefly through the property market, and there was a surge of purchases in the late 1990s. External investment goes to the EU and Brazil. In 2000, 14.5% of the population held shares directly. Average incomes, which were just over half the EU average in 1986, are now three-quarters of the EU average. Mobile phones, mainly on prepayment systems, pioneered in Portugal, are very widespread.

WORLD RANKING

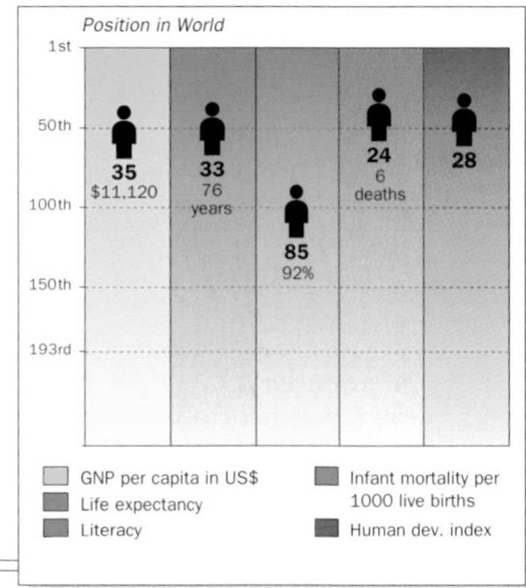
Position in World
1st
50th
100th
150th
193rd
35
$11,120
33
76
years
85
92%
24
6
deaths
28
GNP per capita in US$
Life expectancy
Literacy
Infant mortality per 1000 live births
Human dev. index

P

QATAR

MIDDLE EAST

Bahrain, QATAR, Saudi Arabia, UAE, Asia, Africa

OFFICIAL NAME: State of Qatar **CAPITAL:** Doha
POPULATION: 575,000 **CURRENCY:** Qatar riyal **OFFICIAL LANGUAGE:** Arabic

1971

1971

Sept 3

Q
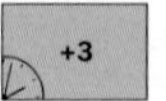
+3
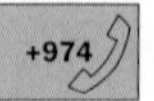
+974

.qa

PROJECTING NORTH FROM the Arabian peninsula into the Gulf, Qatar is mostly flat, semiarid desert. Oil production began in the late 1940s and quickly transformed Qatar from an impoverished pearl producer into a prosperous shaikhdom and a founder member of OPEC. Plentiful oil and gas reserves have made it one of the wealthiest states in the region. Politics is being democratized gradually under the ruling al-Thani clan.

CLIMATE

Hot desert

WEATHER CHART FOR DOHA

■ Average daily temperature — Rainfall

°C/°F J F M A M J J A S O N D cm/in

40/104, 30/86, 20/68, 10/50, 0/32, -10/14, -20/-4 — 40/16, 30/12, 20/8, 10/4, 0

The climate is hot and sultry, with midsummer temperatures reaching 44°C (111°F). Rainfall is rare.

TRANSPORTATION

Drive on right

Doha International
2.83m passengers

64 ships
744,131 grt

THE TRANSPORTATION NETWORK

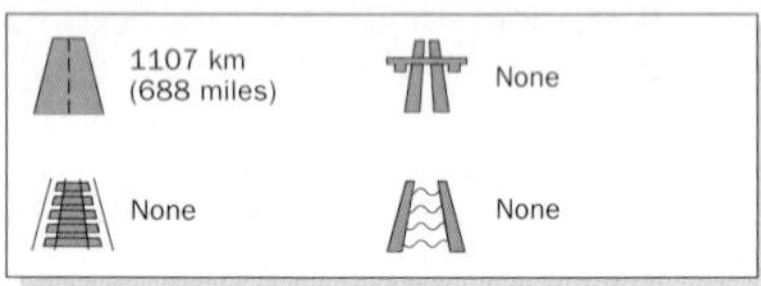
1107 km
(688 miles)
None
None
None

There is no public transportation system to speak of outside Doha, and the rate of car ownership is high.

TOURISM

Visitors : Population 1:1.3

451,000 visitors
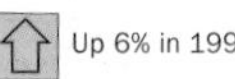
Up 6% in 1998

MAIN TOURIST ARRIVALS

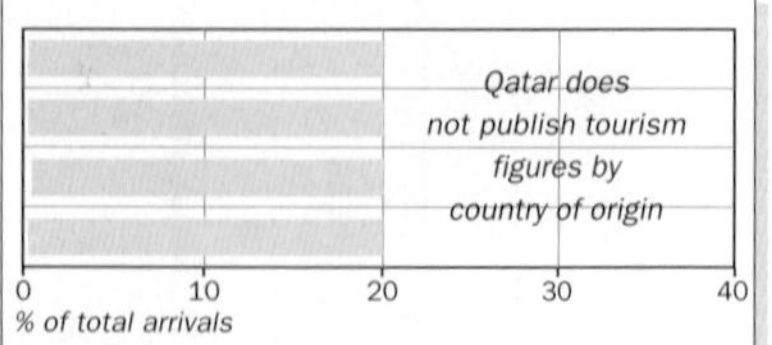
Qatar does not publish tourism figures by country of origin
0
10
20
30
40
% of total arrivals

Tourism is expanding. A government drive to improve Qatar's image as a tourist destination aims to net more than 1.5 million tourists a year by 2010. Attractions include unspoiled beaches, duty-free shopping, and modern hotels.

PEOPLE

Pop. density medium

Arabic

52/km² (135/mi²)

THE URBAN/RURAL POPULATION SPLIT

93% 7%

ETHNIC MAKEUP

Arab 40%, Iranian 10%, Other 14%, Indian 18%, Pakistani 18%

Only one in five inhabitants is native-born. Most are guest workers from the Indian subcontinent, Iran, and north African countries. Western expatriates enjoy a high standard of living and take no part in politics.

Most Qataris are followers of the Wahhabi interpretation of Sunni Islam and espouse conservative religious views. However, women are not obliged to wear a veil and can hold a driving license. Expatriate Christians are allowed freedom to worship but not to promote Christianity.

Since the advent of oil wealth, the Qataris, who were formerly nomadic Bedouins, have become a nation of city dwellers. Almost 90% of the population now inhabit the capital Doha and its suburbs. As a result, northern Qatar is dotted with depopulated and abandoned villages.

Doha, the capital. *Although desert covers the whole country, Qatar now grows most of its own vegetables by tapping groundwater.*

POLITICS

No legislative elections

Not applicable

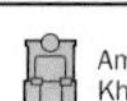
Amir Shaikh Hamad bin Khalifa al-Thani

LEGISLATIVE OR ADVISORY BODIES

Advisory Council 35 seats

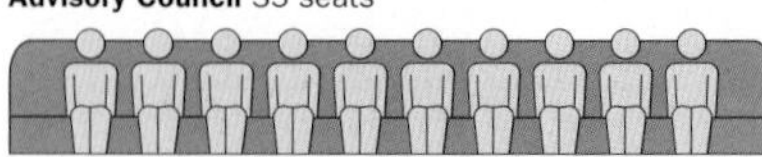

Qatar is an absolute monarchy and has no legislature. The amir rules with the assistance of the Council of Ministers and the Advisory Council.

Qatar is a traditional emirate. The government and religious establishment is dominated by the amir, Shaikh Hamad, who took power from his father, Shaikh Khalifa, in 1995. A failed coup against Hamad in early 1996 was linked with efforts to regain power by Khalifa. The prodemocracy movement has called for reform of the 35-member Advisory Council. Shaikh Hamad responded by authorizing Qatar's first elections, to a new municipal council for Doha, in 1999, in which all adults, including women, were able to vote and stand as candidates. In early 2001 he promised that a directly elected parliament would be established before 2003.

WORLD AFFAIRS

Joined UN in 1971

AL
OIC
GCC
OAPEC
OPEC

Although Qatar was a founder member of the GCC, Shaikh Hamad has adopted a somewhat ambivalent stance toward it. However, Qatar entered into the region's first mutual defense pact, under the aegis of the GCC, in 2000. In 2001 Qatar reached an agreement on the border with Saudi Arabia, but lost its claim to the Hawar Islands when the International Court of Justice ruled in Bahrain's favor.

Although keen to retain strong links with the West, the amir has criticized the US and the UK for their bombing of Iraq and over the campaign in Afghanistan. Qatar supplied liquified natural gas (LNG) to Israel in the late 1990s, but relations, previously the most cordial of the Gulf states, have suffered greatly since the renewal of the Palestinian *intifada* in 2000.

AID

Recipient

No net receipts

Down sharply in 2000

Qatar was a generous aid donor to developing countries during the 1970s and early 1980s. The small amounts of aid received come mainly from France.

Q

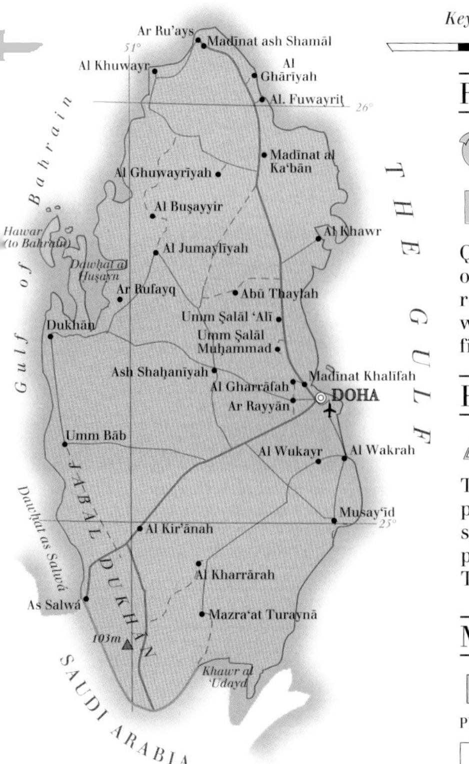

Ar Ru'ays
Madīnat ash Shamāl
Al Khuwayr
Al Ghāriyah
Al Fuwayrit
Madīnat al Ka'bān
Al Ghuwayrīyah
Al Buşayyir
Al Khawr
Al Jumaylīyah
Hawar (to Bahrain)
Dawhat al Huşayn
Ar Rufayq
Abū Thaylah
Umm Şalāl 'Alī
Umm Şalāl Muḥammad
Dukhān
Ash Shaḥanīyah
Madīnat Khalīfah
Al Gharrāfah
DOHA
Ar Rayyān
Umm Bāb
Al Wukayr
Al Wakrah
Musay'id
Al Kir'ānah
Al Kharrārah
As Salwá
Mazra'at Turaynā
103m
Khawr al 'Udayd
Gulf of Bahrain
THE GULF
JABAL DUKHAN
Dawhat as Salwá
SAUDI ARABIA

QATAR

Total Land Area : 11 437 sq. km (4416 sq. miles)

POPULATION
over 100 000
under 10 000

LAND HEIGHT
200m/1640ft
Sea Level

0 20 km
0 20 miles

DEFENSE

No compulsory military service

$1.43bn

Down 3% in 2000

The estimated 12,000-strong armed forces are too small to play a significant role in Qatari affairs, even in the event of political turmoil. A defense agreement with the USA has provided for joint exercises, the stockpiling of US equipment, and US access to bases.

RESOURCES

Electric power 1.9m kw

4207 tonnes

783,000 b/d (reserves 15.2bn barrels)

214,531 sheep, 179,000 goats, 4m chickens

Oil, natural gas

Qatar has the third-smallest reserves of crude oil within OPEC but abundant reserves of gas (the third-largest in the world), including the world's largest field of gas unassociated with oil.

ENVIRONMENT

Not available

None

85.7 tonnes per capita

The desert hinterland supports little plant or animal life. Most native species are extinct in the wild. Oil pollution has damaged marine life. There are salt flats in the south.

MEDIA

TV ownership high

Daily newspaper circulation 161 per 1000 people

PUBLISHING AND BROADCAST MEDIA

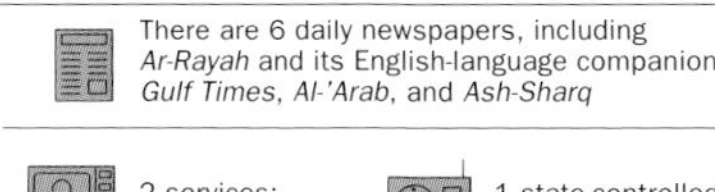
There are 6 daily newspapers, including *Ar-Rayah* and its English-language companion *Gulf Times*, *Al-'Arab*, and *Ash-Sharq*

2 services: 1 state-controlled

1 state-controlled service

Qatari TV is the most independent in the region; Al-Jazeera offers the leading Arab perspective internationally, especially since the events of 2001.

CRIME

Death penalty in use

527 prisoners

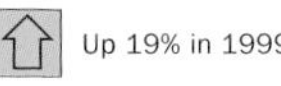
Up 19% in 1999

Traditional Islamic punishments have deterred crime. However, narcotics trafficking is on the increase. The incidence of street crime is low.

CHRONOLOGY

The al-Thanis, related to the Khalifa family of Bahrain, took control of the Qatar peninsula in the 18th century.

- **1971** Sovereignty recognized by UK.
- **1972** Accession of Amir Khalifa.
- **1995** Shaikh Hamad overthrows Shaikh Khalifa.
- **1999** First ever polls, to elect new municipal council for Doha.

EDUCATION

Schooling is free but not compulsory

81%

8475 students

Education is free from primary to university level. The government finances students to study overseas.

HEALTH

Welfare state health benefits

1 per 794 people

Heart, circulatory, and infectious diseases, cancers

Primary health care is free to Qataris. Hospitals operate to Western standards of care and the government also funds treatment abroad.

SPENDING

GDP/cap. increase

CONSUMPTION AND SPENDING

205 per 1000 population

268 per 1000 population

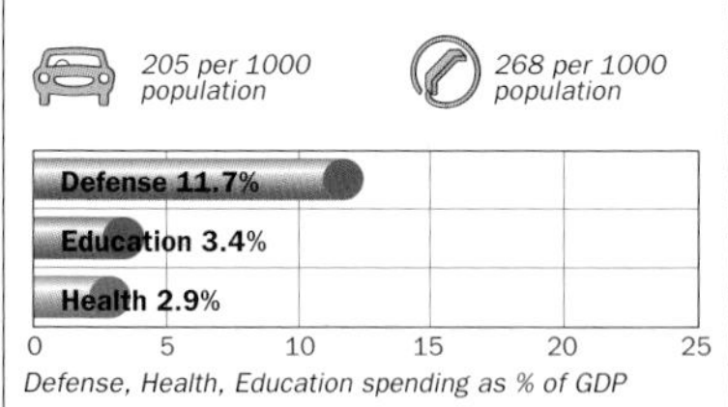

Defense 11.7%
Education 3.4%
Health 2.9%
0
5
10
15
20
25

Defense, Health, Education spending as % of GDP

Qataris have a high per capita income. There is no income tax, public services are free, and the government guarantees jobs for school-leavers. There are no exchange controls.

ECONOMICS

Inflation 2.5% p.a. (1990–2000)

$7.4bn

3.6393–3.6395 Qatar riyals

SCORE CARD

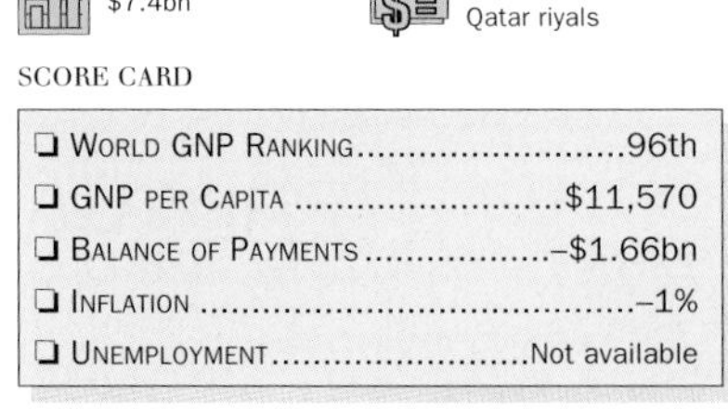
- ❑ WORLD GNP RANKING........................96th
- ❑ GNP PER CAPITA $11,570
- ❑ BALANCE OF PAYMENTS..................–$1.66bn
- ❑ INFLATION ....–1%
- ❑ UNEMPLOYMENT.......................Not available

STRENGTHS

A steady supply of crude oil and huge gas reserves, plus related industries. Soaring world oil prices in 2000. Modern infrastructure. Budget surplus.

WEAKNESSES

Dependence on foreign workforce. All raw materials imported. Virtually all water has to be desalinated. Large foreign reserves, but new industries depend on cementing agreements with foreign partners. Potential threats to security from Iraq and Iran makes some multinationals wary of investment.

EXPORTS

IMPORTS

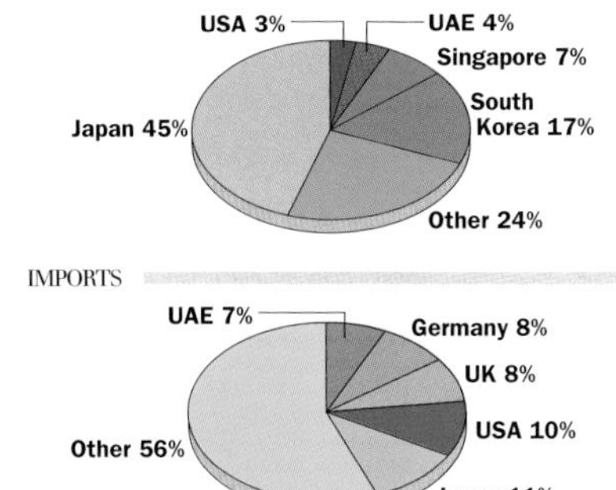

USA 3%
UAE 4%
Singapore 7%
South Korea 17%
Japan 45%
Other 24%
UAE 7%
Germany 8%
UK 8%
USA 10%
Other 56%
Japan 11%

WORLD RANKING

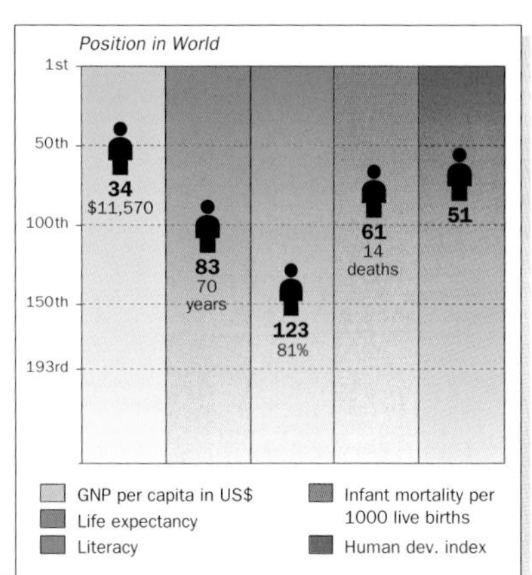

Position in World
1st
50th
100th
150th
193rd
34
$11,570
83
70
years
123
81%
61
14
deaths
51
GNP per capita in US$
Life expectancy
Literacy
Infant mortality per 1000 live births
Human dev. index

Q

Romania

EUROPE
Europe
Ukraine
Moldova
Hungary
ROMANIA
Serb. & Mon.
Bulgaria

Official Name: Romania **Capital:** Bucharest
Population: 21.7 million **Currency:** Romanian leu **Official Language:** Romanian

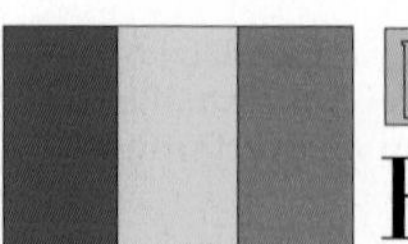

1878

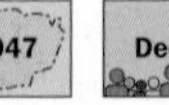
1947

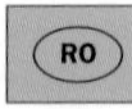
Dec 1

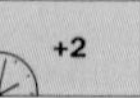
RO

+2

+40

.ro

ROMANIA LIES ON THE Black Sea coast, with the Danube as its southern border. The Carpathian Mountains form an arc across the country, curving around the upland basin of Transylvania. Long dominated by Poles, Hungarians, and Ottomans, Romania became an independent monarchy in 1878. After World War II, this was supplanted by a communist People's Republic, headed from 1965 by Nicolae Ceauşescu. A coup in 1989 resulted in his execution and a limited democracy under Ion Iliescu. Defeated in elections in 1996, Iliescu was returned to office in 2000.

Village in northeastern Romania, in the foothills of the Carpathian Mountains, close to the border with Ukraine. Corn and wheat are Romania's main crops.

CLIMATE

Continental

WEATHER CHART FOR BUCHAREST

Average daily temperature — Rainfall
°C/°F J F M A M J J A S O N D cm/in
40/104 30/86 20/68 10/50 0/32 -10/14 -20/4
40/16 30/12 20/8 10/4 0

Romania has a continental climate with two growing seasons. Rainfall is generally moderate, with most rain falling in spring and early summer. Very heavy spring rains occasionally destroy new crops. Snow is frequent in winter, which can be bitterly cold.

TRANSPORTATION

Drive on right

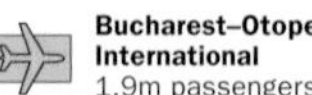

Bucharest–Otopeni International
1.9m passengers

389 ships
2.09m grt

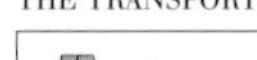

THE TRANSPORTATION NETWORK

103,671 km (64,418 miles)

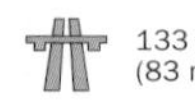

133 km (83 miles)

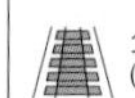

11,364 km (7062 miles)

1724 km (1071 miles)

The road network is inadequate, and traffic levels are rising. EBRD, EU, World Bank, and Japanese funding has focused on the expressway from Bucharest to Hungary, and on improving major roads. Modernization of the port of Constanţa to include a container port and new grain silo is also under way.

TOURISM

Visitors : Population 1:6.6

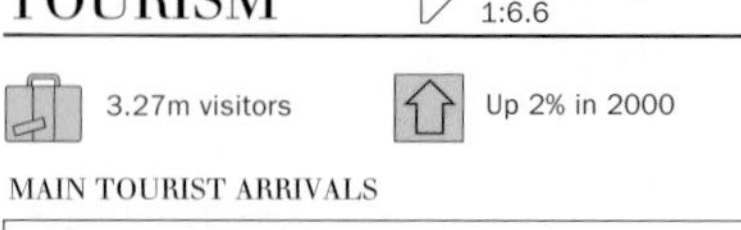

3.27m visitors
Up 2% in 2000

MAIN TOURIST ARRIVALS

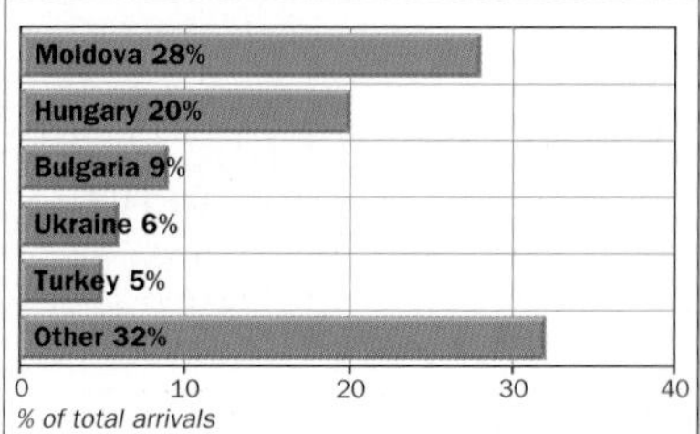
Moldova 28%
Hungary 20%
Bulgaria 9%
Ukraine 6%
Turkey 5%
Other 32%
0
10
20
30
40
% of total arrivals

The Black Sea, Danube delta, and Carpathian Mountains are the primary natural attractions, while Transylvania has a rich historical heritage. A proposal to build a theme park to exploit the Dracula legend was scrapped in 2002 after an international outcry over its location. However, tourist facilities are generally poor. Under Ceauşescu, the need for foreign currency meant that tourist facilities came before Romanians' own housing needs. Today, privatization of property and an acute housing shortage have reduced the accommodation available to visitors.

UKRAINE
MOLDOVA
HUNGARY
SERBIA AND MONTENEGRO (YUGOSLAVIA)
BULGARIA
BLACK SEA
UKRAINE
Satu Mare
Baia Mare
Suceava
Botoşani
Iaşi
Oradea
Zalău
Dej
Bistriţa
Cluj-Napoca
Reghin
Piatra-Neamţ
Bacău
Vaslui
Turda
Târgu Mureş
TRANSYLVANIA
CARPATHIAN MOUNTAINS
CARPAŢII OCCIDENTALI
Arad
Alba Iulia
Sighişoara
Miercurea-Ciuc
Bârlad
Timişoara
Deva
Sfântu Gheorghe
Făgăraş
Tecuci
Lugoj
Sibiu
Focşani
Vârful Moldoveanu 2543m
Braşov
Caransebeş
Petrosani
CARPAŢII MERIDIONALI
Râmnicu Sărat
Galaţi
Reşiţa
Târgu-Jiu
Râmnicu Vâlcea
Buzău
Brăila
Tulcea
Piteşti
Târgovişte
Ploieşti
Lacul Razim
Mouths of the Danube
Drobeta-Turnu Severin
WALACHIA
BUCHAREST
Ialomiţa
Slobozia
Lacul Sinoie
Slatina
Craiova
Caracal
Roşiori de Vede
Călăraşi
Constanţa
Alexandria
Giurgiu
Mangalia
Danube
Mureş
Timiş
Someş
Siret
Prut
Olt
Jiu
48°
46°
44°
22°
24°
26°
28°
N
0 50 km
0 50 miles
ROMANIA
Total Land Area: 237 500 sq. km (91 699 sq. miles)
POPULATION
over 1 000 000
over 100 000
over 50 000
LAND HEIGHT
2000m/6562ft
1000m/3281ft
500m/1640ft
200m/656ft
Sea Level

R

PEOPLE

Pop. density medium

Romanian, Hungarian, German, Romani

94/km² (244/mi²)

THE URBAN/RURAL POPULATION SPLIT

56% 44%

RELIGIOUS PERSUASION

Greek Catholic (Uniate) 1%
Other 2%
Protestant 4%
Greek Orthodox 1%
Roman Catholic 5%
Romanian Orthodox 87%

ETHNIC MAKEUP

Roma 1%
Magyar 9%
Other 1%
Romanian 89%

Since 1989, Romanian nationalism has increased, aggravated by economic austerity measures. The incidence of ethnic violence has also risen, toward Roma and Hungarians in particular. Ethnic Hungarians, or Magyar, who form the largest minority group, are partly protected by the influence of Hungary, whereas the Roma do not have any similar support and tend to suffer greater discrimination.

The population is currently decreasing, due to rising emigration since 1989, mainly for economic reasons, and to a falling birthrate since the early 1990s. The latter trend is in sharp contrast to the 1980s, when the Ceauşescu regime enforced a "pronatalist" policy, banning abortion and contraception; the birthrate rose, but the population as a whole did not grow significantly due to an increase in the mortality rate. Abortion was legalized in 1989; maternal death rates have recently declined. Adoptions by foreigners, spurred partly by shocking conditions in orphanages, were banned for a year in 2001, amid concerns about a "trade in children." Romania, the last country in Europe to lift its ban on homosexuality, did so in 1996, although public acts by homosexuals still receive harsh sentences.

POPULATION AGE BREAKDOWN

Female	Age	Male
1.3%	80+	0.7%
8.8%	60–79	6.9%
12.5%	40–59	11.9%
14.5%	20–39	14.9%
13.9%	0–19	14.6%

% of population by age group

WORLD AFFAIRS

Joined UN in 1955

BSEC

CE

EBRD

OSCE

CEFTA

Romania's priority is building closer links with western Europe. In 1993, it signed an association agreement with the EU, and in 1995 formally applied for membership. Although not one of the front-runners with which the EU opened negotiations in 1998, it was among six "second wave" candidates which began membership negotiations in 2000.

In 1996 Romania signed a treaty of reconciliation and friendship with Hungary, although relations remain tense as Romania has resisted the demands of the Hungarian minority in Transylvania for greater autonomy. In 1997 Romania also signed a treaty with Ukraine recognizing the latter's sovereignty over parts of Bessarabia and Northern Bukovina.

AID

Recipient

$432m (receipts)

Up 12% in 2000

Western aid declined somewhat after the mid-1990s, reflecting uncertainty about the implementation of reform. A World Bank strategy in mid-2001 envisaged a significant increase in annual aid if its suggested reforms were given high priority.

POLITICS

Multiparty elections

L. House 2000/2004
U. House 2000/2004

President Ion Iliescu

AT THE LAST ELECTION

House of Deputies 345 seats

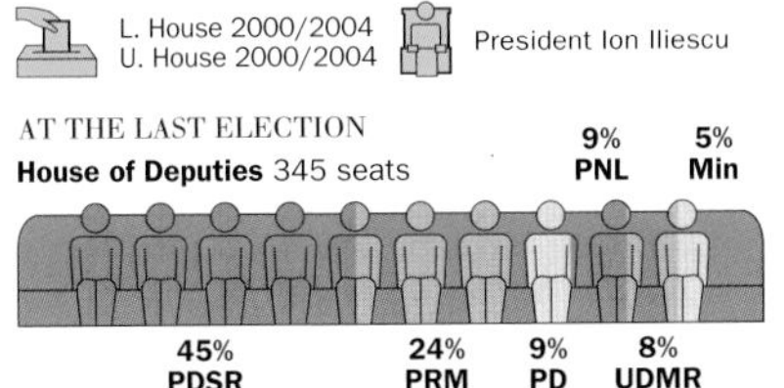
45% PDSR
24% PRM
9% PD
8% UDMR
9% PNL
5% Min

PDSR = Social Democratic Pole of Romania (led by the Social Democracy Party of Romania) **PRM** = Greater Romania Party **PD** = Democratic Party **PNL** = National Liberal Party **UDMR** = Hungarian Democratic Union of Romania **Min** = Minority representatives: 18 seats in the House of Deputies are reserved for national minorities

Senate 140 seats

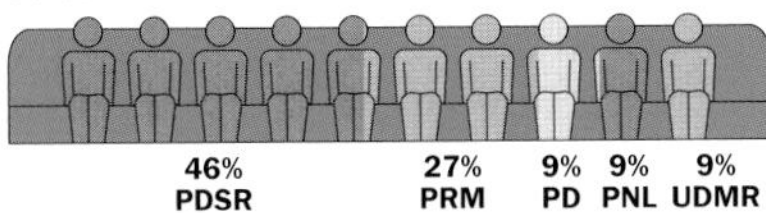
46% PDSR
27% PRM
9% PD
9% PNL
9% UDMR

Romania is a multiparty democracy led by a directly elected president.

PROFILE

The 1989 "revolution" left an old communist elite in power, with no group ready to introduce reform. Only the victory of the center-right in 1996 brought more far-reaching change.

Many of the state assets privatized under the first regime of Ion Iliescu remained closely linked to the ruling clique. Now a declared social democrat, Iliescu retains the support of such conservative groups as miners and rural workers. The coalition parties in power after 1996 came a poor third in the 2000 elections behind Iliescu's PDSR and the extreme nationalist PRM. In mid-2001 the PDSR electoral coalition was formalized in a merger of its main two constituent parties to form the Social Democrat Party (PSD).

MAIN POLITICAL ISSUES

Economic performance

Poverty has grown, while halfhearted attempts at economic reform failed to achieve structural change. Strikes are frequent and protests by miners have been especially forceful, notably against the center-right coalition in power in 1996–2000.

Ethnic tensions

Economic difficulties have led to increased ethnic tensions. The far right has made political gains and nationalism is increasingly accepted. Roma are victims of violent, racially motivated attacks. Benefits given to ethnic Hungarians by the Hungarian government have provoked an outcry from both the Romanian authorities and ordinary Romanians in mixed communities.

Ion Iliescu, Romania's first post-communist president, reelected in 2000.

Adrian Nastase, prime minister since December 2000.

CHRONOLOGY

Many foreign policy tensions stem from Romania's continually redrawn borders. It retains a Hungarian minority in Transylvania. Post-Soviet Moldova opted not to rejoin Romania.

- **1859** Unification of Moldavia and Wallachia forms basis of future Romania.
- **1878** Independence, but at cost of losing eastern Moldavia to Russia.
- **1916–1918** Enters World War I on Allied side. At end of war, gains substantial territory, including Transylvania from Hungary.

R

CHRONOLOGY *continued*

- **1924** Communists banned in unstable political arena. Rise of fascist "Iron Guard."
- **1938** King Carol establishes royal autocracy.
- **1940** Territory forcibly ceded to Soviet Union, Bulgaria, and Hungary. Coup by Iron Guard. King Carol abdicates in favor of son, Michael. Tripartite Pact with Germany.
- **1941** Enters war on Axis side, hoping to recover Moldavian lands.
- **1944** Romania switches sides as Soviet troops reach border.
- **1945** Soviet-backed regime installed. Romanian Communist Party plays an increasing role.
- **1946** Romania regains Transylvania. Eastern Moldavia reverts to USSR, which also demands huge reparations. Communist-led National Democratic Front wins majority in disputed elections.
- **1947** Michael forced to abdicate.
- **1948–1953** Centrally planned economy put in place.
- **1953** Leaders of Jewish community prosecuted for Zionism.
- **1958** Soviet troops withdraw.
- **1964** Prime Minister Gheorghiu-Dej declares national sovereignty. Proposes joint planning by all communist countries to lessen Soviet economic control.
- **1965** Ceauşescu party secretary after death of Gheorghiu-Dej.
- **1968–1980** Condemns Soviet invasion of Czechoslovakia; courts US and European Communities.
- **1982** Ceauşescu vows to pay off foreign debt.
- **1989** Demonstrations; many killed by military. Armed forces join with opposition in National Salvation Front (NSF) to form government. Ion Iliescu declared president. Ceauşescu summarily tried and shot.
- **1990** NSF election victory. Political prisoners freed.
- **1991** New constitution, providing for market reform, approved.
- **1992** Second free elections. NSF splits. Nicolae Vacaroiu forms minority government.
- **1994** General strike demands faster economic reform.
- **1996** Reconciliation treaty with Hungary. Center right wins elections, breaking with communist past; Emil Constantinescu president.
- **1997** Treaty recognizes Ukraine's sovereignty over territory ruled by Romania in 1919–1940.
- **1998** Coalition differences, Prime Minister Victor Ciorbea resigns.
- **2000** December, Ion Iliescu and social democrats win elections.

R

DEFENSE

Compulsory military service

 $809m

 Up 33% in 2000

The military received limited funding under the Ceauşescu regime, and troops were routinely deployed as cheap labor. Romania was the first country to join NATO's Partnerships for Peace program in 1994. Since 1996 the government has actively sought membership of NATO itself. Romanian soldiers were deployed in the US-led "war on terrorism" in Afghanistan from 2002.

ROMANIAN ARMED FORCES

	1373 main battle tanks (821 T-55, 30 T-72, 314 TR-85, 208 TR-580)	52,900 personnel
	1 submarine, 6 frigates, 1 destroyer, and 61 patrol boats	10,200 personnel
	307 combat aircraft (72 IAR-93, 180 MiG-21, 27 MiG-23, 18 MiG-29)	18,900 personnel
	None	

ECONOMICS

Inflation 98% p.a. (1990–2000)

 $37.4bn

 25,925–31,745 Romanian lei

SCORE CARD

World GNP Ranking	53rd
GNP per capita	$1670
Balance of Payments	–$1.36bn
Inflation	45.7%
Unemployment	7%

EXPORTS

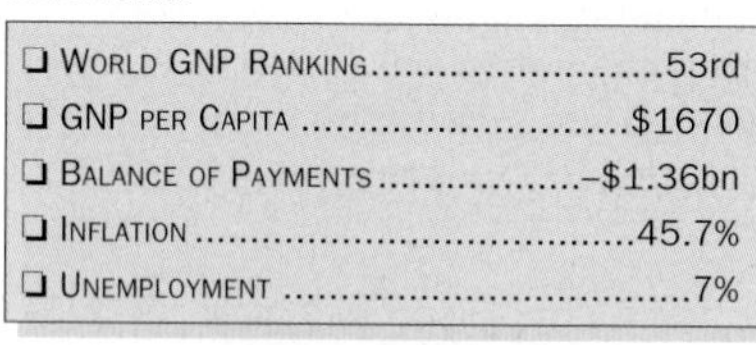
UK 5%
Turkey 6%
France 7%
Other 44%
Germany 16%
Italy 22%

IMPORTS

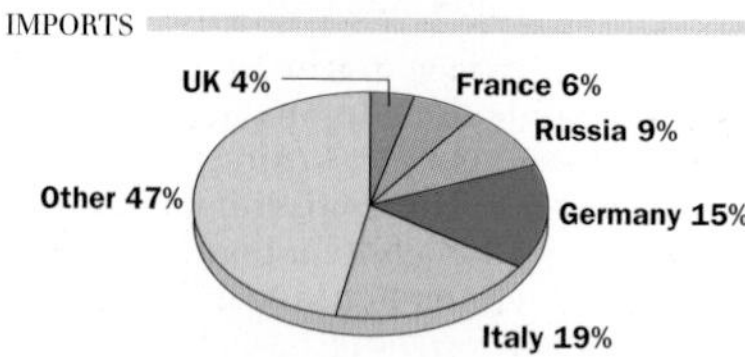
UK 4%
France 6%
Russia 9%
Other 47%
Germany 15%
Italy 19%

ECONOMIC PERFORMANCE INDICATOR

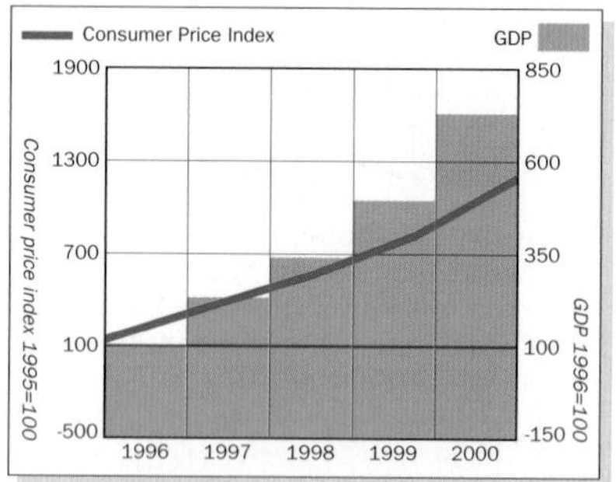
Consumer Price Index
GDP
1900
1300
700
100
-500
850
600
350
100
-150
Consumer price index 1995=100
GDP 1996=100
1996
1997
1998
1999
2000

Strengths

Many foreign joint ventures. Oil reserves. Tourism potential. Fall in inflation.

Weaknesses

Slow transition from centrally planned to market economy. Delays in economic reform. Low foreign investment levels. Large bureaucracy.

Profile

Despite being the first east European country to open its economy to foreign investment, Romania was relatively slow to launch economic reforms, and suffered severe recession for most of the 1990s. The chemical, petrochemical, metal, transportation, and food industries are the priority areas for liberalization and structural overhaul. The reform-oriented government elected in 1996 took tough measures to curb inflation and the budget deficit. Output fell sharply, worsened by disruption to trade during the 1999 Kosovo conflict, and not until 2000 did an export-led recovery appear likely.

Most farmland has been restored to private hands, and a program was launched in mid-2000 to sell off the remaining state farms by the end of 2002. Agriculture, severely undermechanized, still employs over 35% of the workforce.

From 1990 100% foreign ownership of ventures was permitted. Joint ventures, while now numerous, are small in scale; larger investors are put off by bureaucracy and doubts about stability. The privatization of state-owned enterprises, a priority for the center-right government of 1996–2000, was delayed by problems in the sale of larger companies, but by 2000 the private sector accounted for over 60% of GDP.

ROMANIA : MAJOR BUSINESSES

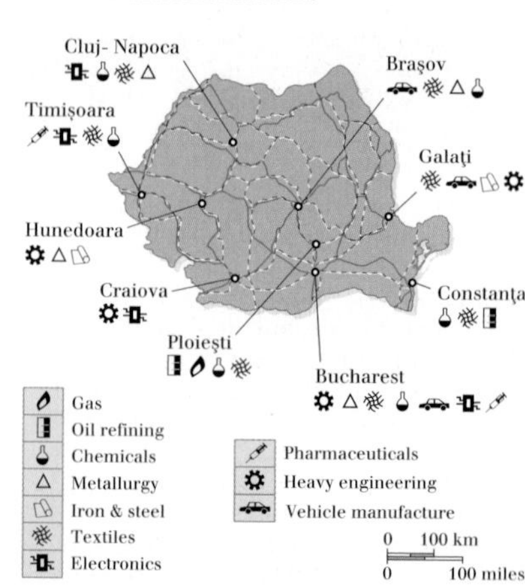
Cluj- Napoca
Braşov
Timişoara
Galaţi
Hunedoara
Craiova
Constanţa
Ploieşti
Bucharest
Gas
Oil refining
Chemicals
Metallurgy
Iron & steel
Textiles
Electronics
Pharmaceuticals
Heavy engineering
Vehicle manufacture
0 100 km
0 100 miles

RESOURCES

Electric power 22.6m kw

16,841 tonnes

130,000 b/d (reserves 1bn barrels)

7.8m sheep, 5.08m pigs, 4m geese, 73.5m chickens

Oil, coal, salt, iron, gas, methane, bauxite, copper, lead, zinc

ELECTRICITY GENERATION

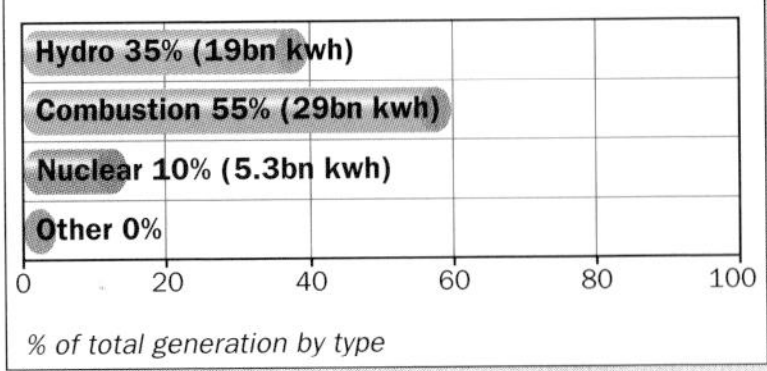
Hydro 35% (19bn kwh)
Combustion 55% (29bn kwh)
Nuclear 10% (5.3bn kwh)
Other 0%
0
20
40
60
80
100
% of total generation by type

Romania has large reserves of oil, but there is little left of proven gas reserves. Oil and gas production from onshore fields has fallen for 25 years and meets barely 40% of domestic demand. Since the mid-1990s efforts have been concentrated on developing offshore reserves in the Black Sea, opening up exploration and processing to foreign investors. Deposits of other minerals are small and contribute little to export earnings. Many coal mines have been shut down.

The electricity industry is outdated but does produce a surplus for export. An agreement in 2000 connected the national grid with that of Bulgaria. The removal of price ceilings since 1997, with a consequent doubling of prices, provides strong incentives to users to improve poor efficiency.

ENVIRONMENT

Sustainability rank: 66th

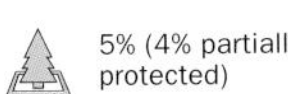
5% (4% partially protected)

4.1 tonnes per capita

ENVIRONMENTAL TREATIES

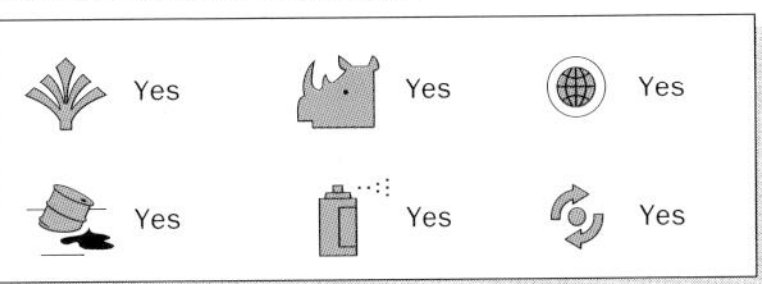
Yes
Yes
Yes
Yes
Yes
Yes

Saddled with the disastrous legacy of its communist-era industry, Romania needs help with a major cleanup. Air pollution, mainly from emissions from cement and power plants but also from exhaust fumes and low-quality coal, is most serious in the south. Cyanide and heavy metal leaked from a gold mine in Baia Mare in 2000 creating a transboundary pollution catastrophe in the Tisza River in Hungary. The Danube delta, despite serious pollution, was designated a UNESCO biosphere reserve in 1998.

MEDIA

TV ownership high

Daily newspaper circulation 298 per 1000 people

PUBLISHING AND BROADCAST MEDIA

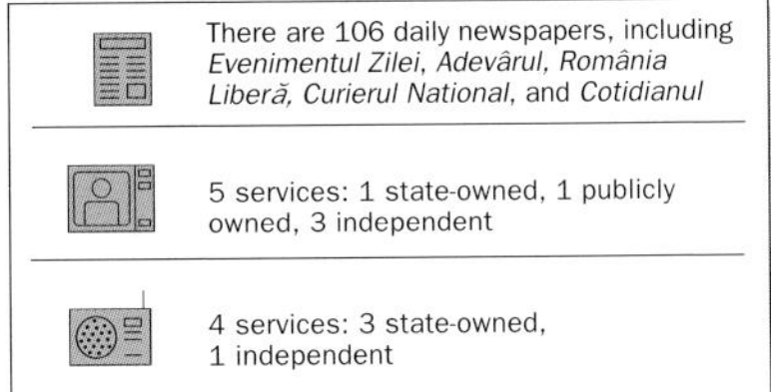
There are 106 daily newspapers, including *Evenimentul Zilei, Adevărul, România Liberă, Curierul Național,* and *Cotidianul*
5 services: 1 state-owned, 1 publicly owned, 3 independent
4 services: 3 state-owned, 1 independent

Many of the newspapers which proliferated after 1989 are now closing, since rising prices mean that fewer readers choose to buy them. In practice the government controls the national TV service. The first satellite TV channel was launched in 1994, and the first exclusively Hungarian radio station began broadcasting in 1999.

CRIME

No death penalty

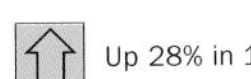
507,710 prisoners

Up 28% in 1999

CRIME RATES

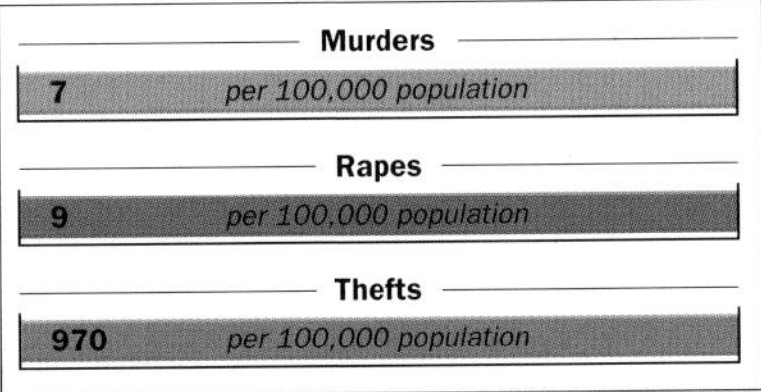
Murders
7
per 100,000 population
Rapes
9
per 100,000 population
Thefts
970
per 100,000 population

The black economy is the main source of income for a third of the population. Levels of tax evasion are extremely high. Romania is a source and transit country for people trafficking.

EDUCATION

School leaving age: 15

98%

452,621 students

THE EDUCATION SYSTEM

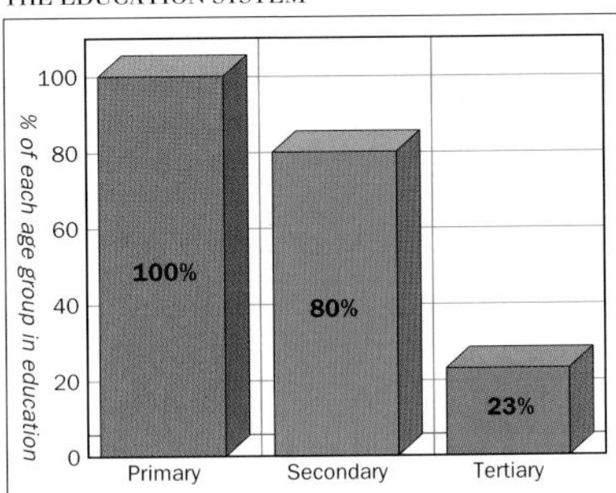
% of each age group in education
100
80
60
40
20
0
100%
80%
23%
Primary
Secondary
Tertiary

Attendance at secondary schools is below the European average. As university enrollment is no longer restricted, the number of tertiary students has risen rapidly. The government increased education expenditure in early 2000, pledging to devote at least 4% of GDP in future years and going some way to meet criticisms over the chronic underfunding of the school system.

ROMANIA : LAND USE

Cropland
Pasture
Forest
Wetlands
Potatoes
Cereals
Sheep

CARPATHIAN MOUNTAINS
Siret
Danube

0 100 km
0 100 miles

HEALTH

Welfare state health benefits

1 per 556 people

Heart & cerebrovascular diseases, cancers, accidents, tuberculosis

Average life expectancy is among the lowest in Europe; in the worst-polluted parts of Transylvania it is as low as 61 years. The incidence of tuberculosis is the highest in Europe. After years of chronic state underfunding, there was a shift in 1999–2001 toward an insurance-based system.

SPENDING

GDP/cap. decrease

CONSUMPTION AND SPENDING

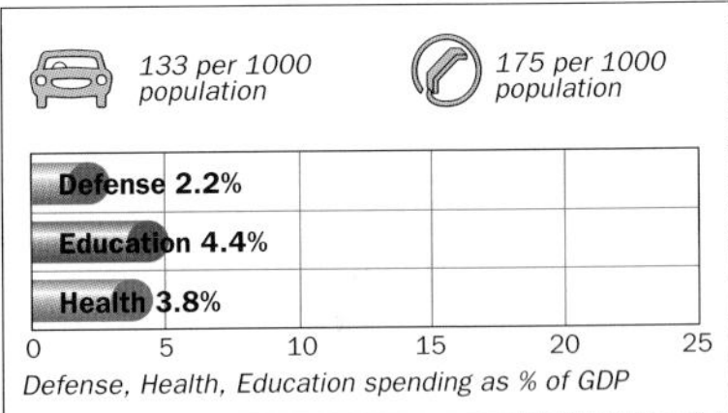
133 per 1000 population
175 per 1000 population
Defense 2.2%
Education 4.4%
Health 3.8%
0
5
10
15
20
25
Defense, Health, Education spending as % of GDP

Real incomes have been hit hard by a decade of economic decline; 40% live below the national poverty line. Most families own their own homes (often overcrowded) and many have small plots of land. Few rural homes have running water or sewerage.

WORLD RANKING

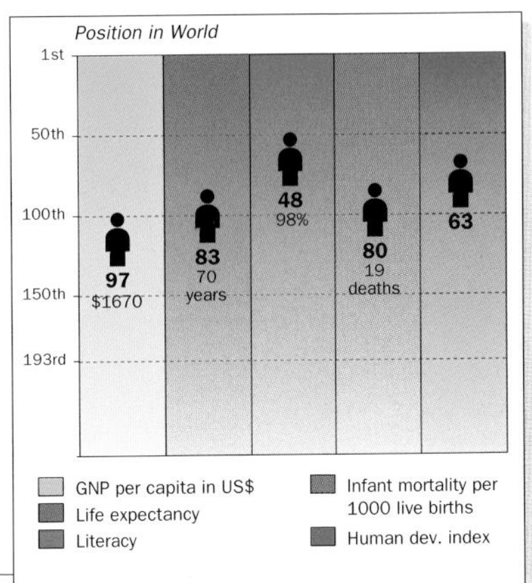
Position in World
1st
50th
100th
150th
193rd
97
$1670
83
70 years
48
98%
80
19 deaths
63
GNP per capita in US$
Life expectancy
Literacy
Infant mortality per 1000 live births
Human dev. index

R

RUSSIAN FEDERATION

ASIA
RUSSIAN FEDERATION
Asia

OFFICIAL NAME: Russian Federation **CAPITAL:** Moscow
POPULATION: 144.7 million **CURRENCY:** Russian rouble **OFFICIAL LANGUAGE:** Russian

1991

1991
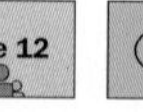
June 12

RUS
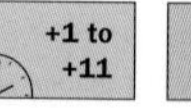
+1 to
+11

+7

.ru

RUSSIA'S TERRITORY, bounded by the Arctic and Pacific Oceans to the north and east, extends over 16 million sq. km (6 million sq. miles). This makes it by far the world's largest state, almost twice as big as either the US or China. With borders established for the federative state in 1954, the Russian Federation's sovereign status dates from 1991 – the dissolution of the USSR. Within the CIS, it maintains a traditionally dominant role in central Asia and the Caucasus. Ethnic Russians make up 82% of the population, but there are around 150 smaller ethnic groups, many with their own national territories within Russia's borders. Regionalism and separatism are major political issues. The situation is complicated by the fact that many of these territories are rich in key resources such as oil, gas, gold, and diamonds.

The Kremlin, Moscow. *Rebuilt in 1475 by Ivan the Great, who commissioned architects from Pskov and Italy, it is enclosed by walls 4 km (1.5 miles) long and lies on the Moscow River.*

CLIMATE

Subarctic/continental/ mountain/steppe

WEATHER CHART FOR MOSCOW

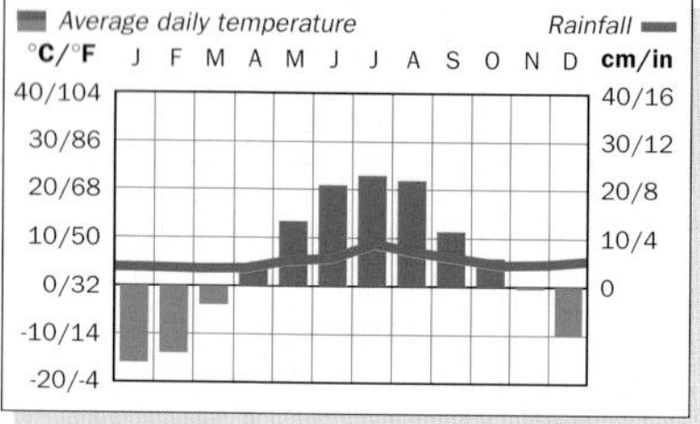
Average daily temperature
Rainfall
°C/°F
J F M A M J J A S O N D
cm/in
40/104
30/86
20/68
10/50
0/32
-10/14
-20/-4
40/16
30/12
20/8
10/4
0

Russia has a cold continental climate, characterized by two widely divergent main seasons. Spring and autumn are very brief periods of transition between warm summers and freezing winters. The country is open to the influences of the Arctic and Atlantic to the north and west. However, mountains to the south and east prevent any warming effects from the Indian and Pacific Oceans filtering across. Severe winters affect most regions. Winter temperatures vary surprisingly little from north to south, but fall sharply in eastern regions. The January temperature of –70°C (–94°F) recorded at Verkhoyansk in Siberia is the world record low outside Antarctica.

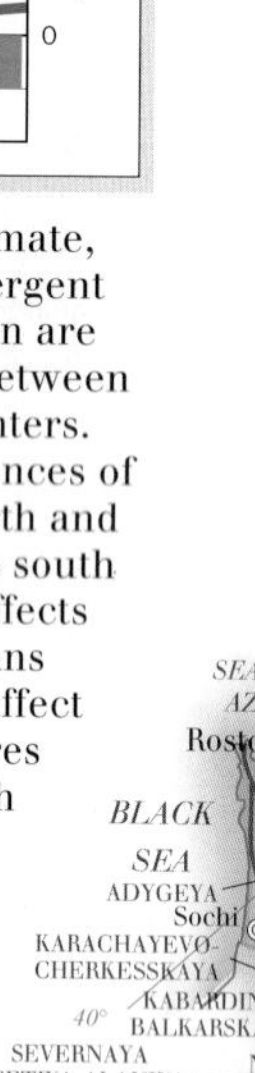
SEA OF AZOV
Rostov-na-Donu
BLACK SEA
ADYGEYA
Sochi
KARACHAYEVO-CHERKESSKAYA
KABARDINO-BALKARSKAYA
SEVERNAYA OSETIYA-ALANIYA
Nal'chik
Vladikavkaz
GEORGIA
AZERBAIJAN

Kaliningrad
POLAND
LITHUANIA
LATVIA
ESTONIA
Gulf of Finland
BELARUS
Smolensk
Bryansk
Orel
Kursk
Belgorod
Voronezh
Lipetsk
Tula
MOSCOW
Tver'
Ryazan'
UKRAINE
Krasnodar
Maykop
Stavropol'
Elista
KALMYKIYA
INGUSHSKAYA
Groznyy
CHECHENSKAYA
DAGESTAN
Makhachkala
Volgograd
Mikhaylovka
Astrakhan'
CASPIAN SEA
Don
Volga

NORWAY
FINLAND
KARELIYA
Murmansk
Apatity
KOL'SKIY POLUOSTROV
BARENTS SEA
BELOYE MORE
Ostrov Kolguyev
Sankt Peterburg
Ladozhskoye Ozero
Pskov
Novgorod
Petrozavodsk
Onezhskoye Ozero
Severodvinsk
Arkhangel'sk
Arctic Circle
Nar'yan-Mar
Cherepovets
Rybinskoye Vdkhr.
Vologda
Yaroslavl'
Kostroma
Ivanovo
Vladimir
Severnaya Dvina
Kotlas
Syktyvkar
K O M I
Pechora
Nizhniy Novgorod
Kirov
MARIY EL
CHAVASH
MORDOVIYA
Kazan'
Tambov
Penza
Ul'yanovsk
Kuybyshevskoye Vdkhr.
TATARSTAN
Saratov
Balakovo
Tol'yatti
Samara
UDMURTSKAYA
Izhevsk
Perm'
Naberezhnyye Chelny
Bereznik
BASHKORTOSTAN
Ufa
Sterlitamak
Orenburg
Ural
URAL MOUNTAINS
Serov
Yekaterinburg
Chelyabinsk
Magnitogorsk
Orsk
Kurgan
Tyumen'
Tobol'sk
Omsk
Irtysh
Vorkuta
Salekhard
Khanty-Mansiysk
Nizhnevartovsk
ZAPAD
SIBIRSK
RAVNI
K A Z A K H S T A
Novaya Zemlya
Rubtsovsk

RUSSIAN FEDERATION

Total Land Area :
17 075 200 sq. km
(6 592 735 sq. miles)

POPULATION
- over 5 000 000
- over 1 000 000
- over 500 000
- over 100 000
- over 50 000
- over 10 000

LAND HEIGHT
- 3000m/9843ft
- 2000m/6562ft
- 1000m/3281ft
- 500m/1640ft
- 200m/656ft
- Sea Level
- -200m/-656ft

0 500 km
0 500 miles

Housing in Moscow. *Living conditions in major cities can be cramped, with families often sharing their small apartments.*

R

TRANSPORTATION

Drive on right

Sheremetyevo-2, Moscow 10.8m passengers

4723 ships 11.1m grt

THE TRANSPORTATION NETWORK

Roads	336,000 km (208,780 miles)	Motorways	None
Railways	86,031 km (53,460 miles)	Waterways	101,000 km (62,758 miles)

Russia has a comprehensive transportation network. However, since 1991, all systems have seen some decline due to lack of funding. Cities are still served by good trolley and bus systems and Moscow has one of the most impressive subway systems in the world. In rural areas, car ownership is low and the population relies on an extensive bus service.

About 20% of the railroad track should be renewed annually owing to frost and other damage. Shortage of funds means that this is no longer done. The railroads are heavily used but seriously overburdened and liable to accidents and delays. New track has been laid for the Sokol (Falcon) high-speed rail link between Moscow and St. Petersburg; the first trains to use it in 2000 cut over an hour off the previous minimum journey time and further dramatic reductions are expected.

Roads in major cities are deteriorating, as are interurban highways. Crime is a problem on railroads – notably the Trans-Siberian – and roads.

The former Aeroflot monopoly of air transportation has been broken up. Aeroflot now competes as Aeroflot Russian Airlines, but hundreds of regional "babyflot" airlines run mainly domestic routes, some with alarming accident records.

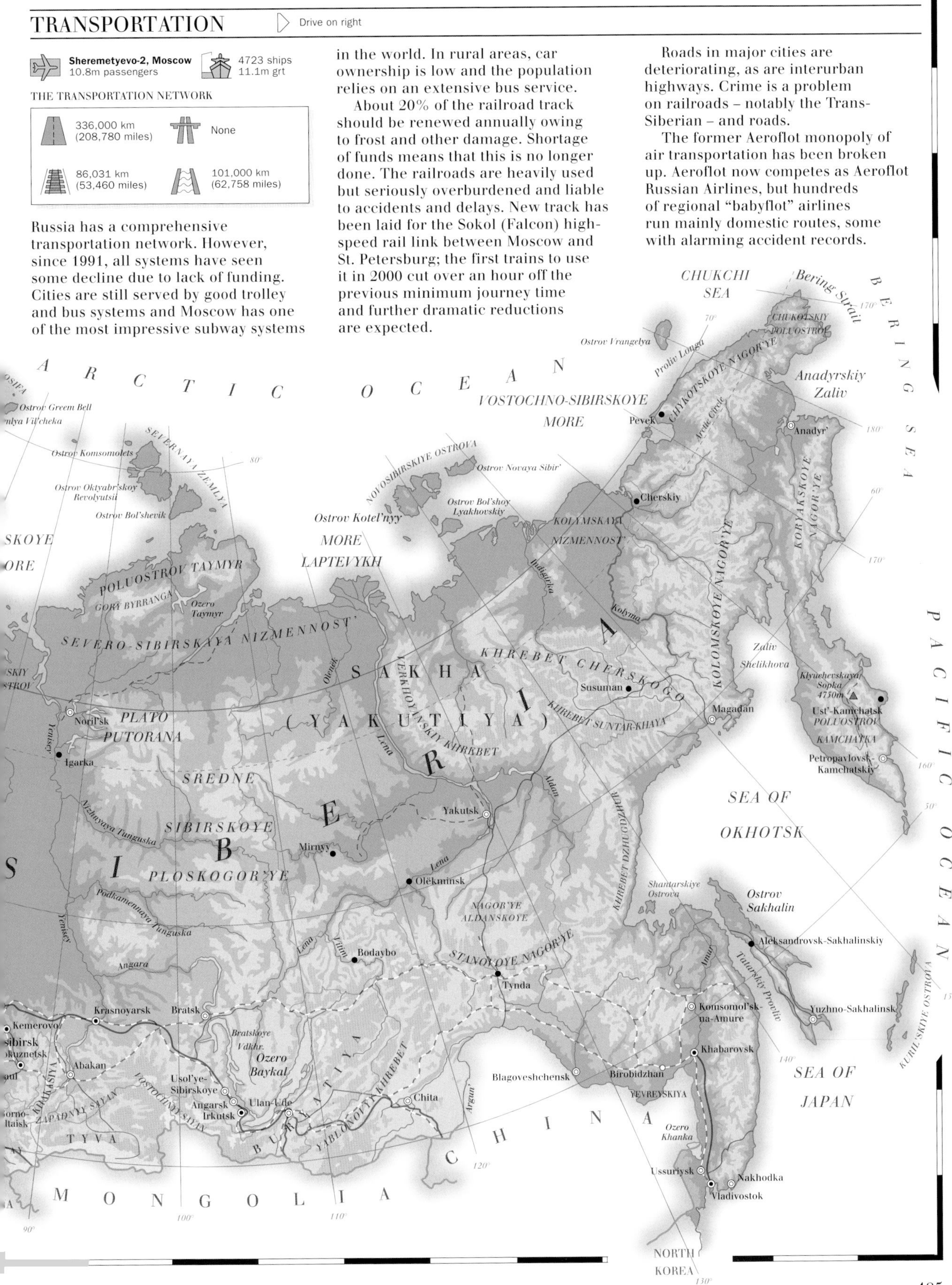
ARCTIC OCEAN
CHUKCHI SEA
Bering Strait
BERING SEA
PACIFIC OCEAN
VOSTOCHNO-SIBIRSKOYE MORE
MORE LAPTEVYKH
SEA OF OKHOTSK
SEA OF JAPAN
Ostrov Vrangelya
Proliv Longa
CHUKOTSKIY POLUOSTROV
CHUKOTSKOYE NAGOR'YE
Anadyrskiy Zaliv
Anadyr'
Pevek
Arctic Circle
Cherskiy
KORYAKSKOYE NAGOR'YE
KOLYMSKAYA NIZMENNOST'
KOLOMSKOYE NAGOR'YE
Kolyma
Indigirka
NOVOSIBIRSKIYE OSTROVA
Ostrov Novaya Sibir'
Ostrov Bol'shoy Lyakhovskiy
Ostrov Kotel'nyy
Ostrov Greem Bell
Ostrov Komsomolets
SEVERNAYA ZEMLYA
Ostrov Oktyabr'skoy Revolyutsii
Ostrov Bol'shevik
POLUOSTROV TAYMYR
GORY BYRRANGA
Ozero Taymyr
SEVERO-SIBIRSKAYA NIZMENNOST'
SAKHA (YAKUTIYA)
SIBERIA
KHREBET CHERSKOGO
KHREBET SUNTAR-KHAYATA
VERKHOYANSKIY KHREBET
Susuman
Magadan
Zaliv Shelikhova
Klyuchevskaya Sopka 4750m
Ust'-Kamchatsk
POLUOSTROV KAMCHATKA
Petropavlovsk-Kamchatskiy
Noril'sk
PLATO PUTORANA
Igarka
Yenisey
Olenëk
Lena
Aldan
SREDNE SIBIRSKOYE PLOSKOGOR'YE
Nizhnyaya Tunguska
Podkamennaya Tunguska
Yakutsk
Mirnyy
Olëkminsk
KHREBET DZHUGDZHUR
NAGOR'YE ALDANSKOYE
Shantarskiye Ostrova
Ostrov Sakhalin
Aleksandrovsk-Sakhalinskiy
Tatarskiy Proliv
Amur
Vitim
Bodaybo
STANOVOYE NAGOR'YE
Tynda
Angara
Krasnoyarsk
Bratsk
Bratskoye Vdkhr.
Ozero Baykal
Kemerovo
Abakan
Usol'ye-Sibirskoye
Angarsk
Irkutsk
Ulan-Ude
Chita
VOSTOCHNYY SAYAN
ZAPADNYY SAYAN
TYVA
BURYATIYA
YABLONOVYY KHREBET
Argun'
Blagoveshchensk
Birobidzhan
YEVREYSKIYA
Komsomol'sk-na-Amure
Khabarovsk
Yuzhno-Sakhalinsk
KURIL'SKIYE OSTROVA
Ozero Khanka
Ussuriysk
Vladivostok
Nakhodka
NORTH KOREA
CHINA
MONGOLIA

TOURISM

Visitors : Population 1:6.8

21.2m visitors

Up 15% in 2000

MAIN TOURIST ARRIVALS

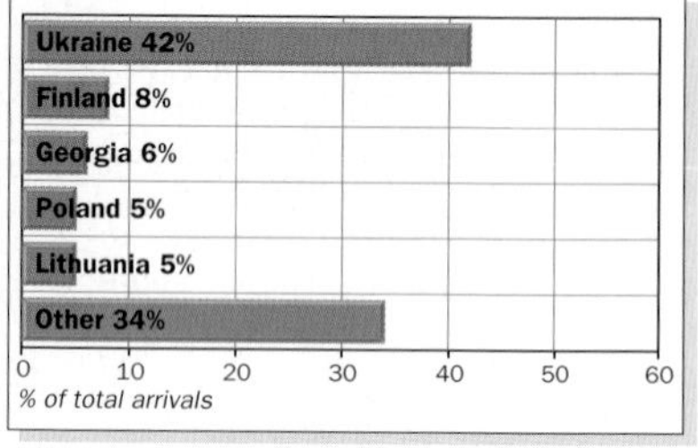

Ukraine 42%
Finland 8%
Georgia 6%
Poland 5%
Lithuania 5%
Other 34%
0 10 20 30 40 50 60
% of total arrivals

The breakup of the previous monopoly tourist agency, Intourist, has led to a vast expansion of tourism opportunities: each region is keen to earn hard currency and to attract rich visitors. By 2000 the total number of tourists a year had reached over 21 million.

At the luxury end of the market, trips from St. Petersburg to Tashkent on former president Brezhnev's official train are now available. River trips down the Volga and visits to medieval monasteries are increasingly popular. Tourists can also explore forests, or fish for salmon in the Kola Peninsula. The defense sector has opened up to tourism and now offers flights in MiG jets, or rides in T-84 Russian tanks. Even the space industry has branched into tourism.

Moscow and St. Petersburg remain favorite destinations, where hotels tend either to be for the well-off or of a basic standard. Near St. Petersburg, Novgorod has many fine churches, and the Pskov area is celebrated as the setting for many of Pushkin's works, including *Eugene Onegin* and *Boris Godunov*.

Many parts of Russia remain inaccessible to most tourists. The communist-era ban on foreigners visiting the Urals has been lifted, but the area still has very few facilities. Resorts such as the subtropical Sochi on the Black Sea, where powerful Russians have *dachas* (country houses), have experienced a building boom.

PEOPLE

Pop. density low

Russian, Tatar, Ukrainian, Chavash, various other national languages

9/km² (22/mi²)

THE URBAN/RURAL POPULATION SPLIT

RELIGIOUS PERSUASION

ETHNIC MAKEUP

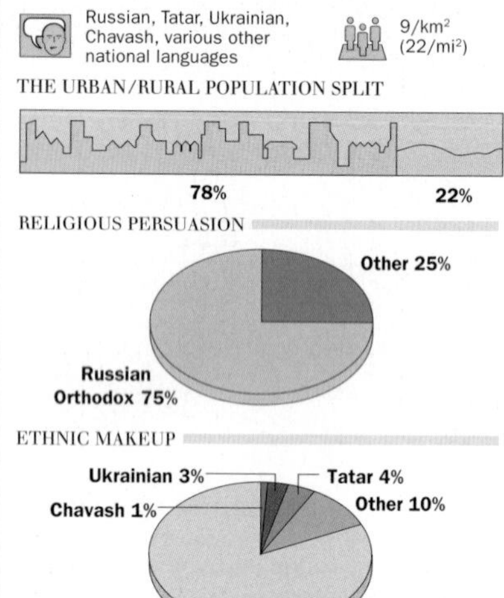

78%
22%
Other 25%
Russian Orthodox 75%
Ukrainian 3%
Tatar 4%
Chavash 1%
Other 10%
Russian 82%

In the former Soviet Union, Russians were just over 50% of the population, but in Russia they are an overwhelming majority. Although significant numbers of Russians still live in some of the neighboring former Soviet republics, notably Ukraine and Latvia, a rise in nationalism has persuaded many to return to Russia.

Within Russia there has also been an increase in ethnic tension, especially in the Caucasus. Hostility toward the Muslim Chechens is particularly acute. There are 57 nationalities with their own territories within the federation and 95 nationalities (but only 6% of the population) without a territory.

Social life in Russia has not changed significantly since the demise of communism. With the lifting of censorship, there has been a greater expression of sexuality and of political and religious views. The expensive rebuilding of Moscow's Church of Christ the Savior symbolized this change. The strong revival of the Russian Orthodox Church is boosted by legal recognition of its "special role" in Russia's history. All religious organizations were required to re-register by 1 January 2000, under conditions which many small minority faiths could not meet, and whereby a church deemed to "violate social order" would be banned.

One marked change of which Russians speak is the growing importance given to money. Mutual support systems of extended friendships are now in decline.

The position of women has changed little since the fall of communism. Many have suffered from the rise in unemployment, but this reflects the demise of part-time or badly paid jobs, rather than gender-motivated social change. Most Russians have very modest living standards and were further impoverished by the collapse of the economy in the late 1990s.

POPULATION AGE BREAKDOWN

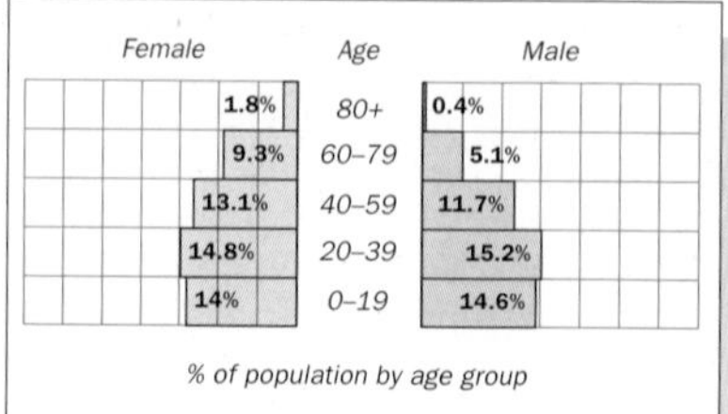

Female
Age
Male
1.8% 80+ 0.4%
9.3% 60–79 5.1%
13.1% 40–59 11.7%
14.8% 20–39 15.2%
14% 0–19 14.6%
% of population by age group

POLITICS

Multiparty elections

L. House 1999/2003
U. House varying

President Vladimir Putin

AT THE LAST ELECTION

State Duma 450 seats

15% **FAR** **5%** **Y** **2%** **Vac**

25% **CP** **23%** **Ind** **16%** **U** **6%** **URF** **4%** **Z** **4%** **Others**

CP = Communist Party **Ind** = Independents **U** = Unity
FAR = Fatherland–All Russia **URF** = Union of Right Forces
Y = Yabloko **Z** = Zhirinovsky's Bloc **Vac** = Vacant

Council of the Federation 178 seats

Each of 89 regions is represented in the Council of the Federation (Soviet Federatsii) by two members chosen by the regional legislature

The government is responsible to the elected parliament (Duma), but executive power lies firmly with the president.

Profile

President Boris Yeltsin's second term (1996–1999) was overshadowed by his health problems, economic crisis, and corruption. Dramatic changes of government personnel were a characteristic of this period, as Yeltsin confronted the Duma in both 1998 and 1999 over his choice of prime minister. The confrontation ended with the appointment of Vladimir Putin, a little-known former head of the Federal Security Service (FSB), who became Yeltsin's favorite to succeed him.

Putin has greatly consolidated his position since coming to power as acting president from the end of 1999. Having stormed to victory in the first round of the 2000 presidential elections, he has tackled the power of the business "oligarchs," and of Russia's 89 regional governors, with his program of centralization. His support within the Duma has been consolidated by the transformation of the Unity bloc (which Putin had formed for the December 1999 legislative elections) into an official party. A merger with the Fatherland

Former president Boris Yeltsin, *renowned for his erratic behavior, stepped down in 1999.*

Mikhail Gorbachev, *whose restructuring ultimately led to the breakup of the Soviet Union.*

POLITICS *continued*

bloc in April 2001 made Unity the largest single party in parliament.

Although the conflict in Chechnya damaged Putin's image internationally, it has by contrast been a key element in his domestic appeal as a strong leader. This was shaken by criticism of his handling of the *Kursk* submarine disaster in August 2000, in which 118 sailors died. However, the marked improvement in the economy during his first year in office stood greatly to his credit in popular opinion, as it also did for Prime Minister Mikhail Kasyanov, appointed in May 2000 with revitalizing the economy as his central task.

The Communist Party, which until April 2001 was the single largest party in the Duma, remains powerful because of its effective organization and its ability to appeal to those who have suffered from the upheavals of the post-Soviet period. However, even with parliamentary allies, it has been unable to take power. Its leader Gennady Zyuganov has three times been beaten to the presidency.

MAIN POLITICAL ISSUES

Living standards

Russians are disillusioned at the failure of politicians to improve their living standards. The fall of communism swept away the securities which used to underpin life – long-term employment, guaranteed housing, and a basic diet – hitting the old particularly hard. Uneven and crisis-prone efforts at introducing a market economy under former president Yeltsin created much insecurity.

Crime and corruption

Crime levels rose alarmingly under the post-Soviet regime, and visitors began to be warned against walking the streets of St. Petersburg or Moscow after dark. Widespread bureaucratic corruption was countered by the power acquired by business tycoons, the so-called "oligarchs," who snapped up privatized industries at bargain prices. President Putin has launched a crusade against these oligarchs, but has shied away from a full-scale review of the privatization process.

Vladimir Putin was handed power by Yeltsin in 1999, and was elected president in 2000.

Mikhail Kasyanov, economic reformist appointed prime minister in 2000.

Regionalism and separatism

Nation-based separatism is brutally suppressed. Nowhere has this been made clearer than by the ferocious military campaign in Chechnya.

Influence accumulated under the Yeltsin regime by Russia's 89 regional governors was reversed by Putin's efforts to concentrate power in the presidency. Control of police and taxation has been centralized in seven huge federal districts, responsible to Putin, and the governors have been stripped of their seats in the upper house, the Federation Council.

Russia's loss of great power status

The collapse of the USSR and Russia's subsequent economic crisis seriously dented its importance on the global stage, where it once counterbalanced the US. Since 2001, however, Putin has had some success in reclaiming Russia's international influence.

WORLD AFFAIRS

Joined UN in 1945

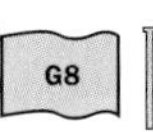

CE | CIS | IAEA | G8 | OSCE

The September 11, 2001 terrorist attacks on the US laid the foundations for a massive change in Russia's relations with the West.

Before then, Russia had been pursuing an independent foreign policy with a renewed skepticism about Western intentions, especially NATO's eastward expansion. The 1999 NATO bombardment of Yugoslavia, Russia's natural coethnic and coreligionist ally, had pushed relations to a low ebb. Russia also felt threatened by US plans for a new missile defense system (NMD). On top of these underlying tensions was heaped Western criticism of the long-running and often bloody Russian campaign against Muslim separatists in Chechnya.

After September 11, by pledging immediate support to the US-led "war on terrorism," President Putin secured unqualified backing for the Chechnya campaign. This was now viewed by the West as a legitimate battle against "terrorism." By agreeing to share intelligence information as part of the new global war, Russia and NATO appeared to bury their grievances, and in 2002 a new NATO–Russia Council was established, effectively making Russia a junior partner in NATO. In return Russia dropped its fundamental objection to the NMD and agreed to a new nuclear disarmament treaty which was widely seen as favoring US interests.

The successor states of the USSR, the "near abroad," retain close links with Russia, particularly Belarus, which is keen on a full reunification.

St. Basil's Cathedral, Moscow. *Commissioned by Ivan the Terrible, it was built in 1555–1561. The exterior domes were decorated in the 1670s.*

AID

Recipient

 $1.57bn (receipts)  Down 20% in 2000

Russia has received billions of dollars in aid from Western countries on several occasions to stave off government debt and to promote economic restructuring. Large-scale IMF credits were obtained during the economic crises of the mid- and late 1990s.

DEFENSE

Phasing out conscription

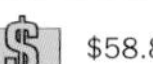 $58.8bn  Up 4% in 2000

RUSSIAN ARMED FORCES

21,820 main battle tanks (T-34, T-55, T-62, T-64, T-72, T-80, T-90)	321,000 personnel
56 submarines, 1 carrier, 17 destroyers, 7 cruisers, 10 frigates, and 108 patrol boats	171,500 personnel
1538 combat aircraft (MiG-25/29/31, Su-24/25/27)	184,600 personnel
776 ICBM, 19 SSBN, 100 ABM	

The loss of the 118-man *Kursk* nuclear submarine in August 2000 symbolized the long-term decline of Russia's military might. Maintaining and using the enormous former communist war machine has proved too expensive. There are plans to axe hundreds of thousands of troops by 2003, and conscription is set to be abolished in 2005. Public anger at conditions within the services has led to pay increases.

Spending on nuclear forces is limited to physical protection of warheads. In 2002, Russia agreed to cut further its nuclear arsenal, expanding on the 2000 Start II Treaty, and slashing the number of warheads by 60%. It also dropped previous objections to the proposed US national missile defense system.

The Northern and Pacific navy fleets are inactive and deteriorating fast. Fourteen admirals were disciplined in 2001 for the "failings" of the Northern fleet.

ECONOMIC UPHEAVAL, REGIONALISM, AND CENTRALISM

THE RUSSIAN ECONOMY surprised many by its recovery from dire crisis in 1998, to grow by 9% in 2000 and by 5% even in the adverse global climate of 2001. Massive problems remain, however, many of them relating to the sprawling and decaying industrial infrastructure, particularly in some remoter regions.

The Yeltsin regime in the 1990s had failed to manage an effective reform and restructuring process. When matters came to a head in 1998, collapse was only narrowly averted. The crisis spotlighted the chronic weakness of central finances. Since then, however, the government has overseen a real improvement in its financial position. The combination of devaluation, an international rescue package, and default on domestic debt was further helped by increased tax revenue and the introduction of a uniform rate of income tax in 2001.

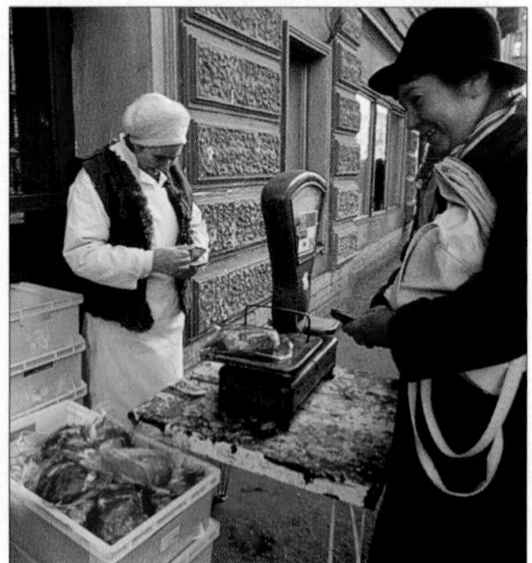

***Residents of St. Petersburg** had to queue in the streets for basic foodstuffs.*

Imports became much more expensive after the 1998 devaluation, and Russian-made goods started appearing in Moscow and St. Petersburg again, beginning with big increases in supplies of such things as bicycles, electric kettles, refrigerators, and washing machines. Middle-class Russians in the cities, who had found ways to survive the crisis, were eager buyers. Gazprom, the giant company monopolizing the gas industry, benefited as an exporter from the low rouble and the improvement in world prices, while such industries as aluminum in Sakha (Yakutia) became even more powerful in the local economy.

Although wages and pensions may now be paid with greater regularity, most Russians are still far from catching up with the consequences of years of inflation. Teachers, for example, went five years without pay increases up to 1999. Pensioners could often be seen in the late 1990s attempting to sell possessions from makeshift stalls on street corners. Even by April 2002, as President Putin conceded, 40 million Russians – 28% of the population – were still living below the official poverty line.

VICTIMS OF CHANGE

For many people in the provinces, the dismantling of the old command economy meant the disappearance of what was often the sole source of employment. Far-flung towns had been built exclusively around a particular heavy industry, such as steel, coal, shipbuilding, machine tools, or arms production. Many of these concerns, no longer with a market, were impossible to privatize. Their workers, put on short time or laid off, have been reduced to living on promises, state handouts, and barter arrangements. In cities in the Arctic north, people leave if they can; Murmansk, the largest, has lost a fifth of its population in a decade. In the far east, the collapse of heating and power systems around Vladivostok left residents freezing; the 2000–2001 winter was the bitterest for 70 years.

The disappearance of proper state welfare provisions in areas such as health care makes the situation even more catastrophic. Rates of alcoholism, drug abuse, and suicide have all risen. Most hard pressed are people with neither work nor land, living with children in small urban settlements. Wherever possible, people grow food to be as self-sufficient as possible, and use what they can for barter. Great reliance is also placed on family support networks. The relevance of the formal economy is thereby much reduced, while rents, transportation, and utility bills are still heavily subsidized. Even when cash changes hands, a large proportion of all purchases involve the black market in one way or another.

***Heavy industries** have declined dramatically, and many factories now lie idle.*

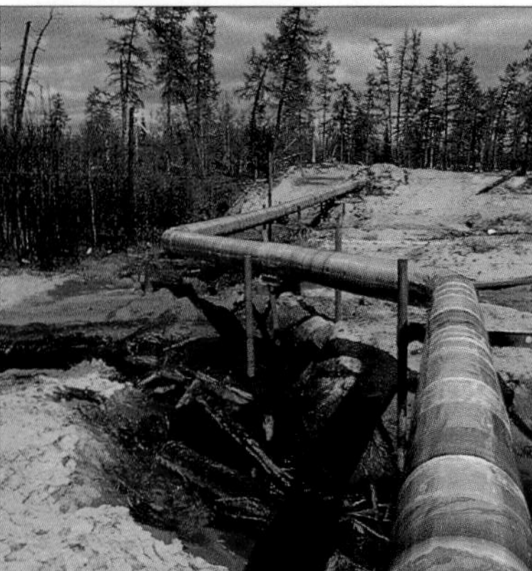

***Poor maintenance** has resulted in cracks and leaks along Russia's network of oil pipelines.*

The running down of heavy industries has left some areas with massive environmental damage. One notorious instance is the pollution of rivers by toxic waste from the metallurgy industry in the Kola peninsula in the north. Problems of disposal of radioactive wastes affect northern coastal areas in particular, although the worst such legacy of the former Soviet Union – the effects of the nuclear accident at Chernobyl – has been to Belarus and Ukraine. Western countries agreed in June 2002 to spend $20 billion over the next decade on destroying discarded nuclear material, on environmental and global security grounds.

REPUBLICS AND CENTRAL CONTROL

Some of the 22 republics within the Russian Federation pressed hard for control over their mineral resources, believing that they could be managed more profitably at regional level, and resenting the outflow of their earnings. Regionalism based on dissatisfaction with central economic management began emerging as a key political issue. Tatarstan, lying on the Volga some 1000 km (625 miles) east of Moscow, notably avoided rushing into privatization, conducted elsewhere in such a way as to allow former managers effectively to appropriate huge assets. Its continuing control of the Tatneft oil company provides both a source of revenue and a means of foreign borrowing on relatively favorable terms.

Tatarstan, along with Chechnya, refused to sign the 1992 treaty between the Russian Federation and its constituent parts. Instead, its president, Mintimer Shaimiev, negotiated better autonomy terms, with detailed revenue-sharing provisions, in a 1994 agreement. Extended for a further five years from 1999, this treaty was to be replaced by a new Tatarstan constitution in 2002. However, in spite of compromises expensively won by the allocation of Russian development funding, this constitution went far enough in its claims on Tatar sovereignty to provoke a crisis over federal acceptance.

President Putin has made no secret of his determination to reverse the fragmentation of the federation. Before becoming president he was already closely identified with "firmness" in relaunching a brutal war in 1999 (ostensibly as an antiterrorist operation) against Chechnya, the only republic actively seeking to secede. After his election in 2000 Putin's first major policy action was to group the federation's 89 regional administrative entities under seven federal districts headed by newly appointed presidential envoys. He asserted powers to disband regional legislatures and dismiss regional governors, and removed the governors from the upper house of parliament, the Federation Council, although they do now sit on a new advisory State Council and have been given the right to stand for more than just two terms of office.

Even in Tatarstan, recognizing the need to compromise, Shaimiev (reelected for a third term in 2001) reduced the emphasis on "Tatarization." Federal control over tax collection and spending became more evident again, as did the presence of federal agencies in Tatarstan's capital, Kazan. A push to reintroduce the Latin alphabet in place of Cyrillic in Tatarstan's schools, launched in 2000, was deferred the following year. The population has at least as many ethnic Russians as Tatars, who – like the Chechens – are mainly Muslims and fiercely proud of their distinctive language and culture.

***New gas pipelines** being laid across the frozen expanses of Siberia.*

ECONOMICS

Inflation 162% p.a. (1990–2000)

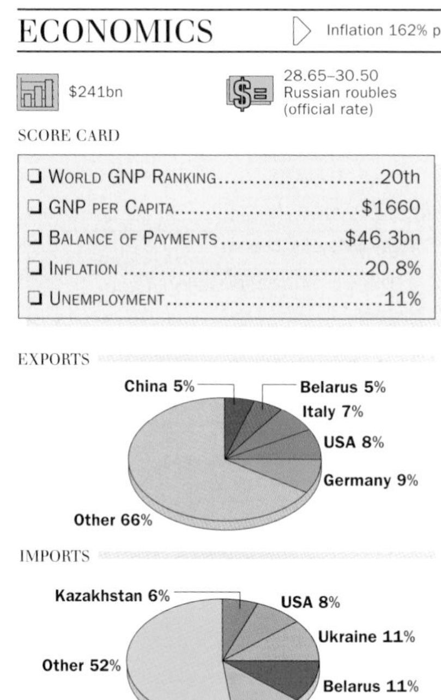
$241bn
28.65–30.50 Russian roubles (official rate)
SCORE CARD
WORLD GNP RANKING....20th
GNP PER CAPITA....$1660
BALANCE OF PAYMENTS....$46.3bn
INFLATION....20.8%
UNEMPLOYMENT....11%
EXPORTS
China 5%
Belarus 5%
Italy 7%
USA 8%
Germany 9%
Other 66%
IMPORTS
Kazakhstan 6%
USA 8%
Ukraine 11%
Belarus 11%
Germany 12%
Other 52%

ECONOMIC PERFORMANCE INDICATOR

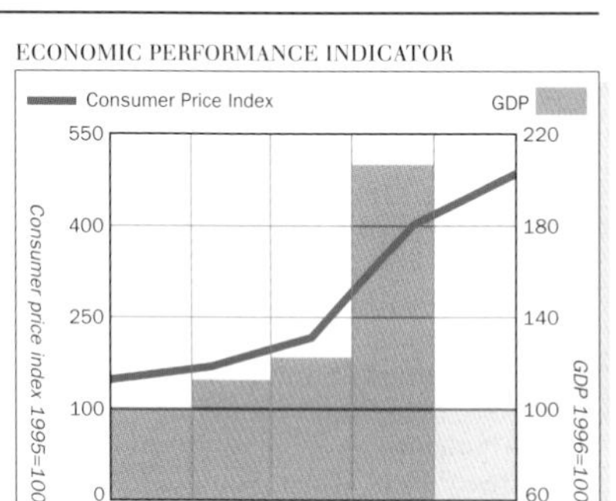
Consumer Price Index
GDP
Consumer price index 1995=100
GDP 1996=100
550
400
250
100
0
220
180
140
100
60
1996
1997
1998
1999
2000

Tax evasion and corruption remain widespread. Rising wages and strengthening rouble have promoted boom in imports, narrowing trade gap, harming local production, and stifling small enterprises.

STRENGTHS

Huge natural resources, in particular hydrocarbons, precious metals, fuel, timber. Potential from future international oil pipelines. Enormous engineering and scientific base. Small enterprises benefited from devaluation of rouble in 1998. Tax reforms increasing government revenue. Lucrative privatizations. Recognized as market economy in 2002, encouraging foreign investment.

WEAKNESSES

Oil profits vulnerable to fluctuating world prices. Crumbling infrastructure. Slow transition to market economy. Privatized companies asset-stripped by former managers. Organized crime controls huge areas of the economy. Regional investment hindered by uneven implementation of federal laws.

PROFILE

The few gains made in the early post-communist era were swept aside in the 1998 economic crisis. Powerful financial "oligarchs" emerged and organized crime moved into most areas of the economy. However, the devaluation of the rouble ironically served to promote a mini-boom in the last years of the 20th century. Real wages fell, encouraging small enterprises to expand, while the devaluation made imports too expensive for the average consumer – promoting local production. Industrial production increased and GDP grew by 7.6% in 2000. Putin has also moved to dismantle the power of the economic elites. The private sale of land was permitted from 2001, and a new 13% flat rate of income tax promised to help reduce widespread tax evasion. However, while the economy weathered the 2001 global slowdown, the rise in real wages and the strengthening of the rouble have begun to offset the previous years' gains.

RUSSIAN FEDERATION : MAJOR BUSINESSES

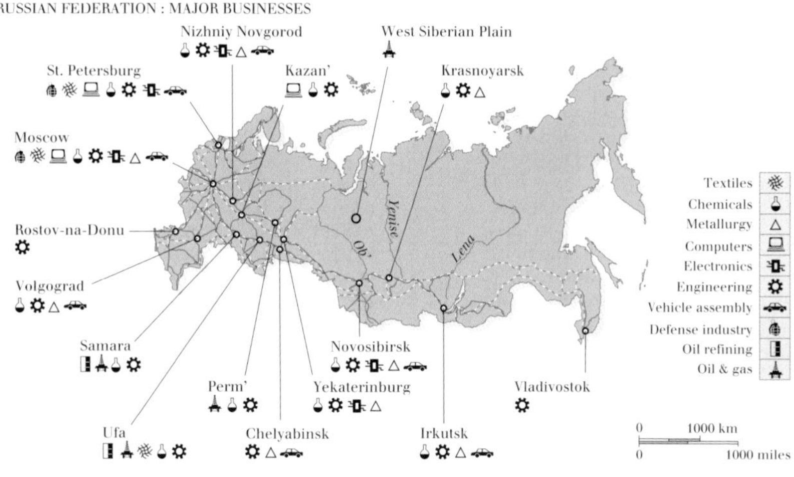
Nizhniy Novgorod
West Siberian Plain
St. Petersburg
Kazan'
Krasnoyarsk
Moscow
Rostov-na-Donu
Volgograd
Samara
Novosibirsk
Perm'
Yekaterinburg
Vladivostok
Ufa
Chelyabinsk
Irkutsk
Ob'
Yenisey
Lena
Textiles
Chemicals
Metallurgy
Computers
Electronics
Engineering
Vehicle assembly
Defense industry
Oil refining
Oil & gas
0 1000 km
0 1000 miles

R

CHRONOLOGY

The first Russian state (Rus) was in present-day Ukraine. Occupation by the Tatars (1240–1480) marked the Russian language and character. From the 17th century, the Romanovs ruled an expanding empire.

- ❑ **1904–1905** Russian war against Japan; ends in defeat for Russia.
- ❑ **1905** "Bloody Sunday" revolution.
- ❑ **1909–1914** Rapid economic expansion.
- ❑ **1914** Enters World War I against Germany.
- ❑ **1917** February Revolution; abdication of Nicholas II. October Revolution; Bolsheviks take over with Lenin as leader.
- ❑ **1918** Nicholas II and family shot.
- ❑ **1918–1920** Civil war.
- ❑ **1921** New Economic Policy; retreat from socialism.
- ❑ **1922** USSR established.
- ❑ **1924** Lenin dies. Leadership struggle eventually won by Stalin.
- ❑ **1928** First Five-Year Plan: forced industrialization and collectivization.
- ❑ **1936–1938** Show trials and campaigns against actual and suspected members of opposition. Millions sent to gulags in Siberia and elsewhere. Purges widespread.
- ❑ **1939** Hitler–Stalin pact gives USSR Baltic states, eastern Poland, and Bessarabia (Moldova).
- ❑ **1941** Germany attacks USSR.
- ❑ **1943** February, tide of war turns with lifting of siege of Stalingrad.
- ❑ **1944–1945** Soviet offensive penetrates Balkans.
- ❑ **1945** Germany defeated. Under Yalta and Potsdam agreements eastern and southeastern Europe are Soviet zone of influence.
- ❑ **1947** Cold War begins; Stalin on defensive and fears penetration of Western capitalist values.
- ❑ **1953** Stalin dies.
- ❑ **1956** Hungarian uprising crushed. Krushchev's "secret speech" attacking Stalin at party congress.
- ❑ **1957** Krushchev consolidates power. *Sputnik* launched.
- ❑ **1961** Yuri Gagarin first man in space.
- ❑ **1962** Cuban missile crisis.
- ❑ **1964** Krushchev ousted in coup, replaced by Leonid Brezhnev.
- ❑ **1975** Helsinki Final Act; confirms European frontiers as at end of World War II. Soviets agree human rights are concern of international community.
- ❑ **1979** Invades Afghanistan. New intensification of Cold War.
- ❑ **1982** Brezhnev dies.
- ❑ **1985** Gorbachev in power. Start of *perestroika*, "restructuring." First of three US–USSR summits resulting in arms

RESOURCES

Electric power 211m kw

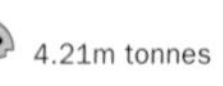

4.21m tonnes

7.06m b/d (reserves 48.6bn barrels)

27.3m cattle, 15.7m pigs, 14m sheep, 325m chickens

Coal, oil, gas, gold, diamonds, iron, aluminum, manganese

ELECTRICITY GENERATION

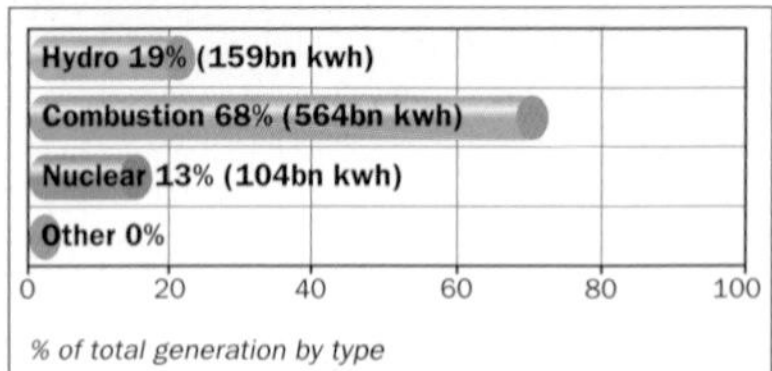
Hydro 19% (159bn kwh)
Combustion 68% (564bn kwh)
Nuclear 13% (104bn kwh)
Other 0%
0
20
40
60
80
100
% of total generation by type

RUSSIAN FEDERATION : LAND USE

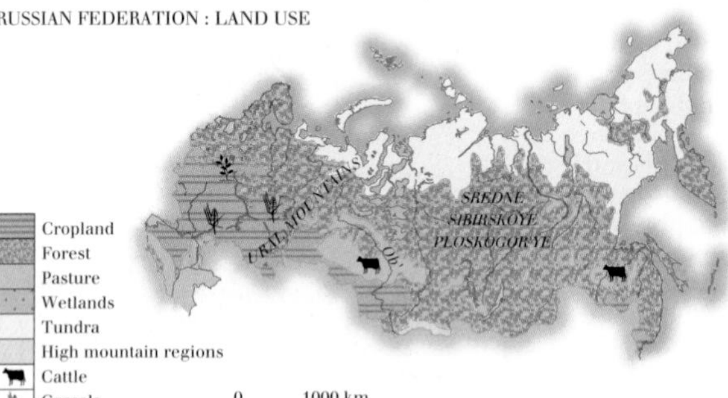
Cropland
Forest
Pasture
Wetlands
Tundra
High mountain regions
Cattle
Cereals
Potatoes
0
1000 km
0
1000 miles
URAL MOUNTAINS
SREDNE SIBIRSKOYE PLOSKOGORYE
Ob'

Russia is a leading world producer of oil, natural gas, and electricity, among other resources. Confirmed reserves make Russia the world's leading country in terms of hydrocarbons, gold, other precious metals, diamonds, and timber.

Unlike some of the other republics of the ex-USSR, Russia has been reluctant to open its resources up to foreign concerns, fearing the loss of control to Western multinationals. The lack of investment and technology is a reason for underexploitation of many of the country's major resources. Geographic remoteness is another. Also, some of the richest energy and mineral deposits are located in national territories such as Tatarstan and Sakha (Yakutia) in Siberia. The desire of these regions for greater autonomy has turned the ownership of these resources into a delicate political issue.

ENVIRONMENT

Sustainability rank: 72nd

3%

9.8 tonnes per capita

ENVIRONMENTAL TREATIES

Yes
Yes
Yes
Yes
No
No

Although awareness of Russia's environmental problems has risen sharply, the resources, political will, and know-how to tackle them are still lacking. While Russia now has an active green movement, it has not as yet won significant support in general elections.

Each region has its own particular problems. The northwest risks contamination from the neglected Soviet-era nuclear submarine fleet and from nuclear waste containers dumped in the Barents Sea. Thousands of tonnes of chemical weapons have been dumped in the Baltic, although their exact location has not been revealed. Many fish species are now extinct in the Volga River in central Russia. In the Urals and the cities of European Russia, many chemical and heavy industrial plants do not treat their effluents at all. In 2001 parliament approved a bill to allow the atomic energy ministry Minatom to earn $2 billion a year from storing and reprocessing foreign nuclear waste.

MEDIA

TV ownership high

Daily newspaper circulation 105 per 1000 people

PUBLISHING AND BROADCAST MEDIA

There are 285 daily newspapers, including *Izvestiya, Rossiiskaya Gazeta, Komsomolskaya Pravda*, and *Trud*

2 main national and regional services, partly state-owned, several independent channels

1 main state-run service, broadcasting 2 channels, 1 foreign broadcast service, several independents

Argumenty i Fakty, founded in 1987, is the best-selling weekly paper, with a circulation of nearly three million. The trade union paper *Trud* is the top-selling daily, whereas the once dominant Communist Party daily *Pravda* now sells barely 70,000 copies. The former Soviet state organ *Izvestiya* survives as an independent paper.

Public Russian Television (ORT), established in 1995 under part private ownership, runs two main TV channels. Bias in TV reporting, rife under Yeltsin, continues under Putin. Reporting events in Chechnya is subject to particular pressure. The NTV network, flagship of Media-Most, the media empire of exiled magnate Vladimir Gusinsky, was taken over in 2001 by the state-owned Gazprom's media division, while critical editorial staff were dismissed from the daily *Sevodnya* and weekly *Itogi*. Many Russians have satellite dishes and tune in to CNN and other Western channels.

CHRONOLOGY *continued*

reduction treaties. Nationality conflicts surface.

- **1988** Law of State Enterprises gives more power to enterprises; inflation and dislocation of economy.
- **1990** Gorbachev becomes Soviet president. First partly freely elected parliament (Supreme Soviet) meets.
- **1991** Boris Yeltsin elected president of Russia. Yeltsin and Muscovites resist hard-line communist coup. Gorbachev sidelined. CIS established; demise of USSR.
- **1992** Economic shock therapy.
- **1993** Yeltsin decrees dissolution of Supreme Soviet and uses force to disband parliament. Elections return conservative State Duma.
- **1994** First Russian military offensive against Chechnya.
- **1995** Communists win elections.
- **1996** Yeltsin reelected despite strong Communist challenge. Peace accord in Chechnya.
- **1998** Economic turmoil forces devaluation of rouble. Severe recession, rampant inflation.
- **1998–1999** Yeltsin repeatedly changes prime minister in successive political crises.
- **1999** December, parliamentary elections. Yeltsin resigns; Prime Minister Putin is acting president.
- **1999–2000** Terrorist violence blamed on Islamic separatists in Dagestan and Chechnya. Military offensive against Chechnya; fall of Chechen capital Grozny to Russian forces.
- **2000** Putin wins presidential election, consolidates power. Attack on "oligarchs" in big business. Improvement in economy. *Kursk* nuclear submarine disaster.
- **2001** April, party mergers make Putin's Unity party the largest grouping in parliament. July, Russian–Chinese friendship treaty.
- **2002** April, Putin declares Chechen war "over." May, Cam Ranh Bay base in Vietnam, last Russian outpost beyond former USSR, closed. Nuclear arsenal cut by 60%. NATO–Russia Council formed.

Tundra in Russia's far east. *Russia has some of the largest uninhabited tracts of land in the world.*

CRIME

Death penalty not used in practice

962,700 prisoners

Up sharply in 1999

CRIME RATES

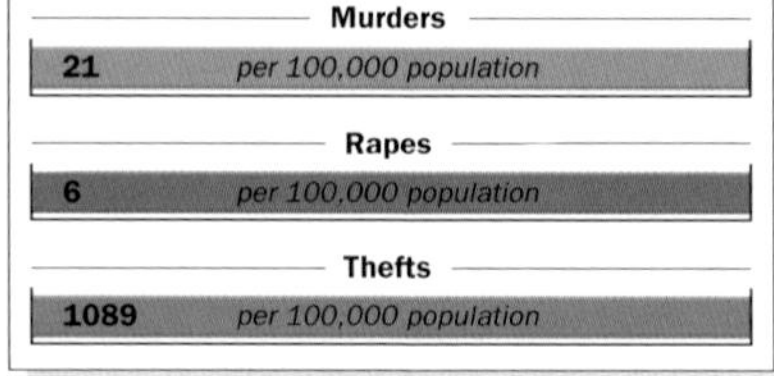
Murders
21 per 100,000 population
Rapes
6 per 100,000 population
Thefts
1089 per 100,000 population

Policing cannot keep pace with the formidable problems of rising crime in Russia. Intergang violence accounts for a sharp rise in the murder rate. Street crime has also increased in the larger cities. Corruption is rife, particularly in the regions. The Russian mafia profits from protection rackets, prostitution, smuggling operations, and narcotics, and is also active in western Europe.

Public fear resulting from the rise in crime contributed to a temporary rise in popularity for authoritarian political platforms.

With around one million inmates, overcrowding, poor conditions, and disease are major problems in prisons.

EDUCATION

School leaving age: 15

99%

4.64m students

THE EDUCATION SYSTEM

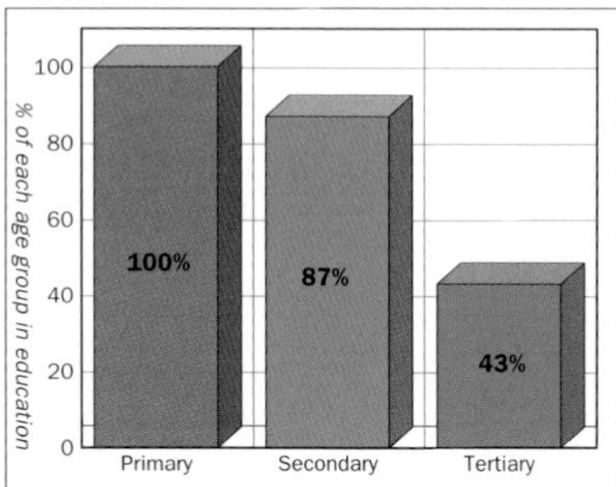
100
80
60
40
20
0
% of each age group in education
100%
87%
43%
Primary
Secondary
Tertiary

Schooling is free, and is compulsory for nine years up to age 15.

Attempts to change the Soviet-based curriculum, still widely in use, are hampered by lack of funds. Hundreds of private lycées, such as those run by the Orthodox Church, offer courses in west European languages. German in particular has made a comeback as a key language for international commerce. The state-subsidized higher education system is seriously underfunded, and some institutions have begun charging students. Prestigious institutions such as the Academy of Sciences have been forced to cut staff and research. Most academics have to rely on extramural earnings.

HEALTH

Welfare state health benefits

1 per 238 people

Heart disease, cancers, accidents, violence, tuberculosis

The health care system is in crisis and medicines are often in short supply.

Until 1991, state enterprises provided considerable health care for employees. Employers should now make payments through the Medical Insurance Fund, but many privatized concerns seek to cut costs. Local authorities lack the resources to take over these responsibilities. Bribing medical staff to obtain treatment is commonplace, and there is a lack of pharmaceutical products and drugs. Hospital patients are normally fed by their relatives.

SPENDING

GDP/cap. decrease

CONSUMPTION AND SPENDING

120 per 1000 population
218 per 1000 population

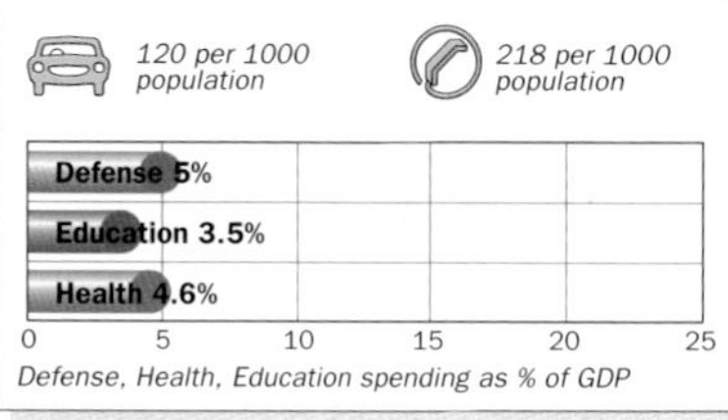
Defense 5%
Education 3.5%
Health 4.6%
0
5
10
15
20
25
Defense, Health, Education spending as % of GDP

Wealth disparities in Russia have increased sharply. A small minority of the population made huge profits from the dismantling of the old Soviet command economy. About 10% are thought to have benefited in some way.

A growing number of dollar millionaires flaunt their wealth, especially in Moscow. The bosses of organized crime are Russian society's wealthiest group. Russia is now the biggest buyer of Rolls Royces, while BMWs, Mercedes, and Volvos are relatively common in Moscow and St. Petersburg. A considerable amount of wealth is deposited abroad. There are thousands of Russian offshore bank accounts; Northern Cyprus is a favorite location.

WORLD RANKING

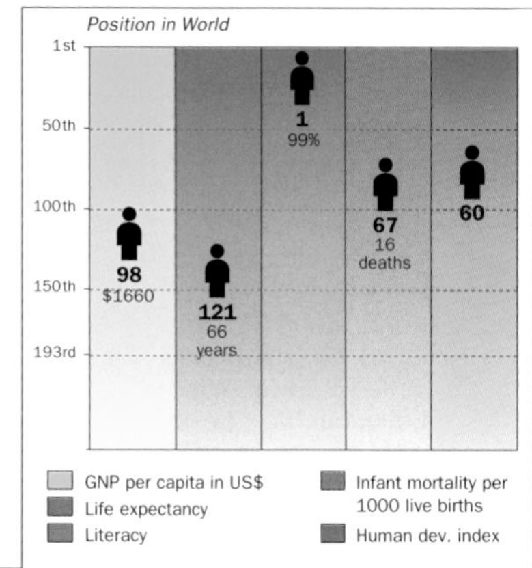
Position in World
1st
50th
100th
150th
193rd
98
$1660
121
66
years
1
99%
67
16
deaths
60
GNP per capita in US$
Life expectancy
Literacy
Infant mortality per 1000 live births
Human dev. index

R

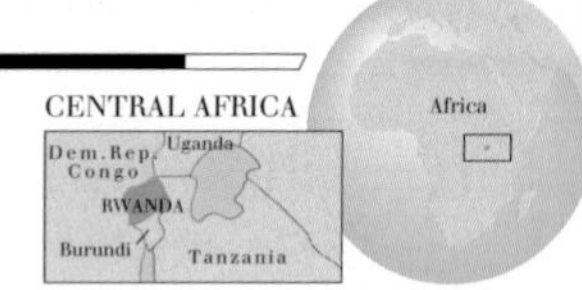
CENTRAL AFRICA
Africa
Dem.Rep. Congo
Uganda
RWANDA
Burundi
Tanzania

RWANDA

OFFICIAL NAME: Republic of Rwanda **CAPITAL:** Kigali **POPULATION:** 7.9 million
CURRENCY: Rwanda franc **OFFICIAL LANGUAGES:** French, English, and Kinyarwanda

1962

1962

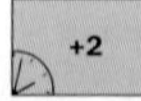
July 1
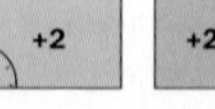

RWA

+2
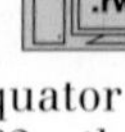
+250

.rw

LANDLOCKED RWANDA lies just south of the equator in east central Africa. Since independence in 1962, ethnic tensions have dominated politics. In 1994, the violent death of the president led to appalling political and ethnic violence. Over half of the surviving population were displaced. The perpetrators of the genocide held sway in desperately overcrowded refugee camps in adjacent countries, greatly complicating the process of eventual repatriation and reintegration.

CLIMATE

Tropical wet & dry

WEATHER CHART FOR KIGALI

Average daily temperature | Rainfall
°C/°F J F M A M J J A S O N D cm/in
40/104 30/86 20/68 10/50 0/32 -10/14 -20/-4
40/16 30/12 20/8 10/4 0

Rwanda's climate is tropical, tempered by altitude. Two wet seasons allow for two harvests each year.

TRANSPORTATION

Drive on right

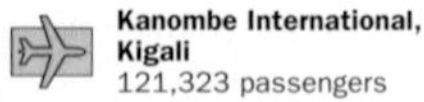
Kanombe International, Kigali
121,323 passengers

Has no fleet

THE TRANSPORTATION NETWORK

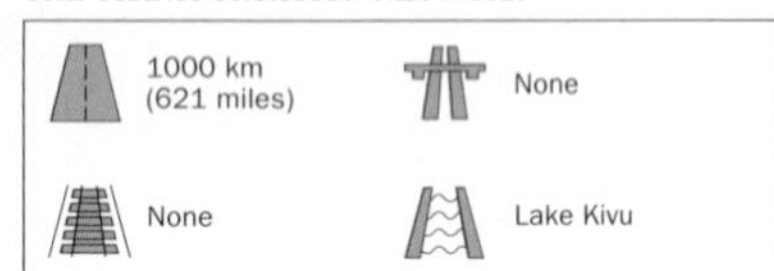
1000 km
(621 miles)
None
None
Lake Kivu

The road network is well developed. The international airport near Kigali was completed in 1986.

TOURISM

Visitors : Population 1:3950

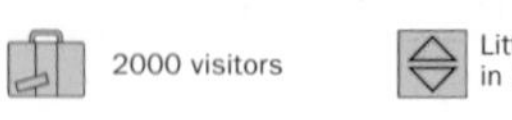
2000 visitors
Little change in 1998

MAIN TOURIST ARRIVALS

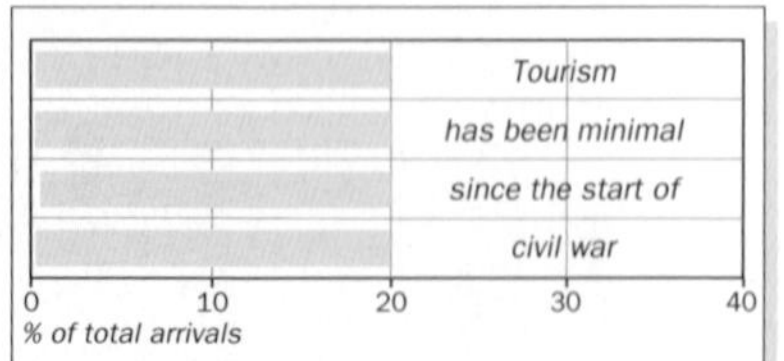
Tourism has been minimal since the start of civil war
0
10
20
30
40
% of total arrivals

Tourism has effectively ceased as a result of the civil war. When peace is secured, Rwanda may be able to regain its status as a destination for wealthy wildlife enthusiasts. Top attractions are the mountain gorillas and Lake Kivu.

PEOPLE

Pop. density high

Kinyarwanda, French, Kiswahili, English

317/km² (820/mi²)

THE URBAN/RURAL POPULATION SPLIT

6% 94%

ETHNIC MAKEUP

Other (including Twa) 1%
Tutsi 9%
Hutu 90%

The Hutu and Tutsi are the main groups; few of the Twa pygmies, the original inhabitants, remain. For more than 500 years, the cattle-owning Tutsi were politically dominant, oppressing the landowning Hutu majority. In 1959, violent revolt led to a reversal of the roles. The two groups have since been waging a spasmodic war. It is estimated that 800,000 people were killed in the violence of the mid-1990s, the majority of them Tutsi victims massacred by Hutus.

POLITICS

In transition

1988/2003

President Paul Kagame

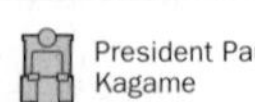

AT THE LAST ELECTION

Transitional National Assembly 74 seats

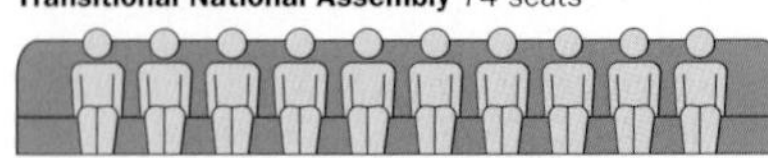

The last legislative election, to the National Development Council, took place in December 1988 when candidates were chosen from members of the National Republican Movement for Development and Democracy (MRND) list.
A Transitional National Assembly was set up in 1994. Polls due in 1999 were postponed until 2003.

In 1993 a peace accord to end the rebellion launched in 1990 by the Tutsi-dominated Rwandan Patriotic Front (FPR) was signed. However, the fragile peace process was halted in 1994 by the death of the president in a plane crash. Genocidal violence was unleashed between the mainly Hutu supporters of the old regime and its mainly, but not exclusively, Tutsi opponents. An estimated 800,000 died and millions fled the conflict, in which the FPR eventually gained control. Hutu were allocated key government posts, including the presidency, but when, in March 2000, the balance was shifted to increase Tutsi representation, President Pasteur Bizimungu resigned. Vice President Paul Kagame, the regime's dominant figure and the FPR leader, was formally elected president in April. Presidential and parliamentary elections are planned for end-2003, following the approval of a constitution.

RWANDA

Total Land Area : 26 338 sq. km (10 169 sq. miles)

POPULATION
over 100 000
over 10 000
under 10 000

LAND HEIGHT
3000m/9843ft
2000m/6562ft
1000m/3281ft

0 40 km
0 40 miles

UGANDA TANZANIA BURUNDI DEM. REP. CONGO
Lac Kivu KIGALI Ruhengeri Gisenyi Kibuye Gitarama Butare Byumba Cyangugu Kanombe Rwamagana Kibungo Rusumo Kirehe Nyanza Ruhango Gikongoro Kabaya Ngororero Kiyumba Kanazi Kigembe Busoro Munini Bugarama Bugumya Rwesero Karaba Nyabisindu Kaduha Rwamatamu Gishyita Birambo Murambi Rushashi Kinihira Busengo Nyarutovu Kirambo Kidaho Ngarama Gisagara
Lac Burera Lac Ruhondo Lac Mugesera Lac Muhazi Lac Rweru Lac Cyohoha Sud Lac Rwanyakizinga Lac Hago Lac Kivumba Lac Ihema Lac Nasho Lac Cyambwe Lac Mpanga
Akagera Nyabarongo Mukungwa Akanyaru Ruzizi Rubyiro
Volcan Karisimbi 4507m

WORLD AFFAIRS

Joined UN in 1962

COMESA

CEPGL

OIF

NAM

AU

Accused of abandoning Rwanda during 1994, the UN in 1995 set up a war crimes tribunal on the genocide.

Relations with neighboring Uganda are particularly tense, especially over the conflict in the DRC.

AID

Recipient

$322m (receipts) — Down 14% in 2000

Large amounts of aid are required, particularly for the agricultural sector, which was severely disrupted by the war. Aid donors in November 2000 urged Rwanda to withdraw its troops from the DRC.

DEFENSE

No compulsory military service

$109m

Down 19% in 2000

The Rwandan army is sufficiently powerful to make it a strong influence in the war-torn region. A national police force was established in 1999.

ECONOMICS

Inflation 15% p.a. (1990–2000)

$1.99bn

359.0–452.9 Rwanda francs

SCORE CARD

- ❑ World GNP Ranking 138th
- ❑ GNP per Capita $230
- ❑ Balance of Payments .... –$7m
- ❑ Inflation 4.3%
- ❑ Unemployment .... Few have formal employment

Strengths

Currently none. Assuming stability, Rwanda could produce coffee and tea. Possible oil and gas reserves. Tourism potential.

Weaknesses

Economic activity completely disrupted by 1994 violence. Lengthy journey to Kenyan and Tanzanian ports means high transportation costs. Few resources.

EXPORTS

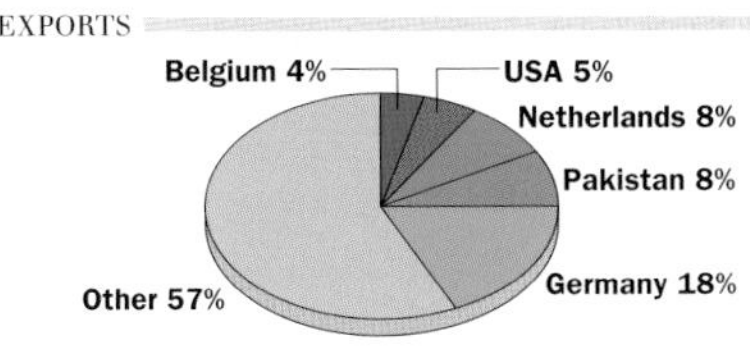
Belgium 4%
USA 5%
Netherlands 8%
Pakistan 8%
Germany 18%
Other 57%

IMPORTS

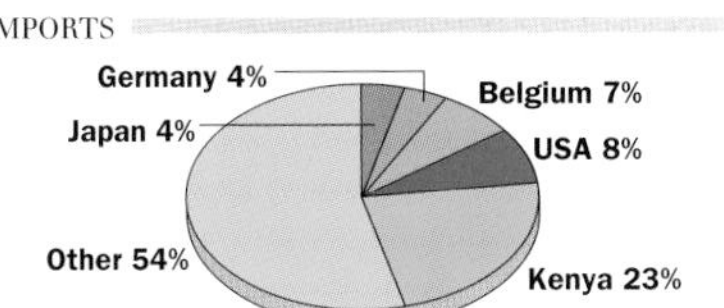
Germany 4%
Belgium 7%
Japan 4%
USA 8%
Other 54%
Kenya 23%

Terraced hillside. *Before the war, Rwanda was the most densely populated country in Africa and its land was intensively cultivated.*

RESOURCES

Electric power 34,000 kw

6733 tonnes

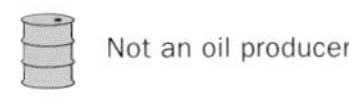
Not an oil producer

800,000 cattle, 700,000 goats, 1.2m chickens

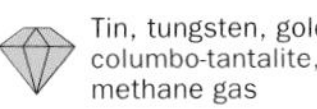
Tin, tungsten, gold, columbo-tantalite, methane gas

Gas deposits in Lake Kivu are likely to be explored with the DRC. Only 20% of urban homes are on the national power grid.

ENVIRONMENT

Sustainability rank: 119th

15%

0.1 tonnes per capita

Apart from the effects of war, soil erosion and forest loss are the major environmental problems. The tourist industry underpinned the preservation of the mountain gorilla.

MEDIA

TV ownership low

Daily newspaper circulation 0.1 per 1000 people

PUBLISHING AND BROADCAST MEDIA

There is 1 daily newspaper. The monthly *Inkingi* and *La Relève* are published in Kinyarwanda and French respectively

1 state-controlled service

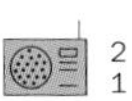
2 services: 1 state-controlled

The media have been used as an important propaganda tool by both sides in the political conflict.

CRIME

Death penalty in use

143,021 prisoners

Crime is rising

In 2001, 125,000 people were awaiting trial on genocide-related charges. A traditional system of participatory justice to address the backlog of cases became operational in 2002.

EDUCATION

School leaving age: 13

67%

5678 students

Schools are run by the state and by Christian missions. Primary education is officially compulsory, but only 78% of children attended in 1997; just 8% go on to secondary schooling.

CHRONOLOGY

The Hutu majority began to arrive in the 14th century, the warrior Tutsi in the 15th. From 1890, German and then Belgian colonizers acted to reinforce Tutsi dominance.

- ❑ **1962** Independence. Hutu-led government.
- ❑ **1960s** Tutsi revolt; massacres by Hutu; thousands of Tutsis in exile.
- ❑ **1973** Coup by Gen. Habyarimana.
- ❑ **1994** Habyarimana dies in plane crash. Genocidal violence unleashed by Hutu extremist regime, ousted by Tutsi-led FPR. Hutu refugee exodus.
- ❑ **1995** Start of war crimes tribunal.
- ❑ **1997** Refugees forcibly repatriated.
- ❑ **2000** Prominent Hutus leave office.
- ❑ **2001–2002** Limited troop withdrawal from DRC.

HEALTH

Welfare state health benefits

1 per 20,000 people

Malaria, measles, diarrheal diseases, violence

Rwanda has a network of 34 hospitals and 188 health centers. At the end of 2001, 9% of the population was estimated to be HIV-positive.

SPENDING

GDP/cap. decrease

CONSUMPTION AND SPENDING

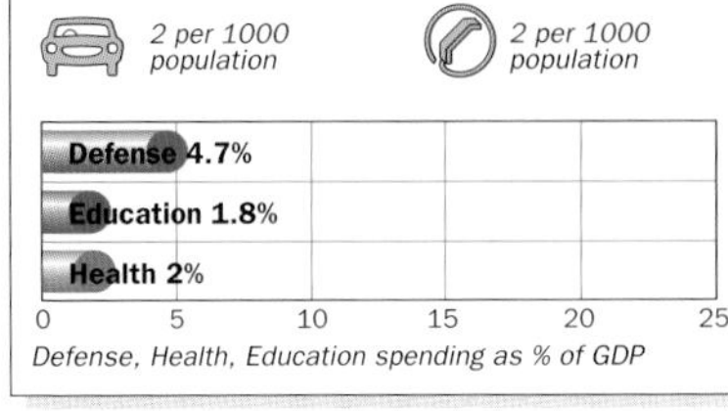
2 per 1000 population
2 per 1000 population
Defense 4.7%
Education 1.8%
Health 2%
0
5
10
15
20
25
Defense, Health, Education spending as % of GDP

Wealth is limited to the country's political elite. Most Rwandans are poor farmers; Twa pygmies and refugees are poorer still.

WORLD RANKING

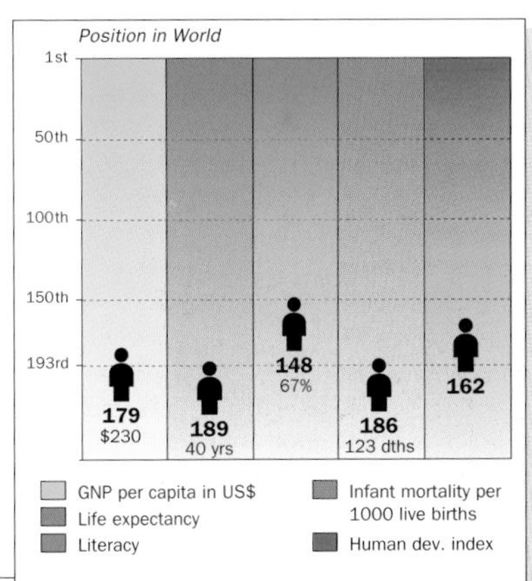
Position in World
1st
50th
100th
150th
193rd
179
$230
189
40 yrs
148
67%
186
123 dths
162
GNP per capita in US$
Life expectancy
Literacy
Infant mortality per 1000 live births
Human dev. index

R

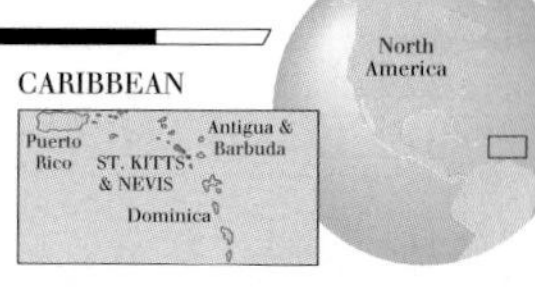
CARIBBEAN
North America
Puerto Rico
ST. KITTS & NEVIS
Antigua & Barbuda
Dominica

St. Kitts & Nevis

Official Name: Federation of Saint Christopher and Nevis **Capital:** Basseterre
Population: 41,000 **Currency:** Eastern Caribbean dollar **Official Language:** English

1983

1983

Sept 19
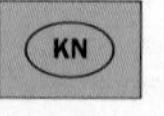
KN
-4

+1869

.kn

One of the Caribbean's most popular tourist destinations, St. Kitts and Nevis, a former British colony, lies at the northern end of the Leeward Islands chain. St. Kitts is of volcanic origin; Mount Liamuiga, a dormant volcano with a crater 227 m (745 ft) deep, is the highest point on the island. Nevis, separated from St. Kitts by a channel 3 km (2 miles) wide, is the lusher but less developed of the two islands. In the 18th century, its famed hot and cold springs gained Nevis the title "the Spa of the Caribbean."

CLIMATE

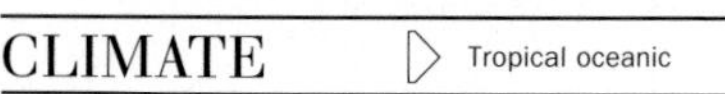
Tropical oceanic

WEATHER CHART FOR BASSETERRE

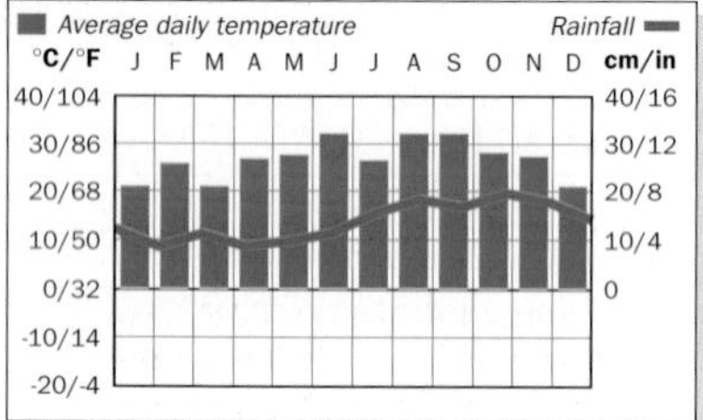
Average daily temperature
Rainfall
°C/°F J F M A M J J A S O N D cm/in
40/104
30/86
20/68
10/50
0/32
-10/14
-20/-4
40/16
30/12
20/8
10/4
0

A combination of high temperatures, trade breezes, and moderate rainfall in summer constitute St. Kitts' typically Caribbean climate.

TRANSPORTATION

Drive on left

Golden Rock International, Basseterre

1 ship 600 dwt

THE TRANSPORTATION NETWORK

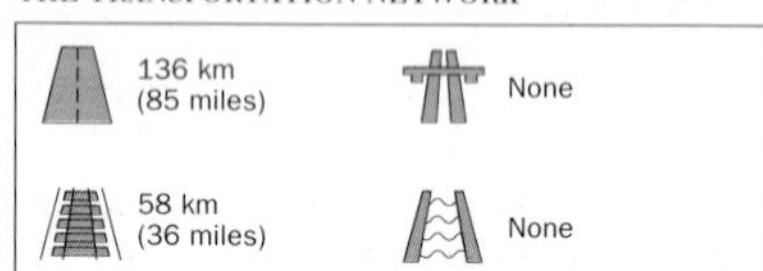
136 km (85 miles)
None
58 km (36 miles)
None

Most roads on the islands follow the coast; just a few cross the interior. Access to the remote southeast peninsula of St. Kitts has been improved. The airport on St. Kitts takes large jets; Nevis airport accepts only light aircraft. Regular ferries connect the islands.

The southeastern peninsula of St. Kitts, looking across to Nevis in the background, on a typical December evening.

TOURISM

Visitors : Population 2:1

84,000 visitors

Down 10% in 1999

MAIN TOURIST ARRIVALS

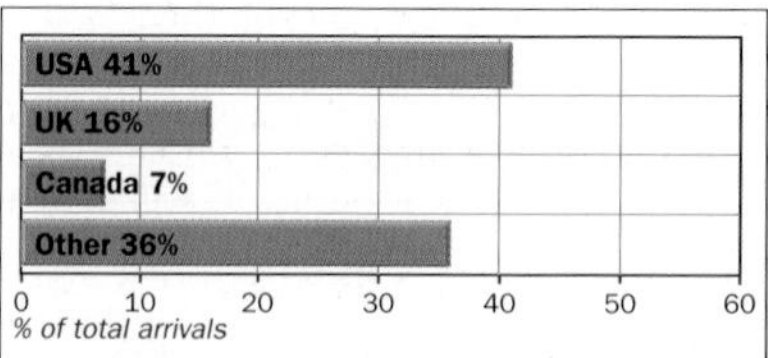
USA 41%
UK 16%
Canada 7%
Other 36%
0 10 20 30 40 50 60
% of total arrivals

St. Kitts has long targeted the mass US tourist market. Improvements to boost tourism include the opening up of the St. Kitts southern peninsula to large-scale tourist developments, and the expansion of the main port to accommodate two cruise ships simultaneously. Most visitors come for sand, sun, and the Caribbean mood, although in recent years safaris inland to see local wildlife and mineral springs have become more popular. On St. Kitts, the old Brimstone Hill fortress has been converted into a museum, as has the Nevis birthplace of Alexander Hamilton, one of the architects of the US constitution.

PEOPLE

Pop. density medium

English, English Creole

114/km² (295/mi²)

THE URBAN/RURAL POPULATION SPLIT

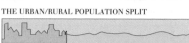

34% 66%

RELIGIOUS PERSUASION

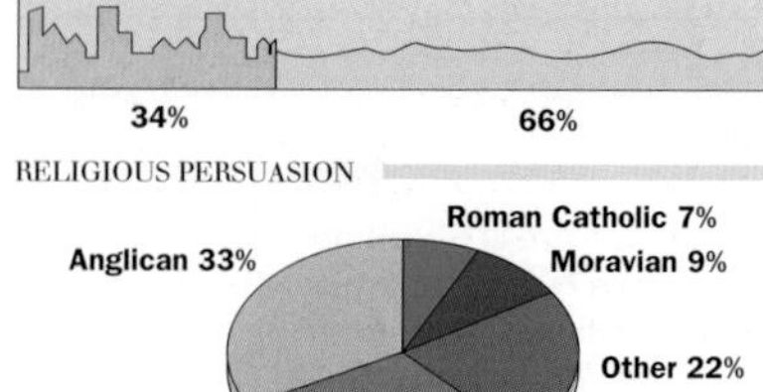
Roman Catholic 7%
Anglican 33%
Moravian 9%
Other 22%
Methodist 29%

Most of the population is descended from Africans brought over in the 17th century; intermarriage has blurred other racial lines. There are small numbers of Europeans and South Asians. High levels of emigration ensure a stable population size and provide a source of foreign currency.

POLITICS

Multiparty elections

2000/2005

H.M. Queen Elizabeth II

AT THE LAST ELECTION

National Assembly 15 seats

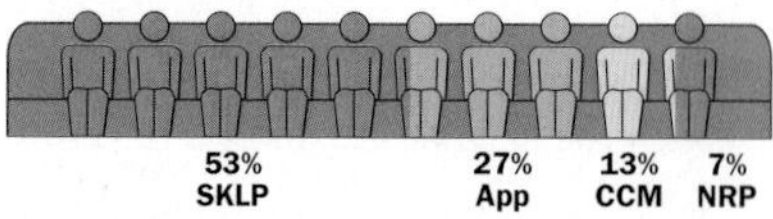
53% SKLP
27% App
13% CCM
7% NRP

SKLP = St. Kitts Labour Party **CCM** = Concerned Citizens' Movement **NRP** = Nevis Reformation Party **App** = Appointed

Nevis has its own executive and legislature, the Nevis Island Assembly, which exercise local power

The center-left SKLP ended 15 years of rule by the right-wing People's Action Movement (PAM) by winning the 1995 general election, and won a second term in 2000. A plan for secession by the Nevis government was narrowly defeated in a referendum held in 1998.

St. Kitts & Nevis

Total Land Area : 261 sq. km (101 sq. miles)

Dieppe Bay Town
Parson's Ground
Sadlers
St. Paul's
Newton Ground
Tabernacle
Mansion
Mt Liamuiga 1156m
Molineux
Lodge
Phillips
Cayon
Sandy Point Town
Brinstone Hill
St. Kitts
Middle Island
Parry's
Old Town Road
St. Peter's
Upper Conaree
Golden Rock Airport
Challengers
Boyd's
Kittitian Village
BASSETERRE
Atlantic Ocean
Caribbean Sea
Great Salt Pond
The Narrows
Newcastle
Cotton Ground
Nevis
Nevis Peak 985m
Charlestown
Bath
Fig Tree
Zion
Market Shop
Brown Hill

LAND HEIGHT

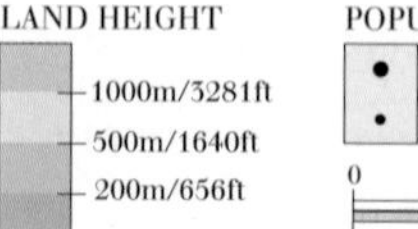
1000m/3281ft
500m/1640ft
200m/656ft
Sea Level

POPULATION

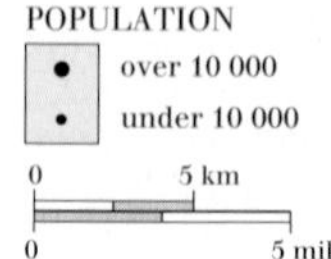
over 10 000
under 10 000
0 5 km
0 5 miles
N

S

WORLD AFFAIRS

Joined UN in 1983

ACS

Comm

Caricom

OAS

OECS

Resumption of hanging in 1998 after a 15-year moratorium drew strong international criticism. Suspected money laundering earned St. Kitts condemnation from G7 in 2000.

AID

Recipient

US$4m (receipts)

Down 20% in 2000

International aid in 1999 supported an economic recovery and relief program, following hurricanes Georges and Lenny, which significantly damaged buildings and infrastructure. Most aid is through the Caribbean Development Bank.

DEFENSE

No compulsory military service

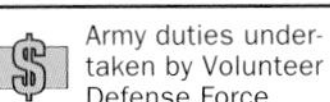
Army duties undertaken by Volunteer Defense Force

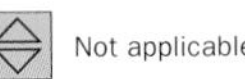
Not applicable

An army existed for six years before it was disbanded to cut government expenditure in 1981. A small paramilitary unit remains within the police; it made a token appearance with US forces during the 1983 invasion of Grenada.

ECONOMICS

Inflation 3% p.a. (1990–2000)

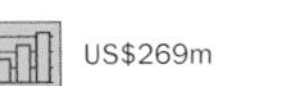
US$269m

2.7 Eastern Caribbean dollars

SCORE CARD

- ❑ WORLD GNP RANKING........................178th
- ❑ GNP PER CAPITA........................US$6570
- ❑ BALANCE OF PAYMENTS..................–US$26m
- ❑ INFLATION..3.9%
- ❑ UNEMPLOYMENT..................................5%

STRENGTHS

Growth in tourism industry, light manufacturing, and financial services. Diversifying agricultural sector.

WEAKNESSES

Tourism and related industries prone to hurricane damage, most recently in 1998–1999. Agricultural exports sensitive to fluctuating world market.

EXPORTS

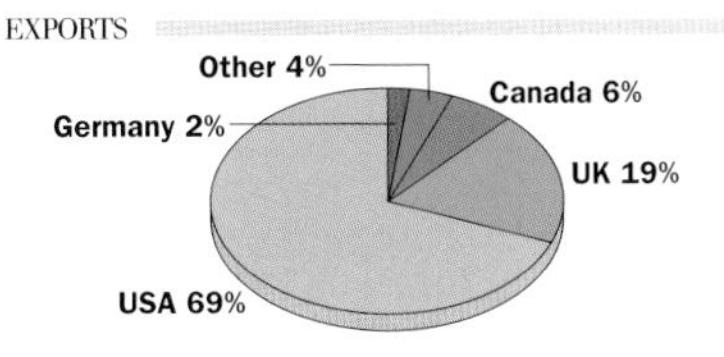
Other 4%
Canada 6%
Germany 2%
UK 19%
USA 69%

IMPORTS

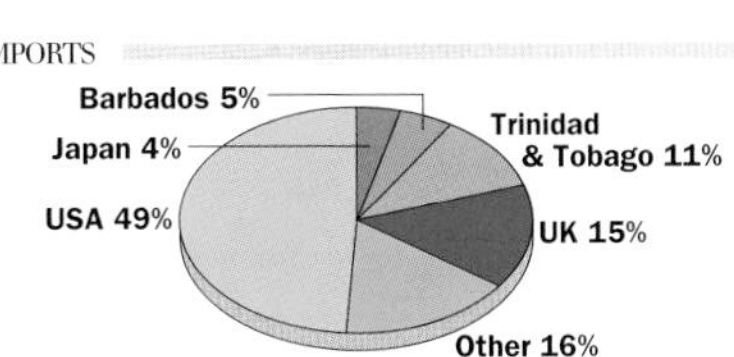
Barbados 5%
Trinidad & Tobago 11%
Japan 4%
USA 49%
UK 15%
Other 16%

RESOURCES

Electric power 16,000 kw

357 tonnes

14,500 goats, 7400 sheep, 3600 cattle, 60,000 chickens

Not an oil producer

None

St. Kitts has no strategic resources. Almost all energy has to be imported, mainly oil from Venezuela and Mexico. Sugar output declined rapidly in the 1990s. New crops, such as Sea Island cotton on Nevis, have been introduced. Offshore fishing has potential.

ENVIRONMENT

Not available

10%

2.5 tonnes per capita

Hurricanes are the greatest environmental threat. Hurricane Georges alone caused damage in 1998 estimated at US$400 million. It was followed in 1999 by Hurricane Lenny. As in the rest of the Caribbean, benefits from encouraging tourism must be set against potential ecological damage. The government has shown sensitivity, with strict preservation orders on the remaining rainforest and on indigenous monkeys.

MEDIA

TV ownership high

There are no daily newspapers

PUBLISHING AND BROADCAST MEDIA

There are no daily newspapers. The two main newspapers are the weekly *The Democrat* and the twice-weekly *Labour Spokesman*
1 state-owned service
4 services: 1 state-owned, 3 independent

Opposition parties have difficulty broadcasting their views on state-owned TV and radio. However, dissent is expressed in the printed press. The two main papers are run by political parties.

CRIME

Death penalty in use

135 prisoners

Down 31% in 1999

The judicial system is based on British common law. The police are UK-trained. Hanging was resumed in 1998, and plans agreed in 2001 to replace the role of the UK Privy Council with a Caribbean Court of Justice raised fears of more executions. Narcotics-related crimes are increasing.

EDUCATION

School leaving age: 12

90%

997 students

Education is free and the government is keen to promote information technology subjects. Students attend the regional University of the West Indies, or go on to colleges in the US and the UK.

CHRONOLOGY

A British colony since 1783, the islands were a part of the Leeward Islands Federation until 1956.

- ❑ **1932** Pro-independence St. Kitts–Nevis–Anguilla Labour Party set up.
- ❑ **1967** Internal self-government.
- ❑ **1980** Anguilla formally separates from St. Kitts and Nevis.
- ❑ **1983** Independence.
- ❑ **1995** Opposition SKLP wins election.
- ❑ **1998** Nevis referendum narrowly rejects secession.
- ❑ **2000** Prime Minister Denzil Douglas and SKLP reelected. Blacklisted by G7 for financial dealings.

HEALTH

Welfare state health benefits

1 per 855 people

Heart and respiratory diseases, cancers

The government-run health service now provides rudimentary care on both St. Kitts and Nevis. The EU and France provided EC$8 million in 1998 for repairs to the main hospital at Basseterre, which was badly damaged by Hurricane Georges.

SPENDING

GDP/cap. increase

CONSUMPTION AND SPENDING

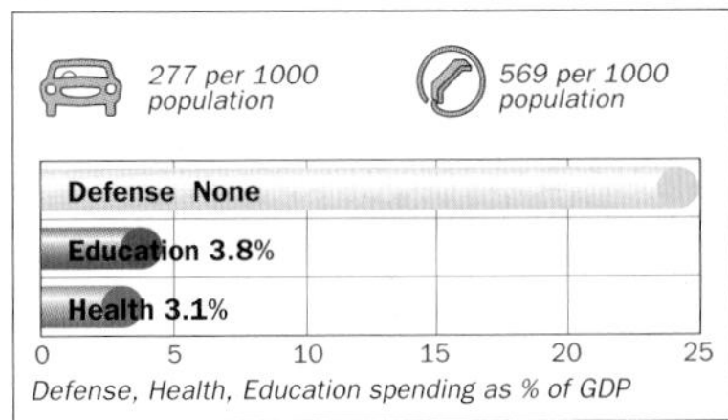
277 per 1000 population
569 per 1000 population
Defense None
Education 3.8%
Health 3.1%
0
5
10
15
20
25
Defense, Health, Education spending as % of GDP

Native professionals and civil servants have replaced expatriates over the past 20 years. They are now the best-paid group, but there are no great extremes of income.

WORLD RANKING

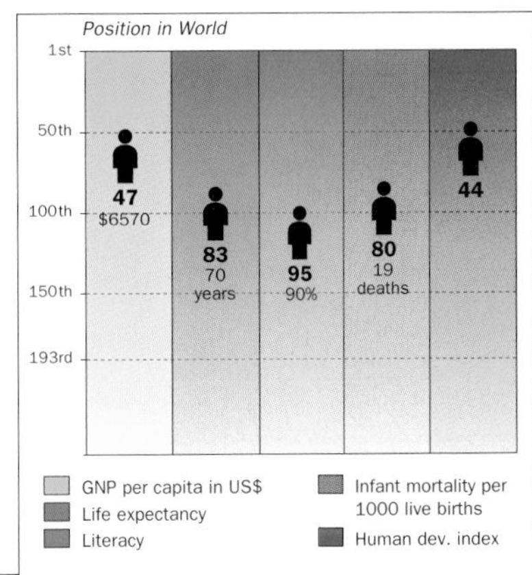
Position in World
1st
50th
100th
150th
193rd
47
$6570
83
70 years
95
90%
80
19 deaths
44
GNP per capita in US$
Life expectancy
Literacy
Infant mortality per 1000 live births
Human dev. index

S

St. Lucia

Official Name: Saint Lucia **Capital:** Castries **Population:** 156,300
Currency: Eastern Caribbean dollar **Official Language:** English

CARIBBEAN
North America
Dominica
ST. LUCIA
St. Vincent & The Grenadines
Barbados

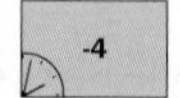

1979
1979
Feb 22
WL
-4
+1758
.lc

ST. LUCIA IS ONE OF THE MOST beautiful islands of the Windward group of the Antilles. The twin Pitons, south of Soufrière, are one of the most striking natural features in the Caribbean. Ruled by the French and the British at different times in its past, St. Lucia retains the character of both. A multiparty democracy, it lives by banana-growing and tourism, with enticing beaches and a rich variety of wildlife in the rainforest.

CLIMATE

Tropical oceanic

WEATHER CHART FOR CASTRIES

Average daily temperature
Rainfall
°C/°F
J F M A M J J A S O N D
cm/in
40/104
30/86
20/68
10/50
0/32
-10/14
-20/-4
40/16
30/12
20/8
10/4
0

The dry season, from January to April, brings intense heat to sheltered parts of St. Lucia. During the rainy season, short warm showers can be expected daily. Rainfall is highest in the mountains.

TRANSPORTATION

Drive on left

Vigie Field, Castries
446,384 passengers
3 ships
911 grt

THE TRANSPORTATION NETWORK

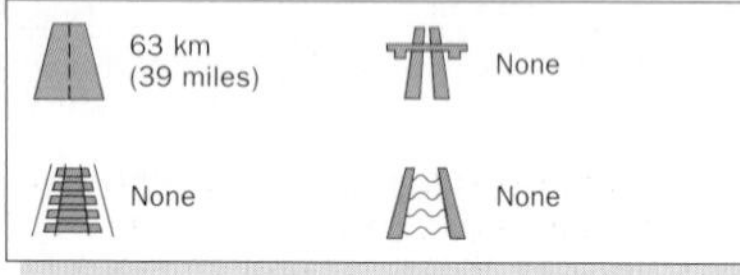
63 km
(39 miles)
None
None
None

Roads are confined to the west and southeast coasts; only half are paved. Flights arrive from major European and North American cities, and other Caribbean locations. Direct passage to South America is largely by sea.

One of the twin Pitons south of Soufrière, marking the entrance to the Jalousie Plantation harbor.

TOURISM

Visitors : Population 1.7:1

259,000 visitors — Down 1% in 2000

MAIN TOURIST ARRIVALS

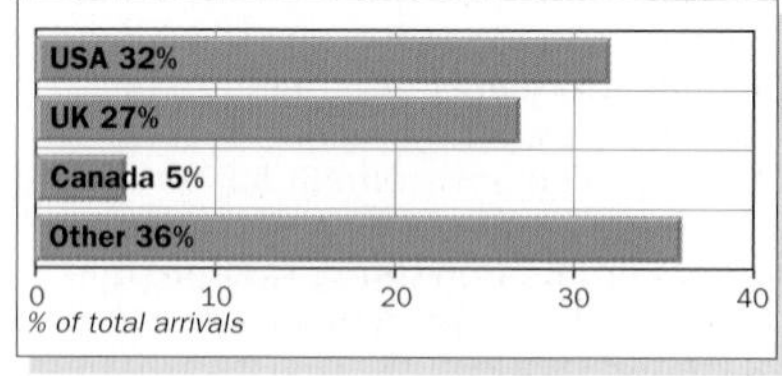
USA 32%
UK 27%
Canada 5%
Other 36%
0
10
20
30
40
% of total arrivals

Tropical beaches and typical Caribbean towns make St. Lucia a favorite destination for cruise ships and stay-over tourists. The number of hotel rooms continues to rise. The pristine rainforest has become the focus of nature tourism, with tours often organized by the National Trust.

PEOPLE

Pop. density high

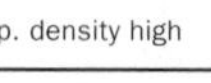

English, French Creole
256/km²
(663/mi²)

THE URBAN/RURAL POPULATION SPLIT

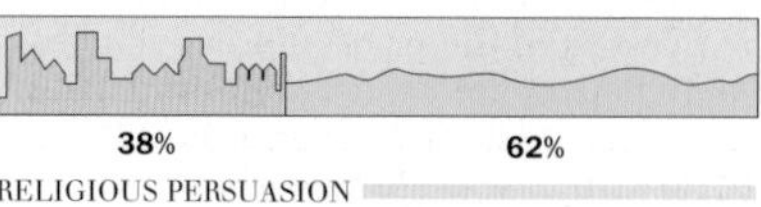
38%
62%

RELIGIOUS PERSUASION

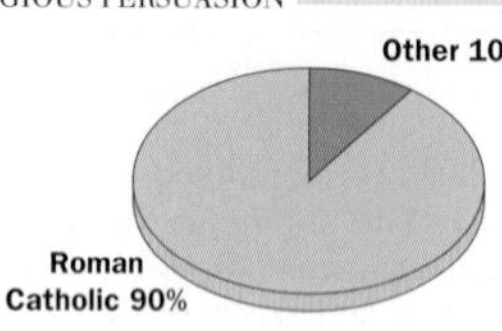
Other 10%
Roman Catholic 90%

St. Lucia has a rich, tension-free racial mix of descendants of Africans, Caribs, and European settlers. Despite relaxed attitudes, family life is central to most St. Lucians, many of whom are practicing Roman Catholics. Small families are the norm. In rural districts, where women run many of the farms, absentee fathers are fairly common. In recent years, women have had greater access to higher education and have moved into professions. A bill to permit the occasional use of Creole in parliament was passed in 1998.

POLITICS

Multiparty elections

L. House 2001/2006
U. House 2001/2006
H.M. Queen Elizabeth II

AT THE LAST ELECTION

House of Assembly 18 seats

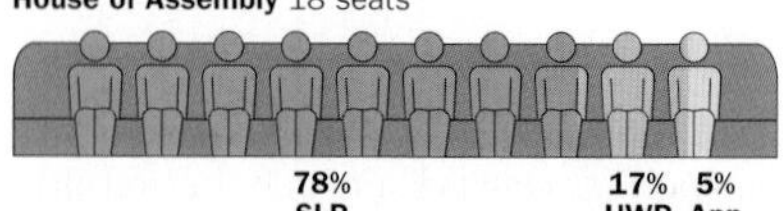
78% SLP
17% UWP
5% App

SLP = St. Lucia Labour Party **UWP** = United Workers' Party **App** = Appointed

Senate 11 seats

6 Senate members are nominated by the government, 3 by the opposition, and 2 by the governor-general on a nonparty basis

Politics revolved around John Compton of the UWP and Julian Hunte of the SLP, until from 1996 a less personalized climate prevailed with the emergence of new political figures. After losing heavily to the SLP in the 1997 general election, the UWP changed its leader three times in as many years, but failed to prevent the SLP's reelection in 2001.

WORLD AFFAIRS

Joined UN in 1979

ACS — Comm — Caricom — OECS — OAS

Relations with the US were strained by the long-running, and ultimately successful, US campaign to end preferential access for Caribbean bananas to the EU. It is feared that from 2006 St. Lucia will be unable to compete with cheaper fruit from Central and South America. St. Lucia supports Japan, an aid donor, in its bid for a permanent seat on the UN Security Council. With the agreement in 2001 to form a Caribbean Court of Justice, one of the last constitutional ties to the UK was set to be cut.

AID

Recipient

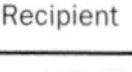
US$11m (receipts)
Down 58% in 2000

The US, the EU, and Japan are the main donors. China has also given aid and grant loans in recent years.

DEFENSE

No compulsory military service

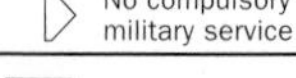

US$5m
Little change

The police force is supported by a small paramilitary unit. Training is provided by the US and the UK.

S

ST. LUCIA

Total Land Area : 620 sq. km (239 sq. miles)

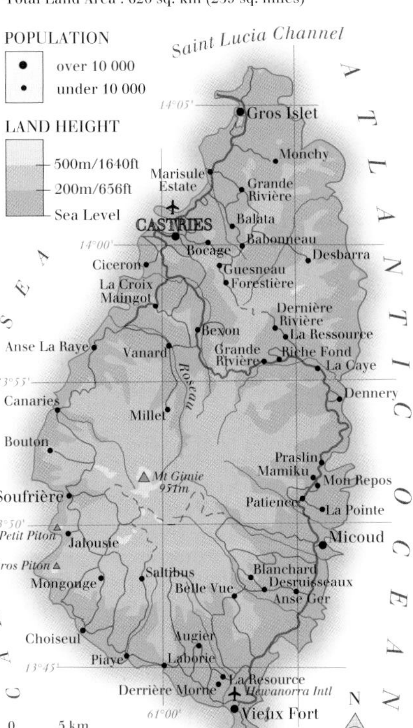
POPULATION
over 10 000
under 10 000
LAND HEIGHT
500m/1640ft
200m/656ft
Sea Level
Saint Lucia Channel
Gros Islet
Monchy
Marisule Estate
Grande Rivière
CASTRIES
Babonneau
Balata
Bocage
Desbarra
Ciceron
La Croix Maingot
Guesneau
Forestière
Dernière Rivière
La Ressource
Bexon
Anse La Raye
Vanard
Grande Rivière
Riche Fond
La Caye
Dennery
Roseau
Canaries
Millet
Bouton
Praslin
Mamiku
Mt Gimie 957m
Mon Repos
Soufrière
Patience
La Pointe
Petit Piton
Jalousie
Micoud
Gros Piton
Blanchard
Saltibus
Desruisseaux
Mongouge
Belle Vue
Anse Ger
Choiseul
Augier
Piaye
Laborie
La Resource
Derrière Morne
Hewanorra Intl
Vieux Fort
CARIBBEAN SEA
ATLANTIC OCEAN
0 5 km
0 5 miles

ECONOMICS

Inflation 2.9% p.a. (1990–2000)

US$642m

2.7 Eastern Caribbean dollars

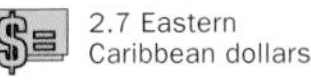

SCORE CARD

- ❑ WORLD GNP RANKING........................165th
- ❑ GNP PER CAPITA.........................US$4120
- ❑ BALANCE OF PAYMENTS..................–US$80m
- ❑ INFLATION...5.4%
- ❑ UNEMPLOYMENT.................................16%

STRENGTHS

Banana industry (privatized in 1998). Tourism, fisheries potential.

WEAKNESSES

Preferential banana trade with EU to end by 2006. Most resorts foreign-owned.

EXPORTS

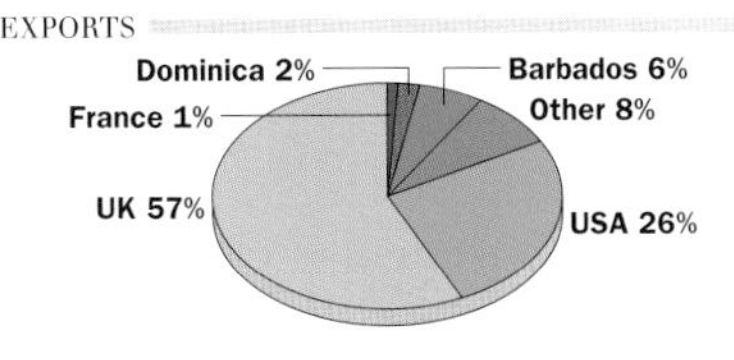
Dominica 2%
Barbados 6%
France 1%
Other 8%
UK 57%
USA 26%

IMPORTS

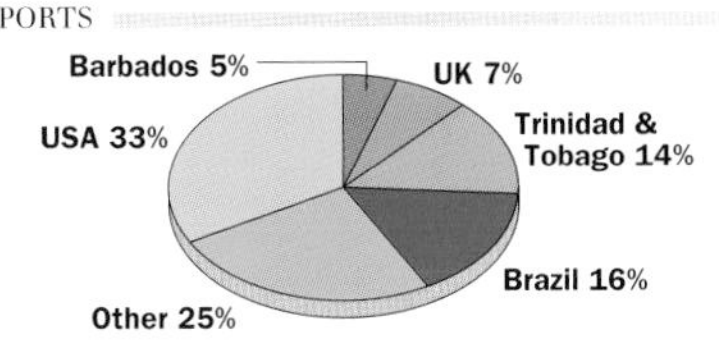
Barbados 5%
UK 7%
USA 33%
Trinidad & Tobago 14%
Brazil 16%
Other 25%

RESOURCES

Electric power 22,000 kw

1719 tonnes

Not an oil producer

14,750 pigs, 12,500 sheep, 210,000 chickens

None

St. Lucia has no mineral resources and imports most of its energy. Plans exist to develop geothermal energy from the hot springs in the volcanic interior.

ENVIRONMENT

Not available

2% partially protected

1.3 tonnes per capita

St. Lucians are proud of their island, and environmental questions arouse fierce debate. In recent years the greatest controversy surrounded the decision to allow a luxury hotel development on the ecologically important Jalousie Plantation, which encompasses the extraordinary twin Pitons and includes an important Amerindian archaeological site. The issue illustrates a key problem in St. Lucia, where business pressures to develop tourism can outweigh vital environmental concerns. One notable conservation success has been the St. Lucia parrot. In 1978, there were 150 birds; strict laws against the trade in parrots ensured that by 2000 numbers had risen to over 800.

MEDIA

TV ownership medium

There are no daily newspapers

PUBLISHING AND BROADCAST MEDIA

There are no daily newspapers. *The Star* and *The Mirror* are published weekly

4 independent services

4 services: 1 state-owned, 3 independent

The privately owned press is free from government intervention. It is possible to receive TV programs from US, Mexican, and some Caribbean stations.

CRIME

Death penalty in use

365 prisoners

Crime is rising

Murder is rare, but narcotics-related deaths are increasing, as is violence in schools. The government has strengthened the police force to combat rising urban crime.

EDUCATION

School leaving age: 15

82%

3881 students

Education is based on the British system. Nobel prizewinners Sir Arthur Lewis (economics) and Derek Walcott (literature) give St. Lucia the world's highest per capita ratio of laureates.

CHRONOLOGY

An excellent naval raiding base in the Caribbean in the 17th and 18th centuries, St. Lucia was fought over by France and Britain. Ownership alternated before it was finally ceded to Britain in 1814. French influence survives in St. Lucian patois and the local cuisine.

- ❑ **1958** Joins West Indies Federation.
- ❑ **1964** Sugar growing ceases.
- ❑ **1979** Gains independence and joins Commonwealth.
- ❑ **1990** Establishes body with Dominica, Grenada, and St. Vincent to discuss forming a Windward Islands Federation.
- ❑ **1997** Hitherto ruling UWP reduced to one seat in general election.
- ❑ **2000** Blacklisted by OECD as international tax haven.

HEALTH

Welfare state health benefits

1 per 2128 people

Heart and respiratory diseases, cancers

The National Insurance Scheme provides free medical cover for all. Work on a new general hospital began in 2001.

SPENDING

GDP/cap. increase

CONSUMPTION AND SPENDING

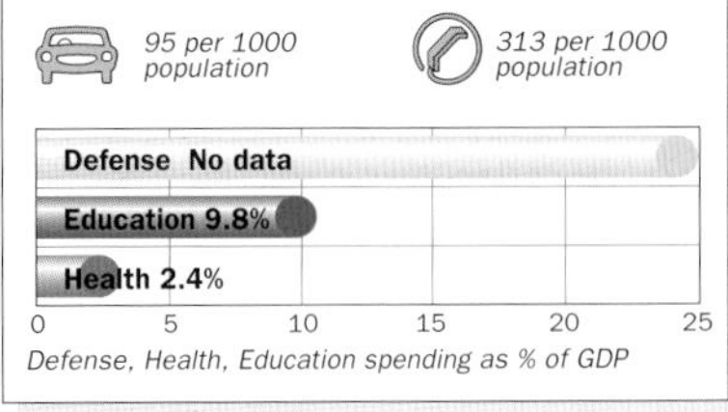
95 per 1000 population
313 per 1000 population
Defense No data
Education 9.8%
Health 2.4%
0
5
10
15
20
25
Defense, Health, Education spending as % of GDP

The island's large-scale banana growers and hotel owners form the richest section of society. Nearly one-fifth of households are considered to be poor.

WORLD RANKING

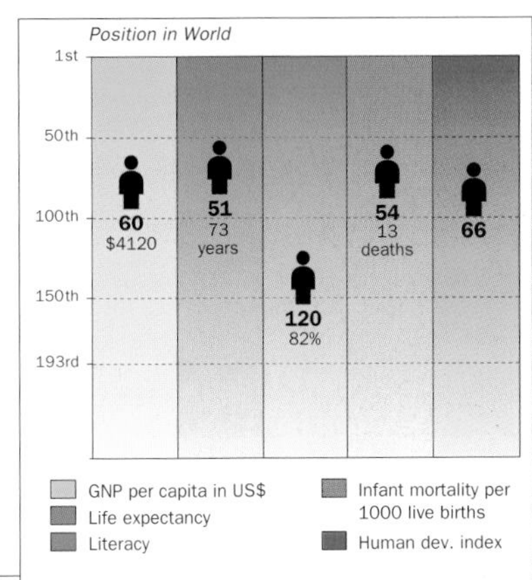
Position in World
1st
50th
100th
150th
193rd
60
$4120
51
73 years
120
82%
54
13 deaths
66
GNP per capita in US$
Life expectancy
Literacy
Infant mortality per 1000 live births
Human dev. index

S

CARIBBEAN

North America

ST. VINCENT & THE GRENADINES
St. Lucia
Barbados
Grenada

ST. VINCENT & THE GRENADINES

OFFICIAL NAME: Saint Vincent and the Grenadines **CAPITAL:** Kingstown
POPULATION: 115,500 **CURRENCY:** Eastern Caribbean dollar **OFFICIAL LANGUAGE:** English

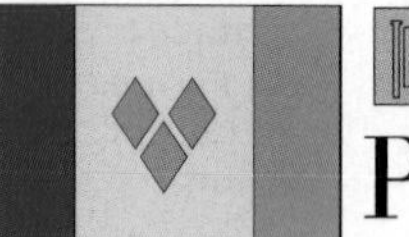

1979

1979

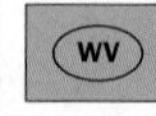
Oct 27

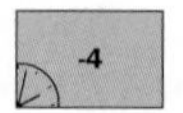
WV

-4

+1784

.vc

PART OF THE WINDWARD ISLANDS group, and bounded by submerged coral reefs, St. Vincent and the Grenadines is the Caribbean playground of the international celebrity circuit. Tourism and bananas are the economic mainstays; St. Vincent is also the world's largest arrowroot producer. It is mostly volcanic; the one remaining active volcano, La Soufrière, last erupted in 1979. The Grenadines are flat, mainly bare, coral islands.

POLITICS

Multiparty elections

2001/2006 H.M. Queen Elizabeth II

AT THE LAST ELECTION

House of Assembly 21 seats

57%
ULP
29%
App
14%
NDP

ULP = Unity Labour Party **App** = Appointed **NDP** = New Democratic Party

Six senators are appointed to the House of Assembly by the governor-general

In March 2001, 17 years of rule by the NDP ended with a crushing electoral defeat. The leader of the long-term opposition ULP, Ralph Gonsalves, was appointed prime minister. A new Caribbean Court of Justice was agreed in 2001, but constitutional ties to the British monarchy remain, despite a strong republican movement.

CLIMATE

Tropical oceanic

WEATHER CHART FOR KINGSTOWN

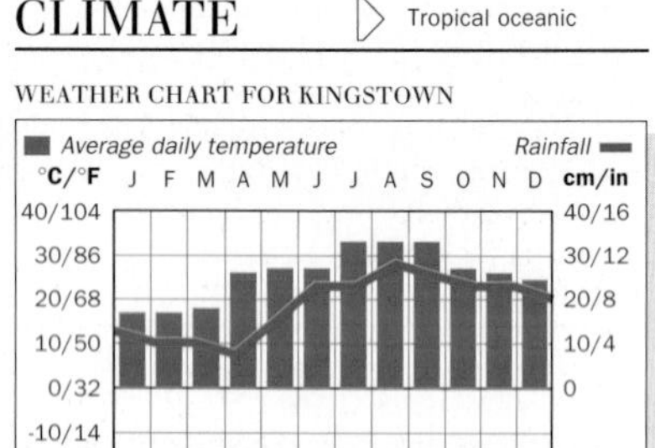
Average daily temperature
Rainfall
°C/°F
J F M A M J J A S O N D
cm/in
40/104
30/86
20/68
10/50
0/32
-10/14
-20/-4
40/16
30/12
20/8
10/4
0

Constant trade winds moderate St. Vincent's tropical climate. Rainfall is heaviest during the summer months. Deep depressions and hurricanes are likely between June and November.

TRANSPORTATION

Drive on left

Arnos Vale, Kingstown 1317 ships 7.88m grt

THE TRANSPORTATION NETWORK

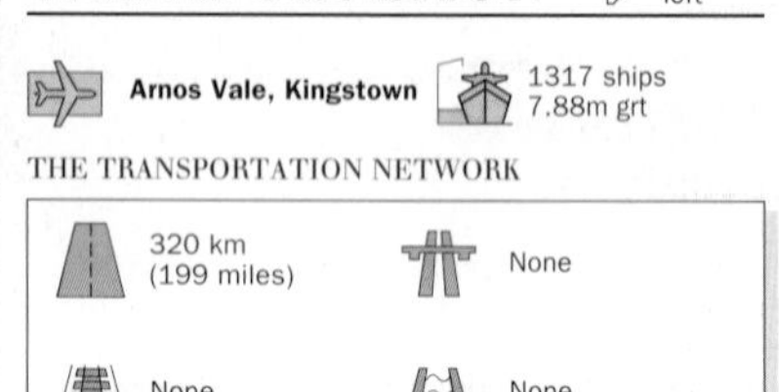
320 km
(199 miles)
None
None
None

Access by air is via neighboring islands. Paved roads encompass most of St. Vincent's coast. Port improvements have been completed in recent years. In 1992, an airport capable of taking executive jets was completed on the island of Bequia.

Aerial view of Union Island in the Grenadines chain. The government is developing the island as a major yachting center.

TOURISM

Visitors : Population 1:1.6

73,000 visitors Up 7% in 2000

MAIN TOURIST ARRIVALS

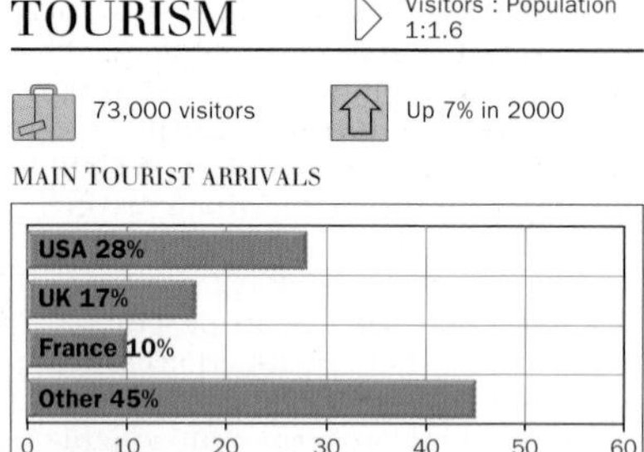
USA 28%
UK 17%
France 10%
Other 45%
0
10
20
30
40
50
60
% of total arrivals

Tourism is targeted at celebrities and cruise ships rather than the mass market, and is concentrated on the Grenadines. Mustique, long associated with Princess Margaret, has a rock music clientele. Union Island draws the yachting rich, and luxury villas, apartments, a golf course, and a casino have been built on Canouan. Layou, on St. Vincent, is the site of pre-Columbian Amerindian petroglyphs.

PEOPLE

Pop. density high

English, English Creole 340/km² (880/mi²)

THE URBAN/RURAL POPULATION SPLIT

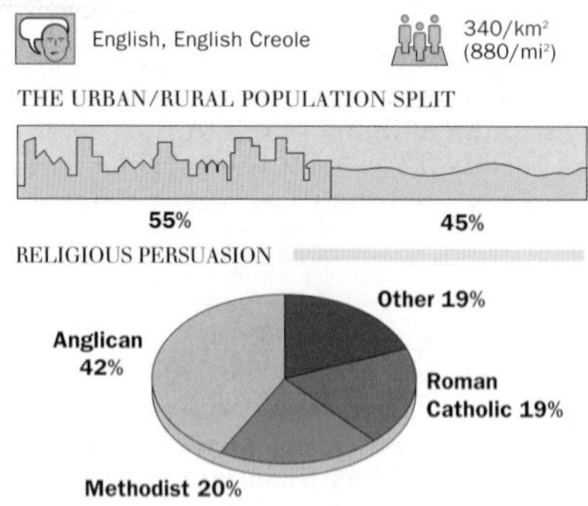
55%
45%
RELIGIOUS PERSUASION
Anglican 42%
Other 19%
Roman Catholic 19%
Methodist 20%

Family life on St. Vincent is heavily influenced by the Anglican Church. Racial tensions are few, and intermarriage has meant that the original communities of descendants of African slaves, Europeans, and the few indigenous Caribs can no longer be distinguished. Many locals fear that traditional island life is being threatened by the expanding tourist industry.

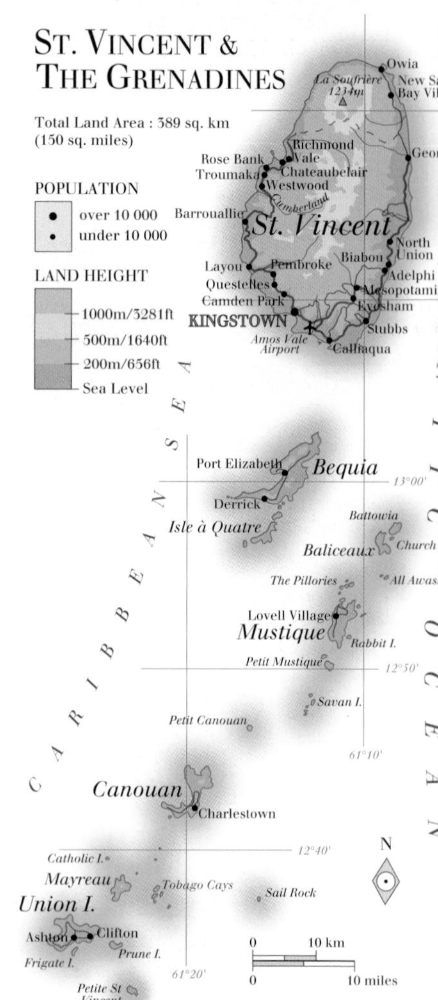
ST. VINCENT & THE GRENADINES
Total Land Area : 389 sq. km (150 sq. miles)
POPULATION
over 10 000
under 10 000
LAND HEIGHT
1000m/3281ft
500m/1640ft
200m/656ft
Sea Level
Owia
La Soufrière 1234m
Rose Bank
Troumaka
Richmond
Vale
Chateaubelair
Westwood
Cumberland
Barrouallie
St. Vincent
North Union
Biabou
Layou
Pembroke
Questelles
Camden Park
Adelphi
Mesopotamia
Evesham
KINGSTOWN
Amos Vale Airport
Stubbs
Calliaqua
Port Elizabeth
Bequia
Derrick
Isle à Quatre
Battowia
Baliceaux
The Pillories
Lovell Village
Mustique
Rabbit I.
Petit Mustique
Savan I.
Petit Canouan
Canouan
Charlestown
Catholic I.
Mayreau
Tobago Cays
Sail Rock
Union I.
Ashton
Clifton
Prune I.
Frigate I.
Petite St Vincent
CARIBBEAN SEA
ATLANTIC OCEAN
13°00'
12°50'
12°40'
61°10'
61°20'
N
0
10 km
10 miles

S

WORLD AFFAIRS

Joined UN in 1980

ACS

Comm

Caricom

OAS

OECS

The successful US bid to end the EU's preferential treatment of Caribbean banana imports has strained relations. St. Vincent supports a united move by Caricom members to promote a presidential system of government in place of the British monarchy.

AID

Recipient

US$6m (receipts)

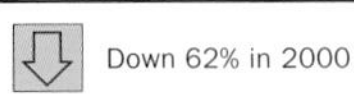
Down 62% in 2000

The EU and Japan are the major sources of development aid. Significant funds also come from the Caribbean Development Bank and Arab countries.

DEFENSE

No compulsory military service

US$3m (estimate)

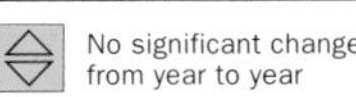
No significant change from year to year

St. Vincent has no army. Its small police force, trained by the US and the UK, is part of the Windward and Leeward Islands' Regional Security System.

ECONOMICS

Inflation 2% p.a. (1990–2000)

US$313m

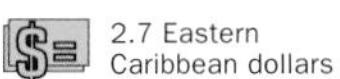
2.7 Eastern Caribbean dollars

SCORE CARD

❑ WORLD GNP RANKING	176th
❑ GNP PER CAPITA	US$2720
❑ BALANCE OF PAYMENTS	–US$66m
❑ INFLATION	0.2%
❑ UNEMPLOYMENT	22%

STRENGTHS
Bananas, but preferential access to EU markets will end in 2006. Great tourist potential. Currency stability. Leading producer of arrowroot starch. Improving infrastructure.

WEAKNESSES
Little diversification. Strong potential competition from Central and South American banana producers.

EXPORTS

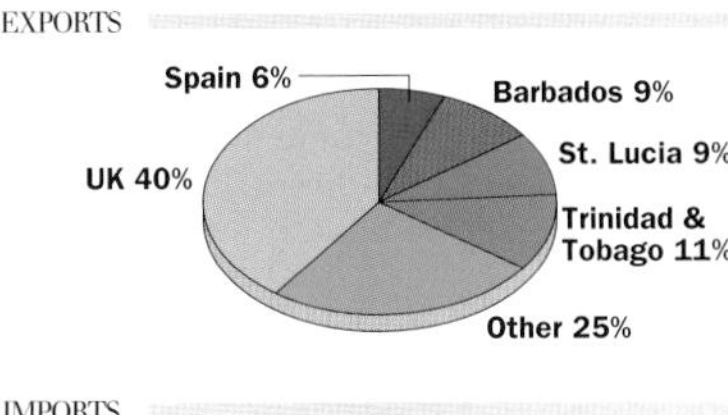
Spain 6%
Barbados 9%
St. Lucia 9%
UK 40%
Trinidad & Tobago 11%
Other 25%

IMPORTS

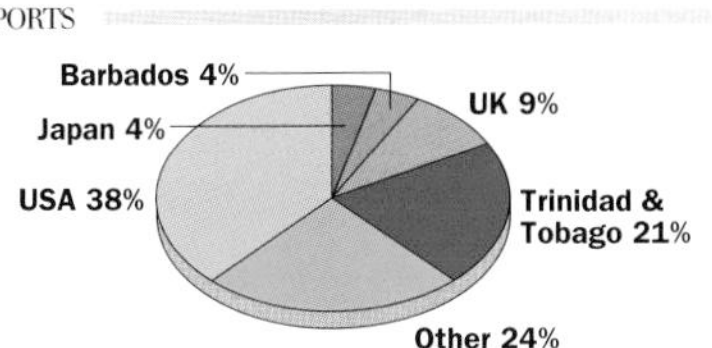
Barbados 4%
UK 9%
Japan 4%
USA 38%
Trinidad & Tobago 21%
Other 24%

RESOURCES

Electric power 16,000 kw

15,573 tonnes

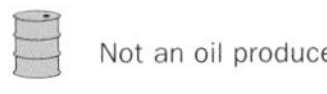
Not an oil producer

13,000 sheep, 9500 pigs, 200,000 chickens

None

There is a hydroelectric plant on the Cumberland River. Virtually all other energy requirements have to be imported. Some of the Grenadines have no fresh water sources.

ENVIRONMENT

Not available

21% (including marine and semi-protected areas)

1.4 tonnes per capita

Hurricanes are the main environmental threat, sometimes destroying as much as 70% of the banana crop. The former inaccessibility of St. Vincent and the Grenadines meant that tourism was a minor environmental threat, and the untouched, idyllic landscape of islands such as Mustique was their attraction. Mustique is reasonably well protected – buildings have been restricted and further development is limited, since fresh water has to be shipped in. On Bequia, the new airport and consequent increase in visitors are seen as a mixed blessing. Schemes to develop Canouan have been opposed by locals.

MEDIA

TV ownership medium

Daily newspaper circulation 9 per 1000 people

PUBLISHING AND BROADCAST MEDIA

There is 1 daily newspaper, *The Herald*. The main weekly newspaper is the independent *The Vincentian*

1 state-owned service

1 state-owned service

Of the many periodicals, three are published by the political parties; the rest are independent. Freedom of the press is written into the constitution.

CRIME

Death penalty in use

405 prisoners

Crime is rising

The incidence of rape and robbery causes most concern, although on the outlying islands both crimes are very rare. St. Vincent is used for narcotics transshipment to the US.

EDUCATION

School leaving age: 15

82%

677 students

State schools follow the former British 11-plus selective system. There are a few private schools. University students go on to the regional University of the West Indies in Jamaica, although increasing numbers are also studying in the US and the UK.

CHRONOLOGY

In 1795, the local Carib population staged a revolt against the British, who deported them, leaving a largely black African population.

- ❑ **1951** Universal suffrage.
- ❑ **1969** Internal self-government.
- ❑ **1972** James Mitchell premier; holds balance of power between People's Political Party (PPP) and St. Vincent Labour Party (SVLP).
- ❑ **1974** PPP–SVLP coalition.
- ❑ **1979** Full independence under Milton Cato of SVLP. La Soufrière volcano erupts.
- ❑ **1984** NDP, founded by Mitchell in 1975, wins first of four terms.
- ❑ **2000** Mitchell resigns premiership.
- ❑ **2001** ULP wins landslide victory. Ralph Gonsalves prime minister.

HEALTH

Welfare state health benefits

1 per 1136 people

Heart and respiratory diseases, cancers

Doctors train at the University of the West Indies. The system is a mixture of state and private hospitals and clinics; facilities are scarcer on the Grenadines.

SPENDING

GDP/cap. increase

CONSUMPTION AND SPENDING

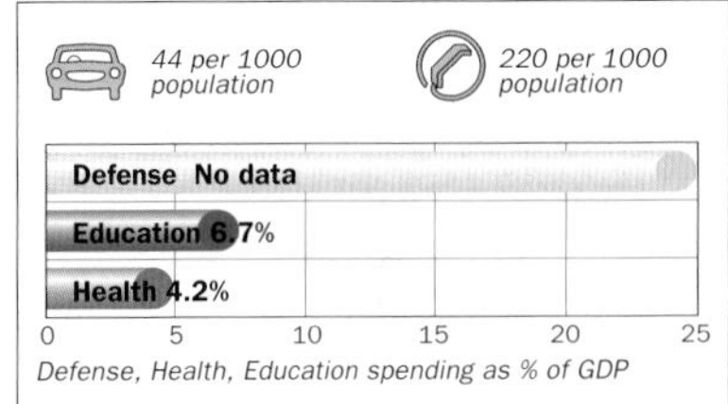
44 per 1000 population
220 per 1000 population
Defense No data
Education 6.7%
Health 4.2%
0
5
10
15
20
25
Defense, Health, Education spending as % of GDP

Jet-set wealth in the islands coexists with the low wages paid to most local workers. Union Island and Mustique in particular attract the wealthy, with their motor yachts and jeeps.

WORLD RANKING

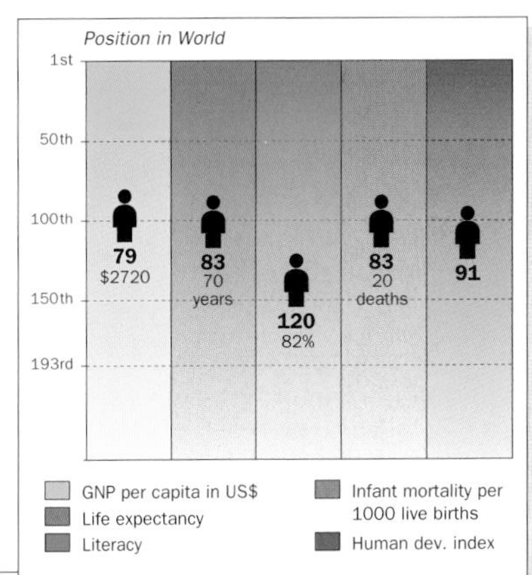
Position in World
1st
50th
100th
150th
193rd
79
$2720
83
70 years
120
82%
83
20 deaths
91
GNP per capita in US$
Life expectancy
Literacy
Infant mortality per 1000 live births
Human dev. index

S

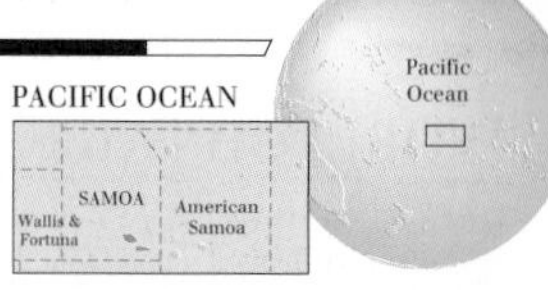
PACIFIC OCEAN
Pacific Ocean
SAMOA
Wallis & Futuna
American Samoa

SAMOA

OFFICIAL NAME: Independent State of Samoa **CAPITAL:** Apia
POPULATION: 159,000 **CURRENCY:** Tala **OFFICIAL LANGUAGES:** Samoan and English

1962

1962
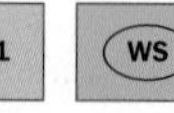
June 1
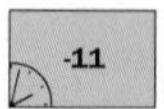
WS

-11

+684

.ws

SAMOA, LYING IN THE HEART of the South Pacific, 2400 km (1500 miles) north of New Zealand, comprises nine volcanic islands. Four are inhabited – Apolima, Manono, Savai'i, and Upolu (where 72% of the population live). Rainforests cloak the mountains; vegetable gardens and coconut plantations thrive around the coasts. A relative boom in the economy in recent years has not lifted Samoa from the ranks of the UN's Least Developed Countries.

CLIMATE

Tropical oceanic

WEATHER CHART FOR APIA

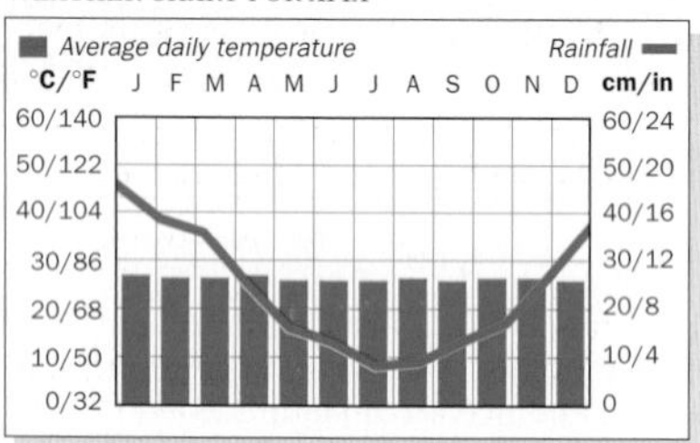
Average daily temperature
Rainfall
°C/°F
J F M A M J J A S O N D
cm/in
60/140
50/122
40/104
30/86
20/68
10/50
0/32
60/24
50/20
40/16
30/12
20/8
10/4
0

The climate is humid and temperatures rarely drop below 25°C (77°F). December to March is the hurricane season.

TRANSPORTATION

Drive on right

Faleolo Apia 160,366 passengers

7 ships 3300 grt

THE TRANSPORTATION NETWORK

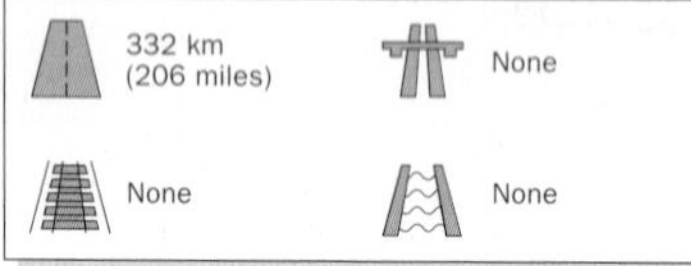
332 km
(206 miles)
None
None
None

Apia port has been improved with Japanese aid. International links are mainly by air. Ferries provide interisland connections.

TOURISM

Visitors : Population 1:1.8

88,000 visitors

Up 4% in 2000

MAIN TOURIST ARRIVALS

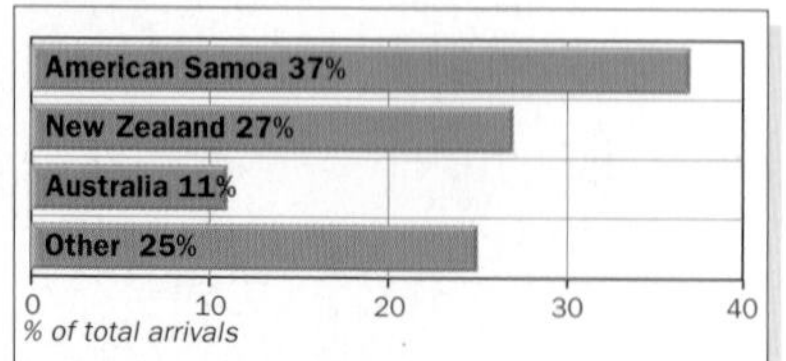
American Samoa 37%
New Zealand 27%
Australia 11%
Other 25%
0
10
20
30
40
% of total arrivals

Tourism is a rapidly growing industry. Small-scale village-based tourism is encouraged. Tourists are attracted by the climate and the easygoing *fa'a Samoa* (Samoan way of life).

PEOPLE

Pop. density medium

Samoan, English

56/km² (146/mi²)

THE URBAN/RURAL POPULATION SPLIT

22% 78%

ETHNIC MAKEUP

Other 1% Euronesian 9% Polynesian 90%

Ethnic Samoans are the world's second-largest Polynesian group, after Maoris. The *fa'a Samoa* – Samoan way of life – is communal and conservative. Extended family groups, in which most people live, own 80% of the land, and cannot sell it. They are each headed by a *matai*, or elected chief, who looks after its social and political interests. Large-scale migration to New Zealand and the US reflects a lack of jobs and the attractions of Western life. Conflict between the *fa'a Samoa* and modern life is strongest among the young, who have a high suicide rate. Euronesians are those of mixed European/Polynesian descent.

SAMOA

Total Land Area : 2860 sq. km (1104 sq. miles)

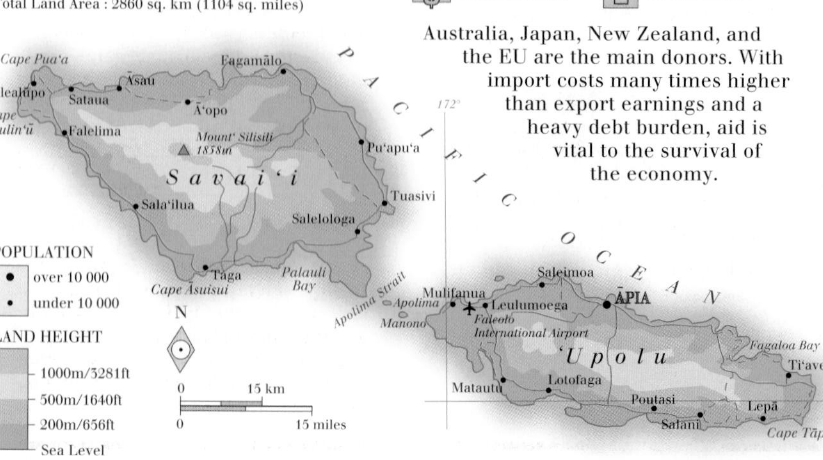
Cape Pua'a
Fagamālo
Falealupo
Āsau
Sataua
A'opo
Cape Mulin'ū
Falelima
Mount Silisili 1858m
Pu'apu'a
S a v a i ' i
Sala'ilua
Tuasivi
Salelologa
Taga
Cape Āsuisui
Palauli Bay
Apolima Strait
Apolima
Manono
Mulifanua
Leulumoega
Faleolo International Airport
Saleimoa
APIA
U p o l u
Fagaloa Bay
Ti'avea
Matautu
Lotofaga
Poutasi
Salani
Lepā
Cape Tāpu
P A C I F I C O C E A N
172°
POPULATION
over 10 000
under 10 000
LAND HEIGHT
1000m/3281ft
500m/1640ft
200m/656ft
Sea Level
N
0 15 km
0 15 miles

POLITICS

Multiparty elections

2001/2006

H.H. Susuga Malietoa Tanumafili II

AT THE LAST ELECTION

Legislative Assembly 49 seats

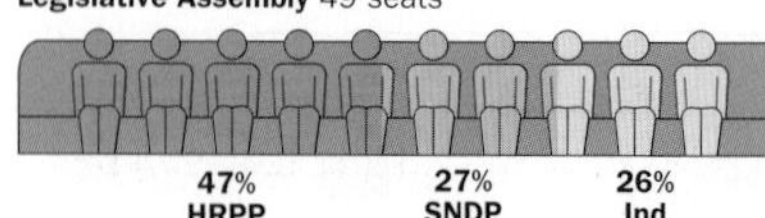
47% HRPP
27% SNDP
26% Ind

HRPP = Human Rights Protection Party **SNDP** = Samoan National Development Party **Ind** = Independents

The conservatism of the *fa'a Samoa* and the Church underpins Samoa's political stability. Allegiance to the two main parties is quite fluid. Until 1990, only the 1800 elected chiefs, or *matai*, could vote for the 47 ethnic Samoan seats in the Assembly; the other two seats are elected by non-Samoans. Universal suffrage was introduced at the 1991 elections. Tofilau Eti Alesana of the HRPP, prime minister twice since 1988, resigned in 1998 amid widespread protest against the government's autocratic style. His successor, Tuilaepa Sailele Malielegaoi, also of the HRPP, was reappointed in 2001.

WORLD AFFAIRS

Joined UN in 1976

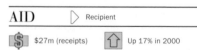

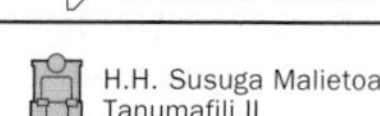
ACP
Comm
IBRD
PC
PIF

Australia is Samoa's main trading partner. New Zealand, the US, Japan, Fiji, American Samoa, and the EU are also important. Relations with China are well established. Samoa has trade links with the Cook Islands, and supports a Polynesian free trade agreement.

AID

Recipient

$27m (receipts)

Up 17% in 2000

Australia, Japan, New Zealand, and the EU are the main donors. With import costs many times higher than export earnings and a heavy debt burden, aid is vital to the survival of the economy.

S

DEFENSE

No compulsory military service

 Samoa has no army and few police

 Not applicable

New Zealand looks after defense under a 1962 treaty. Internal order is mostly maintained by the *matai* (chiefs).

ECONOMICS

Inflation 3.3% p.a. (1990–2000)

 $246m

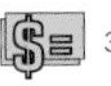 3.339–3.558 tala

SCORE CARD

- ❑ WORLD GNP RANKING........180th
- ❑ GNP PER CAPITA........$1450
- ❑ BALANCE OF PAYMENTS........–$19m
- ❑ INFLATION........1%
- ❑ UNEMPLOYMENT.......Widespread underemployment

STRENGTHS

Light manufacturing expanding, attracting foreign, especially Japanese, firms. Tourism growing rapidly with improved infrastructures. Services expanding rapidly since 1989 launch of offshore banking. Tropical agriculture: taro, coconut cream, cocoa, copra are main exports. Large fishing potential.

WEAKNESSES

Development adversely affected by cyclones. Fluctuating international markets for copra and cocoa. Dependence on aid and expatriate remittances.

EXPORTS

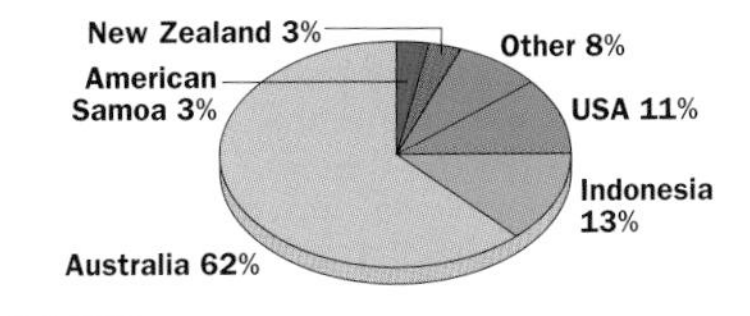
New Zealand 3%
American Samoa 3%
Other 8%
USA 11%
Indonesia 13%
Australia 62%

IMPORTS

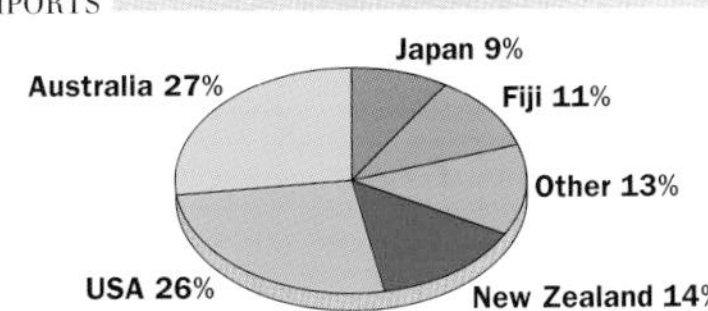
Japan 9%
Australia 27%
Fiji 11%
Other 13%
USA 26%
New Zealand 14%

RESOURCES

Electric power 19,000 kw

 9750 tonnes

 Not an oil producer

 170,000 pigs, 28,000 cattle, 450,000 chickens

 None

With no minerals, Samoa's main resources are its forests and tropical agriculture. The rainforests in lower-lying areas are increasingly exploited for timber. Mahogany and teak plantations are being developed. The volcanic soils, particularly on Upolu, support a wide range of staple and export crops. Two-thirds of the population work in agriculture.

Apia, the capital*, on Upolu, Samoa's second-largest island. It has a central volcanic range of mountains and many rivers.*

ENVIRONMENT

Not available

 None

0.8 tonnes per capita

Strict logging regulations have been introduced to halt irreparable damage to the environment; over 80% of forests have been replaced by plantations. Overhunting and loss of habitat have endangered rare species of fruit bat and pigeon. Samoa is concerned about its marine resources and has taken a firm stance against driftnet fishing.

MEDIA

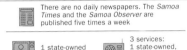

TV ownership medium

 There are no daily newspapers

PUBLISHING AND BROADCAST MEDIA

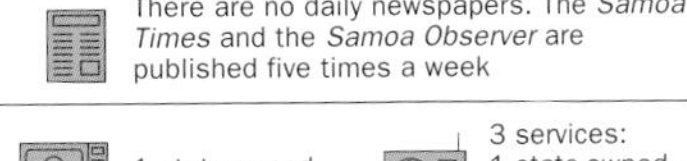

There are no daily newspapers. The *Samoa Times* and the *Samoa Observer* are published five times a week

1 state-owned service

3 services: 1 state-owned, 2 independent

The independent media, notably the *Samoa Observer*, can face strong governmental opposition.

CRIME

Death penalty not used in practice

 176 prisoners

 Crime is rising slowly

Alcohol-related violence is a problem at weekends; otherwise, violent crime is almost unknown. Theft is increasing in urban areas.

EDUCATION

School leaving age: 14

 80%

 810 students

Education is based on the New Zealand system. School attendance is universal. A university was established in Samoa in 1988. There is widespread use of corporal punishment.

HEALTH

No welfare state health benefits

 1 per 2941 people

 Heart and cerebrovascular diseases, pneumonia, suicide

The Samoan preference for being big went well with traditional diets. Diabetes and heart disease are rising as people change to Western-style foods.

CHRONOLOGY

Polynesians settled Samoa in about 1000 BCE. Western rivalry after 1830 led to the 1899 division of the islands into German Western and American Eastern Samoa.

- ❑ **1914** New Zealand occupies Western Samoa.
- ❑ **1962** Becomes first independent Polynesian nation.
- ❑ **1990** Cyclone Ofa leaves 10,000 people homeless.
- ❑ **1991** HRPP retains power in first election under universal adult suffrage.
- ❑ **1996, 2001** HRPP returned to power in elections.
- ❑ **1997** Country's name changed from Western Samoa to Samoa.

SPENDING

GDP/cap. increase

CONSUMPTION AND SPENDING

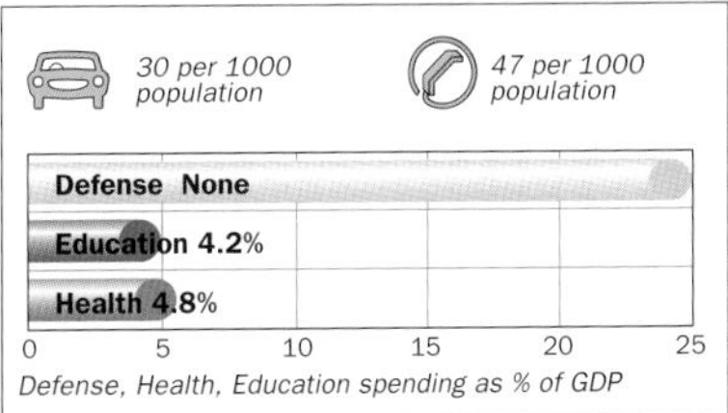
30 per 1000 population
47 per 1000 population
Defense None
Education 4.2%
Health 4.8%
0
5
10
15
20
25
Defense, Health, Education spending as % of GDP

Most of the population depend on subsistence farming and the remittances of relatives for their livelihood. Samoa is classified by the UN as a Least Developed Country. Two-thirds of those with a permanent job work for the government. The prospect of earning higher wages by working in other, wealthier countries in the Pacific region, notably in the tuna canneries of the neighboring US dependency of American Samoa, leads thousands of young Samoans to emigrate every year.

WORLD RANKING

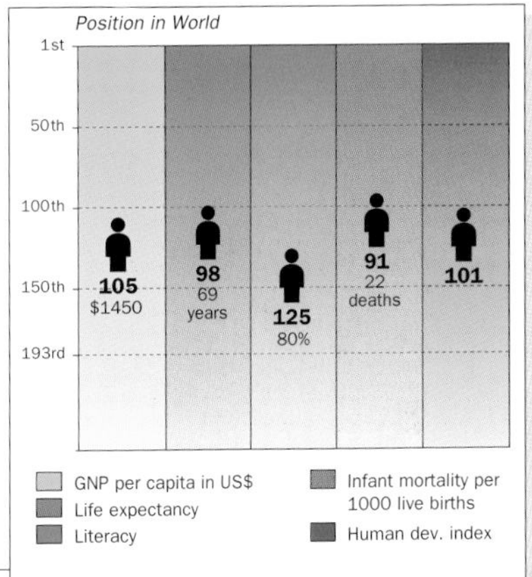
Position in World
1st
50th
100th
150th
193rd
105
$1450
98
69 years
125
80%
91
22 deaths
101
GNP per capita in US$
Life expectancy
Literacy
Infant mortality per 1000 live births
Human dev. index

S

EUROPE

SAN MARINO

Europe

San Marino

Official Name: Republic of San Marino **Capital:** San Marino **Population:** 26,900
Currency: Euro (lira until 2002) **Official Language:** Italian

1631

1631

Sept 3
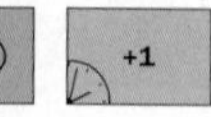
RSM

+1

+378

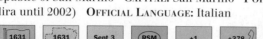

.sm

PERCHED ON THE SLOPES of Mount Titano in the Italian Appennines, tiny San Marino is the world's oldest republic and claims a lineage dating back to the 4th century. The territory is divided into nine castles, or districts. One-third of Sammarinesi live in the northern town of Serravalle. Today San Marino makes its living through agriculture, tourism, and limited industry. Italy effectively controls most of its affairs.

San Marino's second fortress, la Cesta, built in the 13th century, dominates the republic from its pinnacle, 755 m (2477 ft) above sea level.

CLIMATE

Mediterranean

WEATHER CHART FOR SAN MARINO

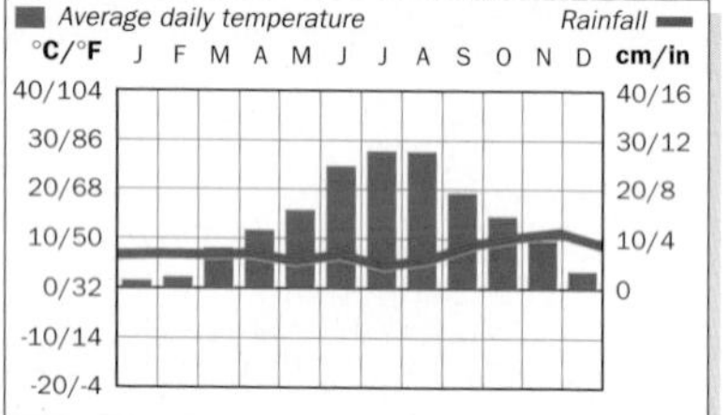
Average daily temperature
Rainfall
°C/°F J F M A M J J A S O N D cm/in
40/104
30/86
20/68
10/50
0/32
-10/14
-20/-4
40/16
30/12
20/8
10/4
0

San Marino's Mediterranean climate is moderated by cool sea breezes and its height above sea level. In summer, temperatures can reach 27°C (81°F), while in winter they fall to 7°C (45°F). There is rarely ever any snow.

TRANSPORTATION

Drive on right

None

Has no fleet

THE TRANSPORTATION NETWORK

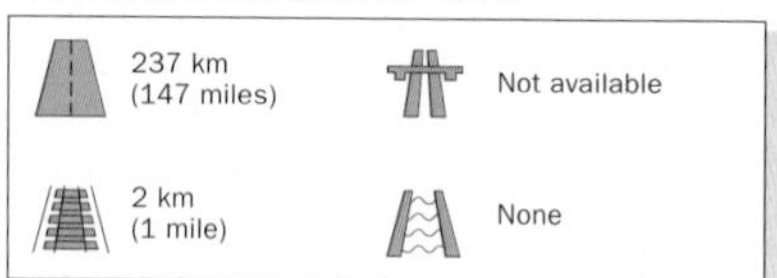
237 km
(147 miles)
Not available
2 km
(1 mile)
None

The 24-km (15-mile) highway to Rimini, which has the nearest airport, is San Marino's most important link. Congestion is a major problem, especially during the annual Mille Miglia car rally. A funicular railroad climbs the east side of Mount Titano. The railroad to Rimini has been closed since World War II.

PEOPLE

Pop. density high

Italian

442/km² (1144/mi²)

THE URBAN/RURAL POPULATION SPLIT

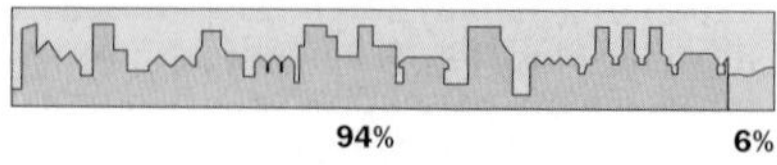

94% 6%

RELIGIOUS PERSUASION

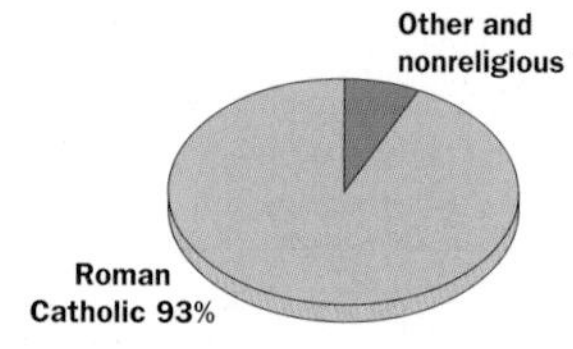
Other and nonreligious 7%
Roman Catholic 93%

Citizenship requires 30 years' residence; it is no longer transmissible by marriage. Women gained the vote in 1960, but could not stand for public office until 1973. Around 20,000 live abroad, mainly in Italy. Some Sammarinesi speak a distinct regional dialect.

TOURISM

Visitors : Population 20:1

532,000 visitors

Little change in 1999

MAIN TOURIST ARRIVALS

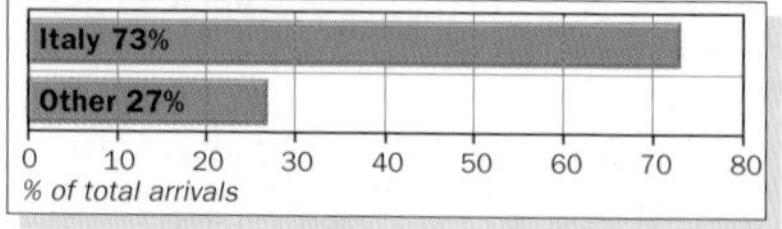
Italy 73%
Other 27%
0 10 20 30 40 50 60 70 80
% of total arrivals

Tourism is the mainstay of San Marino's economy, contributing more than half of government revenue and employment for almost 20% of the workforce. Earnings from tourism are the largest share of GDP. Every year half a million overnight visitors, and a further 2.5 million day visitors, are drawn by its mild climate and contrasting scenery, and come to sample its folklore and museums. The fortresses of Mount Titano – la Rocca, la Cesta, and Montale – built during the Middle Ages, command superb views and are the main attractions, along with the medieval city of San Marino itself. Many visitors to San Marino are day trippers from Italy, although tourism is also boosted by the close proximity of the international airport at Rimini.

The San Marino tourist bureau also attracts thousands of sports enthusiasts to the republic by hosting a series of top international sporting events. In March, both the Rimini–San Marino marathon and the Mille Miglia veteran car meeting are held. May heralds the San Marino Grand Prix, when thousands of Formula One fans descend on the country. June, meanwhile, attracts more motor-racing fans for the World Motocross Championships. A renowned crossbow competition is held to mark San Marino's national day, September 3.

Efforts have been made to attract business meetings and conferences by means of extensive publicity in the Italian media.

***Religious procession.** The official state religion of San Marino is Roman Catholicism, in contrast to Italy, which has no state religion.*

San Marino

Total Land Area : 61 sq. km (23.6 sq. miles)

LAND HEIGHT
500m/1640ft
200m/656ft
above 175m/574ft

POPULATION
under 10 000

Faetano · Dogana · Serravalle · Fiorina · Ventoso · Cailungo · Gualdicciolo · Acquaviva · Domagnano · Borgo Maggiore · SAN MARINO · Monte Titano 739 m · Faetano · Murata · Chiesanuova · Montegiardino · APPENNINES · ITALY · ITALY

0 4 km
0 4 miles

S

POLITICS

Multiparty elections

2001/2006

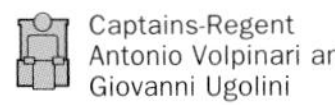

Captains-Regent Antonio Volpinari and Giovanni Ugolini

San Marino is a parliamentary democracy headed by two captains-regent elected every six months. Although the PDCS is the largest single party, it is not guaranteed a place in the ever-shifting coalitions. The most recent government, formed in 2002, is headed by the PSS.

AT THE LAST ELECTION

Great and General Council 60 seats

8% APDS
2% AN
42% PDCS
25% PSS
20% PPDS
3% RC

PDCS = San Marino Christian Democratic Party **PSS** = Socialist Party of San Marino **PPDS** = Progressive Democratic Party **APDS** = Popular Democratic Alliance **RC** = Communist Refoundation **AN** = National Alliance

WORLD AFFAIRS

Joined UN in 1992

CE
OSCE
IBRD
IMF
IWC

Foreign affairs are effectively decided by Italy, on which San Marino is entirely dependent. In 1992, San Marino acquired a seat at the UN.

AID

Neither

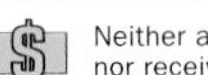

Neither an aid donor nor receiver

Not applicable

San Marino does not receive aid. However, annual subsidies from Italy and free access to the Italian market are essential to the economy.

DEFENSE

No compulsory military service

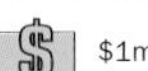

$1m

Little change in 1998

San Marino has a small territorial army and fortification guards. There is no compulsory military service, but males aged 16–55 may be called up in a national emergency.

ECONOMICS

Inflation 5.9% p.a. (1985–1996)

$190m

1.0651–1.1231 euros

SCORE CARD

- ❑ World GNP Ranking 185th
- ❑ GNP per Capita $7830
- ❑ Balance of Payments $11m
- ❑ Inflation 2.2%
- ❑ Unemployment 3%

Strengths

Tourism, providing 60% of government revenue. Light industry, notably mechanical engineering and clothing, with emphasis on sportswear and high-quality prestige lines. Philately.

Weaknesses

Need to import all raw materials.

EXPORTS/IMPORTS

San Marino does not publish independent trade statistics; trade movements are included in the Italian totals.

RESOURCES

Electric power: Included in Italian total

None

Not an oil producer

Small numbers of cattle, pigs, sheep, and horses

None

San Marino has to import all its energy from Italy. It has no exploitable mineral resources now that the stone quarry on Mount Titano has been exhausted.

ENVIRONMENT

Not available

None

Not available

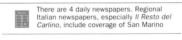

Mount Titano is a unique limestone outcrop in the surrounding Italian plain and so has a very localized ecosystem.

MEDIA

TV ownership high

Daily newspaper circulation 71 per 1000 people

PUBLISHING AND BROADCAST MEDIA

There are 4 daily newspapers. Regional Italian newspapers, especially *Il Resto del Carlino*, include coverage of San Marino

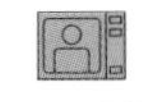

1 independent service

2 independent services

In 1993, a local TV station, San Marino RTV, began broadcasting. Sammarinesi can also receive Italian TV.

CRIME

No death penalty

San Marino does not publish prison figures

Little change from year to year

San Marino has a low crime rate. Justice is mainly administered in Italy. Until mid-1997 homosexuality was illegal.

EDUCATION

School leaving age: 16

99%

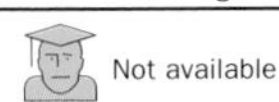

Not available

All teachers are trained abroad, mostly in Italy. Secondary school pupils can go on to Italian universities.

HEALTH

Welfare state health benefits

1 per 375 people

Heart diseases, cancers, accidents

Health care is free and available to all. There is a hospital, but those requiring difficult operations normally go to Rimini for treatment.

CHRONOLOGY

Purportedly founded in the 4th century, the Republic of San Marino became one of many medieval Italian city-states. It was recognized by the papalcy in 1631.

- ❑ **1797** San Marino rejects expansion offered by Napoléon.
- ❑ **1861** Refuses to join unified Italy.
- ❑ **1914–1918** Fights for Italy in World War I.
- ❑ **1940** Supports Axis powers and declares war on the Allies.
- ❑ **1943** Declares neutrality shortly before Italy surrenders.
- ❑ **1960** Women obtain vote.
- ❑ **1978** Coalition of San Marino Communist Party (PCS) and PSS – sole communist-led government in Western Europe.
- ❑ **1986** Financial scandals lead to new PDCS–PCS government.
- ❑ **1988** Joins Council of Europe.
- ❑ **1990** PCS renames itself PPDS.
- ❑ **1992** Joins UN. Collapse of communism in Europe sees PDCS–PPDS alliance replaced by PDCS–PSS coalition.
- ❑ **2002** Adoption of euro.

SPENDING

GDP/cap. increase

CONSUMPTION AND SPENDING

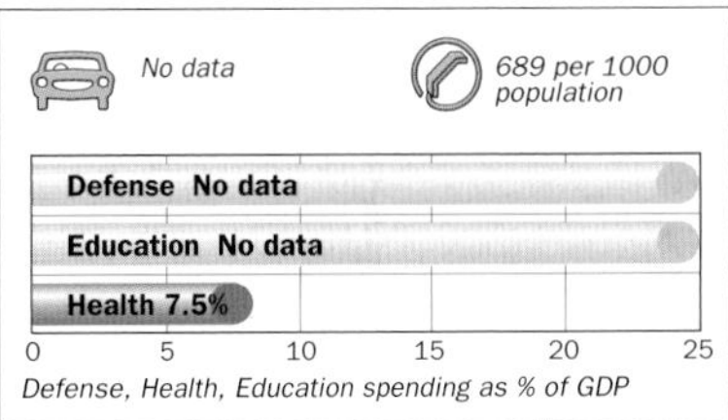
No data
689 per 1000 population
Defense No data
Education No data
Health 7.5%
0
5
10
15
20
25

Defense, Health, Education spending as % of GDP

Living standards are similar to those of northern Italy, while the unemployment rate is well below the Italian average.

WORLD RANKING

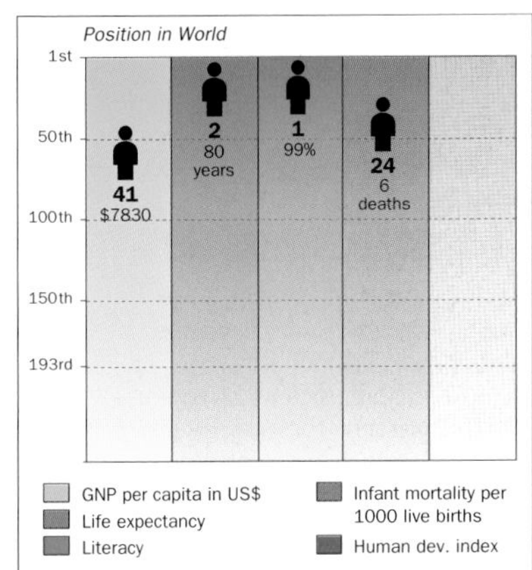
Position in World
1st
50th
100th
150th
193rd
41 $7830
2 80 years
1 99%
24 6 deaths
GNP per capita in US$
Life expectancy
Literacy
Infant mortality per 1000 live births
Human dev. index

S

SÃO TOMÉ & PRÍNCIPE

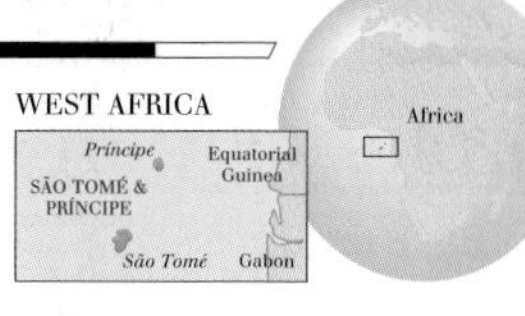
WEST AFRICA
Príncipe
Equatorial Guinea
SÃO TOMÉ & PRÍNCIPE
São Tomé
Gabon
Africa

OFFICIAL NAME: Democratic Republic of São Tomé and Príncipe **CAPITAL:** São Tomé
POPULATION: 159,900 **CURRENCY:** Dobra **OFFICIAL LANGUAGE:** Portuguese

1975
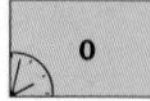
1975

July 12 | STP | 0 | +239 | .st

COMPOSED OF the main islands of São Tomé and Príncipe and surrounding islets, the republic is situated off the western coast of Africa. In 1975, a classic Marxist single-party regime was established following independence from Portugal, but a referendum in 1990 resulted in a 72% vote in favor of democracy. São Tomé's main concerns are relations with Portugal and seeking closer ties with the EU and the US.

CLIMATE

Tropical equatorial

WEATHER CHART FOR SÃO TOMÉ

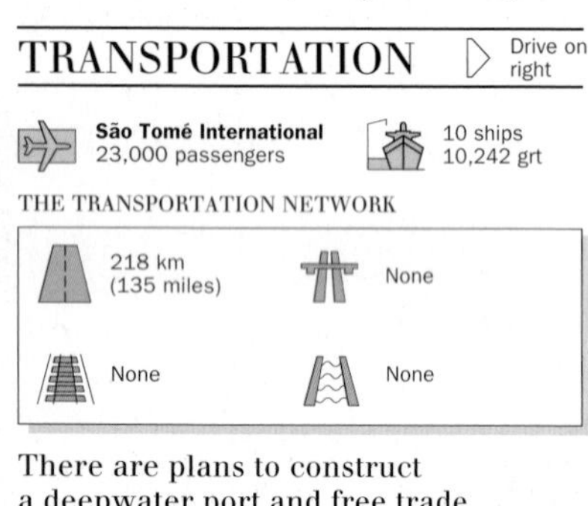
Average daily temperature
Rainfall
°C/°F J F M A M J J A S O N D cm/in
40/104 30/86 20/68 10/50 0/32 -10/14 -20/4
40/16 30/12 20/8 10/4 0

The humid islands straddle the equator. Rainfall averages 500 cm (197 in) on São Tomé and 100 cm (39 in) on Príncipe.

TRANSPORTATION

Drive on right

São Tomé International 23,000 passengers

10 ships 10,242 grt

THE TRANSPORTATION NETWORK

218 km (135 miles)

None

None

None

There are plans to construct a deepwater port and free trade zone at Agulhas Bay on Príncipe.

TOURISM

Visitors : Population 1:32

5000 visitors

Down 17% 1995–1998

MAIN TOURIST ARRIVALS

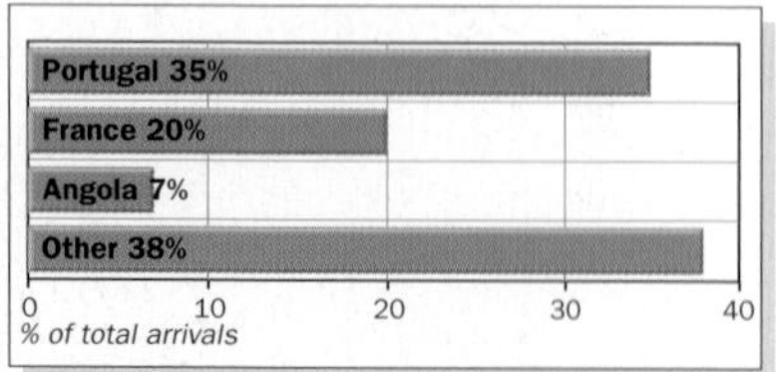
Portugal 35%
France 20%
Angola 7%
Other 38%
0 10 20 30 40
% of total arrivals

Tourism is still small-scale, attracting wealthy Africans and Europeans. Despite recent foreign investment, the islands attract relatively few tourists annually. The first modern hotel opened in 1986.

PEOPLE

Pop. density medium

Portuguese Creole, Portuguese

167/km² (431/mi²)

THE URBAN/RURAL POPULATION SPLIT

47% 53%

ETHNIC MAKEUP

Portuguese and Creole 10%

Black 90%

The population is entirely descended from immigrants, since the islands were uninhabited when the Portuguese arrived in 1470. As the Portuguese settled, they imported Africans as slaves to work the sugar and cocoa plantations. The abolition of slavery in the 19th century, and the departure of 4000 Portuguese at independence, has resulted in a population which is 10% Portuguese and Creole and 90% black African, although Portuguese culture predominates. Blacks run the political parties. Society is well integrated and free of racial tensions. The main conflicts relate to class or differing ideologies. The extended family still offers the best, if not the only, form of social security. Women have a higher status than in most other African states; many have succeeded in attaining prominent positions in the professions.

Lush vegetation on São Tomé. The tropical climate is slightly moderated by the cool Benguela current.

POLITICS

Multiparty elections

2002/2006

President Fradique de Menezes

AT THE LAST ELECTION

National Assembly 55 seats

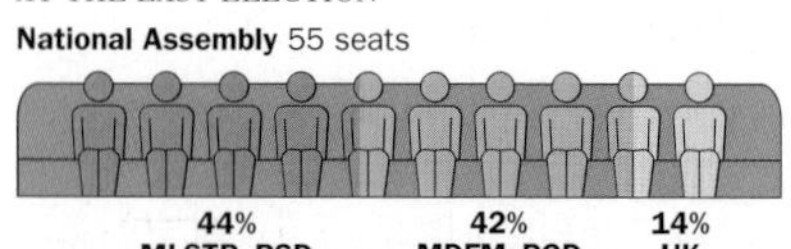

44% MLSTP–PSD | 42% MDFM–PCD | 14% UK

MLSTP–PSD = São Tomé and Príncipe Liberation Movement –Social Democratic Party **MDFM–PCD** = Force for Change Democratic Movement–Democratic Convergence Party **UK** = Ue Kedadji coalition

In 1990, a new multiparty constitution did away with the Marxist single-party state in existence since independence in 1975. The opposition PCD was swept to victory in 1991, and later that year Miguel Trovoada returned from 11 years' exile to be elected as an independent to the presidency.

Elections in 1994 saw the MLSTP, renamed the MLSTP–PSD, return to power. In the 2001 presidential elections, businessman Fradique de Menezes defeated former Marxist president Manuel Pinto da Costa. He was forced to negotiate the formation of a coalition government with the MLSTP–PSD after it won elections in 2002.

Main political concerns are upholding the multiparty system and developing the economy in the light of undersea oil reserves, ownership of which was guaranteed when São Tomé and Nigeria reached agreement in 2000 over their common maritime border.

WORLD AFFAIRS

Joined UN in 1975

CPLP

ACP
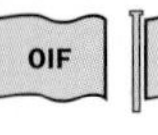
OIF

NAM | AU

São Tomé has achieved rapprochement with Portugal and seeks to maintain links with other former Portuguese colonies, notably Angola. It has always had close ties with Gabon and, while not dropping its ex-communist links, seeks closer relations with other central African states, France, and the US.

AID

Recipient

$35m (receipts)

Up 25% in 2000

São Tomé has one of the highest aid-to-population ratios in Africa. Joining the Lomé Convention in the 1970s has meant that São Tomé has found new sources of aid fairly easily since the demise of communism worldwide. Portugal, France, and the EU are the main donors.

S

DEFENSE

No compulsory military service

 $400,000

 Little change

Since independence, the armed forces have figured prominently in national life. They have put down several attempted coups, notably in 1978, after which 2000 Angolan troops plus Soviet and Cuban advisers were invited in, and in 1988. In 1995, a group of army officers seized temporary control of the country. The national armed forces are believed to number 2000. With the collapse of the Eastern bloc, São Tomé now receives military assistance from the West.

RESOURCES

Electric power 6000 kw

3756 tonnes

 Reserves currently unexploited

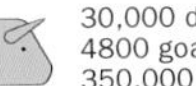 30,000 ducks, 4800 goats, 350,000 chickens

 None

An offshore oil exploration agreement with Nigeria was signed in 2001. There are no mineral resources on the islands. São Tomé is very fertile; cocoa estates are finally back to pre-1975 productivity, and diversification of crops is now a priority. Príncipe has better ports, but its wild scenery makes it more suitable for tourism than farming.

CHRONOLOGY

The entire preindependence history of the islands was as a Portuguese colony exploited by plantation owners.

- **1972–1973** Strikes by plantation workers.
- **1975** Independence as Marxist state. Plantations nationalized.
- **1978** Abortive coup.
- **1990** New democratic constitution.
- **1991–2000** Miguel Trovoada president for two terms.
- **1995** Príncipe granted autonomy.
- **2001** De Menezes wins presidency.

ECONOMICS

Inflation 51% p.a. (1990–2000)

  $43m

 2390–8937 dobras

SCORE CARD

- World GNP Ranking 191st
- GNP per Capita $290
- Balance of Payments .... –$9m
- Inflation 5%
- Unemployment 50%

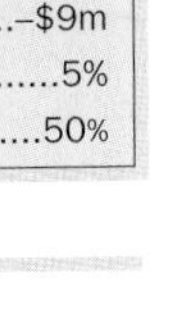

EXPORTS

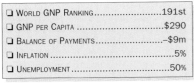

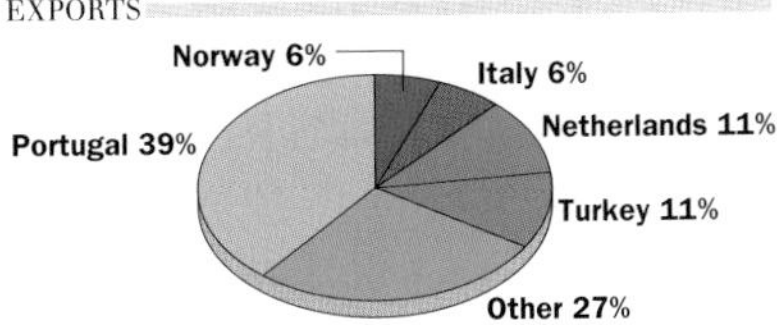
Norway 6%
Italy 6%
Portugal 39%
Netherlands 11%
Turkey 11%
Other 27%

IMPORTS

Belgium–Luxembourg 5%
France 5%
Germany 8%
Portugal 48%
UK 8%
Other 26%

STRENGTHS

Legacy of Portuguese-built infrastructure. Potential for development of fisheries, agriculture, tourism, and oil. Able to attract substantial aid.

WEAKNESSES

Cocoa accounts for 90% of export earnings. Skillful diplomacy has attracted high levels of aid, but mismanagement of these funds has resulted in severe debt. Weak currency.

EDUCATION

School leaving age: 14

75%

 Not available

Education is compulsory for 7–14-year-olds. All staff at the one technical and three secondary schools are foreigners.

HEALTH

No welfare state health benefits

1 per 2128 people

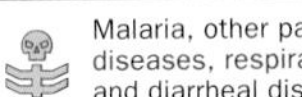 Malaria, other parasitic diseases, respiratory and diarrheal diseases

Although health care is not free, São Tomé has a better system of basic care than other African countries.

SPENDING

GDP/cap. decrease

CONSUMPTION AND SPENDING

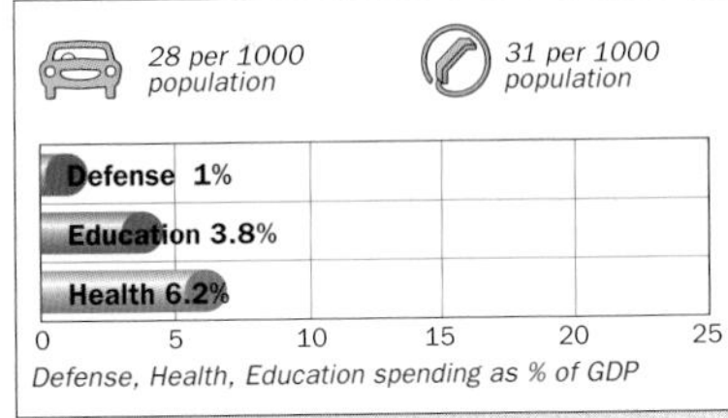
28 per 1000 population
31 per 1000 population
Defense 1%
Education 3.8%
Health 6.2%
0
5
10
15
20
25
Defense, Health, Education spending as % of GDP

Wealth disparities are not conspicuous. There is a growing business class. Cocoa workers form the country's poorest group.

SÃO TOMÉ & PRÍNCIPE

Total Land Area : 1001 sq. km (386 sq. miles)

POPULATION

- over 10 000
- under 10 000

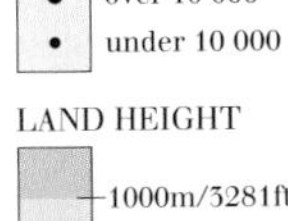

LAND HEIGHT

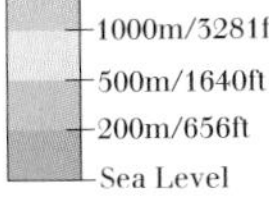
1000m/3281ft
500m/1640ft
200m/656ft
Sea Level

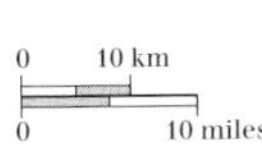
0
10 km
0
10 miles

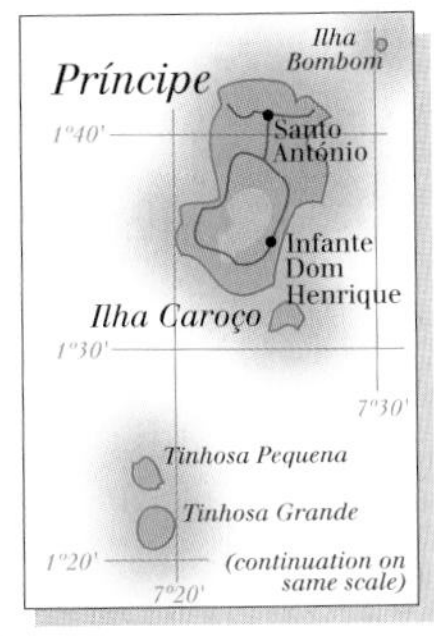
Príncipe
Ilha Bombom
Santo António
Infante Dom Henrique
Ilha Caroço
Tinhosa Pequena
Tinhosa Grande
(continuation on same scale)

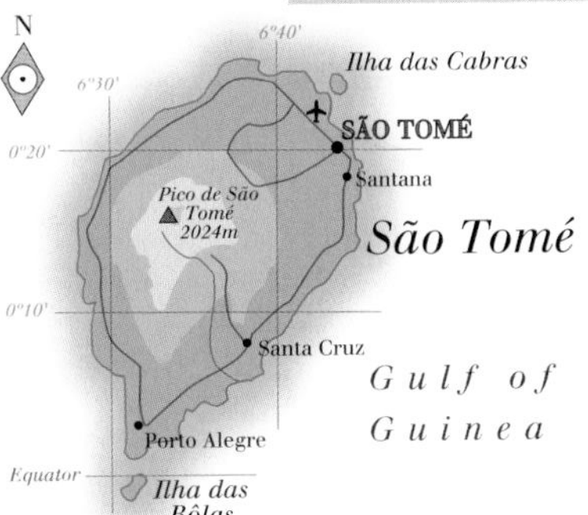
N
Ilha das Cabras
SÃO TOMÉ
Santana
Pico de São Tomé 2024m
São Tomé
Santa Cruz
Gulf of Guinea
Porto Alegre
Equator
Ilha das Rôlas

ENVIRONMENT

Not available

 None

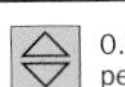 0.5 tonnes per capita

Fish conservation, deforestation for fuelwood, and potential tourism expansion are the major issues.

MEDIA

TV ownership medium

There are no daily newspapers

PUBLISHING AND BROADCAST MEDIA

There are no daily newspapers. *Diário da República* and *Notícias* are published weekly by the government

1 state-controlled service

1 state-controlled service

Freedom of expression is respected. The state controls radio and TV stations. Radio ownership is high for Africa.

CRIME

No death penalty

 99 prisoners

 Crime is decreasing

Crime levels are fairly low owing to the tightly knit nature of the community. Urban robberies are a problem.

WORLD RANKING

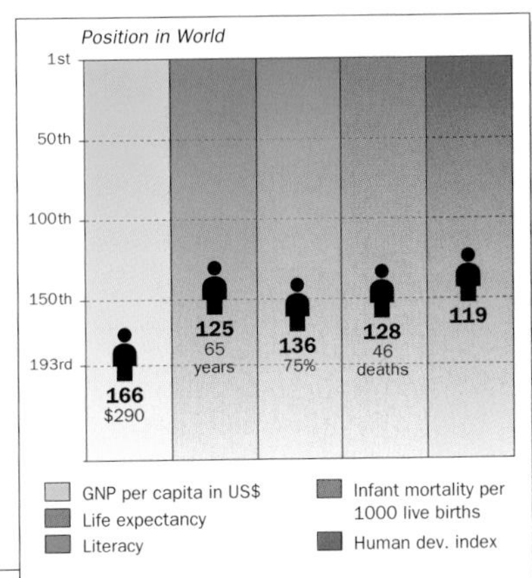
Position in World
1st
50th
100th
150th
193rd
166
$290
125
65 years
136
75%
128
46 deaths
119
GNP per capita in US$
Life expectancy
Literacy
Infant mortality per 1000 live births
Human dev. index

S

SAUDI ARABIA

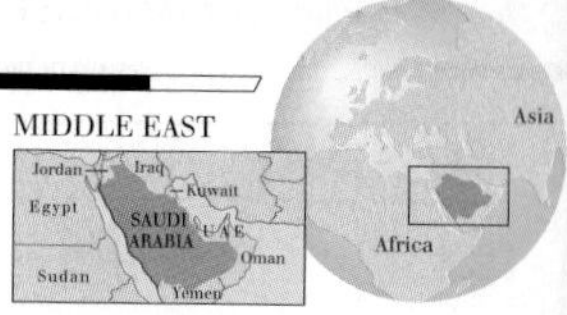
MIDDLE EAST
Jordan
Iraq
Kuwait
Egypt
SAUDI ARABIA
U.A.E.
Oman
Sudan
Yemen
Asia
Africa

OFFICIAL NAME: Kingdom of Saudi Arabia **CAPITAL:** Riyadh; Jiddah (administrative)
POPULATION: 21 million **CURRENCY:** Saudi riyal **OFFICIAL LANGUAGE:** Arabic

1932

1932

Sept 23
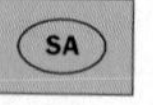
SA
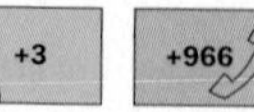
+3

+966

.sa

OCCUPYING MOST OF THE Arabian peninsula, Saudi Arabia covers an area as large as western Europe. Over 95% of its land is desert, with the most arid part, known as the Empty Quarter or Rub al Khali, being in the southeast. Saudi Arabia has the world's largest oil and gas reserves. It includes Islam's holiest cities, Medina and Mecca, visited each year by two million Muslims performing the pilgrimage known as the *haj*. The al-Sa'ud family have been Saudi Arabia's absolutist rulers since 1932. In theory, Islamic *sharia* underpins the constitution.

CLIMATE

Hot desert

WEATHER CHART FOR RIYADH

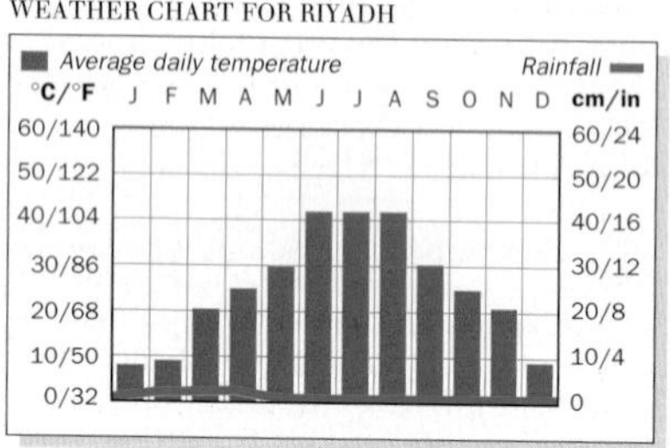
Average daily temperature
Rainfall
°C/°F
J F M A M J J A S O N D
cm/in
60/140
50/122
40/104
30/86
20/68
10/50
0/32
60/24
50/20
40/16
30/12
20/8
10/4
0

The kingdom's only reliable rainfall is in the southern Asir province, making agriculture viable there. The central plateau requires deep artesian wells to water crops. Inland, summer temperatures often soar above 48°C (118°F), but in winter, especially in the northwest, they may fall to freezing point.

TRANSPORTATION

Drive on right

King Abd al-Aziz International, Jiddah 10.6m passengers

279 ships 1.3m grt

THE TRANSPORTATION NETWORK

44,104 km (27,405 miles)

Trans-Arabian Highway

1392 km (865 miles)

None

A modern transportation infrastructure links the main population centers to the Gulf states, Jordan, and Egypt. Saudi Arabian Airlines, founded in 1945, is one of the largest airlines in the world.

TOURISM

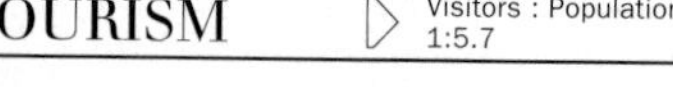
Visitors : Population 1:5.7

3.7m visitors

Up 3% in 1998

MAIN TOURIST ARRIVALS

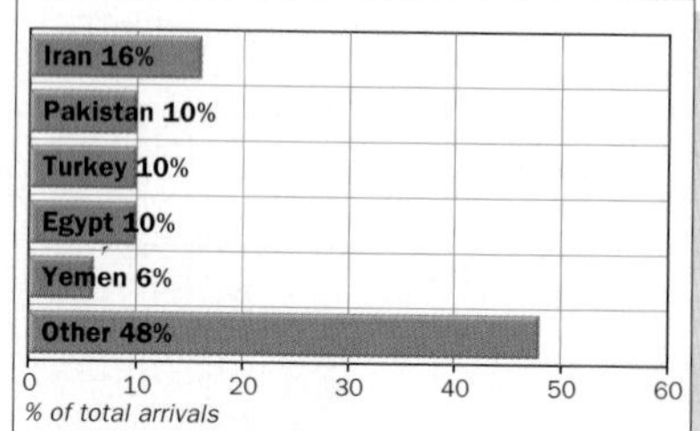
Iran 16%
Pakistan 10%
Turkey 10%
Egypt 10%
Yemen 6%
Other 48%
0
10
20
30
40
50
60
% of total arrivals

Foreign tourism is discouraged. Until a limited relaxation in 2000, only Muslim pilgrims, business people, and foreign workers were permitted entry. Non-Muslims are banned from the holy cities of Mecca and Medina. Although strict quotas have been imposed to avoid overcrowding, stampedes of *haj* pilgrims in 1990, 1997, and 2001 killed or injured thousands. Many choose the port of Jiddah as a base from which to begin the pilgrimage. Also popular is the *umra*, or little pilgrimage, since it can be made at any time of year. An estimated $2.5 billion has been spent on improving *haj* facilities in recent years. Jizan on the Red Sea offers superb scuba diving. The Hejaz railroad and the Nabatean ruins at Medain Salih are of archaeological interest. To escape the summer heat, the government relocates to mountainous Taif, used as a resort by the Saudis.

SAUDI ARABIA

Total Land Area : 1 960 582 sq. km (756 981 sq. miles)

POPULATION

- over 1 000 000
- over 500 000
- over 100 000
- over 50 000
- over 10 000
- under 10 000

LAND HEIGHT

- 3000m/9843ft
- 2000m/6562ft
- 1000m/3281ft
- 500m/1640ft
- Sea Level

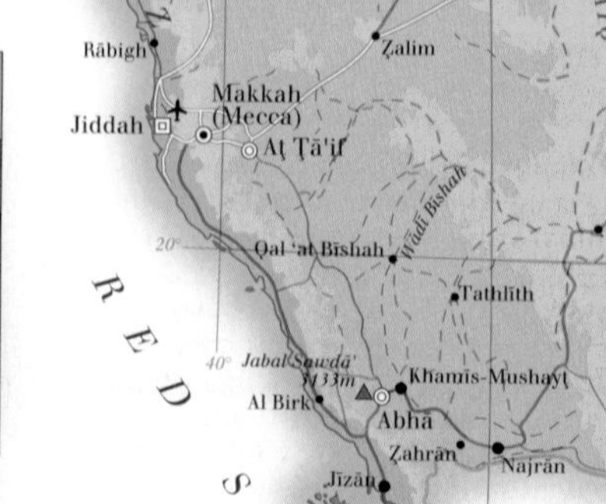
JORDAN
IRAQ
KUWAIT
EGYPT
The Gulf
BAHRAIN
QATAR
UNITED ARAB EMIRATES
OMAN
YEMEN
RED SEA
RUB' AL KHĀLĪ
AN NAFŪD
AD DAHNĀ'
NAJD
HIJAZ
JABAL TUWAYQ
Ţurayf
Al Jawf
Sakākah
Rafḩā'
Tabūk
Ḩafar al Bāţin
Taymā'
Ḩā'il
Al Jubayl
Ras Tannūrah
Ad Dammām
Az Zahran
Al Khubar
Al Wajh
Al 'Ulā
Buraydah
Az Zilfī
Al Majma'ah
'Unayzah
Shaqrā'
Al Mubarraz
Al Hufūf
RIYADH
Al Madīnah (Medina)
Yanbu' al Bahr
Ḩaraḑ
Rābigh
Z̧alim
Jiddah
Makkah (Mecca)
Aţ Ţā'if
Laylá
As Sulayyil
Qal'at Bīshah
Tathlīth
Jabal Sawdā' 3133m
Al Birk
Khamīs-Mushayţ
Abhā
Z̧ahrān
Najrān
Jīzān
Wuday 'ah
Wādī al Bāţin
Wādī ar Rimah
Wādī al Hamḑ
Wādī Bīshah
0
200 km
200 miles
N

Network of modern road junctions spread out across the landscape near Mecca.

PEOPLE

Pop. density low

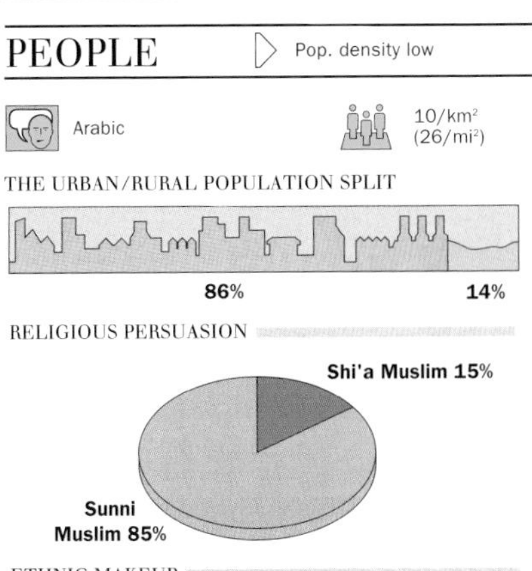
Arabic
10/km²
(26/mi²)
THE URBAN/RURAL POPULATION SPLIT
86%
14%
RELIGIOUS PERSUASION
Shi'a Muslim 15%
Sunni Muslim 85%

ETHNIC MAKEUP

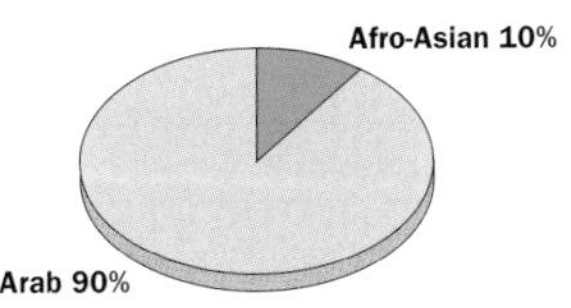
Afro-Asian 10%
Arab 90%

The Saudis, who take their name from the ruling al-Sa'ud family, were united by conquest between 1902 and 1932 by King Abd al-Aziz al-Sa'ud. The vast majority of Saudis are Sunni Muslims who follow the *wahhabi* (puritan) interpretation of Islam and embrace *sharia* (Islamic law). The politically dominant Nejdi tribes from the central plateau around Riyadh are Bedouin in origin. The Hejazi tribes, from the south and west, have a more cosmopolitan, mercantile background, but are largely displaced from politics. In the eastern Al-Hasa province there is a Shi'a minority of some 300,000, many of whom are employed in the oil fields.

Women have to wear the veil, cannot hold a driving license, and have no role in public life. In 2000, however, Saudi Arabia decided to sign the UN convention on women's rights – provided it did not contradict *sharia* – and issued women with identity cards from 2001.

POPULATION AGE BREAKDOWN

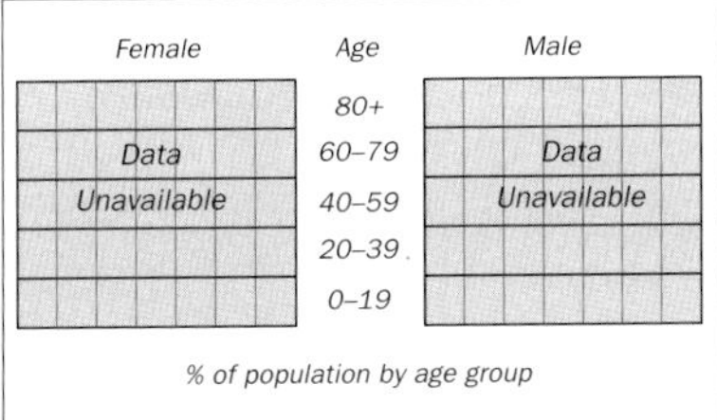
Female
Age
Male
80+
60–79
40–59
20–39
0–19
Data Unavailable
Data Unavailable
% of population by age group

POLITICS

No legislative elections

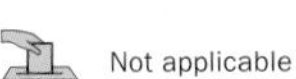
Not applicable

H.M. King Fahd ibn Abd al-Aziz

LEGISLATIVE OR ADVISORY BODIES

Consultative Council 120 seats

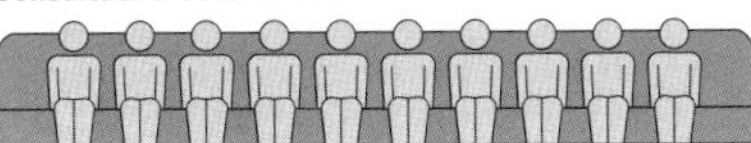

Saudi Arabia is an absolute monarchy. The king rules with the assistance of an appointed Council of Ministers and the Consultative Council

Saudi Arabia is an absolute monarchy. Since 1993 a Consultative Council (Majlis ash-Shoura) has been appointed by the king.

PROFILE

The royal family, the House of Sa'ud, rules by carefully manipulating appointments in all sectors of government. Frequent changes of personnel within the armed forces ensure that officers do not build personal followings. All influential cabinet portfolios, apart from those of oil and religious affairs, are held by members of the royal family.

Absolutist rule means that domestic politics are virtually nonexistent. The regime retains feudal elements: weekly *majlis*, or councils, are held where citizens can present petitions or grievances to leading members of the royal family. Large cash sums are often dispensed at these meetings.

The legitimacy of the regime is built on its adherence to Islamic values, and the backing of the *ulema* (scholars). It is the emphasis on Islam that most colors Saudi life. The 5000-strong *mutawa* (religious police) enforce the five-times-a-day call to prayer, when businesses must close. During Ramadan the *mutawa* are especially active.

MAIN POLITICAL ISSUES

Questioning the ruling family

Following the 1991 Gulf War, a civil rights campaign emerged to challenge the authority of the ruling family, demanding closer adherence to Islamic values. The movement objected to the presence of US troops on Saudi territory and the consequent exposure to corrupt Western culture. The al-Sa'uds moved swiftly to quash the protest but exiled opponents have continued their activities, using faxes and e-mail. The most vociferous denunciations of the royal family have come from Osama bin Laden, formerly part of the inner circle, operating from bases abroad.

The succession issue

The question of succession and the possibility of a future power struggle, rooted in rivalries endemic to the House of Sa'ud, emerged in 1996, when King Fahd suffered a stroke. The management of day-to-day affairs passed briefly to his half-brother, Crown Prince Abdullah, who remains in effective, if no longer formal, control.

WORLD AFFAIRS

Joined UN in 1945

AL

Damasc

GCC

OIC

OPEC

Saudi Arabia's strategic importance is derived largely from its oil reserves and worldwide investments. Relations with the US are particularly close, but frayed when Saudi Arabia refused to back war in Afghanistan in 2001. After Iraq's invasion of Kuwait in 1990, the Saudis had taken a leading role in consolidating the Arab coalition against Iraq. Saudi Arabia sheltered the Kuwaiti royal family, provided military bases to the Western allies, and supplied more troops than any other Arab country. The continued presence of foreign forces then provoked hostility from Islamic militants. Before this gained world attention through statements by Osama bin Laden at the time of the September 2001 terrorist attacks on the US, a bomb attack at a US military complex near Az Zahran in 1996 had killed 19 US personnel.

A pact signed with Yemen in June 2000 ended a simmering border dispute. In early 2002 Crown Prince Abdullah's new peace plan for Israel/Palestine won praise from Arabs and the West – and hinted at a major change in Saudi foreign policy.

AID

Donor

$295m (donations)

Up 59% in 2000

Generous loans and grants from the Saudi Fund for Development are made to other Arab and developing countries, mainly for infrastructure projects. Saudi Arabia promotes Islam through charitable foundations, especially in Africa, Asia, and the former Soviet Union. The royal purse also supports scientific and medical research. Since the liberation of Kuwait in 1991, Saudi Arabia has given large sums to countries that supported the US-led alliance, notably Egypt, Syria, Morocco, and Turkey. In addition, the Saudi government substantially reimbursed the US and the UK for the cost of their expeditionary forces, as well as favoring companies from the allied countries for reconstruction contracts.

***King Fahd ibn Abd al-Aziz** acceded to the Saudi throne in 1982.*

***Crown Prince Abdullah**, commander of the National Guard.*

S

CHRONOLOGY

The unification of Saudi Arabia under King Abd al-Aziz (ibn Sa'ud) was achieved in 1932. The kingdom remains the only country in the world which is named after its royal family.

- ❑ **1937** Oil reserves discovered near Riyadh.
- ❑ **1939** Ceremonial start of oil production at Az Zahran.
- ❑ **1953** King Sa'ud succeeds on the death of his father Abd al-Aziz.
- ❑ **1964** King Sa'ud abdicates in favor of his brother Faisal.
- ❑ **1973** Saudi Arabia imposes oil embargo on Western supporters of Israel.
- ❑ **1975** King Faisal assassinated by a deranged nephew; succeeded by his brother Khalid.
- ❑ **1979** Muslim fundamentalists led by Juhaiman ibn Seif al-Otaibi seize Grand Mosque in Mecca, proclaim a *mahdi* (savior) on first day of Islamic year 1400.
- ❑ **1981** Formation of GCC, with its secretariat in Riyadh.
- ❑ **1982** King Fahd succeeds on the death of his brother King Khalid. Promises to create consultative assembly.
- ❑ **1986** Opening of King Fahd Causeway to Bahrain. Shaikh Yamani sacked as oil minister.
- ❑ **1987** Diplomatic relations with Iran deteriorate after 402 people die in riots involving Islamic fundamentalists at Mecca during the *haj* (pilgrimage).
- ❑ **1989** Saudi Arabia signs nonaggression pact with Iraq. Saudi Arabia brokers political settlement to Lebanese civil war.
- ❑ **1990** Kuwaiti royal family seeks sanctuary in Taif after Iraqi invasion. Many allegedly pro-Iraqi Jordanians and Yemenis expelled.
- ❑ **1990–1991** US, UK, French, Egyptian, and Syrian forces assemble in Saudi Arabia for Operation Desert Storm. Public executions are halted.
- ❑ **1991** Iraqis seize border town of Al Khafji, but are repulsed by Saudi, US, and Qatari forces.
- ❑ **1993** King Fahd appoints Consultative Council (Majlis ash-Shoura).
- ❑ **1996** King Fahd briefly relinquishes control to Crown Prince Abdullah. Bomb attack at US military complex in Az Zahran kills 19 US citizens.
- ❑ **1997, 2001** Consultative Council expanded, first to 90 then to 120 members.
- ❑ **2002** Crown Prince Abdullah unveils Middle East peace plan; endorsed by AL summit in Beirut.

S

DEFENSE

No compulsory military service

$18.3bn

Down 16% in 2000

Saudi Arabia's substantial military contribution to the 1991 Gulf War, at a cost of $55 billion, enhanced its image as a major regional power. Military equipment is purchased mostly from the US, the UK, and France. Weapons systems are advanced and include Patriot missiles and AWACS early warning radar. However, skilled foreign personnel operate many of these: 1000 US air force troops are employed to keep AWACS flying.

The air force is the elite branch of the military. It had one brief period of politicization in 1969 when officers attempted a coup. The paramilitary National Guard is drawn from tribal supporters of the al-Sa'ud regime. Its commander-in-chief is the crown prince rather than the defense minister.

SAUDI ARABIAN ARMED FORCES

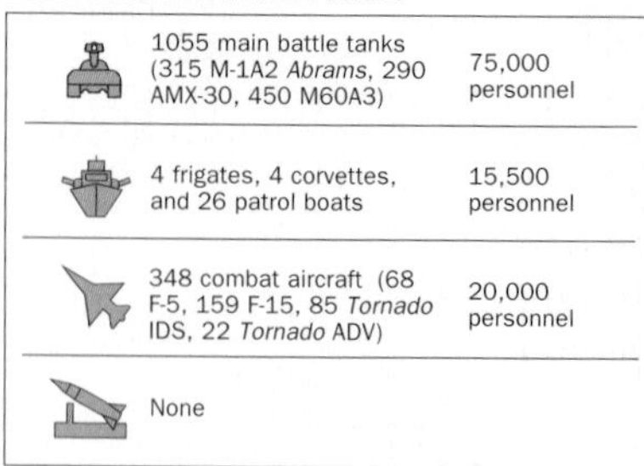
1055 main battle tanks (315 M-1A2 *Abrams*, 290 AMX-30, 450 M60A3)
75,000 personnel
4 frigates, 4 corvettes, and 26 patrol boats
15,500 personnel
348 combat aircraft (68 F-5, 159 F-15, 85 *Tornado* IDS, 22 *Tornado* ADV)
20,000 personnel
None

ECONOMICS

Inflation 2.2% p.a (1990–2000)

$150bn

3.7505–3.7504 Saudi riyals

SCORE CARD

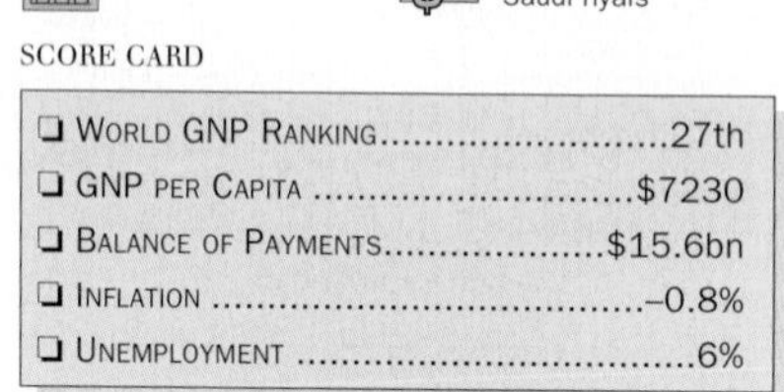

❑ World GNP Ranking	27th
❑ GNP per Capita	$7230
❑ Balance of Payments	$15.6bn
❑ Inflation	–0.8%
❑ Unemployment	6%

EXPORTS

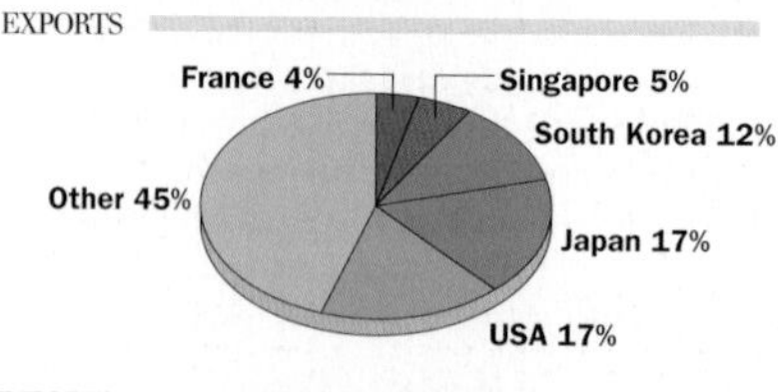
France 4%
Singapore 5%
South Korea 12%
Other 45%
Japan 17%
USA 17%

IMPORTS

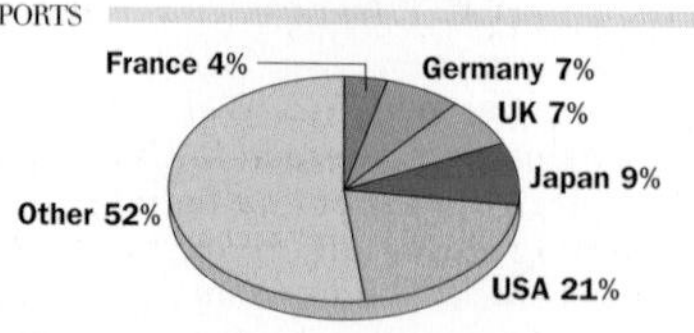
France 4%
Germany 7%
UK 7%
Japan 9%
Other 52%
USA 21%

Strengths

Vast oil and gas reserves. Soaring world oil prices in 2000 signified recovery from 1986 collapse. World-class associated industries. Accumulated surpluses and steady current income. Large income from two million pilgrims to Mecca annually.

Weaknesses

Lack of indigenous skilled workers. Food production requires heavy subsidy. Most consumer items and industrial raw materials imported. Up to 20% youth unemployment. Large national debt. Royal family technically owns most of national wealth.

Profile

Since the 1970s, great efforts have been made to reduce dependence on oil exports and to provide employment for young Saudis. By late 2000, Saudi Arabia had steered OPEC toward controlling oil prices to avoid recession in industrialized countries or spurring a drive to develop alternative energy. Since April 2000, approved foreigners have been allowed complete ownership of Saudi businesses and rights to property. Large sums have been spent on creating a US-standard infrastructure to provide the basis for a manufacturing economy. It remains, however, dependent on foreign workers.

ECONOMIC PERFORMANCE INDICATOR

Consumer Price Index
GDP
160
140
120
100
80
Consumer price index 1995=100
GDP 1996=100
1996
1997
1998
1999
2000

SAUDI ARABIA : MAJOR BUSINESSES

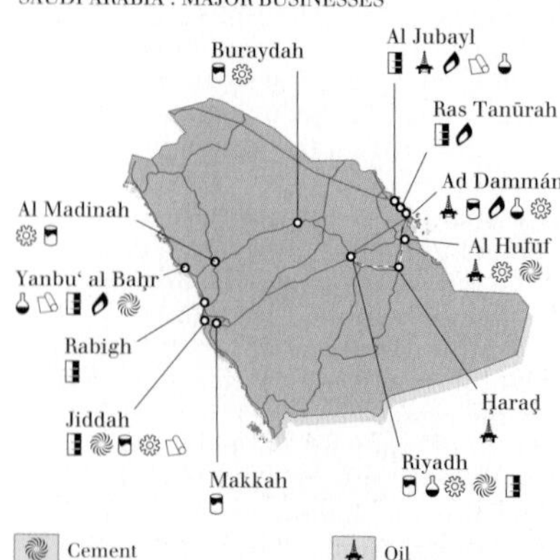
Buraydah
Al Jubayl
Ras Tanūrah
Ad Dammān
Al Madinah
Al Hufūf
Yanbu' al Baḩr
Rabigh
Jiddah
Ḩaraḑ
Riyadh
Makkah
Cement
Steel
Food processing
Light engineering
Petrochemicals/fertilizers
Oil
Oil refining
Gas refining
0
400 km
0
400 miles

RESOURCES

Electric power 22.6m kw

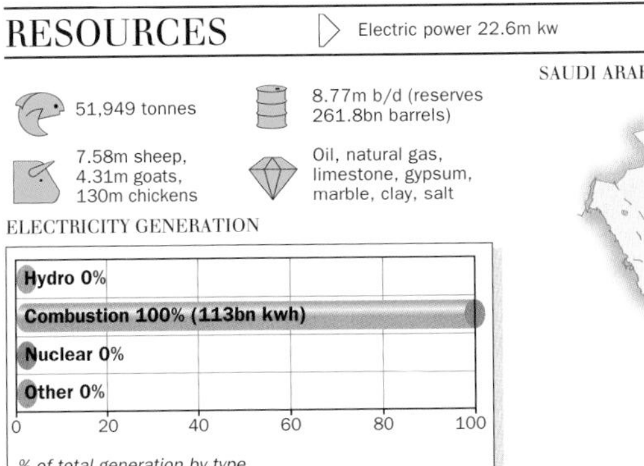
51,949 tonnes
8.77m b/d (reserves 261.8bn barrels)
7.58m sheep, 4.31m goats, 130m chickens
Oil, natural gas, limestone, gypsum, marble, clay, salt
ELECTRICITY GENERATION
Hydro 0%
Combustion 100% (113bn kwh)
Nuclear 0%
Other 0%
0
20
40
60
80
100
% of total generation by type

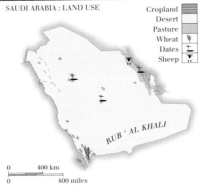

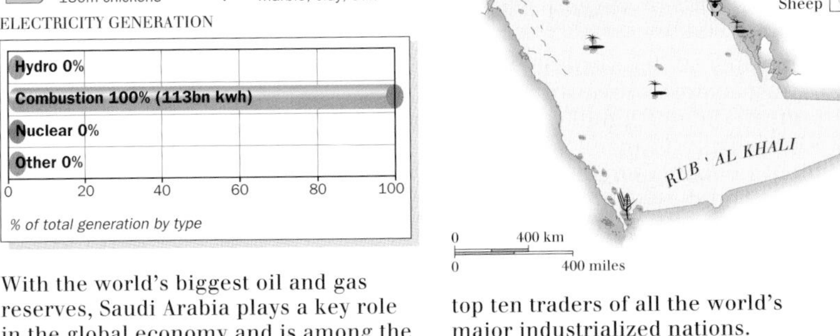
SAUDI ARABIA : LAND USE
Cropland
Desert
Pasture
Wheat
Dates
Sheep
RUB ' AL KHALI
0
400 km
0
400 miles

With the world's biggest oil and gas reserves, Saudi Arabia plays a key role in the global economy and is among the top ten traders of all the world's major industrialized nations.

ENVIRONMENT

Sustainability rank: 138th

2%

14.4 tonnes per capita

ENVIRONMENTAL TREATIES

No	Yes	Yes
Yes	Yes	No

Pollution in the Gulf and Red Sea has threatened some wildlife and their habitats, as have hunters using high-velocity rifles and off-road vehicles. The government has taken steps to confine manufacturing to industrial estates. Environmental legislation is, nevertheless, poorly developed, although planning controls apply in the major cities.

MEDIA

TV ownership high

Daily newspaper circulation 59 per 1000 people

PUBLISHING AND BROADCAST MEDIA

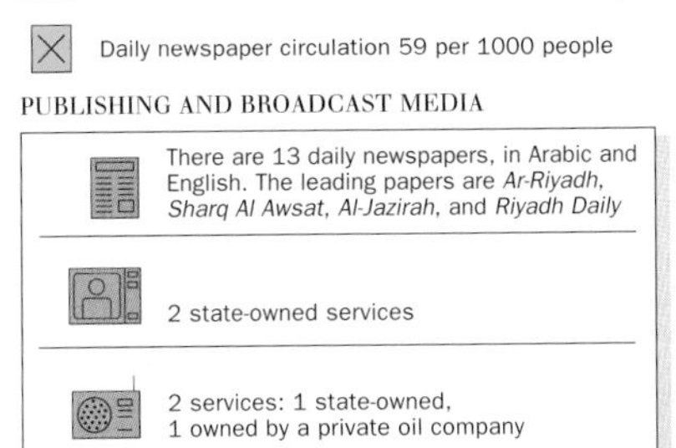
There are 13 daily newspapers, in Arabic and English. The leading papers are *Ar-Riyadh*, *Sharq Al Awsat*, *Al-Jazirah*, and *Riyadh Daily*
2 state-owned services
2 services: 1 state-owned, 1 owned by a private oil company

The government imposes total press censorship and insists on strict morality. In 1994, private citizens were banned from owning satellite dishes, but the authorities turn a blind eye to their use. No allowance is made for Arab satellite broadcasts, which have been criticized for covering anti-Islamic views. The international *Sharq Al Awsat* is a leading Arabic daily. In 2001 the government announced strict rules regarding state and religion on the Internet.

CRIME

Death penalty in use

18,000 prisoners

Down 13% 1996–1998

CRIME RATES

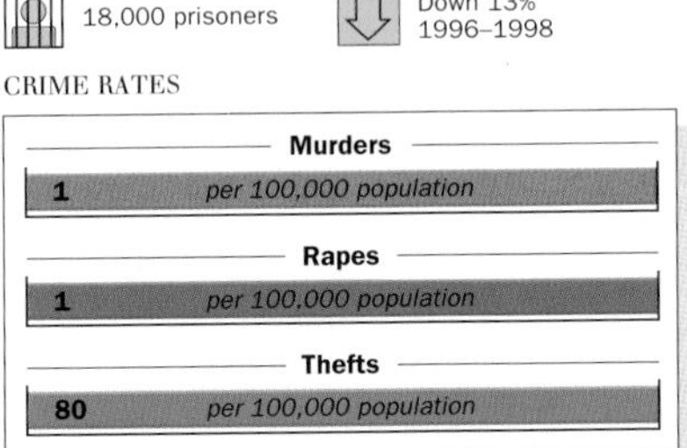
Murders
1 per 100,000 population
Rapes
1 per 100,000 population
Thefts
80 per 100,000 population

Strict Islamic punishments – stoning, amputation, and beheading – are enforced. Criticism for human rights abuses has increased, with an Amnesty International campaign in 2000.

EDUCATION

Schooling is free but not compulsory

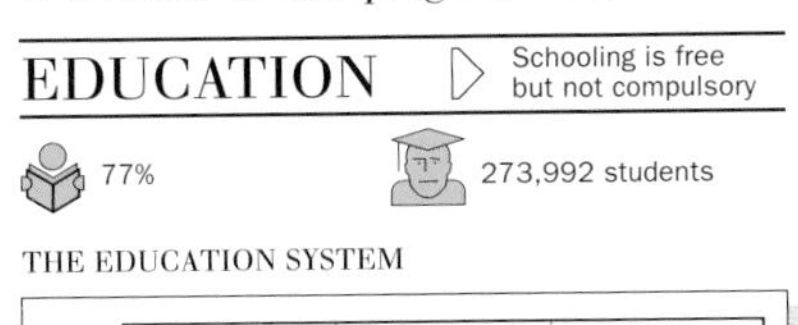
77%
273,992 students

THE EDUCATION SYSTEM

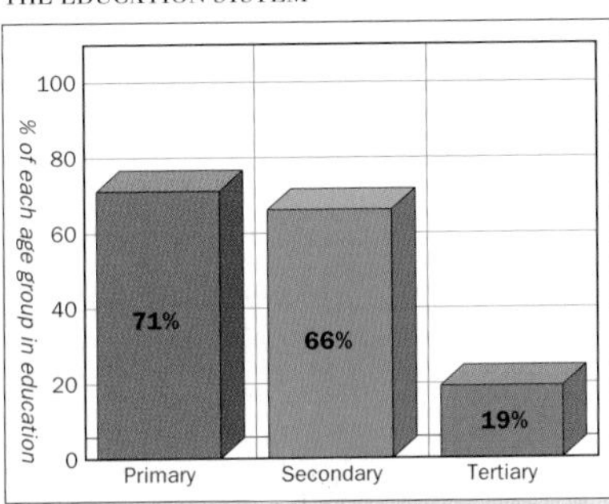
100
80
60
40
20
0
% of each age group in education
71%
66%
19%
Primary
Secondary
Tertiary

Growing numbers of Western-educated Saudis have intensified pressure for social and political change. In the 1950s, the religious establishment was persuaded to give women equal opportunities in education. Economic analysts have criticized Saudi universities for turning out many theology graduates, but not enough engineers and technocrats. Many Saudis still travel abroad to complete their studies.

HEALTH

Welfare state health benefits

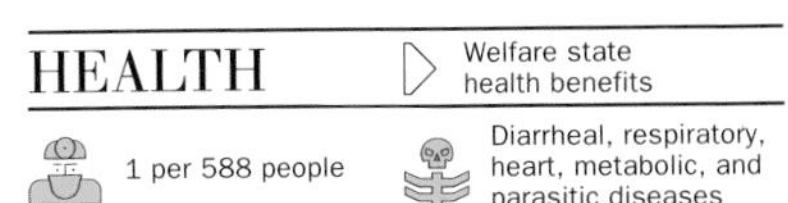
1 per 588 people
Diarrheal, respiratory, heart, metabolic, and parasitic diseases

Infant mortality has dropped and endemic disease has been nearly eradicted. Health care outside major centers such as Riyadh and Jiddah still remains relatively undeveloped, given Saudi Arabia's huge economic resources. However, large sums have been spent on employing Western expertise. Many Saudis are still sent overseas by the government for treatment, especially for transplant operations, which pose some ethical problems for religious leaders. The private sector has also been encouraged.

SPENDING

GDP/cap. increase

CONSUMPTION AND SPENDING

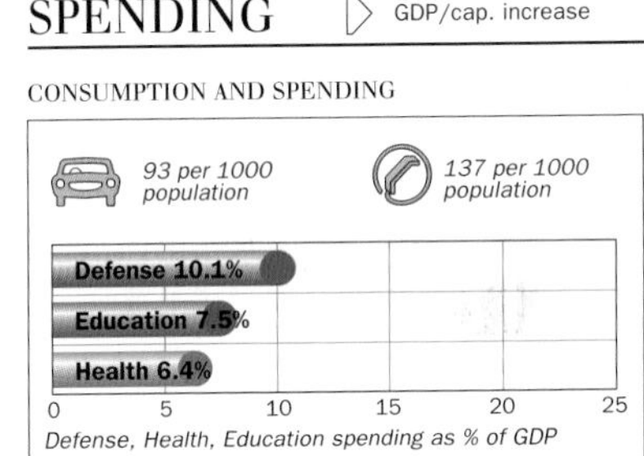
93 per 1000 population
137 per 1000 population
Defense 10.1%
Education 7.5%
Health 6.4%
0
5
10
15
20
25
Defense, Health, Education spending as % of GDP

Saudi citizens are among the most prosperous in the world. Non-Saudi citizens, especially the guest workers from the Indian subcontinent and the Philippines, are much poorer. The al-Sa'uds have used their wealth to create a cradle-to-grave welfare system. Ownership of TVs, telephones, and VCRs is among the region's highest. The distribution of wealth is carefully controlled by the royal family through the *majlis* system. There is no stock market, although shares in public companies are traded privately. Many Saudis refuse for religious reasons to accept interest on deposits with banks, but Islamic banks offer profit-sharing investment schemes as an alternative.

S

WORLD RANKING

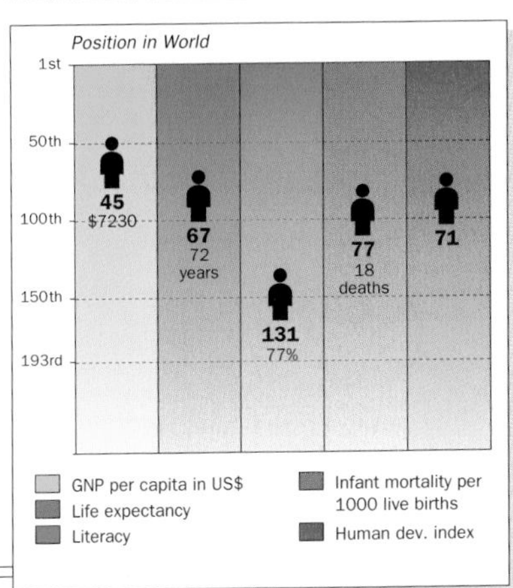
Position in World
1st
50th
100th
150th
193rd
45
$7230
67
72 years
131
77%
77
18 deaths
71
GNP per capita in US$
Life expectancy
Literacy
Infant mortality per 1000 live births
Human dev. index

SENEGAL

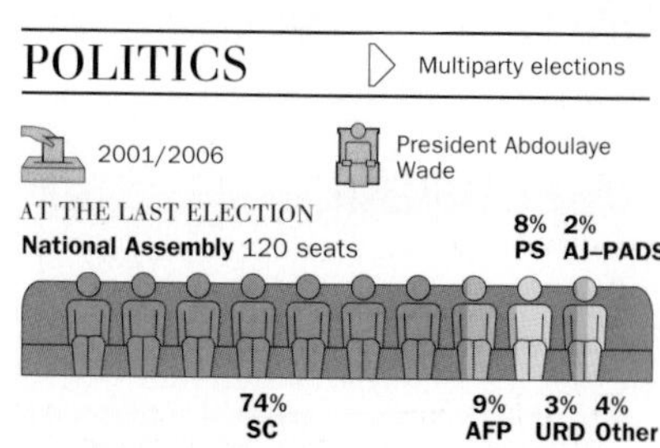
WEST AFRICA
Africa
Mauritania
SENEGAL
Gambia
Mali
Guinea-Bissau
Guinea

OFFICIAL NAME: Republic of Senegal **CAPITAL:** Dakar
POPULATION: 9.7 million **CURRENCY:** CFA franc **OFFICIAL LANGUAGE:** French

1960

1960

April 4
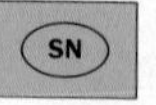
SN
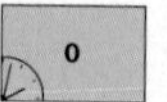
0
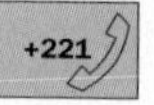
+221

.sn

SENEGAL'S CAPITAL, Dakar, lies on the westernmost cape of Africa. The country is mostly low, with open savanna and semidesert in the north and thicker savanna in the south. After independence from France in 1960, Senegal was ruled until 1981 by President Léopold Senghor. He was succeeded by his prime minister, Abdou Diouf, who held power for almost 20 years until his election defeat in March 2000.

CLIMATE

Steppe/tropical

WEATHER CHART FOR DAKAR

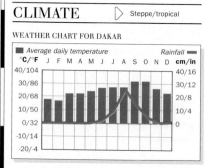

■ Average daily temperature — Rainfall

°C/°F J F M A M J J A S O N D cm/in

40/104 30/86 20/68 10/50 0/32 -10/14 -20/-4 — 40/16 30/12 20/8 10/4 0

The coastal regions, which project into the path of the northern trade winds, are remarkably cool given their latitude.

TRANSPORTATION

Drive on right

Dakar–Yoff International 1.1m passengers

198 ships 51,000 grt

THE TRANSPORTATION NETWORK

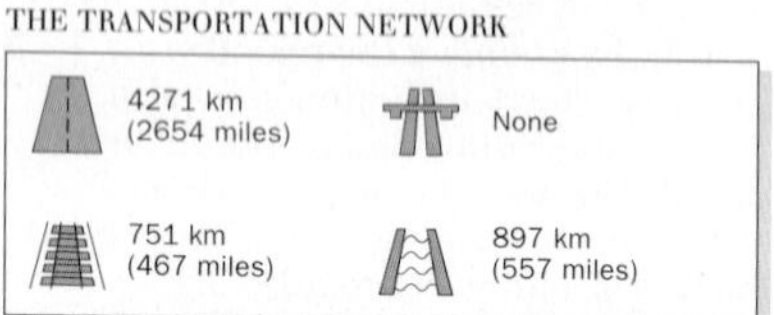
4271 km
(2654 miles)
None
751 km
(467 miles)
897 km
(557 miles)

Dakar is an important west African port, serving Senegal itself, Guinea, the hinterland of Mali, and southern Mauritania. The key rail link to Bamako, Mali's capital, was built in the 1920s.

TOURISM

Visitors : Population 1:26

369,000 visitors

Up 5% in 1999

MAIN TOURIST ARRIVALS

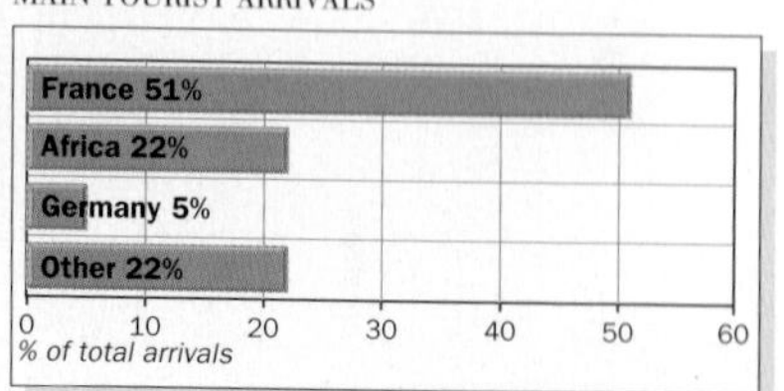
France 51%
Africa 22%
Germany 5%
Other 22%
0
10
20
30
40
50
60
% of total arrivals

In addition to French package tours to coastal resorts, tours for African-Americans to Gorée, an old slave island, are increasingly popular.

PEOPLE

Pop. density medium

Wolof, Fulani, Serer, Diola, Malinke, Soninke, Arabic, French

50/km² (130/mi²)

THE URBAN/RURAL POPULATION SPLIT

47% 53%

RELIGIOUS PERSUASION

Traditional beliefs 5%
Christian (mainly Roman Catholic) 5%
Sunni Muslim 90%

Senegal has a fairly well-developed sense of nationhood, and intermarriage between groups has reduced ethnic tensions. Groups can still be identified regionally, however. Dakar is a Wolof area, the Senegal River is dominated by the Toucouleur, the Malinke mostly live in the east, and the Diola in Casamance. The Diola have felt excluded from politics, and this has led to a long-running rebellion in Casamance. A French-influenced class system is still prevalent. The 2001 constitution gave women property rights for the first time.

POLITICS

Multiparty elections

2001/2006

President Abdoulaye Wade

AT THE LAST ELECTION

National Assembly 120 seats

8% PS, 2% AJ–PADS

74% SC, 9% AFP, 3% URD, 4% Other

SC = Sopi (Change) coalition (led by the Senegalese Democratic Party – **PDS**) **AFP** = Alliance of Progressive Forces **PS** = Senegalese Socialist Party **URD** = Union for Democratic Renewal **AJ–PADS** = And Jëf – African Party for Democracy and Socialism

Senegal has been a multiparty democracy since 1981, when, under the then new president Abdou Diouf, the constitution was amended to allow more than four political parties. However, the PS held power from the 1950s until 2000, and its influence has been pervasive. Presidential elections in 2000 marked a political watershed. Diouf was defeated by Abdoulaye Wade of the liberal democratic PDS, the dominant party in the "Sopi" (Change) coalition which went on to win a landslide victory in the 2001 legislative elections.

A new constitution, approved in 2001 by referendum, abolished the Senate and restricts the president to two terms.

WORLD AFFAIRS

Joined UN in 1960

CILSS ECOWAS FZ OIC OMVG

Maintaining good relations with France, Senegal's main ally and aid donor, is the major foreign affairs concern. Relations with neighboring Gambia, Mauritania, and Guinea-Bissau continue to be a constant preoccupation.

S

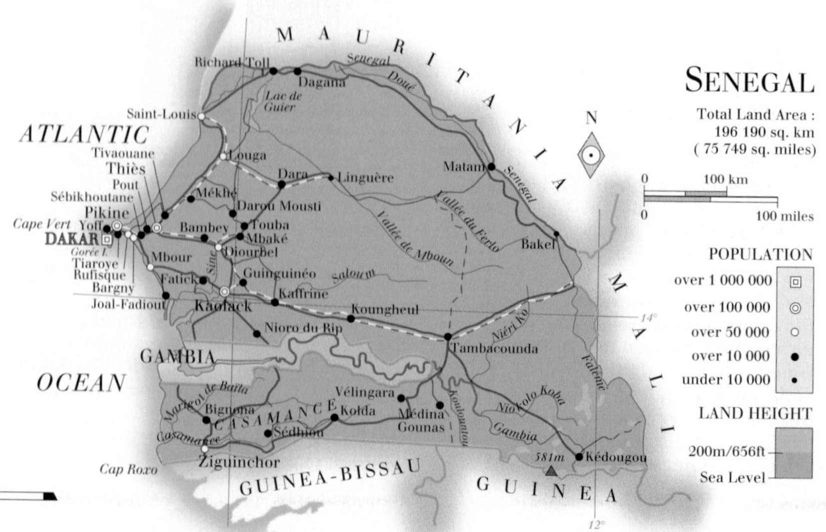
MAURITANIA
MALI
GAMBIA
GUINEA-BISSAU
GUINEA
ATLANTIC
OCEAN
CASAMANCE
Richard-Toll
Dagana
Lac de Guier
Saint-Louis
Louga
Tivaouane
Thiès
Pout
Sébikhoutane
Pikine
Cape Vert
Yoff
DAKAR
Gorée I.
Tiaroye
Rufisque
Bargny
Joal-Fadiout
Mékhé
Dara
Linguère
Darou Mousti
Bambey
Touba
Mbaké
Diourbel
Mbour
Fatick
Kaolack
Guinguinéo
Kaffrine
Koungheul
Nioro du Rip
Matam
Bakel
Tambacounda
Vélingara
Kolda
Médina Gounas
Bignona
Sédhiou
Ziguinchor
Cap Roxo
Kédougou
581m
Senegal
Saloum
Sine
Gambia
Falémé
Niéri Ko
Niokolo Koba
Vallée du Ferlo
Vallée de M'boun
N
SENEGAL
Total Land Area : 196 190 sq. km (75 749 sq. miles)
0 100 km
0 100 miles
POPULATION
over 1 000 000
over 100 000
over 50 000
over 10 000
under 10 000
LAND HEIGHT
200m/656ft
Sea Level
14°
16°
12°

AID

Recipient

$423m (receipts)

Down 21% in 2000

Senegal is one of the highest recipients of aid per capita in Africa, mostly from France, the World Bank, and Japan. Aid is used to import 400,000 tonnes of rice annually, but also helps to finance a sizable civil service, now being cut back. An important structural adjustment program, backed by the IMF, was renewed for a further year in April 2001.

***The mosque in Touba**, religious capital of the Muslim Mouride sect, which was founded in 1887 in Senegal's groundnut-growing district.*

CHRONOLOGY

France colonized Senegal, a major entrepôt from the 15th century, in 1890. Dakar was the capital of French West Africa.

- ❑ **1885** Gambia split off from Senegal.
- ❑ **1960** Independence under Senghor.
- ❑ **1966–1976** One-party state.
- ❑ **1981** Full multipartyism restored.
- ❑ **2000** Presidency won by Abdoulaye Wade in first ever defeat for PS.
- ❑ **2001** Referendum approves new constitution.

DEFENSE

Compulsory military service

$68m

Down 16% in 2000

France maintains an important naval base at Dakar. The armed forces total 9400, plus a paramilitary force of 5800, but the military has never intervened in politics. Senegalese troops took part in Operation Desert Storm in 1991, and intervened in conflicts in Liberia, Rwanda, and the Central African Republic. They also helped to quell revolts in Gambia and Guinea-Bissau.

ECONOMICS

Inflation 4.6% p.a. (1990–2000)

$4.71bn

698.7–736.7 CFA francs

SCORE CARD

❑ WORLD GNP RANKING	114th
❑ GNP PER CAPITA	$490
❑ BALANCE OF PAYMENTS	–$310m
❑ INFLATION	0.7%
❑ UNEMPLOYMENT	4%

EXPORTS

Spain 5%
Mali 6%
Italy 9%
France 16%
India 18%
Other 46%

IMPORTS

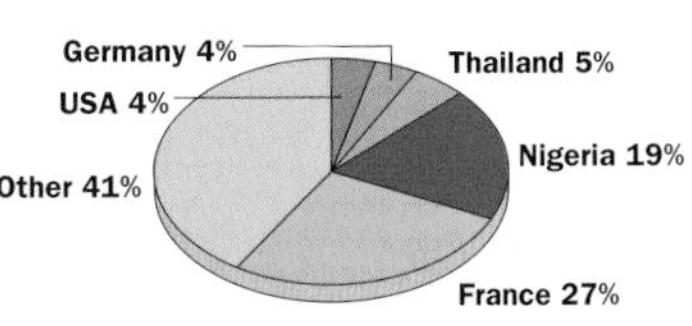
Germany 4%
USA 4%
Thailand 5%
Nigeria 19%
Other 41%
France 27%

STRENGTHS

Good infrastructure. Relatively strong industrial sector. First west African country with international credit rating. Revenue from sale of fishing rights.

WEAKNESSES

Few natural resources exploited, other than groundnuts, phosphates, and fish.

Access to oil potential of Casamance region hampered by rebellion and poor transportation links.

EDUCATION

School leaving age: 13

37%

29,303 students

Illiteracy is Senegal's major educational challenge. There are universities at Dakar and St.-Louis.

HEALTH

No welfare state health benefits

1 per 10,000 people

Malaria, diarrheal diseases

The state health system is rudimentary. A successful education campaign helps to contain the incidence of HIV/AIDS.

SPENDING

GDP/cap. increase

CONSUMPTION AND SPENDING

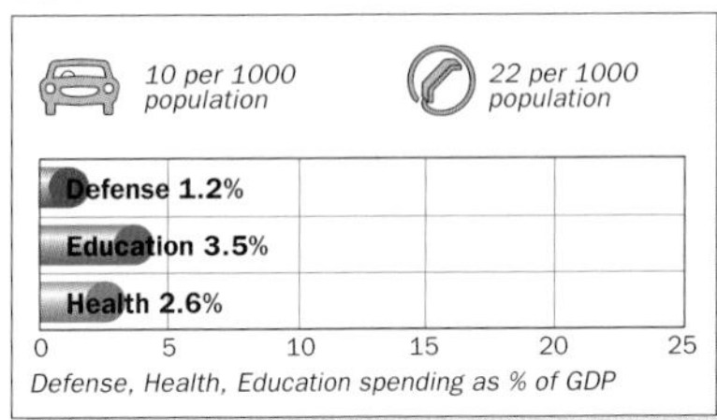
10 per 1000 population
22 per 1000 population
Defense 1.2%
Education 3.5%
Health 2.6%
0
5
10
15
20
25
Defense, Health, Education spending as % of GDP

Wealth disparities are considerable in Senegal, and poverty is widespread. Members of the former ruling PS are the wealthiest group.

RESOURCES

Electric power 235,000 kw

418,280 tonnes

Not an oil producer

4.82m sheep, 4m goats, 45m chickens

Phosphates, bauxite, salt, natural gas, marble, iron, copper

Senegal's electricity capacity is largely dependent on imported fuel; cheaper supplies are expected to become available soon from the Manantali Dam in Mali. Initial explorations suggest that oil reserves may exist off Casamance.

MEDIA

TV ownership low

Daily newspaper circulation 5 per 1000 people

PUBLISHING AND BROADCAST MEDIA

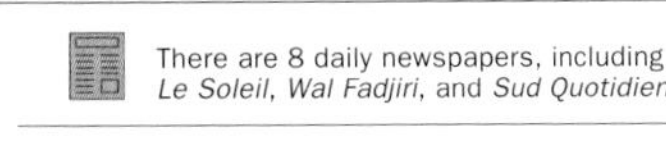
There are 8 daily newspapers, including *Le Soleil, Wal Fadjiri,* and *Sud Quotidien*
2 services: 1 state-owned, 1 private
5 services: 1 state-owned, 4 independent

The independent media flourished with multipartyism. Senegal had the first satirical journal in Africa with the founding of *Le Politicien* in 1978.

WORLD RANKING

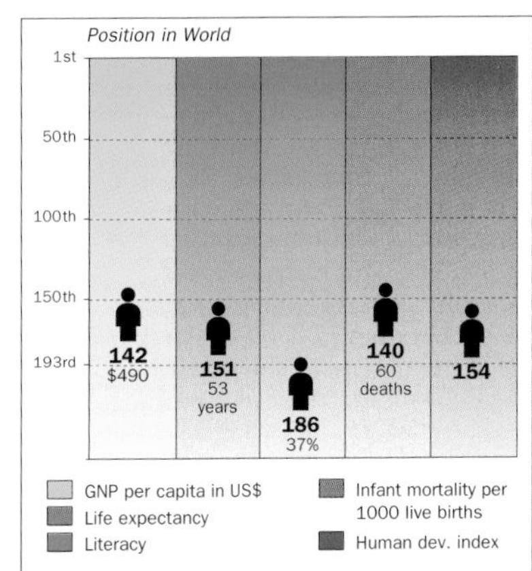
Position in World
1st
50th
100th
150th
193rd
142
$490
151
53 years
186
37%
140
60 deaths
154
GNP per capita in US$
Life expectancy
Literacy
Infant mortality per 1000 live births
Human dev. index

ENVIRONMENT

Sustainability rank: 81st

11% (6% partially protected)

0.4 tonnes per capita

The damming of the Senegal River has caused concern that traditional farming practices, which rely on seasonal floods, may be disrupted. Two major droughts in 1973 and 1983 led to the advance of the Sahara Desert in the west of the country.

CRIME

Death penalty not used in practice

4653 prisoners

Up 8% in 1999

Senegal has comparatively low crime rates, though levels are now rising in Dakar and the surrounding shanty towns, where gangs are based.

S

EUROPE
Europe
Croatia
Romania
B-H
SERBIA & MONTENEGRO
Bulgaria
Albania
Maced.

SERBIA & MONTENEGRO

OFFICIAL NAME: Serbia and Montengro **CAPITAL:** Belgrade **POPULATION:** 10.5 million
CURRENCY: Dinar (euro widely used in Montenegro) **OFFICIAL LANGUAGE:** Serbo-Croat

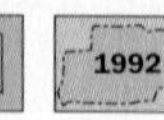
1992

1992

Nov 29
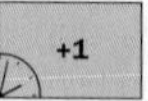
YU

+1
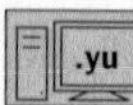
+381
.yu

THE LOOSE CONFEDERATION of Serbia and Montenegro emerged in 2002, when the troubled Federal Republic of Yugoslavia was finally dissolved after 84 years in existence. Literally a "union of south Slavs," Yugoslavia was created after World War I. A maverick communist state after 1945, it was torn apart by nationalist tensions in a violent civil war in 1991–1995, spawning four new Balkan republics. After the ousting in 2000 of the discredited Serbian nationalist Slobodan Milošević as federal president, a more democratic successor regime ended the country's international pariah status.

CLIMATE

Continental

WEATHER CHART FOR BELGRADE

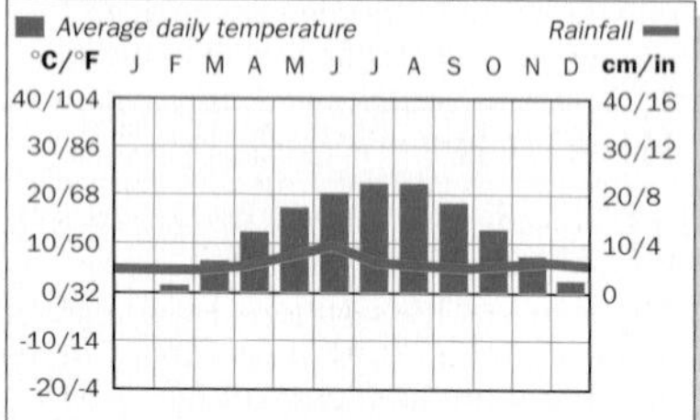
Average daily temperature
Rainfall
°C/°F
J F M A M J J A S O N D
cm/in
40/104
30/86
20/68
10/50
0/32
-10/14
-20/-4
40/16
30/12
20/8
10/4
0

There are three climate zones. The northern plains are characteristically continental: rainy springs, warm summers, and cold winters. While the southern highlands have colder winters with heavy snowfalls, the Adriatic coast boasts hot summers and milder winters.

TRANSPORTATION

Drive on right

Surcin, Belgrade
1.28m passengers

9 ships
4700 grt

THE TRANSPORTATION NETWORK

28,822 km (17,909 miles)

560 km (348 miles)

4059 km (2522 miles)

587 km (365 miles)

About one-third of railroads are electrified. The rail link to Greece is one of Serbia's main trading routes, and lines through Serbia remain the best link between Budapest and Sofia. For internal travel, trains are cheaper, but slower, than buses. Two flights each day link Belgrade with the airports in Montenegro.

Bridges and railroads were specifically targeted during the NATO bombing in 1999. The bombing of the bridge over the Danube at Novi Sad closed the river as a major regional trading artery, and its reopening was made a priority for international assistance once the Milošević regime had fallen.

The country's stunning scenery is a potential draw for tourists. Before the 1990s it attracted millions every year.

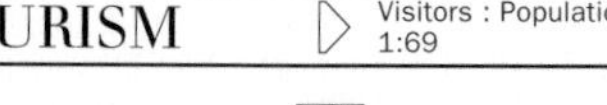
TOURISM
Visitors : Population 1:69

152,000 visitors

Down 46% in 1999

MAIN TOURIST ARRIVALS

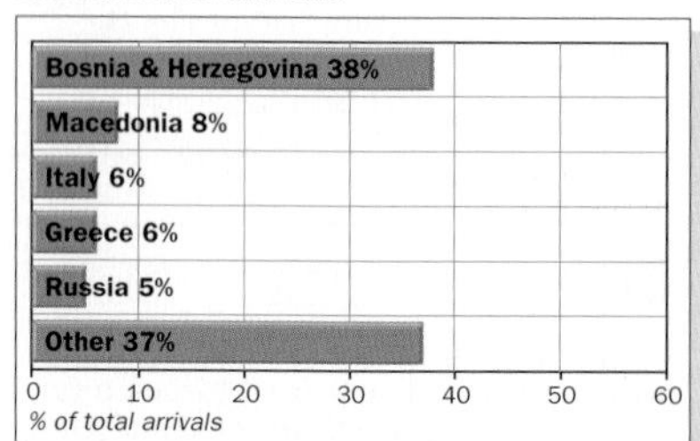
Bosnia & Herzegovina 38%
Macedonia 8%
Italy 6%
Greece 6%
Russia 5%
Other 37%
0
10
20
30
40
50
60
% of total arrivals

Instability, the impact of UN sanctions, and the Kosovo conflict meant that foreign tourism ceased in the 1990s. Serbia has never been a center of tourism. The Montenegrin coast, however, has renowned beaches. In the 1990s they were monopolized by Serbians, particularly by political and criminal elements of the Serbian elite. Hyperinflation and recession kept the average vacationer away.

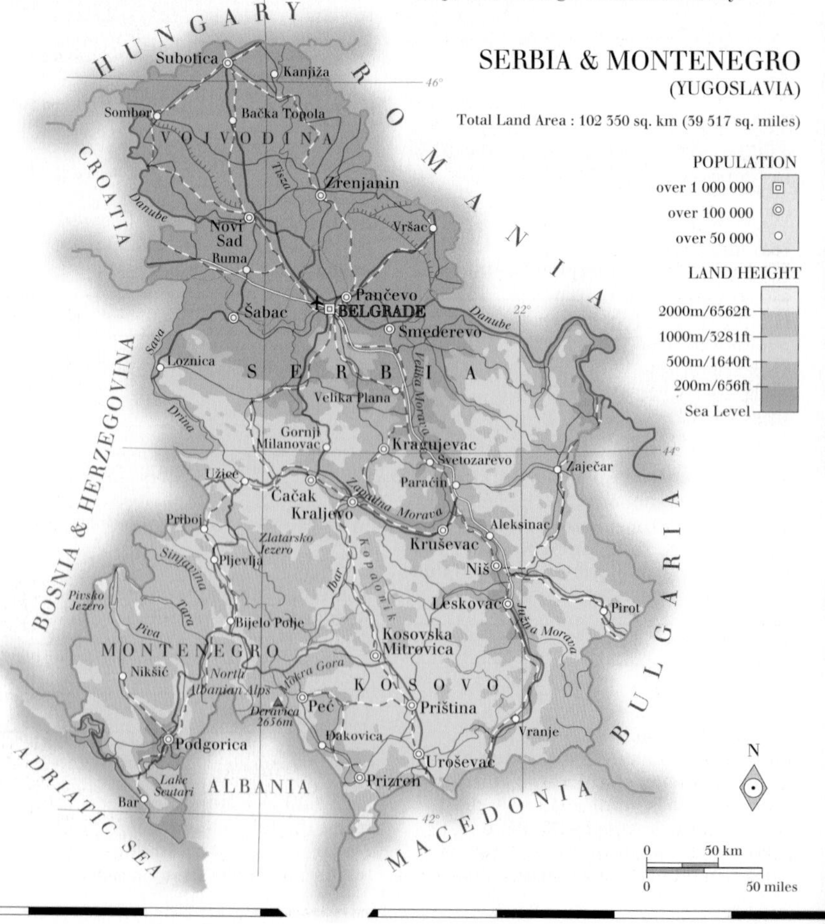

SERBIA & MONTENEGRO (YUGOSLAVIA)

Total Land Area : 102 350 sq. km (39 517 sq. miles)

POPULATION
over 1 000 000
over 100 000
over 50 000

LAND HEIGHT
2000m/6562ft
1000m/3281ft
500m/1640ft
200m/656ft
Sea Level

0 50 km
0 50 miles

PEOPLE

Pop. density medium

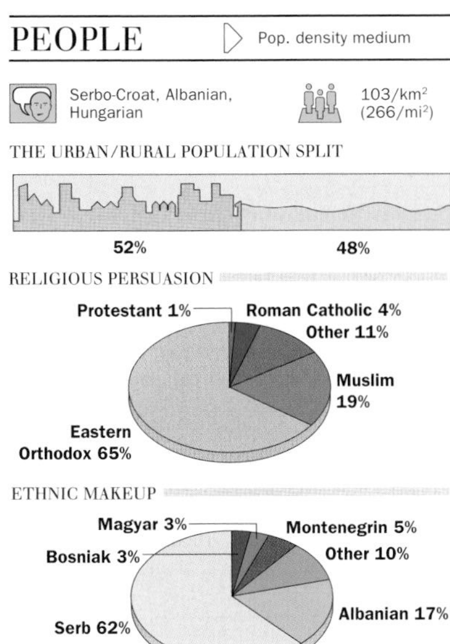

Serbo-Croat, Albanian, Hungarian
103/km² (266/mi²)
THE URBAN/RURAL POPULATION SPLIT
52%
48%
RELIGIOUS PERSUASION
Protestant 1%
Roman Catholic 4%
Other 11%
Muslim 19%
Eastern Orthodox 65%
ETHNIC MAKEUP
Magyar 3%
Montenegrin 5%
Bosniak 3%
Other 10%
Albanian 17%
Serb 62%

The social order has been devastated by the decade of conflict in the region. There is a depressingly high suicide rate among the old living in cities. In Kosovo, 900,000 ethnic Albanians fled (mainly to Albania and Macedonia) during the 1999 conflict. The 1999 agreement opened the way for most to return, since which time many of the province's embattled Serb minority have gone (or been driven out) themselves, fearing both vengeance and the longer-term prospect of Kosovan independence.

The Hungarian (and mainly Roman Catholic) minority is concentrated in, but does not dominate, the relatively prosperous northern Serbian province of Vojvodina. Limited autonomy was restored in 2002, after economically motivated pressure from within the region. Orthodox Christian Serbs who converted to Islam during Ottoman rule are considered to be an ethnically separate group.

POPULATION AGE BREAKDOWN

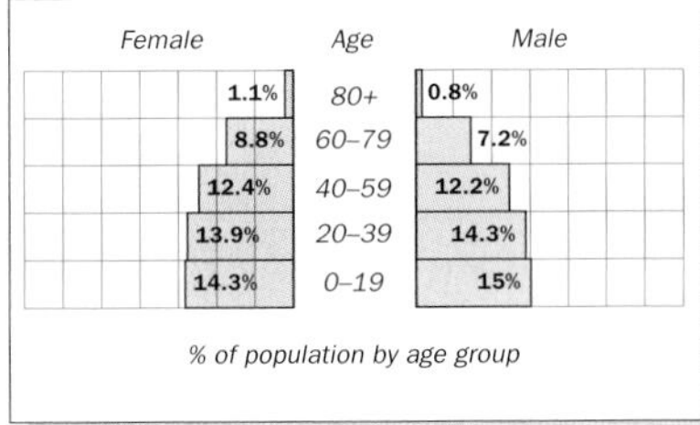

Female
Age
Male
1.1%
80+
0.8%
8.8%
60–79
7.2%
12.4%
40–59
12.2%
13.9%
20–39
14.3%
14.3%
0–19
15%
% of population by age group

POLITICS

Multiparty elections

L. House 2000/2004
U. House 2000/2004
President Vojislav Kostunica

AT THE LAST ELECTION

Chamber of Citizens 138 seats

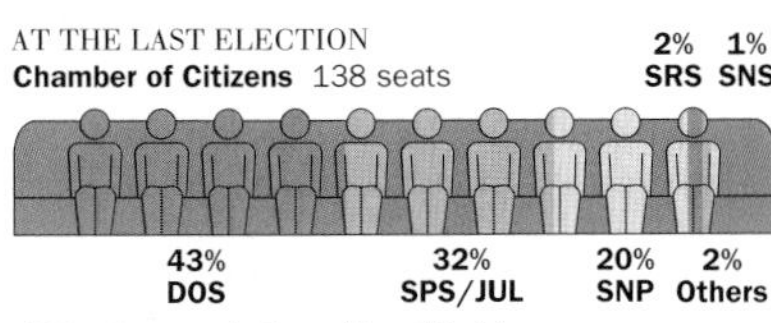

2% SRS
1% SNS
43% DOS
32% SPS/JUL
20% SNP
2% Others

DOS = Democratic Opposition of Serbia
SPS/JUL = Socialist Party of Serbia/Yugoslav Left
SNP = Socialist People's Party **SRS** = Serbian Radical Party
SNS = Serbian People's Party
SPO = Serbian Renewal Movement

Chamber of Republics 40 seats

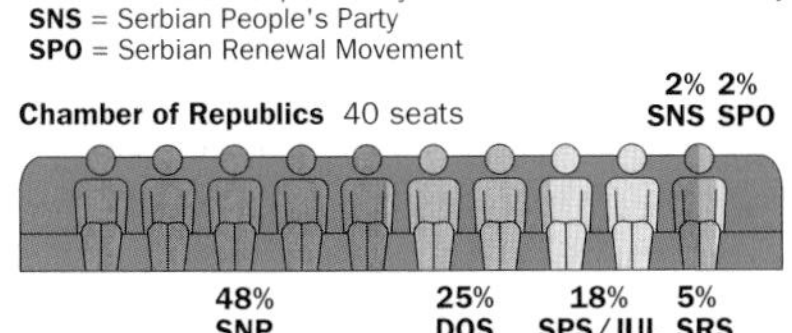

2% SNS
2% SPO
48% SNP
25% DOS
18% SPS/JUL
5% SRS

20 delegates are elected to the Chamber of Republics by each of the two republics of Serbia and Montenegro

Serbia and Montenegro, each with a separate parliament and president, are represented in the Federal Assembly.

PROFILE

Slobodan Milošević, Serbian president from 1989 and federal president from 1997, dominated politics for a decade. Conflict from 1991 to 1995, then again over Kosovo, created a permanent sense of siege. The tightening of UN sanctions in 1999, rather than the NATO bombing, sealed Milošević's fate. After he was ousted in the October 2000 popular uprising, the reformists consolidated their position, dominating elections in December. However, Milošević's extradition in 2001 affronted nationalists in the DOS coalition, including President Vojislav Kostunica; Kostunica has since withdrawn his party.

MAIN POLITICAL ISSUES

Kosovo

The withdrawal from Kosovo of the federal army ended NATO bombing in 1999. The ethnic Albanian majority awaits a promised referendum on its future. International administrators in 2001 oversaw the election of a Kosovo assembly. Local Serbs took part, but are wary of Albanian calls for independence.

Montenegro

Although less subservient to Serbia from 1997, calls for Montenegro's independence were subdued in 2002 when President Milo Djukanović agreed to a trial period in a loose confederation.

Vojislav Kostunica, *federal president since the ousting of Milošević in 2000.*

Slobodan Milošević, *disgraced former president charged with war crimes.*

WORLD AFFAIRS

Joined UN in 1945

CEI
IAEA
EBRD
IMF
OSCE

In 1995, mutual recognition among the countries which had constituted the Socialist Federal Republic of Yugoslavia paved the way for the normalization of relations.

The pariah status of the Milošević regime in the late 1990s was underlined by his indictment by the International Criminal Tribunal for the former Yugoslavia (ICTY) in 1999, and the military action by NATO forces over the situation in Kosovo. Russia, long an ally of Serbia, shares its Orthodox Christianity and its Slavic ethnicity, and strongly opposed the NATO military action over Kosovo, but backed the settlement proposals required to end it. The inauguration of the Kostunica government in 2000 was welcomed enthusiastically by the West and Russia. The new regime was quickly invited to take up its vacant seat in the UN. Agreement to cooperate with the ICTY has reopened aid channels.

AID

Recipient

$1.14bn (receipts)

Up 77% in 2000

Milošević's removal was an explicit condition of large-scale Western aid, urgently needed to rebuild the damaged economy. The country rejoined the World Bank in May 2001 and received pledges of $1.3 billion in aid immediately upon the extradition of Milošević.

CHRONOLOGY

The Serbs were defeated by the Turks at the Battle of Kosovo in 1389. Parts of the region were later ruled by the Austro-Hungarian Empire.

- **1878** Independence gained by Serbia and Montenegro at Congress of Berlin.
- **1918** Joint Kingdom of Serbs, Croats, and Slovenes created.
- **1929** King Alexander of Serbia assumes absolute powers over state, which changes name to Yugoslavia.
- **1941** Germans launch surprise attack. Rival resistance groups: Chetniks (Serb royalist) and Partisans (communist, under Tito).
- **1945** Federal People's Republic of Yugoslavia founded with Tito as prime minister (and president from 1953).
- **1948** Tito breaks with Stalin.
- **1951** Farmers permitted to sell products on free market.
- **1955** Yugoslav–Soviet détente.

S

CHRONOLOGY *continued*

- **1963** Third postwar constitution adopts name Socialist Federal Republic of Yugoslavia (SFRY).
- **1973** Economic cooperation agreement with West Germany. Agreement of noninterference signed with USSR.
- **1974** New constitution decentralizes government. Vojvodina and Kosovo given greater autonomy.
- **1980** Tito dies. Succeeded by collective presidency.
- **1981** Unrest among Kosovo Albanians; state of emergency.
- **1985** Serbian intellectuals publish memorandum listing Serb grievances within Yugoslavia.
- **1986** Slobodan Milošević becomes leader of Communist (later Socialist) Party of Serbia.
- **1987** Wage freeze to combat inflation. Banking system crisis.
- **1988** Belgrade protests against economic austerity. Government brought down over budget failure.
- **1989** 600th anniversary of Battle of Kosovo. Kosovo Albanians protest against Serb police unit; crackdown ends Kosovo's autonomy. Milošević elected president of Serbia.
- **1990** SPS wins elections in Serbia. Communists win presidency and dominate Montenegro elections.
- **1992** EU recognizes breakaway republics of Croatia, Slovenia, and Bosnia and Herzegovina. Bosnian war begins. UN sanctions imposed. Ibrahim Rugova elected president of self-declared republic of Kosovo. Milošević reelected president of Serbia, but SPS loses majority.
- **1995** Milošević signs Bosnian peace accord.
- **1996** UN sanctions formally lifted.
- **1997** Concessions after big protests, acknowledging malpractice in municipal elections. Milošević becomes federal president.
- **1998** Conflict in Kosovo escalates.
- **1999** March, Kosovo talks break down; "ethnic cleansing" precipitates mass exodus. NATO aerial bombing. June, withdrawal of Serbian forces and police from Kosovo, and entry of international force, KFOR.
- **2000** September, defeat of Milošević in first round of presidential election. October, opposition candidate Vojislav Kostunica swept to power by massive anti-Milošević protests. December, DOS dominates elections for Serbian parliament.
- **2001** April, arrest of Milošević. June, Milošević extradited to face war crimes tribunal in The Hague.
- **2002** Serbia and Montenegro agree to dissolve Yugoslavia.

DEFENSE

Compulsory military service

$1.79bn

Up 8% in 2000

SERBIA & MONTENEGRO ARMED FORCES

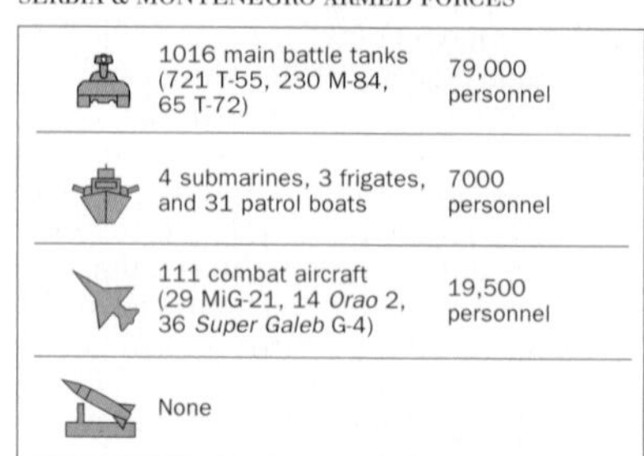

1016 main battle tanks (721 T-55, 230 M-84, 65 T-72)	79,000 personnel
4 submarines, 3 frigates, and 31 patrol boats	7000 personnel
111 combat aircraft (29 MiG-21, 14 *Orao* 2, 36 *Super Galeb* G-4)	19,500 personnel
None	

Military capability was specifically targeted for "degrading" by the NATO air strikes in 1999. The impact on antiaircraft defenses, heavy weaponry, logistics capacity, and infrastructure was less severe than had first been claimed, however.

The Serbian military had previously played a major role in the conflicts in the former Yugoslavia in the early 1990s. Traditionally the center of Yugoslav armaments manufacture, Serbia was able to arm itself – although the need to create money to pay for domestically produced weapons was a major factor in the crippling hyperinflation of that era.

In 2002 the country made its first (modest) contribution to a UN peacekeeping force, in East Timor.

ECONOMICS

Not available

$10bn

13.65–66.48 dinars

SCORE CARD

WORLD GNP RANKING	83rd
GNP PER CAPITA	$940
BALANCE OF PAYMENTS	–$1.42bn
INFLATION	42%
UNEMPLOYMENT	30%

EXPORTS

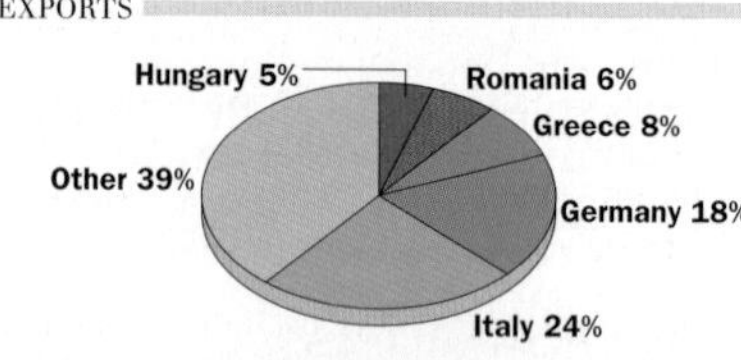
Hungary 5%
Romania 6%
Greece 8%
Germany 18%
Italy 24%
Other 39%

IMPORTS

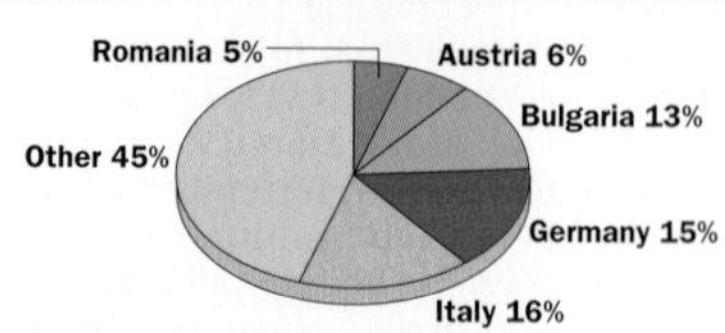
Romania 5%
Austria 6%
Bulgaria 13%
Germany 15%
Italy 16%
Other 45%

ECONOMIC PERFORMANCE INDICATOR

Consumer Price Index
GDP
460
340
220
100
-20
Consumer price index 1995=100
GDP unavailable
1995
1996
1997
1998
1999

STRENGTHS

Return of international aid and investment in 2000–2001. Economic potential of Danube.

WEAKNESSES

Severe damage caused by sanctions and 1999 bombings. Low hard-currency reserves. Outflow of skilled professionals.

PROFILE

The living standards in the former Yugoslavia were among the most advanced of the socialist countries. It has now all but collapsed, although as much as 50% of all activity goes on within the resilient informal sector. The conflicts in the 1990s effectively stalled much-needed economic reform. Sanctions, maintained until 1996 and reimposed more fully in 1999, stifled trade and decimated both the emerging private and the state sectors. Hyperinflation had already pushed the economy to virtual collapse. NATO bombing in 1999 caused extensive damage to infrastructure; the EBRD estimated reconstruction costs at $20 billion over three years. Sanctions were lifted shortly after Milošević's downfall, and investment prospects were boosted by his extradition and the agreement to cooperate with the ICTY.

SERBIA & MONTENEGRO : MAJOR BUSINESSES

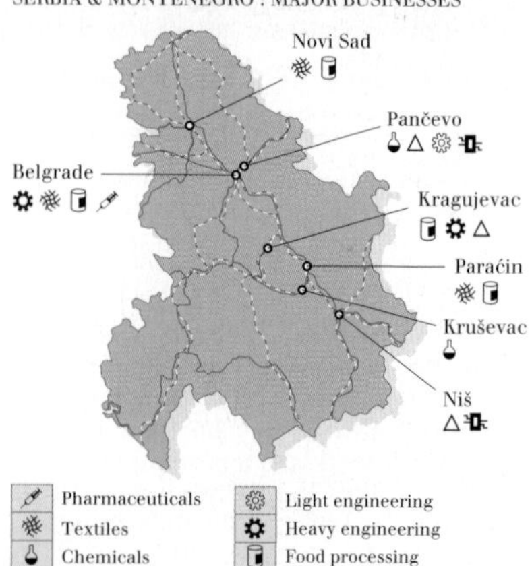
Novi Sad
Pančevo
Belgrade
Kragujevac
Paraćin
Kruševac
Niš
Pharmaceuticals
Textiles
Chemicals
Metallurgy
Electronics
Light engineering
Heavy engineering
Food processing
0
100 km
0
100 miles

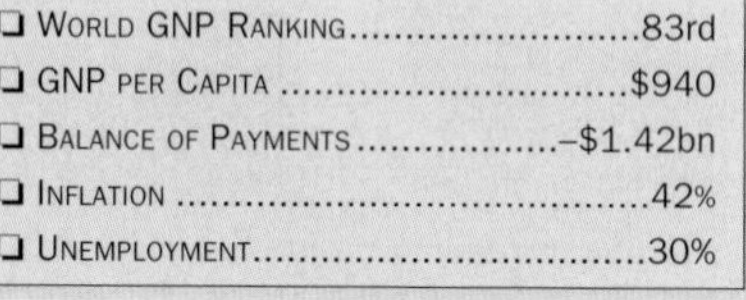

S

RESOURCES

Electric power 11.8m kw

9940 tonnes

18,305 b/d (reserves 80m barrels)

4.37m pigs, 1.92m sheep, 1.83m cattle, 21.1m chickens

Coal, bauxite, iron, lead, copper, zinc

ELECTRICITY GENERATION

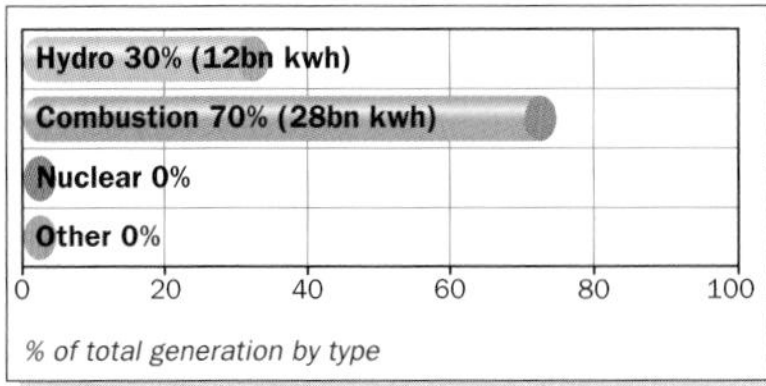
Hydro 30% (12bn kwh)
Combustion 70% (28bn kwh)
Nuclear 0%
Other 0%
0
20
40
60
80
100
% of total generation by type

The country is self-sufficient in coal and electricity production. Vojvodina's oil industry could cater for one-third of its needs, but was badly hit by NATO bombing in March–June 1999.

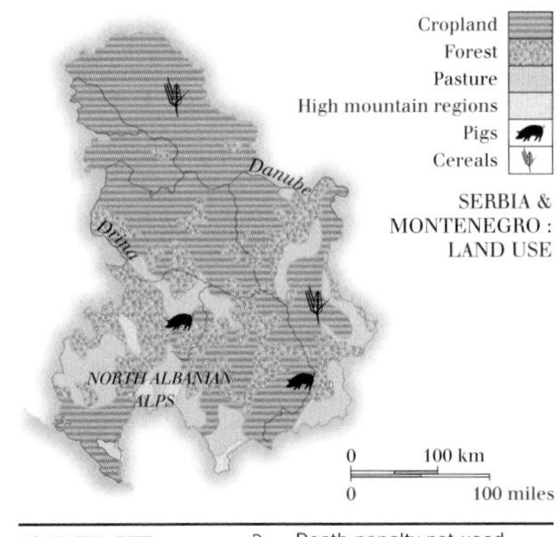
Cropland
Forest
Pasture
High mountain regions
Pigs
Cereals
SERBIA & MONTENEGRO : LAND USE
Danube
Drina
NORTH ALBANIAN ALPS
0
100 km
0
100 miles

ENVIRONMENT

Not available

3%

Not available

ENVIRONMENTAL TREATIES

Yes
Yes
Yes
Yes
No
No

Ecological awareness peaked in the late 1980s, when the Ecological Forum was active. NATO bombing of Serbia in 1999 caused extensive pollution of the Danube and raised fears of contamination from dioxins. Depleted uranium from NATO munitions has also aroused serious concerns both in Serbia proper and in Kosovo.

CRIME

Death penalty not used in practice

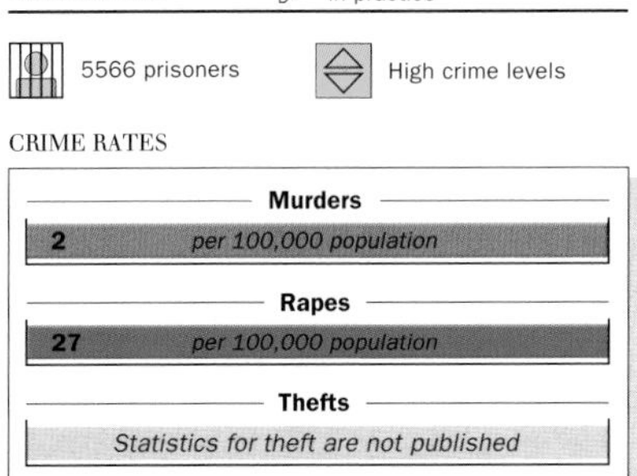
5566 prisoners
High crime levels
CRIME RATES
Murders
2 per 100,000 population
Rapes
27 per 100,000 population
Thefts
Statistics for theft are not published

Crime was rife under Milošević – from currency trading and black marketing to narcotics and extortion. The country has fully cooperated with the ICTY since 2001, and began its own war crimes trials in Serbia in 2002. The death penalty is set to be abolished.

MEDIA

TV ownership high

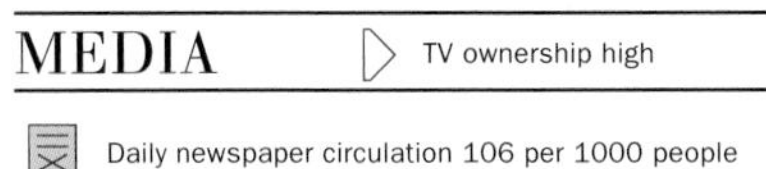
Daily newspaper circulation 106 per 1000 people

PUBLISHING AND BROADCAST MEDIA

There are 18 daily newspapers. Politika has the largest circulation
2 state-controlled, several independent services
3 state-controlled, several independent services

The state broadcasting center was a prime target in the October 2000 popular uprising against the Milošević regime; it had used the broadcast media to control public opinion, while the press was held in check by government control of newsprint. Many journalists supported the opposition. The B92 radio station, when not forced off air, was a beacon of independent reporting. Free media have flourished under the post-Milošević authorities.

EDUCATION

School leaving age: 15

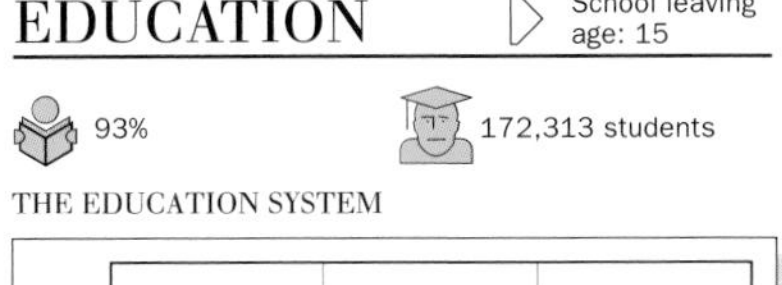
93%
172,313 students

THE EDUCATION SYSTEM

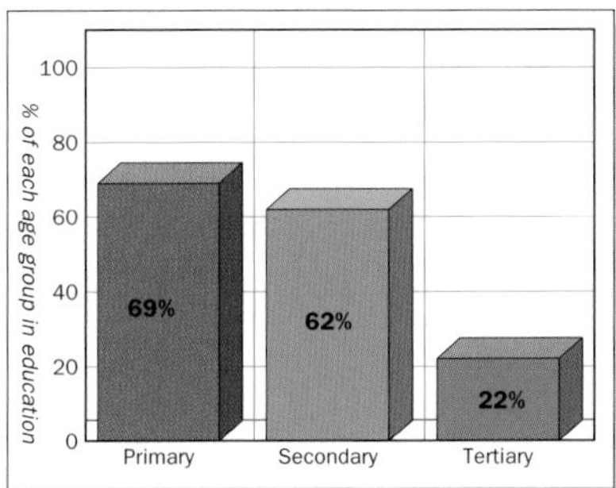
100
80
60
40
20
0
% of each age group in education
69%
62%
22%
Primary
Secondary
Tertiary

Schooling was totally disrupted by the Kosovo conflict in 1999, leaving the education system in crisis; the wealthy go abroad for their education. Literacy rates in Kosovo, where ethnic Albanian schools were closed in 1990, were low even before the conflict. Rebuilding the basic education system is key to reconstruction and reconciliation.

HEALTH

Welfare state health benefits

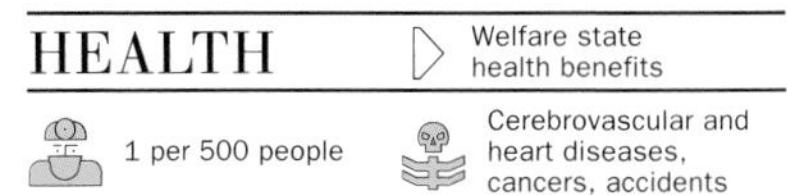
1 per 500 people
Cerebrovascular and heart diseases, cancers, accidents

Isolation from former trading partners has affected the quality of the health service, despite the exemption of medicines and medical supplies from sanctions. Social insurance is obligatory for those in employment, but medicines are scarce and costly, and death rates among infants and the elderly have risen dramatically. Health problems can be aggravated by bitingly cold winters.

SPENDING

GDP/cap. increase

CONSUMPTION AND SPENDING

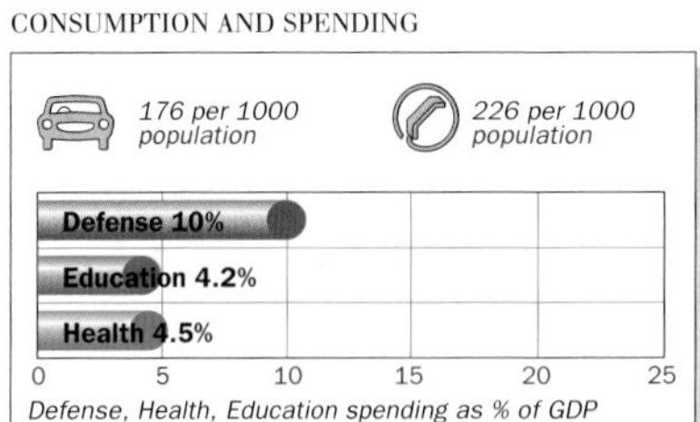
176 per 1000 population
226 per 1000 population
Defense 10%
Education 4.2%
Health 4.5%
0
5
10
15
20
25
Defense, Health, Education spending as % of GDP

The country as a whole was seriously impoverished by sanctions; real incomes fell dramatically, yet food prices remained higher than in much of western Europe. Bank collapses in 1992 and continuing hyperinflation in 1992–1994 wiped out dinar savings. One business that did expand was the illegal import of sanctions-busting goods for the few who could afford them – black marketeers and those close to Milošević.

The lifting of sanctions in 1995–1996 had hardly begun to be reflected in improvements in living conditions when the Kosovo conflict brought further dislocation and hardship in 1999. Even before it erupted, an estimated two-thirds of the population were living below subsistence level. Apart from the desperate situation of refugees and internally displaced families, unsupported pensioners fare worst.

S

WORLD RANKING

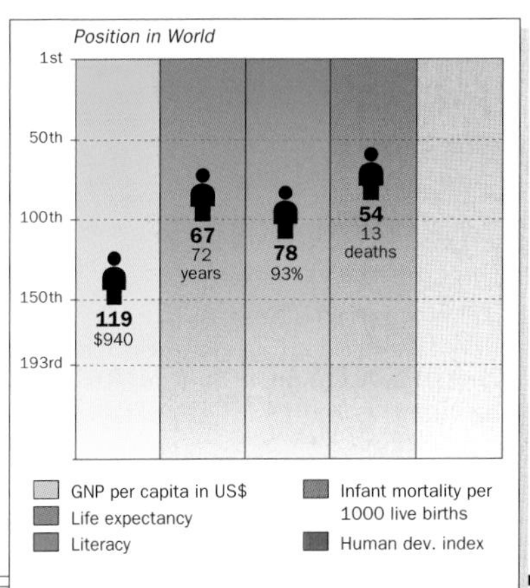
Position in World
1st
50th
100th
150th
193rd
119
$940
67
72
years
78
93%
54
13
deaths
GNP per capita in US$
Life expectancy
Literacy
Infant mortality per 1000 live births
Human dev. index

SEYCHELLES

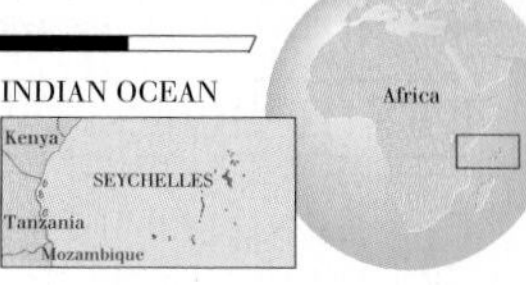
INDIAN OCEAN
Africa
Kenya
SEYCHELLES
Tanzania
Mozambique

OFFICIAL NAME: Republic of the Seychelles **CAPITAL:** Victoria
POPULATION: 79,300 **CURRENCY:** Seychelles rupee **OFFICIAL LANGUAGE:** Seselwa (French Creole)

1976

1976

June 18

SY
+4

+248

.sc

THE 115 ISLANDS of the Seychelles, lying in the Indian Ocean, support unique flora and fauna, including the giant tortoise and the world's largest seed, the *coco-de-mer*. Formerly a UK colony and then under one-party rule for 16 years, the Seychelles became a multiparty democracy in 1993. The economy relies on tourism.

CLIMATE

Tropical oceanic

WEATHER CHART FOR VICTORIA

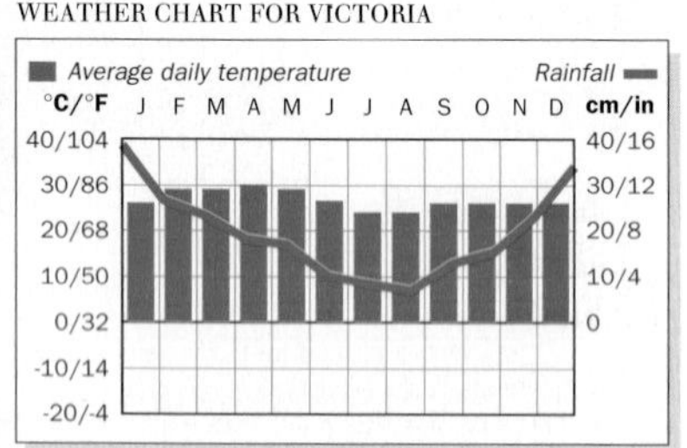
Average daily temperature
Rainfall
°C/°F J F M A M J J A S O N D cm/in
40/104
30/86
20/68
10/50
0/32
-10/14
-20/-4
40/16
30/12
20/8
10/4
0

The islands have a tropical oceanic climate, with only small temperature variations throughout the year.

TRANSPORTATION

Drive on left

Pointe Larue International, Mahé
155,000 passengers
16 ships
17,699 grt

THE TRANSPORTATION NETWORK

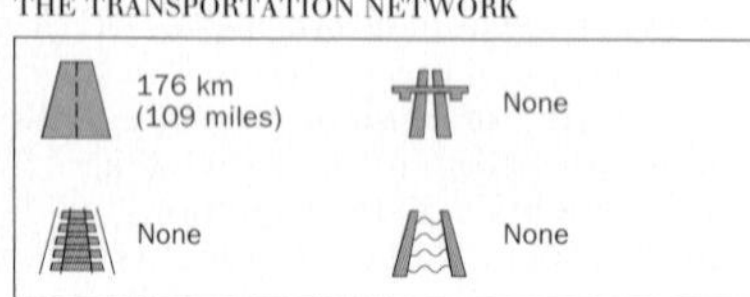
176 km (109 miles)
None
None
None

Nine islands have airstrips. The public transportation fleet and roads are being renewed. Victoria's deep-sea harbor is one of the best run in the region.

TOURISM

Visitors : Population 1.6:1

130,000 visitors

Up 4% in 2000

MAIN TOURIST ARRIVALS

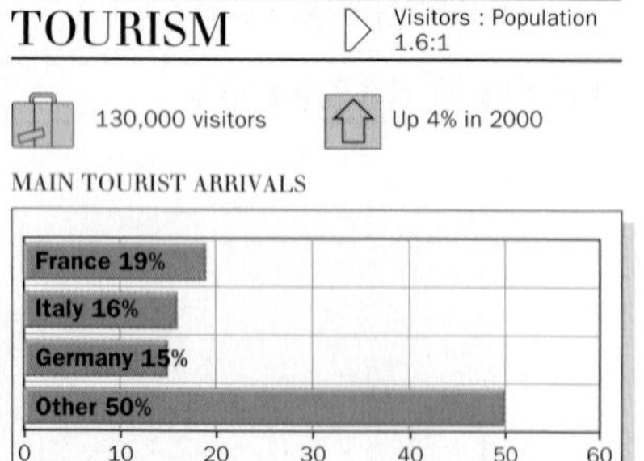
France 19%
Italy 16%
Germany 15%
Other 50%
0 10 20 30 40 50 60
% of total arrivals

Since the international airport was opened at Mahé in 1971, tourism has become the mainstay of the economy, and employs 30% of the workforce. New hotels must comply with laws to protect the islands' beauty and unique wildlife. There is substantial foreign investment.

PEOPLE

Pop. density high

French Creole, English, French

294/km² (761/mi²)

THE URBAN/RURAL POPULATION SPLIT

64% 36%

RELIGIOUS PERSUASION

Other (including Muslim) 2%
Anglican 8%
Roman Catholic 90%

The Seychelles islands were uninhabited before French settlers arrived in the 1770s. Today, the population is markedly homogeneous as a result of intermarriage between different ethnic groups. The Creoles are the descendants of the French settlers and of the Africans who were settled on the islands by British administrators. There are small Chinese and Indian minorities.

Almost 90% of Seychellois live on Mahé. Population growth has been very low, as about 1000 people a year have been emigrating. The transition to democracy may yet prove to reverse this trend.

POLITICS

Multiparty elections

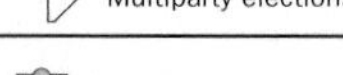
1998/2003
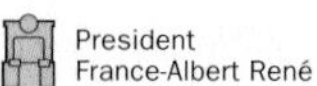
President France-Albert René

AT THE LAST ELECTION

National Assembly 34 seats

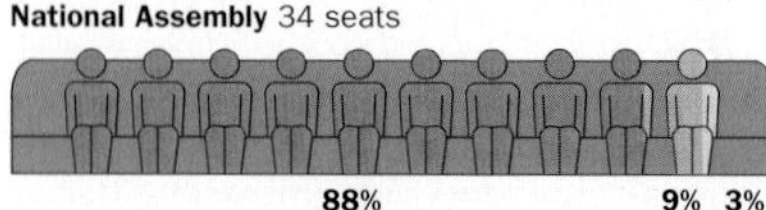

88% SPPF 9% UO 3% DP

SPPF = Seychelles People's Progressive Front **UO** = United Opposition **DP** = Democratic Party

In 1993, the Seychelles returned to democracy after 16 years of one-party socialist rule under President France-Albert René, who had seized power soon after independence. Opposition divisions in the 1993 elections allowed René to retain the presidency. His SPPF kept its majority in the 1998 polls – having dramatically abandoned its leftist ideology, and adopted wide-ranging reforms, encouraging privatization and legalizing trade unions. The highlight is a scheme to develop Seychelles as an International Trading Zone, with free-port facilities and new industry.

WORLD AFFAIRS

Joined UN in 1976

OIF | Comm | COI | NAM | AU

The Seychelles is nonaligned, but its strategic location encourages competing world powers to seek its friendship. It claims the Chagos Archipelago from the UK. Trade accords exist with other Indian Ocean states.

SEYCHELLES

Total Land Area : 455 sq. km (176 sq. miles)

POPULATION
over 10 000
under 10 000
LAND HEIGHT
500m/1640ft
200m/656ft
Sea Level
N
0 100 km
0 100 miles
INNER ISLANDS
Île aux Vaches
Île Denis
Île Aride
Grand'Anse
Praslin
La Digue
Silhouette
Frégate
Mahé
Inset area
AMIRANTE ISLANDS
Île Desroches
Poivre Atoll
Île Platte
INDIAN OCEAN
OUTER ISLANDS
Alphonse Group
Coëtivy
Aldabra Atoll
Grande Terre
Assomption
Cosmolédo Atoll
Astove
ALDABRA GROUP
St. Pierre
Providence Atoll
FARQUHAR GROUP
Farquhar Atoll
Mahé
Anse Étoile
Bel Ombre
VICTORIA
Morne Seychellois 905m
Cascade
Anse Boileau
Anse Royale
Takamaka
Quatre Bornes
0 50 km
0 50 miles

S

AID

Recipient

$18m (receipts) — Up 38% in 2000

Multilateral agencies, notably the EU and the Arab Development Fund, support a range of development projects. In 1996, $13 million was spent on improving transportation links, protecting the environment, and rehabilitating Victoria Market. Bilateral aid comes from the EU, Arab donors, France, and Japan.

DEFENSE

No compulsory military service

$10m

Down 9% in 2000

The Seychelles has a 200-strong army, and a paramilitary guard. The latter includes the coast guard made up of air and sea forces. The army, set up in 1977, was initially trained by Tanzania, and Tanzanian troops were brought in for three years after a coup attempt in 1981. North Korea provided advisers until 1989.

ECONOMICS

Inflation 3.3% p.a. (1990–2000)

$573m

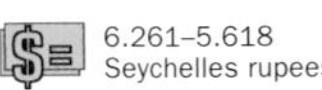

6.261–5.618 Seychelles rupees

SCORE CARD

- World GNP Ranking........168th
- GNP per Capita........$7050
- Balance of Payments........–$114m
- Inflation........6.3%
- Unemployment........9%

Strengths

Tourism. Fish exports, especially shrimp and tuna. Profitable re-export trade. International Trading Zone attracting foreign industrial interest. Copra, cinnamon, tea.

Weaknesses

Growing deficits in early 1990s, caused by drop in tourism following 1991 Gulf War, spending on hosting 1993 Indian Ocean Games, and cost of four elections. High debt-servicing costs. Reliance on food imports, especially for tourist industry. Copra production declining. Reliance on expatriate labor.

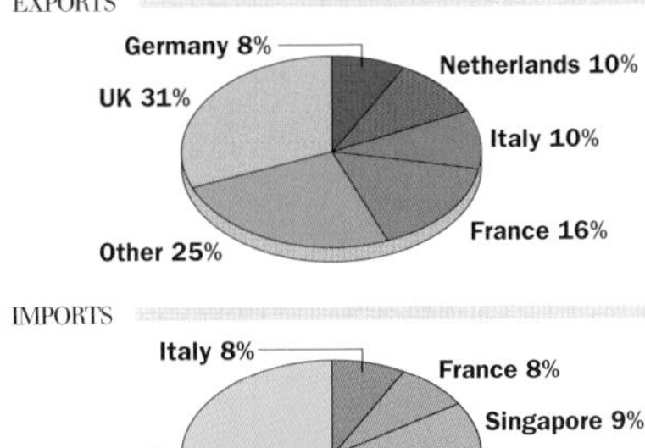
EXPORTS
Germany 8%
Netherlands 10%
UK 31%
Italy 10%
France 16%
Other 25%
IMPORTS
Italy 8%
France 8%
Singapore 9%
Other 52%
South Africa 10%
Saudi Arabia 13%

One of the Inner Islands, which are home to most of the population. Unlike any other mid-ocean islands, all but two are granitic.

RESOURCES

Electric power 28,000 kw

37,992 tonnes

Not an oil producer

18,400 pigs, 5250 goats, 560,000 chickens

Phosphates (guano), salt, granite, natural gas

There are virtually no mineral resources. All fuel is imported; only three islands have electricity. Offshore discoveries of natural gas have spurred a search for oil. Natural habitat and free trade environment are tremendous assets.

ENVIRONMENT

Not available

95%

2.5 tonnes per capita

The Seychelles has been praised for its commitment to conservation. It has two natural World Heritage sites, and helped promote the idea of whale sanctuaries.

MEDIA

TV ownership medium

Daily newspaper circulation 46 per 1000 people

PUBLISHING AND BROADCAST MEDIA

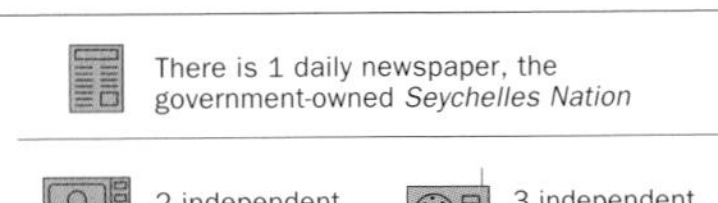

There is 1 daily newspaper, the government-owned *Seychelles Nation*

2 independent services

3 independent services

The state broadcasting company has been reorganized and is now ostensibly free of government control. Privately owned periodicals are now permitted.

CRIME

No death penalty

157 prisoners

Up 72% 1997–1999

Violent crime is rare in the Seychelles. The main concern is the increasing rate of petty theft.

EDUCATION

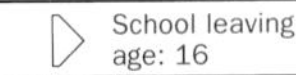

School leaving age: 16

84%

1682 students

The 1995–2008 Educational and Training Plan places special emphasis on increasing levels of female enrollment. National Youth Service is mandatory for entry to higher education.

CHRONOLOGY

The French claimed the islands in 1756. Franco-British rivalry for control ended when France ceded them to Britain in 1815.

- **1952** Political parties formed, led by F. A. René (pro-independence) and James Mancham (pro-UK rule).
- **1965** UK returns Desroches, Aldabra, and Farquhar islands, which are leased to US until 1976.
- **1976** Independence. Coalition with Mancham president, René premier.
- **1977** René takes over in coup.
- **1979** One-party socialist state.
- **1979–1987** Several coup attempts.
- **1992** Politicians return from exile.
- **1993** Democratic elections.
- **2001** René reelected in early presidential elections.

HEALTH

Welfare state health benefits

1 per 758 people

Heart and cerebrovascular diseases, cancers

State health care is free. Private medicine is allowed under new social legislation. Life expectancy is over 70 years.

SPENDING

GDP/cap. increase

CONSUMPTION AND SPENDING

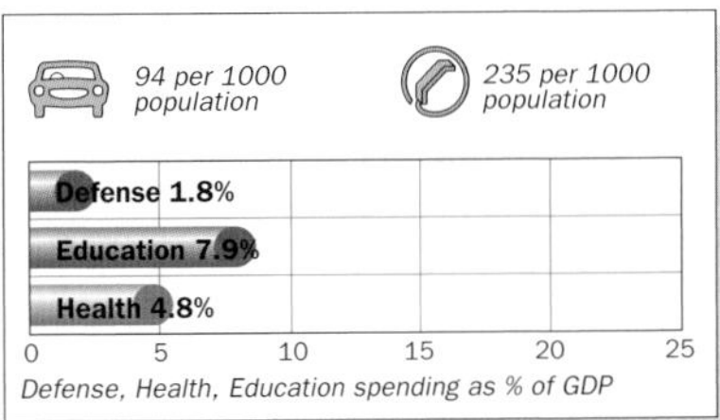
94 per 1000 population
235 per 1000 population
Defense 1.8%
Education 7.9%
Health 4.8%
0
5
10
15
20
25
Defense, Health, Education spending as % of GDP

Living standards are the highest among AU states. There are no slums in the Seychelles, and the state welfare system caters for all.

WORLD RANKING

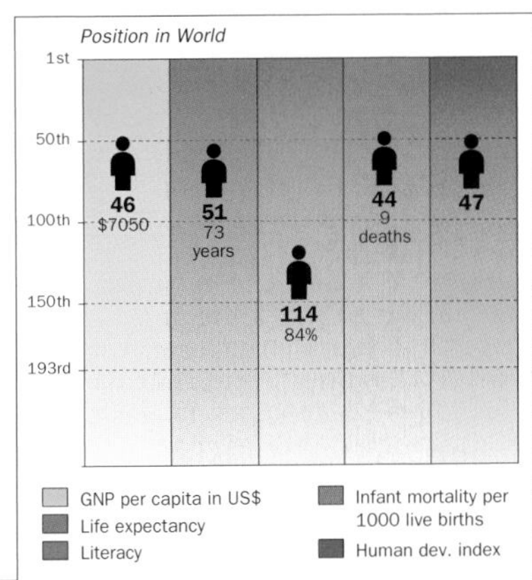
Position in World
1st
50th
100th
150th
193rd
46
$7050
51
73
years
114
84%
44
9
deaths
47
GNP per capita in US$
Life expectancy
Literacy
Infant mortality per 1000 live births
Human dev. index

S

SIERRA LEONE

OFFICIAL NAME: Republic of Sierra Leone **CAPITAL:** Freetown
POPULATION: 4.6 million **CURRENCY:** Leone **OFFICIAL LANGUAGE:** English

WEST AFRICA

Guinea, SIERRA LEONE, Liberia, Ivory Coast (Côte d'Ivoire), Africa

1961
1961

April 27

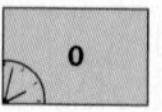
WAL

0

+232

.sl

THE WEST AFRICAN state of Sierra Leone was founded by the British in 1787 for Africans freed from slavery. The terrain rises from coastal lowlands to mountains in the northeast. A democratic government took office in 1996 against a background of bloody rebellion. Sierra Leone soon plunged back into a savage civil war. Although a 1999 peace agreement was short-lived, an ECOWAS-brokered accord signed in late 2000 seems to be holding.

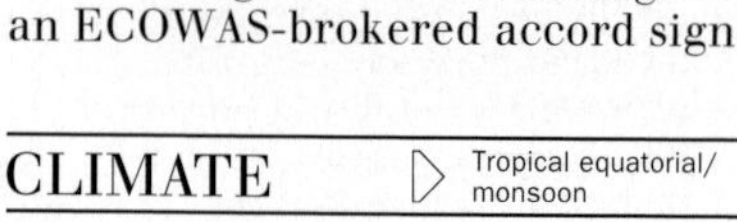

CLIMATE

Tropical equatorial/ monsoon

WEATHER CHART FOR FREETOWN

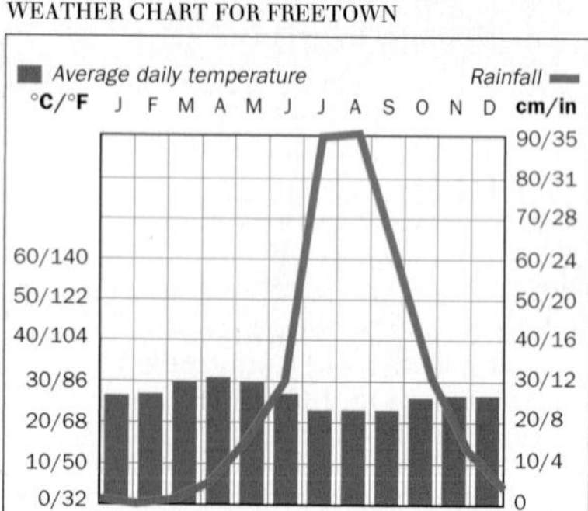
Average daily temperature
Rainfall
°C/°F
J F M A M J J A S O N D
cm/in
60/140
50/122
40/104
30/86
20/68
10/50
0/32
90/35
80/31
70/28
60/24
50/20
40/16
30/12
20/8
10/4
0

Coastal rainfall can be as high as 500 cm (197 in) a year, making Sierra Leone one of the wettest places in coastal west Africa. Humidity is consistently high – about 80% – during the rainy season. The dusty, northeasterly *harmattan* wind often blows during the hotter dry season from November to April. The northeastern savannas are drier, with 190–250 cm (75–98 in) of rain, and are one of the hottest areas.

TRANSPORTATION

Drive on right

Lungi International, Freetown 84,547 passengers

52 ships 18,792 grt

THE TRANSPORTATION NETWORK

904 km (562 miles)

None

84 km (52 miles)

800 km (497 miles)

Little progress has been made in improving Sierra Leone's roads. The 300-km (190-mile) narrow-gauge railroad was abandoned in 1971 as uneconomic, although 84 km (52 miles) of track still runs to the closed iron ore mines at Marampa. Having failed in 1987, Sierra Leone's national airline resumed flights – to Paris only – in 1991. A limited ferry service across the estuary is the only link between Freetown and the airport.

TOURISM

Visitors : Population 1:460

10,000 visitors

Up 67% in 2000

MAIN TOURIST ARRIVALS

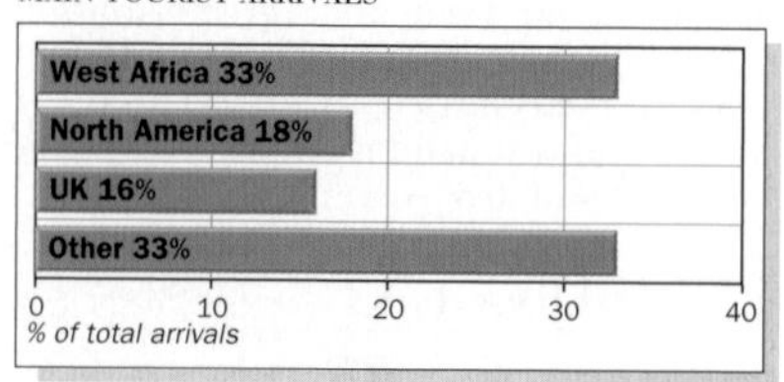
West Africa 33%
North America 18%
UK 16%
Other 33%
0
10
20
30
40
% of total arrivals

Sierra Leone has never attracted many tourists, apart from occasional cruise ship calls. Years of civil war have prevented the development of tourism. Among the chief potential attractions are the beaches along the Freetown peninsula, at present virtually undeveloped.

SIERRA LEONE

Total Land Area : 71 740 sq. km (27 698 sq. miles)

LAND HEIGHT

1000m/3281ft
500m/1640ft
200m/656ft
Sea Level

POPULATION

over 100 000
over 10 000
under 10 000

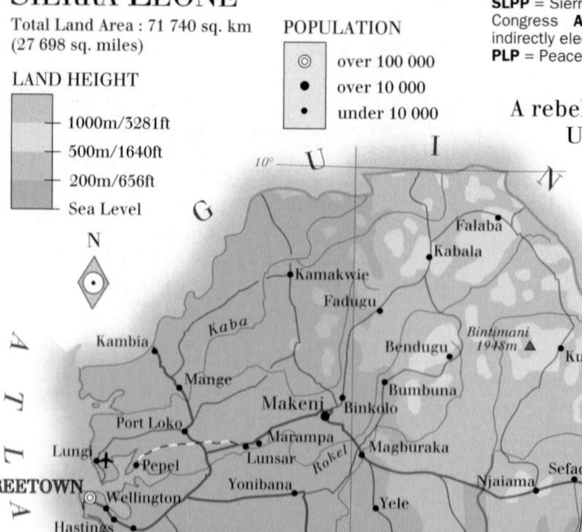
GUINEA
ATLANTIC OCEAN
LIBERIA
Falaba
Kabala
Kamakwie
Fadugu
Bintimani 1948m
Bendugu
Kurubonla
Kambia
Kaba
Mange
Bumbuna
Makeni
Binkolo
Port Loko
Marampa
Lunsar
Rokel
Magburaka
Lungi
Pepel
FREETOWN
Wellington
Yonibana
Njaiama
Sefadu
Koindu
Yele
Hastings
Waterloo
Kailahun
Rotifunk
Banana Is
Taiama
Boajibu
Panguma
Moyamba
Taia
Pendembu
Daru
Sembehun
Mano
Bo
Blama
Kenema
Gbangbatok
Turtle Is
Koribundu
Moa
Bonthe
Sumbuya
Sherbro I.
Sewa
Potoru
Mano
Pujehun
Zimmi
0
50 km
50 miles

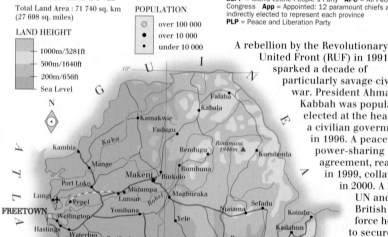

PEOPLE

Pop. density medium

Mende, Temne, Krio, English

64/km² (166/mi²)

THE URBAN/RURAL POPULATION SPLIT

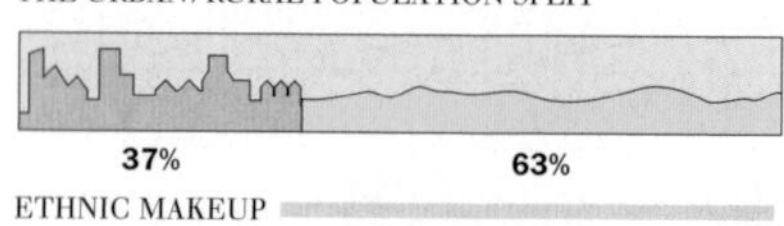
37%
63%

ETHNIC MAKEUP

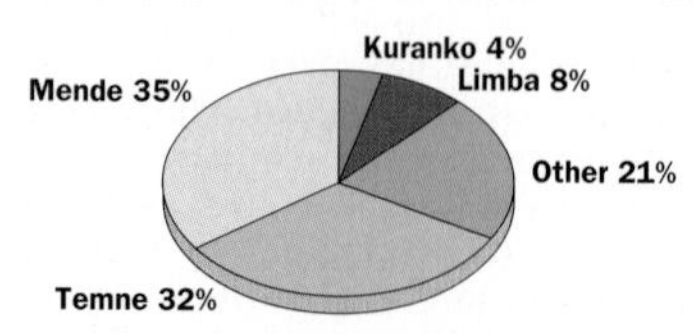
Kuranko 4%
Limba 8%
Mende 35%
Other 21%
Temne 32%

Freetown was founded as a settlement for people freed from slavery. Its citizens' British and North American origins account for Sierra Leone's strongly anglicized Creole culture. An estimated two million people were displaced by the civil war.

POLITICS

Multiparty elections

2002/2007

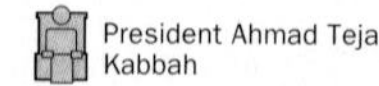

President Ahmad Tejan Kabbah

AT THE LAST ELECTION

National Assembly 124 seats

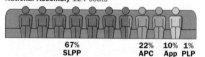

67% SLPP **22% APC** **10% App** **1% PLP**

SLPP = Sierra Leone People's Party **APC** = All People's Congress **App** = Appointed: 12 paramount chiefs are indirectly elected to represent each province
PLP = Peace and Liberation Party

A rebellion by the Revolutionary United Front (RUF) in 1991 sparked a decade of particularly savage civil war. President Ahmad Kabbah was popularly elected at the head of a civilian government in 1996. A peace and power-sharing agreement, reached in 1999, collapsed in 2000. A large UN and British force helped to secure a new cease-fire later that year. Remodeled as a political party, the RUF failed to win a single seat in the 2002 elections, which resulted in a convincing victory for Kabbah and his SLPP.

S

WORLD AFFAIRS

Joined UN in 1961

Comm

ECOWAS

MRU

AU

OIC

UN peacekeepers and British forces assisted government forces during renewed fighting in 2000.

AID

Recipient

 $182m (receipts)

 Up sharply in 2000

The IMF and the World Bank agreed in 2002 to drop 80% of Sierra Leone's debt in return for key reforms.
Aid funds efforts to cope with the humanitarian needs of refugees from Liberia, internal migrants displaced by the civil war, and the near-collapse of public services.

DEFENSE

No compulsory military service

 $9m

Down 18% in 2000

The army is an ineffectual fighting force, prompting assistance from the UK with training in 2000. Both sides in the civil war exploited child fighters.

ECONOMICS

Inflation 29% p.a. (1990–2000)

$647m

 1899–2091 leones

SCORE CARD

- ❏ WORLD GNP RANKING........................164th
- ❏ GNP PER CAPITA $130
- ❏ BALANCE OF PAYMENTS....................–$127m
- ❏ INFLATION ....–0.8%
- ❏ UNEMPLOYMENT ....Widespread

STRENGTHS

Diamonds, although much of the output is smuggled; official exports resumed in late 2000 under a UN certification scheme. Some bauxite and rutile production.

WEAKNESSES

Years of instability affected the most productive areas, including diamond fields, with severe disruption of agricultural and mining sectors.

EXPORTS

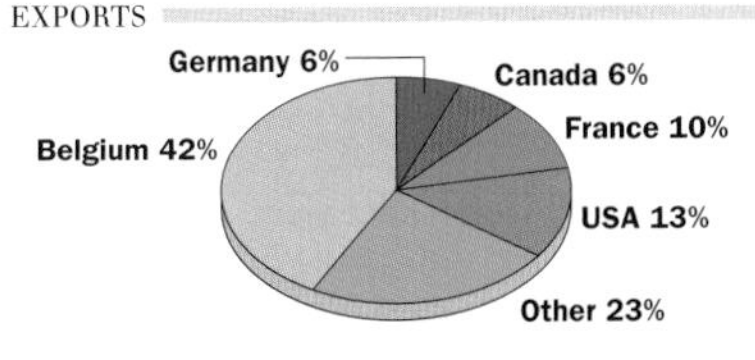
Germany 6%
Canada 6%
France 10%
Belgium 42%
USA 13%
Other 23%

IMPORTS

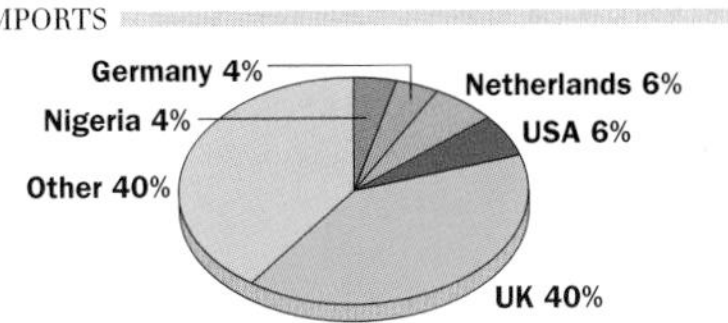
Germany 4%
Netherlands 6%
Nigeria 4%
USA 6%
Other 40%
UK 40%

RESOURCES

Electric power 126,000 kw

 59,437 tonnes

 Not an oil producer

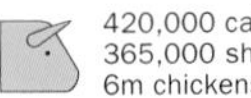 420,000 cattle, 365,000 sheep, 6m chickens

 Diamonds, rutile, bauxite, gold, titanium

The large diamond deposits need fresh investment as areas currently being mined become depleted. The southeast is the most fertile region.

ENVIRONMENT

Sustainability rank: 134th

 1%

 0.1 tonnes per capita

Population pressures and the neglect resulting from years of civil war have depleted the land's productivity.

MEDIA

TV ownership low

 Daily newspaper circulation 5 per 1000 people

PUBLISHING AND BROADCAST MEDIA

 There is 1 daily newspaper, the *Daily Mail*, published by the government

 1 state-controlled service

 1 state-controlled service

A broad range of periodicals is available. The government has promised press freedom, but the Internet is heavily censored.

CRIME

Death penalty in use

 Sierra Leone does not publish prison figures

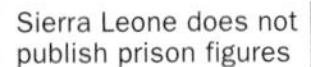

 Crime is rising

The civil war resulted in savage atrocities and the mass looting of resources. As a consequence, the UN has instigated a war crimes tribunal. International restrictions on trade in diamonds from war-torn areas cut RUF revenues, but illegal diamond mining and smuggling remain lucrative crimes.

EDUCATION

Schooling is not compulsory

  36%

 6744 students

Freetown has a long tradition of education, and its university, Fourahbay College, became affiliated with Durham University in the UK in 1876. In recent times, its students have often been active in political dissent. Educational provision has inevitably deteriorated over the past decade.

HEALTH

No welfare state health benefits

 1 per 10,000 people

 Communicable diseases, malaria, malnutrition

Only traditional care is available outside the capital. WHO has ranked Sierra Leone's health care bottom in the world in terms of attainment and efficiency.

The main street, Kabala. *Sierra Leone is consistently near the bottom of the UN's Human Development Index.*

CHRONOLOGY

Freetown was founded in 1787 and became a British colony in 1808; the interior was annexed in 1896.

- ❏ **1961** Independence.
- ❏ **1978** Single-party republic.
- ❏ **1991** RUF rebellion starts.
- ❏ **1996** Civilian rule restored after 1992 army coup; Kabbah president.
- ❏ **1997** Coup ousts Kabbah for a year.
- ❏ **1999–2000** Power-sharing attempt.
- ❏ **2001** RUF ends insurgency.
- ❏ **2002** Government and UN agree to set up war crimes court. Kabbah and SLPP reelected.

SPENDING

GDP/cap. decrease

CONSUMPTION AND SPENDING

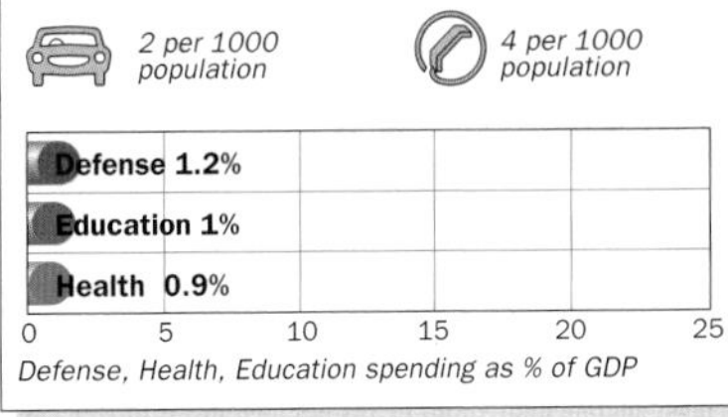
2 per 1000 population
4 per 1000 population
Defense 1.2%
Education 1%
Health 0.9%
0
5
10
15
20
25
Defense, Health, Education spending as % of GDP

In terms of quality of life, the UN has repeatedly ranked Sierra Leoneans as the world's poorest people. Any wealth is associated with political power.

S

WORLD RANKING

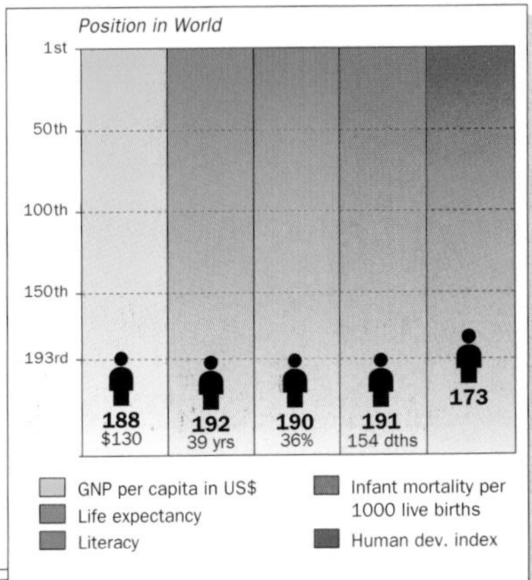
Position in World
1st
50th
100th
150th
193rd
188
$130
192
39 yrs
190
36%
191
154 dths
173
GNP per capita in US$
Life expectancy
Literacy
Infant mortality per 1000 live births
Human dev. index

SINGAPORE

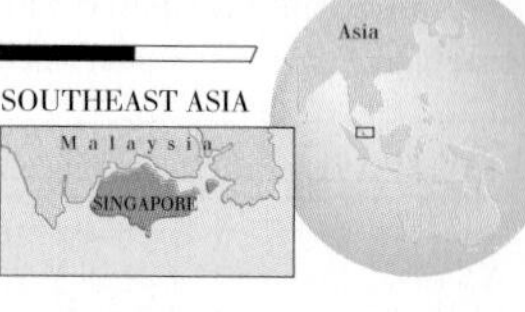
Asia
SOUTHEAST ASIA
Malaysia
SINGAPORE

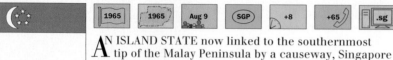

OFFICIAL NAME: Republic of Singapore **CAPITAL:** Singapore **POPULATION:** 4.1 million
CURRENCY: Singapore dollar **OFFICIAL LANGUAGES:** Malay, English, Mandarin, and Tamil

1965

1965

Aug 9
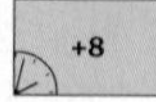
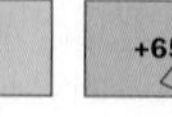
SGP

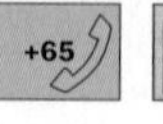
+8

+65

.sg

AN ISLAND STATE now linked to the southernmost tip of the Malay Peninsula by a causeway, Singapore ("lion city") was largely uninhabited until the 19th century. In 1819, an official of the British East India Company, Stamford Raffles, recognized the island's strategic position on key trade routes, and established a trading settlement. Today, Singapore remains one of the most important entrepôts in Asia.

CLIMATE

Tropical equatorial

WEATHER CHART FOR SINGAPORE

Average daily temperature / Rainfall
°C/°F J F M A M J J A S O N D cm/in
40/104 40/16
30/86 30/12
20/68 20/8
10/50 10/4
0/32 0
-10/14
-20/4

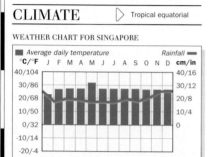

The only variations in the hot, wet, and humid climate are the airless months of September and March, when the trade winds change direction.

TRANSPORTATION

Drive on left

Changi International
28.6m passengers

1677 ships
20.4m grt

THE TRANSPORTATION NETWORK

3038 km (1888 miles)

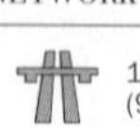

150 km (93 miles)

26 km (16 miles)

None

The Mass Rapid Transit System (subway), completed in 1991, is among the world's most efficient. Space for new roads has run out and monthly auctions are held to sell certificates entitling people to buy from a quota of new cars. The massive port at Pasir Panjang is being expanded on reclaimed land.

The financial center. *More than a quarter of Singapore's GDP is generated by financial and business services.*

TOURISM

Visitors : Population 1.5:1

6.26m visitors

Up 11% in 2000

MAIN TOURIST ARRIVALS

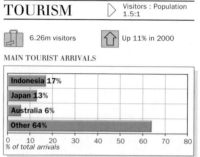

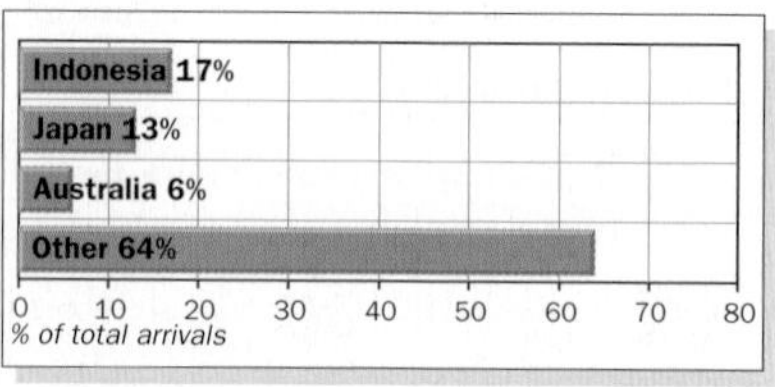
Indonesia 17%
Japan 13%
Australia 6%
Other 64%
0 10 20 30 40 50 60 70 80
% of total arrivals

The buildings of Chinatown, recognized as a picturesque tourist asset, are being restored. A Singaporean consortium is involved in developing a resort on Indonesia's Bintan island, some 45 km (28 miles) across the Strait of Singapore.

PEOPLE

Pop. density high

Mandarin, Malay, Tamil, English

6721/km² (17,408/mi²)

THE URBAN/RURAL POPULATION SPLIT

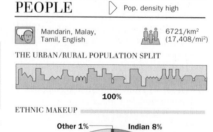

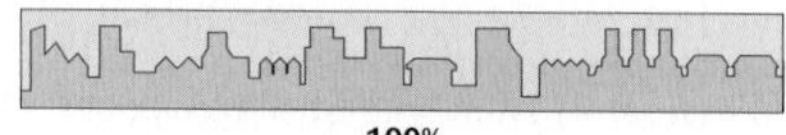

100%

ETHNIC MAKEUP

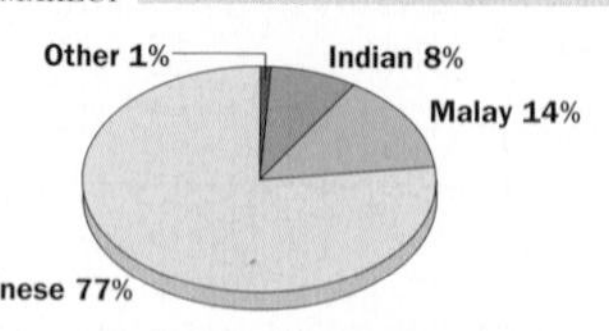
Other 1%
Indian 8%
Malay 14%
Chinese 77%

Singapore is dominated by the Chinese – the old-established English-speaking Straits Chinese and newer Mandarin speakers – who make up almost 80% of the community. Indigenous Malays are generally the poorest group, but today there is little overt ethnic tension. There is also a significant foreign workforce in Singapore.

Long-term plans to stabilize the population structure included the announcement in mid-2000 of cash bonuses for families with more than one child. Society is highly regulated and government campaigns to improve public behavior are frequent.

POLITICS

Multiparty elections

2001/2006

President S. R. Nathan

AT THE LAST ELECTION
Parliament 94 seats

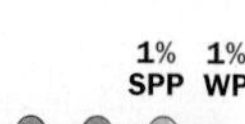
1% 1%
SPP WP

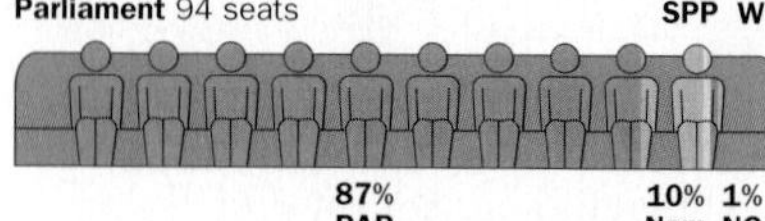
87% PAP
10% Nom
1% NC

PAP = People's Action Party **Nom** = Nominated
SPP = Singapore People's Party **NC** = Nonconstituency member **WP** = Workers' Party

In addition to the 84 elected members, up to six "nonconstituency" members may be nominated from the losers with the most votes and nine members may be nominated to ensure a wider representation in Parliament

Singapore is a multiparty democracy, although the ruling PAP effectively controls all parts of the political process and much of the economy. There are plans to create a national ideology ("shared values") based on Confucian traditions. In 1993, Ong Teng Cheong became the first president to be directly elected. The veteran politician, Lee Kuan Yew, prime minister for more than 30 years until his resignation in 1990, still exercises influence.

The PAP retains its grip on power, having given Singapore one of the highest living standards in the world, based on a free-market economy. The first antigovernment rally was permitted in 2001, but even before the ballot the PAP was sure of reelection, fewer than half of the seats being contested. To improve parliamentary debate, the PAP formed an internal opposition faction.

WORLD AFFAIRS

Joined UN in 1965

APEC
ASEAN
Comm
NAM
WTO

Singapore has established diplomatic relations with China, while continuing to maintain close economic ties with Taiwan. It became the beneficiary in 2001 of Japan's first free trade agreement.

AID

Recipient

US$1m (receipts)

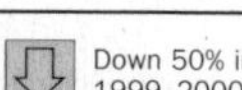

Down 50% in 1999–2000

Aid is not an important issue in Singapore. The state does not provide aid to any states in southeast Asia.

DEFENSE

Compulsory military service

US$4.71bn

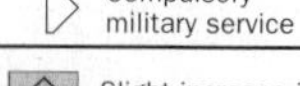

Slight increase in 2000

Despite Singapore's small size, its armed forces have a total strength of over 60,000.

S

ECONOMICS

Inflation 1.3% p.a. (1990–2000)

US$99.4bn

1.734–1.847 Singapore dollars

SCORE CARD

- ❑ WORLD GNP RANKING..........................37th
- ❑ GNP PER CAPITA ....US$24,740
- ❑ BALANCE OF PAYMENTS ....US$21.8bn
- ❑ INFLATION 1.4%
- ❑ UNEMPLOYMENT 3%

EXPORTS

- Taiwan 6%
- Japan 8%
- Hong Kong 8%
- USA 17%
- Malaysia 18%
- Other 43%

IMPORTS

- Taiwan 4%
- China 5%
- USA 15%
- Malaysia 17%
- Japan 17%
- Other 42%

STRENGTHS

Massive accumulated wealth is derived from success as an entrepôt and as a center of high-tech industries. Singapore is a major producer of computer disk drives. Huge state enterprises, such as TAMESEK, with over 450 companies, have proved highly flexible in responding to market conditions. World leader in new biotechnologies.

WEAKNESSES

Dependence on Malaysia for water. Almost all food and energy imported. Skills shortages, notably in engineering. Fluctuations in world electronics market caused recession in 2001. Lack of land.

CHRONOLOGY

In 1819, Sir Stamford Raffles set up a trading post in the village of Singapore; by 1826 it was Britain's colonial center in southeast Asia.

- ❑ **1959** PAP becomes ruling party.
- ❑ **1965** Independence.
- ❑ **1990** Lee Kuan Yew resigns as prime minister.
- ❑ **1993** Ong Teng Cheong first directly elected president.
- ❑ **2001** PAP, in power since independence, wins elections.

RESOURCES

Electric power 5.7m kw

9081 tonnes

Not an oil producer; refines 1.03m b/d

600,000 ducks,
190,000 pigs,
2m chickens

None

Singapore has no strategic resources and has to import almost all the energy and food it needs. Its main resources, on which its wealth as a center of commerce has been built, are its strategic position and its people.

ENVIRONMENT

Not available

5%

21 tonnes per capita

There is a small green belt around the causeway. Singapore sees itself as a world leader in providing the perfect urban environment. There is no litter, thanks to instant heavy fines; chewing gum is banned by law.

MEDIA

TV ownership high

Daily newspaper circulation 324 per 1000 people

PUBLISHING AND BROADCAST MEDIA

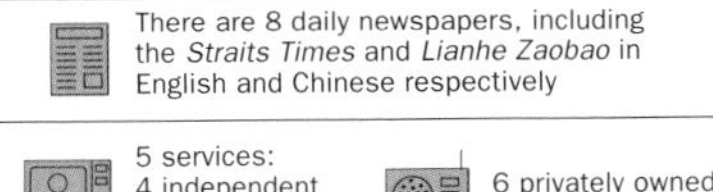
There are 8 daily newspapers, including the *Straits Times* and *Lianhe Zaobao* in English and Chinese respectively

5 services: 4 independent, 1 US-controlled

6 privately owned services

The government is very sensitive to any criticism. However, it has declared its intention partially to liberalize the local media, as a part of "constructive competition." Foreigners may not own newspapers or broadcasting stations.

CRIME

Death penalty in use

13,791 prisoners

Down 21% in 1999

Crime is limited and punishment can be severe. The Triads are no longer a problem; the main issue is intellectual piracy.

EDUCATION

Schooling is not compulsory

92%

124,180 students

Schooling is not compulsory, but attendance is high. Education is seen as the key to a good salary, especially among the Chinese community.

HEALTH

Welfare state health benefits

1 per 625 people

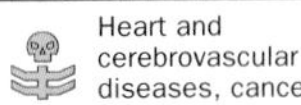
Heart and cerebrovascular diseases, cancers

Singapore has an efficient modern health system. There are incentives to ensure the continuation of the extended family, so that the elderly are cared for at home.

SPENDING

GDP/cap. increase

CONSUMPTION AND SPENDING

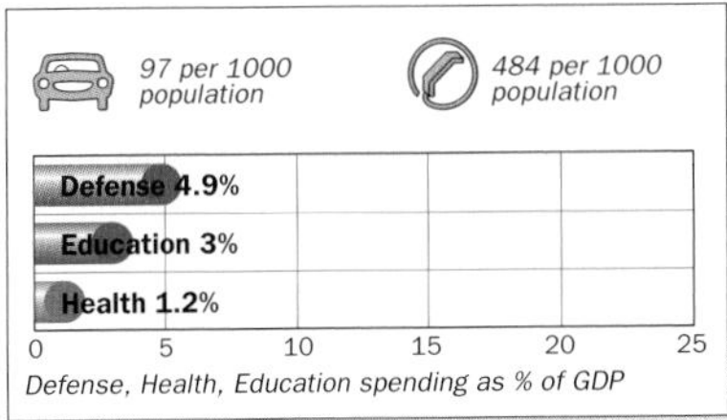
97 per 1000 population
484 per 1000 population
Defense 4.9%
Education 3%
Health 1.2%
0
5
10
15
20
25
Defense, Health, Education spending as % of GDP

The 2001 "Singapore Share" scheme promises to give Singaporeans a share in the country's economy.

WORLD RANKING

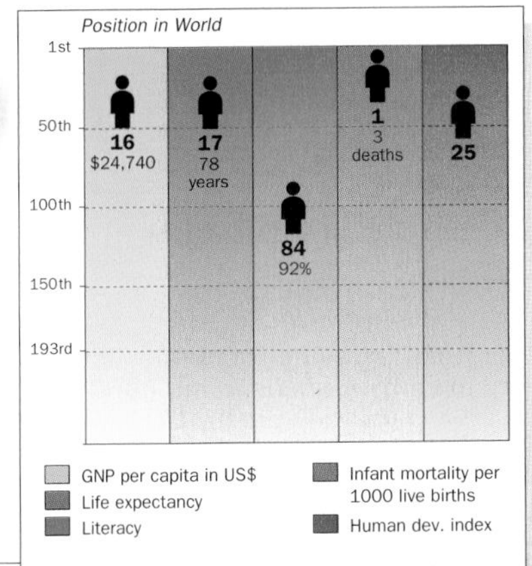
Position in World
1st
50th
100th
150th
193rd
16
$24,740
17
78 years
84
92%
1
3 deaths
25
GNP per capita in US$
Life expectancy
Literacy
Infant mortality per 1000 live births
Human dev. index

SINGAPORE

Total Land Area : 648 sq. km (250 sq. miles)

MALAYSIA
Johore Strait
Causeway
Sembawang
0
5 km
5 miles
N
Sarimbun Res.
Lim Chu Kang
Murai Res.
Kranji Res.
Seletar Res.
Sungai Seletar Res.
Pulau Ubin
Pulau Tekong Kechil
Pulau Tekong Res.
Pulau Tekong
Punggol
Serangoon Harbour
Changi
Changi International Airport
Bukit Panjang
Serangoon
Poyan Res.
Choa Chu Kang
Bukit Timah 176m
Upper Peirce Res.
MacRitchie Res.
Kallang
Bedok Res.
Jurong
Jurong Lake
Bukit Timah
Tengeh Res.
Jurong Town
1°20'
Jurong Industrial Estate
Pandan Res.
Queenstown
City
Geylang
Katong
104°
SOUTH CHINA SEA
Pasir Panjang
Singapore
Telok Blangah
Keppel Harbour
Pulau Pesek
Pulau Pesek Kechil
Pulau Ayer Chawan
Pulau Merlimau
Pulau Seraya
Pulau Retan Laut
Pulau Ayer Merbau
Pulau Sakra
Pulau Brani
Sentosa
Selat Pandan
Pulau Busing
Pulau Bukum
Pulau Hantu
Pulau Sudong
Pulau Tembakul
Pulau Sakijang Pelepah
Pulau Sakijang Bendera
Pulau Pawai
Pulau Semakau
Pulau Sebarok
103°40'
103°50'
Pulau Senang
Strait of Singapore
Urban Areas
Open Areas
Nature Reserve

S

SLOVAKIA

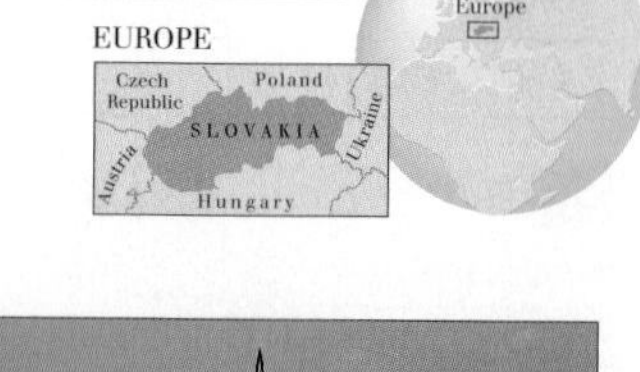
EUROPE
Europe
Czech Republic
Poland
SLOVAKIA
Austria
Ukraine
Hungary

OFFICAL NAME: Slovak Republic **CAPITAL:** Bratislava
POPULATION: 5.4 million **CURRENCY:** Slovak Koruna **OFFICAL LANGUAGE:** Slovak

1993

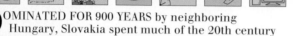

1993

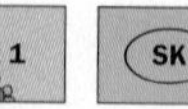
Oct 1

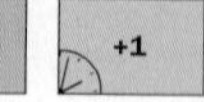
SK

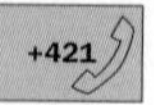
+1

+421

.sk

DOMINATED FOR 900 YEARS by neighboring Hungary, Slovakia spent much of the 20th century as the less developed half of communist Czechoslovakia. An independent democracy since 1993, Slovakia has struggled to create a modern market-led economy, but is now among the "second wave" of states conducting membership negotiations with the EU.

***Levoča, in northeastern Slovakia,** dates from the 13th century and still retains its medieval street plan and town walls.*

CLIMATE

Continental

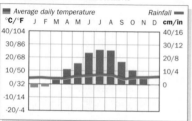

WEATHER CHART FOR BRATISLAVA

Slovakia has a continental climate. Snowfalls are heavy in winter, while summers are moderately warm.

TRANSPORTATION

Drive on right

Milan Rastislav Stefanik, Bratislava 292,515 passengers

Has no fleet

THE TRANSPORTATION NETWORK

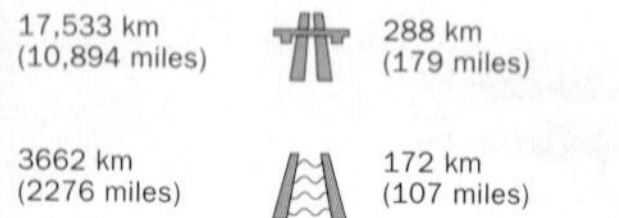
17,533 km
(10,894 miles)
288 km
(179 miles)
3662 km
(2276 miles)
172 km
(107 miles)

The Danube River is a vital artery. Trains are cheap and efficient. Buses and trams are the mainstay of urban transport.

TOURISM

Visitors : Population 1:5.1

1.05m visitors

Up 8% in 2000

MAIN TOURIST ARRIVALS

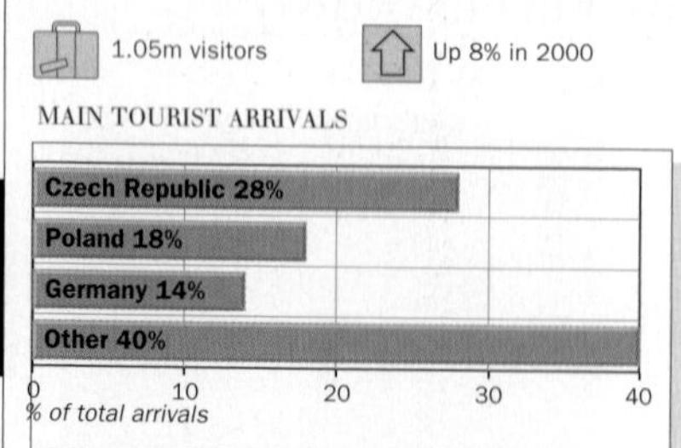
Czech Republic 28%
Poland 18%
Germany 14%
Other 40%
0
10
20
30
40
% of total arrivals

The High Tatras, one of the smallest high-mountain ranges in the world, and the Vrátna valley in the Little Tatras draw hikers and skiers. Tourists also visit Bratislava's castle and old city, and the many thermal-spring health spas.

PEOPLE

Pop. density medium

Slovak, Hungarian, Czech

110/km² (285/mi²)

THE URBAN/RURAL POPULATION SPLIT

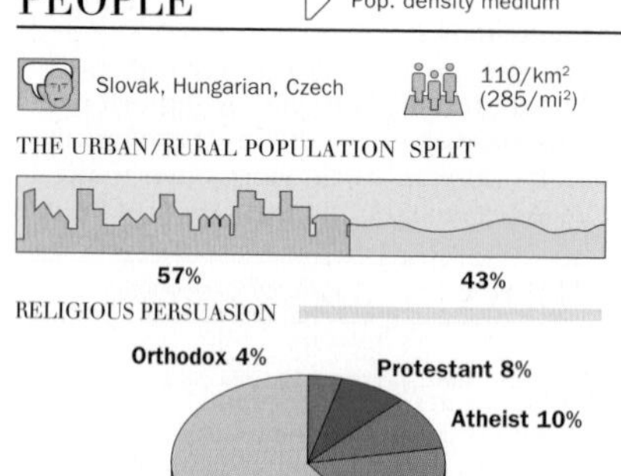
57%
43%
RELIGIOUS PERSUASION
Orthodox 4%
Protestant 8%
Atheist 10%
Other 18%
Roman Catholic 60%

Slovaks dominate society, but 11% of the population is Hungarian, and there is a significant Roma minority which faces discrimination. The Hungarian community, backed by Hungary, seeks protection for its language and culture. Tensions lessened in 1998 when its main political voice, the Hungarian Coalition Pary, joined Dzurinda's government. There were 300,000 Slovaks living in Czech lands in 1993. Dual citizenship is now permitted.

POLITICS

Multiparty elections

2002/2006

President Rudolf Schuster

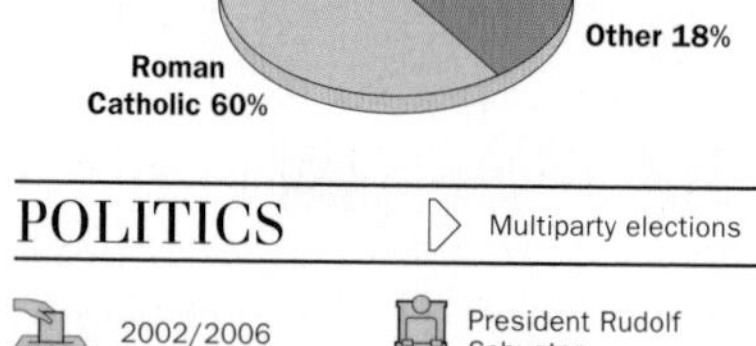

AT THE LAST ELECTION

National Council 150 seats

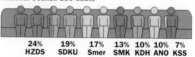

24%	19%	17%	13%	10%	10%	7%
HZDS	SDKU	Smer	SMK	KDH	ANO	KSS

HZDS = Movement for a Democratic Slovakia
SDKU = Slovak Democratic and Christian Union
Smer = Direction **SMK** = Hungarian Coalition Party
KDH = Christian Democratic Movement **ANO** = New Civic Alliance **KSS** = Slovak Communist Party

The move which led to the separation of the two halves of the former Czechoslovakia in 1993 gathered momentum when Slovak leader Vladimir Meciar was tempted by independence as a way of enhancing his power base. The populist Meciar dominated Slovak politics until 1998, but clashed repeatedly with President Michal Kovac, both of them members of the HZDS.

A broad coalition led by center-right politician Mikulas Dzurinda won power in elections in 1998 and retained its position in 2002. Slovakia's first direct presidential election in May 1999 was won by the pro-Western Rudolf Schuster, frustrating Meciar's bid for this office.

The Hungarian minority has its own parties, whereas the Roma have no official representation.

WORLD AFFAIRS

Joined UN in 1993

CE

CEFTA

OECD

EAPC
OSCE

The 1998 change of government reversed the pro-Russian Meciar years. Slovakia was among six "second wave" candidates which began EU membership talks in March 2000. It also became a member of the OECD in 2000.

AID

Recipient

$113m (receipts)

Down 65% in 2000

Foreign aid fell after the mid-1990s, but EU programs are now in place in preparation for EU membership.

DEFENSE

Compulsory military service

$340m

Up 11% in 2000

The Slovak armed forces include some 13,600 conscripts at any one time. Prime Minister Dzurinda, in office since late 1998, reversed Meciar's pro-Russian defense policies.

RESOURCES

Electric power 7.8m kw

2263 tonnes

1203 b/d (reserves 7.5m barrels)

1.49m pigs, 646,100 cattle, 13.6m chickens

Coal, lignite, gas, oil, antimony, copper, iron, mercury, zinc

44% of electricity was nuclear-generated, even before the Mochovce nuclear plant began operations in 1998.

S

SLOVAKIA

Total Land Area : 48 845 sq. km (18 859 sq. miles)

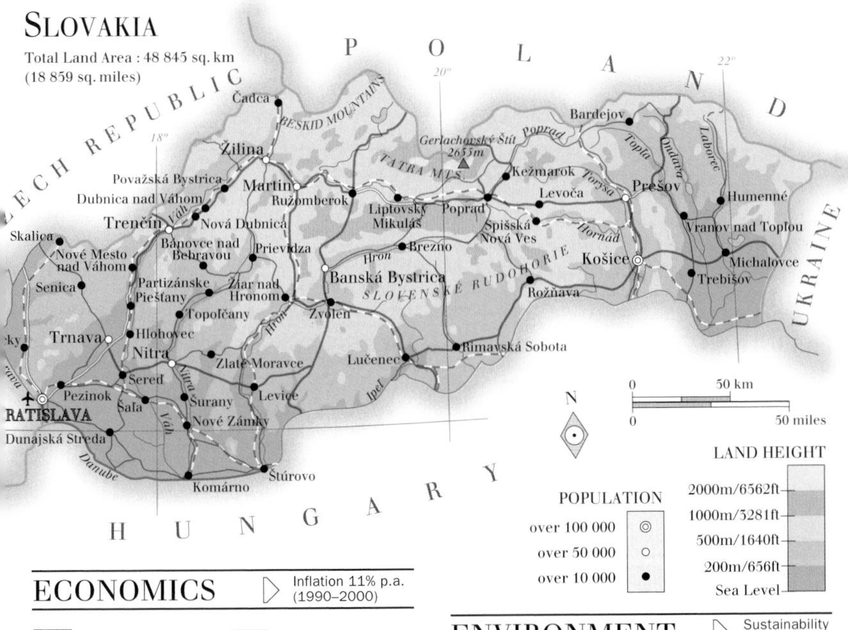
P O L A N D
H U N G A R Y
UKRAINE
CZECH REPUBLIC
BESKID MOUNTAINS
TATRA MTS.
Gerlachovský Štít 2655m
SLOVENSKÉ RUDOHORIE
Čadca
Žilina
Považská Bystrica
Dubnica nad Váhom
Martin
Ružomberok
Trenčín
Nová Dubnica
Bánovce nad Bebravou
Prievidza
Skalica
Nové Mesto nad Váhom
Senica
Partizánske
Piešťany
Žiar nad Hronom
Topoľčany
Zvolen
Banská Bystrica
Brezno
Liptovský Mikuláš
Poprad
Kežmarok
Levoča
Spišská Nová Ves
Bardejov
Prešov
Humenné
Vranov nad Topľou
Košice
Michalovce
Trebišov
Rožňava
Rimavská Sobota
Lučenec
Trnava
Hlohovec
Nitra
Zlaté Moravce
Sereď
Pezinok
Šaľa
Surany
Levice
Nové Zámky
BRATISLAVA
Dunajská Streda
Komárno
Štúrovo
Váh
Hron
Nitra
Ipeľ
Danube
Poprad
Topľa
Ondava
Laborec
Torysa
Hornád
N
0 50 km
0 50 miles
LAND HEIGHT
2000m/6562ft
1000m/3281ft
500m/1640ft
200m/656ft
Sea Level
POPULATION
over 100 000
over 50 000
over 10 000

CHRONOLOGY

Once part of the Austro-Hungarian Empire, Slovakia and the Czech Lands formed the Republic of Czechoslovakia in 1918.

- ❑ **1939–1945** Separate Slovak state under pro-Nazi Jozef Tiso.
- ❑ **1945** Czechoslovak state restored.
- ❑ **1947** Communists seize power.
- ❑ **1968** "Prague Spring" ended by Warsaw pact invasion.
- ❑ **1989** "Velvet Revolution."
- ❑ **1990** Free multiparty elections.
- ❑ **1993** 1st January, separate Slovak and Czech states established.
- ❑ **1994** HZDS election victory.
- ❑ **1998** Broad-based coalition under Dzurinda wins general election.
- ❑ **1999** Rudolf Schuster defeats Meciar in direct presidential poll.
- ❑ **2002** Dzurinda coalition reelected.

ECONOMICS

Inflation 11% p.a. (1990–2000)

$20bn
46.88–48.01 Slovak koruny

SCORE CARD

❑ World GNP Ranking	66th
❑ GNP Per Capita	$3700
❑ Balance of Payments	–$694m
❑ Inflation	12%
❑ Unemployment	19%

STRENGTHS

Increase in manufacturing, especially in Bratislava and surrounding area. Recent progress in cutting budget deficits and restructuring public and private sector. Growth in exports to EU. Potential for tourism, particularly skiing in the Tatra Mountains.

WEAKNESSES

High foreign indebtedness. Dependence on foreign trade makes the economy vulnerable to global recession. Heavy industry has found some new markets in the West, but struggles with poor productivity. Slow to attract foreign investment until 2000. Much poorer eastern region. Growing unemployment.

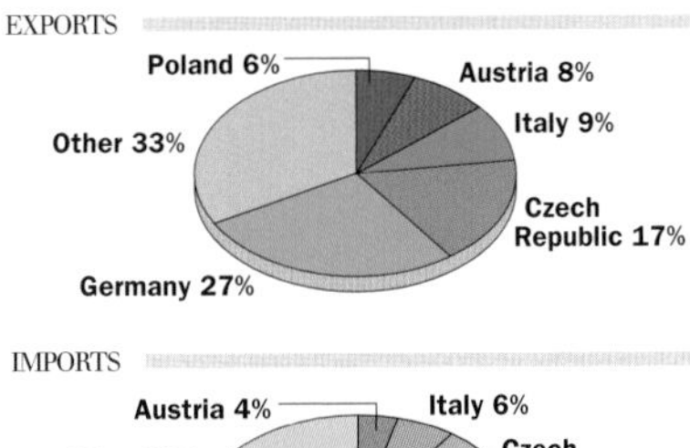
EXPORTS
Poland 6%
Austria 8%
Italy 9%
Other 33%
Czech Republic 17%
Germany 27%
IMPORTS
Austria 4%
Italy 6%
Czech Republic 15%
Other 33%
Russia 17%
Germany 25%

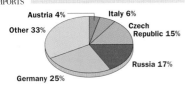

ENVIRONMENT

Sustainability rank: 14th

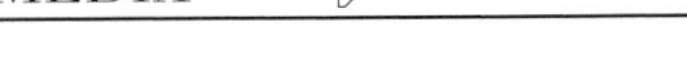
23%
7.1 tonnes per capita

The Gabcikovo Dam and the Bohunice nuclear reactors, now scheduled for partial closure, have provoked criticism.

MEDIA

TV ownership high

Daily newspaper circulation 184 per 1000 people

PUBLISHING AND BROADCAST MEDIA

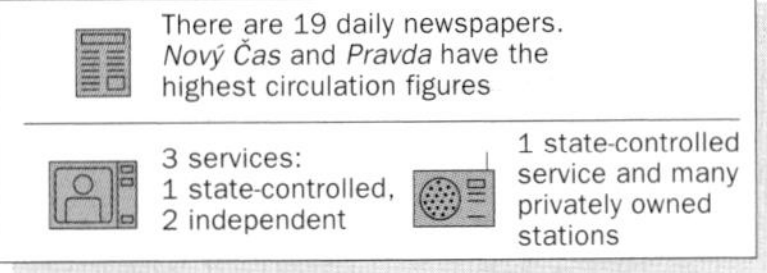
There are 19 daily newspapers. Nový Čas and Pravda have the highest circulation figures
3 services: 1 state-controlled, 2 independent
1 state-controlled service and many privately owned stations

The state news agency TASR, accused of lacking objectivity and depending on government funding, resisted the emergence of independent rival SITA.

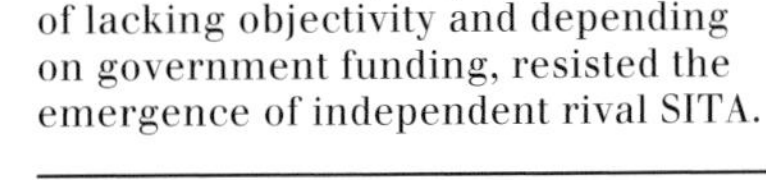

CRIME

No death penalty

7285 prisoners
Down 6% 1996–1998

Organized crime has increased rapidly in recent years, as has "white collar crime" such as business fraud. A new law to control money laundering took effect in 2001. A former economics minister accused of embezzlement was murdered in 1999.

EDUCATION

School leaving age: 15

99%
101,764 students

Schooling now draws on pre-1939 Slovak traditions but it is not adequately resourced, especially in rural areas. There is a modern university in Bratislava.

HEALTH

Welfare state health benefits

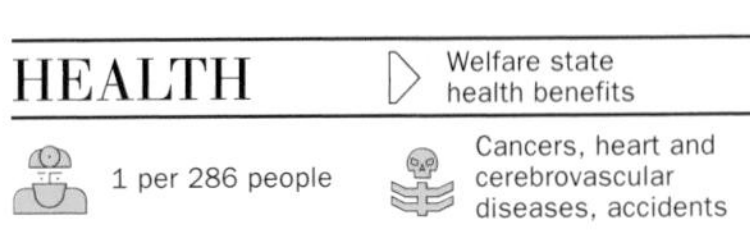
1 per 286 people
Cancers, heart and cerebrovascular diseases, accidents

Rising demand and costs are straining the health service severely. Restoring viability is now a government priority.

SPENDING

GDP/cap. increase

CONSUMPTION AND SPENDING

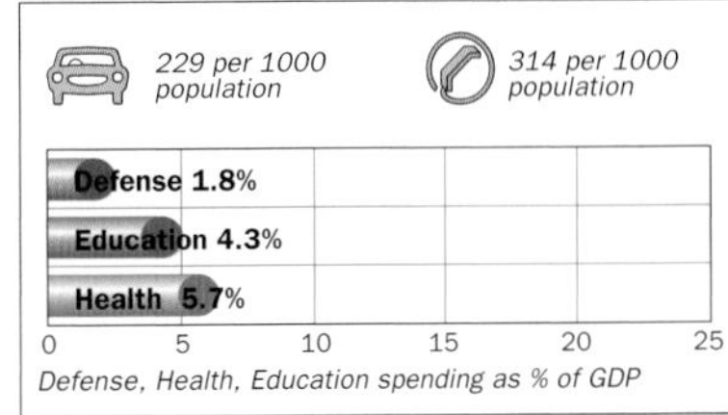
229 per 1000 population
314 per 1000 population
Defense 1.8%
Education 4.3%
Health 5.7%
0 5 10 15 20 25
Defense, Health, Education spending as % of GDP

A new elite is increasing demand for Western goods. Rural workers, Roma, and those living in the east are the poorest.

WORLD RANKING

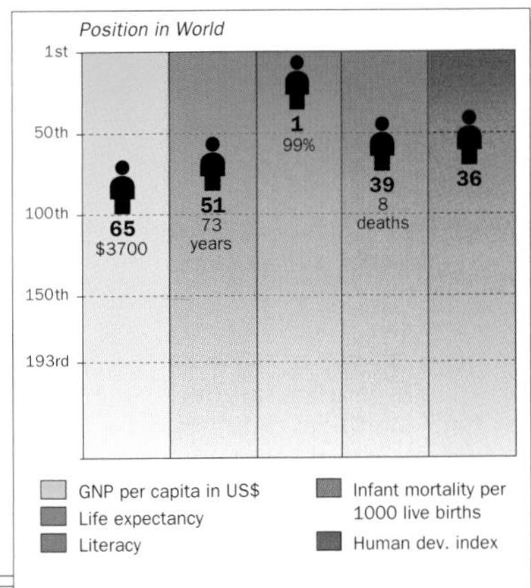
Position in World
1st
50th
100th
150th
193rd
65 $3700
51 73 years
1 99%
39 8 deaths
36
GNP per capita in US$
Life expectancy
Literacy
Infant mortality per 1000 live births
Human dev. index

S

SLOVENIA

EUROPE
Austria
Hungary
Italy
SLOVENIA
Croatia
Europe

OFFICIAL NAME: Republic of Slovenia **CAPITAL:** Ljubljana
POPULATION: 2 million **CURRENCY:** Tolar **OFFICIAL LANGUAGE:** Slovene

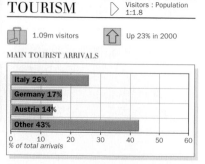

1991

1991

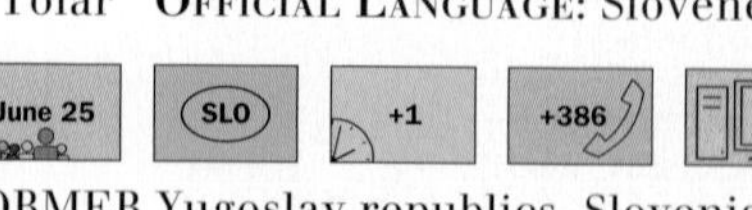
June 25
SLO
+1
+386

.si

OF ALL THE FORMER Yugoslav republics, Slovenia has the closest links with western Europe. Located at the northeastern end of the Adriatic Sea, this small, Alpine country controls some of Europe's major transit routes. Slovenia's transition to independence in 1991 avoided the violence of the breakup of Yugoslavia. The most prosperous of the former communist European states, it is the only former Yugoslav republic on the "fast track" to EU membership.

CLIMATE

Continental/ Mediterranean

WEATHER CHART FOR LJUBLJANA

Average daily temperature — Rainfall
°C/°F J F M A M J J A S O N D cm/in
40/104 40/16
30/86 30/12
20/68 20/8
10/50 10/4
0/32 0
-10/14
-20/-4

Slovenia's interior has a continental climate. Its small coastal region has a mild Mediterranean climate.

TRANSPORTATION

Drive on right

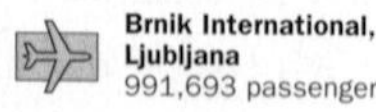

Brnik International, Ljubljana
991,693 passengers

10 ships
1767 grt

THE TRANSPORTATION NETWORK

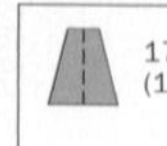

17,745 km
(11,026 miles)

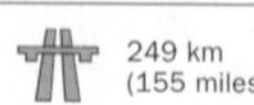

249 km
(155 miles)

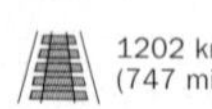

1202 km
(747 miles)

None

Slovenia is strategically situated at some of Europe's major crossroads. In addition, its Adriatic ports provide Austria with its main maritime outlet.

TOURISM

Visitors : Population 1:1.8

1.09m visitors

Up 23% in 2000

MAIN TOURIST ARRIVALS

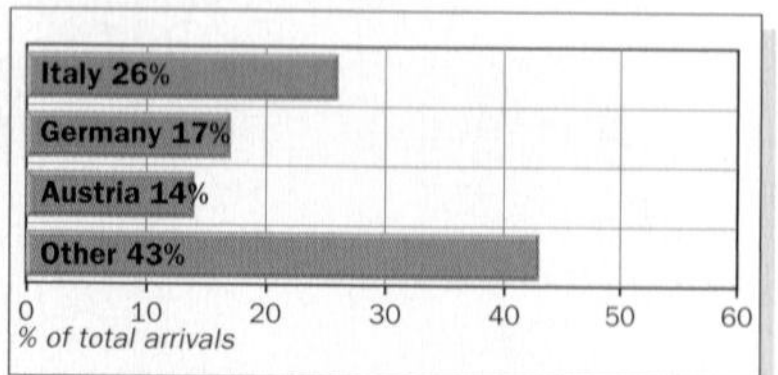
Italy 26%
Germany 17%
Austria 14%
Other 43%
0 10 20 30 40 50 60
% of total arrivals

A revival in tourism has been helped by Slovenia's political stability. Particular attractions include skiing in the Julian Alps, picturesque Ljubljana, and the wine-growing region around Ptuj.

PEOPLE

Pop. density medium

Slovene, Serbo-Croat

99/km²
(256/mi²)

THE URBAN/RURAL POPULATION SPLIT

50% 50%

ETHNIC MAKEUP

Serb 2%
Croat 3%
Muslim 1%
Other 6%
Slovene 88%

Slovenes are ethnically very similar to the neighboring Croats and, like them, are predominantly Roman Catholic. However, Slovenia's long historical association with western Europe, and particularly with Austria, created a distinct Slovene identity. This enabled a smooth transition to independence in 1991. The major non-Slavic minorities are small communities of Hungarians in the east and Italians in the southwestern Istrian region; tensions are few.

Women are not heavily disadvantaged in Slovenian society.

POLITICS

Multiparty elections

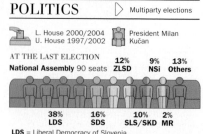

L. House 2000/2004
U. House 1997/2002

President Milan Kučan

AT THE LAST ELECTION

National Assembly 90 seats

12% ZLSD
9% NSi
13% Others
38% LDS
16% SDS
10% SLS/SKD
2% MR

LDS = Liberal Democracy of Slovenia
SDS = Social Democratic Party of Slovenia
ZLSD = United List of Social Democrats
SLS/SKD = Slovene People's Party/Christian Democrats of Slovenia **NSi** = New Slovenia – Christian People's Party
MR = Two seats are reserved for Italian and Hungarian minority representatives

National Council 40 seats

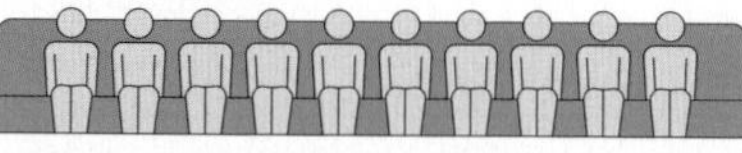

22 members of the National Council, which has an advisory role, are indirectly elected, and 18 are chosen by an electoral college to represent various interests

Slovenia has been strikingly stable since independence, with Milan Kučan as president from 1990. Fragmented party politics makes coalitions essential. Janez Drnovsek, leader of the center-left LDS, has been prime minister almost continuously since 1992, apart from a six-month period in 2000 when the more right-wing SLS pulled out of the coalition. Andrej Bajuk, as SLS leader, then became prime minister, leaving his party after a failed attempt at electoral reform. Drnovsek returned to office following elections in 2000.

Many former communist officials still occupy top posts. Ownership of denationalized property remains an issue and there is still bitterness over the killing of opponents of Tito's partisans in the 1940s.

SLOVENIA

Total Land Area : 20 253 sq. km
(7820 sq. miles)

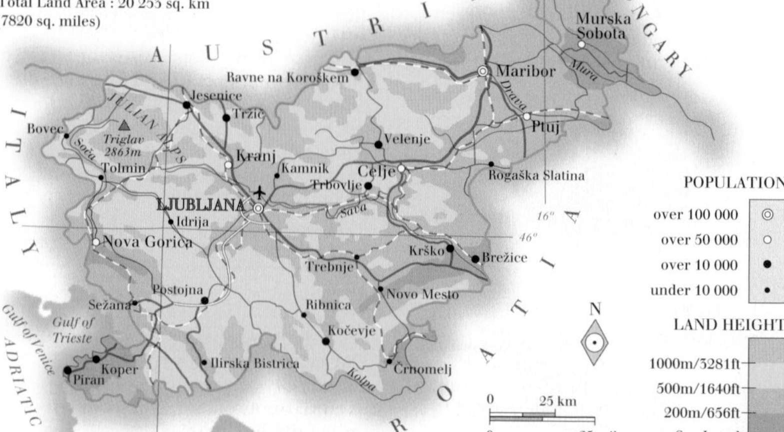
AUSTRIA
HUNGARY
ITALY
CROATIA
Murska Sobota
Maribor
Mura
Drava
Ptuj
Ravne na Koroškem
Jesenice
Tržič
JULIAN ALPS
Triglav 2863m
Bovec
Soča
Tolmin
Kranj
Velenje
Kamnik
Celje
Trbovlje
Rogaška Slatina
Sava
LJUBLJANA
Idrija
Nova Gorica
Krško
Brežice
Trebnje
Postojna
Novo Mesto
Sežana
Ribnica
Kočevje
Gulf of Trieste
Gulf of Venice
ADRIATIC SEA
Koper
Piran
Ilirska Bistrica
Kolpa
Črnomelj
16°
46°
14°
N
0 25 km
0 25 miles
POPULATION
over 100 000
over 50 000
over 10 000
under 10 000
LAND HEIGHT
1000m/3281ft
500m/1640ft
200m/656ft
Sea Level

S

WORLD AFFAIRS

Joined UN in 1992

CE

EAPC

OSCE

PfP

WTO

Slovenia is on the "fast track" to EU membership, probably in 2004. It also aims to be part of the next round of NATO enlargement.

AID

Recipient

 $61m (receipts)

 Up 97% in 2000

EU and World Bank aid focuses on infrastructure, the environment, and agricultural reform, in preparation for EU membership.

DEFENSE

Phasing out conscription

 $223m

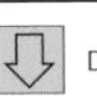 Down 34% in 2000

Troops staved off Yugoslav forces after secession in 1991. It was announced in 2002 that military conscription would be phased out by 2004.

ECONOMICS

Inflation 20% p.a. (1993–2000)

$20bn

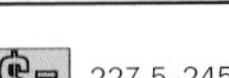 227.5–245.6 tolars

SCORE CARD

- World GNP Ranking......65th
- GNP per Capita......$10,050
- Balance of Payments......–$594m
- Inflation......10.8%
- Unemployment......7%

EXPORTS

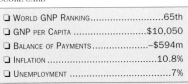

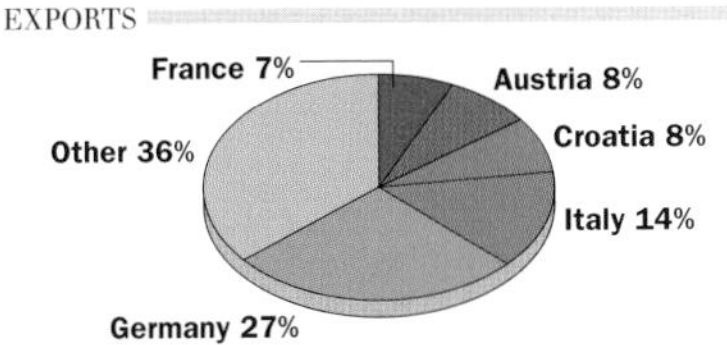
France 7%
Austria 8%
Other 36%
Croatia 8%
Italy 14%
Germany 27%

IMPORTS

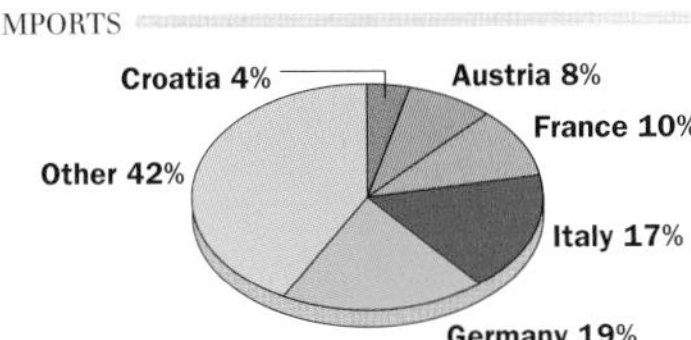
Croatia 4%
Austria 8%
France 10%
Other 42%
Italy 17%
Germany 19%

STRENGTHS

Stability. Competitive manufacturing industry. EU and former Yugoslav export markets. Revoz car plant very productive, making Renault Clios. Competitive port at Koper. Free trade pact with Bosnia from 2001. Least indebted of central and eastern European states.

WEAKNESSES

Economy only partly liberalized. Capital market controls. Slower growth rates than other emerging markets. Slow pace of privatization.

Lake Bled in the Julian Alps, which lie astride the Slovenian–Italian border. The lake is a popular tourist destination.

RESOURCES

Electric power 2.6m kw

 3215 tonnes

 20 b/d

 603,594 pigs, 493,670 cattle, 7.15m chickens

 Coal, lignite, lead, zinc, uranium, silver, mercury, oil

Slovenia has come under pressure from Austria to close the nuclear plant at Krško, which provides one-third of Slovenia's power. There are deposits of brown coal and lignite, but they are difficult to extract and of poor quality.

ENVIRONMENT

Sustainability rank: 23rd

 6%

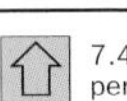 7.4 tonnes per capita

Protecting the country's alpine ecology is a priority. Pollution comes mainly from smelting, the chemicals industry, and burning brown coal and lignite.

MEDIA

TV ownership high

 Daily newspaper circulation 199 per 1000 people

PUBLISHING AND BROADCAST MEDIA

There are 7 daily newspapers. *Dnevnik* is independently owned

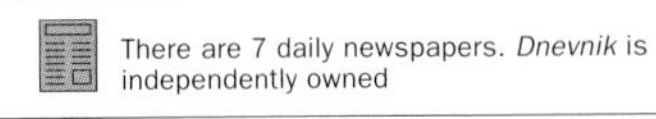
4 services:
1 state-controlled,
3 independent
4 services and
many regional
stations

A free and critical press has developed. State broadcasters have a new ethical code, protecting journalists' sources. POP TV is a commercial success.

CRIME

No death penalty

 1136 prisoners

 Up 11% in 1999

Slovenia's prison population is proportionately among the lowest in Europe. Smuggling people into western Europe is overtaking narcotics smuggling as the focus of organized crime.

EDUCATION

School leaving age: 15

 99%

 74,148 students

School is compulsory from seven to 15 years of age, and standards are high. The university at Ljubljana was founded in 1595.

CHRONOLOGY

Slovenia was part of the Austro-Hungarian Empire until 1918, when it joined the Kingdom of the Serbs, Croats, and Slovenes (Yugoslavia).

- **1949** Tito's break with Moscow.
- **1989** Parliament confirms right to secede. Calls multiparty elections.
- **1990** Control over army asserted, referendum approves secession.
- **1991** Independence declared; first republic to secede. Yugoslav federal army repelled.
- **1992** First multiparty elections. Milan Kučan president, Janez Drnovsek prime minister.
- **1993** Joins IMF and IBRD.
- **1998** EU membership talks begin.
- **2000** Drnovsek briefly ousted; returns to office after elections.

HEALTH

Welfare state health benefits

 1 per 435 people

 Cerebrovascular and heart diseases, cancers, accidents,

National health care in Slovenia uses health centers and outpatient clinics to increase accessibility for patients.

SPENDING

GDP/cap. increase

CONSUMPTION AND SPENDING

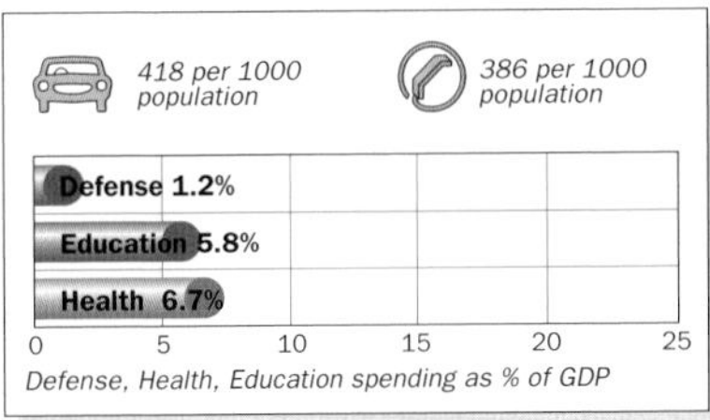
418 per 1000 population
386 per 1000 population
Defense 1.2%
Education 5.8%
Health 6.7%
0
5
10
15
20
25
Defense, Health, Education spending as % of GDP

Slovenia has the highest standard of living of all the central and eastern European states of the former Soviet bloc.

WORLD RANKING

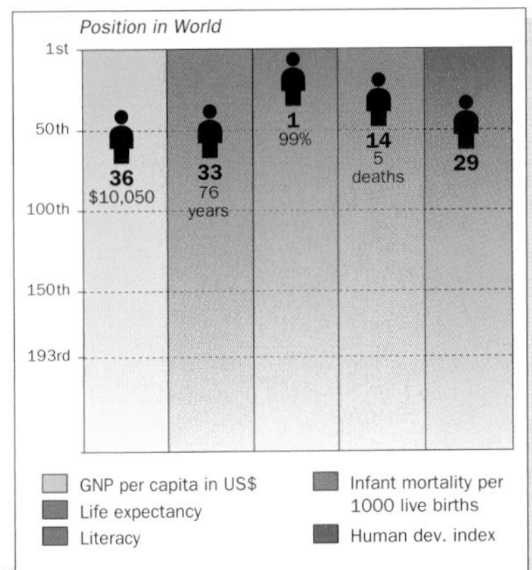
Position in World
1st
50th
100th
150th
193rd
36
$10,050
33
76 years
1
99%
14
5 deaths
29
GNP per capita in US$
Life expectancy
Literacy
Infant mortality per 1000 live births
Human dev. index

S

SOLOMON ISLANDS

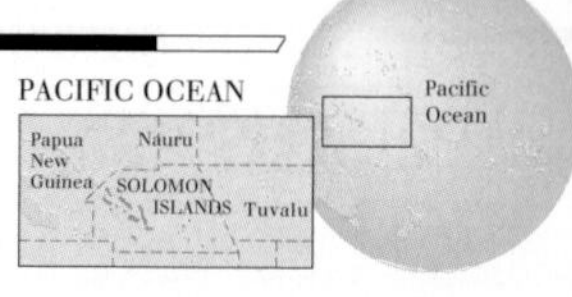

PACIFIC OCEAN
Papua New Guinea
Nauru
SOLOMON ISLANDS
Tuvalu
Pacific Ocean

OFFICIAL NAME: Solomon Islands **CAPITAL:** Honiara
POPULATION: 463,000 **CURRENCY:** Solomon Islands dollar **OFFICIAL LANGUAGE:** English

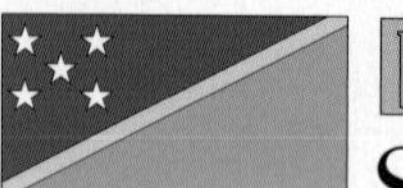

1978

1978

July 7
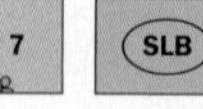

SLB
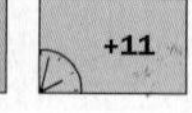

+11

+677
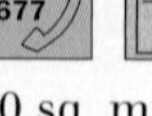

.sb

SCATTERED OVER 645,000 sq. km (250,000 sq. miles), the Solomons archipelago has several hundred islands, but most people live on the six largest – Guadalcanal, Malaita, New Georgia, Makira, Santa Isabel, and Choiseul. The Solomons have been settled since at least 1000 BCE; the Spanish arrived in 1568. Ethnic conflict between rival islanders ravaged the country from 1998 to 2000. Most of the Solomons are coral reefs. Just 1% of the land area is cultivable.

CLIMATE

Tropical equatorial

WEATHER CHART FOR HONIARA

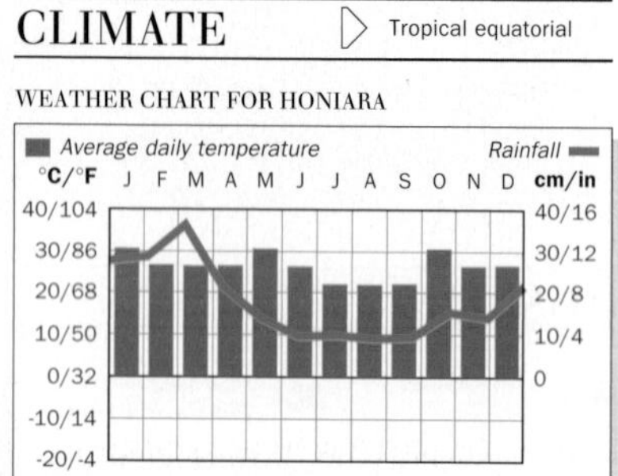

Average daily temperature
Rainfall
°C/°F
J F M A M J J A S O N D
cm/in
40/104
30/86
20/68
10/50
0/32
-10/14
-20/-4
40/16
30/12
20/8
10/4
0

There is little temperature variation in the humid climate, but ferocious cyclones can occur in the rainy season.

TRANSPORTATION

Drive on left

Henderson, Honiara
22,000 passengers

31 ships
1400 grt

THE TRANSPORTATION NETWORK

34 km (21 miles)

None

None

None

International flights from the principal airport, 13 km (8 miles) outside Honiara, were resumed in late 2000 after the ending of open hostilities.

Unloading seed coconuts near Munda on New Georgia in the Solomons' northern chain of islands. Coconuts are by far the largest and most commercially important crop.

TOURISM

Visitors : Population 1:22

21,000 visitors

Up 62% in 1999

MAIN TOURIST ARRIVALS

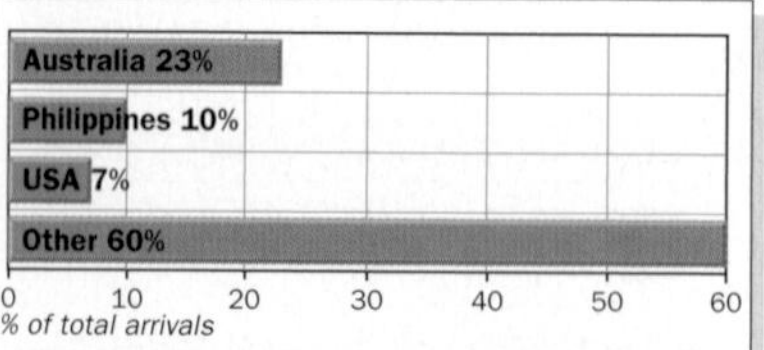

Australia 23%
Philippines 10%
USA 7%
Other 60%
0
10
20
30
40
50
60
% of total arrivals

The importance of Guadalcanal during World War II and the tranquility of the outer islands used to attract tourists. However, ethnic conflict all but destroyed tourism in 1998 and again when fighting intensified in 2000. Lack of funding hampers recovery.

PEOPLE

Pop. density low

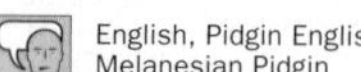

English, Pidgin English, Melanesian Pidgin

17/km² (43/mi²)

THE URBAN/RURAL POPULATION SPLIT

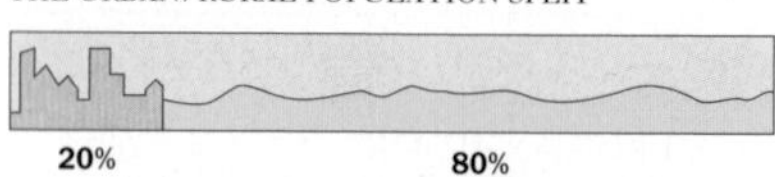

20%
80%

RELIGIOUS PERSUASION

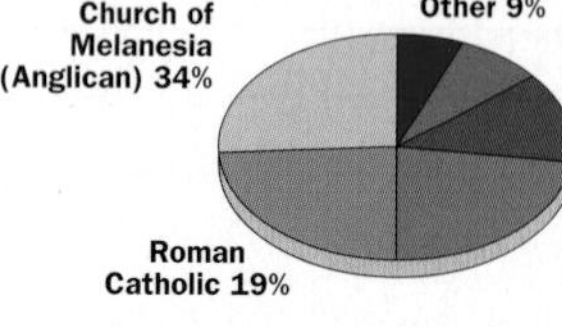

Church of Melanesia (Anglican) 34%
Other 9%
Seventh-day Adventist 10%
Methodist 11%
South Seas Evangelical Church 17%
Roman Catholic 19%

Almost all Solomon Islanders are Melanesian; relations between islands are tense. During the 1998–2000 conflict, 20,000 Malaitans were forced from their homes on Guadalcanal by native (Isatabu) militias. Authorities in outlying islands have pressed for greater autonomy. There are small communities of Micronesians who are descended from I-Kiribati temporarily relocated in 1957. More than 50 dialects are spoken. Although the islanders are nominally Christian, animist beliefs are widespread.

POLITICS

Multiparty elections

2001/2005

H.M. Queen Elizabeth II

AT THE LAST ELECTION

National Parliament 50 seats

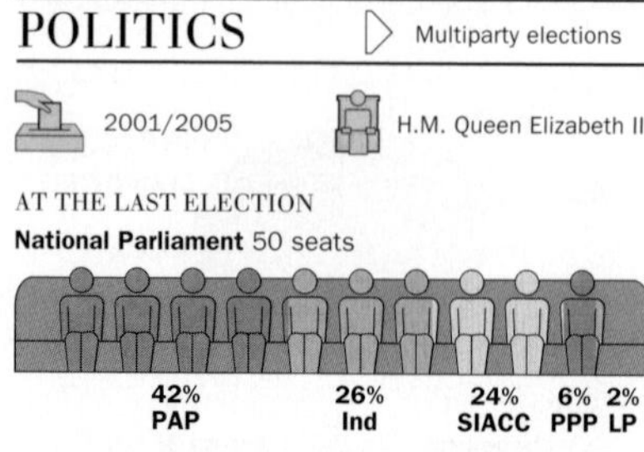

42% PAP
26% Ind
24% SIACC
6% PPP
2% LP

PAP = People's Alliance Party **Ind** = Independents
SIACC = Solomon Islands Alliance for Change Coalition
PPP = People's Progressive Party **LP** = Labour Party

Parliament is based on the Westminster model, but with no dominant political class, prominent local figures – known as "big men" – stand as candidates and turnover of members is high, leading to fluid and unstable coalitions. The civil conflict on Guadalcanal briefly ousted the government in 2000.

A semblance of stability has since been restored with a new devolved "state system," which gives greater regional autonomy.

Rebuilding the economy is now the major issue: "compensation" claims from the recent conflict quickly outstripped aid funds.

SOLOMON ISLANDS

Total Land Area : 28 450 sq. km (10 985 sq. miles)

POPULATION: over 10 000; under 10 000

LAND HEIGHT: 1000m/3280ft; 500m/1640ft; Sea Level

S

WORLD AFFAIRS

Joined UN in 1978

ACP | Comm | IBRD | PC | PIF

The intensification of violence in 2000 caused great concern around the Pacific, and increased mediation efforts, in which Australia was particularly involved. An Australian warship was the scene of key events during the resolution of the conflict, including the prime ministerial election and ratification of the Townsville peace accord.

AID

Recipient

US$68m (receipts)

Up 70% in 2000

Aid has focused very specifically on restoring stability and rebuilding infrastructure after two years of brutal conflict. Regional powers Australia, New Zealand, and Taiwan are key in aiding recovery. Improved relations with Papua New Guinea brought aid of US$23 million between 1998 and 2001.

DEFENSE

No compulsory military service

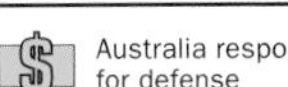
Australia responsible for defense

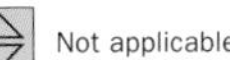
Not applicable

The Peace Plan 2000 includes the creation of a panethnic security force. Under the Townsville peace accord security was overseen by unarmed peacekeepers from neighboring Pacific states. The rival militias are effectively in control on Guadalcanal and Malaita.

ECONOMICS

Inflation 8.5% p.a. (1990–2000)

US$278m

5.163–5.640 Solomon Islands dollars

SCORE CARD

- World GNP Ranking........................177th
- GNP per Capita..........................US$620
- Balance of Payments..................US$21m
- Inflation..8.3%
- Unemployment...........Some underemployment

EXPORTS

UK 7%
Philippines 11%
China 13%
South Korea 17%
Japan 22%
Other 30%

IMPORTS

USA 5%
New Zealand 5%
Japan 5%
Singapore 25%
Other 33%
Australia 27%

Strengths

Good mineral and agricultural resources. Influx of international aid.

Weaknesses

Economy near collapse after ethnic conflict. Destruction of infrastructure. Key gold mine shut by militias. Grossly inflated compensation claims from conflict. Revenue from copra, gold, fish, and palm oil dried up. Social insecurity deters investment.

RESOURCES

Electric power 12,000 kw

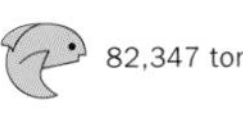
82,347 tonnes

Not an oil producer

67,000 pigs, 12,500 cattle, 200,000 chickens

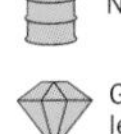
Gold, copper, bauxite, lead, zinc, silver, cobalt, phosphates

Bauxite deposits have been discovered on Rennell Island. In addition, there are traces of gold and copper on Guadalcanal. There is increasing concern over the exploitation of forest and marine resources.

ENVIRONMENT

Not available

None

0.4 tonnes per capita

The environmental movement is strong. Depletion of forest and marine resources are a major concern. In 1998 a sustainable forest-harvesting policy was introduced, but the need to restore the economy puts pressure on environmentally sensitive areas.

MEDIA

TV ownership low

Daily newspaper circulation 16 per 1000 people

PUBLISHING AND BROADCAST MEDIA

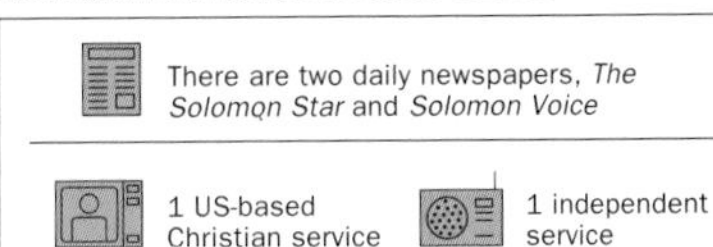
There are two daily newspapers, The Solomon Star and Solomon Voice
1 US-based Christian service
1 independent service

The one radio station broadcasts in English and Pidgin. The first TV service, a Christian channel, began in 2002.

CRIME

No death penalty

134 prisoners

Crime is rising

Under the armed militias, extortion and gang-related violence have flourished, notably on Guadalcanal and Malaita.

CHRONOLOGY

Settled since before 1000 BCE by Melanesian peoples, the Solomons became a British colony in 1893.

- **1900** Britain acquires northern Solomons from Germany.
- **1942–1943** Japanese occupation.
- **1978** Independence from UK.
- **1983** Diplomatic relations with Taiwan established.
- **1998–2000** Civil conflict between Guadalcanal and Malaita islanders.

EDUCATION

School leaving age: 14

62%

Not available

Education is modeled on the British system. Tertiary students go to the University of the South Pacific in Fiji.

HEALTH

Welfare state health benefits

1 per 7143 people

Not available

The main hospital has seriously reduced services; local patients are now expected to provide their own food. Gang violence has spread to hospital wards.

SPENDING

GDP/cap. decrease

CONSUMPTION AND SPENDING

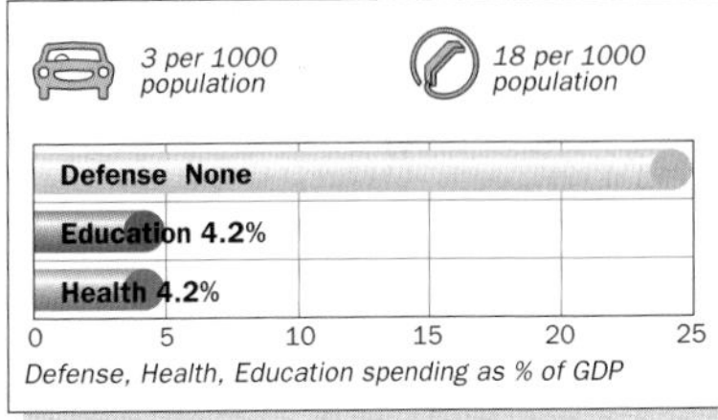
3 per 1000 population
18 per 1000 population
Defense None
Education 4.2%
Health 4.2%
0
5
10
15
20
25
Defense, Health, Education spending as % of GDP

Solomon Islanders in government jobs are the wealthiest group. Inhabitants of the outlying islands are extremely poor.

S

WORLD RANKING

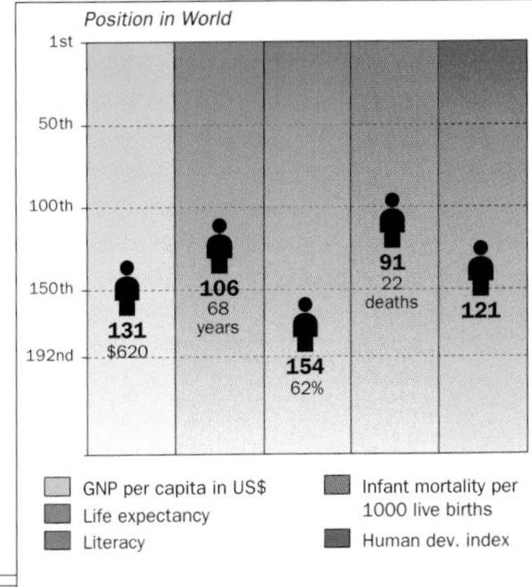
Position in World
1st
50th
100th
150th
192nd
131
$620
106
68 years
154
62%
91
22 deaths
121
GNP per capita in US$
Life expectancy
Literacy
Infant mortality per 1000 live births
Human dev. index

Somalia

EAST AFRICA
Djibouti
Ethiopia
Kenya
SOMALIA
Africa

Official Name: Somalia **Capital:** Mogadishu **Population:** 9.2 million
Currency: Somali shilling **Official Languages:** Somali and Arabic

1960

1960

Oct 21

SO
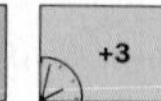
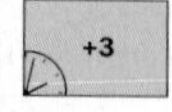
+3

+252

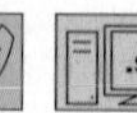
.so

OCCUPYING THE HORN of Africa, Italian Somaliland and British Somaliland joined in 1960 to form an independent Somalia. Except in the more fertile south, the land is semiarid. Years of clan-based civil war have resulted in the collapse of central government, the frustration of US and UN intervention initiatives aimed at easing a huge refugee crisis, and mass starvation.

CLIMATE

Hot desert/steppe

WEATHER CHART FOR MOGADISHU

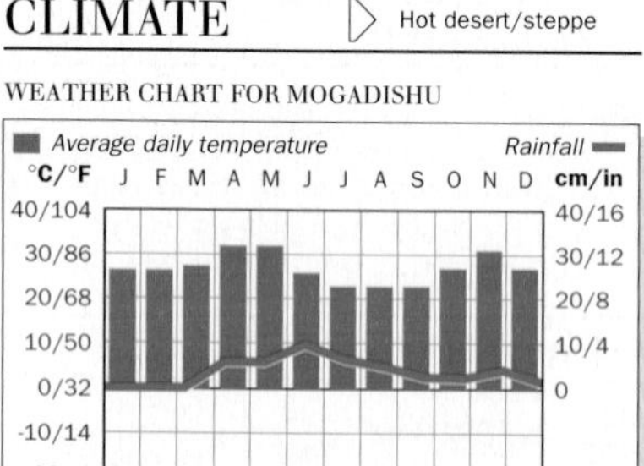

Average daily temperature
Rainfall
°C/°F
J F M A M J J A S O N D
cm/in
40/104
30/86
20/68
10/50
0/32
-10/14
-20/-4
40/16
30/12
20/8
10/4
0

Somalia is very dry. The northern coast is very hot and humid, the eastern less so. The interior has some of the world's highest mean yearly temperatures.

TRANSPORTATION

Drive on left

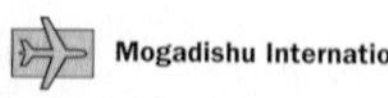
Mogadishu International

22 ships 11,400 grt

THE TRANSPORTATION NETWORK

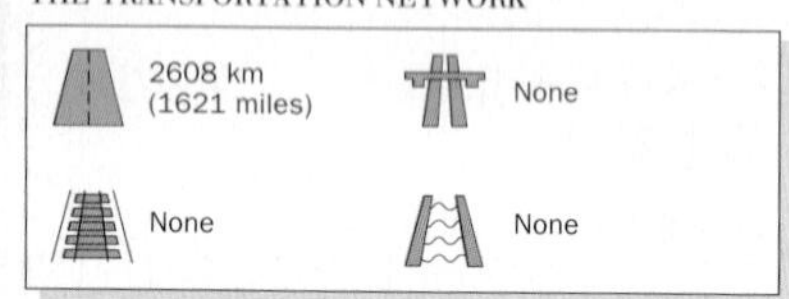

2608 km (1621 miles)
None
None
None

About 50% of Somalis are nomads for whom the camel is the principal means of transportation. In 1990, the IDA agreed to repair the road network, but work on the seven-year project has not yet begun.

TOURISM

Visitors : Population 1:920

10,000 visitors

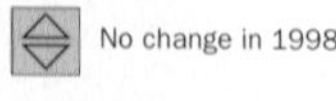
No change in 1998

MAIN TOURIST ARRIVALS

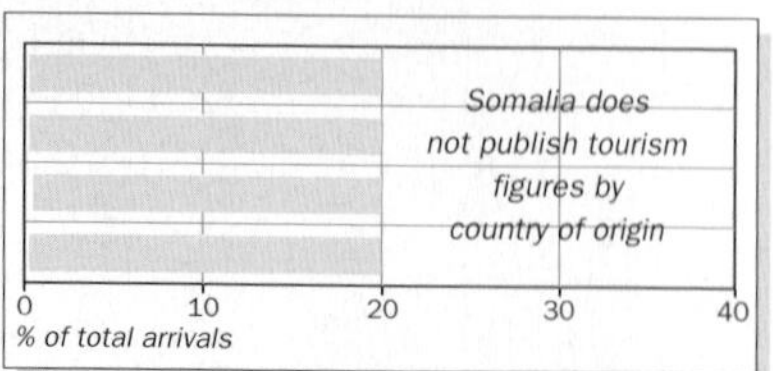

Somalia does not publish tourism figures by country of origin
0
10
20
30
40
% of total arrivals

Aid workers and foreign journalists are the only visitors. Land mines are a major hazard.

Baydhabo market. *Subsistence farming supports most people, despite chaos created by the fighting.*

PEOPLE

Pop. density low

Somali, Arabic, English, Italian

15/km² (38/mi²)

THE URBAN/RURAL POPULATION SPLIT

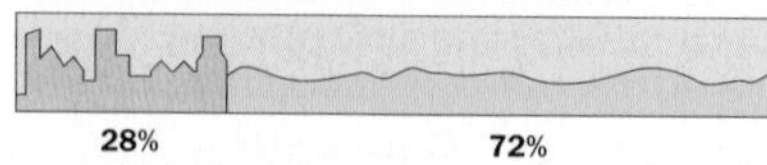

28%
72%

RELIGIOUS PERSUASION

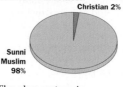

Christian 2%
Sunni Muslim 98%

The clan system is fundamental to Somalia. Shifting allegiances characterize its structure – a tendency stifled by Siad Barre's dictatorship but revived after his fall in 1991. His undermining of the traditional brokers of justice, the elders, contributed to the power vacuum that resulted in civil war, and his persecution of the Issaqs led to Somaliland's declaration of secession in 1991. However, the entire population is ethnic Somali, and national identity remains strong, shown by the widespread opposition to the UN's peacekeeping force.

POLITICS

No legislative elections

1984/Uncertain

No internationally recognized head of state

AT THE LAST ELECTION

National Assembly (suspended)

There has been no prospect of organizing new elections since the overthrow of Siad Barre. A transitional assembly was formed in 2000 on a nonparty basis.

Somalia has remained in anarchy since the former dictator President Siad Barre fled in 1991. The unified state dissolved amid conflict in the south and separatism in the north.

The US led a UN peacekeeping force to the south in 1992, but failed to loosen the grip of the warring factions. Throughout the 1990s rival warlords, including the powerful Gen. Aideed, contended for supreme power. A National Salvation Council sank without trace in 1997.

A conference of businessmen and influential figures, held in neighboring Djibouti in 2000, established a transitional assembly and appointed former Barre minister Abdulkassim Salat Hassan as president. The new government, although warmly received in Mogadishu, was immediately rejected by most of the warlords and by the northern separatist authorities in "Somaliland" and "Puntland."

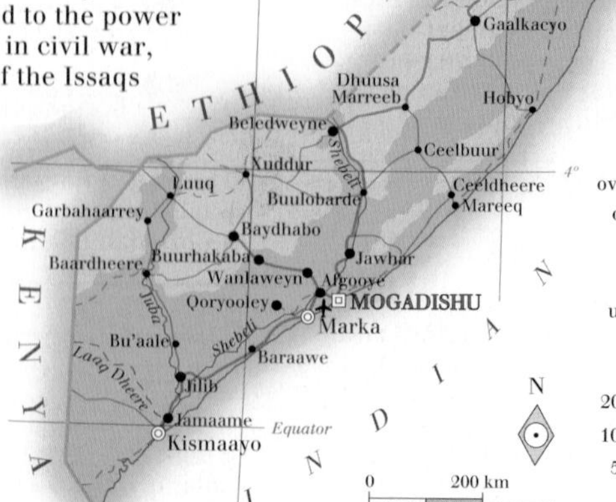

DJIBOUTI
Gulf of Aden
Saylac
Silil
Ceel Gaal
Berbera
Ceerigaabo
Shimbiris 2407m
Boosaaso
Bereeda
Raas Caseyr
PUNTLAND (self proclaimed)
SOMALILAND REPUBLIC (self proclaimed)
Hargeysa
Burco
Laas Dawaco
Raas Xaafuun
Qardho
Bender Beyla
Garoowe
Eyl
Jirriiban
Gaalkacyo
ETHIOPIA
Dhuusa Marreeb
Hobyo
Beledweyne
Ceelbuur
Xuddur
Luuq
Garbahaarrey
Buulobarde
Ceeldheere
Mareeq
Baydhabo
Baardheere
Buurhakaba
Wanlaweyn
Jawhar
Afgooye
MOGADISHU
Qoryooley
Marka
Bu'aale
Baraawe
Jilib
Jamaame
Kismaayo
Equator
KENYA
INDIAN OCEAN
SOMALIA
Total Land Area : 637 657 sq. km (246 199 sq. miles)
POPULATION
over 1 000 000
over 100 000
over 50 000
over 10 000
under 10 000
LAND HEIGHT
2000m/6562ft
1000m/3281ft
500m/1640ft
200m/656ft
Sea Level
0
200 km
200 miles

S

WORLD AFFAIRS

Joined UN in 1960

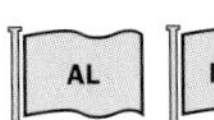

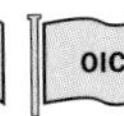
AL
IGAD
NAM
AU
OIC

After the UN force withdrew in 1995, the international community appeared to abandon Somalia until it gave support to the transitional parliament from 2000. Relations with Ethiopia are particularly tense; the government accuses Addis Ababa of sending troops to assist opposition warlords. Belief in the existence of terrorist training camps in Somalia has eroded relations with the US since September 11, 2001.

AID

Recipient

 $104m (receipts)

 Down 10% in 2000

Mass starvation among the Somali population in 1991 finally prompted the UN to launch a large-scale humanitarian aid effort. In this the UN was largely effective, averting widescale starvation and restoring food security.

DEFENSE

No compulsory military service

 $39m

 Down 2% in 2000

Former soldiers have been urged to reenlist. Efforts to demobilize the estimated 75,000 militia began in 2000.

ECONOMICS

Not available

 $835m

2620 Somali shillings

SCORE CARD

❑ WORLD GNP RANKING	157th
❑ GNP PER CAPITA	$100
❑ BALANCE OF PAYMENTS	–$157m
❑ INFLATION	Over 100%
❑ UNEMPLOYMENT	Widespread underemployment

STRENGTHS

Very few. Export of livestock to Arabian peninsula resumed in the north. Inflow of money from Somalis abroad. Growing market in stolen food aid.

WEAKNESSES

Every commodity, except arms, in extremely short supply. Little economic potential in the south. Destruction by drought of livestock. Mogadishu port closed by civil unrest until October 2000.

EXPORTS

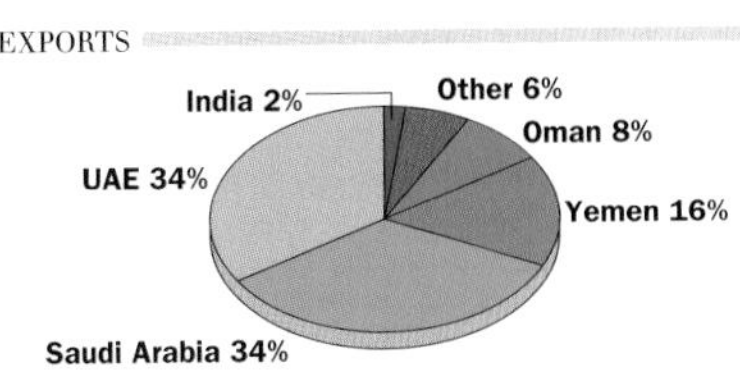
India 2%
Other 6%
Oman 8%
UAE 34%
Yemen 16%
Saudi Arabia 34%

IMPORTS

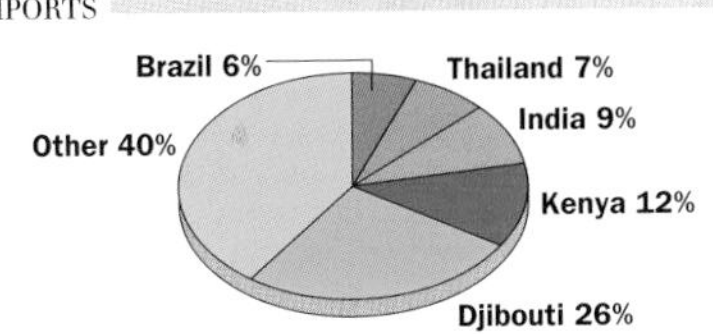
Brazil 6%
Thailand 7%
India 9%
Other 40%
Kenya 12%
Djibouti 26%

RESOURCES

Electric power 79,000 kw

 20,250 tonnes

 Not an oil producer

 13.2m sheep, 12.5m goats, 6.2m camels

 Salt, tin, zinc, copper, gypsum, manganese, uranium, iron

Commercially exploitable minerals remain untapped. An oil exploration agreement was signed with a French oil group in February 2001.

ENVIRONMENT

Sustainability rank: 132nd

0.3%

 Negligible emissions per capita

Human deprivation and starvation caused by the effects of drought and war on land and livestock outweigh all other ecological considerations.

MEDIA

TV ownership low

 Daily newspaper circulation 1 per 1000 people

PUBLISHING AND BROADCAST MEDIA

There are 5 daily newspapers, including *Jamhuuriya*, *Qaran*, and *Xiddigta Oktobar*

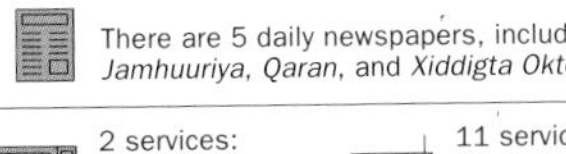

2 services: limited to the Mogadishu area

11 services: mostly political or religious

In Mogadishu there are three faction-run radio stations. Somali Television Network, an independent multichannel, multilingual service, began broadcasting in 1999. There are few newspapers, paper being in very short supply.

CRIME

Death penalty in use

 Somalia does not publish prison figures

 Widespread breakdown of law and order since 1991

Armed clan factions (some, in remoter regions, engaged in family feuds rather than the war) and bandits rule large areas. In Mogadishu a "national" police force has been established, and possession of firearms outlawed. *Sharia* (Islamic law), now the de facto system, is run in a makeshift fashion by elders.

EDUCATION

School leaving age: 14

 24%

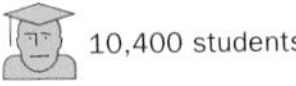 10,400 students

The system collapsed during the civil war. There were reports of improvised open-air schools starting up again in urban areas in 1993. Somali has been a written language only since 1972.

CHRONOLOGY

The lands of the Somalis became British and Italian colonies in the 1880s. Most were unified as Somalia at independence in 1960.

- ❑ **1964–1987** Conflict with Ethiopia over Somali-inhabited Ogaden.
- ❑ **1969** Gen. Siad Barre takes power.
- ❑ **1991** Siad Barre ousted. Civil war and clan chaos. Mass starvation. Somaliland declares secession.
- ❑ **1992** Abortive US intervention.
- ❑ **1995** UN force withdrawn.
- ❑ **1997** Accord signed by 26 clan factions.
- ❑ **2000** National reconciliation conference appoints government; warlords dispute its authority.
- ❑ **2001** Somali Reconciliation and Restoration Council set up, with support from southern clan leaders.

HEALTH

No welfare state health benefits

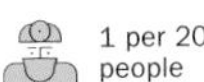 1 per 20,000 people

 Diarrheal, communicable, and parasitic diseases

The state-run system has collapsed entirely. A few very rudimentary facilities are run by foreign workers.

SPENDING

GDP/cap. decrease

CONSUMPTION AND SPENDING

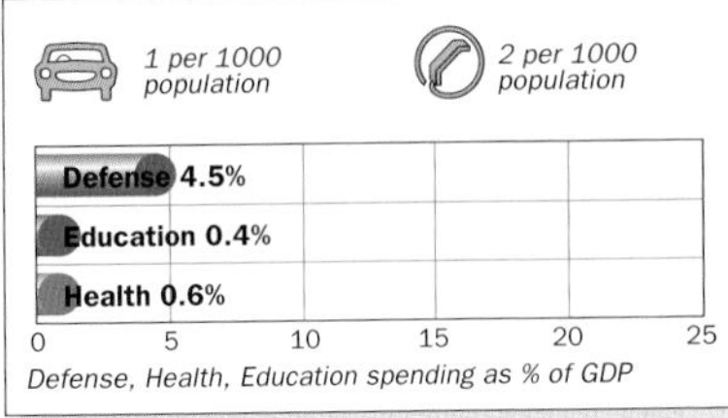
1 per 1000 population
2 per 1000 population
Defense 4.5%
Education 0.4%
Health 0.6%
0
5
10
15
20
25
Defense, Health, Education spending as % of GDP

Bandits and warlords gained rich pickings. Money sent by relatives living overseas is the main income for some people.

WORLD RANKING

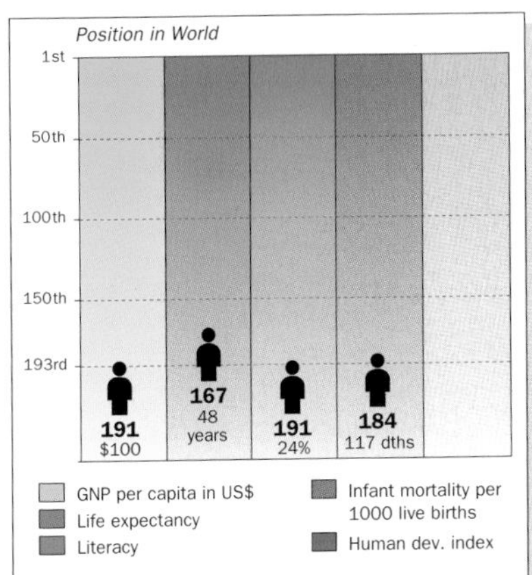
Position in World
1st
50th
100th
150th
193rd
191
$100
167
48 years
191
24%
184
117 dths
GNP per capita in US$
Life expectancy
Literacy
Infant mortality per 1000 live births
Human dev. index

S

SOUTH AFRICA

SOUTHERN AFRICA — Namibia, Botswana, Swaziland, Lesotho, SOUTH AFRICA — Africa

OFFICIAL NAME: Republic of South Africa **CAPITALS:** Pretoria; Cape Town; Bloemfontein
POPULATION: 43.8 million **CURRENCY:** Rand **OFFICIAL LANGUAGES:** Afrikaans, English, and 9 African languages

1934

1994

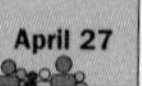
April 27

ZA

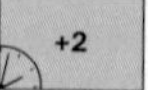
+2

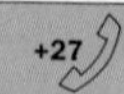
+27

.za

RICH IN NATURAL RESOURCES, South Africa comprises a central plateau, or *veld*, bordered to the south and east by the Drakensberg Mountains. After eight decades of white minority rule, with racial segregation under the apartheid policy since 1948, South Africa held its first multiracial elections in 1994. The revolution in South Africa's politics began in 1990, when black freedom groups were legalized and the dismantling of apartheid began. The African National Congress (ANC), under Nelson Mandela and his successor Thabo Mbeki, is now the leading political movement.

Nelson Mandela, *who became president of South Africa in April 1994.*

Thabo Mbeki, *elected president in 1999 to succeed Mandela.*

CLIMATE

Desert/subtropical/ Mediterranean

WEATHER CHART FOR PRETORIA

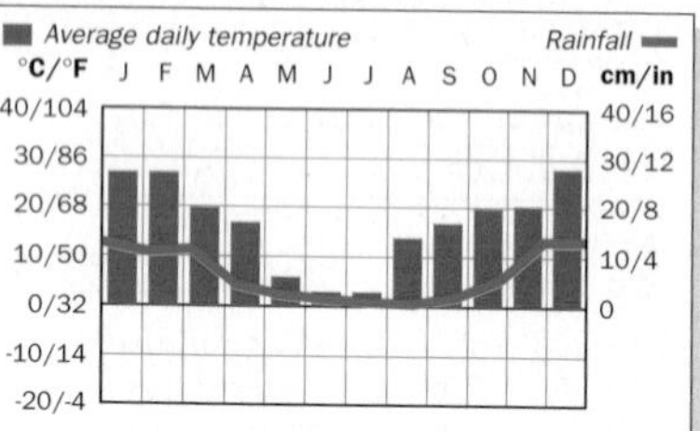
Average daily temperature
Rainfall
°C/°F
J F M A M J J A S O N D
cm/in
40/104
30/86
20/68
10/50
0/32
-10/14
-20/-4
40/16
30/12
20/8
10/4
0

Despite the moderating effects of oceans on three sides, the warm temperate climate is dry; 65% of the country has less than 50 cm (20 in) of rain a year. Drought is a periodic hazard.

TRANSPORTATION

Drive on left

Jan Smuts International, Johannesburg
11.7m passengers

192 ships
383,700 grt

THE TRANSPORTATION NETWORK

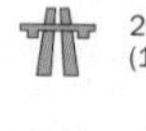

63,027 km (39,163 miles)	2032 km (1263 miles)
22,686 km (14,097 miles)	None

Priorities include expanding port capacity and cross-border rail networks. Public transportation is limited and expensive, but there is an extensive informal network of minibuses and taxis.

Cape Town, *set on a peninsula ending at the Cape of Good Hope, where the Indian and Atlantic oceans meet.*

TOURISM

Visitors : Population 1:7.3

6m visitors

No change in 2000

MAIN TOURIST ARRIVALS

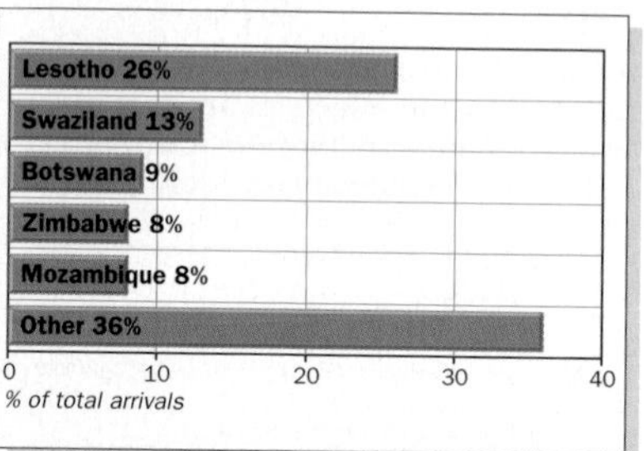
Lesotho 26%
Swaziland 13%
Botswana 9%
Zimbabwe 8%
Mozambique 8%
Other 36%
0
10
20
30
40
% of total arrivals

South Africa has huge tourist potential, with attractions ranging from beaches to mountains, from prizewinning vineyards to world renowned wildlife reserves. The enormous Kruger National Park boasts 137 mammal species and 450 bird species. Visitor numbers increased throughout the 1990s, but tourism is still recovering from the country's isolation during the apartheid era. Today, the key constraint on growth is rising crime. Studies suggest that by 2005 tourism could create an additional 450,000 jobs and contribute 10% toward GDP (compared with 4% in 1995).

PEOPLE

Pop. density low

English, Afrikaans, Zulu, Xhosa, Ndebele, Setswana, Siswati, North Sotho, South Sotho, Tsongo, Venda

36/km² (93/mi²)

THE URBAN/RURAL POPULATION SPLIT

55% 45%

RELIGIOUS PERSUASION

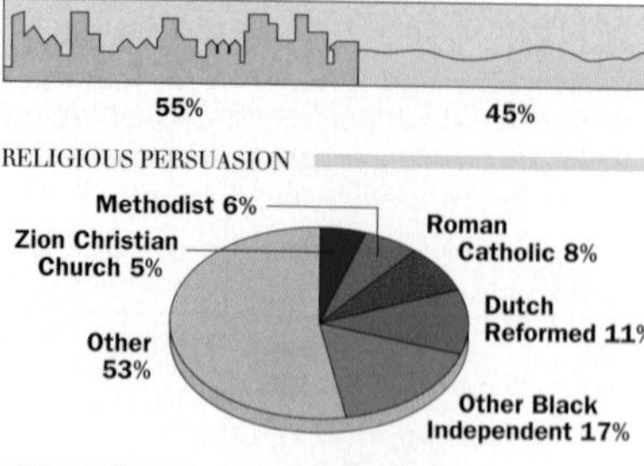
Methodist 6%
Roman Catholic 8%
Zion Christian Church 5%
Dutch Reformed 11%
Other 53%
Other Black Independent 17%

ETHNIC MAKEUP

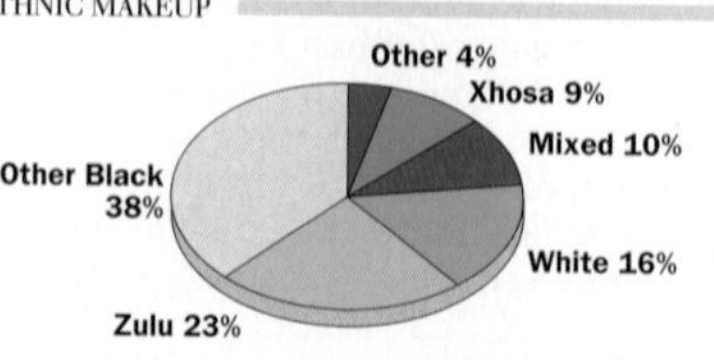
Other 4%
Xhosa 9%
Mixed 10%
Other Black 38%
White 16%
Zulu 23%

Under apartheid, South Africans were divided into racial categories: whites (Afrikaners and English-speakers), and three black groups (Coloreds, people whose descent was deemed mixed; Asians, mainly Indians; and Africans). Each category had different political, economic, and social rights, with whites enjoying the most privileges and Africans the fewest. While blacks now dominate politics, English-speaking whites continue to control the economy.

The extended family has been undermined by regulations forcing men to migrate for work, leaving their wives and children in the rural areas. A small black middle class has developed, but most black South Africans are underemployed.

The expected postapartheid ethnic conflict failed to materialize, although Inkatha has exploited feelings of Zulu identity in its quest for greater political power. An area of the Kalahari Desert was returned to a Khomani San tribe (Bushmen) in 1999.

Many women are now prominent in public life. The new constitution guarantees equality of the sexes.

POPULATION AGE BREAKDOWN

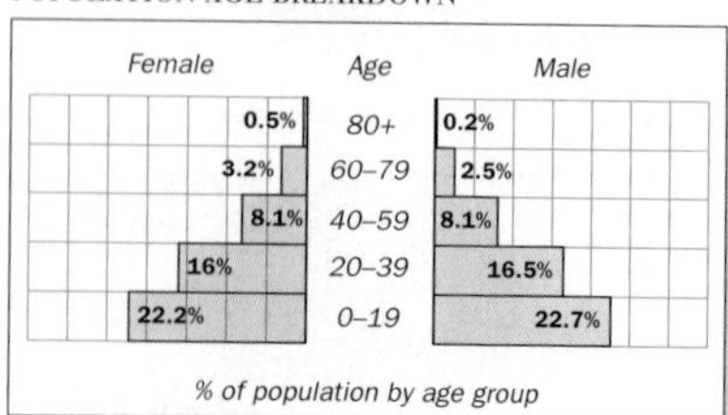
Female
Age
Male
0.5%
80+
0.2%
3.2%
60–79
2.5%
8.1%
40–59
8.1%
16%
20–39
16.5%
22.2%
0–19
22.7%
% of population by age group

S

POLITICS

Multiparty elections

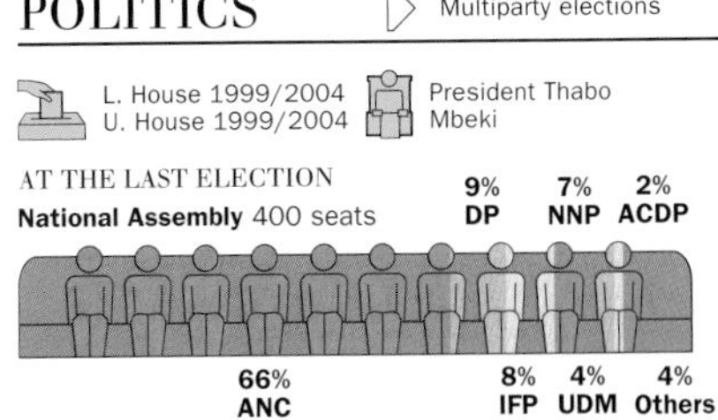
L. House 1999/2004
U. House 1999/2004
President Thabo Mbeki
AT THE LAST ELECTION
National Assembly 400 seats
9% DP
7% NNP
2% ACDP
66% ANC
8% IFP
4% UDM
4% Others

ANC = African National Congress **DP** = Democratic Party **IFP** = Inkatha Freedom Party **NNP** = New National Party **UDM** = United Democratic Movement **ACDP** = African Christian Democratic Party

National Council of Provinces 90 seats

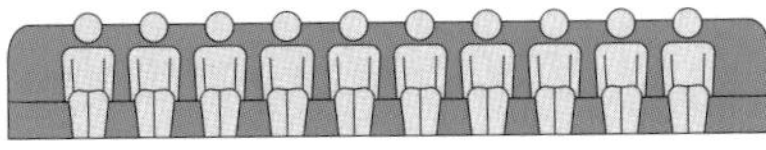

10 members are elected to the National Council of Provinces by each of the nine provincial legislatures

South Africa became a multiracial democracy following elections in 1994.

PROFILE

The 1994 elections ended 45 years of apartheid, and Nelson Mandela was elected president. A new liberal constitution was adopted in 1996. The 1999 elections saw the ANC slightly increase its majority; Thabo Mbeki succeeded Mandela as president. The DP overtook the strife-ridden NNP to become the official opposition, and in 2000 the two organizations merged briefly to form an opposition Democratic Alliance (DA). However, in 2001, the NNP allied itself with the ANC.

MAIN POLITICAL ISSUES

Maintaining unity

In April 1994, South Africa confounded the proponents of violence and ethnic division by holding peaceful elections. The ANC-dominated government, while heeding the aspirations of the black majority, has tried to avoid marginalizing minorities. To deal with the apartheid past, it mandated an innovative Truth and Reconciliation Commission (TRC). Two years of painful and sometimes controversial hearings culminated in a final report in 1998. Initial praise for the report has been modified more recently by allegations of corruption and mishandling of compensation payments for the victims of apartheid.

Reconstruction and development

The costly Reconstruction and Development Program (RDP) aims to improve health, housing, and education, and boost employment. Implementation is too slow for many blacks, who had hoped for immediate benefits from democratic rule, while some members of the minorities see the RDP's affirmative action as reverse discrimination.

WORLD AFFAIRS

Joined UN in 1945

Comm
WTO
G24
AU
SADC

After several decades of political isolation and economic sanctions, South Africa has been welcomed back to the international fold, and has rejoined the UN and the Commonwealth. It now hopes to reattract international investors. South Africa has joined the SADC and leads continental opinion on regional issues. Former President Mandela often intervened to help resolve foreign conflicts, and continues to act as a regional mediator, most notably in Burundi. His successor, Thabo Mbeki, has continued this role, notably in pushing forward the peace process in the Democratic Republic of the Congo. However, Mbeki has been criticized internationally for his support of Robert Mugabe's regime in neighboring Zimbabwe, and for his controversial opinions about HIV and AIDS.

SOUTH AFRICA

Total Land Area : 1 219 912 sq. km (471 008 sq. miles)

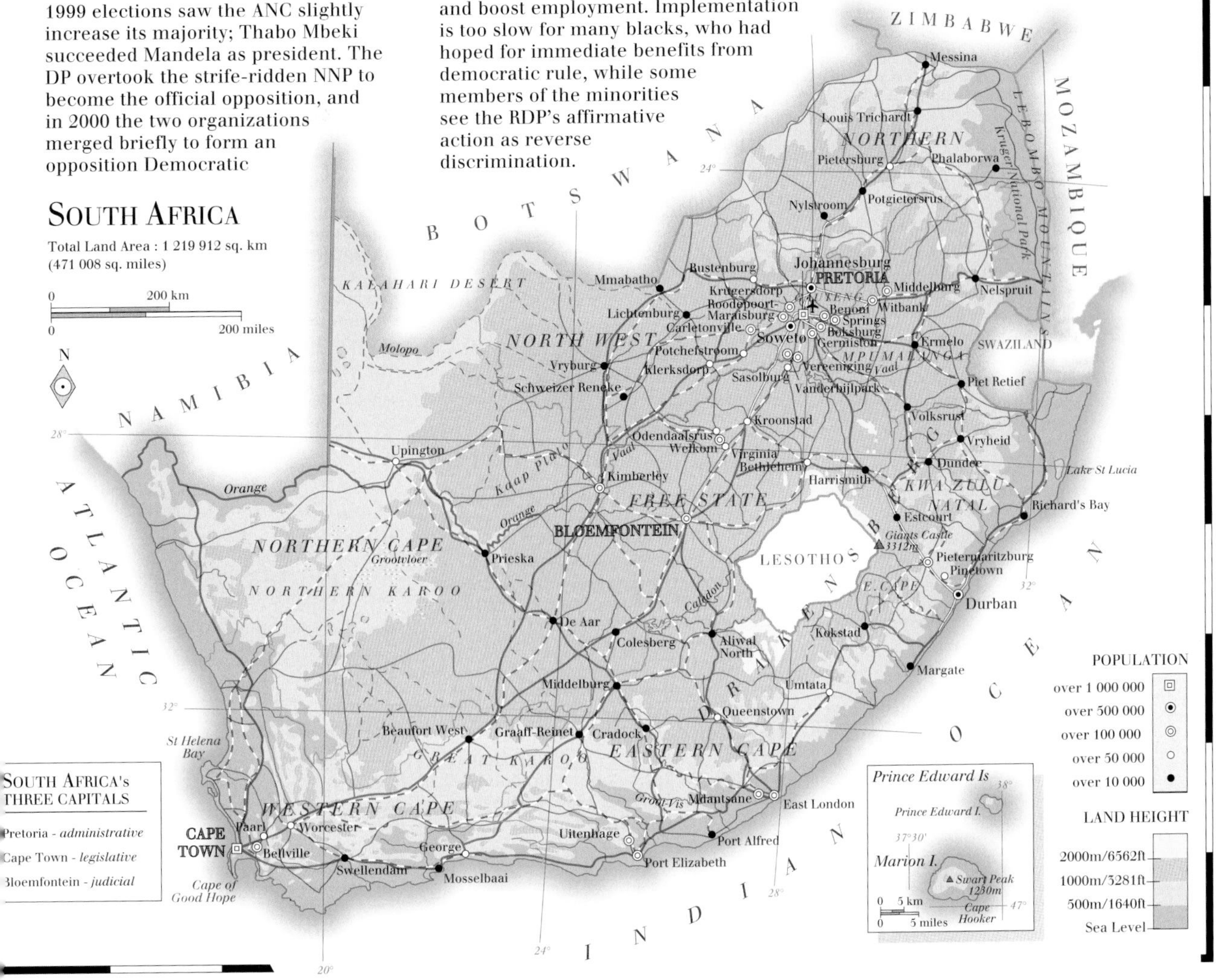

0 200 km
0 200 miles

SOUTH AFRICA'S THREE CAPITALS

Pretoria - *administrative*
Cape Town - *legislative*
Bloemfontein - *judicial*

POPULATION
over 1 000 000
over 500 000
over 100 000
over 50 000
over 10 000

LAND HEIGHT
2000m/6562ft
1000m/3281ft
500m/1640ft
Sea Level

S

AID

Recipient

$488m (receipts)

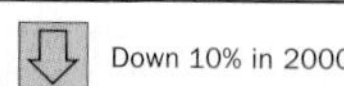
Down 10% in 2000

Apartheid-era South Africa was denied aid, particularly from the World Bank and the IMF. It now seeks financial assistance for massive reconstruction programs. As part of NEPAD – Africa's "Marshall Plan" launched in 2002 – President Mbeki stressed the importance of ending reliance on foreign aid.

CHRONOLOGY

Until 1652, what is now South Africa was peopled by Bantu-speaking groups and Bushmen. Then Dutch settlers arrived. British colonizers followed in the 18th century.

- ❑ **1910** Union of South Africa set up as British dominion; white monopoly of power formalized.
- ❑ **1912** ANC formed.
- ❑ **1934** Independence.
- ❑ **1948** NP takes power; apartheid segregationist policy introduced.
- ❑ **1958–1966** Hendrik Verwoerd prime minister. "Grand Apartheid" policy implemented.
- ❑ **1959** Pan-Africanist Congress (PAC) formed.
- ❑ **1960** Sharpeville massacre. ANC, PAC banned.
- ❑ **1961** South Africa becomes republic; leaves Commonwealth.
- ❑ **1964** Senior ANC leader Nelson Mandela jailed.
- ❑ **1976** Soweto uprisings by black students; hundreds killed.
- ❑ **1978** P. W. Botha in office.
- ❑ **1984** New constitution: Indians and Coloreds get some representation. Growing black opposition.
- ❑ **1985** State of emergency introduced. International sanctions.
- ❑ **1989** F. W. De Klerk replaces Botha as president.
- ❑ **1990** De Klerk legalizes ANC and PAC; frees Nelson Mandela.
- ❑ **1990–1993** International sanctions gradually withdrawn.
- ❑ **1991** Convention for a Democratic South Africa (CODESA) starts work.
- ❑ **1993** Mandela and De Klerk win Nobel Peace Prize.
- ❑ **1994** Multiracial elections won by ANC; Mandela president.
- ❑ **1996** TRC begins work.
- ❑ **1997** New constitution takes effect.
- ❑ **1998** TRC report condemns both apartheid crimes and ANC excesses.
- ❑ **1999** ANC election victory; Thabo Mbeki succeeds Mandela.
- ❑ **2000** DA wins nearly 25% of votes in local elections.
- ❑ **2001** Rand hits record low.
- ❑ **2002** August, world summit on sustainable development held in Johannesburg.

DEFENSE

No compulsory military service

$1.91bn

Up 9% in 2000

SOUTH AFRICAN ARMED FORCES

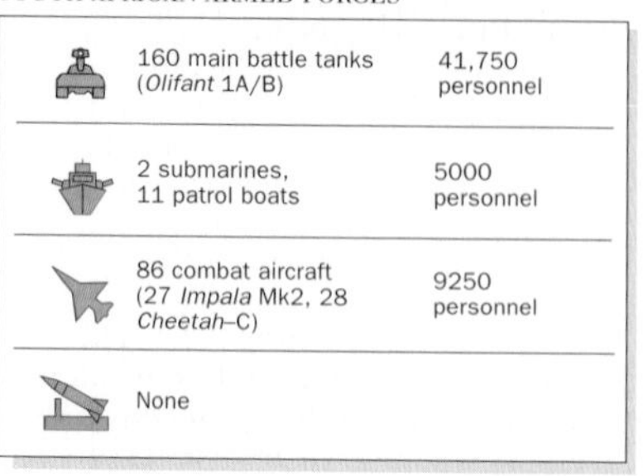

160 main battle tanks (*Olifant* 1A/B)	41,750 personnel
2 submarines, 11 patrol boats	5000 personnel
86 combat aircraft (27 *Impala* Mk2, 28 *Cheetah*–C)	9250 personnel
None	

The creation by postapartheid South Africa of a truly national defense force seems almost miraculous, as it fuses together once bitter enemies: soldiers from the old white-run army, and guerrillas from the liberation groups.

However, doubts have been raised over the army's ability to operate effectively. A freeze on recruitment since 1994 has raised the average age of troops and created a glut of higher-ranking officers, while the incidence of AIDS is increasing. A large arms procurement program, announced in late 1998, failed to overcome the effects of previous swingeing cuts in spending. Few tanks are operational and the air force tends to run out of fuel toward the end of each financial year.

The legacy of years of sanctions is a major arms industry.

ECONOMICS

Inflation 9.6% p.a. (1990–2000)

$129bn

7.570–11.995 rand

SCORE CARD

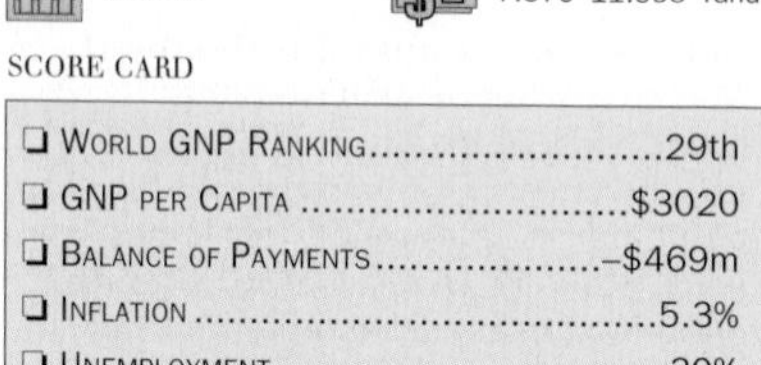

❑ World GNP Ranking	29th
❑ GNP per Capita	$3020
❑ Balance of Payments	–$469m
❑ Inflation	5.3%
❑ Unemployment	30%

EXPORTS

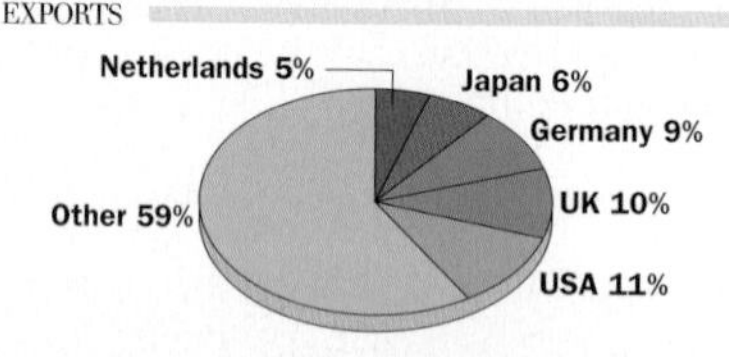

Netherlands 5%
Japan 6%
Germany 9%
UK 10%
USA 11%
Other 59%

IMPORTS

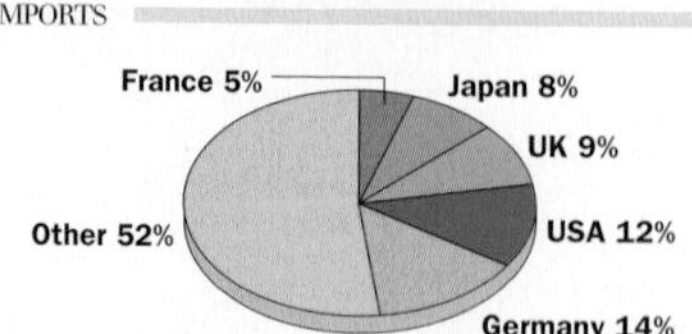

France 5%
Japan 8%
UK 9%
USA 12%
Germany 14%
Other 52%

ECONOMIC PERFORMANCE INDICATOR

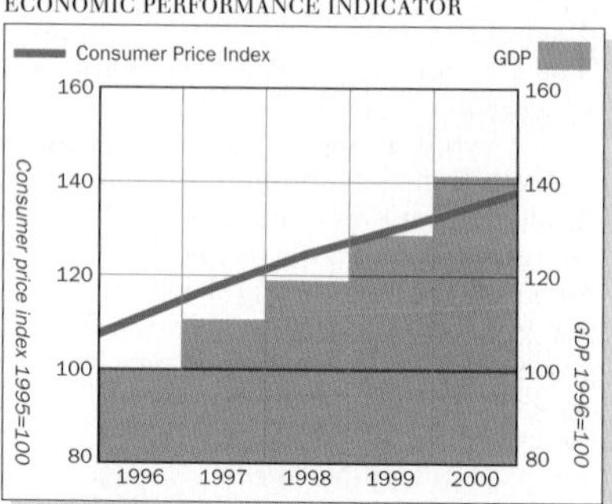

Consumer Price Index
GDP
Consumer price index 1995=100
GDP 1996=100
160
140
120
100
80
1996
1997
1998
1999
2000

Strengths
Africa's largest and most developed economy; highly diversified with modern infrastructure. Strong financial sector for mobilizing investment. Growing manufacturing sector. Varied resource base, particularly of strategically important minerals.

Weaknesses
Foreign investors deterred by political fears. Growth too low to overcome deprivation. Black unemployment growing by 2.5% a year. Emigration of skilled workers. Population boom. Rand at record low in 2001. Falling gold price undermines many sectors.

Profile
South Africa has a large and diverse private sector, much of it controlled by multinationals. International sanctions forced the government to play a central economic role through state corporations in the 1980s. This is now being reduced in a series of privatizations. The ANC has declared its intention to work with big business in order to revivify the economy and develop the townships.

SOUTH AFRICA : MAJOR BUSINESSES

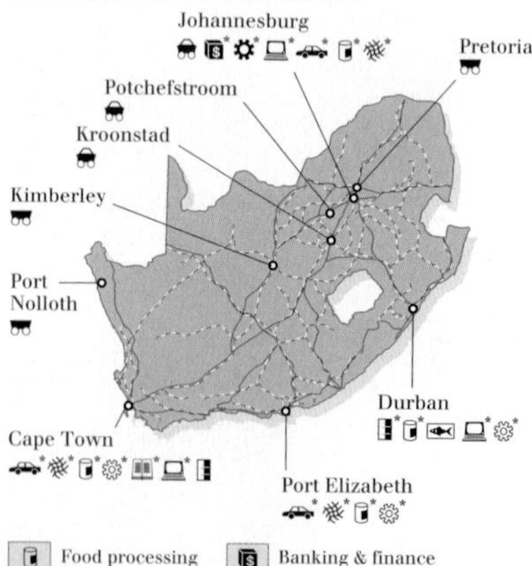

Johannesburg
Pretoria
Potchefstroom
Kroonstad
Kimberley
Port Nolloth
Durban
Cape Town
Port Elizabeth

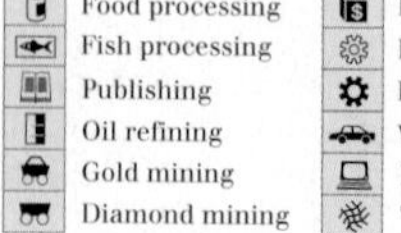

Food processing
Fish processing
Publishing
Oil refining
Gold mining
Diamond mining
Banking & finance
Light engineering
Heavy engineering
Vehicle manufacture
Hi-tech
Textiles
0
500 km
0
500 miles

* significant multinational ownership

S

RESOURCES

Electric power 35.9m kw

592,144 tonnes

146,365 b/d (reserves 107m barrels)

28.8m sheep, 13.7m cattle, 119m chickens

Gold, coal, vanadium, vermiciline, diamonds, chromium, manganese, uranium, nickel

ELECTRICITY GENERATION

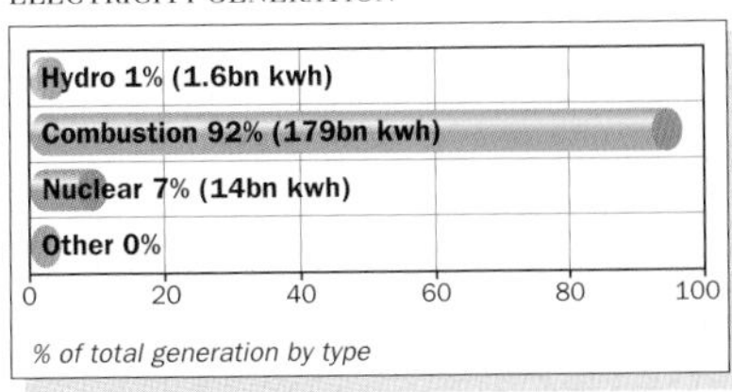
Hydro 1% (1.6bn kwh)
Combustion 92% (179bn kwh)
Nuclear 7% (14bn kwh)
Other 0%
0
20
40
60
80
100
% of total generation by type

South Africa has some of the continent's richest natural resources, in particular minerals. Its dominance of the world market in gold and diamonds helped it survive sanctions during apartheid. The falling price of gold in 2000 meant that for the first time sales of platinum group metals outstripped those of gold. South Africa is the largest single producer of manganese, chrome ore, vanadium, and vermiciline. It also produces uranium.

With no oil, South Africa pioneered the transformation of coal into oil, and otherwise uses its huge coal reserves to generate electricity. About 80% of black homes lack electricity, and the government is considering non-grid options. Agriculture is varied and provides lucrative export earnings.

SOUTH AFRICA : LAND USE

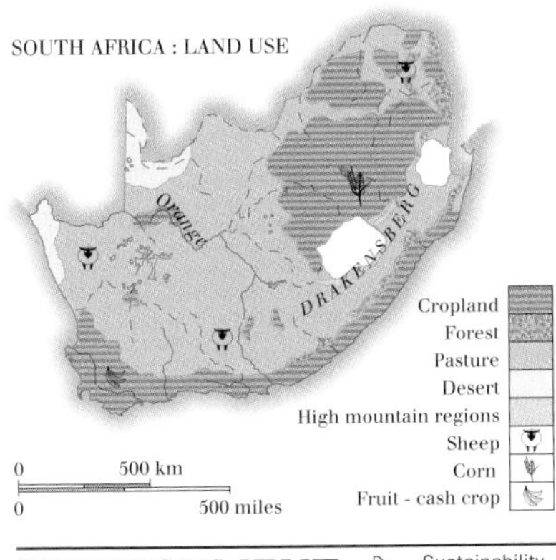
Orange
DRAKENSBERG
Cropland
Forest
Pasture
Desert
High mountain regions
Sheep
Corn
Fruit - cash crop
0
500 km
0
500 miles

ENVIRONMENT

Sustainability rank: 77th

5% (4% partially protected)

8.3 tonnes per capita

ENVIRONMENTAL TREATIES

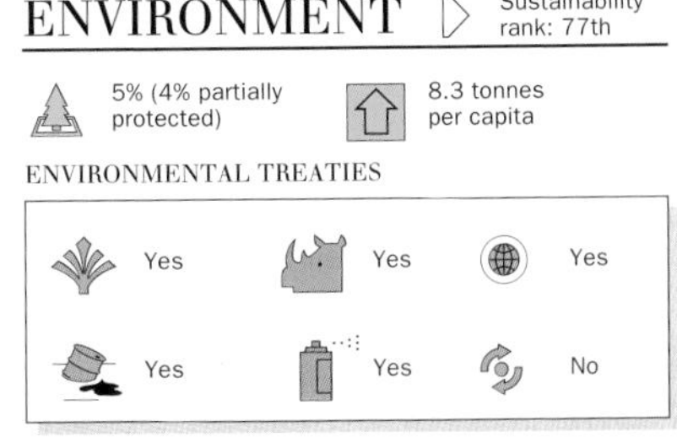
Yes
Yes
Yes
Yes
Yes
No

Floods and drought are familiar hazards. The main concern is protecting animal species. The creation of the world's largest game park, in cooperation with the governments of Zimbabwe and Mozambique, began in 2001.

MEDIA

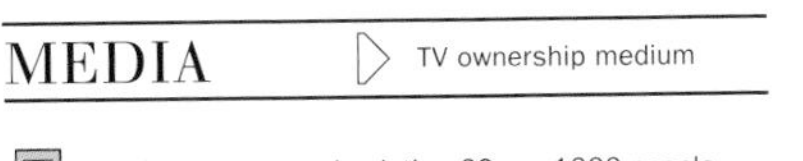
TV ownership medium

Daily newspaper circulation 29 per 1000 people

PUBLISHING AND BROADCAST MEDIA

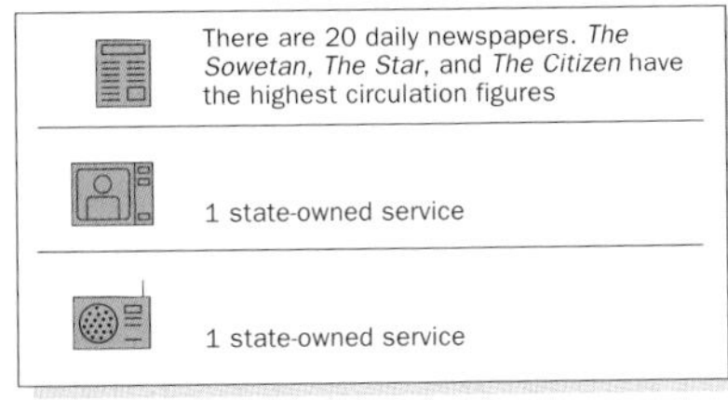
There are 20 daily newspapers. The Sowetan, The Star, and The Citizen have the highest circulation figures
1 state-owned service
1 state-owned service

A drive to combat racial stereotyping in the media was launched following a report on the subject to the Human Rights Commission in early 2000.

CRIME

No death penalty

166,334 prisoners

Up 6% in 1999

CRIME RATES

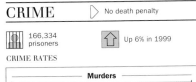

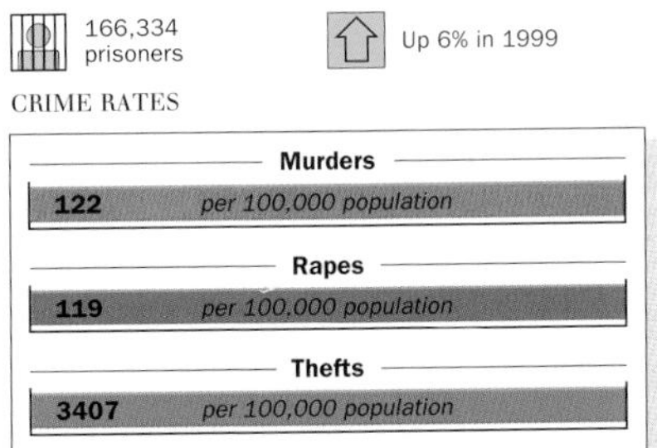
Murders
122 per 100,000 population
Rapes
119 per 100,000 population
Thefts
3407 per 100,000 population

South Africa is a dangerous country and crime rates are rising: murders occur with extreme frequency, and rape, armed robberies, and muggings are rife. Vigilantism is a huge problem in the Cape. The government introduced new gun laws in 2000. The death penalty was abolished in 1997.

EDUCATION

School leaving age: 15

85%

633,918 students

THE EDUCATION SYSTEM

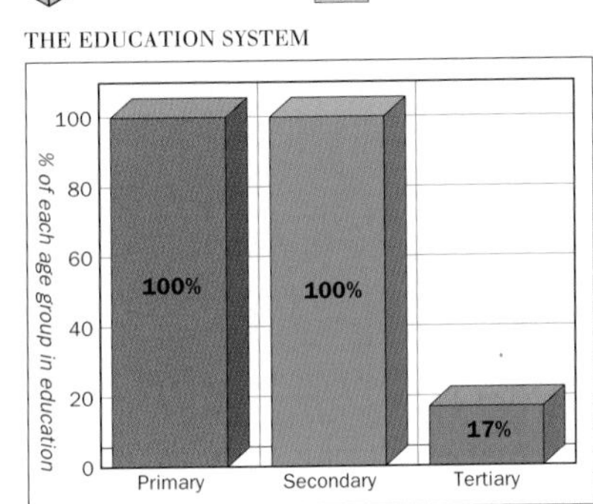
100
80
60
40
20
0
% of each age group in education
100%
100%
17%
Primary
Secondary
Tertiary

Education reform is a central task of the postapartheid government. Progress has been made in improving national literacy, and access to education has been widened through the Tirisano (working together) education program, launched in 2000. Long-established universities continue to be white-dominated.

HEALTH

Welfare state health benefits

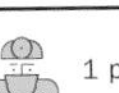
1 per 1667 people

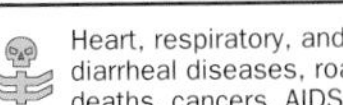
Heart, respiratory, and diarrheal diseases, road deaths, cancers, AIDS

Health services were desegregated formally in 1990, but equal access to care is still a distant goal. Statistics on medical provision hide a strong bias toward whites and urban areas, where 80% of doctors work; 20% of children in rural areas die before the age of five, which is a rate considerably higher than the sub-Saharan average. South Africa has five million AIDS sufferers, more than any other country. The government has won the right to buy cheaper generic drugs for AIDS sufferers and has increased spending, but balked at the potential cost of widespread provision.

SPENDING

GDP/cap. increase

CONSUMPTION AND SPENDING

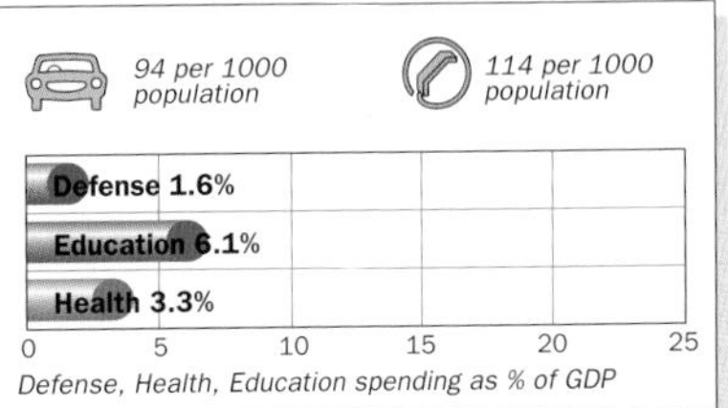
94 per 1000 population
114 per 1000 population
Defense 1.6%
Education 6.1%
Health 3.3%
0
5
10
15
20
25
Defense, Health, Education spending as % of GDP

In South Africa, the black majority forms the poorest group in society. Wealth disparities are marked. At the top, the white elite enjoys living standards similar to those of Californians. In contrast, black living conditions are among Africa's poorest. Nearly half of black adults are unemployed. In between are the mixed race and Asian communities, who enjoyed more privileges under apartheid's strict racial hierarchy. However, a small black middle class is growing slowly, with some black-owned firms doing well on the stock market.

WORLD RANKING

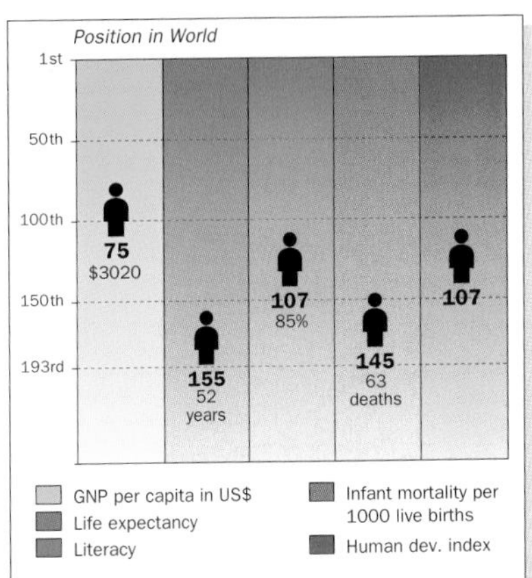
Position in World
1st
50th
100th
150th
193rd
75
$3020
155
52
years
107
85%
145
63
deaths
107
GNP per capita in US$
Life expectancy
Literacy
Infant mortality per 1000 live births
Human dev. index

S

SPAIN

EUROPE

France
SPAIN
Portugal
Andorra
Morocco
Europe

OFFICIAL NAME: Kingdom of Spain **CAPITAL:** Madrid **POPULATION:** 39.9 million
CURRENCY: Euro (peseta until 2002) **OFFICIAL LANGUAGES:** Spanish, Galician, Basque, and Catalan

1492
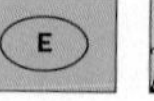
1713

Oct 12
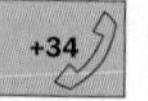
E

+1

+34

.es

OCCUPYING THE MAJOR PART of the Iberian peninsula in southwest Europe, Spain has both an Atlantic and a Mediterranean coast, and is dominated by a central plateau. After the death of Gen. Franco in 1975, the country managed a rapid and relatively peaceful transition to democracy under the supervision of King Juan Carlos I. Since EU membership in 1986, there has been an increasing devolution of power to the regions. For just over 13 years from 1982, Spain had a center-left government, but the right-of-center Popular Party has dominated since 1996.

Alcaudete, Jaén Province, *in the Andalusian mountains between the Guadalquivir River and Granada. The ruined castle is Moorish.*

CLIMATE

Mediterranean/ maritime/mountain

WEATHER CHART FOR MADRID

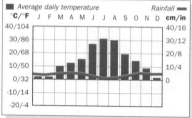

Average daily temperature — Rainfall
°C/°F J F M A M J J A S O N D cm/in
40/104 30/86 20/68 10/50 0/32 -10/14 -20/-4
40/16 30/12 20/8 10/4 0

The central plateau, or *meseta*, endures an extreme climate. Coastal areas are milder, and wetter in the north than in the south.

TRANSPORTATION

Drive on right

Barajas, Madrid 32.9m passengers

1570 ships 1.8m grt

THE TRANSPORTATION NETWORK

343,389 km (213,372 miles)

9063 km (5631 miles)

13,878 km (8624 miles)

1045 km (649 miles)

The AVE high-speed train links Madrid and Seville; more routes are planned. The state-run rail company RENFE is to be privatized.

PEOPLE

Pop. density medium

Spanish, Catalan, Galician, Basque

80/km² (207/mi²)

THE URBAN/RURAL POPULATION SPLIT

78% 22%

RELIGIOUS PERSUASION

Other 4%
Roman Catholic 96%

ETHNIC MAKEUP

Other 2%
Basque 2%
Roma 1%
Galician 6%
Catalan 17%
Castilian Spanish 72%

A vigorous regionalism, suppressed under Franco, now flourishes. Catalonia is an example, with Barcelona its vibrant capital. In the Basque region, the ETA separatists who fight for independence by waging a high-profile terror campaign remain in a minority.

Spain today has one of the lowest birthrates in Europe, just half that of 1975. The influence of the Roman Catholic Church on personal behavior has declined, and attitudes to sexuality are now relaxed. The divorce rate is very low, and family ties remain strong; men often live at home until their late 20s.

POPULATION AGE BREAKDOWN

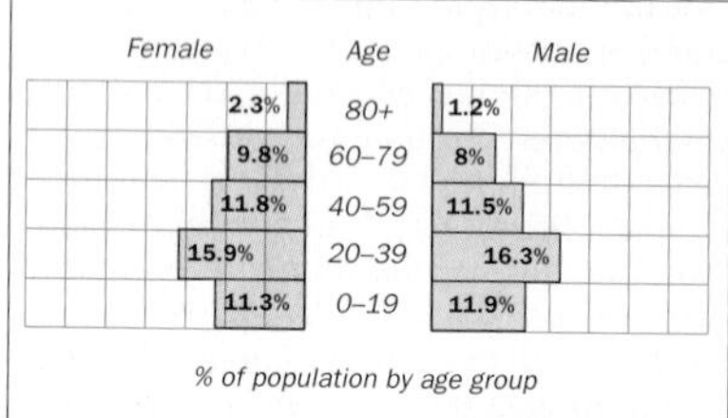
Female
Age
Male
2.3%
80+
1.2%
9.8%
60–79
8%
11.8%
40–59
11.5%
15.9%
20–39
16.3%
11.3%
0–19
11.9%
% of population by age group

Economic growth from the 1970s led to a change in the composition of society. Migration from poor rural regions to the coast was associated with the arrival of job-seeking immigrants from Latin America and – especially – north Africa. A rise in racial tensions and racism has resulted from the subsequent economic downturn and competition for scarce jobs.

Spanish women are increasingly emancipated and more influential in public life, making up 27% of the deputies and senators in the Spanish parliament, and heading 30% of businesses.

TOURISM

Visitors : Population 1.2:1

49.5m visitors

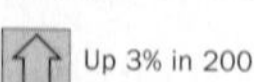
Up 3% in 2001

MAIN TOURIST ARRIVALS

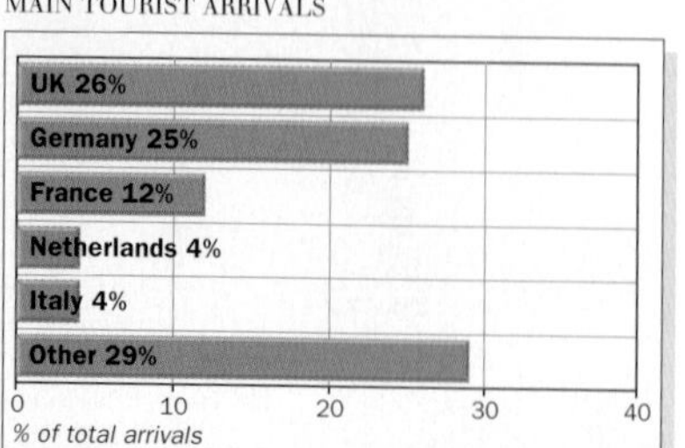
UK 26%
Germany 25%
France 12%
Netherlands 4%
Italy 4%
Other 29%
0
10
20
30
40
% of total arrivals

Tourism earnings in 2001 topped $33 billion, with Germany and the UK still accounting for around 50% of all arrivals. Long dominant in the vacation-package sector, Spain has recently adopted marketing strategies to boost additional cultural, historical, and environmental tourism. Several areas began levying an environmental tax on tourist arrivals in 2001. The cut-price package industry has benefited from political turbulence in potential competitor countries in the Mediterranean.

S

POLITICS

Multiparty elections

L. House 2000/2004
U. House 2000/2004

H.R.H. King Juan Carlos I de Borbón

AT THE LAST ELECTION

Congress of Deputies 350 seats

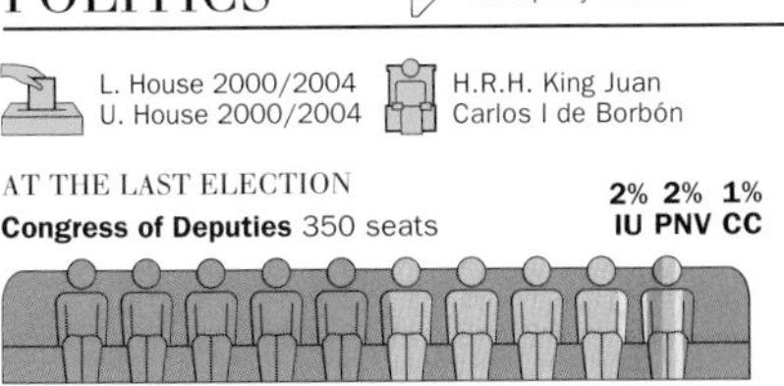

2% IU, 2% PNV, 1% CC

52% PP, 36% PSOE, 5% CiU, 2% Others

PP = Popular Party **PSOE** = Spanish Socialist Workers' Party **CiU** = Convergence and Union **IU** = United Left **PNV** = Basque Nationalist Party **CC** = Canary Islands Coalition **App** = Appointed

Senate 259 seats

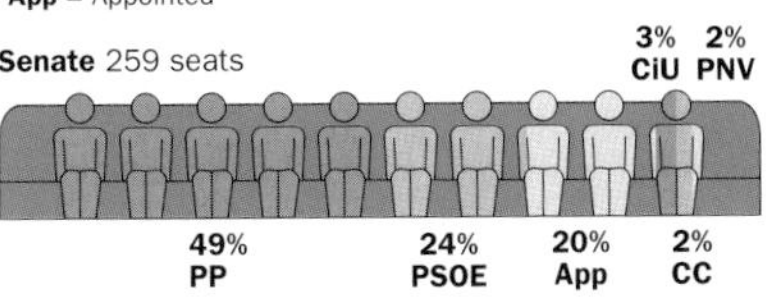

3% CiU, 2% PNV

49% PP, 24% PSOE, 20% App, 2% CC

208 members are directly elected to the Senate, and 51 appointed by autonomous communities

Since 1978, Spain has been a semifederal multiparty parliamentary monarchy. Each region has a legislative assembly.

PROFILE

The PSOE's long period in power blurred the boundaries between party and state. The Cortes (parliament) failed to check executive power, and political disputes were often left to the judiciary, while political corruption undermined voters' faith in Spain's political system. The PP government, led by José María Aznar from 1996, benefited from being seen as a fresh start. Its relative success in running the economy helped it win a second term in 2000, despite an electoral pact between the PSOE and the United Left. Ideological issues no longer sharply divide the main parties, which hold similar views on economic policy and EU membership.

MAIN POLITICAL ISSUES

Increasing regionalism

Spain's 17 autonomous regions all vie for greater funds or independence from Madrid. Many have bypassed central government to borrow funds on the international money markets, and have come close to breaching their legal debt limits. In 1996 the PP government approved a new model of financing for the regions which gave them new powers for raising tax revenue. The Basque separatist movement ETA has, with intermittent cease-fire announcements, waged a protracted violent struggle for independence, prompting large-scale demonstrations against violence. The Basque country, Catalonia, and Galicia each use their own language alongside Castilian Spanish.

Clean government

Initially the PP government suffered less from the corruption scandals which dogged the last years of its PSOE predecessor, but in 2001 its reputation was tarnished by major allegations relating to the Telefonica telecoms group and the Gescarta stockbroking house.

King Juan Carlos, *who became head of state on the death of Franco in 1975.*

José María Aznar, *of the PP, was appointed prime minister in 1996.*

WORLD AFFAIRS

Joined UN in 1955

CE
NATO

OECD

OSCE

EU

Spain, an enthusiastic EU member, has been chary of enlargement to include central Europe, which it sees as a threat to its direct financial benefit. Relations with Morocco, strained over illegal immigrants, were jeopardized in 2002 over sovereignty of rocky islets in the Mediterranean. There is a long-standing dispute with the UK over the status of Gibraltar; Spain and the UK have recently been willing to negotiate, but the colony itself opposes any change. Elsewhere, Spain has sponsored an Ibero-American Community of Nations (a Hispanic Commonwealth). Spain's first contribution to a UN peacekeeping force was for operations in the Balkans.

SPAIN
Total Land Area : 504 782 sq. km (194 896 sq. miles)
POPULATION
over 1 000 000
over 500 000
over 100 000
over 50 000
over 10 000
LAND HEIGHT
3000m/9843ft
2000m/6562ft
1000m/3281ft
500m/1640ft
Sea Level
0 100 km
0 100 miles
FRANCE
PYRENEES
ANDORRA
MEDITERRANEAN SEA
Golfo de Valencia
Islas Baleares
Menorca
Mallorca
Palma
Eivissa
Formentera
Santander
Bilbao
Irún
Donostia-San Sebastián
Baracaldo
Vitoria
Pamplona (Iruña)
Miranda de Ebro
Burgos
Logroño
Huesca
Soria
Zaragoza
Lleida
Figueres
Girona (Gerona)
Terrassa
Sabadell
Mataró
Badalona
Barcelona
Tarragona
Reus
Tortosa
Teruel
Segovia
Alcalá de Henares
Getafe
Cuenca
Castellón de la Plana
Paterna
Valencia
Albacete
Gandia
Benidorm
Alicante
Elda
Elche
Cieza
Murcia
Linares
Jaén
Alcaudete
Lorca
Cartagena
Almería
Motril
Málaga
Costa del Sol
Costa Brava
Costa Dorada
Costa del Azahar
Costa Blanca
Costa Calida
Cerro de Mulhacén 3481m
Melilla (SPAIN)
Islas Canarias
La Palma
Santa Cruz de Tenerife
La Laguna
Gomera
Tenerife
Las Palmas
Hierro
Gran Canaria
Lanzarote
Fuerteventura
0 100 km
0 100 miles

CHRONOLOGY

United under Ferdinand and Isabella in 1492, Spain became a dominant force. A long period of economic and political decline followed, however, and by the mid-19th century, Spain lagged behind many other European countries in stability and prosperity.

- **1874** Constitutional monarchy restored under Alfonso XII.
- **1879** PSOE founded.
- **1881** Trade unions legalized.
- **1885** Death of Alfonso XII.
- **1898** Defeat in war with US results in loss of Cuba, Puerto Rico, and the Philippines.
- **1914–1918** Spain neutral in World War I.
- **1921** Spanish army routed by Berbers in Spanish Morocco.
- **1923** Coup by Gen. Primo de Rivera accepted by King Alfonso XIII. Military dictatorship.
- **1930** Primo de Rivera dismissed by monarchy.
- **1931** Second Republic proclaimed. Alfonso XIII flees Spain.
- **1933** Center-right coalition wins general election.

➪

CHRONOLOGY *continued*

- **1934** Asturias uprising quashed by army. Failure of attempt to form Catalan state.
- **1936** Popular Front wins elections. Right-wing military uprising against Republic. Gen. Francisco Franco subsequently appointed leader.
- **1939** Franco wins civil war which claims 300,000 lives.
- **1940** Franco meets Hitler, but does not enter World War II.
- **1946** UN condemns Franco regime.
- **1948** Spain excluded from the Marshall Plan.
- **1950** UN lifts veto.
- **1953** Concordat with Vatican. Spain grants US military bases.
- **1955** Spain joins UN.
- **1959** Stabilization Plan is basis for 1960s rapid economic growth.
- **1962** Franco government applies for eventual membership of EEC.
- **1969** Gen. Franco names Juan Carlos, grandson of Alfonso XIII, his successor.
- **1970** Spain signs preferential trade agreement with EEC.
- **1973** Basque separatists assassinate Prime Minister Carrero Blanco; replaced by Arias Navarro.
- **1975** Death of Franco. Proclamation of King Juan Carlos I.
- **1976** King appoints Adolfo Suárez as prime minister.
- **1977** First democratic elections since 1936 won by Suárez's Democratic Center Union.
- **1978** New constitution declares Spain a parliamentary monarchy.
- **1981** Leopoldo Calvo Sotelo replaces Suárez. King foils military coup. Calvo takes Spain into NATO.
- **1982** Felipe González wins landslide victory for PSOE.
- **1986** Joins European Communities. González wins referendum on keeping Spain in NATO.
- **1992** Olympic Games held in Barcelona, Expo '92 in Seville.
- **1996** PSOE loses election; José María Aznar of PP prime minister.
- **1998** Former PSOE minister found guilty of involvement in Basque kidnappings. September, ETA cease-fire; holds until December 1999.
- **2000** Aznar and PP win elections.
- **2002** Euro fully adopted.

AID

Donor

$1.2bn (donations) — Down 12% in 2000

Spain has taken steps to increase grant aid after criticism that Spanish aid was of poor quality and tied to the acquisition of goods and services. Aid in 2000 represented 0.2% of GNP.

S

DEFENSE

No compulsory military service

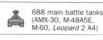

 $7.05bn  Down 2% in 2000

A substantial, largely state-owned and commercially nonviable defense industry is subsidized for strategic reasons. Full integration of NATO military structures was approved in 1997. Defense spending has fallen in recent years, and is now well below the NATO average. National service has been abolished; the last conscripts worked out their terms of duty in the course of 2002.

SPANISH ARMED FORCES

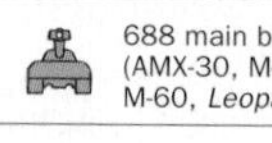	688 main battle tanks (AMX-30, M-48A5E, M-60, *Leopard* 2 A4)	92,000 personnel
	8 submarines, 1 carrier, 15 frigates, 37 patrol boats	26,950 personnel
	211 combat aircraft (F-5B, EF/A-18 A/B, RF-4C, *Mirage* F-1CF/BE/EE)	24,500 personnel
	None	

ECONOMICS

Inflation 3.9% p.a. (1990–2000)

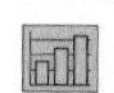 $595bn  1.0651–1.1231 euros

SCORE CARD

- World GNP Ranking 10th
- GNP per Capita $15,080
- Balance of Payments .... –$17.3bn
- Inflation 3.4%
- Unemployment 14%

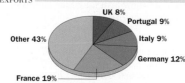

EXPORTS

UK 8%
Portugal 9%
Italy 9%
Germany 12%
France 19%
Other 43%

IMPORTS

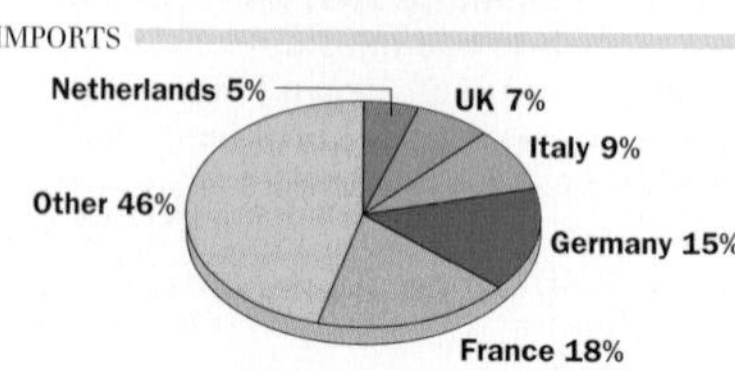
Netherlands 5%
UK 7%
Italy 9%
Germany 15%
France 18%
Other 46%

Strengths
One of the fastest-growing OECD economies. Well-qualified labor force with relatively low labor costs. Privatization has introduced greater competition into gas, oil-refining, electricity, and telecommunications sectors. Recent major investment and expansion into Latin American market.

Weaknesses
Massive foreign penetration of economy and absence of Spanish multinationals. Low investment in research and development, concentration in declining industries, and low productivity – notably in agriculture. Persistent high unemployment.

Profile
Real convergence with the major European economies first became a realistic objective in the late 1980s, as Spain posted the highest investment-led output growth in the OECD. By 1991, GDP per capita stood at almost 80% of the EU average. Recession in the early 1990s was turned around in mid-decade, and growth averaged 4% over 1997–2000, with public debt brought below 60% of GDP by 2001. The economy slowed markedly through 2001, and the annualized growth rate dropped to around 2%. Spain succeeded in meeting the economic convergence criteria necessary for European economic and monetary union and was among the 12 EU countries to adopt the euro fully in January 2002.

ECONOMIC PERFORMANCE INDICATOR

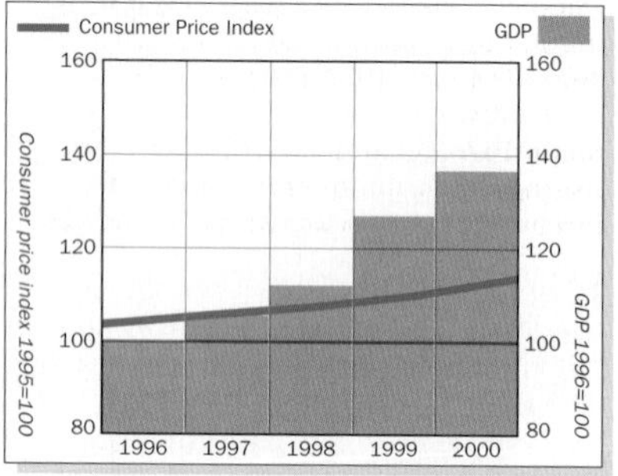
Consumer Price Index
GDP
160
140
120
100
80
1996
1997
1998
1999
2000
Consumer price index 1995=100
GDP 1996=100

SPAIN : MAJOR BUSINESSES

La Coruña
Bilbao
Zaragoza
Barcelona
Vigo
Madrid
Valencia
Huelva
Sevilla
Málaga
Cartagena
Textiles
Agribusiness
Chemicals
Shipbuilding
Vehicle manufacture
Heavy engineering
Light engineering
Fish processing
0 200 km
0 200 miles
* significant multinational ownership

RESOURCES

Electric power 50.1m kw

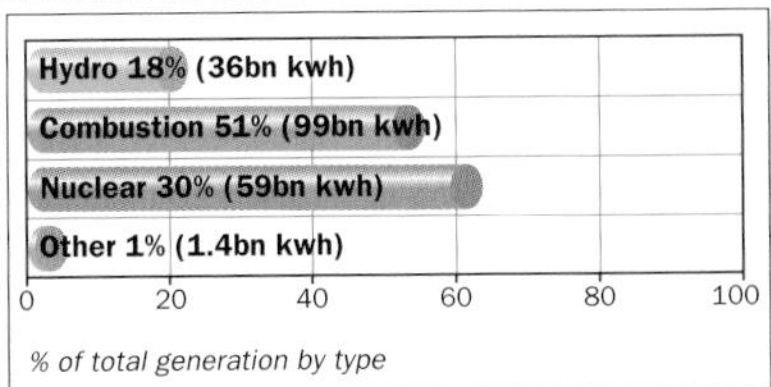
1.49m tonnes
10,606 b/d (reserves 7.7m barrels)
24.4m sheep, 23.3m pigs, 128m chickens
Coal, oil, iron, uranium, mercury, fluorite, gypsum

ELECTRICITY GENERATION

Hydro	18% (36bn kwh)
Combustion	51% (99bn kwh)
Nuclear	30% (59bn kwh)
Other	1% (1.4bn kwh)

% of total generation by type

Spain lacks natural resources, especially water, and is heavily dependent on imported oil and gas. Coal, mined mainly to generate industry, is a declining but still subsidized sector, concentrated in the Asturias region. Spain has one of the world's largest fishing fleets, but EU restrictions have forced cuts in catches since the 1990s.

SPAIN : LAND USE

PYRENEES
Río Duero
Forest
Pasture
Cropland
Wetlands
High mountain regions
Sheep
Olives - cash crop
Citrus fruits
Vineyards
0 200 km
0 200 miles

ENVIRONMENT

Sustainability rank: 44th

9%

6.3 tonnes per capita

ENVIRONMENTAL TREATIES

Yes Yes Yes

Yes Yes Yes

Public opinion is becoming increasingly vocal about environmental matters. Renewable energy, although still tiny in extent, is becoming more visible, particularly with the growth of wind farms. The benefits of a national tree-planting scheme to reduce soil erosion have been offset by increasingly frequent intentional forest fires. More land has national park status than in any other country in Europe. While dams and canals are necessary to counter desertification in the south, environmentalists oppose their construction. A project to bring water from the Ebro River to the parched south is causing controversy, particularly in areas further downstream.

MEDIA

TV ownership high

Daily newspaper circulation 99 per 1000 people

PUBLISHING AND BROADCAST MEDIA

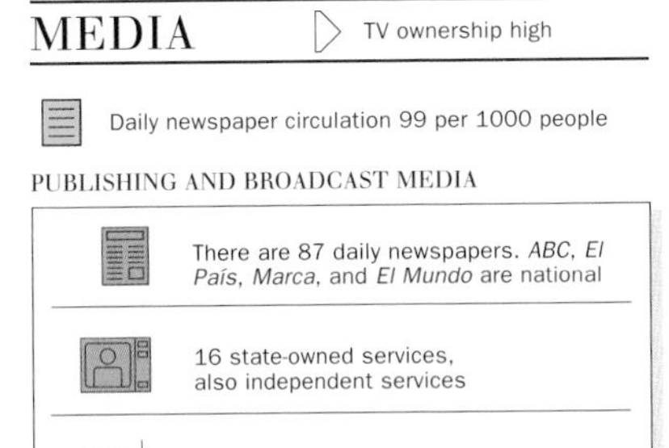
There are 87 daily newspapers. *ABC*, *El País*, *Marca*, and *El Mundo* are national
16 state-owned services, also independent services
363 services: 13 state-owned, 350 independent

Despite the large number of daily newspapers, readership is among the lowest in Europe. Both public and private TV are popular. Radio is of a generally high standard.

CRIME

No death penalty

45,633 prisoners

Up 92% in 1999

CRIME RATES

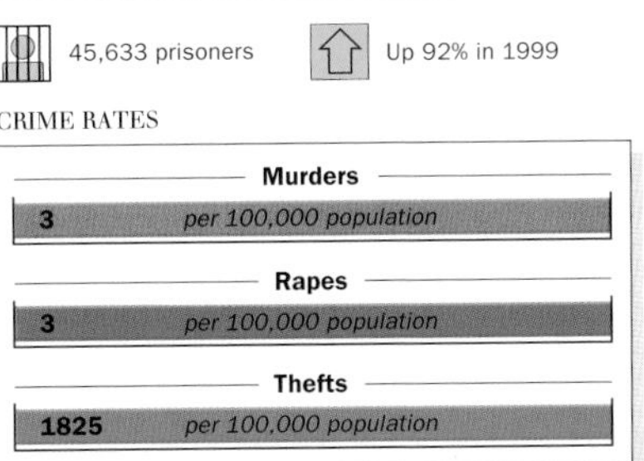
Murders
3 per 100,000 population
Rapes
3 per 100,000 population
Thefts
1825 per 100,000 population

Spain is a major crossroads in the world narcotics trade, and drugs-related crime is rising. Illegal immigration soared in 2000, with authorities in the south unable to cope with the influx.

EDUCATION

School leaving age: 16

98%

1.79m students

THE EDUCATION SYSTEM

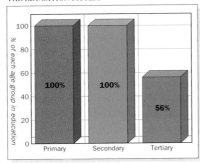

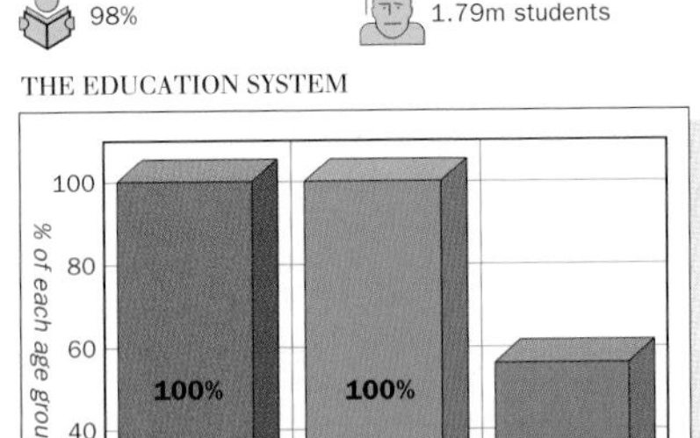
% of each age group in education
100
80
60
40
20
0
100%
100%
56%
Primary
Secondary
Tertiary

The school leaving age has risen since 1990 from 14 to 16. The latest secondary education reforms, announced in 2000, offer a number of additional subjects, improvements in mathematics, philosophy, and languages, and increasing attention to information technology. Autonomous regions regulate by decree the teaching of languages other than Castilian Spanish, such as Basque or Catalan.

HEALTH

Welfare state health benefits

1 per 323 people

Heart and circulatory diseases, cancers, accidents

Public health care is of high quality and readily available. Public hospitals, although widely considered to be superior, are outnumbered by private ones. In spite of very high tobacco and alcohol consumption, Spain has a healthy population, possibly due to its Mediterranean diet. The incidence of AIDS has risen alarmingly, however, to become one of the highest in western Europe.

SPENDING

GDP/cap. increase

CONSUMPTION AND SPENDING

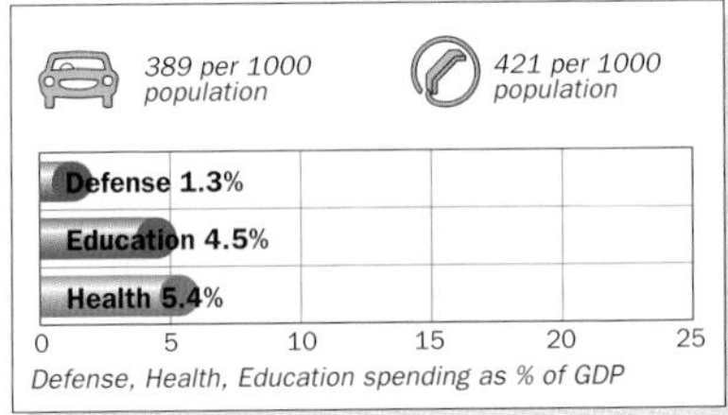
389 per 1000 population
421 per 1000 population
Defense 1.3%
Education 4.5%
Health 5.4%
0
5
10
15
20
25
Defense, Health, Education spending as % of GDP

In the late 1980s, it became fashionable in Spain to compete openly, make money, and acquire consumer goods. Rapid economic growth at this time greatly enriched the professional and managerial classes. The latter became the best-paid, in real terms, in Europe, and Spain quickly became an important market for luxury cars and yachts. In the early 1990s this ostentatious affluence waned in the face of recession and as the unemployment rate soared to become one of the highest in Europe. The boom of 1997–2000 boosted the number of jobs, but also gave rise to anxiety over inflation. By 2001 this boom too had subsided, and there was less financial security, with about one-third of employees on only temporary contracts.

WORLD RANKING

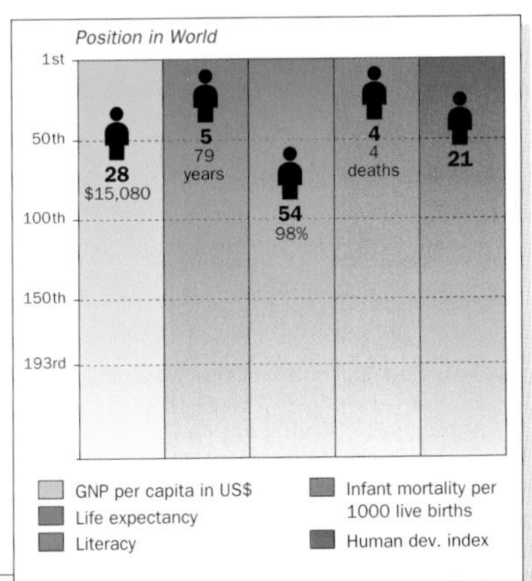
Position in World
1st
50th
100th
150th
193rd
28
$15,080
5
79 years
54
98%
4
4 deaths
21
GNP per capita in US$
Life expectancy
Literacy
Infant mortality per 1000 live births
Human dev. index

S

Sri Lanka

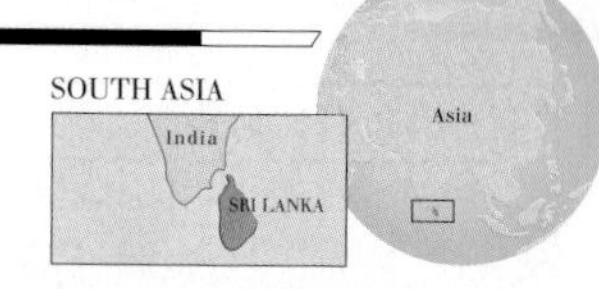
SOUTH ASIA
India
SRI LANKA
Asia

Official Name: Democratic Socialist Republic of Sri Lanka **Capital:** Colombo
Population: 19.1 million **Currency:** Sri Lanka rupee **Official Languages:** Sinhala, Tamil, and English

1948

1948

Feb 4
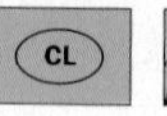
CL
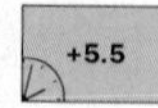
+5.5

+94

.lk

The TEARDROP-SHAPED island of Sri Lanka is separated from India by the Palk Strait. Rugged central uplands give way to fertile plains in the north. The majority Sinhalese, an Indo-Aryan people originating in northern India, have a Buddhism-based identity. Independent since 1948, Sri Lanka suffered from 1983 from a protracted civil war involving the attempted secession of the minority (and mainly Hindu) Tamils in the north and east.

CLIMATE

Tropical monsoon/ equatorial

WEATHER CHART FOR COLOMBO

Average daily temperature — Rainfall
°C/°F J F M A M J J A S O N D cm/in
40/104, 30/86, 20/68, 10/50, 0/32, -10/14, -20/-4
40/16, 30/12, 20/8, 10/4, 0

The climate is tropical, with afternoon breezes on the coast and cooler air in the highlands. The northeast is driest.

TRANSPORTATION

Drive on left

Bandaranaike, Katunayake 2.88m passengers

60 ships 189,200 grt

THE TRANSPORTATION NETWORK

10,721 km (6662 miles)

None

1463 km (909 miles)

430 km (267 miles)

Main roads are crowded and slow. Travel to the Tamil-dominated north and east is being reopened.

TOURISM

Visitors : Population 1:48

400,000 visitors

Down 8% in 2000

MAIN TOURIST ARRIVALS

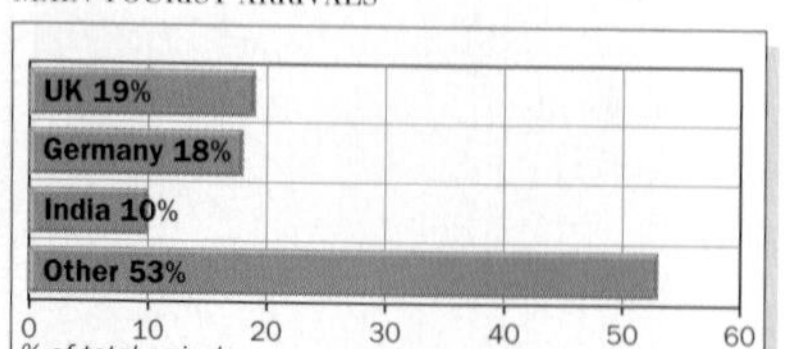
UK 19%
Germany 18%
India 10%
Other 53%
0 10 20 30 40 50 60
% of total arrivals

Sri Lanka has remained a popular destination, despite the civil war. A Tamil attack in 2001 on the military air base next to Colombo's Bandaranaike international airport threatened the lives of dozens of tourists.

PEOPLE

Pop. density high

Sinhalese, Tamil, Sinhalese-Tamil, English

295/km² (764/mi²)

THE URBAN/RURAL POPULATION SPLIT

24% 76%

ETHNIC MAKEUP

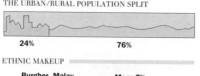

Burgher, Malay, and Veddha 1%
Moor 7%
Tamil 18%
Sinhalese 74%

Ethnic tensions focus on the 19-year conflict between the minority, Hindu, Tamils and majority, Buddhist, Sinhalese. Favored by the British colonial administration, after independence the Tamils were subject to attempts by the Sinhalese to redress the balance. The Tamil Tigers seek a Tamil homeland in the north. A tentative peace was reached in 2002.

The Moors are the Muslim descendants of Arab traders. A few indigenous forest-dwelling Veddhas survive in the remote east of the island.

POLITICS

Multiparty elections

2001/2007

President Chandrika Bandaranaike Kumaratunga

AT THE LAST ELECTION

Parliament 225 seats

48% UNP, 34% PA, 7% JVP, 2% SLMC, 1% DPLF, 7% TULF, 1% EPDP

UNP = United National Party **PA** = People's Alliance, dominated by the Sri Lanka Freedom Party – **SLFP**
JVP = People's Liberation Front **TULF** = Tamil United Liberation Front **SLMC** = Sri Lanka Muslim Congress
EPDP = Eelam People's Democratic Party
DPLF = Democratic People's Liberation Front

Politics has been indelibly marked by the 19-year civil war which has claimed more than 50,000 lives. Breakthrough came in 2001 when the ruling left-wing PA was soundly beaten in elections by the right-of-center UNP. Under Prime Minister Ranil Wickremasinghe, the UNP sought to bring a swift end to the conflict. A permanent cease-fire was agreed in 2002, and the Liberation Tigers of Tamil Eelam (LTTE or Tamil Tigers) hinted that they were willing to drop their long-standing demand for an independent state. The conservative Buddhist clergy remain influential, helping to preserve Sinhalese nationalism, and opposing compromise with the Tamils.

Sri Lanka

Total Land Area : 65 610 sq. km (25 332 sq. miles)

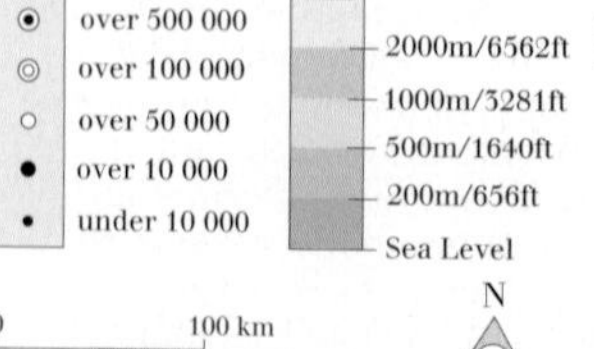
POPULATION
over 500 000
over 100 000
over 50 000
over 10 000
under 10 000
LAND HEIGHT
2000m/6562ft
1000m/3281ft
500m/1640ft
200m/656ft
Sea Level
0 100 km
0 100 miles
N

INDIA
Palk Strait
Gulf of Mannar
Bay of Bengal
INDIAN OCEAN
Kankesanturai
Jaffna
Mannar I.
Kayts
Punkudutivu
Delft
Jaffna Lagoon
Pooneryn
Kilinochchi
Nanthi Kadal Lagoon
Mullaittivu
Karaitivu
Adam's Bridge
Mankulam
Mannar
Talladi
Vavuniya
Horowupotana
Trincomalee
Koddiyar Bay
Medawachchiya
Anuradhapura
Yan
Upaar Lagoon
Puttalam Lagoon
Puttalam
Habarana
Polonnaruwa
Dambulla
Mundal Lagoon
Dedura
Mahaweli Ganga
Chenkaladi
Batticaloa
Chilaw
Matale
Kurunegala
Mahiyangana
Ampara
Negombo
Kandy
Kegalla
Senanayake Samudra
Akkaraipattu
Gampaha
Pidurutalagala 2524m
COLOMBO
Badulla
Dehiwala-Mount Lavinia
Talawakele
Namunukula
Sri Jayawardenapura
Nuwara Eliya
Pottuvil
Moratuwa
Monaragala
Panadura
Kalu
Beragala
Ratnapura
Wellawaya
Kalutara
Pelmadulla
Elpitiya
Ambalangoda
Hambantota
Galle
Weligama
Matara
10°
8°
6°
80°
82°

WORLD AFFAIRS

Joined UN in 1955

Comm

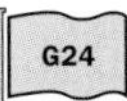
G24

NAM | SAARC | WTO

Relations with India are paramount. However, India's role as peacemaker under the 1987 Indo-Sri Lankan accords was fiercely resisted by the Tamil Tigers, and India was forced to withdraw its peacekeeping troops. In recent years Norway has taken on the role of mediator in the conflict.

AID

Recipient

$276m (receipts)

Up 5% in 2000

The president has responded positively to Western aid donors seeking improvements in Sri Lanka's human rights record.

DEFENSE

No compulsory military service

$862m

Up 7% in 2000

Recent drives to recruit 5000 more soldiers have shown a marked shift in policy, from simply enlarging the army to modernizing it.

ECONOMICS

Inflation 9.1% p.a. (1990–2000)

$16.4bn

82.70–93.16 Sri Lanka rupees

SCORE CARD

- ❏ WORLD GNP RANKING........................73rd
- ❏ GNP PER CAPITA..............................$850
- ❏ BALANCE OF PAYMENTS...................–$1.04bn
- ❏ INFLATION...6.2%
- ❏ UNEMPLOYMENT..................................9%

STRENGTHS

World's largest tea exporter. Foreign investment attracted by privatization, despite left-wing stance of President Kumaratunga.

WEAKNESSES

Civil war has severely drained government funds, and lingering tensions continue to deter investors and many tourists. High unemployment.

EXPORTS

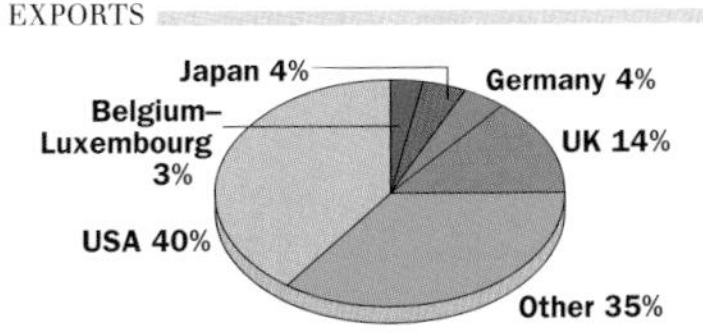
Japan 4%
Germany 4%
Belgium–Luxembourg 3%
UK 14%
USA 40%
Other 35%

IMPORTS

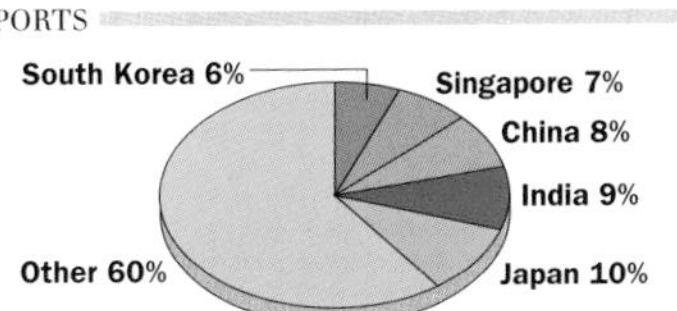
South Korea 6%
Singapore 7%
China 8%
India 9%
Other 60%
Japan 10%

Adam's Peak in mountainous central Sri Lanka is a famous religious site with a Buddhist shrine at the summit.

RESOURCES

Electric power 1.7m kw

279,900 tonnes

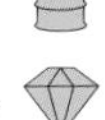
Not an oil producer; refines 50,000 b/d

1.56m cattle, 690,000 buffaloes, 11m chickens

Gemstones, graphite, iron, monazite, uranium, ilmenite, clay

Sri Lanka has to import all its oil. Hydropower supplies around 70% of electricity; droughts are frequent and supplies can be erratic. Sri Lanka is keen to diversify power sources and is turning to coal-powered generation.

ENVIRONMENT

Sustainability rank: 55th

14% (4% partially protected)

0.4 tonnes per capita

Sri Lanka has successfully promoted national parks. Their development is opposed by the Veddha people, who have traditionally occupied such land.

MEDIA

TV ownership medium

Daily newspaper circulation 29 per 1000 people

PUBLISHING AND BROADCAST MEDIA

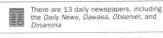

There are 13 daily newspapers, including the *Daily News*, *Dawasa*, *Observer*, and *Dinamina*

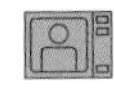
4 independent services

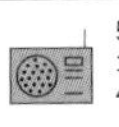
5 services: 1 state-owned, 4 independent

Press censorship was imposed in 1995 to control war reporting by local news agencies. Since 1998 foreign journalists have faced similar restrictions.

CRIME

Death penalty not used in practice

16, 850 prisoners

Up 52% 1992–1996

Extrajudicial killings and other human rights abuses increased the toll of deaths and disappearances during the civil war. New laws are being introduced to combat sex tourism.

EDUCATION

School leaving age: 14

92%

86,931 students

Sri Lanka has a high literacy rate for a developing nation. Many Sri Lankans attend US universities.

CHRONOLOGY

Sri Lanka was inhabited by Tamils and Sinhalese since before the 6th century. Named Ceylon by the British, it became independent in 1948.

- ❏ **1948** Indian Tamil workers stripped of suffrage and citizenship rights.
- ❏ **1956** SLFP wins election, promotes Sinhalese language.
- ❏ **1972** Renamed Sri Lanka.
- ❏ **1983** Tamil Tigers begin civil war.
- ❏ **1993** President Premadasa killed.
- ❏ **1994** Left-wing PA wins election; Chandrika Kumaratunga president.
- ❏ **1995–1996** Collapse of peace talks.
- ❏ **1999** Kumaratunga reelected.
- ❏ **2000** Sirimavo Bandaranaike, world's first woman prime minister, dies.
- ❏ **2001** UNP wins early elections.
- ❏ **2002** Comprehensive cease-fire promises end to civil war.

HEALTH

Welfare state health benefits

1 per 2500 people

Heart attacks, cancers, pneumonia, strokes

Years of high spending on health have resulted in an accessible, fee-free system. Ayurvedic medicine is popular.

SPENDING

GDP/cap. increase

CONSUMPTION AND SPENDING

15 per 1000 population

41 per 1000 population

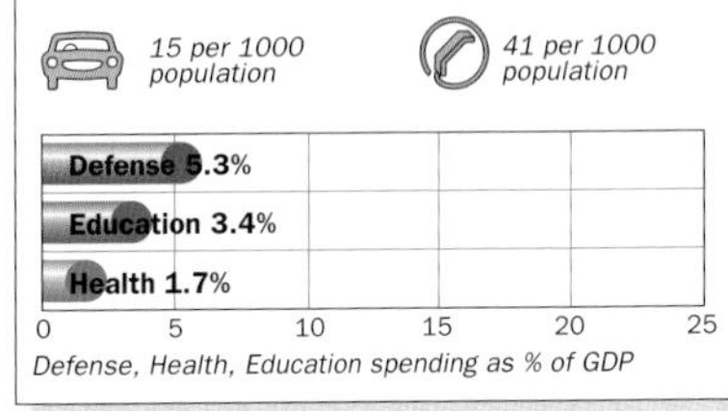
Defense 5.3%
Education 3.4%
Health 1.7%
0
5
10
15
20
25
Defense, Health, Education spending as % of GDP

Economic growth has created a new class of wealthy Sinhalese. Tamil tea workers are the poorest group.

WORLD RANKING

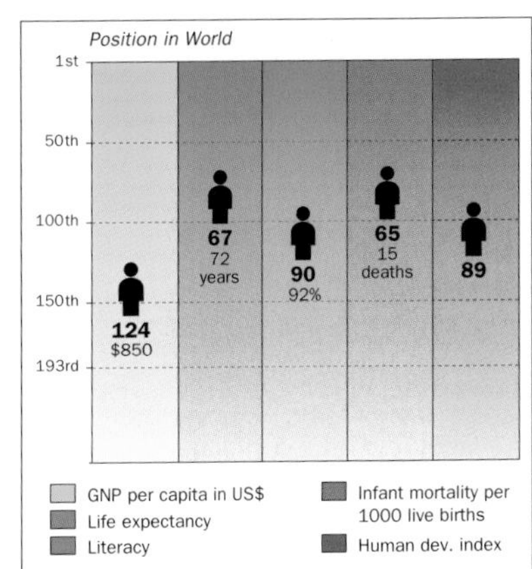
Position in World
1st
50th
100th
150th
193rd
124
$850
67
72
years
90
92%
65
15
deaths
89
GNP per capita in US$
Life expectancy
Literacy
Infant mortality per 1000 live births
Human dev. index

S

SUDAN

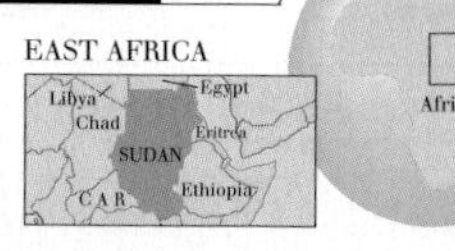
EAST AFRICA
Libya
Chad
Egypt
Eritrea
SUDAN
C.A.R.
Ethiopia
Africa

OFFICIAL NAME: Republic of the Sudan **CAPITAL:** Khartoum **POPULATION:** 31.8 million
CURRENCY: Sudanese pound or dinar **OFFICIAL LANGUAGE:** Arabic

1956 | 1956 | Jan 1 | SUD | +2 | +249 | .sd

BORDERED BY THE RED SEA, Sudan is the largest country in Africa. Its landscape changes from desert in the north to lush tropical in the south, with grassy plains and swamps in the center. Since independence from British and Egyptian rule in 1956, tensions between the Arab north and African south have led to two civil wars. The second of these conflicts remains unresolved. In 1989, an army coup installed a military Islamic fundamentalist regime.

Camel caravan in the dry north. *Periodic drought coupled with war disruption mean that Sudan requires large amounts of food aid.*

CLIMATE

Hot desert/steppe/tropical

WEATHER CHART FOR KHARTOUM

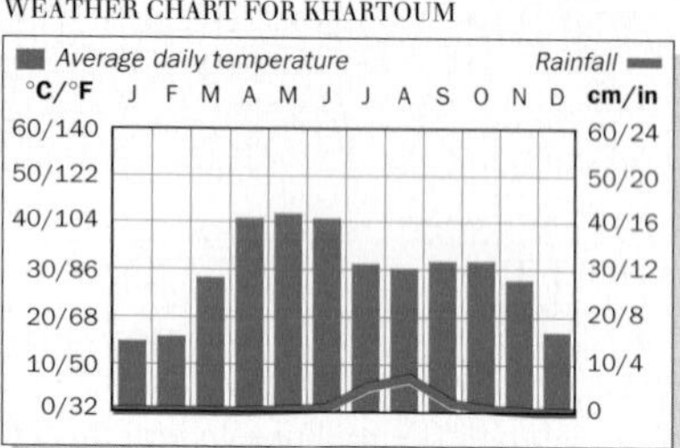
Average daily temperature
Rainfall
°C/°F
J F M A M J J A S O N D
cm/in
60/140
50/122
40/104
30/86
20/68
10/50
0/32
60/24
50/20
40/16
30/12
20/8
10/4
0

Sudan's northern half is hot arid desert with constant dry winds. The rest has a rainy season varying from two months in the center to eight in the south.

TRANSPORTATION

Drive on right

Khartoum International

19 ships 43,078 grt

THE TRANSPORTATION NETWORK

4320 km (2684 miles)

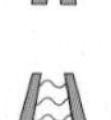

None

4595 km (2855 miles)

5310 km (3299 miles)

The Port Sudan–Khartoum railroad and road are Sudan's most important links. There are few other roads, but Iran is financing a north–south highway. Civil war has stopped all Nile shipping.

TOURISM

Visitors : Population 1:636

50,000 visitors

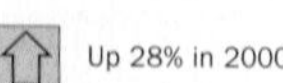

Up 28% in 2000

MAIN TOURIST ARRIVALS

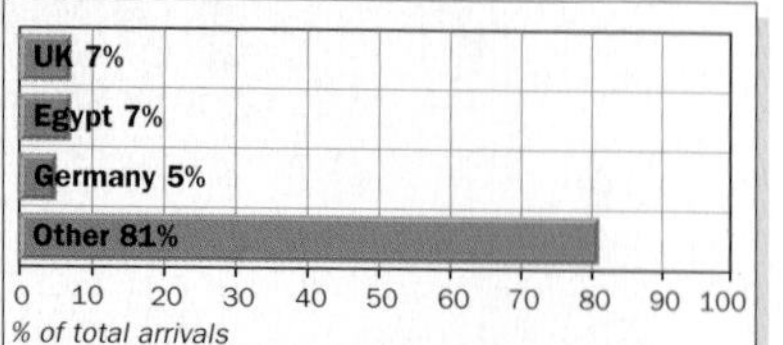
UK 7%
Egypt 7%
Germany 5%
Other 81%
0 10 20 30 40 50 60 70 80 90 100
% of total arrivals

The civil war means that Sudan has very few tourists. Visitors are mostly aid workers or on business.

PEOPLE

Pop. density low

Arabic, Dinka, Nuer, Nubian, Beja, Zande, Bari, Fur, Shilluk, Lotuko

13/km² (35/mi²)

THE URBAN/RURAL POPULATION SPLIT

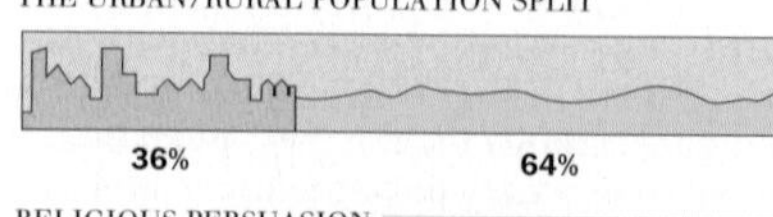
36%
64%

RELIGIOUS PERSUASION

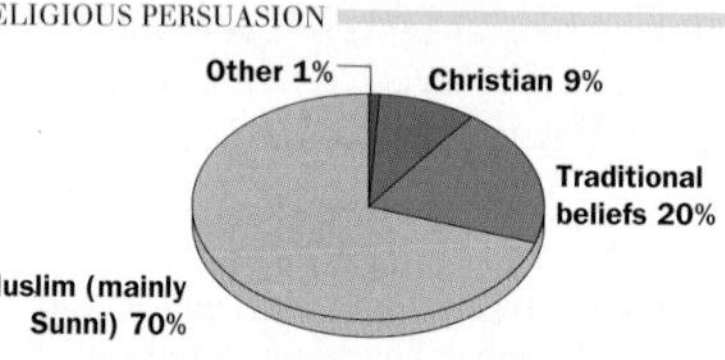
Other 1%
Christian 9%
Traditional beliefs 20%
Muslim (mainly Sunni) 70%

Sudan has a large number of ethnic and linguistic groups. About two million Sudanese are nomads. The major social division, however, is between the Arabized Muslims in the north and the mostly African, largely animist or Christian population in the south. Attempts to impose Arab and Islamic values throughout Sudan have been the root cause of the civil war that has ravaged the south since 1983. There are some non-Arab groups in the north and in the densely populated Darfur region. Women not wearing Islamic dress can suffer harassment or even public flogging. In 2002 the UN reported that thousands of people have been abducted into slavery within Sudan over the last 20 years.

POLITICS

No multiparty elections

2000/2004

President Omar Hassan Ahmad al-Bashir

AT THE LAST ELECTION

National Assembly 400 seats

89% NC
10% Vac
1% Ind

NC = National Congress supporters **Vac** = Vacant **Ind** = Independents

Elections in 2000, under the 1999 constitution allowing "political associations," were boycotted by the opposition, and returned Gen. Omar al-Bashir and his NC bloc to power. The willingness of the Sudanese People's Liberation Army (SPLA) and its former rival, the Sudan People's Defense Force (SPDF), to enter peace talks raised the possibility of an end to nearly 20 years of civil conflict between Muslim north and Christian south. However, increasing hostilities in the oil-rich Upper Nile region in spring 2002 put the peace process at risk.

SUDAN

Total Land Area : 2 505 810 sq. km (967 493 sq. miles)

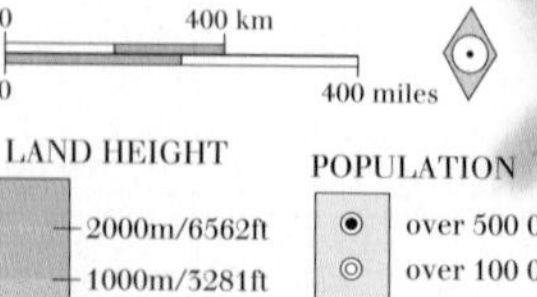
0
400 km
0
400 miles
N
LAND HEIGHT
2000m/6562ft
1000m/3281ft
500m/1640ft
200m/656ft
Sea Level
POPULATION
over 500 000
over 100 000
over 50 000
over 10 000
under 10 000

EGYPT
LIBYA
RED SEA
NUBIAN DESERT
Wadi Halfa
Port Sudan
Red Sea Hills
Argo
Dongola
Nile
LIBYAN DESERT
BAYUDHA DESERT
Atbara
Tokar
Ed Damer
ERITREA
CHAD
Teiga Plateau
Shendi
Khartoum North
Omdurman
KHARTOUM
Kassala
Halfa el Gadida
Khashm el Girba
Sodiri
El Hasaheisa
Wad Medani
Ed Dueim
Gedaref
Geneina
DARFUR
El Fasher
El Obeid
Sennar
Suki
Qala' en Nahl
Zalingei
Kosti
Er Rahad
Umm Ruwaba
Singa
Nyala
Dilling
Ed Damazin
Blue Nile
White Nile
Kadugli
Kurmuk
ETHIOPIA
Malakal
Sobat
Ghazal
Arab
Lol
SUDD
Jonglei Canal
Pongo
Wau
CENTRAL AFRICAN REPUBLIC
Rumbek
Yambio
Juba
Imatong Mountains
Kinyeti 3187m
DEM. REP. CONGO
UGANDA
KENYA

WORLD AFFAIRS

Joined UN in 1956

AL

OIC

COMESA

IGAD

AU

Sudan's support for Iraq in the Gulf War and suspicion that it sponsors terrorism have led to its increasing isolation from the West and the Arab world. Only Iran, Yemen, and Libya maintain friendly relations.

AID

Recipient

 $225m (receipts)

 Down 7% in 2000

The only substantial bilateral aid comes from Iran. IMF funding ceased in 1990. Sudan depends on food aid.

DEFENSE

Compulsory military service

 $568m

 Up 34% in 2000

The NC controls the military and police and has its own paramilitary militia. Sudan's 112,500-strong army was for many years engaged in fighting the two southern factions – the SPLA and SPDF – with forces of a similar size.

ECONOMICS

Inflation 61% p.a. (1990–2000)

 $9.6bn

 258.70 Sudanese dinars

SCORE CARD

- ❑ WORLD GNP RANKING........84th
- ❑ GNP PER CAPITA........$310
- ❑ BALANCE OF PAYMENTS........-$557m
- ❑ INFLATION........16%
- ❑ UNEMPLOYMENT........4%

STRENGTHS
Oil, gas, cotton, gum arabic, sesame, sugar. Some gold mining.

WEAKNESSES
Low industrialization. Lack of foreign exchange for importing energy and spare parts for industry. Drought. Little transportation infrastructure. Huge distances between towns. Civil war delayed exploitation of oil reserves. Alienation of Arab donors and investors.

EXPORTS

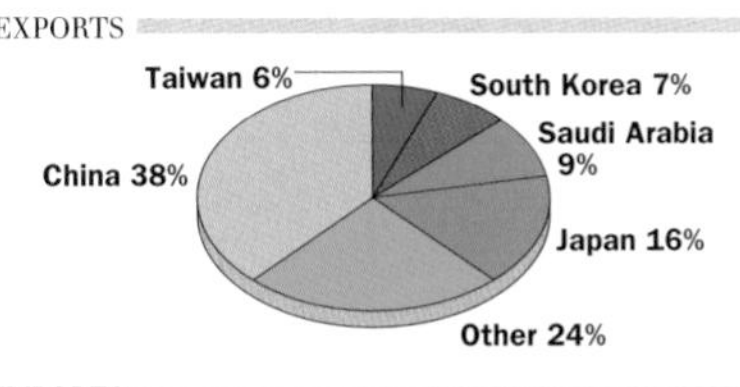
Taiwan 6%
South Korea 7%
Saudi Arabia 9%
China 38%
Japan 16%
Other 24%

IMPORTS

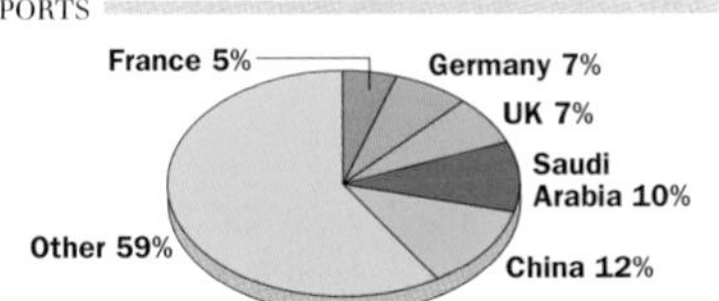
France 5%
Germany 7%
UK 7%
Saudi Arabia 10%
Other 59%
China 12%

RESOURCES

Electric power 606,000 kw

 50,500 tonnes

 200,000 b/d (reserves 300m barrels)

47m sheep, 39m goats, 38.3m cattle, 37.5m chickens

 Oil, gas, gold, copper, gypsum, marble, mica, silver, chromium, zinc

Large oil and gas reserves were found in the south in the 1980s; oil exports started in 1999. The half-thermal, half-hydroelectric generating capacity is insufficient, and weeklong power cuts are frequent. Gold mining has the potential for expansion.

ENVIRONMENT

Sustainability rank: 102nd

4% (0.3% partially protected)

0.1 tonnes per capita

The Jonglei canal project, halted by rebel attacks when only 70% complete, could still devastate the Sudd, the world's largest swamp and a rich wetland habitat, fed by the White Nile.

MEDIA

TV ownership high

 Daily newspaper circulation 27 per 1000 people

PUBLISHING AND BROADCAST MEDIA

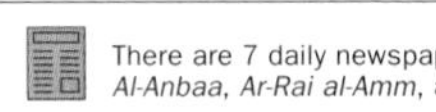 There are 7 daily newspapers, including *Al-Anbaa*, *Ar-Rai al-Amm*, and *Al-Nasr*

 1 state-controlled service

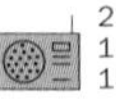 2 services: 1 state-controlled, 1 rebel-controlled

The media are controlled either by the government or the army, and are heavily censored.

CRIME

Death penalty in use

 32,000 prisoners

 Down 7% 1992–1994

Antigovernment dissent is often suppressed by violence, and torture by the security forces is widespread. The UN has condemned Sudan's poor human rights record.

EDUCATION

School leaving age: 14

 58%

 200,538 students

In 1991, measures were introduced to Islamize education. Primary school children must have two years of Islamic religious instruction, and men wishing to enter university must first serve for a year in the People's Militia.

HEALTH

Welfare state health benefits

1 per 10,000 people

 Infectious and parasitic diseases, malnutrition

Health service standards in rural areas are basic. The civil war has led to an increase in communicable diseases. The parasitic infection leishmaniasis is prevalent.

CHRONOLOGY

Northern Sudan was taken by Egypt in 1821, the south by Britain in 1877.

- ❑ **1882** British invade Egypt.
- ❑ **1883** Muslim revolt in Sudan led by Muhammad Ahmed, the Mahdi.
- ❑ **1898** Mahdists defeated. Anglo-Egyptian condominium set up.
- ❑ **1954** Becomes self-governing.
- ❑ **1955** Rebellion in south starts 17 years of civil war.
- ❑ **1956** Independence as republic.
- ❑ **1958–1964** Military rule.
- ❑ **1965** Civilian revolution, elections.
- ❑ **1969** Coup led by Col. Jaafar Nimeiri.
- ❑ **1972** South gets limited autonomy.
- ❑ **1973** Socialist Union sole party.
- ❑ **1983** Southern rebellion resumes. *Sharia* (Islamic law) imposed.
- ❑ **1984** Devastating drought.
- ❑ **1986** Army coup.
- ❑ **1989** Gen. Bashir takes over.
- ❑ **1991** *Sharia* penal code instituted. Pro-Iraq stance in Gulf War.
- ❑ **2000** Bashir ousts fundamentalist al-Turabi from leadership of NC.
- ❑ **2002** SPLA signs cease-fire. Fighting between rival factions over southern oil reserves escalates.

SPENDING

GDP/cap. increase

CONSUMPTION AND SPENDING

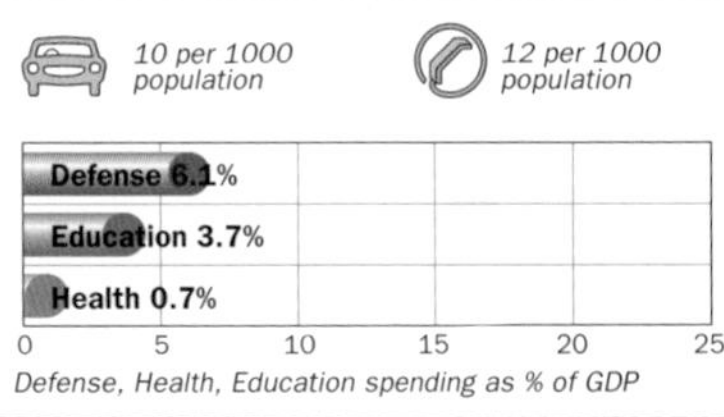
10 per 1000 population
12 per 1000 population
Defense 6.1%
Education 3.7%
Health 0.7%
0
5
10
15
20
25

Defense, Health, Education spending as % of GDP

There are large disparities between rich and poor. Wealth is limited to the NC and southern rebel elites. Most of the population struggles to survive.

WORLD RANKING

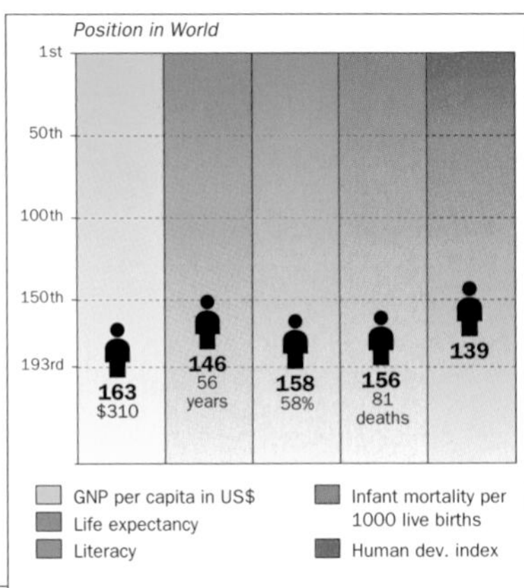
Position in World
1st
50th
100th
150th
193rd
163
$310
146
56 years
158
58%
156
81 deaths
139
GNP per capita in US$
Life expectancy
Literacy
Infant mortality per 1000 live births
Human dev. index

S

SURINAME

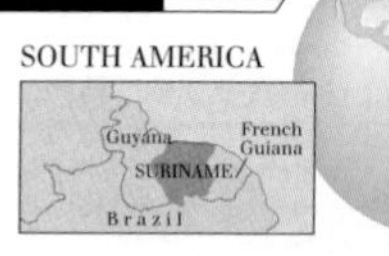
SOUTH AMERICA
Guyana
French Guiana
SURINAME
Brazil

South America

OFFICIAL NAME: Republic of Suriname **CAPITAL:** Paramaribo
POPULATION: 419,000 **CURRENCY:** Suriname guilder or florin **OFFICIAL LANGUAGE:** Dutch

1975

1975

Nov 25

SME

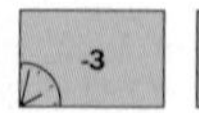
-3

+597

.sr

BOUNDED EAST AND WEST by rivers, Suriname sits on the north coast of South America in the center of the "Guyana Plateau." The interior is rainforested highlands; most people live near the coast. In 1975, after almost 300 years of Dutch rule, Suriname became independent. The Netherlands is still its main aid supplier, and is home to one-third of Surinamese. Multiparty democracy was restored in 1991, after almost 11 years of military rule.

Congested street in Paramaribo. It boasts 18th- and 19th-century Dutch architecture. The large mosque is next to a Jewish synagogue.

CLIMATE

Tropical equatorial

WEATHER CHART FOR PARAMARIBO

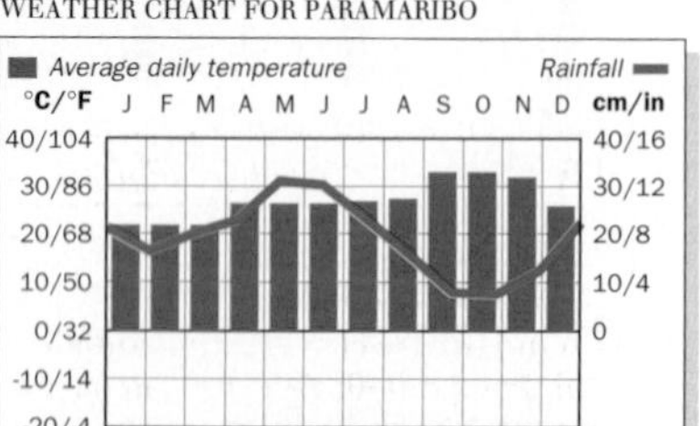
Average daily temperature
Rainfall
°C/°F
J F M A M J J A S O N D
cm/in
40/104
30/86
20/68
10/50
0/32
-10/14
-20/-4
40/16
30/12
20/8
10/4
0

Suriname's tropical climate is cooled by the trade winds. Annual rainfall varies from 150 to 300 cm (60 to 120 in) between coast and interior.

TRANSPORTATION

Drive on left

Johann Pengel International, Paramaribo 175,000 passengers

17 ships 6154 grt

THE TRANSPORTATION NETWORK

1178 km (732 miles)

None

157 km (98 miles)

1200 km (746 miles)

Rivers provide the main north–south links, and the vast interior relies on water or air transportation. The road network runs east–west and focuses on the coast and its immediate hinterland.

TOURISM

Visitors : Population 1:7.4

57,000 visitors

Up 4% in 1999

MAIN TOURIST ARRIVALS

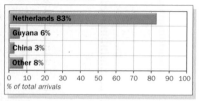

	% of total arrivals
Netherlands	83%
Guyana	6%
China	3%
Other	8%

0 10 20 30 40 50 60 70 80 90 100
% of total arrivals

Tourism is undeveloped. Travelers outside Paramaribo are advised to carry their own hammock and food.

PEOPLE

Pop. density low

Sranan, Dutch, Javanese, Sarnami Hindi, Saramaccan, Chinese, Carib

3/km² (7/mi²)

THE URBAN/RURAL POPULATION SPLIT

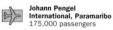

74% 26%

ETHNIC MAKEUP

South Asian 34%
Creole 34%
Javanese 18%
Black 9%
Other 5%

About 250,000 Surinamese have emigrated since 1975. Of those who remain, 90% live near the coast, while the rest live in scattered rainforest communities. Tension between the Creole-dominated government, *bosnegers* (the descendants of runaway slaves), and Amerindians spilled over into armed rebellion in the 1980s. Many South Asians and Javanese work in farming. Christianity, Hinduism, and Islam are the dominant religions.

POLITICS

Multiparty elections

2000/2005

President Ronald Venetiaan

AT THE LAST ELECTION

National Assembly 51 seats

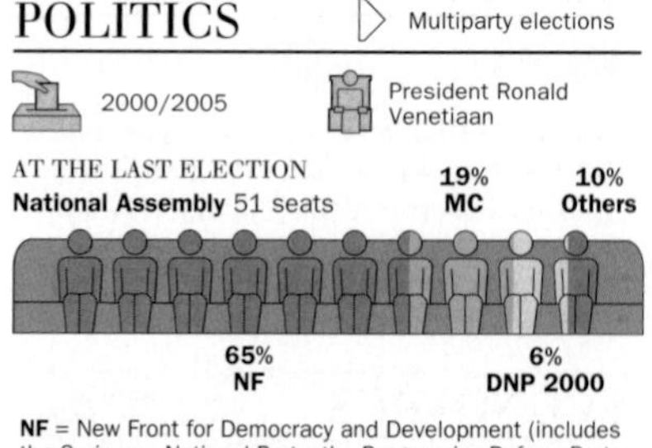
19% MC
10% Others
65% NF
6% DNP 2000

NF = New Front for Democracy and Development (includes the Suriname National Party, the Progressive Reform Party, the Suriname Labor Party, and Pertjajah Luhur)
MC = Millennium Combination (includes the National Democratic Party (**NDP**), the Democratic Alternative, and the Party for Unity and Harmony)
DNP 2000 = Democratic National Platform 2000

A coalition government representing Creoles, South Asians, and Javanese took power under Ronald Venetiaan in 1991. Five years later it was defeated by the NDP, controlled by Desi Bouterse, the military dictator from 1980 to 1988 and the man behind the 1990 coup which ended Suriname's first attempt to return to democracy. Between 1996 and 2000 President Jules Wijdenbosch of the NDP withstood the efforts of opponents in the National Assembly to replace him. In the 2000 legislative elections, however, the NDP was massively defeated by the opposition NF. The new Assembly went on to elect NF leader Venetiaan as president.

ATLANTIC OCEAN
GUYANA
FRENCH GUIANA
BRAZIL
PARAMARIBO
Nieuw Nickerie
Totness
Groningen
Calcutta
Boskamp
Wageningen
Jenny
Nieuw Amsterdam
Galibi
Lelydorp
Onverwacht
Republiek
Paranam
Moengo
Albina
Pilgrimkondre
Apoera
Kwakoegron
Joden Savanne
Zanderij
Bitagron
Berg en Dal
Brokopondo
Kaaimanston
Brownsweg
Afobaka
Njoeng Jacobkondre
W.J. van Blommesteinmeer
Langatabbetje
Pokigron
Stoelmanseiland
Boti-Pasi
Ladoeyani
Poeketi
Asidonhopo
Cottica
Juliana Top 1230m
Teboe Top
Intelewa
Appikalo
Nickerie
Coppename
Corantijn
Kabalebo
Lucie
Coeroeni
Zuid
Gran
Tapanahony
Marowijne
Lawa
Litani
Sipaliwini
N
0 50 km
0 50 miles

SURINAME

Total Land Area : 163 270 sq. km (63 039 sq. miles)

LAND HEIGHT
1000m/3281ft
500m/1640ft
200m/1640ft
Sea Level

POPULATIO

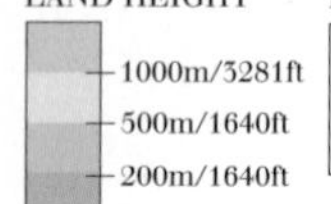
over 1
over 1
under

WORLD AFFAIRS

Joined UN in 1975

ACS

AmCC

Caricom

OAS

OIC

Relations with the Netherlands and the US, Suriname's key aid and trading partners, have been weakened over charges of official connivance in narcotics trafficking. Although ties with Suriname's immediate neighbors are damaged by border disputes, greater regional integration, particularly with the Caribbean, is a priority.

AID

Recipient

 $34m (receipts)

 Down 6% in 2000

The Netherlands is the largest donor, but it has on occasion suspended aid amid deteriorating relations. The IDB and European Investment Bank have granted loans for agricultural and industrial development.

DEFENSE

No compulsory military service

 $11m

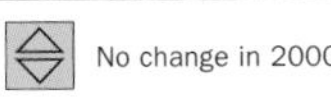 No change in 2000

The army was politically dominant in the 1980s under Lt. Col. Desi Bouterse. A six-year war with *bosneger* rebels ended in 1992. Aid and training have been provided in recent years by both the US and China.

ECONOMICS

Inflation 88% p.a. (1990–2000)

$788m

981–2179 Suriname guilders

SCORE CARD

❑ WORLD GNP RANKING	159th
❑ GNP PER CAPITA	$1890
❑ BALANCE OF PAYMENTS	$32m
❑ INFLATION	64.3%
❑ UNEMPLOYMENT	11%

STRENGTHS

Bauxite. Gold. Timber potential. Oil. Agricultural exports: rice, bananas, citrus fruits. Shrimp exports.

WEAKNESSES

Overdependence on declining bauxite reserves and decreased foreign aid. Weak currency. Severe shortage of foreign exchange. Banana industry damaged by ending of preferential EU access. Net food importer.

EXPORTS

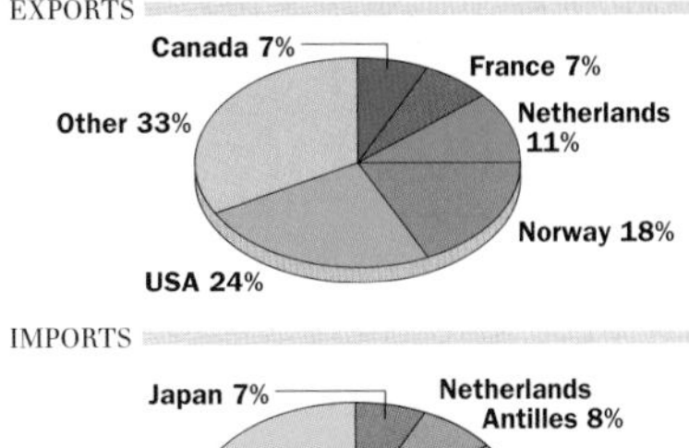
Canada 7%
France 7%
Netherlands 11%
Norway 18%
USA 24%
Other 33%

IMPORTS

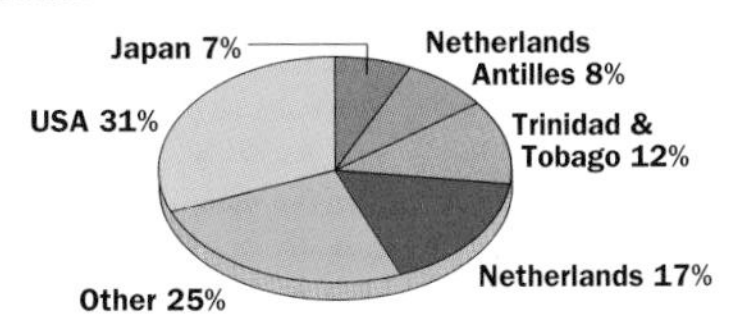
Japan 7%
Netherlands Antilles 8%
Trinidad & Tobago 12%
Netherlands 17%
Other 25%
USA 31%

RESOURCES

Electric power 425,000 kw

13,066 tonnes

 4972 b/d (reserves 87m barrels)

135,000 cattle, 66,000 ducks, 2.9m chickens

Bauxite, iron, gold, manganese, copper, nickel, platinum, oil

Suriname is a major exporter of aluminum and bauxite, but the minerals sector is affected by poor world prices, as is raw gold production. Oil consumption is almost double the level of oil production. Exploitation of the rainforests has begun. Rice and fruit are Suriname's key agricultural products.

ENVIRONMENT

Not available

10% partially protected

5.2 tonnes per capita

In 1998 the government declared some 16,000 sq. km (6150 sq. miles) of rainforest – almost 10% of the country – to be a natural reserve barred to logging but its exploitation for economic gain is still of real concern to environmentalists.

MEDIA

TV ownership medium

 Daily newspaper circulation 122 per 1000 people

PUBLISHING AND BROADCAST MEDIA

 There are 2 daily newspapers, *De Ware Tijd* and *De West*

 2 state-owned services

 10 services: 1 state-owned, 9 independent

There are radio broadcasts in a number of languages. Dutch is used by the daily newspapers and for most TV programs.

CRIME

Death penalty not used in practice

 1933 prisoners

 Relatively high crime levels

Human rights abuses associated with the former military regime have largely ended. Rival armed factions remain in some regions in the interior. Narcotics trafficking and money laundering are a problem, as is urban street crime.

EDUCATION

School leaving age: 12

94%

 4037 students

Education is free and includes adult literacy programs. There is a long tradition of higher education, but most graduates now live in the Netherlands.

CHRONOLOGY

Dutch rule began in 1667, after an Anglo-Dutch treaty whose terms included Britain ceding its colony in Suriname to the Dutch but gaining Nieuw Amsterdam (New York).

- ❑ **1975** Independence.
- ❑ **1980** Coup. Rule by Lt. Col. Desi Bouterse.
- ❑ **1982** Opponents executed. Dutch suspend aid for six years.
- ❑ **1986–1992** *Bosneger* rebel war.
- ❑ **1988–1991** Elections, coup, and new elections. Ronald Venetiaan elected president.
- ❑ **1992** Bouterse quits as army head.
- ❑ **1996** Pro-Bouterse NDP wins polls.
- ❑ **2000** NF defeats NDP. Venetiaan elected president again.

HEALTH

Welfare state health benefits

 1 per 4000 people

 Heart attacks, cancers, malaria, malnutrition, tuberculosis

Urban medical facilities in Suriname are relatively good; Paramaribo has several hospitals. However, provision in the interior is basic.

SPENDING

GDP/cap. increase

CONSUMPTION AND SPENDING

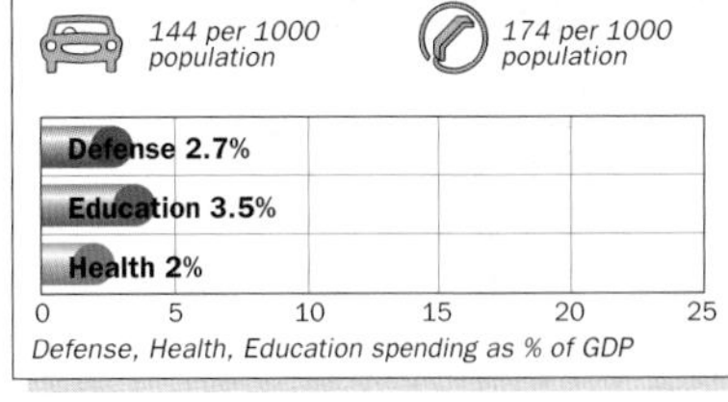
144 per 1000 population
174 per 1000 population
Defense 2.7%
Education 3.5%
Health 2%
0
5
10
15
20
25

Defense, Health, Education spending as % of GDP

Living standards have fallen since 1982, due to the effects of civil war and to aid and loan suspension. Urban Creoles dominate the rich elite. Amerindians and *bosnegers* are the poorest groups.

S

WORLD RANKING

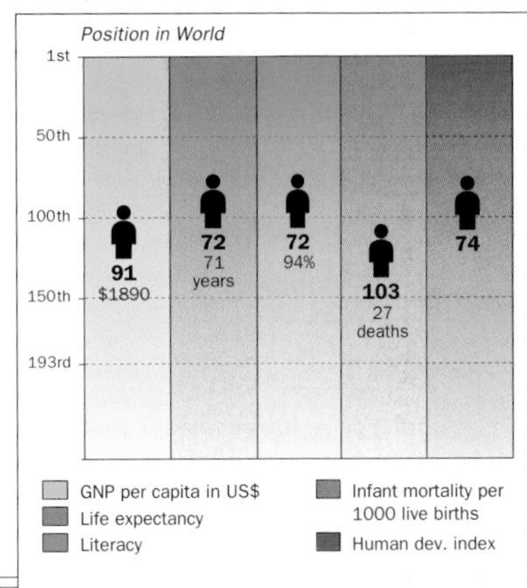
Position in World
1st
50th
100th
150th
193rd
91
$1890
72
71 years
72
94%
103
27 deaths
74
GNP per capita in US$
Life expectancy
Literacy
Infant mortality per 1000 live births
Human dev. index

SWAZILAND

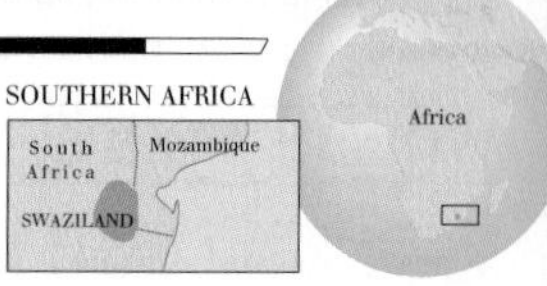

SOUTHERN AFRICA
South Africa
Mozambique
SWAZILAND
Africa

OFFICIAL NAME: Kingdom of Swaziland **CAPITAL:** Mbabane
POPULATION: 938,000 **CURRENCY:** Lilangeni **OFFICIAL LANGUAGES:** English and siSwati

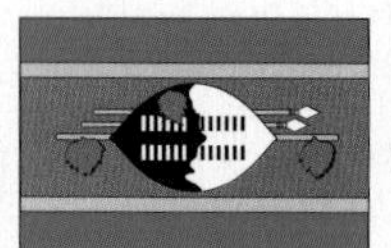

1968

1968

Sept 6
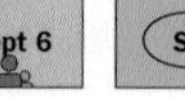
SD

+2
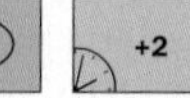
+268

.sz
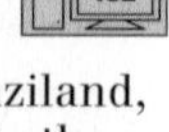

THE TINY SOUTHERN African kingdom of Swaziland, bordered on three sides by South Africa and to the east by Mozambique, comprises mainly upland plateaus and mountains. Governed by a strong hereditary monarchy, Swaziland is a country in which tradition is being challenged by demands for modern multiparty government. King Mswati III, crowned in 1986, has overhauled the electoral process, but has still to legalize party politics.

CLIMATE

Subtropical

WEATHER CHART FOR MBABANE

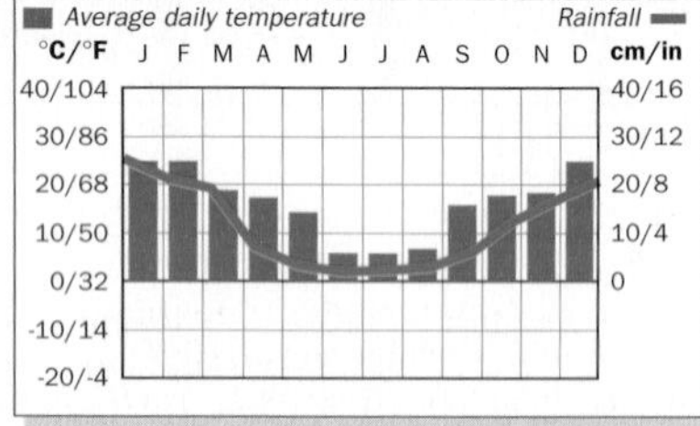

Average daily temperature
Rainfall
°C/°F
J F M A M J J A S O N D
cm/in
40/104
30/86
20/68
10/50
0/32
-10/14
-20/-4
40/16
30/12
20/8
10/4
0

Swaziland is temperate. Temperatures rise and rainfall declines as the land descends eastward, from high to low *veld*. The Low Veld is prone to drought.

TRANSPORTATION

Drive on left

Matsapha, Manzini
93,000 passengers

Has no fleet

THE TRANSPORTATION NETWORK

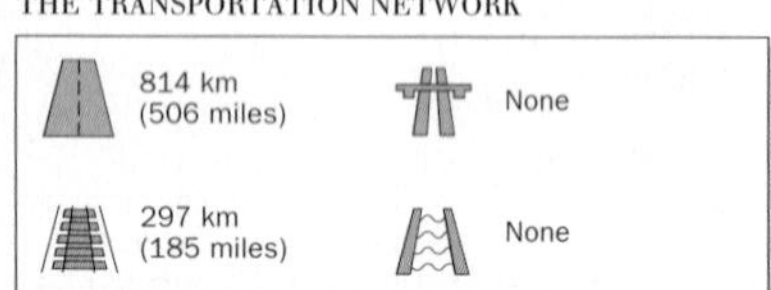

814 km
(506 miles)
None
297 km
(185 miles)
None

A sharp rise in road traffic has necessitated road improvements. The railroad, running to Mozambique and South Africa, mainly carries freight.

TOURISM

Visitors : Population 1:2.9

319,000 visitors

Down 1% in 1998

MAIN TOURIST ARRIVALS

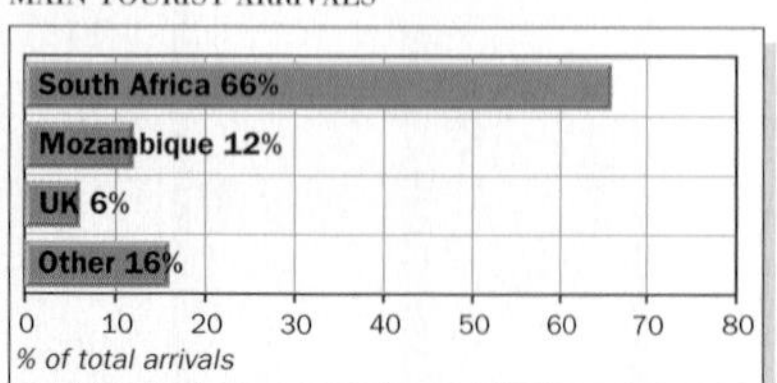

South Africa 66%
Mozambique 12%
UK 6%
Other 16%
0 10 20 30 40 50 60 70 80
% of total arrivals

Swaziland's attractions are its game reserves, mountain scenery, and, for the South Africans who make up two-thirds of tourists, its casinos.

The outskirts of Mbabane. *It lies on the High Veld, where traditional cattle farming has become more difficult owing to overgrazing.*

PEOPLE

Pop. density medium

English, siSwati, Zulu, Tsonga

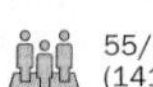
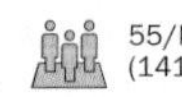
55/km^2 (141/mi^2)

THE URBAN/RURAL POPULATION SPLIT

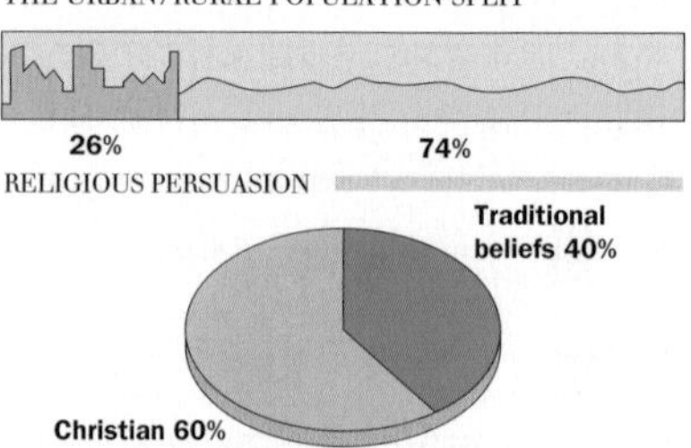

26%
74%
RELIGIOUS PERSUASION
Traditional beliefs 40%
Christian 60%

Over 95% of the population belong to the Swazi ethnic group, making Swaziland one of Africa's most homogeneous states. It is also very conservative, but is now facing pressure from urban-based modernizers. The powerful monarchy dominates politics. Ancient traditions, such as *incwala*, the rainy season's annual movable feast, remain popular. Society is patriarchal and focused around the clan. Chiefs own much "national land," and wield authority through local consultations, called *tindkhundla*. Polygamy is tolerated. Women farm and may vote, but lack economic or political power. The exception is the Queen Mother, the "Great She Elephant," whose power as regent was clear during the mid-1980s.

POLITICS

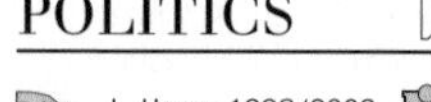

No multiparty elections

L. House 1998/2003
U. House 1998/2003

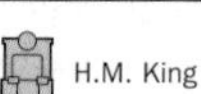
H.M. King Mswati III

AT THE LAST ELECTION

House of Assembly 65 seats

There are no political parties. 10 members of the House of Assembly are appointed by the king

Senate 30 seats

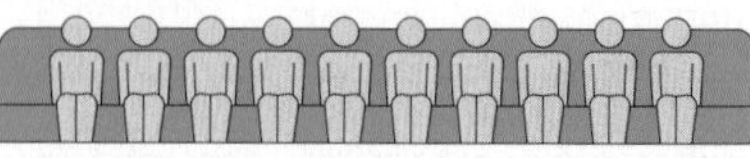

20 members of the Senate are appointed by the king and 10 elected by the House of Assembly

Politics is dominated by a strong executive monarchy and rivalries within the ruling Dlamini clan. Royal advisers complement a nominated cabinet. Direct elections were held in 1993, but parties remain banned. Responding to mounting popular unrest, in 1996 the king appointed Sibusiso Dlamini as prime minister and set up a commission to review the political system. Mass antigovernment protests were staged by prodemocracy activists in 2000.

SWAZILAND

Total Land Area : 17 363 sq. km (6704 sq. miles)

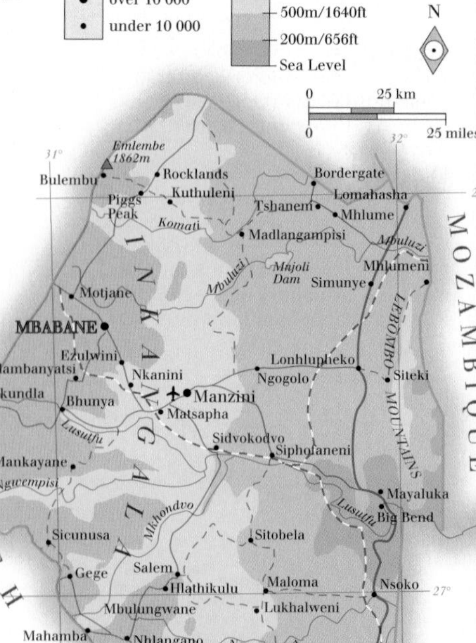

POPULATION
over 50 000
over 10 000
under 10 000
LAND HEIGHT
1000m/3281ft
500m/1640ft
200m/656ft
Sea Level
N
0 25 km
0 25 miles
Emlembe 1862m
Bulembu
Rocklands
Piggs Peak
Kuthuleni
Bordergate
Lomahasha
Tshaneni
Mhlume
Komati
Madlangampisi
Mbuluzi
Mnjoli Dam
Simunye
Mhlumeni
Motjane
MBABANE
Ezulwini
Mhlambanyatsi
Nkanini
Lonhlupheko
Ngogolo
Siteki
Nkundla
Bhunya
Manzini
Matsapha
Lusutfu
Sidvokodvo
Siphofaneni
Mankayane
Ngwempisi
Mayaluka
Big Bend
Mkhondvo
Sicunusa
Sitobela
Gege
Salem
Hlathikulu
Maloma
Nsoko
Mbulungwane
Lukhalweni
Mahamba
Nhlangano
Ngwavuma
Mhlosheni
Salitje
Lavumisa
Jozini Dam
INKANGALA
LEBOMBO MOUNTAINS
MOZAMBIQUE
SOUTH AFRICA

WORLD AFFAIRS

Joined UN in 1968

ACP
Comm
NAM
AU
SADC

Swaziland's membership of the SACU reinforces its traditional dependence on its giant neighbor, South Africa. Having welcomed the election of an ANC-led government there, King Mswati has expressed concern over its support for Swazi prodemocracy campaigners. Peace in Mozambique has meant the return there of 134,000 refugees.

AID

Recipient

 $13m (receipts)

 Down 55% in 2000

Aid helps the balance of payments, and funds the development of the Matsapha industrial estate, roads, and social projects. Donors include Germany, Japan, the UK, and the ADB. EU aid mainly targets "microprojects," such as schools, and supports constitutional reform.

DEFENSE

No compulsory military service

 $23m

 Up 15% in 2000

The Swaziland Defense Force numbers just 3000 troops. Although it does not play an overt political role, its loyalty is to the monarch and the status quo.

ECONOMICS

Inflation 13% p.a. (1990–2000)

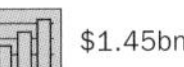 $1.45bn

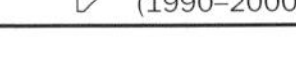 7.570–11.995 emalangeni

SCORE CARD

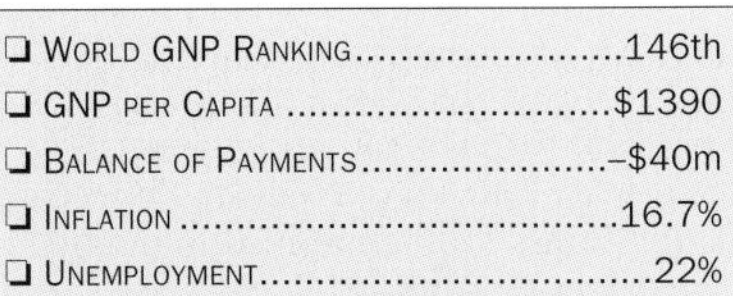

❑ World GNP Ranking	146th
❑ GNP per Capita	$1390
❑ Balance of Payments	–$40m
❑ Inflation	16.7%
❑ Unemployment	22%

STRENGTHS

Economy quite diversified and buoyant. Manufacturing 32% of GDP. Investment rules attractive. Sugar 33% of export earnings. Wood pulp. Debt service low: only 3.8% of export earnings in 1993. Renewed regional stability has reduced risk to exports.

WEAKNESSES

Sugar is vulnerable to world price fluctuations. Dependence on South Africa for jobs, revenue, investment, electricity, and imported goods. Small plots of land and lack of land title hinder farm modernization. High population growth.

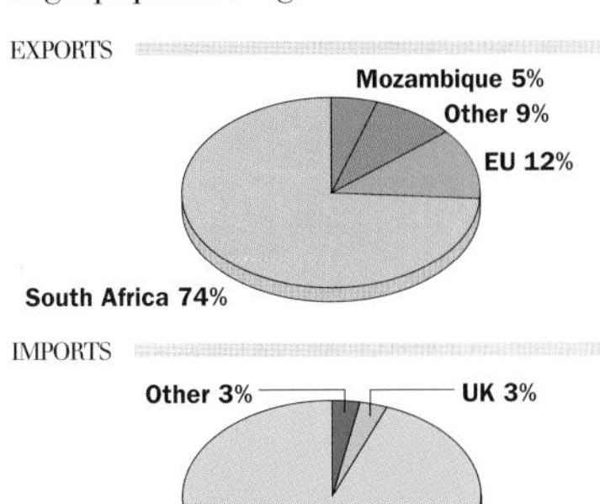
EXPORTS
Mozambique 5%
Other 9%
EU 12%
South Africa 74%
IMPORTS
Other 3%
UK 3%
South Africa 94%

RESOURCES

Electric power: Included in South African total

 131 tonnes

 Not an oil producer

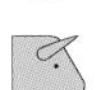 615,000 cattle, 445,000 goats, 3.2m chickens

Coal, diamonds, gold, asbestos, cassiterite, iron, tin

Swaziland's main export is sugarcane, followed by wood pulp, coal, and asbestos. The development of hydroelectric power plants has cut energy imports from South Africa.

ENVIRONMENT

Not available

 2% (3% partially protected)

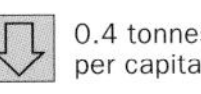 0.4 tonnes per capita

In 1998 Swaziland, Mozambique, and South Africa began an ecological project on the world's largest wetlands – the foothills of the Lebombo Mountains.

MEDIA

TV ownership medium

 Daily newspaper circulation 27 per 1000 people

PUBLISHING AND BROADCAST MEDIA

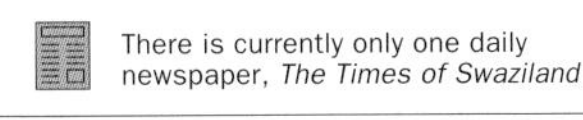 There is currently only one daily newspaper, *The Times of Swaziland*

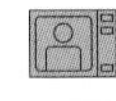 1 state-owned service

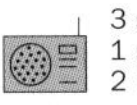 3 services: 1 state-owned, 2 independent

The king, responding to pressure, rapidly reversed his 2001 decree which had ensured the continued closure of the independent *Swaziland Observer*.

CRIME

Death penalty in use

 2221 prisoners

Down 1% in 1999

The crime rate is low. The numbers of illegal weapons brought in by refugees have boosted armed crime.

EDUCATION

School leaving age: 13

 80%

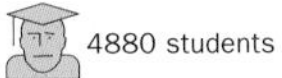 4880 students

Education is compulsory. Parents pay fees at all levels; even so, primary enrollment is about 93%. Drop-out rates at secondary level are high.

CHRONOLOGY

Swaziland became a British protectorate in 1903.

- ❑ **1968** Independence.
- ❑ **1973** King bans political activity, repeals constitution.
- ❑ **1978** New constitution confirms king's executive, legislative control.
- ❑ **1982** King Sobhuza dies. Queen Mother becomes regent for Prince Makhosetive. Power struggle between modernists and traditionalists in royal Dlamini clan.
- ❑ **1986** Makhosetive crowned King Mswati III at the age of 18.
- ❑ **1992** Limited electoral reforms; parties still banned.
- ❑ **1993** Elections under new system.
- ❑ **1996** Review of political system.
- ❑ **1998** Poor turnout at elections.
- ❑ **2000** Mass prodemocracy protests.

HEALTH

No welfare state health benefits

 1 per 5000 people

 Diarrheal and respiratory diseases, AIDS

Health facilities are rudimentary. About a third of the population aged 15 to 49 is estimated to be living with HIV/AIDS.

SPENDING

GDP/cap. increase

CONSUMPTION AND SPENDING

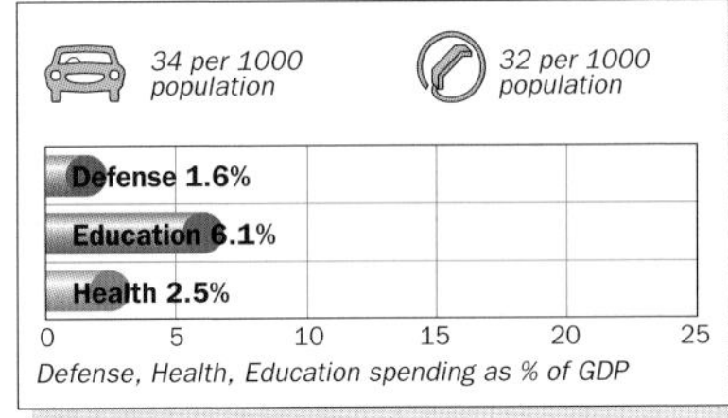
34 per 1000 population
32 per 1000 population
Defense 1.6%
Education 6.1%
Health 2.5%
0
5
10
15
20
25
Defense, Health, Education spending as % of GDP

About 50% of Swazis live below the UN poverty line. The royal Dlamini clan enjoys Western luxuries and travel.

WORLD RANKING

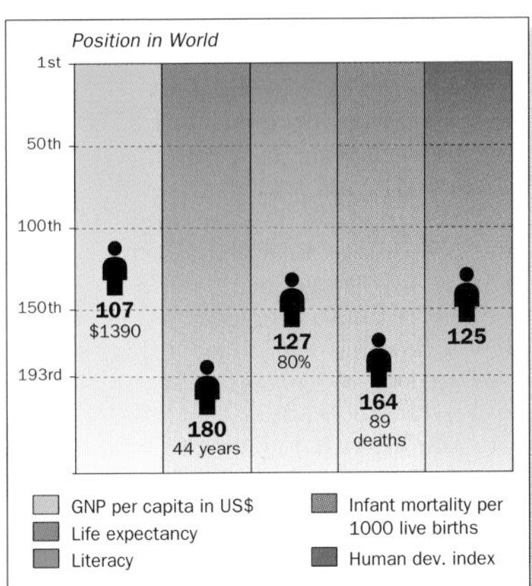
Position in World
1st
50th
100th
150th
193rd
107
$1390
180
44 years
127
80%
164
89 deaths
125
GNP per capita in US$
Life expectancy
Literacy
Infant mortality per 1000 live births
Human dev. index

S

SWEDEN

EUROPE

Europe

Norway Sweden Finland Denmark

OFFICIAL NAME: Kingdom of Sweden **CAPITAL:** Stockholm
POPULATION: 8.8 million **CURRENCY:** Swedish krona **OFFICIAL LANGUAGE:** Swedish

1523
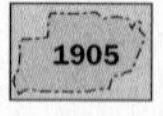
1905

June 6
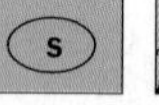
S
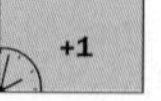
+1
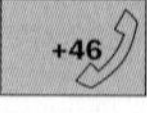
+46

.se

SITUATED ON THE SCANDINAVIAN peninsula with Norway to its west, Sweden is a densely forested country with numerous lakes. The north of Sweden falls within the Arctic Circle; much of the south is fertile and widely cultivated. Sweden has one of the most extensive welfare systems in the world, and is among the world's leading proponents of equal rights for women. Its economic strengths include high-tech industries, such as Ericsson, and car production, most notably Volvo and Saab. Unlike neighboring Norway, it is an EU member, having joined in January 1995.

CLIMATE

Subarctic/continental

WEATHER CHART FOR STOCKHOLM

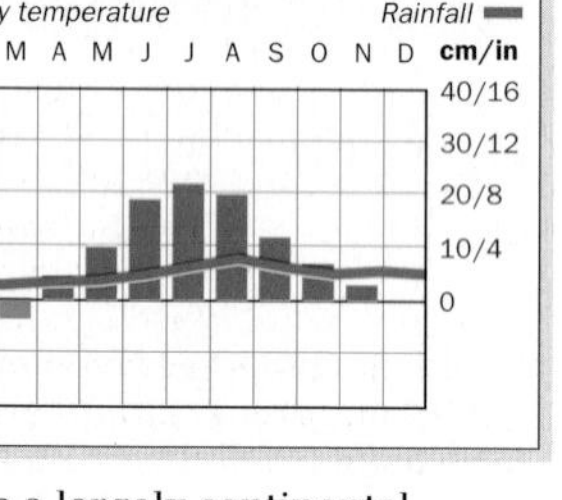

Average daily temperature
Rainfall
°C/°F J F M A M J J A S O N D cm/in
40/104 30/86 20/68 10/50 0/32 -10/14 -20/-4
40/16 30/12 20/8 10/4 0

Sweden has a largely continental climate. The Baltic Sea often freezes in winter, making the east coast much colder than western regions. Summers are mild everywhere, with temperatures varying surprisingly little between northern and southern regions.

TRANSPORTATION

Drive on right

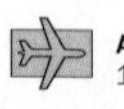
Arlanda, Stockholm 18.4m passengers

412 ships 2.86m grt

THE TRANSPORTATION NETWORK

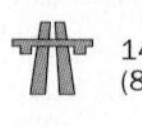
163,453 km (101,565 miles)

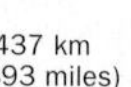
1437 km (893 miles)

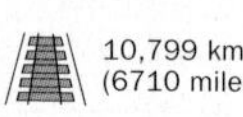
10,799 km (6710 miles)

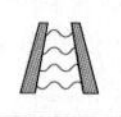
2052 km (1275 miles)

Maintaining and improving transportation links are of prime concern in what is Europe's fifth-largest country. Swedish governments have traditionally spent large sums on infrastructure, as a way of boosting the economy as a whole.

A 16-km (10-mile) Øresund road and rail link by bridge and tunnel connecting Malmö with Copenhagen opened in 2000, providing Sweden with a new road and rail link with Denmark and the rest of Europe. A new rail link between Arlanda airport and Stockholm is also planned. By law, cars must travel with their headlights on at all times, and there are very strict regulations against drink-driving.

TOURISM

Visitors : Population 1:3.2

2.75m visitors

Up 6% in 2000

MAIN TOURIST ARRIVALS

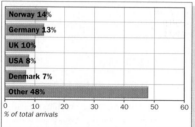

	% of total arrivals
Norway	14%
Germany	13%
UK	10%
USA	8%
Denmark	7%
Other	48%

0 10 20 30 40 50 60
% of total arrivals

Sweden expanded rapidly as a tourist destination in the 1970s and 1980s. Stockholm is renowned for the beauty of its setting, its Old Town, and the *Vasa*, a magnificent 17th-century warship raised from the harbor bed in the 1960s.

Sweden has fewer lakes than Finland, and lacks Norway's dramatic scenery, but it has many natural attractions. The mountains of the "Midnight Sun" lie north of the Arctic Circle, while the southern coast has many white sandy beaches. The vast tracts of deserted landscape and the simple country communal living also attract visitors, but the cost of travel to Sweden means that Norwegians and Germans top the list of visitors.

A crofter's holding in Dalarna, *central Sweden, an area which is still mainly forested. The timber and paper industries play a major role in Sweden's economy.*

PEOPLE

Pop. density low

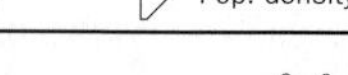
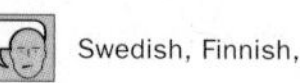
Swedish, Finnish, Sami

21/km² (55/mi²)

THE URBAN/RURAL POPULATION SPLIT

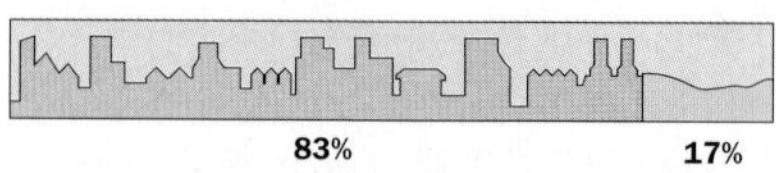

83% 17%

RELIGIOUS PERSUASION

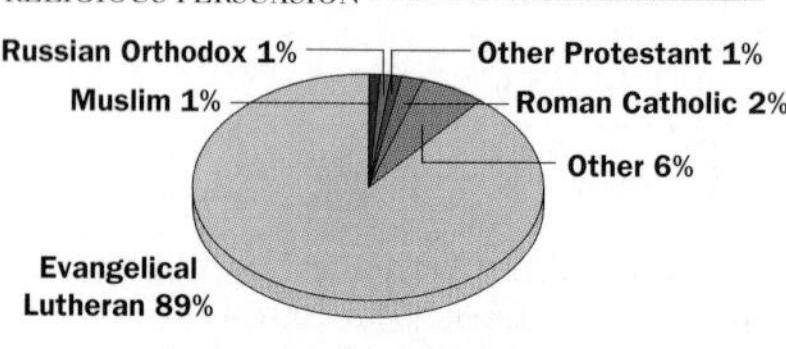

Russian Orthodox 1%
Other Protestant 1%
Muslim 1%
Roman Catholic 2%
Other 6%
Evangelical Lutheran 89%

ETHNIC MAKEUP

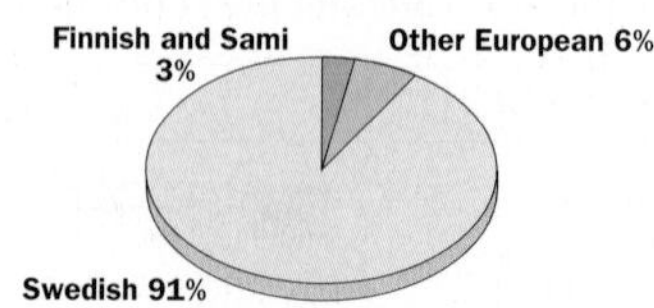

Finnish and Sami 3%
Other European 6%
Swedish 91%

As in all of Scandinavia, the nuclear family forms the basis of society. The birthrate is low with, on average, fewer than two children per family. Marriage is declining, and cohabitation outside marriage is common.

Swedish society has an egalitarian tradition. The role of the state is seen as providing conditions allowing each person to gain economic independence through employment. The welfare system is one of the most extensive in the world. However, in the early 1990s, recession reduced benefits; mothers in particular face problems with the closure of child-care facilities. Women make up nearly half the workforce. Over 45% of MPs are women, the highest percentage in the world. In 1999 the Swedish cabinet became the first in the world to have a majority of women ministers.

Sweden has generous asylum laws, and immigrants comprise about 10% of the population. A 17,000-strong minority of Sami live in northern Sweden. Their traditional way of life is protected.

The Evangelical Lutheran Church was disestablished in 2000.

POPULATION AGE BREAKDOWN

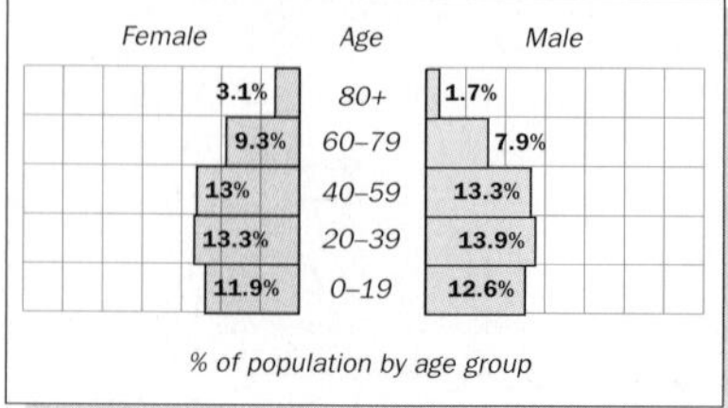

Female Age Male
3.1% 80+ 1.7%
9.3% 60–79 7.9%
13% 40–59 13.3%
13.3% 20–39 13.9%
11.9% 0–19 12.6%
% of population by age group

S

POLITICS

Multiparty elections

2002/2006 | King Carl XVI Gustaf

AT THE LAST ELECTION

Parliament (Riksdag) 349 seats

41% SAP	16% M	14% FP	9% Kd	9% VP	6% CP	5% MpG

SAP = Social Democratic Labor Party **M** = Moderate Party
FP = Liberal Party **Kd** = Christian Democratic Party
VP = Left Party **CP** = Center Party **MpG** = Green Party

Sweden is a constitutional monarchy with an elected parliament under the leadership of the prime minister.

PROFILE

Politics has traditionally been split between the SAP and trade unions on the left, and a host of moderate center and right-wing parties. Since the 1930s, the SAP has governed every term, except in 1976–1982 and 1991–1994. A shift to the right in 1991 was reversed in the 1994 elections. Ingvar Carlsson, SAP leader, formed a minority government but resigned in 1996 and was replaced by Göran Persson. The SAP lost ground in elections in 1998, increasing its independence on Left Party and Green support. A further shift in the 2002 elections left the Greens holding the balance of power.

MAIN POLITICAL ISSUES

EU membership

Sweden joined the EU in 1995 with Austria and Finland, but, like the UK and Denmark, has opted out of joining the eurozone. It supports EU membership for central European states.

The high cost of the welfare state

The cost of the welfare system contributed to enormous budget deficits in the late 1980s and early 1990s. While this has been brought under control, social security pressures remain from the relatively high level of unemployment and the growing number of pensioners.

SWEDEN
Total Land Area : 449 964 sq. km (173 731 sq. miles)
POPULATION
over 1 000 000
over 100 000
over 50 000
over 10 000
LAND HEIGHT
1000m/3281ft
500m/1640ft
200m/656ft
Sea Level
N
0 100 km
0 100 miles
NORWAY
FINLAND
LAPLAND
Gulf of Bothnia
BALTIC SEA
DENMARK
Skagerrak
Kattegat
The Sound
Hanöbukten
Arctic Circle
Kebnekaise 2111m
Kiruna
Gällivare
Luleå
Piteå
Skellefteå
Lycksele
Umeå
Örnsköldsvik
Östersund
Ånge
Sundsvall
Malung
Falun
Gävle
Borlänge
Uppsala
Arvika
Västerås
Täby
STOCKHOLM
Karlstad
Örebro
Eskilstuna
Södertälje
Mälaren
Hjälmaren
Vänern
Vättern
Nyköping
Mellerud
Skövde
Norrköping
Trollhättan
Linköping
Göteborg
Borås
Jönköping
Västervik
Visby
Mölndal
Kungsbacka
Gotland
Växjö
Borgholm
Kalmar
Öland
Halmstad
Helsingborg
Karlskrona
Kristianstad
Lund
Malmö
Muonioälv
Lainioälven
Torneälv
Kalixälv
Luleälv
Skellefteälv
Umeälv
Ljungan
Ljusnan
Dalälven
Klarälven

Carl XVI Gustaf, ascended the throne in 1973. His role is purely ceremonial.

Göran Persson of the SAP became prime minister in 1996.

WORLD AFFAIRS

Joined UN in 1946

EU | CE | NC | OECD | OSCE

Sweden's main recent foreign policy concern has been adjustment to EU membership, which it achieved in 1995. In 1998 parliament voted to join the Schengen passport-free zone linking nine EU states. In the 1980s, Sweden was a vociferous critic of the antagonistic policy used by the US toward the USSR. Since the collapse of the Soviet Union, and more recently the September 11, 2001 attack on the US, Sweden has altered its traditionally neutral stance. Sweden has WEU observer status, and participates in NATO exercises and several UN peacekeeping operations.

AID

Donor

$1.8bn (donations) | Up 10% in 2000

Sweden is one of the few countries to exceed the UN target of 0.7% of GNP in development aid and has declared its intention to increase its allocation.

CHRONOLOGY

Sweden's history has been closely linked to the control of the Baltic Sea and its highly profitable trade routes. Under the house of Vasa, Sweden became a major power, controlling much of the Baltic region. By the 18th century, however, Sweden's position had been eroded by its regional rivals, particularly Russia.

- **1814–1815** Congress of Vienna. Sweden cedes territory to Russia and Denmark. Period of prolonged peace begins.
- **1865–1866** Riksdag (parliament) reformed into a bicameral structure.
- **1905** Norway gains independence from Sweden.
- **1911** First Liberal government comes to power.
- **1914** Government resigns over defense policy.

S

CHRONOLOGY *continued*

- **1914–1917** Sweden remains neutral during WWI but supplies Germany. Allied blockade.
- **1917** Food shortages. Conservative government falls. Nils Edén forms a Liberal government: limits exports contributing to German war effort.
- **1919** Universal adult suffrage.
- **1921** Finland gains Åland Islands as retribution for Sweden's war role.
- **1932** Severe recession. Social Democrat government under Per Albin Hansson elected.
- **1939–1945** Sweden neutral. Grants transit rights to German forces.
- **1945–1976** Continuing Social Democratic rule under Tage Erlander establishes Sweden as world's most advanced welfare state, and one of the most affluent.
- **1950** Gustav VI Adolf becomes king.
- **1953** Joins Nordic Council.
- **1959** Founder member of EFTA.
- **1969** Erlander succeeded by Olof Palme as prime minister.
- **1973** Carl XVI Gustaf on throne.
- **1975** Major constitutional reform. Riksdag (parliament) becomes unicameral with a three-year term. Role of monarchy reduced to ceremonial functions.
- **1976** SAP loses power. Nonsocialist coalition led by Thorbjörn Fälldin in government.
- **1978** Fälldin resigns over issue of nuclear power. Ola Ullsten prime minister.
- **1979** Fälldin prime minister again.
- **1982** Elections. SAP forms minority government. Palme returns as prime minister.
- **1986** Palme shot dead. His deputy, Ingvar Carlsson, succeeds him. Police fail to find killer.
- **1990** Carlsson introduces moderate austerity package, cuts government spending, raises indirect taxes.
- **1991** Sweden applies to join EU. SAP wins election but is unable to form government; Carlsson resigns. Carl Bildt, leader of MS, forms coalition of nonsocialist parties in middle of serious recession.
- **1992** Austerity measures succeed in reducing inflation but SAP refuses to support further spending cuts.
- **1994** EU membership terms agreed. Elections return SAP to power. Referendum favors joining EU.
- **1995** Joins EU.
- **1996** Carlsson resigns; replaced by Göran Persson.
- **1998** Persson remains in office, despite SAP losses in elections; dependent on Left and Greens for parliamentary majority.
- **2001** Defense reform program.
- **2002** Elections: SAP still largest party but lacking majority.

S

DEFENSE

Compulsory military service

$5.19bn

Down 1% in 2000

SWEDISH ARMED FORCES

368 main battle tanks (70 *Centurion*, 40 Strv-103B, 160 Strv-121, 98 Strv-122)	19,100 personnel
7 submarines, 45 patrol boats	7100 personnel
206 combat aircraft (20 SAAB AJSH-37, 95 SAAB JAS-9, 91 SAAB JA-37)	7700 personnel
None	

Sweden's sophisticated and powerful military force is supplied with weaponry manufactured by its advanced home defense industry, including Saab fighter jets and Bofors antiaircraft guns. However, with the ending of the Cold War, strategic priorities have changed. Sweden feels less bound to maintain its neutral stance; it has participated in NATO's Partnerships for Peace program since 1994 and has WEU observer status. In 1999 spending cuts were announced, foreshadowing halving the size of the armed forces, because of the reduced military threat in the Scandinavian and Baltic region. A ten-year reform program began in 2001, offering personnel regular office hours and canceling large-scale exercises.

ECONOMICS

Inflation 2.1% p.a. (1990–2000)

$241bn

9.436–10.490 Swedish kronor

SCORE CARD

- World GNP Ranking 21st
- GNP per Capita $27,140
- Balance of Payments $6.62bn
- Inflation 1%
- Unemployment 5%

EXPORTS

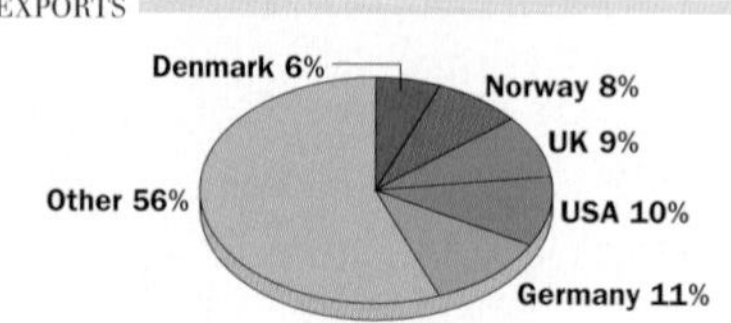

Denmark 6%
Norway 8%
UK 9%
USA 10%
Germany 11%
Other 56%

IMPORTS

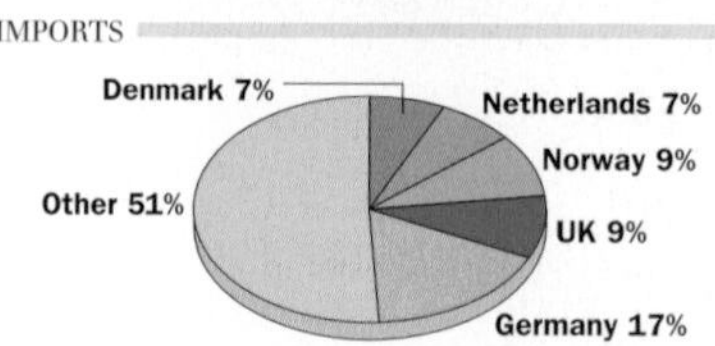

Denmark 7%
Netherlands 7%
Norway 9%
UK 9%
Germany 17%
Other 51%

Strengths

Companies of global importance, including Saab, Volvo, Electrolux and SKF, the world's biggest roller bearing manufacturer. Highly developed and constantly updated infrastructure. Sophisticated technology. Skilled labor force virtually bilingual in English.

Weaknesses

Labor costs remain uncompetitive. Highest taxation of OECD states, accounting for over 60% of GDP. Peripheral location, raising costs for producers and exporters. Major losses by Ericsson, giant telecoms company.

Profile

The state plays a significant role in the economy, particularly the services sector and infrastructure. Sweden's industrial giants have mostly been private-sector companies. The early 1990s saw a shift in economic policy to favor business, but greater growth did not follow, and unemployment and welfare costs drove up the budget deficit in 1994 to one of the OECD's highest. Growth has now resumed, the deficit has been cut back, and unemployment halved, but inflation has started to rise. The world downturn in 2001 brought the krona under pressure. Sweden opted not to introduce the euro from 1999, but the ruling SAP adopted a pro-euro stance in March 2000; the issue is to be decided by national referendum.

ECONOMIC PERFORMANCE INDICATOR

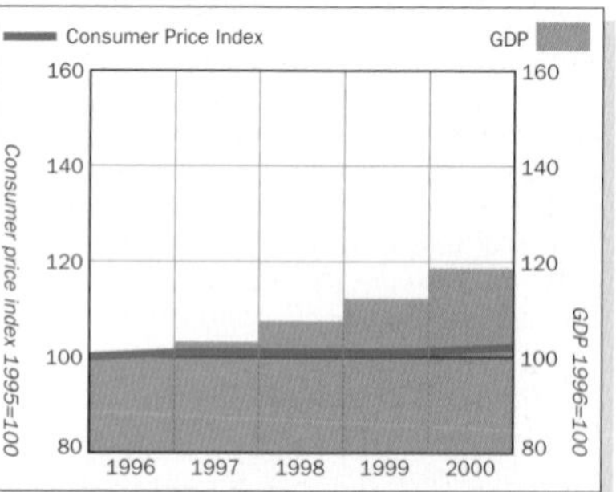

Consumer Price Index
GDP
160
140
120
100
80
Consumer price index 1995=100
GDP 1996=100
1996
1997
1998
1999
2000

SWEDEN : MAJOR BUSINESSES

0 200 km
0 200 miles

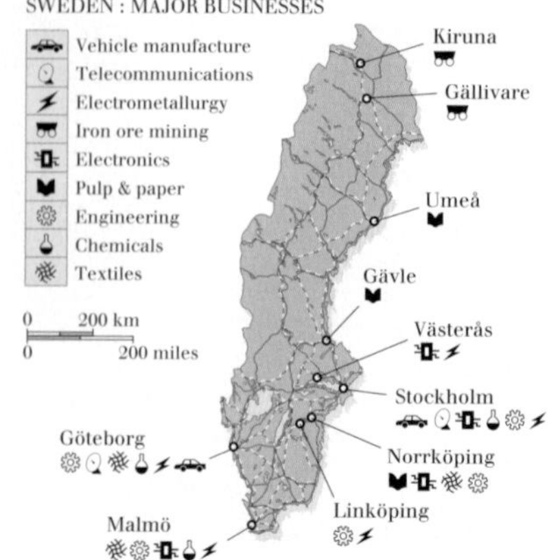

Vehicle manufacture
Telecommunications
Electrometallurgy
Iron ore mining
Electronics
Pulp & paper
Engineering
Chemicals
Textiles
Kiruna
Gällivare
Umeå
Gävle
Västerås
Stockholm
Norrköping
Linköping
Göteborg
Malmö

RESOURCES

Electric power 33.9m kw

357,317 tonnes

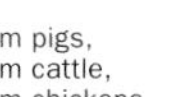

1.89m pigs, 1.65m cattle, 7.41m chickens

Low levels of oil production

Iron, uranium, copper, lead, zinc, silver

ELECTRICITY GENERATION

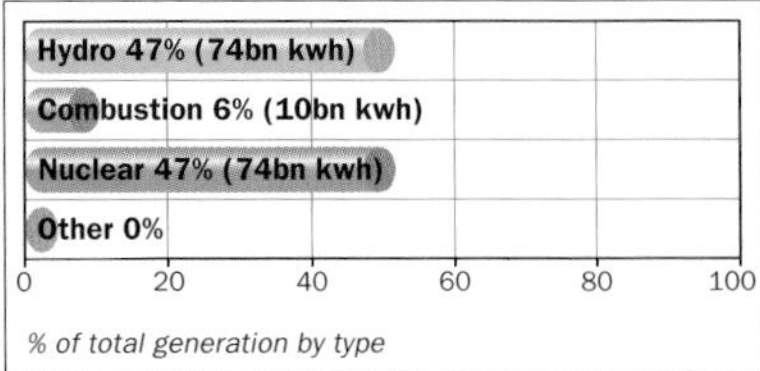

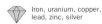

Hydro 47% (74bn kwh)
Combustion 6% (10bn kwh)
Nuclear 47% (74bn kwh)
Other 0%
0
20
40
60
80
100
% of total generation by type

Sweden is rich in minerals, including iron, copper, and silver. While mining and quarrying account for only 0.3% of GDP, they underpin other industrial sectors. In a referendum in 1980 Sweden decided, on environmental grounds, to abandon nuclear power – which in the 1990s accounted for about half of electricity generation – by the year 2010. However, problems in securing sufficient new energy supplies and in reducing consumption meant that, by the end of 2000, only one of the country's four nuclear reactors had been closed down.

SWEDEN : LAND USE

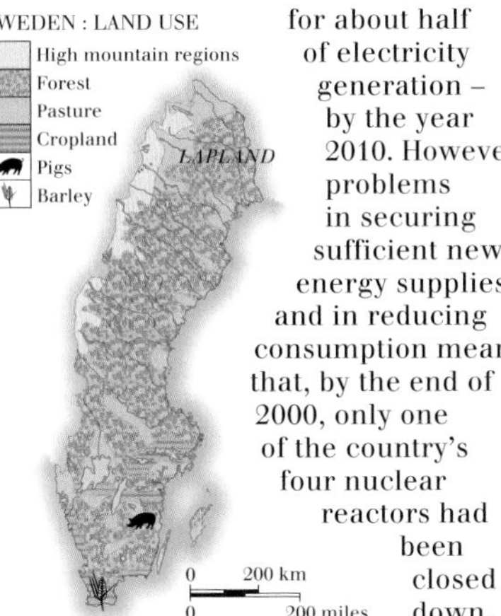

High mountain regions
Forest
Pasture
Cropland
Pigs
Barley
LAPLAND
0 200 km
0 200 miles

ENVIRONMENT

Sustainability rank: 3rd

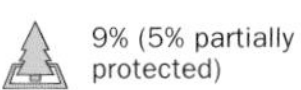

9% (5% partially protected)

5.5 tonnes per capita

ENVIRONMENTAL TREATIES

Yes
Yes
Yes
Yes
Yes
Yes

Since Sweden's pioneering Environment Protection Act in 1969, it has invested heavily in environmental protection measures. It blames acid rain damage to forests and lakes on airborne sulfur dioxide from factories in western Europe. Swedish nuclear reactors are said to be very safe, with filtered venting systems designed to retain 90% of all radioactivity released if there were a core meltdown. Nonetheless, all reactors are due to be shut down by 2010.

MEDIA

TV ownership high

Daily newspaper circulation 420 per 1000 people

PUBLISHING AND BROADCAST MEDIA

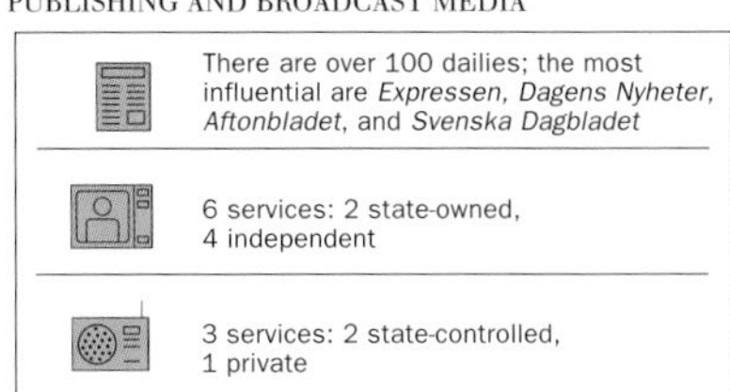

There are over 100 dailies; the most influential are *Expressen, Dagens Nyheter, Aftonbladet*, and *Svenska Dagbladet*
6 services: 2 state-owned, 4 independent
3 services: 2 state-controlled, 1 private

Press freedom is strongly entrenched, although radical views are rarely expressed. The influence of the major daily newspapers is largely confined to Stockholm: the provinces have their own strong press.

Two-thirds of the population have cable or satellite TV.

CRIME

No death penalty

5678 prisoners

Little change in 1999

CRIME RATES

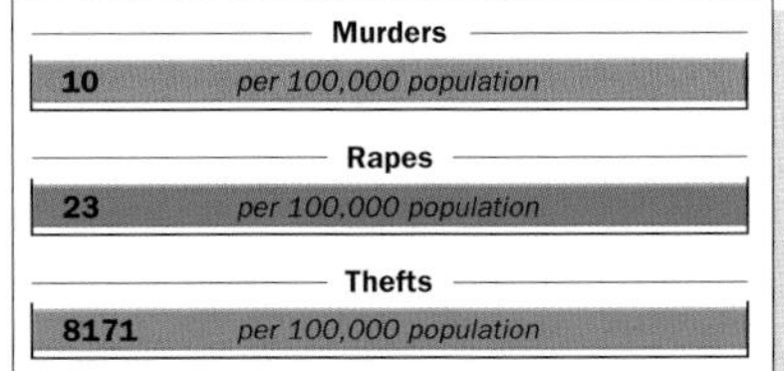

Murders
10 per 100,000 population
Rapes
23 per 100,000 population
Thefts
8171 per 100,000 population

Crime rates are below the European average, although they are the highest among Scandinavian countries. Assault, rape, and theft are growing problems, especially in the cities.

EDUCATION

School leaving age: 16

99%

275,217 students

THE EDUCATION SYSTEM

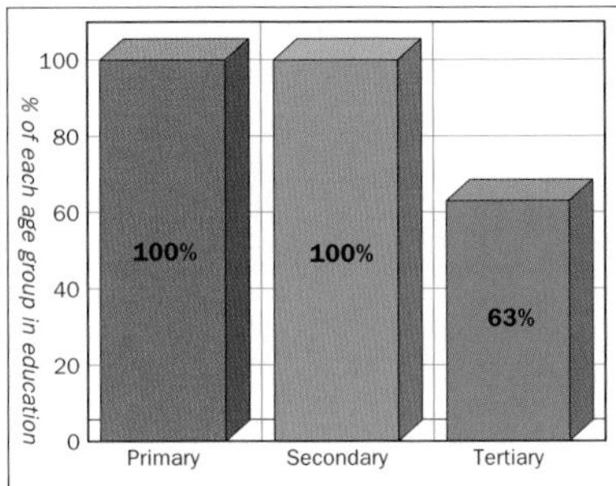

100
80
60
40
20
0
% of each age group in education
100%
100%
63%
Primary
Secondary
Tertiary

Education spending (public and private) is among the OECD's highest as a percentage of GDP.

Coeducational comprehensive schools are the norm. The higher education system is freely available to most of the population, and many adults return to college to do further courses.

HEALTH

Welfare state health benefits

1 per 323 people

Heart and cerebro-vascular diseases, cancers, accidents

Sweden's health care system is comprehensive and of a universally high standard. Spending fell by an average of 2% in real terms in the 1990s, but the trend is now being reversed. Savings have been made by increasing outpatient care, reducing the number of hospital beds, and cutting jobs. Since 1994 individuals have had the right to choose their own doctor, while doctors and specialists can now set up private practices. In 1999 the government agreed to compensate more than 60,000 people subjected to enforced sterilization in 1935–1975.

SPENDING

GDP/cap. increase

CONSUMPTION AND SPENDING

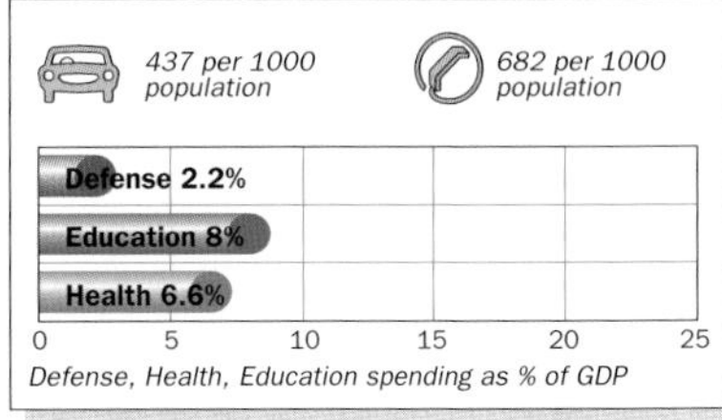

437 per 1000 population
682 per 1000 population
Defense 2.2%
Education 8%
Health 6.6%
0
5
10
15
20
25
Defense, Health, Education spending as % of GDP

Sweden has small income differentials, and Swedish executives are paid less than some of their European counterparts. Compared with other European states or the US, social competition and a sense of hierarchy are limited. Despite some cuts in services, the welfare system still rates highly in Europe.

Swedes are keen overseas property buyers, particularly of villas in Italy and France. Net overseas per capita investment remains among the highest in the world and about two-thirds of households own shares. PC ownership is high, and over 50% of Swedes had Internet access by 2001.

S

WORLD RANKING

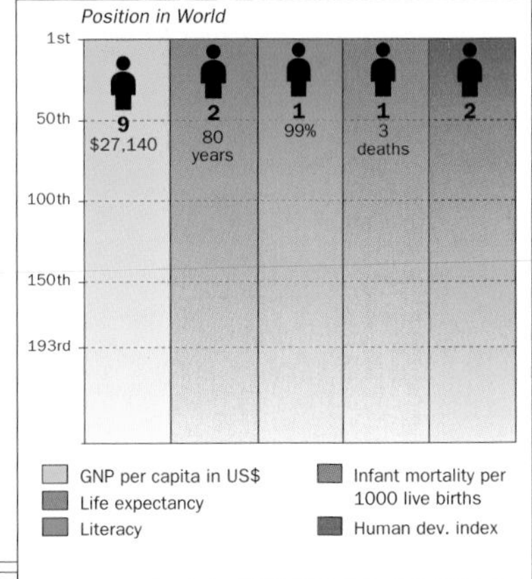

Position in World
1st
50th
100th
150th
193rd
9
$27,140
2
80 years
1
99%
1
3 deaths
2
GNP per capita in US$
Life expectancy
Literacy
Infant mortality per 1000 live births
Human dev. index

Europe
EUROPE
Germany
France
Austria
SWITZERLAND
Liechtenstein
Italy

SWITZERLAND

OFFICIAL NAME: Swiss Confederation **CAPITAL:** Bern
POPULATION: 7.2 million **CURRENCY:** Swiss franc **OFFICIAL LANGUAGES:** French, German, and Italian

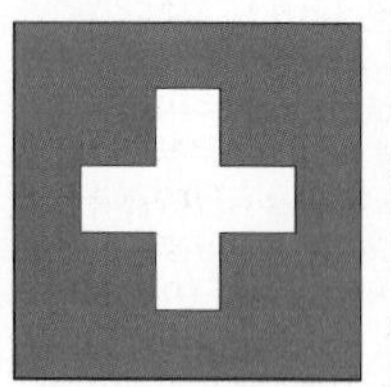

1291

1857
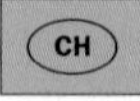
Aug 1

CH
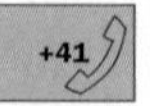
+1

+41

.ch

SWITZERLAND LIES at the center of western Europe geographically, but outside it politically. Sometimes called Europe's water tower, it is the source of all four of the region's major river systems: the Po, the Rhône, the Rhine, and the Inn–Danube. Switzerland has built one of the world's most prosperous economies, aided by the fact that it has retained its neutral status through every major European conflict since 1815. The process of European integration has been the latest and strongest challenge to Swiss neutralism, but it remains outside the EU.

***The Eiger in the Berner Oberland.** In 1994, a referendum voted to ban all truck transit traffic from the Swiss Alps from 2004.*

CLIMATE

Mountain/continental

WEATHER CHART FOR BERN

Average daily temperature — Rainfall
°C/°F J F M A M J J A S O N D cm/in
40/104 — 40/16
30/86 — 30/12
20/68 — 20/8
10/50 — 10/4
0/32 — 0
-10/14
-20/-4

Temperature and weather fluctuate enormously, not only seasonally but because of the huge variations in altitude and the country's location in central Europe. On the plateau north of the Alps, where most of the population lives, summers are warm and winters dry, cool, and often foggy. South of the Alps, it is warmer and sunnier. Strong southerly winds, or *föhn*, can bring summerlike weather even in winter. Avalanches have been a problem in recent years.

TRANSPORTATION

Drive on right

Kloten, Zürich
22.6m passengers

16 ships
368,000 grt

THE TRANSPORTATION NETWORK

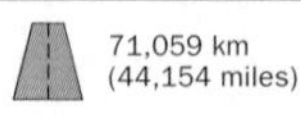
71,059 km
(44,154 miles)

1270 km
(789 miles)

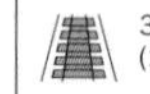
3143 km
(1953 miles)

1208 km
(751 miles)

Switzerland is a major freight transit route. Pollution and safety are major concerns, especially after 11 died in a fire in 2001 in the Gotthard road tunnel. The NEAT project, begun in 1996, will provide a high-speed rail link between Basel and Milan, whose trains will carry trucks. Swissair, bankrupt in 2001, has been replaced by a new airline, "swiss."

TOURISM

Visitors : Population 1.5:1

11m visitors

Up 3% in 2000

MAIN TOURIST ARRIVALS

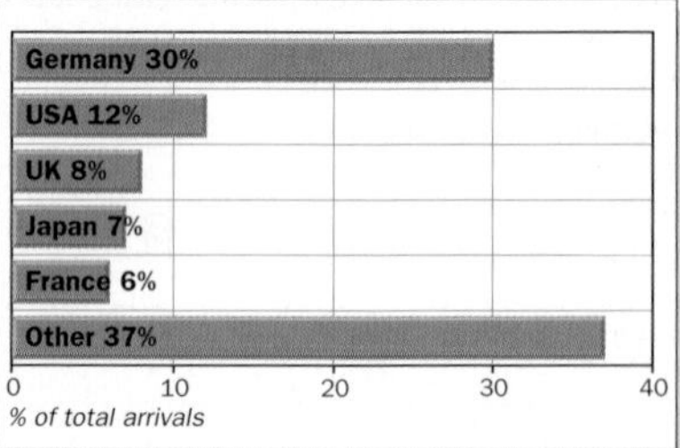
Germany 30%
USA 12%
UK 8%
Japan 7%
France 6%
Other 37%
0
10
20
30
40
% of total arrivals

Tourism is Switzerland's third-largest industry. About 330,000 Swiss earn their living from it, and tourism accounts for around 3% of GDP. The Alps are the main attraction, drawing winter and summer tourists from around the world, and Chillon Castle on Lake Geneva continues to be the country's most popular tourist site. In recent years, warmer winters have resulted in a shorter skiing season. The strong Swiss franc makes Switzerland an expensive destination, and Austria offers tough competition.

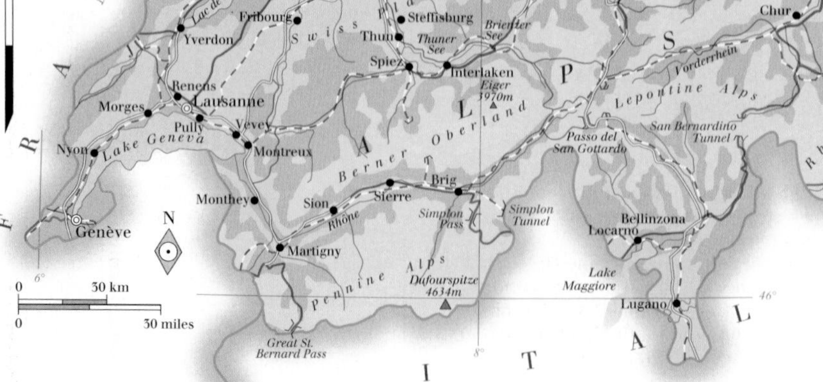
GERMANY
FRANCE
AUSTRIA
ITALY
LIECHTENSTEIN
Schaffhausen
Rhine
Basel
Liestal
Baden
Kreuzlingen
Lake Constance
Frauenfeld
Winterthur
Arbon
Rorschach
Kloten
Wil
Sankt Gallen
Herisau
Aarau
Delémont
Olten
Zürich
Uster
Zürichsee
Horgen
Stäfa
Wädenswil
Jona
Rapperswil
Baar
Zug
Solothurn
Grenchen
Langenthal
Biel
Aare
Doubs
La Chaux-de-Fonds
Bieler See
Burgdorf
Luzern
Einsiedeln
Schwyz
Neuchâtel
Bolligen
BERN
Muri
Vierwaldstätter See
Köniz
Lac de Neuchâtel
Fribourg
Steffisburg
Brienzer See
Thun
Thuner See
Yverdon
Spiez
Interlaken
Eiger 3970m
Chur
Davos
Flüelapass
Vorderrhein
Lepontine Alps
Renens
Lausanne
Morges
Pully
Vevey
Lake Geneva
Nyon
Montreux
Berner Oberland
Passo del San Gottardo
San Bernardino Tunnel
Rhaetian Alps
St. Moritz
Passo del Bernina
Brig
Monthey
Sion
Sierre
Rhône
Simplon Pass
Simplon Tunnel
Genève
Bellinzona
Locarno
Martigny
Pennine Alps
Dufourspitze 4634m
Lake Maggiore
Lugano
Great St. Bernard Pass
47°
46°
6°
8°
10°
N
0 30 km
0 30 miles

SWITZERLAND

Total Land Area : 41 290 sq. km (15 942 sq. miles)

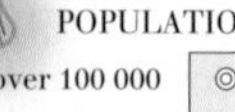
POPULATION
over 100 000
over 50 000
over 10 000

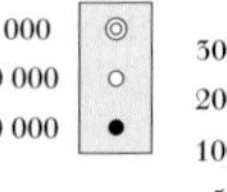
LAND HEIGHT
3000m/9843ft
2000m/6562ft
1000m/3281ft
500m/1640ft
200m/656ft

S

PEOPLE

Pop. density medium

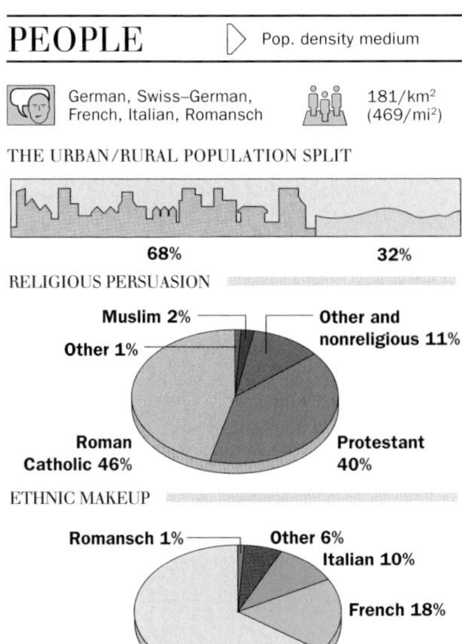
German, Swiss-German, French, Italian, Romansch
181/km²
(469/mi²)
THE URBAN/RURAL POPULATION SPLIT
68%
32%
RELIGIOUS PERSUASION
Muslim 2%
Other 1%
Other and nonreligious 11%
Roman Catholic 46%
Protestant 40%
ETHNIC MAKEUP
Romansch 1%
Other 6%
Italian 10%
French 18%
German 65%

Switzerland is composed of distinct German-Swiss, French-Swiss, and Italian-Swiss linguistic groups. About 40,000 in the eastern canton of Grisons speak Romansch. The German-Swiss, in the majority, are a tightly knit community, with a dialect that is impenetrable to most outsiders. In recent years, the three groups have grown further apart. The French-Swiss, in favor of joining the EU, are opposed by the German-Swiss. In Italian-speaking Ticino, a political party has emerged to champion Italian-Swiss interests. Tension between Swiss and immigrant workers gave rise to vocal support for strict limits on the number of foreigners, but in 2000 this was heavily defeated as a referendum proposal.

Society retains strong conservative elements. Two half-cantons granted women the vote at regional level only in 1989 and 1990. Marriage rates are high and divorce is less common than in most other European states.

POPULATION AGE BREAKDOWN

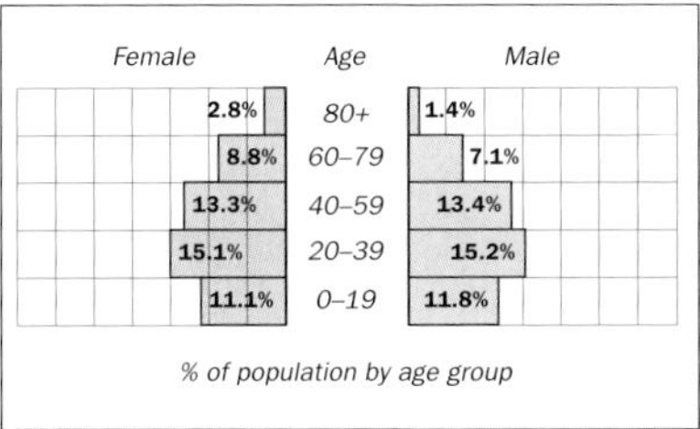
Female
Age
Male
2.8% 80+ 1.4%
8.8% 60–79 7.1%
13.3% 40–59 13.4%
15.1% 20–39 15.2%
11.1% 0–19 11.8%
% of population by age group

WORLD AFFAIRS

Joined UN in 2002

CE
G10

OECD

OSCE

PfP

The basis of Switzerland's foreign policy remains its neutrality. Geneva has retained its position as a center for many international organizations, including the ICRC and the European headquarters of the UN (although Switzerland only voted to join the UN in 2002). The city has often hosted diplomatic negotiations: those for the START nuclear reduction treaties and peace talks for the former Yugoslavia took place there.

Switzerland has so far not joined the process of closer European integration. In 1992 voters rejected EEA membership, widely seen as the first step toward EU membership. Many advocates of joining the EU believe that the economy will suffer without closer integration. Opponents argue that Switzerland's seeming isolation will enhance its role as an international tax haven. In 2001 a proposal to apply for EU membership was overwhelmingly rejected, although a series of bilateral cooperation agreements with the EU entered into force in 2002.

POLITICS

Multiparty elections

L. House 1999/2003
U. House 1999/2003

President Kaspar Villiger

AT THE LAST ELECTION

National Council 200 seats

26% SP/PS · 22% SVP/UDC · 22% FDP/PRD · 17% CVP/PDC · 4% GPS/PES · 9% Others

SP/PS = Social Democratic Party **SVP/UDC** = Swiss People's Party **FDP/PRD** = Radical Democratic Party **CVP/PDC** = Christian Democratic People's Party **GPS/PES** = Green Party of Switzerland

Council of States 46 seats

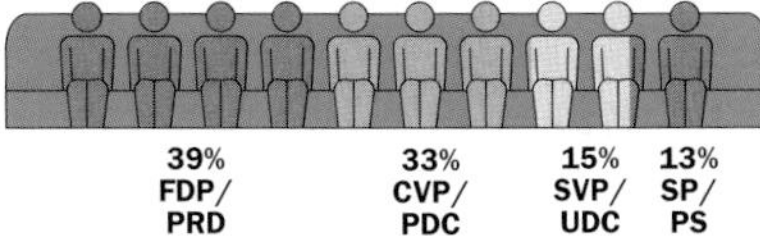
39% FDP/PRD
33% CVP/PDC
15% SVP/UDC
13% SP/PS

Switzerland is a federal democratic republic with 26 autonomous cantons. The presidency rotates every year.

PROFILE

The same four-party coalition has been in power in Switzerland since 1959. Domestic and foreign policies have changed little. Politics has recently become more contentious, however, with voting patterns becoming more polarized. Divisive issues are those of narcotics and of membership of the EU. Both right-wing and green minority parties have recently gained more seats in parliament. The right-wing SVP/UDC in particular capitalized on growing hostility to immigration to perform strongly in the 1999 elections.

Switzerland's political system is unique in Europe, in that taking important decisions depends on the results of referendums. A petition of more than 100,000 signatures can force a referendum on any issue.

MAIN POLITICAL ISSUES

European integration

Almost all politicians and business leaders favor joining the EU, or at least the European Economic Area (EEA), but voters remain sharply divided. The Swiss fear that their decentralized style of government would be lost within the EU. There are also worries that in a barrier-free Europe, Switzerland's high standards of living would fall because of a large influx of immigrants.

Switzerland and World War II

In 2002 the Bergier report criticized Swiss refusal to admit many Jewish refugees during World War II, but found that the banking system was not built on the assets of Holocaust victims. In 1998 the two largest Swiss banks had agreed to pay $1.25 billion to 31,500 victims and their families in return for agreement that there would be no future claims against Swiss banks or the government. A government Fund for Needy Holocaust Victims was wound up in 2002 after paying $179 million to 309,000 survivors.

Ruth Dreifuss, of the CVP, first woman president, held office in 1999.

Kaspar Villiger, of the FDP, who held the one-year presidency in 2002.

CHRONOLOGY

The autonomy of the Swiss cantons was curtailed by the Habsburgs in the 11th century. In 1291, the three cantons of Unterwalden, Schwyz, and Uri set up the Perpetual League to pursue Swiss liberty. Joined by other cantons, they succeeded in 1499 in gaining virtual independence. The Habsburgs retained a titular role.

- **1648** Peace of Westphalia ending Thirty Years' War, in which Switzerland played no active part, recognizes full Swiss independence.
- **1798** Invaded by French.
- **1815** Congress of Vienna after Napoléon's defeat confirms Swiss independence and establishes its neutrality. Geneva and Valais join Swiss Confederation.

⇨

S

CHRONOLOGY *continued*

- **1848** New constitution after brief civil war (1847) – central government given more powers, but cantons' powers guaranteed.
- **1857** Neuchâtel joins confederation.
- **1863** Henri Dunant founds ICRC in Geneva.
- **1874** Referendum established as important decision-making tool.
- **1914–1918** Plays humanitarian role in World War I.
- **1919** Proportional representation ensures future political stability.
- **1920** Joins League of Nations.
- **1939–1945** Neutral in World War II. Refuses to join UN in 1945.
- **1959** Founder member of EFTA. Present four-party coalition comes to power, taking over FDP/PRD dominance of government.
- **1967** Right-wing groups make electoral gains, campaigning to restrict entry of foreign workers.
- **1971** Most women granted right to vote in federal elections.
- **1984** Parliament approves application for UN membership. Elisabeth Kopp is first woman minister (justice minister).
- **1986** Referendum opposes joining UN. Immigrant numbers restricted.
- **1988** Kopp resigns over allegedly violating secrecy of information laws.
- **1990** Kopp acquitted. Case revealed public prosecutor's office held secret files on 200,000 people. Violent protests. State security laws amended.
- **1991** Large increase in attacks on asylum-seekers' hostels.
- **1992** Joins IMF and World Bank. Referendum vetoes joining EEA.
- **1994** Referendum approves new antiracism law and tighter laws against narcotics traffickers and illegal immigrants.
- **1998** $1.25 billion compensation for Holocaust victims whose funds were deposited in Swiss banks.
- **1999** Ruth Dreifuss first woman president.
- **2002** Third referendum on joining UN gives approval. Legalization of abortions as currently carried out also approved by referendum.

AID

Donor

$890m (donations)

Down 10% in 2000

With total disbursements amounting to 0.34% of GNP in 2000, Switzerland ranks above the (OECD) average of 0.22% as an aid donor, although the level of assistance remains below its target of 0.4% of GNP, set in 1994. Good governance and promoting investment are current priorities.

DEFENSE

Compulsory military service

$2.9bn

Down 7% in 2000

SWISS ARMED FORCES

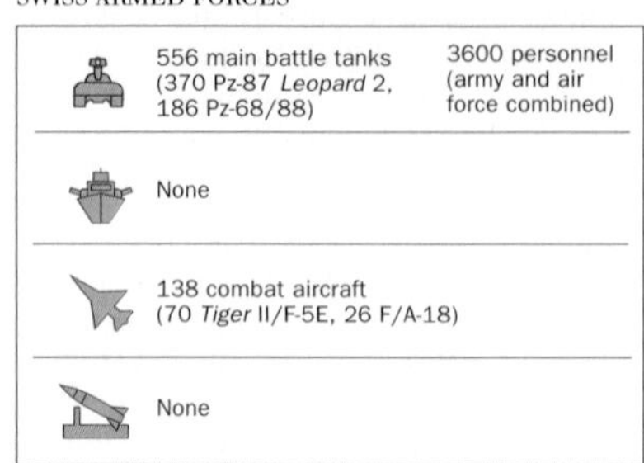
556 main battle tanks
(370 Pz-87 *Leopard* 2, 186 Pz-68/88)
3600 personnel
(army and air force combined)
None
138 combat aircraft
(70 *Tiger* II/F-5E, 26 F/A-18)
None

The army is, in one sense, among the largest in Europe. It is organized so that almost 400,000 conscripts can be called up and armed in a few hours; it still uses skis, bicycles, and horses to protect the Alps. Bridges and tunnels are mined in accordance with a defense strategy drafted in the early 1900s. Military service and further training at intervals are compulsory for males up to the age of 50.

As in the rest of Europe, force numbers are being cut in response to the end of the Cold War. In 1995, legislation allowing civilian service in place of military service was passed. Voters approved in 2001 a referendum proposal allowing Swiss soldiers to bear arms when on international peacekeeping operations.

ECONOMICS

Inflation 1.3% p.a. (1990–2000)

$274bn

1.6205–1.6603 Swiss francs

SCORE CARD

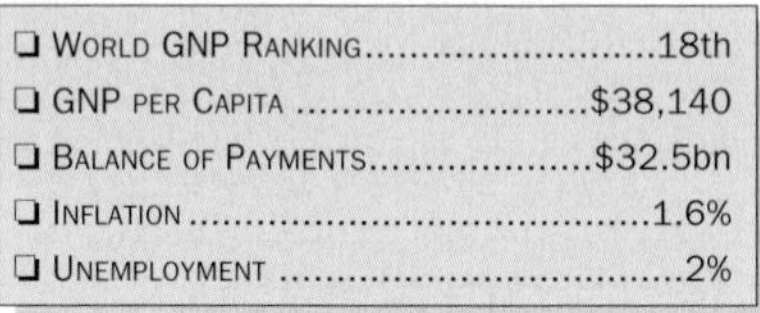

World GNP Ranking	18th
GNP per Capita	$38,140
Balance of Payments	$32.5bn
Inflation	1.6%
Unemployment	2%

EXPORTS

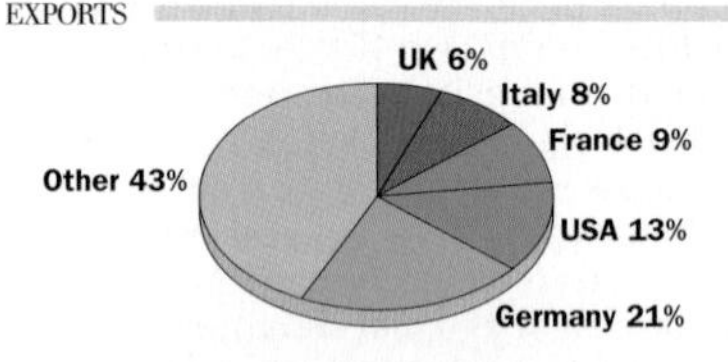
UK 6%
Italy 8%
France 9%
USA 13%
Germany 21%
Other 43%

IMPORTS

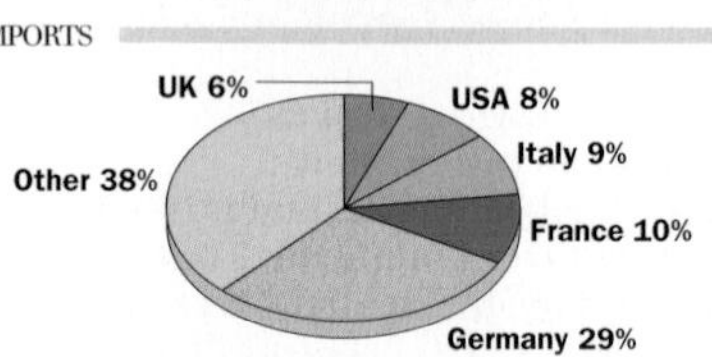
UK 6%
USA 8%
Italy 9%
France 10%
Germany 29%
Other 38%

ECONOMIC PERFORMANCE INDICATOR

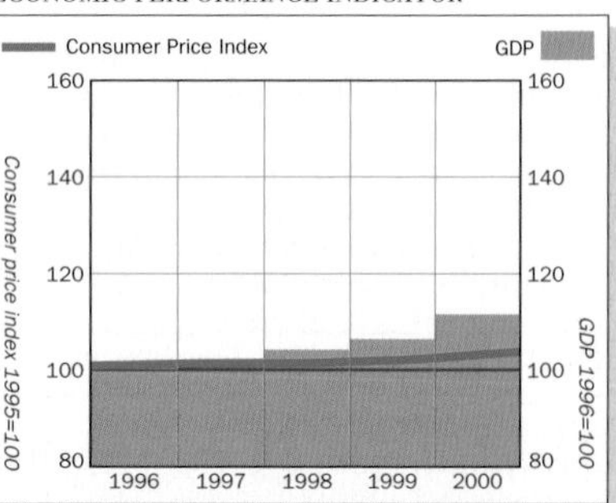
Consumer Price Index
GDP
160
140
120
100
80
Consumer price index 1995=100
GDP 1996=100
1996
1997
1998
1999
2000

Strengths
Highly skilled workforce. Reliable service provider. Major machine tool and precision engineering industries. Powerful chemical, pharmaceutical, and banking multinationals. Banking secrecy laws attract foreign capital; banking sector contributes 9% of GNP. Ability to innovate to capture mass markets, typified by Swatch watch and Swatch-designed Smart car.

Weaknesses
Protected cartels result in many overpriced goods. Highly subsidized agricultural sector.

Profile
The economy is widely diversified, with 62% of GDP coming from services and 34% from industry. There are several large multinational enterprises and, notably, a highly successful banking sector managing around one-third of the world's offshore private wealth. During most of the 1990s the economy grew only very slowly, if at all, but performance picked up in 2000 and 2001. In the long term, government finances face pressure to cater for the needs of an aging population.

SWITZERLAND : MAJOR BUSINESSES

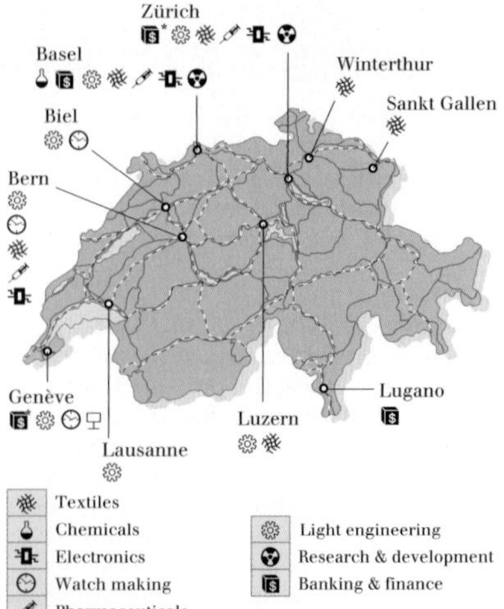
Zürich
Basel
Winterthur
Sankt Gallen
Biel
Bern
Genève
Lugano
Luzern
Lausanne
Textiles
Chemicals
Electronics
Watch making
Pharmaceuticals
Consumer goods
Light engineering
Research & development
Banking & finance
0 50 km
0 50 miles
* significant multinational ownership

S

RESOURCES

Electric power 16.7m kw

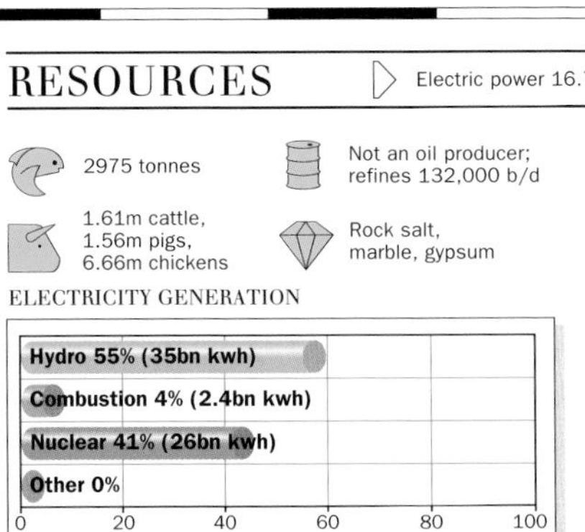

2975 tonnes

Not an oil producer; refines 132,000 b/d

1.61m cattle, 1.56m pigs, 6.66m chickens

Rock salt, marble, gypsum

ELECTRICITY GENERATION

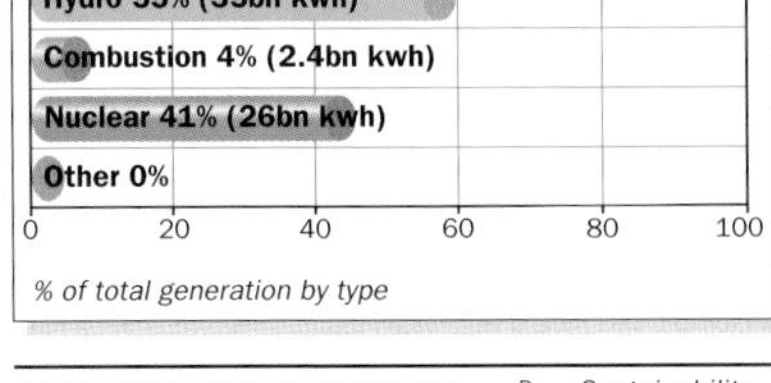
Hydro 55% (35bn kwh)
Combustion 4% (2.4bn kwh)
Nuclear 41% (26bn kwh)
Other 0%
0
20
40
60
80
100
% of total generation by type

Switzerland is poor in natural resources, having no valuable minerals in commercially exploitable quantities. Over half of its electricity comes from hydropower, while five nuclear plants supply most of the rest, so that spending on imported oil and coal is kept to a minimum – they account for less than 4% of the total import bill. Large-scale antinuclear power demonstrations in the 1980s led to the cancellation of plans for a new nuclear power plant, but a referendum approved continued use of existing plants.

SWITZERLAND : LAND USE

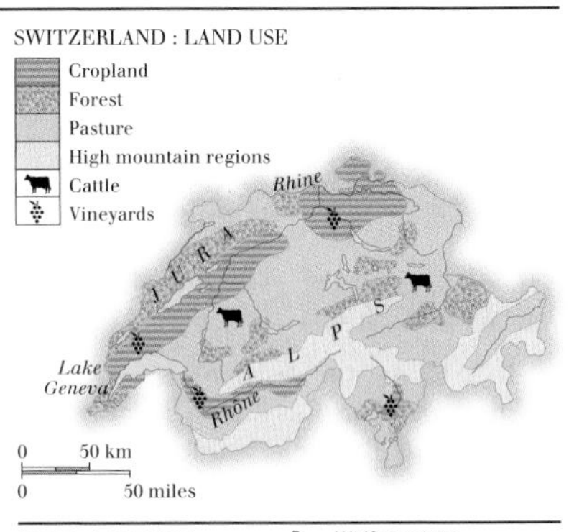
Cropland
Forest
Pasture
High mountain regions
Cattle
Vineyards
Rhine
JURA
ALPS
Lake Geneva
Rhône
0
50 km
0
50 miles

ENVIRONMENT

Sustainability rank: 5th

27%

5.9 tonnes per capita

ENVIRONMENTAL TREATIES

Yes
Yes
Yes
Yes
Yes
No

The Swiss are among the most environmentally conscious people in the world and are willing to back their convictions with money. The Basel–Milan link, designed to carry trucks by rail through a tunnel instead of having them cross the Alps by road, was approved by referendum, despite the estimated $13.3bn cost. The planners' aim is a total ban on carrying freight by road through Switzerland by 2004; this may not be necessary if big time savings make the tunnel sufficiently attractive. The Swiss are keen recyclers and taxation is used to encourage this.

MEDIA

TV ownership high

Daily newspaper circulation 376 per 1000 people

PUBLISHING AND BROADCAST MEDIA

There are 84 daily newspapers. The largest circulations are held by *Tages Anzeiger* and the Zürich-based tabloid *Blick*

3 independent services broadcasting in German, Romansch, French, and Italian

3 independent services broadcasting in German, Romansch, French, and Italian

The Swiss media are organized broadly along regional lines, and reflect the country's linguistic divisions. The German-, Romansch-, French-, and Italian-language TV and radio stations tend to focus on the interests of their specific communities. German, Italian, and French satellite TV is widely available. Few newspapers are distributed throughout the country. *Tribune de Genève* and *Neue Zürcher Zeitung* are exceptions.

CRIME

No death penalty

6390 prisoners

Up 30% in 1999

CRIME RATES

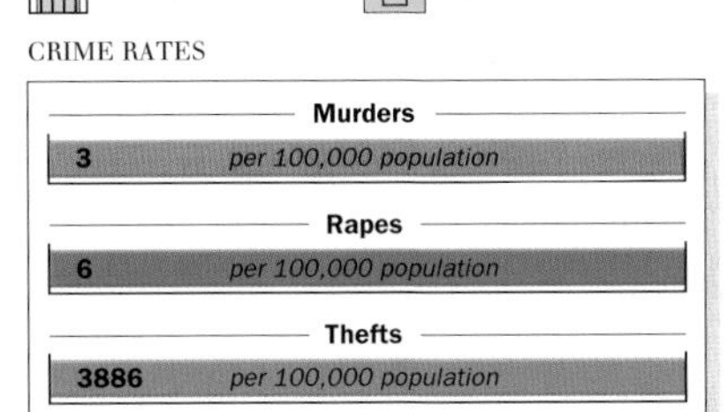
Murders
3 per 100,000 population
Rapes
6 per 100,000 population
Thefts
3886 per 100,000 population

Crime rates are low by international standards. Muggings and burglaries are on the increase and are often related to narcotics. More cases of banking secrecy laws attracting laundered funds are coming to light.

EDUCATION

School leaving age: 16

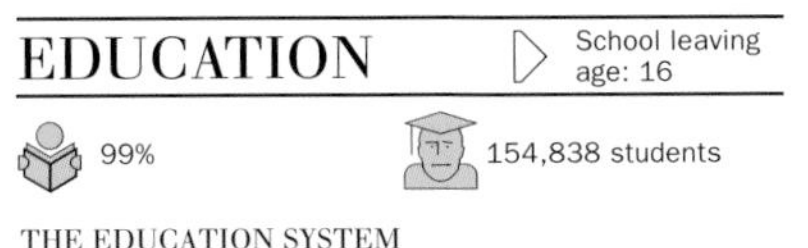
99%
154,838 students

THE EDUCATION SYSTEM

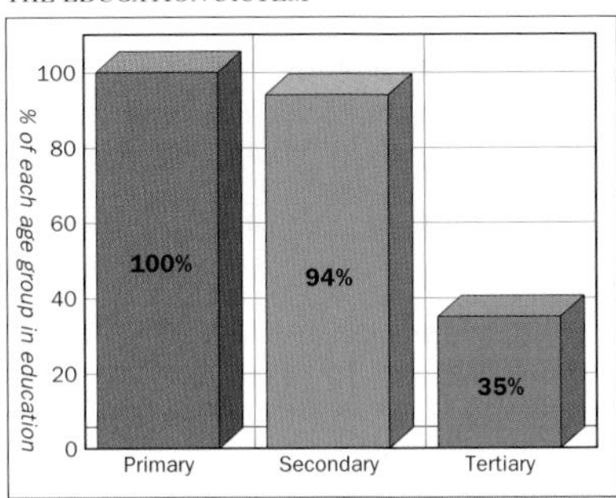
100
80
60
40
20
0
% of each age group in education
100%
94%
35%
Primary
Secondary
Tertiary

Of total government expenditure, around 16% is spent on education. Primary and secondary education are controlled by the cantons, so that there are 26 different systems in operation. Most students after the age of 16 are encouraged to take up vocational studies. Training is thorough and is usually combined with three or four years' apprenticeship in the student's chosen field. The Federal Technological Institute in Zürich has gained an international reputation for its computer programming research.

HEALTH

Welfare state health benefits

1 per 294 people

Heart and cerebrovascular diseases, cancers, accidents

The health system is among the most efficient and pioneering in the world, and is ranked by WHO second, after Japan, for attainment. High health costs are covered by compulsory insurance schemes.

SPENDING

GDP/cap. increase

CONSUMPTION AND SPENDING

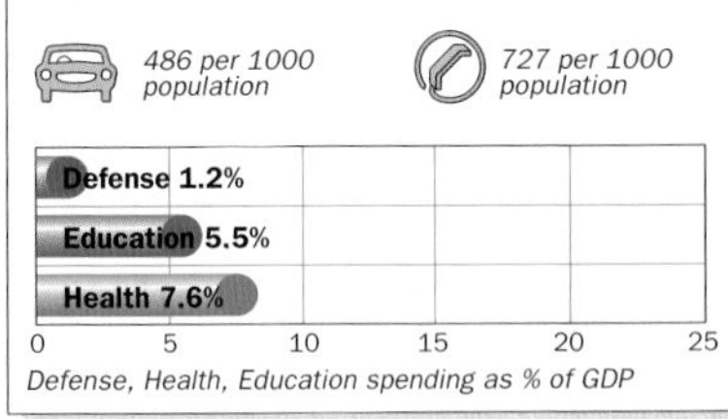
486 per 1000 population
727 per 1000 population
Defense 1.2%
Education 5.5%
Health 7.6%
0
5
10
15
20
25
Defense, Health, Education spending as % of GDP

Immigrant workers do most low-paid and menial jobs. Wages in office jobs are relatively high, although the cost of living is also well above the European average. Many workers choose to live in France and commute across the border. The land market is highly regulated.

WORLD RANKING

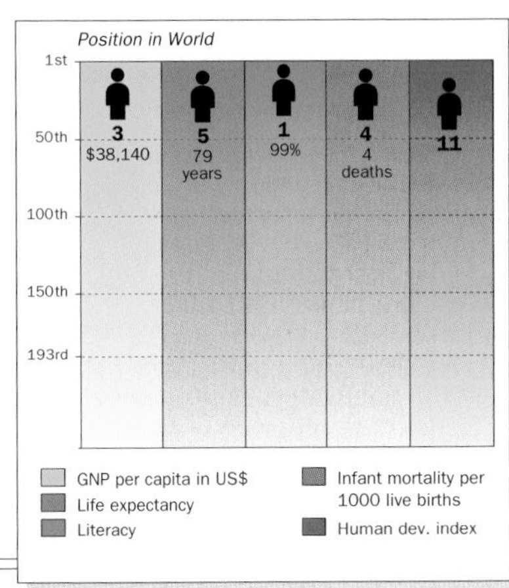
Position in World
1st
50th
100th
150th
193rd
3
$38,140
5
79 years
1
99%
4
4 deaths
11
GNP per capita in US$
Life expectancy
Literacy
Infant mortality per 1000 live births
Human dev. index

S

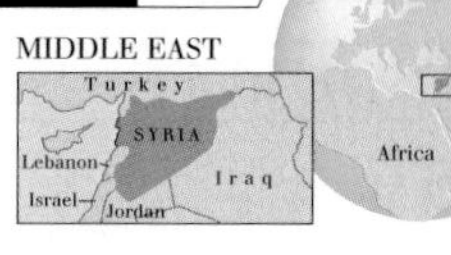
MIDDLE EAST
Turkey
SYRIA
Lebanon
Israel
Jordan
Iraq
Africa
Asia

SYRIA

OFFICIAL NAME: Syrian Arab Republic **CAPITAL:** Damascus
POPULATION: 16.6 million **CURRENCY:** Syrian pound **OFFICIAL LANGUAGE:** Arabic

1941

1967

April 17
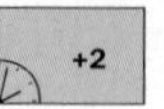
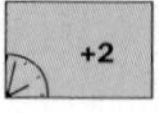
SYR

+2

+963
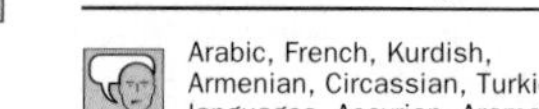

.sy

SYRIA IS REGARDED by many of its people as an artificial creation of French-mandated rule, which lasted from 1920 to independence. They identify instead with a Greater Syria, a successor to the medieval Ummayad caliphate, encompassing Lebanon, Jordan, and Palestine. Since independence, Syria's foreign relations have been turbulent, but the authoritarian Ba'athist regime of Hafez al-Assad (1970–2000) brought a measure of internal stability.

CLIMATE

Steppe/hot desert/ Mediterranean

WEATHER CHART FOR DAMASCUS

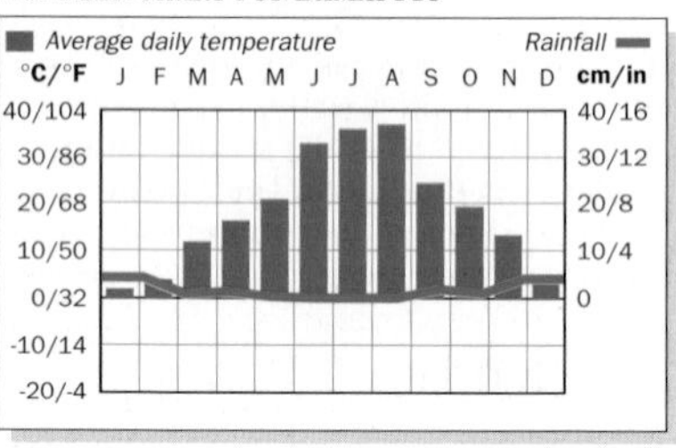
Average daily temperature
Rainfall
°C/°F J F M A M J J A S O N D cm/in
40/104
30/86
20/68
10/50
0/32
-10/14
-20/-4
40/16
30/12
20/8
10/4
0

The coastal climate is Mediterranean, with mild, wet winters and dry, hot summers. Most of the country gets fewer than 25 cm (10 in) of rainfall a year. Away from the coast, rainfall is very unpredictable and the country becomes increasingly arid, with some desert areas. In the mountains, snow is common in winter.

TRANSPORTATION

Drive on right

Damascus International 1.5m passengers

220 ships 427,500 grt

THE TRANSPORTATION NETWORK

26,299 km (16,341 miles)

877 km (545 miles)

1525 km (948 miles)

870 km (541 miles)

The road network is unreliable in rural areas, especially during the winter after rain. Bus services operate to most towns from Damascus and Aleppo. Roads are integrated with the railroads, which carry over four million passengers a year and are vital to freight transportation. The rail link from Aleppo to Mosul, in Iraq, reopened in mid-2000. Damascus is the main international airport and Latakia the main port.

TOURISM

Visitors : Population 1:18

916,000 visitors

Up 24% in 1999

MAIN TOURIST ARRIVALS

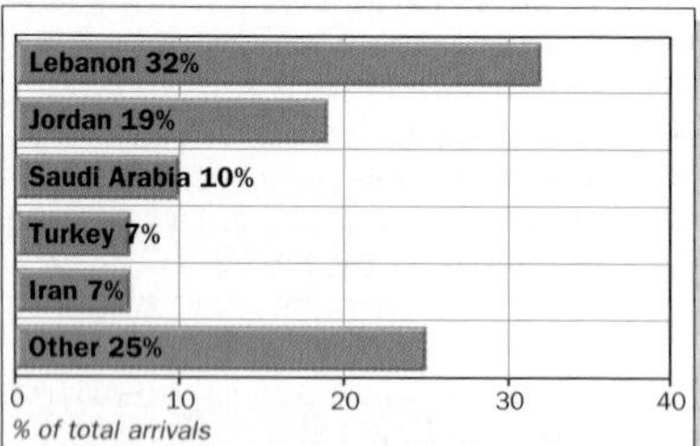
Lebanon 32%
Jordan 19%
Saudi Arabia 10%
Turkey 7%
Iran 7%
Other 25%
0 10 20 30 40
% of total arrivals

Years of political turbulence, allegations of human rights abuses committed under Hafez al-Assad's regime, and strict, complex travel regulations retarded the development of tourism. However, just before the 1990–1991 Gulf War, Syria began to compete with other Middle Eastern states as a tourist destination. Modern hotels were built in most cities and facilities improved to cater for growing numbers of Western visitors. Following the war, tourist numbers dropped sharply, but are now gradually recovering. Syria's main attractions are its antiquities and historic cities, with their covered markets (*soukhs*), baths, and mosques – Damascus, said to be the oldest inhabited city in the world; the ruined desert city of Palmyra; and Aleppo, with its citadel. Syria has a wealth of castles dating back to the Crusades and sites associated with the advent of Islam. In addition, there are as many as 3500 as yet unexcavated archaeological sites. Syria's coastline on the Mediterranean has fine beaches, and there are mountain resorts in Latakia.

The ancient city of Palmyra, *in Syria's central region, was once the capital of the kingdom of Queen Zenobia.*

PEOPLE

Pop. density medium

Arabic, French, Kurdish, Armenian, Circassian, Turkic languages, Assyrian, Aramaic

90/km² (234/mi²)

THE URBAN/RURAL POPULATION SPLIT

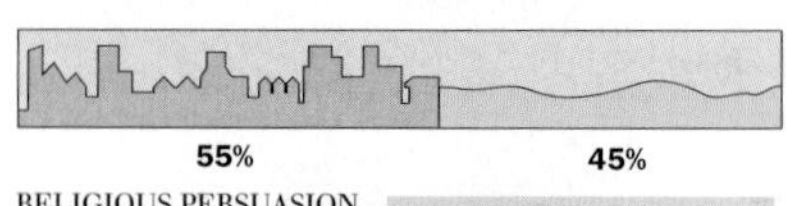
55%
45%

RELIGIOUS PERSUASION

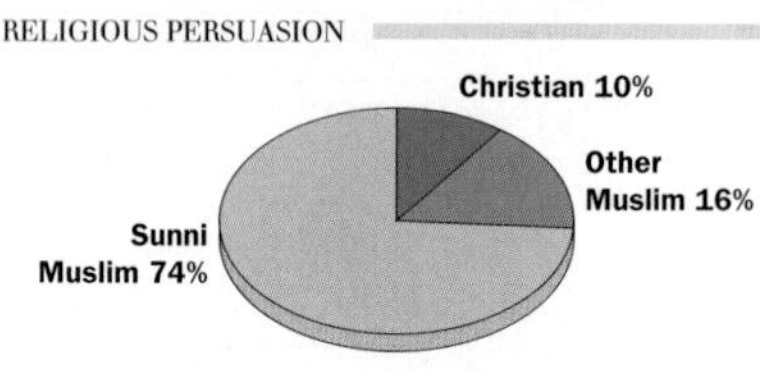
Christian 10%
Other Muslim 16%
Sunni Muslim 74%

ETHNIC MAKEUP

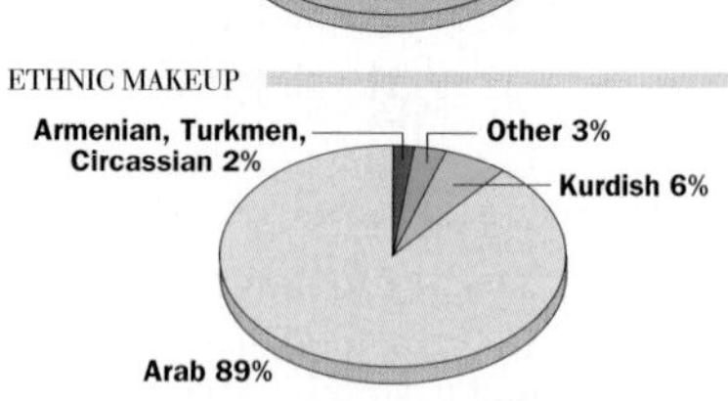
Armenian, Turkmen, Circassian 2%
Other 3%
Kurdish 6%
Arab 89%

Most Syrians live in the west, where the largest cities are sited. About 90% are Muslim. They include the politically dominant Alawis, a heterodox offshoot of Shi'a, comprising 12% of the population, based in Latakia and Tartous provinces. There is also a sizable Christian minority. In the west and north a mosaic of groups includes Kurds, Turkic-speakers, and Armenians, the latter based in cities. Damascus, Al Qamishli, and Aleppo have small Jewish communities, and there are three villages where Aramaic is spoken. In addition, some 300,000 Palestinian refugees have settled in Syria. Minorities were initially attracted to the ruling Ba'ath Party because of its emphasis on the state over sectarian interests. However, disputes between factions led to the Alawis taking control, creating resentment among the Sunni Muslim majority.

The emancipation of women, promoted initially in the late 1960s, was carried forward under President Hafez al-Assad, whose first woman cabinet minister was appointed in 1976.

POPULATION AGE BREAKDOWN

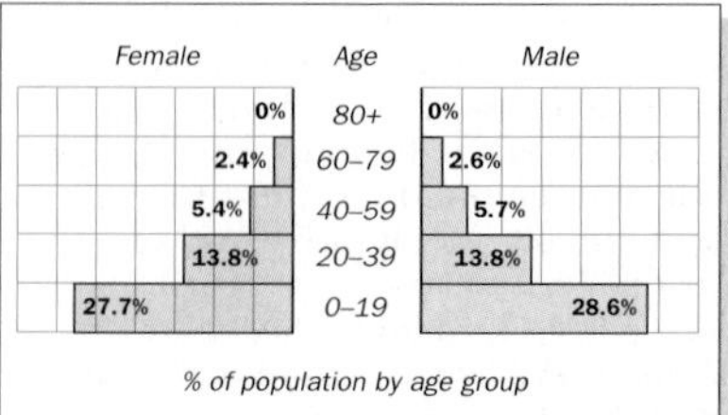
Female
Age
Male
0% 80+ 0%
2.4% 60–79 2.6%
5.4% 40–59 5.7%
13.8% 20–39 13.8%
27.7% 0–19 28.6%
% of population by age group

S

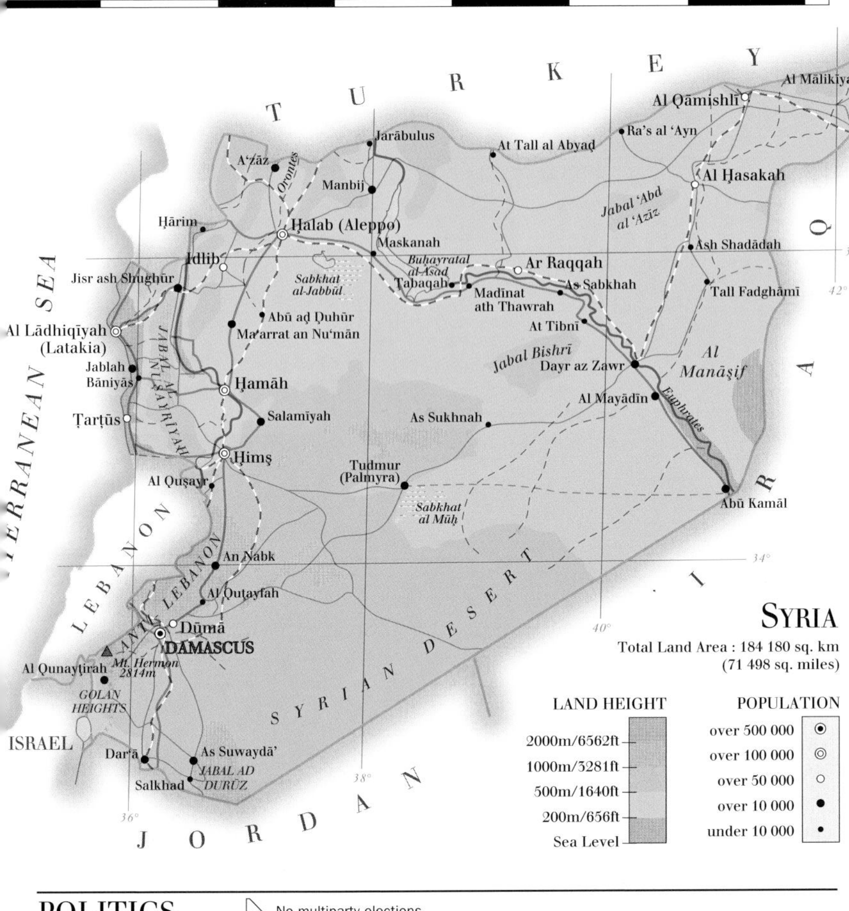
TURKEY
IRAQ
JORDAN
LEBANON
ISRAEL
MEDITERRANEAN SEA
SYRIAN DESERT
Al Mālikīyah
Al Qāmishlī
Ra's al 'Ayn
At Tall al Abyad
Jarābulus
A'zāz
Manbij
Ḩārim
Ḩalab (Aleppo)
Maskanah
Idlib
Jisr ash Shughūr
Buḩayrat al-Asad
Ţabaqah
Madīnat ath Thawrah
Ar Raqqah
As Sabkhah
Al Ḩasakah
Ash Shadādah
Tall Fadghāmī
Jabal 'Abd al 'Azīz
Sabkhat al-Jabbūl
Abū aḑ Ḑuhūr
Ma'arrat an Nu'mān
Al Lādhiqīyah (Latakia)
Jablah
Bāniyās
Ţarţūs
Ḩamāh
Salamīyah
Ḩimş
Al Quşayr
At Tibnī
Jabal Bishrī
Dayr az Zawr
Al Mayādīn
Al Manāşif
Euphrates
As Sukhnah
Tudmur (Palmyra)
Sabkhat al Mūḩ
Abū Kamāl
An Nabk
Al Quţayfah
Dūmā
DAMASCUS
Mt. Hermon 2814m
Al Qunayţirah
GOLAN HEIGHTS
Dar'ā
As Suwaydā'
JABAL AD DURŪZ
Salkhad
Orontes
SYRIA
Total Land Area : 184 180 sq. km (71 498 sq. miles)
LAND HEIGHT
2000m/6562ft
1000m/3281ft
500m/1640ft
200m/656ft
Sea Level
POPULATION
over 500 000
over 100 000
over 50 000
over 10 000
under 10 000

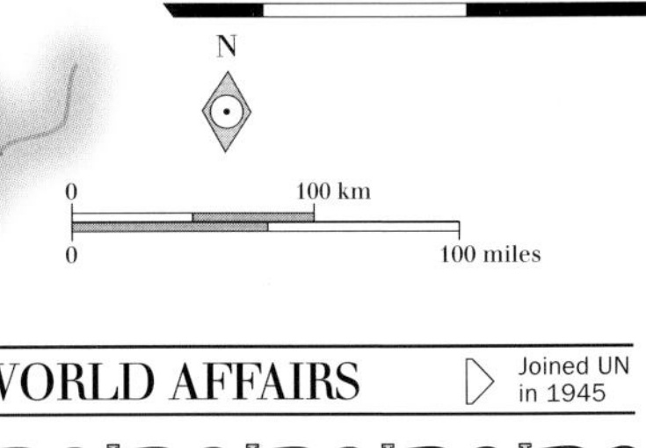
N
0
100 km
0
100 miles

POLITICS

No multiparty elections

1998/2002

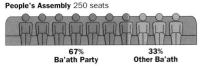

President Bashar al-Assad

AT THE LAST ELECTION

People's Assembly 250 seats

67% Ba'ath Party	33% Other Ba'ath

Ba'ath Party = Members of the National Progressive Front (allies of the Ba'ath Party) **Other Ba'ath** = Parties allied to the Ba'ath Party

Syria is in effect a single-party state. Its military-backed leader from 1970 to 2000 was Hafez al-Assad, a lifelong Ba'ath Party militant. His personal dominance ensured the succession of his son Bashar after his death in June 2000.

PROFILE

The Ba'athist military swept to power in 1963 with a vision of uniting all Arab nations under a single Syrian-dominated socialist system. The coup ended the power of city elites and promoted citizens from rural areas.

Assad consolidated the Ba'ath Party as the major political force. Unrest among Islamic militants was crushed, notably in the army-led massacre at Hamah in 1982. Assad focused on foreign affairs, bidding to make Syria a major power, but plans to unite with fellow Ba'athist Iraq ended in 1981 amid mutual recriminations.

MAIN POLITICAL ISSUES

Human rights

Martial law has not been rescinded since 1963, but the regime has improved its human rights record in recent years. Political prisoners are released under frequent amnesties, and in 1994 all members of the Jewish minority were granted exit visas to travel abroad.

Political pluralism

Assad dominated Syrian politics for 30 years. His military-backed regime, drawn mainly from his own Alawi minority grouping, kept a tight hold on power, although in his last decade Sunnis gained high political posts. Despite promises made under international pressure, Assad never permitted genuine multipartyism. Shortly before his death, the modernizing Mohammed Miro replaced the long-serving prime minister Mahmoud az-Zoubi, who was forced from office. Assad's death was followed immediately by his son's election to the party leadership, which was approved overwhelmingly by referendum. Initial tentative reforms – such as the spread of "discussion clubs" and a more varied cabinet – were later curtailed.

WORLD AFFAIRS

Joined UN in 1945

AL | Damasc | G24 | NAM | OIC

Since Egypt's 1979 accord with Israel, Syria sees itself as the major barrier to Israel's regional dominance. Syria extended its control over Lebanon and hosts radical Palestinian factions, as well as seeking alliances with north African states. The biggest issue with Israel remains the strategically vital Golan Heights, seized by Israel during the Six-Day War in 1967. Peace negotiations foundered when Ariel Sharon became prime minister of Israel in 2001.

There are enduring tensions with Turkey over attitudes to Israel and to Turkish Kurdish guerrillas, access to water, and Syria's desire for the return of its former Alexandretta Province.

Syria, alone among Arab states, backed Iran in the 1980s Iran–Iraq War. Facing international isolation over its alleged backing of terrorists, Syria regained a measure of respectability by securing the release of Western hostages of Shi'a militants in Lebanon. Assad backed the US-led allies in the 1990–1991 Gulf War, and by contributing troops legitimized the action in the eyes of the Arab world. In 2000, however, Syria sent humanitarian aid to Iraq, in defiance of the UN blockade, and in 2002 was accused by the US of continuing to support international "terrorists."

AID

Recipient

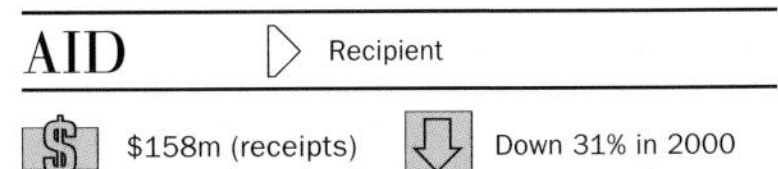

$158m (receipts) | Down 31% in 2000

Syria has historically received little aid owing to its human rights record and self-sufficiency in oil. It received one-off payments totaling $2 billion in 1992 and $1.2 billion in 1993 after the Gulf War, mainly from Saudi Arabia and the Gulf states, but with contributions from the West and Japan.

Hafez al-Assad, *ruled for three decades until his death in 2000.*

Bashar al-Assad, *succeeded his father as president.*

S

CHRONOLOGY

Under French mandate from 1920, Syria declared independence in 1941, and achieved full autonomy in 1946. From 1958 to 1961 Syria merged with Egypt to form the United Arab Republic.

- ❑ **1963** Ba'athist military junta seizes power. Maj. Gen. Amin al-Hafez president.
- ❑ **1966** Hafez ousted by military coup supported by radical Ba'ath Party members.
- ❑ **1967** Israel overruns Syrian positions above Lake Tiberias, seizes Golan Heights, and occupies Quneitra. Syria boycotts Arab summit and rejects compromise with Israel.
- ❑ **1970** Hafez al-Assad seizes power in "corrective coup."
- ❑ **1971–1999** Assad elected president; reelected four times.
- ❑ **1973** New constitution confirms dominance of Ba'ath Party. War launched with Egypt against Israel to regain territory lost in 1967. More territory temporarily lost to Israel.
- ❑ **1976** With peacekeeping mandate from Arab League, Syria intervenes to quell fighting in Lebanon.
- ❑ **1977** Relations broken off with Egypt after Egyptian president Sadat's visit to Jerusalem.
- ❑ **1978** National charter signed with Iraq for union.
- ❑ **1980** Membership of Muslim Brotherhood made capital offense. Treaty of Friendship with USSR.
- ❑ **1981** Israel formally annexes Golan Heights. Charter with Iraq collapses.
- ❑ **1982** Islamic extremist uprising in Hamah crushed. Israel invades Lebanon; Syrian missiles in Bekaa Valley destroyed.
- ❑ **1985** US claims Syrian links to Rome and Vienna airport bombings.
- ❑ **1986** Alleged Syrian complicity in planting of bomb on Israeli airliner in London. EU states, except Greece, impose sanctions and arms embargo.
- ❑ **1989** Diplomatic relations reestablished with Egypt.
- ❑ **1990–1991** Syrian forces crush renegade Gen. Aoun in Beirut. Steers Taif Accords over Lebanon.
- ❑ **1991** Troops take part in Operation Desert Storm. Damascus Declaration aid and defense pact signed with Egypt, Saudi Arabia, Kuwait, UAE, Qatar, Bahrain, and Oman.
- ❑ **1995** Inconclusive talks with Israel.
- ❑ **2000** Forced resignation after 13 years and subsequent suicide of prime minister, Mahmoud az-Zoubi. Death of Hafez al-Assad; succession of son Bashar.
- ❑ **2001** Israel bombs Syrian radar base in Lebanon.

S

DEFENSE

Compulsory military service

$760m

Down 23% in 2000

Syria sees its extensive military capability as a significant deterrent to Israel's territorial expansion. It has fought four wars against Israel since 1948, and is the Arab world's strongest military power after Egypt. Most of its military equipment was obtained from the former Soviet Union.

Throughout 2000, increasing numbers of Lebanese protested against the continuing presence of Syrian troops on their soil – in apparent contravention of the 1989 Taif Accords – and in 2001, Syrian troops were withdrawn from Beirut. In May 2002 a US State Department official accused Syria of harboring weapons of mass destruction.

SYRIAN ARMED FORCES

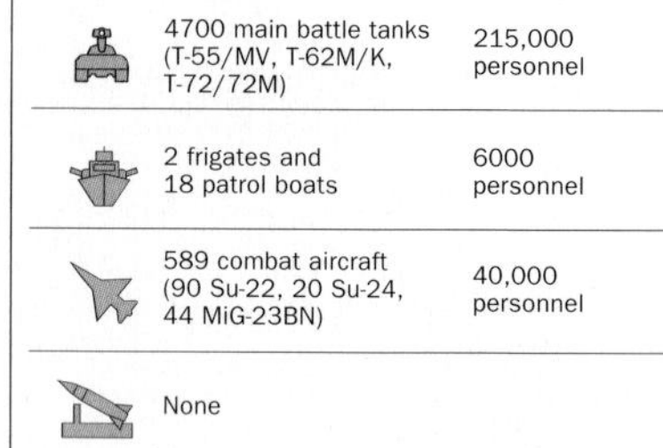
4700 main battle tanks (T-55/MV, T-62M/K, T-72/72M)
215,000 personnel
2 frigates and 18 patrol boats
6000 personnel
589 combat aircraft (90 Su-22, 20 Su-24, 44 MiG-23BN)
40,000 personnel
None

ECONOMICS

Inflation 6.7% p.a. (1990–2000)

$15.1bn

53.5–45.9 Syrian pounds

SCORE CARD

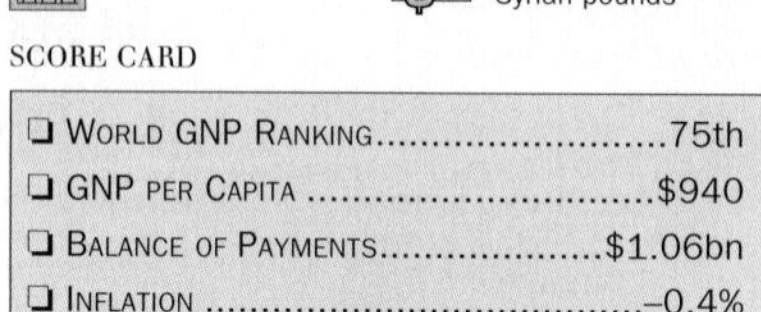

❑ World GNP Ranking	75th
❑ GNP per Capita	$940
❑ Balance of Payments	$1.06bn
❑ Inflation	–0.4%
❑ Unemployment	20%

EXPORTS

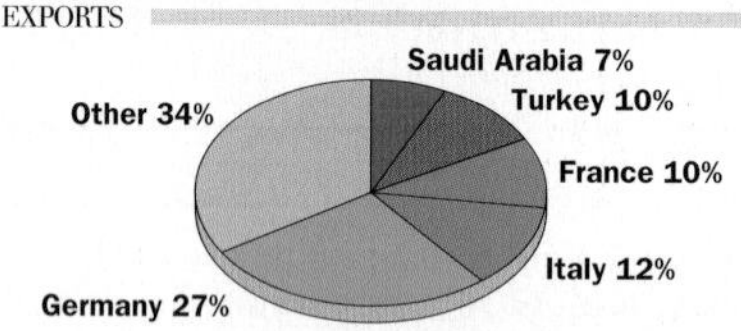
Saudi Arabia 7%
Turkey 10%
France 10%
Italy 12%
Germany 27%
Other 34%

IMPORTS

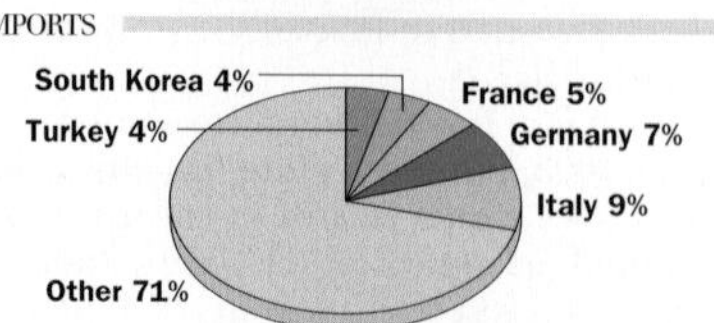
South Korea 4%
France 5%
Turkey 4%
Germany 7%
Italy 9%
Other 71%

Strengths

Exporter of crude oil – production increasing as a result of new oil strikes. Manufacturing base has grown. Thriving agricultural sector. Low inflation.

Weaknesses

High defense spending a major drain. Large black market. Domination of inefficient state-run companies. Lack of foreign investment; foreign currency accounts banned. High population growth and unemployment. Vulnerable water supply. Slow implementation of reforms.

Profile

Billions of dollars flowed into the economy from the West and Gulf states after the 1990–1991 Gulf War. This cash injection, along with increased oil revenue, led to rapid growth. Diversion of water from the Euphrates toward fertile plains, rather than poorer land, led to a rise in agricultural output. However, long-term economic prospects remain uncertain. The public sector employs 20% of the workforce, and state controls inhibit private enterprise and investment, and have created a booming black market. Businessmen often channel funds through the freer Lebanese economy. An economic reform package in 2000 created a stock exchange and permitted private banks.

ECONOMIC PERFORMANCE INDICATOR

Consumer Price Index
GDP
160
140
120
100
80
Consumer price index 1995=100
GDP 1996=100
1996
1997
1998
1999
2000

SYRIA : MAJOR BUSINESSES

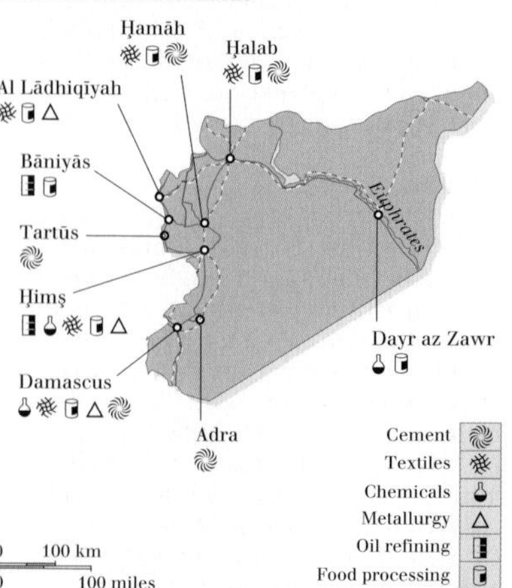
Ḩamāh
Ḩalab
Al Lādhiqīyah
Bāniyās
Tartūs
Ḩimş
Damascus
Adra
Dayr az Zawr
Euphrates
Cement
Textiles
Chemicals
Metallurgy
Oil refining
Food processing
0 100 km
0 100 miles

RESOURCES

Electric power 4.5m kw

14,024 tonnes

551,000 b/d (reserves 2.5bn barrels)

12.4m sheep, 979,325 goats, 21m chickens

Phosphates, oil, natural gas, iron

ELECTRICITY GENERATION

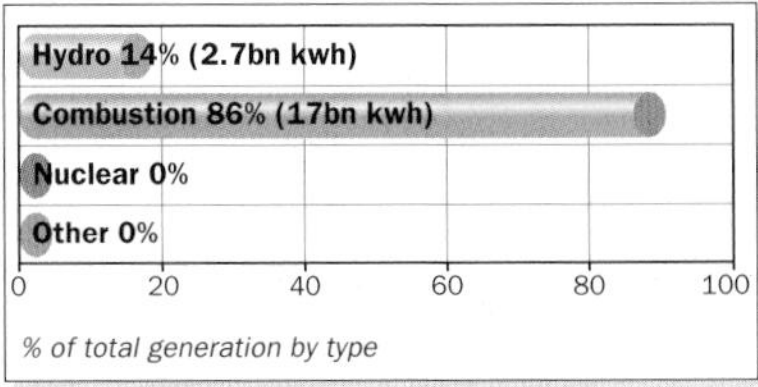
Hydro 14% (2.7bn kwh)
Combustion 86% (17bn kwh)
Nuclear 0%
Other 0%
0
20
40
60
80
100
% of total generation by type

Oil, mostly good-quality light crude, was discovered along the Euphrates in the 1980s. Gas was found in substantial quantities near Palmyra. Other important minerals are phosphates and iron ore. Hydroelectric power provides a small proportion of electricity. In 2001, a joint project was launched to link the electricity grids of Syria, Jordan, and Egypt. Manufacturing is largely in oil-derived industries, textiles, and foodstuffs. Cotton is the main cash crop; fruit and vegetables are also grown. Livestock, especially sheep and goats, support the rural economy.

SYRIA : LAND USE

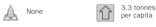

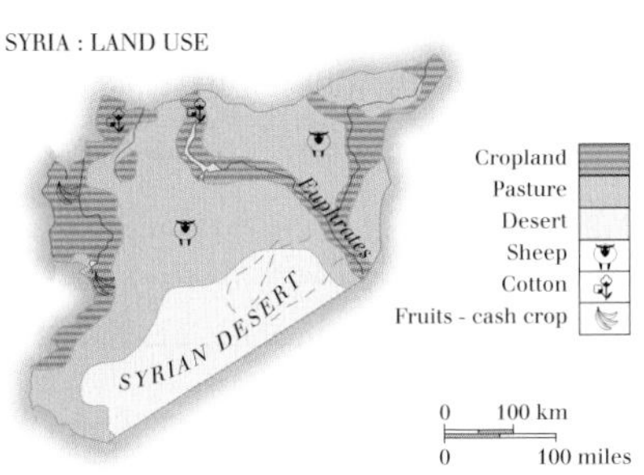
Euphrates
SYRIAN DESERT
Cropland
Pasture
Desert
Sheep
Cotton
Fruits - cash crop
0 100 km
0 100 miles

ENVIRONMENT

Sustainability rank: 107th

None

3.3 tonnes per capita

ENVIRONMENTAL TREATIES

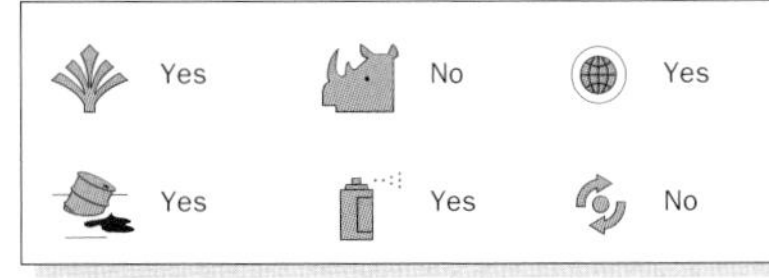
Yes
No
Yes
Yes
Yes
No

The Assad regime's most expensive and controversial environmental project has been the Euphrates Dam, power plant, and irrigation network at Tabaqah. The dam's vast man-made reservoir, Buhayratal al-Assad, engulfed some 300 villages and destroyed 25,000 hectares (62,000 acres) of fertile farmland. A giant cement factory, built by former East Germany at Tartus in the mid-1970s, has increased pollution along a stretch of Syria's Mediterranean coastline. However, its interior remains relatively unspoilt.

MEDIA

TV ownership medium

Daily newspaper circulation 20 per 1000 people

Information became freer after Jordanian papers were again allowed into Syria in 1999, and with the advent of satellite TV, widely watched. Bashar al-Assad encourages Internet use, and in 2001 paved the way for independent publications, including *Ad-domari*, the first satirical journal in 38 years.

PUBLISHING AND BROADCAST MEDIA

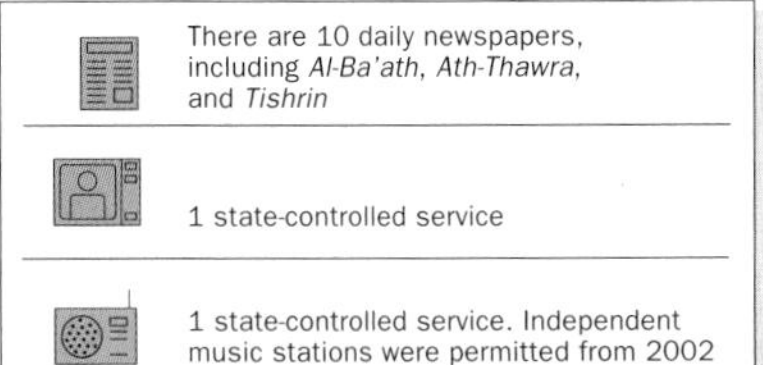

There are 10 daily newspapers, including *Al-Ba'ath, Ath-Thawra*, and *Tishrin*

1 state-controlled service

1 state-controlled service. Independent music stations were permitted from 2002

CRIME

Death penalty in use

14,000 prisoners

Up 164% 1995–1999

CRIME RATES

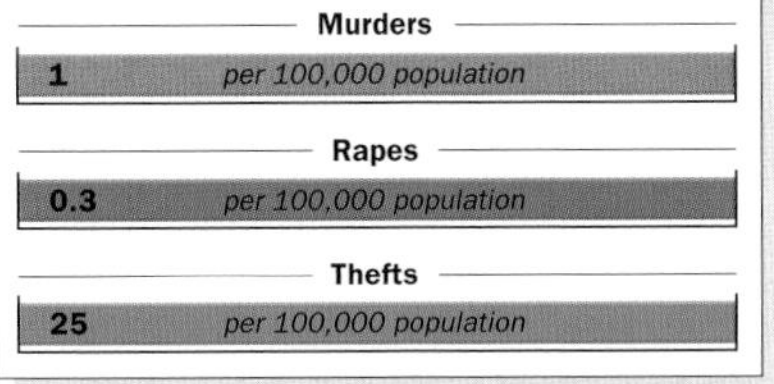
Murders
1 per 100,000 population
Rapes
0.3 per 100,000 population
Thefts
25 per 100,000 population

There is no truly independent judiciary. The powerful security services exercise arbitrary powers of arrest and detention. There are widespread reports of torture in custody. Most politicians removed by President Hafez al-Assad in the 1970s have been released from prison in Damascus.

EDUCATION

School leaving age: 12

74%

227,926 students

THE EDUCATION SYSTEM

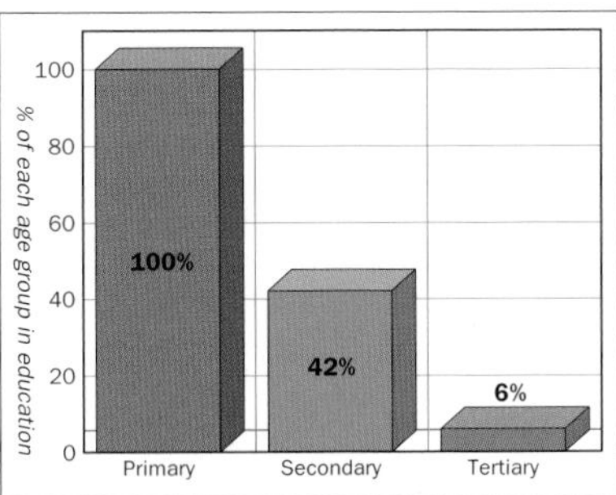
100
80
60
40
20
0
% of each age group in education
100%
42%
6%
Primary
Secondary
Tertiary

A modern and universally accessible system of education remains an important objective. Free and compulsory primary education for all was a priority of the Ba'ath Party when it came to power. Coeducation began in the cities and spread to rural areas under the Assad regime.There are seven state universities, notably at Damascus, Aleppo, Tishrin, and Homs. Private universities were allowed from 2001. Education ranks second, though is far behind defense, in government expenditure. UK and US politicians protest at incitement to racism in Syrian school textbooks.

HEALTH

Welfare state health benefits

1 per 769 people

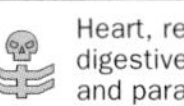

Heart, respiratory, digestive, infectious, and parasitic diseases

An adequate system of primary health care has been set up since the Ba'ath Party came to power. Treatment is free for those unable to pay. However, hospitals often lack modern equipment and medical services are in need of further investment. Rural areas in particular need assistance to combat the spread of heart, respiratory, and infectious diseases.

SPENDING

GDP/cap. increase

CONSUMPTION AND SPENDING

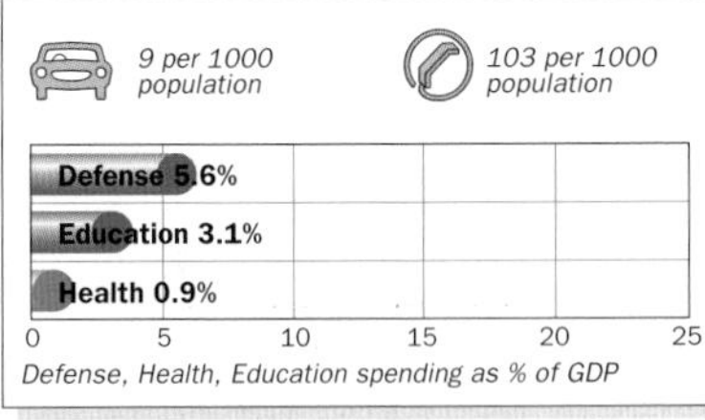
9 per 1000 population
103 per 1000 population
Defense 5.6%
Education 3.1%
Health 0.9%
0
5
10
15
20
25
Defense, Health, Education spending as % of GDP

Syria is far from the equitable society that early Ba'ath Party thinkers envisioned. The gulf between Syria's rich and poor is widening. The political elite, many of whom live in the West Malki suburb of Damascus, is more numerous and richer than ever before. Palestinian refugees and the urban unemployed make up the poorest groups.

WORLD RANKING

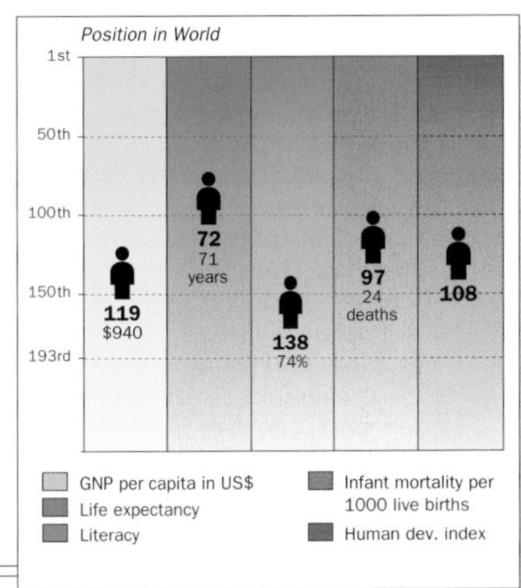
Position in World
1st
50th
100th
150th
193rd
119
$940
72
71 years
138
74%
97
24 deaths
108
GNP per capita in US$
Life expectancy
Literacy
Infant mortality per 1000 live births
Human dev. index

S

TAIWAN

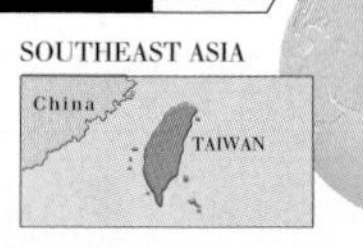
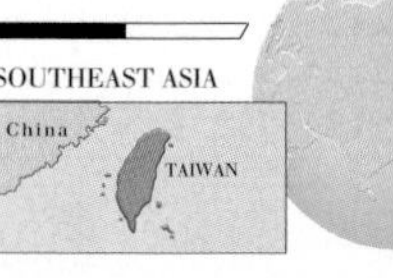
SOUTHEAST ASIA
Asia
China
TAIWAN

OFFICIAL NAME: Republic of China (ROC) **CAPITAL:** Taipei
POPULATION: 22.2 million **CURRENCY:** Taiwan dollar **OFFICIAL LANGUAGE:** Mandarin Chinese

1949

1949
Oct 10
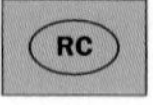
RC
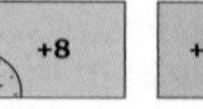
+8

+886

.tw

THE ISLAND OF TAIWAN, formerly known as Formosa, lies off the southeast coast of mainland China. Mountains running north to south cover two-thirds of the island. The lowlands are highly fertile, cultivated mostly with rice, and densely populated. In 1949, when the Chinese Communists ousted Chiang Kai-shek's nationalist Kuomintang (KMT) from power on the mainland, he established the Republic of China government on the island. De facto military rule has been democratized progressively since 1986. Mainland China still considers Taiwan a renegade province, and only a few countries now give official recognition to the regime there.

Wen Wu Temple, *on the shores of Sun Moon Lake in the mountains of central Taiwan – a region famous for its many temples. Nearly the whole population is Buddhist.*

CLIMATE

Tropical monsoon

WEATHER CHART FOR TAIPEI

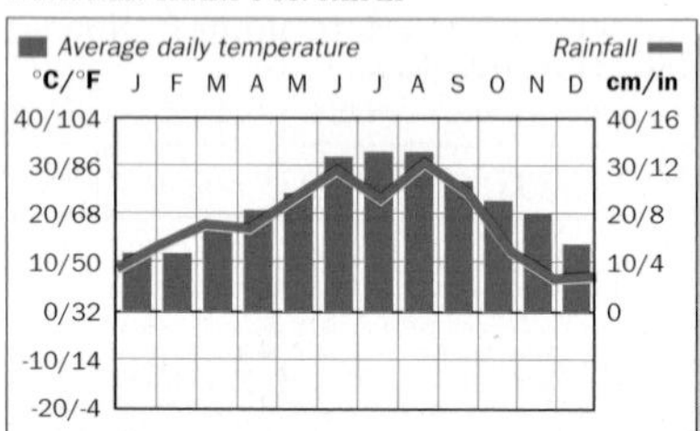
Average daily temperature
Rainfall
°C/°F J F M A M J J A S O N D cm/in
40/104 30/86 20/68 10/50 0/32 -10/14 -20/-4
40/16 30/12 20/8 10/4 0

Taiwan has a tropical monsoon climate similar to that of the southern Chinese mainland. Typhoons from the South China Sea between June and October bring the heaviest rains.

TRANSPORTATION

Drive on right

Chiang Kai-shek International, Taoyuan 18.7m passengers

680 ships
5.09m grt

THE TRANSPORTATION NETWORK

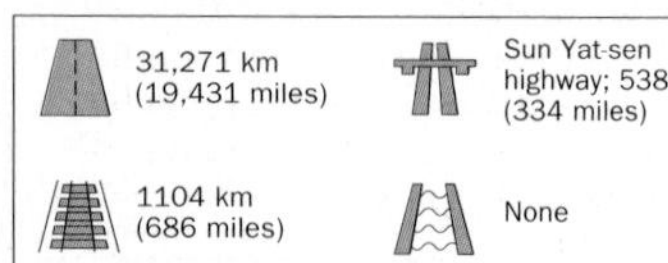
31,271 km (19,431 miles)
Sun Yat-sen highway; 538 km (334 miles)
1104 km (686 miles)
None

A railroad encircles the island. A US$16 billion high-speed rail link along the west coast, linking Taipei to Kaohsiung, is under construction with the help of the Japanese Shinkansen Corporation. It is due for completion in 2005. Car ownership is 148 times higher than the 1971 level. The bicycle is not as popular in Taiwan as in mainland China. However, Taiwan is the world's biggest bicycle producer, exporting mostly to Europe and the US.

Access to mainland China has increased as relations slowly improve. By boat, passengers have to travel from the tiny offshore islands of Matsu and Quemoy. The number of visitors is restricted both ways by yearly quotas.

TOURISM

Visitors : Population 1:8.5

2.62m visitors

Up 9% in 2000

MAIN TOURIST ARRIVALS

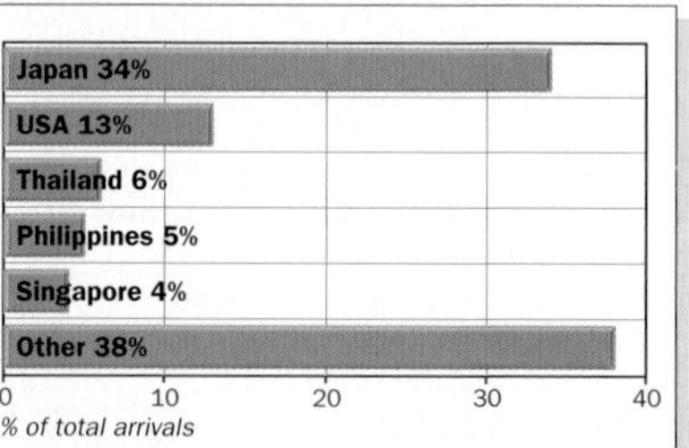
Japan 34%
USA 13%
Thailand 6%
Philippines 5%
Singapore 4%
Other 38%
0 10 20 30 40
% of total arrivals

Taiwan is not a major tourist destination, and it has only recently begun promoting itself in the US and Japan. Restrictions on Chinese tourists are being eased. Successive Six-Year Plans focus on upgrading hotels and improving tourist faciities at international airports. The major attraction is the Palace Museum in Taipei, which includes the massive treasure looted by the nationalists from Beijing. Only 5% can be shown at any one time. Sex tourism is an important business in Taipei. Sex establishments often masquerade as barbershops.

PEOPLE

Pop. density high

Amoy Chinese, Mandarin Chinese, Hakka Chinese

688/km² (1782/mi²)

THE URBAN/RURAL POPULATION SPLIT

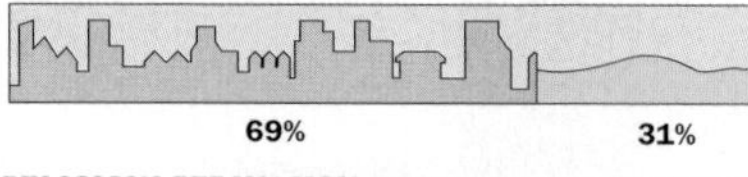
69%
31%

RELIGIOUS PERSUASION

Other 2%
Christian 5%
Buddhist, Confucian, Taoist 93%

ETHNIC MAKEUP

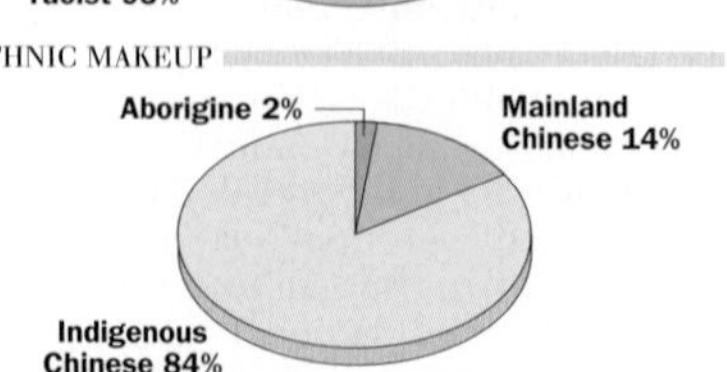
Aborigine 2%
Mainland Chinese 14%
Indigenous Chinese 84%

Most Taiwanese are Han Chinese, descendants of the 1644 migration of the Ming dynasty from mainland China. The 100,000 nationalists who arrived in 1949 established themselves as a ruling class and monopolized the most prestigious jobs in the civil service.

This caused resentment among the local inhabitants, but as the generation elected on the mainland in 1947 has aged, so local Taiwanese have entered the political process.

There is little ethnic tension in Taiwan, although the indigenous minorities who live in the eastern hills do suffer considerable discrimination.

As in the rest of southeast Asia, the extended family is still important and provides a social security net for the elderly. However, the trend is toward European-style nuclear families. Housing shortages are a major issue.

Women are not well represented in the political process, but are prominent in business and the civil service.

POPULATION AGE BREAKDOWN

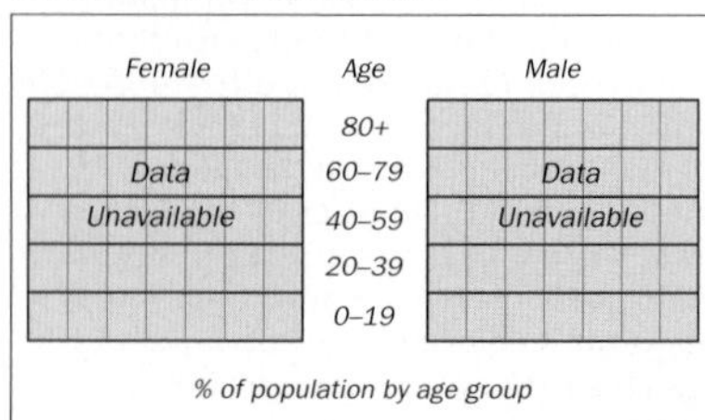
Female
Age
Male
80+
60–79
40–59
20–39
0–19
Data Unavailable
Data Unavailable
% of population by age group

T

POLITICS

Multiparty elections

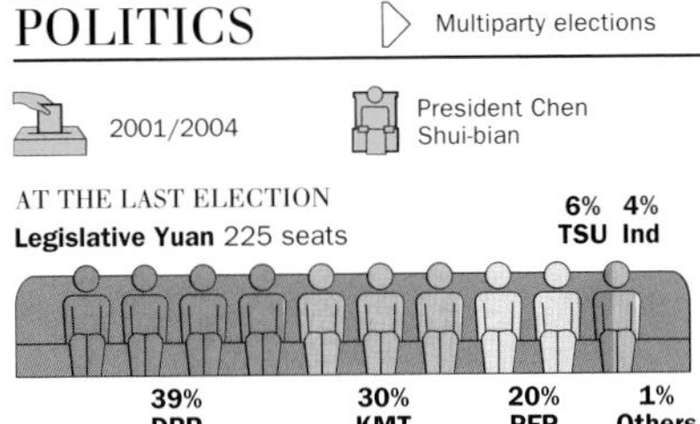
2001/2004
President Chen Shui-bian
AT THE LAST ELECTION
Legislative Yuan 225 seats
6% TSU
4% Ind
39% DPP
30% KMT
20% PFP
1% Others

DPP = Democratic Progressive Party **KMT** = National Party of China (Kuomintang) **PFP** = People First Party **TSU** = Taiwan Solidarity Union **Ind** = Independents

From 1986 Taiwan has been a multiparty democracy, though electoral domination by the KMT only ended in 2000. The government is headed by the president, who is answerable to the National Assembly, an additional body which is convened only when constitutional affairs arise.

Profile

For nearly four decades from the establishment of the Republic of China government on Taiwan in 1949, Chiang Kai-shek's KMT monopolized political power and ruled by strict martial law. In 1986, Gen. Chiang Ching-kuo, Chiang Kai-shek's son and successor, decided to open the way for more democracy. Later that year the first free multiparty elections were held. In March 1996 Lee Teng-hui became the country's first directly elected president.

The KMT retained power in the 1998 legislative elections, winning an absolute majority of seats. Its 50-year political monopoly came to an end with the presidential elections in March 2000 which saw the KMT candidate coming a poor third. Chen Shui-bian of the pro-independence DPP took office in May 2000. The KMT also lost support in the 2001 legislative elections. The DPP became the largest party in the Yuan but did not obtain a majority.

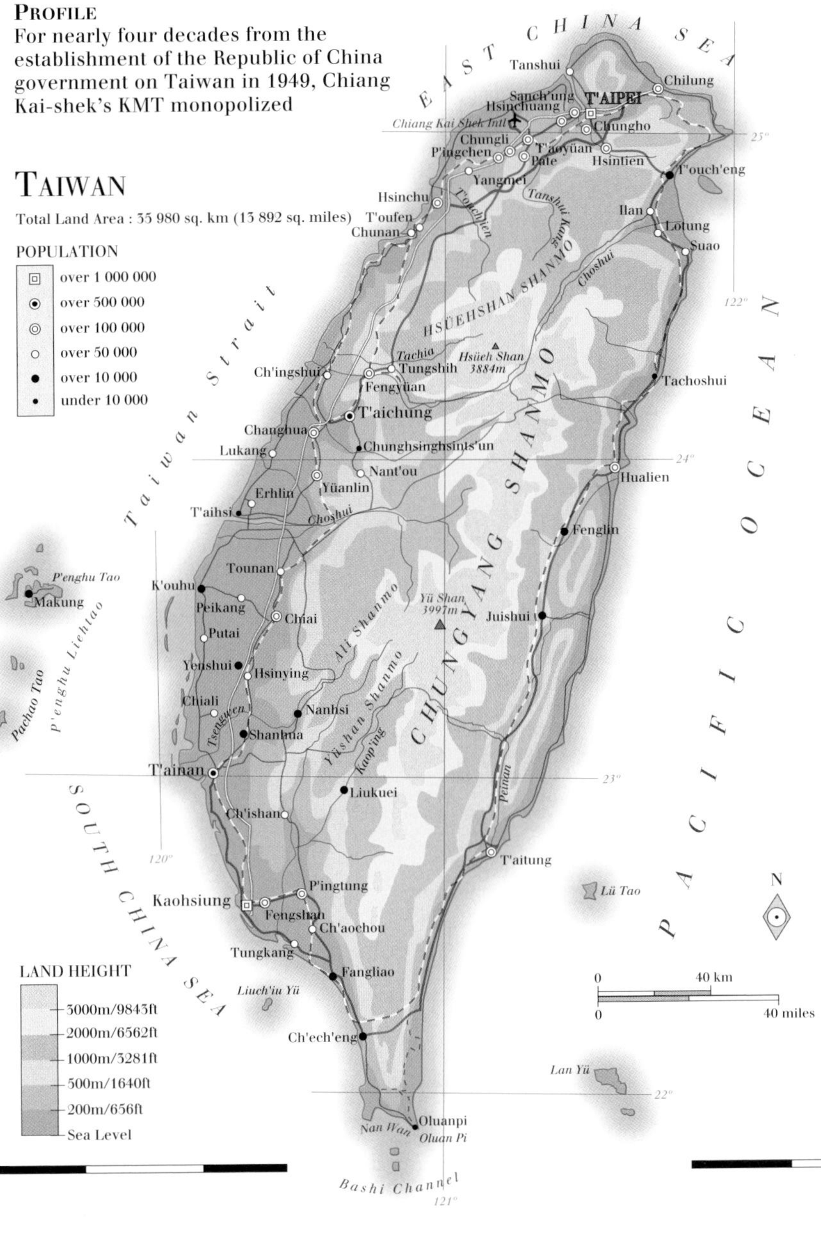

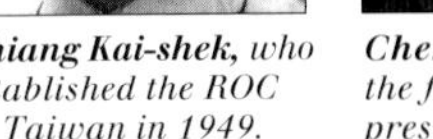

Chiang Kai-shek, who established the ROC in Taiwan in 1949.

Chen Shui-bian, the first non-KMT president since 1949.

Main Political Issues

Relations with China

The DPP has advocated independence from China, despite Chinese threats of military action. Adopting a more prudent stance since his poll victory, Chen Shui-bian promised to make no independence declaration during his term of office. The KMT, officially committed to eventual reunification with China, now favors a more flexible arrangement that presupposes the recognition of a separate Taiwanese national identity.

Political stability

The KMT, in power ever since the Chinese civil war brought Taiwan into existence, has recently been displaced by the DPP. This has meant an unprecedented degree of change within Taiwanese political life. The preservation of political and economic stability, however, remains essential as China intensifies its sovereignty claims on Taiwan after the return of Hong Kong in 1997.

TAIWAN

Total Land Area : 35 980 sq. km (13 892 sq. miles)

POPULATION
- over 1 000 000
- over 500 000
- over 100 000
- over 50 000
- over 10 000
- under 10 000

LAND HEIGHT
- 3000m/9843ft
- 2000m/6562ft
- 1000m/3281ft
- 500m/1640ft
- 200m/656ft
- Sea Level

0 40 km
0 40 miles

EAST CHINA SEA
Taiwan Strait
SOUTH CHINA SEA
PACIFIC OCEAN
Bashi Channel
T'AIPEI
Tanshui
Chilung
Sanch'ung
Hsinchuang
Chiang Kai Shek Intl
Chungho
Chungli
T'aoyüan
P'ingchen
Pate
Hsintien
Yangmei
T'ouch'eng
Hsinchu
Ilan
T'oufen
Chunan
Lotung
Suao
HSÜEHSHAN SHANMO
Ch'ingshui
Tachia
Tungshih
Hsüeh Shan 3884m
Fengyüan
T'aichung
Tachoshui
Changhua
Chunghsinghsints'un
Lukang
Nant'ou
Hualien
Erhlin
Yüanlin
T'aihsi
Choshui
Fenglin
P'enghu Tao
Makung
Tounan
K'ouhu
Peikang
Yü Shan 3997m
Chiai
Juishui
Putai
Yenshui
Hsinying
Chiali
Nanhsi
Shanhua
P'enghu Liehtao
Pachao Tao
T'ainan
Liukuei
Ch'ishan
T'aitung
Kaohsiung
P'ingtung
Lü Tao
Fengshan
Ch'aochou
Tungkang
Fangliao
Liuch'iu Yü
Ch'ech'eng
Lan Yü
Oluanpi
Nan Wan
Oluan Pi
CHUNGYANG SHANMO
Ali Shanmo
Yüshan Shanmo

WORLD AFFAIRS

Not a UN member

APEC ADB WTO

China rejects Taiwan's sovereignty claims, regarding it as a renegade province, and Taiwan conducts overseas relations via trade delegations. It lost its seat on the UN Security Council in 1971, and has since been unable to rejoin the UN. The handover to China of Hong Kong in 1997 deepened fears that China would intensify pressure on countries to sever links with Taiwan. China mounted military exercises in 1999 after President Lee spoke of Taiwan as a separate state. China sought unsuccessfully in 2000 to frighten Taiwanese voters away from backing pro-independence candidate Chen Shui-bian. However, each depends on the other economically; Taiwan is a major foreign investor in China and has recently permitted direct trade. After US rapprochement with China in 1972, Taiwan ceased effectively to be a US client state. US security guarantees are vague, but in practice there are strong bilateral ties with the US and Japan.

AID

Donor

 US$500m (donations)

 Up in 2001

Taiwan has a large aid fund devoted to states which have granted it diplomatic recognition. These include the Pacific states of Kiribati and Tonga, which have looked out for Taiwan's interests in the UN since 1971, when it lost its seat to the People's Republic of China. In 1998 Taiwan donated more than US$2 million to seven Central American supporters, including Panama, to promote literacy.

DEFENSE

Compulsory military service

 US$17.2bn

 Up 15% in 2000

China remains the main defense threat, given the recurring tensions over the issue of independence or reunification. Taiwan has the fifth-largest navy in the world and a sizable army, in order to face a possible Chinese invasion: there are over 1.5 million reservists, and military service lasts for two years. Worries about US loyalty have resulted in the purchase of French *Mirage* fighters in addition to US F-16s.

TAIWANESE ARMED FORCES

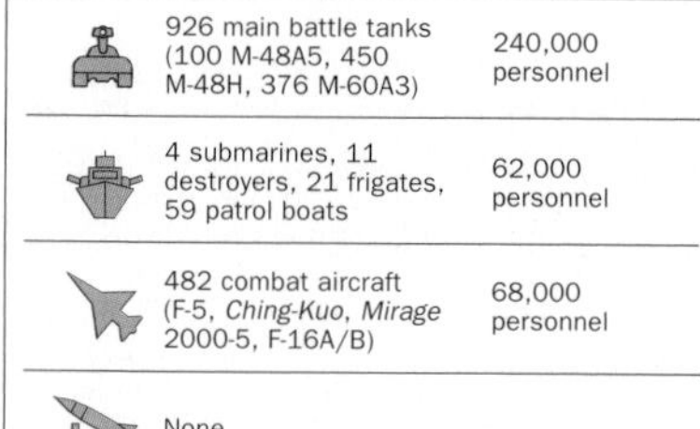

926 main battle tanks (100 M-48A5, 450 M-48H, 376 M-60A3)	240,000 personnel
4 submarines, 11 destroyers, 21 frigates, 59 patrol boats	62,000 personnel
482 combat aircraft (F-5, *Ching-Kuo*, *Mirage* 2000-5, F-16A/B)	68,000 personnel
None	

CHRONOLOGY

Following the 1949 communist revolution in China, Gen. Chiang Kai-shek's nationalist KMT party sought refuge in the island province of Taiwan. The KMT saw the revolution as illegal and itself as the sole rightful Chinese government.

- **1971** People's Republic of China replaces Taiwan at UN, including on UN Security Council.
- **1973** Taipei's KMT regime rejects Beijing's offer of secret talks on reunification of China.
- **1975** President Chiang Kai-shek dies. His son Gen. Chiang Ching-kuo becomes KMT leader. Yen Chia-kan succeeds as president.
- **1978** Chiang Ching-kuo elected president.
- **1979** US severs relations with Taiwan and formally recognizes People's Republic of China.
- **1984** President Chiang reelected.
- **1986** Political reforms: KMT allows multiparty democracy, ends martial law, and permits visits to Chinese mainland for "humanitarian" purposes for first time in 38 years. From 1988, mainland Chinese are allowed to visit Taiwan on same basis.
- **1988** Lee Teng-hui president.
- **1990** KMT formally ends state of war with People's Republic of China.
- **1991** DPP draft constitution for Taiwan independence opposed by ruling KMT and Beijing. KMT reelected with large majority.
- **1995–1996** Legislative elections. KMT majority reduced.
- **1996** Lee Teng-hui wins first direct presidential elections.
- **1998** KMT secures absolute majority in elections to Legislative Yuan.
- **1999** Chinese threats over reference to "separate states" status. Thousands die in earthquake.
- **2000** Chen Shui-bian of DPP wins presidency; ends KMT dominance.
- **2001** Elections: DPP largest single party. Taiwan admitted to WTO.

ECONOMICS

Inflation 5% p.a. (1985–1996)

 US$283bn

 33.08–34.99 Taiwan dollars

SCORE CARD

World GNP Ranking	16th
GNP per Capita	US$13,450
Balance of Payments	US$9.32bn
Inflation	1.3%
Unemployment	3%

EXPORTS

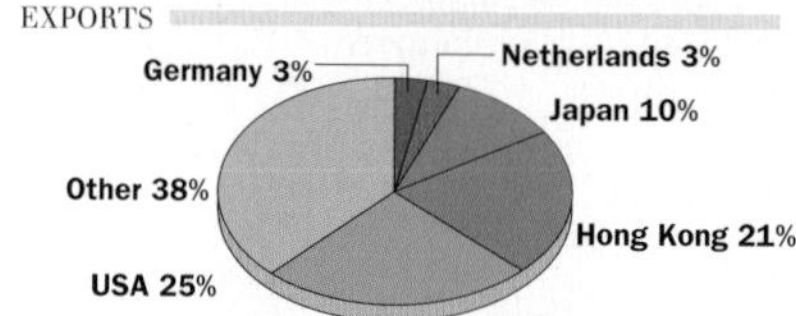
Germany 3%
Netherlands 3%
Japan 10%
Other 38%
Hong Kong 21%
USA 25%

IMPORTS

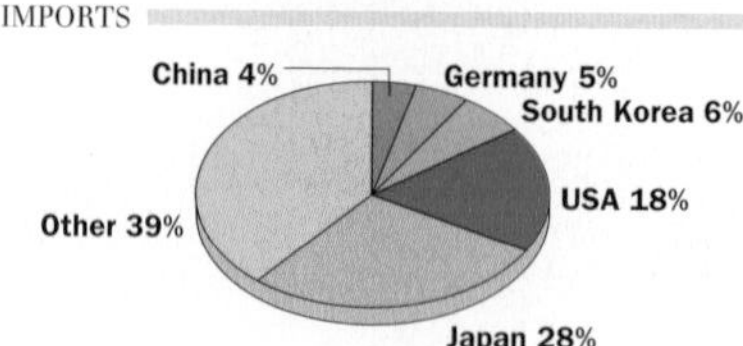
China 4%
Germany 5%
South Korea 6%
USA 18%
Other 39%
Japan 28%

Strengths

Highly educated and ambitious workforce, many US-trained and educated, with an inside knowledge of the US market. Manufacturing economy based on small companies which have proved extremely adaptable to changing market conditions. Track record of capturing major markets. Successively the world's biggest TV, watch, PC, and track shoe manufacturer. Economy in strong surplus, allowing it to invest in other southeast Asian economies.

Weaknesses

Small economic units lack the muscle of Western multinationals, and are unable to follow predatory pricing policies. Weak research and development: economy has no tradition of generating new products or creating new markets. Unresponsive banking system.

Profile

Taiwan's economy has proved resilient. Double-digit growth ended with the Asian financial crisis of 1997–1998, but

ECONOMIC PERFORMANCE INDICATOR

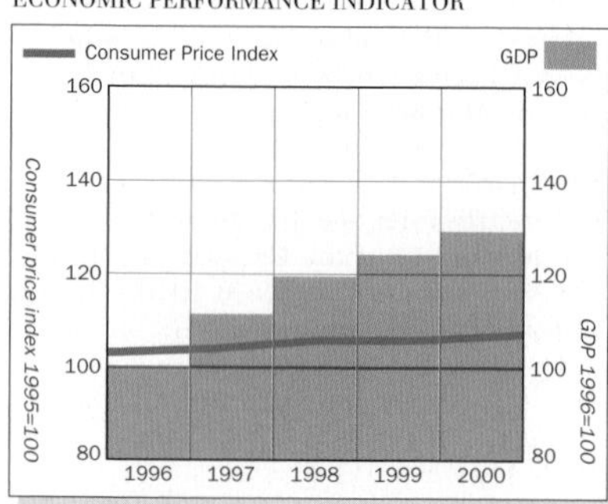
Consumer Price Index
GDP
160
140
120
100
80
Consumer price index 1995=100
GDP 1996=100
1996
1997
1998
1999
2000

Taiwan emerged relatively unscathed, and global downturn caused only a brief recession in 2001. Competition from underdeveloped countries with low production costs is dictating a difficult transition toward service industries. This will entail moving from labor-intensive to capital- and technology-intensive industries. Comprehensive Six-Year Plans reflect a strong element of state direction. Heavy investment abroad includes over 60% of inward investment into China since 1990. Taiwan was admitted, along with China, to the WTO in 2001.

TAIWAN : MAJOR BUSINESSES

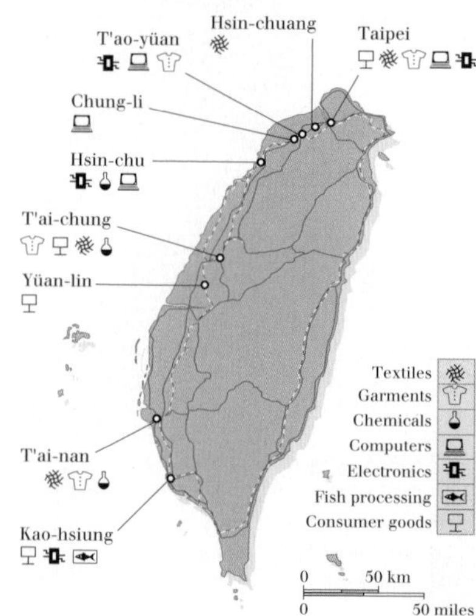
T'ao-yüan
Hsin-chuang
Taipei
Chung-li
Hsin-chu
T'ai-chung
Yüan-lin
T'ai-nan
Kao-hsiung
Textiles
Garments
Chemicals
Computers
Electronics
Fish processing
Consumer goods
0
50 km
0
50 miles

T

RESOURCES

Not available

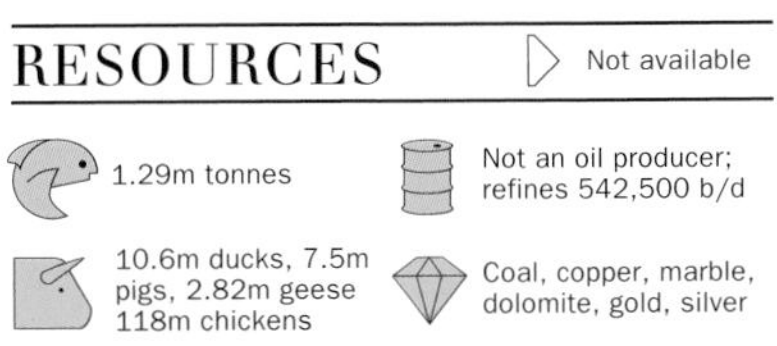
1.29m tonnes
Not an oil producer; refines 542,500 b/d
10.6m ducks, 7.5m pigs, 2.82m geese 118m chickens
Coal, copper, marble, dolomite, gold, silver

ELECTRICITY GENERATION

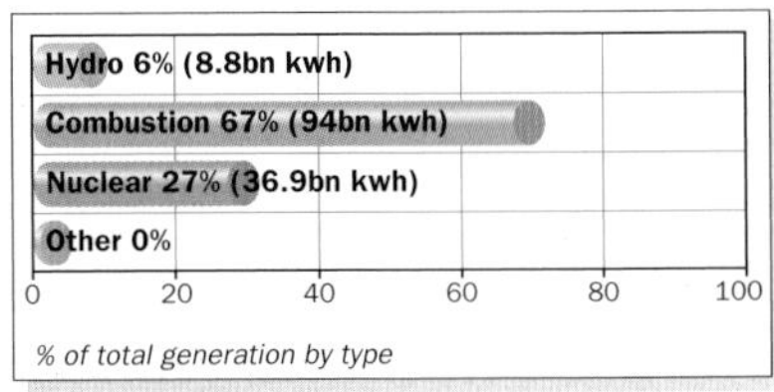
Hydro 6% (8.8bn kwh)
Combustion 67% (94bn kwh)
Nuclear 27% (36.9bn kwh)
Other 0%
0
20
40
60
80
100
% of total generation by type

Taiwan has few strategic resources and its minerals industry is not a major foreign exchange earner. Oil is imported. Taiwan is a major buyer of South African uranium, but proposals to increase reliance on nuclear power have met strong opposition on safety and waste disposal grounds. However, hydroelectric power has been largely exploited already and combustion is also a controversial option.

Fishing is highly successful, and Taiwan is a major supplier to the huge Japanese market. The fleet is often accused of plundering Atlantic stocks.

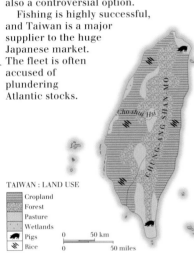

Cho-shui Hsi
CHUNG-YANG SHAN MO

TAIWAN : LAND USE
Cropland
Forest
Pasture
Wetlands
Pigs
Rice
0 50 km
0 50 miles

ENVIRONMENT

Not available

8% (3% partially protected)
8.1 tonnes per capita

ENVIRONMENTAL TREATIES

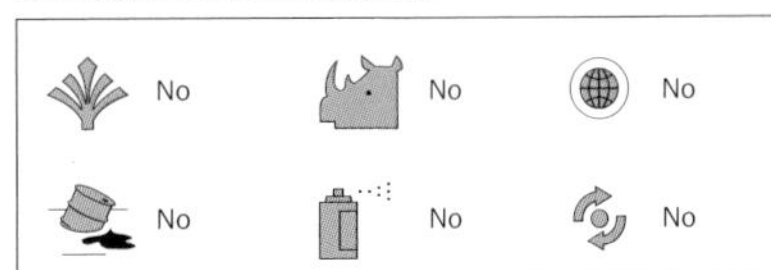
No
No
No
No
No
No

The dash for growth meant the absence of city planning or pollution laws. There is growing opposition to a fourth nuclear power plant, set for completion in 2006, and concern over coal-fired thermal power. Taiwan's fishing industry has been criticized for using longline techniques which trap dolphins, and for plundering other countries' fishing grounds without regard to stock levels.

MEDIA

TV ownership medium

Daily newspaper circulation figures not available

PUBLISHING AND BROADCAST MEDIA

There are 35 daily newspapers. The independent *Lienho Pao* has the largest circulation

1 public-service station, 64 cable companies, and 129 satellite broadcasting channels

110 independent corporations

The rigid state control which used to exist over the media has been relaxed. Opposition parties now have access to the state media. Before the 1990s, press with simplified Chinese characters was banned, thus excluding all publications from the mainland. Taiwan has a large domestic TV and film industry.

CRIME

Death penalty in use

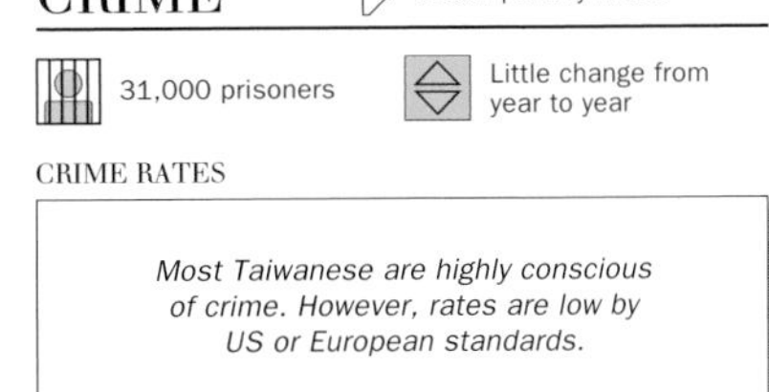
31,000 prisoners
Little change from year to year
CRIME RATES
Most Taiwanese are highly conscious of crime. However, rates are low by US or European standards.

Since the end of martial law in 1986, most political prisoners have been released. Taiwan does not suffer from organized crime to the extent found in Hong Kong or Japan. Multimedia pirating is a major problem.

EDUCATION

School leaving age: 15

94%

1.09m students

THE EDUCATION SYSTEM

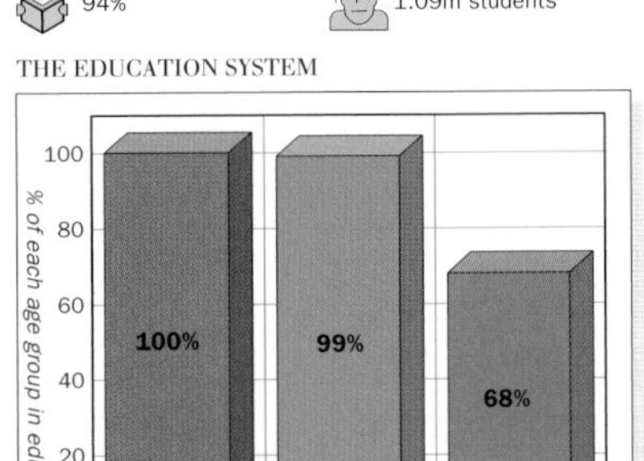
100
80
60
40
20
0
% of each age group in education
100%
99%
68%
Primary
Secondary
Tertiary

Since 2000, government funding has given less priority to higher education and more to reforming and improving the antiquated school system, which is rigid and heavily exam-oriented. Free schooling is available from the age of six to 15 and there are also a number of private schools.

Enrollment levels at tertiary and vocational institutions are among the highest in the world.

HEALTH

No welfare state health benefits

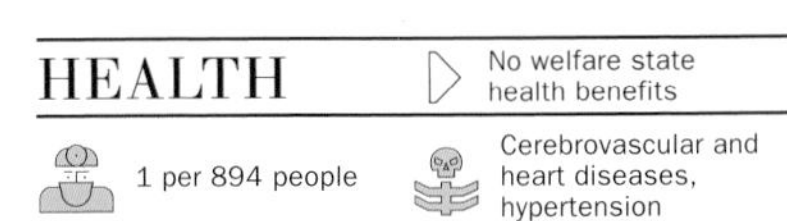
1 per 894 people
Cerebrovascular and heart diseases, hypertension

Most health provision in Taiwan is in the private sector. Taiwanese take out elaborate health insurance schemes and it is essential to prove cover before treatment is provided. Health facilities are on a par with the best in the world, and the Taiwanese enjoy a high life expectancy, similar to that in the US. The incidence of AIDS is low, but the infection rate is increasing. An enteroviral epidemic swept Taiwan in 1998, killing scores of babies and affecting thousands of young children.

SPENDING

GDP/cap. increase

CONSUMPTION AND SPENDING

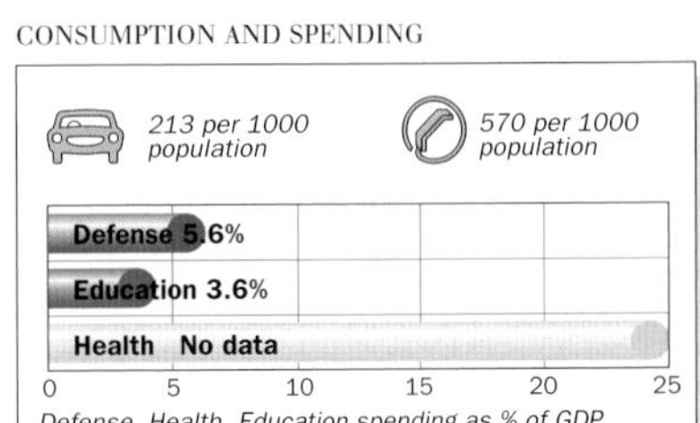
213 per 1000 population
570 per 1000 population
Defense 5.6%
Education 3.6%
Health No data
0
5
10
15
20
25
Defense, Health, Education spending as % of GDP

Export-led growth over a long period enabled Taiwan to build up large cash reserves. The Taiwanese people share much of the benefit of this success, and average living standards are among the highest in Asia. Inequalities of income distribution are comparatively small, and a high degree of social cohesion has been achieved. In part, this is the result of the land reforms of the 1950s, which gave agricultural workers control of the land while compensating landowners and encouraging them to set up business in the cities. Today, the great majority of Taiwanese would describe themselves as middle class. Consumer goods are widely available, and conspicuous consumption is celebrated.

WORLD RANKING

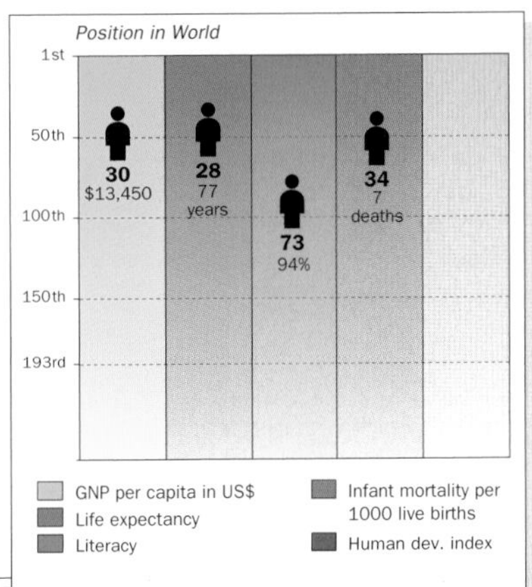
Position in World
1st
50th
100th
150th
193rd
30
$13,450
28
77 years
73
94%
34
7 deaths
GNP per capita in US$
Life expectancy
Literacy
Infant mortality per 1000 live births
Human dev. index

T

TAJIKISTAN

CENTRAL ASIA

OFFICIAL NAME: Republic of Tajikistan **CAPITAL:** Dushanbe
POPULATION: 6.1 million **CURRENCY:** Somoni **OFFICIAL LANGUAGE:** Tajik

1991

1991
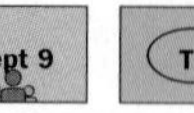
Sept 9

TJ

+5

+992

.tj

TAJIKISTAN LIES ON the western slopes of the Pamirs in central Asia. Language and traditions are similar to those of Iran rather than those of its northern Turkic neighbors. Tajikistan decided on independence only when neighboring Soviet republics declared theirs in late 1991. Fighting between communist government forces and Islamist rebels, which erupted shortly afterward, has been contained since 1997 by a fragile peace agreement.

A herd of goats in the Varzob Gorge, north of Dushanbe. Livestock drives the rural economy.

CLIMATE

Mountain

WEATHER CHART FOR DUSHANBE

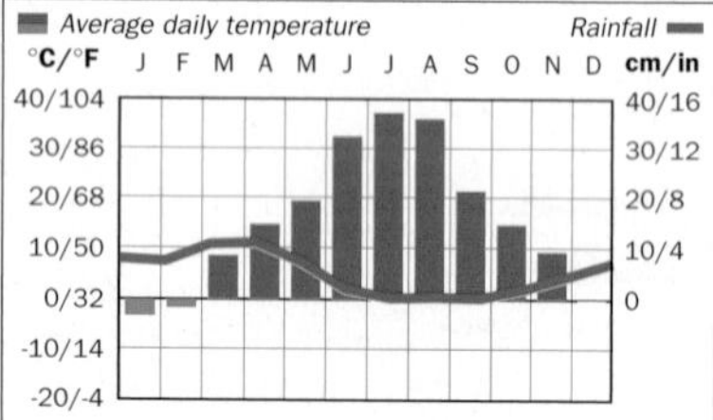
Average daily temperature
Rainfall
°C/°F
J F M A M J J A S O N D
cm/in
40/104
30/86
20/68
10/50
0/32
-10/14
-20/4
40/16
30/12
20/8
10/4
0

Rainfall is low in the valleys. Winter temperatures can fall below –45°C (–49°F) in mountainous areas.

TRANSPORTATION

Drive on right

Dushanbe International
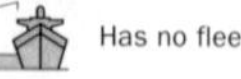
Has no fleet

THE TRANSPORTATION NETWORK

11,330 km (7040 miles)

None

547 km (340 miles)

200 km (124 miles)

Tajikistan has good cross-border roads and well-maintained airfields, the result of its use by Soviet and US forces during conflicts in neighboring Afghanistan. The best way to reach the mountainous interior is by air.

TOURISM

Visitors : Population 1:12

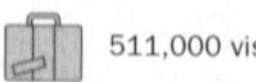
511,000 visitors
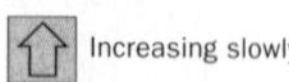
Increasing slowly

MAIN TOURIST ARRIVALS

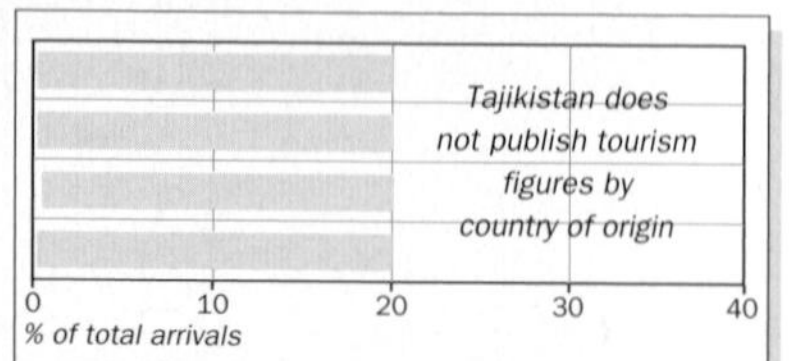
Tajikistan does not publish tourism figures by country of origin
0
10
20
30
40
% of total arrivals

Tourism is virtually nonexistent, and insecurity and poverty make it unlikely that it will be developed soon.

PEOPLE

Pop. density low

Tajik, Russian

43/km² (110/mi²)

THE URBAN/RURAL POPULATION SPLIT

28% 72%

RELIGIOUS PERSUASION

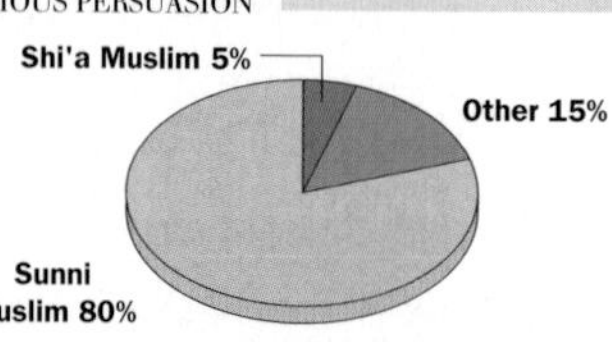
Shi'a Muslim 5%
Other 15%
Sunni Muslim 80%

Unlike the other former Soviet -stans, Tajikistan is dominated by a people of Persian (Iranian), rather than Turkic, origin. Ethnic Tajiks make up around 65% of the population. The main ethnic conflict is with the Turkic Uzbeks. As in neighboring Uzbekistan, Russians are discriminated against, and their numbers have thinned from 400,000 in 1989 to fewer than 200,000. By 1990, the 35,000-strong German minority had left. The struggle between Dushanbe-based communists and the Islamist militants in the central and eastern regions displaced more than 50,000 refugees into Afghanistan.

POLITICS

Multiparty elections

L. House 2000/2005
U. House 2000/2005

President Imomali Rakhmanov

AT THE LAST ELECTION

Assembly of Representatives 63 seats

3% IRP, 3% Others

71% PDPT, 21% CPT, 2% Vac

PDPT = People's Democratic Party of Tajikistan and allies **CPT** = Communist Party of Tajikistan **IRP** = Islamic Revival Party **Vac** = Vacant

National Assembly 33 seats

Five deputies are elected by each regional assembly, and a further eight deputies are appointed by the president

The lull in fighting between government forces and Islamist rebels, aided by a 1997 peace accord, has consolidated the regime of former communists led by President Rakhmanov. In 1998, the Islamist United Tajik Opposition (UTO) joined the government in conformity with the accord, which provided for a National Reconciliation Commission along with parliamentary elections. In the 2000 elections the pro-Rakhmanov PDPT, which headed the poll, claimed some support from former UTO members.

TAJIKISTAN

Total Land Area : 143 100 sq. km (55 251 sq. miles)

POPULATION

- over 500 000
- over 100 000
- over 50 000
- over 10 000
- under 10 000

LAND HEIGHT

- 4000m/13 124ft
- 3000m/9843ft
- 2000m/6562ft
- 1000m/3281ft
- 500m/1640ft
- 200m/656ft

0 100 km
0 100 miles

WORLD AFFAIRS

Joined UN in 1992

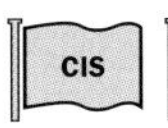

CIS SCO EAPC OIC OSCE

Tajikistan remains heavily dependent on Russia, particularly for economic and military assistance. The introduction of the somoni in 1995 enabled the government to wrest economic control back from Russia. A joint operation with Uzbekistan and Kyrgyzstan was launched in 2000 to combat the Islamic Movement of Uzbekistan, based in northern Tajikistan. Tajik airfields were crucial during the US-led action in Afghanistan in 2001, and in 2002 Tajikistan became the last of the former Soviet Union countries to join NATO's Partnerships for Peace program.

AID

Recipient

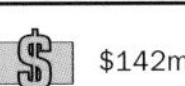
$142m (receipts)

Up 15% in 2000

Russia and Uzbekistan provide military aid. Food aid was required in 2000 and 2001 after severe droughts.

DEFENSE

Compulsory military service

$80m

Down 13% in 2000

Security along the winding southern border with Afghanistan is bolstered by the presence of some 20,000 international (mostly Russian) peacekeepers. The Tajik army numbers only around 6000. There is no navy or air force.

ECONOMICS

Inflation 235% p.a. (1990–2000)

$1.11bn

2.40–2.55 somoni

SCORE CARD

❑ World GNP Ranking	151st
❑ GNP per Capita	$180
❑ Balance of Payments	–$61m
❑ Inflation	33%
❑ Unemployment	6%

Strengths

Few, though Tajikistan has 14% of known world uranium reserves. Hydroelectric power has considerable potential. Carpet-making.

Weaknesses

Formal economy precarious. Dependence on barter economy. No central planning. Little diversification in agriculture; only 6% of land is arable. Exodus of skilled Russians. Production in all sectors in decline.

EXPORTS

IMPORTS

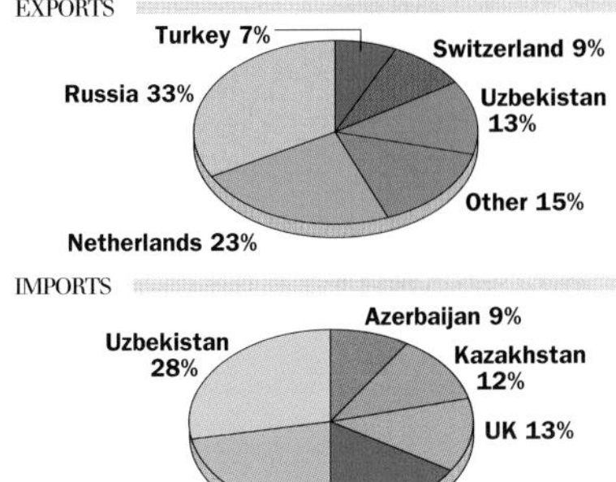
Turkey 7%
Switzerland 9%
Russia 33%
Uzbekistan 13%
Other 15%
Netherlands 23%
Azerbaijan 9%
Uzbekistan 28%
Kazakhstan 12%
UK 13%
Other 22%
Russia 16%

RESOURCES

Electric power 4.4m kw

80 tonnes

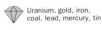

381 b/d (reserves 15m barrels)

1.36m sheep, 1.05m cattle, 900,000 chickens

Uranium, gold, iron, coal, lead, mercury, tin

Tajikistan has one key resource – uranium – which accounted for 30% of the USSR's total production before 1990. The end of the nuclear arms race has reduced its value, however. Most of Tajikistan is bare mountain, and just 6% of the land can be used for agriculture. Industry is concentrated in the Fergana Valley, close to the Uzbek border.

ENVIRONMENT

Sustainability rank: 110th

4%

0.8 tonnes per capita

Landslides, particularly on the lower slopes of the Pamirs, are a serious problem. They are caused as much by the natural geography and by earthquakes as by human activity.

MEDIA

TV ownership high

Daily newspaper circulation 21 per 1000 people

PUBLISHING AND BROADCAST MEDIA

There are 3 daily newspapers, but the weeklies *Djavononi Todjikiston* and *Tochikiston ovozi* are more influential

3 state-controlled services

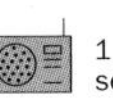
1 state-controlled service

Communist control over the media was tightened in early 1994 with the takeover by President Rakhmanov of the press and broadcast media.

CRIME

Death penalty in use

11,000 prisoners

Crime has been rising dramatically

Only remote areas escape the violence perpetrated by armed gangs. Narcotics smuggling along the border with Afghanistan continues to increase.

EDUCATION

School leaving age: 17

99%

108,203 students

Literacy rates are high. The university at Dushanbe has been weakened by the departure of its Russian academics.

CHRONOLOGY

In the 19th century, Tajikistan was a collection of semi-independent principalities, some under Russian control, others under the influence of the Emirate of Bukhara.

- ❑ **1925** Soviets take over Tajikistan.
- ❑ **1940** Cyrillic script introduced.
- ❑ **1989** Tajik becomes official language.
- ❑ **1991** Independence from Moscow.
- ❑ **1994, 1999** Imomali Rakhmanov reelected president.
- ❑ **1995** Legislative elections. Somoni introduced.
- ❑ **1997–1998** Peace accord with rebels; UTO joins government.
- ❑ **2000** Pro-Rakhmanov PDPT wins legislative elections.

HEALTH

Welfare state health benefits

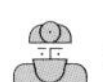
1 per 500 people

Heart, cerebrovascular, respiratory, infectious, and parasitic diseases

Theoretically health provision is free of charge; however, massive underfunding means that patients invariably have to pay to get treatement.

SPENDING

GDP/cap. decrease

CONSUMPTION AND SPENDING

1 per 1000 population

36 per 1000 population

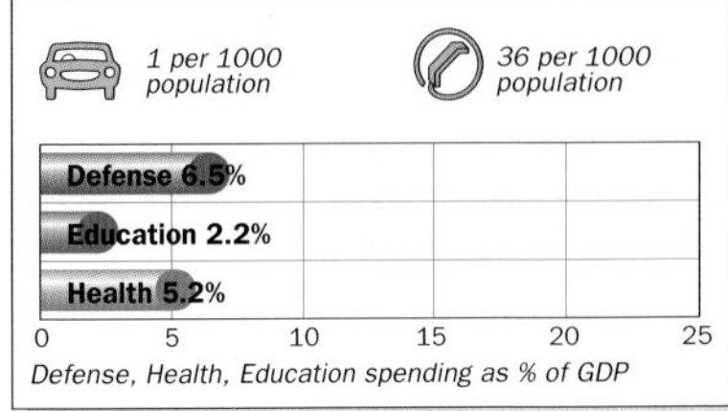
Defense 6.5%
Education 2.2%
Health 5.2%
0
5
10
15
20
25

Defense, Health, Education spending as % of GDP

More than 80% of Tajik people live below the poverty line; the war against the Islamist rebels worsened conditions. The former communist bureaucrats continue to be the wealthiest group.

WORLD RANKING

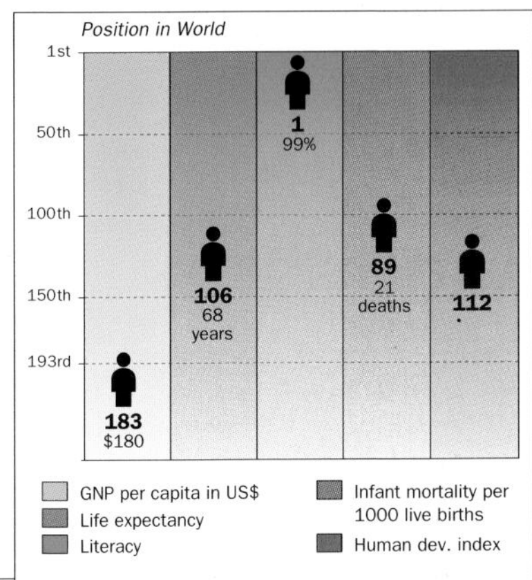
Position in World
1st
50th
100th
150th
193rd
183
$180
106
68
years
1
99%
89
21
deaths
112
GNP per capita in US$
Life expectancy
Literacy
Infant mortality per 1000 live births
Human dev. index

T

TANZANIA

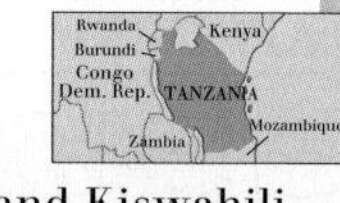
EAST AFRICA
Rwanda
Burundi
Congo
Dem. Rep.
Kenya
TANZANIA
Zambia
Mozambique

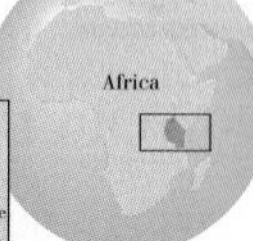
Africa

OFFICIAL NAME: United Republic of Tanzania **CAPITAL:** Dodoma
POPULATION: 36 million **CURRENCY:** Tanzanian shilling **OFFICIAL LANGUAGES:** English and Kiswahili

1964

1964

April 26

EAT

+3

+255

.tz

TANZANIA LIES BETWEEN Kenya and Mozambique on the east African coast. Formed by the union of Tanganyika and the Zanzibar islands, Tanzania comprises a coastal lowland, volcanic highlands, and the Great Rift Valley. It includes Mount Kilimanjaro, Africa's highest peak. Tanzania was led by the socialist Julius Nyerere from 1962 until 1985. His Revolutionary Party of Tanzania (CCM) has won multiparty elections held in 1995 and 2000.

Arusha National Park. *Lying within the Ngurdoto volcanic crater, the park has herds of buffalo, rhino, elephant, and giraffe.*

CLIMATE

Tropical/mountain

WEATHER CHART FOR DODOMA

Average daily temperature — Rainfall
°C/°F J F M A M J J A S O N D cm/in
40/104 — 40/16
30/86 — 30/12
20/68 — 20/8
10/50 — 10/4
0/32 — 0
-10/14
-20/-4

The coast and Zanzibar are tropical. The central plateau is semiarid and the highlands are semitemperate.

TRANSPORTATION

Drive on left

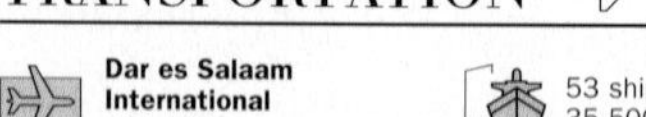
Dar es Salaam International
621,513 passengers
53 ships
35,500 grt

THE TRANSPORTATION NETWORK

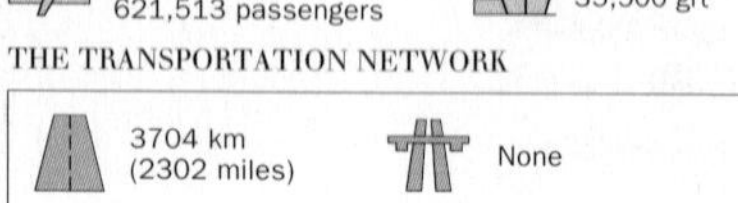
3704 km
(2302 miles)
None

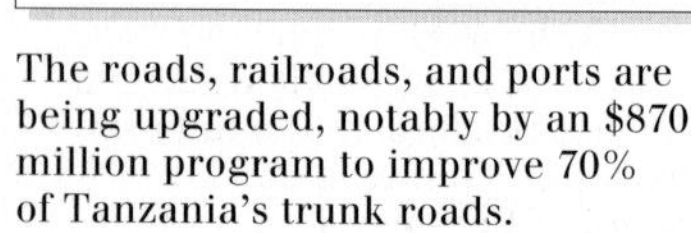
2721 km
(1691 miles)
Lakes Tanganyika, Victoria, and Nyasa are navigable

The roads, railroads, and ports are being upgraded, notably by an $870 million program to improve 70% of Tanzania's trunk roads.

TOURISM

Visitors : Population 1:78

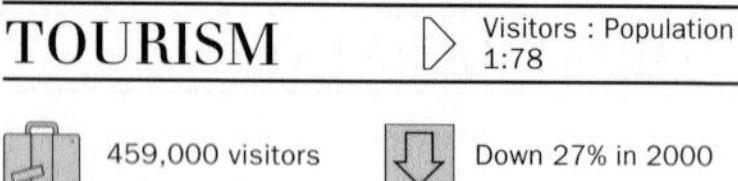
459,000 visitors
Down 27% in 2000

MAIN TOURIST ARRIVALS

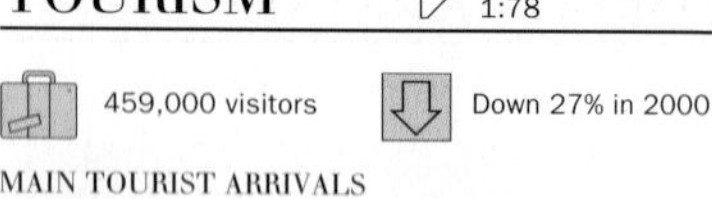

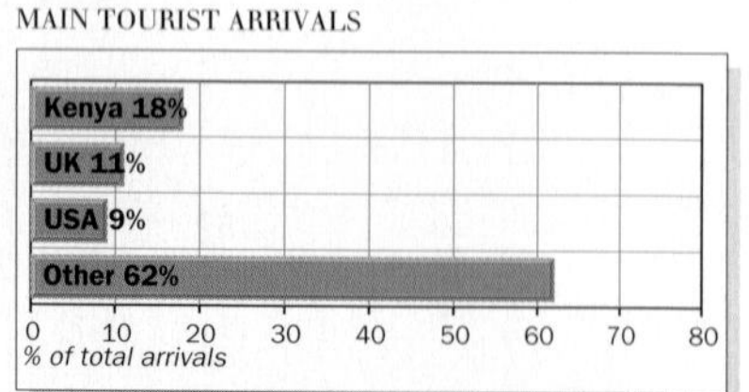
Kenya 18%
UK 11%
USA 9%
Other 62%
0 10 20 30 40 50 60 70 80
% of total arrivals

One-third of Tanzania is national park or game reserve. The Ngorongoro Crater and the Serengeti Plain are top attractions. Tourist numbers increased dramatically in the 1990s.

PEOPLE

Pop. density low

Kiswahili, Sukuma, Chagga, Nyamwezi, Hehe, Makonde, Yao, Sandawe, English

41/km² (105/mi²)

THE URBAN/RURAL POPULATION SPLIT

33% 67%

RELIGIOUS PERSUASION

Other 4%
Muslim 33%
Traditional beliefs 30%
Christian 33%

For many Tanzanians the family is the focus of traditional rural life. About 99% belong to one of 120 small ethnic Bantu groups. The remainder comprise Arab, Asian, and European minorities. The use of Kiswahili as a *lingua franca* has helped make ethnic rivalries almost nonexistent.

POLITICS

Multiparty elections

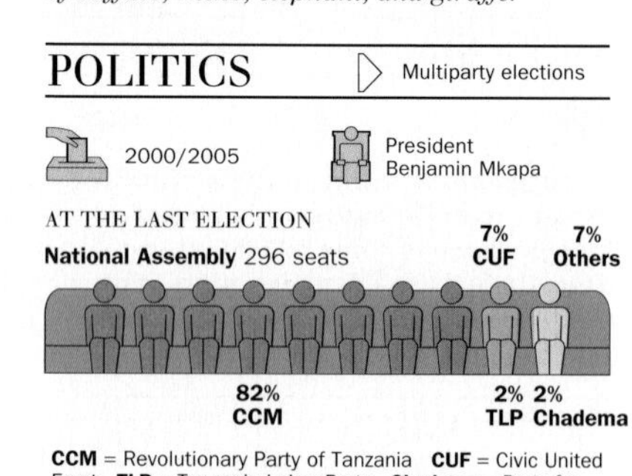
2000/2005
President
Benjamin Mkapa
AT THE LAST ELECTION
National Assembly 296 seats
7% CUF
7% Others
82% CCM
2% TLP
2% Chadema

CCM = Revolutionary Party of Tanzania **CUF** = Civic United Front **TLP** = Tanzania Labor Party **Chadema** = Party for Democracy and Progress
Others include 5 members chosen by the Zanzibar House of Representatives, 10 appointed by the president, and the attorney general who has a seat ex-officio

Julius Nyerere was the dominant force in Tanzanian politics for over two decades. He founded the ruling party, the CCM, and his philosophy of African socialism guided Tanzania's development. Ali Hassan Mwinyi succeeded Nyerere as president in 1985, introducing a transition to multiparty democracy. Mwinyi stood down in 1995, and Benjamin Mkapa was elected president. Separatism in Zanzibar is a key issue, flaring into violent protests in 2001.

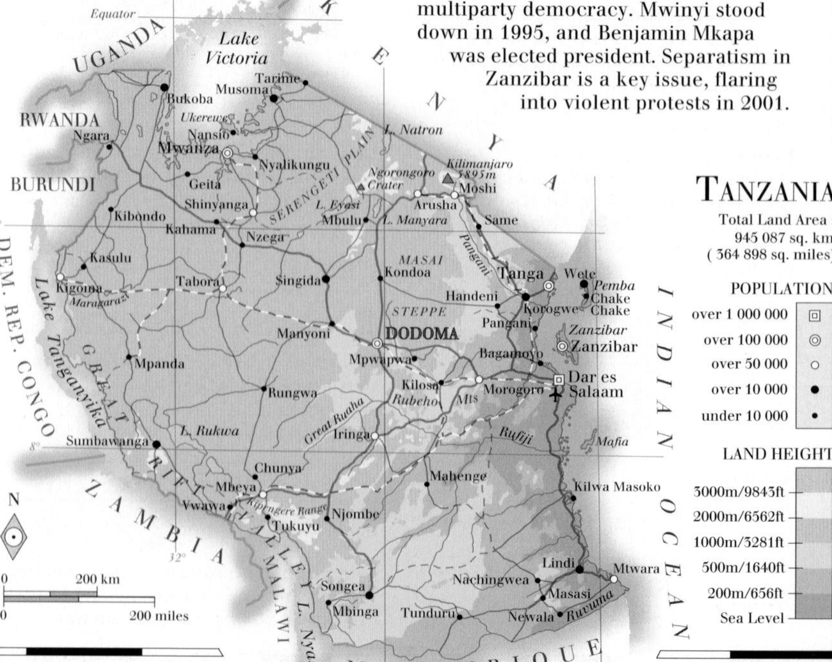
Equator
UGANDA
Lake Victoria
KENYA
RWANDA
BURUNDI
DEM. REP. CONGO
Lake Tanganyika
ZAMBIA
MALAWI
L. Nyasa
MOZAMBIQUE
INDIAN OCEAN
Bukoba
Musoma
Tarime
Ukerewe
Nansio
Ngara
Mwanza
Nyalikungu
SERENGETI PLAIN
L. Natron
Ngorongoro Crater
Kilimanjaro 5895m
Moshi
Geita
Shinyanga
L. Eyasi
Mbulu
Arusha
L. Manyara
Same
Kibondo
Kahama
Nzega
Pangani
Kasulu
MASAI
Kondoa
Tanga
Wete
Pemba
Chake Chake
Kigoma
Malagarasi
Tabora
Singida
STEPPE
Handeni
Korogwe
Pangani
Zanzibar
Manyoni
DODOMA
Mpwapwa
Bagamoyo
Dar es Salaam
GREAT RIFT
Mpanda
Kilosa
Rubeho Mts
Morogoro
Rungwa
Great Ruaha
Rufiji
Mafia
L. Rukwa
Sumbawanga
Iringa
Chunya
Mahenge
Mbeya
Vwawa
Kipengere Range
Njombe
Tukuyu
Kilwa Masoko
Lindi
Mtwara
Nachingwea
Masasi
Songea
Mbinga
Tunduru
Newala
Ruvuma
N
0 200 km
0 200 miles

TANZANIA

Total Land Area : 945 087 sq. km (364 898 sq. miles)

POPULATION
over 1 000 000
over 100 000
over 50 000
over 10 000
under 10 000

LAND HEIGHT
3000m/9843ft
2000m/6562ft
1000m/3281ft
500m/1640ft
200m/656ft
Sea Level

T

WORLD AFFAIRS

Joined UN in 1961

Comm
ACP
EAC
AU
SADC

The instability of Tanzania's central African neighbors is a concern. It accepted over half a million Rwandan and Burundian refugees in the 1990s, but instigated mass repatriations in 1996. Improved relations with Uganda and Kenya since 1985 led to the rebirth in 2001 of the East African Community.

AID

Recipient

 $1.05bn (receipts)  Up 6% in 2000

Tanzania is heavily dependent on aid to help offset a severe balance-of-payments deficit. The World Bank, Japan, and the UK are the major donors. Most aid is now linked to an IMF-backed economic reform program. In 2001 $3 billion of debt was canceled.

DEFENSE

Compulsory military service

 $141m 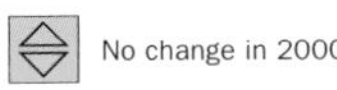 No change in 2000

Defense accounts for 3.5% of budget spending. The armed forces are closely linked with the ruling CCM. There is an 80,000-strong citizens' reserve force.

ECONOMICS

Inflation 22% p.a. (1990–2000)

$9.01bn

805–917 Tanzanian shillings

SCORE CARD

- WORLD GNP RANKING.........88th
- GNP PER CAPITA.........$270
- BALANCE OF PAYMENTS.........–$517m
- INFLATION.........5.9%
- UNEMPLOYMENT.........Not available

STRENGTHS

Coffee, cotton, sisal, tea, cashew nuts. Zanzibar a major producer of cloves. Diamonds, gold. State commitment to effective reforms. Expansion in nontraditional exports. Rise in inward investment. Return to positive growth.

WEAKNESSES

Growth still too low to increase per capita income. Shortage of foreign exchange. Poor credit and equipment limit agricultural development.

EXPORTS

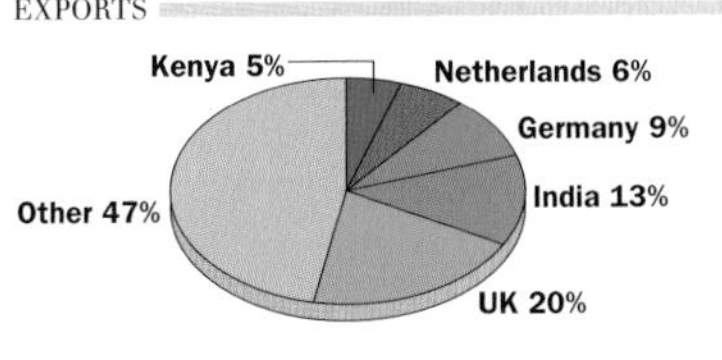
Kenya 5%
Netherlands 6%
Germany 9%
India 13%
UK 20%
Other 47%

IMPORTS

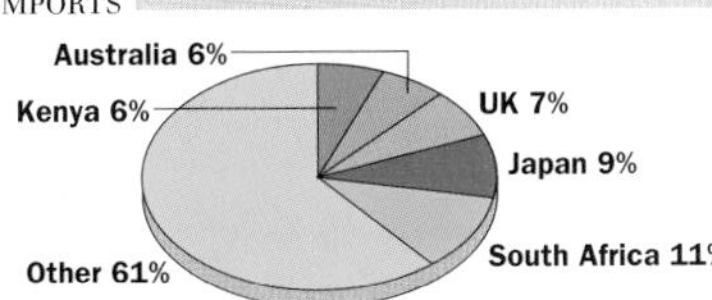
Australia 6%
Kenya 6%
UK 7%
Japan 9%
South Africa 11%
Other 61%

RESOURCES

Electric power 543,000 kw

310,270 tonnes  Limited oil production

 14.4m cattle, 10m goats, 4.25m sheep, 30m chickens  Natural gas, oil, iron, diamonds, gold, salt, phosphates, coal

Agriculture accounts for half of GDP and 80% of employment and exports. Forests cover 50% of land. More than 90% of energy demand is met from wood and charcoal. Hydropower provides 70% of electricity and is being expanded. Oil imports take 40% of export earnings, but offshore gas at Songo Songo is being exploited. Oil has been discovered off Pemba Island. The opening of a gold mine near Mwanza makes Tanzania Africa's fourth-largest gold producer.

ENVIRONMENT

Sustainability rank: 80th

 16% 0.1 tonnes per capita

The demand for fuelwood is a threat to forests. Tourism's demands have to be carefully balanced with those of fragile wildlife environments such as the Ngorongoro Crater and the Serengeti.

MEDIA

TV ownership low

Daily newspaper circulation 4 per 1000 people

PUBLISHING AND BROADCAST MEDIA

 There are 9 daily newspapers, including the *Daily News*, *Uhuru*, and *Kipanga*

3 independent services

5 services: 2 state-owned, 3 independent

The press has grown rapidly since the mid-1990s, with several independent papers. There is no national TV station.

CRIME

Death penalty in use

 45,611 prisoners  Down 1% in 1999

Crime levels are low, although theft in Dar es Salaam has risen. Tanzania's human rights record is good.

EDUCATION

School leaving age: 14

 76%  18,867 students

Primary education, which begins at seven and lasts for seven years, is free; secondary students pay fees. In 1999, enrollment was 75% at primary level, but only 6% for secondary education.

CHRONOLOGY

The mainland became the German colony of Tanganyika in 1884. The Sultanate of Zanzibar became a British protectorate in 1890.

- **1918** Tanganyika British mandate.
- **1961** Tanganyika independent.
- **1962** Nyerere becomes president.
- **1963** Zanzibar independent.
- **1964** Zanzibar signs union with Tanganyika to form Tanzania.
- **1985** President Mwinyi begins relaxation of socialist policies.
- **1992** Political parties allowed.
- **1995** Multiparty elections. Benjamin Mkapa becomes president.
- **1999** Death of Nyerere.
- **2000** Mkapa elected for second term.
- **2001** Increasing unrest among Zanzibar separatists.

HEALTH

Welfare state health benefits

 1 per 20,000 people  Diarrheal and respiratory diseases, malaria

There is a national campaign against HIV/AIDS, which affects 7.8% of the adult population. An immunization program for under-twos was introduced in 2001.

SPENDING

GDP/cap. increase

CONSUMPTION AND SPENDING

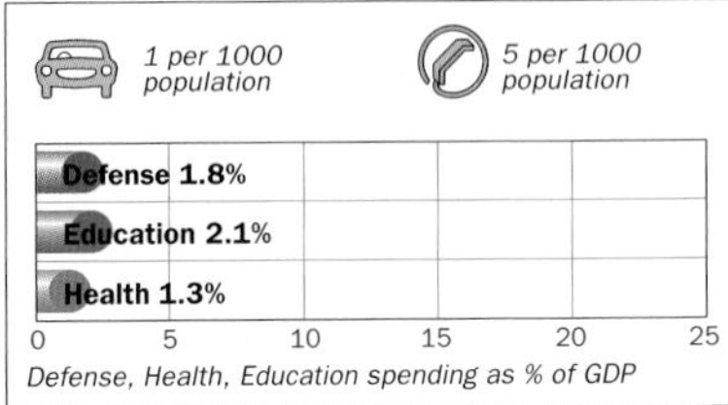
1 per 1000 population
5 per 1000 population
Defense 1.8%
Education 2.1%
Health 1.3%
0
5
10
15
20
25
Defense, Health, Education spending as % of GDP

The majority of Tanzanians are subsistence farmers. The small wealthy elite is composed mainly of Asian and Arab business families.

WORLD RANKING

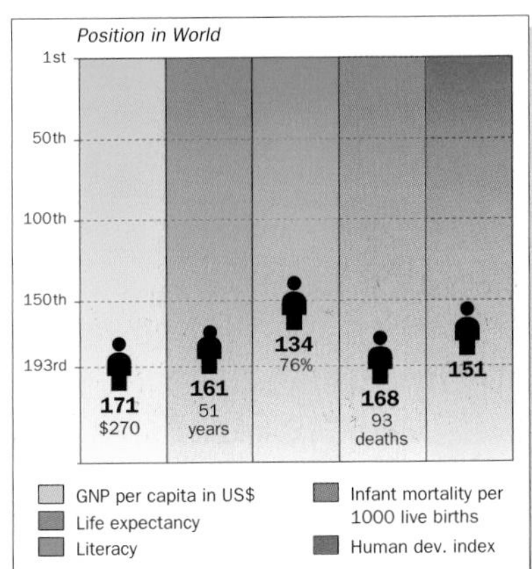
Position in World
1st
50th
100th
150th
193rd
171
$270
161
51
years
134
76%
168
93
deaths
151
GNP per capita in US$
Life expectancy
Literacy
Infant mortality per 1000 live births
Human dev. index

T

SOUTHEAST ASIA
Asia
Burma (Myanmar) Laos THAILAND Cambodia

THAILAND

OFFICIAL NAME: Kingdom of Thailand **CAPITAL:** Bangkok
POPULATION: 63.6 million **CURRENCY:** Baht **OFFICIAL LANGUAGE:** Thai

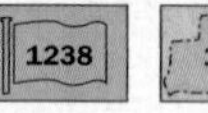
1238

1907

Dec 5
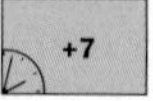
T

+7

+66
.th

THAILAND LIES IN THE HEART of southeast Asia. The north, the border with Burma, and the long Isthmus of Kra between the Andaman Sea and the Gulf of Thailand are mountainous. The central plain is the most fertile and densely populated area, while the low northeastern plateau is the poorest region. Thailand has been an independent kingdom for most of its history, and since 1932 a constitutional monarchy, though with frequent periods of military government. Continuing rapid industrialization results in massive congestion in Bangkok and a serious depletion of natural resources.

CLIMATE

Tropical equatorial/ monsoon

WEATHER CHART FOR BANGKOK

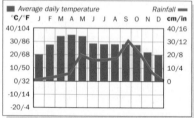

Average daily temperature · Rainfall
°C/°F J F M A M J J A S O N D cm/in
40/104 30/86 20/68 10/50 0/32 -10/14 -20/-4
40/16 30/12 20/8 10/4 0

Thailand's tropical monsoon climate has three seasons – a hot sultry period, rains from May to October, and a dry, cooler season from November to March.

TRANSPORTATION

Drive on left

Bangkok International
29.6m passengers

552 ships
2m grt

THE TRANSPORTATION NETWORK

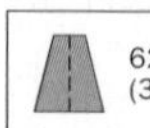
62,985 km
(39,137 miles)

None

4623 km
(2873 miles)

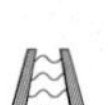
3701 km
(2300 miles)

Bangkok suffers from huge traffic jams. Its first mass transit system became operational in 1999: as well as the elevated railroad, subway lines are planned. A road corridor through Laos into Vietnam was agreed in 2001.

Island in the Andaman Sea. Overdevelopment at Thailand's best-known resorts is pushing tourism into new, remoter locations.

TOURISM

Visitors : Population
1:6.6

9.58m visitors

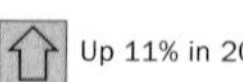
Up 11% in 2000

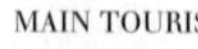
MAIN TOURIST ARRIVALS

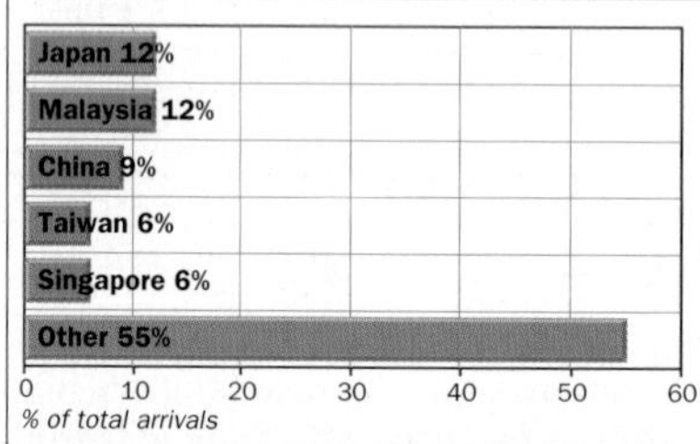
Japan 12%
Malaysia 12%
China 9%
Taiwan 6%
Singapore 6%
Other 55%
0 10 20 30 40 50 60
% of total arrivals

Tourism is an important contributor to the Thai economy. Tourist numbers fell in the early 1990s as a result of both the worldwide recession and local overdevelopment during the 1980s boom. Although the number of arrivals has since recovered, visitors are tending to seek the less-developed resorts. Bangkok's hotel occupancy rates continue to fall as yet more hotels are built. Pattaya beach resort, opposite Phetchaburi has seen such uncontrolled development that sea pollution is now a serious problem, while opposition to the intrusion of large numbers of tourists is growing among northern hill tribes.

Although prostitution is illegal, Bangkok and Pattaya are centers for sex tourism, which thrives despite the state's embarrassment at its effect on Thailand's image. Japanese and German men are among the main clients, while Burmese girls are increasingly recruited as prostitutes. Child prostitution is also a major problem.

There has been a boom in golf tourism, especially among the Japanese. The large number of new golf courses have made Thailand one of the largest golf destinations in Asia. The vast amounts of water needed to maintain the courses is aggravating Thailand's serious water shortage.

PEOPLE

Pop. density medium

Thai, Chinese, Malay, Khmer, Mon, Karen, Miao

124/km² (322/mi²)

THE URBAN/RURAL POPULATION SPLIT

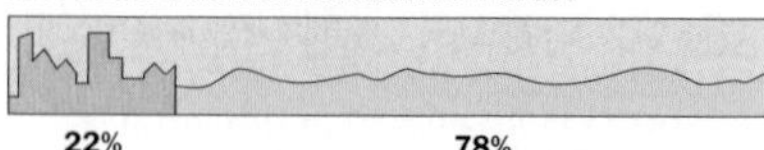
22%
78%

RELIGIOUS PERSUASION

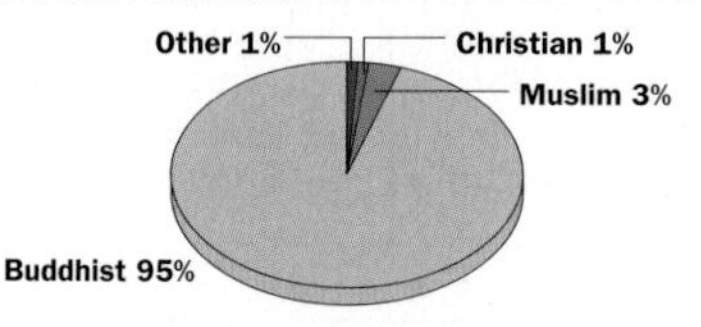
Other 1%
Christian 1%
Muslim 3%
Buddhist 95%

ETHNIC MAKEUP

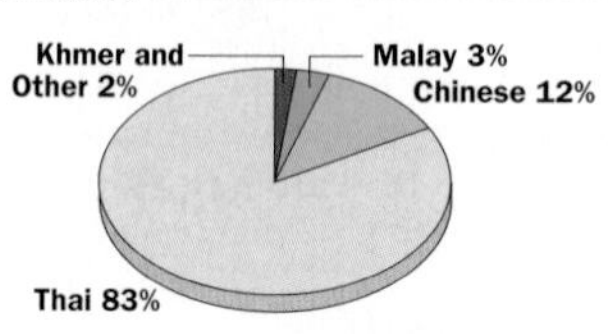
Khmer and Other 2%
Malay 3%
Chinese 12%
Thai 83%

There is little ethnic tension in Thailand, and Buddhism is a great binding force. The majority of Thais follow Theravada Buddhism, although the reformist Asoke Santi Buddhist sect, which advocates a new moral austerity, is gaining influence. Its principles have been espoused in particular by the Palang Dharma (PD), which seeks to clean up politics.

The far north and northeast hills are home to about 600,000 tribespeople with their own languages, and to permanently settled refugees from Laos, mostly of the Hmong tribal group.

The large Chinese community is the most assimilated in southeast Asia. Sino-Thais are particularly dominant in agricultural marketing. Most of Thailand's one million Muslim Malays live in southern Thailand, bordering Malaysia. They feel stronger affinity with Muslims in Malaysia than with Thai culture, and this has given rise to a secessionist movement.

Women are important in business, but their involvement in national politics is limited.

POPULATION AGE BREAKDOWN

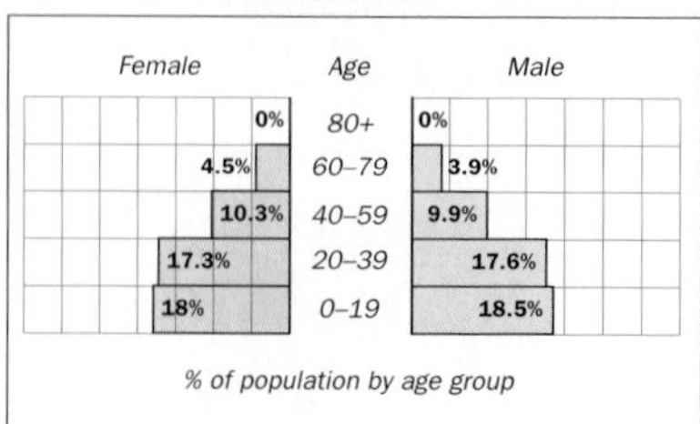
Female
Age
Male
0% 80+ 0%
4.5% 60–79 3.9%
10.3% 40–59 9.9%
17.3% 20–39 17.6%
18% 0–19 18.5%
% of population by age group

T

POLITICS

Multiparty elections

L. House 2001/2005
U. House 2000/2006

H.M. King Bhumibol Adulyadej (Rama IX)

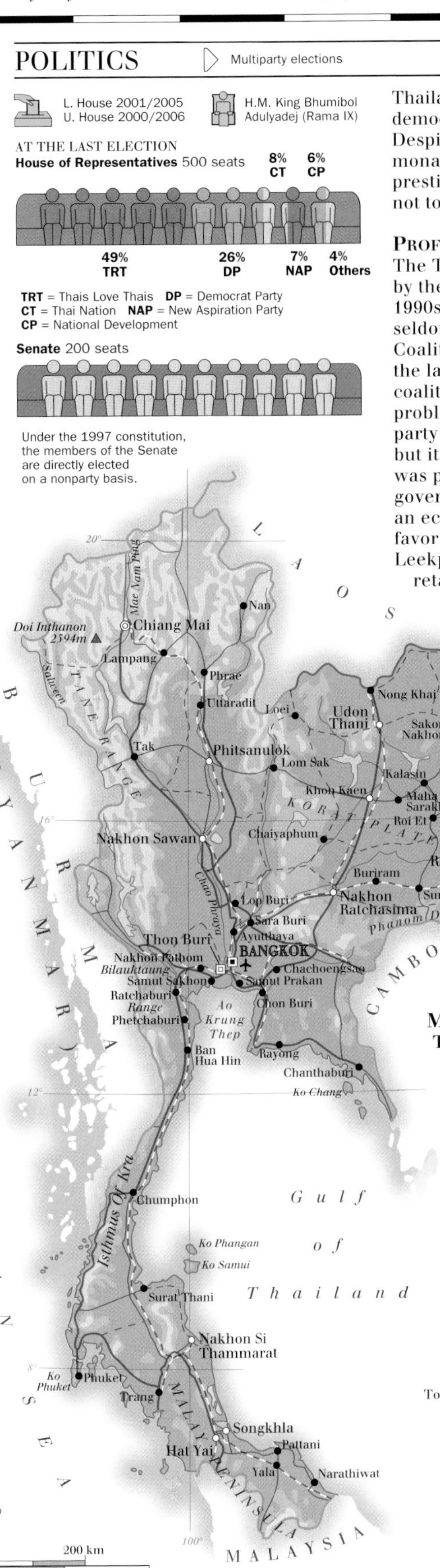

AT THE LAST ELECTION
House of Representatives 500 seats
8% CT
6% CP
49% TRT
26% DP
7% NAP
4% Others
TRT = Thais Love Thais DP = Democrat Party
CT = Thai Nation NAP = New Aspiration Party
CP = National Development
Senate 200 seats
Under the 1997 constitution, the members of the Senate are directly elected on a nonparty basis.
L A O S
Doi Inthanon 2594m
Chiang Mai
Nan
Lampang
Phrae
Uttaradit
Loei
Udon Thani
Nong Khai
Nakhon Phanom
Sakon Nakhon
Tak
Phitsanulok
Lom Sak
Kalasin
Khon Kaen
Maha Sarakham
Roi Et
Nakhon Sawan
Chaiyaphum
Ubon Ratchathani
Buriram
Nakhon Ratchasima
Surin
Lop Buri
Sara Buri
Ayutthaya
Thon Buri
BANGKOK
Nakhon Pathom
Chachoengsao
Samut Sakhon
Samut Prakan
Ratchaburi
Chon Buri
Phetchaburi
Ao Krung Thep
Ban Hua Hin
Rayong
Chanthaburi
Ko Chang
C A M B O D I A
B U R M A
M Y A N M A R
Chumphon
Ko Phangan
Ko Samui
Surat Thani
Nakhon Si Thammarat
G u l f o f T h a i l a n d
Phuket
Trang
Songkhla
Hat Yai
Pattani
Yala
Narathiwat
M A L A Y S I A
200 km
200 miles

Thailand is a multiparty parliamentary democracy. The king is the head of state. Despite his position as a constitutional monarch, he has immense personal prestige. Criticism of the king is not tolerated.

PROFILE

The Thai political process, dominated by the military for decades until the 1990s, is highly personalized. Parties seldom have strong ideologies. Coalitions are often unstable, while the lack of coordination between coalition partners is a recurring problem. The NAP was the largest party in the parliament elected in 1996, but its leader Chaovalit Yongchaiyuth was prime minister only for a year. His government, blamed for mismanaging an economic crisis, was then ousted in favor of another coalition under Chuan Leekpai of the DP, who succeeded in retaining office for four years despite the volatility of coalition politics. However, in the 2001 elections the new populist TRT triumphed, winning just short of a majority of seats. TRT leader and former deputy prime minister Thaksin Shinawatra formed a three-party government. The NAP and TRT merged in 2002.

MAIN POLITICAL ISSUES

The military–democratic cycle

Thailand has been ruled by alternating military and civilian governments since 1932. In 1992 promilitary parties chose an unelected army general as prime minister. Violent demonstrations erupted which forced the resignation of the prime minister and precipitated a constitutional amendment obliging future prime ministers to be elected members of the legislature. The military

King Bhumibol. *On the throne since 1946, he is the world's longest-serving ruler.*

Thaksin Shinawatra, *a controversial billionaire, elected premier in 2001.*

has since maintained a low profile during general elections, and has refrained from interfering in government changes.

Congestion in Bangkok

A major issue is the concentration of industry and commerce in the Bangkok area. Uncontrolled development has left it with traffic congestion which not only is among the world's worst but is also a serious hindrance to economic activity. In December 1999 the first stage of a mass transit system – an elevated railroad – was formally opened.

Since 1993, the government has offered incentives for relocating industry to the provinces. This is also intended to help distribute wealth more evenly – up to 60% of GDP is generated in the Bangkok area.

Water

The national water shortage, caused by rapid industrialization, is so acute that it is affecting farm and industrial output.

WORLD AFFAIRS

Joined UN in 1946

APEC
ASEAN
Mekong River
NAM
WTO

Thailand has friendly relations with China. However, relations with neighboring Burma have been strained over Burma's alleged support for Thai ethnic guerrillas operating along the Thai–Burmese border. Many Thai logging companies, often run by the military, have been active in Burma since Thailand's 1988 logging ban at home. Relations with Vietnam are cordial, despite Thailand's opposition to the Vietnamese regime in Cambodia in the 1980s.

Thailand, Indonesia, and Malaysia have liberalized trade to promote development in each country in regions which are distant from their respective capitals.

Thailand maintains close relations with the US, despite some tension over intellectual property rights and minor trade issues, but no longer has any US military bases on its territory.

THAILAND

Total Land Area : 514 000 sq. km (198 455 sq. miles)

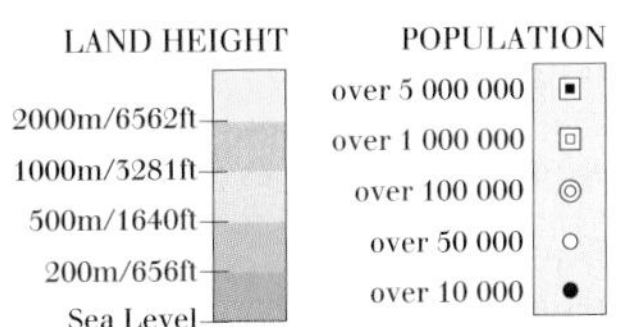

LAND HEIGHT
2000m/6562ft
1000m/3281ft
500m/1640ft
200m/656ft
Sea Level
POPULATION
over 5 000 000
over 1 000 000
over 100 000
over 50 000
over 10 000

T

AID

Recipient

$641m (receipts)

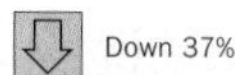

Down 37% in 2000

Japan and the World Bank are the largest aid donors. Thailand has imposed a ceiling on foreign borrowing to keep its debt stable.

CHRONOLOGY

Thailand emerged as a kingdom in the 13th century, and by the late 17th century its then capital, Ayutthya, was the largest city in southeast Asia. In 1782, the present Chakri dynasty and a new capital, Bangkok, were founded.

- ❑ **1855** King Mongut signs Bowring trade treaty with British – Thailand never colonized by Europeans.
- ❑ **1868–1910** King Chulalongkorn westernizes Thailand.
- ❑ **1907** Thailand cedes western Khmer (Cambodia) to France.
- ❑ **1925** King Prajadhipok begins absolute rule.
- ❑ **1932** Bloodless military–civilian coup. Constitutional monarchy.
- ❑ **1933** Military takes control.
- ❑ **1941** Japanese invade. Government collaborates.
- ❑ **1944** Pro-Japanese prime minister and prewar military dictator Phibun voted out of office.
- ❑ **1945** Exiled King Ananda returns.
- ❑ **1946** Ananda assassinated. King Bhumibol accedes.
- ❑ **1947** Military coup. Phibun back.
- ❑ **1957** Military coup. Constitution abolished.
- ❑ **1965** Thailand allows US to use Thai bases in Vietnam War.
- ❑ **1969** New constitution endorses elected parliament.
- ❑ **1971** Army suspends constitution.
- ❑ **1973–1976** Student riots lead to interlude of democracy.
- ❑ **1976** Military takeover.
- ❑ **1980–1988** Gen. Prem Tinsulanond prime minister. Partial democracy.
- ❑ **1988** Elections. Gen. Chatichai Choonhaven, right-wing CT leader, named prime minister.
- ❑ **1991** Military coup. Civilian Anand Panyarachun caretaker premier.
- ❑ **1992** Elections. Gen. Suchinda named premier. Demonstrations. King forces Suchinda to step down and reinstalls Anand. Moderates win new elections.
- ❑ **1995** CT wins general election.
- ❑ **1996** Early elections; Chaovalit Yongchaiyuth of NAP becomes prime minister.
- ❑ **1997** Financial and economic crisis; Chaovalit government falls; DP's Chuan Leekpai prime minister.
- ❑ **2001** TRT, led by Thaksin Shinawatra, wins elections.

DEFENSE

No compulsory military service

$2.46bn

Down 7% in 2000

THAI ARMED FORCES

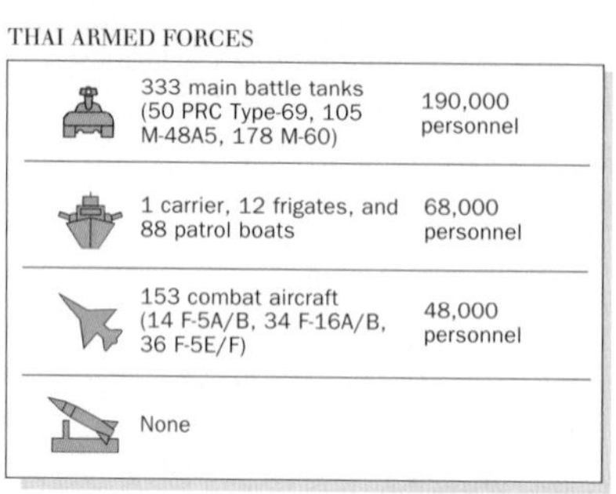

333 main battle tanks (50 PRC Type-69, 105 M-48A5, 178 M-60)	190,000 personnel
1 carrier, 12 frigates, and 88 patrol boats	68,000 personnel
153 combat aircraft (14 F-5A/B, 34 F-16A/B, 36 F-5E/F)	48,000 personnel
None	

The military either ruled Thailand, or played a prominent role in politics, for over half a century from 1932. In 1996, its role in the appointed Senate – hitherto a military stronghold – was reduced. Retired military figures are, however, prominent in the major political parties.

Since 1986, defense spending has tended to focus on the navy and air force. Naval vessels have been bought from China, Germany, and Spain, and aircraft from the UK, the US, and Russia.

Thailand's main defense concerns are border disputes with Cambodia, Burma, and Laos; the Muslim secessionist movement in the south; and piracy and fishing disputes in the South China Sea.

ECONOMICS

Inflation 4.2% p.a. (1990–2000)

$122bn

43.38–44.23 baht

SCORE CARD

- ❑ World GNP Ranking 31st
- ❑ GNP per Capita $2000
- ❑ Balance of Payments $9.2bn
- ❑ Inflation 1.5%
- ❑ Unemployment 4%

EXPORTS

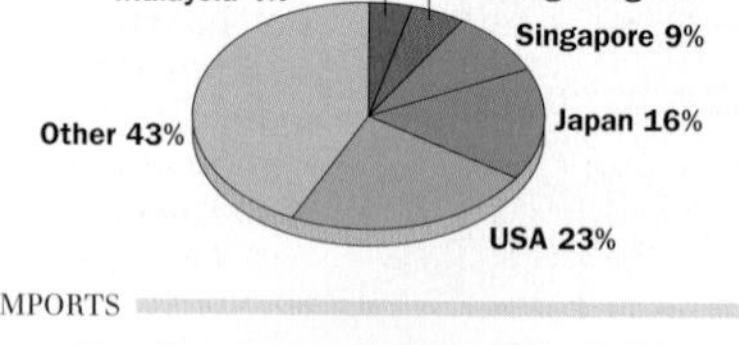
Malaysia 4%
Hong Kong 5%
Singapore 9%
Japan 16%
USA 23%
Other 43%

IMPORTS

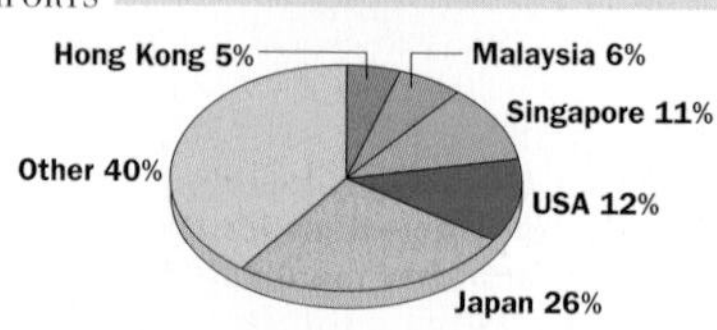
Hong Kong 5%
Malaysia 6%
Singapore 11%
USA 12%
Japan 26%
Other 40%

Strengths

Success of export-based and import-substituting manufacturing. Rapid economic growth. Natural gas. Tourism. Major world exporter of rice and rubber.

Weaknesses

Concentration of economic activity in congested Bangkok area. Inadequate water storage facilities. Rapid growth of foreign debt. 60% of population in low-profit farming.

Profile

Until the late 1990s, the economy grew at over 9% a year for a decade, driven by a rise in manufacturing and huge overseas investments, especially from Japan. However, as domestic wages rose, Thailand faced stiff competition from China and Vietnam. Thailand also lacked a skilled labor force to develop high-tech production, though it is a big producer of electronics goods.

In 1997 mounting foreign debt and the sharp depreciation of the baht made necessary an IMF-led rescue package. Massive retrenchment and stringent austerity measures followed. By mid-2000 the IMF had ended its direct involvement and was optimistic for future expansion. GDP grew by 4.4% in the following year.

ECONOMIC PERFORMANCE INDICATOR

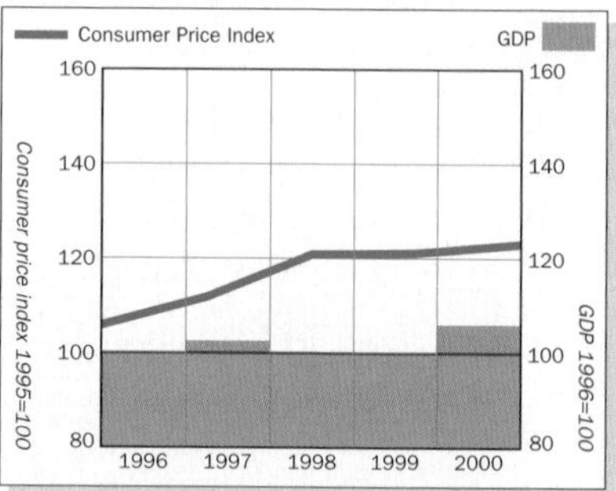
Consumer Price Index
GDP
160
140
120
100
80
1996
1997
1998
1999
2000
Consumer price index 1995=100
GDP 1996=100

THAILAND : MAJOR BUSINESSES

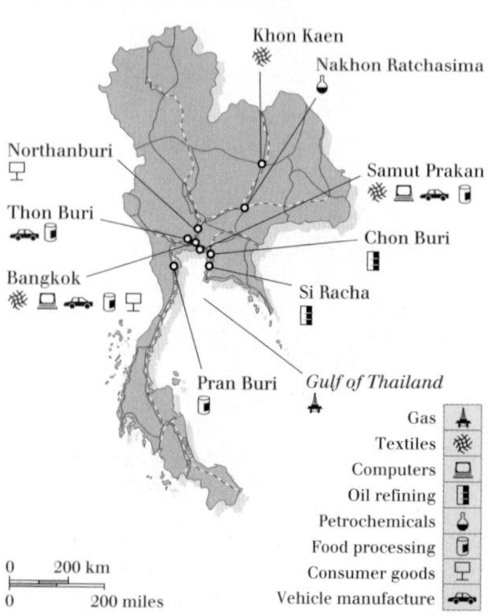
Khon Kaen
Nakhon Ratchasima
Northanburi
Samut Prakan
Thon Buri
Chon Buri
Bangkok
Si Racha
Pran Buri
Gulf of Thailand
Gas
Textiles
Computers
Oil refining
Petrochemicals
Food processing
Consumer goods
Vehicle manufacture
0 200 km
0 200 miles

RESOURCES

Electric power 21.5m kw

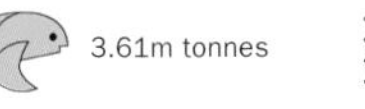
3.61m tonnes

178,000 b/d (reserves 500m barrels)

23m ducks, 8.3m pigs, 6.3m cattle, 190m chickens

Tin, lignite, gas, gems, oil, tungsten, lead, zinc, antimony, gold, copper

ELECTRICITY GENERATION

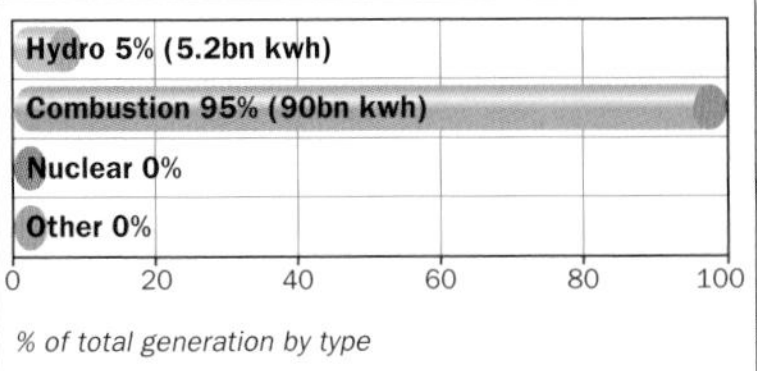
Hydro 5% (5.2bn kwh)
Combustion 95% (90bn kwh)
Nuclear 0%
Other 0%
0
20
40
60
80
100
% of total generation by type

Thailand has minimal crude oil and has rejected the nuclear option in favor of speeding up development of its large natural gas fields. It also has significant lignite deposits for power generation. World demand for Thailand's tin has declined, but recent gold and copper finds offer new potential. Thailand has valuable gemstone deposits. It is the world's fourth-largest producer of shrimp.

THAILAND : LAND USE

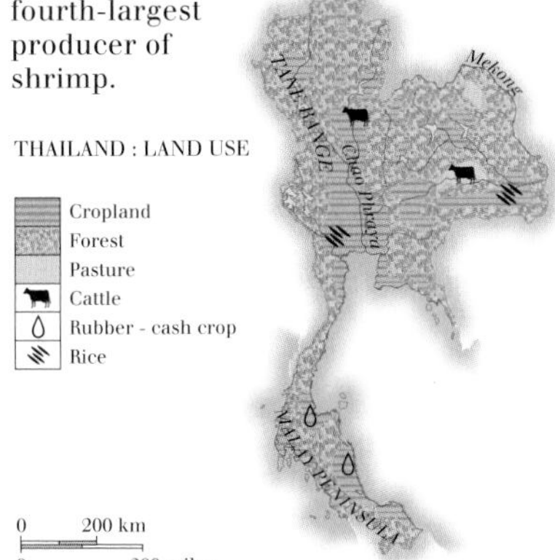
TANEN RANGE
Mekong
Chao Phraya
MALAY PENINSULA
Cropland
Forest
Pasture
Cattle
Rubber - cash crop
Rice
0 200 km
0 200 miles

ENVIRONMENT

Sustainability rank: 54th

14%

3.2 tonnes per capita

ENVIRONMENTAL TREATIES

Yes
Yes
No
Yes
Yes
No

Deforestation, especially of the watersheds in the north, has led to the increasing severity of both floods and droughts. Particularly serious flooding in the south resulted in a total logging ban in 1988. Illegal logging continues, however. Reafforestation projects, some criticized for using single quick-growing species, will not solve the national water shortage. Intensive inland prawn farming leaves a legacy of salination. Mass tourism brings pollution problems to resorts, while backpackers lead the way in exposing remoter locations.

MEDIA

TV ownership high

Daily newspaper circulation 64 per 1000 people

PUBLISHING AND BROADCAST MEDIA

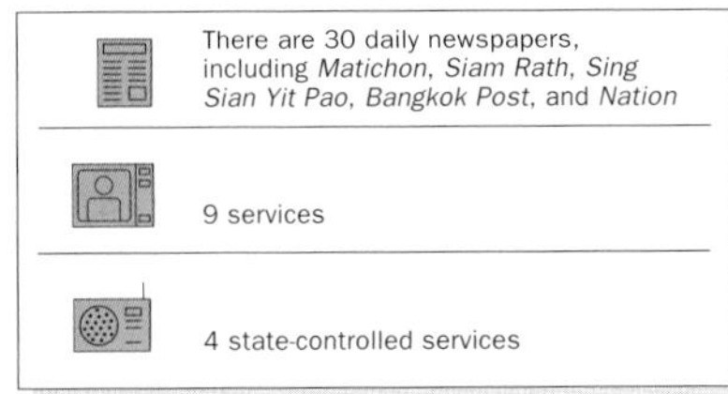
There are 30 daily newspapers, including *Matichon, Siam Rath, Sing Sian Yit Pao, Bangkok Post*, and *Nation*
9 services
4 state-controlled services

Newspapers enjoy a high level of freedom in political reporting. Two of the TV stations are run by the military. There are 60 radio stations around Bangkok, mainly state-run.

CRIME

Death penalty in use

206,011 prisoners

Up 95% 1992–1996

CRIME RATES

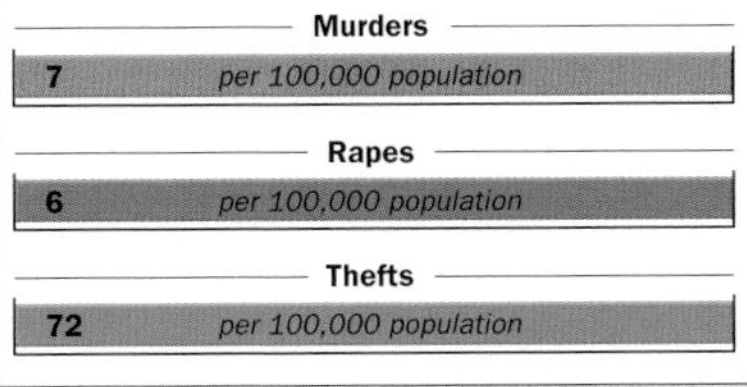
Murders
7 per 100,000 population
Rapes
6 per 100,000 population
Thefts
72 per 100,000 population

Political imprisonment is now extremely rare. There is some police involvement in crime, including extrajudicial killings and ill-treatment of prisoners in detention. The king has inspired an opium-substitution crop program. In the south, drug addiction is a major problem and crime is rising. The government has cracked down on music, software, and video piracy.

EDUCATION

School leaving age: 16

96%

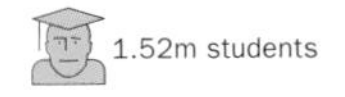
1.52m students

THE EDUCATION SYSTEM

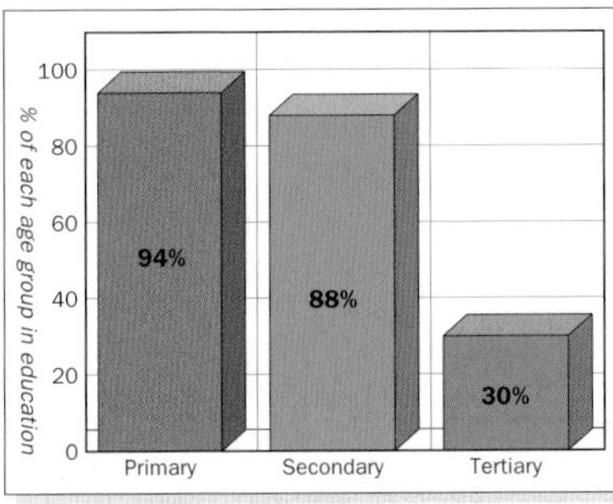
100
80
60
40
20
0
% of each age group in education
94%
88%
30%
Primary
Secondary
Tertiary

A poorly developed education system has led to a shortage of skills needed for the expansion of high-tech industries. The Education for All program aims to rectify this situation, and also to improve adult literacy levels.

HEALTH

Welfare state health benefits

1 per 2500 people

Heart diseases, gastroenteritis

High-quality health care is heavily concentrated in Bangkok. Most of the rural population has access to primary health care, and a new scheme to provide care for the poor for just $1 per person was launched in 2001. Trained personnel are aided by village health volunteers, monks, teachers, and traditional healers.

However, estimates suggest that only 30% of users can afford to pay. The poor can apply annually for a certificate entitling them to free health care.

High-profile family planning programs are slowing population growth. An effective AIDS prevention campaign has helped reduce the number of new infections, although the government has been in conflict with international drug companies over its right to produce cheaper generic drugs for AIDS sufferers. Prostitutes have benefited from an extensive sex education program.

SPENDING

GDP/cap. increase

CONSUMPTION AND SPENDING

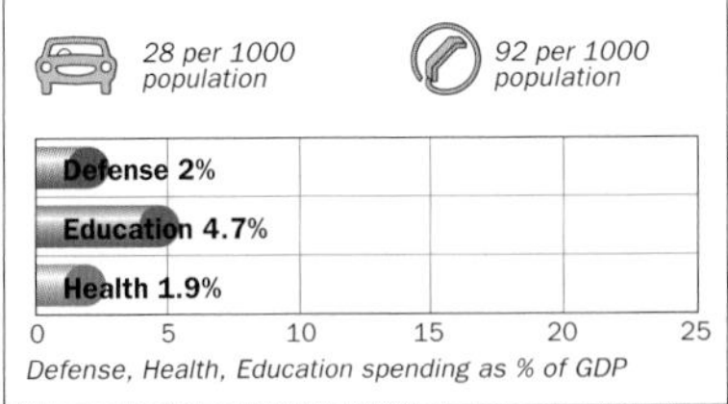
28 per 1000 population
92 per 1000 population
Defense 2%
Education 4.7%
Health 1.9%
0
5
10
15
20
25
Defense, Health, Education spending as % of GDP

The government is trying to diffuse to the provinces the people and wealth currently concentrated to a very great extent in Bangkok. The northeast in particular is very poor. The gap between rich and poor is greater in Thailand than in other industrializing southeast Asian states.

WORLD RANKING

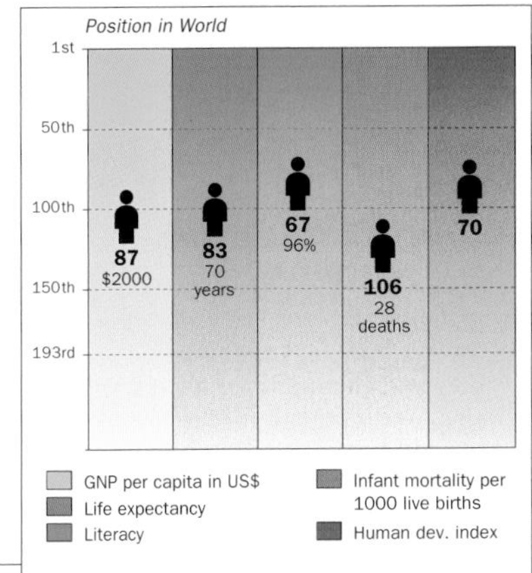
Position in World
1st
50th
100th
150th
193rd
87 $2000
83 70 years
67 96%
106 28 deaths
70
GNP per capita in US$
Life expectancy
Literacy
Infant mortality per 1000 live births
Human dev. index

T

Togo

WEST AFRICA

OFFICIAL NAME: Republic of Togo **CAPITAL:** Lomé **POPULATION:** 4.7 million
CURRENCY: CFA franc **OFFICIAL LANGUAGE:** French

1960

1960
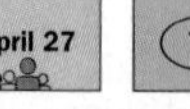
April 27
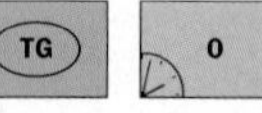
TG

0
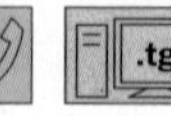
+228

.tg

TOGO IS SANDWICHED between Ghana and Benin in west Africa. A central forested region is bounded by savanna lands to the north and south. The port of Lomé is an important entrepôt for west African trade. Togo's president, Gen. Gnassingbé Eyadéma, has been in power since 1967.

CLIMATE

Tropical equatorial/wet and dry

WEATHER CHART FOR LOMÉ

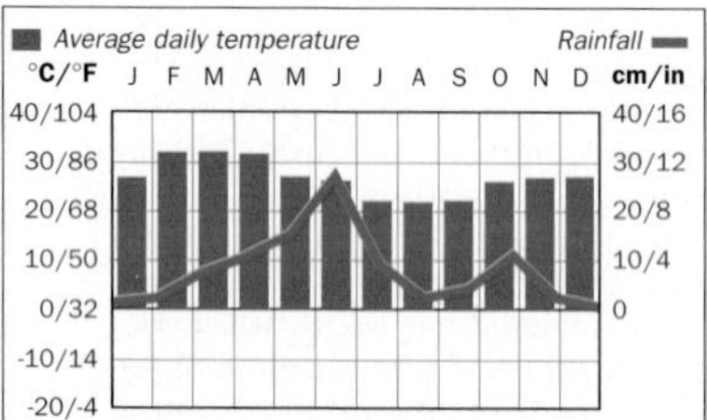
Average daily temperature
Rainfall
°C/°F J F M A M J J A S O N D cm/in
40/104 30/86 20/68 10/50 0/32 -10/14 -20/-4
40/16 30/12 20/8 10/4 0

Togo has a typical Gulf of Guinea climate – very hot and humid on the coast, and drier inland.

TRANSPORTATION

Drive on right

Tokoin, Lomé
232,578 passengers
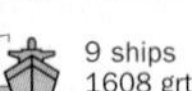
9 ships
1608 grt

THE TRANSPORTATION NETWORK

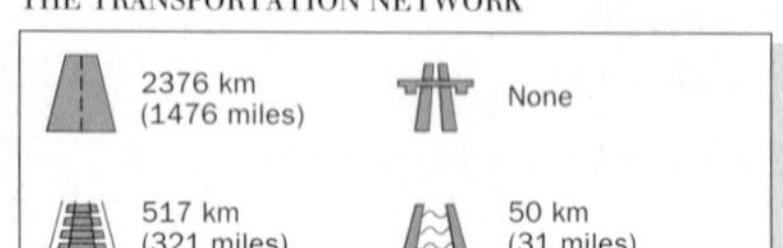
2376 km (1476 miles)
None
517 km (321 miles)
50 km (31 miles)

Improving the already good road network and Lomé's port facilities are priorities, given Togo's role as an entrepôt. Air and rail links to the interior, however, are limited.

TOURISM

Visitors : Population 1:78

60,000 visitors

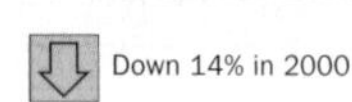
Down 14% in 2000

MAIN TOURIST ARRIVALS

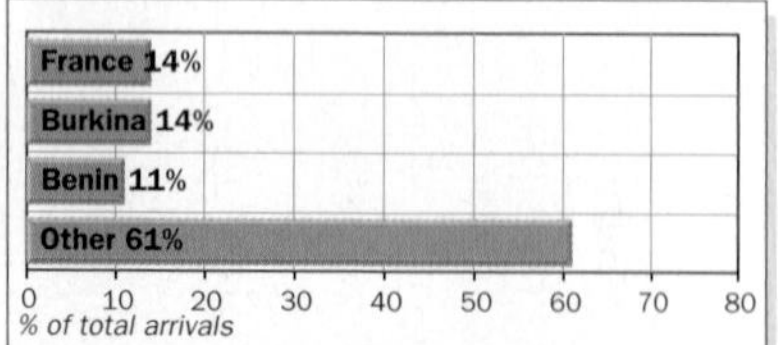
France 14%
Burkina 14%
Benin 11%
Other 61%
0 10 20 30 40 50 60 70 80
% of total arrivals

There is some package tourism, mainly French and German, to coastal tourist villages and hotels built during the expansion program of the 1980s. Tourists have been deterred by the political uncertainty since 1990.

PEOPLE

Pop. density medium

Ewe, Kabye, Gurma, French

86/km² (224/mi²)

THE URBAN/RURAL POPULATION SPLIT

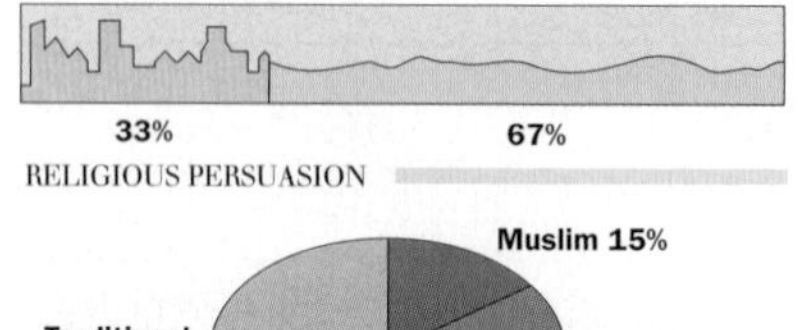
33%
67%

RELIGIOUS PERSUASION

Muslim 15%
Traditional beliefs 50%
Christian 35%

A bitter divide has existed between north and south since before independence. Most southern resentment is directed toward a northern minority, the Kabye people from the Kabye plateau, because of their domination of the military. The Kabye and other northerners in turn resent their own underdevelopment in contrast to the high development, especially educationally, of all southerners. The dominant southern group is the Ewe, who make up more than 40% of the population.

As elsewhere in Africa, the extended family is important and tribalism and nepotism are key factors in everyday life. Some Togolese ethnic groups, such as the Mina, have matriarchal societies. The "Nana Benz," the market-women of Lomé, control the retail trade and have considerable private money. Politics, however, remains a male preserve.

Kabye cultivations near Kara, *in northern Togo. The main food crops grown are cassava, yams, and maize.*

POLITICS

Multiparty elections

1999/2001 (postponed repeatedly)
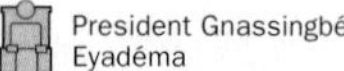
President Gnassingbé Eyadéma

AT THE LAST ELECTION

National Assembly 81 seats

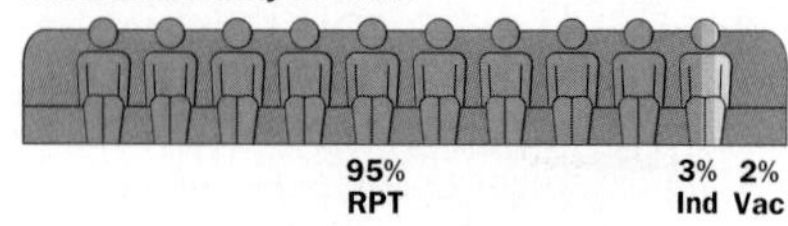
95% RPT
3% Ind
2% Vac

RPT = Rally of the Togolese People
Ind = Independents
Vac = Vacant: results of two seats were annulled

Politics has been dominated for over three decades by Gen. Gnassingbé Eyadéma, in power since 1967 and Africa's longest-serving leader.

A democracy movement has been gathering momentum since 1990. Multiparty presidential elections held in 1993 were won by Eyadéma, although some opposition candidates boycotted the poll over the exclusion of Gilchrist Olympio, son of a former president. Eyadéma claimed victory over Olympio in presidential elections in 1998, amid accusations of malpractice and of the killing of hundreds of opposition supporters immediately afterward in the runup to the 1999 Assembly election. (Serious human rights violations were later confirmed by a UN/OAU report.) During subsequent negotiations, the opposition accepted the election results, Eyadéma stated that he would not stand for reelection in 2003, and an accord provided for a new independent electoral body and a political code of conduct. New Assembly elections, set for late 2001, were repeatedly postponed, however, as the deadlock between Eyadéma and the opposition continued.

WORLD AFFAIRS

Joined UN in 1960

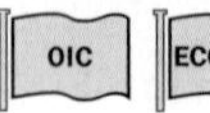
OIC

ECOWAS

FZ

AU
UEMOA

The priority is maintaining traditional links, especially with France. In 1998 President Eyadéma became chair of ECOWAS, acting as mediator in the Guinea-Bissau conflict and hosting talks in Sierra Leone. He was also president of the OAU for 2000/2001.

AID

Recipient

$70m (receipts)

Down 1% in 2000

Development projects and the health of the economy overall have suffered from aid suspensions in the 1990s by donors including the US and the EU.

T

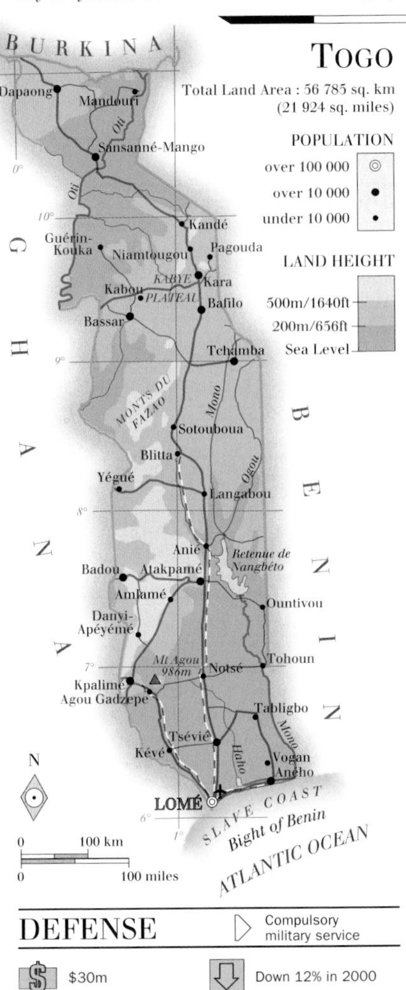
BURKINA
GHANA
BENIN
Dapaong
Mandouri
Oti
Sansanné-Mango
Guérin-Kouka
Niamtougou
Kandé
Pagouda
Kara
KABYE PLATEAU
Kabou
Bafilo
Bassar
Tchamba
MONTS DU FAZAO
Mono
Sotouboua
Blitta
Ogou
Yégué
Langabou
Anié
Retenue de Nangbéto
Badou
Atakpamé
Amlamé
Danyi-Apéyémé
Ountivou
Mt Agou 986m
Notsé
Tohoun
Kpalimé
Agou Gadzepe
Tabligbo
Tsévié
Kévé
Haho
Vogan
Aneho
LOMÉ
SLAVE COAST
Bight of Benin
ATLANTIC OCEAN
N
0 100 km
0 100 miles

TOGO

Total Land Area : 56 785 sq. km (21 924 sq. miles)

POPULATION

over 100 000

over 10 000

under 10 000

LAND HEIGHT

500m/1640ft

200m/656ft

Sea Level

DEFENSE

Compulsory military service

$30m

Down 12% in 2000

The military has an important role in Togo, and spending on defense is quite high. The army's senior ranks are dominated by loyalists from President Eyadéma's northern Kabré tribe. France guarantees Togo's security through a defense accord, and supplies most military equipment and training.

RESOURCES

Electric power 34,000 kw

23,074 tonnes

Not an oil producer

1.42m goats, 1m sheep, 289,200 pigs, 8.5m chickens

Phosphates, iron, chromite, bauxite, marble, dolomite

Phosphates are Togo's most important resource. Offshore oil and gas deposits were found in 1999. The Nangbeto Dam, constructed jointly with Benin and opened in 1988, has reduced dependence on Ghana for energy.

ENVIRONMENT

Sustainability rank: 105th

8%

0.2 tonnes per capita

Ecologists have been critical of the transformation of nature reserves into hunting grounds for the military elite. Other problems include coastal erosion around Aneho and desertification.

MEDIA

TV ownership low

Daily newspaper circulation 4 per 1000 people

PUBLISHING AND BROADCAST MEDIA

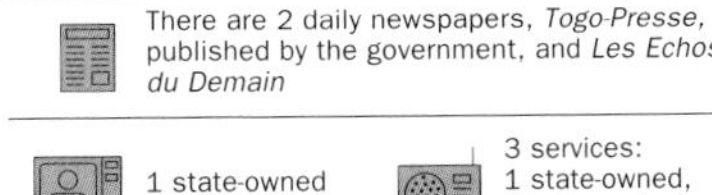

There are 2 daily newspapers, *Togo-Presse*, published by the government, and *Les Echos du Demain*

1 state-owned service

3 services: 1 state-owned, 2 independent

Opposition papers now challenge the government daily *Togo-Presse*, despite some official harassment.

CRIME

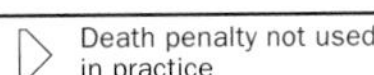

Death penalty not used in practice

2043 prisoners

Crime is rising

Togo is normally relatively peaceable, but urban crime generally increased during the 1990s, particularly during periods of political unrest in the capital.

CHRONOLOGY

After colonization by Germany in 1894, Togoland was divided between France and the UK in 1922.

- ❑ **1960** French sector independent as Togo (UK part joined to Ghana).
- ❑ **1967** Eyadéma takes power.
- ❑ **1991–1992** General strike; repression.
- ❑ **1993** Eyadéma elected president.
- ❑ **1998** Eyadéma claims victory in disputed election.

EDUCATION

School leaving age: 15

57%

15,028 students

Schooling is based on the French model. The University of Bénin in Lomé has more than 4000 students.

HEALTH

No welfare state health benefits

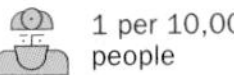

1 per 10,000 people

Malaria, diarrheal, infectious, and parasitic diseases

Health care suffers from a lack of resources. Around 6% of adults were HIV positive by 2000.

SPENDING

GDP/cap. increase

CONSUMPTION AND SPENDING

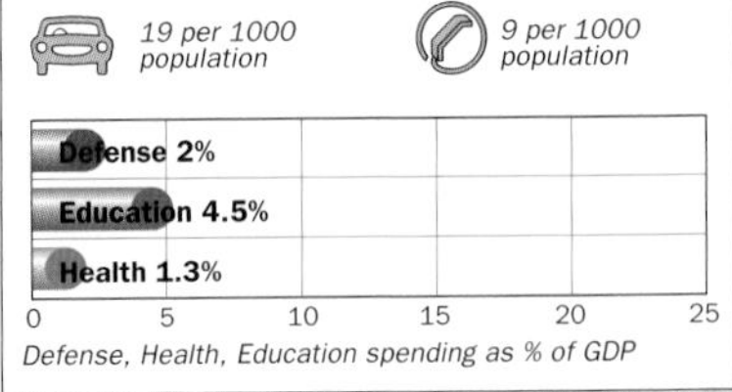
19 per 1000 population
9 per 1000 population
Defense 2%
Education 4.5%
Health 1.3%
0
5
10
15
20
25
Defense, Health, Education spending as % of GDP

Considerable wealth disparities exist between those who work the land and the country's political and business classes. The urban class has been hit by an economic downturn in the late 1990s.

ECONOMICS

Inflation 7.1% p.a. (1990–2000)

$1.32bn

698.7–736.7 CFA francs

SCORE CARD

- ❑ WORLD GNP RANKING........................149th
- ❑ GNP PER CAPITA $290
- ❑ BALANCE OF PAYMENTS...................–$106m
- ❑ INFLATION 1.9%
- ❑ UNEMPLOYMENT ....Not available

EXPORTS

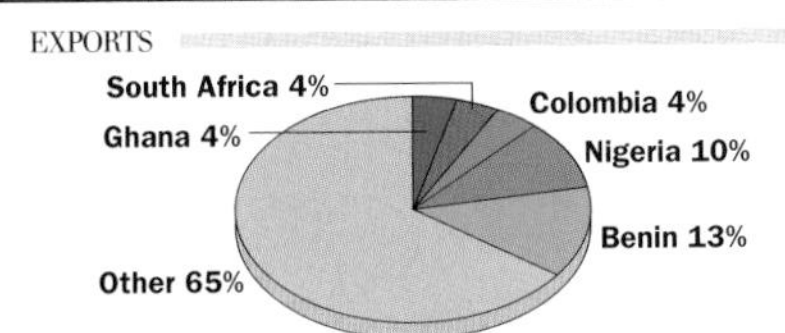
South Africa 4%
Colombia 4%
Ghana 4%
Nigeria 10%
Benin 13%
Other 65%

IMPORTS

Hong Kong 5%
China 7%
Ivory Coast 8%
Other 44%
France 11%
Ghana 25%

STRENGTHS

Efficient civil service. Ideal location for role as entrepôt, based on Lomé port. Resourceful entrepreneurs, notably market-women. Proceeds of widespread smuggling. Phosphate deposits have the world's highest mineral content. Self-sufficient in basic foodstuffs.

WEAKNESSES

Political pariah status led to aid reductions in the 1990s. Low world prices for phosphates. Hydropower generation is vulnerable to drought.

WORLD RANKING

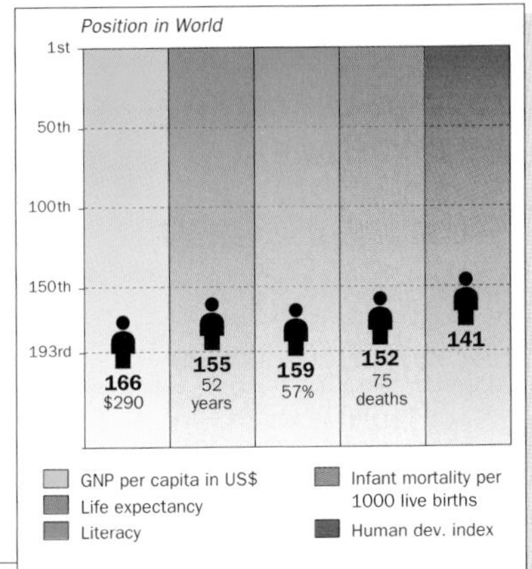
Position in World
1st
50th
100th
150th
193rd
166 $290
155 52 years
159 57%
152 75 deaths
141
GNP per capita in US$
Life expectancy
Literacy
Infant mortality per 1000 live births
Human dev. index

TONGA

PACIFIC OCEAN

OFFICIAL NAME: Kingdom of Tonga **CAPITAL:** Nuku'alofa **POPULATION:** 102,200
CURRENCY: Pa'anga (Tongan dollar) **OFFICIAL LANGUAGES:** English and Tongan

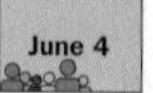

1970

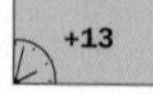
1970
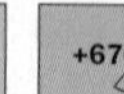
June 4

TO

+13

+676
.to

LOCATED IN THE SOUTH PACIFIC northeast of New Zealand, Tonga is an archipelago of 170 islands. These are divided into three main groups, Vava'u, Ha'apai, and Tongatapu. Tonga's easterly islands are generally low and fertile. Those in the west are higher and volcanic in origin. Tonga's economy is based on agriculture, especially coconut, cassava, and passion fruit production. Politics is effectively controlled by the king.

CLIMATE

Tropical oceanic

WEATHER CHART FOR NUKU'ALOFA

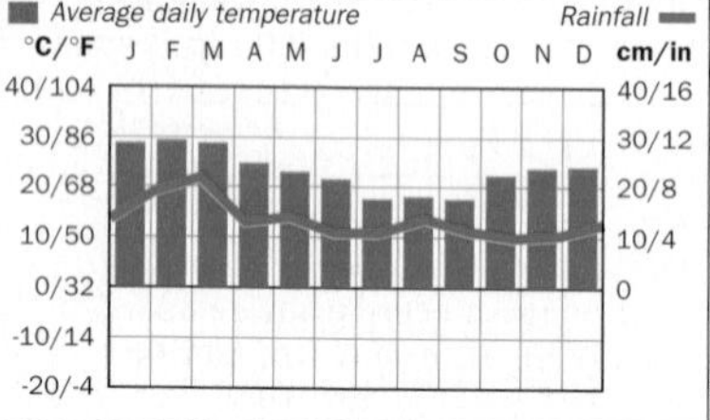
Average daily temperature
Rainfall
°C/°F
J F M A M J J A S O N D
cm/in
40/104
30/86
20/68
10/50
0/32
-10/14
-20/-4
40/16
30/12
20/8
10/4
0

Tonga has a tropical oceanic climate, with year-round temperatures ranging between 17°C (63°F) and 30°C (86°F).

TRANSPORTATION

Drive on left

Fua'amotu International, Tongatabu 67,000 passengers

18 ships 22,200 grt

THE TRANSPORTATION NETWORK

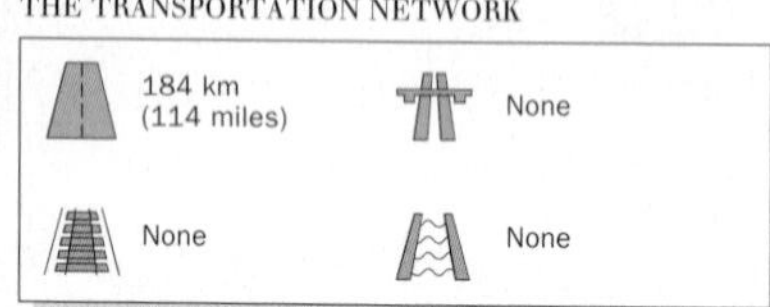
184 km
(114 miles)
None
None
None

Improvements at Fua'amotu Airport have led to an increase in flights to and from Tonga.

TOURISM

Visitors : Population 1:2.9

35,000 visitors

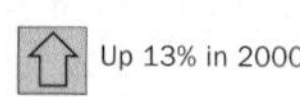
Up 13% in 2000

MAIN TOURIST ARRIVALS

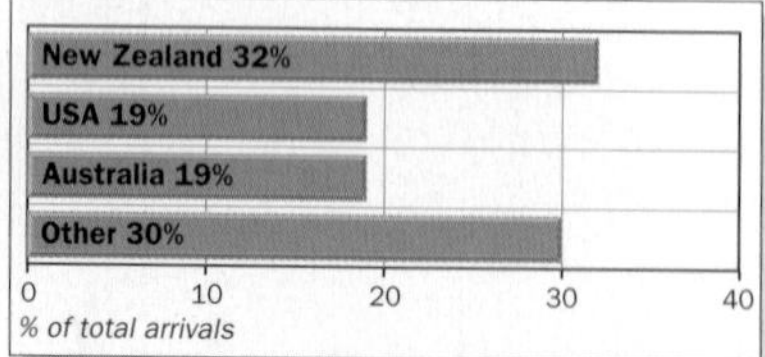
New Zealand 32%
USA 19%
Australia 19%
Other 30%
0
10
20
30
40
% of total arrivals

Tonga's main attractions are its tropical beaches. Flagging tourism has been boosted by political insecurity in the Solomon Islands and Fiji. Fears have been expressed that too many visitors may erode traditional Tongan culture.

Mountainous scenery typical of the westerly islands. Tonga's 170 islands are scattered over a wide expanse of the South Pacific. Only 45 are inhabited.

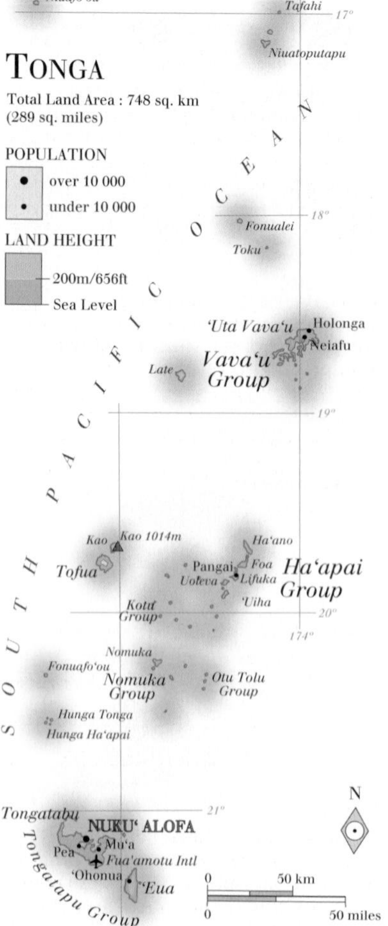
TONGA
Total Land Area : 748 sq. km
(289 sq. miles)
POPULATION
over 10 000
under 10 000
LAND HEIGHT
200m/656ft
Sea Level
SOUTH PACIFIC OCEAN
Niuafo'ou
Tafahi
Niuatoputapu
Fonualei
Toku
'Uta Vava'u
Holonga
Neiafu
Late
Vava'u Group
Kao
Kao 1014m
Tofua
Ha'ano
Pangai
Foa
Uoleva
Lifuka
Ha'apai Group
'Uiha
Kotu Group
Nomuka
Fonuafo'ou
Nomuka Group
Otu Tolu Group
Hunga Tonga
Hunga Ha'apai
Tongatabu
NUKU' ALOFA
Mu'a
Pea
Fua'amotu Intl
'Ohonua
'Eua
Tongatapu Group
Ata
N
0
50 km
50 miles
17°
18°
19°
20°
21°
174°
175°

PEOPLE

Pop. density medium

English, Tongan

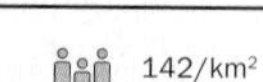
142/km² (368/mi²)

THE URBAN/RURAL POPULATION SPLIT

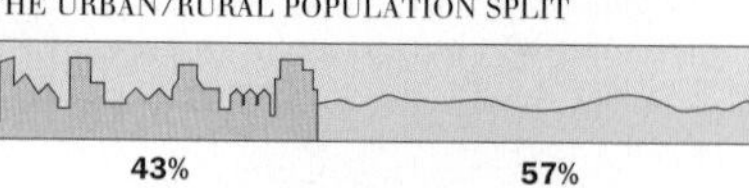
43%
57%

RELIGIOUS PERSUASION

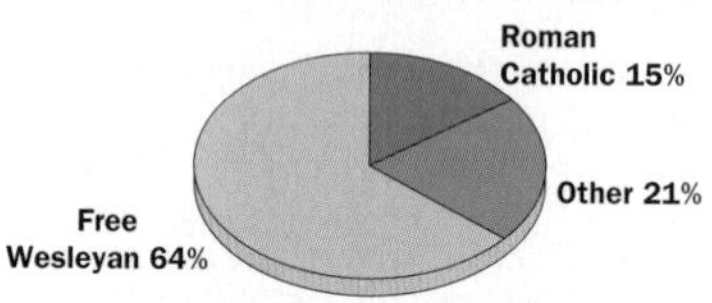
Roman Catholic 15%
Other 21%
Free Wesleyan 64%

Tonga has strong ethnic ties with eastern Fiji, and there has traditionally been considerable population movement between the two states. Tongans tend to see themselves as unique among Pacific islanders, retaining their monarchy and never having been fully colonized.

Respect for traditional values and institutions remains high. Tongans are strong churchgoers; the Wesleyan, Roman Catholic, and Mormon churches are influential and often fund education. A new generation of Western-educated Tongans is querying some traditional attitudes.

POLITICS

No multiparty elections

2002/2005

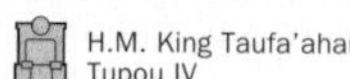
H.M. King Taufa'ahau Tupou IV

AT THE LAST ELECTION

Legislative Assembly 30 seats

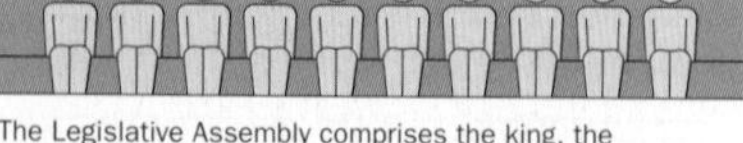

The Legislative Assembly comprises the king, the 11 members of the Privy Council, 9 members indirectly elected by nobles, and 9 directly elected members. There are no political parties: of those elected in 2002, 7 were part of the Human Rights and Democracy Movement.

The main power brokers in Tongan politics are the king, the noble establishment, and the landowners. King Taufa'ahau, on the throne since 1965, effectively heads his government, frequently exercising kingly powers. The Legislative Assembly defers to his judgment and the king has instigated several development projects which have been undertaken without reference to the government.

The king's resistance to growing calls for greater democracy was highlighted in 2000, when he passed over his reformist eldest son and appointed his conservative third son, Prince Ulukalala Lavaka Ata, as premier for life.

WORLD AFFAIRS

Joined UN in 1999

ACP
Comm
PC
PIF
ADB

In 1996, after accepting the Pacific nuclear interests of the US and France in the preceding decade, Tonga finally acceded to the South Pacific Nuclear-free Zone Treaty. It broke its ties with Taiwan in 1998 in return for closer relations with China.

AID

Recipient

 $19m (receipts) 
Down 10% in 2000

Aid finances major infrastructure projects; Japan, Australia, and New Zealand are primary donors. After losing aid in 1997, Tonga fought to reclaim its Least Developed Country status, which guarantees funds.

DEFENSE

No compulsory military service

 $2m (estimate)  No significant change

Tonga has a small defense force, which includes both regulars and reserves. Tongan police assisted in security efforts in the Solomon Islands in 2000.

ECONOMICS

Inflation 1.7% p.a. (1990–2000)

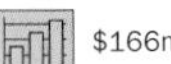 $166m

1.7997–1.9535 pa'anga

SCORE CARD

- ❑ WORLD GNP RANKING........186th
- ❑ GNP PER CAPITA........$1660
- ❑ BALANCE OF PAYMENTS........Zero
- ❑ INFLATION........5.9%
- ❑ UNEMPLOYMENT........13%

STRENGTHS

Agriculture contributes largest percentage of GDP. Tourism main source of hard currency earnings.

WEAKNESSES

Off main shipping routes. Aid-dependent. Importer of food. High youth unemployment. Corruption scandal in royal court in 2001.

EXPORTS

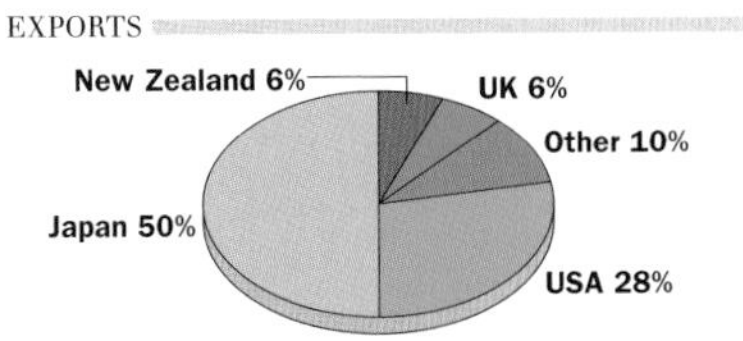
New Zealand 6%
UK 6%
Other 10%
Japan 50%
USA 28%

IMPORTS

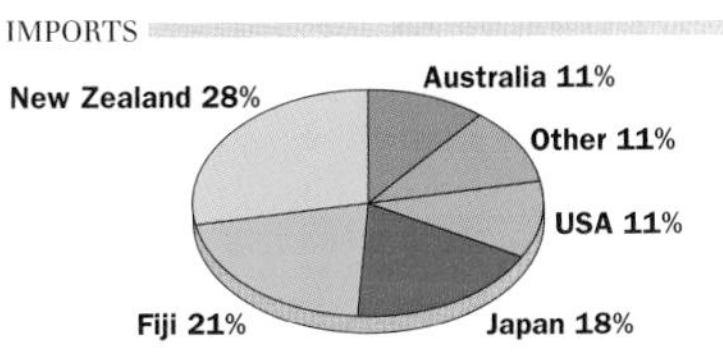
New Zealand 28%
Australia 11%
Other 11%
USA 11%
Fiji 21%
Japan 18%

RESOURCES

Electric power 7000 kw

 3663 tonnes  Not an oil producer

80,853 pigs, 12,500 goats, 300,000 chickens

None

Tonga has no strategic or mineral resources. Electricity is entirely generated from imported fuel. Recent exploration has failed to identify any oil reserves. Tongan waters contain large numbers of tuna.

ENVIRONMENT

Not available

None

1.2 tonnes per capita

Tonga does not suffer from serious environmental problems, although it is occasionally afflicted by natural disasters, such as the recurring El Niño effect. Commercial activity has made little impact on the environment.

MEDIA

TV ownership low

 Daily newspaper circulation 72 per 1000 people

PUBLISHING AND BROADCAST MEDIA

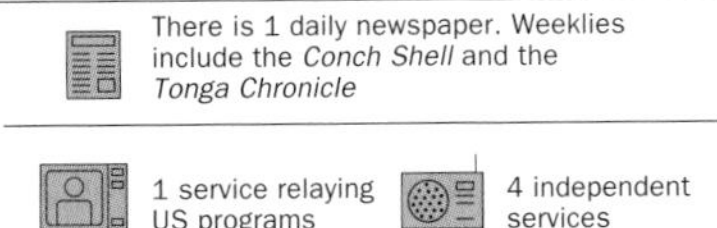
There is 1 daily newspaper. Weeklies include the *Conch Shell* and the *Tonga Chronicle*
1 service relaying US programs
4 independent services

In 2000 Cable and Wireless relinquished control of Tonga's telecommunications services, which it had run since 1978, to a local company.

CRIME

Death penalty not used in practice

 68 prisoners 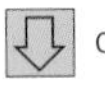 Crime is falling

Crime rates are generally low, partly due to the strong influence of the family. However, offenses such as breaking and entering have increased, along with rising unemployment levels, among young Tongans.

EDUCATION

School leaving age: 14

 98% 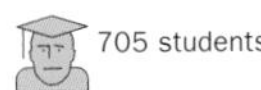
705 students

Church participation in schools is high. Plans for a new national university were approved in 2000 following violence in Fiji, where Tongan students attend the University of the South Pacific.

HEALTH

No welfare state health benefits

 1 per 2176 people  Cerebrovascular, heart, and diarrheal diseases

Tonga has some modern health care facilities. However, patients have to be flown out to Australia or New Zealand for sophisticated surgery.

CHRONOLOGY

Originally discovered by the Polynesians, Tonga was visited by the Dutch in the 17th century and Capt. Cook in the 18th century. In the latter half of the 19th century, during the reign of King George Tupou I, the islands became a unified state after a period of civil war.

- ❑ **1875** First constitution established.
- ❑ **1900** Concern over German ambitions in region; Treaty of Friendship and Protection with UK.
- ❑ **1918–1965** Reign of Queen Salote Tupou III.
- ❑ **1958** Greater autonomy from UK enshrined in Friendship Treaty.
- ❑ **1965** King Taufa'ahau Tupou IV accedes on his mother's death.
- ❑ **1970** Full independence within British Commonwealth.
- ❑ **1988** Treaty allows US nuclear warships right of transit.
- ❑ **2000** King appoints third son as prime minister.
- ❑ **2001** Court jester steals US$20 million of state funds.
- ❑ **2002** Election sees strong showing by prodemocracy candidates, as in 1996 and 1999.

SPENDING

GDP/cap. increase

CONSUMPTION AND SPENDING

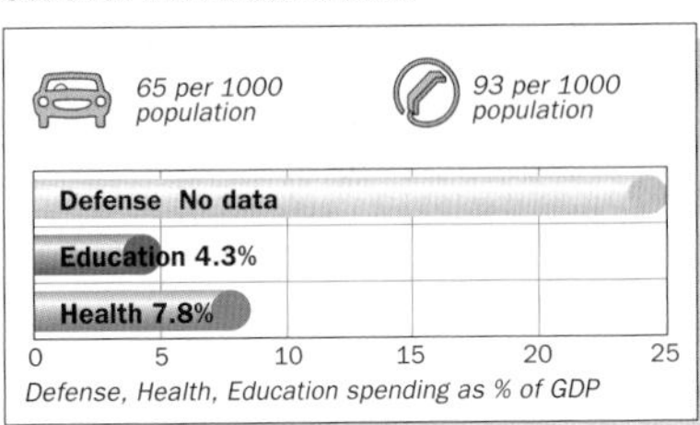
65 per 1000 population
93 per 1000 population
Defense No data
Education 4.3%
Health 7.8%
0
5
10
15
20
25
Defense, Health, Education spending as % of GDP

Tongans indulge in few ostentatious displays of wealth. The well-off provide financial support for relatives.

WORLD RANKING

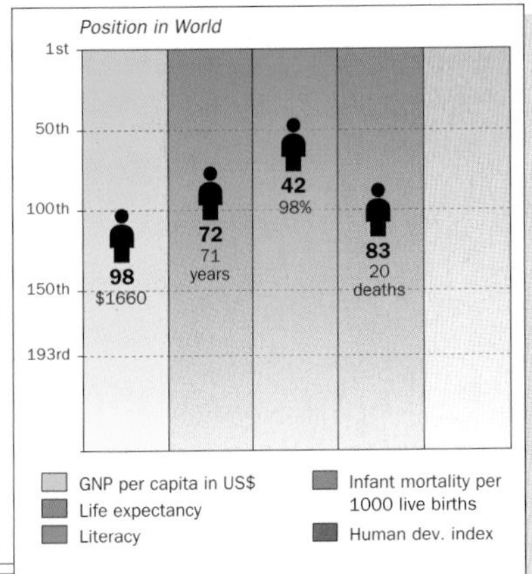
Position in World
1st
50th
100th
150th
193rd
98
$1660
72
71 years
42
98%
83
20 deaths
GNP per capita in US$
Life expectancy
Literacy
Infant mortality per 1000 live births
Human dev. index

T

TRINIDAD & TOBAGO

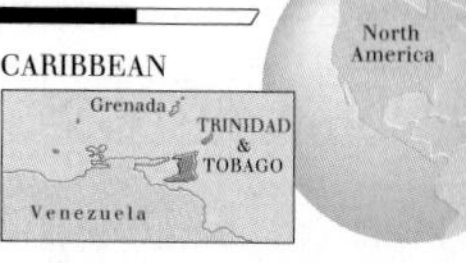

CARIBBEAN
Grenada
TRINIDAD & TOBAGO
Venezuela
North America

OFFICIAL NAME: Republic of Trinidad and Tobago **CAPITAL:** Port-of-Spain
POPULATION: 1.3 million **CURRENCY:** Trinidad and Tobago dollar **OFFICIAL LANGUAGE:** English

1962

1962

Aug 31
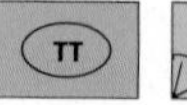
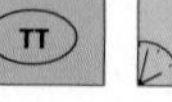
TT
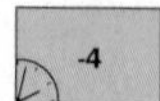
-4

+1868

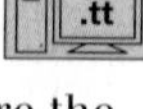
.tt

THE TWO ISLANDS OF Trinidad and Tobago are the most southerly of the Caribbean Windward Islands and lie just 15 km (9 miles) off the Venezuelan coast. They gained joint independence from Britain in 1962, and Tobago was given internal autonomy in 1987. The spectacular mountain ranges and large swamps are rich in tropical flora and fauna. Pitch Lake in Trinidad is the world's largest natural reservoir of asphalt.

CLIMATE

Tropical oceanic

WEATHER CHART FOR PORT-OF-SPAIN

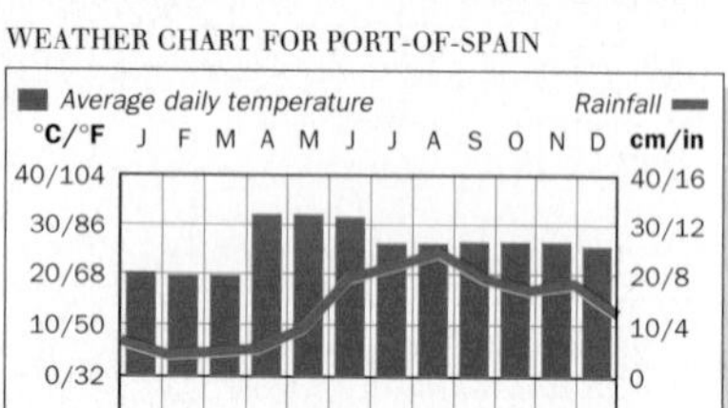

Average daily temperature
Rainfall
°C/°F J F M A M J J A S O N D cm/in
40/104
30/86
20/68
10/50
0/32
-10/14
-20/-4
40/16
30/12
20/8
10/4
0

The islands are a little warmer than others in the Caribbean and escape the hurricanes, which pass by to the north.

TRANSPORTATION

Drive on left

Piarco International, Port-of-Spain
1.93m passengers

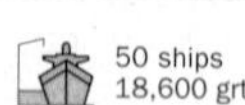

50 ships
18,600 grt

THE TRANSPORTATION NETWORK

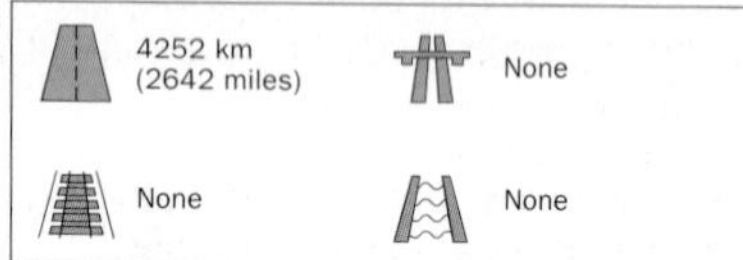

4252 km (2642 miles)
None
None
None

The road network is well developed; there are taxis or minibuses for set routes. National carrier BWIA and Air Caribbean operate Trinidad–Tobago routes. BWIA flies to the US.

TOURISM

Visitors : Population 1:3.9

336,000 visitors

Down 3% in 1999

MAIN TOURIST ARRIVALS

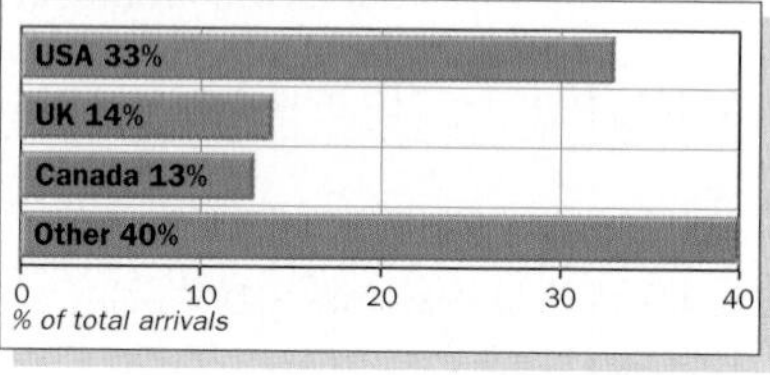

USA 33%
UK 14%
Canada 13%
Other 40%
0
10
20
30
40
% of total arrivals

Oil revenue meant that Trinidad was one of the last Caribbean states to develop tourism. Most is centered on Tobago (said to be the model for the island in *Robinson Crusoe*), famous for its huge variety of South American wildlife, including 210 species of tropical bird.

TRINIDAD & TOBAGO

Total Land Area : 5128 sq. km (1980 sq. miles)

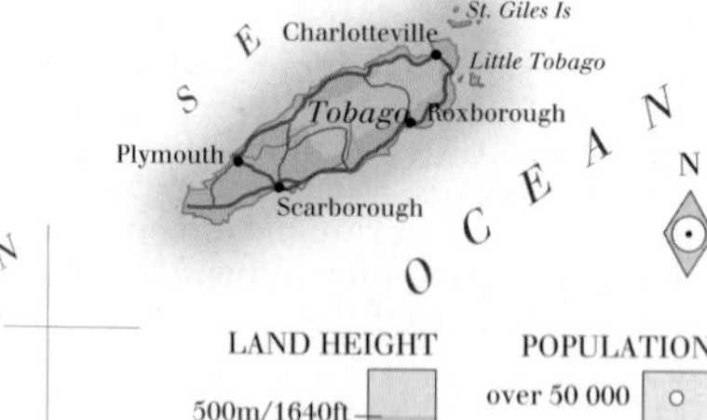

St. Giles Is
Charlotteville
Little Tobago
Tobago
Roxborough
Plymouth
Scarborough
N
LAND HEIGHT
500m/1640ft
200m/656ft
Sea Level
POPULATION
over 50 000
over 10 000
under 10 000
0 30 km
0 30 miles

CARIBBEAN SEA
ATLANTIC OCEAN
Matelot
Blanchisseuse
Redhead
Chacachacare
Monos
Carenage
Chaguaramas
Cantaro
Mount Aripo 940m
Salibea
Four Roads
Morvant
Arima
PORT-OF-SPAIN
Valsayn
Tunapuna
Guaico
Caroni Swamp
Curepe
Caroni
Sangre Grande
Chaguanas
The Dragon's Mouths
Gulf of Paria
Couva
Trinidad
Tabaquite
Rio Claro
St. Joseph
San Fernando
La Brea
Pitch Lake
Princes Town
Ortoire
Point Fortin
Siparia
Moruga
Fullarton
The Serpent's Mouth
Columbus Channel

PEOPLE

Pop. density high

English Creole, English, Hindi, French, Spanish

253/km² (656/mi²)

THE URBAN/RURAL POPULATION SPLIT

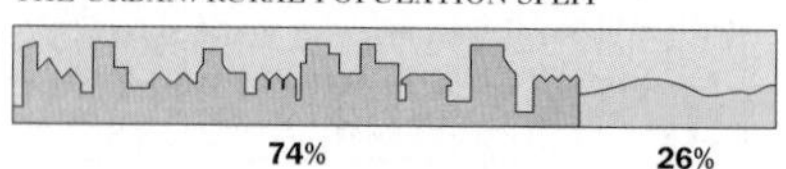

74%
26%

ETHNIC MAKEUP

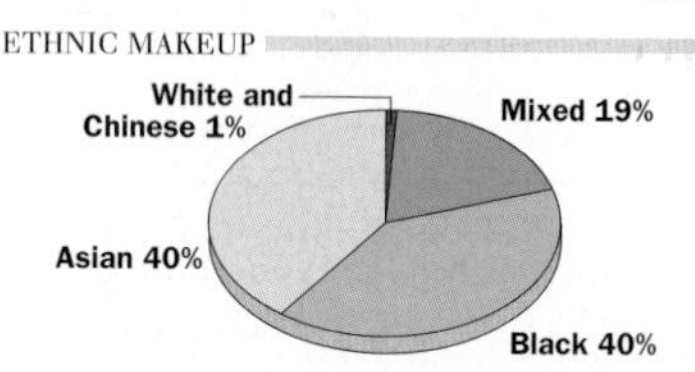

White and Chinese 1%
Mixed 19%
Asian 40%
Black 40%

Trinidad's south Asian community is the largest in the Caribbean and holds on to its Muslim and Hindu inheritance. The open discussion of racial issues goes some way to dissipating latent tensions with the predominantly Christian black Trinidadians.

POLITICS

Multiparty elections

L. House 2001/2002
U. House 2002/2002

President Arthur N. Robinson

AT THE LAST ELECTION

House of Representatives 36 seats

50% PNM **50% UNC**

PNM = People's National Movement
UNC = United National Congress

Senate 31 seats

Senators are appointed by the president, including 16 nominated by the prime minister and 6 by the leader of the opposition

The increasingly right-wing PNM dominated politics from independence in 1962 to the 1990s, leading to political fragmentation and an attempted coup by Muslim extremists in 1990. The UNC's Basdeo Panday became the first ethnic Asian prime minister in 1995. He pledged his coalition government to reducing unemployment, crime, and racial discrimination. The UNC's majority was reduced to a tie with the PNM in the 2001 elections. President Arthur Robinson favored PNM leader Patrick Manning as prime minister, but the political stalemate led to new elections being called for late 2002.

Tobago's white sand beaches, *verdant landscape, and natural anchorages have enabled it to develop a thriving tourist industry.*

WORLD AFFAIRS

Joined UN in 1962

ACS
Caricom
Comm
NAM
OAS

Trinidad withdrew in 1998 and 1999 from the Inter-American and UN Commissions on Human Rights respectively, over appeals against death sentences. New trade agreements include those with Costa Rica, Panama, the Dominican Republic, and Mexico. Sea border disputes with Venezuela relate to fishing and marine oil rights.

AID

Recipient

US$26m (receipts)

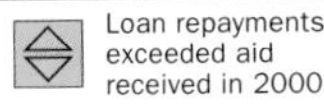
Loan repayments exceeded aid received in 2000

Aid is modest: China provided an interest-free loan of US$20 million in 2000 to help small businesses.

DEFENSE

No compulsory military service

US$35m

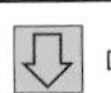
Down 12% in 2000

Defense forces comprise a land army and a coast guard (with air wing), used to patrol fishing grounds.

ECONOMICS

Inflation 5.5% p.a. (1990–2000)

US$6.42bn

6.240–6.115 Trinidad and Tobago dollars

SCORE CARD

- ❑ WORLD GNP RANKING....................103rd
- ❑ GNP PER CAPITA......................US$4930
- ❑ BALANCE OF PAYMENTS.............–US$644m
- ❑ INFLATION.........................3.6%
- ❑ UNEMPLOYMENT......................13%

STRENGTHS

Oil, which accounts for 70% of export earnings. Gas increasingly exploited to support new industries. Methanol, ammonia, iron, and steel exports. Tourism, especially on Tobago.

WEAKNESSES

Insufficiently diversified economy. Reliance on oil – highly sensitive to world price movements. High unemployment.

EXPORTS

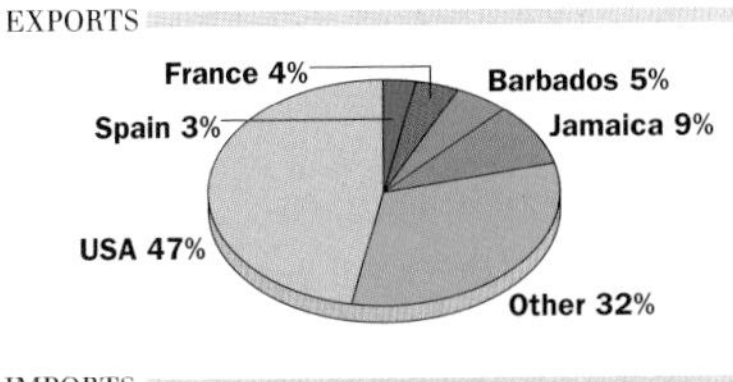
France 4%
Barbados 5%
Spain 3%
Jamaica 9%
USA 47%
Other 32%

IMPORTS

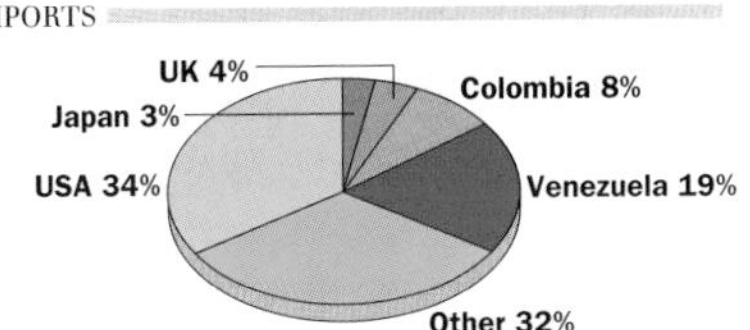
UK 4%
Colombia 8%
Japan 3%
USA 34%
Venezuela 19%
Other 32%

RESOURCES

Electric power 1.2m kw

15,027 tonnes

135,000 b/d (reserves 700m barrels)

59,000 goats, 41,000 pigs, 10m chickens

Oil, natural gas, asphalt, iron

Oil and gas are major resources. Big offshore gas and oil finds in 1998 included the largest discovery of crude oil in 25 years.

ENVIRONMENT

Sustainability rank: 121st

6%

17.4 tonnes per capita

Spillages from oil tankers threaten coastal conservation areas such as the Caroni Swamp, with its many species of butterflies. Forest fires due to periodic drought, and traffic-related pollution and congestion are serious concerns.

MEDIA

TV ownership high

Daily newspaper circulation 123 per 1000 people

PUBLISHING AND BROADCAST MEDIA

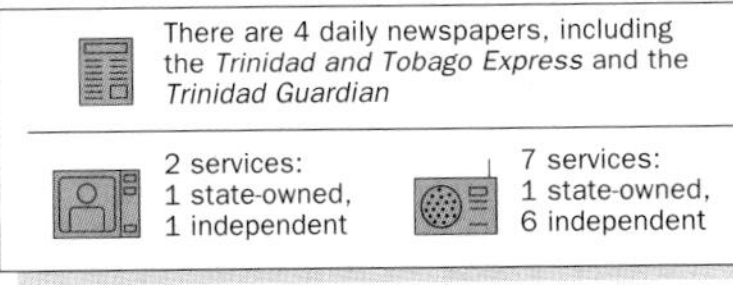
There are 4 daily newspapers, including the *Trinidad and Tobago Express* and the *Trinidad Guardian*

2 services: 1 state-owned, 1 independent

7 services: 1 state-owned, 6 independent

Television schedules are dominated by US programs, while around carnival time radio stations play little but calypso.

CRIME

Death penalty in use

4794 prisoners

Down 63% 1992–1998

Narcotics-related crime increases the murder rate. The country was party to the decision taken in 2001 to replace the authority of the UK's Privy Council with a Caribbean Court of Justice.

EDUCATION

School leaving age: 12

94%

6860 students

Education is based on the former British system. Most students go on to the University of the West Indies; Trinidad hosts the St. Augustine campus. Wealthy Trinidadians, however, go to universities in the US.

HEALTH

Welfare state health benefits

1 per 1250 people

Heart disease, cancers, diabetes, accidents, violence

Oil wealth has given Trinidad a better public health service than most Caribbean states and more private clinics, mainly serving the expatriate community. However, treatment delays are a problem. The spread of HIV/AIDS is of particular concern on Tobago.

CHRONOLOGY

Britain seized Trinidad from Spain in 1797 and Tobago from France in 1802. They were unified in 1888.

- ❑ **1956** Eric Williams founds PNM and wins general election, mainly with support from blacks.
- ❑ **1958–1961** Member of West Indian Federation.
- ❑ **1962** Independence.
- ❑ **1970** Black Power demonstrations.
- ❑ **1980** Tobago gets own House of Assembly; internal autonomy 1987.
- ❑ **1990–1991** Premier taken hostage in failed fundamentalist coup. PNM returned to power.
- ❑ **1995** UNC's Basdeo Panday is first ethnic Asian prime minister.
- ❑ **1998–1999** Trinidad withdraws from international human rights bodies over death sentences.
- ❑ **2001** Elections result in tie. PNM appointed to government.

SPENDING

GDP/cap. increase

CONSUMPTION AND SPENDING

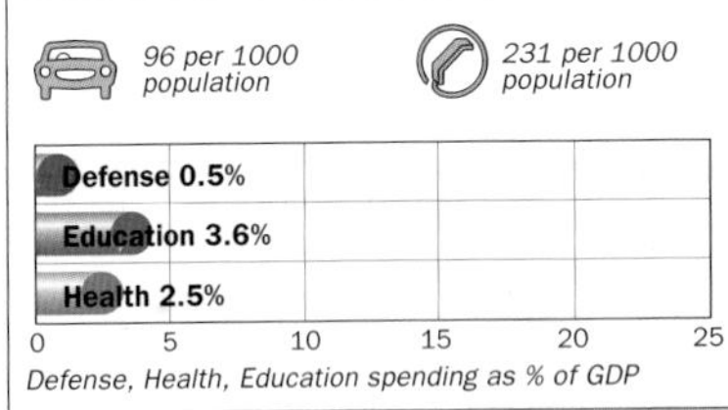
96 per 1000 population
231 per 1000 population
Defense 0.5%
Education 3.6%
Health 2.5%
0
5
10
15
20
25
Defense, Health, Education spending as % of GDP

In Trinidad wealth disparities between the affluent oil-rich business elite, many of whom are expatriate, and farm laborers are particularly marked. Service workers in Tobago's high-value tourism sector are poorly paid. Rural poverty in the interior, particularly among south Asian Trinidadian farmers, is a serious problem.

WORLD RANKING

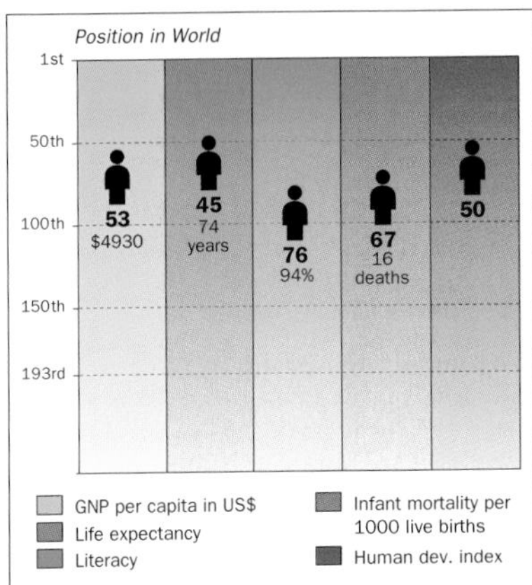
Position in World
1st
50th
100th
150th
193rd
53
$4930
45
74 years
76
94%
67
16 deaths
50
GNP per capita in US$
Life expectancy
Literacy
Infant mortality per 1000 live births
Human dev. index

T

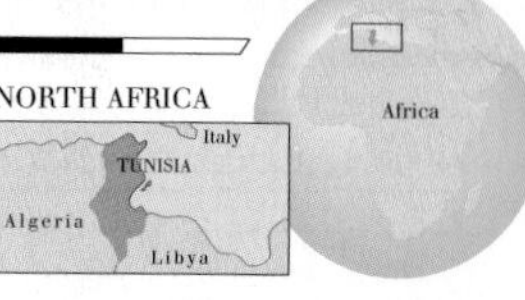
NORTH AFRICA
Africa
Italy
TUNISIA
Algeria
Libya

TUNISIA

OFFICIAL NAME: Republic of Tunisia **CAPITAL:** Tunis
POPULATION: 9.6 million **CURRENCY:** Tunisian dinar **OFFICIAL LANGUAGE:** Arabic

1956

1956
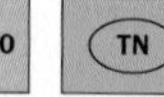
March 20
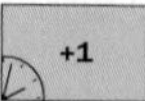
TN

+1

+216

.tn

NORTH AFRICA'S SMALLEST country, Tunisia lies sandwiched between Libya and Algeria. The populous north is mountainous, fertile in places and has a long Mediterranean coastline. The south is largely desert. Habib Bourguiba ruled the country from independence in 1956 until a bloodless coup in 1987. Under President Ben Ali, the government has moved toward multiparty democracy, but is challenged by Islamic fundamentalists. Closer ties with the EU, whose members include Tunisia's main trading partners, were strengthened through the first Euro-Mediterranean conference held in 1995. Manufacturing and tourism are expanding.

CLIMATE

Mediterranean/ hot desert

WEATHER CHART FOR TUNIS

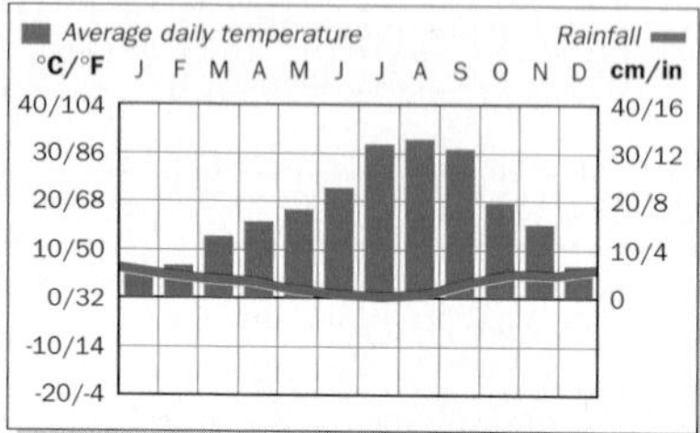
Average daily temperature
Rainfall
°C/°F
J F M A M J J A S O N D
cm/in
40/104
30/86
20/68
10/50
0/32
-10/14
-20/-4
40/16
30/12
20/8
10/4
0

Tunisia is hot in summer. The north is often wet and windy in winter. The far south is arid. The spring brings the dry, dusty *chili* wind from the Sahara.

TRANSPORTATION

Drive on right

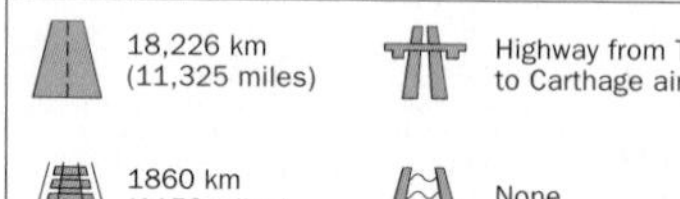
Skanes Airport, Monastir
3.92m passengers
78 ships
193,500 grt

THE TRANSPORTATION NETWORK

18,226 km (11,325 miles)

Highway from Tunis to Carthage airport

1860 km (1156 miles)

None

Tunisia has six international airports. A highway from Tunis to Carthage airport opened in 1993. A light metro in Tunis and a rail link from Gafsa to Gabès are being built. The southern third of the country has few roads.

TOURISM

Visitors : Population 1:1.9

5.1m visitors

Up 5% in 2000

MAIN TOURIST ARRIVALS

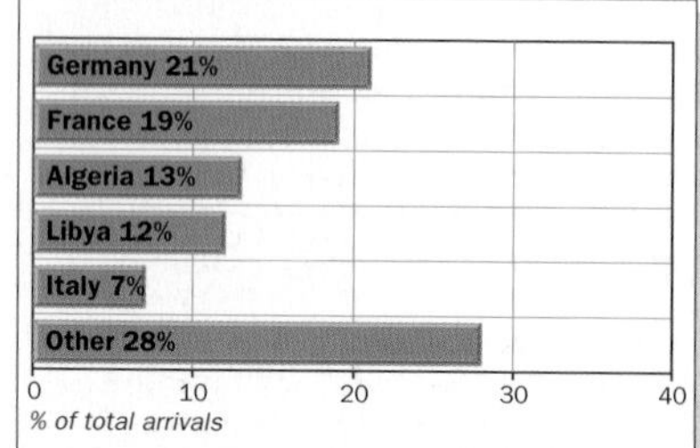
Germany 21%
France 19%
Algeria 13%
Libya 12%
Italy 7%
Other 28%
0
10
20
30
40
% of total arrivals

Tourists have flocked to Tunisia since the 1960s, attracted by its winter sunshine, beaches, desert, and archaeological remains. One of the Mediterranean's cheapest package destinations, Tunisia attracts almost three million European visitors a year, although some tourists are deterred by the fear of attacks by Islamists. In 2002, 14 German tourists were killed by a suicide bombing on the island of Jerba.

Tourism employs more than 200,000 people and is a focus of investment. However, concern about its environmental impact is growing.

PEOPLE

Pop. density medium

Arabic, French

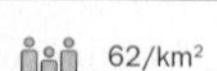
62/km² (160/mi²)

THE URBAN/RURAL POPULATION SPLIT

66% 34%

RELIGIOUS PERSUASION

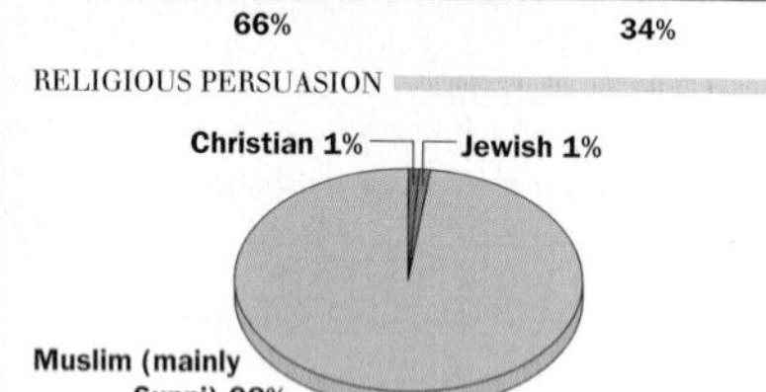
Christian 1%
Jewish 1%
Muslim (mainly Sunni) 98%

ETHNIC MAKEUP

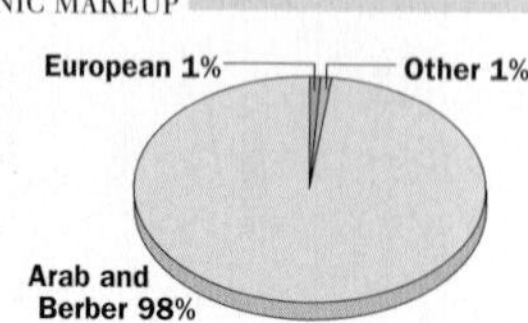
European 1%
Other 1%
Arab and Berber 98%

The population is almost entirely Muslim, of Arab and Berber descent, although there are Jewish and Christian minorities. Many Tunisians still live in extended family groups, in which three or four generations are represented.

Tunisia has traditionally been one of the most liberal Arab states. The 1956 Personal Statutes Code of President Bourguiba gave women fuller rights than in any other Arab country. Further legislation has since given women the right to custody of children in divorce cases, made family violence against women punishable by law, and helped divorced women to get alimony. Family planning and contraception have been freely available since the early 1960s. Tunisia's population growth rate has halved since the 1980s. Women make up 31% of the total workforce and 35% of the industrial workforce. Company ownership by women is steadily increasing; politics, however, remains an exclusively male preserve.

These freedoms are threatened by the growth in recent years of Islamic fundamentalism, which also worries the mainly French-speaking political and business elite who wish to strengthen links with Europe.

The Ben Ali regime has been criticized for its actions against Islamist activists. Despite continuing condemnation of its human rights record, the government's efforts to foster democracy were praised in 2002 when it received the "Mediterranean Award," given by European human rights leagues.

POPULATION AGE BREAKDOWN

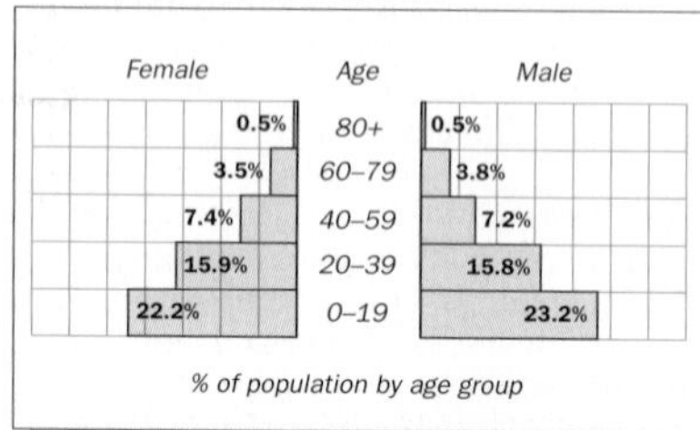
Female
Age
Male
0.5% 80+ 0.5%
3.5% 60–79 3.8%
7.4% 40–59 7.2%
15.9% 20–39 15.8%
22.2% 0–19 23.2%
% of population by age group

***Roman remains** in the western Tozeur region. Diverse archeological remains can be found throughout Tunisia.*

T

POLITICS

Multiparty elections

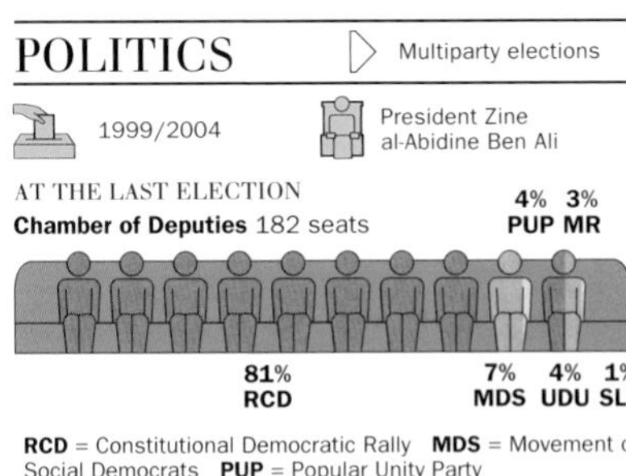
1999/2004
President Zine al-Abidine Ben Ali
AT THE LAST ELECTION
Chamber of Deputies 182 seats
4% PUP
3% MR
81% RCD
7% MDS
4% UDU
1% SLP

RCD = Constitutional Democratic Rally **MDS** = Movement of Social Democrats **PUP** = Popular Unity Party **UDU** = Unionist Democratic Union **MR** = Movement for Renewal **SLP** = Social Liberal Party

Formally a multiparty democracy since 1988, Tunisia is still dominated by the RCD and President Ben Ali.

***President Ben Ali** became head of state in 1987.*

***Mohammed Ghannouchi,** prime minister since 1999.*

PROFILE

President Ben Ali has made some effort to liberalize the political system. The life presidency has been abolished, and political parties and press freedom are encouraged. While allowing for a degree of political plurality, a complex proportional representation system ensured that there was an overwhelming victory for the RCD in the 1994 and 1999 elections. Since 1994 there has been evidence of a renewed crackdown against the left-wing opposition. Measures to promote human rights and further democratization were overshadowed in 2002 when a referendum on constitutional change gave Ben Ali a mandate to stand for a further two terms.

MAIN POLITICAL ISSUES

Fundamentalism

The RCD has clamped down on Islamic fundamentalists, in particular the outlawed Al-Nahda, or Renewal Party. In 1991, 500 Al-Nahda members were arrested following a failed coup, thought to be inspired by fundamentalists. Its leader, Rachid Ghannouchi, is in exile.

Human rights

The RCD has been under increasing attack over its human rights record. The Tunisia League of Human Rights' activities were suspended in January 2001. The RCD is committed to promoting women's rights.

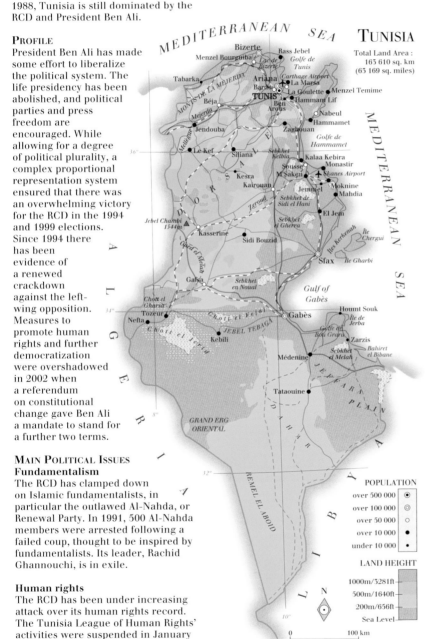

TUNISIA

Total Land Area : 163 610 sq. km (63 169 sq. miles)

POPULATION: over 500 000; over 100 000; over 50 000; over 10 000; under 10 000

LAND HEIGHT: 1000m/3281ft; 500m/1640ft; 200m/656ft; Sea Level

0 – 100 km; 0 – 100 miles

WORLD AFFAIRS

Joined UN in 1956

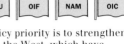

AL
AMU

OIF

NAM
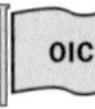
OIC

A foreign policy priority is to strengthen contacts with the West, which have generally been good because of Tunisia's liberal economic and social policies. Attention is focused especially on the EU, Tunisia's main export market. Tunisia played an important role in the run-up to the first Euro-Mediterranean conference, which was held in 1995.

Tunis was host to the Palestine Liberation Organization after it was expelled from Lebanon. Relations with other Arab states, particularly Kuwait and Saudi Arabia, were soured by Tunisia's support for Iraq in the Gulf War. The government regards the political impact of Islamic fundamentalism in neighboring Algeria with concern. Relations with Libya are improving, helped by the fact that Tunisia turned a blind eye to sanctions busters operating through its territory.

CHRONOLOGY

Tunisia has been home to the Zenata Berbers since earliest times and its history is linked to the rise and fall of the Mediterranean-centered empires. Carthage (near present-day Tunis), founded by the Phoenicians in the 9th century BCE, became the hub of a 1000-year trading empire which linked European and African trading networks. Tunisia was then ruled by the Romans, Byzantines, Egyptians, Ottomans, and, finally, the French.

- **1883** La Marsa Treaty makes Tunisia a French protectorate, ending its semi-independence. Bey of Tunis remains monarch.
- **1900** Influx of French and Italians.
- **1920** Destour (Constitution) Party formed; calls for self-government.
- **1935** Habib Bourguiba forms Neo-Destour (New Constitution) Party.
- **1943** Defeat of Axis powers by British troops restores French rule.
- **1955** Internal autonomy. Bourguiba returns from exile.
- **1956** Independence. Bourguiba elected prime minister. Personal Statutes Code gives rights to women. Family planning introduced.
- **1957** Bey is deposed. Tunisia becomes republic with Bourguiba as first president.
- **1964** Neo-Destour made sole legal party; renamed Destour Socialist Party (PSD). Moderate socialist economic program is introduced.
- **1969** Agricultural collectivization program, begun 1964, abandoned.

T

CHRONOLOGY *continued*

- **1974** Bourguiba elected president-for-life by National Assembly.
- **1974–1976** Hundreds imprisoned for belonging to "illegal organizations."
- **1978** Trade union movement, UGTT, holds 24-hour general strike; more than 50 killed in clashes. UGTT leadership replaced with PSD loyalists.
- **1980** New prime minister Muhammed Mazli ushers in greater political tolerance.
- **1981** Elections. Opposition groups allege electoral malpractice.
- **1984** Widespread riots after food price increases.
- **1986** Gen. Zine al-Abidine Ben Ali becomes interior minister. Four Muslim fundamentalists sentenced to death.
- **1987** Fundamentalist leader Rachid Ghannouchi arrested. Ben Ali becomes prime minister; takes over presidency after doctors certify Bourguiba senile. PSD renamed RCD.
- **1988** Most political prisoners released. Constitutional reforms introduce multiparty system and abolish life presidency. Two opposition parties legalized.
- **1989** Elections: RCD wins all seats, Ben Ali president. Fundamentalists take 13% of vote.
- **1990** Tunisia backs Iraq in Gulf War. Clampdown on fundamentalists intensifies.
- **1991** Abortive coup blamed on Al-Nahda; over 500 arrests.
- **1993** Multiparty agreement on electoral reform.
- **1994** Presidential and legislative elections. Ben Ali, sole candidate, is reelected.
- **1996** MDS leader Mohammed Moada imprisoned.
- **1999** Ben Ali and RCD win elections.
- **2002** Referendum allows Ben Ali to stand for office after 2004. Suicide bomb attack in Jerba kills 14 German tourists.

AID

Recipient

$223m (receipts)

Down 12% in 2000

The EU and France are the largest donors, providing over half of all bilateral aid. Japan, Germany, Italy, Belgium, and the Netherlands are other important sources of assistance. Oil-rich Arab states, including Saudi Arabia and Kuwait, suspended their aid programs to Tunisia after 1990 because of its pro-Iraq stance in the Gulf War. Tunisia's total external debt is estimated at over half of GNP.

DEFENSE

Compulsory military service

$350m

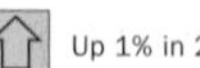
Up 1% in 2000

Despite its small size – 35,000 troops, of which around two-thirds are conscripts – the military is an important political force, armed mainly with US weapons. Officer training is carried out in the US and France, as well as in Tunisia. Border security with Algeria was tightened in 1995 after Algerian Islamists attacked Tunisian border guards in protest against Tunisian support for Algerian security forces.

TUNISIAN ARMED FORCES

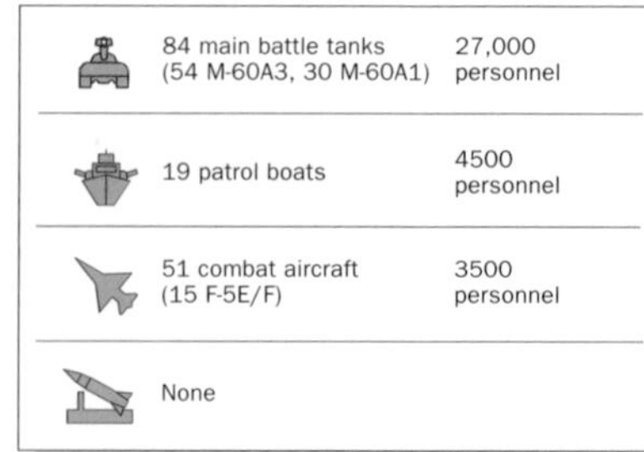

84 main battle tanks (54 M-60A3, 30 M-60A1)	27,000 personnel
19 patrol boats	4500 personnel
51 combat aircraft (15 F-5E/F)	3500 personnel
None	

ECONOMICS

Inflation 4.5% p.a. (1990–2000)

$20.1bn

1.384–1.461 Tunisian dinars

SCORE CARD

World GNP Ranking	62nd
GNP per Capita	$2100
Balance of Payments	–$821m
Inflation	2.9%
Unemployment	16%

EXPORTS

Spain 5%, Belgium 5%, Germany 12%, Italy 22%, France 26%, Other 30%

IMPORTS

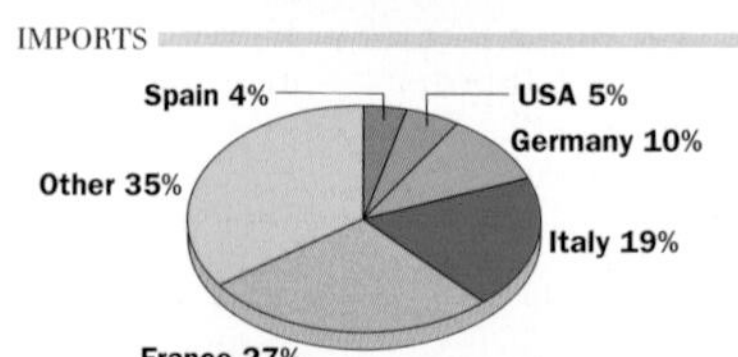

Spain 4%
USA 5%
Germany 10%
Other 35%
Italy 19%
France 27%

Strengths

Well-diversified economy, despite limited resources. Tourism. Oil and gas exports, also agricultural exports: olive oil, olives, citrus fruit, dates. Expanding manufacturing sector, average annual increase was 5.4% in 1990–1998; important sectors are textiles, construction materials, machinery, chemicals. European investment. Ranked as most competitive economy in Africa in World Economic Forum's 2000–2001 report.

Weaknesses

Dependence on growth of drought-prone agricultural sector. Growing domestic energy demand on oil and gas resources.

Profile

Since it began a process of structural adjustment in 1988, supported by the IMF and the World Bank, Tunisia has become an increasingly open, market-oriented economy. Real GDP growth has averaged 5% since 1987, rising to 6% in 2001. Annual inflation remained stable, at just below 3% in 2000, despite higher food and energy prices. Prices have been freed, most state companies privatized, and import barriers reduced.

The balance of payments relies on fluctuating tourism receipts to offset a trade deficit. The government must also balance growth with better social provisions. The member states of the EU are Tunisia's main trading partners, accounting for well over 70% of its imports and nearly 80% of its exports; trade has increased significantly since 1999.

ECONOMIC PERFORMANCE INDICATOR

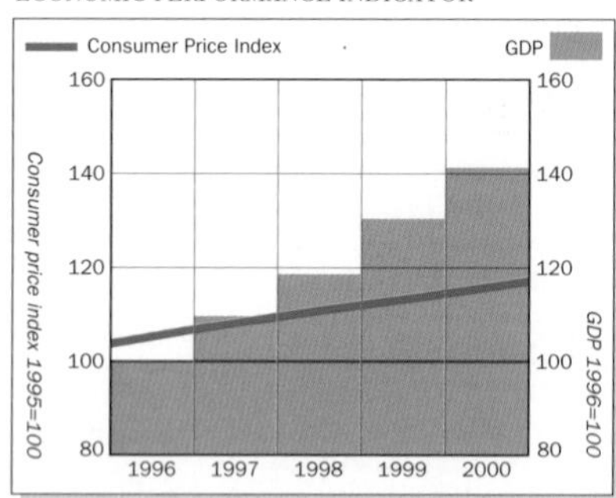

Consumer Price Index
GDP
Consumer price index 1995=100
GDP 1996=100
160
140
120
100
80
1996
1997
1998
1999
2000

TUNISIA : MAJOR BUSINESSES

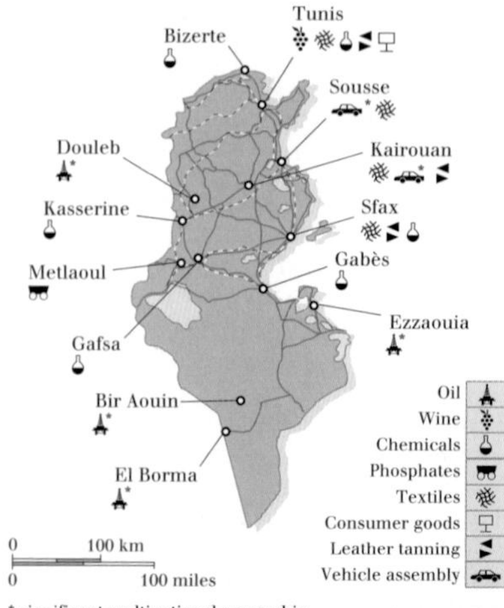

Tunis
Bizerte
Sousse
Douleb
Kairouan
Kasserine
Sfax
Metlaoui
Gabès
Ezzaouia
Gafsa
Bir Aouin
El Borma
Oil
Wine
Chemicals
Phosphates
Textiles
Consumer goods
Leather tanning
Vehicle assembly
0
100 km
0
100 miles
* significant multinational ownership

RESOURCES

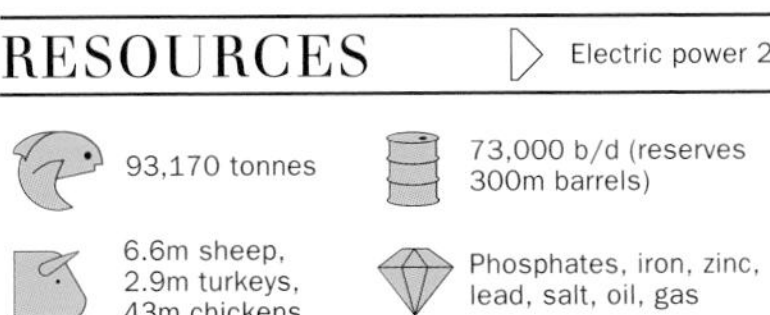
Electric power 2.3m kw
93,170 tonnes
73,000 b/d (reserves 300m barrels)

6.6m sheep, 2.9m turkeys, 43m chickens

Phosphates, iron, zinc, lead, salt, oil, gas

ELECTRICITY GENERATION

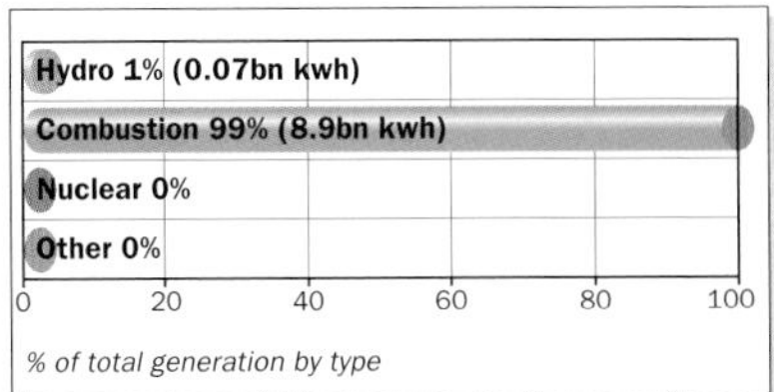
Hydro 1% (0.07bn kwh)
Combustion 99% (8.9bn kwh)
Nuclear 0%
Other 0%
0
20
40
60
80
100
% of total generation by type

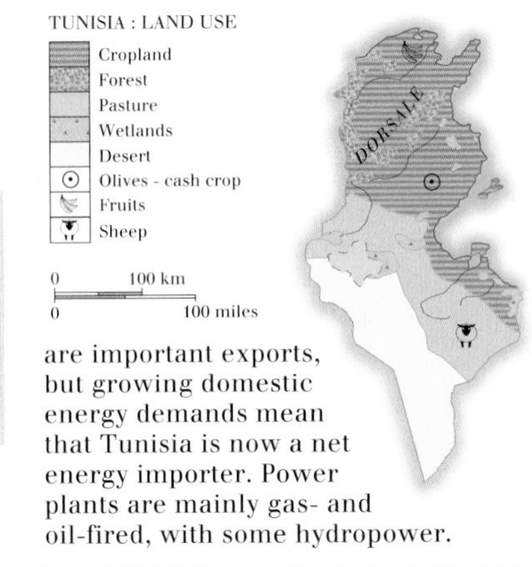
TUNISIA : LAND USE
Cropland
Forest
Pasture
Wetlands
Desert
Olives - cash crop
Fruits
Sheep
0 100 km
0 100 miles
DORSALE

Tunisia is a leading producer of phosphates for fertilizers, mainly from mines near Gafsa. Oil and gas are important exports, but growing domestic energy demands mean that Tunisia is now a net energy importer. Power plants are mainly gas- and oil-fired, with some hydropower.

HEALTH

Welfare state health benefits

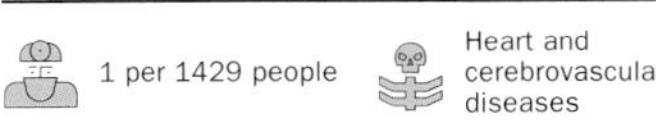
1 per 1429 people
Heart and cerebrovascular diseases

Well-developed family-planning facilities have almost halved Tunisia's birthrate over the past 30 years. The population growth rate has dropped from 3.2% to 1.9% – the lowest in the region. The mortality rate has been halved, to 5.7 per 1000 population, reflecting the extension of free medical services to over 70% of the population. While services lack sophistication, an umbrella of primary care facilities covers all but the most isolated rural communities. Regional committees organize care for the old, needy, and orphaned.

ENVIRONMENT

Sustainability rank: 61st

0.3%
2.4 tonnes per capita
ENVIRONMENTAL TREATIES
Yes
Yes
Yes

Yes · Yes · No

Desertification is a serious problem in the largely arid central and southern regions. However, the dominant environmental issue is the rapid expansion of tourism since the 1980s. Large, insensitively designed hotel and resort developments, which do not fit in with the local architecture, are spoiling coastal areas such as the island of Jerba and Hammamet (although building height restrictions are applied here). Tourism is also making an impact on the fragile desert ecology of the south.

CRIME

Death penalty in use

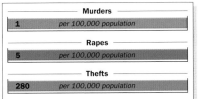

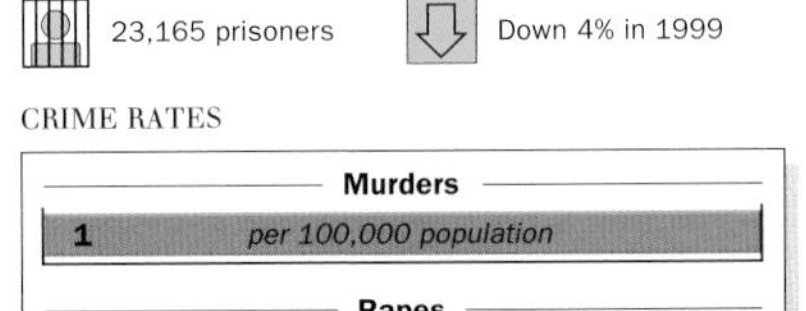
23,165 prisoners
Down 4% in 1999
CRIME RATES
Murders
1 per 100,000 population

Rapes: 5 per 100,000 population

Thefts: 280 per 100,000 population

Street crime is unusual. However, Tunisia's poor human rights record has prompted criticism of its maltreatment of political and other detainees. Arbitrary arrests and torture while in police custody, especially of suspected Islamist activists, are routine.

SPENDING

GDP/cap. increase

CONSUMPTION AND SPENDING

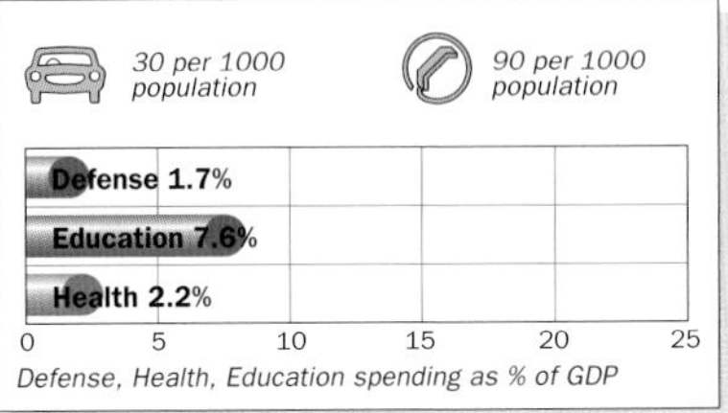
30 per 1000 population
90 per 1000 population
Defense 1.7%
Education 7.6%
Health 2.2%
0
5
10
15
20
25
Defense, Health, Education spending as % of GDP

Today 6% of Tunisians are estimated to live in absolute poverty. In 1970, the figure was 30%. The poorest in the community tend to live in the urban shanty towns, or *bidonvilles*. The Western-oriented elite has links with government or business. Social security covers sickness, old age, and maternity, but not unemployment. The government is concerned that the lack of jobs is encouraging the spread of Islamic fundamentalism; economic growth is its medium-term solution to the problem. Special projects are being set up in the most deprived urban areas to offset the worst effects of poverty.

MEDIA

TV ownership medium

Daily newspaper circulation 31 per 1000 people

PUBLISHING AND BROADCAST MEDIA

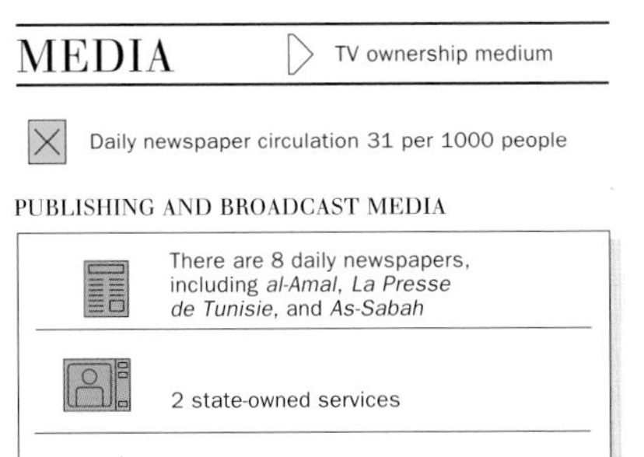
There are 8 daily newspapers, including *al-Amal*, *La Presse de Tunisie*, and *As-Sabah*
2 state-owned services
1 state-owned service

Reforms since the late 1980s have in theory increased press freedom in Tunisia, traditionally considered a source of liberal ideas in the Arab world. In practice, government restrictions remain. The foreign press is also occasionally banned, but the arrival of satellite TV from Europe has enabled people to receive a wide range of programs. The Internet is heavily censored.

EDUCATION

School leaving age: 16

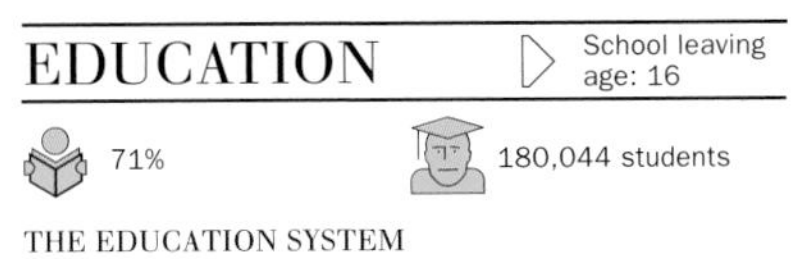
71%
180,044 students

THE EDUCATION SYSTEM

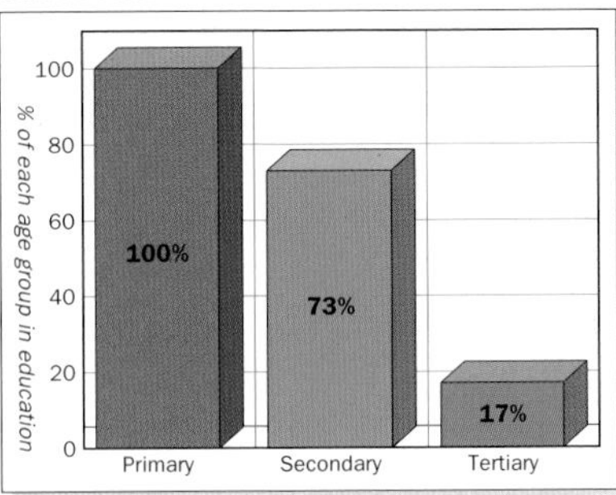
100
80
60
40
20
0
% of each age group in education
100%
73%
17%
Primary
Secondary
Tertiary

Education is compulsory for nine years between the ages of six and 16, with secondary education beginning at the age of 12. Arabic is the first language in schools, but French is also taught, and is used almost exclusively in higher education. There are seven universities; student enrollment has doubled since 1995.

WORLD RANKING

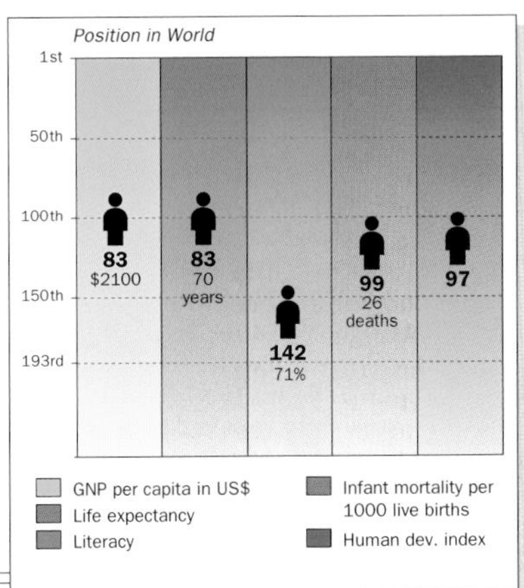
Position in World
1st
50th
100th
150th
193rd
83
$2100
83
70 years
142
71%
99
26 deaths
97
GNP per capita in US$
Life expectancy
Literacy
Infant mortality per 1000 live births
Human dev. index

T

TURKEY

EUROPE
Europe
Bulgaria
Georgia
TURKEY
Greece
Armenia
Syria
Iraq
Iran

OFFICIAL NAME: Republic of Turkey **CAPITAL:** Ankara
POPULATION: 67.6 million **CURRENCY:** Turkish lira **OFFICIAL LANGUAGE:** Turkish

1923

1939

Oct 29

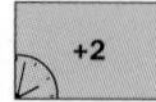
TR

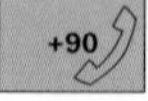
+2

+90

.tr

TURKEY, MAINLY IN WESTERN ASIA, also includes the region of Eastern Thrace in Europe. It thus controls the entrance to the Black Sea, which is straddled by Turkey's largest city, Istanbul. Most Turks live in the western half of the country. The eastern and southeastern reaches of the Anatolia Plateau are Kurdish regions. Turkey's location gives it great strategic influence in the Black Sea, the Mediterranean, and the Middle East. Lying on a major earthquake fault line, many Turkish towns are vulnerable to earthquakes such as the one which devastated Izmit in 1999.

***The Church of the Holy Cross,** on Akdamar Island in Lake Van, was built in the 10th century when Christianity was dominant in the region.*

CLIMATE

Mountain/ Mediterranean

WEATHER CHART FOR ANKARA

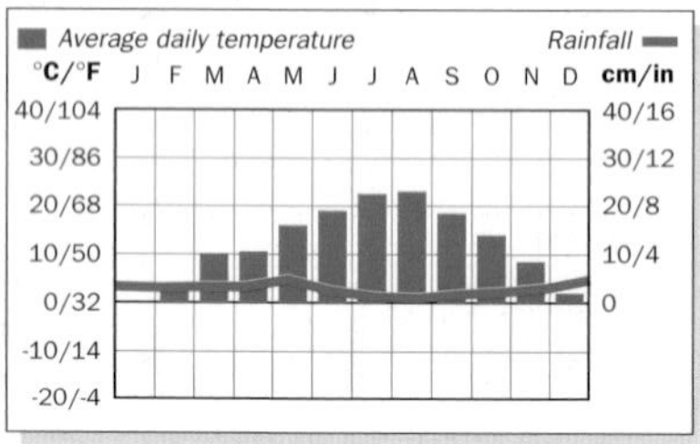

Average daily temperature — Rainfall
°C/°F J F M A M J J A S O N D cm/in
40/104 — 40/16
30/86 — 30/12
20/68 — 20/8
10/50 — 10/4
0/32 — 0
-10/14
-20/4

Coastal regions have a Mediterranean climate. The interior has cold, snowy winters and hot, dry summers.

TRANSPORTATION

Drive on right

Atatürk International, Istanbul
16m passengers

1135 ships
6.25m grt

THE TRANSPORTATION NETWORK

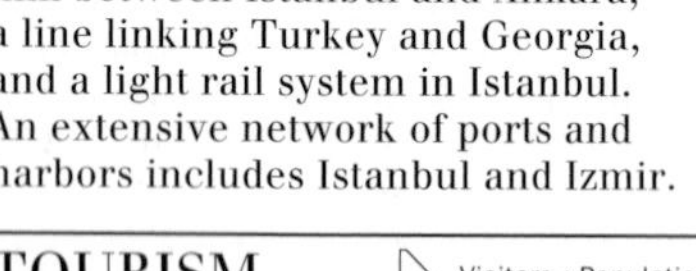
95,599 km
(59,402 miles)
1774 km
(1102 miles)

8682 km
(5395 miles)

1200 km
(746 miles)

Recent rail projects include a high-speed link between Istanbul and Ankara, a line linking Turkey and Georgia, and a light rail system in Istanbul. An extensive network of ports and harbors includes Istanbul and Izmir.

PEOPLE

Pop. density medium

Turkish, Kurdish, Arabic, Circassian, Armenian, Greek, Georgian, Ladino

88/km² (227/mi²)

THE URBAN/RURAL POPULATION SPLIT

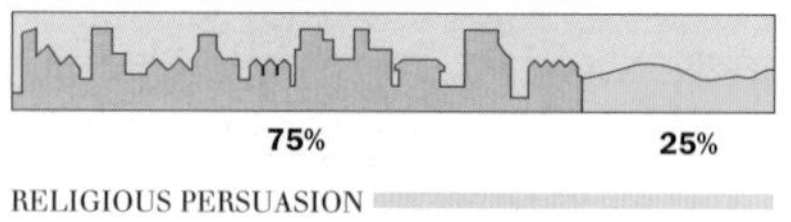
75%
25%

RELIGIOUS PERSUASION

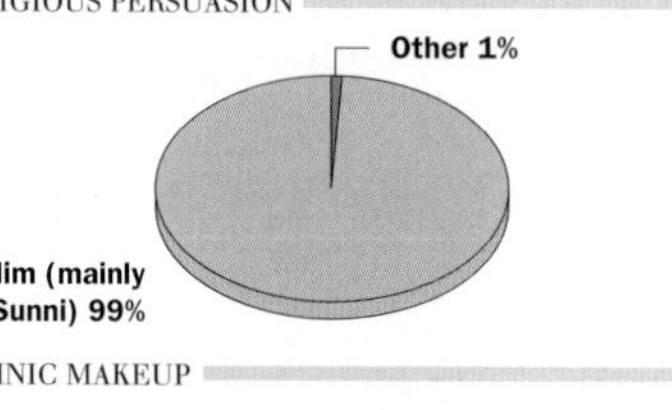
Other 1%
Muslim (mainly Sunni) 99%

ETHNIC MAKEUP

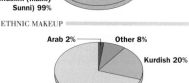

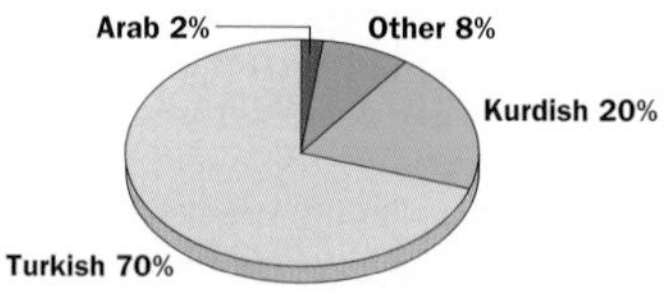

Arab 2%
Other 8%
Kurdish 20%
Turkish 70%

POPULATION AGE BREAKDOWN

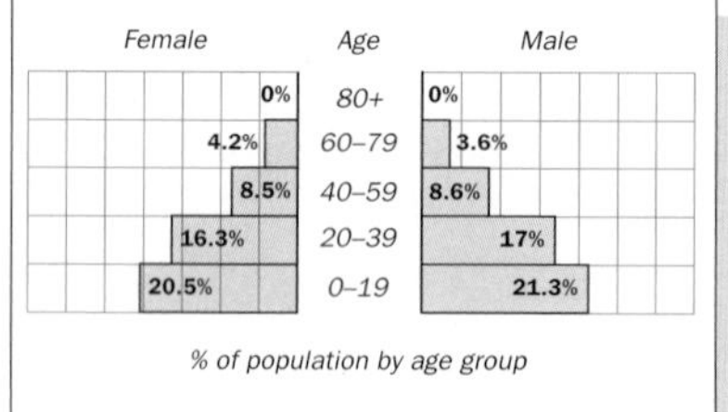
Female
Age
Male
0%
80+
0%
4.2%
60–79
3.6%
8.5%
40–59
8.6%
16.3%
20–39
17%
20.5%
0–19
21.3%
% of population by age group

The Turks are racially diverse. Many are the descendants of refugees, often from the Balkans, but a strong sense of national identity is rooted in a shared language and religion. Most are Sunni Muslim, although a Shi'a community, including the heterodox Alawite sect, is growing fast. The largest minority are the Kurds, and there are some 500,000 Arabic speakers. While women have equal rights in law, men dominate political and even family life. In 2002 women gained the right to an equal portion in the case of divorce.

With a birthrate of 2%, Turkey is projected to have a larger population than any EU country by 2020.

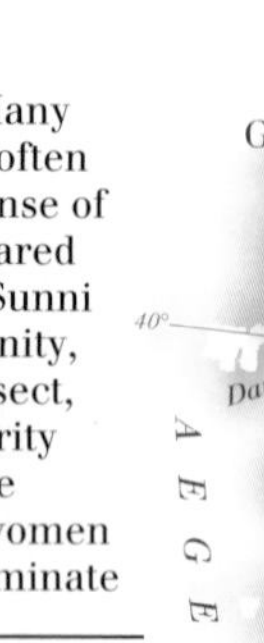

TOURISM

Visitors : Population 1:7.1

9.59m visitors

Up 39% in 2000

Visitors are attracted by fine beaches, classical sites such as Ephesus and Troy, and antiquities from the prehistoric to the Ottoman periods. Attacks on foreigners by Kurdish militants in 1994 hit the tourist trade, but business then recovered, and in 2000 tourists spent $7.6 billion. Visitor numbers dropped after the September 2001 attack on the US.

MAIN TOURIST ARRIVALS

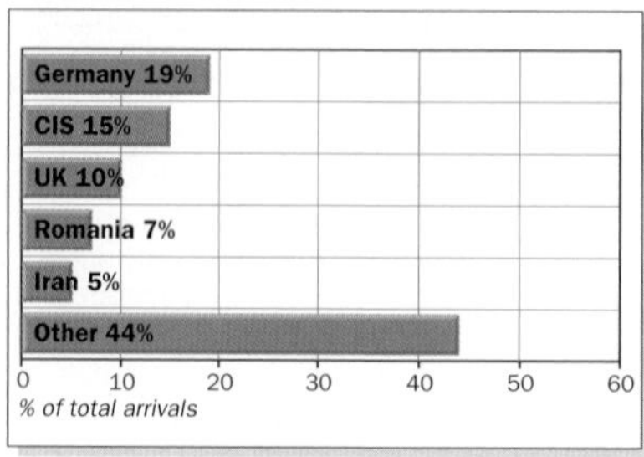
Germany 19%
CIS 15%
UK 10%
Romania 7%
Iran 5%
Other 44%
0
10
20
30
40
50
60
% of total arrivals

BULGARIA
GREECE
Edirne
Kırklareli
İstanbul Boğazi
Tekirdağ
Kâğıthane
İstanbul
Marmara Denizi
İzmit
Yalova
Bandırma
Dardenelles
Çanakkale
Bursa
Bilecik
Balıkesir
Eskişehir
Kütahya
AEGEAN SEA
Manisa
Akhisar
İzmir
Afyon
Turgutlu
Uşak
Aydın
Nazilli
Söke
Büyükmenderes
Denizli
Burdur
Milas
Muğla
Bodrum
Marmaris
Fethiye
Finike
Kaş
Antalya
MEDITERRA

T

POLITICS

Multiparty elections

1999/2002

President Ahmet Necdet Sezer

AT THE LAST ELECTION

Turkish Grand National Assembly 550 seats

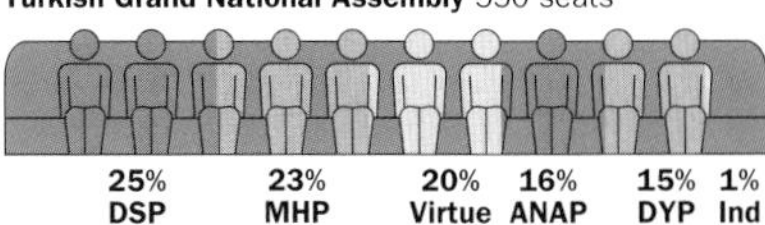

25% DSP
23% MHP
20% Virtue
16% ANAP
15% DYP
1% Ind

DSP = Democratic Left Party **MHP** = Nationalist Action Party **Virtue** = Virtue Party **ANAP** = Motherland Party **DYP** = True Path Party **Ind** = Independents

Turkey is a multiparty republic. The president, who serves a seven-year term, appoints the prime minister.

PROFILE

The main ideological division in politics is Islamic–secular. In 2002 the powers of the National Security Council were restricted and its membership altered to give civilian dominance.

Bulent Ecevit became prime minister in 1999 after the DSP performed well in the election. He formed a new coalition with the far-right MHP and with ANAP (backed by Istanbul's metropolitan interests), which survived Ecevit's failed attempt to enable President Süleyman Demirel to stand for reelection in 2000. Demirel was replaced by nonparty candidate Ahmet Necdet Sezer. Ecevit's refusal to resign because of ill health prompted a crisis in 2002, when key ministers quit their posts in protest.

MAIN POLITICAL ISSUES

Islamic fundamentalism

Modern Turkey's identity as a secular state was profoundly challenged by the Islamist agenda of the Welfare Party (RP) – the largest parliamentary party after the 1995 election. A secular coalition, formed to keep the RP out of power, disintegrated in mid-1996. RP leader Necmettin Erbakan then forged a coalition with the True Path Party. Ousted in mid-1997, the RP was banned in 1998. Many RP Assembly members allied themselves with the Virtue Party, which in turn was banned in 2001.

Kurdish separatism

Thousands of Kurds have been killed since 1984 in a bitter civil war in southeast Turkey. The secessionist Kurdistan Workers' Party (PKK) has proclaimed three cease-fires since 1992. In 1999 its leader Abdullah Ocalan was sentenced to death. However, he remained in prison as, in April 2002, the PKK disbanded and regrouped as a political organization committed to seeking equal rights for Kurds within Turkey. EU pressure led in 2002 to constitutional amendments which lifted the ban on broadcasting and education in Kurdish which had previously been rigorously enforced.

Human rights

Turkey's human rights record has been subject to intense international criticism. Concerns have focused on the curtailment of civil liberties, illegal executions, and the treatment of Kurds. Recent reforms have sought to bring Turkey in line with EU standards. Civil liberties were improved in 1995, and in 2002 executions were banned except in time of war.

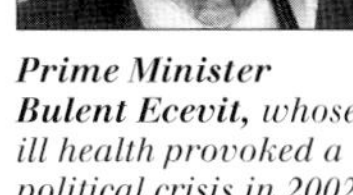

Prime Minister Bulent Ecevit, *whose ill health provoked a political crisis in 2002.*

President Ahmet Necdet Sezer, *has frequently clashed with Ecevit.*

TURKEY

Total Land Area : 780 580 sq. km (301 382 sq. miles)

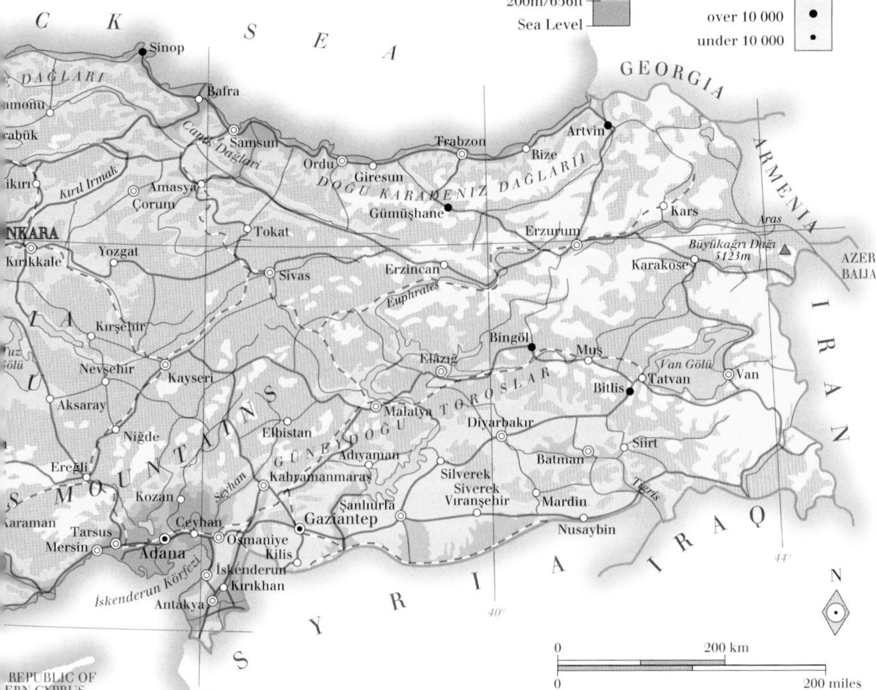

LAND HEIGHT
3000m/9843ft
2000m/6562ft
1000m/3281ft
500m/1640ft
200m/656ft
Sea Level
POPULATION
over 5 000 000
over 1 000 000
over 500 000
over 100 000
over 50 000
over 10 000
under 10 000
GEORGIA
ARMENIA
AZER-BAIJAN
IRAN
IRAQ
SYRIA
Sinop
Bafra
Samsun
Trabzon
Rize
Artvin
Ordu
Giresun
Amasya
Çorum
Gümüşhane
Kars
Tokat
Erzurum
Yozgat
Sivas
Erzincan
Karakose
Büyükağrı Dağı 5123m
Aras
Euphrates
Kırşehir
Bingöl
Muş
Elâzığ
Van Gölü
Tatvan
Van
Nevşehir
Kayseri
Bitlis
Aksaray
Malatya
Diyarbakır
Niğde
Elbistan
Siirt
Adıyaman
Batman
Kahramanmaraş
Silverek
Siverek
Viranşehir
Mardin
Ereğli
Kozan
Şanlıurfa
Gaziantep
Nusaybin
Tarsus
Mersin
Adana
Ceyhan
Osmaniye
Kilis
İskenderun
Kırıkhan
Antakya
İskenderun Körfezi
Tigris
DOĞU KARADENIZ DAĞLARI
GÜNEYDOĞU TOROSLAR
TAURUS MOUNTAINS
Canik Dağları
Kızıl Irmak
Seyhan
Kırıkkale
REPUBLIC OF ... ERN CYPRUS
0 200 km
0 200 miles
N

WORLD AFFAIRS

Joined UN in 1945

CE | NATO | OECD | OIC | OSCE

Turkey had great strategic significance during the Cold War as part of NATO's first line of defense. It now has closer ties with former communist neighbors, particularly Bulgaria and Georgia, and with Turkic-speaking central Asian states. It has joined the BSEC, and has tried to mediate between Armenia and Azerbaijan. In 2000 Turkey agreed to send arms to Uzbekistan – its first military involvement in central Asia since the collapse of the Soviet Union. Before 1991 Turkey's chief trading partner was Iraq, and relations were improving until September 2001, when Turkey became the first Muslim country to join the "war on terrorism." In 2002 it took over leadership of the peacekeeping force in Afghanistan.

Negotiations on joining the EU are aided by recently improving relations with Greece, although the Turkish-backed partition of Cyprus remains an obstacle, as are lingering EU concerns over human rights.

AID

Recipient

$325m (receipts)

Up sharply in 1999–2000

Turkey is a net recipient of aid (especially from the Gulf War allies), despite US suspension of aid in 1994 in protest at the treatment of Kurds. The acute economic crisis in 2000–2001 prompted a $10 billion loan from the IMF. Greece, a traditional adversary, offered humanitarian aid after the 1999 Izmit earthquake.

T

CHRONOLOGY

Following the collapse of the Ottoman Empire and Turkey's defeat in World War I, nationalist Mustafa Kemal Atatürk deposed the ruling sultan in 1922, declaring Turkey a republic in 1923.

- ❑ **1924** Religious courts abolished.
- ❑ **1928** Islam no longer state religion.
- ❑ **1934** Women given the vote.
- ❑ **1938** President Atatürk dies. Succeeded by Ismet Inonu.
- ❑ **1945** Turkey declares war on Germany. Joins UN.
- ❑ **1952** Joins CE and NATO.
- ❑ **1960** Military coup; National Assembly suspended.
- ❑ **1961** New constitution.
- ❑ **1963** Association agreement with European Economic Community.
- ❑ **1974** Invades northern Cyprus.
- ❑ **1980** Military coup; martial law.
- ❑ **1982** New constitution.
- ❑ **1983** General election won by Turgut Özal's ANAP.
- ❑ **1984** Turkey recognizes "Turkish Republic of Northern Cyprus." Kurdish separatist PKK launches guerrilla war in southeast.
- ❑ **1987** Applies to join European Communities.
- ❑ **1990** US-led coalition launches air strikes on Iraq from Turkish bases.
- ❑ **1991** Elections won by DYP. Süleyman Demirel premier.
- ❑ **1992** Joins Black Sea alliance.
- ❑ **1993** Demirel elected president. Tansu Çiller becomes DYP leader and heads coalition.
- ❑ **1995** Major anti-Kurdish offensive. Voting age lowered to 18. Çiller coalition collapses. Pro-Islamic RP wins election, but center-right DYP–ANAP coalition takes office. Customs union with EU.
- ❑ **1996–1997** RP leader Necmettin Erbakan heads first pro-Islamic government since 1923.
- ❑ **1997** Mesut Yilmaz reappointed to head minority ANAP government.
- ❑ **1998** RP banned. Yilmaz resigns amid corruption allegations; replaced by Bulent Ecevit of DSP.
- ❑ **1999** Ecevit heads right-wing coalition after elections. Kurdish leader Abdullah Ocalan sentenced to death. Izmit earthquake kills 14,000.
- ❑ **2000** National Assembly refuses to endorse Demirel's reelection. He is replaced by Ahmet Necdet Sezer.
- ❑ **2001** Acute financial crisis. Hunger strikes in high-security prisons. Virtue Party banned.
- ❑ **2002** Democracy and human rights strengthened in constitution. April, PKK replaced by Kurdistan Freedom and Democracy Congress. July, key ministers resign. Early elections promised.

T

DEFENSE

Compulsory military service

$10.6bn

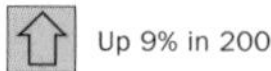
Up 9% in 2000

TURKISH ARMED FORCES

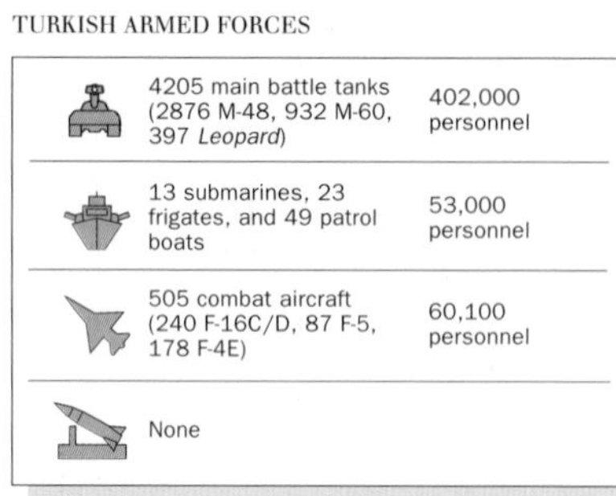

4205 main battle tanks (2876 M-48, 932 M-60, 397 *Leopard*)	402,000 personnel
13 submarines, 23 frigates, and 49 patrol boats	53,000 personnel
505 combat aircraft (240 F-16C/D, 87 F-5, 178 F-4E)	60,100 personnel
None	

Turkey's armed forces are the second-largest in NATO, which it joined in 1952. Turkey is a sizable military power, and it spends a higher percentage of GDP on defense than any other NATO country. NATO membership gives Turkey easy access to Western arms suppliers, although campaigners oppose sales to Turkey on human rights grounds. Israel is an important source for arms. Offensives against Kurdish separatists in northern Iraq and in Turkey's own southeastern provinces have involved over 50,000 troops and repeated incursions into Iraqi territory.

The great majority of Turkey's armed forces personnel are conscripts; 18 months' service is compulsory for all males at the age of 20.

ECONOMICS

Inflation 76% p.a. (1990–2000)

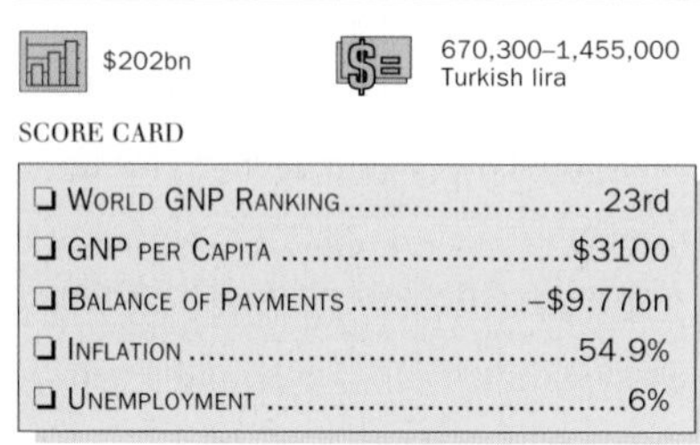

$202bn

670,300–1,455,000 Turkish lira

SCORE CARD

❑ World GNP Ranking	23rd
❑ GNP per Capita	$3100
❑ Balance of Payments	–$9.77bn
❑ Inflation	54.9%
❑ Unemployment	6%

EXPORTS

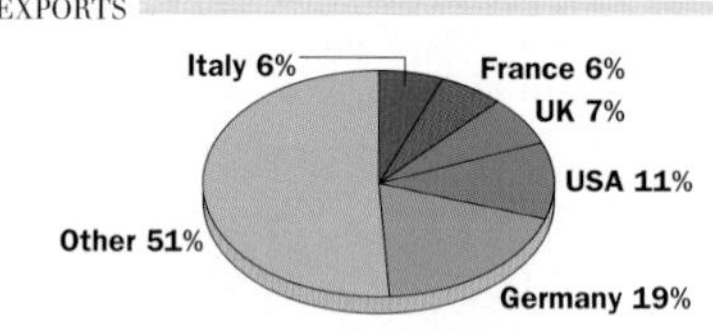
Italy 6%
France 6%
UK 7%
USA 11%
Other 51%
Germany 19%

IMPORTS

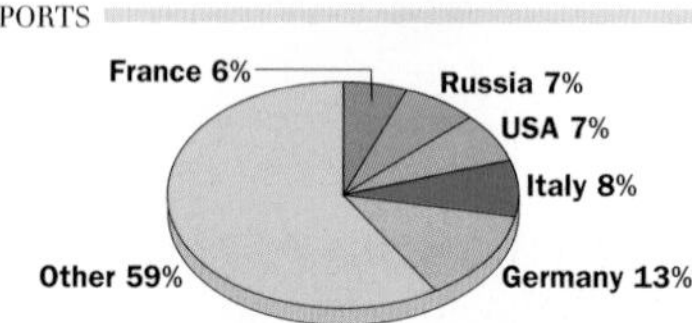
France 6%
Russia 7%
USA 7%
Italy 8%
Other 59%
Germany 13%

ECONOMIC PERFORMANCE INDICATOR

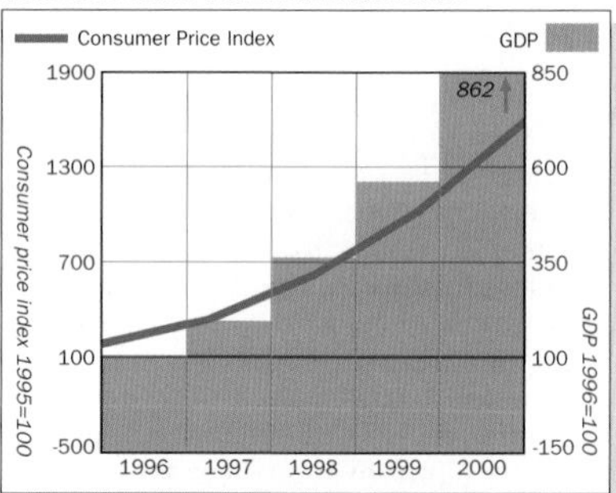
Consumer Price Index
GDP
1900
1300
700
100
-500
850
600
350
100
-150
862
Consumer price index 1995=100
GDP 1996=100
1996
1997
1998
1999
2000

Strengths
Liberalization resulted in strong growth in the 1990s. Textiles, manufacturing, and construction sectors competitive on world markets. Tourism industry. Self-sufficient in agriculture. Dynamic private sector. Skilled labor force. Customs union with EU since 1995.

Weaknesses
Persistently high inflation. Unsound public finances. Large government bureaucracy. Uneven privatization program. Ailing banking sector. Influence of organized crime. High cost of military action against Kurds.

Profile
Turkey has one of the oldest and most advanced of the emerging market economies. In the 1990s it grew strongly, but continued to suffer from persistently high inflation. Despite structural reforms the economy reached crisis in February 2001, with a collapse of the banking system and the loss of over 500,000 jobs. This was compounded by the economic impact of the September 11 attack on the US. IMF rescue packages were agreed in February, May, and November in return for banking reform and privatizing debt-laden state companies.

TURKEY : MAJOR BUSINESSES

Istanbul
Ankara
Bursa
İzmit
Kırıkkale
Erzurum
Sivas
İsparta
Adana
Diyarbakır
İzmir
Mersin
Cement
Textiles
Chemicals
Electronics
Oil refining
Iron & steel
Food processing
Vehicle manufacture
0 200 km
0 200 miles
* significant multinational ownership

RESOURCES

Electric power 23.4m kw

638,097 tonnes

64,621 b/d (reserves 260m barrels)

29.4m sheep, 10.8m cattle, 8.06m goats, 220m chickens

Chromium, oil, copper, borax, coal, gas, bauxite, iron

ELECTRICITY GENERATION

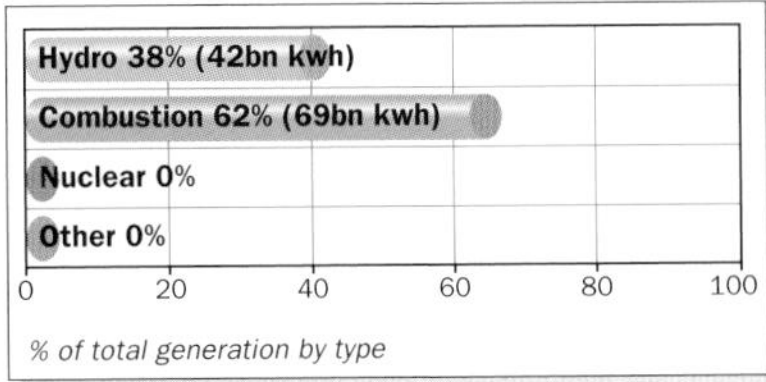
Hydro 38% (42bn kwh)
Combustion 62% (69bn kwh)
Nuclear 0%
Other 0%
0
20
40
60
80
100
% of total generation by type

ENVIRONMENT

Sustainability rank: 62nd

1%

3.2 tonnes per capita

ENVIRONMENTAL TREATIES

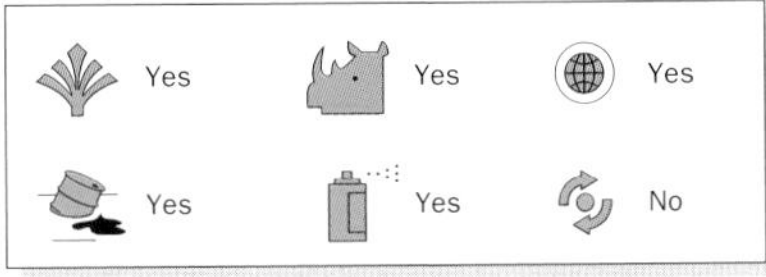
Yes
Yes
Yes
Yes
Yes
No

Turkey's program of dam building on the Tigris and Euphrates has met with condemnation, particularly from Syria and Iraq, whose rivers will suffer reduced flow rates. Plans for the Ilisu Dam were shelved in 2001. Concern has also been expressed at proposals to build a nuclear power plant. There has been uncontrolled tourist development along the western coast.

MEDIA

TV ownership high

Daily newspaper circulation 64 per 1000 people

PUBLISHING AND BROADCAST MEDIA

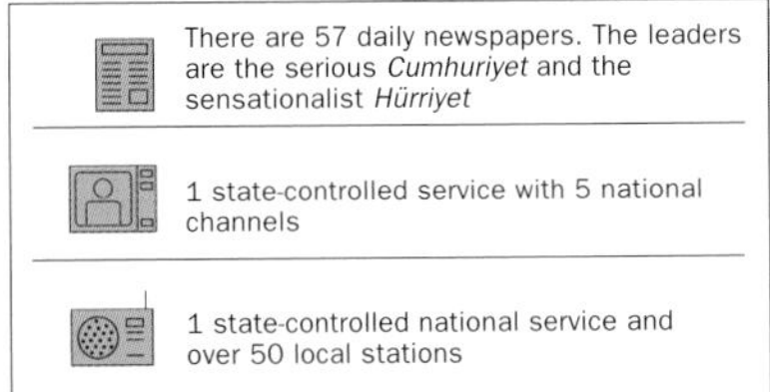
There are 57 daily newspapers. The leaders are the serious *Cumhuriyet* and the sensationalist *Hürriyet*
1 state-controlled service with 5 national channels
1 state-controlled national service and over 50 local stations

The Turkish press is diverse, vigorous, and largely privately owned. Almost all Istanbul newspapers are printed in Ankara and Izmir on the same day. Foreign satellite or cable broadcasts are available, as well as the five national channels of the state Turkish Radio and Television Corporation. The National Assembly has amended censorship laws to ease restrictions on non-Turkish-language broadcasting, but still bans broadcasts in Kurdish and censors Kurdish-related programming. A particularly high number of journalists are imprisoned, and state control over Internet content was increased in 2002.

Under the controversial Southeastern Anatolian Project (GAP) launched in the mid-1980s, Turkey is building 22 dams on the Tigris and Euphrates rivers. In 1999 controversy focused on the Ilisu Dam on the Tigris, which will flood 15 towns and 52 villages, and in 2001 on the Birecik Dam on the Euphrates, which has engulfed the ancient Roman town of Zeugma. Turkey produces oil around Raman, on the Tigris. Eastern provinces are rich in minerals, such as chromium, of which Turkey is a leading producer.

CRIME

Death penalty not used in practice

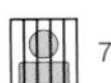
71,860 prisoners

Up 28% in 1999

CRIME RATES

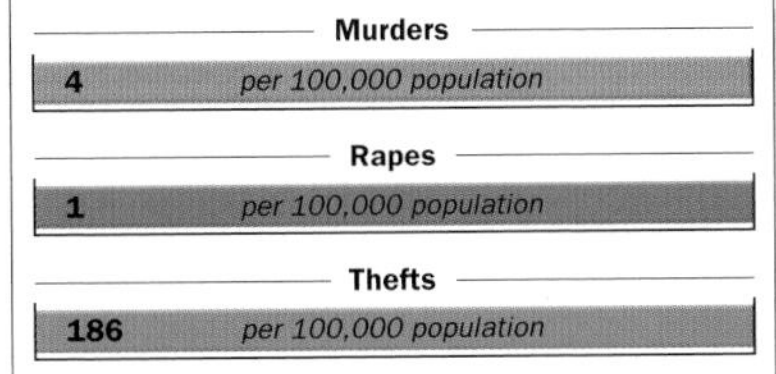
Murders
4 per 100,000 population
Rapes
1 per 100,000 population
Thefts
186 per 100,000 population

The routine use of torture and rape by the police, and the deaths of prisoners in custody, cause concern among human rights groups worldwide. In 2002 the imposition of the death penalty was limited to "times of war" and terrorism.

EDUCATION

School leaving age: 14

85%

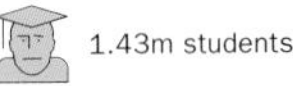
1.43m students

THE EDUCATION SYSTEM

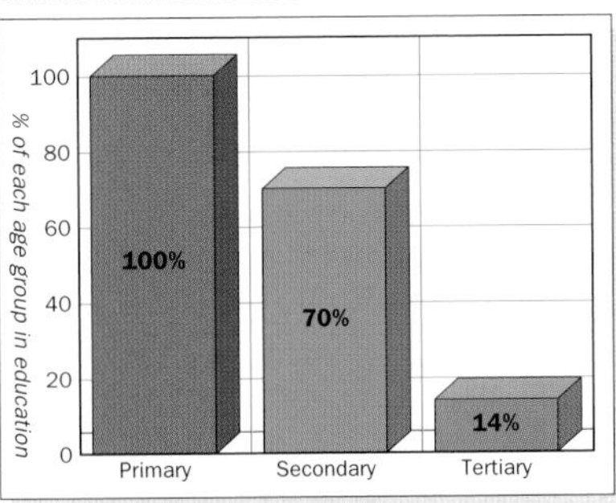
100
80
60
40
20
0
% of each age group in education
100%
70%
14%
Primary
Secondary
Tertiary

After 1923, educational establishments were nationalized. In 1928, a Turkish alphabet with Latin characters was introduced.

In 1997, compulsory education was extended from five to eight years, raising the age for entry into Islamic schools from 11 to 14, in a move seen as designed to reduce attendance at such schools. State schools are coeducational and free. Engineering is usually the strongest faculty in Turkey's many universities.

TURKEY : LAND USE

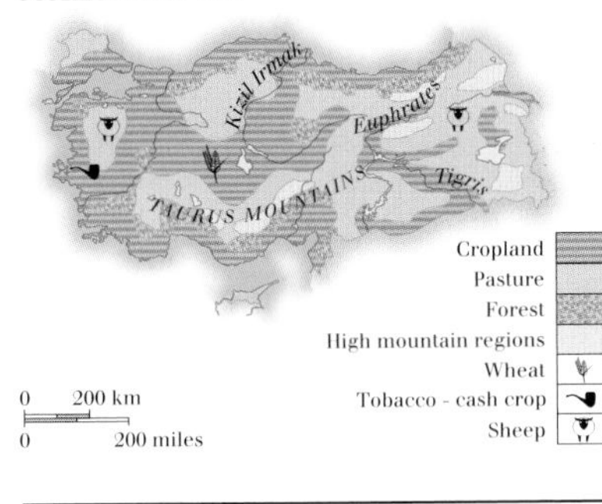
Kizil Irmak
Euphrates
Tigris
TAURUS MOUNTAINS
Cropland
Pasture
Forest
High mountain regions
Wheat
Tobacco - cash crop
Sheep
0 200 km
0 200 miles

HEALTH

Welfare state health benefits

1 per 833 people

Cerebrovascular, heart, respiratory, and digestive diseases

Turkey possesses an adequate national system of primary health care. By Western standards, however, hospitals are underequipped. There are fewer doctors per head than in any European country.

SPENDING

GDP/cap. increase

CONSUMPTION AND SPENDING

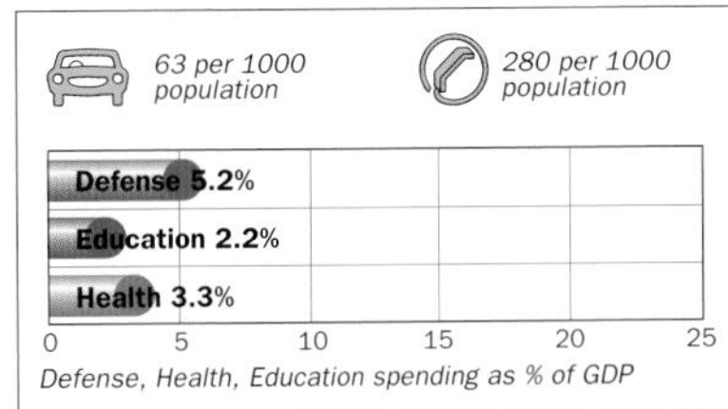
63 per 1000 population
280 per 1000 population
Defense 5.2%
Education 2.2%
Health 3.3%
0
5
10
15
20
25
Defense, Health, Education spending as % of GDP

The economic expansion of the 1980s has created a new class of wealthy entrepreneurs. Urban/rural differences remain pronounced. High inflation in the 1990s eroded earnings of those on fixed incomes, and income inequality has grown. Many Turks take jobs as *Gastarbeiter* (guest workers) in Germany and the Netherlands.

WORLD RANKING

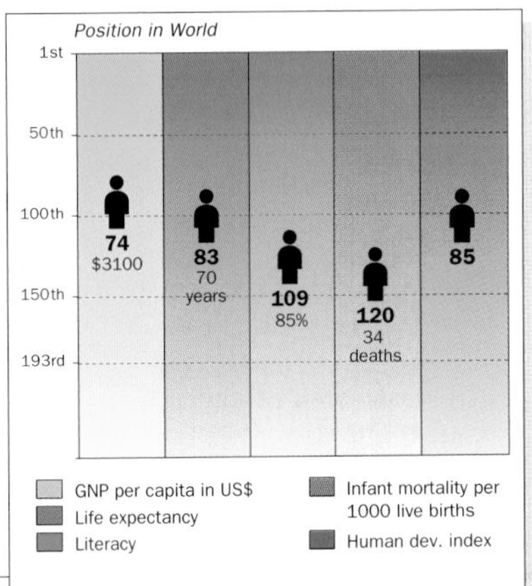
Position in World
1st
50th
100th
150th
193rd
74
$3100
83
70 years
109
85%
120
34 deaths
85
GNP per capita in US$
Life expectancy
Literacy
Infant mortality per 1000 live births
Human dev. index

T

TURKMENISTAN

CENTRAL ASIA

OFFICIAL NAME: Turkmenistan **CAPITAL:** Ashgabat
POPULATION: 4.8 million **CURRENCY:** Manat **OFFICIAL LANGUAGE:** Turkmen

1991

1991

Oct 27
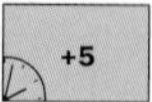
TM

+5

+993
.tm

ONCE THE POOREST of the former Soviet republics, Turkmenistan has adjusted better than most to independence, exploiting the market value of its abundant natural gas supplies. A largely Sunni Muslim area, Turkmenistan is part of the former Turkestan, the last expanse of central Asia incorporated into czarist Russia. Much of life is still based on tribal relationships. Turkmenistan is isolated – telephones are rare and other communications limited.

CLIMATE

Desert/steppe

WEATHER CHART FOR ASHGABAT

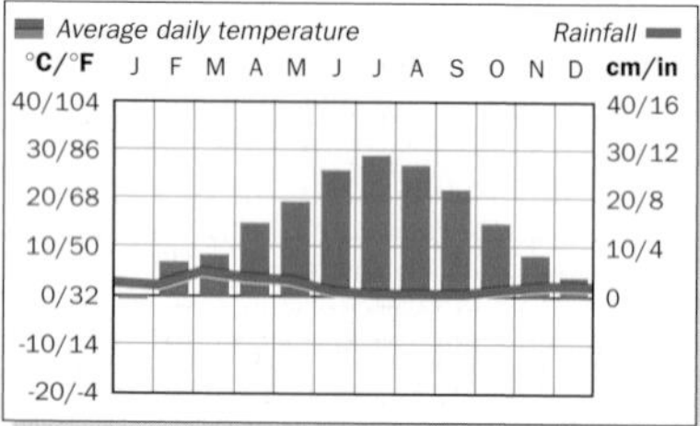
Average daily temperature
Rainfall
°C/°F
J F M A M J J A S O N D
cm/in
40/104
30/86
20/68
10/50
0/32
-10/14
-20/-4
40/16
30/12
20/8
10/4
0

Most of Turkmenistan is arid desert, so that only 2% of the total land area is suitable for agriculture.

TRANSPORTATION

Drive on right

Turkmenistan International, Ashgabat

38 ships 38,400 grt

THE TRANSPORTATION NETWORK

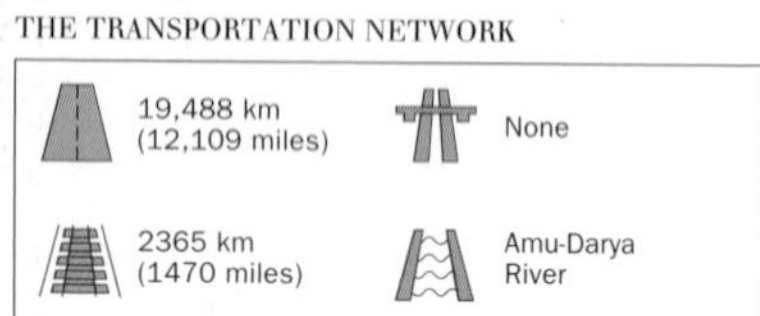
19,488 km (12,109 miles)
None
2365 km (1470 miles)
Amu-Darya River

Trains are cramped and chaotic, and buses unreliable. The border with Iran is officially closed to "foreigners."

TOURISM

Visitors : Population 1:16

300,000 visitors

Up 26% in 1998

MAIN TOURIST ARRIVALS

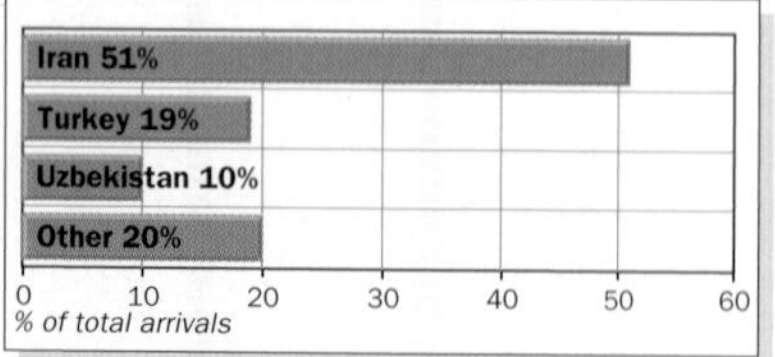
Iran 51%
Turkey 19%
Uzbekistan 10%
Other 20%
0
10
20
30
40
50
60
% of total arrivals

Most visitors are businessmen attracted by Turkmenistan's stability under President Niyazov. Turkmenistan may become a popular tourist destination in future; traditional Turkmen Muslim monuments are slowly being restored.

Karakum Canal zone: *salt flats and the Kopetdag Mountains on the Iranian border. The Karakumy is Turkmenistan's largest desert.*

PEOPLE

Pop. density low

Turkmen, Uzbek, Russian

10/km² (25/mi²)

THE URBAN/RURAL POPULATION SPLIT

45% 55%

RELIGIOUS PERSUASION

Other 2%
Eastern Orthodox 11%
Sunni Muslim 87%

Before czarist Russia annexed Turkmenistan in 1884, the Turkmen were a largely nomadic tribal people. The tribal unit remains strong – the largest tribes are the Tekke in the center, the Ersary on the eastern Afghan border, and the Yomud in the west. Tribal conflicts, rather than tensions with the two main minorities – Russian and Uzbek – are a source of strife. Paradoxically, this has meant that since 1991 nationalism has been less virulent than in other former Soviet republics. Since 1989, Turkmenistan has been rehabilitating its traditional language and culture, as well as reassessing its history. Islam is again central, although few perform the *haj* (pilgrimage) to Mecca and many continue to maintain a cult of ancestors.

Russian remains the language of bureaucracy.

POLITICS

No multiparty elections

L. House 1999/2004
U. House 1998/2003

President Saparmurad Niyazov

AT THE LAST ELECTION

Parliament 50 seats

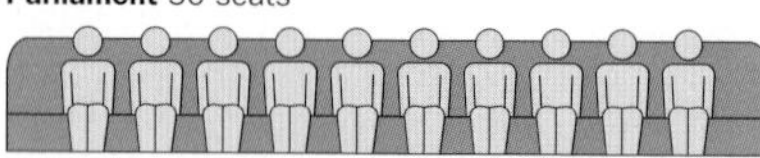

In elections to the Parliament in 1999 all of the seats were won by supporters of the ruling Democratic Party of Turkmenistan (**DPT**) – the only registered party.

People's Council 110 seats

The People's Council has 50 directly elected members, the 50 members of the Parliament, 10 appointed regional members and a varying number of ex-officio members.

Officially, Turkmenistan became a multiparty democracy at independence, but President Saparmurad Niyazov has banned the formation of new parties. As in some other ex-Soviet states, former communists, regrouped as the DPT, still dominate the political process, harboring the traditional communist suspicion of Islamic fundamentalism. A main political concern is to prevent the social and nationalistic conflicts that have blighted other CIS republics.

President Niyazov has encouraged an elaborate personality cult, adopting the title of Turkmenbashi (father of all Turkmen), and there are golden statues of him throughout Ashgabat. His spiritual guide to living, *Ruhnama*, has been adopted as a national code, and he has announced his intention officially to rename all 12 months of the year in commemoration of the country's heroes and most potent national symbols.

WORLD AFFAIRS

Joined UN in 1992

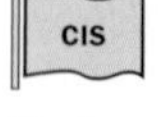
CIS

ECO

EAPC

OIC

OSCE

Turkmenistan is concentrating on establishing good relations with Iran and Turkey. It needs investment from both countries, but is wary of Islamic fundamentalism. President Niyazov opposes economic union with the CIS, and has also expressed caution about closer political union with other Turkic-speaking central Asian states.

AID

Recipient

$32m (receipts)

Up 33% in 2000

Aid is mostly concentrated in the oil and gas industries. Turkmenistan became a member of the ADB in 2000.

T

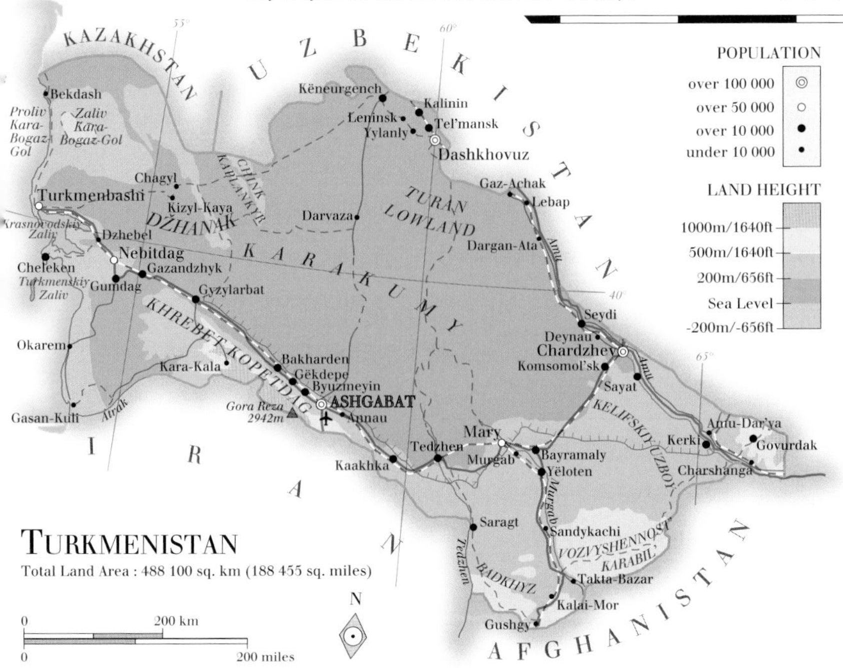
KAZAKHSTAN
UZBEKISTAN
Bekdash
Proliv Kara-Bogaz Gol
Zaliv Kara-Bogaz-Gol
Këneurgench
Kalinin
Leninsk
Tel'mansk
Ylanly
Dashkhovuz
Chagyl
Turkmenbashi
Kizyl-Kaya
DZHANAK
Darvaza
TURAN LOWLAND
Gaz-Achak
Lebap
Krasnovodskiy Zaliv
Dzhebel
Nebitdag
Gazandzhyk
KARAKUMY
Dargan-Ata
Amu
Cheleken
Turkmenskiy Zaliv
Gumdag
Gyzylarbat
KHREBET KOPETDAG
Seydi
Deynau
Chardzhev
Komsomol'sk
Okarem
Bakharden
Gëkdepe
Kara-Kala
Byuzmeyin
ASHGABAT
Annau
Gora Reza 2942m
Sayat
Gasan-Kuli
Atrek
IRAN
Mary
Tedzhen
Murgab
Bayramaly
Yëloten
Kaakhka
KELIFSKIY UZBOY
Kerki
Amu-Dar'ya
Govurdak
Charshanga
Saragt
Sandykachi
VOZVYSHENNOST KARABIL
Takta-Bazar
BADKHYZ
Kalai-Mor
Gushgy
AFGHANISTAN
POPULATION
over 100 000
over 50 000
over 10 000
under 10 000
LAND HEIGHT
1000m/1640ft
500m/1640ft
200m/656ft
Sea Level
-200m/-656ft
N
0
200 km
200 miles

TURKMENISTAN

Total Land Area : 488 100 sq. km (188 455 sq. miles)

CHRONOLOGY

The nomadic peoples of western Turkestan came under Russian imperial control from the 1850s.

- ❑ **1924** Creation of Turkmenistan.
- ❑ **1991** Independence from USSR. Niyazov retains power, becoming president.
- ❑ **1994** Former communists win first elections.
- ❑ **1999** Niyazov's term extended indefinitely by parliament.

DEFENSE

▷ Compulsory military sevice

 $173m

Up 20% in 2000

Turkmenistan relies on Russia for defense. Pilots are trained in Pakistan.

ECONOMICS

▷ Inflation 408% p.a (1990–2000)

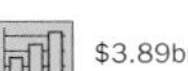 $3.89bn

 5200 manats

SCORE CARD

❑ WORLD GNP RANKING	119th
❑ GNP PER CAPITA	$750
❑ BALANCE OF PAYMENTS	$412m
❑ INFLATION	14%
❑ UNEMPLOYMENT	2%

STRENGTHS

Cotton and gas. Decision to abolish collective farms gradually encouraging private initiative and enterprise.

WEAKNESSES

Cotton monoculture has forced rising food imports. Thriving black market threatens value of manat.

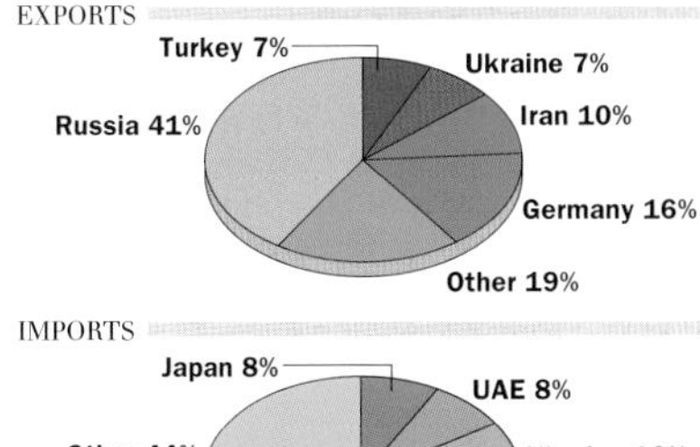
EXPORTS
Turkey 7%
Ukraine 7%
Russia 41%
Iran 10%
Germany 16%
Other 19%
IMPORTS
Japan 8%
UAE 8%
Other 44%
Ukraine 12%
Russia 14%
Turkey 14%

RESOURCES

▷ Electric power 3.9m kw

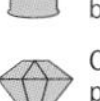 9292 tonnes

162,000 b/d (reserves 500m barrels)

6m sheep, 860,000 cattle, 4.8m chickens

Oil, natural gas, potassium, sulfur, sodium sulfate

During the Soviet years most Turkmen agriculture was turned over to cotton – seen by Moscow as a strategic crop.

ENVIRONMENT

▷ Sustainability rank: 131st

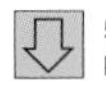 4%

5.7 tonnes per capita

The building of the Karakum Canal has reduced the flow of water to the Aral Sea by 35%. Plans were announced in 2000 for the construction of a large artificial lake in the Karakumy Desert.

MEDIA

▷ TV ownership medium

Daily newspaper circulation figures are not available

PUBLISHING AND BROADCAST MEDIA

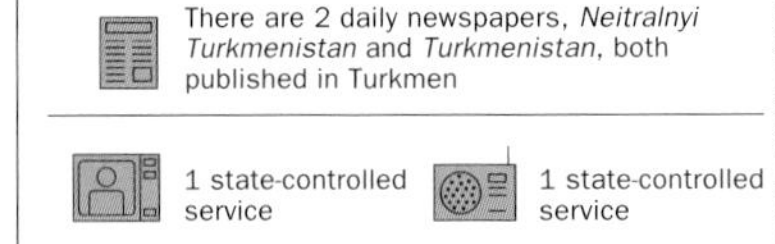

There are 2 daily newspapers, *Neitralnyi Turkmenistan* and *Turkmenistan*, both published in Turkmen

1 state-controlled service

1 state-controlled service

Iranian and Afghan radio stations, beaming in Islamic programs, are popular. TV is only available in cities.

CRIME

▷ No death penalty

 22,000 prisoners

 Increasing levels of theft

Levels of crime are generally low. The death penalty was finally abolished in December 1999.

EDUCATION

▷ School leaving age: 17

 98%

 76,000 students

University students must know the life and works of President Niyazov. They now have to work for two years after their first two years of study.

HEALTH

▷ Welfare state health benefits

 1 per 333 people

Cerebrovascular, heart, and respiratory diseases

Spending is now high for the region, but Turkmenistan's health indicators remain among the worst.

SPENDING

▷ GDP/cap. decrease

CONSUMPTION AND SPENDING

No data

82 per 1000 population

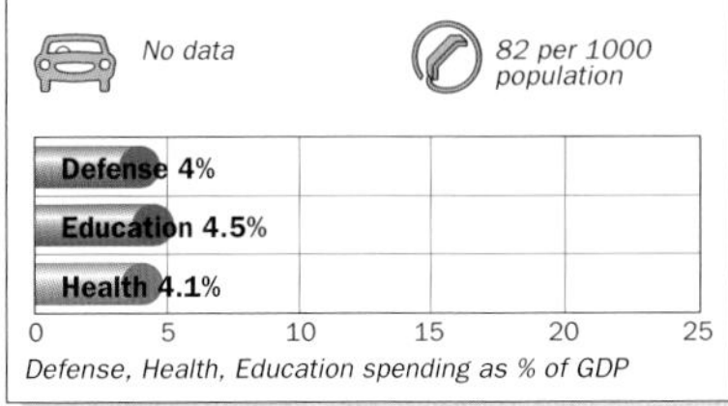
Defense 4%
Education 4.5%
Health 4.1%
0
5
10
15
20
25
Defense, Health, Education spending as % of GDP

The wealthiest group is connected to the government. Transition to a liberal economy has attracted high spending.

WORLD RANKING

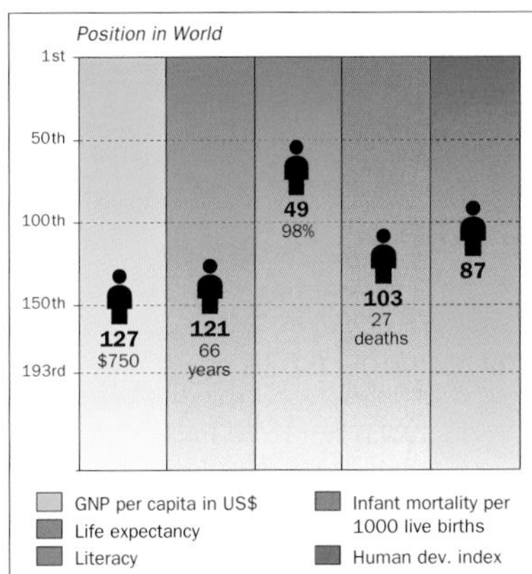
Position in World
1st
50th
100th
150th
193rd
127
$750
121
66
years
49
98%
103
27
deaths
87
GNP per capita in US$
Life expectancy
Literacy
Infant mortality per 1000 live births
Human dev. index

T

TUVALU

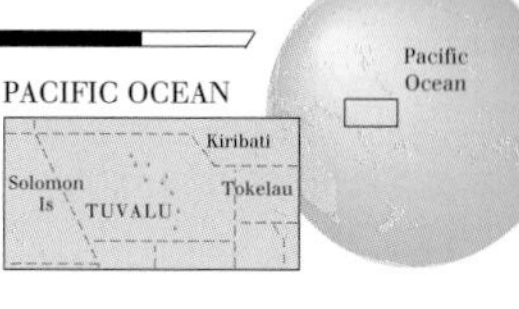
Pacific Ocean
PACIFIC OCEAN
Kiribati
Solomon Is
Tokelau
TUVALU

OFFICIAL NAME: Tuvalu **CAPITAL:** Fongafale, on Funafuti Atoll **POPULATION:** 10,800
CURRENCIES: Australian dollar and Tuvaluan dollar **OFFICIAL LANGUAGE:** English

1978

1978
Oct 1

TUV
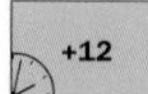
+12

+688

.tv

ONE OF THE WORLD'S smallest, most isolated states, Tuvalu lies 1050 km (650 miles) north of Fiji in the Pacific. A chain of nine coral atolls, it has a land area of just 26 sq. km (10 sq. miles). As the Ellice Islands, it was linked to the Gilbert Islands (now Kiribati) as a British colony until independence in 1978. Politically and socially conservative, Tuvaluans live by subsistence farming and fishing.

CLIMATE

Tropical oceanic

WEATHER CHART FOR FONGAFALE

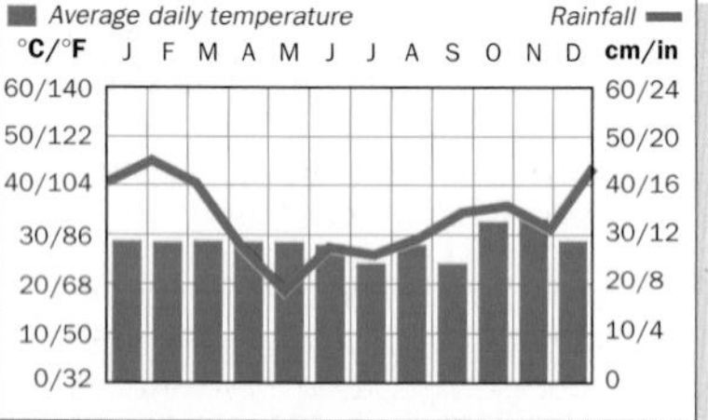
Average daily temperature
Rainfall
°C/°F
J F M A M J J A S O N D
cm/in
60/140
50/122
40/104
30/86
20/68
10/50
0/32
60/24
50/20
40/16
30/12
20/8
10/4
0

Although average humidity exceeds 90%, the climate is pleasantly warm. The mean annual temperature is 29°C (84°F). The October–March hurricane season brings many violent storms.

TRANSPORTATION

Drive on right

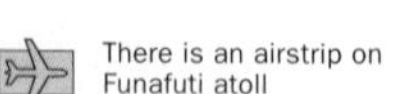
There is an airstrip on Funafuti atoll

49,000 grt

THE TRANSPORTATION NETWORK

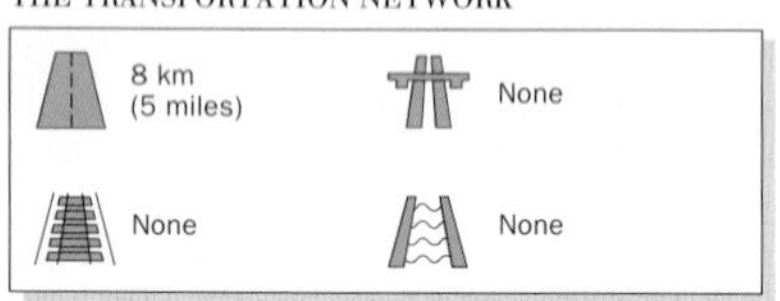
8 km (5 miles)
None
None
None

A ferry links the atolls. There are air links with Kiribati and Fiji. Funafuti and Nukufetau have deepwater berths.

TOURISM

Visitors : Population 1:11

1000 visitors

No change in 2000

MAIN TOURIST ARRIVALS

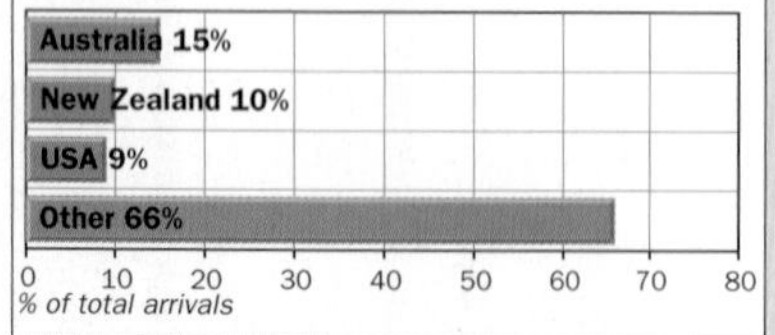
Australia 15%
New Zealand 10%
USA 9%
Other 66%
0 10 20 30 40 50 60 70 80
% of total arrivals

Unspoiled and lapped by some of the world's warmest waters, the remote islands of Tuvalu have surprisingly few visitors. The islands' only paved airstrip and sole hotel are to be found on Funafuti.

PEOPLE

Pop. density high

Tuvaluan, Kiribati, English
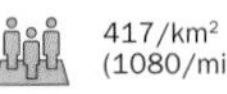
417/km² (1080/mi²)

THE URBAN/RURAL POPULATION SPLIT

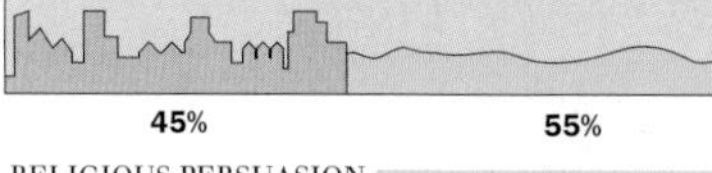
45%
55%

RELIGIOUS PERSUASION

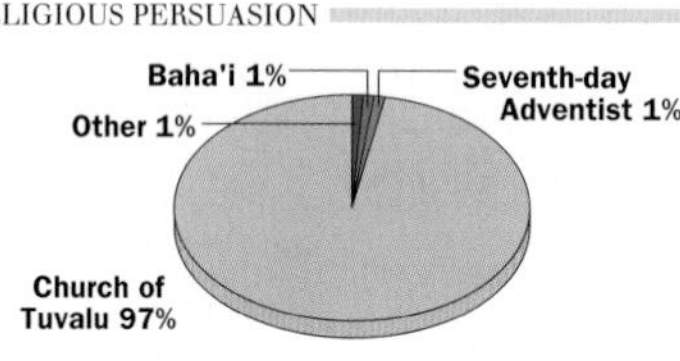
Baha'i 1%
Seventh-day Adventist 1%
Other 1%
Church of Tuvalu 97%

Around 95% of Tuvaluans are ethnically Polynesian. Their ancestors came from Tonga and Samoa 2000 years ago. Nui atoll has Micronesian influences. There is an I-Kiribati community on Funafuti; many Tuvaluans who worked in Kiribati took local wives. Over 40% of the population now live on Funafuti, pushing its population density to almost 1600 per sq. km (4000 per sq. mile). Life is still communal, traditional, and hard. Droughts are common and fresh water is precious. Around two-thirds of people depend on subsistence farming, digging special pits out of the coral to grow most of the islands' limited range of crops. Fishing is also important, and Tuvaluans have a reputation as excellent sailors. About 2000 Tuvaluans work overseas, some in Nauru's phosphate mines, others as merchant seamen.

***Tuvalu's soil** is porous, but sufficiently fertile to support coconut palms, pandanus, and salt-tolerant plants. Fresh water supply is limited.*

POLITICS

Nonparty elections

2002/2006

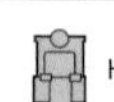
H.M. Queen Elizabeth II

AT THE LAST ELECTION

Parliament of Tuvalu 15 seats

There are no political parties. All members are independent candidates

The 15 MPs, elected every four years, are independents who work in loose political associations. The prime minister, an MP elected by parliament, works with a cabinet of up to four other MPs. Day-to-day administration lies in the hands of an elected council on each of Tuvalu's islands. From independence until 1998 politics was dominated by Tomasi Puapua (now governor-general) and Bikenibeu Paeniu. Parliamentary defections are unchecked, and brought down the short-lived government of Prime Minister Faimalaga Luka in 2001. Saufatu Sopo'aga was appointed prime minister after elections in 2002.

WORLD AFFAIRS

Joined UN in 2000

ACP

Comm

PC

PIF
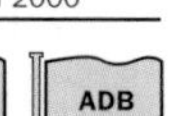
ADB

Tuvalu was admitted as a full member of both the UN and the Commonwealth in 2000, indicating international recognition of its growing involvement in world affairs. Agreements exist with Taiwan, Korea, and the US, whose vessels may exploit Tuvalu's fish-rich territorial waters, which extend for 8.3 million sq. km (3.2 million sq. miles).

AID

Recipient

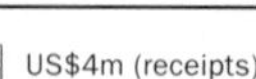
US$4m (receipts)

Down 43% in 2000

With a visible trade deficit, aid is crucial to Tuvalu. Most importantly, in 1987 a trust fund was set up, with US$29 million in grants from Australia, New Zealand, and the UK. While support from the UK has reduced, aid from Taiwan and Japan grows. Tuvalu plans to reduce its reliance on aid through public-sector reform and privatization.

DEFENSE

No compulsory military service

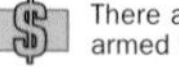
There are no armed forces
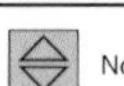
Not applicable

Tuvalu has no military. Internal security is the responsibility of the small police force.

T

ECONOMICS

Not available

US$3m

1.7997–1.9535 Australian dollars

SCORE CARD

- ❑ WORLD GNP RANKING 192nd
- ❑ GNP PER CAPITA ....US$330
- ❑ BALANCE OF PAYMENTS ....Not available
- ❑ INFLATION 7%
- ❑ UNEMPLOYMENT ....Low

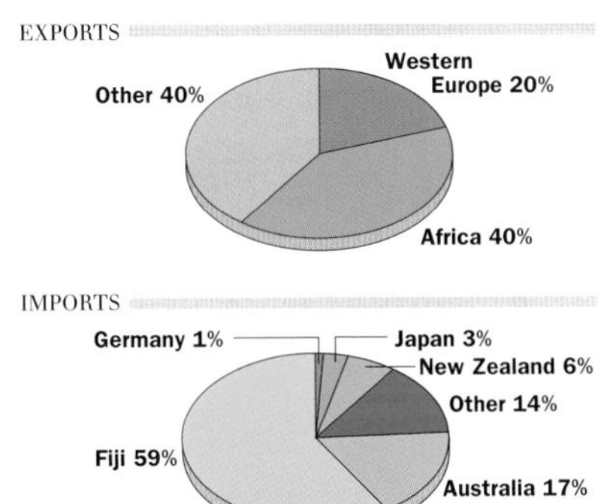
EXPORTS
Other 40%
Western Europe 20%
Africa 40%
IMPORTS
Germany 1%
Japan 3%
New Zealand 6%
Other 14%
Fiji 59%
Australia 17%

STRENGTHS

Sustainable subsistence economy. EEZ: source of jobs and income. Income from trust fund. US$175 million a year from Internet deal for use of .tv suffix.

WEAKNESSES

World's smallest economy. Physical isolation. Few exports: copra, stamps, garments. Dependence on imports and aid. Poor-quality soil. Remittances from overseas workers set to fall as Nauru phosphate mines run out.

RESOURCES

Not available

400 tonnes

Not an oil producer

13,200 pigs, 10,000 ducks, 40,000 chickens

None

Tuvalu's resource potential lies in the waters of its 8.3 million-sq. km (3.2 million-sq. mile) EEZ. Its rich fish stocks are being exploited mainly by foreign boats in return for licensing fees. The salination of soil is increasing, leaving less and less cultivable land. Solar energy is being developed to cut the use of gasoline for power generation. Fuel accounts for about 14% of import costs. The rights to Tuvalu's Internet suffix have been sold to a Canadian media company for up to US$175 million a year.

TUVALU

Total Land Area : 26 sq. km (10 sq. miles)

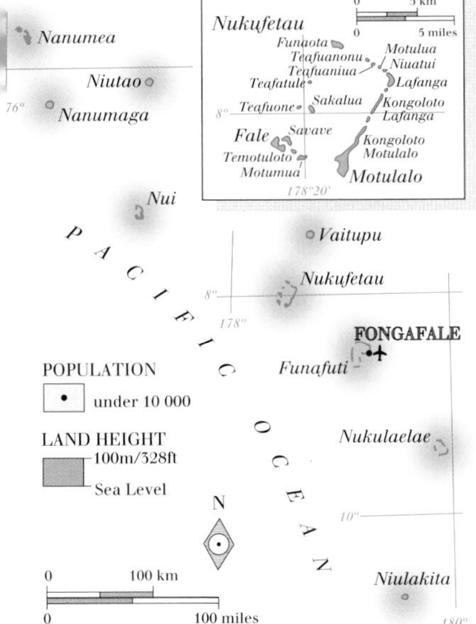
Nukufetau
Funafota
Teafuanonu
Teafuaniua
Teafatule
Teafuone
Sakalua
Motulua
Niuatui
Lafanga
Kongoloto
Fale
Savave
Temotuloto
Motumua
Motulalo
0 5 km
0 5 miles
Nanumea
Niutao
Nanumaga
Nui
Vaitupu
Nukufetau
FONGAFALE
Funafuti
Nukulaelae
Niulakita
PACIFIC OCEAN
POPULATION
under 10 000
LAND HEIGHT
100m/328ft
Sea Level
N
0 100 km
0 100 miles

ENVIRONMENT

Not available

None

Not available

Efforts to protect the environmentally fragile atolls include reafforestation and solar energy projects. On Funafuti, population pressure is leading to overfishing in the atoll lagoon. The "greenhouse effect" is a major concern, since climate changes attributed to it are blamed for a steep rise in cyclone frequency. Any rise in sea levels induced by global warming would quickly submerge the atolls.

MEDIA

TV ownership low

There are no daily newspapers

PUBLISHING AND BROADCAST MEDIA

There are no daily newspapers. The English-language *Tuvalu Echoes* is published biweekly
No TV service
1 state-owned service

Tuvalu Echoes and its Tuvaluan version, *Sikuleo o Tuvalu*, are published by the government.

CRIME

No death penalty

Tuvalu does not publish prison figures

Little change from year to year

Crime is minimal and the result mainly of alcohol-related violence, particularly at the weekends.

EDUCATION

School leaving age: 14

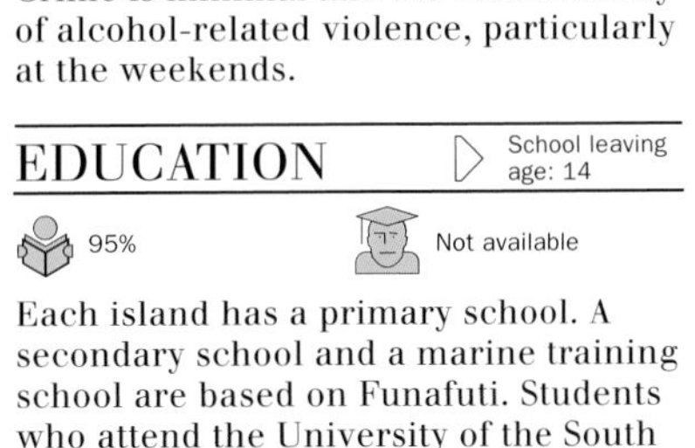
95%
Not available

Each island has a primary school. A secondary school and a marine training school are based on Funafuti. Students who attend the University of the South Pacific in Fiji are state-funded.

CHRONOLOGY

The former Ellice Islands, together with the Gilbert Islands, were annexed by the UK in 1892.

- ❑ **1974** Ellice Islanders vote to separate from Gilbertese.
- ❑ **1978** Independence as Tuvalu.
- ❑ **1987** Tuvalu Trust Fund set up.
- ❑ **1996–1998** Bikenibeu Paeniu prime minister.
- ❑ **2000** Joins UN as 189th member. Admission to Commonwealth as full member.

HEALTH

No welfare state health benefits

1 per 1125 people
Malaria, diarrheal, infectious and parasitic diseases

Concerted efforts since independence to improve health care facilities and programs have cut the incidence of communicable diseases. Serious cases of illness or injury are referred to better-equipped hospitals in Australia or New Zealand.

SPENDING

GDP/cap. increase

CONSUMPTION AND SPENDING

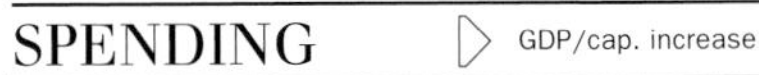
No data
90 per 1000 population

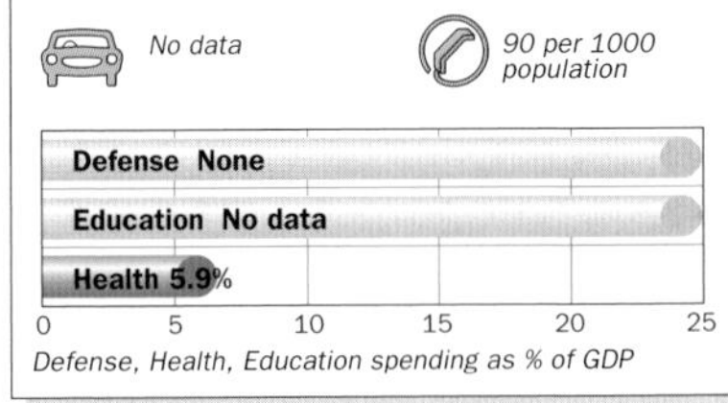
Defense None
Education No data
Health 5.9%
0 5 10 15 20 25
Defense, Health, Education spending as % of GDP

Although living standards are very low, traditional social support systems mean that extreme poverty is rare. Most people rely on subsistence agriculture and fishing, supplemented by remittances from expatriate Tuvaluans.

WORLD RANKING

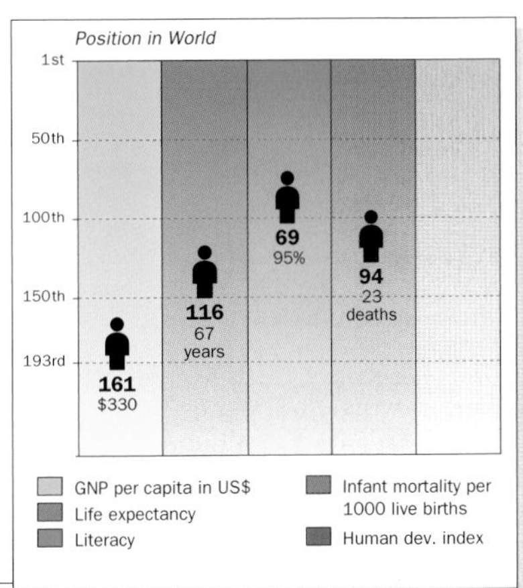
Position in World
1st
50th
100th
150th
193rd
161
$330
116
67 years
69
95%
94
23 deaths
GNP per capita in US$
Life expectancy
Literacy
Infant mortality per 1000 live births
Human dev. index

T

UGANDA

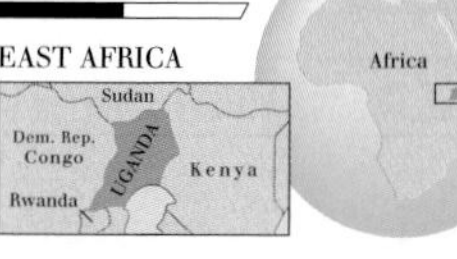
EAST AFRICA
Sudan
Dem. Rep. Congo
UGANDA
Kenya
Rwanda
Africa

OFFICIAL NAME: Republic of Uganda **CAPITAL:** Kampala **POPULATION:** 24 million
CURRENCY: New Uganda shilling **OFFICIAL LANGUAGE:** English

1962

1962

Oct 9
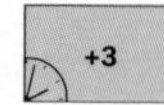
EAU

+3

+256

.ug

AN EAST AFRICAN COUNTRY of fertile upland plateaus and mountains, Uganda has outlets to the sea through Kenya and Tanzania. Its history from independence in 1962 until 1986 was one of ethnic strife. Since 1986, under President Museveni, peace has been restored and steps have been taken to rebuild the economy and democracy.

Kampala, Uganda's capital. Only a tiny proportion of the city's households are supplied with running water

CLIMATE

Tropical wet & dry

WEATHER CHART FOR KAMPALA

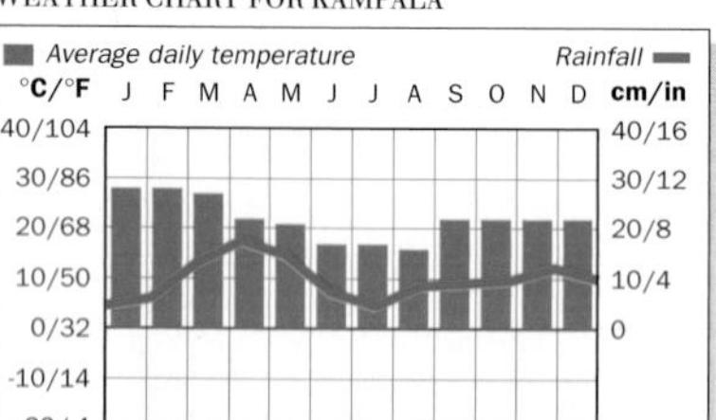
Average daily temperature
Rainfall
°C/°F
J F M A M J J A S O N D
cm/in
40/104
30/86
20/68
10/50
0/32
-10/14
-20/-4
40/16
30/12
20/8
10/4
0

Altitude and the influence of Lake Victoria moderate Uganda's climate. March–May is the wettest period.

TRANSPORTATION

Drive on left

Entebbe International 436,506 passengers

2 ships 5900 dwt

THE TRANSPORTATION NETWORK

1800 km (1118 miles)

None

261 km (162 miles)

Lake Victoria and other lakes are navigable

The government is rebuilding the transportation infrastructure with the help of international aid.

TOURISM

Visitors : Population 1:159

151,000 visitors

Down 37% 1999–2000

MAIN TOURIST ARRIVALS

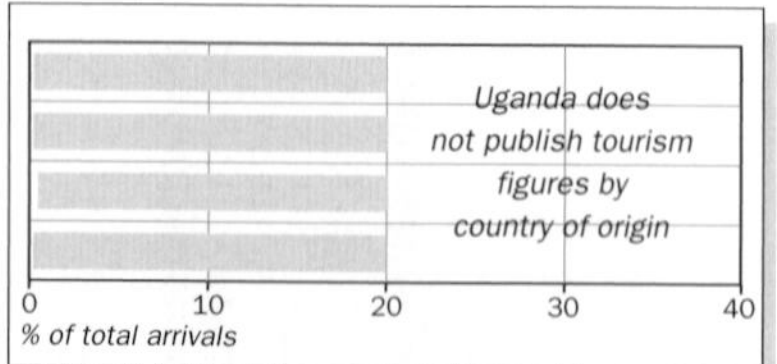
Uganda does not publish tourism figures by country of origin
0
10
20
30
40
% of total arrivals

Major attractions are Uganda's lakes and mountains, notably the rugged Ruwenzori range – the Mountains of the Moon. The brutal murder of eight foreign tourists by Rwandan fighters at the Bwindi national park in March 1999 was a severe setback for Uganda's recovery as a tourist destination.

PEOPLE

Pop. density medium

Luganda, Nkole, Chiga, Lango, Acholi, Teso, Lugbara, English

120/km² (312/mi²)

THE URBAN/RURAL POPULATION SPLIT

14% 86%

RELIGIOUS PERSUASION

Muslim (mainly Sunni) 5%
Other (including Hindu) 11%
Traditional beliefs 13%
Protestant 33%
Roman Catholic 38%

The predominantly rural population consists of 13 main ethnic groups. Traditional animosities were manipulated by ex-rulers Amin and Obote. Since 1986 President Museveni has worked hard for reconciliation, in 1993 allowing the restoration of the four historical monarchies. Uganda now has one of the best human rights records in Africa.

POLITICS

Nonparty elections

2001/2006

President Yoweri Kaguta Museveni

AT THE LAST ELECTION

Parliament 276 members

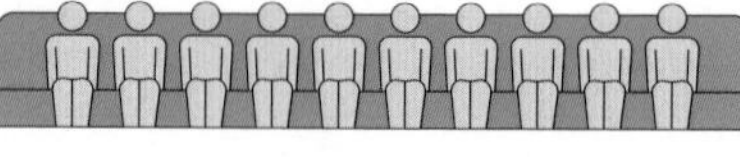

Elections to the Parliament took place on a "no-party" basis in June 2001

Since 1986, President Museveni has run a "no-party democracy," with political parties represented in a broadly based government, but banned from campaigning. Continuing to overcome ethnic tension is the main issue after its catastrophic effect in the 1970s and 1980s. Rebel insurgencies in the north and west led to the deaths, kidnapping, and displacement of tens of thousands of people, and destroyed Uganda's economy. A referendum in 2000 on a return to full multipartyism resulted in an overwhelming vote of support for the "no-party" system. In 2001, Museveni won another term in office, taking 69% of the vote, and his supporters maintained a clear majority in the legislative elections.

UGANDA

Total Land Area : 236 040 sq. km (91 135 sq. miles)

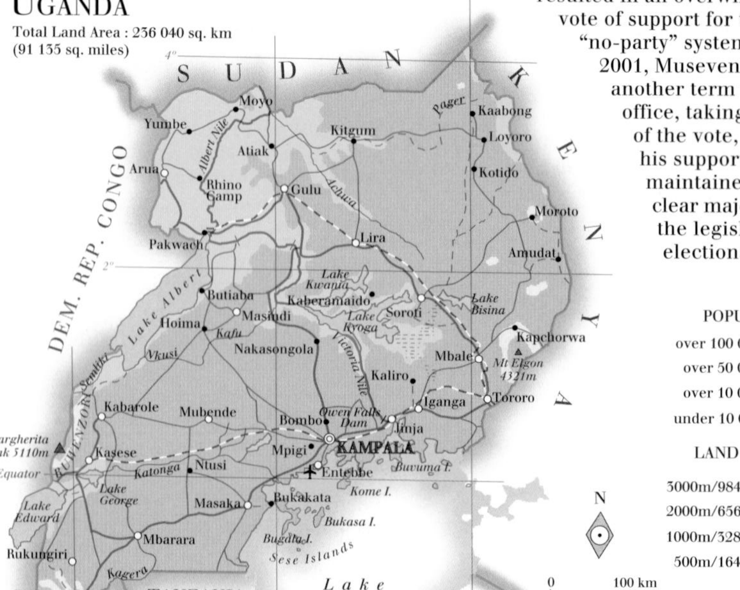
SUDAN
KENYA
DEM. REP. CONGO
TANZANIA
RWANDA
Moyo
Yumbe
Kitgum
Kaabong
Loyoro
Kotido
Atiak
Arua
Rhino Camp
Gulu
Albert Nile
Achwa
Pager
Moroto
Lira
Pakwach
Amudat
Lake Kwania
Kaberamaido
Lake Albert
Butiaba
Hoima
Masindi
Lake Kyoga
Soroti
Lake Bisina
Kapchorwa
Kafu
Nakasongola
Victoria Nile
Mbale
Mt Elgon 4321m
Kaliro
Iganga
Tororo
Kabarole
Mubende
Bombo
Owen Falls Dam
Jinja
Margherita Peak 5110m
Kasese
KAMPALA
Mpigi
Entebbe
Buvuma I.
Equator
Katonga
Ntusi
Lake George
Lake Edward
Masaka
Bukakata
Kome I.
Bukasa I.
Bugala I.
Sese Islands
Mbarara
Rukungiri
Kagera
Kisoro
Kabale
Lake Victoria
Ruwenzori
Semliki
Vkusi
POPULATION
over 100 000
over 50 000
over 10 000
under 10 000
LAND HEIGHT
3000m/9843ft
2000m/6562ft
1000m/3281ft
500m/1640ft
N
0 100 km
0 100 miles

WORLD AFFAIRS

Joined UN in 1962

Comm

COMESA

IGAD

EAC

OIC

Relations with Sudan and the DRC are strained. Conflicts in these countries and in Rwanda have caused a large influx of refugees into Uganda. Uganda's support for antigovernment rebels in Sudan appeared to end during 1999. Ugandan forces embroiled in the fighting in the DRC were withdrawn from the front line in 2001.

AID

Recipient

 $819m (receipts)

 Up 39% in 2000

Aid, mainly from the World Bank and the UK, has risen, its donors encouraged by Uganda's adoption of economic liberalization and private-sector investment policies. Aid has focused on balance-of-payments support, the rehabilitation of the key transportation sector, and the fight against AIDS.

DEFENSE

No compulsory military service

 $247m

 Up 24% in 2000

The pre-1986 army was responsible for many atrocities under Amin's rule. Since the 1990s, Uganda was preoccupied with conflicts in neighboring countries, in particular the DRC, where it supported antigovernment rebels. The army has also been deployed to suppress internal rebellions.

ECONOMICS

Inflation 12% p.a. (1990–2000)

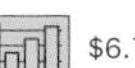 $6.7bn

 1767.5–1727.5 new Uganda shillings

SCORE CARD

- ❑ WORLD GNP RANKING........................100th
- ❑ GNP PER CAPITA $300
- ❑ BALANCE OF PAYMENTS....................–$860m
- ❑ INFLATION 2.8%
- ❑ UNEMPLOYMENT ....Widespread

STRENGTHS

Agriculture. Coffee brings in 93% of export earnings. Potential for more export crops. Road system is being repaired. Proinvestment policies.

WEAKNESSES

Lack of skilled workforce. Instability in the subregion affects confidence. World coffee price fluctuations. High transportation costs.

EXPORTS

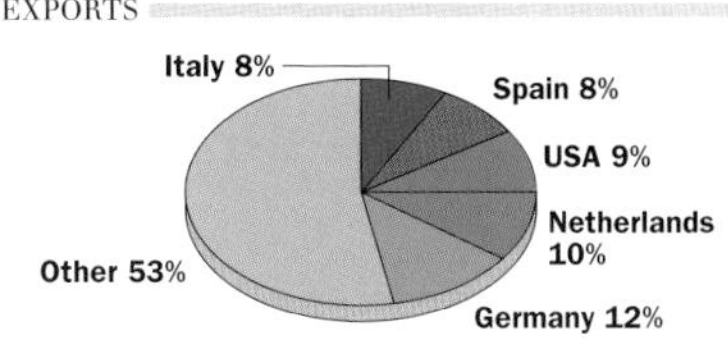
Italy 8%
Spain 8%
USA 9%
Netherlands 10%
Germany 12%
Other 53%

IMPORTS

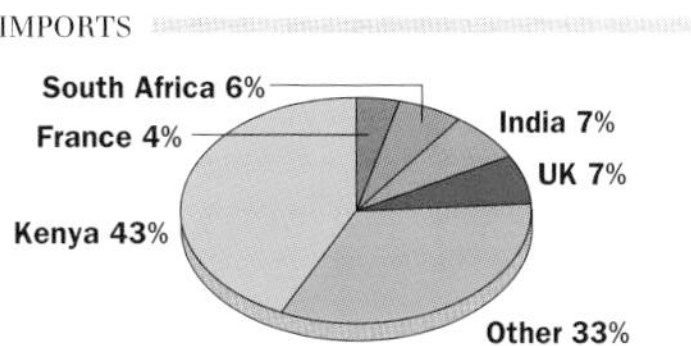
South Africa 6%
France 4%
India 7%
UK 7%
Kenya 43%
Other 33%

RESOURCES

Electric power 189,000 kw

 226,457 tonnes

 Not an oil producer

 6.2m goats, 5.9m cattle, 1.55m pigs, 25.5m chickens

Copper, cobalt, tin, apatite, magnetite, tungsten, gold

Mineral resources are varied but barely exploited. Uganda has sizable copper deposits. The mines, closed under Obote, are now being reopened. Gold and cobalt mining are also due to resume and oil exploration is under way. Hydroelectric output is being expanded, notably at Owen Falls, with the aim of replacing 50% of oil imports.

ENVIRONMENT

Sustainability rank: 76th

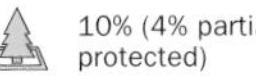 10% (4% partially protected)

 0.1 tonnes per capita

Uganda's priority is economic reconstruction, but ecological issues are not ignored. The construction of a huge hydroelectric power plant at the Kabalega (Murchison) Falls, above Lake Albert, was canceled, following widespread environmental objections.

MEDIA

TV ownership low

 Daily newspaper circulation 2 per 1000 people

PUBLISHING AND BROADCAST MEDIA

There are 5 daily newspapers, *New Vision*, *The Star*, *The Monitor*, *Ngabo*, and *Taifa Uganda Empya*

2 services: 1 state-controlled, 1 independent

4 services: 1 state-controlled, 3 independent

The 13 daily and weekly papers cover the political and religious spectrum; eight are published in English. Only *New Vision* is government-controlled.

CRIME

Death penalty in use

  21,971 prisoners

 Up 83% in 1999

Crime levels are low, though theft in Kampala is a growing problem. In 2000, the remains were discovered of 780 followers of the cult of the Restoration of the Ten Commandments of God.

CHRONOLOGY

Uganda's ancient kingdoms were ruled under a British protectorate from 1893 until independence in 1962.

- ❑ **1962–1971** Milton Obote in power.
- ❑ **1971–1986** Ethnic strife, economic collapse under Idi Amin and Obote.
- ❑ **1986** President Museveni in power.
- ❑ **1996** Museveni wins first presidential elections.
- ❑ **2000** Referendum endorses "no-party" system.

EDUCATION

Schooling is not compulsory

 67%

 40,591 students

All schools charge fees. Enrollment in primary schools is 94%, but only 11% of pupils go on to secondary school.

HEALTH

Welfare state health benefits

 1 per 20,000 people

 Malaria, respiratory, and diarrheal diseases, measles

A successful education and prevention campaign has reduced the prevalence of HIV/AIDS to about 5% of adults, from a peak of 14% in the early 1990s.

SPENDING

GDP/cap. increase

CONSUMPTION AND SPENDING

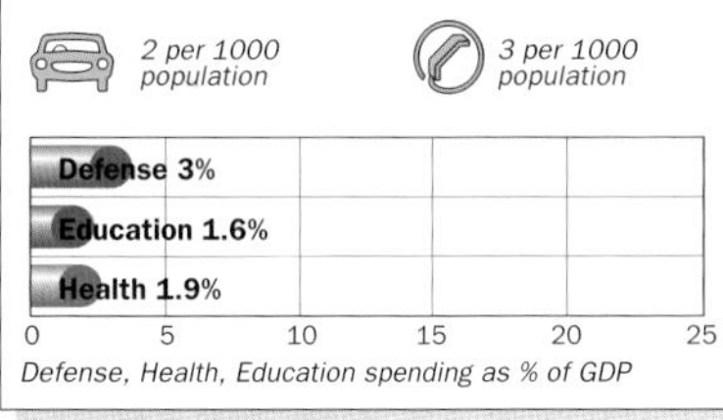
2 per 1000 population
3 per 1000 population
Defense 3%
Education 1.6%
Health 1.9%
0
5
10
15
20
25
Defense, Health, Education spending as % of GDP

Uganda has a small but growing middle class. Those close to the government form the wealthiest group. 44% of the population live below the poverty line.

WORLD RANKING

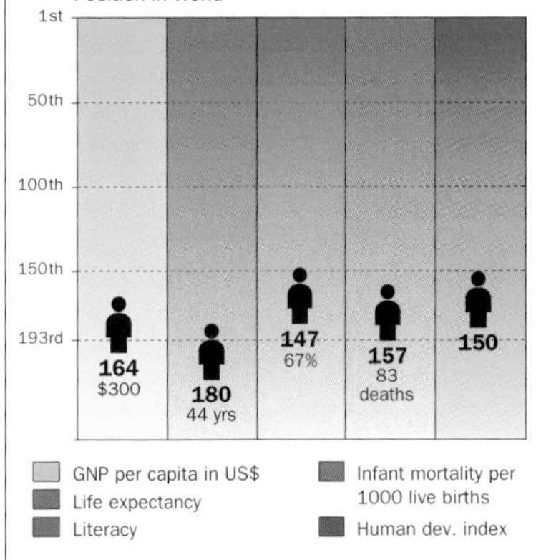
Position in World
1st
50th
100th
150th
193rd
164
$300
180
44 yrs
147
67%
157
83
deaths
150
GNP per capita in US$
Life expectancy
Literacy
Infant mortality per 1000 live births
Human dev. index

U

UKRAINE

EUROPE
Europe
Poland
Russian Fed.
UKRAINE
Romania
Moldova

OFFICIAL NAME: Ukraine **CAPITAL:** Kiev
POPULATION: 48.4 million **CURRENCY:** Hryvna **OFFICIAL LANGUAGE:** Ukrainian

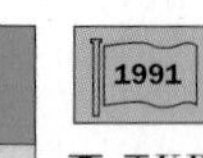
1991

1991

Aug 24

UA

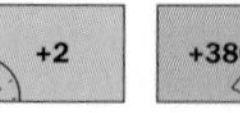
+2

+380

.ua

UKRAINE IS BORDERED by seven states; to the south it lies on the Black Sea and the Sea of Azov. An independent Ukrainian state was established in 1918, but was overrun in the same year by Soviet forces from the east and Polish forces from the west. In 1991, Ukraine again became an independent state. The country has historically been divided between the nationally conscious and Ukrainian-speaking west (including areas which were part of Poland until World War II) and the east, which has a large ethnic Russian population.

View toward the Cathedral of the Assumption in Kharkiv. Many Ukrainian cities are equipped with elaborate trolley networks.

CLIMATE

Continental/steppe/ Mediterranean

WEATHER CHART FOR KIEV

Average daily temperature
Rainfall

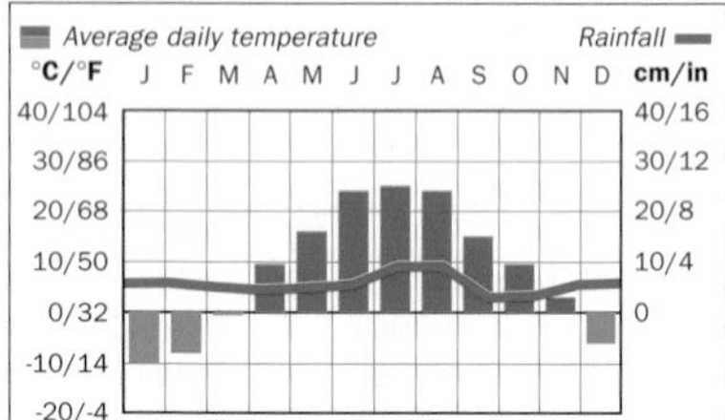
°C/°F J F M A M J J A S O N D cm/in
40/104 30/86 20/68 10/50 0/32 -10/14 -20/-4
40/16 30/12 20/8 10/4 0

Ukraine has a continental climate, with the exception of the southern coast of Crimea, which has a Mediterranean climate. There are four distinct seasons.

TRANSPORTATION

Drive on right

Boryspiel International, Kiev
1.4m passengers

966 ships
2.03m grt

THE TRANSPORTATION NETWORK

170,139 km (105,719 miles)	1770 km (1100 miles)
22,473 km (13,965 miles)	4400 km (2734 miles)

There are Soviet-style subways and trolley networks in the major cities. The rail system and the main highway linking Kiev and Lviv are being upgraded. Part of a former submarine port at Sevastopol has been opened to commercial shipping.

TOURISM

Visitors : Population 1:11

4.23m visitors

Down 32% in 1999

MAIN TOURIST ARRIVALS

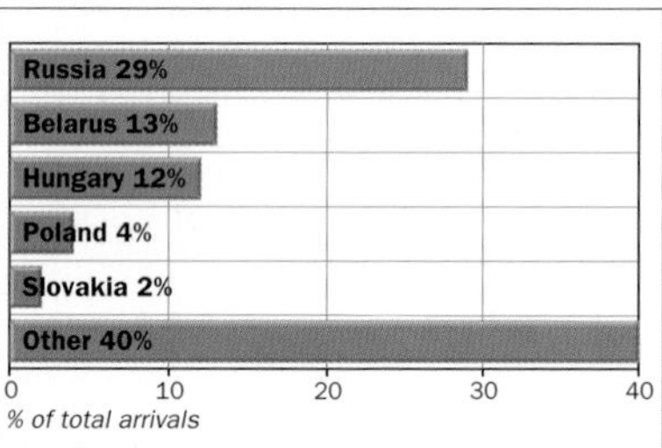
Russia 29%
Belarus 13%
Hungary 12%
Poland 4%
Slovakia 2%
Other 40%
0 10 20 30 40
% of total arrivals

Among potential tourist attractions are warm resort areas in Crimea and the south, and the Carpathian Mountains. The government has maintained a highly regulated system of managing tourism, and bureaucratic hurdles have held up the development of Western-style hotels.

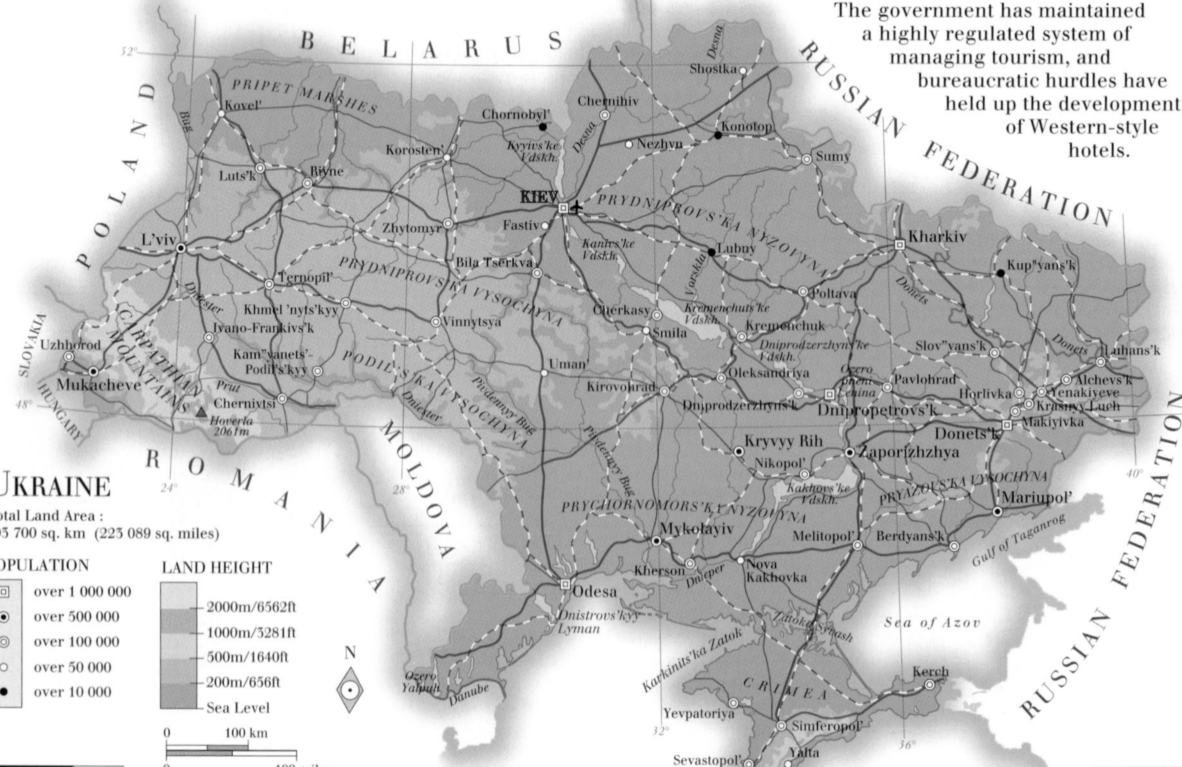
BELARUS
RUSSIAN FEDERATION
POLAND
SLOVAKIA
HUNGARY
ROMANIA
MOLDOVA
BLACK SEA
Sea of Azov
Gulf of Taganrog
PRIPET MARSHES
PRYDNIPROVS'KA NYZOVYNA
PRYDNIPROVS'KA VYSOCHYNA
PODIL'S'KA VYSOCHYNA
PRYCHORNOMORS'KA NYZOVYNA
PRYAZOVS'KA VYSOCHYNA
CARPATHIAN MOUNTAINS
CRIMEA
Hoverla 2061m
KIEV
Kovel'
Luts'k
Rivne
L'viv
Ternopil'
Khmel'nyts'kyy
Ivano-Frankivs'k
Kam"yanets'-Podil's'kyy
Chernivtsi
Uzhhorod
Mukacheve
Korosten'
Zhytomyr
Chornobyl'
Chernihiv
Shostka
Konotop
Nezhyn
Sumy
Fastiv
Bila Tserkva
Vinnytsya
Uman'
Cherkasy
Smila
Lubny
Poltava
Kremenchuk
Kharkiv
Kup"yans'k
Luhans'k
Alchevs'k
Yenakiyeve
Krasnyy Luch
Makiyivka
Horlivka
Slov"yans'k
Pavlohrad
Dnipropetrovs'k
Dniprodzerzhyns'k
Oleksandriya
Kirovohrad
Kryvyy Rih
Nikopol'
Zaporizhzhya
Donets'k
Mariupol'
Berdyans'k
Melitopol'
Mykolayiv
Kherson
Nova Kakhovka
Odesa
Yevpatoriya
Simferopol'
Sevastopol'
Yalta
Kerch
Kyyivs'ke Vdskh.
Kanivs'ke Vdskh.
Kremenchuts'ke Vdskh.
Dniprodzerzhyns'ke Vdskh.
Kakhovs'ke Vdskh.
Ozero Lenina
Ozero Yalpuh
Dnistrovs'kyy Lyman
Karkinits'ka Zatoka
Zatoka Syvash
Desna
Bug
Dniester
Prut
Dnieper
Danube
Donets
Vorskla
Pivdennyy Buh
52°
48°
24°
28°
32°
36°
40°
N

UKRAINE
Total Land Area :
603 700 sq. km (233 089 sq. miles)

POPULATION
- over 1 000 000
- over 500 000
- over 100 000
- over 50 000
- over 10 000

LAND HEIGHT
- 2000m/6562ft
- 1000m/3281ft
- 500m/1640ft
- 200m/656ft
- Sea Level

0 100 km
0 100 miles

U

PEOPLE

Population density medium

Ukrainian, Russian, Tatar

80/km² (208/mi²)

THE URBAN/RURAL POPULATION SPLIT

68% 32%

RELIGIOUS PERSUASION

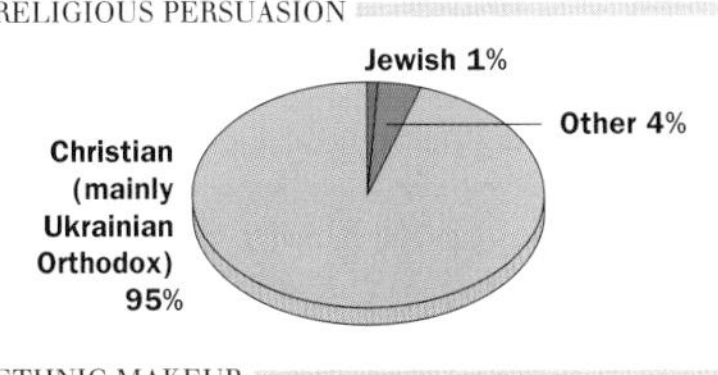

Jewish 1%
Other 4%
Christian (mainly Ukrainian Orthodox) 95%

ETHNIC MAKEUP

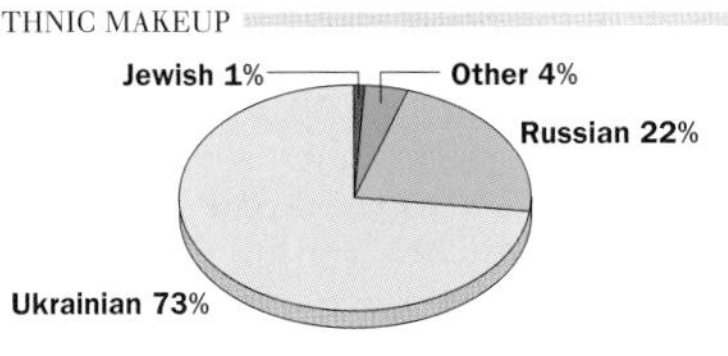

Jewish 1%
Other 4%
Russian 22%
Ukrainian 73%

In the cities and countryside of western Ukraine, Ukrainians make up the vast majority of the population. Russian is spoken by 60% of the people, and in several of the large cities of the east and south, ethnic Russians form a majority, a legacy of 19th-century industrialization and of more recent migration in the Soviet era. At independence, most Russians accepted Ukrainian sovereignty, though tensions remain.

The central government is wary of separatist tendencies in Crimea, however, where Russians make up two-thirds of the population. The Crimea's other main minority, besides ethnic Ukrainians, is the Turkic-speaking Tatar people. Deported en masse to the eastern USSR in 1944 under Stalin, the Tatars have been returning to the Crimea since 1990 and now make up 12% of its population. There is a Romanian-speaking minority in the Odessa region.

POPULATION AGE BREAKDOWN

Female	Age	Male
2%	80+	0.6%
10%	60–79	5.7%
13.6%	40–59	11.8%
14.4%	20–39	14.3%
13.5%	0–19	14.1%

% of population by age group

Leonid Kuchma, *who became president in 1994.*

Viktor Yushchenko, *prime minister in 1999–2001, now leads the opposition.*

POLITICS

Multiparty elections

2002/2006

President Leonid Kuchma

AT THE LAST ELECTION

Parliament 450 seats

25% OU	23% FUU	21% Ind	15% CPU	5% SPU	5% YT	5% SDPU–O	1% Others

OU = Our Ukraine (Viktor Yushchenko bloc)
FUU = For United Ukraine **Ind** = Independents
CPU = Communist Party **SPU** = Socialist Party of Ukraine
SDPU–O = United Social-Democratic Party of Ukraine
YT = Yulia Timoshenko Election Bloc

Ukraine introduced a multiparty system in 1991.

PROFILE

Ex-premier Leonid Kuchma, who had defeated Ukrainian nationalist Leonid Kravchuk in the 1994 presidential elections, gained increased powers under constitutional changes in 1996. Multiparty legislative elections were first held in 1994, and the strong position of the communists and allied pro-Kuchma groups was confirmed four years later. Kuchma's own reelection in 1999 was fiercely contested, however, with opponents claiming fraud and violent confrontations taking place in parliament and on the streets. The boost which Kuchma gained in early 2000 from a referendum backing electoral changes, and Western enthusiasm for his new pro-reform government, was dissipated later that year by a scandal linking him with the murder of a journalist and the uncovering of massive financial frauds. His opponents included growing numbers of discarded former ministers, and the pro-Kuchma For United Ukraine was pushed into second place by the opposition Our Ukraine bloc in legislative elections in 2002.

MAIN POLITICAL ISSUES

Economic reform and corruption

There is generally more support for economic reform in western and central Ukraine than in eastern regions. There have been spasmodic efforts to promote the transition to a market economy, notably during 2000 under the then prime minister Viktor Yushchenko. Such radicalism has foundered on lack of sufficient backing from Kuchma, opposition in parliament, and high-level corruption as powerful business and political factions vie to gain advantage.

Separatism among ethnic Russians

The Crimean parliament's declaration of independence in 1994 has been the strongest threat so far to Ukraine's integrity. It was resisted and later rescinded, although Kiev reaffirmed Crimea's status as an autonomous republic. A new Crimean constitution was approved in 1999. Donbass has also pressed for more autonomy, and the Donetsk region has voted to make Russian a joint official language.

WORLD AFFAIRS

Joined UN in 1945

BSEC  CE  CIS  IAEA OSCE

In the immediate postindependence period Western countries viewed Ukraine as a convenient buffer state between Russia and the rest of Europe. However, their support for Ukraine did not make it strong enough to offset its economic dependence on Russia, especially for fuel. Over time this unqualified backing has been eroded by the West's own relations with Russia, and by distrust of President Kuchma's antidemocratic tendencies. Conversely, Ukraine's ties with Russia, initially seen as a potential aggressor, have been strengthened under Kuchma: a friendship treaty was signed in 1997 and a ten-year economic cooperation accord in 1998. Ukraine signed a trade agreement with the EU in 1995 and was admitted to the Council of Europe in the same year. Despite an ongoing territorial dispute with Romania, a friendship and cooperation treaty was signed in 1997.

CHRONOLOGY

In 1240, Kiev was conquered by the Mongols. The Ukrainian Cossacks later came under the domination of Lithuania, Poland, and Russia.

- **1918** Independent Ukrainian state after collapse of Russian and Austro-Hungarian Empires. Brest-Litovsk Treaty signed with Germany.
- **1919** Red Army invades. Ukrainian Soviet Socialist Republic proclaimed.
- **1920** Poland invades; western Ukraine under Polish occupation.
- **1922** USSR, and Ukrainian SSR, formed.
- **1922–1930** Cultural revival under Lenin's "Ukrainianization" policy to pacify national sentiment.
- **1932–1933** "Ukrainianization" policy reversed. Stalin uses famine to eliminate Ukraine as source of opposition; seven million die.

U

CHRONOLOGY *continued*

- ❑ **1939** Soviet Union invades Poland and incorporates its ethnic Ukrainian territories into the Ukrainian SSR.
- ❑ **1941** Germany invades USSR. 7.5 million Ukrainians die by 1945.
- ❑ **1942** Nationalists form Ukrainian Insurgent Army, which wages war against both Germans and Soviets.
- ❑ **1954** Crimea ceded to Ukrainian SSR.
- ❑ **1972** Widespread arrests of intellectuals and dissidents by Soviet state. Vladimir Shcherbitsky, a Brezhnevite, replaces moderate reformer Petr Shelest as head of Communist Party of Ukraine (CPU).
- ❑ **1986** World's worst nuclear disaster at Chernobyl power plant.
- ❑ **1989** First major coalminers' strike in Donbass. Pro-Gorbachev Volodymyr Ivashko heads CPU.
- ❑ **1990** Ukrainian parliament declares Ukrainian SSR a sovereign state. Leonid Kravchuk replaces Ivashko.
- ❑ **1991** Full independence declared, conditional on approval by referendum, supported by 90% of voters. CPU banned. Crimea becomes an autonomous republic within Ukrainian SSR.
- ❑ **1993** Major strike in Donbass results in costly settlement, which exacerbates budget deficit and stimulates hyperinflation. CPU reestablished at Donetsk congress.
- ❑ **1994** Crimea elects Yuri Meshkov as its first president. Leonid Kuchma defeats Kravchuk to become first democratically elected president of Ukraine.
- ❑ **1996** Hryvna replaces karbovanets as national currency. New constitution comes into force.
- ❑ **1997** Friendship treaty signed with Russia. Accord on Black Sea fleet.
- ❑ **1998** Ten-year cooperation agreement with Russia. CPU wins largest number of seats in election.
- ❑ **1999** Reelection of Kuchma. Opposition claims of fraud. Kuchma appoints pro-reform government.
- ❑ **2000** Chernobyl site closed.
- ❑ **2001** Kuchma linked with murder of journalist. Reformist premier Viktor Yushchenko replaced after parliamentary defeat.
- ❑ **2002** Opposition parties make large gains in legislative elections.

AID

Recipient

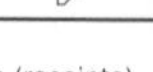
$541m (receipts)

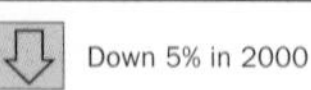
Down 5% in 2000

US assistance in the 1990s was the world's fourth-largest aid program, while EU assistance totaled $3.5 billion in 1991–1999. In 2000 lenders pressed for better controls on fund allocations.

DEFENSE

Phasing out conscription

$1.08bn

Down 25% in 2000

UKRAINIAN ARMED FORCES

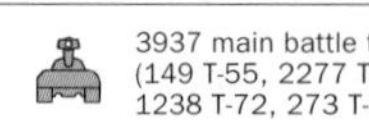

	3937 main battle tanks (149 T-55, 2277 T-64, 1238 T-72, 273 T-80)	151,200 personnel
	Black Sea fleet: 1 submarine, 1 cruiser, 2 frigates, 8 patrol boats	13,000 personnel
	543 combat aircraft (MiG-23, MiG-29, Su-27, Su-25, Su-24)	96,000 personnel
	None	

Ukraine was a center for arms manufacture under the old Soviet system, and now has a major weapons export trade. A member of the CIS, Ukraine finally resolved in 1997 its long-smoldering dispute with Russia over control of the Black Sea fleet, with agreement on the division of the fleet and a 20-year Russian lease on port facilities in Sevastopol. Meanwhile, the Ukrainian parliament had ratified the START-I nuclear disarmament treaty, and Ukraine's nuclear warheads were transferred to Russia under a trilateral weapons dismantling accord also involving substantial US aid.

In 2000 it was decided that compulsory military service, which lasts 18 months, should be ended by 2015. The armed forces have been slimmed down, with further cuts planned to result in a total strength of 285,000 (including central and administrative staff) by 2005.

Ukraine joined NATO's Partnerships for Peace in 1995, signed a security pact with the alliance in 1997, and applied in 2002 for full NATO membership.

ECONOMICS

Inflation 271% p.a. (1990–2000)

$34.6bn

5.435–5.314 hryvnas

SCORE CARD

❑ World GNP Ranking	55th
❑ GNP per Capita	$700
❑ Balance of Payments	$1.48bn
❑ Inflation	22.7%
❑ Unemployment	12%

EXPORTS

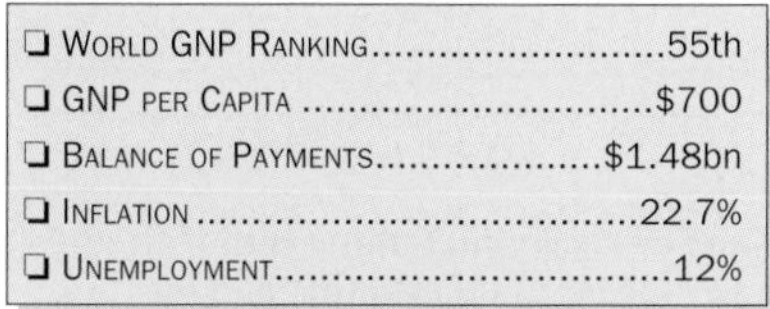

Germany 5%
Italy 4%
USA 5%
Turkey 6%
Other 56%
Russia 24%

IMPORTS

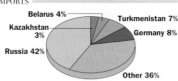

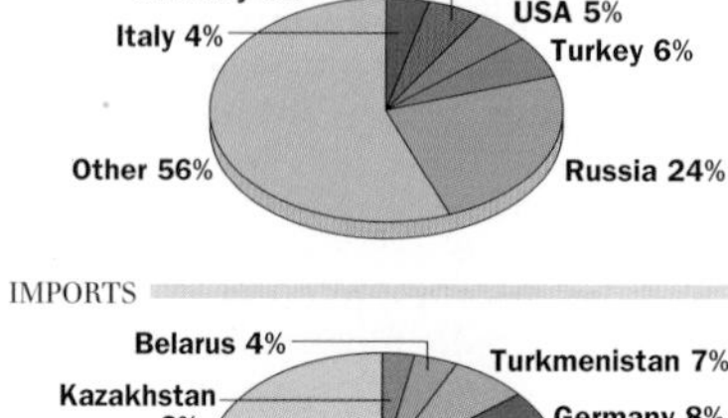

Belarus 4%
Kazakhstan 3%
Turkmenistan 7%
Germany 8%
Russia 42%
Other 36%

ECONOMIC PERFORMANCE INDICATOR

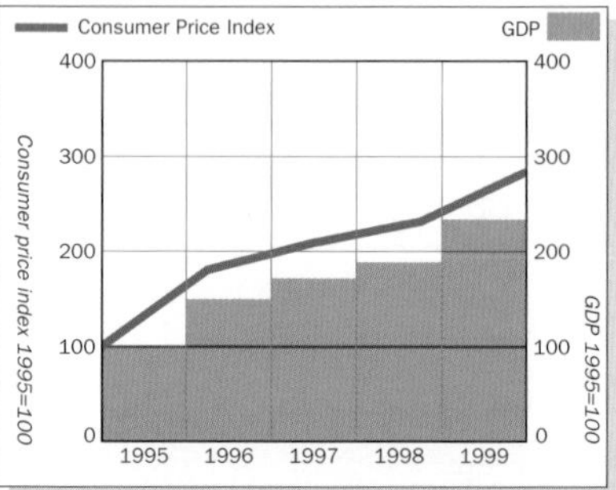

Consumer Price Index
GDP
400
300
200
100
0
Consumer price index 1995=100
GDP 1995=100
1995
1996
1997
1998
1999

Strengths

Well-educated workforce. Good urban transportation infrastructure. Potential for grain and food export. Mineral reserves. Technological potential, especially in aerospace and computers.

Weaknesses

Failure to reform centrally planned economy. High inflation. Weak currency. Huge debt. Antireform political elites. Inefficient, subsidized manufacturing industries. Corruption.

Profile

Real growth was recorded at last in 2000, the economy having contracted by over half over ten years. Privatization of large enterprises has barely begun, and bureaucracy stifles private enterprise and investment. Lack of land reform holds back agriculture in the "bread basket" of Europe. Controversial legislation passed in 2001 paved the way for the sale of farmland after 2004.

UKRAINE : MAJOR BUSINESSES

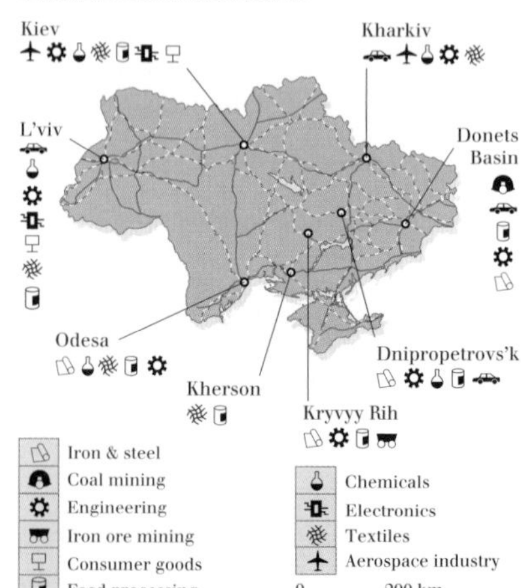

Kiev
Kharkiv
L'viv
Donets Basin
Odesa
Dnipropetrovs'k
Kherson
Kryvyy Rih
Iron & steel
Coal mining
Engineering
Iron ore mining
Consumer goods
Food processing
Vehicle manufacture
Chemicals
Electronics
Textiles
Aerospace industry
0
200 km
0
200 miles

U

RESOURCES

Electric power 53.9m kw

441,672 tonnes

78,195 b/d (reserves 1.6bn barrels)

20m ducks, 9.91m cattle, 9.08m pigs, 108m chickens

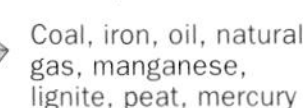
Coal, iron, oil, natural gas, manganese, lignite, peat, mercury

ELECTRICITY GENERATION

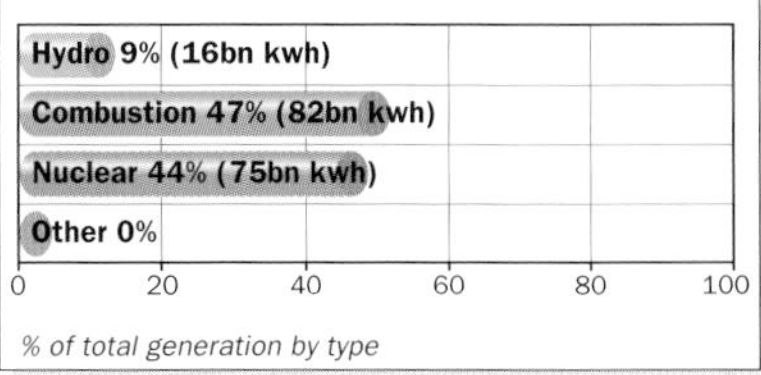
Hydro 9% (16bn kwh)
Combustion 47% (82bn kwh)
Nuclear 44% (75bn kwh)
Other 0%
0
20
40
60
80
100
% of total generation by type

Ukraine imports 80% of its oil and gas (its main source of energy), mostly from Russia. Some gas is in lieu of transit fees for pipelines carrying Russian gas, but it repeatedly fails to keep up payments for the rest. Yet Ukraine has oil and gas reserves of its own. Coal is mined, in appalling safety conditions, mainly in the Donbass–Donetsk region. Just under half of electricity is nuclear-generated.

Ukraine has 5% of global mineral reserves, including the largest titanium reserves, the third-largest deposits of iron ore, and 30% of global manganese ore. There is also mercury, uranium, nickel, and some gold. The metal industry accounted for nearly 20% of GDP in 1997 and 28% of exports. The steel industry has now begun to grow again after seven years of decline.

UKRAINE : LAND USE

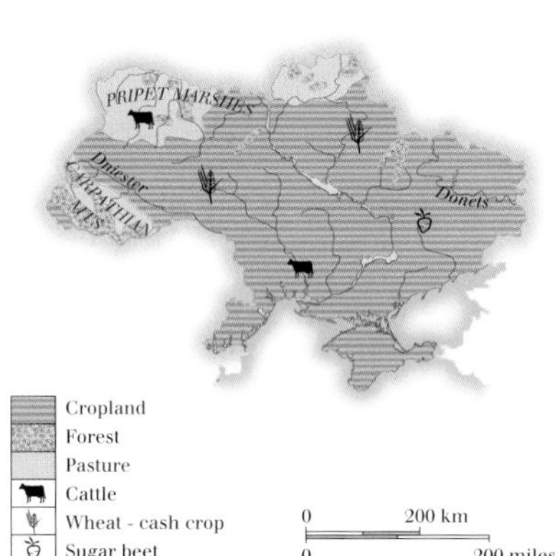
PRIPET MARSHES
Dniester
CARPATHIAN MTS
Donets
Cropland
Forest
Pasture
Cattle
Wheat - cash crop
Sugar beet
0
200 km
0
200 miles

ENVIRONMENT

Sustainability rank: 136th

2%

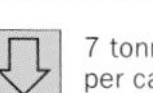
7 tonnes per capita

ENVIRONMENTAL TREATIES

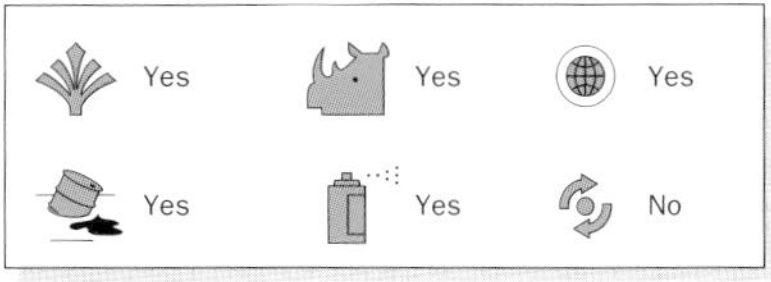
Yes
Yes
Yes
Yes
Yes
No

As a result of the Chernobyl nuclear disaster in 1986 – the world's worst nuclear accident – over three million Ukrainians live in dangerously radioactive areas and 12% of arable land is contaminated. The last working reactor at Chernobyl closed at the end of 2000, under agreements in which Western countries provided large-scale financial assistance. However, nuclear production continues elsewhere because of the cost of Russian oil and gas imports. Coal-fired power plants are old-fashioned, highly polluting, and inefficient. Industrial pollution is widespread, especially from steel and chemical works in the Donbass region, contributing to the acute problem of low air quality in eight major cities.

MEDIA

TV ownership high

Daily newspaper circulation 84 per 1000 people

PUBLISHING AND BROADCAST MEDIA

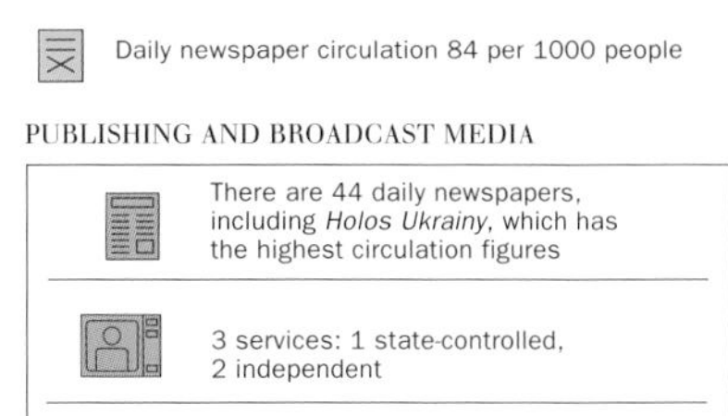
There are 44 daily newspapers, including *Holos Ukrainy*, which has the highest circulation figures

3 services: 1 state-controlled, 2 independent

3 services: 2 state-controlled, 1 independent

Independent, mass-circulation newspapers are published, mainly in Russian. Local TV stations reflect regional political differences.

CRIME

No death penalty

219,555 prisoners

Down 2% in 1999

CRIME RATES

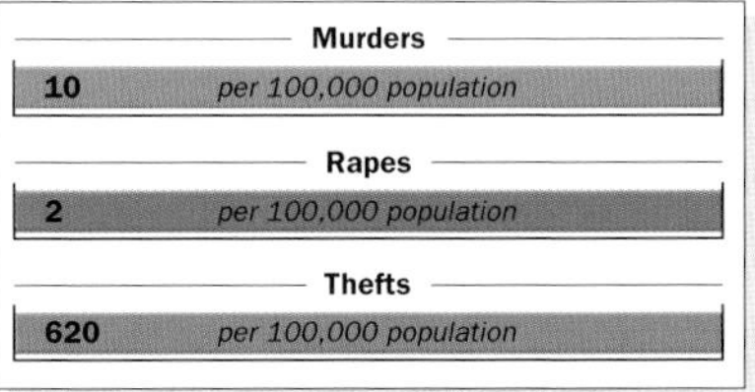

Murders	
10	*per 100,000 population*
Rapes	
2	*per 100,000 population*
Thefts	
620	*per 100,000 population*

Street crime, robberies, violence, and carjackings have increased sharply. The death penalty was abolished in 2000. The country is a major source for people trafficking to the West. Corruption is rampant across the economy. Political killings make headlines, as did the murder in 2000 of journalist Gyorgy Gongadze.

EDUCATION

School leaving age: 15

99%

1.54m students

THE EDUCATION SYSTEM

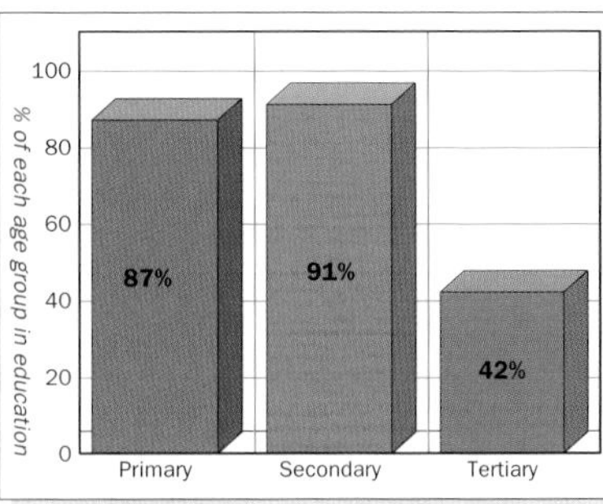
100
80
60
40
20
0
% of each age group in education
87%
91%
42%
Primary
Secondary
Tertiary

Using Ukrainian in schools is the main element in the drive to promote the once-banned language. Some schools in the west no longer teach Russian. Most university teaching is in Russian in eastern regions, and in Ukrainian in those in the west.

HEALTH

Welfare state health benefits

1 per 333 people

Heart disease, cancers, accidents, violence, tuberculosis

Health care, supposedly free to all, has declined significantly in the post-Soviet period. A $2 million UN program provides treatment and preventive care for the 350,000 people who dealt with the Chernobyl disaster. By 2001, 1% of the population was HIV positive.

SPENDING

GDP/cap. decrease

CONSUMPTION AND SPENDING

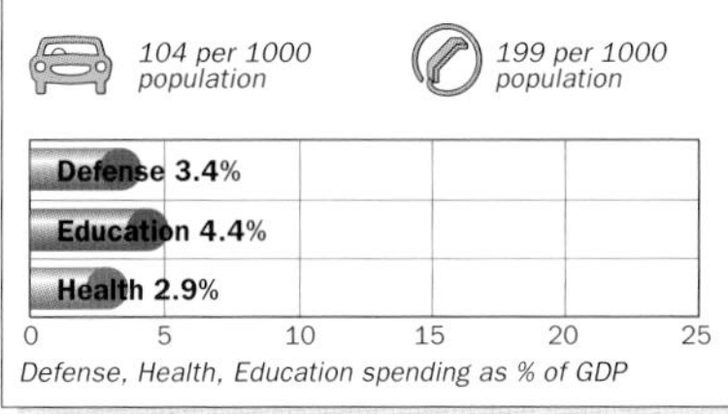
104 per 1000 population
199 per 1000 population
Defense 3.4%
Education 4.4%
Health 2.9%
0
5
10
15
20
25
Defense, Health, Education spending as % of GDP

The gap between rich and poor has widened significantly. Wage arrears – and massive hidden unemployment – are major problems.

WORLD RANKING

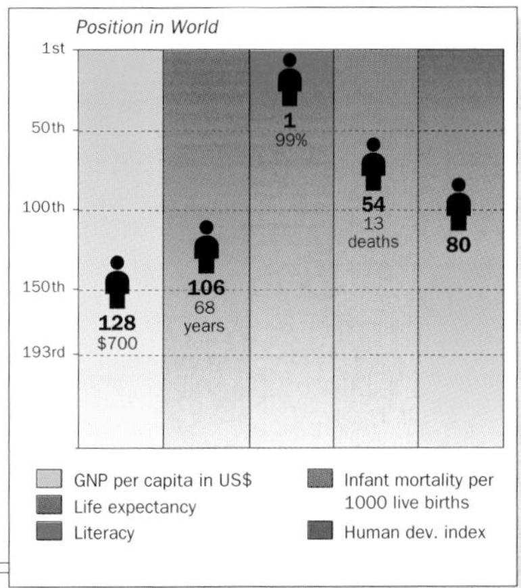
Position in World
1st
50th
100th
150th
193rd
128
$700
106
68
years
1
99%
54
13
deaths
80
GNP per capita in US$
Life expectancy
Literacy
Infant mortality per 1000 live births
Human dev. index

U

United Arab Emirates

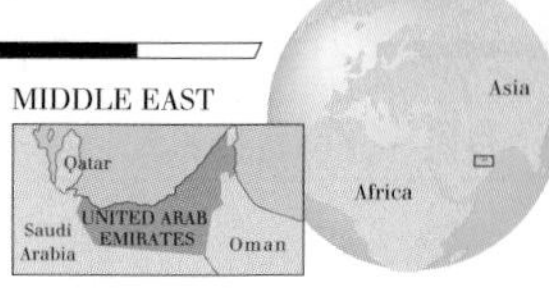
MIDDLE EAST
Qatar
Saudi Arabia
UNITED ARAB EMIRATES
Oman
Asia
Africa

Official Name: United Arab Emirates **Capital:** Abu Dhabi
Population: 2.7 million **Currency:** UAE dirham **Official Language:** Arabic

1971

1972

Dec 2

UAE

+4

+971
.ae

THE UNITED ARAB EMIRATES (UAE), created in 1971, is the Arab world's only working federation. Six of its seven emirates cluster around the northeastern corner of the country, while Abu Dhabi has a larger hinterland of semiarid desert relieved by occasional oases. The cities, watered by extensive irrigation systems, have lavish greenery. Prosperity once relied on pearls, but the UAE is now a sizable gas and oil exporter, and has a growing services sector.

CLIMATE

Hot desert

WEATHER CHART FOR ABU DHABI

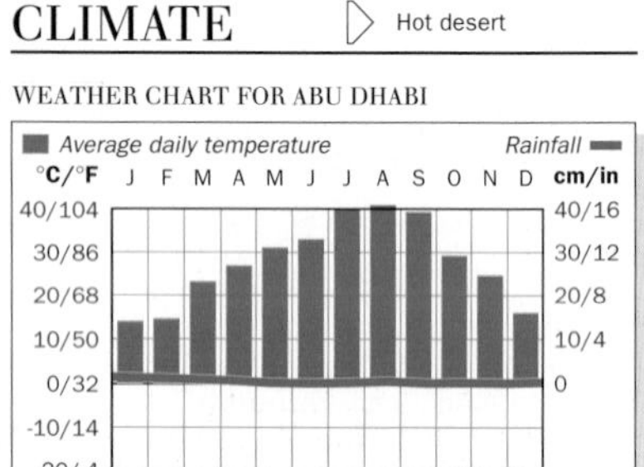
Average daily temperature
Rainfall
°C/°F J F M A M J J A S O N D cm/in
40/104
30/86
20/68
10/50
0/32
-10/14
-20/4
40/16
30/12
20/8
10/4
0

Although rainfall is minimal, summers are humid. Sand-laden *shamal* winds often blow in winter and spring.

TRANSPORTATION

Drive on right

Dubai International 12.3m passengers

332 ships 933,000 grt

THE TRANSPORTATION NETWORK

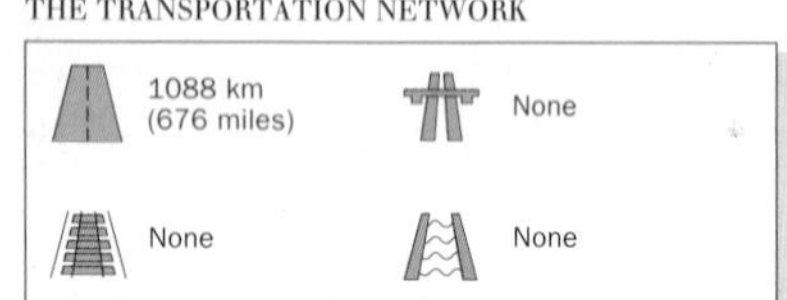
1088 km (676 miles)
None
None
None

The roads are good. Five of the seven emirates have international airports, of which the busiest is Dubai International.

TOURISM

Visitors : Population 1.3:1

3.42m visitors

Up 15% in 2000

MAIN TOURIST ARRIVALS

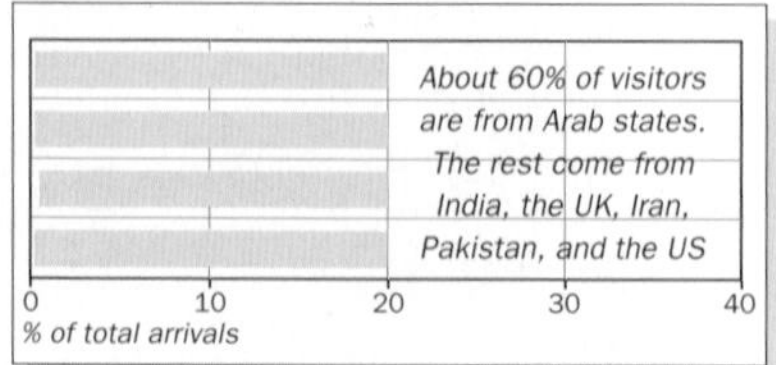
About 60% of visitors are from Arab states. The rest come from India, the UK, Iran, Pakistan, and the US
0
10
20
30
40
% of total arrivals

Until the mid-1980s, tourism was minimal. Led by Dubai, the UAE has now launched initiatives to attract Western visitors during the winter for sunshine, heritage, water sports, desert safaris, and duty-free shopping.

PEOPLE

Pop. density low

Arabic, Farsi, Indian and Pakistani languages, English

32/km² (84/mi²)

THE URBAN/RURAL POPULATION SPLIT

86% 14%

ETHNIC MAKEUP

European 3%
Other Arab 12%
Emirian 25%
Asian 60%

UAE nationals are largely city dwellers, with Abu Dhabi and Dubai the main centers. They are greatly outnumbered by expatriates who arrived in the 1970s during the oil boom, and the Western expatriate community is permitted a virtually unrestricted lifestyle. Most UAE nationals are conservative Sunni Muslims of Bedouin descent, although there is a Shi'a community in Dubai with links to Iran. Islamic fundamentalism is a growing force among the young.

Poverty is rare in the UAE, where the government remains the biggest employer. Women in theory enjoy equal rights with men. A Presidential Marriage Fund discourages UAE men from taking foreign wives.

POLITICS

No legislative elections

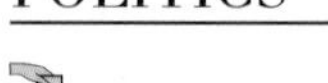

Not applicable

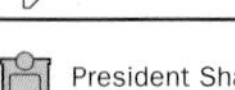

President Shaikh Zayed bin Sultan al-Nahyan

LEGISLATIVE OR ADVISORY BODIES

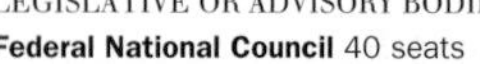

Federal National Council 40 seats

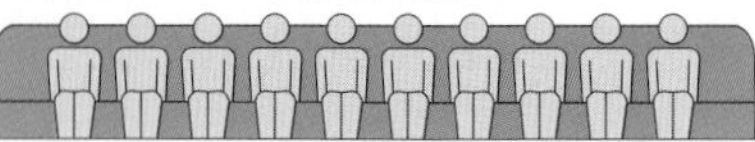

There are no political parties. The method of appointment of members of the Federal National Council is determined individually by each of the seven members of the Federation.

The UAE's seven emirates – Abu Dhabi, Dubai, Sharjah, Ras al Khaimah, Ajman, Umm al Qaiwain, and Fujairah – are dominated by their ruling families. The main personalities are the ruler of Abu Dhabi, Shaikh Zayed, who holds the UAE presidency, and the four al-Maktoum brothers who control Dubai. The eldest, Shaikh Maktoum al-Maktoum, is ruler of Dubai as well as vice president and prime minister of the UAE.

President Zayed has relaunched the advisory Federal National Council in response to criticism of the lack of democracy. The growth of Islamic fundamentalism is also a concern. The freedoms granted to Westerners have aroused some anger but, for economic reasons, they are unlikely to be withdrawn.

United Arab Emirates

Total Land Area : 82 880 sq. km (32 000 sq. miles)

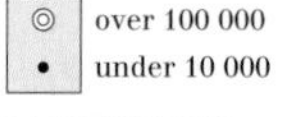
POPULATION
over 100 000
under 10 000

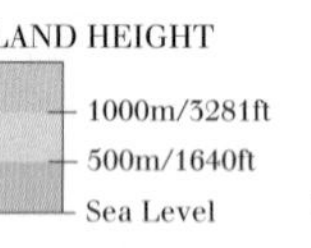
LAND HEIGHT
1000m/3281ft
500m/1640ft
Sea Level

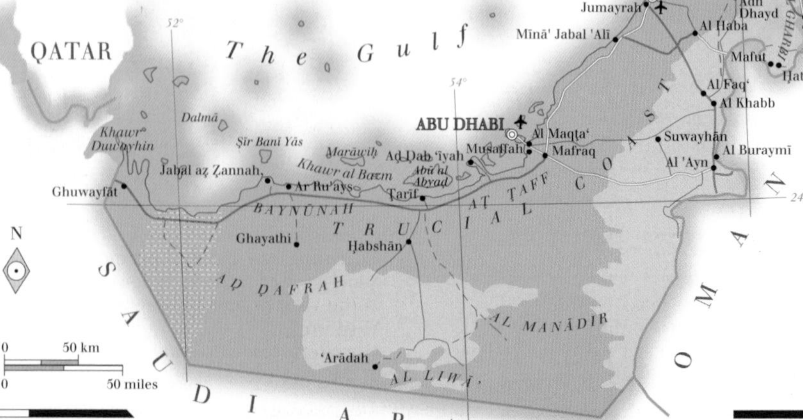
QATAR
The Gulf
Strait of Hormuz
OMAN
Ar Rams
Marid
Ras al Khaimah
Umm al Qaiwain
Al Hamriyah
Sharjah
Ajman
Dubai
Dayrah
Jumayrah
Mīnā' Jabal 'Alī
Dibā
Nabgha
Adh Dhayd
Al Haba
Mafut
Hattā
Al Faq'
Al Khabb
ABU DHABI
Al Maqta'
Mafraq
Musaffah
Ad Dab'iyah
Suwayhān
Al Buraymī
Al 'Ayn
Dalmā
Khawr Duwayhin
Şir Bani Yās
Marawih
Jabal az Zannah
Khawr al Bazm
Ar Ru'ays
Tarif
Ghuwayfat
BAYNUNAH
Ghayathi
Habshān
TRUCIAL COAST
AD DAFRAH
AL MANĀDIR
'Arādah
AL LIWA
SAUDI ARABIA
OMAN
N
0 50 km
0 50 miles
52°
54°
56°
24°

WORLD AFFAIRS

Joined UN in 1971

AL | OIC | GCC | OAPEC | OPEC

The UAE is well known as an advocate of moderation within the Arab world. It maintains close links with most OECD states, especially the UK and the US. In 1992, conflict flared when Iran seized control of three islands in the Strait of Hormuz. Attempts are still being made to settle the dispute through diplomacy.

AID

Donor

 $150m (donations)

 Up 63% in 2000

Once a generous donor to developing countries, the UAE's contributions have fluctuated with energy prices.

DEFENSE

No compulsory military sevice

 $3.34bn

Up 5% in 2000

Training of UAE forces is limited, and personnel are mainly drawn from other Arab states and the Indian subcontinent. US air bases in the UAE supply refueling craft and were used in the 1991 Gulf War, subsequent air patrols over southern Iraq, and during the 2001 air strikes in Afghanistan. The countries of the GCC signed their first defense pact in 2000.

ECONOMICS

Inflation 2.3% p.a. (1990–1999)

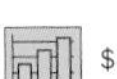 $48.7bn

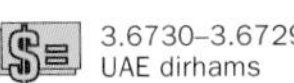 3.6730–3.6729 UAE dirhams

SCORE CARD

❑ WORLD GNP RANKING	49th
❑ GNP PER CAPITA	$17,870
❑ BALANCE OF PAYMENTS	$24.6bn
❑ INFLATION	4.5%
❑ UNEMPLOYMENT	None

STRENGTHS

Oil and gas reserves are third-biggest in OPEC. Soaring oil prices in 2000. Development of service industries and manufacturing sector. Since 2000, regional tax-free base for e-commerce.

WEAKNESSES

Lack of skilled manpower. Most raw materials and foodstuffs have to be imported. Water resources scarce.

EXPORTS

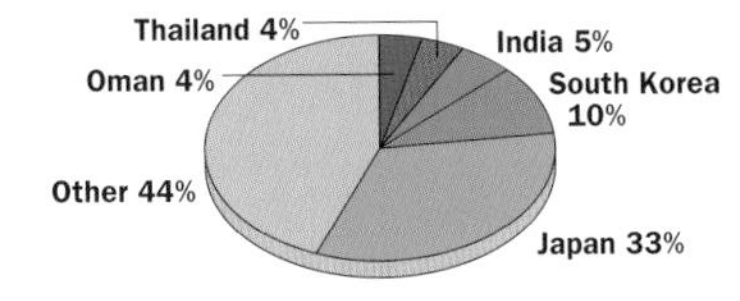
Thailand 4%
Oman 4%
India 5%
South Korea 10%
Other 44%
Japan 33%

IMPORTS

France 6%
USA 6%
Germany 7%
Japan 7%
UK 7%
Other 67%

RESOURCES

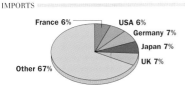

Electric power 5.7m kw

 117,607 tonnes

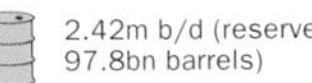 2.42m b/d (reserves 97.8bn barrels)

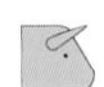 1.2m goats, 467,281 sheep, 14.7m chickens

 Oil, natural gas

The UAE is a major exporter of crude oil and natural gas. Oil production accounts for a great part of export revenue. Mina Jabal Ali in Dubai is the world's largest man-made port and has attracted companies from more than 50 countries. Saadiyat Island off Abu Dhabi is being developed as a financial resort.

ENVIRONMENT

Sustainability rank: 141st

 None

 32.4 tonnes per capita

Despite the harsh desert climate, there is a rich variety of plants and animals. Shaikh Zayed champions conservation parks to avert the threat from hunting.

MEDIA

TV ownership high

 Daily newspaper circulation 170 per 1000 people

PUBLISHING AND BROADCAST MEDIA

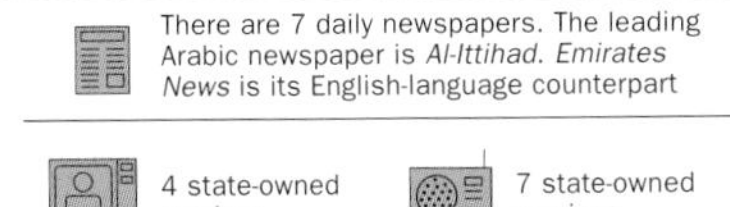

There are 7 daily newspapers. The leading Arabic newspaper is *Al-Ittihad*. *Emirates News* is its English-language counterpart

4 state-owned services

7 state-owned services

Satellite TV is unrestricted. Dubai Media City, opened in 2001, promotes greater press freedom.

CRIME

Death penalty in use

 6000 prisoners

 Up sharply in 1999

Street crime and muggings are rare. Dubai is reputed to be a transit point for the smuggling of narcotics and caviar.

An oasis village, inland from Fujairah, now accessible through a well-developed network of new roads.

CHRONOLOGY

The UAE was influenced by the Portuguese and the Ottomans, but British control became dominant in the 19th century.

- ❑ **1971** The UK withdraws as protecting power and the UAE federation is formed.
- ❑ **1991** UAE offers bases to Western forces after Kuwait is invaded.
- ❑ **2000** GCC defense pact signed.

EDUCATION

School leaving age: 12

 76%

 16,213 students

UAE citizens enjoy completely free education. Zayed University was set up in three emirates in 1998.

HEALTH

Welfare state health benefits

 1 per 556 people

 Circulatory and respiratory diseases, cancers

A high-quality system of primary health care is in place for all UAE citizens, with hospitals able to carry out most operations.

SPENDING

GDP/cap. increase

CONSUMPTION AND SPENDING

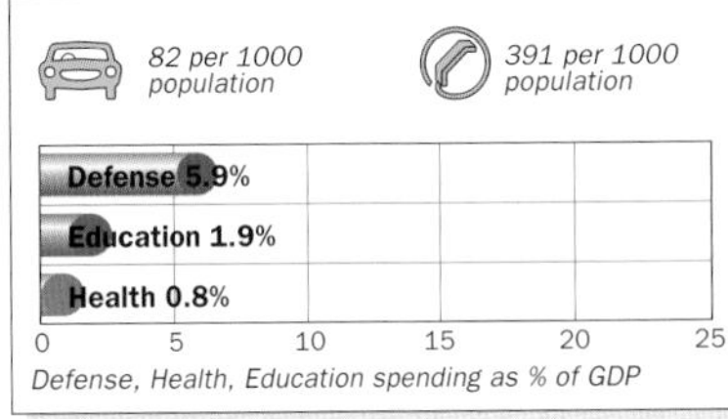
82 per 1000 population
391 per 1000 population
Defense 5.9%
Education 1.9%
Health 0.8%
0
5
10
15
20
25
Defense, Health, Education spending as % of GDP

UAE nationals enjoy one of the highest per capita incomes in the Arab world. There is no income tax, and oil revenues subsidize public services. Entrepreneurship is encouraged.

WORLD RANKING

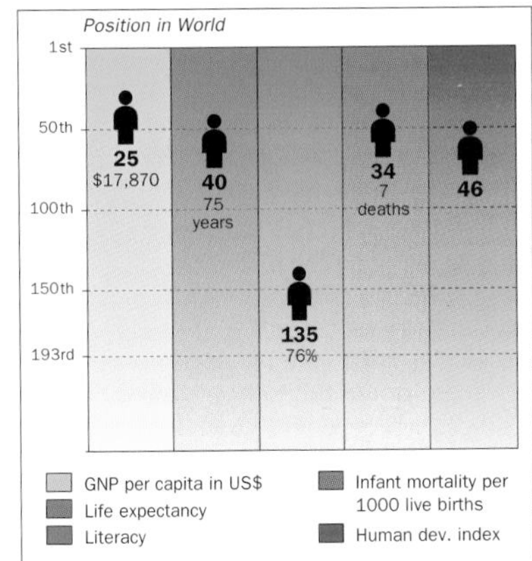
Position in World
1st
50th
100th
150th
193rd
25
$17,870
40
75 years
135
76%
34
7 deaths
46
GNP per capita in US$
Life expectancy
Literacy
Infant mortality per 1000 live births
Human dev. index

U

UNITED KINGDOM

EUROPE

OFFICIAL NAME: United Kingdom of Great Britain and Northern Ireland **CAPITAL:** London
POPULATION: 59.5 million **CURRENCY:** Pound sterling **OFFICIAL LANGUAGES:** English, Welsh (in Wales)

1707 | 1922 | None | GB | 0 | +44 | .uk

LYING IN NORTHWESTERN Europe, the United Kingdom (UK) occupies the major portion of the British Isles. It includes the countries of England, Scotland, and Wales, the constitutionally distinct region of Northern Ireland, and several outlying islands. Its only land border is with the Irish republic. The UK is separated from the European mainland by the English Channel and the North Sea. To the west lies the Atlantic Ocean. Most of the population live in towns and cities and, in England, is fairly well scattered. The most densely populated region is the southeast. Scotland is the wildest region, with the Highlands less populated today than in the 18th century. The UK joined the European Communities (EC – later the EU) in 1973, and most of its trade is now with its European partners. Membership of the UN Security Council also gives the UK a prominent role in international politics.

CLIMATE

Maritime

WEATHER CHART FOR LONDON

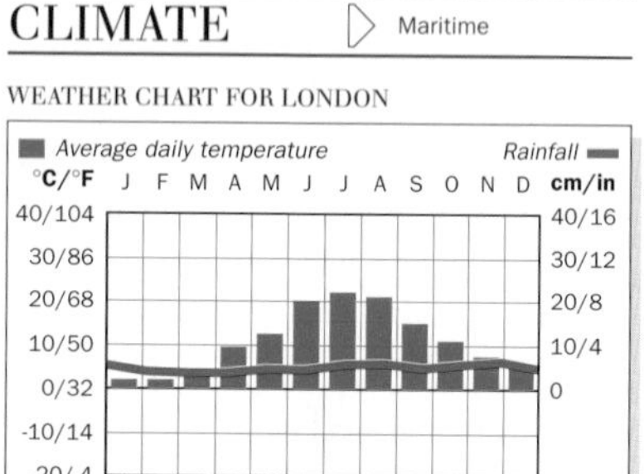

Average daily temperature
Rainfall
°C/°F J F M A M J J A S O N D cm/in
40/104 30/86 20/68 10/50 0/32 -10/14 -20/4
40/16 30/12 20/8 10/4 0

The UK has a generally mild, temperate, and highly changeable climate. Rain, regarded as synonymous with Britain's weather, is fairly well distributed throughout the year, but recently unusually long dry or wet spells have caused water shortages in some areas, and flooding in others. The west is generally wetter than the east, and the south warmer than the north.

TRANSPORTATION

Drive on left

Heathrow, London
64.6m passengers

1421 ships
4.09m grt

THE TRANSPORTATION NETWORK

371,603 km (230,903 miles)

3453 km (2146 miles)

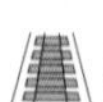
17,064 km (10,604 miles)

5700 km (3542 miles)

The government has not fulfilled its 1997 campaign promise of a more integrated policy, faced with congestion, pollution, and motorists' resentment of high fuel taxes. The rail system, after the rushed privatization of the 1990s, suffers from underinvestment, maintenance problems, and fragmented services. There is still no high-speed rail link to the Channel Tunnel, which opened in 1994.

TOURISM

Visitors : Population 1:2.5

23.4m visitors

Down 7% in 2001

MAIN TOURIST ARRIVALS

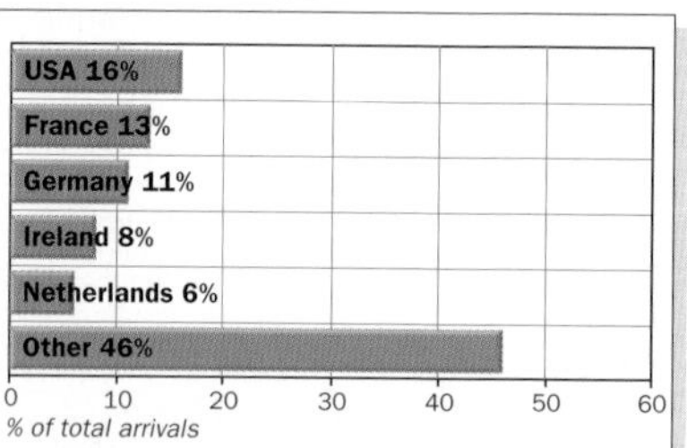

USA 16%
France 13%
Germany 11%
Ireland 8%
Netherlands 6%
Other 46%
0 10 20 30 40 50 60
% of total arrivals

The UK ranks sixth in the world as a tourist destination. Tourism is among the country's most important industries and a growing source of employment. Heritage is the principal selling point, North Americans, French, and Germans are the main visitors, and London, with its art galleries, theaters, and historic buildings, remains the major destination. Visitors also head for the Roman splendors of Bath, Shakespeare's Stratford-upon-Avon, the medieval buildings of Oxford, Cambridge, and York, and Scotland, where the Highlands are a particular attraction.

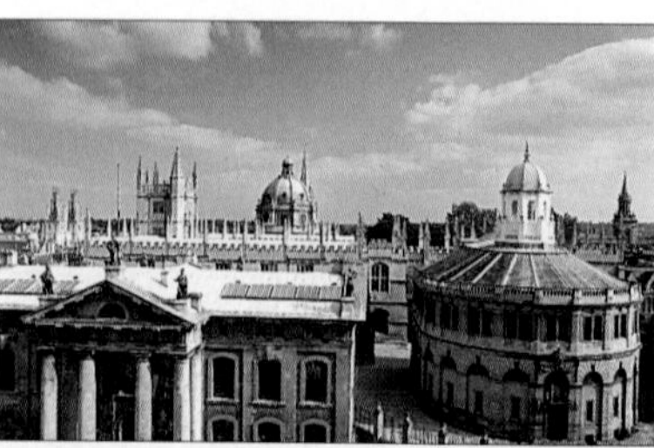

***Oxford,** home to the oldest university in the UK. Teaching began in 1096; the first college was founded in 1249. One of the city's finest buildings is the 17th-century semi-oval Sheldonian Theatre.*

PEOPLE

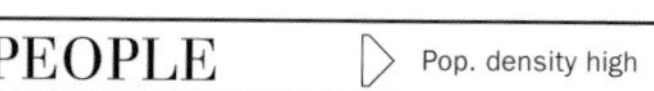

Pop. density high

English, Welsh, Scottish Gaelic, Irish Gaelic
246/km² (638/mi²)

THE URBAN/RURAL POPULATION SPLIT

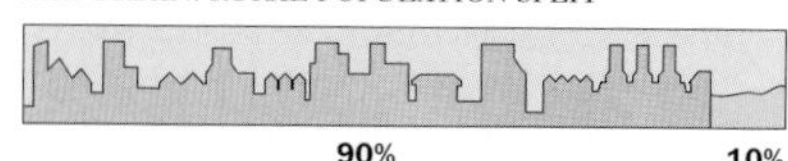

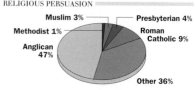

90%
10%

RELIGIOUS PERSUASION

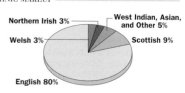

Muslim 3%
Presbyterian 4%
Methodist 1%
Roman Catholic 9%
Anglican 47%
Other 36%

ETHNIC MAKEUP

Northern Irish 3%
West Indian, Asian, and Other 5%
Welsh 3%
Scottish 9%
English 80%

The Scottish and Welsh nations remain recognizably distinct, despite forming part of a unified state. Scotland retains its own legal and educational systems, and recent steps in devolution have given both countries greater autonomy.

Britain's ethnic minorities account for 5% of the total population; over 50% of their members were born in Britain. Ethnic minority communities are generally concentrated in the inner cities, where they face problems of deprivation and social stress, and may also suffer from isolation, particularly among women. Although there is little support for overt racist politics, multiethnic recruitment has made little progress in key areas such as policing, and prejudices persist. The level of institutionalized racism was criticized by the UN in 2000.

Two-fifths of all births occur outside marriage, compared with 12% in 1980, but most of these are to cohabiting couples. Single-parent households account for one-fifth of all families with children under 18.

POPULATION AGE BREAKDOWN

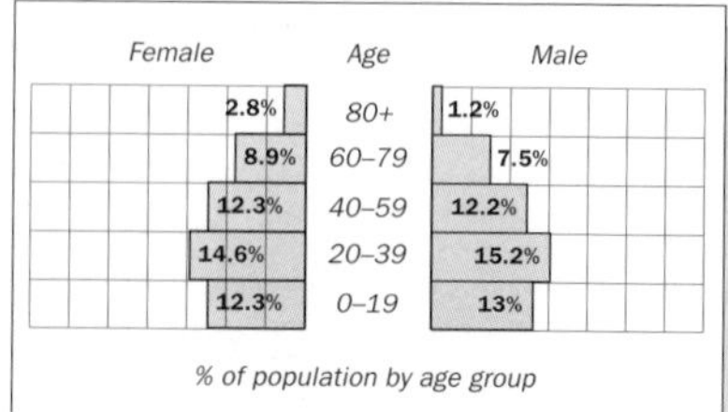

Female
Age
Male
2.8% 80+ 1.2%
8.9% 60–79 7.5%
12.3% 40–59 12.2%
14.6% 20–39 15.2%
12.3% 0–19 13%
% of population by age group

U

Black Mount, Rannoch Moor, *in the Scottish Highlands. The Highlands are one of the UK's wildest regions.*

CHRONOLOGY

Great Britain began the 20th century as one of the world's most advanced economies, backed by a massive trading empire.

- **1906** Reformist Liberal government.
- **1914** World War I begins.
- **1918** Armistice signals end of war. Cost to Britain: 750,000 dead.
- **1922** Irish Free State given dominion status.
- **1926** General Strike.
- **1929** World stock market crash. Widespread unemployment.
- **1931** UK leaves gold standard and devalues pound.
- **1936** Edward VIII abdicates over marriage to Wallis Simpson.
- **1938** Prime Minister Neville Chamberlain meets Hitler in Munich over Czech crisis, says threat of war with Germany averted.
- **1939** Germany invades Poland. UK declares war on Germany. Start of World War II.
- **1940** Winston Churchill prime minister. Battle of Britain.
- **1944** 6th June, D-Day invasion of German-occupied France.
- **1945** End of World War II, costing 330,000 British lives. Labour government elected on social welfare platform.
- **1946** Nationalization of Bank of England, railroads, coal, utilities. ⇨

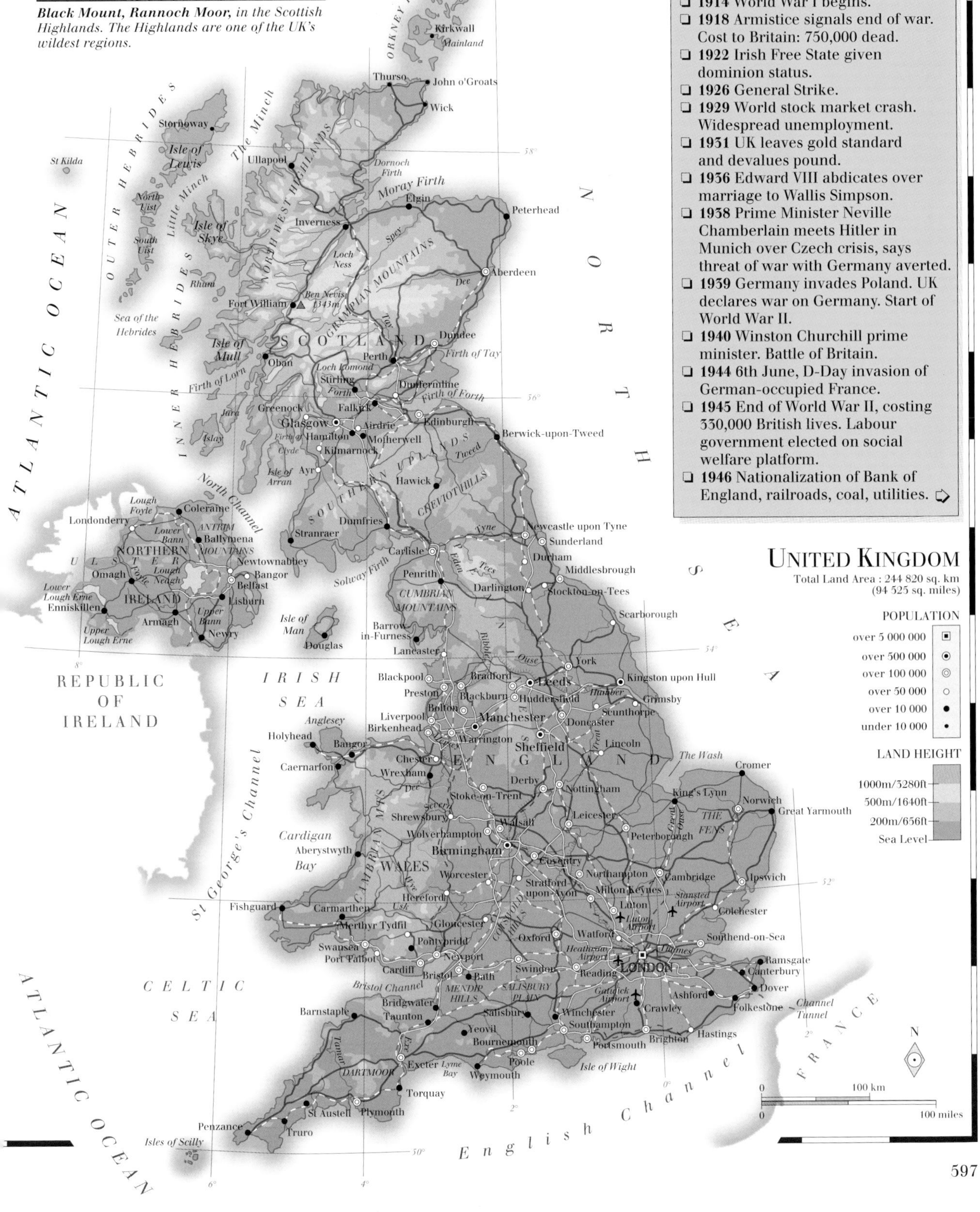

UNITED KINGDOM

Total Land Area : 244 820 sq. km (94 525 sq. miles)

POPULATION

- over 5 000 000
- over 500 000
- over 100 000
- over 50 000
- over 10 000
- under 10 000

LAND HEIGHT

- 1000m/3280ft
- 500m/1640ft
- 200m/656ft
- Sea Level

U

CHRONOLOGY *continued*

- **1947** Indian independence.
- **1948** NHS established.
- **1949** Founder member of NATO.
- **1956** Suez crisis: UK intervenes in Canal Zone, withdraws under US pressure.
- **1957** US nuclear missiles accepted on UK soil.
- **1961** UK application to EC rejected by French president de Gaulle.
- **1968** Abortion and homosexuality are legalized.
- **1969** British troops sent into Northern Ireland.
- **1973** Joins EC under Conservative Edward Heath. Oil crisis. Industry on three-day week following strikes by power workers and miners.
- **1974** Labour government, under Harold Wilson, concedes miners' demands. High inflation.
- **1975** Referendum ratifies EC membership. First North Sea oil pipeline in operation.
- **1979–1997** Conservative rule, until 1990 under Margaret Thatcher.
- **1980** Anti-US Cruise missiles protests. Rising unemployment. Inner-city riots.
- **1981** Privatization program begun.
- **1982** Argentina invades Falklands. Islands retaken by UK task force.
- **1983** Tax-cutting policies.
- **1986** Financial services market deregularized ("Big Bang").
- **1990** John Major replaces Thatcher.
- **1991** UK in Gulf War.
- **1992** Conservatives win fourth consecutive election.
- **1996** Dunblane primary school massacre; tightening of gun control laws. Health crisis linking "mad cow" disease (BSE) with fatal variant Creutzfeldt-Jakob disease (vCJD).
- **1997** May, landslide election victory for Labour under Tony Blair. August, Diana, Princess of Wales, killed in car crash in Paris. September, Scottish and Welsh referendums approve creation of own assemblies.
- **1998–1999** "Good Friday" agreement on political settlement in Northern Ireland, endorsed by referendum but held up by disputes over decommissioning weapons.
- **1999** March–June, involvement in NATO air war with Yugoslavia over Kosovo crisis. May, Scottish Parliament and Welsh Assembly elected, inaugurated. December, devolution to power-sharing executive in Northern Ireland.
- **2001** Foot-and-mouth epidemic and mass livestock culling. June, Labour reelected by huge majority. October, military participation in US-led "war on terrorism" in Afghanistan.
- **2002** 50th anniversary of accession to throne of Queen Elizabeth II.

U

POLITICS

Multiparty elections

 2001/2006

 H.M. Queen Elizabeth II

AT THE LAST ELECTION

House of Commons 659 seats

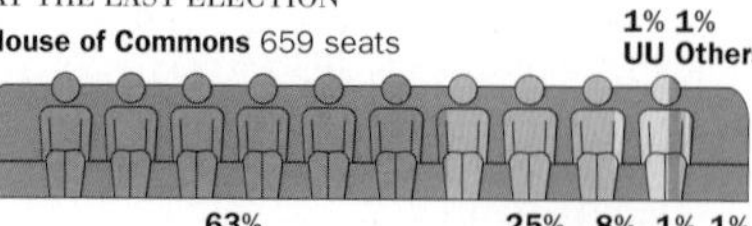
1% 1%
UU Others
63%
Lab
25%
Con
8%
LD
1%
SNP
1%
PC

Lab = Labour Party **Con** = Conservative and Unionist Party **LD** = Liberal Democrats **UU** = Ulster Unionist parties – (official) Ulster Unionist Party, Democratic Unionist Party, UK Unionist **SNP** = Scottish National Party **PC** = Plaid Cymru

House of Lords 679 seats

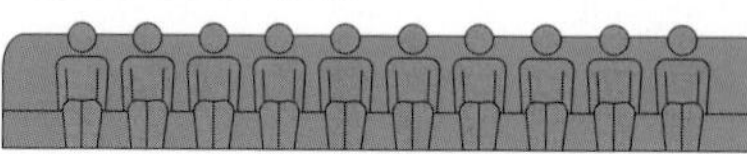

The House of Lords is an unelected body of just under 100 hereditary peers, 26 spiritual peers (bishops), and over 500 life peers (including lords of appeal – judges), appointed by the monarch

The UK is a multiparty democracy. The monarch's power is largely ceremonial.

Profile

Margaret Thatcher's 1979 election victory ushered in almost 18 years of Conservative rule, and monetarist and privatization policies. The Labour Party won back power in 1997 as the Conservatives lost impetus and popularity. Tony Blair's "New Labour" government, occupying the political center, retained a massive majority in the June 2001 election. The low turnout, however, reflected a growing perception that it had become "out of touch." The Conservative Party has so far failed to build a credible challenge around opposition to the euro and pledges to reduce taxes.

Main Political Issues

Europe

Labour plans a referendum on joining the euro single currency. The Conservatives, in opposition, have become increasingly "eurosceptic," considering that membership of the EU erodes national sovereignty.

Constitutional change

Major changes were made to the UK's system of government in the late 1990s. A separate Scottish Parliament, with substantial devolved powers, was approved by referendum and elected in 1999, as was a new Welsh Assembly. The House of Lords was substantially changed by the abolition of voting rights for the great majority of hereditary peers, pending its complete overhaul; 15 "people's peers" were chosen in 2001. London gained greater autonomy in 2000 with the election both of its own assembly and of a mayor.

In Northern Ireland, the 1998 Good Friday agreement brought unionists and Irish republicans into a power-sharing government. Progress has stalled over the decommissioning of paramilitaries' weapons, and tensions recur when Protestant parades attempt to march through Catholic areas in the summer "Marching Season." The joint executive formed in 1999 has on occasion been suspended, and direct rule reimposed from London. However, in 2001 the IRA agreed to put its weapons definitively "beyond use", and in 2002 apologized for the death of "non-combatants" during its violent campaign.

Queen Elizabeth II, head of state since 1952 and head of the Commonwealth.

Tony Blair, prime minister since 1997, and leader of the Labour Party.

***Chancellor of the Exchequer Gordon Brown,** known as the "Iron Chancellor."*

***Baroness Thatcher,** the country's only female prime minister (1979–1990).*

The economy

Fundamental alternatives on running the economy are no longer argued within mainstream politics. Labour no longer believes in renationalizing privatized industries, and is wary of increasing taxes. Its belief in using private finance and management within publicly owned services, such as health and education, has alienated some traditional Labour supporters, and the financial benefits of this policy are questioned by others.

***Canary Wharf,** the centerpiece of the London Docklands development.*

WORLD AFFAIRS

Joined UN in 1945

Comm

EU

G8

NATO

OECD

In 2002 prime minister Blair claimed that the UK, if no longer a "great power," could still play a "pivotal role" in world affairs. It holds a permanent seat on the UN Security Council and was a founder member of NATO. The UK joined the EC only in 1973, and remains wary of full integration in Europe. Generally following a pro-US line during and since the Cold War, the UK has supported US-led action against Iraq, and the 1999 NATO bombing to compel Yugoslav forces to pull out of Kosovo. The question of cooperating in a new US missile defense system promised to be a key test of the UK–US relationship, but in September 2001 both gave top priority to waging a "war on terrorism" worldwide.

AID

Donor

$4.5bn (donations)

Up 31% in 2000

UK foreign aid fell between 1980 and 1997 to below the European average, and well below the nominal target of 0.7% of GNP for industrialized countries. After 1997 the government moved to end the decline, although the figure in 2000 was only 0.31% of GNP. More significant was its concentration on the poorest countries and on partnership with NGOs, building on a change of emphasis already introduced in 1996, when 85% of bilateral aid was directed at 20 states in sub-Saharan Africa and south Asia. The "trade for aid" provision, by which much of the aid budget was tied to contracts for British firms, has been abolished. The program's aims include encouraging good government, widening opportunities for women, and protecting the environment.

DEFENSE

No compulsory military service

$33.9bn

Down 7% in 2000

BRITISH ARMED FORCES

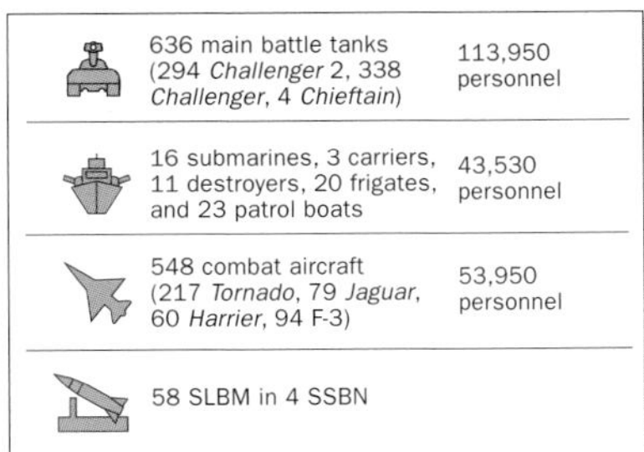

636 main battle tanks (294 *Challenger* 2, 338 *Challenger*, 4 *Chieftain*)	113,950 personnel
16 submarines, 3 carriers, 11 destroyers, 20 frigates, and 23 patrol boats	43,530 personnel
548 combat aircraft (217 *Tornado*, 79 *Jaguar*, 60 *Harrier*, 94 F-3)	53,950 personnel
58 SLBM in 4 SSBN	

The post-Cold War program Options for Change, implemented in 1993, made significant cuts in army and navy personnel and equipment orders. The UK's independent nuclear deterrent was scaled down. Despite such cuts, defense spending is high, and rose in 2002. Troops remain stationed in Northern Ireland, but the strategic emphasis has increasingly been on developing military capabilities relevant for rapid reaction. UK forces have been prominent in peacekeeping in Afghanistan, the Balkans, Sierra Leone, and elsewhere. Reservists were mobilized for the "war on terrorism."

The UK is a leading arms exporter. Major buyers include Middle Eastern and southeast Asian countries.

ECONOMICS

Inflation 2.9% p.a. (1990–2000)

$1460bn

0.6694–0.6871 pounds sterling

SCORE CARD

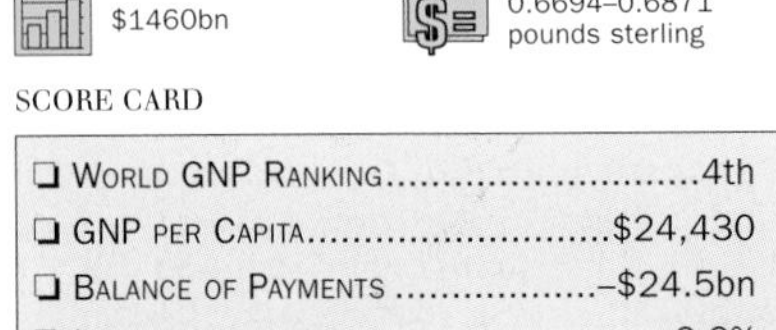

❑ WORLD GNP RANKING	4th
❑ GNP PER CAPITA	$24,430
❑ BALANCE OF PAYMENTS	–$24.5bn
❑ INFLATION	2.9%
❑ UNEMPLOYMENT	4%

EXPORTS

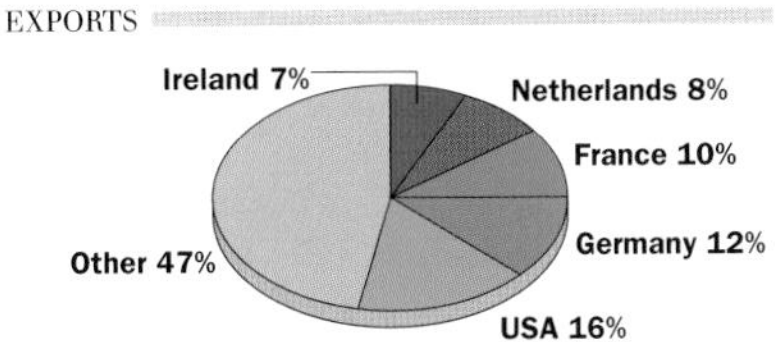
Ireland 7%
Netherlands 8%
France 10%
Germany 12%
USA 16%
Other 47%

IMPORTS

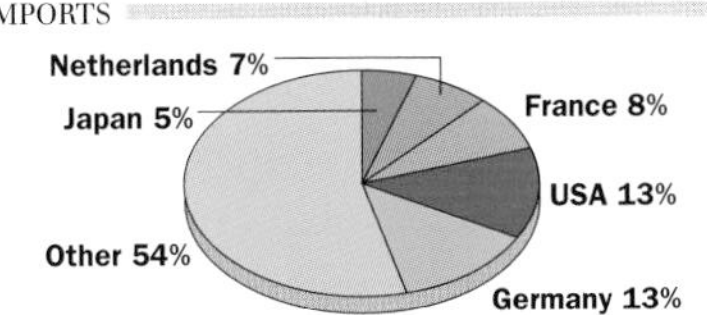
Netherlands 7%
Japan 5%
France 8%
USA 13%
Germany 13%
Other 54%

STRENGTHS

World leader in financial services, pharmaceuticals, and defense industries. Strong multinationals. Precision engineering and high-tech industries, including telecommunications and biotechnology. Energy sector based on North Sea oil and gas production. Innovative in computer software development. Flexible working practices. Success in controlling inflationary tendencies.

WEAKNESSES

Decline of manufacturing sector since 1970s, particularly heavy industries, car manufacturing. Nonparticipation in euro threatens former status as EU's largest recipient of inward investment, has prompted some major investors to close UK factories. Quick-return mentality of many investment decisions.

PROFILE

Manufacturing has been in long-term decline, particularly since the 1980s, when sectors such as financial services expanded rapidly. After sharp recession in 1991, revival was sluggish, but by the late 1990s growth in the UK was faster than that of its European competitors. The government received a £22 billion windfall from the sale in 2000 of "third generation" mobile phone licenses. The rural economy was hit hard by a foot-and-mouth epidemic in 2001 while the wider economy feared recession in 2001–2002 after the US economic downturn and the collapse of the Internet "dot.com" boom. Interest rates were cut to try to boost domestic spending, while the government in 2002 massively increased its own spending on education, health, and defense.

ECONOMIC PERFORMANCE INDICATOR

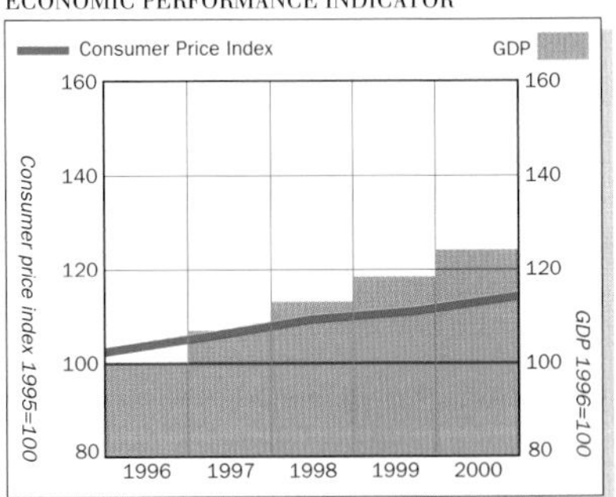
Consumer Price Index
GDP
160
140
120
100
80
Consumer price index 1995=100
GDP 1996=100
1996
1997
1998
1999
2000

UNITED KINGDOM : MAJOR BUSINESSES

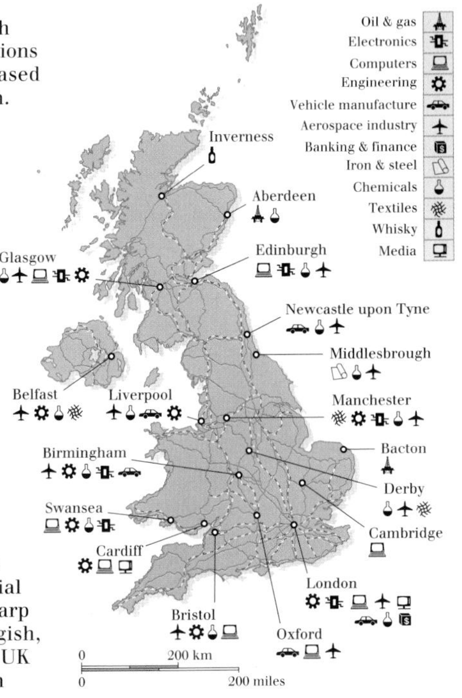
Oil & gas
Electronics
Computers
Engineering
Vehicle manufacture
Aerospace industry
Banking & finance
Iron & steel
Chemicals
Textiles
Whisky
Media
Inverness
Aberdeen
Glasgow
Edinburgh
Newcastle upon Tyne
Middlesbrough
Belfast
Liverpool
Manchester
Birmingham
Bacton
Derby
Swansea
Cambridge
Cardiff
London
Bristol
Oxford
0
200 km
0
200 miles

RESOURCES

Electric power 73.4m kw

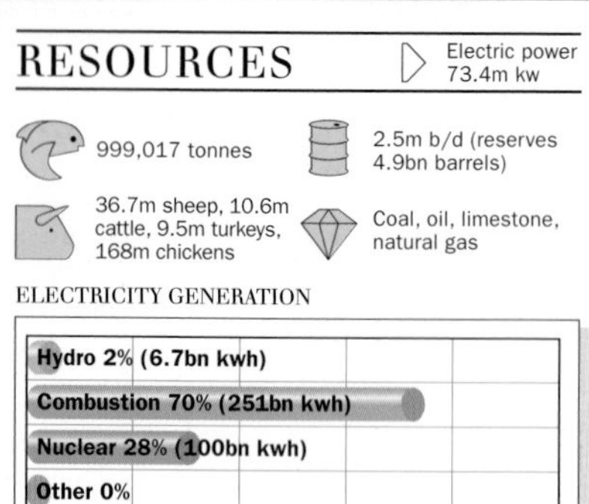
999,017 tonnes
2.5m b/d (reserves 4.9bn barrels)
36.7m sheep, 10.6m cattle, 9.5m turkeys, 168m chickens
Coal, oil, limestone, natural gas

ELECTRICITY GENERATION

Hydro 2% (6.7bn kwh)
Combustion 70% (251bn kwh)
Nuclear 28% (100bn kwh)
Other 0%

0 20 40 60 80 100

% of total generation by type

The UK has the largest energy resources of any EU state, with substantial oil and gas reserves offshore on the continental shelf in the North Sea, and fresh fields in the north Atlantic. Drilled under difficult conditions, North Sea oil is of a high grade. Revenues from taxes on oil companies have been a major contributor to government finances, averaging around $12 billion a year.

Coal reserves are also sizable, but all but a handful of pits have closed, faced with cheap imports and falling demand. Privatization of the electricity industry, and pressure to cut pollution, encouraged the switch from coal- to gas-fired power plants, prompting emergency government measures to save the coal industry in the late 1990s and efforts to boost the role of "cleaner coal" technology. The UK produces few other minerals in significant quantities. Cornwall's last tin mines teeter between closure and rescue. Some very small-scale gold mining survives in Wales and Scotland.

UNITED KINGDOM : LAND USE

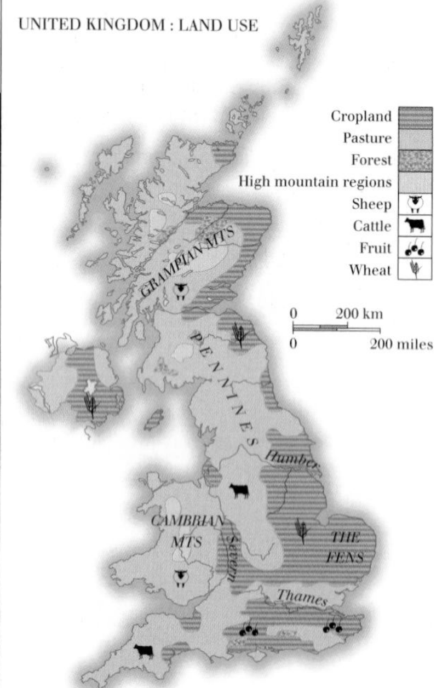
Cropland
Pasture
Forest
High mountain regions
Sheep
Cattle
Fruit
Wheat
0 200 km
0 200 miles
GRAMPIAN MTS
PENNINES
Humber
CAMBRIAN MTS
Severn
THE FENS
Thames

ENVIRONMENT

Sustainability rank: 91st

21%

9.2 tonnes per capita

ENVIRONMENTAL TREATIES

Yes Yes Yes

Yes Yes Yes

Apart from destruction of rural environments by road building and sprawling development, the most important issues are health-related. Urban air pollution from traffic is a major focus, as are nuclear safety issues. Food scares have gripped the public since BSE ("mad cow" disease) was linked in the mid-1990s with human deaths. Opposition to genetically modified (GM) foods is widespread, and GM crop trials have been disrupted for fear that modified genes contaminate other species.

MEDIA

TV ownership high

Daily newspaper circulation 331 per 1000 people

PUBLISHING AND BROADCAST MEDIA

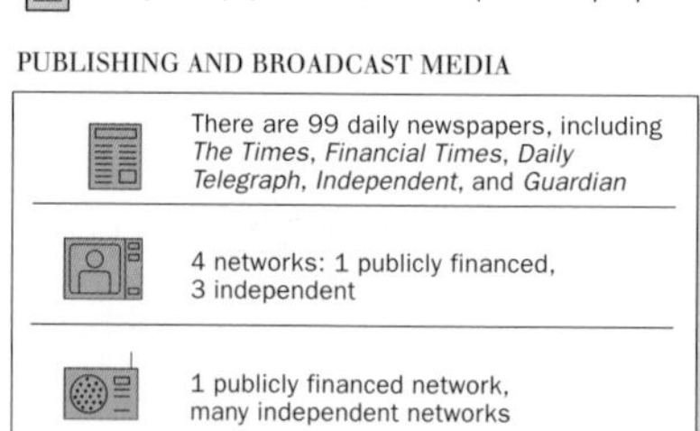
There are 99 daily newspapers, including *The Times, Financial Times, Daily Telegraph, Independent,* and *Guardian*

4 networks: 1 publicly financed, 3 independent

1 publicly financed network, many independent networks

Newspapers are owned mostly by large media corporations. Many publish Internet editions. Criticized for invasions of privacy, the press presents self-regulation as preferable to legislation. Publication deemed contrary to "national interests" may be banned. Satellite TV and digital terrestrial broadcasting have increased competition with the BBC. The BBC's World Service, despite cutbacks, remains an influential news source internationally.

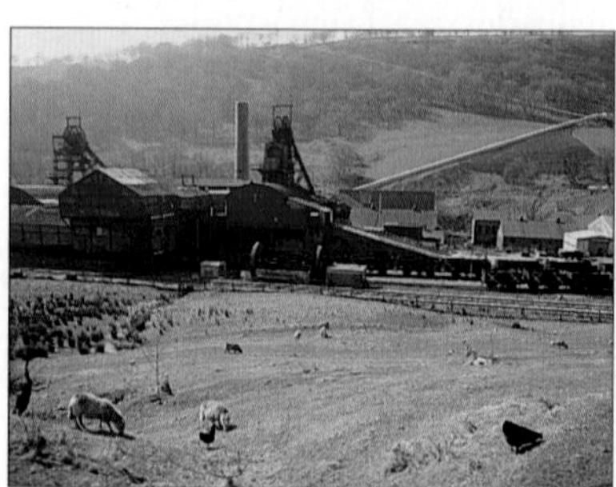

***The Welsh coal industry** has virtually disappeared. Wales now has the highest percentage of small business start-ups, relative to the population, of any part of the UK.*

CRIME

No death penalty

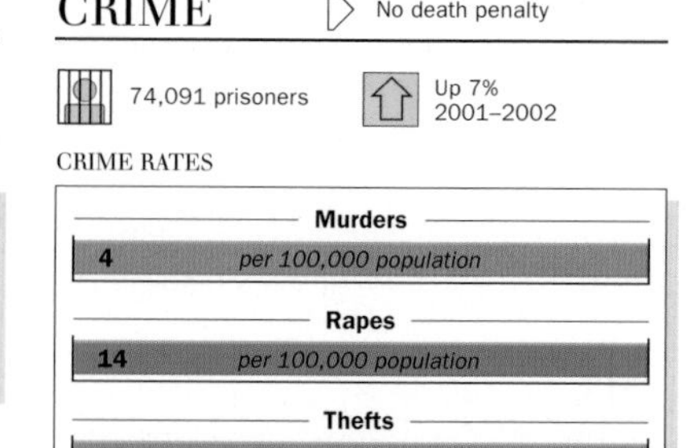
74,091 prisoners
Up 7% 2001–2002
CRIME RATES
Murders
4 per 100,000 population
Rapes
14 per 100,000 population
Thefts
6062 per 100,000 population

Violent crime and domestic abuse are growing problems. Inner-city violence is partly fueled by narcotics dependency and trafficking. Since 1997 the Labour government has maintained a "tough on crime" stance, but sentencing policies place the penal system under serious strain. It has moved toward toleration of marijuana in order to focus on "harder" narcotics. An "antiterrorism" bill introduced in October 2001 gave police extra powers.

EDUCATION

School leaving age: 16

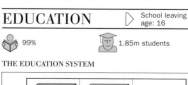

99%

1.85m students

THE EDUCATION SYSTEM

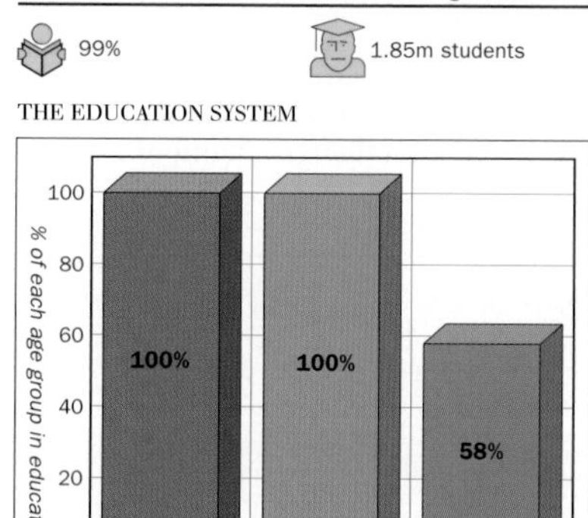
100
80
60
40
20
0
% of each age group in education
100%
100%
58%
Primary
Secondary
Tertiary

The state system is used by 94% of children. Fee-paying private schools include the traditional elite institutions confusingly known as public schools. There is now a new enthusiasm in government for "faith-based" schools.

From the 1960s onward, a two-tier state school system based on academic selection at age 11 was to a great extent replaced by mixed ability comprehensive schools. The 1988 education reforms introduced a national curriculum and weakened the role of local education authorities. The Labour government has focused on testing, assessing teaching standards, and tackling "failing" schools. Public spending targets in 2002 promised significantly increased education expenditure within four years, to 5.6% of GDP.

More colleges were given university status in the 1990s but established centers, particularly Oxford and Cambridge, continue to be the most prestigious and best resourced.

REPRESENTATION AND DEVOLUTION

THE 1801 ACT OF UNION unified Great Britain – England, Wales, and Scotland – and Ireland. The principality of Wales had been joined with England in 1536. The English and Scottish crowns had been united in 1603 (James VI of Scotland becoming James I of England), and the 1707 Act of Union united their parliaments. The present full title of the UK – the United Kingdom of Great Britain and Northern Ireland – dates from 1922, when the Irish Free State was created.

The rise of Welsh and particularly Scottish nationalism prompted an abortive initiative to devolve powers to regional governments in the 1970s. The post-1997 Labour government, giving devolution fresh impetus, created new assemblies in Northern Ireland, Scotland, and Wales. It also raised, but has done little to breathe life into, the idea of English regional assemblies.

SCOTLAND
Glasgow
Edinburgh
Newcastle upon Tyne
NORTHERN IRELAND
Belfast
Leeds
Liverpool
Manchester
ENGLAND
Birmingham
WALES
Cardiff
Bristol
LONDON

NORTHERN IRELAND

In Northern Ireland (pop. 1.65 million, area 14,120 sq. km/5450 sq. miles, capital Belfast) the majority Protestant community dominated a "home rule" parliament at Stormont throughout its 50-year existence, until 1972. After that the troubled province was mainly under direct rule from London, interspersed with attempts to create power-sharing institutions, until the 1998 Good Friday agreement. This created a 108-member Northern Ireland Assembly, elected that June by proportional representation, and a 12-member power-sharing executive. However, implementation of the agreement has repeatedly been held up by disputes, mainly about disarming rival Irish republican and unionist "loyalist" paramilitaries.

***The Scottish Parliament** building, Edinburgh.*

SCOTLAND AND WALES

The Scottish National Party's electoral breakthrough in 1974 was fueled by the perception that independence could be viable, if Scotland controlled the oil wealth of the North Sea. The UK government offered devolution to both Scotland and Wales, but in referendums in 1979 the majority in Scotland was insufficient to carry the proposals, while Wales actually voted no. The devolution proposals of 1997, on the other hand, won endorsement both in Scotland and (narrowly) in Wales.

Scotland (pop. 5.13 million, area 78,742 sq. km/30,394 sq. miles, capital Edinburgh) now has a Scottish Parliament of 129 MSPs, elected in 1999. The Scottish Executive, consisting of a First Minister and an 11-member cabinet, is responsible to the parliament, whose powers notably cover education and changing tax rates to generate revenue for Scottish expenditure. One of its first distinctive initiatives was to reject the imposition of university tuition fees, so controversial elsewhere in the UK.

In Wales (pop. 2.9 million, area 20,761 sq. km/8041 sq. miles, capital Cardiff), a 60-member Welsh Assembly was also elected in 1999. It has fewer powers, not including tax-raising, exercised by an eight-member administration also headed by a First Minister.

***The extensive upland areas** of both Scotland and Wales are used extensively for sheep farming and forestry.*

HEALTH

Welfare state health benefits

 1 per 556 people

 Heart, cerebrovascular, and respiratory diseases, cancers

The National Health Service (NHS) offers universal free health care, but financial pressures have led to shortages, hospital closures, and charges in some areas. In response the government has raised the health budget since 2000, promising $58 billion to the NHS in 2002. Plans involve greater involvement of the private sector.

Recent crises have focused on food safety, from *E. coli* outbreaks to fatal brain disease attributed to eating beef from cattle with "mad cow" disease.

SPENDING

GDP/cap. increase

CONSUMPTION AND SPENDING

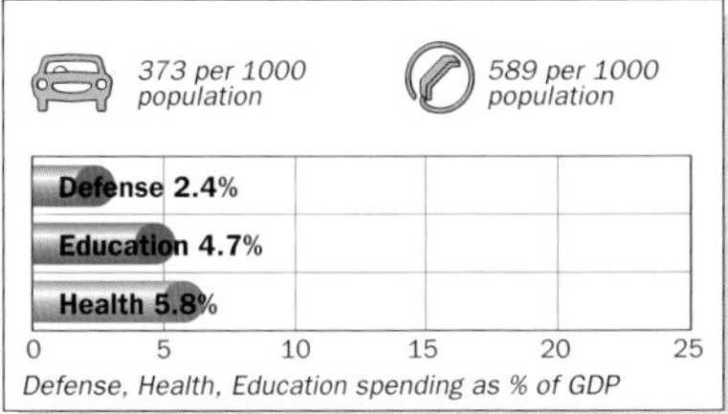
373 per 1000 population
589 per 1000 population
Defense 2.4%
Education 4.7%
Health 5.8%
0
5
10
15
20
25
Defense, Health, Education spending as % of GDP

Income inequality in the UK is now greater than in 1884, when records first began. The UN in 2002 recorded 13.4% living below its "poverty" line, half of average income. Average wages for manufacturing workers in 2001 were $29,000 a year, only 4% of the average received by chief executives in large companies. Under Conservative governments in the 1980s and early 1990s, taxation for higher earners was cut, whereas the value of state benefits and pensions fell. Since the mid-1990s, economic growth has helped to bring unemployment down. Labour's 1997 election promises precluded raising income tax. This limited any scope for redistributive action, leaving antipoverty strategies dependent on better targeting of welfare benefits.

WORLD RANKING

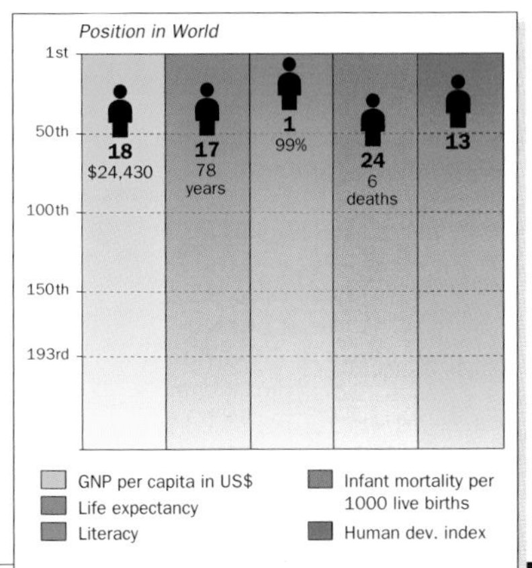
Position in World
1st
50th
100th
150th
193rd
18
$24,430
17
78 years
1
99%
24
6 deaths
13
GNP per capita in US$
Life expectancy
Literacy
Infant mortality per 1000 live births
Human dev. index

U

UNITED STATES

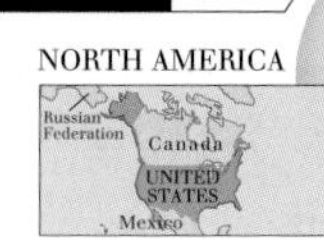
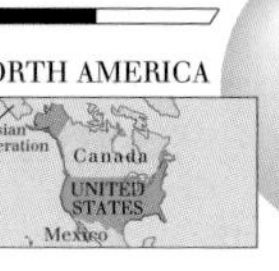
NORTH AMERICA
North America
Russian Federation
Canada
UNITED STATES
Mexico

OFFICIAL NAME: United States of America **CAPITAL:** Washington D.C.
POPULATION: 281.4 million **CURRENCY:** US dollar **OFFICIAL LANGUAGE:** English

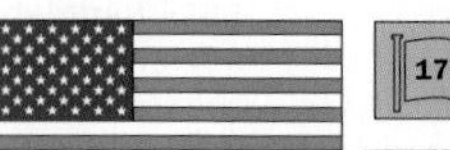

1776

1959

July 4
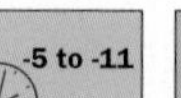
USA
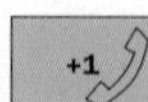
-5 to -11

+1 n/a

THE WORLD'S FOURTH-LARGEST country, the United States is neither overpopulated (like China) nor in the main subject to extremes of climate (like much of Russia and Canada). Its main landmass, bounded by Canada and Mexico, contains 48 of its 50 states. The two others, Alaska at the northwest tip of the Americas and Hawaii in the Pacific, became states in 1959. The US was not built on ethnic identity but on a concept of nationhood intimately bound up with the 18th-century founding fathers' ideas of democracy and liberty – still powerful touchstones in both a political and an economic sense. Since the breakup of the Soviet Union, the US holds a unique position – but arouses extreme hatreds – as the sole global superpower.

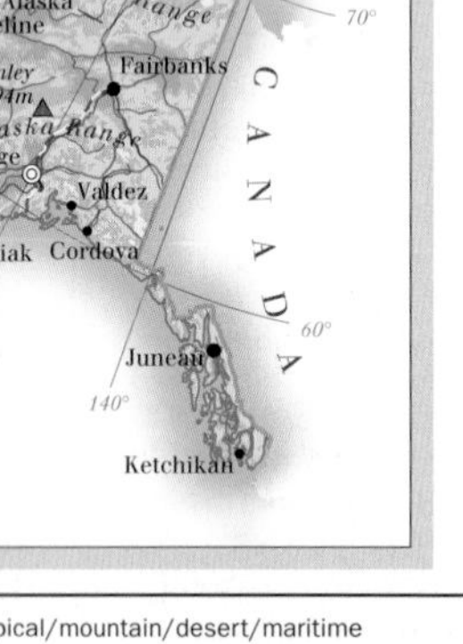
RUSSIAN FEDERATION
Bering Strait
ARCTIC OCEAN
Prudhoe bay
Brooks Range
Trans-Alaska pipeline
ALEUTIAN ISLANDS
Mount McKinley 6194m
Fairbanks
Alaska Range
Anchorage
Valdez
Kodiak
Cordova
CANADA
PACIFIC OCEAN
Juneau
Ketchikan
ALASKA
0 500 km
0 500 miles

CLIMATE

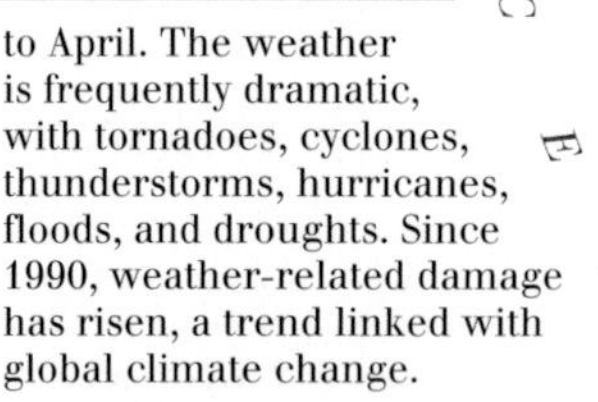

Continental/subtropical/mountain/desert/maritime

WEATHER CHART FOR WASHINGTON D.C.

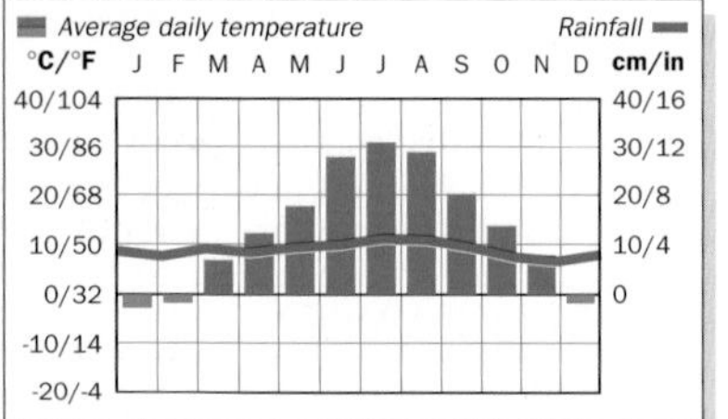
Average daily temperature
Rainfall
°C/°F J F M A M J J A S O N D cm/in
40/104
30/86
20/68
10/50
0/32
-10/14
-20/-4
40/16
30/12
20/8
10/4
0

Spanning a continent and extending far into the Pacific Ocean, the US displays a wide range of climatic conditions. Mean annual temperatures range from 29°C (84°F) in Florida to –13°C (9°F) in Alaska. Except for New England, Alaska, and the Pacific northwest, summer temperatures are higher than in Europe. Southern summers are humid; in the southwest they are dry. Winters are particularly severe in the western mountains and plains and in the Midwest – where the Great Lakes can freeze. The northeast can have heavy snow from November to April. The weather is frequently dramatic, with tornadoes, cyclones, thunderstorms, hurricanes, floods, and droughts. Since 1990, weather-related damage has risen, a trend linked with global climate change.

The Chippendale Block, New York, a notable example of postmodern architecture by the influential US architect Philip Johnson.

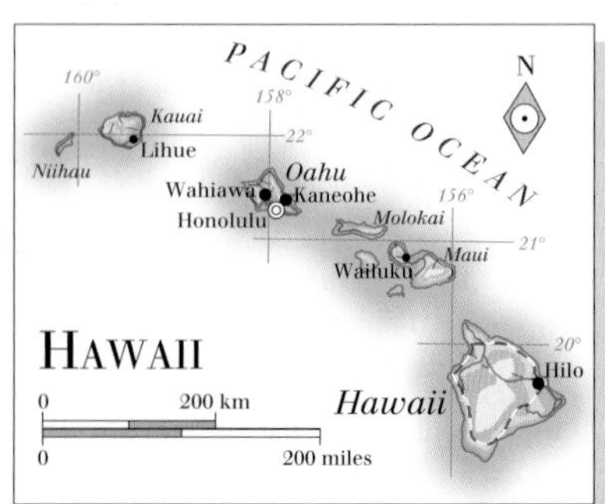
PACIFIC OCEAN
N
Kauai
Lihue
Niihau
Oahu
Wahiawa
Kaneohe
Honolulu
Molokai
Maui
Wailuku
HAWAII
Hawaii
Hilo
0 200 km
0 200 miles

U

TRANSPORTATION

Drive on right

Atlanta, Georgia
80.2m passengers

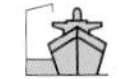
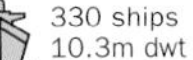
330 ships
10.3m dwt

THE TRANSPORTATION NETWORK

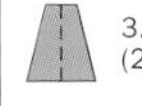
3.73m km
(2.32m miles)

88,727 km
(55,132 miles)

234,131 km
(145,489 miles)

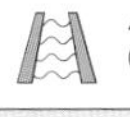
41,009 km
(25,482 miles)

The Mississippi–Missouri river system provided the first transportation network in the US. Today, the US has the world's cheapest, most extensive internal air network and a good system of interstate highways. Railroads, comparatively neglected for years, mainly carry freight, although modern high-speed trains are starting to attract passengers back. Americans have been wedded to the car since Henry Ford began mass production in 1908. By 1919 there were nine million cars in the US. Today the total tops 210 million, including pickups and the ubiquitous "sports utes" (SUVs). Americans make more than half of the world's car journeys. Cheap gasoline underpinned the rise of the car, but problems of congestion and pollution, and the environmental costs of ever more oil production, mean that its role in society needs reviewing.

The Mittens, Monument Valley, Arizona. *These striking natural rock formations are created by erosion of red sandstone. The valley is in the Navajo National Monument.*

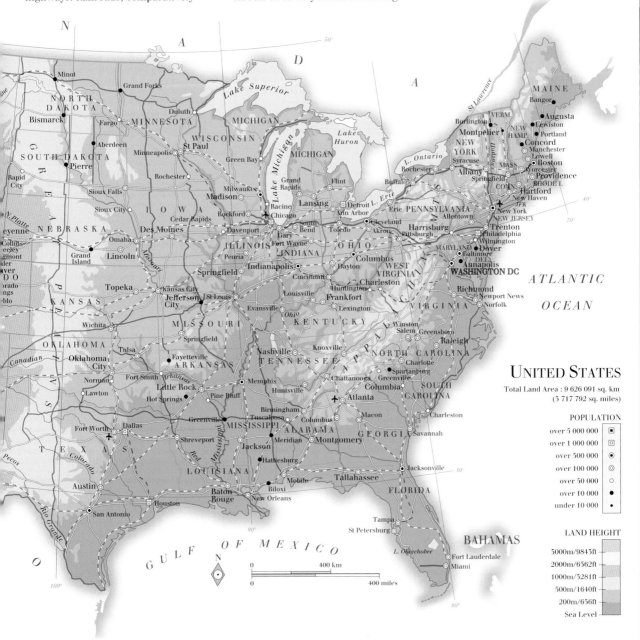

UNITED STATES

Total Land Area : 9 626 091 sq. km
(3 717 792 sq. miles)

POPULATION

over 5 000 000
over 1 000 000
over 500 000
over 100 000
over 50 000
over 10 000
under 10 000

LAND HEIGHT

3000m/9843ft
2000m/6562ft
1000m/3281ft
500m/1640ft
200m/656ft
Sea Level

0 — 400 km
0 — 400 miles

ATLANTIC OCEAN

GULF OF MEXICO

TOURISM

Visitors : Population 1:6.2

 45.5m visitors  Down 11% in 2001

MAIN TOURIST ARRIVALS

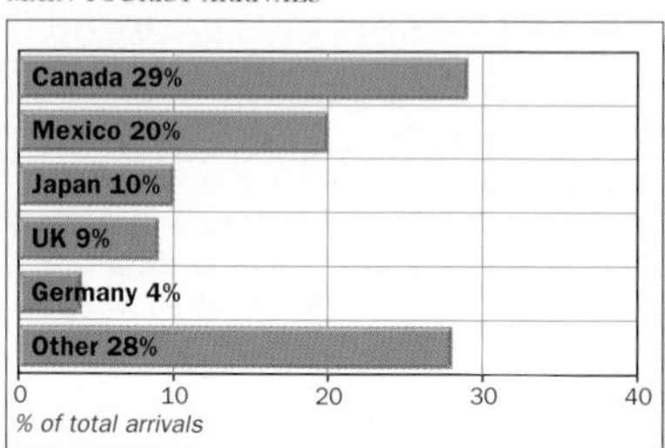
Canada 29%
Mexico 20%
Japan 10%
UK 9%
Germany 4%
Other 28%
0
10
20
30
40
% of total arrivals

The US as a destination for international tourism benefited greatly from the deregulation of air fares. Domestic tourism expanded just as rapidly, along with the rise in real incomes. The impact of the September 11, 2001 attacks was complex. While confidence in air travel took time to be rebuilt, the fact that US tourists put safety first meant that over 80% vacationed within the US in 2002. All the states have their attractions, and most court tourists. Top tourist destinations include Florida's Disney World and Disneyland in California, Niagara Falls, Las Vegas, New York, San Francisco, Los Angeles and Hollywood, the Grand Canyon, Death Valley, New Orleans, Atlantic City, and Washington D.C.

Tourism's rapid expansion has also brought some problems. The parks and sites run by the National Parks Service (NPS) have been particular casualties; visitor numbers rocketed in the three decades after 1970. To try and reduce pressure on the most popular areas, there has been a significant expansion in the area of protected land under NPS management since the mid-1970s. Even so, Yellowstone Park has a continuing traffic management crisis, bumper-to-bumper cars plague other high-profile attractions, and those wanting to take a raft ride down the Grand Canyon are likely to spend many months on a waiting list.

PEOPLE

Pop. density low

 English, Spanish, German, Arabic, Polish, Korean, Chinese, Armenian, French, Italian, Amerindian languages  31/km² (80/mi²)

THE URBAN/RURAL POPULATION SPLIT

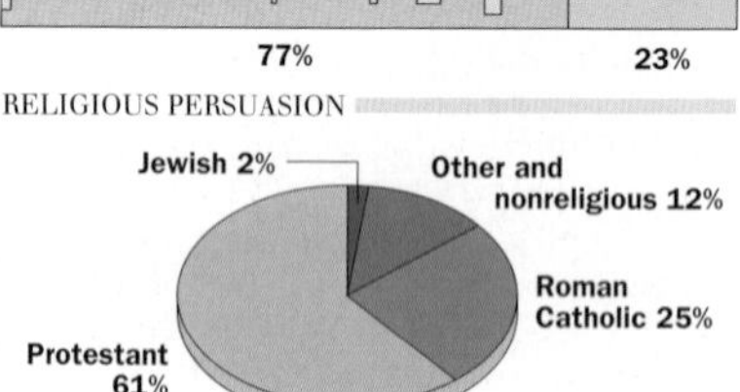
77%
23%

RELIGIOUS PERSUASION

Jewish 2%
Other and nonreligious 12%
Roman Catholic 25%
Protestant 61%

ETHNIC MAKEUP

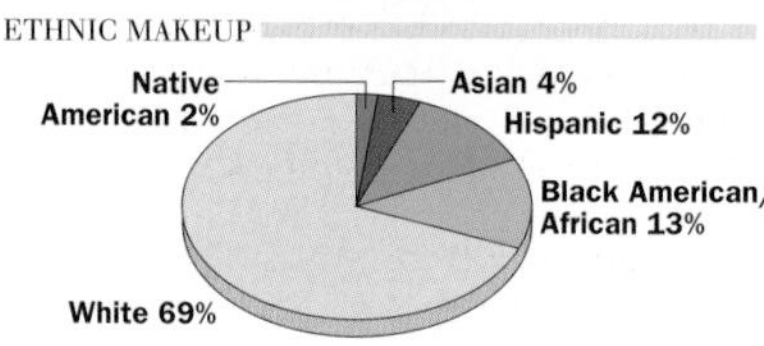
Native American 2%
Asian 4%
Hispanic 12%
Black American/ African 13%
White 69%

Native Americans, the sole inhabitants before white settlers arrived, were dispossessed in the 19th century, and now make up around 2% of the total population. Some of the worst poverty and deprivation in the US can be found in their reservations.

The nation's population increased by some 25 million residents during the 1990s. An immigration boom peaked in the late 1980s and early 1990s, but since then numbers have fallen below a million a year. The new immigrants are disproportionately drawn from Asia and Latin America, with more than two million from Mexico alone in the 1980s.

Census Bureau projections for the year 2050 suggest that 53% of the population will be white non-Hispanics, and just under 14% black; as many as 25% will be Hispanics, while Asians (and Pacific islanders) will total nearly 9%. Blacks increasingly find that they have to compete politically and economically with the newer immigrants. In some communities, such as Los Angeles, this has been a source of tension leading to inner-city riots.

Despite the growth of a black business leadership class, only two black people – television personality Oprah Winfrey and Robert L. Johnson, Black Entertainment Television founder – make the list of the 400 richest Americans. In the mid-1990s, the O. J. Simpson trial, when a black sports hero was acquitted of murdering his white estranged wife, exposed how far the country was divided by race in its perceptions of the justice system. The Nation of Islam, led by the controversial and often provocative Louis Farrakhan, is prominent in emphasizing self-discipline, self-improvement, and community in the quest for empowerment among African-Americans.

POPULATION AGE BREAKDOWN

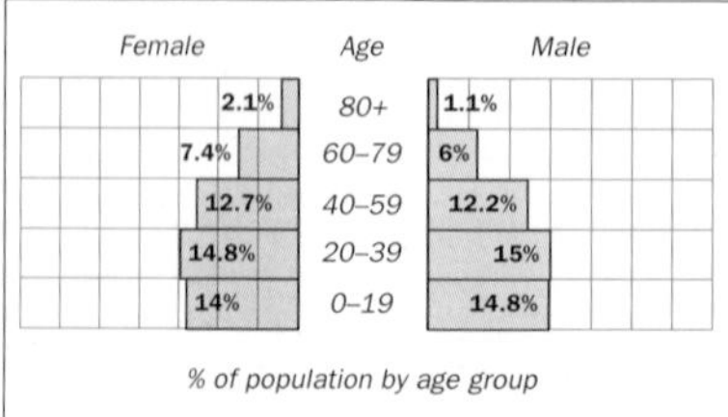
Female
Age
Male
2.1% 80+ 1.1%
7.4% 60–79 6%
12.7% 40–59 12.2%
14.8% 20–39 15%
14% 0–19 14.8%
% of population by age group

POLITICS

Multiparty elections

 L. House 2000/2002 U. House 2000/2002  President George W. Bush

AT THE LAST ELECTION

House of Representatives 435 seats

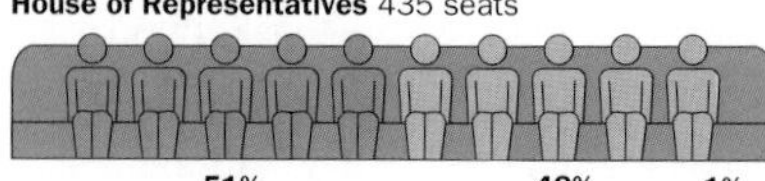
51% Rep
48% Dem
1% Ind

Rep = Republican Party **Dem** = Democratic Party **Ind** = Independents

Senate 100 seats

50% Rep
50% Dem

Presidential elections take place every 4 years, House elections every 2 years. One-third of the senators are elected every 2 years for six-year terms.

The US is a democracy with a federal system of government. Many issues are dealt with by the 50 individual states.

Profile

Despite a "big business" affiliation, controversial policies (notably on energy), and deep divisions when he was first elected in 2000, Republican president George W. Bush gained overwhelming popular support in September 2001, when a wave of patriotic emotion rallied the nation behind him and his declaration of a "war on terrorism," after the attack on the World Trade Center and the Pentagon. Republicans narrowly control the House but Bush still lacks, by the narrowest of margins in the Senate, a complete congressional majority. His predecessors, whether Democrat (Clinton, elected in 1992) or Republican, had struggled to get major initiatives enacted by a hostile Congress.

Main Political Issues

The limits of government

The US has a strong tradition of resisting the extension of government powers. The vigorous defense of constitutional liberties and the rights of citizens, such as freedom of speech or the right to bear arms, is sometimes taken to lengths which appear extreme to other societies. States similarly resist the arrogation of powers by the federal authorities. In areas such as health care

George W. Bush, took office in 2001 after a controversial presidential election.

Gen. Colin Powell, who became the first African-American Secretary of State.

U

POLITICS *continued*

and education, conservatives oppose as interference what others see as the proper concern of government with social welfare. "Big government" is also denounced in the economic sphere. Opponents of environmental controls, for example, portray them as obstructing free enterprise and wealth creation.

The prestige of the presidency

When George W. Bush took office in 2001, the presidency faced a threefold challenge. The tendentious manner of his election, with fewer votes nationwide than his Democrat rival Al Gore, and controversy over the conduct of polling in Florida, left questions as to the legitimacy of his victory. The narrow Republican majority in the House of Representatives, and a "hung" Senate, with an effective Democrat majority after one elected Republican switched to independent, promised problems. Thirdly, the moral authority of the office of president had suffered serious damage under the outgoing administration of Bill Clinton, beset by personal, financial, and political scandals, and hamstrung toward the end by impeachment proceedings. Bush, initially derided by opponents as an intellectual lightweight, rose in public esteem to appear unassailable within his first year in office. His simple messages of belief in "the American way" rallied a nation shocked by the implications of the terrorist attacks of September 11, 2001.

Energy and the environment

Controversial energy policies introduced by Bush in 2001, at the height of a crisis over Californian electricity shortages, offered much freer rein for the US oil industry to exploit reserves in Alaska. Wilderness conservation activists were appalled. The expansion of nuclear power was also revived as an issue under the Bush energy plan, while the US, alone, chose to repudiate the international Kyoto agreement on cutting carbon dioxide emissions.

Crime, race and poverty

Efforts to regenerate depressed urban areas have relied on new economic opportunities and programs that empower the poor (such as self-management of public housing projects). Both rates of criminality and crime victims are higher in the black community than in any other. Tough anticrime policies in cities such as New York have had a real impact in reducing the level of violence, but have also been accused of unfairly targeting minority ethnic groups; there remains the prospect of a permanently disaffected urban underclass.

Hillary Clinton, *senator for New York and wife of former president Bill Clinton.*

Alan Greenspan, *chairman of the Federal Reserve since 1987.*

WORLD AFFAIRS

Joined UN in 1945

G8 | NATO | NAFTA | OAS | OECD

Isolated by two great oceans, the US has for much of its history been able to choose the extent of its involvement in world affairs. Only reluctantly involved in the two world wars, after 1945 it swapped isolationism for involvement. The US took its seat on the Security Council of the UN, based in New York. It helped to set up NATO, although for the US the Cold War was most immediate – and costly – in the Korean and Vietnam wars. The death toll and shock of defeat in Vietnam in the 1970s kept the US out of military involvement overseas for over a decade. Instead, it focused on diplomacy – particularly with China and in the Middle East – and on supporting the opponents of left-wing regimes in developing countries including Nicaragua, Cuba, and Angola.

The collapse of the Eastern bloc after 1989 meant that the US had to redetermine the scope of its foreign responsibilities as the only remaining superpower. It led the intervention in the 1991 Gulf War, but a fiasco in Somalia and a lack of clear policy on Bosnia and Herzegovina and Haiti showed its uncertainty about a role as world policeman. Increasingly isolated in actions against Iraq, the US has shown a preference for air power rather than committing ground forces. Even before the September 11, 2001 attacks, the US had used air strikes to hit back at anti-US terrorism and its alleged sponsors.

The subsequent "war on terrorism" has focused on Islamic extremists, with rumors of imminent attacks on "sponsors" of terrorism, principally Iraq. The US also redefined relations with Russia, prompting further dramatic cuts in both countries' nuclear arsenals and leading Russia to drop its previous objections to the highly controversial national missile defense system (NMD). China, however, remains wary of the NMD, and under Bush the US stance toward China has been notably more confrontational.

9/11

At 08:48 local time on September 11, 2001, the hijacked American Airlines (AA) flight 11 flew into the North Tower of the World Trade Center (WTC) in New York. The world's most devastating terrorist attack had begun. By 10:29 that morning over 3000 people were dead as three more hijacked planes had been crashed; one into the WTC's South Tower, another into the Pentagon military headquarters, and one in rural Pennsylvania after passengers overpowered the hijackers. The finger of blame was immediately pointed at Islamic terrorists of the extremist al-Qaida network, led by Osama bin Laden.

The attacks, caught on camera and broadcast live around the world, are now known as 9/11. Their impact was profound. The US economy shuddered and the country was effectively placed on a permanent state of alert. Warnings are frequently issued on the basis of security intelligence. The government has responded by restructuring the heavily criticized civil and national defense systems. A new Department of Homeland Security was created and President Bush urged reform of the Central Intelligence Agency (CIA) to refocus on potential terrorist threats.

Perhaps the most significant reaction to 9/11 was the launch of Bush's "war on terrorism," beginning with a massive assault on the *taliban* regime in Afghanistan which was sheltering bin Laden. The "war" has been strongly condemned by those in the Islamic world, who feel that it has been unduly targeted, and point out that little has been done to address the motivation of the attackers: US cultural hegemony and in particular its historic support for the Israeli government.

Designs for a memorial *on the WTC site – Ground Zero – have proved controversial. The site's owners have insisted that 1.1 million sq. m (11.6 million sq. ft) of any designs be reserved for commercial purposes.*

THE IMPACT OF INFORMATION TECHNOLOGY

THE UNITED STATES has been, and remains, at the forefront of the so-called information revolution. US society is undergoing profound changes at the beginning of the 21st century as a result of the spread of the personal computer and connectivity through access to electronic mail (e-mail) and the World Wide Web. The Web itself first emerged as a phenomenon of the 1990s, though the Internet on which it is based originated in the 1960s. The Net, developed to allow vital defense communications to be maintained in the event of an attack on a central control site, was first popularized as a medium for exchange within and between universities.

Access to virtual space

Mass ownership of the motor car and cheap fuel had conditioned the attitudes of previous generations of Americans to distance, and helped determine the pattern of development of cities and towns. Information technology, the concept of "cyberspace," and the availability of instant "virtual" interaction, are now sometimes described as having abolished distance.

High-speed "broadband" connections to the Internet are becoming increasingly popular, even for individual users. For most people (80% of home users in 2001), however, the link is still by dial-up on an ordinary telephone line to the local number of their Internet service provider. One consequence has been an explosion in phone line use, particularly because the recently deregulated telephone industry offers most Americans the chance to pay for their local calls on a single flat rate rather than by the minute. Computer modems can thus be left connected for long periods.

As with the automobile, part of society is excluded. In 1998, PC ownership stood at some 43% of the population.

***Access to the Internet** remains out of reach to those who cannot afford a personal computer, such as residents of some of the poverty-stricken inner-city areas like the Bronx in New York.*

***The Microsoft Campus** in Redmond, Washington State. Microsoft is the world's biggest computer software manufacturer; its founder, Bill Gates, is the world's richest man.*

By September 2001, 54% of people had Internet access at home. Special initiatives to spread access among disadvantaged groups have had only limited impact.

Security

Activists of different kinds have made increasing use of the Internet to spread their views and maintain contact with supporters. The use of this tool not only by mainstream political organizations and lobbyists but by fringe groups, terrorists, and the pornography and sex industries give rise to serious security concerns. Such concerns underlay the battle in the late 1990s over the use of encryption software to ensure the privacy of electronic communications. The security authorities fought to ensure that all such software should have a key allowing them to decipher messages. Unfortunately, it is notoriously difficult to prevent the rapid spread of software which is made available on the Internet, and impossible to pin down the location of such programs in the usual geographic sense. The policing of a national jurisdiction is compromised by a communications network which recognizes no national boundaries. The same is true with computer viruses, which can infect the programs on computers that come into contact with them and affect the way they function, in some cases very destructively.

Commerce

The shopping mall, which has had a major influence in reshaping consumer shopping habits since the 1960s, and the mail-order catalog on which many US households rely so heavily for clothes and similar purchases, are now under challenge from the online "virtual mall." Many companies had to take a trial-and-error approach to developing e-commerce. Initial predictions about its rapid growth have proved overstated. However, it is thought that the number of people in the US who regularly buy necessities such as groceries and related goods online will reach 20 million by 2007. Completing business transactions via the Internet became more feasible with legislation making "e-signatures" legally binding, with effect from 2000.

Traffic and Mobility

The take-up of e-commerce on this scale would replace at least half of Americans' average 17 shopping trips a month made to grocery stores and related outlets. It has the added attraction of shielding shoppers from the perceived dangers of being assaulted or robbed.

Information technology is affecting patterns of personal transportation and mobility in other ways too. At one level, online information services and selling techniques make using the Net a more convenient and efficient way of booking airline tickets. Improvements in communications make it increasingly possible and productive for people to work at home or in decentralized facilities away from company headquarters. So-called "knowledge workers" in particular have had to learn new ways of networking electronically to replace the face-to-face contact of the office environment. The use of special-interest discussion facilities and so-called newsgroups on the Internet has mushroomed accordingly. "Telecommuters" continue to form only a small fraction of the total US working population, but even this is significant in terms of managing peak-time commuter traffic. However, videoconferencing is struggling to make headway as an alternative to traveling long distances to hold business meetings, despite the active encouragement of some gurus of the environmental lobby.

***Malls, a typical feature** of the suburban landscape, are under threat from Web use.*

U

AID

Donor

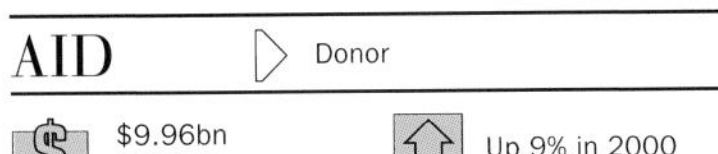

$9.96bn (donations)

Up 9% in 2000

The US gives only 0.1% of GNP in foreign aid, and aid allocations are often held hostage to special pleading in Congress. Egypt and Israel are the major recipients. The strategic aims of the "war on terrorism" prompted an unusual pledge in 2002 of $5 billion for the world's poorest countries.

DEFENSE

No compulsory military service

$295bn

Up 1% in 2000

AMERICAN ARMED FORCES

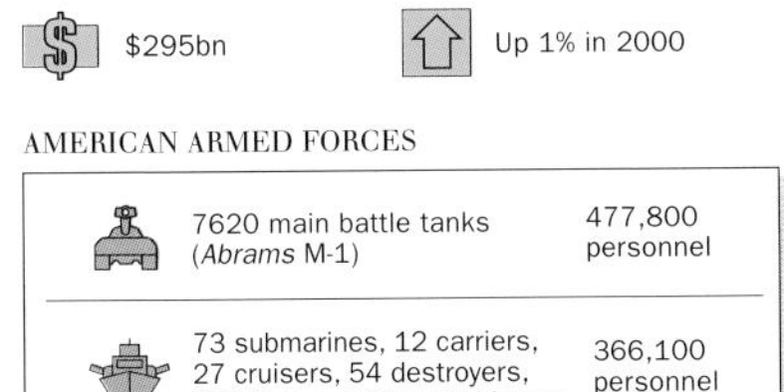

7620 main battle tanks (*Abrams* M-1)	477,800 personnel
73 submarines, 12 carriers, 27 cruisers, 54 destroyers, 35 frigates, & 21 patrol boats	366,100 personnel
4147 combat aircraft (B52H, B-1B, F-4, F-15, F-16, A-10A, F-117, OA-10A)	352,500 personnel
432 SLBM in 18 SSBN, 550 ICBM	

The 9/11 attacks have been popularly cast as a second Pearl Harbor: catching US defenses unprepared and dragging the country into an already simmering global conflict. This time the enemy has been identified as international terrorism. Even before the attacks, emphasis had been shifting away from strategic nuclear deterrence and large warships to "smart" missile systems and "long-range power projection," with rapid intervention capabilities built around air power. Despite setbacks in early tests, a project to create a national missile defense "shield" system was enthusiastically taken up by the incoming Bush administration in 2001.

The enormous US military–industrial complex dates only from the close of the Second World War. In the 1990s, the end of the Cold War and the need to cut the budget deficit combined to slash the defense funds to their lowest level in real terms since 1945. Nuclear weapons tests were superceded by computerized "virtual" tests after the creation of the powerful ASCI White computer.

However, the Republican administration inaugurated in 2001 has steadily increased defense spending. The 2002/2003 budget was up by 11% on the previous year to $350 billion, amounting to more than the combined defense budgets of the world's next nine largest military spenders.

Overseas peacekeeping missions have been put at risk by the US government's objection to the new International Criminal Court.

ECONOMICS

Inflation 2.1% p.a. (1990–2000)

$9602bn

Currency is US dollar

SCORE CARD

❑ WORLD GNP RANKING	1st
❑ GNP PER CAPITA	$34,100
❑ BALANCE OF PAYMENTS	–$445bn
❑ INFLATION	3.4%
❑ UNEMPLOYMENT	4%

ECONOMIC PERFORMANCE INDICATOR

Consumer Price Index — GDP

Consumer price index 1995=100 — GDP 1996=100

1996 1997 1998 1999 2000

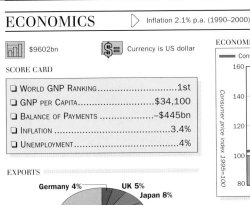

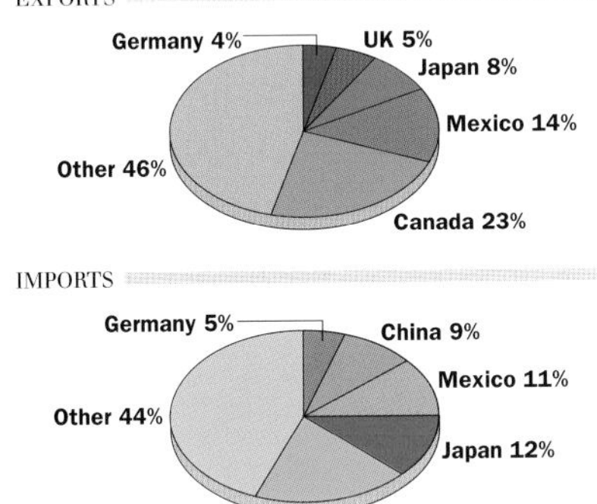
EXPORTS
Germany 4%
UK 5%
Japan 8%
Mexico 14%
Canada 23%
Other 46%
IMPORTS
Germany 5%
China 9%
Mexico 11%
Japan 12%
Canada 19%
Other 44%

STRENGTHS

World's largest economy. Wealth of natural resources: energy, raw materials, and food. Strong high-tech base; world-leading research and development. Global leader in computer software. Advanced and competitive manufacturing industry; sophisticated service sector. Entrepreneurial business ethic. World-class multinationals. High quality of postgraduate education, especially in high-tech business. Global dominance of US culture a major boost to US manufacturers. Subsidized crops and favorable tariffs for domestic industries.

WEAKNESSES

Dramatic fall in manufacturing employment over recent decades. Globalization, problem of job losses to lower-wage economies. Tough competition from Asia and EU in leading-edge technologies. Lower savings rate than many competitors. Volatile market values driven by speculation. Weak business regulation and short-termism. Corporate collapses provoking crisis over accounting standards and scope for fraud.

PROFILE

In 1945, the US accounted for about 50% of world output; by the 1990s its share was down to about 25%. That is not, as Americans often think, a sign of failure, but a clear indication that the 1940s and 1950s were unusual periods. A share of 25% is about the same as was claimed by the US in 1914, when it was already the world's biggest economy.

The US has become a great exporter, and continues to have both a stable political system and a uniquely strong combination of skilled labor and natural resources. In 2001, although still ranked the world's most competitive economy, the US entered a downturn after a nine-year boom, its longest ever. The collapse of the massive Enron corporation in 2001, followed by that of WorldCom in 2002, threatened confidence in business values.

UNITED STATES : MAJOR BUSINESSES

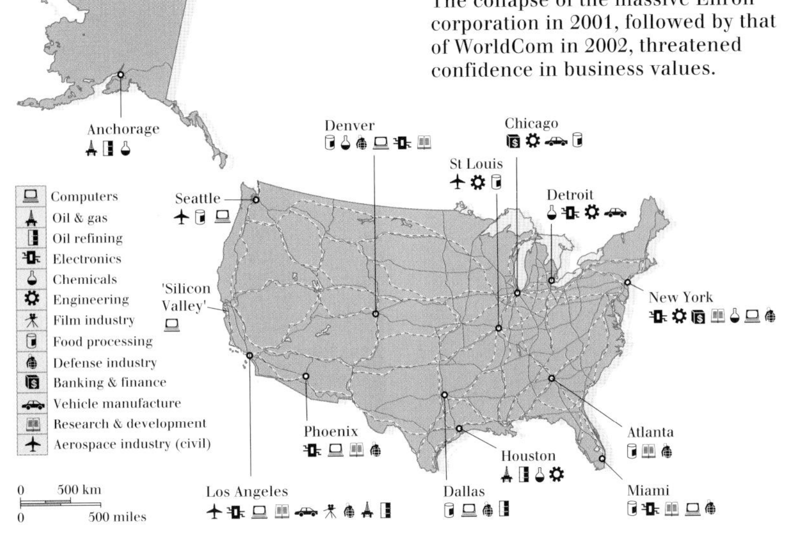
Anchorage
Denver
Chicago
St Louis
Seattle
Detroit
'Silicon Valley'
New York
Phoenix
Houston
Atlanta
Los Angeles
Dallas
Miami
Computers
Oil & gas
Oil refining
Electronics
Chemicals
Engineering
Film industry
Food processing
Defense industry
Banking & finance
Vehicle manufacture
Research & development
Aerospace industry (civil)
0 500 km
0 500 miles

U

CHRONOLOGY

The original 13 colonies, first established by British settlers on the eastern seaboard in the 17th century, joined to wage a war for independence, 1775–1781, which Britain recognized in 1783. The 1776 Declaration of Independence was followed by the writing of the world's first constitution. A century of westward expansion began. Following the victory of the northern states in the 1861–1865 Civil War, slavery was ended throughout the US, but Native Americans were dispossessed of their land in a series of conflicts.

- **1917** US enters World War I.
- **1929** New York stock market collapse; economic depression.
- **1941** Japanese attack on Pearl Harbor; US enters World War II.
- **1950–1953** Korean War.
- **1954** Supreme Court rules racial segregation in schools is unconstitutional. Blacks, seeking constitutional rights, start campaign of civil disobedience.
- **1959** Alaska, Hawaii become states.
- **1961** John F. Kennedy president. Promises aid to South Vietnam. US-backed invasion of Cuba defeated at Bay of Pigs.
- **1962** Soviet missile bases found on Cuba; resulting threat of nuclear war narrowly averted.
- **1963** Kennedy assassinated. Lyndon Baines Johnson president.
- **1964** US involvement in Vietnam stepped up. Civil Rights Act gives blacks constitutional equality.
- **1968** Martin Luther King is assassinated.
- **1969** Republican Richard Nixon takes office as president. Growing public opposition to Vietnam War.
- **1972** Nixon reelected. Makes historic visit to China.
- **1973** Withdrawal of US troops from Vietnam; 58,000 US troops dead by end of war.
- **1974** August, Nixon resigns following Watergate scandal over break-in to Democrat headquarters. Gerald Ford president.
- **1976** Democrat Jimmy Carter elected president.
- **1978** US-sponsored Camp David accord between Egypt and Israel.
- **1979** Seizure of US hostages in Tehran, Iran.
- **1980** Ronald Reagan wins elections for Republicans. Adopts tough anticommunist foreign policy.
- **1983** Military invasion of Grenada.
- **1985** Air strikes against Libyan cities. Relations with USSR improve; first of three summits held.
- **1986** Iran-Contra affair revealed. ⇨

RESOURCES

Electric power 793m kw

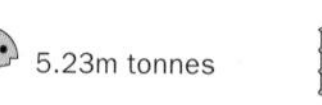

5.23m tonnes

7.72m b/d (reserves 30.4bn barrels)

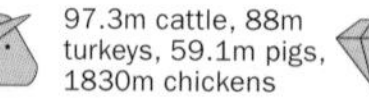

97.3m cattle, 88m turkeys, 59.1m pigs, 1830m chickens

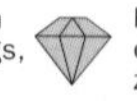

Phosphates, gypsum, oil, coal, sulfur, lead, zinc, copper, gold

ELECTRICITY GENERATION

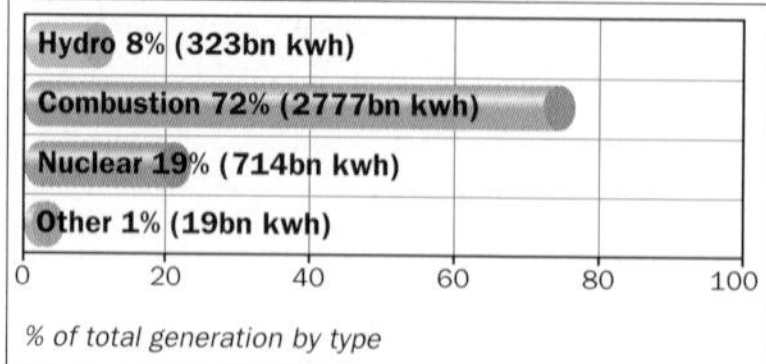
Hydro 8% (323bn kwh)
Combustion 72% (2777bn kwh)
Nuclear 19% (714bn kwh)
Other 1% (19bn kwh)
0
20
40
60
80
100
% of total generation by type

The US has an abundance of natural resources, including oil. The 2001 energy plan aimed to step up oil exploration and output, reducing the need for imports. There are massive deposits of coal in the western states – where almost all mining is open-cast – and substantial mineral deposits in the mountains and intramontane basins.

Environmental concerns halted the development of nuclear power after the 1979 accident at Three Mile Island, but expansion is now being considered.

The timber industry, forced to retreat by conservationists in the Pacific northwest, especially Washington State, has moved to the south, where great stands of pine are harvested as if they were fields of wheat. The US has harnessed hydroelectric power in the past; today, imports of hydropower from Canada are commonplace.

In comparison with western Europe, the US is not intensively farmed. The huge size of farms in the Midwest and west has allowed both arable and livestock farming to be based on a low-input for low-output model.

UNITED STATES : LAND USE

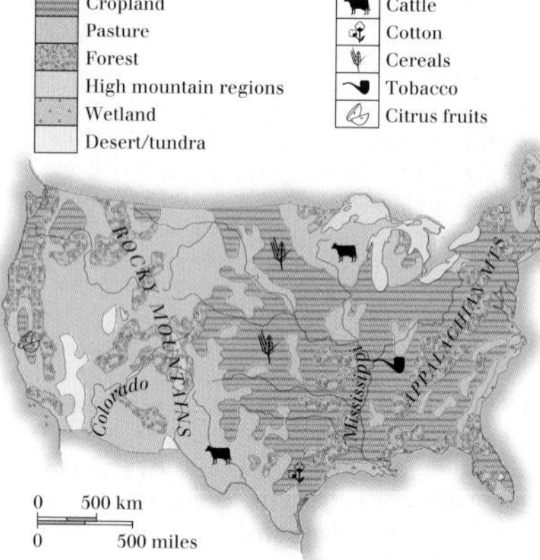
Cropland
Pasture
Forest
High mountain regions
Wetland
Desert/tundra
Cattle
Cotton
Cereals
Tobacco
Citrus fruits
ROCKY MOUNTAINS
APPALACHIAN MTS
Colorado
Mississippi
0 500 km
0 500 miles

ENVIRONMENT

Sustainability rank: 45th

13% (6% partially protected)

19.8 tonnes per capita

ENVIRONMENTAL TREATIES

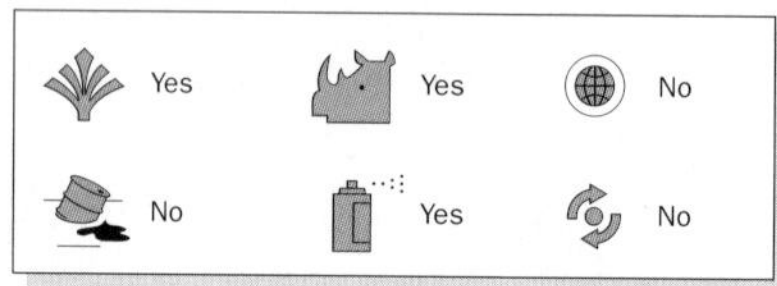
Yes
Yes
No
No
Yes
No

The US lags far behind other Western countries on environmental issues. The international commitment made at the 1997 Kyoto conference on cutting carbon dioxide emissions was scrapped by President Bush in 2001. The Rockies are a battleground between those who want to maintain their beauty, and those who advocate "wise use" – in practice this often means giving ranchers and miners free rein. In 2002, Congress approved plans to dump nuclear waste in Mt. Yucca, Nevada. Similar issues surround the arguments over extending oil drilling in the Alaskan wilderness. The US is leading the field in genetically modified (GM) food. Huge acreages have been planted with GM cereals, and by 2001 over 60% of soybean production was GM. A consumer backlash, especially in Europe, has worried many farmers.

MEDIA

TV ownership high

Daily newspaper circulation 201 per 1000 people

PUBLISHING AND BROADCAST MEDIA

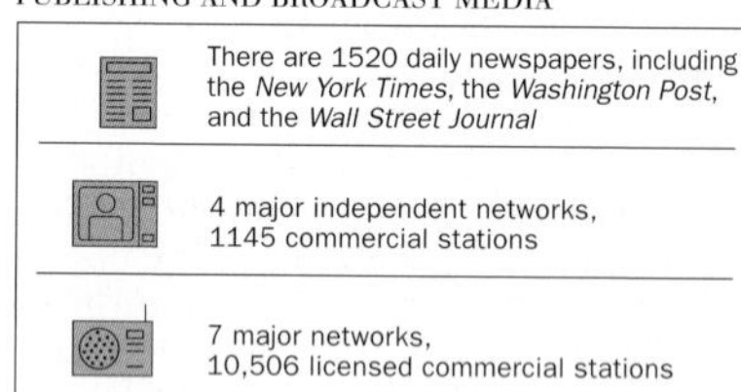
There are 1520 daily newspapers, including the *New York Times*, the *Washington Post*, and the *Wall Street Journal*
4 major independent networks, 1145 commercial stations
7 major networks, 10,506 licensed commercial stations

Mass media as a phenomenon was born in the US. No other society has ever had anything quite like US network TV, or moved so easily into the world of multichannel TV; homes with 50 or more channels are commonplace. The Internet, the most recent in the series of nationwide communication revolutions, is now used regularly by the majority of the population. Newspapers, mostly local rather than national, tend to have very low cover prices, and gain most of their revenue from advertising. They are under increasing threat from cable TV and other outlets. Companies exploring multimedia and ways of providing online news and services are still trying to recover from the crash in their stock values in 2000.

CHRONOLOGY *continued*

- **1987** Intermediate Nuclear Forces Treaty signed by US and USSR.
- **1988** Republican George Bush Sr. wins presidency.
- **1989** US overthrows Gen. Noriega of Panama, arrested on drug charges.
- **1991** January–February, Gulf War against Iraq. US and USSR sign START arms reduction treaty.
- **1992** Black youths riot in Los Angeles and other cities. Bush–Yeltsin summit agrees further arms reductions. Democrat Bill Clinton defeats Bush in presidential election.
- **1994** Health care reform legislation defeated in Congress. Midterm elections, Republican majorities in both houses of Congress.
- **1995** Oklahoma bombing by Timothy McVeigh: over 160 die.
- **1996** Clinton reelected.
- **1998** Scandal over Clinton's affair with White House intern leads to impeachment proceedings. August, bombing of US embassies in Kenya and Tanzania; revenge air strikes on Sudan and Afghanistan. December, air strikes against Iraq.
- **1999** February, Clinton acquitted in Senate impeachment trial. April, Columbine High School shootings by two students. March–June, NATO involvement to end Kosovo conflict, bombardment of Yugoslavia.
- **2000** Democrat Al Gore concedes tightest presidential election ever to Republican George W. Bush.
- **2001** January, President Bush takes office. September, world's worst terrorist attack kills thousands as hijacked planes destroy World Trade Center, damage Pentagon. October, US-led military action in "war on terrorism" begins with intensive aerial bombing campaign in Afghanistan. December, accounting scandal at Enron.
- **2002** July, WorldCom bankruptcy is biggest ever corporate collapse.

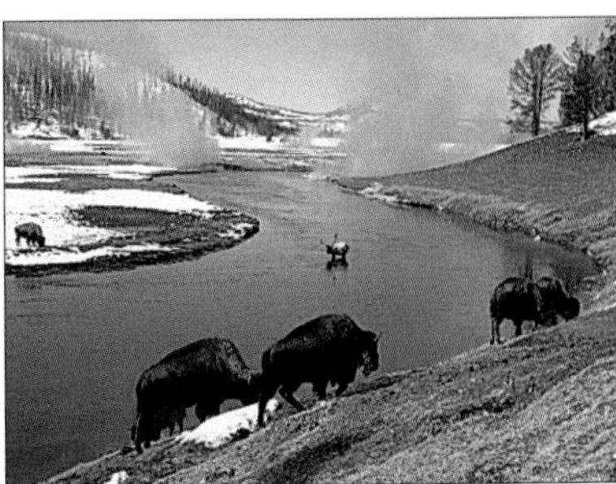

Bison in Yellowstone National Park. *The park's ecosystem is under severe strain due to the number of visitors it attracts.*

CRIME

Death penalty in use

2m prisoners

Crime is decreasing

CRIME RATES

Murders	6	per 100,000 population
Rapes	36	per 100,000 population
Thefts	1145	per 100,000 population

Violent crime – especially murder – is much more common than in other developed countries, even in relatively well-off areas. However, the murder rate has fallen; six deaths per 100,000 people in 2000 was the lowest for over 30 years. Mass shootings have made gun control a major issue, but a powerful lobby opposes restrictions, basing its arguments on the constitution and the defense of individual liberties.

Imprisonment for narcotics crimes in the US is much more widespread than in most Western countries. Capital punishment has increased since the 1980s, especially in the south. Texas carries out most executions. There are two million people in prison in the US, a quarter of the world total.

EDUCATION

School leaving age: 16

99%

14.6m students

THE EDUCATION SYSTEM

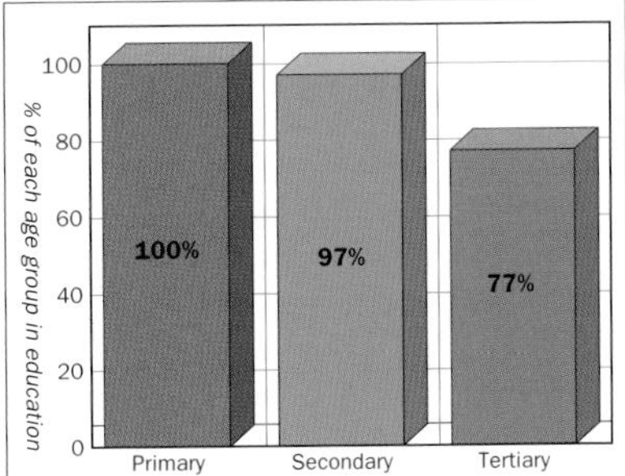

% of each age group in education
100
80
60
40
20
0
100%
97%
77%
Primary
Secondary
Tertiary

Education in the US is primarily the responsibility of the state governments.

Recent reports critical of standards in US high schools cite problems of discipline, poor structural maintenance, and lack of resources in many areas as driving people away from the public education sector. Private education at secondary level continues to develop rapidly. While the number of Catholic private schools has shrunk, more nondenominational fee-paying schools have been founded.

Four out of every five high school students now go on to some form of tertiary college. The leading US universities are internationally recognized as being of world class.

HEALTH

Limited welfare state health benefits

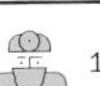

1 per 370 people

Heart and cerebrovascular diseases, cancers, accidents

There are enormous disparities in US health provision, but a proposed reform of the health care system, high on the Democrats' political agenda in the early 1990s, was blocked by Congress. Sophisticated techniques are available to those with insurance (which they typically receive from their employer); the Texas Medical Center, in Houston, for example, has a budget equivalent to that of some small countries. On the other hand, costs have skyrocketed, and facilities for those dependent on state medical care and aid are woefully underfunded. Preventive care fails to reach all sections of society, and infant mortality statistics in some areas are at near-African levels.

Nearly one in five of the population is clinically obese, and one in every two adults is overweight.

SPENDING

GDP/cap. increase

CONSUMPTION AND SPENDING

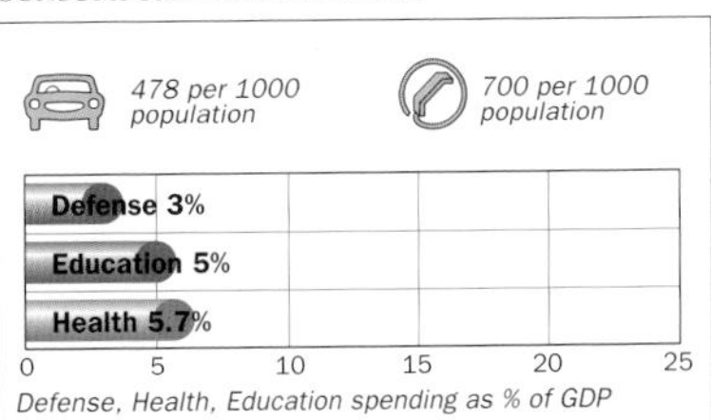

478 per 1000 population
700 per 1000 population
Defense 3%
Education 5%
Health 5.7%
0
5
10
15
20
25
Defense, Health, Education spending as % of GDP

Between 1945 and 1973, most Americans got richer. Since then, however, living standards have gone on rising only among those who finish high school. This "education effect" has led to noticeable class divisions, despite the long economic boom of the 1990s. The top 20% had average household incomes of $137,500 by 2000, whereas the incomes of the poorest 20% averaged only $13,000 – and were lower in real terms than in 1980.

WORLD RANKING

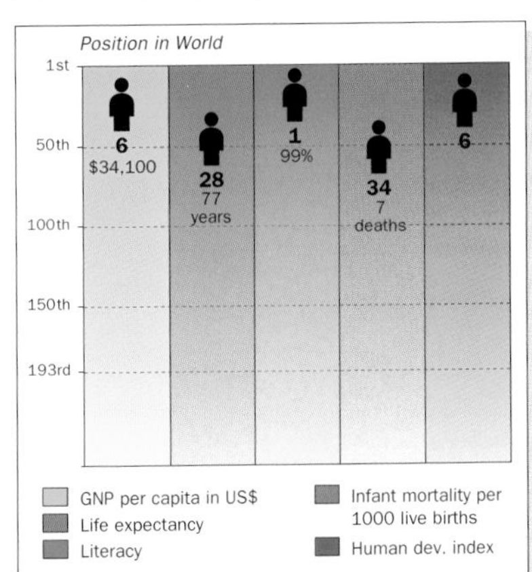

Position in World
1st
50th
100th
150th
193rd
6
$34,100
28
77
years
1
99%
34
7
deaths
6
GNP per capita in US$
Life expectancy
Literacy
Infant mortality per 1000 live births
Human dev. index

U

URUGUAY

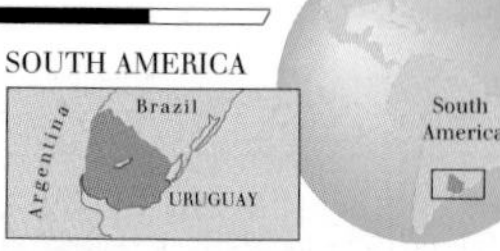
SOUTH AMERICA
Brazil
Argentina
URUGUAY
South America

OFFICIAL NAME: Eastern Republic of Uruguay **CAPITAL:** Montevideo
POPULATION: 3.4 million **CURRENCY:** Uruguayan peso **OFFICIAL LANGUAGE:** Spanish

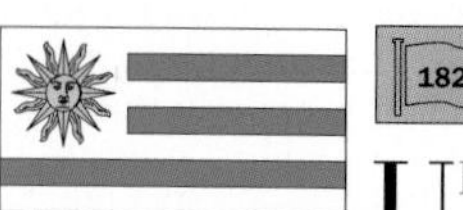

1828

1828

Aug 25
ROU
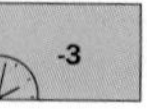
-3

+598

.uy

URUGUAY IS SITUATED IN the southeast of South America, sandwiched between its larger neighbors Brazil and Argentina. Its capital, Montevideo, is an Atlantic port on the River Plate, lying on the opposite bank to the Argentine capital Buenos Aires. Uruguay became independent in 1828, after nearly 150 years of Spanish and Portuguese control. Decades of liberal government ended in 1973 with a military coup that was to result in 12 years of dictatorship, during which 400,000 people emigrated. Most have since returned. Almost the entire low-lying landscape is devoted to the rearing of livestock, especially cattle and sheep; Uruguay is a leading wool exporter. Tourism and offshore banking now bring in substantial foreign earnings.

Uruguayan grasslands. *Rich pasture covers three-quarters of the country, ideal for cattle and sheep. Animals and animal products account for over one-third of export earnings.*

CLIMATE

Subtropical

WEATHER CHART FOR MONTEVIDEO

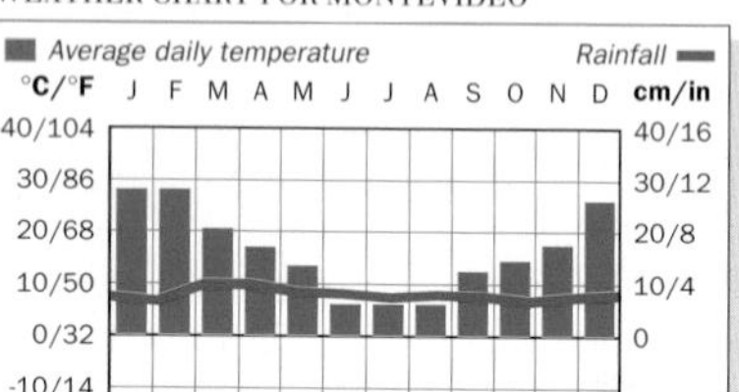
Average daily temperature
Rainfall
°C/°F
J F M A M J J A S O N D
cm/in
40/104
30/86
20/68
10/50
0/32
-10/14
-20/-4
40/16
30/12
20/8
10/4
0

Uruguay has one of the most benign climates in the world. It is uniformly temperate over the whole country. Winters are mild, frost is rare, and it never snows. Summers are generally cool for these latitudes and rarely tropically hot. The moderate rainfall tends to fall in heavy showers, leaving most days sunny.

TRANSPORTATION

Drive on right

Carrasco, Montevideo
1.17m passengers

91 ships
106,900 grt

THE TRANSPORTATION NETWORK

8085 km (5024 miles)

8683 km (5395 miles)

3002 km (1865 miles)

1600 km (994 miles)

The government has sold off its share in the national bus industry – there are extensive internal and international coach and bus services – and has closed down all passenger railroad services. In 1998 the Senate gave the go-ahead for a $1 billion, 45-km (30-mile) road bridge across the River Plate from Colonia to Buenos Aires. Raising international finance for the project will be difficult.

TOURISM

Visitors : Population 1:1.7

2m visitors

Down 5% in 2000

MAIN TOURIST ARRIVALS

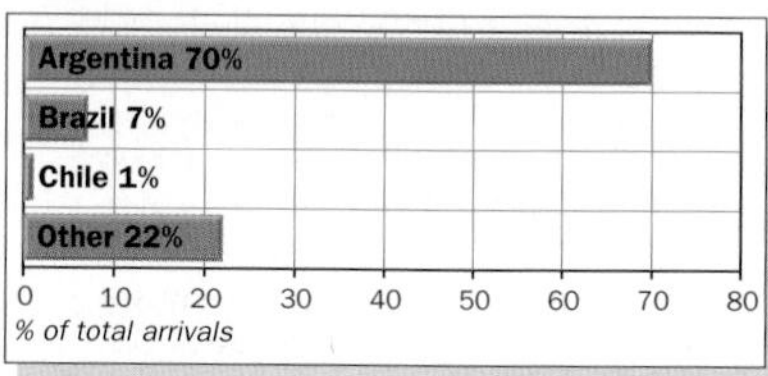
Argentina 70%
Brazil 7%
Chile 1%
Other 22%
0 10 20 30 40 50 60 70 80
% of total arrivals

Sandy beaches near the River Plate estuary are a major attraction. The old Spanish fortifications of Montevideo have been destroyed, but the city retains a colonial atmosphere. Punta del Este, 138 km (86 miles) east of the capital, is the main beach resort. Argentinians account for the majority of visitors.

PEOPLE

Population density low

Spanish

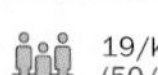

19/km² (50/mi²)

THE URBAN/RURAL POPULATION SPLIT

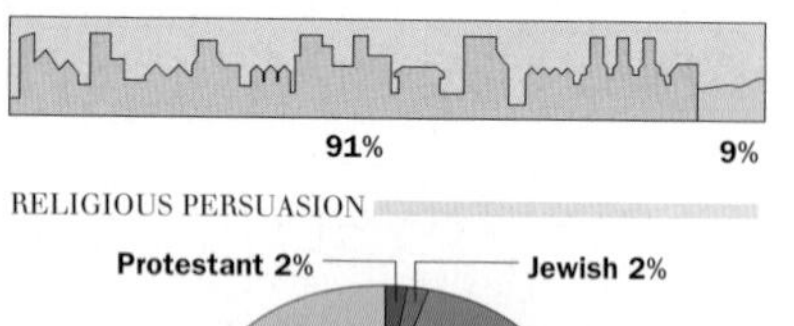
91%
9%

RELIGIOUS PERSUASION

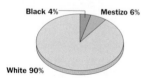

Protestant 2%
Jewish 2%
Nonreligious 30%
Roman Catholic 66%

ETHNIC MAKEUP

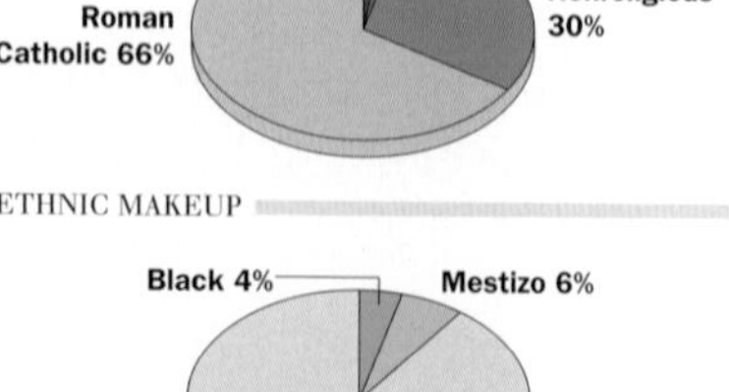
Black 4%
Mestizo 6%
White 90%

Most Uruguayans are second- or third-generation European, mostly of Spanish or Italian descent. There are also some *mestizos* (of mixed blood) and a small minority of people descended from Africans or immigrants from Brazil, who live near the Brazilian border, or in or around Montevideo. All indigenous Amerindian groups became integrated in the *mestizo* population by the mid-19th century. More recent immigrants include Jews, Armenians, and Lebanese. Historically, ethnic tensions have been few. The birthrate is low for Latin America.

The considerable prosperity derived from cattle ranching allowed Uruguay to become a welfare state long before any other Latin American country. In spite of Uruguay's serious economic decline since the end of the 1950s, there is still a sizable, if less prosperous, middle class. A clear sign of the country's economic and social deterioration during the years of military dictatorship was the unprecedented growth of shanty towns around Montevideo.

Although Uruguay is a Roman Catholic country, it is liberal in its attitude to religion, and all forms are tolerated. Divorce is legal. Women, who gained the vote in 1932, are regarded as equal to men.

POPULATION AGE BREAKDOWN

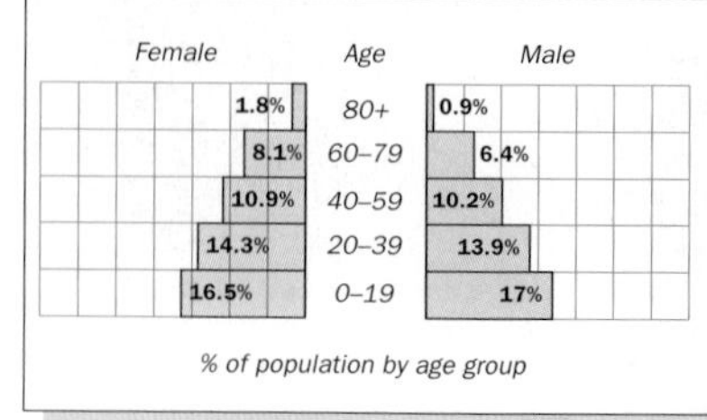
Female
Age
Male
1.8% 80+ 0.9%
8.1% 60–79 6.4%
10.9% 40–59 10.2%
14.3% 20–39 13.9%
16.5% 0–19 17%
% of population by age group

POLITICS

Multiparty elections

L. House 1999/2004
U. House 1999/2004

President Jorge Batlle Ibáñez

AT THE LAST ELECTION

Chamber of Representatives 99 seats

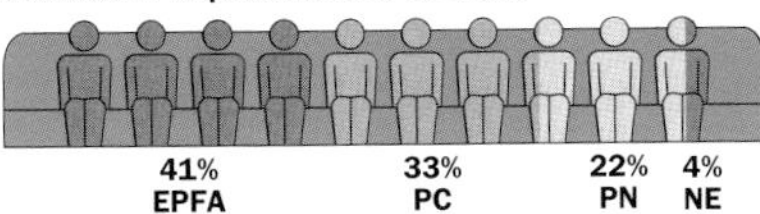
41% EPFA
33% PC
22% PN
4% NE

EPFA = Progressive Broad Front **PC** = Colorado Party (Colorados) **PN** = National Party (Blancos) **NE** = New Space **Res** = Reserved for the vice president

Senate 31 seats

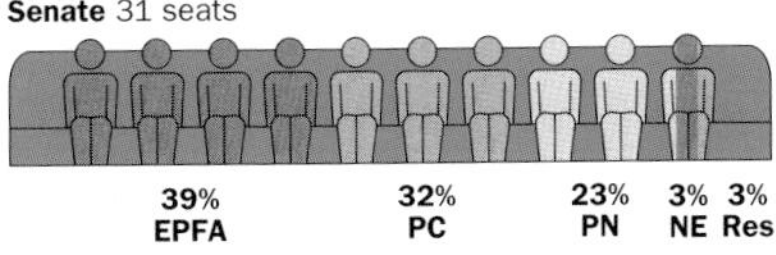
39% EPFA
32% PC
23% PN
3% NE
3% Res

Uruguay is a presidential multiparty democracy.

PROFILE

Since the return to democracy in 1984, the main Colorado (PC) and Blanco (PN) parties have monopolized power, either alone or in coalitions, despite being traditional opponents. The left-wing Broad Front has been the effective opposition, frequently in alliance with trade unions fighting austerity measures and reform of the social security system. Despite the crowded electoral calendar in 1999, and infighting among Blanco factions, there was broad political consensus on the need for continuing economic reform. In the 1999 elections, Colorado candidate Jorge Batlle won the presidency in the face of an unusually strong left-wing challenge.

MAIN POLITICAL ISSUES

Economic crisis

Recession has left the Batlle government little room for maneuver. The unions oppose tough fiscal targets agreed with the IMF. The currency came under pressure in 2002 due to the turmoil in Uruguay's main single export market, Argentina, coming on top of a foot-and-mouth outbreak which hit beef exports.

The disappeared

An official inquiry into the fate of 160 people who "disappeared" under military rule (1973–1985) aims to give Uruguay a "moral and ethical answer."

***Luis Alberto Lacalle Herrera**, president from 1990–1995.*

***Jorge Batlle Ibáñez**, who took office as president in 2000.*

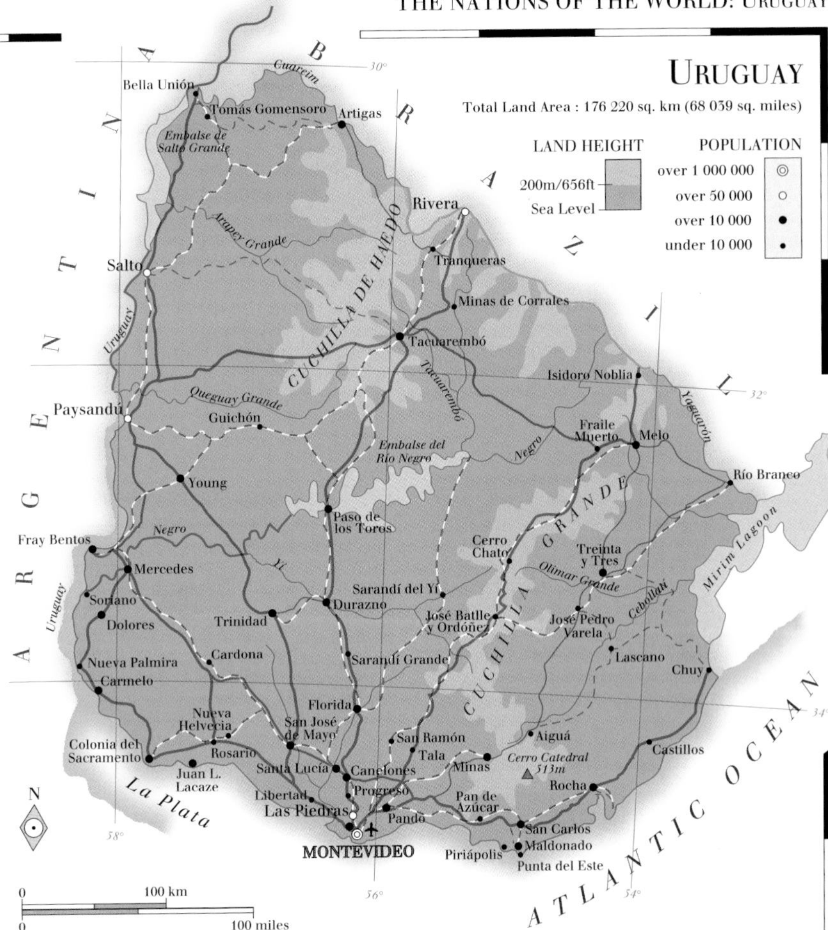
URUGUAY
Total Land Area : 176 220 sq. km (68 039 sq. miles)
LAND HEIGHT
200m/656ft
Sea Level
POPULATION
over 1 000 000
over 50 000
over 10 000
under 10 000
BRAZIL
ARGENTINA
ATLANTIC OCEAN
La Plata
Bella Unión
Tomás Gomensoro
Artigas
Cuareim
Embalse de Salto Grande
Rivera
Tranqueras
Arapey Grande
Salto
Uruguay
Minas de Corrales
CUCHILLA DE HAEDO
Tacuarembó
Isidoro Noblía
Queguay Grande
Paysandú
Guichón
Embalse del Río Negro
Negro
Yaguarón
Fraile Muerto
Melo
Río Branco
Young
Paso de los Toros
CUCHILLA GRANDE
Mirim Lagoon
Fray Bentos
Cerro Chato
Treinta y Tres
Olimar Grande
Mercedes
Yí
Sarandí del Yí
Soriano
Durazno
Dolores
Trinidad
José Batlle y Ordóñez
José Pedro Varela
Cebollatí
Cardona
Sarandí Grande
Lascano
Nueva Palmira
Carmelo
Chuy
Florida
Nueva Helvecia
San José de Mayo
San Ramón
Aiguá
Colonia del Sacramento
Rosario
Tala
Cerro Catedral 513m
Castillos
Juan L. Lacaze
Santa Lucía
Canelones
Minas
Rocha
Libertad
Progreso
Pan de Azúcar
Las Piedras
Pando
San Carlos
MONTEVIDEO
Piriápolis
Maldonado
Punta del Este
N
0 100 km
0 100 miles

WORLD AFFAIRS

Joined UN in 1945

Geplac | Mercsr | IBRD | OAS | RG

Regional integration is a major focus, but President Batlle's determination to clinch a bilateral trade deal with the US risked a row with the Brazilian government, which said that it went against the official policy of Mercosur to negotiate trade agreements as a bloc. Argentina also took exception to plans to impose protective tariffs to cushion Uruguayan industry from the effects of Argentine devaluation. Such difficulties threatened diplomatic moves to strengthen Mercosur as a negotiating bloc in upcoming talks to establish a Free Trade Area of the Americas (FTAA).

Uruguay and the US have agreed a legal assistance treaty to allow easier access to bank accounts of those suspected of laundering the proceeds from narcotics trafficking.

AID

Recipient

$17m (receipts) — Down 23% in 2000

Uruguay received an IMF standby loan of $1.5 billion for 2002–2003, but aid remains otherwise modest.

CHRONOLOGY

The Spaniards were the first to colonize the area north of the River Plate. In 1680, the Portuguese also founded a colony there, at Colonia del Sacramento, so starting 150 years of rivalry between the colonial powers for control of the territory.

- **1726** Spaniards found Montevideo. By end of the century, whole country is divided into large cattle ranches.
- **1808** Montevideo declares independence from Buenos Aires.
- **1811** Patriotic rancher and local caudillo, José Gervasio Artigas, fends off Brazilian attack.
- **1812–1820** Uruguayans, known as Orientales ("Easterners," from the eastern side of the River Plate), fight wars against Argentinian and Brazilian invaders. Brazil finally takes Montevideo.
- **1827** Gen. Lavalleja defeats Brazilians with Argentine help.
- **1828** Seeing trade benefits that an independent Uruguay would bring as a buffer state between Argentina and Brazil, Britain mediates and secures Uruguayan independence.
- **1836** Start of large-scale European immigration.

U

CHRONOLOGY *continued*

- **1838–1865** La Guerra Grande civil war between Blancos (Whites, future conservative party) and Colorados (Reds, future liberals).
- **1865–1870** President Venancio Flores of Colorados takes Uruguay into War of Triple Alliance against Paraguay.
- **1872** Peace under military rule. Blancos strong in country, Colorados in cities.
- **1890s** Violent strikes by immigrant trade unionists against landed elite enriched by massive European investment in ranching.
- **1903–1907** Reformist Colorado, José Batllé y Ordóñez, president.
- **1911–1915** Batllé serves second term in office. Batllismo creates the only welfare state in Latin America with pensions, social security, and free education and health service; also nationalizations, disestablishment of Church, abolition of death penalty.
- **1933** Military coup. Opposition groups excluded from politics.
- **1942** President Alfredo Baldomir dismisses government and tries to bring back proper representation.
- **1939–1945** Neutral in World War II.
- **1951** New constitution replaces president with nine-member council. Decade of great prosperity follows until world agricultural prices plummet. Sharp drop in foreign investment.
- **1958** Blancos win elections for first time in 93 years.
- **1962** Tupamaros urban guerrillas founded. Its guerrilla campaign lasts until 1973.
- **1966** Presidency reinstated. Colorados back in power.
- **1967** Jorge Pacheco president. Tries to stifle opposition to tough anti-inflation policies.
- **1973** Military coup. Promises to encourage foreign investment counteracted by denial of political freedom and brutal repression of the left; 400,000 emigrate.
- **1984–1985** Military step down. Elections. Julio Sanguinetti (Colorado) president.
- **1986** Those guilty of human rights abuse granted amnesty.
- **1989** Referendum endorses amnesty in interests of stability. Elections won by Lacalle Herrera and Blancos.
- **1994–1995** Sanguinetti reelected, forms coalition government. Mercosur membership.
- **1999** October, presidential election won by Colorado Jorge Batlle.
- **2002** Uruguay loses investment grade status due to impact of Argentine crisis.

U

DEFENSE

No compulsory military service

$356m

Up 12% in 2000

The military withdrew from power in 1985 and has since respected civilian rule. "Lodges" operate within the army to promote officers' interests and have displayed opposition to the government's replacements and promotions within the military hierarchy. A 1986 law virtually blocked investigations into killings, torture, and "disappearances" during the dictatorship, but there is still pressure to bring guilty officers to justice. A presidential decree in 1997 granted amnesty to officers punished for political offenses under military rule.

URUGUAYAN ARMED FORCES

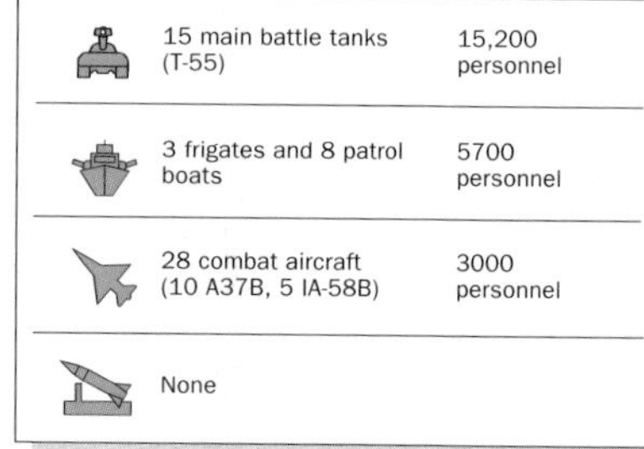
15 main battle tanks (T-55)
15,200 personnel
3 frigates and 8 patrol boats
5700 personnel
28 combat aircraft (10 A37B, 5 IA-58B)
3000 personnel
None

Uruguay's defense budget is modest; the majority of equipment is bought from the US.

ECONOMICS

Inflation 31% p.a. (1990–2000)

$20bn

12.51–14.77 Uruguayan pesos

SCORE CARD

World GNP Ranking	63rd
GNP per Capita	$6000
Balance of Payments	–$593m
Inflation	4.8%
Unemployment	14%

EXPORTS

Mexico 4%
Germany 4%
USA 8%
Argentina 18%
Brazil 23%
Other 43%

IMPORTS

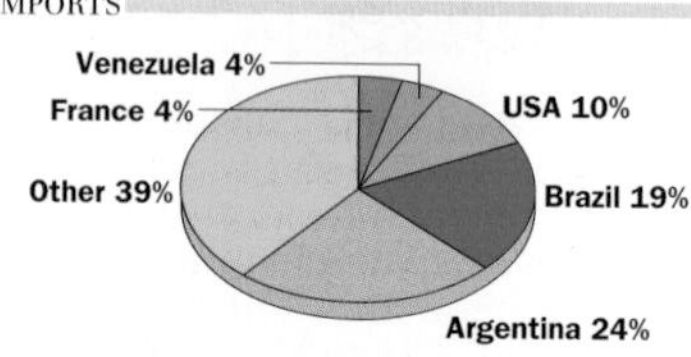
Venezuela 4%
France 4%
USA 10%
Brazil 19%
Argentina 24%
Other 39%

Strengths

Substantial earnings from offshore banking. Tourism. Fertile grasslands. World's second-largest wool exporter. Beef-meat products.

Weaknesses

Few natural resources. Dependence on Brazil and Argentine markets. Modest industry. Large public sector deficit. Sluggish economic growth. Loss of investment grade status.

Profile

Traditionally an agricultural economy, three-quarters of the country is rich pasture, supporting livestock. Much of the rest is given over to crops. Farming, which formerly brought great wealth, still employs about 15% of the labor force, accounting for some 19% of GDP. Livestock and animal products, especially meat and wool, account for over one-third of export earnings. Manufacturing, accounting for some 18% of GDP, is farm-based. Tourism is increasingly important. Most economic activity – and half the population – is concentrated in Montevideo. Recently growth has been sluggish, making it difficult to achieve GDP and fiscal targets agreed with the IMF. Unions resist spending cuts, and, with public opinion hostile to major privatizations, there has been little progress in necessary structural reforms.

ECONOMIC PERFORMANCE INDICATOR

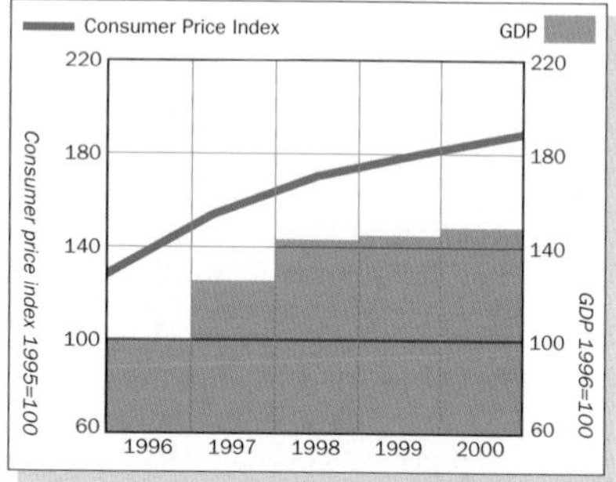
Consumer Price Index
GDP
220
180
140
100
60
Consumer price index 1995=100
GDP 1996=100
1996
1997
1998
1999
2000

URUGUAY : MAJOR BUSINESSES

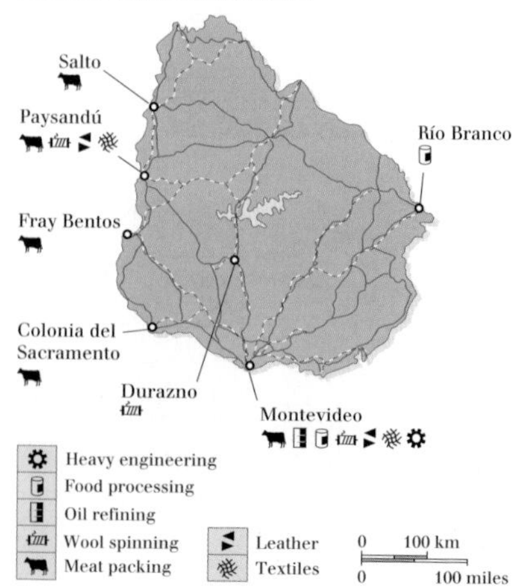
Salto
Paysandú
Río Branco
Fray Bentos
Colonia del Sacramento
Durazno
Montevideo
Heavy engineering
Food processing
Oil refining
Wool spinning
Meat packing
Leather
Textiles
0 100 km
0 100 miles

RESOURCES

Electric power 2.2m kw

103,043 tonnes

Not an oil producer

13m sheep, 10.8m cattle, 13m chickens

Gold, iron, gemstones, copper, zinc, lead, manganese

Most of Uruguay is farmland, much of it given over to cattle and sheep. Rice is the country's only other significant export. There are no known oil or natural gas resources. Considerable potential is believed to exist for the mining sector, but only small quantities of building materials and jewelry-quality agate and amethysts are so far extracted. The mining of gold deposits is currently being developed. Hydropower generates most of the electricity.

ELECTRICITY GENERATION

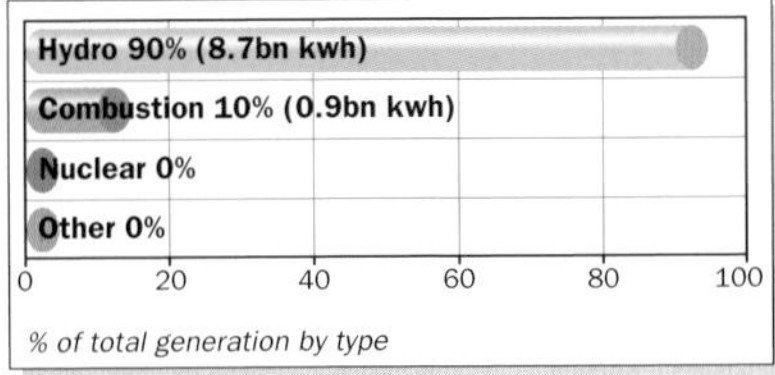
Hydro 90% (8.7bn kwh)
Combustion 10% (0.9bn kwh)
Nuclear 0%
Other 0%
0 20 40 60 80 100
% of total generation by type

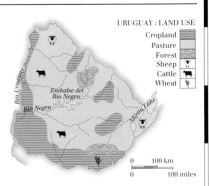

URUGUAY : LAND USE

Cropland
Pasture
Forest
Sheep
Cattle
Wheat

Embalse del Río Negro
Río Uruguay
Río Negro
Mirim Lake
0 100 km
0 100 miles

ENVIRONMENT

Sustainability rank: 6th

0.3% (0.1% partially protected)

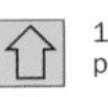
1.8 tonnes per capita

ENVIRONMENTAL TREATIES

Yes
Yes
Yes
Yes
Yes
Yes

Pollution of the main Uruguay and Plate Rivers is a concern, as is traffic density in Montevideo.

MEDIA

TV ownership high

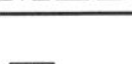
Daily newspaper circulation 293 per 1000 people

PUBLISHING AND BROADCAST MEDIA

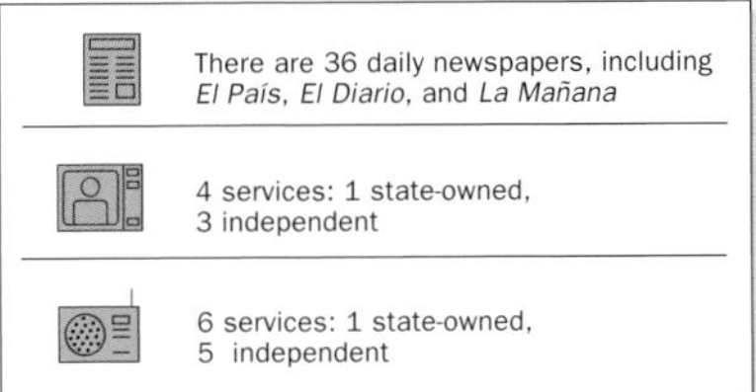
There are 36 daily newspapers, including *El País*, *El Diario*, and *La Mañana*
4 services: 1 state-owned, 3 independent
6 services: 1 state-owned, 5 independent

The press is relatively free. *El País* supports the Blancos (PN), while *La Mañana* backs the Colorados (PC).

CRIME

No death penalty

4012 prisoners

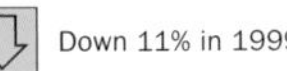
Down 11% in 1999

CRIME RATES

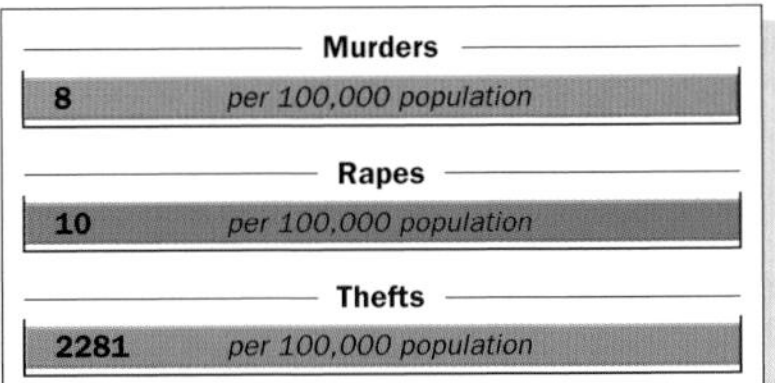
Murders
8 per 100,000 population
Rapes
10 per 100,000 population
Thefts
2281 per 100,000 population

Crime levels in Uruguay are low, particularly compared with its neighbors Brazil and Argentina. Domestic theft is the main problem. Bribery is uncommon.

EDUCATION

School leaving age: 14

98%

77,454 students

THE EDUCATION SYSTEM

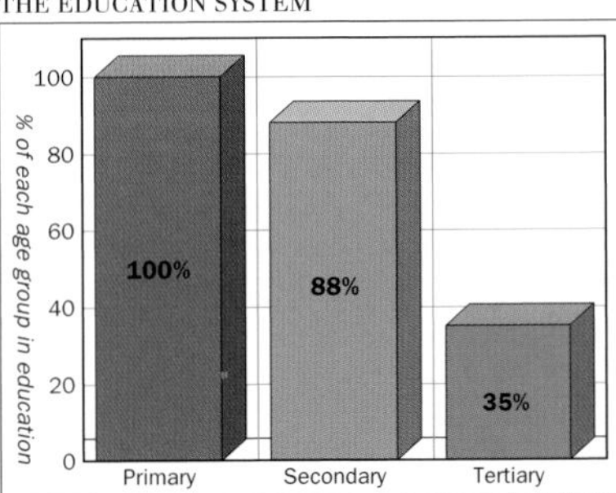
100
80
60
40
20
0
% of each age group in education
100%
88%
35%
Primary
Secondary
Tertiary

Education, inspired by the French *lycée* system, is state-funded for 12 years up to secondary level and is compulsory for all children between the ages of six and 14. Both state and private schools follow the same curriculum; private schools are monitored by the government. Facilities are rudimentary in rural areas. Uruguay has two state-funded universities. The children of wealthy Uruguayans tend to complete their studies in the US. Resistance to tax increases and pressure to reduce the fiscal deficit have both placed serious constraints on education spending. Secondary school students continue to stage protests against the resulting effects on the system.

HEALTH

Welfare state health benefits

1 per 270 people

Cerebrovascular and heart diseases, cancers, accidents

Most Uruguayans have easy access to health services. Average life expectancy is high. Public services provide for 40% of the population, while the private sector caters for the remaining 60%. Despite opposition, the government has privatized some of the state medical establishments.

Health spending has in recent years been a victim of the budget cuts and social welfare reforms aimed at controlling the fiscal deficit.

SPENDING

GDP/cap. increase

CONSUMPTION AND SPENDING

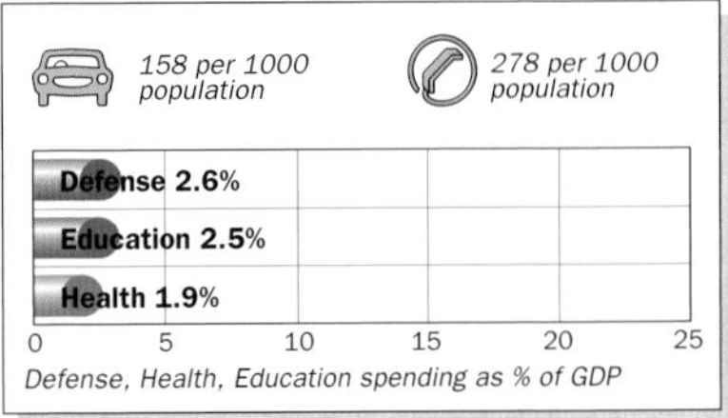
158 per 1000 population
278 per 1000 population
Defense 2.6%
Education 2.5%
Health 1.9%
0 5 10 15 20 25
Defense, Health, Education spending as % of GDP

Uruguay possesses the social mobility which is typical of countries created through decades of large-scale immigration. Many professionals come from modest backgrounds. A 1999 report by the IDB exempted Uruguay (along with Costa Rica and Jamaica) from the regional trend of serious income inequality.

The wealthy tend either to be landowners or are employed in the financial sector. They have traditionally looked toward Europe, rather than the US, for luxury goods.

The most deprived sections of Uruguayan society are the urban poor of Montevideo, a large proportion of whom are of mixed African and European descent, and the rural poor, who own little or no land.

WORLD RANKING

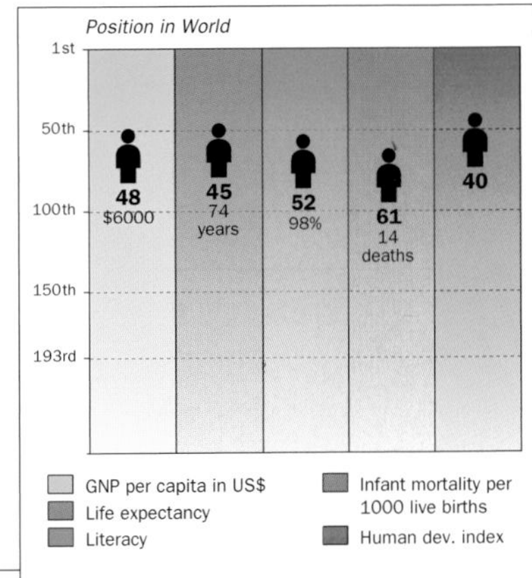
Position in World
1st
50th
100th
150th
193rd
48 $6000
45 74 years
52 98%
61 14 deaths
40
GNP per capita in US$
Life expectancy
Literacy
Infant mortality per 1000 live births
Human dev. index

U

UZBEKISTAN

OFFICIAL NAME: Republic of Uzbekistan **CAPITAL:** Tashkent
POPULATION: 25.3 million **CURRENCY:** Som **OFFICIAL LANGUAGE:** Uzbek

CENTRAL ASIA
Asia
Kazakhstan
Kyrgyzstan
UZBEKISTAN
Turkmenistan
Tajikistan
Afghanistan

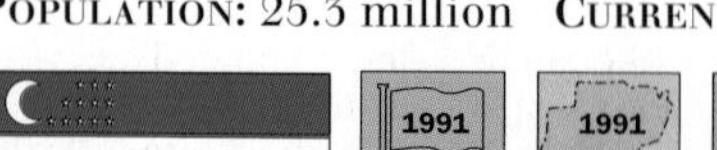

1991

1991
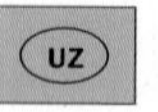
Sept 1
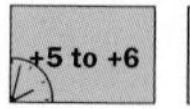
UZ

+5 to +6

+998

.uz

SHARING THE ARAL SEA coastline with its northern neighbor, Kazakhstan, Uzbekistan contains the ancient cities of Samarqand, Bukhara (Bukhoro), Khiva, and Tashkent. It is the most populous central Asian republic and has considerable natural resources. The dictatorship of President Karimov has prevented the spread of Islamic fundamentalism.

CLIMATE

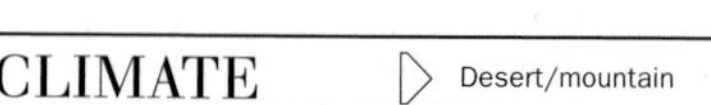
Desert/mountain

WEATHER CHART FOR TASHKENT

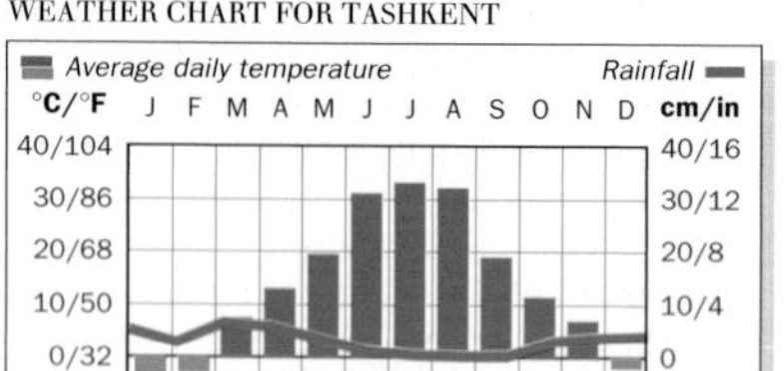

Average daily temperature
Rainfall
°C/°F
J F M A M J J A S O N D
cm/in
40/104
30/86
20/68
10/50
0/32
-10/14
-20/-4
40/16
30/12
20/8
10/4
0

Uzbekistan has a harsh continental climate. Summers can be extremely hot and dry. Large areas of the country are desert.

TRANSPORTATION

Drive on right

Tashkent International 1.93m passengers

Has no fleet

THE TRANSPORTATION NETWORK

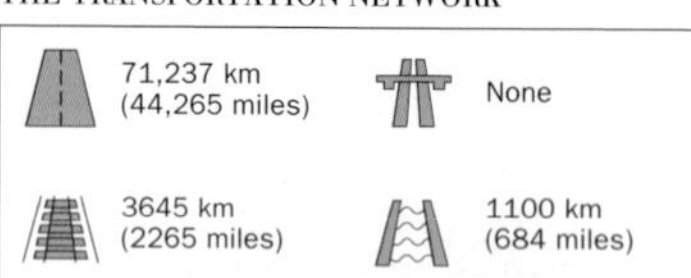

71,237 km (44,265 miles)
None
3645 km (2265 miles)
1100 km (684 miles)

Uzbekistan has a well-developed transportation system. An extensive network of buses serves country areas, while good Soviet-style systems of trolley buses and trams operate in the major cities. Tashkent's subway system was the first in central Asia. Road and rail networks have deteriorated since 1991, however, and are concentrated in the south and east.

TOURISM

Visitors : Population 1:93

272,000 visitors

Up 196% 1995–1998

MAIN TOURIST ARRIVALS

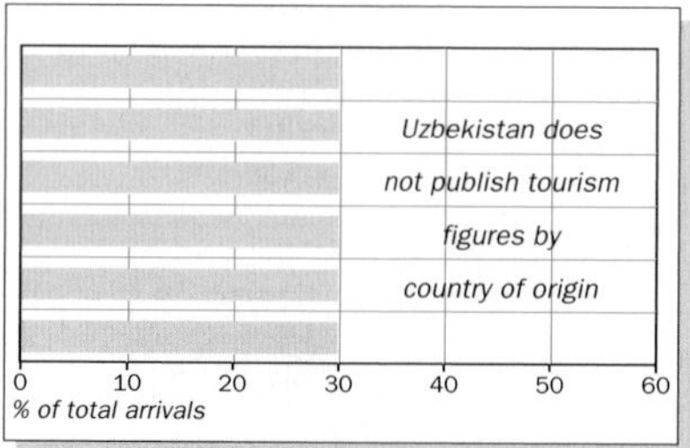

Uzbekistan does not publish tourism figures by country of origin
0 10 20 30 40 50 60
% of total arrivals

Uzbekistan has considerable tourist potential. Bukhara, once a trading center on the silk route, is famous worldwide for its architecture and carpet-making. It has great religious significance for Muslims, who are encouraged to make at least one pilgrimage to its holy shrines. Bukhara's Kalyan Mosque is famous for its minaret built of unbaked bricks. The city of Samarqand was expanded in the 14th century by Timur, and contains the monumental gateway of the Shir Dar Madrasa, one of the most beautiful buildings in the Islamic world.

UZBEKISTAN

Total Land Area : 447 400 sq. km (172 741 sq. miles)

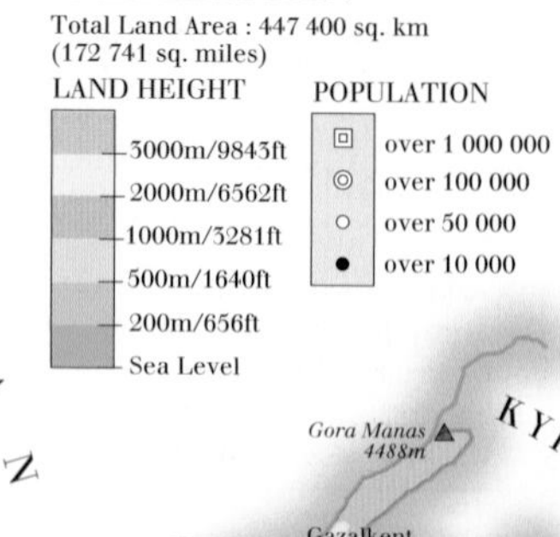

LAND HEIGHT
3000m/9843ft
2000m/6562ft
1000m/3281ft
500m/1640ft
200m/656ft
Sea Level
POPULATION
over 1 000 000
over 100 000
over 50 000
over 10 000
KAZAKHSTAN
ARAL SEA
USTYURT PLATEAU
TURAN LOWLAND
KYZYL KUM
TURKMENISTAN
KYRGYZSTAN
TAJIKISTAN
AFGHANISTAN
Mŭynoq
Chimboy
Takhtakŭpir
Qŭnghirot
Amu Darya
Nukus
Khujayli
Takhiatosh
Beruniy
Urganch
Khiwa
Tŭrtkŭl
Uchquduq
Zarafshon
Gora Manas 4488m
Chirchiq
Gazalkent
Parkent
TASHKENT
Yangiyŭl
CHATKAL RANGE
Chust
Namangan
Andijon
Angren
Olmaliq
Shahrikhon
Asaka
Qŭqon
Syrdar'ya
Syr Darya
Guliston
Yangiyer
Bekobod
Marghilon
Farghona
Aydarkŭl
Nurota
Langar
NUROTA TIZMASI
Nawoiy
Kattaqŭrghon
Jizzakh
Ghijduwon
Gazli
Bukhoro
Kogon
Oqtosh
Bulunghur
Samarqand
Urgut
Qorakŭl
Kitob
Shahrisabz
Kasan
Qarshi
GISSAR RANGE
Denow
Boysun
Jarqŭrghon
Termiz
N
0 100 km
0 100 miles

Mosque in Samarqand. *The city remained an Islamic stronghold, despite communist attempts at suppression, when Uzbekistan formed part of the Soviet Union.*

PEOPLE

Pop. density medium

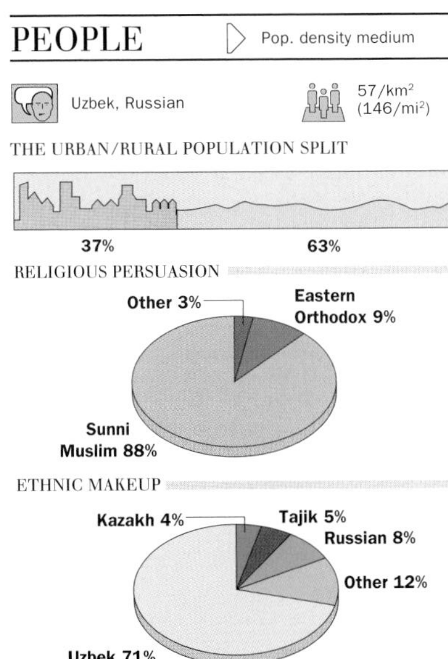
Uzbek, Russian
57/km²
(146/mi²)
THE URBAN/RURAL POPULATION SPLIT
37%
63%
RELIGIOUS PERSUASION
Other 3%
Eastern Orthodox 9%
Sunni Muslim 88%
ETHNIC MAKEUP
Kazakh 4%
Tajik 5%
Russian 8%
Other 12%
Uzbek 71%

Among the former Soviet republics, Uzbekistan has a relatively complex makeup. In addition to the Uzbeks, Russians, Tajiks, and Kazakhs, there are small minorities of Tatars and Karakalpaks. The proportion of Russians has been declining since the 1970s, when net emigration of Russians began. Tensions among ethnic groups have the potential to create regional and racial conflict. The authoritarian nature of the Karimov leadership has so far prevented these antagonisms from becoming violent. Incidents such as the 1989 and 1990 clashes between Meskhetian Turks and Uzbeks are rare. The removal of the dominance of the Communist Party has meant that Uzbek society has reverted to traditional social patterns based on family, religion, clan, and region, rather than on membership of the party. Independence has done little to alter the minor role of women in politics. Arranged marriages are still the custom in the countryside.

POPULATION AGE BREAKDOWN

Female
Age
Male
0.6% 80+ 0.3%
3.4% 60–79 2.2%
6.1% 40–59 6%
15.3% 20–39 15.1%
25.2% 0–19 25.8%
% of population by age group

POLITICS

Multiparty elections

1999/2004

President Islam Karimov

AT THE LAST ELECTION

National Parliament 250 seats

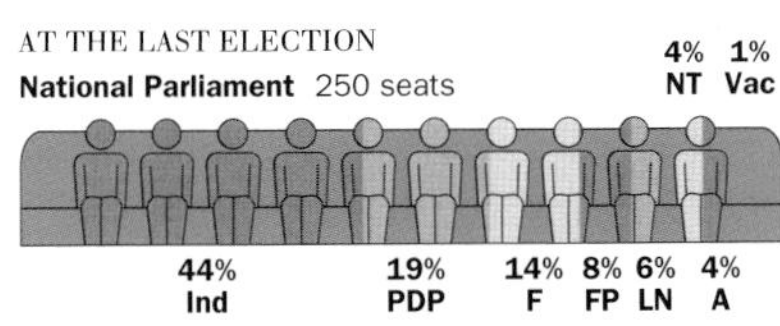
4% 1%
NT Vac
44% Ind
19% PDP
14% F
8% FP
6% LN
4% A

Ind = Independents **PDP** = People's Democratic Party
F = Fidorkorlar **FP** = Fatherland Progress Party
LN = Local nominees **A** = Adolat
NT = National Renaissance **Vac** = Vacant

Uzbekistan is effectively run by a presidential dictatorship. In 2002 a referendum approved the creation of a second legislative chamber and lengthened the presidential term.

PROFILE

President Islam Karimov's PDP has not been willing to devolve or share power. The 1992 constitution appeared to endorse multipartyism, but Karimov took advantage of greater powers granted to his office by banning several opposition parties, including the nationalist Birlik (Unity) movement and the Islamic Renaissance Party. Erk (Will), the only legal opposition party, was proscribed in 1993, and in 1995, a group of its activists, found guilty of political subversion, received stiff sentences. Opposition is now entirely underground. The intimidation and arbitrary imprisonment of dissidents are common, and repression has increased since bomb attacks in Tashkent in 1999. Karimov has kept the support of the Russian minority by avoiding nationalist rhetoric.

MAIN POLITICAL ISSUES

Islamic fundamentalism

Civil war in Tajikistan and the rise of the *taliban* regime in Afghanistan raised fears about Islamic fundamentalism. The constitution stipulates the separation of Islam and the state. A joint operation with Kyrgyzstan and Tajikistan against the pan-regional Islamic Movement of Uzbekistan (IMU) took place in 2000, as cross-border attacks rose. IMU leader Juma Namangani was reported killed alongside *taliban* fighters in Afghanistan in late 2001. A clampdown against the Hizb-ut Tahrir group continues.

Regionalism

The high birthrate puts pressure on limited agricultural resources. Migration from poorer areas have led to calls for secession from some regions. In the densely populated eastern Fergana Valley there has been a number of violent incidents.

***Islam Karimov,** first elected president in 1990 and Uzbekistan's sole leader since independence.*

WORLD AFFAIRS

Joined UN in 1992

CIS
SCO

OIC

NAM

OSCE

Unlike neighboring Turkmenistan, Kyrgyzstan, and Tajikistan, Uzbekistan has the resources to allow it to follow a relatively independent foreign policy. The Karimov leadership has used this to promote Uzbekistan as the leading central Asian state, a role for which it vies with Kazakhstan. It is a member of the Georgia–Ukraine–Uzbekistan–Azerbaijan–Moldova (GUUAM) group of ex-Soviet states, signifying a wish to maintain a certain distance in relations with Russia, although it suspended its membership in 2002. It has border disputes with its immediate neighbors, in particular over the laying of land mines on its borders with Tajikistan and Kyrgyzstan to prevent terrorist incursions.

Ties with the US were strengthened when Uzbekistan played host to US forces during the war in Afghanistan in 2001. Relations with Turkey are also developing. While Western companies have difficulty in sealing contracts in Uzbekistan, Turkish companies have been commissioned to build vital installations such as telecommunications.

CHRONOLOGY

Part of the great Mongol Empire, present-day Uzbekistan was incorporated into the Russian Empire between 1865 and 1876. Russification of the area was superficial, and it was not until Soviet rule that significant Slav immigration occurred. A further influx of Slavs occurred during Stalin's program of forced collectivization.

- **1917** Soviet power established in Tashkent.
- **1918** Turkestan Autonomous Soviet Socialist Republic (ASSR), incorporating present-day Uzbekistan, proclaimed.
- **1923–1941** Alphabet changes, from Arabic to Latin to Iranized Tashkent, finally to Cyrillic.
- **1924** Basmachi rebels who resisted Soviet rule crushed. Uzbek SSR founded (which, until 1929, included the Tajik ASSR).
- **1925** Anti-Islamic campaign bans schools and closes mosques.
- **1936** Karakalpak ASSR (formerly part of the Russian Soviet Federative Socialist Republic) incorporated into the Uzbek SSR.
- **1937** Uzbek communist leadership is purged by Stalin.

U

CHRONOLOGY *continued*

- **1941–1945** Industrial boom.
- **1959** Sharaf Rashidov becomes first secretary of Communist Party of Uzbekistan (CPUz). Retains position until 1983.
- **1966** Tashkent razed by earthquake. Rebuilding brings in large number of Russian and other non-Uzbek migrants.
- **1982–1983** Yuri Andropov becomes leader in Moscow. His anticorruption purge results in emergence of a new generation of central Asian officials.
- **1989** First noncommunist political movement, Unity (Birlik), formed but not officially registered. June, clashes erupt between Meskhetian Turks and indigenous Uzbek population of Fergana Valley resulting in more than 100 deaths. October, Birlik campaign leads to Uzbek being declared the official language.
- **1990** Islam Karimov becomes executive president of the new Uzbek Supreme Soviet. Further interethnic fighting in Fergana Valley; 320 killed.
- **1991** August, independence is proclaimed and Republic of Uzbekistan is adopted as official name. October, Uzbekistan signs treaty establishing economic community with seven other former Soviet republics. November, CPUz restructured as the People's Democratic Party of Uzbekistan (PDP); Karimov remains its leader. December, Karimov confirmed in post of president. Uzbekistan joins the CIS.
- **1992** Price liberalization provokes student riots in Tashkent. New post-Soviet constitution adopted along Western democratic lines. All religious parties banned. September, Uzbekistan sends troops to Tajikistan to suppress violence and strengthen border controls.
- **1993** Growing harassment of opposition political parties, Erk and Birlik.
- **1994** Introduction of som.
- **1995** January, Karimov's PDP wins legislative elections. March, referendum extends Karimov's presidential term until 2000. December, Utkur Sultanov replaces Abdulashim Mutalov as prime minister.
- **1999** Bomb attacks by Islamic terrorists lead to crackdown and arrests of hundreds of opposition activists. Legislative elections.
- **2000** Karimov reelected.
- **2002** Referendum extends president's term to seven years.

U

AID

Recipient

 $186m (receipts)  Up 20% in 2000

Ukraine's human rights violations and lack of commitment to economic stabilization have generally deterred bilateral aid donors. The World Bank is the largest donor, allocating millions of dollars to support economic reform and a range of agricultural, financial, and social projects.

DEFENSE

Compulsory military service

 $1.48bn  Up 20% in 2000

Uzbekistan has a large standing army, mainly conscripts, as well as over 17,000 internal security troops and a 1000-strong National Guard. A military restructuring is under way, with a view to full professionalization of the armed forces.

A policy of mining the land borders with Kyrgyzstan and Uzbekistan to thwart incursions by Islamic militants has drawn international criticism.

UZBEK ARMED FORCES

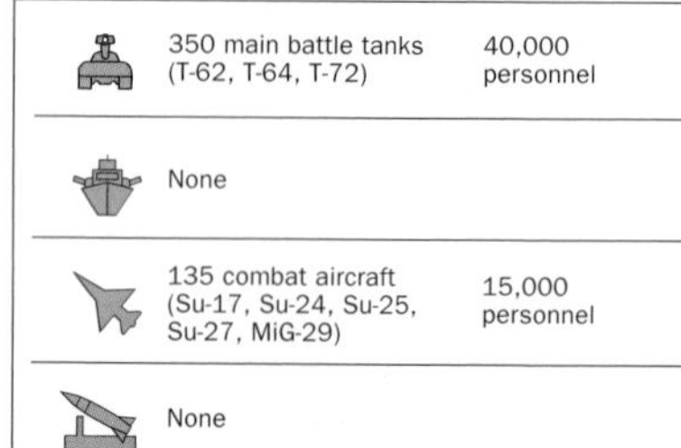
350 main battle tanks (T-62, T-64, T-72)
40,000 personnel
None
135 combat aircraft (Su-17, Su-24, Su-25, Su-27, MiG-29)
15,000 personnel
None

ECONOMICS

Inflation 247% p.a. (1990–2000)

 $8.84bn  322.8–686.9 som

SCORE CARD

World GNP Ranking	89th
GNP per Capita	$360
Balance of Payments	$184m
Inflation	40%
Unemployment	10%

EXPORTS

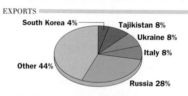

South Korea 4%
Tajikistan 8%
Ukraine 8%
Italy 8%
Other 44%
Russia 28%

IMPORTS

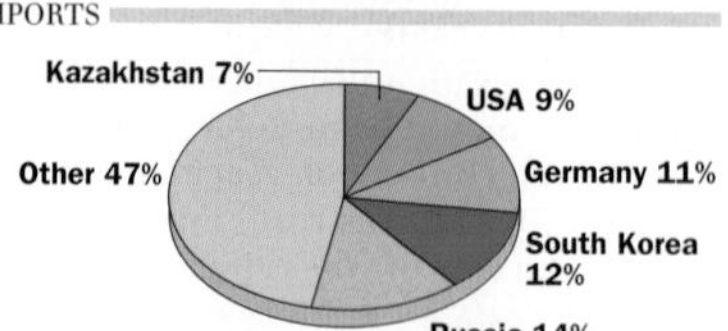
Kazakhstan 7%
USA 9%
Other 47%
Germany 11%
South Korea 12%
Russia 14%

ECONOMIC PERFORMANCE INDICATOR

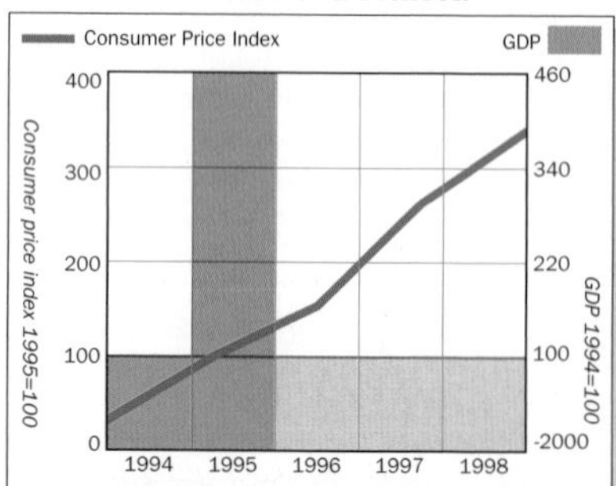
Consumer Price Index
GDP
Consumer price index 1995=100
GDP 1994=100
400
300
200
100
0
460
340
220
100
-2000
1994
1995
1996
1997
1998

Strengths
Gold. Well-developed cotton market. Considerable unexploited deposits of oil and natural gas. Current production of natural gas makes significant contribution to electricity generation. Manufacturing tradition includes agricultural machinery and central Asia's only aviation factory.

Weaknesses
Dependent on grain imports, as domestic production meets only 25% of needs. Very limited economic reform. High inflation. Environmentally damaging irrigation scheme for cotton production.

Profile
With the exception of Tashkent, which became an industrial area during World War II, Uzbekistan's economy is predominantly agricultural. Pro-market reforms have been slow, despite fresh assistance from the World Bank to increase the efficiency of privatized companies. The gold sector has attracted investment from US companies. Energy resources are still to be fully exploited. The som was devalued by 50% in 2001.

UZBEKISTAN : MAJOR BUSINESSES

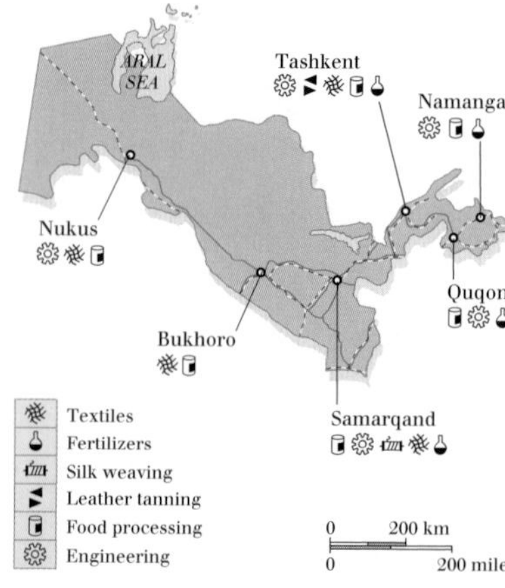
ARAL SEA
Tashkent
Namangan
Nukus
Quqon
Bukhoro
Samarqand
Textiles
Fertilizers
Silk weaving
Leather tanning
Food processing
Engineering
0 200 km
0 200 miles

RESOURCES

Electric power 11.7m kw

 8536 tonnes

 172,000 b/d (reserves 600m barrels)

 8.1m sheep, 5.34m cattle, 14.4m chickens

 Natural gas, coal, oil, gold, uranium, copper, tungsten, aluminum

ELECTRICITY GENERATION

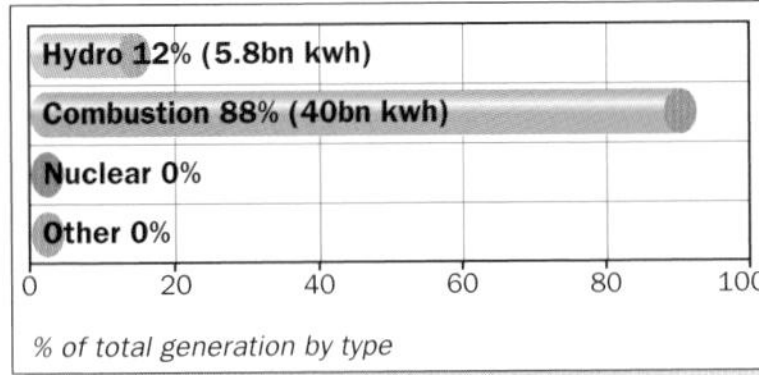
Hydro 12% (5.8bn kwh)
Combustion 88% (40bn kwh)
Nuclear 0%
Other 0%
0
20
40
60
80
100
% of total generation by type

UZBEKISTAN : LAND USE

ARAL SEA

KYZYL KUM

Cropland

Pasture

Forest

High mountain regions

Desert

Wetlands

Sheep

Cotton - cash crop

0 200 km

0 200 miles

Uzbekistan has the world's largest gold mine, at Murantau in the Kyzyl Kum Desert, and also large deposits of natural gas, petroleum, coal, and uranium. An important oil field was discovered in 1992 in the Namangan region and production will rise with further investment. Most gas produced is currently used domestically, but it could also become a strong export.

Cotton is the main focus of agriculture: Uzbekistan is the world's fourth-largest producer. A decision after independence to diversify was reversed when the value of cotton as a commodity on the world market became clear. Fruit, silk cocoons, and vegetables for Russian markets are also of rising importance.

ENVIRONMENT

Sustainability rank: 118th

 2%

 4.5 tonnes per capita

ENVIRONMENTAL TREATIES

Yes
Yes
Yes
Yes
Yes
No

The irrigation schemes required to sustain the cotton industry have wreaked considerable environmental damage. Soil salination is now a major problem. The Aral Sea has also been seriously depleted. From 61,836 sq. km (23,875 sq. miles) in 1974, it had shrunk to less than half that area by 1997. In 1998 the World Bank approved more than $11 million to save the Aral Sea region. The indiscriminate use of fertilizers and pesticides to raise production has polluted many rivers.

MEDIA

TV ownership high

 Daily newspaper circulation 3 per 1000 people

PUBLISHING AND BROADCAST MEDIA

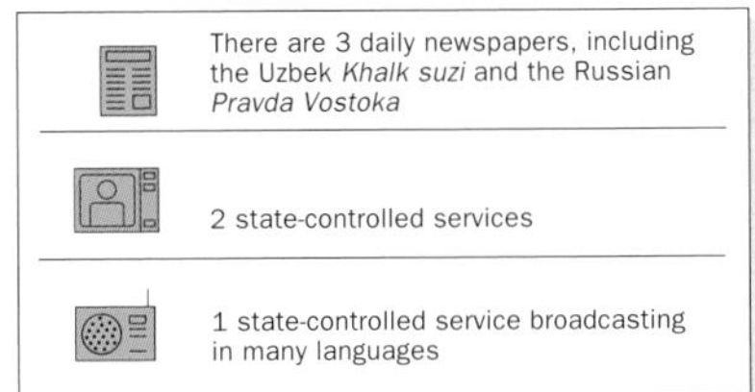
There are 3 daily newspapers, including the Uzbek *Khalk suzi* and the Russian *Pravda Vostoka*
2 state-controlled services
1 state-controlled service broadcasting in many languages

Uzbekistan's restrictions on independent publications are designed to encourage the promotion of the personality cult and policies of Karimov, and manifest themselves both in overt censorship and self-censorship by media outlets. Independent journalists face harassment. The expression of Islamic and nationalist opinion is forbidden.

CRIME

Death penalty in use

 63,900 prisoners

 Up 9% in 1999

CRIME RATES

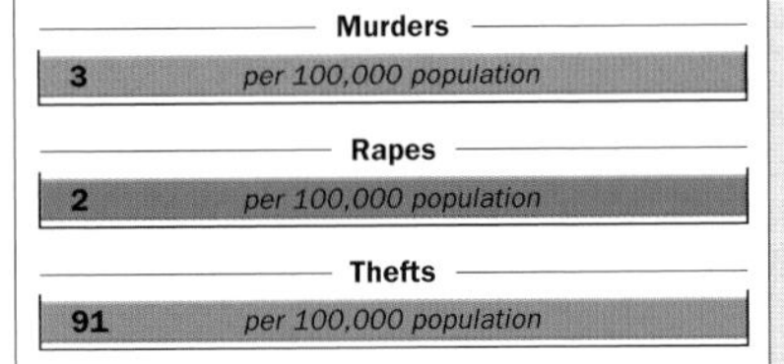
Murders
3 per 100,000 population
Rapes
2 per 100,000 population
Thefts
91 per 100,000 population

Crime has risen as living standards have declined. In 2001 the number of crimes carrying the death penalty was reduced to four. Unofficial Islamic courts in the Fergana Valley are a sign of opposition to the government.

EDUCATION

School leaving age: 18

 89%

 638,200 students

THE EDUCATION SYSTEM

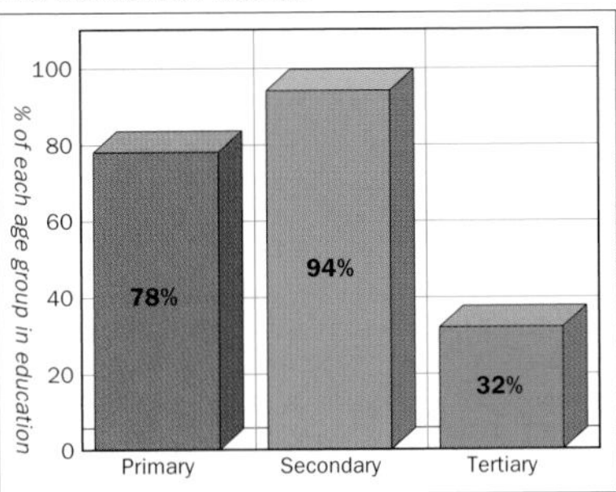
100
80
60
40
20
0
% of each age group in education
78%
94%
32%
Primary
Secondary
Tertiary

The state system still follows the Soviet model, though some instruction is in Uzbek. In the late 1980s, there were a few ethnic Tajik schools and a university in Samarqand. These were closed down in 1992 as relations deteriorated between Uzbekistan and Tajikistan. The rise in Islamic consciousness has led to a growing number of *madaris* – schools attached to mosques. In 1999 the establishment of Tashkent Islamic University was agreed.

HEALTH

Welfare state health benefits

 1 per 323 people

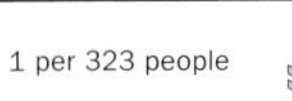 Circulatory and respiratory diseases, accidents, cancers

The health service has been in decline since the dissolution of the USSR. Some rural areas are not served at all. In 1998 a $69.7 million project to improve health services was announced, with the World Bank providing a loan of some $30 million. Serious respiratory diseases among cotton growers are increasing.

SPENDING

GDP/cap. decrease

CONSUMPTION AND SPENDING

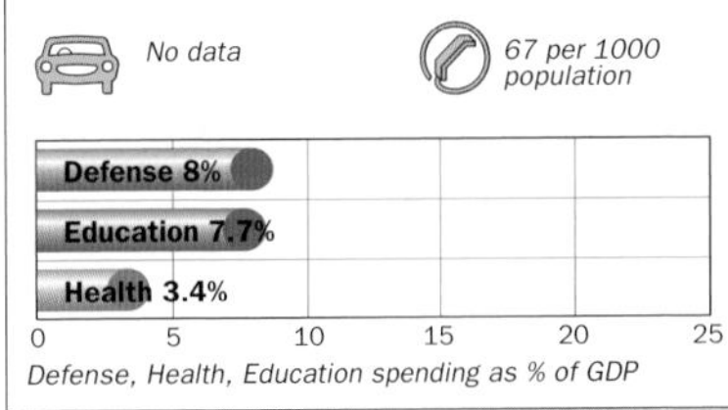
No data
67 per 1000 population
Defense 8%
Education 7.7%
Health 3.4%
0
5
10
15
20
25
Defense, Health, Education spending as % of GDP

Former communists are still the wealthiest group, since they retain control of the economy. Many rural poor live below the poverty line.

WORLD RANKING

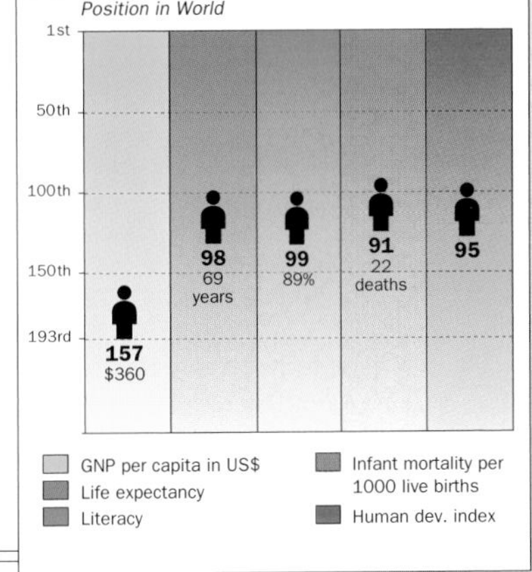
Position in World
1st
50th
100th
150th
193rd
157
$360
98
69 years
99
89%
91
22 deaths
95
GNP per capita in US$
Life expectancy
Literacy
Infant mortality per 1000 live births
Human dev. index

U

VANUATU

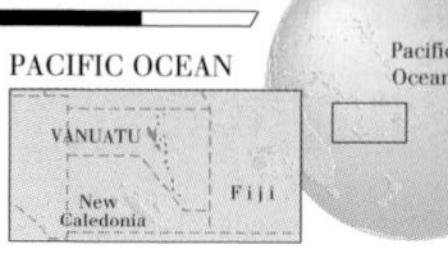
PACIFIC OCEAN
VANUATU
New Caledonia
Fiji
Pacific Ocean

OFFICIAL NAME: Republic of Vanuatu **CAPITAL:** Port Vila
POPULATION: 200,000 **CURRENCY:** Vatu **OFFICIAL LANGUAGES:** Bislama, English, and French

1980

1980

July 30
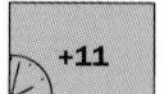
VU

+11

+678
.vu

THE ARCHIPELAGO OF Vanuatu stretches over 1300 km (800 miles) in the South Pacific. Mountainous and volcanic in origin, only 12 of the 82 islands are a significant size – Espiritu Santo and Malekula are the largest. Formerly known as the New Hebrides – ruled jointly by France and Britain from 1906 – Vanuatu became independent in 1980. Politics since then has been democratic but volatile.

CLIMATE

Tropical oceanic

WEATHER CHART FOR PORT VILA

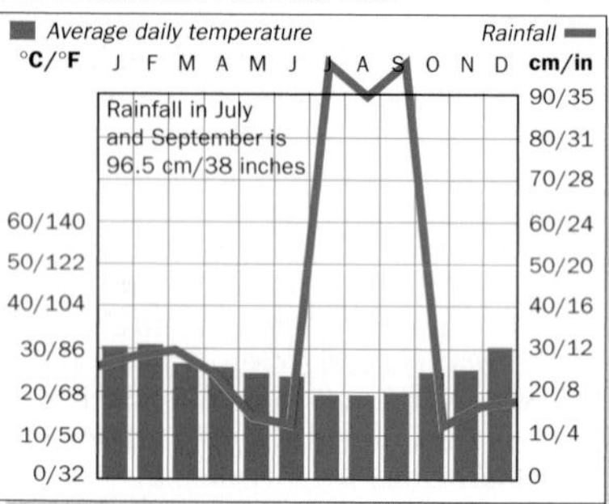
Average daily temperature
Rainfall
°C/°F
J F M A M J J A S O N D
cm/in
Rainfall in July and September is 96.5 cm/38 inches
60/140
50/122
40/104
30/86
20/68
10/50
0/32
90/35
80/31
70/28
60/24
50/20
40/16
30/12
20/8
10/4
0

The climate is tropical and hot. Rainfall and temperatures decrease north to south. Cyclones occur November–April.

TRANSPORTATION

Drive on right

Bauerfield, Port Vila

293 ships
1.6m grt

THE TRANSPORTATION NETWORK

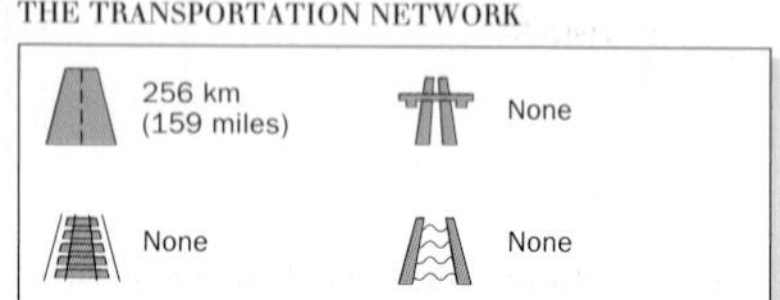
256 km (159 miles)
None
None
None

Road quality is generally poor, with routes on some remote islands impassable in the wet season.

TOURISM

Visitors : Population 1:3.5

57,000 visitors

Up 14% in 2000

MAIN TOURIST ARRIVALS

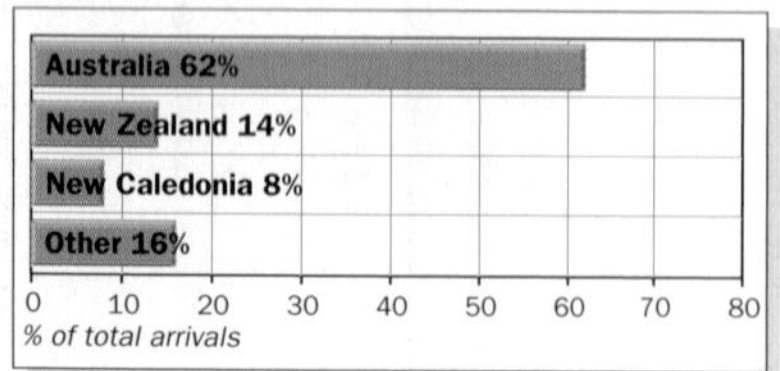
Australia 62%
New Zealand 14%
New Caledonia 8%
Other 16%
0 10 20 30 40 50 60 70 80
% of total arrivals

Tourism in Vanuatu is facing stiff competition from cheaper regional rivals. Organized tours include sea fishing, sailing, kayaking, and diving.

PEOPLE

Pop. density low

Bislama (Melanesian pidgin), English, French

16/km² (42/mi²)

THE URBAN/RURAL POPULATION SPLIT

20% 80%

RELIGIOUS PERSUASION

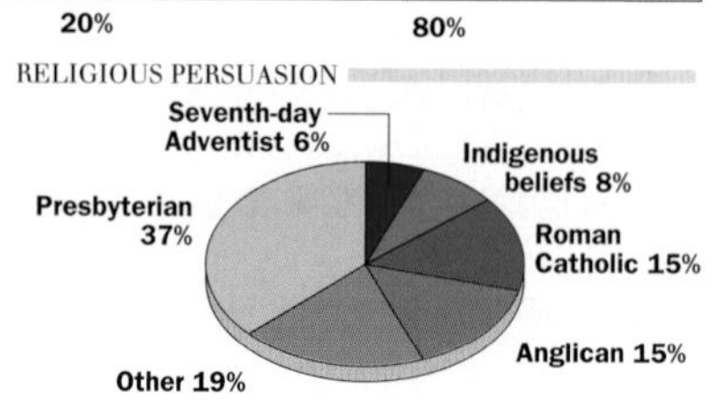
Seventh-day Adventist 6%
Indigenous beliefs 8%
Roman Catholic 15%
Anglican 15%
Other 19%
Presbyterian 37%

Indigenous Melanesians – ni-Vanuatu – comprise 94% of the population. Of Vanuatu's 82 islands, 67 are inhabited, but 80% of people live on 12 main islands. The population is becoming more urbanized as one in eight ni-Vanuatu now lives in Port Vila. However, 75% of the population still live by subsistence agriculture.

Vanuatu is home to some of the Pacific's most traditional peoples, and local social and religious customs are strong. With 105 indigenous languages, Vanuatu boasts the world's highest per capita density of languages. Bislama pidgin is the lingua franca.

Women have lower social status than men, and bride price is still commonly paid. Many educated women refuse to marry because of loss of property rights. To boost equality, primary schools are encouraged to take girls.

Vanuatu's unspoilt beaches *are one of the reasons for the upsurge in the tourist industry.*

POLITICS

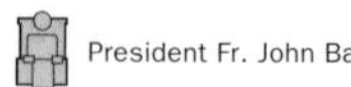
Multiparty elections

2002/2006

President Fr. John Bani

AT THE LAST ELECTION
Parliament 52 seats

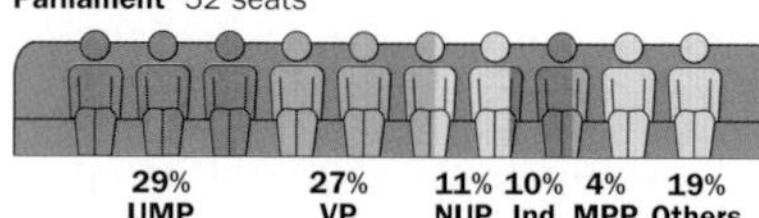
29% UMP
27% VP
11% NUP
10% Ind
4% MPP
19% Others

UMP = Union of Moderate Parties **VP** = Vanua'aku Pati
NUP = National United Party **Ind** = Independents
MPP = Melanesian Progressive Party

Ni-Vanuatu politics is best described as anarchic. Political allegiances are swiftly changed and governments often toppled. Instability was behind the reluctance of France – not shared by Britain – to grant independence, achieved in 1980, and led to alleged French support, reinforced by the anti-French stance of the VP, for an abortive bid for independent status by Espiritu Santo that year.

The VP ruled under the charismatic Fr. Walter Lini from 1980 to 1991, when he was dismissed as party leader and premier. He defected and formed the NUP, and continued to be a dominant figure until his death in 1999. The UMP headed a succession of coalitions from 1991 until 1998 when the VP returned to power. Its government was based on patched-up relations with Lini, but within months it was back in opposition, unable to command a stable majority. In 2001 a broad-based anti-VP coalition was ousted, bringing back the VP, under Edward Natapei, in coalition with the UMP. Natapei continued to head the coalition after the 2002 polls, despite the UMP's marginal victory.

WORLD AFFAIRS

Joined UN in 1981

OIF

Comm

NAM
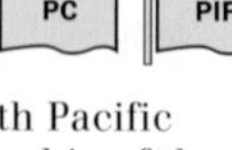
PC
PIF

Vanuatu was the first South Pacific nation to gain full membership of the Non-Aligned Movement. It was censured by the OECD in 2002 for failing to improve financial transparency.

AID

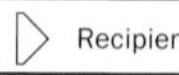
Recipient

$46m (receipts)

Up 24% in 2000

Vanuatu is heavily dependent on aid. Leading donors are Australia, the ADB, the World Bank, France, Japan, New Zealand, and the EU. Vanuatu is classed by the UN as a least developed country (LDC), and as such receives guaranteed aid support.

V

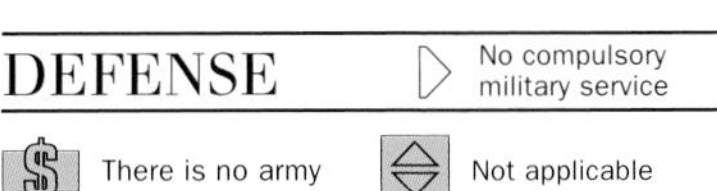

DEFENSE

No compulsory military service

There is no army

Not applicable

There is no army. A small paramilitary force receives training from the US. Papua New Guinean troops helped to end the 1980 secessionist movement on Espiritu Santo.

ECONOMICS

Inflation 2.1% p.a. (1990–2000)

$226m

143.01–146.80 vatu

SCORE CARD

❑ WORLD GNP RANKING	182nd
❑ GNP PER CAPITA	$1150
❑ BALANCE OF PAYMENTS	–$3m
❑ INFLATION	2%
❑ UNEMPLOYMENT	Low

STRENGTHS

Expanding services sector, including tourism and offshore finance. Major economic reforms instituted, including the introduction of a value-added tax and resizing of the public service in return for assistance from the ADB. Large fishing potential.

WEAKNESSES

Large trade and budget deficits. Growth stagnant in recent years and negative in 1999. Dependence on agricultural sector, vulnerable to adverse weather and fluctuating market prices. Shortage of skilled indigenous labor.

EXPORTS

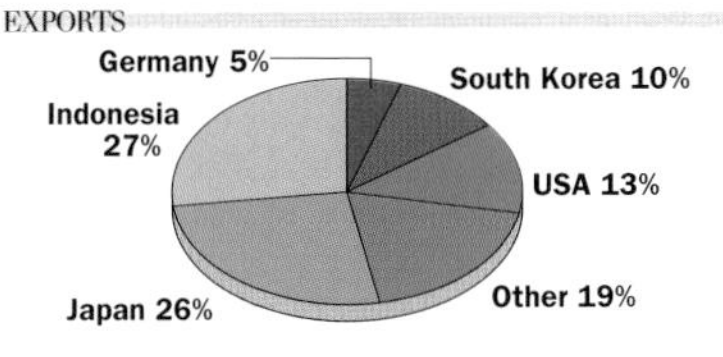
Germany 5%
South Korea 10%
Indonesia 27%
USA 13%
Other 19%
Japan 26%

IMPORTS

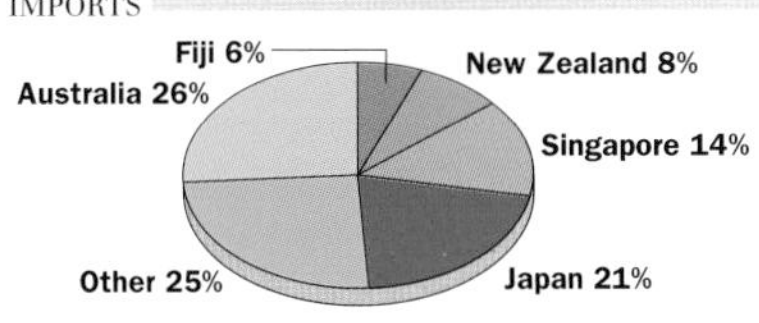
Fiji 6%
New Zealand 8%
Australia 26%
Singapore 14%
Other 25%
Japan 21%

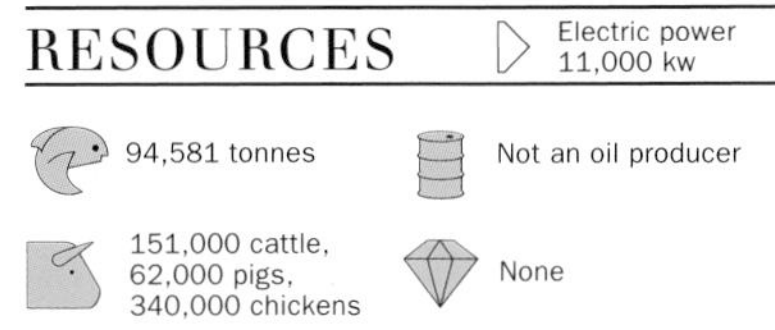

RESOURCES

Electric power 11,000 kw

94,581 tonnes

Not an oil producer

151,000 cattle, 62,000 pigs, 340,000 chickens

None

Vanuatu's main resources are its arable land – only partly utilized – and its forests and waters. These could be exploited by the tourist, timber, and fishing industries. New export crops are being explored to offset declining copra and cocoa exports. Beef is of growing importance. Nuclear power development was banned under 1983 legislation.

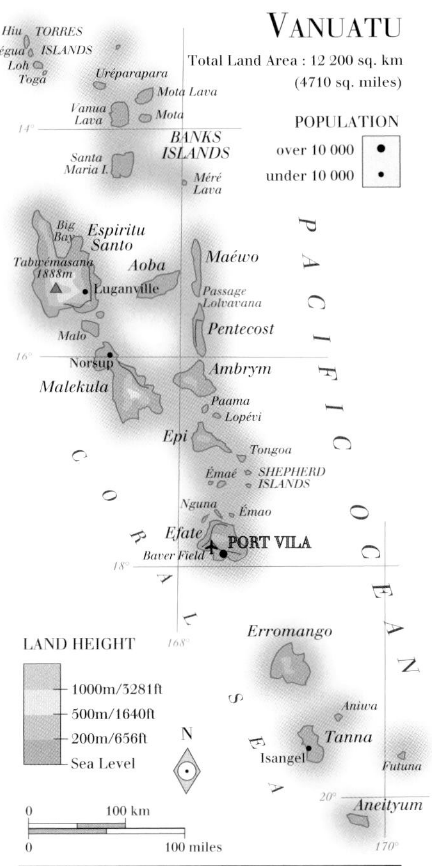
VANUATU
Total Land Area : 12 200 sq. km (4710 sq. miles)
POPULATION
over 10 000
under 10 000
Hiu
TORRES ISLANDS
Tégua
Loh
Toga
Uréparapara
Mota Lava
Vanua Lava
Mota
BANKS ISLANDS
Santa Maria I.
Méré Lava
Big Bay
Espiritu Santo
Tabwémasana 1888m
Luganville
Aoba
Maéwo
Passage Lolvavana
Pentecost
Malo
Norsup
Malekula
Ambrym
Paama
Lopévi
Epi
Tongoa
Émaé
SHEPHERD ISLANDS
Nguna
Émao
Efate
PORT VILA
Bauer Field
PACIFIC OCEAN
CORAL SEA
Erromango
Aniwa
Tanna
Isangel
Futuna
Aneityum
LAND HEIGHT
1000m/3281ft
500m/1640ft
200m/656ft
Sea Level
N
0 100 km
0 100 miles

ENVIRONMENT

Not available

None

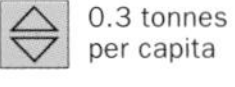

0.3 tonnes per capita

Logging is increasing, but most of the rainforest remains intact, and round wood exports are banned.

Population growth is high, at nearly 3% a year, but is falling. A majority of the population does not have access to a potable and reliable water supply.

MEDIA

TV ownership low

Daily newspaper circulation is very low

PUBLISHING AND BROADCAST MEDIA

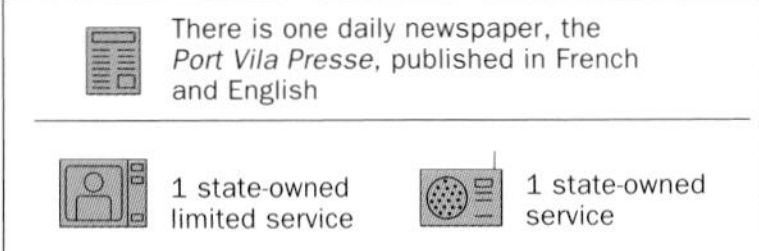

There is one daily newspaper, the *Port Vila Presse*, published in French and English

1 state-owned limited service

1 state-owned service

The *Vanuatu Weekly* appears in each official language. Television Blong Vanuatu broadcasts four hours a day.

CRIME

No death penalty

46 prisoners

Little change from year to year

Domestic violence apart, Vanuatu is almost crime-free, with little alcohol-related violence.

CHRONOLOGY

In 1906, Britain and France set up the New Hebrides under joint rule.

- ❑ **1980** Independence; Walter Lini of VP prime minister. Secession bid by Espiritu Santo.
- ❑ **1991** UMP coalition with NUP set up by Lini after his expulsion from VP.
- ❑ **1999** Tidal wave causes extensive damage. Lini dies.
- ❑ **2001** VP–UMP coalition; Edward Natapei prime minister.

EDUCATION

Schooling is not compulsory

64%

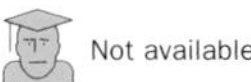

Not available

The abolition of fees has helped to boost primary enrollment. Illiteracy is a major concern.

HEALTH

Welfare state health benefits

1 per 8333 people

Heart diseases, cancers, malaria

A network of rural clinics and village health workers has helped to improve health levels. Nominal fees are charged.

SPENDING

GDP/cap. decrease

CONSUMPTION AND SPENDING

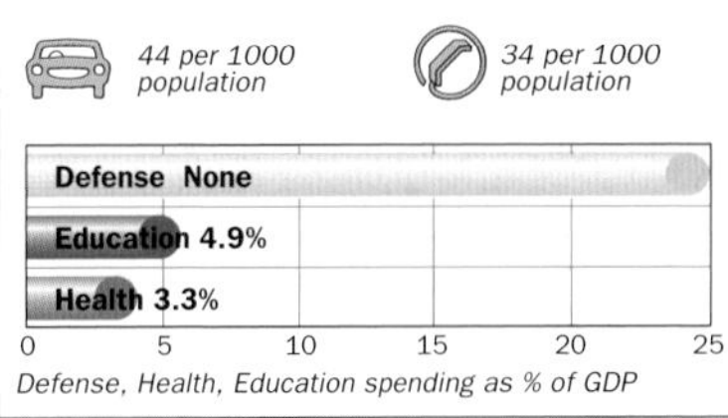
44 per 1000 population
34 per 1000 population
Defense None
Education 4.9%
Health 3.3%
0
5
10
15
20
25
Defense, Health, Education spending as % of GDP

The dominance of subsistence farming and small-scale cash cropping has helped to prevent extreme poverty. Most of the rich are not ni-Vanuatu.

WORLD RANKING

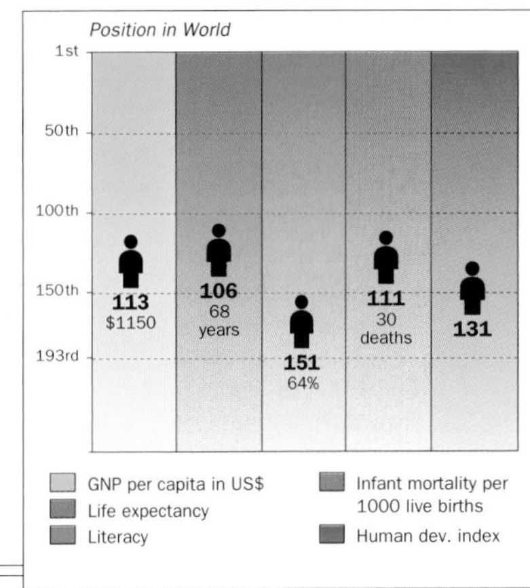
Position in World
1st
50th
100th
150th
193rd
113
$1150
106
68 years
151
64%
111
30 deaths
131
GNP per capita in US$
Life expectancy
Literacy
Infant mortality per 1000 live births
Human dev. index

V

VATICAN CITY

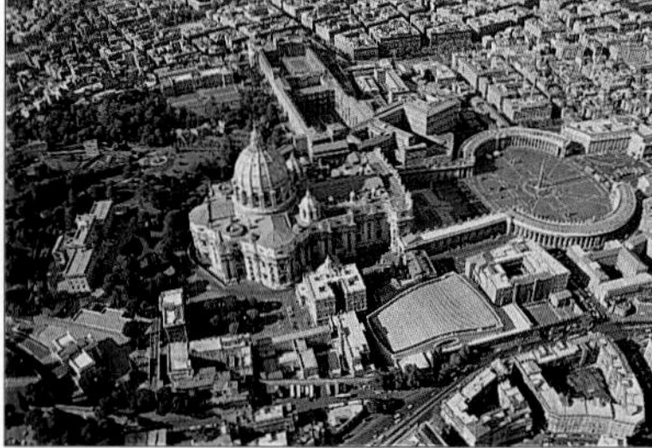
EUROPE
Europe
Italy
VATICAN CITY

OFFICIAL NAME: State of the Vatican City **CAPITAL:** Vatican City **POPULATION:** 524
CURRENCY: Euro (lira until 2002) **OFFICIAL LANGUAGES:** Italian and Latin

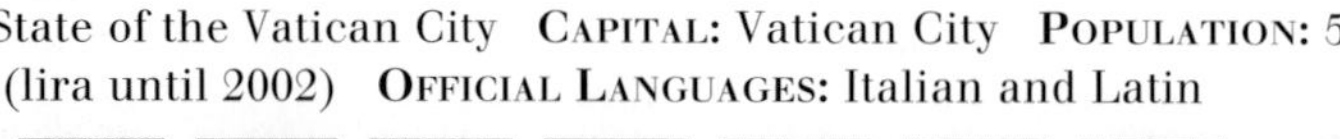

1929

1929

Oct 22
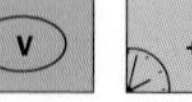
V

+1

+39
.va

THE VATICAN CITY lies close to the Tiber in central Rome and is the world's smallest independent state. It also includes ten other buildings in Rome and the pope's residence at Castel Gandolfo. As the Holy See it is the seat of the Roman Catholic Church, deriving its income from investments and voluntary contributions known as Peter's Pence.

The buildings and gardens of the Vatican City. St. Peter's Basilica was built from 1506–1626 on the traditional site of St. Peter's tomb.

CLIMATE

Mediterranean

WEATHER CHART FOR THE VATICAN CITY

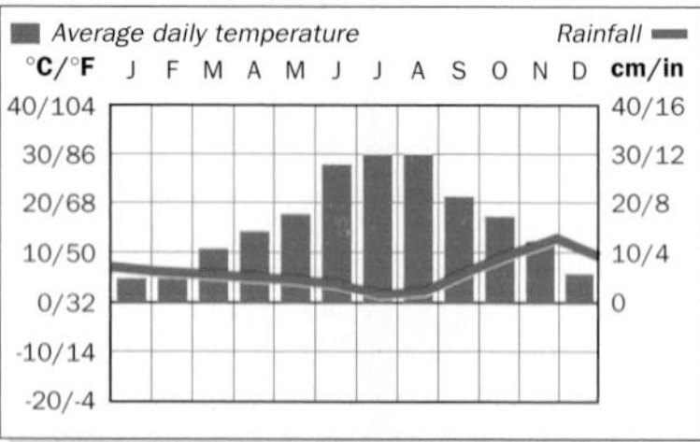
Average daily temperature
Rainfall
°C/°F J F M A M J J A S O N D cm/in
40/104
30/86
20/68
10/50
0/32
-10/14
-20/-4
40/16
30/12
20/8
10/4
0

Summers are hot; winters are mild. November is particularly rainy.

TRANSPORTATION

Drive on right

Heliport for official visitors
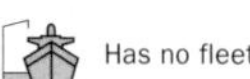
Has no fleet

THE TRANSPORTATION NETWORK

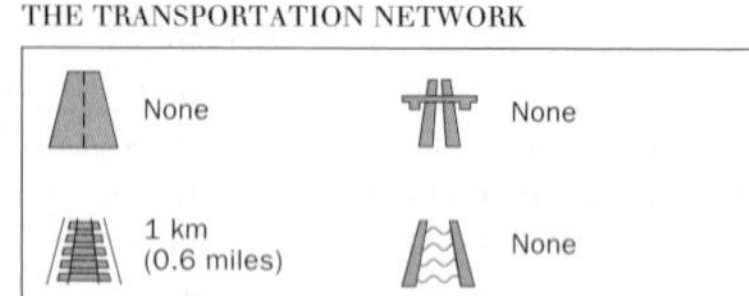
None
None
1 km
(0.6 miles)
None

The railroad is only used for carrying freight. Official visitors are transferred from Rome airport by helicopter.

TOURISM

Not available

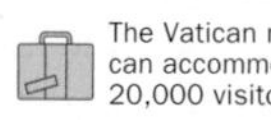
The Vatican museums can accommodate 20,000 visitors daily
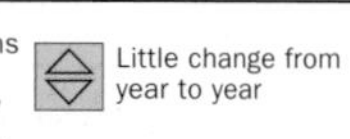
Little change from year to year

MAIN TOURIST ARRIVALS

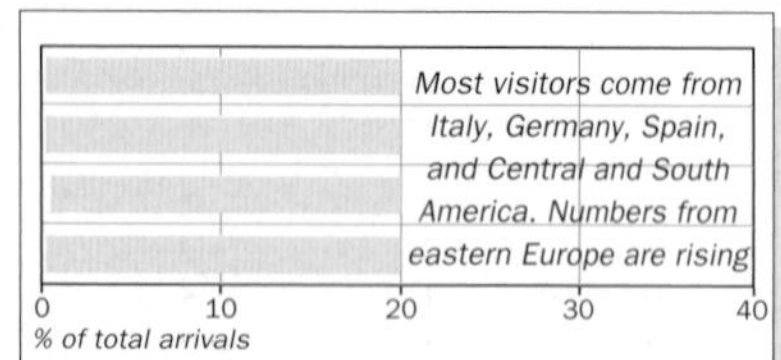
Most visitors come from Italy, Germany, Spain, and Central and South America. Numbers from eastern Europe are rising
0
10
20
30
40
% of total arrivals

Almost all tourists who visit Rome go to the Vatican, while others come as pilgrims. Up to 100,000 hear the pope's annual Easter Message in St. Peter's Square. The Vatican's art collections are among the greatest in the world. Years of restoration work on the Sistine Chapel frescoes were completed in 1999.

PEOPLE

Pop. density high

Italian, Latin

1191/km² (3082/mi²)

THE URBAN/RURAL POPULATION SPLIT

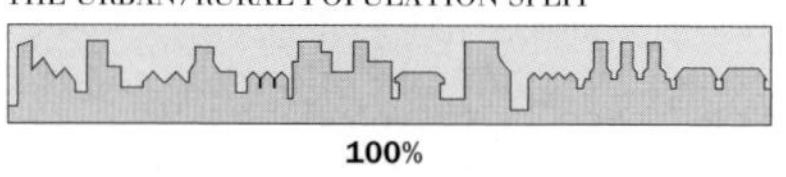

100%

RELIGIOUS PERSUASION

Roman Catholic 100%

The Vatican has about 500 citizens, including over 100 lay persons. Several hundred more lay staff are employed in the city-state. Citizenship can be acquired through stable residence and holding an office or job within the City. A citizen's family can gain residence only by authorization.

The pope is the spiritual leader of around 17% of the world's population. The countries with the largest number of Roman Catholics are Brazil, Mexico, Italy, the US, and the Philippines.

POLITICS

No legislative elections

On death of reigning pope

His Holiness Pope John Paul II

LEGISLATIVE OR ADVISORY BODIES

College of Cardinals 120 seats

Cardinals under the age of 80 are eligible to elect a new pope. There are no political parties.

The Vatican City operates in the manner of an elected monarchy, as the reigning pope has supreme executive, legislative, and judicial powers, and holds office for life. He is elected by the College of Cardinals, who vote until one candidate for the position of Supreme Pontiff achieves a two-thirds majority.

The administration of the Vatican City State, of which the pope is temporal head, is conducted by the Pontifical Commission. The Holy See, which is the governing body of the Roman Catholic Church worldwide and of which the pope is spiritual head, is governed by the Roman Curia, the Church's administrative network. It is the Holy See that maintains diplomatic relations abroad. Pope John Paul II, elected in 1978, was the first non-Italian pope since 1523. Now in his 80s, he continues to fulfill his duties despite suffering from Parkinson's disease and arthritis.

VATICAN CITY

Total Land Area : 0.44 sq. km (0.17 sq. miles)

Main Entrance
Pigna Courtyard
Vatican Museums
Vatican Gardens
Monte Vaticano
ROME
Radio Vatican
Belvedere Courtyard
Raphael Stanza
Sistine Chapel
Papal apartments
Papal Heliport
Saint Peter's Basilica
St. Peter's Square
Vatican Railway Station
ROME
0 200 metres
0 800 feet
N

WORLD AFFAIRS

Not a member of the UN

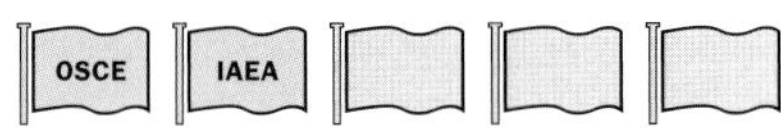
OSCE
IAEA

The Vatican maintains a neutral stance in world affairs and has observer status in many international organizations. It has mediated in many conflicts, notably achieving the 1993 peace agreement in Mozambique. Pope John Paul II has traveled more extensively than any other pope. His main aims have been to promote political dialogue and bridge religious divides, both within and outside the Catholic Church. In 2000, while on a major tour of the Holy Land, he made an unprecedented apology for the Church's 2000 years of anti-Judaism. In May 2001 he became the first pope to enter (and pray in) a mosque – in Syria – and the first to visit Orthodox Greece.

AID

Donor

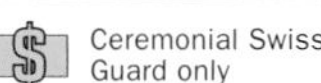 Undisclosed

 Undisclosed

Aid is donated through the pope's charities (such as the Holy Childhood Association, which distributes around $15 million a year to children's causes), through funds donated for use at the pope's discretion, and through religious orders acting under papal charter.

DEFENSE

No compulsory military service

Ceremonial Swiss Guard only

Not applicable

The Vatican is strictly neutral territory. Under the 1954 Hague Convention, it is recognized as "a moral, artistic, and cultural patrimony worthy of being respected as a treasure for all mankind."

ECONOMICS

Not applicable

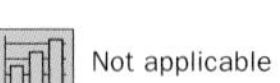 Not applicable

 1.0651–1.1231 euros

SCORE CARD

- ❑ WORLD GNP RANKING ....*The Vatican*
- ❑ GNP PER CAPITA......................*does not have*
- ❑ BALANCE OF PAYMENTS....................*a national*
- ❑ INFLATION ....*economy in the*
- ❑ UNEMPLOYMENT..........................*usual sense*

STRENGTHS

Istituto per le Opere di Religione has assets of $3–4 billion. Voluntary contributions from Catholics worldwide (Peter's Pence). Interest on investments. Gold reserves in Fort Knox, US. Stamp and coin issues. Budgetary deficit of 23 years reversed to modest surplus in 1993. Receipts from tourists.

WEAKNESSES

Losses incurred by Radio Vaticana and *L'Osservatore Romano*. Cost of foreign papal visits, buildings maintenance, and diplomatic missions.

EXPORTS/IMPORTS

The Vatican produces no goods for export. All commodities are imported, mainly from Italy.

RESOURCES

Electric power: None

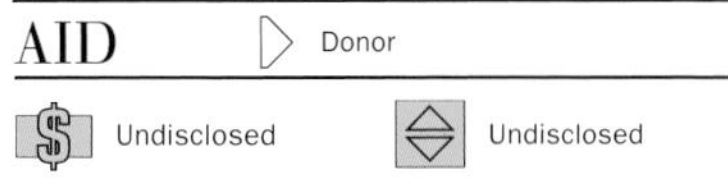

 None

None

 None

None

The Vatican imports all its energy. It has no farmland; its area is restricted to buildings and their formal gardens.

ENVIRONMENT

Not available

 None

Not available

The Vatican is increasingly concerned about the need to balance development and conservation. In 1993, the pope urged a gathering of scientists to press colleagues worldwide to inform people on the need to protect the environment.

MEDIA

TV ownership medium

 Daily newspaper circulation figures not available

PUBLISHING AND BROADCAST MEDIA

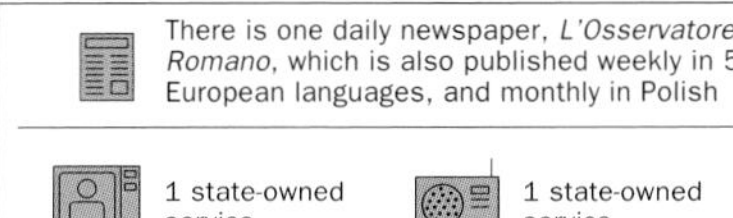

There is one daily newspaper, *L'Osservatore Romano*, which is also published weekly in 5 European languages, and monthly in Polish

1 state-owned service

1 state-owned service

Radio Vaticana's longwave broadcasts are continuing after a brief suspension caused by concern over radiation.

CRIME

No death penalty

 There are no prisons in the Vatican City

 Minimal crime levels

The reputation of the 105-strong Swiss Guard was shaken in 1998, when a young guard shot dead his commandant and the latter's wife and then committed suicide. Three Vatican Bank officials were earlier alleged to have been involved in the Banco Ambrosiano affair.

EDUCATION

Not applicable

 99%

 14,403 students

The university, founded by Gregory XIII, is renowned for its theological and philosophical learning. There are more than 110,000 primary and secondary Catholic schools around the world.

CHRONOLOGY

The Vatican is located in Rome because tradition held that St. Peter was buried on the site of the Church of Constantine, which was pulled down in the Renaissance to make way for the building of St. Peter's Basilica. The Vatican has been the pope's usual residence since 1417, when the pontiffs returned from Avignon in France at the end of the 39 years of Great Schism.

- ❑ **1870** Italy annexes Papal States.
- ❑ **1929** Lateran Treaty – Fascist Italy accepts Vatican City as independent state.
- ❑ **1978** Cardinal Karol Wojtyla pope.
- ❑ **1981–1982** Attempts on pope's life.
- ❑ **1984** Catholicism disestablished as Italian state religion.
- ❑ **1985** Catholic Catechism revised for first time since 1566.
- ❑ **1994–1995** Opposition to abortion and contraception reiterated at UN conferences in Cairo and Beijing.
- ❑ **1998** Statement repenting Catholic passivity during Nazi Holocaust.
- ❑ **2000** Jubilee Year. Papal apology for Catholic violence and oppression over two millennia.
- ❑ **2001** John Paul II becomes first pope to enter a mosque.

HEALTH

Welfare state health benefits

Pope's own doctor is in permanent residence at Vatican

Heart and cardiovascular diseases, cancers

Pope John Paul II's strong opposition to abortion and contraception has prompted criticism from around the world, and from within the Church.

SPENDING

Not applicable

CONSUMPTION AND SPENDING

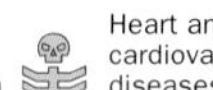
No data

No data

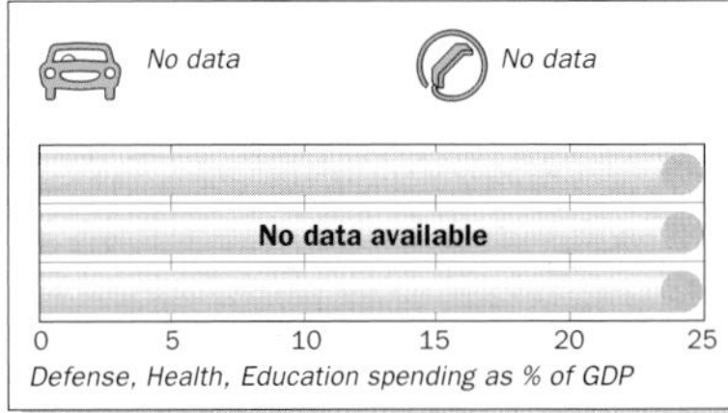
No data available
0
5
10
15
20
25

Defense, Health, Education spending as % of GDP

The wealth of the Vatican is primarily that of the Catholic Church. Its art treasures may not be sold. It is not known how much personal wealth its citizens have.

WORLD RANKING

The Pope and his Vatican staff enjoy one of the highest standards of living in the world

V

VENEZUELA

SOUTH AMERICA

VENEZUELA Guyana Colombia Brazil

South America

OFFICIAL NAME: Bolivarian Republic of Venezuela **CAPITAL:** Caracas
POPULATION: 24.6 million **CURRENCY:** Bolívar **OFFICIAL LANGUAGE:** Spanish

1830

1830

July 5

YV

-4

+58

.ve

LOCATED IN THE NORTH of South America, with a long Caribbean coastline, Venezuela has a vast central plain drained by the Orinoco, while the Guiana Highlands dominate the southwest of the country. A Spanish colony until 1811, Venezuela was lauded as Latin America's most stable democracy until its recent political upheavals. Although the country has some of the largest known oil deposits outside the Middle East, much of Venezuela's population still lives in shanty-town squalor.

President Hugo Chávez *changed the name of the country as part of his Bolivarian Revolution.*

CLIMATE

Tropical wet & dry/ equatorial

WEATHER CHART FOR CARACAS

Average daily temperature / Rainfall
°C/°F J F M A M J J A S O N D cm/in
40/104, 30/86, 20/68, 10/50, 0/32, -10/14, -20/4
40/16, 30/12, 20/8, 10/4, 0

The hot Maracaibo coast is surprisingly dry; the Orinoco Llanos are alternately parched or flooded. Uplands are cold.

TRANSPORTATION

Drive on right

Simón Bolívar International, Caracas 6.66m passengers

246 ships 665,300 grt

THE TRANSPORTATION NETWORK

32,308 km (20,075 miles)

2690 km (1671 miles)

336 km (209 miles)

7100 km (4412 miles)

Massive road building from the 1960s onward has benefited the oil and aluminum industries.The French-designed Caracas subway was completed in 1995. Work on the Centro-Occidental highway and other major roads continues, as does rail development.

The Orinoco. *Its huge Llanos (plains) are grazed by five million cattle, which are herded close to the river in the dry season.*

TOURISM

Visitors : Population 1:52

469,000 visitors

Down 20% in 2000

MAIN TOURIST ARRIVALS

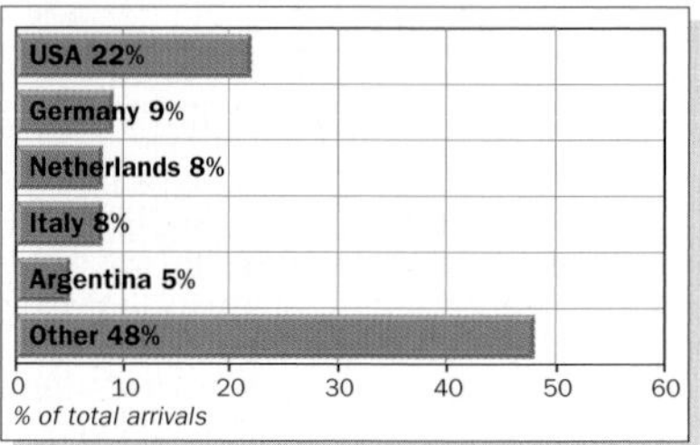
USA 22%
Germany 9%
Netherlands 8%
Italy 8%
Argentina 5%
Other 48%
0 10 20 30 40 50 60
% of total arrivals

Tourism is still a relatively minor industry in Venezuela, but one with enormous potential. Venezuela has many beaches that are the equal of any Caribbean island's, and a fascinating jungle interior which is a target for more adventurous tourists. For many years, the high value of the bolívar made Venezuela an expensive destination, but, after recent devaluations, it has become one of the cheapest in the Caribbean. Privatizing state-run hotels was part of a drive to attract foreign investment.

PEOPLE

Pop. density low

Spanish, Amerindian languages

28/km² (72/mi²)

THE URBAN/RURAL POPULATION SPLIT

87% 13%

RELIGIOUS PERSUASION

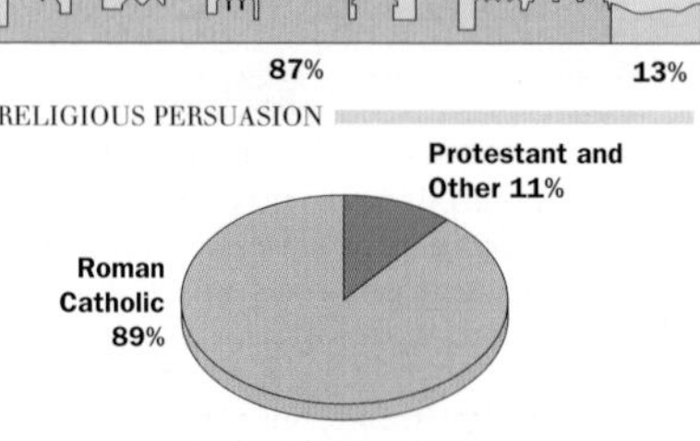
Protestant and Other 11%
Roman Catholic 89%

ETHNIC MAKEUP

Amerindian 2%
Black 9%
White 20%
Mestizo 69%

Venezuela is one of the most highly urbanized societies in Latin America, with most of its population living in cities, mainly in the north. A historic "melting pot," it has experienced large-scale immigration from Italy, Portugal, Spain, and all over Latin America. There remains little of the white Hispanic aristocracy that survives in Colombia and Ecuador. The small number of native Amerindians, such as the Yanomami and Pemón, live in remote regions now threatened by illegal settlers. Most of the black population, who are descended from Africans brought over to work in the cacao industry in the 19th century, live along the Caribbean coast.

Oil wealth has brought comparative prosperity, but life in the *barrios* (shanty towns) which sprawl over the hillsides around Caracas is one of extreme poverty. Discontent peaked in the food riots of 1989 and 1991, which left scores dead along with the country's reputation for being a model democracy. The oil boom accelerated change for women, who today are to be found in all the professions. Politics, however, remains a largely masculine preserve. Oil wealth has also brought a measure of Americanization – boxing and baseball are among the most popular sports.

POPULATION AGE BREAKDOWN

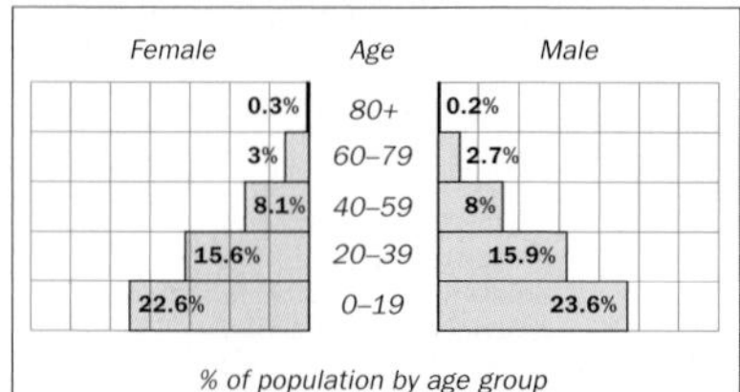
Female Age Male
0.3% 80+ 0.2%
3% 60–79 2.7%
8.1% 40–59 8%
15.6% 20–39 15.9%
22.6% 0–19 23.6%
% of population by age group

V

VENEZUELA

Total Land Area : 912 050 sq. km (352 143 sq. miles)

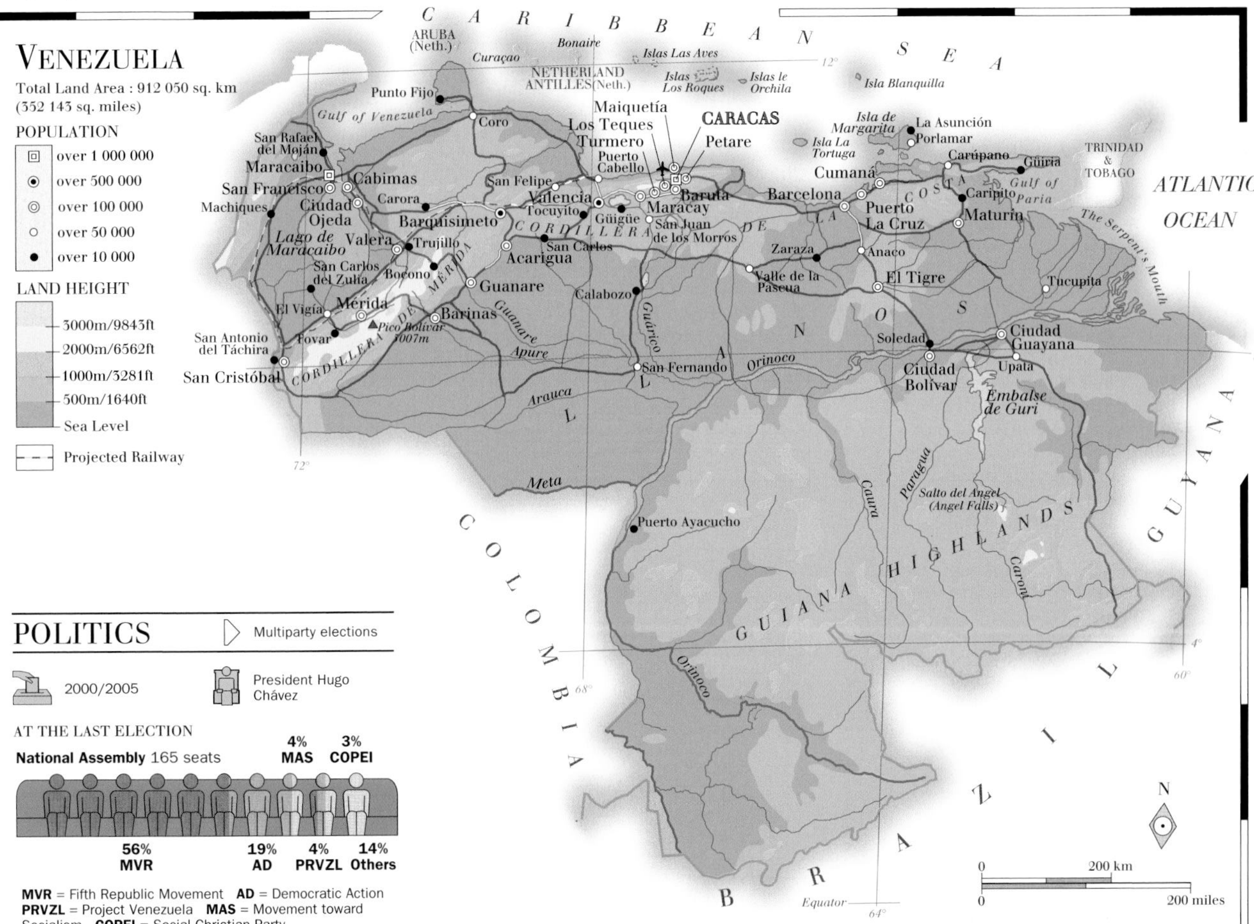
POPULATION
over 1 000 000
over 500 000
over 100 000
over 50 000
over 10 000
LAND HEIGHT
3000m/9843ft
2000m/6562ft
1000m/3281ft
500m/1640ft
Sea Level
Projected Railway
CARIBBEAN SEA
ARUBA (Neth.)
Curaçao
Bonaire
NETHERLAND ANTILLES (Neth.)
Islas Las Aves
Islas Los Roques
Islas le Orchila
Isla Blanquilla
Punto Fijo
Gulf of Venezuela
Coro
San Rafael del Moján
Maracaibo
San Francisco
Cabimas
Machiques
Ciudad Ojeda
Lago de Maracaibo
Carora
Valera
Trujillo
San Carlos del Zulia
Boconó
El Vigía
Mérida
Pico Bolívar 5007m
Tovar
San Antonio del Táchira
San Cristóbal
CORDILLERA DE MÉRIDA
Barinas
Guanare
Acarigua
Barquisimeto
San Felipe
Valencia
Tocuyito
Maiquetía
Los Teques
Turmero
Puerto Cabello
CARACAS
Petare
Maracay
Güigüe
Baruta
San Juan de los Morros
San Carlos
CORDILLERA DE LA COSTA
Calabozo
Guárico
Apure
Arauca
Meta
LLANOS
San Fernando
Orinoco
Isla de Margarita
La Asunción
Porlamar
Isla La Tortuga
Cumaná
Carúpano
Güiria
Barcelona
Puerto La Cruz
Caripito
Gulf of Paria
Maturín
TRINIDAD & TOBAGO
ATLANTIC OCEAN
The Serpent's Mouth
Zaraza
Anaco
Valle de la Pascua
El Tigre
Tucupita
Soledad
Ciudad Guayana
Ciudad Bolívar
Upata
Embalse de Guri
Caura
Paragua
Salto del Angel (Angel Falls)
Caroní
GUIANA HIGHLANDS
Puerto Ayacucho
COLOMBIA
BRAZIL
GUYANA
Equator
N
0 200 km
0 200 miles

POLITICS

Multiparty elections

2000/2005

President Hugo Chávez

AT THE LAST ELECTION

National Assembly 165 seats

56% MVR	19% AD	4% PRVZL	4% MAS	3% COPEI	14% Others

MVR = Fifth Republic Movement **AD** = Democratic Action **PRVZL** = Project Venezuela **MAS** = Movement toward Socialism **COPEI** = Social Christian Party
The Patriotic Front (**PP**) comprises the MVR, the MAS, and some smaller parties

Venezuela is a multiparty democracy.

PROFILE

Corruption, austerity, rising poverty, and anti-price-rise riots in Caracas were the backdrop for two coup attempts in 1992, one led by Hugo Chávez, whose election as president in 1998 broke the stranglehold on power of the traditional parties. His increasingly autocratic style has attracted growing opposition.

MAIN POLITICAL ISSUES

Political stability

President Chávez's fiery rhetoric and populist stance won him the support of the poor as he embarked on root and branch reform, calling elections to a new Constituent Assembly which temporarily replaced the existing Congress. A new "Bolivarian" constitution, approved by referendum, allowed fresh elections in 2000 which gave Chávez a renewed mandate. His command-style running of the country increased political opposition, however, and deterred foreign investment. A military coup in April 2002 installed business leader Pedro Carmona as president for a day, but middle-ranking officers changed tack following mass demonstrations and the generally unfavorable international response. Restored to office, Chávez offered a conciliatory response to his opponents, recognizing the danger of more upheavals.

Oil policy

Support of oil prices and quotas within OPEC remain central, despite Chávez's wish to use oil revenue to build a broader economy.

WORLD AFFAIRS

Joined UN in 1945

ACS AP OAS OPEC RG

A traditionally pro-US orientation was challenged by Chávez, with his personal friendship with Cuba's Fidel Castro, visits to Libya and Iraq, and opposition to "Plan Colombia." The Bush administration acknowledged the new interim government during the April 2002 military coup, but denied helping to install it, although the main coup participants had made very public visits to the US State Department in preceding months. The OAS announced opposition to the coup, as did all of Venezuela's neighbors. Better regional economic integration with the Caribbean and Central and South America are important.

CHRONOLOGY

Venezuela was the first of the Spanish imperial colonies to repudiate Madrid's authority under the guidance of the revolutionary, Simón Bolívar, in 1811.

- **1821** Battle of Carabobo finally overthrows Spanish rule and leads to consolidation of independence within Gran Colombia (Venezuela, Colombia, and Ecuador).
- **1830** Gran Colombia collapses. José Antonio Páez rules Venezuela; coffee planters effectively in control.
- **1870** Guzmán Blanco in power. Rail system constructed.
- **1908** Gen. Juan Vicente Gómez dictator; oil industry developed.
- **1935** Gómez falls from power. Increasing mass participation in political process.
- **1945** Military coup. Rómulo Betancourt of AD takes power as leader of a civilian–military junta.
- **1948** AD wins elections, with novelist Rómulo Gallegos as presidential candidate. Military coup. Marcos Pérez Jiménez forms government, with US and military backing.

➪

CHRONOLOGY *continued*

- **1958** General strike. Adm. Larrázabal leads military coup. Free elections. Betancourt, newly returned from exile, wins presidential election for AD. Anticommunist campaign mounted. A few state welfare programs introduced.
- **1960** Movement of the Revolutionary Left (MIR) splits from AD, begins antigovernment activities.
- **1961** Founder member of OPEC.
- **1962** Communist-backed guerrilla warfare attempts repetition of Cuban revolution in Venezuela; fails to gain popular support.
- **1963** Raúl Leoni (AD) elected president – first democratic transference of power. Antiguerrilla campaign continues.
- **1966** Unsuccessful coup attempt by supporters of former president, Pérez Jiménez.
- **1969** Rafael Caldera Rodríguez of COPEI becomes president. Continues Leoni policies.
- **1973** Oil and steel industries nationalized. World oil crisis. Venezuelan currency peaks in value against the US dollar.
- **1978** Elections won by COPEI's Luis Herrera Campíns. Disastrous economic programs.
- **1983** AD election victory under Jaime Lusinchi. Fall in world oil prices leads to unrest and cuts in state welfare.
- **1988** Carlos Andrés Pérez of AD wins presidency.
- **1989** Caracas food riots; 1500 dead.
- **1993–1995** Andrés Pérez ousted on charges of corruption; Caldera Rodríguez reelected. More social unrest.
- **1998–2000** Hugo Chávez embarks on radical politcal reform after his Patriotic Front coalition defeats COPEI-led coalition in elections. New controversial Constituent Assembly approves new constitution, later endorsed by referendum. Chávez mandate confirmed by presidential elections. New unicameral National Assembly.
- **2002** April, Chávez ousted in military coup. Business leader Pedro Carmona president for a day. Chávez reinstalled a day later after foreign and domestic protests.

AID

Recipient

 $77m (receipts)

 Up 75% in 2000

The IDB is supporting the long-term reform of Venezuela's social security system.

DEFENSE

Compulsory military service

 $1.38bn

 Up 4% in 2000

Chávez led officers opposed to austerity and corruption in a 1992 coup attempt. Military leaders claimed in April 2002 that civil unrest forced them to oust him, and they backed interim president Pedro Carmona, who dissolved the National Assembly and Supreme Court by decree. However, middle-ranking officers got cold feet and reinstalled Chávez after protests.

VENEZUELAN ARMED FORCES

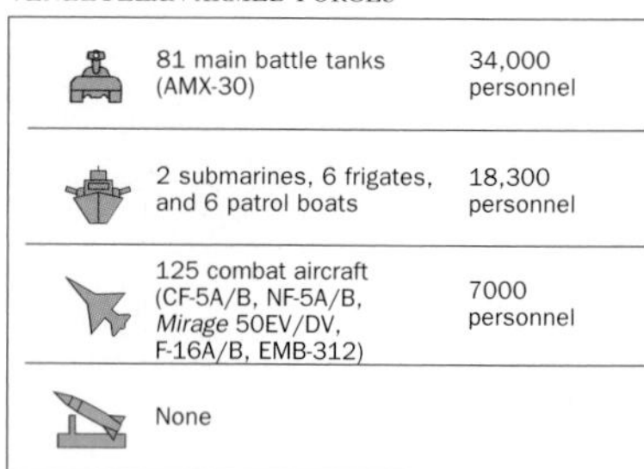
81 main battle tanks (AMX-30)
34,000 personnel
2 submarines, 6 frigates, and 6 patrol boats
18,300 personnel
125 combat aircraft (CF-5A/B, NF-5A/B, *Mirage* 50EV/DV, F-16A/B, EMB-312)
7000 personnel
None

ECONOMICS

Inflation 46% p.a. (1990–2000)

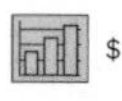 $104bn

 699.8–757.5 bolívares

SCORE CARD

❑ World GNP Ranking	36th
❑ GNP per Capita	$4310
❑ Balance of Payments	$13.2bn
❑ Inflation	16.2%
❑ Unemployment	14%

EXPORTS

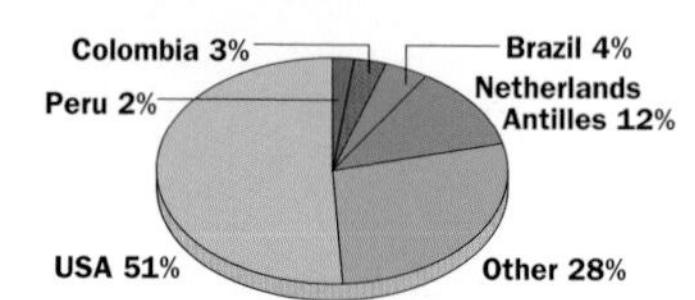
Colombia 3%
Brazil 4%
Peru 2%
Netherlands Antilles 12%
USA 51%
Other 28%

IMPORTS

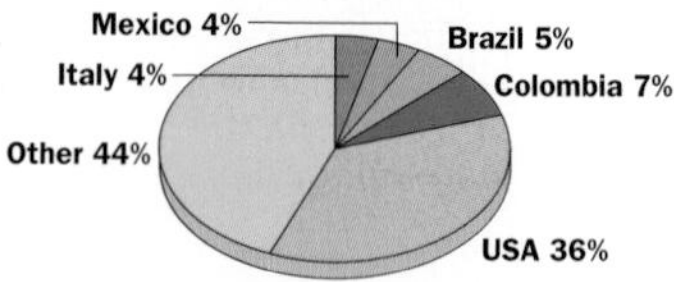
Mexico 4%
Brazil 5%
Italy 4%
Colombia 7%
Other 44%
USA 36%

Strengths

Largest proven oil deposits outside the Middle East and the CIS. Massive reserves of coal, bauxite, iron, and gold, and successful development of new bitumen fuel which has attracted considerable foreign investment. New foreign investment, including telecommunications, banking, iron, and steel. Producer of high-grade aluminum. Labor market becoming more flexible.

Weaknesses

Huge, cumbersome state sector; despite some privatization, large areas of state sector still overmanned, inefficient, and subject to widespread corruption. Poor public services which, despite Venezuela's wealth during the oil-boom years, have been badly maintained. Drop in world oil prices in 2001–2002. Major infrastructure renewal now long overdue. Widespread tax evasion. Weak currency.

ECONOMIC PERFORMANCE INDICATOR

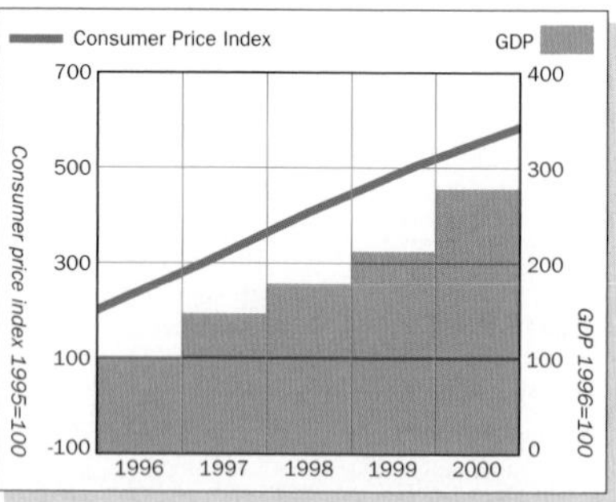
Consumer Price Index
GDP
700
500
300
100
-100
400
300
200
100
0
1996
1997
1998
1999
2000
Consumer price index 1995=100
GDP 1996=100

Profile

Venezuela is an economic paradox. While it has one of the strongest economies in Latin America, its government finances have habitually been in crisis due to a culture of nonaccountability and politically motivated patronage in state-owned industries and government bureaucracies. To date, privatizations and government cuts have failed to solve the problem. Promises by President Chávez to deal with excesses and diversify the economy, by lowering dependence on crude oil exports and promoting domestic processing industries, have received a mixed response from investors, who favor more market-oriented reforms.

VENEZUELA : MAJOR BUSINESSES

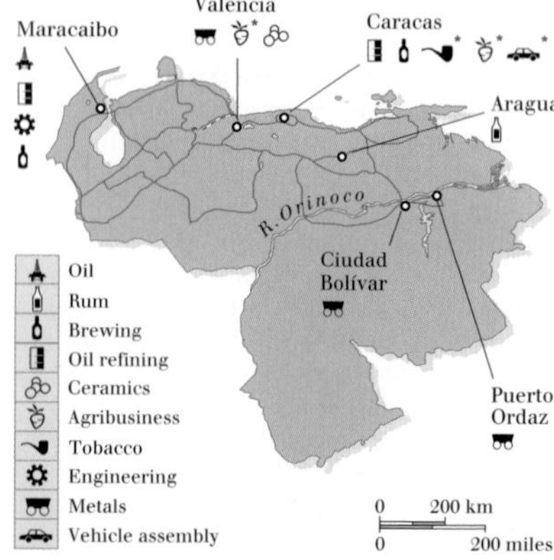
Maracaibo
Valencia
Caracas
Aragua
R. Orinoco
Ciudad Bolívar
Puerto Ordaz
Oil
Rum
Brewing
Oil refining
Ceramics
Agribusiness
Tobacco
Engineering
Metals
Vehicle assembly
0 200 km
0 200 miles
* significant multinational ownership

RESOURCES

Electric power 21.3m kw

422,607 tonnes

3.42m b/d (reserves 77.7bn barrels)

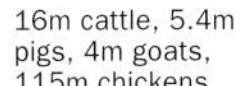

16m cattle, 5.4m pigs, 4m goats, 115m chickens

Oil, bauxite, iron, natural gas, coal, gold, diamonds, aluminum

ELECTRICITY GENERATION

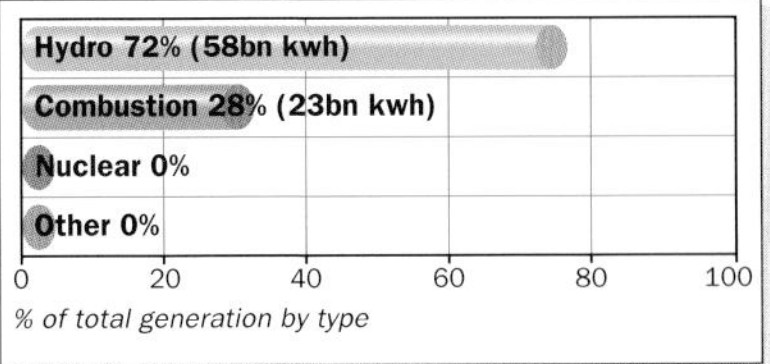
Hydro 72% (58bn kwh)
Combustion 28% (23bn kwh)
Nuclear 0%
Other 0%
0
20
40
60
80
100
% of total generation by type

ENVIRONMENT

Sustainability rank: 48th

37%

6.7 tonnes per capita

ENVIRONMENTAL TREATIES

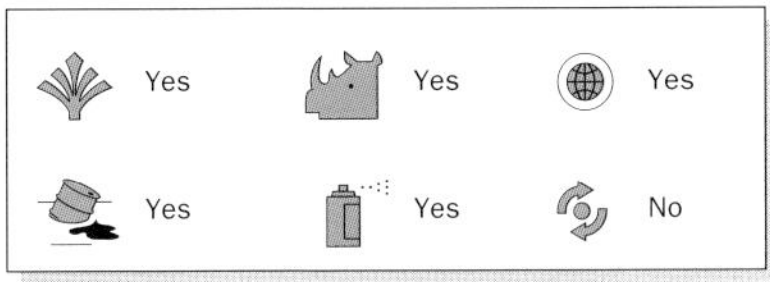
Yes
Yes
Yes
Yes
Yes
No

Flooding and mudflows, exacerbated by overdevelopment of the coastal strip, caused thousands of deaths in late 1999.

MEDIA

TV ownership medium

Daily newspaper circulation 206 per 1000 people

PUBLISHING AND BROADCAST MEDIA

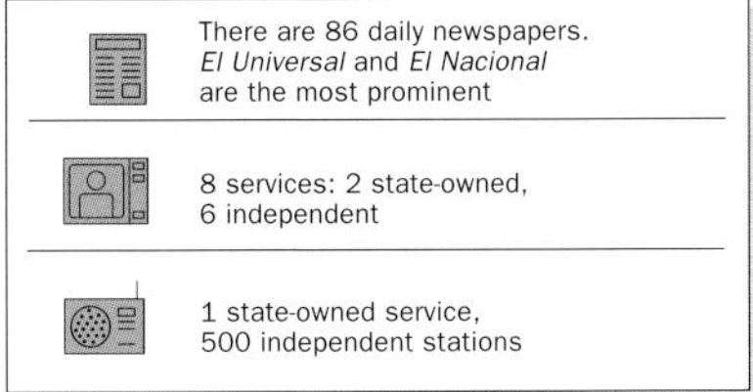
There are 86 daily newspapers. *El Universal* and *El Nacional* are the most prominent
8 services: 2 state-owned, 6 independent
1 state-owned service, 500 independent stations

Private TV stations and newspapers, mainly anti-Chávez, played a major role in the April 2002 coup by working with the military.

CRIME

No death penalty

23,147 prisoners

Little change 1992–1996

CRIME RATES

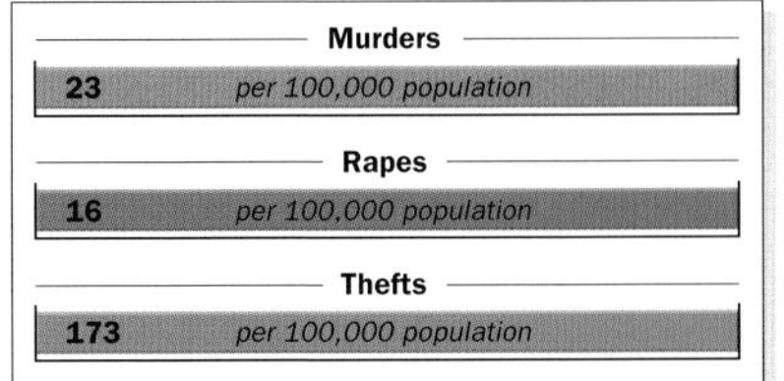
Murders
23 per 100,000 population
Rapes
16 per 100,000 population
Thefts
173 per 100,000 population

Urban robberies and violence involving young delinquents are major problems, as is narcotics-related crime. Cattle smuggling to Colombia is rife.

Venezuela has a remarkable diversity of resources. It has the world's sixth-largest proven oil reserves, vast quantities of coal, iron ore, bauxite, and gold, and cheap hydroelectric power. Huge investment programs are currently under way to raise production in all these sectors as well as in oil-refining capacity. However, the Chávez government wants to cut the investment budget of the state oil company, PDVSA, reduce its output, and increase its contributions to the exchequer. Such uncertainty has deterred private investors. Venezuela has begun exploitation of Orimulsion, a new bitumen-based fuel from the Orinoco; commercially exploitable reserves are estimated at 270 billion barrels. Venezuela's aim to be the world's largest aluminum producer is threatened after difficulties associated with privatizing the sector.

EDUCATION

School leaving age: 14

93%

550,783 students

THE EDUCATION SYSTEM

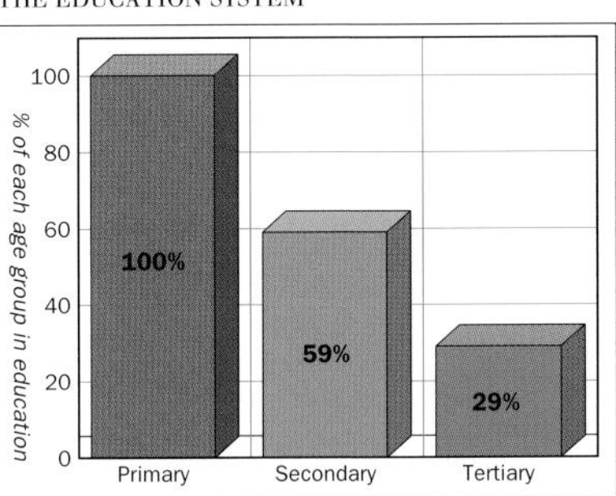
100
80
60
40
20
0
% of each age group in education
100%
59%
29%
Primary
Secondary
Tertiary

An extra $1 billion in social spending approved in 2000 includes raising entitlement in the state sector. Education is characterized by teacher shortages and a high drop-out rate; the quality of education at state universities is low. The private sector is growing.

HEALTH

Welfare state health benefits

1 per 417 people

Heart disease, accidents, violence, cancers

The health service suffered along with other public services from poor management in the 1970s and severe cuts in the 1980s and 1990s. Most health care is concentrated in the towns, and people from indigenous communities often have to travel long distances to receive treatment. Medicines, which have to be paid for, are expensive, and preventable diseases are recurring. Hospitals need modernization.

An additional $1 billion in social spending approved in 2000 includes spending on health.

VENEZUELA : LAND USE

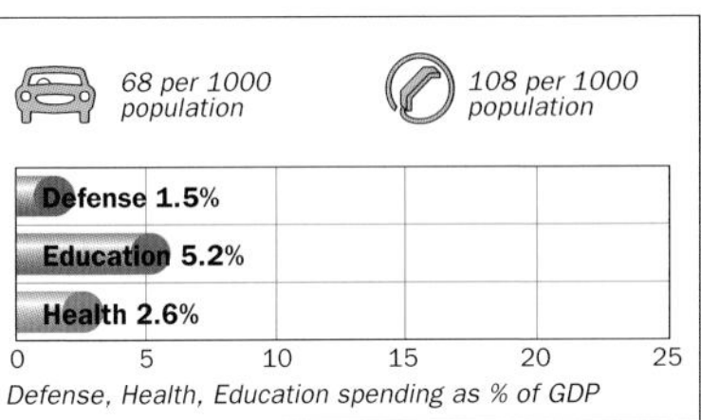
LLANOS
R. Orinoco
GUIANA HIGHLANDS
Cropland
Pasture
Forest
Coffee - cash crop
Cattle
0
200 km
0
200 miles

SPENDING

GDP/cap. increase

CONSUMPTION AND SPENDING

68 per 1000 population

108 per 1000 population

Defense	1.5%
Education	5.2%
Health	2.6%

Defense, Health, Education spending as % of GDP

The oil boom years of the 1970s largely benefited those already rich, with middle-income consumers doing well out of government-sponsored improvements in health and education and subsidized goods, largely at the expense of the poor.

The collapse of world oil prices, economic austerity measures, high inflation, and the devaluation of the bolívar in the 1980s and 1990s has squeezed the middle class and in addition seriously eroded the living standards of working-class households. In 2001, 20.7% of people were living in extreme poverty, according to official figures. Two million were unemployed and 59% of those working were in the informal sector.

WORLD RANKING

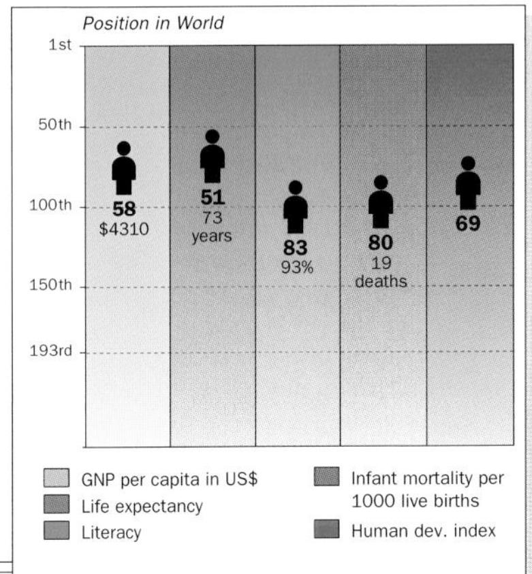
Position in World
1st
50th
100th
150th
193rd
58
$4310
51
73
years
83
93%
80
19
deaths
69
GNP per capita in US$
Life expectancy
Literacy
Infant mortality per 1000 live births
Human dev. index

V

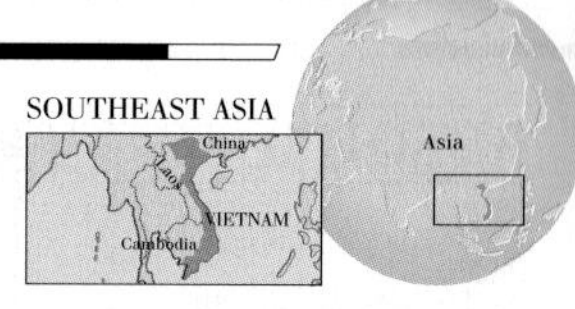
SOUTHEAST ASIA
Asia

VIETNAM

OFFICIAL NAME: Socialist Republic of Vietnam **CAPITAL:** Hanoi
POPULATION: 79.2 million **CURRENCY:** Dông **OFFICIAL LANGUAGE:** Vietnamese

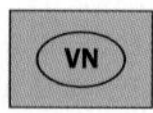
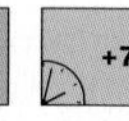
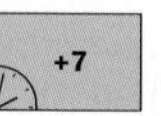
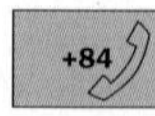
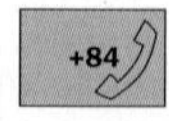
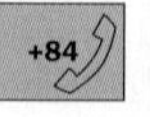

1976
1976
Sept 2
VN
+7
+84
.vn

VIETNAM LIES ON the eastern coast of the Indochinese peninsula. Over half the country is dominated by the heavily forested mountain range, the Chaîne Annamitique. The most populated areas, which are also the most intensively cultivated, are along the Red and Mekong Rivers. Partitioned after World War II, Vietnam was not reunited until 1976, a few months after the communist north finally defeated the southern regime and its US allies in the Vietnam War. Vietnam is now a single-party state ruled by the Communist Party. Since 1986, the regime has pursued a liberal economic policy known as *doi moi* ("renovation").

CLIMATE

Tropical monsoon

WEATHER CHART FOR HANOI

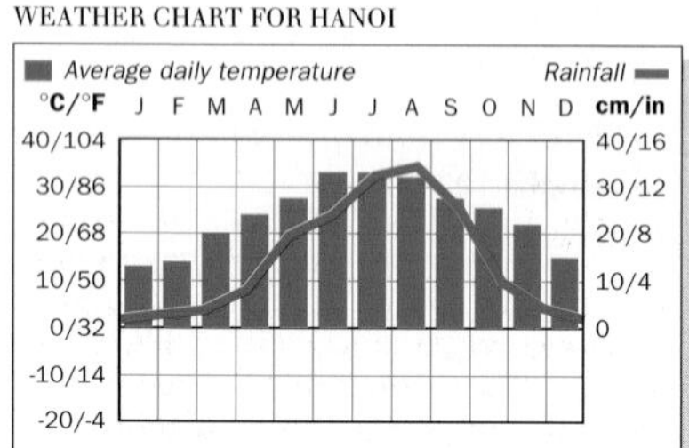
Average daily temperature
Rainfall
°C/°F J F M A M J J A S O N D cm/in
40/104 40/16
30/86 30/12
20/68 20/8
10/50 10/4
0/32 0
-10/14
-20/-4

Vietnam's geography means that there are sharp local contrasts in the climate. The north has cool winters, while the south is tropical, with even temperatures all year round. The central provinces are affected by typhoons. The northern Red River delta is subject to drought, while the Mekong delta in the south suffers heavy flooding.

TRANSPORTATION

Drive on right

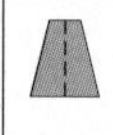

Tan Son Naht International, Ho Chi Minh City

629 ships 784,000 grt

THE TRANSPORTATION NETWORK

23,418 km (14,551 miles)

430 km (267 miles)

2632 km (1636 miles)

17,702 km (10,999 miles)

Rebuilding infrastructure is still a priority. The flagship project, the four-lane Ho Chi Minh Highway linking Hanoi and the south, is unlikely to meet the target of completion by 2003. A major port development plan is under way. Trains travel slowly, with an average speed of around 15 km/h (9 mph), and Hanoi to Ho Chi Minh City takes three days. The bus network is extensive but journeys are also time-consuming. Taxis and cycles provide cheap local transportation. Hanoi has plans for an elevated metro line.

TOURISM

Visitors : Population 1:57

1.38m visitors

Up 14% in 2000

MAIN TOURIST ARRIVALS

	% of total arrivals
China	27%
Taiwan	10%
Japan	6%
Cambodia	4%
France	4%
Other	49%

Until the government opened the way to large-scale tourism in the 1990s, Russians, eastern Europeans, and backpackers from the West made up the bulk of visitors. Other travelers were either on business, or overseas Vietnamese, *Viet Kie*, who were visiting relatives.

Under a "master plan" adopted in 1995, massive investment was channeled into hotels, with an official target of three million tourists a year by 2000. However, arrivals failed to reach even half that figure. Poor transportation infrastructure remains a problem. Vietnam's appeal lies in its unspoiled Asian way of life and in the areas of spectacular natural beauty such as Ha Long Bay on the Red River delta.

Boats moored near Nha Trang. *With 3444 km (2140 miles) of coastline, use of the sea, for transportation and fishing, is vital to Vietnam.*

PEOPLE

Pop. density high

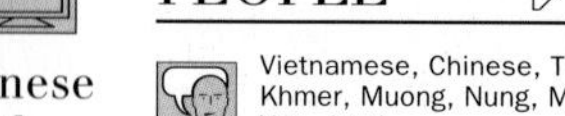

Vietnamese, Chinese, Thai, Khmer, Muong, Nung, Miao, Yao, Jarai

243/km² (630/mi²)

THE URBAN/RURAL POPULATION SPLIT

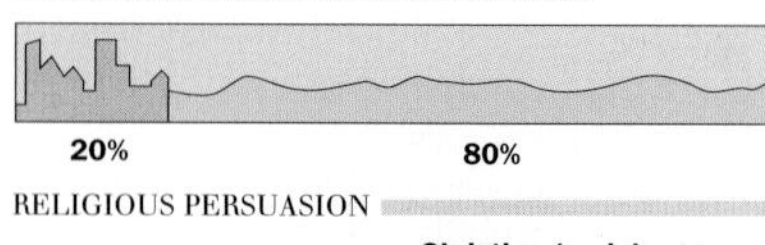
20%
80%

RELIGIOUS PERSUASION

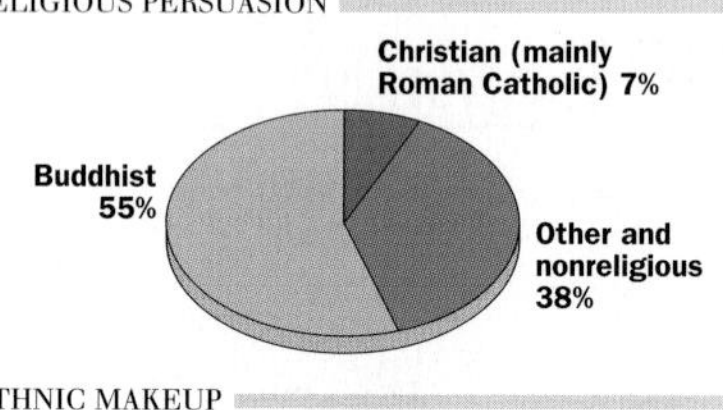
Christian (mainly Roman Catholic) 7%
Buddhist 55%
Other and nonreligious 38%

ETHNIC MAKEUP

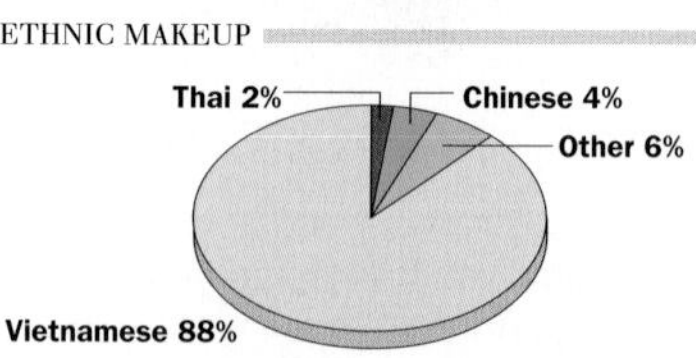
Thai 2%
Chinese 4%
Other 6%
Vietnamese 88%

Family life is strong and is based on kinship groups within village clans. A pronounced north–south cultural split remains evident in the cities. Chinese are the largest minority group. When the victorious communists reunited north and south Vietnam in 1976, they viewed the Saigon Chinese (in what was renamed Ho Chi Minh City), with their Taiwanese links, as a corrupt bourgeoisie. The northern Mountain Chinese were also suspect as a possible fifth column for China's ambitions in Vietnam. Various other mountain minorities (*montagnards*), with a history of collaboration with the French and Americans, were also sidelined by the regime in Hanoi. *Montagnard* resentment over the resettling of lowlanders in mountain regions sparked violent protests in early 2001.

War deaths cause older generations of women to outnumber men. Women form a high proportion of the workforce, and are starting to gain greater political prominence, most notably Vice President Nguyen Thi Binh. Female conscription was reinstated in 2001.

POPULATION AGE BREAKDOWN

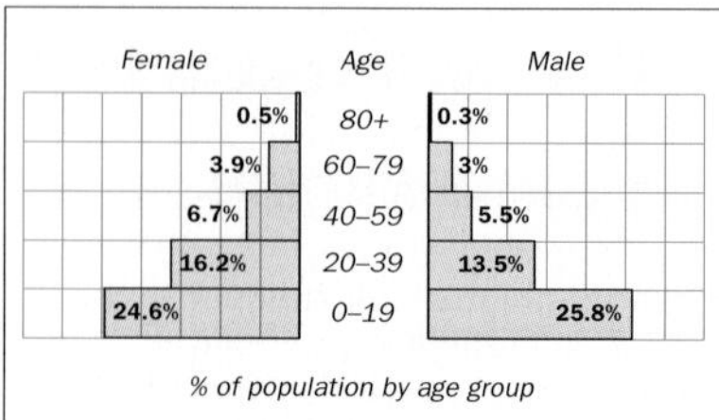
Female
Age
Male
0.5% 80+ 0.3%
3.9% 60–79 3%
6.7% 40–59 5.5%
16.2% 20–39 13.5%
24.6% 0–19 25.8%
% of population by age group

V

Tran Duc Luong, elected president in 1997.

Nong Duc Manh, powerful general secretary of the CPV.

POLITICS

No multiparty elections

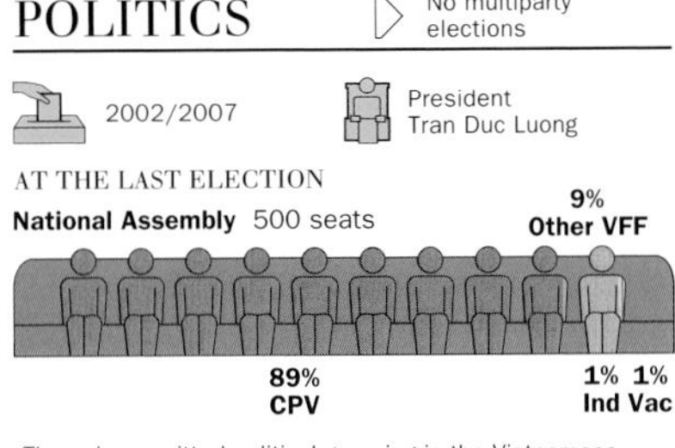
2002/2007
President Tran Duc Luong
AT THE LAST ELECTION
National Assembly 500 seats
9% Other VFF
89% CPV
1% 1% Ind Vac

The sole permitted political grouping is the Vietnamese Fatherland Front (**VFF**), which is dominated by the Communist Party of Vietnam (**CPV**)
Ind = Independents **Vac** = Vacant

Vietnam is effectively a single-party communist state.

PROFILE

A traditional communist system is still in place, with a powerful politburo elected by the party central committee. Heading the secretariat and politburo, the party general secretary wields much power, alongside the prime minister and president. Changes in senior posts in 1997 left reformers such as new prime minister Phan Van Khai still outnumbered by conservatives. The 2001 party congress balanced the theme of greater democracy with a renewed commitment to socialism.

MAIN POLITICAL ISSUE
Economic reform

Vietnam is trying to move to a market economy without political liberalization. The veteran CPV leaders of the 1950s–1970s resisted democratizing the political process and were slow to transfer power. The hand of the economic reformers was strengthened by Nong Duc Manh's elevation to the party leadership at the 2001 congress, but there remains real concern that opening collective farming and state enterprises to competition will undermine stability, encourage "individualism," and weaken the party's monopoly of power.

VIETNAM

Total Land Area : 329 560 sq. km (127 243 sq. miles)

POPULATION
- over 1 000 000
- over 500 000
- over 100 000
- over 50 000
- over 10 000
- under 10 000

LAND HEIGHT
- 2000m/6562ft
- 1000m/3281ft
- 500m/1640ft
- 200m/656ft
- Sea Level

0 100 km
0 100 miles

WORLD AFFAIRS

Joined UN in 1977

ASEAN

IAEA

OIF

Mekong River

NAM

Economic liberalization has improved relations with the US, with lifting of the aid and trade embargo in 1993, full diplomatic relations in 1995, and a landmark bilateral trade agreement in 2000, ahead of a visit by US president Bill Clinton that November.

Vietnam joined ASEAN in 1995, in the wake of the settlement of the Cambodia issue. Trade and economic cooperation links with Japan have been strengthened. Tension with China was reduced by an agreement in 1999 over their mutual land border, although competing claims to the Spratly Islands remain a source of friction.

AID

Recipient

$1.7bn (receipts)

Up 19% in 2000

Vietnam's invasion of Cambodia in 1978 halted all aid from China, Japan, and the West (except for Scandinavian countries), leaving it mostly dependent on the USSR. Western donors resumed assistance in the early 1990s. Their aid rapidly became the main source of capital for improving infrastructure, although foreign investment fell away significantly in the late 1990s.

CHRONOLOGY

From 1825, the brutal persecution of the Catholic community, originally converted by French priests in the 17th century, gave France the excuse to colonize Cochin-China, Annam, and Tonkin, and then merge them with Laos and Cambodia.

- **1920** *Quoc ngu* (Roman script) replaces Chinese script.
- **1930** Ho Chi Minh founds Indo-China Communist Party.
- **1940** Japanese invasion.
- **1941** Viet Minh resistance founded in exile in China.
- **1945** Viet Minh take Saigon and Hanoi. Emperor abdicates. Republic proclaimed with Ho Chi Minh as president.
- **1946** French reenter. First Indochina war.
- **1954** French defeated at Dien Bien Phu. Vietnam divided at 17°N. USSR supports North; US arms South.
- **1960** Groups opposed to southern regime unite as Viet Cong.
- **1961** US pours in military advisers.
- **1964** US Congress approves war.
- **1965** Gen. Nguyen Van Thieu takes over military government of South. First US combat troops arrive.
- **1965–1968** Operation Rolling

V

CHRONOLOGY *continued*

Thunder – intense bombing of North by South and US.

- **1967** Antiwar protests start in US and elsewhere.
- **1968** Tet (New Year) Offensive – 105 towns attacked simultaneously in South with infiltrated arms. Viet Cong suffer serious losses. Peace talks begin. US eases bombing and starts withdrawing troops.
- **1969** Ho Chi Minh dies. Succeeded by Le Duan. War intensifies in spite of talks.
- **1970** US begins secret attacks in Laos and Cambodia and new mass bombing of North to try to stop arms reaching Viet Cong.
- **1972** 11-day Christmas Campaign is heaviest US bombing of war.
- **1973** Paris Peace Agreements signed, but fighting continues.
- **1975** Fall of Saigon to combined forces of North and Provisional Revolutionary (Viet Cong) Government of South. One million flee after end of war.
- **1976** Vietnam united as Socialist Republic of Vietnam. Saigon renamed Ho Chi Minh City.
- **1978** Invasion of Cambodia to oust Pol Pot regime (by January 1979).
- **1979** Nine-Day War with China. Chinese troops pushed back after destroying everything for 40 km (25 miles) inside Vietnam. "Boat people" crisis. At UN conference, Vietnam agrees to allow legal emigration, but exodus continues.
- **1986** Death of Le Duan. Nguyen Van Linh, new Communist Party general secretary, initiates liberal economic policy of *doi moi* (renovation).
- **1987** Fighting in Thailand as Vietnam pursues Kampuchean resistance fighters across border.
- **1989** Troops leave Cambodia.
- **1991** Open anticommunist dissent made a criminal offense.
- **1992** Revised constitution allows foreign investment, but essential role of Communist Party is unchanged.
- **1995** US–Vietnamese relations normalized. Vietnam joins ASEAN.
- **1997** Legislative elections. Tran Duc Luong president, Phan Van Khai prime minister.
- **1998** Asian financial crisis dampens economic boom.
- **1999** Signing of border treaty with China.
- **2000** Worst flooding along Mekong for 40 years. November, visit by US president Bill Clinton.
- **2001** March, visit by Russian president Vladimir Putin. April, ninth party congress. Nong Duc Manh becomes general secretary.

DEFENSE

Compulsory military service

 $931m

 Up 5% in 2000

Vietnam has large and well-equipped armed forces, notably the world's seventh-largest army. Military service is compulsory, with conscripts serving a two-year term. The army's role in preserving both stability and socialism was reaffirmed in 2001. Increased defense spending on the navy reflects the rising tensions in the South China Sea, where there are disputed claims to the Spratly and Paracel Islands.

VIETNAMESE ARMED FORCES

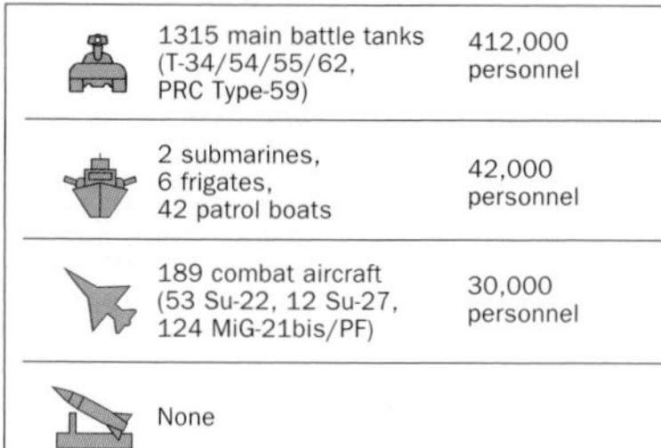

1315 main battle tanks (T-34/54/55/62, PRC Type-59)	412,000 personnel
2 submarines, 6 frigates, 42 patrol boats	42,000 personnel
189 combat aircraft (53 Su-22, 12 Su-27, 124 MiG-21bis/PF)	30,000 personnel
None	

ECONOMICS

Inflation 15% p.a. (1990–2000)

$30.4bn

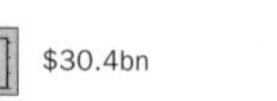 14,514–15,083 dông

SCORE CARD

World GNP Ranking	58th
GNP per Capita	$390
Balance of Payments	$507m
Inflation	–0.6%
Unemployment	25%

EXPORTS 

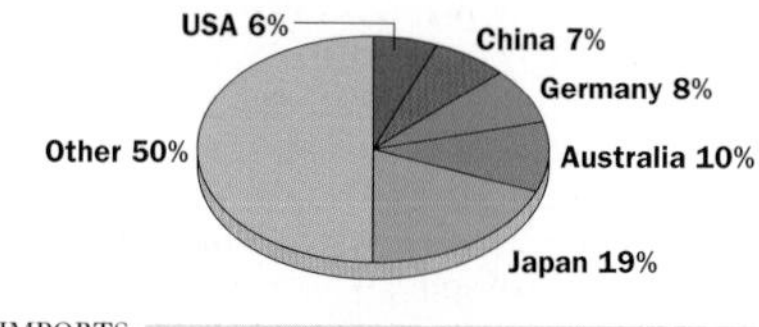
USA 6%
China 7%
Germany 8%
Australia 10%
Japan 19%
Other 50%

IMPORTS

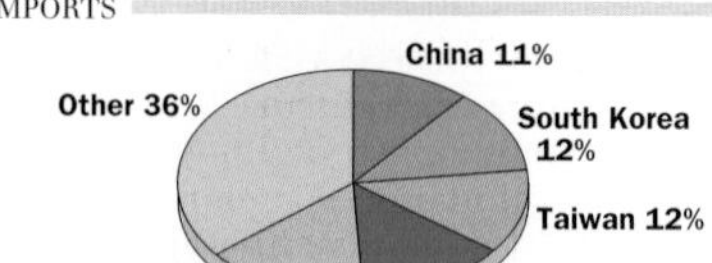
China 11%
South Korea 12%
Taiwan 12%
Japan 14%
Singapore 15%
Other 36%

ECONOMIC PERFORMANCE INDICATOR

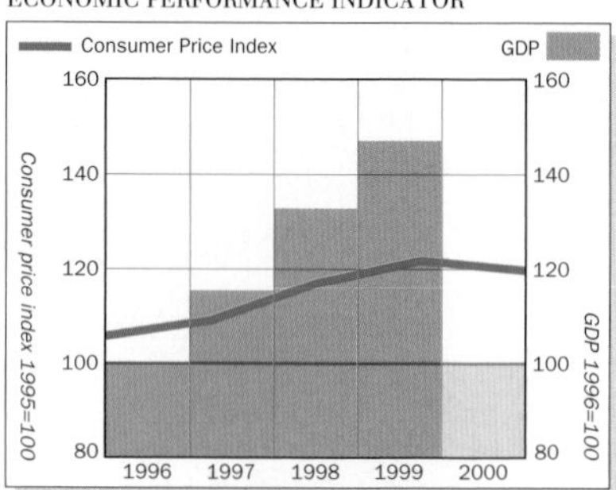
Consumer Price Index
GDP
160
140
120
100
80
Consumer price index 1995=100
GDP 1996=100
1996
1997
1998
1999
2000

Strengths

Diverse resource base; unexploited gas reserves. Young, literate, low-cost labor force. Strong light industrial and handicraft export industries.

Weaknesses

Weak economic institutions. Weight of bureaucracy. Heavy dependence on aid for reconstruction. Enduring suspicion of entrepreneurial southern attitudes and "individualism." Corruption.

Profile

The encouragement of private enterprise began in 1988. Touted by some commentators in the mid-1990s as the next Asian "tiger," Vietnam has aimed at more moderate growth since the Asian crisis of 1997–1998. Annual GDP increases, over 5% again by 2000 and predicted to reach 7% by 2003, are impressive. Inflation, once a huge problem, was held down in the 1990s, and is now under firm control. Increased rice production has boosted incomes, and domestic demand helped the post-1998 upswing. The government promised massive investment in agriculture from 2002.

Attracting foreign investment through reform is essential; government policies regarding state-owned enterprises promise more scope for joint ventures, and new laws on trade-licensing and investment were passed in 2000. Also that year the country's first stock exchange opened. The government has set a target of doubling GDP in the next decade. The potential certainly exists, based on an educated and highly motivated young labor force and on mineral resources, located mostly in the north.

VIETNAM : MAJOR BUSINESSES

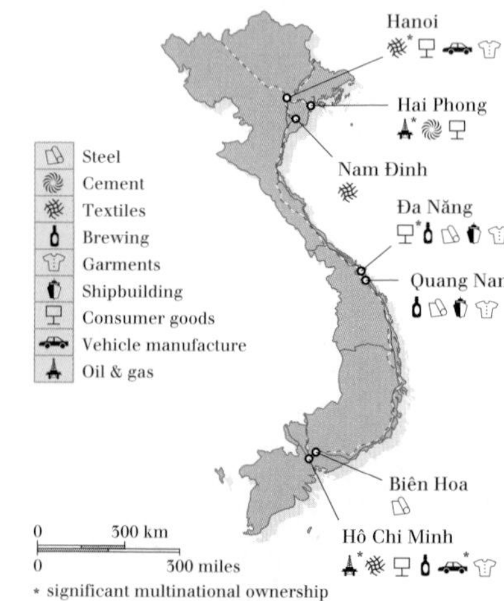
Hanoi
Hai Phong
Nam Đinh
Đa Nǎng
Quang Nam
Biên Hoa
Hô Chi Minh
Steel
Cement
Textiles
Brewing
Garments
Shipbuilding
Consumer goods
Vehicle manufacture
Oil & gas
0 300 km
0 300 miles
* significant multinational ownership

V

RESOURCES

Electric power 5m kw

1.79m tonnes

350,000 b/d (reserves 600m barrels)

20.2m pigs, 57m ducks, 4.2m cattle, 150m chickens

Coal, oil, tin, zinc, iron, antimony, apatite, salt, bauxite

ELECTRICITY GENERATION

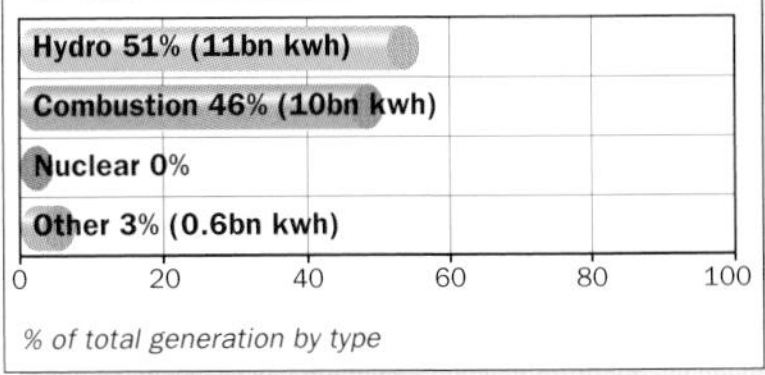
Hydro 51% (11bn kwh)
Combustion 46% (10bn kwh)
Nuclear 0%
Other 3% (0.6bn kwh)
0
20
40
60
80
100
% of total generation by type

Vietnam is among the world's top ten coffee producers and is the third-largest exporter of rice – to the detriment of domestic stocks. Oil production, small by world standards, is sufficient to make it Vietnam's biggest export earner. The Oil and Gas Corporation of Vietnam (PetroVietnam) is involved in joint ventures with international oil firms. Vietnam has unexploited gas reserves in the South China Sea; gas from the only producing field has to be flared off. Timber exports have been banned since 1997 to preserve forests. Northern Vietnam has a surplus of electricity, mainly from hydroelectric schemes.

ENVIRONMENT

Sustainability rank: 94th

3%

0.6 tonnes per capita

ENVIRONMENTAL TREATIES

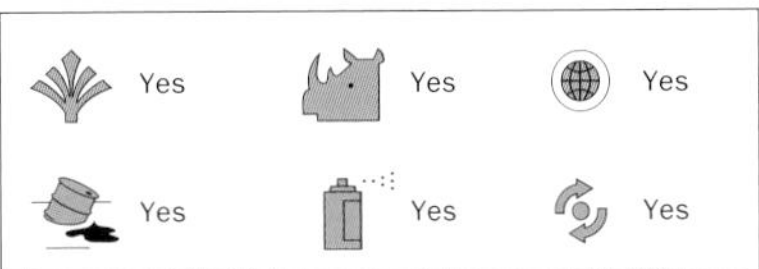
Yes
Yes
Yes
Yes
Yes
Yes

In the Vietnam War, seven million tonnes of bombs were dropped, and the defoliant chemical Agent Orange was sprayed over vast areas; a "census" of the continuing health impact was announced in 1999. Half of Vietnam's forests were seriously damaged and some 5% destroyed. Deforestation continued into the 1990s due to logging and expansion of coffee-growing, causing soil erosion and flooding. Floods along the Mekong in 2000 were the worst for 40 years.

MEDIA

TV ownership medium

Daily newspaper circulation 4 per 1000 people

PUBLISHING AND BROADCAST MEDIA

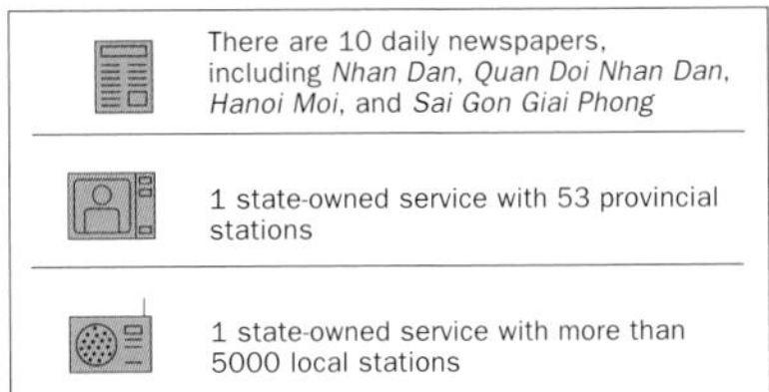
There are 10 daily newspapers, including *Nhan Dan*, *Quan Doi Nhan Dan*, *Hanoi Moi*, and *Sai Gon Giai Phong*
1 state-owned service with 53 provincial stations
1 state-owned service with more than 5000 local stations

The media are tightly regulated. TV is the dominant medium. All editors have to be Party members, but criticism of the authorities is still possible. Even *Nhan Dan*, the CPV newspaper, has been known to expose laxity in the system, especially in the judiciary. However, in 2002 the authorities sought out and destroyed books by proscribed authors.

CRIME

Death penalty in use

55,000 prisoners

Down 15% 1997–1999

CRIME RATES

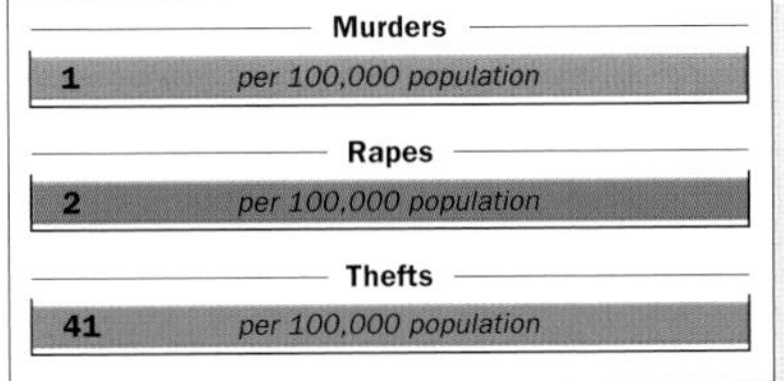
Murders
1 per 100,000 population
Rapes
2 per 100,000 population
Thefts
41 per 100,000 population

The judicial system is based on the Soviet model. The education camps established after liberation have now closed, but religious and political dissidents are still held without trial.

Corruption has risen sharply since economic liberalization, as has the illegal drift of young people to urban areas, where they are blamed for increasing petty crime and "social evils" such as begging, prostitution, and drug-taking. Theft from foreigners is a problem in major cities.

EDUCATION

School leaving age: 15

93%

509,300 students

THE EDUCATION SYSTEM

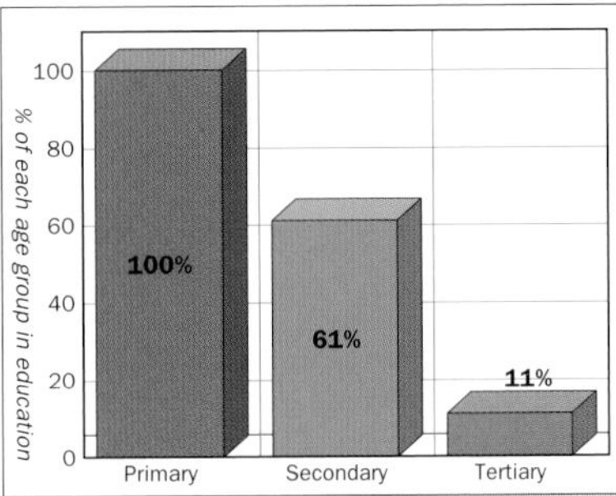
% of each age group in education
100
80
60
40
20
0
100%
61%
11%
Primary
Secondary
Tertiary

Private sponsorship helps fund education. Vietnamese universities have a strong liberal arts tradition. Social pressure to obtain a degree leads to high levels of cheating among university applicants.

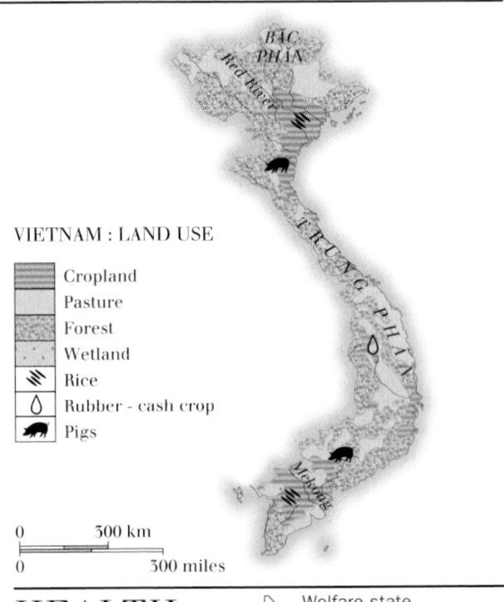
BAC PHAN
Red River
TRUNG PHAN
Mekong
VIETNAM : LAND USE
Cropland
Pasture
Forest
Wetland
Rice
Rubber - cash crop
Pigs
0 300 km
0 300 miles

HEALTH

Welfare state health benefits

1 per 2000 people

Heart disease, cancers, malaria

Vietnam's medical achievements include developing a vaccine for hepatitis B, and extracting an antimalarial drug, artemisinin, from the indigenous thanh hao tree. An extensive campaign is under way to combat the spread of AIDS.

SPENDING

GDP/cap. increase

CONSUMPTION AND SPENDING

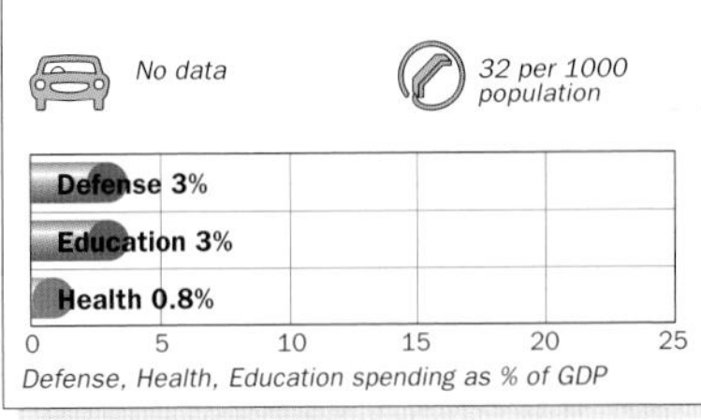
No data
32 per 1000 population
Defense 3%
Education 3%
Health 0.8%
0
5
10
15
20
25
Defense, Health, Education spending as % of GDP

Ostentatious consumerism is rising despite official disapproval, but is beyond most people's reach. Wealth disparities are growing, with rural areas falling deep into poverty.

WORLD RANKING

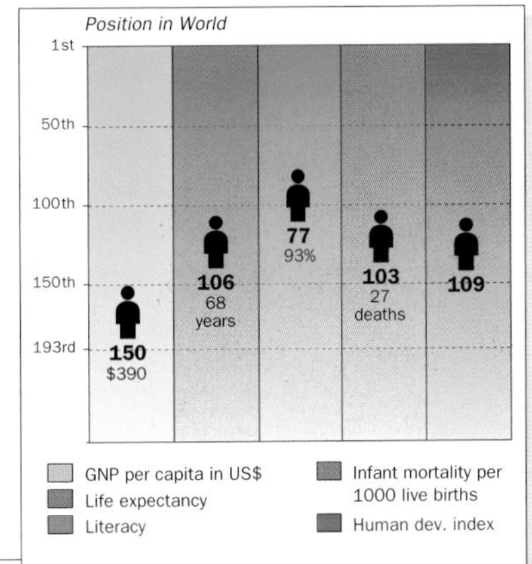
Position in World
1st
50th
100th
150th
193rd
150
$390
106
68 years
77
93%
103
27 deaths
109
GNP per capita in US$
Life expectancy
Literacy
Infant mortality per 1000 live births
Human dev. index

V

YEMEN

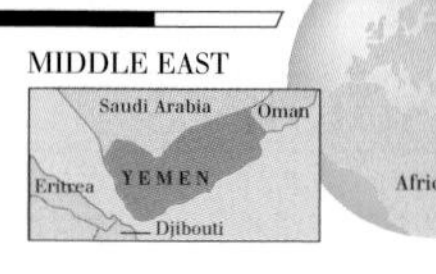
MIDDLE EAST
Saudi Arabia
Oman
YEMEN
Eritrea
Djibouti
Asia
Africa

OFFICIAL NAME: Republic of Yemen **CAPITAL:** Sana
POPULATION: 19.1 million **CURRENCY:** Yemeni rial **OFFICIAL LANGUAGE:** Arabic

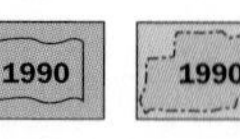
1990
1990

May 22
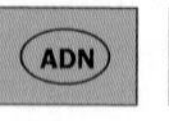
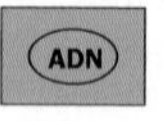
ADN
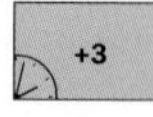

+3

+967
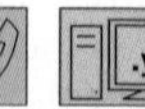
.ye

YEMEN IS LOCATED in southern Arabia. The west is mountainous, with a fertile strip along the Red Sea. The center and south are largely arid mountains and desert. Until 1990 Yemen was two countries, the Yemen Arab Republic (YAR) in the west and the People's Democratic Republic of Yemen (PDRY) in the south. The YAR was run by successive military regimes; the poorer PDRY was the Arab world's only Marxist state. Post-unification conflict between the two ruling hierarchies, nominally in coalition, led to a two-month civil war in 1994, the ousting of the former Marxists, and a new constitution.

CLIMATE

Hot desert/mountain

WEATHER CHART FOR SANA

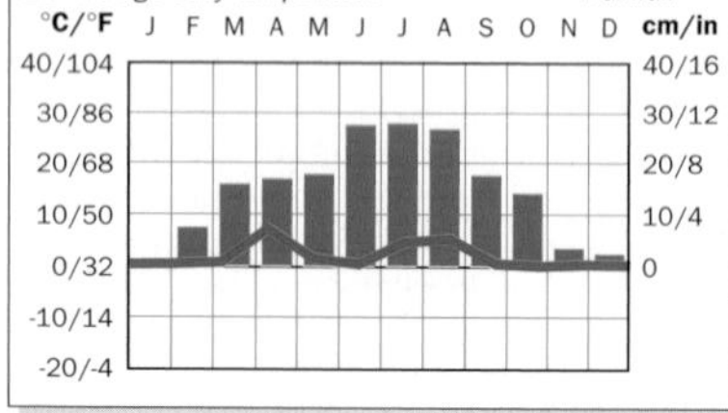
Average daily temperature
Rainfall
°C/°F J F M A M J J A S O N D cm/in
40/104
30/86
20/68
10/50
0/32
-10/14
-20/-4
40/16
30/12
20/8
10/4
0

The desert climate is modified by altitude, which affects temperatures by as much as 12°C (54°F). Rainfall increases in northwest and central Yemen.

TRANSPORTATION

Drive on right

Sana International
858,843 passengers

45 ships
25,300 grt

THE TRANSPORTATION NETWORK

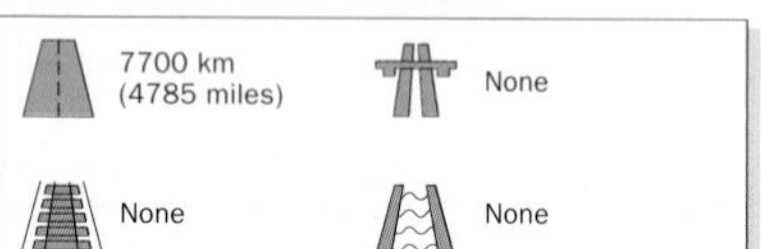
7700 km
(4785 miles)
None
None
None

Aden's history as a port stretches back 3000 years. Adequate roads link the main cities, but many rural areas are inaccessible. International airlines, including the modern fleet of Yemenia, serve Sana and Aden.

***Hilltop village** in northern Yemen, showing traditionally decorated, multistorey houses built from unbaked mud bricks.*

TOURISM

Visitors : Population 1:262

73,000 visitors

Down 13% 1999–2000

MAIN TOURIST ARRIVALS

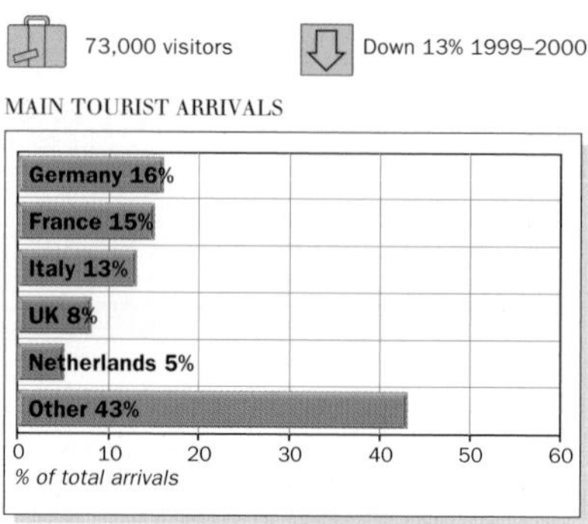
Germany 16%
France 15%
Italy 13%
UK 8%
Netherlands 5%
Other 43%
0 10 20 30 40 50 60
% of total arrivals

Believed to be the home of the legendary Queen of Sheba, Yemen attracts tourists interested in Arab society, architecture, archeology, and historical remains. The Romans called Yemen *Arabia Felix* because of its fertile farmlands and dominance in the frankincense trade. Yemen was the second country, after Saudi Arabia, to convert to Islam.

Southern Yemen has been open to Western visitors only since 1990. Its run-down infrastructure and lack of hotels, especially on the coast, have hindered tourism. Sana, a walled medieval city, is the more interesting center for tourists. It has impressive architecture, particularly tall stone and mud-brick Arab houses, and the palaces of the former imamate. Despite being over 100 km (62 miles) from the capital, the Marib Dam, built in ancient times, is another major attraction.

German and French tourists were the first to travel to northwest Yemen during the 1980s. Hopes of a major rise in tourism following the end of the 1994 civil war were dashed in 1998 after tribesmen kidnapped and killed four tourists.

Tourists are subject to a ban on the consumption of alcohol, except in five-star hotels. Whisky and beer are available on the black market, which operates out of Djibouti.

PEOPLE

Pop. density low

Arabic

34/km²
(88/mi²)

THE URBAN/RURAL POPULATION SPLIT

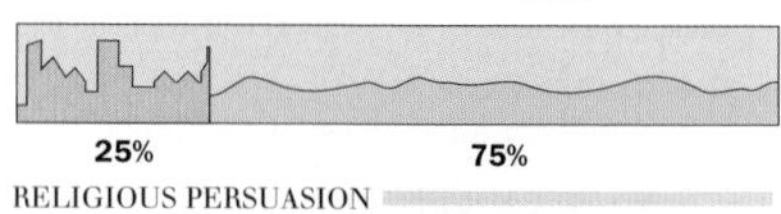
25%
75%

RELIGIOUS PERSUASION

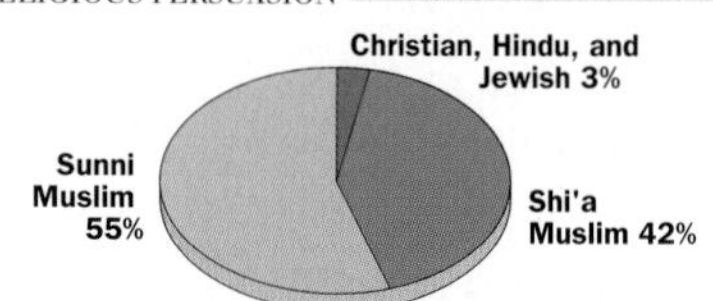
Christian, Hindu, and Jewish 3%
Sunni Muslim 55%
Shi'a Muslim 42%

ETHNIC MAKEUP

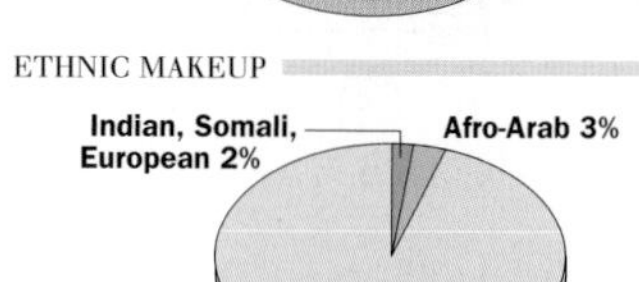
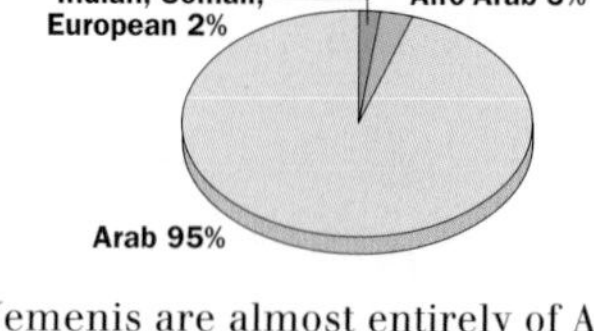
Indian, Somali, European 2%
Afro-Arab 3%
Arab 95%

Yemenis are almost entirely of Arab and Bedouin descent, though there is a small, dwindling, Jewish minority and people of mixed African and Arab descent along the south coast. The majority are Sunni Muslims, of the Shafi sect. However, Zaydi Shi'a are strong in the north, where many people have close family in Saudi Arabia. Many Yemenis consider Saudi Arabia's Asir province to be part of Yemen.

Agriculture supports more than half the population. Many Yemenis went to work in Saudi Arabia and the Gulf states during the 1970s oil boom. More than a million worked in Saudi Arabia, and their expulsion, due to Yemen's support for Iraq's invasion of Kuwait in 1990, has increased domestic unemployment.

In rural areas and in the western highlands, Islamic orthodoxy is strong and most women wear the veil. In the south, however, women still claim the freedoms they had under the Marxist regime, especially in urban areas.

Tension continues to exist between the south, led by cosmopolitan Aden, and the more conservative west, leading in 1994 to civil war.

POPULATION AGE BREAKDOWN

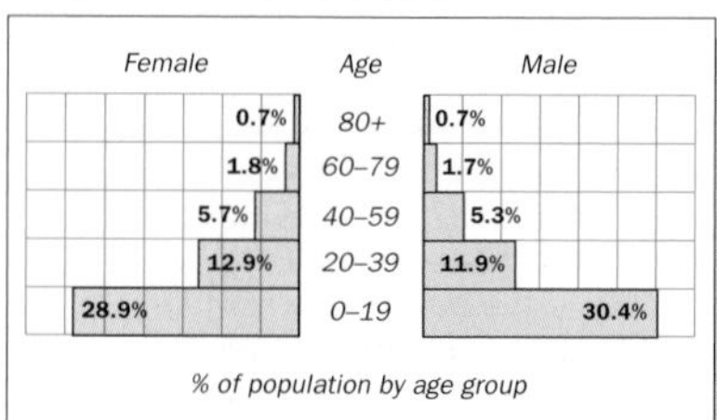
Female
Age
Male
0.7% 80+ 0.7%
1.8% 60–79 1.7%
5.7% 40–59 5.3%
12.9% 20–39 11.9%
28.9% 0–19 30.4%
% of population by age group

Y

YEMEN

Total Land Area : 527 970 sq. km
(203 849 sq. miles)

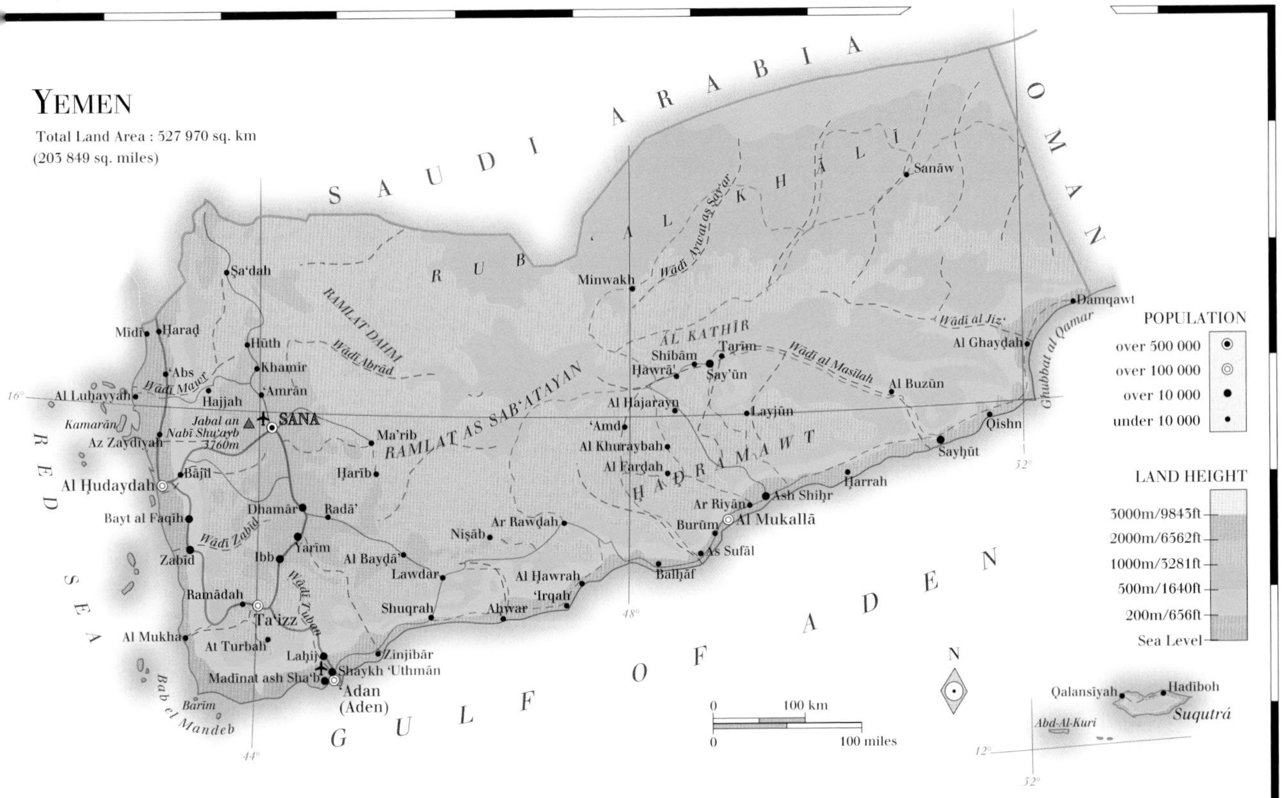

POPULATION
over 500 000
over 100 000
over 10 000
under 10 000

LAND HEIGHT
3000m/9843ft
2000m/6562ft
1000m/3281ft
500m/1640ft
200m/656ft
Sea Level

POLITICS

Multiparty elections

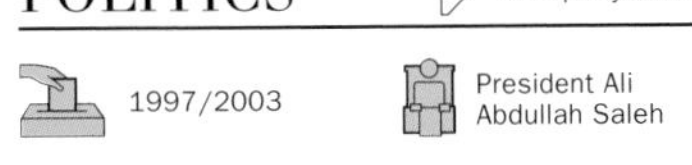

1997/2003

President Ali Abdullah Saleh

AT THE LAST ELECTION

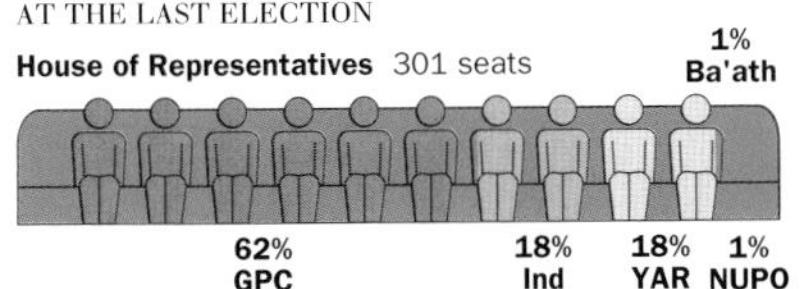
House of Representatives 301 seats
1% Ba'ath
62% GPC
18% Ind
18% YAR
1% NUPO

GPC = General People's Congress **Ind** = Independents
YAR = Yemeni Alliance for Reform (al-Islah)
NUPO = Nasserite Unionist Popular Organization
Ba'ath = Arab Socialist Ba'ath Party

Yemen is a multiparty democracy. The president retains executive power.

PROFILE

The merger of the YAR and the PDRY in 1990 united Yemenis under one ruler for the first time since 1735; free elections were held in 1993. President Ali Saleh initially maintained unity. Then, in 1994, a bloody civil war erupted, fueling a secessionist movement in the south. By mid-1994, the southerners, led by ex-vice president Ali Salem al-Baidh, were crushed. In elections held in 1997, Saleh's GPC won an absolute majority, in part due to a boycott by the former southern-ruling Yemen Socialist Party, and in 1999 he won the region's first direct election of a head of state. Saleh's regime still faces threats from tribesmen angry at levels of poverty in their oil-rich country. Since 1992, tribesmen have kidnapped more than 100 foreigners, including diplomats and tourists.

MAIN POLITICAL ISSUES

Relations with Saudi Arabia

Relations have long been strained – over oil exploration rights, Yemeni claims on Asir, and accusations that Riyadh funds insurgent tribesmen. The two sides clashed violently in 1998 over 1600 km (1000 miles) of disputed border, despite a 1995 memorandum of understanding. In early 2001, both sides withdrew border troops under a pact reached the previous June.

Instability

Since the end of the civil war, stability has been threatened by the border dispute with Saudi Arabia, growing tribal insurgency, and mounting popular discontent with President Saleh's government. The Islamist al-Islah party, formerly a junior partner of the GPC, now forms the chief opposition.

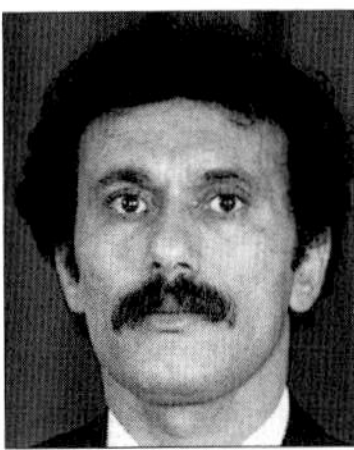

Ali Abdullah Saleh, *former YAR president, now leader of the unified Yemen.*

Ali Salem al-Baidh, *former vice president, ousted when south's resecession failed.*

CHRONOLOGY

From the 9th century, the Zaydi dynasty ruled Yemen until their defeat by the Ottoman Turks in 1517. The Turks were expelled by the Zaydi imams in 1636.

- **1839** Britain occupies Aden.
- **1918** Western Yemen independent.
- **1937** Aden made a crown colony, hinterland a protectorate.
- **1962** Army coup in west. Imam deposed, Yemen Arab Republic (YAR) declared. Civil war.
- **1963** Aden and protectorate united to form Federation of South Arabia.
- **1967** South Arabia independent as People's Republic of South Yemen. British troops leave Aden.
- **1970** South Yemen renamed People's Democratic Republic of Yemen (PDRY). Republican victory in YAR civil war.
- **1972** War between YAR and PDRY ends in peace settlement.
- **1974** Army coup in YAR.
- **1978** Lt. Col. Ali Saleh YAR president. Coup in PDRY. Radical Abdalfattah Ismail in power.
- **1979** PDRY signs 20-year treaty with USSR.
- **1980** Ismail replaced by moderate Ali Muhammed.
- **1982** PDRY peace treaty with Oman. Major earthquake kills 3000.
- **1984** YAR signs 20-year cooperation treaty with USSR.

CHRONOLOGY *continued*

- ❑ **1986** Coup attempt in PDRY leads to civil war. Rebels take control of Aden. New PDRY president meets YAR counterpart.
- ❑ **1987** Oil production starts in YAR.
- ❑ **1988** YAR holds elections for consultative council; Muslim Brotherhood gains influence.
- ❑ **1989** Speeding-up of unification process. PDRY publishes a program of free-market reforms. YAR and PDRY sign unification agreement. Constitution of unified Yemen published.
- ❑ **1990** Restrictions on travel between YAR and PDRY lifted. Ali Saleh becomes president of Republic of Yemen. May, formal unification. Pro-Islamic groups oppose secular constitution.
- ❑ **1991** Yemeni guest workers expelled by Saudi Arabia in retaliation for Yemen's position over Iraqi invasion of Kuwait. Arab states boycott independence celebrations.
- ❑ **1992** Assassinations, food riots, and political unrest delay elections.
- ❑ **1994** Southern secessionists defeated in civil war. Amended constitution adopted.
- ❑ **1997** Saleh's GPC wins absolute majority in general election.
- ❑ **1998–1999** Violent border dispute with Saudi Arabia. Kidnapping of tourists, four killed; three members of Islamic Army of Aden (IAA) sentenced to death.
- ❑ **1999** Saleh reelected.
- ❑ **2000** Yemen agrees border with Saudi Arabia after 66-year dispute. October, terror attacks on US naval vessel and UK embassy.
- ❑ **2001** Referendum approves extension of presidential term to seven years.
- ❑ **2002** Government targets suspected al-Qaida allies in tribal areas, expels 100 foreign "scholars."

WORLD AFFAIRS

Joined UN in 1947/1967

AL
AMF
IBRD
NAM
OIC

Yemen is internationally regarded as a base for Islamic terrorists and was shunned for its support of Iraq in the Gulf War. The government has tried to break this isolation with active support for the US-led "war on terrorism."

AID

Recipient

$265m (receipts)

Down 42% in 2000

Since the receipt of $700 million for economic reforms in 1996, aid to Yemen has decreased. The World Bank is the largest donor.

DEFENSE

Compulsory military service

$489m

Up 14% in 2000

Following unification in 1990, mutual suspicion hampered the integration of the two separate defense forces. Sporadic, bitter clashes have taken place.

The main domestic security concern is insurgent tribesmen and, internationally, anti-Western terrorist activity such as the sinking of USS *Cole* off Aden in late 2000. US military aid has increased accordingly.

YEMENI ARMED FORCES

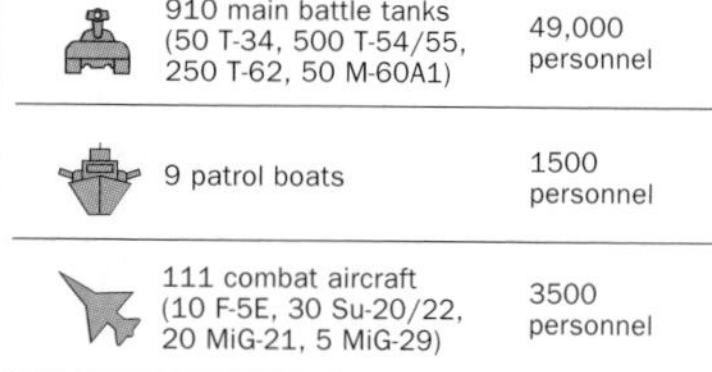

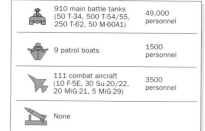

910 main battle tanks (50 T-34, 500 T-54/55, 250 T-62, 50 M-60A1)	49,000 personnel
9 patrol boats	1500 personnel
111 combat aircraft (10 F-5E, 30 Su-20/22, 20 MiG-21, 5 MiG-29)	3500 personnel
None	

ECONOMICS

Inflation 22% p.a. (1990–2000)

$6.55bn

164.39–170.70 Yemeni rials

SCORE CARD

- ❑ World GNP Ranking 102nd
- ❑ GNP per Capita $370
- ❑ Balance of Payments $2.06bn
- ❑ Inflation 7.9%
- ❑ Unemployment 30%

EXPORTS

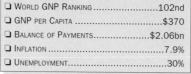

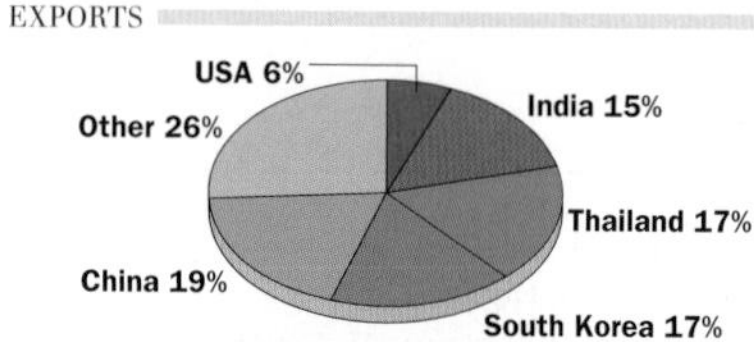
USA 6%
India 15%
Other 26%
Thailand 17%
China 19%
South Korea 17%

IMPORTS

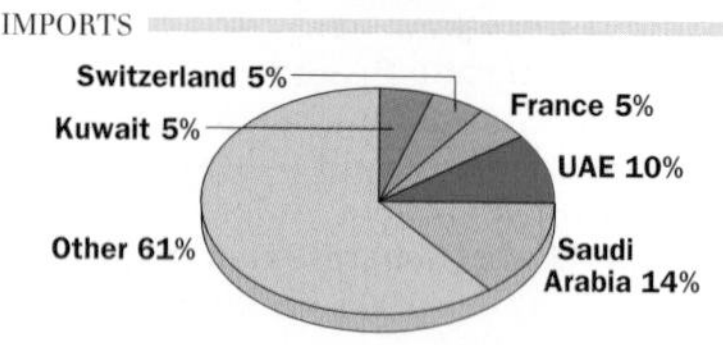
Switzerland 5%
France 5%
Kuwait 5%
UAE 10%
Other 61%
Saudi Arabia 14%

Strengths

Rising oil production. Salt mining. Deposits of copper, gold, lead, zinc, and molybdenum. Industries include oil refining, chemicals, foodstuffs, cement, leather. Improving private sector.

Weaknesses

Political instability deters foreign investment. Heavy economic damage caused by civil war. Well-organized black market undermines tax base. Large balance-of-payments deficit. Dependence on subsistence agriculture. Burden of high population growth, leading to unemployment.

Profile

Unification in 1990 aimed to transform the economy, particularly through the exploitation of large oil and gas reserves, discovered in 1984; exports of oil began in 1987. Industrial investment around Aden was planned. These policies were severely affected by the 1990–1991 Gulf War. In addition, the expulsion of over one million Yemeni guest workers from Saudi Arabia imposed a huge burden on the economy and ended the flow of workers' remittances.

The 1994 civil war inflicted further serious damage on the economy – oil refineries, communications centers, and water systems were destroyed. Economic crisis forced the government to reduce expenditure and subsidies on certain staple foods. This provoked widespread civil unrest (there were particularly violent demonstrations in 1998) and encouraged many farmers to switch from food crops, such as wheat, to growing the more profitable narcotic plant *qat*. As a result, Yemen has increasingly had to import food supplies. Lack of central control, poor integration, and the persistence of patronage politics are hampering economic revival.

ECONOMIC PERFORMANCE INDICATOR

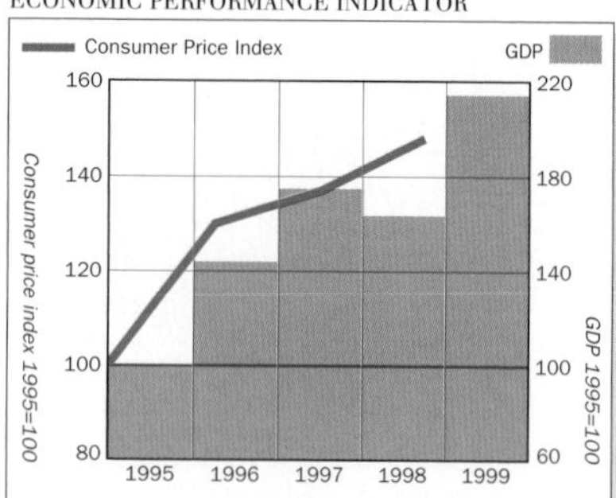
Consumer Price Index
GDP
160
140
120
100
80
220
180
140
100
60
Consumer price index 1995=100
GDP 1995=100
1995
1996
1997
1998
1999

YEMEN : MAJOR BUSINESSES

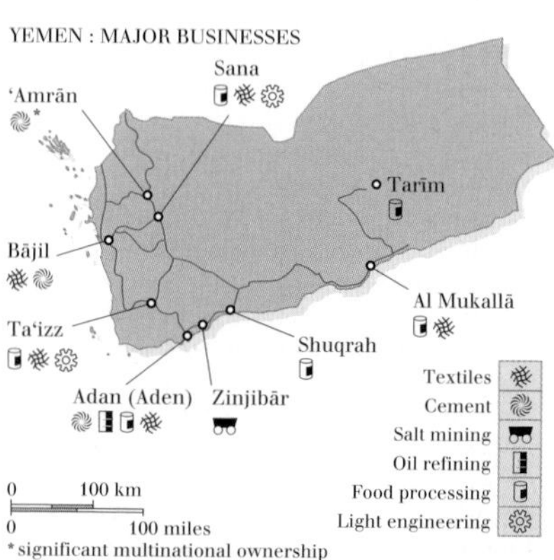
Sana
'Amrān
Tarīm
Bājil
Al Mukallā
Ta'izz
Shuqrah
Adan (Aden)
Zinjibār
Textiles
Cement
Salt mining
Oil refining
Food processing
Light engineering
0 100 km
0 100 miles
* significant multinational ownership

RESOURCES

Electric power 810,000 kw

123,252 tonnes

458,000 b/d (reserves 4bn barrels)

4.8m sheep, 4.25m goats, 1.34m cattle, 29.6m chickens

Oil, natural gas, salt, copper, gold, lead, zinc, molybdenum

ELECTRICITY GENERATION

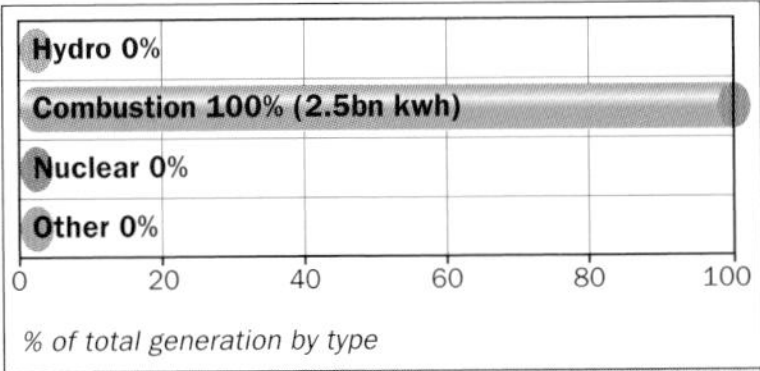
Hydro 0%
Combustion 100% (2.5bn kwh)
Nuclear 0%
Other 0%
0
20
40
60
80
100
% of total generation by type

Oil reserves are considerable, though initial estimates were exaggerated. The 2000 border agreement with Saudi Arabia promises better Yemeni access to oil fields. Salt is the only other mineral to be commercially exploited, and its production continues to grow steadily.

The agricultural sector employs well over half the working population. Cotton is a cash crop. Livestock and livestock products, including dairy products and hides, are mainstays of the north. Yemen's rich fishing grounds in the Arabian Sea now provide a major source of earnings, despite poor equipment. Ambitious plans for an Aden free port, first revealed in 1994, have yet to be realized.

Yemen's population growth, averaging around 3.5 per cent a year and among the highest in the world, is putting severe strain on the country's natural resources, especially water.

YEMEN : LAND USE

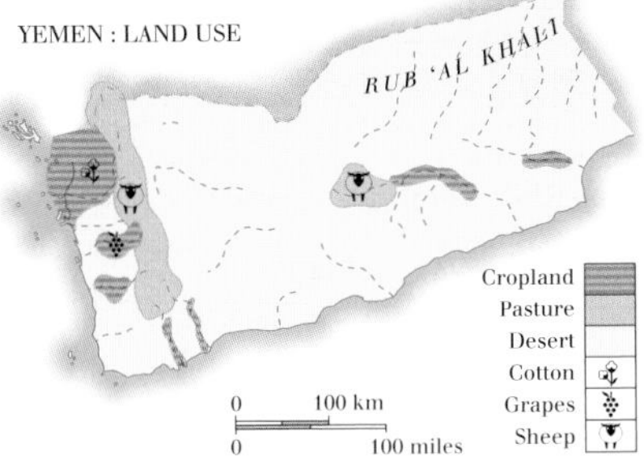
RUB 'AL KHALI
Cropland
Pasture
Desert
Cotton
Grapes
Sheep
0
100 km
0
100 miles

ENVIRONMENT

Not available

None
0.9 tonnes per capita

ENVIRONMENTAL TREATIES

No
Yes
Yes
Yes
Yes
No

Large areas remain untouched by development, preserving habitats for rare birds. Problems include water scarcity, overgrazing, and soil erosion.

MEDIA

TV ownership low

Daily newspaper circulation 15 per 1000 people

PUBLISHING AND BROADCAST MEDIA

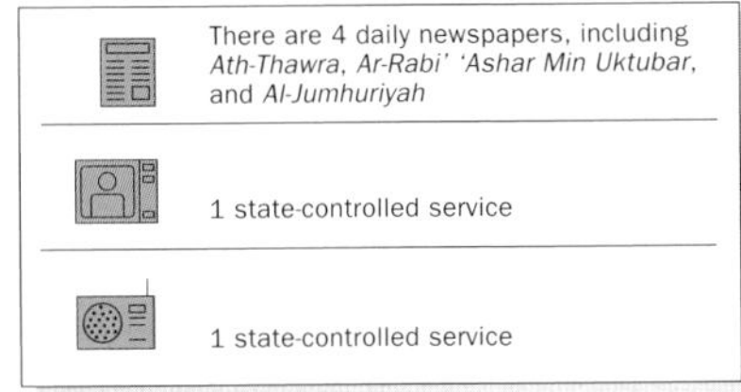
There are 4 daily newspapers, including *Ath-Thawra, Ar-Rabi' 'Ashar Min Uktubar,* and *Al-Jumhuriyah*
1 state-controlled service
1 state-controlled service

Yemen has a long, distinguished tradition of intellectual debate, and legislation embodies freedom of the press, which, however, is poorly developed. The government keeps tight control of the media and vets the entry of foreign journalists. TV and radio are state-controlled and have a limited range around the principal cities. Satellite TV is not generally available. Ownership of radio and TV receivers is low, with only a tiny minority owning a television set.

CRIME

Death penalty in use

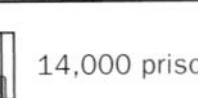
14,000 prisoners

Up 6% 1997–1999

CRIME RATES

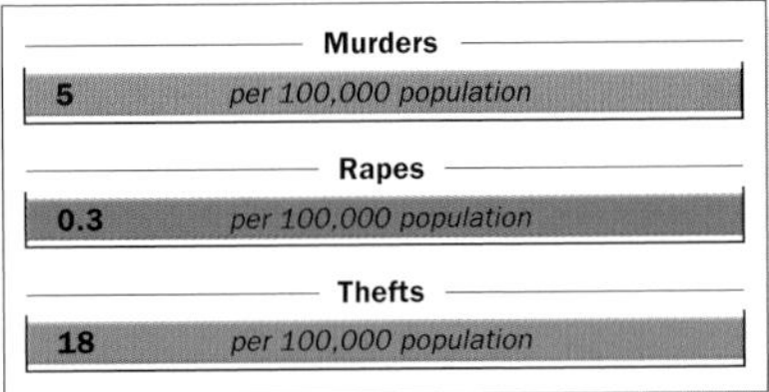
Murders
5
per 100,000 population
Rapes
0.3
per 100,000 population
Thefts
18
per 100,000 population

Political assassinations continue to threaten political stability. There is little formal law enforcement outside the main cities; foreign companies risk kidnappings and theft by Bedouin raiders. There is a proliferation of illicit weapons; the number of firearms has been estimated at 50 million – nearly three times the population size. Some blame lawlessness on the narcotic, *qat*.

EDUCATION

School leaving age: 15

46%

65,675 students

THE EDUCATION SYSTEM

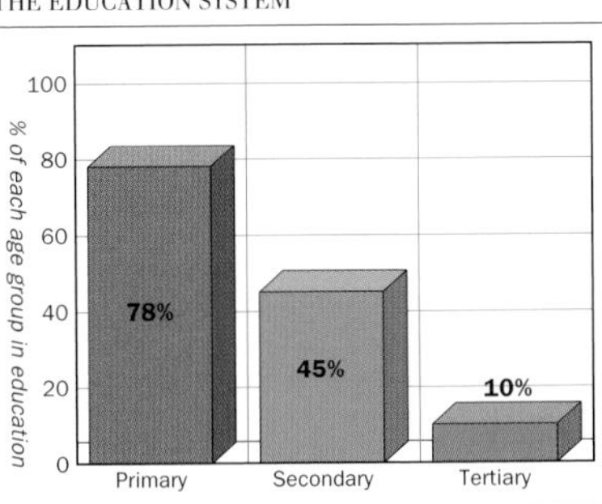
% of each age group in education
100
80
60
40
20
0
78%
45%
10%
Primary
Secondary
Tertiary

Some 80% of the population have had no formal classroom education. Schooling barely extends into the rural areas. Illiteracy is especially high among women: 75% cannot read or write. Only 13% of students at Yemen's two universities – Sana and Aden – are female. Yemen also has some technical colleges. The government's unpopular economic policies have encouraged student activism.

HEALTH

No welfare state health benefits

1 per 5000 people

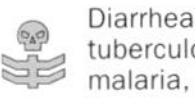
Diarrheal diseases, tuberculosis, malaria, bilharzia

The major cities have an adequate primary health care system. A new 300-bed hospital in Sana is due to be completed in 2004. Rural areas are less well served. Health services are under threat from tribal gangs. In 2002 the World Bank approved a $27 million credit for a Health Reform Support Project to increase health provisions in poor communities.

SPENDING

GDP/cap. increase

CONSUMPTION AND SPENDING

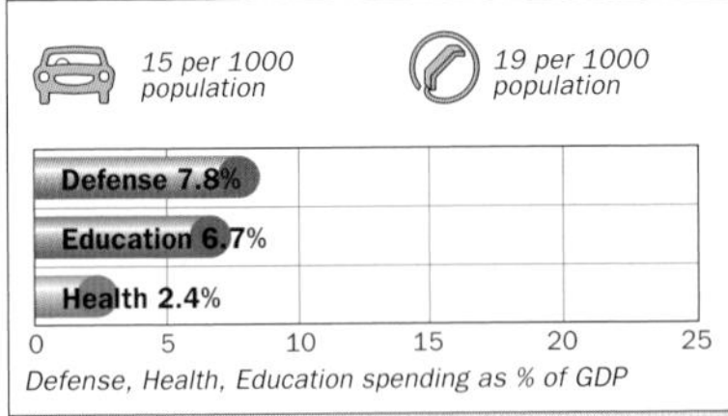
15 per 1000 population
19 per 1000 population
Defense 7.8%
Education 6.7%
Health 2.4%
0
5
10
15
20
25
Defense, Health, Education spending as % of GDP

Most Yemenis suffered a fall in living standards after Saudi Arabia expelled its Yemeni workers. A lack of jobs in other Gulf states has fueled unemployment, estimated at around 30%. Except for a small elite, the ownership of consumer goods is low.

WORLD RANKING

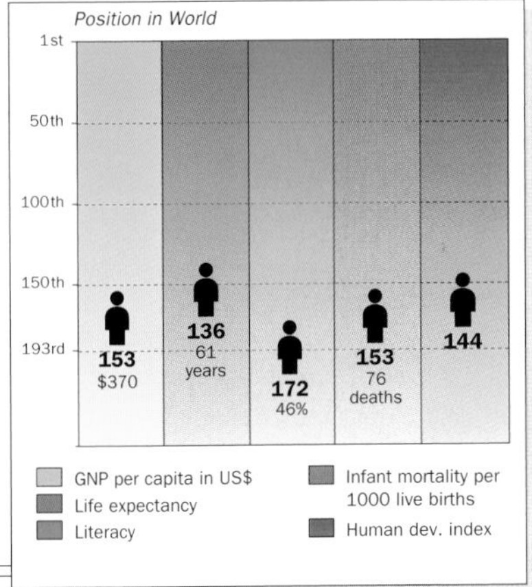
Position in World
1st
50th
100th
150th
193rd
153
$370
136
61
years
172
46%
153
76
deaths
144
GNP per capita in US$
Life expectancy
Literacy
Infant mortality per 1000 live births
Human dev. index

Y

ZAMBIA

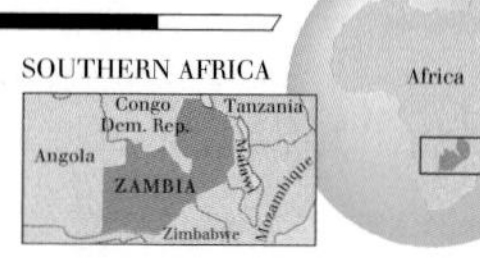
SOUTHERN AFRICA
Congo
Dem. Rep.
Tanzania
Angola
ZAMBIA
Malawi
Mozambique
Zimbabwe
Africa

OFFICIAL NAME: Republic of Zambia **CAPITAL:** Lusaka
POPULATION: 10.6 million **CURRENCY:** Zambian kwacha **OFFICIAL LANGUAGE:** English

1964

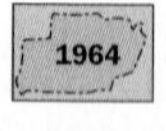
1964

Oct 24
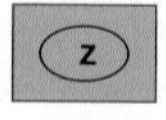
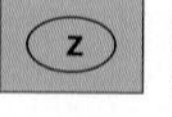
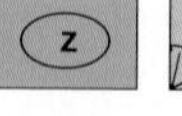
Z
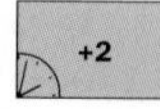
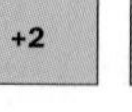
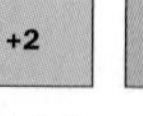
+2

+260

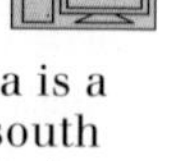
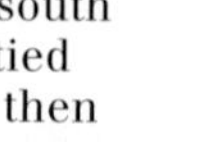
.zm

LYING IN THE HEART of southern Africa, Zambia is a country of upland plateaus, bordered to the south by the Zambezi River. Its economic fortunes are tied to the copper industry. Falling copper prices in the late 1970s, and then the growing inaccessibility of remaining reserves, have led to a severe decline in the economy. In 1991, Zambia achieved a peaceful transition from single-party rule to multiparty democracy.

CLIMATE

Tropical wet & dry

WEATHER CHART FOR LUSAKA

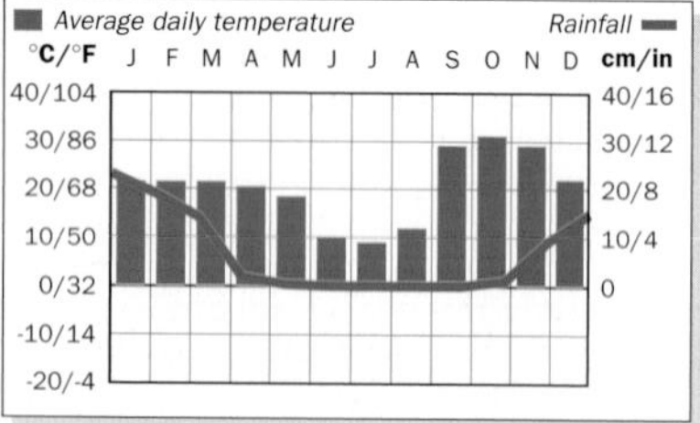
Average daily temperature
Rainfall
°C/°F
J F M A M J J A S O N D
cm/in
40/104
30/86
20/68
10/50
0/32
-10/14
-20/-4
40/16
30/12
20/8
10/4
0

Zambia has a tropical climate, with rains from November to April. The southwest is prone to drought.

TRANSPORTATION

Drive on left

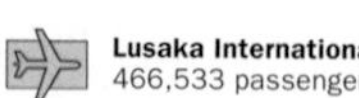
Lusaka International 466,533 passengers

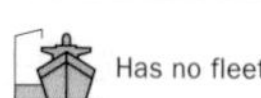
Has no fleet

THE TRANSPORTATION NETWORK

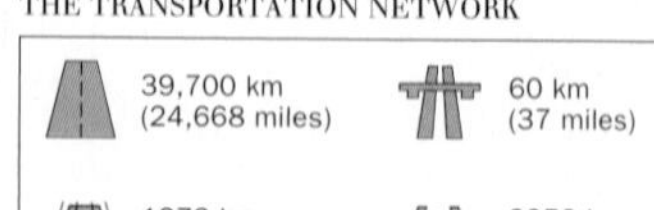
39,700 km (24,668 miles)

60 km (37 miles)

1273 km (791 miles)

2250 km (1398 miles)

The poor rail and road networks, in need of urgent rehabilitation, could sabotage economic recovery. Zambian Airways was liquidated in 1994, and private airlines are now in operation.

TOURISM

Visitors : Population 1:18

574,000 visitors

Up 26% in 2000

MAIN TOURIST ARRIVALS

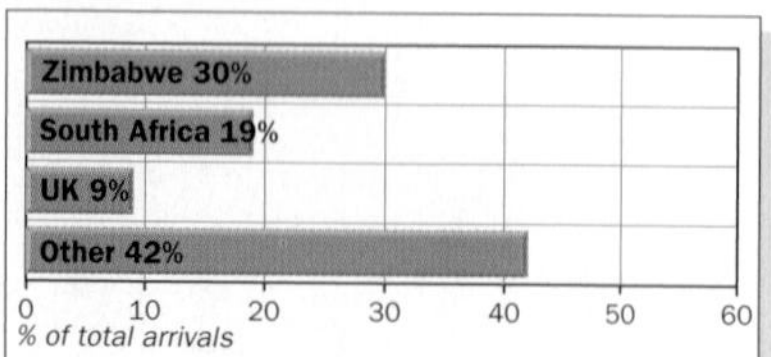
Zimbabwe 30%
South Africa 19%
UK 9%
Other 42%
0 10 20 30 40 50 60
% of total arrivals

Wildlife, the Victoria Falls, and white-water rafting are major attractions. Recent increases in tourism have been at the expense of neighboring Zimbabwe.

PEOPLE

Pop. density low

Bemba, Nyanja, Tonga, Kaonde, Lunda, Luvale, Lozi, English

14/km² (37/mi²)

THE URBAN/RURAL POPULATION SPLIT

45% 55%

RELIGIOUS PERSUASION

Muslim and Hindu communities 1%

Indigenous beliefs 36%

Christian 63%

Although ethnically heterogeneous, with more than 70 different groups, Zambia has been less affected by ethnic tension than many African states. The largest group, about 34% of the population, is the Bemba, who live in the northeast and also predominate in the central Copperbelt. Other major groups are the southern Tonga, the eastern Nyanja, and the Lozi in the west. There are also some 225,000 refugees, mainly from Angola.

Zambia's main urban area is the Copperbelt, where many third- and fourth-generation town dwellers live. Some half a million children are employed there in hazardous conditions. The rural population lives mainly by subsistence farming.

A National Gender Policy was issued in October 2000 to redress inequalities between the sexes.

Musi-o-Tunya (The Smoke That Thunders), known in English as Victoria Falls. Spray from the falls can be seen 30 km (20 miles) away.

POLITICS

Multiparty elections

2001/2006

President Levy Mwanawasa

AT THE LAST ELECTION

National Assembly 158 seats

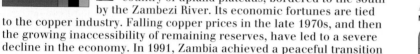

43% MMD · 30% UPND · 8% UNIP · 5% App · 5% Others · 8% FDD · 1% Vacant

MMD = Movement for Multiparty Democracy **UPND** = United Party for National Development **UNIP** = United National Independence Party **FDD** = Forum for Democracy and Development **App** = Appointed

Up to eight members are appointed by the president, and the speaker is also a member

The 1991 defeat of Kenneth Kaunda, in the first multiparty elections for 19 years, expressed popular discontent with the ailing economy and official corruption. President Frederick Chiluba and the MMD government made little headway in revitalizing the economy, despite socially painful reforms. Chiluba himself, once widely respected as a rallying point for democratic opposition to Kaunda, became much criticized for arbitrary and authoritarian rule, and for failing to address poverty and HIV/AIDS. In 2001 Chiluba purged the MMD of key opponents, prompting the formation of new opposition parties, and made way for his chosen successor, Levy Mwanawasa, to win disputed elections.

WORLD AFFAIRS

Joined UN in 1964

Comm

ACP

NAM

AU

SADC

Zambia led Africa's opposition to apartheid South Africa and now enjoys close links with Pretoria. It also has a significant role as a mediator in neighboring conflicts.

AID

Recipient

$795m (receipts)

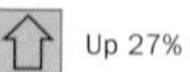
Up 27% in 2000

Regional drought increased the need for aid in 2002. International donors pledged $1.3 billion, returning aid to the yearly levels seen before a freeze in 1997 prompted by state corruption. The IMF agreed a $3.8 billion debt service relief package in 2000.

DEFENSE

No compulsory military service

$65m

Down 26% in 2000

Despite the relatively small budget, the 21,600-strong armed forces are well equipped. Security along the Angolan border is a main concern.

ECONOMICS

Inflation 51% p.a. (1990–2000)

$3.03bn

4500–3810 Zambian kwacha

SCORE CARD

- ❑ WORLD GNP RANKING........................131st
- ❑ GNP PER CAPITA $300
- ❑ BALANCE OF PAYMENTS....................–$269m
- ❑ INFLATION 27.3%
- ❑ UNEMPLOYMENT..................................50%

STRENGTHS

Potential for self-sufficiency in food. Boom in new export crops, such as cotton and flowers. Minerals, notably copper, cobalt, and coal. Market-oriented reforms and privatization drive attracting foreign private investors.

WEAKNESSES

Falling value of copper, on which Zambia depends for 90% of export earnings. Domestic reserves declining. Shortage of finance for restructuring, exacerbated by aid donors' boycott. High inflation, negative growth, and serious drought in 1998. Arable land underutilized. Delays in privatizing state-owned copper consortium.

EXPORTS

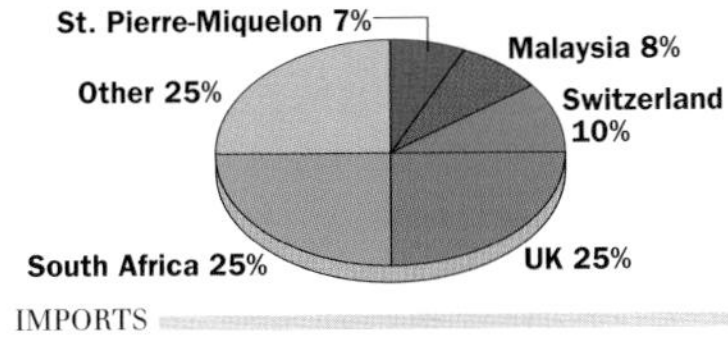
St. Pierre-Miquelon 7%
Malaysia 8%
Other 25%
Switzerland 10%
South Africa 25%
UK 25%

IMPORTS

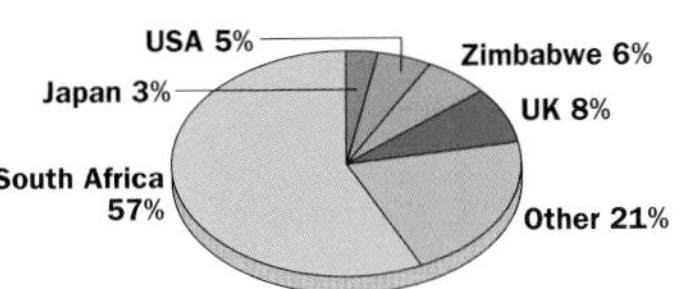
USA 5%
Zimbabwe 6%
Japan 3%
UK 8%
South Africa 57%
Other 21%

RESOURCES

Electric power 2.3m kw

71,507 tonnes

Not an oil producer

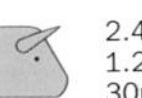

2.4m cattle, 1.27m goats, 30m chickens

Copper, cobalt, coal, zinc, lead, gold, emeralds, amethysts

Despite declining reserves, copper is still the key resource; Zambia remains one of the world's leading producers. It also has rich hydropower potential.

ENVIRONMENT

Sustainability rank: 69th

9%

0.2 tonnes per capita

Drought is a recurrent hazard. Rhinos are almost extinct as a result of poaching. Revenues from legal hunting are being channeled into villages to encourage support for conservation.

MEDIA

TV ownership medium

Daily newspaper circulation 14 per 1000 people

PUBLISHING AND BROADCAST MEDIA

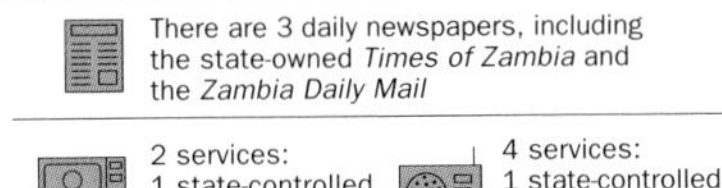

There are 3 daily newspapers, including the state-owned *Times of Zambia* and the *Zambia Daily Mail*

2 services: 1 state-controlled, 1 educational

4 services: 1 state-controlled, 3 independent

Broadcasting is dominated by the government. Opposition journalists have been accused of treason.

CRIME

Death penalty in use

12,500 prisoners

Crime is rising

Cases of violent crime, burglary, and rape are rising rapidly. In 1998 Zambia promised to overhaul its prison and police services.

CHRONOLOGY

Northern Rhodesia was developed by Britain solely for its copper. The UNIP, led by Kenneth Kaunda, took power at Zambian independence in 1964.

- ❑ **1972** UNIP one-party government.
- ❑ **1982–1991** Austerity measures and corruption: pressure for democracy.
- ❑ **1991** MMD government elected; Frederick Chiluba defeats Kaunda.
- ❑ **1996** Controversial elections.
- ❑ **2002** Levy Mwanawasa president.

EDUCATION

School leaving age: 14

78%

22,701 students

Primary education is compulsory. Fees for secondary students have affected the already very low attendance rate.

HEALTH

Welfare state health benefits

1 per 10,000 people

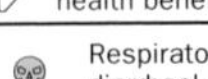

Respiratory infections, diarrheal diseases, AIDS, malaria

An HIV prevention program is reducing rates of infection in Lusaka. More than 25% of town dwellers are HIV-positive.

SPENDING

GDP/cap. increase

CONSUMPTION AND SPENDING

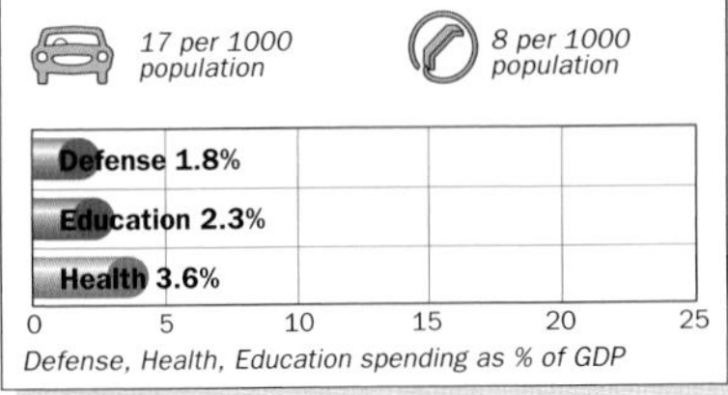
17 per 1000 population
8 per 1000 population
Defense 1.8%
Education 2.3%
Health 3.6%
0
5
10
15
20
25
Defense, Health, Education spending as % of GDP

Standards of living for most Zambians are now lower in real terms than at independence in 1964. Many people lack basic nutrition.

WORLD RANKING

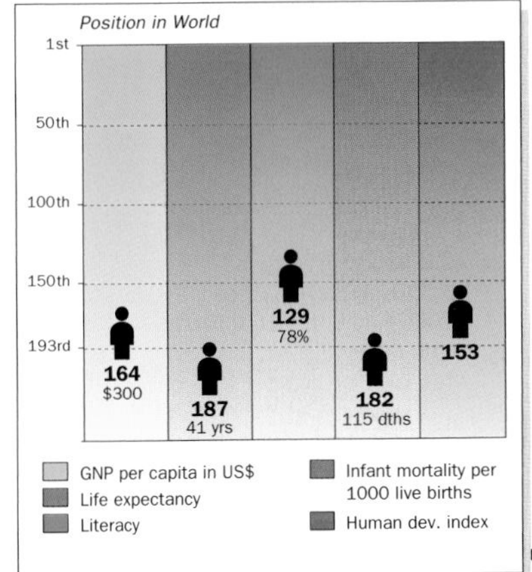
Position in World
1st
50th
100th
150th
193rd
164
$300
187
41 yrs
129
78%
182
115 dths
153
GNP per capita in US$
Life expectancy
Literacy
Infant mortality per 1000 live births
Human dev. index

ZAMBIA

Total Land Area : 752 614 sq. km (290 584 sq. miles)

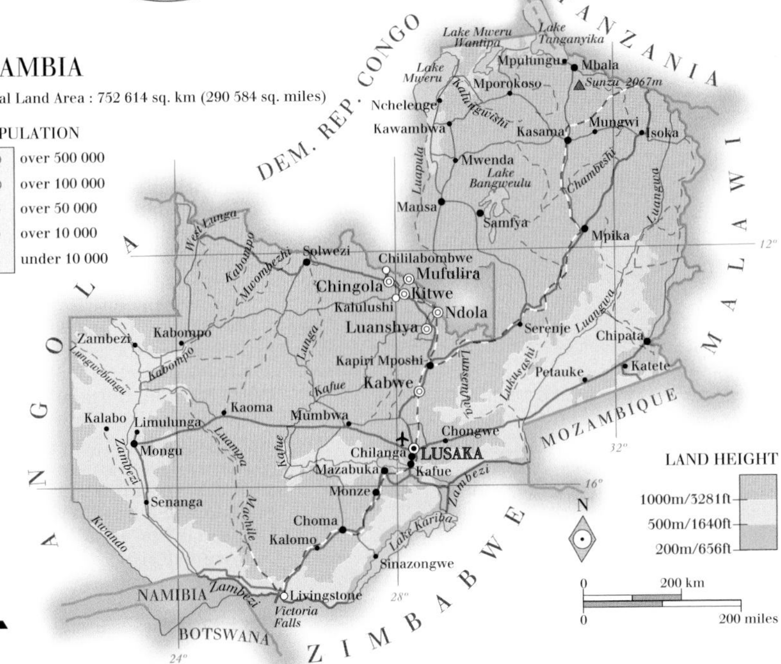
POPULATION
over 500 000
over 100 000
over 50 000
over 10 000
under 10 000
DEM. REP. CONGO
TANZANIA
MALAWI
MOZAMBIQUE
ZIMBABWE
ANGOLA
NAMIBIA
BOTSWANA
Lake Mweru Wantipa
Lake Tanganyika
Lake Mweru
Mpulungu
Mbala
Sunzu 2067m
Mporokoso
Nchelenge
Kawambwa
Kasama
Mungwi
Isoka
Mwenda
Lake Bangweulu
Mansa
Samfya
Mpika
Solwezi
Chililabombwe
Mufulira
Chingola
Kitwe
Kalulushi
Ndola
Luanshya
Serenje
Chipata
Katete
Zambezi
Kabompo
Kapiri Mposhi
Kabwe
Petauke
Kaoma
Kalabo
Limulunga
Mongu
Mumbwa
Chongwe
Chilanga
LUSAKA
Kafue
Mazabuka
Monze
Senanga
Choma
Kalomo
Lake Kariba
Siavonga
Sinazongwe
Livingstone
Victoria Falls
LAND HEIGHT
1000m/3281ft
500m/1640ft
200m/656ft
200 km
200 miles

ZIMBABWE

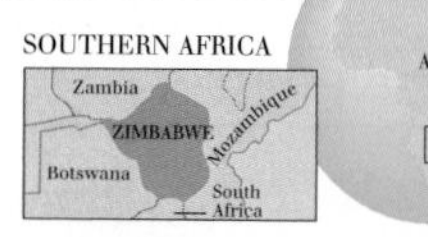
SOUTHERN AFRICA
Zambia
ZIMBABWE
Mozambique
Botswana
South Africa
Africa

OFFICIAL NAME: Republic of Zimbabwe **CAPITAL:** Harare
POPULATION: 12.9 million **CURRENCY:** Zimbabwe dollar **OFFICIAL LANGUAGE:** English

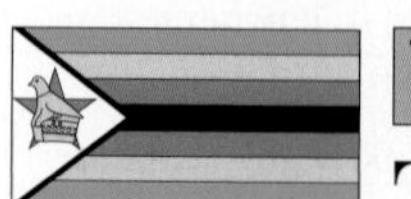

1980

1980

April 18
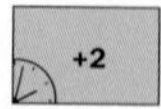
ZW

+2

+263

.ZW

ZIMBABWE IS SITUATED in southern Africa. Its upland center is crisscrossed by rivers flowing into Lake Kariba and the Zambezi River, on which lies the region's most spectacular natural feature, the Victoria Falls (Musi-o-Tunya). Attempts to preserve white rule in the former British colony led to a long guerrilla war before independence in 1980. Robert Mugabe, the country's leader since then and its president since 1987, has become increasingly authoritarian and divisive. Violent seizure of white-owned farmland and severe drought have contributed to virtual economic collapse.

The Kariba Dam, which has created the vast Lake Kariba on the Zambezi River, lies on Zimbabwe's northwest border with Zambia.

CLIMATE

Tropical wet & dry/ steppe

WEATHER CHART FOR HARARE

■ *Average daily temperature* — *Rainfall*

°C/°F J F M A M J J A S O N D cm/in

40/104, 30/86, 20/68, 10/50, 0/32, -10/14, -20/-4 — 40/16, 30/12, 20/8, 10/4, 0

Because of its altitude, Zimbabwe is comparatively temperate for a country in the tropics; humidity is also low. The rainy season occurs between November and March but, with the exception of the eastern highlands, rainfall is erratic and drought is common. Annual rainfall ranges from 140 cm (55 in) in the eastern highlands to 40 cm (16 in) in the Limpopo valley.

TRANSPORTATION

Drive on left

Harare International
1.02m passengers

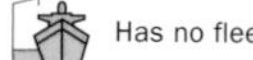
Has no fleet

THE TRANSPORTATION NETWORK

Roads	8692 km (5401 miles)	Motorways	None
Railways	2592 km (1611 miles)	Waterways	Lake Kariba

The number of international air links is being increased. Zimbabwe's rail network, among the densest in sub-Saharan Africa, is being updated.

TOURISM

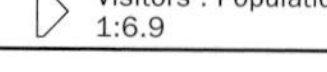
Visitors : Population 1:6.9

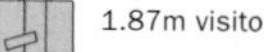
1.87m visitors

Down 11% in 2000

MAIN TOURIST ARRIVALS

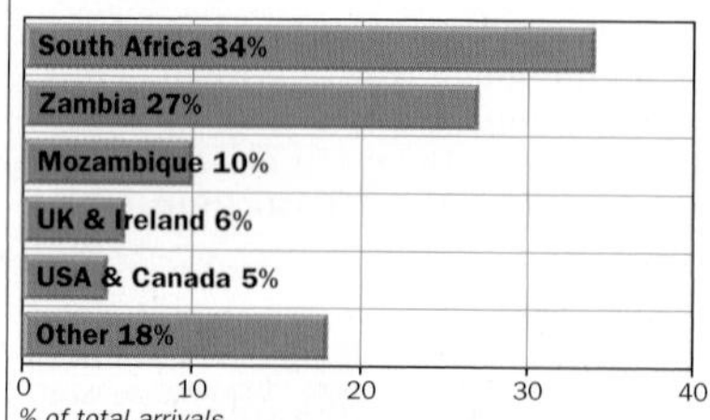
South Africa 34%
Zambia 27%
Mozambique 10%
UK & Ireland 6%
USA & Canada 5%
Other 18%
0
10
20
30
40
% of total arrivals

Zimbabwe's principal attractions are the Victoria Falls, the Kariba Dam, numerous national parks, the Great Zimbabwe ruins near Masvingo, and World's View in the Matopo Hills.

Invasions led by "war veterans" of large commercial farms, and the violence in the run-up to the parliamentary elections in 2000, put Zimbabwe on the list of unsafe destinations for many visitors. Fuel and foreign currency shortages have further undermined the tourism sector.

In addition to these factors, Zimbabwe is wary that mass-market tourism might seriously damage the environment. However, the lure of foreign exchange has encouraged the development in Harare of conference facilities and vacation complexes, such as Elephant Hills, around Victoria Falls. State law requires that there be 30% local ownership of tourist ventures.

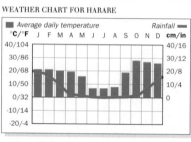

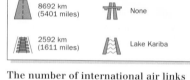

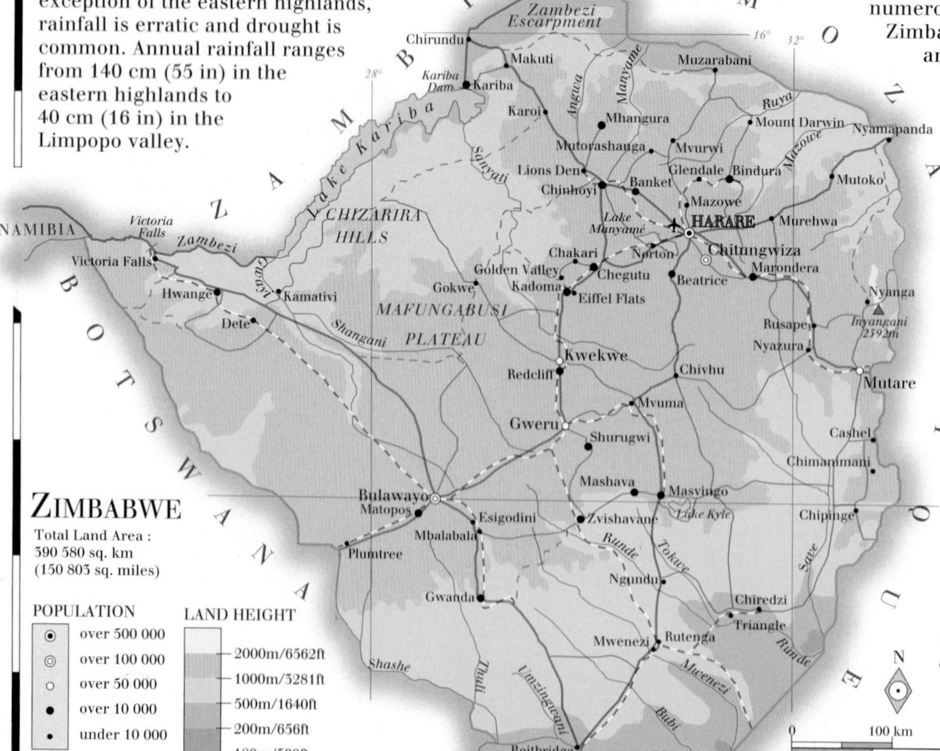
ZAMBIA
MOZAMBIQUE
BOTSWANA
NAMIBIA
SOUTH AFRICA
Zambezi
Zambezi Escarpment
Chirundu
Makuti
Kariba Dam
Kariba
Lake Kariba
Karoi
Muzarabani
Mhangura
Mutorashanga
Mvurwi
Glendale
Bindura
Mount Darwin
Nyamapanda
Mutoko
Lions Den
Chinhoyi
Banket
Mazowe
HARARE
Murehwa
Chitungwiza
Marondera
Lake Manyame
Norton
Chakari
Chegutu
Golden Valley
Kadoma
Eiffel Flats
Beatrice
Gokwe
Victoria Falls
Hwange
Kamativi
Dete
CHIZARIRA HILLS
MAFUNGABUSI PLATEAU
Shangani
Kwekwe
Redcliff
Chivhu
Rusape
Nyazura
Nyanga
Inyangani 2592m
Mutare
Mvuma
Gweru
Shurugwi
Cashel
Chimanimani
Mashava
Masvingo
Lake Kyle
Chipinge
Bulawayo
Matopos
Esigodini
Zvishavane
Mbalabala
Plumtree
Ngundu
Gwanda
Chiredzi
Triangle
Mwenezi
Rutenga
Beitbridge
Limpopo
Shashe
Thuli
Umzingwani
Bubi
Mwenezi
Runde
Tokwe
Save
Ruya
Mazowe
Sanyati
Angwa
Manyame
Gwayi
16°
20°
28°
32°
100 km
100 miles
N

ZIMBABWE
Total Land Area : 390 580 sq. km (150 803 sq. miles)

POPULATION
- over 500 000
- over 100 000
- over 50 000
- over 10 000
- under 10 000

LAND HEIGHT
- 2000m/6562ft
- 1000m/3281ft
- 500m/1640ft
- 200m/656ft
- 180m/590ft

PEOPLE

Pop. density low

Shona, Ndebele, English

33/km² (86/mi²)

THE URBAN/RURAL POPULATION SPLIT

35% 65%

RELIGIOUS PERSUASION

Other (including Muslim) 1%

Traditional beliefs 24%

Syncretic (Christian/ Traditional beliefs) 50%

Christian 25%

ETHNIC MAKEUP

White 1%

Other African 11%

Asian 1%

Ndebele 16%

Shona 71%

There are two main ethnic groups, the majority Shona in the north and the Ndebele in the south. Europeans and Asians comprise 2% of the population.

Ethnic tensions plagued the 1980s. 1500 Ndebele were massacred by the army in 1983 alone as the ruling, Shona-dominated, ZANU–PF attempted to suppress the predominantly Ndebele Zimbabwe African People's Union (PF–ZAPU). A Unity Accord in 1987 eased the conflict, and ZAPU leader Joshua Nkomo was appointed vice president in 1990.

As a legacy of colonial rule, whites remain generally far more affluent than blacks. This imbalance has been redressed in part by policies to improve black education and increase office employment. Land redistribution, which was previously slow and dogged by accusations of corruption, was stepped up in 2000, when the government backed a movement to seize white-owned farms, raising tensions between blacks and whites.

Families are large, and almost half the population is under 15. Zimbabwean society is traditionally patriarchal. In 1999 a Supreme Court ruling provoked protest by according only "junior male" status to black women, especially those marrying under traditional law.

POPULATION AGE BREAKDOWN

Female
Age
Male
0.6%
80+
0.5%
1.6%
60–79
1.6%
5.8%
40–59
5.3%
14.9%
20–39
13.6%
28.3%
0–19
27.8%
% of population by age group

POLITICS

Multiparty elections

2000/2005

President Robert Gabriel Mugabe

AT THE LAST ELECTION

Parliament 150 seats

1% ZANU–Ndonga

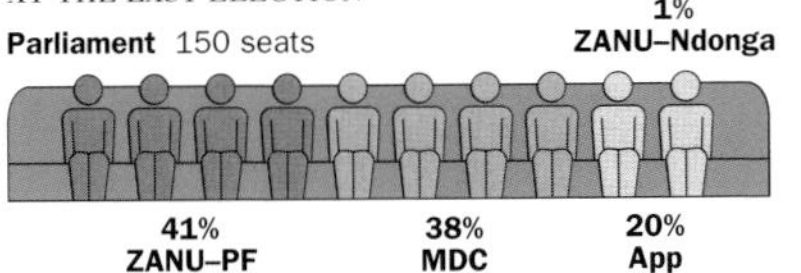

41% ZANU–PF **38% MDC** **20% App**

ZANU–PF = Zimbabwe African National Union
MDC = Movement for Democratic Change **App** = Appointed
ZANU–Ndonga = Zimbabwe African National Union – Ndonga

30 seats are set aside for presidential appointments and traditional chiefs

80% of MPs are elected. The president is directly elected every six years.

PROFILE

ZANU–PF emerged victorious in 1980 from the long struggle with Ian Smith's white-supremacist government. In 1987, rivalries with ZAPU were resolved, and Robert Mugabe led the country in his socialist experiment. In reality repression flourished, and the civil service is closed to non ZANU–PF members.

The influence of ZANU–PF was rocked in 2000 when the MDC won a convincing share of votes despite widespread pre-electoral intimidation. The MDC and its supporters have since been the target of government thugs.

MAIN POLITICAL ISSUES

The rule of President Mugabe

Robert Mugabe was appointed president in 1987. His attempts to create a one-party socialist state were called off in 1991 as opposition to white South Africa and support for the Soviet bloc became meaningless. His ZANU–PF has since lost support in the face of economic collapse and a reinvigorated opposition led by the Movement for Democratic Change (MDC). Mugabe's plans to strengthen his position through a referendum were defeated and ZANU–PF only just held on to power in elections in 2000, making his position increasingly precarious.

Land redistribution

Though it has been generally agreed that the distribution of farmland unfairly favors the white minority, the speed and method of what has been largely belated land redistribution has provoked protest. White-owned farms have been confiscated without compensation, and Mugabe gave his full support to the violent occupations led by self-styled "war veterans." The seizure of white-owned farms has been blamed for the huge drop in grain production in 2001–2002.

Robert Mugabe, *elected prime minister in 1980 and president in 1987.*

Morgan Tsvangirai, *leader of the opposition MDC.*

AID

Recipient

US$178m (receipts)

Down 27% in 2000

Bilateral aid fell by a third between 1994 and 1996, after donors including the UK, Denmark, the US, France, and Germany learned that aid intended for small farmers and indigenous enterprises was siphoned off to large industrial projects. In 1998 the IMF approved US$175 million in standby credit. However, political violence since 2000 has prompted most international aid donors to suspend financial support. The UNDP and USAID agreed additional food aid in 2002 to combat massive food shortages.

WORLD AFFAIRS

Joined UN in 1980

WTO
G15
NAM
AU
SADC

Zimbabwe is an active member of the SADC and the Preferential Trade Area for East and South Africa. Relations with postapartheid South Africa are particularly strong. Zimbabwean troops have been active in the DRC since 1998.

This strong involvement in African affairs and Mugabe's own role as an anticolonial champion have led to regional support for the confrontational policy against white commercial farmers. However, his increasingly antidemocratic stance has brought near-total isolation from the wider international community. The EU and the UK have been particularly vociferous in their condemnation of Mugabe's regime, and sanctions and aid suspensions have contributed to the perilous state of the economy. In 2002 South Africa and Nigeria were participants in the decision to suspend Zimbabwe from the Commonwealth, in a rare display of African displeasure at Mugabe's "unfree and unfair" reelection.

CHRONOLOGY

In 1953, the British colony of Southern Rhodesia became part of the Federation of Rhodesia and Nyasaland with Northern Rhodesia (now Zambia) and Nyasaland (now Malawi).

- ❑ **1961** Joshua Nkomo forms ZAPU.
- ❑ **1962** ZAPU banned. Segregationist Rhodesian Front (RF) wins polls.
- ❑ **1963** African nationalists in Northern Rhodesia and Nyasaland demand dissolution of Federation. ZANU, offshoot of ZAPU, formed by Rev. Sithole and Robert Mugabe.
- ❑ **1964** New RF prime minister Ian Smith rejects British demands for majority rule. ZANU banned.
- ❑ **1965** May, RF reelected. November, state of emergency declared (renewed until 1990). Smith's unilateral declaration of independence. UK imposes economic sanctions. ANC, ZANU, and ZAPU begin guerrilla war.
- ❑ **1974** RF regime agrees cease-fire terms with African nationalists.
- ❑ **1976** ZANU and ZAPU unite as Patriotic Front (PF).
- ❑ **1977** PF backed by "frontline" African states: Mozambique, Tanzania, Botswana, and Zambia.
- ❑ **1979** Lancaster House talks produce agreement on constitution.
- ❑ **1980** Independence as Zimbabwe. Following violent election campaign, Mugabe becomes prime minister of ZANU–PF/ ZAPU–PF coalition. Relations severed with South Africa.
- ❑ **1983–1984** Unrest in Matabeleland, ZAPU–PF's power base.
- ❑ **1985** Elections return ZANU–PF, with manifesto to create one-party state. Many ZAPU–PF members arrested.
- ❑ **1987** September, provision for white seats in parliament abolished. December, ZANU–PF and ZAPU–PF sign unity agreement (merge in 1989). Mugabe elected president.
- ❑ **1990** Elections won by ZANU–PF. Mugabe reelected president.
- ❑ **1991** Mugabe abandons plan for one-party state.
- ❑ **1999** Death of Vice President Nkomo. Opposition forms MDC.
- ❑ **2000** Government loses referendum on new constitution. Expropriations of white-owned farmland by squatters. Strong MDC performance in polls. ZANU–PF accused of using intimidation to retain majority.
- ❑ **2002** Mugabe reelected in flawed poll. Commonwealth membership suspended. Government deadline for white farmers to leave land. Threat of mass starvation and economic collapse.

DEFENSE

No compulsory military service

 US$394m  Down 6% in 2000

Nationalist guerrillas were the heroes of independence in 1980. By the late 1990s, however, resentment grew when ex-combatants demanded enormous pensions.

Although formally nonaligned, Zimbabwe supported the Mozambican regime against RENAMO guerrillas and backed the US-led operation in Somalia in 1992–1995. The withdrawal of troops from the Democratic Republic of the Congo (DRC), dispatched there in 1998 to help President Laurent Kabila fight rebels, began in April 2001, following the Lusaka peace accord. Although the withdrawal was officially completed in 2002, troops remain in the DRC.

Zimbabwe has in the past received military aid and training from the UK and South Korea.

ZIMBABWEAN ARMED FORCES

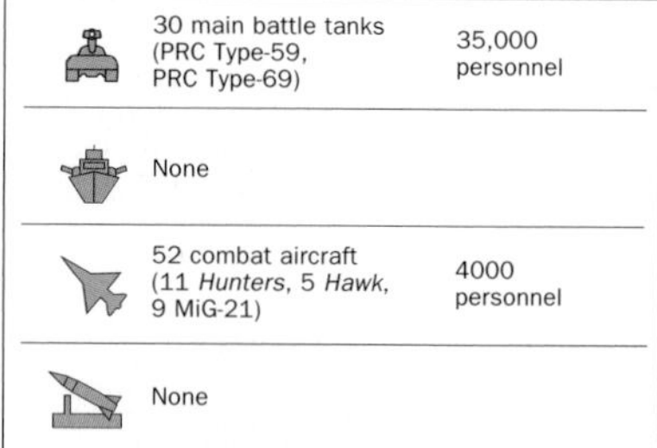
30 main battle tanks (PRC Type-59, PRC Type-69)
35,000 personnel
None
52 combat aircraft (11 *Hunters*, 5 *Hawk*, 9 MiG-21)
4000 personnel
None

ECONOMICS

Inflation 26% p.a. (1990–2000)

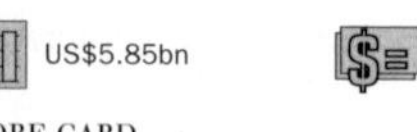
US$5.85bn
55.10–55.45 Zimbabwe dollars

SCORE CARD

❑ World GNP Ranking	105th
❑ GNP per Capita	US$460
❑ Balance of Payments	–US$425m
❑ Inflation	58.5%
❑ Unemployment	50%

EXPORTS

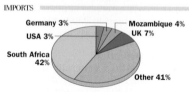

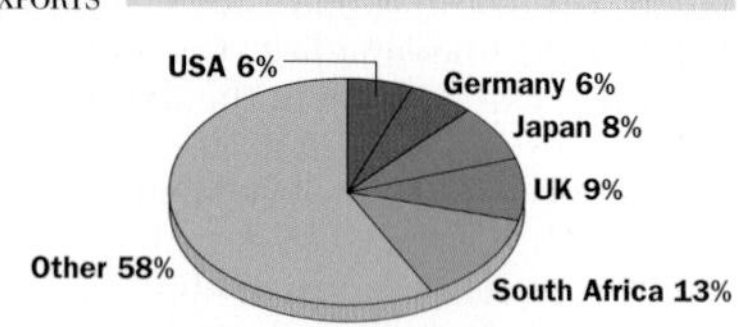
USA 6%
Germany 6%
Japan 8%
UK 9%
Other 58%
South Africa 13%

IMPORTS

Germany 3%
Mozambique 4%
USA 3%
UK 7%
South Africa 42%
Other 41%

ECONOMIC PERFORMANCE INDICATOR

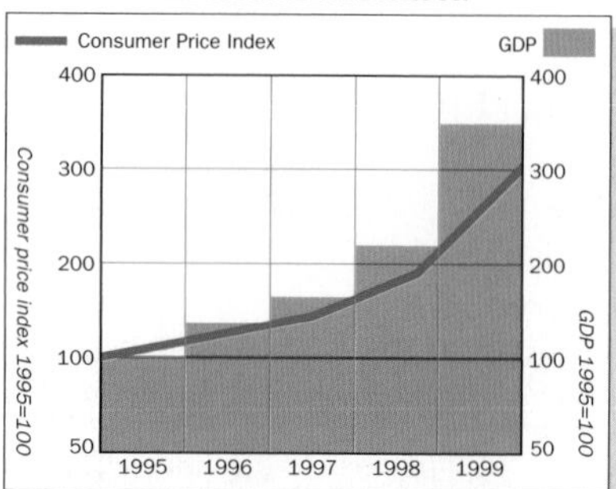
Consumer Price Index
GDP
400
300
200
100
50
Consumer price index 1995=100
GDP 1995=100
1995
1996
1997
1998
1999

Strengths

Most broadly based African economy after South Africa. Sound infrastructure. Virtual self-sufficiency in energy. Founder member of regional free trade area. Gold, coal, tobacco, horticulture.

Weaknesses

Agricultural and hydroelectric output affected by drought. Large budget deficits. High unemployment and inflation. Labor unrest, bank collapses, food price riots. Currency value halved in 1998; devalued again in 2000. Cheap imports damage local industries. Political violence and economic collapse in 2000 scared off investors. Massive food shortages.

Profile

The socialist policies of the 1980s were superseded by a more market-oriented economy, introduced in 1991, which increased unemployment and inflation. Prospects for the mining industry appear particularly bleak: privatization of state copper interests was repeatedly delayed, while collapsing mineral prices have forced the closure of diamond, gold, platinum, and chromium mines. The cost of living and inflation soared during the economic chaos of 2000. Prices for basic provisions were fixed in 2001.

ZIMBABWE : MAJOR BUSINESSES

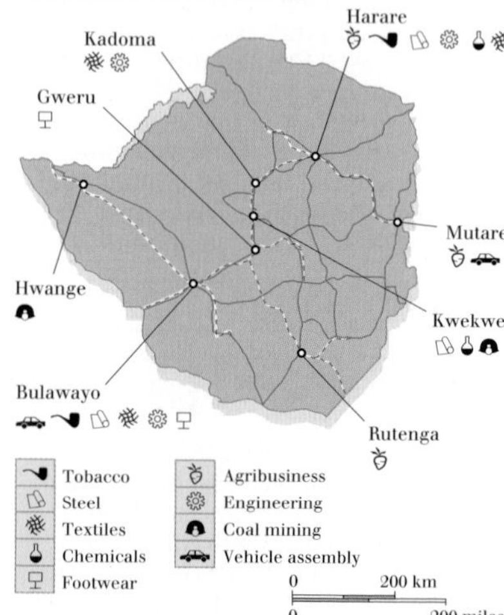
Harare
Kadoma
Gweru
Mutare
Hwange
Kwekwe
Bulawayo
Rutenga
Tobacco
Steel
Textiles
Chemicals
Footwear
Agribusiness
Engineering
Coal mining
Vehicle assembly
0
200 km
0
200 miles

Z

RESOURCES

Electric power 2m kw

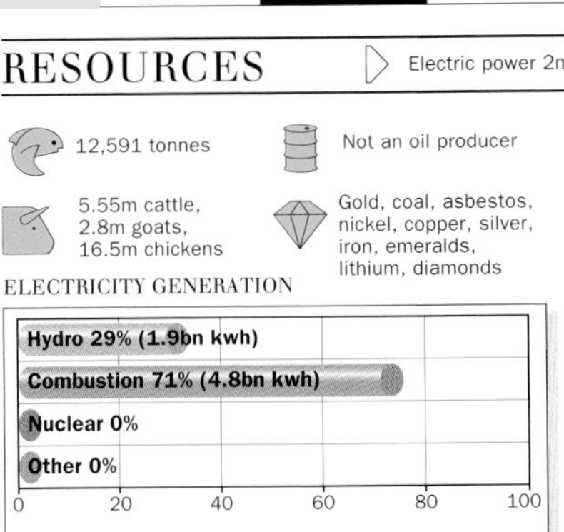
12,591 tonnes
Not an oil producer
5.55m cattle, 2.8m goats, 16.5m chickens
Gold, coal, asbestos, nickel, copper, silver, iron, emeralds, lithium, diamonds
ELECTRICITY GENERATION
Hydro 29% (1.9bn kwh)
Combustion 71% (4.8bn kwh)
Nuclear 0%
Other 0%
0
20
40
60
80
100
% of total generation by type

Almost 30% of Zimbabwe's electricity needs are met by hydropower, notably from the Kariba Dam, jointly owned with Zambia. The state power company is seeking to maximize capacity. In 1991, the government agreed to build an extension facility at Kariba South, and with Zambia a joint HEP plant at Bartoka Gorge. An oil pipeline from Beira, Mozambique, to Mutare is being extended to Harare. Coal mining is expanding at Hwange, where Malaysian investments are helping to exploit deposits of 400 million tonnes.

ZIMBABWE : LAND USE

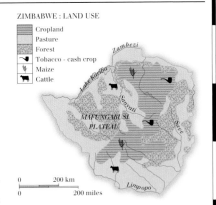

Cropland
Pasture
Forest
Tobacco - cash crop
Maize
Cattle

Zambezi
Lake Kariba
Sanyati
MAFUNGABUSI PLATEAU
Save
Limpopo
0 200 km
0 200 miles

ENVIRONMENT

Sustainability rank: 46th

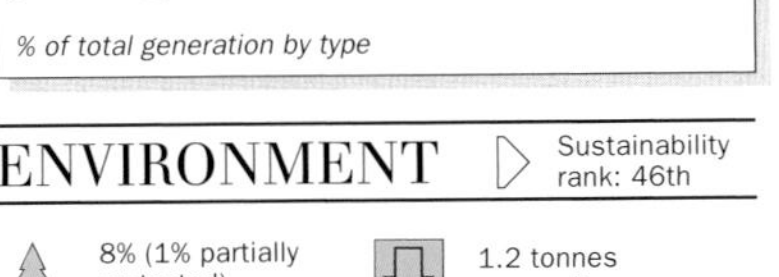
8% (1% partially protected)
1.2 tonnes per capita

ENVIRONMENTAL TREATIES

No
Yes
Yes
No
Yes
No

The 1991–1992 drought left half the population in need of drought relief, and used up 20% of public spending.

In communal areas, the land is suffering from overpopulation and overstocking. Deforestation, soil erosion, and deterioration of wildlife and water resources are widespread.

Measures have been taken to protect the black rhinoceros, including moving animals to safer areas and combating poaching – patrols have killed 150 poachers since 1986. The government also supports a scheme for dehorning rhinos – the horn is the poachers' main target. In 1997 Zimbabwe led the move at the Convention on International Trade in Endangered Species to allow a limited resumption of international trade in ivory. An increase in ivory poaching since 1999 has led to calls for more protection for elephants.

MEDIA

TV ownership low

Daily newspaper circulation 19 per 1000 people

PUBLISHING AND BROADCAST MEDIA

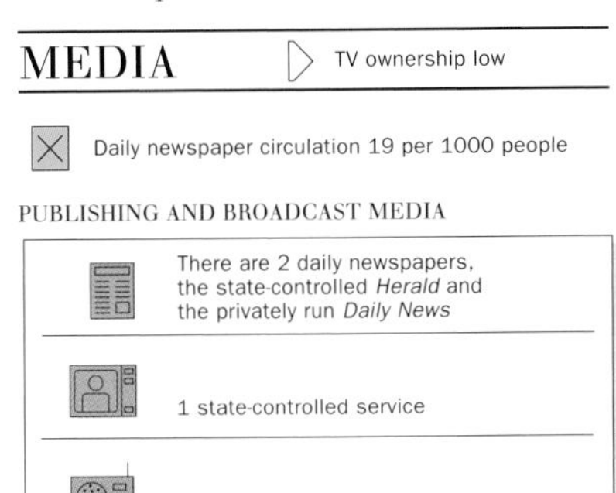
There are 2 daily newspapers, the state-controlled *Herald* and the privately run *Daily News*
1 state-controlled service
1 state-controlled service

The state has a controlling interest in the *Herald* newspaper. Persecution of journalists has increased since a new press law was passed in March 2002.

CRIME

Death penalty in use

19,376 prisoners

Up 8% in 1999

CRIME RATES

Murders
9 per 100,000 population

Rapes
34 per 100,000 population

Thefts
1828 per 100,000 population

Murder and narcotics-related offenses are rife in urban areas. The illegal occupation of white-owned farms, supported by the government, and electoral violence have led to many deaths. The secret service and the army have been criticized for human rights abuses.

EDUCATION

School leaving age: 12

89%

40,850 students

THE EDUCATION SYSTEM

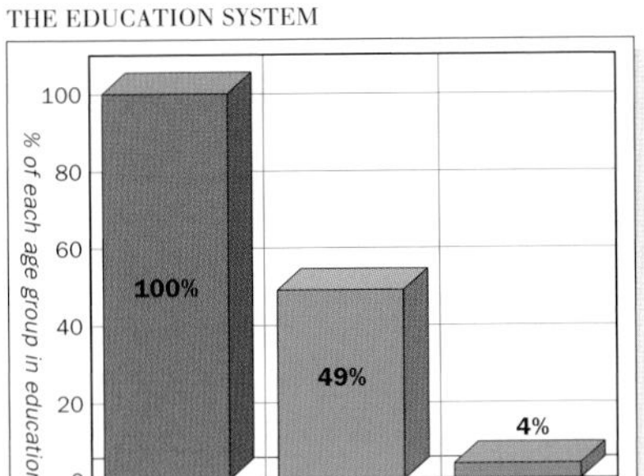
100
80
60
40
20
0
% of each age group in education
100%
49%
4%
Primary
Secondary
Tertiary

Improving education has been one of ZANU–PF's great successes. In barely ten years, primary school attendance rose from 820,000 to some 2.3 million. Education is compulsory and instruction is in English. Fees were introduced after 1992. The government has built two new universities, at Bulawayo and Mutare; it encourages vocational training to create a workforce with skills in agriculture, medicine, and engineering.

HEALTH

Welfare state health benefits

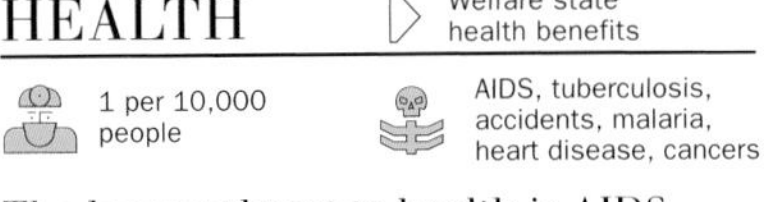
1 per 10,000 people
AIDS, tuberculosis, accidents, malaria, heart disease, cancers

The largest threat to health is AIDS. It has reduced average life expectancy to just 43 years, created some 800,000 orphans, and kills 700 people a week. A belated AIDS program, offering generic drugs, is now in place. Malaria and tuberculosis account for many other deaths. The beleaguered health system is free for the poor.

SPENDING

GDP/cap. increase

CONSUMPTION AND SPENDING

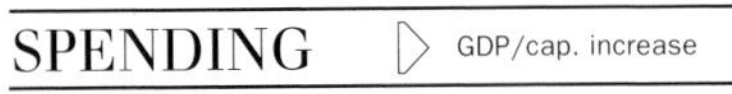
29 per 1000 population
18 per 1000 population

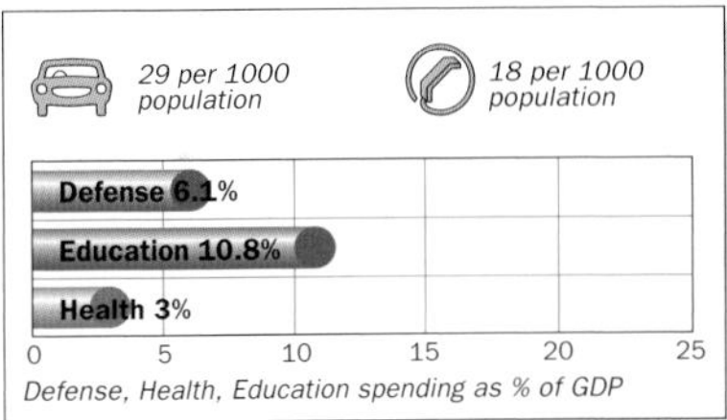
Defense 6.1%
Education 10.8%
Health 3%
0
5
10
15
20
25
Defense, Health, Education spending as % of GDP

Socialist policies in the 1980s lessened the gap between blacks and whites. But currency depreciation and inflation have since greatly reduced real wages.

WORLD RANKING

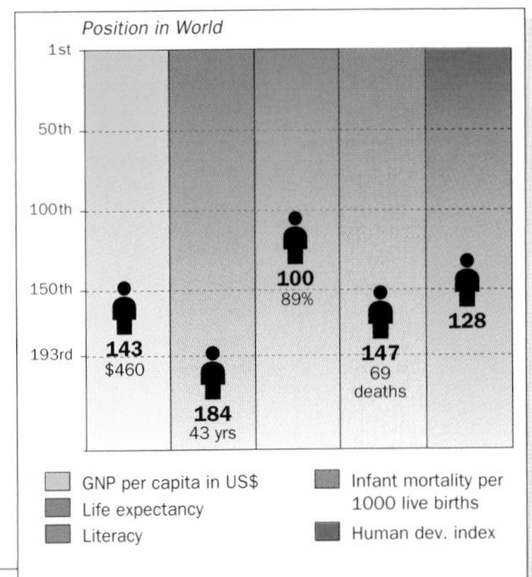
Position in World
1st
50th
100th
150th
193rd
143
$460
184
43 yrs
100
89%
147
69 deaths
128
GNP per capita in US$
Life expectancy
Literacy
Infant mortality per 1000 live births
Human dev. index

Z

OVERSEAS TERRITORIES & DEPENDENCIE

DESPITE THE RAPID process of decolonization since 1945 (pages 52–55), roughly seven million people around the world still live in nonsovereign territories under the protection of the UK, the US, France, Netherlands, Denmark, Norway, Australia, or New Zealand. These remnants of former colonial empires may have persisted for economic, strategic, or political reasons. Hong Kong and Macao reverted to Chinese control in the late 1990s. Others await political developments, such as referendums, which will determine their future status. Finally, there is a large group of territories that are considered too small, remote, or weak to be able to survive as independent states.

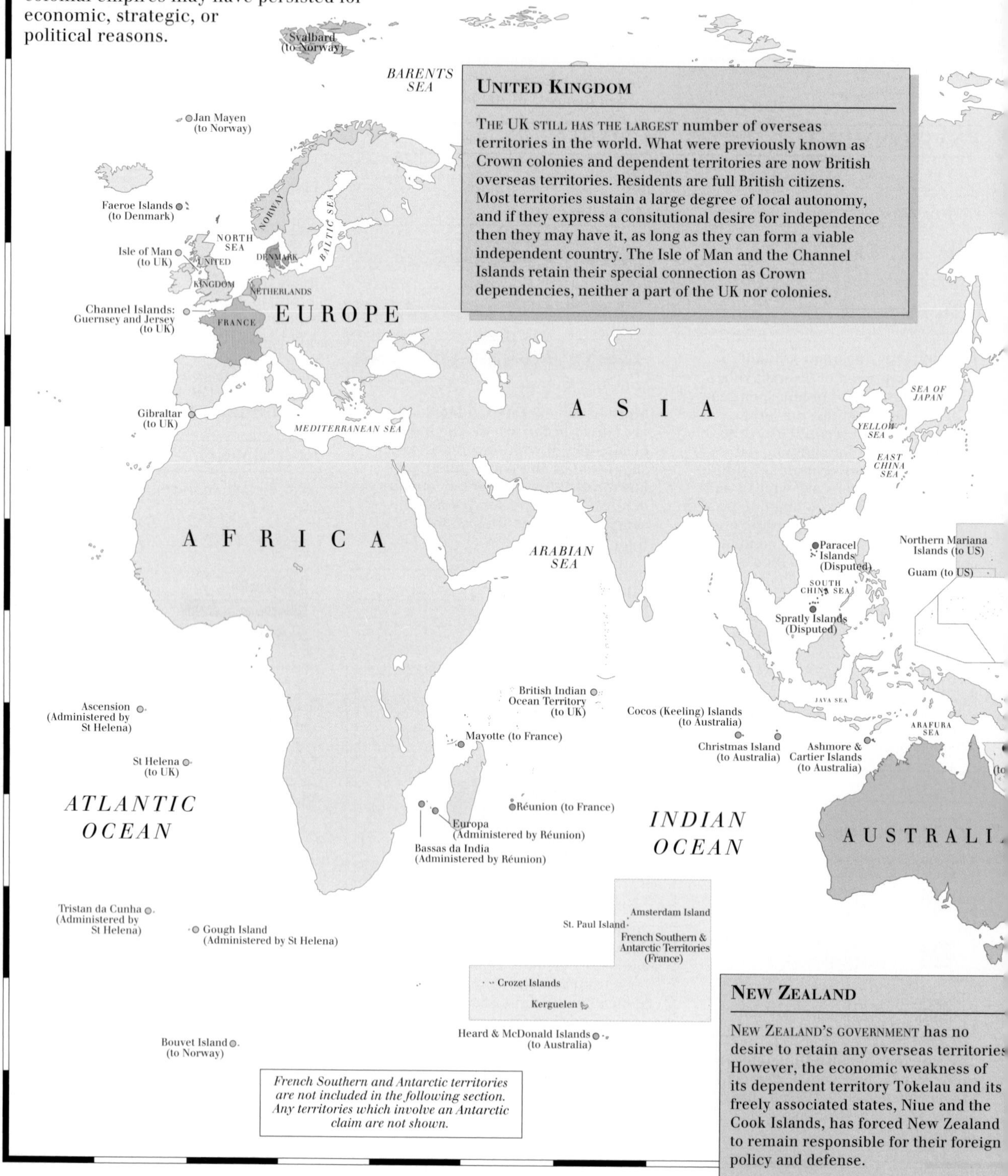

UNITED KINGDOM

THE UK STILL HAS THE LARGEST number of overseas territories in the world. What were previously known as Crown colonies and dependent territories are now British overseas territories. Residents are full British citizens. Most territories sustain a large degree of local autonomy, and if they express a consitutional desire for independence then they may have it, as long as they can form a viable independent country. The Isle of Man and the Channel Islands retain their special connection as Crown dependencies, neither a part of the UK nor colonies.

Svalbard (to Norway)
BARENTS SEA
Jan Mayen (to Norway)
Faeroe Islands (to Denmark)
Isle of Man (to UK)
NORTH SEA
NORWAY
BALTIC SEA
DENMARK
UNITED KINGDOM
NETHERLANDS
Channel Islands: Guernsey and Jersey (to UK)
FRANCE
EUROPE
Gibraltar (to UK)
MEDITERRANEAN SEA
ASIA
SEA OF JAPAN
YELLOW SEA
EAST CHINA SEA
AFRICA
ARABIAN SEA
Paracel Islands (Disputed)
Northern Mariana Islands (to US)
Guam (to US)
SOUTH CHINA SEA
Spratly Islands (Disputed)
British Indian Ocean Territory (to UK)
Cocos (Keeling) Islands (to Australia)
JAVA SEA
ARAFURA SEA
Ascension (Administered by St Helena)
Mayotte (to France)
Christmas Island (to Australia)
Ashmore & Cartier Islands (to Australia)
St Helena (to UK)
ATLANTIC OCEAN
Réunion (to France)
Europa (Administered by Réunion)
Bassas da India (Administered by Réunion)
INDIAN OCEAN
AUSTRALI
Tristan da Cunha (Administered by St Helena)
Gough Island (Administered by St Helena)
Amsterdam Island
St. Paul Island
French Southern & Antarctic Territories (France)
Crozet Islands
Kerguelen
Bouvet Island (to Norway)
Heard & McDonald Islands (to Australia)

French Southern and Antarctic territories are not included in the following section. Any territories which involve an Antarctic claim are not shown.

NEW ZEALAND

NEW ZEALAND'S GOVERNMENT has no desire to retain any overseas territories However, the economic weakness of its dependent territory Tokelau and its freely associated states, Niue and the Cook Islands, has forced New Zealand to remain responsible for their foreign policy and defense.

OVERSEAS TERRITORIES AND DEPENDENCIES
Australia
New Zealand
United Kingdom
United States
France
Denmark
Netherlands
Norway
Disputed
ARCTIC OCEAN
BEAUFORT SEA
BERING SEA
Greenland (to Denmark)
ATLANTIC OCEAN
NORTH AMERICA
St Pierre & Miquelon (to France)
UNITED STATES OF AMERICA
Turks & Caicos Islands (to UK)
Puerto Rico (to US)
British Virgin Islands (to UK)
Anguilla (to UK)
Bermuda (to UK)
Virgin Islands (to US)
Guadeloupe (to France)
CARIBBEAN SEA
Netherlands Antilles (to Neth.)
Montserrat (to UK)
Aruba (to Neth.)
Martinique (to France)
Midway Islands (to US)
Gulf of Mexico
Cayman Islands (to UK)
Navassa Island (to US)
CARIBBEAN SEA
Johnston Atoll (to US)
PACIFIC OCEAN
Clipperton Island (Administered by French Polynesia)
Kingman Reef (to US)
Palmyra Atoll (to US)
Jarvis Island (to US)
French Guiana (to France)
Tokelau (to NZ)
Cook Islands (to NZ)
SOUTH AMERICA
American Samoa (to US)
Niue (to NZ)
French Polynesia (to France)
Pitcairn Islands (to UK)
Norfolk Island (to Australia)
NEW ZEALAND
Falkland Islands (to UK)
South Georgia and South Sandwich Islands (to UK)

NITED STATES OF AMERICA

E OVERSEAS TERRITORIES OF THE US have been seen as ategically useful, if expensive, parts of its backyard. e US has, in most cases, given the local population ay in deciding its own status. Thus, full sovereignty s been granted to three successor states of the mer US-administered UN Trust Territory of the cific Islands. A US commonwealth territory, such as erto Rico, has a greater level of independence than at of a US unincorporated or external territory.

FRANCE

FRENCH *TERRITOIRES D'OUTRE-MER* are considered an indivisible part of the French Republic. As a result, France has developed economic ties, and stressed the advantage of interdependence over independence, with its overseas territories. A distinct hierarchy has been developed. Overseas *départements*, officially part of France, have their own governments. Territorial *collectivités* are administered by a French-appointed commissioner and a locally elected council, while overseas *territoires* have varying degrees of autonomy.

AMERICAN SAMOA

STATUS: Unincorporated territory of the USA **CLAIMED:** 1900
CAPITAL: Pago Pago **POP.:** 57,291 **DENSITY:** 294/km² (764/mi²)

COMPRISING THE EASTERN half of the Samoan islands, American Samoa sits on the edge of Polynesia in the South Pacific Ocean. Although Christianity, introduced in the 19th century, has taken a very firm hold – Samoa is known as the Bible belt of the Pacific – the traditional and conservative *fa'a Samoa* (Samoan way of life) continues to dominate the islands' culture. At its base is the extended family, the *aiga*, while traditional chiefs, or *matai*, retain their central role in government. Samoa came under the control of the US in 1900, and life there remained largely unchanged until a US-led drive for modernization in the 1960s. Along with better health care and industrial development, *fa'a Amerika* also meant unemployment, pollution, and rising petty crime fueled by alcohol. Tuna processed by Pago Pago's canneries represent 95% of American Samoa's exports. Efforts to diversify include the development of other light industries and tourism.

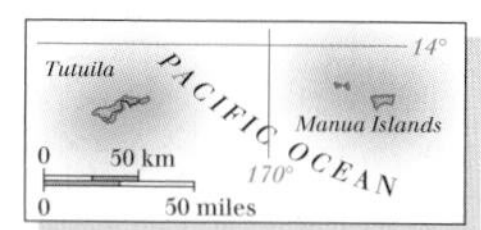

Tutuila
PACIFIC OCEAN
Manua Islands
14°
170°
0 50 km
0 50 miles

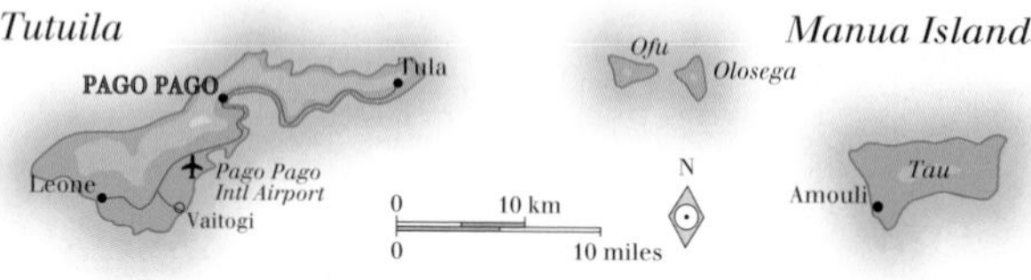

Tutuila
PAGO PAGO
Tula
Leone
Pago Pago Intl Airport
Vaitogi
0 10 km
0 10 miles
N
Manua Islands
Ofu
Olosega
Tau
Amouli

ANGUILLA

STATUS: British overseas territory **CLAIMED:** 1650
CAPITAL: The Valley **POP.:** 11,561 **DENSITY:** 120/km² (312/mi²)

LYING IN THE CENTER of the Leeward Islands, in the Caribbean, Anguilla has a subtropical climate, the heat and humidity being tempered by trade winds. Lumped by the UK into a joint colonial administration with St. Kitts and Nevis, Anguillans took up arms to protect their dependent status in 1967 when St. Kitts and Nevis were awarded internal self-government. Lacking any major industry, Anguilla relied on the economic stability that came with being an overseas territory. In the 1980s the island's government resolved to introduce a tourist industry and targeted the luxury end of the market. Tourism is now Anguilla's main source of income. Almost 50,000 tourists, around 60% from the US, take their vacations in Anguilla every year.

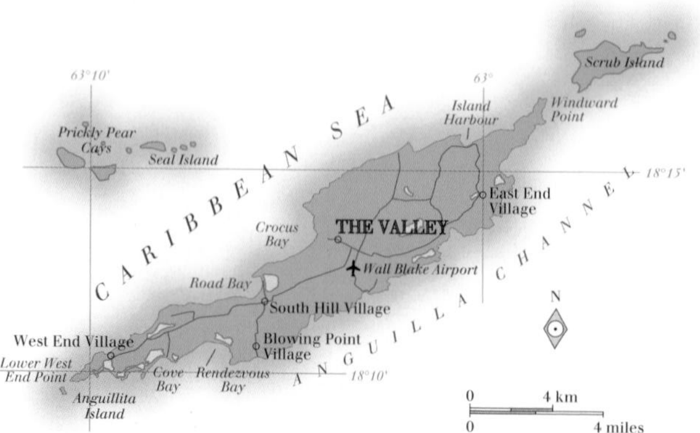

63°10'
63°
Scrub Island
Island Harbour
Windward Point
Prickly Pear Cays
Seal Island
CARIBBEAN SEA
18°15'
East End Village
Crocus Bay
THE VALLEY
Wall Blake Airport
Road Bay
South Hill Village
West End Village
Blowing Point Village
Lower West End Point
Cove Bay
Rendezvous Bay
18°10'
Anguillita Island
ANGUILLA CHANNEL
N
0 4 km
0 4 miles

ARUBA

STATUS: Autonomous part of the Netherlands **CLAIMED:** 1643
CAPITAL: Oranjestad **POP.:** 101,000 **DENSITY:** 523/km² (1347/mi²)

THE MOST DEVELOPED island among the Dutch Caribbean territories, Aruba lies 25 km (15 miles) off the coast of Venezuela. Its tropical climate is moderated by constant trade winds sweeping in from the Atlantic. Formerly the richest island in the Netherlands Antilles, Aruba became a separate dependency of the Netherlands in 1986. Transition to full independence, expected in 1996, was halted in 1994 after an agreement was reached between the governments of the Netherlands, Aruba, and the Netherlands Antilles. The Netherlands voiced concern over the island's security and the danger of its becoming a base for narcotics trafficking, and the Aruban government questioned the desirability of full independence, citing high unemployment and economic instability.

Since 1986, the economy of Aruba, formerly dependent on oil refining, has diversified. Tourism and offshore finance have become the most important sectors of the economy, and there are now more than 700,000 visitors annually; nearly 60% of them come from the US. However, the rapid expansion of tourism has put considerable strain on Aruba's infrastructure, and some attempt has been made to restrict the number of visitors. At the same time facilities have been improved to encourage the growth of a data-processing industry.

70°
CARIBBEAN SEA
Palm Beach
Noord
ORANJESTAD
Queen Beatrix Intl Airport
12°30'
San Nicolas
Seru Colorado
N
0 10 km
0 10 miles

Palm Beach, Aruba, also known as the Turquoise Coast, lies on the western side of the island. The beach stretches for 10 km (6 miles) and is the site of a low-rise beach resort.

***Oranjestad,** Aruba's capital, contains many Dutch colonial-style buildings. Although first claimed by the Spanish in 1499, Aruba was colonized by the Dutch in the 17th century.*

Aruba's cooperation with the US in the region includes support for its actions against narcotics trafficking from South America, and since the closure of the US base in Panama in 1999, US aircraft have used bases on the island to launch reconnaissance flights. Those who oppose this cooperation fear that it could drag Aruba unnecessarily into the civil conflict in Colombia.

LAND HEIGHT: above Sea Level | 200m/656ft | 500m/1640ft | 1000m/3281ft | 1500m/4572ft | above 2000m/6562ft

BERMUDA

STATUS: British overseas territory **CLAIMED:** 1612
CAPITAL: Hamilton **POP.:** 61,688 **DENSITY:** 1164/km² (3084/mi²)

SITUATED MORE THAN 1000 km (650 miles) off the coast of the US, Bermuda consists of a chain of more than 150 coral islands. The Gulf Stream, flowing between Bermuda and the eastern seaboard of the US, keeps the climate mild and humid. Bermuda is racially mixed; some 60% of the population are of mostly European extraction. Racial tension has lessened since the 1960s and 1970s. A more representative electoral system was established after a Royal Commission visited Bermuda in 1978.

For 30 years after the first general election, held in 1968, Bermuda was ruled by the conservative United Bermuda Party (UBP). Its veteran leader, Sir John Swan, resigned as prime minister and party leader in 1995, when a referendum decisively rejected his campaign for independence from the UK. In a general election in November 1998 the UBP, now under the leadership of Pamela Gordon, was heavily defeated by the Progressive Labour Party, led by Jennifer Smith, who said that she had no plans to pursue her party's own pro-independence aspirations. Major issues are the social and economic challenges posed by the withdrawal in 1995 of both the US naval base and the British military base, environmental issues, and narcotics trafficking.

Bermuda is overwhelmingly a service economy. Lilies are grown for export, but few other agricultural products are grown in sufficient quantity, and the islands are heavily dependent on food imports.

Tourist figures have been falling steadily, but tourism is still a significant industry, most visitors coming from the US. However, financial services have become the most important sector of the economy, helping to maintain one of the highest per capita incomes in the world. The government has attempted to head off international criticism of its financial environment through a series of reforms. Bermuda also operates one of the world's largest flag-of-convenience shipping fleets.

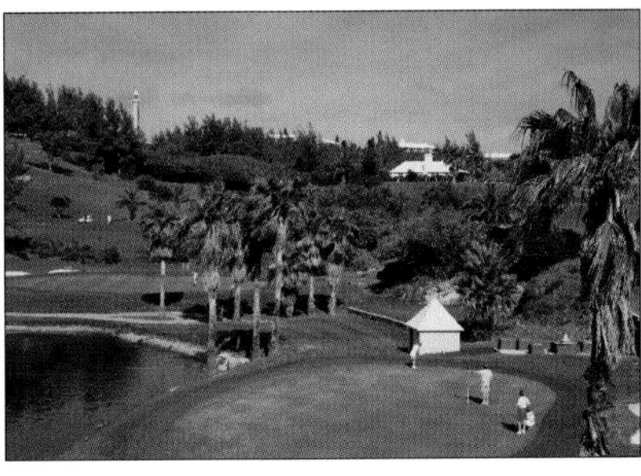

***Bermuda** has one of the highest densities of golf courses in the world. Eight courses have now been developed.*

BRITISH INDIAN OCEAN TERRITORY

STATUS: British overseas territory **CLAIMED:** 1814
CAPITAL: Diego Garcia **POP.:** 3200 **DENSITY:** 53/km² (139/mi²)

THE BRITISH Indian Ocean Territory, or Chagos Islands, lies in the middle of the Indian Ocean. The coral atolls are uninhabited, except for the US–UK military base on Diego Garcia, and the UK has undertaken to cede the islands to Mauritius when they are no longer required. In 2000 the Ilois people, evicted by the UK from the islands in 1968, won the right in the UK High Court to return, but face strong resistance from the US.

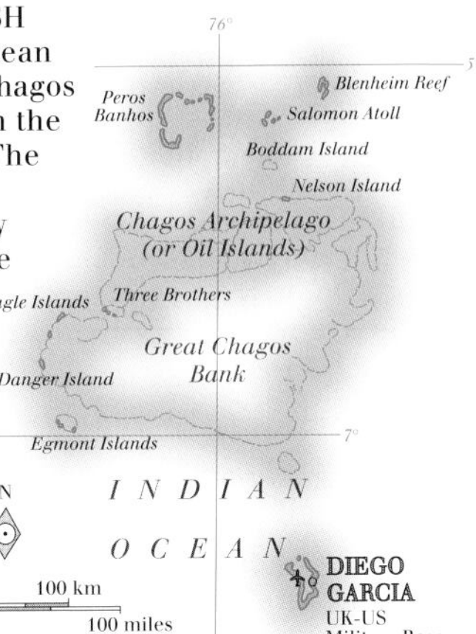
Peros Banhos
Blenheim Reef
Salomon Atoll
Boddam Island
Nelson Island
Chagos Archipelago (or Oil Islands)
Eagle Islands
Three Brothers
Great Chagos Bank
Danger Island
Egmont Islands
INDIAN OCEAN
DIEGO GARCIA
UK-US Military Base
N
0 100 km
0 100 miles

BRITISH VIRGIN ISLANDS

STATUS: British overseas territory **CLAIMED:** 1672
CAPITAL: Road Town **POP.:** 19,864 **DENSITY:** 130/km² (337/mi²)

AN ARCHIPELAGO of 60 Caribbean islands, 15 of them inhabited, the British Virgin Islands lie at the northwestern end of the Leeward Islands chain. Tourism, now a major economic activity, is suited to the tropical climate, but there is concern about its effect on the environment. There are also fears that traditional place-names are being altered to be more tourist-friendly. The offshore finance sector is important, and has been more tightly regulated since 1990, following scandals involving foreign companies registered in the territory.

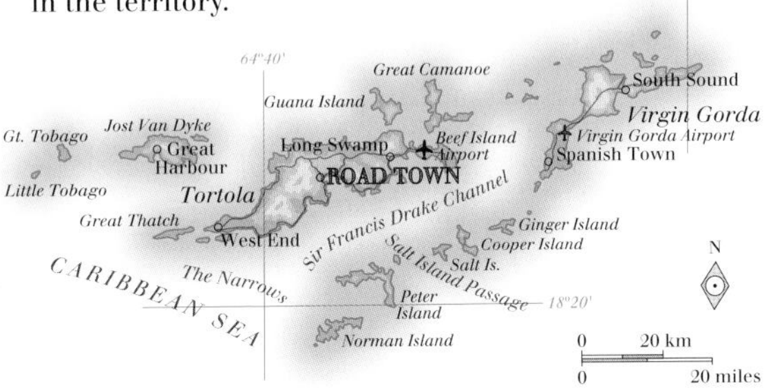
Anegada
The Settlement
ATLANTIC OCEAN
Great Camanoe
Guana Island
South Sound
Virgin Gorda
Virgin Gorda Airport
Spanish Town
Jost Van Dyke
Great Harbour
Gt. Tobago
Little Tobago
Long Swamp
Beef Island Airport
ROAD TOWN
Tortola
Great Thatch
West End
Sir Francis Drake Channel
Ginger Island
Cooper Island
Salt Island Passage
Salt Is.
The Narrows
Peter Island
CARIBBEAN SEA
Norman Island
N
0 20 km
0 20 miles

POPULATION ○ under 5000 ● under 10 000 ● over 10 000 ○ over 50 000 ◎ over 100 000 Urban areas AIRPORTS International Local

CAYMAN ISLANDS

STATUS: British overseas territory **CLAIMED:** 1670
CAPITAL: George Town **POP.:** 39,410 **DENSITY:** 152/km² (394/mi²)

THE LARGEST OF BRITAIN'S territories in the Caribbean, the Cayman Islands lie 225 km (140 miles) west of Jamaica and south of Cuba. The abundance of exotic wildlife, especially marine life, is a powerful draw for tourists. Grand Cayman is credited as the home of modern scuba diving, the first ever specialist shop opening there in 1957. The islanders have rejected greater autonomy, persuaded that their economic stability is linked to their status as an overseas territory. Thanks to the absence of tax and foreign-exchange controls, the islands form one of the world's largest offshore financial centers, but tourism continues to underpin the economy.

20° Little Cayman CARIBBEAN SEA Cayman Brac Grand Cayman 80°
Grand Cayman West Bay North Sound GEORGE TOWN Owen Roberts Airport East End Bodden Town
Little Cayman Cayman Brac Spot Bay South Town Gerard Smith Airport West End
N 0 20 km 0 20 miles

CHRISTMAS ISLAND

STATUS: Australian external territory **CLAIMED:** 1958
CAPITAL: Flying Fish Cove **POP.:** 2771 **DENSITY:** 21/km² (53/mi²)

SO NAMED because it was sighted on Christmas Day in 1643, Christmas Island lies in the Indian Ocean, 380 km (240 miles) south of Java. Its population is mostly Malay and Chinese, descended from laborers imported to mine rich phosphate deposits. A national park covers some 70% of the island. In 2001 Australia agreed with Russia to begin construction of a major rocket-launching site on the island.

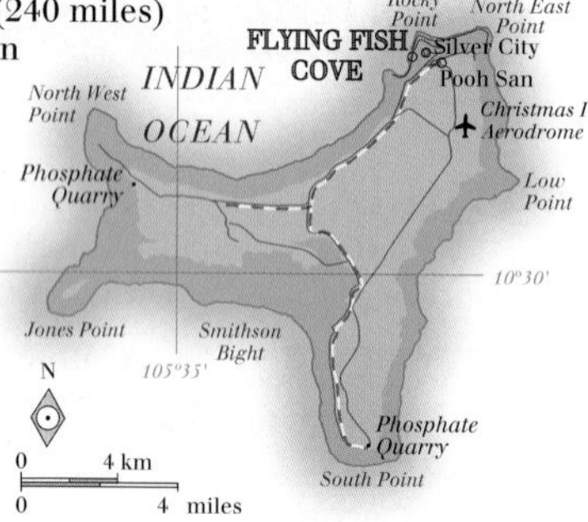
Rocky Point
North East Point
FLYING FISH COVE
Silver City
Poon San
North West Point
INDIAN OCEAN
Christmas I Aerodrome
Phosphate Quarry
Low Point
10°30'
Jones Point
Smithson Bight
105°35'
N
Phosphate Quarry
South Point
0 4 km
0 4 miles

COCOS (KEELING) ISLANDS

STATUS: Australian external territory **CLAIMED:** 1955
CAPITAL: West Island **POP.:** 633 **DENSITY:** 45/km² (127/mi²)

IN ALL, 27 coral atolls make up the Cocos (Keeling) Islands. Situated in the Indian Ocean, roughly halfway between Australia and Sri Lanka, they have been part of the Northern Territory electoral district since 1992. The inhabited islands are the European-dominated West Island and Home Island, which has a mainly Cocos Malay community. Coconuts are the sole cash crop.

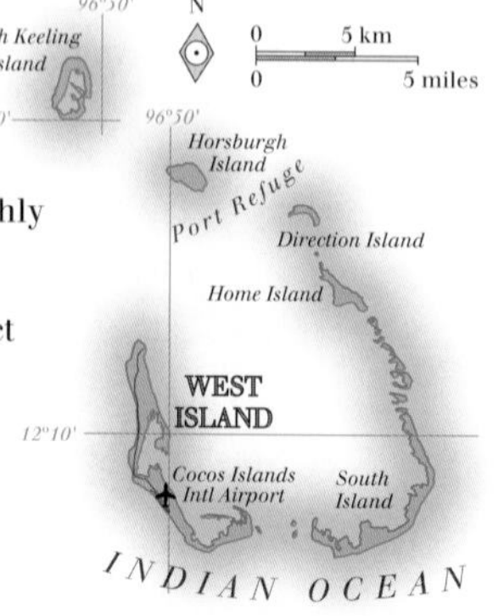
96°50'
North Keeling Island
11°50'
N
0 5 km
0 5 miles
96°50'
Horsburgh Island
Port Refuge
Direction Island
Home Island
WEST ISLAND
12°10'
Cocos Islands Intl Airport
South Island
INDIAN OCEAN

COOK ISLANDS

STATUS: Territory in free association with New Zealand **CLAIMED:** 1
CAPITAL: Avarua **POP.:** 12,900 **DENSITY:** 54/km² (140/mi²)

LYING IN THE MIDDLE of the South Pacific 3000 km (1900 miles) from New Zealand, the Cook Islands are a combination of 24 coral atolls and volcanic islands. Achieving self-government in 1965, they have adopted a diversified economy focusing primarily on tourism and banking, but with significant trade in giant clams and pearls.

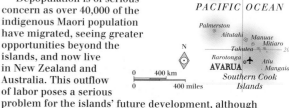

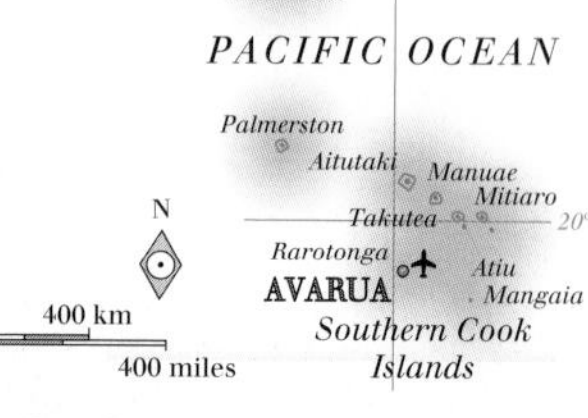
165°
Penrhyn
Rakahanga
10°
Pukapuka
Manihiki
Nassau
Northern Cook Islands
Suwarrow
PACIFIC OCEAN
Palmerston
Aitutaki
Manuae
Mitiaro
N
Takutea
20°
Rarotonga
Atiu
AVARUA
Mangaia
0 400 km
0 400 miles
Southern Cook Islands

Depopulation is of serious concern as over 40,000 of the indigenous Maori population have migrated, seeing greater opportunities beyond the islands, and now live in New Zealand and Australia. This outflow of labor poses a serious problem for the islands' future development, although remittances to relatives form an important source of income.

The government of the Cook Islands, headed since 2002 by centrist Prime Minister Robert Woonton, is advised by a traditional council known as the House of Ariki. It is seeking a program of political reform including devolving administration to the outer islands and cutting back an expensive bureaucracy.

FAEROE ISLANDS

STATUS: Self-governing territory of Denmark **CLAIMED:** 1380
CAPITAL: Tórshavn **POP.:** 46,196 **DENSITY:** 33/km² (86/mi²)

MIDWAY BETWEEN Scotland and Iceland in the North Atlantic, the Faeroe Islands have a moderate climate for their latitude – a result of the warm Gulf Stream current. Home rule since 1948 has given the Faeroese a strong sense of national identity – they voted against joining the European Communities with Denmark in 1973, but now have favorable terms of trade with most EU members. Fishing is the dominant industry, providing over 90% of exports. In the face of international criticism, the Faeroese have continued their traditional cull of pilot whales and bottle-nosed dolphins. Sheep farming is important, and there is a small textile industry which exports traditional woolens, and puffin and eider-duck feathers.

Denmark's moves toward ever closer European integration have strengthened calls in the Faeroes for full independence. Negotiations to establish a "sovereign nation" under the Danish monarchy began in 1998. However, the Danish government's threat in 2001 to suspend subsidies to the islands quashed calls for a referendum.

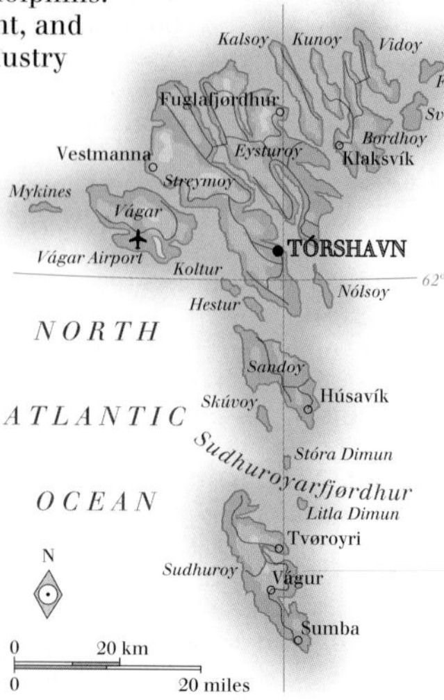
7°
Kalsoy
Kunoy
Vidoy
Fugloy
Svínoy
Fuglafjørdhur
Bordhoy
Vestmanna
Eysturoy
Klaksvík
Mykines
Streymoy
Vágar
TÓRSHAVN
Vágar Airport
Koltur
62°
Nólsoy
Hestur
NORTH
Sandoy
ATLANTIC
Skúvoy
Húsavík
Sudhuroyarfjørdhur
Stóra Dimun
OCEAN
Lítla Dimun
Tvøroyri
N
Sudhuroy
Vágur
Sumba
0 20 km
0 20 miles

LAND HEIGHT above Sea Level 200m/656ft 500m/1640ft 1000m/3281ft 1500m/4572ft above 2000m/6562ft

FALKLAND ISLANDS

STATUS: British overseas territory **CLAIMED:** 1832
CAPITAL: Port Stanley **POP.:** 2913 **DENSITY:** 0.24/km² (0.6/mi²)

SITUATED IN THE South Atlantic Ocean, over 12,000 km (7440 miles) from the UK, the Falkland Islands are influenced by the cold Antarctic current. The main islands of East and West Falkland and the hundreds of outlying islands have a cool, temperate climate with frequent strong winds.

The islands gained international attention with the Argentine invasion, and subsequent British recapture, in 1982. Since then, the UK government has invested heavily in a "Fortress Falklands" policy. A new runway and a military base to house an enlarged garrison were built at Mount Pleasant. The islanders, for their part, are determined to maintain the political status quo, but in 1999 improving relations led to the restoration of scheduled air connections with Argentina. Since the Falklands War, the economy of the islands has prospered. Falklanders invested heavily in schools, roads, and tourism in a fresh drive for a strong identity. By 1987, the Falklands had become financially solvent through the sale of fishing licenses. Although sales by Argentina of cheaper, less restrictive licenses caused a fall in fishing revenues, fishing (mostly of squid) is still the major source of income and employment. Depressed wool prices caused a slump in the fortunes of the sheep-farming industry. The UK and Argentina reached agreement in 1995 on oil exploration, and the discovery of oil reserves in the Falklands' territorial waters is revolutionizing prospects for the economy. Tourism, attracting birdwatchers, photographers, and military historians, is steadily growing.

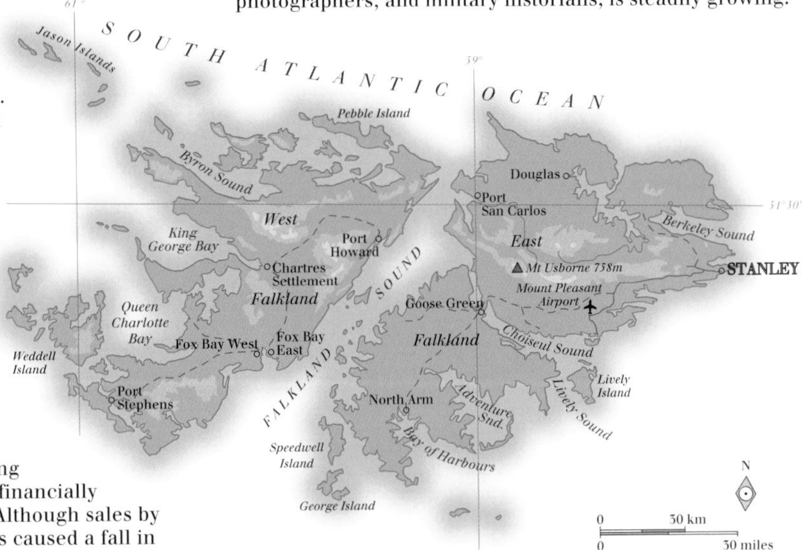
SOUTH ATLANTIC OCEAN
Jason Islands
Pebble Island
Byron Sound
Douglas
Port San Carlos
Berkeley Sound
West
King George Bay
Port Howard
East
Chartres Settlement
Falkland
Mt Usborne 738m
STANLEY
Mount Pleasant Airport
Queen Charlotte Bay
Goose Green
Fox Bay West
Fox Bay East
Falkland
Choiseul Sound
Weddell Island
FALKLAND SOUND
Port Stephens
North Arm
Adventure Snd.
Lively Island
Lively Sound
Bay of Harbours
Speedwell Island
George Island
N
0 30 km
0 30 miles

FRENCH GUIANA

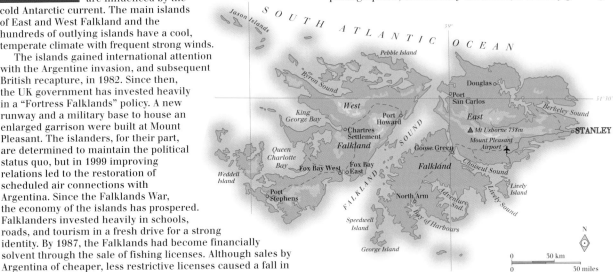

STATUS: French overseas department **CLAIMED:** 1817
CAPITAL: Cayenne **POP.:** 172,605 **DENSITY:** 2/km² (5/mi²)

SANDWICHED BETWEEN Brazil and Suriname, French Guiana is the only remaining colony in South America. A belt of coastal marsh, and an interior of equatorial jungle, combine in a location which was, for years, notorious for the offshore penal colony, Devil's Island. The rainforest, which covers 90% of the territory, is particularly rich in flora and fauna. It harbors over 400,000 species, including more different species of bird than in the whole of Europe.

Concentrated near the coast, the population is ethnically mixed. While 40% are creoles, there are some 20,000 Amerindians and a village of 1000 Hmong who fled civil war in Laos in the 1980s.

***Kourou** was selected for the launch of the* Ariane *rocket because of its equatorial site. The town has grown from 800 to 15,000 people.*

A campaign for greater autonomy in the late 1970s and early 1980s led to limited decentralization of power to a regional council. The grip on local power by the Guianese Socialist Party (PSG) has been undermined since 1993 by a more unified opposition, but it is still the largest party in the regional council.

During the 1990s the people have become increasingly vocal in their condemnation of the French government's perceived indifference to their country's problems, and there were riots in 1996 and 1997 over the education system. The PSG has accordingly campaigned for greater autonomy. As an overseas *département* of metropolitan France, French Guiana is also a region of the EU, but it is heavily dependent on France itself for aid, food, and manufactured goods. It has a number of valuable natural resources, including gold, fishing, and forestry, and also has potential for increased tourism, but these are yet to be fully exploited because of a lack of skilled labor and investment and an underdeveloped infrastructure. The Guiana Space Center, which is situated on the coast at Kourou, has been operational since 1964. From there the Ariane rockets of the European Space Agency (ESA) are launched.

ATLANTIC OCEAN
Mana
le Maroni
St.-Laurent-du-Maroni
Sinnamary
Devil's I.
Apatou
Cambrouze
Kourou
CAYENNE
Saint-Élie
Montsinéry
Rochambeau
Roura
Citron
Délices
MONTAGNES DE LA TRINITÉ
Cacao
SURINAME
Régina
Grande-Santi
la Lawa
la Mana
MONTAGNE TORTUE
Ouanary
l'Approuague
St- Georges
Maripasoula
Saül
l'Oyapok
Bienvenue
Camopi
MONTS APITIRI
Camopi
la Litani
Claimed by Suriname
Siminiout
BRAZIL
N
0 60 km
0 60 miles

French Polynesia

Status: French overseas territory **Claimed:** 1843

Capital: Papeete **Pop.:** 235,400 **Density:** 67/km² (173/mi²)

A SCATTER OF 130 South Pacific islands and coral atolls over an area the size of Europe combine to form French Polynesia. The average annual temperature varies between 20°C (68°F) and 29°C (84°F), with rainfall of over 150 cm (58 in). Nearly 75% of the population live on the main island of Tahiti. The French administration has developed the islands with little regard for local wishes, and the 70% West Polynesian (Mahoi) majority have seen their simple, self-sufficient economy transformed into one dependent on the French military and tourism. Nuclear testing on Mururoa atoll created many jobs, but there was growing opposition, and a final series of tests, held in 1995-1996 despite widespread international protests, provoked local demonstrations and riots in Papeete.

The Polynesian majority has called increasingly for more autonomy, reduced tourism, and the rebuilding of indigenous trade. Future hopes rest largely on new tuna fishing ventures.

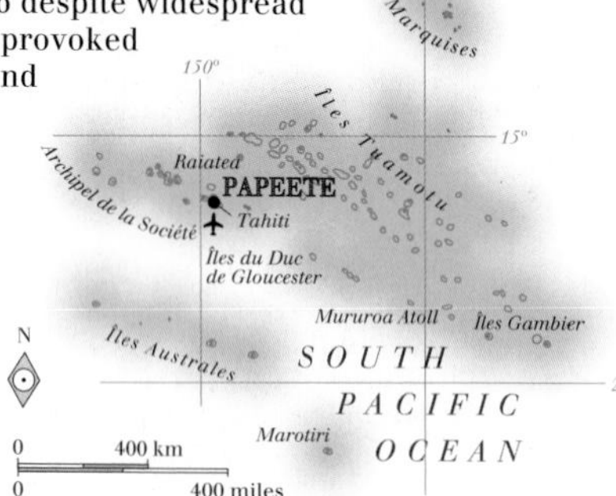
140°
Îles Marquises
150°
Îles Tuamotu
15°
Archipel de la Société
Raiatea
PAPEETE
Tahiti
Îles du Duc de Gloucester
Mururoa Atoll
Îles Gambier
Îles Australes
SOUTH
PACIFIC
OCEAN
Marotiri
25°
N
0 400 km
0 400 miles

Gibraltar

Status: British overseas territory **Claimed:** 1713

Capital: Gibraltar **Pop.:** 27,025 **Density:** 3861/km² (9008/mi²)

GUARDING THE western entrance to the Mediterranean, Gibraltar has survived on military and marine revenues. However, as Britain has cut defense spending, so its military presence on the Rock has declined. In response Gibraltarians have developed a vibrant offshore banking industry. Strict antismuggling legislation, in force since 1995, has curbed extensive smuggling from north Africa into Spain. Gibraltar's relationship with Britain and Spain remains contentious. However, the two governments have undertaken to resolve the territory's status by December 2002, prompting mass protests from the colony's inhabitants, who have felt sidelined in the discussions.

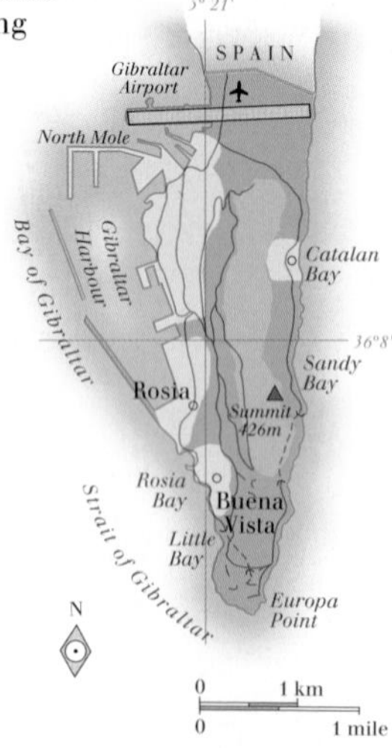
5° 21'
SPAIN
Gibraltar Airport
North Mole
Bay of Gibraltar
Gibraltar Harbour
Catalan Bay
36°8'
Sandy Bay
Rosia
Summit 426m
Rosia Bay
Buena Vista
Little Bay
Strait of Gibraltar
Europa Point
N
0 1 km
0 1 mile

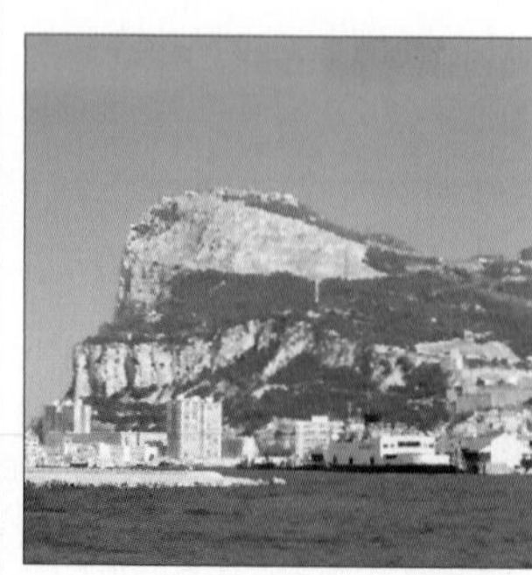

The Rock of Gibraltar. The British built 143 caves, 50 km (30 miles) of roads, and as many km of tunnels, for defensive purposes.

Greenland

Status: Self-governing territory of Denmark **Claimed:** 1380

Capital: Nuuk **Pop.:** 56,569 **Density:** 0.03/km² (0.07/mi²)

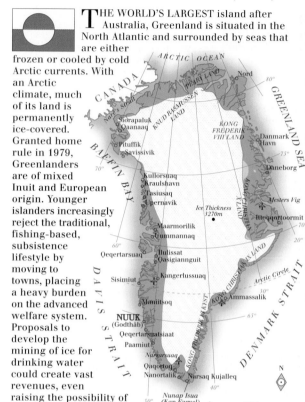

THE WORLD'S LARGEST island after Australia, Greenland is situated in the North Atlantic and surrounded by seas that are either frozen or cooled by cold Arctic currents. With an Arctic climate, much of its land is permanently ice-covered. Granted home rule in 1979, Greenlanders are of mixed Inuit and European origin. Younger islanders increasingly reject the traditional, fishing-based, subsistence lifestyle by moving to towns, placing a heavy burden on the advanced welfare system. Proposals to develop the mining of ice for drinking water could create vast revenues, even raising the possibility of economic independence.

Guadeloupe

Status: French overseas department **Claimed:** 1635

Capital: Basse-Terre **Pop.:** 431,000 **Density:** 242/km² (627/mi²)

GUADELOUPE lies at the southern end of the Leeward Islands in the Caribbean and includes the north of St. Martin island and St. Barthélémy island near Anguilla. The movement for independence from France has been pronounced and intermittently violent since the 1960s. The economy is largely based on tourism and agriculture, with sugar, rum, and bananas the main exports, but the islands depend on French and EU regional aid. The vulnerability of the banana industry to hurricanes and global trade pacts has prompted efforts to expand sugar production and develop tourism.

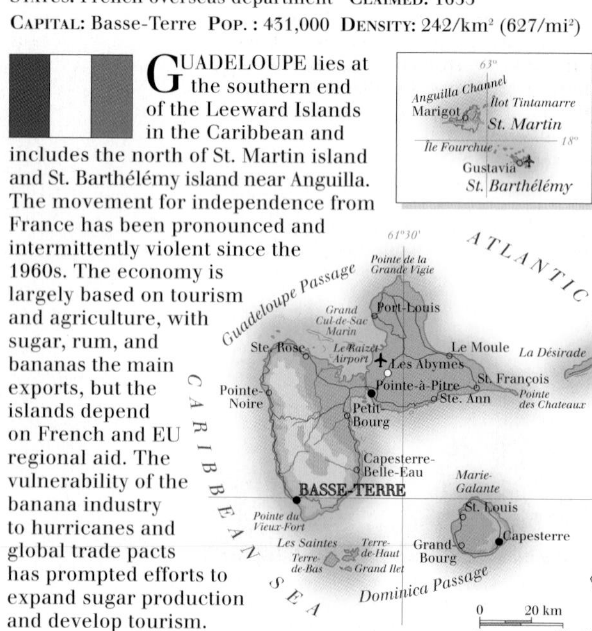
63°
Anguilla Channel
Marigot
Îlot Tintamarre
St. Martin
Île Fourchue
18°
Gustavia
St. Barthélémy
61°30'
ATLANTIC
Guadeloupe Passage
Pointe de la Grande Vigie
Port-Louis
Grand Cul-de-Sac Marin
Ste. Rose
Le Raizet Airport
Les Abymes
Le Moule
La Désirade
Pointe-Noire
Pointe-à-Pitre
St. François
Ste. Ann
Pointe des Chateaux
Petit-Bourg
CARIBBEAN SEA
Capesterre-Belle-Eau
BASSE-TERRE
Marie-Galante
Pointe du Vieux-Fort
St. Louis
Les Saintes
Terre-de-Haut
Terre-de-Bas
Grand Ilet
Grand-Bourg
Capesterre
Dominica Passage
0 20 km
0 20 miles

LAND HEIGHT above Sea Level 200m/656ft 500m/1640ft 1000m/3281ft 1500m/4572ft above 2000m/6562ft

GUERNSEY

STATUS: British Crown dependency **CLAIMED:** 1066
CAPITAL: St. Peter Port **POP.:** 59,807 **DENSITY:** 920/km² (2392/mi²)

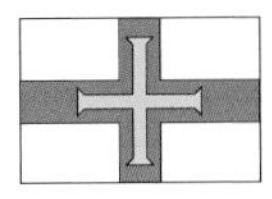

LYING 47 KM (29 miles) off the coast of France, Guernsey and its dependencies form the northwestern part of the Channel Islands, historically part of the Duchy of Normandy. English is the language most commonly used, but the Norman patois is spoken in some villages, and some formal business of the legislature is conducted in French. Travel to France is easier than to the UK; Alderney is only 13 km (8 miles) from the French mainland. Residents on the smaller islands have no need for cars, and life continues in an unhurried manner that has changed little through the centuries. The islanders guard this lifestyle with strict residence laws. Guernsey's mild climate has encouraged the development of tourism and market gardening as major industries. Tomatoes and flowers are produced mainly for the UK market. The low tax system, independent of the UK, has led to a substantial and profitable financial services industry. Many international banks have Guernsey subsidiaries.

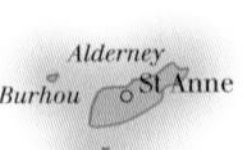
Alderney
Burhou
St Anne

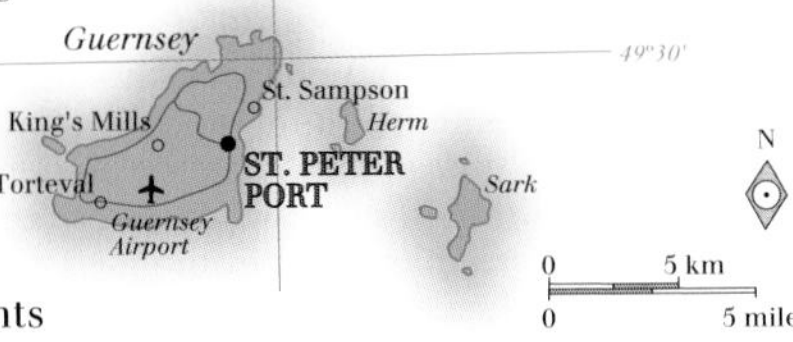
ENGLISH CHANNEL
Guernsey
St. Sampson
Herm
King's Mills
Torteval
ST. PETER PORT
Sark
Guernsey Airport
0 5 km
0 5 miles

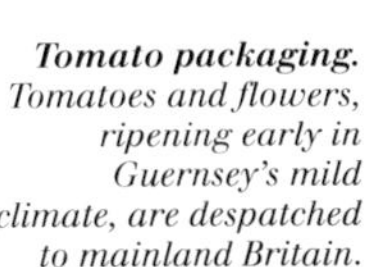

Tomato packaging. *Tomatoes and flowers, ripening early in Guernsey's mild climate, are despatched to mainland Britain.*

GUAM

STATUS: Unincorporated territory of the USA **CLAIMED:** 1898
CAPITAL: Hågatña **POP.:** 158,000 **DENSITY:** 288/km² (745/mi²)

THE VOLCANIC island of Guam lies at the southern end of the Mariana archipelago in the Pacific. Its tropical climate has encouraged tourism, although it lies in a region where typhoons are common. Guam's indigenous Chamorro people, who comprise just under half the population, dominate the island's political and social life. They are famous for a set of facial expressions, called "eyebrow," which virtually constitutes a language of its own. Although English is the official language, Chamorro is commonly spoken, and in 1998 the spelling of the capital was changed from Agaña to the Chamorran Hagåtña. The US military base, covering one-third of the island, has made Guam strategically important to the US. Military spending and tourism revenues have failed to benefit all islanders and 23% live below the poverty line. The influx of US culture has also threatened to upset Guam's social stability. Greater independence has been an issue since the early 1980s, with a series of referendums since 1982. A draft Commonwealth Act was rejected by the US Congress after nearly 15 years of deliberation.

ISLE OF MAN

STATUS: British Crown dependency **CLAIMED:** 1765
CAPITAL: Douglas **POP.:** 76,315 **DENSITY:** 133/km² (345/mi²)

LYING HALFWAY BETWEEN England and Northern Ireland in the Irish Sea, the Isle of Man has been inhabited for centuries by the Celtic Manx people. Established by the Vikings in the 9th century, the Manx parliament, the Tynwold, has autonomy from the UK in a number of matters, including taxation, and the death penalty was only officially abolished in 1993. The islanders have used this independence to establish a thriving financial and business sector, which has aided employment as the traditional industries of agriculture and fishing decline. There is still a shellfish industry, specializing in scallops. Tourism is also important: there are more than 200,000 visitors each year. Numbers have been boosted by the growing number of films being made on the island. Manx culture received a boost in 1993, when the local language, in danger of dying out, began to be taught in the island's schools once more. The Calf of Man, a small uninhabited island, is administered as a nature reserve.

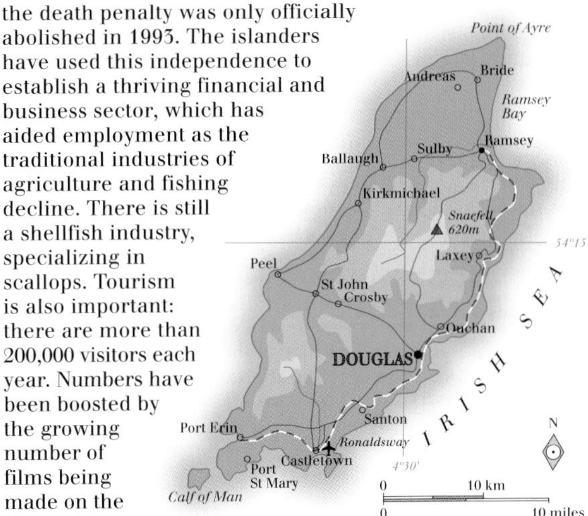
Point of Ayre
Andreas
Bride
Ramsey Bay
Ballaugh
Sulby
Ramsey
Kirkmichael
Snaefell 620m
Laxey
Peel
St John
Crosby
Onchan
DOUGLAS
IRISH SEA
Port Erin
Santon
Ronaldsway
Port St Mary
Castletown
Calf of Man
0 10 km
0 10 miles

The annual TT motorbike race *on the Isle of Man. Thousands of people come each year to watch the island's famous Touring Trophy race.*

Jersey

Status: British Crown dependency **Claimed:** 1066
Capital: St. Helier **Pop.:** 87,186 **Density:** 752/km² (1937/mi²)

THE BAILIWICK OF JERSEY, the largest of the Channel Islands, lies some 22 km (14 miles) from the coast of Normandy in France. The official language (since 1960) is English, but French is still used in the courts. The island has a mild climate owing to the Gulf Stream, fine beaches, and more sunshine than anywhere in the British Isles. Jersey has its own legislative and taxation systems which are a blend of the French and British versions. The Jersey States Assembly is one of the oldest legislative bodies in the world. Members stand as independents, rather than for political parties. It is considered a "Peculiar" of the UK monarchy, and has the right to reject "unacceptable" UK laws.

Historically, agriculture has been Jersey's most important industry, with dairy cows its most famous export, closely followed by early-harvested potatoes, tomatoes, and flowers. By the end of the 20th century, however, farming had been eclipsed by the rise of offshore finance and tourism. The growth of these sectors, and rigid controls on the rights of residence, have ensured high living standards for most of the inhabitants. Jersey is also host to a large Portuguese community which works in the island's tourist industry.

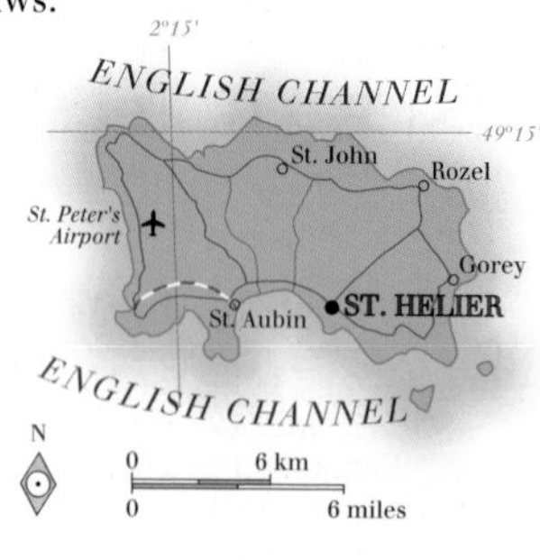
2°15'
ENGLISH CHANNEL
49°15'
St. John
Rozel
St. Peter's Airport
Gorey
St. Aubin
ST. HELIER
ENGLISH CHANNEL
N
0 6 km
0 6 miles

Johnston Atoll

Status: Unincorporated territory of the USA **Claimed:** 1858
Capital: *Not applicable* **Pop.:** *Not applicable*

JOHNSTON ATOLL LIES 1150 km (714 miles) southwest of Hawaii. The atoll consists of a coral reef, two highly modified natural islands, Johnston and Sand, and two completely man-made islands, Akau (North) and Hikina (East). The islands, which were used by the US for nuclear weapons tests, were seriously contaminated with plutonium in 1962, when a nuclear missile exploded during testing. Regular tests began in 1971, and until 2000 the islands were also used for the storage of nuclear material and the destruction of chemical and biological weapons, including sarin nerve gas and the defoliant Agent Orange. Cleanup operations began in 2000. The only inhabitants left are US government personnel and civilian contractors who maintain the plant. The islands have also been designated by the US as a wildlife refuge, a breeding place for seabirds and green turtles. The US army intends to evacuate the base entirely by 2003.

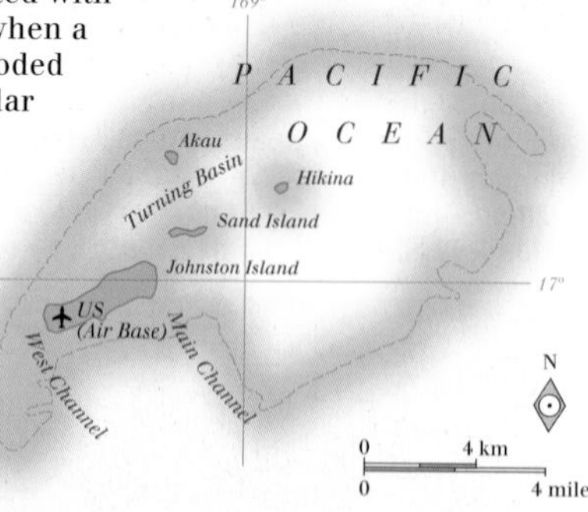
169°
PACIFIC
OCEAN
Akau
Turning Basin
Hikina
Sand Island
Johnston Island
17°
US (Air Base)
West Channel
Main Channel
N
0 4 km
0 4 miles

Martinique

Status: French overseas department **Claimed:** 1635
Capital: Fort-de-France **Pop.:** 381,427 **Density:** 338/km² (875/mi²)

CHRISTOPHER COLUMBUS described Martinique as "the most beautiful country in the world." It lies in the middle of the Caribbean Windward Islands but has retained remarkably close links both culturally and economically to mainland France. The island is dominated by the Montagne Pelée volcano, which violently erupted and engulfed the old capital, St. Pierre, in 1902. Situated in the Caribbean's hurricane belt, Martinique suffers an average of one natural disaster every five years.

Its long association with France and its status as an overseas *département* have left Martinique with a distinctly French feel; nonetheless nearly 90% of the population are of African or mixed ethnicity, and this influence has created a vibrant Caribbean tradition, particularly in music. Some of Martinique's more famous children include Joséphine Bonaparte (Napoléon's first wife) and Frantz Fanon, the black revolutionary who influenced anticolonial movements in the 20th century.

Economic power remains in the hands of the *Bekes* (descendants of white colonial settlers), who own most of the agricultural land. This situation has led in the past to outbreaks of violence and calls for greater autonomy. However, high living standards depend on French subsidies and a French-style social welfare system. The traditionally agricultural economy, based on the cultivation of sugarcane and bananas, has been forced to diversify as EU subsidy cuts come into effect, and high-class tourism is now the biggest source of income and the largest provider of employment. Almost 80% of the half-million annual visitors come from France. Since the late 1980s unemployment and emigration have been high, with the result that over 30% of Martiniquais nationals are resident in metropolitan France.

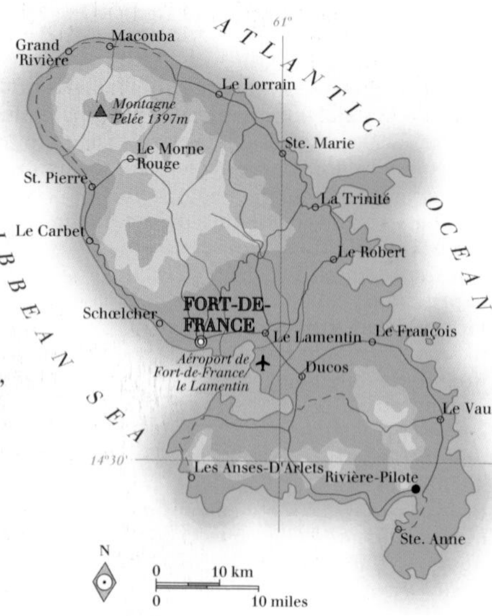
61°
ATLANTIC OCEAN
CARIBBEAN SEA
Grand 'Rivière
Macouba
Le Lorrain
Montagne Pelée 1397m
Ste. Marie
Le Morne Rouge
St. Pierre
La Trinité
Le Carbet
Le Robert
Schœlcher
FORT-DE-FRANCE
Le Lamentin
Le François
Aéroport de Fort-de-France/ le Lamentin
Ducos
Le Vauclin
14°30'
Les Anses-D'Arlets
Rivière-Pilote
Ste. Anne
N
0 10 km
0 10 miles

Martinique. *Tourists are attracted to the island's beaches, its mountainous interior, and the historic towns of Fort-de-France and Saint Pierre.*

LAND HEIGHT | above Sea Level | 200m/656ft | 500m/1640ft | 1000m/3281ft | 1500m/4572ft | above 2000m/6562ft

MAYOTTE

STATUS: French territorial collectivity **CLAIMED:** 1843
CAPITAL: Mamoudzou **POP.:** 142,000 **DENSITY:** 380/km² (986/mi²)

PART OF THE COMOROS archipelago, Mayotte lies about 8000 km (5000 miles) from France, between Madagascar and the east African coast. It was the only island in the archipelago to vote against independence from France in a 1974 referendum. The other islands declared unilateral independence in 1975 and laid claim to Mayotte. Despite widespread poverty, endemic unemployment, and a cost of living twice that of France, the Mahorais voted again in 1976 to maintain the link. The main political movement has since unsuccessfully demanded that Mayotte be given the status of a French *département*, hoping that this would bring more aid to develop their largely agricultural economy. France opposes this idea because of the expense involved, but did grant the island *département*-style autonomy in 2000.

The economy is still largely agricultural, producing crops both for internal consumption and for export. However, large quantities of foodstuffs are also imported. France has invested in an airport and port, but tourism has been slow to develop. Nevertheless, the relative prosperity of Mayotte has encouraged separatist movements on the two other small Comoros islands to seek closer relations with France.

MIDWAY ISLANDS

STATUS: Unincorporated territory of the USA **CLAIMED:** 1867
CAPITAL: *Not applicable* **POP.:** 453 **DENSITY:** *Not applicable*

NAMED BECAUSE of its position between California and Japan, Midway is a coral atoll at the western end of the Hawaiian islands; there have been moves to make it part of Hawaii. The site of a major World War II battle, the atoll comprises two large islands, totaling over 4 sq. km (1.5 sq. miles), and several smaller ones. It functions as a naval air base and wildlife refuge. The population is limited to military personnel and civilian contractors, but some tourism is permitted, mainly connected with the wildlife.

MONTSERRAT

STATUS: British overseas territory **CLAIMED:** 1632
CAPITAL: Plymouth **POP.:** 5000 **DENSITY:** 49/km² (128/mi²)

MONTSERRAT IS ONE of the Leeward Islands chain in the eastern Caribbean. It has been devastated by volcanic eruptions which began in 1995 and culminated in massive explosions of the Soufrière Hills volcano in 1997 and 1998. As a result, the southern two-thirds of the island, where Plymouth and Blackburne airport are located, have become uninhabitable and it is illegal to enter the volcano "exclusion zone." Over half of the 10,000 population left permanently for neighboring islands or the UK. Calls for independence, based on a tourist boom in the 1980s, have been largely dropped, as the island is now dependent on UK aid. The disaster soured relations, setting off a bitter dispute over the cost of resettlement and reconstruction. A new capital, tentatively named Port Diana, is planned for the "safe" northern coast. The tourism industry is struggling to rebuild itself, but is hindered by the closure of both the airport and seaport by the eruption. Montserrat can now only be reached via neighboring Antigua.

Montserrat. *Known as the Caribbean's "emerald isle" because of its luxuriant flora and Irish heritage.*

NETHERLANDS ANTILLES

STATUS: Autonomous part of the Netherlands **CLAIMED:** 1816
CAPITAL: Willemstad **POP.:** 217,000 **DENSITY:** 271/km² (702/mi²)

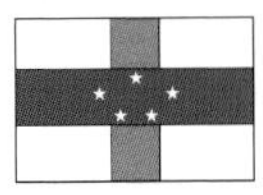

THE NETHERLANDS Antilles are composed of two Caribbean island groups. Curaçao – the richest and wealthiest island – and Bonaire lie just off the Venezuelan coast, while Saba, St. Eustatius, and Sint Maarten – whose northern half is part of Guadeloupe – lie 800 km (500 miles) to the north. Financial scandals, political instability, and the issue of the federation's future, among other things, have strained relations with the Dutch government, the major aid provider. Refining petroleum using oil from Venezuela is the islands' principal industrial activity.

Curaçao
Bonaire
CARIBBEAN SEA
Noordpunt
Westpunt
Hato Intl Airport
St. Willibrordus
WILLEMSTAD
Santa Caterina
Malmok
Rincon
Klein Bonaire
Kralendijk
Flaming Field Airport
Lacre Punt

New Caledonia

Status: French overseas territory **Claimed:** 1853
Capital: Nouméa **Pop.:** 200,000 **Density:** 10/km² (27/mi²)

NEW CALEDONIA, or, as it is known to the indigenous Kanaks, Kanaky, is an island group 400 km (250 miles) west of Vanuatu and 1350 km (840 miles) off the coast of eastern Australia. Tension over socioeconomic inequalities and independence between the Melanesian Kanaks, who form over half of the population, and the influential expatriate *Caldoches*, resulted in a long history of political violence. Under the 1988 Matignon Accord, France imposed a year of direct rule as the prelude to a new constitutional structure which attempted to address Kanak grievances by providing greater provincial autonomy. Although some racial violence continued after 1988, it has not again reached the same level.The Nouméa accord, signed in 1998, set out a 15-year program for gradual autonomy which would end in a vote on self-determination.

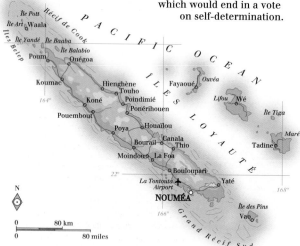

Nickel mining is the territory's most valuable export industry, at over 90% of export income. New Caledonia has over 20% of world reserves, and is the fourth-largest producer in the world, but the industry employs relatively few people, and is vulnerable to fluctuations in the world price. It was seriously affected by the Asian financial crisis of 1997–1998, but recovered on the back of high world prices in 2000. Tourism and agriculture are bigger employers, although less than 1% of total land area is cultivated. Corn, yams, sweet potatoes, and coconuts have traditionally been the main crops, and since the 1990s large numbers of melons have been exported to Japan. Fishing is important, the main products being tuna and shrimps, most of which are also exported to Japan. A project for farming giant clams started in 1996. Unemployment nevertheless remains high among young Kanaks.

A nickel mine, *New Caledonia. The importance of the nickel industry to the territory's economy has made the control of reserves a dominant issue in politics, and in negotiations over the island's independence from France.*

Niue

Status: Territory in free association with New Zealand **Claimed:** 1
Capital: Alofi **Pop.:** 1857 **Density:** 7/km² (18/mi²)

THE WORLD'S LARGEST coral island, Niue lies 2400 km (1500 miles) northeast of New Zealand. The subsistence economy produces tropical fruits, while tourism and the sale of postage stamps provide foreign currency. The island's financial services have recently received warnings from international monitors. Mass emigration has seen the Niuean community in New Zealand grow to over 20,000. In an effort to stem the tide, New Zealand has invested heavily in the Niuean economy. Despite this boost, growth is slow.

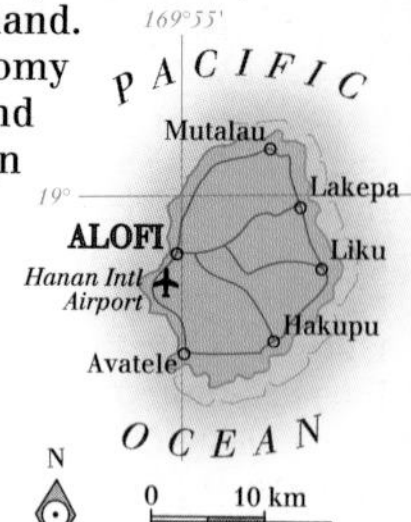
PACIFIC OCEAN
Mutalau
Lakepa
Liku
ALOFI
Hanan Intl Airport
Hakupu
Avatele
N
0 10 km
0 10 miles

Norfolk Island

Status: Australian external territory **Claimed:** 1774
Capital: Kingston **Pop.:** 2601 **Density:** 74/km² (200/mi²)

INHABITED by Australian migrants and descendants of the mutineers of HMS *Bounty*, Norfolk Island lies 1400 km (869 miles) east of Australia. Islanders speak a hybrid language, mixing Westcountry English, Gaelic, and ancient Tahitian. They enjoy substantial autonomy, and in 1991 rejected a plan to become part of the Australian federal state. Tourists, attracted by the climate and unique flora, have brought islanders a relatively high standard of living.

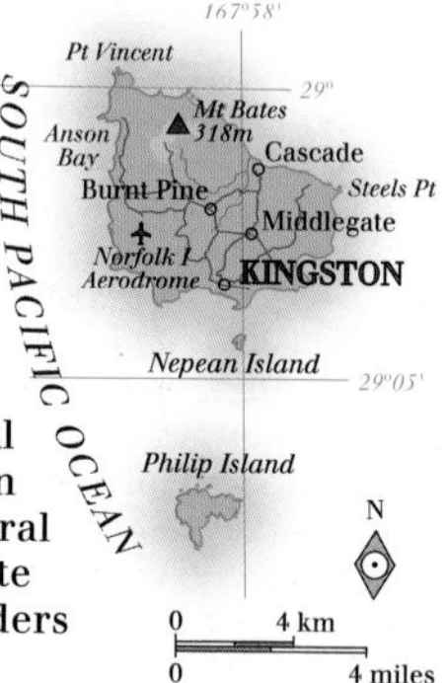
SOUTH PACIFIC OCEAN
Pt Vincent
Anson Bay
Mt Bates 318m
Cascade
Steels Pt
Burnt Pine
Middlegate
Norfolk I. Aerodrome
KINGSTON
Nepean Island
Philip Island
N
0 4 km
0 4 miles

Northern Mariana Is.

Status: Commonwealth territory of the USA **Claimed:** 1947
Capital: Saipan **Pop.:** 69,221 **Density:** 151/km² (393/mi²)

A FORMER UN trust territory, the Northern Marianas preferred in 1987 to retain links with the US rather than opt for independence. However, local politicians have questioned their current status. US aid fueled a boom during the 1980s, but it depended on immigrant workers who by the early 1990s outnumbered the local Chamorro population. In addition, tourism has speeded the decline of the traditional subsistence economy.

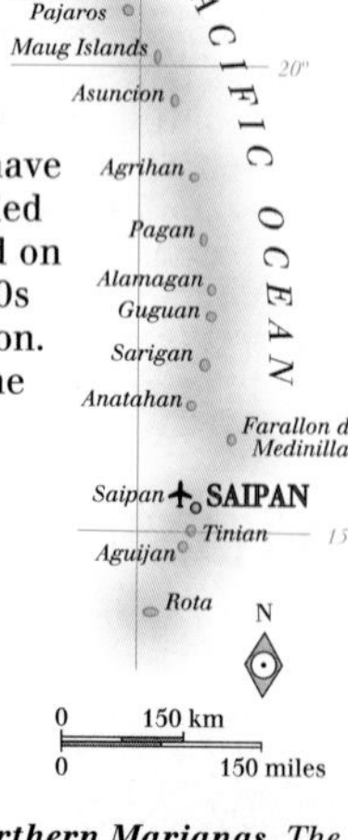
PACIFIC OCEAN
Farallon de Pajaros
Maug Islands
Asuncion
Agrihan
Pagan
Alamagan
Guguan
Sarigan
Anatahan
Farallon de Medinilla
Saipan
SAIPAN
Tinian
Aguijan
Rota
N
0 150 km
0 150 miles

Rota, Northern Marianas. *The limestone outcrop of Wedding Cake Mountain overlooks the small village of Songsong.*

 LAND HEIGHT above Sea Level 200m/656ft 500m/1640ft 1000m/3281ft 1500m/4572ft above 2000m/6562ft

PARACEL ISLANDS

STATUS: *Disputed* **CLAIMED:** *Not applicable*
CAPITAL: *Not applicable* **POPULATION:** *Unknown*

OCCUPIED BY CHINESE FORCES (who call them the Xisha islands), but also claimed by Taiwan and Vietnam, the Paracel Islands are a small collection of coral atolls situated some 325 km (200 miles) east of Vietnam, in the South China Sea. Subject to frequent typhoons and with a tropical climate, the Paracels are at the center of a regional dispute over the vast reserves of oil and natural gas which are believed to lie beneath their territorial waters. China has built port facilities and an airport on Woody Island to support its claim.

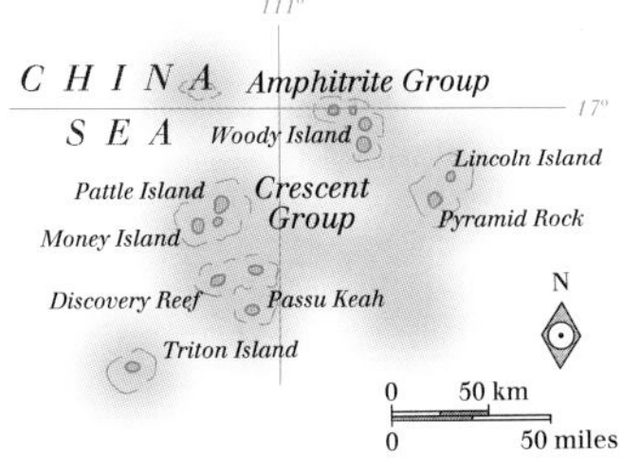
CHINA SEA
Amphitrite Group
Woody Island
Lincoln Island
Pattle Island
Crescent Group
Pyramid Rock
Money Island
Discovery Reef
Passu Keah
Triton Island
N
0 50 km
0 50 miles

PITCAIRN ISLANDS

STATUS: British overseas territory **CLAIMED:** 1887
CAPITAL: Adamstown **POP.:** 43 **DENSITY:** 1.2/km² (3/mi²)

A GROUP OF volcanic South Pacific islands, Pitcairn is Britain's most isolated dependency. Pitcairn Island was the last refuge for the mutineers from HMS *Bounty*. The economy operates by barter, fishing, and subsistence farming, and is reliant on regular airdrops from New Zealand and periodic visits by supply vessels. Postage stamp sales provide foreign currency earnings. In 2002, 20 Pitcairners were charged with sexual assault.

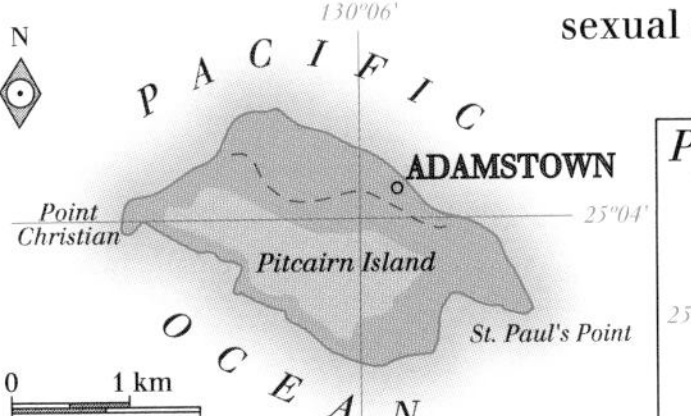
PACIFIC OCEAN
ADAMSTOWN
Point Christian
Pitcairn Island
St. Paul's Point
N
0 1 km
0 1 miles

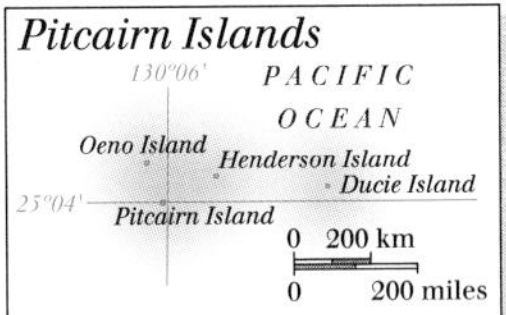
Pitcairn Islands
PACIFIC OCEAN
Oeno Island
Henderson Island
Ducie Island
Pitcairn Island
0 200 km
0 200 miles

PUERTO RICO

STATUS: Commonwealth territory of the USA **CLAIMED:** 1898
CAPITAL: San Juan **POP.:** 4 million **DENSITY:** 446/km² (1156/mi²)

PUERTO RICO, a US territory since its invasion in 1898, is by far the most populous nonindependent territory. It is the easternmost of the Greater Antilles chain in the Caribbean. The population density, highest around San Juan, is comparable with the Netherlands and is higher than in any US state. The tropical climate attracts growing numbers of tourists, 80% from the US, and there have been major efforts to expand hotel and resort facilities.

Puerto Rico was granted its current commonwealth status in 1952, four years after an abortive pro-independence uprising. The inhabitants have US citizenship but only limited self-government. In three plebiscites, in 1967, 1993, and 1998, the islanders endorsed continued commonwealth status rather than opting for either US statehood or independence. The most recent of these votes was extremely close, but the pro-statehood governor who called the 1993 and 1998 votes, Pedro Rossello, was replaced by the anti-statehood Sila Calderón – the first female governor of Puerto Rico – in 2001.

Although thousands of the mostly Spanish-speaking Puerto Ricans have migrated to the US mainland in search of higher wages, the islanders have one of the highest living standards in the region. Tax relief, cheap labor, and the island's role as an export-processing zone, mainly for the US market, attracted many businesses. Clothing, electronics, petrochemical, and pharmaceutical industries traditionally dominated, but the decision to phase out tax exemptions for companies reinvesting in the island caused a slump in 1996, and more emphasis is now being placed on the service sector. New industries include health care and clinical testing, biotechnology, and other knowledge-based areas.

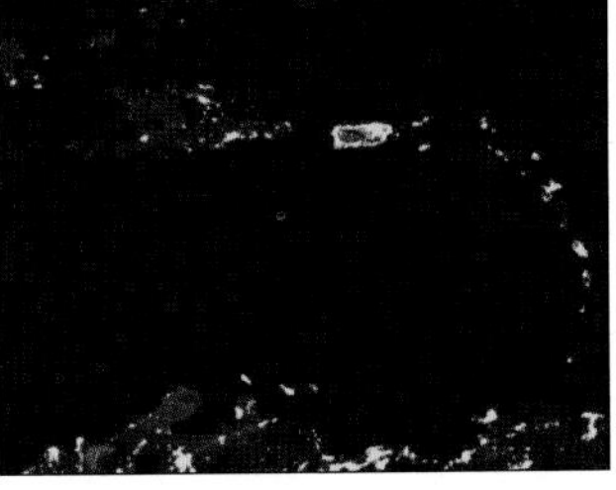

***At night,** the bright lights of Puerto Rico's well-developed roads, settlements, and busy ports are in sharp contrast to the rest of the Caribbean – notably the dark outline of Haiti, just to the west.*

Governor Calderón spearheaded the campaign to stop the US navy from using the populated eastern island of Vieques for bombing practice. In 2000 an invasion of the bombing range by protestors led to some high-profile arrests, including that of Robert Kennedy Jr. A year later newly elected US president George W. Bush announced that the navy would not use the island after 2003.

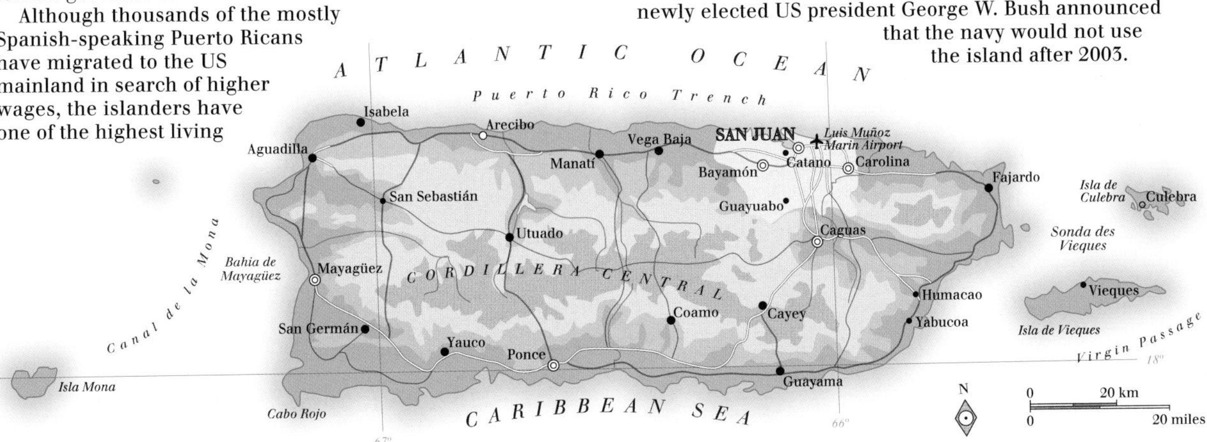
ATLANTIC OCEAN
Puerto Rico Trench
Isabela
Arecibo
Aguadilla
Vega Baja
SAN JUAN
Luis Muñoz Marín Airport
Manatí
Bayamón
Cataño
Carolina
Fajardo
San Sebastián
Guayuabo
Isla de Culebra
Culebra
Utuado
Caguas
Sonda des Vieques
Bahía de Mayagüez
Mayagüez
CORDILLERA CENTRAL
Humacao
Vieques
Canal de la Mona
Coamo
Cayey
Yabucoa
San Germán
Isla de Vieques
Yauco
Ponce
Virgin Passage
Isla Mona
Guayama
Cabo Rojo
CARIBBEAN SEA
N
0 20 km
0 20 miles

POPULATION  under 5000 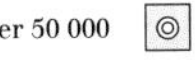 under 10 000 • over 10 000 ○ over 50 000 ◎ over 100 000 ▢ Urban areas AIRPORTS International Local

Réunion

Status: French overseas department **Claimed:** 1638

Capital: Saint-Denis **Pop.:** 728,400 **Density:** 290/km² (751/mi²)

THE LARGE VOLCANIC ISLAND of Réunion, 800 km (500 miles) east of Madagascar, provides France with an important strategic presence – and a large military base – in the Indian Ocean. Its mountainous interior has forced the majority of the population to live along the coast. Tensions still exist between the very poor black community and the wealthy Indian and European groups, although the violence of 1991 has not been repeated. Despite the introduction of measures applicable to all French overseas *départements*, intended to improve social and economic standards, unemployment remains high and the cost of living very expensive. Réunion's main crop is sugarcane, although Cyclone Dina devastated production in 2002.

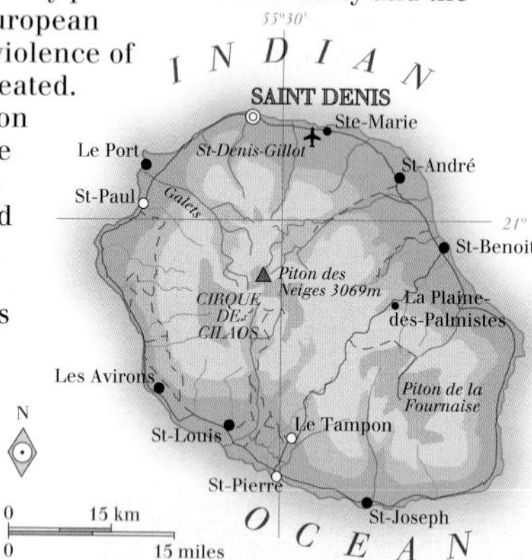
INDIAN
SAINT DENIS
Ste-Marie
St-Denis-Gillot
Le Port
St-André
St-Paul
Galets
St-Benoit
Piton des Neiges 3069m
CIRQUE DE CILAOS
La Plaine-des-Palmistes
Les Avirons
Piton de la Fournaise
St-Louis
Le Tampon
St-Pierre
St-Joseph
OCEAN
N
0 15 km
0 15 miles

St. Pierre & Miquelon

Status: French territorial collectivity **Claimed:** 1604

Capital: St. Pierre **Pop.:** 6316 **Density:** 26/km² (68/mi²)

ST. PIERRE & Miquelon is a group of barren islands lying just off the south coast of Newfoundland, Canada. The islands are surrounded by some of the world's richest fishing grounds. Their inhabitants have traditionally earned a living from fishing, and from servicing foreign trawler fleets off the coast. A long-running and sometimes bitter dispute between Canada and France over fishing and mineral rights was settled in 1992. The ruling, which was generally deemed to be in Canada's favor, has led the French authorities to diversify the economy by developing port facilities and encouraging tourism.

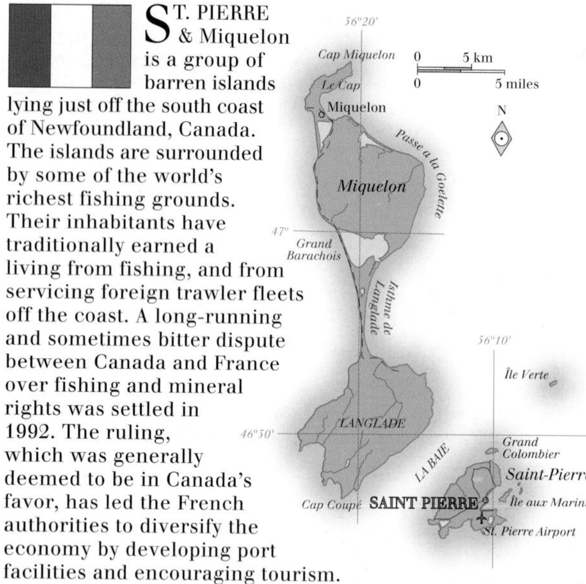
Cap Miquelon
Le Cap
Miquelon
Passe a la Goelette
Miquelon
Grand Barachois
Isthme de Langlade
Île Verte
LANGLADE
LA BAIE
Grand Colombier
Saint-Pierre
Cap Coupé
SAINT PIERRE
Île aux Marins
St. Pierre Airport
0 5 km
0 5 miles
N

St. Helena & Dependencies

Status: British overseas territory **Claimed:** 1673

Capital: Jamestown **Pop.:** 4647 **Density:** 38/km² (99/mi²)

TOGETHER, the islands of St. Helena, Tristan da Cunha, and Ascension form Britain's main dependency in the south Atlantic. St. Helena is famed for being the final place of exile for Napoléon. Its main economic activities – fishing, livestock farming, and the sale of handicrafts – are unable to support the population; as a result, underemployment on the island is a major problem. Many St. Helenians have been forced to seek work on Ascension, which has no resident population and is operated as a military base and communications center, although civilian flights have been permitted since 1998. Tristan da Cunha, a volcanic island 2000 km (1240 miles) south of St. Helena, is inhabited by a small, closely knit farming community. It was badly hit by severe winter storms in 2001. The 2002 British Overseas Territories Act granted all overseas citizens full British citizenship, assuaging a source of local resentment.

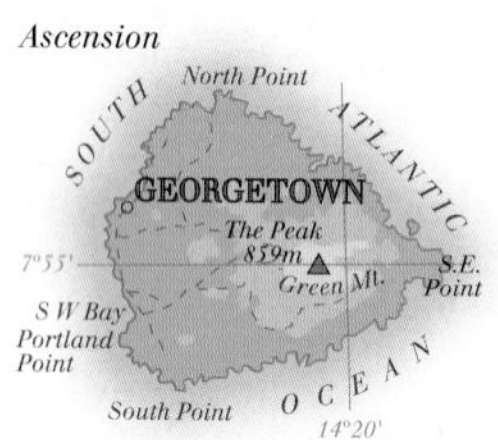
Ascension
SOUTH ATLANTIC OCEAN
North Point
GEORGETOWN
The Peak 859m
Green Mt.
S.E. Point
S W Bay
Portland Point
South Point

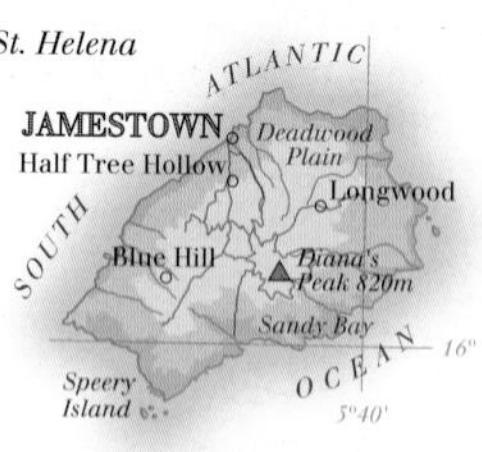
St. Helena
SOUTH ATLANTIC OCEAN
JAMESTOWN
Deadwood Plain
Half Tree Hollow
Longwood
Blue Hill
Diana's Peak 820m
Sandy Bay
Speery Island

Tristan da Cunha
SOUTH ATLANTIC OCEAN
Edinburgh
Queen Mary's Peak 2060m
Anchorstock Point
Sandy Point
Deadman's Bay
Stonyhill Point

N
0 8 km
0 8 miles

Spratly Islands

Status: *Disputed* **Claimed:** *Not applicable*

Capital: *Not applicable* **Population:** *Unknown*

SCATTERED ACROSS a large area of the South China Sea, the reefs, islands, and atolls that make up the Spratly Islands have become one of South Asia's most serious security issues. Claimed, all or in part, by China, Taiwan, Vietnam, Brunei, Malaysia, and the Philippines, more than 40 of the larger islands now have garrisons from some of the claimant states. The reasons for this interest, and the occasional skirmish, are twofold. Strategically, the islands control some of the world's most important shipping lanes. In addition, surveys suggest that some of the largest oil and gas reserves yet found lie in the Spratlys' territorial waters.

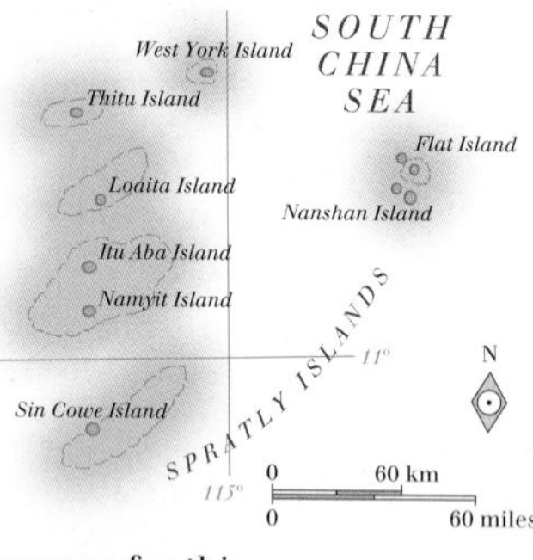
SOUTH CHINA SEA
West York Island
Thitu Island
Flat Island
Loaita Island
Nanshan Island
Itu Aba Island
Namyit Island
SPRATLY ISLANDS
Sin Cowe Island
N
0 60 km
0 60 miles

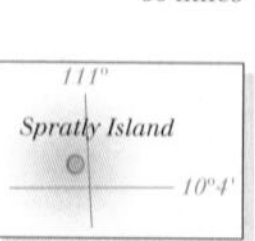
Spratly Island

The isolated Chinese occupying force on one of the Spratly Islands.

LAND HEIGHT above Sea Level 200m/656ft 500m/1640ft 1000m/3281ft 1500m/4572ft above 2000m/6562ft

SVALBARD

STATUS: Norwegian dependency **CLAIMED:** 1920
CAPITAL: Longyearbyen **POP.:** 2515 **DENSITY:** 0.04/km² (0.1/mi²)

MORE THAN 150 ice-covered Arctic islands 650 km (400 miles) north of Norway make up Svalbard. In accordance with the Spitsbergen Treaty of 1920, nationals of the treaty powers have equal rights to exploit the coal deposits, subject to regulation by Norway. The only companies still mining are Russian and Norwegian. There has been conflict with Iceland over fishing rights. Over half of the area of the islands is designated as environmentally protected.

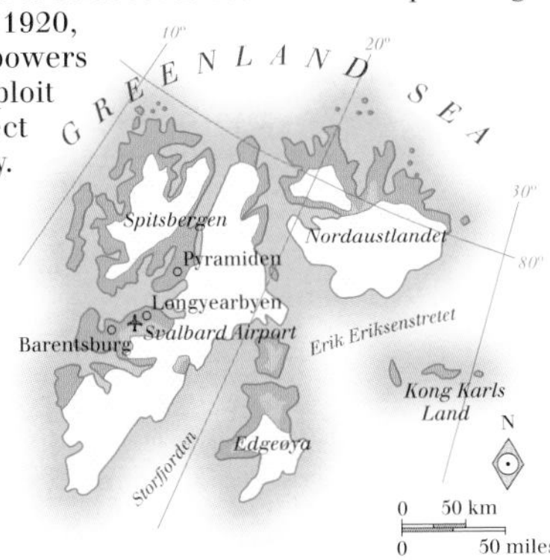
GREENLAND SEA
10°
20°
30°
80°
Spitsbergen
Pyramiden
Nordaustlandet
Longyearbyen
Svalbard Airport
Barentsburg
Erik Eriksenstretet
Kong Karls Land
Storfjorden
Edgeøya
N
0 50 km
0 50 miles

TOKELAU

STATUS: New Zealand dependent territory **CLAIMED:** 1926
CAPITAL: *Not applicable* **POP.:** 1487 **DENSITY:** 149/km² (372/mi²)

A 1989 UN REPORT states that in the 21st century these islands in the South Pacific will disappear under the sea, unless action is taken to stop global warming. The economy depends on a tuna cannery and the sale of fishing licenses, postage stamps, and coins; a catamaran link between the atolls has increased tourist potential. Tokelau's small size and economic fragility make independence unlikely, but in 1996 it gained the right to enact its own internal legislation, and since 2001 the local authorities have had full control of the island's public services. Nearly 4000 Tokelauans live in New Zealand.

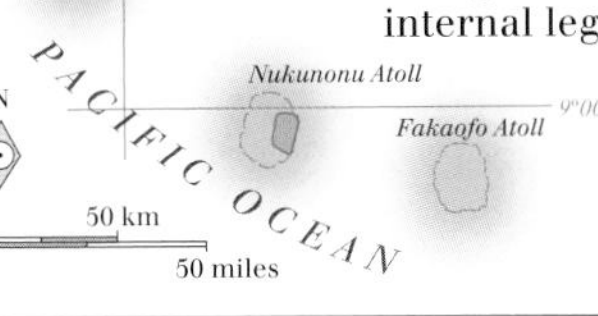
172° 30'
Atafu Atoll
Nukunonu Atoll
Fakaofo Atoll
9°00'
PACIFIC OCEAN
N
0 50 km
0 50 miles

TURKS & CAICOS ISLANDS

STATUS: British overseas territory **CLAIMED:** 1766
CAPITAL: Cockburn Town **POP.:** 16,863 **DENSITY:** 39/km² (102/mi²)

SITUATED 40 km (25 miles) south of the Bahamas, the Turks and Caicos Islands is a group of 30 low-lying islands, eight of which are inhabited. A traditional salt-based economy was exhausted in 1964, leading to two decades of stagnation. Since the 1980s, however, tourism and offshore banking have led a dramatic turnaround in the islands' fortunes.

ATLANTIC OCEAN
Caicos Passage
22°
North Caicos
Bottle Creek
Providenciales
Kew
Bambarra
Blue Hills
Grand Caicos
West Caicos
East Caicos
Caicos Bank
South Caicos
Grand Turk Island
Cockburn Harbour
Cockburn Town
Grand Turk Airport
Turks Islands
Mouchoir Passage
Ambergris Cays
N
0 30 km
0 30 miles
72°

VIRGIN ISLANDS (US)

STATUS: Unincorporated territory of the USA **CLAIMED:** 1917
CAPITAL: Charlotte Amalie **POP.:** 108,612 **DENS.:** 313/km² (811/mi²)

THE US VIRGIN ISLANDS are a collection of 53 volcanic islands, just to the east of Puerto Rico. Most of the population – a mix of African and European ethnic groups – live on the main islands of St. John, St. Thomas, and St. Croix. Tourism is the principal activity, although St. Croix has also used federal aid to develop industry. It has one of the world's largest oil refineries.

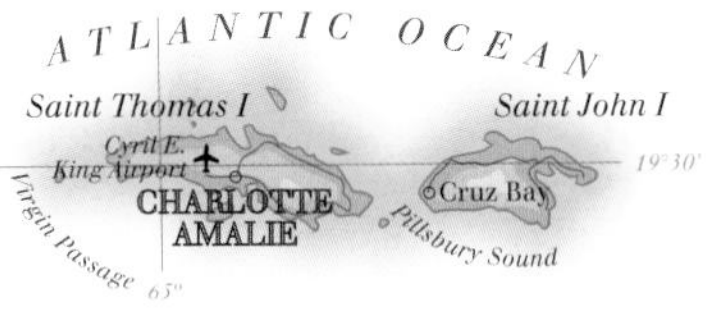
ATLANTIC OCEAN
Saint Thomas I
Saint John I
Cyril E. King Airport
CHARLOTTE AMALIE
Cruz Bay
19°30'
Virgin Passage
Pillsbury Sound
65°

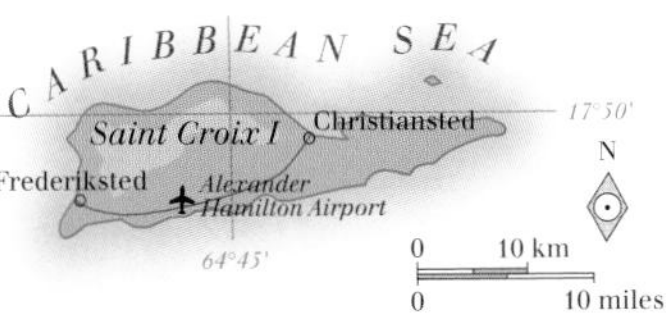
CARIBBEAN SEA
Saint Croix I
Christiansted
17°50'
Frederiksted
Alexander Hamilton Airport
64°45'
N
0 10 km
0 10 miles

St. Thomas, US Virgin Islands, *is a major stop-off for Caribbean cruise ships. Tourists are attracted by the island's duty-free shopping.*

WAKE ISLAND

STATUS: Unincorporated territory of the USA **CLAIMED:** 1898
CAPITAL: *Not applicable* **POP.:** *Not applicable*

WAKE ISLAND, in fact three islands that form the rim of an extinct volcano, has a US air base, whose airstrip can be used in emergencies by trans-Pacific flights. After widespread condemnation a 1998 proposal to store nuclear waste there was dropped. It is claimed by the Marshall Islands.

WALLIS & FUTUNA

STATUS: French overseas territory **CLAIMED:** 1842
CAPITAL: Matá'Utu **POP.:** 14,600 **DENSITY:** 53/km² (138/mi²)

UNLIKE FRANCE'S other South Pacific overseas territories, the inhabitants of Wallis and Futuna have little desire for greater autonomy. The islands' subsistence economy produces a variety of tropical crops, while expatriate remittances and the sale of licenses to Japanese and South Korean fishing fleets provide foreign exchange. Deforestation, leading to soil erosion, is of great concern.

178°10'
PACIFIC OCEAN
Pointe Fatua
Île Futuna
Nuku
Leava
Kolia
Pointe Vele
Taoa
14°20'
Pointe Mafa'a
Île Alofi
Pointe Sauma
N
0 10 km
0 10 miles

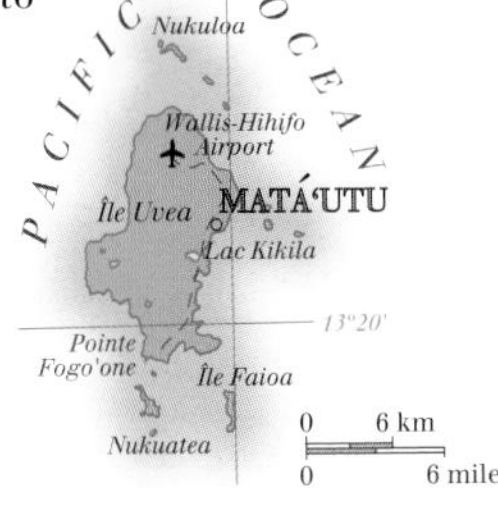
176°10'
PACIFIC OCEAN
Nukuloa
Wallis-Hihifo Airport
Île Uvea
MATÁ'UTU
Lac Kikila
13°20'
Pointe Fogo'one
Île Faioa
Nukuatea
0 6 km
0 6 miles

Glossary of Geographical Terms

The glossary following lists all geographical terms occurring on the maps and in main-entry names in the Index~Gazetteer. These terms may precede, follow, or be run together with the proper element of the name; where they precede it the term is reversed for indexing purposes – thus Poluostrov Yamal is indexed as Yamal, Poluostrov.

KEY
Geographical term *Language*, Term

A

Å *Danish, Norwegian,* River
Alpen *German,* Alps
Altiplanicie *Spanish,* Plateau
Älv(en) *Swedish,* River
Anse *French,* Bay
Archipiélago *Spanish,* Archipelago
Arcipelago *Italian,* Archipelago
Arquipélago *Portuguese,* Archipelago
Aukštuma *Lithuanian,* Upland

B

Bahía *Spanish,* Bay
Baía *Portuguese,* Bay
Baḩr *Arabic,* River
Baie *French,* Bay
Bandao *Chinese,* Peninsula
Banjaran *Malay,* Mountain range
Batang *Malay,* Stream
-berg *Afrikaans, Norwegian,* Mountain
Birket *Arabic* , Lake
Boğazı *Turkish,* Lake
Bucht *German,* Bay
Bugten *Danish,* Bay
Buḩayrat *Arabic,* Lake, reservoir
Buḩeiret *Arabic,* Lake
Bukit *Malay,* Mountain
-bukta *Norwegian,* Bay
bukten *Swedish,* Bay
Burnu *Turkish,* Cape, point
Buuraha *Somali,* Mountains

C

Cabo *Portuguese,* Cape
Cap *French,* Cape
Cascada *Portuguese,* Waterfall
Cerro *Spanish,* Mountain
Chaîne *French,* Mountain range
Chau *Cantonese,* Island
Chāy *Turkish,* River
Chhâk *Cambodian,* Bay
Chhu *Tibetan,* River
-chŏsuji *Korean,* Reservoir
Chott *Arabic,* Salt lake, depression
Ch'ün-tao *Chinese,* Island group
Chuôr Phnum *Cambodian,* Mountains
Cordillera *Spanish,* Mountain range
Costa *Spanish,* Coast
Côte *French,* Coast
Cuchilla *Spanish,* Mountains

D

Dağı *Azerbaijani, Turkish,* Mountain
Dağları *Azerbaijani, Turkish,* Mountains
-dake *Japanese,* Peak
Danau *Indonesian,* Lake
Đao *Vietnamese,* Island
Daryā *Persian,* River
Daryācheh *Persian,* Lake
Dasht *Persian,* Plain, desert
Dawḩat *Arabic,* Bay
Dere *Turkish,* Stream
Dili *Azerbaijani,* Spit
-do *Korean,* Island
Dooxo *Somali,* Valley
Düzü *Azerbaijani,* Steppe
-dwīp *Bengali,* Island

E

Embalse *Spanish,* Reservoir
Erg *Arabic,* Dunes
Estany *Catalan,* Lake
Estrecho *Spanish,* Strait
-ey *Icelandic,* Island
Ezero *Bulgarian, Macedonian,* Lake

F

Fjord *Danish,* Fjord
-fjorden *Norwegian,* Fjord
-fjørdhur *Faeroese,* Fjord
Fleuve *French,* River
Fliegu *Maltese,* Channel
-fljór *Icelandic,* River

G

-gang *Korean,* River
Ganga *Nepali, Sinhala,* River
Gaoyuan *Chinese,* Plateau
-gawa *Japanese,* River
Gebel *Arabic,* Mountain
-gebirge *German,* Mountains
Ghubbat *Arabic,* Bay
Gjiri *Albanian,* Bay
Gol *Mongolian,* River
Golfe *French,* Gulf
Golfo *Italian, Spanish,* Gulf
Gora *Russian, Serbian,* Mountain
Gory *Russian,* Mountains
Guba *Russian,* Bay
Gunung *Malay,* Mountain

H

Ḩadd *Arabic,* Spit
-haehyŏp *Korean,* Strait
Haff *German,* Lagoon
Hai *Chinese,* Sea, bay
Ḩammādat *Arabic,* Plateau
Hāmūn *Persian,* Lake
Hawr *Arabic,* Lake
Hāyk' *Amharic,* Lake
He *Chinese,* River
Helodrano *Malagasy,* Bay
-hegység *Hungarian,* Mountain range
Hka *Burmese,* River
-ho *Korean,* Lake
Hô *Korean,* Reservoir
Ḫolot *Hebrew,* Dunes
Hora *Belarussian,* Mountain
Hrada *Belarussian,* Mountains, ridge
Hsi *Chinese,* River
Hu *Chinese,* Lake

I

Île(s) *French,* Island(s)
Ilha(s) *Portuguese,* Island(s)
Ilhéu(s) *Portuguese,* Islet(s)
Irmak *Turkish,* River
Isla(s) *Spanish,* Island(s)
Isola (Isole) *Italian,* Island(s)

J

Jabal *Arabic,* Mountain
Jāl *Arabic,* Ridge
-järvi *Finnish,* Lake
Jazīrat *Arabic,* Island
Jazīreh *Persian,* Island
Jebel *Arabic,* Mountain
Jezero *Serbo-Croat,* Lake
Jiang *Chinese,* River
-joki *Finnish,* River
-jökull *Icelandic,* Glacier
Juzur *Arabic,* Islands

K

Kaikyō *Japanese,* Strait
-kaise *Lappish,* Mountain
Kali *Nepali,* River
Kalnas *Lithuanian,* Mountain
Kalns *Latvian,* Mountain
Kang *Chinese,* Harbor
Kangri *Tibetan,* Mountain(s)
Kaôh *Cambodian,* Island
Kapp *Norwegian,* Cape
Kavīr *Persian,* Desert
K'edi *Georgian,* Mountain range
Kediet *Arabic,* Mountain
Kepulauan *Indonesian, Malay,* Island group
Khalîg, Khalīj *Arabic,* Gulf
Khawr *Arabic,* Inlet
Khola *Nepali,* River
Khrebet *Russian,* Mountain range
Ko *Thai,* Island
Kolpos *Greek,* Bay
-kopf *German,* Peak
Körfäzi *Azerbaijani,* Bay
Körfezi *Turkish,* Bay

Kõrgustik *Estonian,* Upland
Koshi *Nepali,* River
Kowtal *Persian,* Pass
Kūh(hā) *Persian,* Mountain(s)
-kundo *Korean,* Island group
-kysten *Norwegian,* Coast
Kyun *Burmese,* Island

L

Laaq *Somali,* Watercourse
Lac *French,* Lake
Lacul *Romanian,* Lake
Lago *Italian, Portuguese, Spanish,* Lake
Laguna *Spanish,* Lagoon, Lake
Laht *Estonian,* Bay
Laut *Indonesian,* Sea
Lembalemba *Malagasy,* Plateau
Lerr *Armenian,* Mountain
Lerrnashght'a *Armenian,* Mountain range
Les *Czech,* Forest
Lich *Armenian,* Lake
Liqeni *Albanian,* Lake
Lumi *Albanian,* River
Lyman *Ukrainian,* Estuary

M

Mae Nam *Thai,* River
-mägi *Estonian,* Hill
Maja *Albanian,* Mountain
-man *Korean,* Bay
Marios *Lithuanian,* Lake
-meer *Dutch,* Lake
Melkosopochnik *Russian,* Plain
-meri *Estonian,* Sea
Mifraz *Hebrew,* Bay
Monkhafad *Arabic,* Depression
Mont(s) *French,* Mountain(s)
Monte *Italian, Portuguese,* Mountain
More *Russian,* Sea
Mörön *Mongolian,* River

N

Nagor'ye *Russian,* Upland
Nahal *Hebrew,* River
Nahr *Arabic,* River
Nam *Laotian,* River
Nehri *Turkish,* River
Nevado *Spanish,* Mountain (snow-capped)
Nisoi *Greek,* Islands
Nizmennost' *Russian,* Lowland, plain
Nosy *Malagasy,* Island
Nur *Mongolian,* Lake
Nuruu *Mongolian,* Mountains
Nuur *Mongolian,* Lake
Nyzovyna *Ukrainian,* Lowland, plain

O

Ostrov(a) *Russian,* Island(s)
Oued *Arabic,* Watercourse
-oy *Faeroese,* Island
-øy(a) *Norwegian,* Island
Oya *Sinhala,* River
Ozero *Russian, Ukrainian,* Lake

P

Passo *Italian,* Pass
Pegunungan *Indonesian, Malay,* Mountain range
Pelagos *Greek,* Sea
Penisola *Italian,* Peninsula
Peski *Russian,* Sands
Phanom *Thai,* Mountain
Phou *Laotian,* Mountain
Pi *Chinese,* Point
Pic *Catalan,* Peak
Pico *Portuguese, Spanish,* Peak
Pik *Russian,* Peak
Planalto *Portuguese,* Plateau
Planina, Planini *Bulgarian, Macedonian, Serbo-Croat,* Mountain range
Ploskogor'ye *Russian,* Upland
Poluostrov *Russian,* Peninsula
Potamos *Greek,* River
Proliv *Russian,* Strait
Pulau *Indonesian, Malay,* Island
Pulu *Malay,* Island
Punta *Portuguese, Spanish,* Point

Q

Qā' *Arabic,* Depression
Qolleh *Persian,* Mountain

R

Raas *Somali,* Cape
-rags *Latvian,* Cape
Ramlat *Arabic,* Sands
Ra's *Arabic,* Cape, point, headland
Ravnina *Bulgarian, Russian,* Plain
Récif *French,* Reef
Represa (Rep.) *Spanish, Portuguese,* Reservoir
-rettō *Japanese,* Island chain
Riacho *Spanish,* Stream
Riban' *Malagasy,* Mountains
Rio *Portuguese,* River
Río *Spanish,* River
Riu *Catalan,* River
Rivier *Dutch,* River
Rivière *French,* River
Rowd *Pashtu,* River
Rūd *Persian,* River
Rudohorie *Slovak,* Mountains
Ruisseau *French,* Stream

S

Sabkhat *Arabic,* Salt marsh
Şaḩrā' *Arabic,* Desert
Samudra *Sinhala,* Reservoir
-san *Japanese, Korean,* Mountain
-sanchi *Japanese,* Mountains
-sanmaek *Korean,*
Sarīr *Arabic,* Desert
Sebkha, Sebkhet *Arabic,* Salt marsh, depression
See *German,* Lake
Selat *Indonesian,* Strait
-selkä *Finnish,* Ridge
Selseleh *Persian,* Mountain range
Serra *Portuguese,* Mountain
Serranía *Spanish,* Mountain
Sha'īb *Arabic,* Watercourse
Shamo *Chinese,* Desert
Shan *Chinese,* Mountain(s)
Shan-mo *Chinese,* Mountain range
Shaţţ *Arabic,* Distributary
-shima *Japanese,* Island
Shiqqat *Arabic,* Depression
Shui-tao *Chinese,* Channel
Sierra *Spanish,* Mountains
Sơn *Vietnamese,* Mountain
Sông *Vietnamese,* River
-spitze *German,* Peak
Štít *Slovak,* Peak
Stoeng *Cambodian,* River
Stretto *Italian,* Strait
Su Anbarı *Azerbaijani,* Reservoir
Sungai *Indonesian, Malay,* River
Suu *Turkish,* River

T

Tal *Mongolian,* Plain
Tandavan' *Malagasy,* Mountain range
Tangorombohitr' *Malagasy,* Mountain massif
Tao *Chinese,* Island
Tassili *Berber,* Plateau, mountain
Tau *Russian,* Mountain(s)
Taungdan *Burmese,* Mountain range
Teluk *Indonesian, Malay,* Bay
Terara *Amharic,* Mountain
Tog *Somali,* Valley
Tônlé *Cambodian,* Lake
Top *Dutch,* Peak
-tunturi *Finnish,* Mountain
Tur'at *Arabic,* Channel

V

Väin *Estonian,* Strait
-vatn *Icelandic,* Lake
-vesi *Finnish,* Lake
Vinh *Vietnamese,* Bay
Vodokhranilishche (Vdkhr.) *Russian,* Reservoir
Vodoskhovyshche (Vdskh.) *Ukrainian,* Reservoir
Volcán *Spanish,* Volcano
Vozvyshennost' *Russian,* Upland, plateau
Vrh *Macedonian,* Peak
Vysochyna *Ukrainian,* Upland
Vysočina *Czech,* Upland

W

Waadi *Somali,* Watercourse
Wādī *Arabic,* Watercourse
Wāḩat, Wâhat *Arabic,* Oasis
Wald *German,* Forest
Wan *Chinese,* Bay
Wyżyna *Polish,* Upland

X

Xé *Laotian,* River

Y

Yarımadası *Azerbaijani,* Peninsula
Yazovir *Bulgarian,* Reservoir
Yoma *Burmese,* Mountains
Yü *Chinese,* Island

Z

Zaliv *Bulgarian, Russian,* Bay
Zatoka *Ukrainian,* Bay
Zemlya *Russian,* Bay

Glossary of Abbreviations

This glossary provides a comprehensive guide to the abbreviations used.

A

abbrev. abbreviated
ABM antiballistic missile(s)
Afr. Afrikaans
AIDS acquired immunodeficiency syndrome
Alb. Albanian
ALCM air-launched Cruise missile(s)
Amh. Amharic
ANC African National Congress
anc. ancient
APC armored personnel carrier(s)
approx. approximately
Ar. Arabic
Arm. Armenian
ASSR Autonomous Soviet Socialist Republic
Aust. Australian
Az. Azerbaijani
Azerb. Azerbaijan

B

Basq. Basque
BBC British Broadcasting Corporation
BCE Before Common Era
b/d barrels per day
Bel. Belarussian
Ben. Bengali
Ber. Berber
B-H Bosnia and Herzegovina
bn billion (1000 million)
BP British Petroleum
Bret. Breton
Brig. Brigadier
Brit. British
BSE bovine spongiform encephalopathy
Bul. Bulgarian
Bur. Burmese

C

C central
C. Cape
°C degrees (Centigrade)
Cam. Cambodian
Cant. Cantonese
Capt. Captain
CAR Central African Republic
Cast. Castilian
Cat. Catalan
CE Common Era
Chin. Chinese
CIA Central Intelligence Agency
cm centimeter(s)
Cmdr. Commander
CNN Cable News Network
Col. Colonel
Cro. Croat
Cz. Czech
Czech Rep. Czech Republic

D E

Dan. Danish
dept. department
dev. development
Dom. Rep. Dominican Republic
Dr. Doctor
DRC Democratic Republic of Congo
Dut. Dutch
dwt dead weight tonnage
E east
EEC/EC European Community
EEZ Exclusive Economic Zone
ECU European Currency Unit
EMS European Monetary System
Eng. English
est. estimated
Est. Estonian

F G

°F degrees (Fahrenheit)
Faer. Faeroese
Fij. Fijian
Fin. Finnish
Fr. Father
Fr. French
Franc Francophone
Fris. Frisian
ft foot/feet
FYRM Former Yugoslav Republic of Macedonia
FZ Franc Zone
Gael. Gaelic
Gal. Galician
GATT General Agreement on Tariffs and Trade
GDP Gross Domestic Product (the total value of goods and services produced by a country excluding income from foreign countries)
Gen. General
Geor. Georgian
Ger. German
Gk Greek
GNP Gross National Product (the total value of goods and services produced by a country)

H I

Heb. Hebrew
HEP hydroelectric power
HH His/Her Highness
Hind. Hindi
HIPC heavily indebted poor country(ies)
hist. historical
HIV human immunodeficiency virus
H.M. His/Her Majesty
HMS His/Her Majesty's ship
H.R.H His/Her Royal Highness
H.S.H His/Her Serene Highness
Hung. Hungarian
I. Island
ICBM intercontinental ballistic missile(s)
Icel. Icelandic
in inch(es)
In. Inuit (Eskimo)
Ind. Indonesian
Intl International
Ir. Irish
IRBM intermediate-range ballistic missile(s)
Is Islands
It. Italian

J K L

Jap. Japanese
Kaz. Kazakh
kg kilogram(s)
Kir. Kirghiz
km kilometer(s)
km² square kilometer (singular)
Kor. Korean
Kurd. Kurdish
kw kilowatt(s)
kwh kilowatt hour(s)
L. Lake
Lao. Laotian
Lat. Latin
Latv. Latvian
Liech. Liechtenstein
Lith. Lithuanian
LNG liquefied natural gas
Lt. Lieutenant
Lusoph Lusophone
Lux. Luxembourg

M N

m million/meter(s)
Mac. Macedonian
Maced. Macedonia
Maj. Major
Mal. Malay
Malg. Malagasy
Malt. Maltese
MBA Master of Business Administration
mi. mile(s)
mi² square mile(s)
Mong. Mongolian
MP Member of Parliament
MSP Member of Scottish Parliament
Mt. Mountain/Mount
Mts Mountains
MW megawatt(s)
N north
NASA National Aeronautics and Space Administration
Nep. Nepali
Nepad New Partnership for Africa's Development
Neth. Netherlands
NGO Nongovernmental Organization
NIC Newly Industrialized Country
Nic. Nicaraguan
Nor. Norwegian
NPT Non-Proliferation Treaty
NZ New Zealand

P Q R

Pash. Pashtu
Per. Persian
PLO Palestine Liberation Organization
PNG Papua New Guinea
Pol. Polish
Poly. Polynesian
Port. Portuguese
POW prisoner of war
prev. previously
Rep. Represa (Spanish, Portuguese for reservoir)
Rep. Republic
Res. Reservoir
Rev. Reverend
Rmsch. Romansch
Rom. Romanian
Rus. Russian
Russ. Fed. Russian Federation

S

S south
S. & Mon. Serbia & Montenegro
SALT Strategic Arms Limitation Treaty
SCr. Serbo-Croatian
Serb. Serbian
Sinh. Sinhala
SLBM submarine-launched ballistic missile(s)
Slvk. Slovak
Slvn. Slovene
Som. Somali
Sp. Spanish
sq. square
SSBN nuclear-fuelled ballistic-missile submarine(s)
SSM surface-to-surface missile(s)
St. Saint
START Strategic Arms Reduction Treaty
Strs. Straits
Swa. Swahili
Swe. Swedish
Switz. Switzerland

T U

Taj. Tajik
TGV *train à grande vitesse*
Th. Thai
Thai. Thailand
Tib. Tibetan
Turk. Turkish
Turkm. Turkmenistan
UAE United Arab Emirates
Uigh. Uighur
UK United Kingdom
Ukr. Ukrainian
UN United Nations
Urd. Urdu
US/USA United States of America
USS United States ship
USSR Union of Soviet Socialist Republics
Uzb. Uzbek

V W X Y

var. variant
VCR video cassette recorder
Vdkhr. Vodokhranilishche (Russian for reservoir)
Vdskh. Vodoskhovyshche (Ukrainian for reservoir)
Vtn. Vietnamese
W west
Wel. Welsh

GEOGRAPHICAL PLACE-NAMES

THE CHOICES confronting a map-maker when deciding which place-name style to use on a map are surprisingly varied. The criteria adopted may be affected by a range of factors: the existence of foreign and native language forms of a place name (London, Londres, Londra), variant spellings used within the country itself (Gent, Gand), and the existence of completely different language forms for international features (the English Channel, La Manche).

In addition to these, political expedience, simple clarity, and the use to which the published map may be put are all factors that need consideration.

The revision of place-name forms and spellings, which is a continuing administrative activity worldwide, adds a further dimension of complexity to the subject. Since the collapse of Soviet communism, for instance, place-names in Russia have been altered to expunge traces of communist ideology (the most famous being the 1991 reversion of Leningrad to its pre-1914 name, St. Petersburg). In many former Soviet republics, Russian names have been replaced with native language forms (notably in Ukraine, Belarus, Georgia, and Armenia).

Standardized Arabic names have been hindered by the persistent use of French forms in practice.

THE MAPS

The maps in the Nations of the World section of this book have used the most up-to-date reference sources available to provide local name forms and spellings, that is to say those used within the country. In an age when international travel, on holiday or on business, is commonplace, this criterion seems the most appropriate.

English conventional forms have been used for all international features (such as sea areas between countries, and cross-border mountain ranges); for all country names (the Index-Gazetteer provides local forms and spellings, while commonly used alternative names, such as Burma/Myanmar, are also made clear in the national A–Z entry); and for all capital cities. The Index-Gazetteer provides a fully cross-referenced system that will guide the reader the short distance from the English conventional "Florence" to the local "Firenze," as used on the maps.

English conventional forms also appear on all the maps in the World Factfile. These maps have not been indexed, as all contemporary places featured are most usefully and accurately identified on the national maps.

THE INDEX-GAZETTEER

The Index-Gazetteer lists all names that appear on the maps in the Nations of the World section of the book. Physical features are defined as such, as are countries and those administrative or regional names included on the maps; all other names are those of population centers. Location is given by page number, then country, and is narrowed down by positional reference as N(orth), S(outh), E(ast), W(est), or C(entral), or combinations of these as appropriate.

Following each main entry name are given: variant spellings of the name most commonly found; its previous name or names; and such foreign-language forms of the name as are pertinent to modern history since 1940. This is the cut-off date generally adopted, permitting the inclusion of all place-name changes made during or after World War II. Exceptionally, name changes made in Russia and other countries of the former Soviet Union before 1940 are given, since many old names in these countries are now being restored.

INDEX

PAGE NUMBERS SHOWN IN ITALICS

Geographical Terms and Abbreviations on pages 654-657 / ❖ = *capital*

A

Aa see Gauja
Aabenraa see Åbenrå
Aachen 266 Fr. Aix-la-Chapelle, Dut. Aken. W Germany
Aalborg see Ålborg
Aalesund see Ålesund
Aaley 364 var. Ālayh, Aley. C Lebanon
Aalsmeer 429 W Netherlands
Aalst 127 Fr. Alost. C Belgium
Aanaarjävri see Inarijärvi
Aanjar 364 C Lebanon
Aarau 550 N Switzerland
Aare 550 var. Aar. River of W Switzerland
Aarhus see Århus
Aarlen see Arlon
Aarschot 127 C Belgium
Aassi, Nahr el see Orantes
Aba 440 S Nigeria
Abaco Island 112 island of N Bahamas
Ābādān 309 W Iran
Abai see Blue Nile
Abaiang 348 island of the Gilbert Is, W Kiribati
Abakan 485 prev. Khakassk, Ust'-Abakanskoye. C Russian Federation
Abancay 463 SE Peru
Abariringa see Kanton
Ābaya Hāyk' 245 It. Abbaia, Eng. Lake Margherita. Lake of SW Ethiopia
Abay Wenz see Blue Nile
Abbeville 255 N France
'Abd al 'Azīz, Jabal 555 mountains of NE Syria
'Abdalī 356 S Kuwait
Abd-Al-Kuri 631 island of SE Yemen, off the Horn of Africa
Abéché 180 var. Abécher. E Chad
Abemama 348 island of the Gilbert Is, W Kiribati
Abengourou 328 E Ivory Coast
Åbenrå 218 var. Aabenraa, Ger. Apenrade. Jylland, SW Denmark
Abeokuta 440 SW Nigeria
Abercorn see Mbala
Aberdeen 597 NE Scotland, UK
Aberdeen 603 South Dakota, NC USA
Abergwaun see Fishguard
Abersee see Wolfgangsee
Abertawe see Swansea
Aberystwyth 597 W Wales, UK
Abhā 506 S Saudi Arabia
Abhe, Lake 222, 245 Amh. Ābhē Bid Hāyk'. Lake of Djibouti and Ethiopia
Abidjan 328 S Ivory Coast
Åbo see Turku
Aboisso 328 SE Ivory Coast
Abo, Massif d' 180 mountain range of N Chad
Abomey 132 S Benin
Abong Mbang 168 SE Cameroon
Abou-Déïa 180 S Chad
Aboudouhour see Abū aḍ Ḍuhūr
Abou Kémal see Abū Kamāl
Aboumi 260 E Gabon
Abovyan 98 var. Abovjan. C Armenia
Abra 466 river of Luzon, N Philippines
Abrād, Wādī 631 seasonal river of NW Yemen
Abraham Bay see Carlton, The
Abruzzese, Appennino 323 mountain range of C Italy
'Abs 631 var. Sūq 'Abs. W Yemen
Abşeron Yarımadası 110 Rus. Apsheronskiy Poluostrov. Oil-rich peninsula of E Azerbaijan
Abū aḍ Ḍuhūr 555 Fr. Aboudouhour. NW Syria
Abū al Abyaḍ 594 island of N United Arab Emirates
Abū al Jirfān, Sha'īb 356 var. Sh'ib Abu Jarfan. Dry watercourse of N Kuwait
Abū al Khaṣīb 312 var. Abul Khasib. SE Iraq
Abu al Mawj, Ra's 115 cape of W Bahrain
Abu Dhabi 594 Ar. Abū Ẓaby, var. Abū Ẓabī. ❖ of United Arab Emirates
Abuja 440 ❖ of Nigeria, C Nigeria
Abū Jarjūr, Ra's 115 cape of E Bahrain
Abū Kamāl 555 Fr. Abou Kémal. E Syria
Abul Khasib see Abū al Khaṣīb
Abuná 136 river of Bolivia and Brazil
Abū Thaylah 479 NE Qatar
Abū Ẓabī see Abu Dhabi
Abū Ẓaby see Abu Dhabi
Abyaḍ, Baḥr al see White Nile
Åbybro 218 N Denmark
Abyssinia see Ethiopia
Acaill see Achill Island
Acajutla 237 W El Salvador
Acapulco 403 var. Acapulco de Juárez. S Mexico
Acaraí, Serra 144 Eng. Acarai Mountains. Mountain range of Brazil and Guyana
Acarigua 623 NW Venezuela
Accra 272 ❖ of Ghana, SE Ghana
Achacachi 136 W Bolivia
Acharnés 275 prev. Akharnaí. SE Greece
Acheloos 275 var. Aspropotamos, prev. Akhelóös. River of W Greece
Achénouma 439 NE Niger
Achill Island 316 Ir. Acaill. Island of W Ireland
Achna see Athna
Achorstock Point 652 headland of W Tristan da Cunha
Achwa 588 var. Aswa. River of N Uganda
Acireale 323 Sicilia, S Italy
Acklins Island 112 island of S Bahamas
Aconcagua, Cerro 95 mountain of W Argentina
Açores see Azores
Açores, Arquipélago dos see Azores
A Coruña 534 Cast. La Coruña. NW Spain
Acoua 649 NW Mayotte
Acquaviva 502 NW San Marino
Acre see 'Akko
Acurnam 238 var. Acurenan, Akurenan. S Río Muni, Equatorial Guinea
Adalia see Antalya
Adalia, Gulf of see Antalya Körfezi
Adam 448 N Oman
Adama see Nazrēt
Adamaoua, Massif d' 168 Eng. Adamawa. Plateau of West Africa
Adam-jo-Tando see Tando Ādam
Adam's Bridge 538 chain of shoals to the NW of Sri Lanka
Adamstown 651 ❖ of Pitcairn Islands, NE Pitcairn Island, Pitcairn Islands
'Adan 631 Eng. Aden. SW Yemen
Adana 581 var. Seyhan. S Turkey
Adapazarı 580 var. Sakarya. NW Turkey
Aḍ Ḍab 'iyah 594 C United Arab Emirates
Aḍ Ḍafrah 594 desert region of W United Arab Emirates
Ad Dahnā' 506 N Saudi Arabia.
Ad Dakhla 414 W Western Sahara
Ad Dalanj see Dilling
Ad Dammām 506 desert region of NE Saudi Arabia
Ad Dawḥah see Doha
Ad Dibdibah 356 mountain of W Kuwait
Addīgrat 245 N Ethiopia
Ad Dirāz 115 NW Bahrain
Addis Ababa 245 Amh. Ādīs Ābeba. ❖ of Ethiopia, C Ethiopia
Ad Dīwānīyah 312 var. Diwaniya. C Iraq
Addu Atoll 390 atoll of S Maldives
Adelaide 101 S Australia
Adelphi 498 E St Vincent, St Vincent & the Grenadines
Adelsberg see Postojna
Aden see 'Adan
Aden, Gulf of 222, 528, 631 var. Badyarada 'Adméd. Gulf connecting the Indian Ocean and Red Sea
Adh Dhayd 594 var. Deira. NE United Arab Emirates
Adh Dhirā' 338 W Jordan
Ādī Ārk'ay 245 var. Addi Arkay. N Ethiopia
Adige 323 Ger. Etsch. River of N Italy
Adi Keyih 240 var. Adi Keyah. SE Eritrea
Adi Kwala 240 S Eritrea
Ādīs Ābeba see Addis Ababa
Ādīs Zemen 245 NW Ethiopia
Adi Tekelezan 240 C Eritrea
Adıyaman 581 SE Turkey
Admiralty Islands 458 island group of N Papua New Guinea
Ado-Ekiti 440 SW Nigeria
Adola see Kibre Mengist
Ado-Odo 440 var. Ado. SW Nigeria
Adour 255 river of SW France
Adrar des Ifôghas 392 mountainous region of C Sahara, NE Mali
Adriatic Sea 323, 512 It. Mare Adriatico, Slvn. Jadransko Morje, SCr. Jadransko More, Alb. Deti Adriatik. Area of the Mediterranean Sea, between Italy and SE Europe
Adriatik, Deti see Adriatic Sea
Adventure Sound 645 bay of the South Atlantic Ocean, E Falkland Islands
Ādwa 245 var. Adowa, It. Adua. N Ethiopia
Adygeya, Respublika 484 autonomous republic of SW Russian Federation
Adzopé 328 SE Ivory Coast
Aeankan District 396 district of Majuro, SE Marshall Islands
Aegean Sea 275, 580 Gk Aigaío Pélagos, Turk. Ege Denizi. Area of the Mediterranean Sea
Aeolian Islands see Eolie, Isole
Ærø 218 Ger. Arrö. Island of S Denmark
Afadjado 272 var. Afadjato, Afadjoto. Mountain of SE Ghana
'Afak 312 C Iraq
Afar Depression see Danakil Desert
Afghanistan 76-79 officially Islamic State of Afghanistan, prev. Republic of Afghanistan. Country of C Asia divided into 30 admin. units (velayats). ❖ Kābul
Afgooye 528 It. Afgoi. S Somalia
Afikpo 440 S Nigeria
Afobaka 542 NE Suriname
'Afula 319 N Israel
Afyon 580 prev. Afyonkarahisar. W Turkey
Agadez 439 prev. Agadès. C Niger
Agadir 414 SW Morocco
Agaña see Hagåtña
Aga Point 647 headland on the S coast of Guam
Āgaro 245 W Ethiopia
Agat 647 W Guam
Agboville 328 SE Ivory Coast
Ağcabädi 110 Rus. Agdzhabedi, var. Agdžabedi. C Azerbaijan
Agdam 110 SW Azerbaijan
Agedabia see Ajdābiyā
Agen 255 SW France
Agere Hiywet see Hāgere Hiywet
Agia Napa see Ayia Napa
Agigialousa see Yenierenköy
Agios Ioannis see Ayios Ioannis
Agona Swedru 272 var. Swedru. SE Ghana
Agordat see Akordat
Agou Gadzepe 571 SW Togo
Agou, Mont 571 prev. Pic Baumann. Mountain of SW Togo
Āgra 298 N India
Agram see Zagreb
Agrigento 323 prev. Girgenti. Sicilia, S Italy
Agrihan 650 island of N Northern Mariana Islands
Agrínio 275 prev. Agrínion. W Greece
Aguachica 195 N Colombia
Aguadilla 651 NW Puerto Rico
Aguadulce 456 S Panama
Aguán 290 river of N Honduras
Aguarico 230 river of Ecuador and Peru
Aguascalientes 403 C Mexico
Aguijan 650 island of S Northern Mariana Islands
Agusan 466 river of Mindanao, S Philippines
Ahaggar 83 var. Hoggar. Mountain range of SE Algeria
Ahja 242 var. Ahja Jõgi. River of SE Estonia
Ahmadābād 298 var. Ahmedabad. W India
Ahmadpur East 451 E Pakistan
Ahuachapán 237 W El Salvador
Ahvāz 309 var. Ahwāz. W Iran
Ahvenanmaa see Åland
Aḥwar 631 SW Yemen
Aibak see Āybak
Aigaío Pélagos see Aegean Sea
Aiguá 611 S Uruguay
Ai-hun see Heihe
Ailigandí 456 E Panama
Ailinginae 396 island of NW Marshall Islands
Ailinglaplap 396 prev. Ailinglapalap. Island of S Marshall Islands
Ailuk 396 island of NE Marshall Islands
'Aïn Ben Tili 398 N Mauritania
Aïn Oussera 83 var. Aïn Wessara. N Algeria
Aintab see Gaziantep
Aïoun el Atroûss see 'Ayoûn el 'Atroûs
Aiquile 136 C Bolivia
Airai 455 C Palau
Airdrie 170 SW Canada
Airdrie 597 C Scotland, UK
Airlalang see Rokan
Aitape 458 var. Eitape. NW Papua New Guinea
Aitos see Aytos
Aitutaki 644 island of Southern Cook Islands, S Cook Islands
Aix-en-Provence 255 SE France
Aix-la-Chapelle see Aachen
Aizu-Wakamatsu 332 Honshū, N Japan
Ajaccio 255 Corse, SE France
Ajdābiyā 371 var. Ajdābiyah, Agedabia. NE Libya
Ajeltake District 396 district of Majuro, SE Marshall Islands
Ajjinena see Geneina
Ajka 292 W Hungary
Ajman 594 Ar. 'Ajmān, var. 'Ujmān. NE United Arab Emirates
Ajtos see Aytos
Akaba see Al 'Aqabah
Akagera 162, 492, 588 var. Kagera. River of E Africa

B

Bonthe *518* Sherbro I, Sierra Leone
Boosaaso *528 var.* Bosaso, Bender Qaasim, *It.* Bender Cassim. N Somalia
Boothia Felix *see* Boothia Peninsula
Boothia, Gulf of *170* gulf of the Arctic Ocean between the Boothia Peninsula and Melville Peninsula, N Canada
Boothia Peninsula *170 prev.* Boothia Felix. Peninsula of N Canada
Booué *260* NE Gabon
Bophuthatswana Bantustan 'self-governing homeland'; abolished in 1994
Bopolu *368* NW Liberia
Boquete *456 var.* Bajo Boquete. W Panama
Boraha, Nosy *see* Sainte Marie, Nosy
Borås *547* SW Sweden
Borborema, Planalto da *145* plateau of NE Brazil
Bordeaux *255* SW France
Borden Peninsula *171* peninsula of Baffin Island, N Canada
Bordergate *544* NE Swaziland
Bordhoy *644 var.* Bordø. Island of NE Faeroe Islands
Bordj Bou Arreridj *83* N Algeria
Borgarnes *296* W Iceland
Borgholm *547* Öland, S Sweden
Borgo Maggiore *502* NW San Marino
Borisov *see* Barysaw
Bor, Lagh *344 var.* Lak Bor. Dry watercourse of NE Kenya
Borlänge *547* C Sweden
Borneo *304, 386-387* island of SE Asia divided between Brunei, two Malaysian states and part of Indonesia
Bornholm *218* island of E Denmark
Boromo *157* SW Burkina
Borongo *see* Black Volta
Borovo *209* NE Croatia
Borujerd *309 var.* Burujird. W Iran
Börzsöny *292* mountain range of N Hungary
Bosanska Gradiška *140* N Bosnia & Herzegovina
Bosanski Brod *140* N Bosnia & Herzegovina
Bosanski Šamac *140* N Bosnia & Herzegovina
Boscobelle *121* NE Barbados
Bösing *see* Pezinok
Boskamp *542* N Suriname
Boskovice *216* SE Czech Republic
Bosna *140* river of N Bosnia & Herzegovina
Bosna I Hercegovina, Federacija *140 republic* Bosnia & Herzegovina
Bosnia & Herzegovina *140-141* officially The Republic of Bosnia and Herzegovina. Country of SE Europe. ❖ Sarajevo
Bosporus *see* İstanbul Boğazi
Bossangoa *179* C Central African Republic
Bossembélé *179* C Central African Republic
Boston *603* Massachusetts, NE USA
Boteti *142 var.* Botletle. River of C Botswana
Botevgrad *152 prev.* Orkhanie, Orkhaniye. W Bulgaria
Bothnia, Gulf of *251, 547 Fin.* Pohjanlahti, *Swe.* Bottniska Viken. Gulf of the Baltic Sea, between Finland and Sweden
Boti-Pasi *542* C Suriname
Botna *408* river of E Moldova
Botoşani *480* NE Romania
Botrange *127* mountain of E Belgium
Botswana *142-143* officially Republic of Botswana. Country of southern Africa divided into 10 admin. units (districts). ❖ Gaborone
Bottle Creek *653* North Caicos, N Turks and Caicos Islands
Bottniska Viken *see* Bothnia, Gulf of
Bottom, The *650* Saba, N Netherlands Antilles
Bouaflé *328* C Ivory Coast
Bouaké *328 var.* Bwake. C Ivory Coast
Bouar *179* W Central African Republic
Bouca *179* C Central African Republic
Bouéni *649* SW Mayotte
Bouéni, Baie de *649 var.* Bonéni Bay. Bay of the Mozambique Channel on the SW coast of Mayotte
Bouenza *200* river of S Congo
Boufarik *83* N Algeria
Bougainville Island *458* island of E Papua New Guinea
Bougie *see* Béjaïa
Bougouni *392* SW Mali
Bougouriba *157* river of SW Burkina
Boujdour *414* W Western Sahara
Boukoumbé *132 var.* Boukombé. NW Benin
Boukra *414* C Western Sahara
Boulder *603* Colorado, SW USA
Boulogne-sur-Mer *255 var.* Boulogne. N France
Bouloupari *650* S New Caledonia
Bouma *248* Taveuni, N Fiji
Boumango *260* SE Gabon
Boumba *168* river of SE Cameroon
Boumbé II *179* river of Cameroon and Central African Republic
Boûmdeïd *398* S Mauritania
Bouna *328* NE Ivory Coast
Boundiali *328* N Ivory Coast
Boungou *179* river of C Central African Republic
Bourail *650* C New Caledonia
Bourgas *see* Burgas
Bourg-en-Bresse *255* E France
Bourges *255* C France
Bourgogne *255 Eng.* Burgundy. Cultural region of E France
Bourke *101* E Australia
Bournemouth *597* S England, UK
Boussé *157* C Burkina
Boussouma *157* C Burkina
Boutilimit *398* SW Mauritania
Bouton *497* W St Lucia
Bovec *524 Ger.* Flitsch, *It.* Plezzo. NW Slovenia
Boven Kapuas, Pegunungan *386* mountain range of SW Borneo, Malaysia
Boyd's *494* S St Kitts, St Kitts & Nevis
Boyne *316 Ir.* An Bhóinn. River of E Ireland
Boysun *614 Rus.* Baysun. SE Uzbekistan
Bozen *see* Bolzano
Bozoum *179* W Central African Republic
Brač *209 var.* Brach, *It.* Brazza. Island of S Croatia
Bradford *597* N England, UK
Braga *474* NW Portugal
Bragado *95* E Argentina
Bragança *474 Eng.* Braganza. NE Portugal
Braganza *see* Bragança
Brahestad *see* Raahe
Brāhmanbāria *117* E Bangladesh
Brahmaputra *117, 186, 299 var.* Tsangpo, *Ben.* Jamuna. River of S Asia
Brăila *480* E Romania
Braine-l'Alleud *127* C Belgium
Branco *144* river of NW Brazil
Branco, Ilhéu *176* island of N Cape Verde
Brandberg *423* mountain of NW Namibia
Brande *218* Jylland, W Denmark
Brandenburg *267* C Germany
Brandon *170* S Canada
Brani, Pulau *521* island of S Singapore
Brasília *145* ❖ of Brazil, C Brazil
Braslaw *122* N Belarus
Braşov *480 prev.* Stalin, *Ger.* Kronstadt, *Hung.* Brassó. C Romania
Brasschaat *127* N Belgium
Brassey, Banjaran *387 var.* Brassey Range. Mountain range of E Borneo, Malaysia
Brassey Range *see* Brassey, Banjaran
Bratislava *523 Ger.* Pressburg, *Hung.* Pozsony. ❖ of Slovakia, SW Slovakia
Bratsk *485* C Russian Federation
Bratskoye Vodokhranilishche *485 Eng.* Bratsk Reservoir. C Russian Federation
Braunau am Inn *106 var.* Braunau. N Austria
Braunschweig *267 Eng.* Brunswick. N Germany
Brava *see* Baraawe
Brava *176* island of SW Cape Verde
Brava, Costa *535* coastal region of E Spain
Bravo Del Norte *see* Grande, Río
Bray *316 Ir.* Bri Chuallan. E Ireland
Brazil *146-149* officially Federative Republic of Brazil. Country of South America divided into 28 admin. units (26 states, 1 territory, 1 federal district). ❖ Brasília
Brazos *603* river of SW USA
Brazza *see* Brač
Brazzaville *200* ❖ of Congo, S Congo
Brčko *140* NE Bosnia & Herzegovina
Breda *429* SW Netherlands
Bregalnica *381* river of E FYR Macedonia
Bregenz *106* W Austria
Breidhafjördhur *296* bay of the Denmark Strait, on the coast of W Iceland
Bremen *266* NW Germany
Bremerhaven *266* NW Germany
Bremersdorp *see* Manzini
Brennero, Passo del *see* Brenner Pass
Brenner Pass *106, 323 It.* Passo del Brennero, *Ger.* Brennerpass, *var.* Brenner Sattel. Mountain pass of Austria and Italy
Brerton *121* C Barbados
Brescia *322* N Italy
Breslau *see* Wrocław
Brest *122 Pol.* Brześć nad Bugiem, *prev.* Brześć Litewski, *Rus.* Brest-Litovsk. SW Belarus
Brest *254* NW France
Bretagne *254 Eng.* Brittany. Cultural region of NW France
Brewerville *368* W Liberia
Brezhnev *see* Naberezhnyye
Brežice *524 Ger.* Rann. E Slovenia
Brezno *523 prev.* Brezno nad Hronom, *Ger.* Bries, *var.* Briesen, *Hung.* Breznóbánya. C Slovakia
Bria *179* C Central African Republic
Bribrí *206* E Costa Rica
Bri Chuallan *see* Bray
Bride *647* N Isle of Man
Bridgetown *100* SW Australia
Bridgetown *121* ❖ of Barbados, SW Barbados
Bridgwater *597* SW England, UK
Brienzer See *550* lake of SW Switzerland
Brig *550* SW Switzerland
Brighton *597* SE England, UK
Brikama *263* W Gambia
Brindisi *323* S Italy
Brinstone Hill *494* NW St Kitts, St Kitts & Nevis
Brisbane *101* E Australia
Bristol *597* SW England, UK
Bristol Channel *597* inlet of the Atlantic Ocean, SW England, UK
British Columbia *170* province of SW Canada
British Guiana *see* Guyana
British Indian Ocean Territory *643* British dependent territory of the Indian Ocean. ❖ Diego Garcia
British North America *see* Canada
British Virgin Islands *643* British dependent territory of the Caribbean Sea. ❖ Road Town.
Brittany *see* Bretagne
Brizan *278* W Grenada island, Grenada
Brno *216 Ger.* Brünn. SE Czech Republic
Bród *see* Slavonski Brod
Brodeur Peninsula *170* peninsula of Baffin Island, N Canada
Broken Hill *101* S Australia
Brokopondo *542* NE Suriname
Bromberg *see* Bydgoszcz
Brønderslev *218* Jylland, N Denmark
Brooks Range *602* mountain range of Alaska, USA
Broome *100* NW Australia
Brorup *218* Jylland, W Denmark
Broughton Bay *see* Tongjosŏn-man
Brown Hill *494* S Nevis, St Kitts & Nevis
Browns Town *331* N Jamaica
Brownsweg *542* C Suriname
Bruck an der Mur *106 var.* Bruck. C Austria
Brufut *263* W Gambia
Brugge *127 Fr.* Bruges. NW Belgium
Bruit, Pulau *386* island of W Borneo, Malaysia
Brunei *150-151, 386* officially Sultanate of Brunei, *Mal.* Negara Brunei Darussalam. Country on island of Borneo, SE Asia, divided into 4 admin. units (districts). ❖ Bandar Seri Begawan
Brunei Bay *386 var.* Teluk Brunei. Bay of the South China Sea
Brunei Town *see* Bandar Seri Begawan
Brünn *see* Brno
Brunswick *see* Braunschweig
Brusa *see* Bursa
Brussels *127 Fr.* Bruxelles, *Dut.* Brussel, *Ger.* Brüssel. ❖ of Belgium, C Belgium
Brüx *see* Most
Bryansk *484* W Russian Federation
Brześć Litewski *see* Brest
Brześć nad Bugiem *see* Brest
Bsharrī *see* Bcharré
Bua *385* river of C Malawi
Bu'aale *528* SW Somalia
Buada Lagoon *424* lagoon on the coast of Nauru
Buala *526* SE Santa Isabel, Solomon Islands
Bū al Ḩīdān, Wādī *371* dry watercourse of C Libya
Buba *284* S Guinea-Bissau
Bubanza *162* NW Burundi
Bubaque *284* Ilha de Bubaque, Guinea-Bissau
Bubaque, Ilha de *284* island of SW Guinea-Bissau
Bubi *636 var.* Bubye. River of S Zimbabwe
Būbīyan, Jazirat *356* island of NE Kuwait
Buca *248 prev.* Mbutha. Vanua Levu, N Fiji
Bucaramanga *195* N Colombia
Buchanan *368 prev.* Grand Bassa. SW Liberia
Bucharest *480 Rom.* Bucureşti. ❖ of Romania, S Romania
Budapest *292* ❖ of Hungary, N Hungary
Búdhardalur *296* W Iceland
Budweis *see* České Budějovice
Buea *168* SW Cameroon
Buenaventura *195* W Colombia
Buena Vista *130* N Belize
Buena Vista *136* C Bolivia
Buena Vista *646* S Gibraltar
Buenos Aires *95* ❖ of Argentina, E Argentina
Buenos Aires *206* SE Costa Rica
Buenos Aires, Lago *95, 183 Sp.* Lago General Carrera. Lake of Argentina and Chile
Buffalo *603* New York, NE USA
Buff Bay *331* E Jamaica
Bug *471, 590 Ukr.* Zakhidnyy Buh, *Rus.* Zapadnyy Bug. River of E Europe
Buga *195* W Colombia
Bugala Island *588* island of Sese Islands, S Uganda
Buganda *162* NW Burundi
Bugarama *492* SW Rwanda
Bughotu *see* Santa Isabel
Buġibba *395* N Malta
Bugojno *140* C Bosnia & Herzegovina
Bugumya *492* SW Rwanda
Buin *458* Bougainville I, Papua New Guinea
Buitenzorg *see* Bogor

Bujumbura *162 prev.* Usumbura. ❖ of Burundi, W Burundi
Buka Island *458* island of E Papua New Guinea
Bukakata *588* S Uganda
Bukasa Island *588* island of Sese Islands, S Uganda
Bukavu *203 prev.* Costermansville. E Dem. Rep. Congo
Bukeye *162 var.* Bukaye. C Burundi
Bukhoro *614 var.* Bokhara, *Rus.* Bukhara. S Uzbekistan
Bukit Mertajam *386* NW Peninsular Malaysia
Bukit Panjang *521* area of C Singapore
Bukit Timah *521* area of C Singapore
Bükk *292* mountain range of NE Hungary
Bukoba *564* NW Tanzania
Bukum Kechil, Pulau *521* island of SW Singapore
Bukum, Pulau *521* island of S Singapore
Bula *284* W Guinea-Bissau
Bulawayo *636 var.* Buluwayo. SW Zimbabwe
Bulembu *544* NW Swaziland
Bulgan *412* N Mongolia
Bulgaria *152-155* officially Republic of Bulgaria, *prev.* People's Republic of Bulgaria. Country of E Europe divided into 8 admin. units (regions). ❖ Sofia
Buller *433* river of N South Island, New Zealand
Bulolo *458* SE Papua New Guinea
Bulunghur *614 prev.* Krasnogvardeysk, *Rus.* Bulungur. SE Uzbekistan
Bumba *203* N Dem. Rep. Congo
Bumbah, Khalīj *371 var.* Kalīj Bumbah. Gulf of the Mediterranean Sea, on the N coast of Libya
Bumbuna *518* C Sierra Leone
Bunbury *100* SW Australia
Bundaberg *101* E Australia
Bundoran *316* N Ireland
Bungoma *344* W Kenya
Bunia *203* NE Dem. Rep. Congo
Bunya *see* Bhunya
Buôn Ma Thuột *627 var.* Lac Giao. S Vietnam
Bur Acaba *see* Buurhakaba
Buraimi *see* Al Buraymī
Buraydah *506 var.* Buraida. C Saudi Arabia
Burco *528 var.* Burao, Bur'o. NW Somalia
Burdur *580* SW Turkey
Burē *245* NW Ethiopia
Burera, Lac *492* lake of N Rwanda
Būrewāla *see* Mandi Būrewāla
Burgas *152 var.* Bourgas. E Bulgaria
Burgaski Zaliv *152* bay of the Black Sea on the E coast of Bulgaria
Burgdorf *550* NW Switzerland
Burgos *535* N Spain
Burgundy *see* Bourgogne
Burhou *647* island of N Guernsey
Burias Island *466* island of C Philippines
Buriram *567* E Thailand
Burkina *156-157* officially Burkina Faso, *prev.* Upper Volta. Country of West Africa divided into 30 admin. units (provinces). ❖ Ouagadougou
Burlington *603* Vermont, NE USA
Burma *158-161* officially Union of Myanmar, *var.* Myanmar. Country of SE Asia divided into 14 admin. units (7 states, 7 divisions). ❖ Rangoon
Burnie *101* Tasmania, SE Australia
Burnt Pine *650* C Norfolk Island
Burrel *81 var.* Burreli. C Albania
Burrell Boom *130* E Belize
Bursa *580 prev.* Brusa, *var.* Brusa. NW Turkey
Būr Sa'īd *see* Port Said
Būr Sūdān *see* Port Sudan
Burtnieku Ezers *362* lake of N Latvia
Buru *304 prev.* Boeroe. Island of Maluku, E Indonesia
Burujird *see* Borujerd
Burūm *631* SE Yemen
Burundi *162-163* officially Republic of Burundi, *prev.* Kingdom of Burundi. Country of Central Africa divided into 15 admin. units (provinces). ❖ Bujumbura
Bururi *162* SW Burundi
Buryatiya, Respublika *485* autonomous republic of C Russian Federation
Busan *see* Pusan
Busengo *492* NW Rwanda
Buševa Planina *see* Baba
Bushire *see* Bandar-e Bushehr
Busia *344* W Kenya
Busing, Pulau *521* island of SW Singapore
Busira *203* river of NW Dem. Rep. Congo
Busoro *492* S Rwanda
Busuanga Island *466* island of W Philippines
Buta *203* N Dem. Rep. Congo
Butare *492 prev.* Astrida. S Rwanda
Butaritari *348* island of the Gilbert Is, W Kiribati
Butawal *see* Butwal
Butembo *203* NE Dem. Rep. Congo
Butha Buthe *366* N Lesotho
Butiaba *588* NW Uganda
Butterworth *386* NW Peninsular Malaysia
Butuan *466* Mindanao, S Philippines
Butung, Pulau *304 var.* Buton, *prev.* Boetoeng. Island to the SE of Celebes, C Indonesia
Butwal *427 var.* Butawal. C Nepal
Buurgplaatz *378* mountain of N Luxembourg
Buurhakaba *528 var.* Bur Hakkaba, *It.* Bur Acaba. SW Somalia
Buvuma Island *588* island of S Uganda
Büyükağrı Dağı *581 Eng.* Mount Ararat, *var.* Great Ararat. Mountain of E Turkey
Büyükmenderes *580* river of SW Turkey
Büyükzap Suyu *see* Great Zab
Buzău *480 Hung.* Bodza. SE Romania
Buzi *419* river of C Mozambique
Bwake *see* Bouaké
Bwiam *263* SW Gambia
Byahoml' *122* N Belarus
Byala Slatina *152 var.* Bjala Slatina. NW Bulgaria
Byaroza *122* SW Belarus
Bydgoszcz *471 Ger.* Bromberg. N Poland
Byelruskaya Hrada *122 Rus.* Belorusskaya Gryada. Hilly region of N Belarus
Byerazino *122 Rus.* Berezina. River of C Belarus
Byerazino *122 Rus.* Berezino. C Belarus
Bykhaw *122 Rus.* Bykhov. E Belarus
Byron Sound *645* area of the South Atlantic Ocean, NW Falkland Islands
Byrranga, Gory *485* mountain range of N Russian Federation
Bystrovka *see* Kemin
Bytom *471 Ger.* Beuthen. S Poland
Byumba *492 var.* Biumba. N Rwanda
Byuzmeyin *585 prev.* Bezmein. S Turkmenistan

C

Caacupé *460* S Paraguay
Caaguazú *460* SE Paraguay
Caála *88 var.* Kaala, *Port.* Vila Robert Williams. C Angola
Caatinga, Serra da *145* mountain range of E Brazil
Caazapá *460* S Paraguay
Cabaiguán *210* C Cuba
Cabanatuan City *466* Luzon, Philippines
Cabháin *see* Cavan
Cabimas *623* NW Venezuela
Cabinda *88 var.* Kabinda. Non-contiguous region, NW Angola
Cabinda *88 var.* Kabinda. NW Angola
Cabora Bassa, Lake *see* Cahora Bassa, Lago de
Cabot Strait *171* strait connecting the Atlantic Ocean and Gulf of St. Lawrence, between Newfoundland & Labrador and Nova Scotia
Cape Verde, Ilhas de *see* Cape Verde
Cabras, Ilha das *505* island to the N of São Tomé, Sao Tome & Principe
Cabras Island *647* island, W Guam
Čačak *512* C Serbia, Serbia & Montenegro (Yugo.)
Cacao *645* NE French Guiana
Cáceres *535* W Spain
Cachacrou *see* Scotts Head Village
Cacheu *284 var.* Cacheo. W Guinea-Bissau
Cacheu *284* river of NW Guinea-Bissau
Cachimbo, Serra do *145* mountain range of C Brazil
Cacine *284* S Guinea-Bissau
Cacine *284* river of S Guinea-Bissau
Caconda *88* C Angola
Čadca *523 Hung.* Csaca. N Slovakia
Cadiz *466* Negros, Philippines
Cádiz *534* SW Spain
Cádiz, Golfo de *474, 535* area of the Atlantic Ocean, on the SW coast of Spain
Caen *255* NW France
Caerdydd *see* Cardiff
Caerfyrddin *see* Carmarthen
Caergybi *see* Holyhead
Caernarfon *597 var.* Caernarvon. N Wales, UK
Cagayan *466* river of Luzon, Philippines
Cagayan de Oro *466* Mindanao, Philippines
Cagayan Islands *466* island group of C Philippines
Cagliari *322* Sardegna, W Italy
Caguas *651* E Puerto Rico
Cahabón *280* river of C Guatemala
Cahora Bassa Dam *419* dam of NW Mozambique
Cahora Bassa, Lago de *419 var.* Lake Cabora Bassa. Reservoir of NW Mozambique
Cahors *255* S France
Cahul *408 Rus.* Kagul. S Moldova
Caia *419* C Mozambique
Caibarién *210* C Cuba
Caicos Bank *653* undersea feature of the Atlantic Ocean
Caicos Passage *112, 653* strait of the Atlantic Ocean, between the Bahamas and Caicos Islands
Cai Lan *627* N Vietnam
Caille Island *278* island to the N of Grenada island, Grenada
Cailungo *502* N San Marino
Caió *284* W Guinea-Bissau
Cairns *101* NE Australia
Cairo *232 Ar.* Al Qāhirah, *var.* El Qâhira. N Egypt
Cajamarca *463 prev.* Caxamarca. NW Peru
Cajón, Represa el *290* reservoir of W Honduras
Çakilli Dere *see* Yialias
Čakovec *209 Hung.* Csáktornya, *Ger.* Csakathurn, *prev.* Tschakathurn. N Croatia
Calabar *440* S Nigeria
Calabozo *623* C Venezuela
Calabrai *327.* S Italy
Calagua Islands *466* island group of N Philippines
Calais *255* N France
Calama *183* N Chile
Calamian Group *466* island group of W Philippines
Calandula *88 var.* Kalandula. NW Angola
Calanscio Sand Sea *see* Kalanshiyū, Sarīr
Călăraşi *480* SE Romania
Călăraşi *408* var. Călăras, Kalarash. C Moldova
Calbayog *466* Samar, W Philippines
Calceta *230* W Ecuador
Calcutta *130* N Belize
Calcutta *299 var.* Kolkata. E India
Calcutta *542* N Suriname
Caldas da Rainha *474* W Portugal
Caledon *366, 531* river of Lesotho and South Africa
Caleta Olivia *95* SE Argentina
Calf of Man *647* island of S Isle of Man
Calgary *170* SW Canada
Calhau *176* São Vincente, N Cape Verde
Cali *195* W Colombia
Calibishie *224* N Dominica
Calicut *298 var.* Kozhikode. S India
Calida, Costa *535* coastal region of SE Spain
California *602* state of W USA
California, Golfo de *403* gulf of the Pacific Ocean
Cälilabad *110 Rus.* Dzhalilabad, *var.* Džalilabad, *prev.* Astrakhan-Bazar. S Azerbaijan
Callao *463* W Peru
Calliaqua *498* S St Vincent, St Vincent & the Grenadines
Caltagirone *323* Sicilia, S Italy
Caltanissetta *323* Sicilia, S Italy
Camá *226* river of C Dominican Republic
Camabatela *88* NW Angola
Camacupa *88 Port.* General Machado. C Angola
Camagüey *210 prev.* Puerto Príncipe. C Cuba
Camagüey, Archipiélago de *210* island group of N Cuba
Ca Mau *627 var.* Quan Long. S Vietnam
Cambay, Gulf of *see* Khambhat, Gulf of
Cambodia *164-167* officially State of Cambodia, *Cam.* Kampuchea, *prev.* People's Democratic Republic of Kampuchea. Country of SE Asia divided into 20 admin. units (provinces). ❖ Phnom Penh
Cambrian Mountains *597* mountain range of Wales, UK
Cambridge *331* NW Jamaica
Cambridge *433* C North Island, New Zealand
Cambridge *597* E England, UK
Cambridge Bay *170* Victoria Island, NW Canada
Cambrouze *645* N French Guiana
Cambulo *88* NE Angola
Camden Park *498* SW St Vincent, St Vincent & the Grenadines
Cameron Highlands *386* highlands of C Peninsular Malaysia
Cameroon *168-169* officially Republic of Cameroon, *Fr.* Cameroun. Country of W Africa divided into 10 admin. units (provinces). ❖ Yaoundé
Camiri *136* S Bolivia
Camopi *645* river of SE French Guiana
Camopi *645* E French Guiana
Camotes Islands *466* island group of E Philippines
Campana *95* E Argentina
Campbell River *170* Vancouver Island, SW Canada
Campeche *403* SE Mexico
Campeche, Bahía de *403 var.* Gulf of Campeche. Bay of the Gulf of Mexico
Campeche, Gulf of *see* Campeche, Bahía de
Câm Pha *627* N Vietnam
Campina Grande *145* E Brazil
Campinas *145* S Brazil
Campine *see* Kempen
Campobasso *323* S Italy
Campo Grande *145* SW Brazil
Campos *145 var.* Campo dos Goitacazes. SE Brazil
Campossa *284* river of E Guinea-Bissau
Cam Ranh *627* SE Vietnam

Camrose *170* SW Canada
Canada *170-175* *prev.* British North America. Country of North America, divided into 12 admin. units (10 provinces, 2 territories). ❖ Ottawa
Cañada de Gómez *95* C Argentina
Canadian *603* river of SW USA
Çanakkale *580* W Turkey
Çanakkale Boğazi *580* *Eng.* Dardanelles. Strait connecting Marmara Denizi and Aegean Sea
Canala *650* C New Caledonia
Canarias, Islas *535* *Eng.* Canary Islands. Islands of the Atlantic Ocean, part of Spain.
Canaries *497* W St Lucia
Canarreos, Archipiélago de los *210* island group of W Cuba
Canary Islands *see* Canarias, Islas
Cañas *206* NW Costa Rica
Canberra *101* ❖ of Australia, SE Australia
Canchungo *284* *prev.* Teixeira Pinto. W Guinea-Bissau
Cancuén *see* Santa Isabel
Candia *see* Irákleio
Canea *see* Chaniá
Canelones *611* *var.* Guadalupe. S Uruguay
Canguzo *see* Cankuzo
Canik Dağları *581* mountain range of N Turkey
Canillo *86* C Andorra
Canjambari *284* river of N Guinea-Bissau
Çankırı *581* N Turkey
Cankuzo *162* *var.* Canguzo. E Burundi
Cannes *255* SE France
Canoas *145* S Brazil
Cano, Pico do *176* mountain of Fogo, S Cape Verde
Canot, Rivière *288* river of C Haiti
Canouan *498* island of S St Vincent & the Grenadines
Cantabria *535* autonomous community of N Spain
Cantábrica, Cordillera *534-535* mountains of N Spain
Cantaro *574* N Trinidad, Trinidad & Tobago
Canterbury *597* SE England, UK
Canterbury Bight *433* area of the Pacific Ocean, SE New Zealand
Canterbury Plains *433* plain of C South Island, New Zealand
Cân Thơ *627* S Vietnam
Canton *see* Guangzhou
Canton Island *see* Kanton
Cao Băng *627* N Vietnam
Cape Breton Island *171* *Fr.* Île du Cap-Breton. Island of SE Canada
Cape Coast *272* *prev.* Cape Coast Castle. S Ghana
Cape Domesnes *see* Kolkasrags
Capellen *378* SW Luxembourg
Cape Palmas *see* Harper
Capesterre *646* Marie-Galante, S Guadeloupe
Capesterre-Belle-Eau *646* S Guadeloupe
Cape Town *531* *Afr.* Kaapstad. Legislative capital of South Africa, Western Cape, SW South Africa
Cape Verde *176-177* officially Republic of Cape Verde, *Port.* Ilhas do Cabo Verde. Country of volcanic islands in the Atlantic Ocean divided into 14 admin. units (districts). ❖ Cidade de Praia
Cape York *101* N Australia
Cape York Peninsula *101* peninsula of N Australia
Cap-Haïtien *288* *var.* Le Cap. N Haiti
Capitán Arturo Prat *90* Chilian research station of South Shetland Islands, Antarctica
Capitán Bado *460* E Paraguay
Capitán Pablo Lagerenza *460* *var.* Mayor Pablo Lagerenza. N Paraguay
Capodistria *see* Koper
Capri, Isola di *323* island of S Italy
Caprivi Concession *see* Caprivi Strip
Caprivi Strip *142, 423* *prev.* Caprivi Concession, *Ger.* Caprivizipfel. Finger of territory of NE Namibia
Caprivizipfel *see* Caprivi Strip
Cap Saint-Jacques *see* Vung Tau
Caquetá *195* river of S Colombia
CAR *see* Central African Republic
Caracal *480* S Romania
Caracas *623* ❖ of Venezuela, N Venezuela
Carache, Ilha de *284* island of W Guinea-Bissau
Caracollo *136* W Bolivia
Caransebes *480* W Romania
Caráquez *230* *prev.* Bahía de Caráquez. W Ecuador
Caratasca, Laguna de *290* lagoon on the E coast of Honduras
Caravela *284* Ilha Caravela, Guinea-Bissau
Caravela, Ilha *284* island of Arquipélago dos Bijagós, W Guinea-Bissau
Carbonia *322* SW Sardegna, Italy
Carcassonne *255* S France
Cardamom, Chaîne des *see* Krâvanh, Chuŏr Phmun
Cardamom Mountains *see* Krâvanh, Chuŏr Phmun
Cárdenas *210* NW Cuba
Cardiff *597* *Wel.* Caerdydd. S Wales, UK
Cardigan Bay *597* bay of W Wales, UK
Cardona *611* SW Uruguay
Cardonagh *316* N Ireland
Čardžou *see* Chardzhev
Carenage *574* NW Trinidad, Trinidad & Tobago
Caribbean Sea *206, 288, 642, 643* arm of the Atlantic Ocean, Central America
Caripito *623* NE Venezuela
Carlton, The *112* *var.* Abraham Bay. Mayaguana, Bahamas
Carletonville *531* North West, N South Africa
Carlisle *597* NW England, UK
Carlow *316* SE Ireland
Carlsruhe *see* Karlsruhe
Carmarthen *597* *Wel.* Caerfyrddin. W Wales, UK
Carmelo *611* *var.* Carmelo del Este. SW Uruguay
Carmona *206* W Costa Rica
Carmona *see* Uíge
Carnavon *100* W Australia
Carnegie, Lake *100* salt lake of W Australia
Carnot *179* W Central African Republic
Caroço, Ilha *505* island to the S of Príncipe, Sao Tome & Principe
Carolina *651* NE Puerto Rico
Caroline Island *see* Millennium island
Caroline Islands *406* island group of Micronesia
Caroni *574* river of N Trinidad, Trinidad & Tobago
Caroní *623* river of E Venezuela
Carora *623* NW Venezuela
Carpathian Mountains *216, 471, 480, 590* *var.* Carpathians, *Ger.* Karpaten, *Cz./Pol.* Karpaty. Mountain range of E Europe
Carpaţii Meridionali *480* *var.* South Carpathians, *Eng.* Transylvanian Alps. Mountain range of C Romania
Carpaţii Occidentali *480* mountain range of W Romania
Carpentaria, Gulf of *101* gulf of the Arafura Sea, on the coast of N Australia
Carrantual *see* Carrauntoohill
Carrara *322* N Italy
Carrauntoohill *316* *var.* Carrauntohil, *Ir.* Carrantual. Mountain of SW Ireland
Carriacou *278* island to the north of Grenada, Grenada
Carrick on Shannon *316* N Ireland
Carson City *602* Nevada, W USA
Carstensz, Puntjak *see* Jaya, Puncak
Cartagena *195* NW Colombia
Cartagena *535* SE Spain
Cartago *195* W Colombia
Cartago *206* C Costa Rica
Carúpano *623* NE Venezuela
Casablanca *414* *Ar.* Dar el Beida. NW Morocco
Casamance *510* cultural region of SW Senegal
Casamance *510* river of SW Senegal
Cascade *650* NE Norfolk Island
Cascade *516* Mahé, Seychelles
Cascade Range *602* mountain range of Canada and USA
Casey *90* Australian research station of Greater Antarctica, Antarctica
Caseyr, Raas *528* *var.* Ras Aser, *prev.* Cape Guardafui. Cape on the NE coast of Somalia
Cashel *636* E Zimbabwe
Casilda *95* C Argentina
Ca *627* river of Laos and Vietnam
Casper *603* Wyoming, NW USA
Caspian Sea *110* *Az.* Xäzär Dänizi, *Rus.* Kaspiyskoye More, *Per.* Daryā-ye Khazar, Baḩr-e Khazar, *Kaz.* Kaspiy Tengizi. Shallow inland sea tween W Asia and E Europe
Cassacatiza *419* NW Mozambique
Cassai *see* Kasai
Cassel *see* Kassel
Cassiar Mountains *170* mountain range of W Canada
Castellamare di Stabia *323* S Italy
Castellón de la Plana *535* *Cat.* Castelló de la Plana. E Spain
Castelo Branco *474* C Portugal
Castelvetrano *323* Sicilia, S Italy
Castilhiano *176* São Nicolau, N Cape Verde
Castilla *463* NW Peru
Castilla-La Mancha *535* autonomous community of NE Spain
Castillos *611* SE Uruguay
Castlebar *316* W Ireland
Castle Bruce *224* NE Dominica
Castle Harbour *643* inlet of the North Atlantic Ocean, E Bermuda
Castletown *647* SE Isle of Man
Castries *497* ❖ of St Lucia, N St Lucia
Castro *183* Isla de Chiloé, W Chile
Castrovillari *323* SE Italy
Catacamas *290* C Honduras
Catacaos *463* NW Peru
Catalan Bay *646* *var.* Caleta. Bay of the Mediterranean Sea on the E coast of Gibraltar
Cataluña *535* autonomous community of N Spain
Catamarca *see* San Fernando del Valle de Catamarca
Catanduanes Island *466* island of E Philippines
Catania *323* Sicilia, S Italy
Catano *651* NE Puerto Rico
Catanzaro *323* SE Italy
Catedral, Cerro *611* mountain of S Uruguay
Catherine, Mount *see* Katherina, Gebel
Catholic Island *498* island of SW St Vincent & the Grenadines
Catió *284* S Guinea-Bissau
Cat Island *112* island of C Bahamas
Catuane *419* S Mozambique
Catumbela *88* W Angola
Cauca *195* river of N Colombia
Caucasia *195* NW Colombia
Caucasus *264, 484* *Rus.* Kavkaz. Mountain range of SE Europe and SW Asia, commonly taken as the boundary between Asia and Europe
Caura *623* river of E Venezuela
Căuşeni *408* *Rus.* Kaushany. E Moldova
Causse du Kelifely *see* Kôsin' i Kelifely
Cauto *210* river of SE Cuba
Cavaia *see* Kavajë
Cavalla *see* Cavally
Cavally *328, 368* *var.* Cavalla. River of Ivory Coast and Liberia
Cavan *316* *Ir.* Cabháin. NE Ireland
Cave Hill *121* W Barbados
Caviana de Fora, Ilha *145* *var.* Ilha Cavinana. Island of N Brazil
Cawnpore *see* Kānpur
Caxamarca *see* Cajamarca
Caxias do Sul *145* S Brazil
Caxito *88* NW Angola
Cayambe *230* N Ecuador
Cayenne *645* ❖ of French Guiana, NE French Guiana
Cayey *651* SE Puerto Rico
Çayırova *215* *var.* Ayios Theodhoros. NE Cyprus
Cayman Brac *644* island, E Cayman Islands
Cayman Islands *644* British dependent territory of the Caribbean Sea. ❖ George Town.
Cayo *see* San Ignacio
Cayon *494* E St Kitts, St Kitts & Nevis
Cayes *288* *var.* Les Cayes. SW Haiti
Cay Sal *112* islet of W Bahamas
Cazin *140* NW Bosnia & Herzegovina
Ceadâr-Lunga *see* Ciadîr-Lunga
Ceanannus Mór *316* NE Ireland
Ceará *see* Fortaleza
Cébaco, Isla *456* island of SW Panama
Cebollatí *611* river of E Uruguay
Cebu *466* Cebu, C Philippines
Cebu *466* island of C Philippines
Cedar Grove *92* N Antigua, Antigua & Barbuda
Cedar Rapids *603* Iowa, C USA
Cedros, Isla *403* island of NW Mexico
Ceelbuur *528* *It.* El Bur. C Somalia
Ceeldheere *528* *var.* Ceel Dheere, *It.* El Dere. E Somalia
Ceel Gaal *528* *var.* El Gal. NW Somalia
Ceel Xamurre *see* Jirriiban
Ceerigaabo *528* *var.* Erigabo, Erigavo. N Somalia
Cegléd *292* *prev.* Czegléd. C Hungary
Celaya *403* C Mexico
Celebes *304* *Ind.* Sulawesi. Island of C Indonesia
Celebes Sea *304, 387, 466* *Ind.* Laut Sulawesi. Sea of the Pacific Ocean, between N Indonesia and S Philippines
Čeleken *see* Cheleken
Celje *524* *Ger.* Cilli. C Slovenia
Celtic Sea *316, 597* *Ir.* An Mhuir Cheilteach. Sea of the Atlantic Ocean, between S Ireland and SW England, UK
Cenderawasih, Teluk *305* bay of the Pacific Ocean on the NW coast of Irian Jaya, E Indonesia
Central African Republic *178-179* abbrev. CAR. Country of Central Africa divided into 16 admin. units (prefectures). ❖ Bangui
Central Brāhui Range *451* mountain range of W Pakistan
Central China *191*
Central, Cordillera *195* range of the Andes in W Colombia
Central, Cordillera *226* mountain range of C Dominican Republic
Central, Cordillera *466* mountain range of Luzon, N Philippines
Central, Cordillera *651* mountain range of C Puerto Rico
Central Group *see* Inner Islands
Central Island *344* island in Lake Rudolf, NW Kenya
Central Makrān Range *450* mountain range of W Pakistan
Central Range *458* mountain range of C Papua New Guinea
Central Siberian Plateau *see* Srednesibirskoye Ploskogor'ye
Centre de Flacq *400* E Mauritius
Centre Hills *649* mountain range, C Montserrat
Cephalonia *see* Kefallinía
Ceram *see* Seram
Ceram Sea *see* Seram, Laut
Cerfs, Ile aux *400* *var.* Isle aux Cerfs. Island of E Mauritius
Cernăuţi *see* Chernivtsi
Cërrik *81* *var.* Cerriku. C Albania

Columbia *602* river of NW USA
Columbia *603* South Carolina, SE USA
Columbia, District of *603* federal district of NE USA
Columbus *603* Georgia, SE USA
Columbus *603* Ohio, NE USA
Columbus Channel *574* channel connecting the Atlantic Ocean and Gulf of Paria
Colville Channel *433* channel linking the Bay of Plenty and Hauraki Gulf, N of North Island, New Zealand
Comandante Ferraz *90* Brazilian research station of Antarctic Peninsula, Antarctica
Comarapa *136* C Bolivia
Comas *463* W Peru
Comayagua *290* W Honduras
Comendador *226 prev.* Elías Piña. W Dominican Republic
Comer *see* Como, Lago di
Comilla *117 Ben.* Kumillā. E Bangladesh
Commissioner's Point *643* headland of Ireland Island North, W Bermuda
Communism Peak *see* Garmo, Qullai
Como *322* N Italy
Comodoro Rivadavia *95* SE Argentina
Como, Lago di *322 var.* Lario, *Eng.* Lake Como, *Ger.* Comer See. Lake of N Italy
Comoros *198-199* officially Federal Islamic Republic of the Comoros. Island group of the Indian Ocean, between Madagascar and the African mainland, divided into 3 admin. units (districts). ❖ Moroni
Comrat *408 Rus.* Komrat. S Moldova
Conakry *283* ❖ of Guinea, SW Guinea
Concepción *136* E Bolivia
Concepción *183* C Chile
Concepción *460 var.* Villa Concepción. C Paraguay
Concepción *see* Riaba
Concepción de La Vega *see* La Vega
Conchos *403* river of NW Mexico
Concord *278* W Grenada island, Grenada
Concord *603* New Hampshire, NE USA
Concordia *95* E Argentina
Condado *210* C Cuba
Côn Dao *627 var.* Con Son. Island of S Vietnam
Condroz *127* physical region of SE Belgium
Congo *88, 200, 203 var.* Zaire, Kongo, Lualaba. River of C Africa
Congo *200-201* officially The Republic of the Congo. Country of C Africa divided into 9 admin. units (regions). ❖ Brazzaville
Congo Basin *203* drainage basin of C Africa
Congo, Democratic Republic of *202-205* officially Democratic Republic of Congo, *prev.* Congo (Kinshasa), Belgian Congo, Zaire. Country of C Africa divided into 11 admin. units (10 provinces, 1 city). ❖ Kinshasa
Con, Loch *see* Conn, Lough
Connaught *316* province of W Ireland
Connecticut *603* state of NE USA
Conn, Lough *316 Ir.* Loch Con. Lake of NW Ireland
Consolación del Sur *210* W Cuba
Constance, Lake *106, 266, 550 Ger.* Bodensee. Lake of C Europe
Constanţa *480 Ger.* Konstantza, *Turk.* Köstence, *var.* Küstendje, *Eng.* Constanza. SE Romania
Constantine *83 Ar.* Qoussantîna, *var.* Qacentina, NE Algeria
Constantine *278* SW Grenada island, Grenada
Constantinople *see* İstanbul
Constant Spring *331* SE Jamaica
Constanza *226* C Dominican Republic
Contagem *145* SE Brazil
Contuboel *284* NE Guinea-Bissau
Cook Islands *644* territory in free association with New Zealand, Pacific Ocean. ❖ Avarua.
Cook, Mount *433 prev.* Aorangi. Mountain of W South Island, New Zealand
Cook, Récif de *650* reef of the Pacific Ocean, N New Caledonia
Cook Strait *433 var.* Raukawa. Strait between North and South Islands of New Zealand, connecting the South Pacific Ocean and Tasman Sea
Cooktown *101* NE Australia
Cooper Creek *101 var.* Barcoo, Cooper's Creek. River of C Australia
Cooper Island *643* island of SE British Virgin Islands
Copacabana *136* W Bolivia
Copenhagen *218 Dan.* København. ❖ of Denmark, Sjælland, E Denmark
Copiapó *183* N Chile
Coppename *542 var.* Koppename. River of C Suriname
Coppermine *see* Kugluktuk
Coquilhatville *see* Mbandaka
Coquimbo *183* N Chile
Corail *288* SW Haiti
Coral Harbour *171* Southampton Island, NE Canada
Coral Sea *101, 619, 458* sea of the Pacific Ocean between Australia and Papua New Guinea
Corantijn *286, 542 var.* Coeroeni, Corentyne, Courantyne. River of Guyana and Suriname
Córdoba *95* C Argentina
Córdoba *535 var.* Cordoba, *Eng.* Cordova. SW Spain
Cordova *602* Alaska, USA
Corentyne *see* Corantijn
Corfu *see* Kérkyra
Corinth *see* Kórinthos
Corinth *278* SE Grenada island, Grenada
Corinth, Gulf of *see* Korinthiakós Kólpos
Corinth, Isthmus of *see* Korínthou, Isthmós
Corinto *436* W Nicaragua
Coriole *see* Qoryooley
Corisco, Isla de *238* Island of SW Equatorial Guinea
Cork *316 Ir.* Corcaigh. S Ireland
Cork Hill *649* W Montserrat
Corleone *323* Sicilia, S Italy
Corner Brook *171* Newfoundland & Labrador, E Canada
Corn Exchange *366* NW Lesotho
Corn Islands *see* Maíz, Islas
Cornwallis Island *170* island of Parry Islands, N Canada
Coro *623 var.* Santa Ana de Coro. NW Venezuela
Corocoro *136* W Bolivia
Coromandel Peninsula *433* peninsula of NE North Island, New Zealand
Coronel Bogado *460* S Paraguay
Coronel Oviedo *460* SE Paraguay
Çorovodë *81 var.* Çorovoda, Corovoda. SE Albania
Corozal *130* N Belize
Corrib, Lough *316 Ir.* Loch Corrib. Lake of W Ireland
Corrientes *95* NE Argentina
Corriverton *286* E Guyana
Corriza *see* Korçë
Corse *255 Eng.* Corsica. Island of SE France
Corsica *see* Corse
Cortés *206* SE Costa Rica
Corubal *284 var.* Cocoli, Rio Grande. River of W Africa
Çorum *581* N Turkey
Corvallis *602* Oregon, NW USA
Corvo *474 var.* Ilha do Corvo. Island of the Azores, Portugal
Cosenza *323* S Italy
Cosmolédo Atoll *516* atoll of the Aldabra Group, SW Seychelles
Cospicua *395* E Malta
Costa, Cordillera de la *623 var.* Cordillera de Venezuela. Mountain range of N Venezuela
Costa Rica *206-207* officially Republic of Costa Rica. Country of Central America divided into 7 admin. units (provinces). ❖ San José
Costermansville *see* Bukavu
Cotagaita *136* S Bolivia
Côte d'Ivoire *see* Ivory Coast
Côtière, Chaîne *see* Coast Mountains
Cotonou *132 var.* Kotonu. S Benin
Cotopaxi *230* active volcano of N Ecuador
Cotswold Hills *597* hills of W England, UK
Cottbus *267 prev.* Kottbus. E Germany
Cottica *542* E Suriname
Cotton Ground *494* NW Nevis, St Kitts & Nevis
Cotuí *226* C Dominican Republic
Coulibistri *224* W Dominica
Coupe, Cap *652* cape of the Atlantic Ocean on the coast of Miquelon, S Saint Pierre and Miquelon
Courantyne *see* Corantijn
Courcelles *127* S Belgium
Courland *see* Kurzeme
Courtrai *see* Kortrijk
Couva *574* W Trinidad, Trinidad & Tobago
Couvin *127* S Belgium
Cova Figueira *176* Fogo, S Cape Verde
Cove Bay *642* bay of the Caribbean Sea on the S coast of Anguilla
Coventry *597* C England, UK
Covilhã *474* E Portugal
Cox's Bāzār *117* S Bangladesh
Coyah *283* SW Guinea
Coyhaique *see* Coihaique
Cozumel, Isla de *403* island of SE Mexico
Cracow *see* Kraków
Cradock *531* Eastern Cape, S South Africa
Craiova *480* SW Romania
Cranbrook *170* SW Canada
Crane, The *121* SE Barbados
Crawley *597* SE England, UK
Créoles, River des *400* river of SE Mauritius
Cres *209 It.* Cherso. Island of W Croatia
Crescent Group *651* island group of C Paracel Islands
Crete *see* Kríti
Créteil *255* N France
Crete, Sea of *275 Gk* Kritikó Pélagos. Area of the Mediterranean Sea, SE Greece
Creuse *255* river of C France
Crikvenica *209 It.* Cirquenizza. NW Croatia
Crimea *see* Krym
Cristóbal *456* N Panama
Cristóbal Colón, Pico *195* peak of N Colombia
Crna Gora *see* Montenegro
Crna reka *381* river of S FYR Macedonia
Crni Drim *see* Drinit të Zi
Črnomelj *524 Ger.* Tschernembl. S Slovenia
Croatia *238 -209* officially Republic of Croatia, *SCr.* Hrvatska. Country of S Europe divided into 21 admin. units (provinces). ❖ Zagreb
Crochu *278* SE Grenada island, Grenada
Crocker, Banjaran *387 var.* Crocker Range. Mountain range of C Borneo, Malaysia
Crocodile *see* Limpopo
Crocus Bay *642* bay of the Caribbean Sea on the W coast of Anguilla
Croia *see* Krujë
Cromer *597* E England, UK
Crooked Island *112* island of the E Bahamas
Crooked Island Passage *112* passage of the Atlantic Ocean between Crooked Island and Long Island, Bahamas
Crooked Tree *130* NE Belize
Crosby *647* C Isle of Man
Cross *440* river of Cameroon and Nigeria
Crossroads *316* N Ireland
Crotone *323* S Italy
Crucero Contramaestre *210* S Cuba
Cruz Bay *653* Saint John, E Virgin Islands (US)
Cruz del Eje *95* C Argentina
Csaca *see* Čadca
Csakathurn *see* Čakovec
Csáktornya *see* Čakovec
Csíkszereda *see* Miercurea-Ciuc
Cuamba *419* N Mozambique
Cuan Dhun Dealgan *see* Dundalk Bay
Cuando *88, 635 var.* Kwando. River of southern Africa
Cuango *see* Kwango
Cuan na Gaillimhe *see* Galway Bay
Cuanza *88 var.* Kwanza. River of Angola
Cuareim *611 Port.* Quaraí. River of Brazil and Uruguay
Cuba *210-213* officially Republic of Cuba. Country of the West Indies divided into 15 admin. units (14 provinces and 1 special municipality). ❖ Havana
Cubal *88* W Angola
Cubango *88 var.* Kuvango, *Port.* Vila Artuur de Paiva, Vila da Ponte. C Angola
Cubango *88, 142, 423 var.* Kavango, Kavengo, Kubango, Okavango, Okavanggo. River of Southern Africa
Cúcuta *195 var.* San José de Cúcuta. N Colombia
Cuddapah *298* S India
Cudjoehead *649* NW Montserrat
Cuenca *230* S Ecuador
Cuenca *535* C Spain
Cueno *322* N Italy
Cuera *see* Chur
Cuernavaca *403* S Mexico
Cueto *210* SE Cuba
Cufra *see* Al Kufrah
Cuiabá *145 prev.* Cuyabá. SW Brazil
Cuilapa *280* S Guatemala
Cuito *88 var.* Kwito. River of Angola
Cuito Cuanquale *88* S Angola
Culebra *651* Isla de Culebra, NE Puerto Rico
Culebra, Isla de *651* island of NE Puerto Rico
Culiacán *403* W Mexico
Culion Island *466* island of Calamian Group, W Philippines
Cumaná *623* NE Venezuela
Cumberland *498* river of W St Vincent, St Vincent & the Grenadines
Cumberland Peninsula *171* peninsula of Baffin Island, NE Canada
Cumbrian Mountains *597 var.* Lake District. Mountain range of NW England, UK
Cunene *88, 423 var.* Kunene. River of Angola and Namibia
Cunnamulla *101* E Australia
Curaçao *623, 650* island of the Netherland Antilles, Caribbean Sea
Curanilahue *183* C Chile
Curaray *230* river of Ecuador and Peru
Curepe *574* NW Trinidad, Trinidad & Tobago
Curepipe *400* C Mauritius
Curicó *183* C Chile
Curitiba *145 prev.* Curytiba. S Brazil
Curral Velho *176* Boa Vista, Cape Verde
Current *112* Eleuthera Island, Bahamas
Curuguaty *460* E Paraguay
Curzola *see* Korčula
Cusco *463 var.* Cuzco. SE Peru
Čust *see* Chust
Cutch, Gulf of *see* Kachch, Gulf of
Cuttack *298* E India
Cutting Camp *366* SW Lesotho
Cuxhaven *266* NW Germany
Cuyabá *see* Cuiabá
Cuyo East Passage *466* passage of the Sulu Sea between Cuyo Islands and Panay, Philippines
Cuyo Islands *466* islands of C Philippines
Cuyo West Passage *466* passage of the Sulu Sea between Cuyo Islands and Palawan, Philippines

Cuyuni *286* river of Guyana and Venezuela
Cuzco *see* Cusco
Čvrsnica *140* mountain range of SW Bosnia & Herzegovina
Cyambwe, Lac *492* lake of E Rwanda
Cyangugu *492* SW Rwanda
Cyclades *see* Kyklades
Cymru *see* Wales
Cyohoha Sud, Lac *162, 492* *var.* Cohoha, Lac Tshohoha Sud. Lake of Burundi and Rwanda
Cyprus *214-215* officially Republic of Cyprus, *Gk* Kypros, *Turk.* Kıbrıs, Kıbrıs Cumhuriyeti. Country of E Mediterranean divided into 5 admin. units (districts). ❖ Nicosia. Following Turkish invasion of 1974, northern sector became self-proclaimed state, officially Turkish Republic of Northern Cyprus (TRNC). ❖ Nicosia
Cyrenaica *see* Barqah
Czechoslovakia *see* Czech Republic and Slovakia
Czech Republic *216-217* *Cz.* Česká Republika, *prev.* constituent republic of Czechoslovakia. Country of C Europe divided into 7 admin. units (regions). ❖ Prague
Czegléd *see* Cegléd
Czernowitz *see* Chernivtsi
Częstochowa *471* *Ger.* Tschenstochau. S Poland

D

Dabakala *328* NE Ivory Coast
Dabola *283* C Guinea
Dabou *328* S Ivory Coast
Dąbrowa Górnicza *471* S Poland
Dacca *see* Dhaka
Dachau *267* S Germany
Dadanawa *286* SW Guyana
Dadda 'to *222* N Djibouti
Dadeldhura *427* *var.* Dandeldhura. W Nepal
Dadnah *594* *var.* Dhadnah. NE United Arab Emirates
Daegu *see* Taegu
Daga *134* S Bhutan
Dagana *510* N Senegal
Dagden *see* Hiiumaa
Dagestan, Respublika *484* autonomous republic of SW Russian Federation
Dagö *see* Hiiumaa
Dagupan *466* Luzon, N Philippines
Dahar *577* physical region of S Tunisia
Da Hinggan Ling *187* *Eng.* Great Khingan Range. Mountain range of Nei Mongol Zizhiqu, NE China
Dahlak Archipelago *240* island group of E Eritrea
Dahlak Island *240* island of Dahlak Archipelago, E Eritrea
Dahm, Ramlat *631* desert region of NW Yemen
Dahomey *see* Benin
Dahra *see* Dara
Dahūk *312* *var.* Dohuk, *Kurd.* Dihōk. N Iraq
Dai Island *526* island of E Solomon Is
Dailekh *427* W Nepal
Daingin, Bá an *see* Dingle Bay
Dajabón *226* NW Dominican Republic
Dakar *510* ❖ of Senegal, W Senegal
Dakoro *439* SW Niger
Đakovica *512* *var.* Djakovica, *Alb.* Gjakovë. S Serbia, Serbia & Montenegro (Yugo.)
Đakovo *209* *var.* Djakovo, *Hung.* Diakovár. NE Croatia
Dakshin *see* Deccan
Dalaba *283* W Guinea
Dalai Nor *see* Hulun Nur
Dalälven *547* river of SE Sweden
Dalandzadgad *412* S Mongolia
Dalap District *396* district of Majuro, SE Marshall Islands
Đa Lat *627* S Vietnam
Dali *see* Dhali
Dalian *187* *var.* Jay Dairen, Ta-lien, *Rus.* Dalny. Liaoning, NE China
Dallas *603* Texas, SC USA
Dallol Bosso *439* seasonal watercourse of W Niger
Dalmā *594* island of W United Arab Emirates
Dalmacija *209* *Eng.* Dalmatia. Cultural region of S Croatia
Dalmatia *see* Dalmacija
Dalny *see* Dalian
Daloa *328* C Ivory Coast
Dalvík *296* N Iceland
Damanhûr *232* *var.* Damanhûr. N Egypt
Damar, Kepulauan *304* *var.* Kepulauan Barat Daya. Island group to the E of Nusa Tenggara, C Indonesia
Damara *179* S Central African Republic
Damasak *441* NE Nigeria
Damascus *555* *var.* Esh Sham, *Fr.* Damas, *Ar.* Dimashq. ❖ of Syria, SW Syria
Damāvand, Qolleh-ye *309* mountain of N Iran
Dambulla *538* C Sri Lanka
Dame-Marie *288* SW Haiti
Damêrdjôg *222* E Djibouti
Damietta *see* Dumyât
Damongo *272* NW Ghana
Damoûr *364* *var.* Ad Dāmūr. W Lebanon
Damphu *134* S Bhutan
Damqawt *631* *var.* Damqut. E Yemen
Dâmrei, Chuŏr Phmun *165* *Fr.* Chaîne de l'Éléphant. Mountain range of SW Cambodia
Danakil Desert *240, 245* *var.* Danakil Plain, Afar Depression. Desert region of Eritrea and Ethiopia
Danané *328* W Ivory Coast
Đă Nâng *627* *prev.* Tourane. C Vietnam
Dandeldhura *see* Dadeldhura
Dandong *187* *var.* Tan-tung, *prev.* An-tung. Liaoning, NE China
Daneborg *646* *var.* Danborg. E Greenland
Dänew *see* Deynau
Dangal *240* SE Eritrea
Dangara *see* Danghara
Danger Island *643* island of W British Indian Ocean Territory
Danghara *562* *Rus.* Dangara. W Tajikistan
Dānglā *245* *var.* Dangila. NW Ethiopia
Dangme *134* river of S Bhutan
Dang Raek, Phanom *165, 567* *var.* Phanom Dong Rak, *Cam.* Chuor Phmum Dângrêk, *Fr.* Chaîne des Dangrek. Mountain range of Cambodia and Thailand
Dangrak, Chaîne des *see* Dang Raek, Phanom
Dângrêk, Chuŏr Phmun *see* Dang Raek, Phanom
Dangriga *130* *var.* Stann Creek. SE Belize
Dank *448* *var.* Dhank. NW Oman
Danlí *290* S Honduras
Danmark *see* Denmark
Danmark Havn *646* E Greenland
Danmarksstraedet *see* Denmark Strait
Danube *106, 152, 209, 267, Bul.* Danav, *Hung.* Duna, *Cz.* Dunaj, *Ger.* Donau, *Rom.* Dunărea. River of C Europe
Danube, Mouths of the *480* *Rom.* Delta Dunării. Delta of Romania and Ukraine
Danubian Plain *see* Dunavska Ravnina
Danyi-Apéyémé *571* *prev.* Apéyémé. W Togo
Danzig *see* Gdańsk
Danzig, Gulf of *471* *var.* Gulf of Gdańsk,*Gk* Danziger Bucht, *Pol.* Zatoka Gdańska, *Rus.* Gdan'skaya Bukhta. Gulf of the Baltic Sea, N Pola
Dapaong *571* N Togo
Dara *510* *var.* Dahra. NW Senegal
Dar'ā *555* *var.* Der'a, *Fr.* Déraa. SW Syria
Da Rang *627* *var.* Ba. River of S Vietnam
Dardanelles *see* Canakkale Boğazi
Dar el Beida *see* Casablanca
Dar es Salaam *564* E Tanzania
Darfur *540* *var.* Darfur Massif. Mountain range of W Sudan
Dargan-Ata *585* *var.* Darganata. E Turkmenistan
Dargaville *433* NW North Island, New Zealand
Dargol *439* W Niger
Darhan *412* N Mongolia
Darien, Isthmus of *see* Panamá, Istmo de
Darién, Serranía del *456* mountain range of Colombia and Panama
Darjiling *299* *prev.* Darjeeling. NE India
Darling *101* river of E Australia
Darling Range *100* mountain range of SW Australia
Darlington *597* N England, UK
Darmstadt *266* SW Germany
Darnah *371* *var.* Derna. NE Libya
Daroot-Korgon *359* *var.* Daraut-Kurgan. SW Kyrgyzstan
Darou Mousti *510* NW Senegal
Darrell Island *643* island of W Bermuda
Dartmoor *597* moorland of SW England, UK
Dartmouth *171* SE Canada
Daru *458* SW Papua New Guinea
Daru *518* SE Sierra Leone
Darvaza *585* *Turkm.* Derweze. C Turkmenistan
Darvel Bay *see* Lahad Datu, Telukan
Darvel, Teluk *see* Lahad Datu, Telukan
Darvos, Qatorkûhi *562* *Rus.* Darvazskiy Khrebet. Mountain range of C Tajikistan
Darwin *101* N Australia
Darwin, Isla *230* island of the NW Galapagos Is, Ecuador
Dashkhovuz *585* *prev.* Tashauz, *var.* Tašauz, *Turkm.* Dashhowuz. N Turkmenistan
Dasht Kaur *450* river of SW Pakistan
Daska *451* NE Pakistan
Da, Sông *see* Black River
Dassa *132* *var.* Dassa-Zoumé. S Benin
Datong *187* *var.* Ta-t'ung. Shanxi, N China
Datu, Teluk *386* bay of the South China Sea, on the coast of Borneo, E Malaysia
Daua *344* *Amh.* Dawa Wenz. River of E Africa
Daugava *see* Western Dvina
Daugavpils *362* *Ger.* Dünaburg, *Rus.* Dvinsk. SE Latvia
Daule *230* W Ecuador
Daule *230* river of W Ecuador
Daurada, Costa *534* *var.* Costa Dorada. Coastal region of E Spain
Davao *466* Mindanao, S Philippines
Davao Gulf *466* gulf of the Pacific Ocean
Davenport *603* Iowa, C USA
David *456* W Panama
Davis *90* Australian research station of Greater Antarctica, Antarctica
Davis Strait *171, 646* strait connecting the Atlantic Ocean and Baffin Bay, NE Canada between Baffin Island and Greenland
Davos *550* E Switzerland
Davyd-Haradok *122* *Rus.* David Gorodok, *Pol.* Dawidgródek. S Belarus
Dawa Wenz *see* Daua
Dawei *see* Tavoy
Dawidgródek *see* Davyd-Haradok
Dawra *414* NW Western Sahara
Dawson *170* NW Canada
Dawwah *448* *var.* Dauwa. E Oman
Dayrah *594* *var.* Deira. NE United Arab Emirates
Dayr az Zawr *555* *var.* Deir ez Zor. E Syria
Dayrīk *see* Al Mālikīyah
Dayton *603* Ohio, NE USA
De Aar *531* Northern Cape, C South Africa
Deadman's Bay *652* bay of the South Atlantic Ocean on the SW coast of Tristan da Cunha
Dead Sea *319, 320, 338* *Ar.* Al Baḩr al Mayyit, Baḩrat Lūt, *Heb.* Yam HaMelaḥ Salt lake of SW Asia
Deadwood Plain *652* plain of N St Helena
Debar *381* W FYR Macedonia
Débo, Lac *392* lake of C Mali
Debre Birhan *245* *var.* Debra Birhan. C Ethiopia
Debrecen *292* *prev.* Debreczen, *Ger.* Debreczin. E Hungary
Debre Mark'os *245* NW Ethiopia
Debre Tabor *245* NW Ethiopia
Debre Zebīt *245* N Ethiopia
Debre Zeyit *245* *var.* Debre Zeyt, *prev.* Bishoftu, *It.* Biscoftù. C Ethiopia
Deccan *298* *Hind.* Dakshin. Plateau of C India
Děčín *216* *Ger.* Tetschen. NW Czech Republic
Dedeagach *see* Alexandroúpoli
Dededo *647* N Guam
Dedegaç *see* Alexandroúpoli
Dedoplistsqaro *264* *Rus.* Dedoplis-Tskaro, *prev.* Tsiteli-Tskaro. SE Georgia
Dédougou *157* W Burkina
Deduru Oya *538* C Sri Lanka
Dedza *385* SW Malawi
Dee *597* river of NE Scotland, UK
Dee *597* *Wel.* Dyfrdwy. River of N Wales, UK
Deep Bay *see* Chilumba
Deep Water Bay *see* Hau Hoi Wan
Değirmenlik *215* *var.* Kythrea. N Cyprus
Dehiwala-Mount Lavinia *538* SW Sri Lanka
Deinze *127* W Belgium
Deira *see* Adh Dhayd
Deir el Balaḥ *319, 320* C Gaza Strip
Deir el-Bahri *232* E Egypt
Deir ez Zo *see* Dayr az Zawr
Deirgeirt, Loch *see* Derg, Lough
Dej *480* NW Romania
Dekemhare *240* S Eritrea
Dékoa *179* C Central African Republic
Delagoa Bay *see* Maputo, Baía de
Delārām *77* SW Afghanistan
Delaware *603* state of E USA
Delčevo *381* NE FYR Macedonia
Delcommune, Lac *see* Nzilo, Lac
Delémont *550* *Ger.* Delsberg. NW Switzerland
Delft *429* W Netherlands
Delft *538* island of NW Sri Lanka
Delfzijl *429* NE Netherlands
Delhi *298* *Hind.* Dilli. N India
Délices *224* SE Dominica
Délices *645* C French Guiana
Delsberg *see* Delémont
Delvinë *81* *var.* Delvina, *It.* Delvino. S Albania
Delvino *see* Delvinë
Demba *203* C Dem. Rep. Congo
Dembéni *649* E Mayotte
Dembéni *198* S Grande Comore, Comoros
Dembī Dolo *245* *var.* Dembidollo. W Ethiopia
Demerara *286* river of N Guyana
Denau *see* Denow
Dender *127* *Fr.* Dendre. River of W Belgium
Dendre *see* Dender
Den Haag *see* 's-Gravenhage
Den Helder *429* NW Netherlands
Denis, Île *516* island of the Inner Islands, NE Seychelles
Denizli *580* SW Turkey
Denmark *218-221* officially Kingdom of Denmark, *Dan.* Danmark. Country of W Europe, divided into 14 admin. units (counties). ❖ Copenhagen
Denmark Strait *296, 646* *var.* Danmarksstraedet. Strait between Greenland and Iceland

Dennery 497 E St Lucia
Denow 614 *Rus.* Denau. SE Uzbekistan
Denpasar 304 *prev.* Paloe. Bali, C Indonesia
D'Entrecasteaux Islands 458 island group of SE Papua New Guinea
Denver 603 Colorado, SW USA
Der'a *see* Dar'ā
Deraa *see* Dar'ā
Dera Ghāzi Khān 451 C Pakistan
Dera Ismāīl Khān 451 N Pakistan
Đeravica 512 *var.* Durmitor. Mountain of S Serbia & Montenegro (Yugo.)
Derby 597 C England, UK
Derg, Lough 316 *Ir.* Loch Deirgeirt. Lake of C Ireland
Dernière Rivière 497 NE St Lucia
Derrick 498 Bequia, St Vincent & the Grenadines
Derrière Morne 497 S St Lucia
Derrubado 176 N Boa Vista, E Cape Verde
Derry *see* Londonderry
Derventa 140 N Bosnia & Herzegovina
Derweze *see* Darvaza
Deryneia *see* Dherinia
Desaguadero 136 river of Bolivia and Peru
Desbarra 497 NE St Lucia
Desdunes 288 W Haiti
Desē 245 *var.* Desse, *It.* Dessie. N Ethiopia
Desertas, Ilhas 474 island group of the Madeira Is, Portugal
Des Moines 603 Iowa, C USA
Desna 590 river of N Ukraine
Despoto Planina *see* Rhodope Mountains
Desroches, Île 516 *var.* Desroches. Island of the Amirante Islands, C Seychelles
Desruisseaux 497 SE St Lucia
Dessau 267 C Germany
Dessie *see* Desē
Destêrro *see* Florianópolis
Dete 636 *prev.* Dett. W Zimbabwe
Detroit 603 Michigan, NC USA
Deutsch-Brod *see* Havlíčkův Brod
Deutsche Bucht *see* German Bight
Deutschendorf *see* Poprad
Deutschland *see* Germany
Deva 480 *Ger.* Diemrich, *Hung.* Déva. W Romania
Đevđelija *see* Gevgelija
Deventer 429 E Netherlands
Devil's Island 645 island of N French Guiana
Devollit 81 *var.* Devoll. River of SE Albania
Devon Island 170-171 *prev.* North Devon Island. Island of Parry Islands, N Canada
Devonport 101 Tasmania, Australia
Deynau 585 *var.* Dyanev, *Turkm.* Dänew. E Turkmenistan
Dezfūl 309 *var.* Dizful. W Iran
Dhaalu Atoll *see* South Nilandhe Atoll
Dhahran *see* Az̧ Z̧ahrān
Dhaka 117 *var.* Dacca. ❖ of Bangladesh, C Bangladesh
Dhali 215 *var.* Dali. C Cyprus
Dhamar 631 W Yemen
Dhanbād 298 NE India
Dhankuta 427 E Nepal
Dharan 427 *var.* Dharan Bazar. E Nepal
Dheere Laaq 528 *It.* Lach Dera. Seasonal river of SW Somalia
Dhekelia 215 *var.* Dekeleia. UK Air base, SE Cyprus
Dherinia 215 *var.* Deryneia. E Cyprus
Dhiarizos 215 *var.* Diarizos. River of SW Cyprus
Dhībān 338 NW Jordan
Dhofar *see* Z̧ufār
Dhráma *see* Dráma
Dhún na nGall, Bá *see* Donegal Bay
Dhuudo 528 *It.* Uadi Dudo. Seasonal river of NE Somalia
Dhuusa Marreeb 528 *var.* Dusa Marreb, *It.* Dusa Mareb. C Somalia
Diafarabé 392 C Mali
Diakovár *see* Đakovo
Diamond Island 278 island to the N of Grenada island, Grenada
Diana's Peak 652 mountain of C St Helena
Diangounté-Kamara 392 W Mali
Diapaga 157 E Burkina
Diarizos *see* Dhiarizos
Dibā 594 *var.* Dibā al Ḩişn, Dibba. NE United Arab Emirates
Dicle *see* Tigris
Diébougou 157 SW Burkina
Diego Garcia 643 island of S British Indian Ocean Territory
Diego Garcia 643 ❖ of British Indian Ocean Territory, Diego Garcia, S British Indian Ocean Territory
Diégo-Suarez *see* Antsirañana
Diekirch 378 C Luxembourg
Diéma 392 W Mali
Diemrich *see* Deva
Điện Biên 627 *var.* Dien Bien Phu. NW Vietnam
Dieppe 255 N France
Dieppe Bay Town 494 N St Kitts, St Kitts & Nevis
Diffa 439 SE Niger
Differdange 378 SW Luxembourg
Digne 255 SE France
Digoin 255 C France
Digul 305 *prev.* Digoel. River of Irian Jaya, E Indonesia
Dihōk *see* Dahūk
Dijlah *see* Tigris
Dijon 255 NE France
Dikhil 222 SW Djibouti
Dīla 245 *var.* Dilla. S Ethiopia
Dilbeek 127 C Belgium
Dili 228 ❖ of East Timor, N East Timor
Dilia 439 *var.* Dillia. Seasonal river of SE Niger
Dilijan 98 *Rus.* Dilizhan, *var.* Diližan. NE Armenia
Di Linh 627 S Vietnam
Dilli *see* Delhi
Dillia *see* Dilia
Dillikot 427 W Nepal
Dilling 540 *var.* Ad Dalanj. C Sudan
Dilolo 203 S Dem. Rep. Congo
Dimashq *see* Damascus
Dimbokro 328 E Ivory Coast
Dimitrovgrad 152 S Bulgaria
Dimitrovo *see* Pernik
Dimlang 440 *var.* Vogel Peak. Mountain of E Nigeria
Dimona 319 S Israel
Dinagat Island 466 island of E Philippines
Dinājpur 117 NW Bangladesh
Dinan 254 NW France
Dinant 127 S Belgium
Dinara 209 mountain of S Croatia
Dinara 140, 209 mountain range of Bosnia & Herzegovina and Croatia
Dinaric Alps 209 mountain range of Bosnia & Herzegovina and Croatia
Dindigal 298 S India
Dingle Bay 316 *Ir.* Bá an Daingin. Bay of the Atlantic Ocean, SW Ireland
Dingli 395 SW Malta
Dingli Cliffs 395 cliffs of SW Malta
Dinguiraye 283 C Guinea
Dion 283 river of S Guinea
Dioro 392 *var.* Dyero. SW Mali
Diourbel 510 W Senegal
Dipayal 427 W Nepal
Dipkarpaz 215 *var.* Rizokarpaso. NE Cyprus
Direction Island 644 *var.* Pulu Tikus. Island of E Cocos Islands
Dirē Dawa 245 E Ethiopia
Diriamba 436 SW Nicaragua
Dirj 371 NW Libya
Dirk Hartog Island 100 island of W Australia
Disappointment, Lake 100 salt lake of W Australia
Discovery Reef 651 reef of the China Sea, W Paracel Islands
Disna *see* Drysa
Disûq 232 *var.* Disūq. N Egypt
Diu 298 W India
Diuata Mountains 466 mountain range of Mindanao, S Philippines
Diva *see* Piva
Divinópolis 145 SE Brazil
Divisa 456 S Panama
Divo 328 S Ivory Coast
Diwaniya *see* Ad Dīwanīyah
Diyālá 312 *Per.* Rūdkhāneh-ye Sīrvān,irwan. River of Iran and Iraq
Diyarbakır 581 SE Turkey
Dizful *see* Dezfūl
Dja 168 river of SE Cameroon
Djailolo *see* Halmahera
Djakarta *see* Jakarta
Djakovica *see* Đakovica
Djakovo *see* Đakovo
Djamâa 83 NE Algeria
Djambala 200 W Congo
Djambi *see* Hari
Djambi *see* Jambi
Djanet 83 *prev.* Fort Charlet. SE Algeria
Djawa *see* Java
Djéblé *see* Jablah
Djelfa 83 *var.* El Djelfa. N Algeria
Djéma 179 E Central African Republic
Djember *see* Jember
Djénné 392 *var.* Jenné. C Mali
Djérablous *see* Jarābulus
Djerba *see* Jerba, Île de
Djerba *see* Houmt Souk
Djerem 168 *var.* Djérem. River of C Cameroon
Djevdjelija *see* Gevgelija
Djibo 157 N Burkina
Djibouti 222 *var.* Jibuti. ❖ of Djibouti, E Djibouti
Djibouti 222-223 officially Republic of Djibouti, *var.* Jibuti, *prev.* French Territory of the Afars and Issas 1967-77, French Somaliland -1967. Country of East Africa divided into 5 admin. units (districts). ❖ Djibouti
Djidjel *see* Jijel
Djidjelli *see* Jijel
Djiguéni 398 SE Mauritania
Djirataoua 439 S Niger
Djisr el Choghour *see* Jisr ashShughúr
Djoua 200 river of Congo and Gabon
Djoué 200 river of S Congo
Djougou 132 W Benin
Djúpivogur 296 SE Iceland
Dmitriyevsk *see* Makiyivka
Dnieper 590 *Bel.* Dnyapro, *Ukr.* Dnipro, *Rus.* Dnepr. River of E Europe
Dniester 408, 590 *Rom.* Nistru, *Rus.* Dnestr, *Ukr.* Dnister. River of Moldova and Ukraine
Dniprodzerzhyns'k 590 *Rus.* Dneprodzerzhinsk, *prev.* Kamenskoye. E Ukraine
Dniprodzerzhyns'ke Vodoskhovyshche 590 *Rus.* Dneprodzerzhinskoye Vodokhranilische. Reservoir of C Ukraine
Dnipropetrovs'k 590 *Rus.* Dnepropetrovsk, *prev.* Ekaterinoslav. E Ukraine
Dnistrovs'kyy Lyman 590 *Rus.* Dnestrovskiy Liman. Inlet of the Black Sea, SW Ukraine
Dnyapro *see* Dnieper
Doba 180 S Chad
Dobele 362 *Ger.* Doblen. W Latvia
Doberai, Jazirah 304 *Dut.* Vogelkop. Region of Irian Jaya, E Indonesia
Doboj 140 N Bosnia & Herzegovina
Dobrich 152 *var.* Dobrič, *prev.* Tolbukhin, *Rom.* Bazargic. NE Bulgaria
Dobrush 122 SE Belarus
Doctor Pedro P. Peña 460 W Paraguay
Dodekánisos 275 *prev.* Dhodhekánisos, *var.* Noties Sporádes, *Eng.* Dodecanese. Island group of SE Greece
Dodoma 564 ❖ of Tanzania, C Tanzania
Dodona 275 site of ancient city, N Greece
Dodwekon 368 *var.* Dudwiokahn. SE Liberia
Doetinchem 429 SE Netherlands
Dogana 502 NE San Marino
Dogondoutchi 439 SW Niger
Doğu Karadeniz Dağlarıı 581 *var.* Anadolu Dağları. Mountain range of NE Turkey
Doha 479 *Ar.* Ad Dawḩah. ❖ of Qatar, E Qatar
Dohuk *see* Dahuk
Doko 283 NE Guinea
Dolisie *see* Loubomo
Dolobil 368 C Liberia
Dolomites *see* Dolomitiche, Alpi
Dolomitiche, Alpi 323 *var.* Dolomiti, *Eng.* Dolomites. Mountain range of N Italy
Dolo Odo 245 *var.* Dollo Odo, Dolo. S Ethiopia
Dolores 280 N Guatemala
Dolores 611 W Uruguay
Domagnano 502 NE San Marino
Domel Island *see* Letsok-aw I
Dominica 224-225 officially Commonwealth of Dominica. Country of the West Indies divided into 10 admin. units (parishes). ❖ Roseau
Dominica Channel *see* Martinique Passage
Dominican Republic 226-227 Country of the West Indies divided into 30 admin. units (1 national district and 29 provinces). ❖ Santo Domingo
Dominica Passage 646 passage of the Caribbean Sea, N Guadeloupe
Domoni 198 SE Anjouan, Comoros
Don 484 river of W Russian Federation
Donau *see* Danube
Doncaster 597 N England, UK
Dondo 88 NW Angola
Dondo 419 C Mozambique
Donegal 316 N Ireland
Donegal Bay 316 *Ir.* Bá Dhún na nGall. Bay of the Atlantic Ocean, to the N of Ireland
Donets 590 *Rus.* Severskiy Donets, *Ukr.* Sivers'kyy Donets'. River of E Ukraine
Donets'k 590 *prev.* Stalino, *Rus.* Donetsk. E Ukraine
Donga 168 river of Cameroon and Nigeria
Đông Ha 627 C Vietnam
Đông Hơi 627 C Vietnam
Dong Nai 627 *var.* Donnai, Dong-nai, Dong Noi. River of S Vietnam
Dongola 540 *var.* Dunqulah, Donqola N Sudan
Dongou 200 NE Congo
Dong Rak, Phanom *see* Dang Rak, Phanom
Dongting Hu 187 *var.* Tung-t'ing Hu. Lake of SE China
Donostia-San Sebastián 535 N Spai
Doornik *see* Tournai
Dorada, Costa *see* Daurada, Costa
Dordogne 255 river of SW France
Dordrecht 429 SW Netherlands
Dori 157 N Burkina
Dornbirn 106 W Austria
Dornoch Firth 597 estuary of the river Dornoch, NE Scotland, UK
Dorpat *see* Tartu
Dorra 222 NW Djibouti
Dorsale 577 mountain range of N Tunisia
Dortmund 266 W Germany
Dos D'Âne 224 N Dominica
Dospad Dagh *see* Rhodope Mountains
Dos Puntas, Cabo 238 cape on the W coast of Río Muni, Equatorial Guinea
Dosso 439 SW Niger
Dostuk 359 C Kyrgyzstan
Douai 255 N France
Douala 168 *var.* Duala. SW Cameroon
Double Headed Shot Cays 112 islets of W Bahamas
Doubs 550 *var.* Le Doubs. River of France and Switzerland
Doudoub Bololé 222 S Djibouti
Doué 510 river of N Senegal
Douglas 645 East Falkland, NE Falkland Islands
Douglas 597, 647 ❖ of Isle of Man, SE Isle of Man

Douma *see* Dūmā
Douro *see* Duero
Dover *597 Fr.* Douvres.SE England,UK
Dover *603* Delaware, E USA
Dover, Strait of *255 var.* Straits of Dover, *Fr.* Pas de Calais. Strait connecting the English Channel and North Sea between England and France
Dovrefjell *444* mountain of SW Norway
Dowa *385* C Malawi
Dōzen *332* island to the N of Honshū, W Japan
Drâa *414* seasonal river of S Morocco
Drac *see* Durrës
Draç *see* Durrës
Dragon's Mouths, The *574 Sp.* Bocas del Dragón. Strait connecting the Caribbean Sea and Gulf of Paria
Dra, Hamada du *83 var.* Haut Plateau du Dra, Hammada du Drâa. Desert region of W Algeria
Drakensberg *366, 531* mountain range of Lesotho and South Africa
Drake Passage *183* passage connecting Pacific Ocean and Atlantic Ocean between South America and Antarctica
Dráma *275 var.* Dhráma. NE Greece
Drammen *444* S Norway
Drangajökull *296* glacier of NW Iceland
Drava *106, 209, 292, 524 Eng.* Drave, *Hung.* Dráva, *Ger.* Drau, *SCr.* Drava. River of C Europe
Dresden *267* E Germany
Drina *140, 512* river of Bosnia & Herzegovina and Serbia & Montenegro (Yugo.)
Drin Gulf *see* Drinit, Gjiri i
Drinit *81 var.* Drin. River of NW Albania
Drinit, Gjiri i *81 var.* Pellg i Drinit, Drin Gulf. Gulf of the Adriatic Sea, NW Albania
Drinit të Zi *81, 381 var.* Drin i Zi, *Eng.* Black Drin, *SCr.* Crni Drim. River of Albania and FYR Macedonia
Drin i Zi *see* Drinit të Zi
Drinos *81* river of S Albania
Drissa *122* river of Belarus and Russian Federation
Drobeta-Turnu Severin *480 prev.* Turnu Severin. SW Romania
Drochia *408 Rus.* Drokiya. N Moldova
Drogheda *316 Ir.* Droichead Átha. E Ireland
Drontheim *see* Trondheim
Druskininkai *376 Pol.* Druskienniki. S Lithuania
Drysa *122 Rus.* Disna. River of Belarus and Lithuania
Dschang *168* W Cameroon
Duala *see* Douala
Duarte, Pico *226* mountain of C Dominican Republic
Dubai *594 Ar.* Dubayy. NE United Arab Emirates
Dubăsari *408 Rus.* Dubossary. NE Moldova
Dubăsari Reservoir *408* reservoir of NE Moldova
Dubawnt *170* river of C Canada
Dubbo *101* E Australia
Dublanc *224* NW Dominica
Dublin *316 Ir.* Baile Átha Cliath. ❖ of Ireland, E Ireland
Dubnica nad Váhom *523 Hung.* Máriatölgyes, *prev.* Dubnicz. NW Slovakia
Dubnicz *see* Dubnica nad Váhom
Dubossary *see* Dubăsari
Dubréka *283* SW Guinea
Dubrovnik *209 It.* Ragusa. SE Croatia
Duc de Gloucester, Îles du *646* island group of C French Polynesia
Ducie Island *651* island of E Pitcairn Islands
Ducos *649* C Martinique
Dudelange *378* S Luxembourg
Dudo, Uadi *see* Dhuudo
Dudwiokahn *see* Dodwekon
Duékoué *328* W Ivory Coast
Duero *474, 534-535 Port.* Douro. River of Portugal and Spain
Duesseldorf *see* Düsseldorf
Duff Islands *526* small island group within Santa Cruz Is, Solomon Is
Dufourspitze *550* mountain of S Switzerland
Dugi Otok *209 It.* Isola Lunga. Island of W Croatia
Duinkerden *see* Dunkerque
Duisburg *266* W Germany
Duitama *195* C Colombia
Duitse Bocht *see* German Bight
Dukhān *479* W Qatar
Dukhan Heights *see* Dukhān, Jabal
Dukhān, Jabal *479 var.* Dukhan Heights. Hilly region of SW Qatar
Dukhan, Jabal *see* Dukhān, Jabal ad
Dukhān, Jabal ad *115 var.* Dukhan Heights, Jabal Dukhan. Mountain of C Bahrain
Dukou *see* Panzhihua
Dulce, Golfo *see* Izabal, Lago de
Dulce Nombre de Culmí *290* Honduras
Dulit, Banjaran *386 var.* Dulit Range. Mountain range of W Borneo, Malaysia
Duluth *603* Minnesota, NC USA
Dūmā *555 Fr.* Douma. SW Syria
Dumfries *597* SW Scotland, UK
Dumistān *115* NW Bahrain
Dumont d'Urville *90* French research station of Greater Antarctica, Antarctica
Dumyât *232 Eng.* Damietta. N Egypt
Düna *see* Western Dvina
Dünaburg *see* Daugavpils
Dunai *427* W Nepal
Dunaj *see* Danube
Dunaj *see* Vienna
Dunajská Streda *523 Hung.* Dunaszerdahely. SW Slovakia
Dunapentele *see* Dunaújváros
Dunărea *see* Danube
Dunării, Delta *see* Danube, Mouths of the
Dunaszerdahely *see* Dunajská Streda
Dunaújváros *292 prev.* Sztálinváros, *prev.* Dunapentele. C Hungary
Dunav *see* Danube
Dunavska Ravnina *152 Eng.* Danubian Plain. Lowland region of N Bulgaria
Dundalk *316 Ir.* Dún Dealgan. NE Ireland
Dundalk Bay *316 Ir.* Cuan Dhun Dealgan. Bay of the Irish Sea, to the NE of Ireland
Dundas *see* Pituffik
Dundee *531* Kwazulu Natal, E South Africa
Dundee *597* E Scotland, UK
Dunedin *433* S South Island, New Zealand
Dunfermline *597* E Scotland, UK
Dungarvan *316 Ir.* Dun Garbhain. S Ireland
Dunkerque *255 Eng.* Dunkirk, *Dut.* Duinkerken. N France
Dunkirk *see* Dunkerque
Dunkwa *272* SW Ghana
Dún Laoghaire *316 prev.* Kingstown. E Ireland
Dunqulah *see* Dongola
Dupnitsa *152 prev.* Stanke Dimitrov, *prev.* Marek. W Bulgaria
Duqm *448 var.* Daqm. E Oman
Duque de Caxias *145* SE Brazil
Durán *see* Eloy Alfaro
Durance *255* river of SE France
Durango *403* W Mexico
Durazno *611 var.* San Pedro del Durazno. C Uruguay
Durazzo *see* Durrës
Durazzo, Gulf of *see* Durrësit, Gjiri i
Durban *531* Kwazulu Natal, E South Africa
Durbe *362 Ger.* Durben. W Latvia
Durben *see* Durbe
Durdur *528* seasonal river of NW Somalia
Durham *597* NE England, UK
Durmitor *see* Đeravica
Durrës *81 var.* Durrësi, Dursi, *It.* Durazzo, *SCr.* Drač, *Turk.* Draç. W Albania
Durrësit, Gjiri i *81 var.* Gulf of Durazzo. Gulf of the Adriatic Sea, W Albania
Durūz, Jabal ad *555* mountain range of SW Syria
D'Urville Island *433* island to the NE of South Island, New Zealand
Dusa Mareb *see* Dhuusa Marreeb
Dushanbe *562 var.* Dušanbe, Dyushambe, *prev.* Stalinabad. ❖ of Tajikistan, W Tajikistan
Düsseldorf *266 var.* Duesseldorf. W Germany
Dŭstí *562* SW Tajikistan
Dutch New Guinea *see* Irian Jaya
Dutch East Indies *see* Indonesia
Dutch West Indies *see* Netherlands Antilles
Düzce *see* Athna
Dvinsk *see* Daugavpils
Dyanev *see* Deynau
Dyero *see* Dioro
Dyfrdwy *see* Dee
Dyushambe *see* Dushanbe
Džalilabad *see* Cälilabad
Dzaoudzi *649* Petite-Terre, E Mayotte
Džarkurgan *see* Dzharkurgan
Dzaudzhikau *see* Vladikavkaz
Dzavhan *412* river of W Mongolia
Džebel *see* Dzhebel
Dzerzhinskiy *see* Nar'yan-Mar
Dzhalal-Abad *359 var.* Džalal-Abad, *Kir.* Jalal-Abad. SW Kyrgyzstan
Dzhalilabad *see* Cälilabad
Dzhambul *see* Taraz
Dzhanak *585* region of W Turkmenistan
Dzharkurgan *see* Jarqŭrghon
Dzhebel *585 var.* Džebel, *Turkm.* Jebel. W Turkmenistan
Dzhelandy *562* SE Tajikistan
Dzhergalan *359 var.* Džergalan, *Kir.* Jyrgalan. NE Kyrgyzstan
Dzhermuk *see* Jermuk
Dzhezkazgan *see* Zhezkazgan
Dzhetygara *see* Zhitikara
Dzhirgatal' *see* Jirgatol
Dzhizak *see* Jizzakh
Dzhugdzhur, Khrebet *485* mountain range of E Russian Federation
Dzhusaly *340* SW Kazakhstan
Dzongsa *134* SW Bhutan
Dzunmod *412* C Mongolia
Dzüünharaa *412* N Mongolia
Dzvina *see* Western Dvina
Dzyarzhynskaya, Hora *122* mountain of C Belarus

E

Eagle Islands *643* island group of W British Indian Ocean Territory
Eagle Passage *645* passage connecting Falkland Sound and Atlantic Ocean, S Falkland Islands
East Caicos *653* island of N Turks and Caicos Islands
East China Sea *559 Chin.* Nan Hai. Sea of Pacific Ocean, off E Asia
East End *644* Grand Cayman, W Cayman Islands
East End Village *642* E Anguilla
Easter Island *183* Pacific island of Chile
Eastern Cape *531* province of SE South Africa
Eastern Desert *see* Sharqîya, Sahara el
Eastern Ghats *298* mountains of SE India
Eastern Sayans *see* Vostochnyy Sayan
Eastern Scheldt *see* Oosterschelde
Eastern Sierra Madre *see* Sierra Madre Oriental
Eastern Transvaal *see* Mpumalanga
East Falkland *645* island of E Falkland Islands
East Fayu Island *406* island of C Micronesia
East Frisian Islands *see* Ostfriesische Inseln
East London *531 Afr.* Oos-Londen. Eastern Cape, S South Africa
East Pakistan *see* Bangladesh
East Malaysia *386-387* eastern part of Malaysia situated on N Borneo
East Siberian Sea *see* Vostochno-Sibirskoye More
East Timor *228-229* Country of SE Asia, ❖ Dili
Eauripik *406* atoll of C Micronesia
Ebebiyin *238* NE Río Muni, Equatorial Guinea
Ebeltoft *218* Jylland, C Denmark
Ebetsu *333* Hokkaidō, N Japan
Ebinayon *see* Evinayong
Eblana *see* Dublin
Ebolowa *168* S Cameroon
Ebon *396* island of S Marshall Islands
Ébrié, Lagune *328* lake of SW Ivory Coast
Ebro *535* river of NE Spain
Ech Cheliff *see* Chlef
Echmiadzin *see* Ejmiadzin
Echternach *378* E Luxembourg
Écija *535* SW Spain
Ečmiadzin *see* Ejmiadzin
Ecuador *230-231* officially Republic of Ecuador. Country of NW South America divided into 20 admin. units (provinces). ❖ Quito
Ed *240* SE Eritrea
Ed Damazin *540 var.* Ad Damazīn. E Sudan
Ed Damer *540 var.* Ad Dāmir, Ad Damar. NE Sudan
Ed Dueim *540 var.* Ad Duwaym, Ad Duwēm. C Sudan
Ede *429* C Netherlands
Ede *440* W Nigeria
Edéa *168* SW Cameroon
Eden *597* river of NW England, UK
Edfu *see* Idfu
Edgeøya *653* island of S Svalbard
Edina *368* SW Liberia
Edinburgh *652* N Tristan da Cunha
Edinburgh *597* E Scotland, UK
Edineţ *408 var.* Edineţi, *Rus.* Yedintsy. NW Moldova
Edirne *580* NW Turkey
Edmonton *170* SW Canada
Edward, Lake *588, 203 var.* Lake Rutanzige, Edward Nyanza, Albert Edward Nyanza, Lac Idi Amin. Lake of Uganda and Dem. Rep. Congo
Eems *see* Ems
Eesti Vabariik *see* Estonia
Efate *619 Fr.* Vaté, *prev.* Sandwich Islands. Island group of C Vanuatu
Egadi, Isole *323* island group to the W of Sicilia, S Italy
Ege Denizi *see* Aegean Sea
Eger *292 Ger.* Erlau. NE Hungary
Eger *see* Cheb
Eger *see* Ohře
Egersund *444* SW Norway
Egilsstadhir *296* E Iceland
Egmont, Cape *433* cape of SW North Island, New Zealand
Egmont Islands *643* island group of W British Indian Ocean Territory
Egmont, Mount *see* Taranaki, Mount
Egypt *232-235* officially Arab Republic of Egypt, *prev.* United Arab Republic. Country of NE Africa divided into 26 admin. units (governorates). ❖ Cairo
Eidsvoll *444* S Norway
Eifel *266* plateau of W Germany
Eiffel Flats *656* C Zimbabwe
Eight Degree Channel *390* channel of the Indian Ocean between N Maldives and Lakshadweep, SW India
Eil *see* Eyl
Eilat *see* Elat
Eil Malk *455* island of S Palau
Eindhoven *429* S Netherlands
Einsiedeln *550* NE Switzerland
Eipel *see* Ipeľ, Ipoly
Eire *see* Ireland, Republic of
Eisen *see* Yŏngch'ŏn
Eisenstadt *106* E Austria
Eishū *see* Yŏngju
Eitape *see* Aitape

Eivissa *535 Cast.* Ibiza, *var.* Iviza. Island of the Islas Baleares, E Spain
Ejmiadzin *98 Rus.* Echmiadzin, Ečmiadzin, Etchmiadzin. W Armenia
Ekaterinoslav *see* Dnipropetrovs'k
Ekerem *see* Okarem
Ekeren *127* N Belgium
Ekibastuz *340* NE Kazakhstan
El Alto *136* W Bolivia
El Araïche *see* Larache
El 'Arîsh *232 var.* Al Arīsh. NE Egypt
El Asnam *see* Chlef
Elat *319 var.* Elath, Eilat. S Israel
Elat, Gulf of *see* Aqaba, Gulf of,
Elato *406* atoll of C Micronesia
Elâzığ *581 var.* Elâziz, Elâziğ. E Turkey
Elba, Isola d' *322* island of C Italy
Elbasan *81 var.* Elbasani. C Albania
Elbe *216, 267 Cz.* Labe. River of Czech Republic and Germany
El Beqaa *364 var.* Al Biqā', Bekaa Valley. Valley of E Lebanon
Elbing *see* Elbląg
Elbistan *581* S Turkey
Elbląg *471 Ger.* Elbing. N Poland
El Boulaida *see* Blida
Elbow Bay *643* bay of the North Atlantic Ocean, C Bermuda
El Bur *see* Ceelbuur
Elburgon *344* W Kenya
Elburz Mountains *see* Alborz, Reshteh-ye Kuhhā-ye
El Carmen de Bolívar *195* NW Colombia
El Cayo *see* San Ignacio
El Cerro del Aripo *see* Aripo, Mount
Elche *535 Cat.* Elx. SE Spain
El Congo *237* W El Salvador
El Copé *456* C Panama
Elda *535* SE Spain
El Dere *see* Ceeldheere
El Djazaïr *see* Algiers
El Djelfa *see* Djelfa
Eldoret *344* W Kenya
Eleja *362* C Latvia
Elemi Triangle *245, 344* disputed region of Kenya and Sudan
Éléphant, Chaîne de l' *see* Dâmrei, Chuŏr Phmun
Elephant Island *90* island of South Shetland Islands, Antarctica
Eleuthera Island *112* island of C Bahamas
El Faiyûm *232 var.* Al Fayyūm. N Egypt
El Fasher *540 var.* Al Fāshir. W Sudan
El Ferrol del Caudillo *see* Ferrol
El Gal *see* Ceel Gaal
El Gedaref *see* Gedaref
Elgin *597* NE Scotland, UK
El Gîza *232 var.* Gîza, Al Jīzah. N Egypt
El Goléa *83* C Algeria
Elgon, Mount *588* mountain of E Uganda
El Hasaheisa *540 var.* Al Ḩuşayḩişah, Al Hasahisa, Hasaheisa. C Sudan
El Haseke *see* Al Ḩasakah
Elías Piña *see* Comendador
Elimina *272* S Ghana
Élisabethville *see* Lubumbashi
El Iskandarîyah *see* Alexandria
Elizabeth *101* S Australia
El Jadida *414 prev.* Mazagan. W Morocco
El Jafr *see* Qā' al Jafr
El Jem *577 var.* Al Djem, Al Jamm. NE Tunisia
El Kef *see* Le Kef
El Kelâa des Srarhna *414* C Morocco
El Khârga *232 var.* Al Khārijah. C Egypt
El Khiyam *364 var.* Khiam. S Lebanon
Elkhovo *152 var.* Elhovo. SE Bulgaria
Ellás *see* Greece
Ellef Ringnes Island *170* island of Sverdrup Islands, N Canada
Ellerton *121* C Barbados
Ellesmere Island *170* island of Queen Elizabeth Islands, N Canada
Ellice Islands *see* Tuvalu
Ellsworth Land *90* physical region of Lesser Antarctica, Antarctica
El Mahalla el Kubra *232 var.* Maḩallah al Kubrá. N Egypt
El Manşûra *232 var.* Al Manşūrah. N Egypt
El Mediyya *see* Médéa
El Mina *364 var.* Al Mīnā'. N Lebanon
El Minya *232 var.* Al Minyā. C Egypt
El Mohammaidia *see* Mohammadia
Elmshorn *266* N Germany
El Nebk *see* An Nabk
El Obeid *540 var.* Al Ubayyiḑ, Al Obayyid. C Sudan
El Oued *83 var.* El Ouâdi, El Wad. NE Algeria
Eloy Alfaro *230 var.* Durán. SW Ecuador
El Palmar *136* E Bolivia
El Paraíso *290* S Honduras
el Pas de las Casa *86* E Andorra
El Paso *602* Texas, SC USA
Elpitiya *538* S Sri Lanka
El Porvenir *456* NE Panama
El Progreso *280 var.* Guastatoya. C Guatemala
El Progreso *290* NW Honduras
El Qâhira *see* Cairo
Elqui *183* river of N Chile
El Quneitra *see* Al Qunayţirah
El Quseir *see* Al Quşayr
El Quweira *see* Al Quwayrah
El Rama *436* SE Nicaragua
El Salvador *211* SE Cuba
El Salvador *236-237* officially Republic of El Salvador. Country of Central America divided into 14 admin. units (13 departments, 1 metropolitan area). ❖ San Salvador
El Seibo *226 var.* Santa Cruz de El Seibo. E Dominican Republic
el Serrat *86* N Andorra
Elsinore *see* Helsingør
El Suweida *see* As Suwaydā'
el Tarter *86* NE Andorra
El Tigre *623* NE Venezuela
El Tigre de San Lorenzo *456* SW Panama
El Triunfo *237* SE El Salvador
Elva *242 Ger.* Elwa. SE Estonia
Elvas *474* C Portugal
Elverum *444* S Norway
El Vigía *623* W Venezuela
El Wad *see* El Oued
El Wâhât el Khârga *232 var.* Al Wāḩāt al Khārijah.S Egypt
Elx *see* Elche
Émaé *619* island of C Vanuatu
Emajõgi *242 var.* Emajygi, Ema Jõgi. River of SE Estonia
Emao *619* island of C Vanuatu
Emba *340* W Kazakhstan
Emba *340* river of W Kazakhstan
Embangweni *385 var.* Ephangweni. NW Malawi
Emborcação, Represa da *145* reservoir of S Brazil
Embu *344* C Kenya
Emden *266* NW Germany
Emerald *101* E Australia
Emi Koussi *180* mountain of N Chad
Emlembe *544* mountain of South Africa and Swaziland
Emmen *429* NE Netherlands
Empada *284* SW Guinea-Bissau
Empty Quarter *see* Rub'al Khali
Ems *266, 429 Dut.* Eems. River of Germany and Netherlands
Enaretträsk *see* Inarijärvi
Encamp *86* C Andorra
Encarnación *460* S Paraguay
Enchi *272* SW Ghana
Endeavour *121* NW Barbados
Endelave *218* island of C Denmark
Enderbury Island *348* island of the Phoenix Is, C Kiribati
Enderby Land *90* physical region of Greater Antarctica, Antarctica
Enewetak *396 var.* Eniwetok. Atoll of W Marshall Islands
Enghershatu *240* mountain of N Eritrea
England *597* national region of UK
English Channel *255, 597, 647 var.* The Channel, *Fr.* La Manche. Channel connecting the North Sea and Atlantic Ocean between England and France
English Harbour Town *92* S Antigua, Antigua & Barbuda
Engordany *86* C Andorra
Engures Ezers *362* lake of NW Latvia
Enguri *264 Rus.* Inguri. River of NW Georgia
Eniwetok *see* Enewetak
Enkeldoorn *see* Chivhu
Enkhuizen *429* NW Netherlands
Enna *323* Sicilia, S Italy
En Nâqaoûra *364 var.* An Nāqūrah. SW Lebanon
Ennedi *180* plateau of E Chad
Ennis *316 Ir.* Inis. W Ireland
Enniscorthy *316* E Ireland
Enniskillen *597* Northern Ireland, UK
Enns *106* river of C Austria
Enriquillo, Lago *226* lake of SW Dominican Republic
Enschede *429* E Netherlands
Ensenada *403* NW Mexico
Entebbe *588* S Uganda
Enugu *440* S Nigeria
Enu, Ilha de *284* island of SW Guinea-Bissau
Eolie, Isole *323 var.* Isole Lipari, *Eng.* Lipari Islands, *var.* Aeolian Islands. Island group of S Italy
Epéna *200* NE Congo
Eperies *see* Prešov
Eperjes *see* Prešov
Ephangweni *see* Embangweni
Epi *619* island of C Vanuatu
Épinal *255* NE France
Epira *286* E Guyana
Episkopi *215* SW Cyprus
Episkopi Bay *215* bay of the Mediterranean Sea, on the SE coast of Cyprus
Epping Forest *278* S Grenada island, Grenada
Equatorial Channel *390* channel of the Indian Ocean between the atolls of Fuammulah and South Huvadhu Atoll, S Maldives
Equatorial Guinea *238-239* officially Republic of Equatorial Guinea. Country of W Africa divided into 7 admin. units (provinces). ❖ Malabo
Erbil *see* Arbīl
Erdenet *412* N Mongolia
Erdi Ma *180* desert region of NE Chad
Ereğli *581* S Turkey
Erenköy *see* Kokkina
Erenköy *215* W Cyprus
Erevan *see* Yerevan
Erfurt *267* C Germany
Erg Chech *83* desert region of SW Algeria
Erg du Djourab *180* desert region of C Chad
Erguig *180* river of SW Chad
Erhlin *559 Jap.* Nirin. W Taiwan
Erie *603* Pennsylvania, NE USA
Erie, Lake *171, 603 Fr.* Lac Érié. Lake of Canada and USA
Erigabo *see* Ceerigaabo
Erik Erikenstretet *653* strait of the Arctic Ocean, E Svalbard
Erikub *396* island of C Marshall Islands
Eritrea *240-241* officially State of Eritrea. Country of E Africa divided into 10 admin. units (regions). ❖ Asmara
Erivan *see* Yerevan
Erlangen *267* S Germany
Erlau *see* Eger
Ermak *see* Yermak
Ermelo *531* Eastern Transvaal, NE South Africa
Er Rachidia *414* E Morocco
Er Rahad *540 var.* Ar Rahad. C Sudan
Er Rif *see* Rif
Erromango *619* island of S Vanuatu
Ersekë *81 var.* Erseka, Kolonjë. SE Albania
Érsekújvár *see* Nové Zámky
Ertis *see* Irtysh
Erzenit *81 var.* Erzen. River of C Albania
Erzgebirge *216, 267 Cz.* Krušné Hory, *Eng.* Ore Mountains. Mountain range of Czech Republic and Germany
Erzincan *581* E Turkey
Erzurum *581 prev.* Erzerum. NE Turkey
Esa'ala *458* Normanby I, Papua New Guinea
Esbjerg *218* Jylland, W Denmark
Esbo *see* Espoo
Escaut *see* Scheldt
Eschen *374* NW Liechtenstein
Esch-sur-Alzette *378* S Luxembourg
Escuintla *280* S Guatemala
Eséka *168* SW Cameroon
Esenguly *see* Gasan-Kuli
Esentepe *215 var.* Ayios Amvrosios. N Cyprus
Eşfahān *309 var.* Ispahan. W Iran
Esh Sham *see* Damascus
Esh Sharā *see* Ash Sharāh
Esigodini *636 prev.* Essexvale. SW Zimbabwe
Esil *see* Ishim
Eski Dzhumaya *see* Turgovishte
Eskifjördhur *296* E Iceland
Eskije *see* Xánthi
Eskilstuna *547* S Sweden
Eskişehir *580* W Turkey
Esla, Embalse de *535 var.* Embalse de Elsa. Reservoir of NW Spain
Esmeralda *210* E Cuba
Esmeraldas *230* N Ecuador
Esna *see* Isna
Esneux *127* E Belgium
España *see* Spain
Española, Isla *230* island of SE Galapagos Is, Ecuador
Esparta *290* NW Honduras
Esperance *100* SW Australia
Esperanza *90* Argentinian research station of Antarctic Peninsula, Antarctica
Esperanza *226* N Dominican Republic
Espinal *195* C Colombia
Espinhaço, Serra do *145* mountain range of SE Brazil
Espiritu Santo *619 var.* Santo. Island of C Vanuatu
Espoo *251 Swe.* Esbo. SW Finland
Espungabera *419* SW Mozambique
Esquel *95* SW Argentina
Esquipulas *280* SE Guatemala
Êssalou *222* river of N Djibouti
Essaouira *414 prev.* Mogador. W Morocco
Essau *263* W Gambia
Esseg *see* Osijek
Essen *266* W Germany
Essequibo *286* river of Guyana
Essequibo Islands *286* islands of N Guyana
Essexvale *see* Esigodini
Estados, Isla de los *95 Eng.* Staten Island. Island to the E of Tierra del Fuego, Argentina
Estanyó, Pic de l' *86* mountain of N Andorra
Estcourt *531* Kwazulu Natal, E South Africa
Estelí *436* W Nicaragua
Estevan *170* S Canada
Est, Ile de l' *400* E Mauritius
Estonia *242-243* officially Republic of Estonia, *Est.* Eesti Vabariik, *prev.* Estonian SSR, *Rus.* Estonskaya SSR. Country of E Europe divided into 15 admin. units (counties).❖ Tallinn
Estoril *474* W Portugal
Estrela *474* mountain of C Portugal
Estrela, Serra de *474* mountain range of C Portugal
Eszék *see* Osijek
Esztergom *292 Ger.* Gran. N Hungary
Etchmiadzin *see* Ejmiadzin
Etembue *238* SW Río Muni, Equatorial Guinea

Ethiopia *244-247 prev.* Abyssinia. Country of E Africa divided into 14 admin. units (regions). ❖ Addis Ababa
Ethiopian Plateau *245* plateau of W Ethiopia
Etna, Mount *323* volcano of Sicilia, S Italy
Etosha Pan *423* salt basin of N Namibia
Etoumbi *200* NW Congo
Etrek *see* Atrak
Etsch *see* Adige
Ettelbrück *378* C Luxembourg
'Eua *572 prev.* Middleburg Island. Island of Tongatapu Group, Tonga
Eugene *602* Oregon, NW USA
Euphrates *312, 555, 580* River of SW Asia
Eureka *170* Ellesmere Island, N Canada
Europa Point *646* headland on the S coast of Gibraltar
Euthini *385 var.* Eutini. NW Malawi
Euxine Sea *see* Black Sea
Evansville *603* Indiana, C USA
Everest, Mount *186, 427 Chin.* Qomolangma Feng, *Nep.* Sagarmatha. Mountain of China and Nepal
Evesham *498* SE St Vincent, St Vincent & the Grenadines
Evinayong *238 var.* Evinayoung, Ebinayon. S Río Muni, Equatorial Guinea
Évora *474* C Portugal
Évreux *255* N France
Évry *255* N France
Evrykhou *215 var.* Evrychou. W Cyprus
Évvoia *275* Island of E Greece
Ewarton *331* C Jamaica
Ewaso Ngiro *344* river of C Kenya
Ewo *200* W Congo
Exe *597* river of SW England, UK
Exeter *597* SW England, UK
Exuma Cays *112* islets of C Bahamas
Exuma Sound *112* stretch of water between Cat I and Exuma Cays, Bahamas
Eyasi, Lake *564* lake of N Tanzania
Eyl *528 It.* Eil. E Somalia
Eyre North, Lake *101* salt lake of C Australia
Eyre Peninsula *101* peninsula of S Australia
Eyre South, Lake *101* salt lake of C Australia
Eysturoy *644 var.* Østerø. Island of N Faeroe Islands
Extremadura *534* autonomous community of W Spain
Ezulwini *544* W Swaziland

F

Faadhippolhu Atoll *390 var.* Fadiffolu, Lhaviyani Atoll. Atoll of N Maldives
Faafu Atoll *see* North Nilandhe Atoll
Fåborg *218* Fyn, S Denmark
Fabriano *323* C Italy
Facpi Point *647* headland on the SW coast of Guam
Fada *180* E Chad
Fada-N'gourma *157 var.* Fadan-Gourma. E Burkina
Fadghāmī *see* Tall Fadghāmī
Fadiffolu *see* Faadhippolhu Atoll
Fadugu *518* N Sierra Leone
Færingehavn *646 var.* Kangerluarsoruseq, S Greenland
Faeroe Islands *644 Faer.* Fóroyar, *Dan.* Færøerne. Self-governing territory of Denmark, North Atlantic Ocean. ❖ Tórshavn
Faetano *502* E San Marino
Fagaloa Bay *500* bay of the Pacific Ocean on Upolu, SE Samoa
Fagamālo *500* Savai'i, Samoa
Făgăraş *480* C Romania
Faguibine, Lac *392 var.* Lake Fagibina. Lake of NW Mali
Fahaheel *see* Al Fuḩayḩīl
Faial *474 var.* Ilha do Faial. Island of the Azores, Portugal
Faifo *see* Hôi An
Failaka Island *see* Faylakah
Faioa, Île *653* island of Île Uvea, S Wallis & Futuna
Fairbanks *602* Alaska, USA
Fair Isle *597* island of N Scotland, UK
Fais *406* island of W Micronesia
Faisalābād *451 prev.* Lyallpur. NE Pakistan
Faizabad *see* Feyzābād
Fajã *176* Brava, S Cape Verde
Fajara *263* W Gambia
Fajardo *651* NE Puerto Rico
Fajãzinha *176* Fogo, S Cape Verde
Fakaofo Atoll *653* island of SE Tokelau
Fako *168* active volcano of W Cameroon
Falaba *518* N Sierra Leone
Falam *159* NW Burma
Falciano *502* NE San Marino
Fale *587* islet of Nukufetau, Tuvalu
Faleālupo *500* Savai'i, Samoa
Falelima *500* Savai'i, Samoa
Falémé *392, 510* river of W Africa
Faleshty *see* Fălești
Făleşti *408 Rus.* Faleshty. NW Moldova
Falkat *240* seasonal river of N Eritrea
Falkirk *597* C Scotland, UK
Falkland Islands *95, 645 Sp.* Islas Malvinas. British dependent territory of the South Atlantic Ocean. ❖ Stanley.
Falkland Sound *645* strait of the South Atlantic Ocean between East Falkland and West Falkland, Falkland Islands
Falluja *see* Al Fallūjah
Falmouth *92* S Antigua, Antigua & Barbuda
Falmouth *331* NW Jamaica
Falster *218* island of SE Denmark
Falun *547* C Sweden
Famagusta *see* Gazimağusa
Famagusta Bay *215 var.* Ammochostos Bay, Gazimağusa Körfezi. Bay of the Mediterranean Sea, on the E coast of Cyprus
Fandriana *382* C Madagascar
Fangliao *559* SW Taiwan
Fanning Island *see* Tabuaeran
Fano *323* N Italy
Fan Si Pan *627* mountain of NW Vietnam
Faraday *90* UK research station of Antarctic Peninsula, Antarctica
Faradofay *see* Tôlañaro
Farafangana *382* SE Madagascar
Farafenni *263* NW Gambia
Farāh *77* W Afghanistan
Farāh Rūd *77* river of W Afghanistan
Farallon de Medinilla *650* island of C Northern Mariana Islands
Farallon de Pajaros *650* island of N Northern Mariana Islands
Faranah *283* S Guinea
Faraulep *406* atoll of C Micronesia
Farghona *614 Rus.* Fergana, *prev.* Novyy Margilan. E Uzbekistan
Fargo *603* North Dakota, NC USA
Farīdābād *298* N India
Farīdpur *117* C Bangladesh
Farim *284* NW Guinea-Bissau
Farkhor *562 Rus.* Parkhar. SW Tajikistan
Farm *649* river of E Montserrat
Farmington *602* New Mexico, SW USA
Faro *168* river of Cameroon and Nigeria
Faro *474* S Portugal
Farquhar Atoll *516* atoll of the Farquhar Group, S Seychelles
Farquhar Group *516* island group of S Seychelles
Fars, Khalīj-e *see* The Gulf
Farvel, Kap *see* Nunap Isua
Fass *263* W Gambia
Fastiv *590* NW Ukraine
Fatala *283* river of W Guinea
Fatick *510* W Senegal
Fátima *474* W Portugal
Fatoto *263* E Gambia
Fatua, Pointe *653 var.* Pointe Nord. Headland of Île Futuna, N Wallis & Futuna
Fatutaka *526 var.* Mitre I. Island of E Solomon Islands
Faxaflói *296* bay of North Atlantic Ocean, on SW coast of Iceland
Faya *180* N Chad
Fayaoué *650* Ouvéa, Îles Loyauté, N New Caledonia
Fayetteville *603* Arkansas, SC USA
Faylakah *356 var.* Failaka Island. Island of E Kuwait
Fazao, Monts du *571* mountain range of W Togo
Fazzān *371 Eng.* Fezzan. Cultural region of W Libya
Fdérik *398 prev.* Fort-Gouraud. NW Mauritania
Feabhai, Loch *see* Foyle, Lough
Fédala *see* Mohammedia
Fehmarn *267* island of N Germany
Fehmarnbelt *267* strait connecting Kieler Bucht and Mecklenburger Bucht, between Denmark and Germany
Feira de Santana *145 prev.* Feira. E Brazil
Feistritz *106* river of SE Austria
Fejaj, Chott el *577* salt lake of C Tunisia
Feldkirch *106* W Austria
Félegyháza *see* Kiskunfélegyháza
Felidhu Atoll *390* atoll of C Maldives
Fénérive *see* Fenoarivo Atsinanana
Fenglin *559 Jap.* Hōrin. E Taiwan
Fengshan *559 Jap.* Hōzan. SW Taiwan
Fengtien *see* Shenyang
Fengtien *see* Liaoning
Fengyüan *559 var.* Toyohara, *Jap.* Hōgen. W Taiwan
Feni *117* E Bangladesh
Fennern *see* Vändra
Fenoarivo Atsinanana *382 prev.* Fénérive. NE Madagascar
Fens, The *597* wetlands of E England, UK
Ferdinand *see* Montana
Fergana *see* Farghona
Fergana Valley *562* physical region of C Asia
Fergusson Island *458 var.* Kaluwawa. Island of SE Papua New Guinea
Ferizaj *see* Uroševac
Ferkessédougou *328* N Ivory Coast
Ferlo *see* Vallée du Ferlo
Fernandina, Isla *230* island of W Galapagos Is, Ecuador
Fernando de la Mora *460* S Paraguay
Fernando de Noronha *145* island of E Brazil
Fernando Po *see* Bioko
Ferrara *323* N Italy
Ferrol *534 prev.* El Ferrol del Caudillo. NW Spain
Ferryville *see* Menzel Bourguiba
Ferto-tó *see* Neusiedler See
Fès *414 Eng.* Fez. N Morocco
Fethiye *580* SW Turkey
Feyzābād *77 var.* Faizabad. NE Afghanistan
Fezzan *see* Fazzān
Fianarantsoa *382* C Madagascar
Fianga *180* SW Chad
Fichē *245 It.* Ficce. C Ethiopia
Fielding *433* S North Island, New Zealand
Fier *81 var.* Fieri. SW Albania
Fierzës, Liqeni i *81* lake of N Albania
Fig Tree *494* S Nevis, St Kitts & Nevis
Figueira da Foz *474* W Portugal
Figueres *535* E Spain
Figuig *414 var.* Figig. E Morocco
Fiji *248-249* officially Republic of Fiji, *Fij.* Viti. Country of the Pacific Ocean divided into 4 admin. units (divisions). ❖ Suva
Filadelfia *206* W Costa Rica
Filingué *439* W Niger
Fimi *203* river of W Dem. Rep. Congo
Finike *580* SW Turkey
Finland *250-253* officially Republic of Finland, *Fin.* Suomen Tasavalta. Country of N Europe divided into 12 admin. units (11 provinces and 1 autonomous region). ❖ Helsinki
Finland, Gulf of *242, 251, 484 Fin.* Suomenlahti, *Swe.* Finska Viken, *Est.* Soome Laht, *Rus.* Finskiy Zaliv. Gulf of the Baltic Sea, NE Europe
Finnmarksvidda *444* physical region of NE Norway
Fins *448* NE Oman
Finskiy Zaliv *see* Finland, Gulf of
Fiorina *502* NE San Marino
Firenze *322 Eng.* Florence. NW Italy
Fischbacher Alpen *106* mountain range of E Austria
Fish *423 Afr.* Vis. River of S Namibia
Fishguard *597 Wel.* Abergwaun. W Wales, UK
Fiskeræsset *646 var.* Qeqertarsuatsiaat, S Greenland
Fiume *see* Rijeka
Five Islands Village *92* W Antigua, Antigua & Barbuda
Fizuli *see* Füzuli
Fjerritslev *218* Jylland, NW Denmark
Fläming *267* hill region of NE Germany
Flanders *127 Dut.* Vlaanderen, *Fr.* Flandres. Cultural region of W Belgium
Flat Island *400 var.* Île Plate. Island of N Mauritius
Flat Island *652* island of NE Spratly Islands
Flatts Village *643* C Bermuda
Flensburg *266* N Germany
Flessingue *see* Vlissingen
Flinders *101* river of N Australia
Flinders Island *101* island of SE Australia
Flinders Ranges *101* mountain range of S Australia
Flin Flon *170* SW Canada
Flint *603* Michigan, NC USA
Flint Island *348* island of the Line Is, E Kiribati
Flitsch *see* Bovec
Floréal *400* C Mauritius
Florence *see* Firenze
Florencia *195* SW Colombia
Flores *280* N Guatemala
Flores *304* island of Nusa Tenggara, C Indonesia
Flores *474* island of the Azores, Portugal
Flores, Lago de *see* Petén Itza, Lago
Flores, Laut *304 Eng.* Flores Sea. Sea of the Pacific Ocean, C Indonesia
Floreşti *408 Rus.* Floreshty. N Moldova
Florianópolis *145 prev.* Destêrro. S Brazil
Florida *210* SE Cuba
Florida *290* W Honduras
Florida *603, 607* state of SE USA
Florida *611* S Uruguay
Floridablanca *195* NE Colombia
Florida Islands *526* group of islands of C Solomon Is
Florida, Straits of *112, 210* strait connecting the Atlantic Ocean and Gulf of Mexico
Flórina *275 var.* Phlórina. N Greece
Flüelapass *550* mountain pass of E Switzerland
Flushing *see* Vlissingen
Fly *458* river of Indonesia and Papua New Guinea
Flying Fish Cove *644* ❖ of Christmas Island
Fnjóská *296* river of C Iceland
Foa *572* island of Ha'apai Group, Tonga
Foča *140* SE Bosnia & Herzegovina
Focşani *480* E Romania
Foggia *323* S Italy
Fogo *176* island of SW Cape Verde
Fogo'one, Pointe *653* headland of Île Uvea, S Wallis & Futuna
Foix *255* S France
Folkestone *597* SE England, UK
Fomboni *198* N Mohéli, Comoros

Fon *283* mountainous region of E Guinea
Fond St. Jean *224* S Dominica
Fongafale *587* *var.* Funafuti. ❖ of Tuvalu, Funafuti, Tuvalu
Fonseca, Gulf of *237, 290, 436* gulf of the Pacific Ocean, on the W coast of Central America
Fontvieille *410* SW Monaco
Fonuafo'ou *572* island of Nomuka Group, W Tonga
Fonualei *572* island of N Tonga
Foochow *see* Fuzhou
Forécariah *283* SW Guinea
Forestière *497* N St Lucia
Forlì *323* N Italy
Formentera *535* island of the Islas Baleares, E Spain
Formosa *95* NE Argentina
Formosa Bay *see* Ungama Bay
Formosa, Ilha *284* island of Arquipélago dos Bijagós, SW Guinea-Bissau
Formosa, Serra *145* mountains of C Brazil
Formosa Strait *see* Taiwan Strait
Forssa *251* SW Finland
Fortaleza *136* N Bolivia
Fortaleza *145* *prev.* Ceará. NE Brazil
Fort-Archambault *see* Sarh
Fort-Bayard *see* Zhanjiang
Fort-Cappolani *see* Tidjikja
Fort Charlet *see* Djanet
Fort Collins *603* Colorado, SW USA
Fort-Crampel *see* Kaga Bandoro
Fort-Dauphin *see* Tôlañaro
Fort-de-France *649* ❖ of Martinique, W Martinique
Fort-Foureau *see* Kousséri
Fort George *see* Grande Rivière, La
Fort-Gouraud *see* Fdérik
Forth *597* river of C Scotland, UK
Fort Hall *see* Murang'a
Forth, Firth of *597* estuary of the river Forth, E Scotland, UK
Fort Hill *see* Chitipa
Fortín General Diaz *460* W Paraguay
Fortín General Eugenio Garay *see* General Eugenio A. Garay
Fort Jameson *see* Chipata
Fort Johnson *see* Mangochi
Fort-Lamy *see* N'Djamena
Fort Lauderdale *603* Florida, SE USA
Fort-Liberté *288* N Haiti
Fort Manning *see* Mchinji
Fort McMurray *170* W Canada
Fort-Repoux *see* Akjoujt
Fort Rosebery *see* Mansa
Fort Saint John *170* W Canada
Fort Saskatchewan *170* SW Canada
Fort-Shevchenko *340* W Kazakhstan
Fort-Sibut *see* Sibut
Fort Smith *170* W Canada
Fort Smith *603* Arkansas, SC USA
Fort-Trinquet *see* Bir Mogreïn
Fort Victoria *see* Masvingo
Fort Wayne *603* Indiana, C USA
Fort William *597* W Scotland, UK
Fort Worth *603* Texas, SC USA
Fougamou *260* C Gabon
Foulenzem *260* NW Gabon
Foumban *168* NW Cameroon
Foumbouni *198* S Grande Comore, Comoros
Fourchue, Île *646* island of N Guadeloupe
Fournaise, Piton de la *652* mountain of SE Réunion
Four Roads *574* NW Trinidad, Trinidad & Tobago
Fouta Djallon *283* *var.* Futa Jallon. Mountainous region of W Guinea
Foveaux Strait *433* strait between South Island and Stewart Island, New Zealand
Fox Bay East *645* West Falkland, W Falkland Islands
Fox Bay West *645* West Falkland, W Falkland Islands
Foxe Peninsula *171* peninsula of Baffin Island, NE Canada
Foyle *597* river of Ireland and UK
Foyle, Lough *316, 597* *Ir.* Loch Feabhai. Inlet of the Atlantic Ocean, Ireland and UK
Fraile Muerto *611* E Uruguay
Frakštát *see* Hlohovec
France *254-259* officially The French Republic. Country of Europe divided into 22 admin. units (regions, comprising 96 departments). ❖ Paris
Franceville *see* Massoukou
Francistown *142* NE Botswana
Franconian Jura *see* Fränkische Alb
Frankfort *603* Kentucky, C USA
Frankfurt am Main *266* *Eng.* Frankfort on the Main, Frankfurt. SW Germany
Frankfurt an der Oder *267* E Germany
Fränkische Alb *267* *Eng.* Franconian Jura. Mountain range of S Germany
Frantsa-Iosifa, Zemlya *454-485* *Eng.* Franz Josef Land. Island group of N Russian Federation
Fraser *170* river of SW Canada
Fraser Island *101* *var.* Great Sandy Island. Island of E Australia
Frauenburg *see* Saldus
Frauenfeld *550* NE Switzerland
Fray Bentos *611* W Uruguay
Fredericia *218* Jylland, SW Denmark
Fredericton *171* SE Canada
Frederiksdal *see* Narsaq Kujalleq
Frederikshåb *see* Paamiut
Frederikshavn *218* Jylland, N Denmark
Frederiksted *653* Saint Croix, S Virgin Islands
Fredrikshald *see* Halden
Fredrikstad *444* S Norway
Freemans *92* C Antigua, Antigua & Barbuda
Freeport *112* *var.* Freeport-Lucaya. Bahamas
Freetown *92* SE Antigua, Antigua & Barbuda
Free State *531* *prev.* Orange Free State. Province of C South Africa
Freetown *518* ❖ of Sierra Leone, W Sierra Leone
Frégate *516* island of the Inner Islands, NE Seychelles
Freiburg *see* Fribourg
Freiburg im Breisgau *266* *var.* Freiburg. SW Germany
Freistadtl *see* Hlohovec
Fremantle *100* SW Australia
French Guiana *645* *var* Guyane. French overseas department of N South America. ❖ Cayenne.
French Polynesia *646* French overseas possession of the Pacific Ocean.❖ Papeete.
French Somaliland *see* Djibouti
French Sudan *see* Mali
French Territory of the Afars and Issas *see* Djibouti
French Togoland *see* Togo
Fria *283* W Guinea
Fribourg *550* *Ger.* Freiburg. W Switzerland
Friedek-Mistek *see* Frýdek-Místek
Friedrichshafen *266* S Germany
Friendly Islands *see* Tonga
Frigate Island *278* island to the S of Carriacou, Grenada
Frigate Island *498* island of SW St Vincent & the Grenadines
Frisches Haff *see* Vistula Lagoon
Frobisher Bay *see* Iqaluit
Frome, Lake *101* salt lake of S Australia
Front Range *see* Maluti
Frunze *see* Bishkek
Frýdek - Místek *216* *Ger.* Friedek - Mistek. SE Czech Republic
Fuammulah *390* *var.* Gnaviyani Atoll. Atoll of S Maldives
Fu-chien *see* Fujian
Fu-chou *see* Fuzhou
Fucht *414* W Western Sahara
Fuenlabrada *535* C Spain
Fuerte Olimpo *460* NE Paraguay
Fuerteventura *535* island of Islas Canarias, SW Spain
Fuglafjørdhur *644* *var.* Fuglefjord. Eysturoy, N Faeroe Islands
Fugloy *644* *var.* Fuglø. Island of NE Faeroe Islands
Fu-hsin *see* Fuxin
Fujairah *594* *Ar.* Al Fujayrah. NE United Arab Emirates
Fuji *332* Honshū, SE Japan
Fujian *187* *var.* Fukien, Fu-chien. Province of SE China
Fuji-san *332* mountain of Honshū, SE Japan
Fujisawa *332* Honshū, SE Japan
Fukuchiyama *332* Honshū, C Japan
Fukue *332* island of Gotō-rettō, SW Japan
Fukue *332* Gotō-rettō, SW Japan
Fukui *332* Honshū, C Japan
Fukuoka *332* Kyūshū, SW Japan
Fukushima *332* Honshū, N Japan
Fukuyama *332* Honshū, W Japan
Fulacunda *284* C Guinea-Bissau
Fulaga *248* island of the Lau Group, E Fiji
Fulda *266* C Germany
Fullarton *574* SW Trinidad, Trinidad & Tobago
Funabashi *332* Honshū, SE Japan
Funafuti *see* Fongafale
Funafuti *587* coral atoll of C Tuvalu
Funaota *587* islet of Nukufetau, Tuvalu
Funchal *474* Madeira, Madeira Islands, Portugal
Fünen *see* Fyn
Fünfkirchen *see* Pécs
Funhalouro *419* SE Mozambique
Furna *176* Brava, S Cape Verde
Furnas, Represa de *145* reservoir of SE Brazil
Furneaux Group *101* island group of SE Australia
Fusan *see* Pusan
Fushun *187* Liaoning, NE China
Futa Jallon *see* Fouta Djallon
Futuna *619* island of S Vanuatu
Futuna, Île *653* island of N Wallis & Futuna
Fuwairet *see* Al Fuwayriţ
Fuxin *187* *var.* Fu-hsin, Fusin. Liaoning, NE China
Fuzhou *187* *var.* Foochow, Fu-chou. Fujian, SE China
Füzuli *110* *Rus.* Fizuli SW Azerbaijan
Fyn *218* *Ger.* Fünen. Island of C Denmark

G

Gaafu Alifu Atoll *see* North Huvadhu Atoll
Gaafu Dhaalu Atoll *see* South Huvadhu Atoll
Gaalkacyo *528* *var.* Galka'yo, *It.* Galcaio. C Somalia
Gabela *88* W Angola
Gabès *577* *var.* Qābis. C Tunisia
Gabès, Gulf of *577* gulf of the Mediterranean Sea to the E of Tunisia
Gablonz an der Neisse *see* Jablonec nad Nisou
Gabon *260-261* officially The Gabonese Republic. Country of West Africa divided into 9 admin. units (provinces). ❖ Libreville
Gaborone *142* *prev.* Gaberones. ❖ of Botswana, SE Botswana
Gabriel, Ilot *400* *Eng.* Gabriel Island. Island of N Mauritius
Gabrovo *152* C Bulgaria
Gabú *284* *prev.* Nova Lamego. E Guinea-Bissau
Gaeta, Golfo di *323* *var.* Gulf of Gaeta. Gulf of the Tyrrhenian Sea, on the W coast of Italy
Gaferut *406* island of C Micronesia
Gafsa *577* *var.* Qafşah. W Tunisia
Gagnoa *328* C Ivory Coast
Gagra *264* NW Georgia
Gaherré *222* NE Djibouti
Gahnpa *see* Ganta
Gaibānda *117* NW Bangladesh
Gaillimh *see* Galway
Gailtaler Alpen *106* mountain range of S Austria
Gairdner, Lake *101* salt lake of S Australia
Gaiziņ *see* Gaizina Kalns
Gaizina Kalns *362* *var.* Gaiziņ. Mountain of E Latvia
Gâlâfi *222* W Djibouti
Galana *344* river of SE Kenya
Galapagos Islands *230* *var.* Tortoise Islands, *Sp.* Archipiélago de Colón. Island group of W Ecuador in the Pacific Ocean
Galaţi *480* *Ger.* Galatz. E Romania
Galaymor *see* Kalai-Mor
Galcaio *see* Gaalkacyo
Gales Point *130* E Belize
Galets *652* river of NW Réunion
Galgóc *see* Hlohovec
Galibi *542* NE Suriname
Galicia *535* NW Spain
Galilee, Sea of *see* Tiberias, Lake
Galle *538* *prev.* Point de Galle. SW Sri Lanka
Gällivare *547* N Sweden
Gâlma *see* Guelma
Galomaro *284* C Guinea-Bissau
Galway *316* *Ir.* Gaillimh. W Ireland
Galway Bay *316* *Ir.* Cuan na Gaillimhe. Bay of the Atlantic Ocean, to the W of Ireland
Gamamudo *284* NE Guinea-Bissau
Gamba *260* SW Gabon
Gambia *262-263* officially Republic of the Gambia. Country of W Africa divided into 6 admin. units (divisions). ❖ Banjul
Gambia *263, 283, 510* *Fr.* Gambie. River of W Africa
Gambier, Îles *646* island group of E French Polynesia
Gambissara *263* E Gambia
Gamboma *200* E Congo
Gamboula *179* SW Central African Republic
Gamgadhi *427* *var.* Gum. W Nepal
Gamlakarleby *see* Kokkola
Gammouda *see* Sidi Bouzid
Gampaha *538* W Sri Lanka
Gamprin *374* NW Liechtenstein
Gâm *627* river of N Vietnam
Gan *390* C Maldives
Ganaane *see* Juba
Gäncä *110* *Rus.* Gyandzha, *prev.* Kirovabad, Yelisavetpol. W Azerbaijan
Gand *see* Gent
Gandajika *203* S Dem. Rep. Congo
Gandía *535* E Spain
Ganges *117, 298-299* *Ben.* Padma, *Hind.* Ganga. River of S Asia
Ganges, Mouths of the *117, 299* large delta area of Bangladesh and India
Gansu *187* *var.* Kansu. Province of NW China
Ganta *368* *var.* Gahnpa. NE Liberia
Gao *392* E Mali
Gaoua *157* SW Burkina
Gaoual *283* N Guinea
Gap *255* E France
Garabogazköl Bogazy *see* Kara-Bogaz-Gol, Proliv
Garagum *see* Karakumy
Garagum Kanaly *see* Karakumskiy Kanal
Garam *see* Hron
Garamszentkereszt *see* Žiar nad Hronom
Garango *157* C Burkina
Garbahaarrey *528* *It.* Garba Harre. SW Somalia
Garda, Lago di *322* *var.* Benaco, *Eng.* Lake Garda. Lake of N Italy
Gardēz *77* *var.* Gardeyz. E Afghanistan
Gardner Island *see* Nikumaroro
Gardo *see* Qardho
Garissa *344* E Kenya
Garm *see* Gharm

Goettingen *see* Göttingen
Gogounou *132* N Benin
Goiânia *145 prev.* Goyania. S Brazil
Gökdepe *see* Gëkdepe
Gokwe *636* NW Zimbabwe
Gol *444* S Norway
Golan Heights *319* disputed territory of SW Syria
Gold Coast *101* coastal region of E Australia
Gold Coast *272* coastal region of W Africa
Golden Bay *433* bay on the coast of N South Island, New Zealand
Golden Valley *636* N Zimbabwe
Goldingen *see* Kuldīga
Golfito *206* SE Costa Rica
Gollel *see* Lavumisa
Golmud *186 var.* Golmo, *Chin.* Ko-erh-mu. Qinghai, W China
Golungo Alto *88* NW Angola
Goma *203* NE Dem. Rep. Congo
Gombe *440* E Nigeria
Gomel' *see* Homyel'
Gomera *535* island of Islas Canarias, SW Spain
Gómez Palacio *403* NW Mexico
Gonaïves *288* W Haiti
Gonâve, Canal de la *288 var.* Canal de Sud. Channel of the Caribbean Sea between Île de la Gonâve and Haiti
Gonâve, Golfe de la *288* gulf of the Caribbean Sea to the W of Haiti
Gonâve, Île de la *288* island of W Haiti
Gonder *245 var.* Gondar. NW Ethiopia
Gondomar *474* NW Portugal
Goodenough Island *458 var.* Morata. Island of SE Papua New Guinea
Good Hope *224* E Dominica
Good Hope, Cape of *531 Afr.* Kaap die Gooie Hoop. Coastal feature of SW South Africa
Goodlands *400* NE Mauritius
Goose Bay *see* Happy Valley-Goose Bay
Goose Green *645* East Falkland, C Falkland Islands
Gopālpur *117* N Bangladesh
Gorakhpur *298* NE India
Gorce Island *510* island of W Senegal
Gore *433* S South Island,New Zealand
Goré *180* S Chad
Gorē *245* W Ethiopia
Gorey *648* E Jersey
Gorgān *309 var.* Gurgan. N Iran
Gori *264* C Georgia
Goris *98* SE Armenia
Gorki *see* Horki
Gor'kiy *see* Nizhniy Novgorod
Görlitz *267* E Germany
Gorlovka *see* Horlivka
Gorna Dzhumaya *see* Blagoevgrad
Gorna Oryakhovitsa *152 var.* Gorna Orjahovica. N Bulgaria
Gornji Milanovac *512* C Serbia, Serbia & Montenegro (Yugo.)
Gorno Altaysk *485* S Russian Federation
Goroka *458* C Papua New Guinea
Gorontalo, Teluk *see* Tomini, Teluk
Gorzów Wielkopolski *471 Ger.* Landsberg. W Poland
Gosford *101* E Australia
Goshogawara *332* Honshū, N Japan
Gospić *209* C Croatia
Gostivar *381* W FYR Macedonia
Göteborg *547 Eng.* Gothenburg. SW Sweden
Gotland *547* island of SE Sweden
Gotō-rettō *332* island group to the W of Kyūshū, SW Japan
Gotse-Delchev *152 var.* Goce Delčev. SW Bulgaria
Gottardo, San Passo del *550 Eng.* St. Gotthard Pass. Mountain pass of S Switzerland
Göttingen *267 var.* Goettingen. C Germany
Gottschee *see* Kočevje
Gottwaldov *see* Zlín
Goubétto *222* SE Djibouti
Gouda *429* SW Netherlands
Goulburn *101* SE Australia
Goundam *392* NW Mali
Gourcy *157* NW Burkina
Gouré *439* SE Niger
Gourma *157* cultural region of E Burkina
Gouyave *278 var.* Charlotte Town. W Grenada island, Grenada
Goverla, Gora *see* Hoverla
Governador Valadares *145* SE Brazil
Governor's Harbour *112* Eleuthera Island, Bahamas
Govurdak *585 prev.* Guardak, *Turkm.* Gowurdak. SE Turkmenistan
Goya *95* NE Argentina
Goyania *see* Goiânia
Göyçay *110 Rus.* Geokchay. C Azerbaijan
Goz-Beïda *180* SE Chad
Gozo *395 var.* Ghawdex. Island of NW Malta
Graaff-Reinet *531* Eastern Cape, S South Africa
Gračanica *140* NE Bosnia & Herzegovina
Gracias *290* W Honduras
Graciosa *474 var.* Ilha Graciosa. Island of the Azores, Portugal
Gradačac *140* N Bosnia & Herzegovina
Gradaús, Serra dos *145* mountain range of C Brazil
Gradsko *381* C FYR Macedonia
Grafton *101* E Australia
Graham Bell Island *see* Greem Bell, Ostrov
Grain Coast *368* coastal region of Liberia
Grampian Mountains *597* mountain range of C Scotland, UK
Gramsh *81 var.* Gramshi. C Albania
Gran *see* Hron
Gran *see* Esztergom
Gran *542* river of C Suriname
Granada *436* S Nicaragua
Granada *535* S Spain
Gran Canaria *535 Eng.* Grand Canary. Island of Islas Canarias, SW Spain
Gran Chaco *95, 136, 460 var.* Chaco. Lowland plain of C South America
Grand Anse *278* SW Grenada island, Grenada
Grand'Anse *516* Praslin, Seychelles
Grand Bahama Island *112* island of N Bahamas
Grand Baie *400 var.* Grande Baie. NW Mauritius
Grand Barachois *652* inlet of the Atlantic Ocean on the coast of Miquelon, N Saint Pierre and Miquelon
Grand Bassa *see* Buchanan
Grand-Bassam *328 var.* Bassam. SE Ivory Coast
Grand Bay *278* E Carriacou, Grenada
Grand Bourg *646* Marie-Galante, S Guadeloupe
Grand Caicos *653 var.* Middle Caicos. Island of N Turks and Caicos Islands
Grand Canary *see* Gran Canaria
Grand Canyon *602* canyon of SW USA
Grand Cayman *644* island of W Cayman Islands
Grand Cess *368* SE Liberia
Grand Colombier *652* island of SE Saint Pierre and Miquelon
Grand Cul-de-Sac Marin *646* bay of the Caribbean Sea, N Guadeloupe
Grande-Anse *see* Portsmouth
Grande, Bahía *95* bay of the Atlantic Ocean, to the SE of Argentina
Grande Comore *198 var.* Njazidja. Island of Comoros
Grande, Cuchilla *611* mountain range of E Uruguay
Grande de Buba, Rio *284* river of S Guinea-Bissau
Grande de Chiloé, Isla *see* Chiloé, Isla de
Grande de Gurupá, Ilha *145* island of N Brazil
Grande de Matagalpa, Río *436* river of C Nicaragua
Grande de Santiago, Río *403 var.* Santiago. River of SW Mexico
Grande de Tárcoles, Río *206* river of C Costa Rica
Grande de Térraba, Río *206* river of SE Costa Rica
Grande Prairie *170* W Canada
Grand Erg Occidental *83* desert region of W Algeria
Grand Erg Oriental *83, 577* desert region of Algeria and Tunisia
Grande, Río *403, 602-603 Sp.* Bravo Del Norte. River of Mexico and USA
Grande Rivière *497* N St Lucia
Grande Rivière *497* C St Lucia
Grande-Rivière-du-Nord *288* N Haiti
Grande Rivière, La *171 var.* Fort George River. River of SE Canada
Grande-Saline *288* W Haiti
Grande-Santi *645* W French Guiana
Grande Terre *516* island of the Aldabra Group, SW Seychelles
Grande Vigie, Pointe de la *646* headland of N Guadeloupe
Grand Falls *344* waterfall of C Kenya
Grand Forks *603* North Dakota, NC USA
Grand Goâve *288* S Haiti
Grand Harbour *395* port of Valletta, E Malta
Grand Ilet *646* bay of the Caribbean Sea, S Guadeloupe
Grand Island *603* Nebraska, C USA
Grand Junction *602* Colorado, SW USA
Grand Lac de l'Ours *see* Great Bear Lake
Grand Lac des Esclaves *see* Great Slave Lake
Grand Montagne *400* Rodrigues, Mauritius
Grand Paradis *see* Gran Paradiso
Grand-Popo *132* S Benin
Grand Rapids *603* Michigan, NC USA
Grand Récif Sud *650* reef of the Pacific Ocean, S New Caledonia
Grand River South East *400* river of E Mauritius
Grand 'Rivière *649* N Martinique
Grand Roy *278* W Grenada island, Grenada
Grand Turk Island *653* island of SE Turks and Caicos Islands
Granges *see* Grenchen
Gran Lago *see* Nicaragua, Lago de
Gran Paradiso *322 Fr.* Grand Paradis. Mountain of NW Italy
Gran San Bernardo, Passo di *see* Great St Bernard Pass
Grape Bay *643* bay of the North Atlantic Ocean, E Bermuda
Grassy Bay *643* bay of the North Atlantic Ocean, W Bermuda
Gråsten *218* Jylland, SW Denmark
Graudenz *see* Grudziądz
Grau Roig *86* E Andorra
Graz *106* SE Austria
Great Abaco *112* island of N Bahamas
Great Admiralty Island *see* Manus Island
Great Ararat *see* Büyükağrı Dağı
Great Artesian Basin *101* lowlands of C Australia
Great Australian Bight *101* large bay of the Indian Ocean, S Australia
Great Barrier Island *433* island to the NE of North Island, New Zealand
Great Barrier Reef *101, 458* coral reef to the NE of Australia, and the largest in the world
Great Basin *602* physical region of W USA
Great Bear Lake *170 Fr.* Grand Lac de l'Ours. Lake of NW Canada
Great Belt *see* Storebælt
Great Camanoe *643* island of N British Virgin Islands
Great Chagos Bank *643* undersea feature of the Indian Ocean, C British Indian Ocean Territory
Great Coco Island *159* island of SW Burma
Great Dividing Range *101* mountain range of E Australia
Greater Antarctica *90* physical region of Antarctica
Greater Caucasus *110 Rus.* Bol'shoy Kavkaz. Mountain range of SW Asia and SE Europe
Great Exhibition Bay *433* inlet of the Pacific Ocean, on the NE coast of North Island, New Zealand
Great Exuma Island *112* island of C Bahamas
Great Falls *602* Montana, NW USA
Great Fish *see* Groot-Vis
Great Harbour *643* Jost Van Dyke, W British Virgin Islands
Great Hungarian Plain *see* Alföld
Great Inagua *112* island of S Bahamas
Great Indian Desert *see* Thar Desert
Great Karoo *531 var.* Great Karroo, *Afr.* Groot Karoo. Plateau region of S South Africa
Great Khingan Range *see* Da Hinggan Ling
Great Lake *see* Tônlé Sap
Great Lakes, the *606 see* also Erie, Huron, Michigan, Ontario, Superior
Great Mercury Island *433* island to the NE of North Island, New Zealand
Great Nicobar *298* island of Nicobar Islands, to the SE of India
Great Ouse *597 var.* Ouse. River of E England, UK
Great Plains *603* plains of Canada and USA
Great Rift Valley *245, 344, 385, 564 var.* Rift Valley. Depression of E Africa and SW Asia
Great Ruaha *564* river of C Tanzania
Great Saint Bernard Pass *550 Fr.* Col du Grand-Saint-Bernard, *It.* Passo di Gran San Bernardo. Mountain pass of SW Switzerland
Great Salt Lake *602* salt lake of SW USA
Great Salt Pond *494* SE St Kitts, St Kitts & Nevis
Great Sand Sea. *232, 371* desert of Egypt and Libya
Great Sandy Desert *100* desert region of W Australia
Great Sandy Desert *see* Rub 'al Khali
Great Sandy Island *see* Fraser Island
Great Slave Lake *170 Fr.* Grand Lac des Esclaves. Lake of W Canada
Great Sound *643* bay of W Bermuda
Great Thatch *643* island of W British Virgin Islands
Great Tobago *643* island of W British Virgin Islands
Great Usutu *see* Lusutfu
Great Victoria Desert *100-101* desert region of S Australia
Great Wall *90* Chinese research station of South Shetland Islands, Antarctica
Great Zab *312 Ar.* Az Zāb Al Kabīr, *Turk.* Büyükzap Suyu, *Kurd.* Zē-i Bādīnān. River of Iraq and Turkey
Greco, Cape *215 var.* Cape Gkreko. Cape of E Cyprus
Greece *274-277* officially Hellenic Republic, *Gk* Ellás. Country of SE Europe divided into 51 admin. units (nomos). ❖ Athens
Greeley *603* Colorado, SW USA
Greem Bell, Ostrov *485 Eng.* Graham Bell Island. Island of Zemlya Frantsa-Iosifa, N Russian Federation
Green Bay *603* Wisconsin, NC USA
Green Island *92* island to the E of Antigua, Antigua & Barbuda
Green Island *278* island to the N of Grenada island, Grenada
Green Islands *458 var.* Nissan Is. Island group of E Papua New Guinea
Greenland *646 var.* Grønland. Self governing territory of Denmark, North Atlantic Ocean. ❖ Nuuk.
Greenland Sea *646* sea of the Arctic Ocean, NE Greenland

Green Mountain *652* mountain of C Ascension Island
Greenock *597* W Scotland, UK
Greensboro *603* North Carolina, SE USA
Greenville *368* *var.* Sino, Sinoe. SE Liberia
Greenville *603* Mississippi, SE USA
Greenville *603* South Carolina, SE USA
Greifswald *267* NE Germany
Grenå *218* NJylland, Denmark
Grenada *278* island which, with the southern Grenadines, comprises the independent state of Grenada
Grenada *278-279* island state of the West Indies, comprising of Grenada island and the southern Grenadine islands to the north. Divided into 6 admin. units (parishes). ❖ St George's
Grenadines *278, 498* group of over 500 small islands between Grenada and St Vincent in the Caribbean Sea. The southern islands are administered by Grenada, while the northern islands form, with St Vincent, the independent state of St Vincent the Grenadines
Grenchen *550* *Fr.* Granges. NW Switzerland
Grenoble *255* E France
Grenville *278* E Grenada island, Grenada
Grevelingen *429* inlet of the North Sea, on the coast of SW Netherlands
Grevenmacher *378* E Luxembourg
Greymouth *433* W South Island, New Zealand
Grey Range *101* mountain range of E Australia
Gribingui *see* Nana-Grébizi
Griffith *101* SE Australia
Grijalva *403* *var.* Tabasco. River of SE Mexico
Grimari *179* C Central African Republic
Grimbergen *127* C Belgium
Grimsby *597* NE England, UK
Grímsey Island *296* *var.* Grimsey. Island of N Iceland
Grímsstadhir *296* E Iceland
Grindavík *296* W Iceland
Grindsted *218* Jylland, W Denmark
Grise Fiord *170* *var.* Ausuitoq. Ellesmere Island, N Canada
Grmeč *140* mountain range of NW Bosnia & Herzegovina
Grobiņa *362* W Latvia
Grodno *see* Hrodna
Groningen *429* NE Netherlands
Groningen *542* N Suriname
Groote Eylandt *101* island of N Australia
Grootfontein *423* N Namibia
Groot Karoo *see* Great Karoo
Groot-Vis *531* *Eng.* Great Fish. River of S South Africa
Grootvloer *531* salt flat of Northern Cape, W South Africa
Gros Islet *497* N St Lucia
Gros-Morne *288* N Haiti
Gros Piton *497* mountain of SW St Lucia
Grossbetschkerek *see* Zrenjanin
Grosseto *322* NW Italy
Grosskanizsa *see* Nagykanizsa
Grossmichel *see* Michalovce
Gross-Sankt-Johannis *see* Suure-Jaani
Gross-Steffelsdorf *see* Rimavská Sobota
Grosswardein *see* Oradea
Groznyy *484* SW Russian Federation
Großglockner *106* mountain of W Austria
Grudziądz *471* *Ger.* Graudenz. N Poland
Grünau *423* S Namibia
Grünberg in Schlesien *see* Zielona Góra
Gruziya *see* Georgia
Guabalá *456* SW Panama
Guacanayabo, Golfo de *210* gulf of the Caribbean Sea, on the S coast of Cuba
Guácimo *206* NE Costa Rica
Guadalajara *403* SW Mexico
Guadalcanal *526* island of C Solomon Is
Guadalquivir *534-535* river of W Spain
Guadalupe *456* C Panama
Guadalupe *see* Canelones
Guadeloupe *646* French overseas department of the Caribbean Sea. ❖ Basse-Terre
Guadeloupe Passage *92, 224* passage connecting the Atlantic Ocean and Caribbean Sea between Guadeloupe and Dominica
Guadiana *474, 534-535* river of Portugal and Spain
Guaico *574* NE Trinidad, Trinidad & Tobago
Guáimero *210* S Cuba
Gualaco *290* C Honduras
Gualán *280* C Guatemala
Gualdicciolo *502* NW San Marino
Gualeguaychú *95* E Argentina
Guam *647* unincorporated territory of the USA, Pacific Ocean. ❖ Hagåtña
Guanabacoa *210* NW Cuba
Guanabo *210* NW Cuba
Guana Island *643* island of N British Virgin Islands
Guanaja *290* Islas de la Bahía, Honduras
Guanajay *210* NW Cuba
Guanajuato *403* C Mexico
Guanare *623* NW Venezuela
Guanare *623* river of W Venezuela
Guangdong *187* *var.* Kwangtung, Kuang-tung. Province of SE China
Guangju *see* Kwangju
Guangxi *187* *Chin.* Guangxi Zhuangzu Zizhiqu, *Eng.* Kwangsi Chuang Autonomous Region, *var.* Kuang-hsi. Autonomous region of S China
Guangyuan *187* *var.* Kuang-yuan, Kwangyuan. Sichuan, SW China
Guangzhou *187* *var.* Kwangchow, Kuang-chou, *Eng.* Canton. Guangdong, SE China
Guantánamo *211* SE Cuba
Guantanamo Bay *211* US military base, E Cuba
Guaporé *136, 144* *var.* Iténez. River of Bolivia and Brazil
Guaranda *230* C Ecuador
Guardafui, Cape *see* Caseyr, Raas
Guardak *see* Govurdak
Guárico *623* river of C Venezuela
Guasave *403* W Mexico
Guastatoya *see* El Progreso
Guatemala *280-281* officially Republic of Guatemala. Country of Central America divided into 22 admin. units (departments). ❖ Guatemala City
Guatemala City *280* *Sp.* Ciudad de Guatemala. ❖ of Guatemala, C Guatemala
Guaviare *195* river of E Colombia
Guayama *651* SE Puerto Rico
Guayaquil *230* *var.* Santiago de Guayaquil. SW Ecuador
Guayaquil, Gulf of *230, 463* gulf of the Pacific Ocean to the W of Ecuador and Peru
Guayaramerín *136* N Bolivia
Guayuabo *651* N Puerto Rico
Gudara *see* Ghŭdara
Gudenå *218* river of Jylland, NW Denmark
Guékédou *283* *var.* Guéckédou. S Guinea
Guélîlé *222* S Djibouti
Guelma *83* *var.* Gâlma. NE Algeria
Guelta Zemmur *414* E Western Sahara
Guéné *132* NE Benin
Güera *see* Guérou
Guéret *255* C France
Guerguerat *414* SW Western Sahara
Guérin-Kouka *571* NW Togo
Guernsey *254, 647* *var.* Bailiwick of Guernsey. British Crown dependency, Channel Islands, English Channel. ❖ St Peter Port.
Guérou *398* *var.* Güera. S Mauritania
Guesneau *497* N St Lucia
Guguan *650* island of C Northern Mariana Islands
Guiana Highlands *144-145, 286, 145, 286, 623* *Port.* Planalto das Guianas. Mountain range of N South America
Guiana Island *92* island to the NE of Antigua, Antigua & Barbuda
Guiba *see* Juba
Guichón *611* NW Uruguay
Guider *168* *var.* Guidder. N Cameroon
Guidimouni *439* S Niger
Guier, Lac de *510* *var.* Lac de Guiers. Lake of N Senegal
Guietsou *260* SW Gabon
Guiglo *328* W Ivory Coast
Güigüe *623* N Venezuela
Güija, Lago de *237* lake of El Salvador and Guatemala
Guillet *224* NW Dominica
Guimarães *474* N Portugal
Guinea *282-283* officially Republic of Guinea, *prev.* People's Revolutionary Republic of Guinea. Country of West Africa divided into 4 admin. units (regions). ❖ Conakry
Guinea-Bissau *284-285* officially Republic of Guinea-Bissau, *Port.* Guiné-Bissau. Independent state of West Africa divided into 8 admin. units (regions). ❖ Bissau
Guinea, Gulf of *168, 238, 272, 328* *Fr.* Golfe de Guinée. Gulf of the Atlantic Ocean, on the W coast of Africa
Guinguinéo *510* W Senegal
Güiria *623* NE Venezuela
Guiyang *187* *var.* Kuei-Yang, Kweiyang. Guizhou, S China
Guizhou *187* *var.* Kweichow, Kuei-chou. Province of S China
Gujrānwāla *451* NE Pakistan
Gujrāt *451* NE Pakistan
Gukasyan *see* Ashots'k
Gulariya *427* *var.* Gularia. W Nepal
Gulbene *362* *Ger.* Alt-Schwanenburg. NE Latvia
Gul'cha *359* *var.* Gul'ča, *Kir.* Gülchö. SW Kyrgyzstan
Gülchö *see* Gul'cha
Gulf, The *115, 312, 356, 479, 506* *var.* Persian Gulf, *Ar.* Khalīj al 'Arabī, *Per.* Khalīj-e Fars. Gulf of the Arabian Sea between the Arabian Peninsula and Iran
Guliston *614* *Rus.* Gulistan. E Uzbekistan
Gulja *see* Yining
Gulu *588* N Uganda
Gum *see* Gamgadhi
Gumal *451* river of Afghanistan and Pakistan
Gumdag *585* *prev.* Kum-Dag. W Turkmenistan
Gümülcine *see* Komotiní
Gümüljina *see* Komotiní
Gümüşhane *581* NE Turkey
Gund *562* *Rus.* Gunt. River of S Tajikistan
Güneydoğu Toroslar *581* mountain range of SE Turkey
Gunjur *263* W Gambia
Gunner's Quoin *400* *var.* Coin de Mire. Island of N Mauritius
Gunsan *see* Kunsan
Guntūr *298* SE India
Gunzan *see* Kunsan
Gurgan *see* Gorgān
Guri, Embalse de *623* reservoir of E Venezuela
Gurjaani *264* *Rus.* Gurdzhaani. E Georgia
Gurk *106* river of S Austria
Gurkfeld *see* Krško
Gurktaler Alpen *106* mountain range of S Austria
Gur'yev *see* Atyrau
Gusau *440* N Nigeria
Gushgy *585* *prev.* Kushka. S Turkmenistan
Gustavia *646* St. Barthélémy, N Guadeloupe
Gutenstein *see* Ravne na Koroškem
Gutland *378* physical region of S Luxembourg
Guwāhāti *299* *prev.* Gauhāti. NE India
Guyana *286-287* officially Co-operative Republic of Guyana, *prev.* British Guiana. Country of Central America divided into 10 admin. units (regions). ❖ Georgetown
Guyane *see* French Guiana
Güzelyurt *215* *var.* Morfou, Morphou. W Cyprus
Güzelyurt Körfezi *see* Morphou Bay
Gwādar *450* SW Pakistan
Gwalior *298* N India
Gwanda *636* SW Zimbabwe
Gwayi *636* river of W Zimbabwe
Gweru *636* *prev.* Gwelo. C Zimbabwe
Gwy *see* Wye
Gyandzha *see* Gäncä
Gydanskiy Poluostrov *484* *Eng.* Gyda Peninsula. Peninsula of N Russian Federation
Gympie *101* E Australia
Gyöngyös *292* NE Hungary
Győr *292* *Ger.* Raab. NW Hungary
Gyula *292* SE Hungary
Gyulafehérvár *see* Alba Iulia
Gyumri *98* *Rus.* Kumayri, *prev.* Leninakan, Aleksandropol'. W Armenia
Gyzylarbat *585* *prev.* Kizyl-Arvat. W Turkmenistan
Gyzylgaya *see* Kizyl-Kaya
Ġzira *395* N Malta

H

Ha *134* W Bhutan
Ha'ano *572* island of the Ha'apai Group, Tonga
Ha'apai Group *572* *var.* Haabai. Island group of C Tonga
Haapsalu *242* *Ger.* Hapsal. W Estonia
Haarlem *429* W Netherlands
Hab *451* river of S Pakistan
Habarana *538* C Sri Lanka
Ḩabbānīyah, Buḩayrat al *312* *var.* Hawr al Ḩabbānīyah. Lake of C Iraq
Habiganj *117* NE Bangladesh
Habomai Islands *333* disputed island group of SE Russian Federation
Ḩabshān *594* SW United Arab Emirates
Hachijō-jima *332* island to the SE of Honshū, SE Japan
Hachinohe *332* Honshū, N Japan
Hačmas *see* Xaçmaz
Hadama *see* Nazrēt
Ḩadd al Jamal *115* *var.* Ra's al Yaman. Cape of E Bahrain
Haddummati Atoll *see* Hadhdhunmathi Atoll
Hadejia *440* N Nigeria
Hadejia *440* river of N Nigeria
Ḩadera *319* C Israel
Haderslev *218* *Ger.* Hadersleben. Jylland, SW Denmark
Hadhdhunmathi Atoll *390* *var.* Haddummati Atoll, Laamu Atoll. Atoll of S Maldives
Hadhramaut *see* Ḩaḑramawt
Hadíboh *631* NE Suqutra, Yemen
Ha Đông *627* N Vietnam
Ḩaḑramawt *631* *Eng.* Hadhramaut. Mountain range of S Yemen
Haedo, Cuchilla de *611* mountain range of NW Uruguay
Haeju *351* S North Korea
Ha-erh-pin *see* Harbin
Ḩafar al Bāţin *506* N Saudi Arabia
Hafnarfjördhur *296* W Iceland
Hafnarhreppur *296* S Iceland
Hafren *see* Severn
Hagar Nish Plateau *240* plateau of N Eritrea
Hagåtña *647* *var.* Agaña. ❖ of Guam, NW Guam
Hagen *266* W Germany
Hägere Hiywet *245* *var.* Agere Hiywet, Ambo. C Ethiopia
Hagi *332* Honshū, SW Japan
Ha Giang *627* NW Vietnam
Hago, Lac *492* lake of NE Rwanda

I

İstanbul Boğazį *581* Karadeniz Boğazi, *Eng.* Bosporus. Strait connecting rmara Denizi and Black Sea
Istra *209* *Eng.* Istria. Peninsula of SE Europe
Istria *see* Istra
Itabuna *145* E Brazil
Itagüí *195* NW Colombia
Itaipú, Represa de *145, 460* reservoir of Brazil and Paraguay
Italy *322-327* officially Italian Republic, *It.* Italia, Repubblica Italiana. Country of S Europe divided into 20 admin. units (regions). ❖ Rome
Itany *see* Litani
Itassi *see* Vieille Case
Iténez *see* Guaporé
Itonamas *136* river of NE Bolivia
Itremo *382* *var.* Massif de l'Itremo. Mountain range of C Madagascar
Itsamia *198* S Mohéli, Comoros
Itsandra *198* W Grande Comore, Comoros
Ittoqqortoormiit *646* *Dan.* Scoresbysund. E Greenland
Itu Aba Island *652* island of W Spratly Islands
Ituni *286* C Guyana
Iturup *333* disputed island of Kurile Islands, SE Russian Federation
Ivakoany *382* *var.* Massif de l'Ivakoany. Mountain range of SE Madagascar
Ivalojoki *251* river of N Finland
Ivano-Frankivs'k *590* *Rus.* Ivano-Frankovsk, *prev.* Stanislav, *Pol.* Stanisławów, *Ger.* Stanislau. W Ukraine
Ivanovo *484* W Russian Federation
Ivatsevichy *122* SW Belarus
Ivindo *260* river of C Africa
Iviza *see* Eivissa
Ivoire, Côte d' *see* Ivory Coast
Ivory Coast *328* *Fr.* Côte d'Ivoire. Coastal region of S Ivory Coast
Ivory Coast *328-329* officially Republic of the Côte d'Ivoire, *Fr.* Côte d'Ivoire. Country of W Africa divided into 34 admin. units (departments). ❖ Yamoussoukro
Ivujivik *171* NE Canada
Iwakuni *332* Honshū, W Japan
Iwaki *332* Honshū, N Japan
Iwo *440* SW Nigeria
Iwŏn *351* E North Korea
Ixcán *280* river of Guatemala and Mexico
Izabal, Lago de *280* *prev.* Golfo Dulce. Lake of E Guatemala
Izhevsk *484* *prev.* Ustinov. W Russian Federation
Izkī *448* N Oman
İzmir *581* *prev.* Smyrna. W Turkey
İzmit *581* *var.* Kocaeli. NW Turkey
Izuhara *332* Tsushima, W Japan
Izumo *332* Honshū, W Japan
Izu-shotō *332* island group to the SE of Honshū, SE Japan

J

Jabal aẓ Ẓannah *594* *var.* Jebel Dhanna. W United Arab Emirates
Jabāliya *319, 320* NE Gaza Strip
Jabalpur *298* *prev.* Jubbulpore. C India
Jabat *396* *var.* Jabwot Island. island of S Marshall Islands
Jabbul, Sabkhat al *555* salt-flat of NW Syria
Jablah *555* *var.* Jeble, *Fr.* Djéblé. W Syria
Jablanica *81* mountain range of E Albania
Jablonec nad Nisou *216* *Ger.* Gablonz an der Neisse. N Czech Republic
Jaboatão *145* E Brazil
Jabwot Island *see* Jabat
Jaceel *528* *It.* Uadi Giahel. Seasonal river of NE Somalia
Jackson *603* Mississippi, SE USA
Jacksonville *603* Florida, SE USA
Jacmel *288* *var.* Jaquemel. S Haiti
Jacó *206* SW Costa Rica
Jacob *see* Nkayi
Jacobābād *451* SW Pakistan
Jadotville *see* Likasi
Jadransko More *see* Adriatic Sea
Jādū *371* NW Libya
Jaén *535* SW Spain
Jaffna *538* N Sri Lanka
Jaffna Lagoon *538* lagoon of N Sri Lanka
Jafr, Qā' al *338* *var.* El Jafr. Salt pan of S Jordan
Jägala *242* *var.* Jägala Jõgi. River of N Estonia
Jägerndorf *see* Krnov
Jagodina *see* Svetozarevo
Jaguarão *see* Yaguarón
Jailolo *see* Halmahera
Jaipur *298* *prev.* Jeypore. N India
Jaipur Hāt *117* NW Bangladesh
Jajce *140* W Bosnia & Herzegovina
Jakar *134* C Bhutan
Jakarta *304* *prev.* Djakarta, *Dut.* Batavia. ❖ of Indonesia, Java, C Indonesia
Jakobshavn *see* Ilulissat
Jakobstad *see* Pietersaari. W Finland
Jakobstadt *see* Jēkabpils
Jalal-Abad *see* Dzhalal-Abad
Jalālābād *77* E Afghanistan
Jalandhar *298* *prev.* Jullundur. N India
Jalapa *403* *var.* Jalapa Enríquez, *prev.* Xalapa. SE Mexico
Jalapa *280* C Guatemala
Jalousie *497* SW St Lucia
Jālū *371* NE Libya
Jaluit *396* island of S Marshall Islands
Jamaame *528* *It.* Giamame. S Somalia
Jamaare *440* river of NE Nigeria
Jamaica *330-331* island state of the West Indies, divided into 14 admin. units (parishes). ❖ Kingston
Jamaica Channel *288, 331* channel of the Caribbean Sea between Haiti and Jamaica
Jamālpur *117* N Bangladesh
Jambi *304* *prev.* Djambi, *var.* Telanaipura. Sumatra, W Indonesia
Jambol *see* Yambol
Jamdena *see* Yamdena, Pulau
James Bay *171* inlet of Hudson Bay, C Canada
Jamestown *652* ❖ of St Helena, N St Helena
Jamestown *see* Holetown
Jammāl *see* Jemmel
Jammerbugten *218* bay to the NW of Denmark
Jammu *298* N India
Jāmnagar *298* *prev.* Navangar. W India
Jämsä *251* S Finland
Jamshedpur *298* E India
Jamuna *117* lower course of the Brahmaputra, N Bangladesh
Jamundá *see* Nhamundá
Janakpur *427* E Nepal
Janela *176* Santo Antão, N Cape Verde
Jangijul *see* Yangiyŭl
Janīn *see* Jenin
Janina *see* Ioánnina
Janow *see* Jonava
Jantra *see* Yambol
Janzūr *371* NW Libya
Japan *332-337* country of E Asia, divided into 47 admin. units (prefectures). ❖ Tokyo
Japan, Sea of *332, 351, 485, 352* *Rus.* Yapanskoye More. Sea of Pacific Ocean, between E Asia and Japan
Jappeni *263* C Gambia
Japurá *144* *var.* Yapurá. River of Brazil
Jaquemel *see* Jacmel
Jarabacoa *226* C Dominican Republic
Jarābulus *555* *var.* Jerablus, *Fr.* Djérablous. N Syria
Jarash *338* *var.* Jerash. NW Jordan
Jarbah, Jazīrat *see* Jerba, Île de
Jardines de la Reina, Archipiélago de los *210* island group of S Cuba
Jarej District *396* district of Majuro, SE Marshall Islands
Jari *145* *var.* Jary. River of N Brazil
Jarīd, Shaṭṭ al *see* Jerid, Chott el
Jaroměř *216* NE Czech Republic
Jarqŭrghon *614* *Rus.* Dzharkurgan. SE Uzbekistan
Jars, Plain of *see* Xiangkhoang, Plateau de
Järvenpää *251* *Swe.* Träskända. S Finland
Jason Islands *645* island group of NW Falkland Islands
Jassy *see* Iaşi
Jastrzębie Zdrój *471* S Poland
Jászberény *292* NE Hungary
Jauf *see* Al Jawf
Jaunpiebalga *362* NE Latvia
Java *304* *var.* Jawa, *prev.* Djawa. Island of C Indonesia
Javari *144-145* *var.* Yavarí. River of Brazil and Peru
Java Sea *see* Jawa, Laut
Jawa *see* Java
Jawa, Laut *304* *Eng.* Java Sea. Sea of the Pacific Ocean, C Indonesia
Jawhar *528* *var.* Jowhar, *It.* Giohar. S Somalia
Jayapura *305* *prev.* Sukarnapura, *Dut.* Hollandia. Irian Jaya, E Indonesia
Jay Dairen *see* Dalian
Jaya, Puncak *305* *prev.* Puntjak Sukarno, Puntjak Carstensz. Mountain of Irian Jaya, E Indonesia
Jazā'ir, Ra's al *115* cape of SW Bahrain
Jaz Murian, Hamun-e *309* lake of SE Iran
Jazzīn *see* Jezzine
Jbaïl *364* *var.* Jubayl. W Lebanon
Jdiriya *414* NE Western Sahara
Jebba *440* W Nigeria
Jebel *see* Dzhebel
Jebel, Bahr el *see* White Nile
Jebel Dhanna *see* Jabal aẓ Ẓannah
Jeble *see* Jablah
Jedda *see* Jiddah
Jeffara Plain *371, 577* *var.* Gefara, Al Jifārah. Plain of Libya and Tunisia
Jefferson City *603* Missouri, C USA
Jega *440* NW Nigeria
Jehegnadzor *see* Yeghegnadzor
Jēkabpils *362* *Ger.* Jakobstadt. SE Latvia
Jelenia Góra *471* *Ger.* Hirschberg in Riesengebirge. SW Poland
Jelgava *362* *Ger.* Mitau. C Latvia
Jember *304* *prev.* Djember. Java, C Indonesia
Jemmel *577* *var.* Jammāl. N Tunisia
Jemo *396* island of C Marshall Islands
Jena *267* C Germany
Jendouba *577* *var.* Jundūbah. NW Tunisia
Jenin *319, 320* *var.* Janīn, *Ar.* Jinīn. N West Bank
Jenné *see* Djenné
Jennings *92* W Antigua, Antigua & Barbuda
Jenny *542* N Suriname
Jequitinhonha *145* river of E Brazil
Jerablus *see* Jarābulus
Jerada *414* NE Morocco
Jerash *see* Jarash
Jerba *see* Houmt Souk
Jerba, Île de *577* *var.* Djerba, Jazīrat Jarbah. Island of E Tunisia
Jérémie *288* SW Haiti
Jerevan *see* Yerevan
Jerez de la Frontera *534* SW Spain
Jericho *319, 320* *Heb.* Yeriḥo, *Ar.* Arīḥā. E West Bank
Jerid, Chott el *577* *var.* Shaṭṭ al Jarīd. Salt lake of SW Tunisia
Jermuk *98* *Rus.* Dzhermuk. SE Armenia
Jersey *254, 648* British Crown dependency of the English Channel. ❖ St Helier.
Jerusalem *319, 320* *Ar.* Al Quds, *Heb.* Yerushalayim. ❖ of Israel, Israel and West Bank
Jesenice *524* *Ger.* Assling. NW Slovenia
Jesselton *see* Kota Kinabalu
Jessore *117* W Bangladesh
Jesús Menéndez *210* SE Cuba
Jeta, Ilha de *284* island of W Guinea-Bissau
Jevlah *see* Yevlax
Jeypore *see* Jaipur
Jezercës, Maja e *81* *var.* Jezerce. Mountain of N Albania
Jezzine *364* *var.* Jazzīn. S Lebanon
Jhālakāti *117* S Bangladesh
Jhang *451* *var.* Jhang Sadar, Jhang Sadr. NE Pakistan
Jhelum *451* NE Pakistan
Jhelum *451* river of India and Pakistan
Jhenida *117* W Bangladesh
Jiamusi *187* *var.* Chia-mu-ssu, Kiamusze. Heilongjiang, NE China
Jiangsu *187* *var.* Kiangsu, Chiang-su. Province of E China
Jiangxi *187* *var.* Kiangsi, Chiang-hsi. Province of SE China
Jibuti *see* Djibouti
Jičín *216* N Czech Republic
Jiddah *506* *Eng.* Jedda. W Saudi Arabia
Jiddah *115* island of NW Bahrain
Jidd Ḥafş *115* *var.* Judd Ḥafş. N Bahrain
Jiftilik Post *320* E West Bank
Jiguaní *210* SE Cuba
Jihlava *216* *Ger.* Iglau. S Czech Republic
Jijel *83* *var.* Djidjel, *prev.* Djidjelli. NE Algeria
Jijiga *245* *It.* Giggiga. E Ethiopia
Jilf al Kabīr, Haḍabat al *see* Gilf Kebir Plateau
Jilib *528* *It.* Gelib. S Somalia
Jilin *187* *var.* Kirin, Chi-lin, *prev.* Yungki. Jilin, NE China
Jilin *187* *var.* Kirin, Chi-lin. Province of NE China
Jīma *245* *var.* Jimma, Ft. Gimma. SW Ethiopia
Jimaní *226* W Dominican Republic
Jinan *187* *var.* Chinan, Tsinan. Shandong, E China
Jinīn *see* Jenin
Jinja *588* S Uganda
Jinotega *436* C Nicaragua
Jinotepe *436* S Nicaragua
Jinsen *see* Inch'ŏn
Jintotlolo Channel *466* channel connecting Mindoro Strait and Visayan Sea
Jinzhou *187* *var.* Chin-chou, Chinchow, *prev.* Chinhsien. Liaoning, NE China
Jipijapa *230* W Ecuador
Jiquilisco *237* S El Salvador
Jiquilisco, Bahia de *237* bay of the Pacific Ocean to the S of El Salvador
Jirgatol *562* *Rus.* Dzhirgatal'. C Tajikistan
Jirriiban *528* *prev.* Ceel Xamurre, *It.* El Hamurre. E Somalia
Jisr ash Shughūr *555* *var.* Djisr el Choghour. NW Syria
Jiu *480* *Ger.* Schyl, *Hung.* Zsily. River of S Romania
Jiulong *see* Kowloon
Jixi *187* *var.* Chi-hsi. Heilongjiang, NE China
Jīzān *506* *var.* Qīzān. S Saudi Arabia
Jizuka *332* Kyūshū, SW Japan
Jiz', Wādī al *631* dry watercourse of E Yemen
Jizzakh *614* *Rus.* Dzhizak. SE Uzbekistan
Jleeb, Shagat Al *see* Qalīb, Shiqqat al
Jleeb al Shuyoukh *see* Qalīb ash Shuyūkh
Joal-Fadiout *510* *prev.* Joal. W Senegal
João Barrosa *176* Boa Vista, E Cape Verde
João Pessoa *145* *prev.* Paraíba. E Brazil
Jo-ch'iang *see* Ruoqiang
Joden Savanne *542* NE Suriname
Jodhpur *298* NW India
Joel's Drift *366* N Lesotho
Joensuu *251* SE Finland
Jõgeva *242* *Ger.* Laisholm. C Estonia
Jogjakarta *see* Yogyakarta
Johannesburg *531* Pretoria-Witwatersrand-Vereeniging, NE South Africa

John o'Groats *597* N Scotland, UK
Johnsons Point *92* SW Antigua, Antigua & Barbuda
Johnson, Rapides *203* rapids of Dem. Rep. Congo and Zambia
Johnston Atoll *648* unincorporated territory of the USA, Pacific Ocean
Johnston Island *648* island of S Johnston Atoll
Johor Bahru *386* SE Peninsular Malaysia
Johore Strait *521* strait connecting Strait of Malacca and South China Sea
Joinville *145* *var.* Joinvile. S Brazil
Jolo Group *466* island group of Sulu Archipelago, SW Philippines
Jolo Island *466* island of Jolo Group, SW Philippines
Jomsom *427* W Nepal
Jona *550* NE Switzerland
Jonava *376* *Ger.* Janow. C Lithuania
Jones Point *644* headland on the W coast of Christmas Island
Jonglei Canal *540* canal of S Sudan
Jönköping *547* S Sweden
Jonquière *171* SE Canada
Jordan *319, 320, 338* *Ar.* Urdunn, *Heb.* HaYarden. River of SW Asia
Jordan *338-339* officially Hashemite Kingdom of Jordan, *Ar.* Al Urdunn. Country of SW Asia divided into 8 admin. units (governorates). ❖ Amman
Jos *440* C Nigeria
José Batlle y Ordóñez *611* C Uruguay
José E. Bisanó *226* N Dominican Republic
José Pedro Varela *611* SE Uruguay
Joseph Bonaparte Gulf *101* gulf of Timor Sea on the coast of NW Australia
Jos Plateau *440* plateau of C Nigeria
Jos Sudarso *see* Yos Sudarso, Pulau
Jost Van Dyke *643* island of W British Virgin Islands
Jotunheimen *444* mountains of SW Norway
Joûnié *364* *var.* Jūniyah, Juniye. W Lebanon
Jovellanos *210* NW Cuba
Jozini Dam *544* reservoir of South Africa and Swaziland
Jsahaya *332* Kyūshū, SW Japan
Juan Fernández Islands *183* island group of W Chile
Juan L. Lacaze *611* *prev.* Sauce. SW Uruguay
Juarzon *368* *var.* Juazohn. SE Liberia
Juazeiro do Norte *145* E Brazil
Juazohn *see* Juarzon
Juba *245, 528* *Som.* Jubba, *var.* Ganaane, *Amh.* Genale Wenz, *It.* Guiba. River of Ethiopia and Somalia
Juba *540* *var.* Jūbā. S Sudan
Jubany *90* Argentinian research station of Antarctic Peninsula, Antarctica
Jubba *see* Juba
Jubbulpore *see* Jabalpur
Júcar *535* river of C Spain
Juclà, Estany de *86* lake of NE Andorra
Judd Ḩafş *see* Jidd Ḩafş
Judenburg *106* C Austria
Juigalpa *436* S Nicaragua
Juishui *559* E Taiwan
Juiz de Fora *145* SE Brazil
Jujuy *see* San Salvador de Jujuy
Juliaca *463* SE Peru
Julian Alps *524* *Ger.* Julische Alpen, *It.* Alpi Giulie, *Slvn.* Julijske Alpe. Mountains of NW Slovenia
Juliana Top *542* mountain of C Suriname
Julianehåb *see* Qaqortoq
Jullundur *see* Jalandhar
Jumayrah *594* *var.* Jumeirah. NE United Arab Emirates
Jumla *427* E Nepal
Jumna *see* Yamuna
Jundūbah *see* Jendouba
Juneau *602* Alaska, USA
Jungbunzlau *see* Mladá Boleslav
Junín *95* E Argentina
Junk Bay *see* Tseung Kwan O
Juntas *206* W Costa Rica
Junten *see* Sunch'ŏn
Juozapinės Kalnas *376* mountain of SE Lithuania
Jupiá, Represa de *145* reservoir of S Brazil
Jura *597* island of Inner Hebrides, W Scotland, UK
Jura *255, 550* *var.* Jura Mountains. Mountain range of France and Switzerland
Juraguá *210* C Cuba
Jura Mountains *see* Jura
Jurbarkas *376* *Ger.* Jurburg, *var.* Georgenburg. W Lithuania
Jūrmala *362* NW Latvia
Jurong Lake *521* lake of W Singapore
Jurong Town *521* W Singapore
Juruá *144* river of Brazil and Peru
Juruena *144* river of W Brazil
Jutiapa *280* S Guatemala
Juticalpa *290* C Honduras
Jutland *see* Jylland
Juventud, Isla de la *210* *var.* Isla de Pinos, *Eng.* Isle of Pines. Island of W Cuba
Južna Morava *512* river of SE Serbia, Serbia & Montenegro (Yugo.)
Jwaneng *142* S Botswana
Jylland *218* *Eng.* Jutland. Island of W Denmark
Jyrgalan *see* Dzhergalan
Jyväskylä *251* S Finland

K

K2 *451* *Eng.* Mount Godwin Austen. Mountain of China and Pakistan
Kaabong *588* NE Uganda
Kaafu Atoll *see* Male' Atoll
Kaaimanston *542* NW Suriname
Kaakhka *585* *var.* Kaachka, Kaka. S Turkmenistan
Kaala *see* Caála
Kaapstad *see* Cape Town
Kaba *518* *var.* Little Scarcies. River of Guinea and Sierra Leone
Kabakama *263* E Gambia
Kabala *518* N Sierra Leone
Kabale *588* SW Uganda
Kabalega Falls *see* Murchison Falls
Kabara *248* *prev.* Kambara. Island of the Lau Group, E Fiji
Kabardino-Balkarskaya, Respublika *484* autonomous republic of SW Russian Federation
Kabarnet *344* W Kenya
Kabarole *588* W Uganda
Kabaya *492* NW Rwanda
Kaberamaido *588* C Uganda
Kabinda *203* SE Dem. Rep. Congo
Kabinda *see* Cabinda
Kābol *see* Kābul
Kabompo *635* W Zambia
Kabompo *635* river of W Zambia
Kabou *571* N Togo
Kābul *77* *Per.* Kābol. ❖ of Afghanistan, E Afghanistan
Kabul *77, 451* river of Afghanistan and Pakistan
Kabwe *635* C Zambia
Kabye Plateau *571* plateau of E Togo
Kachch, Gulf of *298* *var.* Gulf of Cutch, Gulf of Kutch. Gulf of Arabian Sea to the W of India
Kachhi *451* lowland region of C Pakistan
Kadan Island *159* *prev.* King I. Island of S Burma
Kadavu *248* *prev.* Kandavu. Island to the S of Viti Levu, SW Fiji
Kadavu Passage *248* channel of the Pacific Ocean between Kadavu and Vitu Levu, Fiji
Kadéï *168, 179* river of Cameroon and Central African Republic
Kadoma *636* *prev.* Gatooma. C Zimbabwe
Kadugli *540* *var.* Kāduqlī. S Sudan
Kaduha *492* SW Rwanda
Kaduna *440* C Nigeria
Kaduna *440* river of N Nigeria
Kadzharan *see* K'ajaran
Kadzhi-Say *359* *Kir.* Kajisay. NE Kyrgyzstan
Kaédi *398* S Mauritania
Kaélé *168* N Cameroon
Kaesŏng *351* S North Korea
Kaewieng *see* Kavieng
Kafan *see* Kapan
Kāfar Jar Ghar *77* mountain range of C Afghanistan
Kaffrine *510* C Senegal
Kafr el Dauwâr *232* *var.* Kafr ad Dawwār. N Egypt
Kafr el Sheikh *232* *var.* Kafr ash Shaykh. N Egypt
Kafu *588* *var.* Kafo. River of W Uganda
Kafue *635* river of C Zambia
Kafue *635* SE Zambia
Kaga Bandoro *179* *prev.* Fort-Crampel. C Central African Republic
Kagan *see* Kogon
Kaganovichabad *see* Kolkhozobod
Kagera *see* Akagera
Kagi *see* Chiai
Kâgıthane *581* NW Turkey
Kagoshima *332* Kyūshū, SW Japan
Kagul *see* Cahul
Kahama *564* NW Tanzania
Kahayan *304* river of Borneo, C Indonesia
Kahnple *368* NE Liberia
Kahnwia *368* SE Liberia
Kahramanmaraş *581* *var.* Marash, Maraş. S Turkey
Kaiaf *263* S Gambia
Kaieteur Falls *286* waterfall of C Guyana
Kaifeng *187* Henan, C China
Kai, Kepulauan *304* *prev.* Kei Islands. Island group of Maluku, E Indonesia
Kaikoura *433* NE South Island, New Zealand
Kailahun *518* E Sierra Leone
Kainan *332* Honshū, C Japan
Kainji Reservoir *440* reservoir of W Nigeria
Kaipara Harbour *433* harbour of NW North Island, New Zealand
Kairouan *577* *var.* Al Qayrawān. N Tunisia
Kaiserslautern *266* SW Germany
Kaitaia *433* NW North Island, New Zealand
Kajaani *251* *Swe.* Kajana. C Finland
Kajana *see* Kajaani
Kajang *386* W Peninsular Malaysia
K'ajaran *98* *Rus.* Kadzharan, *var.* Kadžaran. SE Armenia
Kajisay *see* Kadzhi-Say
Kaka *see* Kaakhaa
Kakamega *344* W Kenya
Kakata *368* C Liberia
Kakhovs'ke Vodokhovyshche *590* *Rus.* Kakhovskoye Vodokhranilische. Reservoir of SE Ukraine
Kakia *see* Khakhea
Kakogawa *332* Honshū, C Japan
Kakshaal-Too, Khrebet *see* Kokshaal-Tau
Kalaa Kebira *577* *var.* Al Qal'ah al Kubrá. N Tunisia
Kalabo *635* W Zambia
Kalahari Desert *142, 423, 531* desert region of southern Africa
Kalaikhum *see* Qal'aikhum
Kalai-Mor *585* *Turkm.* Galaymor. SE Turkmenistan
Kalamáki *275* *prev.* Kalmákion. SE Greece
Kalamariá *275* *prev.* Kalamaria. N Greece
Kalámata *275* *prev.* Kalámai. S Greece
Kalandula *see* Calandula
Kalang *see* Kallang
Kalanshiyū, Sarīr *371* *var.* Calanscio Sand Sea. Desert region of E Libya
Kalarash *see* Călăraşi
Kalasin *567* *var.* Muang Kalasin. NE Thailand
Kalāt *77* *var.* Qalāt. S Afghanistan
Kālat *451* *var.* Kelat. W Pakistan
Kalbā *594* *var.* Kalba, NE United Arab Emirates
Kaldakvísl *296* river of C Iceland
Kalemie *203* *prev.* Albertville. SE Dem. Rep. Congo
Kalgan *see* Zhangjiakou
Kalgoorlie *100* SW Australia
Kali Gandaki *427* river of C Nepal
Kalima *203* E Dem. Rep. Congo
Kalimantan *304* *Eng.* Indonesian Borneo. Region of Borneo, administered by Indonesia
Kalinin *585* N Turkmenistan
Kalininabad *see* Kalininobod
Kaliningrad *484* W Russian Federation
Kalinino *see* Tashir
Kalininobod *562* *Rus.* Kalininabad. SW Tajikistan
Kalinkavichy *122* *Rus.* Kalinkovichi. SE Belarus
Kaliro *588* SE Uganda
Kalisz *471* *Ger.* Kalisch. C Poland
Kalixälv *547* river of NE Sweden
Kalkandelen *see* Tetovo
Kalkfeld *423* NW Namibia
Kallang *521* *var.* Kalang. River of C Singapore
Kallaste *242* *Ger.* Krasnogor. E Estonia
Kallavesi *251* lake of SE Finland
Kalmar *547* S Sweden
Kalmykiya, Respublika *484* autonomous republic of SW Russian Federation
Kalomo *635* S Zambia
Kalpeni Island *298* island of Lakshadweep, SW India
Kalsoy *644* *var.* Kalsø. Island of N Faeroe Islands
Kalu Ganga *538* river of S Sri Lanka
Kalulushi *635* C Zambia
Kalundborg *218* Sjælland, C Denmark
Kalungwishi *635* river of N Zambia
Kalutara *538* SW Sri Lanka
Kaluwawa *see* Fergusson Island
Kalyān *298* W India
Kálymnas *275* island of SE Greece
Kama *203* E Dem. Rep. Congo
Kamai *180* N Chad
Kamaishi *333* Honshū, N Japan
Kamakwie *518* NW Sierra Leone
Kamālia *451* NE Pakistan
Kamanjab *423* NW Namibia
Kamarān *631* island of W Yemen
Kamarang *286* W Guyana
Kamativi *636* W Zimbabwe
Kambar *451* *var.* Qambar. SW Pakistan
Kambing, Pulau *307* island of E Timor
Kambara *see* Kabara
Kamchatka, Poluostrov *485* *Eng.* Kamchatka Peninsula. Peninsula of NE Russian Federation
Kamchiya *152* *var.* Kamčija. River of E Bulgaria
Kamenets-Podol'sk *see* Kam"yanets'-Podil's'kyy
Kamenets-Podol'skiy *see* Kam"yanets'-Podil's'kyy
Kamenica *381* NE FYR Macedonia
Kamenskoye *see* Dniprodzerzhyns'k
Kamina *203* S Dem. Rep. Congo
Kamishli *see* Al Qāmishlī
Kamloops *170* SW Canada
Kammersee *see* Attersee
Kamnik *524* *Ger.* Stein. C Slovenia
Kamo *98* C Armenia
Kamp *106* river of N Austria
Kampala *588* ❖ of Uganda, S Uganda
Kampar *386* W Peninsular Malaysia
Kampar *304* river of Sumatra, W Indonesia
Kampo *see* Ntem
Kampong Batang Duri *150* NE Brunei
Kampong Benutan *150* C Brunei
Kampong Bukit Sawat *150* C Brunei
Kampong Bunut *150* N Brunei
Kâmpóng Cham *165* *prev.* Kompong Cham. S Cambodia
Kâmpóng Chhnăng *165* C Cambodia
Kampong Jerudong *150* N Brunei

Kâmpóng Khleăng *165 prev.* Kompong Kleang. NW Cambodia
Kampong Kuala Abang *150* C Brunei
Kampong Kuala Balai *150* SW Brunei
Kampong Labi *150* S Brunei
Kampong Labu *150* NE Brunei
Kampong Lumut *150* W Brunei
Kampong Paring *150* N Brunei
Kampong Parit *150* N Brunei
Kâmpóng Saôm *165 var.* Kompong Som, *prev.* Sihanoukville. SW Cambodia
Kâmpóng Saôm, Chhâk *165 Fr.* Baie de Kompong Som. Bay of the Gulf of Thailand on the SW coast of Cambodia
Kâmpóng Spoe *165 prev.* Kompong Speu. S Cambodia
Kampong Sukang *150* S Brunei
Kampong Tanajor *150* C Brunei
Kampong Teraja *150* S Brunei
Kâmpóng Thum *165 prev.* Kompong Thom. C Cambodia
Kâmpôt *165* S Cambodia
Kampuchea *see* Cambodia
Kamsar *283* W Guinea
Kam"yanets'-Podil's'kyy *590 Rus.* Kamenets-Podol'skiy, *prev.* Kamenets-Podol'sk. W Ukraine
Kanacea *248 prev.* Kanathea. Taveuni, N Fiji
Kanacea *248* island of the Lau Group, E Fiji
Kananga *203 prev.* Luluabourg. SW Dem. Rep. Congo
Kanazawa *332* Honshū, C Japan
Kanazi *492* SE Rwanda
Kandahār *77 var.* Qandahār. S Afghanistan
Kandavu *see* Kadavu
Kandé *571* NE Togo
Kandi *132* N Benin
Kandrian *458* New Britain, E Papua New Guinea
Kandy *538* C Sri Lanka
Kaneohe *602* Oahu, Hawaii, USA
Kanevskoye Vodokhranilische *see* Kanivs'ke Vodoskhovyshche
Kang *142* C Botswana
Kangar *386* NW Peninsular Malaysia
Kangaroo Island *101* island of S Australia
Kangaruma *286* C Guyana
Kangchenjunga *299 var.* Kanchenjunga. Mountain of NE India
Kangerlussuaq *646 Dan.* Søndre Strømfjord. SW Greenland
Kanggye *351* N North Korea
Kanghwa-do *352 Jap.* Kōka-tō. Island of NW South Korea
Kangnŭng *352 Jap.* Kōryō. NE South Korea
Kango *260* NW Gabon
Kanibadam *562* N Tajikistan
Kani, Baie de *649 var.* Kani Bay. Bay of the Mozambique Channel on the SW coast of Mayotte
Kanivs'ke Vodoskhovyshche *590 Rus.* Kanevskoye Vodokhranilische. Reservoir of C Ukraine
Kanjiža *512 prev.* Stara Kanjiža, *Ger.* Altkanischa, *Hung.* Magyarkanizsa, Ókanizsa. N Serbia, Serbia & Montenegro (Yugo.)
Kankan *283* E Guinea
Kankesanturai *538* N Sri Lanka
Kan-kō *see* Han
Kankossa *398* S Mauritania
Kanli Dere *see* Pedhieos
Kanmaw Island *159 var.* Kettharin I, Kisseraing. Island of S Burma
Kano *440* N Nigeria
Kanombe *492* C Rwanda
Kanoya *332* Kyūshū, SW Japan
Kānpur *298 prev.* Cawnpore. N India
Kanra-san *see* Halla-san
Kansas *603* state of C USA
Kansas City *603* Kansas, C USA
Kant *359* C Kyrgyzstan
Kantipur *see* Kathmandu
Kanton *348 var.* Abariringa, Canton I, *prev.* Mary I. Island of the Phoenix Is, C Kiribati
Kanyaru *see* Akanyaru
Kanye *142* S Botswana
Kao *572* island of W Tonga
Kao *572* mountain of Kao, Tonga
Kaôh Nhêk *165* E Cambodia
Kaohsiung *559 var.* Kaohiung, *Jap.* Takao. SW Taiwan
Kaolack *510 var.* Kaolak. W Senegal
Kaolak *see* Kaolack
Kaolan *see* Lanzhou
Kaoma *635* W Zambia
Kaop'ing Hsi *559* river of C Taiwan
Kapan *98 var.* Ghap'an. *Rus.* Kafan. SE Armenia
Kapchorwa *588* E Uganda
Kapenguria *344* W Kenya
Kapfenberg *106* C Austria
Kapingamarangi *406* atoll of S Micronesia
Kapiri Mposhi *635* C Zambia
Kapiti Island *433* island to the S of North Island, New Zealand
Kapka, Massif du *180* mountains of E Chad
Kaposvár *292* SW Hungary
Kaproncza *see* Koprivnica
Kapsabet *344* W Kenya
Kapsukas *see* Marijampolė
Kapuas *304 prev.* Kapoeas. River of Borneo, C Indonesia
Kapuas Mountains *304, 386 Ind.* Pegunungan Kapuas Hulu. Mountain range of Indonesia and Malaysia
Kara *571 var.* Lama-Kara. NE Togo
Karaba *492* SW Rwanda
Kara-Balta *359* NW Kyrgyzstan
Karabil', Vozvyshennost' *585* region of SE Turkmenistan
Kara-Bogaz-Gol, Zaliv *585* NW Turkmenistan
Kara-Bogaz-Gol, Proliv *585 Turkm.* Garabogazköl Bogazy. Strait of the Caspian Sea, on the NW coast of Turkmenistan
Karabük *581* N Turkey
Karachayevo-Cherkesskaya SSR *484* autonomous republic of SW Russian Federation
Karāchi *451* S Pakistan
Karadeniz *see* Black Sea
Karadeniz Boğazi *see* İstanbul Boğazi
Karaferiye *see* Véroia
Karaganda *340 Kaz.* Qaraghandy. C Kazakhstan
Karaitivu *538* N Sri Lanka
Karaj *309* NW Iran
Karak *see* Al Karak
Kara-Kala *585 var.* Garrygala. SW Turkmenistan
Karaklin *see* Vanadzar
Karakol *359 var.* Karakolka. E Kyrgyzstan
Karakol *359 prev.* Przheval'sk, *var.* Prževalsk. NE Kyrgyzstan
Karakoram Range *298, 451* mountain range of C Asia
Karakose *581* NE Turkey
Kara-Kul' *359 Kir.* Kara-Köl. W Kyrgyzstan
Karakul' *see* Qorakŭl
Karakul' *see* Qarokŭl
Karakul', Ozero *see* Qarokŭl
Karakumskiy Kanal *585 Turkm.* Garagum Kanaly. Canal of SE Turkmenistan
Karakumy *585 Eng.* Kara Kum, *Turkm.* Garagum, *var.* Qara Qum. Desert region of C Turkmenistan
Karaman *581* S Turkey
Karamay *186 var.* Karamai, *Chin.* K'o-la-ma-i. Xinjiang Uygur Zizhiqu, NW China
Karamea Bight *433* area of the Tasman Sea, on the NW coast of South Island, New Zealand
Kara-Say *359* E Kyrgyzstan
Karasburg *423* S Namibia
Kara Sea *see* Karskoye More
Karasjok *444* NE Norway
Kara Su *see* Mesta, Néstos
Karatau *340 Kaz.* Qaratau. S Kazakhstan
Karatsu *332* Kyūshū, SW Japan
Karavastasë, Laguna e *81 var.* Kënet' e Karavastas, Kravasta Lagoon. Lagoon of W Albania
Karawang *304 prev.* Krawang. Java, C Indonesia
Karawanken *106 Slvn.* Karavanke. Mountain range of C Europe
Karbalā' *312 var.* Kerbala. C Iraq
Kardítsa *275* C Greece
Kärdla *242 Ger.* Kertel. Hiiumaa, Estonia
Kareliya, Respublika *484* autonomous republic of NW Russian Federation
Karen *see* Hualien
Kari *see* Chiali
Kariba *636* N Zimbabwe
Kariba Dam *636* dam at NE end of Lake Kariba, on Zambezi river, NW Zimbabwe
Kariba, Lake *635, 636* reservoir of Zambia and Zimbabwe
Karibib *423* C Namibia
Karimama *132* N Benin
Karimata, Selat *304* strait connecting Laut Jawa and the South China Sea, E Indonesia
Karisimbi, Volcan *492 var.* Mount Karisimbi. Mountain of Rwanda and Dem. Rep. Congo
Karkaralinsk *340* E Kazakhstan
Karkar Island *458* island of NE Papua New Guinea
Karkinits'ka Zatoka *590 Rus.* Karkinitskiy Zaliv. Gulf of the Black Sea, S Ukraine
Karleby *see* Kokkola
Karl-Marx-Stadt *see* Chemnitz
Karlö *see* Hailuoto
Karlovac *209 Ger.* Karlstadt, *Hung.* Károlyváros. N Croatia
Karlovo *152 prev.* Levskigrad. C Bulgaria
Karlovy Vary *216 Ger.* Karlsbad, *var.* Carlsbad. W Czech Republic
Karlsbad *see* Karlovy Vary
Karlskrona *547* S Sweden
Karlsruhe *266 var.* Carlsruhe. SW Germany
Karlstad *547* SW Sweden
Karlstadt *see* Karlovac
Karmi 'él *319* N Israel
Karnali *427 var.* Kauriala. River of W Nepal
Karnobat *152* E Bulgaria
Karoi *636* N Zimbabwe
Károlyváros *see* Karlovac
Karonga *385* N Malawi
Karonje, Mount *162* mountain of W Burundi
Karpasia *215 var.* Karpas Peninsula. Peninsular of NE Cyprus
Karpaten *see* Carpathian Mountains
Kárpathos *275* island of SE Greece
Karpaty *see* Carpathian Mountains
Karrānah *115* N Bahrain
Kars *581* NE Turkey
Karshi *see* Qarshi
Karskoye More *485 Eng.* Kara Sea. Sea of Arctic Ocean, bordering N Russian Federation
Karumba *101* NE Australia
Kartung *263* W Gambia
Kārūn *309* river of W Iran
Karungu Bay *344* bay of Lake Victoria, to the SW of Kenya
Karuzi *162* C Burundi
Karviná *216 Ger.* Karwin. E Czech Republic
Karzakkān *115* NW Bahrain
Kas *581* SW Turkey
Kasai *88, 203 var.* Kassai, Cassai. River of Angola and Dem. Rep. Congo
Kasama *635* N Zambia
Kasan *see* Koson
Kasane *142* N Botswana
Kasari *242* river of W Estonia
Kasbegi *see* Qazbegi
Kaschau *see* Korice
Kasese *588* SW Uganda
Kashaf Rūd *309* river of NE Iran
Kāshān *309* NW Iran
Kashgar *see* Kashi
Kashi *186 Uigh.* Kashgar. Xinjiang Uygur Zizhiqu, NW China
Kashiwa *332* Honshū, SE Japan
Kashiwazaki *332* Honshū, N Japan
Käsmark *see* Kežmarok
Kasongo *203* E Dem. Rep. Congo
Kaspi *264* C Georgia
Kaspiyskoye More *see* Caspian Sea
Kaspiy Tengizi *see* Caspian Sea
Kassa *see* Košice
Kassai *see* Kasai
Kassala *540 var.* Kassalā, Kasala. E Sudan
Kassándra *275* peninsula of NE Greece
Kassel *266 prev.* Cassel. C Germany
Kasserine *577 var.* Al-Qaṣrayn. W Tunisia
Kassikaityu *286* river of S Guyana
Kastamonu *581* N Turkey
Kastsyukovichy *122* E Belarus
Kasugai *332* Honshū, C Japan
Kasulu *564* W Tanzania
Kasumiga-ura *332* lake of Honshū, SE Japan
Kasungu *385* C Malawi
Kasupe *see* Machinga
Katchang *263* C Gambia
Kateríni *275* N Greece
Katete *635* E Zambia
Katha *159* N Burma
Katherina, Gebel *232 var.* Jabal Katrīnah, *Eng.* Mt. Catherine. Mountain of NE Egypt
Katherine *101* N Australia
Kathmandu *427 prev.* Kantipur. ❖ of Nepal, C Nepal
Kati *392* SW Mali
Katima Mulilo *423 var.* Ngweze. NE Namibia
Katiola *328* C Ivory Coast
Katonga *588* river of SW Uganda
Katowice *471 Ger.* Kattowitz. S Poland
Katrīnah, Jabal *see* Katherina, Gebel
Katsina *440* N Nigeria
Kattaqŭrghon *614 Rus.* Kattakurgan. SE Uzbekistan
Kattegat *218, 547* strait between Denmark and Sweden
Katumbi *385* NW Malawi
Katwijk aan Zee *429* W Netherlands
Kauai *602* island of Hawaii, USA, C Pacific
Kaufbeuren *267* S Germany
Kaunas *376 Ger.* Kauen, *Pol.* Kowno, *Rus.* Kovno. C Lithuania
Kauno Marios *376* reservoir of S Lithuania
Kauriala *see* Karnali
Kaushany *see* Căuşeni
Kau-Ur *263* N Gambia
Kavadarci *381* S FYR Macedonia
Kavajë *81 It.* Cavaia. W Albania
Kavála *275 prev.* Kaválla. NE Greece
Kavango *see* Cubango
Kavaratti Island *298* island of Lakshadweep, SW India
Kavengo *see* Cubango
Kavieng *458 var.* Kaewieng. New Ireland I, Papua New Guinea
Kavīr, Dasht-e *309* desert region of N Iran
Kavirondo Gulf *see* Winam Gulf
Kavkaz *see* Caucasus
Kawagoe *332* Honshū, SE Japan
Kawambwa *635* N Zambia
Kawasaki *332* Honshū, SE Japan
Kaya *157* C Burkina
Kayagangiri, Mont *179* mountain of W Central African Republic
Kayan *304* river of Borneo, C Indonesia
Kayan *159* S Burma
Kayangel Islands *455* island group of N Palau
Kayanza *162* N Burundi
Kayes *392* W Mali
Kayl *378* S Luxembourg
Kayogoro *162* S Burundi
Kayokwe *162* C Burundi
Kayrakkumskoye Vodokhranilishche *see* Qayrokkum, Obanbori
Kayseri *581* C Turkey

Kayts *538* island of N Sri Lanka
Kazakh *see* Qazax
Kazakhskiy Melkosopochnik *340 Eng.* Kazakh Uplands. Uplands of C Kazakhstan
Kazakhstan *340-343* officially Republic of Kazakhstan, *Kaz.* Qazaqstan, *prev.* Kazakh SSR. *Rus.* Kazakhskay SSR. Country of C Asia divided into 19 admin. units (provinces). ❖ Astana
Kazakh Uplands *see* Kazakhskiy Melkosopochnik
Kazan' *484* W Russian Federation
Kazandzhik *see* Gazandzhyk
Kazanlŭk *152 var.* Kazanlăk, Kazanlik. C Bulgaria
Kazan-rettō *332 Eng.* Volcano Islands. Island group to the SE of Honshū, SE Japan
Kazarman *359* C Kyrgyzstan
Kazbek *264* mountain of N Georgia
Kazi Magomed *see* Qazimämmäd
Kazincbarcika *292* NE Hungary
Kazvin *see* Qazvin
Kéa *275* island of SE Greece
Kéamu *see* Aneityum
Kebili *577 var.* Qibilī. C Tunisia
Kebnekaise *547* mountain of N Sweden
Kecskemét *292* C Hungary
Kėdainiai *376* C Lithuania
Kediet ej Jill *398 var.* Kediet Ijill, Kédia d'Idjil. Mountain of NW Mauritania
Kediri *304* Java, C Indonesia
Kédougou *510* SE Senegal
Keeling Islands *see* Cocos Islands
Keelung *see* Chilung
Keetmanshoop *423* S Namibia
Kefallinía *275 Eng.* Cephalonia. Island of W Greece
Kefar Sava *319* C Israel
Keflavík *296 var.* Iceflavik. W Iceland
Kegalla *538 var.* Kegalle. C Sri Lanka
Kegel *see* Keila
Kei Islands *see* Kai, Kepulauan
Keijō *see* Seoul
Keila *242 Ger.* Kegel. NW Estonia
Keila *242 var.* Keila Jōgi. River of NW Estonia
Keishū *see* Kyŏngju
Kéita *180 var.* Doka. River of S Chad
Keïta *439* SW Niger
Keitele *251* lake of C Finland
Këk-Art *359 prev.* Alaykel'. SW Kyrgyzstan
Kékes *292* mountain of N Hungary
Kelang *386 var.* Klang, *prev.* Port Swettenham. W Peninsular Malaysia
Kelantan *386* river of N Peninsular Malaysia
Kelbia, Sebkhet *577 var.* Sabkhat Kalbīyah. Salt flat of NE Tunisia
Këlcyrë *81 var.* Këlcyra. S Albania
Kelifskiy Uzboy *585* region of SE Turkmenistan
Kéllé *200* W Congo
Kelmė *376* NW Lithuania
Kélo *180* SW Chad
Kelowna *170* SW Canada
Keluang *386 var.* Kluang. SE Peninsular Malaysia
Kembolcha *245 var.* Kombolcha. N Ethiopia
Kemerovo *485 prev.* Shcheglovsk. C Russian Federation
Kemi *251* NW Finland
Kemijärvi *251* N Finland
Kemijoki *251* river of NW Finland
Kemin *359 prev.* Bystrovka. N Kyrgyzstan
Kemiö *see* Kimito
Kemmuna *395* island of NW Malta
Kemmunett *395* island of NW Malta
Kempen *127 Fr.* Campine, *Ger.* Kempenland. Heathland of NE Belgium
Kempten *267* S Germany
Kenema *518* SE Sierra Leone
Këneurgench *585 prev.* Kunya-Urgench, Kunja-Urgenč, *Turkm.* Köneür gench. N Turkmenistan
Kénitra *414 prev.* Port Lyautey. NW Morocco
Kenmare *316* SW Ireland
Kentau *340* S Kazakhstan
Kentucky *603* state of C USA
Kenya *344-347* officially Republic of Kenya. Country of E Africa divided into 7 admin. units (provinces). ❖ Nairobi
Kenya, Mount *see* Kirinyaga
Keppel Harbour *521* harbour, S Singapore
Keppel Island *see* Niuatoputapu
Kerava *251 Swe.* Kervo. S Finland
Kerch *590 Rus.* Kerch'. SE Ukraine
Kerema *458* S Papua New Guinea
Keren *240 var.* Cheren. C Eritrea
Kerewan *263* W Gambia
Kericho *344* W Kenya
Kerio *344* river of W Kenya
Kerkenah, Îles *577 var.* Kerkenna Islands, *Ar.* Juzur Qarqannah. Island group of E Tunisia
Kerki *585* SE Turkmenistan
Kerkrade *429* S Netherlands
Kérkyra *275 prev.* Kérkira, *Eng.* Corfu. Island of W Greece
Kérkyra *275 Eng.* Corfu, *prev.* Kérkira. W Greece
Kermān *309 var.* Kirman. SE Iran
Kermānshāh *see* Bākhtarān
Kerora *240* N Eritrea
Kérouané *283* SE Guinea
Kertel *see* Kärdla
Kerulen *412 var.* Herlen Gol. River of China and Mongolia
Kervo *see* Kerava
Keryneia *see* Girne
Kesen'-numa *333* Honshū, N Japan
Késmárk *see* Kežmarok
Kesra *577 var.* Kisrah. NW Tunisia
Keta *272* SE Ghana
Ketchikan *602* Alaska, USA
Kete-Krachi *272 var.* Kete Krakye. E Ghana
Kétou *132* SE Benin
Kettharin Island *see* Kanmaw Island
Keur Massène *398* SW Mauritania
Kévé *571* SW Togo
Kew *653* North Caicos, NW Turks and Caicos Islands
Kežmarok *523 Ger.* Käsmark, *Hung.* Késmárk. NE Slovakia
Khabarovsk *485* SE Russian Federation
Khabura *see* Al Khaburah
Khachmas *see* Xaçmaz
Khairpur *451* S Pakistan
Khakasiya, Respublika *485* autonomous republic of C Russian Federation
Khakassk *see* Abakan
Khakhea *142 var.* Kakia. S Botswana
Khalándrion *see* Chalándri
Khalkidhikí *see* Chalkidikí
Khalkís *see* Chalkída
Khalūf *448 var.* Al Khaluf. S Oman
Khambhat, Gulf of *298 Eng.* Gulf of Cambay. Gulf of Arabian Sea to the W of India
Khamir *631 var.* Khamr. W Yemen
Khamīs Mushayṭ *506* S Saudi Arabia
Khānābād *77* NE Afghanistan
Khānaqīn *312* E Iraq
Khānewāl *451* NE Pakistan
Khanh Hung *see* Soc Trăng
Khanka, Lake *187, 485 var.* Lake Hanka, *Rus.* Ozero Khanka, *Chin.* Xingkai Hu, Hsing-K'ai Hu. Lake of China and Russian Federation
Khanka, Ozero *see* Khanka, Lake
Khankendy *see* Xankändi
Khānpur *451* SE Pakistan
Khanty-Mansiysk *484 prev.* Ostyako-Voguls'k. C Russian Federation
Khān Yūnis *319, 320 Ar.* Khan Yunus. Gaza Strip
Kharâb, Ghoubbet el *222* bay at the head of Golfe de Tadjoura, E of Djibouti
Kharanah *see* Al Kir'ānah
Khāriān *451* NE Pakistan
Kharîṭ, Wâdi el *232 var.* Wādī al Kharīṭ. Dry watercourse SE Egypt
Kharkiv *590 Rus.* Khar'kov. NE Ukraine
Kharmanli *152 var.* Harmanli. S Bulgaria
Khartoum *540 var.* Al Khurṭūm. ❖ of Sudan, C Sudan
Khartoum North *540 var.* Al Khurṭūm al Baḥrī. E Sudan
Khasab *see* Al Khaṣab
Khāsh Rūd *77* river of W Afghanistan
Khashuri *264* C Georgia
Khaskovo *152 var.* Haskovo. S Bulgaria
Khatt *see* Al Khaṭṭ
Khawr al Bazm *594 var.* Khor al Bizm. Inlet of the The Gulf, on the coast of United Arab Emirates
Khawr al 'Udayd *479 var.* Khor al Udeid. Inlet of the The Gulf on the coast of SE Qatar
Duwayhin, Khawr *594* inlet of the The Gulf, on the coast of United Arab Emirates
Khawr Fakkān *594 var.* Khor Fakkan. NE United Arab Emirates
Khaydarkan *359 var.* Khaydarken, Hajdarken. SW Kyrgyzstan
Khazar, Baḥr-e *see* Caspian Sea
Khazar, Daryā-ye *see* Caspian Sea
Khenchela *83 var.* Khenchla. NE Algeria
Khénifra *414* C Morocco
Kherson *590 var.* Cherson. S Ukraine
Khezqazghan *see* Zhezkazgan
Khíos *see* Chíos
Khiwa *614 Uzb.* Khiwa. W Uzbekistan
Khmel 'nyts'kyy *590 Rus.* Khmel'nitskiy, *prev.* Proskurov. W Ukraine
Khodzhent *see* Khujand
Khodzheyli *see* Khujayli
Khoi *see* Khvoy
Khojend *see* Khujand
Kholm *77* N Afghanistan
Khomeynīshahr *309 prev.* Homāyūnshahr. W Iran
Khoms *see* Al Khums
Khong Sedone *see* Muang Khôngxédôn
Khon Kaen *567 var.* Muang Khon Kaen. N Thailand
Khor al Udeid *see* Khawr al 'Udayd
Khôr 'Angar *222* NE Djibouti
Khorixas *423* NW Namibia
Khorramābād *309* W Iran
Khorramshahr *309 prev.* Khūnīnshahr. W Iran
Khorugh *562 var.* Horug, *Rus.* Khorog. S Tajikistan
Khotan *see* Hotan
Khouribga *414* C Morocco
Khowst *77* E Afghanistan
Khoyniki *122* SE Belarus
Khrysokhou Bay *215 var.* Chrysochou Bay. Bay of the Mediterranean Sea, on the NW coast of Cyprus
Khujand *562 prev.* Leninabad, Khodzhent, Khojend. NW Tajikistan
Khujayli *614 Rus.* Khodzheyli. W Uzbekistan
Khulna *117* SW Bangladesh
Khūnīnshahr *see* Khorramshahr
Khurīyā Murīyā, Jazā'ir *see* Ḥalāniyāt, Juzur al
Khurramshahr *see* Khorramshahr
Khushāb *451* NE Pakistan
Khvoy *309 var.* Khoi. NW Iran
Khyber Pass *77, 451* mountain pass connecting Afghanistan with Pakistan
Kia *526* SW Santa Isabel, Solomon Is
Kiamusze *see* Jiamusi
Kiangsi *see* Jiangxi
Kiangsu *see* Jiangsu
Kiayi *see* Chiai
Kibondo *564* NW Tanzania
Kibre Mengist *245 var.* Adola. S Ethiopia
Kibungo *492 var.* Kibungu. SE Rwanda
Kibuye *492* W Rwanda
Kičevo *381* W FYR Macedonia
Kidaho *492* NW Rwanda
Kiel *267* N Germany
Kiel Bay *218, 267 Ger.* Kieler Bucht. Bay of the Baltic Sea
Kielce *471* S Poland
Kieler Bucht *see* Kiel Bay
Kieta *458* Bougainville I, Papua New Guinea
Kiev *590 Ukr.* Kyyiv, *Rus.* Kiyev. ❖ of Ukraine, N Ukraine
Kiffa *398* S Mauritania
Kigali *492* ❖ of Rwanda, C Rwanda
Kigembe *492* S Rwanda
Kigoma *564* W Tanzania
Kigwena *162* SW Burundi
Kikila, Lac *653* lake of Île Uvea, S Wallis & Futuna
Kihnu *242* island of SW Estonia
Kikládhes *see* Kyklades
Kikori *458* river of C Papua New Guinea
Kikwit *203* W Dem. Rep. Congo
Kilchu *351* NE North Korea
Kili *396* island of S Marshall Islands
Kilien Mountains *see* Qilian Shan
Kilifi *344* SE Kenya
Kilimanjaro *564* mountain of NE Tanzania
Kilingi-Nõmme *242 Ger.* Kurkund. S Estonia
Kilinochchi *538* N Sri Lanka
Kilis *581* S Turkey
Kilkee *316* W Ireland
Kilkenny *316 Ir.* Cill Choinnigh. SE Ireland
Kilkís *275* N Greece
Kilkoch *316* E Ireland
Killarney *316 Ir.* Cill Airne. SW Ireland
Kilmarnock *597* W Scotland, UK
Kilosa *564* C Tanzania
Kilwa Masoko *564* SE Tanzania
Kimbe *458* New Britain, Papua New Guinea
Kimberley *531* Northern Cape, C South Africa
Kimberley Plateau *100* plateau of NW Australia
Kimch'aek *351 prev.* Sŏngjin. E North Korea
Kimch'ŏn *352* C South Korea
Kimhae *352* SE South Korea
Kimito *251 Swe.* Kemiö. Island of SW Finland
Kimje *352* SW South Korea
Kinabatangan *386* river of NE Borneo, Malaysia
Kinabalu, Gunung *386* mountain of N Borneo, Malaysia
Kindamba *200* S Congo
Kindia *283* SW Guinea
Kindu *203* C Dem. Rep. Congo
King George Bay *645* bay of the South Atlantic Ocean, W Falkland Islands
King George Land *90* island of South Shetland Islands, Antarctica
King Island *101* island of SE Australia
King Island *see* Kadan I
Kingissepp *see* Kuressaare
King Sejong *90* South Korea research station of Antarctic Peninsula, Antarctica
King's Lynn *597* E England, UK
King's Mills *647* SW Guernsey
Kingston *171* SE Canada
Kingston *331* ❖ of Jamaica, E Jamaica
Kingston *650* ❖ of Norfolk Island, S Norfolk Island
Kingston upon Hull *597 var.* Hull. NE England, UK
Kingstown *498* ❖ of St Vincent & the Grenadines, SW St Vincent
King William Island *170 var.* King William. Island of N Canada
Kinihira *492* N Rwanda
Kinkala *200* S Congo
Kinneret-Negev Conduit *319* canal of S Israel
Kinsale *649* SW Montserrat
Kinshasa *203 prev.* Léopoldville. ❖ of Dem. Rep. Congo, W Dem. Rep. Congo
Kintampo *272* C Ghana

Kinyeti *540* mountain of S Sudan
Kinyinya *162* SE Burundi
Kioa *248* island to the E of Vanua Levu, N Fiji
Kipengere Range *564* SW Tanzania
Kipushi *203* SE Dem. Rep. Congo
Kirakira *526* San Cristobal I, Solomon Islands
Kirambo *492* N Rwanda
Kirdzhali *see* Kŭrdzhali
Kirehe *492* SE Rwanda
Kirghizia *see* Kyrgyzstan
Kirghiz Range *359 Rus.* Kirgizskiy Khrebet, *prev.* Alexander Range. Mountain range of Kazakhstan and Kyrgyzstan
Kirghiz Steppe *340* plain of W Kazakhstan
Kiribati *348-349* officially Republic of Kiribati, *prev.* Gilbert Islands, Phoenix Islands, Line Islands. Country of the SC Pacific Ocean. ❖ Bairiki
Kırıkhan *581* S Turkey
Kırıkkale *581* C Turkey
Kirin *see* Jilin
Kirinyaga *344 var.* Mount Kenya. Extinct volcano of C Kenya
Kiritimati *348 var.* Christmas I. Island of the Line Is, E Kiribati
Kiriwina Islands *458 var.* Trobriand Is. Island group of SE Papua New Guinea
Kirkenes *444* NE Norway
Kirklareli *581* NW Turkey
Kirkmichael *647* W Isle of Man
Kirkpatrick, Mount *90* mountain of Greater Antarctica, Antarctica
Kirkūk *312 var.* Karkūk. N Iraq
Kirkwall *597* Orkney Islands, N Scotland, UK
Kirman *see* Kermān
Kirov *484 prev.* Vyatka. W Russian Federation
Kirovabad *see* Gäncä
Kirovakan *see* Vanadzor
Kirovohrad *590 prev.* Kirovo, Zinov'yevsk, Yelizavetgrad. C Ukraine
Kirşehir *581* C Turkey
Kīrthar Range *451* mountain range of S Pakistan
Kirun *see* Chi-lung
Kiruna *547* N Sweden
Kirundo *162 var.* Kirundu. N Burundi
Kiryū *332* Honshū, SE Japan
Kisangani *203 prev.* Stanleyville. NE Dem. Rep. Congo
Kishan *see* Ch'ishan
Kishinev *see* Chişinău
Kishiwada *332* Honshū, C Japan
Kishorganj *117* NE Bangladesh
Kisii *344* SW Kenya
Kiskörei-víztároló *292* reservoir of E Hungary
Kiskunfélegyháza *292 prev.* Félegyháza. C Hungary
Kiskunhalas *292 prev.* Halas. S Hungary
Kismaayo *528 var.* Kismayu, Chisimayu, *It.* Chisimaio. S Somalia
Kisoro *588* SW Uganda
Kisseraing *see* Kanmaw Island
Kissidougou *283* S Guinea
Kistna *see* Krishna
Kisumu *344 prev.* Port Florence. W Kenya
Kisuru *see* Gisuru
Kita *392* W Mali
Kitakami *332* Honshū, N Japan
Kitakyūshū *332* Kyūshū, SW Japan
Kitale *344* W Kenya
Kitami *333* Hokkaidō, N Japan
Kitchener *171* S Canada
Kitega *see* Gitega
Kitgum *588* N Uganda
Kitinen *251* river of N Finland
Kitob *614 Rus.* Kitab. SE Uzbekistan
Kit Stoddart's *494* SE St Kitts, St Kitts & Nevis
Kittitian Village *494* SE St Kitts, St Kitts & Nevis
Kitui *344* S Kenya
Kitwe *635 var.* Kitwe-Nkana. C Zambia
Kitzbühel *106* W Austria
Kitzbüheler Alpen *106* mountain range of W Austria
Kiunga *458* W Papua New Guinea
Kivalo *251* ridge of C Finland
Kiviõli *242* NE Estonia
Kivu, Lac *see* Kivu, Lake
Kivu, Lake *492, 203 Fr.* Lac Kivu. Lake of Rwanda and Dem. Rep. Congo
Kivumba, Lac *492* lake of E Rwanda
Kiyev *see* Kiev
Kiyevskoy Vodokhranilische *see* Kyyivs'ke Vodoskhovyshche
Kiyomizu *see* Ch'ing-shui
Kiyumba *492* C Rwanda
Kizan *see* Ch'ishan
Kizyl-Arvat *see* Gyzylarbat
Kizyl-Kaya *585 var.* Kizyl-Kaja, Gyzylgaya. NW Turkmenistan
Kjølen *see* Kölen
Kladno *216* NW Czech Republic
Klagenfurt *106* S Austria
Klaipėda *376 Ger.* Memel. NW Lithuania
Klaksvík *644* Bordhoy, N Faeroe Islands
Klang *see* Kelang
Klarälven *547* river of SW Sweden
Klatovy *216* W Czech Republic
Klausenburg *see* Cluj-Napoca
Klein Bonaire *650* island to the W of Bonaire, S Netherlands Antilles
Klerksdorp *531* North West, N South Africa
Klirou *215* W Cyprus
Ključ *140* NW Bosnia & Herzegovina
Klosterneuburg *106* NE Austria
Kloten *550* N Switzerland
Kluang *see* Keluang
Klyuchevskaya Sopka *485* Mountain of NE Russian Federation
Knezha *152 var.* Kneža. NW Bulgaria
Knin *209* S Croatia
Knox Atoll *see* Narikrik
Knoxville *603* Tennessee, SE USA
Knud Rasmussen Land *646* physical region of N Greenland
Kōbe *332* Honshū, C Japan
København *see* Copenhagen
Kobenni *398* S Mauritania
Koblenz *266* W Germany
Kobryn *122 Rus.* Kobrin. SW Belarus
Kocaeli *see* İzmit
Kočani *381* NE FYR Macedonia
Kočevje *524 Ger.* Gottschee. S Slovenia
Kōchi *see* Cochin
Kochi *332* Shikoku, SW Japan
Kochkor *see* Kochkorka
Kochkorka *359 Kir.* Kochkor. NE Kyrgyzstan
Koddiyar Bay *538* bay of the Indian Ocean, on the NE coast of Sri Lanka
Kodiak *602* Alaska, USA
Koedoes *see* Kudus
Koeln *see* Köln
Ko-erh-mu *see* Golmud
Koes *423* SE Namibia
Koetai *see* Mahakam
Kofarnihon *562 prev.* Ordzhonikidzeabad. W Tajikistan
Kofinou *see* Kouklia
Koforidua *272* SE Ghana
Køge *218* Sjælland, E Denmark
Kogil'nik *see* Cogîlnic
Kogo *see* Cogo
Kogon *283* river of W Guinea
Kogon *614 Rus.* Kagan. S Uzbekistan
Kōgŭm-do *352* island of S South Korea
Kohāt *451* N Pakistan
Kohtla-Järve *242* NE Estonia
Kohŭng *352* S South Korea
Koilabas *427* W Nepal
Koimbani *198* E Grande Comore, Comoros
Koindu *518* E Sierra Leone
Koi Sanjaq *312 var.* Kūysanjaq, Koysanjaq. N Iraq
Kōje-do *352 Jap.* Kyōsai-tō. Island of S South Korea
Kokand *see* Qŭqon
Kōka-tō *see* Kanghwa-do
Kokemäenjoki *251* river of SW Finland
Kök-Janggak *see* Kok-Yangak
Kokkina *215 var.* Erenköy. W Cyprus
Kokkola *251 Swe.* Karleby, *prev.* Gamlakarleby. W Finland
Koko Nor *see* Qinghai Hu
Kokshaal-Tau *359 Rus.* Khrebet Kakshaal-Too. Mountain range of China and Kyrgyzstan
Kokshetau *340* N Kazakhstan
Kokstad *531* Kwazulu Natal, E South Africa
Kok-Yangak *359 var.* Kok-Jangak, *Kir.* Kök-Janggak. SW Kyrgyzstan
Kolahun *368* N Liberia
K'o-la-ma-i *see* Karamay
Kola Peninsula *see* Kol'skiy Poluostrov
Kolda *510* S Senegal
Kolding *218* Jylland, W Denmark
Kölen *547 Nor.* Kjølen. Mountains of N Sweden
Kolenté *283* river of Guinea and Sierra Leone
Kolga Laht *242* bay of the Gulf of Finland, on the coast of N Estonia
Kolguyev, Ostrov *484* Island of NW Russian Federation
Kolhāpur *298* SW India
Kolhumadulu Atoll *390 var.* Kolumadulu Atoll, Thaa Atoll. Atoll of S Maldives
Kolia *653* Île Futuna, N Wallis & Futuna
Koliba *283* river of NW Guinea
Kolín *216 Ger.* Kolin. C Czech Republic
Kolkasrags *362 prev.* Cape Domesnes. Cape of NW Latvia
Kolkata *see* Calcutta
Kolkhozobod *562 var.* Kaganovichabad, Tugalan; *Rus.* Kolkhozabad. SW Tajikistan
Kolmar *see* Colmar
Köln *266 var.* Koeln, *prev.* Cöln, *Eng.* Cologne. W Germany
Kolokani *392* W Mali
Kolombangara *526 var.* Nduke. New Georgia Is, Solomon Islands
Kolomskoye Nagor'ye *485 Eng.* Kolyma Range. Mountain range of NE Russian Federation
Kolonia *406 var.* Colonia. Pohnpei, Micronesia
Kolonjë *see* Ersekë
Kolonyama *366* NW Lesotho
Kolozsvár *see* Cluj-Napoca
Kolpa *524 SCr.* Kupa, *Ger.* Kulpa. River of S Slovenia
Kol'skiy Poluostrov *484 Eng.* Kola Peninsula. Peninsula of NW Russian Federation
Koltur *644* island of C Faeroe Islands
Kolumadulu Atoll *see* Kolhumadulu Atoll
Kolwezi *203* S Dem. Rep. Congo
Kolyma *485* river of NE Russian Federation
Kolyma Lowland *see* Kolymskaya Nizmennost'
Kolymskaya Nizmennost' *485 Eng.* Kolyma Lowland. Lowland region of NE Russian Federation
Komanit, Liqeni i *81* lake of N Albania
Komárno *523 Ger.* Komorn, *Hung.* Komárom. SW Slovakia
Komárom *see* Komárno
Komati *544* river of SE Africa
Komatsu *332* Honshū, C Japan
Kombissiri *157 var.* Kombissiguiri. C Burkina
Kome Island *588* island of S Uganda
Komi, Respublika *484* autonomous republic of NW Russian Federation
Komló *292* SW Hungary
Kommunarsk *see* Alchevs'k
Komoé *328* river of E Ivory Coast
Komono *200* SW Congo
Komorn *see* Komárno
Komotau *see* Chomutov
Komotiní *275 Turk.* Gümülcine, Gümüljina. NE Greece
Kompong Kleang *see* Kâmpóng Khleang
Kompong Som *see* Kâmpóng Saôm
Kompong Speu *see* Kâmpóng Spoe
Kompong Thom *see* Kâmpóng Thum
Komsomol *see* Komsomol'sk
Komsomolets, Ostrov *485* island of Severnaya Zemlya, N Russian Federation
Komsomol'sk *585 Turkm.* Komsomol. SE Turkmenistan
Komsomol'sk-na-Amure *485* SE Russian Federation
Komusan *351* NE North Korea
Kondoa *564* C Tanzania
Koné *650* W New Caledonia
Köneürgench *see* Këneurgench
Kŏng *165, 360* river of Cambodia and Laos
Kŏng, Kaôh *165 prev.* Kas Kong. Island of SW Cambodia
Kong Christian IX Land *646* physical region of SE Greenland
Kong Christian X Land *646* physical region of E Greenland
Kong Frederik VIII Land *646* physical region of NE Greenland
Kong Frederik VI Kyst *646* physical region of SE Greenland
Kongju *352 Jap.* Kōshū. W South Korea
Kong Karls Land *653* island group of SE Svalbard
Kongo *see* Congo
Kongoloto Lafanga *587* islet of Nukufetau, Tuvalu
Kongoloto Motulalo *587* islet of Nukufetau, Tuvalu
Kongoussi *157* N Burkina
Kongsberg *444* S Norway
Kongsvinger *444* S Norway
Königgrätz *see* Hradec Králové
Königshütte *see* Chorzów
Konispol *81 var.* Konispoli. S Albania
Köniz *550* W Switzerland
Konjic *140* S Bosnia & Herzegovina
Konkämäälv *see* Konkämäeno
Konkämäeno *251 Swe.* Konkämäälv. River of N Europe
Konkouré *283* river of W Guinea
Konotop *590* NE Ukraine
Konrei *455* N Palau
Konsankoro *283* SE Guinea
Konstantza *see* Constanţa
Konstanz *266* S Germany
Kontagora *440* NW Nigeria
Kon Tum *627 var.* Kontum. S Vietnam
Konya *581 prev.* Konia. C Turkey
Kopaonik *512* mountain range of C Serbia, Serbia & Montenegro (Yugo.)
Kópavogur *296* W Iceland
Koper *524 It.* Capodistria. SW Slovenia
Kopetdag, Khrebet *585 Turkm.* Kopetdag Gershi, *Per.* Koppeh Dāgh. Mountain range of W Turkmenistan
Kophinou *215* S Cyprus
Kophinou *see* Kouklia
Koplik *81* N Albania
Koppeh Dāgh *see* Kopetdag Khrebet
Koppename *see* Coppename
Koprivnica *209 Ger.* Kopreinitz, *Hung.* Kaproncza. NE Croatia
Köprülü *see* Titov Veles
Köprülü Rezevuar *see* Kouklia Reservoir
Korat *see* Nakhon Ratchasima
Korat Plateau *567* plateau of NE Thailand
Korçë *81 var.* Korça, *prev.* Koritsa, *It.* Corriza, *Gk* Korytsa. SE Albania
Korčula *209 It.* Curzola. Island of S Croatia
Korea *see* North Korea, South Korea
Korea Bay *187, 351* bay of Yellow Sea, off the coast of E Asia
Korea Strait *352 Kor.* Taehan-haehyŏp, *Jap.* Chōsen-kaikyō. Channel connecting the East China Sea and e Sea of Japan, E Asia
Korhogo *328* N Ivory Coast

L

Liban, Jebel *364* *Eng.* Lebanon, Mount Lebanon, *Ar.* Jabal Lubnān. Mountain range of C Lebanon

Libau *see* Liepāja

Liberec *216* *Ger.* Reichenberg. N Czech Republic

Liberia *206* NW Costa Rica

Liberia *368-369* officially The Republic of Liberia. Country of West Africa, divided into 9 admin. units (counties). ❖ Monrovia

Liberta *92* S Antigua, Antigua & Barbuda

Libertad *611* S Uruguay

Libertad *130* *prev.* Pembroke Hall. N Belize

Librazhd *81* *var.* Librazhdi. E Albania

Libreville *260* ❖ of Gabon, NW Gabon

Libya *370-373* officially The Great Socialist People's Libyan Arab Jamahiriya, *prev.* Libyan Arab Republic. Country of N Africa, the current administrative structure is not clear. ❖ Tripoli

Libyan Desert *232, 371, 540* *Ar.* Aş Şahrā' al Lībīyah. Desert of N Africa

Libyan Plateau *232, 371* *Ar.* Aḍ Ḍiffah. Plateau of Egypt and Libya

Licata *323* Sicilia, S Italy

Lichinga *419* N Mozambique

Lichtenburg *531* North West, N South Africa

Lida *122* W Belarus

Lido di Ostia *323* C Italy

Liechtenstein *374-375* officially Principality of Liechtenstein. Country of C Europe divided into 11 admin. units (communes). ❖ Vaduz

Liège *127* *Dut.* Luik, *Ger.* Lüttich. E Belgium

Liegnitz *see* Legnica

Lieksa *251* E Finland

Lien-yun *see* Liangyungang

Lienz *106* W Austria

Liepāja *362* *Ger.* Libau. W Latvia

Lier *127* *Fr.* Lierre. N Belgium

Liestal *550* N Switzerland

Lievenhof *see* Līvāni

Liezen *106* C Austria

Lifford *316* N Ireland

Lifou *650* island, Îles Loyauté, E New Caledonia

Lifuka *572* island of Ha'apai Group, Tonga

Līgatne *362* NE Latvia

Ligure, Appennino *322* mountain range of N Italy

Ligure, Mar *see* Ligurian Sea

Ligurian Sea *255, 322* *It.* Mar Ligure, *Fr.* Mer Ligurienne. Area of the Mediterranean Sea, between France and Italy

Lihue *602* Kauai, Hawaii, USA

Lihula *242* *Ger.* Leal. W Estonia

Liivi Laht *see* Riga, Gulf of

Likasi *203* *prev.* Jadotville. SE Dem. Rep. Congo

Likiep *396* island of C Marshall Islands

Likouala *200* river of NW Congo

Likouala aux Herbes *200* river of E Congo

Liku *650* E Niue

Lille *255* *Dut.* Rijssel. N France

Lillebælt *218* *Eng.* Little Belt, *var.* Lille Bælt. Straits between Fyn and Jylland, SW Denmark

Lillehammer *444* S Norway

Lillestrøm *444* S Norway

Lilongwe *385* ❖ of Malawi, W Malawi

Lilongwe *385* river of W Malawi

Lima *463* ❖ of Peru, W Peru

Limassol *215* *var.* Lemesos. SW Cyprus

Limbe *385* S Malawi

Limbe *168* *prev.* Victoria. SW Cameroon

Limbé *288* N Haiti

Lim Chu Kang *521* area of NW Singapore

Limerick *316* *Ir.* Luimneach. SW Ireland

Limfjorden *218* fjord of Jylland, NW Denmark

Límni Megáli Préspa *see* Prespa, Lake

Limni Prespa *see* Prespa, Lake

Límnos *275* *var.* Lemnos. Island of E Greece

Limoges *255* C France

Limón *206* E Costa Rica

Limón *290* NE Honduras

Limon, Mont *400* mountain of Rodrigues, Mauritius

Limousin *255* cultural region of C France

Limpopo *142, 419, 423, 636* *var.* Crocodile. River of southern Africa

Limulunga *635* W Zambia

Linakeng *366* E Lesotho

Linares *183* C Chile

Linares *535* S Spain

Lincoln *597* E England, UK

Lincoln *603* Nebraska, C USA

Lincoln Island *651* island of E Paracel Islands

Linden *286* E Guyana

Lindi *203* river of NE Dem. Rep. Congo

Lindi *564* SE Tanzania

Line Islands *348* island group of E Kiribati

Lingayen Gulf *466* gulf of the South China Sea, N Philippines

Lingga, Kepulauan *304* island group to the E of Sumatra, W Indonesia

Linguère *510* N Senegal

Linköping *547* S Sweden

Linyanti *142* river of Botswana and Namibia

Linz *106* N Austria

Lion, Golfe du *255* *Eng.* Gulf of Lions. Gulf of the Mediterranean Sea to the S of France

Lions Den *636* N Zimbabwe

Lipa *466* Luzon, N Philippines

Lipari Islands *see* Eolie, Isole

Lipari, Isola *323* island of S Italy

Lipari, Isole *see* Eolie, Isole

Lipetsk *484* W Russian Federation

Lipsk *see* Leipzig

Liptovský Mikuláš *523* *Ger.* Liptau-Sankt-Nikolaus, *Hung.* Liptószentmiklós. C Slovakia

Lira *588* N Uganda

Liranga *200* E Congo

Liri *323* river of C Italy

Lisala *203* N Dem. Rep. Congo

Lisbon *474* *Port.* Lisboa. ❖ of Portugal, W Portugal

Lisburn *597* Northern Ireland, UK

Lisieux *255* NW France

Lismore *101* E Australia

Lissa *see* Vis

Litani *542, 645* *var.* Itany. River of French Guiana and Suriname

Litani *364* river of C Lebanon

Litaven *see* Lithuania

Litavra *see* Lithuania

Lithgow *101* SE Australia

Lithuania *376-377* officially Republic of Lithuania, *Lith.* Lietuva, *Ger.* Litauen, *Pol.* Litwa, *Rus.* Litva, *prev.* Lithuanian SSR, *Rus.* Litovskaya SSR. Country of E Europe divided into 44 admin. units (disicts). ❖ Vilnius

Litla Dimun *644* island of S Faeroe Islands

Little Abaco *112* island of N Bahamas

Little Alföld *292* plain of Hungary and Slovakia

Little Andaman *298* island of Andaman Islands to the SE of India

Little Barrier Island *433* island to the N of North Island, New Zealand

Little Belt *see* Lillebælt

Little Cayman *644* island of C Cayman Islands

Little Coco Island *159* island of SW Burma

Little Inagua *112* island of S Bahamas

Little Minch *597* strait of the Atlantic Ocean, NW Scotland, UK

Little Rock *603* Arkansas, SC USA

Little Scarcies *see* Kaba

Little Sound *643* bay of the North Atlantic Ocean, W Bermuda

Little Tobago *643* island of W British Virgin Islands

Little Tobago *574* *var.* Bird of Paradise Island. Island to the E of Tobago, Trinidad & Tobago

Little Zab *312* *Ar.* Zāb aş Şaghīr, *Kurd.* Zē-i Kya. River of Iran and Iraq

Litva *see* Lithuania

Litwa *see* Lithuania

Liuch'iu Yü *559* island of SW Taiwan

Liukuei *559* S Taiwan

Liuzhou *187* *var.* Liu-chou, Liuchow. Guangxi, S China

Livadhi *215* *var.* Leivadi. River of W Cyprus

Līvāni *362* *Ger.* Lievenhof. SE Latvia

Lively Island *645* island of E Falkland Islands

Lively Sound *645* area of the South Atlantic Ocean, E Falkland Islands

Liverpool *597* NW England, UK

Lívingston *280* E Guatemala

Livingstone *635* *var.* Maramba. S Zambia

Livingstonia *385* N Malawi

Livno *140* SW Bosnia & Herzegovina

Livojoki *251* river of C Finland

Livonia *see* Vidzeme

Livorno *322* *Eng.* Leghorn. C Italy

Liwonde *385* S Malawi

Liyāḩ, Jāl al *356* ridge of NW Kuwait

Ljouwert *see* Leeuwarden

Ljubelj *see* Loibl Pass

Ljubljana *524* *var.* Lyublyana, *Ger.* Laibach, *It.* Lubiana. ❖ of Slovenia, C Slovenia

Ljubrlj *see* Loibl Pass

Ljungan *547* river of C Sweden

Ljusnan *547* river of C Sweden

Llallagua *136* SW Bolivia

Lleida *535* *Cast.* Lérida. NE Spain

Llolleo *183* C Chile

Lloydminster *170* SW Canada

Lô *627* river of China and Vietnam

Loaita Island *652* island of W Spratly Islands

Loangwa *see* Luangwa

Lobatse *142* *var.* Lobatsi. S Botswana

Lobaye *179* river of SW Central African Republic

Lobito *88* W Angola

Lob Nor *see* Lop Nur

Locarno *550* *Ger.* Luggarus. S Switzerland

Lôc Ninh *627* SW Vietnam

Locri *323* S Italy

Lőcse *see* Levoča

Lod *319* *var.* Lydda. C Israel

Lodge *494* NE St Kitts, St Kitts & Nevis

Lodja *203* C Dem. Rep. Congo

Lodwar *344* NW Kenya

Łódź *471* *Rus.* Lodz. C Poland

Loei *567* *var.* Muang Loei. N Thailand

Lofa *368* *var.* Loffa. River of Guinea and Liberia

Lofoten *444* *var.* Lofoten Islands. Islands of NE Norway

Loga *439* W Niger

Logan, Mount *170* mountain of NW Canada

Logone *168, 180* *var.* Lagone. River of Cameroon and Chad

Logroño *535* N Spain

Lögurinn *296* *var.* Lagarfljót. Lake of E Iceland

Loh *619* Torres Islands, N Vanuatu

Loibl Pass *106* *var.* Ljubelj, *Ger.* Loiblpass, *Slvn.* Ljubrlj. Mountain pass of Austria and Slovenia

Loikaw *159* E Burma

Loir *255* river of NW France

Loire *255* river of C France

Loita Hills *344* hilly region of SW Kenya

Loja *230* S Ecuador

Lökbatan *110* *Rus.* Lokbatan. E Azerbaijan

Lokeren *127* NW Belgium

Lokitaung *344* NW Kenya

Løkken *218* Jylland, NW Denmark

Lokoja *440* C Nigeria

Lokossa *132* S Benin

Loksa *242* *Ger.* Loxa. N Estonia

Lol *540* river of S Sudan

Lola *283* SE Guinea

Lolland *218* *prev.* Laaland. Island of S Denmark

Lolotique *237* SE El Salvador

Lolvavana, Passage *619* strait between Maewo and Pentecost, C Seychelles

Lom *152* *prev.* Lom-Palanka. NW Bulgaria

Lom *168* river of Cameroon and Central African Republic

Lomahasha *544* NE Swaziland

Lomami *203* river of C Dem. Rep. Congo

Lomas de Zamora *95* E Argentina

Lombok *304* island of Nusa Tenggara, C Indonesia

Lomé *571* ❖ of Togo, S Togo

Lomond, Loch *597* lake of C Scotland, UK

Lom Sak *567* *var.* Muang Lom Sak. N Thailand

Londiani *344* W Kenya

London *597,* ❖ of United Kingdom

London *171* S Canada

London *348* Kiritimati, E Kiribati

London Bridge *278* island to the N of Grenada island, Grenada

Londonderry *597* *var.* Derry. Northern Ireland, UK

Londrina *145* S Brazil

Longa, Proliv *485* *Eng.* Long Strait. Strait connecting Chukchi Sea and East Siberian Sea, between NE Asia and NW North America

Long Bay *643* bay of the North Atlantic Ocean, E Bermuda

Long Bay *643* bay of the North Atlantic Ocean, W Bermuda

Longford *316* *Ir.* Longphort. C Ireland

Long Island *643* island of W Bermuda

Long Island *see* Arop Island

Long Island *112* island of C Bahamas

Long Island *92* island to the N of Antigua, Antigua & Barbuda

Longmont *603* Colorado, SW USA

Longoni, Baie de *649* *var.* Longoni Bay. Bay of the Indian Ocean on the N coast of Mayotte

Longphort *see* Longford

Longreach *101* E Australia

Long Strait *see* Longa, Proliv

Long Swamp *643* Tortola, C British Virgin Islands

Longwood *652* E St Helena

Long Xuyên *627* SW Vietnam

Longyearbyen *653* ❖ of Svalbard, Spitsbergen, W Svalbard

Lonhlupheko *544* E Swaziland

Lons-le-Saunier *255* E France

Loop Head *316* promontory on the W coast of Ireland

Lop Buri *567* C Thailand

Lopévi *619* island of C Vanuatu

Lopez, Cap *260* W Gabon

Lop Nur *186* *var.* Lop Nor, Lob Nor, *Chin.* Lo-pu Po. Lake of Xinjiang Uygur Zizhiqu, NW China

Lo-pu Po *see* Lop Nur

Lora, Hāmūn-i- *450* salt marsh of W Pakistan

Lord Howe Island island of E Australia

Lord Howe Island *see* Ontong Java Atoll

Lorca *535* S Spain

Lorengau *458* *var.* Lorungau, Manus I, Papua New Guinea

Lorentz *305* river of Irian Jaya, E Indonesia

Loreto *460* C Paraguay

Lorian Swamp *344* swamp E Kenya

Lorient *254* W France

Lorn, Firth of *597* inlet of Atlantic Ocean, W Scotland, UK

Lorraine *255* cultural region of NE France

Los Amates *280* E Guatemala

Los Andes *183* C Chile

Los Angeles *602* California, W USA

Los Ángeles *183* C Chile

Losap *406* atoll of C Micronesia

Los Chiles *206* NW Costa Rica

Los, Îles de 283 Island group to the SW of Guinea
Lošinj 209 It. Lussino. Island of W Croatia
Loslau see Wodzisław Śląski
Los Mochis 403 W Mexico
Losonc see Lučenec
Losontz see Lučenec
Lospalos 228 E East Timor
Los Roques, Islas 623 island group of N Venezuela
Los Teques 623 N Venezuela
Lot 255 river of S France
Lotofaga 500 Upolu, Samoa
Lo-tung 559 Jap. Ratō. NE Taiwan
Louang Namtha 360 var. Luong Nam Tha. N Laos
Louangphrabang 360 var. Luang Prabang. C Laos
Loubiere 224 SW Dominica
Loubomo 200 prev. Dolisie. S Congo
Loudima 200 S Congo
Louéssé 200 river of SW Congo
Louga 510 NW Senegal
Loughrea 316 W Ireland
Louis Gentil see Youssoufia
Louisiade Archipelago 458 island group of SE Papua New Guinea
Louisiana 603 state of SC USA
Louis Trichardt 531 Northern Transvaal, NE South Africa
Louisville 603 Kentucky, C USA
Louisville 130 N Belize
Loukoléla 200 E Congo
Loum 168 W Cameroon
Louna 200 river of SE Congo
Louny 216 NW Czech Republic
Lourenço Marques see Maputo
Lourenço Marques, Baía de see Maputo, Baía de
Louvain see Leuven
Lovech 152 var. Loveč. NW Bulgaria
Lovell Village 498 Mustique, St Vincent & the Grenadines
Lóvua 88 N Angola
Lowell 603 Massachusetts, NE USA
Löwen see Leuven
Lower Bann 597 river of Northern Ireland, UK
Lower California see Baja California
Lower Carlton 121 NW Barbados
Lower Hutt 433 S North Island, New Zealand
Lower Lough Erne 597 lake of Northern Ireland, UK
Lower Mortlocks 406 island group of C Micronesia
Lower Rhine see Neder-Rijn
Lower Tunguska see Nizhnyaya Tunguska
Lower West End Point 642 headland on the SW coast of Anguilla
Low Point 644 headland on the E coast of Christmas Island
Loyada 222 E Djibouti
l'Oyapok see Oiapoque
Loyauté, Îles 650 island group of E New Caledonia
Loyoro 588 NE Uganda
Loznica 512 W Serbia, Serbia & Montenegro (Yugo.)
Lualaba 203 var. Zaire, Fr. Loualaba. River of Dem. Rep. Congo
Luampa 635 river of W Zambia
Luanda 88 ❖ of Angola, NW Angola
Luang Prabang see Louangphrabang
Luang Prabang Range 360 mountain range of W Laos
Luangwa 635 Port. Aruângua. River of Mozambique and Zambia
Luanshya 635 C Zambia
Luapula 203, 635 river of Dem. Rep. Congo and Zambia
Luar, Pulu see Horsburgh Island
Luba 238 prev. San Carlos. W Bioko, Equatorial Guinea
Lubānas Ezers 362 lake of E Latvia
Lubang Island 466 island of N Philippines
Lubango 88 Port. Sá da Bandeira. SW Angola
Lubao 203 SE Dem. Rep. Congo
Lübeck 267 N Germany
Lübecker Bucht 267 bay of the Baltic Sea, on the N coast of Germany
Lubelska, Wyżyna 471 plateau of SE Poland
Lubiana see Ljubljana
Lublin 471 Rus. Lyublin. E Poland
Lubnān, Jabal see Liban, Jebel
Lubny 590 C Ukraine
Lubumbashi 203 prev. Élisabethville. SE Dem. Rep. Congo
Luca 136 SW Bolivia
Lucala 88 NW Angola
Lucan 316 Ir. Leamhcán. E Ireland
Lucano, Appennino 323 mountain range of S Italy
Lucapa 88 var. Lukapa. NE Angola
Lucea 331 NW Jamaica
Lucena 466 Luzon, N Philippines
Lučenec 523 Hung. Losonc, Ger. Losontz. C Slovakia
Lucerne see Luzern
Lucerne, Lake of see Vierwaldstätter See
Luchow see Hefei
Lucie 542 Suriname
Łuck see Luts'k
Lucknow 298 Hind. Lakhnau. N India
Lüderitz 423 prev. Angra Pequena. SW Namibia
Ludhiāna 298 N India
Ludwigshafen 266 var. Ludwigshafen am Rhein. SW Germany
Luebo 203 SW Dem. Rep. Congo
Luena 88 Port. Luso. E Angola
Lufira, Lac de Retenue de la 203 var. Lac Tshangalele. Lake of SE Dem. Rep. Congo
Lugano 550 Ger. Lauis. S Switzerland
Luganville 619 Espiritu Santo, Vanuatu
Lugards Falls 344 waterfall of SE Kenya
Lugenda 419 river of N Mozambique
Luggarus see Locarno
Lugh Ganana see Luuq
Lugo 535 NW Spain
Lugoj 480 W Romania
Lugusi see Rugusye
Luhans'k 590 Rus. Lugansk, prev. Voroshilovgrad. E Ukraine
Luiana 88 river of SE Angola
Luichow Peninsula see Leizhou Bandao
Luik see Liège
Luimneach see Limerick
Luján 95 C Argentina
Lukang 559 var. Lu-chiang, Jap. Rokkō. W Taiwan
Lukapa see Lucapa
Lukenie 203 river of C Dem. Rep. Congo
Lukhalweni 544 S Swaziland
Lukusashi 635 river of C Zambia
Luleå 547 NE Sweden
Luleälv 547 river of NE Sweden
Lulonga 203 river of NW Dem. Rep. Congo
Lulua 203 river of S Dem. Rep. Congo
Luluabourg see Kananga
Lumbo 419 NE Mozambique
Lumi 458 NW Papua New Guinea
Lumphăt 165 prev. Lomphat. NE Cambodia
Lumpungu see Rumpungwe
Lund 547 S Sweden
Lunga 635 river of Zambia
Lunga, Isola see Dugi Otok
Lungi 518 W Sierra Leone
Lungkiang see Qiqihar
Lungwebungu 635 river of Angola and Zambia
Luninyets 122 Rus. Luninets. SW Belarus
Lunsar 518 W Sierra Leone
Lunsemfwa 635 river C Zambia
Luong Nam Tha see Louang Namtha
Luoyang 187 Henan, C China
Luque 460 S Paraguay
Lúrio 419 NE Mozambique
Lúrio 419 river of NE Mozambique
Lusaka 635 ❖ of Zambia, SE Zambia
Lushnjë 81 var. Lushnja. C Albania
Luso see Luena
Lussino see Lošinj
Lusutfu 544 var. Usutu, Great Usutu. River of southern Africa
Lü Tao 559 island of SE Taiwan
Lūt, Baḩrat see Dead Sea
Lūt, Dasht-e 309 var. Kavīr-e Lūt. Desert region of E Iran
Luton 597 C England, UK
Luts'k 590 Rus. Lutsk, Pol. Łuck. NW Ukraine
Lüttich see Liège
Luud, Waadi 528 It. Uadi Lhut. Seasonal river of N Somalia
Luuq 528 It. Lugh Ganana. SW Somalia
Luvironza see Ruvyironza
Luxembourg 378 ❖ of Luxembourg
Luxembourg 378-379 officially Grand Duchy of Luxembourg, var. Lëtzebuerg. Country of W Europe divided into 3 admin. units (districts). ❖ Luxembourg
Luxor 232 Ar. Al Uqşur. E Egypt
Luz, Costa de la 535 coastal region of SW Spain
Luzern 550 Fr. Lucerne. C Switzerland
Luzon 466 island of N Philippines
L'viv 590 Rus. L'vov, Pol. Lwów, Ger. Lemberg. W Ukraine
Lyallpur see Faisalābād
Lyangar see Langar
Lycksele 547 N Sweden
Lydda see Lod
Lyepyel' 122 Rus. Lepel'. N Belarus
Lyme Bay 597 bay of S England, UK
Lyon 255 Eng. Lyons E France
Lys see Leie
Lysi 215 var. Akdoğan. C Cyprus
Lyublin see Lublin
Lyublyana see Ljubljana

M

Ma'ān 338 SW Jordan
Maanselkä 251 mountain range of NE Finland
Maardu 242 Ger. Maart. N Estonia
Maarianhamina see Mariehamn
Maarmorilik 646 W Greenland
Ma'arrat an Nu'mān 555 NW Syria
Maarssen 429 C Netherlands
Maart see Maardu
Maas see Meuse
Maasmechelen 127 NE Belgium
Maastricht 429 S Netherlands
Maaza Plateau 232 plateau of NE Egypt
Mabalane 419 S Mozambique
Mabanda 260 S Gabon
Mabayi 162 NW Burundi
Mabouya Island 278 island to the W of Carriacou, Grenada
Macao 187 Chin. Aomen. SE China
Macapá 145 N Brazil
Macará 230 S Ecuador
Macarsca see Makarska
MacArthur see Ormoc
Macas 230 SE Ecuador
Macassar see Ujungpandang
Macdonnell Ranges 101 mountain range of C Australia
Macedonia 380-381 officially the Former Yugoslav Republic of Macedonia, abbrev. FYR Macedonia, FYROM, Mac. Makedonija. Country of SE Europe divided into 34 admin. units (opcine). ❖ Skopje
Maceió 145 E Brazil
Macenta 283 SE Guinea
Macgillicuddy's Reeks 316 var. Macgillicuddy's Reeks Mountains, Ir. Na Cruacha Dubha. Mountain range of SW Ireland
Machakos 344 S Kenya
Machala 230 SW Ecuador
Machaneng 142 SE Botswana
Machanga 419 SE Mozambique
Machaze 419 SW Mozambique
Machile 635 var. Machili. River of SW Zambia
Machinga 385 var. Kasupe, Kasupi. S Malawi
Machiques 623 NW Venezuela
Machu Picchu 463 S Peru
Macia 419 var. Vila de Macia. S Mozambique
Macías Nguema Biyogo see Bioko
Mackay 101 NE Australia
Mackay, Lake 101 salt lake of C Australia
Mackenzie 170 river of NW Canada
Mackenzie Mountains 170 mountain range of NW Canada
Macleod, Lake 100 lake of W Australia
Macomer 322 Sardegna, S Italy
Macon 603 Georgia, SE USA
Mâcon 255 E France
Macouba 649 N Martinique
MacRitchie Reservoir 521 reservoir of C Singapore
Macroom 316 SW Ireland
Macuelizo 290 W Honduras
Ma'dabā 338 var. Mādabā. NW Jordan
Madagascar 382-383 officially Democratic Republic of Madagascar, Malg. Madagasikara, prev. Malagasy Republic. Country of SE Africa divided into 6 admin. units (provinces). ❖ Antananarivo
Madan 152 S Bulgaria
Madang 458 E Papua New Guinea
Madanīyīn see Médenine
Mādārīpur 117 S Bangladesh
Madeira 474 var. Ilha de Madeira. Madeira Is, Portugal
Madeira 144 river of Bolivia and Brazil
Madeira, Arquipélago da see Madeira Islands
Madeira Islands 474 Port. Arquipélago da Madeira. Island group to the SW of Portugal
Madina do Boé see Boé
Madīnat al Abyār 371 NE Libya
Madīnat al Ka'bān 479 var. Al Ka'aban. N Qatar
Madīnat ash Sha'b 631 prev. Al Ittiḩād. SW Yemen
Madīnat ash Shamāl 479 var. Madinat el Shamal. N Qatar
Madīnat ath Thawrah 555 N Syria
Madinat el Shamal see Madīnat ash Shamāl
Madīnat Ḩamad 115 var. Hamada Town, Hamad. W Bahrain
Madīnat 'Īsá 115 var. Isa Town. N Bahrain
Madīnat Khalīfa 479 E Qatar
Madingo-Kayes 200 S Congo
Madingou 200 S Congo
Madison 603 Wisconsin, NC USA
Madiun 304 prev. Madioen. Java, C Indonesia
Madlangampisi 544 N Swaziland
Madona 362 Ger. Modohn. E Latvia
Madras see Chennai
Madre de Dios 136, 463 river of Bolivia and Peru
Madrid 535 ❖ of Spain, C Spain
Madriu, Riu 86 river of S Andorra
Madura 304 prev. Madoera. Island to NE of Java, C Indonesia
Madurai 298 prev. Madura. S India
Maebashi 332 Honshū, SE Japan
Mae Name Khong see Mekong
Mae Nam Khong see Mekong
Mae Nam Moi see Thaungyin
Mae Nam Ping 567 river of NW Thailand
Maéwo 619 prev. Aurora. Island of C Vanuatu
Mafa'a, Pointe 653 headland of Île Alofi, N Wallis & Futuna
Mafeteng 366 W Lesotho
Mafia 564 island of E Tanzania
Mafou 283 river of C Guinea
Mafraq see Al Mafraq
Mafraq 594 C United Arab Emirates
Mafungabusi Plateau 636 plateau of C Zimbabwe
Mafut 594 NE United Arab Emirates
Magadan 485 NE Russian Federation
Magadi 344 SW Kenya
Magadi, Lake 344 lake of SW Kenya

Mania *382* river of C Madagascar
Manica *419 var.* Vila de Manica. W Mozambique
Manihiki *644* island of Northern Cook Islands, N Cook Islands
Maniitsoq *646 Dan.* Sukkertoppen. SW Greenland
Manikaraku *526* E Guadalcanal, Solomon Is
Manikganj *117* C Bangladesh
Manila *466 var.* Manila City. ❖ of the Philippines, Luzon, N Philippines
Manisa *580 prev.* Saruhan. W Turkey
Man, Isle of *597, 647* British Crown dependency of the Irish Sea. ❖ Douglas
Manitoba *170* province of S Canada
Manizales *195* W Colombia
Manjimup *100* SW Australia
Mankayane *544 var.* Mankaiana. W Swaziland
Mankono *328* C Ivory Coast
Mankulam *538* N Sri Lanka
Mannar *538 var.* Manar. NW Sri Lanka
Mannar, Gulf of *298, 538* gulf of Indian Ocean, to the S of India
Mannar Island *538* island to the N of Sri Lanka
Mannheim *266* SW Germany
Mano *518* SW Sierra Leone
Mano *518* river of Liberia and Sierra Leone
Manombo Atsimo *382 var.* Manombo. SW Madagascar
Manono *500* Upolu, Samoa
Manono *203* SE Dem. Rep. Congo
Manorhamilton *316* N Ireland
Manp'o *351 var.* Manp'ojin. NW North Korea
Manra *348 var.* Sydney I. Island of the Phoenix Is, C Kiribati
Mansa *635 prev.* Fort Rosebery. N Zambia
Mansabá *284* NW Guinea-Bissau
Mansajang Kunda *263* E Gambia
Mansa Konko *263* C Gambia
Mansion *494* NE St Kitts, St Kitts & Nevis
Mansôa *284* W Guinea-Bissau
Mansôa *284* river of W Guinea-Bissau
Manta *230* W Ecuador
Mantes-la-Jolie *255 prev.* Mantes-sur-Seine, Mantes-Gassicourt. N France
Mantova *322 Eng.* Mantua, *Fr.* Mantoue. N Italy
Mantsonyane *366* C Lesotho
Manuae *644* island of Southern Cook Islands, S Cook Islands
Manua Islands *642* island group of E American Samoa
Manukau Harbour *433* harbour of W North Island, New Zealand
Manurewa *433* N North Island, New Zealand
Manus Island *458 var.* Great Admiralty I. Island of NE Papua New Guinea
Manyame *636 var.* Panhame, *prev.* Hunyani. River of Mozambique and Zimbabwe
Manyame, Lake *636 prev.* Robertson, Lake. Reservoir of N Zimbabwe
Manyara, Lake *564* lake of NE Tanzania
Manyoni *564* C Tanzania
Manzanillo *210* SE Cuba
Manzhouli *187 var.* Man-chou-li. Nei Mongol Zizhiqu, NE China
Manzil Bū Ruqaybah *see* Menzel Bourguiba
Manzil Tamīm *see* Menzel Temime
Manzini *544 prev.* Bremersdorp. C Swaziland
Mao *180* W Chad
Mao *226* NW Dominican Republic
Maoke, Pegunungan *305 Dut.* Sneeuw-gebergte, *Eng.* Snow Mountains. Mountain range of Irian Jaya, E Indonesia
Mapoteng *366* NW Lesotho
Mapou *400* N Mauritius
Maputo *419 prev.* Lourenço Marques. ❖ of Mozambique, S Mozambique
Maputo, Baía de *419 var.* Baía de Lourenço Marques, *Eng.* Delagoa Bay. Bay on the coast of Mozambique
Mara *286* E Guyana
Marabá *145* NE Brazil
Maracaibo *623* NW Venezuela
Maracaibo, Lago de *623* inlet of Caribbean Sea, NW Venezuela
Maracay *623* N Venezuela
Marada *371* N Libya
Maradi *439* S Niger
Maragarazi *162, 564 var.* Muragarazi. River of Burundi and Tanzania
Marāgheh *309 var.* Maragha. NW Iran
Marahoué *see* Bandama Rouge
Marajó, Baía de *145* N Brazil
Marajó, Ilha de *145* island of N Brazil
Marakabei *366 var.* Marakabeis. C Lesotho
Marakei *348* island of the Gilbert Is, W Kiribati
Maralal *344* C Kenya
Maralik *98* W Armenia
Maramasike *526* island of E Solomon Is
Maramba *see* Livingstone
Marambio *90* Argentinian research station near Antarctic Peninsula, Antarctica
Maramvya *162* SW Burundi
Marandellas *see* Marondera
Marañón *463* river of N Peru
Marash *see* Kahramanmaraş
Maravovo *526* W Guadalcanal, Solomon Is
Marāwiḥ *594 var.* Merawwah. Island of W United Arab Emirates
Marbella *535* S Spain
Marburg *see* Maribor
Marburg an der Lahn *266* W Germany
Marcal *292* river of W Hungary
Marche *255* cultural region of C France
Marche-en-Famenne *127* SE Belgium
Marchena, Isla *230* island of N Galapagos Is, Ecuador
Marchfield *121* SE Barbados
Mar Chiquita, Lago *95* lake of C Argentina
Marcounda *see* Markounda
Marcovia *290* S Honduras
Mardān *451* N Pakistan
Mar del Plata *95* E Argentina
Mardin *581* SE Turkey
Maré *650* island, Îles Loyauté, E New Caledonia
Mareeq *528 var.* Mereeg, *It.* Meregh. E Somalia
Marek *see* Dupnitsa
Marfa Ridge *395* ridge of NW Malta
Margarita, Isla de *623* island of N Venezuela
Margate *531* Kwazulu Natal, SE South Africa
Margherita, Lake *see* Ābaya Hāyk'
Margherita Peak *588, 203* mountain of Uganda and Dem. Rep. Congo
Marghilon *614 var.* Margelan, *Rus.* Margilan. E Uzbekistan
Mārgow, Dasht-e- *77* desert of SW Afghanistan
Mari *215* S Cyprus
Marianao *210* NW Cuba
Marías, Islas *403* Island of W Mexico
Maria-Theresiopel *see* Subotica
Máriatölgyes *see* Dubnica nad Váhom
Mar'ib *631* W Yemen
Maribo *218* Lolland, S Denmark
Maribor *524 Ger.* Marburg. NE Slovenia
Marid *594* NE United Arab Emirates
Marie Byrd Land *90* physical region of Greater Antarctica, Antarctica
Marie-Galante *646* island of SE Guadeloupe
Mariehamn *251 var.* Maarianhamina. Aland, Finland
Mariel *210* NW Cuba
Marienburg *see* Alūksne
Mariental *423* S Namibia
Marigot *646* St. Martin, N Guadeloupe
Marigot *224* NE Dominica
Marigot de Baïla *510* river of SW Senegal
Mariguana *see* Mayaguana
Marijampolė *376 prev.* Kapsukas. S Lithuania
Marília *145* S Brazil
Marinduque Island *466* island of C Philippines
Maringá *145* S Brazil
Marins, Île aux *652* island of SE Saint Pierre and Miquelon
Marion Island *531* island of Prince Edward Islands, S South Africa
Ionio, Mar *see* Ionian Sea
Maripasoula *645* W French Guiana
Mariscal Estigarribia *460* NW Paraguay
Marisule Estate *497* N St Lucia
Maritsa *152, 275 var.* Marica, *Gk* Évros, *Turk.* Meriç. River of SE Europe
Mariupol' *590 prev.* Zhdanov. SE Ukraine
Mariy El, Respublika *484* autonomous republic of W Russian Federation
Märjamaa *242 Ger.* Merjama. W Estonia
Marjayoun *364 var.* Marj 'Uyūn. S Lebanon
Marka *528 var.* Merca. S Somalia
Marka *385* S Malawi
Market Shop *494* SE Nevis, St Kitts & Nevis
Markounda *179 var.* Marcounda. NW Central African Republic
Marlánské Lázně *216* W Czech Republic
Marmara Denizi *580 Eng.* Sea of Marmara. Sea to the NW of Turkey
Marmaris *580* SW Turkey
Marne *255* river of NE France
Marneuli *264* S Georgia
Maro *180* S Chad
Maroantsetra *382* NE Madagascar
Maromokotro *382* mountain of N Madagascar
Marondera *636 var.* Marandellas. NE Zimbabwe
Maroni *542, 645 Dut.* Marowijne. River of French Guiana and Suriname
Maros *see* Mureş
Marosvásárhely *see* Târgu Mureş
Marotiri *646* island group of S French Polynesia
Maroua *168* N Cameroon
Marovoay *382* NW Madagascar
Marowijne *see* Maroni
Marqūbān *115* NE Bahrain
Marquises, Îles *646* island group of N French Polynesia
Marrakech *414 var.* Marakesh, *Eng.* Marrakesh, *prev.* Morocco. W Morocco
Marrupa *419* N Mozambique
Marsa *395* C Malta
Marsá al Burayqah *see* Al Burayqah
Marsabit *344* N Kenya
Marsala *323* Sicilia, S Italy
Marsaxlokk *395* SE Malta
Marsaxlokk Bay *395* inlet on the SW coast of Malta
Marseille *255 prev. Eng.* Marseilles. SE France
Marshall *368* W Liberia
Marshall Islands *396-397* officially Republic of the Marshall Islands. Country of the Pacific Ocean divided into 33 admin. units (districts). ❖ Majuro
Marsh Harbour *112* Great Abaco, Bahamas
Martaban *159* SE Burma
Martadi *427 var.* Bajura. W Nepal
Martigny *550* SW Switzerland
Martigues *255* SE France
Martin *523 prev.* Turčiansky Svätý Martin, *Ger.* Sankt Martin, *Hung.* Turócszentmárton. NW Slovakia
Martinique *649* French overseas department of the Caribbean Sea. ❖ Fort-de-France.
Martinique Passage *224 var.* Dominica Channel, Martinique Channel. Passage connecting the Atlantic Ocean and Caribbean Sea between Dominica and Martinique
Martuni *98* E Armenia
Marungu *203* mountain range of SE Dem. Rep. Congo
Mary *585 prev.* Merv. SE Turkmenistan
Maryborough *101* E Australia
Mary Island *see* Kanton
Maryland *603* state of E USA
Marzūq *see* Murzuq
Masai Steppe *564* grassland of NW Tanzania
Masaka *588* SW Uganda
Masākin *see* M'saken
Masally *see* Massilı
Masampo *see* Masan
Masan *352 prev.* Masampo. S South Korea
Masasi *564* SE Tanzania
Masatepe *436* SW Nicaragua
Masaya *436* S Nicaragua
Masbate *466* island of C Philippines
Mascara *83 var.* Mouaskar. NW Algeria
Maseru *366* ❖ of Lesotho, W Lesotho
Mas-ha *320* W West Bank
Mashava *636 prev.* Mashaba. SE Zimbabwe
Mashhad *309 var.* Meshed. NE Iran
Māshkel *309, 450 var.* Rūd-i Māshkel, Māshkīd. River of Iran and Pakistan
Māshkel, Hāmūn-i *450* salt marsh of Iran and Pakistan
Māshkīd *see* Māshkel
Mashtagi *see* Mastağı
Masīlah, Wādī al *631* dry watercourse of E Yemen
Masindi *588* W Uganda
Masinga Reservoir *344* reservoir of C Kenya
Masirah, Gulf of *see* Maşīrah, Khalīj
Maşīrah, Jazīrat *448 var.* Masirah, Masira. Island of E Oman
Maşīrah, Khalīj *448 var.* Gulf of Masirah. Bay of the Arabian Sea, E Oman
Masis *98* SW Armenia
Masjed Soleymān *309 var.* Masjed-e Soleymān, Masjid-i Sulaiman. W Iran
Maskall *130* NE Belize
Maskanah *555 var.* Meskene. N Syria
Maskin *448 var.* Miskin. N Oman
Mask, Lough *316 Ir.* Loch Measca. Lake of W Ireland
Ma *627* river of Laos and Vietnam
Massa *322* N Italy
Massachusetts *603* state of NE USA
Massacre *224* W Dominica
Massawa *240 Amh.* Mits'iwa. E Eritrea
Massawa Channel *240* channel of the Red Sea between Dahlak Archipelago and mainland Eritrea
Massenya *180* SW Chad
Massif Central *255* plateau region of C France
Massilı *110 Rus.* Masally. S Azerbaijan
Massoukou *260 var.* Masuku, *prev.* Franceville. E Gabon
Maştağa *110 Rus.* Mastaga, *var.* Maštaga, Mashtagi. E Azerbaijan
Masterton *433* S North Island, New Zealand
Masuda *332* Honshū, W Japan
Masunga *142* NE Botswana
Masvingo *636 prev.* Nyanda, *prev.* Fort Victoria. SE Zimbabwe
Mât *652* river of NE Réunion
Matacawa Levu *248* island of the Yasawa Group, NW Fiji
Matadi *203* W Dem. Rep. Congo
Matagalpa *436* C Nicaragua
Matale *538* C Sri Lanka
Matam *510* NE Senegal
Matamoros *403* E Mexico
Matana *162* C Burundi
Matanzas *210* NW Cuba
Matara *538* S Sri Lanka
Mataró *535* E Spain

Meta Incognita Peninsula *171* peninsula of Baffin Island, NE Canada
Metangula *419* N Mozambique
Metapán *237* NW El Salvador
Metema *245* NW Ethiopia
Meterlam *see* Mehtarlām
Methariam *see* Mehtarlām
Metković *209* SE Croatia
Metu *245* *var.* Mattu, Mettu. W Ethiopia
Metz *255* NE France
Meuse *127, 255, 429* *var.* Maas. River of W Europe
Mexcala *see* Balsas
Mexiana, Ilha *145* island of N Brazil
Mexicali *403* NW Mexico
Mexicana, Altiplanicie *403* *Eng.* Plateau of Mexico, Mexican Plateau. Plateau of N Mexico
Mexico *402-405* officially United States of Mexico, *Sp.* Estados Unidos Mexicanos, Méjico. Country of North or Central America divided into 31 admin. units (states). ❖ Mexico City
Mexico City *403* *Sp.* Ciudad de México. of Mexico, C Mexico
Mexico, Gulf of *210, 403* *Sp.* Golfo de México. Gulf of the Atlantic Ocean, on the SE coast of North America
Mexico, Plateau of *see* Mexicana, Altiplanicie
Meyadine *see* Al Mayādīn
Meymaneh *77* *var.* Maimana. NW Afghanistan
Mezdra *152* NW Bulgaria
Mfanganu Island *344* *var.* Mfangano Island. Island of Lake Victoria, SW Kenya
Mfouati *200* S Congo
Mhangura *636* *var.* Mangula. N Zimbabwe
Mhlambanyatsi *544* W Swaziland
Mhlosheni *544* S Swaziland
Mhlume *544* NE Swaziland
Mhlumeni *544* NE Swaziland
Miami *603* Florida, SE USA
Miānwāli *451* NE Pakistan
Michalovce *523* *Ger.* Grossmichel, *Hung.* Nagymihály. E Slovakia
Michigan *603* state of NC USA
Michigan, Lake *171, 603* Lake of NC USA
Micomeseng *see* Mikomeseng
Micoud *497* SE St Lucia
Micronesia *406-407* officially Federated States of Micronesia, *prev.* Caroline Islands. Country of the Pacific Ocean divided into 4 admin. units (states). ❖ Palikir
Middelburg *429* SW Netherlands
Middelburg *531* Eastern Cape, S South Africa
Middelburg *531* Eastern Transvaal, NE South Africa
Middelfart *218* Fyn, SW Denmark
Middle Andaman *298* island of Andaman Islands to the SE of India
Middle Atlas *see* Moyen Atlas
Middlegate *650* C Norfolk Island
Middle Island *494* W St Kitts, St Kitts & Nevis
Middlesbrough *597* NE England, UK
Middlesex *130* E Belize
Mīdī *631* *var.* Maydī. NW Yemen
Miercurea-Ciuc *480* *Hung.* Csíkszereda. C Romania
Mieres del Camino *535* NW Spain
Mi'eso *245* *var.* Miesso, Meheso. C Ethiopia
Mikhaylovgrad *see* Montana
Mikhaylovka *484* W Russian Federation
Mikkeli *251* *Swe.* Sankt Michel. S Finland
Mikomeseng *238* *var.* Micomeseng. NE Río Muni, Equatorial Guinea
Mikuni-sammyaku *332* mountain range of Honshū, N Japan
Milagro *230* SW Ecuador
Milange *419* N Mozambique
Milano *322, 327.* *Eng.* Milan, *Ger.* Mailand. N Italy
Milas *580* SW Turkey
Mildura *101* SE Australia
Mil Düzü *110* *Rus.* Mil'skaya Step'. Physical region of C Azerbaijan
Milgis *344* *var.* Malgis. River of C Kenya
Mili *396* island of SE Marshall Islands
Milḥ, Baḥr al *see* Razāzah, Buḥayrat ar
Milḥ, Wādī al *see* Melah, Oued el
Millennium Island *348* *prev.* Caroline Island. Island of the Line Is, E Kiribati
Millet *497* C St Lucia
Millstätter See *106* lake of S Austria
Milo *283* river of E Guinea
Milondo, Mont *260* mountain of C Gabon
Mílos *275* island of SE Greece
Mil'skaya Step' *see* Mil Düzü
Milton Keynes *597* C England, UK
Milwaukee *603* Wisconsin, NC USA
Milyang *see* Miryang
Mimongo *260* C Gabon
Mīnā' 'Abd Allāh *356* *var.* Mina Abdulla. E Kuwait
Mīnā' al Aḥmadī *356* *var.* Mina Ahmadi. E Kuwait
Mīnā' Jabal 'Alī *594* NE United Arab Emirates
Minas *611* S Uruguay
Mīnā' Sa'ūd *356* *var.* Mīnā' Su'ud. SE Kuwait
Minas de Corrales *611* N Uruguay
Minas de Matahambre *210* W Cuba
Minatitlán *403* SE Mexico
Minbu *159* W Burma
Minch, The *597* strait of the Atlantic Ocean, between Outer Hebrides and Scotland
Mincivan *110* *Rus.* Mindzhivan. SW Azerbaijan
Mindanao *466* island of S Philippines
Mindanao Sea *see* Bohol Sea
Mindelo *176* *var.* Porto Grande. São Vincente, N Cape Verde
Mindoro *466* island of C Philippines
Mindoro Strait *466* strait connecting South China Sea and Sulu Sea
Mindouli *200* S Congo
Mindživan *see* Mincivan
Mingäçevir *110* *Rus.* Mingechaur *var.* Mingečaur. C Azerbaijan
Mingäçevir Su Anbarı *110* *Rus.* Mingechaurskoye Vodokhranilishche. Reservoir of NW Azerbaijan
Mingala *179* SE Central African Republic
Mingāora *451* *var.* Mingora, Mongora. N Pakistan
Mingechaurskoye Vodokhranilishche *see* Mingäçevir Su Anbarı
Ming-Kush *see* Min-Kush
Minho *see* Miño
Minicoy Island *298* island of Lakshadweep, SW India
Min-Kush *359* *Kir.* Ming-Kush. C Kyrgyzstan
Minna *440* C Nigeria
Minneapolis *603* Minnesota, NC USA
Minnesota *603* state of NC USA
Miño *474, 534* *Port.* Minho. River of Portugal and Spain
Minorca *see* Menorca
Minot *603* North Dakota, NC USA
Minsk *122* ❖ of Belarus, C Belarus
Minto Reef *406* atoll of C Micronesia
Minvoul *260* N Gabon
Minwakh *631* N Yemen
Miquelon *652* N Saint Pierre and Miquelon
Miquelon *652* island of N Saint Pierre and Miquelon
Miquelon, Cap *652* cape of the Atlantic Ocean on the coast of Miquelon, N Saint Pierre and Miquelon
Miragoâne *288* SW Haiti
Miranda de Ebro *535* N Spain
Mirbāṭ *448* *var.* Marbat. SW Oman
Mirebalais *288* C Haiti
Miri *386* NW Borneo, Malaysia
Mirim Lagoon *145, 611* *var.* Lake Mirim. Lagoon of Brazil and Uruguay
Mirim, Lake *see* Mirim Lagoon
Mirny *90* CIS research station of Greater Antarctica, Antarctica
Mirnyy *485* C Russian Federation
Mīrpur *see* New Mīrpur
Mirtóo Pelagos *275* *Eng.* Mirtoan Sea. Area of the Mediterranean Sea, S Greece
Miryang *352* *var.* Milyang *Jap.* Mitsuō. SE South Korea
Misery, Mount *see* Liamuiga, Mount
Miskito Coast *see* Mosquito Coast
Miskitos, Cayos *436* island group of NE Nicaragua
Miskolc *292* NE Hungary
Misool, Pulau *304* island of Maluku, E Indonesia
Miṣrātah *371* *var.* Misurata. N Libya
Mississippi *603* river of C USA
Mississippi *603* state of SE USA
Missoula *602* Montana, NW USA
Missouri *603* river of NC USA
Missouri *603* state of C USA
Misurata *see* Miṣrātah
Mitau *see* Jelgava
Mitchell *101* river of NE Australia
Mitèmboni *see* Mitemele, Río
Mitemele, Río *238* *var.* Mitèmboni, Temboni, Utamboni. River of Equatorial Guinea and Gabon
Mitiaro *644* island of Southern Cook Islands, S Cook Islands
Mito *332* Honshū, SE Japan
Mitre Island *see* Fatutaka
Mitrovica *see* Kosovska Mitrovica
Mitrovicë *see* Kosovska Mitrovica
Mitsamiouli *198* N Grande Comore, Comoros
Mits'iwa *see* Massawa
Mitsoudjé *198* SW Grande Comore, Comoros
Mitsuyō *see* Miryang
Mitú *195* SE Colombia
Mitumba, Monts *203* *var.* Chaîne des Mitumba, Mitumba Range. Mountain range of E Dem. Rep. Congo
Mitzic *260* N Gabon
Miyako *333* Honshū, N Japan
Miyako-jima *332* island of Sakishima-shotō, SW Japan
Miyakonojō *332* Kyūshū, SW Japan
Miyazaki *332* Kyūshū, SW Japan
Miyoshi *332* Honshū, W Japan
Mizdah *371* *var.* Mizda. NW Libya
Mjøsa *444* *var.* Mjøsen. Lake of SE Norway
Mkhondvo *544* *var.* Mkondo. River of South Africa and Swaziland
Mladá Boleslav *216* *Ger.* Jungbunzlau. N Czech Republic
Mlanje *see* Mulanje
Mljet *209* *It.* Meleda. Island of S Croatia
Mmabatho *531* North West, N South Africa
Mmathethe *142* S Botswana
Mnjoli Dam *544* reservoir of NE Swaziland
Mo *444* NE Norway
Moa *518* river of W Africa
Moa *211* SE Cuba
Moabi *260* SW Gabon
Moala *248* island to the SE of Viti Levu, S Fiji
Moamba *419* SW Mozambique
Moanda *260* SE Gabon
Moba *203* E Dem. Rep. Congo
Mobaye *179* S Central African Republic
Mobile *603* Alabama, SE USA
Moca *226* N Dominican Republic
Moçambique *419* island and settlement of NE Mozambique
Moçâmedes *see* Namibe
Moce *248* island of the Lau Group, E Fiji
Mocha *see* Al Mukhā
Mochudi *142* S Botswana
Mocímboa da Praia *419* *var.* Vila de Mocímboa da Praia. N Mozambique
Môco *88* *var.* Serra Môco, Morro de Môco. Mountain of W Angola
Mocoa *195* SW Colombia
Mocuba *419* E Mozambique
Modena *322* NW Italy
Mödling *106* NE Austria
Modohn *see* Madona
Modriča *140* N Bosnia & Herzegovina
Moe *101* SE Australia
Moen *see* Weno
Möen *see* Møn
Moena *see* Muna, Pulau
Moengo *542* NE Suriname
Moesi *see* Musi
Moeskroen *see* Mouscron
Mogadishu *528* *Som.* Muqdisho, *It.* Mogadiscio. ❖ of Somalia, S Somalia
Mogador *see* Essaouira
Mogilëv *see* Mahilyow
Mogotón, Pico *436* mountain of NW Nicaragua
Mohales Hoek *366* SW Lesotho
Mohammadia *83* *var.* El Mohammaidia. NW Algeria
Mohammedia *414* *prev.* Fédala. NW Morocco
Moharek *see* Al Muḥarraq
Mohéli *198* *var.* Mwali. Island of Comoros
Mohn *see* Muhu
Moindou *650* C New Caledonia
Mõisaküla *242* *Ger.* Moiseküll. S Estonia
Moïssala *180* S Chad
Mokhotlong *366* NE Lesotho
Mokil *406* atoll of E Micronesia
Moknine *577* *var.* Al Muknīn. NE Tunisia
Mokp'o *352* *Jap.* Moppo. SW South Korea
Mokra Gora *512* mountain range of SW Serbia, Serbia & Montenegro (Yugo.)
Mokwa *440* W Nigeria
Moldau *see* Vltava
Moldova *408-409* officially Republic of Moldova, *var.* Moldova, *prev.* Moldavian SSR, *Rus.* Moldavskaya SSR. Country of E Europe divided into 40 admin. units (districts). ❖ Chişinău
Molde *444* SW Norway
Moldo-Too, Khrebet *359* mountain range of C Kyrgyzstan
Moldova *see* Moldova
Molepolole *142* S Botswana
Môle-St-Nicolas *288* NW Haiti
Molineux *494* NE St Kitts, St Kitts & Nevis
Möll *106* river of S Austria
Mölndal *547* SW Sweden
Molodechno *see* Maladzyechna
Molodeczno *see* Maladzyechna
Molodezhnaya *90* CIS research station of Greater Antarctica, Antarctica
Molokai *602* island of Hawaii, USA, C Pacific
Molopo *142, 531* seasonal river of southern Africa
Molotov *see* Severodvinsk
Molotov *see* Perm'
Moloundou *168* SE Cameroon
Moluccas *see* Maluku
Molucca Sea *see* Maluku, Laut
Mombasa *344* SE Kenya
Môn *see* Anglesey
Møn *218* *prev.* Möen. Island of SE Denmark
Mona, Canal de la *226, 651* channel connecting the Atlantic Ocean and Caribbean Sea, between Dominican Republica and Puerto Rico
Monaco *410-411* officially Principality of Monaco. Country of W Europe divided into 4 admin. units (quarters). ❖ Monaco
Monaco *see* München
Monaghan *316* *Ir.* Muineachán. NE Ireland
Monagrillo *456* S Panama
Mona, Isla *651* island of SW Puerto Rico
Monapo *419* NE Mozambique
Monaragala *538* SE Sri Lanka
Monastir *577* *var.* Al Munastīr. NE Tunisia
Monastir *see* Bitola

Mönchengladbach *266* *prev.* München-Gladbach. W Germany
Monchy *497* N St Lucia
Monclova *403* N Mexico
Moncton *171* SE Canada
Mondego *474* river of N Portugal
Mondsee *106* lake of N Austria
Money Island *651* island of W Paracel Islands
Monfalcone *323* N Italy
Mongar *134* E Bhutan
Mongo *180* C Chad
Mongolia *412-413* country of NE Asia divided into 21 admin. units (18 provinces, 3 cities). ❖ Ulan Bator
Mongomo *238* E Río Muni, Equatorial Guinea
Mongora *see* Mingāora
Mongos, Chaîne des *see* Bongo, Massif des
Mongouge *497* SW St Lucia
Mongoumba *179* SW Central African Republic
Mongu *635* W Zambia
Mönh Hayrhan Uul *412* mountain of W Mongolia
Moni *215* S Cyprus
Monkey Bay *385* SE Malawi
Monkey River Town *130* SE Belize
Mono *132, 571* river of Benin and Togo
Monopoi *323* E Italy
Monos *574* island to the NW of Trinidad, Trinidad & Tobago
Mon Repos *497* E St Lucia
Monrovia *368* ❖ of Liberia, W Liberia
Mons *127* *Dut.* Bergen. SW Belgium
Montana *602-603* state of NW USA
Montana *152* *prev.* Mikhaylovgrad, *var.* Mihajlovgrad, Mikhailovgrad, *prev.* Ferdinand. NW Bulgaria
Montauban *255* S France
Montbéliard *255* NE France
Mont Blanc *255, 322* *It.* Monte Bianco. Mountain of France and Italy
Mont-de-Marsan *255* SW France
Monteagudo *136* S Bolivia
Monte-Carlo *410* NE Monaco
Monte Cristi *226* NW Dominican Republic
Monte Croce Carnico, Passo di *see* Plöcken
Montegiardino *502* SE San Marino
Montego Bay *331* NW Jamaica
Montelindo *460* river of C Paraguay
Montenegro *512* *Serb.* Crna Gora. Republic of Serbia & Montenegro (Yugo.)
Monte Plata *226* C Dominican Republic
Montepuez *419* N Mozambique
Montería *195* NW Colombia
Montero *136* C Bolivia
Monterrey *403* N Mexico
Montes Claros *145* SE Brazil
Montevideo *611* ❖ of Uruguay, S Uruguay
Montgomery *603* Alabama, SE USA
Montgomery *see* Sāhīwāl
Monthey *550* SW Switzerland
Montijo *474* W Portugal
Montpelier *603* Vermont, NE USA
Montpellier *255* S France
Montréal *171* *Eng.* Montreal. SE Canada
Montreux *550* SW Switzerland
Montserrat *649* British dependent territory of the Caribbean Sea. ❖ Plymouth
Montsinéry *645* NE French Guiana
Monywa *159* NW Burma
Monza *322* N Italy
Monze *635* S Zambia
Moora *100* W Australia
Moose Jaw *170* SW Canada
Moosonee *171* SE Canada
Moppo *see* Mokp'o
Mopti *392* C Mali
Moqor *77* SE Afghanistan
Moquegua *463* SE Peru
Morales *280* E Guatemala
Moramanga *382* E Madagascar
Morant Bay *331* E Jamaica
Morata *see* Goodenough Island
Moratuwa *538* SW Sri Lanka
Morava *see* Velika Morava
Morava *216, 523* *Ger.* March. River of C Europe
Moravia *216* cultural region of E Czech Republic
Moravská Ostrava *see* Ostrava
Morawhanna *286* N Guyana
Moray Firth *597* inlet of Atlantic Ocean, NE Scotland, UK
Mordoviya SSR *484* autonomous republic of W Russian Federation
Moree *101* E Australia
Morehead *458* SW Papua New Guinea
Morelia *403* C Mexico
Moreno *136* N Bolivia
Morfou *see* Güzelyurt
Morfou Bay *see* Morphou Bay
Morges *550* SW Switzerland
Morghāb, Daryā-ye *77* river of Afghanistan and Turkmenistan
Morija *366* W Lesotho
Morioka *332* Honshū, N Japan
Morne Aux Frégates *224* E Dominica
Morne Diablotins *224* mountain of N Dominica
Morne Raquette *224* W Dominica
Morne Seychellois *516* mountain of Mahé, NE Seychelles
Morne Trois Pitons *224* mountain of C Dominica
Morocco *see* Marrakech
Morocco *414-417* officially Kingdom of Morocco, *Ar.* Al Mamlakah. Country of N Africa divided into 37 admin. units (provinces and prefectures). ❖ Rabat
Morogoro *564* E Tanzania
Moro Gulf *466* area of the Sulu Sea
Morombe *382* SW Madagascar
Morón *210* C Cuba
Morón *95* E Argentina
Mörön *412* N Mongolia
Morondava *382* W Madagascar
Moroni *198* ❖ of Comoros, W Grande Comore, Comoros
Morotai, Pulau *304* island of Maluku, E Indonesia
Moroto *588* NE Uganda
Morphou *see* Güzelyurt
Morphou Bay *215* *var.* Morfou Bay, Güzelyurt Körfezi. Bay of the Mediterranean Sea, on the NW coast of Cyprus
Morris *649* S Montserrat
Morro de Môco *see* Môco
Mors *218* island of NW Denmark
Moruga *574* S Trinidad, Trinidad & Tobago
Morvan *255* physical region of NE France
Morvant *574* NW Trinidad, Trinidad & Tobago
Morwell *101* SE Australia
Moscow *484, 488* *Rus.* Moskva. ❖ of Russia, W Russian Federation
Mosel *255, 266, 378* *Fr.* Moselle. River of W Europe
Moselle *see* Mosel
Moshi *564* NE Tanzania
Mosjøen *444* C Norway
Moskva *562* *prev.* Chubek; *Rus.* Moskovskiy. SW Tajikistan
Moson *see* Mosonmagyaróvár
Mosoni-Duna *292* river of NW Hungary
Mosonmagyaróvár *292* *prev.* Moson, Magyaróvár, *Ger.* Wieselburg. NW Hungary
Mosquito Coast *290, 436* *var.* Miskito Coast, *Sp.* La Mosquitia. Coastal region of E Central America
Mosquitos, Golfo de *456* gulf of the Caribbean Sea to the N of Panama
Moss *444* S Norway
Mossâmedes *see* Namibe
Mosselbaai *531* *Eng.* Mossel Bay. Western Cape S South Africa
Mossendjo *200* SW Congo
Mossoró *145* NE Brazil
Most *216* *Ger.* Brüx. NW Czech Republic
Mosta *395* *var.* Musta. C Malta
Mostaganem *83* *var.* Mestghanem. NW Algeria
Mostar *140* S Bosnia & Herzegovina
Mosul *see* Al Mawşil
Mota *619* Banks Islands, N Vanuatu
Motaba *200* river of N Congo
Motagua *280* river of Guatemala and Honduras
Mota Lava *619* Banks Islands, N Vanuatu
Mothae *366* NE Lesotho
Motherwell *597* C Scotland, UK
Motjane *544* NW Swaziland
Motril *535* S Spain
Motueka *433* river of NE of South Island, New Zealand
Motulalo *587* islet of Nukufetau, Tuvalu
Motulua *587* islet of Nukufetau, Tuvalu
Motumua *587* islet of Nukufetau, Tuvalu
Mouaskar *see* Mascara
Moucha, Îles *222* islands in the Gulf of Aden, of E Djibouti
Mouchoir Passage *653* passage of the Pacific Ocean, SE Turks and Caicos Islands
Moudjéria *398* SW Mauritania
Mouila *260* C Gabon
Mould Bay *170* Prince Patrick Island, N Canada
Moulèngui Binza *260* S Gabon
Moulhoulé *222* N Djibouti
Moulins *255* C France
Moulmein *159* *var.* Mawlamyine. SE Burma
Moulmeingyun *159* SW Burma
Mouloud *222* SW Djibouti
Moulouya *414* seasonal river of NE Morocco
Moun Hou *see* Black Volta
Mounana *260* S Gabon
Moundou *180* SW Chad
Moŭng Roessi *165* *var.* Moung. W Cambodia
Mount Barclay *368* W Liberia
Mount Darwin *636* NE Zimbabwe
Mount Friendship *121* SW Barbados
Mount Gambier *101* SE Australia
Mount Hagen *458* C Papua New Guinea
Mount Isa *101* C Australia
Mount Lebanon *see* Liban, Jebel
Mount Magnet *100* W Australia
Mount Maunganui *433* C North Island, New Zealand
Mount Moorosi *366* SW Lesotho
Mount Moritz *278* SW Grenada island, Grenada
Mount Rose *278* NE Grenada island, Grenada
Mourdi, Dépression de *180* desert lowland of E Chad
Mouscron *127* *Dut.* Moeskroen. W Belgium
Moussa 'Ali *222* mountain of NW Djibouti
Moussoro *180* W Chad
Moutsamoudou *198* NW Anjouan, Comoros
Moya *198* SW Anjouan Comoros
Moyale *245* S Ethiopia
Moyale *344* N Kenya
Moyamba *518* W Sierra Leone
Moyen Atlas *414* *Eng.* Middle Atlas. Mountain range of N Morocco
Moyeni *366* *var.* Quthing. SW Lesotho
Moyo *588* NW Uganda
Moyobamba *463* NW Peru
Moyuta *280* S Guatemala
Mozambique *418-421* officially Republic of Mozambique, *prev.* People's Republic of Mozambique, Portuguese East Africa. Country of East Africa divided into 10 admin. units (provinces). ❖ Maputo
Mozambique Channel *198, 382* *var.* Canal de Mozambique, Canal de Moçambique, Lakandranon' i Mozambika. Strait of the Indian Ocean, between Mozambique and Madagascar
Mozambique Island *see* Moçambique
Mozyr' *see* Mazyr
Mpama *200* river of C Congo
Mpanda *564* W Tanzania
Mpanga, Lac *492* lake of E Rwanda
Mphaki *366* S Lesotho
Mpigi *588* S Uganda
Mpika *635* NE Zambia
Mporokoso *635* N Zambia
Mpulungu *635* N Zambia
Mpumalanga *531* *prev.* Eastern Transvaal. Province of NE South Africa
Mpwapwa *564* C Tanzania
Mqocha *385* NW Malawi
Mrémani *198* S Anjouan, Comoros
M'Saken *577* *var.* Masakin. N Tunisia
Msida *395* C Malta
M 'Sila *83* N Algeria
Mtoko *see* Mutoko
Mtsamboro *649* N Mayotte
Mtsamgamouji *649* W Mayotte
Mtsapa *see* Matsapha
Mtwara *564* SE Tanzania
Mu'a *572* Tongatabu, Tongatapu Group, Tonga
Mualama *419* E Mozambique
Mualo *see* Messalo
Muang Ham *360* NE Laos
Muang Hinboun *360* S Laos
Muang Kalasin *see* Kalasin
Muang Khammouan *360* *var.* Thakhek. S Laos
Muang Không *360* S Laos
Muang Khôngxédôn *360* *var.* Khong Sedone. S Laos
Muang Khon Kaen *see* Khon Kaen
Muang Lampang *see* Lampang
Muang Loei *see* Loei
Muang Lom Sak *see* Lom Sak
Muang Nakhon Sawan *see* Nakhon Sawan
Muang Namo *360* N Laos
Muang Nan *see* Nan
Muang Pakxan *360* *var.* Pak Sane. C Laos
Muang Phalan *360* S Laos
Muang Phin *360* S Laos
Muang Phitsanulok *see* Phitsanulok
Muang Phôn-Hông *360* C Laos
Muang Phrae *see* Phrae
Muang Roi Et *see* Roi Et
Muang Samut Prakan *see* Samut Prakan
Muang Sing *360* NW Laos
Muang Vangviang *360* C Laos
Muang Xaignabouri *360* *var.* Sayaboury. W Laos
Muang Xay *360* *var.* Muong Sai. N Laos
Muang Xépôn *360* *var.* Sepone. SE Laos
Muar *386* river of S Peninsular Malaysia
Muar *386* *prev.* Bandar Maharani. S Peninsular Malaysia
Muara Besar, Pulau *150* island of N Brunei
Mubende *588* SW Uganda
Mucojo *419* N Mozambique
Mudanjiang *187* *var.* Mu-tan-chiang. Heilongjiang, NE China
Mudon *159* SE Burma
Mueda *419* NE Mozambique
Muenster *see* Münster
Mufulira *635* C Zambia
Muğan Düzü *110* *Rus.* Muganskaya Step'. Physical region of S Azerbaijan
Mugera *162* C Burundi
Mugesera, Lac *492* lake of E Rwanda
Mughsu *562* *Rus.* Muksu. River of NE Tajikistan
Mugi *332* Shikoku, SW Japan
Mugla *580* SW Turkey
Muharraq *see* Al Muḩarraq
Muḩarraq, Jazīrat al *115* island of N Bahrain
Muhazi, Lac *492* lake of E Rwanda
Muhinga *see* Muyinga
Muh, Sabkhat al *555* salt-flat of S Syria

N

Nangen *see* Namwŏn
Nan Hai *see* East China Sea and South China Sea
Nanhsi *559* SW Taiwan
Nanjing *187 var.* Nanking, Nan-ching. Jiangsu, E China
Nankai-tō *see* Namhae-do
Nanning *187 prev.* Yung-ning. Guangxi, S China
Nanortalik *646* S Greenland
Nansei-shotō *332* island group to the SW of Kyūshū, SW Japan
Nanshan Island *652* island of E Spratly Islands
Nansio *564* NW Tanzania
Nanterre *255* N France
Nantes *254* W France
Nanthi Kadal Lagoon *538* lagoon of N Sri Lanka
Nant'ou *559* W Taiwan
Nanuku Passage *248* channel of the Pacific Ocean between the Lau Group and Taveuni, NE Fiji
Nanumaga *587 prev.* Nanumanga. Coral atoll of NW Tuvalu
Nanumea *587* coral atoll of NW Tuvalu
Nan Wan *559* bay of the South China Sea, S Taiwan
Nanyang *187* Henan, C China
Nanyuki *344* C Kenya
Naogaon *117* NW Bangladesh
Napier *433* SE North Island, New Zealand
Naples *see* Napoli
Napo *230, 463* river of Ecuador and Peru
Napoli *323 Eng.* Naples, *Ger.* Neapel. S Italy
Nāra *451* irrigation canal of S Pakistan
Nara *332* Honshū, C Japan
Narathiwat *567* S Thailand
Narayani *427* river of C Nepal
Narbada *see* Narmada
Narbonne *255* S France
Nare's Strait *646* strait of NW Greenland
Narew *471* river of E Poland
Narganá *456* NE Panama
Narikrik *396 prev.* Knox Atoll. Atoll of SE Marshall Islands
Narmada *298 var.* Narbada. River of C India
Narok *344* SW Kenya
Närpes *251 Swe.* Närpiö. SW Finland
Narrows, The *494* channel connecting the Atlantic Ocean and Caribbean Sea, between Nevis and St Kitts
Narsaq Kujalleq *646 Dan.* Frederiksdal. S Greenland
Narsingdi *117* C Bangladesh
Nartës, Gjol i *see* Nartës, Liqeni i
Nartës, Liqeni i *81 var.* Gjol i Nartës. Lake of SW Albania
Naruto *332* Shikoku, SW Japan
Narva *242 prev.* Narova. River of Estonia and Russian Federation
Narva *242* NE Estonia
Narva Bay *242 Est.* Narva Laht, *Rus.* Narviskiy Zaliv. Bay of the Gulf of Finland
Narva Reservoir *242 Est.* Narva Veehoidla. Reservoir of Estonia and Russian Federation
Narvik *444* NE Norway
Nar'yan-Mar *484 prev.* Dzerzhinskiy, *prev.* Beloshchel'ye. NW Russian Federation
Naryn *359* E Kyrgyzstan
Naryn *359* river of Kyrgyzstan and Uzbekistan
Nasau *248* Koro, C Fiji
Nāshik *298 prev.* Nāsik. W India
Nasho, Lac *492* lake of E Rwanda
Nashville *603* Tennessee, SE USA
Näsijärvi *251* lake of SW Finland
Nasirābād *see* Mymensingh
Nâşir, Buḩeirat *232 var.* Buḩayrat Nāşir, *Eng.* Lake Nser. Lake of Egypt and Sudan
Nasiriya *see* An Nāşirīyah
Nás Na Ríogh *see* Naas
Nassau *112* ❖ of Bahamas, New Providence, Bahamas
Nassau *644* island of Northern Cook Islands, N Cook Islands
Nasser, Lake *see* Nâsir, Buheiret
Nata *142* NE Botswana
Natal *145* E Brazil
Nathula *see* Nacula
Natitingou *132* NW Benin
Natl *338 var.* Nitil. NW Jordan
Nator *117* W Bangladesh
Natron, Lake *564* lake of Kenya and Tanzania
Natuna Besar, Pulau *304* island of Kepulauan Natuna, W Indonesia
Natuna, Kepulauan *304* island group to the NW of Borneo, W Indonesia
Nau *see* Nov
Naujoji Akmenė *376* NW Lithuania
Nā'ūr *338* NW Jordan
Nauru *424-425* officially The Republic of Nauru, *prev.* Pleasant Island. Island country of the Pacific Ocean divided into 14 admin. units (districts)
Naushahra *see* Nowshera
Nausori *248* Viti Levu, Fiji
Navabad *see* Navobod
Navaga *248* W Koro, W Fiji
Navahrudak *122 Rus.* Novogrudok, *Pol.* Nowogródek. W Belarus
Navangar *see* Jāmnagar
Navapolatsk *122 Rus.* Novopolotsk. N Belarus
Navarra *535* autonomous community of N Spain
Naviti *248* island of the Yasawa Group, NW Fiji
Navoalevu *248* NE Vanua Levu, N Fiji
Navobod *562 Rus.* Navabad. W Tajikistan
Navoi *see.* Nawoiy
Navua *248* Viti Levu, W Fiji
Nawābganj *117* NW Bangladesh
Nawābshāh *451* S Pakistan
Nawmah, Ra's *115 var.* Ra's Noma. Cape of SW Bahrain
Nawoiy *614 Rus.* Navoi. S Uzbekistan
Naxçıvan *110 Rus.* Nakhichevan', *var.* Nahičevan'. SW Azerbaijan
Náxos *275* island of SE Greece
Nayau *248* island of the Lau Group, E Fiji
Nazareth *see* Nazerat
Nazca *463* S Peru
Naze *332* Nansei-shotō, SW Japan
Nazerat *319 Eng.* Nazareth. N Israel
Nazerat 'Illit *319* N Israel
Nazilli *580* SW Turkey
Nazran' *484* SW Russian Federation
Nazrēt *245 var.* Adama, Hadama. C Ethiopia
Nazwá *448* N Oman
Nchelenge *635* N Zambia
Ncheu *see* Ntcheu
Nchisi *see* Ntchisi
Ncue *238* N Río Muni, EquatorialGuinea
Ndaghamcha, Sebkra de *see* Te-n-Dghâmcha, Sebkhet
N'Dalatando *88 Port.* Vila Salazar. NW Angola
Ndali *132* C Benin
Ndélé *179* N Central African Republic
Ndendé *260* S Gabon
Ndeni *see* Nendö
Ndindi *260* S Gabon
N'Djamena *180 var.* Njamena, *prev.* Fort-Lamy. ❖ of Chad, W Chad
Ndjolé *260* C Gabon
Ndoki *200* river of N Congo
Ndola *635* C Zambia
Ndora *162* NW Burundi
Ndréméani *198* S Mohéli, Comoros
Ndrhamcha, Sebkha de *see* Te-n-Dghâmcha, Sebkhet
Nduindui *526* S Guadalcanal, Solomon Is
Nduke *see* Kolombangara
Neagh, Lough *597* lake of Northern Ireland, UK
Neapel *see* Napoli
Nébeck *see* An Nabk
Nebitdag *585* W Turkmenistan
Nebk *see* An Nabk
Neblina, Pico da *144* mountain of NW Brazil
Nebraska *603* state of C USA
Neckar *266* river of SW Germany
Necochea *95* E Argentina
Nederland *see* Netherlands
Neder-Rijn *429 Eng.* Lower Rhine. River of C Netherlands
Nefasit *240* C Eritrea
Nefta *577 var.* Naftah. W Tunisia
Neftezavodsk *see* Seydi
Negara Brunei Darussalam *see* Brunei
Negēlē *245 var.* Negelli, *It.* Neghelli. S Ethiopia
Negev *see* HaNegev
Neghelli *see* Negēlē
Negomane *419 var.* Negomano. N Mozambique
Negombo *538* SW Sri Lanka
Negotino *381* C FYR Macedonia
Negril *331* W Jamaica
Negro, Rio *144, 195* river of N South America
Negro, Río *see* Sico
Negro, Río *611* river of Brazil and Uruguay
Negro, Río *see* Chixoy
Negros *466* island of C Philippines
Neiafu *572* Uta Vava'u, Vava'u Group, Tonga
Neiba *226* SW Dominican Republic
Neiges, Piton des *652* mountain of C Réunion
Neily *206* SE Costa Rica
Nei Mongol Zizhiqu *187 Eng.* Inner Mongolian Autonomous Region, *prev.* Nei Monggol Zizhiqu. Autonomous region of N China
Neiva *195* W Colombia
Nek'emtē *245 var.* Nakamti, Lakamti, Lekemti. W Ethiopia
Nelson *170* river of C Canada
Nelson *433* N South Island, New Zealand
Nelson Island *643* island of N British Indian Ocean Territory
Nelspruit *531* Eastern Transvaal, NE South Africa
Néma *398* SE Mauritania
Neman *122, 376 Bel.* Nyoman, *Lith.* Nemunas, *Ger.* Memel, *Pol.* Niemen. River of NE Europe
Německý Brod *see* Havlíčkův Brod
Nemunas *see* Neman
Nenagh *316* S Ireland
Nendeln *374* C Liechtenstein
Nendö *526 var.* Ndeni. Santa Cruz Is, Solomon Islands
Nepal *426-427* officially Kingdom of Nepal. Country of Asia divided into 5 admin. units (regions). ❖ kathmandu
Nepalganj *427* W Nepal
Nepean Island *650* island of C Norfolk Island
Neretva *140* river of S Bosnia & Herzegovina
Neris *376 Bel.* Viliya, *Pol.* Wilja. River of Belarus and Lithuania
Neskaupstadhur *296* E Iceland
Ness, Loch *597* lake of N Scotland, UK
Néstos *152, 275 Turk.* Kara Su, *Bul.* Mesta. River of Bulgaria and Greece
Netanya *319* C Israel
Netherlands *430-431* officially Kingdom of the Netherlands, *var.* Holland, *Dut.* Nederland. Country of W Europe divided into 12 admin. units (provinces). ❖ Amsterdam, The Hague
Netherlands Antilles *623, 650 prev.* Dutch West Indies. Autonomous part of the Netherlands, Caribbean Sea. ❖ Willemstad
Netherlands East Indies *see* Indonesia
Netrakona *117* N Bangladesh
Netze *see* Noteć
Neubrandenburg *267* NE Germany
Neuchâtel *550 Ger.* Neuenburg. W Switzerland
Neuchâtel, Lac de *550 Ger.* Neuenburger See. Lake of W Switzerland
Neuenburger See *see* Neuchâtel, Lac de
Neugradiska *see* Nova Gradiška
Neuhäusl *see* Nové Zámky
Neumarkt *see* Târgu Mures
Neumarktl *see* Tržič
Neumünster *267* N Germany
Neunkirchen *106* E Austria
Neuquén *95* SE Argentina
Neusatz *see* Novi Sad
Neusiedler See *106, 292 Hung.* Fertő-tó. Lake of Austria and Hungary
Neusohl *see* Banská Bystrica
Neustadt *see* Baia Mare
Neustadtl *see* Novo Mesto
Neutra *see* Nitra
Neu-Ulm *267* S Germany
Nevada *602* state of W USA
Nevers *255* C France
Nevis *494* island of the Lesser Antilles which, with St Kitts, forms the independent state of St Kitts & Nevis
Nevis Peak *494* mountain peak of C Nevis, St Kitts & Nevis
Nevşehir *581* C Turkey
Newala *564* SE Tanzania
New Amsterdam *286* E Guyana
New Britain *458* island of E Papua New Guinea
New Brunswick *171* province of SE Canada
New Bussa *440* W Nigeria
New Caledonia *650* French overseas territory of the Pacific Ocean ❖ Nouméa
Newcastle *101* E Australia
Newcastle *494* N Nevis, St Kitts & Nevis
Newcastle upon Tyne *597* NE England, UK
New Delhi *298* ❖ of India, N India
Newfield *92* SE Antigua, Antigua & Barbuda
Newfoundland & Labrador *171 Fr.* Terre-Neuve. Island of S E Canada
Newfoundland & Labrador *171* province of E Canada
New Georgia *526* island of the New Georgia Is, W Solomon Is
New Georgia Islands *526* island group of W Solomon Is
New Guinea *305, 458 Dut.* Nieuw Guinea, *Ind.* Irian. Large island of W Pacific Ocean, divided administratively into the Indonesian state of Irian Jaya and the independent country of Papua New Guinea
New Halfa *see* Halfa el Gadida
New Hampshire *603* state of NE USA
New Haven *603* Connecticut, NE USA
New Hebrides *see* Vanuatu
New Ireland *458* island of NE Papua New Guinea
New Jersey *603* state of E USA
Newman *100* W Australia
New Mexico *602-603* state of SW USA
New Mīrpur *451 prev.* Mīrpur. NE Pakistan
New Orleans *603* Louisiana, SC USA
New Plymouth *433* SW North Island, New Zealand
Newport *597* S Wales, UK
Newport News *603* Virginia, E USA
New Providence *112* island of C Bahamas
New River *286* river of SE Guyana
New River *130* river of N Belize
New Ross *316* SE Ireland
Newry *597* Northern Ireland, UK
New Sandy Bay Village *498* N St Vincent, St Vincent & the Grenadines
New Siberian Islands *see* Novosibirskiye Ostrova
New South Wales *101* state of SE Australia
Newton Ground *494* NW St Kitts, St Kitts & Nevis
Newtownabbey *597* Northern Ireland, UK
New Winthorpes *92* N Antigua, Antigua & Barbuda
New York *603* state of NE USA
New York *603, 607* New York, NE USA

New Zealand *432-435* officially The Dominion of New Zealand. Country of the Pacific Ocean, divided into 14 admin. units (regions). ❖ Wellington
Nezhyn *590* N Ukraine
Ngabé *200* SE Congo
Ngadda *441* river of NE Nigeria
Ngala *441* NE Nigeria
Ngangerabeli Plain *344* plain of SE Kenya
Ngaoundéré *168* *var.* N'Gaoundéré, N'Gaundere. N Cameroon
Ngara *564* NW Tanzania
Ngarama *492* N Rwanda
Ngardmau *455* C Palau
Ngaruangl *455* island of N Palau
Ngatik *406* atoll of E Micronesia
Ngau *see* Gau
N'Gaundere *see* Ngaoundéré
Nggamea *see* Qamea
Nggatokae *526* island of the New Georgia Islands, W Solomon Islands
Nggwavuma *544* *var.* Ingwavuma. River of South Africa and Swaziland
N'Giva *88* *var.* Ondjiva *Port.* Vila Pereira de Eça. S Angola
Ngo *200* SE Congo
Ngogolo *544* C Swaziland
Ngoko *168, 200* river of Cameroon and Congo
Ngorongoro Crater *564* crater and conservation area of N Tanzania
Ngororero *492* W Rwanda
Ngounié *260* river of Congo and Gabon
Ngouoni *260* E Gabon
Ngourti *439* E Niger
Ngozi *162* N Burundi
Nguigmi *439* SE Niger
Ngulu *406* atoll of W Micronesia
Ngum *360* river of C Laos
Nguna *619* island of C Vanuatu
Ngundu *636* S Zimbabwe
N'Gunza *see* Sumbe
Nguru *440* NE Nigeria
Ngwempisi *544* river of South Africa and Swaziland
Ngweze *see* Katima Mulilo
Nhacra *284* W Guinea-Bissau
Nhamundá *144* *var.* Yamundá, Jamundá. River of NW Brazil
Nha Trang *627* SE Vietnam
Nhlangano *544* *prev.* Goedgegun. SW Swaziland
Niagara Falls *171* SE Canada
Niagassola *283* *var.* Nyagassola. NE Guinea
Niamey *439* ❖ of Niger, SW Niger
Niamtougou *571* N Togo
Niandan *283* river of E Guinea
Niangay, Lac *392* lake of E Mali
Nianija Bolon *263* river of Gambia and Senegal
Niantanina *283* E Guinea
Niari *200* river of S Congo
Nias, Pulau *304* island to the W of Sumatra, W Indonesia
Niassa, Lago *see* Nyasa, Lake
Nicaragua *436-437* officially Republic of Nicaragua. Country of Central America divided into 16 admin. units (departments). ❖ Managua
Nicaragua, Lago de *436* *var.* Cocibolca, Gran Lago. Lake of S Nicaragua
Nicastro *323* S Italy
Nice *255* *It.* Nizza. SE France
Nicholls Town *112* Andros I, Bahamas
Nickerie *542* river of NW Suriname
Nicobar Islands *298* island group to the SE of India
Nicosia *215* *var.* Lefkosia, *Turk.* Lefkoşa. ❖ of Cyprus, C Cyprus
Nicoya *206* W Costa Rica
Nicoya, Península de *206* peninsula of W Costa Rica
Nictheroy *see* Niterói
Nidaros *see* Trondheim
Niedere Tauern *106* mountain range of C Austria
Niefang *238* *var.* Sevilla de Niefang. NW Río Muni, Equatorial Guinea
Niemen *see* Neman
Niéri Ko *510* river of SE Senegal
Nieuw Amsterdam *542* N Suriname
Nieuwegein *429* C Netherlands
Nieuwkoop *429* W Netherlands
Nieuw Nickerie *542* NW Suriname
Niğde *581* C Turkey
Niger *132, 283, 392, 439* river of W Africa
Niger *438-439* officially Republic of Niger. Country of West Africa divided into 7 admin. units (departments). ❖ Niamey
Nigeria *440-441* officially Federal Republic of Nigeria. Country of West Africa divided into 20 admin. units (19 states and 1 federal capital Territory). ❖ Abuja
Niger, Mouths of the *440* delta of the river Niger, on the S coast of Nigeria
Niigata *332* Honshū, N Japan
Niihama *332* Shikoku, SW Japan
Niihau *602* island of Hawaii, USA, C Pacific
Niimi *332* Honshū, W Japan
Nijmegen *429* *Ger.* Nimwegen. SE Netherlands
Nikki *132* E Benin
Nikolainkaupunki *see* Vaasa
Nikolayev *see* Mykolayiv
Nikol'sk-Ussuriyskiy *see* Ussuriysk
Nikol'skiy *see* Satpayev
Nikopol' *590* SE Ukraine
Nikšić *512* W Montenegro, Serbia & Montenegro (Yugo.)
Nikumaroro *348* *var.* Gardner I. Island of the Phoenix Is, C Kiribati
Nikunau *348* island of the Gilbert Is, W Kiribati
Nile *232, 540* *Ar.* Nahr an Nīl. River of N Africa
Nile Delta *232* delta of N Egypt
Nīl, Nahr an *see* Nile
Nilphāmāri *117* NW Bangladesh
Nimba, Monts *283* *var.* Nimba Mountains. Mountain range of W Africa
Nimba, Mount *328, 368* mountain of W Africa
Nimba Mountains *see* Nimba, Monts
Nîmes *255* SE France
Nimwegen *see* Nijmegen
Ningbo *187* *var.* Ning-po, *prev.* Ninghsien. Zhejiang, E China
Ning-hsia *see* Ningxia
Ninghsien *see* Ningbo
Ning-po *see* Ningbo
Ningxia *187* *Chin.* Ningxia Huizu Zizhiqu, *Eng.* Ningsia Hui Autonomous Region, *var.* Ning-hsia. Autonomous region of N China
Ninotsminda *264* *prev.* Bogdanovka. S Georgia
Ninove *127* C Belgium
Niokolo Koba *510* river of SE Senegal
Niono *392* C Mali
Nioro *392* *var.* Nioro du Sahel. W Mali
Nioro du Rip *510* SW Senegal
Niort *255* W France
Nippon-kai *see* Japan, Sea of
Niquero *210* S Cuba
Nirin *see* Erhlin
Niš *512* *Eng.* Nish. E Serbia, Serbia & Montenegro (Yugo.)
Nişāb *631* *var.* Anşāb. SW Yemen
Nisporeni *408* *Rus.* Nisporeny. W Moldova
Nissan Islands *see* Green Islands
Nissum Bredning *218* inlet of North Sea on the NW coast of Denmark
Nistru *see* Dniester
Niterói *145* *prev.* Nictheroy. SE Brazil
Nitil *see* Natl
Nitra *523* *Ger.* Neutra, *Hung.* Nyitra. River of SW Slovakia
Nitra *523* *Ger.* Neutra, *Hung.* Nyitra. SW Slovakia
Niuafo'ou *572* *var.* Niuafoo. Island of NW Tonga
Niuatoputapu *572* *var.* Niuatobutabu, *prev.* Keppel Island. Island of N Tonga
Niuatui *587* islet of Nukufetau, Tuvalu
Niue *650* territory in free association with New Zealand, Pacific Ocean. ❖ Alofi
Niulakita *587* *var.* Nurakita. Coral atoll of S Tuvalu
Niutao *587* coral atoll of NW Tuvalu
Nizāmābād *298* C India
Nizhnevartovsk *484* C Russian Federation
Nizhniy Novgorod *484* *prev.* Gor'kiy. W Russian Federation
Nizhniy Pyandzh *see* Panji Poyon
Nizhnyaya Tunguska *485* *Eng.* Lower Tunguska. River of C Russian Federation
Nizza *see* Nice
Njaba *see* Nja Kunda
Njaiama *518* E Sierra Leone
Nja Kunda *263* *var.* Njaba. NW Gambia
Njamena *see* N'Djamena
Njardhvík *296* SW Iceland
Njazidja *see* Grande Comore
Njoeng Jacobkondre *542* C Suriname
Njombe *564* S Tanzania
Njoro *344* W Kenya
Nkanini *544* W Swaziland
Nkata Bay *see* Nkhata Bay
Nkayi *200* *var.* N'Kayi, *prev.* Jacob. S Congo
Nkhata Bay *385* *var.* Nkata Bay. N Malawi
Nkhotakota *385* *var.* Kota Kota, Nkota Kota. C Malawi
Nkonfap *440* S Nigeria
Nkongsamba *168* *var.* N'Kongsamba. W Cameroon
Nkumekie *238* C Río Muni, Equatorial Guinea
Nkundla *544* W Swaziland
Nkusi *588* river of W Uganda
Nmai Hka *159* *var.* Me Hka. River of N Burma forming a headstream of the Irrawaddy river
Noākhāli *117* *prev.* Sudharam. S Bangladesh
Nobeoka *332* Kyūshū, W Japan
Noboribetsu *332* Hokkaidō, N Japan
Nogal, Uadi *see* Nugaal
Noire, Rivière *400* river of SW Mauritius
Noire, Rivière *see* Black River
Noirmoutier, Île de *254* island of W France
Nokia *251* SW Finland
Nokou *180* W Chad
Nokoué, Lac *132* lake of S Benin
Nola *179* SW Central African Republic
Nólsoy *644* island of E Faeroe Islands
Noma, Ra's *see* Nawmah, Ra's
Nomuka *572* island of the Nomuka Group, Tonga
Nomuka Group *572* island group of W Tonga
Nomwin *406* atoll of C Micronesia
Nông Hèt *360* E Laos
Nong Khai *567* NE Thailand
Nonouti *348* island of the Gilbert Is, W Kiribati
Nonsan *352* *Jap.* Ronzan. W South Korea
Nonsuch Island *643* island of E Bermuda
Noord *642* N Aruba
Noord-Beveland *429* island of SW Netherlands
Noordoewer *423* S Namibia
Noordpunt *650* headland of Curaçao, W Netherlands Antilles
Noordzee *see* North Sea
Nor Achin *see* Nor Hachn
Nor Ačin *see* Nor Hachn
Norak *562* *Rus.* Nurek. W Tajikistan
Nord *646* N Greenland
Nordaustlandet *653* island of NE Svalbard
Norddeutsches Tiefland *266-267* *Eng.* North German Plain. Plain of N Germany
Nordfriesische Inseln *266* *Eng.* North Frisian Islands. Island group of NW Germany
Nordhausen *267* C Germany
Nordishavet *see* Arctic Ocean
Nord, Massif du *288* mountainous region of Haiti
Nord, Mer du *see* North Sea
Nord-Pas de Calais *258* administrative region of N France
Nordsee *see* North Sea
Nordsjøen *see* North Sea
Nordsøen *see* North Sea
Nordtiroler Kalkalpen *106* mountain range of W Austria
Nore *316* *Ir.* An Fheoir. River of SE Ireland
Norfolk *603* Virginia, E USA
Norfolk Island *650* Australian external territory of the South Pacific Ocean. ❖ Kingston
Norge *see* Norway
Nor Hachn *98* *var.* Nor Hachyn, *Rus.* Nor Achin, *var.* Nor Ačcin. C Armenia
Nor Hachyn *see* Nor Hachn
Noril'sk *485* N Russian Federation
Norman *603* Oklahoma, SC USA
Normanby Island *458* island of SE Papua New Guinea
Normandie *255* *Eng.* Normandy. Cultural region of N France
Normandie, Collines de *255* hilly region of NW France
Norman Island *643* island of S British Virgin Islands
Norrköping *547* S Sweden
Norseman *100* SW Australia
Norskehavet *see* Norwegian Sea
Norsup *619* Malekula, Vanuatu
North Albanian Alps *81, 512* *SCr.* Prokletije, *Alb.* Bjeshkët e Nemuna. Mountain range of Albania and Serbia & Montenegro (Yugo.)
Northam *100* SW Australia
Northampton *597* C England, UK
North Andaman *159, 298* island of the Andaman Is, E India
North Battleford *170* SW Canada
North Bay *171* SE Canada
North Caicos *653* island of NW Turks and Caicos Islands
North Carolina *603* state of SE USA
North Channel *597* strait of Atlantic Ocean, between Northern Ireland and Scotland, UK
North Comino Channel *see* Għawdex, Il-Fliegu ta'
North Dakota *603* state of NC USA
North Devon Island *see* Devon Island
Northern Territory *101* territory of N Australia
North East China *191*
North East Point *644* headland on the NE coast of Christmas Island
Northeast Providence Channel *112* channel between Eleuthera I and Great Abaco I, Bahamas
Northern *531* *prev.* Northern Transvaal. Province of NE South Africa
Northern Cape *531* province of W South Africa
Northern Cook Islands *644* island group of N Cook Islands
Northern Cyprus, Republic of *see* Cyprus
Northern Dvina *see* Severnaya Dvina
Northern Forest Reserve *224* nature reserve of N Dominica
Northern Ireland *597* *var.* the Six Counties. Political division of UK
Northern Karoo *531* *var.* High Veld, *Afr.* Hoë Karoo. Plateau region of W South Africa
Northern Mariana Islands *650* Commonwealth territory of the USA, Pacific Ocean. ❖ Saipan
Northern Rhodesia *see* Zambia
Northern Sporades *see* Vor eioi Sporades
Northern Transvaal *see* Northern
North Frisian Islands *see* Nordfriesische Inseln
North German Plain *see* Norddeutsches Tiefland
North Huvadhu Atoll *390* *var.* Gaafu Alifu Atoll. Atoll of S Maldives
North Island *433* northernmost of the two main islands that comprise New Zealand
North Keeling Island *644* island of NW Cocos Islands

North Korea 350-351 officially Democratic People's Republic of Korea. Country of E Asia divided into 12 admin. units (9 provinces, 3 independent municipalities). ❖ Pyongyang

North Maalhosmadulu Atoll 390 var. North Malosmadulu Atoll, Raa Atoll. Atoll of N Maldives

North Miladummadulu Atoll 390 atoll of N Maldives

North Nilandhe Atoll 390 var. Faafu Atoll. Atoll of C Maldives

North Point 652 headland on the N coast of Ascension Island

North Saskatchewan 170 river of SW Canada

North Sea 266, 597 Dan. Nordsøen, Nor. Nordsjøen, Fr. Mer du Nord, Dut. Noordzee, Ger. Nordsee, prev. German Ocean. Sea of the Atlantic Ocean, between mainland Europe and Britain

North Siberian Lowland see Severo-Sibirskaya Nizmennost'

North Siberian Plain see Severo-Sibirskaya Nizmennost'

North Sound 644 area of the Caribbean Sea, Grand Cayman, W Cayman Islands

North Taranaki Bight 433 area of the Tasman Sea, to the W of North Island, New Zealand

North Uist 597 island of Outer Hebrides, NW Scotland, UK

North Union 498 E St Vincent, St Vincent & the Grenadines

North West 531 province of N South Africa

North West Bluff 649 headland on the N coast of Montserrat

North West Highlands 597 mountain range of N Scotland, UK

North West Point 644 headland on the NW coast of Christmas Island

Northwest Providence Channel 112 channel of Atlantic Ocean, between Grand Bahama Island and Bimini Islands, Bahamas

Northwest Territories 170-171, 175 territory of N Canada

Norton 636 NE Zimbabwe

Norway 444-447 officially Kingdom of Norway, Nor. Norge. Country of N Europe divided into 19 admin. units (counties). ❖ Oslo

Norwegian Sea 296, 444 var. Norske Havet, Norskehavet. Area of Arctic Ocean between Iceland, Greenland and Norway

Norwich 597 E England, UK

Nösen see Bistriţa

Noshiro 332 Honshū, N Japan

Nosop 142, 423 var. Nossob, Nossop. River of southern Africa

Nossob see Nosop

Nossop see Nosop

Noteć 471 Ger. Netze. River of NW Poland

Noties Sporádes see Dodekánisos

Notodden 444 S Norway

Notsé 571 S Togo

Nottingham 597 C England, UK

Nouâdhibou 398 prev. Port Étienne. W Mauritania

Nouakchott 398 ❖ of Mauritania, SW Mauritania

Noual, Sebkhet en 577 var. Sabkhat an Nawāl. Salt flat of C Tunisia

Nouméa 650 ❖ of New Caledonia, S New Caledonia

Nouna 157 W Burkina

Nouvelle France 400 S Mauritius

Nov 562 Rus. Nau. NW Tajikistan

Nová Dubnica 523 W Slovakia

Nova Gorica 524 W Slovenia

Nova Gradiška 209 Ger. Neugradiska, Hung. Újgradiska. NE Croatia

Nova Iguaçu 145 SE Brazil

Nova Kakhovka 590 SE Ukraine

Nova Lamego see Gabú

Nova Lisboa see Huambo

Novara 322 N Italy

Nova Scotia 171 province of SE Canada

Novaya Sibir', Ostrov 485 island of Novosibirskiye Ostrovo, N Russian Federation

Novaya Zemlya 484 island group of N Russian Federation

Nova Zagora 152 C Bulgaria

Nové Mesto nad Váhom 523 Ger. Waagneustadtl, Hung. Vágújhely. W Slovakia

Nové Zámky 523 Ger. Neuhäusl, Hung. Érsekújvár. SW Slovakia

Novgorod 484 NW Russian Federation

Novi Sad 512 Ger. Neusatz, Hung. Újvidék. N Serbia, Serbia & Montenegro (Yugo.)

Novogrudok see Navahrudak

Novokuznetsk 485 prev. Stalinsk. C Russian Federation

Novolazarevskaya 90 CIS research station of Greater Antarctica, Antarctica

Novo Mesto 524 Ger. Rudolfswert, prev. Neustadtl. SE Slovenia

Novopolotsk see Navapolatsk

Novo Redondo see Sumbe

Novosibirsk 485 C Russian Federation

Novosibirskiye Ostrova 485 Eng. New Siberian Islands. Island Group of N Russian Federation

Novo Urgench see Urganch

Nový Jičín 216 SE Czech Republic

Novyy Margilan see Farghona

Novyy Uzen' see Zhanaozen

Nowogródek see Navahrudak

Nowshera 451 var. Naushahra. NE Pakistan

Nowy Sącz 471 S Poland

Noyemberyan 98 N Armenia

Nsanje 385 S Malawi

Nsawam 272 SE Ghana

Nsoc 238 var. Nsork. SE Río Muni, Equatorial Guinea

Nsoc-Nsomo 238 E Río Muni, Equatorial Guinea

Nsoko 544 SE Swaziland

Nsuta 272 SW Ghana

Ntcheu 385 var. Ncheu. S Malawi

Ntchisi 385 var. Nchisi. C Malawi

Ntega 162 N Burundi

Ntem 168, 238 prev. Campo, var. Kampo. River of Cameroon and Equatorial Guinea

Ntomba, Lac 203 var. Lac Tumba. Lake of NW Dem. Rep. Congo

Ntoum 260 NW Gabon

NTsaouéni 198 NW Grande Comore, Comoros

Ntusi 588 SW Uganda

Nuanetsi see Mwenezi

Nubian Desert 540 desert of NE Sudan

Nu Chiang see Salween

Nu'eima 320 E West Bank

Nueva Caceres see Naga

Nueva Concepción 280 SW Guatemala

Nueva Gerona 210 Isla de la Juventud, Cuba

Nueva Helvecia 611 SW Uruguay

Nueva Loja 230 NE Ecuador

Nueva Ocotepeque 290 W Honduras

Nueva Palmira 611 SW Uruguay

Nueva San Salvador 237 prev. Santa Tecla. SW El Salvador

Nuevitas 210 E Cuba

Nuevo Laredo 403 N Mexico

Nuevo Rocafuerte 230 E Ecuador

Nugaal 528 It. Uadi Nogal. Seasonal river of N Somalia

Nui 587 coral atoll of W Tuvalu

Nu Jiang see Salween

Nukha see Şäki

Nuku 653 Île Futuna, N Wallis & Futuna

Nuku'alofa 572 ❖ of Tonga, Tongatabu, Tongatapu Group, Tonga

Nukuatea 653 island of Île Uvea, S Wallis & Futuna

Nukufetau 587 coral atoll of C Tuvalu

Nukulaelae 587 var. Nukulailai. Coral atoll of E Tuvalu

Nukuloa 653 island of Île Uvea, S Wallis & Futuna

Nukunonu Atoll 653 island of C Tokelau

Nukuoro 406 atoll of SE Micronesia

Nukus 614 W Uzbekistan

Nullarbor Plain 100-101 plateau of S Australia

Numan 441 E Nigeria

Numazu 332 Honshū, SE Japan

Nunap Isua 646 var. Uummannarsuaq. Dan. Kap Farvel, Eng. Cape Farewell. Cape on the S coast of Greenland

Nunavut 170-171, 175 territory of N Canada

Nuoro 322 Sardegna, W Italy

Nupani 526 Santa Cruz Is, Solomon Is

Nuquí 195 W Colombia

Nurakita see Niulakita

Nurata see Nurota

Nuratau, Khrebet see Nurota Tizmasi

Nurek see Norak

Nurmes 251 E Finland

Nürnberg 267 Eng. Nuremberg. S Germany

Nurota 614 Rus. Nurata. SE Uzbekistan

Nurota Tizmasi 614 Rus. Khrebet Muratau. Mountains of E Uzbekistan

Nusa Tenggara 304 Eng. Lesser Sunda Islands. Island group of C Indonesia

Nusaybin 581 SE Turkey

Nuşayrīyah, Jabal al 555 mountains of W Syria

Nuuk 646 Dan. Godthåb. ❖ Greenland, SW Greenland

Nuwara Eliya 538 S Sri Lanka

Nyaake 368 SE Liberia

Nyabarongo 492 river of W Rwanda

Nyabisindu 492 SW Rwanda

Nyabugogo 492 river of C Rwanda

Nyagassola see Niagassola

Nyahururu 344 W Kenya

Nyainqêntanglha Shan 186 mountain range of Xizang Zizhiqu, W China

Nyala 540 W Sudan

Nyalikungu 564 N Tanzania

Nyamapanda 636 NE Zimbabwe

Nyanda see Masvingo

Nyanga 636 var. Inyanga. E Zimbabwe

Nyanga 200, 260 river of Congo and Gabon

Nyanza-Lac 162 S Burundi

Nyarutovu 492 NW Rwanda

Nyasa, Lake 385, 419, 564 var. Lake Malawi, Port. Lago Niassa, prev. Lago Nyassa. Great lake of E Africa

Nyasaland see Malawi

Nyasvizh 122 W Belarus

Nyaunglebin 159 S Burma

Nyazura 636 prev. Inyazura. E Zimbabwe

Nyenyen 368 C Liberia

Nyeri 344 C Kenya

Nyika 344 plain of SE Kenya

Nyika Plateau 385 var. Nyika Uplands. Plateau of N Malawi

Nyíregyháza 292 NE Hungary

Nyiri Desert 344 desert region of SE Kenya

Nyitra see Nitra

Nykøbing 218 Sjælland, E Denmark

Nykøbing-Falster 218 var. Nykøbing. Falster, S Denmark

Nyköping 547 SE Sweden

Nylstroom 531 Northern Transvaal, NE South Africa

Nyoman see Neman

Nyon 550 SW Switzerland

Nyong 168 river of SW Cameroon

Nyslott see Savonlinna

Nzambi 200 S Congo

Nzega 564 NW Tanzania

Nzérékoré 283 SE Guinea

Nzi 328 river of C Ivory Coast

Nzilo, Lac 203 var. Barrage de Nzilo, prev. Lac Delcommune. Lake of SE Dem. Rep. Congo

Nzwani see Anjouan

Oahu 602 island of Hawaii, USA, C Pacific

Oakland 602 California, W USA

Oamaru 433 S South Island, New Zealand

Oaxaca 403 var. Oaxaca de Juárez. S Mexico

Ob' 454-485 river of C Russian Federation

Oban 597 W Scotland, UK

Obando see Puerto Inírida

Obbia see Hobyo

Obdorsk see Salekhard

Oberhausen 266 W Germany

Oberpahlen see Põltsamaa

Obihiro 333 Hokkaidō, N Japan

Obo 179 E Central African Republic

Obock 222 E Djibouti

Obskaya Guba 485 bay of Karskoye More, N Russian Federation

Obuasi 272 C Ghana

Ocaña 195 N Colombia

Occidental, Cordillera 136 range of the Andes in Bolivia and Chile

Occidental, Cordillera 195 range of the Andes in W Colombia

Occidental, Cordillera 463 range of the Andes in W Peru

Ocean Falls 170 W Canada

Ocean Island see Banaba

Och'amch'ire 264 Rus. Ochamchira. W Georgia

Ocho Rios 331 N Jamaica

Ocotal 436 NW Nicaragua

October Revolution Island see Oktyabr'skoy Revolyutsii, Ostrov

Oda 272 SE Ghana

Ōdate 332 Honshū, N Japan

Odawara 332 Honshū, SE Japan

Oddur see Xuddur

Ödenburg see Sopron

Odendaalsrus 531 Orange Free State, C South Africa

Odenpäh see Otepää

Odense 218 Fyn, S Denmark

Oder 216, 267, 471 Cz./Pol. Odra. River of C Europe

Oderhaff 267, 471 var. Stettiner Haff, Pol. Zalew Szczeciński. Bay of the Baltic Sea, on the N coast of Germany and Poland

Odesa 590 Rus. Odessa. S Ukraine

Odienné 328 NW Ivory Coast

Ôdôngk 165 S Cambodia

Oeno Island 651 island of NW Pitcairn Islands

Ofanto 323 river of S Italy

Offa 440 W Nigeria

Offenbach 266 var. Offenbach am Main. SW Germany

Offenbach am Main see Offenbach

Offenburg 266 SW Germany

Ofu 642 island of Manua Islands, E American Samoa

Ogaden 245 arid plateau of SE Ethiopia

Ōgaki 332 Honshū, C Japan

Ogasawara-shotō 332 Eng. Bonin Islands. Island group to the SE of Honshū, SE Japan

Ogbomosho 440 W Nigeria

Ogden 602 Utah, SW USA

Ogea Driki 248 island of the Lau Group, E Fiji

Ogea Levu 248 island of the Lau Group, E Fiji

Ogooué 200, 260 river of Congo and Gabon

Ogou 571 river of E Togo

Ogražden 381 mountain range of Bulgaria and FYR Macedonia

Ogre 362 Ger. Oger. C Latvia

Ogulin 209 N Croatia

Ohau, Lake 433 W South Island, New Zealand

Ohio 603 river of NC USA

Ohio 603 state of NE USA

Ohobela 366 N Lesotho

Ouahigouya *157* NW Burkina
Ouahran *see* Oran
Ouaka *179* river of C Central African Republic
Oualâta *398* SE Mauritania
Ouallam *439* *var.* Oualam. W Niger
Ouanary *645* E French Guiana
Ouanda Djallé *179* NE Central African Republic
Ouani *198* N Anjouan, Comoros
Ouara *179* river of E Central African Republic
Ouargla *83* *var.* Wargla. NE Algeria
Ouarkziz *414* seasonal river of SW Morocco
Ouarzazate *414* S Morocco
Ouazzane *414* N Morocco
Oubangui *see* Ubangi
Ouchan *647* E Isle of Man
Oued Zem *414* C Morocco
Ouégoa *650* N New Caledonia
Ouéléssébougou *392* *var.* Ouolossébougou. SW Mali
Ouémé *132* river of C Benin
Ouessant, Île d' *254* *Eng.* Ushant. Island of NW France
Ouèssè *132* *var.* Ouéssé. E Benin
Ouésso *200* NW Congo
Ouham *179, 180* river of Central African Republic and Chad
Ouidah *132* *Eng.* Whydah, *var.* Wida. S Benin
Oujda *414* NE Morocco
Oujeft *398* C Mauritania
Ould Yenjé *398* S Mauritania
Ouled Djellal *83* *var.* Awled Djellal. N Algeria
Oulu *251* *Swe.* Uleåborg. C Finland
Oulujärvi *251* *Swe.* Uleträsk. Lake of C Finland
Oulujoki *251* *Swe.* Uleälv. River of C Finland
Oumé *328* C Ivory Coast
Oum er Rbia *414* river of C Morocco
Oumm ed Droûs Telli, Sebkhet *398* salt lake of N Mauritania
Ounasjoki *251* river of N Finland
Ounianga Kébir *180* NE Chad
Ountivou *571* E Togo
Ouolossébougou *see* Ouéléssébougou
Our *378* river of W Europe
Ourense *534* *Cast.* Orense. NW Spain
Ourthe *127* river of E Belgium
Ouse *see* Great Ouse
Ouse *597* river of N England, UK
Outaouais *see* Ottawa
Outer Hebrides *597* *var.* Western Isles. Island group of NW Scotland, UK
Outer Islands *516* island group of C and SW Seychelles
Outjo *423* N Namibia
Ouvéa *650* island of Îles Loyauté, NE New Caledonia
Ovalau *248* island to the NE of Viti Levu, C Fiji
Ovalle *183* N Chile
Ovan *260* NE Gabon
Overflakkee *429* island of SW Netherlands
Overhalla *444* C Norway
Ovgos *215* river of NW Cyprus
Oviedo *534* NW Spain
Owando *200* C Congo
Owen Falls Dam *588* dam of S Uganda
Owen Stanley Range *458* mountain range of SE Papua New Guinea
Owerri *440* S Nigeria
Owia *498* N St Vincent, St Vincent & the Grenadines
Owo *440* SW Nigeria
Öxarfjördhur *296* *var.* Axarfjördhur. Fjord of NE Iceland
Oxbow *366* N Lesotho
Oxford *597* C England, UK
Oyama *332* Honshū, N Japan
Oyem *260* N Gabon
Oyo *440* W Nigeria
Oyo *200* C Congo
Oyster Island *159* island of W Burma
Ozama *226* river of S Dominican Republic
Ózd *292* NE Hungary
Özgön *see* Uzgen
Ozurget'i *264* *prev.* Makharadze. W Georgia

P

Paama *619* island of C Vanuatu
Paamiut *646* *Dan.* Frederikshåb. SW Greenland
Paarl *531* Western Cape, SW South Africa
Pābna *117* W Bangladesh
Pacaraima, Serra *145, 286* *var.* Pakaraima Mountains. Mountain range of N South America
Pachao Tao *559* island group of W Taiwan
Pachna *see* Pakhna
Pachuca *403* *var.* Pachuca de Soto. C Mexico
Pacific Ocean *90, 283, 332-333, 642, 647* world's largest ocean bounded by Asia and Australia to the W, the Americas to the E and Antarctica to the S
Padang *304* Sumatra, W Indonesia
Paderborn *266* NW Germany
Padma *117* name of the Ganges in Bangladesh, *see* Ganges
Padova *323* *Eng.* Padua. N Italy
Paektu-san *351* *Chin.* Baitou Shan. Mountain of China and North Korea
Pafos *see* Paphos
Pag *209* *It.* Pago. Island of C Croatia
Pagan *650* island of C Northern Mariana Islands
Pager *588* river of NE Uganda
Paget Island *643* island of E Bermuda
Pago *see* Pag
Pagon, Bukit *150* mountain of SE Brunei
Pago Pago *642* ❖ of American Samoa, Tutuila, W American Samoa
Pagouda *571* *var.* Kpagouda. NE Togo
Pahang *386* *var.* Syngei Pahang. River of C Peninsular Malaysia
Pai-ch'eng *see* Baicheng
Paide *242* *Ger.* Weissenstein. C Estonia
Päijänne *251* lake of S Finland
Pailĭn *165* W Cambodia
Paine, Cerro *183* mountain of S Chile
Paisance *see* Piacenza
País Valenciano *535* *Cat.* València, *Eng.* Valencia. Autonomous community of NE Spain
País Vasco *535* autonomous community of N Spain
Pakambaru *304* Sumatra, W Indonesia
Pakaraima Mountains *see* Pacaraima, Serra
Pakch'ŏn *351* W North Korea
Pakhna *215* *var.* Pachna. SW Cyprus
Pakin *406* atoll of E Micronesia
Pakistan *450-453* officially Islamic Republic of Pakistan. Country of Asia divided into 4 admin. units (provinces). v Islāmābād
Pak Lay *360* W Laos
Pakokku *159* W Burma
Pākpattan *451* E Pakistan
Pak Sane *see* Muang Pakxan
Pāksey *117* W Bangladesh
Pakwach *588* NW Uganda
Pakxé *360* *var.* Pakse. S Laos
Pal *86* W Andorra
Pala *180* SW Chad
Palapye *142* SE Botswana
Palau *454-455* *var.* Belau. Country of the Pacific Ocean. ❖ Koror
Palauli Bay *500* bay of Pacific Ocean off Sava'i, SW Samoa
Palawan *466* island of W Philippines
Palawan Passage *466* passage of the South China Sea, between Spratly Islands and Palawan, Philippines
Paldiski *242* *prev.* Baltiski, *Eng.* Baltic Port, *Ger.* Baltischport. NW Estonia
Palembang *304* Sumatra, W Indonesia
Palencia *535* NW Spain
Palermo *323* *Fr.* Palerme. Sicilia, S Italy
Palikir *406* ❖ of Micronesia, Pohnpei, Micronesia
Palimé *see* Kpalimé
Palk Strait *298, 538* strait connecting the Bay of Bengal and Gulf of Mannar, between India and Sri Lanka
Palma *535* *var.* Palma de Mallorca. Mallorca, E Spain
Palma *419* N Mozambique
Palmar Norte *206* SE Costa Rica
Palma Soriano *210* SE Cuba
Palm Beach *642* NW Aruba
Palmeira *176* Sal, NE Cape Verde
Palmer *90* US research station of Antarctic Peninsula, Antarctica
Palmerston *644* island of Southern Cook Islands, S Cook Islands
Palmerston North *433* S North Island, New Zealand
Palmetto Point *92* SW Barbuda, Antigua & Barbuda
Palmira *195* W Colombia
Palmyra *see* Tudmur
Paloe *see* Denpasar
Palu *304* Celebes, C Indonesia
Pamandzi *649* Petite-Terre, E Mayotte
Pamir *562* river of Afghanistan, Pakistan and Tajikistan
Pamirs *562* mountain range of E Tajikistan
Pampa Aullagas, Lago *see* Poopó, Lago
Pampas *95* flatlands of South America
Pampeluna *see* Pamplona
Pamplemousses *400* NW Mauritius
Pamplona *535* *var.* Pampeluna, *Basq.* Iruña. N Spain
Pamplona *195* NE Colombia
Pana *260* S Gabon
Panadura *538* SW Sri Lanka
Panagyurishte *152* *var.* Panagjurište. W Bulgaria
Panama *456-457* officially Republic of Panama. Country of Central America divided into 10 admin. units (9 provinces, and 1 special territory). ❖ Panama City
Panamá, Bahía de *456* bay to the S of Panama
Panama Canal *456* shipping canal linking the Caribbean Sea to the Pacific Ocean, passing through C Panama
Panama City *456* *Sp.* Panamá, *var.* Ciudad de Panama. ❖ of Panama, C Panama
Panamá, Golfo de *456* gulf of the Pacific Ocean to the S of Panama
Panamá, Istmo de *456* *prev.* Isthmus of Darien, *Eng.* Isthmus of Panama. Narrow strip of land, between North America and South America
Panay *466* island of C Philippines
Panay Gulf *466* gulf of the Sulu Sea
Pančevo *512* *Ger.* Pantschowa, *Hung.* Pancsova. N Serbia, Serbia & Montenegro (Yugo.)
Panda *419* S Mozambique
Pandan, Selat *521* strait connecting Strait of Malacca and South China Sea
Pandan Reservoir *521* reservoir of SW Singapore
Pandaruan *150* river of NE Brunei
Pan de Azúcar *611* S Uruguay
Pandėlys *376* *var.* Pandelis. NE Lithuania
Pandivere Kõrgustik *242* *var.* Pandivere Kõrgendik. Plateau of NW and NE Estonia
Pando *611* S Uruguay
Panevėžys *376* NE Lithuania
Panfilov *340* SE Kazakhstan
Pangai *572* Lifuka, Hai'pai Group, Tonga
Pangani *564* E Tanzania
Pangani *564* river of NE Tanzania
Pangar *168* river of C Cameroon
Pangkalpinang *304* Pulau Bangka, W Indonesia
Panguma *518* E Sierra Leone
Panguna *458* Bougainville I, Papua New Guinea
Pangutaran Group *466* island group of Sulu Archipelago, SW Philippines
Panhame *see* Manyame
Paniai, Danau *304* lake of Irian Jaya, E Indonesia
Panj *562* *Rus.* Pyandzh. SW Tajikistan
Panj *77, 562* *Rus.* Pyandzh. River of Afghanistan and Tajikistan
Panjakent *562* *Rus.* Pendzhikent. W Tajikistan
Panjang, Pulu *see* West Island
Pānji *298* SW India
Panji Poyon *562* *Rus.* Nizhniy Pyandzh. SW Tajikistan
Pano Lefkara *215* S Cyprus
Pano Panayia *215* *var.* Pano Panagia. W Cyprus
Pano Platres *215* SW Cyprus
Pantanal *144, 149* swamp region of SW Brazil
Pantelleria *323* island to the SW of Sicilia, S Italy
Pante Makasar *228* W East Timor
Pantschowa *see* Pančevo
Pánuco *403* river of C Mexico
Panzhihua *187* *prev.* Dukou *var.* Tu-k'ou. Sichuan, SW China
Panzós *280* E Guatemala
Pao-chi *see* Baoji
Paoki *see* Baoji
Paola *395* E Malta
Paola *323* S Italy
Pao-shan *see* Baoshan
Pao-ting *see* Baoding
Pao-t'ou *see* Baotou
Paotow *see* Baotou
Pápa *292* W Hungary
Papakura *433* N North Island, New Zealand
Papatoetoe *433* NW North Island, New Zealand
Papayes, River *400* river of W Mauritius
Papeete *646* ❖ of French Polynesia, Tahiti, W French Polynesia
Paphos *215* *var.* Pafos. W Cyprus
Papua, Gulf of *458* gulf of the Coral Sea, to the S of Papua New Guinea
Papua New Guinea *458-459* officially Independent State of Papua New Guinea, *prev.* Territory of Papua and New Guinea. Country of the SW Pacific divided into 19 admin. units (provinces). ❖ Port Moresby
Papuk *209* mountain range of NE Croatia
Paquera *206* W Costa Rica
Pará *see* Belém
Paraburdoo *100* W Australia
Paracel Islands *651* disputed island group of the South China Sea. ❖ Woody Island
Paraćin *512* C Serbia, Serbia & Montenegro (Yugo.)
Paradise *278* E Grenada island, Grenada
Paraguá *136* river of NE Bolivia
Paragua *623* river of SE Venezuela
Paraguaçu *145* *var.* Paraguassú. River of E Brazil
Paraguai *see* Paraguay
Paraguarí *460* S Paraguay
Paraguassú *see* Paraguaçu
Paraguay *95, 144, 460* *Port.* Paraguai. River of C South America
Paraguay *460-461* officially Republic of Paraguay. Country of South America divided into 20 admin. units (19 departments and 1 province). ❖ Asunción
Paraíba *see* Joao Pessoa
Paraíso *206* C Costa Rica
Parakou *132* C Benin
Paralimni *215* E Cyprus
Paramaribo *542* ❖ of Suriname, N Suriname
Paraná *95* E Argentina
Paraná *95, 145, 460* *var.* Alto Paraná. River of C South America
Paranam *542* N Suriname
Paraparaumu *433* S North Island, New Zealand
Pardubice *216* *Ger.* Pardubitz, C Czech Republic

Pardubitz *see* Pardubice
Parecis, Chapada dos *144 var.* Serra dos Parecis. Mountain range of W Brazil
Pares *92* E Antigua, Antigua & Barbuda
Parham *92* NE Antigua, Antigua & Barbuda
Paria, Gulf of *574, 623* gulf of the Atlantic Ocean, between Trinidad and Venezuela
Parika *286* NE Guyana
Parima, Serra *144* mountain range of Brazil and Venezuela
Paris *255, 258* ❖ of France, N France
Paris *348* Kiritimati, E Kiribati
Parita, Bahía de *456* bay of the Gulf of Panama
Parkan *see* Štúrovo
Párkány *see* Štúrovo
Parkent *614* E Uzbekistan
Parkhar *see* Farkhor
Parma *322* N Italy
Parnaíba *145* river of NE Brazil
Pärnu *242 Rus.* Pyarnu, *prev.* Pernov, *Ger.* Pernau. SW Estonia
Pärnu *242 var.* Pärnu Jõgi, *Ger.* Pernau. River of SW Estonia
Pärnu Laht *242* bay of the Gulf of Riga, on the SW coast of Estonia
Paro *134* W Bhutan
P'aro-ho *352 var.* Hwach'ŏn-chŏsuji. Reservoir of N South Korea
Páros *275* island of SE Greece
Parral *183* C Chile
Parrita *206* S Costa Rica
Parry Islands *170* island group of N Canada
Parry's *494* SE St Kitts, St Kitts & Nevis
Parson's Ground *494* N St Kitts, St Kitts & Nevis
Partizánske *523 prev.* Šimonovany, *Hung.* Simony. W Slovakia
Pasaje *230* SW Ecuador
Pasaquina *237* E El Salvador
Pas de Calais *see* Dover, Strait of
Pa-shih Hai-hsia *see* Bashi Channel
Pashmakli *see* Smolyan
Pasión *280* river of N Guatemala
Pasir Mas *386* N Peninsular Malaysia
Pasir Panjang *521* reservoir of SW Singapore
Paso de los Toros *611* C Uruguay
Passau *267* SE Germany
Passo Fundo *145* S Brazil
Passu Keah *651* island of S Paracel Islands
Pastavy *122 Rus.* Postavy, *Pol.* Postawy. NW Belarus
Pastaza *230, 463* river of Ecuador and Peru
Pasto *195* SW Colombia
Patagonia *95* semi-arid region of S South America
Patchchacan *130* N Belize
Pate *559* N Taiwan
Pate Island *344 var.* Patta Island. Island of SE Kenya
Paterna *535* E Spain
Pathein *see* Bassein
Patía *195* river of SW Colombia
Patience *497* E St Lucia
Pati Point *647* headland on the NE coast of Guam
Patlong *366* S Lesotho
Patna *298 var.* Azimabad. NE India
Patos *81 var.* Patosi. SW Albania
Patos, Lagoa dos *145* lagoon of S Brazil
Pátra *275 var.* Patras, *prev.* Pátrai. S Greece
Patta Island *see* Pate Island
Pattani *567* S Thailand
Pattle Island *651* island of W Paracel Islands
Patuākhāli *117* S Bangladesh
Patuca *290* river of E Honduras
Pau *255* SW France
Paungde *159* SW Burma
Pāvilosta *362* W Latvia
Pavlodar *340* NE Kazakhstan
Pavlohrad *590 Rus.* Pavlograd. E Ukraine
Pavuvu *526* island of C Solomon Islands
Pawai, Pulau *521* island of SW Singapore
Paysandú *611* NW Uruguay
Paz *237* river of Guatemala and El Salvador
Pazardzhik *152 var.* Pazardžik, *prev.* Tatar Pazardzhik. SW Bulgaria
Pazin *209* NW Croatia
Pčinja *381* river of N FYR Macedonia
Pea *572* Tongatabu, Tongatapu Group, Tonga
Peace *170* river of W Canada
Peak, The *652* mountain of C Ascension Island
Pearl Islands *see* Perlas, Archipiélago de las
Pearl Lagoon *see* Perlas, Laguna de
Peary Land *646* physical region of N Greenland
Pebble Island *645* island of N Falkland Islands
Peć *512* S Serbia, Serbia & Montenegro (Yugo.)
Pechora *484* River of NW Russian Federation
Pecixe, Ilha de *284* island of W Guinea-Bissau
Pecos *603* river of SW USA
Pécs *292 Ger.* Fünfkirchen. SW Hungary
Pedernales *226* SW Dominican Republic
Pedhieos *215 var.* Kanli Dere. River of NE Cyprus
Pedhoulas *215* W Cyprus
Pedja *242 var.* Pedja Jõgi. River of C Estonia
Pedoulas *see* Pedhoulas
Pedra Lume *176* Sal, NE Cape Verde
Pedregal *456* W Panama
Pedro Juan Caballero *456* E Paraguay
Pedro Santana *226* W Dominican Republic
Peel *647* W Isle of Man
Pegasus Bay *433* bay of the South Pacific Ocean, on the E coast of South Island, New Zealand
Pegeia *see* Peyia
Pegu *159 var.* Bago. S Burma
Péhonko *132* NW Benin
Pei-ching *see* Beijing
Peikang *559 var.* Pei-chiang, *Jap.* Hokkō. W Taiwan
Peinan Hsi *559* river of C Taiwan
Peineville *224* N Dominica
Peipsi Järv *see* Peipus, Lake
Peipus, Lake *242 Est.* Peipsi Järv, *Rus.* Chudskoye Ozero. Lake of Estonia and Russian Federation
Peiraías *275 prev.* Piraiévs, *Eng.* Piraeus. SE Greece
Peka *366* NW Lesotho
Pekalongan *304* Java, C Indonesia
Pekan Muara *150* N Brunei
Pekan Seria *150* W Brunei
Peking *see* Beijing
Pelée, Montagne *649* mountain of N Martinique
Peleliu *455* island of S Palau
Péligre, Lac de *288* C Haiti
Pelly Bay *170* N Canada
Pelmadulla *538* S Sri Lanka
Pélmonostor *see* Beli Manastir
Pelopónnisos *275 Eng.* Peloponnese. Peninsula of S Greece
Pelotas *145* S Brazil
Pemagatsel *134* SE Bhutan
Pematangsiantar *304* Sumatra, W Indonesia
Pemba *564* island of E Tanzania
Pemba *419 prev.* Porto Amélia. NE Mozambique
Pembroke *498* SW St Vincent, St Vincent & the Grenadines
Pembroke Hall *see* Libertad
Penambo, Banjaran *387 var.* Penambo Range, Banjaran Tama Abu. Mountain range of Borneo, Malaysia and Indonesia
Penambo Range *see* Penambo, Banjaran
Penang *see* George Town
Peñas Blancas *436* S Nicaragua
Pen-ch'i *see* Benxi
Pendé *179* river of Central African Republic and Chad
Pendembu *518* E Sierra Leone
Pendjari *132, 157* river of Benin and Burkina
Pendzhikent *see* Panjakent
P'enghu Liehtao *559 Eng.*Pescadores Islands. Island group of W Taiwan
P'eng-hu Shui-tao *559 Eng.* Pescadores Channel. Channel connecting South China Sea and Taiwan Strait
P'enghu Tao *559* island of W Taiwan
Peng-pu *see* Bengbu
Penibético, Sistema *535 Eng.* Baetic Cordillera, Baetic Mountains. Mountain range of S Spain
Peniche *474* W Portugal
Peninsular Malaysia *386 var.* Malaya, *prev.* West Malaysia. Western partof Malaysia situated on S Malay Peninsula
Penki *see* Benxi
Pennine Alps *550 var.* Alpes Penninae, *Fr.* Alpes Pennines, *It.* Alpi Pennine. Mountain range of SW Switzerland
Pennines *597 var.* Pennine Chain. Mountain range of N England, UK
Pennsylvania *603* state of NE USA
Penong *101* S Australia
Penonomé *456* C Panama
Penrhyn *644* island of Northern Cook Islands, N Cook Islands
Penrith *597* NW England, UK
Pentaschoinos *see* Yermasoyia
Pentecost *619 Fr.* Pentecôte. Island of C Vanuatu
Penticton *170* SW Canada
Penza *484* W Russian Federation
Penzance *597* SW England, UK
Peoria *603* Illinois, C USA
Pepel *518* W Sierra Leone
Pereira *195* W Colombia
Pergamino *95* E Argentina
Perico *210* NW Cuba
Périgueux *255* SW France
Perim *see* Barīm
Peringat *386* N Peninsular Malaysia
Perkhemahan Berakas *150* N Brunei
Perlas, Archipiélago de las *456 var.* Pearl Islands. Island group of SE Panama
Perlas, Laguna de *436 var.* Pearl Lagoon. Lagoon of the Caribbean Sea on the E coast of Nicaragua
Perlepe *see* Prilep
Perm' *484 prev.* Molotov. W Russian Federation
Përmet *81 var.* Permeti, Premet. S Albania
Pernambuco *see* Recife
Pernau *see* Pärnu
Pernik *152 prev.* Dimitrovo. W Bulgaria
Pernov *see* Pärnu
Peros Banhos *643* island of N British Indian Ocean Territory
Pérouse *see* Perugia
Perpignan *255* S France
Perquín *237* E El Salvador
Përrenjas *81 var.* Prenjasi, Prenjas. E Albania
Persian Gulf *see* The Gulf
Persia *see* Iran
Perth *100* SW Australia
Perth *597* N Scotland, UK
Peru *462-465* officially Republic of Peru. Country of South America divided into 25 admin. units (24 departments and 1 constitutional province). ❖ Lima
Perugia *323 Fr.* Pérouse. C Italy
Perugia, Lake of *see* Trasimeno, Lago
Pesaro *323* N Italy
Pescadores Channel *see* P'enghu Liehtao
Pescadores Islands *see* P'eng-hu Ch'ü-tao
Pescara *323* C Italy
Pesek Kechil, Pulau *521* island of SW Singapore
Pesek, Pulau *521* island of SW Singapore
Peshāwar *451* N Pakistan
Peshkopi *81 var.* Peshkopia, Peshkopija. NE Albania
Pessons, Pic dels *86* mountain of SE Andorra
Petah Tiqwa *319* C Israel
Petaling Jaya *386* W Peninsular Malaysia
Pétange *378* SW Luxembourg
Petani *386 var.* Patani. NW Peninsular Malaysia
Petare *623* N Venezuela
Petauke *635* E Zambia
Petén Itzá, Lago *280 var.* Lago de Flores. Lake of N Guatemala
Peterborough *597* E England, UK
Peterborough *171* SE Canada
Peterhead *597* NE Scotland, UK
Peter Island *643* island of S British Virgin Islands
Peters Mine *286* NW Guyana
Pétionville *288* S Haiti
Petit-Bourg *646* C Guadeloupe
Petit Canouan *498* island of S St Vincent & the Grenadines
Petite Butte *400* Rodrigues, Mauritius
Petite Côte *510* coastal region of W Senegal
Petite Dominique *278* island to the NE of Carriacou, Grenada
Petite Martinique *278* island to the NE of Carriacou, Grenada
Petite-Rivière-de-l'Artibonite *288* C Haiti
Petite-Rivière Noire, Piton de la *400* mountain range of SW Mauritius
Petite Savane *224* S Dominica
Petite Soufrière *224* E Dominica
Petite-Terre *649* island, E Mayotte
Petit-Goâve *288* S Haiti
Petitjean *see* Sidi Kacem
Petit Mustique *498* island of C St Vincent & the Grenadines
Petit Piton *497* mountain of SW St Lucia
Petit-Popo *see* Aného
Petit St. Vincent Island *278* island to the NE of Carriacou, Grenada
Petra *338* archaeological site of W Jordan
Petre Bay *433* bay of the South Pacific Ocean, on the coast of Chatham Island, New Zealand
Petrich *152 var.* Petrič. SW Bulgaria
Petrinja *209* N Croatia
Petroaleksandrovsk *see* Türtkŭl
Petropavlovsk *340* N Kazakhstan
Petropavlovsk-Kamchatskiy *485* NE Russian Federation
Petrópolis *145* SE Brazil
Petrosani *480* W Romania
Petrovgrad *see* Zrenjanin
Petrovsk-Port *see* Makhachkala
Petrozavodsk *484 Fin.* Petroskoi. NW Russian Federation
Pettau *see* Ptuj
Pevek *485* NE Russian Federation
Peyia *215 var.* Pegeia. SW Cyprus
Pezinok *523 Ger.* Bösing, *Hung.* Bazin. SW Slovakia
Phalaborwa *531* Northern Transvaal, NE South Africa
Phangan, Ko *567* island of S Thailand
Phan Rang-Thap Cham *627* SE Vietnam
Phan Thiêt *627* S Vietnam
Phet Buri *see* Phetchaburi
Phetchaburi *567 var.* Phet Buri. C Thailand
Philadelphia *603* Pennsylvania, NE USA
Philip Island *650* island of S Norfolk Island
Philippeville *see* Skikda
Philippines *466-469* officially Republic of the Philippines. Country of SE Asia divided into 14 admin. units (regions). ❖ Manila
Philippine Sea *406, 466* sea of the Pacific Ocean to the E of the Philippines
Philipsburg *650* St Martin, N Netherlands Antilles
Phillips *494* NE St Kitts, St Kitts & Nevis

Port Loko *518* W Sierra Leone
Port-Louis *646* N Guadeloupe
Port Louis *400* *var.* Port-Louis. ❖ of Mauritius, NW Mauritius
Port Lyautey *see* Kénitra
Port Macquarie *101* E Australia
Port Maria *331* N Jamaica
Port Mathurin *400* Rodrigues, Mauritius
Port Morant *331* E Jamaica
Portmore *331* SE Jamaica
Port Moresby *458* ❖ of Papua New Guinea, SE Papua New Guinea
Porto *474* *Eng.* Oporto. NW Portugal
Porto Alegre *145* *prev.* Pôrto Alegre. S Brazil
Porto Alegre *505* S São Tomé, Sao Tome & Principe
Porto Alexandre *see* Tombua
Porto Amélia *see* Pemba
Portobelo *456* *var.* Porto Bello, Puerto Bello. N Panama
Porto Edda *see* Sarandë
Port-of-Spain *574* ❖ of Trinidad & Tobago, NW Trinidad, Trinidad & Tobago
Porto Gole *284* C Guinea-Bissau
Porto Grande *see* Mindelo
Porto-Novo *132* ❖ of Benin, S Benin
Porto Santo *474* *var.* Ilha do Porto Santo. Island of the Madeira Is, Portugal
Porto Torres *323* Sardegna, W Italy
Porto Velho *144* *prev.* Pôrto Velho. W Brazil
Portoviejo *230* *var.* Puertoviejo. W Ecuador
Port Pirie *101* S Australia
Port Refuge *644* strait of the Indian Ocean between Horsburgh Island and Direction Island, C Cocos Islands
Port Royal *331* SE Jamaica
Port Said *232* *Ar.* Bur Sa'íd. N Egypt
Port St Mary *647* S Isle of Man
Portsmouth *597* S England, UK
Portsmouth *224* *var.* Grande-Anse. NW Dominica
Port Stanley *see* Stanley
Port Stephens *645* West Falkland, W Falkland Islands
Port Sudan *540* *var.* Būr Sūdān. NE Sudan
Port Swettenham *see* Kelang
Port Talbot *597* S Wales, UK
Portugal *474-477* officially Republic of Portugal. Country of W Europe divided into 18 admin. units (districts). ❖ Lisbon
Portuguese East Africa *see* Mozambique
Port-Vila *619* *var.* Vila. ❖ of Vanuatu, Éfate, Vanuatu
Porvenir *183* Tierra del Fuego, Chile
Porvenir *136* NW Bolivia
Posadas *95* NE Argentina
Posen *see* Poznań
Posŏng *352* river of S South Korea
Postojna *524* *Ger.* Adelsberg, *It.* Postumia. SW Slovenia
Pöstyén *see* Piešťany
Potaro *286* river of C Guyana
Potchefstroom *531* North West, N South Africa
Potenza *323* S Italy
Potgietersrus *531* Northern Transvaal, NE South Africa
Pot House *121* E Barbados
P'ot'i *264* W Georgia
Potiskum *440* NE Nigeria
Potoru *518* S Sierra Leone
Potosí *136* S Bolivia
Potsdam *267* NE Germany
Potters Village *92* C Antigua, Antigua & Barbuda
Pott, Île *650* island of Îles Belep, W New Caledonia
Pottuvil *538* SE Sri Lanka
Potwar Plateau *450* plateau of NE Pakistan
Poudre d'Or *400* NE Mauritius
Pouembout *653* W New Caledonia
Poum *650* W New Caledonia
Pout *510* W Senegal
Poutasi *500* Upolu, Samoa
Poŭthĭsăt *165* *var.* Pursat. River of W Cambodia
Poŭthĭsăt *165* *prev.* Pursat. W Cambodia
Po Valley *322* valley of N Italy
Považská Bystrica *523* *Ger.* Waagbistritz, *Hung.* Vágbeszterce. NW Slovakia
Póvoa de Varzim *474* NW Portugal
Powell, Lake *602* reservoir of SW USA
Poya *650* C New Caledonia
Poyang Hu *187* lake of E China
Poyan Reservoir *521* reservoir of W Singapore
Poza Rica *403* *var.* Poza Rica de Hidalgo. C Mexico
Poznań *471* *Ger.* Posen. W Poland
Pozo Colorado *460* C Paraguay
Pozsega *see* Slavonska Požega
Pozsony *see* Bratislava
Prábis *284* W Guinea-Bissau
Præstø *218* Sjælland, SE Denmark
Prague *216* *Cz.* Praha, *Ger.* Prag. ❖ of Czech Republic, NW Czech Republic
Praia *176* ❖ of Cape Verde, Santiago, S Cape Verde
Praslin *516* island of the Inner Islands, NE Seychelles
Praslin *497* E St Lucia
Prato *322* N Italy
Preguiça *176* São Nicolau, N Cape Verde
Prenjas *see* Përrenjas
Preparis Island *159* island of SW Burma
Přerov *216* *Ger.* Prerau. SE Czech Republic
Presidente Prudente *145* S Brazil
Prešov *523* *Ger.* Eperies, *var.* Preschau, *Hung.* Eperjes. NE Slovakia
Prespa, Lake *81, 275, 381* *Alb.* Liqen i Prespës, *Mac.* Prespansko Ezero, *Gk* Límni Megáli Préspa, *var.* Limni Prespa. Lake of SE Europe
Prespës, Liqen i *see* Prespa, Lake
Pressburg *see* Bratislava
Prestea *272* SW Ghana
Preston *597* NW England, UK
Pretoria *531* ❖ of South Africa. Pretoria-Witwatersrand-Vereeniging, NE South Africa
Pretoria-Witwatersrand-Vereeniging *531* province of NE South Africa
Préveza *275* W Greece
Prey Vêng *165* S Cambodia
Priboj *512* W Serbia, Serbia & Montenegro (Yugo.)
Příbram *216* W Czech Republic
Prickly Pear Cays *642* island group of NW Anguilla
Prieska *531* Northern Cape, C South Africa
Prievidza *523* C Slovakia
Prijedor *140* NW Bosnia & Herzegovina
Prilep *381* *Turk.* Perlepe. S FYR Macedonia
Prince Albert *170* SW Canada
Prince Edward Island *171* province and island of SE Canada
Prince Edward Island *531* island of the Prince Edward Islands, S South Africa
Prince Edward Islands *531* island group of S South Africa
Prince George *170* W Canada
Prince Island *see* Príncipe
Prince of Wales Island *see* Pinang, Pulau
Prince of Wales Island *170* island of N Canada
Prince Patrick Island *170* island of Parry Islands, N Canada
Prince Rupert *170* W Canada
Prince Rupert Bay *224* bay of the Caribbean Sea, to the NW of Dominica
Princes Town *574* SW Trinidad, Trinidad & Tobago
Príncipe *505* *var.* Príncipe Island, *Eng.* Prince Island. Island to the N of São Tomé, Sao Tome & Principe
Pripet *122* river of S Belarus
Pripet Marshes *122, 590* forested and swampy region of Belarus and Ukraine
Priština *512* S Serbia, Serbia & Montenegro (Yugo.)
Privas *255* SE France
Privigye *see* Prievidza
Priwitz *see* Prievidza
Prizren *512* *Alb.* Prizreni. S Serbia, Serbia & Montenegro (Yugo.)
Probištip *381* N FYR Macedonia
Probolinggo *304* Java, C Indonesia
Progreso *611* S Uruguay
Prome *159* *var.* Pyè. SW Burma
Promissão, Represa de *145* reservoir of S Brazil
Proskurov *see* Khmel 'nyts'kyy
Prostějov *216* *Ger.* Prossnitz. SE Czech Republic
Provadiya *152* *var.* Provadija. E Bulgaria
Provence *255* cultural region of SE France
Providence *603* Rhode Island, NE USA
Providence *121* S Barbados
Providence Atoll *516* *var.* Providence. Atoll of the Farquhar Group, S Seychelles
Providenciales *653* island of NW Turks and Caicos Islands
Provo *602* Utah, SW USA
Prudhoe Bay *602* Alaska, USA
Prune Island *498* island of SW St Vincent & the Grenadines
Prut *408, 480, 590* *Ger.* Pruth. River of E Europe
Pruth *see* Prut
Pruzhany *122* SW Belarus
Pryazova'ks Vysochyna *590* mountain range of SE Ukraine
Prychornomors'ka Nyzovyna *590* mountain range of S Ukraine
Prydniprovs'ka Nyzovyna *590* mountain range of NE Ukraine
Prydniprovs'ka Vysochnya *590* mountain range of NW Ukraine
Przemyśl *471* SE Poland
Przheval'sk *see* Karakol
Pskov *484* *Ger.* Pleskau. W Russian Federation
Pskov, Lake *242* *Est.* Pihkva Järv, *Rus.* Pskovskoye Ozero. Lake of Estonia and Russian Federation
Ptsich *122* *Rus.* Ptich'. River of C Belarus
Ptuj *524* *Ger.* Pettau. NE Slovenia
Pua'a, Cape *500* cape on the coast of Savai'i, NW Samoa
Pu'apu'a *500* Savali'i, Samoa
Pucallpa *463* C Peru
Puch'ŏn *352* *prev.* Punwŏn. NW South Korea
Pudasjärvi *251* C Finland
Puebla *403* *var.* Puebla de Zaragoza. S Mexico
Pueblo *603* Colorado, SW USA
Pueblo Nuevo Tiquisate *280* *var.* Tiquisate. SW Guatemala
Puente Alto *183* C Chile
Puerto Acosta *136* W Bolivia
Puerto Aisén *183* S Chile
Puerto Armuelles *456* W Panama
Puerto Ayacucho *623* SW Venezuela
Puerto Bahía Negra *460* N Paraguay
Puerto Baquerizo Moreno *230* San Cristobal I, Galapagos Is.
Puerto Barrios *280* E Guatemala
Puerto Bello *see* Portobelo
Puerto Berrío *195* N Colombia
Puerto Busch *136* *var.* Puerto General Busch. SE Bolivia
Puerto Cabello *623* N Venezuela
Puerto Cabezas *436* *var.* Bilwi. NE Nicaragua
Puerto Carreño *195* E Colombia
Puerto Casado *460* C Paraguay
Puerto Cooper *460* C Paraguay
Puerto Cortés *290* NW Honduras
Puerto El Carmen de Putumayo *230* *var.* Putumayo. NW Ecuador
Puerto el Triunfo *237* S El Salvador
Puerto Inírida *195* *var.* Obando. E Colombia
Puerto Iradier *see* Cogo
Puerto La Cruz *623* NE Venezuela
Puerto Lempira *290* E Honduras
Puertolland *535* SW Spain
Puerto Maldonado *463* E Peru
Puerto México *see* Coatzacoalcos
Puerto Montt *183* C Chile
Puerto Natales *183* S Chile
Puerto Padre *210* SE Cuba
Puerto Pinasco *460* C Paraguay
Puerto Plata *226* *var.* San Felipe de Puerto Plata. N Dominican Republic
Puerto Presidente Stroessner *see* Ciudad del Este
Puerto Princesa *466* Palawan, W Philippines
Puerto Príncipe *see* Camagüey
Puerto Rico *651* Commonwealth territory of the USA, Caribbean Sea. ❖ San Juan
Puerto Rico Trench *651* undersea feature of the Caribbean Sea, N Puerto Rico
Puerto San José *280* *var.* San José. S Guatemala
Puerto Suárez *136* E Bolivia
Puerto Vallarta *403* W Mexico
Puerto Varas *183* C Chile
Puerto Viejo *206* NE Costa Rica
Puertoviejo *see* Portoviejo
Pujehun *518* S Sierra Leone
Pukaki, Lake *433* lake of C South Island, New Zealand
Pukapuka *644* island of Northern Cook Islands, N Cook Islands
Pukch'ŏng *351* E North Korea
Pukë *81* *var.* Puka. N Albania
Pukekohe *433* NW North Island, New Zealand
Pukhan *352* river of North Korea and South Korea
Pula *209* *prev.* Pulj, *It.* Pola. W Croatia
Pulangi *466* river of Mindanao, S Philippines
Pulap *406* atoll of C Micronesia
Pulau *305* river of Irian Jaya, E Indonesia
Pulau Tekong Reservoir *521* reservoir of E Singapore
Pul-i-Khumri *see* Pol-e Khomrī
Pully *550* SW Switzerland
Pulusuk *406* island of C Micronesia
Puluwat *406* atoll of C Micronesia
Puná, Isla *230* island to the SW of Ecuador, in the Gulf of Guayaquil
Punakha *134* C Bhutan
Punata *136* C Bolivia
Pune *298* *prev.* Poona. W India
Punggol *521* area of NE Singapore
Púngoè *419* *var.* Pungue, Pungwe. River of C Mozambique
Punkudutivu *538* island of N Sri Lanka
Puno *463* SE Peru
Punta Arenas *183* *prev.* Magallanes. S Chile
Punta Chame *456* C Panama
Punta del Este *611* S Uruguay
Punta Gorda *130* S Belize
Puntarenas *206* W Costa Rica
Punta Santiago *238* S Bioko, Equatorial Guinea
Punto Fijo *623* NW Venezuela
Punwŏn *see* Puch'ŏn
Purari *458* river of C Papua New Guinea
Puri *298* E India
Purmerend *429* NW Netherlands
Pursat *see* Poŭthĭsăt
Purus *144* river of Brazil and Peru
Pusan *352* *var.* Busan, *Jap.* Fusan. SE South Korea
Pusat Gayo, Pegunungan *304* mountain range of Sumatra, W Indonesia
Pushkino *see* Biläsuvar
Putai *559* W Taiwan
Putorana, Plato *485* mountain range of N Russian Federation
Puttalam *538* W Sri Lanka
Puttalam Lagoon *538* lagoon of W Sri Lanka
Putumayo *195, 463* river of NW South America

Putumayo *see* Puerto El Carmen de Putumayo
Puyo *230* C Ecuador
Pyandzh *see* Panj
Pyapon *159* S Burma
Pyarnu *see* Pärnu
Pyinmana *159* C Burma
Pyltsamaa *see* Põltsamaa
P'yŏngt'aek *352* NW South Korea
Pyongyang *351* *Kor.* P'yŏngyang. ❖ of North Korea, SW North Korea
Pyramiden *653* Spitsbergen, W Svalbard
Pyramid Rock *651* island of E Paracel Islands
Pyrenees *86, 254-255, 535* *Sp.* Pirineos, *Fr.* Pyrénées. Mountain range of SW Europe
Pyu *159* S Burma
Pyuntaza *159* S Burma
Pyuthan *427* W Nepal

Q

Qaanaaq *646* *Dan.* Thule. NW Greenland
Qabātiya *320* N West Bank
Qābis *see* Gabès
Qacentina *see* Constantine
Qafşah *see* Gafsa
Qahremānshahr *see* Bākhtarān
Qala' en Nahl *540* *var.* Qala' an Naḩl. E Sudan
Qal'aikhum *562* *Rus.* Kalaikhum. C Tajikistan
Qalali *115* Jazirat al Muharraq, Bahrain
Qalansīyah *631* NW Suqutra, Yemen
Qalāt *see* Kalāt
Qal 'at Bīshah *506* SW Saudi Arabia
Qal'eh-ye Now *77* *var.* Qala Nau. NW Afghanistan
Qalīb ash Shuyūkh *356* *var.* Jleeb al Shuyoukh. C Kuwait
Qalīb, Shiqqat al *356* *var.* Shagat Al Jleeb. Desert region of NW Kuwait
Qalqīlya *320* NW West Bank
Qamar, Ghubbat al *631* bay of Arabian Sea, E Yemen
Qamar, Jabal al *448* mountain range of SW Oman
Qambar *see* Kambar
Qamea *248* *prev.* Nggamea. Island to the E of Taveuni, N Fiji
Qandahār *see* Kandahār
Qaqortoq *646* *Dan.* Julianehåb. S Greenland
Qaraghandy *see* Karaganda
Qaraoun, Lac de *364* *var.* Buḩayrat al Qir'awn. Lake of S Lebanon
Qara Qum *see* Karakumy
Qarataū *see* Karatau
Qardho *528* *It.* Gardo. N Somalia
Qareh Chāy *309* river of NW Iran
Qarkilik *see* Ruoqiang
Qarokŭl *562* *Rus.* Karakul'. E Tajikistan
Qarokŭl *562* *Rus.* Ozero Karakul'. Lake of E Tajikistan
Qarshi *614* *Rus.* Karshi, *prev.* Bek-Budi. S Uzbekistan
Qartaba *364* *var.* Qarţabā. N Lebanon
Qasigianguit *646* *Dan.* Christianshåb. W Greenland
Qatar *478-479* officially State of Qatar. Country of SW Asia divided into 9 admin. units (municipalities). ❖ Doha
Qaţţâra, Monkhafad el *232* *var.* Munkhafaḑ al Qaţţārah, *Eng.* Qattara Depression. Arid desert bin of NW Egypt
Qayrokkum, Obanbori *562* *Rus.* Kayrakkumskoye Vodolkhranilishche. Reservoir of NW Tajikistan
Qazaqstan *see* Kazakhstan
Qazax *110* *Rus.* Kazakh. W Azerbaijan
Qazbegi *264* *Rus.* Kazbegi. NE Georgia
Qazimämmäd *110* *Rus.* Kazi-Magomed. SE Azerbaijan
Qazvīn *309* *var.* Kazvin. NW Iran
Qena *232* *var.* Qina. E Egypt
Qena, Wâdi *232* *var.* Wādī Qinā. Seasonal river of E Egypt
Qeqertarsuaq *646* *Dan.* Godhavn. W Greenland
Qeqertarsuatsiaat *646* *Dan.* Fiskenæsset. SW Greenland
Qeshm *309* *var.* Jazīreh-ye Qeshm, Qeshm Island. Island of S Iran
Qezel Owzan *309* river of NW Iran
Qibilī *see* Kebili
Qilian Shan *186* *var.* Kilien Mountains. Mountain range of W China
Qingdao *187* *var.* Tsintao, Ching-Tao, Ch'ing-tao. Shandong, E China
Qinghai *186* *var.* Chinghai, Tsinghai. Province of W China
Qinghai Hu *186* *var.* Ch'ing Hai *Mong.* Koko Nor. Lake of W China
Qing-Zang Gaoyuan *186* *Eng.* Plateau of Tibet. Plateau of Xizang Zizhiqu, W China
Qiqihar *187* *prev.* Lungkiang, *var.* Tsitsihar, Ch'i-ch'i-ha-erh. Heilongjiang, NE China
Qir'awn, Buḩayrat al *see* Qaraoun, Lac de
Qirba, Khashim Al *see* Girba, Khashm el
Qiryat Ata *319* N Israel
Qiryat Gat *319* C Israel
Qiryat Motzkin *319* N Israel
Qiryat Shemona *319* N Israel
Qishn *631* SE Yemen
Qishon, Naḩal *319* river of N Israel
Qīzān *see* Jīzān
Qizilqum *see* Kyzyl Kum
Qizilrabot *562* *Rus.* Kyzylrabot. E Tajikistan
Qom *309* *var.* Qum, Kum. NW Iran
Qomolangma Feng *see* Everest, Mount
Qomul *see* Hami
Qondūz *see* Kunduz
Qondūz, Daryā-ye *77* seasonal river of NE Afghanistan
Qorakŭl *614* *Rus.* Karakul'. S Uzbekistan
Qormi *395* C Malta
Qornet es Saouda *364* mountain of NE Lebanon
Qoryooley *528* *It.* Coriole. SW Somalia
Qostanay *see* Kostanay
Qoubaïyât *364* *var.* Al Qubayyāt. NE Lebanon
Qoussantîna *see* Constantine
Quang Ngai *627* E Vietnam
Quan Long *see* Ca Mau
Quanzhou *187* *var.* Ch'uan-chou, *prev.* Chin-chiang, *var.* Tsinkiang. Fujian, SE China
Quaraí *see* Cuareim
Quarles, Pegunungan *304* mountain range of Celebes, W Indonesia
Quarnero *see* Kvarner
Quartier Militaire *400* C Mauritius
Quatre Bornes *400* W Mauritius
Quatre Bornes *516* Mahé, Seychelles
Quatre, Isle à *498* island of C St Vincent & the Grenadines
Quba *110* *Rus.* Kuba. N Azerbaijan
Queanbeyan *101* SE Australia
Québec *171* SE Canada
Quebec *171, 175* province of SE Canada
Quebo *284* S Guinea-Bissau
Queen Charlotte Bay *645* bay of the South Atlantic Ocean, W Falkland Islands
Queen Charlotte Islands *170* *Fr.* Îles de la Reine-Charlotte. Island group of SW Canada
Queen Charlotte Sound *170* area of the Pacific Ocean between the Queen Charlotte Islands and Vancouver Island, SW Canada
Queen Elizabeth Islands *170* *Fr.* Îles de la Reine-Élisabeth. Island group of N Canada
Queen Mary's Peak *652* mountain of C Tristan da Cunha
Queen Maud Gulf *170* gulf of the Arctic Ocean on the coast of N Canada
Queen Maud Land *90* physical region of Greater Antarctica, Antarctica
Queensland *101, 105* state of N Australia
Queenstown *433* S South Island, New Zealand
Queenstown *521* area of S Singapore
Queenstown *531* Eastern Cape, S South Africa
Queguay Grande *611* river of W Uruguay
Quelimane *419* E Mozambique
Quelpart *see* Cheju-do
Queluz *474* W Portugal
Quepos *206* S Costa Rica
Que Que *see* Kwekwe
Querétaro *403* C Mexico
Quesada *206* N Costa Rica
Questelles *498* SW St Vincent, St Vincent & the Grenadines
Quetta *451* NW Pakistan
Quezaltenango *280* *var.* Quetzaltenango. W Guatemala
Quezaltepeque *237* C El Salvador
Quibdó *195* W Colombia
Quillacollo *136* C Bolivia
Quilpué *183* C Chile
Quimper *254* W France
Quinhámel *284* W Guinea-Bissau
Quiniluban Group *466* island group of C Philippines
Quissico *419* S Mozambique
Quito *230* ❖ of Ecuador, N Ecuador
Qum *see* Qom
Qŭnghirot *614* *Rus.* Kungrad. NW Uzbekistan
Qŭqon *614* *var.* Khokand, *Rus.* Kokand. E Uzbekistan
Qurayn, Ra's al *115* cape of SE Bahrain
Qurayyāt *448* *var.* Qurayat, Quraiyat. NE Oman
Qŭrghonteppa *562* *Rus.* Kurgan-Tyube. W Tajikistan
Qurlurtuuq *see* Kugluktuk
Qus *232* *var.* Qūş. E Egypt
Quthing *see* Moyeni
Quy Nhơn *627* *var.* Qui Nhon, Quinhon. SE Vietnam
Qvareli *264* *Rus.* Kvareli. E Georgia
Qyteti Stalin *see* Kuçovë
Qyzylorda *see* Kzyylorda

R

Raab *see* Rába
Raab *see* Győr
Raahe *251* *Swe.* Brahestad. W Finland
Ra'ananna *319* C Israel
Raas Xaatuun *528* *It.* Ras Hafun. NE Somalia
Rába *106, 292* *Ger.* Raab. River of Austria and Hungary
Rabat *395* W Malta
Rabat *414* ❖ of Morocco, NW Morocco
Rabaul *458* New Britain, NE Papua New Guinea
Rabbit Island *498* island of SE St Vincent & the Grenadines
Rábca *292* river of NW Hungary
Rabi *248* *prev.* Rambi. Island to the E of Vanua Levu, N Fiji
Rābigh *506* W Saudi Arabia
Rabinal *280* C Guatemala
Râbniţa *see* Rîbniţa
Rabyānah, Ramlat *371* *var.* Şaḩrā' Rabyāh. Desert of SE Libya
Rachaïya *364* *var.* Rāshayyā. S Lebanon
Rach Gia *627* SW Vietnam
Rach Gia, Vinh *627* bay of the Gulf of Thailand on the SW coast of Vietnam
Racine *603* Wisconsin, NC USA
Radā *631* *var.* Ridā. W Yemen
Radom *471* C Poland
Radoviš *381* *var.* Radovište. E FYR Macedonia
Radviliškis *376* N Lithuania
Rafaela *95* E Argentina
Rafah *319, 320* *Heb.* Rafiaḩ. SW Gaza Strip
Rafḩā' *506* N Saudi Arabia
Ragged Island Range *112* island group of S Bahamas
Ragusa *323* Sicilia, S Italy
Ragusa *see* Dubrovnik
Rahachow *122* *Rus.* Rogachëv. E Belarus
Rahaeng *see* Tak
Rahīmyār Khān *451* SE Pakistan
Raiatea *646* island of W French Polynesia
Raipur *298* C India
Rairok District *396* district of Majuro, SE Marshall Islands
Rājahmundry *298* SE India
Rajang *386* river of SW Borneo, Malaysia
Rājbāri *117* C Bangladesh
Rajbiraj *427* E Nepal
Rājkot *298* W India
Rājshāhi *117* *prev.* Rāmpur Boalia. W Bangladesh
Rakahanga *644* island of Northern Cook Islands, N Cook Islands
Rakaia *433* river of C South Island, New Zealand
Rakhshān *450* river of W Pakistan
Rakiraki *248* N Viti Levu, W Fiji
Rakka *see* Ar Raqqah
Rakutō-kō *see* Naktong
Rakvere *242* *Ger.* Wesenberg. N Estonia
Raleigh *603* North Carolina, SE USA
Ralik Chain *396* island group of W Marshall Islands
Ramādah *631* W Yemen
Ramallah *319, 320* C West Bank
Ramat Gan *319* C Israel
Ramatlabama *142* S Botswana
Rambi *see* Rabi
Ramechhap *427* C Nepal
Ramier Island *see* Glover Island
Ramla *319* C Israel
Ramm, Jabal *338* mountain of SW Jordan
Râmnicu Sarat *480* E Romania
Râmnicu Vâlcea *480* *prev.* Rîmnicu-Vîlcea. C Romania
Ramotswa *142* S Botswana
Rampur Boalia *see* Rajshahi
Ramree Island *159* island of W Burma
Ramsey *647* NE Isle of Man
Ramsey Bay *647* bay of the Irish Sea on the NE coast of Isle of Man
Ramsgate *597* SE England, UK
Ramu *458* river of NE Papua New Guinea
Rancagua *183* C Chile
Rānchi *298* E India
Randa *222* C Djibouti
Randers *218* Jylland, N Denmark
Rāngāmāti *117* SE Bangladesh
Rangiora *433* E South Island, New Zealand
Rangitaiki *433* river of E North Island, New Zealand
Rangitata *433* river of C South Island, New Zealand
Rangitikei *433* river of S North Island, New Zealand
Rangoon *159* *var.* Yangon. ❖ of Burma, S Burma
Rangpur *117* N Bangladesh
Rankin Inlet *170* C Canada
Rankovićevo *see* Kraljevo
Rann *see* Brežice
Rann of Kachch *298* *var.* Rann of Cutch, Rann of Kutch. Salt marsh of India and Pakistan
Ranongga *526* *var.* Ghanongga. New Georgia Is, Solomon Islands
Rantau, Puala *see* Tebingtinggi, Pulau
Rapallo *322* N Italy
Rapid City *603* South Dakota, NC USA
Räpina *242* *Ger.* Rappin. SE Estonia
Rapla *242* *Ger.* Rappel. NW Estonia
Rapperswil *550* NW Switzerland
Rappin *see* Räpina
Rarotonga *644* island of Southern Cook Islands, S Cook Islands
Ra's al 'Ayn *555* N Syria
Ras al Hadd *see* Al Ḩadd
Ras al Khaimah *594* NE United Arab Emirates
Ra's an Naqb *338* SW Jordan
Ras Dashen Terara *245* mountain of N Ethiopia
Rasdu Atoll *390* atoll of C Maldives
Raseiniai *376* W Lithuania
Rashîd *232* *Eng.* Rosetta. N Egypt
Rasht *309* *var.* Resht. NW Iran

Rashū *see* Kŭmsong
Raso, Ilhéu *176* island of NW Cape Verde
Rass Jebel *577 var.* Ra's al Jabal. N Tunisia
Ras Tannūrah *506* E Saudi Arabia
Ratak Chain *396* island group of E Marshall Islands
Ratchaburi *567 var.* Rat Buri. C Thailand
Rathkeale *316* SW Ireland
Rätische Alpen *see* Rhaetian Alps
Ratnapura *538* S Sri Lanka
Ratō *see* Lotung
Raub *386* C Peninsular Malaysia
Raufarhöfn *296* NE Iceland
Raukawa *see* Cook Strait
Rauma *251 Swe.* Raumo. SW Finland
Rauna *362* NE Latvia
Răuţel *408 var.* Reuţel. River of N Moldova
Ravenna *323* N Italy
Ravensthorpe *100* SW Australia
Rāvi *451* river of India and Pakistan
Ravne na Koroškem *524 Ger.* Gutenstein. N Slovenia
Rawaki *348 var.* Phoenix Island. Island of Phoenix Islands, C Kiribati
Rāwalpindi *451* NE Pakistan
Rawson *95* SE Argentina
Rayak *364 var.* Riyāq. E Lebanon
Rayong *567* C Thailand
Raysūt *448* SW Oman
Razāzah, Buḩayrat ar *312 var.* Baḩr al Milḩ. Lake of C Iraq
Razdan *see* Hrazdan
Razim, Lacul *480 prev.* Lacul Rezelm. Lagoon of E Romania
Reading *597* SE England, UK
Reăng Kései *165* W Cambodia
Rebun-tō *332* island to the NW of Hokkaidō, N Japan
Rechytsa *122 Rus.* Rechitsa. SE Belarus
Recife *145 prev.* Pernambuco. E Brazil
Recklinghausen *266* W Germany
Reconquista *95* NE Argentina
Redange *378* W Luxembourg
Redcliff *636* C Zimbabwe
Red Deer *170* SW Canada
Redhead *574* NE Trinidad, Trinidad & Tobago
Redon *254* NW France
Red River *170* river of Canada and USA
Red River *603* river of SC USA
Red River *627 var.* Sông Coi, *Chin.* Yuan Jiang. River of China and Vietnam
Red Sea *240, 631* sea of Indian Ocean, between the Arabian Peninsula and NE Africa
Red Sea Hills *540* hilly region of NE Sudan
Red Volta *157 Fr.* Volta Rouge. River of Burkina and Ghana
Ree, Lough *316 Ir.* Loch Ri. Lake of C Ireland
Reefton *433* N South Island, New Zealand
Regar *see* Tursunzode
Regensburg *267* SE Germany
Reggane *83* C Algeria
Reggio di Calabria *323 var.* Reggio Calabria. S Italy
Reggio nell' Emilia *322 var.* Reggio Emilia. N Italy
Reghin *480* N Romania
Regina *170* S Canada
Régina *645* E French Guiana
Rehoboth *423* C Namibia
Reẖovot *319* C Israel
Reichenberg *see* Liberec
Reifnitz *see* Ribnica
Ré, Île de *254* island of W France
Reims *255 Eng.* Rheims. NE France
Reine-Charlotte, Îles de la *see* Queen Charlotte Islands
Reine-Élisabeth, Îles de la *see* Queen Elizabeth Islands
Reisduoddarhalde *see* Haltiatunturi
Reisui *see* Yŏsu
Reka *see* Rijeka
Relizane *83 var.* Ghilizane, Ghelîzâne. NW Algeria
Remel el Abiod *577* desert region of S Tunisia
Remich *378* SE Luxembourg
Rendezvous Bay *642* bay of the Caribbean Sea on the S coast of Anguilla
Rendova *526* island of the New Georgia Is, W Solomon Is
Renens *550* SW Switzerland
Rengo *183* C Chile
Rennell *526 var.* Mu Nggava. Island of S Solomon Islands
Rennes *254 Bret.* Roazon. NW France
Reno *602* Nevada, W USA
Réo *157* W Burkina
Republiek *542* N Suriname
Rere *526* E Guadalcanal, Solomon Is
Resen *381* SW FYR Macedonia
Resistencia *95* NE Argentina
Reşiţa *480 Hung.* Resicabánya, *Ger.* Reschiza. W Romania
Resolute *170* Cornwallis Island, N Canada
Resolution Island *433* island to the SW of South Island, New Zealand
Retalhuleu *280* SW Guatemala
Retan Laut, Pulau *521* island SW Singapore
Retiche, Alpi *see* Rhaetian Alps
Réunion *652* French overseas department of the Indian Ocean. ❖ St Denis
Reus *535* E Spain
Reutlingen *266* S Germany
Reval *see* Tallinn
Rewa *286* river of S Guyana
Rey *309 var.* Shahr Rey. NW Iran
Reyes *136* NW Bolivia
Rey, Isla del *456* island of SE Panama
Reykjahlídh *296* NE Iceland
Reykjavík *296* ❖ of Iceland, W Iceland
Reynosa *403* N Mexico
Reza, Gora *585 var.* Gora Riza. Mountain of SW Turkmenistan
Rezā 'īyeh *see* Orūmīyeh
Rezā'īyeh, Daryācheh-ye *see* Orūmīeh, Daryācheh-ye
Rēzekne *362 Ger.* Rositten, *Rus.* Rezhitsa. E Latvia
Rezina *408* NE Moldova
Rēznas Ezers *362* lake of SE Latvia
Rhadames *see* Ghadamis
Rhaetian Alps *550 Ger.* Rätische Alpen, *Fr.* Alpes Rhétiques, *It.* Alpi Retiche. Mountain range of E Switzerland
Rheden *429* SE Netherlands
Rhein *see* Rhine
Rheinisches Schiefergebirge *266 Eng.* Rhenish Slate Mountains. Mountains of W Germany
Rhenish Slate Mountains *see* Rheinisches Schiefergebirge
Rhétiques, Alpes *see* Rhaetian Alps
Rhine *255, 266, 374, 550 Ger.* Rhein, *Fr.* Rhin, *Dut.* Rijn. River of W Europe
Rhino Camp *588* NW Uganda
Rhode Island *603* state of NE USA
Rhodes *see* Rodos
Rhodesia *see* Zimbabwe
Rhodope Mountains *152, 275 Gk* Orosirá Rodópis, *Bul.* Despoto Planina, *Turk.* Dospad Dagh. Mountain range of Bulgaria and Greece
Rhône *255, 550* river of France and Switzerland
Rhum *597 var.* Rum. Island of Inner Hebrides, W Scotland, UK
Riaba *238 prev.* Concepción. S Bioko, Equatorial Guinea
Riau, Kepulauan *304 var.* Riau Archipelago, *Dut.* Riouw Archipel. Island group to the E of Sumatra, W Indonesia
Riban i Manamby *382* S Madagascar
Ribáuè *419* NE Mozambique
Ribble *597* river of NW England, UK
Ribe *218* Jylland, SW Denmark
Ribeira da Barça *176* Santiago, S Cape Verde
Ribeira Funda *176* São Nicolau, N Cape Verde
Ribeira Grande *176* Santo Antão, N Cape Verde
Ribeirão Preto *145* S Brazil
Riberalta *136* N Bolivia
Ribnica *524 Ger.* Reifnitz. S Slovenia
Rîbniţa *408 var.* Râbniţa, *Rus.* Rybnitsa. NE Moldova
Richard's Bay *531* Kwazulu Natal, E South Africa
Richard Toll *510* N Senegal
Riche Fond *497* E St Lucia
Richmond *603* Virginia, E USA
Richmond Vale *498* NW St Vincent, St Vincent & the Grenadines
Ridā *see* Radāa
Ridderkerk *429* SW Netherlands
Rif *414 var.* Riff, Er Rif. Mountain range of N Morocco
Rift Valley *see* Great Rift Valley
Riga *362 Latv.* Rīga. ❖ of Latvia, C Latvia
Riga, Gulf of *242, 362 Est.* Liivi Laht, *prev.* Riia Laht, *Rus.* Rizhskiy Zaliv, *Latv.* Rīgas Jūras Līci Gulf of the Baltic Sea, on the coasts of Estonia and Latvia
Rīgestān *77 var.* Registan. Desert region of S Afghanistan
Riihimäki *251* SW Finland
Rijeka *209 Slvn.* Reka, *Ger.* Sankt Veit am Flaum, *It.* Fiume. NW Croatia
Rijn *see* Rhine
Rijssel *see* Lille
Ri, Loch *see* Ree, Lough
Rimah, Wādī ar *506* dry watercourse of C Saudi Arabia
Rimaszombat *see* Rimavská Sobota
Rimavská Sobota *523 Ger.* Gross-Steffelsdorf, *Hung.* Rimaszombat. SE Slovakia
Rimini *323* N Italy
Rincon *650* Bonaire, S Netherlands Antilles
Ringe *218* Fyn, S Denmark
Ringkøbing *218* Jylland, W Denmark
Ringkøbing Fjord *218* fjord of Jylland, W Denmark
Ringsted *218* Sjælland, SE Denmark
Ringvassøya *444* island of NE Norway
Riobamba *230* C Ecuador
Rio Branco *144* W Brazil
Río Branco *611* E Uruguay
Rio Claro *574* SE Trinidad, Trinidad & Tobago
Río Cuarto *95* C Argentina
Rio de Janeiro *145* SE Brazil
Río Gallegos *95 var.* Puerto Gallegos, Gallegos. S Argentina
Rio Grande *145 var.* São Pedro do Rio Grande do Sul. S Brazil
Ríohacha *195* N Colombia
Río Muni *238* mainland region of Equatorial Guinea
Río Negro, Embalse del *611 var.* Lago Artificial de Rincón del Bonete. Reservoir of C Uruguay
Rioni *264* river of W Georgia
Río Sereno *456* W Panama
Riouw Archipel *see* Riau, Kepulauan
Riri *see* Iri
Rîşcani *408 var.* Râşcani. NW Moldova
Rishiri-tō *332* island to the NW of Hokkaidō, N Japan
Rishon Le Ẕiyyon *319* C Israel
Ritidian Point *647* headland on the N coast of Guam
Rivadavia *95* W Argentina
Rivas *436* S Nicaragua
Rivera *611* N Uruguay
Rivercess *see* Cess
River Sallee *278* NE Grenada island, Grenada
Rivière des Anguilles *400* S Mauritius
Rivière-Pilote *649* SE Martinique
Rivne *590 Pol.* Równe, *Rus.* Rovno. NW Ukraine
Riyadh *506 var.* Ar Riyāḑ. ❖ of Saudi Arabia, C Saudi Arabia
Riyāq *see* Rayak
Rize *581* NE Turkey
Rizhskiy Zaliv *see* Riga, Gulf of
Rizokarpaso *see* Dipkarpaz
Rkîz, Lac *398* lake of SW Mauritania
Road Bay *642* bay of the Caribbean Sea on the W coast of Anguilla
Road Town *643* ❖ of British Virgin Islands, Tortola, C British Virgin Islands
Roanne *255* E France
Roaring Creek *130* C Belize
Roatán *290* Islas de la Bahía, Honduras
Roazon *see* Rennes
Robertson, Lake *see* Manyame, Lake
Robertsport *368* W Liberia
Robinson Crusoe, Isla *183* island of Juan Fernández Islands, W Chile
Rocas, Atol das *145* island of E Brazil
Rocha *611* SE Uruguay
Rochambeau *645* NE French Guiana
Rochester *603* Minnesota, NC USA
Rochester *603* New York, NE USA
Rocheuses, Montagnes *see* Rocky Mountains
Rock, The *646* E Gibraltar
Rockford *603* Illinois, C USA
Rockhampton *101* E Australia
Rockies *see* Rocky Mountains
Rockingham *100* SW Australia
Rocklands *544* NW Swaziland
Rock Sound *112* Eleuthera I, Bahamas
Rock Springs *602* Wyoming, NW USA
Rockstone *286* E Guyana
Rocky Mountains *170, 602 var.* Rockies, *Fr.* Montagnes Rocheuses. Mountain range of NW America
Rocky Point *644* headland on the N coast of Christmas Island
Rodez *255* S France
Ródhos *see* Rodos
Rodi *see* Rodos
Rodi Gargancio *323* C Italy
Rodonit, Gjiri i *81* gulf of the Adriatic Sea, NW Albania
Rodópis, Orosirá *see* Rhodope Mountains
Rodos *275 Eng.* Rhodes, *It.* Rodi, *prev.* Ródhos. Island of SE Greece
Ródos *275 Eng.* Rhodes, *It.* Rodi, *prev.* Ródhos. Ródos, SE Greece
Rodosto *see* Tekirdağ
Rodrigues *400 var.* Rodriquez. Island of E Mauritius
Roermond *429* S Netherlands
Roeselare *127 Fr.* Roulers, *prev.* Rousselaere. W Belgium
Rogachëv *see* Rahachow
Rogaška Slatina *524 prev.* Rogatec-Slatina, *Ger.* Rohitsch-Sauerbrunn. E Slovenia
Rogatec-Slatina *see* Rogaška Slatina
Roger *224* W Dominica
Rogozhina *see* Rrogozhinë
Rohitsch-Sauerbrunn *see* Rogaška Slatina
Roi Et *567 var.* Muang Roi Et. NE Thailand
Roja *362* NW Latvia
Rojo, Cabo *651* cape on the SW coast of Puerto Rico
Rokan *304 var.* Airlalang. River of Sumatra, W Indonesia
Rokel *518 var.* Seli. River of C Sierra Leone
Rokkō *see* Lukang
Rôlas, Ilha das *505* island to the S of São Tomé, Sao Tome & Principe
Roma *see* Rome
Roma *366* W Lesotho
Romang Strait *304* strait connecting the Arafura Sea and Laut Banda, E Indonesia
Romania *480-483 prev.* Socialist Republic of Romania, *var.* Rumania. Country of SE Europe, divided into 40 admin. units (judeţ). ❖ Bucharest
Romano, Cayo *210* island of NE Cuba
Rombo, Ilhéus do *176 var.* Ilhéus Secos. Island of S Cape Verde
Rome *323 It.* Roma. ❖ of Italy, C Italy
Rømø *218* island of SW Denmark
Ronde, Île *see* Round Island
Ronde Island *278* island to the N of Grenada island, Grenada
Rongelap *396* island of NW Marshall Islands

S

St. George's *278* ❖ of Grenada, SW Grenada
St. George's Channel *316, 597* *Ir.* Muir Bhreatan. Channel connecting the Celtic Sea and Irish Sea
St. George's Harbour *643* bay of E Bermuda
St. George's Island *643* island of E Bermuda
St. Giles Islands *574* *prev.* Melville Islands. Islands to the NE of Tobago, Trinidad & Tobago
St Helena *652* British dependent territory of the South Atlantic Ocean ❖ Jamestown
St. Helena Bay *531* bay of Atlantic Ocean, of coast of W South Africa
St Helier *648* ❖ of Jersey, S Jersey
St John *648* N Jersey
St John *647* C Isle of Man
Saint John *171* SE Canada
St. John *368* river of Guinea and Liberia
Saint John Island *653* island of NE Virgin Islands
St John's *649* N Montserrat
Saint John's *171* Newfoundland & Labrador, E Canada
St. John's *92* ❖ of Antigua & Barbuda, NW Antigua
St. John's Island *see* Sakijang Bendera, Pulau
St. Johnston Village *92* C Antigua, Antigua & Barbuda
St-Joseph *652* S Réunion
St Joseph *224* W Dominica
St. Joseph *574* SE Trinidad, Trinidad & Tobago
St Julian's *395* N Malta
St Kilda *597* island of NW Scotland, UK
Saint Kitts *494* island of the Lesser Antilles, which, with Nevis, forms the independent state of St Kitts & Nevis
Saint Kitts and Nevis *494-495* officially Federation of Saint Christopher and Nevis. Country of the West Indies. ❖ Basseterre
Saint-Laurent, Golfe du *see* Saint Lawrence, Gulf of
St-Laurent-du-Maroni *645* NW French Guiana
Saint Lawrence *171* *Fr.* Fleuve Saint-Laurent. River of SE Canada
Saint Lawrence, Gulf of *170* Gulf of the Atlantic Ocean, SE Canada
St-Lô *254* NW France
St-Louis *652* SW Réunion
St Louis *646* Marie-Galante, S Guadeloupe
St Louis *603* Missouri, C USA
Saint-Louis *510* NW Senegal
St-Louis-du-Nord *288* *var.* St-Luis du Nord. N Haiti
Saint Lucia *496-497* independent island state of the Caribbean. ❖ Castries
Saint Lucia Channel *497* channel connecting the Atlantic Ocean and Caribbean Sea
St. Lucia, Lake *531* lake of Kwazulu Natal, E South Africa
St-Malo *254* NW France
St-Malo, Golfe de *254* gulf of the English Channel to the NW of France
St-Marc *288* W Haiti
St-Marc, Canal de *288* channel of the Caribbean Sea between Île de la Gonâve and W Haiti
Sainte Marie, Nosy *382* *var.* Nosy Boraha. Island of NE Madagascar
Ste. Marie *649* NE Martinique
Ste-Marie *652* NE Réunion
St Martin *650* island of N Netherlands Antilles
St. Martin *646* island of N Guadeloupe
St. Martins *121* SE Barbados
St.Moritz *550* *Ger.* Sankt Mortiz, *Rmsch.* San Murezzan. SE Switzerland
St-Nazaire *254* W France
St-Nicolas *see* Sint-Niklaas
St. Patricks *121* S Barbados
St-Paul *652* NW Réunion
St Paul *603* Minnesota, NC USA
St. Paul *368* river of Guinea and Liberia
St. Paul's *494* NW St Kitts, St Kitts & Nevis
Saint Paul's Bay *see* San Pawl il Baħar
St Paul's Point *651* headland of Pitcairn Island, S Pitcairn Islands
St Peter Port *647* ❖ of Guernsey, C Guernsey
St Peters *494* SE St Kitts, St Kitts & Nevis
St Petersburg *603* Florida, SE USA
St. Philips *92* SE Antigua, Antigua & Barbuda
Saint-Pierre *652* ❖ of Saint Pierre and Miquelon, SE Saint Pierre
Saint Pierre *652* island of SE Saint Pierre and Miquelon
St-Pierre *652* SW Réunion
St Pierre *649* NW Martinique
St. Pierre *516* island of the Farquhar Group, Seychelles
St. Pierre and Miquelon *652* French territorial collectivity of the Atlantic Ocean ❖ Saint-Pierre
St. Sampson *647* S Guernsey
Saint Sauveur *224* E Dominica
Saint Thomas Island *653* island of W Virgin Islands
Saint Thomas Island *see* São Tomé
St-Trond *see* Sint-Truiden
Saint Vincent *498* island of the Lesser Antilles which, with the Northern Grenadines forms the independent state of St Vincent & the Grenadines
Saint Vincent and the Grenadines *498-499* country of the West Indies. ❖ Kingstown
St Willibrordus *650* Curaçao, S Netherlands Antilles
Saipan *650* island of S Northern Mariana Islands
Saipan *650* ❖ of Northern Mariana Islands, Saipan, S Northern Mariana Islands
Saishū *see* Cheju
Sajama, Nevado *136* mountain of W Bolivia
Sakaide *332* Shikoku, SW Japan
Sakākah *506* N Saudi Arabia
Sakalua *587* islet of Nukufetau, Tuvalu
Sakanthit *see* Saganthit Island
Sakarya *see* Adapazarı
Sakarya *580* river of NW Turkey
Sakata *332* Honshū, N Japan
Sakchu *351* W North Korea
Sakété *132* S Benin
Sakhalin, Ostrov *485* island of SE Russian Federation
Sakha, Respublika *484* *var.* Respublika Yakutiya. Autonomous republic of E Russian Federation
Şäki *110* *Rus.* Sheki, *var.* Šeki, *prev.* Nukha. NW Azerbaijan
Sakijang Bendera, Pulau *521* *prev.* St. John's Island S Singapore
Sakijang Pelepah, Pulau *521* *prev.* Lazarus Island S Singapore
Sakis-Adasi *see* Chíos
Sakishima-shotō *332* island group of Nansei-shotō, SW Japan
Sakon Nakhon *567* NE Thailand
Sakra, Pulau *521* island of SW Singapore
Sakskøbing *218* Lolland, SE Denmark
Sal *176* island of NE Cape Verde
Šaľa *523* *Hung.* Sellye. SW Slovakia
Salacgrīva *362* N Latvia
Salado *210* river of SE Cuba
Salaga *272* C Ghana
Sala'ilua *500* Savai'i, Samoa
Salala *368* C Liberia
Şalalah *448* SW Oman
Salamá *280* C Guatemala
Salamanca *403* C Mexico
Salamanca *534* NW Spain
Salamat *180* river of S Chad
Salamīyah *555* *var.* Selemia. W Syria
Salani *500* Upolu, Samoa
Salcedo *226* N Dominican Republic
Šalčininkai *376* SE Lithuania
Saldus *362* *Ger.* Frauenburg. W Latvia
Sale *101* SE Australia
Salé *414* NW Morocco
Sale'imou *500* Upolu, Samoa
Salekhard *484* *prev.* Obdorsk. N Russian Federation
Salelologa *500* Savai'i, Samoa
Salem *649* W Montserrat
Salem *298* S India
Salem *544* S Swaziland
Salem *602* Oregon, NW USA
Salemy *see* As Salimi
Salentina, Penisola *323* peninsula of S Italy
Salerno *323* S Italy
Salerno, Golfo di *323* gulf of the Tyrrhenian Sea, on the W coast of Italy
Salgótarján *292* N Hungary
Salibea *574* NE Trinidad, Trinidad & Tobago
Salibia *224* NE Dominica
Salihorsk *122* *Rus.* Soligorsk. S Belarus
Salikene *263* W Gambia
Salima *385* C Malawi
Salinas *see* Chixoy
Salinas *230* W Ecuador
Saline Island *278* island to the S of Carriacou, Grenada
Salisbury *597* S England, UK
Salisbury *224* *var.* Baroui. W Dominica
Salisbury *see* Harare
Salisbury, Lake *see* Bisina, Lake
Salisbury Plain *597* plain of S England, UK
Salitje *544* S Swaziland
Saljani *see* Salyan
Salkhad *555* SW Syria
Salla *251* NE Finland
Sallūm, Khalīj as *see* Sollum, Gulf of
Sallyana *see* Salyan
Salomon Atoll *643* atoll of N British Indian Ocean Territory
Salona *see* Solin
Salon-de-Provence *255* SE France
Salonica *see* Thessaloníki
Saloum *510* river of C Senegal
Salpausselkä *251* physical region of S Finland
Sal Rei *176* *var.* Vila de Sal Rei. Boa Vista, E Cape Verde
Salt *see* As Salţ
Salta *95* N Argentina
Saltholm *218* island of E Denmark
Saltibus *497* S St Lucia
Saltillo *403* N Mexico
Salt Island *643* island of SE British Virgin Islands
Salt Island Passage *643* passage of the Caribbean Sea between Peter Island and Salt Island, S British Virgin Islands
Salt Lake City *602* Utah, SW USA
Salto *611* NW Uruguay
Salto del Guairá *460* E Paraguay
Salto Grande, Embalse de *611* reservoir of Argentina and Uruguay
Saltpond *272* SE Ghana
Salvador *145* *prev.* São Salvador. E Brazil
Salvaleón de Higüey *see* Higüey
Salwa *see* As Salwá
Salwá, Dawḩat as *479* *var.* Dawḩat Salwah. Inlet of the Gulf of Bahrain on the coast of SW Qatar
Salwah *see* As Salwá
Salween *159, 186, 567* *Chin.* Nu Chiang, *var.* Nu Jiang, *Bur.* Thanlwin. River of SE Asia
Salyan *110* *Rus.* Sal'yany, *var.* Saljani. SE Azerbaijan
Salyan *427* *var.* Sallyana. W Nepal
Salzburg *106* N Austria
Salzburg Alps *106* *Ger.* Salzburger Kalkalpen. Mountain range of C Austria
Salzgitter *267* *prev.* Watenstedt-Salzgitter. C Germany
Samāhīj *115* Jazirat al Muharraq, Bahrain
Samā'il *448* *var.* Sumail. NE Oman
Samales Group *466* island group of Sulu Archipelago, SW Philippines
Samaná *226* NE Dominican Republic
Samana Cay *112* island of SE Bahamas
Samar *466* island of E Philippines
Samar *338* NW Jordan
Samara *484* *prev.* Kuybyshev. W Russian Federation
Samarai *458* SE Papua New Guinea
Samarinda *304* Borneo, C Indonesia
Samarkandski *see* Temirtau
Samarqand *614* *Rus.* Samarkand. SE Uzbekistan
Sāmarrā' *312* C Iraq
Samawa *see* As Samāwah
Şamaxı *110* *Rus.* Shemakha. C Azerbaijan
Sambava *382* NE Madagascar
Sambre *127* river of Belgium and France
Samchi *134* SW Bhutan
Samch'ŏk *352* *Jap.* Sanchoku. NE South Korea
Samch'ŏnpŏ *352* *Jap.* Sansenhō. S South Korea
Samdrup Jongkhar *134* SE Bhutan
Same *564* NE Tanzania
Samfya *635* *var.* Samfya Mission. N Zambia
Sam Hall's Bay *643* bay of the North Atlantic Ocean, E Bermuda
Samina *374* river of Austria and Liechtenstein
Saminatal Valley *374* valley of Austria and Liechtenstein
Sam Neua *see* Xam Nua
Samoa *500-501* officially Independent State of Samoa, Sam. Samoa i Sisfo. Country of the Pacific Ocean divided into 11 admin. units (districts). ❖ Apia
Samobor *209* N Croatia
Samokov *152* W Bulgaria
Sámos *275* island of SE Greece
Samosch *see* Someş
Samothráki *275* *Eng.* Samothrace. Island of NE Greece
Samsø *218* island of C Denmark
Samsun *581* N Turkey
Samtredia *264* W Georgia
Samui, Ko *567* island of S Thailand
Samur *110* river of Azerbaijan and Russian Federation
Samut Prakan *567* *var.* Muang Samut Prakan. C Thailand
Samut Sakhon *567* C Thailand
San *165* *var.* Se San. River of Cambodia and Vietnam
San *392* C Mali
San *471* river of SE Poland
Sana *140* river of NW Bosnia & Herzegovina
Sana *631* *Ar.* Şan'ā'. ❖ of Yemen, W Yemen
Saña *463* NW Peru
Sanābis *115* N Bahrain
Sanae *90* South African research station of Greater Antarctica, Antarctica
Sanaga *168* river of C Cameroon
Sanandaj *309* NW Iran
San Antonio *130* SW Belize
San Antonio *603* Texas, SC USA
San Antonio, Cabo *210* cape of W Cuba
San Antonio de Cortés *290* W Honduras
San Antonio del Táchira *623* W Venezuela
San Antonio de Ureca *see* Ureca
Sanāw *631* *var.* Sanaw. NE Yemen
San Bernardo *183* C Chile
San Blas, Cordillera de *456* mountain range of NE Panama
San Carlos *645* East Falkland, N Falkland Islands
San Carlos *183* C Chile
San Carlos *466* Luzon, N Philippines
San Carlos *206* river of N Costa Rica
San Carlos *623* C Venezuela
San Carlos *see* Luba

São Tomé *505* ❖ of Sao Tome & Principe, NE São Tomé
Sao Tome & Principe *504-505* officially Democratic Republic of Sao Tome and Principe, *Port.* São Tomé e Príncipe. Country of W Africa divided into 7 admin. units (districts). ❖ São Tomé
São Tomé, Pico de *505* mountain of São Tomé, Sao Tome & Principe
São Vicente *145* S Brazil
São Vicente *176* island of N Cape Verde
Sapele *440* S Nigeria
Sapitwa *385* mountain of S Malawi
Saponé *157* C Burkina
Sappemeer *429* NE Netherlands
Sapporo *332* Hokkaidō, N Japan
Sapta Koshi *427* river of India and Nepal
Sār *115* NW Bahrain
Šara *381* mountain range of FYR Macedonia and Serbia & Montenegro (Yugo.)
Sara Buri *567* C Thailand
Saragossa *see* Zaragoza
Saragt *585 prev.* Serakhs, *var.* Serahs. S Turkmenistan
Sarajevo *140* ❖ of Bosnia & Herzegovina, SE Bosnia & Herzegovina
Saran' *340* C Kazakhstan
Sarandë *81 var.* Saranda, *It.* Porto Edda, *prev.* Santi Quaranta. S Albania
Sarandí del Yí *611* C Uruguay
Sarandí Grande *611* S Uruguay
Sarangani Islands *466* island group of SE Philippines
Saratov *484* W Russian Federation
Saravan *360 var.* Saravane. SE Laos
Sarbhang *134* S Bhutan
Sardegna *322 Eng.* Sardinia. Island of W Italy
Sardinia *see* Sardegna
Sargodha *451* NE Pakistan
Sarh *180 prev.* Fort-Archambault. S Chad
Sārī *309* N Iran
Sarigan *650* island of C Northern Mariana Islands
Sarikol Range *562 Rus.* Sarykol'skiy Khrebet. Mountain range of China and Tajikistan
Sarimbun Reservoir *521* NW Singapore
Sariwŏn *351* SW North Korea
Sark *647* island of SE Guernsey
Sarpsborg *444* S Norway
Sarrebruck *see* Saarbrücken
Sarstoon *130, 280 var.* Sarstún. River of Belize and Guatemala
Sarstún *see* Sarstoon
Sarthe *255* river of NW France
Saruhan *see* Manisa
Sarykol'skiy Khrebet *see* Sarikol Range
Sary-Tash *359 var.* Sary-Ta#. SW Kyrgyzstan
Sasebo *332* Kyūshū, SW Japan
Saseno *see* Sazan
Saskatchewan *170* river of C Canada
Saskatchewan *170* province of C Canada
Saskatoon *170* SW Canada
Sasolburg *531* Orange Free State, C South Africa
Sassandra *328* S Ivory Coast
Sassandra *328 var.* Ibo. River of S Ivory Coast
Sassari *322* Sardegna, W Italy
Sassnitz *267* NE Germany
Sasstown *368* SE Liberia
Sataua *500* Savai'i, Samoa
Satawal *406* island of C Micronesia
Satawan *406* atoll of C Micronesia
Sātkhira *117* SW Bangladesh
Satpayev *340 prev.* Nikol'skiy. C Kazakhstan
Satpura Range *298* mountains of C India
Satu Mare *480 Hung.* Szatmárnémeti. NW Romania
Satunan-shotō *332* island group of Nansei-shotō, SW Japan
Sau *see* Sava
Sauce *see* Juan L. Lacaze
Saudhárkrókur *296* N Iceland
Saudi Arabia *506-509* officially Kingdom of Saudi Arabia. Country of SW Asia divided into 13 admin. units (provinces). ❖ Riyadh
Saül *645* C French Guiana
Saulkrasti *362* N Latvia
Sault Sainte Marie *171* S Canada
Sauma, Pointe *653* headland of Île Alofi, N Wallis & Futuna
Saumâtre, Étang *288* lake of SE Haiti
Saurimo *88 Port.* Vila Henrique de Carvalho. NE Angola
Sauteurs *278* N Grenada island, Grenada
Sava *140, 209, 524, 512 Eng.* Save, *Hung.* Száva, *Ger.* Sau. River of SE Europe
Savai'i *500* island of NW Samoa
Savalou *132* S Benin
Savan Island *498* island of S St Vincent & the Grenadines
Savannah *603* Georgia, SE USA
Savannakhét *360* S Laos
Savanna-La-Mar *331* W Jamaica
Savave *587* islet of Nukufetau, Tuvalu
Save *419, 636 var.* Sabi. River of Mozambique and Zimbabwe
Savè *132* SE Benin
Savissivik *646* NW Greenland
Savona *322* N Italy
Savonlinna *251 Swe.* Nyslott. SE Finland
Savusavu *248* Vanua Levu, N Fiji
Savu Sea *see* Sawu, Laut
Savute *142* river of N Botswana
Sawdā', Jabal *506* mountain of SW Saudi Arabia
Sawdā, Jabal as *371* mountain range of C Libya
Sawdirī *see* Sodiri
Sawhaj *see* Sohâg
Şawqirah *448* SE Oman
Şawqirah, Ghubbat *see* Suqrah Bay
Sawu, Laut *304 Eng.* Savu Sea. Sea of the Indian Ocean, C Indonesia
Say *439* SW Niger
Sayaboury *see* Muang Xaignabouri
Sayat *585* E Turkmenistan
Sayhūt *631* E Yemen
Saylac *528 var.* Zeila. NW Somalia
Saynshand *412* S Mongolia
Say 'ūn *631 var.* Saywūn. C Yemen
Sazan *81 It.* Saseno. Island of SW Albania
Scaldis *see* Scheldt
Scarborough *597* NE England, UK
Scarborough *574* S Tobago, Trinidad & Tobago
Scarborough *121* S Barbados
Scebeli *see* Shebeli
Schaan *374* W Liechtenstein
Schaanwald *374* NE Liechtenstein
Schaffhausen *550* N Switzerland
Schaulen *see* Šiauliai
Schefferville *171* E Canada
Scheldt *127 Dut.* Schelde, *Fr.* Escaut. River of W Europe
Schellenberg *374* N Liechtenstein
Schiedam *429* SW Netherlands
Schiermonnikoog *429* island of Waddeneilanden, N Netherlands
Schifflange *378* S Luxembourg
Schneekoppe *see* Sněžka
Schneidemühl *see* Pila
Schœlcher *649* W Martinique
Schoten *127* N Belgium
Schouwen *429* island of SW Netherlands
Schwäbische Alb *266 Eng.* Swabian Jura. Mountain range of SW Germany
Schwarzwald *see* Black Forest
Schwaz *106* W Austria
Schweizer Mittelland *see* Swiss Plateau
Schweizer Reneke *531* North West, N South Africa
Schwerin *267* N Germany
Schweriner See *266* lake of N Germany
Schwyz *550* C Switzerland
Schyl *see* Jiu
Sciacca *323* Sicilia, S Italy
Sciasciamana *see* Shashemenee
Scio *see* Chíos
Scoresbysund *see* Ittoqqortoormiit
Scotland *597, 597* national region of UK divided into 12 admin. units (9 regions, 3 island authorities)
Scott Base *90* New Zealand research station near Ross Shelf, Antarctica
Scott Island *90* island to the N of Ross Ice Shelf, Antarctica
Scotts Head Village *224 var.* Cachacrou. S Dominica
Scrub Island *642* island of NE Anguilla
Scunthorpe *597* NE England, UK
Scutari *see* Shkodër
Scutari, Lake *81, 512 Alb.* Liqeni i Shkodrës, *SCr.* Skadarsko Jezero. Lake of Albania and Serbia & Montenegro (Yugo.)
Seal Island *642* island of NW Anguilla
Seatons *92* E Antigua, Antigua & Barbuda
Seattle *602, 607* Washington, NW USA
Sébaco *436* C Nicaragua
Sebaiera *414* C Western Sahara
Sebapala *366* SW Lesotho
Sebarok, Pulau *521* island S Singapore
Sebastían Vizcaíno, Bahía *403* bay of the Pacific Ocean, on the NW coast of Mexico
Sebastopol *see* Sevastopol'
Sebenico *see* Šibenik
Sébikhoutane *510* W Senegal
Sebou *414* river of N Morocco
Secos, Ilhéus *see* Rombo, Ilhéus de
Sedberat *240* W Eritrea
Sédhiou *510* SW Senegal
Seeheim Noord *423* S Namibia
Seeland *see* Sjælland
Sefadu *518* E Sierra Leone
Sefrou *414* N Morocco
Segamat *386* S of Peninsular Malaysia
Ségbana *132* NE Benin
Segewold *see* Sigulda
Segna *see* Senj
Ségou *392 var.* Segu C Mali
Segovia *535* C Spain
Segovia *see* Coco
Segu *see* Ségou
Séguédine *439* NE Niger
Séguéla *328* W Ivory Coast
Séguénéga *157* NW Burkina
Segura *535* river of S Spain
Sehlabathebe *366* E Lesotho
Seinäjoki *251 Swe.* Östermyra. SW Finland
Seine *255* river of N France
Seine, Baie de la *254-255* bay of the English Channel to the NW of France
Seiyū *see* Chŏngju
Sejerø *218* island of C Denmark
Šeki *see* Şäki
Sekoma *142* S Botswana
Sekondi-Takoradi *272* S Ghana
Selânik *see* Thessaloníki
Selemia *see* Salamīyah
Selenge *412* river of Mongolia and Russian Federation
Seletar Reservoir *521* reservoir of C Singapore
Selfoss *296* SW Iceland
Seli *see* Rokel
Sélibabi *398* S Mauritania
Selibi Phikwe *142* E Botswana
Sélingué, Lac de *392* reservoir of S Mali
Selle, Massif de la *288* mountain range of S Haiti
Selle, Pic la *288 var.* La Selle. Mountain of S Haiti
Sellore Island *see* Saganthit Island
Sellye *see* Skalica
Sellye *see* Šaľa
Sélouma *283* C Guinea
Selukwe *see* Shurugwi
Selvagens, Ilhas *474* island group of the Madeira Is, Portugal
Semakau, Pulau *521* island S Singapore
Semanit *81 var.* Seman. River of W Albania
Semara *414* N Western Sahara
Semarang *304* Java, C Indonesia
Sembawang *521* area of N Singapore
Sembé *200* NW Congo
Sembehun *518* SW Sierra Leone
Semberong *386* river of SE Peninsular Malaysia
Semendria *see* Smederevo
Semipalatinsk *340 Kaz.* Semey. E Kazakhstan
Semirara Islands *466* island group of C Philippines
Semliki *588* river of W Uganda
Sên *165 var.* Sen. River of C Cambodia
Sena *see* Vila de Sena
Senafe *240* SE Eritrea
Senanayake Samudra *538* lake of E Sri Lanka
Senanga *635* SW Zambia
Senang, Pulau *521* island of S Singapore
Sendai *332* Kyūshū, SW Japan
Sendai *332* Honshū, N Japan
Senegal *510-511* officially Republic of Senegal. *Fr.* Sénégal. Country of West Africa divided into 10 admin. units (regions). ❖ Dakar
Senegal *398, 392, 510 Fr.* Sénégal. River of W Africa
Senica *523 Ger.* Senitz, *Hung.* Szenice. W Slovakia
Senigallia *323* C Italy
Senj *209 Ger.* Zengg, Ital. Segna. NW Croatia
Senja *444 prev.* Senjen. Island of NW Norway
Senkaku-shotō *332* island group of Nansei-shotō, SW Japan
Senmonorom *165* E Cambodia
Sennar *540 var.* Sannār. C Sudan
Senne *127 Dut.* Zenne. River of C Belgium
Senqunyane *366* river of C Lesotho
Senshin-kō *see* Sŏmjin
Sensuntepeque *237* NE El Salvador
Sentery *203* SE Dem. Rep. Congo
Sentosa *521* island S Singapore
Senye *238* W Río Muni, Equatorial Guinea
Seongnam *see* Sŏngnam
Seoul *352 Kor.* Sŏul, *prev.* Kyŏngsŏng, *Jap.* Keijō. of South Kore NW South Korea
Sepik *458* river of Indonesia and Papua New Guinea
Sepone *see* Muang Xéphôn
Sept-Iles *170* E Canada
Serahs *see* Saragt
Seraing *127* E Belgium
Serakhis *see* Serrakhis
Serakhs *see* Saragt
Seram *304 var.* Serang, *Eng.* Ceram. Island of Maluku, E Indonesia
Seram, Laut *304 Eng.* Ceram Sea. Sea of the Pacific Ocean, E Indonesia
Serang *304* Java, C Indonesia
Serangoon Harbour *521* harbour, E Singapore
Serasan, Selat *304, 386* strait of the South China Sea between Borneo and Kepulauan Natuna, W Indonesia
Seraya, Pulau *521* island of SW Singapore
Serbia *512 Serb.* Srbija. Republic of Serbia & Montenegro (Yugo.)
Sereď *523 Hung.* Szered. SW Slovakia
Serekunda *263* W Gambia
Seremban *386* W Peninsular Malaysia
Serengeti Plain *564* plain of N Tanzania
Serenje *635* E Zambia
Sereth *see* Siret
Sérifos *275* island of SE Greece
Serov *484* C Russian Federation
Serowe *142* SE Botswana
Serpa Pinto *see* Menongue
Serpent's Mouth, The *574 Sp.* Boca de la Serpiente. Strait connecting the Colombus Channel and the Gulf of Paria
Serrakhis *215 var.* Serrachis, Serakhis. River of NW Cyprus

Serravalle *502* N San Marino
Sérres *275* *prev.* Sérrai. NE Greece
Serule *142* E Botswana
Se San *see* Tônlé San
Sesana *see* Sežana
Sese Islands *588* island group of S Uganda
Sesvete *209* N Croatia
Seti *427* river of W Nepal
Sétif *83* *var.* Stif. N Algeria
Settat *414* W Morocco
Setté Cama *260* SW Gabon
Settlement, The *643* Anegada, N British Virgin Islands
Setúbal *474* W Portugal
Setúbal, Baía de *474* bay of the Atlantic Ocean, to the SW of Portugal
Sevan *98* C Armenia
Sevana Lich *98* *Eng.* Lake Sevan. Lake of E Armenia
Sevani Lerrnashght'a *see* Shakh-Dag
Sevan, Lake *see* Sevana Lich
Sévaré *392* C Mali
Sevastopol' *590* *Eng.* Sebastopol. S Ukraine
Severn *597* *Wel.* Hafren. River of England and Wales, UK
Severn *170* river of S Canada
Severnaya Dvina *484* *Eng.* Northern Dvina. River of NW Russian Federation
Severnaya Osetiya-Alaniya *484* autonomous republic of SW Russian Federation
Severnaya Zemlya *485* island group of N Russian Federation
Severnyy Ledovityy Okean *see* Arctic Ocean
Severodvinsk *484* *prev.* Molotov, *prev.* Sudostroy. NW Russian Federation
Severo-Sibirskaya Nizmennost' *485* *Eng.* North Siberian Lowland, *var.* North Siberian Plain. Lowland region of N Russian Federation
Severskiy Donets *see* Donets
Sevilla *534* *Eng.* Seville. SW Spain
Sevilla de Niefang *see* Niefang
Sevlievo *152* C Bulgaria
Sewa *518* river E Sierra Leone
Seychelles *516-517* officially Republic of the Seychelles. Country of the Indian Ocean divided into 23 admin. units (districts). ❖ Victoria
Seydhisfjördhur *296* E Iceland
Seydi *585* *prev.* Neftezavodsk. E Turkmenistan
Seyhan *581* river of S Turkey
Seyhan *see* Adana
Sežana *524* *It.* Sesana. SW Slovenia
Sfântu Gheorghe *480* *prev.* Sfîntu Gheorghe. C Romania
Sfax *577* *var.* Safāqis. E Tunisia
's-Gravenhage *429* *var.* Den Haag, *Eng.* The Hague, *Fr.* La Haye. Seat of government, W Netherlands
Shaanxi *187* *var.* Shensi, Shan-hsi. Province of C China
Shabani *see* Zvishavane
Shabeelle, Webi *see* Shebeli
Shaddādī *see* Ash Shadādah
Shah Alam *386* W Peninsular Malaysia
Shāhbāzpur *117* river of S Bangladesh
Shāhdādkot *451* SW Pakistan
Shaḩḩāt *371* NE Libya
Shahrikhon *614* *Rus.* Shakhrikhan. E Uzbekistan
Shahrisabz *614* *Rus.* Shakhrisabz. SE Uzbekistan
Shahr Rey *see* Rey
Shahrtuz *562* *Rus.* Shaartuz. W Tajikistan
Shahzadpur *117* W Bangladesh
Shakawe *142* NW Botswana
Shakh-Dag *98* *Arm.* Sevani Lerrnashght'a, *Rus.* Shakhdagskiy Khrebet. Mountain range of Armenia and Azerbaijan
Shakhdagskiy Khrebet *see* Shakh-Dag
Shakhrisabz *see* Shahrisabz
Shakhtinsk *340* C Kazakhstan
Shaki *440* W Nigeria
Shām, Bādiyat ash *see* Syrian Desert
Shām, Jabal ash *448* *var.* Jebel Sham. Mountain of N Oman
Shandī *see* Shendi
Shandong *187* *var.* Shantung. Province of E China
Shandong Bandao *187* *var.* Shantung Peninsula. Peninsula of E China
Shangani *636* river of W Zimbabwe
Shanghai *187, 190* city and municipality of E China
Shan-hsi *see* Shaanxi
Shan-hsi *see* Shanxi
Shanhua *559* SW Taiwan
Shannon *316* *Ir.* An tSionainn. River of C Ireland
Shansi *see* Shanxi
Shantar Islands *see* Shantarskiye Ostrova
Shantarskiye Ostrova *485* *Eng.* Shantar Islands. Island group of SE Russian Federation
Shantou *187* *var.* Swatow. Guangdong, SE China
Shantung *see* Shandong
Shantung Peninsula *see* Shandong Bandao
Shanxi *187* *var.* Shansi, Shan-hsi. Province of NE China
Shaoguan *187* *var.* Shao-kuan, *prev.* Ch'u-chiang, *Cant.* Kukong. Guangdong, SE China
Shao-kuan *see* Shaoguan
Shaqrā' *506* C Saudi Arabia
Shaqrā *see* Shuqrah
Shar *340* *var.* Charsk. E Kazakhstan
Sharasume *see* Altay
Shari *see* Chari
Sharïn Gol *412* N Mongolia
Sharjah *594* NE United Arab Emirates
Shark Bay *100* bay to the W of Australia
Sharon, Plain of *see* HaSharon
Sharqī, Jabal ash *see* Anti-Lebanon
Sharqī, Jazīrat ash *see* Chergui, Île
Sharqī, Jebel esh *see* Anti-Lebanon
Shashe *142, 636* *var.* Shashi. River of Botswana and Zimbabwe
Shashemenē *245* *var.* Shashemenne, Shashhamana, *It.* Sciasciamana. S Ethiopia
Shashi *187* *var.* Sha-shih, Shasi. Hubei, C China
Shāṭi', Wādī ash *371* dry watercourse of W Libya
Shaykh, Jabal ash *see* Hermon, Mount
Shaykh 'Uthmān *631* SW Yemen
Shcheglovsk *see* Kemerovo
Shchuchinsk *340* N Kazakhstan
Shea *286* S Guyana
Shebeli *528* *Som.* Webi Shabeelle, *Amh.* Shebele Wenz, *It.* Scebeli. River of Ethiopia and Somalia
Sheberghān *77* *var.* Shibarghan. N Afghanistan
Shedadi *see* Ash Shadādah
Shefar 'am *319* N Israel
Sheffield *597* N England, UK
Shekhem *see* Nāblus
Shekhūpura *451* NE Pakistan
Sheki *see* Şäki
Shelikhova, Zaliv *485* *Eng.* Shelekhov Gulf. Gulf of Sea of Okhotsk, bordering NE Russian Federation
Shemakha *see* Şamaxı
Shemgang *134* C Bhutan
Shendi *540* *var.* Shandī. NE Sudan
Shengking *see* Liaoning
Shensi *see* Shaanxi
Shenyang *187* *prev.* Fengtien, *Eng.* Mukden. Liaoning, NE China
Shepherd Islands *619* islands to the C of Vanuatu
Shepparton *101* SE Australia
Sherbro Island *518* island of SW Sierra Leone
Sherbrooke *171* SE Canada
Sheridan *602* Wyoming, NW USA
Sherpur *117* N Bangladesh
's-Hertogenbosch *429* *Ger.* Herzogenbusch, *Fr.* Bois-le-Duc. S Netherlands
Sherwood Ranch *142* SE Botswana
Shetland *597* islands of NE Scotland,UK
Shevchenko *see* Aktau
Shibām *631* C Yemen
Shibarghan *see* Sheberghān
Shibata *332* Honshū, N Japan
Shibh Jazīrat Sīnā' *see* Sinai
Shibîn el Kôm *232* *var.* Shibīn al Kawm. N Egypt
Shihmen *see* Shijiazhuang
Shijak *81* *var.* Shijaku. W Albania
Shijiazhuang *187* *var.* Shihkiachwang, Shih-chia-chuang, *prev.* Shihmen. Hebei, NE China
Shikārpur *451* S Pakistan
Shikoku *332* island of SW Japan
Shiliguri *298* *prev.* Siliguri. NE India
Shimbiris *528* *var.* Shimbir Berris. Mountain of N Somalia
Shimizu *332* Honshū, SE Japan
Shimonoseki *332* Honshū, W Japan
Shimonoseki-kaikyō *332* strait connecting the Sea of Japan and Inland Sea, between Honshū and Kyūsh ū, W Japan
Shinano *332* river of Honshū, N Japan
Shināş *448* NW Oman
Shīndand *77* W Afghanistan
Shinei *see* Hsinying
Shinshō *see* Hsinchuang
Shinshū *see* Chinju
Shinten *see* Hsintien
Shinyanga *564* NW Tanzania
Shiogama *332* Honshū, N Japan
Shīrāz *309* SW Iran
Shire *385* *Port.* Chire. River of Malawi and Mozambique
Shire Highlands *385* hilly region of S Malawi
Shirvanskaya Step' *see* Şirvan Düzü
Shirwa, Lake *see* Chilwa, Lake
Shizuoka *332* Honshū, SE Japan
Shkodër *81* *var.* Shkodra, *It.* Scutari, *SCr.* Skadar. NW Albania
Shkodrës, Liqeni i *see* Scutari, Lake
Shkubinit *81* *var.* Shkumbî, Shkumbin. River of C Albania
Shoe Rock *649* headland on the S coast of Montserrat
Shōka *see* Changhua
Sholāpur *see* Solāpur
Shorkot *451* NE Pakistan
Shortland Island *526* *var.* Alu. Island of the Shortland Is, W Solomon Islands
Shortland Islands *526* island group of the W Solomon Islands
Shostka *590* N Ukraine
Shreveport *603* Louisiana, SC USA
Shrewsbury *597* C England, UK
Shū *340* *var.* Chu. SE Kazakhstan
Shu'aybah *356* *var.* Shuaiba. E Kuwait
Shubrâ el Kheima *232* *var.* Shubrā al Khaymah. N Egypt
Shūlgareh *77* N Afghanistan
Shumen *152* *var.* Šumen. E Bulgaria
Shunsen *see* Ch'unch'ŏn
Shuqrah *631* *var.* Shaqrā. SW Yemen
Shurugwi *636* *prev.* Selukwe. C Zimbabwe
Shwebo *159* N Burma
Shweli *159* river of Burma and China
Shymkent *340* *prev.* Chimkent. S Kazakhstan
Shyashchytsy *122* C Belarus
Siāhān Range *450* mountain range of W Pakistan
Siāh Kūh *77* mountain range of W Afghanistan
Siālkot *451* NE Pakistan
Siam *see* Thailand
Siam, Gulf of *see* Thailand, Gulf of
Sian *see* Xi'an
Siangtan *see* Xiangtan
Siargao Island *466* island of E Philippines
Šiauliai *376* *Ger.* Schaulen. NW Lithuania
Siazan' *see* Siyäzän
Šibenik *209* *It.* Sebenico. S Croatia
Siberut, Pulau *304* island of Kepulauan Mentawai, W Indonesia
Sibi *451* C Pakistan
Sibiti *200* S Congo
Sibiu *480* *Ger.* Hermannstadt, *Hung.* Nagyszeben. C Romania
Sibu *386* W Borneo, Malaysia
Sibut *179* *prev.* Fort-Sibut. C Central African Republic
Sibutu Passage *386* passage connecting Celebes Sea and Sulu Sea
Sibuyan Island *466* island of C Philippines
Sibuyan Sea *466* sea of the Pacific Ocean
Sichuan *187* *var.* Szechuan, Ssu-ch'uan. Province of SW China
Sicilia *323* *Eng.* Sicily. Island of S Italy
Sicily *see* Sicilia
Sico *290* *var.* Tinto, Río Negro. River of NE Honduras
Sicunusa *544* SW Swaziland
Siders *see* Sierre
Sidi Bel Abbès *83* NW Algeria
Sidi Bouzid *577* *var.* Sīdī bū Zayd, Gammouda. C Tunisia
Sidi el Hani, Sebkhet de *577* *var.* Sabkhat Sīd' al Hānī'. Salt flat of NE Tunisia
Sidi Kacem *414* *prev.* Petitjean. N Morocco
Sidra *see* Surt
Sidra, Gulf of *see* Surt, Khalīj
Sidvokodvo *544* C Swaziland
Siegen *266* W Germany
Sielo *368* N Liberia
Siĕmréab *165* *prev.* Siem Reap. NW Cambodia
Siena *323* *Fr.* Sienne. C Italy
Sienne *see* Siena
Sierra de Guadarrama *535* mountains of C Spain
Sierra Leone *518-519* officially Republic of Sierra Leone. Country of W Africa divided into 4 admin. units (provinces). ❖ Freetown
Sierra Madre *466* mountain range of Luzon, N Philippines
Sierra Madre *280, 403* mountain range of Guatemala and Mexico
Sierra Madre del Sur *403* mountain range of S Mexico
Sierra Madre Occidental *403* *var.* Western Sierra Madre. Mountain range of NW Mexico
Sierra Madre Oriental *403* *var.* Eastern Sierra Madre. Mountain range of N Mexico
Sierra Maestra *210* mountain range of SE Cuba
Sierra Morena *534-535* mountain range of SW Spain
Sierra Nevada *602* mountain range of W USA
Sierra Nevada de Mérida *see* Mérida, Cordillera de
Sierre *550* *Ger.* Siders. SW Switzerland
Sigatoka *248* *prev.* Singatoka. Viti Levu, W Fiji
Siġġiewi *395* S Malta
Sighişoara *480* C Romania
Siglufjördhur *296* N Iceland
Signy *90* UK research station of South Orkney Islands, Antarctica
Sigsig *230* S Ecuador
Siguatepeque *290* W Honduras
Siguiri *283* NE Guinea
Sigulda *362* *Ger.* Segewold. NE Latvia
Sihanoukville *see* Kâmpóng Saôm
Siirt *581* SE Turkey
Sikasso *392* S Mali
Sikwane *142* S Botswana
Silay *466* Negros, C Philippines
Silesia *471* region of SW Poland
Silgadhi *427* *var.* Silgarhi. W Nepal
Silhouette *516* island of the Inner Islands, SE Seychelles
Siliana *577* *var.* Silyānah. NW Tunisia
Silicon Valley *606* business region of SW USA
Siliguri *see* Shiliguri
Silil *528* *var.* Silel. Seasonal river of NW Somalia
Silinhot *see* Xilinhot
Silisili, Mount *500* *var.* Mauga Silisili. Mountain of NW Samoa

Silistra *152 var.* Silistria. NE Bulgaria
Silkeborg *218* Jylland, W Denmark
Sillamäe *242 Ger.* Sillamäggi. NE Estonia
Sillein *see* Žilina
Šilutė *376 var.* Šilute. W Lithuania
Silva Porto *see* Kuito
Silver City *644* NE Christmas Island
Silverek *581* SE Turkey
Sima *198* W Anjouan, Comoros
Simanggang *see* Bandar Sri Aman
Simbirsk *see* Ul'yanovsk
Simeto *323* river of Sicilia, S Italy
Simeulue, Pulau *304* island to the NW of Sumatra, W Indonesia
Simferopol *590* S Ukraine
Simikot *427* W Nepal
Siminiout *645* S French Guiana
Šimonovany *see* Partizánske
Simony *see* Partizánske
Simplon Pass *550* mountain pass of S Switzerland
Simplon Tunnel *550* tunnel of Italy and Switzerland
Simpson Desert *101* desert region of C Australia
Simunye *544* NE Swaziland
Sinai *232 Ar.* Shibh Jazīrat Sīnā'. Desert region of NE Egypt
Sinazongwe *635* S Zambia
Sincelejo *195* NW Colombia
Sinchwang *see* Hsinchuang
Sin Cowe Island *652* island of SW Spratly Islands
Sindh *451* administrative region of SE Pakistan
Sindhulimadi *427* C Nepal
Sindi *242* SW Estonia
Sine *510* river of W Senegal
Sinendé *132* N Benin
Sines *474* S Portugal
Sinfra *328* C Ivory Coast
Singa *540 var.* Sinjah, Sinja. E Sudan
Singapore *521* river of S Singapore
Singapore *520-521* officially Republic of Singapore. Country of SE Asia divided into 5 admin. units (districts). ❖ Singapore City
Singapore Strait *386, 521 var.* Strait of Singapore. Strait connecting Strait of Malacca and South China Sea
Singatoka *see* Sigatoka
Sîngerei *408 var.* Sângerei, *prev.* Lazovsk. N Moldova
Singida *564* C Tanzania
Singora *see* Songkhla
Sining *see* Xining
Sinj *209* SE Croatia
Sinjavina *512 var.* Sinjajevina. Mountain range of N Montenegro, Serbia & Montenegro (Yugo.)
Sinkiang Uighur Autonomous Region *see* Xinjiang Uygur Zizhiqu
Sinnamary *645* N French Guiana
Sinnûris *232 var.* Sinnūris. N Egypt
Sino *see* Greenville
Sinoe *see* Greenville
Sinoia *see* Chinhoyi
Sinoie, Lacul *480 prev.* Lacul Sinoe. Lagoon of E Romania
Sinop *581* N Turkey
Sinp'o *351* E North Korea
Sintien *see* Hsintien
Sint-Niklaas *127 Fr.* St.-Nicolas. N Belgium
Sintra *474 prev.* Cintra. W Portugal
Sint-Truiden *127 Fr.* St.-Trond. E Belgium
Sinŭiju *351* W North Korea
Sinyang *see* Xinyang
Sió *292* river of W Hungary
Sion *550 Ger.* Sitten. SW Switzerland
Siorapaluk *646* NW Greenland
Sioux City *603* Iowa, C USA
Sioux Falls *603* South Dakota, NC USA
Sipaliwini *542* river of S Suriname
Siparia *574* SW Trinidad, Trinidad & Tobago
Siphofaneni *544 var.* Sipofaneni. C Swaziland
Siping *187 var.* Ssu-p'ing, Szeping, *prev.* Ssu-p'ing-chieh. Jilin, NE China
Siple *90* US research station of South Orkney Islands, Antarctica
Siput *386 var.* Sungei Siput. NW Peninsular Malaysia
Siquirres *206* E Costa Rica
Siracusa *323 Eng.* Syracuse. Sicilia, S Italy
Sirājganj *117* N Bangladesh
Şīr Banī Yās *594* island of W United Arab Emirates
Sirdaryo *see* Syr Darya
Sir Edward Pellew Group *101* island group of N Australia
Siret *480 var.* Siretul, *Ger.* Sereth. River of Romania and Ukraine
Sir Francis Drake Channel *643* channel connecting the Atlantic Ocean and Caribbean Sea, C British Virgin Islands
Sirte *see* Surt
Sirte, Gulf of *see* Surt, Khalij
Şirvan Düzü *110 Rus.* Shirvanskaya Step'. Mountain range of C Azerbaijan
Sirwan *see* Diyālá
Sisak *209 Hung.* Sziszek, *Ger.* Sissek. N Croatia
Sisian *98* SE Armenia
Sisimiut *646 var.* Holsteinsborg. SW Greenland
Sisŏphŏn *165* NW Cambodia
Sissek *see* Sisak
Sīstān, Daryācheh-ye *309 var.* Hāmūn-e Şāberī, Daryācheh-ye Hāmūn. Lake of E Iran
Sisters, The *278* islands N of Grenada island, Grenada
Siteki *544 var.* Stegi. E Swaziland
Sithoniá *275* peninsula of NE Greece
Sitobela *544* S Swaziland
Sitona *240* SW Eritrea
Sitrah *115 var.* Sitra. Island of NE Bahrain
Sittang *159 var.* Sittoung. River of C Burma
Sittard *429* S Netherlands
Sitten *see* Sion
Sittwe *159 prev.* Akyab. W Burma
Siuna *436* NE Nicaragua
Sivas *581* C Turkey
Sivers'kyy Donets' *see* Donets
Six Counties, the *see* Northern Ireland
Siyäzän *110 Rus.* Siazan'. NE Azerbaijan
Sjælland *218 Ger.* Seeland, *Eng.* Zealand. Island of E Denmark
Skadar *see* Shkodër
Skadarsko Jezero *see* Scutari, Lake
Skagaströnd *296 prev.* Höfdhakaupstadhur. N Iceland
Skagen *218* Jylland, N Denmark
Skagerrak *218, 444, 547 var.* Skagerak. Area of the Baltic Sea
Skalica *523 Hung.* Szakolca. W Slovakia
Skeleton Coast *423* coastal region of NW Namibia
Skellefteå *547* NE Sweden
Skellefteälv *547* river of N Sweden
Skien *444* S Norway
Skikda *83 prev.* Philippeville. NE Algeria
Skive *218* Jylland, NW Denmark
Skjálfandafljót *296* river of C Iceland
Skjern *218* Jylland, W Denmark
Skjern Å *218* river of W Denmark
Skon *165* S Cambodia
Skopje *381 prev.* Skoplje, *Turk.* Üsküb. ❖ of FYR Macedonia, N FYR Macedonia
Skoplje *see* Skopje
Skövde *547* S Sweden
Skrunda *362* W Latvia
Skúvoy *644* island of C Faeroe Islands
Skye, Isle of *597* island of W Scotland, UK
Skýros *275* island of E Greece
Slagelse *218* Sjælland, SE Denmark
Slaney *316 Ir.* An tSláine. River of SE Ireland
Slatina *209 prev.* Podravska Slatina, *Hung.* Szlatina. NE Croatia
Slatina *480* S Romania
Slave Coast *571* coastal region of W Africa, Atlantic Ocean
Slavonska Požega *209 prev.* Požega, *Hung.* Pozsega. NE Croatia
Slavonski Brod *209 prev.* Brod, *Hung.* Bród. E Croatia
Slavyansk *see* Slov"yans'k
Sléibhte Chill Mhantáin *see* Wicklow Mountains
Slēmānī *see* As Sulaymānīyah
Sliema *395* N Malta
Sligo *316 Ir.* Sligeach. N Ireland
Sliven *152 var.* Slivno. E Bulgaria
Slobozia *480* SE Romania
Slobozia *408 Rus.* Slobodzeya. E Moldova
Slonim *122 Rus.* Slonin. W Belarus
Slovakia *522-523* officially Slovak Republika, *prev.* constituent republic of Czechoslovakia. Country of C Europe divided into 4 admin. regions (kraj). ❖ Bratislava
Slovenia *524-525* officially Republic of Slovenia, *Slvn.* Slovenija. Country divided into 86 admin. units (občina). ❖ Ljubljana
Slovenské Rudohorie *523 Ger.* Slowakisches Erzgebirge, *var.* Ungarisches Erzgebirge. Mountain range of C Slovakia
Slov'yans'k *590 Rus.* Slavyansk. E Ukraine
Słupsk *471 Ger.* Stolp. N Poland
Slutsk *122* C Belarus
Smallwood Reservoir *171* lake of S Canada
Smarhon' *122* NW Belarus
Smederevo *512 Ger.* Semendria. N Serbia, Serbia & Montenegro (Yugo.)
Smila *590* C Ukraine
Smith's Island *643* island of E Bermuda
Smithson Bight *644* bay of the Indian Ocean on the S coast of Christmas Island
Smolensk *484* W Russian Federation
Smolyan *152 var.* Smoljan, *prev.* Pashmakli. SW Bulgaria
Smyrna *see* İzmir
Snaefell *647* mountain of C Isle of Man
Snake *602* river of NW USA
Sneeuw-gebergte *see* Maoke, Pegunungan
Sněžka *216 Ger.* Schneekoppe. Mountain of N Czech Republic
Snow Mountains *see* Maoke, Pegunungan
Snug Corner *278* SW Grenada island, Grenada
Snuŏl *165* E Cambodia
Soacha *195* C Colombia
Sobaek-sanmaek *352* mountain range of S South Korea
Sobat *540* river of Ethiopia and Sudan
Sobradinho, Represa de *145 var.* Barragem de Sobradinho. Reservoir of E Brazil
Soča *323, 524 It.* Isonzo. River of Italy and Slovenia
Socabaya *463* SE Peru
Sochi *484* SW Russian Federation
Société, Archipel de la *646* island group of W French Polynesia
Socotra *see* Suquţrá
Soc Trăng *627 var.* Khanh, *Hung.* S Vietnam
Sodankylä *251* N Finland
Södertälje *547* SE Sweden
Sodiri *540 var.* Sawdirī, Sodari. C Sudan
Sodo *245 var.* Soddo, Soddu. SW Ethiopia
Soekaboemi *see* Sukabumi
Soela Väin *242* strait of Baltic Sea, between the islands of Hiiumaa and Saaremaa, W Estonia
Soembawa *see* Sumbawa
Soerabaja *see* Surabaya
Soerakarta *see* Surakarta
Sofala, Baía de *419* Bay of Indian Ocean, off Mozambique
Sofia *382* seasonal river of NW Madagascar
Sofia *152 var.* Sofija, *Bul.* Sofiya. ❖ of Bulgaria, W Bulgaria
Sogamoso *195* C Colombia
Sognefjorden *444* fjord of SW Norway
Sohâg *232 var.* Sawhaj. C Egypt
Sŏjosŏn-man *351* inlet of Korea Bay, on W coast of N Korea
Sokch'o *352* N South Korea
Söke *581* SW Turkey
Sokhumi *264 Rus.* Sukhumi. NW Georgia
Sokodé *571* C Togo
Sokolov *216* NW Czech Republic
Sokoto *440* NW Nigeria
Sokoto *440* river of NW Nigeria
Sola *444* SW Norway
Solapur *298 var.* Sholapur. SW India
Sol, Costa del *535* coastal region of S Spain
Soldeu *86* NE Andorra
Soledad *623* E Venezuela
Soledad *195* N Colombia
Soleure *see* Solothurn
Soligorsk *see* Salihorsk
Solimões *145* local name for a stretch of the Amazon river, NW Brazil
Solin *209 It.* Salona. S Croatia
Solingen *266* W Germany
Sollum, Gulf of *232 Ar.* Khalīj as Sallūm. Gulf of the Mediterranean Sea, NW Egypt
Sololá *280* W Guatemala
Solomon Islands *526-527 prev.* British Solomon Islands Protectorate. Country of the South Pacific Ocean divided into 7 admin. units (provinces). ❖ Honiara
Solomon Sea *458, 526* sea of the Pacific Ocean, to the E of Papua New Guinea
Solothurn *550 Fr.* Soleure. NW Switzerland
Solun *see* Thessaloníki
Solway Firth *597* arm of the Irish Sea, W UK
Solwezi *635* NW Zambia
Soma *263* C Gambia
Somalia *528-529* officially Somali Democratic Republic, *prev.* Somaliland Protectorate, Italian Somaliland. Country of E Africa divided into 16 admin. units (regions). ❖ Mogadishu
Sombor *512 Hung.* Zombor. NW Serbia, Serbia & Montenegro (Yugo.)
Somerset *643* Somerset Island, W Bermuda
Somerset Island *643* island of W Bermuda
Somerset Island *170* island of N Canada
Somerset Nile *see* Victoria Nile
Someş *292, 480 Hung.* Szamos, *Ger.* Samosch. River of Hungary and Romania
Sŏmjin *352 Jap.* Senshin-kō. River of S South Korea
Somme *255* river of N France
Somosomo *248* Taveuni, N Fiji
Somotillo *436* W Nicaragua
Somoto *436* NW Nicaragua
Soná *456* SW Panama
Sonaco *284* NE Guinea-Bissau
Sonda des Vieques *651* bay of the Caribbean Sea, E Puerto Rico
Sønderborg *218 Ger.* Sonderburg. Als, S Denmark
Søndre Strømfjord *see* Kangerlussuaq
Songea *564* S Tanzania
Songhua Jiang *see* Sungari
Sŏngjin *see* Kimch'aek
Songkhla *567 Mal.* Singora. S Thailand
Sŏngnam *352 var.* Seongnam. NW South Korea
Songnim *351* SW North Korea
Songo *419* NW Mozambique
Sông Tiên Giang *see* Mekong
Songwe *385* river of Malawi and Tanzania
Sonmiāni Bay *451* bay of the Arabian Sea, on the S coast of Pakistan
Sonoran Desert *10* desert Mexico/USA
Sonsonate *237* W El Salvador
Sonsorol Islands *455* island group of Palau

Sue *540* river of S Sudan
Sue Wood Bay *643* bay of the North Atlantic Ocean, C Bermuda
Suez *232* *Ar.* As Suways, *var.* El Suweis. NE Egypt
Suez Canal *232* *Ar.* Qanāt as Suways. Canal of NE Egypt
Suez, Gulf of *232* *Ar.* Khalīj al 'Aqabah. Gulf of the Red Sea, to the NE of Egypt
Sūf *338* NW Jordan
Sugar Loaf *278* *var.* Levera Island. N of Grenada island, Grenada
Şuḩār *448* *var.* Sohar. NW Oman
Sühbaatar *412* N Mongolia
Suigen *see* Suwŏn
Suir *316* *Ir.* An tSiúir. River of S Ireland
Sukabumi *304* *prev.* Soekaboemi. Java, C Indonesia
Sukagawa *332* Honshū, N Japan
Sukarnapura *see* Jayapura
Sukarno, Puntjak *see* Jaya, Puncak
Sukhne *see* As Sukhnah
Sukhumi *see* Sokhumi
Suki *540* E Sudan
Sukkertoppen *see* Maniitsoq
Sukkur *451* S Pakistan
Sukuta *263* W Gambia
Sulaimaniya *see* As Sulaymānīyah
Sulaimān Range *451* mountain range of C Pakistan
Sula, Kepulauan *304* *prev.* Xulla Islands, Soela. Island group to the E of Celebes, E Indonesia
Sulawesi *see* Celebes
Sulawesi, Laut *see* Celebes Sea
Sulby *647* N Isle of Man
Sullana *463* NW Peru
Sullivan Island *see* Lanbi Island
Sultan Alonto, Lake *see* Lanao, Lake
Sulu Archipelago *466* island group of SW Philippines
Sulu Sea *387, 466* sea of the Pacific Ocean, to the NE of Borneo, Malaysia
Sulyukta *359* *Kir.* Sülüktü. SW Kyrgyzstan
Sumatera *see* Sumatra
Sumatra *304* *var.* Sumatera. Island of W Indonesia
Šumava *see* Bohemian Forest
Sumba *644* Sudhuroy, S Faeroe Islands
Sumba *304* *prev.* Soemba, *Eng.* Sandalwood Island. Island of Nusa Tenggara, C Indonesia
Sumba, Selat *304* strait of the Indian Ocean between Sumba and Sumbawa, C Indonesia
Sumbawa *304* *prev.* Soembawa. Island of Nusa Tenggara, C Indonesia
Sumbawanga *564* W Tanzania
Sumbe *88* *Port.* Novo Redondo. W Angola
Sumbuya *518* S Sierra Leone
Šumen *see* Shumen
Sumisu-jima *332* island to the SE of Honshū, SE Japan
Šumperk *216* *Ger.* Mährisch-Schönberg. E Czech Republic
Sumpul *237* river of Honduras and El Salvador
Sumqayıt *110* *Rus.* Sumgait. E Azerbaijan
Sumy *590* NE Ukraine
Sunan *351* SW North Korea
Sunch'ŏn *351* SW North Korea
Sunch'ŏn *352* *Jap.* Junten. S South Korea
Sunda, Selat *304* strait connecting Indian Ocean and Laut Jawa between Java and Sumatra, W Indonesia
Sunderland *597* NE England, UK
Sundsvall *547* C Sweden
Sungai Seletar Reservoir *521* reservoir of N Singapore
Sungari *187* *Chin.* Songhua Jiang. River of NE China
Sun Koshi *427* river of E Nepal
Suntar-Khayata, Khrebet *485* mountain range of NE Russian Federation
Sunyani *272* W Ghana
Sunzu *635* mountain NE Zambia
Suō *see* Suao
Suomenlahti *see* Finland, Gulf of
Suomenselkä *251* physical region of C Finland
Suŏng *165* SE Cambodia
Superior de Tristaina, Estany *86* lake of NW Andorra
Superior, Lake *171, 603* *Fr.* Lac Supérieur. Lake of Canada and USA
Sup'ung-ho *351* reservoir of China and North Korea
Sūq 'Abs *see* 'Abs
Sūq ash Shuyūkh *312* SE Iraq
Suqrah Bay *448* Bay of the Arabian Sea, SE Oman
Suquṭrá *631* *Eng.* Socotra. Island of SE Yemen, off the Horn of Africa
Şūr *448* NE Oman
Surabaya *304* *prev.* Surabaja, Soerabaja. Java, C Indonesia
Surakarta *304* *prev.* Soerakarta. Java, C Indonesia
Šurany *523* *Hung.* Nagysurány. SW Slovakia
Sūrat *298* W India
Surat Thani *567* S Thailand
Sûre *378* river of W Europe
Sure, Lagh *344* dry watercourse of NE Kenya
Surfers Paradise *101* Queensland, E Australia
Surin *567* E Thailand
Surinam *400* S Mauritius
Suriname *542-543* officially Republic of Suriname, *var.* Surinam. Country of Central America divided into 8 admin. units (provinces). ❖ Paramaribo
Surinam *121* E Barbados
Surkhet *see* Birendranagar
Surkhob *562* river of C Tajikistan
Şurmān *371* NW Libya
Surt *371* *var.* Sidra, Sirte. N Libya
Surt, Khalīj *371* *var.* Gulf of Sirte, Gulf of Sidra. Gulf of the Mediterranean Sea, off N coast of Libya
Sūsah *see* Sousse
Susana *284* W Guinea-Bissau
Susuman *485* Ostrov Sakhalin, E Russian Federation
Sutlej *451* river of India and Pakistan
Suure-Jaani *242* *Ger.* Gross-Sankt-Johannis. C Estonia
Suur Munamägi *242* *var.* Munamägi. Mountain of SE Estonia
Suur Väin *242* strait of the Baltic Sea, between the mainland and the island of Muhu, W Estonia
Suva *248* ❖ of Fiji, Viti Levu, W Fiji
Suwa *240* SE Eritrea
Suwarrow *644* island of Northern Cook Islands, N Cook Islands
Suwayhān *594* E United Arab Emirates
Suways, Qanāt as *see* Suez Canal
Suwŏn *352* *var.* Suweon, *Jap.* Suigen. NW South Korea
Suzhou *187* *var.* Soochow, Su-chou, Suchow, *prev.* Wuhsien. Jiangsu, E China
Suzuka *332* Honshū, C Japan
Svalbard *653* Norwegian dependency of the Greenland Sea
Sväty Kríž nad Hronom *see* Žiar nad Hronom
Svay Chék *165* river of Cambodia and Thailand
Svay Riĕng *165* SE Cambodia
Svendborg *218* Fyn, S Denmark
Sverdlovsk *see* Yekaterinburg
Sverdrup Islands *170* island group of N Canada
Sveti Nikole *381* *prev.* Sveti Nikola. C FYR Macedonia
Svetlogorsk *see* Svyetlahorsk
Svetozarevo *512* *prev.* Jagodina. C Serbia, Serbia & Montenegro (Yugo.)
Svilengrad *152* *prev.* Mustafa-Pasha. SE Bulgaria
Svínoy *644* *var.* Svinø. Island of NE Faeroe Islands
Svishtov *152* *var.* Svištov. N Bulgaria
Svitavy *216* E Czech Republic
Svyetlahorsk *122* *Rus.* Svetlogorsk. SE Belarus
Swabian Jura *see* Schwäbische Alb
Swakopmund *423* W Namibia
Swallow Islands *526* small island group within Santa Cruz Is, E Solomon Is
Swan *100* river of SW Australia
Swansea *597* *Wel.* Abertawe. S Wales, UK
Swatow *see* Shantou
Swaziland *544-545* officially Kingdom of Swaziland. Country of southern Africa divided into 4 admin. units (districts).❖ Mbabane
Sweden *546-549* officially Kingdom of Sweden, *Swe.* Sverige. Country of Scandinavia divided into 24 admin. units (läns).❖ Stockholm
Swedru *see* Agona Swedru
Swellendam *531* Western Cape, S South Africa
Swetes *92* S Antigua, Antigua & Barbuda
Swift Current *170* SW Canada
Swindon *597* C England, UK
Swiss Plateau *550* *Ger.* Schweizer Mittelland. Plateau of W Switzerland
Switzerland *550-553* officially Swiss Confederation, *Ger.* Schweiz, *It.* Svizzera. Country of C Europe divided into 26 admin. units (cantons). ❖ Bern
Swords *316* *Ir.* Sórd Choluim Chille. E Ireland
Syaphrubesi *427* *var.* Syabrubensi. C Nepal
Sydney *101, 105* SE Australia
Sydney *171* Cape Breton Island, SE Canada
Sydney Island *see* Manra
Syktyvkar *484* *prev.* Ust'-Sisol'sk. NW Russian Federation
Sylhet *117* NE Bangladesh
Syowa *90* Japanese research station of Greater Antarctica, Antarctica
Syracuse *603* New York, NE USA
Syracuse *see* Siracusa
Syr Darya *340, 562, 614* *Rus.* Syrdar'ya, *Kaz.* Syrdariya, *Uzb.* Sirdaryo. River of C Asia
Syrdar'ya *614* E Uzbekistan
Syria *554-555* officially Syrian Arab Republic, *Ar.* Suriyah. Country divided into 13 admin. units (governorates).❖ Damascus
Syriam *159* S Burma
Syrian Desert *312, 338, 555* *Ar.* Bādiyat ash Shām. Desert of SW Asia
Syvash, Zatoka *590* inlet of the Sea of Azov
Szabadka *see* Subotica
Szamos *see* Someş
Szatmárnémeti *see* Satu Mare
Szczecin *471* *Ger.* Stettin. NW Poland
Szczeciński, Zalew *see* Oderhaff
Szechuan *see* Sichuan
Szeged *292* *Ger.* Szegedin.SE Hungary
Székesfehérvár *292* *Ger.* Stuhlweissenburg. W Hungary
Szekszárd *292* S Hungary
Szenice *see* Senica
Szentes *292* SE Hungary
Szeping *see* Siping
Szered *see* Sereď
Sziszek *see* Sisak
Szlatina *see* Podravska Slatina
Szolnok *292* C Hungary
Szombathely *292* *Ger.* Steinamanger. W Hungary
Sztálinváros *see* Dunaújváros

T

Tabac, River *400* river of S Mauritius
Ţabaqah *555* N Syria
Tabaquite *574* C Trinidad, Trinidad & Tobago
Tabarka *577* *var.* Ţabarqah. NW Tunisia
Tabasará, Serranía de *456* mountain range of W Panama
Tabasco *see* Grijalva
Tabernacle *494* NE St Kitts, St Kitts & Nevis
Tabiteuea *348* island of the Gilbert Is, W Kiribati
Tablas Island *466* island of C Philippines
Table Hill Gordon *92* SE Antigua, Antigua & Barbuda
Tabligbo *571* SE Togo
Tábor *216* SW Czech Republic
Tabora *564* W Tanzania
Tabou *328* *var.* Tabu. S Ivory Coast
Tabrīz *309* NW Iran
Tabuaeran *348* *var.* Fanning Island. Island of the Line Is, E Kiribati
Tabūk *506* NW Saudi Arabia
Tabwémasana *619* mountain of Espiritu Santo, W Vanuatu
Täby *547* SE Sweden
Tachia Hsi *559* river of W Taiwan
Tachoshui *559* E Taiwan
Tacloban *466* Leyte, E Philippines
Tacna *463* SE Peru
Tacoma *602* Washington, NW USA
Tacuarembó *611* N Uruguay
Tacuarembó *611* river of C Uruguay
Tacurupucú *see* Hernandarias
Tademaït, Plateau du *83* plateau of C Algeria
Tadine *650* Maré, Îles Loyauté, E New Caledonia
Tadjoura *222* E Djibouti
Tadjoura, Golfe de *222* inlet of the Gulf of Aden, E of Djibouti
T'aebaek-sanmaek *352* mountain range of South Korea
Taedong *351* river of C North Korea
Taegu *352* *var.* Daegu, *Jap.* Taikyū. SE South Korea
Taehan-haehyŏp *see* Korea Strait
Taejŏn *352* *Jap.* Taiden. C South Korea
Tafahi *572* island of N Tonga
Tafí Viejo *95* NW Argentina
Taftlund *218* Jylland, SW Denmark
Taga *500* Savai'i, Samoa
Taganrog, Gulf of *590* *Ukr.* Tahanroz'ka Zatoka, *Rus.* Taganrogskiy Zaliv. Gulf of the Sea of Azov, SE Ukraine
Tagarzimat *414* W Western Sahara
Tagiura *see* Tājūrā'
Tagliamento *323* river of N Italy
Tagtabazar *see* Takhta-Bazar
Taguasco *210* C Cuba
Taguatinga *145* C Brazil
Tagula Island *458* *prev.* Southeast I. Island of SE Papua New Guinea
Tagum *466* river of Mindanao, S Philippines
Tahanroz'ka Zatoka *see* Taganrog, Gulf of
Tahat *83* mountain of SE Algeria
Tahiti *646* island of W French Polynesia
Tahoua *439* W Niger
Taia *518* river of C Sierra Leone
Taiama *518* C Sierra Leone
T'aichung *559* *Jap.* Taichū. W Taiwan
Taiden *see* Taejŏn
Taieri *433* river of S South Island, New Zealand
Taihoku *see* Taipei
T'aihsi *559* W Taiwan
Tai Hu *187* lake of E China
Taikyū *see* Taegu
T'ainan *559* *Jap.* Tainan. SW Taiwan
Tai Pang Wan *see* Mirs Bay
T"aipei *559* *var.* Taipei, *Jap.* Taihoku. ❖ of Taiwan, N Taiwan
Taiping *386* NW Peninsular Malaysia
T'aitung *559* *Jap.* Taitō. SE Taiwan
Taiwan *558-561* officially Republic of China (Taiwan). Country of E Asia divided into 16 admin. units (counties). ❖ Taipei

U

V

W

Willemstad *650* ❖ Netherlands Antilles, Curaçao, S Netherlands Antilles
Willikies *92* E Antigua, Antigua & Barbuda
Willis *278* SW Grenada island, Grenada
Wilmington *603* Delaware, E USA
Wilna *see* Vilnius
Wilno *see* Vilnius
Wiltz *378* NW Luxembourg
Winam Gulf *344* *var.* Kavirondo Gulf. Gulf of Lake Victoria to the SW of Kenya
Winchester *597* S England, UK
Windau *see* Ventspils
Windhoek *423* *Ger.* Windhuk. ❖ of Namibia, C Namibia
Windsor *171* S Canada
Windward *278* N Carriacou, Grenada
Windward Passage *210, 288* channel connecting the Atlantic Ocean and the Caribbean Sea between Cuba and Haiti
Windward Point *642* headland on the NE coast of Anguilla
Winisk *171* C Canada
Winkowitz *see* Vinkovci
Winneba *272* SE Ghana
Winnipeg *170* S Canada
Winnipeg, Lake *170* lake of S Canada
Winnipegosis, Lake *170* lake of S Canada
Winston-Salem *603* North Carolina, SE USA
Winterthur *550* NE Switzerland
Winton *101* E Australia
Wisby *see* Visby
Wisconsin *603* state of NC USA
Wisła *471* *Ger.* Weichsel, *Eng.* Vistula. River of C Poland
Wiślany, Zalew *see* Vistula Lagoon
Wismar *266* N Germany
Witbank *531* Eastern Transvaal, NE South Africa
Witti, Banjaran *386* *var.* Witti Range. Mountain range of NE Borneo, Malaysia
W.J. van Blommesteinmeer *542* reservoir of NE Suriname
Włocławek *471* C Poland
Wodzisław Śląski *471* *Ger.* Loslau. S Poland
Woja District *396* district of Majuro, SE Marshall Islands
Wojwodina *see* Vojvodina
Woleai *406* atoll of C Micronesia
Woleu *see* Uolo, Río
Wolfgangsee *106* *var.* St Wolfgangsee, Abersee. Lake of N Austria
Wolf, Isla *230* island of NW Galapagos Is, Ecuador
Wolfsberg *106* SE Austria
Wolfsburg *267* N Germany
Wołkowysk *see* Vawkavysk
Wöllan *see* Velenje
Wollaston Peninsula *170* peninsula of Victoria Island, NW Canada
Wollongong *101* SE Australia
Wolmar *see* Valmiera
Wolverhampton *597* C England, UK
Wŏnju *352* *Jap.* Genshū. N South Korea
Wŏnsan *351* SE North Korea
Woodlands Bay *649* bay of the Caribbean Sea on the W coast of Montserrat
Woodlark Island *see* Murua Island
Woods, Lake of the *170* *Fr.* Lac des Bois. Lake of Canada and USA
Woody Island *651* island of Amphitrite Group, N Paracel Islands
Worcester *531* Western Cape, SW South Africa
Worcester *597* C England, UK
Worcester *603* Massachusetts, NE USA
Worms *see* Vormsi
Worthing *121* SW Barbados
Wotho *396* island of W Marshall Islands
Wotje *396* island of E Marshall Islands
Wrangel Island *see* Vrangelya, Ostrov
Wrexham *597* N Wales, UK
Wrocław *471* *Ger.* Breslau. SW Poland
Wu-chou *see* Wuzhou
Wuday 'ah *506* S Saudi Arabia
Wuhan *187* *prev.* Hankow, *var.* Han-k'ou, Hanyang, Wuchang. Hubei, C China
Wu-hsi *see* Wuxi
Wuhsien *see* Suzhou
Wuhu *187* *var.* Wu-na-mu. Anhui, E China
Wukari *440* E Nigeria
Wu-lu-mu-ch'i *see* Ürümqi
Wum *168* W Cameroon
Wu-na-mu *see* Wuhu
Wuppertal *266* W Germany
Würzburg *267* SW Germany
Wusih *see* Wuxi
Wuxi *187* *var.* Wu-hsi, Wusih. Jiangsu, E China
Wuzhou *187* *var.* Wu-chou, Wuchow. Guangxi, S China
Wye *597* *Wel.* Gwy. River of England and Wales, UK
Wyndham *101* N Australia
Wyoming *602-603* state of NW USA
Wysg *see* Usk

X

Xaçmaz *110* *Rus.* Khachmas, *var.* Hačmas. N Azerbaijan
Xaghra *395* Gozo, Malta
Xai-Xai *419* *prev.* Vila de João Belo, *var.* João Belo S Mozambique
Xam Nua *360* *var.* Sam Neua. NE Laos
Xankändi *110* *Rus.* Khankendy, *prev.* Stepanakert. SW Azerbaijan
Xánthi *275* *var.* Eskije, *Turk.* Iskeçe. NE Greece
Xauen *see* Chaouèn
Xäzär Dänizi *see* Caspian Sea
Xi'an *187* *var.* Hsi-an Sian. Shaanxi, C China
Xianggang *see* Hong Kong
Xiangkhoang *360* *var.* Xieng Khouang. E Laos
Xiangkhoang, Plateau de *360* *var.* Plain of Jars. Plateau of C Laos
Xiang Ngeun *360* C Laos
Xiangtan *187* *var.* Hsiang-t'an, Siangtan. Hunan, S China
Xiao Hinggan Ling *187* *Eng.* Lesser Khingan Range. Mountain range of NE China
Xieng Khouang *see* Xiangkhoang
Xi Jiang *187* *var.* Hsi Chiang. *Eng.* West River. River of S China
Xilinhot *187* *var.* Silinhot. Nei Mongol Zizhiqu, NE China
Xingkai Hu *see* Khanka, Lake
Xingu *145* river of C Brazil
Xining *187* *var.* Sining, Hsi-ning. Qinghai province, W China
Xinjiang Uygur Zizhiqu *186* *Eng.* Sinkiang Uighur Autonomous Region. Autonomous region of NW China
Xinpu *see* Liangyungang
Xinyang *187* *var.* Hsin-yang, Sinyang. Henan, C China
Xitole *284* SE Guinea-Bissau
Xixón *see* Gijón.
Xizang Zizhiqu *186* *Eng.* Tibetan Autonomous Region. Autonomous region of W China
Xolotlán *see* Managua, Lago de
Xuddur *528* *It.* Oddur. SW Somalia
Xulla Islands *see* Sula, Kepulauan
Xuzhou *187* *var.* Hsu-chou, Suchow, *prev.* T'ung-shan. Jiangsu, E China
Xylophaghou *215* *var.* Xylofagou. SE Cyprus

Y

Yabassi *168* W Cameroon
Yabēlo *245* S Ethiopia
Yablonovyy Khrebet *485* mountain range of C Russian Federation
Yabucoa *651* SE Puerto Rico
Yacata *248* island of the Lau Group, E Fiji
Yacuiba *136* S Bolivia
Yadua *248* *prev.* Yandua. Island to the W of Vanua Levu, NW Fiji
Yafran *371* NW Libya
Yagasa Cluster *248* islands of the Lau Group, E Fiji
Yagoua *168* NE Cameroon
Yaguarón *611* *var.* Jaguarão. River of Brazil and Uruguay
Yakhegnadzor *see* Yeghegnadzor
Yako *157* W Burkina
Yakutiya, Respublika *see* Sakha, Respublika
Yakutsk *485* E Russian Federation
Yala *567* S Thailand
Yalala, Chute *203* waterfall of W Dem. Rep. Congo
Yalinga *179* C Central African Republic
Yalova *580* NW Turkey
Yalpug *see* Ialpug
Yalpuh, Ozero *590* *Rus.* Ozero Yalpug. Lake of SW Ukraine
Yalta *590* S Ukraine
Yalu *351* *var.* Yalü, Amnok, *Jap.* Oryokko. River of China and North Korea
Yamagata *332* Honshū, N Japan
Yamaguchi *332* Honshū, W Japan
Yamal, Poluostrov *484* Peninsula of N Russian Federation
Yaman, Ra's al *see* Ḩadd al Jamal
Yambio *540* *var.* Yambiyo. S Sudan
Yambol *152* *var.* Jambol, *Turk.* Yanboli. E Bulgaria
Yamdena, Pulau *304* *prev.* Jamdena. Island of Kepulauan Tanimbar, Indonesia
Yamethin *159* C Burma
Yam HaMelah *see* Dead Sea
Yamoussoukro *328* ❖ of Ivory Coast, C Ivory Coast
Yamuna *298* *prev.* Jumna. River of N India
Yamundá *see* Nhamundá
Yanboli *see* Yambol
Yanbu' al Baḩr *506* W Saudi Arabia
Yandé, Île *650* island of Îles Belep, W New Caledonia
Yandua *see* Yadua
Yangambi *203* N Dem. Rep. Congo
Yangdŏk *351* S North Korea
Yanggeta *see* Yaqeta
Yangiyer *614* E Uzbekistan
Yangiyŭl *614* *var.* Jangijul. E Uzbekistan
Yangku *see* Taiyuan
Yangmei *559* N Taiwan
Yangon *see* Rangoon
Yangtze Kiang *see* Chang Jiang
Yannina *see* Ioánnina
Yan Oya *538* river of N Sri Lanka
Yantai *187* *var.* Yan-t'ai, *prev.* Chih-fu, *var.* Chefoo. Shandong, E China
Yantra *152* *var.* Jantra. N Bulgaria
Yaoundé *168* *var.* Yaunde. ❖ of Cameroon, S Cameroon
Yap *406* island of W Micronesia
Yapanskoye More *see* Japan, Sea of
Yapen, Pulau *305* island to the N of Irian Jaya, E Indonesia
Yapurá *see* Japurá
Yaqaga *248* island to the W of Vanua Levu, N Fiji
Yaqeta *248* *prev.* Yanggeta. Island of the Yasawa Group, NW Fiji
Yaque del Norte *226* river of NW Dominican Republic
Yaque del Sur *226* river of SW Dominican Republic
Yaqui *403* river of NW Mexico
Yarīm *631* W Yemen
Yarlung Zangbo Jiang *see* Brahmaputra
Yarmouth *171* SE Canada
Yaroslavl' *484* W Russian Federation
Yarumal *195* NW Colombia
Yasawa *248* island of the Yasawa Group, NW Fiji
Yasawa Group *248* island group of NW Fiji
Yasyel'da *122* river of SW Belarus
Yaté *650* S New Caledonia
Yatsushiro *332* Kyūshū, SW Japan
Yatta Plateau *344* plateau of SE Kenya
Yauco *651* SW Puerto Rico
Yaunde *see* Yaoundé
Yavan *see* Javan
Yavarí *see* Javari
Yaviza *456* SE Panama
Yawatahama *332* Shikoku, SW Japan
Yazd *309* *var.* Yezd. C Iran
Ybbs *106* river of C Austria
Yding Skovhøj *218* hill of Jylland, C Denmark
Ye *159* SE Burma
Yedintsy *see* Edineţ
Yedseram *441* river of E Nigeria
Yeghegnadzor *98* *Rus.* Yakhegnadzor, *var.* Jehegnadzor. SE Armenia
Yégué *571* W Togo
Yekaterinburg *484* *prev.* Sverdlovsk. C Russian Federation
Yekaterinodar *see* Krasnodar
Yekepa *368* NE Liberia
Yele *518* C Sierra Leone
Yelisavetpol *see* Gäncä
Yelizavetgrad *see* Kirovohrad
Yellow Hole *649* bay of the Caribbean Sea on the NE coast of Montserrat
Yellowknife *170* W Canada
Yellow River *see* Huang He
Yellow Sea *352* *Kor.* Hwang-Hae, *Chin.* Huang Hai. Sea of the Pacific Ocean between China and Korea
Yellowstone *602-603* river of NW USA
Yëloteu *585* *prev.* Iolotan, *Turkm.* Yolöten. SE Turkmenistan
Yel'sk *122* SE Belarus
Yelwa *440* E Nigeria
Yelwa *440* W Nigeria
Yemen *630-633* officially Republic of Yemen, *Ar.* Al Yaman, Al Jumhuriyah al Yamaniyah, *prev.* divided into South Yemen and Yemen Arab Republic (North Yemen) *prev.* Federation of South Arabia, Aden Protectorate. Country dividednto 17 admin. units (governorates). ❖ Sana
Yenakiyeve *590* *Rus.* Yenakiyevo, *prev.* Ordzhonikidze, Rykovo. E Ukraine
Yenangyaung *159* W Burma
Yên Bai *627* NW Vietnam
Yendi *272* NE Ghana
Yeniboğaziçi *215* *var.* Ayios Seryios. E Cyprus
Yenierenköy *215* *var.* Agigialousa, Yialousa. NE Cyprus
Yenisey *485* river of C Russian Federation
Yenshui *559* W Taiwan
Yeovil *597* S England, UK
Yerevan *98* *var.* Erevan, Jerevan, *Eng.* Erivan. ❖ of Armenia, C Armenia
Yeriho *see* Jericho
Yermak *340* *Kaz.* Ermak. NE Kazakhstan
Yermasoyia *215* *var.* Pentaschoinos. River of S Cyprus
Yerushalayim *see* Jerusalem
Yeu, Île d' *254* island of W France
Yevlax *110* *Rus.* Yevlakh, *var.* Jevlah. C Azerbaijan
Yevpatoriya *590* S Ukraine
Yerreyskiya *485* *Eng.* Jewish Autonomous Oblast. Autonomous region of SE Russian Federation
Yezd *see* Yazd
Ygatimí *460* *var.* Igatimí. E Paraguay
Yí *611* river of C Uruguay
Yialias *215* *var.* Yalyas, Çakilli Dere. River of C Cyprus
Yialousa *see* Yenierenköy
Yigo *647* NE Guam
Yinchuan *187* *var.* Yinchwan. Ningxia, C China
Yining *186* *var.* I-ning, *Uigh.* Gulja, *var.* Kuldja. Xinjiang Uygur Zizhiqu, NW China
Yirga 'Alem *245* *It.* Irgalem. S Ethiopia
Yisra'el *see* Israel
Yoboki *222* C Djibouti
Yof *510* W Senegal
Yogyakarta *304* *prev.* Jogjakarta. Java, C Indonesia
Yojoa, Lago de *290* lake of W Honduras

Z

ACKNOWLEDGMENTS

ORLING KINDERSLEY would like to express their thanks to the ollowing individuals, companies and institutions for their elp in preparing this atlas:

DDITIONAL CARTOGRAPHY

dvanced Illustration (Congleton, UK)
ndrew Bright
osmographics (Watford, UK)
lalcolm Porter
wanston Publishing (Derby, UK)
ndrew Thompson

ESIGN

oyd Annison, Icon Solutions (Chesham, UK) *for Macintosh onsultancy and chart templates*
runo Maag, Dalton Maag (London, UK) *for font consultancy and roduction*

ESEARCH AND REFERENCE

r D Alkhateeb, Organization of Petroleum Exporting Countries OPEC, Vienna, Austria)
mnesty International (London, UK)
aroline Blunden
NN International (New York, USA)
ATAQUEST EUROPE SA (PARIS, FRANCE)
SL Davies
epartment of Trade and Industry Export Market Information Centre London, UK)
he Flag Institute (Chester, UK)
oreign and Commonwealth Office (London, UK)
lexander Fyges-Walker
hristel Heideloff, Institute of Shipping Economics and Logistics Bremen, Germany)
nternational Bank for Reconstruction and Development World Bank, Washington, DC, USA)
nternational Committee of the Red Cross (ICRC, Geneva, Switzerland)
nternational Civil Aviation Organization (ICAO, Montreal, Canada)
nternational Criminal Police Organization (INTERPOL, Lyon, France)
International Institute for Strategic Studies, for information from *The Military Balance* (London, UK)
International Boundaries Research Unit, University of Durham
Institute of Latin American Studies, University of London (London, UK)
Intermediate Technology Development Group (Rugby, UK)
Chris Joseph, United States Travel and Tourism Administration (USTTA, London, UK)
Latin American Bureau (London, UK)
Patrick Mahaffey, Ohio European Office (Brussels, Belgium)
Peter Mansfield
Robert Minton-Taylor
National Meteorological Library and Archive (Bracknell, UK)
Oil and Gas Journal (Houston, Texas)
Organization for Economic Cooperation and Development (OECD, Paris, France)
Penal Reform International (London, UK)
Matt Ridley
Screen Digest (London, UK)
William Smith, Chicago Sun-Times (Chicago, USA)
Tourism Concern (London, UK)
United Nations Crime Prevention and Criminal Justice Branch (UNCPC, Vienna, Austria)
United Nations Development Programme (UNDP, New York, USA)
United Nations Environment Programme (UNEP, Nairobi, Kenya)
United Nations Food and Agriculture Organization (UNFAO, Rome, Italy)
United Nations International Labour Organization (UNILO, Geneva, Switzerland)
United Nations Population Fund (UNFPA, New York, USA)
Westminster Reference Library (London, UK)
World Conservation Monitoring Centre (Cambridge, UK)
World Health Organization (WHO, Geneva, Switzerland)
World Tourism Organization (Madrid, Spain)

The many embassies, High Commissions, airports, national information and tourist offices in London and around the world.

PICTURE CREDITS

=top, b=below, a=above, l=left, r=right, c=center

gence France Presse: 300crb, 332tr, 337tc, 445cr; Gustavo Amarelle 95crb; Dong-A llbo 53bcr; Victor Drachev 123bcr; Vassil Donev 153bcr; Gonzalo Espinoza 137cr; Eric Feferberg 88br; Francois Guillot 256crb; Martyn Hayhow 256bcr; Doug Kanter 601tr; Attila Kisbenedek 93br; John MacDougall 77tr; Shah Marai 78br; Tatiana Munoz 195tcr; Keld Navntoft 219tcr; nacio Rosa 475bcr; Dibyangshu Sarkar 303tc; Bernd Settnik 269br; Sergei Supinsky 587tr; Weda 29tc; Greg Wood 337bl; **Ancient Art & Architecture Collection:** 44bcr; 45bl; 45cr; 47tc; G ortoli 45tcb; **Arcaid:** P Mauss Esto 598bc; **Art Archive:** 44bl; 49bl; 49cr; 51tcl; **Aspect Picture ibrary:** 202tr; Brian Seed 418c; D Bayes 370bl; Fiona Nichols 304tr; **Associated Press AP:** 45bcr; AFP 415bcr; Aaron Favila 467crl; Humberto Pradera/Agencia Estado 149cb; Bullit arquez 191bl; Jan Bauer 271cb Alexander Zemlianichenko 488cb; **Bridgeman Art Library, ondon/New York:** Hermitage, St Petersburg 46bc; Lauros - Giraudon / Château de almaison 48bcr; National Maritime Museum, London 47br; Private Collection 48bcl; **D Donne ryant Stock Picture Agency:** 461tc; Byron Augustin 436tr; **Dale Buckton:** 594br; **Camera Press:** A Pucciano 95cr; F Goodman 649bca; H Andrews 627bcr; S Smith 245bcl; T Charlier 07bl; **The Canadian Alliance:** 172crb; **The J Allan Cash Photolibrary:** 55tl; 106c; 116bc; 46br; 149tl; 168tr; 171tr; 241tr; 291tc; 308tr; 320br, 339cr; 342bl; 383tc; 386c; 438bc; 440tr; 51bc; 474tr; 566bc; **CDA:** 429tcr; **Bruce Coleman Ltd:** 498bl; B&C Calhoun 174bc; Dr MP Kahl 4tr; F Prenzel 104tc; Gerald Cubitt 88tr; Gerald Cubitt 244bc; Gordon Langsbury 262bc; J Fry 11tc; J Jurka 542bc; K. Maj 470c; Kim Taylor 515tr; L Lee Rue 91tl; LC Marigo 281tc; M Berge 96tr; MPL Fogden 462ca; O Langrand 370tr; P Davey 214bc; S Prato 274bc; **Colorific:** J olleross / M Kreiner 54cl; M Rogers 536tr; Sandro Tucci 159bl; **Comstock:** 194bc, 24tl, 535tc; or Eigeland 399tc; **Corbis UK Ltd:** AFP 387bcl; Bettmann 146car, 233bcr, 555tcr; Bob Krist 42br; Francoise de Mulder 141bc; Jack Fields 648bl; Jack Fields 650bl; James L Amos 647tr; iki Kratsman 486br; Nik Wheeler 175tc; Franz-Marc Frei 261cb; Peter Wilson 327cb; Joseph ohm 603bl; **James Davis Travel Photography:** 144bl; 145tl; 218c; 250bc; 330bc; 379tl; 390tr; 01tc; 441tl; 428bc; 432bc; 493tc; 494bl; 496bl; 502tr; 550bc; 558tr; 586tr; 626bl; 643cr; Prisma 6cla; Prisma / Schwarz 226tr; S Begawan 151tl; S Thingeyjar 297tc; World View - Fototheek msterdam 177tc; **Mary Evans Picture Library:** 47bl, 47trb; **Chris Fairclough Colour Library:** 1cr; 375tl; 572br; **Robert Harding Picture Library:** 76tr; 180c; 225tl; 273tc; 282bl; 293tr; 312tr; 36bl; 358bc; 449tc; 466tr; 480tr; 484tr; 501tc; 513tc; A Woolfitt 114bc; C Martin 266bc; C Rennie 80ca; D Hughes 259bc; Explorer 377tc; Explorer / Roy 110tr; F Dubes 148bl; Frerck / Odyssey 02tr; G Hellier 217tr; G Hellier 518tr; G Roli 444tr; Gascoine 610bl; P Craven 254tr; Photri 22ca; R Rainford 593tl; Rosehaven Management Ltd 653cr; Sassoon 134cb; Sassoon 264cl; **Paul Harris Photography:** Paul Harris Photography 491bl; **Hulton Getty:** 50bcr; 51br; 53br; **Robert Hunt Library:** 53tc; **Hutchison Library:** 136bl; 369tc; 596bc; 636tr; Andrew Hill 634bc; Bernard Gérard 261bc; Christine Pemberton 210tr; J Henderson 286ca; JG Fuller 650bcr; L Taylor 484bl; M Macintyre 568ca; Robert Francis 560tr; Trevor Page 524ca; **Image Bank:** 394bl; A Rippy 186tr; G Jung 576tr; GA Rossi 646bl; M Beebe 209cl; ME Newman 324tr; P Trummer 232tr; T Madison 187br; **Images Colour Library:** 426bl; **Image Select:** Ann Ronan 42bc; **Impact Photos:** A le Garsmeur 412ca; Alain leGarsmeur 175bl; Ben Edwards 450tr; C Penn 584tr; G-J Norman 356tr; J Arthur 526tcr; Mark Henley 188tr; Robin Lubbock 236tr; Ben Edwards 303bl; B Babarov/Vika 489bl; **David King Collection:** 55tcb **Magnum:** H Cartier-Bresson 53tcb; **Panos Pictures:** B Tobiasson 329bc; Chris Stowers 333bl; D Hulcher 360tr; Jeremy Hartley 156bc; Marc French 289tc; Morris Carpenter 365bc; Neil Cooper 130tr; R Giling 538tr; S Sprague 160tr; Sean Sprague 136c; Penny Tweedie 105tc; Michael Harvey 149cl; John Miles 190tr;.JC Tordai 320cl; 488tr; 597cb; 602bl; **Pa News Photo Library:** AFP 419cbl; **Popperfoto:** 51trb; 52br; 55bl; AFP / Armand 54br; David Mercado / Reuters 137crl; Will Burgess / Reuters 105bl; Reuters 271cl; Jeremy Piper / Reuters 102bcr; John Cobb / Reuters 345tr; Official U.S. Air Force Photo 55cr; **Popperfoto / Reuters**: Peter Andrews 637cr; Simon Baker 433tlb; Denis Balibouse 203cl; Russell Boyce 594cra; Andreas Comas 415bcl; Bogdan Cristel 481cr; Jack Dabaghian 551br; Dimitar Dilkoff 153bcl; Larry Downing 601bl; Marcus Gyger 107bcl; Mohamed Hammi 573tcr; Hyungwon Kang 601bcr; Sergei Karpukhin 487bcl; Kamal Kishore 300cbr; Francois Lenoir 127bcr; Havakuk Levison 318bcr; David Loh 623tcl; Christinne Muschi 172cbr; Alexander Natruskin 487bl; Patrick de Noirmont 159bcl; Enny Nuraheni 305bcr; Pilar Olivares 463tcl; Hrvoje Polan 293bcr; Romero Ranoco 467cr; Molly Riley 268bcr; Oswaldo Rivas 555tr; Henry Romero 403bcl; Jayanta Shaw 623tl; Ruben Sprich 547cr; STR 451tr, 577car; Ray Stubblebine 251tcr, 293bcl, 577trb, 607bcl; Eriko Sugita 334cbr; Sukree Sukplang 563tr; Supri 305bcl; Susumu Takahashi 334cbl; Martin Thomas 182tr; Goran Tomasevic 631bcl; Pierre Virot 476bcl; Haydn West 594tcr; Kimberly White 618tcr; Darren Whiteside 165tcr; Rick Wilking 600bcr, 600br; **Reuters:** 123bcl, 146tcr; 507br; **Rex Features:** 116tcbr; 172bcr; 251tr; 256cb; 275tcr; 275tr; 341bcl; 481cbr; 486bcr; 516bl; 526tr; 543tcl; 587tcr; 594tr; 594trb; 611bc; 627bcl; Andy Hernandez 488cl, J Sutton Hibbert 597ct; Paul Brown 602cr; David Hartley 219tr; F Stevens 211bcl; Farnood 309tcr; Ken McKay 531tr; Sipa-Press 77tcr; Sipa-Press 95cbr; Sipa-Press 83bcl; Sipa-Press 127bcl; Sipa-Press 188bcr; Sipa-Press 313c; Sipa-Press 318br; Sipa-Press 429tcl; Sipa-Press 445c; Sipa-Press 563tcr; Sipa-Press 631bcr; Sipa-Press 645bl; Tony Kyriacou 353bcl; Torregano 419cl; **Harry Smith Collection:** 46clb; **South American Pictures:** Jevan Berrange 206br; P Dixon 456tr; T Morrison 606tr; **Sovfoto/Eastfoto:** 652br; **Frank Spooner Pictures:** 53bl; 83bl; 324bc;345tcr; 543tr; A Denize 371tcr; A Sassaku 146tr; Alain Morvan / Gamma 547c; Alexis Duclos 182tcr; Bob Stern / Liason 243tr; C Hires 551bcr; C Poulet 647cr; Eric Vandeville 195tr; Tim Crosby / Gamma - Liaison 602ct; Gamma 471tr; Gamma / B Iverson 233bcl; Gamma / E de Keerle 471tcr; Gamma / F Apesteguy 507bcr; Gamma / Iliona - Figaro Magazine 113tc; Gamma / K Al Arab 573tcl; Gamma / L Chaperon 268cbr; Gamma / N Jallot 408tr; Gamma / Najer 387bcr; Gamma / Xinhua 188bcl; Georges Merillon 441bcl; KJ Eddy 107bcl; L Anticoli 324cbr; Liason / Peterson 441bcr; Liason / Markel 637c; N Sagansky 203cr; P Perrin 313cr; Patrick Piel 268bra; Reglain 165tr; Victoria Brynner 451tcr; **Tony Stone Images:** 179bc; 198tr; 354bl; 444br; A Cassidy 302br; Alan Kearney 182bc; Alan Smith 362c; C Waite 592bc; D Armand 172tr; D Hanson 562bl; D Schultz 50bl; Dennis Stone 344tr; G Allison 605bl; H Kurihara 270bc; J Pragen 616tr; Joe Cornish 317bc; Marcus Brooke 158tr; O Benn 93tc; 279tr; P Chesley 406ca; Penny Tweedie 384bc; R Evans 152tr; R. Everts 415tr; R Smith 102tr; R Smith 231tl; S Egan 530tr; S Egan 546tr; Steven Rothfeld 393tc; **Sygma:** aldev 350ca; R Reuter 502bl; **Telegraph Colour Library:** 100tr; Ford Motor Company Ltd 52clb; **Topham Picturepoint:** 133tc; 238tr; 320br; 327cl; 371tr; 433cal; 478bc; 487tr; 504bc; 642c; Keystone 324cbl; **Trip:** 506bl; G Spenceley 380tr; T Goodman 285tc; V Shuba 122c; V Sidoropolev 340tr; **World Pictures:** 83cla; 322bl; 422ca; 554tr; **Zefa Picture Library**: 80bc; 120bl; 200ca; 526bl; 570bc; 618bl; Everts 630ca; F Lanting 142tr; Streichan 268bl.

NOTES

For the very latest information, visit:
www.dk.com and click on the Maps & Atlases icon

NOTES

For the very latest information, visit:
www.dk.com and click on the Maps & Atlases icon

NOTES

For the very latest information, visit:
www.dk.com and click on the Maps & Atlases icon

NOTES

For the very latest information, visit:
www.dk.com and click on the Maps & Atlases icon

NOTES

For the very latest information, visit:
www.dk.com and click on the Maps & Atlases icon

NOTES

For the very latest information, visit:
www.dk.com and click on the Maps & Atlases icon

NOTES

For the very latest information, visit:
www.dk.com and click on the Maps & Atlases icon